D0768840

REFRIGERATION AND AIR CONDITIONING TECHNOLOGY

REFRIGERATION AND AIR CONDITIONING TECHNOLOGY, 3rd edition

Concepts, Procedures, and Troubleshooting Techniques

WILLIAM C. WHITMAN
WILLIAM M. JOHNSON

Central Piedmont Community College
Charlotte, North Carolina

Delmar Publishers™

I(T)P™ An International Thomson Publishing Company

Albany • Bonn • Boston • Cincinnati • Detroit • London • Madrid • Melbourne
Mexico City • New York • Pacific Grove • Paris • San Francisco • Singapore • Tokyo
Toronto • Washington

NOTICE TO THE READER

Cover photos courtesy of Bill Johnson, Yukon Energy Corporation, York International, Robinair, Carolina Products Inc., Tecumseh Products Company.

DELMAR STAFF
Senior Administrative Editor: Vernon Anthony
Developmental Editor: Lisa Reale / Catherine Eads
Project Editor: Eleanor Isenhart
Production Coordinator: Dianne Jensis
Art / Design Coordinator: Heather Brown

Printed in the United States of America

For more information, contact:

Delmar Publishers
3 Columbia Circle, Box 15015
Albany, New York 12212-5015

International Thomson Publishing Europe
Berkshire House 168-173
High Holborn
London, WC1V7AA
England

Thomas Nelson Australia
102 Dodds Street
South Melbourne 3205
Victoria, Australia

Nelson Canada
1120 Birchmont Road
Scarborough, Ontario
Canada M1K 5G4

International Thomson Editores
Campos Eliseos 385, Piso 7
Col Polanco
11560 Mexico D F Mexico

International Thomson Publishing GmbH
Königswinterer Strasse 418
53227 Bonn
Germany

International Thomson Publishing Asia
221 Henderson Road
#05-10 Henderson Building
Singapore 0315

International Thomson Publishing Japan
Kyowa Building, 3F
2-2-1 Hirakawacho
Chiyoda-ku, Tokyo 102
Japan

6 7 8 9 10 XXX 00 99 98 97

Library of Congress Cataloging-in-Publication Data

Whitman, William C.
Refrigeration and air conditioning technology : concepts, procedures, and troubleshooting techniques / William C. Whitman, William M. Johnson.—3rd ed.
p. cm.
Includes index.
ISBN 0-8273-5646-3
1. Refrigeration and refrigerating machinery. 2. Air conditioning. I. Johnson, William M. II. Title.
TP492.W6 1995
621.5'7—dc20 94-30065
CIP

Brief Contents

Contents

Preface

Refrigeration and Air Conditioning Technology is an instructional text that can be used by students in vocational-technical schools and colleges, in comprehensive community colleges and also for industrial upgrading and apprenticeship programs. The term "air conditioning" includes cooling, heating, cleaning, humidifying, and dehumidifying air. Considerable effort has been made to provide content in a format that is appropriate for students who are attending classes full-time trying to prepare for their first job in this field and for students attending classes part-time preparing for a career change or for those working in the field who wish to increase their knowledge and skills. Emphasis throughout the text is on practical application of the knowledge and skills technicians need to be productive in the refrigeration and air conditioning industry.

Considerable thought and study has been devoted to the organization of this text. Difficult decisions had to be made to provide a text in a format that would meet the needs of varied institutions. Instructors from various areas were asked for their ideas regarding the organization of the instructional content.

The text is organized so that after completing the first three sections, students may concentrate on courses in refrigeration or air conditioning (heating and/or cooling). If the objective is to complete a whole program, the instruction may proceed until the sequence scheduled by the school's curriculum is completed.

Every effort has been made to keep the material at a reading level that most students can easily understand. Terms used commonly by technicians and mechanics have been used throughout to make the text easy to read and to present the material in a practical way.

Illustrations and photos are used extensively throughout the text. Full-color treatment of most photos and illustrations helps amplify most concepts.

FEATURES OF THE TEXT

OBJECTIVES

- ***Objectives*** **are listed at the beginning of each unit. The objective statements are kept clear and simple to give students direction.**

SAFETY CHECKLIST

* A *Safety Checklist* is presented at the beginning of each appropriate unit immediately following the Objectives. This checklist emphasizes the importance of safety and is included in units where "hands-on" activities are discussed.

TROUBLESHOOTING

Practical troubleshooting procedures are an important feature of this text. There are practical component and system troubleshooting suggestions and techniques.

SERVICE CALLS

In many units, practical examples of *service technician calls* are presented in a down-to-earth situational format. These are realistic service situations in which technicians may find themselves. In some instances the solution is provided in the text and in others the reader must decide what the best solution should be.

SAFETY PRECAUTIONS: ****Safety* **is emphasized throughout the text. One entire unit is dedicated to safety. There is a safety checklist at the beginning of most units. Safety precautions and techniques are highlighted in red throughout the text. It would be impossible to include a safety precaution for every conceivable circumstance that may arise, but an attempt has been made to be as thorough as possible. Work safely whether in a school shop or laboratory or on the job. Use common sense.*

PREVENTIVE MAINTENANCE

Preventive maintenance procedures **are contained in many units and relate specifically to the equipment presented in that particular unit. Technicians can provide some routine preventive maintenance service when on other types of service calls as well as when on strictly preventive maintenance calls. The preventive maintenance procedures provide valuable information for the new or aspiring technician as well as those with experience.**

GOLDEN RULES

Golden Rules for the refrigeration and air conditioning technician give advice and practical hints for developing good customer relations. These "golden rules" appear in appropriate units.

RECOVERY/RECYCLING/RECLAIMING

⊛Discussion relating to *recovery, recycling or reclaiming* is highlighted in green throughout the text. In addition, there is one complete unit on refrigerant management.⊛

SUMMARY

- The *Summary* appears near the end of each unit. It can be used to review the unit and to stimulate class discussion.

REVIEW QUESTIONS

1. *Review Questions* follow the Summary in each unit and can help to measure the student's knowledge of the unit.

DIAGNOSTIC CHARTS

Diagnostic Charts are included at the end of many units that include material on troubleshooting and diagnosis.

NEW UNITS IN THIS EDITION

Introduction

The topics include History of Refrigeration and Air Conditioning (Cooling), History of Home and Commercial Heating, Career Opportunities, Certification, and Quality of Work.

Refrigerant Management—Recovery/Recyling/Reclaiming

The topics include Refrigerants and the Environment, CFC Refrigerants, HCFC Refrigerants, HFC Refrigerants, Refrigerant Blends, Regulations, and Recovery/Recycle/Reclaim.

Special Refrigeration Applications

The topics include Transport Refrigeration, Truck Refrigeration Systems, Railway Refrigeration, Extra Low Temperature Refrigeration, Quick Freezing Methods, Marine Refrigeration, and Air Cargo Hauling.

High-Pressure, Low-Pressure and Absorption Chilled-Water Systems

The topics include Reciprocating, Scroll, Rotary Screw, and Centrifugal Compressor Chillers (High and Low Pressure), Absorption Chillers, and Motors and Drives for Chillers.

Cooling Towers and Pumps

The topics include Cooling Tower Function, Types of Cooling Towers, Flow Patterns, Tower Materials, Fans, and Water Pumps.

Operation, Maintenance, and Troubleshooting of Chilled-Water Systems

The topics include Compression Cycle Chiller Start-Up, Large Positive-Displacement Chiller Operation, Absorption Chilled Water System Start-Up, and Absorption Chiller Operation and Maintenance.

SUPPLEMENTS

Refrigeration and Air Conditioning Technology Study Guide/Lab Manual

As the title implies a study guide that includes a brief description for each unit, a list of terms and a series of objective questions has been combined with a lab manual. The lab exercises include a series of practical exercises that the student completes in a "hands-on" lab environment.

Refrigeration and Air Conditioning Technology Instructor's Guide

This guide includes the suggested use of transparency masters included in the Guide and the suggested use of materials from other sources that may be used to enhance student learning. It includes the answers to the review questions in the text, and all questions in the Study Guide/Lab Manual. In addition it includes the "trouble" that should be inserted in the equipment in order to perform certain lab exercises and the diagnosis for those Service Technician Calls that do not include this information in the text. Also included are 150 transparency masters and the printed test bank.

Computerized Test Bank

Computerized Testmaker and Testbank contains a data bank of questions to test the student's knowledge of each unit. This versatile tool enables the instructor to manipulate the data to create original tests.

Introduction

Refrigeration, as used in this text, relates to the cooling of air or liquids, thus providing lower temperatures to preserve food, cool beverages, make ice, and many other applications. Air conditioning includes space cooling, heating, humidification, dehumidification, air filtration, and ventilation to condition the air and improve indoor air quality.

HISTORY OF REFRIGERATION AND AIR CONDITIONING (COOLING)

Most evidence indicates that the Chinese were the first to store natural ice and snow to cool wine and other delicacies. Evidence has been found that ice cellars were used as early as 1000 BC in China. Early Greeks and Romans also used underground pits to store ice, which they covered with straw, weeds, and other materials to provide insulation and preserve it over a long period.

Ancient people of Egypt and India cooled liquids in porous earthen jars. These jars were set in the dry night air and the liquids seeping through the porous walls evaporated to provide the cooling. Some evidence indicates that ice was produced due to the vaporization of water through the walls of these jars, radiating heat into the night air.

In the 18th and 19th centuries, natural ice was cut from lakes and ponds in the winter in northern climates and stored underground for use in the warmer months. Some of this ice was packed in sawdust and transported to southern states to be used for preserving food. In the early 20th century, it was still common in the northern states for ice to be cut from ponds and then stored in open ice houses. This ice was insulated with sawdust and delivered to homes and businesses.

In 1834, Jacob Perkins, an American, developed a closed refrigeration system using expansion of a liquid and compression to produce cooling. He used ether as a refrigerant, a hand-operated compressor, a water-cooled condenser, and an evaporator in a liquid cooler. He was awarded a British patent for this system. In Great Britain during the same year, L. W. Wright produced ice by the expansion of compressed air. Mechanical refrigeration was first designed to produce ice.

During the middle 1800s, other refrigeration systems were designed in the United States, Australia, and England. In the following years many improvements were made in the equipment design and by the 1930s refrigeration was well on its way to being used extensively in American homes and commercial establishments.

HISTORY OF HOME AND COMMERCIAL HEATING

A human being's first exposure to fire was probably when lightning or other natural occurrence such as a volcanic eruption ignited forests or grasslands. After overcoming the fear of fire, early man found that a more comfortable living environment could be created by placing a controlled fire in a cave or other shelter. A fire was often carried from one place to another. Smoke was always a problem, however, and it was soon found that methods needed to be developed for venting it outside. Indians had vents at the peaks of their tepees and some were constructed with a vane that could be adjusted to prevent downdrafts.

Fireplaces were common in Europe and America and were vented through chimneys. Early stoves were found to be more efficient than fireplaces. These early stoves were constructed of a type of fire brick, ceramic materials, or iron.

In the middle 18th century, a jacket for the stove and a duct system were developed. The stove could then be located at the lowest place in a structure; the air in the jacket around the stove was heated and would rise through the duct system and grates into the living area. This was the beginning of the development of circulating warm air heating systems.

Boilers that heated water were developed and this water was circulated through pipes in duct systems. The water heated the air around the pipes. This heated air passed into the rooms to be heated. Radiators were then developed. The heated water circulated by convection through the pipes to the radiators and heat was passed into the room by radiation. These early systems were forerunners of modern hydronic heating systems.

CAREER OPPORTUNITIES

People in the United States expect to be comfortable. In the cold weather they expect to be able to go inside and be warm and in the warmer climates they expect to be able to go inside and be cool. They expect beverages to be cold when they want them and they expect their food to be properly preserved. Many buildings are constructed so that the quality of the air must be controlled by specialized equipment. The condition of the air must be controlled in many manufacturing processes. Heating and air conditioning systems control the temperature, humidity, and total air quality in residential, commercial, industrial, and other buildings. Refrigeration systems are used to store and transport food, medicine, and other perishable items.

Refrigeration and air conditioning technicians install and maintain these systems. These technicians may specialize in installation or service and they may further specialize in one type of equipment.

Employment for service technicians in these fields is generally more stable during economic downturns because the need to service this equipment is present regardless of the economic conditions.

Many contractors and service companies specialize in commercial refrigeration. The installation and service technicians employed by these companies install and service refrigeration equipment in supermarkets, restaurants, hotels/motels, flower shops, and many other types of retail and wholesale commercial businesses.

Other contractors and service companies may specialize in air conditioning. Some may specialize in residential only or commercial only installation and service; others may install and service both residential and commercial equipment up to a specific size. Air conditioning may include cooling, heating, humidifying, dehumidifying, or air cleaning. The heating equipment may include gas, oil, electric, or heat pumps. The number of each type of installation will vary from one part of the country to another, depending on the climate and availability of the heat source. The heating equipment may be a space heating type or a hydronic furnace. The hydronic furnace heats water and pumps it to the space to be heated where the heat is transferred to the air by means of a radiator or other type of heat exchanger. The hydronic type of furnace may also heat water until it turns to steam and the steam forced to a radiator in the space to be heated where heat is given off to the air.

Technicians may specialize in installation or service of this equipment or they may be involved with both. Other technicians may design installations, whereas others work in sales. Salespersons may be in the field selling equipment to contractors or business or home owners; others may work in wholesale supply stores. Other technicians may represent manufacturers, selling equipment to wholesalers and large contractors.

Many opportunities exist for technicians to be employed in industry or by companies owning large buildings. These technicians may be responsible for the operation of air conditioning equipment or they may be involved in the service of this equipment.

Opportunities also exist for employment servicing household refrigeration and room air conditioners. This would include refrigerators, freezers, and window or through-the-wall air conditioners.

Opportunities are available for employment in a field often called transport refrigeration. This includes servicing refrigeration equipment on trucks or on large containers hauled by trucks and ships.

Most modern houses and other buildings are constructed to keep outside air from entering except through planned ventilation. Consequently, the same air is circulated through the building many times. The quality of this air may eventually cause a health problem for people spending many hours in the building. This indoor air quality (IAQ) presents another opportunity for employment in the air conditioning field. Technicians clean filters and ducts, take air measurements, check ventilation systems, and perform other tasks to help ensure healthy air quality.

Other technicians work for manufacturers of air conditioning equipment. These technicians may be employed to assist in equipment design or in the manufacturing process or as equipment salespersons.

CERTIFICATION

⊛**Technicians involved with the transfer of refrigerants into or out of a system must be certified in a program approved by the Environmental Protection Agency (EPA), an agency of the federal government.**⊛ Many city, county, and state governments require technicians to be certified or licensed under certain conditions. You should be aware of the requirements in your area.

QUALITY OF WORK

All technicians should be concerned with the quality of their work. Particularly in recent years, some technicians have provided unsatisfactory service, which has caused doubt among customers as to whether or not they are getting the quality service they are paying for. As professionals, technicians should strive to provide the best workmanship possible. Quality work will prove beneficial to the technicians, to their companies, and to the consumers or customers.

Reviewers

Keith Bilbrey
Cedar Valley Community College
Lancaster, TX

James C. Bledsoe
Houston Community College
Houston, TX

John Brewer
Central Texas College
Killeen, TX

Robert F. Burris
Riverside Community College
Riverside, CA

Ronald A. Hovey
Technical College of the Low County
Beaufort, SC

Bill Johnson
Moraine Valley Community College
Palos Hills, IL

Kevin Joyce
Ranken Technical College
St. Louis, MO

Clyde Perry
Gateway Community College
Phoenix, AZ

William S. Pritchard
Eastfield College
Garland, TX

Edward J. Radigan
West Side Institute of Technology
Cleveland, OH

Greg Skudlarek
Minneapolis Technical College
Minneapolis, MN

Darius Spence
Northern Virginia Community College
Woodbridge, VA

Del Winston
Platte County Area Vocational Technical School
Platte City, MO

SECTION

1 *Theory of Heat*

UNIT 1

Theory

OBJECTIVES

After studying this unit, you should be able to

- define temperature.
- make conversions between Fahrenheit and Celsius scales.
- describe molecular motion at absolute 0.
- define the British thermal unit.
- describe heat flow between substances of different temperatures.
- explain the transfer of heat by conduction, convection, and radiation.
- discuss sensible heat, latent heat, and specific heat.
- state atmospheric pressure at sea level and explain why it varies at different elevations.
- describe two types of barometers.
- explain psig and psia as they apply to pressure measurements.

1.1 TEMPERATURE

Temperature can be thought of as a description of the level of heat and heat can be thought of as energy in the form of molecules in motion. The starting point of temperature is, therefore, the starting point of molecular motion.

Most people know that the freezing point of water is 32 degrees Fahrenheit (32°F) and that the boiling point is 212 degrees Fahrenheit (212°F). These points are commonly indicated on a thermometer, which is an instrument that measures temperature.

Early thermometers were of glass-stem types and operated on the theory that when the substance in the bulb was heated it would expand and rise in the tube, Figure 1–1. Mercury and alcohol are still commonly used today for this application.

We must qualify the statement that water boils at 212°F. Pure water boils at precisely 212°F at sea level when the atmosphere is 70°F. This qualification concerns the relationship of the earth's atmosphere to the boiling point and will be covered in detail later in this section in the discussion on pressure. The statement that water boils at 212°F at sea level when the atmosphere is 70°F is important because these are standard conditions that will be applied to actual practice in later units.

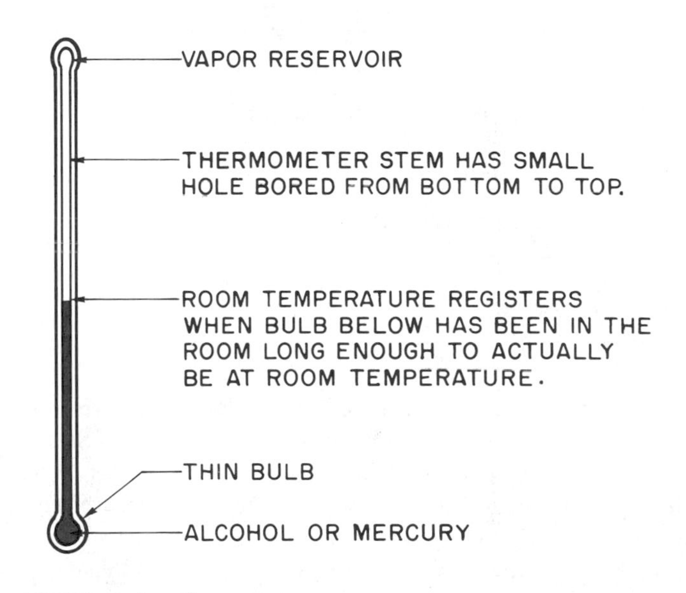

FIGURE 1–1 Thermometer.

Pure water has a freezing point of 32°F. Obviously the temperature can go lower than 32°F, but the question is, how much lower?

The theory is that molecular motion stops at −460°F. This is theoretical because molecular motion has never been totally stopped. The complete stopping of molecular motion is expressed as absolute zero. This has been calculated to be −460°F. Scientists have actually come within a few degrees of causing substances to reach absolute zero. Figure 1–2 is an illustration of some levels of heat (molecular motion) shown on a thermometer scale.

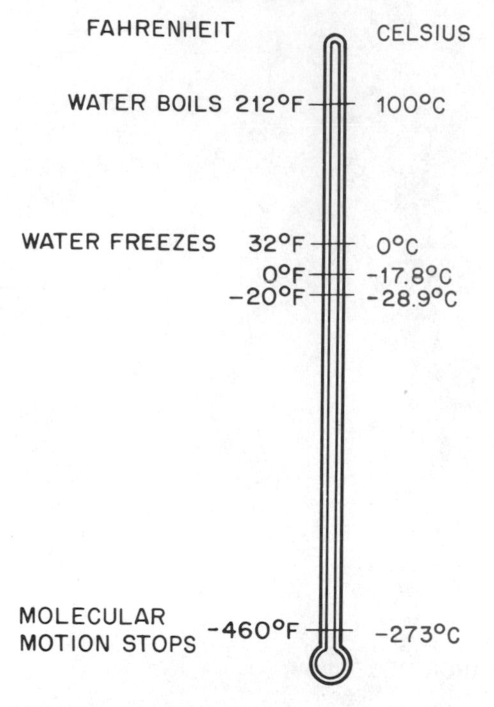

FIGURE 1–2 Fahrenheit scale compared to Celsius scale.

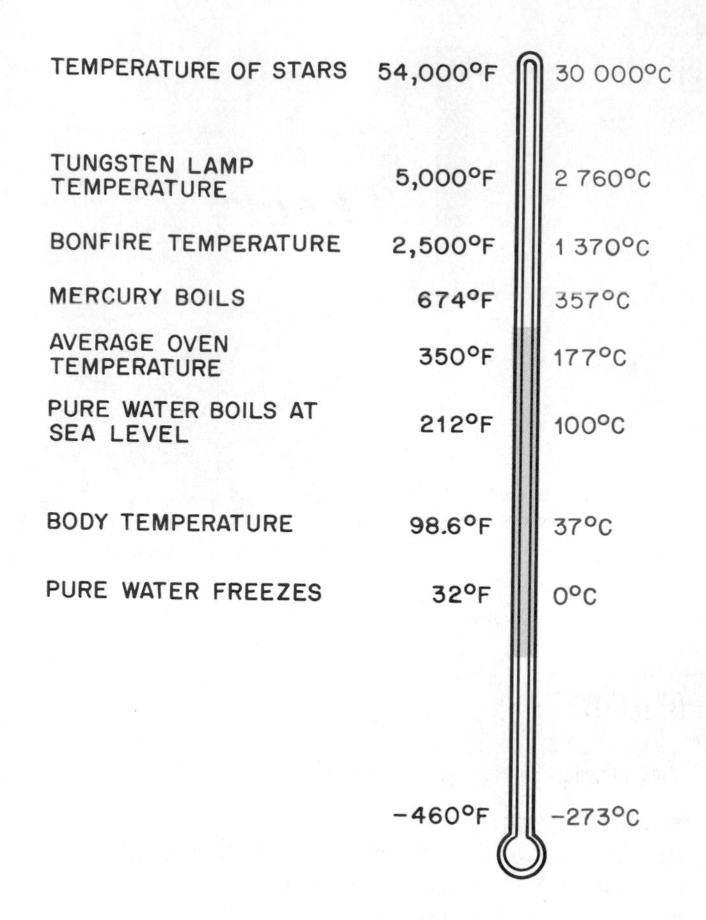

FIGURE 1–3 Civilization is generally exposed to a comparatively small range of temperatures.

Fahrenheit is a system of temperature measurement used in the English measurement system. The United States is one of the few countries in the world that uses this system. Celsius is a term used in the metric measurement system used by most countries.

As the United States develops trade opportunities with the rest of the world, it may become necessary to use the metric system. Figure 1–2 illustrates a thermometer with some important Fahrenheit and Celsius equivalent temperatures. Figure 1–3 illustrates a thermometer showing many more equivalent temperatures. See the Temperature Conversion Table in the appendix for conversion of temperatures. Formulas can also be used to make conversions. Use of the table and formulas is discussed at the end of this unit.

Most of the terms of measurement in this book are in English terms because at this time these are the terms most often used by technicians in this industry in the United States.

Temperature has been expressed in everyday terms up to this point. It is equally important in the air conditioning, heating, and refrigeration industry to describe temperature in terms engineers and scientists use. Performance ratings of equipment are established in terms of *absolute* temperature. Equipment is rated to establish criteria for comparing equipment performance. Using these ratings, different manufacturers can make comparisons with other manufacturers regarding their products. We can use the equipment rating to evaluate these comparisons. The Fahrenheit absolute scale is called the *Rankine* scale (named for its inventor, W. J. M. Rankine), and the Celsius absolute scale is known as the *Kelvin* scale (named for the scientist, Lord Kelvin). Absolute temperature scales begin where molecular motion starts; they use 0 as the starting point. For instance, 0 on the Fahrenheit absolute scale is called absolute zero of 0° Rankine (0°R). Similarly, 0 on the Celsius absolute scale is called absolute zero of 0° Kelvin (0°K), Figure 1–4.

The Fahrenheit/Celsius and the Rankine/Kelvin scales are used interchangeably to describe equipment and fundamentals of this industry. Memorization is not required. A working knowledge of these scales and conversion formulas and a ready reference table are more practical. Figure 1–4 shows how these four scales are related. The world we live in accounts for only a small portion of the total temperature spectrum.

Our earlier statement that temperature describes the level of heat or molecular motion can now be explained. As a substance becomes warmer, its molecular motion, and therefore the temperature, increases, Figure 1–5.

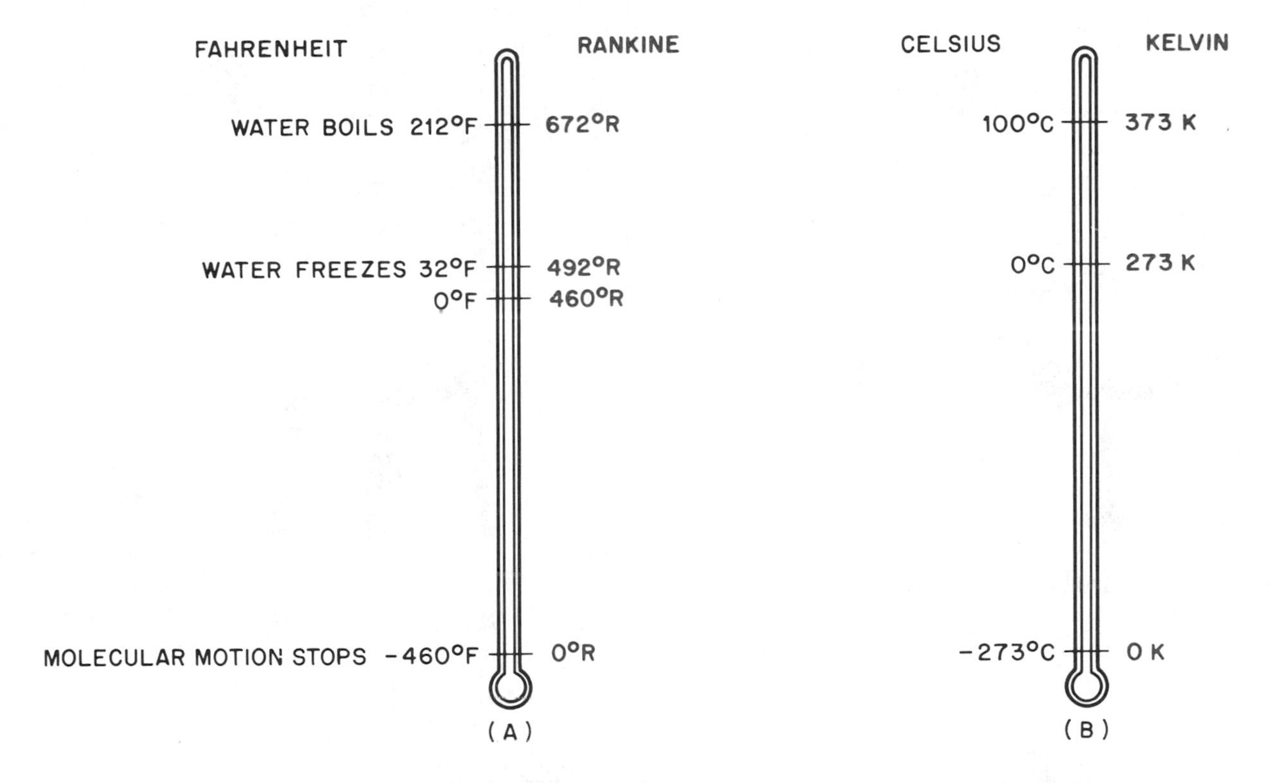

FIGURE 1–4 (A) Fahrenheit and Rankine thermometer. (B) Celsius and Kelvin thermometer.

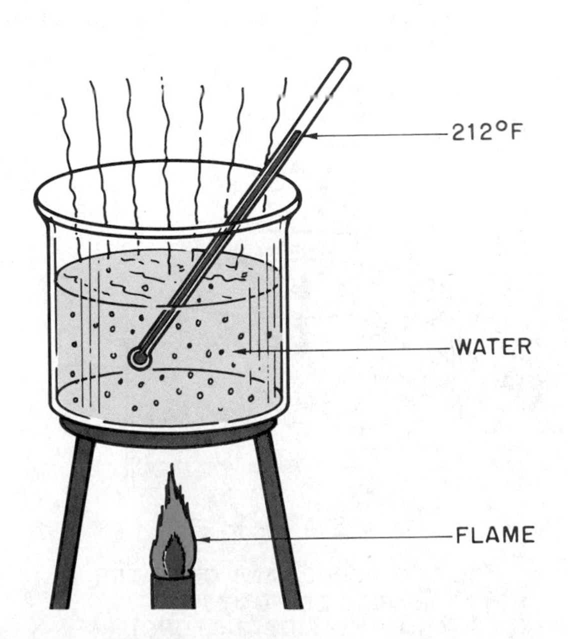

FIGURE 1–5 The water in the pot boils because the molecules move faster when heat is applied.

1.2 INTRODUCTION TO HEAT

The laws of thermodynamics can help us to understand what heat is all about. One of these laws states that heat can neither be created nor destroyed. This means that all of the heat that the world experiences is not created but is merely converted to usable heat from something that is already here. This heat can also be accounted for when it is transferred from one substance to another.

Temperature describes the level of heat with reference to no heat. The term used to describe the quantity of heat is known as the *British thermal unit* (Btu). This term explains how much heat is contained in a substance. The rate of heat consumption can be determined by adding time.

The Btu is defined as the amount of heat required to raise the temperature of 1 lb of water 1°F. For example, when 1 lb of water (about 1 pint) is heated from 68°F to 69°F, 1 Btu of heat energy is absorbed into the water, Figure 1–6. To actually measure how much heat is absorbed in a process like this, we need an instrument of laboratory quality. This instrument is called a *calorimeter.* Notice the similarity to the word "calorie," the food word for energy.

When there is a temperature difference between two substances, heat transfer will occur. Temperature difference is the driving force behind heat transfer. The greater the temperature difference, the greater the heat transfer. Heat flows naturally from a warmer substance to a cooler substance. Rapidly moving molecules in the warmer substance

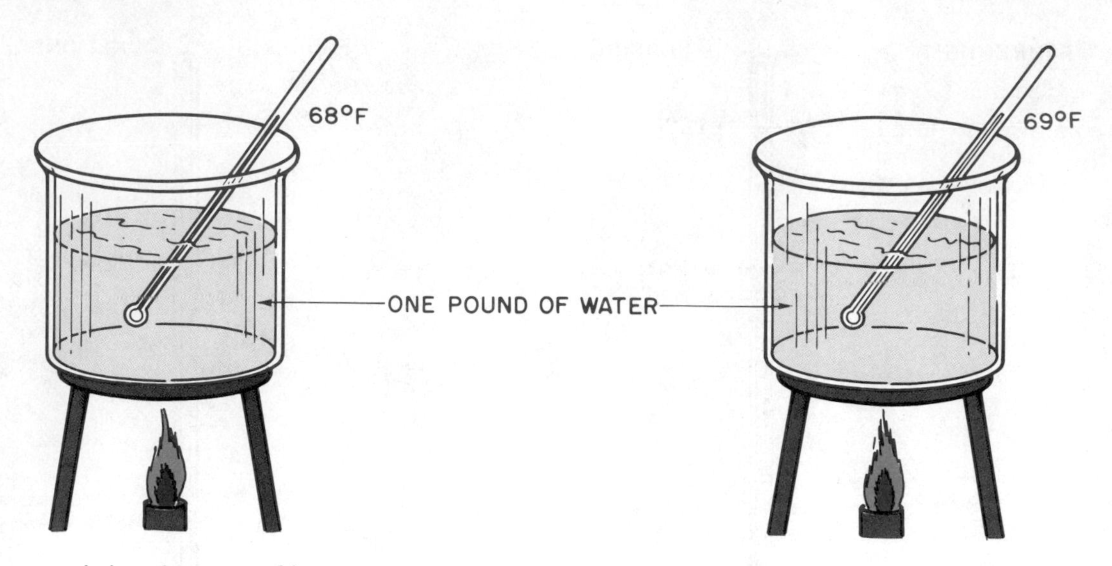

FIGURE 1–6 British thermal unit (Btu) of heat energy is required to heat 1 lb of water from 68°F to 69°F.

give up some of their energy to the slower-moving molecules in the cooler substance. The warmer substance cools because the molecules have slowed. The cooler substance becomes warmer because the molecules are moving faster.

The following example illustrates the difference in the quantity of heat compared to the level of heat. One tank of water weighing 10 lb (slightly more than 1 gallon [gal]) is heated to a temperature level of 200°F. A second tank of water weighing 100,000 lb (slightly more than 12,000 gal) is heated to 175°F. The 10-lb tank will cool to room temperature much faster than the 100,000-lb tank. The temperature difference of 25°F is not much, but the cool-down time is much longer for the 100,000-lb tank due to the quantity of water, Figure 1–7.

A comparison using water may be helpful in showing the level versus the quantity of heat. A 200-ft deep well would not contain nearly as much water as a large lake with a water depth of 25 ft. The depth of water (in feet) tells us the level of water, but it in no way expresses the quantity (gallons) of water.

In practical terms, each piece of heating equipment is rated according to the amount of heat it will produce. If the equipment had no such rating, it would be difficult for a buyer to intelligently choose the correct appliance.

A gas or oil furnace used to heat a home has the rating permanently printed on a nameplate. Either furnace would be rated in Btu per hour, which is a *rate* of energy consumption. Later, this rate will be used to calculate the amount of

ROOM TEMPERATURE (70°F)

10-POUND TANK OF WATER
IS HEATED TO 200°F
(ABOUT 1 GALLON)

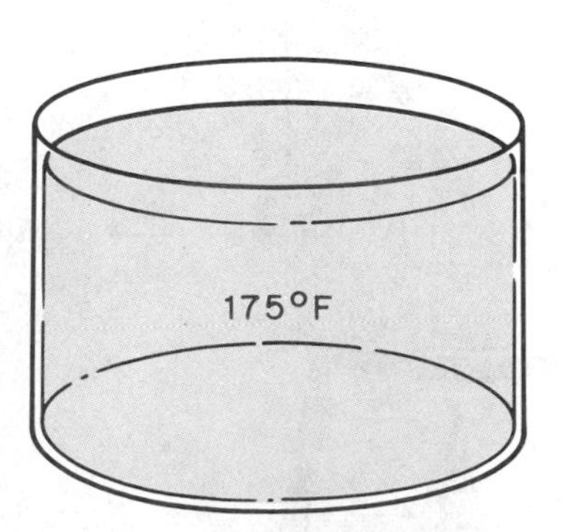

100,000-POUND TANK OF WATER
IS HEATED TO 175°F
(ABOUT 12,000 GALLONS)

FIGURE 1–7 The smaller tank will cool to room temperature first because there is a smaller quantity of heat.

fuel required to heat a house or a structure. For now, it is sufficient to say that if one needs a 75,000-Btu/h furnace to heat a house on the coldest day, a furnace rated at 75,000 Btu/h should be chosen. If not, the house will begin to get cold on any day the temperature falls below the capacity of the furnace.

In the metric or SI (Systems International) system of measurement, the term joule (J) is used to express the quantity of heat. Because a joule is very small, metric units of heat in this industry are usually expressed in kilojoules (kJ) or 1000 joules. One Btu equals 1.055 kJ.

The term gram (g) is used to express weight in the metric system. Again, this is a small quantity of weight so the term kilograms (kg) is often used. One pound equals 0.453 59 kg.

The amount of heat required to raise the temperature of 1 kg of water 1°C is equal to 4.187 kJ.

Cold is a comparative term used to describe lower temperature levels. Because all heat is a positive value in relation to no heat, cold is not a true value. It is really an expression of comparison. When a person says it is cold outside, it is in relation to the normal expected temperature for the time of year or to the inside temperature. Cold has no number value and is used by most people as a basis of comparison only.

FIGURE 1–9 The car fender and the fence post are actually the same temperature, but the fender feels colder because the metal conducts heat away from the hand faster than the wooden fence post.

1.3 CONDUCTION

Conduction heat transfer can be explained as the energy actually traveling from one molecule to another. As a molecule moves faster, it causes others to do the same. For example, if one end of a copper rod is in a fire, the other end gets too hot to handle. The heat travels up the rod from molecule to molecule, Figure 1–8.

Conduction heat transfer is used in many heat transfer applications that are experienced regularly. Heat is transferred by conduction from the hot electric burner on the cookstove to the pan of water. It is then transferred by conduction into the water. Note that there is an orderly explanation for each step.

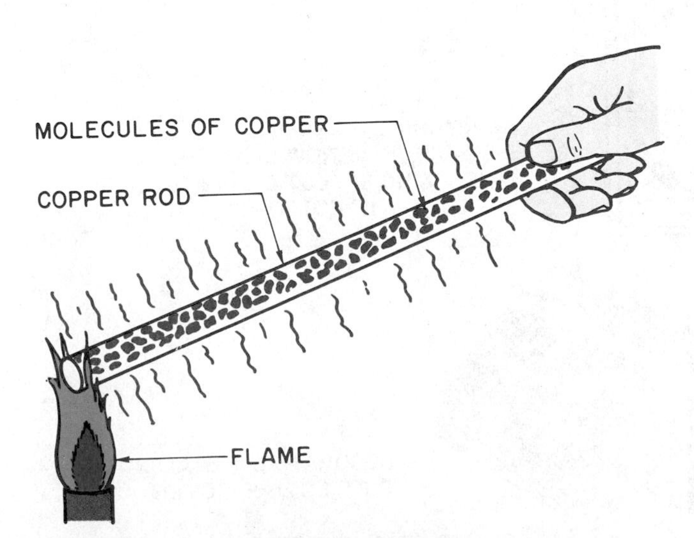

FIGURE 1–8 The copper rod is held in the flame only for a short time before heat is felt at the far end.

Heat does not conduct at the same rate in all materials. Copper, for instance, conducts at a different rate from iron. Glass is a very poor conductor of heat.

Touching a wooden fence post or another piece of wood on a cold morning does not give the same sensation as touching a car fender or another piece of steel. The piece of steel feels much colder. Actually the steel is not colder; it just conducts heat out of the hand faster, Figure 1–9.

The different rates at which various materials conduct heat have an interesting similarity to the conduction of electricity. As a rule, substances that are poor conductors of heat are also poor conductors of electricity. For instance, copper is one of the best conductors of electricity and heat, and glass is one of the poorest conductors of both. Glass is actually used as an insulator of electrical current flow.

1.4 CONVECTION

Convection heat transfer is used to move heat from one location to another. When heat is moved, it is normally transferred into some substance that is readily movable, such as air or water. Many large buildings have a central heating plant where water is heated and pumped throughout the building to the final heated space. Notice the similarity of the words "convection" and "convey" (to carry from one place to another).

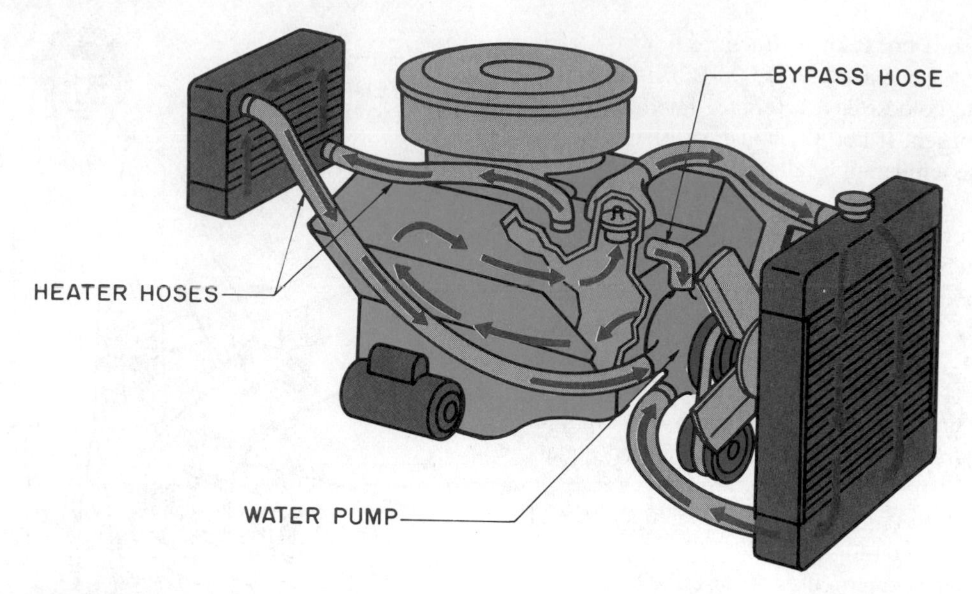

FIGURE 1–10 The engine generates heat that is transferred into the circulating water by conduction. The circulating water is pumped through the heater coil inside the car (forced convection). Air is passed over the coil where it absorbs heat from the coil by conduction. Warm air is then conveyed to the interior by a fan.

The automobile heater is a good example of convection heat. Heat from the engine's combustion process is passed by conduction to the water. Hot water from the engine is then passed through a heater coil. The heat in the water is transferred by conduction from the water in the engine to the heater coil. The heat is transferred through the coil from the water to the air and conveyed to the car's interior by the heater fan. When a fan or pump is used to convey the heat, the process is called *forced* convection, Figure 1–10.

Another example of heat transfer by convection is when air is heated, it rises; this is called *natural* convection. When air is heated, it expands, and the warmer air becomes less dense or lighter than the surrounding unheated air. This principle is applied in many ways in the air conditioning industry. Baseboard heating units are an example. They are normally installed on the outside walls of buildings and use electricity or hot water as the heat source. When the air near the floor is heated, it expands and rises. This heated air is displaced by cooler air around the heater and sets up a natural convection current in the room, Figure 1–11.

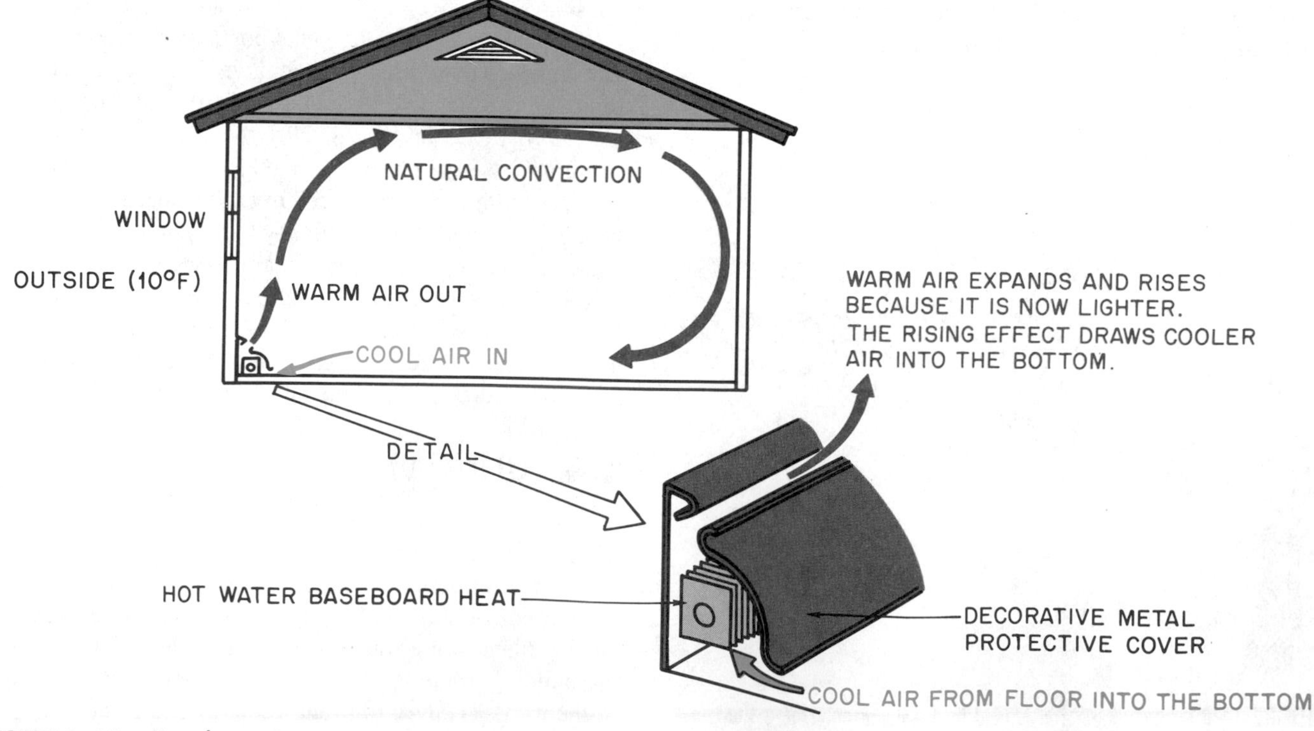

FIGURE 1–11 Natural convection.

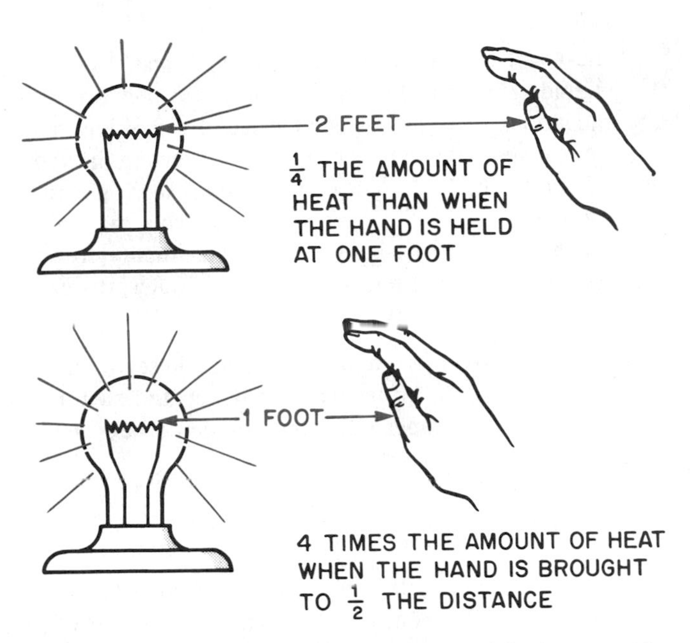

FIGURE 1–12 The intensity of the heat diminishes by the square of the distance.

1.5 RADIATION

Radiation heat transfer can best be explained by using the sun as an example of the source. The sun is approximately 93,000,000 miles from the earth's surface, yet we can feel its intensity. The sun's surface temperature is extremely hot compared to anything on earth. Heat transferred by radiation travels through space without heating the space and is absorbed by the first solid object that it encounters. The earth does not experience the total heat of the sun because heat transferred by radiation diminishes by the square of the distance traveled. In practical terms, this means that every time the distance is doubled, the heat intensity decreases by one fourth. If you hold your hand close to a light bulb, for example, you feel the heat's intensity, but if you move your hand twice the distance away, you feel only one fourth of the heat intensity. Keep in mind that, because of the square-of-the-distance explanation, radiant heat does not transfer the actual temperature value. If it did, the earth would be as hot as the sun, Figure 1–12.

Electric heaters that glow red hot are practical examples of radiant heat. The electric heater coil glows red hot and radiates heat into the room. It does not heat the air, but it warms the solid objects that the heat rays encounter. Any heater that glows has the same effect.

1.6 SENSIBLE HEAT

Heat level can readily be measured when it changes the temperature of a substance (remember the example of changing 1 lb of water from 68°F to 69°F). This change in the heat level can be measured with a thermometer. When a change of temperature can be registered, we know that the level of heat has changed and is called *sensible heat*.

1.7 LATENT HEAT

Another type of heat is called *latent* or *hidden* heat. In this process heat is known to be added but no temperature rise is noticed. An example is heat added to water while it is boiling in an open container. Once water is brought to the boiling point, adding more heat only makes it boil faster; it does not raise the temperature.

The following example describes the sensible heat and latent heat characteristics of water. These are explored from 0°F through the temperature range to above the boiling point. Examine the chart in Figure 1–13 and notice that

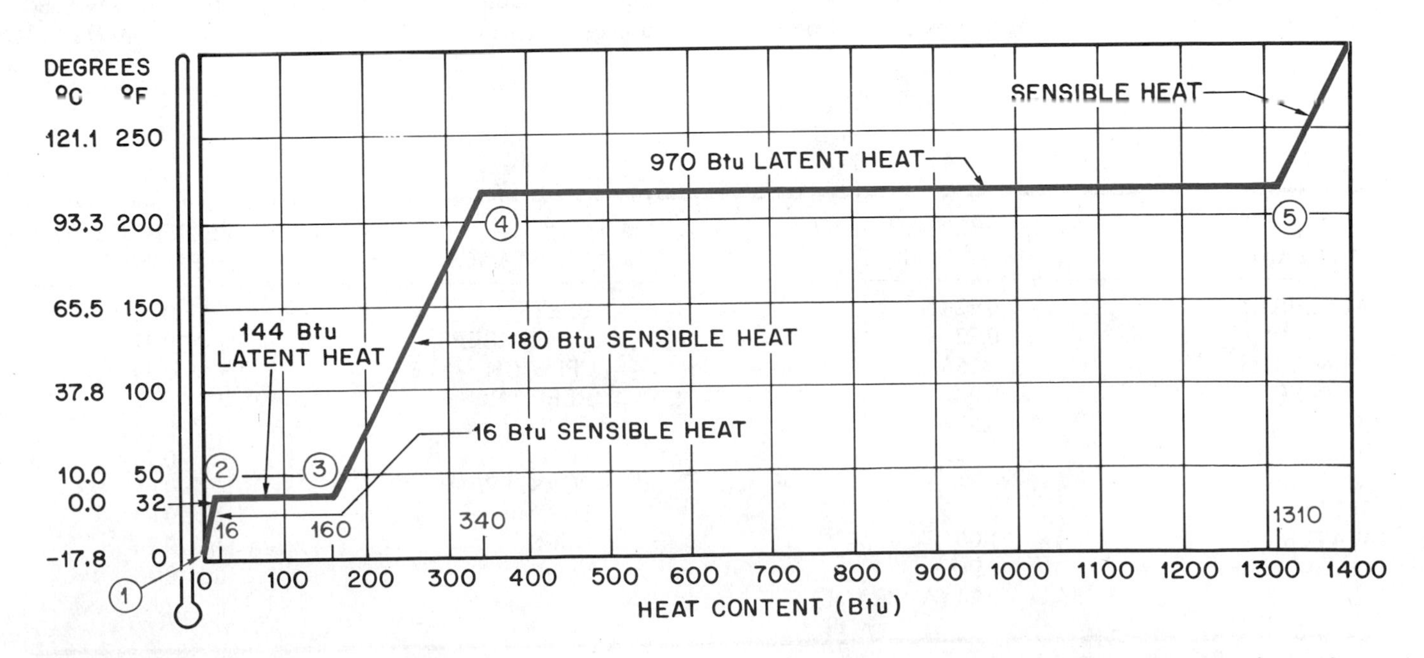

FIGURE 1–13 Heat/temperature graph. An increase in sensible heat causes a rise in temperature. An increase in latent heat causes a change of state.

temperature is plotted on the left margin, and heat content is plotted along the bottom of the chart. We see that as heat is added the temperature will rise except during the latent or hidden-heat process. This chart is interesting because heat can be added without causing a rise in temperature.

The following statements should help you to understand the chart.

1. Water is in the form of ice at point 1 where the example starts. Point 1 is *not* absolute 0. It is 0°F and is used as a point of departure.
2. Heat added from point 1 to point 2 is sensible heat. This is a registered rise in temperature. Note, it only takes 0.5 Btu of heat to raise 1 lb of ice 1°F.
3. When point 2 is reached, the ice is thought of as being saturated with heat. This means that if more heat is added, it will be known as latent heat and will melt the ice but not raise the temperature. Adding 144 Btu of heat will change the 1 lb of ice to 1 lb of water. Removing any heat will cool the ice below 32°F.
4. When point 3 is reached, the substance is now water and is known as a saturated liquid. Adding more heat causes a rise in temperature (this is sensible heat). Removal of any heat at point 3 results in some of the water changing back to ice. This is known as removing latent heat because there is no change in temperature.
5. Heat added from point 3 to point 4 is sensible heat; when point 4 is reached, 180 Btu of heat will have been added: 1 Btu/lb/°F temperature change.
6. Point 4 represents another saturated point. The water is saturated with heat to the point that the removal of any heat causes the liquid to cool off below the boiling point. Heat added is identified as latent heat and causes the water to boil and to start changing to a vapor (steam). Adding 970 Btu makes the 1 lb of liquid boil to point 5 and become a vapor.
7. Point 5 represents another saturated point. The water is now in the vapor state. Heat removed would be latent heat and would change some of the vapor back to a liquid. This is called *condensing the vapor.* Any heat added at point 5 is sensible heat; it raises the vapor temperature above the boiling point. Heating the vapor above the boiling point is called *superheat.* Superheat will be important in future studies. Note that in the vapor state it only takes 0.5 Btu to heat the water vapor (steam) 1°F. The same was true while water was in the ice (solid) state.

SAFETY PRECAUTION: *When examining these principles in practice, be careful because the water and steam are well above body temperature and you could be seriously burned.*

1.8 SPECIFIC HEAT

We now realize that different substances respond differently to heat. When 1 Btu of heat energy is added to 1 lb of water, it changes the temperature 1°F. This only holds true for water. When other substances are heated, different values occur. For instance, we noted that adding 0.5 Btu of heat energy to either ice or steam (water vapor) caused a 1°F rise per pound while in these states. They heated at twice the rate. Adding 1 Btu would cause a 2°F rise. This difference in heat rise is known as *specific heat.*

Specific heat is the amount of heat necessary to raise the temperature of 1 lb of a substance 1°F. Every substance has a different specific heat. Note that the specific heat of water is 1 Btu/lb/°F. See Figure 1–14 for the specific heat of some other substances.

1.9 SIZING HEATING EQUIPMENT

Specific heat is significant because the amount of heat required to change the temperatures of different substances is used to size equipment. Recall the example of the house and furnace earlier in this unit.

SUBSTANCE	SPECIFIC HEAT Btu/lb/°F	SUBSTANCE	SPECIFIC HEAT Btu/lb/°F
ALUMINUM	0.224	BEETS	0.90
BRICK	0.22	CUCUMBERS	0.97
CONCRETE	0.156	SPINACH	0.94
COPPER	0.092	BEEF, FRESH LEAN	0.77
ICE	0.504	FISH	0.76
IRON	0.129	PORK, FRESH	0.68
MARBLE	0.21	SHRIMP	0.83
STEEL	0.116	EGGS	0.76
WATER	1.00	FLOUR	0.38
SEA WATER	0.94		
AIR	0.24 (AVERAGE)		

FIGURE 1–14 Specific heat table.

The following example shows how this would be applied in practice. A manufacturing company may need to buy a piece of heating equipment to heat steel before it can be machined. The steel may be stored outside in the cold at 0°F and need preheating before machining. The temperature desired for the machining is 70°F. How much heat must be added to the steel if the plant wants to machine 1000 lb/h?

The steel is coming into the plant at a fixed rate of 1000 lb/h and that heat has to be added at a steady rate to stay ahead of production. Figure 1–14 gives a specific heat of 0.116 Btu/lb/°F for steel. This means that 0.116 Btu of heat energy must be added to 1 lb of steel to raise its temperature 1°F.

Q = Weight × Specific Heat × Temperature Difference

where Q = quantity of heat needed. Substituting in the formula, we get

$Q = 1000$ lb/h × 0.116 Btu/lb/°F × 70°F

$Q = 8120$ Btu/h required to heat the steel for machining.

The previous example has some known values and a value to be found. The known information is used to find the unknown value with the help of the formula. The formula can be used when adding heat or removing heat and is used often in heat load calculations for sizing both heating and cooling equipment.

1.10 PRESSURE

Pressure is defined as force per unit of area. This is normally expressed in pounds per square inch (psi). Simply stated, when a 1-lb weight rests on an area of 1 square inch (1 in^2), the pressure exerted downward is 1 pound per square inch (psi). Similarly, when a 100-lb weight rests on a 1-in^2 area, 100 psi of pressure is exerted, Figure 1–15.

When you swim below the surface of the water, you feel a pressure pushing inward in your body. This pressure is caused by the weight of the water and is very real. A different sensation is felt when flying in an airplane without a pressurized cabin. Your body is subjected to less pressure instead of more, yet you still feel uncomfortable.

It is easy to understand why the discomfort under water exists. The weight of the water pushes in. In the airplane, the reason is just the reverse. There is less pressure high in the air than down on the ground. The pressure is greater inside your body and is pushing out.

Water weighs 62.4 pounds per cubic foot (lb/ft^3). A cubic foot (7.48 gal) exerts a downward pressure of 62.4 lb/ft^2 when it is in its actual cube shape, Figure 1–16. How much weight is then resting on 1 in^2? The answer is simply calculated. The bottom of the cube has an area of 144 in^2 (12 in. × 12 in.) sharing the weight. Each square inch has a total pressure of 0.433 lb (62.4 ÷ 144) resting on it. Thus, the pressure at the bottom of the cube is 0.433 psi, Figure 1–17.

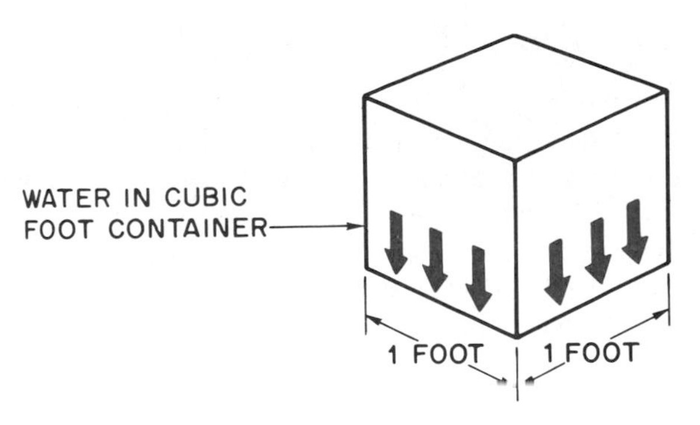

FIGURE 1–16 One cubic foot (1 ft^3) of water (7.48 gal) exerts all of its pressure downward. 1 ft^3 of water weighs 62.4 lb spread over 1 ft^2.

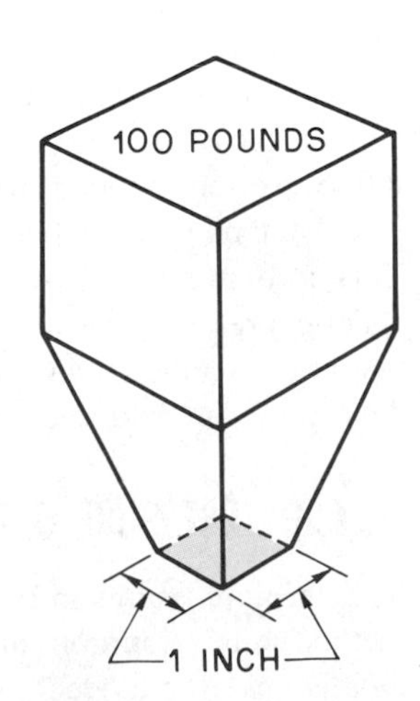

FIGURE 1–15 Both weights are resting on a 1-square inch (in^2) surface. One weight exerts a pressure of 1 psi, the other 100 psi.

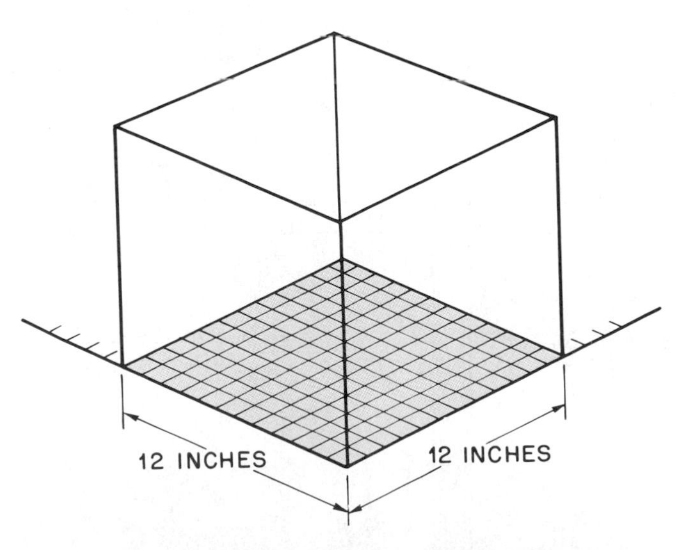

FIGURE 1–17 One cubic foot (1 ft^3) of water exerts a pressure downward of 62.4 lb/ft^2 on the bottom surface area.

1.11 ATMOSPHERIC PRESSURE

The sensation of being underwater and feeling the pressure of the water is familiar to many people. The earth's atmosphere is like an ocean of air that has weight and exerts pressure. The earth's surface can be thought of as being at the bottom of this ocean of air. Different locations are at different depths. For instance, there are sea level locations such as Miami, Florida, or mountainous locations such as Denver, Colorado. The atmospheric pressures at these two locations are different. For now, we will assume that we live at the bottom of this ocean of air.

The atmosphere that we live in has weight just as water does, but not as much. Actually the earth's atmosphere exerts a weight or pressure of 14.696 psi at sea level when the surrounding temperature is 70°F. These are standard conditions.

Atmospheric pressure can be measured with an instrument called a *barometer*. The barometer is a glass tube about 36 in. long that is closed on one end and filled with mercury. It is then inserted open-side down into a puddle of mercury and held upright. The mercury will try to run down into the puddle, but it will not all run out. The atmosphere is pushing down on the puddle, and a vacuum is formed in the top of the tube. The mercury in the tube will fall to 29.92 in. at sea level when the surrounding atmospheric temperature is 70°F, Figure 1–18. This is a standard that is used for comparison in engineering and scientific work. If the barometer is taken up higher, such as on a mountain, the mercury column will start to fall. It will fall about 1 in./1000 ft of altitude. When the barometer is at standard conditions and the mercury drops, it is called a low-pressure system; this means the weather is going to change. Listen closely to the weather report, and the weather forecaster will make these terms more meaningful.

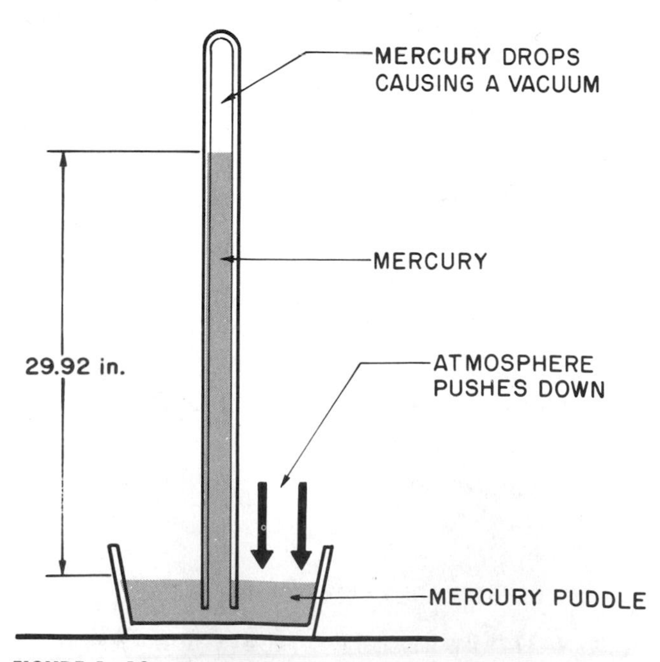

FIGURE 1–18 Mercury (Hg) barometer.

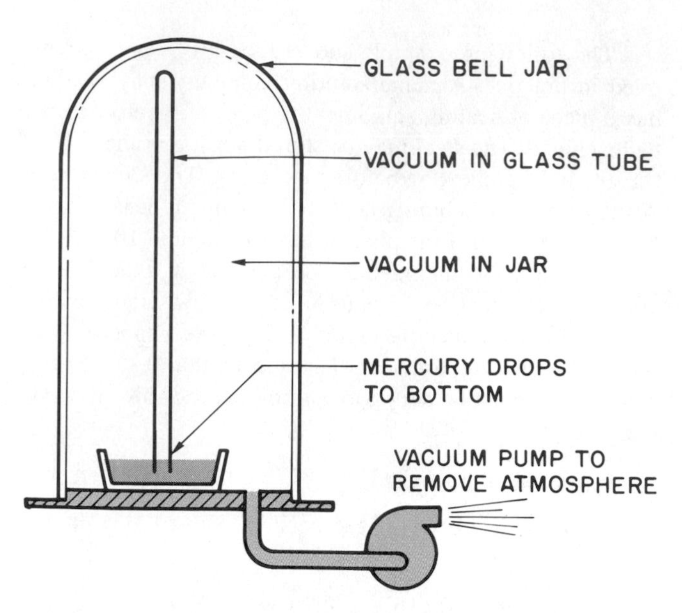

FIGURE 1–19 When the mercury barometer is placed in a closed glass jar (bell jar) and the atmosphere is removed, the pressure at the top of the column is the same as the pressure in the jar; the column of mercury drops to the level of the puddle.

If the barometer is placed inside a closed jar and the atmosphere evacuated, the mercury column falls to a level with the puddle in the bottom, Figure 1–19. When the atmosphere is allowed back into the jar, the mercury again rises because a vacuum exists above the mercury column in the tube.

The mercury in the column has weight and counteracts the atmospheric pressure of 14.696 psi at standard conditions. A pressure of 14.696 psi then is equal to the weight of a column of mercury (Hg) 29.92 in. high. The expression "inches of mercury" thus becomes an expression of pressure and can be converted to pounds per square inch. The conversion factor is 1 psi = 2.036 in. Hg (29.92 ÷ 14.696); 2.036 is often rounded off to 2 (30 in. Hg ÷ 15 psi).

Another type of barometer is the *aneroid* barometer. This is a more practical instrument to transport. Atmospheric pressure has to be measured in many places, so instruments other than the mercury barometer had to be developed for field use, Figure 1–20.

1.12 PRESSURE GAGES

Measuring pressures in a closed system requires a different method—the Bourdon tube, Figure 1–21. The *Bourdon tube* is linked to a needle and can measure pressures above and below atmosphere. A common tool used in the refrigeration industry to take readings in the field or shop is a

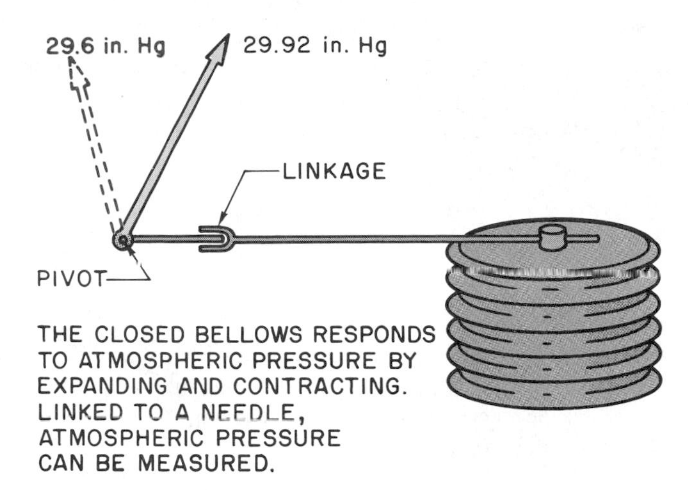

FIGURE 1–20 The aneroid barometer uses a closed bellows that expands and contracts with atmospheric pressure changes.

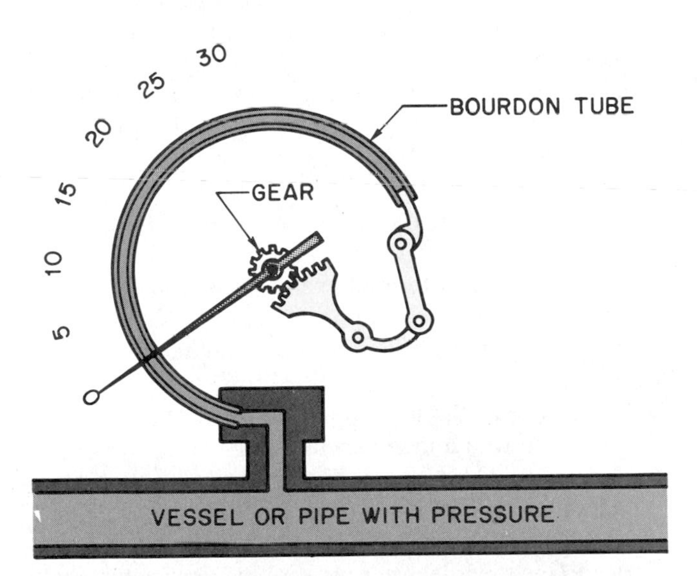

FIGURE 1–21 The Bourdon tube is made of a thin substance such as brass. It is closed on one end and the other end is fastened to the pressure being checked. When an increase in pressure is experienced, the tube tends to straighten out. When attached to a needle linkage, pressure changes are indicated.

combination of a low-pressure gage (called the *low-side gage*) and a high-pressure gage (called the *high-side gage*), Figure 1–22. The gage on the left reads pressures above and below atmospheric pressure. It is called a *compound gage*. The gage on the right will read up to 500 psi and is called the *high-pressure gage*.

These gages read 0 psi when opened to the atmosphere. If they do not, then they should be calibrated to 0 psi. These gages are designed to read pounds-per-square-inch gage pressure (psig). Atmospheric pressure is used as the starting or reference point. If you want to know what the absolute pressure is, you must add the atmospheric pressure to the gage reading. For example, to convert a gage reading of 50 psig to absolute pressure, you must add the atmospheric pressure of 14.696 psi to the gage reading. Let's round off 14.696 to 15 for this example. Then 50 psig + 15 = 65 psia (pounds per square inch absolute), Figure 1–23.

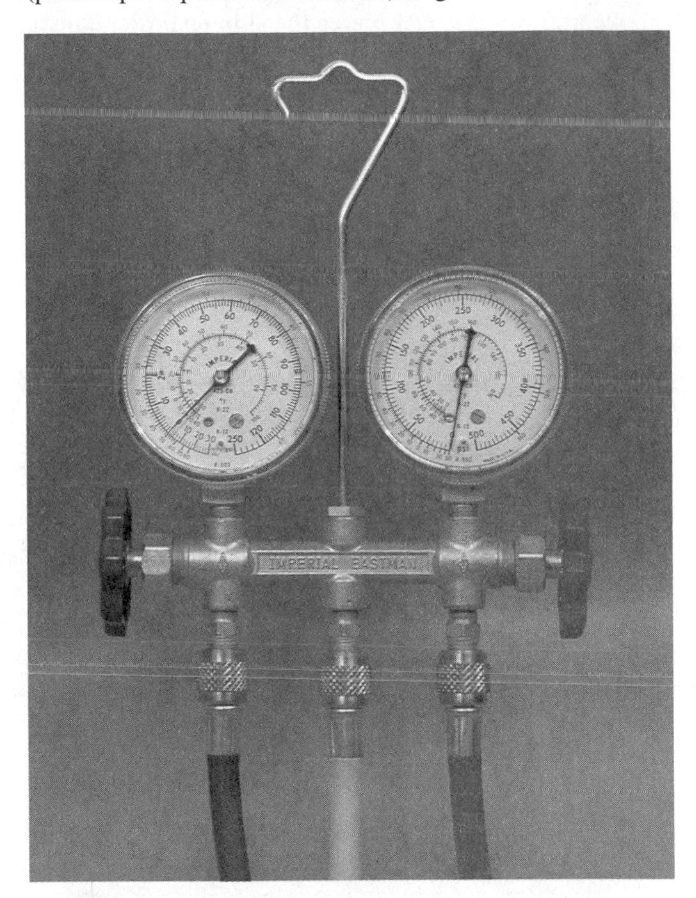

FIGURE 1–22 The gage on the left is called a compound gage because it reads below atmospheric pressure in in. Hg and above atmospheric pressure in psig. The right-hand gage reads high pressure up to 500 psig. *Photo by Bill Johnson*

FIGURE 1–23 This gage reads 50 psig. To convert this gage reading to psia, add the atmospheric pressure, 50 psig + 15 psi atmosphere = 65 psia. *Photo by Bill Johnson*

SAFETY PRECAUTION: *Working with temperatures that are above or below body temperature can cause skin and flesh damage. Proper protection, such as gloves and safety glasses, *must be used.* Pressures that are above or below the atmosphere's can cause bodily injury. A vacuum can cause a blood blister on the skin. Pressures above atmospheric can pierce the skin or inflict damage by flying objects.*

1.13 TEMPERATURE CONVERSION—FAHRENHEIT AND CELSIUS

You may find it necessary to convert specific temperatures from Fahrenheit to Celsius or from Celsius to Fahrenheit. This conversion can be done by using the Temperature Conversion Table in the appendix or by using a formula.

Using the table:

To convert a room temperature of 78°F to degrees Celsius, move down the column labeled Temperature to be Converted until you find 78. Look to the right under the column marked °C and you will find 25.6°C.

To convert 36°C to degrees Fahrenheit, look down the column labeled Temperature to be Converted until you find 36. Look to the left and you will find 96.8°F.

Using formulas:

$$°C = \frac{°F - 32°}{1.8} \qquad °F = 1.8\,(°C) + 32°$$

or or

$$°C = \frac{5}{9}\,(°F - 32°) \qquad °F = \frac{9}{5}\,(°C) + 32°$$

To convert a room temperature of 75°F to degrees C:

$$°C = \frac{75° - 32°}{1.8} = 23.9°$$

$$°C = 23.9°F$$

To convert a room temperature of 25°C to degrees F:

$$°F = 1.8 \times 25° + 32° = 77°$$

$$°F = 77°$$

$$(25°C = 77°F)$$

1.14 PRESSURE MEASURED IN METRIC TERMS

Pressure, like temperature, can be expressed in metric terms. Remember, pressure is an expression of force per unit of area. Several terms have been used to express the measurement of pressure in the past by different countries, but the present standard metric expression for pressure is the term newton per square meter (N/m^2). Pressure in English measurement is expressed in pounds per square inch (psi). It is difficult to compare pounds per square inch and newton per square meter. To make this comparison easier, the newton per square meter has been given the name pascal in honor of the scientist and mathematician Blaise Pascal. The standard metric term for pressure is the kilopascal (kPa) or 1000 pascals. One psi is equal to 6890 pascals, or 6.89 kPa. To convert psi to kPa, simply multiply the number of psi by 6.89.

SUMMARY

- Thermometers measure temperature. Four temperature scales are Fahrenheit, Celsius, Fahrenheit absolute (Rankine), and Celsius absolute (Kelvin).
- Molecules in matter are constantly moving. The higher the temperature, the faster they move.
- The British thermal unit (Btu) describes the quantity of heat in a substance. One Btu is the amount of heat necessary to raise the temperature of 1 lb of water 1°F.
- The transfer of heat by conduction is the transfer of heat from molecule to molecule. As molecules in a substance move faster and with more energy, they cause others near them to do the same.
- The transfer of heat by convection is the actual moving of heat in a fluid (vapor state or liquid state) from one place to another. This can be by natural convection, where heated liquid or air rises naturally, or by forced convection, where liquid or air is moved with a pump or fan.
- Radiant heat is a form of energy transmitted through a medium, such as air, without heating it. Solid objects absorb the energy, become heated, and transfer the heat to the air.
- Difference in the temperature of sensible heat can be measured with a thermometer.
- Latent (or hidden) heat is that heat added to a substance causing a change of state and not registering on a thermometer. For example, heat added to melting ice causes ice to melt but does not increase the temperature.
- Specific heat is the amount of heat (measured in Btu) required to raise the temperature of 1 lb of a substance 1°F. Substances have different specific heats.
- Pressure is the force applied to a specific unit of area. The atmosphere around the earth has weight and therefore exerts pressure. The weight or pressure is greater at sea level (14.696 psi or 29.92 in. Hg at 70°F) than at higher elevations.
- Barometers measure atmospheric pressure in inches of mercury. Two barometers used are the mercury and the aneroid.
- Gages have been developed to measure pressures in enclosed systems. Two common gages used in the air conditioning, heating, and refrigeration industry are the compound gage and the high-pressure gage. The compound gage reads pressures both above and below atmospheric pressure.
- The metric term kilopascal (kPa) is used to express pressure in the refrigeration and air conditioning field.

nt heat?

ssure at sea level under

ssure less when measured

ween a mercury barometer

nches of mercury are equal
t sea level under standard

pressure gage using a Bour-

tween psig and psia.
Celsius.
Fahrenheit.
it is used to express psi?

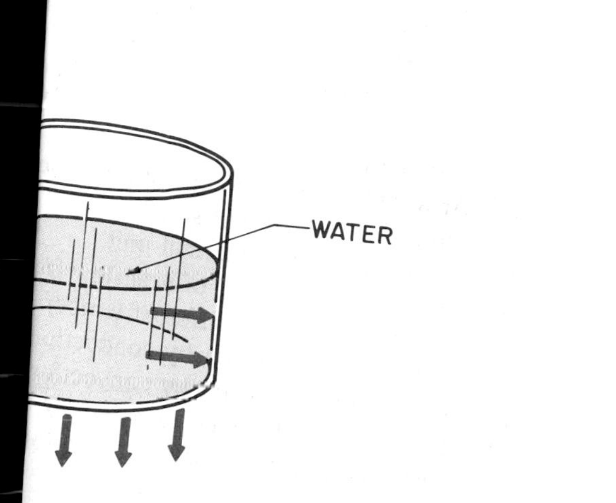

vater in the container exerts pressure outward and
pressure is what makes water seek its own level.
l have a small amount of adhesion to each other.

r is heated above the freezing point, it
a liquid state. The molecular activity is
r molecules have less attraction for each
quid state exerts a pressure outward and
eeks its own level by pushing outward
ure 2–2.

ove the liquid state, 212°F at standard
a vapor. In the vapor state the mole-
attraction for each other and are said to
e vapor exerts pressure more or less in
2–3.

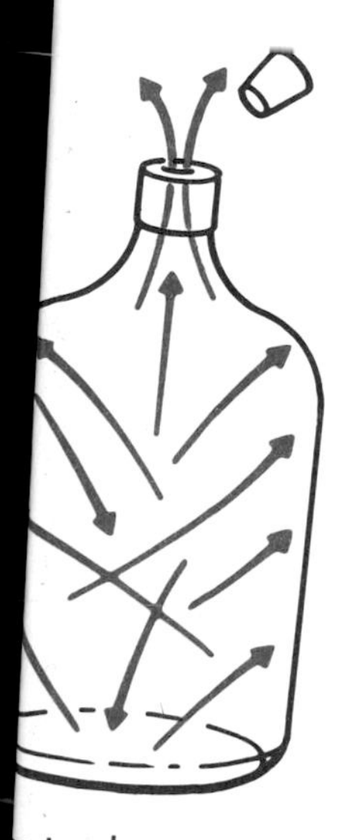

s travel at random. When a container with
e is opened, the gas molecules seem to

UNIT 2

Matter and Energy

OBJECTIVES

After studying this unit, you should be able to

- define matter.
- list the three states in which matter is commonly found.
- define density.
- discuss Boyle's Law.
- state Charles' Law.
- discuss Dalton's Law as it relates to the pressure of different gases.
- define specific gravity and specific volume.
- state two forms of energy important to the air conditioning (heating and cooling) and refrigeration industry.
- describe work and state the formula used to determine the amount of work in a given task.
- define horsepower.
- convert horsepower to watts.
- convert watts to British thermal units.

2.1 MATTER

Matter is commonly explained as a substance that occupies space and has weight. The weight comes from the earth's gravitational pull. Matter is made up of atoms. Atoms are very small parts of a substance and may combine to form molecules. Atoms of one substance may be combined chemically with those of another to form a new substance. When molecules are formed they cannot be broken down any further without changing the chemical content of the substance. Matter also exists in three states: *solids, liquids,* and *gases*. The heat content and pressure determine the state of matter. For instance, water is made up of molecules containing atoms of hydrogen and oxygen. There are two atoms of hydrogen and one atom of oxygen in each molecule of water. The chemical expression of these molecules is H_2O.

Water in the solid state is known as ice. It exerts all of its force downward—it has weight. The molecules of the water are highly attracted to each other, Figure 2–1.

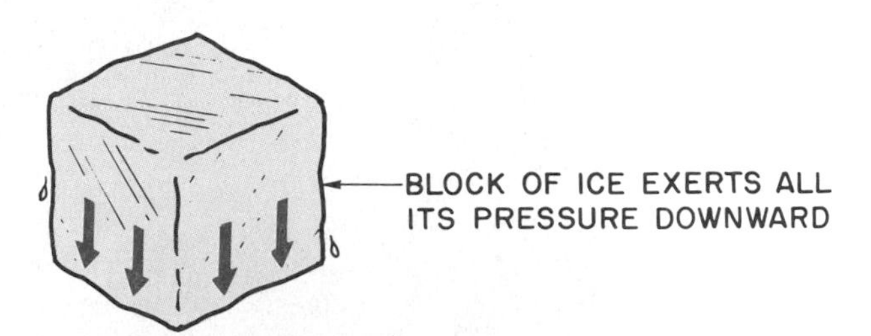
BLOCK OF ICE EXERTS ALL
ITS PRESSURE DOWNWARD

FIGURE 2–1 Solids exert all their pressure downward. The molecules of solid water have a great attraction for each other and hold together.

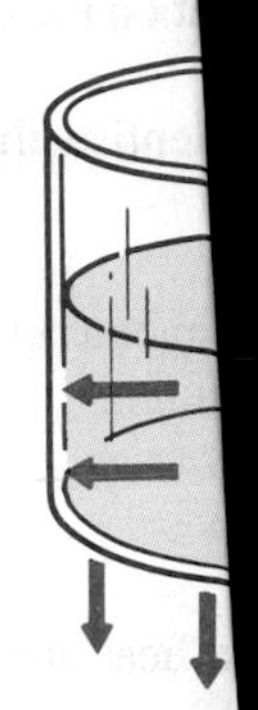

FIGURE 2–2 The w
downward. The outwar
The water molecules sti

When the wat
begins to change to
higher, and the wat
other. Water in the l
downward. It now
and downward, Fig

Water heated a
conditions, becomes
cules have even less
travel at random. Th
all directions, Figure

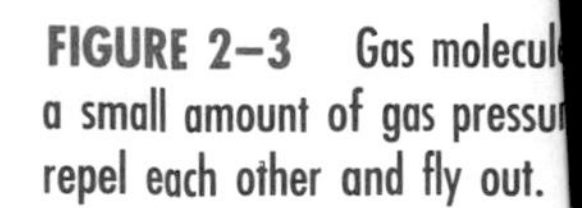

FIGURE 2–3 Gas molecul
a small amount of gas pressu
repel each other and fly out.

The study of matter leads to the study of other terms that help to understand how different substances compare to each other.

2.2 MASS

Mass is a term that is used along with weight to describe matter. The universe is made up of matter that has weight or mass and takes up space. All solid matter has mass. A liquid such as water is said to have mass. The air in the atmosphere has weight or mass. When the atmosphere is evacuated out of a jar, the mass is removed and a vacuum is created.

2.3 DENSITY

The *density* of a substance describes its mass-to-volume relationship. The mass contained in a particular volume is the density of that substance. In the British system of units, volume is measured in cubic feet. Sometimes it is advantageous to compare different substances according to weight per unit volume. Water, for example, has a density of 62.4 lb/ft^3. Wood floats on water because the density (weight per volume) of wood is less than the density of water. In other words, it weighs less per cubic foot. Iron, on the other hand, sinks because it is denser than water. Figure 2–4 lists some typical densities.

2.4 SPECIFIC GRAVITY

Specific gravity compares the densities of various substances. The specific gravity of water is 1. The specific gravity of iron is 7.86. This means that a volume of iron is 7.86 times heavier than an equal volume of water. Since water weighs 62.4 lb/ft^3, a cubic foot of iron would weigh 62.4×7.86 or 490 lb. Aluminum has a specific gravity of 2.7, so it has a density or weight per cubic foot of $2.7 \times 62.4 = 168$ lb.

SUBSTANCE	DENSITY lb/ft^3	SPECIFIC GRAVITY
ALUMINUM	168	2.7
BRASS	536	8.7
COPPER	555	8.92
GOLD	1204	19.3
ICE	57	0.92
IRON	490	7.86
LITHIUM	33	0.53
TUNGSTEN	1186	19.0
MERCURY	845	13.54
WATER	62.4	1

FIGURE 2–4 Table of density and specific gravity.

2.5 SPECIFIC VOLUME

Specific volume compares the densities of gases. It indicates the space (volume) a weight of gas will occupy. One pound of clean dry air has a volume of 13.33 ft^3 at standard conditions. Hydrogen has a density of 179 ft^3/lb under the same conditions. Because there are more cubic feet of hydrogen per pound, it is lighter than air. Although both are gases, the hydrogen has a tendency to rise when mixed with air.

Natural gas is explosive when mixed with air, but it is lighter than air and has a tendency to rise like hydrogen. Propane gas is another frequently used heating gas and has to be treated differently from natural gas because it is heavier than air. Propane has a tendency to fall and collect in low places and to cause potential danger from ignition.

The specific volumes of various gases that are pumped is valuable information that enables the engineer to choose the size of the compressor or vapor pump to do a particular job. The specific volumes for vapors vary according to the pressure the vapor is under. An example is refrigerant-22, which is a common refrigerant used in residential air conditioning units. At 3 psig about 2.5 ft^3 of gas must be pumped to move 1 lb of gas. At the standard design condition of 70 psig, only 0.48 ft^3 of gas needs to be pumped to move a pound of the same gas. A complete breakdown of specific volume can be found in the properties of liquid and saturated vapor tables in engineering manuals for any refrigerant.

2.6 GAS LAWS

It is necessary to have a working knowledge of gases and how they respond to pressure and temperature changes. Several scientists made significant discoveries many years ago. A simple explanation of some of the gas laws developed by these scientists may help you understand the reaction of gases and the pressure/ temperature relationships in various parts of a refrigeration system.

Boyle's Law

Robert Boyle, a citizen of Ireland, developed in the early 1600s what has come to be known as Boyle's Law. He discovered that when pressure is applied to a volume of air that is contained, the volume of air becomes smaller and the pressure greater. Boyle's Law states that *the volume of a gas varies inversely with the absolute pressure, provided the temperature remains constant.* For example, if a cylinder with a piston at the bottom and enclosed at the top was filled with air and the piston moved halfway up the cylinder, the pressure of the air would double, Figure 2–5. That part of the law pertaining to the temperature remaining constant keeps Boyle's Law from being used in practical situations. This is because when a gas is compressed some heat is transferred to the gas from the mechanical compression, and when gas is expanded heat is given up. However, this law, when combined with another, will make it practical to use.

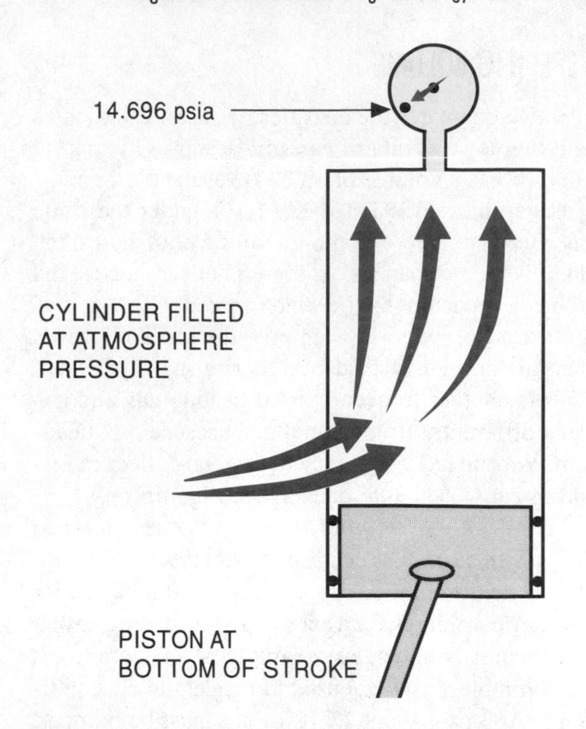

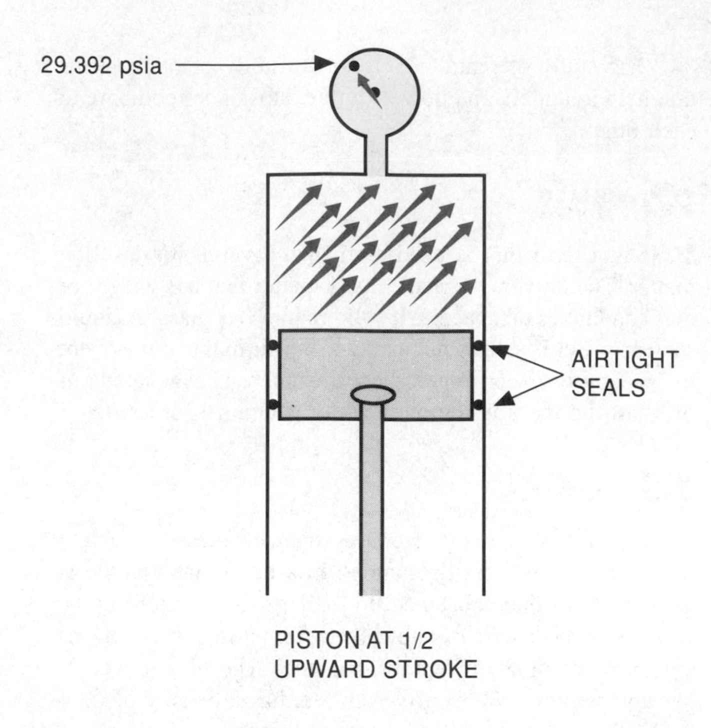

FIGURE 2–5 **Absolute pressure in a cylinder doubles when the volume is reduced by 1/2.**

The formula for Boyle's Law is:

$$P_1 \times V_1 = P_2 \times V_2$$

where P_1 = original absolute pressure
V_1 = original volume
P_2 = new pressure
V_2 = new volume

For example, if the original pressure was 40 psia and the original volume 30 in³, what would the new volume be if the pressure were increased to 50 psia? We are determining the new volume. The formula would have to be rearranged so we could find the new volume.

$$V_2 = \frac{P_1 \times V_1}{P_2}$$

$$V_2 = \frac{40 \times 30}{50}$$

$$V_2 = 24 \text{ in}^3$$

Charles' Law

In the 1800s, a French scientist named Jacques Charles made discoveries regarding the effect of temperature on gases. Charles' Law states that *at a constant pressure, the volume of a gas varies as to the absolute temperature and at a constant volume, the pressure of a gas varies directly with the absolute temperature*. Stated in a different form, when a gas is heated and if it is free to expand, it will do so and the volume will vary directly as to the absolute temperature. If a gas is confined in a container that will not expand and it is heated, the pressure will vary with the absolute temperature.

This law can also be stated with formulae. There are two of them because one part of the law pertains to pressure and temperature and the other part to volume and temperature.

This formula pertains to volume and temperature:

$$\frac{V_1}{T_1} = \frac{V_2}{T_2}$$

where V_1 = original volume
V_2 = new volume
T_1 = original temperature
T_2 = new temperature

If 2000 ft³ of air is passed through a gas furnace and heated from 75°F room temperature to 130°F, what is the volume of the air leaving the heating unit? See Figure 2–6.

V_1 = 2000 ft³
T_1 = 75°F + 460° = 535°R (absolute)
V_2 = unknown
T_2 = 130°F + 460° = 590°R

We must mathematically rearrange the formula so that the unknown is alone on one side of the equation.

$$V_2 = \frac{V_1 \times T_2}{T_1}$$

$$V_2 = \frac{2000 \text{ ft}^3 \times 590°\text{R}}{535°\text{R}}$$

$$V_2 = 2205.6 \text{ ft}^3$$

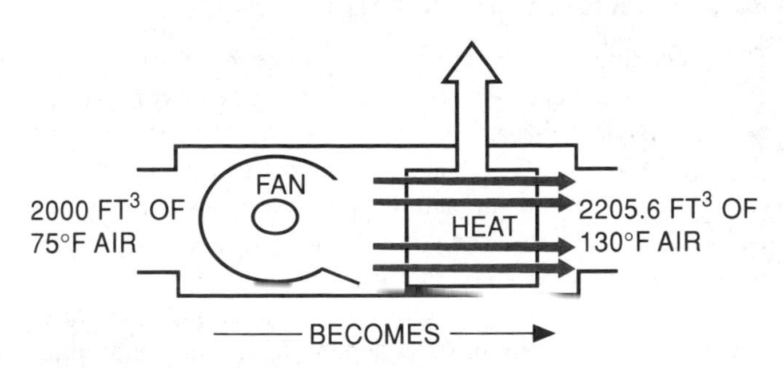

FIGURE 2–6 Air expands when heated.

The air expanded when heated.

The following formula pertains to pressure and temperature:

$$\frac{P_1}{T_1} = \frac{P_2}{T_2}$$

where P_1 = original pressure
T_1 = original temperature
P_2 = new pressure
T_2 = new temperature

If a large natural gas tank holding 500,000 ft³ of gas is stored at 70°F in the spring and the temperature rises to 95°F in the summer, what would the pressure be if the original pressure was 25 psig in the spring?

P_1 = 25 psig + 14.696 (atmospheric pressure) or 39.696 psia
T_1 = 70°F + 460°R or 530°R (absolute)
P_2 = unknown
T_2 = 95°F + 460° or 555°R (absolute)

Again the formula must be rearranged so that the unknown is on one side of the equation by itself.

$$P_2 = \frac{P_1 \times T_2}{T_1}$$

$$P_2 = \frac{39.696 \text{ psia} \times 555°\text{R}}{530°\text{R}}$$

$$P_2 = 41.57 \text{ psia} - 14.696 = 26.87 \text{ psig}$$

General Law of Perfect Gas

A general gas law, often called the General Law of Perfect Gas, is a combination of Boyle's and Charles' laws. This combination law is more practical because it includes temperature, pressure, and volume.

The formula for this law can be stated as follows:

$$\frac{P_1 \times V_1}{T_1} = \frac{P_2 \times V_2}{T_2}$$

where P_1 = original pressure
V_1 = original volume
T_1 = original temperature
P_2 = new pressure
V_2 = new volume
T_2 = new temperature

For example, 20 ft³ of gas is being stored in a container at 100°F and a pressure of 50 psig. This container is connected by pipe to one that will hold 30 ft³ (a total of 50 ft³) and the gas is allowed to equalize between the two containers. The temperature of the gas is lowered to 80°F. What is the pressure in the combined containers?

P_1 = 50 psig + 14.696 or 64.696
V_1 = 20 ft³
T_1 = 100° + 560°R
P_2 = unknown
V_2 = 50 ft³
T_2 = 80°F + 460° or 540°R

The formula is mathematically rearranged to solve for the unknown P_2:

$$P_2 = \frac{P_1 \times V_1 \times T_2}{T_1 \times V_2}$$

$$P_2 = \frac{64.696 \times 20 \times 540}{560 \times 50}$$

$$P_2 = 24.95 - 14.696 = 10.26 \text{ psig}$$

Dalton's Law

In the early 1800s, John Dalton, an English mathematics professor, made the discovery that the atmosphere is made up of several different gases. He found that each gas created its own pressure and that the total pressure was the sum of each. Dalton's Law states that *the total pressure of a confined mixture of gases is the sum of the pressures of each of the gases in the mixture.* For example, when nitrogen and oxygen are placed in a closed container, the pressure on the container will be the total pressure of the nitrogen as if it were in the container by itself added to the oxygen pressure in the container by itself.

2.7 ENERGY

Energy is important because using energy properly to operate equipment is a major goal of the air conditioning and refrigeration industry. Energy in the form of electricity drives the motors; heat energy from the fossil fuels of natural gas, oil, and coal heats homes and industry. What is this energy and how is it used?

The only new energy we get is from the sun heating the earth. Most of the energy we use is converted to usable heat from something already here (eg, fossil fuels). This conversion from fuel to heat can be direct or indirect. An example of direct conversion is a gas furnace, which converts the gas flame to usable heat by combustion. The gas is burned in a combustion chamber, and the heat from combustion is trans-

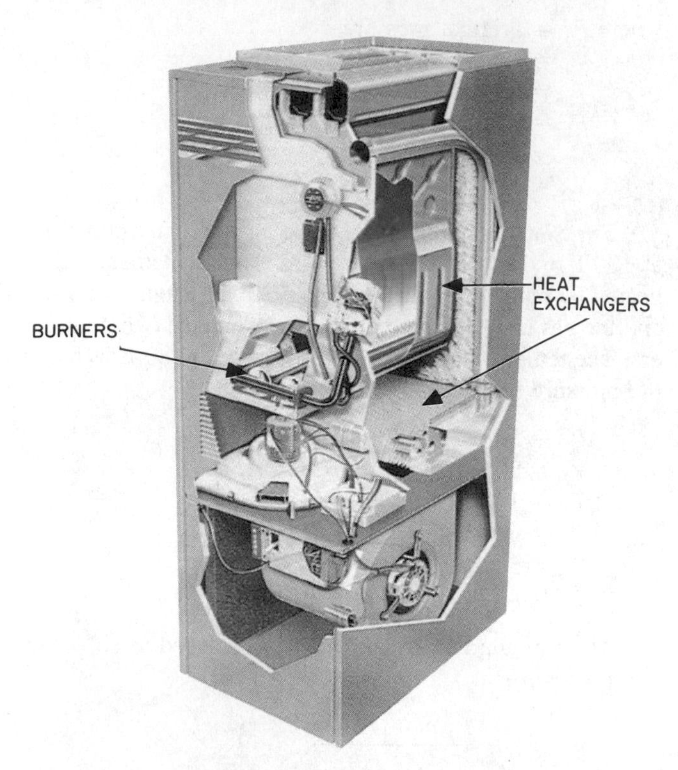

FIGURE 2–7 Cutaway of a high-efficiency condensing gas furnace. *Courtesy Heil-Quaker Corporation*

ferred to circulated air by conduction through the heat exchanger wall of thin steel. The heated air is then distributed throughout the heated space, Figure 2–7.

An example of indirect conversion is a fossil-fuel power plant. Gas may be used in the power plant to produce the steam that turns a steam turbine generator to produce electricity. The electricity is then distributed by the local power company and consumed locally as electric heat, Figure 2–8.

2.8 CONSERVATION OF ENERGY

The preceding leads to the law of conservation of energy. The law states that *energy is neither created or destroyed.* It can then be said that energy can be accounted for.

We have already indicated that most of the energy we use is a result of the sun supporting plant growth for thousands of years. Fossil fuels come from decayed vegetable and animal matter covered by earth and rock during changes in the earth's surface. This decayed matter is in various states, such as gas, oil, or coal, depending on the conditions it was subjected to in the past, Figure 2–9.

2.9 ENERGY CONTAINED IN HEAT

We indicated previously that temperature is a measure of the level of heat and that heat is a form of energy because of the motion of molecules. Because molecular motion does not stop until −460°F, energy is still available in a substance even at very low temperatures. This energy is in relationship to other substances that are at lower temperatures. For example, if two substances at very low temperatures are moved close together, heat will transfer from the warmer substance to the colder one. In Figure 2–10, a substance at −200°F is placed next to a substance at −350°F. As we discussed earlier, the warmer substance gives up heat (energy) to the cooler substance. The energy used by home and industry is not at these low levels.

Electric heat is different from gas heat. Electrons flow through a special wire in the heater section that converts electrical energy to heat. A moving air stream is then passed over the heated wire, allowing heat to be transferred to the air by conduction and moved to the heated space by forced convection (the fan).

2.10 ENERGY IN MAGNETISM

Magnetism is another method of converting electron flow to usable energy. Electron flow is used to develop magnetism

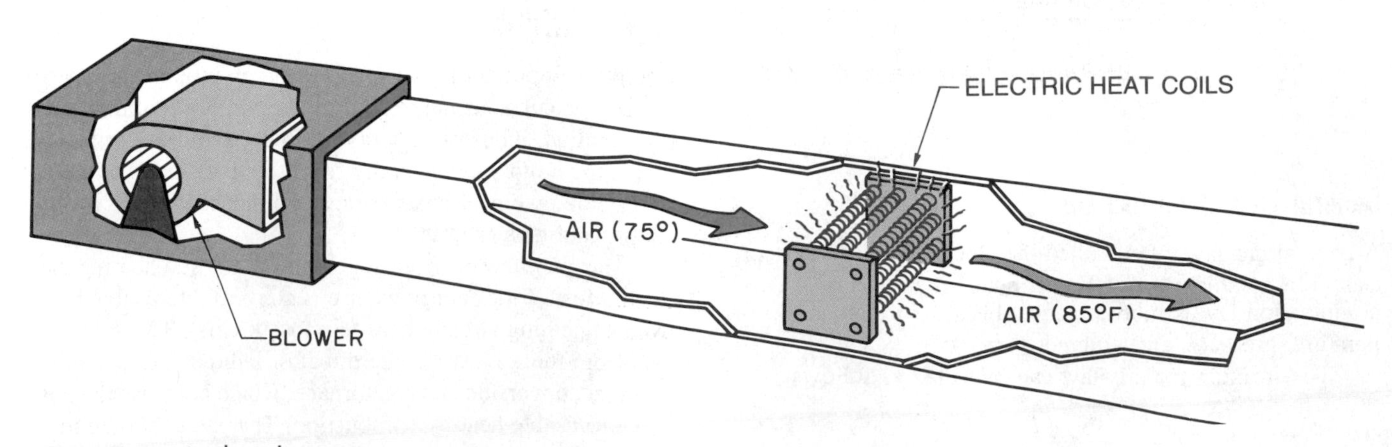

FIGURE 2–8 An electric heat airstream.

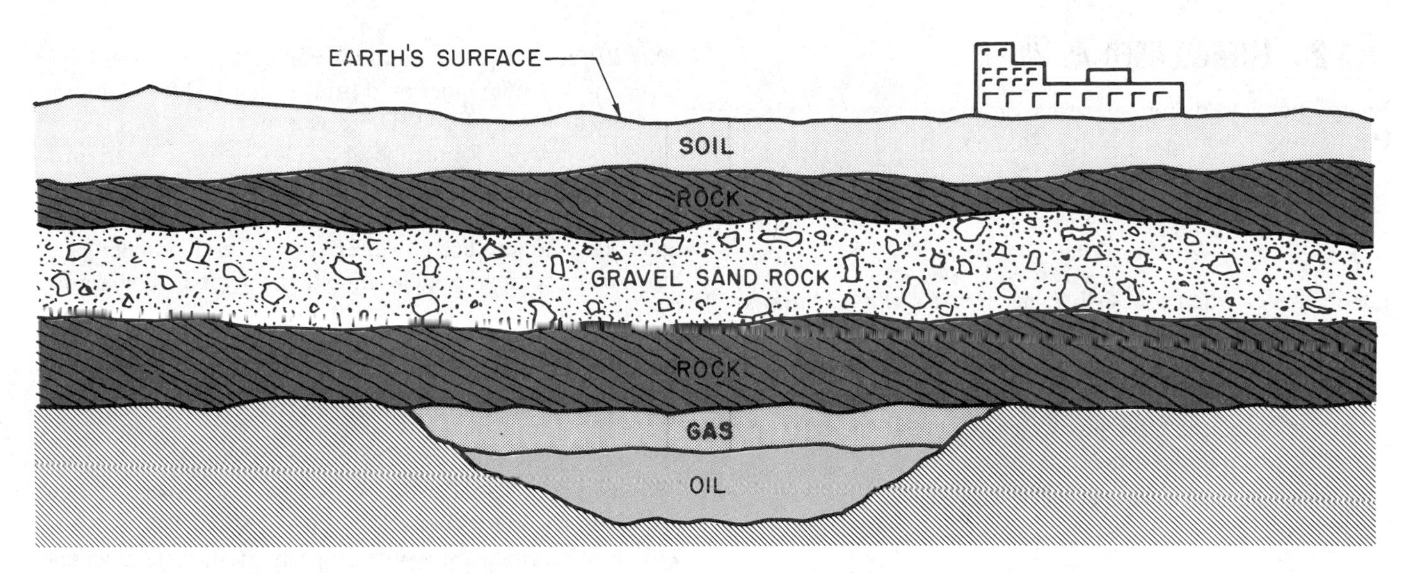

FIGURE 2–9 Gas and oil deposits settle into depressions.

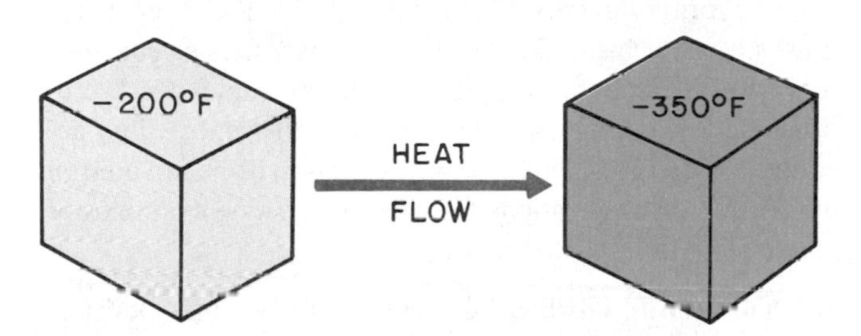

FIGURE 2–10 Heat energy is still available at these low temperatures and will still transfer from the warmer to the colder substance.

to turn motors. The motors turn the prime movers—fans, pumps, compressors—of air, water, and refrigerant. In Figure 2–11 an electric motor turns a water pump to boost the water pressure from 20 to 60 psig. This takes energy. The energy in this example is purchased from the power company.

The preceding examples serve only as an introduction to the concepts of gas heat, electric heat, and magnetism. Each subject will later be covered in detail. For now it is important to realize that any system furnishing heating or cooling uses energy.

2.11 PURCHASE OF ENERGY

Energy must be transferred from one owner to another and accounted for. This energy is purchased as a fossil fuel or as electric power. Energy purchased as a fossil fuel is normally purchased by the unit. Natural gas is an example. Natural gas flows through a meter that measures how many cubic feet have passed during some time span, such as a month. Fuel oil is normally sold by the gallon, coal by the ton. Electrical energy is sold by the kilowatt. The amount of heat each of these units contains is known, so a known amount of heat is purchased. Natural gas, for instance, has a heat content of about 1000 Btu/ft^3, whereas the heat content of coal varies from one type of coal to another.

FIGURE 2–11 Magnetism used in an electric motor is converted to work to boost water pressure to force circulation.

2.12 ENERGY USED AS WORK

Energy purchased from electrical utilities is known as electric power. *Power* is the rate of doing work. *Work* can be explained as a force moving an object in the direction of the force. It is expressed by this formula.

$$\text{Work} = \text{Force} \times \text{Distance}$$

For instance, when a 150-lb man climbs a flight of stairs 100 ft high (about the height of a 10-story building), he performs work. But how much? The amount of work in this example is equivalent to the amount of work necessary to lift this man the same height. We can calculate the work by using the preceding formula.

$$\begin{aligned}\text{Work} &= 150 \text{ lb} \times 100 \text{ ft} \\ &= 15{,}000 \text{ ft-lb}\end{aligned}$$

Notice that no time limit has been added. This example can be accomplished by a healthy man in a few minutes. But if the task were to be accomplished by a machine such as an elevator, more information is necessary. Do we want to take seconds, minutes, or hours to do the job? The faster the job is accomplished, the more power is required.

2.13 HORSEPOWER

Power is the rate of doing work. An expression of power is *horsepower* (hp). Many years ago it was determined that an average good horse could lift the equivalent of 33,000 lb a height of 1 ft in 1 min, which is the same as 33,000 ft-lb/min or 1 hp. This describes a rate of doing work because time has been added. Keep in mind that lifting 330 lb a height of 100 ft in 1 min or 660 lb 50 ft in 1 min is the same amount of work. As a point of reference, the fan motor in the average furnace can be rated at 1/2 hp. See Figure 2–12 for an illustration of the horse lifting 1 hp.

When the horsepower is compared to the man climbing the stairs, the man would have to climb the 100 ft in less than 30 sec to equal 1 hp. That makes the task seen even harder. A 1/2-hp motor could lift the man 100 ft in 1 min if only the man were lifted. The reason is that 15,000 ft-lb of work is required. (Remember that 33,000 ft-lb of work in 1 min equals 1 hp.)

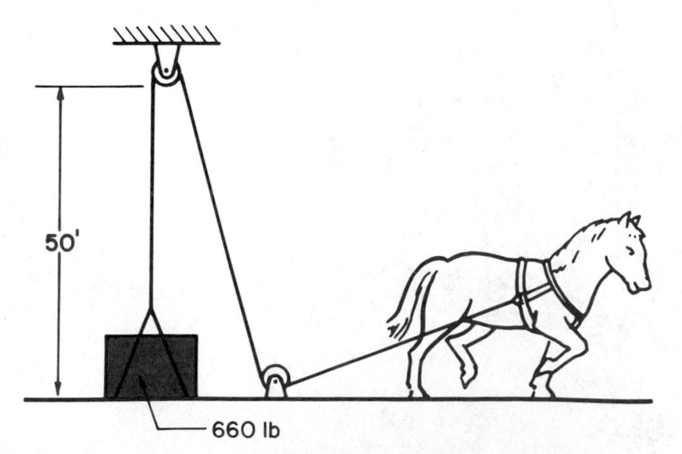

FIGURE 2–12 When a horse can lift the 660 lb a height of 50 ft in 1 min, it has done the equivalent of 33,000 ft-lb of work in 1 min, or 1 hp.

Our purpose in discussing these topics is to help you understand how to use power effectively and to understand how power companies determine their methods of charging for power.

2.14 ELECTRICAL POWER—THE WATT

The unit of measurement for electrical power is the *watt* (W). This is the unit used by the power company. When converted to electrical energy, 1 hp = 746 W; that is, when 746 W of electrical energy is properly used, the equivalent of 1 hp of work has been accomplished.

Fossil-fuel energy can be compared to electrical energy and one form of energy can be converted to the other. There must be some basis, however, for conversion so that one fuel can be compared to another. The examples we use to illustrate this comparison will not take efficiencies into account. Efficiencies for the various fuels will be covered in the section on applications for each fuel. Some examples of conversions follow.

1. **Converting electric heat rated in kilowatts (kW) to the equivalent gas or oil heat rated in Btu.** Suppose we want to know the capacity in Btu for a 20-kW electric heater (a kilowatt is 1000 watts). 1 kW = 3413 Btu.

   $$20 \text{ kW} \times 3413 \text{ Btu/kW} = 68{,}260 \text{ Btu of heat energy}$$

2. **Converting Btu to kW.** Suppose a gas or oil furnace has an output capacity of 100,000 Btu/h. Since 3413 Btu = 1 kW, we have

   $$100{,}000 \text{ Btu} \div 3413 = 29.3 \text{ kW}$$

   In other words, a 29.3-kW electric heat system would be required to replace the 100,000-Btu/h furnace.

Contact the local utility company for rate comparisons between different fuels.

SAFETY PRECAUTION: *Any device that consumes power, such as an electric motor or gas furnace, is potentially dangerous. These devices should only be handled or adjusted by experienced people.*

SUMMARY

- Matter takes up space, has weight, and can be in the form of a solid, a liquid, or a gas.
- The mass of a substance is its weight.
- In the British system of units, density is the weight of a substance per cubic foot.
- Specific gravity is the term used to compare the density of various substances.

- Specific volume is the amount of space a pound of a vapor or a gas will occupy.
- Boyle's Law states that the volume of a gas varies inversely with the absolute pressure, provided the temperature remains constant.
- Charles' Law states that at a constant pressure, the volume of a gas varies as to the absolute temperature, and at a constant volume the pressure of a gas varies directly with the absolute temperature.
- Dalton's Law states that the total pressure of a confined mixture of gases is the sum of the pressures of each of the gases in the mixture.
- Electrical energy and heat energy are two forms of energy used in this industry.
- Fossil fuels are purchased by the unit. Natural gas is metered by the cubic foot, oil is purchased by the gallon, and coal is purchased by the ton. Electricity is purchased from the electric utility company by the kilowatt-hour (kWh).
- Work is the amount of force necessary to move an object: Work = Force × Distance.
- Horsepower is the equivalent of lifting 33,000 lb a height of 1 ft in 1 min, or some combination totaling the same.
- Watts are a measurement of electrical power. One horsepower equals 746 W.
- 3.413 Btu = 1 W. 1 kW (1000 W) = 3413 Btu.

REVIEW QUESTIONS

1. Define matter.
2. What are the three states in which matter is commonly found?
3. What is the term used for water when it is in the solid state?
4. In what direction does a solid exert force?
5. In what direction does a liquid exert force?
6. Describe how vapor exerts pressure.
7. Define density.
8. Define specific gravity.
9. Describe specific volume.
10. Why is information regarding the specific volume of gases important to the designer of air conditioning, heating, and refrigeration equipment?
11. Describe Boyle's Law.
12. At a constant pressure how does a volume of gas vary with respect to the absolute temperature?
13. Describe Dalton's Law as it relates to a confined mixture of gases.
14. What are the two types of energy most frequently used or considered in this industry?
15. How were fossil fuels formed?
16. What is work?
17. State the formula for determining the amount of work accomplished in a particular task.
18. If an air conditioning compressor weighing 300 lb had to be lifted 4 ft to be mounted on a base, how many ft-lb of work must be accomplished?
19. Describe horsepower and list the three quantities needed to determine horsepower.
20. How many watts of electrical energy are equal to 1 hp?
21. How many Btu are in 4000 W (4 kW)?
22. How many Btu would be produced in a 12-kW electric heater?
23. What unit of energy does the power company charge the consumer for?

UNIT 3

Refrigeration and Refrigerants

OBJECTIVES

After studying this unit, you should be able to

- state three reasons why ice melts in ice boxes.
- discuss applications for high-, medium-, and low-temperature refrigeration.
- describe the term ton of refrigeration.
- describe the basic refrigeration cycle.
- explain the relationship between pressure and the boiling point of water or other liquids.
- describe the function of the evaporator or cooling coil.
- explain the purpose of the compressor.
- list the compressors normally used in residential and light commercial buildings.
- discuss the function of the condensing coil.
- state the purpose of the metering device.
- list the three refrigerants commonly used in residential and light commercial refrigeration and air conditioning systems.
- list four characteristics to consider when choosing a refrigerant for a system.
- list the designated colors for refrigerant cylinders for various types of refrigerants.
- describe how refrigerants can be stored or processed while refrigeration systems are being serviced.
- plot a refrigeration cycle on a pressure/enthalpy diagram.

SAFETY CHECKLIST

* Areas where there are potential refrigerant leaks should be properly ventilated.
* Extra precautions should be taken to ensure that no refrigerant leaks occur near an open flame.
* Refrigerants are stored in pressurized containers and should be handled with care. Goggles with side shields and gloves should be worn when checking pressures and when transferring refrigerants from the container to a system or from the system to an approved container.

3.1 INTRODUCTION TO REFRIGERATION

This unit is an introduction to refrigeration and refrigerants. The term refrigeration is used here to include both the cooling process to preserve food and comfort cooling (air conditioning).

Preserving food is one of the most valuable uses of refrigeration. Food spoilage slows down as molecular motion slows. This retards the growth of bacteria that causes food to spoil. Below the frozen hard point, food-spoiling bacteria stop growing. The frozen hard point for most foods is considered to be 0°F. The food temperature range between 35°F and 45°F is known in the industry as medium temperature; below 0°F is considered low temperature. These ranges are used to describe many types of refrigeration equipment and applications.

For many years dairy products and other perishables were stored in the coldest room in the house, the basement, the well, or a spring. In the South, temperatures as low as 55°F could be reached in the summer with underground water. This would add to the time that some foods could be kept. Ice in the North and to some extent in the South was placed in "ice boxes" in kitchens. The ice melted when it absorbed heat from the food in the box, cooling the food, Figure 3–1.

In the early 1900s, ice was manufactured by mechanical refrigeration and sold to people with ice boxes, but still only the wealthy could afford it.

Also in the early 1900s, some companies manufactured the household refrigerator. Like all new items, it took a while to become popular. Now, of course, most houses have a refrigerator with a freezing compartment.

Frozen food was just beginning to become popular about the time World War II began. Because most people did not have a freezer at this time, central frozen food locker plants were established so that a family could have its own locker. Food that is frozen fresh is appealing because it stays fresh. Refrigerated foods, both medium temperature and low temperature, are so common now that most people take them for granted.

The refrigeration process is now used in the comfort cooling of the home and business and in the air conditioning of automobiles. The air conditioning application of refrigeration is known in this industry as high-temperature refrigeration.

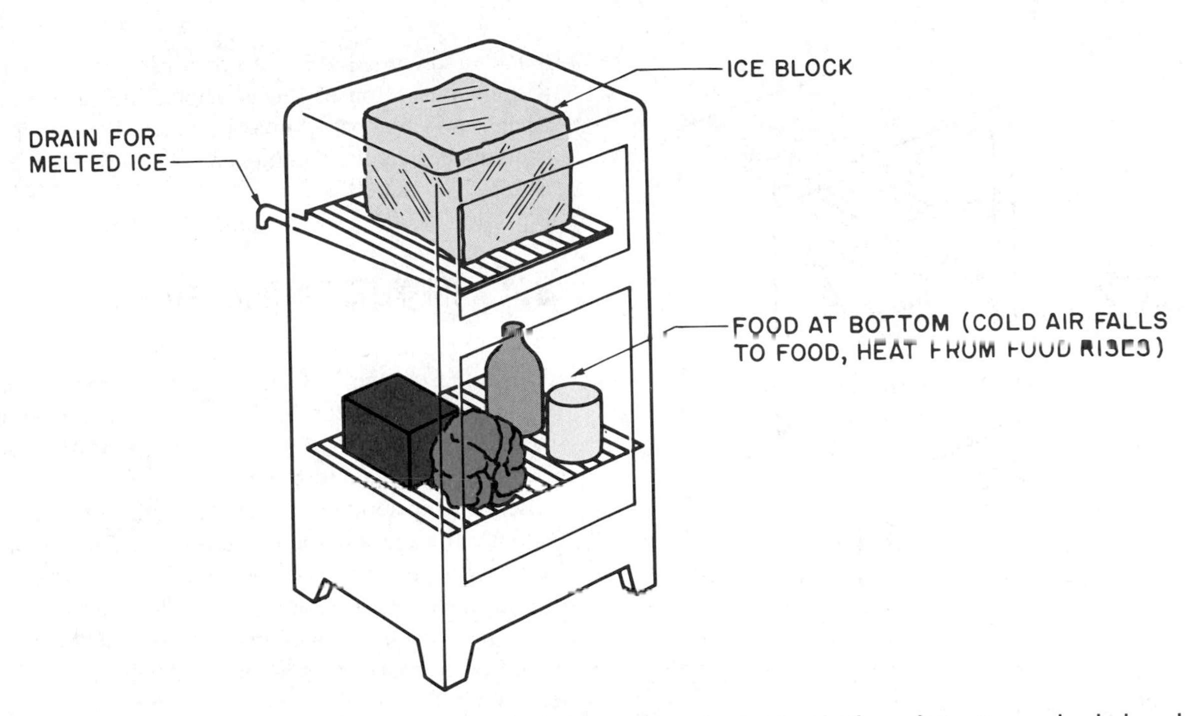
ICE BLOCK
DRAIN FOR MELTED ICE
FOOD AT BOTTOM (COLD AIR FALLS TO FOOD, HEAT FROM FOOD RISES)

FIGURE 3–1 Ice boxes were made of wood at first, then metal. The boxes were insulated with cork. If a cooling unit were placed where the ice is, this would be a refrigerator.

3.2 REFRIGERATION

***Refrigeration* is the process of removing heat from a place where it is not wanted and transferring that heat to a place where it makes little or no difference.** In the average household, the room temperature from summer to winter is normally between 70°F and 90°F. The temperature inside the refrigerator fresh food section should be about 35°F. Heat flows naturally from a warm level to a cold level. Therefore, heat in the room is trying to flow into the refrigerator and it does through the insulated walls, through the door when it is opened, and through warm food placed in the refrigerator, Figures 3–2, 3–3, and 3–4.

WARM AIR REPLACES THE COLD AIR
COLD AIR FALLS OUT BECAUSE IT IS HEAVIER

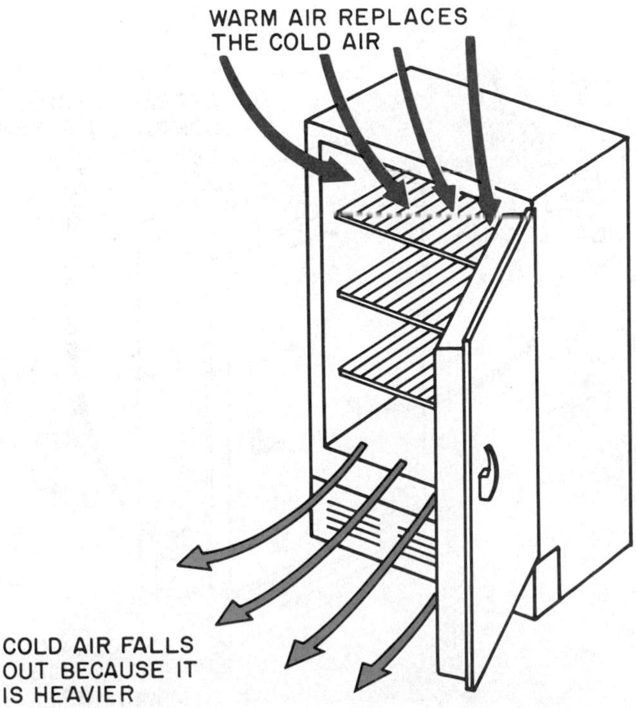

FIGURE 3–2 The colder air falls out of the refrigerator because it is heavier. It is replaced with warmer air from the top. This warm air is a heat leakage.

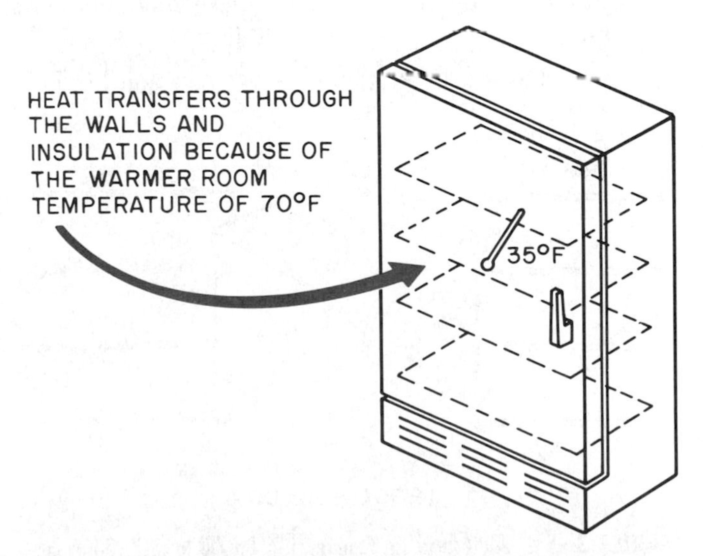
HEAT TRANSFERS THROUGH THE WALLS AND INSULATION BECAUSE OF THE WARMER ROOM TEMPERATURE OF 70°F
35°F

FIGURE 3–3 Heat transfers through the walls into the box by conduction. The walls have insulation, but this does not stop the leakage completely.

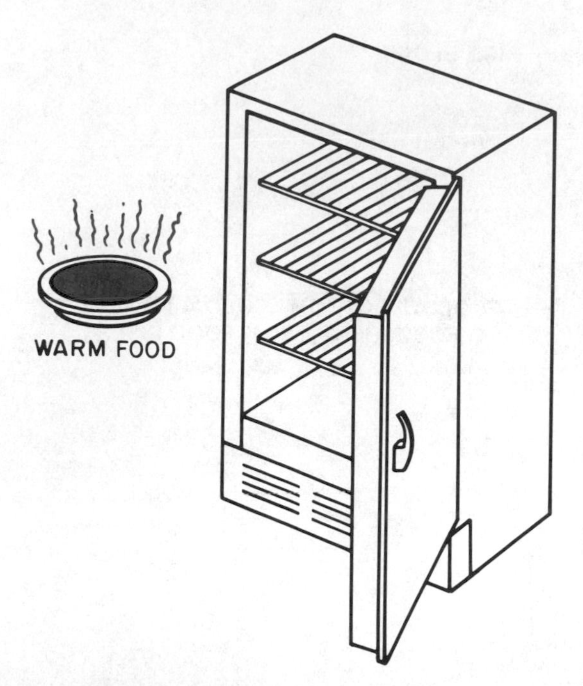

FIGURE 3–4 Warm food that is moved from the room or stove adds heat to the refrigerator and is considered heat leakage. This added heat has to be removed, or the inside temperature will rise.

3.3 RATING REFRIGERATION EQUIPMENT

Refrigeration equipment must also have a rating system so that equipment can be compared. The method for rating refrigeration equipment goes back to the days of using ice as the source for removing heat. It takes 144 Btu of heat energy to melt a pound of ice at 32°F. This same figure is also used in the rating of refrigeration equipment.

The term for this rating is the ton. *One ton of refrigeration* is the amount of heat required to melt 1 ton of ice in a 24-h period. Previously, we saw that it takes 144 Btu of heat to melt a pound of ice. It would then take 2000 times that much heat to melt a ton of ice (2000 lb = 1 ton):

$$144 \text{ Btu/lb} \times 2000 \text{ lb} = 288{,}000 \text{ Btu}$$

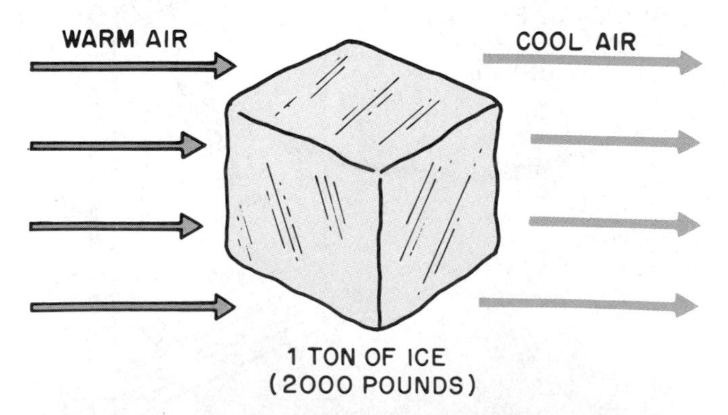

FIGURE 3–5 2000 lb of ice requires 144 Btu/lb to melt. 2000 lb × 144 Btu/lb = 288,000 Btu. When this is accomplished in 24 h, it is known as a work rate of 1 ton of refrigeration. This is the same as 12,000 Btu/h or 200 Btu/min.

to melt a ton of ice. When accomplished in a 24-h period, it is known as 1 ton of refrigeration. The same rules apply when removing heat from a substance. For example, an air conditioner, having a 1-ton capacity, will remove 288,000 Btu/24 h or 12,000 Btu/h (288,000 ÷ 24 = 12,000), or 200 Btu/min (12,000 ÷ 60 = 200), Figure 3–5.

3.4 THE REFRIGERATION PROCESS

The refrigerator pump has to pump the heat up the temperature scale from the 35°F or 0°F refrigeration compartments to the room temperature of 70°F to 90°F. The components of the refrigerator are used to accomplish this task, Figure 3–6. The heat leaking into the refrigerator does not normally raise the temperature of the contents of food an appreciable amount. If it did, the food would spoil. The heat inside the refrigerator rises to a predetermined level; the refrigeration system then comes on and pumps the heat out.

The process of pumping heat out of the refrigerator could be compared to pumping water from a valley to the top of a hill. It takes just as much energy to pump water up the hill as it does to carry it. A water pump with a motor accomplishes work. If a gasoline engine, for instance, were driving the pump, the gasoline would be burned and converted to work energy. An electric motor uses electric power as work energy, Figure 3–7. Refrigeration is the process of moving heat from a lower temperature into a medium with a higher temperature. This takes energy that must be purchased.

Following is an example of using a residential window air conditioning system to explain the basics of refrigeration. Residential air conditioning, whether window unit or central system, is considered to be high-temperature refrig-

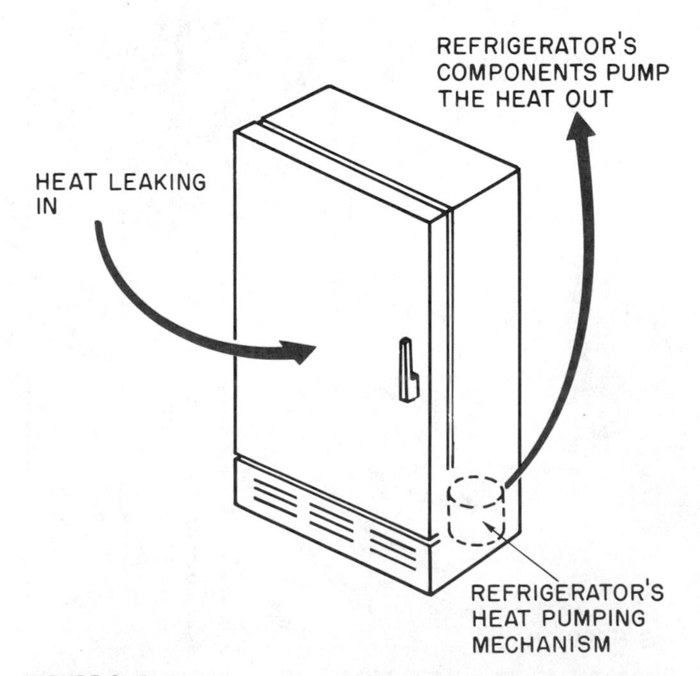

FIGURE 3–6 Heat that leaks into the refrigerator from any source has to be removed by the refrigerator's heat pumping mechanism. The heat has to be pumped from a 35°F box temperature to a 70°F room.

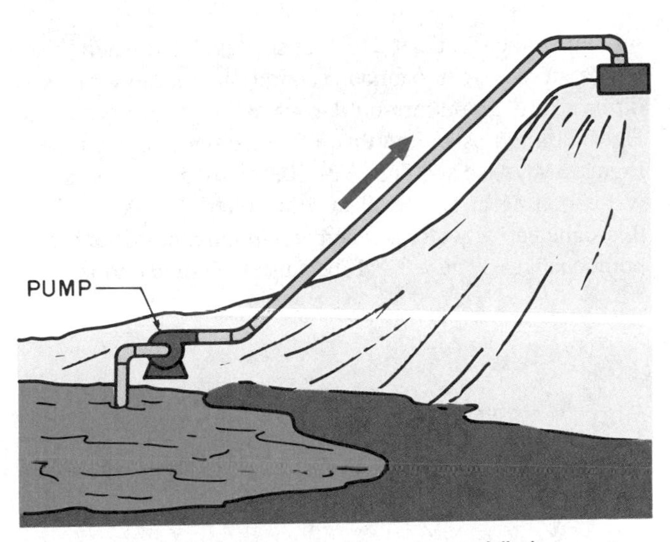

FIGURE 3–7 Power is required to pump water uphill; the same is true for pumping heat up the temperature scale from a 35°F box temperature to a 70°F room temperature.

eration and is used for comfort cooling. The residential system can be seen from the outside, touched, and listened to for the examples that will be given.

The refrigeration concept in the residential air conditioner is the same as in the household refrigerator. It pumps the heat from inside the house to the outside of the house. When hot air is coming into the house, this heat must be exhausted. The hot air is exhausted outside with a part of the system. The cold air in the house is recirculated air. Room air at approximately 75°F from the room goes into the unit, and air at approximately 55°F comes out. This is the same air with some of the heat removed, Figure 3–8.

The following example illustrates this concept. The following statements are also guidelines to some of the design data used throughout the air conditioning field.

1. The outside design temperature is 95°F.
2. The inside desired temperature is 75°F.
3. The cooling coil temperature is 40°F. This coil transfers heat from the room into the refrigeration system. Notice that with a 75°F room temperature and a 40°F cooling coil temperature, heat will transfer from the room air into the coil.
4. This heat transfer makes the air leaving the coil about 55°F.
5. The outside coil temperature is 125°F. This coil transfers heat from the system to the outside air. Notice that when the outside air temperature is 95°F and the coil temperature is 125°F, heat will be transferred from the system to the outside air.

Careful examination of Figure 3–8 shows that heat from the house is transferred into the refrigeration system through the inside coil and transferred to the outside air from the refrigeration system through the outside coil. The air conditioning system is actually pumping the heat out of

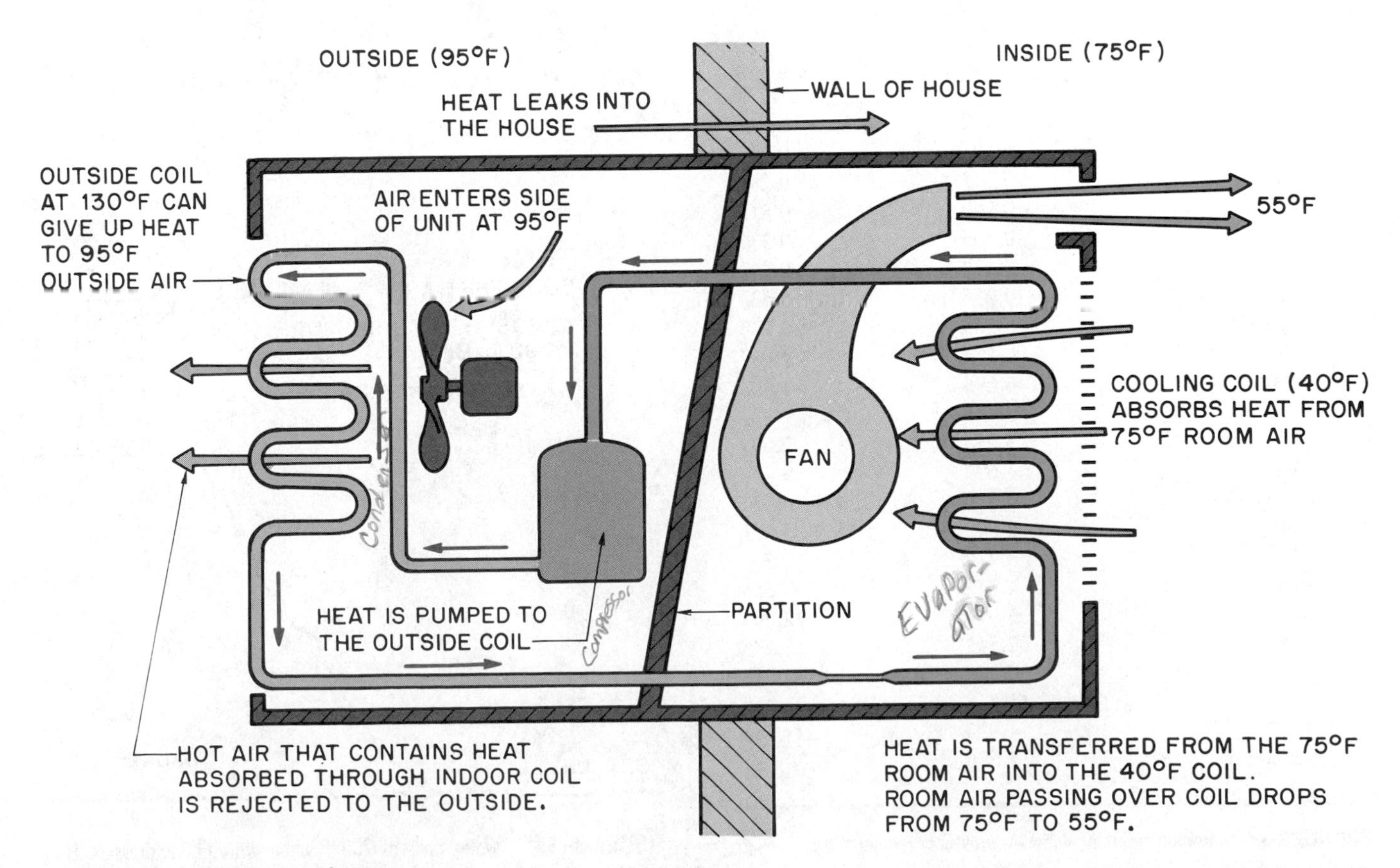

FIGURE 3–8 Window air conditioning unit.

the house. The system capacity must be large enough to keep the heat pumped out of the house so that the occupants will not become uncomfortable.

3.5 PRESSURE AND TEMPERATURE RELATIONSHIP

To understand the refrigeration process, we must go back to Figure 1–13 (heat/temperature graph), where water was changed to steam. Water boils at 212°F at 29.92 in. Hg pressure. This suggests that water has other boiling points. The next statement is one of the most important in this text. You may wish to memorize it. **The boiling point of water can be changed and controlled by controlling the vapor pressure above the water.** Understanding this concept is necessary because water is used as the heat transfer medium in the following example. The next few paragraphs are critically important for understanding refrigeration.

The *pressure/temperature relationship* correlates the vapor pressure and the boiling point of water and is the basis for controlling the system's temperatures.

WATER TEMPERATURE °F	ABSOLUTE PRESSURE lb/in.^{2}	ABSOLUTE PRESSURE in. Hg
10	0.031	0.063
20	0.050	0.103
30	0.081	0.165
32	0.089	0.180
34	0.096	0.195
36	0.104	0.212
38	0.112	0.229
40	0.122	0.248
42	0.131	0.268
44	0.142	0.289
46	0.153	0.312
48	0.165	0.336
50	0.178	0.362
60	0.256	0.522
70	0.363	0.739
80	0.507	1.032
90	0.698	1.422
100	0.950	1.933
110	1.275	2.597
120	1.693	3.448
130	2.224	4.527
140	2.890	5.881
150	3.719	7.573
160	4.742	9.656
170	5.994	12.203
180	7.512	15.295
190	9.340	19.017
200	11.526	23.468
210	14.123	28.754
212	14.696	29.921

FIGURE 3–9 Boiling point of water—pressure/temperature relationship table.

Pure water boils at 212°F at sea level and when the air temperature is 70°F (standard conditions) because this condition exerts a pressure on the water's surface of 29.92 in. Hg (or 14.696 psi). Find this reference point on the table in Figure 3–9. Also see Figure 3–10, showing the container of water boiling at sea level at atmospheric pressure. When this same pan of water is taken to a mountaintop, the boiling point changes, Figure 3–11, because the thinner atmosphere

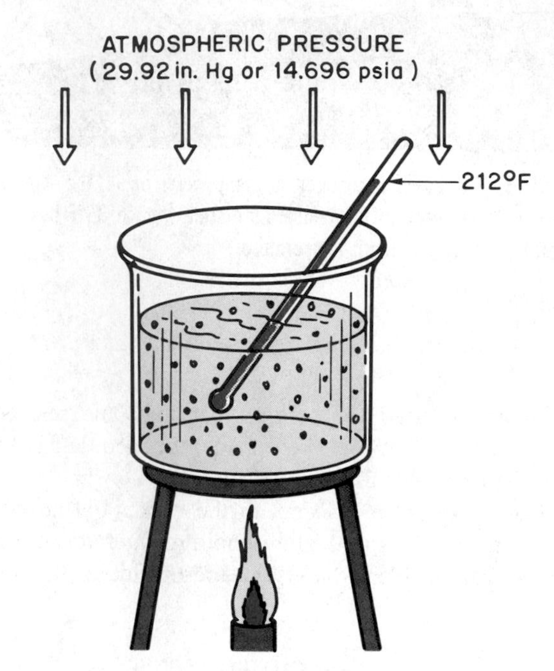

FIGURE 3–10 Water boils at 212°F when atmospheric pressure is 29.92 in. Hg.

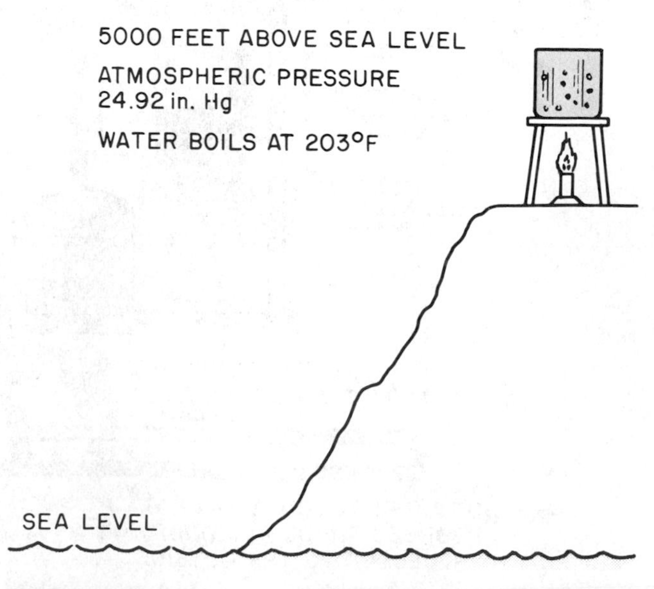

FIGURE 3–11 Water boils at 203°F when atmospheric pressure is 24.92 in. Hg.

causes a reduction in pressure (about 1 in. Hg/1000 ft). In Denver, Colorado, for example, which is about 5000 ft above sea level, the atmospheric pressure is approximately 25 in. Hg. Water boils at 203.4°F at that pressure. This makes cooking foods such as potatoes and dried beans more difficult because they need a higher temperature. But by placing the food in a closed container that can be pressurized, such as a pressure cooker, and allowing the pressure to go up to about 15 psi above atmosphere (or 30 psia), the boiling point can be raised to 250°F, Figure 3–12.

Studying the water pressure/temperature table reveals that whenever the pressure is increased, the boiling point increases, and that whenever the pressure is reduced, the boiling point is reduced. If water were boiled at a temperature low enough to absorb heat out of a room, we could have comfort cooling (air conditioning).

Let's place a thermometer in the pan of pure water, put the pan inside a bell jar with a barometer, and start the vacuum pump. Suppose the water is at room temperature (70°F). When the pressure in the jar reaches the pressure that corresponds to the boiling point of water at 70°F, the water will start to boil and vaporize. This point is 0.739 in. Hg (0.363 psia). Figure 3–13 illustrates the container in the jar.

If we were to lower the pressure in the jar to correspond to a temperature of 40°F, this new pressure of 0.248 in. Hg (0.122 psia) will cause the water to boil at 40°F. The water is not hot even though it is boiling. The thermometer in the pan indicates this. If the jar were opened to the atmosphere, the water would be found to be cold.

Now let's circulate this water boiling at 40°F through a cooling coil. If room air were passed over it, it would absorb heat from the room air. Because this air is giving up heat to the coil, the air leaving the coil is cold. Figure 3–14 illustrates the cooling coil.

When water is used in this way, it is called a *refrigerant*. **A *refrigerant* is a substance that can be changed**

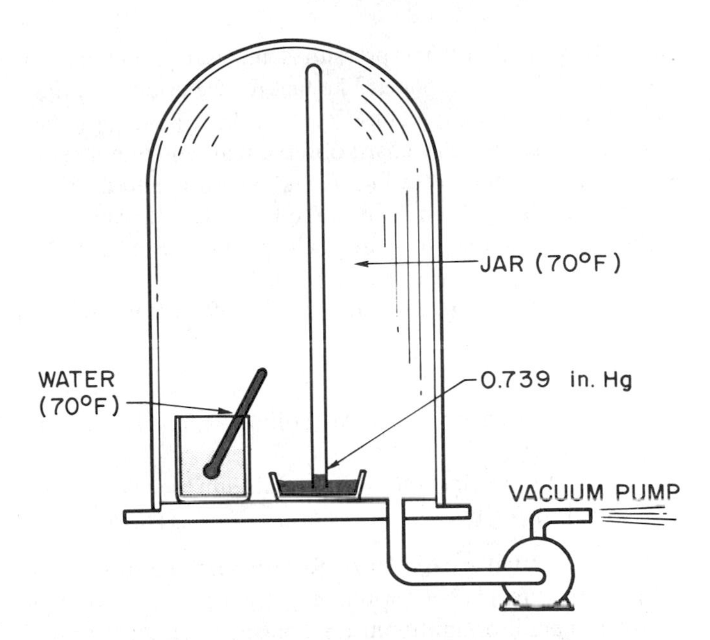

FIGURE 3–13 Pressure in the bell jar is reduced to 0.739 in. Hg, and the boiling temperature of the water is reduced to 70°F because the pressure is 0.739 in. Hg (0.363 psia).

readily to a vapor by boiling it and then to a liquid by condensing it. The refrigerant must be able to make this change repeatedly without altering its characteristics. Water is not normally used as a refrigerant in small applications for reasons that will be discussed later. We used it in this example because most people are familiar with its characteristics.

To explore how a real refrigeration system works, we will use refrigerant-22 (R-22) in the following examples because it is commonly used in residential air conditioning.

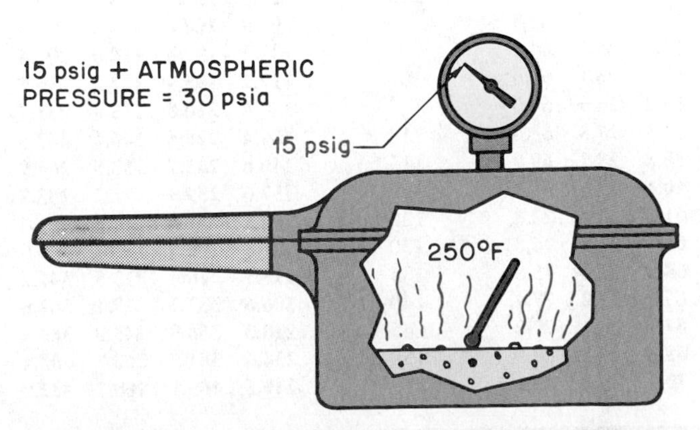

FIGURE 3–12 The water in the pressure cooker boils at 250°F. As heat is added, the water boils to make vapor. The vapor cannot escape, and the vapor pressure rises to 15 psig. The water boils at 250°F because the pressure is 15 psig.

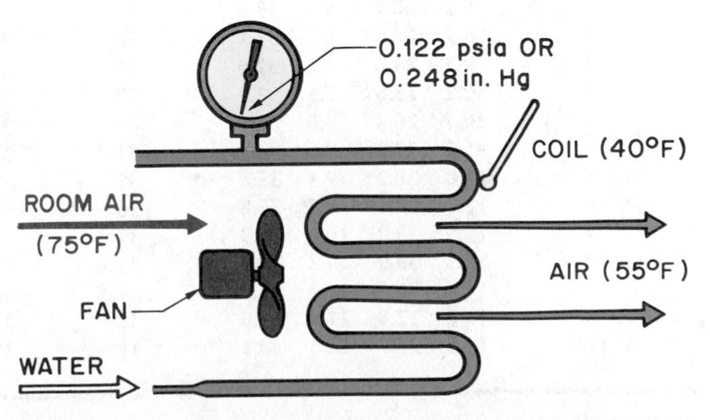

FIGURE 3–14 The water is boiling at 40°F because the pressure is 0.122 psia or 0.248 in. Hg. The room air is 75°F and gives up heat to the 40°F coil.

See Figure 3–15 for the pressure/temperature relationship chart for R-22. This chart is like that for water but at different temperature and pressure levels. Take a moment to become familiar with this chart; observe that temperature is in the left column, expressed in °F, and pressure is to the right expressed in psig. Find 40°F in the left column, read to the right, and notice that the gage reading is 68.5 psig for R-22. What does this mean in usable terms?

The pressure and temperature of a refrigerant will correspond when both liquid and vapor are present under two conditions:

1. When the change of state (boiling or condensing) is occurring.
2. When the refrigerant is at equilibrium (ie, no heat is added or removed).

Suppose that a cylinder of R-22 is allowed to set in a room until it reaches the room temperature of 75°F. It will then be in equilibrium because no outside forces are acting on it. The cylinder and its partial liquid partial vapor contents will now be at the room temperature of 75°F. The pressure and temperature chart indicates a pressure of 132 psig, Figure 3–15.

Suppose that the same cylinder of R-22 is moved into a walk-in cooler and allowed to reach the room temperature of 35°F and attain equilibrium. The cylinder will then reach a new pressure of 61.5 psig because while it is cooling off to 35°F, the vapor inside the cylinder is reacting to the cooling effect by partially condensing; therefore, the pressure drops.

If we move the cylinder (now at 35°F) back into the warmer room (75°F) and allow it to warm up, the liquid inside it reacts to the warming effect by boiling slightly and creating vapor. Thus, the pressure gradually increases to 132 psig, which corresponds to 75°F.

If we move the cylinder (now at 75°F) into a room at 100°F, the liquid again responds to the temperature change by slightly boiling and creating more vapor. As the liquid boils and makes vapor, the pressure steadily increases (according to the pressure/temperature chart) until it corresponds to the liquid temperature. This continues until the contents of the cylinder reach the pressure, 196 psig, corresponding to 100°F, Figures 3–16, 3–17, and 3–18.

TEMPERATURE	REFRIGERANT				TEMPERATURE	REFRIGERANT				TEMPERATURE	REFRIGERANT			
°F	12	22	134a	502	°F	12	22	134a	502	°F	12	22	134a	502
−60	19.0	12.0		7.2	12	15.8	34.7	13.2	43.2	42	38.8	71.4	37.0	83.8
−55	17.3	9.2		3.8	13	16.4	35.7	13.8	44.3	43	39.8	73.0	38.0	85.4
−50	15.4	6.2		0.2	14	17.1	36.7	14.4	45.4	44	40.7	74.5	39.0	87.0
−45	13.3	2.7		1.9	15	17.7	37.7	15.1	46.5	45	41.7	76.0	40.1	88.7
−40	11.0	0.5	14.7	4.1	16	18.4	38.7	15.7	47.7	46	42.6	77.6	41.1	90.4
−35	8.4	2.6	12.4	6.5	17	19.0	39.8	16.4	48.8	47	43.6	79.2	42.2	92.1
−30	5.5	4.9	9.7	9.2	18	19.7	40.8	17.1	50.0	48	44.6	80.8	43.3	93.9
−25	2.3	7.4	6.8	12.1	19	20.4	41.9	17.7	51.2	49	45.7	82.4	44.4	95.6
−20	0.6	10.1	3.6	15.3	20	21.0	43.0	18.4	52.4	50	46.7	84.0	45.5	97.4
−18	1.3	11.3	2.2	16.7	21	21.7	44.1	19.2	53.7	55	52.0	92.6	51.3	106.6
−16	2.0	12.5	0.7	18.1	22	22.4	45.3	19.9	54.9	60	57.7	101.6	57.3	116.4
−14	2.8	13.8	0.3	19.5	23	23.2	46.4	20.6	56.2	65	63.8	111.2	64.1	126.7
−12	3.6	15.1	1.2	21.0	24	23.9	47.6	21.4	57.5	70	70.2	121.4	71.2	137.6
−10	4.5	16.5	2.0	22.6	25	24.6	48.8	22.0	58.8	75	77.0	132.2	78.7	149.1
−8	5.4	17.9	2.8	24.2	26	25.4	49.9	22.9	60.1	80	84.2	143.6	86.8	161.2
−6	6.3	19.3	3.7	25.8	27	26.1	51.2	23.7	61.5	85	91.8	155.7	95.3	174.0
−4	7.2	20.8	4.6	27.5	28	26.9	52.4	24.5	62.8	90	99.8	168.4	104.4	187.4
−2	8.2	22.4	5.5	29.3	29	27.7	53.6	25.3	64.2	95	108.2	181.8	114.0	201.4
0	9.2	24.0	6.5	31.1	30	28.4	54.9	26.1	65.6	100	117.2	195.9	124.2	216.2
1	9.7	24.8	7.0	32.0	31	29.2	56.2	26.9	67.0	105	126.6	210.8	135.0	231.7
2	10.2	25.6	7.5	32.9	32	30.1	57.5	27.8	68.4	110	136.4	226.4	146.4	247.9
3	10.7	26.4	8.0	33.9	33	30.9	58.8	28.7	69.9	115	146.8	242.7	158.5	264.9
4	11.2	27.3	8.6	34.9	34	31.7	60.1	29.5	71.3	120	157.6	259.9	171.2	282.7
5	11.8	28.2	9.1	35.8	35	32.6	61.5	30.4	72.8	125	169.1	277.9	184.6	301.4
6	12.3	29.1	9.7	36.8	36	33.4	62.8	31.3	74.3	130	181.0	296.8	198.7	320.8
7	12.9	30.0	10.2	37.9	37	34.3	64.2	32.2	75.8	135	193.5	316.6	213.5	341.2
8	13.5	30.9	10.8	38.9	38	35.2	65.6	33.2	77.4	140	206.6	337.2	229.1	362.6
9	14.0	31.8	11.4	39.9	39	36.1	67.1	34.1	79.0	145	220.3	358.9	245.5	385.0
10	14.6	32.8	11.9	41.0	40	37.0	68.5	35.1	80.5	150	234.6	381.5	262.7	408.4
11	15.2	33.7	12.5	42.1	41	37.9	70.0	36.0	82.1	155	249.5	405.1	280.7	432.9

Vacuum—Red Figures
Gage Pressure—Bold Figures

FIGURE 3–15 Pressure/temperature relationship chart in inches Hg vacuum or psig.

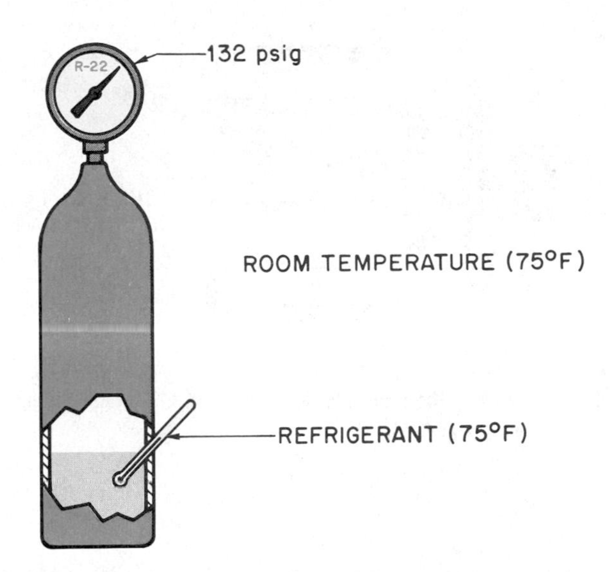

FIGURE 3–16 The cylinder of R-22 is left in a 75°F room until it and its contents are at room temperature. The cylinder contains a partial liquid, partial vapor mixture; when both become room temperature, they are in equilibrium; no more temperature changes will be occuring. At this time the cylinder pressure, 132 psig, will correspond to its temperature of 75°F.

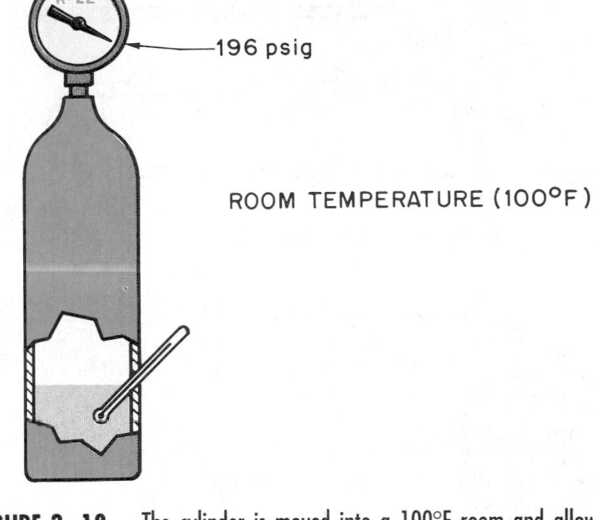

FIGURE 3–18 The cylinder is moved into a 100°F room and allowed to reach the point of equilibrium at 100°F, 196 psig. The pressure rise is due to some of the liquid refrigerant boiling to a vapor and increasing the total cylinder pressure.

Further study of the temperature/pressure chart shows that when the pressure is lowered to atmospheric pressure R-22 boils at about −41°F. ⊛**Do not perform the following exercises because allowing refrigerant to escape to the atmosphere is now against the law. These are stated here for illustration purposes only.**⊛ If the valve on the cylinder of R-22 were opened slowly and the vapor allowed to escape to the atmosphere, the pressure loss of the vapor would cause the liquid remaining in the cylinder to drop in temperature. Soon the pressure in the cylinder would be down to atmospheric pressure, and it would frost over and become −41°F. If a gage line were connected to the liquid port on a cylinder and the valve opened very slowly, holding the end of the gage line in a thick coffee cup, the escaping liquid will cool the cup to −41°F, Figure 3–19.

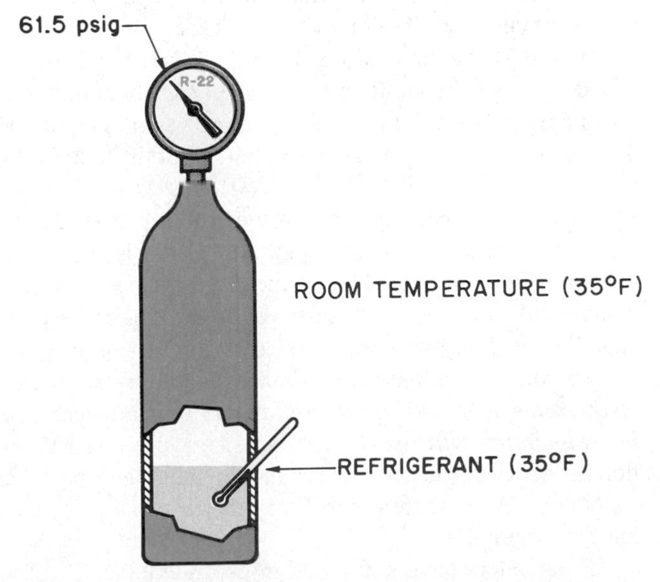

FIGURE 3–17 The cylinder is moved into a walk-in cooler and left until it and its contents become the same as the inside of the cooler, 35°F. Until the cylinder and its contents get to 35°F some of the vapor will be changing to a liquid, reducing the pressure. Soon the cylinder pressure will correspond to the walk-in cooler temperature of 35°F, 61.5 psig.

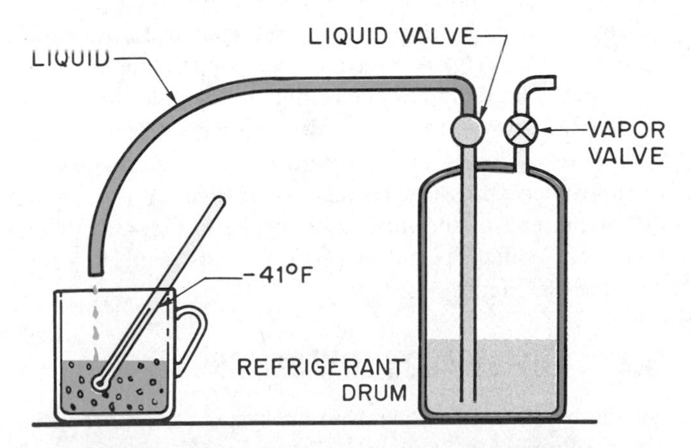

FIGURE 3–19 When a gage line is attached to the liquid line valve on an R-22 refrigeration cylinder and liquid is allowed to trickle out of it into the cup, the liquid will collect in the cup. It will boil at a temperature of −41°F at atmospheric pressure. ⊛**Do not perform this experiment because it is illegal to vent refrigerant to the atmosphere.**⊛

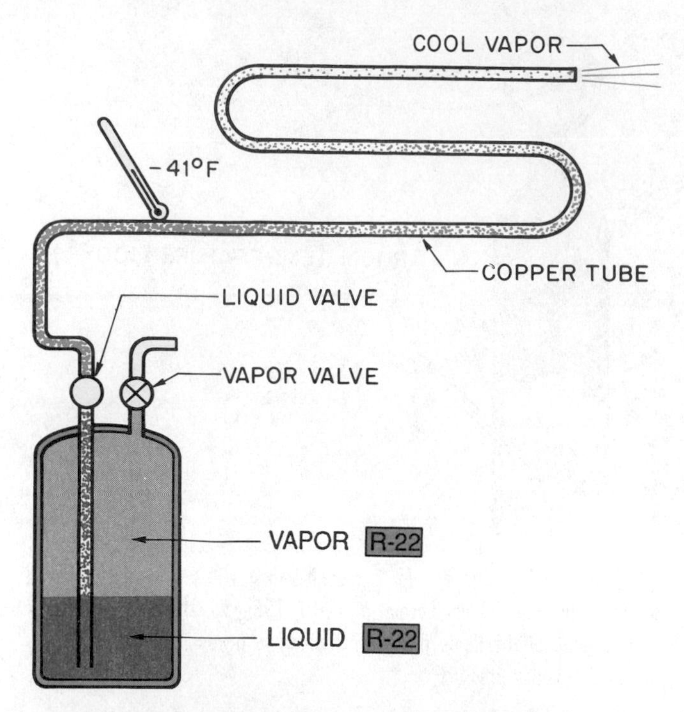

FIGURE 3–20 When the tubing is attached to the liquid valve on an R-22 cylinder and liquid is allowed to trickle into the tubing, the liquid will boil at −41°F at atmospheric pressure. **⊛Do not perform this experiment because it is illegal to vent refrigerant to the atmosphere.⊛**

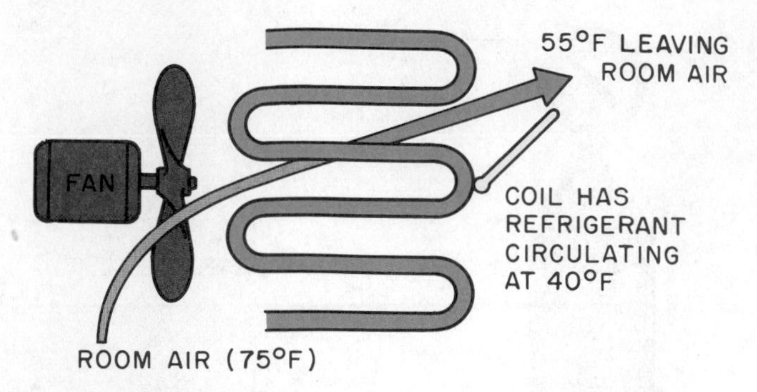

FIGURE 3–21 The evaporator is operated at 40°F to be able to absorb heat from the 75°F air.

The liquid will accumulate in the bottom of the cup and will boil. **⊛Again, do not perform the above experiments.⊛**

A crude but effective demonstration has been used in the past to show how air can be cooled. **⊛Do not perform this experiment because venting refrigerant to the atmosphere is illegal.⊛** A long piece of copper tubing is fastened to the liquid tap on the refrigerant cylinder and liquid refrigerant allowed to trickle into the tube while air passes over it. The tube has a temperature of −41°F, corresponding to atmospheric pressure, because the refrigerant is escaping out of the end of the tube at atmospheric pressure. If the tube were coiled up and placed in an airstream, it would cool the air, Figure 3–20.

3.6 REFRIGERATION COMPONENTS

By adding some components to the system, these problems can be eliminated. Four major components to mechanical refrigeration systems are covered in this book:

1. The evaporator
2. The compressor
3. The condenser
4. The refrigerant metering device

3.7 THE EVAPORATOR

The *evaporator* absorbs heat into the system. When the refrigerant is boiled at a lower temperature than that of the substance to be cooled, it absorbs heat from the substance. The boiling temperature of 40°F was chosen in the previous air conditioning examples because it is the design temperature normally used for air conditioning systems. The reason is that room temperature is close to 75°F, which readily gives up heat to a 40°F coil. The 40°F temperature is also well above the freezing point of the coil. See Figure 3–21 for the coil-to-air relationships.

Let's see what happens as the R-22 refrigerant passes through the evaporator coil. The refrigerant enters the coil as a mixture of about 75% liquid and 25% vapor. The mixture is tumbling and boiling down the tube, with the liquid being turned to vapor all along the coil because heat is being added to the coil from the air, Figure 3–22. About halfway down the coil, the mixture becomes more vapor than liquid. The purpose of the evaporator is to boil all of the liquid into a vapor just before the end of the coil. This occurs approximately 90% of the way through the coil, when all of the liquid is gone, leaving pure vapor. At this precise point we have a saturated vapor. This is the point where the vapor would start to condense if heat were removed, or become superheated if any heat were added. **When a vapor *is superheated, it no longer corresponds to the pressure and temperature relationship; it will take on sensible heat and its temperature will rise.*** Superheat is considered refrigeration insurance because it ensures that no liquid gets past the evaporator. When there is superheat, there is no liquid leaving the evaporator.

Evaporators have many design configurations. But for now just remember that they absorb the heat into the system from the substance to be cooled. The substance may be solid, liquid, or gas, and the evaporator has to be designed to fit the condition. See Figure 3–23 for a typical evaporator. Once absorbed into the system, the heat is now in the refrigerant gas and is drawn into the compressor.

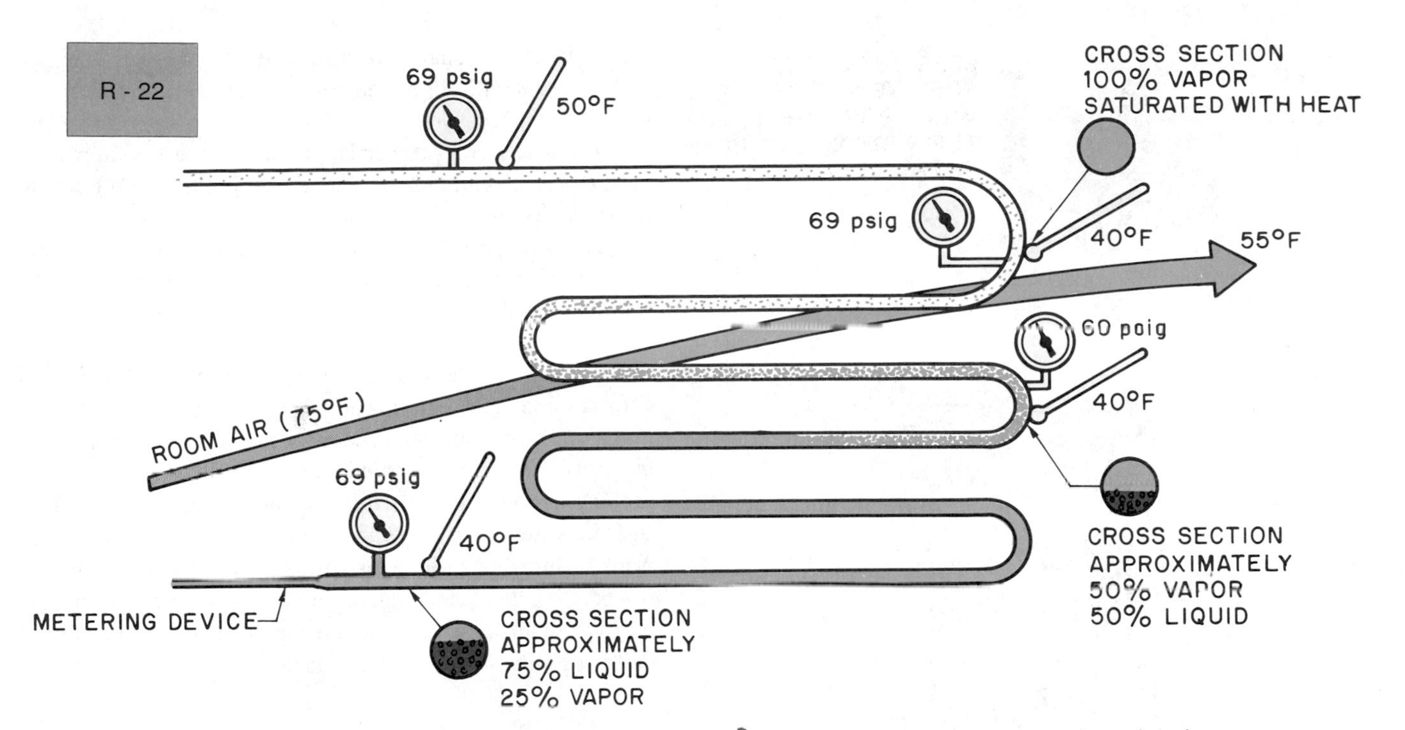

FIGURE 3–22 The evaporator absorbs heat into the refrigeration system by boiling the refrigerant at a temperature lower than the room air passing over it. The 75°F room air readily gives up heat to the 40°F evaporator by conduction.

FIGURE 3–23 Typical evaporator. *Courtesy Larkin Coils, Inc.*

3.8 THE COMPRESSOR

The compressor is the heart of the refrigeration system. It pumps heat through the system in the form of heat-laden refrigerant vapor. A compressor can be considered a vapor pump. It reduces the pressure on the low-pressure side of the system, which includes the evaporator and increases the pressure in the high-pressure side of the system. All compressors in refrigeration systems perform this function by compressing the vapor refrigerant. This compression can be accomplished in several ways with different types of compressors. The most common compressors used in residential and light commercial air conditioning and refrigeration are *the reciprocating, the rotary, and the scroll.*

The reciprocating compressor uses a piston in a cylinder to compress the refrigerant, Figure 3–24. Valves, usually reed or flapper valves, ensure that the refrigerant flows in the correct direction, Figure 3–25. This compressor is known as a positive displacement compressor. When the cylinder is filled with vapor, it must be emptied as the compressor turns or damage will occur. For many years, it was the most commonly used compressor for systems up to 100 hp. Newer and more efficient designs of compressors are now also being used.

The rotary compressor is also a positive displacement compressor and is used for applications that are typically in

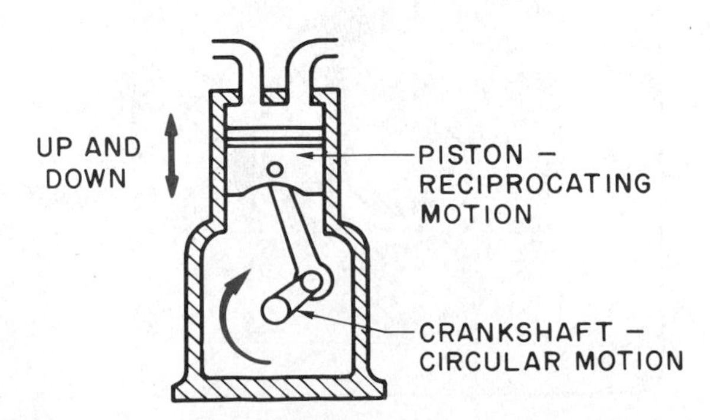

FIGURE 3–24 The crankshaft converts the circular motion of the motor to the reciprocating or back and forth motion of the piston.

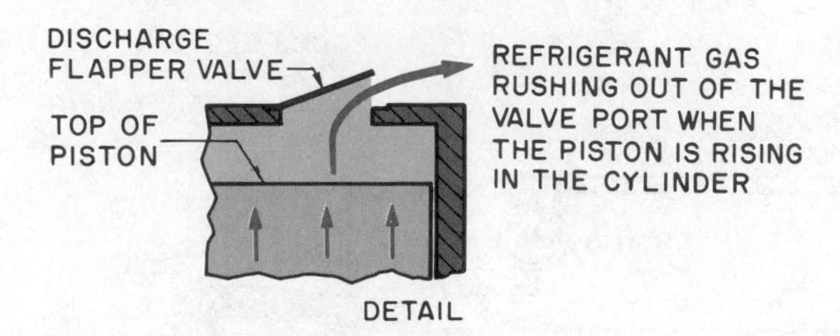

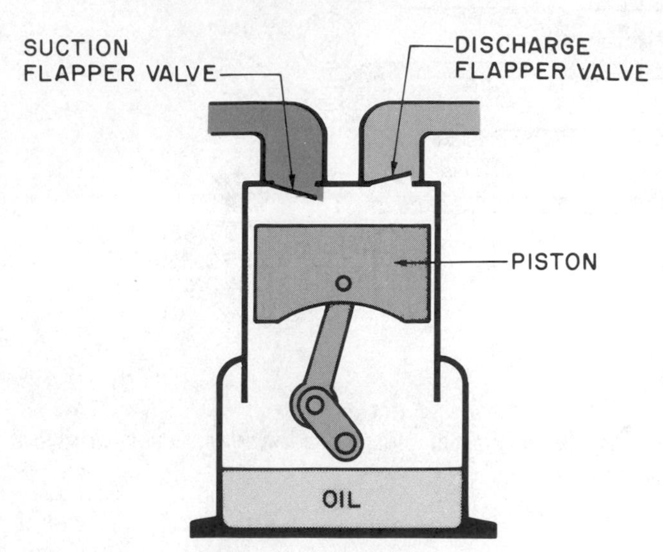

FIGURE 3–25 Flapper valves and compressor components.

the small equipment range, such as window air conditioners, household refrigerators, and some central air conditioning systems. These compressors are extremely efficient and have few moving parts, Figure 3–26. This compressor uses a rotating drumlike piston that squeezes the vapor refrigerant out the discharge port. These compressors are typically very small compared to the same capacity of reciprocating compressors.

The scroll compressor is one of the latest compressors to be developed and has an entirely different working mechanism. It has a stationary part that looks like a coil spring and a moving part that matches and meshes with the stationary part, Figure 3–27. The movable part orbits inside the stationary part and squeezes the vapor from the low-pressure side to the high-pressure side of the system between the movable and stationary parts. Several stages of compression are taking place in the scroll at the same time, making it a very smooth running compressor with few moving parts. The scroll is sealed on the bottom and top with the rubbing action and at the tip with a tip seal. These sealing surfaces prevent refrigerant from the high-pressure side from pushing back to the low-pressure side while running. It is a positive displacement compressor with a limitation. It is positive displacement until too much pressure differential builds up, then the scrolls are capable of moving apart and

BLADE MOVES IN
AND OUT WHEN
ROTOR TURNS
DISCHARGE
SPRING APPLIES
PRESSURE TO BLADE
(OR VALVE)
SUCTION GAS
ROTOR

FIGURE 3–26 Rotary compressor with motion in one direction and no back stroke.

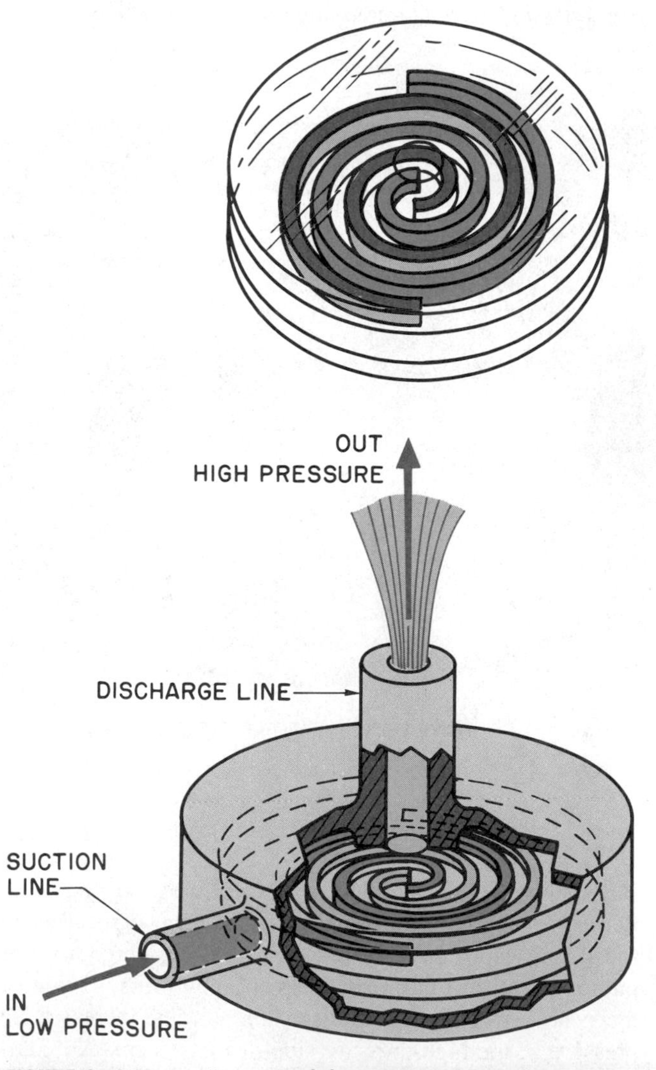

FIGURE 3–27 An illustration of the operation of a scroll compressor mechanism.

high-pressure refrigerant can blow back through the compressor and prevent overload.

Large commercial systems use other types of compressors because they must move much more refrigerant vapor through the system. The centrifugal compressor is used in large air conditioning systems. It is much like a large fan and is not positive displacement, Figure 3–28. The screw compressor is also used for the same reasons as the centrifugal compressor except it is also applied to low-temperature applications in refrigeration. It is a positive displacement compressor, Figure 3–29.

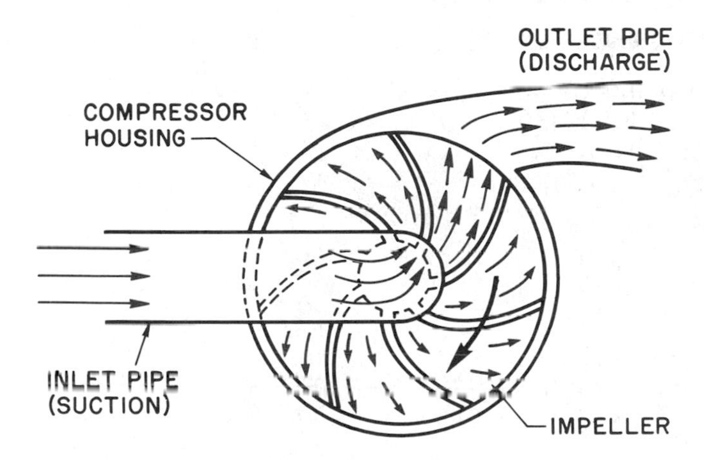
OUTLET PIPE (DISCHARGE)
COMPRESSOR HOUSING
INLET PIPE (SUCTION)
IMPELLER
THE TURNING IMPELLER IMPARTS CENTRIFUGAL FORCE ON THE REFRIGERANT FORCING THE REFRIGERANT TO THE OUTSIDE OF THE IMPELLER. THE COMPRESSOR HOUSING TRAPS THE REFRIGERANT AND FORCES IT TO EXIT INTO THE DISCHARGE LINE. THE REFRIGERANT MOVING TO THE OUTSIDE CREATES A LOW PRESSURE IN THE CENTER OF THE IMPELLER WHERE THE INLET IS CONNECTED.

FIGURE 3-28 An illustration of the operation of a centrifugal compressor mechanism

The important thing for you to remember is that a compressor performs the same function no matter what the type is. For now it can be thought of as a component that increases the pressure in the system and moves the vapor refrigerant from the low-pressure side to the high-pressure side into the condenser.

3.9 THE CONDENSER

The *condenser* rejects heat from the refrigeration system that the evaporator absorbed and the compressor pumped. The condenser receives the hot gas after it leaves the compressor through the short pipe between the compressor and the condenser called the hot gas line, Figure 3–30. The hot gas is forced into the top of the condenser coil by the compressor. The gas is being pushed along at high speed and at high temperature (about 200°F). The gas is not corresponding to the pressure/temperature relationship because the head pressure is 278 psig for R-22. The head pressure for 200°F would be off the pressure/temperature chart. Remember that the temperature at which the change of state would occur is 125°F. This temperature establishes the head pressure of 278 psig.

The gas entering the condenser is so hot compared to the surrounding air that a heat exchange begins to occur immediately in the air. The surrounding air that is being passed over the condenser is 95°F as compared to the near 200°F of the gas entering the condenser. As the gas moves through the condenser, it begins to give up heat to the surrounding air. This causes a drop in gas temperature. The gas keeps cooling off until it reaches the condensing temperature of 125°F, and the change of state begins to occur. The change of state begins slowly at first with small amounts of vapor to liquid and gets faster as the combination gas-liquid mixture moves toward the end of the condenser.

When the condensing refrigerant gets about 90% of the way through the condenser, the refrigerant in the pipe becomes almost, or all, pure liquid. Now more heat can be taken from the liquid. The liquid at the end of the condenser

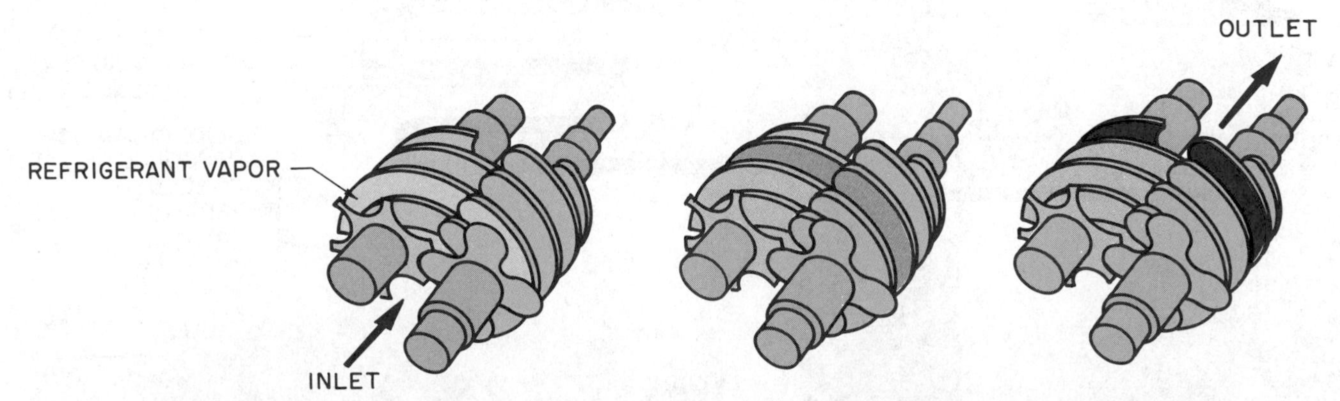
OUTLET
REFRIGERANT VAPOR
INLET
ONE STAGE OF COMPRESSION AS REFRIGERANT MOVES THROUGH SCREW COMPRESSOR

FIGURE 3–29 An illustration of the internal working mechanism of a screw compressor.

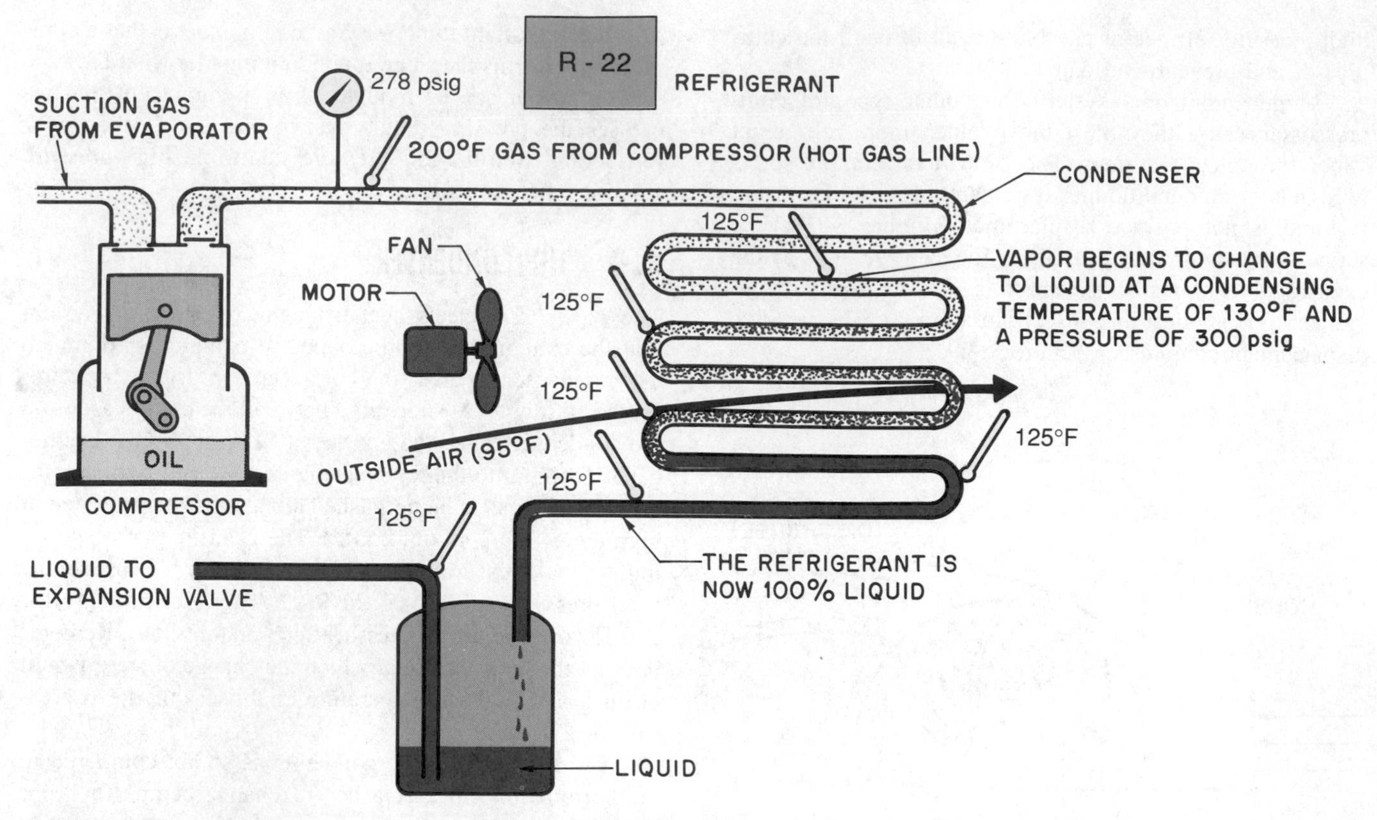

FIGURE 3–30 Vapor inside the condenser changes to liquid refrigerant.

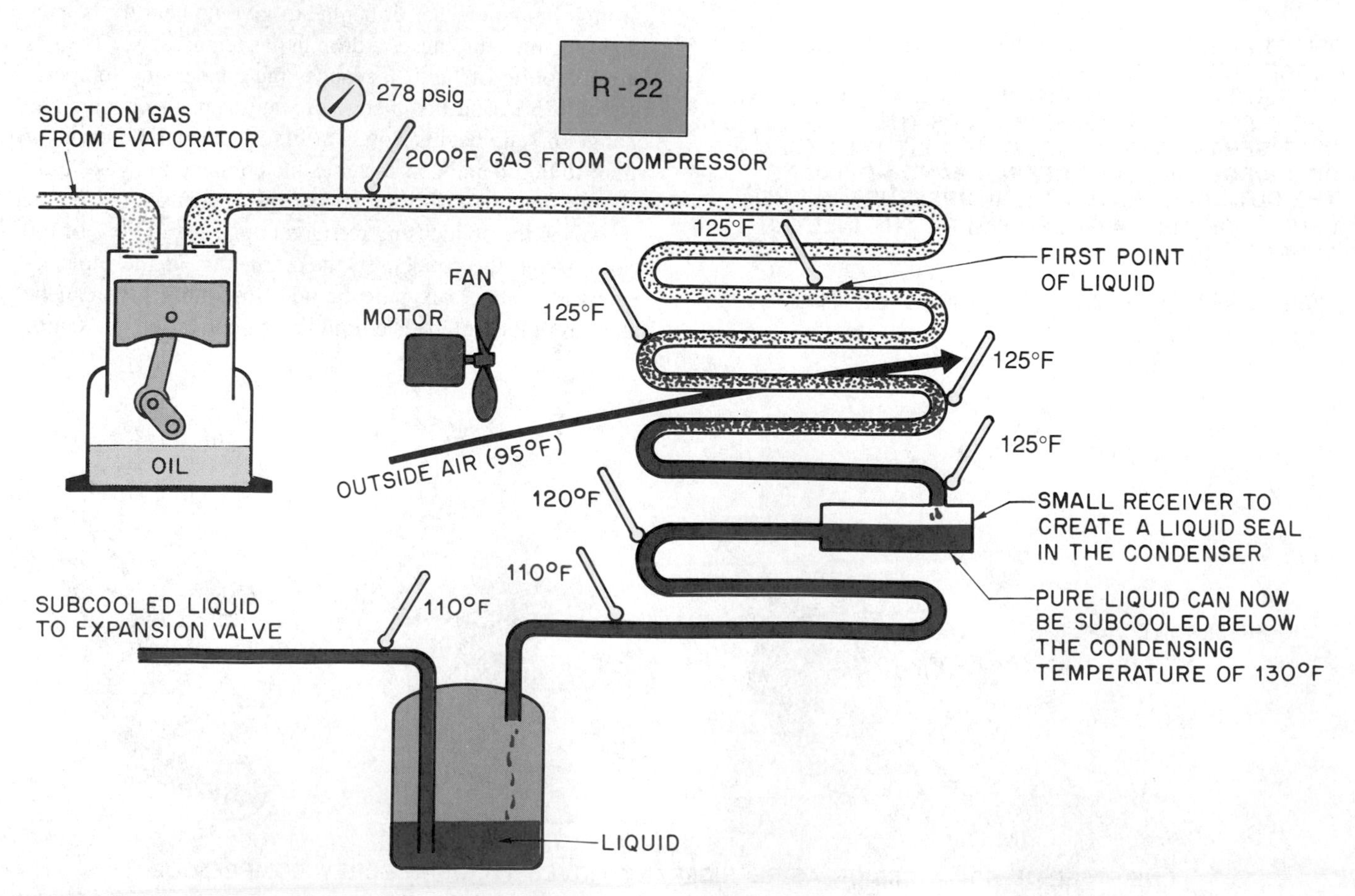

FIGURE 3–31 Condenser with subcooling.

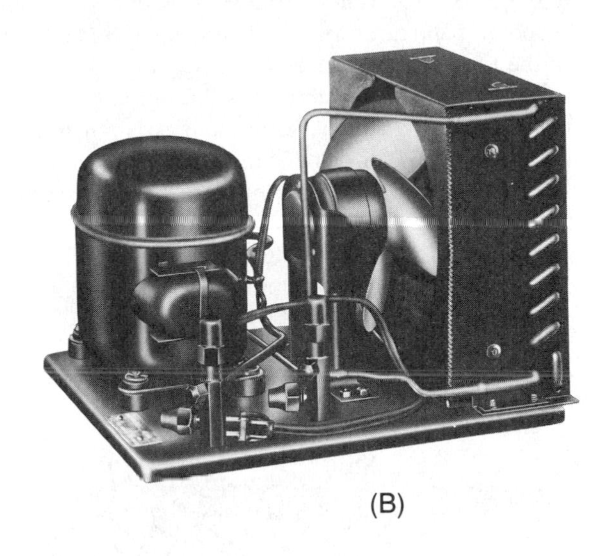

FIGURE 3–32 Typical condensing units. (A) Air-cooled semihermetic condensing unit. (B) Air-cooled hermetic condensing unit. The compressor is shown as part of the condensing unit. *Courtesy (A) Copeland Corporation, (B) Tecumseh Products Company*

is at the condensing temperature of 125°F and can still give up some heat to the surrounding 95°F air. When the liquid at the end of the condenser goes below 125°F, it is called *subcooled,* Figure 3–31.

Three important things may happen to the refrigerant in the condenser.

1. The hot gas from the compressor is de-superheated from the hot discharge temperature to the condensing temperature. Remember, the condensing temperature determines the head pressure.
2. The refrigerant is condensed from a vapor to a liquid.
3. The liquid refrigerant temperature may then be lowered below the condensing temperature, or subcooled. The refrigerant can usually be subcooled to between 10°F and 20°F below the condensing temperature, Figure 3–31.

Many types of condensing devices are available. The condenser is the component that rejects the heat out of the refrigeration system. The heat may have to be rejected into a solid, liquid, or a gas substance, and a condenser can be designed to do the job. Figure 3–32 shows some typical condensing units. The compressor is shown with the condenser.

3.10 THE REFRIGERANT METERING DEVICE

The warm liquid is now moving down the liquid line in the direction of the *metering device*. The liquid temperature is about 115°F and may still give up some heat to the surroundings before reaching the metering device. This line may be routed under a house or through a wall where it may easily reach a new temperature of about 110°F. Any heat given off to the surroundings is helpful because it came from within the system and will help the system capacity.

One type of metering device is a simple fixed-size type known as an *orifice*. It is a small restriction of a fixed size in the line, Figure 3–33. This device holds back the full flow of refrigerant and is the dividing point between the high-pressure and the low-pressure sides of the system. Only pure liquid must enter it. The pipe leading to the orifice may be the size of a pencil, and the precision-drilled hole in the orifice may be the size of a very fine sewing needle. As you can see from the figure, the gas flow is greatly restricted here. The liquid refrigerant entering the orifice is at a pressure of 278 psig; the refrigerant leaving the orifice is a *mixture* of about 75% liquid and 25% vapor at a new pressure of 70 psig and a new temperature of 40°F, Figure 3–33. Two questions usually arise at this time.

1. Why did 25% of the liquid change to a gas?
2. How did the mixture of 100% pure liquid go from 110°F to 40°F in such a short space?

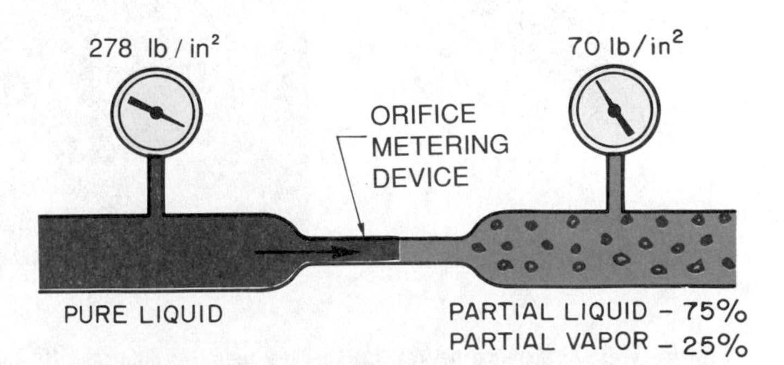

FIGURE 3–33 An orifice metering device.

These questions can be answered by using a garden hose under pressure, in which the water coming out feels cooler, Figure 3–34. The water actually is cooler because some of it evaporates and turns to mist. This evaporation takes heat out of the rest of the water and cools it down. Now when the high-pressure refrigerant passes through the orifice, it does the same thing as the water in the hose; it changes pressure (278 psig to 70 psig), and some of the refrigerant flashes to a vapor (called *flash gas*), which cools the remaining gas to the pressure/temperature relationship of 70 psig, or 40°F. The liquid entering the metering device would heat the liquid in the evaporator if it were warmer when it actually reached the evaporator. This quick drop in pressure in the metering device lowers the boiling point of the liquid leaving the metering device.

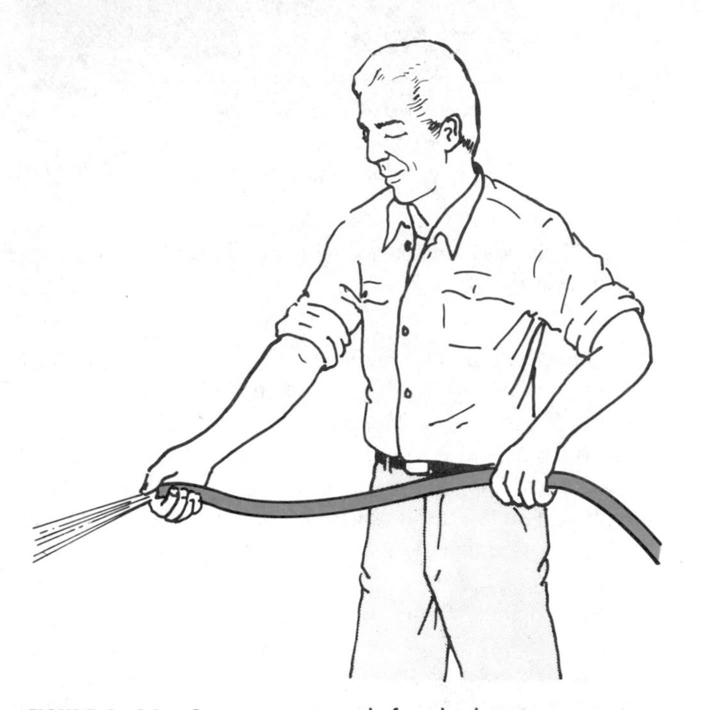

FIGURE 3–34 Person squeezing end of garden hose.

Several types of metering devices are available for many applications. They will be covered in detail in later units. See Figure 3–35 for some examples of the various types of metering devices.

3.11 REFRIGERATION SYSTEM AND COMPONENTS

The basic components of the mechanical compression system have been described according to function. These components must be properly matched for each specific application. For instance, a low-temperature compressor cannot be applied to a high-temperature application because of the pumping characteristics of the compressor. Some equipment can be mismatched successfully by using the manufacturer's data, but only someone with considerable knowledge and experience should do so.

Following is a description of a matched system correctly working at design conditions. Later we will explain malfunctions and adverse operating conditions.

A typical air conditioning system operating at a design temperature of 75°F inside temperature has a humidity (moisture content of the conditioned room air) of 50%. These conditions are to be maintained inside the house. The air in the house gives up heat to the refrigerant. The humidity factor has been brought up at this time because the indoor coil is also responsible for removing some of the moisture from the air to keep the humidity at an acceptable level. This is known as *dehumidifying*.

Moisture removal requires considerable energy. Approximately the same amount (970 Btu) of latent heat removal is required to condense a pound of water from the air as to condense a pound of steam. All air conditioning systems must have a method for dealing with this moisture

(A)

(B)

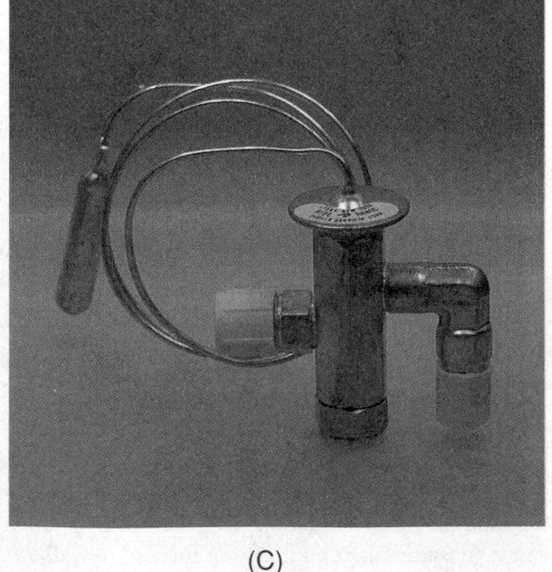

(C)

FIGURE 3–35 Metering devices. (A) Capillary tube. (B) Automatic. (C) Thermostatic. *(A) Courtesy Parker Hannifin Corporation, (B) and (C) Photos by Bill Johnson*

after it has turned to a liquid. Some units drip, some drain the liquid into plumbing waste drains, and some use the liquid at the outdoor coil to help the system capacity by evaporating it at the condenser.

Remember that part of the system is inside the house and part of the system is outside the house. The numbers in the description correspond to the circled numbers in Figure 3–36.

1. A mixture of 75% liquid and 25% vapor leaves the metering device and enters the evaporator.
2. The mixture is R-22 at a pressure of 69 psig, which corresponds to a 40°F boiling point. It is important to remember that *the pressure is 69 psig because the evaporating refrigerant is boiling at 40°F.*
3. The mixture tumbles down the tube in the evaporator with the liquid evaporating as it moves along.
4. When the mixture is about halfway through the coil, it is composed of 50% liquid and 50% vapor and still at the same temperature/pressure relationship because the change of state is taking place.
5. The refrigerant is now 100% vapor. In other words, it has reached the *saturation point* of the vapor. Recall the example using the saturated water table; the water reached various points where it was saturated with heat. We say it is saturated with heat because if any heat is removed at this point, some of the vapor changes back to a liquid and if any heat is added, the vapor rises in temperature. This is called *superheat.* (Superheat is sensible heat.) The saturated vapor is still at 40°F and still able to absorb more heat from the 75°F room air.
6. Pure vapor now exists that is normally superheated about 10°F above the saturation point. Examine the line in Figure 3–36 at this point and you will see that the temperature is about 50°F.

   NOTE: To arrive at the correct superheat reading at this point, take the following steps.

   a. Note the suction pressure reading from the suction gage: 69 psig.
   b. Convert the suction pressure reading to suction temperature using the pressure/temperature chart for R-22: 40°F.
   c. Use a suitable thermometer to record the actual temperature of the suction line: 50°F.
   d. Subtract the saturated suction temperature from the actual suction line temperature: 50°F − 40°F = 10°F of superheat.

   This vapor is said to be heat laden because it contains the heat removed from the room air. The heat was absorbed into the vaporizing refrigerant that boiled off to this vapor in the end of the suction line.
7. The vapor is drawn into the compressor by its pumping action, which creates a low-pressure suction. When the vapor left the evaporator, its temperature was about 50°F with 10°F of superheat above the satu-

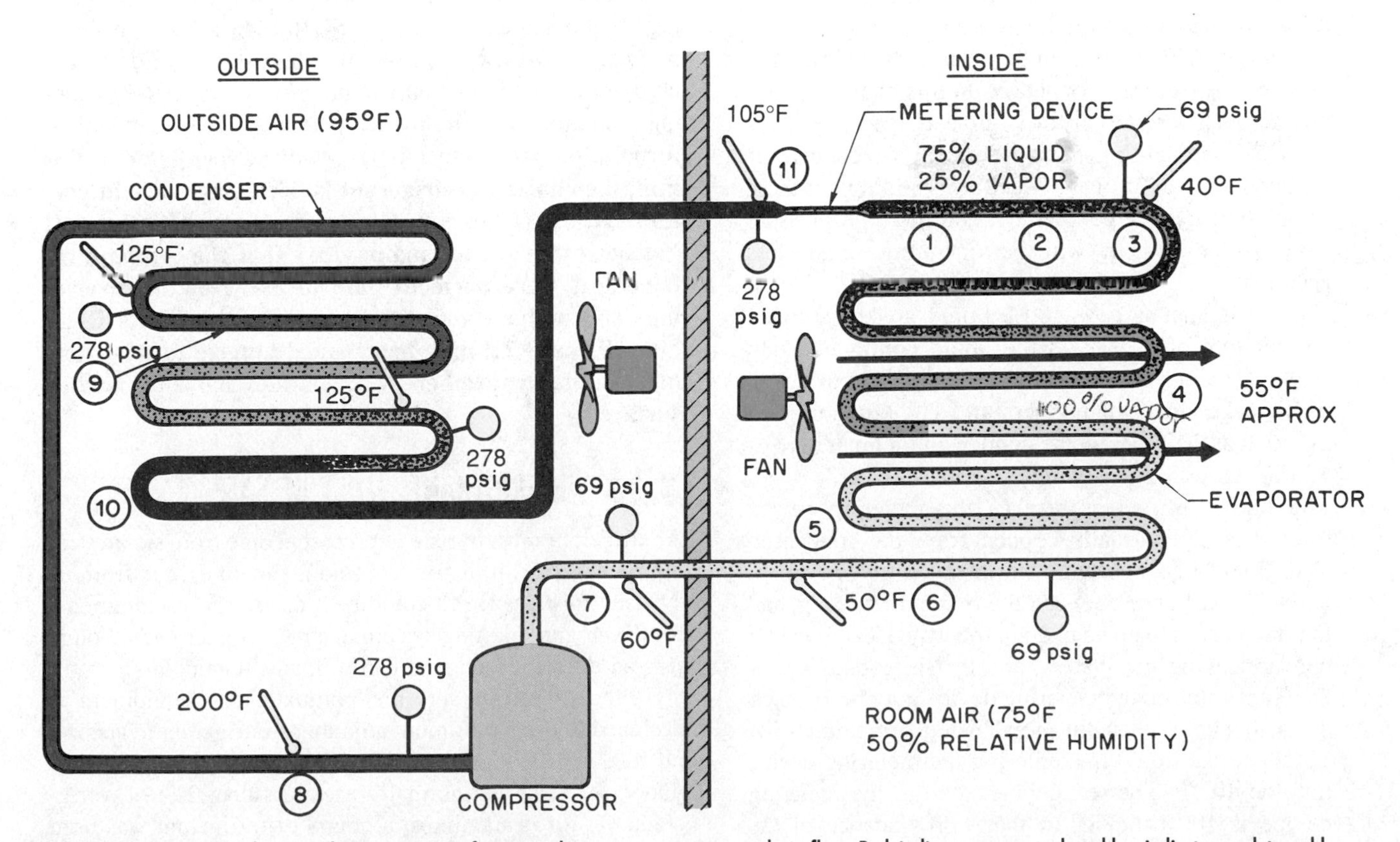

FIGURE 3–36 Typical air conditioning system for R-22 showing temperatures and air flow. Red indicates warm to hot; blue indicates cool to cold.

rated boiling temperature. As the vapor moves along toward the compressor, it is contained in the suction line. This line is usually copper and insulated to keep it from drawing heat into the system from the surroundings; however, it still picks up some heat. Because the suction line carries vapor, any heat that it picks up will quickly raise the temperature. Remember that it does not take much sensible heat to raise the temperature of a vapor. Depending on the length of the line and the quality of the insulation, the suction line temperature may be 60°F at the compressor end.

8. Highly superheated gas leaves the compressor through the *hot gas line* on the high-pressure side of the system. This line normally is very short because the condenser is usually close to the compressor. On a hot day the hot gas line may be close to 200°F with a pressure of 278 psig. Because the saturated temperature corresponding to 278 psig is 125°F, the hot gas line has about 70°F of superheat that must be removed before condensing can occur. Because the line is so hot and a vapor is present, the line will give up heat readily to the surroundings. The surrounding air temperature is 95°F.
9. The superheat has been removed down to the 125°F condensing temperature, and liquid refrigerant is beginning to form. Now notice that the coil temperature is corresponding to the high-side pressure of 278 psig and 125°F. The high-pressure reading of 278 psig is due to the refrigerant condensing at 125°F. The condensing conditions are arrived at by knowing the efficiency of the condenser. In this example we use a standard condenser, which has a condensing temperature about 30°F higher than the surrounding air used to absorb heat from the condenser. In this example, 95°F outside air is used to absorb the heat so 95°F + 30°F = 125°F condensing temperature. Some condensers will condense at 25°F above the surrounding air; these are high-efficiency condensers, and the high-pressure side of the system will be operating under less pressure.
10. The refrigerant is now 100% liquid at the saturated temperature of 125°F. As the liquid continues along the coil, the air continues to cool the liquid to below the actual condensing temperature. The liquid may go as much as 20°F below the condensing temperature of 125°F before it reaches the metering device.
11. The liquid refrigerant reaches the metering device through a pipe, usually copper, from the condenser. This liquid line is often field-installed and not insulated. The distance between the two may be long, and the line may give up heat along the way. Heat given up here is leaving the system, and that is good. The refrigerant entering the metering device may be as much as 20°F cooler than the condensing temperature of 125°F, so the liquid line entering the metering device may be 105°F. The refrigerant entering the metering device is 100% liquid. In the short distance of the metering device's orifice (a pinhole about the size of a small sewing needle), the above liquid is changed to a mixture of about 75% liquid and 25% vapor. The 25% vapor is known as *flash gas* and is used to cool the remaining 75% of the liquid down to 40°F, the boiling temperature of the evaporator.

The refrigerant has now completed the refrigeration cycle and is ready to go around again. It should be evident that a refrigerant does the same thing over and over, changing from a liquid to a vapor in the evaporator and back to liquid form in the condenser. The expansion device meters the flow to the evaporator, and the compressor pumps the refrigerant out of the evaporator.

The following statements briefly summarize the refrigeration cycle.

1. **The evaporator absorbs heat into the system.**
2. **The condenser rejects heat from the system.**
3. **The compressor pumps the heat-laden vapor through the system.**
4. **The expansion device meters the flow of refrigerant.**

3.12 REFRIGERANTS

Previously we have used water and R-22 as examples of refrigerants. Although many products have the characteristics of a refrigerant, only a few will be covered here. Residential and light commercial air conditioning and refrigeration systems commonly use three refrigerants (yet another is still encountered in existing equipment): R-22, used primarily in air conditioning; R-12 used primarily in medium- and high-temperature refrigeration ⊛**(R-134a is being used as a replacement for R-12 in some installations)**⊛ R-502, used primarily in low-temperature refrigeration (R-22 is being used more often in low temperature). R-500 can still be found in older equipment. ⊛**As will be seen later in this unit, the choice of refrigerant is becoming more important because of the pollution factor of the refrigerant. It has been thought for many years that the common refrigerants were perfectly safe to use. New discoveries have shown that some of the common refrigerants, R-12, 500, 502, and 22, may be causing damage to the ozone layer in the stratosphere, 7 to 30 miles above the earth's surface.**⊛

3.13 REFRIGERANTS MUST BE SAFE

A refrigerant must be safe to protect people from sickness or injury, even death, if the refrigerant should escape from its system. For instance, it could be a disaster to use ammonia for the air conditioning system in a public place even though it is an efficient refrigerant from many standpoints.

Modern refrigerants are nontoxic, and equipment is designed to use a minimum amount of refrigerant to accomplish its job. A household refrigerator or window air conditioner, for example, normally uses less than 2 lb of refrigerant, yet for years almost a pound of refrigerant was used as the propellent in a 16-oz aerosol can of hair spray.

SAFETY PRECAUTION: *Because refrigerants are heavier than air, proper ventilation is important. For example, if a leak in a large container of refrigerant should occur in a basement, the oxygen could be displaced by the refrigerant and a person could be overcome. Avoid open flame when a refrigerant is present. When refrigeration equipment or cylinders are located in a room with an open gas flame, such as a pilot light on a gas water heater or furnace, the equipment must be kept leak free. If the refrigerant escapes and gets to the flame, the flame will sometimes burn an off-blue or blue-green. This means the flame is giving off a toxic and corrosive gas that will deteriorate any steel in the vicinity and burn the eyes and nose and severely hamper the breathing of anyone in the room. The refrigerants themselves will not burn.*

3.14 REFRIGERANTS MUST BE DETECTABLE

A good refrigerant must be readily detectable. The first leak detection device that can be used for some large leaks is listening for the hiss of the escaping refrigerant, Figure 3–37(A). This is not the best way in all cases as some leaks

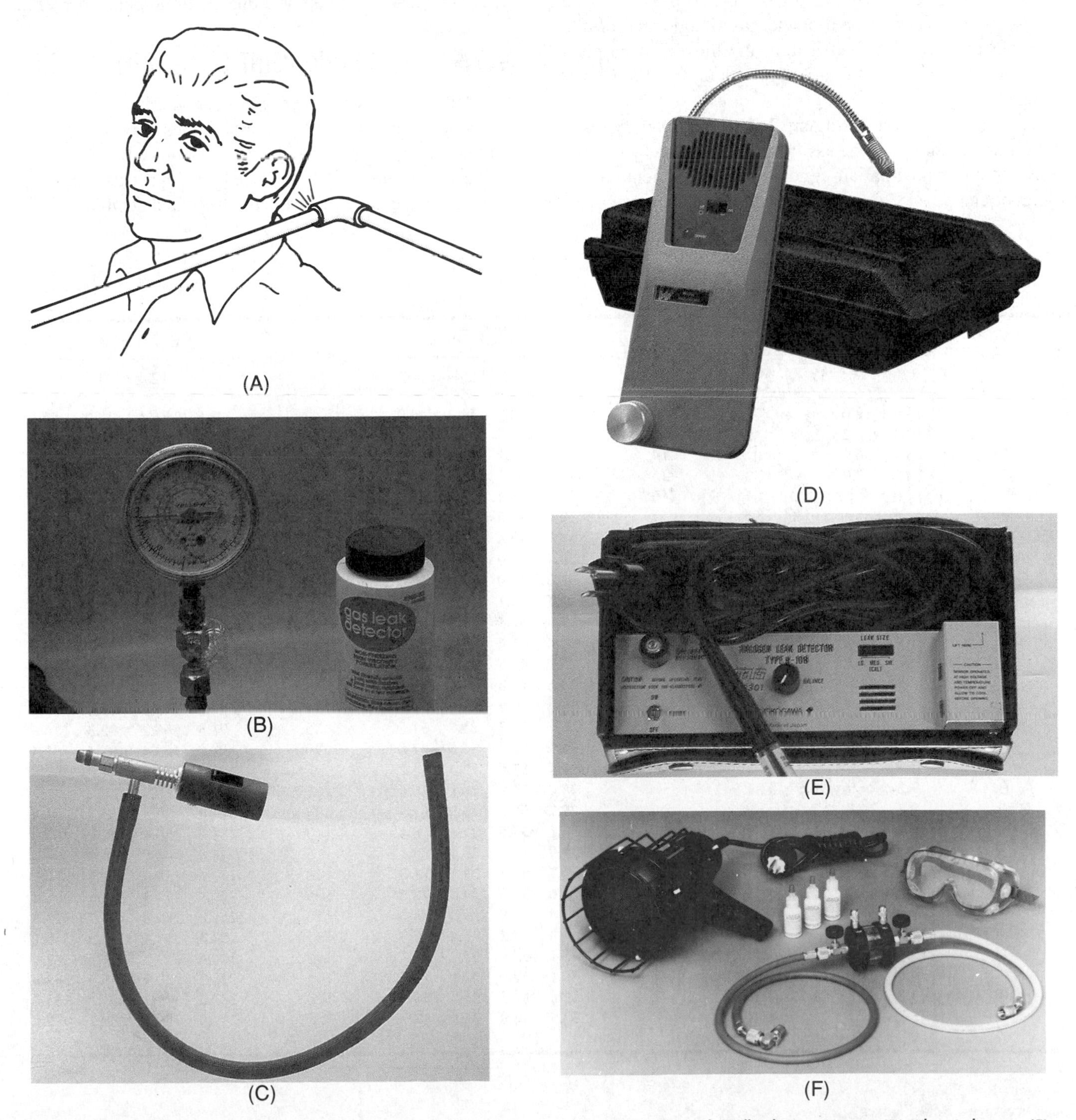

FIGURE 3–37 Methods and equipment for determining refrigerant leaks. *(B), (C), and (E) Photos by Bill Johnson, Courtesy (D) White Industries, (F) Spectronics Corporation.*

may be so small they may not be heard by the human ear. However, many leaks can be found in this way.

Soap bubbles are a practical and yet simple leak detector. Commercially prepared products that blow large elastic types of bubbles are used by many service technicians. These are valuable when it is known that a leak is in a certain area. Soap bubble solution can be applied with a brush to the tubing joint to see exactly where the leak is. Leaking refrigerant will cause bubbles, Figure 3–37(B). At times a piece of equipment can be submerged in water to watch for bubbles. This is effective when it can be used.

The halide leak detector, Figure 3–37(C), is available for use with acetylene or propane gas. It operates on the principle that when the refrigerant is allowed in an open flame in the presence of glowing copper the flame will change color.

The leak detector in Figure 3–37(D) is battery operated, is small enough to be easily carried, and has a flexible probe. Some residential air conditioning equipment has refrigerant charge specifications that call for half-ounce accuracy. The electronic leak detectors are capable of detecting leak rates down to one-half ounce per year, Figure 3–37(E).

Another system uses a high-intensity ultraviolet lamp, Figure 3–37(F). An additive is induced into the refrigerant system. The additive will show as a bright yellow-green glow under the ultraviolet lamp at the source of the leak. The area can be wiped clean with a general purpose cleaner after the leak has been repaired and the area can be reinspected. The additive can remain in the system. Should a new leak be suspected at a later date it will still show the yellow-green color under the ultraviolet light. This system will detect leaks as small as a quarter of an ounce per year.

3.15 THE BOILING POINT OF THE REFRIGERANT

The boiling point of the refrigerant should be low at atmospheric pressure so that low temperatures may be obtained without going into a vacuum. For example, R-502 can be boiled as low as −50°F before the boiling pressure goes into a vacuum, whereas R-12 can only be boiled down to −21°F

TEMPERATURE °F	REFRIGERANT 12	22	134a	502	TEMPERATURE °F	REFRIGERANT 12	22	134a	502	TEMPERATURE °F	REFRIGERANT 12	22	134a	502
−60	19.0	12.0		7.2	12	**15.8**	**34.7**	**13.2**	**43.2**	42	**38.8**	**71.4**	**37.0**	**83.8**
−55	17.3	9.2		3.8	13	**16.4**	**35.7**	**13.8**	**44.3**	43	**39.8**	**73.0**	**38.0**	**85.4**
−50	15.4	6.2		0.2	14	**17.1**	**36.7**	**14.4**	**45.4**	44	**40.7**	**74.5**	**39.0**	**87.0**
−45	13.3	2.7		**1.9**	15	**17.7**	**37.7**	**15.1**	**46.5**	45	**41.7**	**76.0**	**40.1**	**88.7**
−40	11.0	**0.5**	14.7	**4.1**	16	**18.4**	**38.7**	**15.7**	**47.7**	46	**42.6**	**77.6**	**41.1**	**90.4**
−35	8.4	**2.6**	12.4	**6.5**	17	**19.0**	**39.8**	**16.4**	**48.8**	47	**43.6**	**79.2**	**42.2**	**92.1**
−30	5.5	**4.9**	9.7	**9.2**	18	**19.7**	**40.8**	**17.1**	**50.0**	48	**44.6**	**80.8**	**43.3**	**93.9**
−25	2.3	**7.4**	6.8	**12.1**	19	**20.4**	**41.9**	**17.7**	**51.2**	49	**45.7**	**82.4**	**44.4**	**95.6**
−20	**0.6**	**10.1**	3.6	**15.3**	20	**21.0**	**43.0**	**18.4**	**52.4**	50	**46.7**	**84.0**	**45.5**	**97.4**
−18	**1.3**	**11.3**	2.2	**16.7**	21	**21.7**	**44.1**	**19.2**	**53.7**	55	**52.0**	**92.6**	**51.3**	**106.6**
−16	**2.0**	**12.5**	0.7	**18.1**	22	**22.4**	**45.3**	**19.9**	**54.9**	60	**57.7**	**101.6**	**57.3**	**116.4**
−14	**2.8**	**13.8**	**.3**	**19.5**	23	**23.2**	**46.4**	**20.6**	**56.2**	65	**63.8**	**111.2**	**64.1**	**126.7**
−12	**3.6**	**15.1**	**1.2**	**21.0**	24	**23.9**	**47.6**	**21.4**	**57.5**	70	**70.2**	**121.4**	**71.2**	**137.6**
−10	**4.5**	**16.5**	**2.0**	**22.6**	25	**24.6**	**48.8**	**22.0**	**58.8**	75	**77.0**	**132.2**	**78.7**	**149.1**
−8	**5.4**	**17.9**	**2.8**	**24.2**	26	**25.4**	**49.9**	**22.9**	**60.1**	80	**84.2**	**143.6**	**86.8**	**161.2**
−6	**6.3**	**19.3**	**3.7**	**25.8**	27	**26.1**	**51.2**	**23.7**	**61.5**	85	**91.8**	**155.7**	**95.3**	**174.0**
−4	**7.2**	**20.8**	**4.6**	**27.5**	28	**26.9**	**52.4**	**24.5**	**62.8**	90	**99.8**	**168.4**	**104.4**	**187.4**
−2	**8.2**	**22.4**	**5.5**	**29.3**	29	**27.7**	**53.6**	**25.3**	**64.2**	95	**108.2**	**181.8**	**114.0**	**201.4**
0	**9.2**	**24.0**	**6.5**	**31.1**	30	**28.4**	**54.9**	**26.1**	**65.6**	100	**117.2**	**195.9**	**124.2**	**216.2**
1	**9.7**	**24.8**	**7.0**	**32.0**	31	**29.2**	**56.2**	**26.9**	**67.0**	105	**126.6**	**210.8**	**135.0**	**231.7**
2	**10.2**	**25.6**	**7.5**	**32.9**	32	**30.1**	**57.5**	**27.8**	**68.4**	110	**136.4**	**226.4**	**146.4**	**247.9**
3	**10.7**	**26.4**	**8.0**	**33.9**	33	**30.9**	**58.8**	**28.7**	**69.9**	115	**146.8**	**242.7**	**158.5**	**264.9**
4	**11.2**	**27.3**	**8.6**	**34.9**	34	**31.7**	**60.1**	**29.5**	**71.3**	120	**157.6**	**259.9**	**171.2**	**282.7**
5	**11.8**	**28.2**	**9.1**	**35.8**	35	**32.6**	**61.5**	**30.4**	**72.8**	125	**169.1**	**277.9**	**184.6**	**301.4**
6	**12.3**	**29.1**	**9.7**	**36.8**	36	**33.4**	**62.8**	**31.3**	**74.3**	130	**181.0**	**296.8**	**198.7**	**320.8**
7	**12.9**	**30.0**	**10.2**	**37.9**	37	**34.3**	**64.2**	**32.2**	**75.8**	135	**193.5**	**316.6**	**213.5**	**341.2**
8	**13.5**	**30.9**	**10.8**	**38.9**	38	**35.2**	**65.6**	**33.2**	**77.4**	140	**206.6**	**337.2**	**229.1**	**362.6**
9	**14.0**	**31.8**	**11.4**	**39.9**	39	**36.1**	**67.1**	**34.1**	**79.0**	145	**220.3**	**358.9**	**245.5**	**385.0**
10	**14.6**	**32.8**	**11.9**	**41.0**	40	**37.0**	**68.5**	**35.1**	**80.5**	150	**234.6**	**381.5**	**262.7**	**408.4**
11	**15.2**	**33.7**	**12.5**	**42.1**	41	**37.9**	**70.0**	**36.0**	**82.1**	155	**249.5**	**405.1**	**280.7**	**432.9**

Vacuum—Red Figures
Gage Pressure—Bold Figures

FIGURE 3–38 Pressure/temperature relationship chart in inches Hg vacuum or psig.

before it goes into a vacuum. Water would have to be boiled at 29.67 in. Hg vacuum just to boil at 40°F.

NOTE: When using the compound gage below atmospheric pressure, the scale reads in reverse of the inches of mercury absolute scale. It starts at atmospheric pressure and counts down to a perfect vacuum, called inches of mercury vacuum. When possible, design engineers avoid using refrigerants that boil below 0 psig. This is one reason why R-502 is a good choice for a low-temperature system. When a system operates in a vacuum and a leak occurs, the atmosphere is pulled inside the system instead of the refrigerant leaking out of the system.

3.16 PUMPING CHARACTERISTICS

The pumping characteristics have to do with how much refrigerant vapor is pumped per amount of work accomplished. Water was disqualified as a practical refrigerant for small equipment partly for this reason. One pound of water at 40°F has a vapor volume of 2445 ft^3 compared to about 0.6 ft^3 for R-22. Thus, the compressor would have to be very large for a water system.

Modern refrigerants meet all of these requirements better than any of the older types. Figure 3–38 presents the pressure/temperature chart for the refrigerants we have discussed.

3.17 REFRIGERANT CHEMICAL MAKEUP

Each refrigerant has a chemical formula and chemical name. Sometimes this formula or name best indicates the refrigerant used in a particular application. The formulas and names are as follows:

- The chemical formula for R-12 is CCl_2F_2; the chemical name is dichlorodifluoromethane.
- The chemical formula for R-22 is $CHClF_2$; the chemical name is monochlorodifluoromethane.
- The chemical formula for R-134a is CF_3CFH_2; the chemical name is tetrafluoroethane.
- There is no chemical formula or name for R-502. It is an azeotropic mixture of 48.8% R-22 and 51.2% R-115. The chemical formula for R-115 is $CClF_2CF_3$; its chemical name is chloropentafluoroethane. This mixture is produced while the refrigerant is being manufactured. When a refrigerant is mixed as an azeotrope, it will not separate into the different components should a leak occur. This can be important for some refrigerants because one of their components may be flammable.
- R-500 is an azeotropic mixture of 73.8% R-12 and 26.2% R-152a. The chemical formula for R-152a is CH_3CH_2Cl; its chemical name is difluoroethane. R-500 is also mixed during manufacture.

3.18 REFRIGERANT CYLINDER COLOR CODES

Each type of refrigerant is contained in a cylinder that has a designated color. Following are the colors for some of the most frequently used refrigerants.

FIGURE 3-39 Color-coded refrigerant cylinders and drums for some of the newer refrigerants available. *Courtesy of Allied Signal Genetron Products*

R-11	Orange
R-12	White
R-22	Green
R-113	Purple
R-134a	Light Blue
R-114	Dark Blue
R-500	Yellow
R-502	Orchid
R-717	Silver

Some equipment manufacturers color code their compressors to indicate the type of refrigerant used in the system. Figure 3–39 shows refrigerant containers for some newer refrigerants.

3.19 RECOVERY, RECYCLE, OR RECLAIM OF REFRIGERANTS

⊛It is necessary for technicians to recover and/or recycle refrigerants during installation and servicing operations to help reduce emissions of chlorofluorocarbons (CFCs) to the atmosphere. Examples of recovery equipment are shown in Figure 3–40. Many larger systems can be fitted with receivers or dump tanks into which the refrigerant can be pumped and stored while the system is serviced. However, in smaller capacity systems it

FIGURE 3–40 Examples of refrigerant recovery systems. *Courtesy Robinair Division, SPX Corporation*

is not often feasible to provide these components. Recovery units or other storage devices may be necessary. Most recovery and/or recycle units that have been developed to date vary in their technology and capabilities so manufacturers' instructions must be followed carefully when using this equipment. Unit 9 includes a detailed description of the recovery, recycle, and reclaim of refrigerants.⊛

3.20 PLOTTING THE REFRIGERANT CYCLE

A graphic picture of the refrigerant cycle may be plotted on a pressure/enthalpy diagram. *Enthalpy* describes how much heat a substance contains from some starting point. Many people refer to enthalpy as total heat. This is not an absolutely accurate statement. It is a convenient statement because it is the total heat from a starting point. Refer to Figure 1–13, the heat/temperature graph for water. We used 0°F as the starting point of heat for water, knowing that you can really remove more heat from the water (ice) and lower the temperature below 0°F. We described the process as the amount of heat added starting at 0°F. This heat is called enthalpy. A similar diagram is available for all refrigerants; it is called a pressure/enthalpy diagram and is used to plot the refrigerant cycle in a complete loop, Figure 3–41.

The pressure/enthalpy diagram plots pressure on the left hand column and enthalpy or total heat on the bottom of the diagram. The enthalpy readings start at −40°F saturated liquid with a heat content of 0 Btu/lb. Readings below −40°F saturated liquid show as minus values. Notice, temperature corresponds to pressure and is plotted on the inside of the chart along the left and right of the horseshoe-shaped curve.

The horseshoe-shaped curve is the saturation curve with the temperature corresponding to the absolute pressure. Anytime a plot falls on this curve, the refrigerant is saturated with heat. There are two saturation curves; the one on the left is the saturated liquid curve. If heat is added, the refrigerant will start changing state to a vapor. If heat is removed, the liquid will be subcooled. The right hand curve is the saturated vapor curve. If heat is added, the vapor will superheat. If heat is removed, the vapor will start changing state to a liquid. Notice that the saturated liquid and vapor curves touch at the top. This is called the critical temperature, or pressure. Above this point, the refrigerant will not condense. It is a vapor regardless of how much pressure is applied.

The area between the saturated liquid and saturated vapor curve, inside the horseshoe-shaped curve, is where the change of state occurs. Anytime a plot falls between the saturation curves, the refrigerant is in the partial liquid, partial vapor state. The slanted, near vertical lines between the saturated liquid and saturated vapor lines are the constant quality lines and describe the percent of liquid to vapor in the mixture between the saturation points. If the plot is closer to the saturated liquid curve, there is more liquid than vapor. If the plot is closer to the saturated vapor curve, there is more vapor than liquid. For example, find a point on the chart at 40°F (from the left column) and 30 Btu/lb (along the bottom), Figure 3–42. This point is inside the horseshoe-

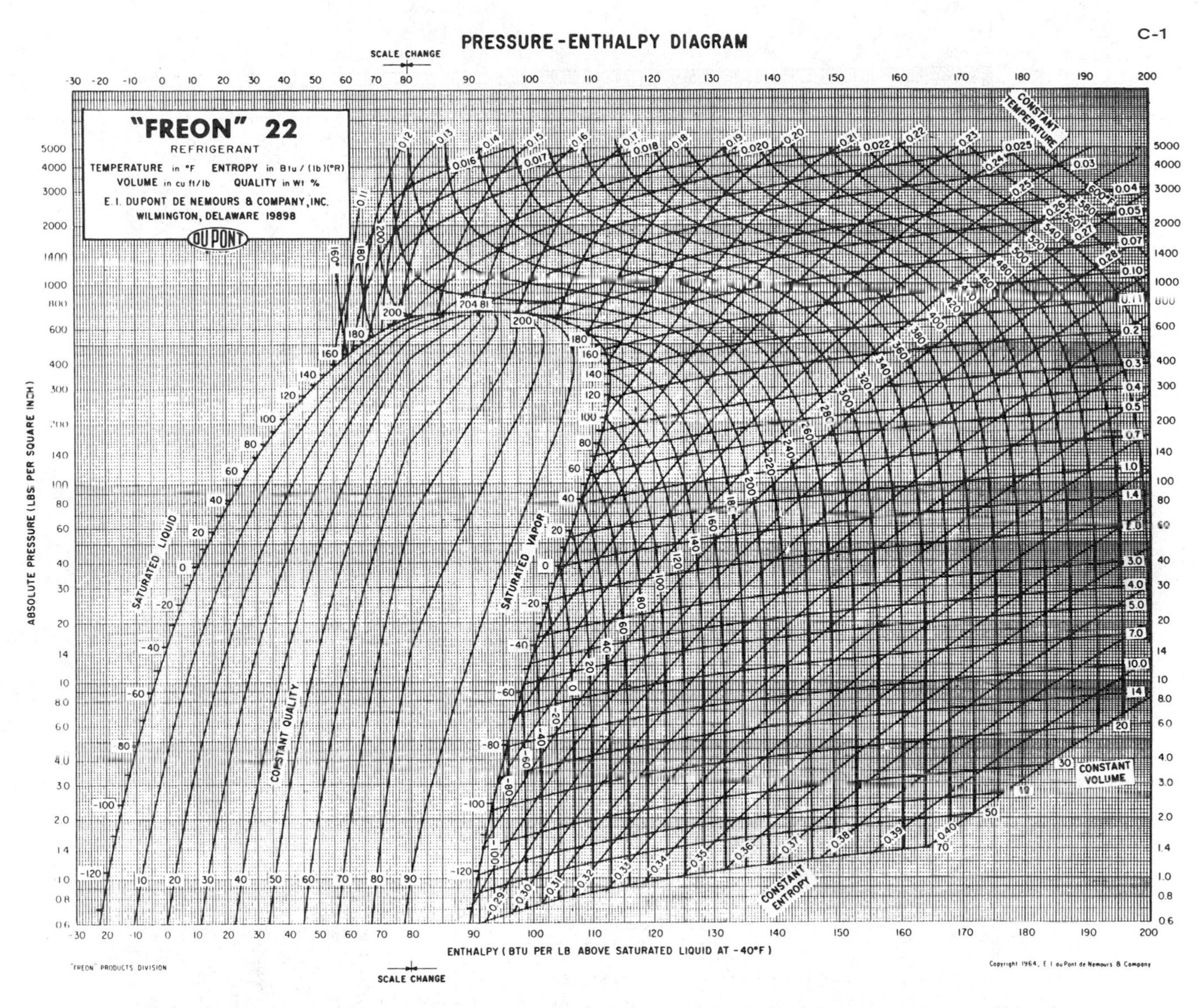

FIGURE 3–41 Pressure/enthalpy chart for R-22 expressed in Btu/lb of refrigerant circulated. The chart uses −40°F saturated liquid as the starting point for heat content. To find this point, find 0 Btu/lb on the bottom line and follow it straight up to the saturated liquid curve at −40°F. *Courtesy E. I. DuPont*

shaped curve and the refrigerant is 90% liquid and 10% vapor.

A refrigeration cycle is plotted in Figure 3–43. The system to be plotted is an air conditioning system using R-22. The system is operating at 130°F condensing temperature (296.8 psig or 311.5 psia discharge pressure) and an evaporating temperature of 40°F (68.5 psig or 83.2 psia suction pressure). The cycle plotted has no subcooling, and 10°F of superheat as the refrigerant leaves the evaporator with another 10°F of superheat absorbed by the suction line in route to the compressor. Notice, the compressor is an air-cooled compressor and the suction gas enters the suction valve adjacent to the cylinders.

1. Refrigerant R-22 enters the expansion device as a saturated liquid at 311.5 psia (296 psig) and 130°F, point A. The heat content is 49 Btu/lb entering the expansion valve and 49 Btu/lb leaving the expansion valve. The temperature of the liquid refrigerant before the valve is 130°F and the temperature leaving the valve is 40°F. The temperature drop can be accounted for by observing that we have 100% liquid entering the valve and about 67% liquid leaving the valve About 33% of the liquid changed to a vapor (called flash gas), lowering the remaining liquid temperature to 40°F.
2. Usable refrigeration starts at point B where the refrigerant has a heat content of 49 Btu/lb. As heat is added

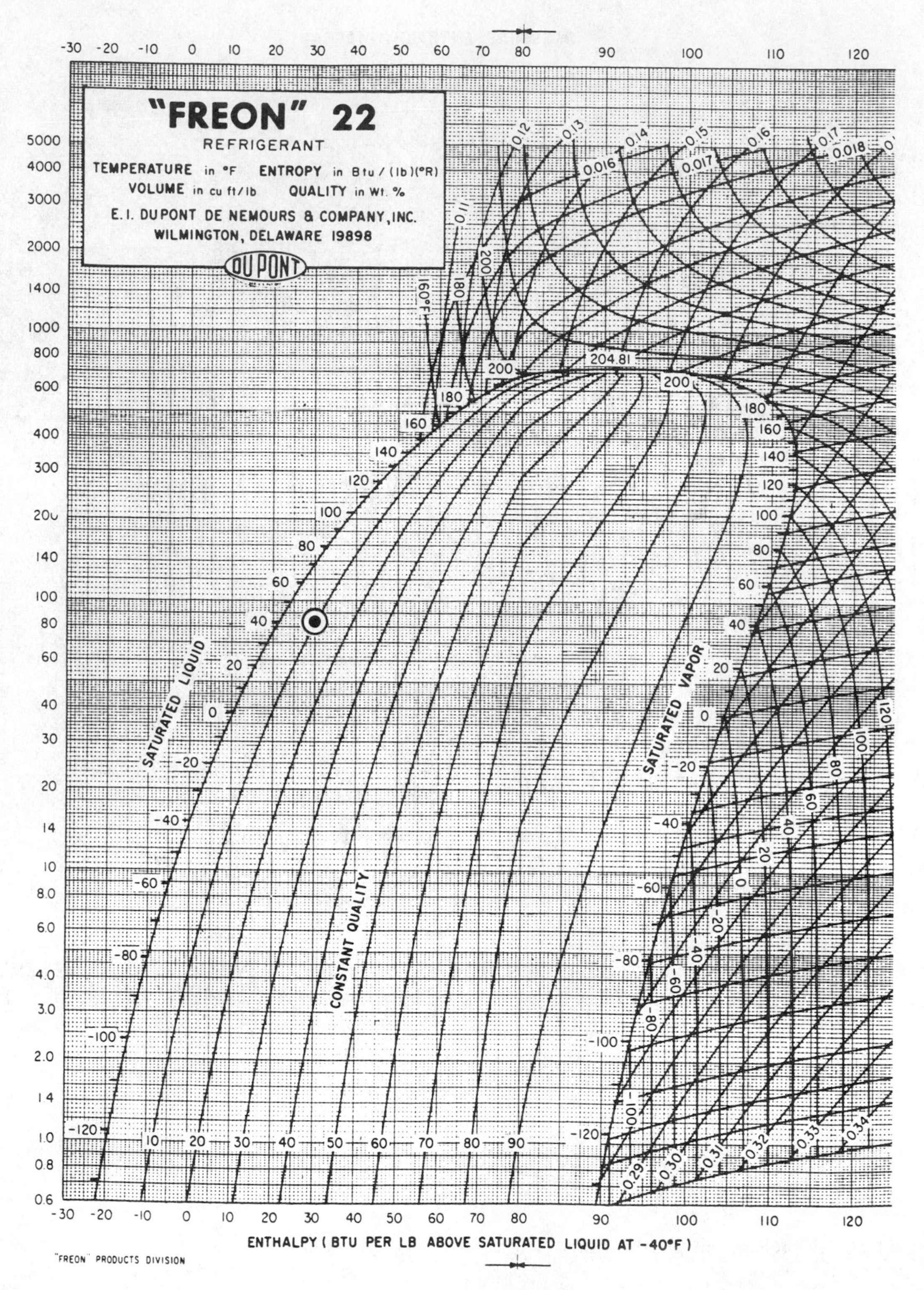

FIGURE 3–42 Follow the 40°F line to the right until it intersects the vertical 30 Btu/lb line and we find the refrigerant mixture to be 90% liquid and 10% vapor. *Courtesy E. I. DePont*

in the evaporator, the refrigerant changes the state to a vapor. All liquid changes to a vapor when it reaches the saturated vapor curve, and a small amount of heat is added to the refrigerant in the form of superheat (10°F). The refrigerant leaves the evaporator at point C, it contains about 110 Btu/lb. This is a net refrigeration effect of 61 Btu/lb (110 Btu/lb − 49 Btu/lb = 61 Btu/lb) of refrigerant circulated. The net refrigeration effect is the same as usable refrigeration, the heat actually extracted from the conditioned space. About 10 more Btu/lb are absorbed into the suction line before reaching the compressor inlet at point D. This is not usable refrigeration

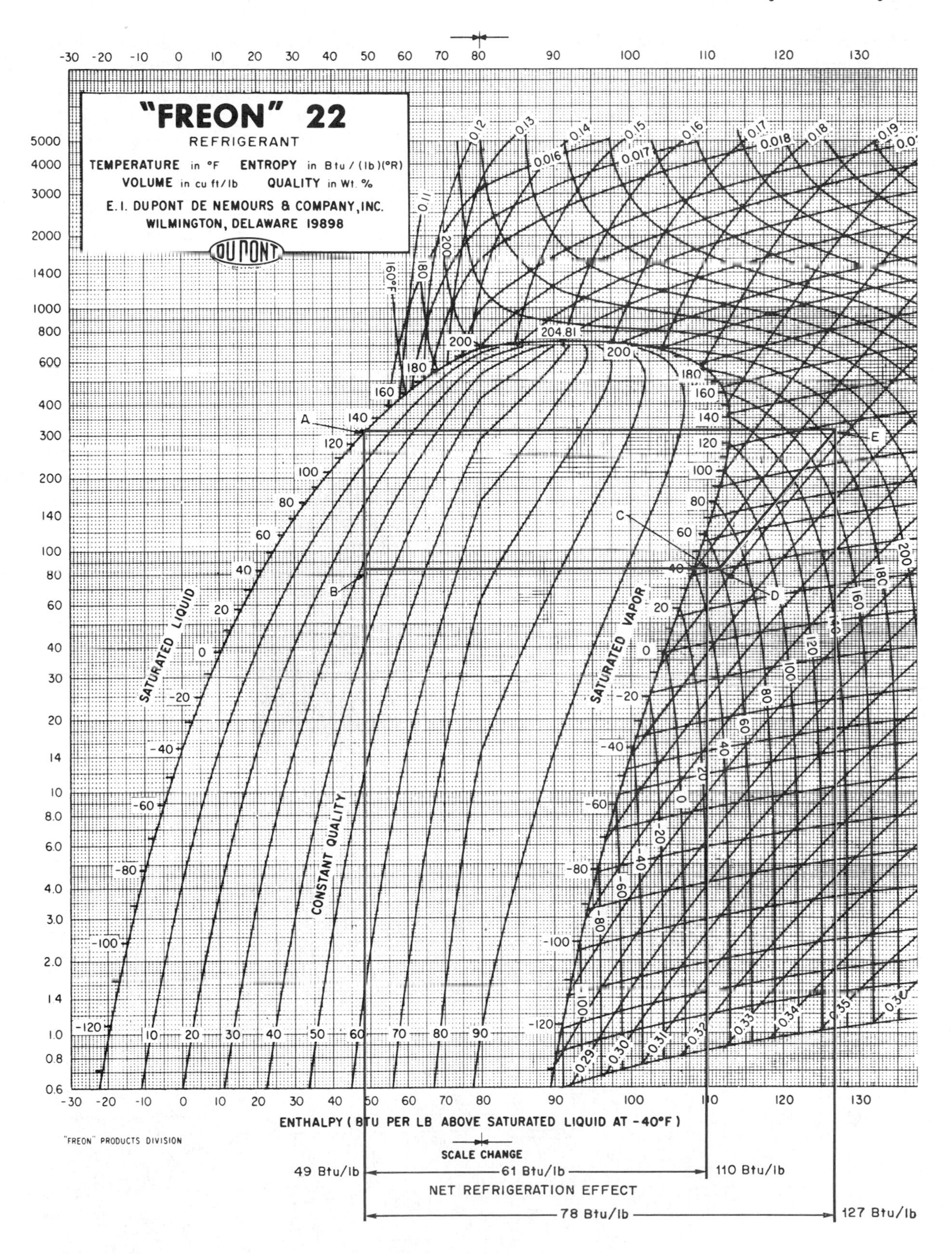

FIGURE 3–43 Refrigeration cycle plotted on the pressure/enthalpy chart. *Courtesy E. I. DuPont*

because the heat does not come from the conditioned space, but it is heat that must be pumped by the compressor and rejected by the condenser.

3. The refrigerant enters the compressor at point D and leaves the compressor at point E. No heat has been added in the compressor except heat of compression,

because the compressor is air cooled. The refrigerant enters the compressor cylinder from the suction line. (A fully hermetic compressor with a suction cooled motor would not plot out just like this. We have no way of knowing how much heat is added by the motor so we don't know what the temperature of the suction gas entering the compressor cylinder would be for a suction cooled motor. Manufacturers obtain their own figures for this using internal thermometers during testing.) The compression process in this compressor is called adiabatic compression, where no heat is added or taken away.

4. The refrigerant leaves the compressor at point E and contains about 127 Btu/lb. This condenser must reject 78 Btu/lb (127/Btu/lb − 49 Btu/lb = 78 Btu/lb), called the heat of rejection. The temperature of the discharge gas is about 190°F (see the constant temperature lines for temperature of superheated gas). When the hot gas leaves the compressor it contains the maximum amount of heat that must be rejected by the condenser.
5. The refrigerant enters the condenser at point E as a highly superheated gas. The refrigerant condensing temperature is 130°F and the hot gas leaving the compressor is 190°F, so it contains 60°F (190°F − 130°F = 60°F) of superheat. The condenser will first remove the superheat down to the condensing temperature, then it will condense the refrigerant to a liquid of 130°F for reentering the expansion device at point A for another trip around the cycle.

The refrigerant cycle in the previous example can be improved by removing some heat from the condensed liquid by subcooling it. This can be seen in Figure 3–44, a scaled up diagram. The same conditions are used in this figure as in Figure 3–43, except the liquid is subcooled 15°F (from 130°F condensing temperature to 115°F liquid). The system then has a net refrigeration effect of 68 Btu/lb instead of 61 Btu/lb. This is an increase in capacity of about 11%. Notice the liquid leaving the expansion valve is only about 23% vapor instead of the 33% vapor in the first example. This is where the capacity is gained. Less capacity is lost to flash gas.

Other conditions may be plotted on the pressure/enthalpy diagram. For example, suppose the head pressure is raised due to a dirty condenser, Figure 3–45. Using the first example of saturated liquid and raising the condensing temperature to 140°F (337.2 psig or 351.9 psia), we see the percent of liquid leaving the expansion valve to be about 64% with a heat content of 53 Btu/lb. Using the same heat content leaving the evaporator, 110 Btu/lb, we have a net refrigeration effect of 57 Btu/lb. This is a net refrigeration effect reduction of about 7% from the original example, which contained 49 Btu/lb at the same point. This shows the importance of keeping condensers clean.

Figure 3–46 shows how increased superheat affects the first system in Figure 3–43. The suction line has not been insulated and absorbs heat. The suction gas may leave the evaporator at 50°F and rise to 80°F before entering the compressor. Notice the high discharge temperature (about

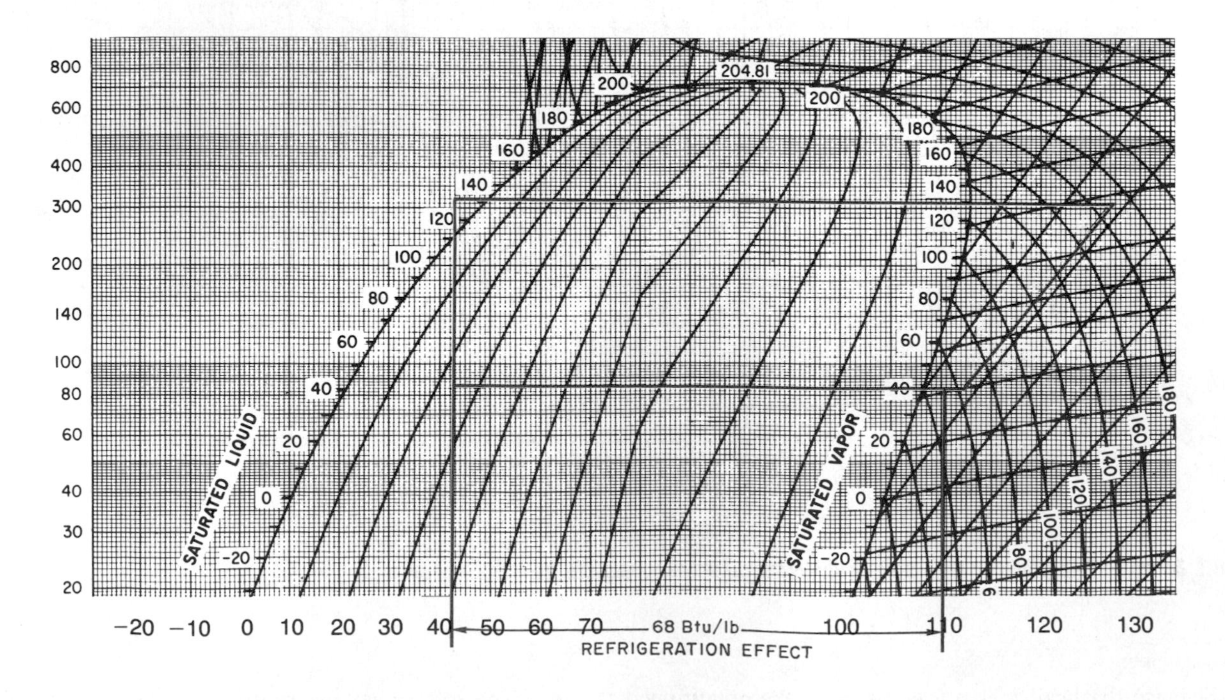

FIGURE 3–44 Subcooling the refrigerant 15°F (compared to Figure 3–43) increases the net refrigeration effect from 61 to 68 Btu/lb. This increases the capacity by about 11%. Only a very slight amount of extra heat must be rejected by the condenser with no added load to the compressor. *Courtesy E. I. DuPont*

Pressure-Enthalpy Diagram for R-22

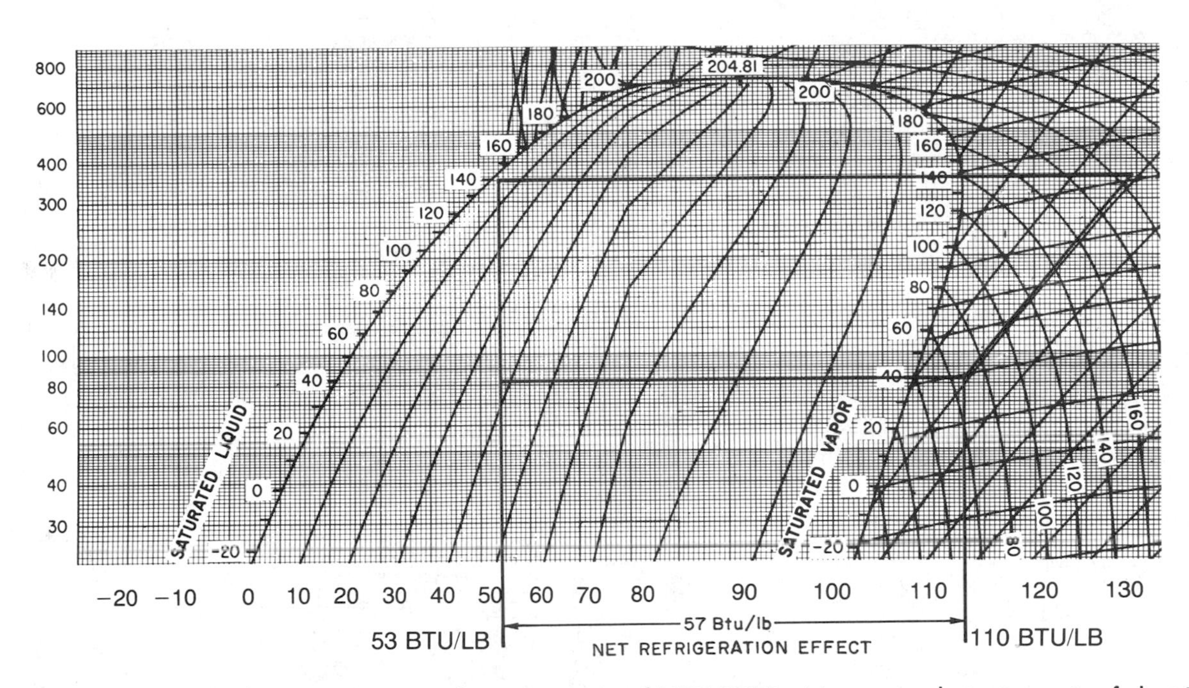

FIGURE 3–45 An increase in head pressure to a new condensing temperature of 140°F (337.2 psig) causes a reduction in capacity of about 7%. *Courtesy E. I. DuPont*

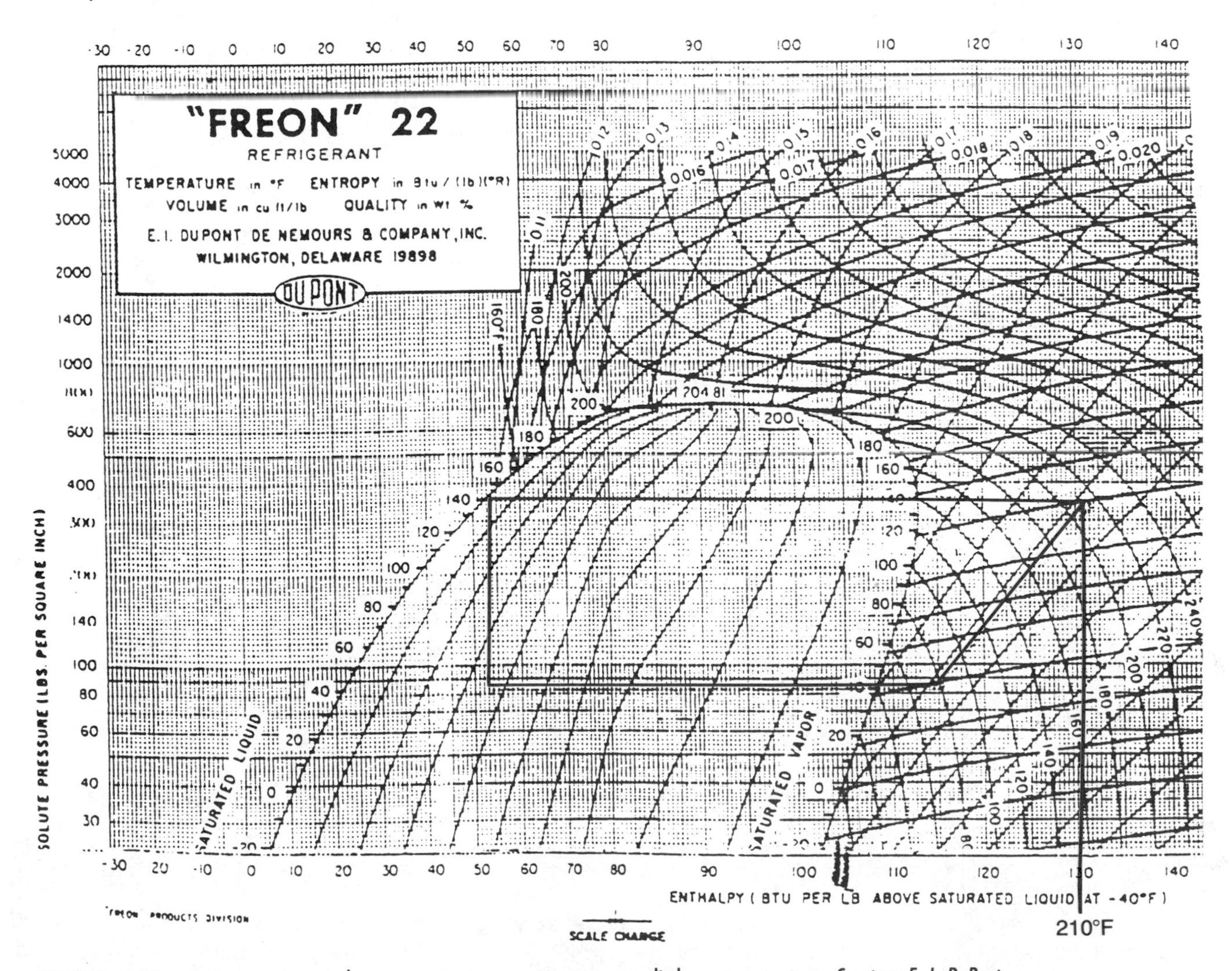

FIGURE 3–46 An increase in superheat causes an increase in compressor discharge temperature. *Courtesy E. I. DuPont*

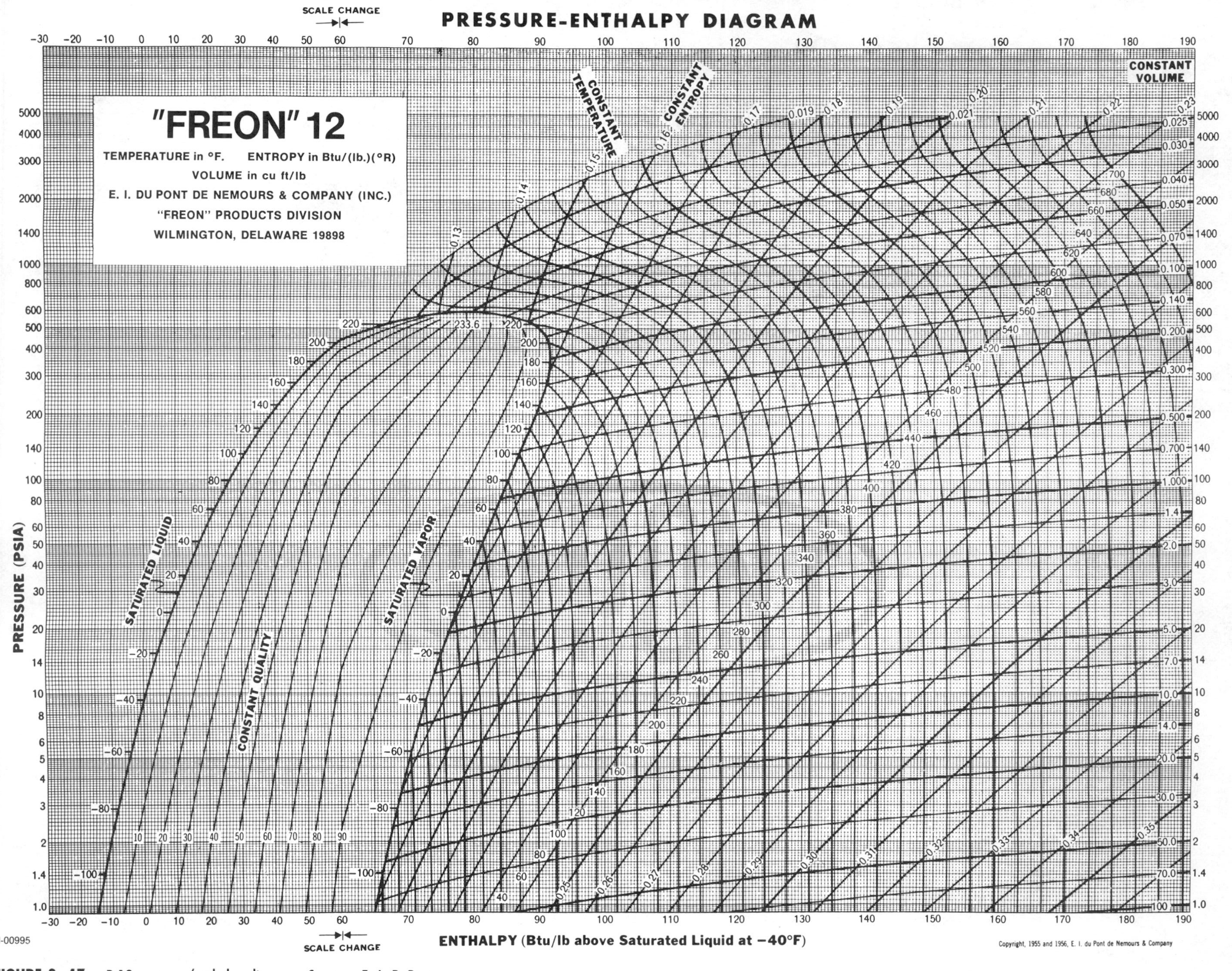

FIGURE 3–47 R-12 pressure/enthalpy diagram. *Courtesy E. I. DuPont*

210°F). This is approaching the temperature that will cause oil to breakdown and form acids in the system. Most compressors must not exceed 250°F. The compressor must pump more refrigerant to accomplish the same refrigeration effect and the condenser must reject more heat.

R-12 has been the most popular refrigerant for medium-temperature applications for many years. The CFC/ozone depletion issue has caused the manufacturers to explore two different refrigerants for the replacement of R-12, namely, R-134a and R-22. R-134a is considered to have no ozone depletion potential, which is discussed in detail in Unit 9, but may have some compatability problems with current equipment and oils in existing systems. Figure 3–47 shows a pressure/enthalpy chart for R-12.

The following sequence shows how R-12 performs in a typical medium-temperature application. We will use an evaporator temperature of 20°F and a condensing temperature of 115°F. This lower condensing temperature is common for medium- and low-temperature applications. Follow this description in the illustration in Figure 3–48. The pressure/enthalpy explanation appears in Figure 3–49.

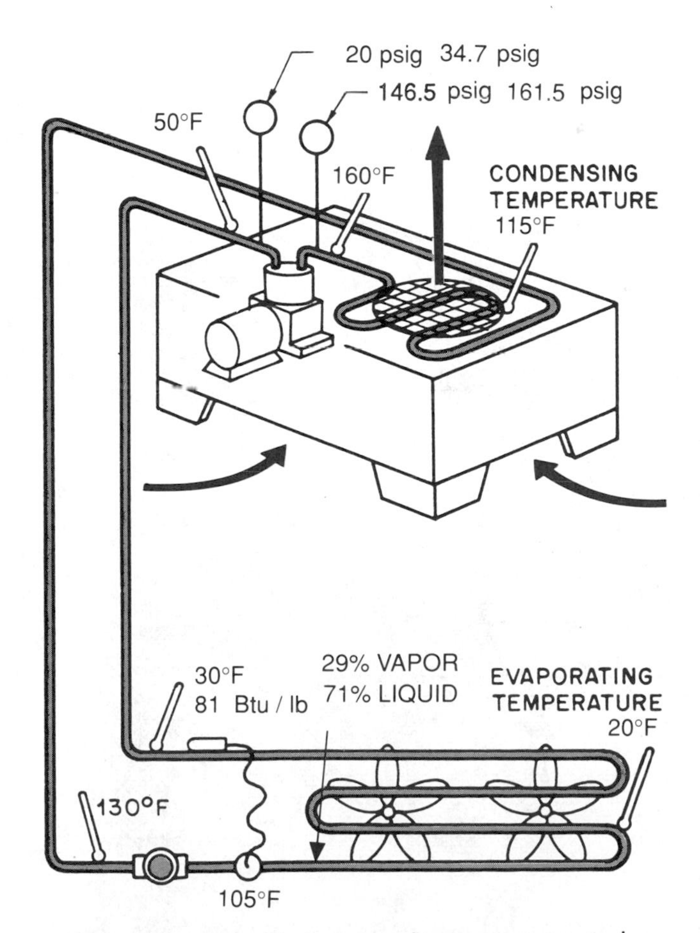

FIGURE 3-48 Air conditioning system showing temperatures and pressures.

1. Liquid enters the expansion valve at point A at 105°F. Note, the liquid is subcooled 10°F from 115°F.
2. Partial liquid and partial vapor leave the expansion valve at point B (29% vapor and 71% liquid). The heat content per pound is 32 Btu/lb.
3. Vapor refrigerant leaves the evaporator at point C with 10° of superheat and a heat content of 81 Btu/lb. Note, the net refrigeration effect is 49 Btu/lb (81 − 32 = 49).
4. The refrigerant enters the compressor at point D at a temperature of 50°F. The refrigerant has a total of 30° of superheat considering what was picked up in the evaporator and the suction line.
5. The refrigerant is compressed along the line between point D and E where it leaves the compressor at a temperature of 160°F. Note that the low temperature is because this is R-12 and we are operating at a lower head pressure to keep the discharge gas temperature down. This is accomplished with a larger condenser.
6. The refrigerant is condensed along the line from point E to A at 115°F.

Following is an example of the same system using R-22, for medium-temperature application. Follow the description in Figure 3–50 and compare with the previous example.

1. Refrigerant enters the expansion valve at point A at 105°F, subcooled 10°F from the condensing temperature of 115°F, just like the example for R-12.
2. The refrigerant leaves the expansion valve at point B at 28% vapor and 72% liquid with a heat content of 41 Btu/lb.
3. The refrigerant leaves the evaporator at point C in the vapor state with 10° of superheat, a heat content of 108 Btu/lb, and a net refrigeration effect of 67 Btu/lb.
4. Refrigerant vapor enters the compressor at point D at 50°F, containing 30° of superheat.
5. The vapor refrigerant is compressed on the line from D to E and leaves the compressor at point E at a temperature of about 185°F. *This is 25°F higher than the temperature for the same conditions for R-12 and is one of the main differences in the refrigerants. R-22 has a much higher discharge temperature than R-12.* As the condensing temperature of the application becomes higher, the temperature rises. At some point, the designer must decide to either use a different refrigerant or change the application.

R-134a, a replacement refrigerant for R-12, would plot out on the pressure/enthalpy chart as follows. Use Figure 3–51 to follow this example.

1. Refrigerant enters the expansion valve at point A at 105°F, subcooled 10° from 115°F.
2. Refrigerant leaves the expansion valve at point B with a heat content of 47 Btu/lb. Note, there are no constant quality lines on this chart to show the percent of liquid and vapor.
3. Vapor refrigerant leaves the evaporator at point C with a heat content of 114 Btu/lb and a net refrigeration effect of 67 Btu/lb (114 − 47 = 67).

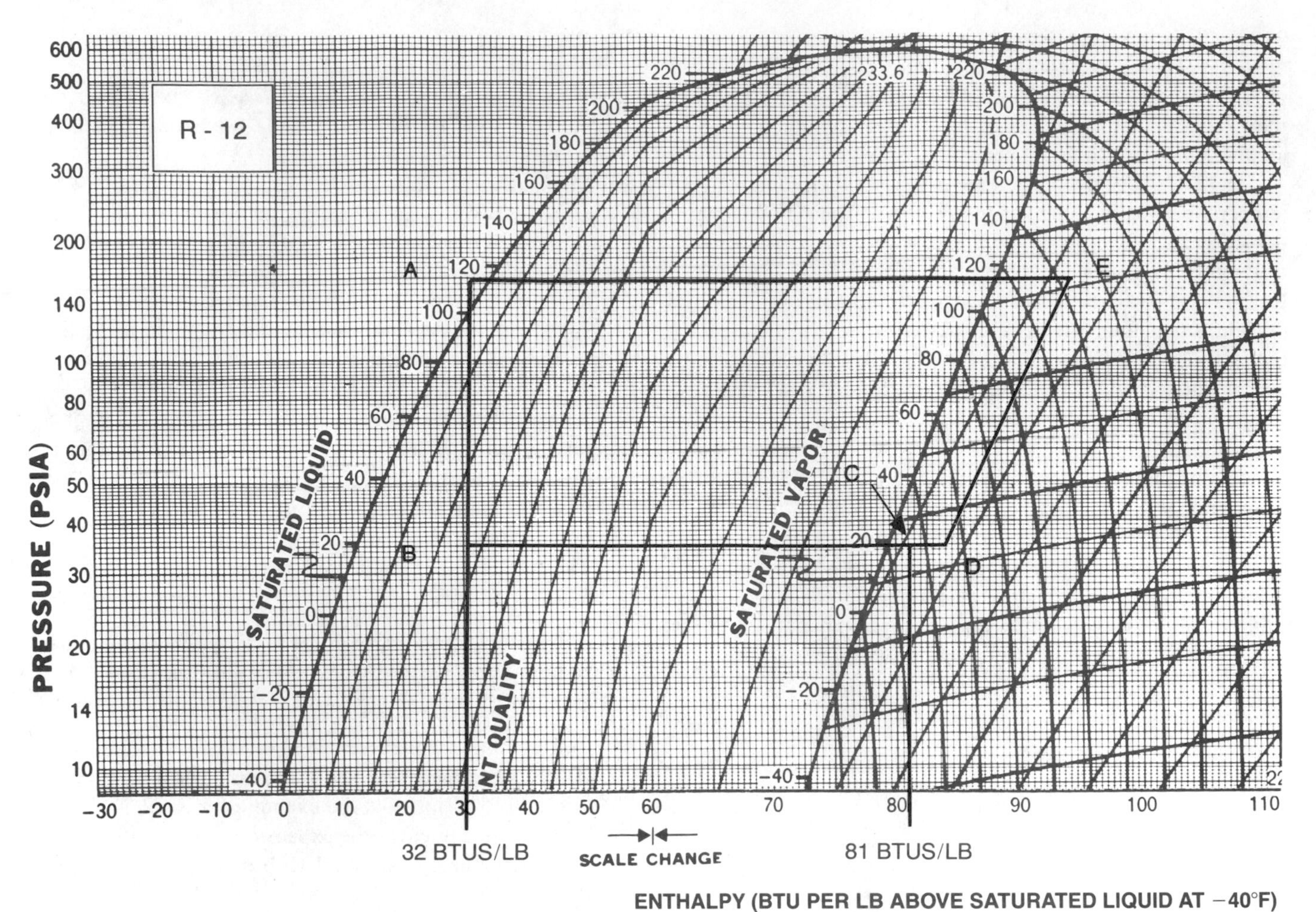

FIGURE 3-49 R-12 medium-temperature system. *Courtesy E. I. DuPont*

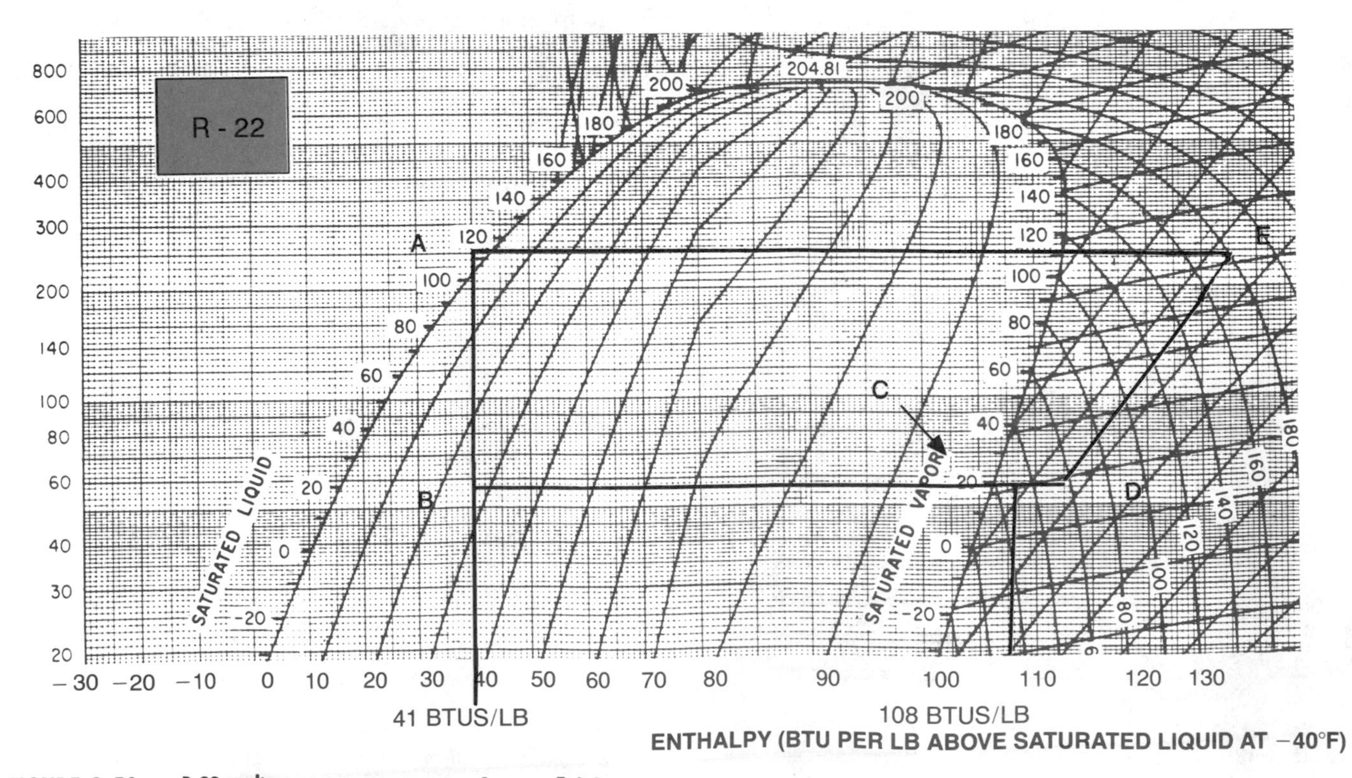

FIGURE 3-50 R-22 medium-temperature system. *Courtesy E. I. DuPont*

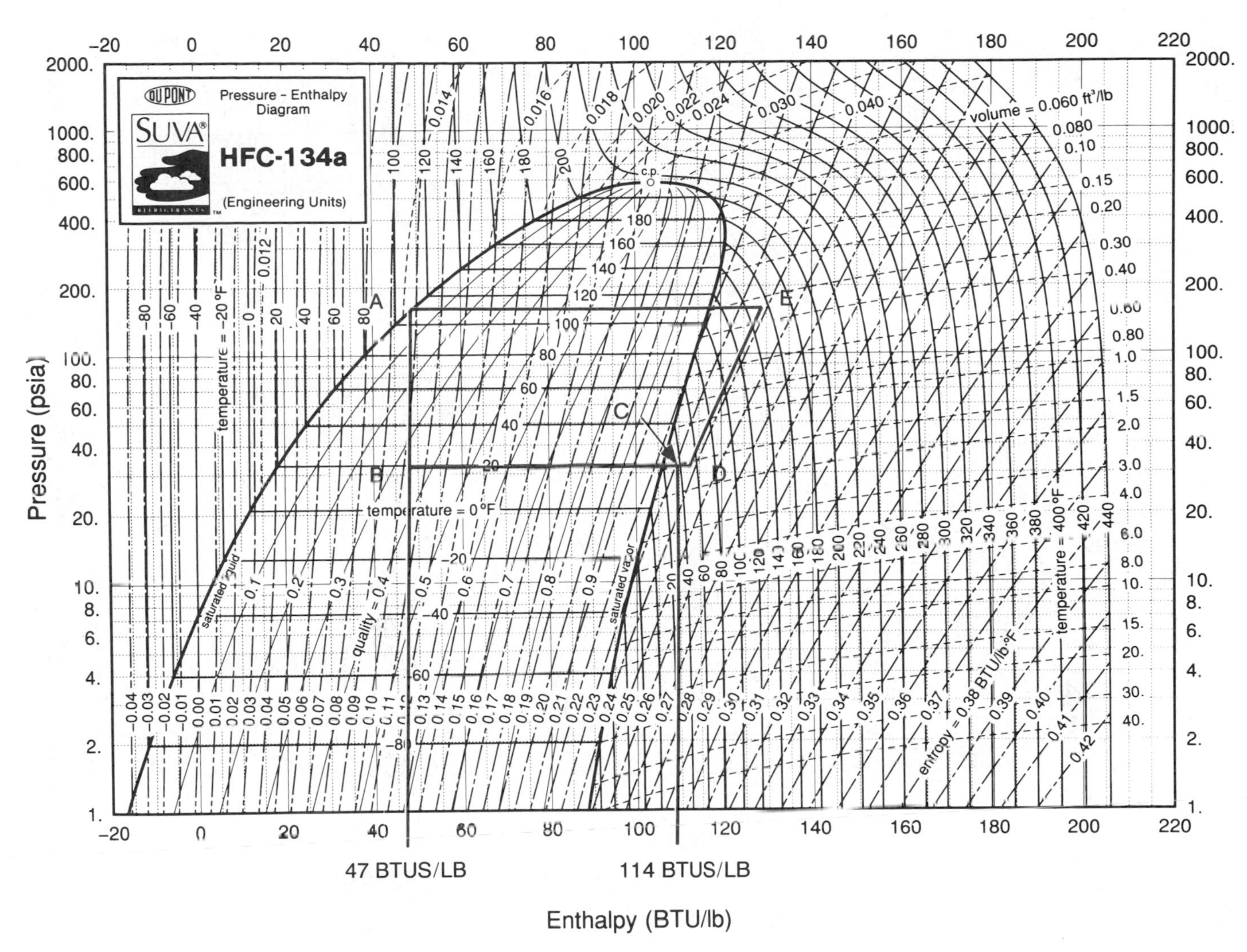

FIGURE 3-51 R-134a pressure/enthalpy diagram. *Courtesy E. I. DuPont*

4. The refrigerant enters the compressor at point D with a superheat of 30° and the vapor is compressed along the line from D to E. *Note the lower discharge temperature of 155°F.*
5. The refrigerant leaves the compressor at point E and is compressed at 115°F between point E and A.

Another comparison of refrigerants may be made using low temperature as the application. Here we will see much higher discharge temperatures and see why some decisions about various refrigerants are made. We will use a condensing temperature of 115°F and an evaporator temperature of −20°F and compare R-12 to R-502, then to R-22.

Figure 3–52 shows a low-temperature R-12 refrigeration system on a pressure/enthalpy chart. Follow the plot below.

1. The refrigerant enters the expansion valve at point A at 105°F, subcooled 10° from 115°F, like the above examples.
2. The refrigerant leaves the expansion valve at point B at -20°F to be boiled to a vapor in the evaporator. *Note, the pressure for R-12 boiling at −20°F is 0.6 psig, very close to atmospheric pressure. If the boiling point were any lower, the low side of the system would be in a vacuum. This is one of the disadvantages of R-12 as a refrigerant for low-temperature application.* For this system to maintain a room temperature of about 0°F, the coil temperature can only be 20°F below the return air temperature (0°F room air temp − (−20°F) temperature difference = coil temp), Figure 3–53. If the thermostat were to be turned down too low, the low-pressure side

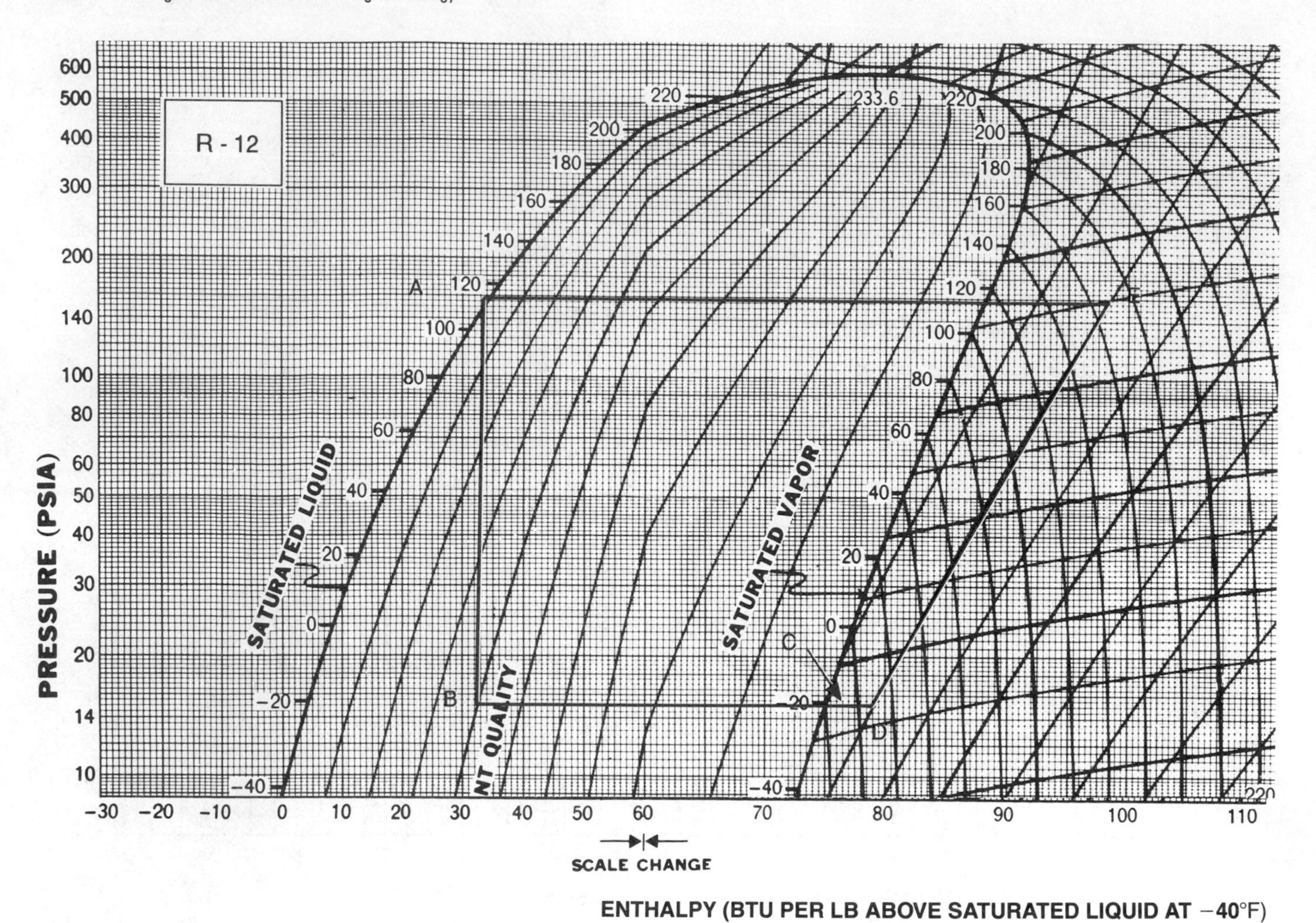

FIGURE 3-52 R-12 low-temperature system. *Courtesy E. I. DuPont*

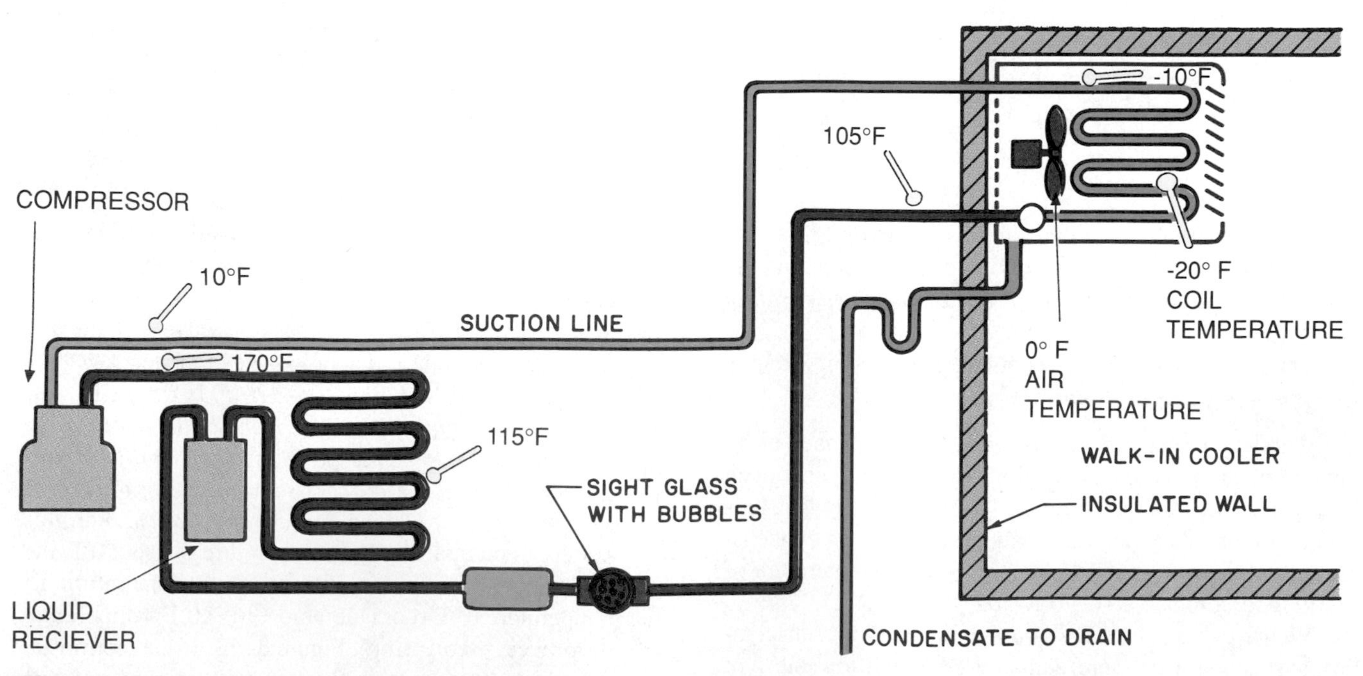

FIGURE 3-53 The refrigerant in the evaporator is boiling at -20°F. The room air temperature is 0°F.

of this system would be in a vacuum. *Note, the percent of liquid to vapor mixture is different for low-temperature than for medium-temperature applications. In this example, we have 39% vapor and 61% liquid leaving the expansion valve. The difference is extra flash gas is required to lower the remaing liquid to the lower temperature of −20°F.*

3. At point C, the refrigerant leaves the evaporator in the vapor state with 10° of superheat and a temperature of -10°F.
4. The vapor refrigerant enters the compressor at point D with 30° of superheat at a temperature of 10°F and is compressed along the line to point E.
5. The refrigerant leaves the compressor at point E to be condensed in the condenser. Notice that the discharge temperature is only 170°F, a very cool discharge temperature.

R-502 has been used for many low-temperature applications to prevent the system from operating in a vacuum on the low-pressure side. Follow the same conditions using R-502 in Figure 3–54.

1. Refrigerant enters the expansion valve at 105°F, subcooled from 115°F and 38% vapor, 62% liquid.
2. The remaining vapor is evaporated and 10° of superheat is added by the time the vapor reaches point C.
3. The vapor enters the compressor at point D and is compressed to point E where the vapor leaves the compressor at 170°F.
4. The vapor is condensed and subcooled from point E to A.
5. Notice that the suction pressure is 15.3 psig for R-502 while boiling at −20°F. R-502 will not go into a vacuum until the temperature is −50°F. This refrigerant is very good for low temperature applications because of this. It also has a very acceptable discharge gas temperature.

R-22 is being used for many low-temperature applications because R-502 is being phased out due to the CFC/ozone depletion issue. R-22 has the problem of the high discharge gas temperature. Figure 3–55 shows a plot of the low-temperature application above using R-22. Follow the sequence below.

1. The refrigerant enters the expansion valve at point A at 105°F, subcooled from 115°F, just as in the previous problem.
2. The refrigerant is evaporated to a vapor at −20°F and enters the compressor at point D with 30° superheat. Notice that R-22 boils at 10.1 psig at −20°F. It is in a positive pressure. Evaporators using R-22 can be operated down to −41°F before the suction pressure goes into a vacuum.

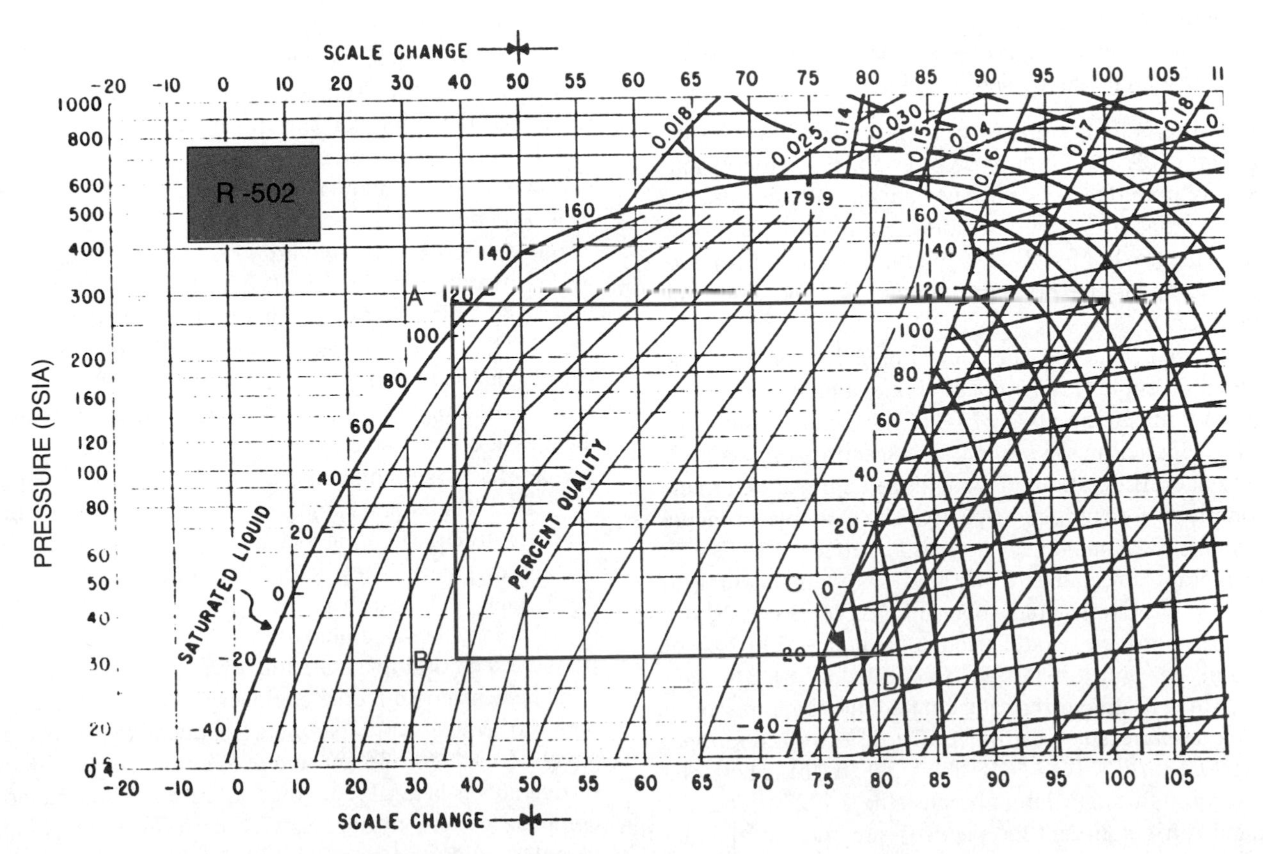

FIGURE 3-54 R-502 low-temperature system. *Courtesy E. I. DuPont*

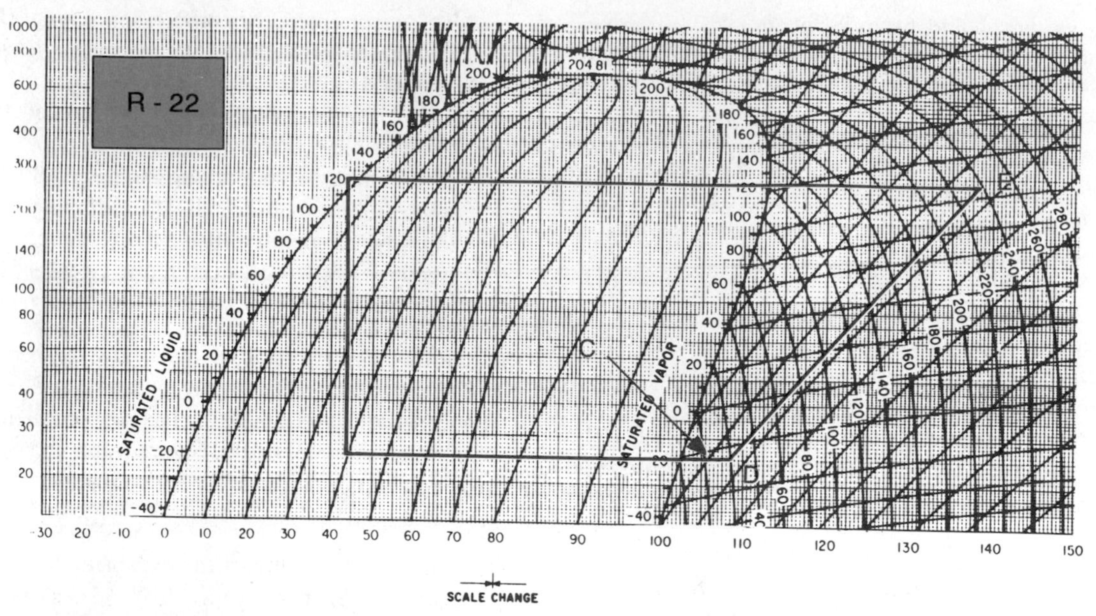

FIGURE 3-55 R-22 low-temperature system. *Courtesy E. I. DuPont*

3. The vapor is compressed along line D to E where it leaves the compressor at point E at a temperature of 235°F. Remember, R-12 had a discharge temperature of 170°F at the same condition and R-502 a discharge temperature of 170°F. R-22 is much hotter. A 235°F discharge temperature can be worked with but is close to being too high.
4. The vapor leaves the compressor at point E where it is condensed and subcooled to point A.

Pressure/enthalpy diagrams are useful in showing the refrigerant cycle for the purpose of establishing the various conditions around the system. They are partially constructed from properties of refrigerant tables. Figure 3–56 is a page from a typical table for R-22. Column 1 is the temperature corresponding to the pressure columns for the saturation temperature.

Column 5 lists the specific volume for the saturated refrigerant in cubic feet per pound. For example, at 60°F, the compressor must pump 0.4727 ft^3 of refrigerant to circulate 1 lb of refrigerant in the system. The specific volume along with the net refrigeration effect help the engineer determine the compressor's pumping capacity. The example in Figure 3–43 using R-22 had a net refrigeration effect of 61 Btu/lb of refrigerant circulated. If we had a system needing to circulate enough refrigerant to absorb 36,000 Btu/h (3 tons of refrigeration) we would need to circulate 590.2 lb of refrigerant per hour (36,000 Btu/h divided by 61 Btu/lb = 590.2 lb/h). If the refrigerant entered the compressor at 60°F, the compressor must move 279 ft^3 of refrigerant per hour (590.2 lb/h × 0.4727 ft^3/lb = 279 ft^3/h). There is a slight error in this calculation because the 0.4727 ft^3/h is for saturated refrigerant and the vapor is superheated entering the compressor. Superheat tables are available, but will only complicate this calculation more and there is very little error. Many compressors are rated in cubic feet per minute so this compressor would need to pump 4.65 ft^3/min (279 ft^3/h ÷ 60 min/h = 4.65 ft^3/min).

The density portion of the table tells the engineer how much a particular volume of liquid refrigerant will weigh at the rated temperature. For example, R-22 weighs 76.773 lb/ft^3 when the liquid temperature is 60°F. This is important in determining the weight of refrigerant in components, such as evaporators, condensers, and receivers.

The enthalpy portion of the table (total heat) lists the heat content of the liquid and vapor and the amount of latent heat required to boil a pound of liquid to a vapor. For example, at 60°F, saturated liquid refrigerant would contain 27.172 Btu/lb compared to 0 Btu/lb at −40°F. It would require 82.54 Btu/lb to boil 1 lb of 60°F saturated liquid to a vapor. The saturated vapor would then contain 109.712 Btu/lb total heat (27.172 + 82.54 = 109.712).

The entropy column is of no practical value except on the pressure/enthalpy chart where it is used to plot the compressor discharge temperature.

These charts and tables are not normally used in the field for troubleshooting but are for engineers to use to design equipment. They help the technician understand the refrigerants and the refrigerant cycle.

Different refrigerants have different temperature/pressure relationships and enthalpy relationships. These all must be considered by the engineer when choosing the correct refrigerant for a particular application. A complete study of each refrigerant and its comparison to all other refrigerants is helpful, but you don't need to understand the complete picture to successfully perform in the field. A complete study and comparison is beyond the scope of this

"FREON" 22 SATURATION PROPERTIES—TEMPERATURE TABLE

TEMP.	PRESSURE		VOLUME cu ft/lb		DENSITY lb/cu ft		ENTHALPY Btu/lb			ENTROPY Btu/(lb)(°R)		TEMP.
°F	PSIA	PSIG	LIQUID v_f	VAPOR v_g	LIQUID $1/v_f$	VAPOR $1/v_g$	LIQUID h_f	LATENT h_{fg}	VAPOR h_g	LIQUID s_f	VAPOR s_g	°F
10	47.464	32.768	0.012088	1.1290	82.724	0.88571	13.104	92.338	105.442	0.02932	0.22592	10
11	48.423	33.727	0.012105	1.1077	82.612	0.90275	13.376	92.102	105.538	0.02990	0.22570	11
12	49.396	34.700	0.012121	1.0869	82.501	0.92005	13.648	91.986	105.633	0.03047	0.22548	12
13	50.384	35.688	0.012138	1.0665	82.389	0.93761	13.920	91.808	105.728	0.03104	0.22527	13
14	51.387	36.691	0.012154	1.0466	82.276	0.95544	14.193	91.630	105.823	0.03161	0.22505	14
15	52.405	37.709	0.012171	1.0272	82.164	0.97352	14.466	91.451	105.917	0.03218	0.22484	15
16	53.438	38.742	0.012188	1.0082	82.051	0.99188	14.739	91.272	106.011	0.03275	0.22463	16
17	54.487	39.791	0.012204	0.98961	81.938	1.0105	15.013	91.091	106.105	0.03332	0.22442	17
18	55.551	40.855	0.012221	0.97144	81.825	1.0294	15.288	90.910	106.198	0.03389	0.22421	18
19	56.631	41.935	0.012238	0.95368	81.711	1.0486	15.562	90.728	106.290	0.03446	0.22400	19
20	57.727	43.031	0.012255	0.93631	81.597	1.0680	15.837	90.545	106.383	0.03503	0.22379	20
21	58.839	44.143	0.012273	0.91932	81.483	1.0878	16.113	90.362	106.475	0.03560	0.22358	21
22	59.967	45.271	0.012290	0.90270	81.368	1.1078	16.389	90.178	106.566	0.03617	0.22338	22
23	61.111	46.415	0.012307	0.88645	81.253	1.1281	16.665	89.993	106.657	0.03674	0.22318	23
24	62.272	47.576	0.012325	0.87055	81.138	1.1487	16.942	89.807	106.748	0.03730	0.22297	24
25	63.450	48.754	0.012342	0.85500	81.023	1.1696	17.219	89.620	106.839	0.03787	0.22277	25
26	64.644	49.948	0.012360	0.83978	80.907	1.1908	17.496	89.433	106.928	0.03844	0.22257	26
27	65.855	51.159	0.012378	0.82488	80.791	1.2123	17.774	89.244	107.018	0.03900	0.22237	27
28	67.083	52.387	0.012395	0.81031	80.675	1.2341	18.052	89.055	107.107	0.03958	0.22217	28
29	68.328	53.632	0.012413	0.79604	80.558	1.2562	18.330	88.865	107.196	0.04013	0.22198	29
30	69.591	54.895	0.012431	0.78208	80.441	1.2786	18.609	88.674	107.284	0.04070	0.22178	30
31	70.871	56.175	0.012450	0.76842	80.324	1.3014	18.889	88.483	107.372	0.04126	0.22158	31
32	72.169	57.473	0.012468	0.75503	80.207	1.3244	19.169	88.290	107.459	0.04182	0.22139	32
33	73.485	58.789	0.012486	0.74194	80.089	1.3478	19.449	88.097	107.546	0.04239	0.22119	33
34	74.818	60.122	0.012505	0.72911	79.971	1.3715	19.729	87.903	107.632	0.04295	0.22100	34
35	76.170	61.474	0.012523	0.71655	79.852	1.3956	20.010	87.708	107.719	0.04351	0.22081	35
36	77.540	62.844	0.012542	0.70425	79.733	1.4199	20.292	87.512	107.804	0.04407	0.22062	36
37	78.929	64.233	0.012561	0.69221	79.614	1.4447	20.574	87.316	107.889	0.04464	0.22043	37
38	80.336	65.640	0.012579	0.68041	79.495	1.4697	20.856	87.118	107.974	0.04520	0.22024	38
39	81.761	67.065	0.012598	0.66885	79.375	1.4951	21.138	86.920	108.058	0.04576	0.22005	39
40	83.206	68.510	0.012618	0.65753	79.255	1.5208	21.422	86.720	108.142	0.04632	0.21986	40
41	84.670	69.974	0.012637	0.64643	79.134	1.5469	21.705	86.520	108.225	0.04688	0.21968	41
42	86.153	71.457	0.012656	0.63557	79.013	1.5734	21.989	86.319	108.308	0.04744	0.21949	42
43	87.655	72.959	0.012676	0.62492	78.892	1.6002	22.273	86.117	108.390	0.04800	0.21931	43
44	89.177	74.481	0.012695	0.61448	78.770	1.6274	22.558	85.914	108.472	0.04855	0.21912	44
45	90.719	76.023	0.012715	0.60425	78.648	1.6549	22.843	85.710	108.553	0.04911	0.21894	45
46	92.280	77.584	0.012735	0.59422	78.526	1.6829	23.129	85.506	108.634	0.04967	0.21876	46
47	93.861	79.165	0.012755	0.58440	78.403	1.7112	23.415	85.300	108.715	0.05023	0.21858	47
48	95.463	80.767	0.012775	0.57476	78.280	1.7398	23.701	85.094	108.795	0.05079	0.21839	48
49	97.085	82.389	0.012795	0.56532	78.157	1.7689	23.988	84.886	108.874	0.05134	0.21821	49
50	98.727	84.031	0.012815	0.55606	78.033	1.7984	24.275	84.678	108.953	0.05190	0.21803	50
51	100.39	85.69	0.012836	0.54698	77.909	1.8282	24.563	84.468	109.031	0.05245	0.21785	51
52	102.07	87.38	0.012856	0.53808	77.784	1.8585	24.851	84.258	109.109	0.05301	0.21768	52
53	103.78	89.08	0.012877	0.52934	77.659	1.8891	25.139	84.047	109.186	0.05357	0.21750	53
54	105.50	90.81	0.012898	0.52078	77.534	1.9202	25.429	83.834	109.263	0.05412	0.21732	54
55	107.25	92.56	0.012919	0.51238	77.408	1.9517	25.718	83.621	109.339	0.05468	0.21714	55
56	109.02	94.32	0.012940	0.50414	77.282	1.9836	26.008	83.407	109.415	0.05523	0.21697	56
57	110.81	96.11	0.012961	0.49606	77.155	2.0159	26.298	83.191	109.490	0.05579	0.21679	57
58	112.62	97.93	0.012982	0.48813	77.028	2.0486	26.589	82.975	109.564	0.05634	0.21662	58
59	114.46	99.76	0.013004	0.48035	76.900	2.0818	26.880	82.758	109.638	0.05689	0.21644	59
60	116.31	101.62	0.013025	0.47272	76.773	2.1154	27.172	82.540	109.712	0.05745	0.21627	60
61	118.19	103.49	0.013047	0.46523	76.644	2.1495	27.464	82.320	109.785	0.05800	0.21610	61
62	120.09	105.39	0.013069	0.45788	76.515	2.1840	27.757	82.100	109.857	0.05855	0.21592	62
63	122.01	107.32	0.013091	0.45066	76.386	2.2190	28.050	81.878	109.929	0.05910	0.21575	63
64	123.96	109.26	0.013114	0.44358	76.257	2.2544	28.344	81.656	110.000	0.05966	0.21558	64

FIGURE 3-56 Portion of properties of R-22 table. *Courtesy E. I. DuPont*

STANDARD REFRIGERANT DESIGNATION	CHEMICAL NAME	BOILING POINT °F	CHEMICAL FORMULA	MOLECULAR WEIGHT
11	Trichlorofluoromethane	74.8	CCl_3F	137.4
12	Dichlorodifluoromethane	−21.6	CCl_2F_2	120.9
13	Chlorotrifluoromethane	−114.6	$CClF_3$	104.5
22	Chlorodifluoromethane	−41.4	$CHClF_2$	86.5
30	Methylene Chloride	105.2	CH_2Cl_2	84.9
40	Methyl Chloride	−10.8	CH_3Cl	50.5
50	Methane	−259.	CH_4	16.0
113	Trichlorotrifluoroethane	117.6	CCl_2FCClF_2	187.4
114	Dichlorotetrafluoroethane	38.4	$CClF_2CClF_2$	170.9
123	Dichlorotrifluoroethane	82.17	CCl_2HCF_3	152.93
134a	Tetrafluoroethane	−15.08	CF_3CFH_2	102.03
170	Ethane	−127.5	CH_3CH_3	30.0
290	Propane	−44.0	$CH_3CH_2CH_3$	44.0
500*	Refrigerants 12/152a; 73.8/26.2% (Wt)	−28.0	CCl_2F_2/CH_3CHF_2	99.3
502*	Refrigerants 22/115; 48.8/51.2% (Wt)	−50.1	$CHClF_2/CClF_2CF_3$	112.0
601	Isobutane	14.0	$CH(CH_3)_3$	58.1
717	Ammonia	−28.0	NH_3	17.0
718	Water	212.0	H_2O	18.0

* Denotes Azeotropic Mixture

FIGURE 3–57 List of refrigerants and some of their characteristics.

book. Figure 3–57 shows some of the characteristics of many refrigerants. Figure 3–58 shows a pressure/temperature comparison graph of many of the available refrigerants.

SAFETY PRECAUTION: *All refrigerants that have been discussed in this text are stored in pressurized containers and should be handled with care. Consult your instructor or supervisor for use and handling of these refrigerants. Goggles and gloves should be worn while transferring the refrigerants from the container to the system.*

SUMMARY

- Bacterial growth that causes food spoilage slows at low temperatures.
- The frozen hard point for most foods is considered to be 0°F. Below this point, food-spoiling bacteria stop growing.
- Product temperatures above 45°F and below room temperature are considered high-temperature refrigeration.
- Product temperatures between 35°F and 45°F are considered medium-temperature refrigeration.
- Product temperatures from 0°F to −10°F are considered low-temperature refrigeration.
- Refrigeration is the process of removing heat from a place where it is not wanted and transferring it to a place where it makes little or no difference.
- One ton of refrigeration is the amount of heat necessary to melt 1 ton of ice in a 24-h period. It takes 288,000 Btu to melt 1 ton of ice in a 24-h period or 12,000 Btu in 1 h or 200 Btu in 1 min.
- The boiling point of liquids can be changed and controlled by controlling the pressure on the refrigerant. The relationship of the vapor pressure and the boiling point is called the pressure/temperature relationship. When the pressure is increased, the boiling point increases. When the pressure is decreased, the boiling point decreases.
- A compressor can be considered a vapor pump. It lowers the pressure in the evaporator to the desired temperature and increases the pressure in the condenser to a level where the vapor may be condensed to a liquid.
- The liquid refrigerant moves from the condenser to the metering device where it again enters the evaporator. The metering device causes part of the liquid to vaporize, and the pressure and temperature are greatly reduced as the refrigerant enters the evaporator and starts the cycle over again.
- Refrigerants have a definite chemical makeup and are usually designated with an "R" and a number for field identification. R-22 is primarily used in air conditioning systems, R-12 in high-, medium-, and some low-temperature applications, and R-502 in low- and extra low-temperature applications.
- A refrigerant must be safe, must be detectable, must have a low boiling point, and must have good pumping characteristics.
- Refrigerant cylinders are color coded to indicate the type of refrigerant they contain.
- Refrigerants should be recovered or stored while a refrigeration system is being serviced, then recycled, if appropriate, or sent to a manufacturer to be reclaimed.
- Pressure/enthalpy diagrams may be used to plot refrigeration cycles.

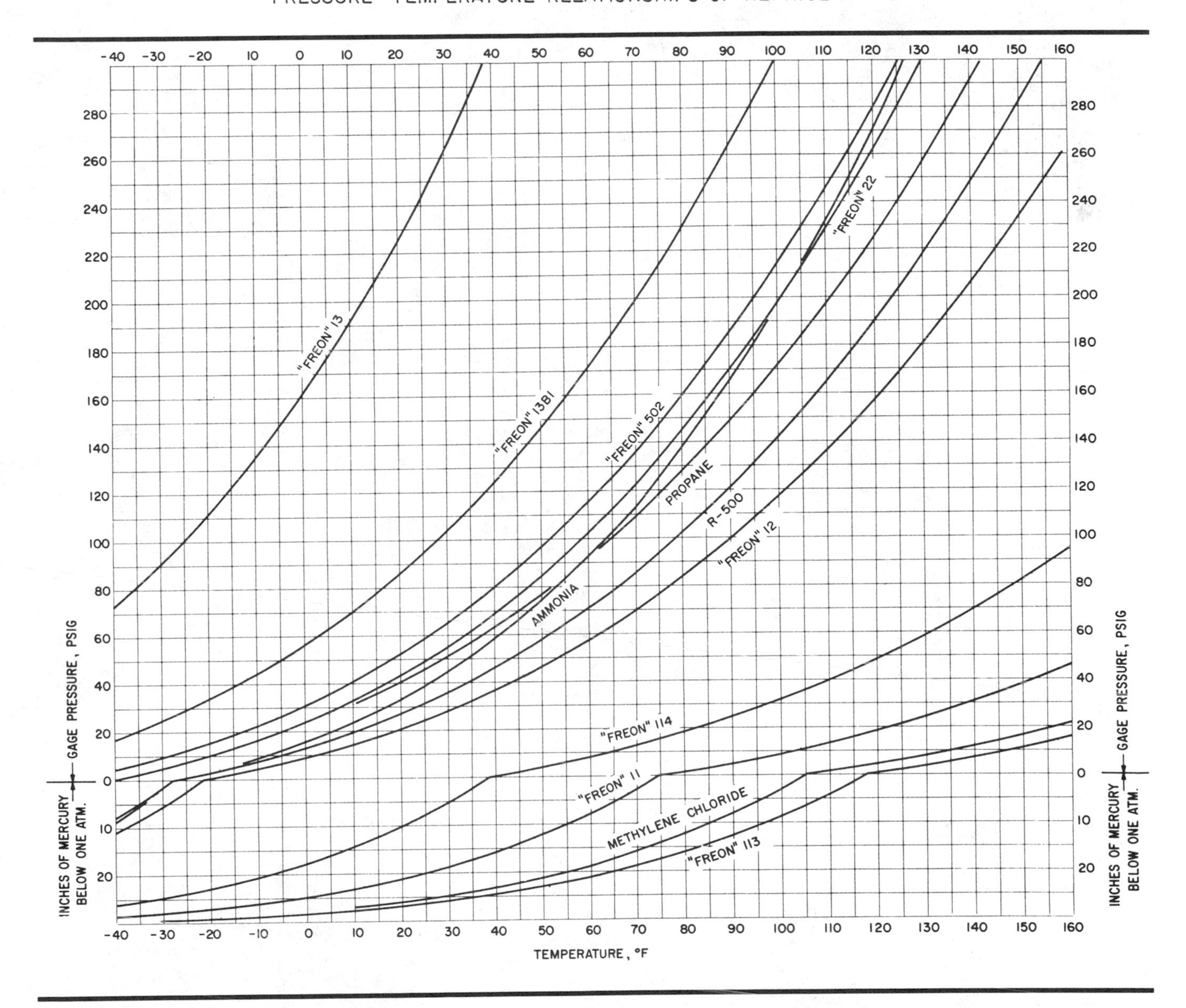

FIGURE 3–58 Pressure/temperature comparison graph for many available refrigerants. *Courtesy E. I. DuPont*

REVIEW QUESTIONS

1. Name three reasons why ice melts in an icebox.
2. What are the approximate temperature ranges for low-, medium-, and high-temperature refrigeration applications?
3. What is a ton of refrigeration?
4. Describe briefly the basic refrigeration cycle.
5. What is the relationship between pressure and the boiling point of liquids?
6. What is the function of the evaporator in a refrigeration or air conditioning system?
7. What does the compressor do in the refrigeration system?
8. What happens to the refrigerant in the condenser?
9. What happens to heat in the refrigerant while in the condenser?
10. What does the metering device do?
11. Describe the difference between a reciprocating compressor and a rotary compressor.
12. Give a simple definition of a compressor.
13. Name two common refrigerants used in refrigeration.
14. What refrigerant is commonly used in air conditioning systems?
15. List four characteristics a manufacturer must consider when choosing a refrigerant.
16. List the cylinder color codes for R-12, R-22, and R-502.
17. State two methods for storing or recovering refrigerants while a refrigeration system is being serviced.
18. Define enthalpy.

SECTION

2 *Safety, Tools and Equipment, Shop Practices*

U N I T 4

* General Safety Practices

OBJECTIVES

After studying this unit, you should be able to

- describe proper procedures for working with pressurized systems and vessels, electrical energy, heat, cold, rotating machinery, chemicals, and moving heavy objects.
- work safely, avoiding safety hazards.

The heating, air conditioning, and refrigeration technician works close to many potentially dangerous situations: liquids and gases under pressure, electrical energy, heat, cold, chemicals, rotating machinery, moving heavy objects, and so on. The job must be completed in a manner that is safe for the technician and the public.

This unit describes some general safety practices and procedures with which all technicians should be familiar. Whether in a school laboratory, a shop, a commercial service shop, or a manufacturing plant, always be familiar with the location of emergency exits and first aid and eye wash stations. When you are on the job, be aware of how you would get out of a building or particular location should an emergency occur. Know the location of first aid kits and first aid equipment. Many other more specific safety practices and tips are given within units where they may be applied. Always use common sense and be prepared.

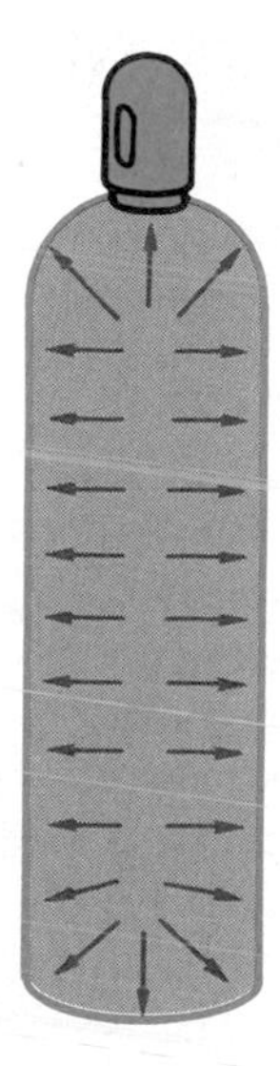

FIGURE 4–1 Pressure exerted across the entire surface area of refrigerant cylinder.

4.1 PRESSURE VESSELS AND PIPING

Pressure vessels and piping are part of many systems that are serviced by refrigeration and air conditioning technicians. For example, a cylinder of R-22 in the back of an open truck with the sun shining on it may have a cylinder temperature of 110°F on a summer day. The temperature/pressure chart indicates that the pressure inside the cylinder at 110° is 226 psig. This pressure reading means that the cylinder has a pressure of 226 lb for each square inch of surface area. A large cylinder may have a total area of 1500 in^2. This gives a total inside pressure (pushing outward) of

$$1500 \text{ in}^2 \times 226 \text{ psi} = 339{,}000 \text{ lb}$$

This is equal to

$$339{,}000 \text{ lb} \div 2000 \text{ lb/ton} = 169.5 \text{ ton}$$

See Figure 4–1. This pressure is well contained and will be safe if the cylinder is protected. Do not drop it. ***Move the cylinder only while a protective cap is on it, if it is designed for one, Figure 4–2.*** Cylinders too large to be carried should be moved chained to a cart, Figure 4–3.

FIGURE 4–2 Refrigerant cylinder with protective cap. *Photo by Bill Johnson*

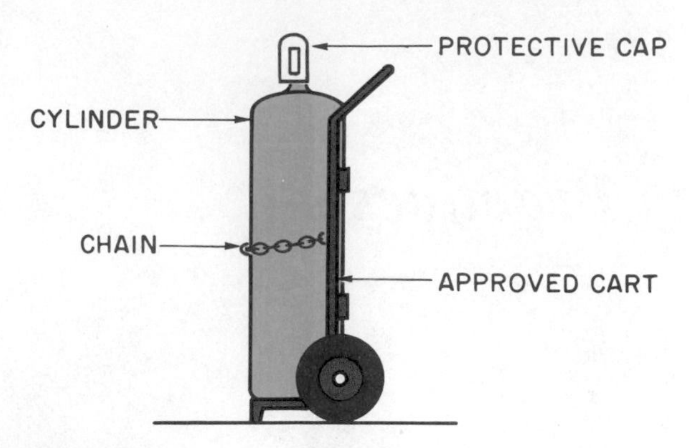

FIGURE 4–3 Pressurized cylinders should be chained to and moved safely on an approved cart. The protective cap must be secured.

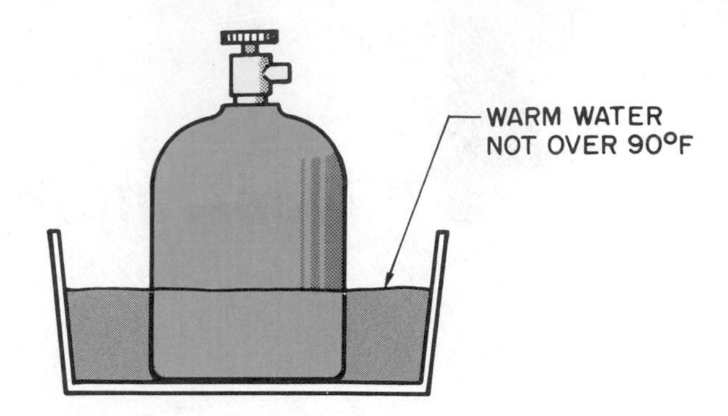

FIGURE 4–5 Refrigerant cylinder in warm water (not warmer than 90°F).

The pressure in this cylinder can be thought of as a potential danger. It will not become dangerous unless it is allowed to escape in an uncontrolled manner. The cylinder has a relief valve in the top in the vapor space. If the pressure builds up to the relief valve setting, it will start relieving vapor. As the vapor pressure is relieved, the liquid in the cylinder will begin to cool and reduce the pressure, Figure 4–4. Relief valve settings are set at values above the worst typical operating conditions and are typically over 400 psig.

The refrigerant cylinder has a fusible plug made of a material with a low melting temperature. The plug will melt and blow out if the cylinder gets too hot. This prevents the cylinder from bursting and injuring personnel and property around it. ***Some technicians may apply heat to refrigerant cylinders while charging a system to keep the pressure from dropping in the cylinder. This is an extremely dangerous practice. It is recommended for the above purpose that the cylinder of refrigerant be set in a container of warm water with a temperature no higher than 90°F, Figure 4–5.***

You will be taking pressure readings on refrigeration and air conditioning systems. Liquid R-22 boils at −41°F when it is released to the atmosphere. If you are careless and get this refrigerant on your skin or in your eyes, it will quickly cause frostbite. Keep your skin and eyes away from any liquid refrigerant. When you attach gages and take refrigerant pressure readings or transfer refrigerant into or out of a system, wear gloves and side shield goggles. Figure 4–6 is a photo of protective eye goggles. ***If a leak develops and refrigerant is escaping, the best thing to do is stand back and look for a valve with which to shut it off. Do not try to stop it with your hands.***

Released refrigerant, and any particles that may become airborne with the blast of vapor, can harm the eyes. Wear approved goggles, which are vented to keep them from fogging over with condensation and which help keep the operator cool, Figure 4–6.

***All cylinders into which refrigerant is transferred should be approved by the Department of Transportation as recovery cylinders. Figure 4–7 shows cylinders**

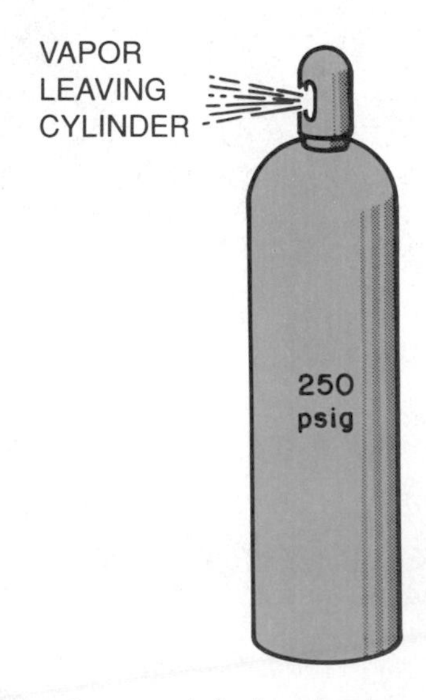

FIGURE 4–4 Pressure relief valve will reduce pressure in the cylinder.

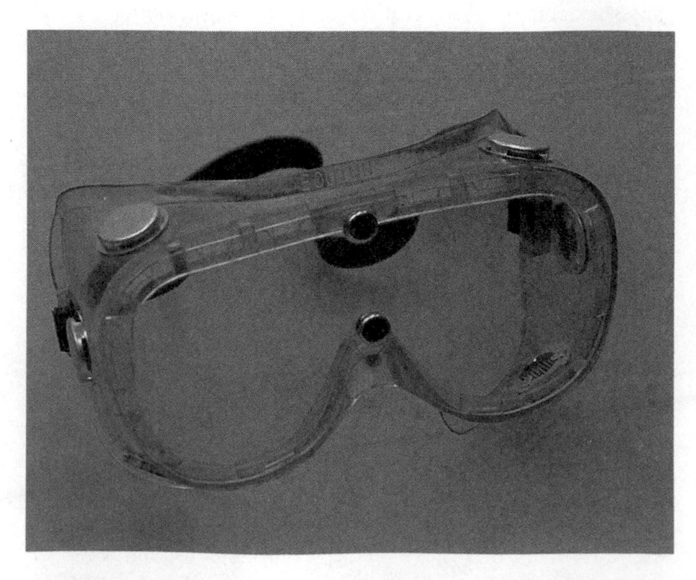

FIGURE 4–6 Protective eye side shield goggles. *Photo by Bill Johnson*

FIGURE 4–7 Color-coded refrigerant recovery cylinders. *Courtesy White Industries*

with the color code indicating that they are approved as recovery cylinders. It is against the law to transport refrigerant in nonapproved cylinders.*

In addition to the pressure potential inside a refrigerant cylinder, there is tremendous pressure potential inside nitrogen and oxygen cylinders. They are shipped at pressures of 2500 psig. These cylinders must not be moved unless the protective cap is in place. They should be chained to and moved on carts designed for this purpose, Figure 4–3. Dropping the cylinder without the protective cap may result in breaking the valve off the cylinder. The pressure inside the cylinder can propel the cylinder like a balloon full of air that is turned loose, Figure 4–8.

Nitrogen must also have its pressure regulated before it can be used. The pressure in the cylinder is too great to be connected to a system. If a person allowed nitrogen under cylinder pressure to enter a refrigeration system, the pressure could burst some weak point in the system. This could be particularly dangerous if it were at the compressor shell.

Because of this high pressure, oxygen also must be regulated. In addition, all oxygen lines must be kept absolutely oil free. Oil residue in an oxygen regulator connection may cause an explosion. The explosion may blow the regulator apart in your hand.

Oxygen is often used with acetylene. Acetylene cylinders are not under the same high pressure as nitrogen and oxygen but must be treated with the same respect because acetylene is highly explosive. A pressure-reducing regulator must be used. Always use an approved cart, chain the cylinder to the cart, and secure the protective cap, Figure 4–3. All stored cylinders must be supported so they will not fall over and they must be separated as required by code, Figure 4–9.

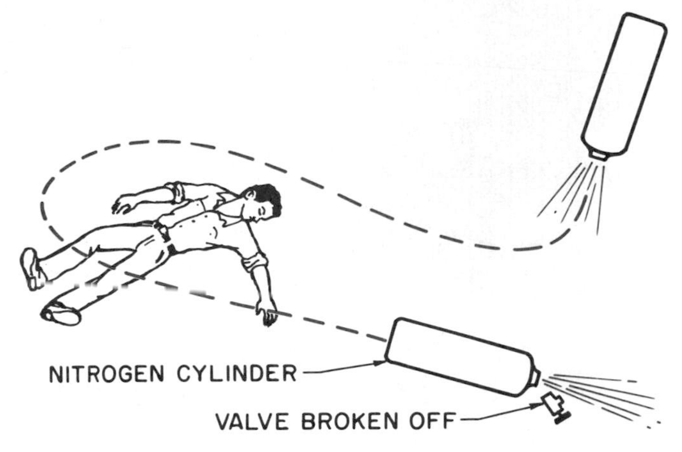

FIGURE 4–8 When a cylinder valve is broken off, the cylinder becomes a projectile until pressure is exhausted.

4.2 ELECTRICAL HAZARDS

Electrical shocks and *burns* are ever-present hazards. It is impossible to troubleshoot all circuits with the power off, so you must learn safe methods for troubleshooting "live" circuits. As long as electricity is contained in the conductors and the devices where it is supposed to function, there is nothing to fear. When uncontrolled electrical flow occurs (eg, if you touch two live wires), you are very likely to get hurt.

SAFETY PRECAUTION: *Electrical power should always be shut off at the distribution or entrance panel and locked out in an approved manner when installing equipment. Whenever possible the power should be shut off and locked out when servicing the equipment. Electrical panels are furnished with a place for a padlock for the purpose of lock out. To avoid someone turning the power on, you should keep the panel locked when you are out of sight of it and keep the only key with you. Don't ever think that you are good enough or smart enough to work with live electrical power when it is not necessary.*

However, at certain times tests have to be made with the power on. Extreme care should be taken when making these tests. Ensure that your hands touch only the meter probes and that your arms and the rest of your body stay clear of all electrical terminals and connections. You should know the voltage in the circuit you are checking. Make sure that the range selector on the test instrument is set properly before using it. ***Don't stand in a wet or damp area when**

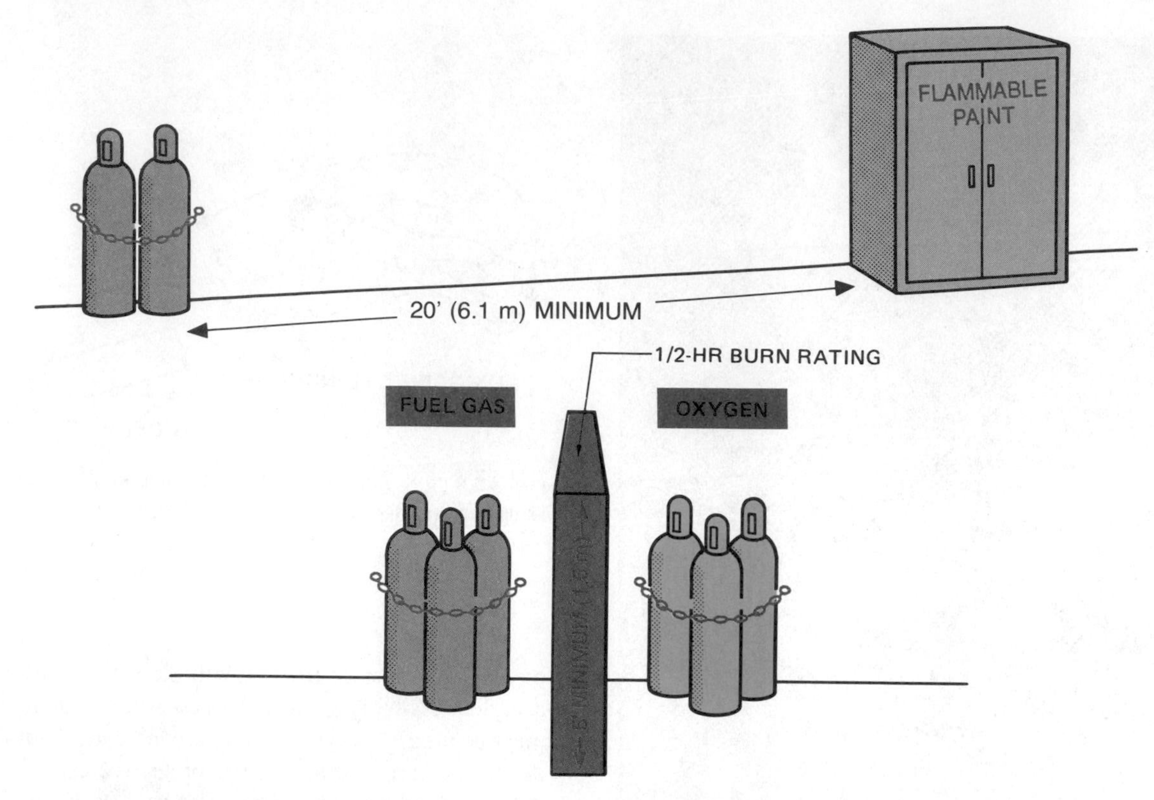

FIGURE 4–9 Pressurized cylinders must be chained when stored. The minimum safe distance between stored fuel gas cylinders and any flammable materials is 20 ft or a wall 5 ft high. From Jeffus & Johnson, *Welding: Principles and Applications, 3E,* © 1992 by Delmar Publishers Inc.

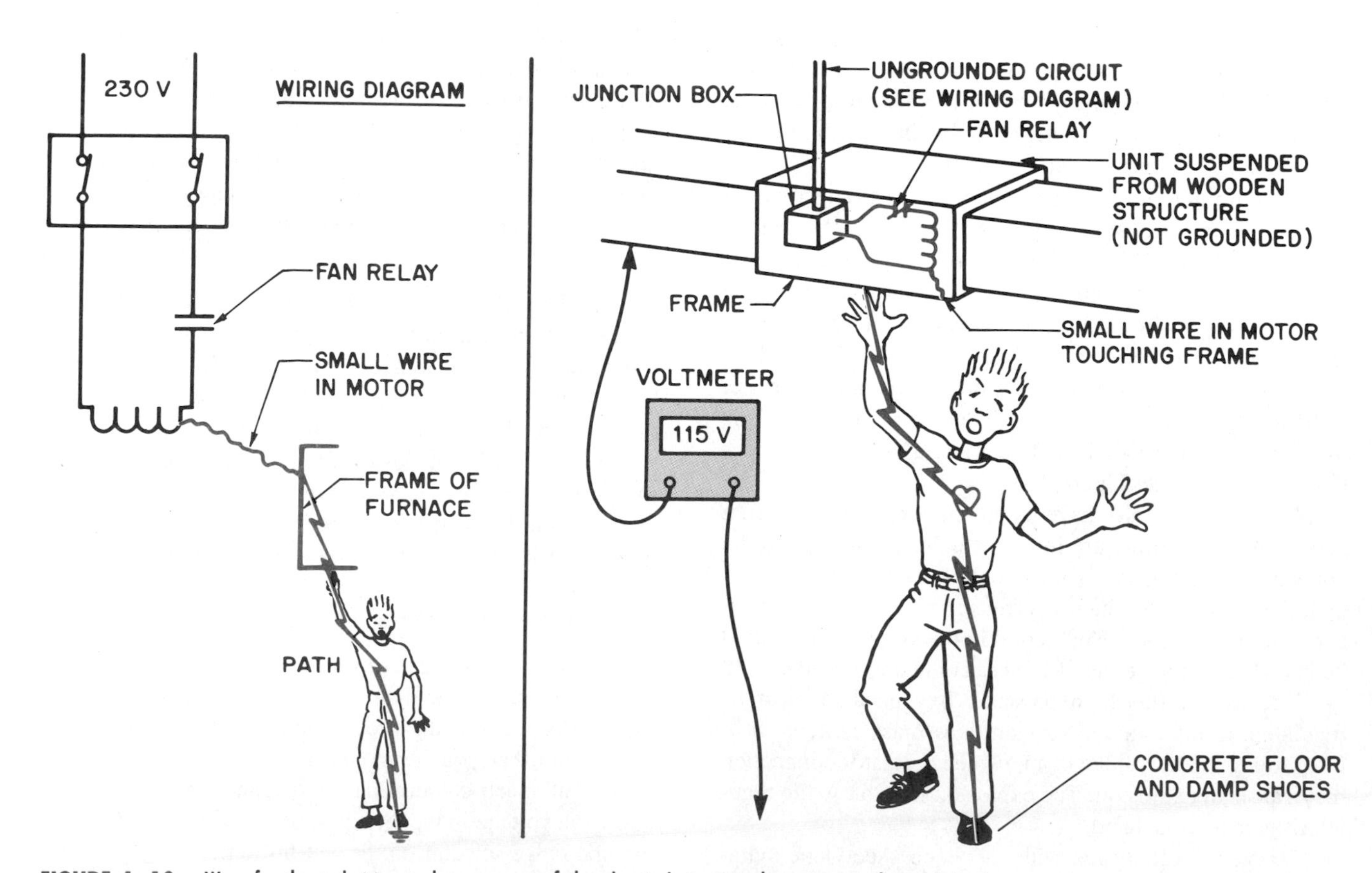

FIGURE 4–10 Ways for the technician to become part of the electrical circuit and receive an electrical shock.

making these checks. Use only proper test equipment, and make sure it is in good condition. Wear heavy shoes with an insulating sole and heel.* Intelligent and competent technicians take all precautions.

Electrical Shock

Electrical shock occurs when you become part of the circuit. Electricity flows through your body and can damage your heart—stop it from pumping, resulting in death if it is not restarted quickly. It is a good idea to take a first aid course that includes cardiopulmonary resusitation methods for lifesaving.

To prevent electrical shock, don't become a conductor between two live wires or a live (hot) wire and ground. The electricity must have a path to flow through. Don't let your body be the path. Figure 4–10 is a wiring diagram showing situations in which the technician is part of a circuit.

The technician should use only properly grounded power tools connected to properly grounded circuits. Caution should be taken when using portable electric tools. These are handheld devices with electrical energy inside just waiting for a path to flow through. Some portable electric tools are constructed with metal frames. These should all have a grounding wire in the power cord. The grounding wire protects the operator. The tool will work without it, but it is not safe. If the motor inside the tool develops a loose connection and the frame of the tool becomes electrically hot, the third wire, rather than your body, will carry the current, and a fuse or breaker will interrupt the circuit, Figure 4–11.

In some instances technicians may use the three- to two-prong adapters at job sites because the wall receptacle may only have two connections and their portable electric tool has a three-wire plug, Figure 4–12. This adapter has a third wire that must be connected to a ground for the circuit

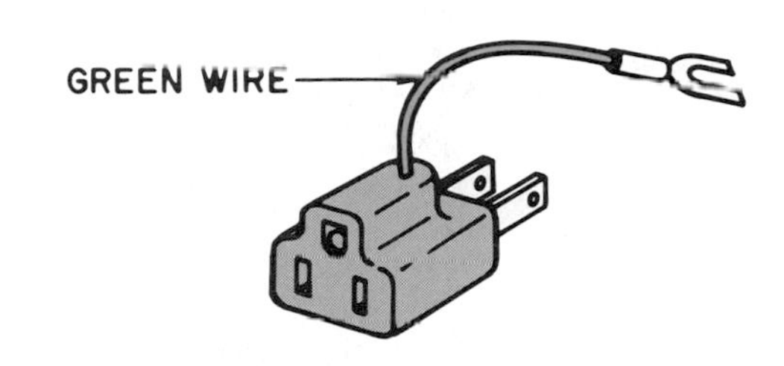

FIGURE 4–12 A three- to two-prong adapter.

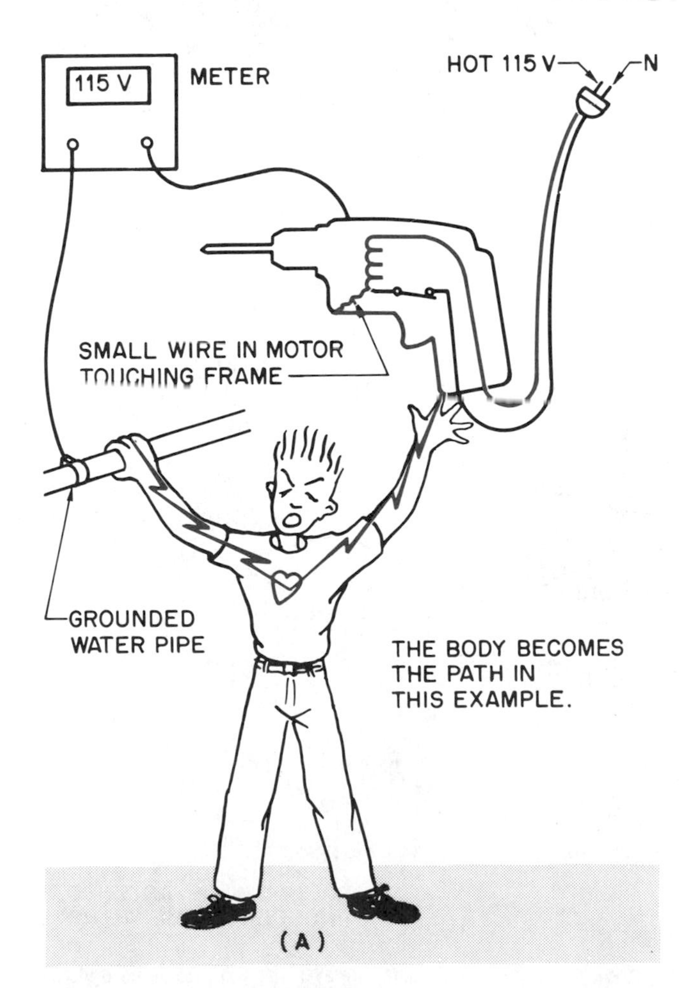

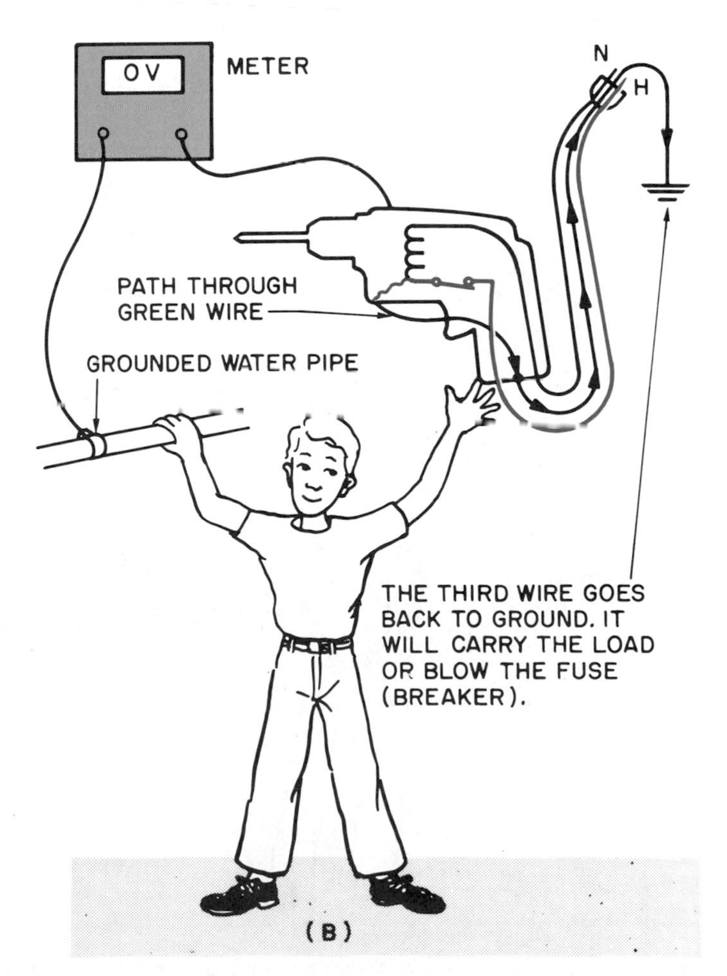

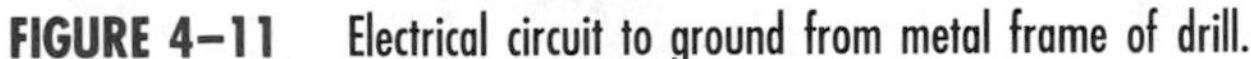

FIGURE 4–11 Electrical circuit to ground from metal frame of drill.

to give protection. If this wire is fastened under the wall plate screw and this screw terminates in an ungrounded box nailed to a wooden wall, you are not protected, Figure 4–13. Ensure that the third or ground wire is properly connected to a ground.

Alternatives to the older style metal portable electric tools are the plastic-cased and the battery-operated tool. In the plastic-cased tool the motor and electrical connections are insulated within the tool. It is called double-insulated and is considered safe, Figure 4–14(A). The battery operated tool uses rechargeable batteries and is very convenient and safe, Figure 4–14(B).

An extension cord with a ground fault circuit interrupter (GFCI) is recommended to be used with portable electric tools, Figure 4–15. These are designed to help protect the operator from shock as they detect very small electrical leaks to ground. A small electrical current leak will cause the GFCI to open the circuit, preventing further current flow.

Nonconducting Ladders

Nonconducting ladders should be used on all jobs and should be the type furnished with service trucks. Two types of nonconducting ladders are those made from wood and fiberglass. Nonconducting ladders work as well as aluminum ladders; they are just heavier. They are also much safer. ***A ladder may be raised into a power line or placed against a live electrical hazard without realizing it. Chances should not be taken. When the technician is standing on a nonconducting ladder it will provide protection from electrical shock to ground. However, it will**

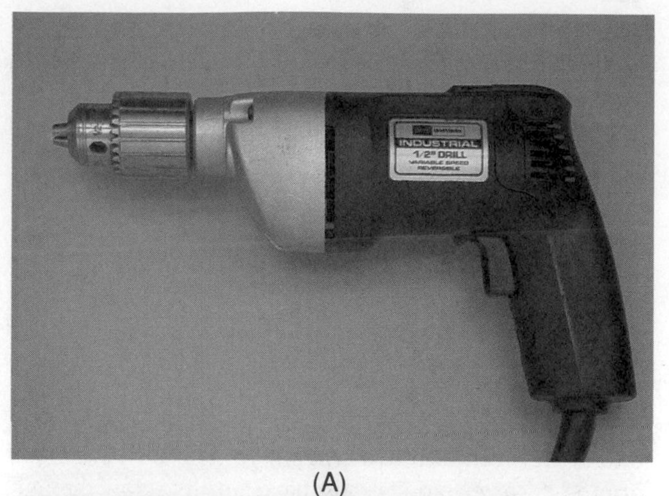

(A)

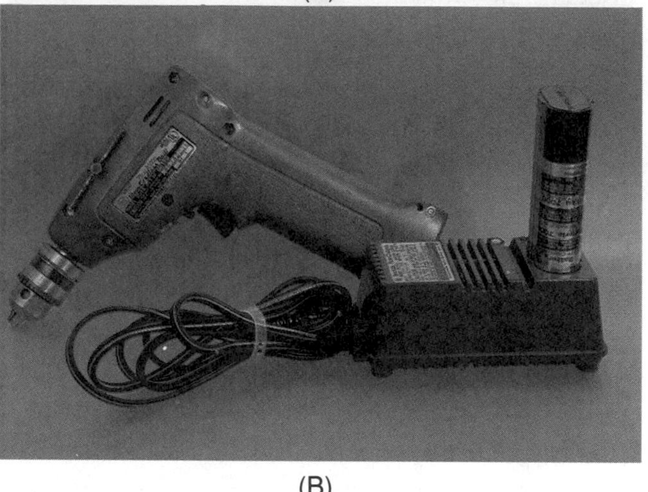

(B)

FIGURE 4–14 (A) Double-insulated electric drill. (B) Battery-operated electric drill. *Photos by Bill Johnson*

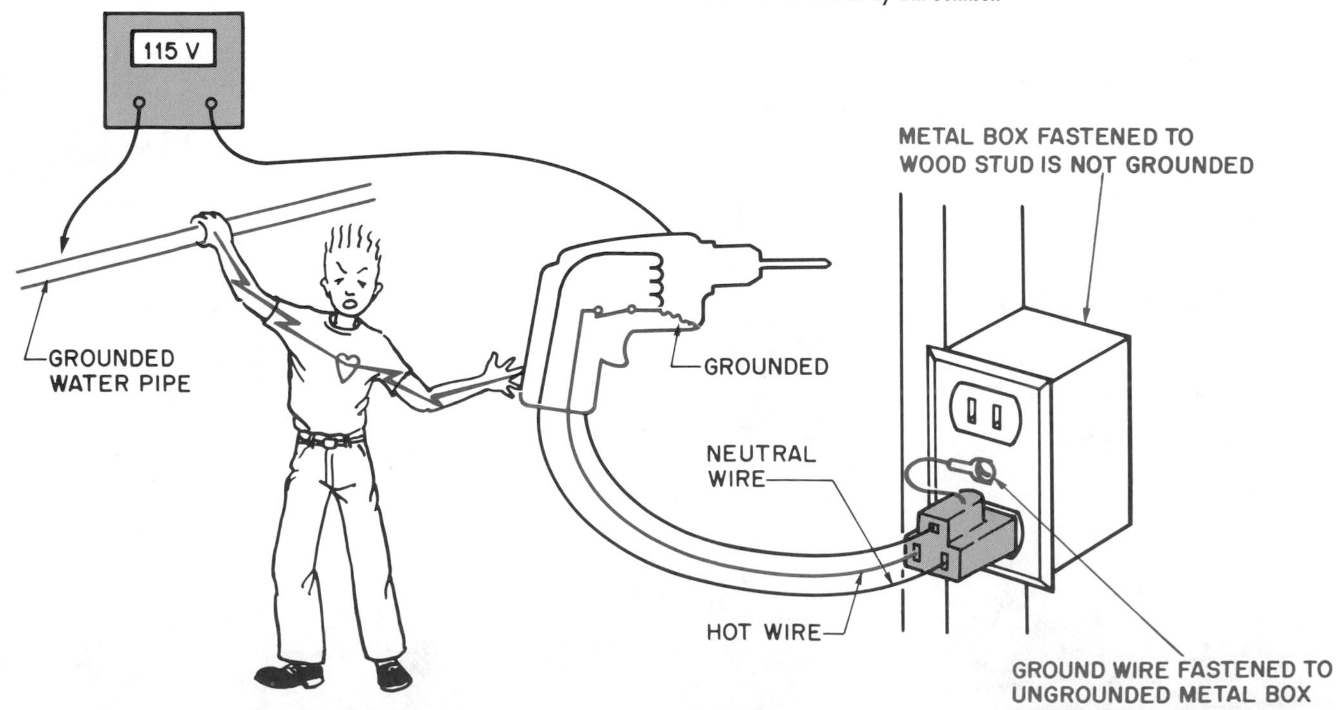

FIGURE 4–13 The wire from the adapter is intended to be fastened under the screw in the duplex wall plate. However, this will provide no protection if the outlet box is not grounded.

FIGURE 4–15 Extension cord with ground fault circuit interrupter receptacle. *Photo by Bill Johnson*

not provide protection between two or more electrical conductors.*

Electrical Burns

Do not wear jewelry (rings and watches) while working on live electrical circuits because they can cause shock and possible burns.

Never use a screwdriver or other tool in an electrical panel when the power is on. Electrical burns can come from an electrical arc, such as in a short circuit to ground when uncontrolled electrical energy flows. For example, if a screwdriver slipped while you were working in a panel and the blade completed a circuit to ground, the potential flow of electrical energy is tremendous. When a circuit has a resistance of 10 Ω and is operated on 120 V, it would, using Ohm's law, have a current flow of

$$I = \frac{E}{R} = \frac{120\ \text{V}}{10\ \Omega} = 12\ \text{A}$$

If this example is calculated again with less resistance, the current will be greater because the voltage is divided by a smaller number. If the resistance is lowered to 1 Ω; the current flow is then

$$I = \frac{E}{R} = \frac{120\ \text{V}}{1\ \Omega} = 120\ \text{A}$$

If the resistance is reduced to 0.1 Ω, the current flow will be 1200 A. By this time the circuit breaker will trip, but you may have already incurred burns or an electrical shock, Figure 4–16. Remember, current flow through the body of 0.015 ampere or less can prove fatal.

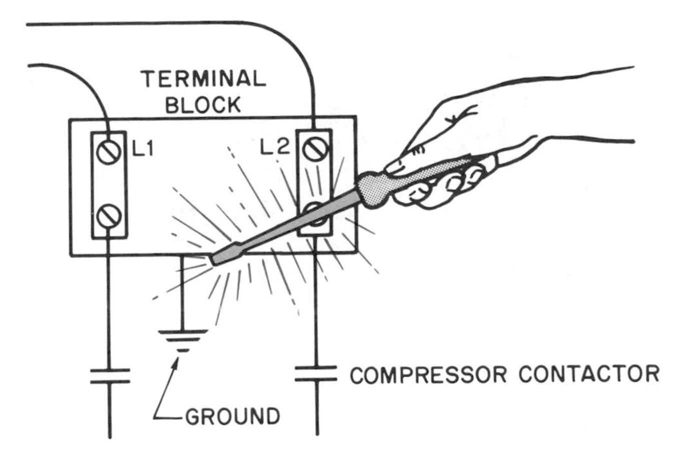

FIGURE 4–16 Wiring illustration showing short circuit caused by a slip of a screwdriver.

4.3 HEAT

The use of heat requires special care. A high concentration of heat comes from torches. Torches are used for many things, including soldering or brazing. Many combustible materials may be in the area where soldering is required. For example, the refrigeration system in a restaurant may need repair. The upholstered restaurant furniture, grease, and other flammable materials must be treated as carefully as possible. ***When soldering or using concentrated heat, a fire extinguisher should always be close by, and you should know exactly where it is and how to use it.*** Learn to use a fire extinguisher *before* the fire occurs. A fire extinguisher should always be included as a part of the service tools and equipment on a service truck, Figure 4–17.

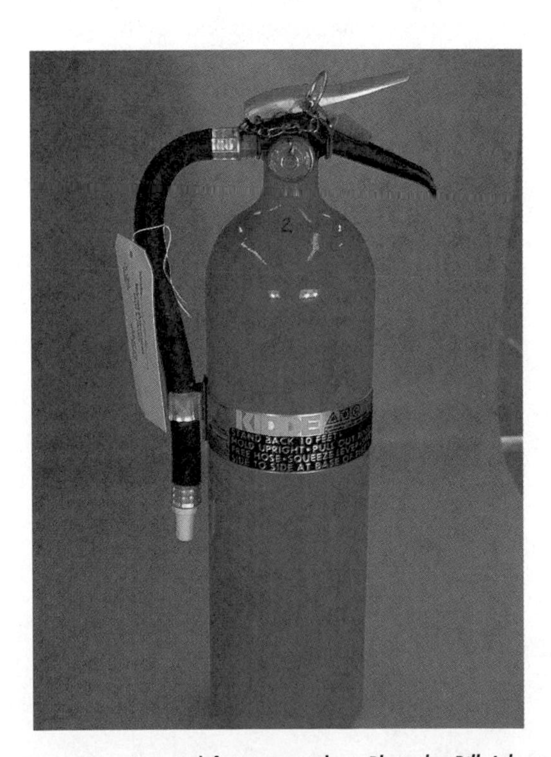

FIGURE 4–17 A typical fire extinguisher. *Photo by Bill Johnson*

FIGURE 4–18 A shield used when soldering. *Courtesy Wengaersheek*

When a solder connection must be made next to combustible materials or a finished surface, use a shield of noncombustible materials for insulation, Figure 4-18. A fire-resistant spray may also be used to decrease the flammability of wood if a torch must be used nearby, Figure 4–19. This spray retardant should be used with an appropriate shield. It is also often necessary to use a shield when soldering within an equipment cabinet, for example, when an ice-maker compressor is changed or when a drier must be soldered in the line.

FIGURE 4–19 Fire-retardant spray.

SAFETY PRECAUTION: *Never solder tubing lines that are sealed. Service valves or Schrader ports should be open before soldering is attempted.*

SAFETY PRECAUTION: *Hot refrigerant lines, hot heat exchangers, and hot motors can burn your skin and leave a permanent scar. Care should be used while handling them.*

4.4 COLD

Cold can be as harmful as heat. Liquid refrigerant can freeze your skin or eyes instantaneously. But long exposure to cold is also harmful. Working in cold weather can cause frostbite. Wear proper clothing and waterproof boots, which also help protect against electrical shock. A cold wet technician will not always make decisions based on logic. Make it a point to stay warm. ***Waterproof boots not only protect your feet from water and cold, but help to protect you from electrical shock hazard when working in wet weather.***

Low-temperature freezers are just as cold in the middle of the summer as in the winter. Cold-weather gear must be used when working inside these freezers. For example, an expansion valve may need changing, and you may be in the freezer for more than an hour. It is a shock to the system to step into a room that is 0°F from the outside where it may be 95° or 100°F. If you are on call for any low-temperature applications, carry a coat and gloves and wear them in cold environments.

4.5 MECHANICAL EQUIPMENT

Rotating equipment can damage body and property. Motors that drive fans, compressors, and pumps are among the most dangerous because they have so much power. ***If a shirt sleeve or coat were caught in a motor drive pulley or coupling, severe injury could occur. Loose clothing should never be worn around rotating machinery, Figure 4–20. Even a small electric hand drill can wind a necktie up before the drill can be shut off.***

FIGURE 4–20 ***Never wear a necktie or loose clothing when using or working around rotating equipment.***

When starting an open motor, stand well to the side of the motor drive mechanism. If the coupling or belt were to fly off the drive, it would fly outward in the direction of rotation of the motor. All set screws or holding mechanisms must be tight before a motor is started, even if the motor is not connected to a load. All wrenches must be away from a coupling or pulley. A wrench or nut thrown from a coupling can be a lethal projectile, Figure 4–21.

When a large motor, such as a fan motor, is coasting to a stop, don't try to stop it. If you try to stop the motor and fan by gripping the belts, the momentum of the fan and motor may pull your hand into the pulley and under the belt, Figure 4–22.

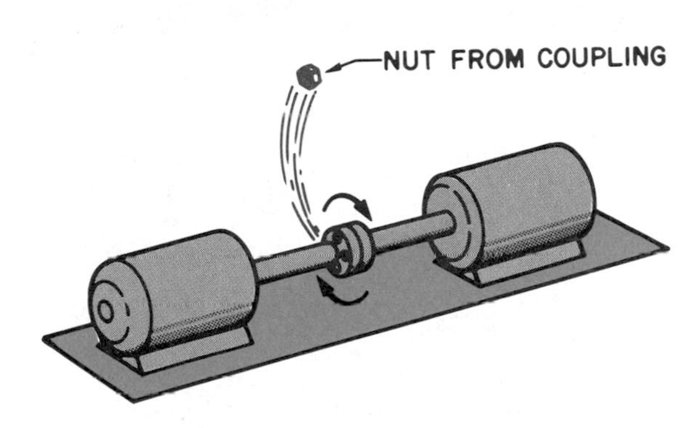

FIGURE 4–21 ***Ensure that all nuts are tight on couplings and other components.***

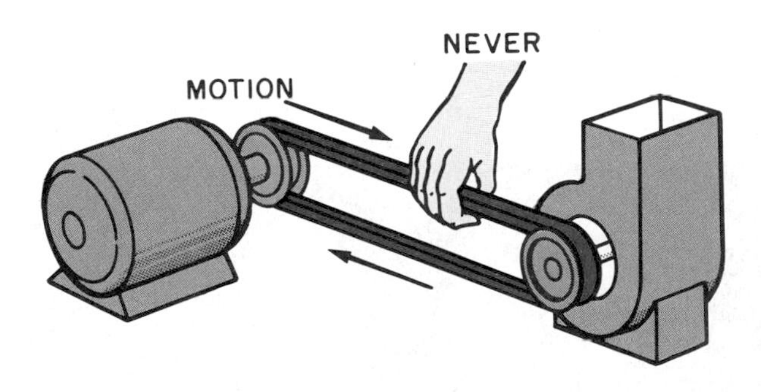

FIGURE 4–22 ***Never attempt to stop a motor or other mechanism by gripping the belt.***

Never wear jewelry while working on a job that requires much movement. A ring may be caught on a nail head, or a bracelet may be caught on the tailgate of a truck as you jump down, Figure 4–23.

When using a grinder to sharpen tools, remove burrs, or for other reasons, use a face shield, Figure 4–24. Most grinding stones are made for grinding ferrous metals, cast iron, steel, stainless steel, and others. However, other stones are made for nonferrous metals such as aluminum, copper,

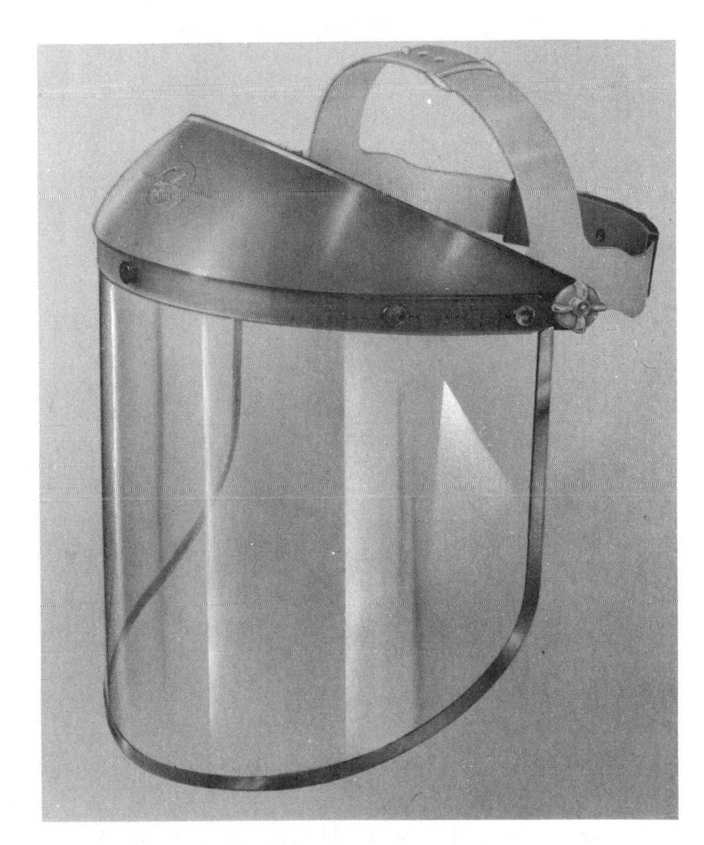

FIGURE 4–24 Use a face shield when grinding. *Courtesy Jackson Products*

FIGURE 4–23 Jewelry can catch on nails or other objects and cause injury.

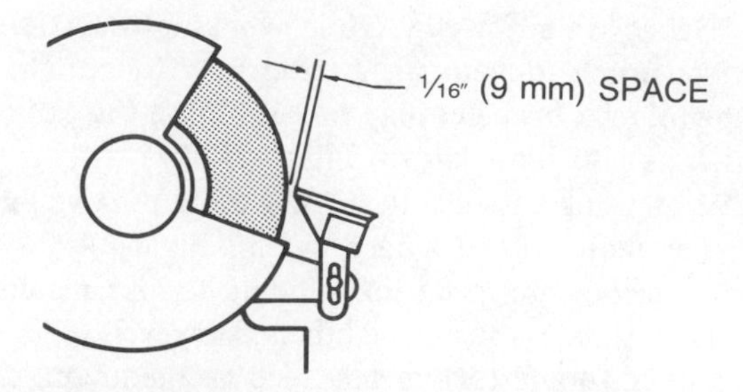

FIGURE 4–25 Keep the tool rest on a grinder adjusted properly.

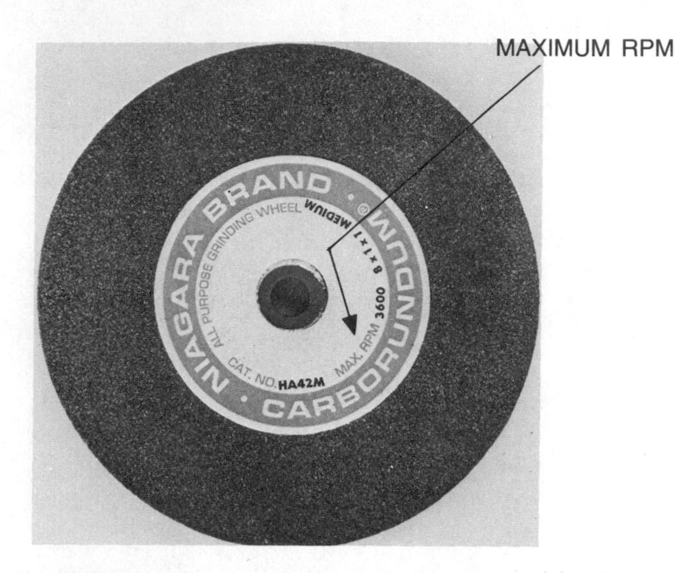

FIGURE 4–26 When installing a grinding stone, ensure that it is compatible with the grinder. From Jeffus & Johnson, *Welding: Principles and Applications, 3E,* © 1992 by Delmar Publishers Inc.

or brass. Use the correct grinding stone for the metal you are grinding. The tool rest should be adjusted to approximately 1/16 in. from the grinding stone, Figure 4–25. As the stone wears down keep the tool rest adjusted to this setting. Grinding stones must not be used on a grinder that turns faster than the stones rated maximum revolutions per minute (rpm) as they may explode, Figure 4–26.

4.6 MOVING HEAVY OBJECTS

Heavy objects must be moved from time to time. Think out the best and safest method to move these objects. Don't just use muscle power. Special tools can help you move equipment. When equipment must be installed on top of a building, a crane or even a helicopter can be used, Figure 4–27. Do not take a chance by trying to lift heavy equipment by yourself; get help from another person and use tools and equipment designed for the purpose. A technician without proper equipment is limited.

When you must lift, use your legs not your back and wear an approved back brace belt, Figure 4–28. Some available tools are a pry bar, a lever truck, a refrigerator hand truck, a lift gate on the pickup truck, and a portable dolly, Figure 4–29.

FIGURE 4–27 Helicopter lifting air conditioning equipment to roof.

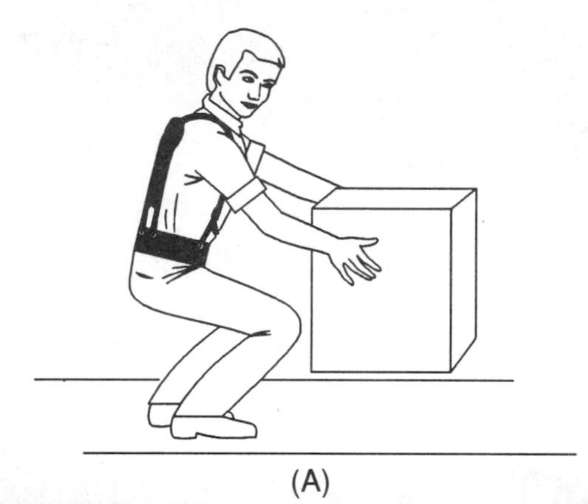

(A)

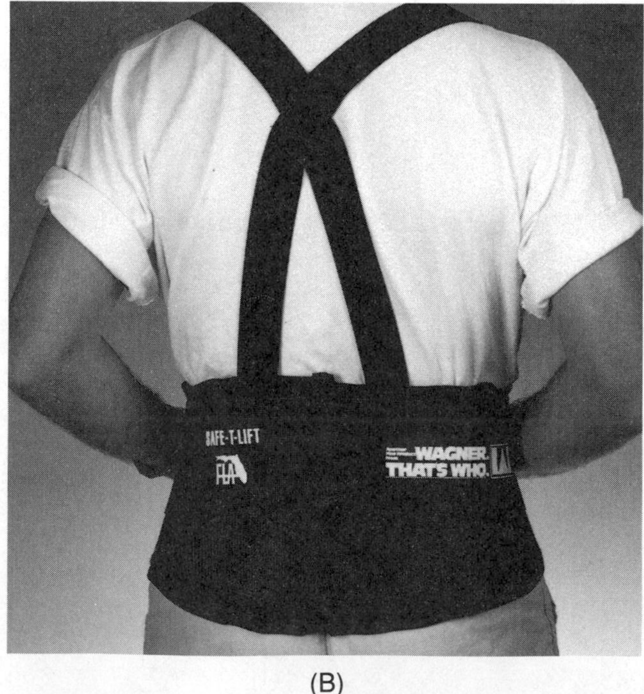

(B)

FIGURE 4–28 ***Use legs, not back, to lift objects. Keep back straight. Use back belt brace.*** *Courtesy Wagner Products Corp.*

FIGURE 4–29 (A) Pry bar. (B) Lever truck. (C) Hand truck. (D) Lift gate on a pickup truck. (E) Portable dolly. *Photos by Bill Johnson*

When moving large equipment across a carpeted or tiled floor, or across a gravel-coated roof, first lay down some plywood. Keep the plywood in front of the equipment as it is moved along. When equipment has a flat bottom, such as a package air conditioner, short lengths of pipe may be used to move the equipment across a solid floor, Figure 4–30.

FIGURE 4–30 Moving equipment using short lengths of pipe as rollers.

4.7 REFRIGERANTS IN YOUR BREATHING SPACE

Fresh refrigerant vapors and many other gases are heavier than air and can displace the oxygen in a closed space. Proper ventilation must be used at all times to prevent being overcome by the the lack of oxygen. Should you be in a close space and the concentration of refrigerant becomes too great, you may not notice it until it is too late. Your symptoms would be a dizzy feeling and your lips may become numb. If you should feel this way, move to a place with fresh air.

Proper ventilation should be set up in advance of starting a job. Fans may be used to push or pull fresh air into a confined space where work must be performed. Cross ventilation can help prevent a buildup of fumes, Figure 4–31.

Refrigerant vapors that have been heated while soldering or passing through an open flame, such as a leak in the presence of a fire, are dangerous. They are toxic and may cause harm. They have a strong odor. ***If you are soldering in a close place, keep your head below the rising fumes and have plenty of ventilation, Figure 4–32.**

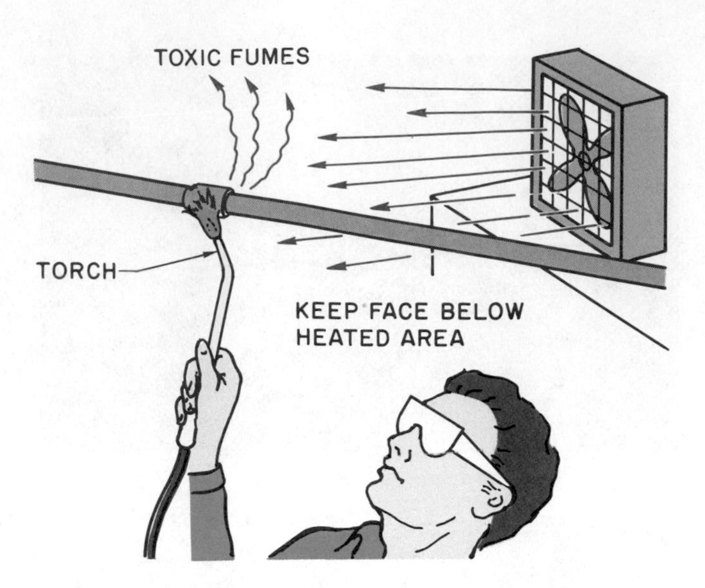

FIGURE 4–32 Keep face below heated area and ensure that area is well ventilated.

4.8 USING CHEMICALS

Chemicals are often used to clean equipment such as air-cooled condensers and evaporators. They are also used for water treatment. The chemicals are normally simple and mild, except for some harsh cleaning products used for water treatment. ***These chemicals should be handled according to the manufacturer's directions. Do not get careless. If you spill chemicals on your skin or splash them in your eyes, follow the manufacturer's directions and go to a doctor. It is a good idea to read the entire label before starting a job. It's hard to read the first aid treatment for eyes after they've been damaged.***

Refrigerant and oil from a motor burnout can be harmful. The contaminated refrigerant and oil may be hazardous to your skin, eyes, and lungs because they contain acid. ***Keep your distance if a line is subject to rupture or any amount of this refrigerant is allowed to escape.***

SUMMARY

- ***The technician must use every precaution when working with pressures, electrical energy, heat, cold, rotating machinery, chemicals, and moving heavy objects.***
- ***Safety situations involving pressure are encountered while working with pressurized systems and vessels.***
- ***Electrical energy is present while troubleshooting energized electrical circuits. Be careful. Turn off**

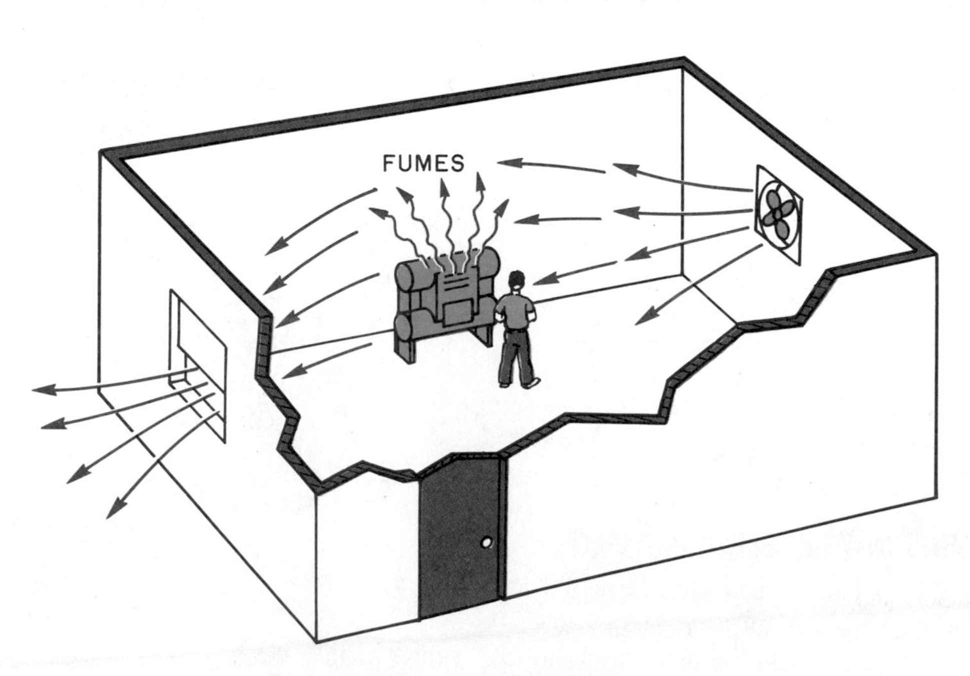

FIGURE 4–31 Cross ventilation with fresh air will help prevent fumes from accumulating.

electrical power, if possible, when working on an electrical component. Lock panel or disconnect box and keep the only key in your possession.***

- ***Heat is encountered while soldering and working on heating systems.***
- ***Liquid R-22 refrigerant boils at −41°F at atmospheric pressure and will cause frostbite.***
- ***Rotating equipment such as fans and pumps can be dangerous and should be treated with caution.***
- ***When moving heavy equipment, use correct techniques, appropriate tools and equipment, and wear a back brace belt.***
- ***Chemicals are used for cleaning and water treatment and must be handled with care.***

REVIEW QUESTIONS

1. Where would a technician encounter freezing temperatures with liquid refrigerant?
2. Should a technician ever refill a refrigerant throwaway cylinder?
3. What can happen to an oxygen or nitrogen cylinder if the valve is broken off?
4. Why must nitrogen be used with a regulator when charging a system?
5. What can happen when oil is mixed with oxygen under pressure?
6. Why should technicians not use their hands to stop liquid refrigerant from escaping?
7. What part of the body does electrical shock harm?
8. What two types of injury are caused by electrical energy?
9. How can the technician prevent electrical shock?
10. How can the technician prevent paint burning on equipment while soldering?
11. What safety precaution should be taken before starting a large motor with the coupling disconnected?
12. What is the third wire used for on a protable electric drill?
13. How can heavy equipment be moved across a rooftop?
14. What special precautions should be taken before using chemicals to clean an air-cooled condenser?
15. How can refrigerant from a motor burnout be harmful?

UNIT 5

Tools and Equipment

OBJECTIVES

After studying this unit, you should be able to

- describe hand tools used by the air conditioning, heating, and refrigeration technician.
- describe equipment used to install and service air conditioning, heating, and refrigeration systems.

SAFETY CHECKLIST

* Tools and equipment should be used only for the job for which they were designed. Other use may damage the tool or equipment and may be unsafe for the technician.

Air conditioning, heating, and refrigeration technicians must be able to properly use hand tools and specialized equipment relating to this field. Technicians must use the tools and equipment intended for the job. This unit contains a brief description of most of the general and specialized tools and equipment that technicians use in this field. Some of these are described in more detail in other units as they apply to specific tasks.

5.1 GENERAL HAND TOOLS

Figures 5–1 through 5–6 illustrate many general hand tools used by service technicians.

PORTABLE ELECTRIC DRILLS. Portable electric drills are used extensively by refrigeration and air conditioning technicians. They are available in cord type (115 V) or cordless (battery operated), Figures 5–6(E) and 5–6(F). If the drill is a cord type and not double-insulated, it must have a three-prong plug and be used only in a grounded receptacle. Drills that have variable speed and reversing options are recommended. Many technicians prefer the battery-operated type for its convenience and safety. Be sure that the batteries are kept charged.

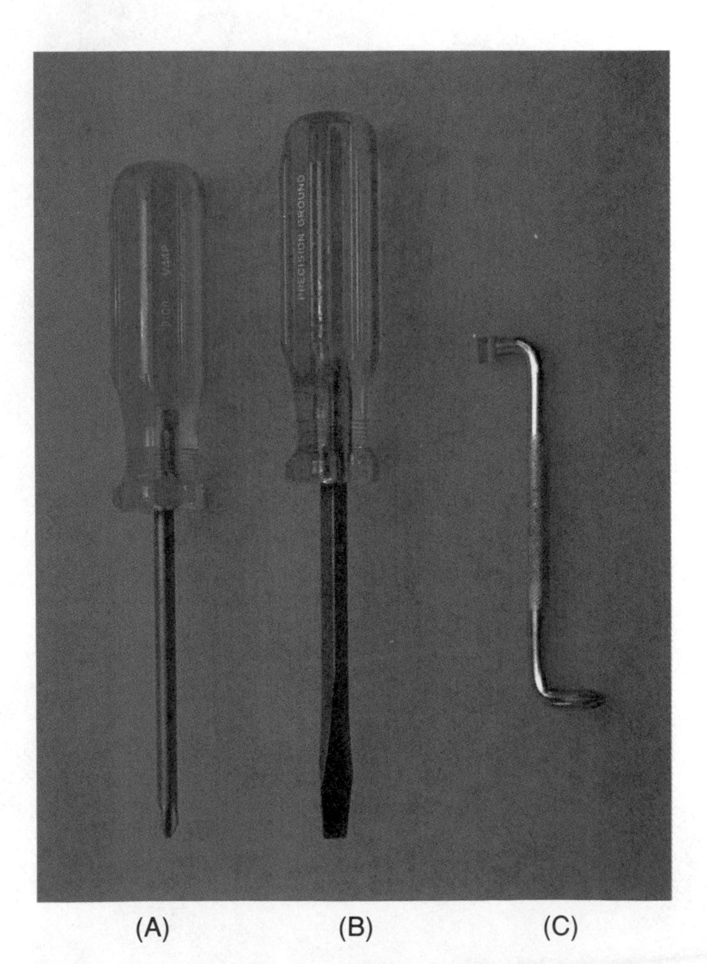

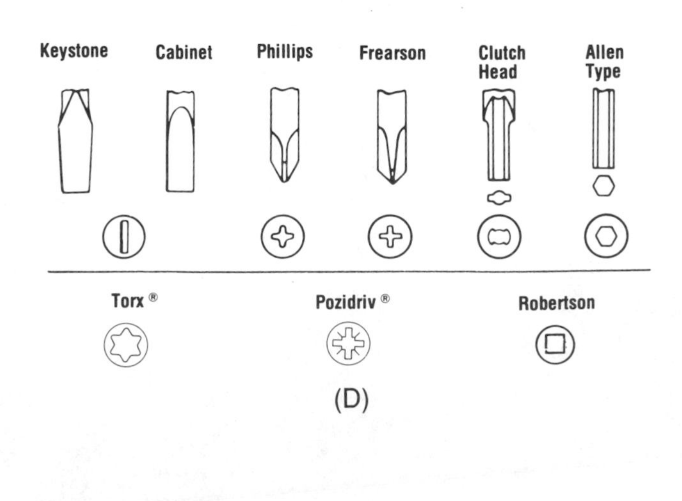

FIGURE 5–1 Screwdrivers. (A) Phillips tip. (B) Straight or slot blade. (C) Offset. (D) Standard screwdriver bit types. *(A), (B), and (C) Photos by Bill Johnson, (D) Courtesy Vaco Products Company*

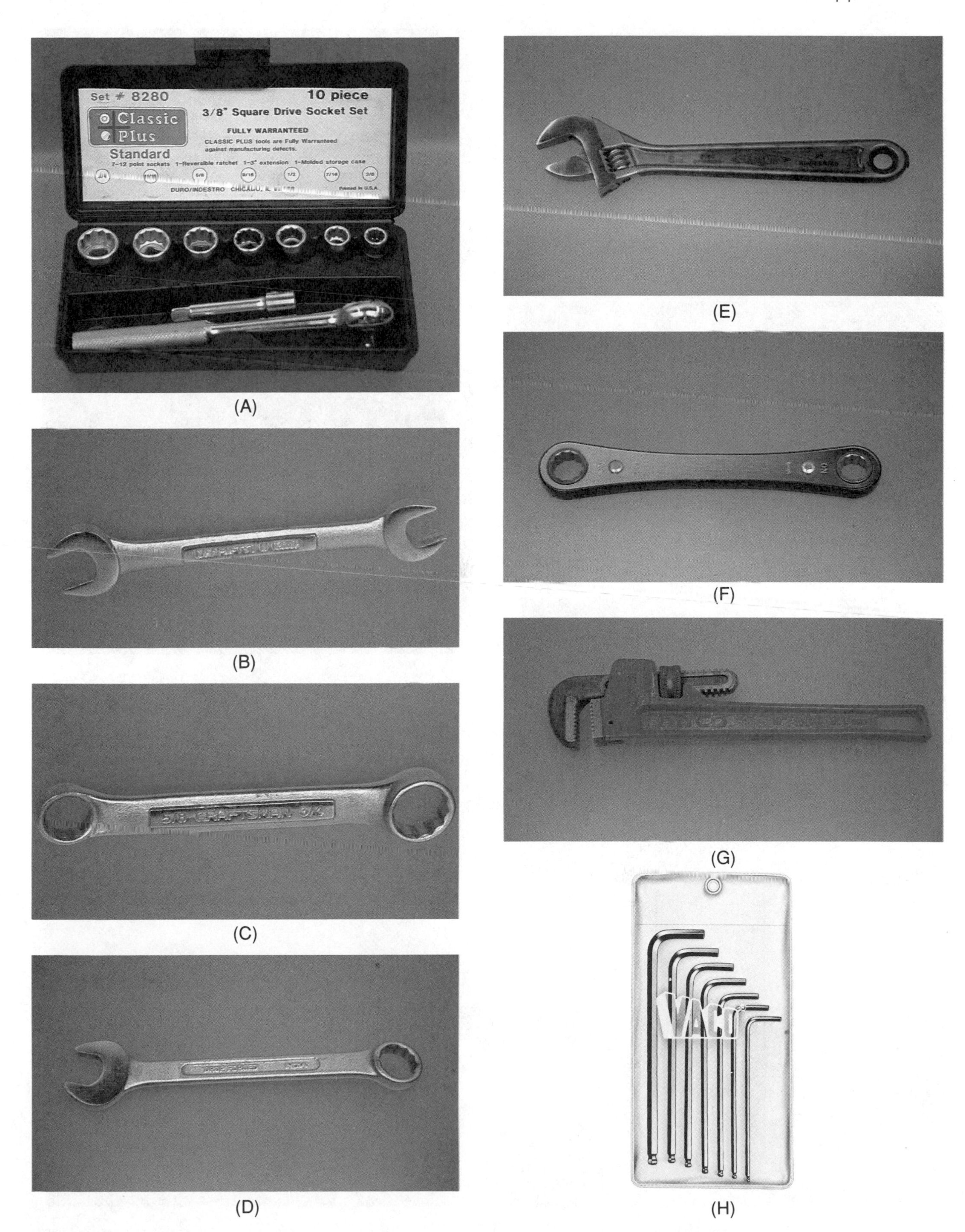

FIGURE 5–2 Wrenches. (A) Socket with ratchet handle. (B) Open end. (C) Box end. (D) Combination. (E) Adjustable open end. (F) Ratchet box. (G) Pipe. (H) Hex key. *(A) through (G) Photos by Bill Johnson, (H) Courtesy Vaco Products Company*

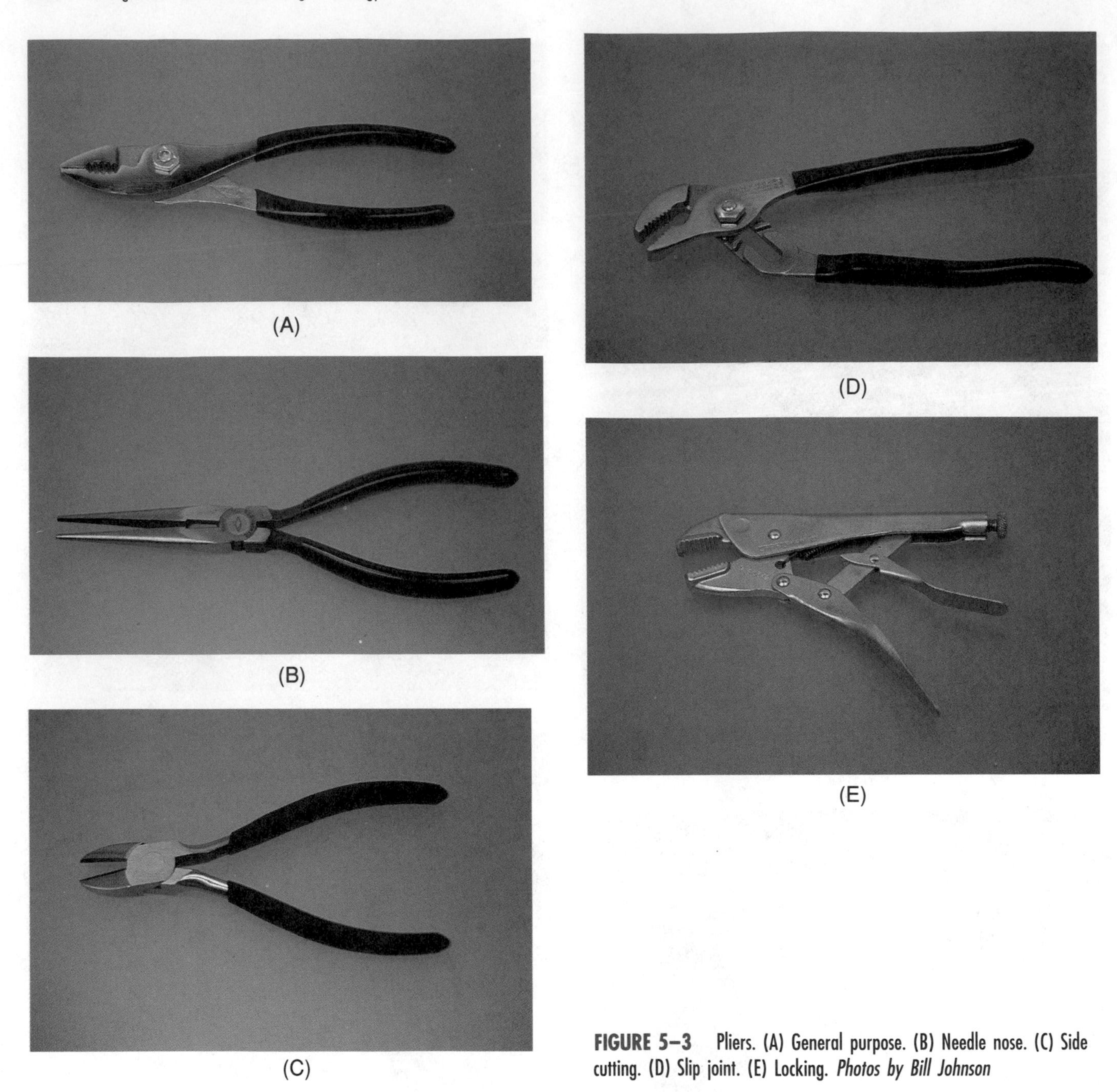

FIGURE 5–3 Pliers. (A) General purpose. (B) Needle nose. (C) Side cutting. (D) Slip joint. (E) Locking. *Photos by Bill Johnson*

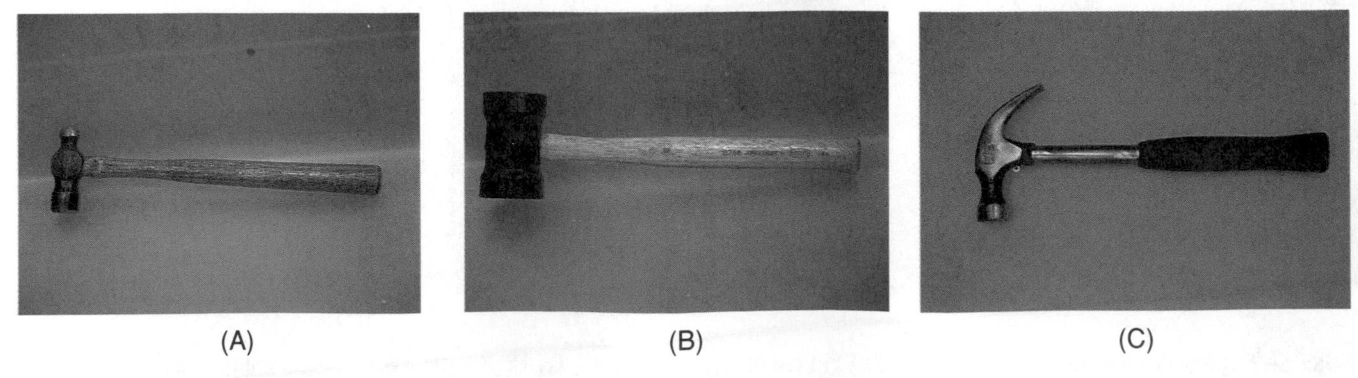

FIGURE 5–4 Hammers. (A) Ball peen. (B) Soft head. (C) Carpenter's claw. *Photos by Bill Johnson*

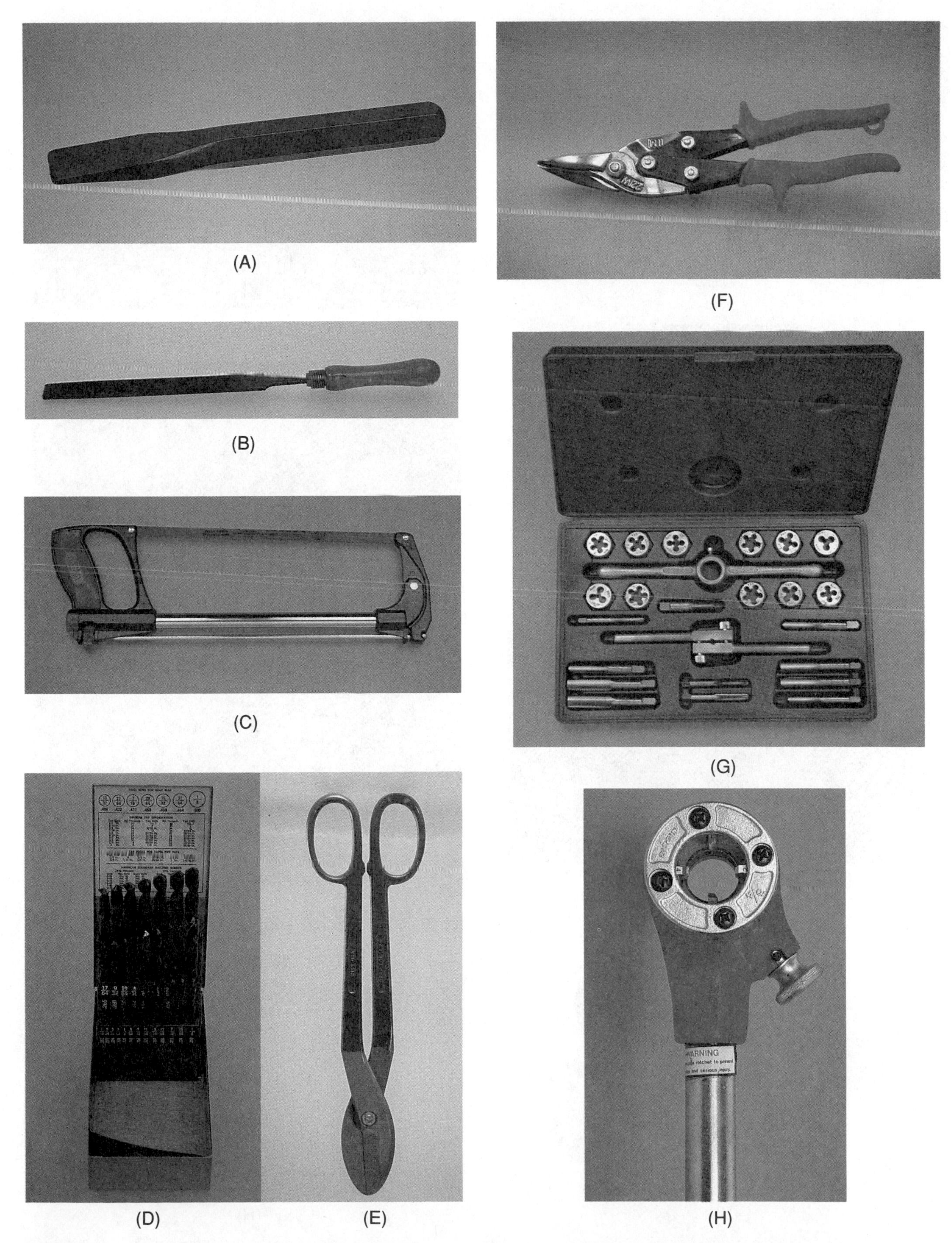

FIGURE 5–5 General metal-cutting tools. (A) Cold chisel. (B) File. (C) Hacksaw. (D) Drill bits. (E) Straight metal snips. (F) Aviation metal snips. (G) Tap and die set. A tap is used to cut an internal thread. A die is used to cut an external thread. (H) Pipe-threading die. *Photos by Bill Johnson*

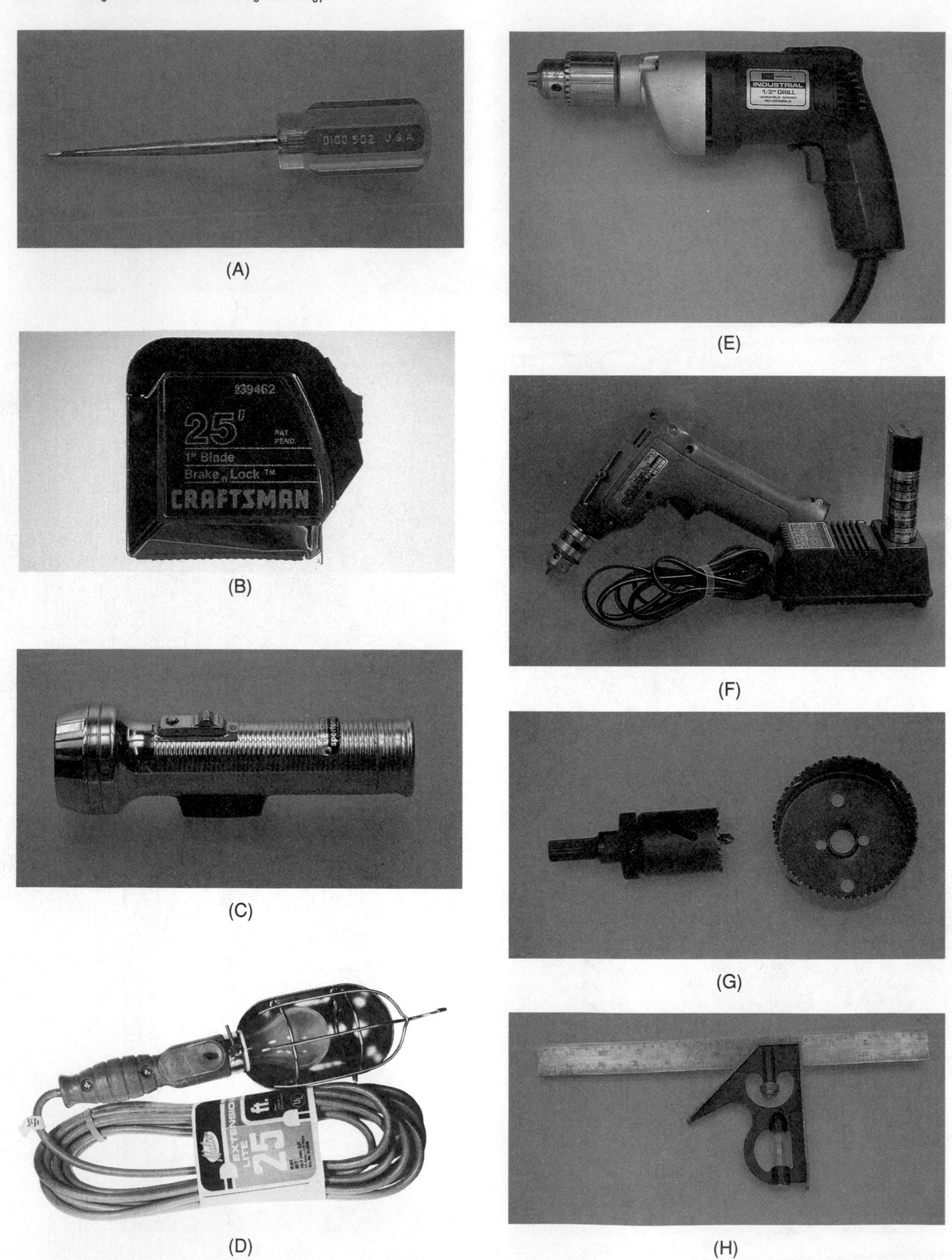

FIGURE 5–6 Other general-purpose tools. (A) Awl. (B) Rule. (C) Flashlight. (D) Extension cord/light. (E) Portable electric drill, cord type. (F) Portable electric drill, cordless. (G) Hole saws. (H) Square.

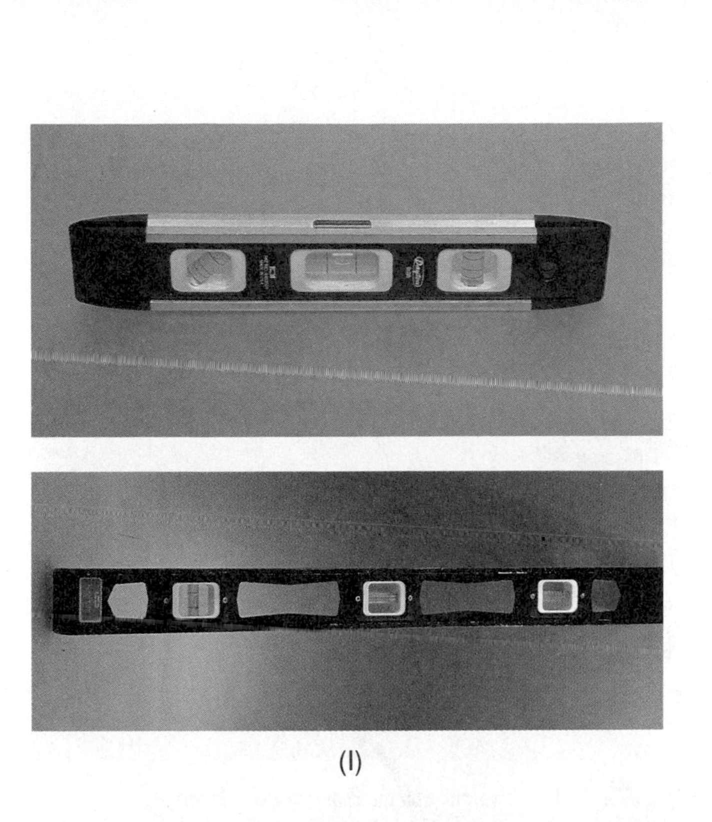

(I)

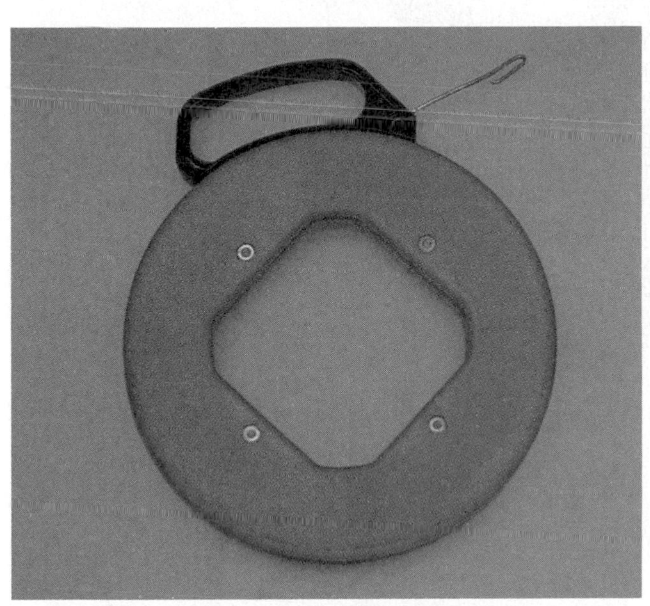

(J)

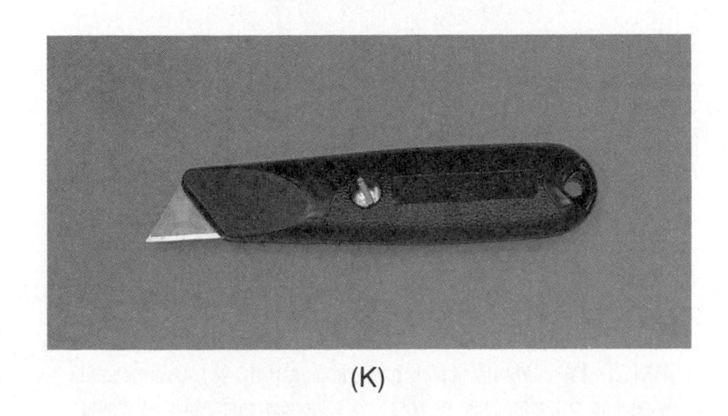

(K)

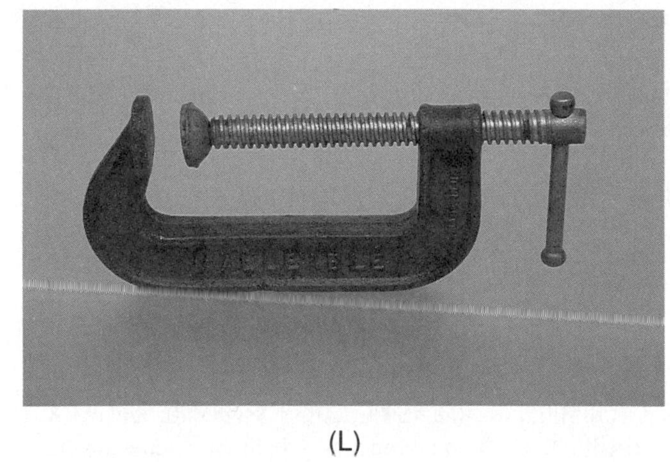

(L)

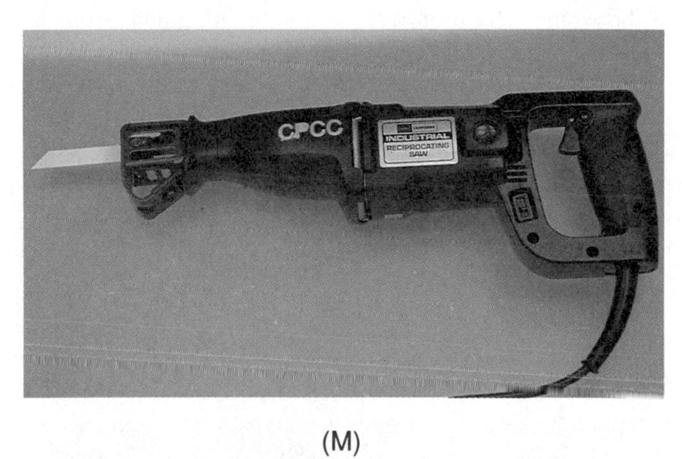

(M)

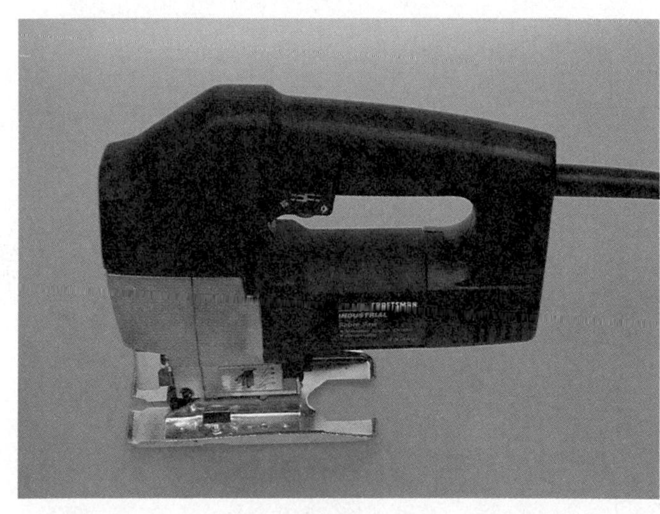

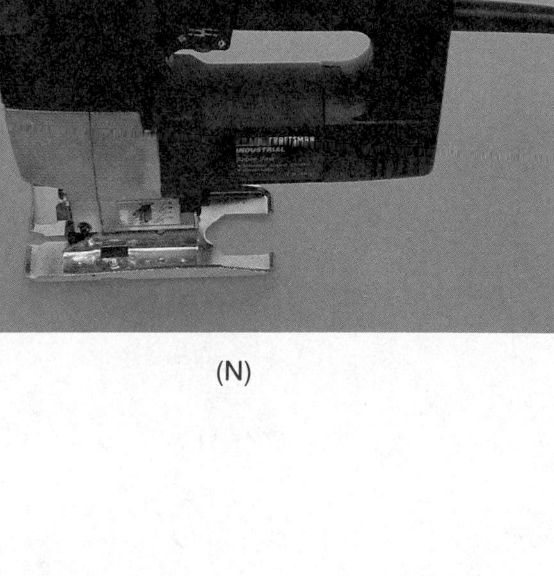

(N)

FIGURE 5–6 (Continued) (I) Levels. (J) Fish tape. (K) Utility knife. (L) C-Clamp. (M) Reciprocating saw. (N) Jig saw. *Photos by Bill Johnson*

5.2 SPECIALIZED HAND TOOLS

The following tools are regularly used by technicians in the air conditioning, heating, and refrigeration field.

NUT DRIVERS. Nut drivers have a socket head and are used primarily to drive hex head screws from panels on air conditioning, heating, and refrigeration cabinets. They are available with hollow shaft, solid shaft, and extra long or stubby shafts, Figure 5–7. The hollow shaft allows the screw to protrude into the shaft when it extends beyond the nut.

AIR CONDITIONING AND REFRIGERATION REVERSIBLE RATCHET BOX WRENCHES. Ratchet box wrenches, where the ratchet direction can be changed by pushing the leverlike button near the end of the wrench, are used with air conditioning and refrigeration valves and fittings. Two openings on each end of the wrench allow it to be used on four sizes of valve stems or fittings, Figure 5–8.

FLARE NUT WRENCH. The flare nut wrench is used like a box end wrench. The opening at the end of the wrench allows it to be slipped over the tubing. The wrench can then be placed over the flare nut to tighten or loosen it, Figure 5–9.

Wrenches should not be used if they are worn because they will round off the corners of fittings made of soft brass.

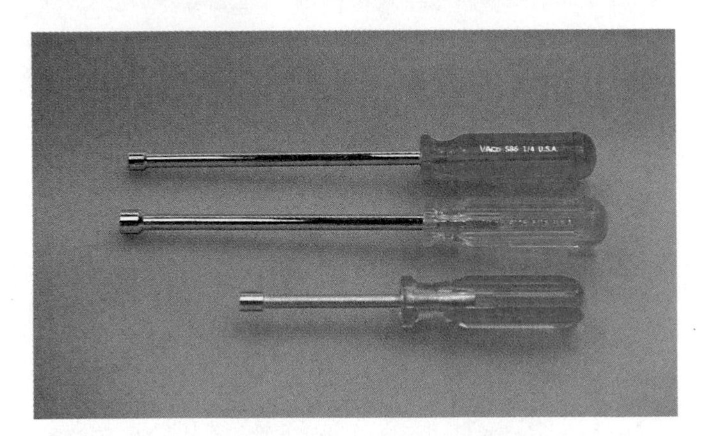

FIGURE 5–7 Assorted nut drivers. *Photo by Bill Johnson*

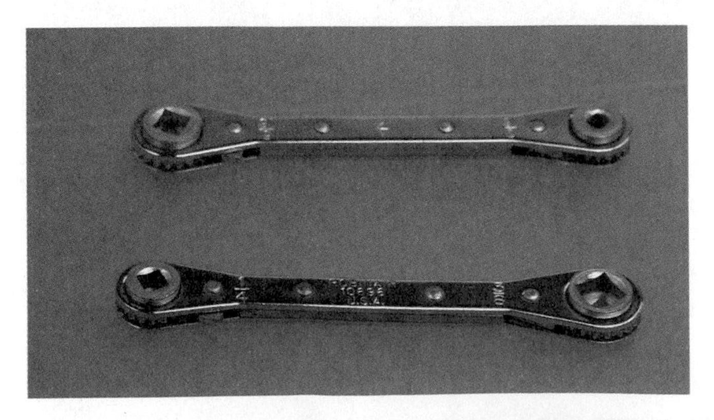

FIGURE 5–8 Air conditioning and refrigeration reversible ratchet box wrenches. *Photo by Bill Johnson*

Do not use a pipe to extend the handle on a wrench for more leverage. Fittings can be loosened by heating or by using a penetrating oil.

WIRING AND CRIMPING TOOLS. Wiring and crimping tools are available in many designs. Figure 5–10 illustrates a combi-

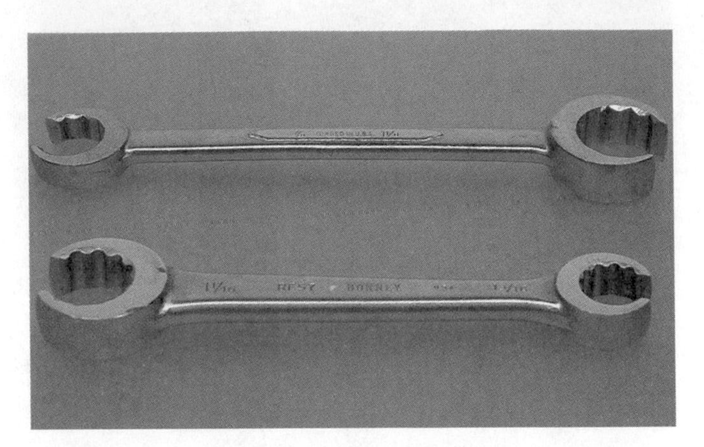

FIGURE 5–9 Flare nut wrench. *Photo by Bill Johnson*

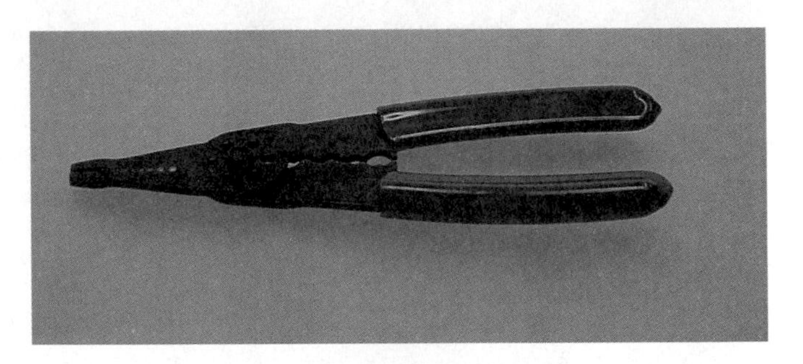

(A)

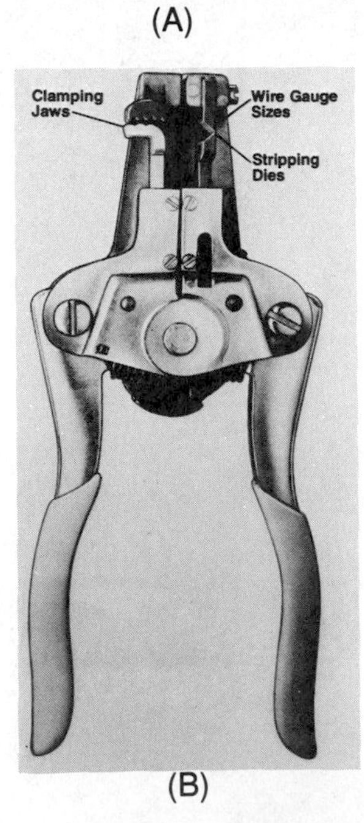

(B)

FIGURE 5–10 Wire stripping and crimping tools. (A) A combination crimping and stripping tool for crimping solderless connectors, stripping wire, cutting wire, and cutting small bolts. (B) Automatic wire stripper *(A) Photo by Bill Johnson, (B) Courtesy Vaco Products Company*

nation tool for crimping solderless connectors, stripping wire, cutting wire, and cutting small bolts. This figure also illustrates an *automatic wire stripper*. To use this tool, insert the wire into the proper strip-die hole. The length of the strip is determined by the amount of wire extending beyond the die away from the tool. Hold the wire in one hand and squeeze the handles with the other. Release the handles and remove the stripped wire.

INSPECTION MIRRORS. Inspection mirrors are usually available in rectangular or round shapes, with fixed or telescoping handles, some over 30 in. long. The mirrors are used to inspect areas or parts in components that are behind or underneath other parts, Figure 5–11.

STAPLING TACKERS. Stapling tackers are used to fasten insulation and other soft materials to wood; some types may be used to install low-voltage wiring, Figure 5–12.

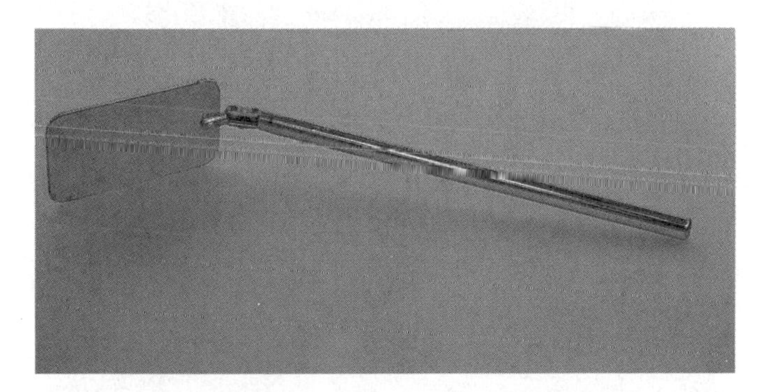

FIGURE 5–11 Inspection mirror. *Photo by Bill Johnson*

FIGURE 5–12 Stapling tacker. *Photo by Bill Johnson*

The following tools are used to install tubing. Most will be described more fully in other units in the book.

TUBE CUTTER. Tube cutters are available in different sizes and styles. The standard tube cutter is shown in Figure 5–13. They are also available with a ratchet feed mechanism. The cutter opens quickly to insert the tubing and slides to the cutting position. Some models have a flare cutoff groove that reduces the tube loss when removing a cracked flare. Many models also include a retractable reamer to remove inside burrs and a filing surface to remove outside burrs. Figure 5–14 illustrates a cutter for use in tight spaces where standard cutters do not fit.

INNER-OUTER REAMERS. Inner-outer reamers use three cutters to ream both the inside and trim the outside edges of tubing, Figure 5–15.

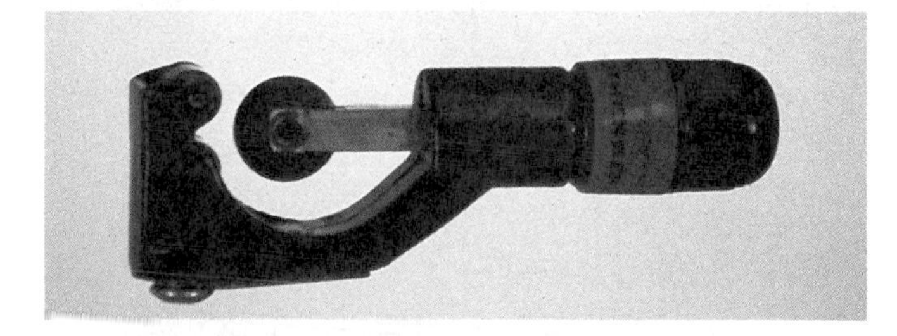

FIGURE 5–13 Tube cutter. *Photo by Bill Johnson*

FIGURE 5–14 Small tube cutter used in tight places. *Photo by Bill Johnson*

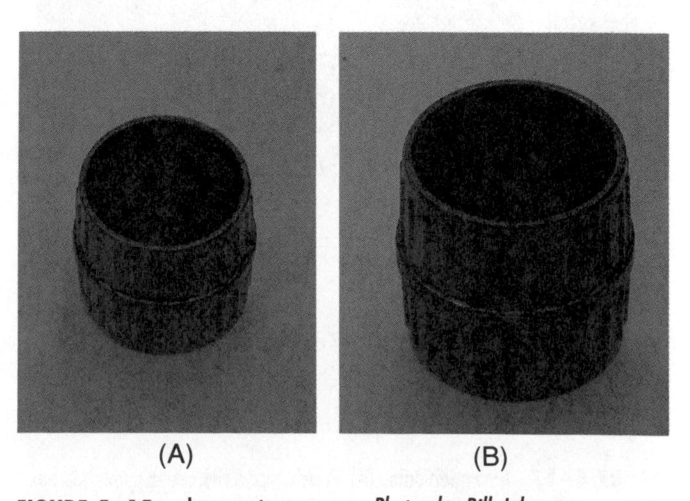

FIGURE 5–15 Inner-outer reamers. *Photos by Bill Johnson*

FLARING TOOLS. The flaring tool has a flaring bar to hold the tubing, a slip-on yoke, and a feed screw with flaring cone and handle. Several sizes of tubing can be flared with this tool, Figure 5–16.

SWAGING TOOLS. Swaging tools are available in punch type and lever type, Figure 5–17.

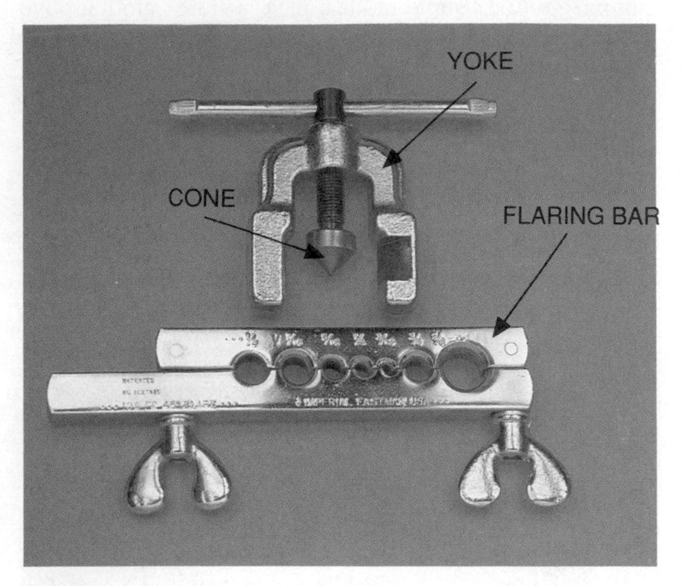

FIGURE 5–16 Flaring tool. This tool has a flaring bar, a yoke and a feed screw with flaring cone. *Photo by Bill Johnson*

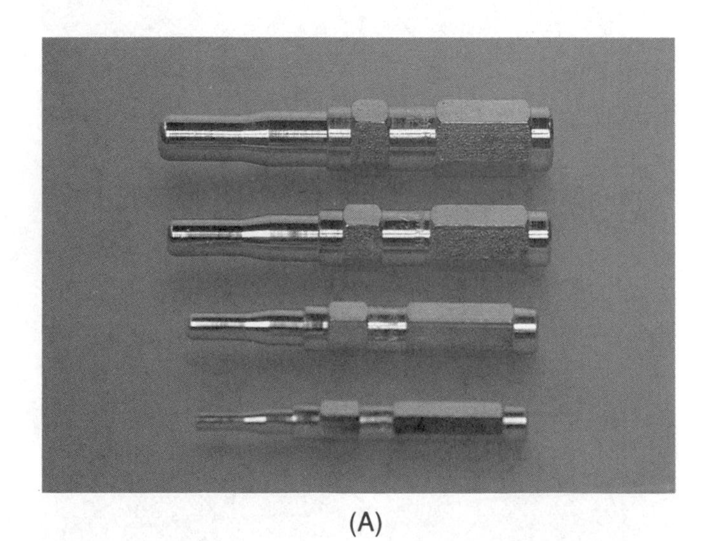

(A)

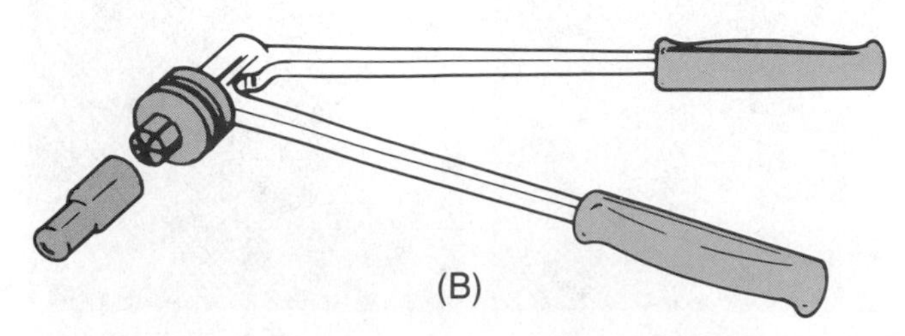

(B)

FIGURE 5–17 Swaging tool. (A) Punch type. (B) Lever type. *(A) Photo by Bill Johnson*

TUBE BENDERS. Three types of tube benders may be used: spring type, lever type, Figure 5–18, and, to a lesser extent, gear type. These tools are used for bending soft copper and aluminum.

TUBE BRUSHES. Tube brushes clean the inside and outside of tubing and the inside of fittings. Some types can be turned by hand or by an electric drill, Figure 5–19.

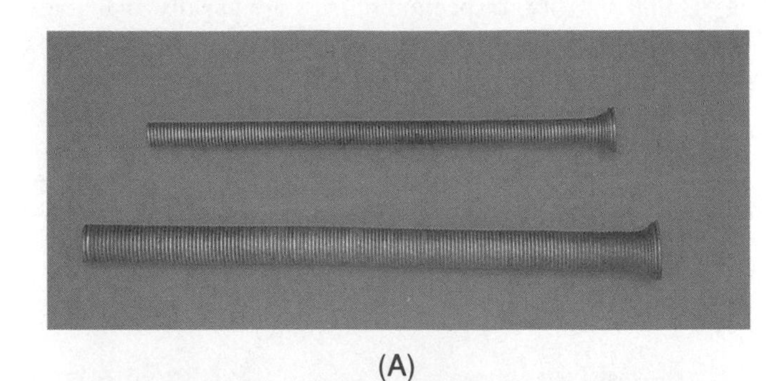

(A)

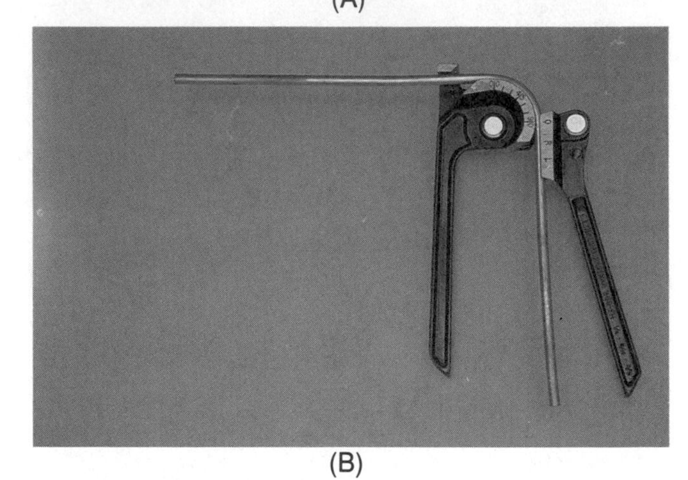

(B)

FIGURE 5–18 Tube benders. (A) Spring type. (B) Lever type. *Photos by Bill Johnson*

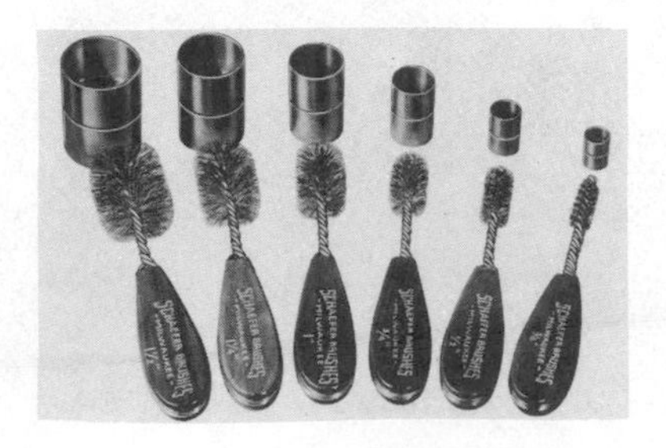

FIGURE 5–19 Tube brushes. *Courtesy Shaefer Brushes*

PLASTIC TUBING SHEAR. A plastic tubing shear cuts plastic tubing and non–wire-reinforced plastic or synthetic hose, Figure 5–20.

TUBING PINCH-OFF TOOL. A tubing pinch-off tool is used to pinch shut the short stub of tubing often provided for service, such as the service stub on a compressor. This tool is used to pinch shut this stub before sealing it by soldering, Figure 5–21.

METALWORKER'S HAMMER. A metalworker's hammer straightens and forms sheet metal for duct work, Figure 5–22.

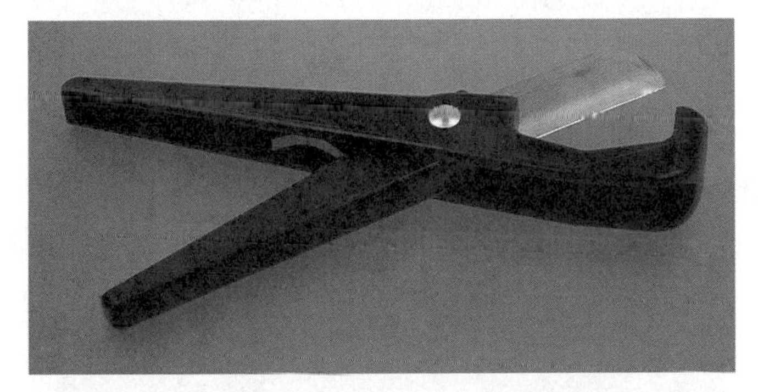

FIGURE 5–20 Plastic tubing shear. *Photo by Bill Johnson*

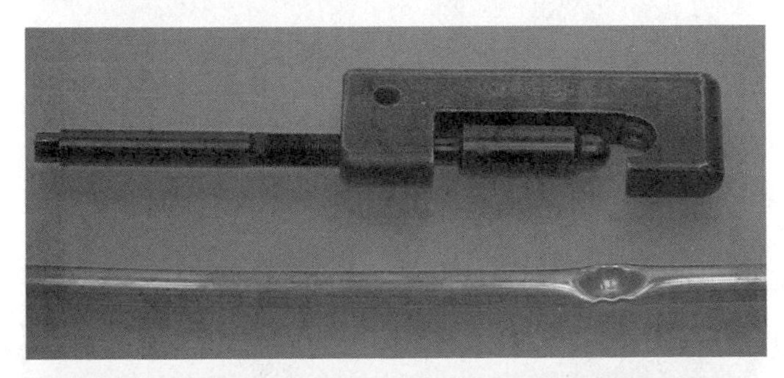

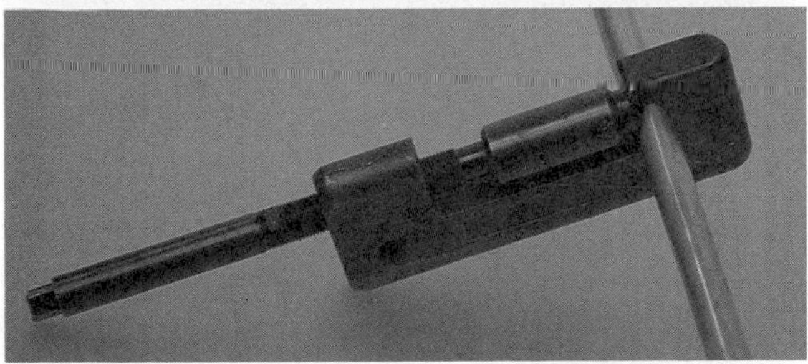

FIGURE 5–21 Tubing pinch-off tool. *Photos by Bill Johnson*

FIGURE 5–22 Metalworker's hammer. *Photo by Bill Johnson*

5.3 SPECIALIZED SERVICE AND INSTALLATION EQUIPMENT

GAGE MANIFOLD. One of the most important of all pieces of refrigeration and air conditioning service equipment is the *gage manifold.* This equipment normally includes the *compound gage* (low-pressure and vacuum), *high-pressure gage,* the *manifold,* valves, and hoses. The gage manifold may be two-valve or four-valve. The four-valve design has separate valves for the vacuum, low-pressure, high-pressure, and refrigerant cylinder connections, Figure 5–23.

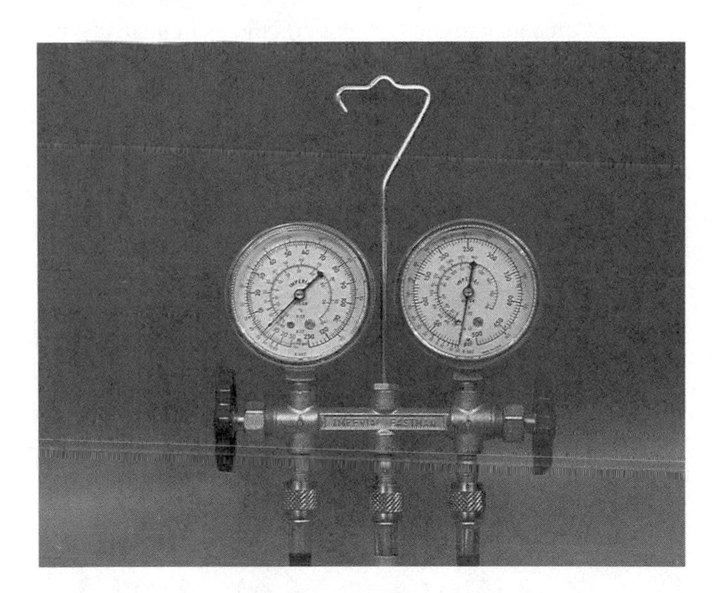

(A)

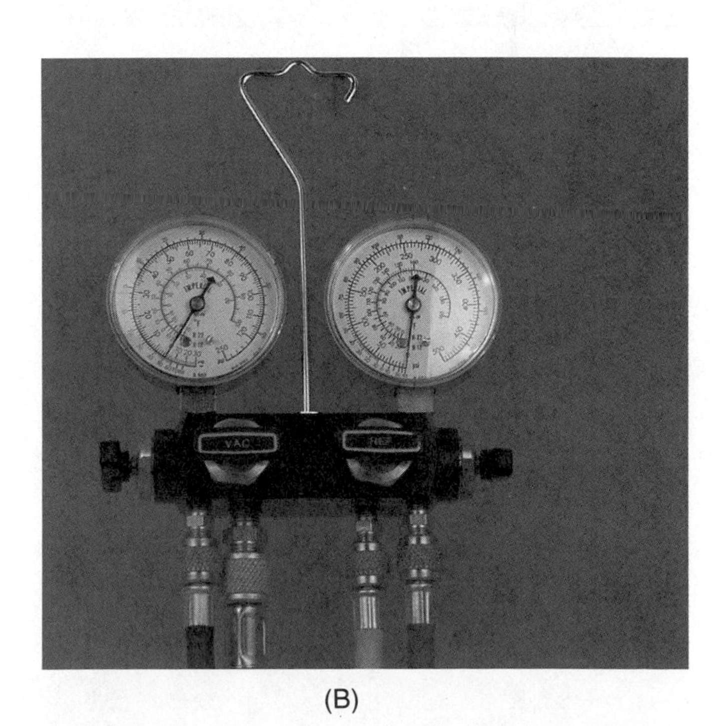

(B)

FIGURE 5–23 Gage manifolds. (A) Two-valve gage manifold. (B) Four-valve gage manifold. *Photos by Bill Johnson*

REFRIGERANT CHARGER. This device allows the technician to accurately charge a system with refrigerant. It can be charged by pressure or by refrigerant weight in 1-oz increments. The particular instrument shown in Figure 5–24 has both digital and analog displays and charges R-12, R-22, R-500, and R-502.

ELECTRONIC CHARGING SCALE. These types of scales will allow a technician to accurately charge refrigerant by weight. This can usually be done manually or automatically. The amount of refrigerant to be charged into the system can be programmed. A microprocessor controls the charge, stopping the process when the programmed weight has been dispensed. Often an audible signal alerts the technician that the proper amount of refrigerant has been charged into the system. Figure 5–25 is an example of an electronic charging scale.

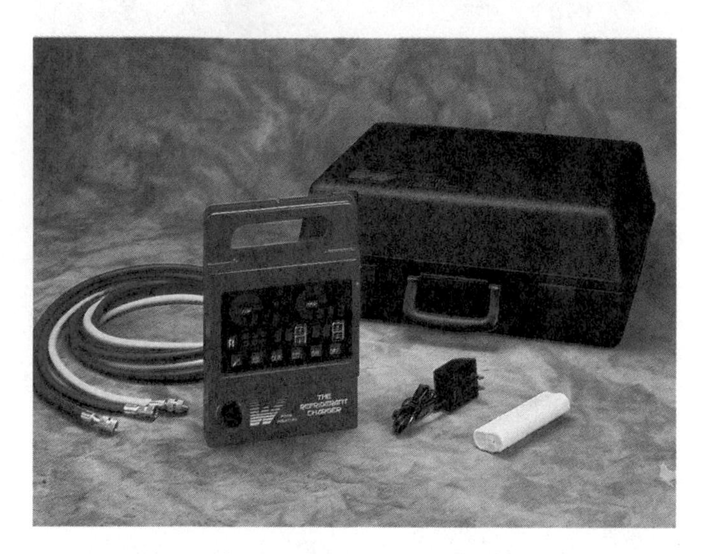

FIGURE 5–24 Refrigerant charger. *Courtesy White Industries*

FIGURE 5-25 Electronic charging scale. *Courtesy Robinair Division, SPX Corporation*

U-TUBE MERCURY MANOMETER. A U-tube mercury manometer shows the level of vacuum while evacuating a refrigeration system. It is accurate to approximately 0.5 mm Hg, Figure 5–26.

ELECTRONIC VACUUM GAGE. An electronic vacuum gage measures the vacuum when evacuating a refrigeration, air conditioning, or heat pump system. It measures a vacuum down to about 1 micron or 0.001 millimeter (mm) (1000 microns = 1 mm), Figure 5–27.

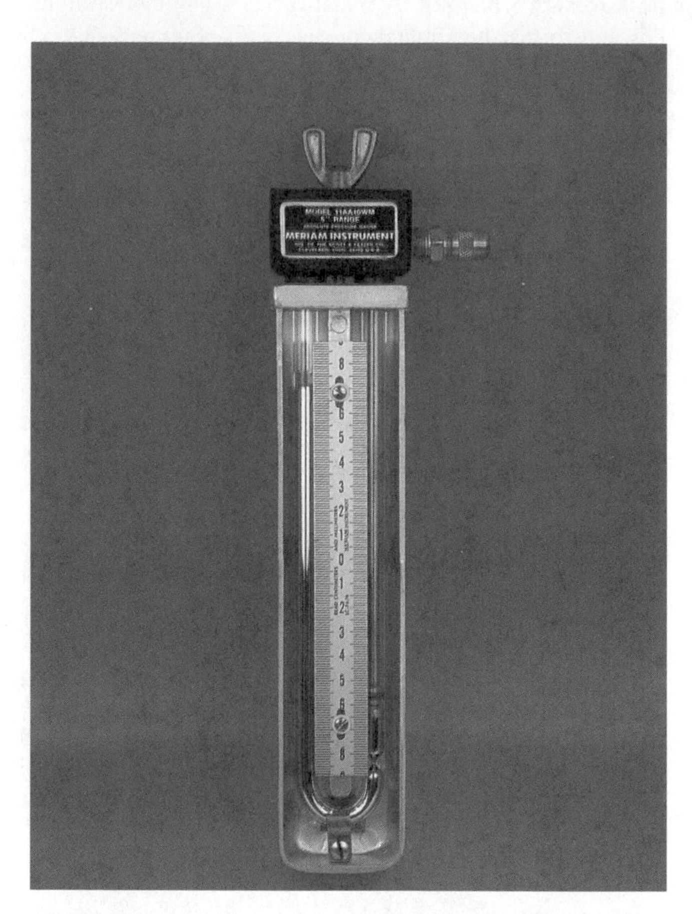

FIGURE 5–26 U-tube mercury manometer. *Photo by Bill Johnson*

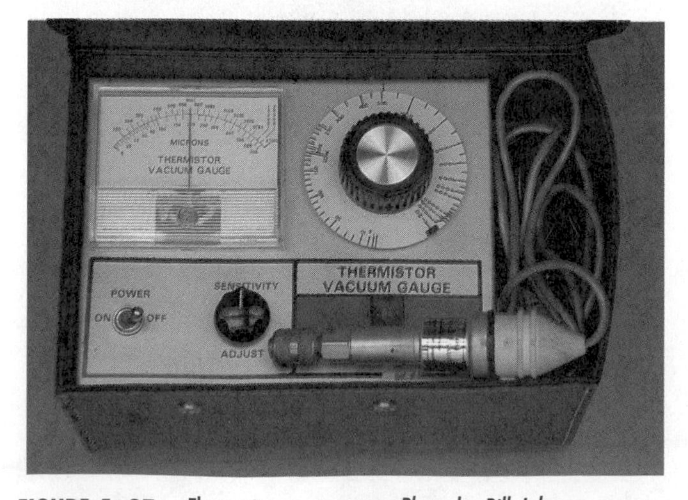

FIGURE 5–27 Electronic vacuum gage. *Photo by Bill Johnson*

VACUUM PUMP. Vacuum pumps designed specifically for servicing air conditioning and refrigeration systems remove the air and noncondensible gases from the system. This is called *evacuating* the system and is necessary because the air and noncondensable gases take up space, contain moisture, and cause excessive pressures. Figure 5–28 shows photos of various vacuum pumps.

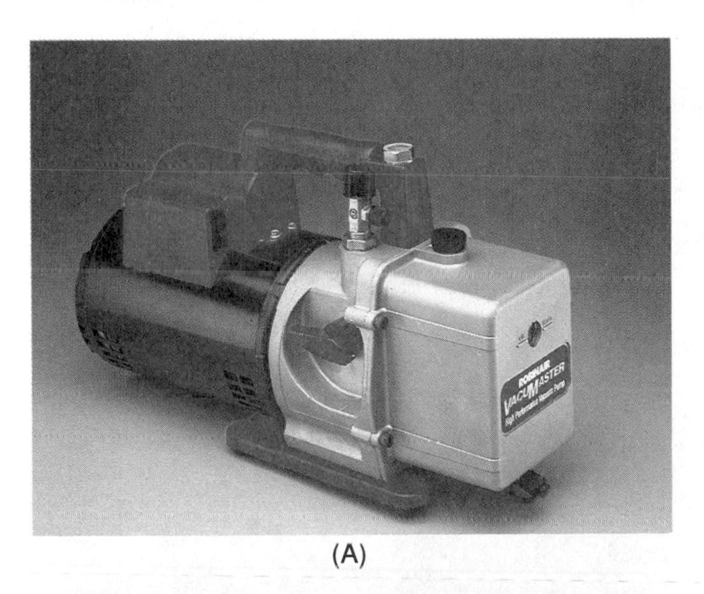

(A)

(B) (C)

FIGURE 5–28 Typical vacuum pumps. *Courtesy (A) Robinair Division, SPX Corporation, (B) and (C) White Industries*

REFRIGERANT RECOVERY RECYCLING STATION. ⊕It is illegal to vent refrigerant to the air. Figure 5–29 is a photo of a typical recovery recycling station. The refrigerant from a refrigeration or air conditioning system is pumped into a cylinder or container at this station where it is stored until it can either be charged back into the system if it meets the requirements or transferred to another approved container for transportation to a refrigerant reclaiming facility.⊕

FIGURE 5–29 Refrigerant recovery recycling station. *Courtesy Robinair Division, Sealed Power Corp.*

HALIDE LEAK DETECTOR. A halide leak detector, Figure 5–30, detects refrigerant leaks. It is used with acetylene or propane gas. When the detector is ignited, the flame heats a copper disc. Air for the combustion is drawn through the attached hose. The end of the hose is passed over or near fittings or other areas where a leak may be suspected. If there is a leak, the refrigerant will be drawn into the hose and contact the copper disc. This breaks down the halogen refrigerants into other compounds and changes the color of the flame. The colors range from green to purple, depending on the size of the leak.

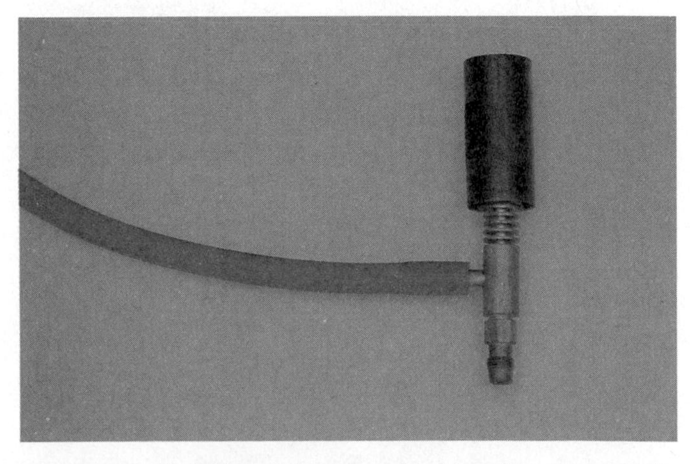

FIGURE 5–30 Halide leak detector used with acetylene. *Photo by Bill Johnson*

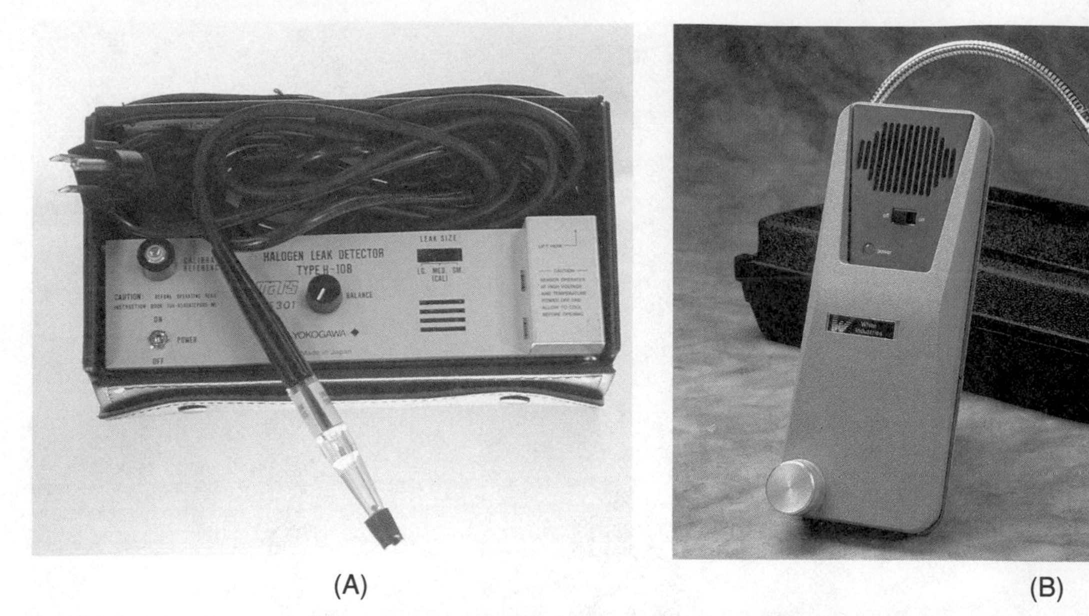

(A) (B)

FIGURE 5–31 Electronic leak detectors. *(A) Photo by Bill Johnson, (B) Courtesy White Industries*

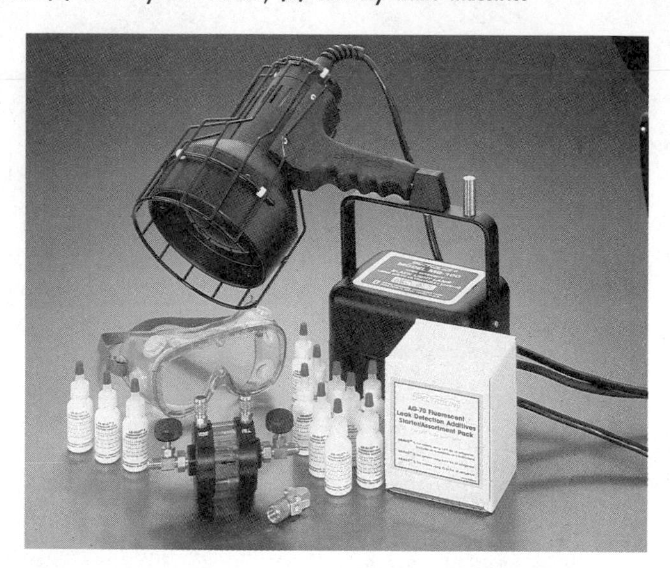

FIGURE 5-32 Fluorescent refrigerant leak detection system using additive with high-intensity ultraviolet lamp. *Courtesy Spectronics Corporation*

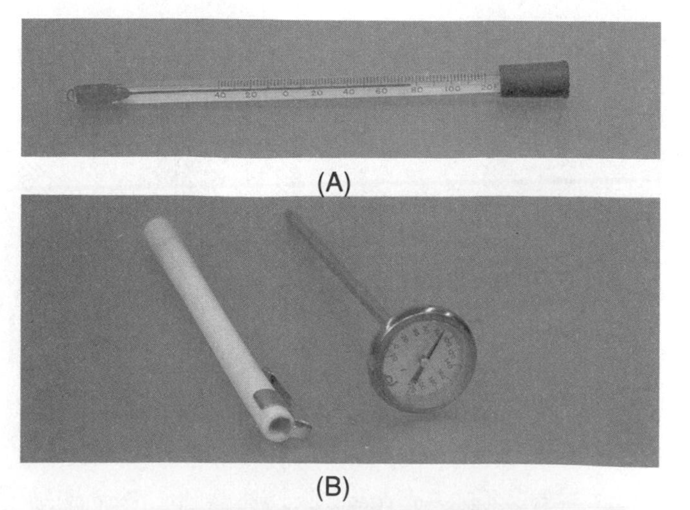

(A)

(B)

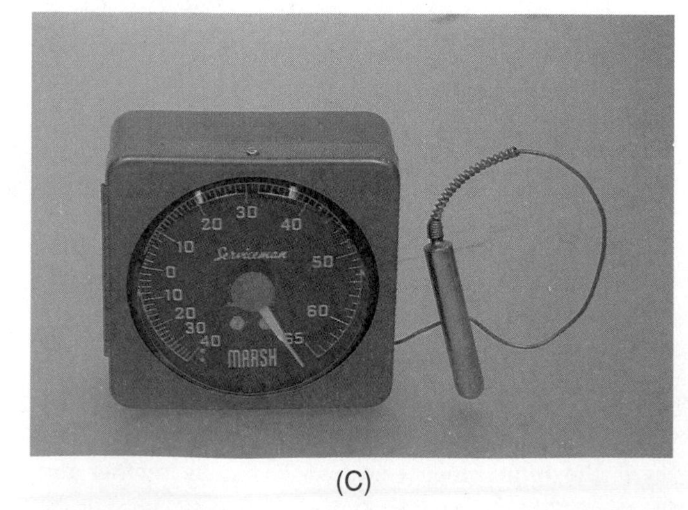

(C)

FIGURE 5–33 Thermometers. (A) Glass stem. (B) Dial indicator. (C) Remote bulb. *Photos by Bill Johnson*

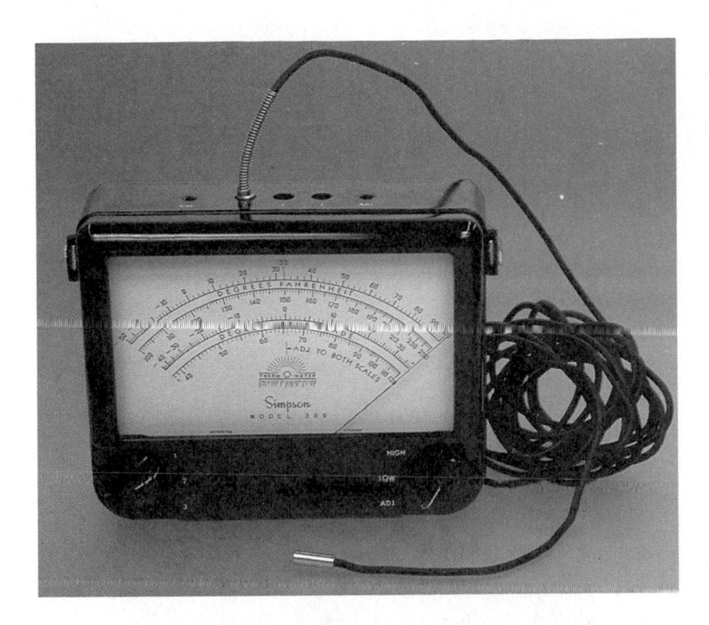

FIGURE 5–34 Electronic thermometer. *Photo by Bill Johnson*

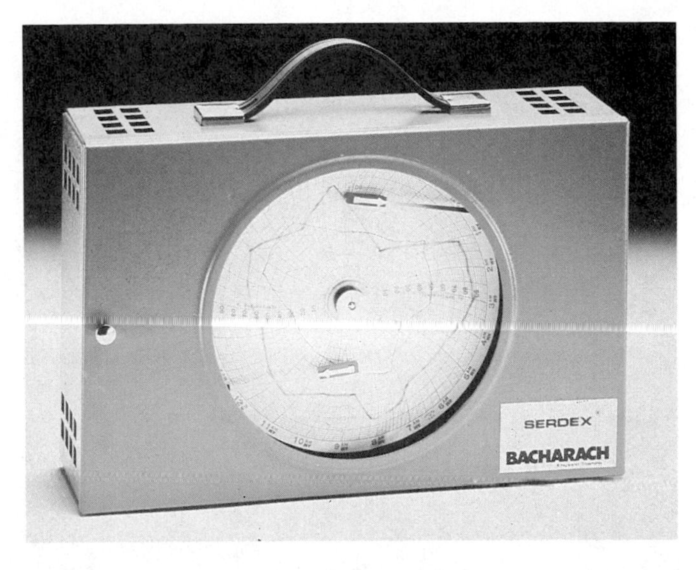

FIGURE 5–35 Recording thermometer. *Courtesy United Technologies Bacharach*

ELECTRONIC LEAK DETECTORS. Electronic leak detectors contain an element sensitive to halogen gases. The device may be battery or AC powered and often has a pump to suck in the gas and air mixture. A ticking signal that increases in frequency and intensity as the probe "homes in" on the leak is used to alert the operator. Many also have varying sensitivity ranges that can be adjusted, Figure 5–31.

ADDITIVE USED WITH A HIGH-INTENSITY ULTRAVIOLET LAMP. With this system, Figure 5–32, an additive is induced into the refrigerant system. This additive shows up as a bright yellow-green glow under the ultraviolet lamp at the source of the leak. The additive can remain in the system should a new leak be suspected at a later date.

THERMOMETERS. Thermometers range from simple pocket styles to electronic and recording types. Pocket-style mercury column and dial-indicator thermometers, Figure 5–33, and remote bulb thermometers are frequently used. For more sophisticated temperature measurements an electronic or recording-type thermometer may be used, Figures 5–34 and 5–35.

FIN STRAIGHTENERS. Fin straighteners are available in different styles. Figure 5–36 is an example of one type capable of straightening condensor and evaporator coil fins having spacing of 8, 9, 10, 12, 14, and 15 fins per inch.

HEAT GUNS. Heat guns are needed in many situations to warm refrigerant, melt ice, and do other tasks. Figure 5–37 is a heat gun carried by many technicians.

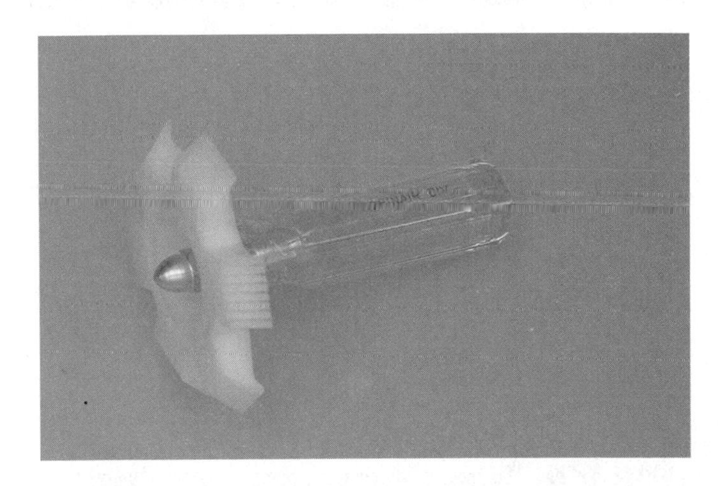

FIGURE 5–36 Condenser or evaporator coil fin straightener. *Photo by Bill Johnson*

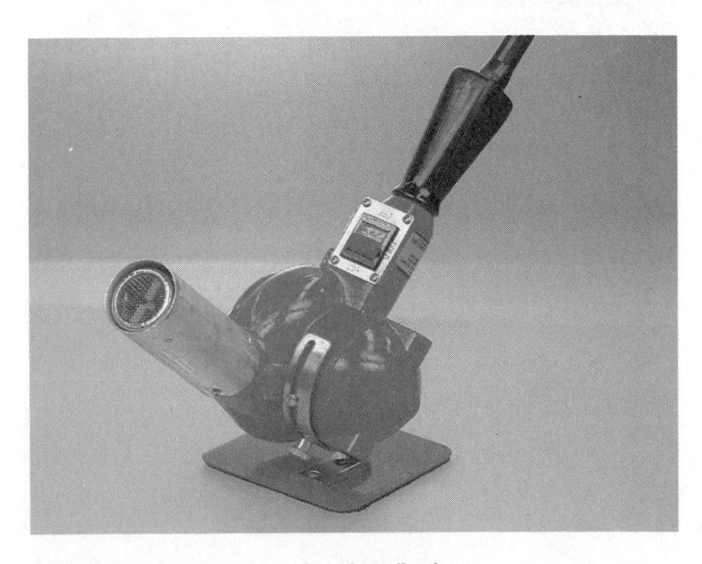

FIGURE 5–37 Heat gun. *Photo by Bill Johnson*

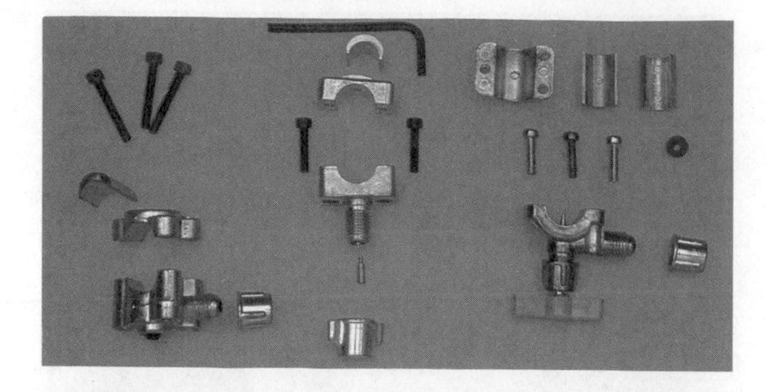

FIGURE 5–38 Tubing piercing valve. *Photos by Bill Johnson*

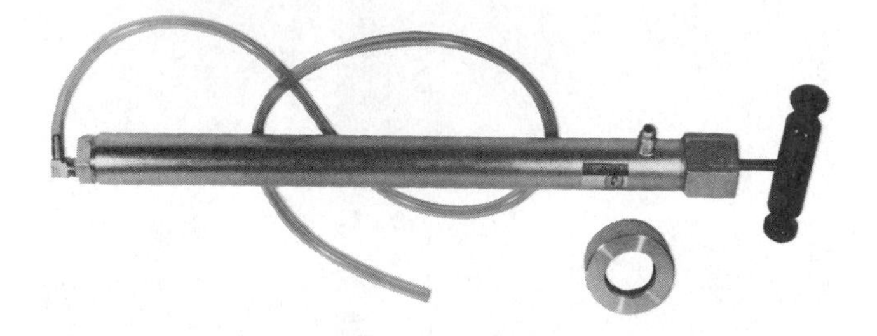

FIGURE 5–39 Compressor oil charging pump. *Courtesy Robinair Division-Sealed Power Corporation*

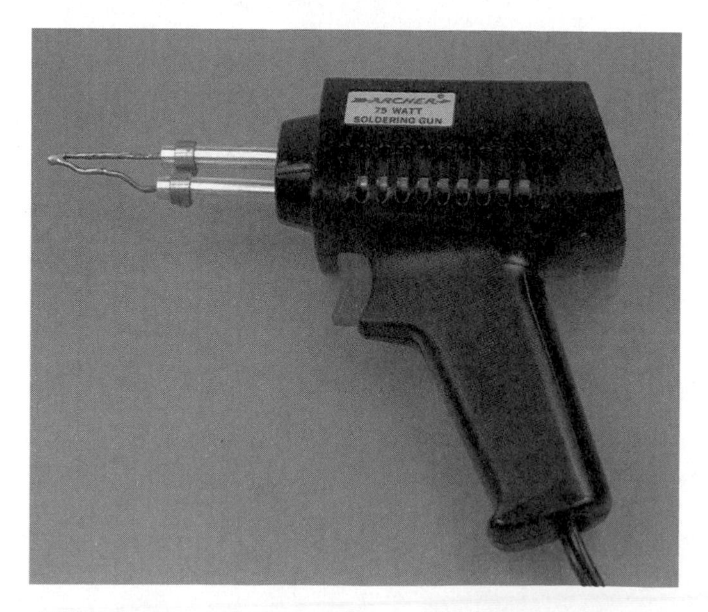

FIGURE 5–40 Soldering gun. *Photo by Bill Johnson*

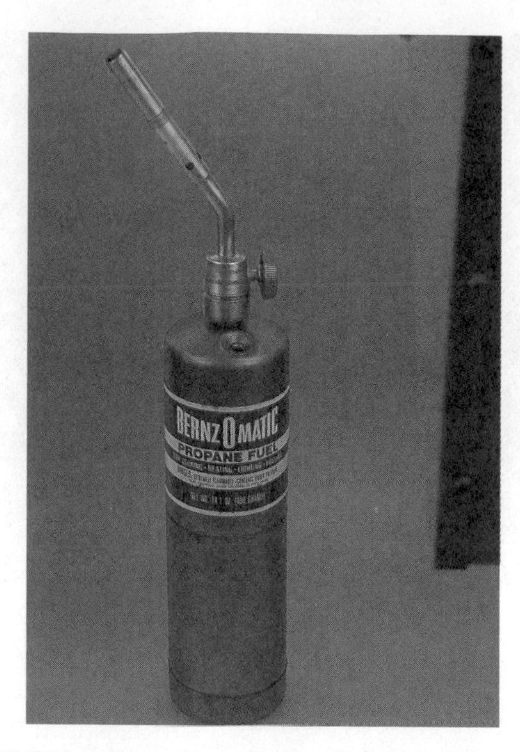

FIGURE 5-41 Propane gas torch with a disposable propane gas tank. *Photo by Bill Johnson*

HERMETIC TUBING PIERCING VALVES. Piercing valves are an economical way to tap a line for charging, testing, or purging hermetically sealed units. A valve such as that in Figure 5–38 is clamped to the line. Often these are designed so that the valve stem is turned until a sharp needle pierces the tubing. Then the stem is backed off so that the service operations can be accomplished. These valves should be installed following manufacturer's directions. The valve can then be left on the tubing for future servicing.

COMPRESSOR OIL CHARGING PUMP. Figure 5–39 is an example of a compressor oil charging pump used for charging refrigeration compressors with oil without pumping the compressor down.

SOLDERING AND WELDING EQUIPMENT. Soldering and welding equipment is available in many styles, sizes, and qualities.

SOLDERING GUN. A soldering gun is used primarily to solder electrical connections. It does not produce enough heat for soldering tubing, Figure 5–40.

PROPANE GAS TORCH. Figure 5–41 shows a propane gas torch with a disposable propane gas tank. This is an easy torch to use. The flame adjusts easily and can be used for many soldering operations.

AIR-ACETYLENE UNIT. Air-acetylene units provide sufficient heat for soldering and brazing. They consist of a torch, which can be fitted with several sizes of tips, a regulator, hoses, and an acetylene tank, Figure 5–42.

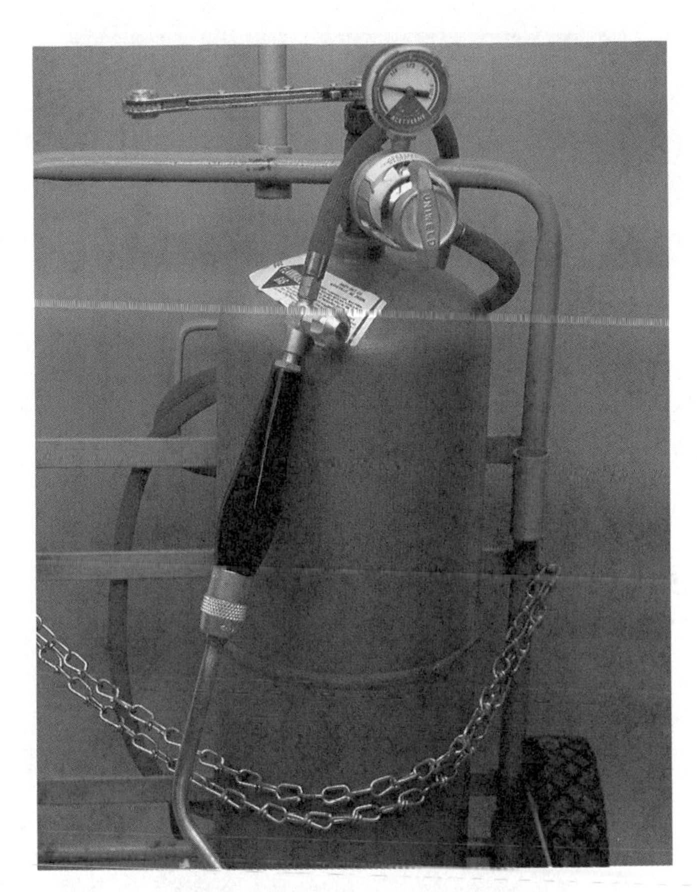

FIGURE 5–42 Air-acetylene unit. *Photo by Bill Johnson*

OXYACETYLENE WELDING UNITS. Oxyacetylene welding units are used for certain soldering, brazing, and welding applications. They consist of a torch, regulators, hoses, oxygen tanks, and acetylene gas tanks. When the oxygen and acetylene gases are mixed in the proper proportion, a very hot flame is produced, Figure 5–43.

SLING PSYCHROMETER. The sling psychrometer uses the wet-bulb/dry-bulb principle to obtain relative humidity readings quickly. Two thermometers, a dry bulb and a wet bulb, are whirled together in the air. Evaporation will occur at the wick of the wet-bulb thermometer, giving it a lower temperature reading. The difference in temperature depends on the humidity in the air. The drier the air, the greater the difference in the temperature readings because the air absorbs more moisture, Figure 5–44. Most manufacturers provide a scale so that the relative humidity can easily be determined. Before you use it, be sure that the wick is clean and wet with pure water, if possible.

MOTORIZED PSYCHROMETER. A motorized psychrometer is often used when many readings are taken over a large area.

NYLON STRAP FASTENER. Figure 5–45 shows the installation tool used to install nylon strap clamps around flexible duct. This tool automatically cuts the clamping strap off flush when a preset tension is reached.

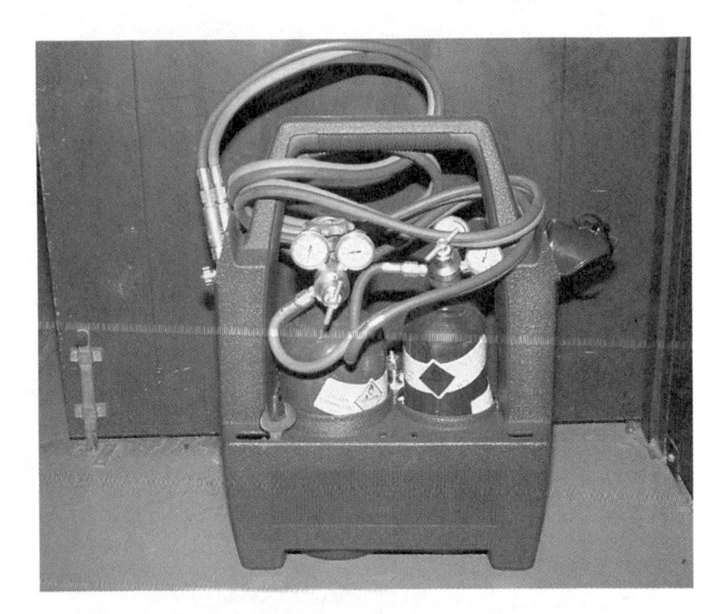

FIGURE 5–43 Oxyacetylene welding unit. *Photo by Bill Johnson*

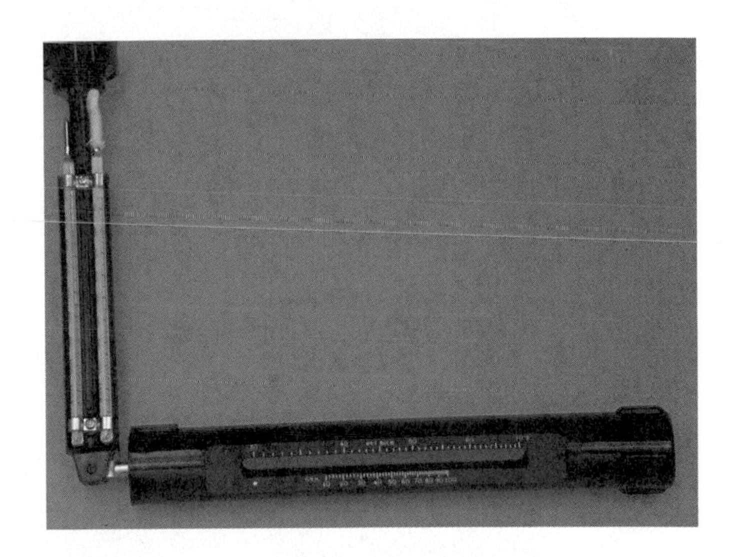

FIGURE 5–44 Sling psychrometer. *Photo by Bill Johnson*

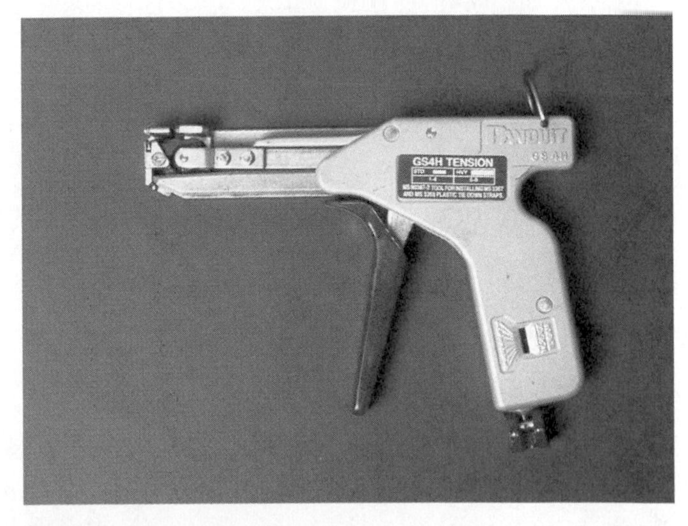

FIGURE 5–45 Tool used to install nylon strap clamps around flexible duct. *Photo by Bill Johnson*

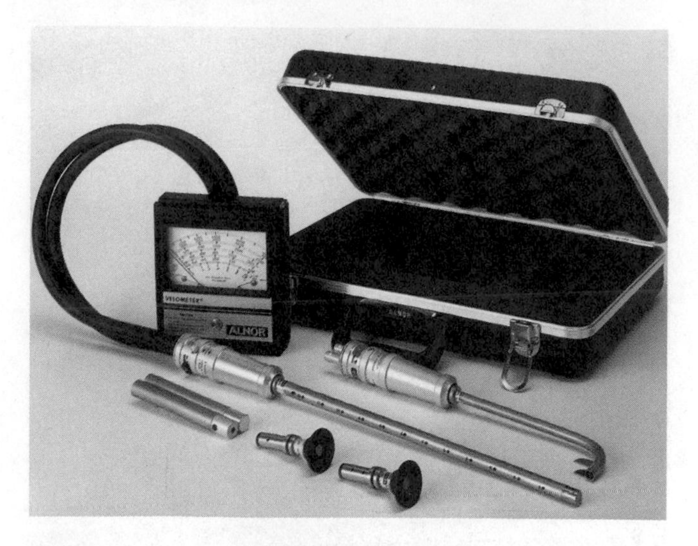

FIGURE 5–46 Air velocity measuring kit. *Courtesy Alnor Instrument Company*

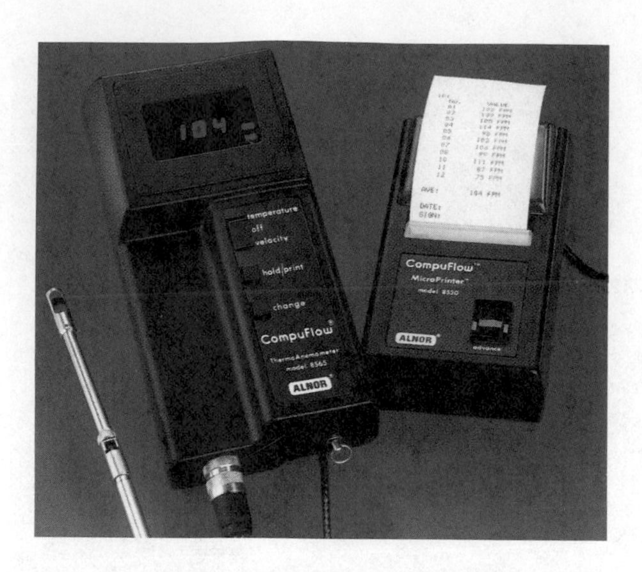

FIGURE 5–48 Air velocity measuring instrument with microprocessor. *Courtesy Alnor Instrument Company*

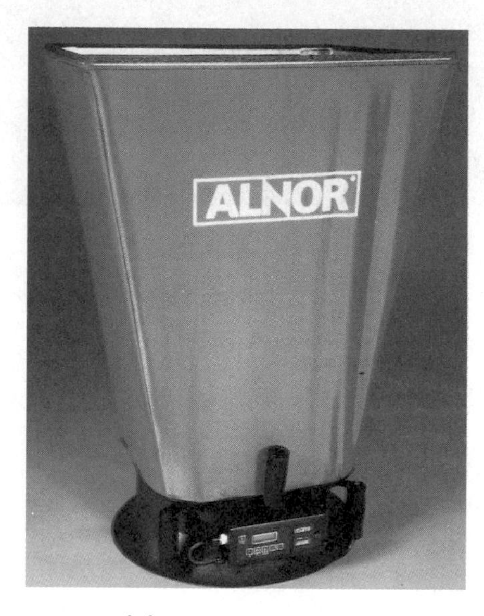

FIGURE 5–49 Air balancing meter. *Courtesy Alnor Instrument Company*

AIR VELOCITY MEASURING INSTRUMENTS. Air velocity measuring instruments are necessary to balance duct systems, check fan and blower characteristics, and make static pressure measurements. They measure air velocity in feet per minute. Figure 5–46 shows an air velocity measuring kit. This type of instrument makes air velocity measurements from 50 to 10,000 ft/min. Figure 5–47 shows a dial face on one of these instruments. Figure 5–48 shows an air velocity instrument incorporating a microprocessor that will take up to 250 readings across hood openings or in a duct and will display the average air velocity and temperature readings when needed.

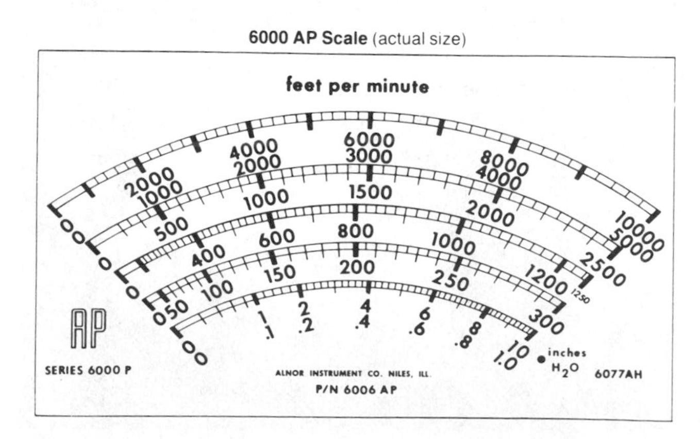

FIGURE 5–47 Dial face on an air velocity measuring instrument. *Courtesy Alnor Instrument Company*

AIR BALANCING METER. The air balancing meter shown in Figure 5–49 eliminates the need to take multiple readings. The meter will make readings directly from exhaust or supply grilles in ceilings, floors, or walls, and read out in cubic feet per minute (cfm).

CARBON DIOXIDE (CO_2) AND OXYGEN (O_2) INDICATORS. CO_2 and O_2 indicators are used to make flue-gas analyses to determine combustion efficiency in gas or oil furnaces, Figure 5–50.

CARBON MONOXIDE (CO) INDICATOR. A CO indicator is used to take flue-gas samples in natural gas furnaces to determine the percentage of CO present, Figure 5–51.

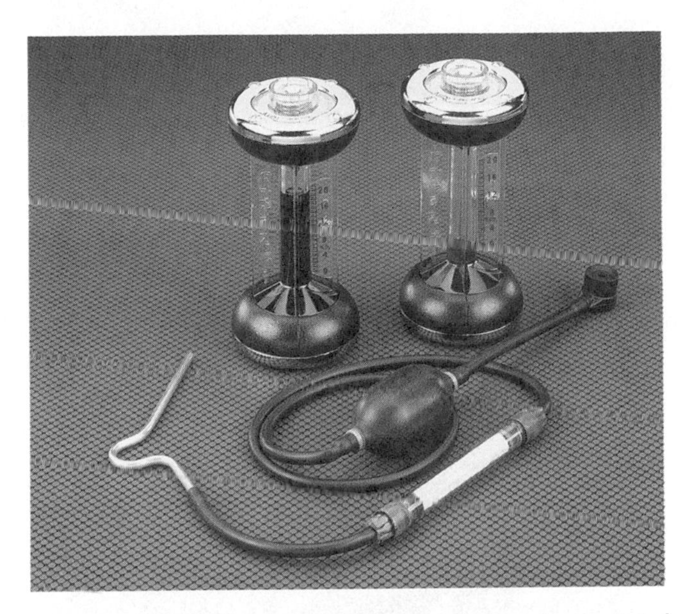

FIGURE 5–50 Carbon dioxide and oxygen indicators. *Courtesy United Technologies Bacharach*

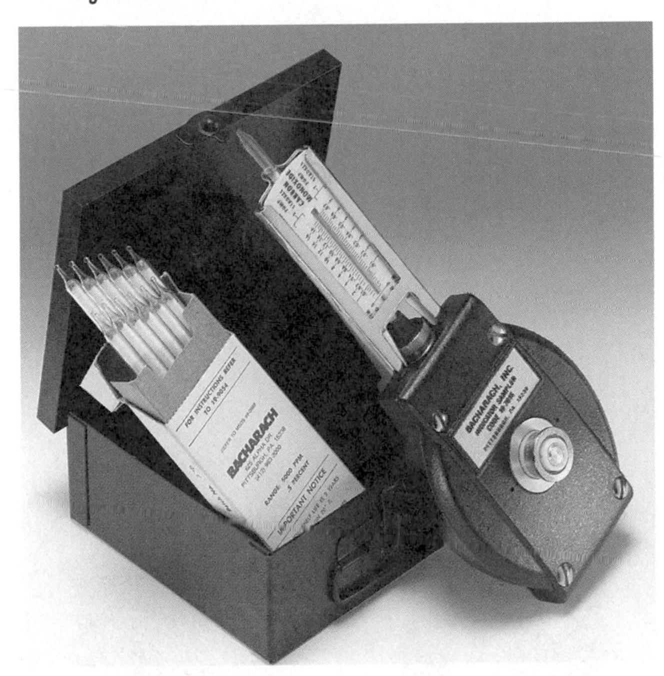

FIGURE 5-51 Carbon monoxide indicator. *Courtesy United Technologies Bacharach*

ELECTRONIC COMBUSTION ANALYZER. The electronic combustion analyzer is used to measure oxygen (O_2) concentrations within flue gases. It indicates flue-gas temperature, tests smoke, and tests for carbon monoxide, Figure 5–52. Other models designed to be used with larger heating equipment have built-in microcomputers that automatically compute the percentage of excess air, the percentage of carbon dioxide, and the combustion efficiency.

DRAFT GAGE. A draft gage is used to check the pressure of the flue gas in gas and oil furnaces to ensure that the flue gases are moving up the flue at a satisfactory speed, Figure 5–53. The flue gas is normally at a slightly negative pressure.

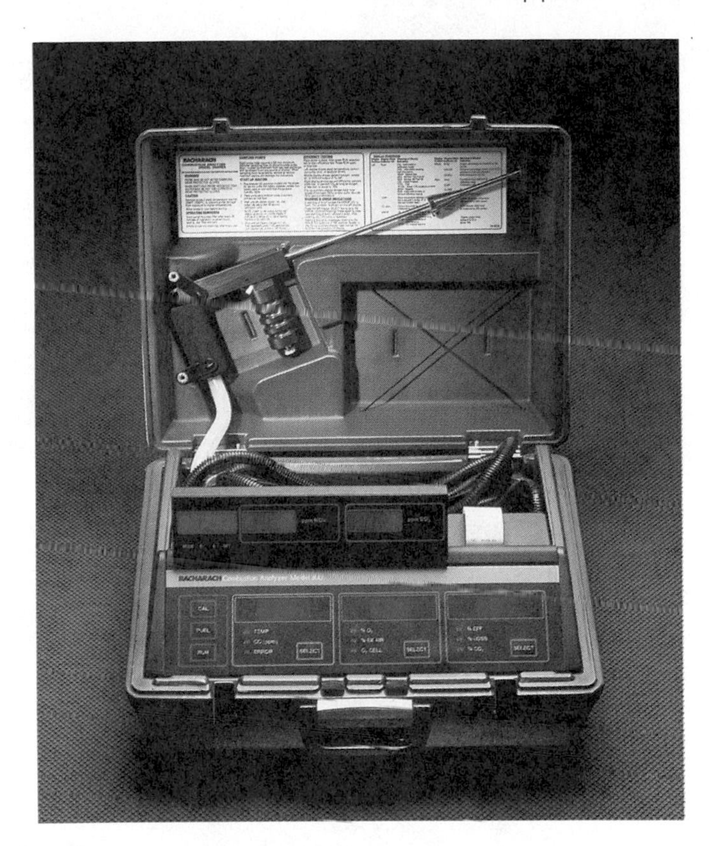

FIGURE 5–52 Electronic combustion analyzer. *Courtesy United Technologies Bacharach*

FIGURE 5–53 Draft gage. *Photo by Bill Johnson*

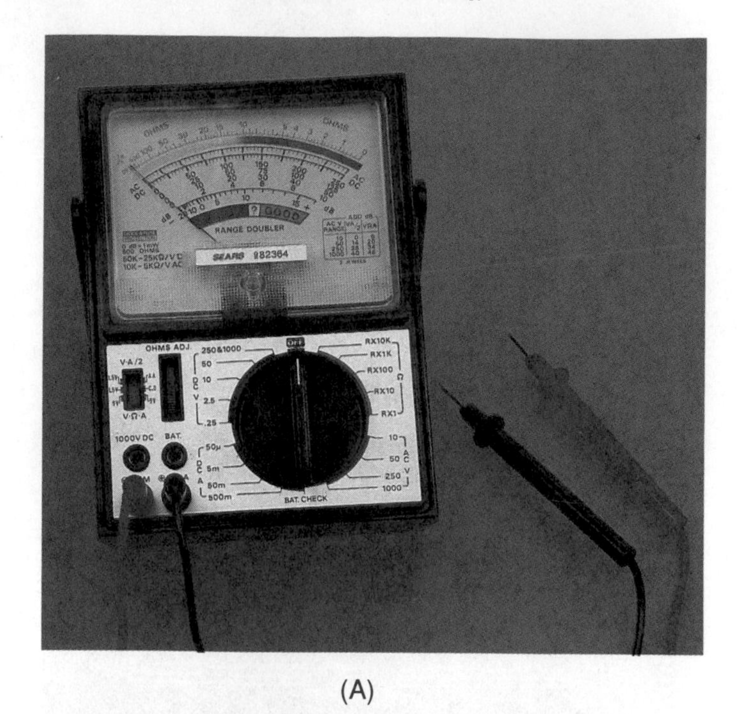

(A)

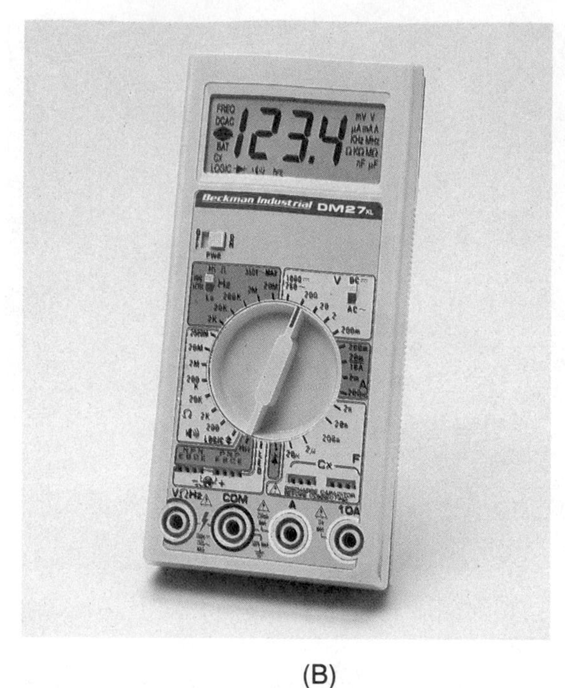

(B)

FIGURE 5–54 Volt-ohm-milliammeter (VOM). (A) Analog. (B) Digital. *(A) Photo by Bill Johnson, (B) Courtesy Beckman Industrial Corporation*

VOLT-OHM-MILLIAMMETER (VOM). A VOM, often referred to as a multimeter, is an electrical instrument that measures voltage (volts), resistance (ohms), and current (milliamperes). These instruments have several ranges in each mode. They are available in many types, ranges, and quality, either with a regular dial (analog) readout or a digital readout. Figure 5–54 illustrates both an analog and digital type. If you purchase one, be sure to select one with the features and ranges used by technicians in this field.

AC CLAMP-ON AMMETER. An AC clamp-on ammeter is a versatile instrument. It is also called clip-on, tang-type, snap-on, or other names. Some can also measure voltage or resistance or both. Unless you have an ammeter like this, you must interrupt the circuit to place the ammeter in the circuit. With this instrument you simply clamp the jaws around a single conductor, Figure 5–55.

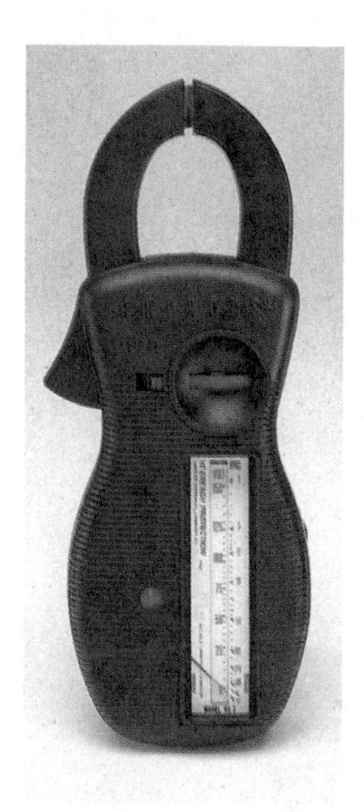

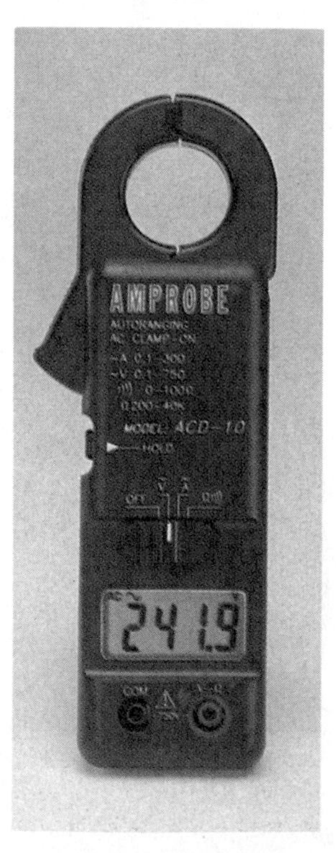

FIGURE 5-55 AC clamp-on ammeter. *Courtesy Amprobe Instrument, Division of Core Industries Inc.*

U-TUBE WATER MANOMETER. A U-tube water manometer displays natural gas and propane gas pressures during servicing or installation of gas furnaces and gas-burning equipment, Figure 5–56(A).

INCLINED WATER MANOMETER. An inclined water manometer determines air pressures in very low-pressure systems, such as to 0.1 in. of water column. These are used to analyze air flow in air conditioning and heating air distribution systems, Figure 5–56(B).

BELT TENSION GAGES. Figure 5–57 shows photos of two types of belt tension gages. These gages are used to accurately adjust compressor and other V belts to proper tension. This helps provide maximum life for the belts and protects the bearings.

A technician will use many other tools and equipment, but the ones covered in this unit are the most common.

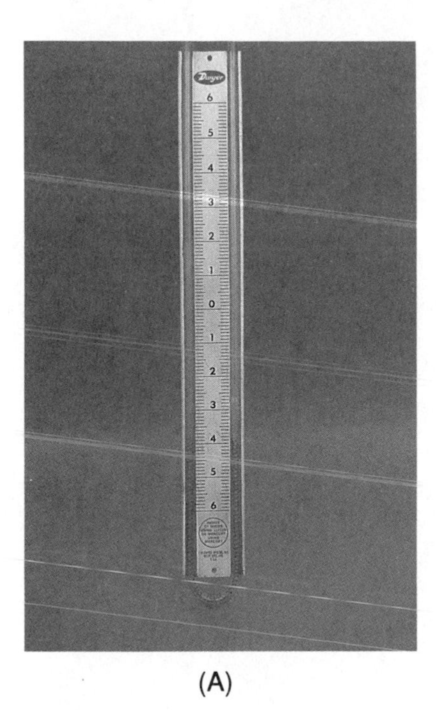

(A)

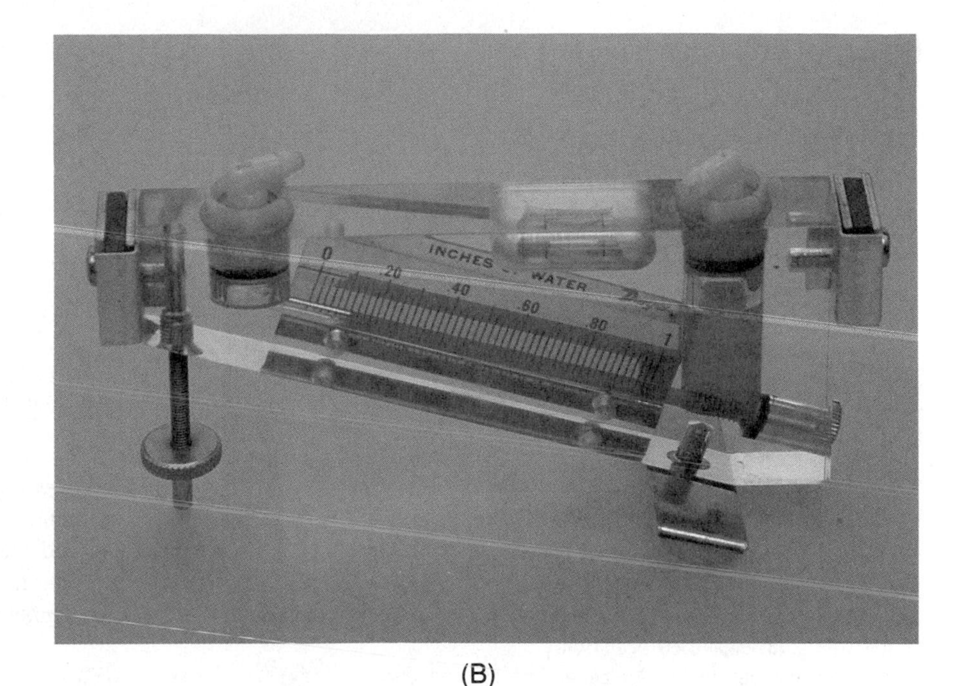

(B)

FIGURE 5–56 Manometers. (A) Water U tube. (B) Inclined water. *Photos by Bill Johnson*

FIGURE 5–57 Belt tension gages. *Courtesy Robinair Division, SPX Corp.*

SUMMARY

- Air conditioning, heating, and refrigeration technicians should be familiar with available hand tools and equipment.
- Technicians should properly use hand tools and specialized equipment.
- ***These tools and this equipment should be used only for the job for which they were designed. Other use may damage the tool or equipment and may be unsafe for the technician.***

REVIEW QUESTIONS

1. List two types of screwdriver tips.
2. Give an example of a use for a nut driver.
3. List five different types of wrenches.
4. List four different types of pliers.
5. What is the name of a hammer used primarily for metal work?
6. What is a hacksaw used for?
7. Describe uses for a wiring and crimping tool.
8. What is a tube cutter used for?
9. What are flaring tools used for?
10. Describe two types of benders for soft tubing.
11. List the components usually associated with a gage manifold unit.
12. What is the purpose of a vacuum pump?
13. What are two types of refrigerant leak detectors?
14. List three types of thermometers.
15. What is a soldering gun normally used for?
16. What is an air acetylene torch used for?
17. What is a sling psychrometer used for?
18. When are air velocity measuring instruments used?
19. Describe the purpose of CO_2 and O_2 indicators.
20. What is the purpose of a draft gage?
21. What does a VOM measure?
22. What does an AC clamp-on ammeter measure?
23. Why is it necessary to recover refrigerant?
24. Why would a technician use an electronic charging scale?
25. What are two reasons for using belt tension gages?

UNIT

6 Fasteners

OBJECTIVES

After studying this unit, you should be able to

- identify common fasteners used with wood.
- identify a common fastener used with sheet metal.
- write and explain a typical tapping screw dimension.
- identify typical machine screw heads.
- write and explain each part of a machine screw thread dimension.
- describe a fastener used in masonry.
- describe hanging devices for piping, tubing, and duct.

SAFETY CHECKLIST

* When staples are used to fasten a wire in place, do not hammer them too tight as they may damage the wire.
* When using fasteners, be sure that all materials are strong enough for the purpose for which they are being used.

As an air conditioning (heating and cooling) and refrigeration technician, you need to know about different types of fasteners and the various fastening systems so that you will use the right fastener or system for the job and securely install all equipment and materials.

6.1 NAILS

Probably the most common fastener used in wood is the nail. Nails are available in many styles and sizes. *Common nails* are large, flat-headed wire nails with a specific diameter for each length, Figure 6–1(A). A *finishing nail*, Figure

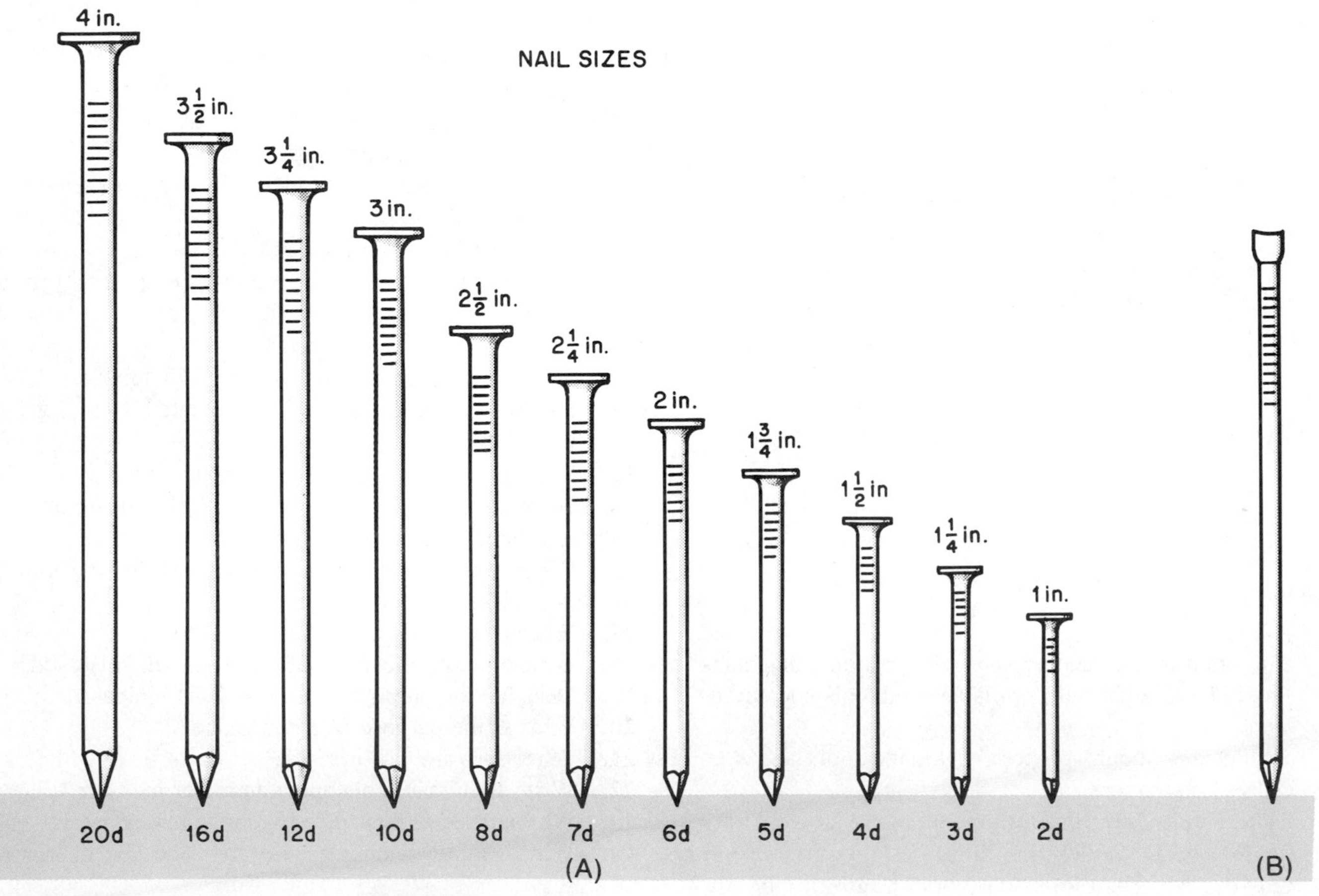

FIGURE 6–1 (A) Common nails. (B) Finishing nail.

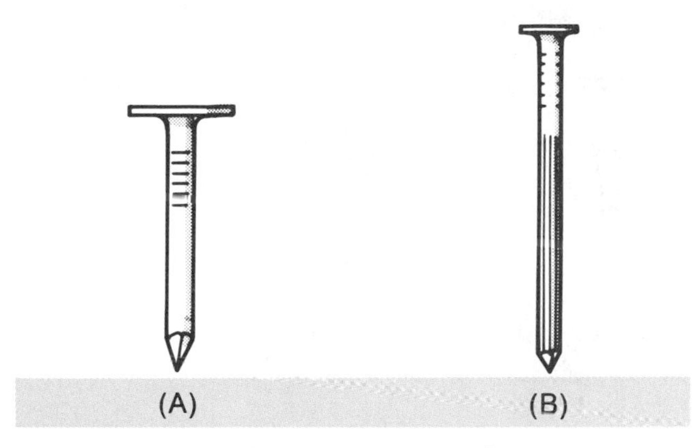

FIGURE 6–2 (A) Roofing nail. (B) Masonry nail.

6–1(B), has a very small head so that it can be driven below the surface of the wood; it is used where a good finish is desired.

These nails are sized by using the term penny, which is abbreviated by the letter *d*, Figure 6–1(A). For instance, an 8d common nail will describe the shape and size. These nails can be purchased plain or with a coating, usually zinc or resin. The coating protects against corrosion and helps to prevent the nail from working out of the wood.

A *roofing nail* is used to fasten shingles and other roofing materials to the roof of a building. It has a large head to keep the shingles from tearing and pulling off. A roofing nail can sometimes be used on strapping to hang duct work or tubing, Figure 6–2(A). *Masonry nails* are made of hardened steel and can be driven into masonry, Figure 6–2(B).

6.2 STAPLES AND RIVETS

STAPLES. A *staple*, Figure 6–3, is a fastener made somewhat like a nail but shaped in a U with a point on each end. Staples are available in different sizes for different uses. One use is to fasten wire in place. The staples are simply hammered in over the wire. ***Never hammer them too tight because they can damage the wire.***

FIGURE 6–3 Staples used to fasten wire. *Photo by Bill Johnson*

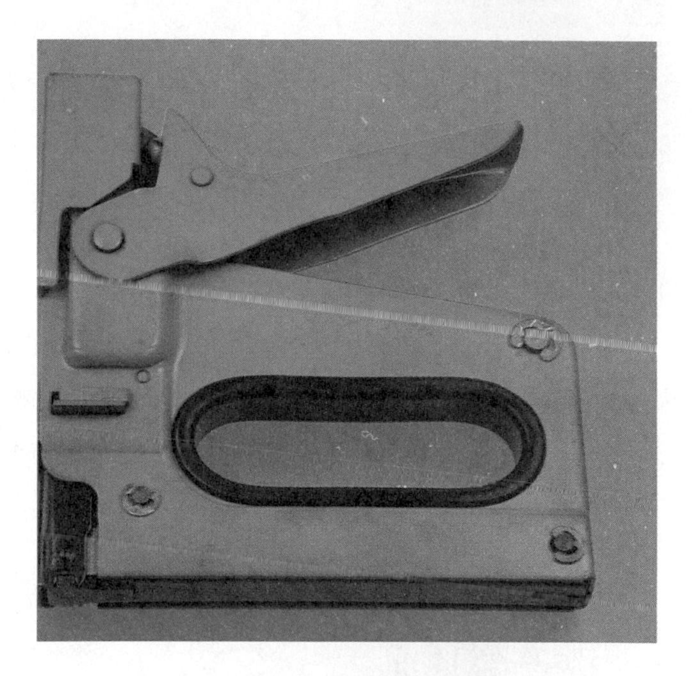

FIGURE 6–4 Stapling tacker. *Photo by Bill Johnson*

FIGURE 6-5 Staple clinched outward.

Other types of staples may be fastened in place with a *stapling tacker,* Figure 6–4. By depressing the handle a staple can be driven through paper, fabric, or insulation into the wood. *Outward clinch tackers* will anchor staples inside soft materials. The staple legs spread outward to form a strong tight clinch, Figure 6–5. They can be used to fasten insulation around heating or cooling pipes and ducts and to install ductboard. Other models of tackers are used to staple low-voltage wiring to wood.

RIVETS. *Pin rivets* or *blind rivets* are used to join two pieces of sheet metal. They are actually hollow rivets assembled on a pin, often called a *mandrel.* Figure 6–6 illustrates a pin rivet assembly, how it is used to join sheet metal, and a riveting tool or gun. The rivets are inserted and set from only one side of the metal. Thus, they are particularly useful when there is no access to the back side of the metal being fastened.

To use a pin rivet, drill a hole in the metal the size of the rivet diameter. Insert the pin rivet in the hole with the pointed end facing out. Then place the nozzle of the riveting tool over the pin and squeeze the handles. Jaws grab the pin and pull the head of the pin into the rivet, expanding it and forming a head on the other side. Continue squeezing the handles until the head on the reverse side forms a tight joint.

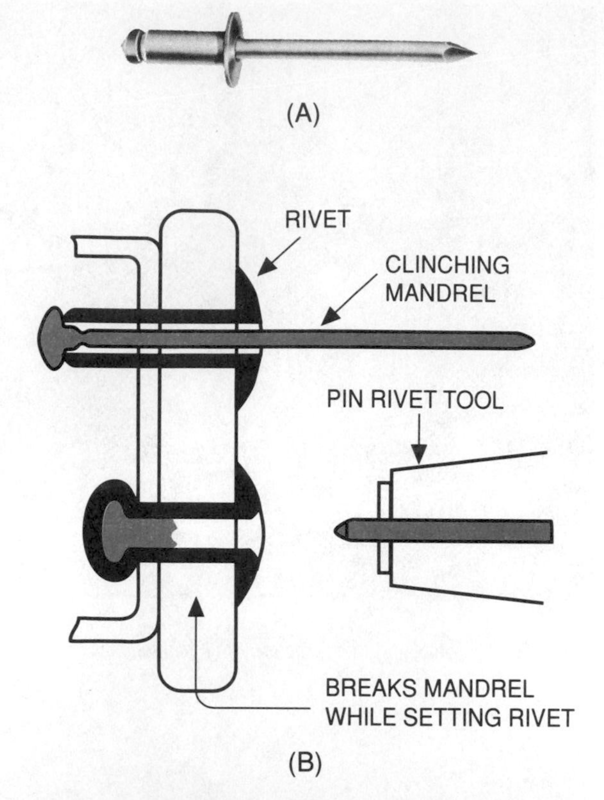

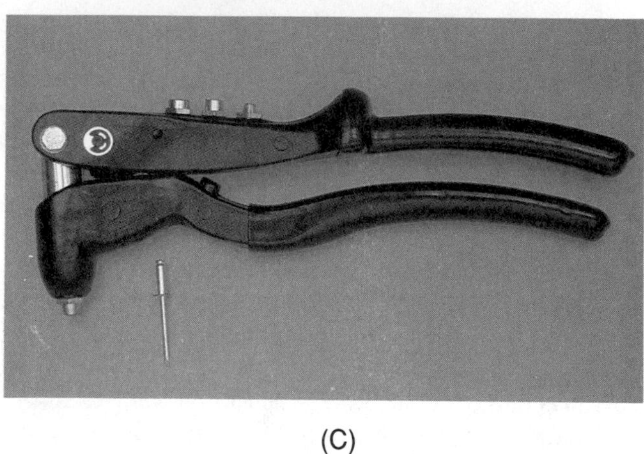

(C)

FIGURE 6–6 (A) Pin rivet assembly. (B) Using pin rivets to fasten sheet metal. (C) Riveting gun. *(A) and (B) Courtesy Duro Dyne Corp., (C) Photo by Bill Johnson*

When this happens, the pin breaks off and is ejected from the riveting tool.

6.3 THREADED FASTENERS

We will describe only a few of the most common threaded fasteners.

WOOD SCREWS. *Wood screws* are used to fasten many types of materials to wood. They generally have a flat, round, or oval head, Figure 6–7. Flat and oval heads have an angle beneath

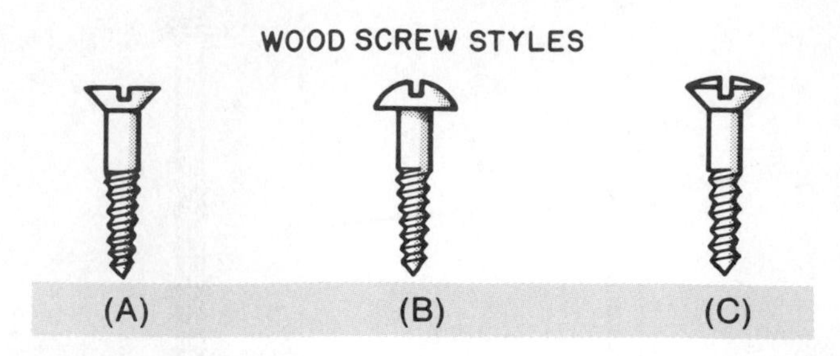

FIGURE 6–7 (A) Flat head wood screw. (B) Round head wood screw. (C) Oval head wood screw.

the head. Holes for these screws should be countersunk so that their angular surface will be recessed. Wood screws can have a straight slotted head or a Phillips recessed head.

Wood screw sizes are specified by length (in inches) and shank diameter (a number from 0 to 24). The larger the number, the larger the diameter or gauge of the shank.

TAPPING SCREWS. *Tapping* or *sheet metal* screws are used extensively by service technicians. These screws may have a straight slot, Phillips head, or a straight slot and hex head, Figure 6–8. To use a tapping screw, drill a hole into the sheet metal the approximate size of the root diameter of the thread. Then turn the screw into the hole with a conventional screwdriver.

Figure 6–9 shows a tapping screw, often called a self-drilling screw, that can be turned into sheet metal with an

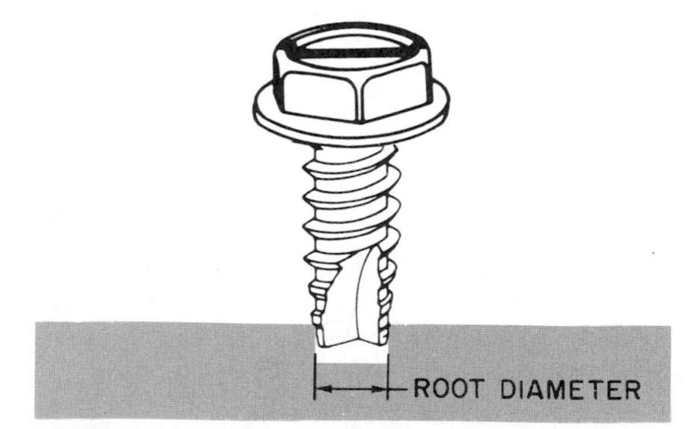

FIGURE 6–8 Tapping screw.

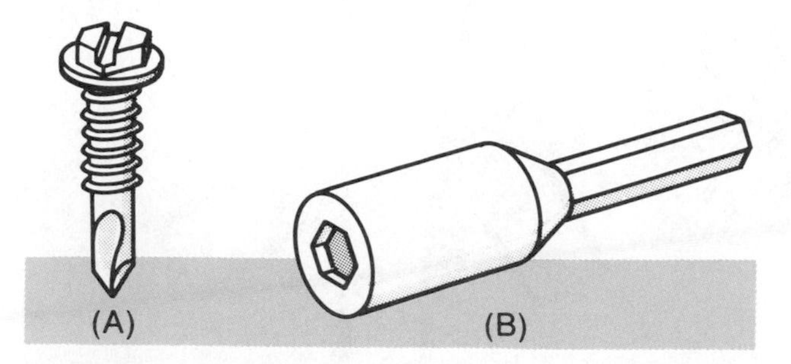

FIGURE 6–9 (A) Self-drilling screw. Note the special point used to start the hole. (B) Drill chuck.

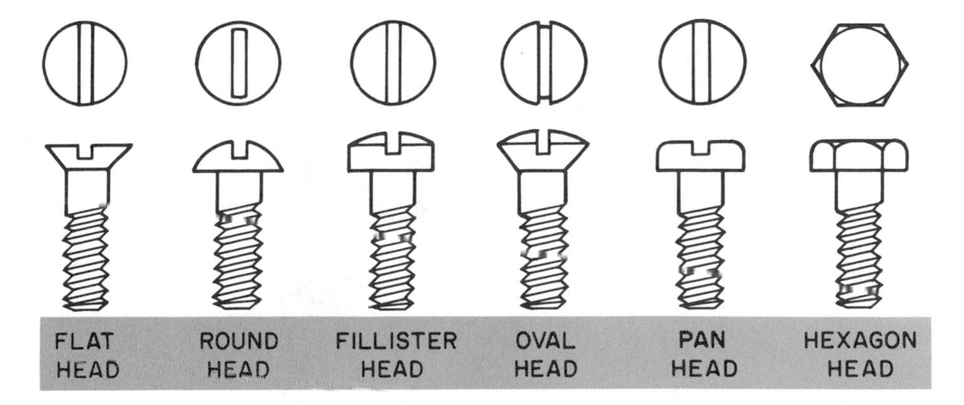

FIGURE 6–10 Machine screws.

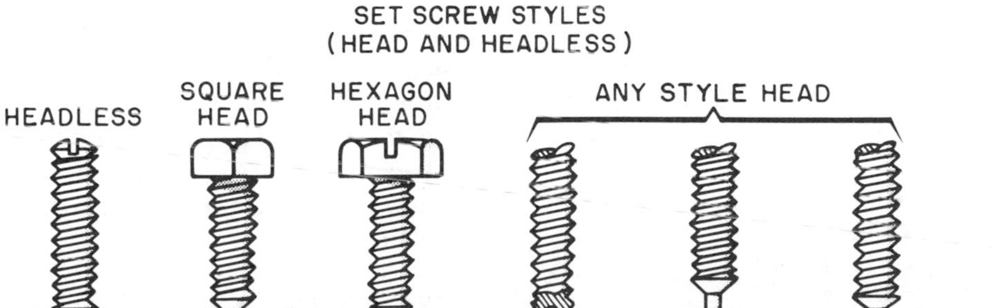

FIGURE 6–11 Styles of set screws.

electric drill using a chuck similar to the one shown. This screw has a special point used to start the hole. It can be used with light to medium gauges of sheet metal.

Tapping screws are identified or labeled in a specific way. Following is a typical identifying label for tapping screws.

6–20 X 1/2 Type AB, Slotted hex
6(0.1380) = Outside thread diameter
20 = Number of threads per inch
1/2 = Length of the screw
Type AB = Type of point
Slotted hex = Type of head

Other items can be included, but they are very technical in nature. A thread dimension may not include all of the features listed. The diameter can be a number, a fraction, or a decimal fraction. The type of point is often not included.

MACHINE SCREWS. There are many types of machine screws. They are normally identified by their head styles. Some of these are indicated in Figure 6–10.

A thread dimension or label is indicated as follows.

5/16-18 UNC – 2
5/16 = Outside thread diameter
18 = Number of threads per inch
UNC = Unified thread series (Unified National Coarse)
2 = Class of fit (the amount of play between the internal and external threads)

SET SCREWS. *Set screws* have points as illustrated in Figure 6–11. They have square heads, hex heads, or no heads. The headless type may be slotted for a screwdriver, or it may have a hexagonal or fluted socket. These screws are used to keep a pulley from turning on a shaft, and other similar applications.

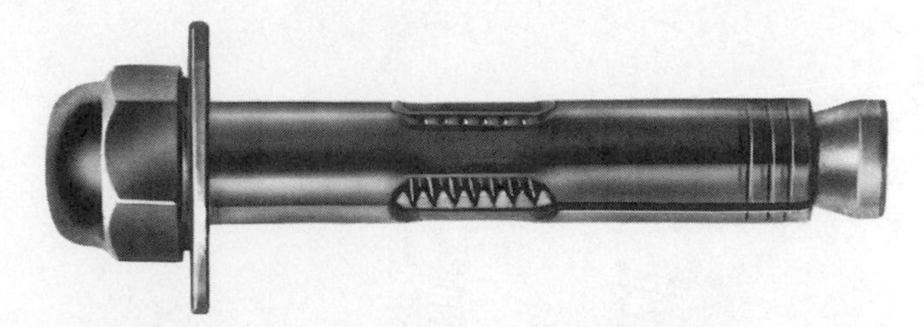

FIGURE 6–12 Anchor shield with screw. *Courtesy Rawlplug Company, Inc.*

ANCHOR SHIELDS. *Anchor shields* with bolts or screws are used to fasten objects to masonry or, in some instances, hollow walls. Figure 6–12 illustrates a multipurpose steel anchor bolt used in masonry material. Drill a hole in the masonry the size of the sleeve. Tap the sleeve and bolt into the hole with a hammer. Turn the bolt head. This expands the sleeve and secures the bolt in the masonry. A variety of head styles are available.

WALL ANCHOR. A *hollow wall anchor,* Figure 6–13, can be used in plaster, wallboard, gypsum board, and similar materials. Once the anchor has been set, the screw may be removed as often as necessary without affecting the anchor.

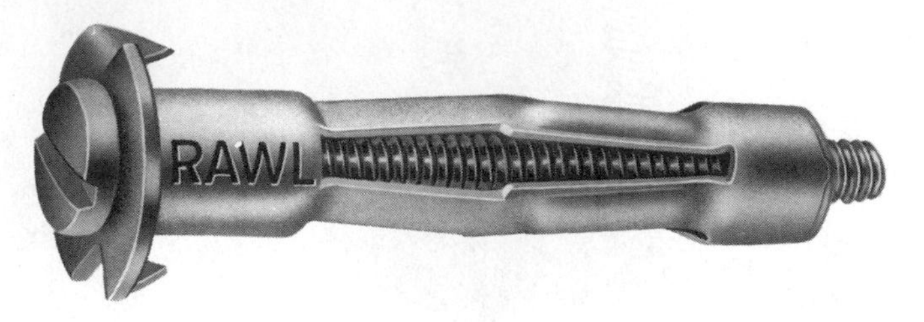

FIGURE 6–13 Hollow wall anchor. *Courtesy Rawlplug Company, Inc.*

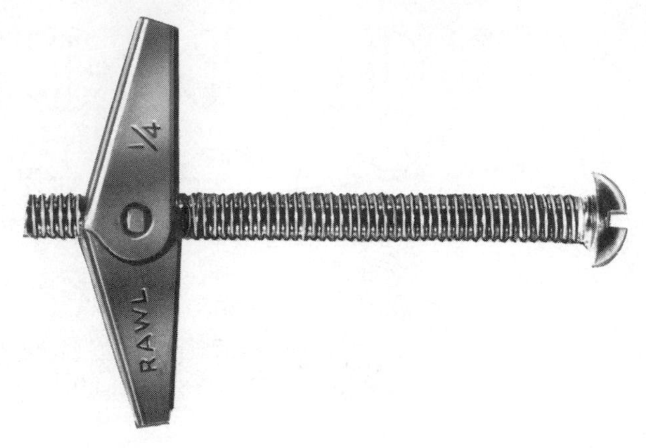

FIGURE 6–14 Toggle bolt. *Courtesy Rawlplug Company, Inc.*

TOGGLE BOLTS. *Toggle bolts* provide a secure anchoring in hollow tile, building block, plaster over lath, and gypsum board. Drill a hole in the wall large enough for the toggle in its folded position to go through. Push the toggle through the hole and use a screwdriver to turn the bolt head. You must maintain tension on the toggle or it will not tighten, Figure 6–14.

THREADED ROD AND ANGLE STEEL. *Threaded rod* and *angle steel* can be used to custom make hangers for pipes or components such as an air handler, Figure 6–15. ***Be sure that all materials are strong enough to adequately support the equipment.***

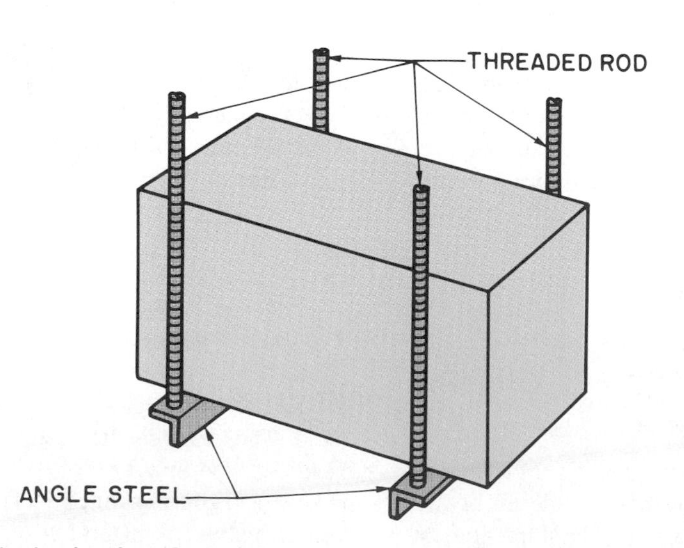

FIGURE 6–15 Using threaded rod and angle steel to make a hanger.

6.4 OTHER FASTENERS

COTTER PINS. *Cotter pins,* Figure 6–16, are used to secure pins. The cotter pin is inserted through the hole in the pin, and the ends spread to retain it.

PIPE HOOK. A *wire pipe hook* is a wire bent into a U with a point at an angle on both ends. The pipe or tubing rests in the bottom of the U, and the pointed ends are driven into wooden joists or other wood supports, Figure 6–17.

PIPE STRAP. A *pipe strap* is used for fastening pipe and tubing to joists, ceilings, or walls, Figure 6–18(A). The arc on the strap should be approximately the same as the outside diameter of the pipe or tubing. These straps are normally fastened with round head screws. Notice that the underside of the head of a round head screw is flat and provides good contact with the strap.

PERFORATED STRAP. A *perforated strap* may also be used to support pipe and tubing. Round head stove bolts with nuts fasten the strap to itself, and round head wood screws fasten the strap to a wood support, Figure 6–18(B).

FIGURE 6–16 A cotter pin.

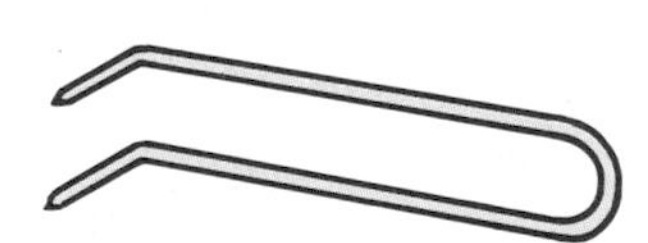

FIGURE 6–17 A wire pipe hook.

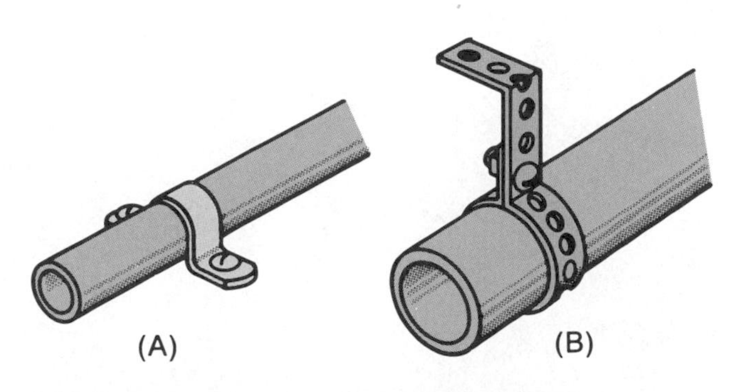

FIGURE 6–18 (A) A pipe strap. (B) A perforated strap.

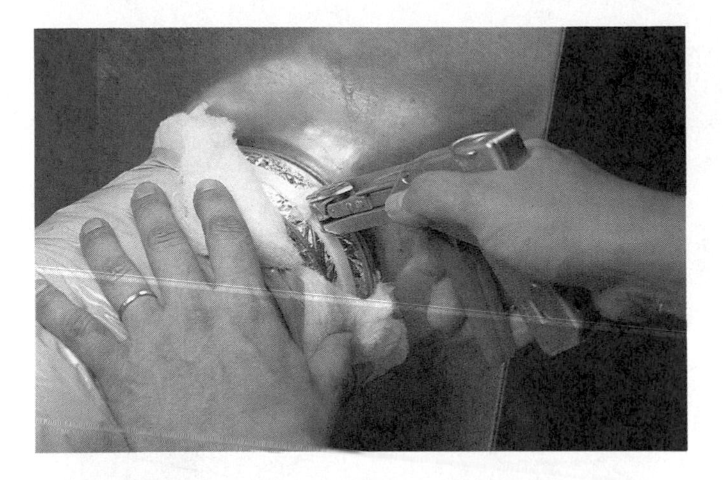

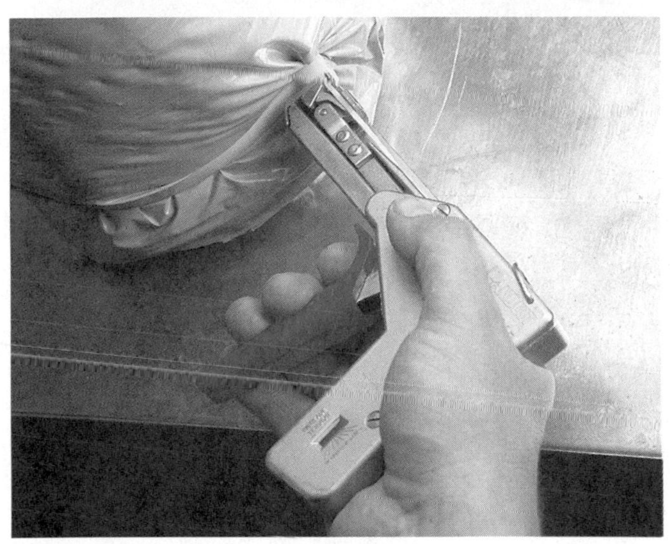

FIGURE 6–19 A system for fastening round flexible duct to a sheet metal collar. *Courtesy Panduit Corporation*

NYLON STRAP. To fasten round flexible duct to a sheet metal collar, apply the inner liner of the duct with a sealer and clamp it with a *nylon strap*. A special tool is manufactured to install this strap, which applies the correct tension and cuts the strap off. Now position the insulation and vapor barrier over the collar, which can then be secured with another nylon strap. Apply duct tape over this strap and duct end to further seal the system. Figure 6–19 illustrates this procedure.

GRILLE CLIPS. *Grille clips* fasten grilles to the ends of fiberglass duct. The clips are bent around the end of the duct, with the points pushed into the sides of the duct, Figure 6–20. Screws fasten the grilles to these clips.

Figure 6–21 illustrates a system to fasten damper regulators, controls, and other components to fiberglass duct. It consists of a drill screw, a head plate, and a backup plate.

Many fasteners not described here are highly specialized. More are developed each year. The intent of this unit is to depict the more common ones and to encourage you to keep up to date by discussing new fastening techniques with your supplier.

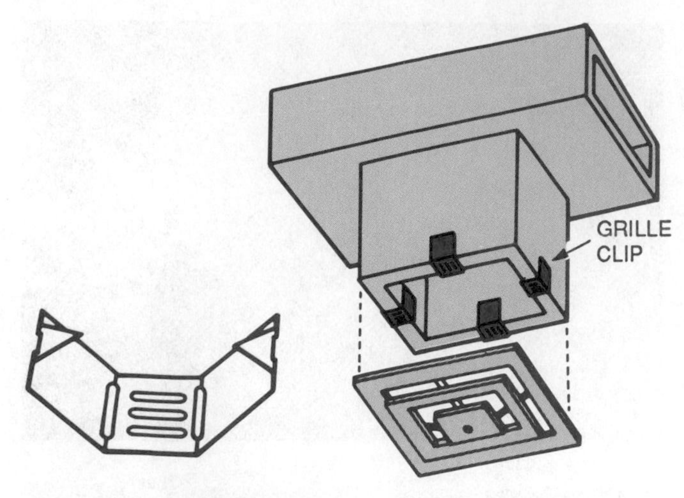

FIGURE 6–20 Fastening a grille to fiberglass duct with a grille clip.

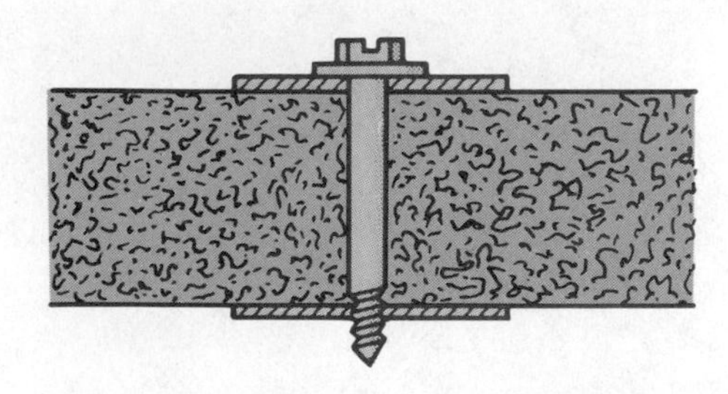

FIGURE 6–21 A system for fastening fiberglass duct.

SUMMARY

- Technicians need to use a broad variety of fasteners.
- Some of these fasteners are common nails, wood screws, masonry nails, staples, tapping screws, and set screws.
- Anchor shields and other devices are often used with screws to secure them in masonry walls or hollow walls.
- Typical hangers used with pipe, tubing, and duct work are wire pipe hooks, pipe straps, perforated steel straps, and custom hangers made from threaded rod and angle steel.
- Other specialty fasteners are available for flexible duct and fiberglass duct.

REVIEW QUESTIONS

1. Name three types of nails.
2. What term is used to describe the size of nails?
3. What is the abbreviation for this term?
4. What are masonry nails made from?
5. Describe two types of staples.
6. Describe the procedure for fastening two pieces of sheet metal together with a pin rivet.
7. Write a typical dimension for a tapping screw. Explain each part of the dimension.
8. Sketch three types of machine screw heads.
9. Write a typical machine screw thread dimension and explain each part of the dimension.
10. Describe the procedure used for fastening two pieces of sheet metal together with a tapping screw.
11. Describe two types of tapping screws.
12. Describe two types of set screws.
13. How are anchor shields used?
14. What is a pipe strap? What is it used for?
15. What is a grille clip used for?

U N I T 7

Tubing and Piping

OBJECTIVES

After studying this unit, you should be able to

- list the different types of tubing used in heating, air conditioning, and refrigeration applications.
- describe two common ways of cutting copper tubing.
- list procedures used for bending tubing.
- discuss procedures used for soldering and brazing tubing.
- describe two methods for making flared joints.
- state procedures for making swaged joints.
- describe procedures for preparing and threading steel pipe ends.
- list four types of plastic pipe and describe uses for each.

SAFETY CHECKLIST

* Use care while reaming tubing; the burr can stick in your hand.
* Be careful while using a hacksaw; the blade is sharp.
* Do not get your skin near the flame of a torch; the 5000°F and higher temperatures can cause severe burns.
* Always wear eye protection while performing any task where particles may be in the air.
* Use extra caution with the oxyacetylene heat source. ***Do not allow any oil around the fittings or to enter the hoses or regulators.***
* Be aware of flammable materials in the area where you are soldering or brazing.
* Pipe cutting and threading creates burrs that can cut your skin and otherwise be dangerous; use caution.
* Do not breathe excessive amounts of the glue used for plastic fittings.

7.1 PURPOSE OF TUBING AND PIPING

The correct size, layout, and installation of tubing, piping, and fittings helps to keep a refrigeration or air conditioning system operating properly and prevents refrigerant loss. The piping system provides passage for the refrigerant to the evaporator, the compressor, the condenser, and the expansion valve. It also provides the way for oil to drain back to the compressor. Tubing, piping, and fittings are used in numerous applications, such as fuel lines for oil and gas burners and water lines for hot water heating. The tubing, piping, and fittings used must be of the correct material and the proper size; the system must be laid out properly and installed correctly.

SAFETY PRECAUTION: ***Careless handling of the tubing and poor soldering or brazing techniques may cause serious damage to system components. You must keep contaminants, including moisture, from air conditioning and refrigeration systems.***

7.2 TYPES AND SIZES OF TUBING

Copper tubing is generally used for plumbing, heating, and refrigerant piping. Steel and wrought iron pipe are used for gas piping and frequently hot water heating. Plastic pipe is used for waste drains, condensate drains, water supplies, water-source heat pumps, and venting high-efficiency gas furnaces.

Copper tubing is available as soft copper or hard-drawn copper tubing. Soft copper may be bent or used with elbows, tees, and other fittings. Hard-drawn tubing is not intended to be bent; use it only with fittings to obtain the necessary configurations. Copper tubing used for refrigeration or air conditioning is called *ACR* (air conditioning and refrigeration) tubing. Copper tubing used in plumbing and heating is available in four standard weights: type K is heavy duty; type L is the standard size and used most frequently; types M and DWV are not used extensively in this industry. The outside diameter (OD) of these four types is approximately 1/8 in. larger than the size indicated; that is, 1/2-in. tubing has an OD of 5/8 in. ACR tubing is sized by its OD; 1/2-in. tubing has an OD of 1/2 in. In general, for plumbing and heating applications use the inside diameter (ID), and for air conditioning and refrigeration applications use the OD, Figure 7–1. Copper tubing is normally available in diameters from 3/16 in. to greater than 6 in.

Soft copper tubing is normally available in 25-ft or 50-ft rolls and in diameters from 3/16 in. to 3/4 in. It can be special ordered in 100-ft lengths. ACR tubing is capped on each end to keep it dry and clean inside and often has a charge of nitrogen to keep it free of contaminants. Proper practice should be used to remove tubing from the coil. Never uncoil the tubing from the side of the roll. Place it on

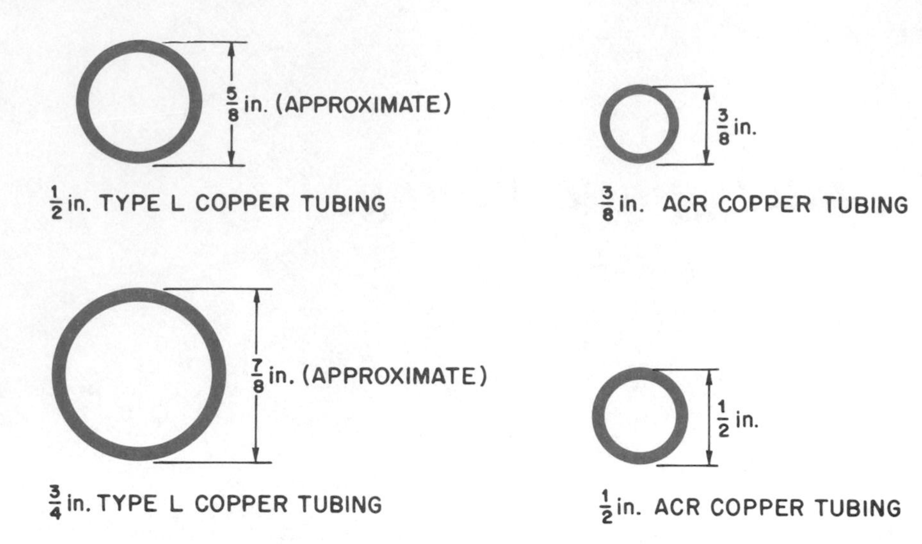

FIGURE 7–1 Tubing used for plumbing and heating is sized by its inside diameter; ACR tubing is sized by its outside diameter.

FIGURE 7–2 Roll of soft tubing. Place on a flat surface and unroll. *Photo by Bill Johnson*

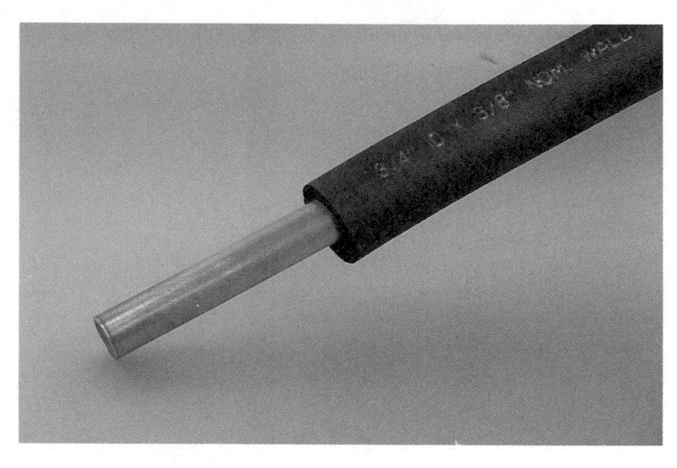

FIGURE 7–3 ACR tubing with insulation. *Photo by Bill Johnson*

a flat surface and unroll it, Figure 7–2. Cut only what you need and recap the ends.

Do not bend or straighten the tubing more than necessary because it will harden. This is called *work hardening*. Work-hardened tubing can be softened by heating and allowing it to cool slowly. This is called *annealing*. When annealing, don't use a high concentrated heat in one area, but use a flared flame over 1 ft at a time. Heat to a cherry red and allow to cool slowly.

Hard-drawn copper tubing is available in 20-ft lengths and in larger diameters than soft copper tubing. Be as careful with hard-drawn copper as with soft copper, and recap the ends when the tubing is not used.

7.3 TUBING INSULATION

ACR tubing is often insulated on the low-pressure side of an air conditioning or refrigeration system between the evaporator and compressor to keep the refrigerant from absorbing heat, Figure 7–3. Insulation also prevents condensation from forming on the lines. The closed-cell structure of this insulation eliminates the need for a vapor barrier. The insulation may be purchased separately from the tubing, or it may be factory installed. If you install the insulation, it is easier, where practical, to apply it to the tubing before assembling the line. The ID of the insulation is usually powdered to allow easy slippage even around most bends. You can buy adhesive to seal the ends of the insulation together, Figure 7–4.

For existing lines, or when it is impractical to insulate before installing the tubing, the insulation can be slit with a sharp utility knife and snapped over the tubing. All seams must be sealed with an adhesive. Do not use tape.

Do not stretch tubing insulation because the wall thickness of the insulation will be reduced and the adhesive may then fail to hold, and the effectiveness of the insulation reduced.

FIGURE 7–4 When joining two ends of tubing insulation, use an adhesive made specifically for this purpose. *Photo by Bill Johnson*

FIGURE 7–5 A typical line set. *Photo by Bill Johnson*

7.4 LINE SETS

Tubing can be purchased as line sets. These sets are charged with refrigerant, sealed on both ends, and may be obtained with the insulation installed. These line sets normally will have fittings on each end for quicker and cleaner field installation, Figure 7–5. Precharging helps to eliminate improper field charging. It also reduces the possibility of contamination in the system, eliminating clogged systems and compressor damage.

7.5 CUTTING TUBING

Tubing is normally cut with a tube cutter or a hacksaw. The tube cutter is most often used with soft tubing and smaller diameter hard-drawn tubing. A hacksaw may be used with larger diameter hard-drawn tubing. To cut the tubing with a tube cutter, follow these steps, as shown in Figure 7–6:

1. Place the tubing in the cutter and align the cutting wheel with the cutting mark on the tube. Tighten the adjusting screw until a moderate pressure is applied to the tubing.
2. Revolve the cutter around the tubing, keeping a moderate pressure applied to the tubing by gradually turning the adjusting screw.
3. Continue until the tubing is cut. *Do not apply excessive pressure because it may break the cutter wheel and constrict the opening in the tubing.*

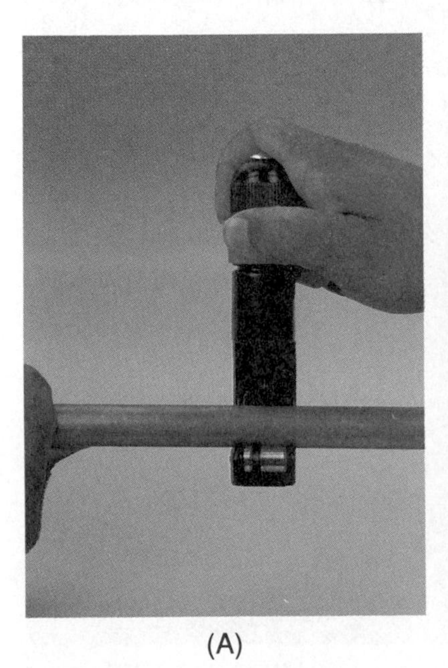

(A)

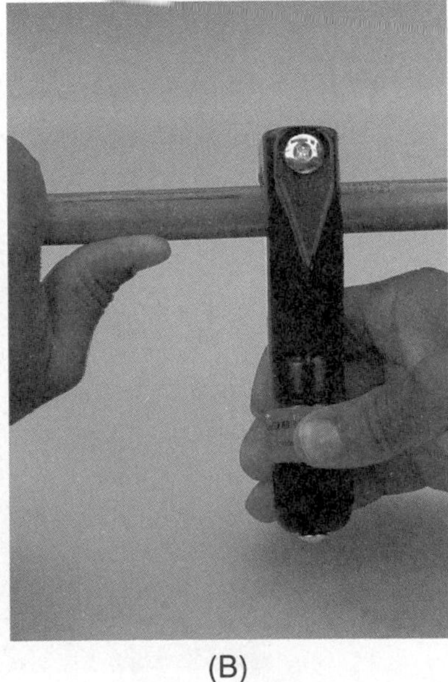

(B)

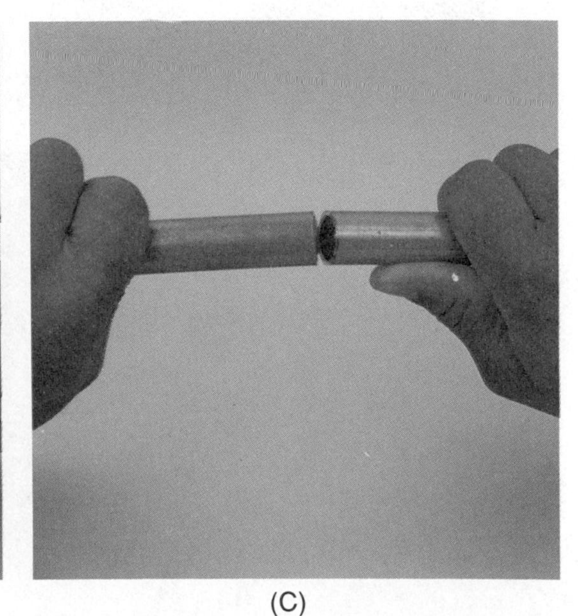

(C)

FIGURE 7–6 Proper procedure for using a tube cutter. *Photos by Bill Johnson*

When the cut is finished, the excess material (called a *burr*) pushed into the pipe by the cutter wheel must be removed, Figure 7–7. Burrs cause turbulence and restrict the fluid or vapor passing through the pipe.

To cut the tubing with a hacksaw, make the cut at a 90° angle to the tubing. A fixture may be used to ensure an accurate cut, Figure 7–8. After cutting, ream the tubing and file the end. Remove all chips and filings, making sure that no debris or metal particles get into the tubing.

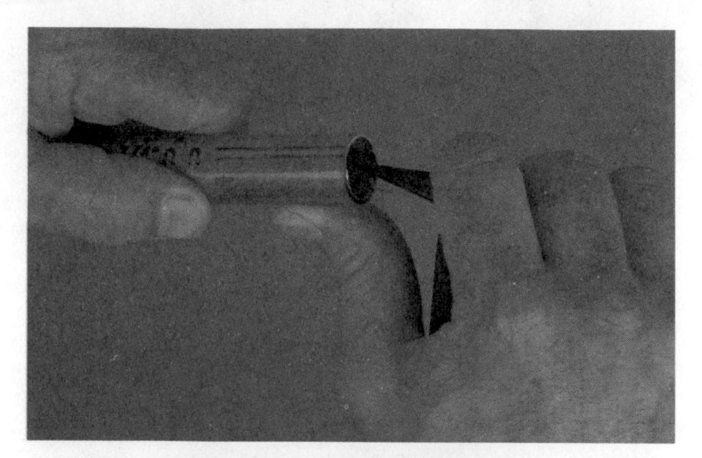

FIGURE 7–7 Removing a burr. *Photo by Bill Johnson*

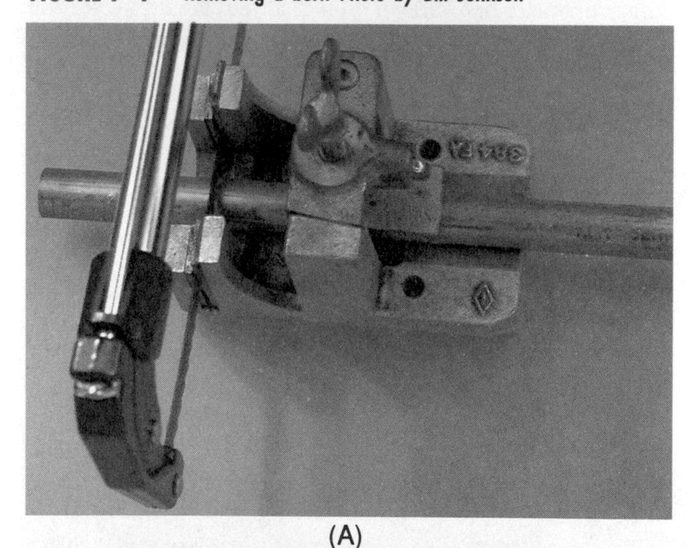

(A)

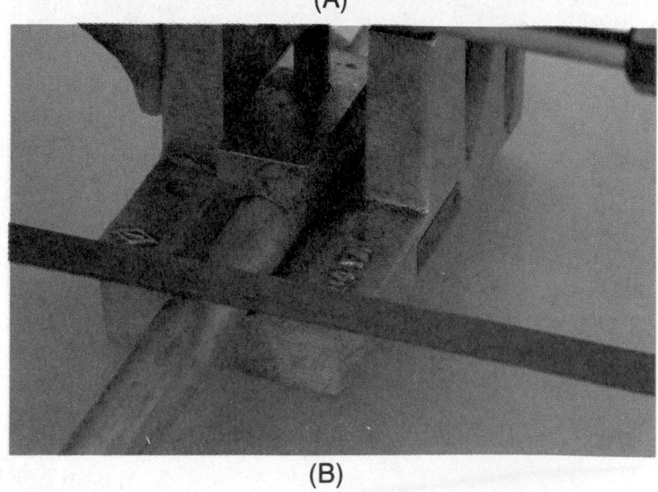

(B)

FIGURE 7–8 Proper procedure for using a hacksaw. *Photos by Bill Johnson*

7.6 BENDING TUBING

Only soft tubing should be bent. Use as large a radius bend as possible, Figure 7–9A. All areas of the tubing must remain round. *Do not allow it to flatten or kink,* Figure 7–9B. Carefully bend the tubing, gradually working around the radius.

Tube bending springs may be used to help make the bend, Figure 7–10. They can be used either inside or outside

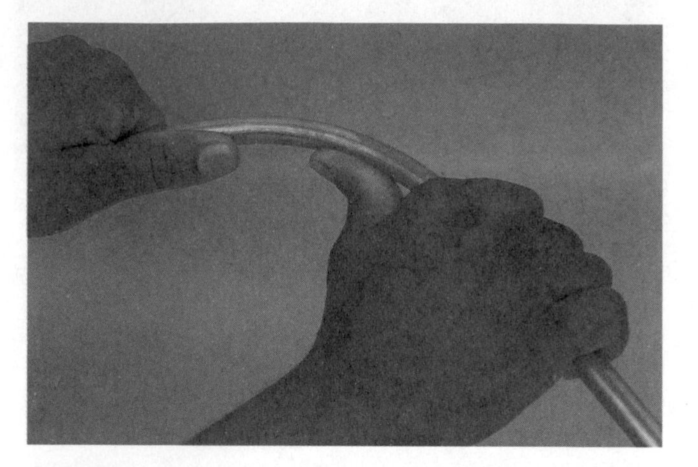

FIGURE 7–9A Tubing bent by hand. Use as large radius as possible. *Photo by Bill Johnson*

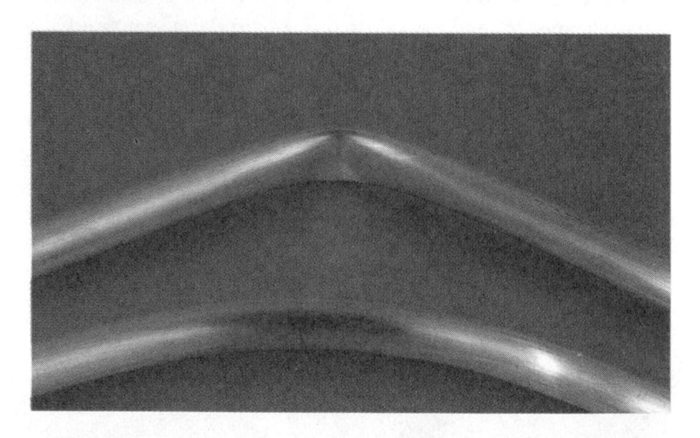

FIGURE 7–9B Do not allow the tubing to flatten or kink when bending. *Photo by Bill Johnson*

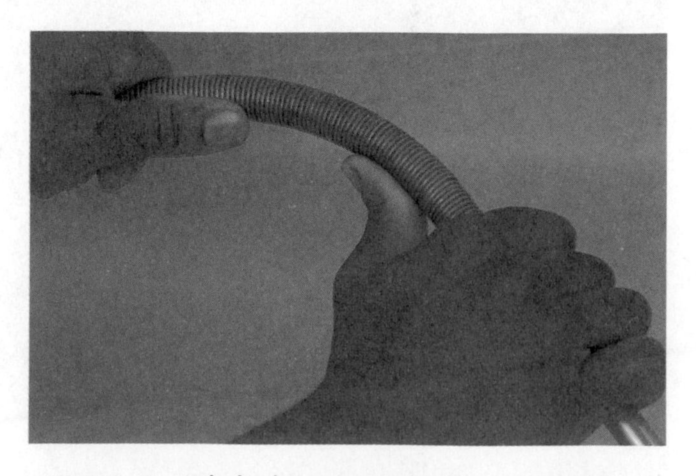

FIGURE 7–10 Tube bending springs used inside or outside of the tubing. Be sure to use the proper size. *Photo by Bill Johnson*

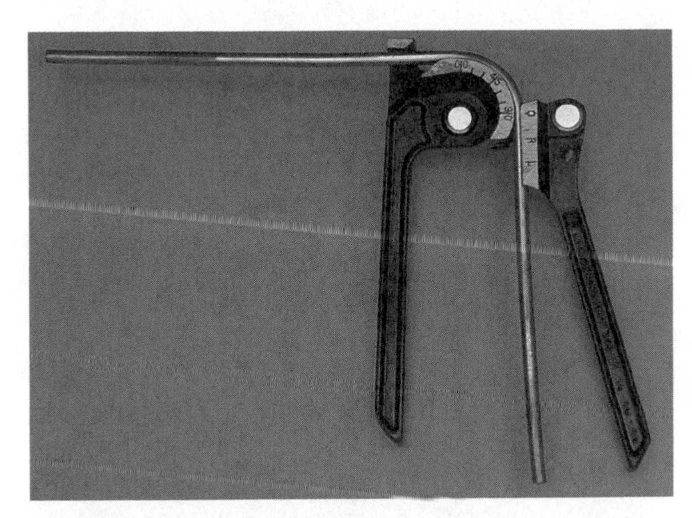

FIGURE 7–11 Use of a lever-type tube bender. *Photo by Bill Johnson*

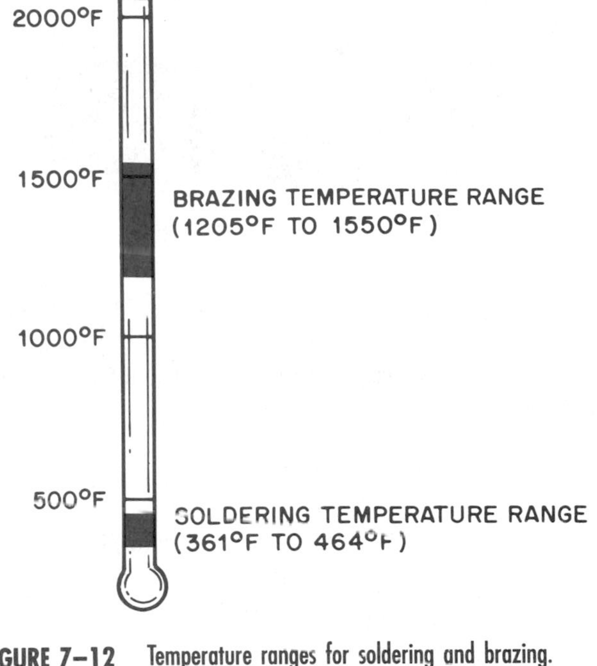

FIGURE 7–12 Temperature ranges for soldering and brazing.

the tube. They are available in different sizes for different diameter tubing. To remove the spring after the bend, you might have to twist it. If you use a spring on the OD, bend the tube before flaring so that the spring may be removed.

Lever-type tube benders, Figure 7–11, which are available in different sizes, are used to bend soft copper and thin-walled steel tubing.

7.7 SOLDERING AND BRAZING PROCESSES

Soldering is a process used to join piping and tubing to fittings. It is used primarily in plumbing and heating systems utilizing copper and brass piping and fittings. Large refrigeration systems also use hard tubing and fittings. Soldering, often called soft soldering, is done at temperatures under 800°F, usually in the 375°F to 500°F range, Figure 7–12.

The 50/50 tin-lead solder is suitable for moderate pressures and temperatures. For higher pressures, or where greater joint strength is required, 95/5 tin-antimony solder can be used.

Brazing, requiring higher temperatures, is often called silver brazing and is similar to soldering. It is used to join tubing and piping in air conditioning and refrigeration systems. Don't confuse this with welding brazing. In brazing processes, temperatures over 800°F are used. The differences in temperature are necessary due to the different combinations of alloys used in the filler metals.

Brazing filler metals suitable for joining copper tubing are alloys containing 15% to 60% silver (BAg), or copper alloys containing phosphorus (BCuP). Brazing filler metals are sometimes referred to as *hard solders* or *silver solders.* These are confusing terms often used by technicians, and it's better to avoid using them.

In soldering and brazing, the base metal (the piping or tubing) is heated to the melting point of the filler material. *The piping and tubing must not melt.* When two close-fitting, clean, smooth metals are heated to the point where the filler metal melts, this molten metal is drawn into the close-fitting space by *capillary attraction* (see Figure 7–13

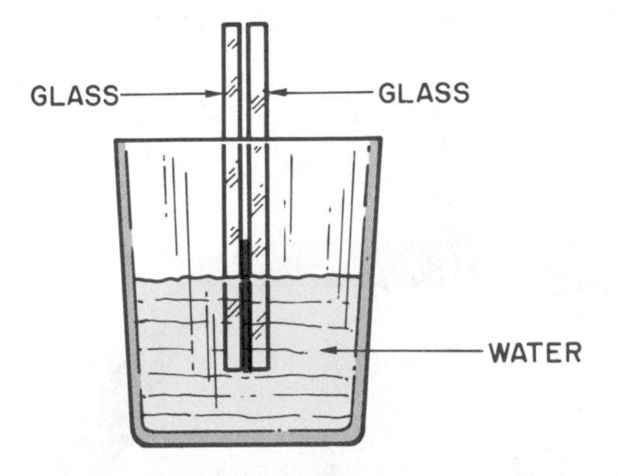

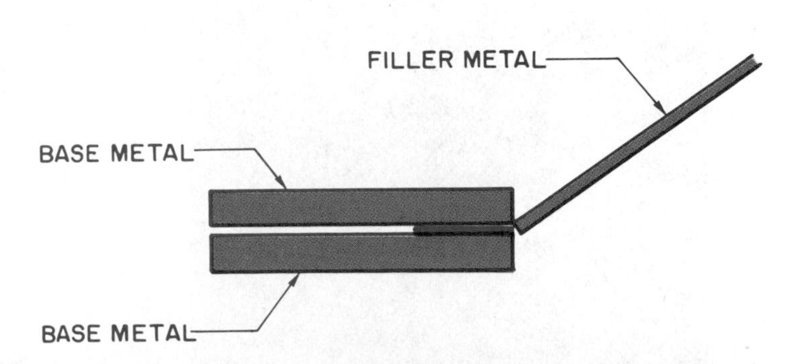

FIGURE 7–13 Two examples of capillary attraction. On the left are two pieces of glass spaced close together. When inserted in water, capillary attraction draws the water into the space between the two pieces of glass. The water molecules have a greater attraction for the glass than for each other. Therefore, they work their way up between the two pieces of glass. On the right is an illustration showing melted filler metal being drawn into the space between the two pieces of base metal. The molecules in the filler metal have a greater attraction to the base metal than they have for each other. These molecules work their way along the joint, first "wetting" the base metal and then filling the joint.

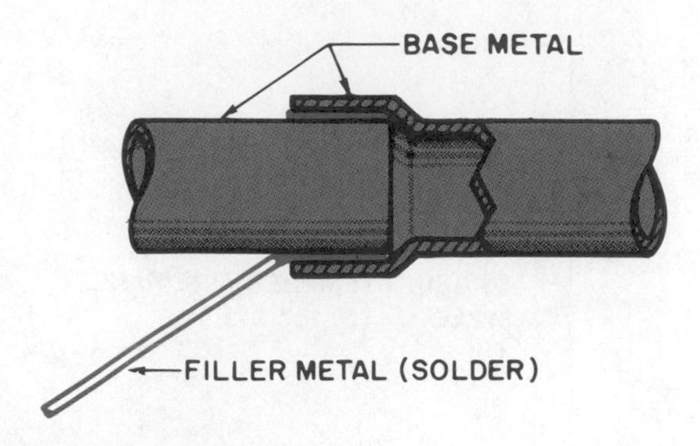

FIGURE 7–14 The molten solder in a soldered joint will be absorbed into the surface pores of the base metal.

for an explanation of this). If the soldering is properly done, the molten solder will be absorbed into the pores of the base metal, adhere to all surfaces, and form a bond, Figure 7–14.

7.8 HEAT SOURCES FOR SOLDERING AND BRAZING

Propane, butane, or air-acetylene torches are the most commonly used sources of heat when soldering or brazing. A *propane* or *butane* torch can be easily ignited and adjusted to the type and size of joint being soldered. Various tips are available, Figure 7–15.

FIGURE 7–15 A propane torch with typical tip for soldering. *Photo by Bill Johnson*

An *air-acetylene* unit is a type of heat source used often by air conditioning and refrigeration technicians. It usually consists of a B tank of acetylene gas, a regulator, a hose, and a torch, Figure 7–16. Various sizes of standard tips are available for a unit like this. The smaller tips are used for small-diameter tubing; larger tips are used for large-diameter tubing and for high-temperature applications. A high-velocity tip may be used to provide more concentration of heat. Figure 7–17 illustrates a popular high-velocity tip.

Figure 7–16 A typical air-acetylene setup. *Photo by Bill Johnson*

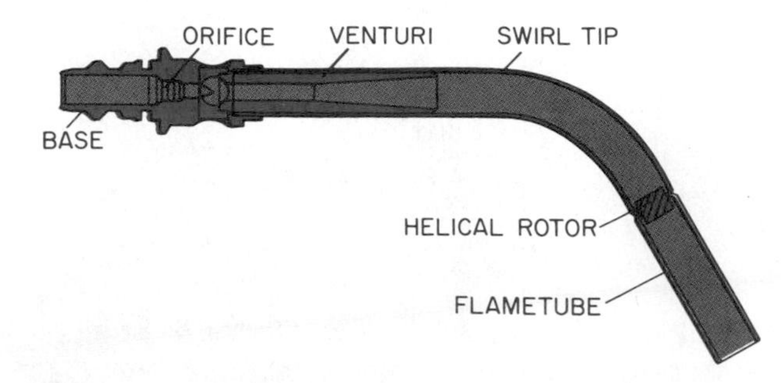

FIGURE 7–17 A high-velocity tip popular with many technicians. *Courtesy Wingaersheek*

Follow these procedures when setting up, igniting, and using an air-acetylene unit, Figure 7–18:

1. Before connecting the regulator to the tank, open the tank valve slightly to blow out any dirt that may be lodged at the valve.
2. Connect the regulator with hose and torch to the tank. Be sure that all connections are tight.
3. Open the tank valve one-half turn.
4. Adjust the regulator valve to about midrange.
5. Open the needle valve on the torch slightly, and ignite the gas with the spark lighter. ***Do not use matches or cigarette lighters.*** Adjust the flame using the needle valve at the handle so that there will be a sharp inner flame and a blue outer flame. After each use, shut off the valve on the tank and bleed off the acetylene in the hose by opening the valve on the torch handle. Bleeding the acetylene from the hoses relieves the pressure when not in use.

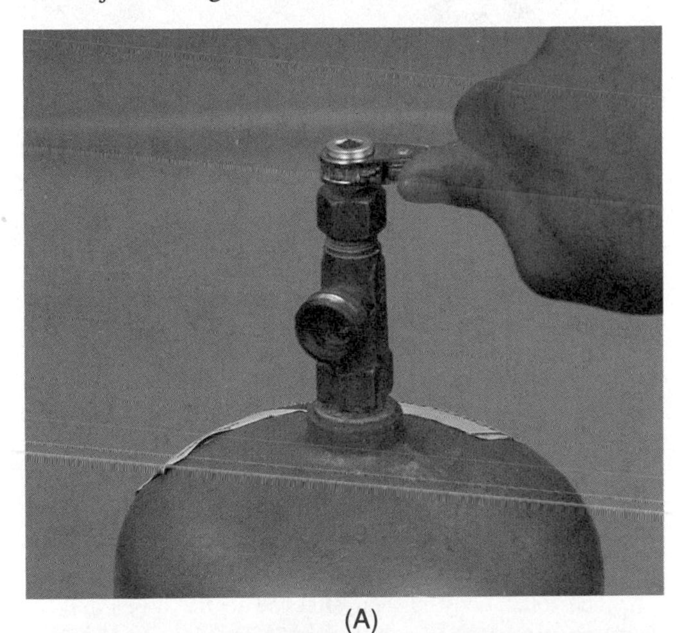

(A)

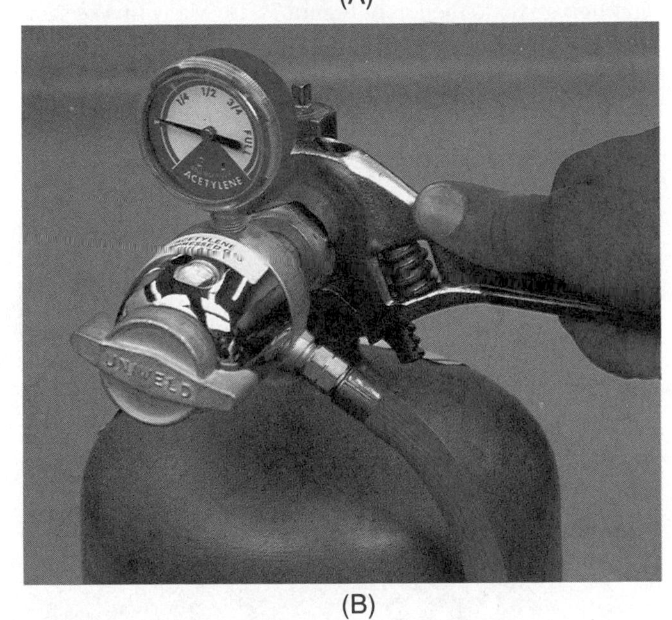

(B)

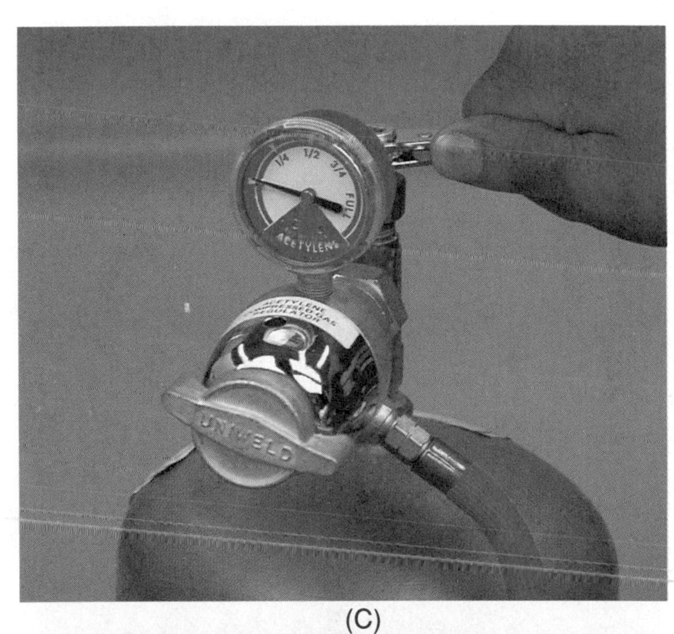

(C)

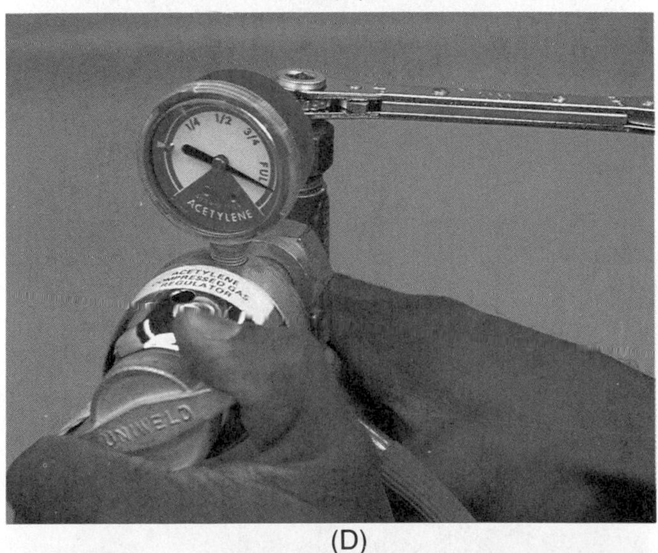

(D)

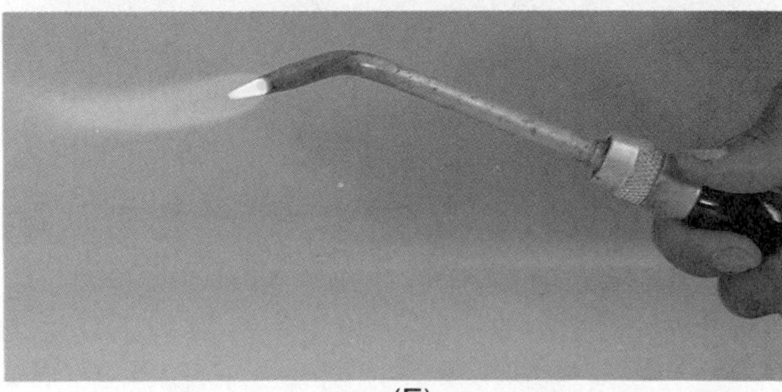

(E)

FIGURE 7–18 Proper procedures for setting up, igniting, and using an air-acetylene unit. *Photos by Bill Johnson*

Oxyacetylene torches may be preferred by some technicians, particularly when brazing large-diameter tubing or other applications requiring higher temperatures. ***This equipment can be extremely dangerous when not used properly. It is necessary that you thoroughly understand proper instructions for using oxyacetylene equipment before attempting to use it. It is also highly recommended that when you first begin to use the equipment, use it only under the close supervision of a qualified person.***

Following is a brief description of oxyacetylene welding and brazing equipment. Oxyacetylene brazing and welding processes use a high-temperature flame. Oxygen is mixed with acetylene gas to produce this high heat. The equipment includes oxygen and acetylene cylinders, oxygen and acetylene pressure regulators, hoses, fittings, safety valves, torches, and tips, Figure 7–19.

The regulators each have two gages, one to register tank pressure and the other to register pressure to the torch. The pressures indicated on these regulators are in pounds per square inch gage (psig). Figure 7–20 is a photo of an oxygen and an acetylene regulator. These regulators must be used only for the gases and service for which they are intended. ***All connections must be free from dirt, dust, grease, and oil. Oxygen can produce an explosion when in contact with grease or oil. A reverse flow valve should be used. These valves allow the gas to flow in one direction only.*** Follow instructions when attaching these valves because some are designed to attach to the hose connection on the torch body and some to the hose connection on the regulator.

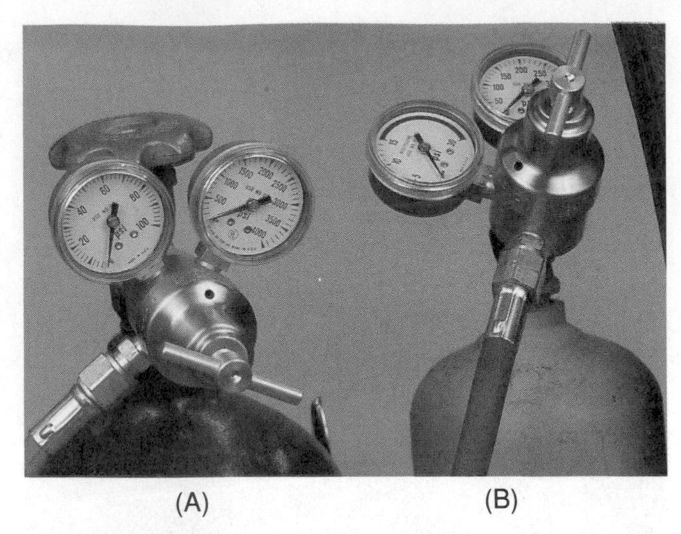

(A) (B)

FIGURE 7–20 (A) Acetylene regulator. (B) Oxygen regulator. *Photo by Bill Johnson*

The red hose is attached to the acetylene regulator with left-handed threads and the green hose to the oxygen regulator with right-handed threads.

The torch body is then attached to the hoses and the appropriate tip to the torch, Figure 7–21. Many tip sizes and

FIGURE 7–19 Oxyacetylene equipment. *Photo by Bill Johnson*

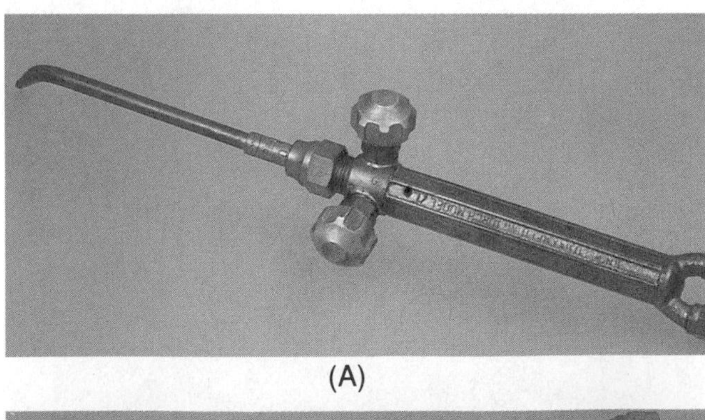

(A)

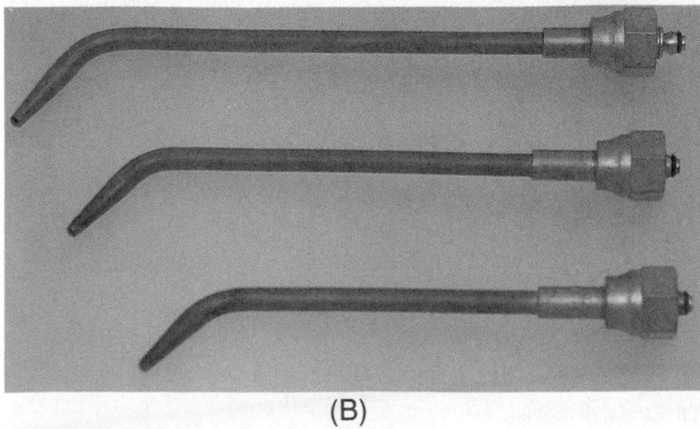

(B)

FIGURE 7–21 (A) Oxyacetylene torch with tip. (B) Assortment of oxyacetylene torch tips. *Photos by Bill Johnson*

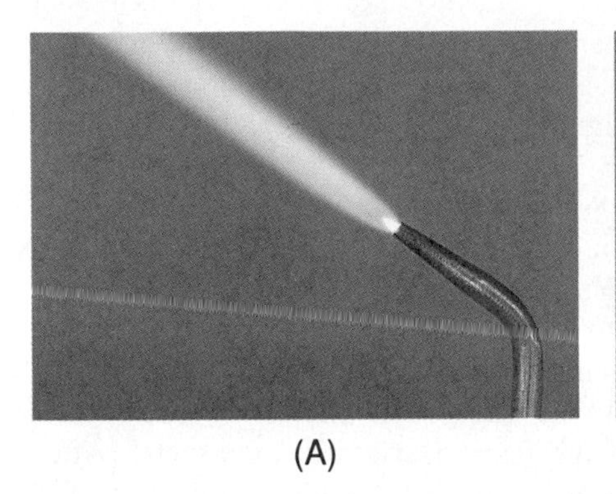
(A)

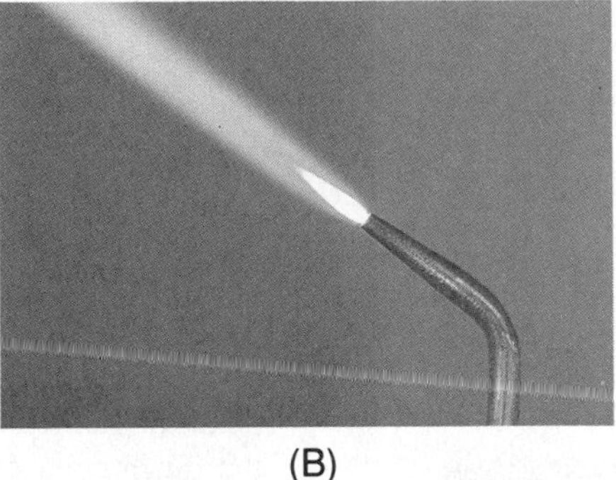
(B)

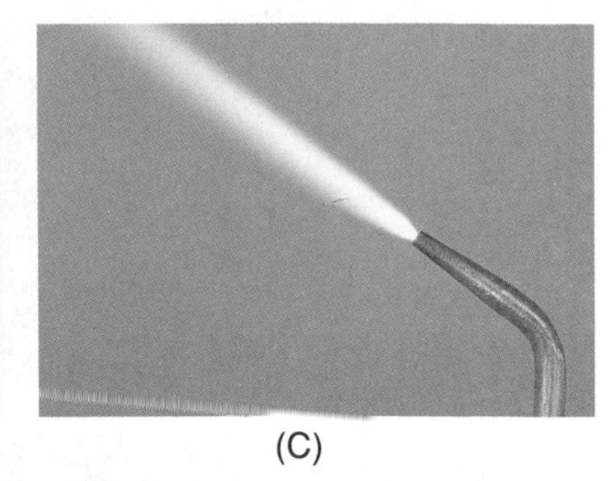
(C)

FIGURE 7–22 (A) Neutral flame, (B) Carbonizing flame, and (C) Oxidizing flame. *Photos by Bill Johnson*

styles are available. You should learn through your specialized training in using this equipment which tips should be used for particular applications.

A neutral flame should be used for these operations. See Figure 7–22 for photos of a neutral flame, carbonizing flame (too much acetylene), and an oxidizing flame (too much oxygen).

As stated previously there are many safety precautions that should be followed when using oxyacetylene equipment. Be sure that you follow all of these precautions. You should become familiar with them during your training using this equipment.

7.9 SOLDERING TECHNIQUES

The mating diameters of tubing and fittings are designed or sized to fit together properly. For good capillary attraction there should be a space between the metals of approximately 0.003 in. After the tubing has been cut to size and deburred, you must do the following for good soldered joints:

1. Clean mating parts of the joint.
2. Apply a flux to the male connection.
3. Assemble the tubing and fitting.
4. Heat the joint and apply the solder.
5. Wipe the joint clean.

CLEANING. The end of the copper tubing and the inside of the fitting must be absolutely clean. Even though these surfaces may look clean, they may contain fingerprints, dust, or oxidation. A fine sand cloth, a cleaning pad, or a special wire brush may be used. When the piping system is for a hermetic compressor, the sand cloth should be a nonconducting approved type, Figure 7–23A.

FLUXING. Apply flux soon after the surfaces are cleaned. For soft soldering, flux may be a paste, jelly, or liquid. Apply the flux with a *clean* brush or applicator. Do *not* use a brush that has been used for any other purpose. Apply the flux only to the area to be joined, and avoid getting it into the piping system. The flux minimizes oxidation while the joint is being heated. It also helps to float dirt or dust out of the joint.

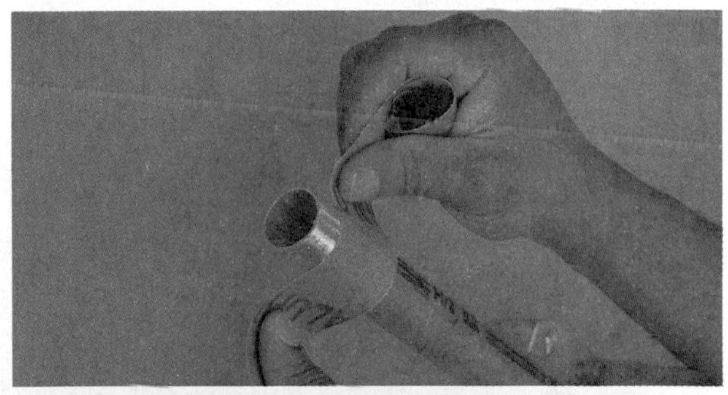

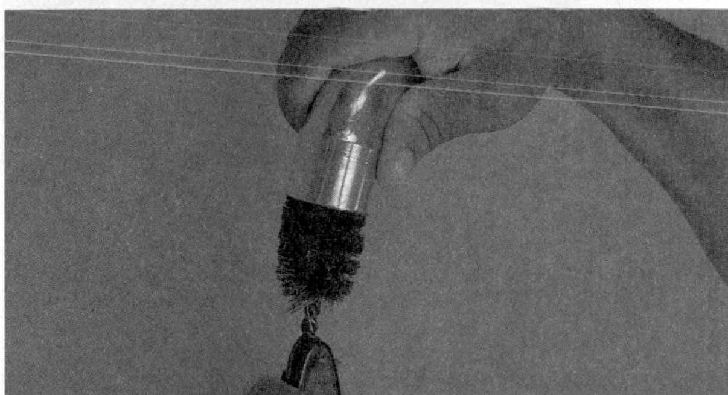

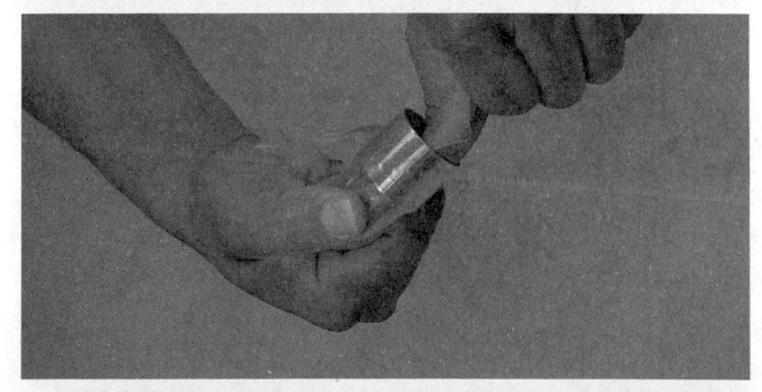

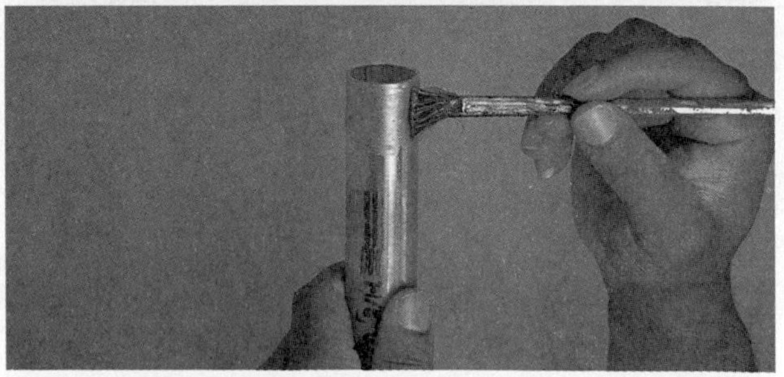

FIGURE 7–23A Cleaning and fluxing. (1) Clean tubing with sand cloth. (2) Clean fitting with a brush. (3) Clean fitting with sand cloth. (4) Apply flux. *Photos by Bill Johnson*

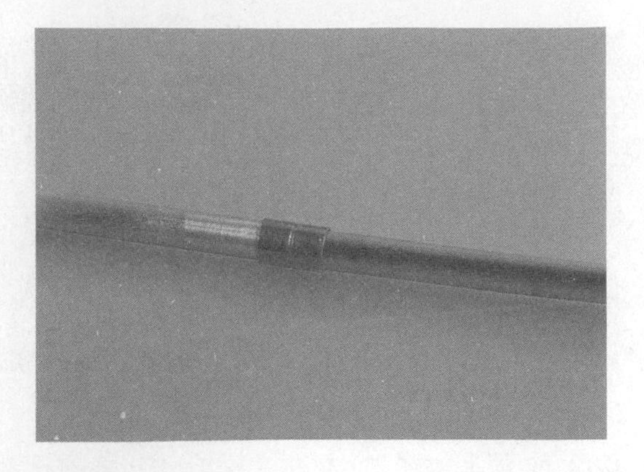

FIGURE 7–23B A properly assembled and supported joint ready to be soldered. *Photo by Bill Johnson*

ASSEMBLY. Soon after the flux is applied, assemble and support the joint so that it is straight and will not move while being soldered, Figure 7–23B.

HEATING AND APPLYING SOLDER. When soldering, heat the tubing near the fitting first for a short time. Then move the torch from the tubing to the fitting. Keep moving the torch from the tubing to the fitting. Keep moving the torch to spread the heat evenly and do not overheat any area. Do not point the flame into the fitting socket. Hold the torch so that the inner cone of the flame just touches the metal. After briefly heating the joint, touch the solder to the joint. If it does not readily melt, remove it and continue heating the joint. Continue to test the heat of the metal with the solder. Do *not* melt the solder with the flame; use the heat in the metal. When the solder flows freely from the heat of the metal, feed enough solder in to fill the joint. Do not use excessive solder. Figure 7–24 is a step-by-step procedure for heating the joint and applying solder.

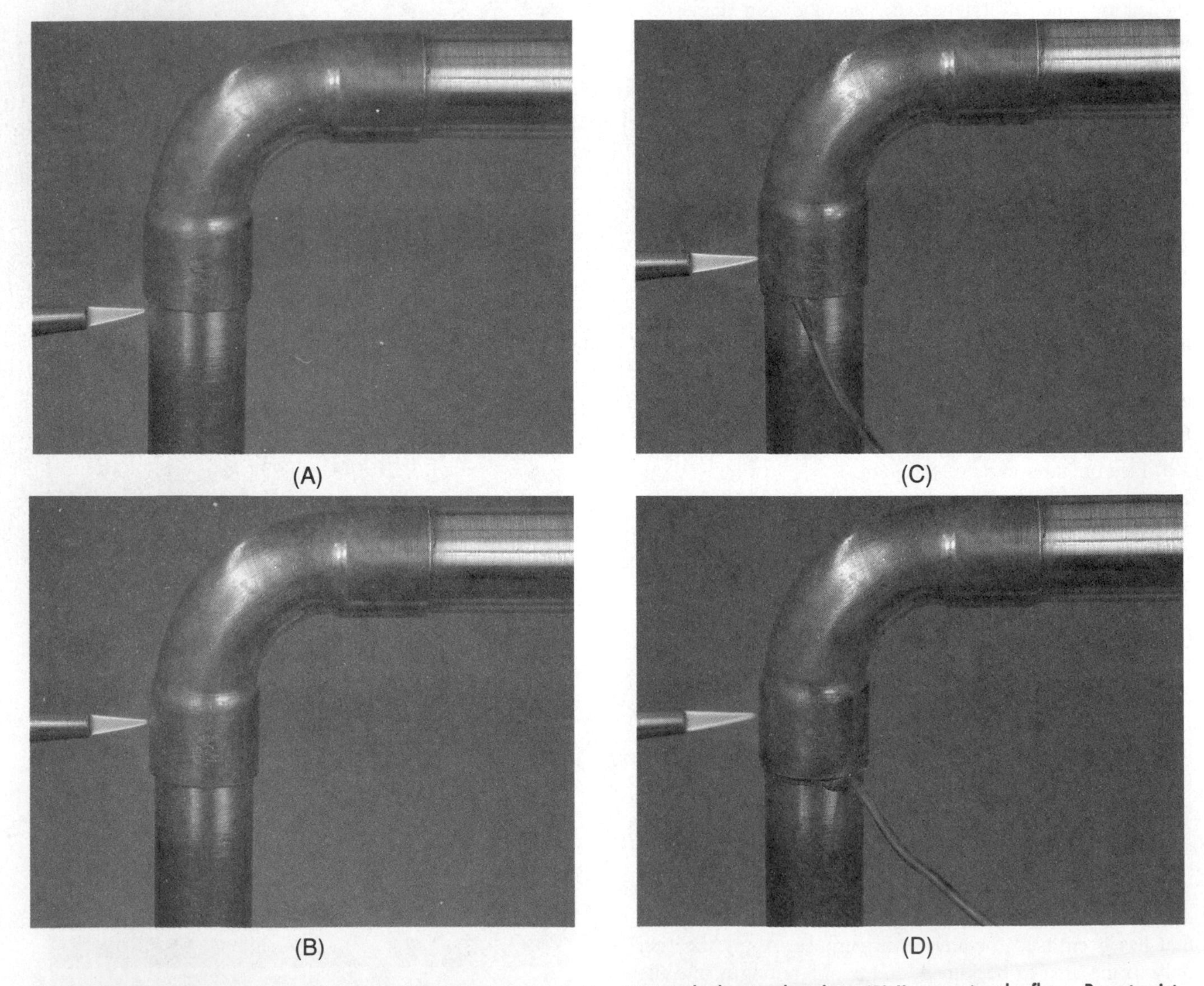

FIGURE 7–24 Proper procedures for heating a joint and applying solder. (A) Start by heating the tubing. (B) Keep moving the flame. Do not point flame into edge of fitting. (C) Touch solder to joint to check for proper heat. Do not melt solder with flame. (D) When joint is hot enough, solder will flow. *Photos by Bill Johnson*

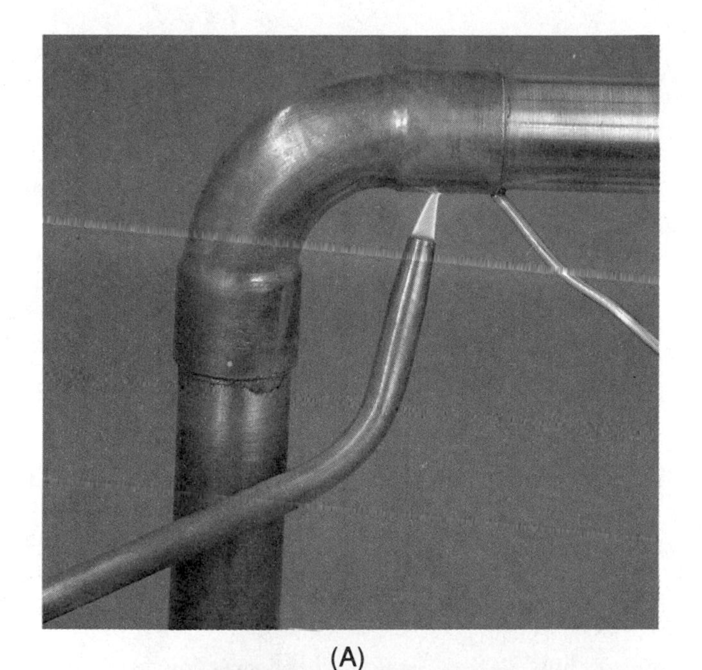

(A)

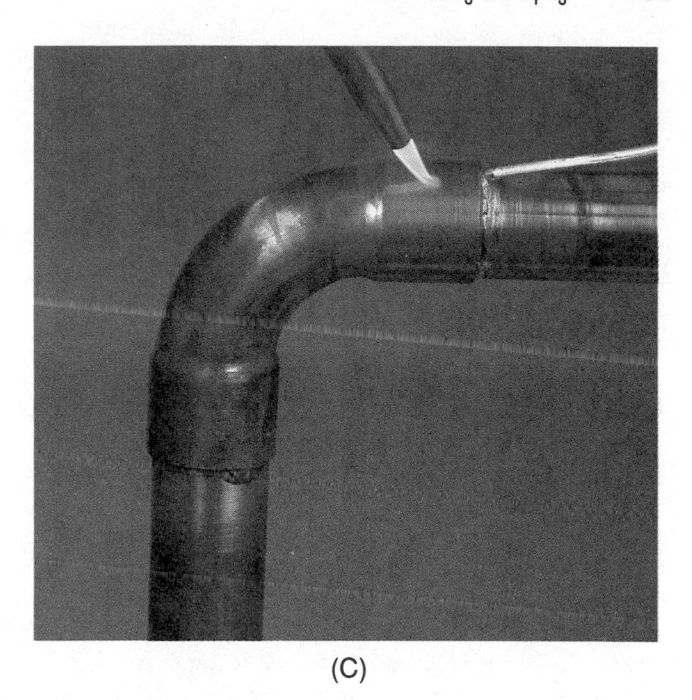

(C)

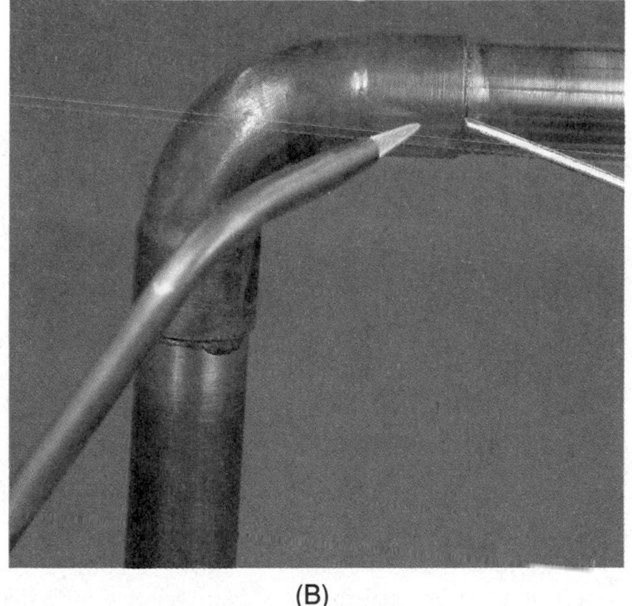

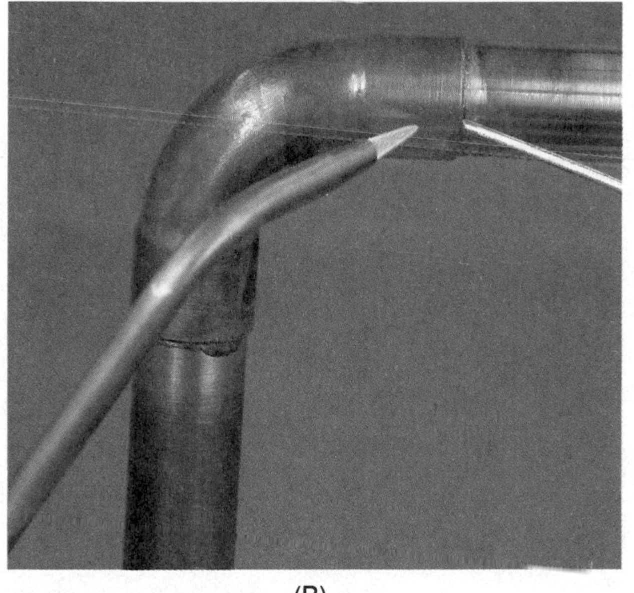

(B)

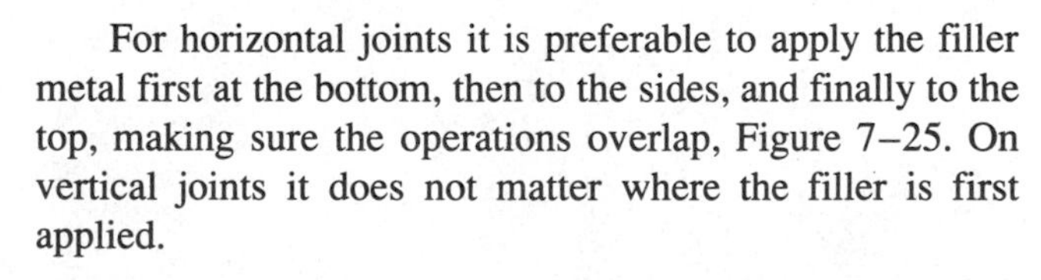

FIGURE 7–25 When making soldered or brazed joints in the horizontal position, (A) apply the filler metal at the bottom, (B) then to the two sides, and (C) finally to the top, making sure the operations overlap. *Photos by Bill Johnson*

For horizontal joints it is preferable to apply the filler metal first at the bottom, then to the sides, and finally to the top, making sure the operations overlap, Figure 7–25. On vertical joints it does not matter where the filler is first applied.

WIPING. While the joint is still hot, you may wipe it with a rag to remove excess solder. This is not necessary for producing a good bond, but it improves the appearance of the joint.

7.10 BRAZING TECHNIQUES

CLEANING. The cleaning procedures for brazing are similar to those for soldering. The brazing flux is applied with a brush to the cleaned area of the tube end. Avoid getting flux inside the piping system. Some brands of a silver-copper-phosphorus alloy do not require extensive cleaning when brazing copper to copper. Follow the instructions from the filler-material manufacturer.

APPLYING HEAT FOR BRAZING. *Before you heat the joint, it is good practice to force nitrogen or carbon dioxide into the system to purge the air and reduce the possibility of oxidation.* Apply heat to the parts to be joined with an air-acetylene or an oxyacetylene torch. Heat the tube first, beginning about 1 in. from the edge of the fitting, sweeping the flame around the tube. It is very important to keep the flame in motion and to not overheat any one area. Then switch the flame to the fitting at the base of the cup. Heat uniformly, sweeping the flame from the fitting to the tube. Apply the filler rod or wire at a point where the tube enters the socket. When the proper temperature is reached, the filler metal will flow readily by capillary attraction, into the space between the tube and the fitting. As in soldering, do not heat the rod or wire itself. The temperature of the metal at the joint should be hot enough to melt the filler metal. When the joint is at the correct temperature, it will be cherry red in color. The procedures are the same as with soldering except for the materials used and the higher heat applied.

The flux used in the brazing process will cause oxidation. When the brazing is done, wash these joints with soap and water.

7.11 PRACTICAL SOLDERING AND BRAZING TIPS

LOW TEMPERATURE. When soldering, the surfaces between the male and female parts being joined must be clean. Clean surfaces are necessary to ensure leak-free connections. It takes much longer to prepare surfaces for soldering than it does to actually make the soldered joint. If a connection to be soldered is cleaned in advance, it may need a touch-up cleaning when it is actually soldered. Copper oxidizes and iron or steel begins to rust immediately. Some fluxes may be applied after cleaning to prevent oxidation and rust until the tubing and fittings are ready for soldering.

Only the best solders should be used for low-temperature solder connections for refrigeration and air conditioning. For many years, systems were soldered successfully with 95/5 solder. If the soldered connection is completed in the correct manner, 95/5 can still be used. It should never be used on the high-pressure side of the system close to the compressor. The high temperature of the discharge line and the vibration will very likely cause it to leak.

A better choice of low-temperature solder than 95/5 would be a solder with a high strength and low temperature. Low-temperature solders with a silver content offer this greater strength with low melting temperatures.

One of the problems with most low-temperature solders is that the melting and flow points are too close together. This is evident when you are trying to use the solder; it flows too fast and you have a hard time keeping it in the clearance in the joint. Some of the silver-type low-temperature solders have a wider melt and flow point and are easier to use. They also have the advantage of being more elastic during the soldering procedure. This allows gaps between fittings to be filled more easily.

HIGH-TEMPERATURE (BRAZING). There are several choices of high-temperature brazing materials. Some include a high silver content (45% silver) and must always be used with a flux. Some high-temperature brazing materials have been developed that do not have high silver content (15% silver). These may not require flux when making copper-to-copper connections; however, flux may be used. Other brazing materials have been developed with no silver content. Your experience will soon help you choose which you like.

DIFFERENT JOINTS. The type of joint dictates the solder or brazing materials used. All connections are not copper-to-copper. Some may be copper-to-steel, copper-to-brass, brass-to-steel. These may be called dissimilar metal connections. Examples follow.

1. A copper suction line to a steel compressor or connection. The logical choice is to use 45% silver content because of the strength of the connection combined with the high melting temperature.
2. A copper suction line to a brass accessory valve. The best choice would be 45% silver from a strength standpoint. Another choice would be a low-melting temperature solder with a silver content that has a high strength and a low melting temperature. The valve body will not have to be heated to the high melting temperature of the 45% silver solder.
3. A copper liquid line to a steel filter drier. Forty-five percent silver brazing material is a good choice but requires a lot of heat. A low-temperature solder with silver content may be the best choice. It also gives you the option of easily removing the drier at a later date for replacement.
4. A large copper suction line connection using hard-drawn tubing. A high-temperature brazing material with low silver content is the choice of many technicians but you may not want to take the temper out of the hard-drawn tubing. A low-temperature solder with a silver content will give the proper strength and the low melting temperature will not take the temper from the pipe.

CHOICE OF HEAT FOR SOLDERING AND BRAZING. Many technicians use the air-acetylene torch combinations because the oxy-acetylene ones are difficult to use and are heavy. It should be pointed out that the air-acetylene torches that use the twist tip method for mixing the air and acetylene may be used for both low- and high-temperature soldering. There are practically no instances where oxyacetylene is necessary while doing general piping work with high-temperature solder until you are installing systems over 15 tons. An air-acetylene torch has a flame temperature of 5589°F. The correct tip must be used for the solder type and pipe size, Figure 7–26.

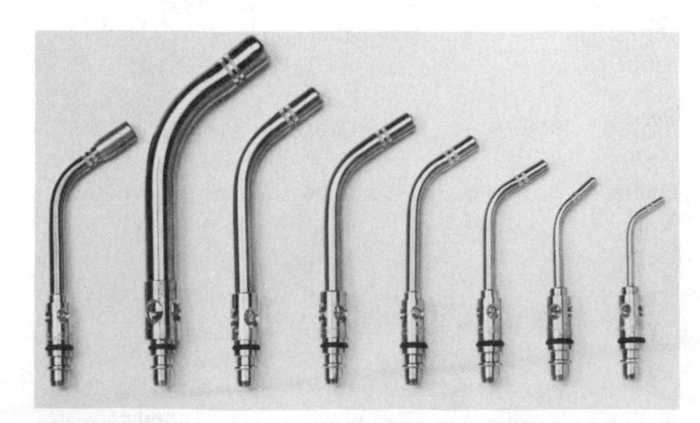

FIGURE 7–26 Different tip sizes may be used for different pipe sizes and solder combinations. *Courtesy Thermadyne Industries, Inc.*

ACETYLENE TORCH TIPS

	Tip Size		Gas Flow		Copper Tubing Size Capacity			
			@ 14 psi	(0.9 Bar)	Soft Solder		Silver Solder	
Tip No.	in.	mm	ft ³/hr	m³/hr	in.	mm	in.	mm
A-2	3/16	4.8	2.0	.17	1/8-1/2	3-15	1/8-1/4	3-10
A-3	1/4	6.4	3.6	.31	1/4-1	5-25	1/8-1/2	3-12
A-5	5/16	7.9	5.7	.48	3/4-1 1/2	20-40	1/4-3/4	10-20
A-8	3/8	9.5	8.3	.71	1-2	25-50	1/2-1	15-30
A-11	7/16	11.1	11.0	.94	1 1/2-3	40-75	7/8-1 5/8	20-40
A-14	1/2	12.7	14.5	1.23	2-3 1/2	50-90	1-2	30-50
A-32*	3/4	19.0	33.2	2.82	4-6	100-150	1 1/2-4	40-100
MSA-8	3/8	9.5	5.8	.50	3/4-3	20-40	1/4-3/4	10-20

*Use with large tank only.

NOTE: For air conditioning, add 1/8 inch for type L tubing.

FIGURE 7–26 continued

Another heat source is MAPP gas. It is a composite gas that is similar in nature to propane and may be used with air. The flame temperature of MAPP gas is 5301°F. It does not get as hot as air-acetylene but is supplied in larger and possibly lighter containers, Figure 7–27.

SOLDERING AND BRAZING TIPS. Examples follow.

1. Clean all surfaces to be soldered or brazed.
2. Keep filings and flux from inside the pipe.
3. When making upright soldered joints, apply heat to the top of the fitting.
4. When soldering or brazing fittings of different weights, such as soldering a copper line to a large brass valve body, most of the heat should be applied to the large mass of metal, the valve.
5. Do not overheat the connections. The heat may be varied by moving the torch closer to or further from the joint. It is recommended that once heat is applied to a connection, that it not be completely removed because air moves in and oxidation occurs.
6. Do not apply low-temperature solders in excess to a connection. It is a good idea to mark the length of solder you intend to use with a bend, Figure 7–28. When you get to the bend, stop or you will overfill the joint and the excess may be in the system.
7. When using flux with high-temperature brazing material, always chip the flux away when finished. ***Wear eye protection.*** It is hard and appears like glass on the brazed connection, Figure 7–29. This hard substance may cover a leak and be blown out later.

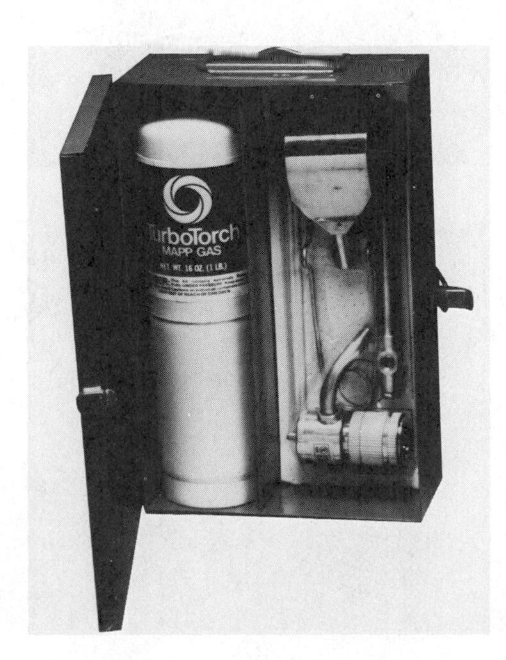

FIGURE 7–27 MAPP gas container. *Courtesy Thermadyne Industries, Inc.*

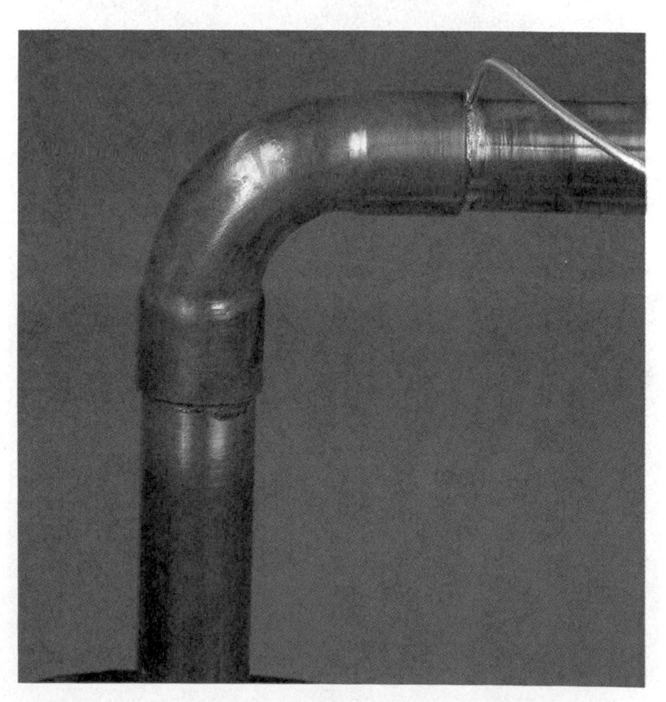

FIGURE 7–28 Make a bend in the end of the solder so you will know when to stop. *Photo by Bill Johnson*

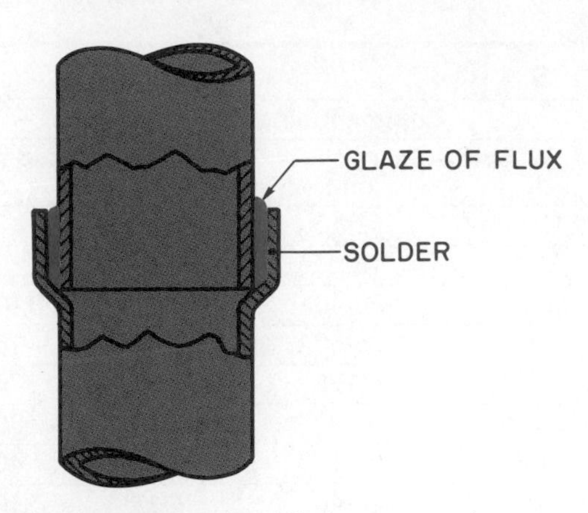

FIGURE 7–29 Flux used with high-temperature brazing materials will form a glaze that looks like glass.

8. When using any flux that will corrode the pipe, such as some fluxes for low-temperature solders, wash the flux off the connection or corrosion will occur. If this is not done it will soon look like a poor job.
9. Talk to the expert at a supply house for your special solder needs.

7.12 MAKING FLARE JOINTS

Another method of joining tubing and fittings is the *flare joint*. This joint uses a flare on the end of the tubing against an angle on a fitting and is secured with a flare nut behind the flare on the tubing, Figure 7–30.

FIGURE 7–30 Components for a flare joint. *Photo by Bill Johnson*

The flare on the tubing can be made with a screw-type flaring tool. To make the flare on the end of the tube use the following procedure.

1. Cut the tube to the right length.
2. Ream to remove all burrs and clean all residue from the tubing.
3. Slip the flare nut or coupling nut over the tubing with the threaded end facing the end of the tubing.
4. Clamp the tube in the flaring block, Figure 7–31(A). Adjust it so that the tube is slightly above the block (about one third of the total height of the flare).

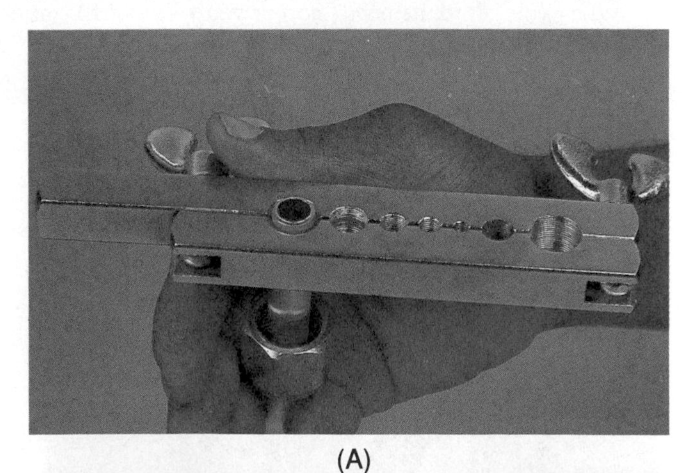

(A)

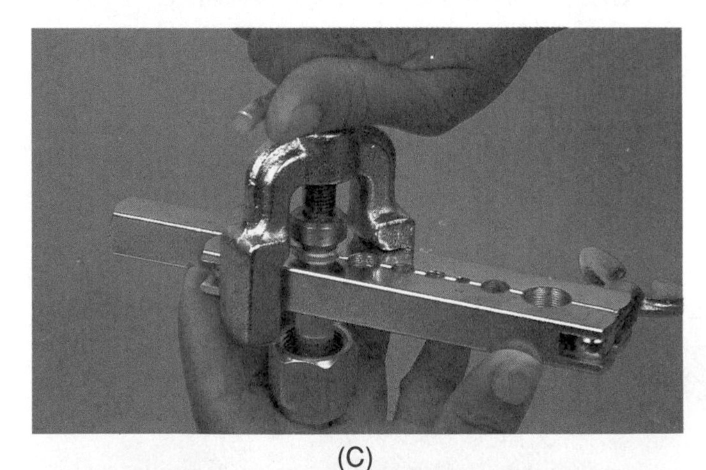

(C)

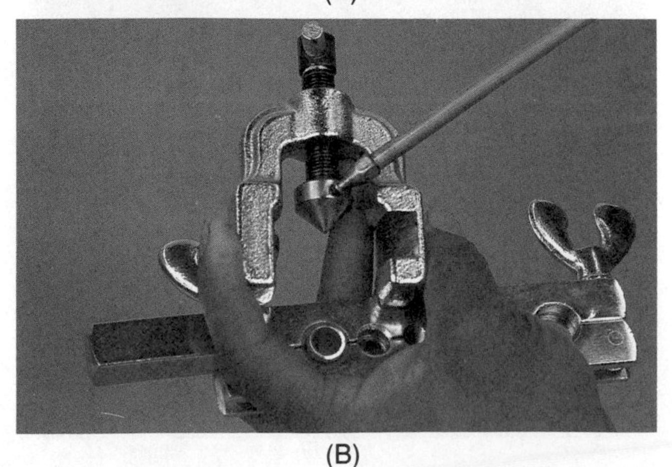

(B)

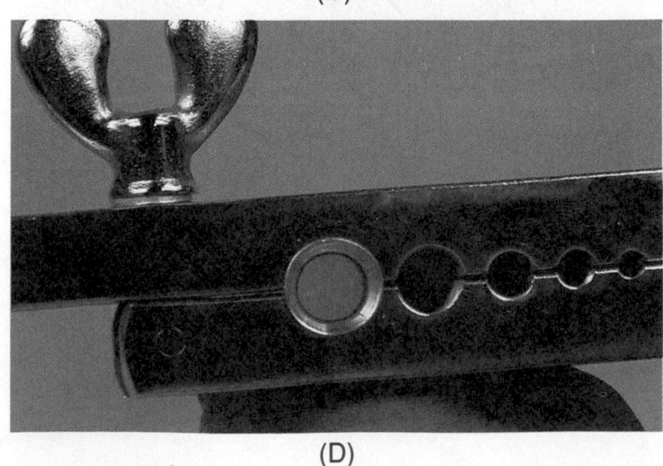

(D)

FIGURE 7–31 Proper procedure for making a flare joint using a screw-type flaring tool. *Photos by Bill Johnson*

5. Place the yoke on the block with the tapered cone over the end of the tube. Many technicians use a drop or two of refrigerant oil to lubricate the inside of the flare while it is being made, Figure 7–31(B).
6. Turn the screw down firmly, Figure 7–31(C). Continue to turn the screw until the flare is completed.
7. Remove the tubing from the block, Figure 7–31(D). Inspect for defects. If you find any, cut off the flare and start over.
8. Assemble the joint.

7.13 MAKING A DOUBLE-THICKNESS FLARE

A double-thickness flare provides more strength at the flare end of the tube. To make the flare is a two-step operation. Either a punch and block or combination flaring tool is used. Figure 7–32 illustrates the procedure for making double-thickness flares with the combination flaring tool.

Many fittings are available to use with a flare joint. Each of the fittings has a 45° angle on the end that fits against the flare on the end of the tube, Figure 7–33.

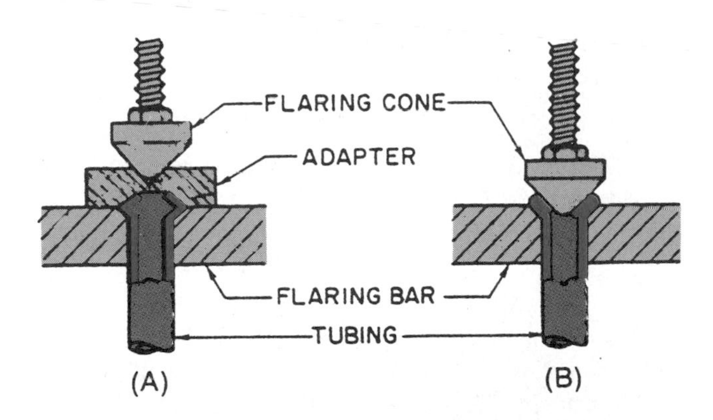

FIGURE 7–32 Procedure for making a double-thickness flare. (A) Place adapter of combination flaring tool over the tube extended above flaring block. Screw down to bell out tubing. (B) Remove adapter, place cone over tube and screw down to form double flare.

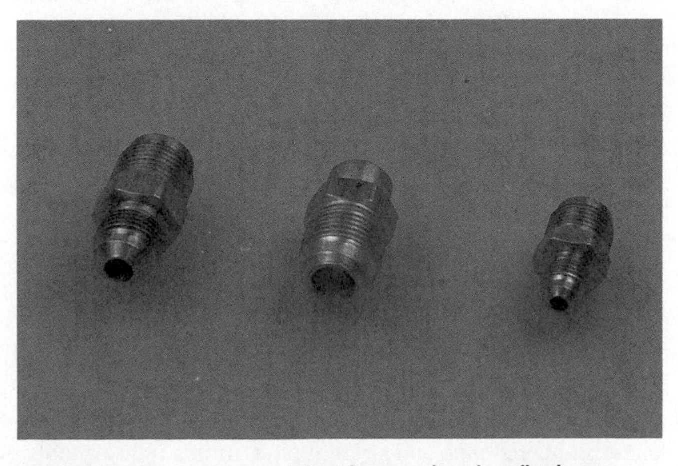

FIGURE 7–33 Examples of flare fittings. *Photo by Bill Johnson*

7.14 SWAGING TECHNIQUES

Swaging is not as common as flaring, but you should know how to make a swaged joint.

Swaging is the joining of two pieces of copper tubing of the same diameter by expanding or stretching the end of one piece to fit over the other so the joint may be soldered or brazed, Figure 7–34. As a general rule, the length of the joint that fits over the other is equal to the approximate OD of the tubing.

You can make a swaged joint by using a punch or a lever-type tool to expand the end of the tubing, Figure 7–35.

Place the tubing in a flare block or an anvil block that has a hole equal to the size of the OD of the tubing. The tube should extend above the block by an amount equal to the

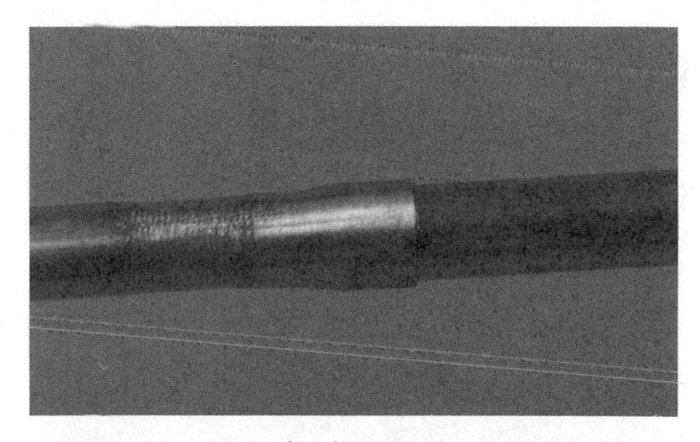

FIGURE 7–34 Joining tubing by a swaged joint. *Photo by Bill Johnson*

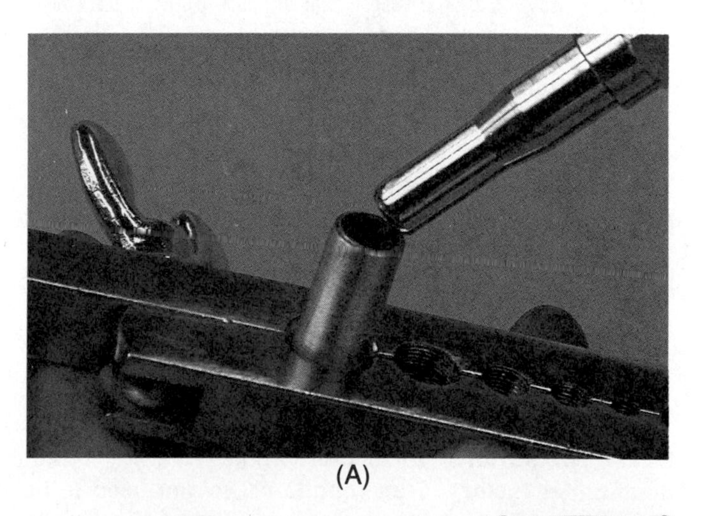

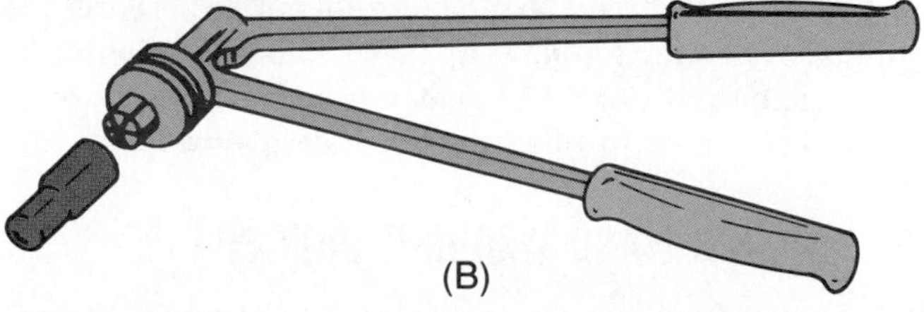

FIGURE 7–35 (A) Swaging punch. (B) Lever-type swaging tool. *(A) Photo by Bill Johnson, (B) Courtesy Lang, Air Conditioning: Procedures and Installation,* © 1982 by Delmar Publishers, Inc.

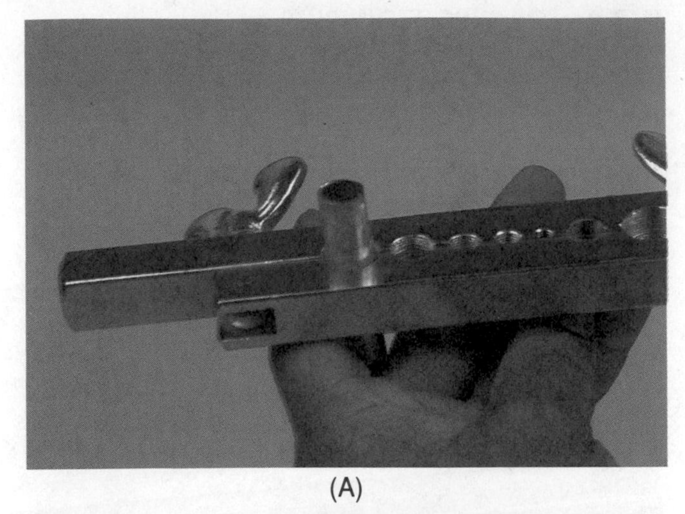

(A)

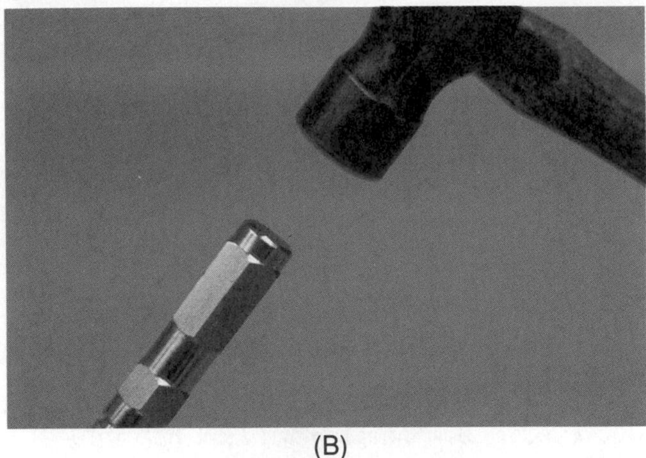

(B)

FIGURE 7–36 **Making a swaged joint. (A) Tube placed in block for swaging. (B) Swaging punch expanding metal.** ***Photos by Bill Johnson***

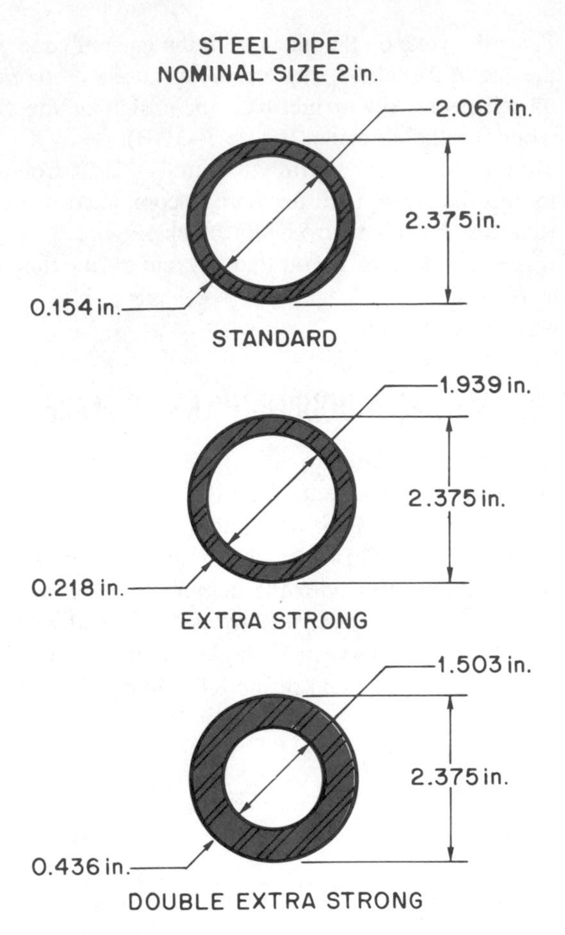

FIGURE 7–37 **Cross section of standard, extra strong, and double extra strong steel pipe.**

OD of the tube plus approximately 1/8 in., Figure 7–36(A). Place the correct size swaging punch in the tube and strike it with a hammer until the proper shape and length of the joint has been obtained, Figure 7–36(B). Follow the same procedure with screw-type or lever-type tools. A drop or two of refrigerant oil on the swaging tool will help but must be cleaned off before soldering. Assemble the joint. The tubing should fit together easily.

Always inspect the tubing after swaging to see if there are cracks or other defects. If any are seen or suspected, cut off the swage and start over.

SAFETY PRECAUTION: *Field fabrication of tubing is not done under factory-clean conditions, so you need to be observant and careful that *no* foreign materials enter the piping. When the piping is applied to air conditioning or refrigeration, it should be remembered that any foreign matter *will* cause problems. Utmost care *must* be taken.*

7.15 STEEL AND WROUGHT IRON PIPE

The terms "steel pipe," "wrought steel," and "wrought iron" pipe are often used interchangeably and incorrectly. When you want wrought iron pipe, specify "genuine wrought iron" to avoid confusion.

When manufactured, the steel pipe is either welded or produced without a seam by drawing hot steel through a forming machine. This pipe may be painted, left black, or coated with zinc (*galvanized*) to help resist rusting.

Steel pipe is often used in plumbing, hydronic (hot water) heating, and gas heating applications. The size of the pipe is referred to as the *nominal* size. For pipe sizes 12 in. or less in diameter the nominal size is approximately the size of the ID of the pipe. For sizes larger than 12 in. in diameter the OD is considered the nominal size. The pipe comes in many wall thicknesses but is normally furnished in standard, extra strong, and double extra strong sizes. Figure 7–37 is a cross section showing the different wall thicknesses of a 2-in. pipe. Tables are published indicating this information for each standard pipe diameter. Steel pipe is normally available in 21-ft lengths.

7.16 JOINING STEEL PIPE

Steel pipe is joined (with fittings) either by welding or by threading the end of the pipe and using threaded fittings. There are two types of American National Standard pipe threads: tapered pipe and straight pipe. In this industry only the tapered threads are used because they produce a tight

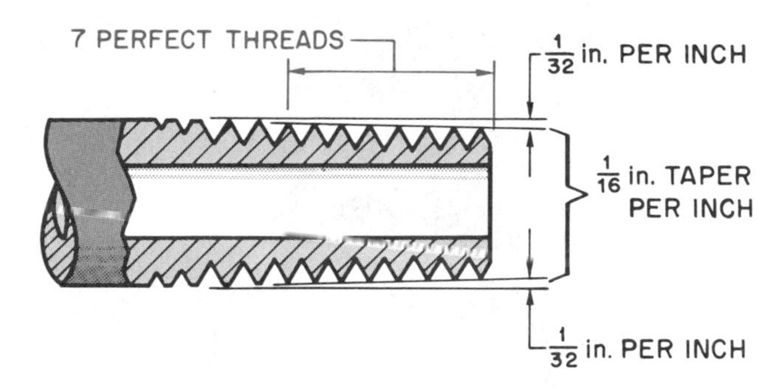

FIGURE 7–38 Cross section of a pipe thread.

PIPE SIZE (INCHES)	THREADS PER INCH
$\frac{1}{8}$	27
$\frac{1}{4}$, $\frac{3}{8}$	18
$\frac{1}{2}$, $\frac{3}{4}$	14
1 to 2	$11\frac{1}{2}$
$2\frac{1}{2}$ to 12	8

FIGURE 7–39 Threads per inch for some pipe sizes.

joint and help prevent the pressurized gas or liquid in the pipe from leaking.

Pipe threads have been standardized. Each thread is V-shaped with an angle of 60°. The diameter of the thread has a taper of 3/4 in./ft or 1/16 in./in. There should be approximately seven perfect threads and two or three imperfect threads for each joint, Figure 7–38. Perfect threads must not be nicked or broken, or leaks may occur.

Thread diameters refer to the approximate ID of the steel pipe. The nominal size then will be smaller than the actual diameter of the thread. Figure 7–39 shows the number of threads per inch for some pipe sizes. A thread dimension is written as follows: first the diameter, then the number of threads per inch, then the letters *NPT,* Figure 7–40.

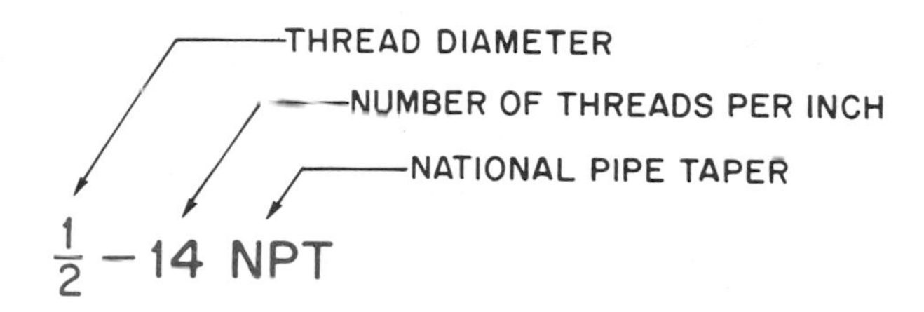

FIGURE 7–40 Thread specification.

You also need to be familiar with various fittings. Some common fittings are illustrated in Figure 7–41.

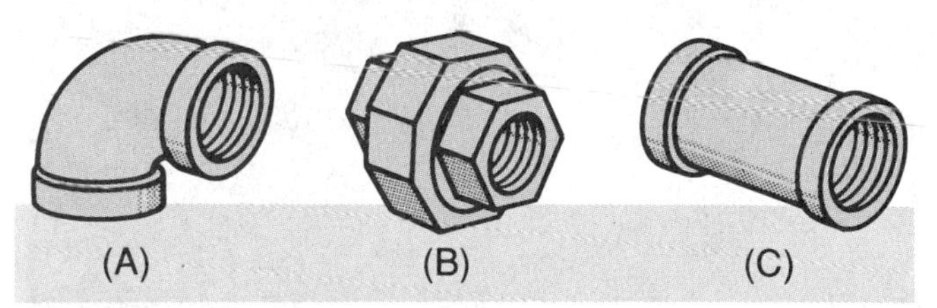

FIGURE 7–41 Steel pipe fittings. (A) 90° elbow. (B) Union. (C) Coupling.

Four tools are needed to cut and thread pipe.

- A *hacksaw* (use one with 18 to 24 teeth per inch) or *pipe cutter* is generally used to cut the pipe, Figure 7–42. A

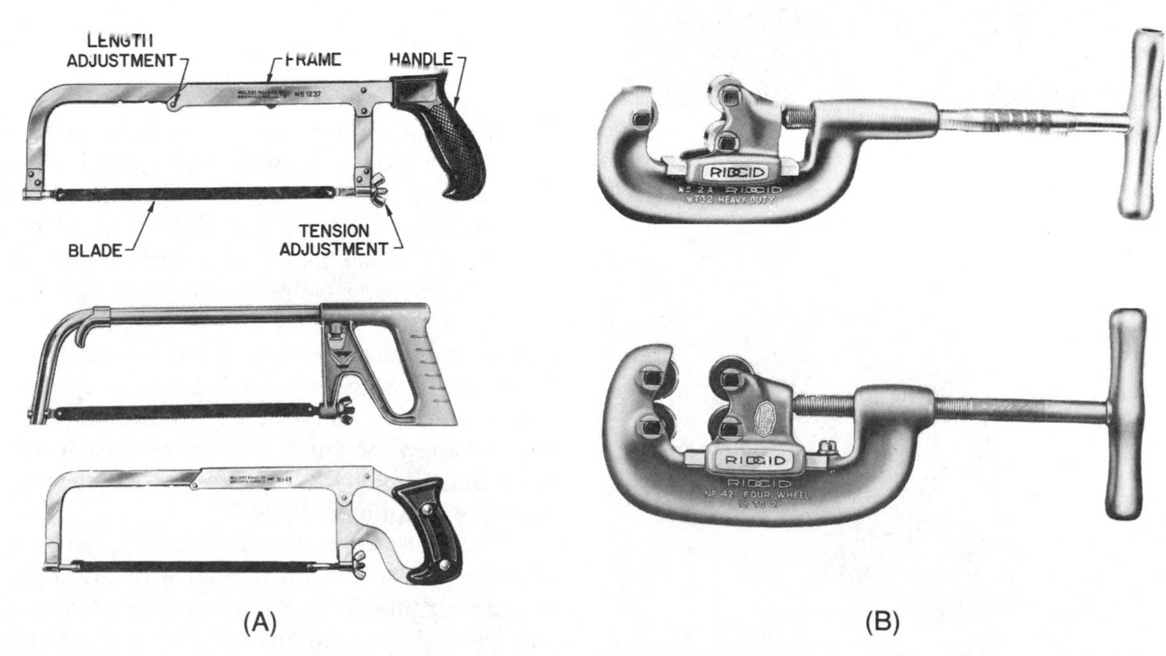

FIGURE 7–42 (A) Standard hacksaws. (B) Pipe cutters. *Courtesy (A) Slater and Smith, Basic Plumbing,* © 1979 by Delmar Publishers, Inc., *(B) Ridge Tool Company, Elyria, Ohio*

pipe cutter is best because it makes a square cut, but there must be room to swing the cutter around the pipe.

- A *reamer* removes burrs from the inside of the pipe after it has been cut. The burrs must be removed because they restrict the flow of the fluid or gas, Figure 7–43.
- *A threader,* also known as a *die*. Most threading devices used in this field are fixed-die threaders, Figure 7–44.
- Holding tools such as the *chain vise, yoke vise,* and *pipe wrench*. Figure 7–45.

When large quantities of pipe are cut and threaded regularly, special machines can be used. These machines are not covered in this text.

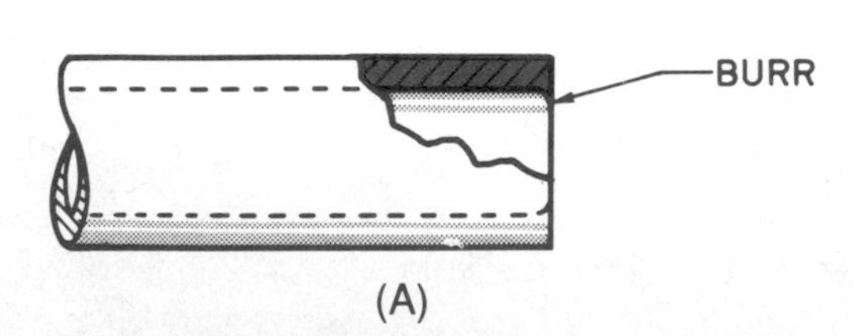

(A)

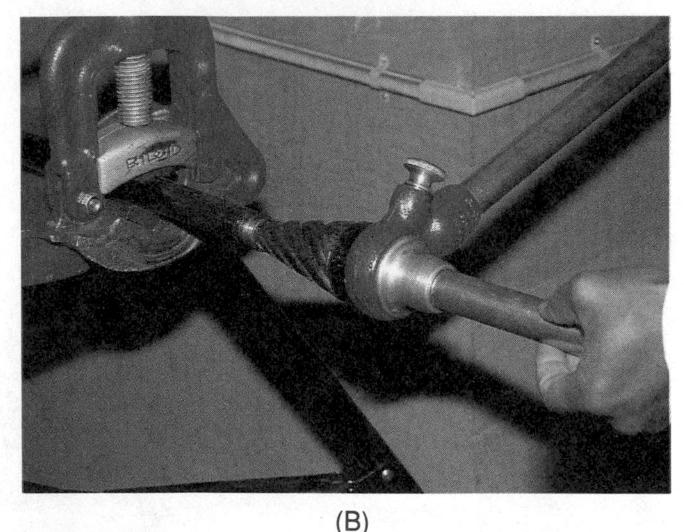

(B)

FIGURE 7–43 (A) Burr inside a pipe. (B) Using a reamer to remove a burr. *Photo by Bill Johnson*

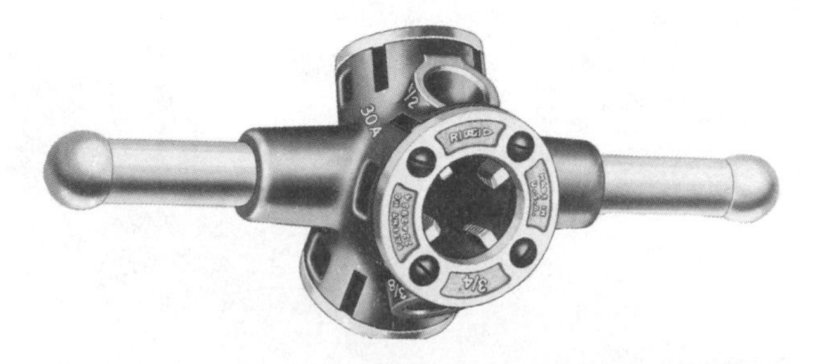

FIGURE 7–44 Fixed die-type pipe threader. *Courtesy Ridge Tool Company, Elyria, Ohio*

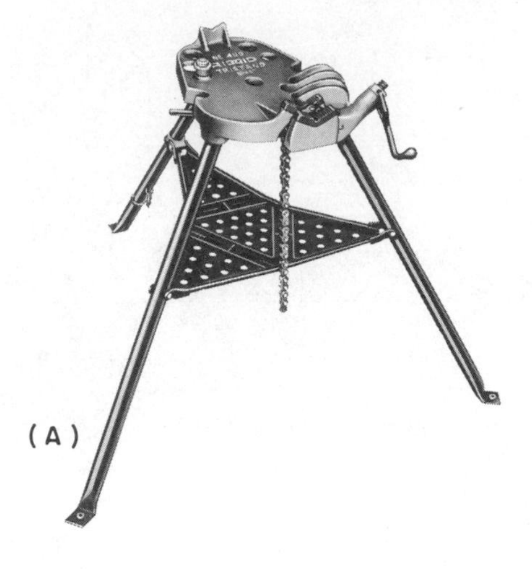

(A)

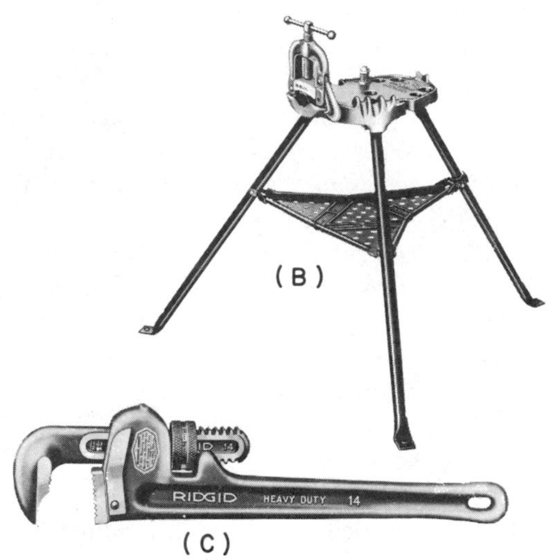

(B)

(C)

FIGURE 7–45 (A) Chain vise. (B) Yoke vise. (C) Pipe wrench. *Courtesy Ridge Tool Company, Elyria, Ohio*

CUTTING. The pipe must be cut square to be threaded properly. If there is room to revolve a pipe cutter around the pipe, you can use a one-wheel cutter. Otherwise, use one with more than one cutting wheel. Hold the pipe in the chain vise or yoke vise if it has not yet been installed. Place the cutting wheel directly over the place where the pipe is to be cut. Adjust the cutter with the T handle until all the rollers or cutters contact the pipe. Apply moderate pressure with the T handle and rotate the cutter around the pipe. Turn the handle about one-quarter turn for each revolution around the pipe. ***Do not apply too much pressure because it will cause a large burr inside the pipe and excessive wear of the cutting wheel, Figure 7–46.***

To use a hacksaw, start the cut gently, using your thumb to guide the blade or use a holding fixture, Figure 7–47. ***Keep your thumb away from the teeth.*** A hacksaw will only cut on the forward stroke. Do not apply pressure on the backstroke. Do not force the hacksaw or apply excessive pressure. Let the saw do the work.

FIGURE 7–46 Pipe cutter. *Courtesy Ridge Tool Company, Elyria, Ohio*

FIGURE 7–47 Cutting steel pipe with a hacksaw and holding fixture. *Photo by Bill Johnson*

REAMING. After the pipe is cut, put the reamer in the end of the pipe. Apply pressure against the reamer and turn clockwise. Ream only until the burr is removed, Figure 7–43(B).

THREADING. To thread the pipe, place the die over the end and make sure it lines up square with the pipe. Apply cutting oil on the pipe and turn the die once or twice. Then reverse the die approximately one-quarter turn. Rotate the die one or two more turns, and reverse again. Continue this procedure and apply cutting oil liberally until the end of the pipe is flush with the far side of the die, Figure 7–48.

7.17 INSTALLING STEEL PIPE

When installing steel pipe, hold or turn the fittings and pipe with pipe wrenches. These wrenches have teeth set at an angle so that the fitting or pipe will be held securely when

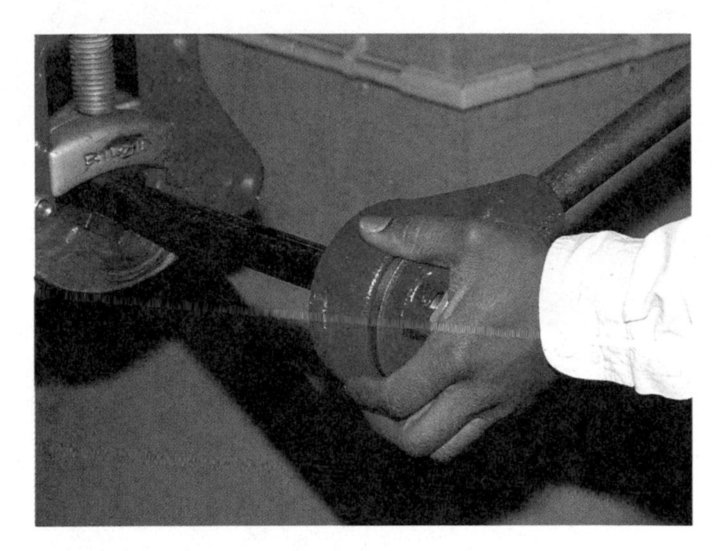

FIGURE 7–48 Threading pipe. *Photos by Bill Johnson*

pressure is applied. Position the wrenches in opposite directions on the pipe and fitting, Figure 7–49.

When assembling the pipe, use the *correct* pipe thread dope on the male threads. Do not apply the dope closer than two threads from the end of the pipe, Figure 7–50, otherwise it might get into the piping system.

All state and local codes must be followed. You should continually familiarize yourself with all applicable codes.

SAFETY PRECAUTION: The technician should also realize the importance of installing the size pipe specified. A

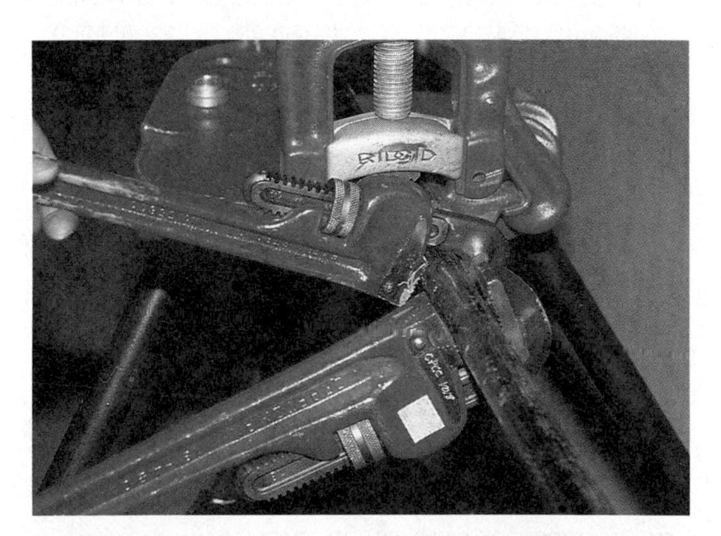

FIGURE 7–49 Holding pipe and turning fitting with pipe wrenches. *Photo by Bill Johnson*

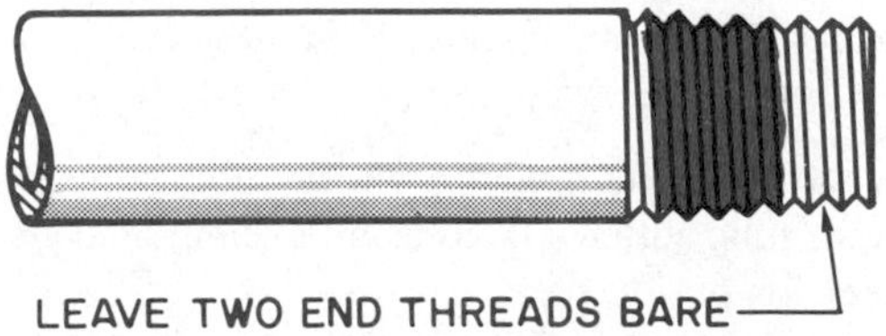

USE MODERATE AMOUNT OF DOPE

LEAVE TWO END THREADS BARE

FIGURE 7–50 Applying pipe dope.

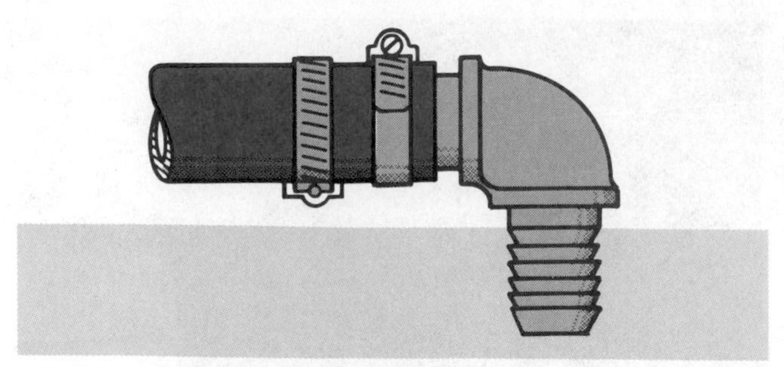

FIGURE 7–51 Position of clamps on PE pipe.

designer has carefully studied the entire system and has indicated the size that will deliver the correct amount of gas or fluid. Pipe sizes other than those specified should never be substituted without permission of the designer.

7.18 PLASTIC PIPE

Plastic pipe is used for many plumbing, venting, and condensate applications. You should be familiar with the following types.

ABS (ACRYLONITRILEBUTADIENE STYRENE). *ABS* is used for water drains, waste, and venting. It can withstand heat to 180°F without pressure. Use a solvent cement to join ABS with ABS; use a transition fitting to join ABS to a metal pipe. ABS is rigid and has good impact strength at low temperatures.

PE (POLYETHYLENE). *PE* is used for water, gas, and irrigation systems. It can be used for water-supply and sprinkler systems and water-source heat pumps. PE is not used with a hot water supply, although it can stand heat with no pressure. It is flexible and has good impact strength at low temperatures. It is normally attached to fittings with two hose clamps. Place the screws of the clamps on opposite sides of the pipe, Figure 7–51.

PVC (POLYVINYL CHLORIDE). *PVC* can be used in high-pressure applications at low temperatures. It can be used for water, gas, sewage, certain industrial processes, and in irrigation systems. It is a rigid pipe and a high impact strength. PVC can be joined to PV fittings with a solvent cement, or it can be threaded and used with a transition fitting for joining to metal pipe.

CPVC (CHLORINATED POLYVINYL CHLORIDE). *CPVC* is similar to PVC except that it can be used with temperatures up to 180°F at 100 psig. It is used for both hot and cold water supplies and is joined to fittings in the same manner as PVC.

The following and Figure 7–52 describe how to prepare PVC or CPVC for joining.

1. Cut the end square with a plastic tubing shear, a hacksaw, or tube cutter. The tube cuter should have a special wheel for plastic pipe.
2. Deburr the pipe inside and out with a knife or a half-round file.

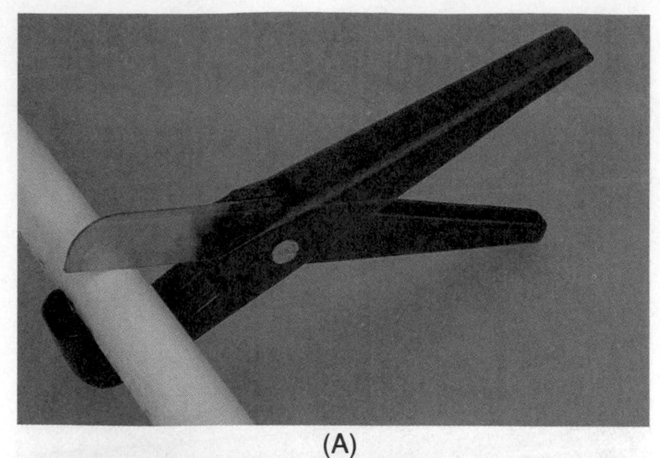

(A)

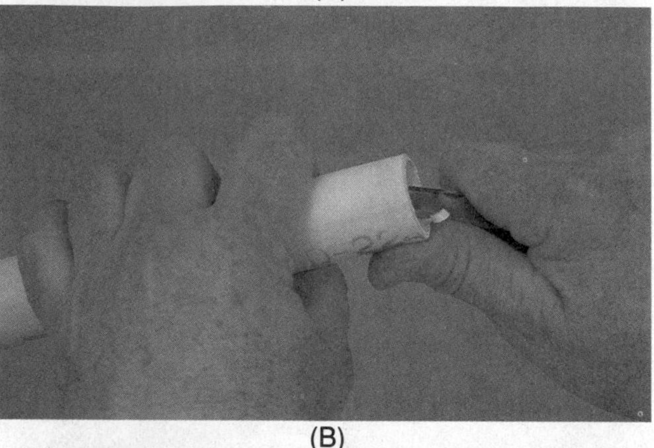

(B)

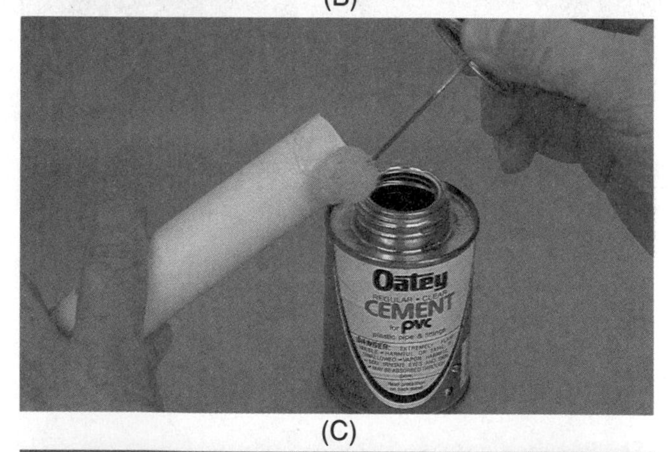

(C)

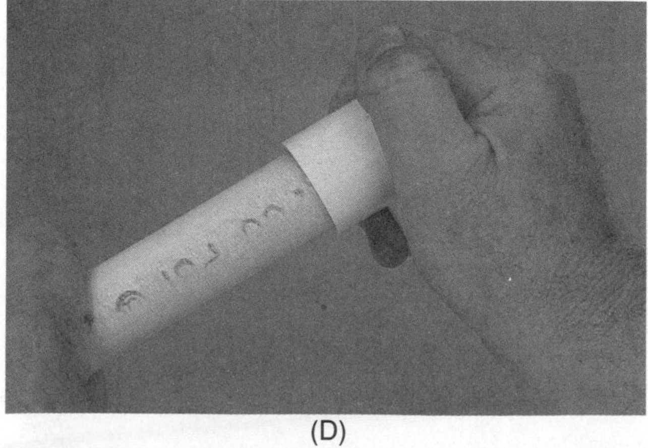

(D)

FIGURE 7–52 Cutting and joining PVC or CPVC pipe. *Photos by Bill Johnson*

3. Clean the pipe end. Apply primer and solvent to both the outside of the pipe and the inside of the fitting. (Follow instructions on primer and cement containers.)
4. Insert the pipe all the way into the fitting. Turn approximately one-quarter turn to spread the cement and allow it to set (dry) for about 1 minute.

PVC and CPVC are prepared for joining similar to ABS except that a primer must be applied before applying the solvent cement. The same cement cannot be used for ABS and PVC or CPVC. Schedule #80 PVC and CPVC can be threaded. A regular pipe thread die can be used. ***Do not use the same die for metal and plastic pipe. The die used for metal will become too dull to be used for plastic. The plastic pipe die must be kept very sharp. Always follow manufacturer's directions when using any plastic pipe and cement.***

SUMMARY

- *The use of correct tubing, piping, and fittings along with the proper installation is necessary for a refrigeration or air conditioning system to operate properly. Careless handling of the tubing and poor soldering or brazing techniques may cause serious damage to the components of the system.*
- Copper tubing is generally used for plumbing, heating, and refrigerant piping.
- Copper tubing is available in soft- or hard-drawn copper. Type L is the standard size used most frequently in plumbing and heating. ACR tubing is used in air conditioning and refrigeration.
- The size of heating and plumbing tubing refers to the ID of the tubing. The size of ACR tubing refers to the OD.
- A proper adhesive should be used to fasten tubing insulation together.
- ACR tubing can be purchased as line sets charged with refrigerant and sealed on both ends.
- Tubing may be cut with a hacksaw or tube cutter.
- Soft tubing may be bent. Tube bending springs or lever-type benders may be used, or the bend can be made by hand.
- Soldering and brazing fasten tubing and fittings together. Temperatures below 800°F are used for soldering; temperatures over 800°F, for brazing.
- Air-acetylene units are frequently used for soldering and brazing.
- Oxyacetylene equipment is also used, particularly for brazing requiring higher temperatures.
- The flare joint is another method of joining tubing and fittings.
- The soldered swaged joint is a method used to fasten two pieces of copper tubing together, but it is not used as much as soldering or brazing fittings to tubing or using flared joints.
- Steel pipe is used in plumbing, hydronic heating, and gas heating applications.
- Steel pipe is joined with fittings by welding or by threaded joints.
- ABS, PE, PVC, and CPVC are four types of plastic pipe; each has a different use.
- All but the PE type are joined to fittings with a solvent cement.

REVIEW QUESTIONS

1. Which type of copper tubing may be bent?
2. What standard type of copper tubing is used most frequently in plumbing and heating?
3. What would the size of 1/2 in. refer to with regard to copper tubing used in plumbing and heating?
4. What would the size 1/2 in. refer to with regard to ACR tubing?
5. In what size rolls is soft copper tubing normally available?
6. In what lengths is hard-drawn copper normally available?
7. Why are some ACR tubing lines insulated?
8. What sealant is best for all tubing insulation seams?
9. Describe a line set.
10. Describe the procedure for cutting tubing with a tube cutter.
11. Describe procedures for bending soft copper tubing.
12. What are the approximate temperature ranges used when soldering?
13. What is the minimum temperature for brazing?
14. What type of solder is suitable for moderate pressures and temperatures?
15. What are elements that make up brazing filler metal alloys?
16. Describe how to make a good soldered joint.
17. What type of equipment is used most frequently in soldering and brazing applications?
18. List the procedures used when setting up, igniting, and using an air-acetylene unit.
19. Describe the procedures used to make a good brazed joint.
20. Describe a flared joint.
21. Describe the proper procedures for making a flared joint.
22. Describe a double-thickness flare.
23. What is swaging?
24. What should you do if you see or suspect a crack in a flared joint?
25. What are two methods of manufacturing steel pipe?
26. What are some uses of steel pipe?
27. What does the nominal pipe size 12 in. or less in diameter refer to?
28. Describe the procedure for preparing and threading the end of steel pipe.
29. Describe the thread dimension 1/4-18 NPT.
30. List four types of plastic pipe.
31. State a use for each type of plastic pipe.
32. Describe the procedure for joining each type to fittings.

UNIT 8 System Evacuation

OBJECTIVES

After studying this unit, you should be able to

- describe a deep vacuum.
- describe two different types of evacuation.
- describe two different types of vacuum measuring instruments.
- choose a proper high-vacuum pump.
- list some of the proper evacuation practices.
- describe a high-vacuum single evacuation.
- describe a triple evacuation.

SAFETY CHECKLIST

* Care should be used while handling any of the products from a contaminated system because acids may be encountered.
* Do not place your hand over any opening that is under high vacuum because the vacuum may cause a blood blister on the skin.
* Wear goggles and gloves while transferring refrigerant.
* Do not allow mercury from any instrument to escape. It is a pollutant.

8.1 PURPOSE OF SYSTEM EVACUATION

Refrigeration systems are designed to operate with only refrigerant and oil circulating inside them. When systems are assembled or serviced, air enters the system. Air contains oxygen, nitrogen, hydrogen, and water vapor, all of which are detrimental to the system. These gases cause two problems. The nitrogen is called a noncondensable gas. It will not condense in the condenser and will occupy condenser space that would normally be used for condensing. This will cause a rise in head pressure. Figure 8–1 illustrates a condenser with noncondensable vapors inside. These other gases also cause chemical reactions that produce acids in the system. Acids in the system cause deterioration of the system's parts, copper plating of the running gear, and break down of motor insulation. Actually, oxygen is the real problem when air is allowed into the system. Air contains 13% oxygen, which will react with the various chemicals that make up the refrigerants. Some of these chemical combinations will create mild forms of hydrofluoric or hydrochloric acid. These acids may remain in the system for years without showing signs of a problem. Then, the motor will burn out or copper will be deposited on the crankshaft from the copper in the system, causing the crankshaft to become slightly oversized and binding will occur. This will cause the rubbing surfaces to score and become worn prematurely.

These noncondensable gases must be removed from the system if it is to have a normal life expectancy. Many systems have been operated for years with small amounts of these products inside them, but they will not last and give the reliability the customer pays for.

Noncondensable gases are removed by vacuum pumps after the system is leak checked. The pressure inside the system is reduced to an almost perfect vacuum.

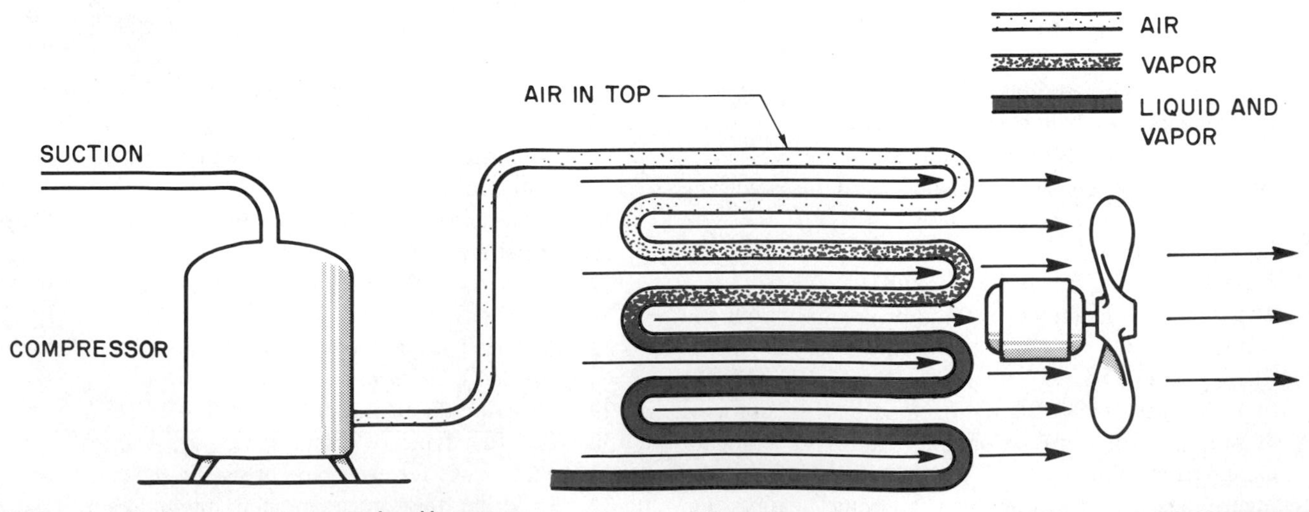

FIGURE 8–1 Condenser containing noncondensable gases.

8.2 THEORY INVOLVED WITH EVACUATION

To *pull a vacuum* means to lower the pressure in a system below the atmosphere's pressure. The atmosphere exerts a pressure of 14.696 psia (29.92 in. Hg) at sea level at 70°F. Vacuum is commonly expressed in millimeters of mercury (mm Hg). The atmosphere will support a column of mercury 760 mm (29.92 in.) high. To pull a vacuum in a refrigeration system, for example, the pressure inside the system must be reduced to 0 psia (29.92 in. Hg vacuum) to remove all of the atmosphere. This would represent a perfect vacuum, which has never been achieved.

A compound gage is often used to indicate the vacuum level. A compound gage starts at 0 in. Hg vacuum and reduces to 30 in. Hg vacuum. When the term "vacuum" is used, it is applied to the compound gage. When the term "in. Hg" is used, it is applied to a manometer or barometer. Gages used for refrigeration have scales graduated in in. Hg. There are methods other than evacuation to remove noncondensables; these methods will be discussed later.

We can use the bell jar in Figure 8–2A to describe a typical system evacuation. A refrigeration system contains a volume of gas like the bell jar. The only difference is that a refrigeration system is composed of many small chambers

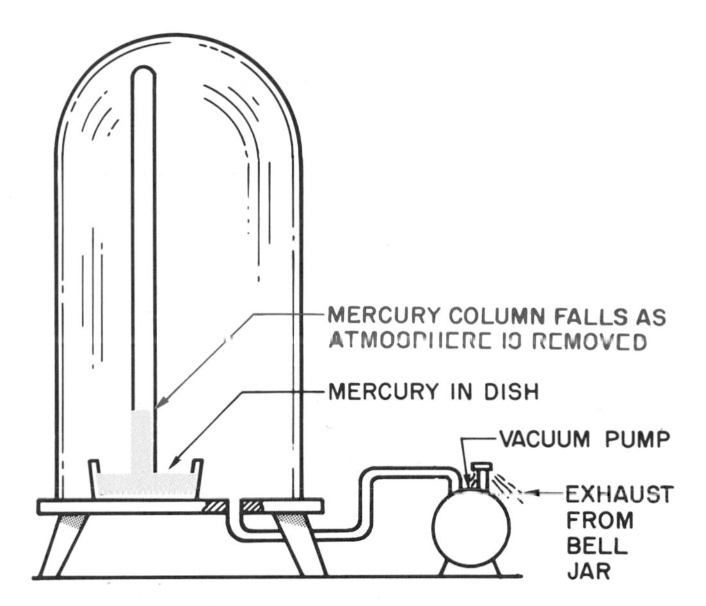

FIGURE 8–2A The mercury barometer in this bell jar illustrates how the atmosphere will support a column of mercury. As the atmosphere is removed from the jar, the column of mercury will begin to fall. If all of the atmosphere could be removed, the mercury would be at the bottom of the column.

ATMOSPHERIC PRESSURES, ABSOLUTE VALUES				COMPOUND GAGE READING	SATURATION POINTS of H_2O (BOILING—CONDENSING)
psia	in. Hg	mm Hg	microns	in. Hg VACUUM	°F
14.696	29.921	759.999	759,999	00.000	212.00
14.000	28.504	724.007	724,007	1.418	209.56
13.000	26.468	672.292	672,292	3.454	205.88
12.000	24.432	620.577	620,577	5.490	201.96
11.000	22.396	568.862	568,862	7.526	197.75
10.000	20.360	517.147	517,147	9.617	193.21
9.000	18.324	465.432	436,432	11.598	188.28
8.000	16.288	413.718	413,718	13.634	182.86
7.000	14.252	362.003	362,003	15.670	176.85
6.000	12.216	310.289	310,289	17.706	170.06
5.000	10.180	258.573	258,573	19.742	162.24
4.000	8.144	206.859	206,859	21.778	152.97
3.000	6.108	155.144	155,144	23.813	141.48
2.000	4.072	103.430	103,430	25.849	126.08
1.000	2.036	51.715	51,715	27.885	101.74
0.900	1.832	46.543	46,543	28.089	98.24
0.800	1.629	41.371	41,371	28.292	94.38
0.700	1.425	36.200	36,200	28.496	90.08
0.600	1.222	31.029	31,029	28.699	85.21
0.500	1.180	25.857	25,857	28.903	79.58
0.400	0.814	20.686	20,686	29.107	72.86
0.300	0.611	15.514	15,514	29.310	64.47
0.200	0.407	10.343	10,343	29.514	53.14
0.100	0.204	5.171	5,171	29.717	35.00
0.000	0.000	0.000	0.000	29.921	—

NOTE: psia × 2.035 966 = in. Hg psia × 51.715 = mm Hg psia × 51,715 = microns

FIGURE 8–2B Pressure and temperature relationships for water below atmospheric pressure.

connected by piping. These chambers include the cylinders of the compressor, which may have a reed valve partially sealing it from the system. When the atmosphere is removed from the bell jar, it is often called *pulling a vacuum.* When the noncondensables are removed from the refrigeration equipment, the process is often called *pulling a vacuum.* As the atmosphere is pulled out of the bell jar, the barometer inside the jar changes, Figure 8–2A. The standing column of mercury begins to drop. When the column drops down to 1 mm, only a small amount of the atmosphere is still in the jar (1/760 of the original volume since 760 mm = 29.92 in.). Figure 8–2B shows comparative scales for the saturation points of water.

8.3 MEASURING THE VACUUM

When the pressure in the bell jar is reduced to 1 mm Hg, the mercury column is hard to see, so another pressure measurement called the micron is used (1000 microns = 1 mm Hg). Microns are measured with electronic instruments.

Accurately measuring and proving a vacuum in the low micron range can be accomplished with an electronic instrument, such as a thermocouple or thermistor vacuum gage. Figure 8–3 shows a typical electronic vacuum gage. Several companies manufacture an electronic vacuum gage, so the choice may be made by asking a reliable supply house which gage seems to be the most popular.

The electronic vacuum gage is attached to the system so it can measure the pressure inside the system. It has a separate sensor that attaches to the system with wires connecting it to the instrument, Figure 8–4. The sensor portion of the instrument must never be exposed to system pressure, so it is advisable to have a valve between it and the system. The sensor may also be installed using gage lines where it is disconnected from the system before refrigerant is added, Figure 8–5. *The sensor should always be located in an upright position so that any oil in the system will not drain into the sensor,* Figure 8–6.

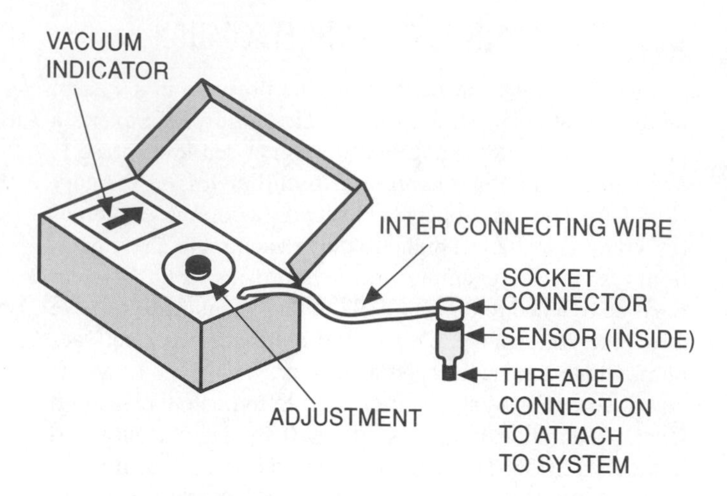

FIGURE 8–4 Components of electronic vacuum gage.

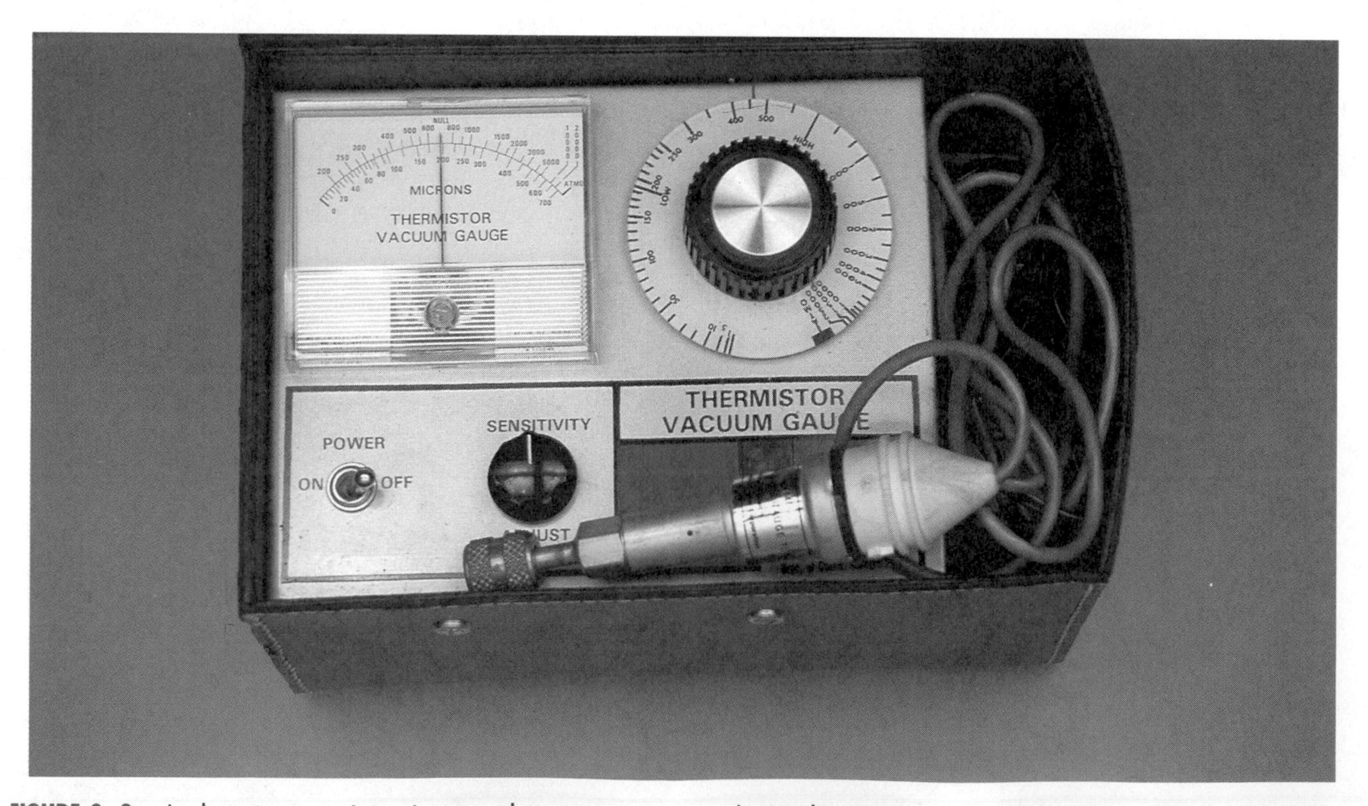

FIGURE 8–3 An electronic vacuum (micron) gage used to measure vacuums in the very low range. *Photo by Bill Johnson*

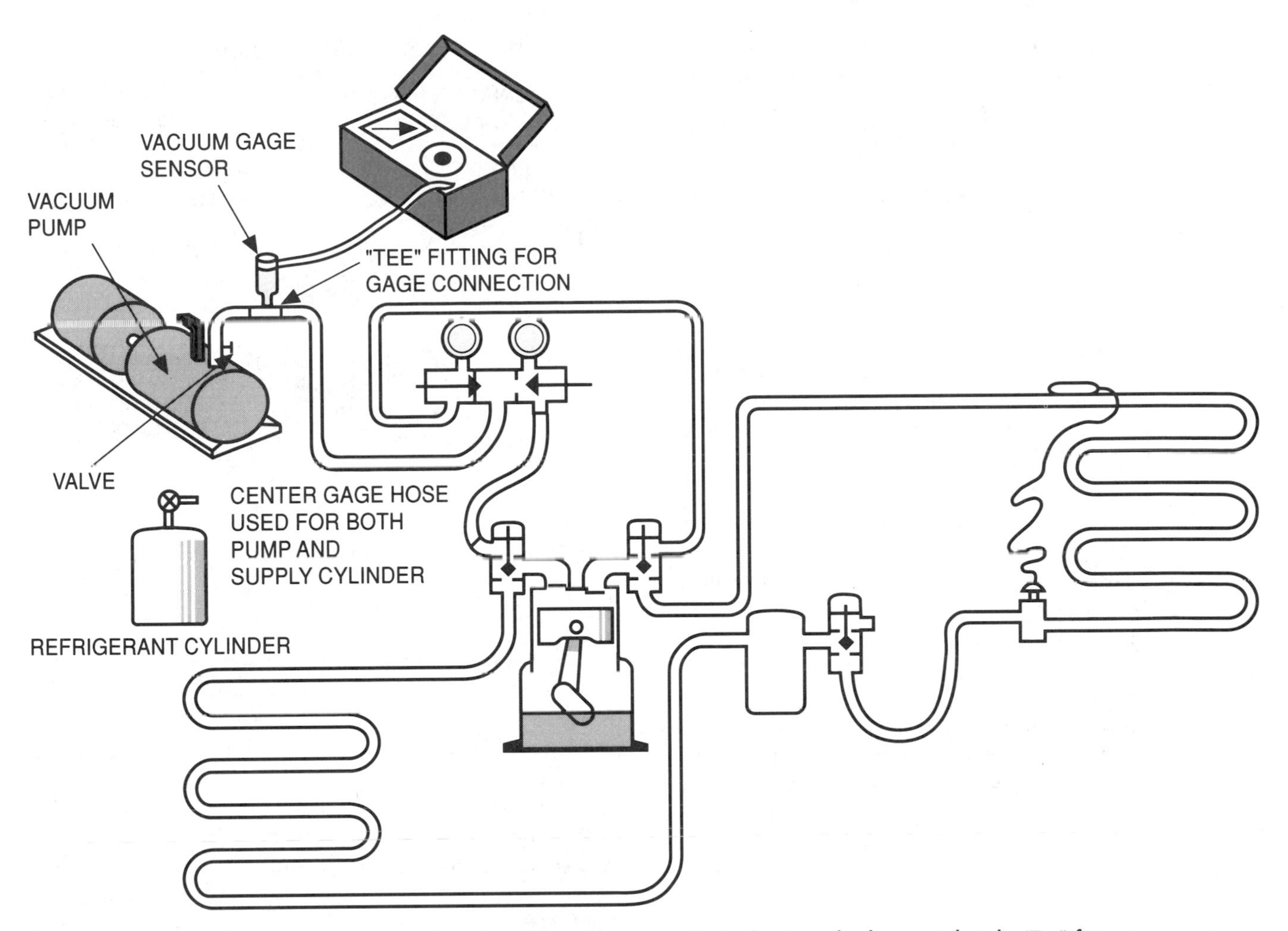

FIGURE 8–5 Electronic vacuum gage sensor is located near the vacuum pump. Notice that it can be disconnected at the "Tee" fitting to prevent pressure from entering the sensor.

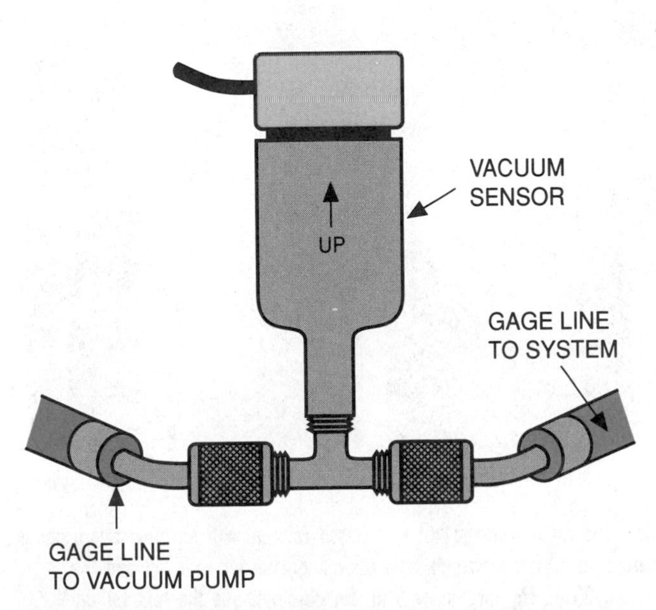

FIGURE 8–6 The electronic vacuum sensor must be mounted upright to prevent oil from entering the sensor.

When the electronic sensor is used, the vacuum pump should be operated for some period of time, then when the gage-manifold gage begins to drop, indicating a deep vacuum, the valve to the sensor should be opened and the instrument turned on. It is pointless to turn the instrument on until a fairly deep vacuum is achieved, some value below 25 in. Hg as indicated on the manifold gage.

When the vacuum gage reaches the vacuum desired, usually about 250 microns, the vacuum pump should then be valved off and the reading marked. The instrument reading may rise for a very short time, about 1 min, then it should stabilize. When the reading stabilizes, the true system reading can be recorded. If it continues to rise, either a leak or moisture is present and boiling to create pressure.

One of the advantages of the electronic vacuum gage is the rapid response to pressure rise. It will respond quickly to very small pressure rises. The smaller the system, the faster the rise. A very large system will take time to reach new pressure levels. These pressure differences may be seen instantly on the electronic instrument. *Be sure you allow for the first rise in pressure mentioned above before deciding a leak is present.*

Another vacuum gage often used in refrigeration work is the U-tube manometer, Figure 8–7(A), a glass gage closed on one side, which uses mercury as an indicator. The two columns of mercury balance each other. The atmosphere has been removed from one side of the mercury column so that the instrument has a standing column of about 5 in. Hg. This device can be used for fairly accurate readings down to about 1 mm. Since the columns of mercury are only about 5 in. different in height, this gage starts indicating at about 25 in. Hg vacuum below atmospheric pressure. When the gage is attached to the system and the vacuum pump is started, the gage will not read until the vacuum reaches about 25 in. Hg vacuum, Figure 8–7(B). Then the gage will gradually fall until the two columns of mercury are equal. At this time the vacuum in the system is between 1 mm Hg and a perfect vacuum. The instrument cannot be read much closer than that because the eye cannot see any better, Figure 8–7(C).

Using a special valve arrangement helps a technician check the system pressure and the vacuum pump's capability. This arrangement allows the technician to isolate the vacuum pump and the sensor by closing the valve closest to the system so just the sensor can be evacuated. If the vacuum pump cannot develop enough vacuum, it can be shown at this point. By closing the valve closest to the vacuum pump, the sensor and the system are isolated to check the system pressure, Figure 8–8. These tests can be useful to determine vacuum pump operation and system pressures for a standing pressure test.

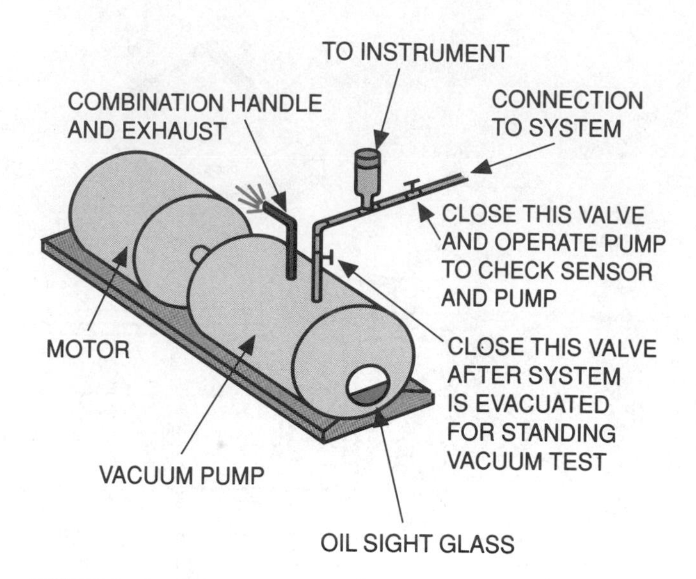

FIGURE 8–8 Valve arrangements for vacuum pump and sensor.

8.4 RECOVERING REFRIGERANT

Before evacuating a system to remove contaminants or noncondensable gases, if there is or has been refrigerant in the system, it must be removed. This must be done with EPA-approved recovery equipment. The amount of vacuum to be achieved in removing this refrigerant depends on the size of the system, the type of refrigerant, and whether or not the recovery equipment was manufactured before or after November 15, 1993. The recovery of refrigerant is discussed in more detail in Unit 9.

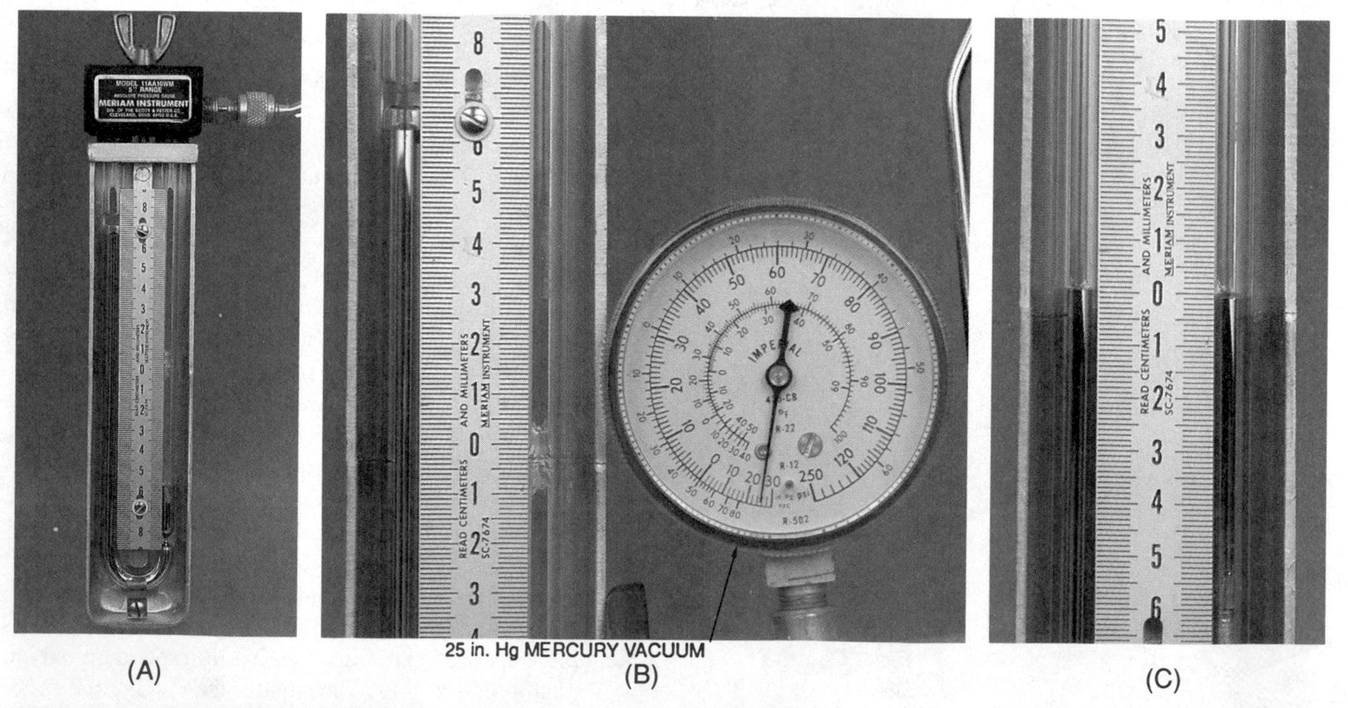

FIGURE 8–7 The mercury U-tube manometer is at various stages of evacuation. The column on the left is a closed column with no atmosphere above it. When the column on the right is connected to a system that is below atmosphere at a very low vacuum, the column on the left will fall and the column on the right will rise. This will not start until the vacuum on the right is about 25 in. Hg vacuum or 5 in. Hg absolute. See the text for an explanation of the difference. As the atmosphere is pulled out of the right hand column, the column rises. It will rise to be exactly parallel to the left hand column at a perfect vacuum. This is very hard to see. *Photos by Bill Johnson*

FIGURE 8–9 Two-stage rotary vacuum pump. *Photo by Bill Johnson*

8.5 THE VACUUM PUMP

A vacuum pump capable of removing the atmosphere down to a very low vacuum is necessary. The vacuum pumps usually used in the refrigeration field are manufactured with rotary compressors. The pumps that produce the lowest vacuums are two-stage rotary vacuum pumps, Figure 8–9. These vacuum pumps are capable of reducing the pressure in a leak-free vessel down to 0.1 micron. It is not practical to pull a vacuum this low in a field-installed system because the refrigerant oil in the system will boil slightly and create a vapor. The usual vacuum required by manufacturers is approximately 250 microns, although some may require a vacuum as low as 50 microns.

When moisture is in a system, a very low vacuum will cause the moisture to boil to a vapor. This vapor will be removed by the vacuum pump and exhausted to the atmosphere. Small amounts of moisture can be removed this way, but it is not practical to remove large amounts with a vacuum pump because of the large amount of vapor produced by the boiling water. For example, 1 lb of water (about a pint) in a system will turn to 867 ft^3 of vapor if boiled at 70°F.

8.6 DEEP VACUUM

The *deep vacuum* method involves reducing the pressure in the system to about 50 to 250 microns. When the vacuum reaches the desired level, the vacuum pump is valved off and the system is allowed to stand for some time period to see if the pressure rises. If the pressure rises and stops at some point, a material such as water is boiling in the system. If this occurs, continue evacuating. If the pressure continues to rise, there is a leak, and the atmosphere is seeping into the system. In this case the system should be pressured and leak checked again.

When a system pressure is reduced to 50 to 250 microns and the pressure remains constant, no noncondensable gas or moisture is left in the system. Reducing the system pressure to 250 microns is a slow process because when the vacuum pump pulls the system pressure to below about 5 mm (5000 microns), the pumping process slows down. The technician should have other work planned and let the vacuum pump run. Most technicians plan to start the vacuum pump as early as possible and finish other work while the vacuum pump does its work.

Some technicians leave the vacuum pump running all night because of the time involved in reaching a deep vacuum. Then the vacuum should be at the desired level the next morning. This is a good practice if some precautions are taken. When the vacuum pump pulls a vacuum, the system becomes a large volume of low pressure with the vacuum pump between this volume and the atmosphere, Figure 8–10. If the vacuum pump shuts off during the night from a power failure, it may lose its lubricating oil to the system by the vacuum in the system. If the power is restored and it starts back up, it will be without adequate lubrication

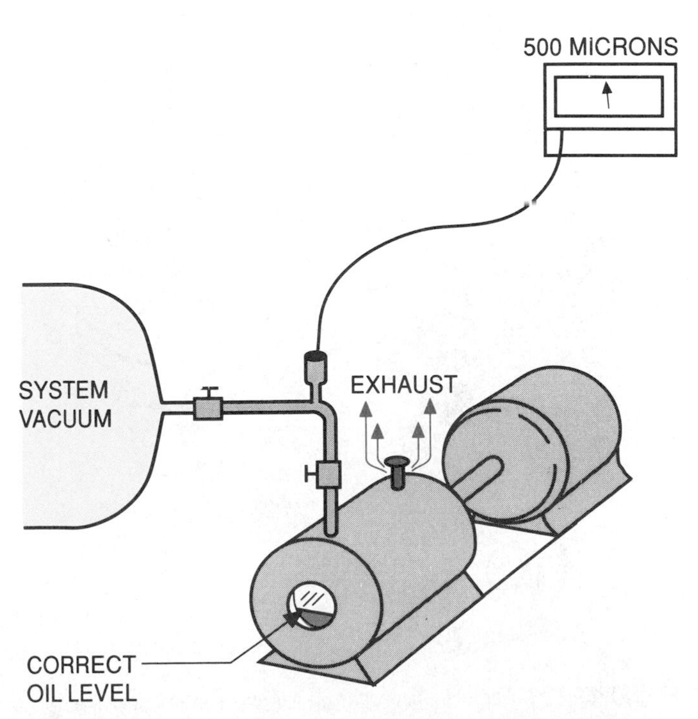

FIGURE 8–10 This vacuum pump has pulled a vacuum on a large system.

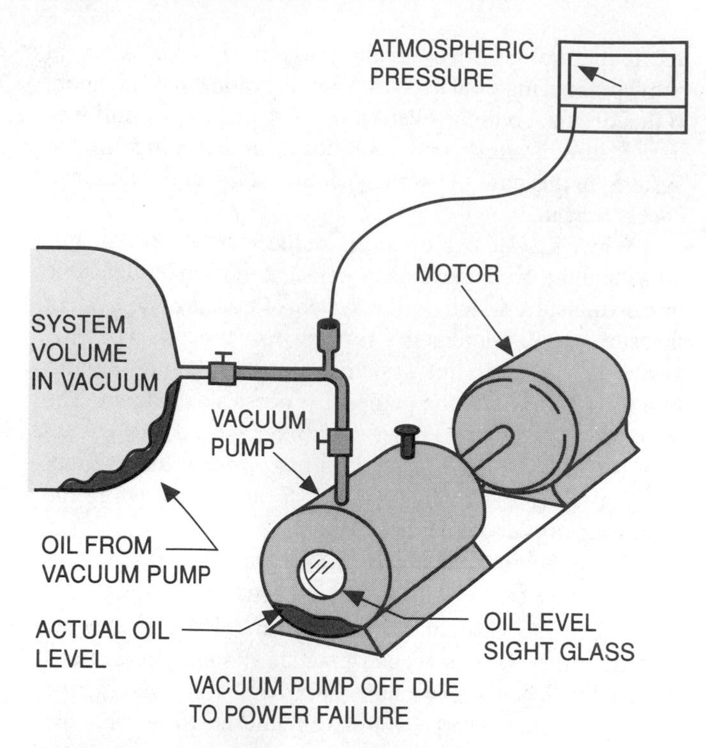

FIGURE 8–11 This system in a vacuum has pulled the oil out of the vacuum pump.

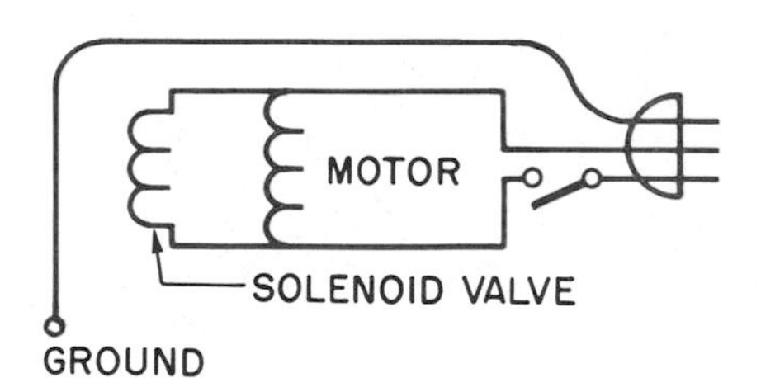

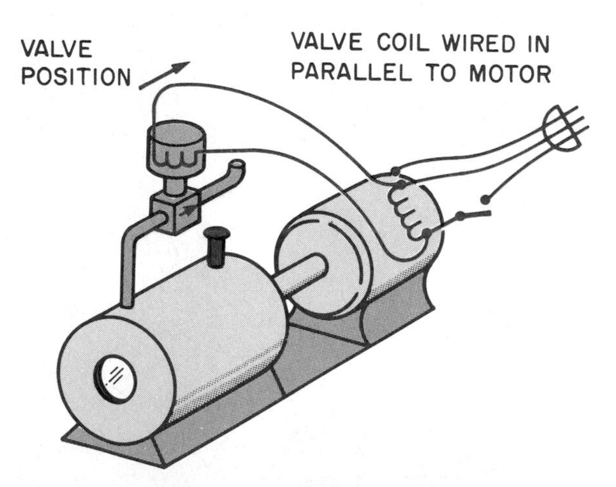

FIGURE 8–12 Vacuum pump with solenoid valve in the inlet line. Note the direction of the arrow on the solenoid valve. It is installed to prevent flow from the pump. It must be installed in this direction.

and could be damaged. The oil is pulled out of the vacuum pump by the large vacuum volume, Figure 8–11. This can be prevented by installing a large solenoid valve in the vacuum line entering the vacuum pump and wiring the solenoid valve coil in parallel with the vacuum pump motor. The solenoid valve should have a large port to keep from restricting the flow, Figure 8–12. This will be discussed in more detail later in this unit. Now, if the power fails, or if someone disconnects the vacuum pump (a good possibility at a construction site), the vacuum will not be lost, and the vacuum pump will not lose its lubrication.

8.7 MULTIPLE EVACUATION

Multiple evacuation is used by many technicians for removing the atmosphere to the lowest level of contamination. Multiple evacuation is normally accomplished by evacuating a system to a low vacuum, about 1 or 2 mm, and then allowing a small amount of refrigerant to bleed into the system. The system is then evacuated until the vacuum is again reduced to 1 mm Hg. The following is a detailed description of a multiple evacuation. This one is performed three times and called a *triple evacuation.* Figure 8–13 is a diagram of the valve arrangements.

1. Attach a U-tube mercury manometer to the system. The best place is as far from the vacuum port as possible. For example, on a refrigeration system, the pump may be attached to the suction and discharge service valves and the U-tube manometer to the liquid receiver valve port. Then start the vacuum pump.
2. Let the vacuum pump run until the manometer reaches 1 mm. The mercury manometer should be positioned vertically to take accurate readings. Lay a straightedge across the manometer to help determine the column heights, Figure 8–14.
3. Allow a small amount of refrigerant to enter the system until the vacuum is about 20 in. Hg. This must be indicated on the manifold gage because the mercury in the mercury gage will rise to the top and give no indication. Figure 8–15 shows the manifold gage reading. This small amount of refrigerant vapor will fill the system and absorb and mix with other vapors.
4. Open the vacuum pump valve and start the vapor moving from the system again. Let the vacuum pump run until the vacuum is again reduced to 1 mm Hg. Then repeat step 3.
5. When the refrigerant has been added to the system the second time, open the vacuum pump valve and again remove the vapor. Operate the vacuum pump for a long time during this third pull down. It is best to operate the vacuum pump until the manometer columns are equal. Some technicians call this *flat out.*
6. When the vacuum has been pulled the third time, allow refrigerant to enter the system until the system is about 5 psig above the atmosphere. Now remove the mercury manometer (it cannot stand system pressure) and charge the system.

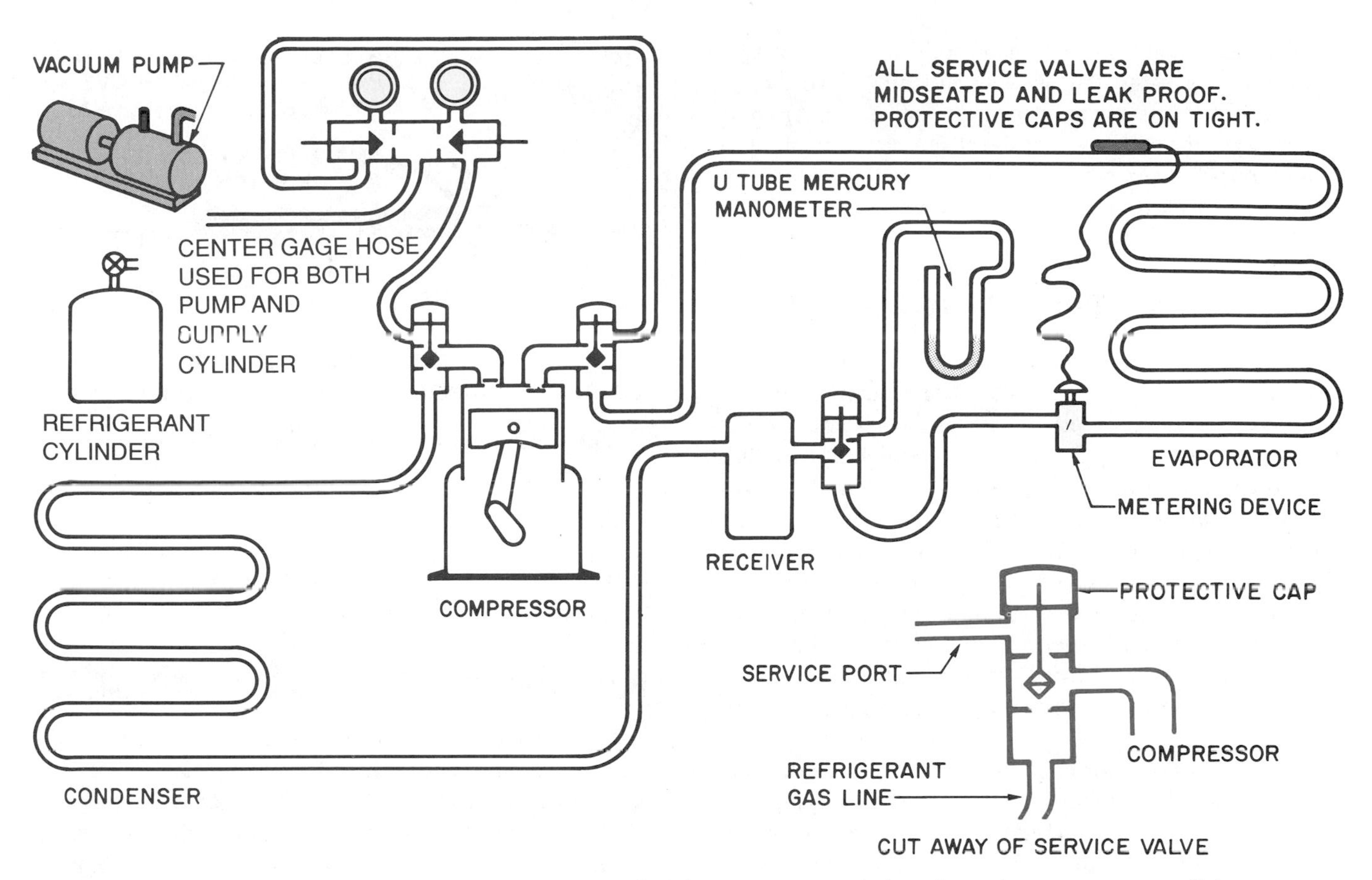

FIGURE 8–13 This system is ready for multiple evacuation. Notice the valve arrangements and where the U-tube manometer is installed.

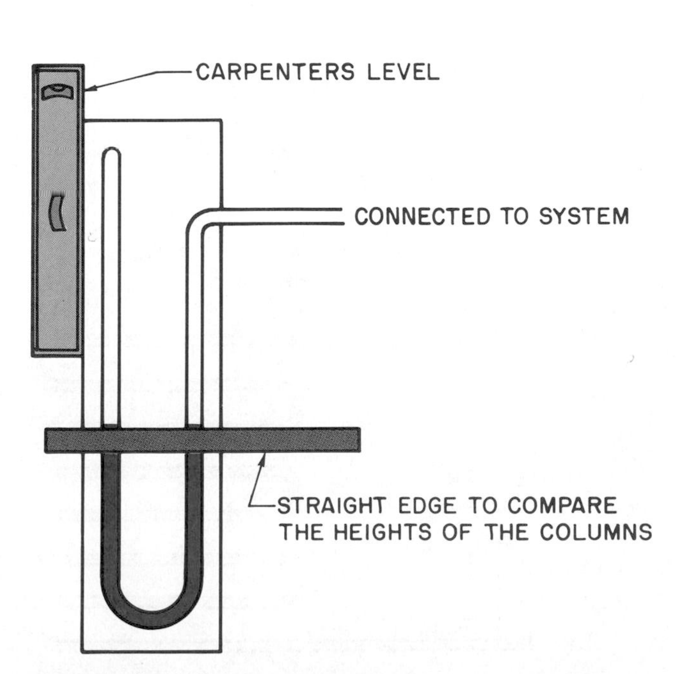

FIGURE 8–14 Mercury U-tube manometer being read as close as possible. The manometer is positioned vertically and a straightedge is used to compare the two columns of mercury.

FIGURE 8–15 Manifold gage reading 20 in. Hg vacuum. *Photo by Bill Johnson*

The electronic vacuum gage may be used for triple evacuation by using the above-mentioned valve arrangement to isolate the system between evacuations. Again, remember that the advantage of the electronic vacuum gage is that it responds to pressure changes in the system very quickly. If a leak is present, it will tell you much faster than the U-tube manometer.

8.8 LEAK DETECTION WHILE IN A VACUUM

We mentioned that if a leak is present in a system, the vacuum gage will start to rise, if the system is still in a vacuum, indicating a pressure rise in the system. The vacuum gage will rise very fast. ***Many technicians use this as an indicator that a leak is still in the system, but this is not a recommended leak test procedure.*** It allows air to enter the system, and the technician cannot determine from the vacuum where the leak is. Also, when a vacuum is used for leak checking it is only proving that the system will not leak under a pressure difference of 14.696 psi. If all of the atmosphere is removed from a system, only the atmosphere under atmospheric pressure is trying to get back into the system.

When checking for a leak using a vacuum, the technician is using a reverse pressure (the atmosphere trying to get into the system) of only 14.696 psi. The system may be operating under an operating pressure of 350 psig + 14.696 psi = 364.696 psia for an R-22 air-cooled condenser on a very hot day when fully loaded, Figure 8–16.

Using a vacuum for leak checking also may hide a leak. For example, if a pin-sized hole is in a solder connection that has a flux buildup on it, the vacuum will tend to pull the flux into the pinhole and may even hide it to the point that a deep vacuum can be achieved, Figure 8–17. Then when pressure is applied to the system, the flux will blow out of the pinhole, and a leak will appear.

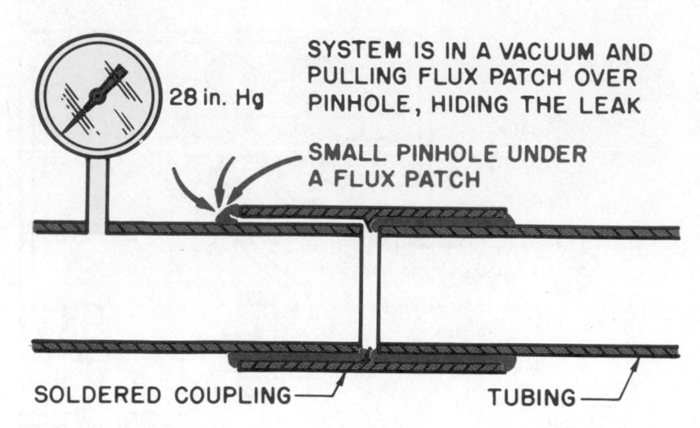

FIGURE 8–17 This system was leak checked under vacuum.

8.9 LEAK DETECTION—STANDING PRESSURE TEST

The best leak-checking procedure is a *standing pressure test* using a pressure source that will not change any appreciable amount with temperature changes. Nitrogen is a good gas to use for this. ***Never use air. Air mixed with some refrigerants under pressure can create an explosive mixture.*** To perform this test, put a small amount of refrigerant in the system for leak-checking purposes (up to about 10 psig) with the nitrogen. Using R-22 as a trace refrigerant for leak-checking purposes is acceptable by the EPA. The small amount of refrigerant can be detected with any common leak detector. ***Do not pressurize any system above its working pressure written on the label of the equipment.***

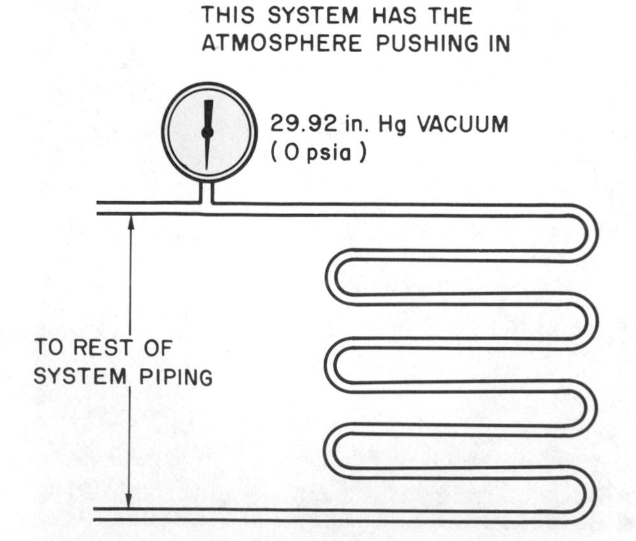

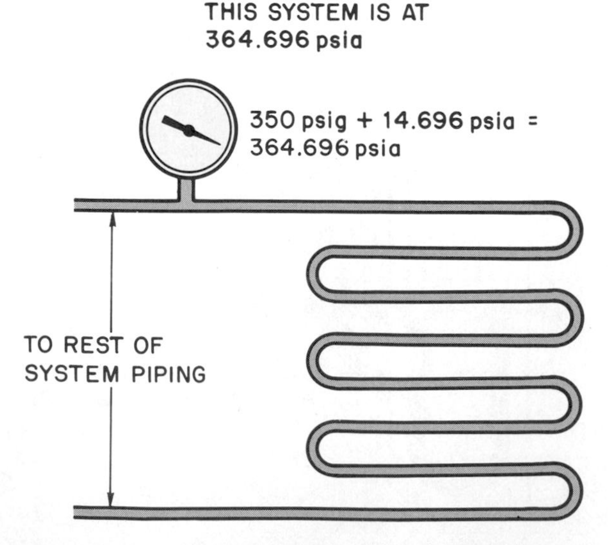

FIGURE 8–16 These two systems are being compared to each other under different pressure situations. One is evacuated and has the atmosphere under atmospheric pressure trying to get into the system. The other has 350 psig + 14.696 = 364.696 psia. The one under the most pressure is under the most stress. Using a vacuum as a leak test does not give the system a proper leak test.

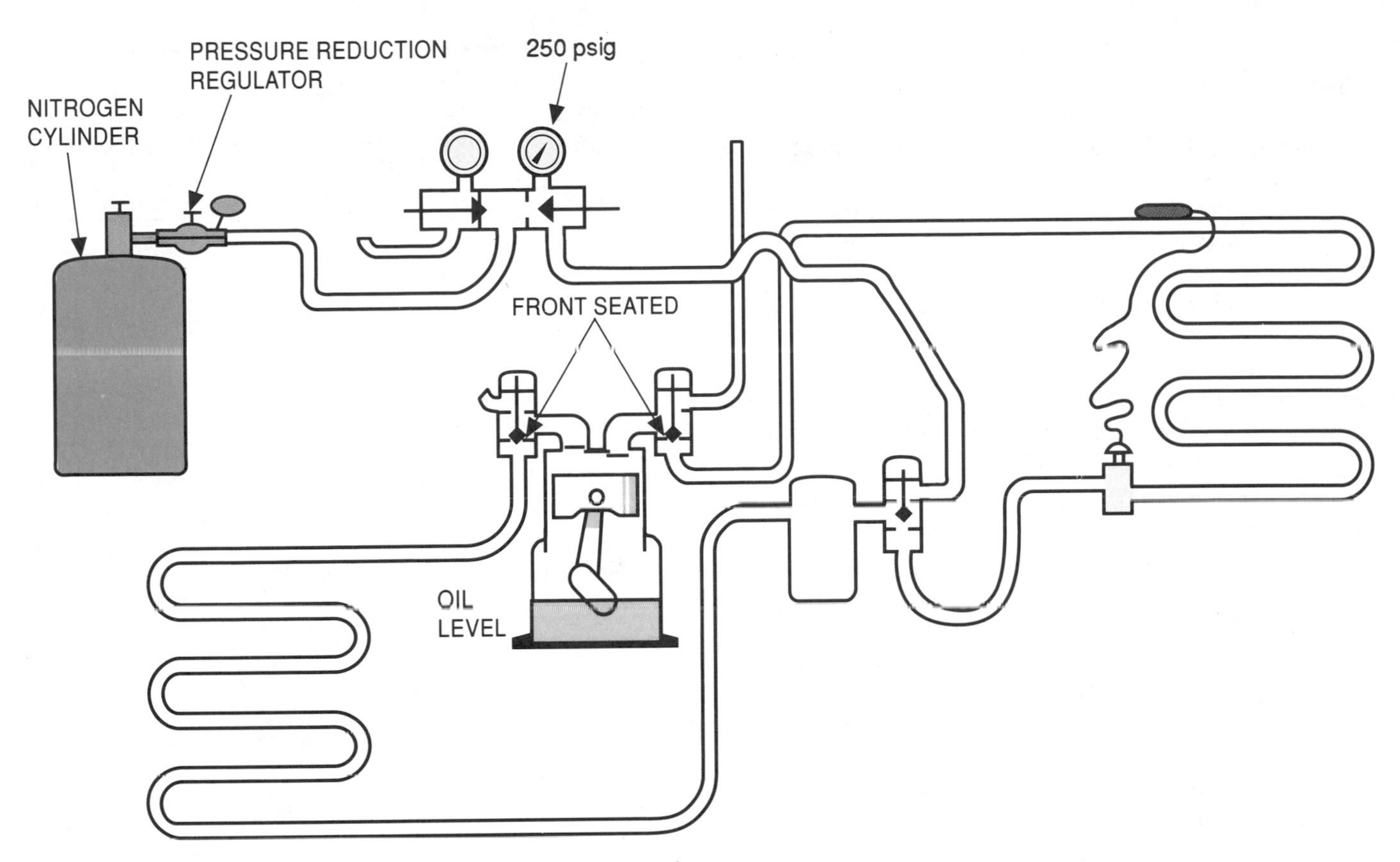

FIGURE 8–18 Isolating the compressor to pressurize the condenser and evaporator.

No manufacturer uses a working pressure lower than 150 psig with the high-pressure refrigerants discussed in this book (R-12, R-22, and R-502). Therefore, we can assume that 150 psig is safe for any system using the common refrigerants. Actually, the compressor has the lowest working pressure of any component in the system. Compressor manufacturers' specifications show a working pressure of 150 psig. If more pressure is desired for a pressure test, the compressor may be isolated and the component may be pressure tested to a higher value, Figure 8–18. The condenser will have a higher working pressure than the evaporator in a typical system. It is often practical to pressure test the piping, condenser, or evaporator to a higher pressure than the compressor.

The common test pressure is 150 psig. When the system pressure is up to 150 psig, tap the gage slightly to make sure the needle is free, and make a mark, Figure 8–19. Let the system stand at this pressure while leak checking. When the leak check is complete, observe the gage reading again. If it has fallen, there is a leak. Do not forget that the gage manifold and connections may leak. When the system is leak checked and no drop is found in the gage, let the system stand for a while. The smaller the system, the shorter the standing time needed. For example, a small beverage cooler may need to stand for only an hour to be sure that the system is leak free, whereas a 20-ton system may need to stand under pressure for 12 h. If the standing time is long, you will have more assurance that no leak is present.

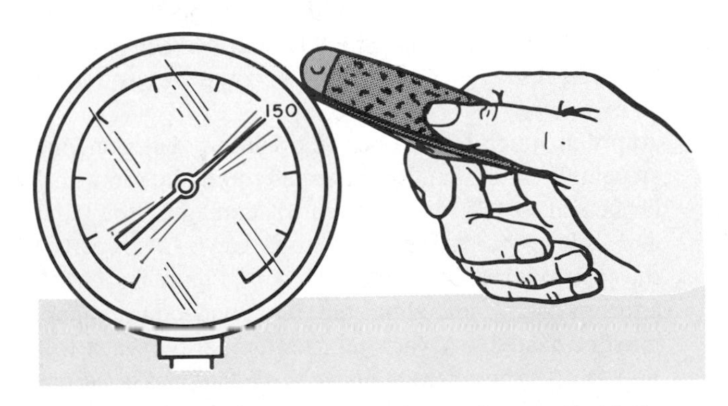

FIGURE 8–19 When using a pressure gage for a standing leak test, tap the gage lightly to make sure the needle is free; then mark the gage.

8.10 REMOVING MOISTURE WITH A VACUUM

Removing moisture with a vacuum is the process of using the vacuum pump to remove moisture from a refrigeration system. There are two kinds of moisture in the system, vapor and liquid. When the moisture is in the vapor state, it is easy to remove. When it is in the liquid state, it is much more difficult to remove. The example earlier in this unit shows that 867 ft³ of vapor at 70°F must be pumped to remove 1 lb of water. This is not a complete explanation because as the vacuum pump begins to remove the moisture, the water will boil and the temperature of the trapped water will drop. For example, if the water temperature drops to 50°F, 1 lb of water will then boil to 1702 ft³ of vapor that must be removed. This is a pressure level in the system of 0.362 in. Hg or 9.2 mm Hg (0.362 X 25.4 mm/in.). The vacuum level is just reaching the low ranges. As the vacuum pump pulls lower, the water will boil more (if the vacuum pump has the capacity to pump this much vapor), and the temperature will decrease to 36°F. The water will now create a vapor volume of 2837 ft³. This is a vapor pressure in the system of 0.212 in. Hg or 5.4 mm Hg (0.212 × 25.4 mm/in.). This illustrates that lowering the pressure level creates more vapor. It takes a large vacuum pump to pull moisture out of a system. (See Figure 8–20 for the relationships between temperature, pressure, and volume.)

If the system pressure is reduced further, the water will turn to ice and be even more difficult to remove. If large amounts of moisture must be removed from a system with a vacuum pump, the following procedure will help.

TEMPERATURE		SPECIFIC VOLUME OF WATER VAPOR	ABSOLUTE PRESSURE		
°C	°F	ft³/lb	lb/in.²	kPa	in. Hg
−12.2	10	9054	0.031	0.214	0.063
−6.7	20	5657	0.050	0.345	0.103
−1.1	30	3606	0.081	0.558	0.165
0.0	32	3302	0.089	0.613	0.180
1.1	34	3059	0.096	0.661	0.195
2.2	36	2837	0.104	0.717	0.212
3.3	38	2632	0.112	0.772	0.229
4.4	40	2444	0.122	0.841	0.248
5.6	42	2270	0.131	0.903	0.268
6.7	44	2111	0.142	0.978	0.289
7.8	46	1964	0.153	1.054	0.312
8.9	48	1828	0.165	1.137	0.336
10.0	50	1702	0.178	1.266	0.362
15.6	60	1206	0.256	1.764	0.522
21.1	70	867	0.363	2.501	0.739
26.7	80	633	0.507	3.493	1.032
32.2	90	468	0.698	4.809	1.422
37.8	100	350	0.950	6.546	1.933
43.3	110	265	1.275	8.785	2.597
48.9	120	203	1.693	11.665	3.448
54.4	130	157	2.224	15.323	4.527
60.0	140	123	2.890	19.912	5.881
65.6	150	97	3.719	25.624	7.573
71.1	160	77	4.742	32.672	9.656
76.7	170	62	5.994	41.299	12.203
82.2	180	50	7.512	51.758	15.295
87.8	190	41	9.340	64.353	19.017
93.3	200	34	11.526	79.414	23.468
98.9	210	28	14.123	97.307	28.754
100.0	212	27	14.696	101.255	29.921

FIGURE 8–20 This partial pressure/temperature relationship table for water shows the specific volume of water vapor that must be removed to remove a pound of water from a system.

1. Use a large vacuum pump. If the system is flooded, for example if a water-cooled condenser pipe ruptures from freezing, a 5-cfm vacuum pump is recommended for systems up to 10 tons. If the system is larger, a larger pump or a second pump should be used.
2. Drain the system in as many low places as possible. Remove the compressor and pour the water and oil from the system. **Do not add the oil back until the system is ready to be started, after evacuation. If you add it earlier, the oil may become wet and hard to evacuate.**
3. Apply as much heat as possible without damaging the system. If the system is in a heated room, the room may be heated to 90°F without fear of damaging the room and its furnishings or the system, Figure 8–21. If part of the system is outside, use a heat lamp, Figure 8–22. The entire system, including the interconnecting piping, must be heated to a warm temperature, or the water will boil to a vapor where the heat is applied and condense where the system is cool. For example, if you know water is in the evaporator inside the structure and you apply heat to the evaporator, the water will boil to a vapor. If it is cool outside, the water vapor may condense outside in the condenser piping. The water is only being moved around.
4. Start the vacuum pump and observe the oil level in it. As moisture is removed, some of it will condense in the vacuum pump crankcase. Some vacuum pumps have a feature called *gas ballast* that introduces some atmosphere between the first and second stages of the two-stage pump. This prevents some of the moisture from condensing in the crankcase. Regardless of the vacuum pump, watch the oil level. ***The water will displace the oil and raise the oil out of the pump. Soon, water may be the only lubricant in the vacuum pump crankcase and damage may occur to the vacuum pump. They are *very* expensive and should be protected.***

8.11 GENERAL EVACUATION PROCEDURES

Some general rules apply to deep vacuum and multiple evacuation procedures. If the system is large enough or if you must evacuate the moisture from several systems, you can construct a cold trap to use in the field. The *cold trap* is a refrigerated volume in the vacuum line between the wet system and the vacuum pump. When the water vapor passes through the cold trap, the moisture freezes to the walls of the trap, which is normally refrigerated with dry ice (CO_2), a commercially available product. The trap is heated, pres-

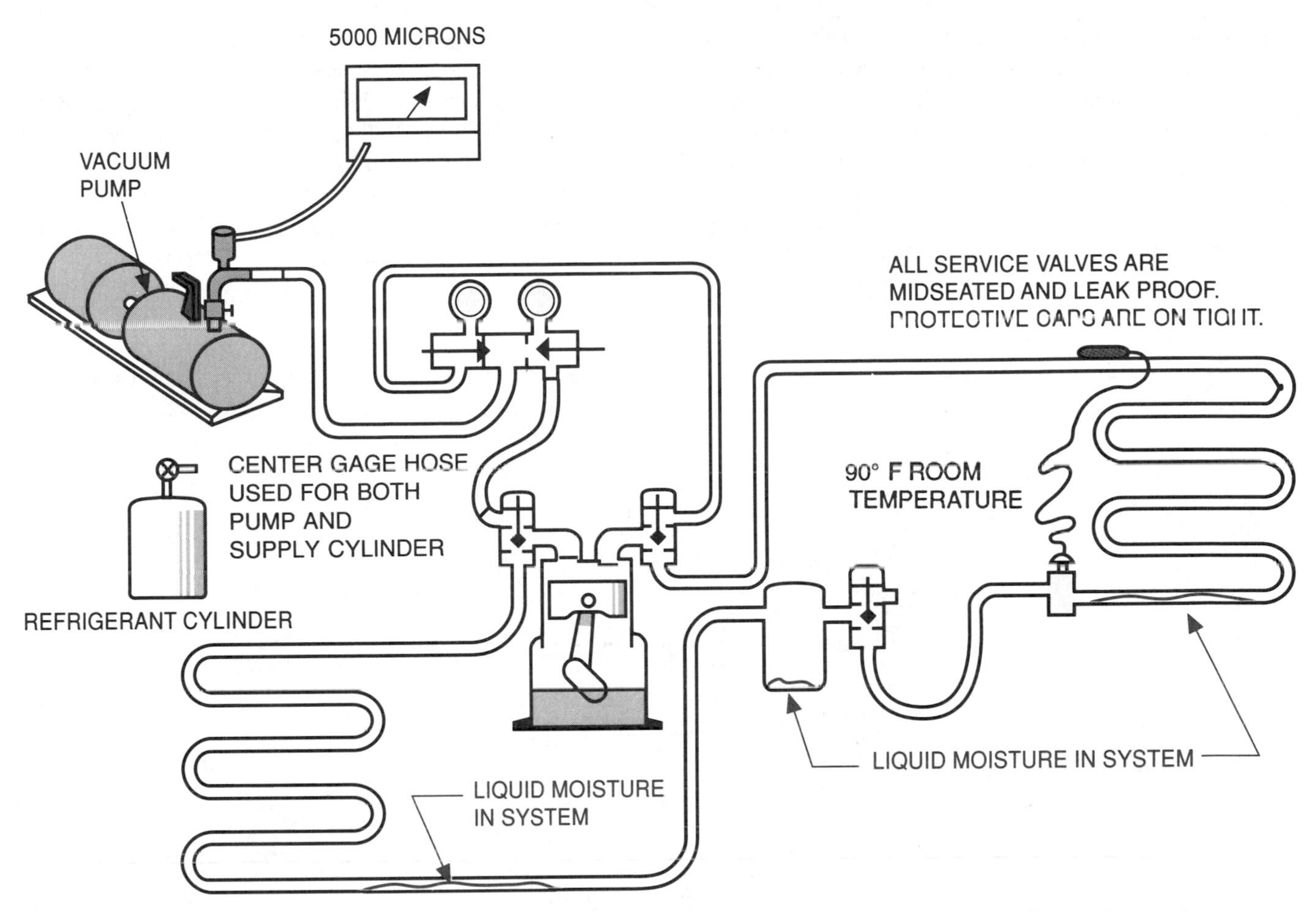

FIGURE 8–21 When a system has moisture in it and is being evacuated, heat may be applied to the system. This will cause the water to turn to vapor, and the vacuum pump will remove it.

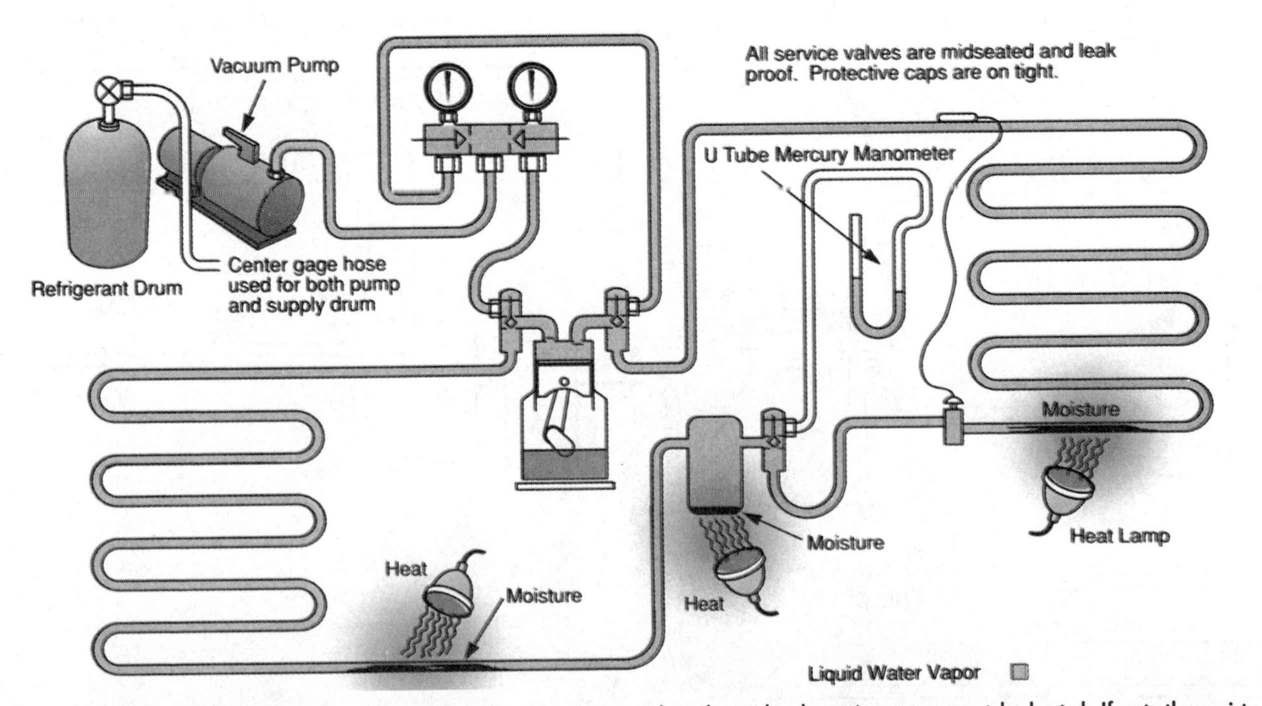

FIGURE 8–22 When heat is supplied to a large system with components inside and outside, the entire system must be heated. If not, the moisture will condense where the system is cool.

surized, and drained periodically to remove the moisture, Figure 8–23. The cold trap can save a vacuum pump.

Noncondensable gases and moisture may be trapped in a compressor and is as difficult to release as a vapor that can be pumped out of the system. A compressor has small chambers, such as cylinders, that may contain air or moisture. Only the flapper valves are setting on top of these chambers, but there is no reason for the air or water to move out of the cylinder while it is under a vacuum. At times it is advisable to start the compressor after a vacuum has been tried. This is easy to do with the triple evacuation method. When the first vacuum has been reached, refrigerant can be charged into the system until it reaches atmospheric pressure. The compressor can then be started for a few seconds. All chambers should be flushed at this time. ***Do not start a hermetic compressor while it is in a deep vacuum. Motor damage may occur.*** Figure 8–24 is an example of vapor trapped in the cylinder of a compressor.

Water can be trapped in a compressor under the oil. The oil has surface tension, and the moisture may stay under it even under a deep vacuum. During a deep vacuum, the oil surface tension can be broken with vibration, such as striking the compressor housing with a soft face hammer. Any kind of movement that causes the oil's surface to shake will work, Figure 8–25. Applying heat to the compressor crankcase will also release the water, Figure 8–26.

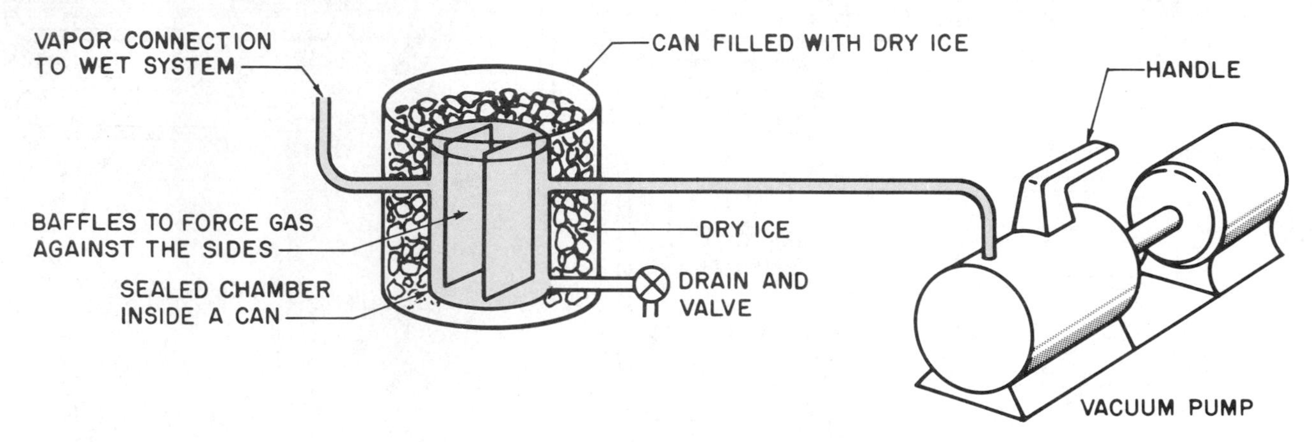

FIGURE 8–23 A cold trap.

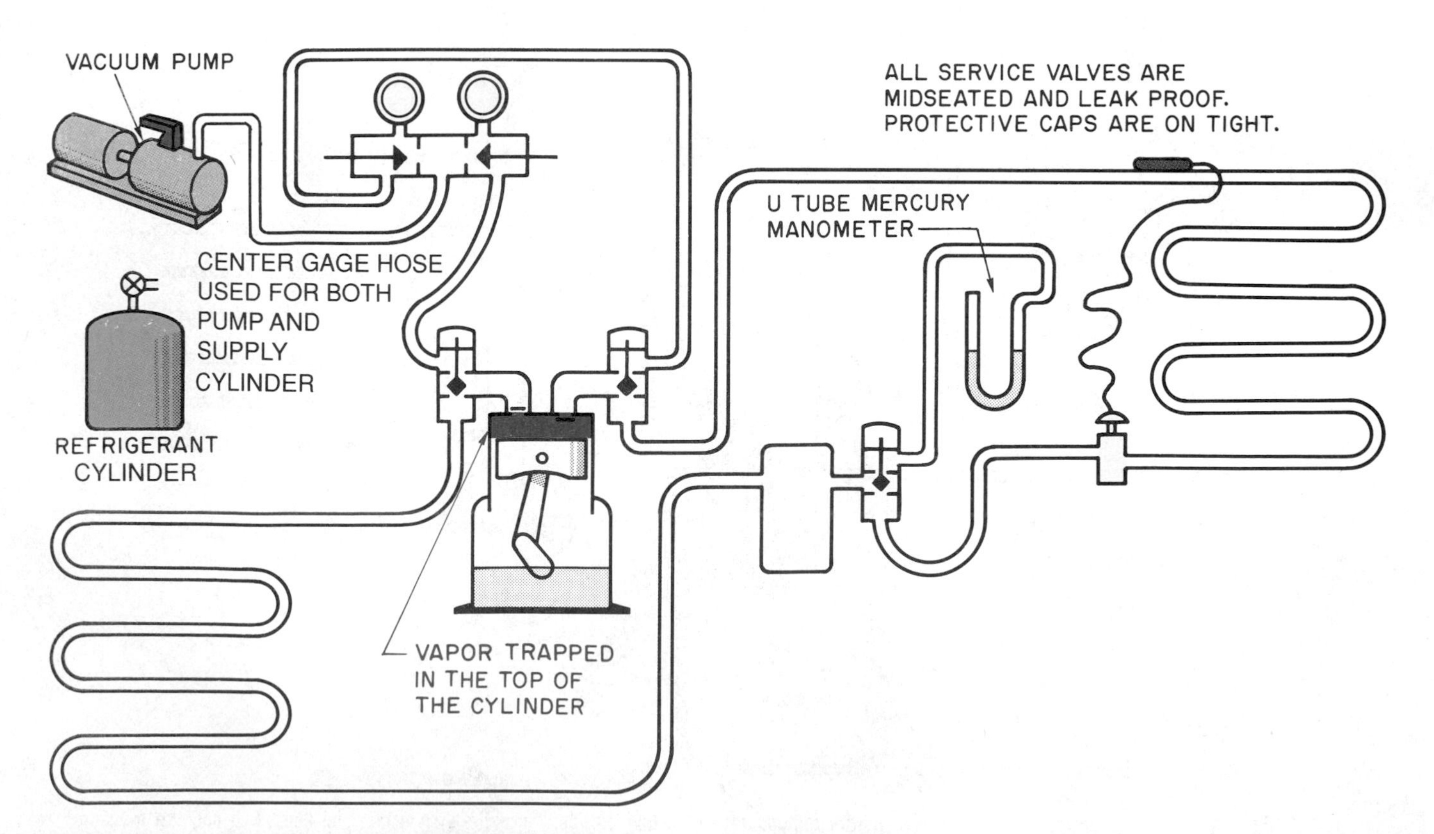

FIGURE 8–24 Vapor trapped in the cylinder of the compressor.

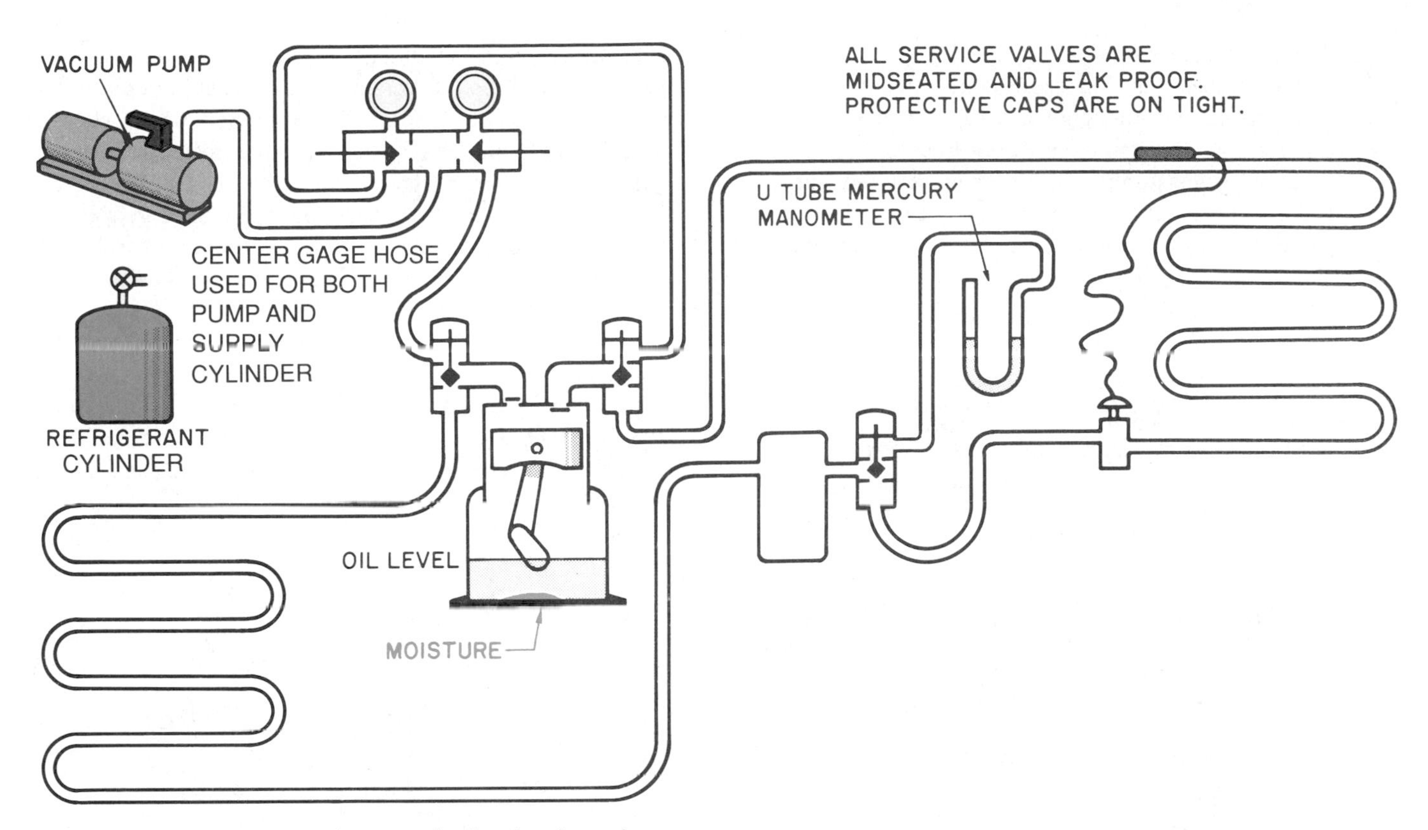

FIGURE 8–25 Compressor with water under the oil in the crankcase.

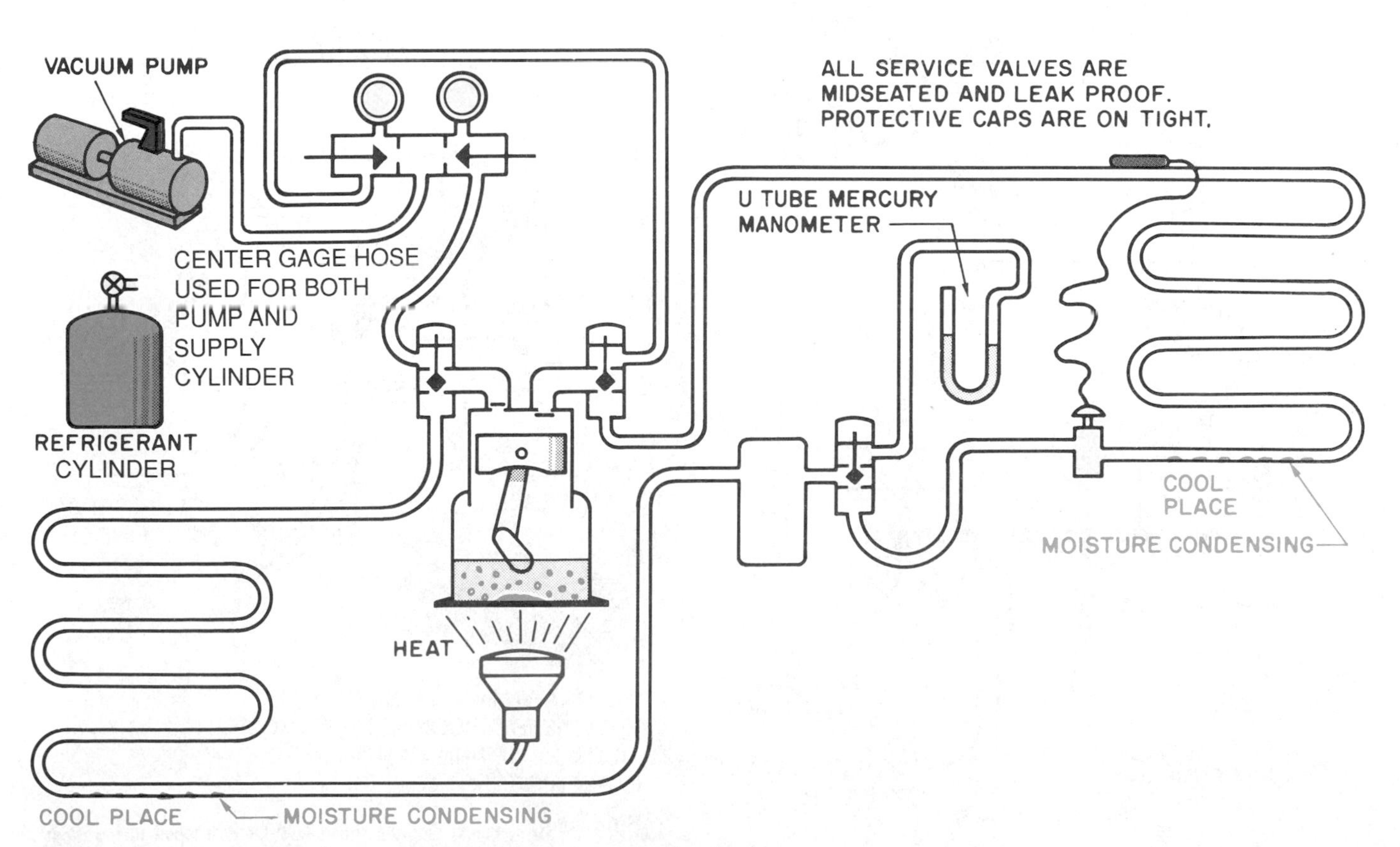

FIGURE 8–26 Heat is applied to compressor to boil the water under the oil.

The technician who evacuates many systems must use timesaving procedures. For example, a typical gage manifold may not be the best choice because it has very small valve ports that slow the evacuation process, Figure 8–27. However, some gage manifolds are manufactured with large valve ports and a special large hose for the vacuum pump connection, Figure 8–28. The gage manifold in Figure 8–29 has four valves and four hoses. The extra two valves are used to control the refrigerant and the vacuum pump lines. When using this manifold, you need not disconnect the vacuum pump and switch the hose line to the refrigerant cylinder to charge refrigerant into the system. When the time comes to stop the evacuation and charge refrigerant into the system, close one valve and open the other, Figure 8–29. This is a much easier and cleaner method of changing from the vacuum line to the refrigerant line.

When a gage line is disconnected from the vacuum pump, air is drawn into the gage hose. This air must be purged from the gage hose at the top, near the manifold. It is impossible to get all of the air out of the manifold because some will be trapped and pushed into the system, Figure 8–30.

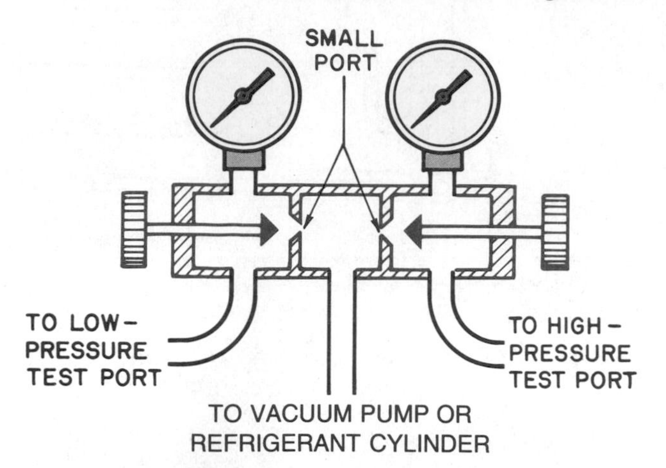

FIGURE 8–27 Gage manifold with small ports.

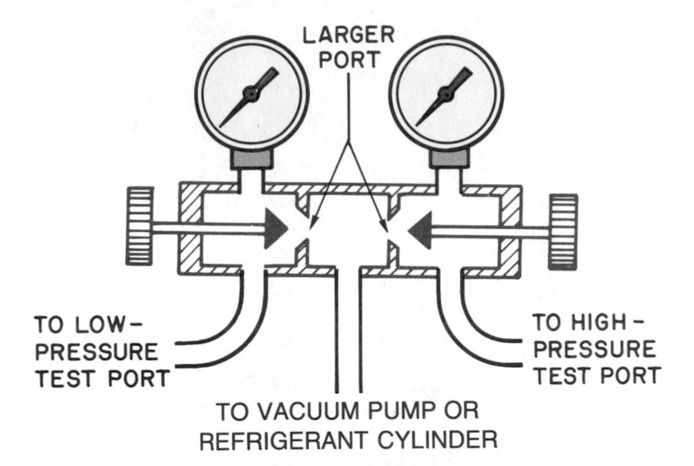

FIGURE 8–28 Gage manifold with large ports.

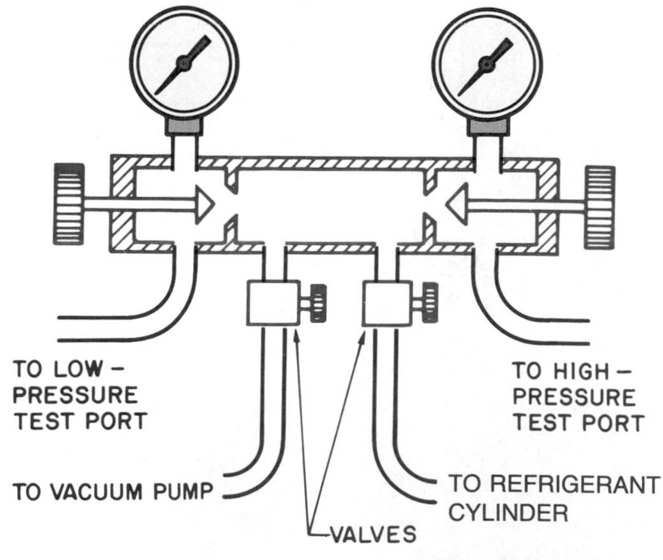

FIGURE 8–29 Manifold with four valves and four gage hoses.

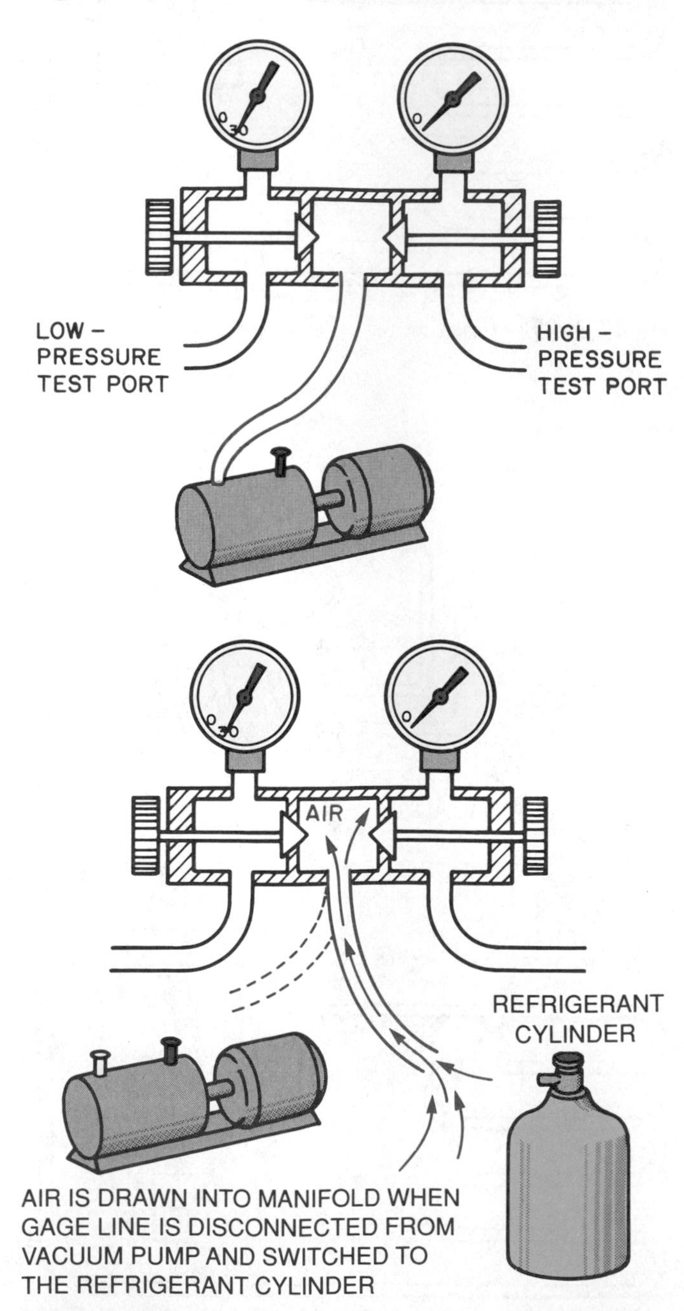

FIGURE 8–30 Piping diagram shows how air is trapped in the gage manifold.

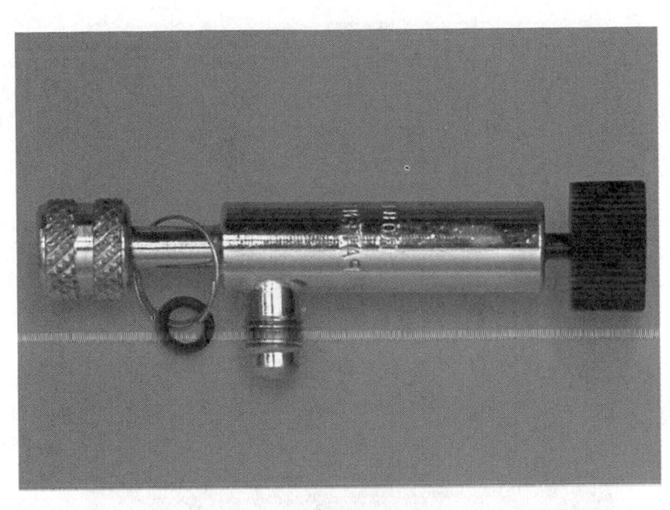

FIGURE 8–31 This gage adapter can be used instead of the gage depressors that are normally in the end of gage lines. The adapters may be used for gage readings. *Photo by Bill Johnson*

FIGURE 8–32 This small valve can also be used for controlled gage readings. When the technician wants to read a pressure in a Schrader port, this adapter valve may be used by turning the valve handle down. *Photo by Bill Johnson*

Most gage manifolds have valve stem depressors in the ends of the gage hoses. The depressors are used for servicing systems with Schrader access valves. These valves are much like the valve and stem on an automobile tire. These depressors are a restriction to the evacuation process. When a vacuum pump pulls down to the very low ranges (1 mm), these valve depressors slow the vacuum process considerably. Many technicians erroneously use oversized vacuum pumps and undersized connectors because they do not realize that the vacuum can be pulled much faster with larger connectors. The valve depressors can be removed from the ends of the gage hoses, and adapters can be used when valve depression is needed. Figure 8–31 shows one of these adapters. Figure 8–32 is a small valve that can be used on the end of a gage hose; it will even give the technician the choice as to when the valve stem is depressed.

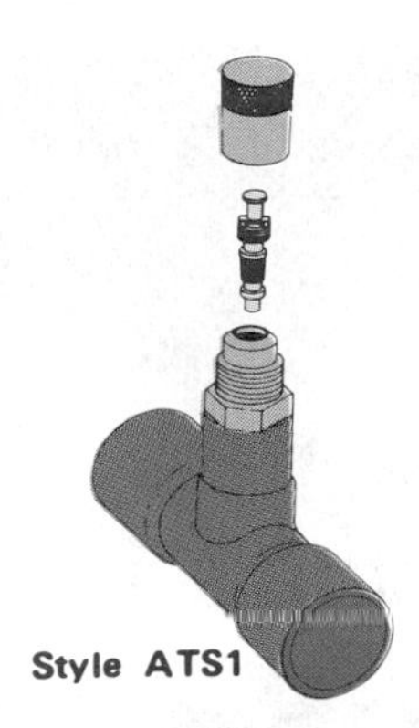

FIGURE 8–33 Schrader valve assembly. *Courtesy J/B Industries*

8.12 SYSTEMS WITH SCHRADER VALVES

A system with Schrader valves for gage ports will take much longer to evacuate than a system with service valves. The reason is that the valve stems and the depressors act as very small restrictions. An alternative is to remove the valve stems during evacuation and replace them when evacuation is finished. A system with water to be removed will take much time to evacuate if there are Schrader valve stems in the service ports. These valve stems are designed to be removed for replacement, so they can also be removed for evacuation, Figure 8–33.

A special tool, called a *field service valve,* can be used to replace Schrader valve stems under pressure, or it can be used as a control valve during evacuation. The tool has a valve arrangement that allows the technician to evacuate a system through it with the stem backed out of the Schrader valve. The stem is replaced when evacuation is completed.

Schrader valves are shipped with a special cap, which is used to cover the valve when it is not in use. This cover has a soft gasket and should be the only cover used for Schrader valves. If a standard brass flare cap is used and overtightened, the Schrader valve top will be distorted and valve stem service will be very difficult, if it can be done at all.

8.13 GAGE MANIFOLD HOSES

The standard gage manifold uses flexible hoses with connectors on the ends. These hoses sometimes get pinhole leaks, usually around the connectors, that may leak while under a vacuum but not be evident when the hose has pressure inside it. The reason is that the hose swells when pressurized. If you have trouble while pulling a vacuum and you

FIGURE 8–34 Gage manifold with copper gage lines used for evacuation. *Photo by Bill Johnson*

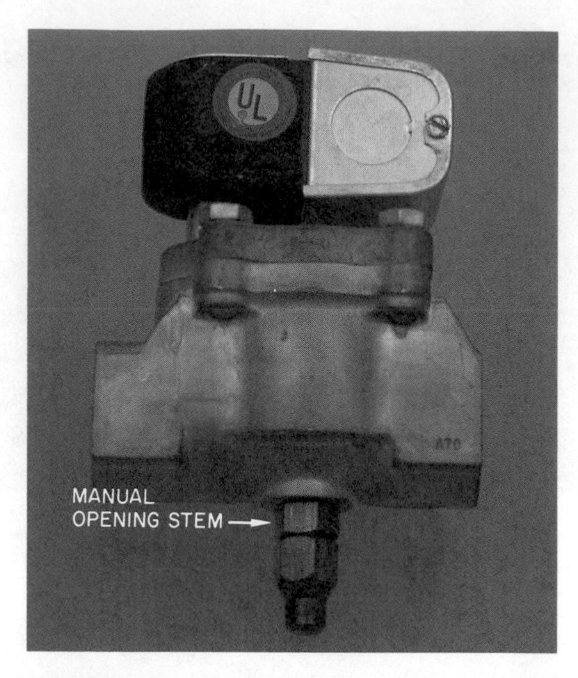

FIGURE 8–36 Manual opening stem for a solenoid valve. *Photo by Bill Johnson*

can't find a leak, substitute soft copper tubing for the gage lines, Figure 8–34.

8.14 SYSTEM VALVES

For a system with many valves and piping runs, perhaps even multiple evaporators, check the system's valves to see if they are open before evacuation. A system may have a closed solenoid valve. The valve may trap air in the liquid line between the expansion valve and the solenoid valve, Figure 8–35. This valve must be opened for complete evacuation. It may even need a temporary power supply to operate its magnetic coil. Some solenoid valves have a screw on the bottom to manually jack the valve open, Figure 8–36.

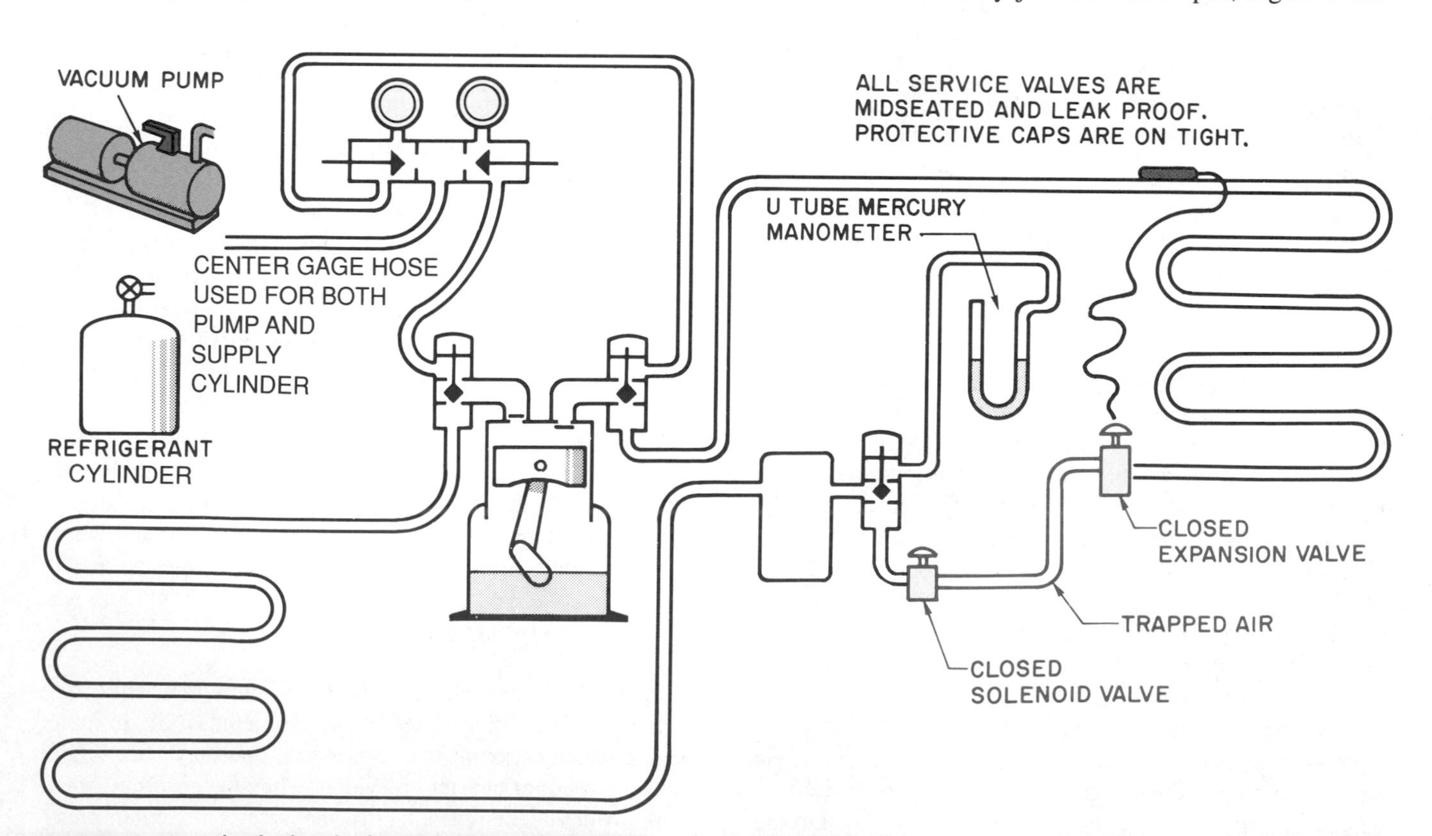

FIGURE 8–35 Closed solenoid valve trapping air in the system liquid line.

8.15 USING DRY NITROGEN

Good workmanship practices while assembling or installing a system can make system evacuation an easier task. When piping is field installed, sweeping dry nitrogen through the refrigerant lines can keep the atmosphere pushed out and clean the pipe. It is relatively inexpensive to use a dry nitrogen setup, and using it saves time and money.

When a system has been open to the atmosphere for some time, it needs evacuation. The task can be quickened by sweeping the system with dry nitrogen before evacuation. Figure 8–37 shows how this is done.

8.16 CLEANING A DIRTY SYSTEM

The technician should be aware of how to use the vacuum to clean a dirty system. Actually several types of contaminants may be in a system. Water and air have been discussed, but they are not the only contaminants that can form in a system. The hermetic motor inside a sealed system is the source of heat in a motor burn circumstance. This heat source can heat the refrigerant and oil to temperatures that will break down the oil and refrigerants to acids, soot (carbon), and sludge that cannot be removed with a vacuum pump. Let's use a bad motor burn example to demonstrate how most manufacturers would expect you to clean a system.

Suppose a 5-ton air conditioning system with a fully hermetic compressor were to have a severe motor burn while running. When a motor burn occurs while the compressor is running, the soot and sludge from the hot oil move into the condenser, Figure 8–38. The following steps will be taken to clean the system.

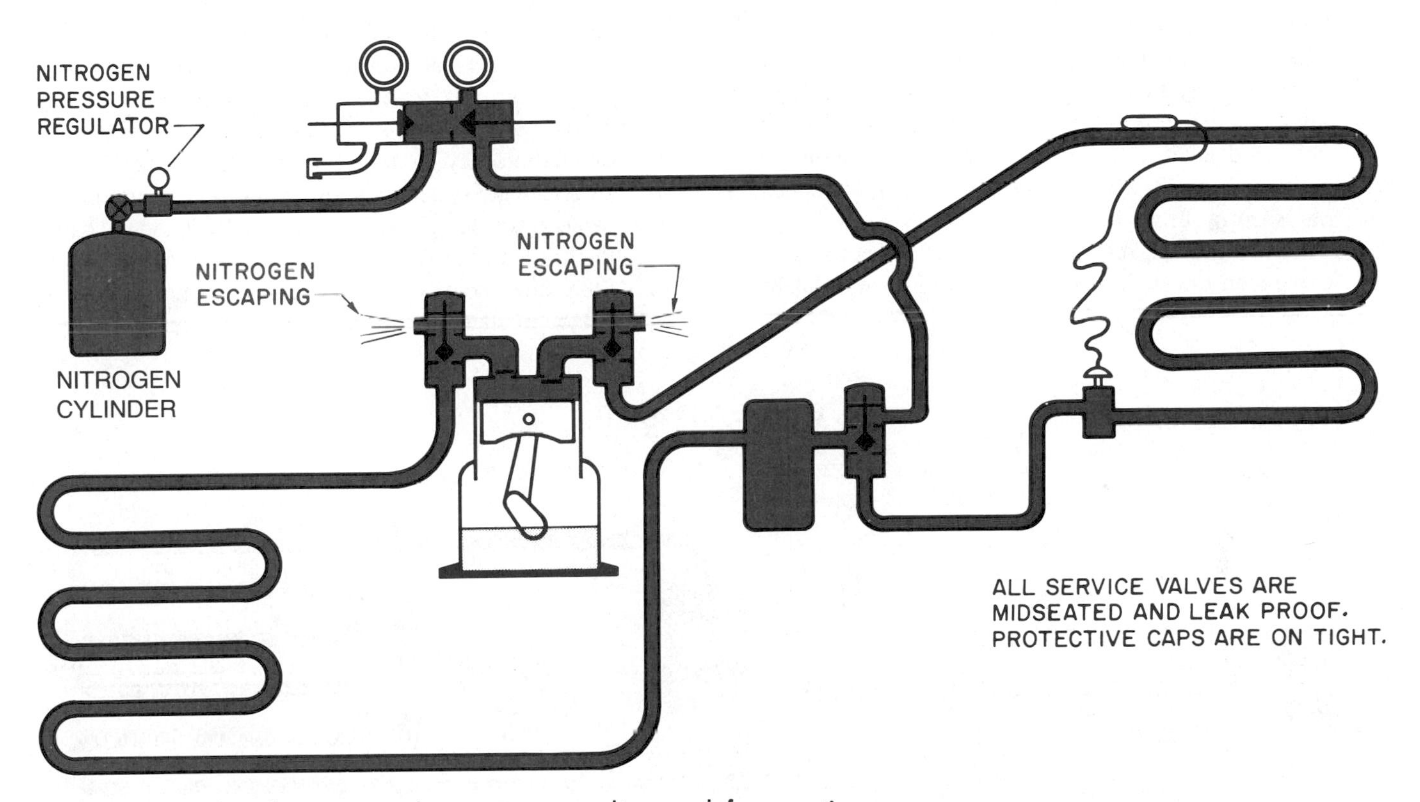

FIGURE 8–37 The technician is using dry nitrogen to sweep this system before evacuation.

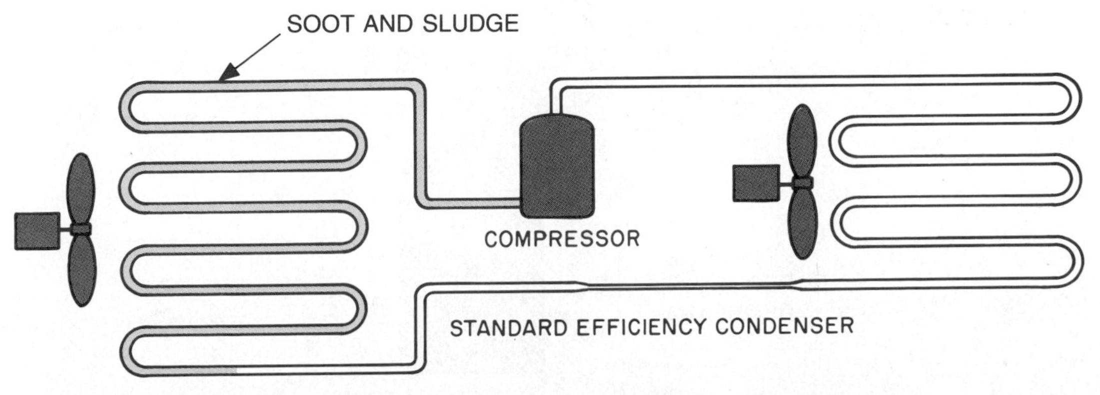

FIGURE 8–38 Soot and sludge move into the condenser when a motor burn occurs while the compressor is running.

1. ⊛*The refrigerant must be recovered from the system.* **This process is discussed in detail in Unit 9.**⊛
2. When the refrigerant is removed from the system, the compressor can then be changed for the new one. It will not be connected until later.
3. As mentioned before, contamination is in the system. It is in the vapor, liquid, and solid states. The vacuum pump can only remove the vapor state substances so other methods may be used to sweep some of the contaminants from the system.
4. Dry nitrogen may be used to push some of the contaminants out of the system by simply attaching the nitrogen regulator to one of the lines and allowing it to blow out the other. ***Do not exceed the system working pressure with the nitrogen. Without the compressor in the system, you can safely use 250 psig for high-pressure refrigerant systems.*** Because the contaminants are known to be in the condenser, the line may be disconnected before the expansion device where a liquid line filter drier can be installed later. Nitrogen may then be purged through the liquid line toward the compressor and discharged out the compressor discharge line before the compressor is connected, Figure 8–39. This will push all loose contaminants out at this point.
5. The nitrogen cylinder may then be connected to the expansion device side of the liquid line, and this line may be purged toward the compressor suction line. It should be noted that the velocity of the refrigerant will be reduced because of the expansion device, but this is all that can be done without disconnecting the expansion device. If it is a capillary tube system, there will be several connections and this will not be practical.
6. The system has been purged as much as possible, but contaminants are still in the system, as solids and liquids (soot and contaminated oil). The compressor is now connected to the system with a suction line filter drier installed just before the compressor. This will prevent any contamination from entering the compressor on start-up. Purging the system with dry nitrogen also ensures maximum capacity from the filter drier because some of the contaminants have already been pushed out of the system.
7. A liquid line filter drier is installed just before the metering device to prevent contamination from restricting the refrigerant flow.
8. The system is leak checked and ready to evacuate with a vacuum pump. If you have a choice of an old used or a new vacuum pump, use the old one, because contamination may be pulled through the pump.
9. Evacuate the system to a low vacuum, 250 microns, or triple evacuate, whichever you usually do, then charge the system with refrigerant.
10. Start the system and keep it running as long as practical to circulate the refrigerant through the filter driers.

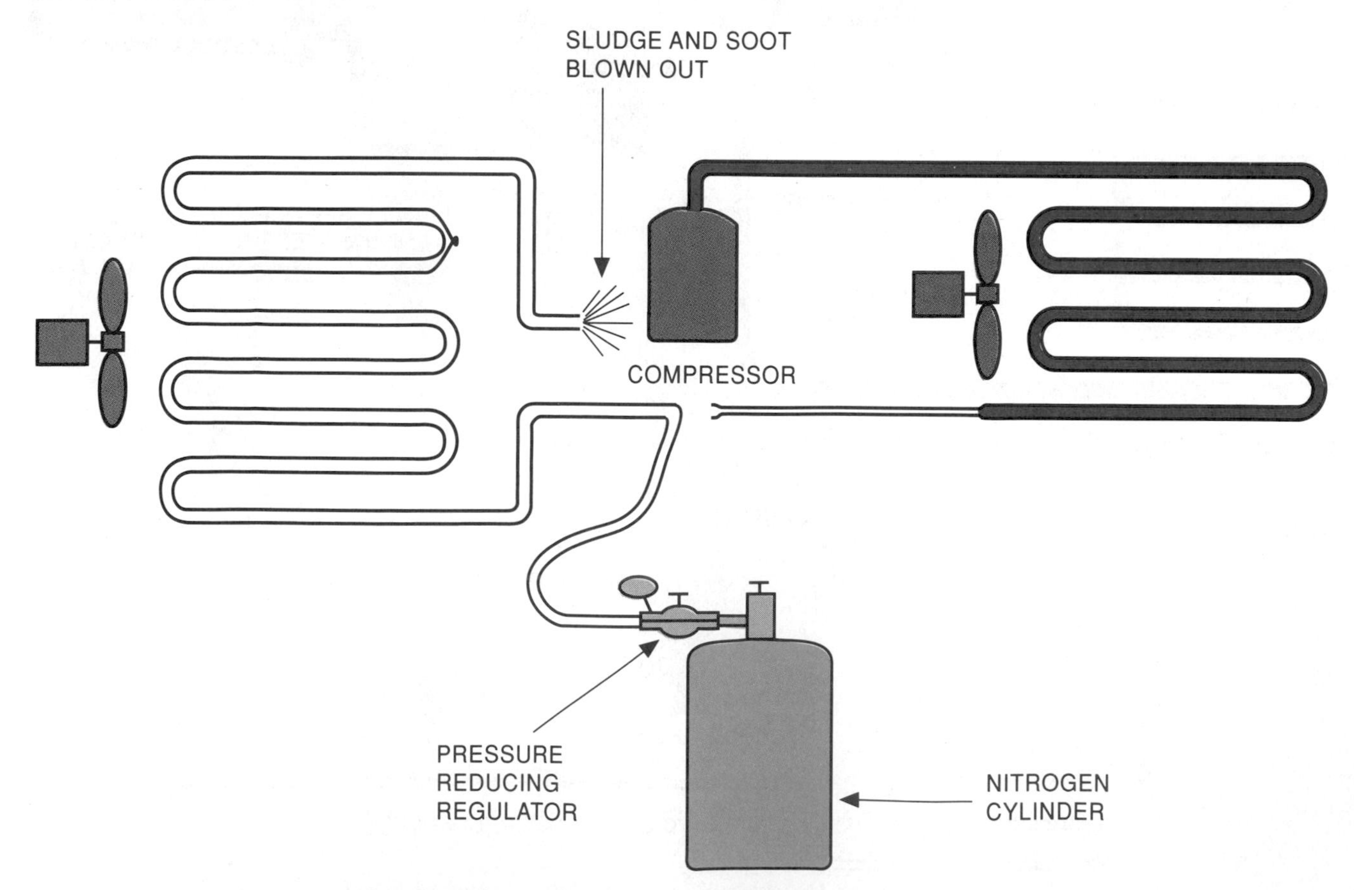

FIGURE 8–39 Purging contaminants from the condenser.

The refrigerant in the system is the best solvent you can find to break loose the contaminants that will be trapped in the filter driers. The pressure drop across the suction line drier may be monitored to see if it is gathering particles and beginning to become restricted, Figure 8–40. The manufacturer's literature will tell you what maximum pressure drop is allowed. If it becomes restricted beyond the manufacturer's recommendation, pump the system down and change the drier. If it does not become restricted, just leave it in the system; it will not hurt anything.

11. Change the oil in the vacuum pump while it is hot. Run it for 30 min and change it again to be sure all contamination has been removed from the pump crankcase. Many technicians will do a great job of cleaning the refrigeration system and neglect the vacuum pump, which is very expensive and just as important as the system compressor.
12. Check the pressure drop across the liquid line filter drier by measuring the temperature in and out of the drier. If there is a temperature drop, there is a pressure drop, Figure 8–41.

Filter drier manufacturers have done a good job of developing filter media that will remove acid, moisture, and carbon sludge. They claim that nothing can be created inside a system that has only oil and refrigerant that the filter driers will not remove.

SUMMARY

- Only two products should be circulating in a refrigeration system: refrigerant and oil.
- ***Noncondensable gases and moisture are common foreign matter that get into systems during assembly and repair. They must be removed.***
- Evacuation using low vacuum levels removes noncondensable gases and involves pumping the system to below atmospheric pressure.
- Vapors will be pumped out by the vacuum pump. Liquids must be boiled to be removed with a vacuum.
- Water makes a large volume of vapor when boiled at low pressure levels. It should be drained from a system if possible.

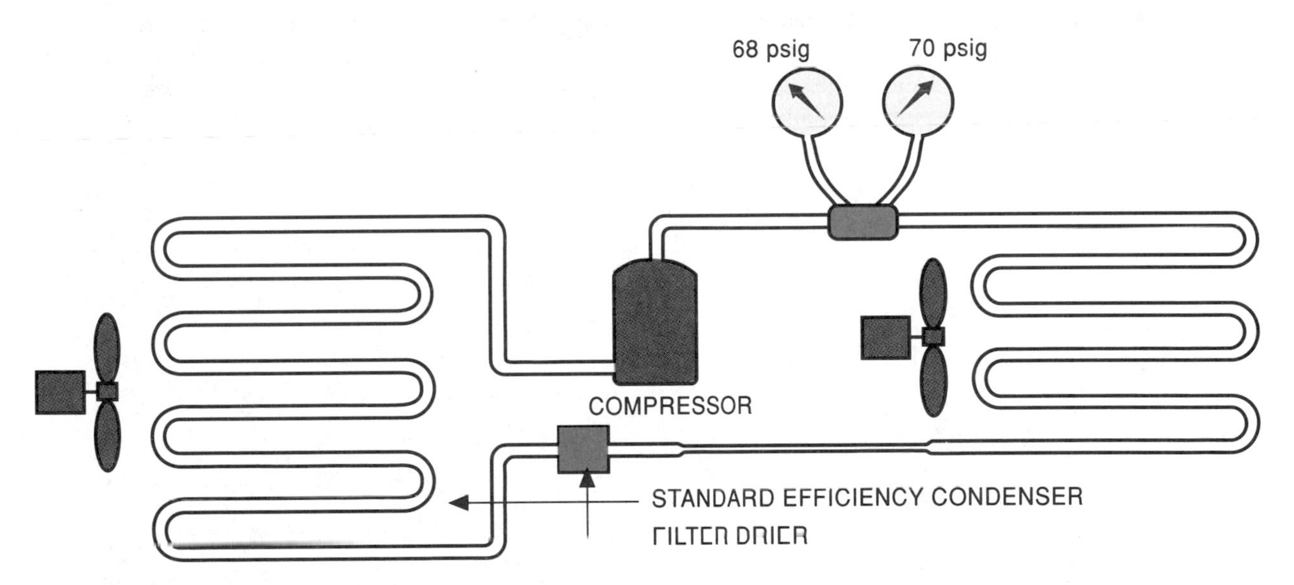

FIGURE 8–40 Checking the pressure drop across a suction line filter drier.

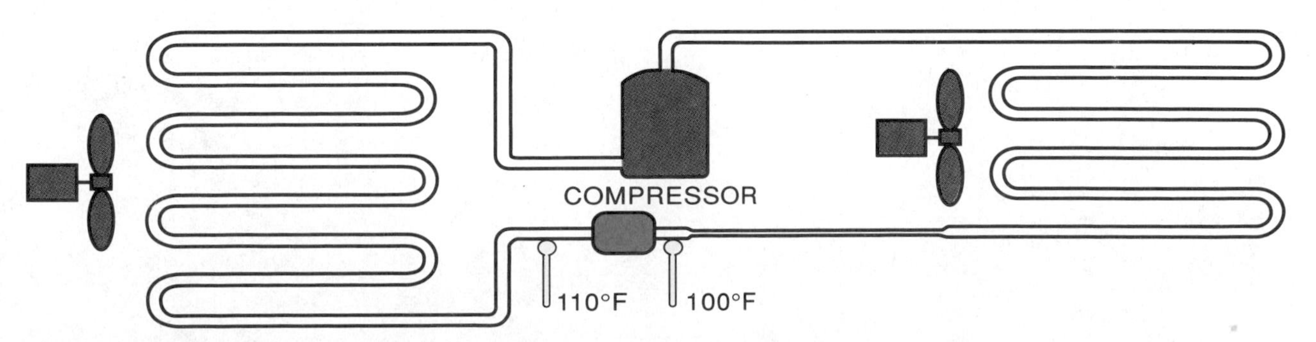

FIGURE 8–41 The 10°F temperature differential indicates a pressure drop; the drier should be changed.

- The two common vacuum gages are the U-tube manometer and the electronic micron gage.
- Pumping a vacuum may be quickened with large unrestricted lines.
- Good workmanship and piping practice along with a dry nitrogen setup will lessen the evacuation time.
- When noncondensables are left in a system, mild acids (hydrochloric and hydrofluoric) will slowly form and deteriorate the system by attacking the motor windings and causing copper plating on the crankshaft.
- The oxygen in air is the real problem when air is allowed to enter a system.
- Nitrogen in a system will cause excess head pressure because it takes up condensing space.
- The best system pressure test is a standing pressure test, normally 150 psig.
- The only advantage in using a vacuum to test for leaks is that vacuum instruments quickly respond to leaks. The vacuum leak test only proves that the system piping will prevent atmosphere from entering the system (14.696 psi).
- When a vacuum pump is allowed to run unattended, the system becomes a large vacuum reservoir and if the vacuum pump were to be shut off, the oil will be pulled into the system. If the vacuum pump is then restarted, it will be operating without lubrication.
- Special valve arrangements allow the technician to check the vacuum gage and pump and also check the system pressure.
- System Schrader valve cores may be removed to quicken the vacuum on a system; then the cores may be replaced before the system is put back in operation.
- Using dry nitrogen to sweep a contaminated system will help in evacuation of a dirty system.
- *Do not forget to clean the vacuum pump after evacuating a contaminated system.*

REVIEW QUESTIONS

1. What are foreign matters in a refrigeration system that must be removed?
2. What is evacuation?
3. What are the only two products that should be circulating in a refrigeration system?
4. Why is it hard to remove water from a refrigeration system with a vacuum pump?
5. Name some things that can be done to help remove water from a refrigeration system.
6. What can be done to get water out from under oil in a compressor crankcase?
7. Name the two evacuation processes.
8. What must be done to solenoid valves and other valves in a system to ensure proper evacuation?
9. Name two common vacuum test gages.
10. How many microns are there in 1 mm Hg?

UNIT 9 Refrigerant Management—Recovery/Recycling/Reclaiming

OBJECTIVES

After studying this unit, you should be able to

- describe ozone.
- discuss how CFCs deplete the earth's ozone layer.
- differentiate between CFCs, HCFCs, and HFCs.
- discuss EPA regulations as they relate to refrigerants.
- define the terms recover, recycle, and reclaim.
- describe methods of recovering refrigerants.
- identify a DOT-approved recovery cylinder.

SAFETY CHECKLIST

* Gloves and goggles should be worn any time a technician transfers refrigerant from one container to another.
* Transfer refrigerants only into DOT-approved containers.

9.1 REFRIGERANTS AND THE ENVIRONMENT

As the earth's population grows and demands for comfort and newer technologies increase, more and more chemicals are produced and used in various combinations. Many of these chemicals reach the earth's atmosphere, producing different types of pollution. Some of these chemicals make up the refrigerants that cool or freeze our food and cool our homes, office buildings, stores, and other buildings in which we live and work. Many of the refrigerants we use were developed in the 1930s.

While contained within a system, these refrigerants are stable, are not pollutants, and produce no harmful effects to the earth's atmosphere. Over the years, however, quantities have leaked from systems and have been allowed to escape by the willful purging or venting from systems during routine service procedures.

Over time these refrigerants are thought to slowly rise to the earth's stratosphere (7 to 30 miles above the earth), mostly with the help of wind currents. It is believed that some of them react with a type of oxygen called ozone. When this happens, new chemical compounds are formed and the ozone no longer exists in its present form.

Ozone is a type of oxygen in which each molecule consists of three atoms of oxygen, written as a chemical symbol O_3. The oxygen molecules we breathe consist of two atoms of oxygen, O_2. Ozone is also present near the earth's surface in the form of pollution. It is caused by ultraviolet radiation from the sun acting on smog and other air pollutants. Much of this pollution can be attributed to automobile exhaust and to industrial wastes in the lower atmosphere. This ozone is harmful to breathe. It is present in greater quantities when certain weather conditions encourage smog to remain near ground level.

Do not confuse the near-ground-level ozone with the ozone layer in the stratosphere, which is formed by ultraviolet rays from the sun acting on oxygen (O_2) molecules. The stratospheric ozone layer acts as a shield by preventing excessive amounts of the sun's ultraviolet rays from reaching the earth. The depletion of this layer will allow more of these harmful rays to reach the earth.

These ultraviolet rays in sufficient quantities can cause damage to human beings, animals, and plants. Depletion of the ozone shield can cause an increase in skin cancer and other health problems. Research has also shown that more of these ultraviolet rays reaching the earth could have an adverse effect on crops and other plant growth.

When released to the earth's atmosphere, refrigerants rise to the stratosphere very slowly. These compounds are thought to stay in the atmosphere for a long time and break down slowly. The refrigerants released many months or even years ago are what is suspected to be affecting the ozone layer today. The refrigerants released today may take months or possibly years to affect the ozone layer.

Many refrigerants contain chlorine atoms. It is these chlorine atoms that are thought to combine with the ozone atoms to cause a chemical change and, consequently, ozone depletion. Most refrigerants in use today are classified as chlorofluorocarbons (CFCs), hydrochlorofluorocarbons (HCFCs), and hydrofluorocarbons (HFCs).

9.2 CFC REFRIGERANTS

The CFCs contain chlorine, fluorine, and carbon and are considered the most damaging because they reach the stratosphere intact. The chlorine atom reacts with the ozone and the sunlight, causing the ozone to break down.

Following is a chart of CFCs with the chemical name and formula for each:

CHLOROFLUOROCARBONS (CFCs)

REFRIGERANT NO.	CHEMICAL NAME	CHEMICAL FORMULA
R-11	Trichlorofluoromethane	CCl_3F
R-12	Dichlorodifluoromethane	CCl_2F_2
R-113	Trichlorotrifluoroethane	CCl_2FCClF_2
R-114	Dichlorotetrafluoroethane	$CClF_2CClF_2$
R-115	Chloropentafluoroethane	$CClF_2CF_3$

All of the refrigerants in the CFC group are on the list to be phased out of manufacturing in 1995. One refrigerant that is important to us is R-12 because it is commonly used for residential and light commercial refrigeration and for centrifugal chillers in some commercial buildings. The other refrigerants are used in other applications. For example, R-11 is used for many centrifugal chillers in office buildings and also as an industrial solvent to clean parts. R-113 is used in smaller commercial chillers in office buildings and also as a cleaning solvent. R-114 has been used in some household refrigerators in the past and in centrifugal chillers for marine applications. R-115 is part of the blend of refrigerants that make up R-502 used for low-temperature refrigeration systems.

9.3 HCFC REFRIGERANTS

The second group of refrigerants in common use is the HCFC group. These refrigerants contain hydrogen, chlorine, fluorine, and carbon. These refrigerants have a small amount of chlorine in them, but also have hydrogen in the compound that makes them less stable in the atmosphere. These refrigerants have much less potential for ozone depletion because they tend to break down in the atmosphere, releasing the chlorine before it reaches and reacts with the ozone in the stratosphere. However, this group is on the list for phaseout by year 2030 because of the chlorine content.

This group of refrigerants includes:

HYDROCHLOROFLUOROCARBONS (HCFCs)

REFRIGERANT NO.	CHEMICAL NAME	CHEMICAL FORMULA
R-22	Chlorodifluoromethane	$CHClF_2$
R-123	Dichlorotrifluoroethane	$CHCl_2CF_3$

9.4 HFC REFRIGERANTS

The third group of refrigerants is the HFC group. These refrigerants have no chlorine atom and are believed to do no harm to the environment. They are considered to be possible replacement refrigerants for the future. The original plan was to replace R-12 with R-134a, but this plan is complicated by the fact that R-134a is not compatible with any oil left in an R-12 system. It is also not compatible with some of the gaskets in the R-12 systems. An R-12 system, however, may be converted to R-134a by modifying the system. This task must be performed by an experienced technician because a complete oil change is necessary. The residual oil left in the system pipes must be removed and all system gaskets must be checked for compatibility. The compressor materials must be checked by contacting the manufacturer to be sure these materials are compatible.

Following is a list of refrigerants in the HFC group:

HYDROFLUOROCARBONS (HFCs)
(Possible replacement refrigerants)

REFRIGERANT NO.	CHEMICAL NAME	CHEMICAL FORMULA
R-124	Chlorotetrafluoroethane	$CHClFCF_3$
R-125	Pentafluoroethane	CHF_2CF_3
R-134a	Tetrafluoroethane	CH_2FCF_3

9.5 REFRIGERANT BLENDS

Other refrigerants are mixtures of some of the existing refrigerants and may be called either blends or azeotropes. Because the refrigerants in the blend may boil at different temperatures, an ordinary blend of two or more refrigerants may separate into the different mixtures if leaked out of the system. Some of these blends may contain methane, which is natural gas and is flammable; the concern is that the refrigerant could become explosive. Over time, more and more blends will be tried as replacements for the existing CFC and HCFC refrigerants.

Azeotrope refrigerants are blends that are combined in the chemical process of manufacturing in such a manner that they will not separate into the individual components. They are thought of as being a specific refrigerant even when they leak into the atmosphere. R-500 is an example of an azeotrope blend of two refrigerants. It is made of 73.8% R-12 + 26.2% R-152 by weight. This refrigerant was used in some air conditioning (cooling) systems many years ago.

The common refrigerants mentioned throughout the text are:

1. R-12 for medium- and high-temperature commercial refrigeration applications and domestic refrigeration.
2. R-502 for low-temperature commercial refrigeration applications. Recently, R-22 has become the choice refrigerant for low-temperature applications because it will probably be on the market longer than R-502.
3. R-22 for residential and light commercial air conditioning and refrigeration and window unit air conditioning applications.
4. R-500 may still be in use in some older air conditioning systems.
5. R-11 is used in low-pressure water chillers for large commercial water chillers in large buildings.
6. R-123 is used as a replacement for R-11 in large chillers.

Manufacturers are currently working on several other refrigerant blends and azeotropic blends that will be studied and used in the future. This will be ongoing for years to come.

9.6 REGULATIONS

The United States, Canada, and more than 30 other countries met in Montreal, Canada in September 1987 to try to solve the problems of released refrigerants. This conference is known as the Montreal Protocol and those attending agreed to reduce by 50% by 1999 the production of refrigerants believed to be harmful ozone-destroying chemicals. This would begin when ratified by at least 11 countries representing 60% of the global production of these chemicals. Additional meetings have been held since 1987 and further reduction of these refrigerants has been agreed on. The use of refrigerants as we have known them has changed to some extent and will change dramatically in the near future.

The United States Clean Air Act Amendments of 1990 regulate the use and disposal of CFCs and HCFCs. The United States Environmental Protection Agency (EPA) is charged with implementing the provisions of this legislation.

This act affects the technician servicing refrigeration and cooling equipment. According to the Clean Air Act of 1990, technicians may not "knowingly vent or otherwise release or dispose of any substance used as a refrigerant in such appliance in a manner which permits such substance to enter the environment." This prohibition became effective July 1, 1992. Severe fines and penalties are provided for, including prison terms. The first level is the power of the EPA to obtain an injunction against the offending party, prohibiting the discharge of refrigerant to the atmosphere. The second level is a $25,000 fine per day and a prison term not exceeding 5 years.

To help police the illegal discharge of refrigerants, a reward may be given to any person who furnishes information leading to the conviction of a person willfully venting refrigerant to the environment.

Each refrigerant has been assigned a number to help identify the rate at which it is thought to deplete the ozone layer. This number is called the ozone depletion potential (ODP). The numbers are based on R-11, which has an ODP of 1. Other ODP numbers are:

OZONE DEPLETION FACTORS

REFRIGERANT NUMBER	OZONE DEPLETION NUMBER
R-11	1
R-12	1
R-22	0.05
R-113	0.8
R-114	1
R-115	0.8
R-123	0.016
R-124	not rated
R-125	not rated
R-134a	0
R-500	not rated
R-502	not rated

Note that R-11 and R-12 are major contributors to ozone depletion and will be among the first to be phased out.

The United States budget for 1990 contained provisions for an excise tax on CFC refrigerants. The CFCs included were R-11, R-12, R-113, R-114, and R-115. This tax is to be used to help industry in the conversion to the use of new refrigerants. This tax applies to newly manufactured refrigerants. Recycled and reclaimed refrigerants are exempt from the tax.

FEDERAL EXCISE TAX ON REFRIGERANTS
(per pound of refrigerant)

1990	$1.37	1995	$3.10
1991	1.37	1996	3.55
1992	1.67	1997	4.00
1993	2.65	1998	4.45
1994	2.65	1999	4.90

9.7 RECOVER, RECYCLE, OR RECLAIM

You should become familiar with these three terms: *recover, recycle,* or *reclaim.* These are the three terms the industry and its technicians must understand.

Technicians will have to contain all refrigerants except those used to purge lines and tools of the trade, such as the gage manifold or any device used to capture and save the refrigerant. The Clean Air Act Amendments of 1990 charged the EPA to issue regulations to reduce the use and emissions of refrigerants to the lowest achievable level. The law states that the technician shall not "knowingly vent or otherwise knowingly release or dispose of any such substance used as a refrigerant in such an appliance in a manner which permits such substance to enter the environment. De minimus releases associated with good faith attempts to recapture and recycle or safely dispose of any such substance shall not be subject to prohibition set forth in the preceding sentence." This prohibition became effective July 1, 1992. *De minimus releases* means the minimum amount possible under the circumstances.

Some situations requiring the removal of refrigerant from a component or system are the following:

When:

1. a compressor motor is burned.
2. a system is being removed to be replaced. It cannot be disposed of at a salvage yard with refrigerant in the system.
3. a repair must be made to a system and the refrigerant cannot be pumped, using the system compressor, and captured in the condenser or receiver.

When refrigerant must be removed, the technician must study the system and determine the best procedure for removal. The system may have several different valve configurations. The compressor may or may not be operable. Sometimes the system compressor may be used to assist in refrigerant recovery. We will consider only the refrigerants and systems covered in this text. The following situations may be encountered by the technician in the field where refrigerant must be removed from a system:

1. The compressor will *not* run and the system has no service valves or access ports, such as when a unit like a domestic refrigerator, freezer, or window unit is being discarded.
2. The compressor will *not* run and the system has no access ports, such as a burnout on a domestic refrigerator, freezer, or window unit.

3. The compressor will run and the unit has service ports (Schrader ports) only, as in a central air conditioner.
4. The compressor will *not* run and the unit has service ports (Schrader ports) only, as in a central air conditioner.
5. The compressor will run and the system has some service valves, such as a residential air conditioning system with liquid and suction line isolation valves in the line set.
6. The compressor will *not* run and the system has some service valves such as a residential air conditioning system with liquid and suction line isolation valves on the line set.
7. The compressor will run and the system has a complete set of service valves, such as a refrigeration system.
8. The compressor will *not* run and the system has a complete set of service valves, such as a refrigeration system.
9. A heat pump with any combination of service valves.

When the technician begins to try to recover the refrigerant in these cases, a certain amount of knowledge and experience is necessary. The condition of the refrigerant in the system must also be considered. It may be contaminated with air, another refrigerant, nitrogen, acid, water, or motor burn particulates.

Cross contamination to other systems must never be allowed because it can cause damage to the system into which the contaminated refrigerant is transferred. This damage may occur slowly and may not be known for long periods of time. Warranties of compressors and systems may be voided if contamination can be proven. Food loss from inoperative refrigeration systems and money loss in places of business when the air conditioning system is off can be a result.

The words recover, recycle, and reclaim are used to mean the same thing by many technicians, but they refer to three entirely different procedures.

RECOVER REFRIGERANT. "To remove refrigerant in any condition from a system and store it in an external container without necessarily testing or processing it in any way." This refrigerant may be contaminated with air, another refrigerant, nitrogen, acid, water, or motor burn particulates. A refrigerant must not be used in another system unless it is known to be clean or cleaned to the Air Conditioning and Refrigeration Institute (ARI) specifications "ARI 700."

If the technician is removing the refrigerant from the system and it is operable, it may be charged back into the system. For example, assume a system has no service valves and a leak occurs. The remaining refrigerant may be recovered and reused in this system only. It may not be sold to another customer without meeting the ARI 700 specification standard. Analysis of the refrigerant to the ARI 700 standard may only be performed by a qualified chemical laboratory. This test is expensive and is usually only performed on larger volumes of refrigerant. Some owners of multiple equipment may reuse refrigerant in another unit they own if they are willing to take the chance with their own equipment. Some owners may be willing to take the chance after the equipment is out of warranty. It is up to the technician to use an approved, clean, refrigerant cylinder for recovery to avoid cross contamination.

RECYCLE REFRIGERANT. "To clean the refrigerant by oil separation and single or multiple passes through devices, such as replaceable core filter driers, which reduce moisture, acidity and particulate matter." This term usually applies to procedures implemented at the job site or at a local service shop. If it is suspected that refrigerant is dirty with certain contaminates, it may be recycled, which means to filter and clean. Filter driers may be used only to remove acid, particles, and moisture from a refrigerant. In some cases, the refrigerant will need to pass through the driers several times before complete cleanup is accomplished. Purging air from the system may only be done when the refrigerant is contained in a separate container or a recycling unit.

A form of recycling has been used for many years. When a motor burn occurred and the compressor had service valves, the valves were front seated, the compressor changed, and the only refrigerant lost was the refrigerant vapor in the compressor. This was only a small amount of vapor. Driers were used to clean the refrigerant while running the new system compressor. Often the system was started, pumped down into the condenser and receiver and additional acid-removing filter driers added to the suction line to further protect the compressor, Figure 9–1. This recycle process occurred within the equipment and normal operation resumed.

A recycling unit removes the refrigerant from the system, cleans it, and returns it to the system. This is accomplished by moving the refrigerant to the recovery cylinder. Then using the two valves on the cylinder the refrigerant is circulated around through the recycle unit and its filtration system. Some units have an air purge that allows any air introduced into the system to be purged before transferring the refrigerant back into the system, Figure 9–2.

RECLAIM REFRIGERANT. "To process refrigerant to new product specifications by means which may include distillation. It will require chemical analysis of the refrigerant to determine that appropriate product specifications are met. This term usually implies the use of processes or procedures available only at a reprocessing or manufacturing facility." This refrigerant is recovered at the job site, stored in approved cylinders, and shipped to the reprocess site where it is chemically analyzed and declared that it can be reprocessed (reclaimed). It is then reprocessed and shipped as reclaimed refrigerant because it meets the ARI 700 standard. At this time there is no way of testing to the ARI standard in the field.

When reclaimed refrigerant is sold, it is not subject to the excise tax as mentioned above because this tax applies to only newly manufactured refrigerant. This tax saving is an incentive to save and reclaim old refrigerant.

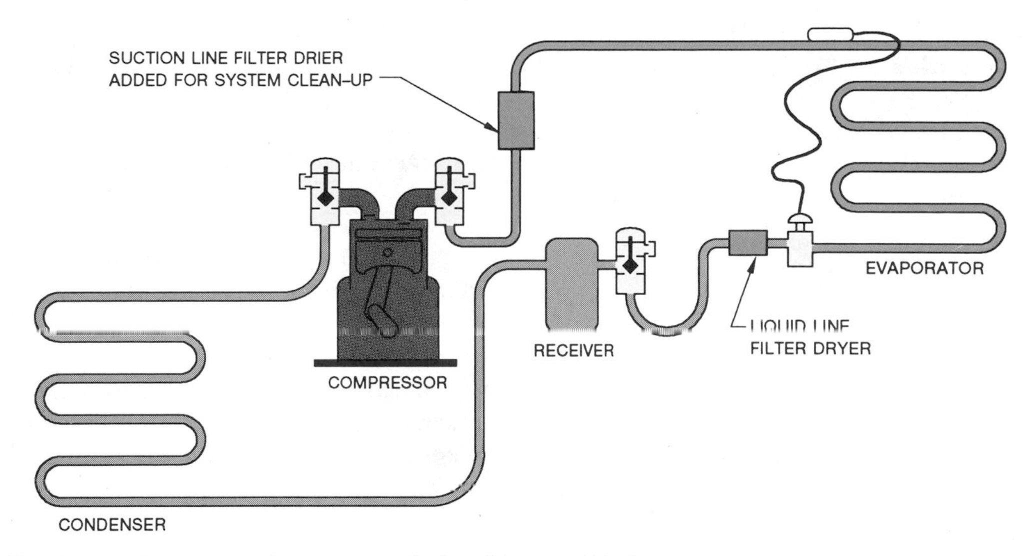

FIGURE 9–1 Filter driers in the system can be used to recycle the refrigerant within the system.

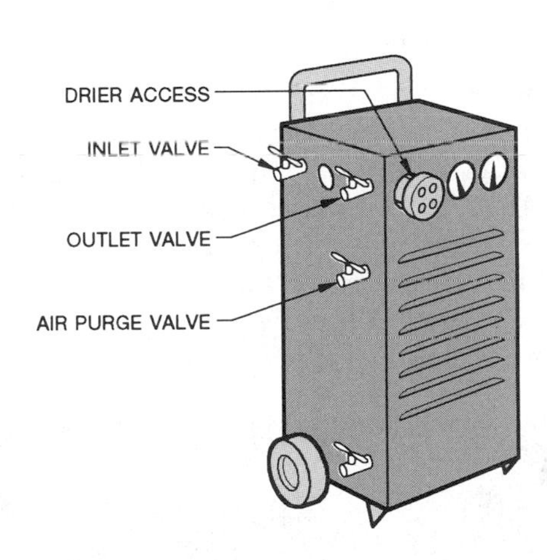

FIGURE 9–2 The recovery/recycle unit has an air purge valve from which accumulated air may be vented.

9.8 METHODS OF RECOVERY

Several methods of recovery for the refrigerants are used in the systems mentioned in this text. All of these refrigerants are thought of as high-pressure refrigerants. They are also called low-boiling point refrigerants. Some of the CFC refrigerants, such as R-11, are low-pressure refrigerants (high boiling point). It boils at 74.9°F at atmospheric pressure. This refrigerant is very hard to remove from a system because of its high boiling point. A very low vacuum must be pulled to boil it all from the system. This is different from R-12, which has the highest boiling point of the commonly used refrigerants in small systems. It boils at -21.62°F at atmospheric pressure. That means that if the temperature of the system is above -21.62°F, the R-12 will boil and can be removed from the system easily. Refrigerant 134a has a boiling point of -15.7°F, but it is not yet a common refrigerant. The boiling points of some refrigerants at atmospheric pressure are:

REFRIGERANT	BOILING POINT (°F)
R-12	−21.62
R-22	−41.36
R-125	−55.3 (to replace R-502)
R-134a	−15.7 (to replace R-12)
R-500	−28.3
R-502	−49.8
R-123	+82.2 (to replace R-11)
R-11	+74.9

The following chart shows that most of the common refrigerants discussed in this text have high vapor pressures and will be relatively easy to remove from the systems. The vapor pressures at room temperature of 70°F would be:

REFRIGERANT	VAPOR PRESSURE AT 70°F
R-12	70 psig
R-22	121 psig
R-125	158 psig (to replace R-502)
R-134a	71 psig (to replace R-12)
R-500	85 psig
R-502	137 psig

The boiling point is important because the Environmental Protection Agency (EPA) has ruled on how low the pressure in a system must be reduced to accomplish the recovery guideline.

Effective August 12, 1993, all companies have had to certify to the EPA that they have recovery equipment to perform adequate recovery of refrigerant for the systems they service.

Equipment manufactured before November 15, 1993 must meet qualifications set for this time period. Equipment that is manufactured after November 15, 1993 must be approved by an ARI-approved third-party laboratory to meet ARI Standard 740-1993.

As of November 15, 1993, the following is the evacuation guide used by the EPA for evacuation of equipment except appliances and motor vehicle air conditioning:

1. HCFC-22 appliance, or isolated component of such appliance, normally containing less than 200 lb of refrigerant, 0 psig.
2. HCFC-22 appliance, or isolated component of such appliance, normally containing more than 200 lb of refrigerant, 10 in. Hg vacuum.
3. Other high-pressure appliance, or isolated component for such appliance, normally containing less than 200 lb of refrigerant (R-12, R-500, R-502, R-114), 15 in. Hg vacuum.
4. Other high-pressure appliance, or isolated component of such appliance, normally containing 200 lb or more of refrigerant (R-12, R-500, R-502, R-114), 15 in. Hg vacuum.
5. Very high-pressure appliance (R-13, R-503), 0 psig.
6. Low-pressure appliance (R-11, R-113, R-123), 25 mm Hg absolute, which is 29 in. Hg vacuum.

The exception to the above would be when the system could not be reduced to these pressure levels, such as when the system has a large leak, there is no need to pull the system full of air.

The rules are different for small appliance repair technicians. Effective August 12, 1993, they must certify to the EPA that the equipment they are using is capable of recovering 80% of the refrigerant in a system or achieving a 4-in. vacuum under conditions of ARI 740-1993 if the equipment is manufactured before November 15, 1993. Recovery equipment manufactured after November 15, 1993 that is used during maintenance, service, or repair of small appliances must be able to recover 90% of the refrigerant or achieve a 4-in. vacuum if the compressor in the appliance is operable. If the compressor is not operable, the recovery machine must be capable of recovering 80% of the refrigerant.

Technicians will be required to be certified to purchase refrigerant after November 14, 1994. There will be four different certification areas for the different types of service work.

TYPE I CERTIFICATION. Small Appliance—Manufactured, *charged and hermetically sealed with 5 lb or less of refrigerant.* Includes refrigerators, freezers, room air conditioners, *package terminal heat pumps,* dehumidifiers, under-the-counter ice makers, vending machines, and drinking water coolers.

TYPE II CERTIFICATION. High-Pressure Appliance—Uses refrigerant with a boiling point between -50°C (-58°F) and 10°C (50°F) at atmospheric pressure. Includes *12, 22, 114, 500,* and *502* refrigerant.

TYPE III CERTIFICATION. Low-Pressure Appliance—Uses refrigerant with a boiling point above 10°C (50°F) at atmospheric pressure. Includes 11, 113, and 123 refrigerants.

UNIVERSAL CERTIFICATION. Certified in all of the above: Type I, II, and III.

It is believed that technician certification should ensure that the technician knows how to handle the refrigerant in a safe manner without exhausting it to the atmosphere.

The refrigerant may be removed from the system as either a vapor or liquid or in the partial liquid and vapor state. It must be remembered that lubricating oil is also circulating in the system with the refrigerant. If the refrigerant is removed in the vapor state, the oil is more likely to stay in the system. This is desirable from two standpoints:

1. If the oil is contaminated, it may have to be handled as a hazardous waste. Much more consideration is necessary and a certified hazardous waste technician must be available.
2. If the oil remains in the system, it will not have to be measured and replaced back to the system. Time may be saved in either case and time is money. The technician should pay close attention to the management of the oil from any system.

Refrigerant must be transferred only into approved refrigerant cylinders. These cylinders are approved by the Department of Transportation (DOT). Never use the DOT 39 disposable cylinders. The approved cylinders are recognizable by the color and valve arrangement. They are yellow on top with gray bodies and a special valve that allows liquid or vapor to be added or removed from the cylinder, Figure 9–3.

FIGURE 9–3 Approved Department of Transportation (DOT) cylinders. *Courtesy White Industries*

This cylinder must be clean and in a deep vacuum before the recovery or recycle process is started. It is suggested that the cylinder be evacuated to 1000 microns (1 mm Hg) before recovery is started. Refrigerant from a previous job will be removed with this vacuum. All lines to the cylinder must be purged of air before connections are made.

9.9 MECHANICAL RECOVERY SYSTEMS

Many mechanical recovery systems are available. Some are sold for the purpose of recovering refrigerant only. Some are designed to recover and recycle and a few sophisticated ones claim to have reclaim capabilities. ARI will be the organization that will certify the equipment specifications and these machines will be expected to perform to the 1993 ARI 740 specifications.

For recovery and recycle equipment to be useful in the field, it must be portable. Some of the equipment is designed to use in the shop where the refrigerant is brought in for recycling. But for the equipment to be useful in the field, it must be able to be easily moved to the rooftop, where many systems are located. This presents a design challenge because the typical unit must be able to both recover and recycle in the field for it to be effective. It must also be small enough to be hauled in the technician's truck. Units for use on rooftops should not weigh more than approximately 50 lb because anything heavier may be too much for the technician to carry up a ladder. A heavy unit may require two technicians where only one would normally be needed.

Some technicians may overcome this weight problem by using long hoses to reach from the rooftop to the unit, which may be left on the truck or on the top floor of a building, Figure 9–4. This is not often used because the technician must recover the refrigerant from the long hose. The long hose also slows the recovery process. Many units have wheels so they may be rolled to the top floor, Figure 9–5.

Manufacturers are meeting this portability challenge in several ways. Some have developed modular units that may be carried to the rooftops or remote jobs as separate components, Figure 9–6. Some are reducing the size of the units by limiting the internal components and using them for recovery only. In this case, the refrigerant may be recovered to a cylinder and taken to the shop for recycling, then returned to the job. However, every time the refrigerant is transferred from one container to another, refrigerant will be lost.

Manufacturers are attempting to provide a unit that is as small as practical, yet will still perform properly. The task is to remove the refrigerant from the system. They are trying to remove vapor, liquid, and vapor/liquid mixtures and deal

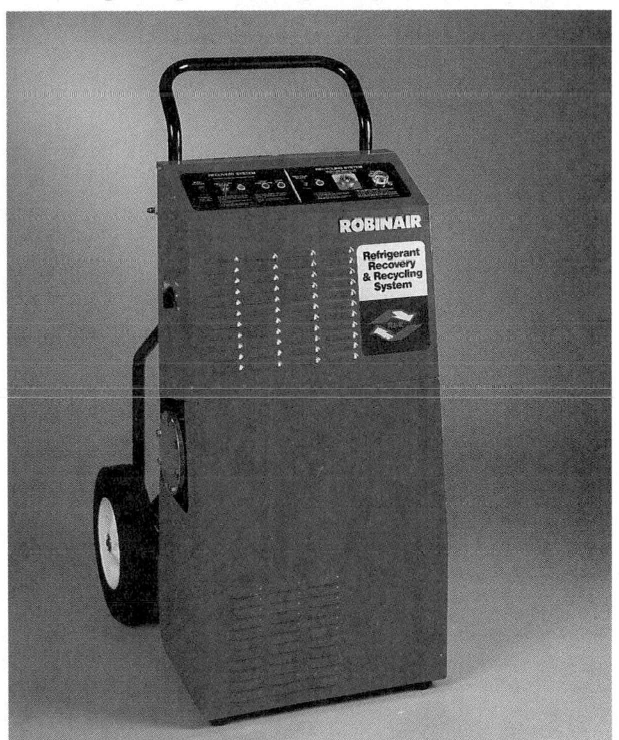

FIGURE 9–5 This recovery/recycle unit is heavy and has wheels to make it easy to move around. *Courtesy of Robinair Division, SPX Corporation*

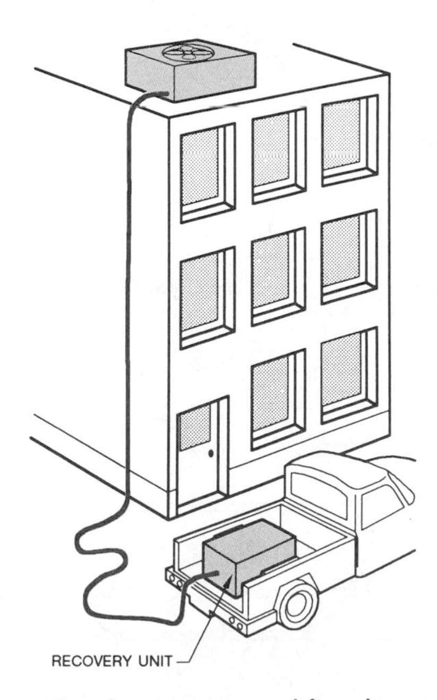

FIGURE 9–4 The refrigerant is recovered from this unit through a long hose to the unit on the truck. This is a slower process.

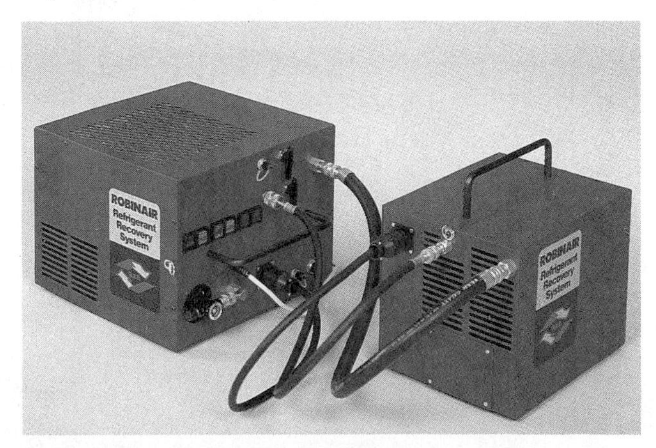

FIGURE 9–6 This modular unit reduces a heavy unit to two manageable pieces. *Courtesy of Robinair Division, SPX Corporation*

with oil that may be suspended in the liquid refrigerant. Many are accomplishing this challenge by using a small hermetic compressor in the unit to pump the refrigerant vapor. As the compressors become larger, they become heavier. Larger condensers are also needed, which adds to the weight. Some manufacturers are using rotary compressors to reduce the weight. Rotary compressors may also pull a deeper vacuum on the system with less effort than a reciprocating compressor. Deeper vacuums may be required in the future for removing more of the system refrigerant. All of these compressors in recovery units also pump oil, so some method of oil return is needed to return the oil to the recovery unit compressor crankcase.

The *fastest* method to remove refrigerant from a system is to take it out in the liquid state. It occupies a smaller volume per pound of refrigerant. If the system is large enough, it may have a liquid receiver where most of the charge is collected. Many systems, however, are small and if the compressor is not operable it cannot pump the refrigerant.

The *slowest* method of removing refrigerant is to remove it in the vapor state. When removed in the vapor state using a recovery/recycle unit, the unit will remove the vapor faster if the hoses and valve ports are not restricted and if a greater pressure difference can be created. The warmer the system is, the warmer and denser the vapor is. The compressor in the recovery/recycle unit will be able to pump more pounds in a minute. As the vapor pressure in the system is reduced, the vapor becomes less dense and the unit capacity is reduced. It will take more time as the system pressure drops. For example, when removing R-12 from a system, if saturated vapor is removed at 70 psig, only 0.482 ft^3 of refrigerant vapor must be removed to remove 1 lb of refrigerant. When the pressure in the system is reduced to 20 psig, 1.147 ft^3 of vapor must be removed to remove 1 lb of refrigerant. When the pressure is reduced to 0 psig, there is still refrigerant in the system, but 2.540 ft^3 of refrigerant vapor must be removed to remove 1 lb of refrigerant, Figure 9–7. The unit recovery rate will slow down as the refrigerant pressure drops because the compressor pumping rate is a constant.

The technician must also be able to rely on having equipment that will operate under the conditions for which it will be needed. It may be 100°F on the rooftop. The unit will be functioning under some extremely hot conditions. The system pressure may be very high due to the ambient temperature, which may tend to overload the unit compressor. A crankcase pressure regulator is sometimes used in recovery/recycle units to prevent overloading the compressor with high suction pressure. The recovery unit will have to condense the refrigerant using the 100°F air across the condenser, Figure 9–8. The condenser on the unit must have enough capacity to condense this refrigerant without overloading the compressor.

Recovery/recycling unit manufacturers give the typical pumping rate for their equipment in their specifications. This is the rate at which the equipment is capable of removing refrigerant. The rate is generally 2 lb/min. The rate will be more constant and faster while removing refrigerant in the liquid state. More factors must be considered when removing vapor. When the system pressure is high, the unit will be able to remove vapor relatively fast. When the system pressure is lowered, the rate of vapor removal will slow down. With a small compressor when the system vapor pressure reaches about 20 psig, the rate of removal will become very slow.

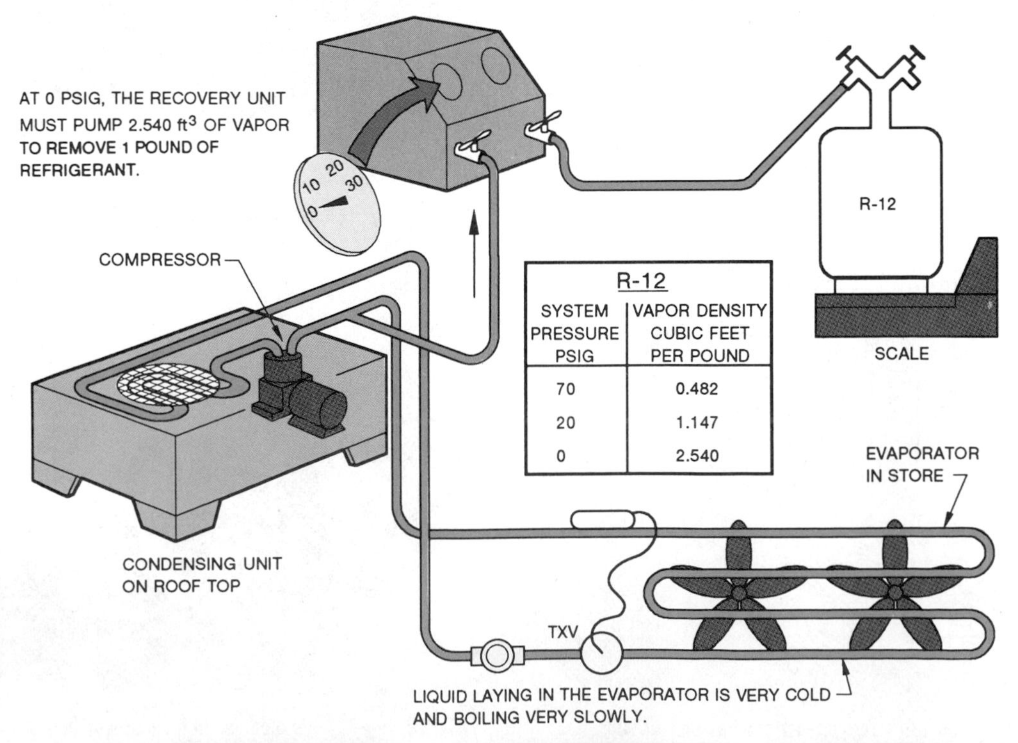
AT 0 PSIG, THE RECOVERY UNIT MUST PUMP 2.540 ft³ OF VAPOR TO REMOVE 1 POUND OF REFRIGERANT.
10
20
30
0
COMPRESSOR
R-12
SCALE
CONDENSING UNIT ON ROOF TOP
EVAPORATOR IN STORE
TXV
LIQUID LAYING IN THE EVAPORATOR IS VERY COLD AND BOILING VERY SLOWLY.

R-12

SYSTEM PRESSURE PSIG	VAPOR DENSITY CUBIC FEET PER POUND
70	0.482
20	1.147
0	2.540

FIGURE 9–7 Refrigerant vapor has less weight per cubic foot as the pressure is reduced.

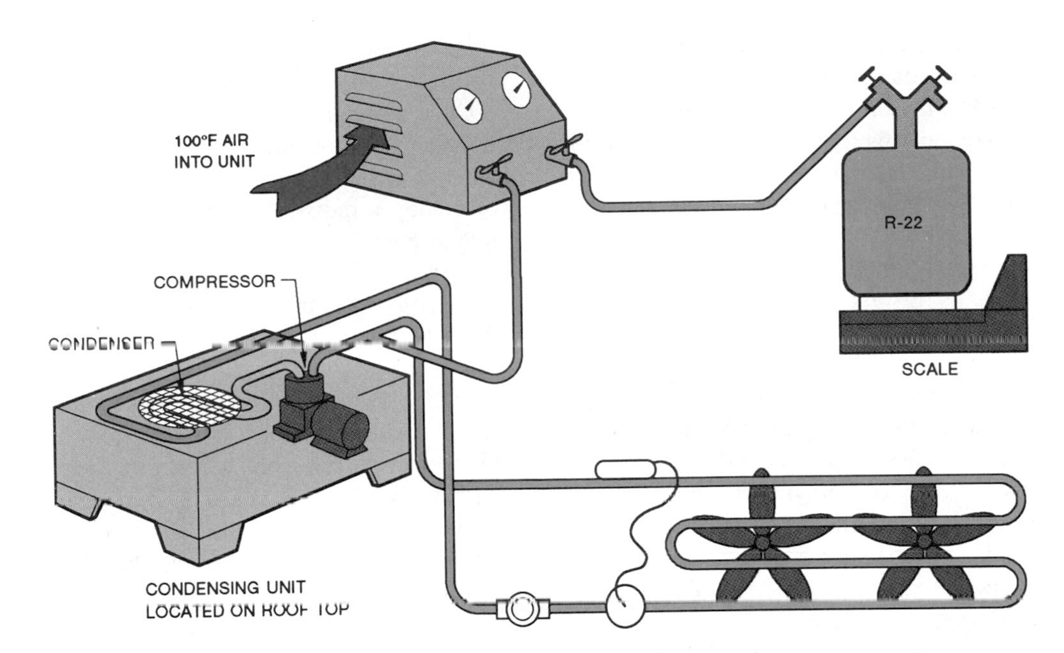

FIGURE 9–8 The unit will have to condense the refrigerant using 100°F air on the rooftop. The condenser must be large enough to keep the pressure low enough so the compressor is not overloaded.

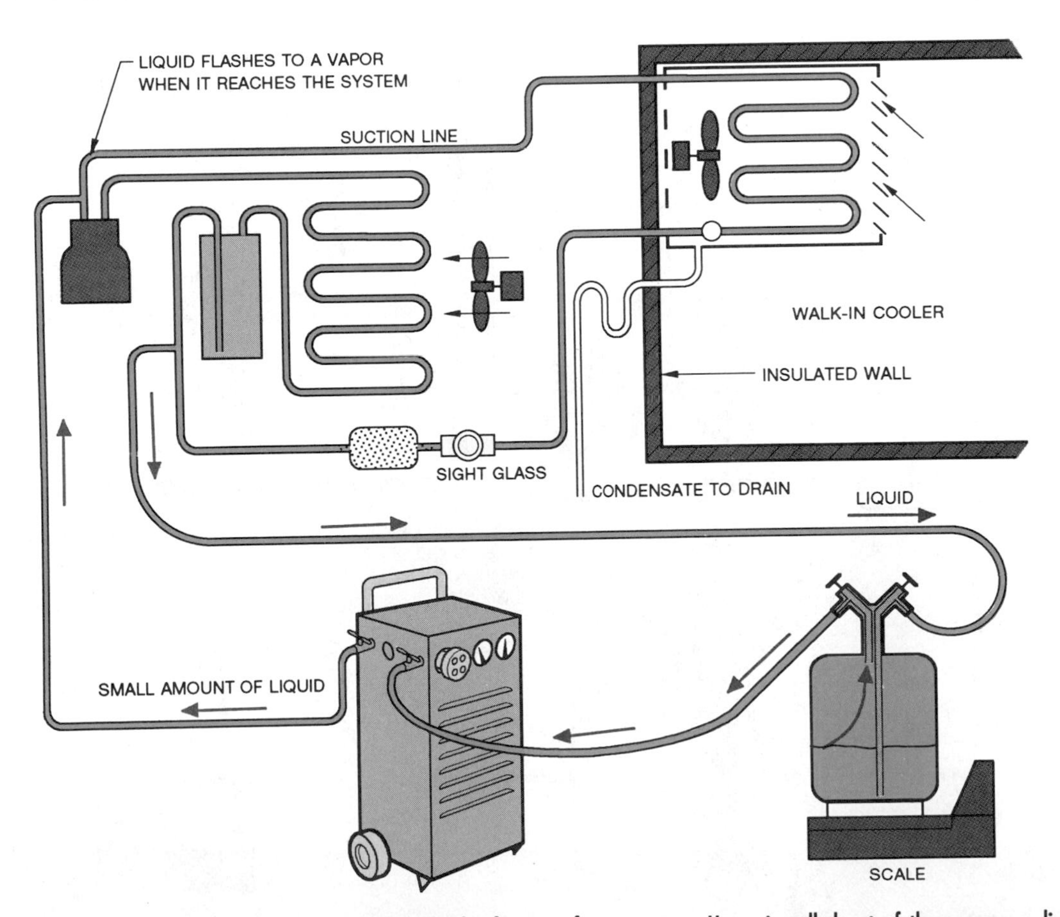

FIGURE 9–9 This unit uses a push-pull method of removing liquid refrigerant from a system. Vapor is pulled out of the recovery cylinder, creating a low pressure in the cylinder. A small amount of condensed liquid is allowed back into the system to build pressure and push liquid into the cylinder.

The compressor in the recovery/recycle unit is a vapor pump. Liquid refrigerant cannot be allowed to enter the compressor because it will cause damage. Different methods may be used to prevent this from occurring. Some manufacturers use a push-pull method of removing refrigerant. This is accomplished by connecting the liquid line fitting on the unit to the liquid line fitting on the cylinder. The suction line on the unit is connected to the vapor fitting on the cylinder. A connection is then made from the unit discharge back to the system. When the unit is started, vapor is pulled out of the recovery cylinder from the vapor port and condensed by the recovery unit. A very small amount of liquid is pushed into the system where it flashes to a vapor to build pressure and push liquid into the receiving cylinder, Figure 9–9. A sight glass is used to monitor the liquid and the recovery cylinder is weighed to prevent overfilling. When it is determined that no more liquid may be removed, the suction line from the recovery unit is reconnected to the vapor portion of the system and the unit discharge is fastened to the refrigerant cylinder, Figure 9–10. The unit is started and the remaining vapor is removed from the system and condensed in the cylinder. A crankcase pressure regulator valve is often installed in the suction line to the compressor in the recovery/recycle unit to protect it from overloading and offers some liquid refrigerant protection, Figure 9–11. When liquid is removed, it contains oil. The oil will be in the refrigerant in the cylinder. When this liquid refrigerant is charged back into the system, the oil will go with it. Oil that may move through the recovery unit with any small amount of liquid will be stopped by the oil separator. Any oil removed must be accounted for to add back to the system, Figure 9–12. An acid check of the system's oil should be performed if there is any question as to its quality, such as after a motor burn, Figure 9–13.

The DOT-approved cylinder used for some of the recovery/recycle systems has a float and switch that will shut the unit off when the liquid in the cylinder reaches a certain level. At this point, the cylinder is 80% full. This switch helps prevent the technician from overfilling the cylinder with refrigerant. It also helps compensate for oil that may be in the refrigerant. When oil and refrigerant are added to a cylinder, the amount of weight that the cylinder can hold changes because oil is much lighter than refrigerant. If a cylinder contains several pounds of oil due to poor practices, it can easily be overfilled with refrigerant if the cylinder is filled only by weight. The float switch prevents this from happening, Figure 9–14. ***CAUTION: The recovery cylinder must be clean, empty, and in a vacuum of about 1000 microns before recovery is started.***

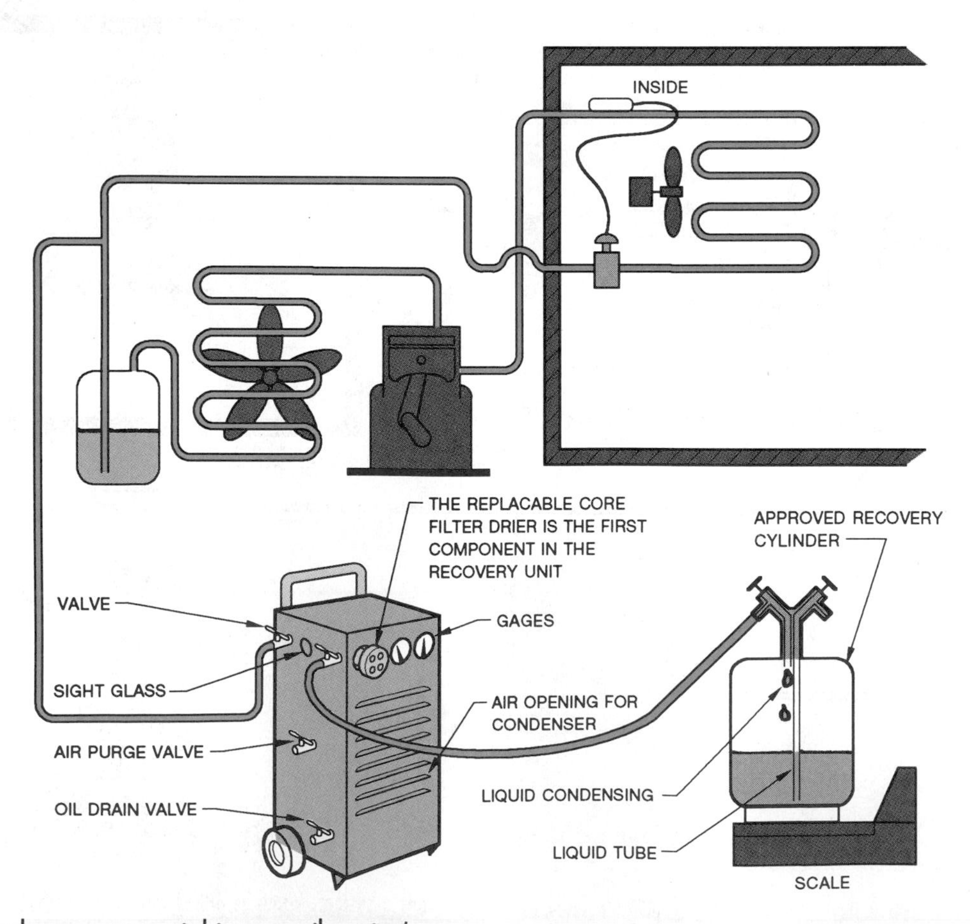

FIGURE 9–10 The hoses are reconnected to remove the system's vapor.

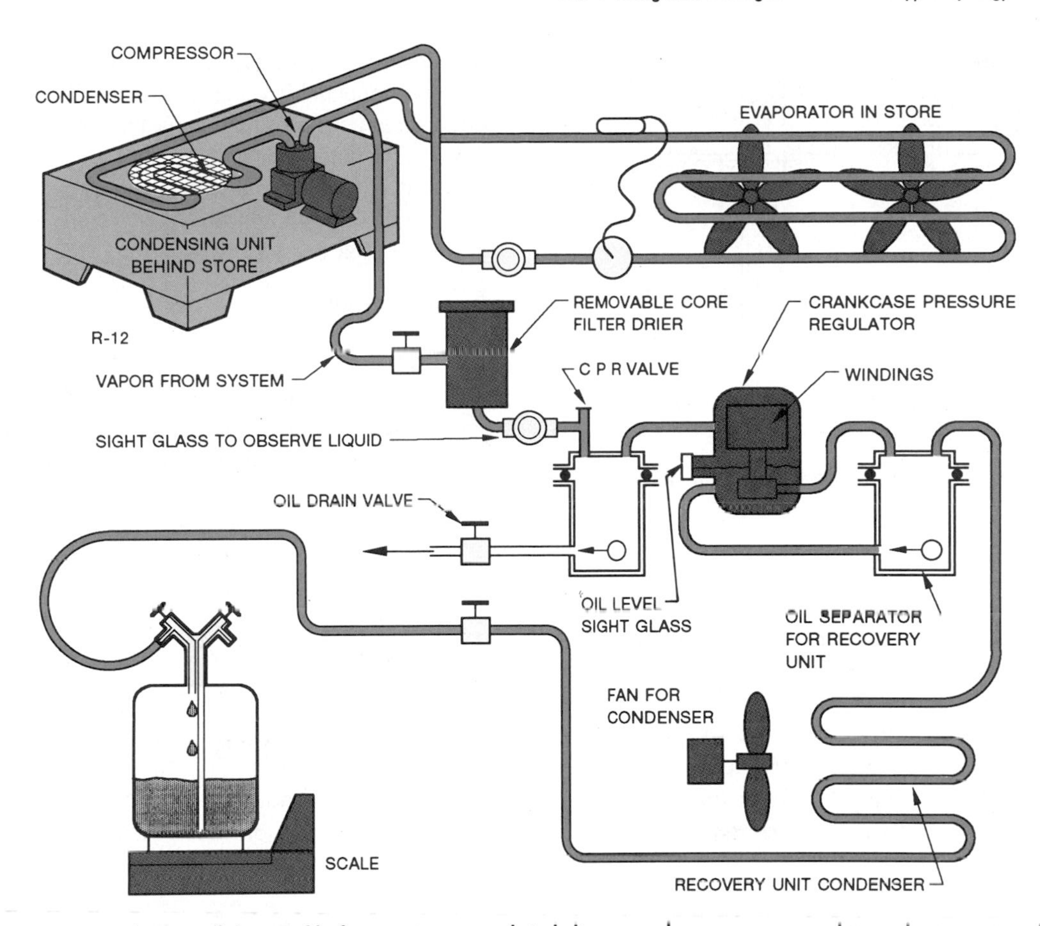

FIGURE 9–11 A system with some of the possible features incorporated, including a crankcase pressure regulator valve to protect the compressor.

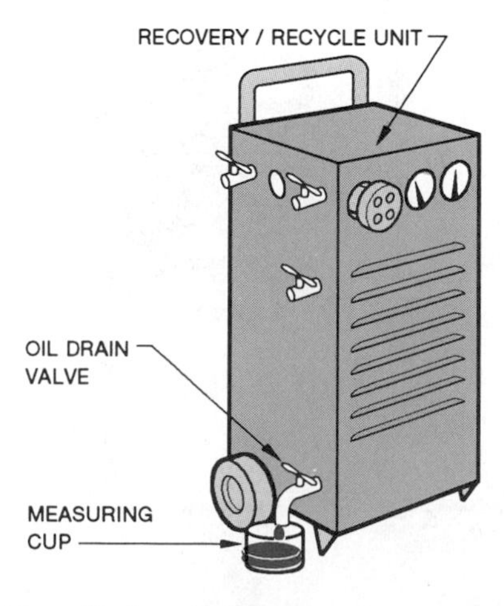

FIGURE 9–12 All oil removed from a system must be accounted for.

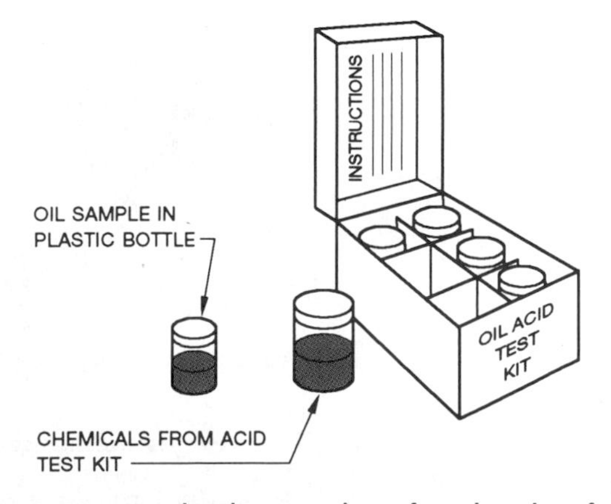

FIGURE 9–13 An oil acid test may be performed on the refrigerant oil in the field.

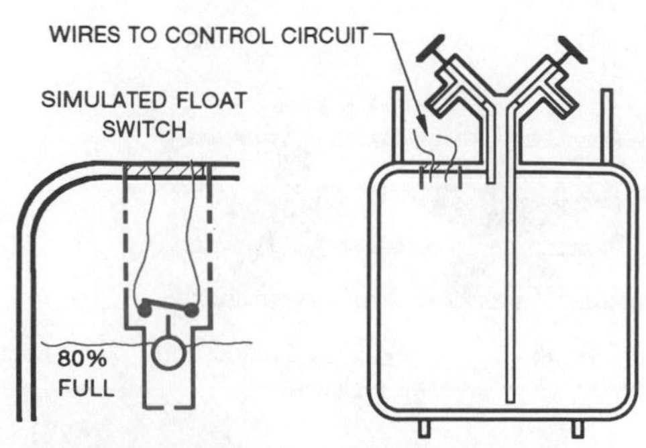

FIGURE 9–14 This DOT-approved cylinder has a float switch to prevent overfilling.

Some manufacturers remove liquid from the system through a metering device and an evaporator that is used to boil the liquid to a vapor. The compressor in the recovery/recycle unit moves the liquid refrigerant and some oil from the system and evaporates the refrigerant from the oil, Figure 9–15. The oil must then be separated out, or it will collect in the unit compressor. Usually an oil separator is used. The oil from the separator must be drained periodically and measured to know how much oil to add back to the system if it is to be put back in service. Oil at this time may be checked for acid content using an acid test kit.

All of the recovery/recycle unit compressors will pump oil out the discharge line and should have an oil separator to direct the oil back to the unit compressor crankcase.

Manufacturers may use other features to accomplish the same things. Keep in mind that manufacturers are trying to incorporate features in their equipment to make it lightweight, dependable, and efficient. Follow the manufacturers' instructions while using their equipment; they know the capabilities of their equipment.

The unit to be purchased should be chosen carefully. Some questions should be asked when determining which unit to purchase.

Recovery Unit

1. Will it run under the conditions needed for the majority of the jobs encountered? For example, will it operate properly on the rooftop in hot or cold weather?
2. Will it recover both liquid and vapor if you need these features?
3. What is the pumping rate in the liquid and vapor state for the refrigerants you need to recover?
4. Are the drier cores standard and can they be purchased locally?
5. Is the unit portability such that you can conveniently perform your service jobs?

Recovery/Recycle Units

1. Will the unit run under the conditions needed for the majority of the jobs encountered?
2. Will it recycle enough refrigerant for the type of jobs encountered?
3. What is the recycle rate and does it correspond to the rate you may need for the size systems you service?
4. Are the drier cores standard and can they be purchased locally?
5. Is the unit portable enough for your jobs and transportation availability?

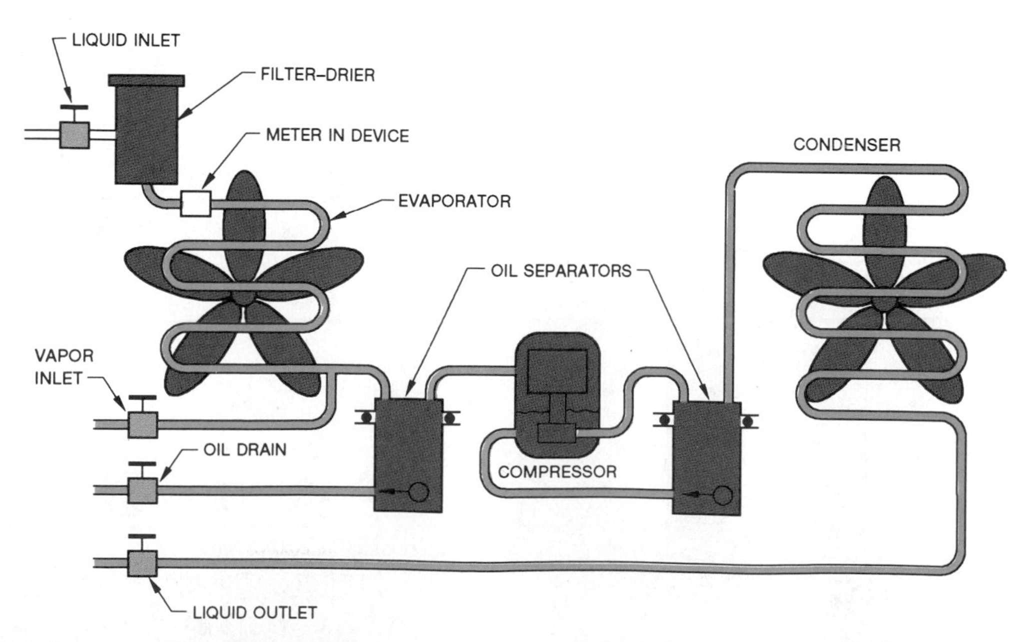

FIGURE 9–15 This system removes either liquid or vapor.

9.10 RECOVERING REFRIGERANT FROM APPLIANCES

Recovering refrigerant from appliances may be an easier job than from larger systems because not as much refrigerant is involved. Most small appliances will contain less than 2 lb of refrigerant. This can probably be removed into an evacuated approved cylinder by reducing the pressure in the cylinder to 1000 microns and connecting it to the appliance and opening the valves. Only 80% to 90% of the charge needs to be removed from appliances. The procedure would be as follows:

1. Evacuate the recovery cylinder to 1000 microns.
2. Fasten a line tap valve on the appliance liquid line leaving the condenser.
3. Fasten the other end of the line to the approved recovery cylinder and purge the line of air.
4. Start the compressor if it will run and open the valve to the cylinder at the same time.
5. Let the refrigerant equalize over to the cylinder until it stops flowing. Place the line to your ear and listen for the flow to stop, Figure 9–16.

The compressor will pull the refrigerant from the low side of the system and move it to the high side of the system and most of the refrigerant charge will flow into the cylinder. If the compressor will not run, the cylinder may be cooled with ice to encourage more refrigerant to move into the cylinder.

Refrigerant bags are available for recovery of refrigerant from small appliances. These bags are plastic and will hold the charge of several refrigerators, Figure 9–17. The technician must have enough room in the service truck to haul the bag. When the bag is full, it may be taken to the shop and the refrigerant transferred into the reclaim cylinder for later pickup by or shipment to the reclaim company.

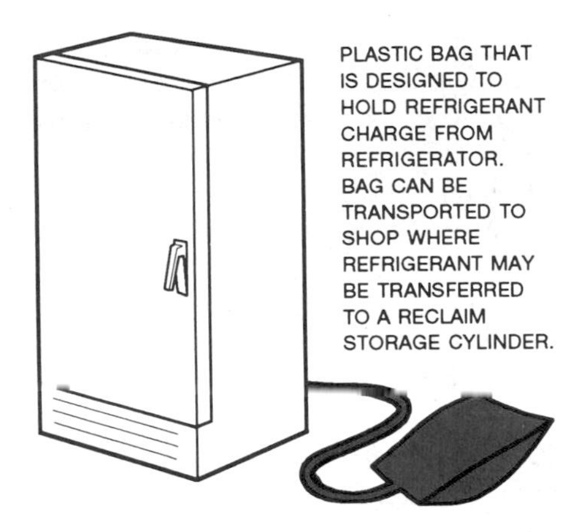

PLASTIC BAG THAT IS DESIGNED TO HOLD REFRIGERANT CHARGE FROM REFRIGERATOR. BAG CAN BE TRANSPORTED TO SHOP WHERE REFRIGERANT MAY BE TRANSFERRED TO A RECLAIM STORAGE CYLINDER.

FIGURE 9–17 Removing refrigerant to special plastic bag.

9.11 RECLAIMING REFRIGERANT

Because refrigerant must meet the ARI 700 standard before it is resold or reused in another customer's equipment, all refrigerant for reclaim will need to be sent to a reclaim company where it can be checked and received (provided it is in good enough condition to be used). Many service companies will not try to recycle refrigerant in the field because of the complications and the possibility of cross contamination of products. Reclaiming the refrigerant is another option. Refrigerant that meets the ARI 700 specification can be purchased from reclaim companies and it does not have the added tax. Since the refrigerant cannot be vented, the only other choice is to recover it to a central location and place it in a cylinder that may be shipped to a reclaim company. The reclaim companies will only accept large quantities of refrigerant so the service shop will have a need for a cylinder of the correct size for each of the refrigerants

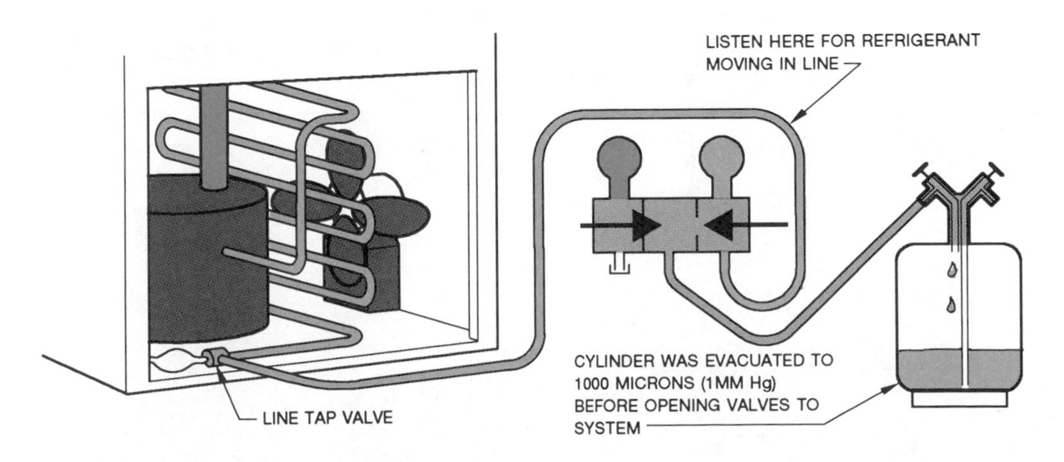

FIGURE 9–16 Listen to determine when the refrigerant flow has stopped.

that will be shipped. Reclaim companies will be accepting refrigerant in cylinders of 100 lb and larger. Some companies are buying recovered refrigerant and accumulating it to be reclaimed.

The refrigerant in the reclaim cylinders must be of one type of refrigerant. The refrigerants cannot be allowed to be mixed or the reclaim company cannot reclaim it. As mentioned before, mixed refrigerants cannot be separated. When the reclaim company receives refrigerant that cannot be reclaimed, it must be destroyed. The method of destroying refrigerant is to incinerate the refrigerant in such a manner that the fluorine in the refrigerant is captured. This is an expensive process, so technicians must be careful not to mix the refrigerants.

9.12 REFRIGERANTS IN THE FUTURE

It will pay the technician of the future to stay up to date on all developments as they occur. You should join local trade organizations that are watching industry trends. For example, in the future there will be changes in the tools used to handle refrigerant. Vacuum pumps will soon be required to be able to pump refrigerant into recovery cylinders so the refrigerant that is removed from a system during evacuation will be captured.

Gage lines that will not leak refrigerant are necessary. Many older gage lines seep refrigerant while connected. Gage lines are also required to have valves to reduce the amount of refrigerant lost when they are disconnected from Schrader connections. These valves are located in the end of the hoses and hold the refrigerant in the hose when disconnected.

Each technician should make an effort to know what the latest information and technology is and practice the best techniques available.

SUMMARY

- CFC refrigerants must not be vented to the atmosphere.
- Ozone is a form of oxygen, with each molecule consisting of three atoms of oxygen (O_3). The oxygen we breathe contains two atoms (O_2).
- The ozone layer, located in the stratosphere 7 to 30 miles above the earth's surface, protects the earth from harmful ultraviolet rays from the sun, which can cause damage to human beings, animals, and plants.
- Chlorofluorocarbons (CFCs) contain chlorine, fluorine, and carbon atoms.
- Hydrochlorofluorocarbons (HCFCs) contain hydrogen, chlorine, fluorine, and carbon atoms.
- Hydrofluorocarbons (HFCs) contain hydrogen, fluorine, and carbon atoms.
- Hydrogen atoms help the HCFCs break down before they reach the stratosphere and react with the ozone.
- The EPA has established many regulations governing the handling of refrigerants.
- To recover refrigerant is "to remove refrigerant in any condition from a system and store it in an external container without necessarily testing or processing it in any way."
- To recycle refrigerant is "to clean the refrigerant by oil separation and single or multiple passes through devices, such as replaceable core filter-driers, which reduce moisture, acidity and particulate matter. This term usually applies to procedures implemented at the job site or at a local service shop."
- To reclaim refrigerant is "to process refrigerant to new product specifications by means which may include distillation. It will require chemical analysis of the refrigerant to determine that approriate product specifications are met. This term usually implies the use of processes or procedures available only at a reprocessing or manufacturing facility."
- Refrigerant can be transferred only into Department of Transportation (DOT)-approved cylinders and tanks.
- Manufacturers have developed many types of equipment to recover and recycle refrigerant.

REVIEW QUESTIONS

1. What is the difference between the oxygen we breathe and the ozone located in the earth's stratosphere?
2. Why are CFCs more harmful to the ozone layer than HCFCs?
3. What is the name of the conference that was held in Canada in 1987 to attempt to solve the problems of released refrigerants?
4. What agency of the federal government is charged with implementing the United States Clean Air Act Amendments of 1990?
5. Are federal excise taxes on reclaimed refrigerants the same as those for new refrigerants?
6. To recover refrigerant means to ______.
7. To recycle refrigerant means to ______.
8. To reclaim refrigerant means to ______.
9. What is the name of the federal agency that must approve containers into which refrigerant may be transferred?
10. State the standard that a refrigerant must meet after it has been reclaimed and before it can be sold.

UNIT 10 System Charging

OBJECTIVES

After studying this unit, you should be able to

- describe how refrigerant is charged into systems in the vapor and the liquid states.
- describe system charging using two different weighing methods.
- state the advantage of using electronic scales for weighing refrigerant into a system.
- describe two types of charging devices.

SAFETY CHECKLIST

* Never use concentrated heat from a torch to apply heat to a refrigerant cylinder. Use a gentle heat such as from a tub of warm water no hotter than 90°F.
* Do not charge liquid refrigerant into the suction line of a system. Liquid refrigerant must not enter the compressor.
* Liquid refrigerant is normally added in the liquid line and then only under the proper conditions.
* If a device is used to flash liquid refrigerant into the suction line, its use must be limited to those systems where the suction gas passes over the windings of the compressor motor.

10.1 CHARGING A REFRIGERATION SYSTEM

Charging a system refers to the adding of refrigerant to a refrigeration system. The correct charge must be added for a refrigeration system to operate as it was designed to, and this is not always easy to do. Each component in the system must have the correct amount of refrigerant. The refrigerant may be added to the system in the vapor or liquid states by weighing, by measuring, or by using operating pressure charts.

This unit describes how to add the refrigerant correctly and safely in the vapor state and the liquid state. The correct charge for a particular system will be discussed in the unit where that system is covered. Heat pump charging, for example, is covered in the unit on heat pumps.

10.2 VAPOR REFRIGERANT CHARGING

Vapor refrigerant charging of a system is accomplished by allowing vapor to move out of the vapor space of a refrigerant cylinder and into the low-pressure side of the refrigerant system. When the system is not operating—for example, when a vacuum has just been pulled, or when the system is out of refrigerant—you can add vapor to the low- and high-pressure sides of the system. When the system is running, refrigerant may normally be added only to the low side of the system because the high side is under more pressure than the refrigerant in the cylinder. For example, an R-12 system may have a head pressure of 169 psig on a 95°F day (this is determined by taking the outside temperature of 95°F and adding 30°F; this gives a condensing temperature of 125°F, or 169 psig for R-12). The cylinder is setting in the same ambient temperature of 95°F but only has a pressure of 108 psig, Figure 10–1.

The low-side pressure in an operating system is much lower than the cylinder pressure if the cylinder is warm. For

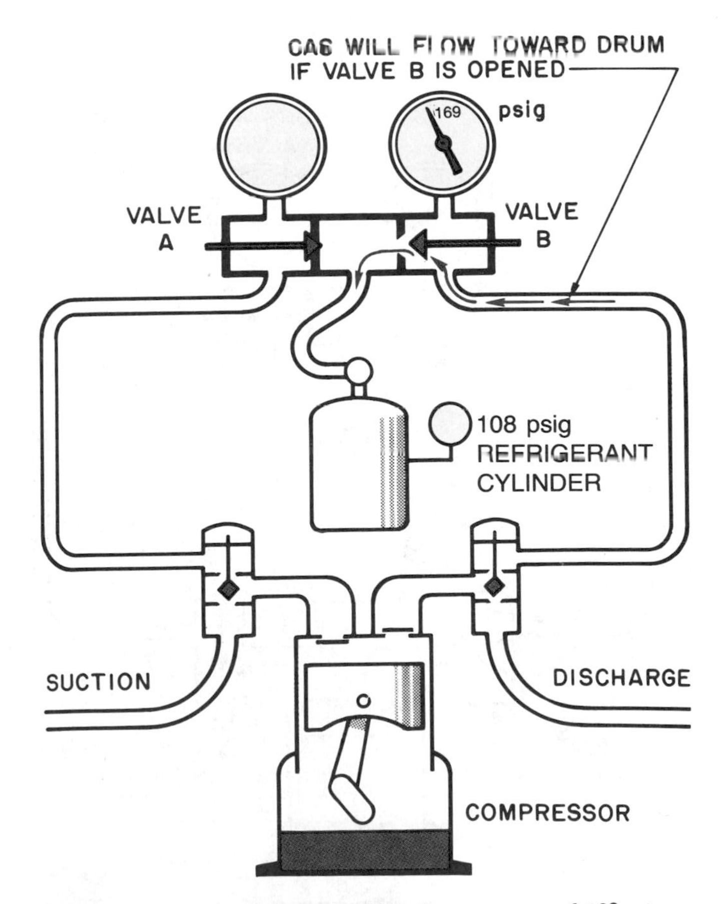

FIGURE 10–1 This refrigerant cylinder has a pressure of 108 psig. The high side of the system has a pressure of 169 psig. Pressure in the system will prevent the refrigerant in the cylinder from moving into the system.

example, on a 95°F day, the cylinder will have a pressure of 108 psig, but the evaporator pressure may be only 20 psig. Refrigerant will easily move into the system from the cylinder, Figure 10–2. In the winter, however, the cylinder may have been in the back of the truck all night and its pressure may be lower than the low side of the system, Figure 10–3. In this case the cylinder will have to be warmed to get refrigerant to move from the cylinder to the system. It is a good idea to have a cylinder of refrigerant stored in the equipment room of large installations. The cylinder will always be there in case you have no refrigerant in the truck, and the cylinder will be at room temperature even in cold weather.

When vapor refrigerant is pulled out of a refrigerant cylinder, the liquid boils to replace the vapor that is leaving. As more and more vapor is released from the cylinder, the liquid in the bottom of the cylinder continues to boil, and its temperature decreases. If enough refrigerant is released, the cylinder pressure will decrease to the low-side pressure of the system. Heat will have to be added to the liquid refrigerant to keep the pressure up. ***But *never* use concentrated heat from a torch. Gentle heat, such as from a tub of warm water, is safer. The water temperature should not exceed 90°F.*** This will maintain a cylinder pressure of 100 psig for R-12 if the refrigerant temperature is kept the same as the water in the tub. Move the refrigerant cylinder around to keep the liquid in the center of the cylinder in touch with the warm outside of the cylinder, Figure 10–4. A water

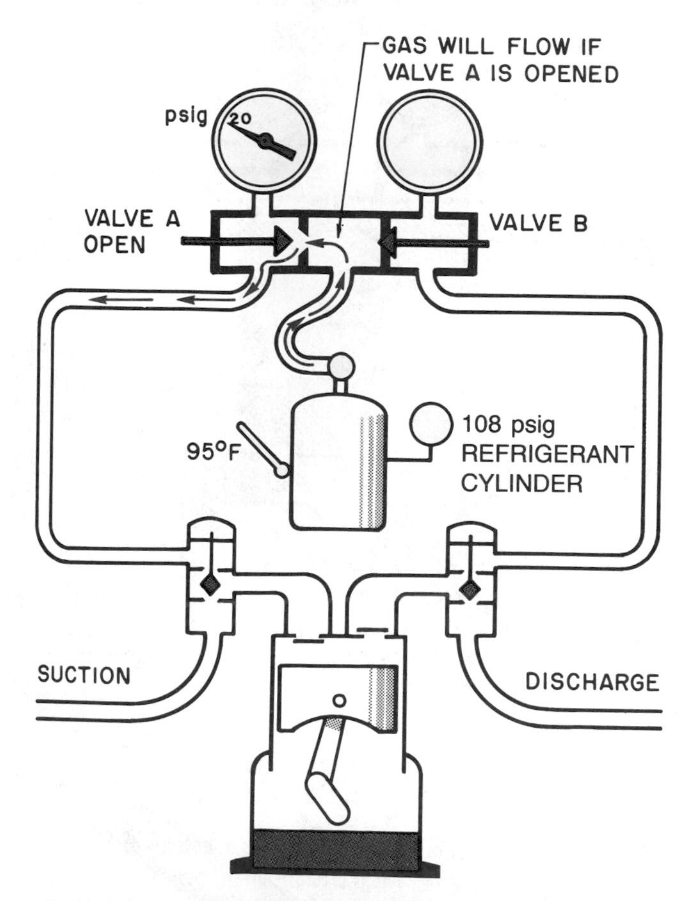

FIGURE 10–2 Temperature of the cylinder is 95°F. The pressure inside the cylinder is 108 psig. The low-side pressure is 20 psig.

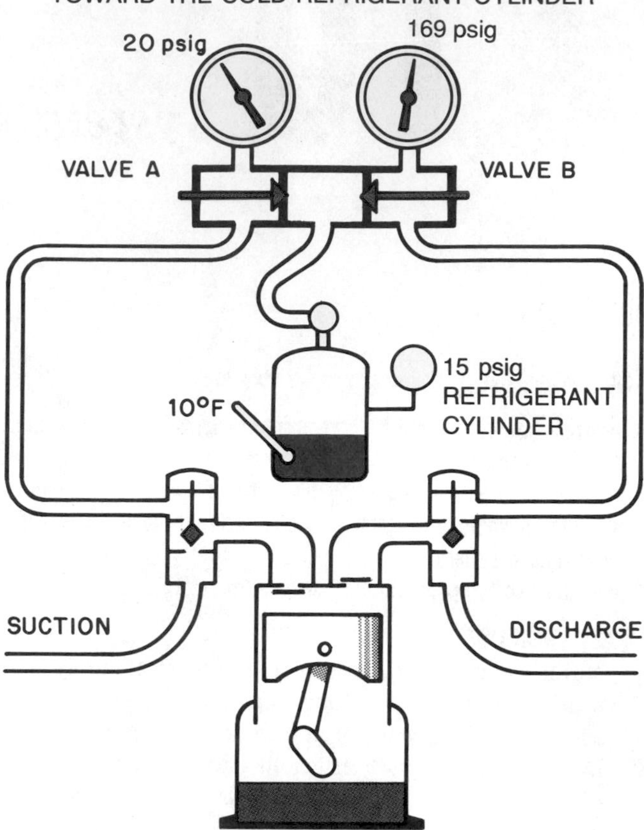

FIGURE 10–3 The refrigerant in this cylinder is at a low temperature and pressure because it has been in the back of a truck all night in cold weather. The cylinder pressure is 15 psig, which corresponds to 10°F. The pressure in the system is 20 psig.

FIGURE 10–4 Refrigerant cylinder is in warm water to keep the pressure up.

temperature of 90°F is a good one to work with because it is the approximate temperature of the human hand. If the water begins to feel warm to the hand, it is getting too hot.

The larger the volume of liquid refrigerant in the bottom of the cylinder, the longer the cylinder will maintain the pressure. When large amounts of refrigerant must be charged into a system, use the largest cylinder available. For example, don't use a 25-lb cylinder to charge 20 lb of refrigerant into a system if a 125-lb cylinder is available.

10.3 LIQUID REFRIGERANT CHARGING

Liquid refrigerant charging of a system is normally accomplished in the liquid line. For example, when a system is out of refrigerant, liquid refrigerant can be charged into the *king* valve on the liquid line. If the system is in a vacuum, you can connect to the liquid valve of a cylinder of refrigerant and liquid refrigerant may be allowed to enter the system until it has nearly stopped. The liquid will enter the system and move toward the evaporator and the condenser. When the system is started, the refrigerant is about equally divided between the evaporator and the condenser, and there is no danger of liquid flooding into the compressor, Figure 10–5. When charging with liquid refrigerant, the cylinder pressure is not reduced. When large amounts of refrigerant are needed, the liquid method is preferable to other methods because it saves time.

When a system has a king valve, it may be *front seated* while the system is operating, and the low-side pressure of the system will drop. Liquid from the cylinder may be charged into the system at this time through an extra charging port. The liquid from the cylinder is actually feeding the expansion device. **Be careful not to overcharge the system,** Figure 10–6. The low-pressure control may have to be bypassed during charging to keep it from shutting the system off. Be sure to remove the electrical bypass when charging is completed, Figure 10–7. ***Every manufacturer cautions against charging liquid refrigerant into the suction line of a compressor. To repeat, liquid refrigerant must not enter the compressor.***

Some commercially available charging devices allow the cylinder liquid line to be attached to the suction line for charging a system that is running. They are orifice-metering devices that are actually a restriction between the gage man-

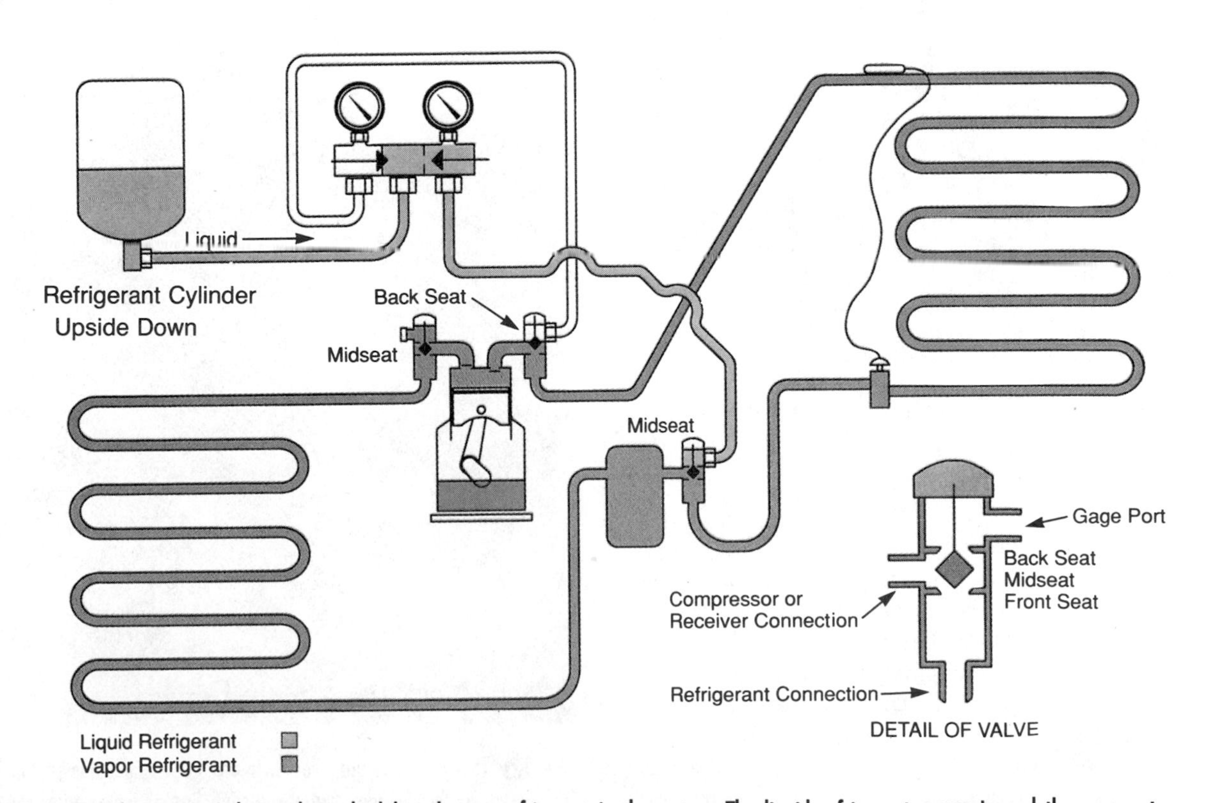

FIGURE 10–5 This system is being charged while it has no refrigerant in the system. The liquid refrigerant moves toward the evaporator and the condenser when doing this. No liquid refrigerant will enter the compressor.

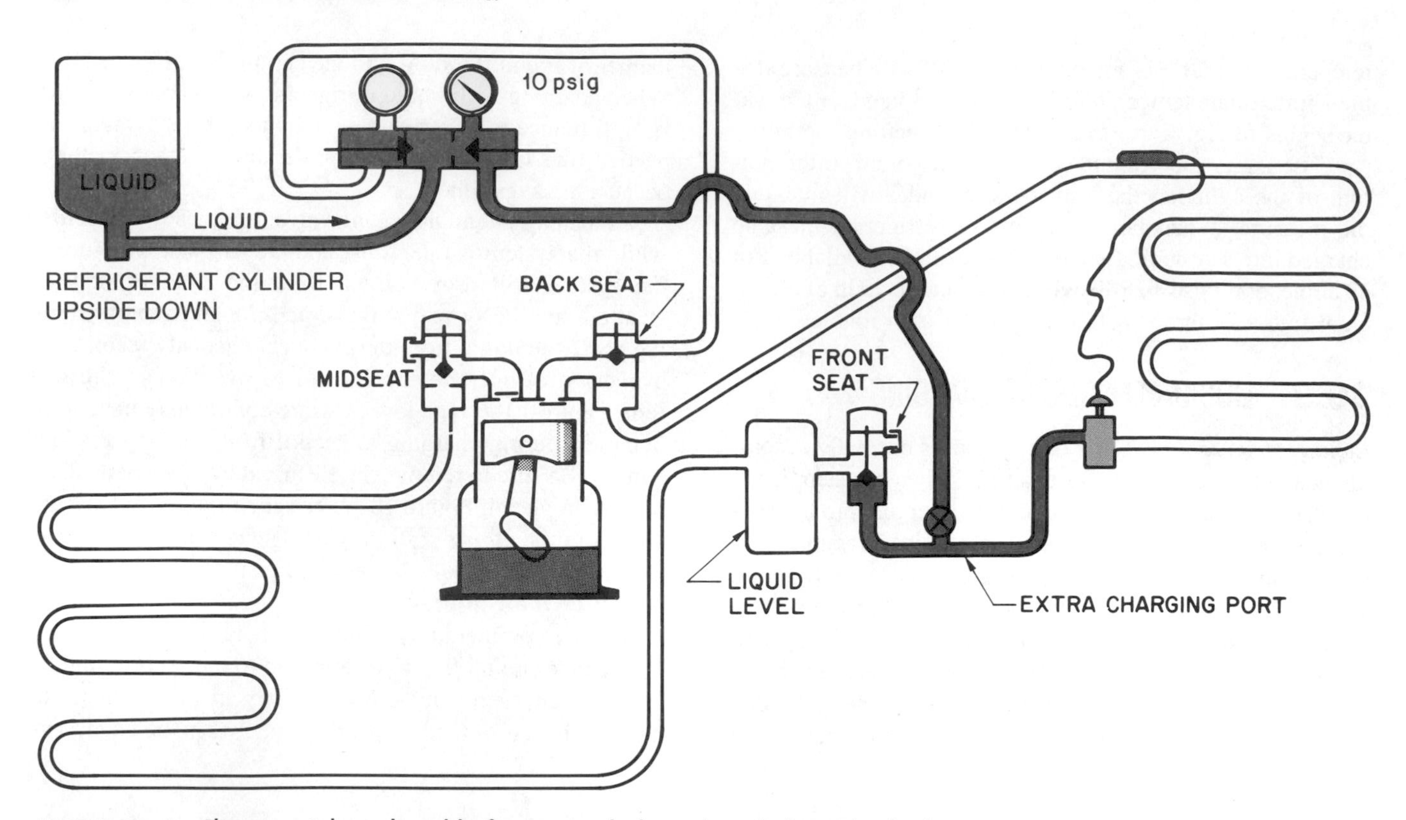

FIGURE 10–6 This system is being charged by front seating the king valve and allowing liquid refrigerant to enter the liquid line.

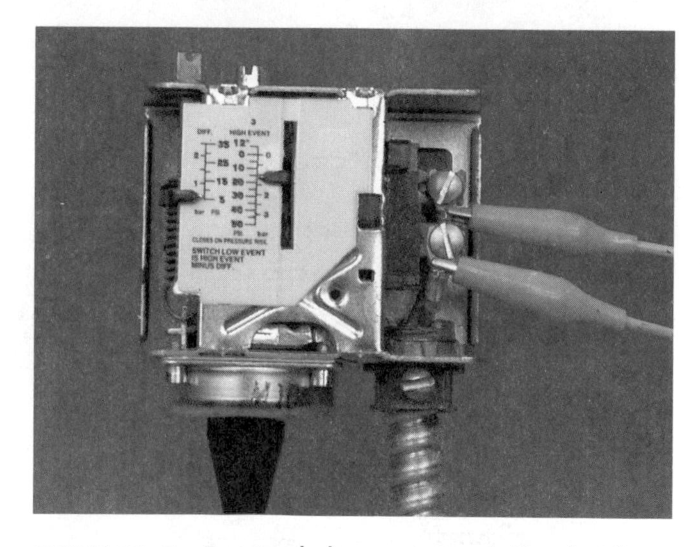

FIGURE 10–7 Bypassing the low-pressure control. *Photo by Bill Johnson*

ifold and the system's suction line, Figure 10–8. They meter liquid refrigerant into the suction line where it flashes into a vapor. The same thing may be accomplished using the gage manifold valve, Figure 10–9. The pressure in the suction line is maintained at a pressure of not more than 10 lb higher than the system suction pressure. This will meter the liquid refrigerant into the suction line as a vapor. ***It should only be used on compressors where the suction gas passes over the motor windings. This will boil any small amounts of liquid refrigerant that may reach the compressor. If**

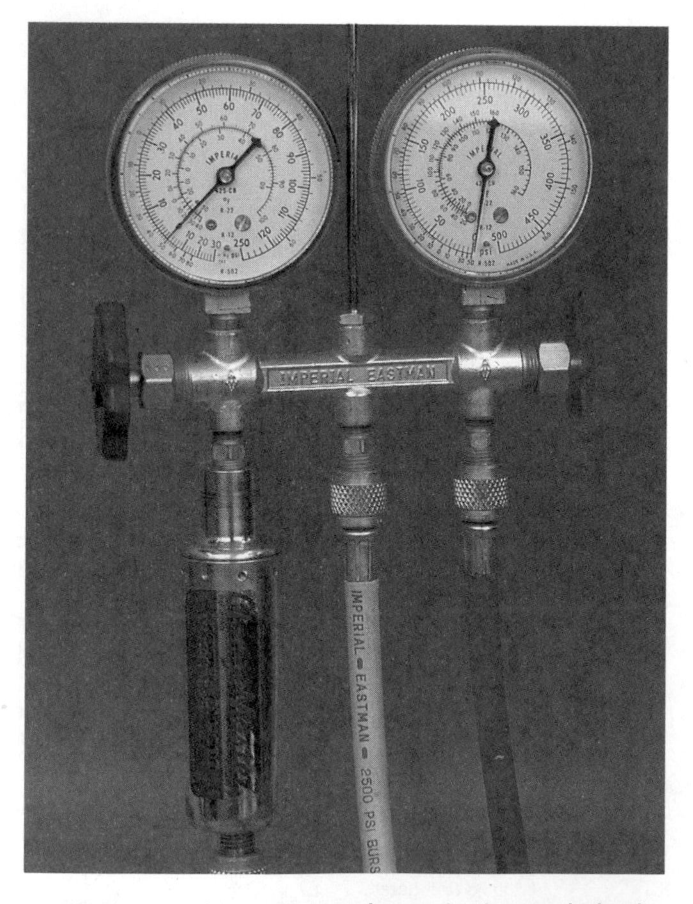

FIGURE 10–8 Charging device in the gage line between the liquid refrigerant in the cylinder and the suction line of the system. *Photo by Bill Johnson*

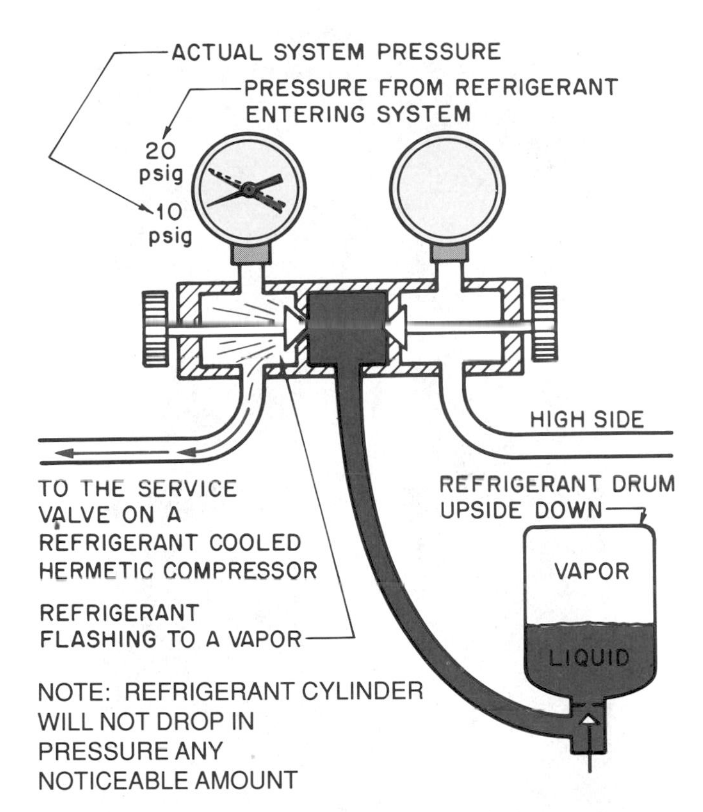

FIGURE 10–9 Gage manifold used to accomplish the same thing as in Figure 10–8.

the lower compressor housing becomes cold, stop adding liquid. This method should only be performed under the supervision of an experienced person.*

When a measured amount of refrigerant must be charged into a system, it may be weighed into the system or measured in by using a graduated charging cylinder. Package systems, such as air conditioners and refrigerated cases will have the recommended charge printed on the nameplate. This charge must be added to the system from a deep vacuum or it will not be correct.

10.4 WEIGHING REFRIGERANT

Weighing refrigerant may be accomplished with various scales. Bathroom and other inaccurate scales should *not* be used. Figure 10–10 shows an accurate dial scale graduated

FIGURE 10–10 Scale for measuring refrigerant. *Photo by Bill Johnson*

in pounds and ounces. Secure the scales (make sure they are portable) in the truck to keep the mechanism from shaking and changing the calibration. Dial scales can be difficult to use, as the next example shows.

Suppose 28 oz of refrigerant is needed. Put the cylinder of refrigerant on a dial scale and it is found to weigh 24 lb 4 oz. As refrigerant runs into the system, cylinder weight decreases. Determining the final cylinder weight is not easy for some technicians.

The calculated final cylinder weight is

$$24 \text{ lb } 4 \text{ oz} - 28 \text{ oz} = 22 \text{ lb } 8 \text{ oz}$$

To determine this, 24 lb 4 oz was converted to ounces:

$$24 \text{ lb} \times 16 \text{ oz/lb} = 384 \text{ oz} + 4 \text{ oz} = 388 \text{ oz}$$

Now subtract 28 oz from the 388 oz:

$$388 \text{ oz} - 28 \text{ oz} = 360 \text{ oz}$$

is the final cylinder weight. Because the scales do not read in ounces, you must convert to pounds and ounces:

$$360 \text{ oz} \div 16 \text{ oz/lb} = 22.5 \text{ lb} = 22 \text{ lb } 8 \text{ oz}$$

It's easy to make a mistake in this calculation, Figure 10–11.

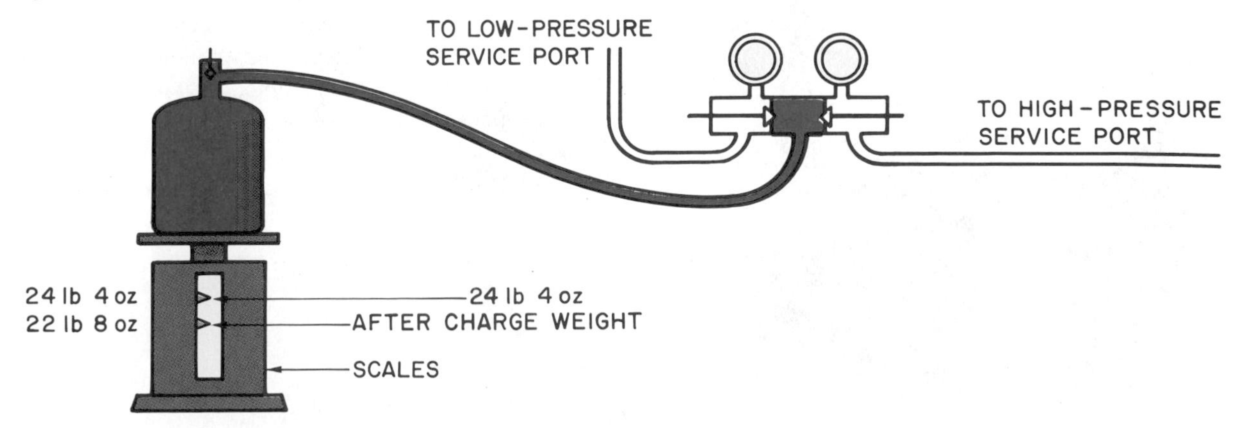

FIGURE 10–11 Using a set of scales to measure refrigerant into a system.

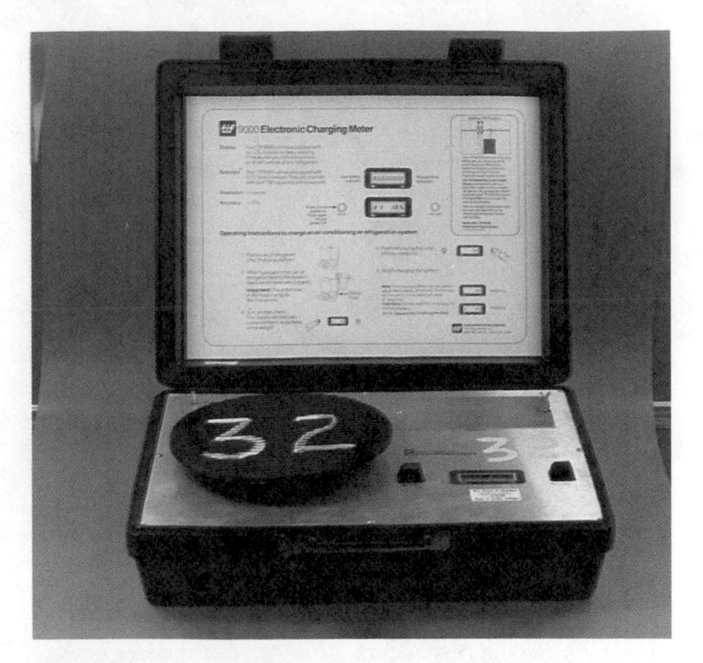

FIGURE 10–12 Electronic scale with adjustable zero feature. *Photo by Bill Johnson*

Electronic scales are often used, Figure 10–12. These are very accurate but more expensive than dial-type scales. These scales can be adjusted to zero with a full cylinder, so as refrigerant is added to the system the scales read a positive value. For example: if 28 oz of refrigerant is needed in a system, put the refrigerant cylinder on the scale and set the scale at 0. As the refrigerant leaves the cylinder, the scale counts upward. When 28 oz is reached, the refrigerant flow can be stopped. This is a timesaving feature that avoids the cumbersome calculations involved with the dial scale. Figure 10–13 shows an electronic scale that can be programmed for the correct amount of charge. A solid state microprocessor controls a solenoid that stops the charging process when the programmed weight has been dispensed.

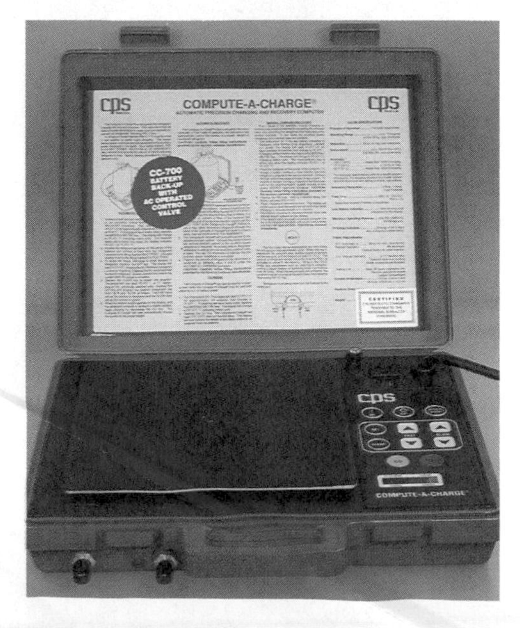

FIGURE 10–13 Programmable electronic scale. *Photo by Bill Johnson*

FIGURE 10–14 Graduated cylinder used to measure refrigerant into a system using volume. *Photo by Bill Johnson*

10.5 USING CHARGING DEVICES

Graduated cylinders are often used to add refrigerant to systems. These cylinders have a visible column of liquid refrigerant, so you can observe the liquid level in the cylinder. Use the pressure gage at the top of the cylinder to determine the temperature of the refrigerant. The liquid refrigerant has a different volume at different temperatures, so the temperature of the refrigerant must be known. This temperature is dialed on the graduated cylinder, Figure 10–14. The final liquid level inside the cylinder must be calculated much like the previous example, but it is not as complicated.

Suppose a graduated cylinder has 4 lb 4 oz of R-12 in the cylinder at 100 psig. Turn the dial to 100 psig and record the level of 4 lb 4 oz. The system charge of 28 oz is subtracted from the 4 lb 4 oz as follows:

$$4 \text{ lb} \times 16 \text{ oz/lb} = 64 \text{ oz} + \text{the remaining } 4 \text{ oz} = 68 \text{ oz}$$

Then

$$68 \text{ oz} - 28 \text{ oz} = 40 \text{ oz}$$

is the final cylinder level.

$$40 \text{ oz} \div 16 \text{ oz/lb} = 2.5 \text{ lb} = 2 \text{ lb } 8 \text{ oz}$$

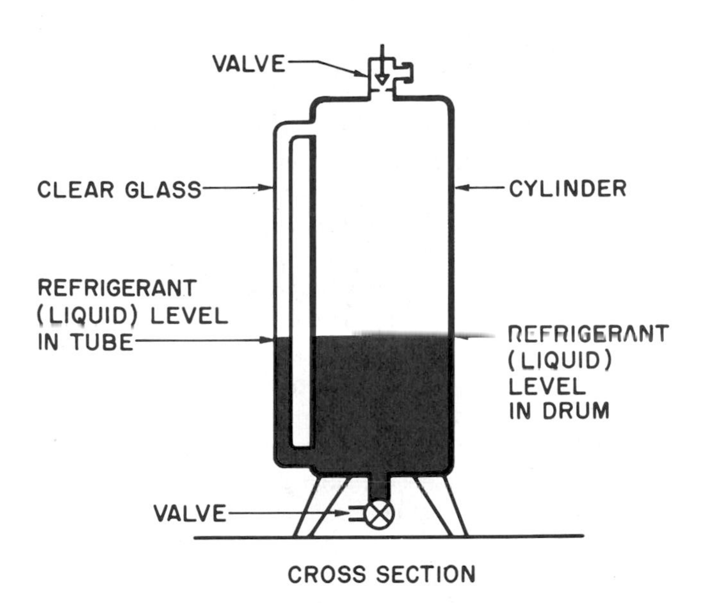

FIGURE 10–15 Cross section of a graduated cylinder. The refrigerant may be seen in the tube as it moves into the system.

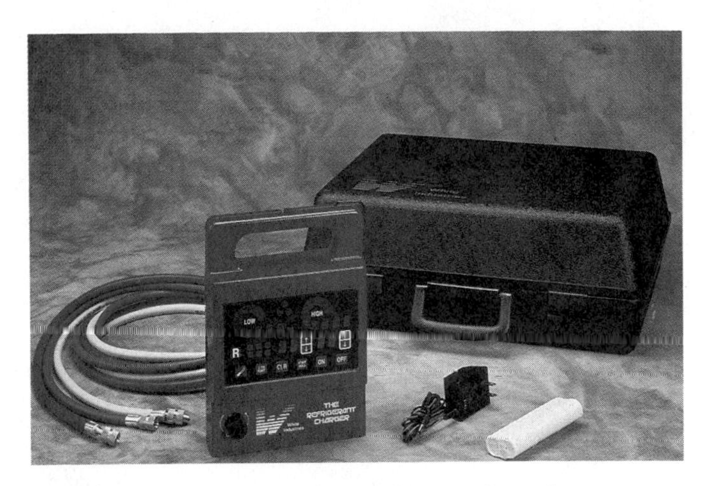

FIGURE 10-16 Refrigerant charger. *Courtesy White Industries*

The advantage of the graduated cylinder is the refrigerant can be seen as the level drops. See Figure 10–15 for an example of this setup.

Some graduated cylinders have heaters in the bottom to keep the refrigerant temperature from dropping when vapor is pulled from the cylinder.

When selecting a graduated cylinder for charging purposes, be sure you select one that is large enough for the systems that you will be working with. It is difficult to use a cylinder twice for one accurate charge. When charging systems with more than one type of refrigerant, you need a charging cylinder for each type of refrigerant. You will also not overcharge the customer or use the wrong amount of refrigerant if you closely follow the above-mentioned methods.

Other available refrigerant charging devices may make it more convenient to charge refrigerant into a system. Figure 10–16 is an example of one of these. This device can be set to charge a system using pressure or weight. A predetermined amount of refrigerant can be charged in pounds and ounces in 1-oz increments. This type of device also has many other features. Be sure to follow the manufacturer's instructions.

Some system manuals give typical operating pressures that may be compared to the gage readings for determining the correct charge. These are called *charging charts*. Figure 10–17 can be used while using the manufacturer's directions. A charging chart must be developed by the manufacturer for each system manufactured so there are no general charts. This is discussed in various units in this text as it applies to a particular system.

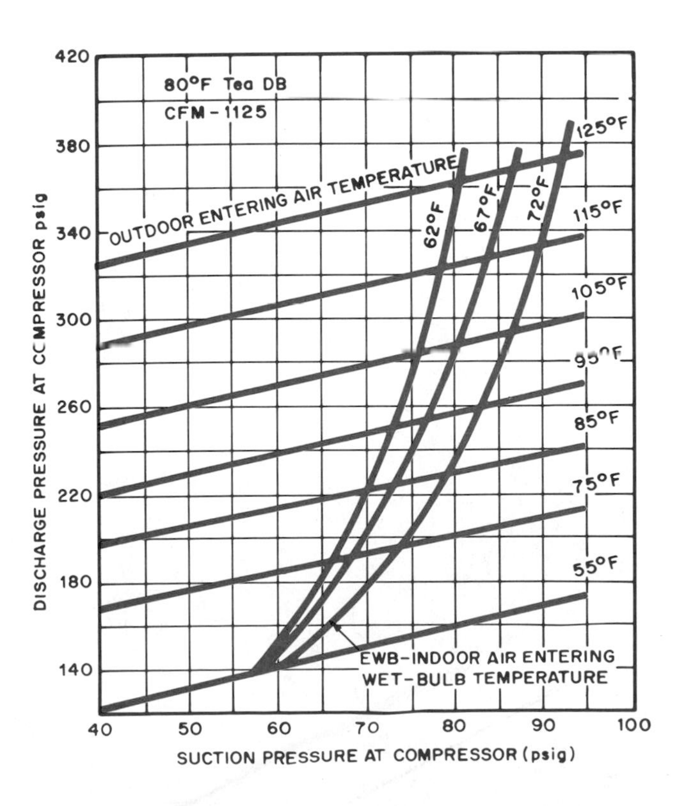

FIGURE 10–17 Typical system charging chart. Pressures and temperatures are plotted to arrive at the correct charge. *Courtesy Carrier Corporation*

SUMMARY

- Refrigerant may be added to the refrigeration system in the vapor state or the liquid state under the proper conditions.
- When refrigerant is added in the vapor state, the refrigerant cylinder will lose pressure as the vapor is pushed out of the cylinder.
- ***Liquid refrigerant is normally added in the liquid line and only under the proper conditions.***
- ***Liquid refrigerant must never be allowed to enter the compressor.***
- Refrigerant is measured into systems using weight and volume.
- It can be difficult to add refrigerant using dial scales because the final cylinder weight must be calculated. The scales are graduated in pounds and ounces.
- Electronic scales may have a cylinder-emptying feature that allows the scales to be adjusted to zero with a full cylinder of refrigerant on the platform.
- Graduated cylinders use the volume of the liquid refrigerant. This volume varies at different temperatures. These may be dialed onto the cylinder for accuracy.

REVIEW QUESTIONS

1. How is liquid refrigerant added to the refrigeration system when the system is out of refrigerant?
2. How is the refrigerant cylinder pressure kept above the system pressure when charging with vapor from a cylinder?
3. Why does the refrigerant pressure decrease in a refrigerant cylinder while charging with vapor?
4. What is the main disadvantage of dial scales?
5. What type of equipment normally has the refrigerant charge printed on the nameplate?
6. What feature of digital electronic scales makes them useful for refrigerant charging?
7. How is refrigerant pressure maintained in a graduated cylinder?
8. How does a graduated cylinder account for the volume change due to temperature changes?
9. What must you remember when purchasing a charging cylinder?
10. What methods besides weighing and measuring are used for charging systems?

UNIT 11

Calibrating Instruments

OBJECTIVES

After studying this unit, you should be able to

- describe instruments used in heating, air conditioning, and refrigeration.
- test and calibrate a basic thermometer at the low- and high-temperature ranges.
- check an ohmmeter for accuracy.
- describe the comparison test for an ammeter and a voltmeter.
- describe procedures for checking pressure instruments above and below atmospheric pressure.
- check flue-gas analysis instruments.

11.1 THE NEED FOR CALIBRATION

The service technician cannot always see or hear what is occurring within a machine or piece of equipment. Instruments such as voltmeters and temperature testers are used to help. Therefore, these instruments must be reliable. Although the instruments should be calibrated when manufactured, this is not always the case. They may need to be checked before you use them. Even if they are perfectly calibrated, the instruments may not remain so due to use and because of conditions such as moisture and vibration. Instruments may be transported in a truck over rough roads to the job site. The instrument stays in the truck through extremes of hot and cold weather. The instrument compartment may sweat (moisture due to condensation of humidity from the air) and cause the instruments to become damp. All of these things cause stress to the instrument. They may not stay in calibration over a long period of time.

The technician should always take proper care of tools and equipment. The technician is very dependent on the tools and equipment used everyday in the performance of a job. There is no substitute for good common sense, careful use, and proper storage of these tools and various pieces of equipment. *Some of these instruments are used to check for voltage to protect the technician from electrical shock. This instrument must function correctly for safety sake.*

11.2 CALIBRATION

Some instruments cannot be calibrated. *Calibration* means to change the instrument's output or reading to correspond to a standard or correct reading. For example, if a speedometer shows 55 mph for an automobile actually traveling at 60 mph, the speedometer is out of calibration. If the speedometer can be changed to read the correct speed, it can be calibrated. If the speedometer cannot be changed to read the actual speed, it cannot be calibrated.

Some instruments are designed for field use and will stay calibrated longer. The new electronic instruments with digital readout features may not be as sensitive to field use as the analog (needle-type) instruments, Figure 11–1.

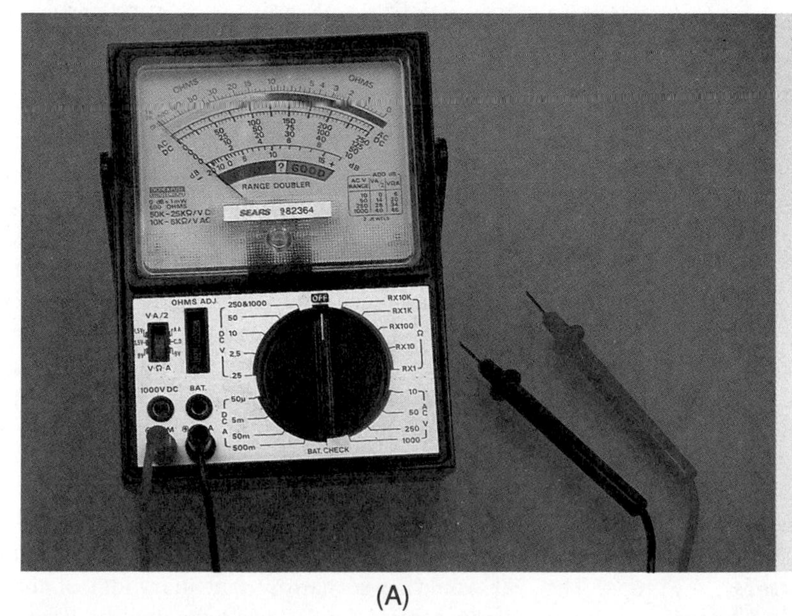

(A)

(B)

FIGURE 11–1 (A) Analog meter. (B) Digital meter. *(A) Photo by Bill Johnson, (B) Courtesy Beckman Industrial Corporation*

This unit deals with the most common instruments used for troubleshooting. These instruments measure temperature, pressure, voltage, amperage, resistance, and refrigerant leaks and make flue-gas analyses. To check and calibrate instruments, you must have reference points. Some instruments can be readily calibrated, some must be returned to the manufacturer for calibration, and some cannot be calibrated. We recommend that whenever you buy an instrument, you check some readings against known values. If the instrument is not within the standards the manufacturer states, return it to the supplier or to the manufacturer. Save the box the instrument came in as well as the directions and warranty. They can save you much time.

11.3 TEMPERATURE-MEASURING INSTRUMENTS

Temperature-measuring instruments measure the temperature of vapors, liquids, and solids. Air, water, and refrigerant in copper lines are the common substances measured for temperature level. Regardless of the medium to be measured, the methods for checking the accuracy of the instruments are similar.

Refrigeration technicians must have thermometers that are accurate from −50°F to 50°F to measure the refrigerant lines and the inside of coolers. Higher temperatures are experienced when measuring ambient temperatures, such as when the operating pressures for the condenser are being examined. Heating and air conditioning technicians must measure air temperatures from 40°F to 150°F, and water temperatures as high as 220°F for normal service. This can require a wide range of instruments. For temperatures above 250°F, for example, flue-gas analysis in gas- and oil-burning equipment, special thermometers are used. The thermometer is included in the flue-gas analysis kit, Figure 11–2.

In the past, most technicians relied on glass-stem mercury or alcohol thermometers. These are easy to use to measure fluid temperature when the thermometer can be

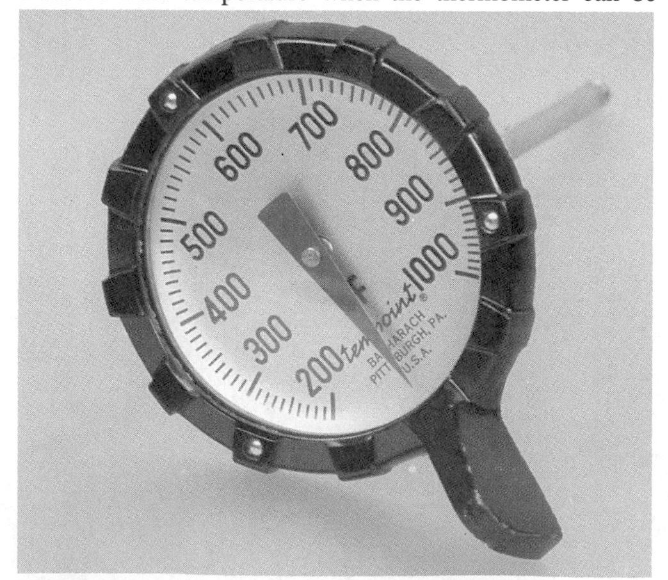

FIGURE 11–2 The thermometer is included in the flue-gas analysis kit. Note the high-temperature range of the thermometer. *Photo by Bill Johnson*

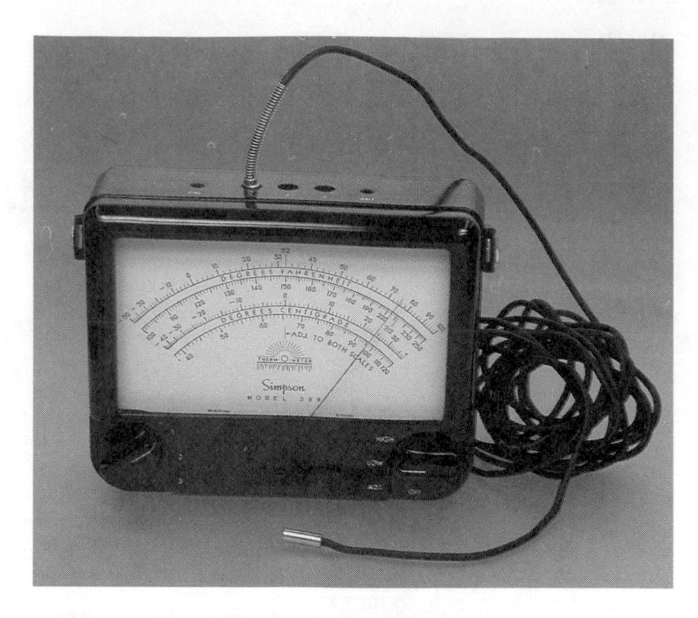

FIGURE 11–3 Analog-type electronic thermometer. *Photo by Bill Johnson*

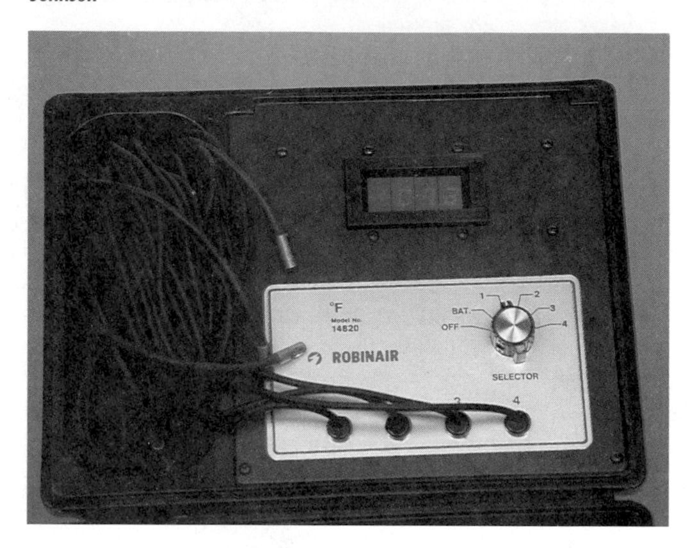

FIGURE 11–4 Digital-type electronic thermometer. *Photo by Bill Johnson*

inserted into the fluid, but they are difficult to use to measure temperature of solids. They are being replaced by the electronic thermometer, which is very popular. Electronic thermometers are simple, economical, and accurate, Figures 11–3 and 11–4. Both the analog and digital versions are adequate. Although the digital instrument costs more, it retains accuracy for a longer time under rough conditions.

The pocket-type dial thermometer is often used for field readings, Figure 11–5. It is not intended to be a laboratory-grade accurate instrument. The scale on this unit goes from 0°F to 220°F in a very short distance. The distance the needle must move to travel from the bottom of the scale to the top of the scale is only about 2.5 in. (the circumference of the dial). This would be like having a speedometer on an automobile with a 1.25 in. travel from 0 to 100 mph. The dial would be so narrow that the width of the needle would cover several miles per hour and the driver could not be sure of the actual speed.

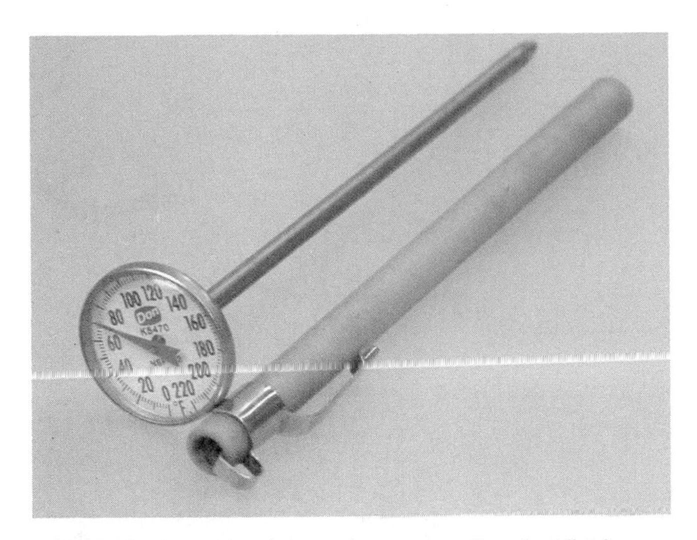

FIGURE 11–5 Pocket dial-type thermometer. *Photo by Bill Johnson*

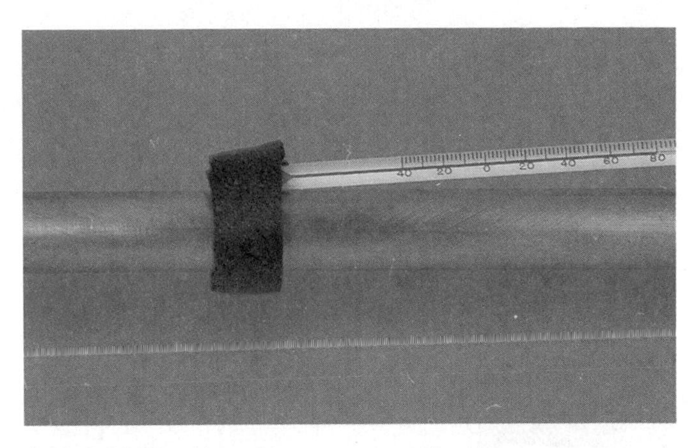

FIGURE 11–7 The technician must remember that a temperature-sensing element indicates the temperature of the sensing element. *Photo by Bill Johnson*

Three reference points are easily obtainable for checking temperature-measuring instruments: 32°F (ice and water), 98.6°F (body temperature), and 212°F (boiling point of water), Figure 11–6. The reference points should be close to the temperature range in which you are working. When using any of these as a reference for checking the accuracy of a temperature-measuring device, remember that a thermometer indicates the temperature of the sensing element. The reason for mentioning this is that many technicians make the mistake of thinking that the sensing element indicates the temperature of the medium being checked. *It does not.* Many inexperienced technicians merely set a thermometer lead on a copper line and read the temperature, but the thermometer sensing element has more contact with the surrounding air than with the copper line, Figure 11–7. It must be in contact with the medium to be measured for a long enough time for the sensor to become the same temperature as the medium for an accurate reading to appear on the instrument.

One method of temperature-instrument checking is to submerge the instrument-sensing element into a known temperature condition (such as ice and water while the change of state is occurring) and allow the sensing element to reach the known temperature. The following method checks an electronic thermometer with four plug-in leads that can be moved from socket to socket.

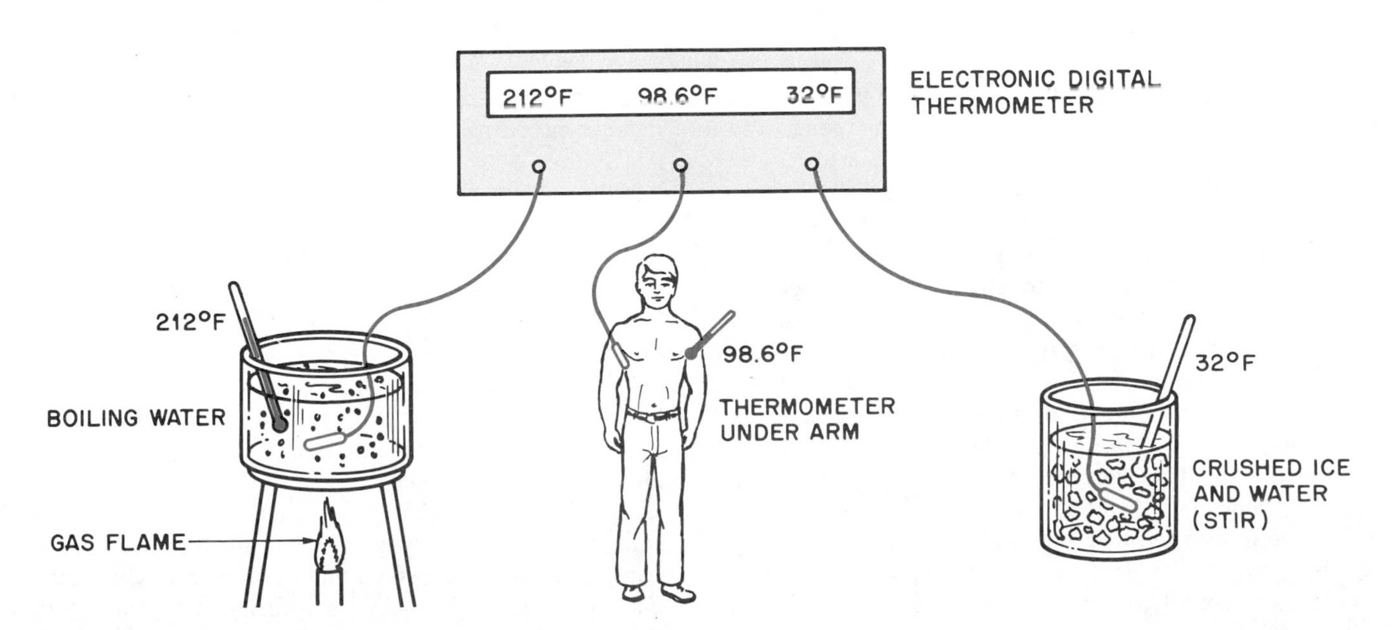

FIGURE 11–6 Three reference points that a service technician may use.

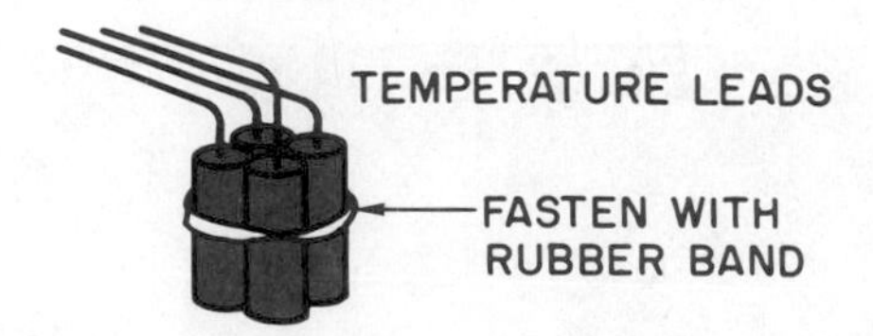

FIGURE 11–8 The four leads to the temperature tester are fastened at the ends so that they can all be submerged in water at the same time.

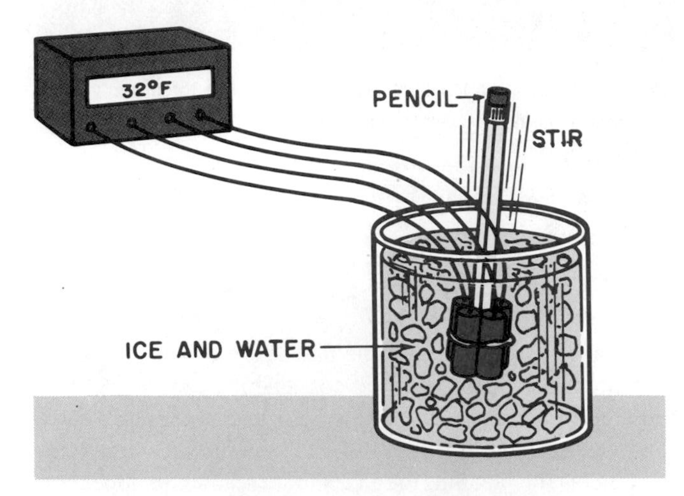

FIGURE 11–9 A pencil is fastened to the group to give the leads some rigidity so that they can be stirred in the ice and water. Note that the ice must reach to the bottom of the pan.

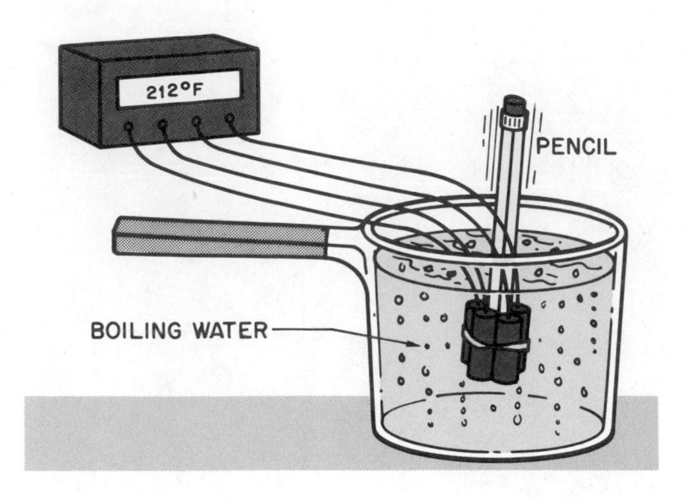

FIGURE 11–10 High-temperature test for accuracy of four temperature leads.

1. Fasten the four leads together as shown in Figures 11–8 and 11–9. Something solid can be fastened with them so that they can be stirred in ice and water.
2. For a low-temperature check, crush about a quart of ice, preferably made from pure water. If pure water is not available, make sure the water has no salt or sugar because either one changes the freezing point. You must crush the ice very fine (wrap it in a towel and pound it with a hammer), or there may be warm spots in the mixture.
3. Pour enough water, pure if possible, over the ice to almost cover the ice. **Do not cover the ice completely with the water,** or it will float and may be warmer on the bottom of the mixture. The ice must reach to the bottom of the vessel.
4. Stir the temperature leads in the mixture of ice and water, where the change of state is taking place, for at least 5 min. The leads must have enough time to reach the temperature of the mixture.
5. If the leads vary, note which leads are in error and by how much. The leads should be numbered, and the temperature differences marked on the instrument case, or mark the leads with their error.
6. For a high-temperature check, put a pan of water on a stove-top heating unit and bring the water to a boil. Make sure the thermometers you are checking indicate up to 212°F. If they do, immerse them in the boiling water. ***Do not let them touch the bottom of the pan or they may conduct heat directly from the pan bottom to the lead, Figure 11–10.*** Stir the thermometers for at least 5 min and check the readings. It is not critical that the thermometers be accurate to a perfect 212°F, because with products at these temperature levels, a degree or two one way or the other does not make a big difference. If any lead reads more than 4°F from 212°F, mark it defective. Remember that water boils at 212°F, at sea level at 70°F only. Any altitude above sea level will make a slight difference. If you are more than 1000 ft above sea level, we highly recommend that you use a laboratory glass thermometer as a standard and that you do not rely on the boiling water temperature being correct.

Accuracy is more important in the lower temperature ranges where small temperature differences are measured. A 1°F error does not sound like much until you have one lead that is off +1°F and another off −1°F and try to take an accurate temperature rise across a water heat exchanger that only has a 10°F rise. You have a built-in 20% error, Figure 11–11.

If a digital thermometer with leads that cannot be moved from socket to socket is used, there may be an adjustment in the back of the instrument for each lead. Figure 11–12 shows how this thermometer can be calibrated.

Glass thermometers often cannot be calibrated because the graduations are etched on the stem. If the graduations are printed on the back of the instrument, the back may be adjustable. A laboratory-grade glass thermometer is certified as to its accuracy, and it may be used as a standard for field instruments. It is a good investment for calibrating electronic thermometers, Figure 11–13.

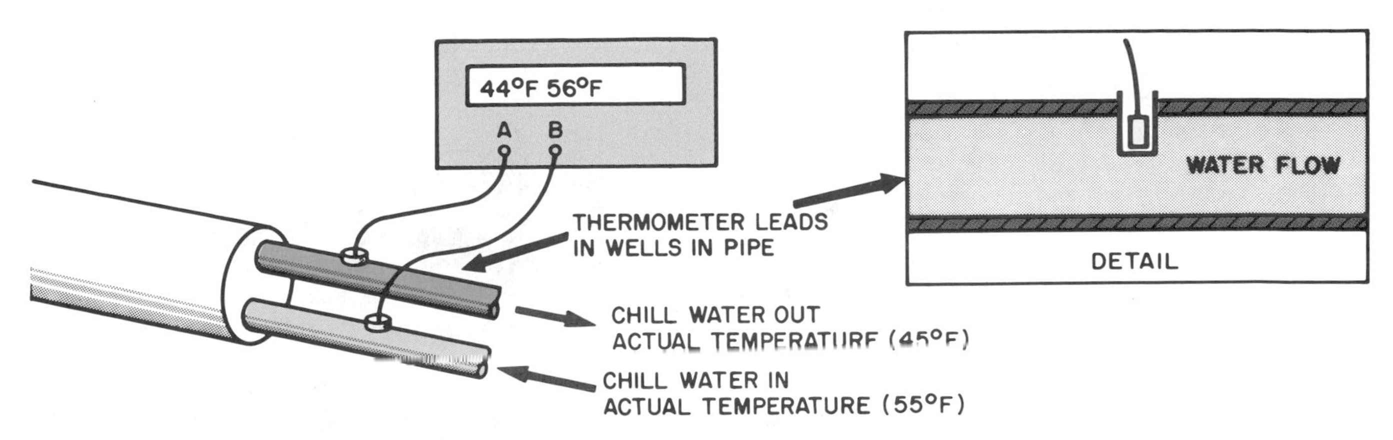

FIGURE 11–11 Thermometer with two leads accurate to within 1°F.

FIGURE 11–12 Calibration of thermometer with leads that cannot be moved or relocated within the instrument. *Photo by Bill Johnson*

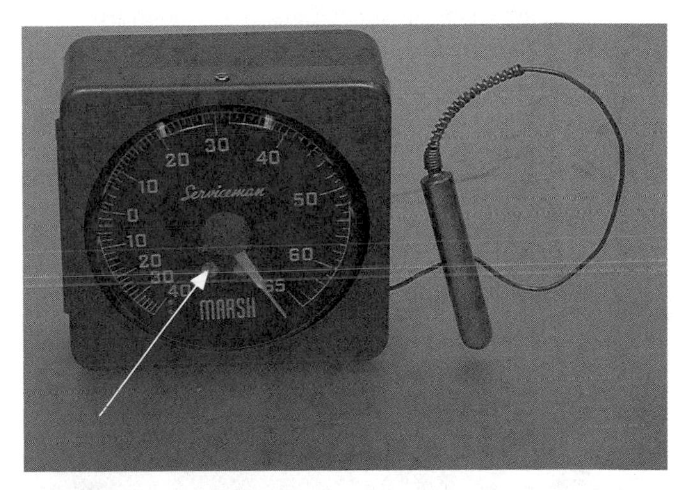

FIGURE 11–14 Large dial-type thermometers may be calibrated with the calibration screw. *Photo by Bill Johnson*

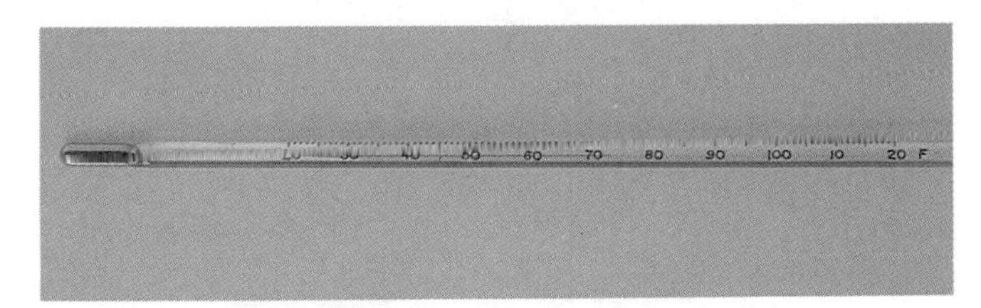

FIGURE 11–13 Laboratory-grade glass thermometer. *Photo by Bill Johnson*

Many dial-type thermometers have built-in means for making adjustments. These instruments may be tested for accuracy as we have described, and calibrated if possible. If not, the dial may need to be marked, Figure 11–14.

Body temperature may also be used as a standard when needed. Remember that the outer extremities, such as the hands, are not at body temperature. The body is 98.6°F in the main blood flow, next to the trunk of the body, Figure 11–15.

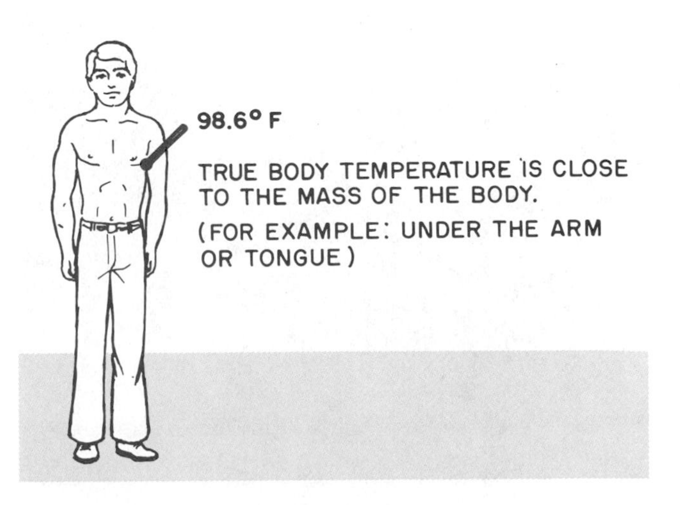

FIGURE 11–15 Using the human body as a standard.

11.4 PRESSURE TEST INSTRUMENTS

Pressure test instruments register pressures above and below atmospheric pressure. The gage manifold and its construction were discussed earlier. The technician must be able to rely on these gages and have some reference points with which to check them periodically. This is particularly necessary when there is reason to doubt gage accuracy. This instrument is used frequently and is subject to considerable abuse, Figure 11–16.

Gage readings can be taken from a cylinder of refrigerant and the pressure/temperature relationship can be compared in the following manner. The gage manifold should be opened to the atmosphere, and both gages should be checked to see that they read 0 psig. *It is impossible to determine a correct gage reading if the gage is not set at 0 at the start of the test.* Connect the gage manifold to a cylinder of fresh new refrigerant that has been in a room with a fixed temperature for a long time. Purge the gage manifold of air. Using an old cylinder may lead to errors due to cylinder pollution of air or another refrigerant. If the cylinder pressure is not correct due to pollution, it *cannot* be used to check the gages. The cylinder pressure is the standard and must be reliable. A 1-lb cylinder may be purchased and kept at a fixed temperature just for the purpose of checking gages, Figure 11–17.

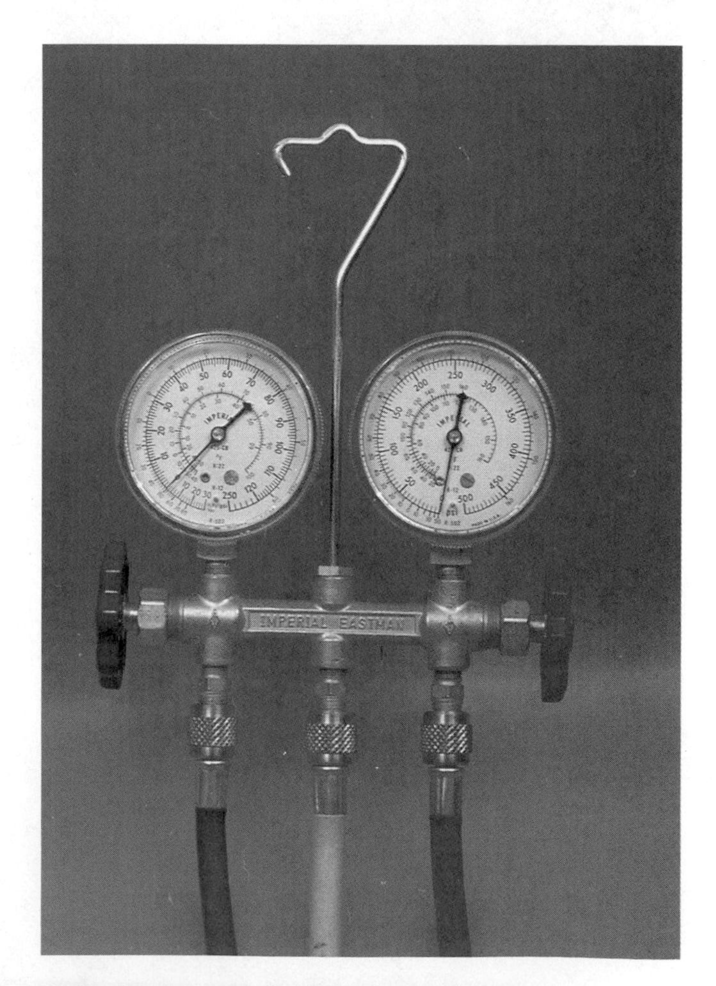

FIGURE 11–16 Gage manifold. *Photo by Bill Johnson*

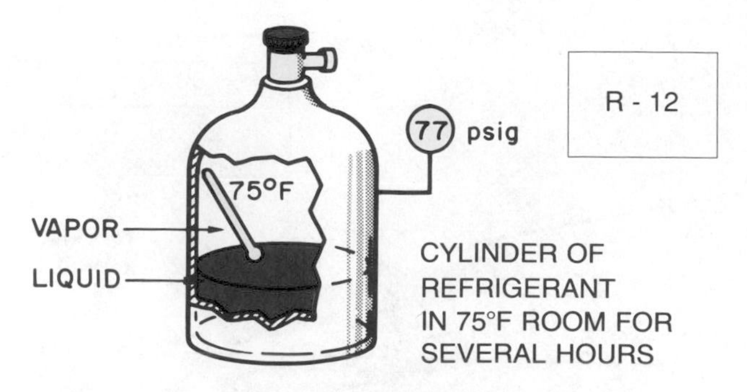

FIGURE 11–17 This refrigerant cylinder has been left at a known temperature for long enough that the temperature of the refrigerant is the same as the known room temperature. When the gage manifold is attached to this cylinder, the pressure inside can be obtained from a pressure/temperature chart.

The refrigerant should have a known pressure if the cylinder temperature is known. Typically, the cylinder is left in a temperature-controlled office all day, and the readings are taken late in the afternoon. Keep the cylinder out of direct sunlight. If the refrigerant is R-12 and the office is 75°F, the cylinder and refrigerant should be 75°F if they have been left in the office for a long enough time. When the gages are connected to the cylinder and purged of any air, the gage reading should compare to 75°F and read 77 psig. (See the pressure/temperature chart for R-12.) The cylinder can be connected in such a manner that both gages (the low and high sides) may be checked at the same time, Figure 11–18. The same test can be performed with R-22, and the reading will be higher. The reading for R-22 at 75°F should be 132 psig. Performing the test with both refrigerants checks the gages at two different pressure ranges.

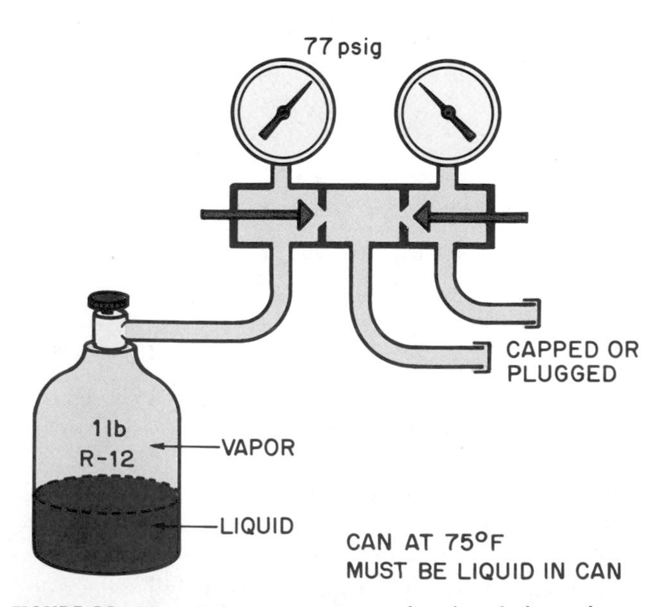

FIGURE 11–18 Both gages are connected to the cylinder so that they can be checked at the same time.

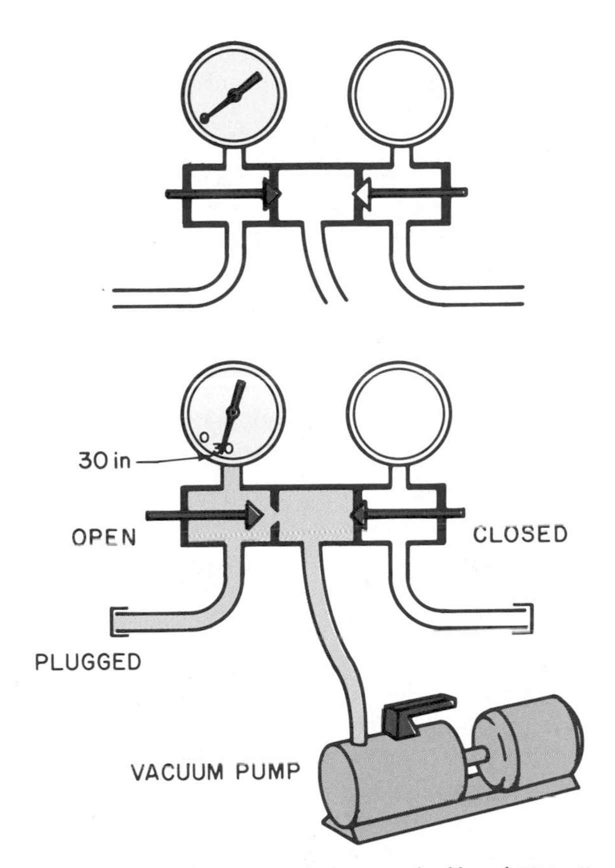

FIGURE 11–19 When evacuated, the gage should read 30 in. Hg vacuum. If this is the case, all points in between 0 psig and 30 in. Hg should be correct.

Checking the low-side gage in a vacuum is not as easy as checking the gages above atmosphere because we have no readily available known vacuum. One method is to open the gage to atmosphere and make sure that it reads 0 psig. Then connect the gage to a two-stage vacuum pump and start the pump. When the pump has reached its lowest vacuum, the gage should read 30 in. Hg (29.92 in. Hg vacuum), Figure 11–19. *Note:* The vacuum pump will not make as

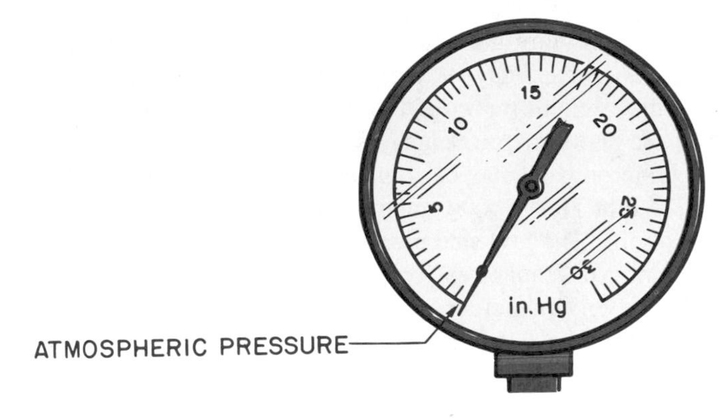

FIGURE 11–20 A large gage is more accurate for monitoring systems that operate in a vacuum because the needle moves farther from 0 psig to 30 in. Hg.

much noise at a low vacuum as it will at a pressure close to atmospheric pressure so an experienced technician can tell from the sound of the pump when it is in a deep vacuum. If the gage is correct at atmospheric pressure and at the bottom end of the scale, you can assume that it is correct in the middle of the scale. If vacuum readings closer than this are needed for monitoring a system that runs in a vacuum, you should buy a larger more accurate vacuum gage, Figure 11–20.

The mercury manometer and the electronic micron gage may be checked in the following manner. This test is a field test and not 100% accurate, but it is sufficient to tell the technician whether the instruments are within a working tolerance or not.

1. Prepare a two-stage vacuum pump for the lowest vacuum that it will pull. Change the oil, if it has been used, to improve the pumping capacity. Connect a gage manifold to the vacuum pump with the mercury manometer and the micron gage as shown in Figure 11–21. The

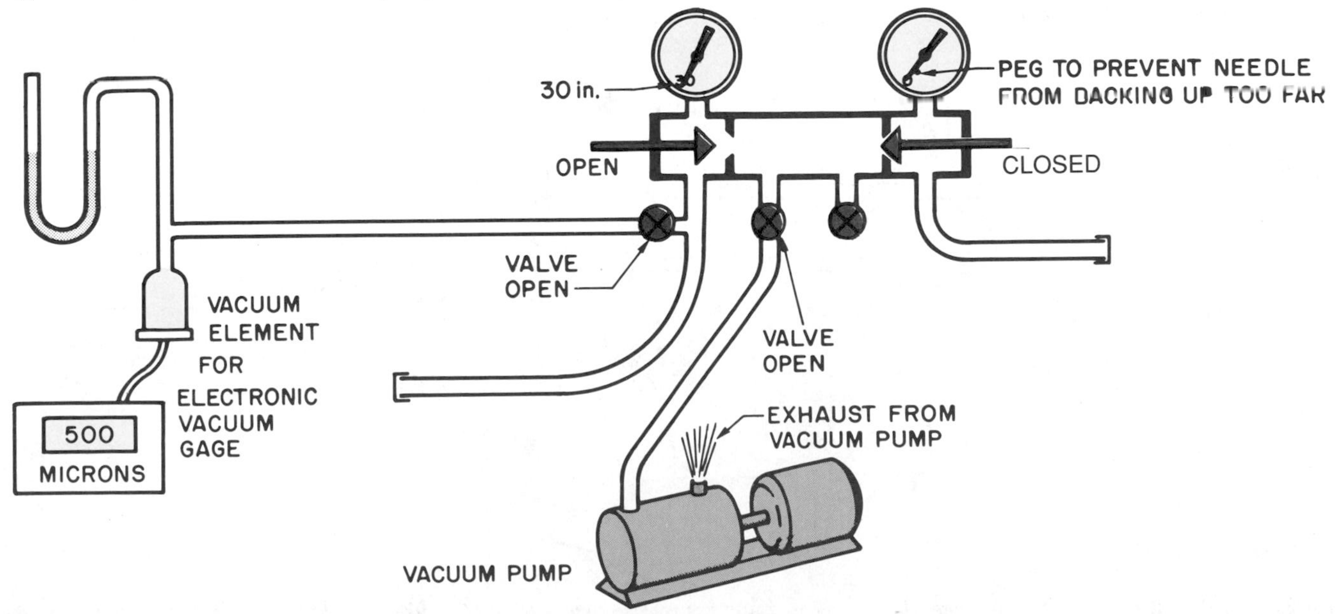

FIGURE 11–21 Setup to check mercury U-tube manometer and an electronic micron gage.

low-pressure gage, the micron gage, and the mercury manometer may be compared at the same time. Start the vacuum pump. If the micron gage has readings in the 5000-micron range, the mercury manometer and the micron gage may be compared at this point. Remember, 1 mm Hg = 1000 microns so 5 mm Hg = 5000 microns. The low-side manifold gage will read 30 in. Hg, and you cannot easily distinguish movement below 1 mm on the mercury manometer.

2. When the vacuum pump has evacuated the manifold and gages, observe the readings. If the mercury manometer is reading *flat out*—both columns of mercury are at the same level when compared (be sure the instrument is perfectly vertical)—the micron gage should read between 0 and 1000 microns.
(Note: if any atmosphere has seeped into the left column of the mercury manometer, then when the vacuum is pulled the right column will rise higher than the left column. This indicates more than a perfect vacuum, which is not possible. Whenever the right column rises higher than the left column, the reading is *wrong*. Check the manometer for a bubble on top of the left column with the instrument at atmospheric pressure.) It is difficult to compare the mercury manometer closer than this because it is hard to compare the columns. If the mercury columns are flat out and the micron gage is still reading high, you should send the micron gage to the manufacturer for calibration.
NOTE: If the vacuum pump will not pull the mercury manometer and the micron gage down to a very low level—flat out on the mercury manometer and 50 microns on the micron gage—either the vacuum pump is not pumping or the connections are leaking. The connections can be replaced with copper lines if you suspect a leak. If the vacuum still will not pull down, connect the micron gage directly to the vacuum pump with the shortest possible connection and see if the vacuum pump will pull the gage down. If it will not, check the gage on another pump to see which is not performing, the gage or the pump.

It is difficult to get instruments to correlate exactly in a vacuum. It is also difficult to determine which instrument is correct and whether the vacuum pump is evacuating the system. If the evacuation happens too quickly to be observed, a volume, such as an empty refrigerant cylinder, can be used to retard the pull-down time, Figure 11–22.

11.5 ELECTRICAL TEST INSTRUMENTS

Electrical test instruments are not as easy to calibrate; however, they may be checked for accuracy. The technician must know that the ohm scale, the volt scale, and the ammeter scale are correct. The milliamp scale on the meter is seldom used and must be checked by the manufacturer or compared to another meter.

There are many grades of electrical test instruments. When electrical testing procedures must be relied on for accuracy, it pays to buy a good quality instrument. If you don't need accuracy from an electrical test instrument, a less expensive one may be satisfactory. You should periodically (at least once a year) check electrical test instruments for accuracy. One way to check these instruments is to compare the instrument reading against known values.

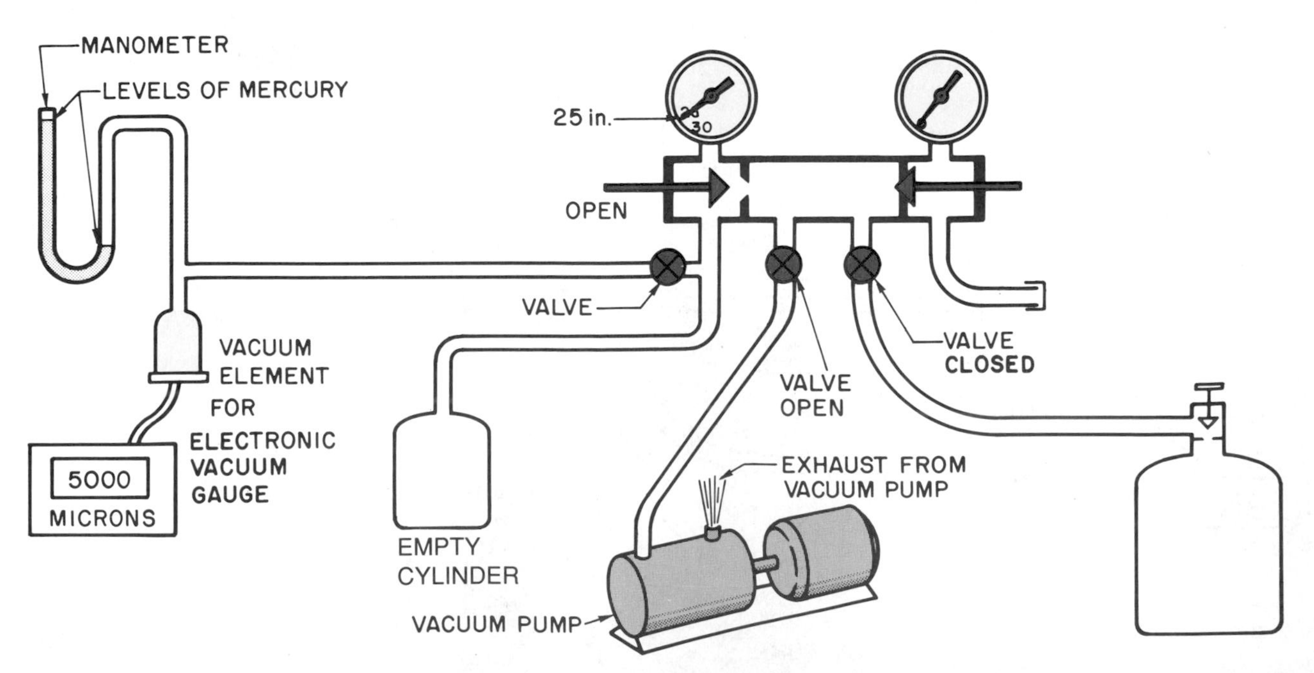

FIGURE 11–22 If the evacuation in Figure 11–21 happens too fast, a volume, such as an empty refrigerant cylinder, may be connected to retard the evacuation.

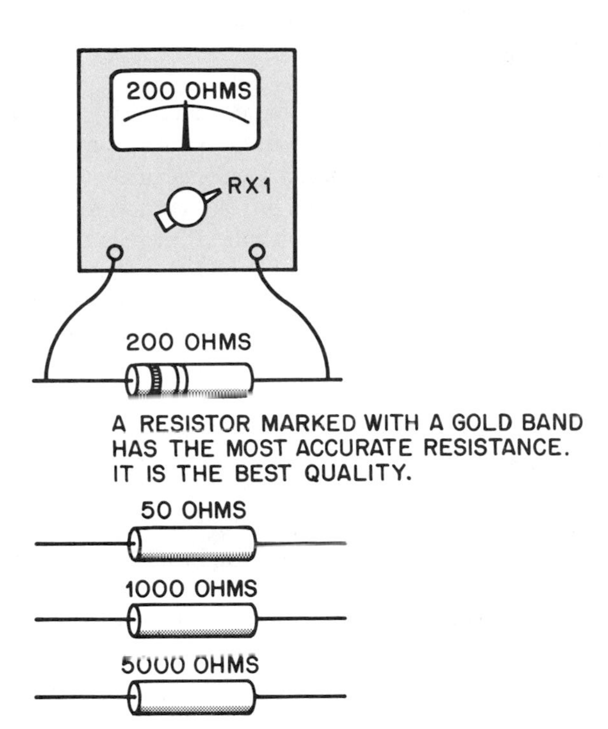

FIGURE 11–23 Checking an ohmmeter at various resistance ranges with known resistances.

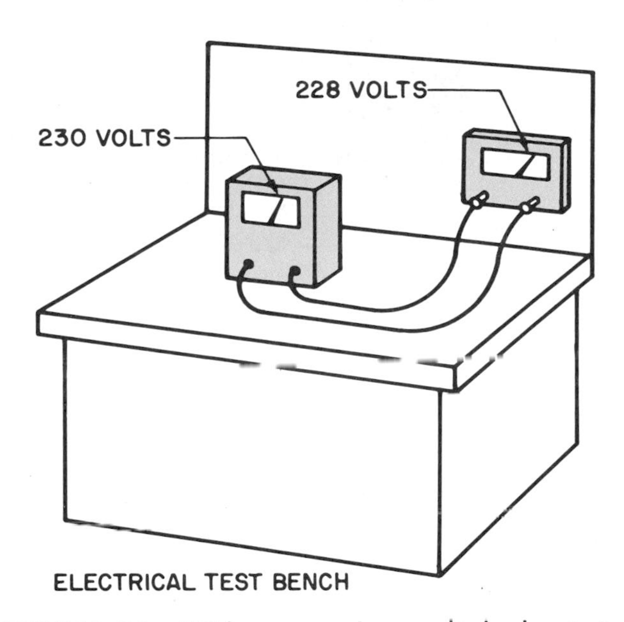

FIGURE 11–24 VOM being compared to a quality bench meter in the voltage mode.

The ohmmeter feature of a volt ohm milliammeter (VOM) can be checked by obtaining several high-quality resistors of a known resistance at an electronic supply house. Get different values of resistors, so you can test the ohmmeter at each end and the midpoint of every scale on the meter, Figure 11–23. Always start each test by a zero adjustment check of the ohmmeter. ***If the meter is out of calibration to the point that it cannot be brought to the zero adjustment, check the batteries and change if necessary. If the instrument will not read zero with fresh batteries, send it to the experts. Do not try to repair it yourself. Be sure to start each test by a zero adjustment check of the ohmmeter.***

The volt scale is not as easy to check as the ohm scale. A friend at the local power company or technical school may allow you to compare your meter to a high-quality bench meter. This is recommended at least once a year or whenever you suspect your meter is incorrect, Figure 11–24. It is satisfying to know your meter is correct when you call the local power company and report that it has low voltage on a particular job.

The clamp-on ammeter is used most frequently for amperage checks. This instrument clamps around one conductor in an electrical circuit. Like the voltmeter, it can be compared to a high-quality bench meter, Figure 11–25. Some amount of checking can be done by using Ohm's law and comparing the ampere reading to a known resistance heater circuit. For example, Ohm's law states that current (I) is equal to voltage (E) divided by resistance (R or Ω for ohms):

$$I = \frac{E}{R}$$

If a heater has a resistance of 109 Ω and an applied line voltage of 228 V, the amperage (current) on this circuit should be

$$I = \frac{228\text{ V}}{10\ \Omega} = 22.8\text{ A}$$

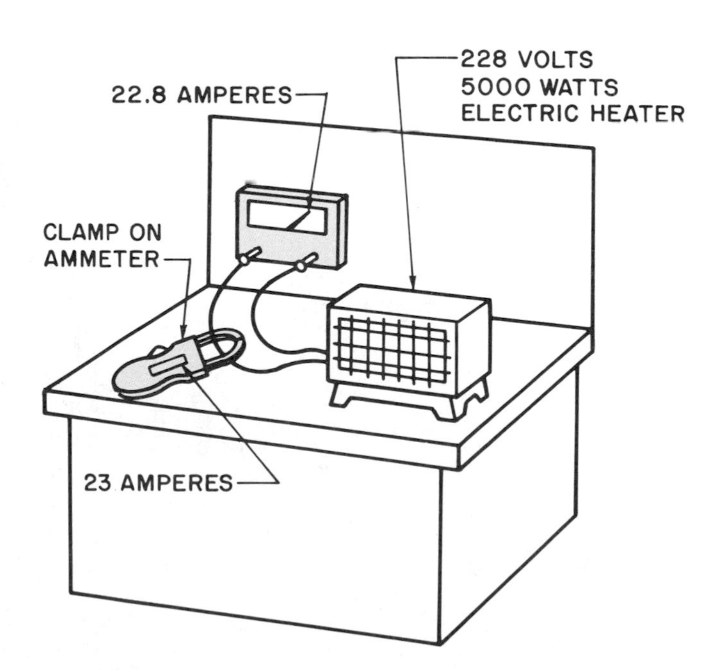

FIGURE 11–25 Clamp-on ammeter being compared to a quality bench ammeter.

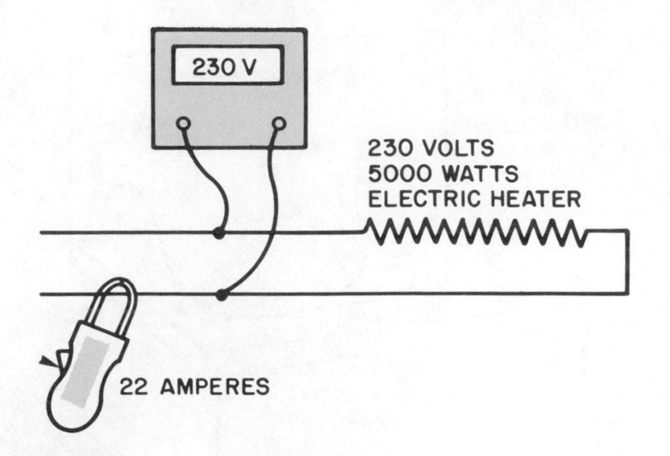

FIGURE 11–26 Using an electric heater to check the calibration of an ammeter.

Remember to read the voltage at the same time as the amperage, Figure 11–26. You will notice small errors because the resistance of the electric heaters will change when they get hot. The resistance will be greater, and the exact ampere reading will not compare precisely to the calculated one. But the purpose is to check the ammeter, and if it is off by more than 10% send it to the repair shop.

11.6 REFRIGERANT LEAK DETECTION DEVICES

Two refrigerant detection devices are commonly used: the halide torch and the electronic leak detector.

Halide Torch

The halide torch cannot be calibrated, but it can be checked to make sure that it will detect leaks. It must be maintained for it to be reliable. It will detect a leak rate of about 6 oz per year. The halide torch uses the primary air port to draw air into the burner through a flexible tube. If there is any refrigerant in this air sample, it passes over a copper element and the color of the flame changes from the typical blue of a gas flame to a green color, Figure 11–27. A large leak will extinguish the flame of the halide torch.

The maintenance on this torch consists of keeping the tube clear of debris and keeping a copper element in the burner head. If the sample tube becomes restricted, the flame may burn yellow. You can place the end of the sample tube close to your ear and hear the rushing sound of air being pulled in the tube. If you can't hear it, or if it burns yellow, clean the tube, Figure 11–28.

The copper element is replaceable. If you cannot find an element, you can make a temporary one out of a piece of copper wire, Figure 11–29.

Electronic Leak Detectors

Electronic leak detectors are much more sensitive than the halide torch (they can detect leak rates of about 1/2 oz per year) and are widely used. The electronic leak detector samples air; if the air contains refrigerant, the detector either sounds an alarm or lights the probe end. These devices are manufactured in both battery- and 120 VAC-powered units. Some units may have a pump to pull the sample across the

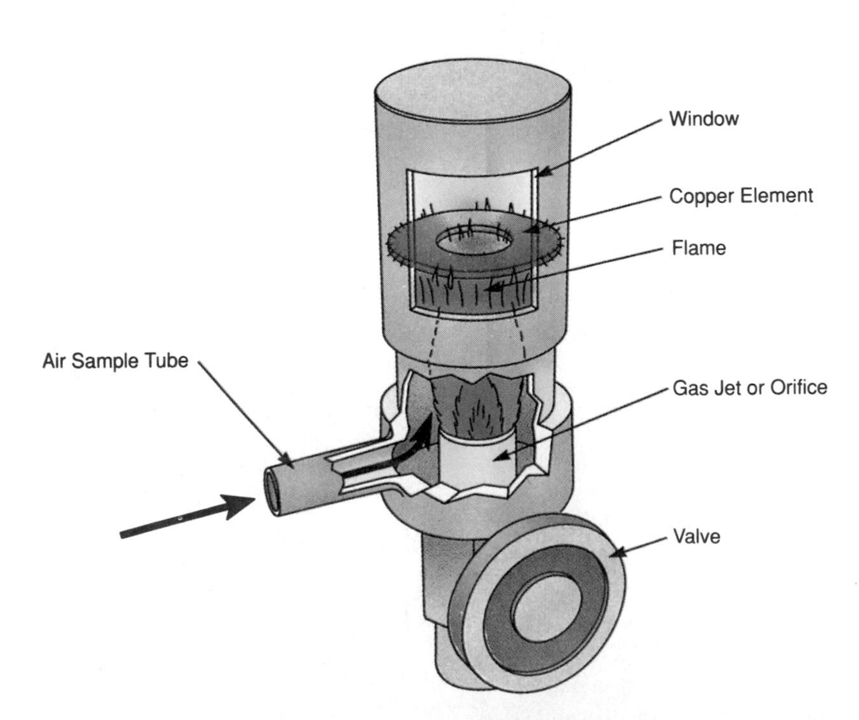

FIGURE 11–27 Halide torch for detecting refrigerant leaks. The halide torch flame turns color from blue to green, depending on the amount of refrigerant in the air sample. The glowing copper element makes this possible.

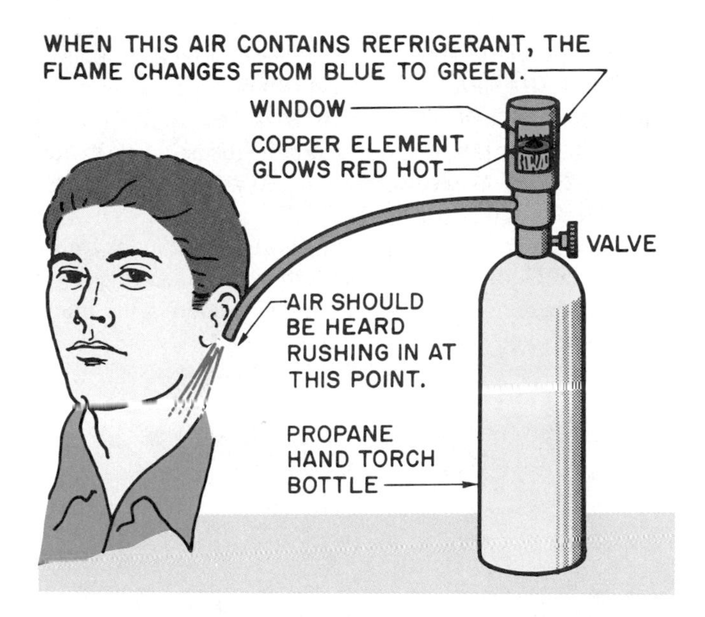
WHEN THIS AIR CONTAINS REFRIGERANT, THE FLAME CHANGES FROM BLUE TO GREEN.
WINDOW
COPPER ELEMENT GLOWS RED HOT
VALVE
AIR SHOULD BE HEARD RUSHING IN AT THIS POINT.
PROPANE HAND TORCH BOTTLE

FIGURE 11–28 A rushing sound may be heard at the end of the sampling tube if the halide torch detector is pulling in air and working correctly.

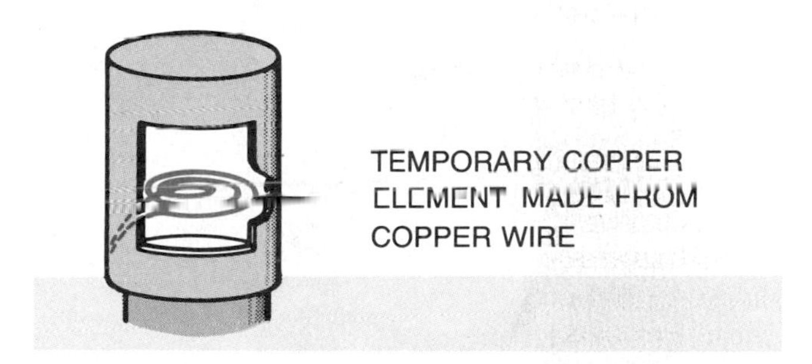
TEMPORARY COPPER ELEMENT MADE FROM COPPER WIRE

FIGURE 11–29 Temporary copper element.

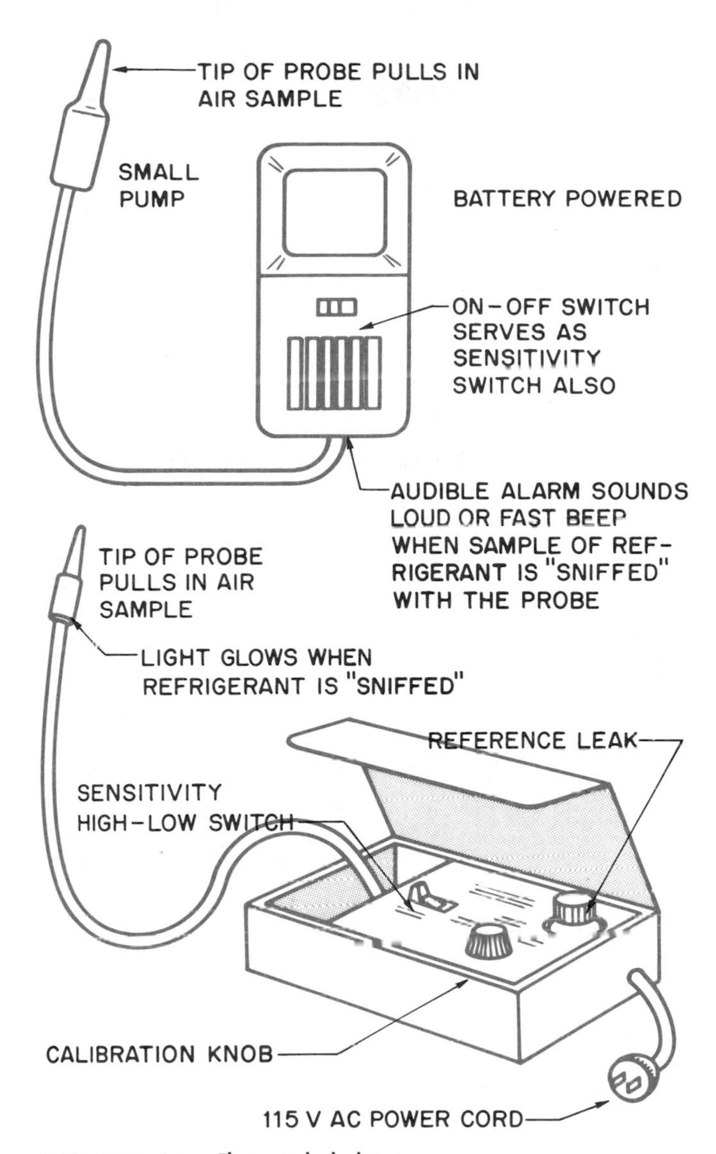
TIP OF PROBE PULLS IN AIR SAMPLE
SMALL PUMP
BATTERY POWERED
ON-OFF SWITCH SERVES AS SENSITIVITY SWITCH ALSO
AUDIBLE ALARM SOUNDS LOUD OR FAST BEEP WHEN SAMPLE OF REFRIGERANT IS "SNIFFED" WITH THE PROBE
TIP OF PROBE PULLS IN AIR SAMPLE
LIGHT GLOWS WHEN REFRIGERANT IS "SNIFFED"
REFERENCE LEAK
SENSITIVITY HIGH-LOW SWITCH
CALIBRATION KNOB
115 V AC POWER CORD

FIGURE 11–30 Electronic leak detectors.

sensing element, and some have the sensing element located in the head of the probe, Figure 11–30.

Some electronic leak detectors have an adjustment that will compensate for background refrigerant. Some equipment rooms may have many small leaks and a small amount of refrigerant may be in the air all the time. The electronic leak detector will indicate all the time unless it has a feature to account for this background refrigerant.

No matter what style of leak detector you use, you must be confident that the detector will actually detect a leak. Remember that the detector only indicates what it samples. If the detector is in the middle of a refrigerant cloud and the sensor is sensing air, it will not sound an alarm or light up. For example, a pinhole leak in a pipe can be passed by with the probe of a leak detector. The sensor is sensing air next to the leak, not the leak itself, Figure 11–31.

Some manufacturers furnish a sample refrigerant container of R-11. The container has a pinhole in the top with

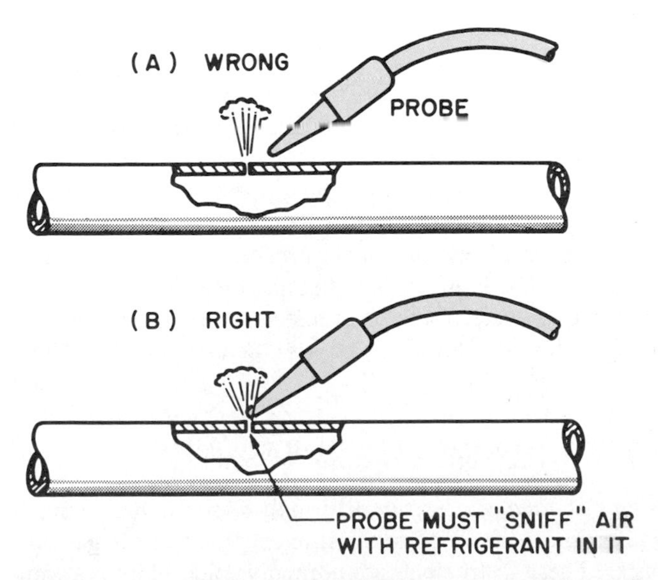
(A) WRONG
PROBE
(B) RIGHT
PROBE MUST "SNIFF" AIR WITH REFRIGERANT IN IT

FIGURE 11–31 (A) Electronic leak detector is not sensing the small pinhole leak because it is spraying past the detector's sensor. (B) Sensor will detect refrigerant leak.

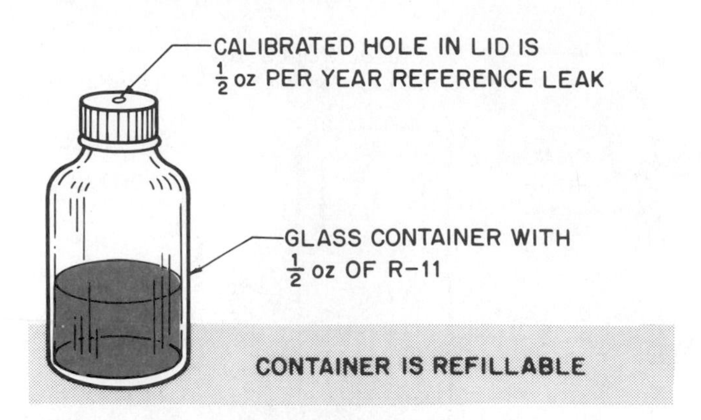

FIGURE 11–32 This is a small refrigerant vial that serves as a reference leak.

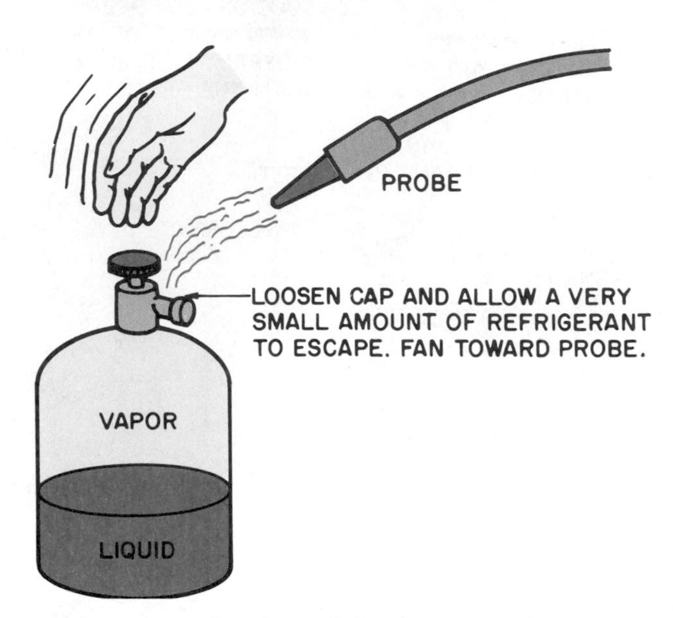

FIGURE 11–33 If a reference leak vial is not available, a small amount of refrigerant may be allowed to escape and to mix with air. This sample may be fanned toward the probe.

a calibrated leak, Figure 11–32. The refrigerant will remain in the container for a long time if the lid is replaced after each use.

Never spray pure refrigerant into the sensing element. Damage will occur. If you do not have a reference leak canister, a gage line under pressure from a refrigerant cylinder can be loosened slightly, and the refrigerant can be fanned to the electronic leak detector sensing element. In doing it this way, air is mixed with the refrigerant to dilute it, Figure 11–33.

11.7 FLUE-GAS ANALYSIS INSTRUMENTS

Flue-gas analysis instruments analyze fossil fuel-burning equipment products of combustion, such as oil and gas furnaces. These instruments are normally sold in kit form with a carrying case. ***There are chemicals in the flue-gas kit that must not be allowed to contact any tools or other instruments. This chemical is intended to stay in the container or the instrument that uses it. The instrument has a valve at the top that is a potential leak source. It should be checked periodically. It is best to store and transport the kit in the upright position so that if the valve does develop a leak the chemical will not run out of the instrument, Figure 11–34.***

These are precision instruments that cost money and deserve care and attention. The draft gage is sensitive to a very fine degree. This kit should not be hauled in the truck except when you intend to use it.

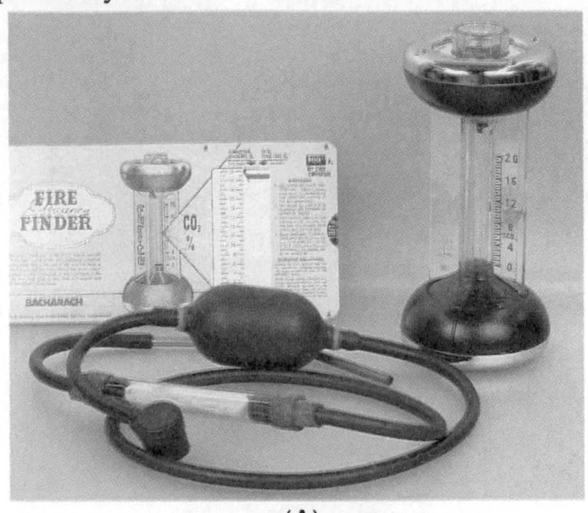

(A)

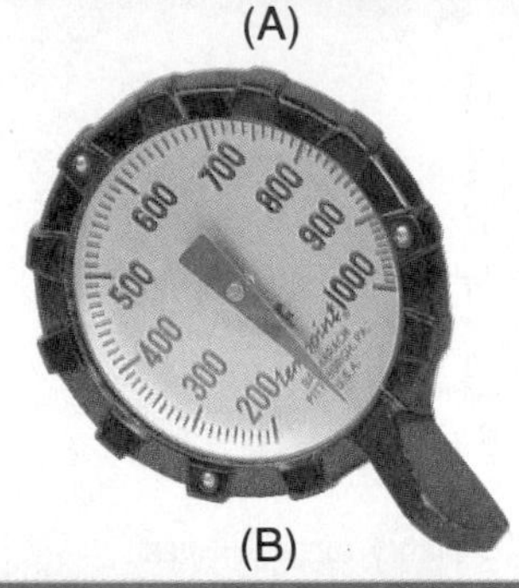

(B)

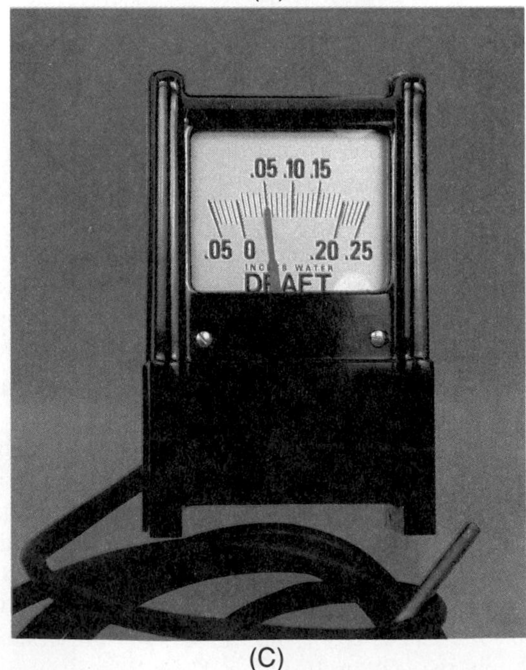

(C)

FIGURE 11–34 Flue-gas analysis kit. (A) CO_2 tester. (B) Thermometer. (C) Draft gage. *Photos by Bill Johnson*

FIGURE 11-35 Zero adjustment on the sliding scale of flue-gas analysis kit. *Photo by Bill Johnson*

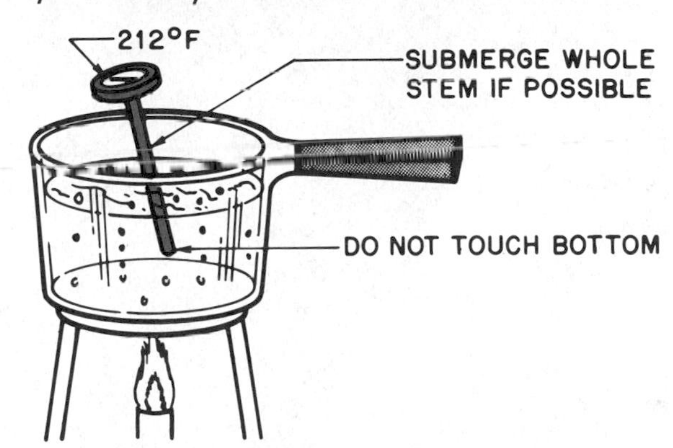

FIGURE 11–36 Flue-gas kit thermometer being checked in boiling water.

A calibration check of these instruments is not necessary. The chemicals in the analyzer should be changed according to the manufacturer's suggestions. These instruments are direct-reading instruments and cannot be calibrated. The only adjustment is the zero adjustment on the sliding scale, Figure 11–35, which is adjusted at the beginning of the test. This is done by venting the sample chamber to the atmosphere. The fluid should fall. If it gets dirty, change it.

The thermometer in the kit is used to register very high temperatures, up to 1000°F. There is no easy reference point except the boiling point of water, 212°F. This may be used as the reference, even though it is near the bottom of the scale, Figure 11–36.

11.8 GENERAL MAINTENANCE

Any instrument with a digital readout will have batteries. They must be maintained. *Buy the best batteries available; inexpensive batteries may cause problems. Good batteries will not leak acid on the instrument components if left unattended and the battery goes dead.*

The instruments you use extend your senses; the instruments therefore must be maintained so that they can be believed. Airplane pilots sometimes have a malady called vertigo. They become dizzy and lose their sensation and relationship with the horizon. Suppose the pilot were in a storm and being tossed around, even upside down at times. The pilot can be upside down and have a sensation that the plane is right side up and climbing. The plane may be diving toward the earth while the pilot thinks it is climbing. Instruments must be believable, and the technician must have reference points to have faith in the instruments.

SUMMARY

- The instruments used by technicians must be reliable.
- Reference points for all instruments should be established to give the technician confidence.
- The three easily obtainable reference points for temperature-measuring instruments are ice and water at 32°F, body temperature at 98.6°F, and boiling temperature at 212°F.
- ***Make sure that the temperature-sensing element reflects the actual temperature of the medium used as the standard.***
- Pressure-measuring instruments must be checked above and below atmospheric pressure. There are no good reference points below the atmosphere, so a vacuum pump pulling a deep vacuum is used as the reference.
- Some electronic leak detectors have reference leak canisters furnished.
- Flue-gas analysis kits need no calibration except for the sliding scale on the sample chamber.
- The thermometers in flue-gas analysis kits may be checked in boiling water.

REVIEW QUESTIONS

1. Name the three reference points for checking temperature-measuring instruments.
2. What does "calibrating an instrument" mean?
3. How can a glass-stem thermometer be used in the calibration of an electronic thermometer?
4. What should you do when an electronic thermometer's leads are slightly out of calibration.
5. Can all thermometers be calibrated?
6. What should a gage manifold reading indicate when opened to the atmosphere?
7. What two reference points are used for checking vacuum for a gage manifold low-side gage?
8. What reference point may be used for checking a pressure gage?
9. How is a flue-gas instrument calibrated?
10. What types of batteries are suggested for instruments?

SECTION

3 Basic Automatic Controls

UNIT 12

Basic Electricity and Magnetism

OBJECTIVES

After studying this unit, you should be able to

- describe the structure of an atom.
- identify atoms with a positive charge and atoms with a negative charge.
- explain the characteristics that make certain materials good conductors.
- describe how magnetism is used to produce electricity.
- state the differences between alternating current and direct current.
- list the units of measurement for electricity.
- explain the differences between series and parallel circuits.
- state Ohm's Law.
- state the formula for determining electrical power.
- describe a solenoid.
- explain inductance.
- describe the construction of a transformer and the way that a current is induced in a secondary circuit.
- describe how a capacitor works.
- state the reasons for using proper wire sizes.
- describe the physical characteristics and the function of several semiconductors.
- describe procedures for making electrical measurements.

SAFETY CHECKLIST

* Do not make any electrical measurements without specific instructions from a qualified person.
* Use only electrical conductors of the proper size to avoid overheating and possibly fire.
* Electrical circuits must be protected from current overloads. These circuits are normally protected with fuses or circuit breakers.
* Extension cords used by technicians to provide electrical power for portable power tools and other devices should be protected with ground fault circuit interrupters.
* When servicing equipment, the electrical service should be shut off at a disconnect panel whenever possible, the disconnect panel locked, and the only key kept by the technician.

12.1 STRUCTURE OF MATTER

To understand the theory of how an electric current flows, you must understand something about the structure of matter. Matter is made up of atoms. Atoms are made up of protons, neutrons, and electrons. Protons and neutrons are located at the center (or nucleus) of the atom. Protons have a positive charge. Neutrons have no charge and have little or no effect as far as electrical characteristics are concerned. Electrons have a negative charge and travel around the nucleus in orbits. The number of electrons in an atom is the same as the number of protons. Electrons in the same orbit are the same distance from the nucleus but do not follow the same orbital paths, Figure 12–1.

The hydrogen atom is a simple atom to illustrate because it has only one proton and one electron, Figure 12–2.

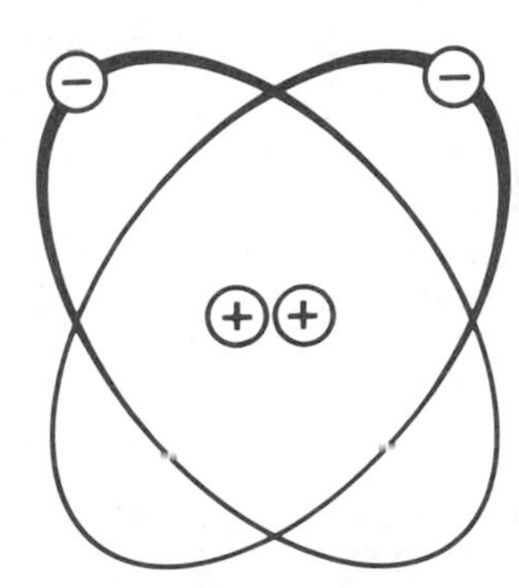

FIGURE 12–1 Orbital paths of electrons.

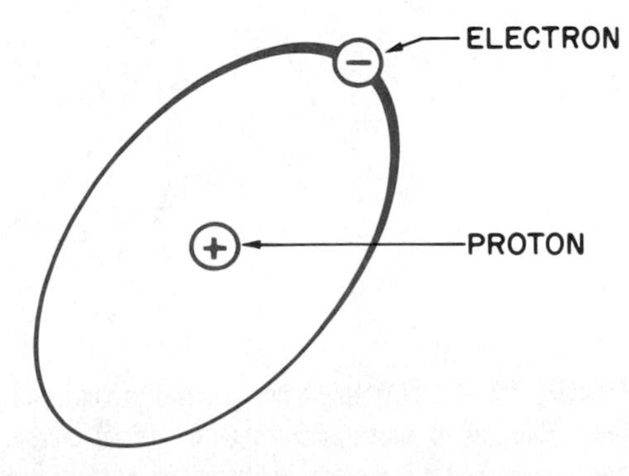

FIGURE 12–2 Hydrogen atom with one electron and one proton.

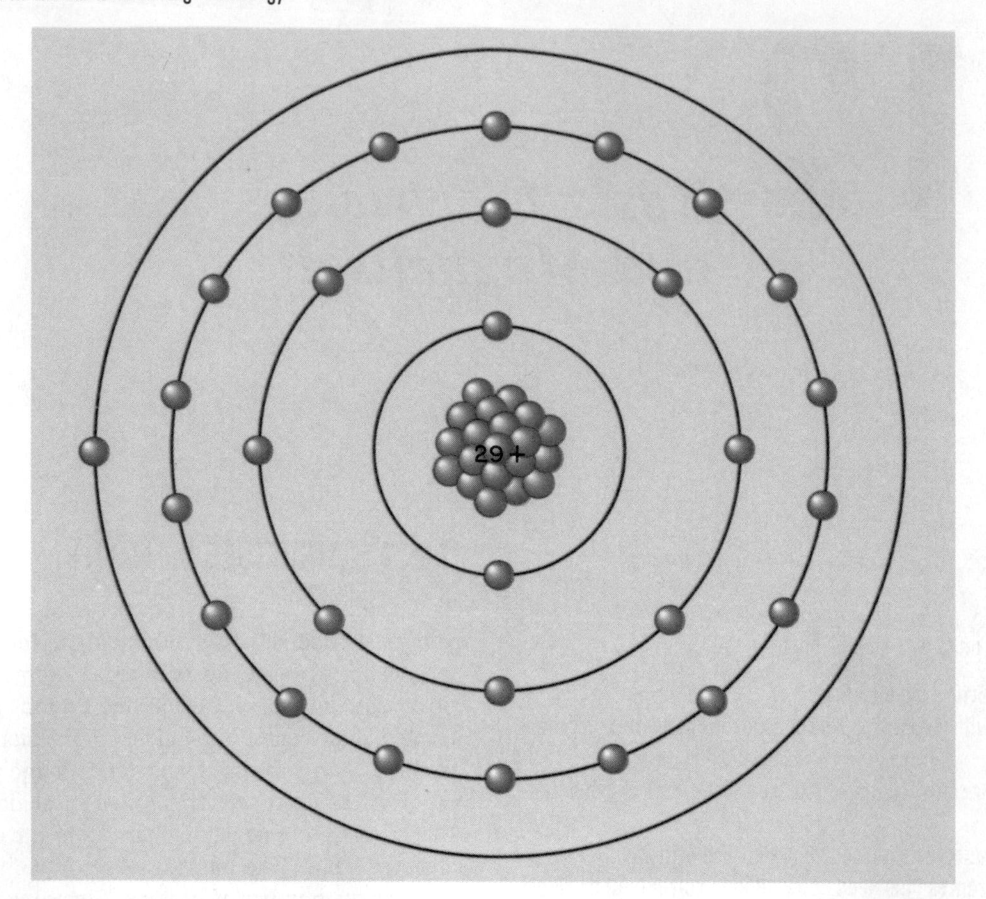

FIGURE 12–3 Copper atom with 29 protons and 29 electrons.

Not all atoms are as simple as the hydrogen atom. Most wiring used to conduct an electrical current is made of copper. Figure 12–3 illustrates a copper atom, which has 29 protons and 29 electrons. Some electron orbits are farther away from the nucleus than others. As can be seen, 2 travel in an inner orbit, 8 in the next, 18 in the next, and 1 in the outer orbit. It is this single electron in the outer orbit that makes copper a good conductor.

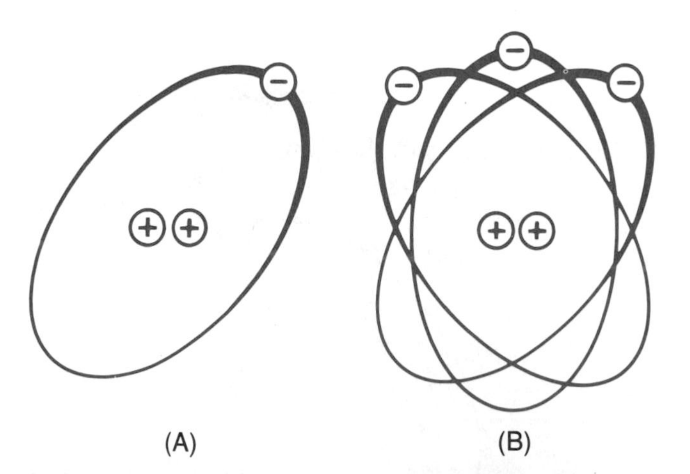

FIGURE 12–4 (A) This atom has two protons and one electron. It has a shortage of electrons and thus a positive charge. (B) This atom has two protons and three electrons. It has an excess of electrons and thus a negative charge.

12.2 MOVEMENT OF ELECTRONS

When sufficient energy or force is applied to an atom, the outer electron (or electrons) becomes free and moves. If it leaves the atom, the atom will contain more protons than electrons. Protons have a positive charge. This means that this atom will have a positive charge, Figure 12–4(A). The atom the electron joins will contain more electrons than protons, so it will have a negative charge, Figure 12–4(B).

Like charges repel each other, and unlike charges attract each other. An electron in an atom with a surplus of electrons (negative charge) will be attracted to an atom with a shortage of electrons (positive charge). An electron entering an orbit with a surplus of electrons will tend to repel an electron already there and cause it to become a free electron.

12.3 CONDUCTORS

Good conductors are those with few electrons in the outer orbit. Three common metals—copper, silver, and gold—are good conductors, and each has one electron in the outer orbit. These are considered to be free electrons because they move easily from one atom to another.

12.4 INSULATORS

Atoms with several electrons in the outer orbit are poor conductors. These electrons are difficult to free, and materials made with these atoms are considered to be insulators. Glass, rubber, and plastic are examples of good insulators.

12.5 ELECTRICITY PRODUCED FROM MAGNETISM

Electricity can be produced in many ways, for example, from chemicals, pressure, light, heat, and magnetism. The electricity that air conditioning and heating technicians are most involved with is produced by a generator using magnetism.

Magnets are common objects with many uses. Magnets have poles usually designated as the north (N) pole and the south (S) pole. They also have fields of force. Figure 12–5 shows the lines of the field of force around a permanent bar magnet. This field causes the like poles of two magnets to repel each other and the unlike poles to attract each other.

If a conductor, such as a copper wire, is passed through this field and cuts these lines of force, the outer electrons in the atoms in the wire are freed and begin to move from atom to atom. They will move in one direction. It does not matter if the wire moves or if the magnetic field moves. It is only necessary that the conductor cut through the lines of force, Figure 12–6.

This movement of electrons in one direction produces the electric current. The current is an impulse transferred from one electron to the next. If you pushed a golf ball into a tube already filled with golf balls, one would be ejected instantly from the other end, Figure 12–7. Electric current travels in a similar manner at a speed of 186,000 miles/sec. The electrons do not travel through the wire at this speed, but the repelling and attracting effect causes the current to do so.

An electrical generator has a large magnetic field and many turns of wire cutting the lines of force. A large magnetic field or one with many turns of wire produces more electricity than a smaller field or a field with few turns of wire. The magnetic force field for generators is usually produced by electromagnets. Electromagnets have similar characteristics to permanent magnets and are discussed later in this unit. Figure 12–8 shows a simple generator.

12.6 DIRECT CURRENT

Direct current (DC) travels in one direction. Because electrons have a negative charge and travel to atoms with a positive charge, DC is considered to flow from negative to positive.

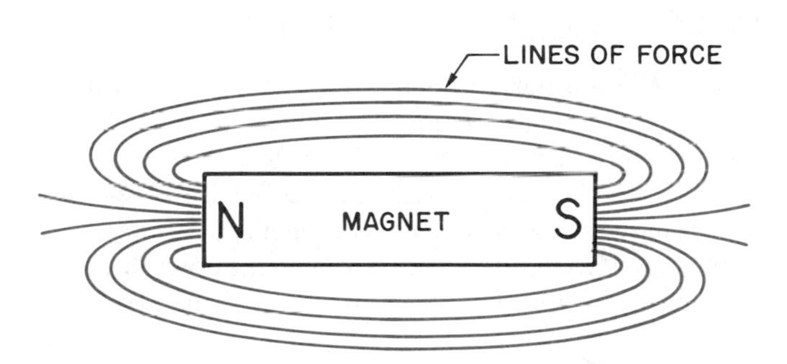

FIGURE 12–5 Permanent magnet with lines of force.

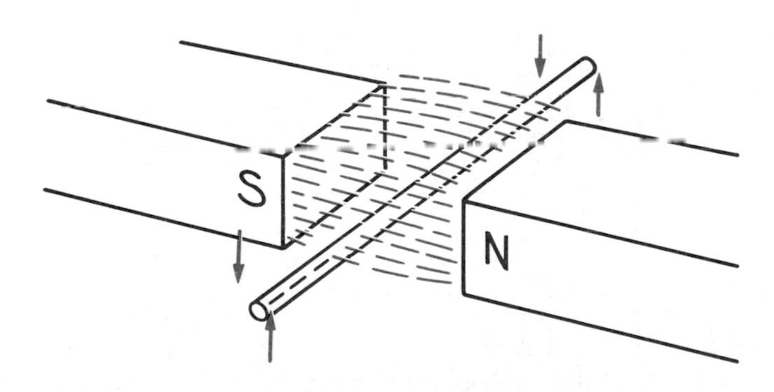

FIGURE 12–6 Movement of wire up and down and cutting the lines of force causes an electric current to flow in the wire.

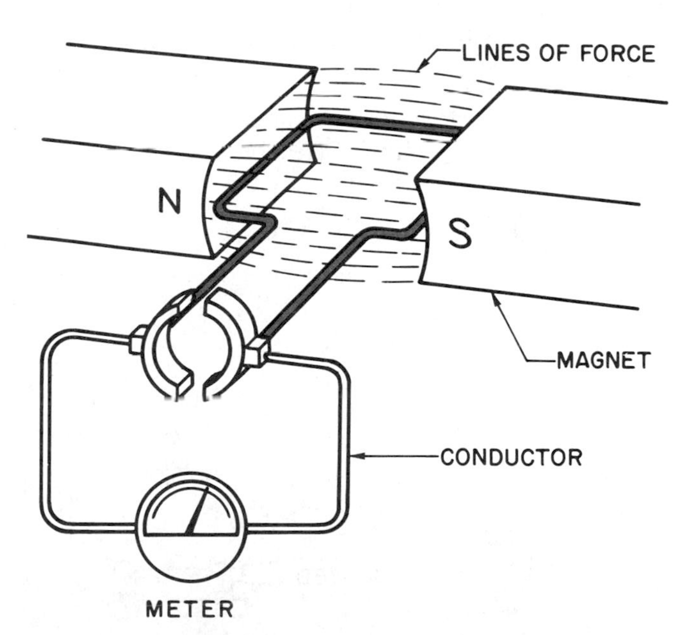

FIGURE 12–8 Simple generator.

FIGURE 12–7 Tube filled with golf balls; when one ball is pushed in, one ball is pushed out.

12.7 ALTERNATING CURRENT

Alternating current (AC) is continually and rapidly reversing. The charge at the power source (generator) is continually changing direction; thus the current continually reverses itself. For several reasons, most electrical energy generated for public use is AC. It is much more economical to transmit electrical energy long distances in the form of AC. The voltage of this type of electrical current flow can be readily changed so that it has many more uses. DC still has many applications, but it is usually obtained by changing AC to DC or by producing the DC locally where it is to be used.

12.8 ELECTRICAL UNITS OF MEASUREMENT

Electromotive force (emf) or voltage (V) is used to indicate the difference of potential in two charges. When an electron surplus builds up on one side of a circuit and a shortage of electrons exists on the other side, a difference of potential or emf is created. The unit used to measure this force is the *volt.*

The *ampere* is the unit used to measure the quantity of electrons moving past a given point in a specific period of time (electron flow).

All materials oppose or resist the flow of an electrical current to some extent. In good conductors this opposition or resistance is low. In poor conductors the resistance is high. The unit used to measure resistance is the *ohm.* A conductor has a resistance of 1 ohm when a force of 1 volt causes a current of 1 ampere to flow.

Volt = Electrical force or pressure (V)
Ampere = Quantity of electron flow (A)
Ohm = Resistance to electron flow (Ω)

12.9 THE ELECTRICAL CIRCUIT

An electrical circuit must have a power source, a conductor to carry the current, and a load or device to use the current. There is also generally a means for turning the electrical current flow on and off. Figure 12–9 shows an electrical generator for the source, a wire for the conductor, a light bulb for the load, and a switch for opening and closing the circuit.

The generator produces the current by passing many turns of wire through a magnetic field. If it is a DC generator, the current will flow in one direction. If it is an AC generator, the current will continually reverse itself. However, the effect on this circuit will generally be the same whether it is AC or DC.

The wire or conductor provides the means for the electricity to flow to the bulb and complete the circuit. The electrical energy is converted to heat and light energy at the bulb element.

The switch is used to open and close the circuit. When the switch is open, no current will flow. When it is closed, the bulb element will produce heat and light.

12.10 MAKING ELECTRICAL MEASUREMENTS

In the circuit illustrated in Figure 12–9 electrical measurements can be made to determine the voltage (emf) and amperes (current). In making the measurements, Figure 12–10, the voltmeter is connected across the terminals of the bulb without interrupting the circuit. The ammeter is connected directly into the circuit so that all the current flows through it. Figure 12–11 illustrates the same circuit using symbols.

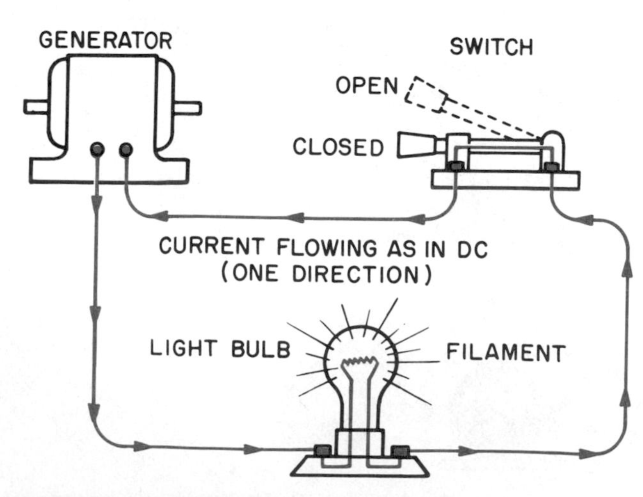

FIGURE 12–9 Electric circuit.

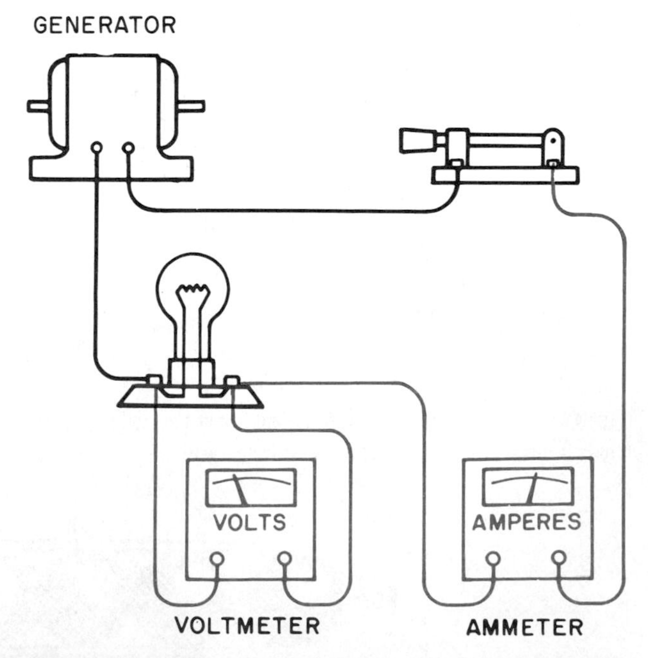

FIGURE 12–10 Voltage is measured across the resistance. Amperage is measured in series.

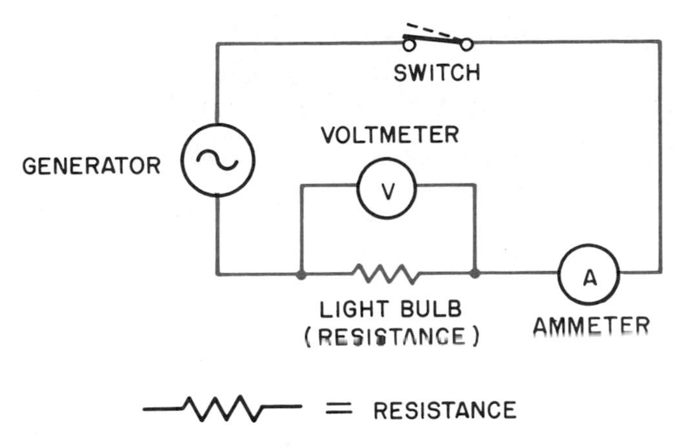

FIGURE 12–11 Same circuit as Figure 12–10, illustrated with symbols.

Often a circuit will contain more than one resistance or load. These resistances may be wired in series or in parallel, depending on the application or use of the circuit. Figure 12–12 shows three loads in series. This is shown pictorially and by symbols. Figure 12–13 illustrates these loads wired in parallel.

In circuits where devices are wired in series, all of the current passes through each load. When two or more loads are wired in parallel, the current is divided among the loads. This is explained in more detail later. Power-passing devices such as switches are wired in series. Most resistance or loads (power-consuming devices) air conditioning and heating technicians work with are wired in parallel.

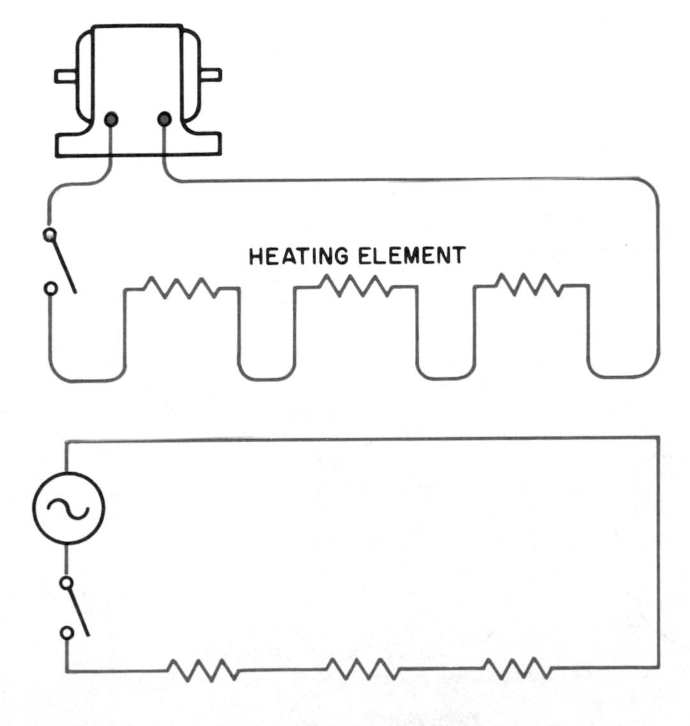

FIGURE 12–12 Multiple resistances (small heating elements) in series.

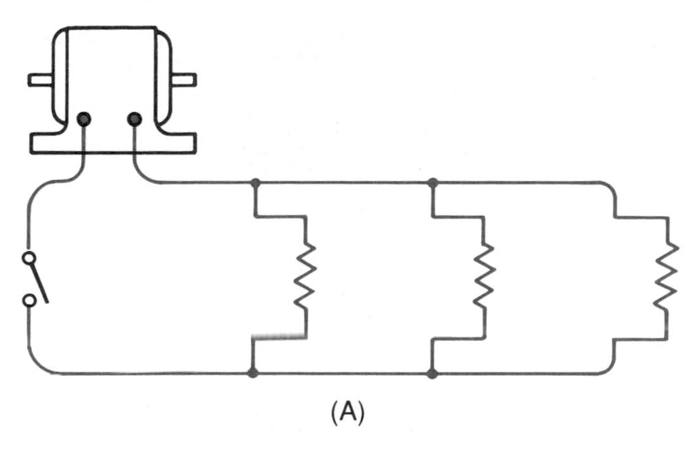

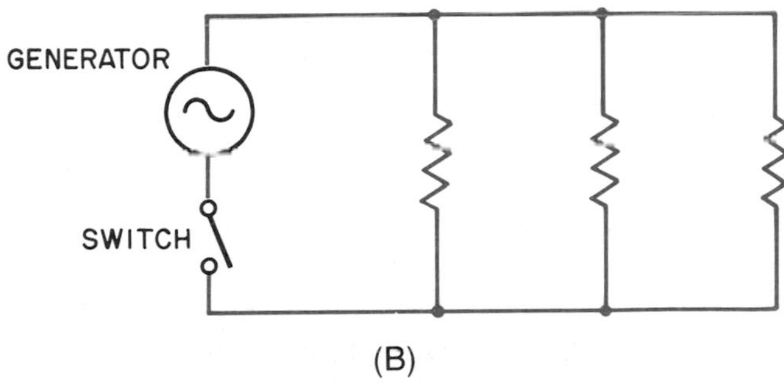

FIGURE 12–13 (A) Multiple resistances (small heating elements) in parallel. (B) Three resistances in parallel using symbols.

Figure 12–14 illustrates how a voltmeter is connected for each of the resistances. The voltmeter is in parallel with each resistance. The ammeter is also shown and is in series in the circuit. An ammeter has been developed that can be clamped around a single conductor to measure amperes, Figure 12–15. This is convenient because it is often difficult to disconnect the circuit to connect the ammeter in series. This type of ammeter, usually called a clamp-on type, is discussed in this unit.

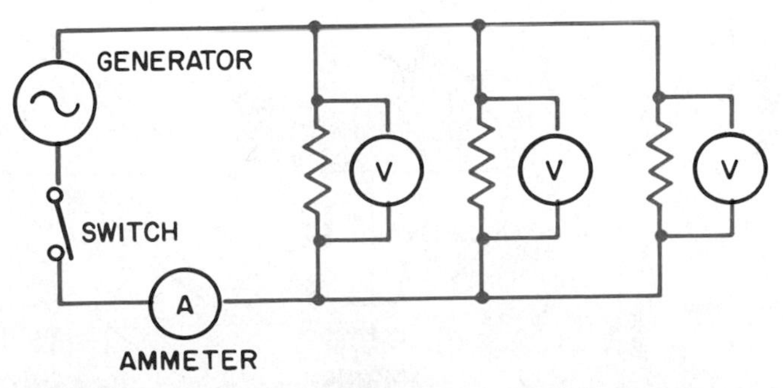

FIGURE 12–14 Voltage readings are taken across the resistances in the circuit.

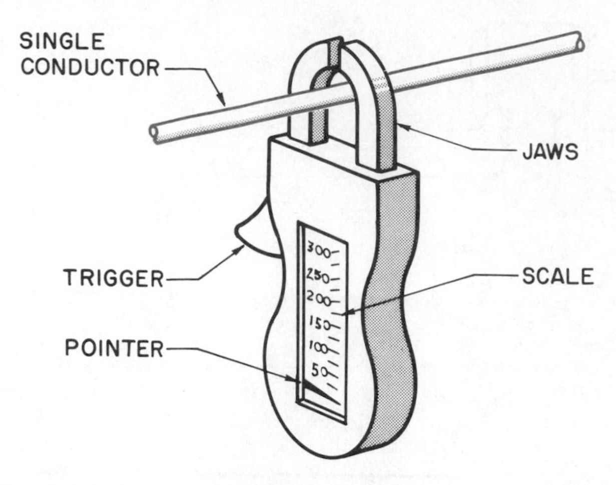

FIGURE 12–15 Clamp-on ammeter. *Courtesy Amprobe Instrument Division, Core Industries*

12.11 OHM'S LAW

During the early 1800s the German scientist Georg S. Ohm did considerable experimentation with electrical circuits and particularly with regard to resistances in these circuits. He determined that there is a relationship between each of the factors in an electrical circuit. This relationship is called Ohm's Law. The following describes this relationship. Letters are used to represent the different electrical factors.

E or V = Voltage (emf)
I = Amperage (current)
R = Resistance (load)

The voltage equals the amperage times the resistance:

$$E = I \times R$$

The amperage equals the voltage divided by the resistance:

$$I = \frac{E}{R}$$

The resistance equals the voltage divided by the amperage:

$$R = \frac{E}{I}$$

Figure 12–16 shows a convenient way to remember these formulae. The symbol for ohms is Ω.

In Figure 12–17 the resistance of the heating element can be determined as follows:

$$R = \frac{E}{I} = \frac{120}{1} = 120\ \Omega$$

In Figure 12–18 the voltage across the resistance can be calculated as follows:

$$E = I \times R = 2 \times 60 = 120\ \text{V}$$

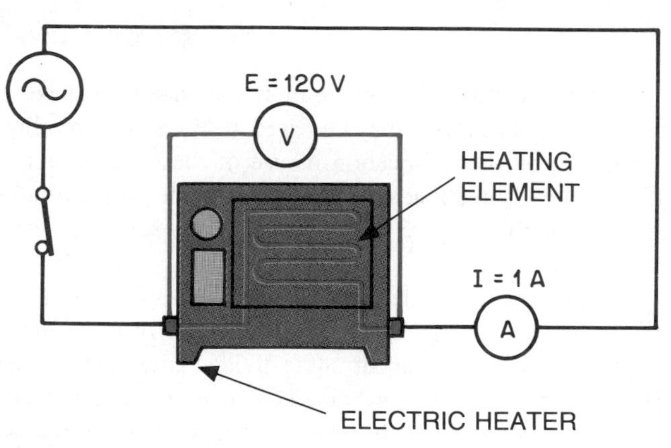

FIGURE 12–17

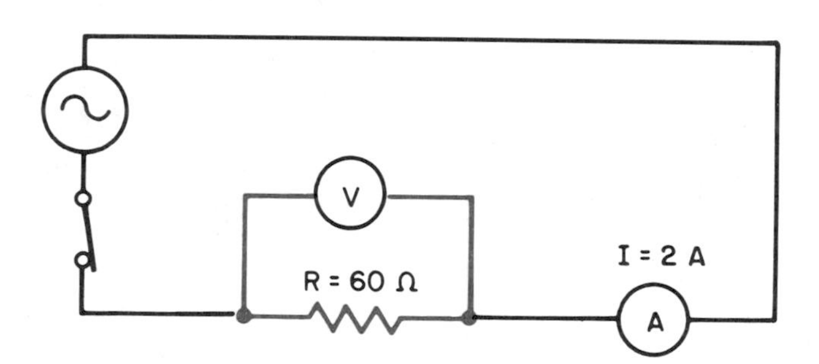

FIGURE 12–18

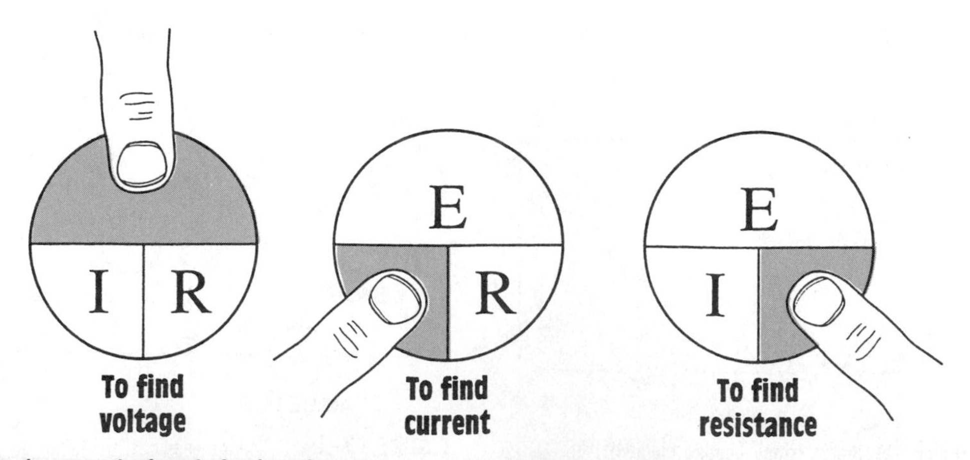

FIGURE 12–16 To determine the formula for the unknown quantity, cover the letter representing the unknown.

Figure 12–19 indicates the voltage to be 120 V and the resistance to be 20 Ω. The formula for determining the current flow is

$$I = \frac{E}{R} = \frac{120}{20} = 6 \text{ A}$$

In series circuits with more than one resistance, simply add the resistances together as if they were one. In Figure 12–20 there is only one path for the current to follow (a series circuit), so the resistance is 40 Ω (20 Ω + 10 Ω + 10 Ω). The amperage in this circuit will be

$$I = \frac{E}{R} = \frac{120}{40} = 3 \text{ A}$$

The ohms in the individual resistances can be determined by disconnecting the resistance to be measured from the circuit and reading the ohms from an ohmmeter, as illustrated in Figure 12–21.

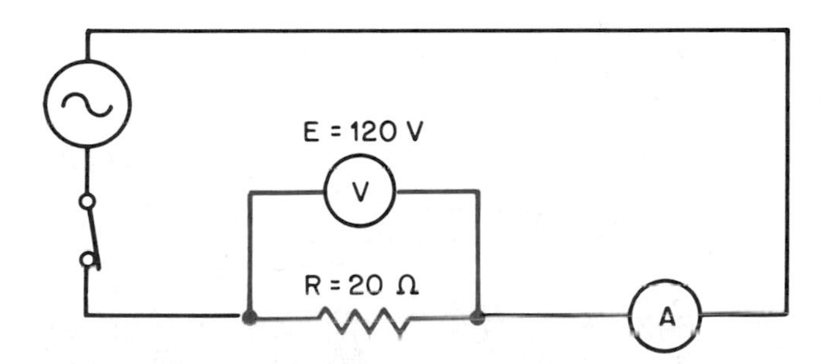

FIGURE 12–19

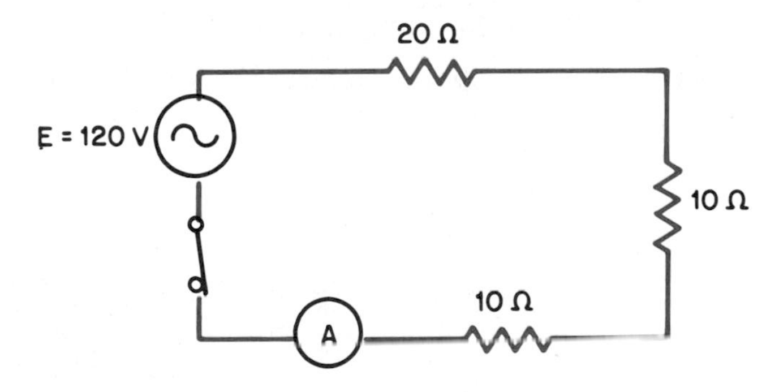

FIGURE 12–20

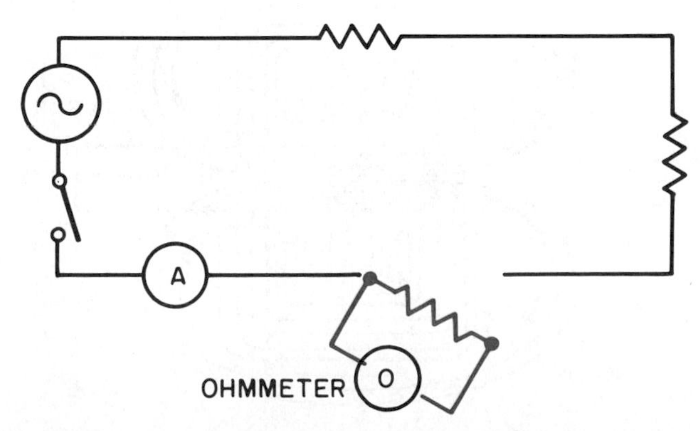

FIGURE 12–21 To determine resistance, disconnect the resistance from the circuit and check with an ohmmeter.

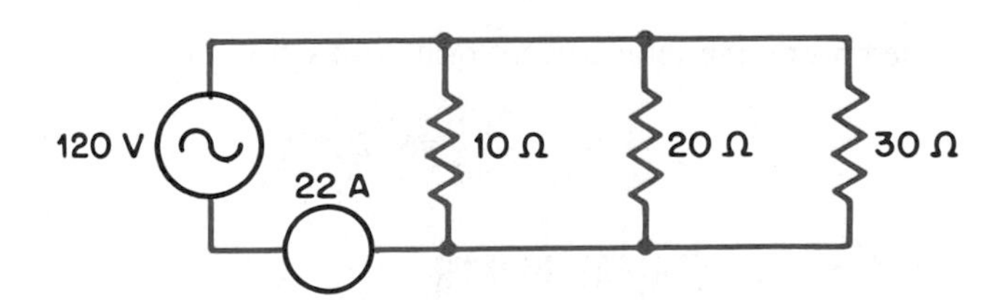

FIGURE 12–22

SAFETY PRECAUTION: *Do not use any electrical measuring instruments without specific instructions from a qualified person. The use of electrical measuring instruments is discussed later in this unit.*

12.12 CHARACTERISTICS OF SERIES CIRCUITS

In series circuits:

- The voltage is divided across the different resistances.
- The total current flows through each resistance or load.
- The resistances are added together to obtain the total resistance, Figure 12–22.

12.13 CHARACTERISTICS OF PARALLEL CIRCUITS

In parallel circuits:

- The total voltage is applied across each resistance.
- The current is divided between the different loads, or the total current is equal to the sum of the currents in each branch.
- The total resistance is less than the value of the smallest resistance.

Calculating the resistances in a parallel circuit requires a different procedure than simply adding them together as in a series circuit. A parallel circuit allows current flow along two or more paths at the same time. This type of circuit applies equal voltage to all loads. The general formula used to determine *total resistance* in a parallel circuit is as follows:

$$R_{total} = \frac{1}{\frac{1}{R_1} + \frac{1}{R_2} + \frac{1}{R_3} + \cdots}$$

The total resistance of the circuit in Figure 12–22 is determined as follows:

$$\begin{aligned} R_{total} &= \frac{1}{\frac{1}{10} + \frac{1}{20} + \frac{1}{30}} \\ &= \frac{1}{0.1 + 0.05 + 0.033} \\ &= \frac{1}{0.183} \\ &= 5.46\ \Omega \end{aligned}$$

To determine the total current draw use Ohm's Law:

$$I = \frac{E}{R} = \frac{120}{5.46} = 22 \text{ A}$$

12.14 ELECTRICAL POWER

Electrical power (P) is measured in watts. A *watt* (W) is the power used when 1 ampere flows with a potential difference of 1 volt. Therefore, power can be determined by multiplying the voltage times the amperes flowing in a circuit.

$$\text{Watts} = \text{Volts} \times \text{Amperes}$$

or

$$P = E \times I$$

The consumer of electrical power pays the electrical utility company according to the number of kilowatts (kW) used. A kilowatt is equal to 1000 W. To determine the power being consumed, divide the number of watts by 1000:

$$P \text{ (in kW)} = \frac{E \times I}{1000}$$

12.15 MAGNETISM

Magnetism was briefly discussed previously in this unit to point out how electrical generators are able to produce electricity. Magnets are classified as either permanent or temporary. Permanent magnets are used in only a few applications that air conditioning and refrigeration technicians would work with. Electromagnets, a form of temporary magnet, are used in many components of air conditioning and refrigeration equipment.

A magnetic field exists around a wire carrying an electrical current, Figure 12–23. If the wire or conductor is formed in a loop, the magnetic field will be increased, Figure 12–24. If the wire is wound into a coil, a stronger magnetic field will be created, Figure 12–25. This coil of wire carrying an electrical current is called a *solenoid.* This solenoid or electromagnet will attract or pull an iron bar into the coil, Figure 12–26.

If an iron bar is inserted permanently in the coil, the strength of the magnetic field will be increased even more.

This magnetic field can be used to generate electricity and to cause electric motors to operate. The magnetic attraction can also cause motion, which is used in many controls

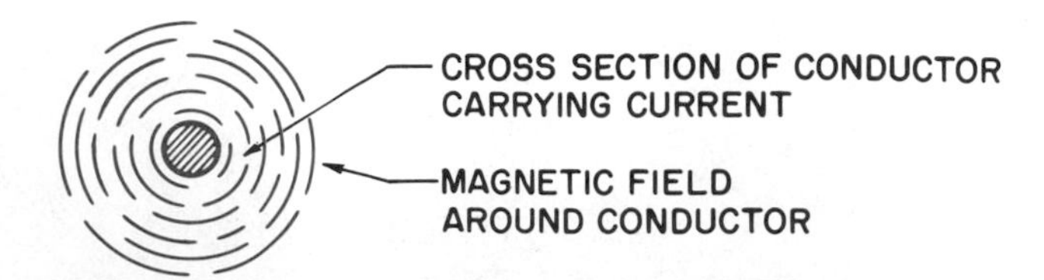

FIGURE 12–23 Cross section of a wire shows magnetic field around conductor.

FIGURE 12–24 Magnetic field around loop of wire. This is a stronger field than that around a straight wire.

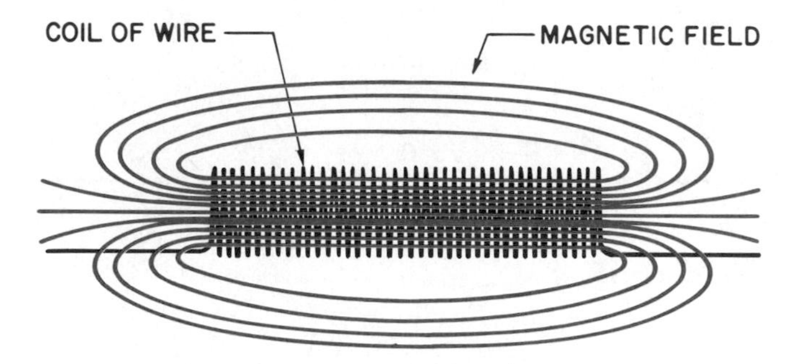

FIGURE 12–25 There is a stronger magnetic field surrounding wire formed into a coil.

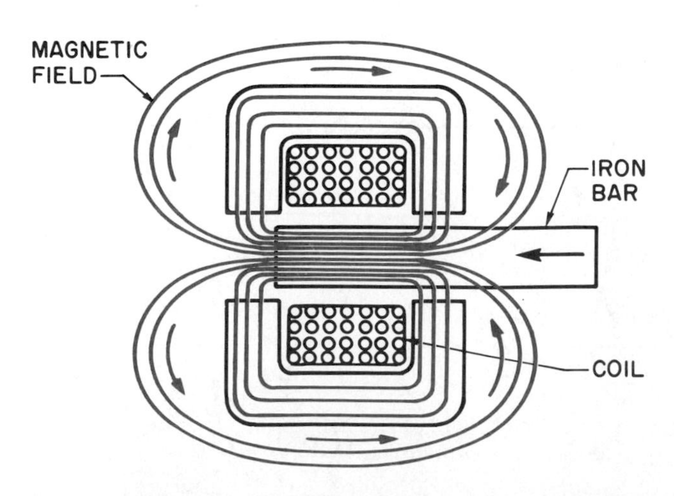

FIGURE 12–26 When current flows through coil, the iron bar will be attracted into it.

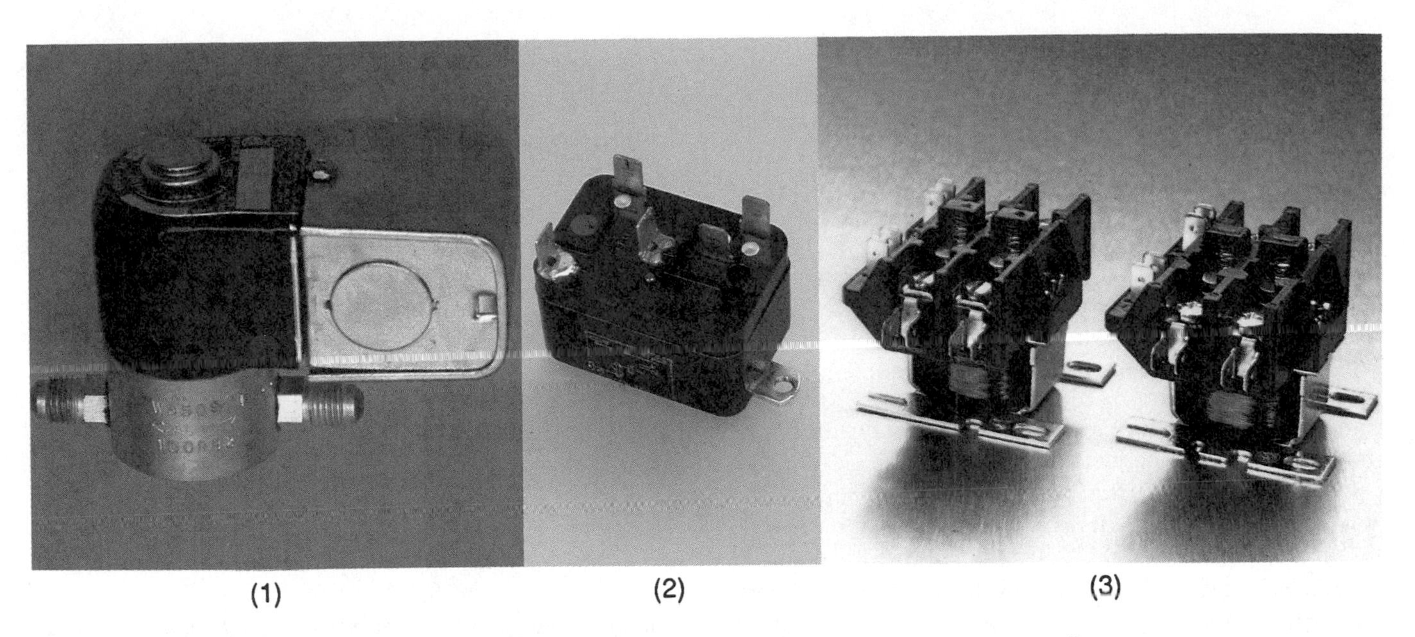

FIGURE 12–27A (1) Solenoid. (2) Relay. (3) Contactor. *Photos (1) and (2) by Bill Johnson, (3) Courtesy Honeywell, Inc.*

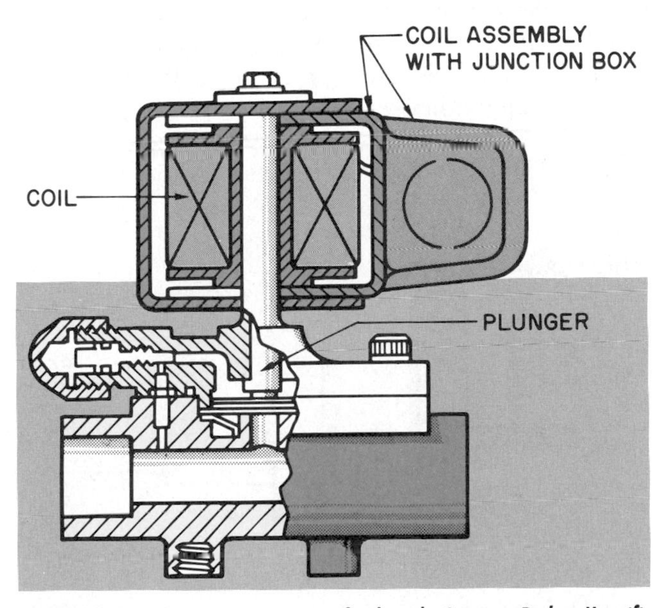

FIGURE 12–27B Cutaway view of solenoid. *Courtesy Parker Hannifin Corporation*

and switching devices, such as solenoids, relays, and contactors, Figure 12–27A. Figure 12–27B is a cutaway view of a solenoid.

12.16 INDUCTANCE

As mentioned previously, when voltage is applied to a conductor and current flows, a magnetic field is produced around the conductor. In an AC circuit the current is continually changing direction. This causes the magnetic field to continually build up and immediately collapse. When these lines of force build up and collapse, they cut through the wire or conductor and produce an emf or voltage. This voltage opposes the existing voltage in the conductor.

In a straight conductor this induced voltage is very small and is usually not considered, Figure 12–28. However, if a conductor is wound into a coil, these lines of force overlap and reinforce each other, Figure 12–29. This does develop an emf or voltage that is strong enough to provide opposition to the existing voltage. This opposition is called *inductive reactance* and is a type of resistance in an AC circuit. Coils, chokes, and transformers are examples of

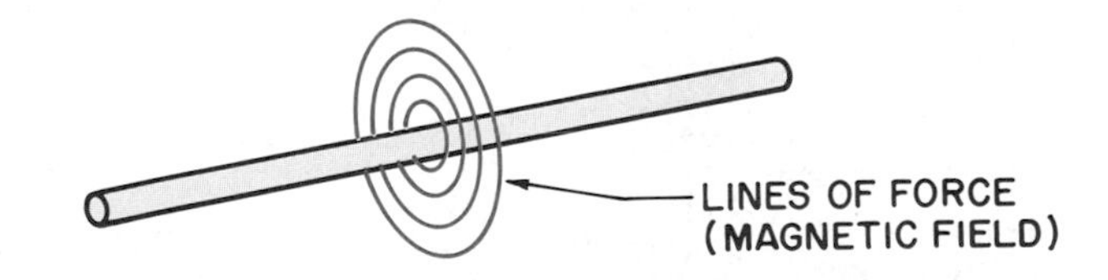

FIGURE 12–28 Straight conductor with magnetic field.

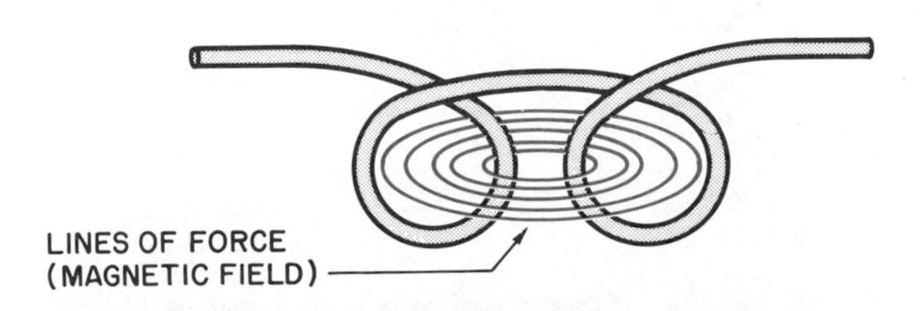

FIGURE 12–29 Conductor formed into coil with lines of force.

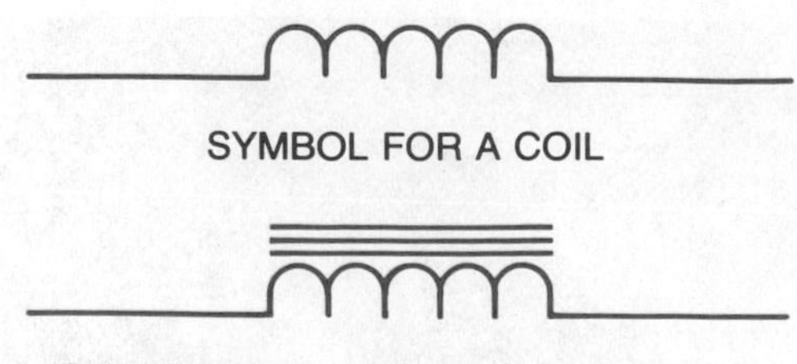

FIGURE 12–30 Symbols for a coil.

components that produce inductive reactance. Figure 12–30 shows symbols for the electrical coil.

12.17 TRANSFORMERS

Transformers are electrical devices that produce an electrical current in a second circuit through electromagnetic induction.

In Figure 12–31A a voltage applied across terminals A-A will produce a magnetic field around the steel or iron core. This is AC causing the magnetic field to continually build up and collapse as the current reverses. This will cause the magnetic field around the core in the second winding to cut across the conductor wound around it. An electrical current is induced in the second winding.

Transformers, Figure 12–31B, have a primary winding, a core usually made of thin plates of steel laminated together, and a secondary winding. There are step-up and step-down transformers. A step-down transformer contains more turns of wire in the primary winding than in the secondary

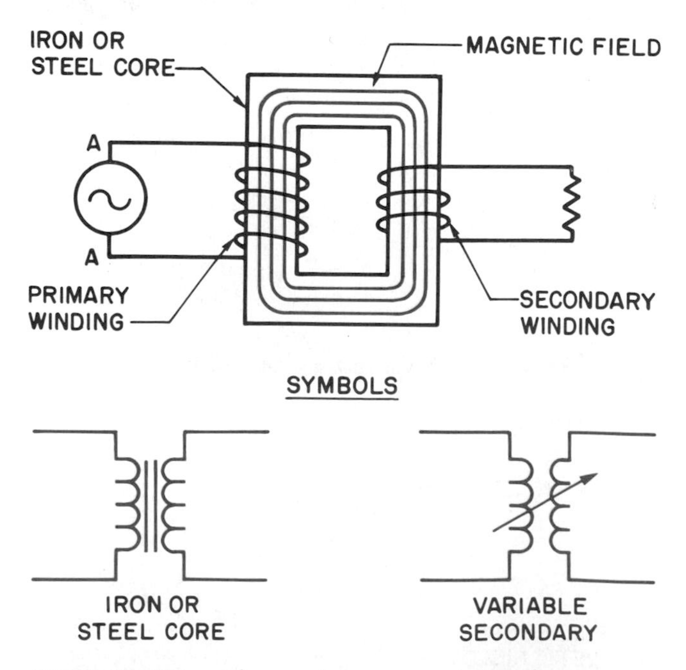

FIGURE 12–31A Voltage applied across terminals produces a magnetic field around an iron or steel core.

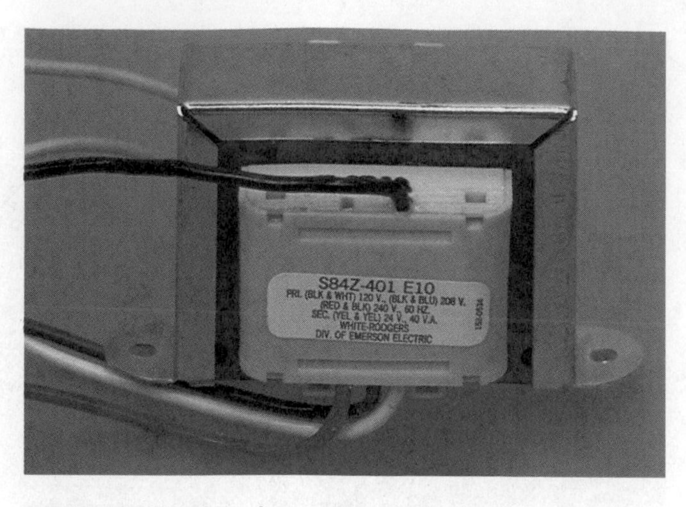

FIGURE 12–31B Transformer. *Photo by Bill Johnson*

winding. The voltage at the secondary is directly proportional to the number of turns of wire in the secondary as compared to the primary windings. For example, Figure 12–32 is a transformer with 1000 turns in the primary and 500 turns in the secondary. A voltage of 120 V is applied, and the voltage induced in the secondary is 60 V. Actually the voltage is slightly less due to some loss into the air of the magnetic field and because of resistance in the wire.

A step-up transformer has more windings in the secondary than in the primary. This causes a larger voltage to be induced into the secondary. In Figure 12–33, with 1000

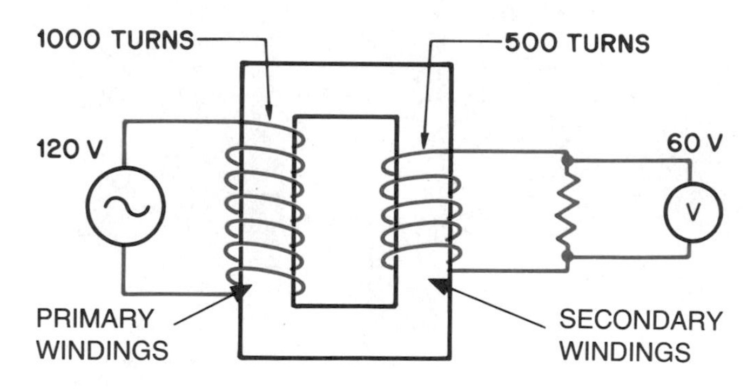

FIGURE 12–32 Step-down transformer.

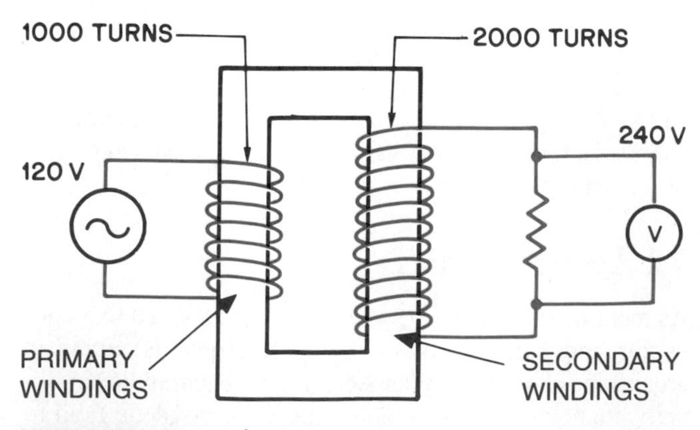

FIGURE 12–33 Step-up transformer.

turns in the primary, 2000 in the secondary, and an applied voltage of 120 V, the voltage induced in the secondary is doubled, or approximately 240 V.

The same power (watts) is available at the secondary as at the primary (except for a slight loss). If the voltage is reduced to one half that at the primary, the current capacity nearly doubles.

Step-up transformers are used at generating stations to increase the voltage to produce more efficiency in delivering the electrical energy over long distances to substations or other distribution centers. At the substation the voltage is reduced for further distribution. To reduce the voltage, a step-down transformer is used. At a residence the voltage may be reduced to 240 V or 120 V. Further step-down transformers may be used with air conditioning and heating equipment to produce the 24 V commonly used in thermostats and other control devices.

12.18 CAPACITANCE

A device in an electrical circuit that allows electrical energy to be stored for later use is called a *capacitor.* A simple capacitor is composed of two plates with insulating material between them, Figure 12–34. The capacitor can store a charge of electrons on one plate. When the plate is fully charged in a DC circuit, no current will flow until there is a path back to the positive plate, Figure 12–35. When this path is available, the electrons will flow to the positive plate until the negative plate no longer has a charge, Figures 12–36 and 12-37. At this point both plates are neutral.

In an AC circuit, the voltage and current are continuously changing direction. As the electrons flow in one direction, the capacitor plate on one side becomes charged. As this current and voltage reverses, the charge on the capacitor becomes greater than the source voltage, and the capacitor begins to discharge. It is discharged through the circuit, and the opposite plate becomes charged. This continues through each AC cycle.

A capacitor has *capacitance,* which is the amount of charge that can be stored. The capacitance is determined by the following physical characteristics of the capacitor:

1. Distance between the plates
2. Surface area of the plates
3. Dielectric material between the plates

Capacitors are rated in farads. However, farads represent such a large amount of capacitance that the term microfarad is normally used. A microfarad is one millionth (0.000001)

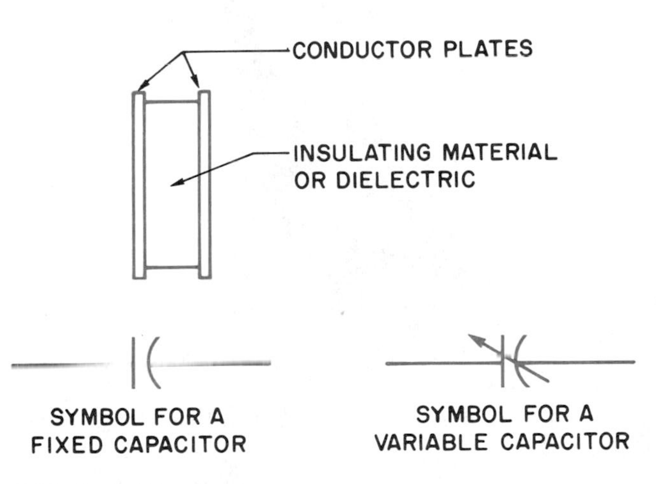

FIGURE 12–34 Capacitor.

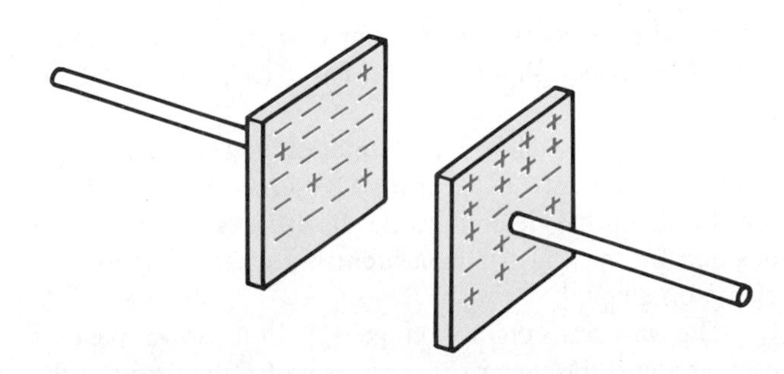

FIGURE 12–35 Charged capacitor.

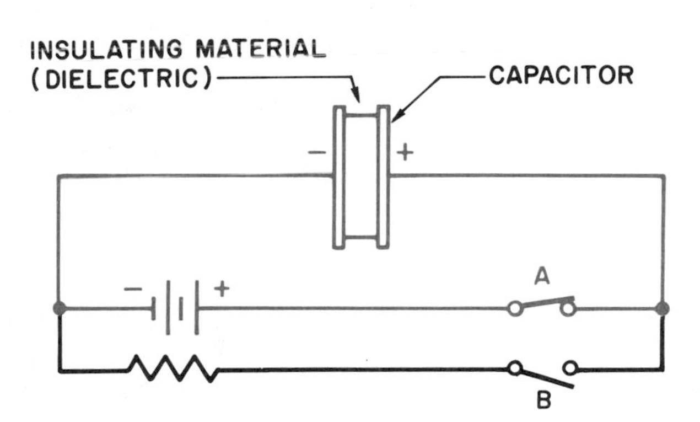

FIGURE 12–36 Electrons will flow to negative plate from battery. Negative plate will charge until capacitor has the same potential difference as the battery.

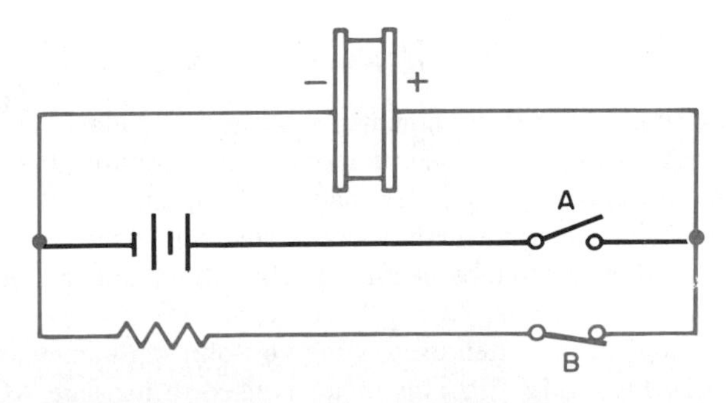

FIGURE 12–37 When capacitor is charged, switch A opened, and switch B closed, the capacitor discharges through the resistor to the positive plate. The capacitor then has no charge.

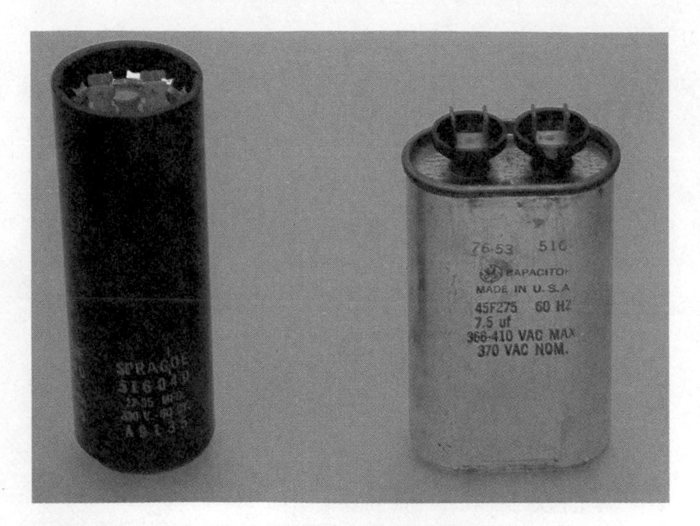

FIGURE 12–38 Run capacitor and start capacitor. *Photo by Bill Johnson*

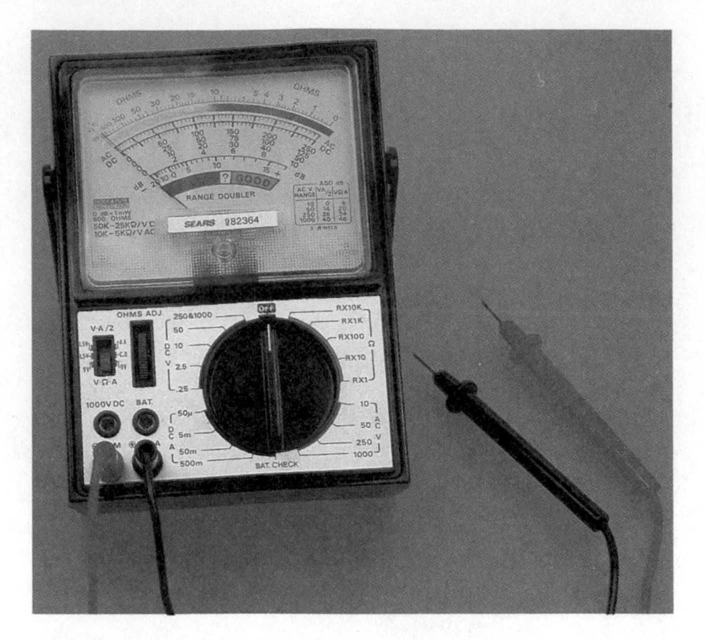

FIGURE 12–39 Volt-ohm-milliammeter (VOM). *Courtesy Simpson Electric Co., Elgin, Illinois*

of a farad. Capacitors can be purchased in ranges up to several hundred microfarads. The symbol for micro is the Greek letter μ (mu) and the symbol for farad is the capital letter F (μF).

A capacitor opposes current flow in an AC circuit similar to a resistor or to inductive reactance. This opposition or type of resistance is called *capacitive reactance*. The capacitive reactance depends on the frequency of the voltage and the capacitance of the capacitor.

Two types of capacitors used frequently in the air conditioning and refrigeration industry are the starting and running capacitors used on electric motors, Figure 12–38.

12.19 IMPEDANCE

We have learned that there are three types of opposition to current flow in an AC circuit. There is pure resistance, inductive reactance, and capacitive reactance. The total effect of these three is called impedance. The voltage and current in a circuit that has only resistance are in phase with each other. The voltage leads the current across an inductor and lags behind the current across a capacitor. Inductive reactance and capacitive reactance can cancel each other. Impedance is a combination of the opposition to current flow produced by these characteristics in a circuit.

FIGURE 12–40 Ammeter adaptor to VOM. *Courtesy Simpson Electric Co., Elgin, Illinois*

12.20 ELECTRICAL MEASURING INSTRUMENTS

A multimeter is an instrument that measures voltage, current, resistance, and, on some models, temperature. It is a combination of several meters and can be used for making AC or DC measurements in several ranges. It is the instrument used most often by heating, refrigeration, and air conditioning technicians.

A multimeter often used is the volt-ohm-milliammeter (VOM), Figure 12–39. This meter is used to measure AC and DC voltages, DC, resistance, and AC amperage when used with an AC ammeter adaptor, Figure 12–40. The meter has two main switches: the *function* switch and the *range* switch. The function switch, located at the left side of the lower front panel, Figure 12–41, has −DC, +DC, and AC positions.

The range switch is in the center of the lower part of the front panel, Figure 12–41. It may be turned in either direction to obtain the desired range. It also selects the proper position for making AC measurements when using the AC clamp-on adapter.

The zero ohms control on the right of the lower panel is used to adjust the meter to compensate for the aging of the meter's batteries, Figure 12–42.

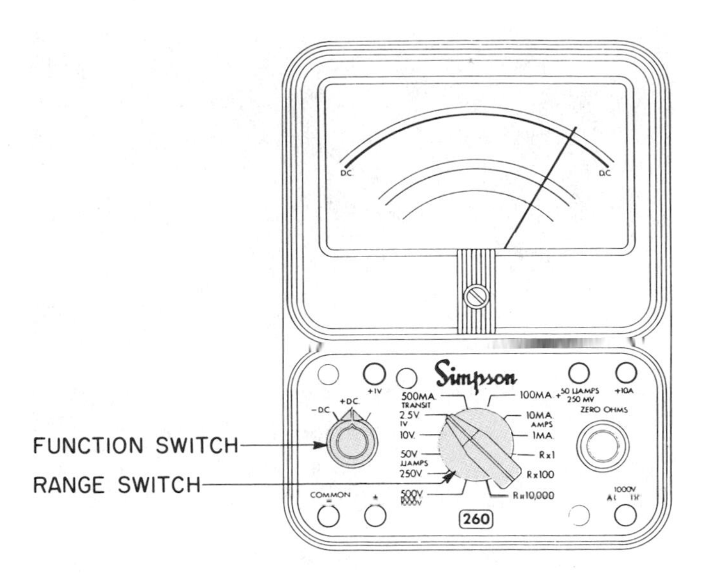

FIGURE 12–41 Function switch and range switch on VOM. *Courtesy Simpson Electric Co., Elgin, Illinois*

When the meter is ready to operate, the pointer must read zero. If the pointer is off zero, use a screwdriver to turn the screw clockwise or counterclockwise until the pointer is set exactly at zero, Figure 12–42.

The test leads may be plugged into any of eight jacks. In this unit the common (−) and the positive (+) jacks will be the only ones used, Figure 12–43. Only a few of the basic measurements are discussed. Other measurements are described in detail in other units.

The following instructions are for familiarization with the meter and procedure only. ***Do not make any measurements without instructions and approval from an instructor or supervisor.*** Insert the black test lead into the common (−) jack. Insert the red test lead into the + jack.

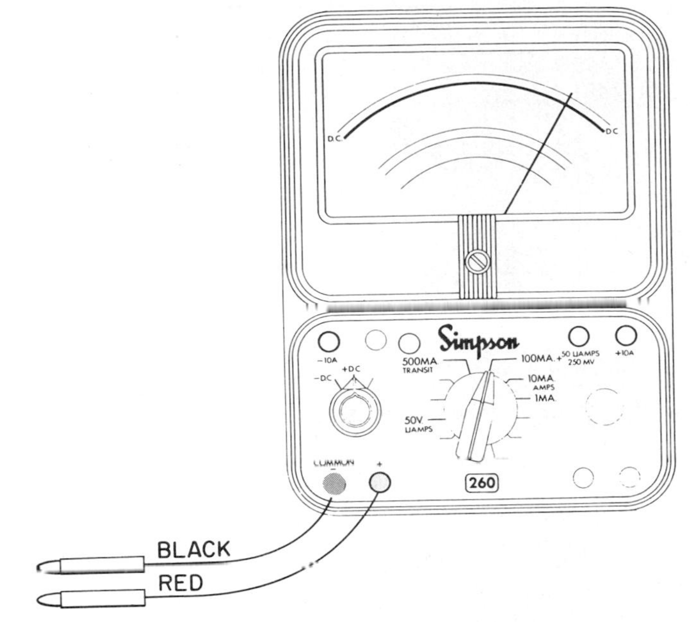

FIGURE 12–43 VOM showing the − and + jacks. *Courtesy Simspson Electric Co., Elgin, Illinois*

Figure 12–44 is a DC circuit with 15 V from the battery power source. To check this voltage with the VOM, set the function switch to +DC. Set the range switch to 50 V, Figure 12–44. If you are not sure about the magnitude of the voltage, always set the range switch to the highest setting. After measuring, you can set the switch to a lower range if necessary to obtain a more accurate reading. Be sure the switch in the circuit is open. Now connect the black test lead to the negative side of the circuit, and connect the red test lead to the positive side, as indicated in Figure 12–44. Note that the meter is connected across the load (in parallel). Close the switch and read the voltage from the DC scale.

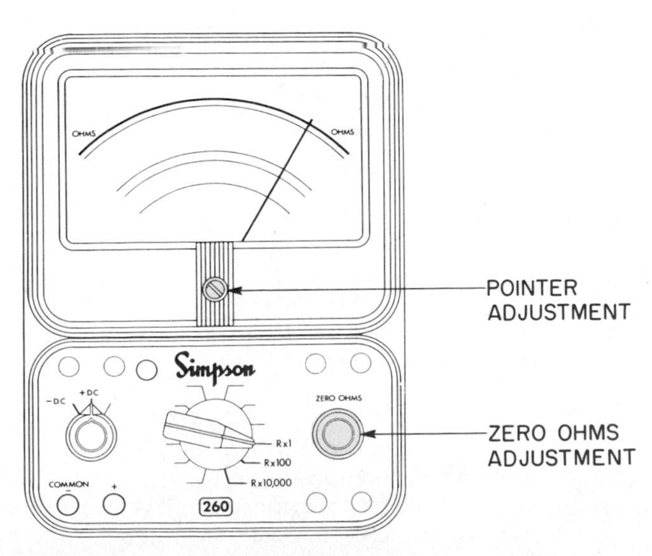

FIGURE 12–42 Zero ohms adjustment and pointer adjustment. *Courtesy Simpson Electric Co., Elgin, Illinois*

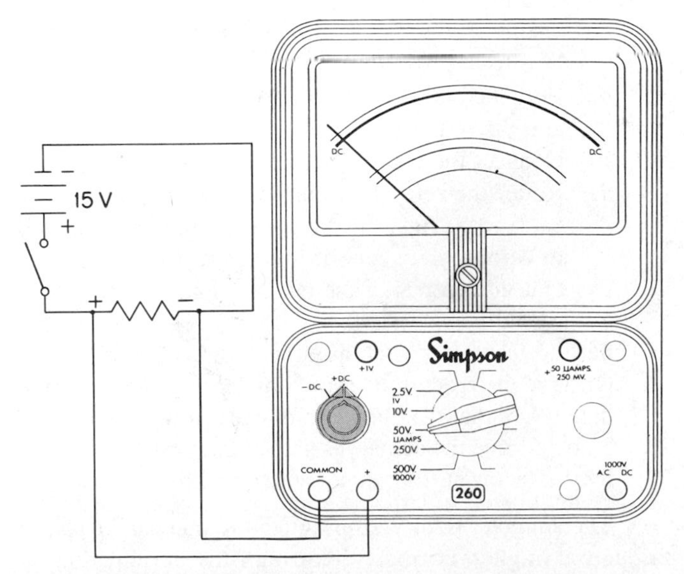

FIGURE 12–44 VOM with function switch set at +DC. Range switch set at 50 V. *Courtesy Simpson Electric Co., Elgin, Illinois*

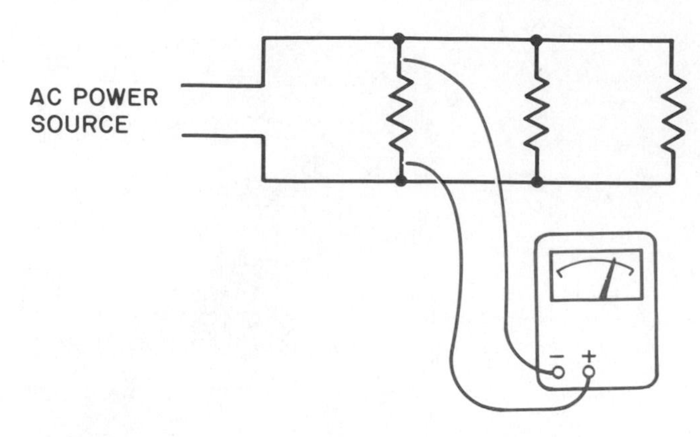

FIGURE 12–45 Connecting test leads across a load.

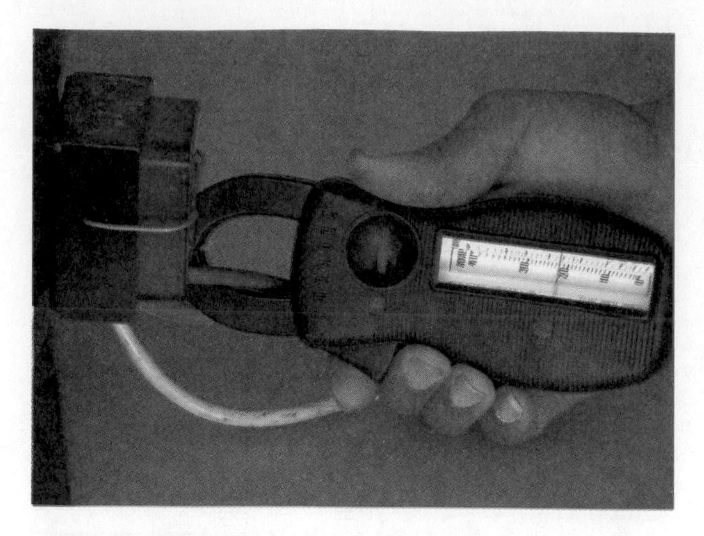

FIGURE 12–46 Measuring amperage by clamping meter around conductor. *Photo by Bill Johnson*

To check the voltage in the AC circuit in Figure 12–45, follow the steps listed:

1. Turn off the power.
2. Set the function switch to AC.
3. Set the range switch to 500 V.
4. Plug the black test lead into the common (−) jack and the red test lead into the (+) jack.
5. Connect the test leads across the load as shown in Figure 12–45.
6. Turn on the power. Read the red scale marked AC. Use the 0 to 50 figures and multiply the reading by 10.
7. Turn the range switch to 250 V. Read the red scale marked AC and use the black figures immediately above the scale.

To determine the resistance of a load, disconnect the load from the circuit. Make sure all power is off while doing this.

1. Make the zero ohms adjustment in the following manner:
   a. Turn the range switch to the desired ohms range.
      Use R × 1 for 0 to 200 Ω
      Use R × 100 for 200 to 20,000 Ω
      Use R × 10,000 for above 20,000 Ω
   b. Plug the black test lead into the common (−) jack and the red test lead into the (+) jack.
   c. Connect the test leads to each other.
   d. Rotate the zero ohms control until the pointer indicates zero ohms. (If the pointer cannot be adjusted to zero, replace one or both batteries.)
2. Disconnect the ends of the test leads and connect them to the load being tested.
3. Set the function switch at either −DC or +DC.
4. Observe the reading on the ohms scale at the top of the dial. (Note that the ohms scale reads from right to left.)
5. To determine the actual resistance, multiply the reading by the factor at the range switch position.

The ammeter has a clamping feature that can be placed around a single wire in a circuit and the current flowing through the wire can be read as amperage from the meter, Figure 12–46.

SAFETY PRECAUTION: *Do not perform any of the above or following tests in this unit without approval from an instructor or supervisor. These instructions are simply a general orientation to meters. Be sure to read the operator's manual for the particular meter available to you.*

It is often necessary to determine voltage or amperage readings to a fraction of a volt or ampere, Figure 12–47.

Many styles and types of meters are available for making electrical measurements. Figure 12–48 shows some of these meters. Many modern meters come with digital readouts, Figure 12–49.

Electrical troubleshooting is taken one step at a time. Figure 12–50 is a partial wiring diagram of an oil burner fan circuit showing the process. The fan motor does not operate because of an open motor winding. The following voltage checks could be made by the technician to determine where the failure is.

1. This is a 120-V circuit. The technician would set the VOM range selector switch to 250 V AC. The neutral (− or common) meter lead is connected to a neutral terminal at the power source.

VOLTAGE
1 volt = 1000 millivolts (m V)
1 volt = 1 000 000 microvolts (μ V)

AMPERAGE
1 ampere = 1000 milliampers (m A)
1 ampere = 1000 000 microamperes (μ A)
Note that the symbol for micro or millions is μ

FIGURE 12–47 Units of voltage and amperage.

(A)

(B)

(C)

FIGURE 12–48 Meters used for electrical measurements. (A) DC millivoltmeter. (B) Multimeter (VOM). (C) Digital clamp-on ammeter. *Photos by Bill Johnson*

FIGURE 12–49 Typical VOM with digital readout. *Courtesy Beckman Industrial Corporation*

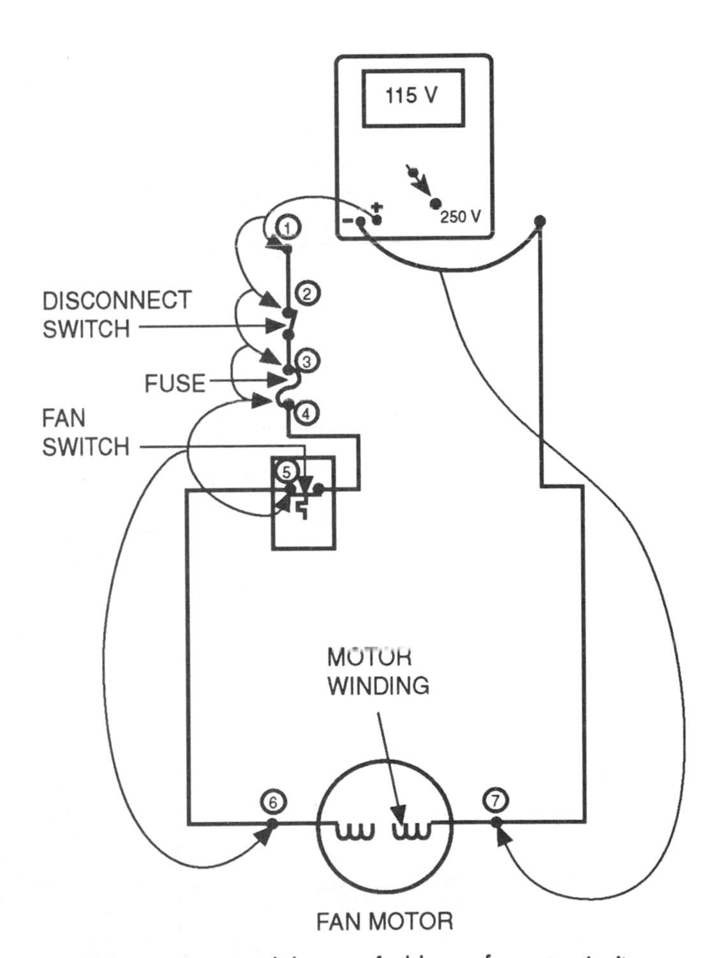

FIGURE 12–50 Partial diagram of oil burner fan motor circuit.

2. The positive (+) meter probe is connected to a line-side power source terminal, Figure 12–50 (1). The meter should read 120 V, indicating that there is power at the source.
3. Then the positive lead is connected to the line-side terminal of the disconnect switch (2). There is power.

4. The positive lead is connected to the line-side of the fuse (3). There is power.
5. The positive lead is connected to the load side of the fuse (4). There is power.
6. The positive lead is connected to the load side of the fan switch (5). There is power.
7. The positive lead is connected to the line side of the motor terminal (6). There is power.
8. To ensure there is power through the conductor and terminal connections on the neutral or ground side, connect the neutral meter probe to the neutral side of the motor (7). Leave the positive probe on the line side of the motor. There should be a 120-V reading, indicating that current is flowing through the neutral conductor.

In all of the above checks, there is 120 V but the motor does not run. It would be appropriate to conclude that the motor is defective.

The motor winding can be checked with an ohmmeter. The motor winding must have a measurable resistance for it to function properly. To check this resistance the technician may do the following, Figure 12–51:

1. Turn off the power source to the circuit.
2. Disconnect one terminal on the motor from the circuit.
3. Set the meter selector switch to ohms R × 1.
4. Touch the meter probes together and adjust the meter to 0 ohms.
5. Touch one meter probe to one motor terminal and the other to the other terminal. The meter reads infinity (∞).

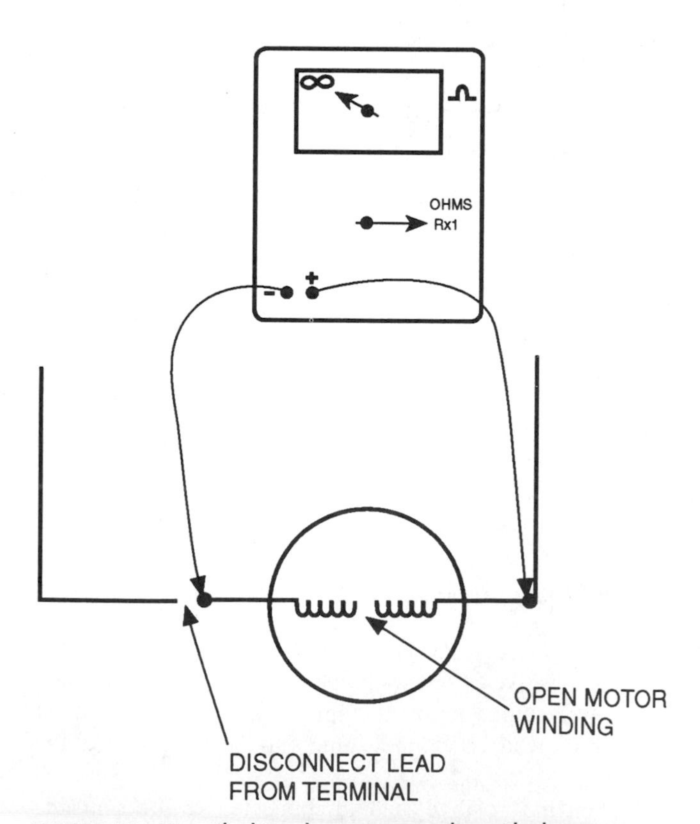

FIGURE 12–51 Checking electric motor winding with ohmmeter.

This is the same reading you would get by holding the meter probes apart in the air. There is no circuit through the windings indicating that they are open.

Most clamp-on type ammeters do not read accurately in the lower amperage ranges such as in the 1 ampere or below range. However, a standard clamp-on ammeter can be modified to produce an accurate reading. For instance transformers are rated in volt-amperes (VA). A 40-VA transformer is often used in the control circuit of combination heating and cooling systems. These transformers produce 24 V and can carry a maximum of 1.66 A. This is determined as follows:

$$\text{Output in amperes} = \frac{\text{VA rating}}{\text{Voltage}}$$

$$I = \frac{40}{24} = 1.66\text{ A}$$

Figure 12–52 illustrates how the clamp-on ammeter can be used with 10 wraps of wire to multiply the amperage reading by 10. To determine the actual amperage divide the amperage indicated on the meter by 10.

12.21 WIRE SIZES

All conductors have some resistance. The resistance depends on the conductor material, the cross-sectional area of the conductor, and the length of the conductor. A conductor with low resistance carries a current more easily than a conductor with high resistance.

Always use the proper wire (conductor) size. The size of a wire is determined by its diameter or cross section, Figure 12–53. A large diameter wire has more current-carrying capacity than a smaller diameter wire. ***If a wire is too small for the current passing through, it will overheat and possibly burn the insulation and could cause a fire.*** Standard copper wire sizes are identified by American Standard Wire Gauge numbers and measured in circular mils. A circular mil is the area of a circle 1/1000 in. in diameter. Temperature is also considered because resistance increases as temperature increases. Increasing wire size numbers indicate *smaller* wire diameters and greater resistance. Check the tables in the National Electrical Code to determine proper wire size.

12.22 CIRCUIT PROTECTION DEVICES

SAFETY PRECAUTION: *Electrical circuits *must* be protected from current overloads. If too much current flows through the circuit, the wires and components will overheat, resulting in damage and possible fire. Circuits are normally protected with fuses or circuit breakers.*

Fuses

A *fuse* is a simple device. Most fuses contain a strip of metal that has a higher resistance than the conductors in the circuit. This strip also has a relatively low melting point. Be-

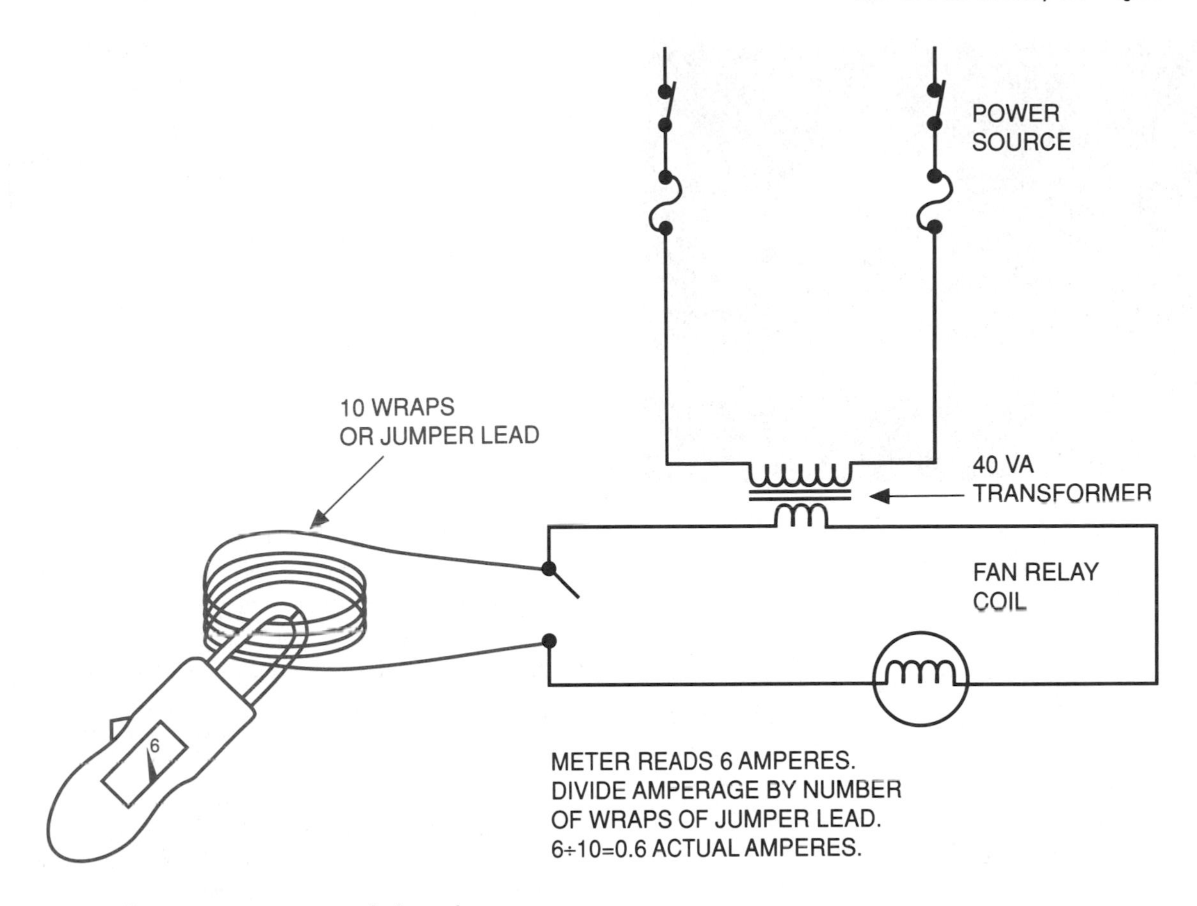

FIGURE 12–52 Illustration using 10-wrap multiplier with ammeter.

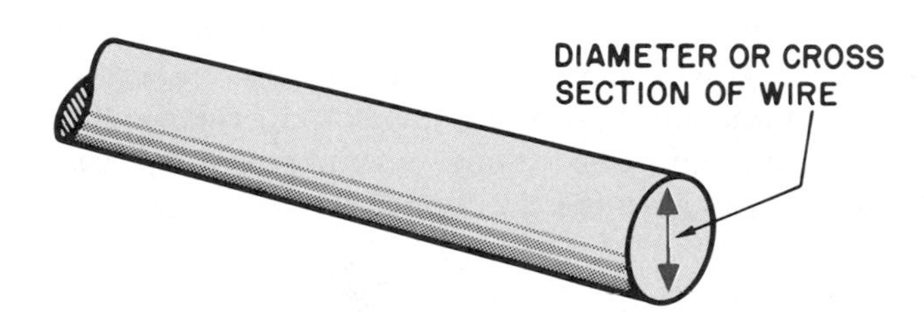

FIGURE 12–53 Cross section of a wire.

cause of its higher resistance, it will heat up faster than the conductor. When the current exceeds the rating on the fuse, the strip melts and opens the circuit.

PLUG FUSES. Plug fuses have either an Edison base or a Type S base, Figure 12–54(A). Edison-base fuses are used in older installations and can be used for replacement only. Type S fuses can be used only in a Type S fuse holder specifically designed for the fuse; otherwise an adapter must be used, Figure 12–54(B). Each adapter is designed for a specific ampere rating, and these fuses cannot be interchanged. The amperage rating determines the size of the adapter. Plug fuses are rated up to 125 volts and 30 A.

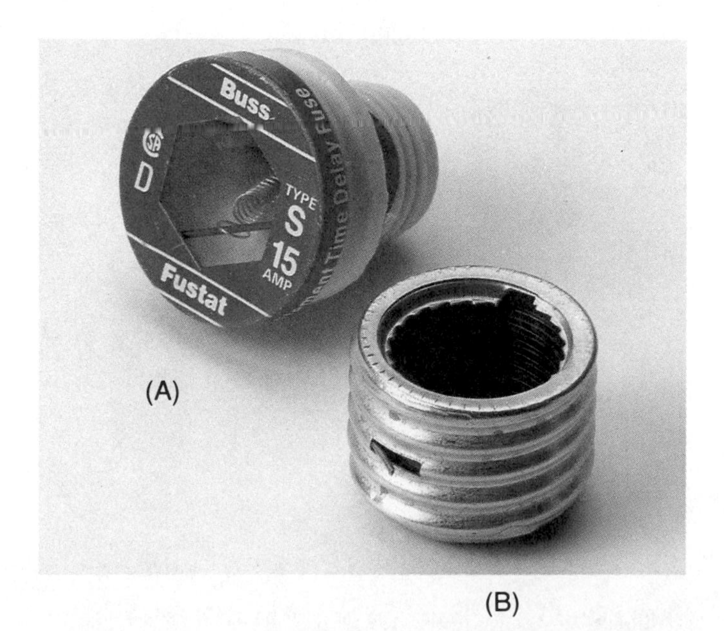

FIGURE 12–54 (A) Type S base plug fuse. (B) Type S fuse adapter. ***Reprinted with permission by Bussmann Division, McGraw-Edison Company***

FIGURE 12–55 Dual-element plug fuse. *Reprinted with permission by Bussmann Division, McGraw-Edison Company*

DUAL-ELEMENT PLUG FUSES. Many circuits have electric motors as the load or part of the load. Motors draw more current when starting and can cause a plain (single element) fuse to burn out or open the circuit. Dual-element fuses are frequently used in this situation, Figure 12–55. One element in the fuse will melt when there is a large overload, such as a short circuit. The other element will melt and open the circuit when there is a smaller current overload lasting more than a few seconds. This allows for the larger starting current of an electric motor.

CARTRIDGE FUSES. For 230-V to 600-V service up to 60 A the ferrule cartridge fuse is used, Figure 12–56(A). From 60 A to 600 A knife-blade cartridge fuses can be used, Figure 12–56(B). A cartridge fuse is sized according to its ampere rating to prevent a fuse with an inadequate rating from being used. Many cartridge fuses have an arc-quenching material around the element to prevent damage from arcing in severe short-circuit situations, Figure 12–57.

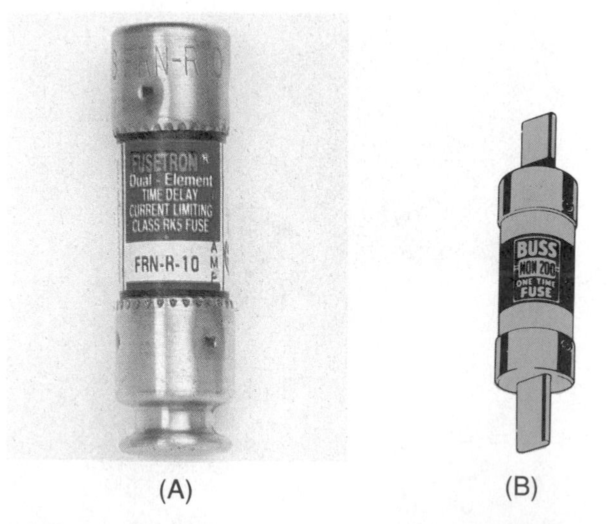

FIGURE 12–56 (A) Ferrule-type cartridge fuses. (B) Knife-blade cartridge fuse. *Reprinted with permission by Bussmann Division, McGraw-Edison Company*

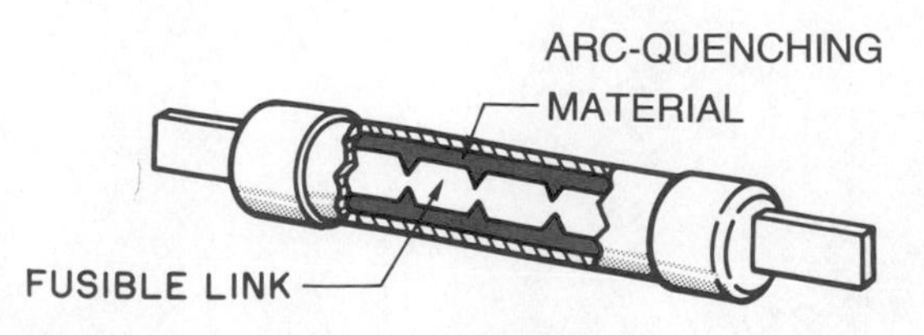

FIGURE 12–57 Knife-blade cartridge fuse with arc-quenching material.

Circuit Breakers

A circuit breaker can function as a switch as well as a means for opening a circuit when a current overload occurs. Most modern installations in houses and many commercial and industrial installations use circuit breakers rather than fuses for circuit protection.

Circuit breakers use two methods to protect the circuit. One is a bimetal strip that heats up with a current overload and trips the breaker, opening the circuit. The other is a magnetic coil that causes the breaker to trip and open the circuit when there is a short circuit or other excessive current overload in a short time, Figure 12–58.

Ground Fault Circuit Interrupters

SAFETY PRECAUTION: *Ground fault circuit interrupters (GFCI) help protect individuals against shock, in addition to providing current overload protection. The GFCI, Figure 12–59, detects even a very small current leak to a ground. Under certain conditions this leak may cause an electrical shock. This small leak, which may not be detected by a conventional circuit breaker, will cause the GFCI to open the circuit.*

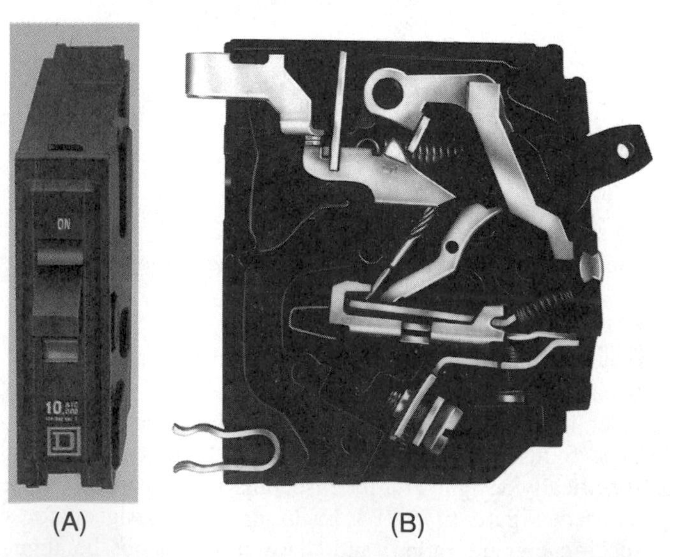

FIGURE 12–58 (A) Circuit breaker. (B) Cutaway. *Courtesy Square D Company*

ON
OFF
PUSH
TO
TEST

FIGURE 12–59 Ground fault circuit interrupter. *Courtesy Square D Company*

Circuit protection is essential to prevent the conductors in the circuit from being overloaded. If one of the circuit power-consuming devices were to cause an overload due to a short circuit within its coil, the circuit protector would stop the current flow before the conductor became overloaded and hot. Remember, a circuit consists of a power supply, the conductor, and the power-consuming device. The conductor must be sized large enough that it does not operate beyond its rated temperature, typically 140°F (60°C) while in an ambient of 86°F (30°C). For example, a circuit may be designed to carry a load of 20 A. As long as the circuit is carrying up to this amperage, overheating is not a potential hazard. If the amperage in the circuit is gradually increased, the conductor will begin to become hot, Figure 12–60. Proper understanding of circuit protection is a lengthy process. More details can be obtained from the National Electrical Code and from further study of electricity.

12.23 SEMICONDUCTORS

The development of what are commonly called semiconductors or solid-state components has caused major changes in the design of electrical devices and controls.

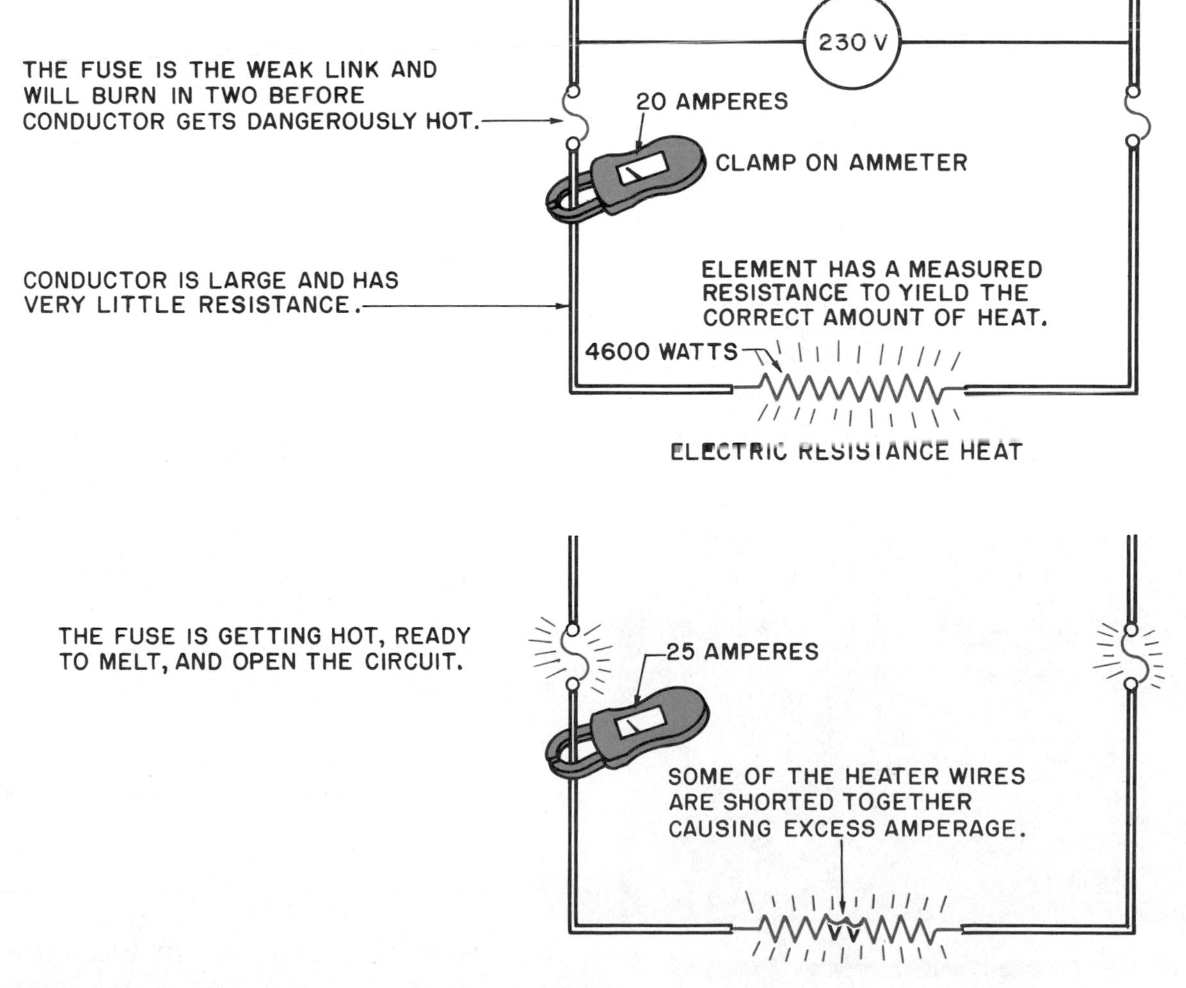
230 V
THE FUSE IS THE WEAK LINK AND WILL BURN IN TWO BEFORE CONDUCTOR GETS DANGEROUSLY HOT.
20 AMPERES
CLAMP ON AMMETER
CONDUCTOR IS LARGE AND HAS VERY LITTLE RESISTANCE.
ELEMENT HAS A MEASURED RESISTANCE TO YIELD THE CORRECT AMOUNT OF HEAT.
4600 WATTS
ELECTRIC RESISTANCE HEAT
THE FUSE IS GETTING HOT, READY TO MELT, AND OPEN THE CIRCUIT.
25 AMPERES
SOME OF THE HEATER WIRES ARE SHORTED TOGETHER CAUSING EXCESS AMPERAGE.

FIGURE 12–60 Fuses protect the circuit.

Semiconductors are generally small and lightweight and can be mounted in circuit boards, Figure 12–61. In this unit we describe some of the individual solid-state devices and some of their uses. Refrigeration and air conditioning technicians do not normally replace solid-state components on circuit boards. They should have some knowledge of these components, however, and should be able to determine when one or more of the circuits in which they are used are defective. In most cases, when a component is defective the entire board will need to be replaced. Often these circuit boards can be returned to the manufacturer or sent to a company that specializes in repairing or rebuilding them.

Semiconductors are usually made of silicon or germanium. Semiconductors in their pure form, and as their name implies, do not conduct electricity well. However, for semiconductors to be of value they must conduct electricity in some controlled manner. To accomplish this an additional substance, often called an impurity, is added to the crystal-like structures of the silicon or germanium. This is called doping. One type of impurity produces a hole in the material where an electron should be. Because the hole replaces an electron (which has a negative charge), it results in the material having fewer electrons or a net result of a positive charge. This is called a P-type material. If a material of a different type is added to the semiconductor, an excess of electrons is produced, the material has a negative charge, and is called an N-type material.

When a voltage is applied to a P-type material, electrons fill these holes and move from one hole to the next still moving from negative to positive. However, this makes it appear that the holes are moving in the opposite direction (from positive to negative) as the electrons move from hole to hole.

N-type material has an excess of electrons that move from negative to positive when a voltage is applied.

Solid-state components are made from a combination of N-type and P-type substances. The manner in which the materials are joined together, the thickness of the materials, and other factors determine the type of solid-state component and its electronic characteristics.

FIGURE 12–61 Circuit board with semiconductors. *Photo by Bill Johnson*

DIODES. Diodes are simple solid-state devices. They consist of P-type and N-type material connected together. When this combination of P- and N-type material is connected to a power source one way, it will allow current to flow and is said to have forward bias. When reversed it is said to have reverse bias, and no current will flow. Figure 12–62 is a drawing of a simple diode. Figure 12–63 is a photo of several types of diodes. One of the connections on the diode is called the cathode and the other the anode, Figure 12–64. If the diode is to be connected to a battery to have forward bias (current flow), the negative terminal on the battery should be connected to the cathode, Figure 12–65. Connec-

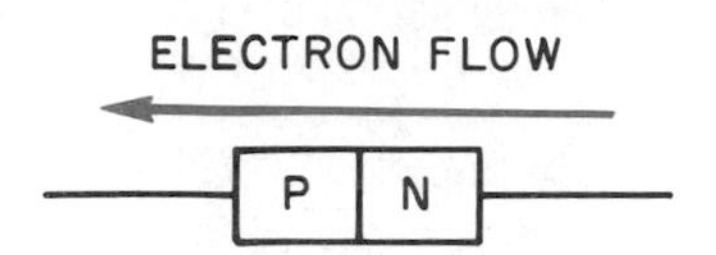

FIGURE 12–62 Pictorial drawing of a diode.

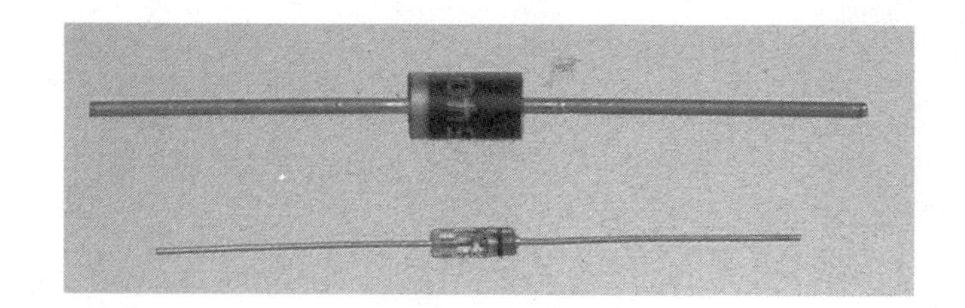

FIGURE 12–63 Typical diodes. *Photos by Bill Johnson*

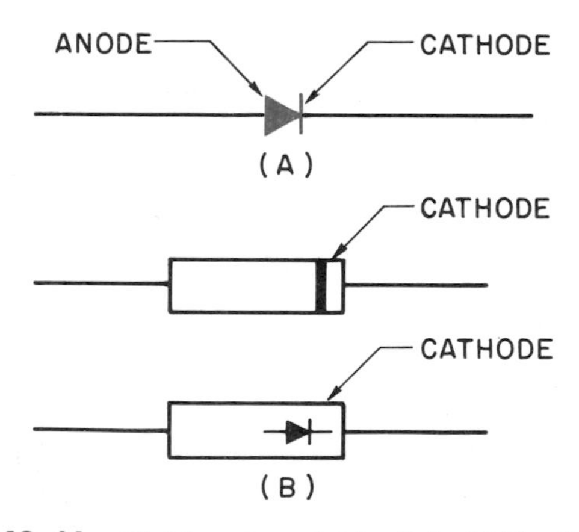

FIGURE 12–64 (A) Schematic symbol for diode. (B) Identifying markings on diode.

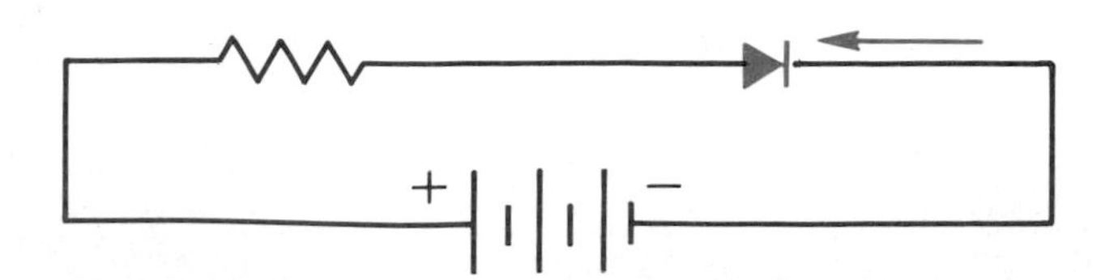

FIGURE 12–65 Simple diagram with diode indicating forward bias.

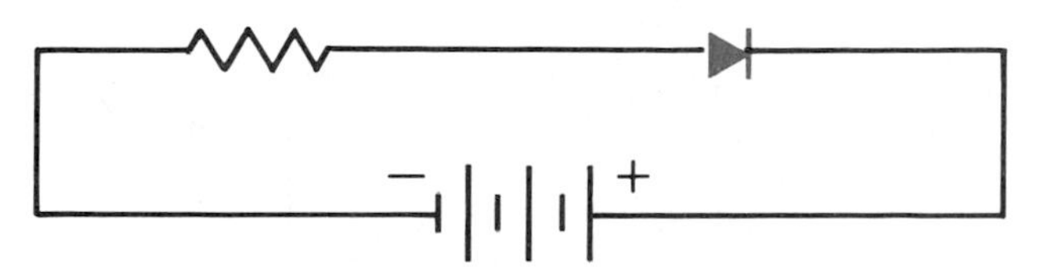

FIGURE 12–66 Circuit with diode indicating reverse bias.

ting the negative terminal of the battery to the anode will produce reverse bias (no current flow), Figure 12–66.

Checking a Diode

A diode may be tested by connecting an ohmmeter across it. The diode must be removed from the circuit. The negative probe should be touched to the cathode and the positive to the anode. With the selector switch on R × 1 the meter should show a small resistance, indicating that there is continuity, Figure 12–67. Reverse the leads. The meter should show infinity, indicating there is no continuity. A diode should show continuity in one direction and not in the other. If it shows continuity in both directions, it is defective. If it does not show continuity in either direction, it is defective.

RECTIFIER. A diode can be used as a solid-state rectifier, changing AC to DC. The term diode is normally used when rated for less than 1 A. A similar component rated above 1 A is called a rectifier. A rectifier allows current to flow in one direction. Remember that AC flows in first one direction and then reverses, Figure 12–68(A). The rectifier allows the AC to flow in one direction but blocks it from reversing, Figure 12–68(B). Therefore, the output of a rectifier circuit is in one direction or direct current. Figure 12–69 illustrates a rectifier circuit. This is called a half wave rectifier because it allows only that part of the AC moving in one direction to pass through. Figure 12–69 point (A) shows the AC before it is rectified and Figure 12–69 point (B) shows the AC after it is rectified. Full wave rectification can be achieved by using a more complicated circuit such as in Figure 12–70. During one half of the AC cycle, D_1 will conduct. Current flows through D_1, through

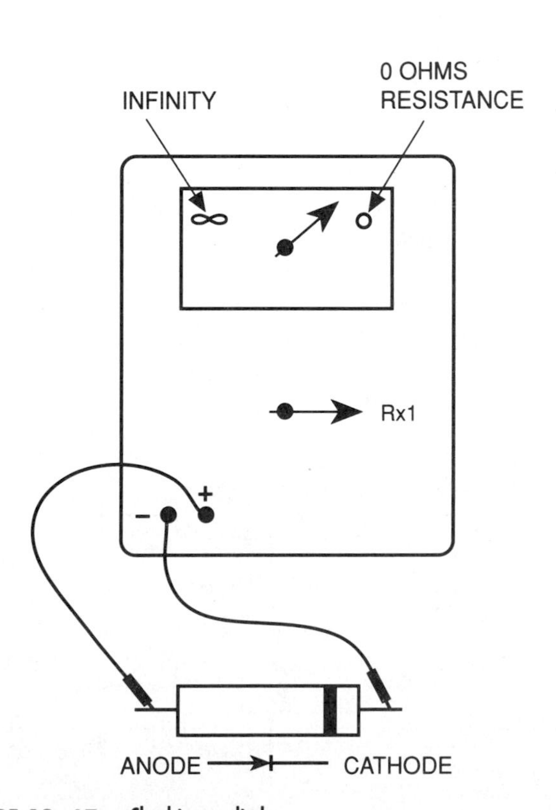

FIGURE 12–67 Checking a diode.

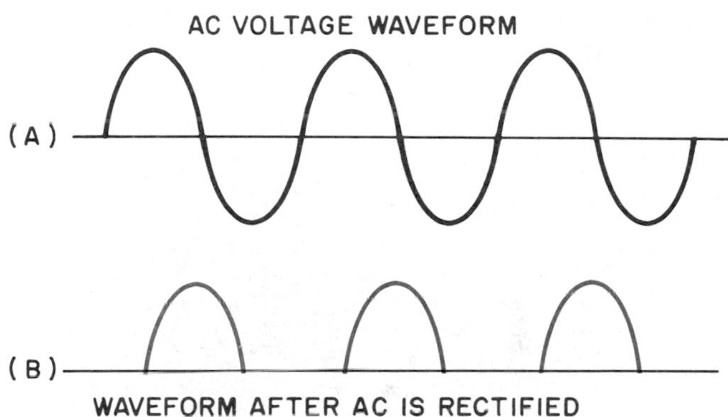

FIGURE 12–68 (A) Full wave AC waveform. (B) Half wave DC waveform.

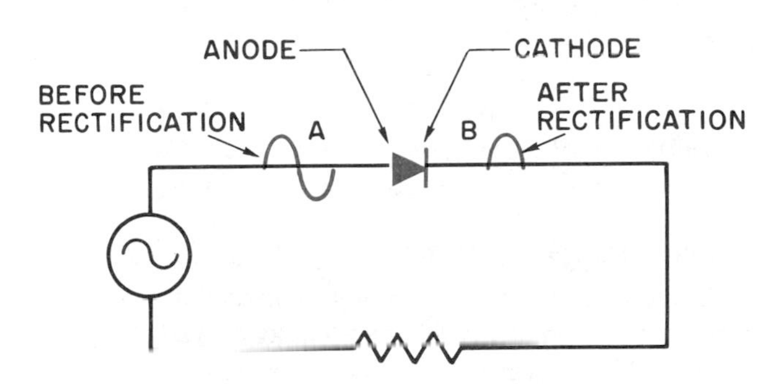

FIGURE 12–69 Diode rectifier circuit.

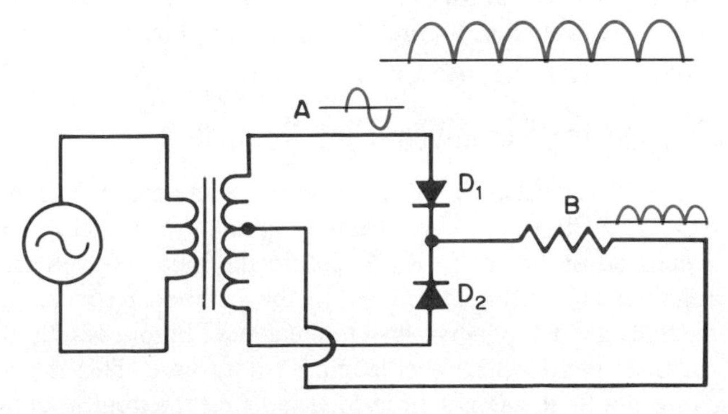

FIGURE 12–70 Full wave rectifier.

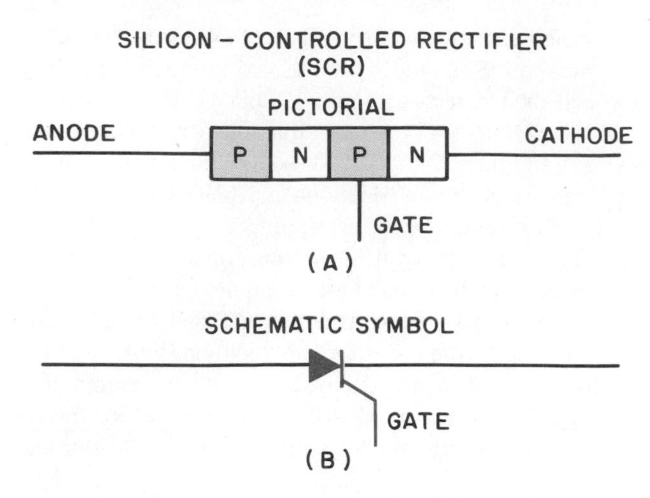

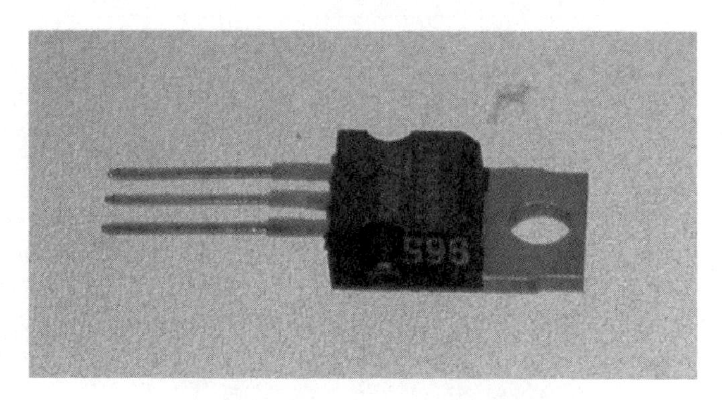

FIGURE 12–71 Pictorial and schematic drawings of a silicon-controlled rectifier.

FIGURE 12–72 Typical silicon-controlled rectifier. *Photo by Bill Johnson*

the resistor, and back to the center tap of the transformer. When the voltage reverses, D_2 conducts; current flows through the resistor and back to the center tap.

SILICON-CONTROLLED RECTIFIER. Silicon-controlled rectifiers (SCRs) consist of four semiconductor materials bonded together. These form a PNPN junction, Figure 12–71(A); Figure 12–71(B) illustrates the schematic symbol. Notice that the schematic is similar to the diode except for the gate. The SCR is used to control devices that may use large amounts of power. The gate is the control for the SCR. These devices may be used to control the speed of motors or to control the brightness of lights. Figure 12–72 shows photos of typical SCRs.

Checking the Silicon-Controlled Rectifier

The SCR can also be checked with an ohmmeter. Ensure that the SCR is removed from a circuit. Set the selector switch on the meter to R × 1. Zero the meter. Fasten the negative lead from the meter to the cathode terminal of the SCR and the positive lead to the anode, Figure 12–73. If the SCR is good, the needle should not move. This is because the SCR has not fired to complete the circuit. Use a jumper to connect the gate terminal to the anode. The meter

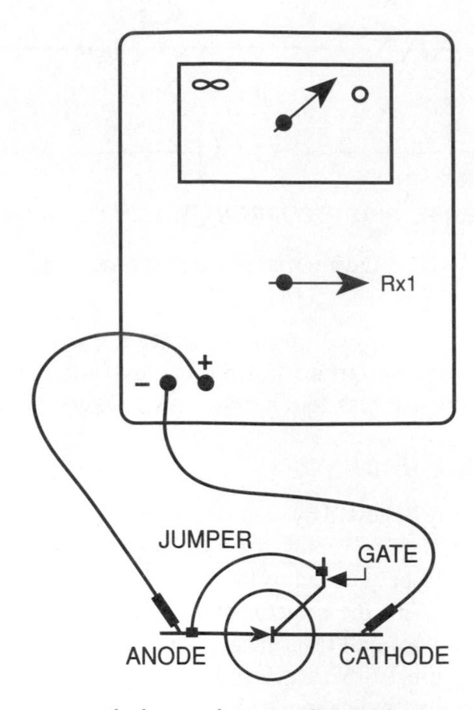

FIGURE 12–73 Checking a silicon-controlled rectifier.

needle should now show continuity. If it does not, you may not have the cathode and anode properly identified. Reverse the leads and change the jumper to the new suspected anode. If it fires, you had the anode and cathode reversed. When the jumper is removed, the SCR will continue to conduct if the meter has enough capacity to keep the gate closed. If the meter were to show current flow without firing the gate, the SCR is defective. If the gate will not fire after the jumper and leads are attached correctly, the SCR is defective.

TRANSISTORS. Transistors are also made of N-type and P-type semiconductor materials. Three pieces of these materials are sandwiched together. Transistors are either NPN or PNP types, Figure 12–74 shows diagrams and Figure 12–75 shows the schematic symbols for the two types. As the symbols show, each transistor has a base, a collector, and an

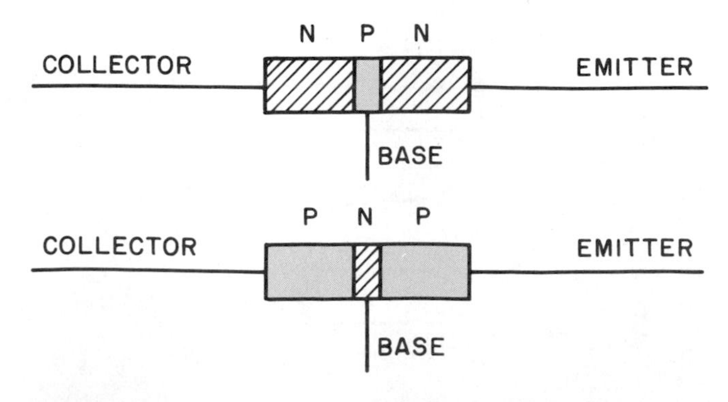

FIGURE 12–74 Pictorial drawings of NPN and PNP transistors.

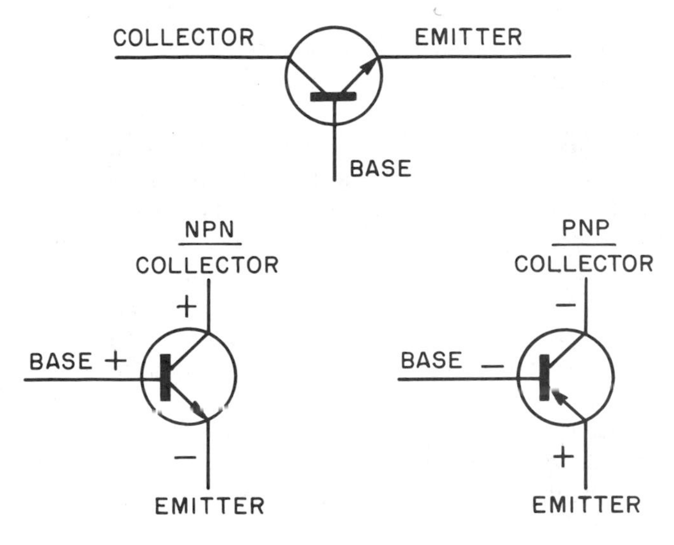

FIGURE 12–75 Schematic drawings of NPN and PNP transistors.

emitter. In the NPN type the collector and the base are connected to the positive, the emitter to the negative. The PNP transistor has a negative base and collector connection and a positive emitter. The base must be connected to the same polarity as the collector to provide forward bias. Figure 12–76 is a photo of typical transistors.

The transistor may be used as a switch or a device to amplify or increase an electrical signal. One application of a transistor used in an air conditioning control circuit would be to amplify a small signal to provide enough current to operate a switch or relay.

Current will flow through the base emitter and through the collector emitter. The base-emitter current is the control and the collector-emitter current produces the action. A very small current passing through the base emitter may allow a much larger current to pass through the collector-emitter junction. A small increase in the base-emitter junction can allow a much larger increase in current flow through the collector emitter.

FIGURE 12–76 Typical transistors. *Photo by Bill Johnson*

THERMISTORS. A thermistor is a type of resistor that is sensitive to temperature, Figure 12–77. The resistance of a thermistor changes with a change in temperature. There are two types of thermistors. A positive temperature coefficient (PTC) thermistor causes the resistance of the thermistor to increase when the temperature increases. A negative temperature coefficient (NTC) causes the resistance to decrease with an increase in temperature. Figure 12–78 illustrates the schematic symbol for a thermistor.

An application of a thermistor is to provide motor overload protection. The thermistor is imbedded in the windings of a motor. When the winding temperature exceeds a predetermined amount, the thermistor changes in resistance. This change in resistance is detected by an electronic circuit that causes the motor circuit to open.

Another application is to provide start assistance in a PSC (permanent split-capacitor) electric motor. This thermistor is known as a positive temperature coefficient device. It allows full voltage to reach the start windings during start-up of the motor. The thermistor heats during start-up and creates resistance, turning off power to the start winding at the appropriate time. This does not give the motor the

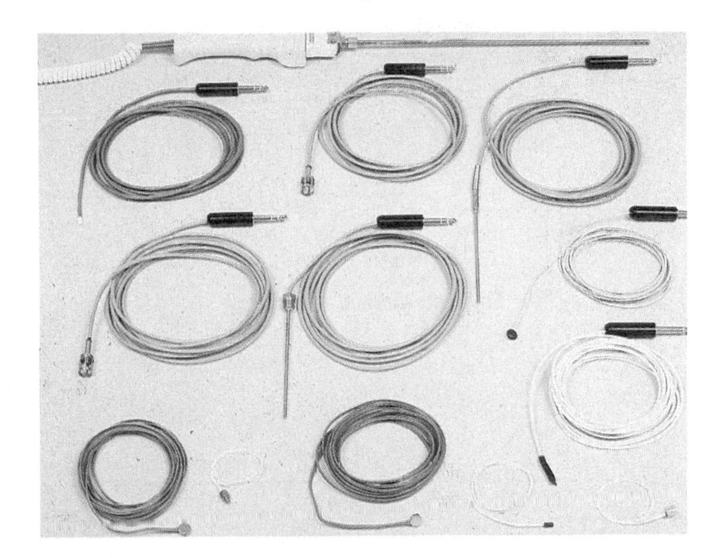

FIGURE 12–77 Typical thermistors. *Courtesy Omega Engineering*

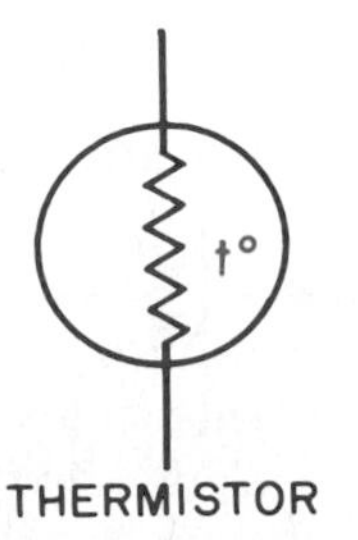

FIGURE 12–78 Schematic symbol for thermistor.

starting torque that a start capacitor does, but it is advantageous in some applications because of its simple construction and lack of moving parts.

DIACS. The diac is a two-directional electronic device. It can operate in an AC circuit and its output is AC. It is a voltage-sensitive switch that operates in both halves of the AC waveform. When a voltage is applied, it will not conduct (or will remain off) until the voltage reaches a predetermined level. Let's assume this predetermined level to be 24 V. When the voltage in the circuit reaches 24 V, the diac will begin to conduct or fire. Once it fires it will continue to conduct even at a lower voltage. Diacs are designed to have a higher cut-in voltage and lower cut-out voltage. If the cut-in voltage is 24 V, let's assume that the cut-out voltage is 12 V. In this case the diac will continue to operate until the voltage drops below 12 V at which time it will cut off.

Figure 12–79 shows two schematic symbols for a diac. Figure 12–80 illustrates a diac in a simple AC circuit. Diacs are often used as switching or control devices for triacs.

TRIACS. A triac is a switching device that will conduct on both halves of the AC waveform. The output of the triac is AC. Figure 12–81 shows the schematic symbol for a triac. Notice that it is similar to the diac but has a gate lead. A pulse supplied to the gate lead can cause the triac to fire or to conduct. Triacs were developed to provide better AC switching. As was mentioned previously, diacs often provide the pulse to the gate of the triac.

HEAT SINKS. Some solid-state devices may appear a little different from others of the same type. This is because of the differing voltages and current they are rated to carry and the purpose for which they are designed and used. Some produce much more heat than others and will only operate at the rating specified if kept within a certain temperature range. Heat that could change the operation of or destroy the device must be dissipated. This is done by adhering the solid-state component to an object called a heat sink that has a much greater surface area, Figure 12–82. Heat will travel from the device to this object with the large surface area allowing the excess heat to be dissipated into the surrounding air.

This has been a short introduction into semiconductors. Brief descriptions in previous paragraphs outlined procedures for checking certain semiconductors. These tests were described to show how semiconductors may be checked in the field. However, it is seldom practical to remove these devices from a printed circuit board and perform the test. Other instruments may be used in the shop or laboratory to check these same components while they are still in the circuit. Refrigeration and air conditioning technicians will be more involved with checking the input and output of circuit board circuits than with checking individual elec-

FIGURE 12–79 Schematic symbols for a diac.

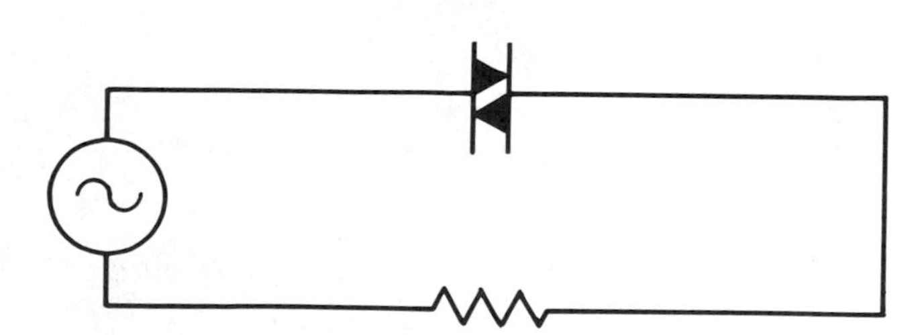

FIGURE 12–80 Simple diac circuit.

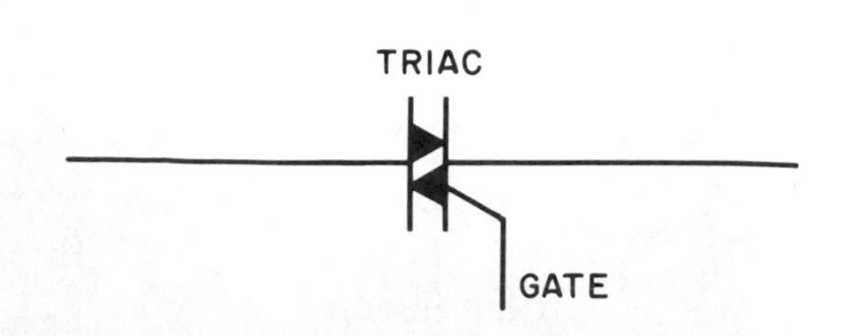

FIGURE 12–81 Schematic symbol for a triac.

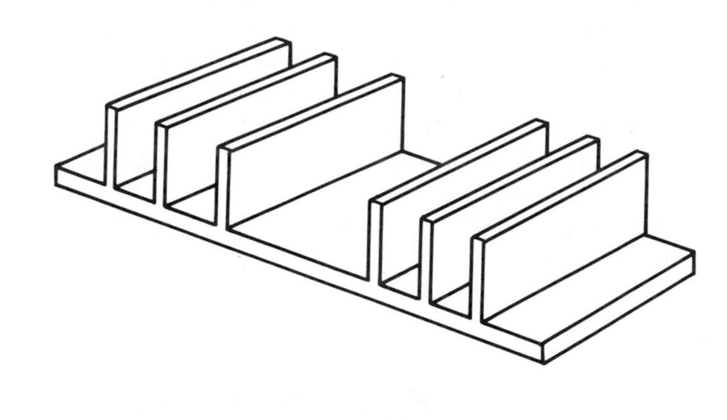

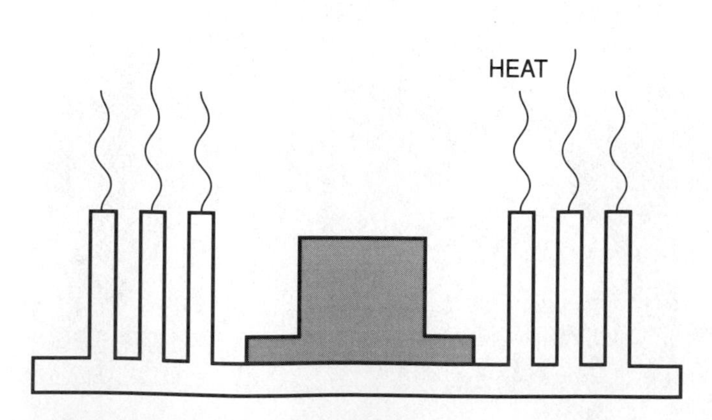

FIGURE 12–82 One type of heat sink.

tronic components. The preceding information has been provided so that you will have some idea of the purpose of these control devices. You are encouraged to pursue the study of these components as more and more solid-state electronics will be used in the future to control the systems you will be working with. Each manufacturer will have a control sequence procedure that you must use to successfully check their controls. Make it a practice to attend seminars and factory schools in your area to increase your knowledge in all segments of this ever-changing field.

SUMMARY

- Matter is made up of atoms.
- Atoms are made up of protons, neutrons, and electrons.
- Protons have a positive charge, and electrons have a negative charge.
- Electrons travel in orbits around the protons and neutrons.
- Electrons in outer orbits travel from one atom to another.
- When there is a surplus of electrons in an atom, it has a negative charge. When there is a deficiency of electrons, the atom has a positive charge.
- Conductors have fewer electrons in the atom's outer orbit. They conduct electricity easily by allowing electrons to move easily from atom to atom.
- Insulators have more electrons in the outer orbits, which makes it difficult for the electrons to move from atom to atom. Insulators are poor conductors of electricity.
- Electricity can be produced by using magnetism. A conductor cutting magnetic lines of force produces electricity.
- Direct current (DC) is an electrical current moving in one direction.
- Alternating current (AC) is an electrical current that is continually reversing.
- Volt = electrical force or pressure.
- Ampere = quantity of electron flow.
- Ohm = resistance to electron flow.
- An electrical circuit must have an electrical source, a conductor to carry the current, and a resistance or load to use the current.
- Resistances or loads may be wired in series or in parallel.
- Voltage (E) = Amperage (I) × Resistance (R). This is Ohm's Law.
- In series circuits the voltage is divided across the resistances, the total current flows through each resistance, and the resistances are added together to obtain the total resistance.
- In parallel circuits the total voltage is applied across each resistance, the current is divided between the resistances, and the total resistance is less than that of the smallest resistance.
- Electrical power is measured in watts, $P = E \times I$.
- Inductive reactance is the resistance caused by the magnetic field surrounding a coil in an AC circuit.
- A coil with an electric current flowing through the loops of wire will cause an iron bar to be attracted into it. Electrical switching devices can be designed to use this action. These switches are called solenoids, relays, and contactors.
- Transformers use the magnetic field to step up or step down the voltage. A step-up transformer increases the voltage and decreases the current. A step-down transformer decreases the voltage and increases the current.
- A capacitor in a DC circuit collects electrons on one plate. These collect until they are equal to the source voltage. When a path is provided for these electrons to discharge, they will do so until the capacitor becomes neutral.
- A capacitor in an AC circuit will continually charge and discharge as the current in the circuit reverses.
- A capacitor has capacitance, which is the amount of charge that can be stored.
- Impedance is the opposition to current flow in an AC circuit from the combination of resistance, inductive reactance, and capacitive reactance.
- A multimeter is a common electrical measuring instrument used by air conditioning, heating, and refrigeration technicians. A multimeter often used is the VOM (volt-ohm-milliammeter).
- ***Properly sized conductors must be used. Larger wire sizes will carry more current than smaller wire sizes without overheating.***
- ***Fuses and circuit breakers are used to interrupt the current flow in a circuit when the current is excessive.***
- Semiconductors are usually made from silicon or germanium. In their pure state they do not conduct electricity well, but when doped with an impurity form an N-type or P-type material that will conduct in one direction.
- Diodes, rectifiers, transistors, thermistors, diacs, and triacs are examples of semiconductors.

REVIEW QUESTIONS

1. Describe the structure of an atom.
2. What is the charge on an electron? a proton? a neutron?
3. Describe the part of an atom that moves from one atom to another.
4. What effect does this movement have on the losing atom? on the gaining atom?
5. Describe the electron structure in a good conductor.
6. Describe the electron structure in an insulator.
7. Describe how electricity is generated through the use of magnetism.
8. State the differences between DC and AC.
9. State the electrical units of measurement and describe each.

10. What components make up an electrical circuit?
11. Describe how a meter would be connected in a circuit to measure the voltage at a light bulb.
12. Describe how a meter would be connected in a circuit to measure the amperage.
13. Describe how an amperage reading would be made using a clamp-on or clamp-around ammeter.
14. Describe how the resistance in a DC circuit is determined.
15. Write Ohm's Law for determining voltage, amperage, and resistance.
16. Sketch three loads wired in parallel in a circuit.
17. Illustrate how three loads would be wired in series.
18. Describe the characteristics of the voltage, amperage, and resistances when there is more than one load in a series circuit.
19. Describe the characteristics of the voltage, amperage, and resistances when there is more than one load in a parallel circuit.
20. What is the formula for the total resistance of three loads in a parallel circuit?
21. What is the unit of measurement for electrical power?
22. What is the formula for determining electrical power?
23. Explain inductance.
24. Explain how a solenoid operates.
25. Describe how a transformer operates.
26. Sketch a step-up transformer.
27. How does a step-down transformer differ from a step-up transformer?
28. Describe how a transformer is constructed.
29. Describe a capacitor.
30. How does a capacitor work in a DC circuit?
31. How does a capacitor work in an AC circuit?
32. What two types of capacitors are used frequently in the air conditioning and refrigeration industry?
33. What are the three types of opposition to current flow that impedance represents?
34. What electrical measurements will a multimeter make?
35. What do the letters VOM stand for when referring to an electrical measuring instrument?
36. What are the two main switches on a VOM?
37. Why is there a zero ohms adjustment on a VOM?
38. Why is it important to use a properly sized wire in a particular circuit?
39. What is a circular mil?
40. Describe two kinds of plug fuses.
41. Describe two reasons for using a circuit breaker.
42. What force opens a circuit breaker?
43. What are the two letters used to represent the types of material in most semiconductors?
44. Briefly describe a diode.
45. What does forward bias on a diode mean?
46. What does reverse bias on a diode mean?
47. List four types of semiconductors and state one use of each.

UNIT 13

Introduction to Automatic Controls

OBJECTIVES

After studying this unit, you should be able to

- define bimetal.
- make general comparisons between different bimetal applications.
- describe the rod and tube.
- describe fluid-filled controls.
- describe partial liquid, partial vapor-filled controls.
- distinguish between the bellows, diaphragm, and Bourdon tube.
- discuss the thermocouple.
- explain the thermistor.

13.1 TYPES OF AUTOMATIC CONTROLS

The heating, air conditioning, and refrigeration field requires many types and designs of automatic controls to stop or start equipment. Modulating controls that vary the speed of a motor or those that open and close valves varying amounts are found less frequently in the residential and light commercial range of equipment. Controls also provide protection to people and equipment.

Controls can be classified in the following categories: electrical, mechanical, electromechanical, and electronic. Pneumatic and hydraulic controls are not discussed because they normally do not apply to residential and light commercial equipment.

Electrical controls are electrically operated and normally control electrical devices. Mechanical controls are operated by pressure and temperature to control fluid flow. Electromechanical controls are driven by pressure or temperature to provide electrical functions, or they are driven by electricity to control fluid flow. Electronic controls use electronic circuits and devices to perform the same functions that electrical and electromechanical controls perform.

The automatic control of a system is intended to maintain stable or constant conditions with a controllable device. This can involve protection of people and equipment. The system must regulate itself within the design boundaries of the equipment. If the system's equipment is allowed to operate outside of its design boundaries, the equipment components may be damaged.

In this industry, the job is to control space or product condition by controlling temperature, humidity, and cleanliness.

13.2 DEVICES THAT RESPOND TO THERMAL CHANGE

Automatic controls in this industry usually provide some method of controlling temperature. Temperature control is used to maintain space or product temperature and to protect equipment from damaging itself. When used to control space or product temperature, the control is called a *thermostat;* when used to protect equipment, it is known as a *safety device.* A good example of both of these applications can be found in a household refrigerator. The refrigerator maintains the space temperature in the fresh food section at about 35°F. When food is placed in the box and stored for a long time, it becomes the same temperature as the space, Figure 13–1. If the space temperature is allowed to go much below 35°F, the food begins to freeze. Foods such as eggs, tomatoes, and lettuce are not good after freezing.

The refrigerator is often able to maintain this fresh food compartment condition for 15 or 20 years without failure. The frozen food compartment is another situation and is discussed later. For now think of what it would be like without automatic controls that keep food cold to preserve it but not cold enough to freeze and ruin it. The owner of the

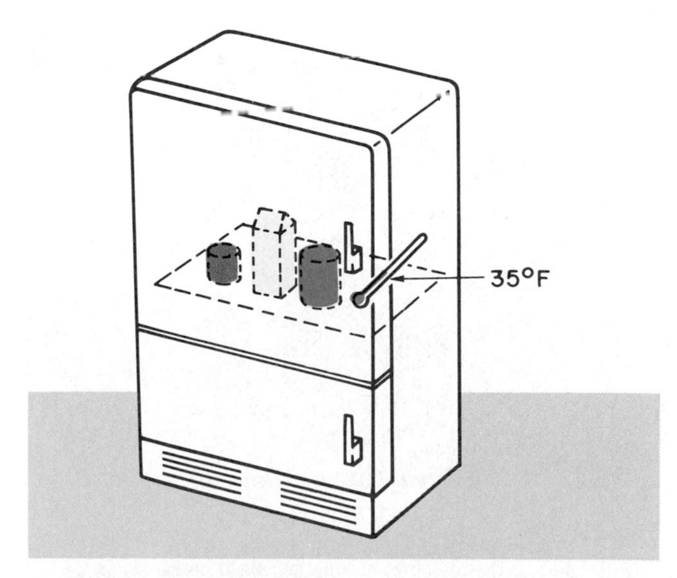

FIGURE 13–1 The household refrigerator maintains a specific temperature inside the box.

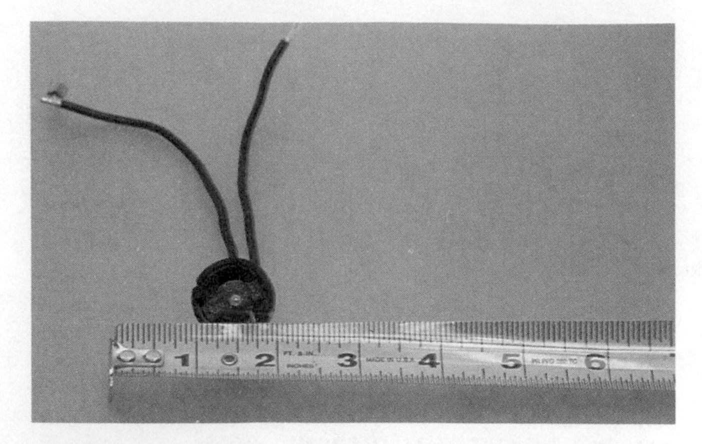

FIGURE 13–2 Compressor overload device. *Photo by Bill Johnson*

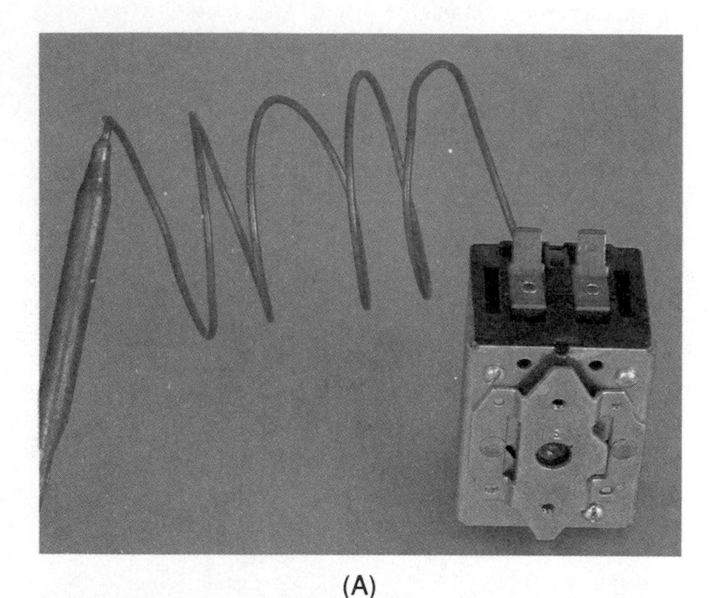

(A)

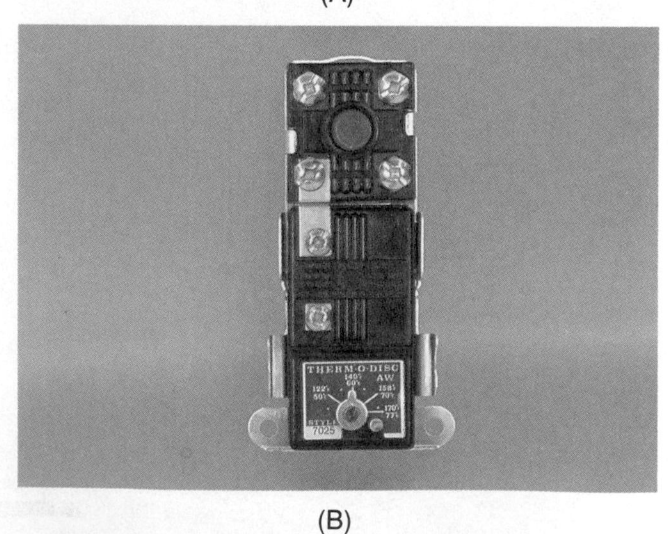

(B)

FIGURE 13–3 These controls operate household appliances. (A) Refrigerator thermostat. (B) Water heater thermostat. *Photos by Bill Johnson*

refrigerator would have to anticipate the temperature in the food compartment and get up in the middle of the night and turn it on or off to maintain the temperature. It is hard to imagine how many times the thermostat stops and starts the refrigeration cycle in 20 years to maintain the proper conditions in the refrigerator.

The refrigerator compressor has a protective device to keep it from overloading and damaging itself, Figure 13–2. This overload is an automatic control designed to function on the rare occasion that an overload or power problem may cause damage to the compressor. One such occasion is when the power goes off and comes right back on while the refrigerator is running. The overload will stop the compressor for a cool-down period until it is ready to go back to work again without overloading and hurting itself.

Some common automatically controlled devices are:

- Household refrigerator's fresh and frozen food compartments
- Residential and office cooling and heating systems
- Water heater temperature control
- Electric oven temperature control
- Garbage disposal overload control
- Fuse and circuit breakers that control current flow in electrical circuits in a home

Figure 13–3 shows two examples of automatic controls.

Automatic controls used in the air conditioning and refrigeration industry are devices that monitor temperature and its changes. Some controls respond to temperature changes and are used to monitor electrical overloads by temperature changes in the wiring circuits. This response is usually a change of dimension or electrical characteristic in the control-sensing element.

13.3 THE BIMETAL DEVICE

The *bimetal* device is probably the most common device used to detect thermal change. In its simplest form the device consists of two unlike metal strips, attached back to back, that have different rates of expansion, Figure 13–4.

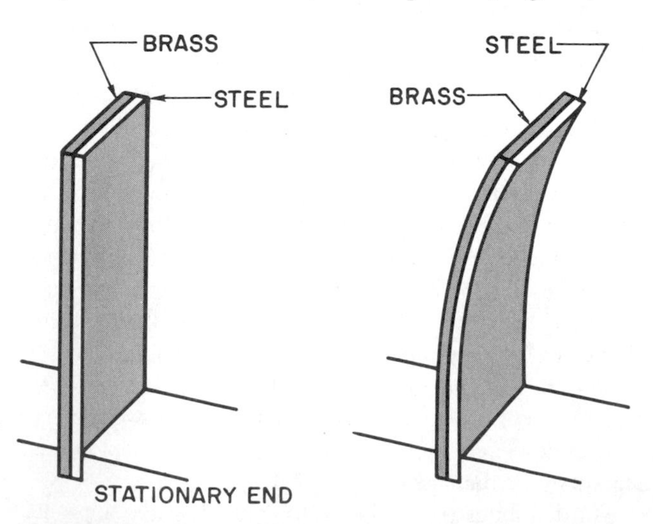

FIGURE 13–4 Basic bimetal strip made of two unlike metals such as brass and steel fastened back to back.

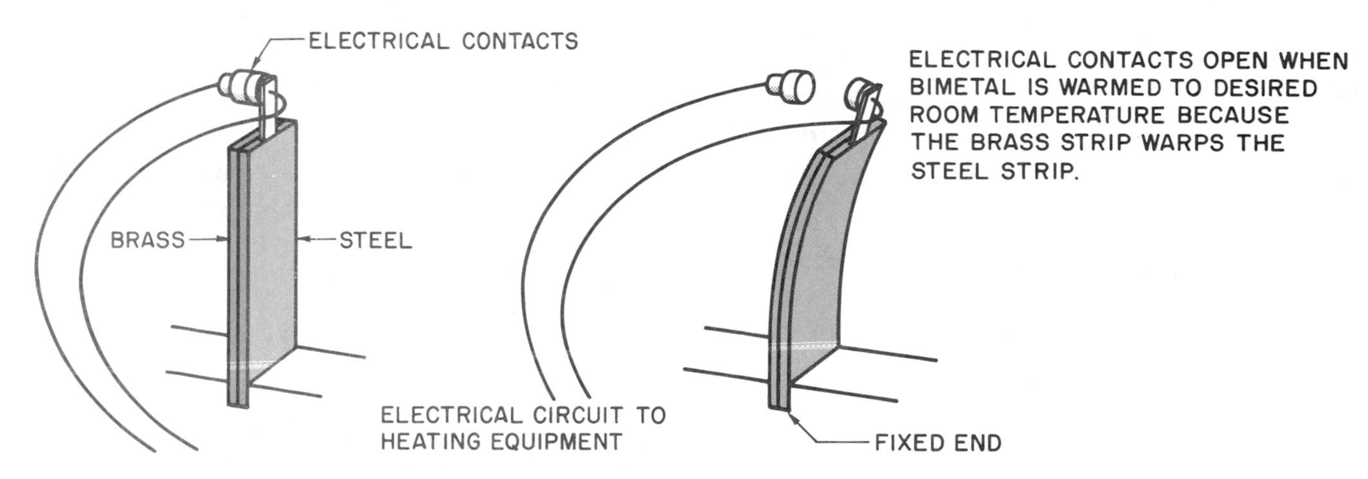

FIGURE 13–5 Basic bimetal strip used for a heating thermostat.

Brass and steel are commonly used. When the device is heated, the brass expands faster than the steel, and the device is warped out of shape. This warping action is a known dimensional change that can be attached to an electrical component or valve to stop, start, or modulate electrical current or fluid flow, Figure 13–5. This control is limited to its application by the amount of warp it can accomplish with a temperature change. For instance, when the bimetal is fixed on one end and heated, the other end moves a certain amount per degree of temperature change, Figure 13–6.

To obtain enough travel to make the bimetal practical over a wider temperature range, add length to the bimetal strip. When adding length, the bimetal strip is normally coiled into a circle, shaped like a hairpin, wound into a helix, or formed into a worm shape, Figure 13–7. The movable end of the coil or helix can be attached to a pointer to indicate temperature, a switch to stop or start current flow,

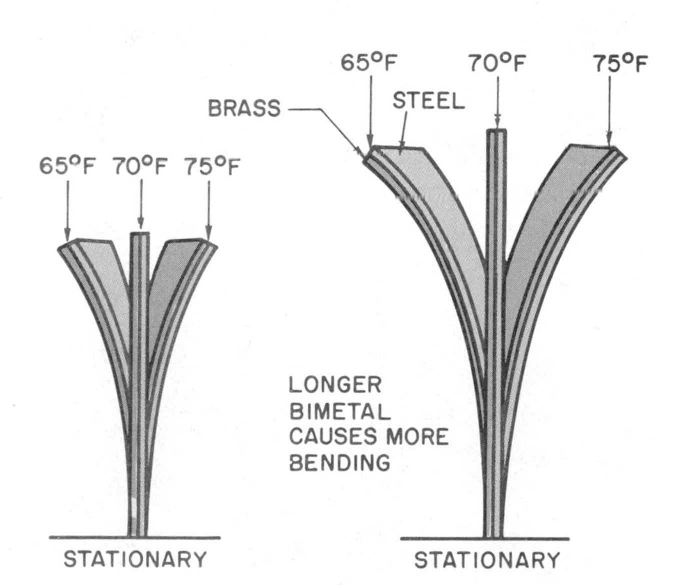

FIGURE 13–6 This bimetal is straight at 70°F. The brass side contracts more than the steel side when cooled, causing a bend to the left. The brass side expands faster than the steel side on a temperature rise, causing a bend to the right. This bend is a predictable amount per degree of temperature change. The longer the strip the more the bend.

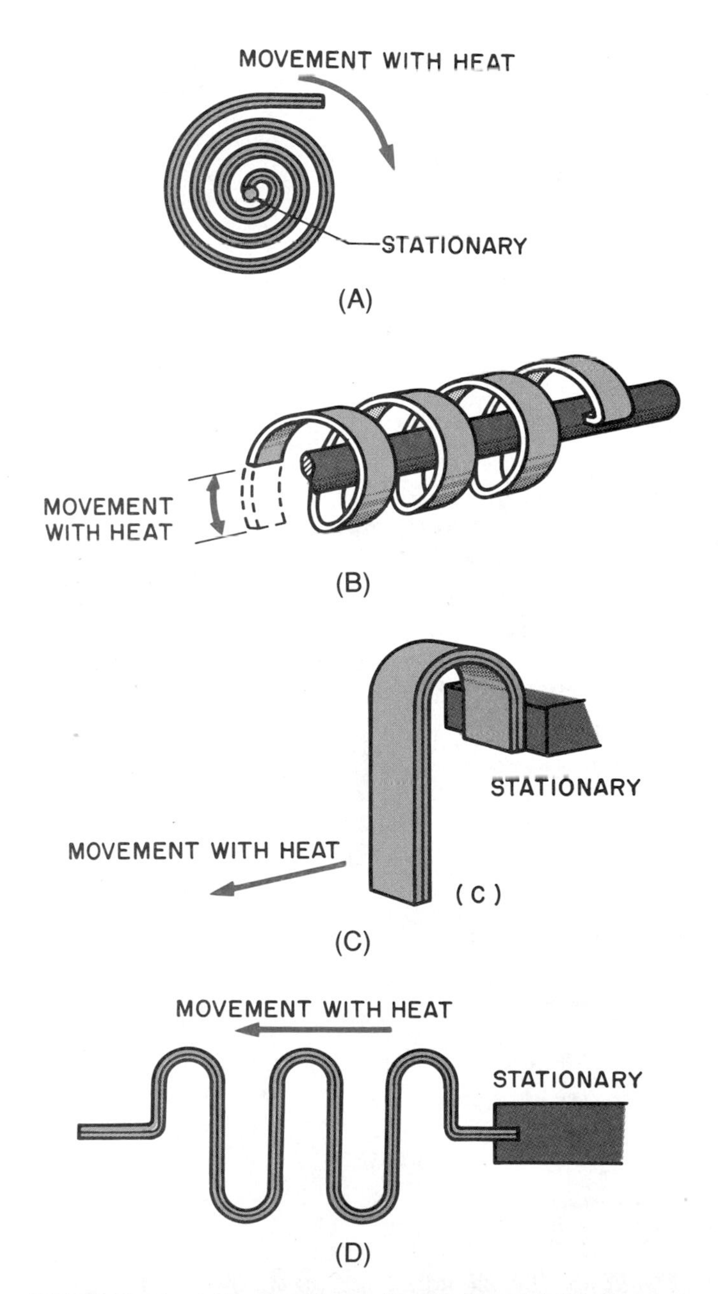

FIGURE 13–7 Adding length to the bimetal. (A) Coiled. (B) Wound into helix. (C) Hairpin shape. (D) Formed into worm shape.

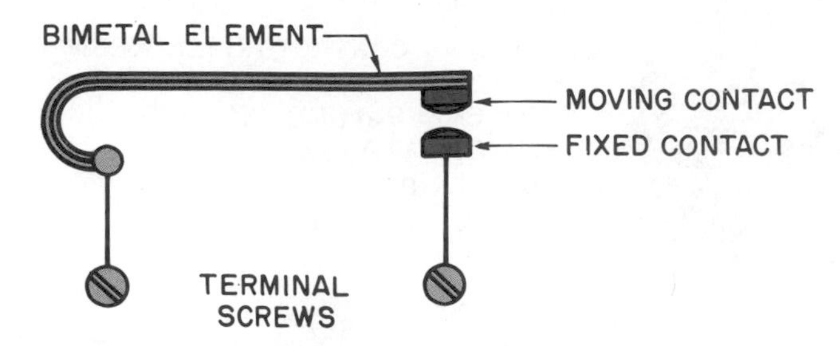

FIGURE 13–8 Movement of the bimetal due to changes in temperature opens or closes electrical contacts.

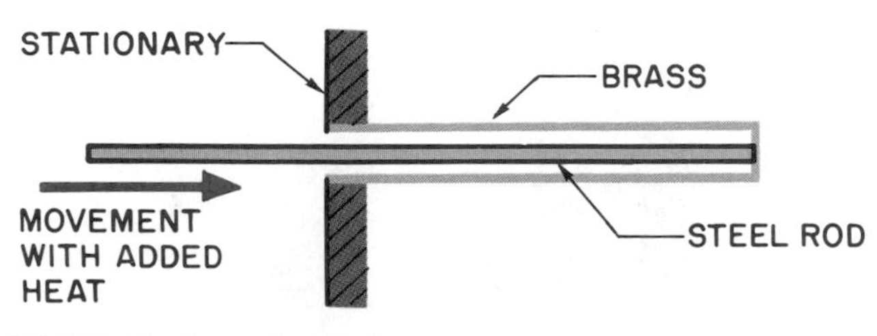

FIGURE 13–9 Rod and tube.

BRASS TUBE INSERTED IN A
HOT WATER TANK TO CONTROL
THE GAS BURNER
STATIONARY
STEEL ROD
WATER
PIVOT
TO BURNER
GAS SUPPLY
PIVOT
TO BURNER
GAS SUPPLY

FIGURE 13–10 The rod and tube type of control consists of two unlike metals with the fastest expanding metal in a tube normally inserted in a fluid such as a hot water tank.

or a valve to modulate fluid flow. One of the basic control applications is shown in Figure 13–8.

ROD AND TUBE. The *rod and tube* is another type of control that uses two unlike metals and the difference in thermal expansion. The rod and tube can be described more accurately as the rod in tube. It has an outer tube of metal with a high expansion rate and an internal rod of metal with a low expansion rate, Figure 13–9. This control has been used for years in the residential gas water heater. The tube is inserted into the tank and provides very accurate sensing of the water temperature. As the tank water temperature changes, the tube pushes the rod and opens or closes the gas valve to start or stop the heat to the water in the tank, Figure 13–10.

SNAP-DISC. The *snap-disc* is another type of bimetal used in some applications to sense temperature changes. This control is treated apart from the bimetal because of its snap characteristic that gives it a quick open-and-close feature. Some sort of snap-action feature has to be incorporated into all controls that stop and start electrical loads, Figure 13–11.

13.4 CONTROL BY FLUID EXPANSION

Fluid expansion is another method of sensing temperature change. Earlier, a mercury thermometer was described as a bulb with a thin tube of mercury rising up a stem. As the

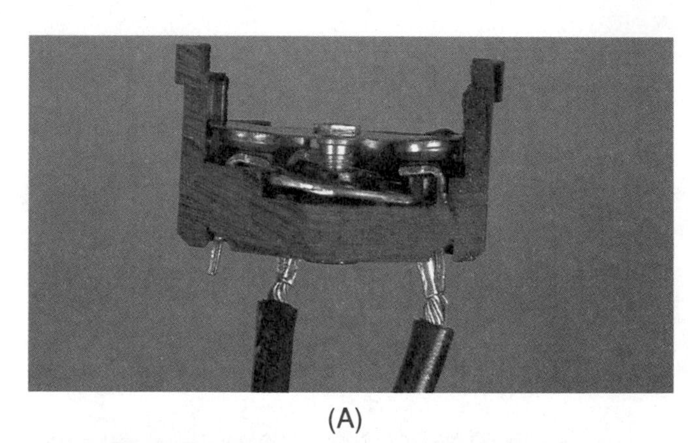

(A)

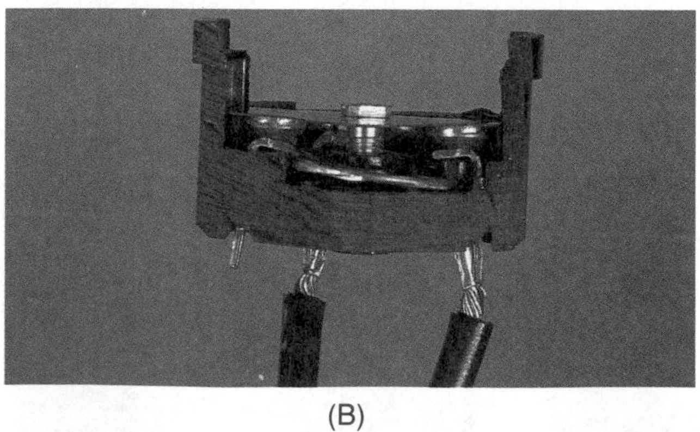

(B)

FIGURE 13–11 The snap-disc is another variation of the bimetal concept. The snap-disc is usually round and fastened on the outside. When heated, the disc snaps to a different position. (A) Open circuit. (B) Closed circuit. *Photos by Bill Johnson*

mercury in the bulb is heated or cooled, it expands or contracts and either rises or falls in the stem of the thermometer. The level of the mercury in the stem is based on the temperature of the mercury in the bulb, Figure 1–1. The reason for the rising and falling is because the mercury in the tube has no place else to go. When the mercury in the bulb is heated and expands, it has to rise up in the tube. When it is cooled, it naturally falls down the tube. This same idea can be used to transmit a signal to a control that a temperature change is occurring.

The liquid rising up the transmitting tube has to act on some device to convert the rising liquid to usable motion. One device used is the diaphragm. A *diaphragm* is a thin, flexible metal disc with a large area. It can move in and out with pressure changes underneath it, Figure 13–12.

When a bulb is filled with a liquid and connected to a diaphragm with piping, the bulb temperature can be transmitted to the diaphragm by the expanding liquid. In Figure 13–13 the bulb is filled with mercury and placed in the pilot light flame on a gas furnace to ensure that a pilot light is present to ignite the gas burner before the main gas valve is opened. The entire mercury-filled tube and mercury-filled diaphragm are sensitive to temperature changes. The pilot light is very hot so only the sensing bulb is located in the pilot light flame.

To maintain more accurate control at the actual bulb location, you can use a bulb partially filled with a liquid that will boil and make a vapor, which is then transmitted to the diaphragm at the control point, Figure 13–14. You should realize that the liquid will respond to temperature change much more than will the vapor, which is used to transmit the pressure.

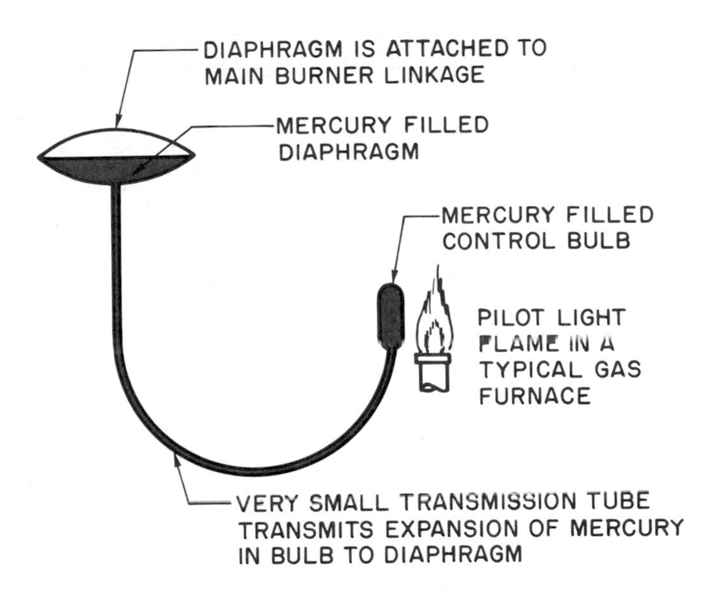

FIGURE 13–13 **The bulb in or near the flame is mercury filled. The heated bulb causes the mercury to expand and move up the transmission tube and flex out the diaphragm. This proves that the pilot flame is present to ignite the main burner.**

The following example using a walk-in refrigerated box with R-12 describes how this can work. Refer to Figure 3–15 for the pressure/temperature chart for R-12. The inside temperature is maintained by cutting the refrigeration system off when the box temperature reaches 35°F and starting it when the box temperature reaches 45°F. A control with a remote bulb is used to regulate the space temperature. The bulb is located inside the box, and the control is located outside the box so it can be adjusted.

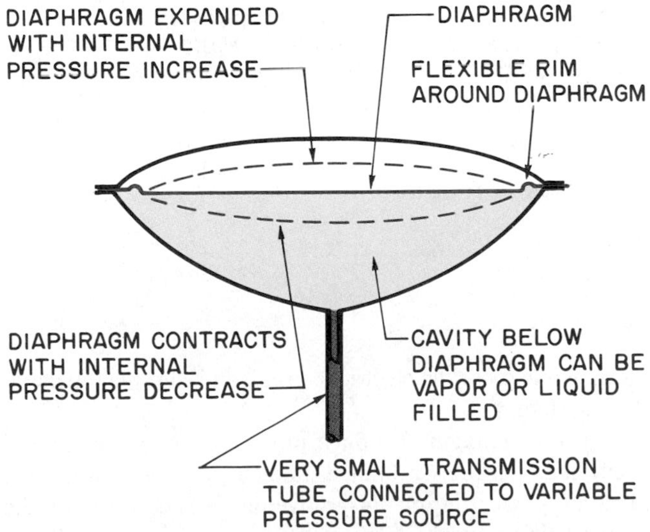

FIGURE 13–12 **The diaphragm is a thin, flexible, movable membrane (brass-steel or other metal) used to convert pressure changes to movement. This movement can stop and start controls or modulate controls.**

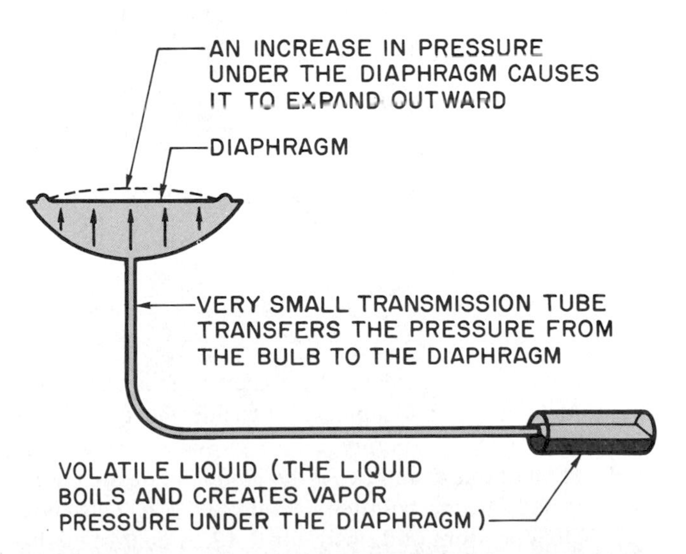

FIGURE 13–14 **A large bulb is partially filled with a volatile liquid, one that boils and creates vapor pressure when heated. This causes an increase in vapor pressure, which forces the diaphragm to move outward. When cooled, the vapor condenses, and the diaphragm moves inward.**

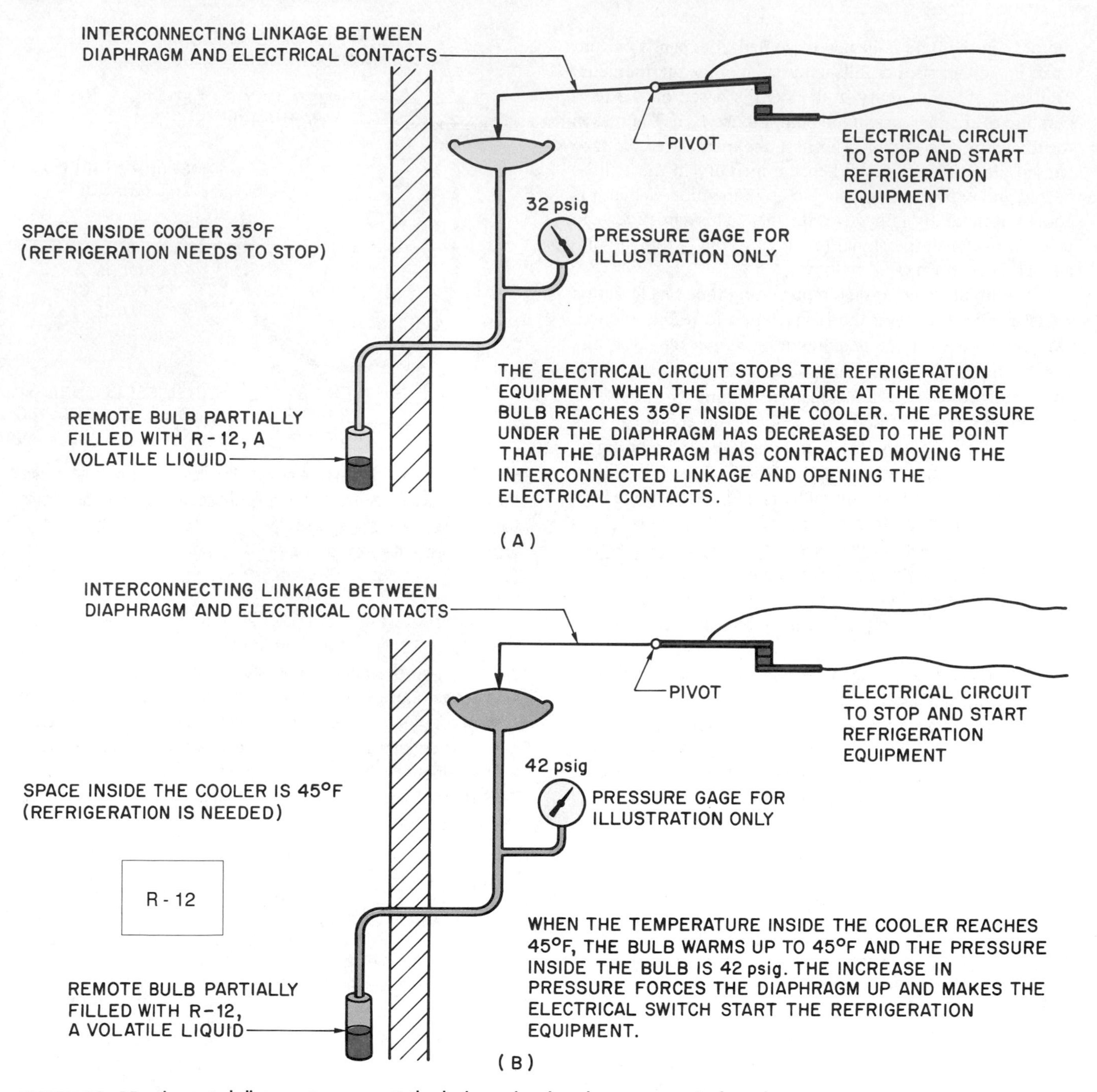
INTERCONNECTING LINKAGE BETWEEN DIAPHRAGM AND ELECTRICAL CONTACTS
PIVOT
ELECTRICAL CIRCUIT TO STOP AND START REFRIGERATION EQUIPMENT
32 psig
SPACE INSIDE COOLER 35°F (REFRIGERATION NEEDS TO STOP)
PRESSURE GAGE FOR ILLUSTRATION ONLY
REMOTE BULB PARTIALLY FILLED WITH R-12, A VOLATILE LIQUID
THE ELECTRICAL CIRCUIT STOPS THE REFRIGERATION EQUIPMENT WHEN THE TEMPERATURE AT THE REMOTE BULB REACHES 35°F INSIDE THE COOLER. THE PRESSURE UNDER THE DIAPHRAGM HAS DECREASED TO THE POINT THAT THE DIAPHRAGM HAS CONTRACTED MOVING THE INTERCONNECTED LINKAGE AND OPENING THE ELECTRICAL CONTACTS.
(A)
INTERCONNECTING LINKAGE BETWEEN DIAPHRAGM AND ELECTRICAL CONTACTS
PIVOT
ELECTRICAL CIRCUIT TO STOP AND START REFRIGERATION EQUIPMENT
42 psig
SPACE INSIDE THE COOLER IS 45°F (REFRIGERATION IS NEEDED)
PRESSURE GAGE FOR ILLUSTRATION ONLY
R-12
REMOTE BULB PARTIALLY FILLED WITH R-12, A VOLATILE LIQUID
WHEN THE TEMPERATURE INSIDE THE COOLER REACHES 45°F, THE BULB WARMS UP TO 45°F AND THE PRESSURE INSIDE THE BULB IS 42 psig. THE INCREASE IN PRESSURE FORCES THE DIAPHRAGM UP AND MAKES THE ELECTRICAL SWITCH START THE REFRIGERATION EQUIPMENT.
(B)

FIGURE 13–15 A remote bulb transmits pressure to the diaphragm based on the temperature in the cooler.

For illustration purposes a pressure gage is installed in the bulb to monitor the pressures inside the bulb as the temperature changes. Figure 13–15 presents a progressive explanation of this example. At the point that the unit needs to be cycled off, the bulb temperature is 35°F. This corresponds to a pressure of 32 psig for R-12. A control mechanism can be designed to open an electrical circuit and stop it at this point. When the cooler temperature rises to 45°F, it is time to restart the unit. At 45°F for R-12, the pressure inside the control is 42 psig, and the same mechanism can be designed to close the electrical circuit and start the refrigeration system.

The diaphragm also must move like the bimetal strip. The diaphragm has a limited amount of travel but a great deal of power during this movement. The travel of the liquid-filled control is limited to the expansion of the liquid in the bulb for the temperature range it is working within. When more travel is needed, another device, called the *bellows,* can be used. The bellows is much like an accordion. It has a lot of internal volume with a lot of travel, Figure

13–16. The bellows is normally used with a vapor instead of a liquid inside it.

The remote bulb partially filled with liquid may also be used to indicate temperature using a Bourdon tube by driving a needle on a calibrated dial, Figure 13–17.

The partially filled bulb control is widely used in this industry because it is reliable, simple, and economical. This control has been in the industry since it began, and it has many configurations. Figure 13–18 illustrates some remote bulb thermostats. The Bourdon tube is often used in the same manner as the diaphragm and the bellows to monitor fluid expansion.

13.5 THE THERMOCOUPLE

The *thermocouple* differs from other methods of controlling with thermal change because it does not use expansion; instead, it uses electrical principles. The thermocouple con-

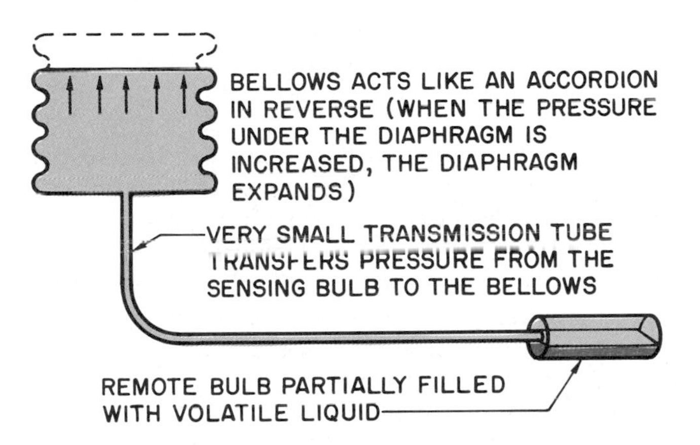
BELLOWS ACTS LIKE AN ACCORDION IN REVERSE (WHEN THE PRESSURE UNDER THE DIAPHRAGM IS INCREASED, THE DIAPHRAGM EXPANDS)
VERY SMALL TRANSMISSION TUBE TRANSFERS PRESSURE FROM THE SENSING BULB TO THE BELLOWS
REMOTE BULB PARTIALLY FILLED WITH VOLATILE LIQUID

FIGURE 13–16 The bellows is applied where more movement per degree is desirable. This control would normally have a partially filled bulb with vapor pushing up in the bellows section.

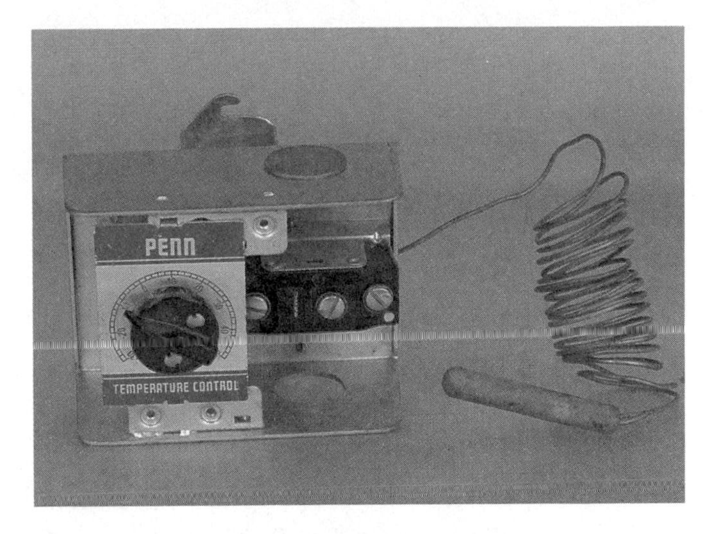
PENN
TEMPERATURE CONTROL

(A)

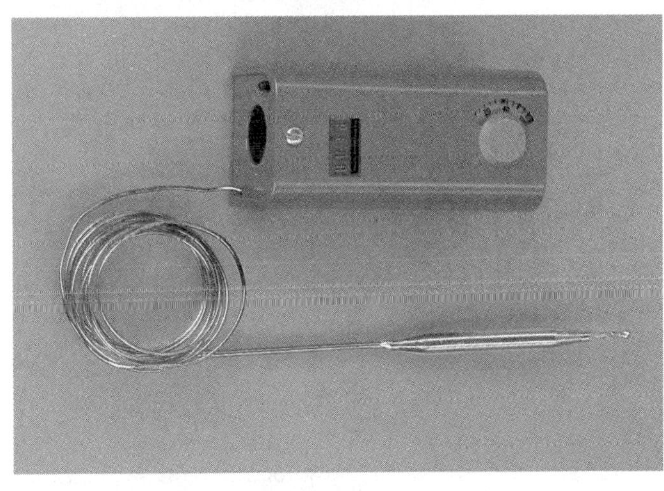

(B)

FIGURE 13–18 Remote bulb refrigeration temperature controls. *Photos by Bill Johnson*

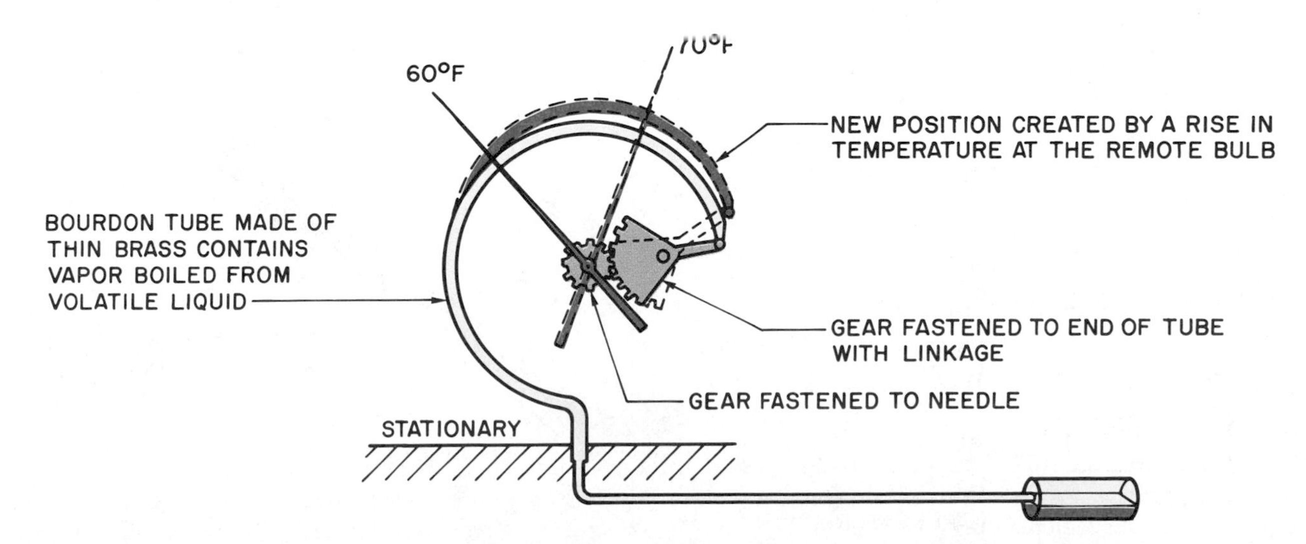
70°F
60°F
NEW POSITION CREATED BY A RISE IN TEMPERATURE AT THE REMOTE BULB
BOURDON TUBE MADE OF THIN BRASS CONTAINS VAPOR BOILED FROM VOLATILE LIQUID
GEAR FASTENED TO END OF TUBE WITH LINKAGE
GEAR FASTENED TO NEEDLE
STATIONARY

FIGURE 13–17 This remote bulb is partially filled with liquid. When heated, the expanded vapor is transmitted to a Bourdon tube that straightens out with an increase in vapor pressure. A decrease in pressure causes the Bourdon tube to curl inward.

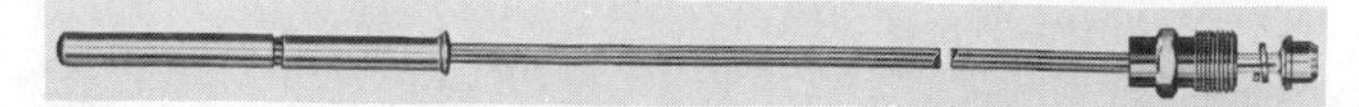

FIGURE 13–19 Thermocouple used to detect a pilot light in a gas-burning appliance. *Courtesy Robertshaw Controls Company*

sists of two unlike metals formed together on one end (usually wire made of unlike metals such as iron and constantan), Figure 13–19. When heated on the fastened end, an electrical current flow is started due to the difference in temperature in the two ends of the device, Figure 13–20. Thermocouples can be made of many different unlike metal combinations, and each one has a different characteristic. Each thermocouple has a *hot* junction and a *cold* junction. The hot junction, as the name implies, is at a higher temperature level than the cold junction. This difference in temperature is what starts the current flowing. Heat will cause an electrical current to flow in one direction in one metal and in the opposite direction in the other. When these metals are connected, they make an electrical circuit, and current will flow when heat is applied to one end of the device.

Tables and graphs for various types of thermocouples show how much current flow can be expected from a thermocouple under different conditions for hot and cold junctions. The current flow in a thermocouple can be monitored by an electronic circuit and used for many temperature-related applications, such as a thermometer, a thermostat to stop or start a process, or a safety control, Figure 13–21.

The thermocouple has been used extensively for years in gas furnaces to detect the pilot light flame for safety purposes. This application is beginning to be phased out because it works best with standing pilot light systems. A gradual change in design using intermittent pilots (these are pilot lights that are extinguished each time the burner goes out and relit each time the room thermostat calls for heat) and other newer types of ignition is causing this phase out. This thermocouple application has an output voltage of about 20 millivolts, all that is needed to control a safety circuit to prove there is flame, Figure 13–22.

FIGURE 13–21 Thermocouple senses whether the gas furnace pilot light is on. *Photo by Bill Johnson*

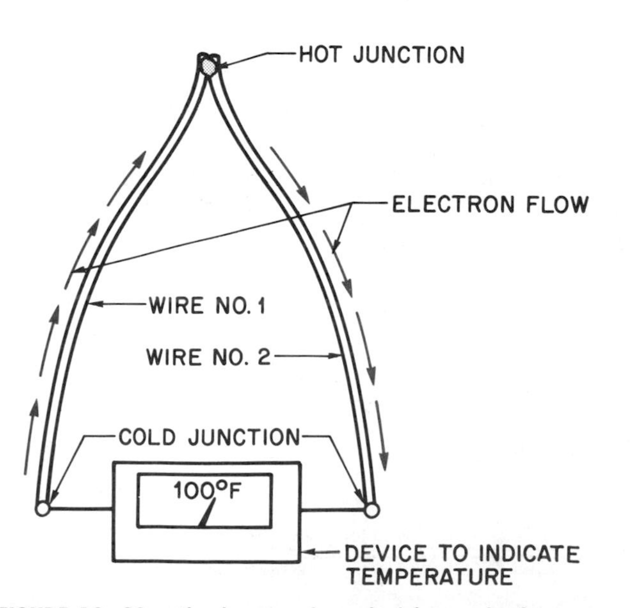

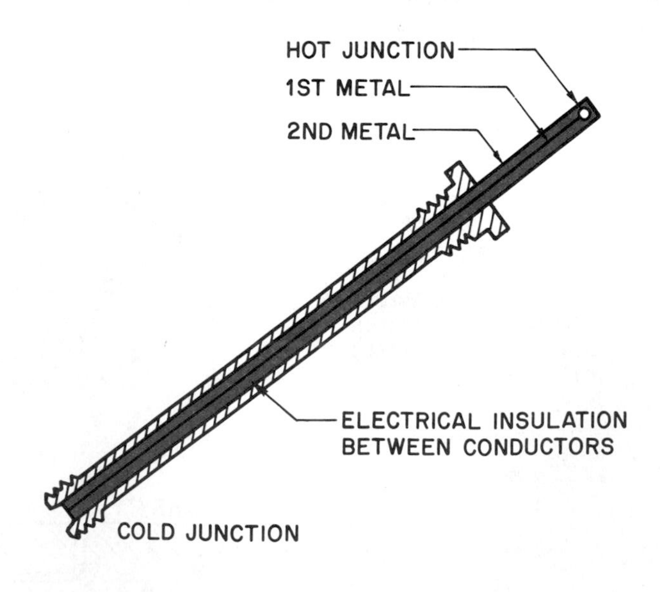

FIGURE 13–20 The thermocouple on the left is made of two wires of different metals welded on one end. It is used to indicate temperature. The thermocouple on the right is a rigid device and used to detect a pilot light.

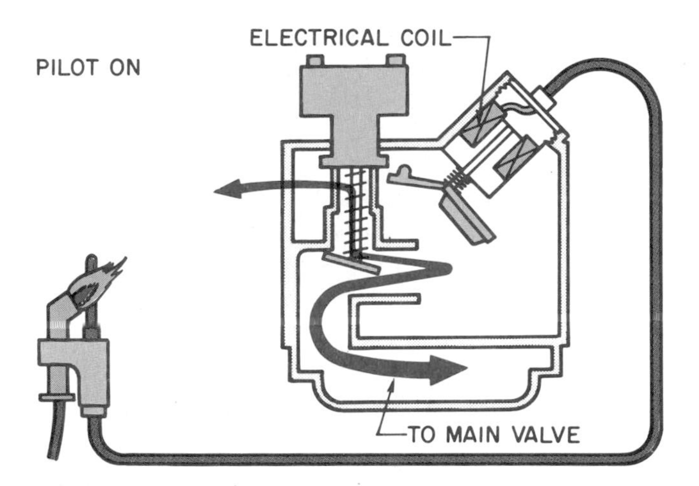

FIGURE 13–22 Thermocouple and control circuit used to detect a gas flame. When the flame is lit, the thermocouple generates an electrical current. This energizes an electromagnet that holds the gas valve open. When the flame is out, the thermocouple stops generating electricity, and the valve closes. Gas is not allowed to flow.

Thermocouples ganged together to give more output are called *thermopiles,* Figure 13–23. The thermopile is used on some gas-burning equipment as the only power source. This type of equipment has no need for power other than the control circuit, so the power supply is very small (about 500 millivolts). The thermopile has also been used to operate radios using the sun or heat from a small fire in remote areas.

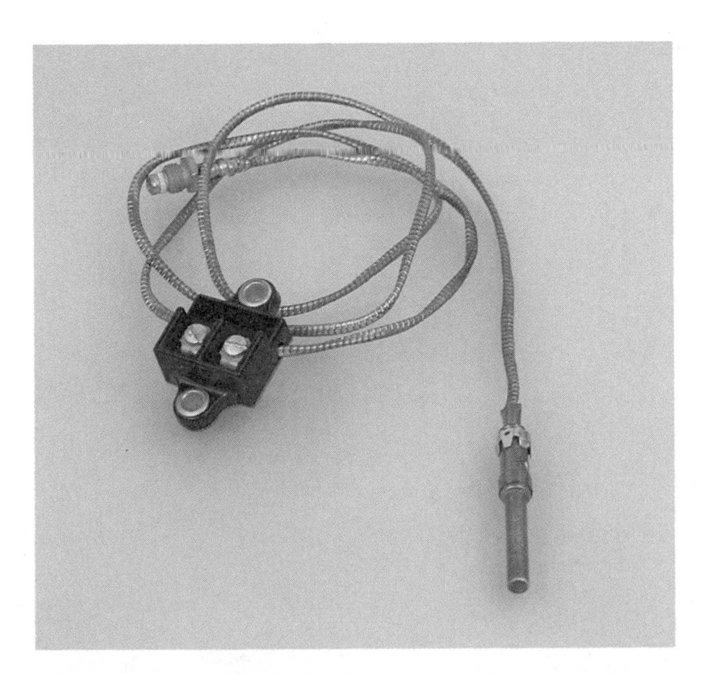

FIGURE 13–23 A thermopile consists of a series of thermocouples in one housing. *Photo by Bill Johnson*

13.6 ELECTRONIC TEMPERATURE-SENSING DEVICES

The *thermistor* is an electronic solid-state device known as a semiconductor and requires an electronic circuit to utilize its capabilities. It varies its resistance to current flow based on its temperature.

The thermistor can be very small and will respond to small temperature changes. The changes in current flow in the device are monitored by special electronic circuits that can stop, start, and modulate machines or provide a temperature readout, Figures 13–24 and 13–25.

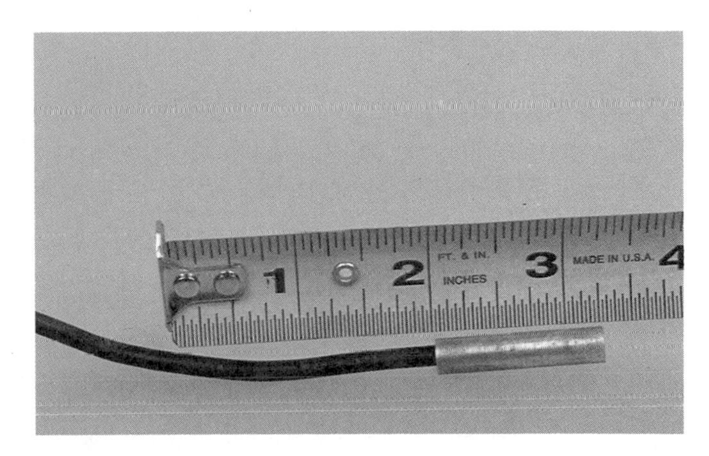

FIGURE 13–24 A thermometer probe using a thermistor to measure temperature. *Photo by Bill Johnson*

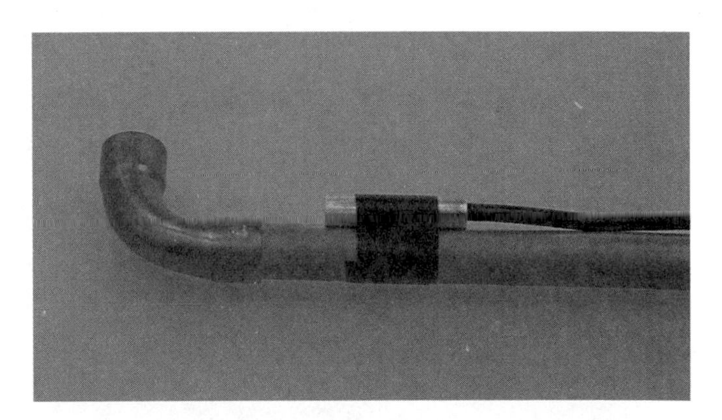

FIGURE 13–25 Thermistor application. *Photo by Bill Johnson*

SUMMARY

- A bimetal element is two unlike metal strips such as brass and steel fastened back to back.
- Bimetal strips warp with temperature changes and can be used to stop, start, or modulate electrical current flow and fluid flow when used with different mechanical, electrical, and electronic helpers.

- The travel of the bimetal can be extended by coiling it. The helix, worm shape, hairpin shape, and coil are the names given on the extended bimetal.
- The rod and tube is another version of the bimetal.
- Fluid expansion is used in the thermometer to indicate temperature and to operate controls that are totally liquid filled.
- The pressure/temperature relationship is applied to some controls that are partially filled with liquid.
- The diaphragm is used to move the control mechanism when either liquid or vapor pressure is applied to it.
- The diaphragm has very little travel but much power.
- The bellows is used for more travel and is normally vapor filled.
- The Bourdon tube is sometimes used like the diaphragm or bellows.
- The thermocouple generates electrical current flow when heated at the hot junction.
- This current flow can be used to monitor temperature changes to stop, start, or modulate electrical circuits.
- The thermistor is an electronic device that varies its resistance to electrical current flow based on temperature changes.

REVIEW QUESTIONS

1. Describe the bimetal strip.
2. What are some applications of the bimetal strip?
3. How can the bimetal strip be extended to have more stroke per degree of temperature change?
4. What is a rod and tube?
5. What two metals can be used in the bimetal or the rod and tube?
6. Define a diaphragm.
7. Name two characteristics of a diaphragm.
8. What fluid can be used in a totally liquid-filled bulb type of control?
9. What is one application for the totally liquid-filled control?
10. Define a bellows.
11. What is the difference between a bellows and a diaphragm?
12. Which type of fluid is normally found in a bellows?
13. How is the pressure/temperature relationship used to understand the partially filled remote bulb type of control?
14. What is a thermocouple?
15. How can a thermocouple be used to verify that a gas flame is present?
16. What is used with a thermocouple to control mechanical devices?
17. What are thermocouples normally made of?
18. Define a thermistor.
19. What must be used with a thermistor to control a mechanical device or machine?
20. What does a thermistor do that makes it different from a thermocouple?

UNIT 14

Automatic Control Components and Applications

OBJECTIVES

After studying this unit, you should be able to

- discuss space temperature control.
- describe the mercury control bulb.
- describe the difference between low- and high-voltage controls.
- name components of low- and high-voltage controls.
- name two ways motors are protected from high temperature.
- describe the difference between a diaphragm and a bellows control.
- state the uses of pressure-sensitive controls.
- describe a high-pressure control.
- describe a low-pressure control.
- describe a pressure relief valve.
- describe the functions of mechanical and electromechanical controls.

SAFETY CHECKLIST

* The subbase for a line voltage thermostat is attached to an electrical outlet box. If an electrical arc due to overload or short circuit occurs, it is enclosed in the conduit or box. Do not consider this as a low-voltage device.
* The best procedure when a hot compressor is encountered is to shut off the compressor with the space temperature thermostat "off" switch and return the next day. This gives it time to cool and keeps the crankcase heat on.
* If the above is not possible or feasible and you must cool the compressor quickly, turn the power off, and cover electrical circuits with plastic before you use water to cool the compressor. Do not stand in this water when you make electrical checks.
* When used as safety controls, pressure controls should be installed in such a manner that they are not subject to being valved off by the service valves.
* The high-pressure control is a method of providing safety for equipment and surroundings. It should not be tampered with.
* A gas pressure switch is a safety control and should never be bypassed except for troubleshooting and then only by experienced service technicians.
* The pressure relief valve is a safety device that is factory set and should not be changed or tampered with.

14.1 RECOGNITION OF CONTROL COMPONENTS

This unit describes how various controls look and operate. Recognizing a control and understanding its function and what component it influences are vitally important and will eliminate confusion when you read a diagram or troubleshoot a system. The ability to see a control on a circuit diagram and then recognize it on the equipment will be easier after a description and illustration are studied.

Fortunately there are some similarities and categories that can make control recognition easier.

14.2 TEMPERATURE CONTROLS

Temperature is controlled in many ways for many reasons. For instance, space temperature is controlled for comfort. The motor-winding temperature in a compressor motor is controlled to prevent overheating and damage to the motor. The motor could overheat and damage could occur from the very power that operates it. Both types of controls require some device that will sense temperature rise with a sensing element and make a known response. The space temperature example uses the control as an operating control, whereas the motor temperature example serves as a safety device.

The space temperature application has two different actions depending on whether winter heating or summer cooling is needed. In winter the control must break a circuit to stop the heat when conditions are satisfied. In summer the conditions are reversed; the control must make a circuit and start the cooling based on a temperature rise. The heating thermostat opens on a rise in temperature, and the cooling thermostat closes on a rise in temperature.

NOTE: In both cases the control was described as *functioning on a rise*. This terminology is important because it is used in this industry.

The motor temperature cutout has the same circuit action as the heating thermostat. It opens the circuit on a rise in motor temperature and stops the motor. The heating ther-

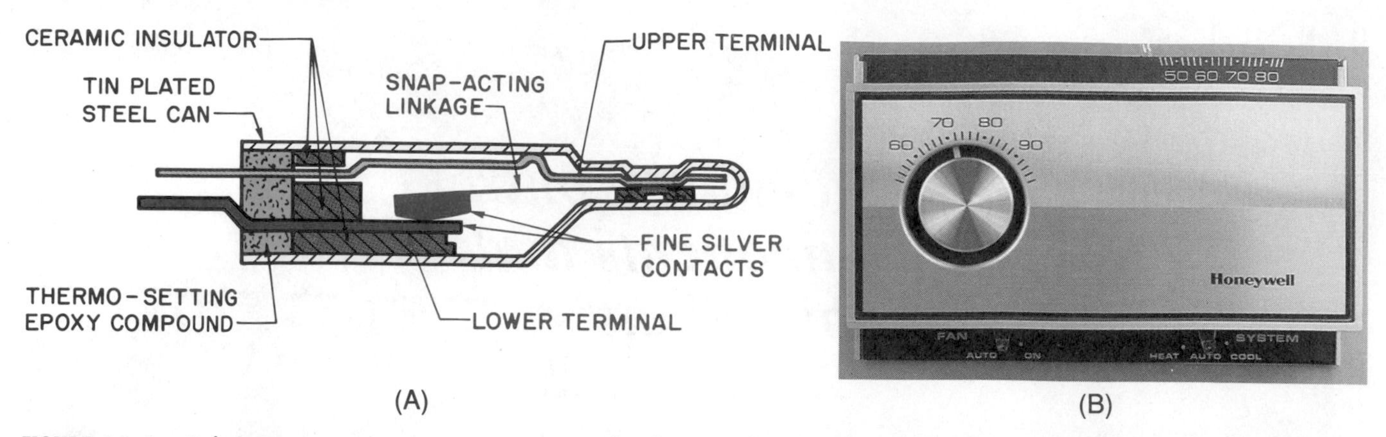

FIGURE 14–1 Both thermostats open on a rise in temperature, but they serve two different purposes. (A) The motor-winding thermostat measures the temperature of iron and copper while in close contact with motor windings. (B) The space thermostat measures air temperature from random air currents. *Photo by Bill Johnson*

mostat and a motor-winding thermostat may make the same move under the same conditions, but they do not physically resemble each other, Figure 14–1.

Another difference between the two thermostats is the medium to be detected. The motor-winding thermostat must be in close contact with the motor winding. It is fastened to the winding itself. The space temperature thermostat is mounted on a wall with the control components suspended in air under the decorative cover. The thermostat relies on random air currents passing over it.

Another important design concept is the current-carrying characteristics of the various controls. In the space temperature application the stopping or starting of a heating system, such as the gas or oil furnace, involves stopping and starting low-voltage (24 V) components and high-voltage (120 V or 230 V) components. The gas or oil furnace normally has a low-voltage gas valve or relay and a high-voltage fan motor.

No firm rule states that one voltage or the other is all that is used in any specific application. However, the stopping and starting of a 3-hp compressor requires a larger switching mechanism than a simple gas valve. A 3-hp compressor could require a running current of 18 A and a starting current of 90 A, whereas a simple gas valve might use only 1/2 A. If the bimetal were large enough to carry the current for a 3-hp compressor, the control would be so large that it would be slow to respond to air temperature changes. This is one reason for using low-voltage controls to stop and start high-voltage components.

Residential systems usually have low-voltage control circuits. There are four reasons for this:

1. Economy
2. Safety
3. More precise control of relatively still air temperature
4. In many states a technician does not need an electrician's license to install and service low-voltage wiring.

The low-voltage thermostat is energized from the residence power supply that is reduced to 24 V with a small transformer usually furnished with the equipment, Figure 14–2.

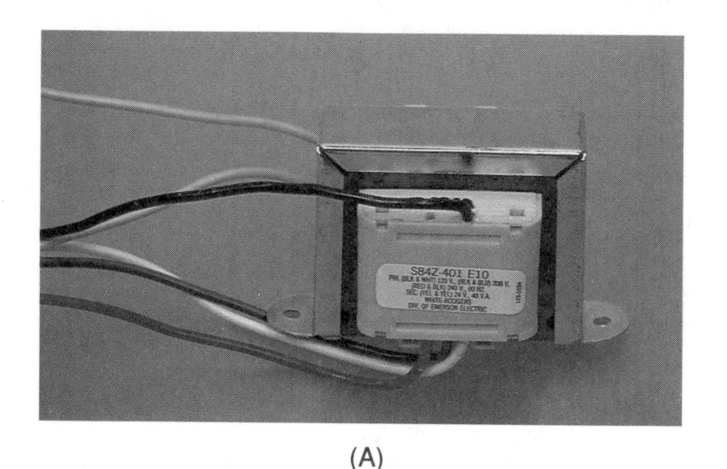

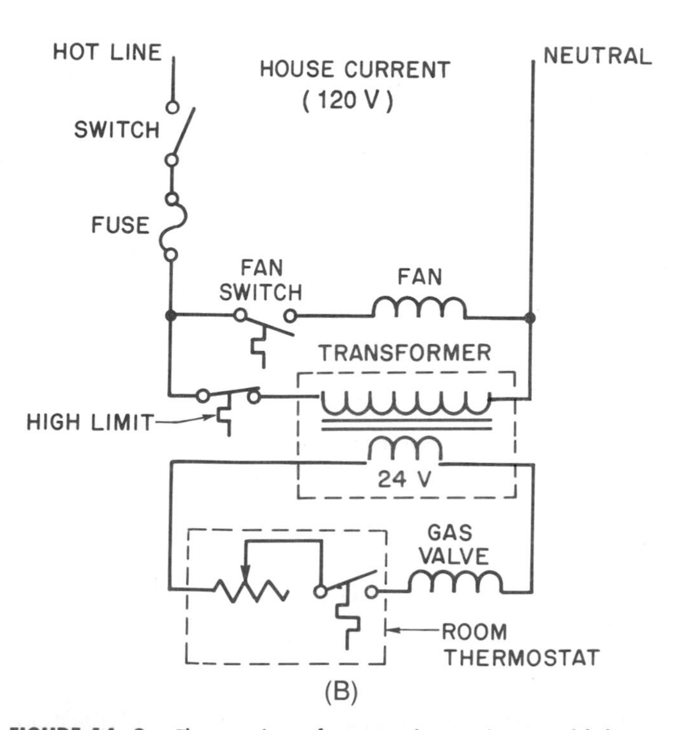

FIGURE 14–2 The typical transformer used in residences and light commercial buildings to change 120 V to 24 V (control voltage). *Photo by Bill Johnson*

14.3 SPACE TEMPERATURE CONTROLS, LOW VOLTAGE

The low-voltage space temperature control (thermostat) normally regulates other controls and does not carry much current—seldom more than 2 A. The thermostat consists of the following components.

ELECTRICAL CONTACT TYPE. The mercury bulb is probably the most popular component used to make and break the electrical circuit in low-voltage thermostats. The mercury bulb is inside the thermostat, Figure 14–3. It consists of a glass bulb filled with an inert gas (a gas that will not support oxidation) with a small puddle of mercury free to move from one end to the other. The principal is to be able to make and break a small electrical current in a controlled atmosphere. When an electrical current is either made or broken, a small arc is present. The arc is hot enough to cause oxidation in the vicinity of the arc. When this arc occurs inside the bulb with an inert gas, where there is no oxygen, there is no oxidation.

The mercury bulb is fastened to the movable end of the bimetal, so it is free to rotate with the movement of the bimetal. The wire that connects the mercury bulb to the electrical circuit is very fine to prevent drag on the movement of the bulb. The mercury cannot be in both ends of the bulb at the same time, so when the bimetal rolls the mercury bulb to a new position, the mercury rapidly makes or breaks the electrical current flow. This is called *snap* or *detent action.*

Two other types of contacts in the low-voltage thermostat use conventional contact surfaces of silver-coated steel contacts. One is simply an open set of contacts, usually with a protective cover, Figure 14–4. The other is a set of silver-coated steel contacts enclosed in a glass bulb. Both of these contacts use a magnet mounted close to the contact to achieve detent or snap action.

WIRE CONNECTOR

(A)

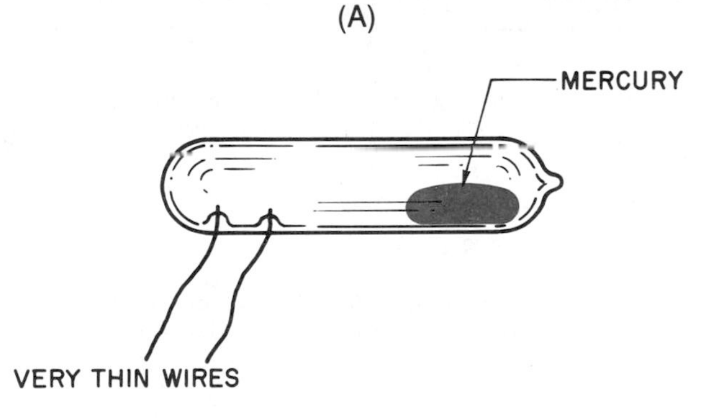

MERCURY BULB

(B)

FIGURE 14–3 (A) A wall thermostat with the cover off and the mercury bulb exposed. (B) A detail of the mercury bulb. Note the very fine wire that connects the bulb to the circuit. The bulb is attached to the movable end of a bimetal coil. When the bimetal tips the bulb, the mercury flows to the other end and closes the circuit by providing the contact between the two wires. The wires are fastened to contacts inside the glass bulb, which is filled with an inert gas. This inert gas helps keep the contacts from pitting and burning up. *Photo by Bill Johnson*

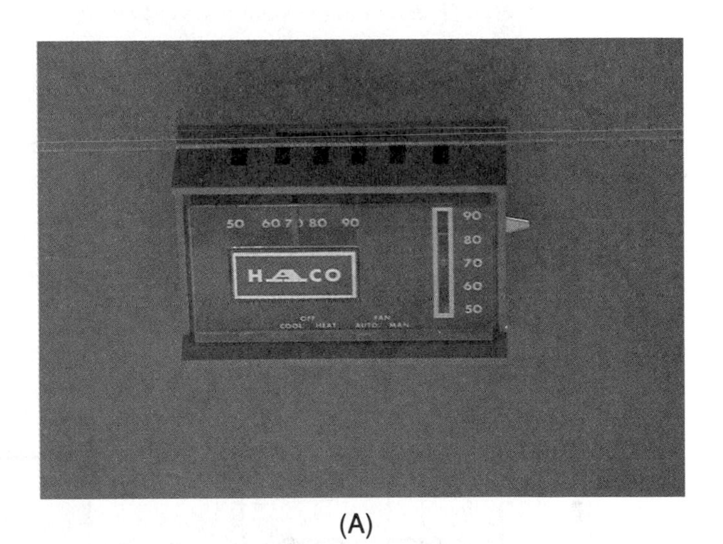

(A)

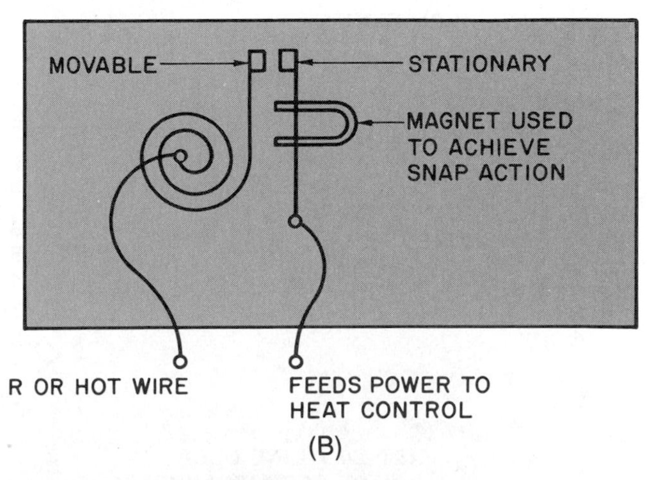

(B)

FIGURE 14–4 (A) Low-voltage thermostat. (B) Low-voltage thermostat illustrated with open contacts. *Photo by Bill Johnson*

HEAT ANTICIPATOR. The *heat anticipator* is a small resistor, usually adjustable, used to cut off the heating equipment prematurely, Figure 14–5. Consider that an oil or a gas furnace has heated a home to just the right cutoff point. The furnace itself could weigh several hundred pounds and be hot from running a long time. When the combustion is stopped in the furnace, it still contains a great amount of heat. The fan is allowed to run to dissipate this heat. The heat left in the furnace is enough to drive the house temperature past the comfort point.

The heat anticipator is located inside the thermostat to cause the furnace to cut off early to dissipate this heat before the space temperature rises to an uncomfortable level, Figure 14–6. The resistor gives off a small amount of heat and causes the bimetal to become warmer than the room temperature.

NOTE: The current flow through the heat anticipator must be accurately matched with the heat anticipator setting. Directions come with the thermostat to show how to do this. The heat anticipator is wired in series with the mercury bulb heating contacts so it is energized anytime the thermostat contacts are closed.

THE COLD ANTICIPATOR. The cooling system needs to be started just a few minutes early to allow the air conditioning system to get up to capacity when needed. If the air conditioning system were not started until it was needed, it would be 5 to 15 min before it would be producing to capacity. This is enough time to cause a temperature rise in the conditioned space. The cold anticipator causes the system to start a few minutes early, so capacity is reached on time. This is nor-

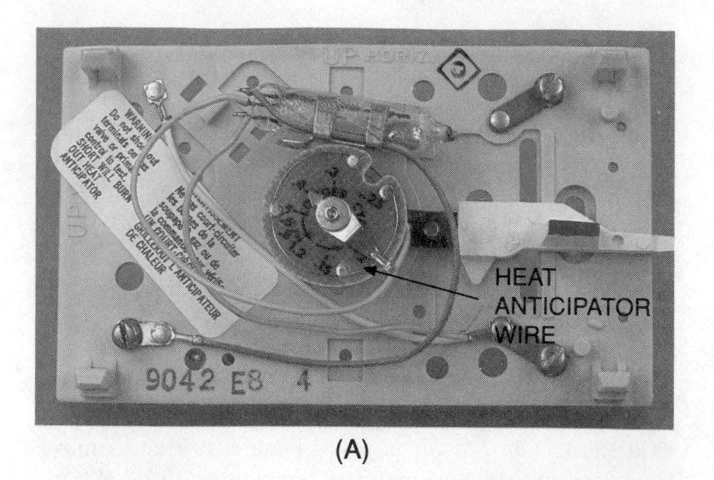

(A)

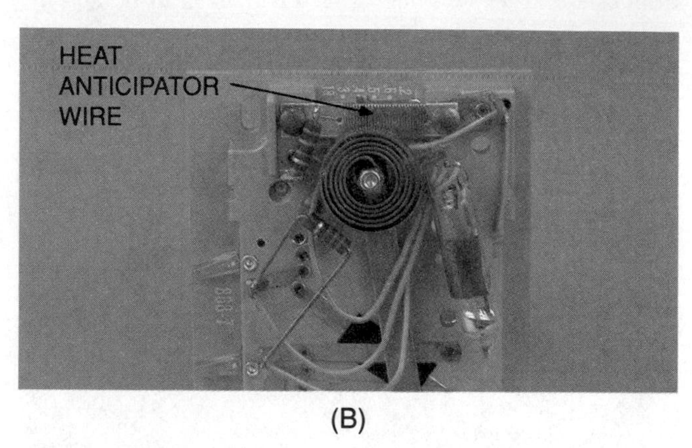

(B)

FIGURE 14–5 Heat anticipators as they appear in actual thermostats. *Photos by Bill Johnson*

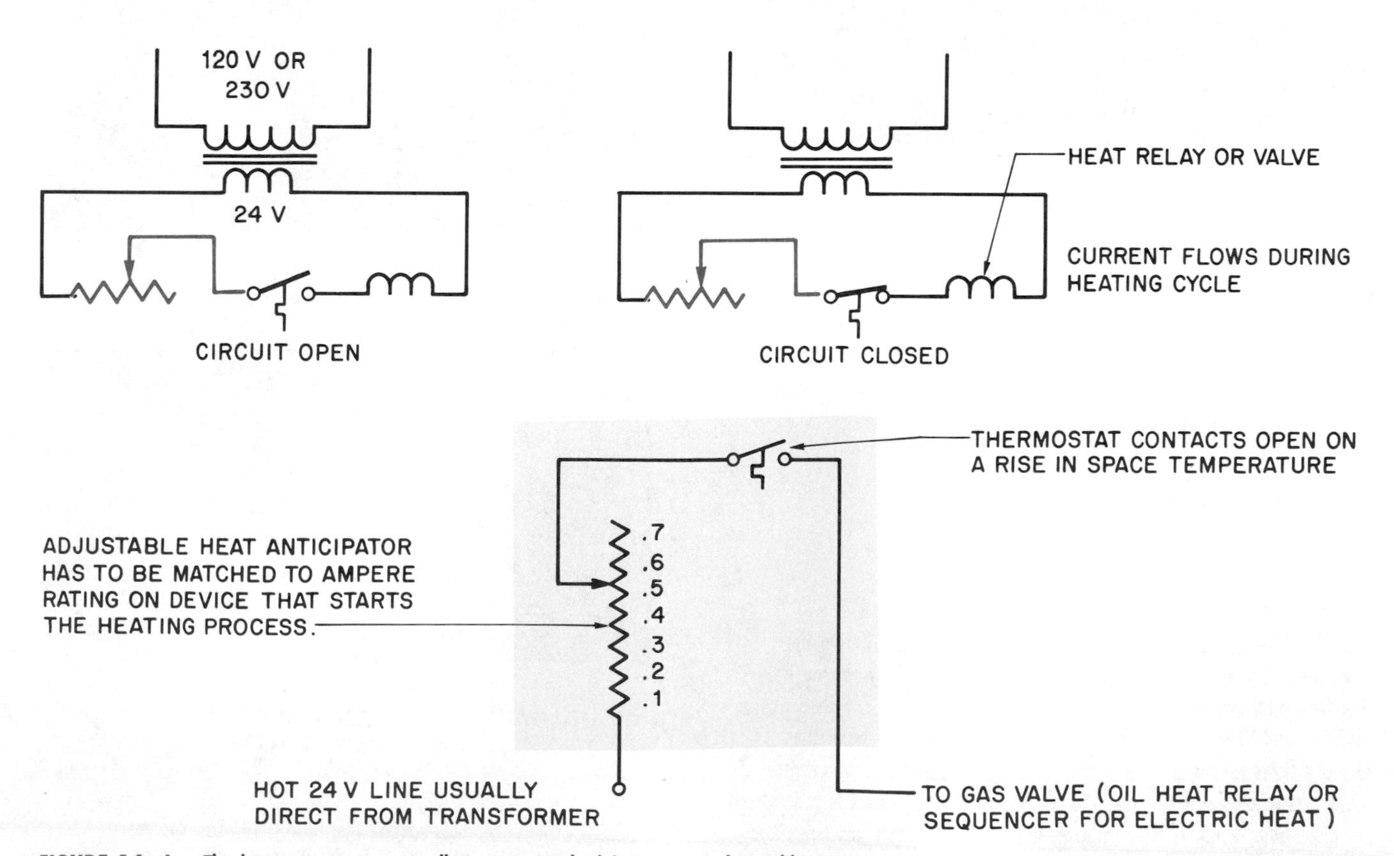

FIGURE 14–6 The heat anticipator is usually a wirewound, slide-bar type of variable resistor.

mally a fixed resistor that is not adjustable in the field, Figure 14–7. The cold anticipator is wired in parallel with the mercury bulb cooling contacts and is energized during the off cycle, Figure 14–8.

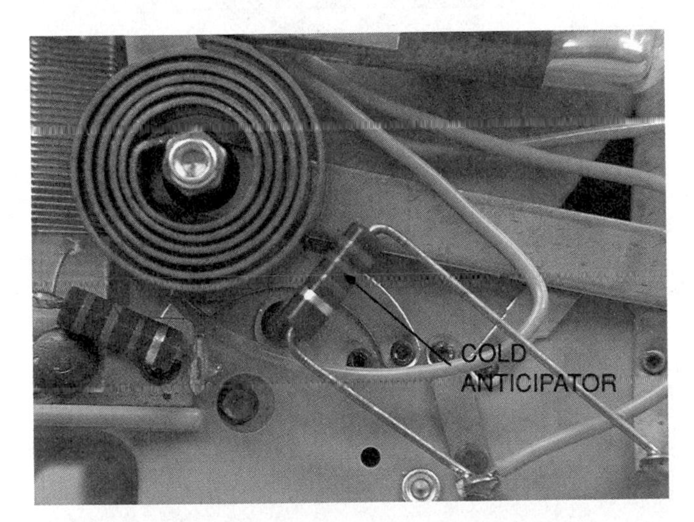

FIGURE 14–7 The cold anticipator is normally a fixed resistor similar to resistors found in electronic circuitry. They are small round devices with colored bands to denote the resistance and wattage. *Photo by Bill Johnson*

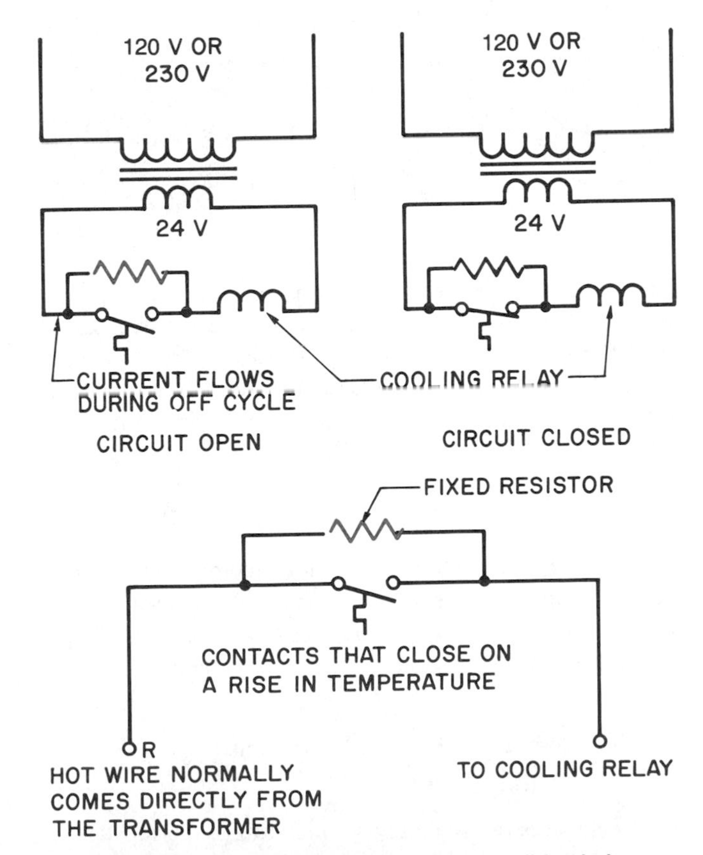

FIGURE 14–8 The cold anticipator is wired in parallel with the cooling contacts on the thermostat. This allows the current to flow through it during the off cycle. This means that the cold anticipator and the relay to start the cooling cycle are in the circuit when the thermostat is open.

THERMOSTAT COVER. The thermostat cover is intended to be decorative and protective. A thermometer is usually mounted on it to indicate the surrounding (ambient) temperature. The thermometer is functionally separate from any of the controls and would serve the same purpose if it were hung on the wall next to the thermostat. Thermostats come in many shapes. Each manufacturer tries to have a distinctive design, Figure 14–9.

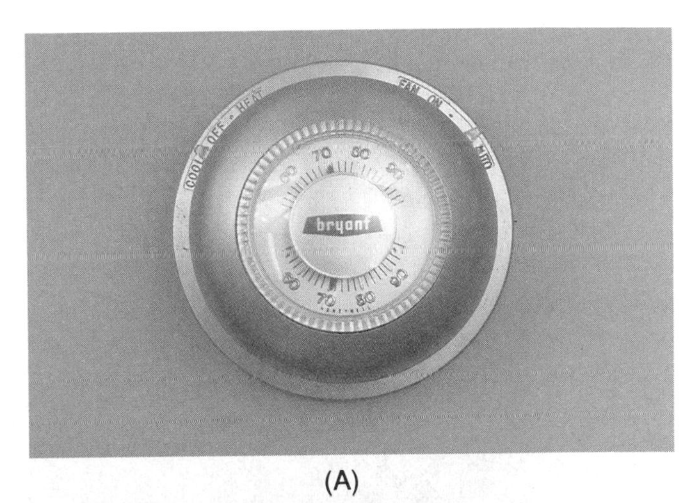

(A)

(B)

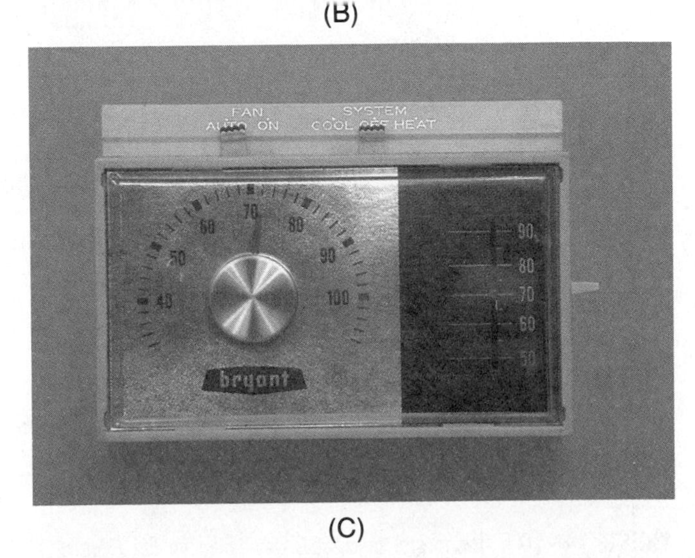

(C)

FIGURE 14–9 Thermostat decorative covers. *Photos by Bill Johnson*

THERMOSTAT ASSEMBLY. The *thermostat assembly* houses the thermostat components already mentioned and is normally mounted on a subbase fastened to the wall. This assembly could be called the brain of the system. In addition to mercury bulbs and anticipators, the thermostat assembly includes the movable levers that adjust the temperature. These levers or indicators normally point to the set point or desired temperature, Figure 14–10. When the thermostat is functioning correctly, the thermometer on the front will read the same as the set point.

THE SUBBASE. The *subbase,* which is usually separate from the thermostat, contains the selector switching levers, such as the FAN-ON switch or the HEAT-OFF-COOL switch. Figure 14–11. The subbase is important because the thermostat mounts on it. The subbase is first mounted on the wall, then the interconnecting wiring is attached to the subbase. When the thermostat is attached to the subbase, the electrical connections are made between the two components.

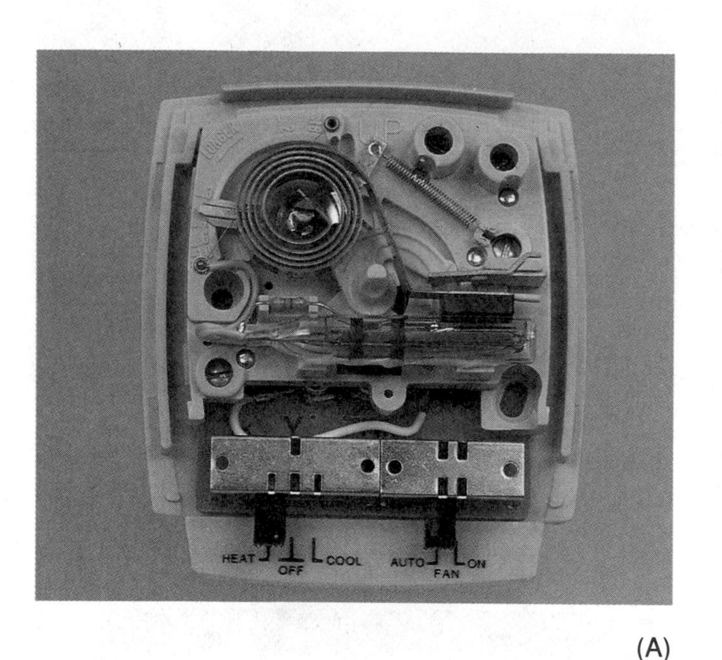

(A)

(B)

FIGURE 14–10 Thermostat assemblies. *(A) Photo by Bill Johnson, (B) Courtesy Robertshaw Controls Company*

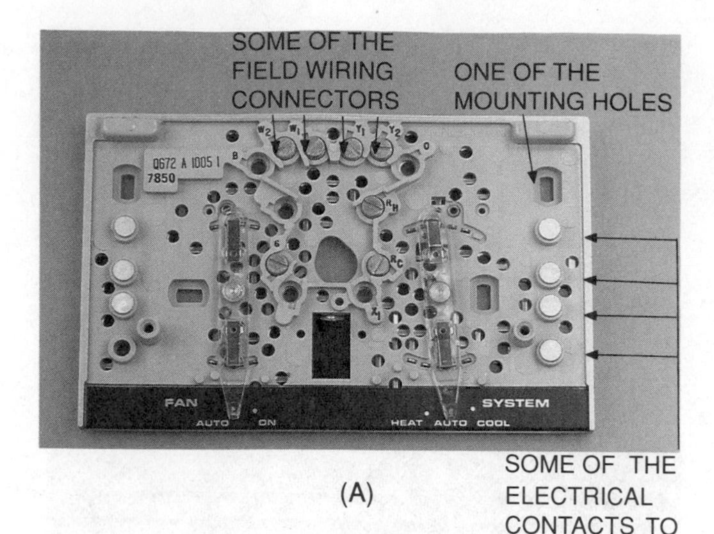

(A)

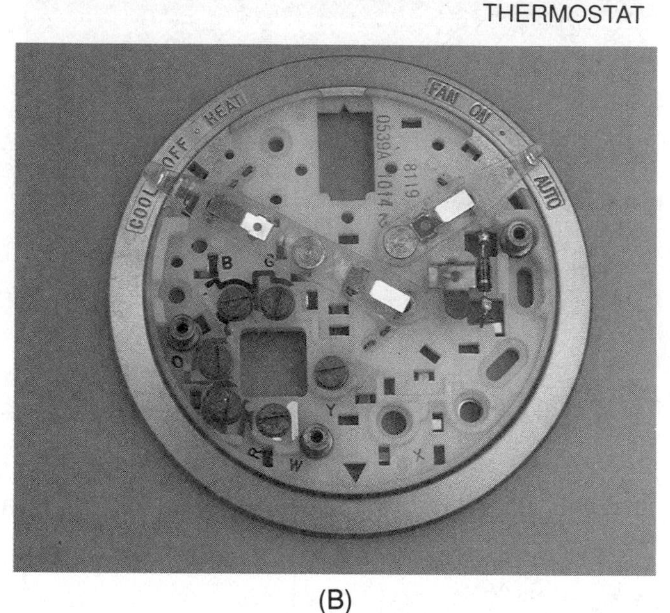

(B)

FIGURE 14–11 The subbase normally mounts on the wall and the wiring is fastened inside the subbase on terminals. These terminals are designed in such a way as to allow easy wire makeup. When the thermostat is screwed down onto the subbase, electrical connections are made between the two. The subbase normally contains the selector switches, such as FAN ON-AUTO and HEAT-OFF-COOL. *Photos by Bill Johnson*

14.4 SPACE TEMPERATURE CONTROLS, HIGH (LINE) VOLTAGE

Sometimes it is desirable to use line-voltage thermostats to stop and start equipment. Some self-contained equipment does not need a remote thermostat. To add a remote thermostat would be an extra expense and would only result in more potential trouble.

The window air conditioner is a good example of this. It is self-contained and needs no remote thermostat. If it had a remote thermostat, it would not be a plug-in device but would require an installation of the thermostat. Note that the remote bulb type of thermostat is normally used with the bulb located in the return airstream, Figure 14–12. The fan usually runs all of the time and keeps a steady stream of room return air passing over the bulb. This gives more sensitivity to this type of application.

The concept of the line-voltage thermostat is used in many types of installations. The household refrigerator, reach-in coolers, and free-standing package air conditioning equipment are just a few examples. All of these have something in common—the thermostat is a heavy-duty type and may not be as sensitive as the low-voltage type. When replacing these thermostats, an exact replacement or one recommended by the equipment supplier should be used.

The line-voltage thermostat must be matched to the voltage and current that the circuit is expected to use. For example, a reach-in cooler for a convenience store has a compressor and fan motor to be controlled (stopped and started). The combined running-current draw for both is 16.2 A at 120 V (15.1 A for the compressor and 1.1 A for the fan motor). The locked-rotor amperage for the compressor is 72 A. (Locked-rotor amperage is the inrush current that the circuit must carry until the motor starts turning.

(A)

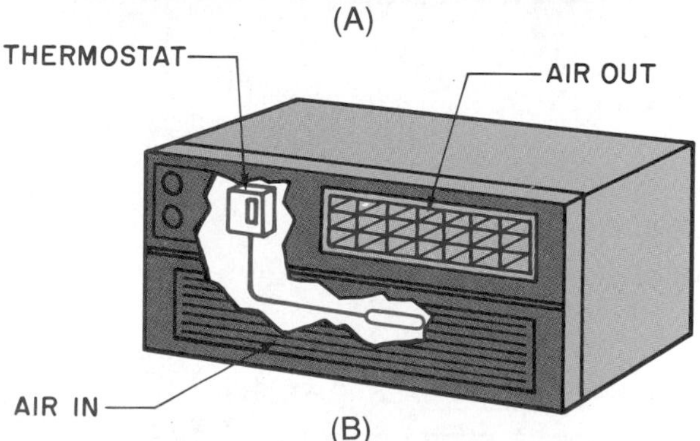

(B)

FIGURE 14–12 The window air conditioner has a line-voltage thermostat that stops and starts the compressor. When the selector switch is turned to cool, the fan comes on and stays on. The thermostat cycles the compressor only. *(A) Courtesy Whirlpool Corporation*

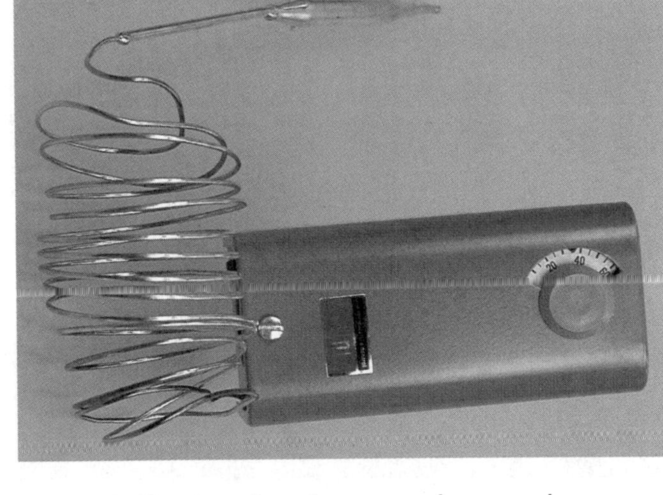

FIGURE 14–13 Line-voltage thermostat used to stop and start a compressor that draws up to 20 A full-load current and has an inrush current up to 80 A. *Photo by Bill Johnson*

The inrush was not considered for the fan because it is so small.) This is a 3/4-hp compressor motor. A reliable, long-lasting thermostat must be chosen to operate this equipment. A control rated at 20 A running current and 80 A locked-rotor current is selected, Figure 14–13.

Usually, thermostats are not rated at more than 25 A because the size becomes prohibitive. This limits the size of the compressor that can be started directly with a line-voltage thermostat to about 1 1/2 hp on 120 V or 3 hp on 230 V. Remember, the same motor would draw exactly half of the current when the voltage is doubled.

When larger current-carrying capacities are needed, a motor starter is normally used with a line-voltage thermostat or a low-voltage thermostat.

Line-voltage thermostats usually consist of the following components.

THE SWITCHING MECHANISM. Some line-voltage thermostats use mercury as the contact surface, but most switches are a silver-coated base metal. The silver helps conduct current at the contact point. This silver contact point takes the real load in the circuit and is the component in the control that wears first, Figure 14–14.

THE SENSING ELEMENT. The sensing element is normally a bimetal, a bellows, or a remote bulb, Figure 14–15, located where it can sense the space temperature. If sensitivity is important, a slight air velocity should cross the element. The levers or knobs used to change the adjustment of the thermostat are attached to the main thermostat.

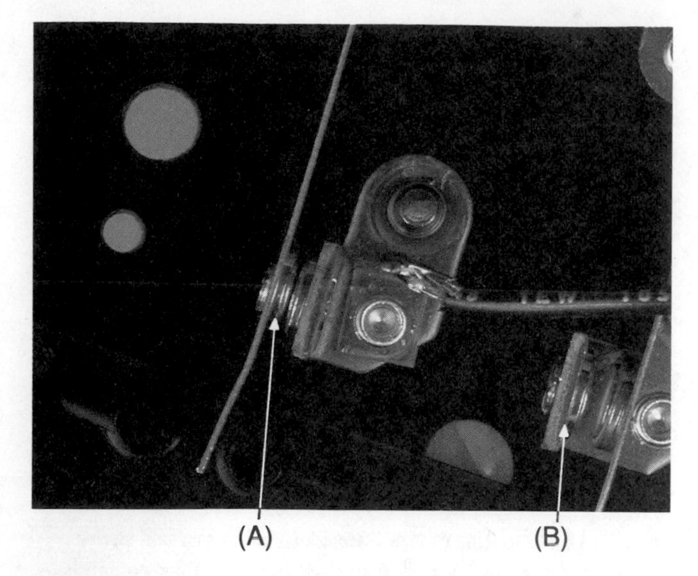

FIGURE 14–14 Silver-contact type line-voltage contacts. (A) Closed. (B) Open. *Photos by Bill Johnson*

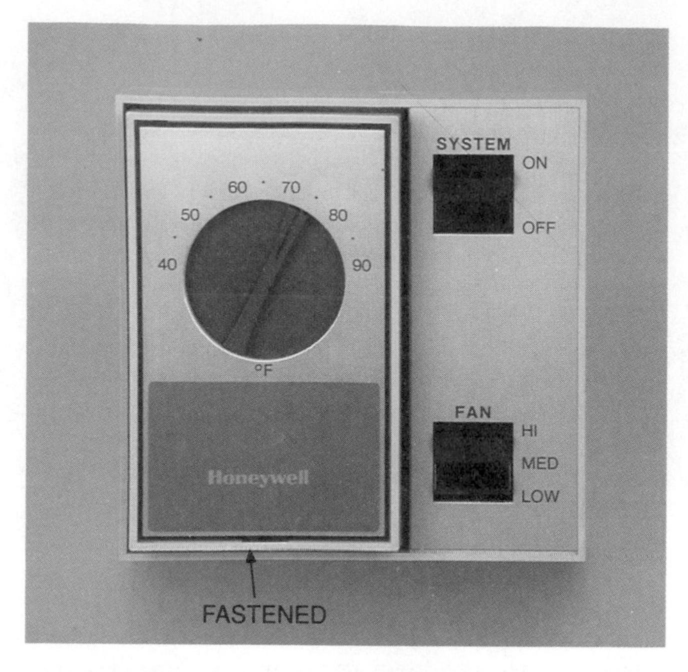

FIGURE 14–16 The cover for a line-voltage thermostat is usually less decorated than the cover on a low-voltage thermostat. *Photo by Bill Johnson*

THE COVER. Because of the line voltage inside, the cover is usually attached with some sort of fastener to discourage easy entrance. Figure 14–16. If the control is applied to room space temperature, such as a building, a thermometer may be mounted on the cover to read the room temperature. If the control is used to control a box space temperature, such as a reach-in cooler, the cover might be just a plain protective cover.

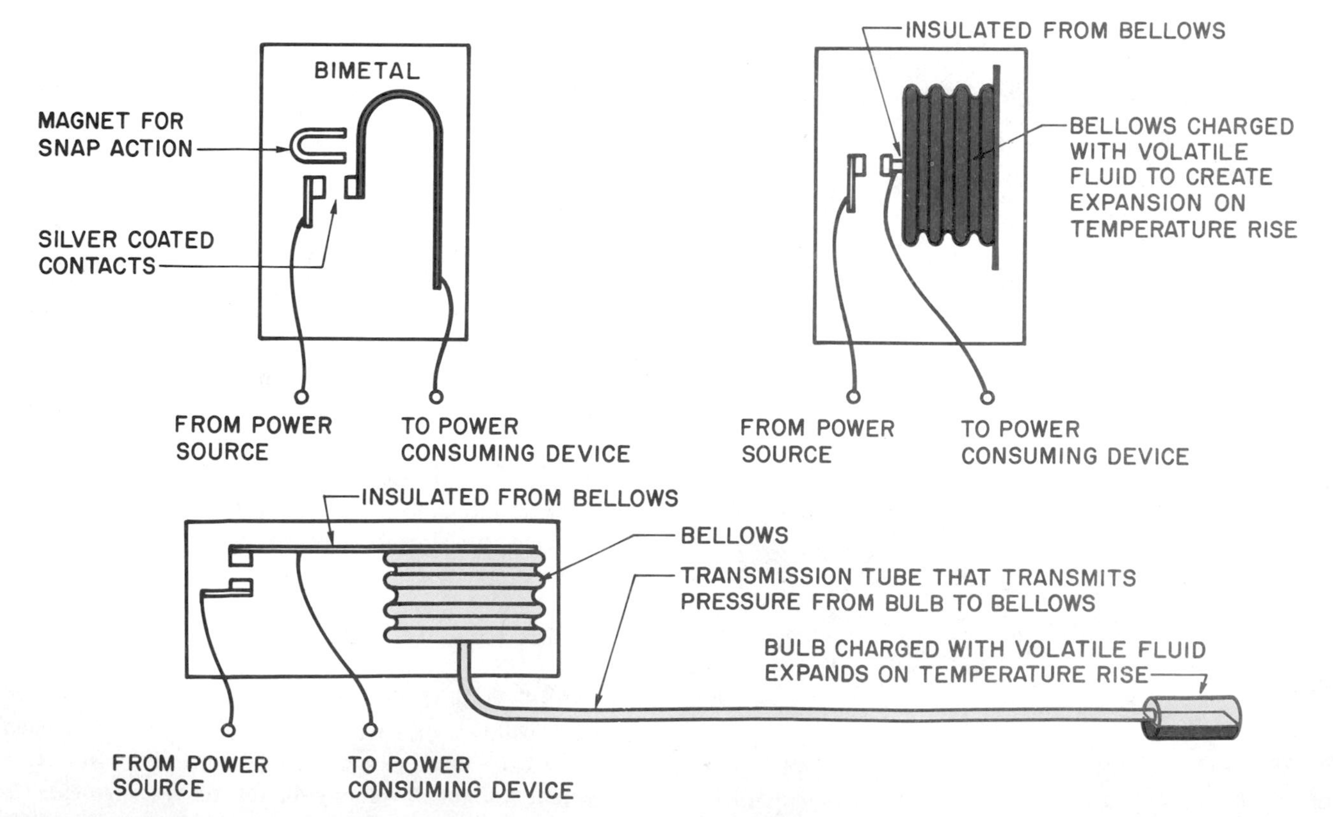

FIGURE 14–15 Bimetal, bellows, and remote bulb sensing elements used in line-voltage thermostats. The levers or knobs used to adjust these controls are arranged to apply more or less pressure to the sensing element to vary the temperature range.

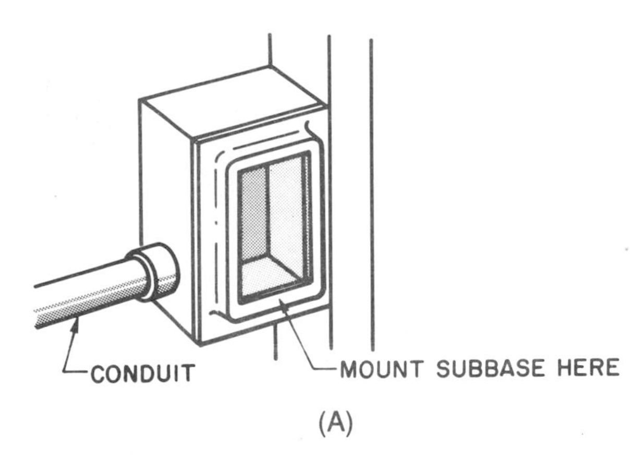

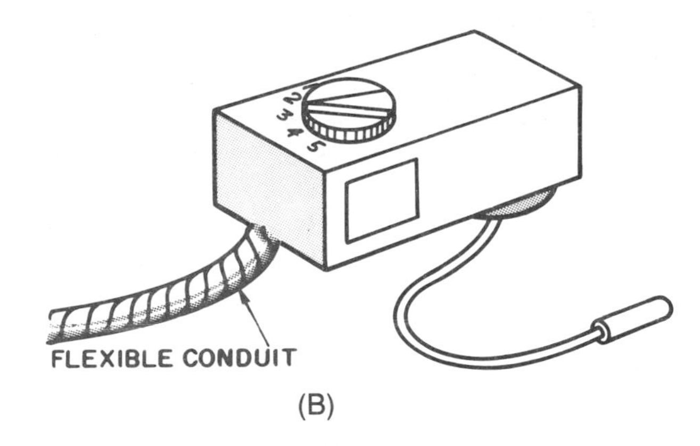

FIGURE 14–17 (A) Box for line-voltage, wall-mounted thermostat. (B) Remote bulb thermostat with flexible conduit. In both cases the interconnected wiring is covered and protected.

SUBBASE. When the thermostat is used for room space temperature control, there must be some way to mount it to the wall. A *subbase* that fits on the electrical outlet box is usually used. The wire leading into a line-voltage control is normally a high-voltage wire routed between points in conduit. The conduit is connected to the box that the thermostat is mounted on. ***If an electrical arc due to overload or short circuit occurs, it is enclosed in the conduit box. This reduces fire hazard, Figure 14–17.***

As mentioned earlier, a room thermostat for the purpose of controlling a space-heating system is the same as a motor temperature thermostat as far as action is concerned. They both open or interrupt the electrical circuit on a rise in temperature. However, the difference in the appearance of the two controls is considerable. One control is designed to sense the temperature of slow-moving air, and the other is designed to sense the temperature of the motor winding. The winding has much more mass, and the control must be in much closer contact to get the response needed.

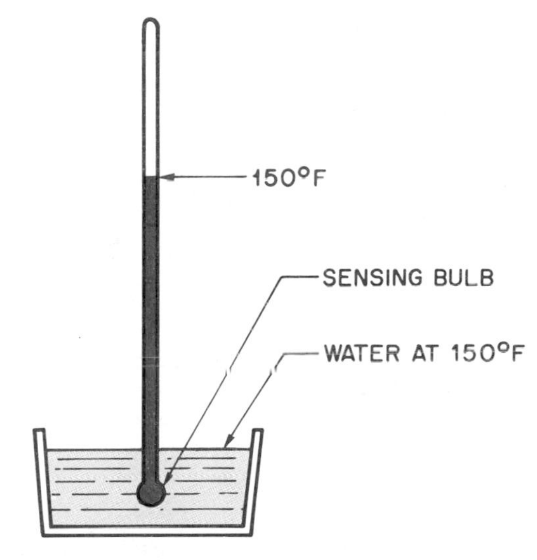

FIGURE 14–18 The mercury thermometer rises or falls based on expansion or contraction of the mercury in the bulb. The bulb is the sensing device.

14.5 SENSING THE TEMPERATURE OF SOLIDS

A key point to remember is that any sensing device indicates or reacts to the temperature of the sensing element. A mercury bulb thermometer indicates the temperature of the bulb at the end of the thermometer, not the temperature of the substance it is submerged in. If it stays in the substance long enough to attain the temperature of the substance, an accurate reading will be achieved. Figure 14–18 illustrates the mercury thermometer.

To accurately determine the temperature of solids, the sensing element must assume the temperature of the substance to be sensed as soon as practical. It can be difficult to get a round mercury bulb close enough to a flat piece of metal so that it senses only the temperature of the metal. Only a fractional part of the bulb will touch the flat metal at any time, Figure 14–19. This leaves most of the area of the bulb exposed to the surrounding (ambient) air.

Placing insulation over the thermometer to hold it tightly on the plate and to shield the bulb from the ambient air helps give a more accurate reading. Sometimes a gum-type substance is used to hold the thermometer's bulb against the

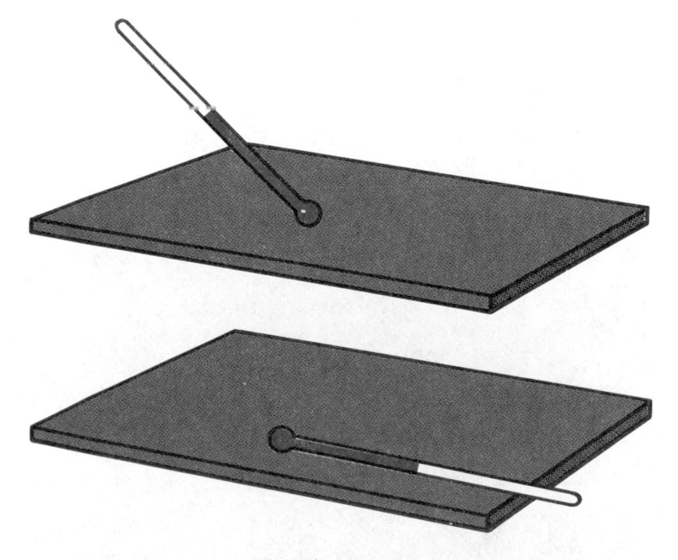

FIGURE 14–19 Only a fractional part of the mercury bulb thermometer can touch the surface of the flat iron plate at one time. This leaves most of the bulb exposed to the surrounding (ambient) temperature.

surface and insulate it from the ambient air, Figure 14–20. A well in which to insert the thermometer may be necessary for a more permanent installation, Figure 14–21.

Most sensing elements will have similar difficulties in reaching the same temperature of the substance to be sensed. Some sensing elements are designed to fit the surface to be sensed. The external motor temperature-sensing element is a good example. It is manufactured flat to fit close to the motor housing, Figure 14–22. One reason it can be made flat is because it is a bimetal. This control is normally mounted inside the terminal box of the motor or compressor to shield it from the ambient temperature.

FIGURE 14–20 Example of sensing temperature often used for field readings. The insulating gum holds the bulb against the metal and insulates it from the ambient air.

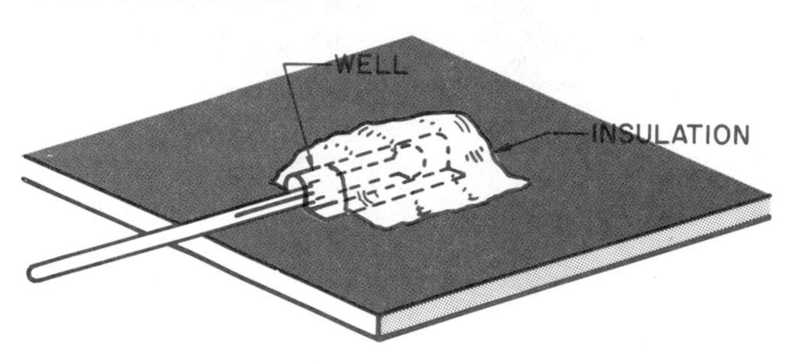

FIGURE 14–21 Example of how a well for a thermometer is designed. The well is fastened to the metal plate so that heat will conduct both into and out of the bulb.

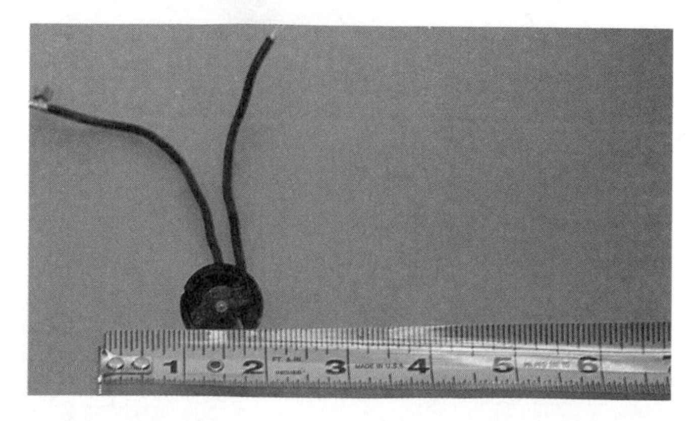

FIGURE 14–22 Motor temperature-sensing thermostat. *Photo by Bill Johnson*

The protection of electrical motors is important in this industry because motors are the prime movers of refrigerants, air, and water. Motors, especially the compressor motor, are the most expensive components in the system. Motors are normally made of steel and copper and need to be protected from heat and from an overload that will cause heat.

All motors build up heat as a normal function of work. The electrical energy that passes to the motor is intended to be converted to magnetism and work, but some of it is converted to heat. All motors have some means of detecting and dissipating this heat under normal or design conditions. If the heat becomes excessive, it must be detected and dealt with. Otherwise, the motor will overheat and be damaged.

Motor high-temperature protection is usually accomplished with some variation of the bimetal or the thermistor. The bimetal device can either be mounted on the outside of thc motor, usually in the terminal box, or embedded in the windings themselves, Figure 14–23. The section on motors covers in detail how these devices actually protect the motors.

The thermistor type is normally embedded in the windings. This close contact with the windings gives fast, accurate response, but it also means that the wires must be brought to the outside of the compressor. This involves extra terminals in the compressor terminal box, Figure 14–24.

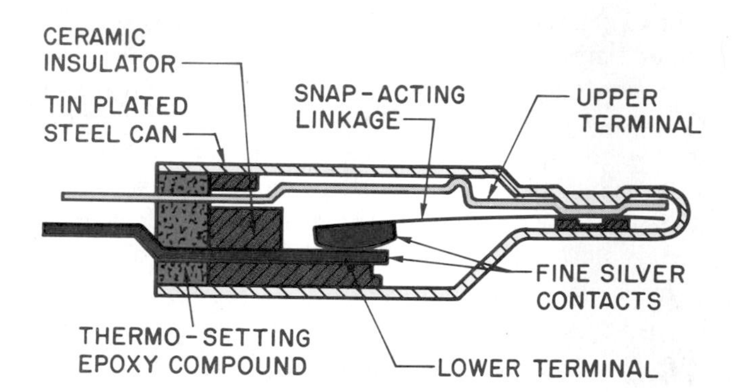

FIGURE 14–23 Bimetal motor temperature protection device.

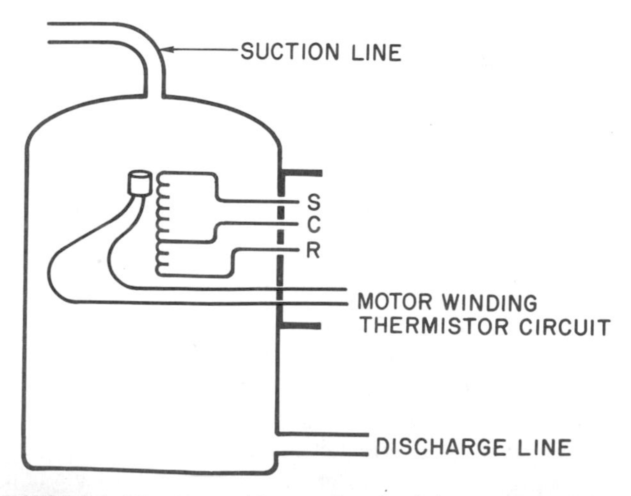

FIGURE 14–24 Terminal box on the side of the compressor. It has more terminals than the normal common, run, and start of the typical compressor. The extra terminals are for the internal motor protection.

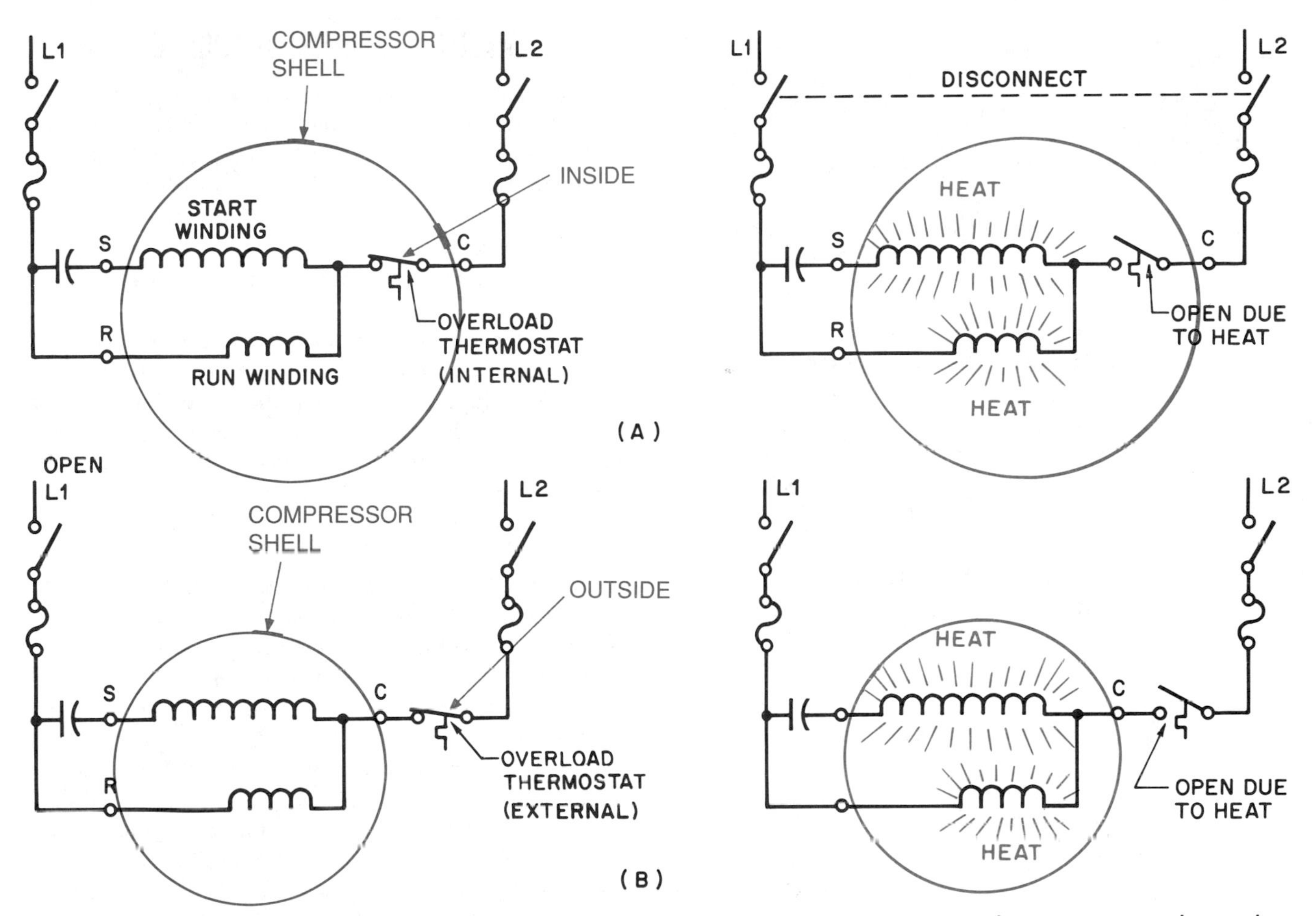

FIGURE 14–25 Line-voltage bimetal sensing device under hot and normal operating conditions. (A) Permanent split capacitor motor with internal protection. Note that a meter would indicate an open circuit if the ohm reading were taken at the C terminal to either start or run if the overload thermostat were to open. There is still a measurable resistance between start and run. (B) The same motor, except the motor protection is on the outside. Note that it is easier to troubleshoot the overload, but it is not as close to the windings for fast response.

Figures 14–25 and 14–26 show wiring diagrams of temperature protection devices.

Because of its size and weight, a motor can take a long time to cool after overheating. If the motor is an open type, a fan or moving airstream can be devised to cool it more quickly. If the motor is inside a compressor shell, it may be suspended from springs inside the shell. This means that the actual motor and the compressor may be hot and hard to cool even though the shell does not feel hot. Figure 14–27 shows a method often used to cool a hot compressor. There is a vapor space between the outside of the shell and the actual heat source, Figure 14–28. ***The unit must have time to cool. If you are in a hurry, set up a fan, or even cool the compressor with water, but don't allow water to get into the electrical circuits. Turn the power off and cover electrical circuits with plastic before using water to cool a hot compressor. Be careful when restarting. Use an ammeter to look for overcurrent and gages to determine the charge level of the equipment.*** Most hermetic compressors are cooled by the suction gas. If there is an undercharge, there will be undercooling (or overheating) of the motor.

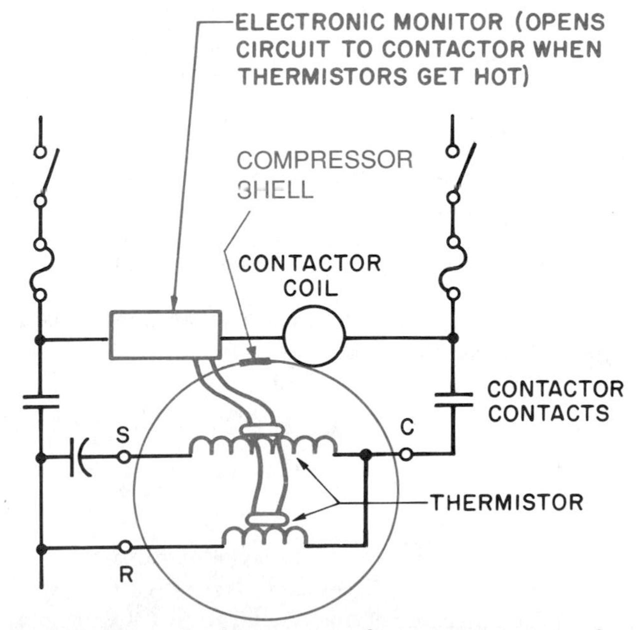

FIGURE 14–26 The thermistor type of temperature-monitoring device uses an electronic monitoring circuit to check the temperature at the thermistor. When the temperature reaches a predetermined high, the monitor interrupts the circuit to the contactor coil and stops the compressor.

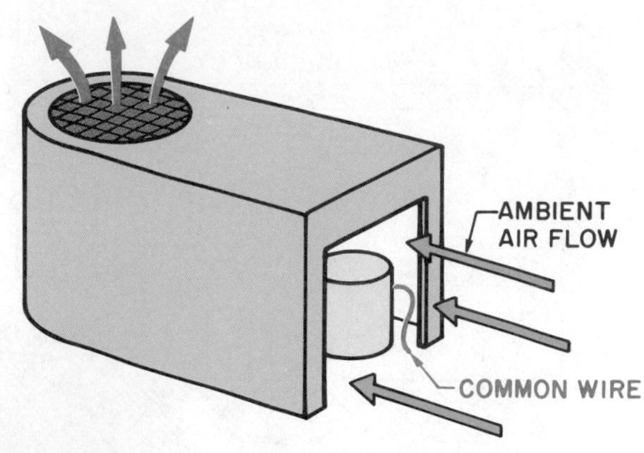

FIGURE 14–27 Ambient air is used to cool the compressor motor when located in fan compartment.

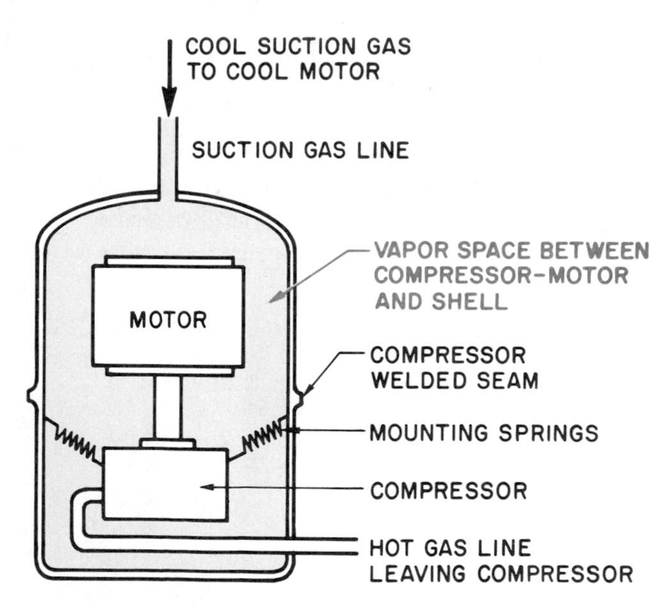

FIGURE 14–28 Compressor and motor suspended in vapor space in compressor shell. The vapor conducts heat slowly.

The best procedure when a hot compressor is encountered is to shut off the compressor with the space temperature thermostat switch and return the next day. This gives it ample time to cool and keeps the crankcase heat on. Many service technicians have diagnosed an open winding in a compressor that was only hot, and they later discovered that the winding was open because of internal thermal protection.

14.6 MEASURING THE TEMPERATURE OF FLUIDS

The term *fluid* applies to both the liquid and vapor states of matter. Liquids are heavy and change temperature very slowly. The sensing element must be able to reach the temperature of the medium to be measured as soon as practical. Because liquids are contained in vessels (or pipes), the measurement can be made either by contact with the vessel or by some kind of immersion. When a temperature is detected from the outside of the vessel by contact, care must be taken that the ambient temperature does not affect the reading.

A good example of this is the sensing bulb used in refrigeration work for measuring the performance of the thermostatic expansion valve. The sensing bulb must sense refrigerant gas temperature accurately to keep liquid from entering the suction line. Often the technician will strap the sensing bulb to the suction line in the correct location but will fasten it to the line incorrectly. The technician may forget to insulate the bulb from the ambient temperature when it needs to be insulated. In this case the bulb is sensing the ambient temperature and averaging it in with the line temperature. Make sure when mounting a sensing bulb to a line for a contact reading that the bulb is in the very best possible contact, Figure 14–29.

When a temperature reading is needed from a larger pipe in a permanent installation, different arrangements are made. A well can be welded into the pipe during installation so that a thermometer or controller-sensing bulb can be inserted into it. The well must be matched to the sensing bulb to get a good contact fit, or you won't get an accurate reading.

Sometimes it is desirable to remove the thermometer from the well and insert an electronic thermometer with a small probe for troubleshooting purposes. The well inside diameter is much larger than the probe. The well can be packed so that the probe is held firm against the well for an accurate reading, Figure 14–30.

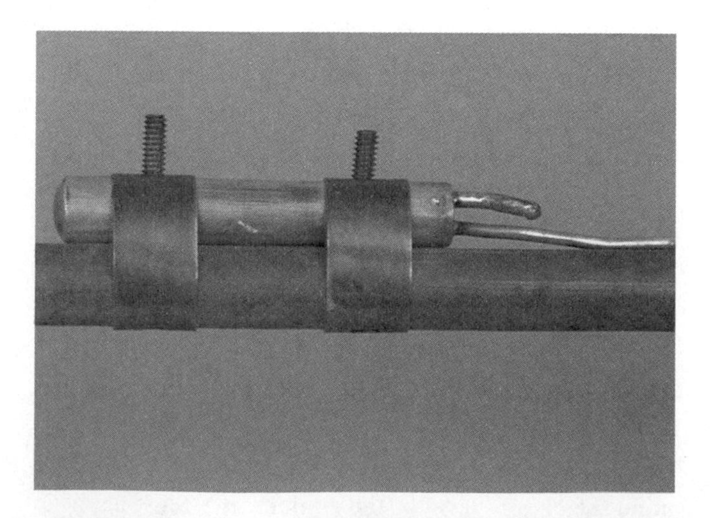

FIGURE 14–29 Correct way to get good contact with a sensing bulb and the suction line. Bulb is mounted on a straight portion of the line with a strap that holds it secure against the line. *Photo by Bill Johnson*

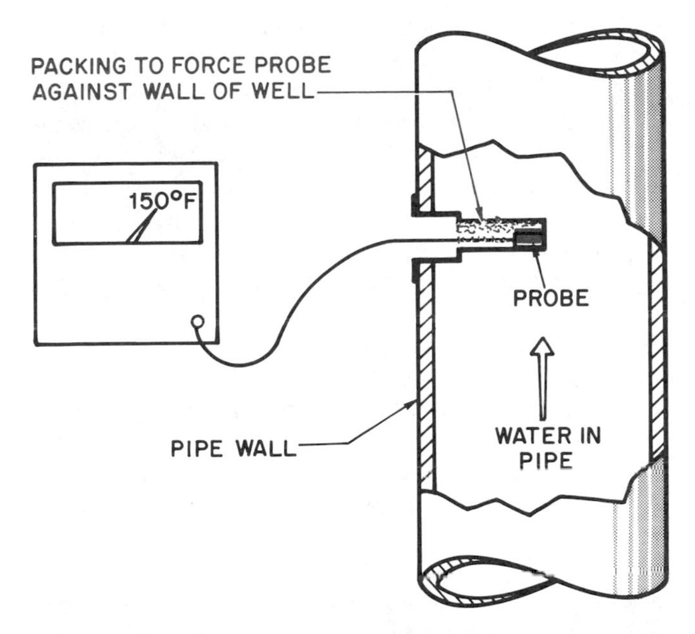

FIGURE 14–30 The well is packed to get the probe against the wall of the well.

Another method for obtaining an accurate temperature reading in a water circuit for test purposes is to use one of the valves in a water line to give constant bleed. For instance, if the leaving-water temperature of a home boiler is needed and the thermometer in the well is questionable, try the following procedure. Place a small container under the drain valve in the leaving-water line. Allow a small amount of water from the system to run continuously into the container. An accurate reading can be obtained from this water. This is not a long-term method for detecting temperature because it requires a constant bleeding of water, but it is an effective field method, Figure 14–31.

FIGURE 14–32 A snap-disc, which can be used to sense air temperatures in duct work and furnaces. *Photo by Bill Johnson*

14.7 SENSING TEMPERATURE IN AN AIRSTREAM

Sensing temperature in fast-moving airstreams such as duct work and furnace heat exchangers is usually done by inserting the sensing element into the actual airstream. The bimetal such as the flat-type snap-disc or the helix coil is usually used. Figure 14–32 shows a snap-disc.

14.8 THINGS TO REMEMBER ABOUT SENSING DEVICES

Sensing devices are not mysterious; something reacts to the temperature change, such as a bimetal device or a thermistor. Look over any temperature-sensing control and study it, and you will usually be able to determine the way it operates. If you are still confused, consult a catalog or the supplier of the control.

14.9 PRESSURE-SENSING DEVICES

Pressure-sensing devices are normally used when measuring or controlling the pressure of refrigerants, air, gas, and water. These are sometimes strictly pressure controls or they

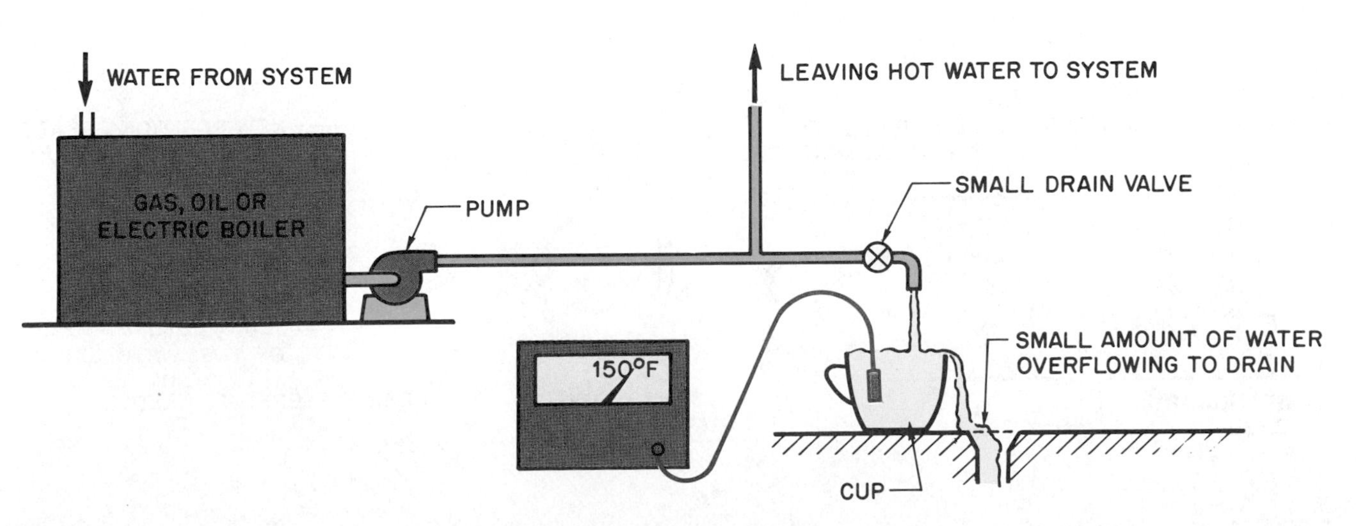

FIGURE 14–31 Method to obtain a leaving-water temperature reading from a small boiler.

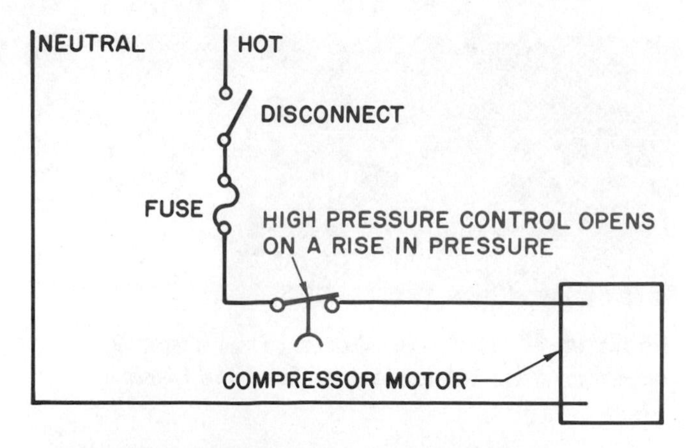

FIGURE 14–33 Electrical circuit of a refrigeration compressor with a high-pressure control. This control has a normally closed contact that opens on a rise in pressure.

can be used to operate electrical switching devices. The terms pressure control or pressure switch are often used interchangeably in the field. The actual application of the component should indicate if the device is used to control a fluid or to operate an electrical switching device.

Some applications for pressure-operated controls:

1. Pressure switches are used to stop and start electrical loads, such as motors, Figure 14–33.
2. Pressure controls contain a bellows, a diaphragm, or a Bourdon tube to create movement when the pressure inside it is changed. Pressure controls may be attached to switches or valves, Figure 14–34.
3. When used as a switch, the bellows, Bourdon tube, or diaphragm is attached to the linkage that operates the electrical contacts. When used as a valve, it is normally attached directly to the valve.
4. The electrical contacts are the component that actually open and close the electrical circuit.
5. The electrical contacts either open or close with snap action on a rise in pressure, Figure 14–35.
6. The pressure control can either open or close on a rise in pressure. This opening and closing action can control water or other fluids, depending on the type.
7. The pressure control can sense a pressure differential and be designed to open or close a set of electrical contacts, Figure 14–36.

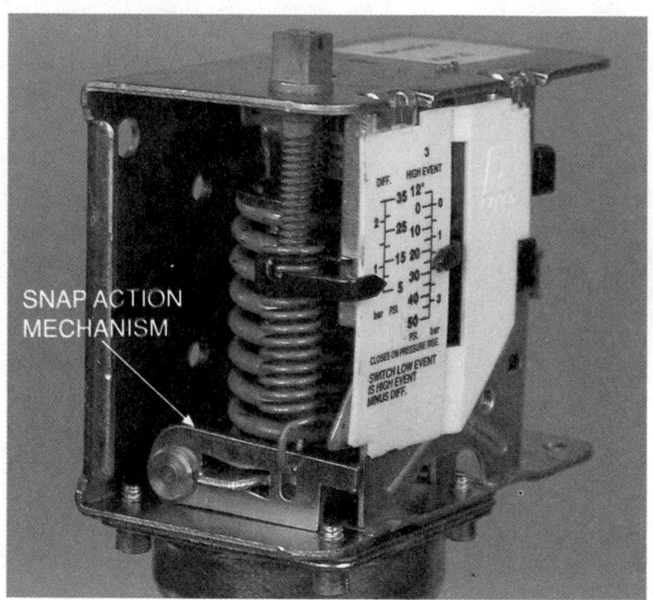

FIGURE 14–35 Snap action over center device. *Photo by Bill Johnson*

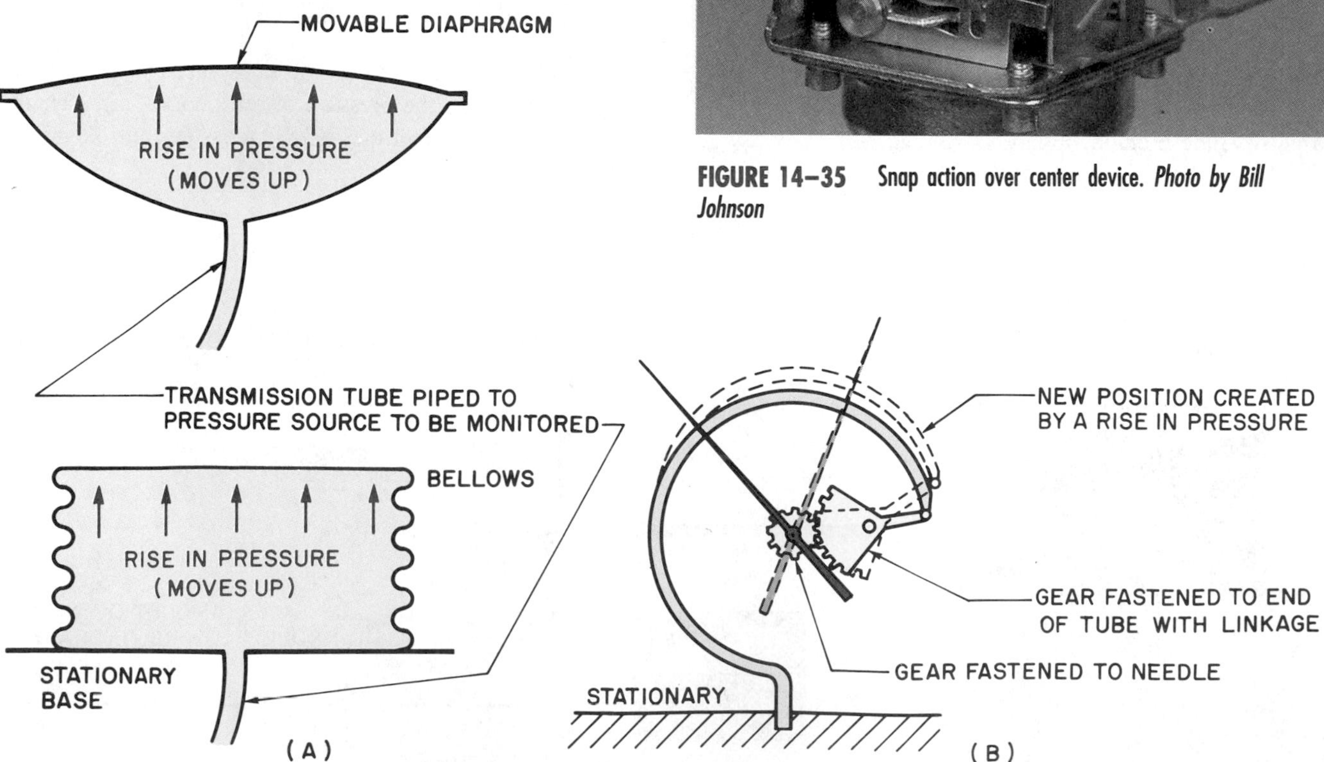

FIGURE 14–34 Moving part of most pressure-type controls.

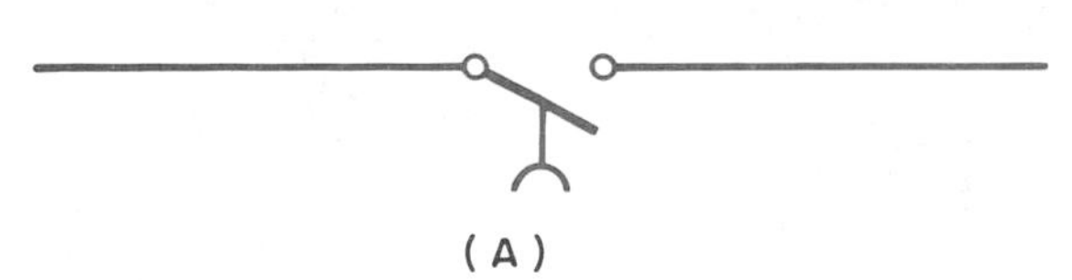

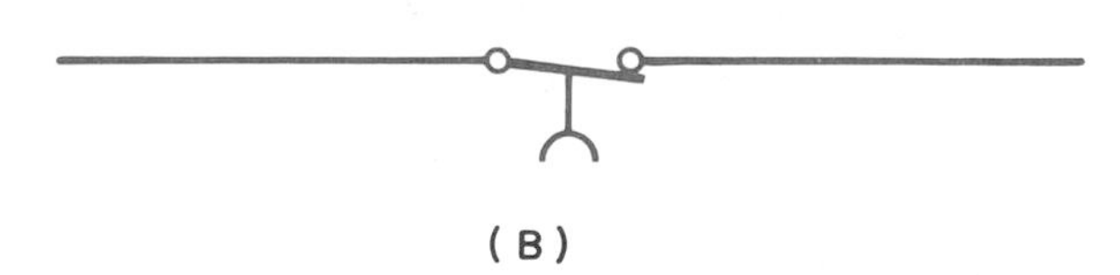

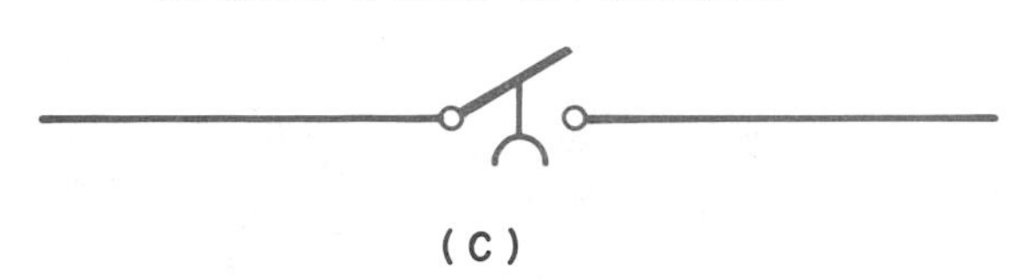

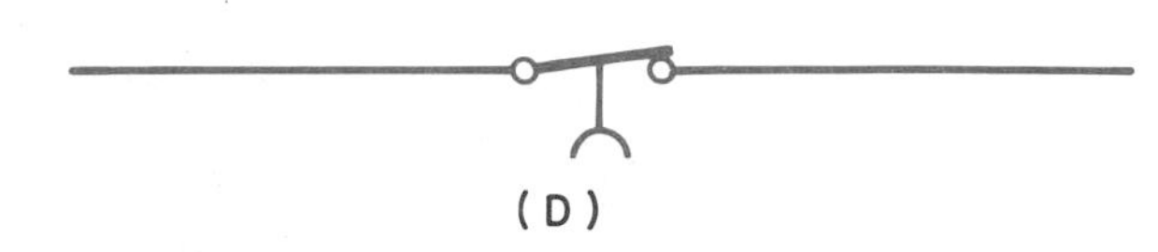

FIGURE 14–36 These symbols show how pressure controls connected to switches appear on control diagrams. The two symbols, A and B, indicate that the circuit will make on a rise in pressure. A indicates the switch is normally open when the machine does not have power to the electrical circuit. B shows that the switch is normally closed without power. The two symbols, C and D, indicate that the circuit will open on a rise in pressure. C is normally open and D is normally closed.

8. The pressure control can be an operating-type control or a safety-type control, Figure 14–37.
9. It can operate either at low pressures (even below atmospheric pressure) or high pressures, depending on the design of the control mechanism.
10. Pressure controls can sometimes be recognized by the small pipe running to them for measuring fluid pressures.
11. Pressure switches are manufactured to handle control voltages or line currents to start a compressor up to about 3 hp maximum. The refrigeration industry is the only industry that uses the high-current draw controls.
12. The high-pressure and the low-pressure controls in refrigeration and air conditioning equipment are the two most widely used pressure controls in this industry.
13. Some pressure switches are adjustable (Figure 14–38), and some are not.
14. Some controls are automatic reset, and some are manual reset, Figure 14–39.

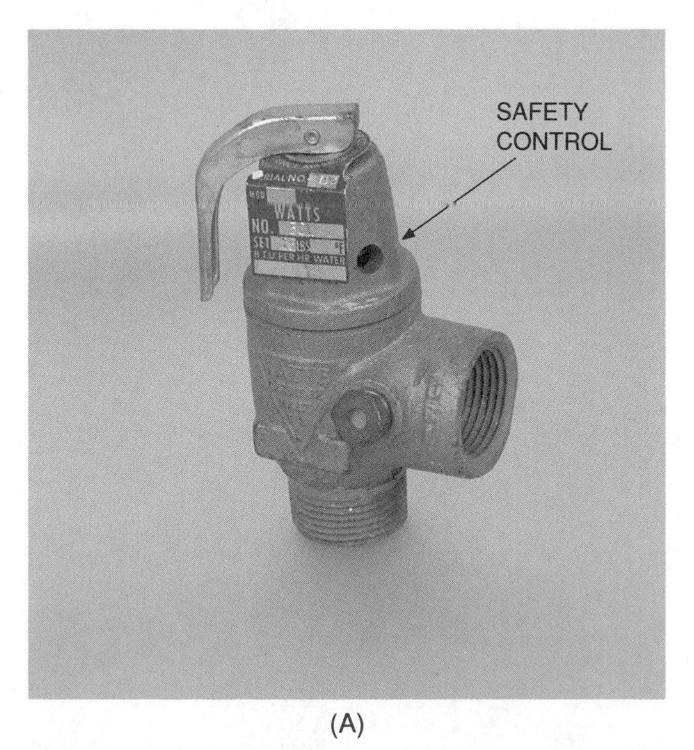

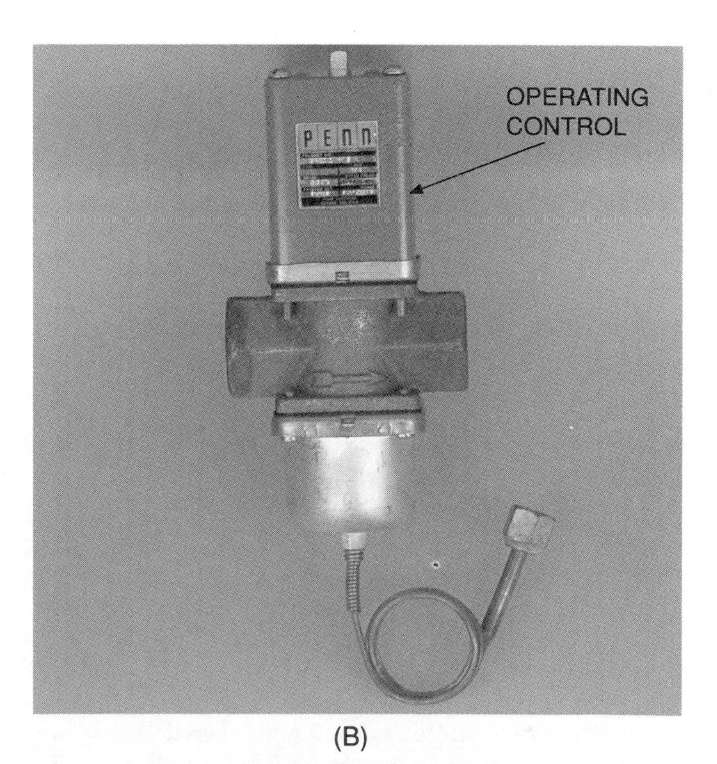

FIGURE 14–37 (A) A safety control (boiler relief valve) and (B) an operating control (water-regulating valve). *Photos by Bill Johnson*

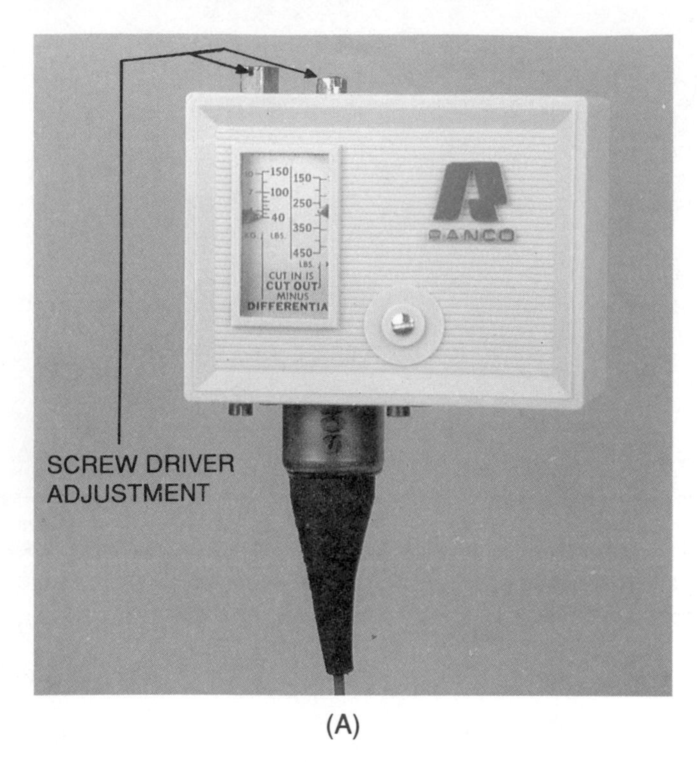

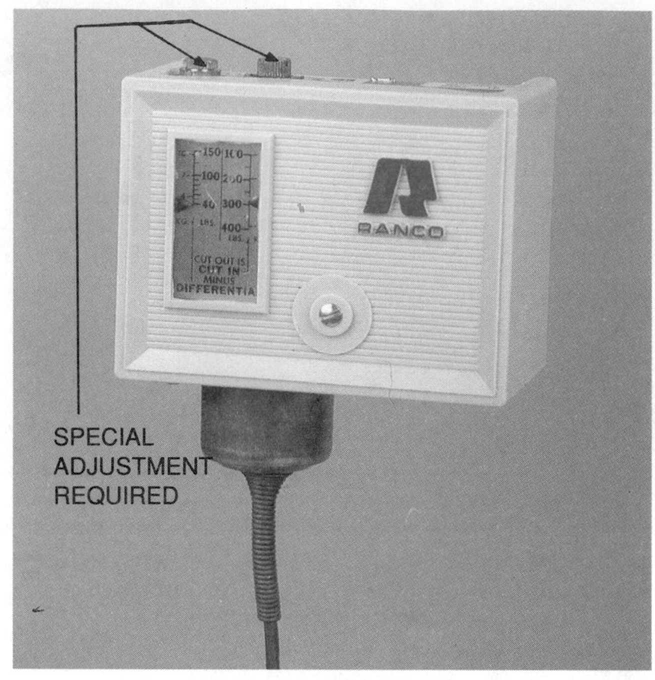

FIGURE 14–38 Commonly used pressure controls. *Photos by Bill Johnson*

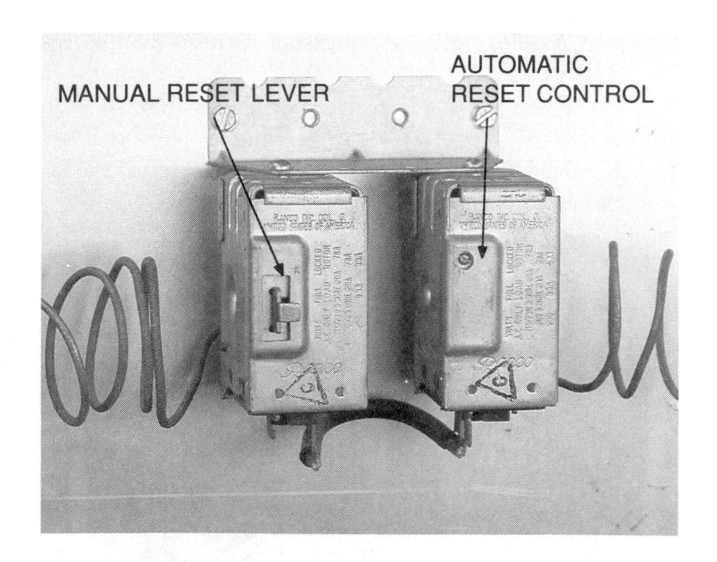

FIGURE 14–39 The control on the left is a manual reset control, the one on the right is automatic reset. Note the push lever on the left hand control. *Photo by Bill Johnson*

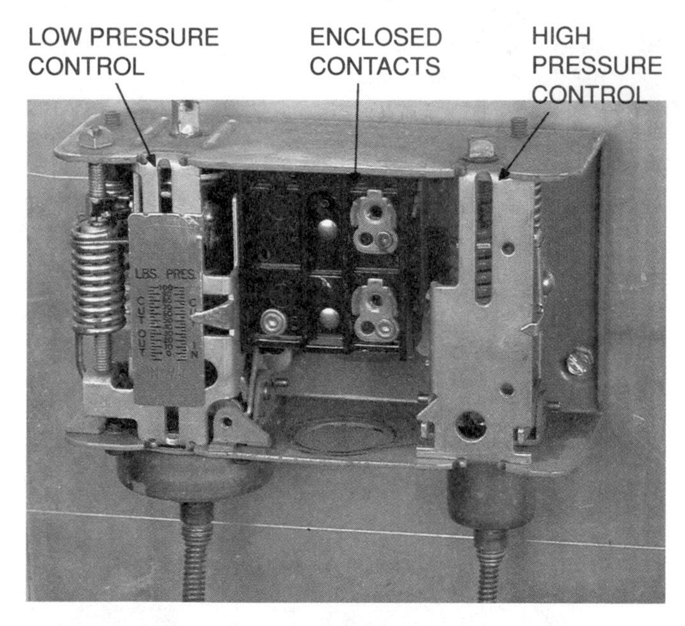

FIGURE 14–40 The control has two bellows acting on one set of contacts. Either control can stop the compressor. *Photo by Bill Johnson*

15. In some pressure controls the high-pressure and the low-pressure controls are built into one housing. They are called *dual-pressure controls,* Figure 14–40.
16. Pressure controls are usually located near the compressor on air conditioning and refrigeration equipment.
17. ***When used as safety controls, pressure controls should be installed in such a manner that they are not subject to being valved off by the service valves.***
18. The point or pressure setting at which the control interrupts the electrical circuit is known as the *cutout.* The point or pressure setting at which the electrical circuit is made is known as the *cut-in.* The difference in the two settings is known as the *differential.*

14.10 HIGH-PRESSURE CONTROLS

The high-pressure control (switch) on an air conditioner stops the compressor if the pressure on the high-pressure side becomes excessive. This control appears in the wiring diagram as a normally closed control that opens on a rise in pressure. The manufacturer may determine the upper limit of operation for a particular piece of equipment and furnish a high-pressure cutout control to ensure that the equipment does not operate above these limits, Figure 14–41.

A compressor is known as a positive displacement device. When it has a cylinder full of vapor, it is going to pump out the vapor or stall. If a condenser fan motor on an air-cooled piece of equipment burns out and the compressor continues to operate, very high pressures will occur. ***The high-pressure control is one method of ensuring safety for the equipment and the surroundings. Some compressors are strong enough to burst a pipe or a container. The overload device in the compressor offers some protection in this event, but it is really a secondary device because it is not directly responding to the pressures. The motor overload device may also be a little slow to respond.***

(A)

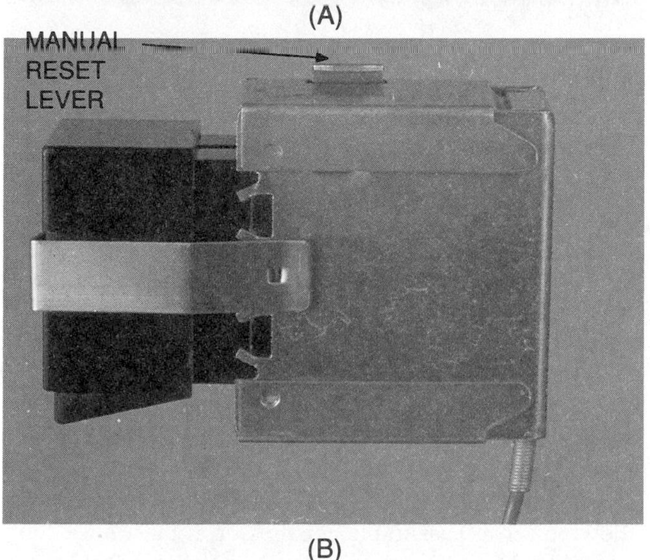

(B)

FIGURE 14–41 High-pressure controls. (A) Automatic reset. (B) Manual reset. *Photos by Bill Johnson*

14.11 LOW-PRESSURE CONTROLS

The low-pressure control (switch) is used in the air conditioning field as a low-charge protection, Figure 14–42. When the equipment loses some of the refrigerant from the system, the pressure on the low-pressure side of the system will fall. Manufacturers may have a minimum pressure under which they will not allow the equipment to operate. This is the point at which the low-pressure control cuts off the compressor. Manufacturers use different settings based on their requirements, so their recommendations should be followed.

The recent popularity of the capillary tube as a metering device has caused the low-pressure control to be reconsidered as a standard control on all equipment. The capillary tube metering device equalizes pressure during the off cycle and causes the low-pressure control to short cycle the compressor if it is not carefully applied. The capillary tube is a fixed-bore metering device and has no shutoff valve action. To prevent this short cycle, some equipment comes with a time-delay circuit that will not allow the compressor to restart for a predetermined time period.

There are two good reasons why it is undesirable to operate a system without an adequate charge.

1. The compressor motors used most commonly in this industry, particularly in the air conditioning field, are cooled by the refrigerant. Without this cooling action, the motor will build up heat when the charge is low. The motor temperature cutout is used to detect this condition. It often takes the place of the low-pressure cutout by sensing motor temperature.
2. If the refrigerant escaped from the system through a leak in the low side of the system, the system can

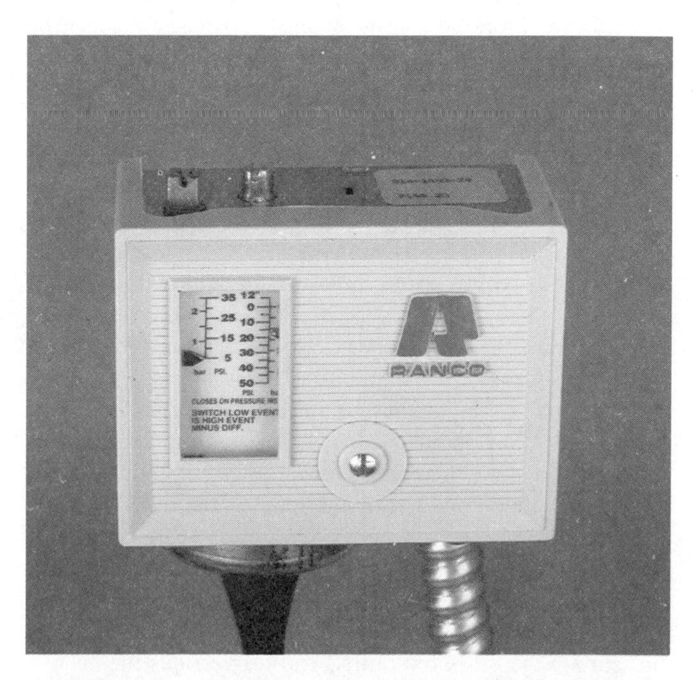

FIGURE 14–42 Low-pressure control. *Photo by Bill Johnson*

operate until it goes into a vacuum. When a vessel is in a vacuum, the atmospheric pressure is greater than the vessel pressure. This causes the atmosphere to be pushed into the system. Technicians often say "pull air into the system." The reference point the technician uses is atmospheric pressure. This air in the system is sometimes hard to detect; if it is not removed, it causes acid to form in the system.

14.12 OIL PRESSURE SAFETY CONTROLS

The oil pressure safety control (switch) is used to ensure that the compressor has oil pressure when operating, Figure 14–43. This control is used on larger compressors and has a different sensing arrangement than the high- and low-pressure controls. The high- and low-pressure controls are single diaphragm or single bellows controls because they are comparing atmospheric pressure to the pressures inside the system. Atmospheric pressure can be considered a constant for any particular locality because it does not vary more than a small amount.

The oil pressure safety control is a pressure differential control. This control measures a difference in pressure to establish that positive oil pressure is present. A study of the compressor will show that the compressor crankcase (this is where the oil pump suction inlet is located) is the same as the compressor suction pressure, Figure 14–44. The suction pressure will vary from the off or standing reading to the actual running reading. For example, when a system is using R-22 as the refrigerant, the pressures may be similar to the following: 125 psig while standing, 70 psig while operating, and 20 psig during a low-charge situation.

A plain low-pressure cutout control would not function at all of these levels, so a control had to be devised that would sensibly monitor pressures at all of these conditions.

Most compressors need at least 30 psig of actual oil pressure for proper lubrication. This means that whatever the suction pressure is, the oil pressure has to be at least 30 lb above the oil pump inlet pressure, because the oil pump inlet pressure is the same as the suction pressure. For example, if the suction pressure is 70 psig, the oil pump outlet pressure must be 100 psig for the bearings to have a net oil pressure of 30 psig. This difference in the suction pressure and the oil pump outlet pressure is called the *net oil pressure*.

The basic low-pressure control has the pressure under the diaphragm or bellows and the atmospheric pressure on the other side of the diaphragm or bellows. The atmospheric pressure is considered a constant because it doesn't vary more than a small amount. The oil pressure control uses a double bellows—one bellows opposing the other to detect the net or actual oil pressure. The pump inlet pressure is under one bellows, and the pump outlet pressure is under the other bellows. These bellows are opposite each other either physically or by linkage. The bellows with the most pressure is the oil pump outlet, and it overrides the bellows with the least amount of pressure. This override reads out in

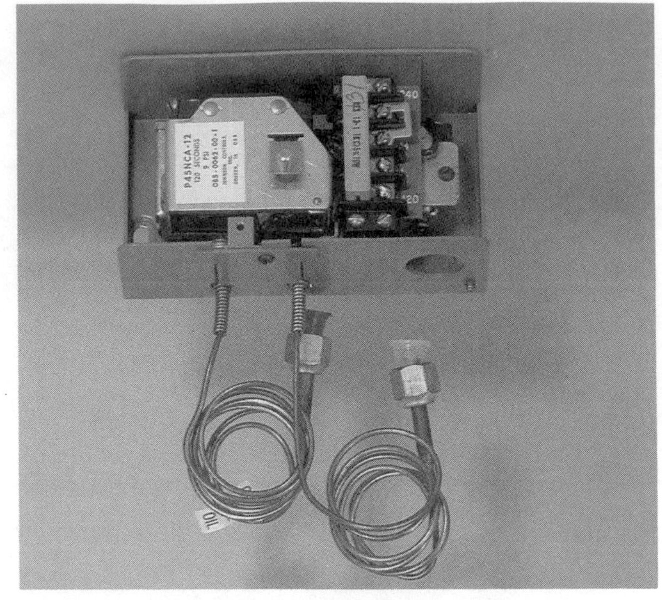

(A)

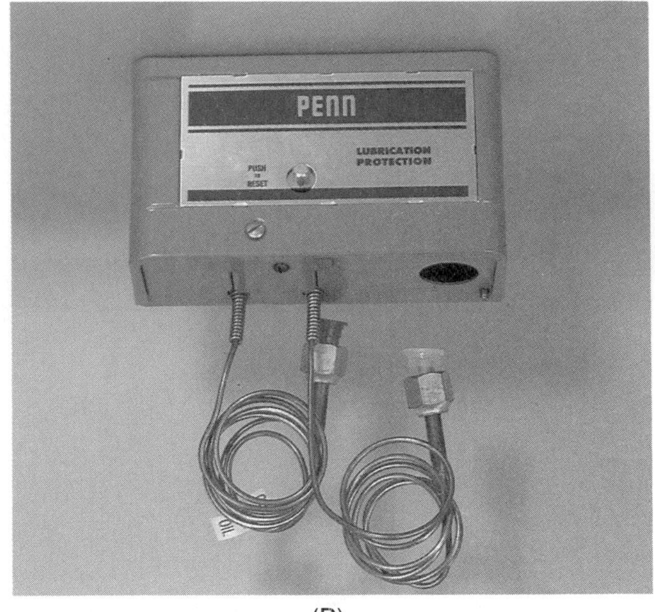

(B)

FIGURE 14–43 Two views of an oil pressure safety control. This control satisfies two requirements: how to measure net oil pressure effectively and how to get the compressor started to build oil pressure. *Photos by Bill Johnson*

net pressure and is attached to a linkage that can stop the compressor when the net pressure drops for a predetermined time.

Because the control needs a differential in pressure to allow power to pass to the compressor, it must have some means to allow the compressor to get started. Remember, there is no pressure differential until the compressor starts to turn, because the oil pump is attached to the compressor

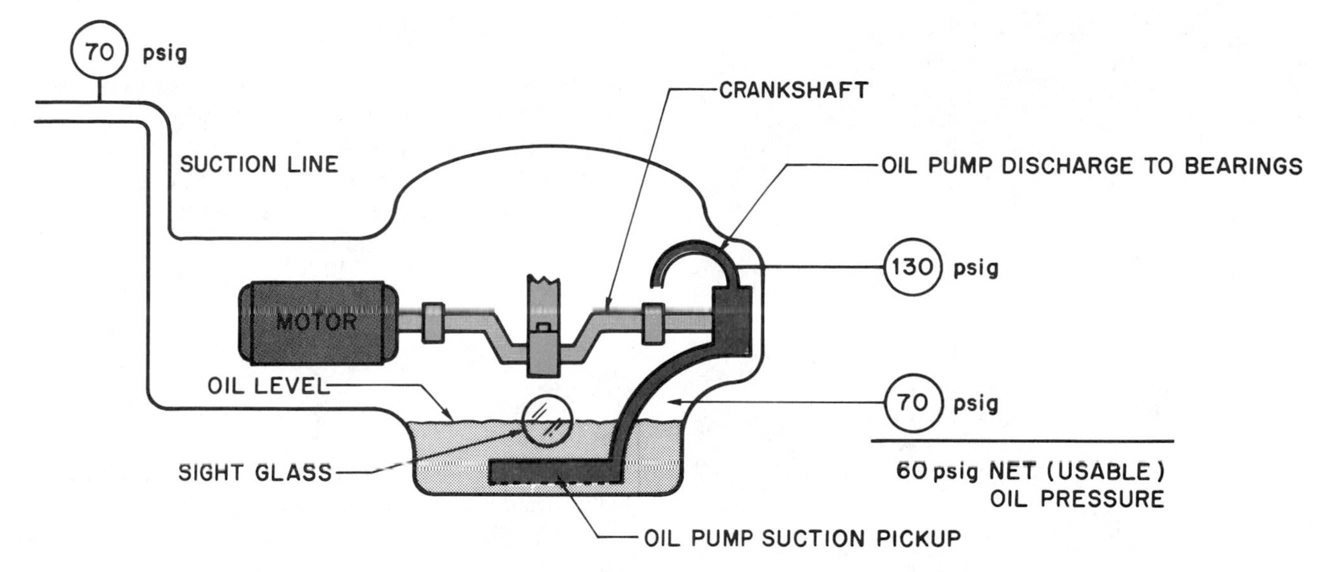

FIGURE 14–44 The oil pump suction is actually the suction pressure of the compressor. This means that the true oil pump pressure is the oil pump discharge pressure less the compressor suction pressure. For example if the oil pump discharge pressure is 130 psig and the compressor suction pressure is 70 psig, the net oil pressure is 60 psig. This is usable oil pressure.

crankshaft. There is a time delay built into the control to allow the compressor to get started and to prevent unneeded cutouts when oil pressure may vary for only a moment. This time delay is normally about 90 sec. It is accomplished with a heater circuit and a bimetal device or electronically. ***The manufacturer's instructions should be consulted when working with any compressor that has an oil safety control, Figure 14–45.***

14.13 AIR PRESSURE CONTROLS

Air pressure controls (switches) are used in the following applications.

1. Heat pumps have air pressure drop across the outdoor coil due to ice buildup. When the coil has a predetermined pressure drop across it, there is an ice buildup that justifies a defrost. When used in this manner, the control is an operating control, Figure 14–46. All manufacturers do not use air pressure as an indicator for defrost.

Air pressure switch sensors need a large diaphragm to sense the very low pressures in air systems. Mounted on the outside of the diaphragm is a small switch (microswitch) capable only of stopping and starting control circuits.

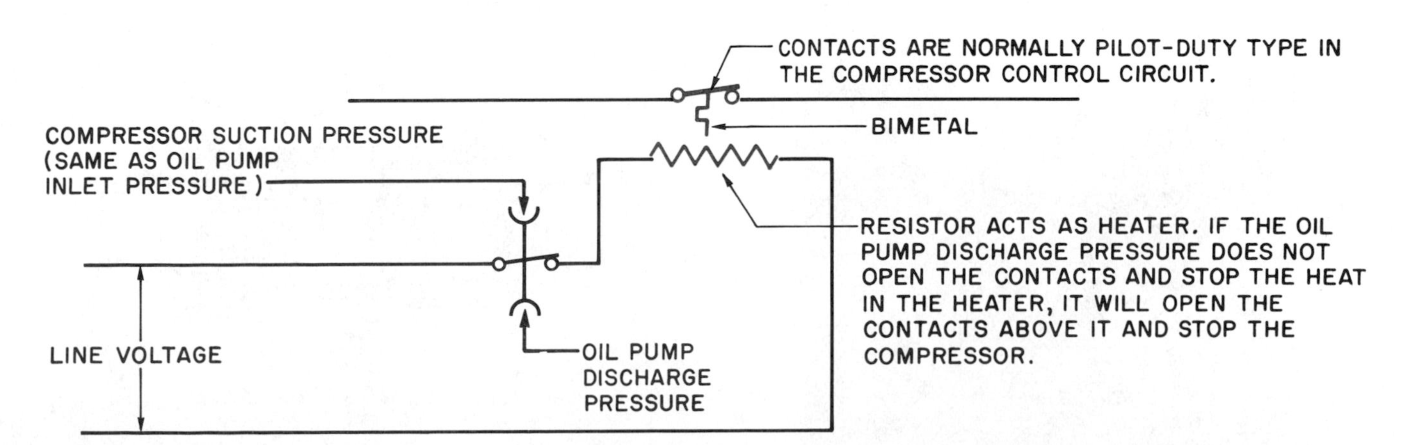

FIGURE 14–45 The oil pressure control is used for lubrication protection for the compressor. The oil pump that lubricates the compressor is driven by the compressor crankshaft. Therefore, a time delay is necessary to allow the compressor to start up and build oil pressure. This time delay is normally 90 sec. The time delay is accomplished either with a heater circuit heating a bimetal or an electronic circuit.

FIGURE 14–46 An air pressure differential switch used to detect ice on a heat pump outdoor coil. *Photo by Bill Johnson*

2. When electric heat is used in a remote duct as terminal heat, an air switch (sail switch) is sometimes used to ensure that the fan is passing air through the duct before the heat is allowed to be energized. This is a safety switch and should be treated as such, Figure 14–47. This switch has a very lightweight and sensitive sail. When air is passing through the duct from the fan, the sail is blown to an approximate horizontal position. In this position it allows the heat source to be activated. When the fan stops, the sail rises ensuring that the heat source is turned off.

FIGURE 14–47 A sail switch detects air movement. *Photo by Bill Johnson*

14.14 GAS PRESSURE SWITCHES

Gas pressure switches are similar to air pressure switches but usually smaller. They detect the presence of gas pressure in gas-burning equipment before burners are allowed to ignite. ***This is a safety control and should never be bypassed except for troubleshooting and then only by experienced service technicians.***

14.15 DEVICES THAT CONTROL FLUID FLOW AND DO NOT CONTAIN SWITCHES

The pressure relief valve can be considered a pressure-sensitive device. It is used to detect excess pressure in any system that contains fluids (water or refrigerant). Boilers, water heaters, hot water systems, refrigerant systems, and gas systems all use pressure relief valves, Figure 14–48. This device can either be pressure and temperature sensitive (called a P & T valve) or just pressure sensitive. The P & T

(A)

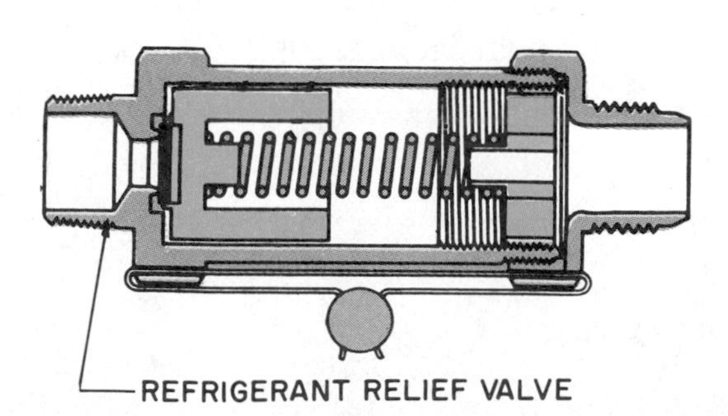

(B)

FIGURE 14–48 Pressure relief valves. *(A) Photo by Bill Johnson, (B) Courtesy Superior Valve Company*

valves are normally associated with water heaters. The pressure-sensitive type will be the only one considered here.

The pressure relief valve can usually be recognized by its location. It can be found on all boilers at the high point on the boiler. ***It is a safety device that must be treated with the utmost respect.*** Some of these valves have levers on top that can be raised to check the valve for flow. Most of the valves have a visible spring that pushes the seat downward toward the system's pressure. A visual inspection will usually reveal that the principle of the valve is for the spring to hold the seat against the valve body. The valve body is connected to the system, so, in effect, the spring is holding the valve seat down against the system pressure. When the system pressure becomes greater than the spring tension, the valve opens and relieves the pressure and then reseats. ***The pressure setting on a relief valve is factory set and should never be tampered with. Normally the valve setting is sealed or marked in some manner so that it can be seen if it is changed.***

Most relief valves are automatically reset. After they relieve, they automatically seat back to the original position. A seeping or leaking relief valve should be replaced. ***Never adjust or plug the valve. Extreme danger could be encountered.***

14.16 WATER PRESSURE REGULATORS

Water pressure regulators are common devices used to control water pressure. Two types of valves are commonly used: the pressure-regulating valve for system pressure, and the pressure-regulating valve for head pressure control on water-cooled refrigeration systems. Air conditioning and refrigeration systems operate at lower pressures and temperatures when water cooled, but the water circuit needs a special kind of maintenance. At one time water-cooled equipment was widely used. Air-cooled equipment has now become the dominant type because it is easy to maintain. Water valves, although still used on equipment, are therefore not as common as they were on air conditioning and refrigeration equipment.

Water pressure-regulating valves are used in two basic applications:

1. They reduce supply water pressure to the operating pressure in a hot water heating system. This type of valve has an adjustment screw on top that increases the tension on the spring regulating the pressure, Figure 14–49. A great many boilers in hydronic systems in homes or businesses use circulating hot water at about 15 psig. The supply water may have a working pressure of 75 psig or more and have to be reduced to the system working pressure. If the supply pressure were to be allowed into the system, the boiler pressure relief valve would open. The water-regulating valve is installed in the supply water makeup line that adds water to the system when some leaks out. Most water-regulating valves have a manual valve arrangement that allows the service technician to remove the valve from the system and service it without stopping the system. A manual feedline is also furnished with most systems to allow the system to be filled by bypassing the water-regulating valve. ***Care must be taken not to overfill the system.***
2. The water-regulating valve controls water flow to the condenser on water-cooled equipment for head pressure control. This valve takes a pressure reading from the high-pressure side of the system and uses this to control the water flow to establish a predetermined head pressure, Figure 14–50. For example, if an ice maker were to be installed in a restaurant where the noise of the fan

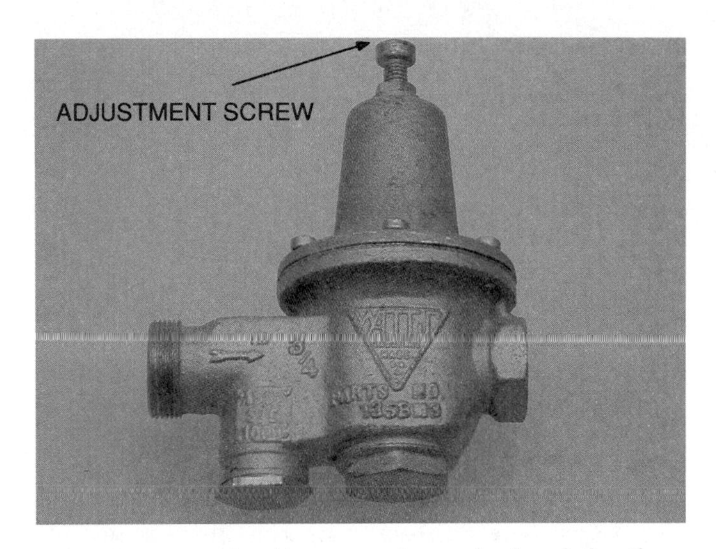

FIGURE 14–49 Adjustable water-regulating valve for a boiler. *Photo by Bill Johnson*

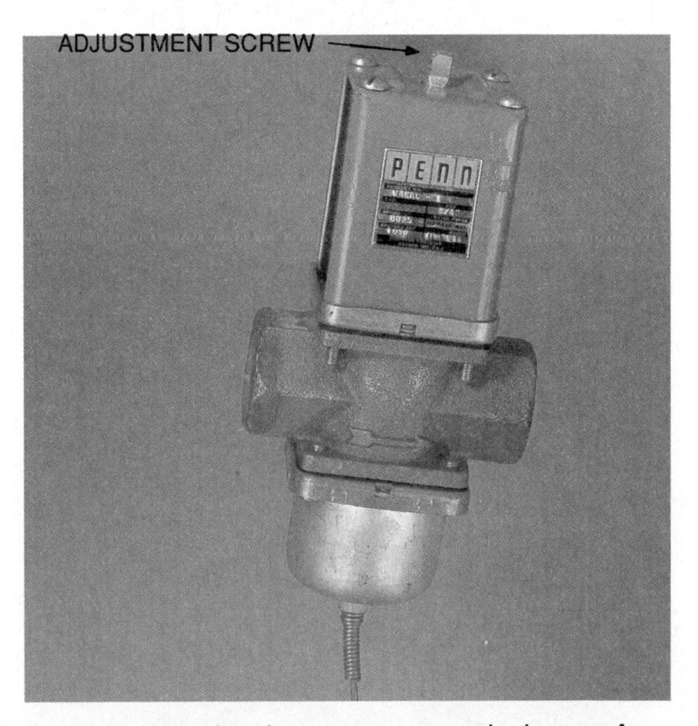

FIGURE 14–50 This valve maintains a constant head pressure for a water-cooled system during changing water temperatures and pressures. *Photo by Bill Johnson*

would be objectionable, a water-cooled ice maker may be used. This might be a "wastewater type of system" where the cooling water is allowed to go down the drain. In the winter the water may be very cool, and in the summer the water may get warm. In other words, the ice maker may have a winter need and a summer need. The water-regulating valve would modulate the water in both cases to maintain the required head pressure. There is an added benefit because when the system is off for any reason, the head pressure is reduced and the water flow is stopped until needed again. All of this is accomplished without an electric solenoid.

14.17 GAS PRESSURE REGULATORS

The gas pressure regulator is used in all gas systems to reduce the gas transmission pressure to usable pressure at the burner. The gas pressure at the street in a natural gas system could be 5 psig, and the burners are manufactured to burn gas at 3.5 in. of water column (this is a pressure-measuring system that indicates how high the gas pressure can push a column of water as the indicator of pressure). ***This pressure must be reduced to the burner's design, and this is done with a pressure-reducing valve that acts like the water pressure-regulating valve in the boiler system, Figure 14–51.*** There is an adjustment on top of the valve for qualified personnel to adjust the valve.

When bottled gas is used, the tank can have as much as 150 psig pressure to be reduced to the burner design pressure of 11 in. of water column. The regulator is normally located at the tank for this pressure reduction, Figure 14–52.

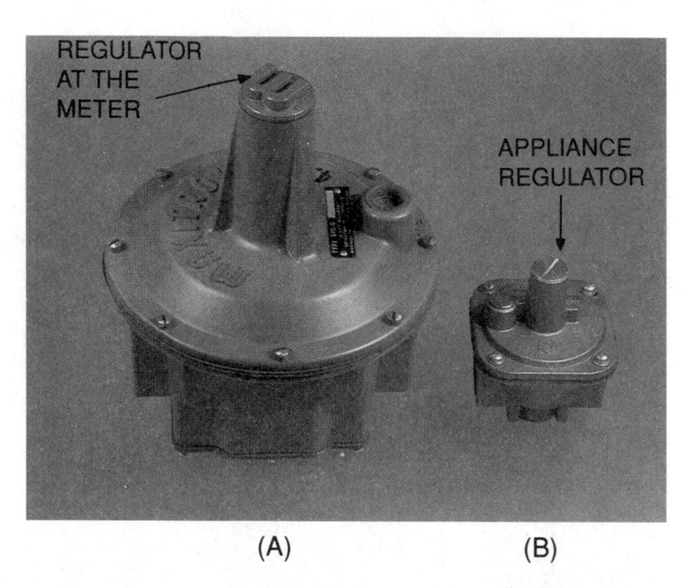

FIGURE 14–51 Gas pressure regulators. (A) Gas pressure regulator at the meter. (B) Gas pressure regulator at the appliance. *Photos by Bill Johnson*

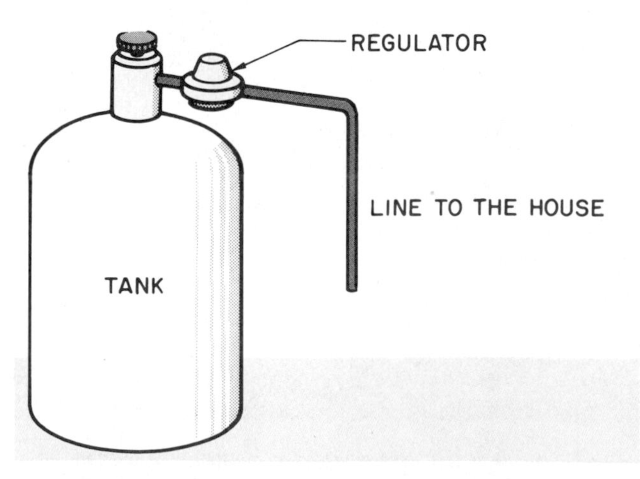

FIGURE 14–52 Gas pressure regulator on a bottled-gas system. This regulator is at the tank.

14.18 MECHANICAL CONTROLS

An example of a mechanical control is a water pressure-regulating valve. This valve is used to maintain a preset water pressure in a boiler circuit and has no electrical contacts or connections, Figure 14–49. It acts independently from any of the other controls, yet the system depends on it as part of the team of controls that make the system function trouble free. The technician must be able to recognize and understand the function of each mechanical control. Doing this is more difficult than for electrical controls because the control diagram does not always describe mechanical controls as well as it does electrical controls.

14.19 ELECTROMECHANICAL CONTROLS

Electromechanical controls convert a mechanical movement into some type of electrical activity. A high-pressure switch is an example of an electromechanical control. The switch contacts are the electrical part, and the bellows or diaphragm is the mechanical part. The mechanical action of the bellows is transferred to the switch to stop a motor when high pressures occur, Figure 14–53. Electromechanical con-

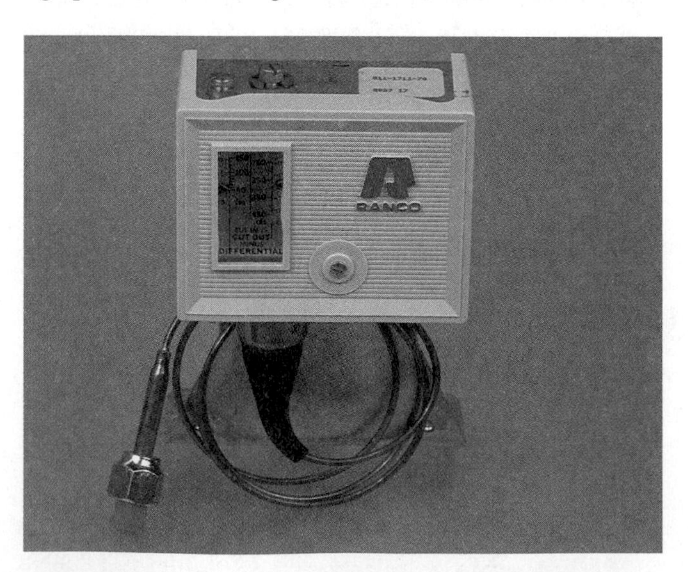

FIGURE 14–53 High-pressure control. It is considered an electromechanical control. *Photo by Bill Johnson*

PRESSURE AND VACUUM SWITCHES		LIQUID LEVEL SWITCHES	
N.O.	N.C.	N.O.	N.C.
FLOW SWITCH (AIR, WATER, ETC)		TIMER CONTACTS ENERGIZED COIL	
N.O.	N.C.	N.O.T.C.	N.C.T.O.

FIGURE 14–54 Table of electromechanical control symbols.

trols normally appear on the electrical print with a symbol adjacent to the electrical contact describing what the control does, Figure 14–54. These symbols are supposed to be standard; however, old equipment, installed long before standardization was considered, is still in service. In some instances you need imagination and experience to understand the intent of the manufacturer.

14.20 MAINTENANCE OF MECHANICAL CONTROLS

Mechanical controls are used to control the flow of fluids such as water, refrigerant liquids and vapors, or natural gas. These fluids flowing through the controls must be contained within the system. The controls typically are designed with diaphragms, bellows, and gaskets that are subject to leakage after being used for long periods of time. A visual check is always wise any time you are near the controls for service or maintenance. Look for any signs of a leak or physical change.

Water is one of the most difficult substances to contain and is likely to leak through the smallest opening. All water-regulating valves should be inspected for leaks by looking for wet spots or rust streaks. Water circulating in a system will also leave mineral deposits in the piping and valve sets or mechanisms. For example, a water-regulating valve for a hot water system, Figure 14–49, is subject to problems that may cause water leaks or control set point drift. Some of these valves are made from brass to prevent corrosion and some are made of cast iron. The cast iron ones are less expensive but may rust and bind the moving parts. The valve also has a flexible diaphragm that moves every time the valve functions. This diaphragm is subject to deterioration from flexing and age. If the diaphragm leaks, water will escape to the outside and drip on the floor. Mineral deposits or rust inside a valve can prevent the valve from moving up and down to feed water. This might overfeed or underfeed the system.

Pressure relief valves on boilers are constructed of material that will not corrode because this control is a safety control. These valves must be rust free to ensure that they will not stick shut and not function. An explosion may occur because of a valve that is stuck shut. Many technicians pull the lever on the top of safety relief valves from time to time and let them relieve a small amount of steam or hot water. This may be considered a good practice because it ensures that the valve port is free from deposits that may stop it from functioning. On some occasions, the valve may not seat back when the lever is released and the valve may seep. Usually pulling the lever again will clear this up by blowing out any trash in the valve seat. If it cannot be stopped from leaking, the valve must be replaced. The fact that this procedure may start a leak is the reason many technicians do not "test relieve" valves.

Pressure gages and boiler valves that may be in contact with the system water or steam must be checked from time to time to make sure they are functioning. Technicians develop their own tests for each system. For example, the pressure-regulating valve may be checked by allowing some water to escape from the system and then observing as the regulating valve refills the system to the correct pressure.

14.21 MAINTENANCE OF ELECTROMECHANICAL CONTROLS

Electromechanical controls have both a mechanical and an electrical action. Many of the same procedures used in the maintenance of the mechanical controls should be followed when water is involved. Inspect for leaks. Electromechanical pressure controls often are connected to the system with small tubing similar to capillary tubes used as metering devices. This tube is usually copper and is the cause for many leaks in refrigeration systems because of misapplication or poor installation practices. For example, a low- or high-pressure control may be mounted close to a compressor and the small control tube routed to the compressor to sense low or high pressure. The control must be mounted securely to the frame and the control tube routed in such a manner that it does not touch the frame or any other component. The vibration of the compressor will vibrate the tube and rub a hole in the control tube or one of the refrigerant lines if they are not kept isolated.

The electrical section of the control will usually have a set of contacts or a mercury switch to stop and start some component in the system. The electrical contacts are often enclosed and cannot be viewed so visual inspection is not possible. You should look for frayed wires or burned wire insulation adjacent to the control when a problem occurs. For example, if the sealed contacts inside a control are becoming burned, the wire and the control terminal have likely been hot and will have burned the wire's insulation.

The mercury in mercury switches may be viewed through the clear glass enclosure. If the mercury becomes dark, the tube is allowing oxygen to enter the space where the mercury makes and breaks the electrical contact and oxidation occurs. In this case the switch should be replaced.

It is possible that the switch will conduct across the oxidation and the switch will then appear to be closed at all times. This could prevent a boiler from shutting off, allowing overheating to occur.

14.22 SERVICE TECHNICIAN CALLS

SERVICE CALL 1

A customer calls indicating that the boiler in the equipment room at a motel has hot water running out and down the drain all the time. Another service company has been performing service at the motel for the last few months. *The problem is the water-regulating valve (boiler water feed) is out of adjustment. Water is seeping from the boiler's pressure relief valve, Figure 14–55.* The technician arrives, goes to the boiler room, and notices as the customer said, water is seeping out the relief valve pipe that terminates in the drain. This is heated water, causing a very inefficient situation. The technician looks at the temperature and pressure on the boiler gage. The needle is in the red; the pressure is 30 psig. No wonder the valve is relieving the pressure; it is too high. The boiler is hot, but the technician cannot tell if it is too hot. The burner is not operating, so the technician decides that this may be a pressure problem instead of a temperature problem.

The technician looks for and finds the water pressure-regulating valve, which is in the supply water line entering the boiler. Actually, this valve keeps the supply water pressure from reaching the boiler and feeds water into the system when there is a loss of water due to a leak. The supply water pressure could easily be 75 psig or more. The boiler working pressure is 30 psig with a relief valve setting of 30 psig. If the supply water were allowed directly into the boiler, it would push the pressure past the design pressure of the system. The design pressure of the system must be less than 30 psig. If the pressure were to be above this, it would cause the pressure relief valve to relieve the pressure and dump water from the system until the pressure was lowered.

The technician shuts off the water valve leading to the boiler makeup system and bleeds some water from the system until the pressure at the boiler is down to 15 psig and then adjusts the water-regulating valve for a lower pressure. The supply water valve is then opened and water is allowed to enter the system. The technician can hear water entering the system through the water-regulating valve. In a few minutes, the water stops entering. The pressure in the system is now adjusted to 18 psig and all is well.

The problem in this example is one in which a mechanical automatic control is causing another control to show that the system has problems. If you did not examine the problem, you might think that the relief valve is defective because water is seeping out of it. This is not the case; the relief valve is doing its job. The water-regulating valve is at fault.

The technician shows the motel manager what the problem was and completes the paperwork before leaving for another call.

SERVICE CALL 2

A customer in a service station calls and says the hot water boiler in the equipment room is blowing water out from time to time. A technician is quickly dispatched to the job because of the potential hazard. *The problem is the gas valve will not close. The boiler is overheating. The relief valve is relieving water periodically.* The season is mild, so the boiler has a small load on it.

Because the relief valve is relieving intermittently, you may believe there is a periodic pressure buildup in the boiler. If the problem were the same as in Service Call 1, where a constant, bleeding relief occurred, the water-regulating valve or the relief valve might be the suspect. This is not the case.

The technician arrives and goes straight to the boiler and reads the gage; it is in the red and about to relieve again. The technician then looks at the flame in the burner section and notices that it is not

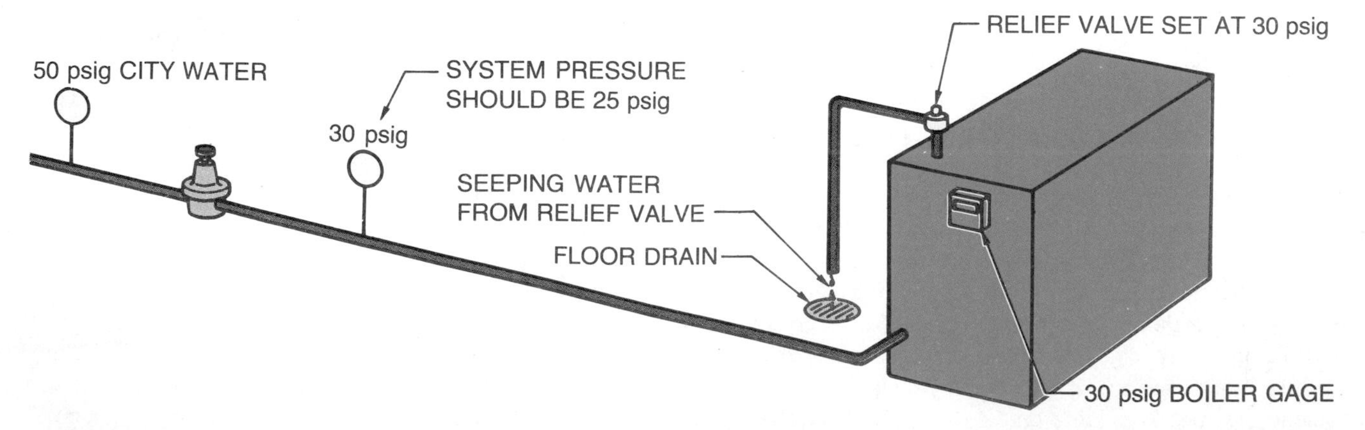

FIGURE 14–55 This illustration indicates how one control can cause another control to look defective. The water pressure-regulating valve adjustment has been changed by mistake, allowing more than operating pressure into the system. The water pressure-regulating valve is designed to regulate the water pressure down to 25 psig. This valve is necessary to allow water to automatically keep the system full should a water loss develop due to a leak.

a full flame; it is small. It is decided that the gas valve may be stuck open, allowing fuel to enter the burner all the time.

The inline gas hand valve is shut off to put out the flame so the technician can discover the problem.

A check with a voltmeter shows no voltage at the gas valve terminals, so the thermostat is trying to shut the gas off. The valve must be stuck open.

The technician obtains a new gas valve and installs it. The boiler is started up and allowed to run. The boiler is cold after being off for some time for the gas valve change so it takes awhile for it to get up to temperature. The set point of the thermostat is 190°F and the gas valve shuts off when this temperature is reached.

These service calls are examples of how a trained technician might approach these service problems. Technicians develop their own methods of approach to problems. The point is that no diagram was consulted. A technician either must have prior knowledge of the system or must be able to look at the system and make accurate deductions based on a knowledge of basic principles.

Each control has distinguishing features to give it a purpose. There are hundreds of controls and dozens of manufacturers (some of whom are out of business). If you find a control with no information as to what it does, consult the manufacturer. If this is impossible, try to find someone who has experience with the equipment. In general, when the control is examined in the proper perspective, the design parameters will help. For example: What is the maximum temperature or pressure that is practical? What is the minimum temperature or pressure that is practical? Ask yourself these questions when you run across a strange control.

The following service calls do not have the solution described. The solution can be found in the Instructor's Guide.

SERVICE CALL 3

A residential customer calls and reports that the heating system was heating until early this morning but for no apparent reason it stopped. The furnace seems to be hot, but no heat is coming out the air outlets. This is a gas furnace located in the basement.

The technician talks to the home owner about the problem and from the description believes the problem may be with the fan. The technician first checks to see if there is power at the furnace and notices that although the furnace is hot to the touch, no heat is coming out of the duct. Further investigation shows that the fan is not running.

The technician knows there is power to the furnace because it is hot and the burner ignites from time to time. The problem must be in the fan circuit.

A voltage check shows that power is leaving the fan switch going to the fan motor, but it is still not running. See the diagram in Figure 14–56 for direction.

What is the problem and the recommended solution?

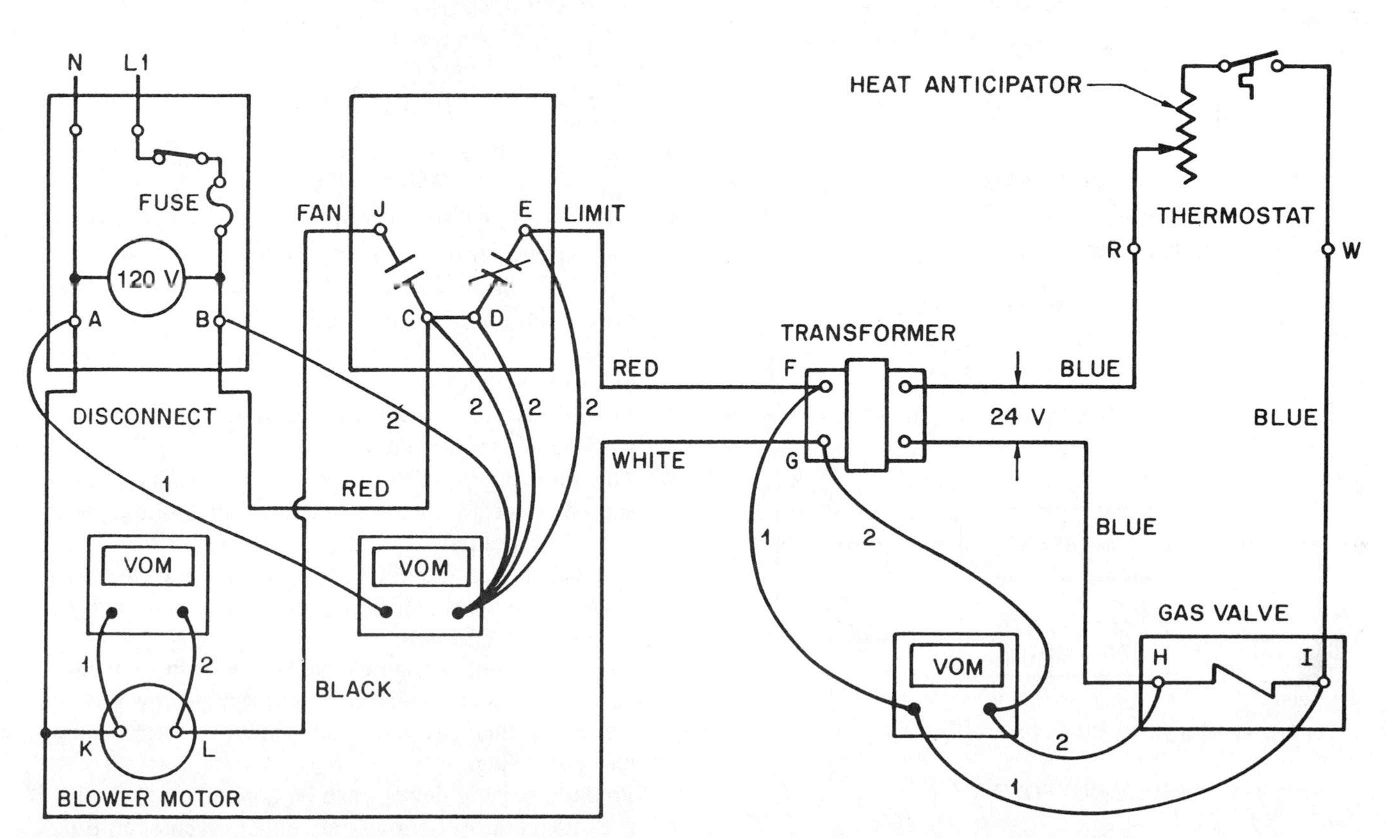

FIGURE 14–56 This diagram can be used for Service Call 3.

SERVICE CALL 4

A customer calls and says the central air conditioning system in their small office building will not cool.

The service technician arrives and turns the thermostat to COOL and turns the thermostat indicator to call for cooling. The indoor fan motor starts. This indicates that the control voltage is operating and will send power on as requested for cooling. Only recirculated air is coming out of the ducts.

The technician checks the outside unit and notices that it is not running.

The control panel cover is removed to see what controls will actually prevent the unit from starting. This will give some indication as to what the possible problem may be. Use the diagram in Figure 14–57.

Pressure gages are applied to determine the system pressures. Most air conditioning systems use R-22. If the system were standing at 80°F, the system should have 144 psig of pressure, according to the pressure/temperature chart for R-22 at 80°F.

NOTE: Care should be used in the standing pressure comparison. Part of the unit is inside, at 75°F air, and part of the unit is outside. Some of the liquid refrigerant will go to the coolest place if the system valves will allow it to.

The gage readings may correspond to a temperature between the indoor and outdoor temperature but not less than the cooler temperature. Many modern systems use fixed-opening metering devices that allow the system pressures to equalize and the refrigerant to migrate to the coolest place in the system.

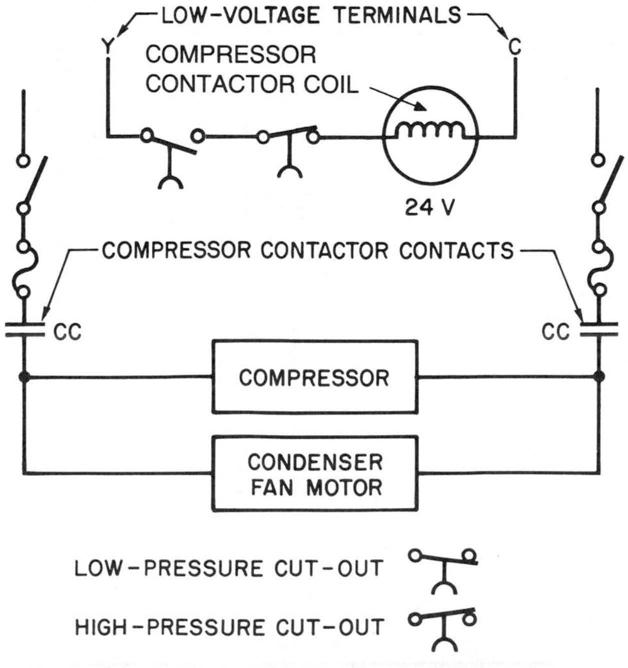

FIGURE 14–57 This diagram can be used for Service Call 4.

In this example the pressure in the system is 60 psig, the refrigerant is R-22, and the temperature is about 80°F. A typical pressure control setup may call for the control to interrupt the compressor when the system pressure gets down to 20 psig and to make the control and start the compressor when the pressure rises back to a temperature corresponding to 70°F.

What is the problem and the recommended solution?

The foregoing examples are used as typical service problems to show how a service technician might arrive at the correct diagnosis, using the information at hand.

Always save and study the manufacturer's literature. There is no substitute for having access to manufacturer's specifications.

SUMMARY

- Temperature controls are either operating controls or safety controls.
- Space temperature controls can either be low-voltage or line-voltage types.
- Low voltage is normally applied to residential heating and cooling controls.
- Some temperature control devices may have the electric contacts enclosed in a bulb and use mercury as the contact surface, or they may have open contacts.
- Heating thermostats normally have a heat anticipator circuit in series with the thermostat contacts.
- Cooling thermostats may have a cooling anticipator in parallel with the thermostat contacts.
- Line-voltage thermostats are normally rated up to 20 A of current because they are used to switch high-voltage current.
- Low-voltage thermostats normally will carry only 2 A of current.
- To get a correct temperature reading of a flat or round surface, the sensing element (either mercury bulb, remote bulb, or bimetal) must be in good contact with the surface and insulated from the ambient air.
- Some installations have wells in the substance to be sensed in which the sensor can be placed.
- Motors have both internal and external types of motor temperature-sensing devices.
- The internal type of motor temperature-sensing device can be either a bimetal or a thermistor inserted inside the motor windings.
- Some sensing elements are inserted into the fluid stream. Examples are the fan or limit switch on a gas, oil, or electric furnace.
- All temperature-sensing elements change in some manner with a change in temperature: the bimetal warps, the thermistor changes resistance, and the thermocouple changes voltage.
- Pressure-sensing devices can be applied to all fluids.
- Pressure controls normally are either diaphragm, Bourdon tube, or bellows operated.

- Pressure controls can operate at high pressures, low pressures, even below atmospheric pressure (in a vacuum), and can detect differential pressures.
- Pressure-sensitive controls can control fluid flow. These are sometimes modulating-type controls.
- ***Gas pressure must be reduced before it enters a house or place of business to be burned in an appliance.***
- Mechanical controls perform mechanical functions without electricity.
- Electromechanical controls have both electrical and mechanical functions.
- Some purely mechanical controls are water-regulating valves, pressure relief valves, and expansion valves.
- Some electromechanical controls are low-pressure cutout, high-pressure cutout, and thermostats.
- Keep and study manufacturer's literature.

REVIEW QUESTIONS

1. Why is it important to be able to recognize the various controls?
2. Why is a low-voltage thermostat normally more accurate than a high-voltage thermostat?
3. Name three kinds of switching mechanisms in a low-voltage thermostat.
4. What is an inert gas?
5. What does a heat anticipator do?
6. What does a cold anticipator do?
7. Describe how a bimetal functions.
8. In a residential system what voltage is considered low voltage?
9. What component steps down the voltage to the low-voltage value?
10. Name four reasons why low voltage is desirable for residential control voltage.
11. What is the maximum amperage usually encountered by a low-voltage thermostat?
12. A heating thermostat ______ on a rise in temperature.
13. A cooling thermostat ______ on a rise in temperature.
14. What two types of switches are normally found in the subbase of a low-voltage thermostat?
15. When is a line-voltage thermostat used?
16. Name three sensing elements that can be used with a line-voltage thermostat?
17. What is the maximum amperage generally encountered in a line-voltage thermostat?
18. Why is a line-voltage thermostat mounted on an electrical box that has conduit connected to it?
19. What method is used to get a mercury thermometer to indicate accurately on a flat or round surface?
20. Why do motors build up heat?
21. How is motor heat dissipated?
22. What are the two types of motor temperature-sensing devices?
23. What is the main precaution for an externally mounted motor temperature protector?
24. What is the principal of operation of most externally mounted overload protection devices?
25. Name a method to speed up the cooling of an open motor.
26. Describe two methods to cool an overheated compressor.
27. How can an electronic thermometer be used in a well to obtain an accurate response?
28. How can the temperature be checked in a water circuit if no temperature well is provided?
29. Name a temperature-sensing device commonly used in measuring airflow.
30. How does the thermistor change with a temperature change?
31. How does a thermocouple change with a temperature change?
32. What is a pressure-sensitive device?
33. Name two methods used to convert pressure changes into action.
34. Name two actions that can be obtained with a pressure change.
35. Can pressures below atmospheric pressure be detected?
36. Name one function of the low-pressure control.
37. Name one function of the high-pressure control.
38. Name two types of water pressure control.
39. Why should a safety control not be adjusted if it has a seal to prevent tampering?
40. How can you find information about a control when there is no description with the unit?

UNIT 15 Troubleshooting Basic Controls

OBJECTIVES

After studying this unit, you should be able to

- descibe and identify power- and non–power-consuming devices.
- describe how a voltmeter is used to troubleshoot electrical circuits.
- identify some typical problems in an electrical circuit.
- describe how an ammeter is used to troubleshoot an electrical circuit.
- recognize the components in a heat-cool electrical circuit.
- follow the sequence of electrical events in a heat-cool electrical circuit.
- differentiate between a pictorial and a line-type electrical wiring diagram.

SAFETY CHECKLIST

* When troubleshooting electrical malfunctions disconnect electrical power unless power is necessary to make appropriate checks. Lock and tag the panel where the disconnect is made and keep the only available key on your person.
* When checking continuity or resistance, turn off power and disconnect at least one lead from component being checked.
* Ensure that meter probes come in contact only with terminals or other contacts intended.

INTRODUCTION TO TROUBLESHOOTING

Each control must be evaluated as to its function (major or minor) in the system. Recognizing the control and its purpose requires that you understand its use in the system. Studying the purpose of a control before you take action will save a great deal of troubleshooting time. Look for pressure lines going to pressure controls and temperature-activated elements on temperature controls. See if the control stops and starts a motor, opens and closes a valve, or provides some other function. As mentioned previously, controls are either electrical, mechanical, or a combination of electrical and mechanical. (Electronic controls are considered the same as electrical controls for now.)

Electrical devices can be considered as power-consuming or non–power-consuming as far as their function in the circuit is concerned. Power-consuming devices use power and have either a magnetic coil or a resistance circuit. They are wired in parallel with the power source. Non–power-consuming devices pass power.

These two terms can be demonstrated with a simple light bulb circuit with a switch. The light bulb in the circuit actually consumes the power, and the switch passes the power to the light bulb. The object is to wire the light bulb to both sides (hot and neutral) of the power supply, Figure 15–1. This will then ensure that the light bulb is in parallel with the power supply, and a complete circuit will be established. The switch is a power-passing device and is wired in series with the light bulb. For any power-consuming device

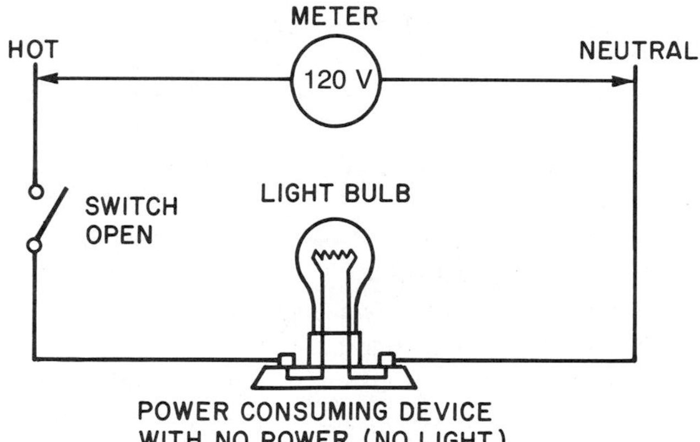

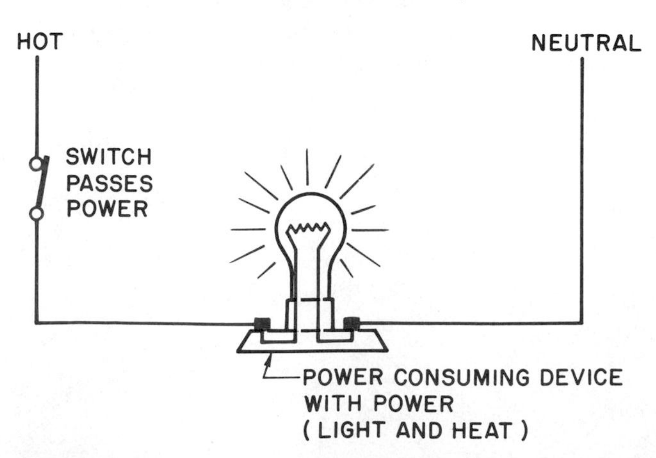

FIGURE 15–1 Power-consuming device (light bulb) and non–power-consuming device (switch).

to consume power it must have a potential voltage. A potential voltage is the voltage indicated on a voltmeter between two power legs (such as line 1 to line 2) of a power supply. This can be thought of as electrical pressure in a home; for example, the light bulb has to be wired from the hot leg (the wire that has a fuse or breaker in it) to the neutral leg (the grounded wire that actually is wired to the earth ground).

The following statements may help you understand electron flow in an electrical circuit. Some liberty is used in explaining the electrical circuits in this unit, but these liberties should help your understanding.

1. It doesn't make any difference which way the electrons are flowing in an alternating current (AC) circuit when the object is to get the electrons (current) to a power-consuming device and complete the circuit.
2. The electrons may pass through many power-passing devices before getting to the power-consuming device(s).
3. Devices that don't consume power pass power.
4. Devices that pass power can be either safety devices or operating devices.

Suppose the light bulb mentioned above was to be used to add heat to a well pump to keep it from freezing in the winter. To prevent the bulb from burning all the time, a thermostat is installed. A fuse must also be installed in the line to protect the circuit, and a switch must be wired to allow the circuit to be serviced, Figure 15–2. Note that the power now must pass through three power-passing devices to reach the light bulb. The fuse is a safety control, and the thermostat turns the light bulb on and off. The switch is not a control but a service convenience device. Notice that all switches are on one side of the circuit. Let's suppose that the light bulb does not light up when it is supposed to. Which component is interrupting the power to the bulb: the switch, the fuse, or the thermostat? Or is the light bulb itself at fault? In this example the thermostat (switch) is open.

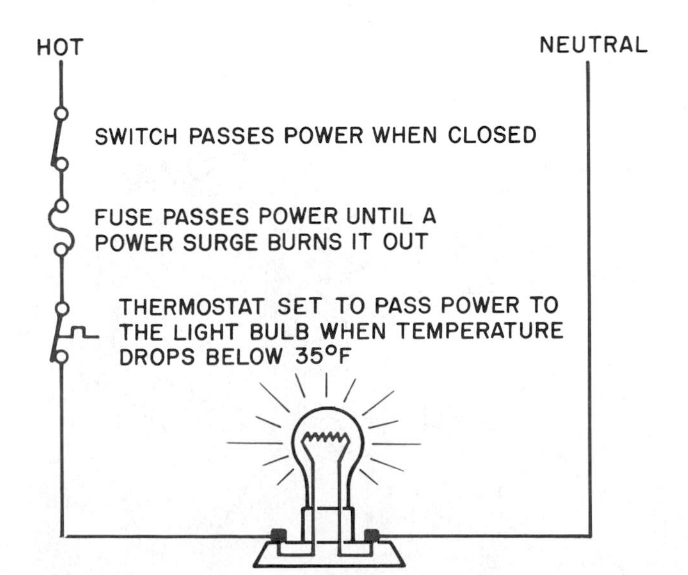

FIGURE 15–2 When closed, the switch passes power to the fuse.

15.2 TROUBLESHOOTING A SIMPLE CIRCUIT

To troubleshoot the circuit in the previous paragraph, a voltmeter may be used in the following manner. Turn the voltmeter selector switch to a voltage setting higher than the voltage supply. In this case the supply voltage should be 120 V, so the 250-V scale is a good choice. Follow the diagram in Figure 15–3 as you read the following.

1. Place the red lead of the voltmeter on the hot line and the black lead on the neutral. The meter will read 120 V.
2. Place the red lead, the lead being used to find and detect power in the hot line, on the load side of the switch. (The "load" side of the switch is the side of the switch that the load is connected to. The other side of the switch, where the line is connected, is the "line" side of the switch.) The black lead should remain in contact with the neutral line. The meter will read 120 V.
3. Place the red lead on the line side of the fuse. The meter will read 120 V.
4. Place the red lead on the load side of the fuse. The meter will read 120 V.
5. Place the red lead on the line side of the thermostat. The meter will read 120 V.

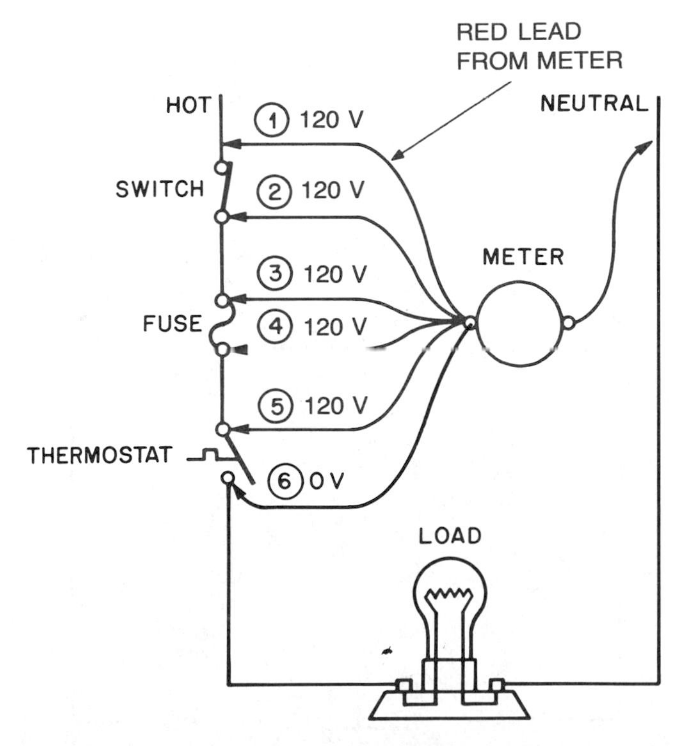

FIGURE 15–3 The troubleshooting procedure is to establish the main power supply of 120 V from the hot line to the neutral line. When this power supply is verified, the lead on the hot side is moved down the circuit toward the light bulb. The voltage is established at all points. When point 6 is reached, no voltage is present. The thermostat contacts are open.

6. Place the red lead on the load side of the thermostat. The meter will read 0 V. There is no power available to energize the bulb. The thermostat contacts are open. Now ask the question, is the room cold enough to cause the thermostat contacts to make? If the room temperature is below 35°F, the circuit should be closed; the contacts should be made.
NOTE: The red lead was the only one moved. It is important to note that if the meter had read 120 V at the light bulb connection when the red lead was moved to this point, then further tests should be made.
Another step is necessary to reach the final conclusion. Let's suppose that the thermostat is good and 120 V is indicated at point 6.
7. The red meter lead can now be moved to the terminal on the light bulb, Figure 15–4. Suppose it reads 120 V. Now, move the black lead to the light bulb terminal on the right. If there is no voltage, the neutral wire is open between the source and the bulb.
If there is voltage at the light bulb and it will not burn, the bulb is defective, Figure 15–5. *When there is a power supply and a path to flow through, current will flow.* The light bulb filament is a measurable resistance and should be the path for the current to flow through.

15.3 TROUBLESHOOTING A COMPLEX CIRCUIT

The following example is a progressive circuit that would be typical of a combination heating-cooling unit. This circuit is not standard because each manufacturer may design its circuits differently. They may vary, but will be similar in nature.

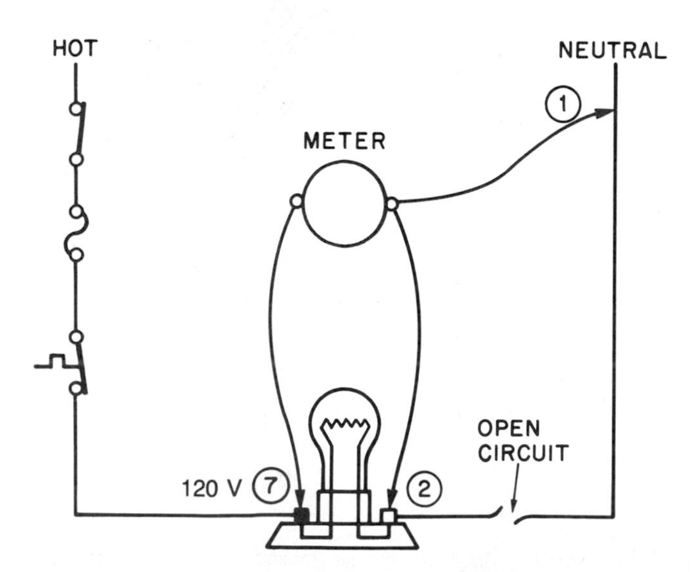

FIGURE 15–4 Thermostat contacts are closed, and there is a measurement of 120 V when one lead is on the neutral and one lead on the light bulb terminal (see position 7). When the black lead is moved to the light bulb in position 2, there is no voltage reading. The neutral line is open.

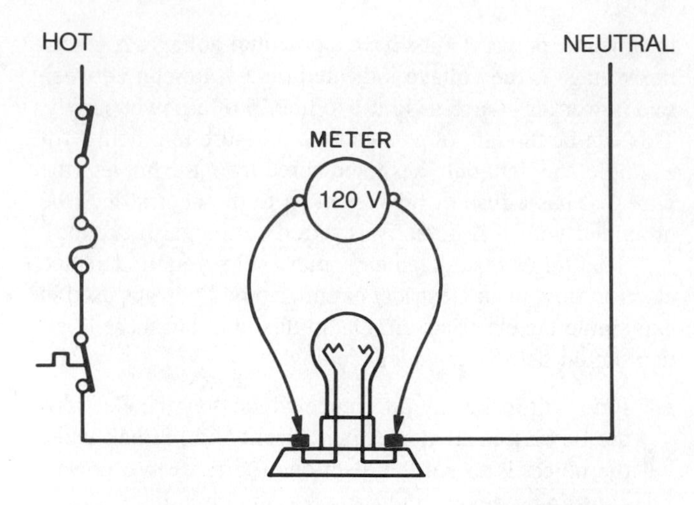

FIGURE 15–5 Power is available at the light bulb. The hot line on the left side and the neutral on the right side complete the circuit. The light bulb filament is burned out, so there is no path through the bulb.

The unit in this example is a package air conditioner with 1 1/2 tons cooling and 5 kW of electric heat. This unit resembles a window air conditioner because all of the components are in the same cabinet. The unit can be installed through a wall or on a roof, and the supply and return air can be ducted to the conditioned space. The reason for using this unit is that it comes in many sizes, from 1 1/2 tons of cooling (18,000 Btu/h) to very large systems. The unit is popular with shopping centers because a store could have several roof units to give good zone control. If one unit were off, other units would help to hold the heating-cooling conditions. The unit also has all of the control components, except the room thermostat, within the unit's cabinet. The thermostat is mounted in the conditioned space. The unit can be serviced without disturbing the conditioned space, Figure 15–6.

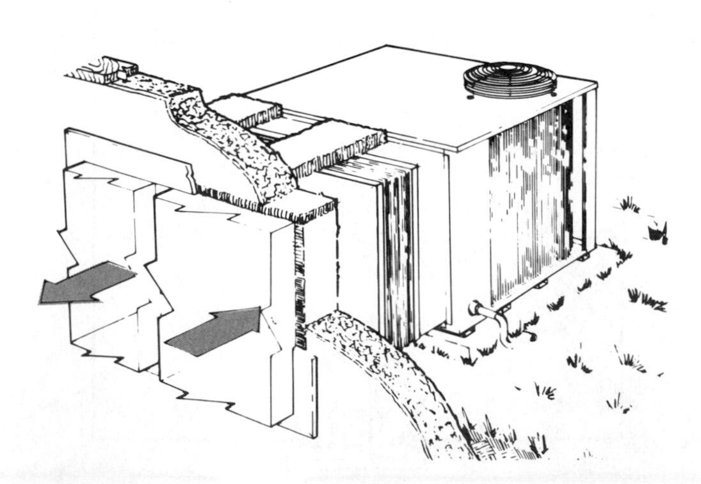

FIGURE 15–6 Small package air conditioner. *Courtesy Climate Control*

The first thing we will consider is the thermostat. The thermostat is not standard but is a version used for illustration purposes. This is a usable circuit designed to illustrate troubleshooting.

1. This thermostat is equipped with a selector switch for either HEAT or COOL. When the selector switch is in the HEAT position and heat is called for, the *electric heat relay* is energized. The fan must run in the heating cycle. It will be started and run in the high-voltage circuit. This is discussed later.
2. When the fan is switched to ON, the indoor fan will run all the time. This thermostat is equipped with a FAN ON or FAN AUTO position, so the fan can be switched on manually for air circulation when desired. When the fan switch is in the AUTO position, the fan will start on a call for cooling.
3. When the selector switch is in the cooling mode, the cooling system will start on a call for cooling. The indoor fan will start through the AUTO mode on the fan selector switch. The outdoor fan must run also, so it is wired in parallel with the compressor.

Figure 15–7 is the beginning of the control explanation. This is the simplest of thermostats, having only a set of heating contacts and a selector switch. Follow the power from the R terminal through the selector switch to the thermostat contacts and on to the W terminal. The W terminal destination is not universal but is common. When these contacts are closed, we can say that the thermostat is calling for heat. These contacts pass power to the heat relay coil. Power for the other side of the heat relay coil comes directly from the control transformer. When this coil is energized (24 V), the heat should come on. This coil is going to close a set of contacts in the high-voltage circuit to pass power to the electric heat element.

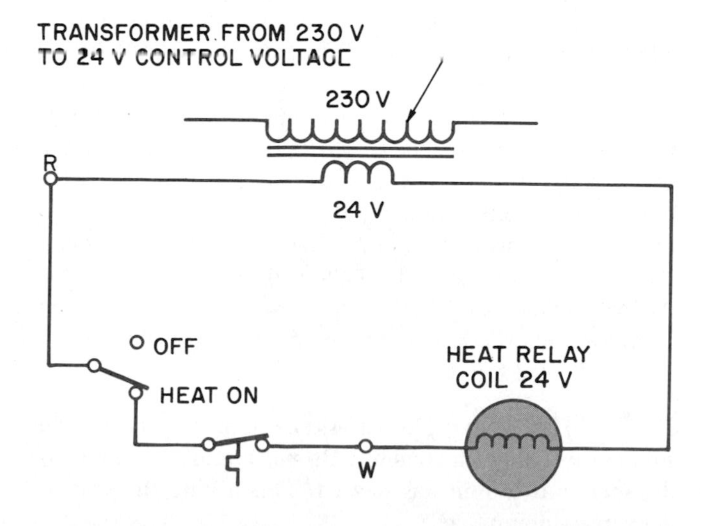

FIGURE 15–7 Simple thermostat with a HEAT OFF/HEAT ON position. There is no heat anticipator. The selector switch is closed, and the thermostat circuit is closed, calling for heating. When the heat relay is energized, the heating system starts.

In Figure 15–8 the heat anticipator is added to the circuit. The heat anticipator is in series with the heat relay coil and has current passing through it at any time the heat relay coil is energized. This current passing through the anticipator creates a small amount of heat in the vicinity of the thermostat bimetal sensing element. This causes the bimetal to break the thermostat contacts early to dissipate the heat in the heater. If the contacts to the heater did not open until the space temperature was actually up to temperature, the extra heat would overheat the space. The fan to move the heat is stopped and started in the high-voltage circuit. This is discussed later.

The heat anticipator is often a variable resistor and must be adjusted to the actual system. The current must be matched to the current used by the heat relay coil. All thermostat manufacturers explain how this is done in the installation instructions for the specific thermostat.

Figure 15–9 illustrates how a fan-starting circuit can be added to a thermostat. Notice the addition of the G terminal. It is used to start the indoor fan. The G terminal designation is not universal but is common. Follow the power from the R terminal through the fan selector switch to the G terminal and on to the indoor fan relay coil. Power is supplied to the other side of the coil directly from the control transformer. When this coil has power (24 V), it closes a set of contacts in the high-voltage circuit and starts the indoor fan.

Figure 15–10 completes the circuitry in this thermostat by adding a Y terminal for cooling. Again, this is not the only letter used for cooling, but it is also common. Follow the power from the R terminal down to the cool side of the

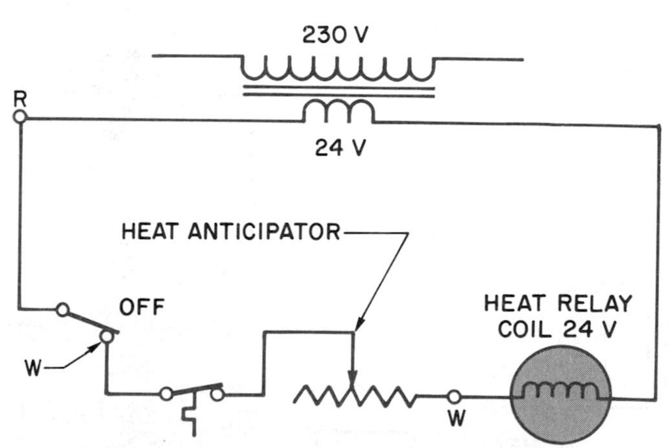

NOTE: THE HEAT ANTICIPATOR IS A VARIABLE RESISTANCE THAT IS DESIGNED TO CREATE A SMALL AMOUNT OF HEAT NEAR THE BIMETAL. THIS FOOLS THE THERMOSTAT INTO CUTTING OFF EARLY. THE FAN WILL CONTINUE TO RUN AND DISSIPATE THE HEAT REMAINING IN THE HEATERS.

FIGURE 15–8 Same diagram as Figure 15–7 with a heat anticipator in series with the thermostat contacts. The selector switch is set for heat, and the thermostat contacts are closed, calling for heat. The fan in this application will be operated with the high-voltage circuit when in the heating mode.

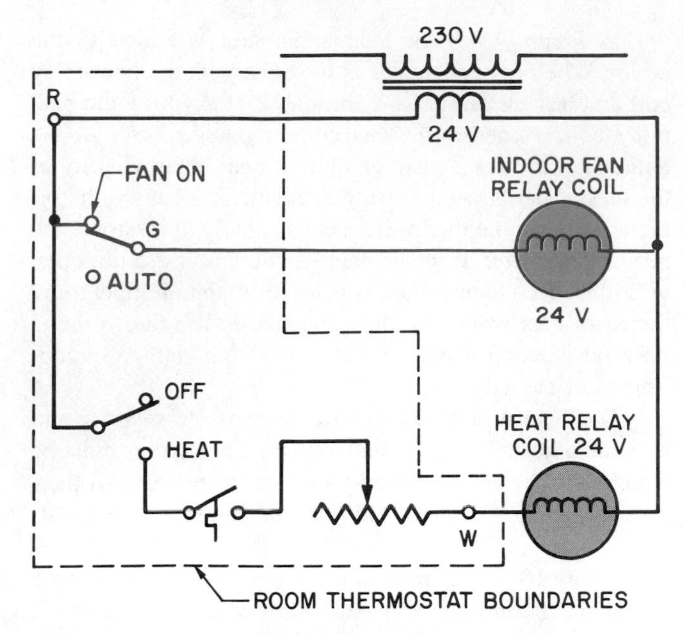

FIGURE 15–9 The addition of the fan relay to the thermostat allows the owner to switch the fan on for continuous operation.

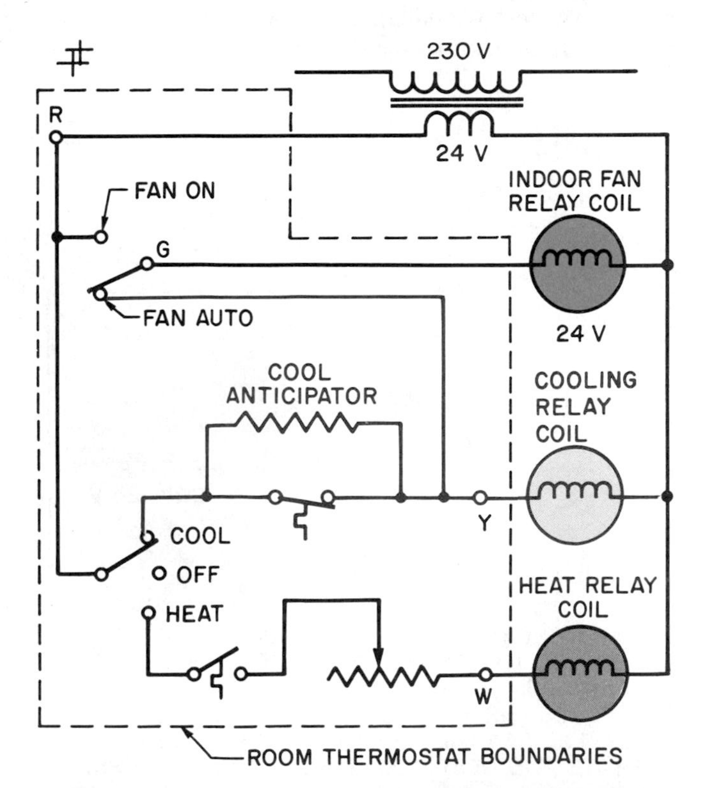

NOTE: THE CIRCUIT THROUGH FAN-AUTO ENSURES THE FAN WILL COME ON DURING COOLING.

FIGURE 15–10 Same diagram as Figure 15–9 with cooling added. The cooling circuit has a cooling anticipator in parallel with the cooling contacts, so current flows through it during the off cycle.

selector switch through to the contacts and on to the Y terminal. When these contacts are closed, power will pass through them and go two ways. One path will be through the fan AUTO switch to the G terminal and on to start the indoor fan. It has to run in the cooling cycle. The other way is straight to the cool relay. When the cool relay is energized (24 V), it closes a set of contacts in the high-voltage circuit to start the cooling cycle.

NOTE: The cooling anticipator is in parallel with the cooling contacts. This means that current will flow through this anticipator when the thermostat is satisfied (or when the thermostat's circuit is open). The cool anticipator is normally a fixed, nonadjustable type.

15.4 TROUBLESHOOTING THE THERMOSTAT

The thermostat is an often misunderstood, frequently suspected component during equipment malfunction. But it does a straightforward job of monitoring temperature and distributing the power leg of the 24-V circuit to the correct component to control the temperature. Service technicians should remind themselves as they approach the job that "one power leg enters the thermostat and it distributes power where called." Every technician needs a technique for checking the thermostat for circuit problems.

One way to troubleshoot a thermostat on a heating and cooling system is to first turn the fan selector switch to the FAN ON position and see if the fan starts. If the fan does not start, the problem may be with the control voltage. Control voltage must be present for the thermostat to operate. Assume for a moment that the thermostat will cause the fan to come on when turned to FAN ON but will not operate the heat or cooling cycles. The next step may be to take the thermostat off the subbase and jump the circuit out manually with an *insulated jumper.* Jump from R to G, and the fan should start. Jump from R to W, and the heat should come on. Jump from R to Y, and the cooling should come on. If the circuit can be made to operate without the thermostat but not with it, the thermostat is defective, Figure 15–11. One does not have to be afraid to jump these circuits out one at a time because only one leg of power comes to the thermostat. If all circuits were jumped out at one time in this thermostat, the only thing that would happen is that the heating and cooling would run at the same time. Of course, this should not be allowed to continue.

SAFETY PRECAUTION: *This should be done only under the supervision of the instructor. Do not restart the air conditioner until 5 min has passed. This allows the system pressures to equalize.*

We have just covered what happens in the basic thermostat. The next step is to move into the high-voltage circuit and see how the thermostat's actions actually control

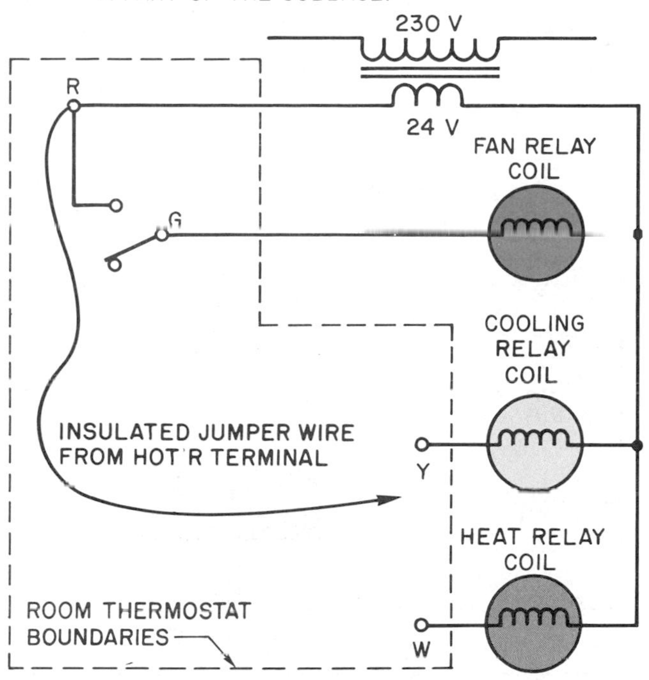

FIGURE 15–11 Terminal designation in a heat-cool thermostat subbase. The letters R, G, Y, and W are common designations. A simple jumper wire from the R terminal to G should start the fan. From R to Y should start the cooling; from R to W should start the heat.

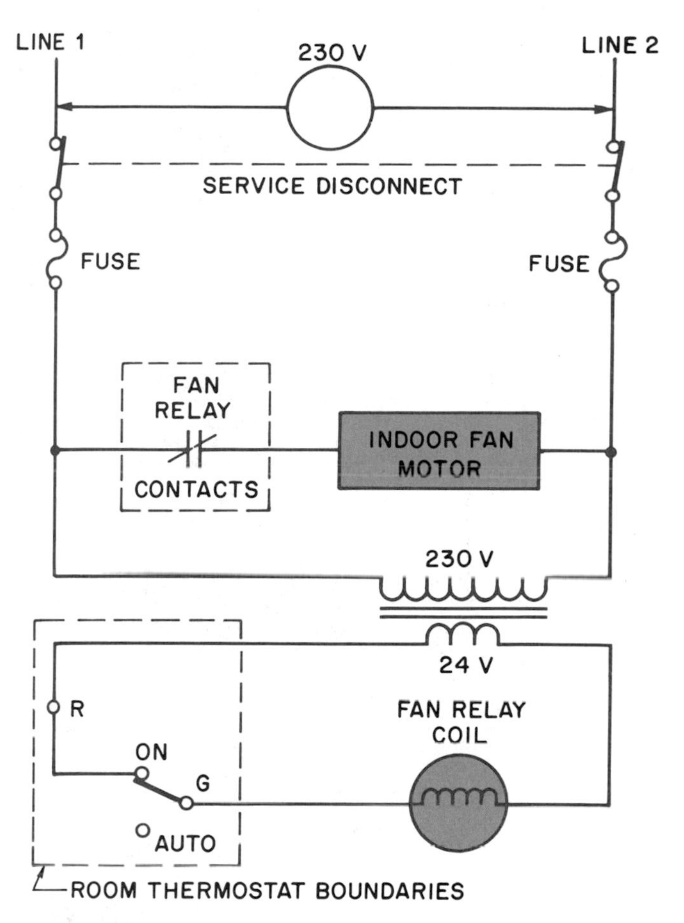

FIGURE 15–12 When the fan switch is turned to the ON position, a circuit is completed to the fan relay coil. When this coil is energized, it closes the fan relay contacts and passes power to the indoor fan motor.

the fan, cooling, and heating. The progressive circuit will again be used.

The first thing to remember is that the high voltage (230 V) is the input to the transformer. Without high voltage there is no low voltage. When the service disconnect is closed, the potential voltage between line 1 and line 2 is the power supply to the primary of the transformer. The primary input then induces power to the secondary.

Figure 15–12 illustrates the high-voltage operation of the fan circuit. The power-consuming device is the fan motor in the high-voltage circuit. The fan relay contact is in the line 1 circuit. It should be evident that the fan relay contacts have to be closed to pass power to the fan motor. These contacts close when the fan relay coil is energized in the low-voltage control circuit.

Figure 15–13 illustrates the addition of the electric heat element and the electric heat relay. The electric heat element and the fan motor are the power-consuming devices in the high-voltage circuit. In the low-voltage circuit when the HEAT-COOL selector switch is moved to the HEAT position, power is passed to the thermostat contacts. When there is a call for heat, these contacts close and pass power through the heat anticipator to the heat relay coil. This closes the two sets of contacts in the high-voltage circuit. One set starts the fan, and the other set passes power to the limit switch, and then through the limit switch to the line 1 side of the heater. The circuit is completed through the fuse link in line 2.

In Figure 15–14 cooling or air conditioning has been added to the circuit. For this illustration the heating circuit was removed to reduce clutter and confusion. Three components have to operate in the cooling mode: the indoor fan, the compressor, and the outdoor fan. The compressor and the outdoor fan are wired in parallel and, for practical purposes, can be thought of as one component.

The line 2 side of the circuit goes directly to the three power-consuming components. Power for the line 1 side of the circuit comes through two different relays. The cooling relay contacts start the compressor and outdoor fan motor; the indoor fan relay starts the indoor fan motor. In both cases, the relays pass power to the components on a call from the thermostat.

When the thermostat selector switch is set to the COOL position, power is passed to the contacts in the thermostat. When these contacts are closed, power passes to the cooling relay coil. When the cooling relay coil is energized, the cooling relay contacts in the high-voltage circuit are closed. This passes power to the compressor and outdoor fan motor.

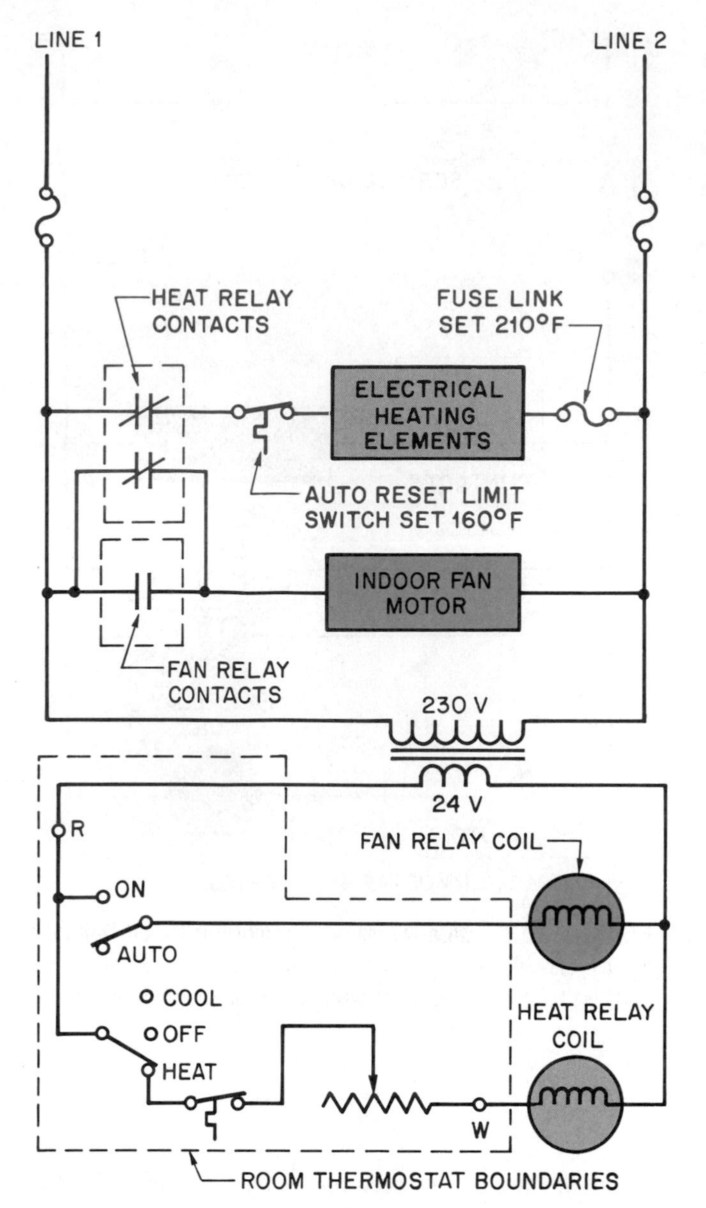

FIGURE 15–13 When the selector switch is in the HEAT position and the thermostat contacts close, the heat relay coil is energized. This action closes two sets of contacts in the heat relay. One set starts the fan because they are wired in parallel with the fan relay contacts. The other set passes power to the auto reset limit switch (set at 160°F). Power reaches the other side of the heat element through the fuse link. When power reaches both sides of the element, the heating system is functioning.

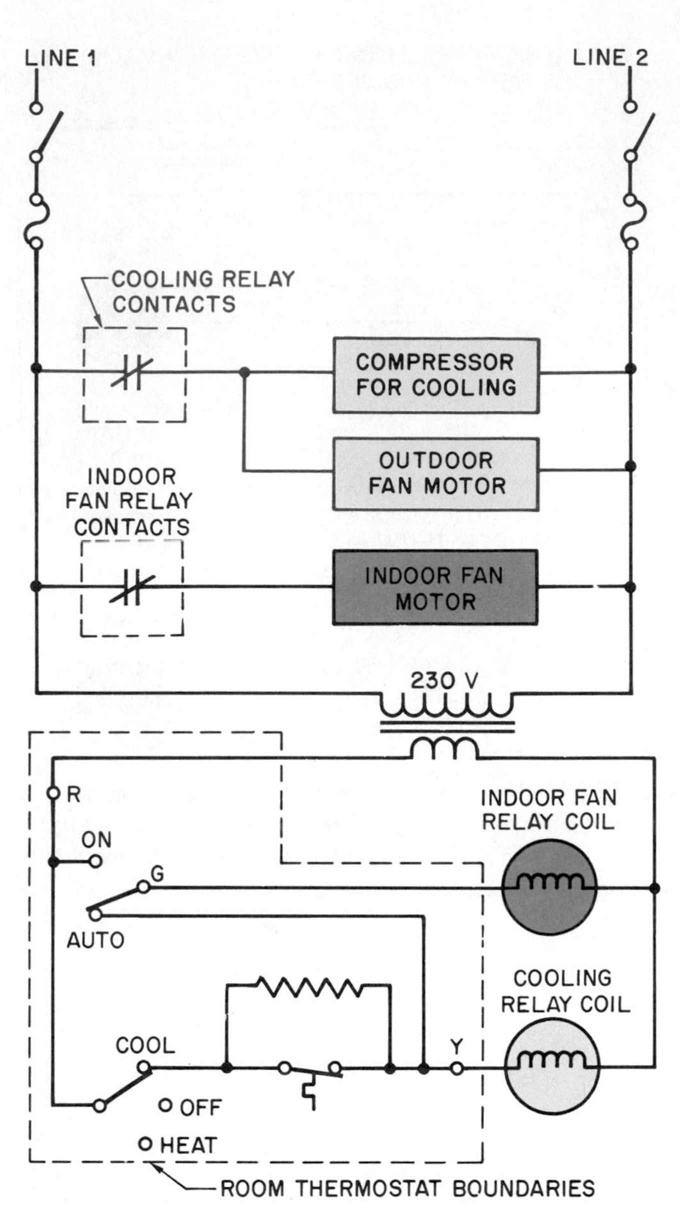

FIGURE 15–14 Addition of cooling to the system. When the thermostat selector switch is set to COOL and the thermal contacts close, the cooling relay coil is energized, closing the high-voltage contacts to the compressor and outdoor fan motor. Notice the circuit in the thermostat that starts the indoor fan through the AUTO switch. If this switch were ON, the fan would have been running.

The indoor fan must run whenever there is a call for cooling. When power passes through the contacts in the thermostat, a circuit is made through the AUTO side of the selector switch to energize the indoor fan relay coil and start the indoor fan. Note that if the fan selector switch is in the FAN ON position, the fan would run all the time and would still be on for cooling.

15.5 TROUBLESHOOTING AMPERAGE IN LOW-VOLTAGE CIRCUITS

Transformers are rated in *volt-amperes,* commonly called VA. This rating can be used to determine if the transformer is underrated or drawing too much current.

For example, it is quite common to use a 40-VA transformer for the low-voltage power source on a combination cooling and heating piece of equipment. This tells the technician that at 24 V, the maximum amperage the transformer can be expected to carry is 1.66 A. This was determined as follows:

$$\frac{40\text{ VA}}{24\text{ V}} = 1.66\text{ A}$$

Clamp-on ammeters will not readily measure such a low current with any degree of accuracy, so arrangements must be made to determine an accurate current reading. Use a jumper wire and coil it 10 times (called a 10-wrap amperage multiplier). Place the ammeter's jaws around the 10-wrap loop and place the jumper in series with the circuit, Figure 15–15. **The reading on the ammeter will have to be divided by 10.** For instance, if the heat relay coil amperage reads 7 A, it is really only carrying 0.7 A.

Some ammeters have attachments to read in ohms and volts. The volt attachments can be helpful for voltage readings, but the ohm attachment should be checked to make sure that it will read in the range of ohms needed. Some of the ohmmeter attachments will not read very high resistance. For instance, it may show an open circuit on a high-voltage coil that has a considerable amount of resistance.

15.6 TROUBLESHOOTING VOLTAGE IN THE LOW-VOLTAGE CIRCUIT

The volt-ohm-milliammeter (VOM) has the capability of checking continuity, milliamps, and volts. The most common applications are checking voltage and continuity. The

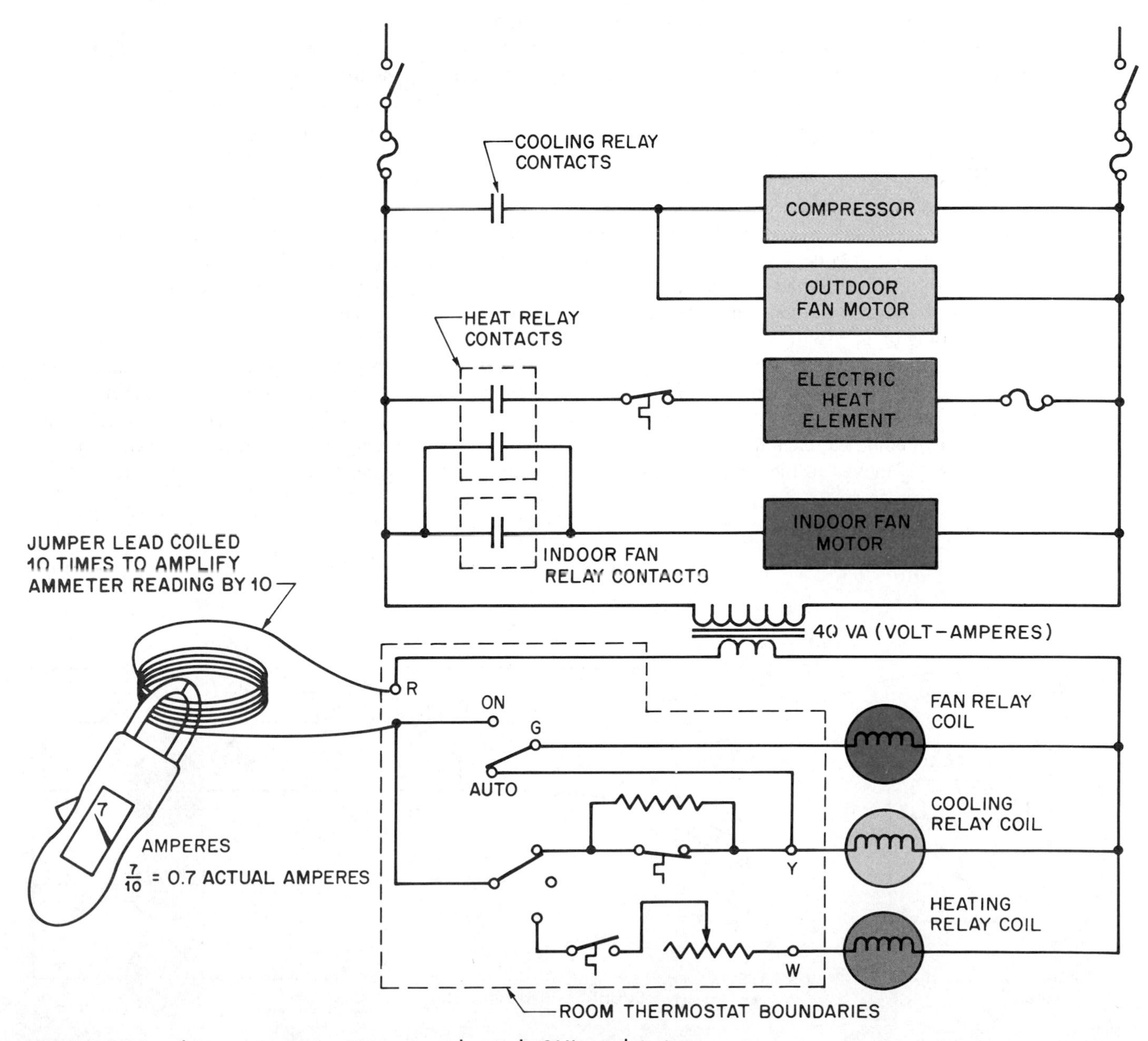

FIGURE 15–15 Clamp-on ammeter to measure current draw in the 24-V control circuit.

volt scale can be used to check for the presence of voltage, and the ohmmeter scale can be used to check for continuity. A high-quality meter will give accurate readings for each measurement.

Remember that any power-consuming device in the circuit must have the correct voltage to operate. The voltmeter can be applied to any power-consuming device for the voltage check by placing one probe on one side of the coil and the other probe on the other side of the coil. When power is present at both points, the coil should function. If not, a check of continuity through the coil is the next step, using the ohmmeter. Again keep in mind that when voltage is present and a path is provided, current will flow. If the flowing current does not cause the coil to do its job, the coil is defective and will have to be changed.

The voltmeter cannot be used to much advantage at the actual thermostat because the thermostat is a closed device. When the thermostat is removed from the subbase, the method discussed earlier of jumping the thermostat terminals is usually more effective. With the thermostat removed from the subbase and the subbase terminals exposed, the voltmeter can be used in the following manner.

Turn the voltmeter selector switch to the scale just higher than 24 V. Attach the voltmeter lead to the hot leg feeding the thermostat, Figure 15–16. This terminal is sometimes labeled R or V; no standard letter or number is used. When the other lead is placed on the other terminals one at a time, the circuits will be verified.

For example, with one lead on the R terminal and one on the G terminal, the terminal normally used for the fan circuit, a voltage reading of 24 V will be read. The meter is reading one side of the line straight from the transformer and the other side of the line through the coil on the fan relay. When the meter probe is moved to the circuit assigned to cooling (this terminal could be lettered Y), 24 V is read. When the meter probe is moved to the circuit assigned to heating (sometimes the W terminal), 24 V again appears on the meter. The fact that the voltage reads through the coil in the respective circuits is evidence that a complete circuit is present.

15.7 PICTORIAL AND LINE DIAGRAMS

The previous examples were all on troubleshooting the low-voltage circuit. This is recommended to start with. The control circuits that use high voltage (120 V or 230 V) work the same way. The same rules apply with more emphasis on safety. The service technician must have a good mental picture of the circuit or a good diagram to work from. There are two distinct types of diagrams furnished with equipment: the *pictorial* diagram and the *line* diagram. Some equipment has only one diagram, some has both.

The pictorial diagram is used to locate the different components in the circuit, Figure 15–17. The example in the figure is a gas furnace printed circuit board, which includes connections for an electronic air cleaner (terminals EAC 1 & 2) and a vent damper shutoff. Also built into the board is a time delay and an electronic pilot light ignition. This diagram is organized just as you would see it with the panel door open. For instance, if the diagram illustrates the control transformer in the upper portion of the picture, it will be in the upper portion of the control panel when the door is opened. This is useful when you don't know what a particular component looks like. Study the diagram until you find the control and then locate it in the corresponding place in the control panel. The diagram also gives the wire color to further verify the component. The general outline of components is the same. In this diagram, the transformer is shown at the top with four prominent wires leading to it.

The line diagram, sometimes called the *ladder* diagram, is the easier diagram to use to follow the circuit, Figure

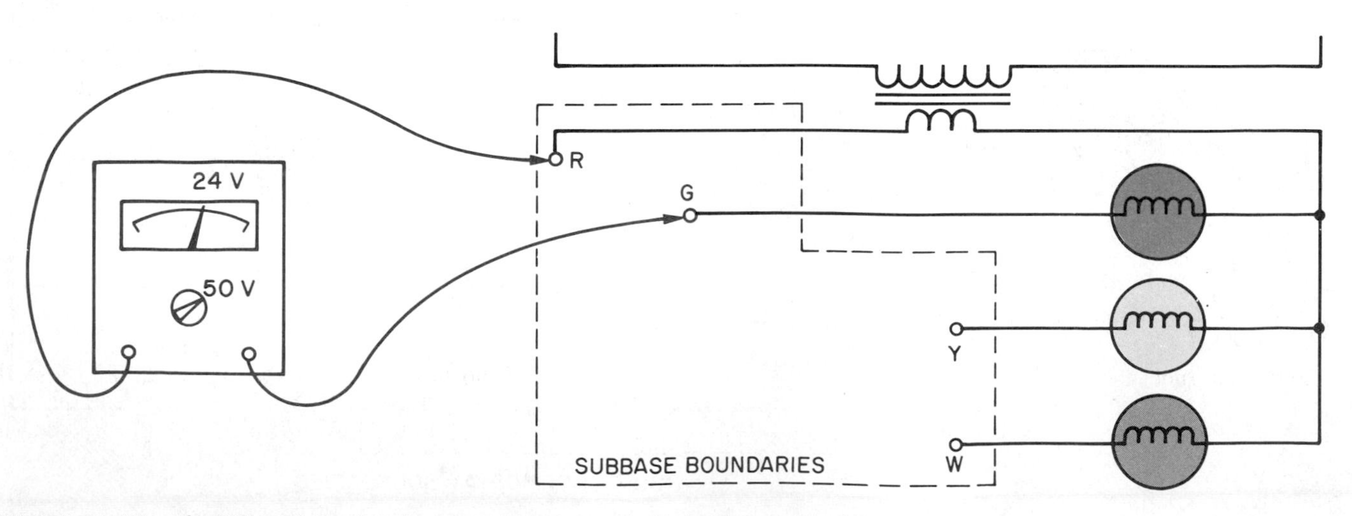

FIGURE 15–16 The VOM can be used at the thermostat location with the thermostat removed from the subbase.

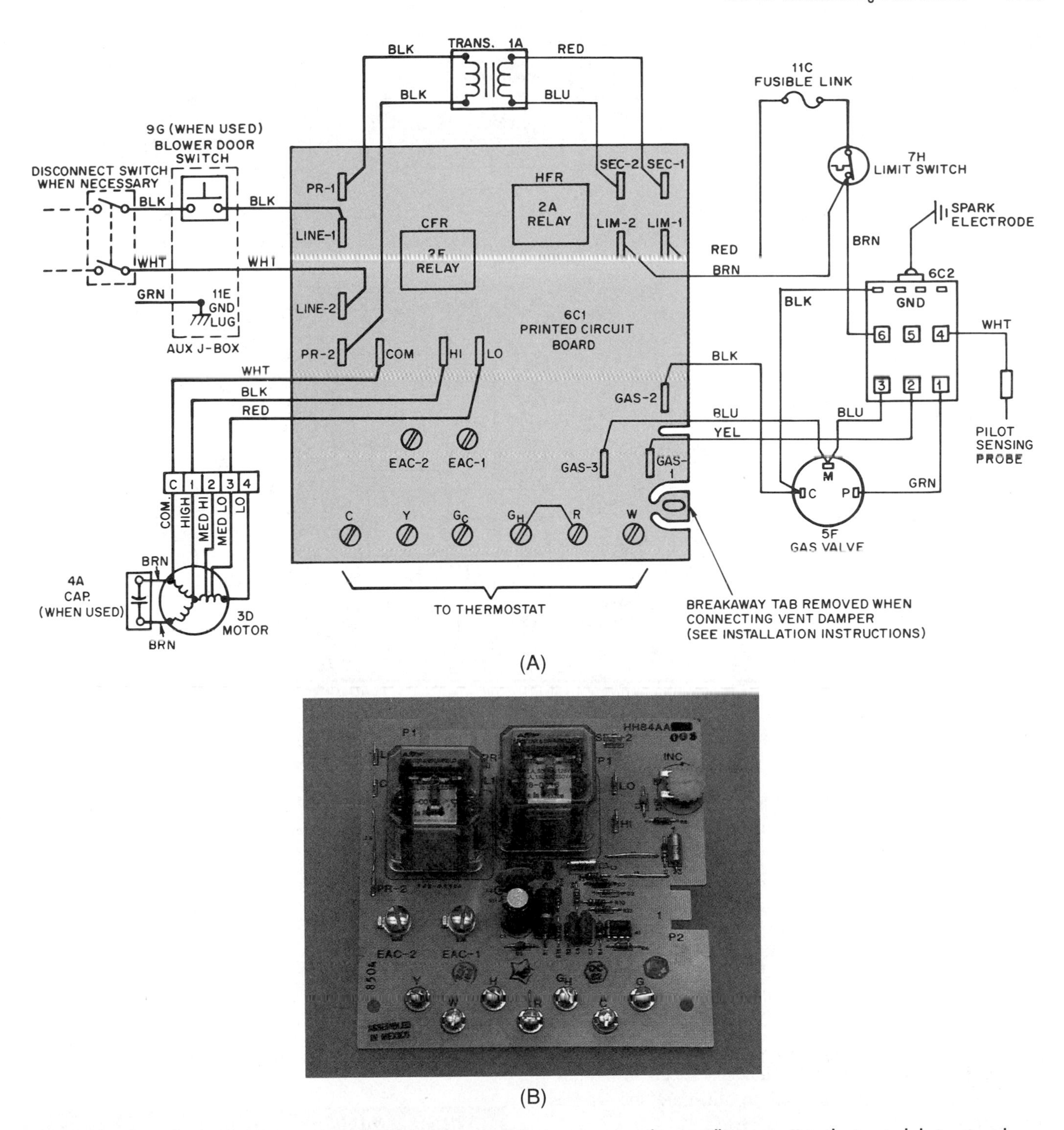

FIGURE 15–17 (A) A pictorial diagram shows the relative positions of the components as they actually appear. Note the terminal designation along the bottom. (B) Photo shows the actual circuit board. *(A) Courtesy BDP Company, a part of United Technologies Corporation, (B) Photo by Bill Johnson*

15–18. This diagram can normally be studied briefly, and the circuit function should become obvious. All power-consuming devices are between the lines. Most manufacturers try to make the right side of the diagram a common line to all power-consuming devices. The right side of the diagram will normally have no switches in it. This can make troubleshooting the circuit more practical.

Notice that the components have to be separated to illustrate them in the line diagram style. For example, the fan relay contacts 2F at the top of the diagram in Figure 15–18 are actually operated by the 2F coil illustrated at the bottom of the diagram; follow the G_c wire to the right.

Pictorial and line diagrams are an example of the way most manufacturers illustrate the wiring in their equipment.

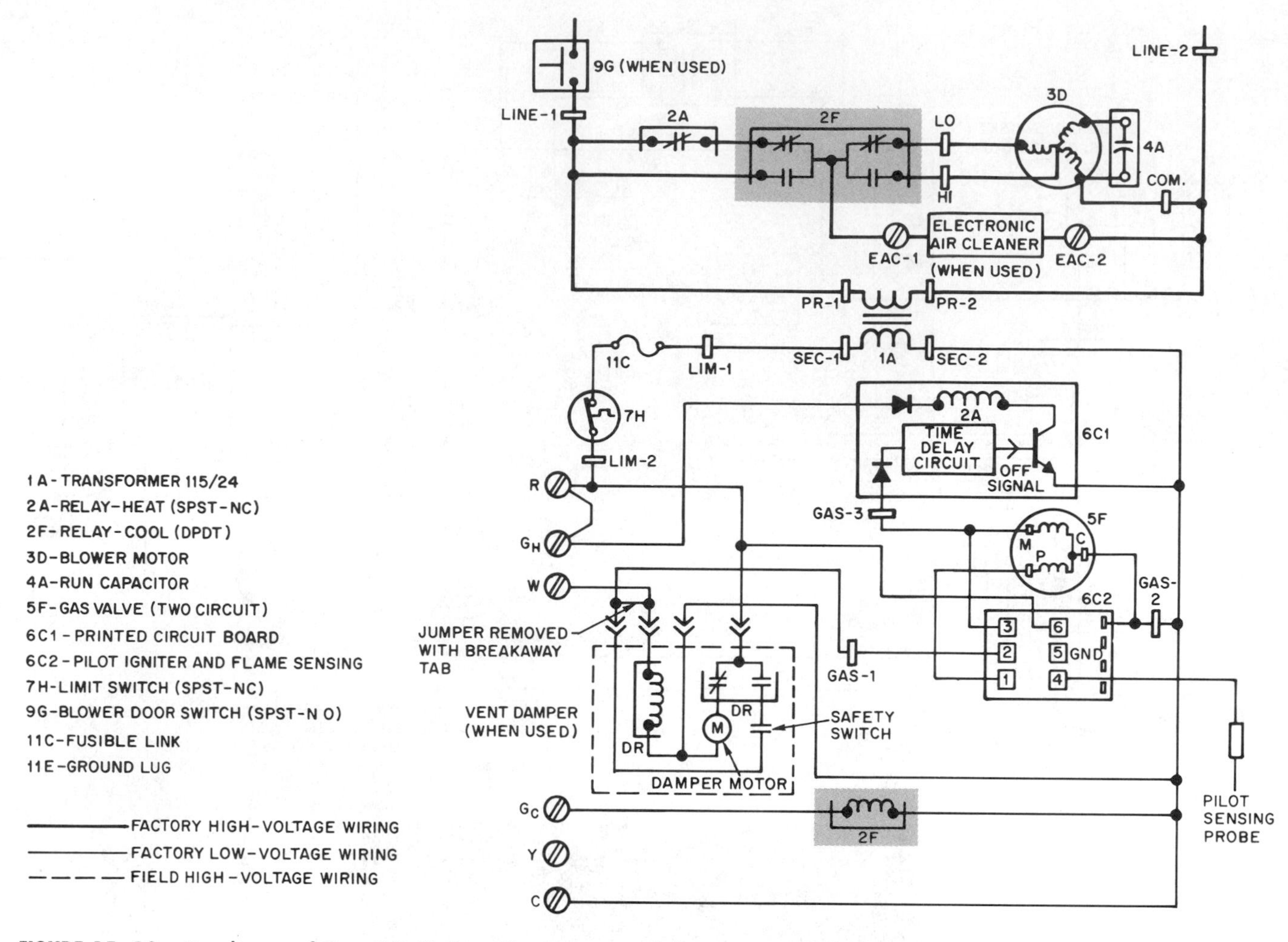

FIGURE 15–18 Line diagram of Figure 15–17. Notice the arrangement of the components. The right side has no switches. The power goes straight to the power-consuming devices. Notice also that the components have to be separated to illustrate them in this manner. *Courtesy BDP Company, a part of United Technologies Corporation*

Each manufacturer has its own way to illustrate points of interest. The only standard that industry seems to have established is the symbols used to illustrate the various components.

Anyone studying electrical circuits could benefit by first using a colored pencil for each circuit. A skilled person can divide every diagram into circuits. The colored pencil will allow an unskilled person to make a start in dividing the circuits into segments.

The following service calls are examples of actual service situations. The problem and solution are stated at the beginning of the first three service calls so that you will have a better understanding of the troubleshooting procedures. The last three service calls describe the troubleshooting procedures up to a point. Use your critical thinking abilities or class discussion to determine the solutions. The solutions can be found in the Instructor's Guide.

15.8 SERVICE TECHNICIAN CALLS

SERVICE CALL 1

A customer with a package air conditioner calls and tells the dispatcher that the unit is not cooling. *The problem is the control transformer has an open circuit in the primary, Figure 15–19.*

The technician arrives and goes to the indoor thermostat and turns the indoor fan switch to the FAN ON position. The fan will not start. This is an indication that possibly there is no control voltage. There must be high voltage before there is low voltage. The technician then goes to the outdoor unit and with the meter range selector switch set on 250 V, a check for high voltage shows there is power.

The cover is removed from the control compartment (usually this compartment can be found because the low-voltage wires enter close

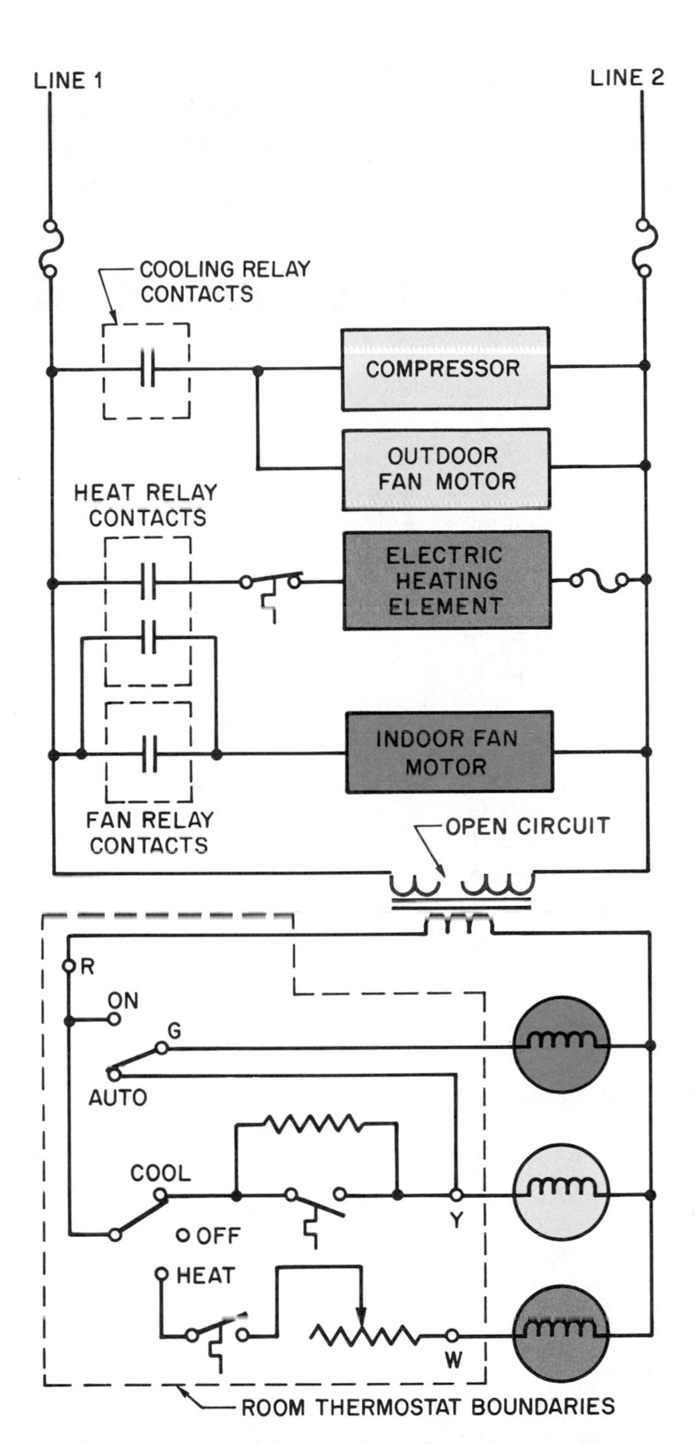

FIGURE 15–19 The control transformer has an open circuit.

by). A check for high voltage at the primary of the transformer shows that there is power.

A check for power at the secondary of the low-voltage transformer indicates no power. The transformer must be defective.

To prove this, the main power supply is turned off. One lead of the low-voltage transformer is removed and the ohmmeter section of the VOM is used to check the transformer for continuity. *Here the technician finds the primary circuit is open.* A new transformer must be installed.

The technician installs the new transformer and turns the power on. The system starts and runs correctly. The paperwork is handled with the customer and the technician goes to the next call.

SERVICE CALL 2

A customer calls with a "no cooling" complaint. *The problem is that the cooling relay coil is open, Figure 15–20.*

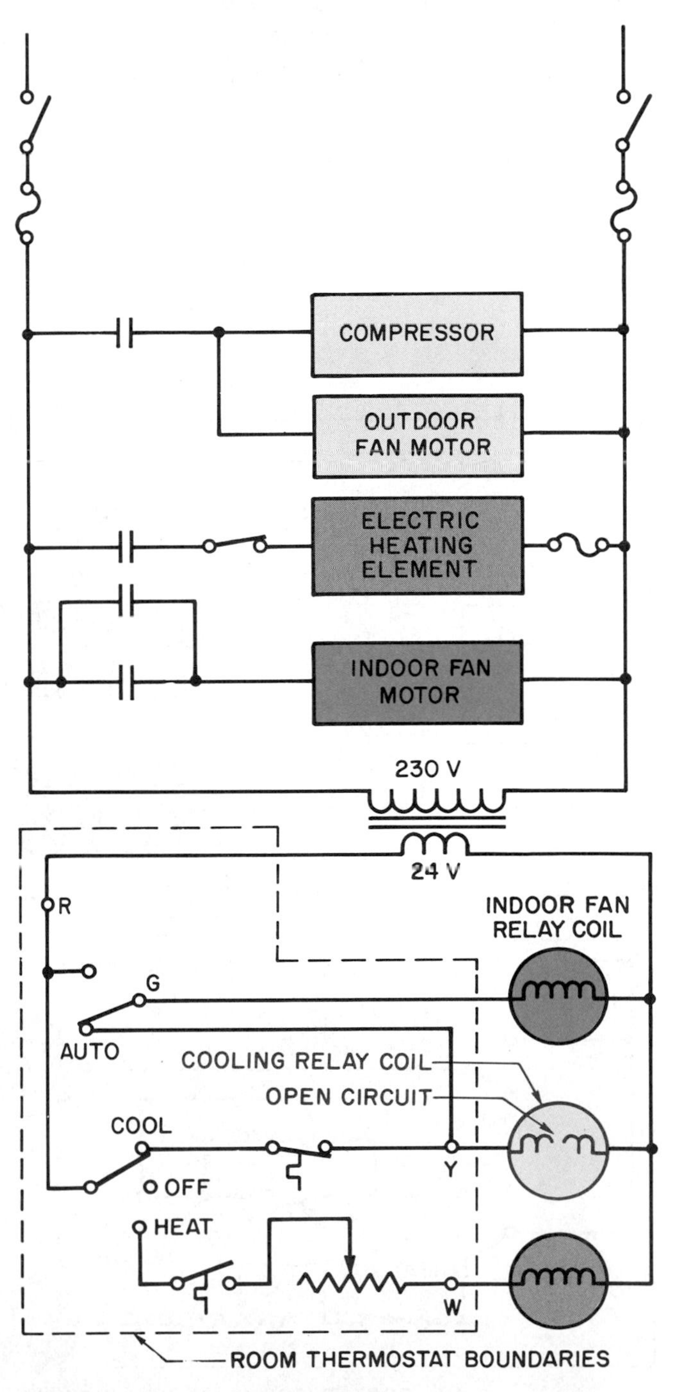

FIGURE 15–20 The cooling relay has an open circuit and will not close the cooling relay contacts. Note that the indoor fan is operable.

The technician arrives, goes to the thermostat, and tries the fan circuit. The fan switch is moved to FAN ON and the indoor fan starts. Then the thermostat is switched to the COOL position and the fan runs. This means that the power is passing through the thermostat, so the problem is probably not there. The thermostat is left in position to call for cooling while the technician goes to the outdoor unit.

Since the primary for the low-voltage transformer comes from the high voltage, the power supply is established.

A look at the diagram shows the technician that the only requirement for the cooling relay to close its contacts is that its coil be energized. A check for 24 V at the cooling relay coil shows 24 volts. The coil must be defective. To be sure, the technician turns off the power and removes one of the leads on the coil. The ohmmeter is used to check the coil. There is no continuity. It is open.

The technician changes the whole contactor (it is less expensive to change the contactor from stock on the truck than to go to a supply house and get a coil), turns the power on, and starts the unit. It runs correctly. All paperwork is handled and the technician leaves.

SERVICE CALL 3

A customer complains of no heat. *The problem is the heat relay has a shorted coil that has overloaded the transformer and burned it out, Figure 15–21.*

The technician arrives, goes to the thermostat and tries the indoor fan. It won't run. There is obviously no current flow to the fan. This could be a problem in either the high- or low-voltage circuits.

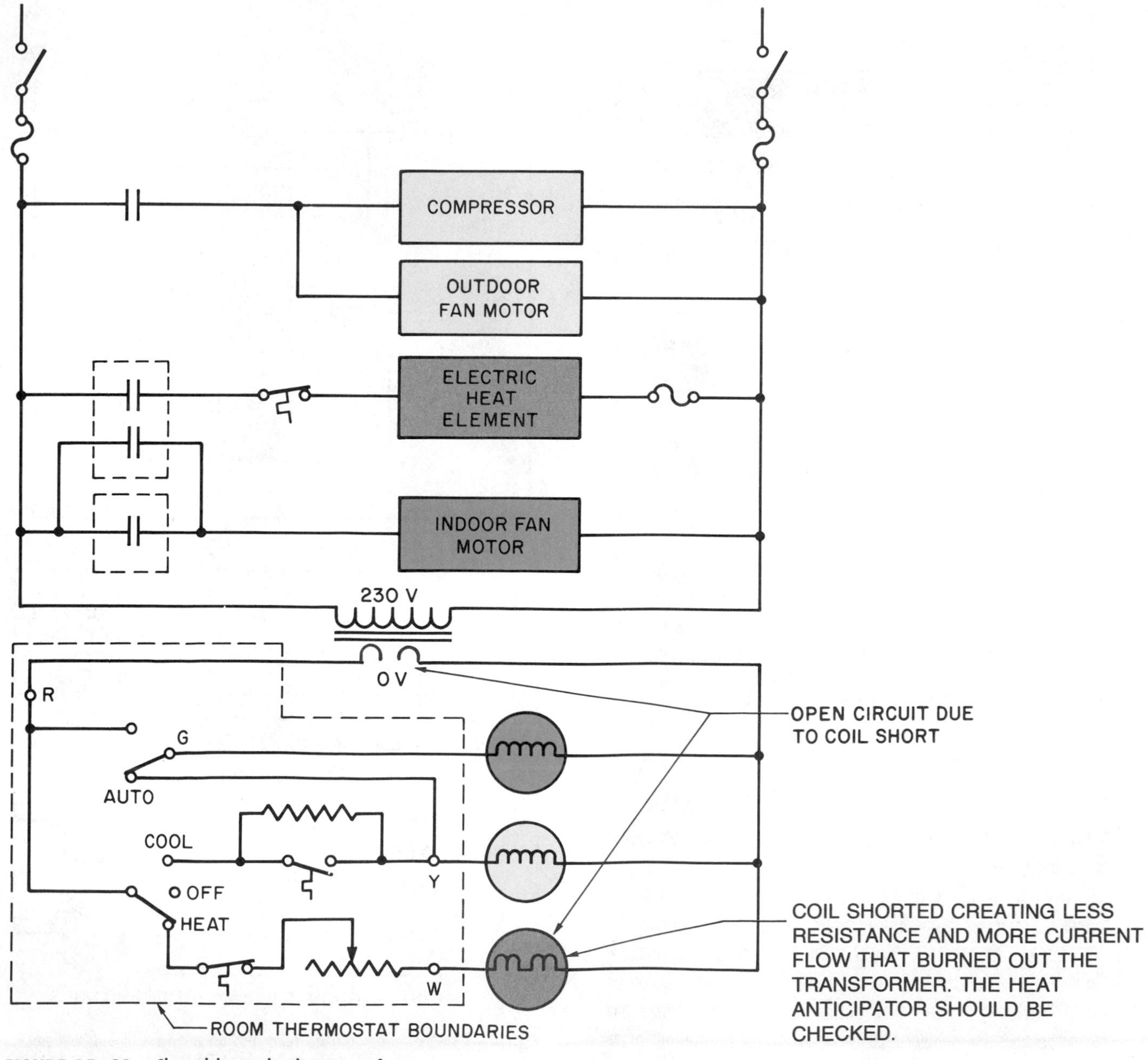

FIGURE 15–21 Shorted heat relay burns transformer.

The technician goes to the outdoor unit to check for high voltage and finds there is power. A check for low voltage shows there is none. The transformer must be defective. It is removed and has a burnt smell. It is checked with the VOM and found to have an open secondary.

When the transformer is changed and the system is switched to HEAT, the system does not come on. The technician notices the transformer is getting hot. It appears to be overloaded. The technician turns off the power before a second transformer is damaged. The problem is to find out which circuit is overloading. Since the system was in the heating mode, the technician assumes for the moment it is the heat relay. The technician realizes that if too much current is allowed to pass through the heat anticipator in the room thermostat, it will burn also and a thermostat will have to be replaced.

The technician goes to the room thermostat and turns it off to take it out of the circuit, then goes to the outdoor unit where the low-voltage terminal block is located. An ammeter is applied to the transformer circuit leaving the transformer. A 10-wrap coil of wire is installed to amplify the current reading. The power is turned on and no amperage is recorded. The problem must not be a short in the wiring.

The technician then jumps from R to W to call for heat. The amperage goes to 25 A; divided by 10 because of the 10 wrap, the real amperage is 2.5 A in the circuit. This is a 40-VA transformer with a rated amperage of 1.67 A (40/24 = 1.666). The heat relay is pulling enough current to overheat the transformer and burn it up in a short period of time. It is lucky that the room thermostat heat anticipator was not burned also.

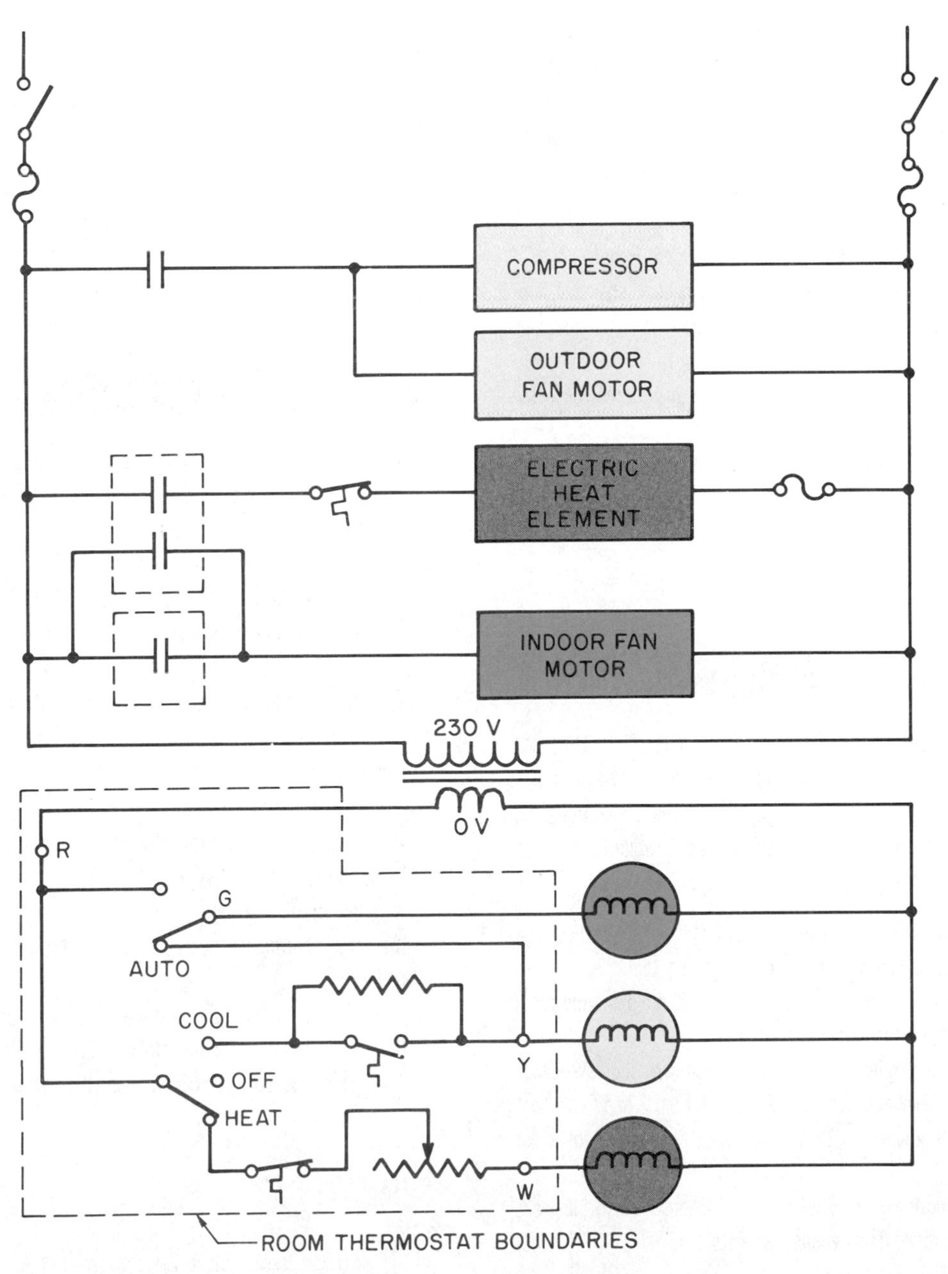

FIGURE 15–22 Use this diagram for discussion of Service Call 4.

The technician changes the heat relay and tries the circuit again and the amperage is only 5 A, divided by 10, or 0.5 A. This is correct for the circuit. The jumper is removed, the panels are put back in place, and the system started with the room thermostat and it operates normally.

The following service calls do not have the solution described.

SERVICE CALL 4

A customer calls and indicates that the air conditioner quit running after the cable TV repairman was under the house installing TV cable.

The technician arrives and enters the house. It is obvious that the air conditioner is not running. The technician can hear the fan running, but it is hot in the house. A check outside shows that the condensing unit is not running. The technician then proceeds through the following checklist. Use the diagram in Figure 15–22 to follow the checkout procedure.

1. Low voltage is checked at the condenser and there is none.
2. The technician then goes to the room thermostat and removes it from the subbase and jumps from R to Y; the condensing unit does not start.
3. Leaving the jumper in place, the technician then goes to the condenser and checks voltage at the contactor and finds no voltage.

What is the problem and the recommended solution?

SERVICE CALL 5

A store manager calls and reports that the heat is off in the shoe store. This store has an electric heat package unit on the roof.

The technician arrives, talks to the customer, and uses the following checkout procedure to determine the problem. You can follow the procedure using the diagram in Figure 15–23.

1. The customer says the system worked well until about an hour ago when the manager noticed the store was getting cold.
2. The technician notices that the set point on the room thermostat is 75°F and the store temperature is 65°F.
3. The technician then goes to the roof and removes the low-voltage control panel. The voltage from C to R is 24 V, from C to W (the heat terminal) is 0 V. There is a 24-V power supply, but the heat is not operating.
4. The room thermostat is removed from the subbase and a jumper is applied from R to W. The heating system starts.

What is the problem and the recommended solution?

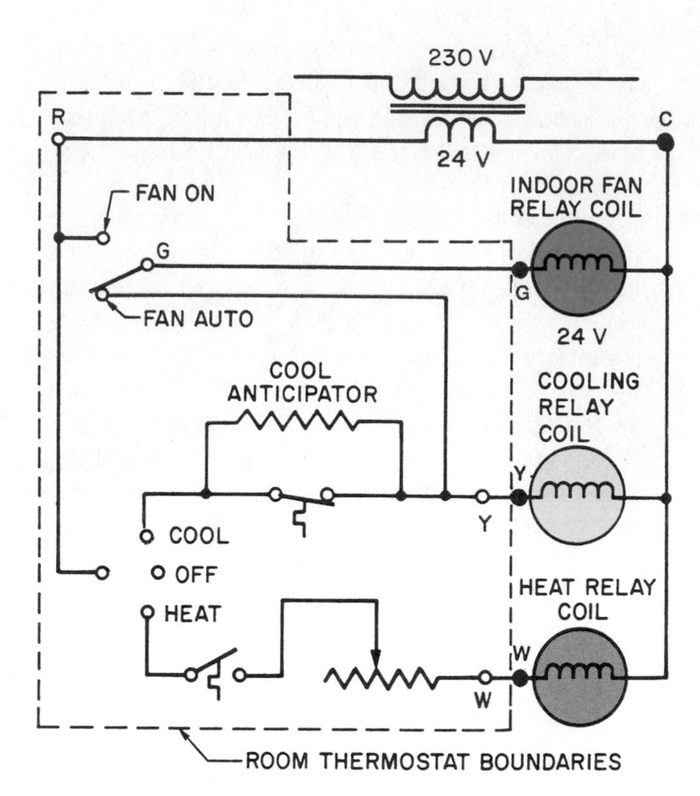

FIGURE 15–23 Use this diagram for discussion of Service Call 5.

SERVICE CALL 6

A customer calls on a very cold day and explains that the gas furnace just shut off and the house is getting cold. The system has a printed circuit board, Figure 15–24.

The technician arrives and checks the room thermostat. It is set at 73°F and the house temperature is 65°F. The thermostat is calling for heat. The fan is running, but there is no heat.

The technician goes to the furnace and removes the front door. This stops the furnace fan because of the door switch (see 9G at the top of the diagram). The door switch is taped down so the technician can troubleshoot the circuit.

The technician places one meter lead on SEC-2 (the common side of the circuit) and the other on SEC-1 and finds that there is 24 V. The meter lead on SEC-1 is then moved to the R terminal and 24 V is found there. The lead is then moved to G_H where the signal from the room thermostat passes power to the circuit board and there is power.

The technician then removes the lead from G_H and places it on the GAS-1 terminal where the technician finds 24 V.

What is the problem and the recommended solution?

SUMMARY

- Each control must be evaluated as to its purpose in a circuit.

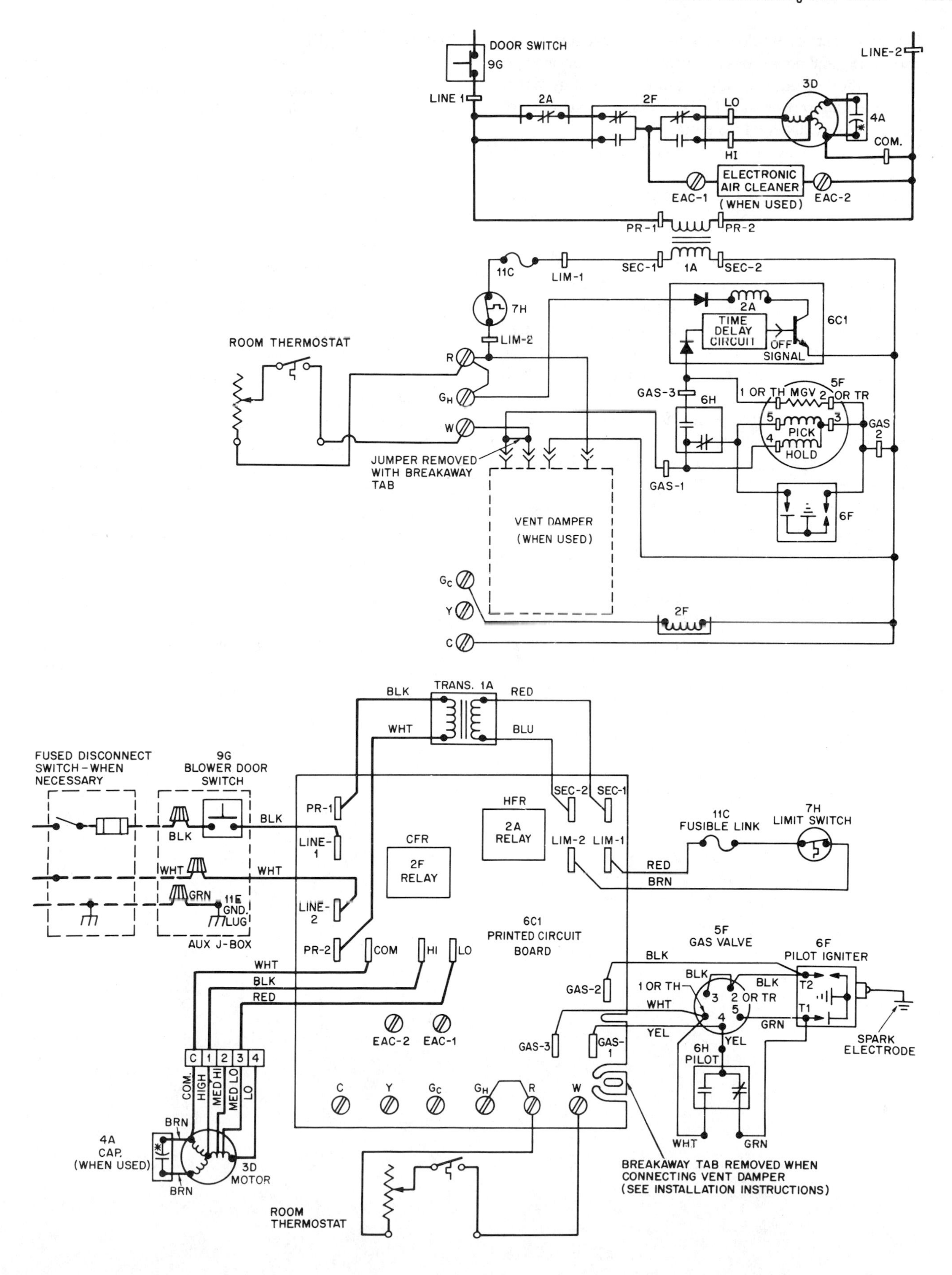

FIGURE 15–24 Use this diagram for discussion of Service Call 6.

- Electrical controls are divided into two categories: power-consuming and power-passing (non–power-consuming).
- One method of understanding a circuit basically is that the potential power supply is between two different power legs of different potential.
- Devices or controls that pass power are known as safety, operating, or control devices.
- The light bulb controlled by a thermostat with a fuse in the circuit is an example of an operating device and a safety device in the same circuit.
- The voltmeter may be used to follow the circuit from the beginning to the power-consuming device.
- The three separate power-consuming circuits in the low-voltage control of a typical heating and cooling fan unit are the heat circuit, cool circuit, and fan circuit. The selector switch for heating and cooling decides which function will operate.
- The fan relay, cooling relay, and heat relay are all power-consuming devices.
- The low-voltage relays start the high-voltage power-consuming devices.
- The voltmeter is used to trace the actual voltage at various points in the circuit.
- The ohmmeter is used to check for continuity in a circuit.
- The ammeter is used to detect current flow.
- The pictorial diagram has wire colors and destinations printed on it. It shows the actual locations of all components.
- The line or ladder diagram is used to trace the circuit to understand its purpose.

REVIEW QUESTIONS

1. Name three types of automatic controls.
2. Name two categories of electric controls.
3. Name the circuit that is energized with a thermostat terminal designated Y.
4. Name the circuit that is energized with a thermostat terminal designated G.
5. Name the circuit that is energized with a thermostat terminal designated W.
6. When a thermostat terminal is designated R, it is known by what name?
7. Is the heat anticipator in parallel or in series with the heat contacts in a thermostat?
8. Is the cool anticipator in parallel or in series with the cooling contacts in a thermostat?
9. Which anticipator is adjustable?
10. True or False: The indoor fan must always run in the cooling cycle.
11. True or False: When the control transformer is not working, the indoor fan will not run in the FAN ON position.
12. What switch should be turned off when working on the compressor circuit?
13. True or False: The pictorial wiring diagram is used to locate components from the diagram to the actual unit.
14. True or False: The line diagram is used to follow and understand the intent of the circuit.
15. How can amperage be measured in a low-voltage circuit with a clamp-on ammeter?

UNIT 16

Electronic and Programmable Controls

OBJECTIVES

After studying this unit, you should be able to

- describe several applications for electronic controls.
- describe why electronic controls are more applicable to some situations than are electromechanical controls.
- describe the electronic control boards used for air conditioning, oil burner, and gas furnace circuits.
- recognize and troubleshoot a basic electronic control circuit board.
- describe programmable thermostats.

SAFETY CHECKLIST

* Use all electrical safety precautions when servicing or troubleshooting electrical/electronic circuits.
* When replacing line voltage components ensure that power is off, panel is locked and tagged, and that you have the only key.

16.1 ELECTRONIC CONTROLS

Electronic controls are being used more frequently for automatic control of equipment. For many years electronic controls have been used to control larger equipment. The development of these controls has now reached the point that they are feasible for use in residential and light commercial equipment. These controls are economical and reliable and provide efficient energy management.

Electronic controls serve the same purposes as many electrical and electromechanical controls. Operating and safety functions are the main uses, but energy management applications are becoming more popular each year. These controls normally come in circuit boards with terminal strips for external circuit connections. Some manufacturers design these circuit boards as though they were individual controls. For troubleshooting purposes this control board can normally be treated as an individual component, like a switch. Another method used by some manufacturers is to furnish a special module that can be plugged into the control circuit to locate problems. When a module is required, the technician is basically tied to this specific manufacturer for an orderly repair.

Following are examples of applications of electronic controls in residential and light commercial systems.

16.2 GAS FURNACE PILOT LIGHTS

As more efficient gas furnaces are being developed, many are equipped with new and different types of ignition rather than with a standing pilot. The standing pilot would normally be left on even when the furnace was not used for long periods. The summer off period can be lengthy in mild climate areas, and the air conditioning system must overcome the pilot light's heat. Many of these new types of ignition systems are controlled with an electronic circuit. The manufacturers furnish the best troubleshooting guide for these controls because each manufacturer has its own method of applying these controls. Figure 16–1 presents an example of an electronic circuit board used in a gas furnace.

The circuit board has some extra features and other components may easily be added to the board, such as controls for an electronic air cleaner, a humidifier, and a vent damper shutoff. The component's wires can be connected to the board without a lot of control circuit wiring. For example, wiring an electronic air cleaner or a humidifier requires that neither component be energized unless the indoor furnace fan is running. This is called *interlocking* the components. The circuit board has the connections that interlock the electronic air cleaner or the humidifier with the furnace fan in both summer (high-speed) and winter (low-speed) applications. No wiring decisions have to be made by the installing technician. The application and troubleshooting of intermittent ignition are covered in more detail in the unit on gas heat.

16.3 OIL FURNACES

The oil furnace industry is currently using an electronic circuit to prove the flame in oil-fired furnaces. This control is known as a *cad cell,* which stands for cadmium sulfide cell. This cell varies in resistance with more or less light, Figure 16–2. This varying resistance is monitored by an electronic circuit to stop the oil from being injected into the furnace when there is no flame to ignite it. Figure 16–3 shows an example of the circuit board for an oil furnace. This control is called a primary control when applied to an oil furnace. The cad cell is not used on gas furnaces because of the quality of light. The oil furnace burns with a yellow-orange flame, and the gas furnace burns with a blue flame. The cad cell does not see blue well. The cad cell is covered in more detail in the unit on oil heat.

16.4 AIR CONDITIONING APPLICATIONS

The electronic control board is used by some manufacturers of air conditioning equipment to provide electrical protection for the equipment. For example, the electronic control

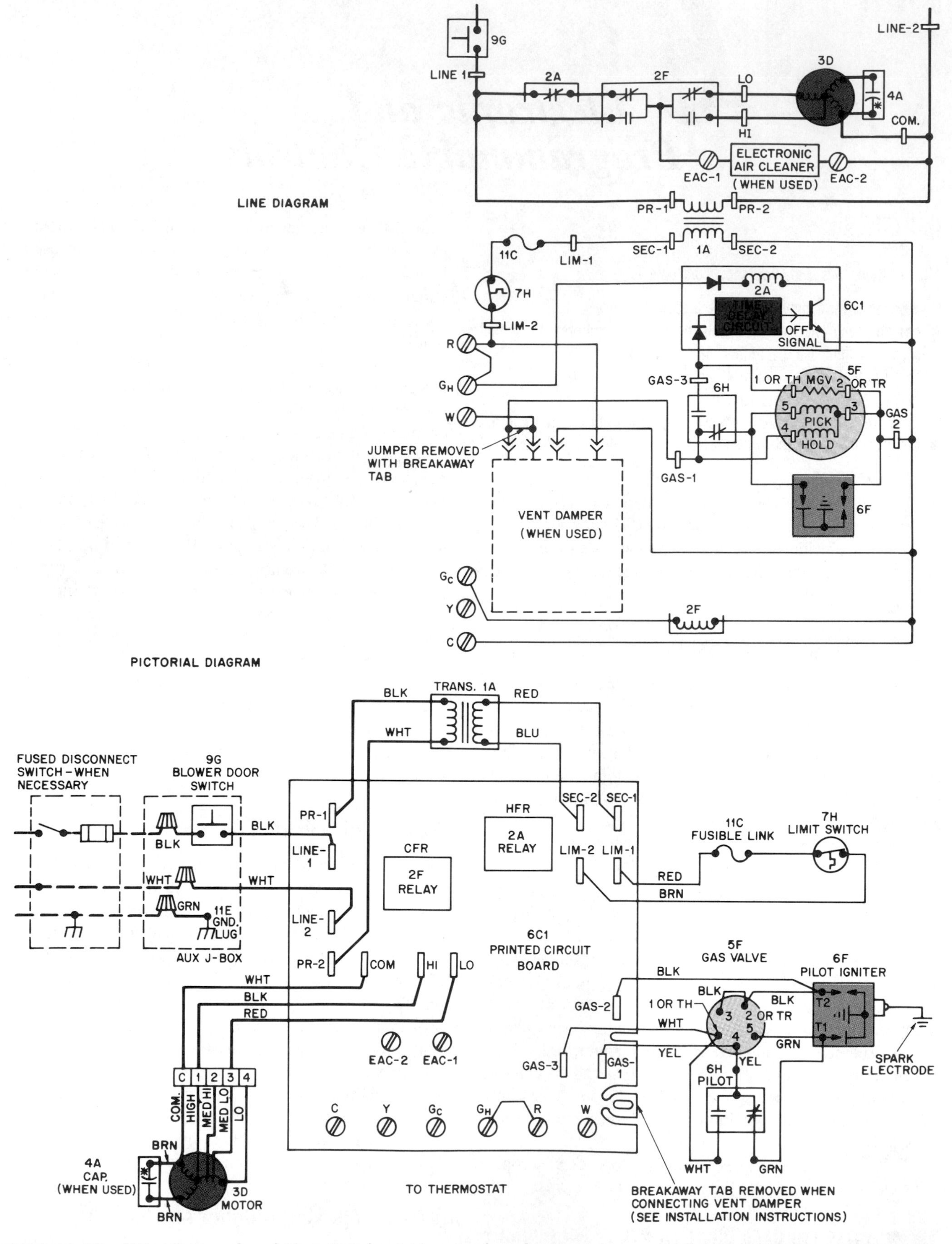

FIGURE 16–1A Wiring diagrams of an electronic circuit board. The terminals on the pictorial diagram are placed exactly like the ones in the actual unit. Different components can be added. This board automatically starts these components at the correct time. *Courtesy BDP Company, a part of United Technologies Corporation*

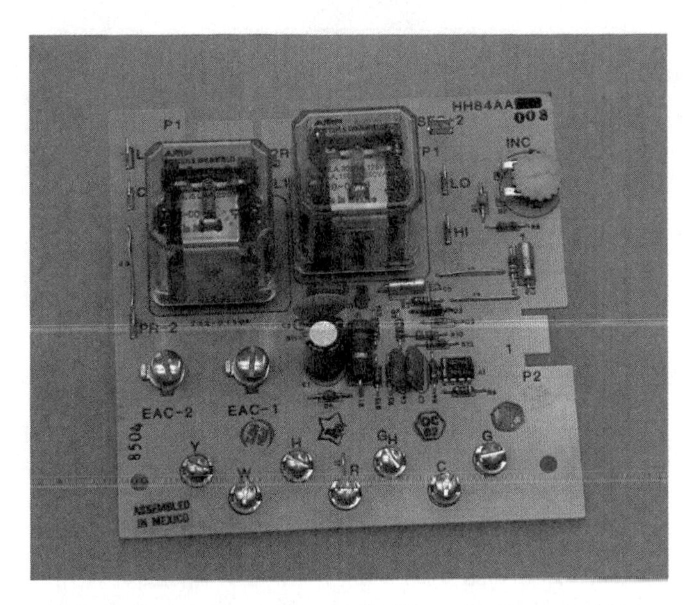

FIGURE 16–1B Photo of the electronic circuit board. *Photo by Bill Johnson*

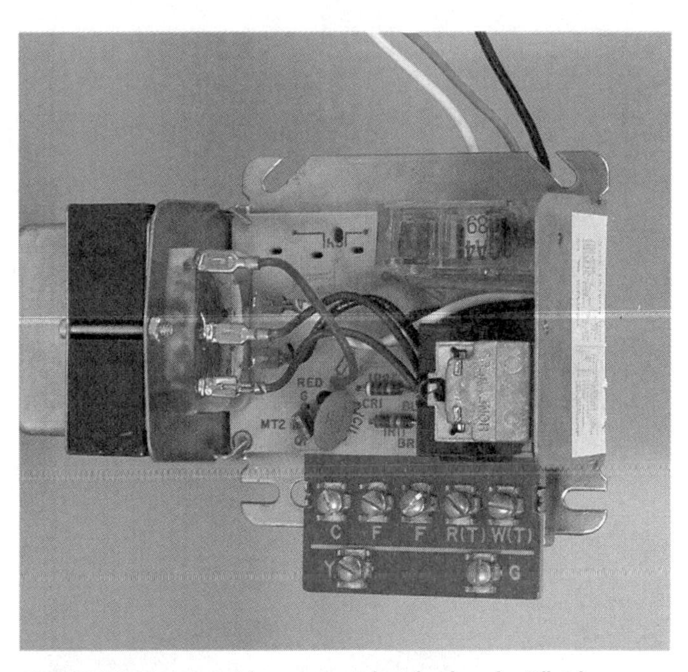

FIGURE 16–3 An oil burner circuit board. *Photo by Bill Johnson*

can monitor the voltage being supplied to a unit. While measuring the voltage, it can measure the current draw of the compressor motor to protect it from overload, Figure 16–4. Overload protection can be readily obtained with electromechanical devices, but low- and high-voltage monitoring and protection cannot.

The electromechanical controls can be connected to the circuit board in such a manner that the time-delay feature is used to keep the unit from short cycling. The time delay before a compressor restart could be approximately 5 min. For example, most residential air conditioning units use fixed-bore metering devices, such as the capillary tube, that

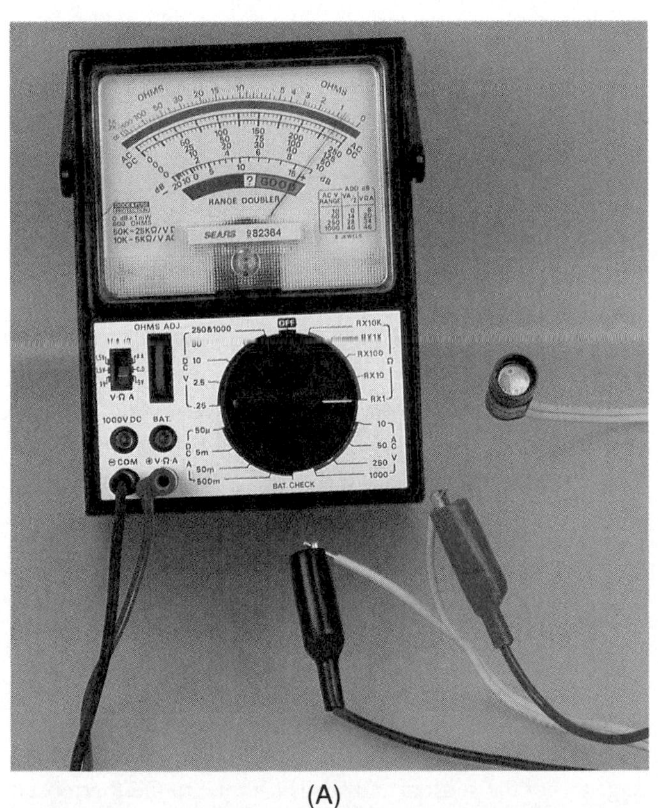

(A)

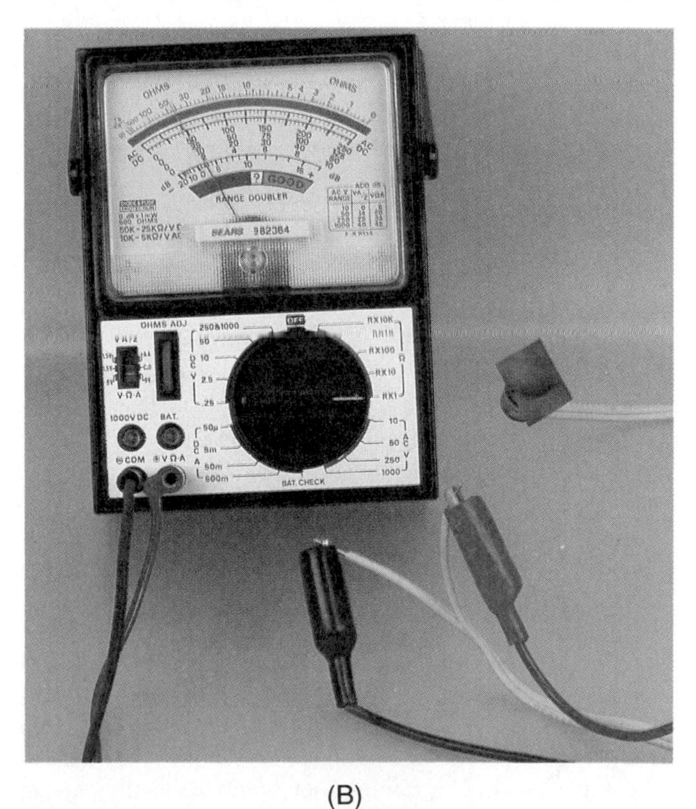

(B)

FIGURE 16–2 The cad cell changes resistance with light changes. (A) The cad cell with light shining in and a resistance of 20 Ω. (Note: The meter is on the R × 1 scale.) (B) The cell eye is covered with electrical tape and the reading changes to 620 Ω. This cad cell is used to "see" the flame in an oil furnace. *Photos by Bill Johnson*

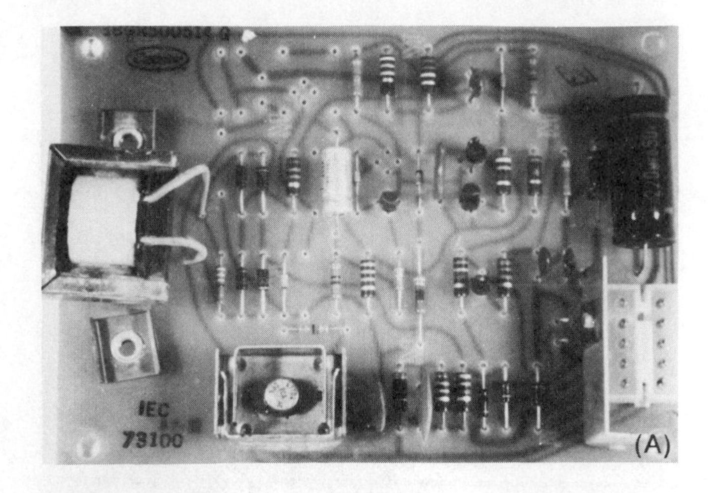

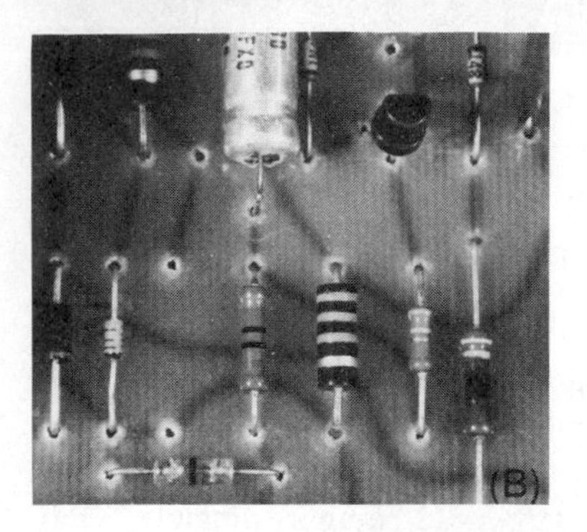

FIGURE 16–4 (A) Circuit board used in residential air conditioning. It has the following features: time delay, low-voltage monitor, high-voltage monitor, and amperage protection. (B) The electronics in this board resemble those in a radio or TV. The service technician cannot repair the components but will have to replace the board if the components fail. (C) Notice the pin connectors for field wiring connections. *Photos by Bill Johnson*

equalize during the off cycle. (The thermostatic expansion valve does not equalize completely during the off cycle, so the short cycle is not as much of a problem when that valve is used exclusively.) The unit could come right back on after a low-pressure cutout if not for the time-delay circuit. The time delay can be easily obtained with electronic circuits.

16.5 ELECTRONIC THERMOSTATS

Reasonably priced electronic thermostats can be manufactured with programs for various needs. This makes them attractive to the small user who wants economy and total control. They can be easily programmed to stop and start the heating and cooling equipment at many predetermined times, Figure 16–5. For example, homeowners may want to lower the heating or cooling set point while they are at work, then turn on the heat or air conditioner before they arrive home. The thermostat can also cut the system back at night and bring the home back to temperature just before the residents get up in the morning. The program schedule can

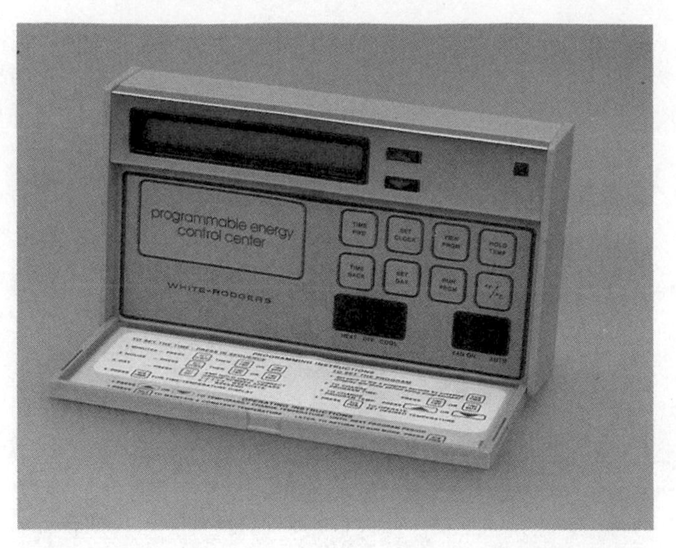

FIGURE 16–5 Thermostat programmable for energy management. It can be used in any 24-V circuit for one-stage cooling and one-stage heating. This is typical of one that could be installed in a residence or small business. *Photo by Bill Johnson*

be altered when necessary. It will automatically return to the original program at the next scheduled change.

There are many features available that concern programming time. The temperature control itself has a faster response time than the bimetal type because of the mass of the bimetal sensing element. It could easily be a thermistor element, which is lighter than the bimetal. From a troubleshooting standpoint, this thermostat is the same as a regular thermostat. The terminal designations in the subbase are basically the same. We can say that the thermostat either works or it doesn't. If it doesn't work, install a new one. The directions furnished by the manufacturer show how to check the program to make sure that the problem is not an operator's error. If the program is correct and the thermostat will not function as it was designed to, it will probably have to be replaced.

16.6 DIAGNOSTIC THERMOSTATS

Many commercial, industrial, and some residential installations use thermostats that are diagnostic message centers. Sensors are mounted in strategic locations on the system to monitor pressures and temperatures. When the system begins to experience difficulty, a computer monitors the messages from the sensors and sends a signal to the thermostat. The thermostat will display the problems that the unit is experiencing. In anticipation of trouble, the technician can call the system computer and get a readout of potential problems.

16.7 TROUBLESHOOTING ELECTRONIC CONTROLS

Troubleshooting electronic circuit boards currently manufactured is much like troubleshooting a single component in the circuit. Figure 16–6 shows an electronic thermostat that is programmable along with the subbase. The low-voltage power supply accomplishes the same thing as in a conventional circuit in that it is still working to supply power to the power-consuming control circuit. The hot leg of the transformer goes directly to the thermostat, and the thermostat distributes this power to the individual components just as in a conventional thermostat. There is still a compressor contactor, a fan relay, and the same heating components to energize.

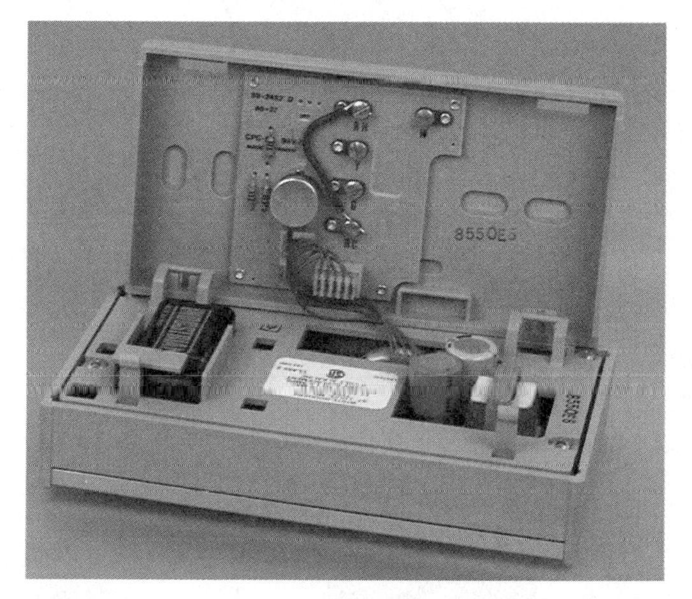

FIGURE 16–6 Photo shows the back of the thermostat. *Photo by Bill Johnson*

When a technician has a problem with a circuit and believes that it may be the circuit board, the first thing to do is to check the manufacturer's troubleshooting procedures. This may be attached to the unit or you may need to place a call to the distributor of the product. If your problem occurs at a time when research is not practical, you may have to proceed on your own. First, look for any warnings on the circuit board or in the compartment. Then determine from the wiring diagram which wire may be carrying the signal to the circuit board and which wire may be used to carry the signal from the circuit board to the power-consuming device. For example, on an air conditioning circuit board the wire feeding the low-voltage signal to the board may be the Y wire, which may feed the circuit through the board's electronic circuit to the compressor contactor coil. In some cases a jumper across this circuit may start the unit. In this case, remove the jumper immediately as you have proven the board to be defective, Figure 16–7.

16.8 TROUBLESHOOTING THE ELECTRONIC THERMOSTAT

The following is an example of how a typical programmable thermostat for combination heating and cooling can be diagnosed for some typical problems. This could be considered a replacement for a standard heat-cool thermostat. A review of the unit on thermostats may be helpful in understanding this example.

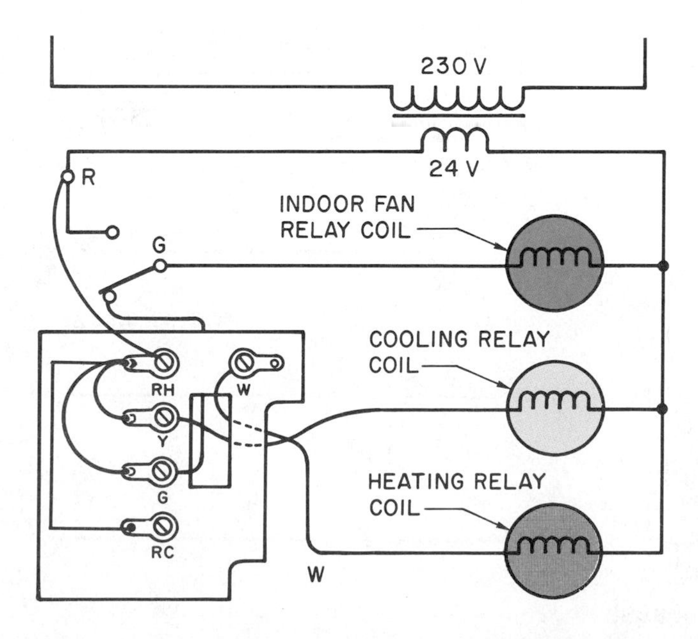

FIGURE 16–7 The 24-V signal leaves the transformer and goes through the room thermostat to the Y terminal on the circuit board. If the signal reaches the Y terminal but will not pass through the board to the terminal for the compressor contactor, the board may be defective.

Remember that the hot terminal feeding the thermostat provides the power to the other terminals. See Figures 16–8, 16–9, and 16–10 for circuit diagrams. As you can see from the manufacturer's literature, the installation is quite simple.

The circuit signal enters the thermostat at the RH terminal, Figure 16–9. If there is a problem with the circuit and the thermostat is suspected, a jumper applied between RH and any other circuit will prove whether or not the thermostat is functioning. If the system will start with the jumper in place and will not start without it, the thermostat is defective. A time delay may be built into the thermostat, so be sure to allow 5 min after energizing the thermostat for the unit to start.

When the existing unit has two different power supplies, the transformers must be *in phase,* or they will oppose each other, Figures 16–10 and 16-11. Isolation relays are often used to prevent stray unwanted electrical feedback that can cause erratic operation. This manufacturer recommends *isolation relays* in certain applications that have elec-

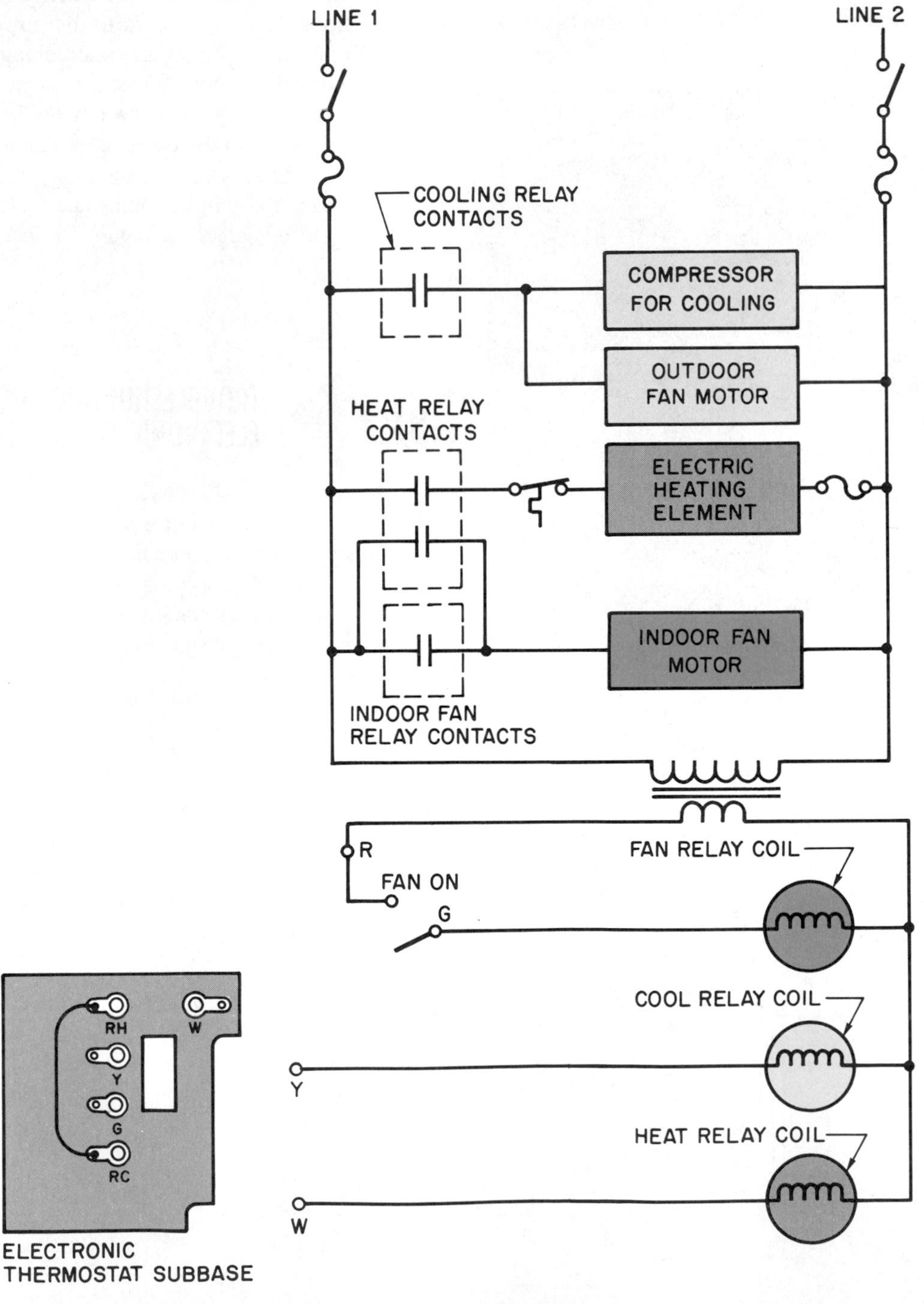

FIGURE 16–8 Same application as was shown in Figure 15–11, only on this diagram there are two R terminals with a jumper between them. This is a split-base thermostat that can use two 24-V power supplies. When only one is used, a jumper is applied.

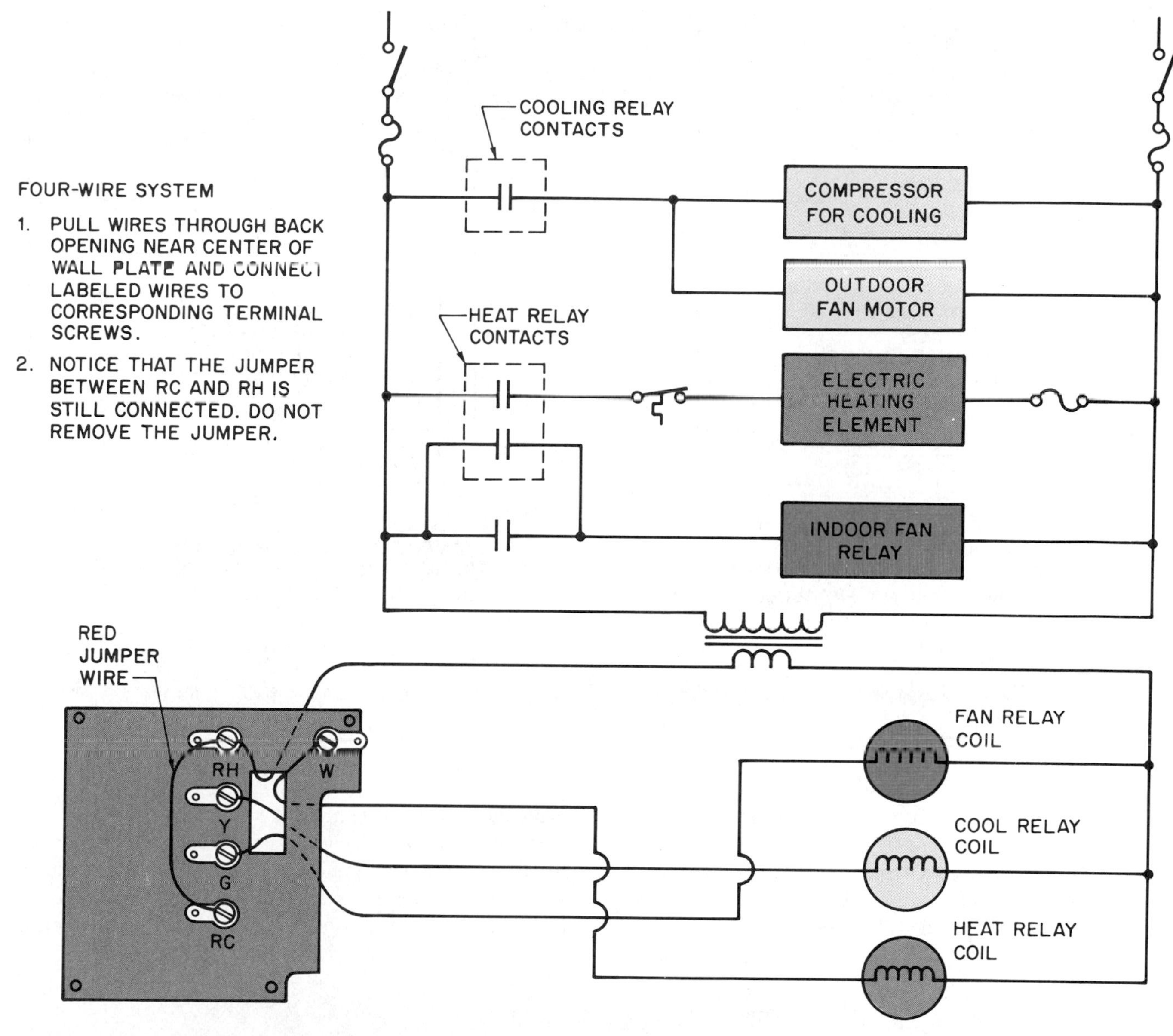

FIGURE 16–9 Standard heat-cool thermostat replaced with a programmable thermostat.

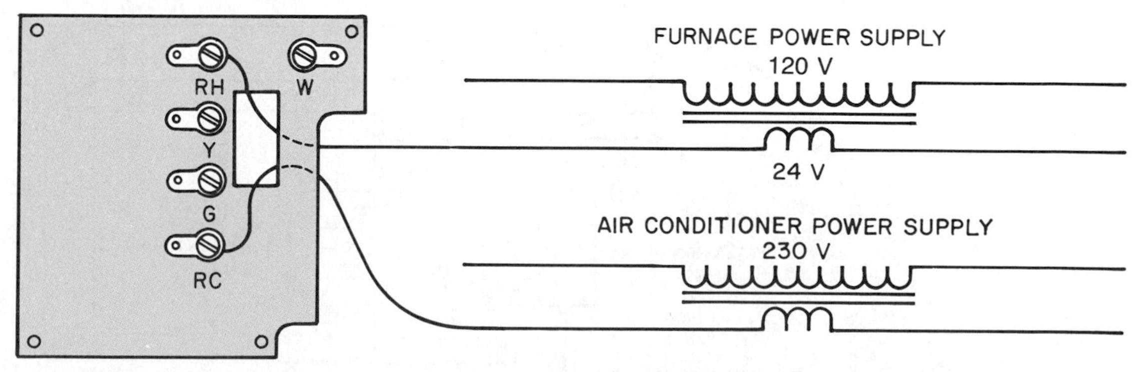

FIVE-WIRE SYSTEM WHERE SEPARATE TRANSFORMERS ARE USED ON HEATING SYSTEM AND COOLING SYSTEM.
1. REMOVE AND DISCARD THE RED JUMPER WIRE THAT CONNECTS TERMINALS RH AND RC ON WALL PLATE.
2. PULL WIRES THROUGH BACK OPENING NEAR CENTER OF WALL PLATE AND CONNECT LABELED WIRES TO CORRESPONDING TERMINAL SCREWS.

FIGURE 16–10 This illustrates an application with two 24-V power supplies. One power supply comes to RH (heat) and the other to RC (cool). The thermostat and subbase keep the two supplies from coming together. (If the two power supplies are phase checked, it is permissible to wire them toget' Phase checking involves making sure that the primaries of the transformers are parallel and that the secondaries are parallel, in phase.)

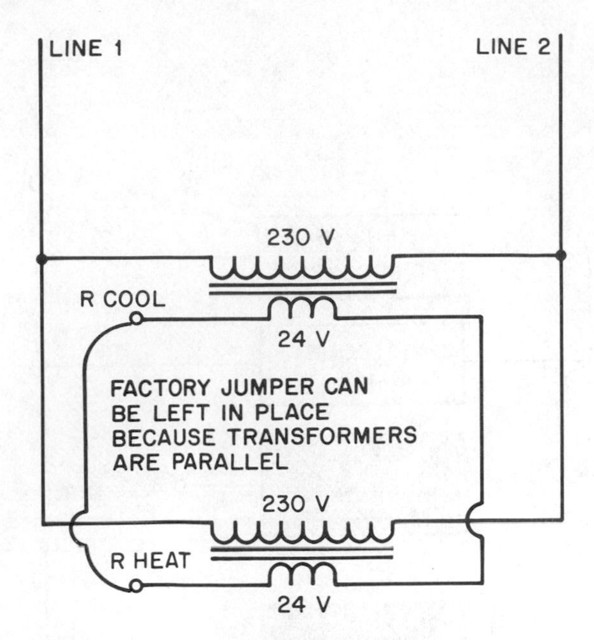

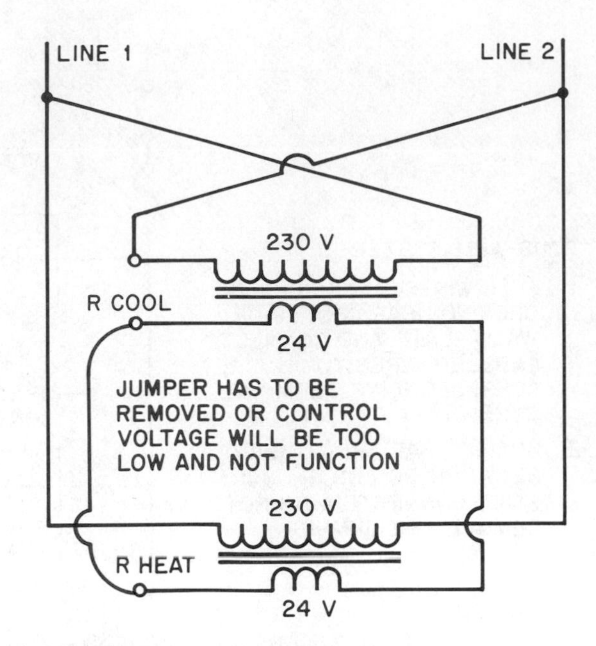

FIGURE 16–11 The second example can be corrected by changing the primary or secondary wiring. Sparks will be emitted if the secondary wiring is wrong when touched together.

tronic spark ignition and a vent damper shutoff, Figure 16–12.

This thermostat has many features that can save money in the cooling and heating seasons by turning on the equipment only when really needed. The conventional thermostat maintains a constant temperature in the conditioned space unless it is manually changed. The electronic thermostat can be set up to maintain minimum or maximum temperatures when no one is present and then automatically change to the correct conditions just before the people return. The typical

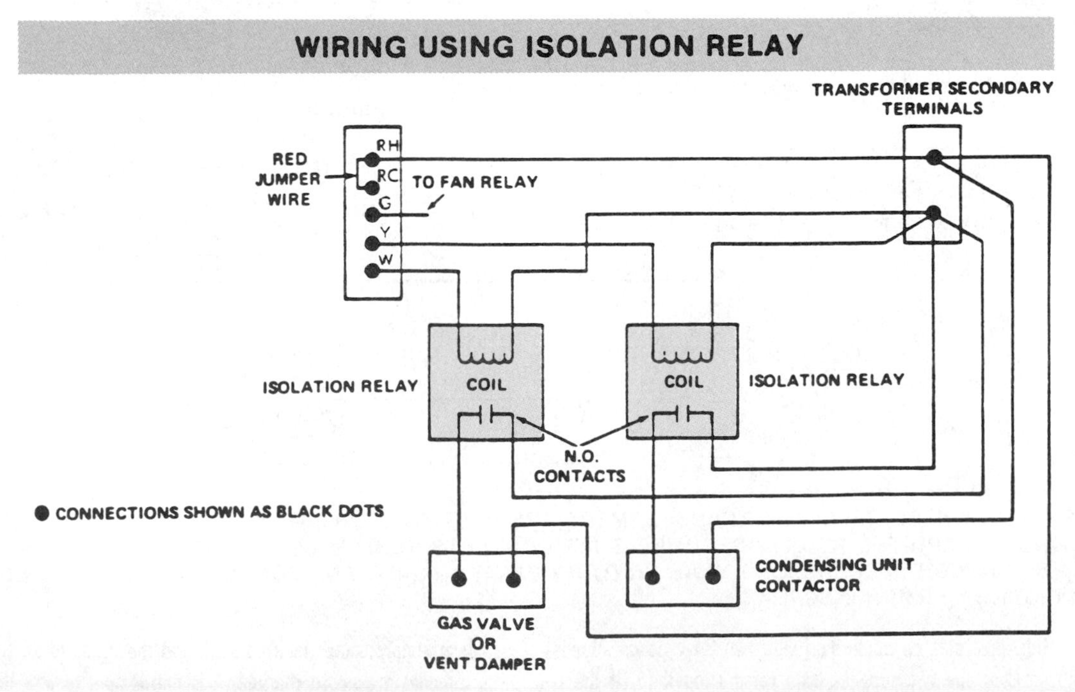

FIGURE 16–12 Wiring diagram shows use of an isolation relay.

homeowner or business manager can also override the program in special cases. When the program has been altered, it will automatically go back to the program at the next scheduled change. See Figure 16–13 for a list of the manufacturer's features and specifications.

16.9 PROGRAMMING THE ELECTRONIC THERMOSTAT

The programming of this thermostat is typical of most thermostats on the market and is intended for the average person to be able to set the basic functions. Each manufacturer has its own method of programming its thermostat but most have similar features and similar methods to achieve them. Most of the thermostats have explicit instructions for the operator to follow. These instructions should be studied and the operator should become aware of the various operating modes. Figure 16–14 gives some of the programming information on this thermostat. It is flexible for the owner who does not have a predictable schedule.

16.10 POWER OUTAGE AND THE ELECTRONIC THERMOSTAT

If the power goes off for a relatively short period, the program is not normally lost. Manufacturers use various methods to prevent this. For example, in Figure 16–15 the thermostat is backed up with a battery for up to 3 days. When a long power failure occurs, the program may be lost. The owner's manual shows how to reset the program. Some thermostats maintain an acceptable set temperature with a default program until the program is reestablished. This protects the premises from overheating or overcooling (maybe below freezing).

16.11 SERVICE TECHNICIAN CALLS

SERVICE CALL 1

A customer calls and tells the company dispatcher that the unit in her house is not cooling. *The problem is that a wire is burned off the compressor contactor due to a bad connection, Figure 16–16.*

The technician arrives and talks to the customer about the problem, then goes to the indoor thermostat, which is a programmable type. The indoor fan is switched to ON. It runs. The technician then turns the thermostat to call for cooling. The indoor fan comes on again. This proves that the control voltage and thermostat are trying to operate the equipment.

The technician then goes to the outdoor unit. The cooling relay is humming. This means that the low-voltage call for cooling is reaching the outdoor unit. When the contactor is humming, it is energized and trying to start the compressor.

The power is turned off and the cover to the compartment and controls is removed. When the covers are removed, the power is turned on. The cooling relay is energized and still humming. Line power has to be reaching the unit because it is the power source for the control transformer. Line power is checked into and out of the

SPECIAL FEATURES

- Separate set-back programming for five day week and two day weekend.
- Four separate time/temperature settings per 24 hour period.
- LCD displays continuous set point, time and room temperature alternately.
- Manual temperature control for overriding program.
- Fahrenheit or Celsius temperature display.
- Sensor for 100° F protection.
- Independently adjustable anticipation for heating and cooling.
- Compressor short cycle protection.
- Pre-programmed thermostat.
- Heating set point resets to 68° F (power loss).
- Cooling set point resets to 78° F (power loss).
- Indicators for "Hold Temp" — "System Cycle".
- Battery backup option.

SPECIFICATIONS

Electrical Rating: 17 to 30 volts 60 Hz
0.15 to 1.5 Amperes
WARNING: **Do not use on circuits exceeding 30 volts. Higher voltage will damage control—could cause shock or fire hazard.**
Temperature Range: 40° F to 90° F
4° C to 32° C
Rated Differential: 0.3° to 1.8° F—Heating
0.8° to 1.8° F—Cooling
Mountings: Wiring wall plate mounts on wall.
Dimensions: 6-3/8" W × 1-3/4" D × 3-1/2" H

FIGURE 16–13 An example of features and specifications as shown by one manufacturer. *Courtesy White-Rodgers Division, Emerson Electric Co.*

Get to know your Thermostat

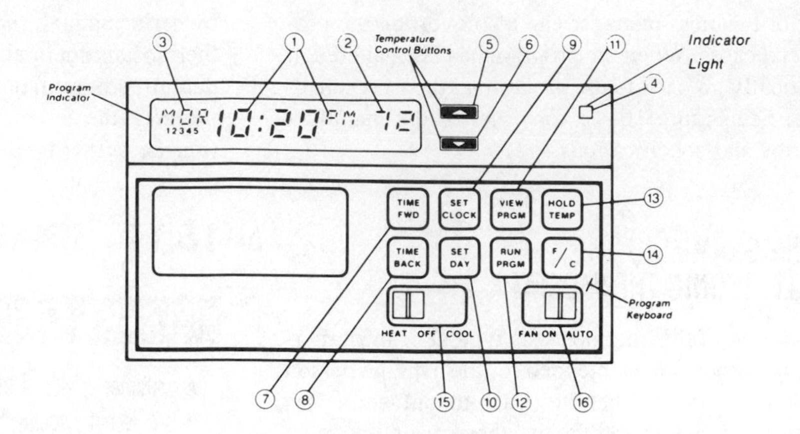

Program Display.	①	Alternately shows time and current room temperature.
	②	Continuously shows programmed temperature setting.
	③	When VIEW PROGRAM button is pressed, display shows time period being programmed; such as MOR, DAY and etc. When HOLD TEMP button is pressed, "HLD" is displayed. This shows the thermostat has been set to continuously hold a fixed temperature.
Indicator Light.	④	Shows heating or cooling system is running.
Temperature Control Buttons.	⑤	Red arrow raises temperature setting. (90°F or 32°C maximum)
	⑥	Blue arrow lowers temperature setting. (40°F or 4°C minimum)
Program Control Buttons.	⑦ TIME FWD	For advancing time to change program setting.
	⑧ TIME BACK	For moving back time to change program setting.
	⑨ SET CLOCK	For setting current time.
	⑩ SET DAY	For setting current day of the week.
	⑪ VIEW PRGM	Press and hold to review weekday/weekend programming for all 8 program time/temperature schedules.
	⑫ RUN PRGM	Push to start your thermostat when programming is complete.
	⑬ HOLD TEMP	To temporarily override all program temperatures without changing the weekly program.
	⑭ F/C	For setting the display to show either Fahrenheit or Celsius temperature.
	⑮ HEAT OFF COOL	Sets the thermostat to control either your furnace or your air conditioner.
	⑯ FAN ON AUTO	For selecting continuous or automatic fan operation.

FIGURE 16–14 Manufacturer's instructions for a programmable thermostat. *Courtesy White-Rodgers Division, Emerson Electric Co.*

Planning Your Personal Schedule

Before you begin to program the thermostat you must plan your program. To do this, fill out the **Personal-Use Chart** on the next page as you answer the following questions:

Now, let's plan your heating season schedule for weekdays:

1. Morning
   a. What time does the first member of your household usually get up in the morning? (We suggest that you program the thermostat 1/2 hour before this time, so the house reaches the temperature you want at the desired time.)
   b. What temperature would you like the house, at this time?
2. Day
   a. What time does the last person usually leave the house for the day (for work, school, etc.)?
   b. What temperature would you like the house, at this time?
   (**NOTE:** If someone stays home all day, you can program this temperature to be the same as in the morning.)
3. Evening
   a. What time does your first family member return home?
   (We suggest that you program the thermostat 1/2 hour before this time, so the house reaches the temperature you want at the desired time.)
   b. What temperature would you like the house, at this time?
   (**NOTE:** If someone stays home all day, you can program this temperature to be the same as in the morning).
4. Night
   a. What time does the last member of your family usually go to bed?
   b. What temperature would you like the house, at this time?

Planning your weekend and cooling season schedules:

WHEN PLANNING YOUR INDIVIDUAL WEEKEND & COOLING SEASON SCHEDULES, SIMPLY REPEAT STEPS 1 THROUGH 4.

Remember, for the greatest energy savings:

—when *heating* your home, program the temperature to be cooler when you are gone.

—when *cooling* your home, program the temperature to be warmer when you are gone.

Pre-Programmed Time and Temperature

Your thermostat is pre-programmed with Time and Temperatures for Heating and Cooling Programs that are typical of the average residential user's lifestyle as follows:

HEATING Program for All Days of the Week:

STEP	TIME	TEMP
MOR	5:00 A.M.	70
DAY	9:00 A.M.	70
EVE	4:00 P.M.	70
NTE	10:00 P.M.	62

NOTE: If "System Switch" is in "HEAT" or "OFF" position, when thermostat is installed, the pre-program shown will be for **Heating Program Only.** (Program will be shown when view program button is pressed.)

COOLING Program for All Days of the Week:

STEP	TIME	TEMP
MOR	5:00 A.M.	78
DAY	9:00 A.M.	82
EVE	4:00 P.M.	78
NTE	10:00 P.M.	78

NOTE: If "System Switch" is in "COOL" position when thermostat is installed, the pre-program shown will be for **Cooling Program Only.** (Program will be shown when view program button is pressed.)

FIGURE 16–14 Continued

PERSONAL-USE CHART		HEATING		COOLING	
		TIME	TEMP	TIME	TEMP
Weekdays (1–5)	1. Morning	a.	b.	a.	b.
	2. Day	a.	b.	a.	b.
	3. Evening	a.	b.	a.	b.
	4. Night	a.	b.	a.	b.
Weekends (6 & 7)	1. Morning	a.	b.	a.	b.
	2. Day	a.	b.	a.	b.
	3. Evening	a.	b.	a.	b.
	4. Night	a.	b.	a.	b.

FIGURE 16–14 Continued

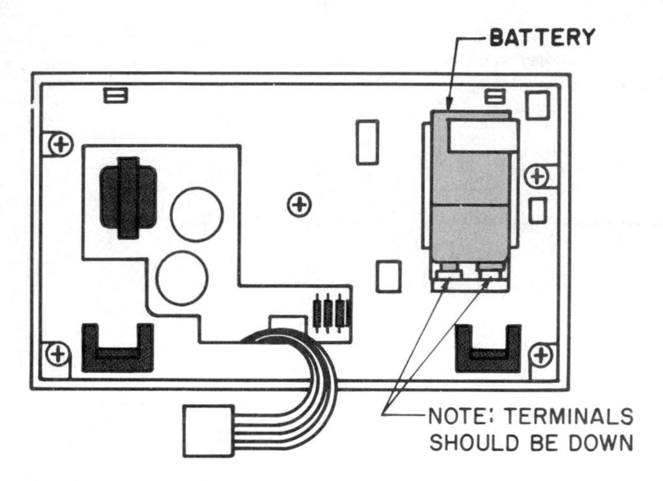

FIGURE 16–15 An optional 9-V battery can be installed as a backup for power failure.

compressor contactor. Power is going in but not coming out. The relay has a bad connection. ***The power is turned off, the panel is locked and tagged, and the connection repaired.**** When the power is turned back on, the unit starts and runs normally.

The technician notifies the customer as to what the problem was, completes the paperwork, and leaves for another job.

SERVICE CALL 2

A customer calls and complains that the cooling is not operating in a small retail craft shop in a shopping center. The system is an electric furnace with a split-system air conditioner; the condenser is behind the store. *The system will not cool because the electronic programmable thermostat will not pass power, Figure 16–17.*

The technician goes to the indoor thermostat and switches on the indoor fan and it starts. This proves that the line voltage is reaching the indoor unit because the 24-V control transformer is located in the indoor unit. The thermostat is switched to call for cooling. Neither the indoor fan nor the outdoor unit starts.

The technician goes to the outdoor unit, removes the panel to the low-voltage terminal block, and checks for voltage at the cooling relay coil. There is no voltage.

The technician returns to the indoor thermostat and checks the program according to the manufacturer's instructions. The program is correct. The thermostat is removed from the subbase. A jumper is placed on the R terminal and the G terminal. The indoor fan starts. The jumper is left on the fan circuit and another is placed between the R terminal and the Y terminal. The outdoor fan and compressor start and run with the indoor fan. The thermostat is not passing power and needs to be replaced.

The technician obtains another thermostat and changes it. The system is then switched to cool, and after a short time delay, the condensing unit starts cooling. While the technician is there with the new thermostat, the complete control circuit is checked by operating the heating system using the new thermostat. It all functions correctly.

The store owner asks the technician to program the thermostat for the typical store hours, 8 AM to 5 PM, Monday through Saturday and off on Sunday.

The following service calls do not describe the solution.

SERVICE CALL 3

A customer in a small retail store has a package unit with electric heat and air conditioning and complains that the unit has stopped heating.

The technician arrives and goes to the room thermostat and notices that there are no figures showing on the face of the thermo-

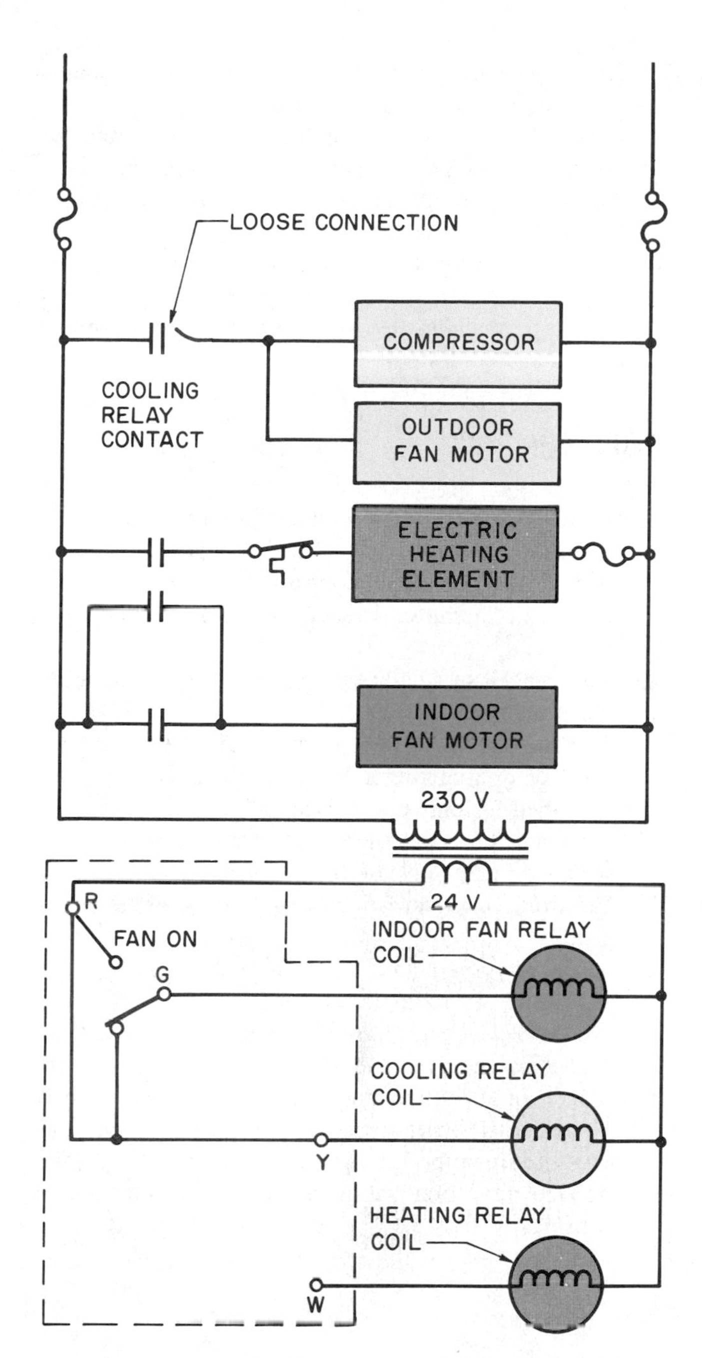

FIGURE 16–16 A loose connection at the cooling relay terminal does not allow power to get to the compressor and outdoor fan motor. The coil is energized because it is humming. This represents what happens inside the electronic circuit on a call for cooling. R passes power to G and Y.

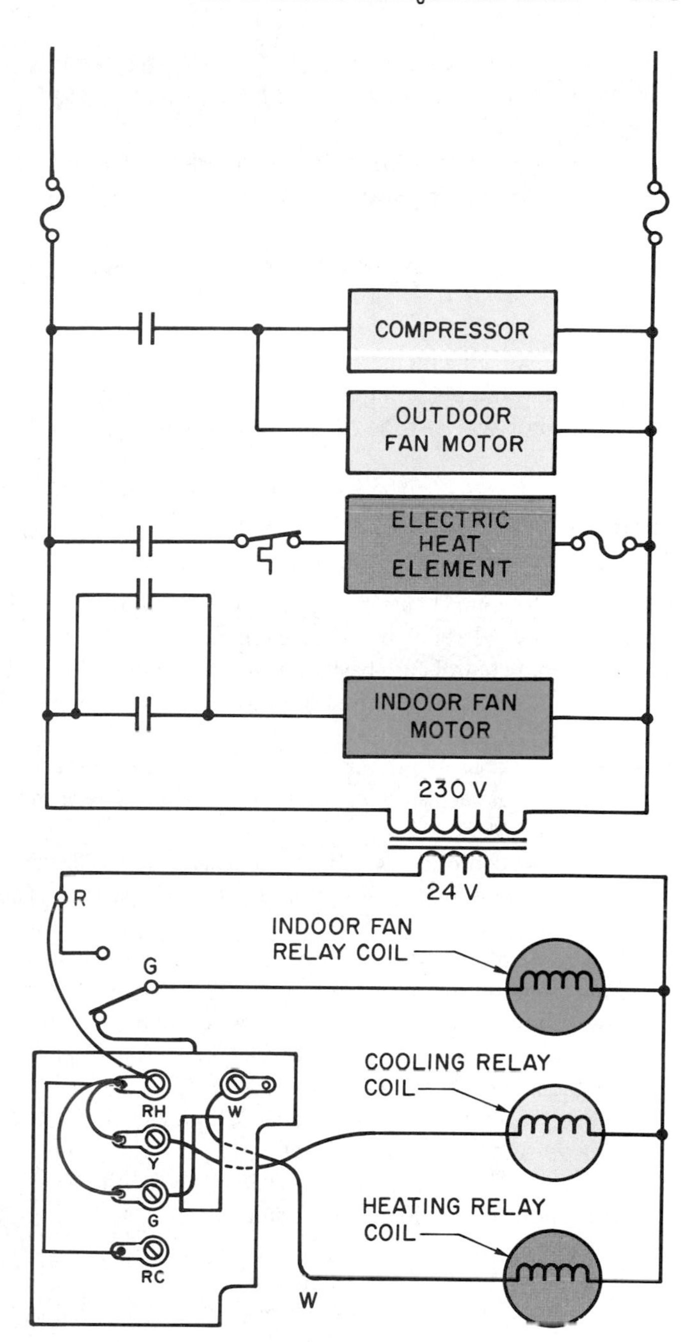

FIGURE 16–17 The 24-V control voltage is going into the thermostat but will not come out on a call for cooling. The thermostat is removed from the subbase, a jumper attached from R to G, and the indoor fan starts. Another jumper is attached from R to Y, and the outdoor unit starts.

stat. There should be some numbers or writing on the face of the digital thermostat. The technician reviews the diagram in Figure 16–17.

The technician removes the thermostat from the subbase and jumps from RH to W, goes to the electric furnace and discovers that there is no power at the heating relay. The technician then goes to the room thermostat and jumps from RC to G and the indoor fan does not start.

What is the problem and the recommended solution?

SERVICE CALL 4

A customer calls to indicate that the cooling is not operating in a small retail store. This is an electric furnace with a split-system air conditioner with the condenser behind the store.

The technician arrives and notices the indoor fan is not running. The evaporator is iced up because the compressor is still running. The technician goes to the furnace and can tell the fan is not trying to run. The door is removed; the fan is free to turn and the motor is not hot.

The technician checks for voltage at the fan and there is no line voltage to the fan motor. The assumption is made that the relay is not functioning.

The technician then returns to the room thermostat and removes it from the subbase and a jumper is applied from RC to G and the fan starts.

What is the problem and the recommended solution?

Get acquainted with the written material provided by the manufacturer before starting any installation or service job. The distributor will also be helpful if you have questions.

SUMMARY

- Electronic controls are being used more frequently because they are reliable and economical.
- Electronic controls are basically used to replace electrical and electromechanical controls where they are applicable.
- Electronic controls are basically used in a safety and operating capacity, with more emphasis on energy management.
- Electronic controls can monitor high and low voltages and easily add time delays and sequence of operation to the control system.
- The residential and light commercial air conditioning control circuit may use electronics to monitor high and low voltages, time delay, and current draw of the compressor. It has the board connections for electromechanical controls where applicable.
- The oil furnace uses an electronic control circuit to achieve flame-proving methods.
- Oil burner equipment uses the cad cell to see the flame and report to the electronic circuit.
- The gas furnace can use electronic controls to monitor gas flame for flame-proving methods.
- The gas furnace uses several different methods to prove the flame and report to the electronic circuit board.
- The electronic programmable thermostat is becoming a popular electronic device.
- Troubleshooting electronic controls is similar to troubleshooting electromechanical controls because each circuit board is normally treated as a single component; current goes in and comes out.

REVIEW QUESTIONS

Answer questions 1 through 9 either True or False.

1. Electronic controls are normally more economical than electromechanical controls.
2. Electronic thermostats react faster than bimetal thermostats.
3. The cad cell sees darkness and shuts off the oil burner when there is no flame.
4. The gas furnace can use electronic flame detection.
5. Electronic programmable thermostats can be programmed for only one program.
6. Electronic programmable thermostats can normally be installed in place of a typical thermostat.
7. Electronic circuits can generally be repaired in the field?
8. Electronic programmable thermostats are designed so the owner can reprogram them.
9. A qualified technician should be called to troubleshoot electronic circuit problems.
10. Which of the following control objectives are best obtained with electronic controls: time delay, low-pressure monitoring, high-voltage protection, low-voltage protection, vibration protection, temperature sensing?

SECTION

4 *Electric Motors*

UNIT

17 Types of Electric Motors

OBJECTIVES

After studying this unit, you should be able to

- describe the different types of open single-phase motors used to drive fans, compressors, and pumps.
- describe the applications of the various types of motors.
- state which motors have high starting torque.
- list the components that cause a motor to have a higher starting torque.
- describe a multispeed permanent split-capacitor motor and indicate how the different speeds are obtained.
- explain the operation of a three-phase motor.
- describe a motor used for a hermetic compressor.
- explain the motor terminal connections in various compressors.
- describe the different types of compressors that use hermetic motors.
- describe the use of variable speed motors.

17.1 USES OF ELECTRIC MOTORS

Electric motors are used to turn the prime movers of air, water, and refrigerant. The prime movers are the fans, pumps, and compressors, Figure 17–1. Several types of motors, each with its particular use, are available. For example, some applications need motors that will start under heavy loads and still develop their rated work horsepower at a continuous running condition, whereas others are used in installations that don't need much starting torque but must develop their rated horsepower under a continuous running condition. Some motors run for years in dirty operating conditions, and others operate in a refrigerant atmosphere. These are a few of the typical applications of motors in this industry. The technician must understand which motor is suitable for each job so that effective troubleshooting can be accomplished and, if necessary, the motor replaced by the proper type. But the basic operating principles of an electric motor must be first understood. Although many types of electric motors are used, most motors operate on similar principles.

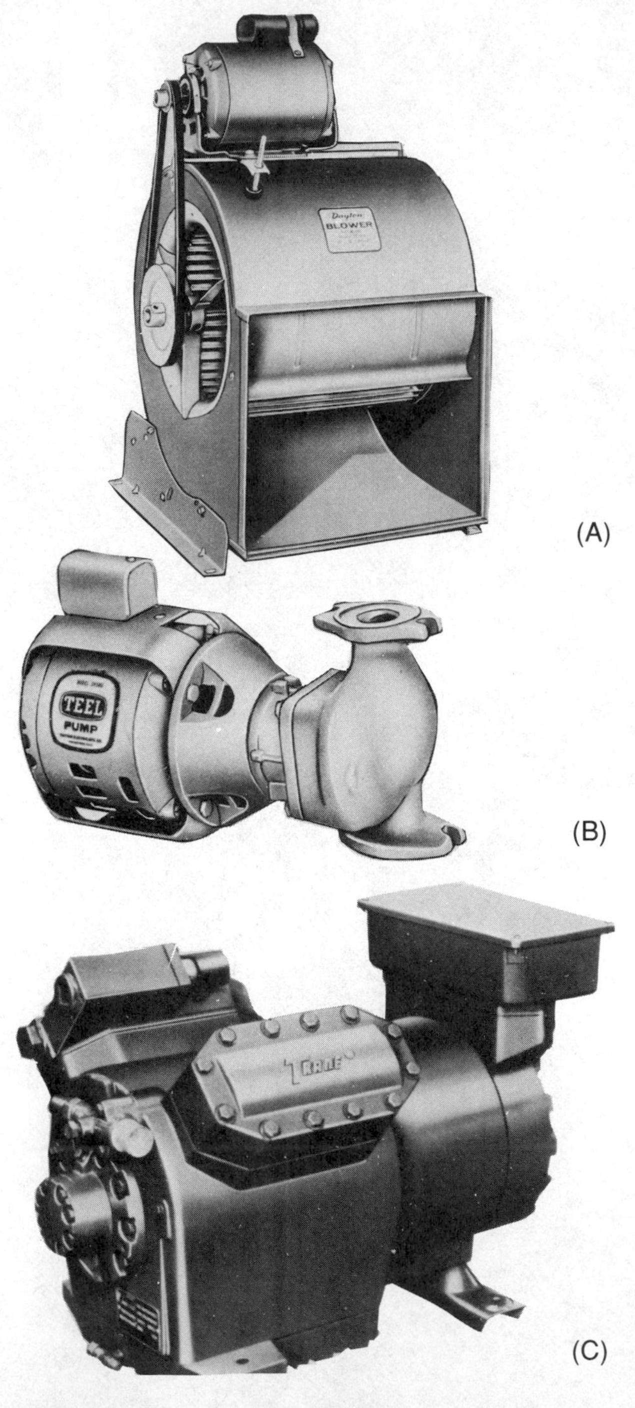

FIGURE 17–1 (A) Fans move air. (B) Pumps move water. (C) Compressors move vapor refrigerant. *Courtesy (A) and (B) W. W. Grainger, Inc., (C) Trane Company*

17.2 PARTS OF AN ELECTRIC MOTOR

Electric motors have a *stator* with windings, a *rotor, bearings, end bells, housing,* and some means to hold these parts in the proper position, Figures 17–2 and 17–3.

FIGURE 17–2 Cutaway of an electric motor. *Courtesy Century Electric, Inc.*

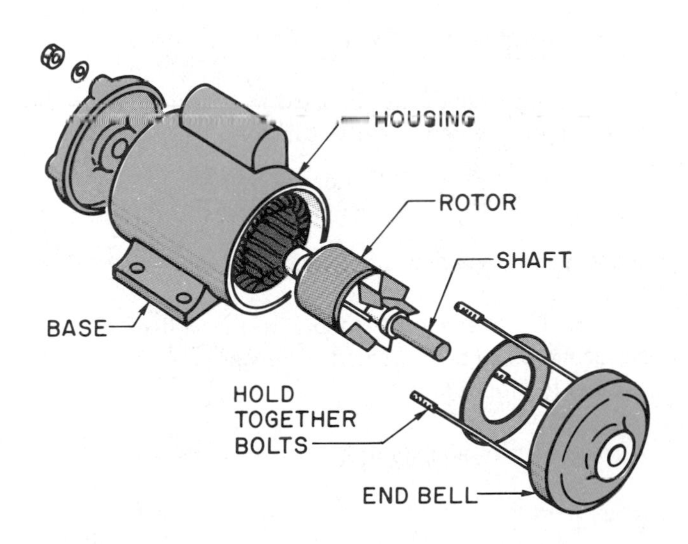

FIGURE 17–3 Individual electric motor parts.

17.3 ELECTRIC MOTORS AND MAGNETISM

Unlike poles of a magnet attract each other, and like poles repel each other. If a stationary horseshoe magnet were placed with its two poles (north and south) at either end of a free-turning magnet as in Figure 17–4, one pole of the free rotating magnet would line up with the opposite pole of the horseshoe magnet. If the horseshoe magnet were an electromagnet and the wires on the battery were reversed, the poles of this magnet would reverse and the poles on the free magnet would be repelled, causing it to rotate until the unlike poles again were lined up. This is the basic principle of electric motor operation. The horseshoe magnet is the stator and the free-rotating magnet the rotor.

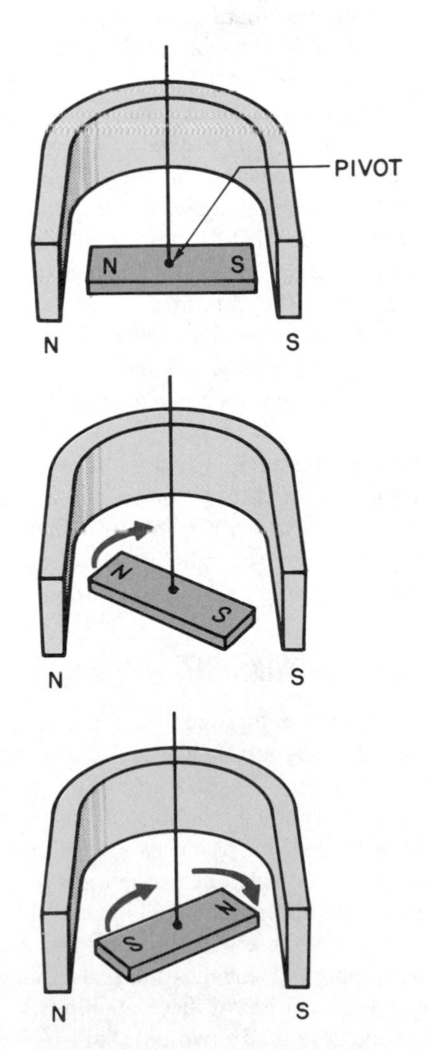

FIGURE 17–4 Poles (north and south) on rotating magnet will line up with opposite poles on stationary magnet.

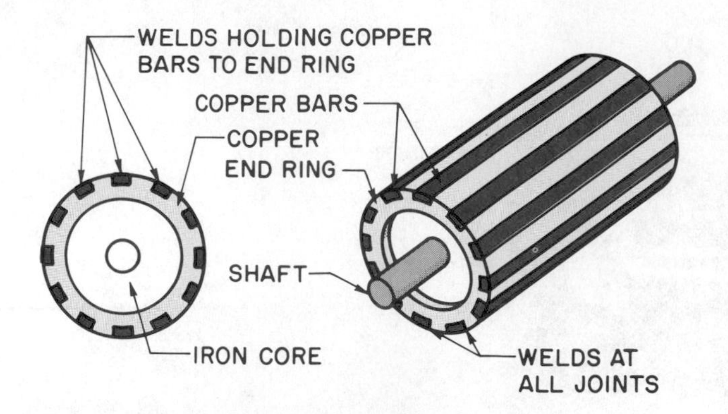

FIGURE 17–5 Simple sketch of squirrel cage rotor.

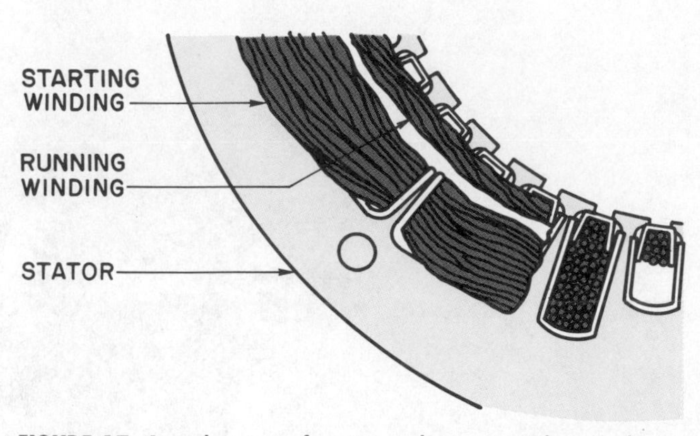

FIGURE 17–6 Placement of starting and running windings inside stator.

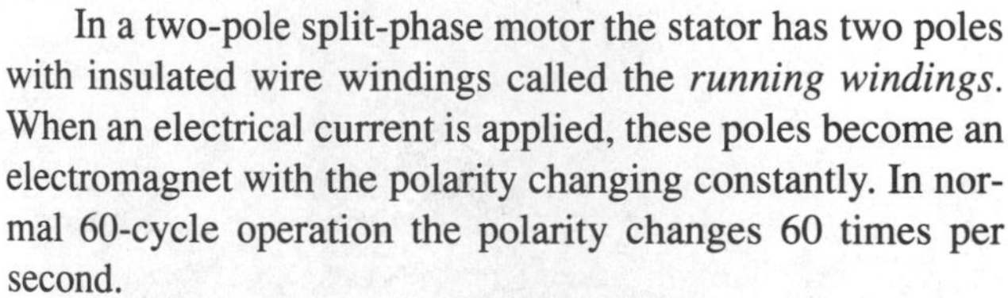

In a two-pole split-phase motor the stator has two poles with insulated wire windings called the *running windings*. When an electrical current is applied, these poles become an electromagnet with the polarity changing constantly. In normal 60-cycle operation the polarity changes 60 times per second.

The rotor may be constructed of bars, Figure 17–5. This type is called a *squirrel cage rotor.* When the rotor shaft is placed in the bearings in the bell type ends, it is positioned between the running windings. When an alternating current (AC) is applied to these windings, a magnetic field is produced in the windings and a magnetic field is also induced in the rotor. The bars in the rotor actually form a coil. This is similar to the field induced in a transformer secondary by the magnetic field in the transformer primary. The field induced in the rotor has a polarity opposite that in the running windings.

The attracting and repelling action between the poles of the running windings and the rotor sets up a rotating magnetic field and causes the rotor to turn. Since this is AC reversing 60 times per second, the rotor turns, in effect "chasing" the changing polarity in the running windings.

17.4 STARTING WINDINGS

Starting windings placed between the running windings ensure that the rotor starts properly and that it turns in the desired direction, Figure 17–6. The starting windings have more turns than the running windings and are wound with a smaller diameter wire. This produces a larger magnetic field and greater resistance, which helps the rotor to start turning and determines the direction in which it will turn. This happens as a result of these windings being located between the running windings. It changes the phase angle between the voltage and the current in these windings.

We have just described a two-pole split-phase *induction* motor, which is rated to run at 3600 revolutions per minute (rpm). It actually turns at a slightly slower speed when running under full load. When the motor reaches approximately 75% of its normal speed, a centrifugal switch opens the circuit to the starting windings and the motor continues to operate on only the running windings. Many split-phase motors have four poles and run at 1800 rpm.

17.5 DETERMINING A MOTOR'S SPEED

The following formula can be used to determine the synchronous speed (without load) of motors.

$$S \text{ (rpm)} = \frac{\text{Frequency} \times 120}{\text{Number of poles}}$$

Frequency is the number of cycles per second (also called *hertz*).

NOTE: The magnetic field builds and collapses twice each second (each time it changes direction), therefore 120 is used in the formula instead of 60.

$$\text{Speed of two-pole split-phase motors} = \frac{60 \times 120}{2} = 3600$$

$$\text{Speed of four-pole split-phase motors} = \frac{60 \times 120}{4} = 1800$$

The speed under load of each motor will be approximately 3450 rpm and 1750 rpm. The difference between synchronous speed and the actual speed is called slip. Slip is caused by the load.

17.6 STARTING AND RUNNING CHARACTERISTICS

Two major considerations of electric motor applications are the starting and running characteristics. A motor applied to a refrigeration compressor must have a high starting torque —it must be able to start under heavy starting loads. A refrigeration compressor may have a head pressure of 155 psig and a suction pressure of 5 psig and still be required to start in systems where the pressures do not equalize. The pressure difference of 150 psig is the same as saying that the compressor has a starting resistance of 150 psi of piston area. If this compressor has a 1-in. diameter piston, the area of the piston is 0.78 in.^{2} ($A = \pi r^2$ or $3.14 \times 0.5 \times 0.5$). This area multiplied by the pressure difference of 150 psi is the starting resistance for the motor (117 lb). This is similar to a 117-lb weight resting on top of the piston when it tries to start, Figure 17–7.

To start a small fan, a motor does not need as much starting torque. The motor must simply overcome the friction needed to start the fan moving. There is no pressure difference because the pressures equalize when the fan is not running, Figure 17–8.

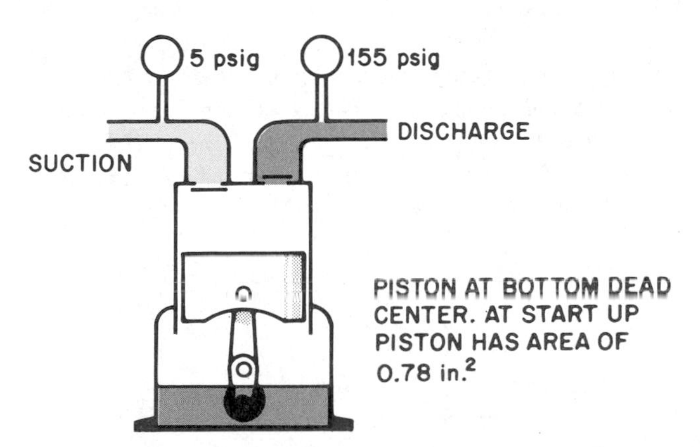

FIGURE 17–7 Compressor with high-side pressure 155 psig and low-side pressure 5 psig.

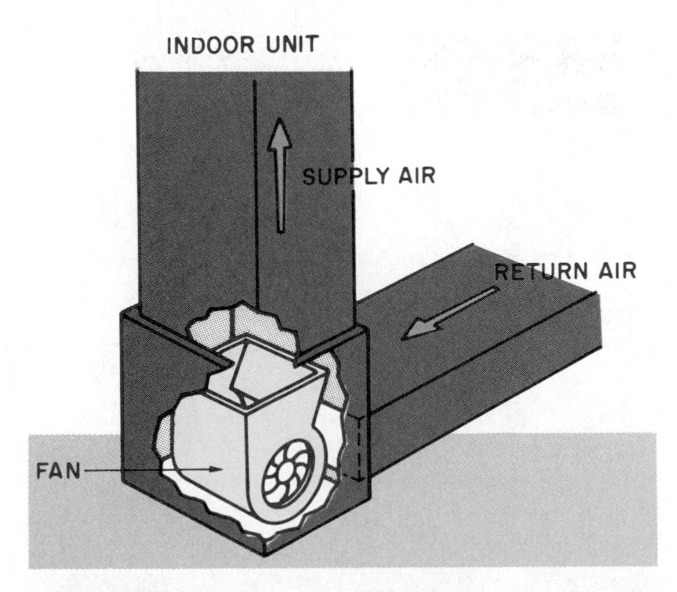

FIGURE 17–8 This fan has no pressure difference to overcome while starting. When it stops, the air pressure equalizes.

17.7 POWER SUPPLIES FOR ELECTRIC MOTORS

One other basic difference in electric motors is the power supply used to operate them. For example, only single-phase power is available to homes. Motors designed to operate on single-phase power must therefore be used. For large loads in factories, for example, single-phase power is inadequate, so three-phase power is used to operate the motors. The difference in the two power sources changes the starting and running characteristics of the motors.

Following is a description of some motors currently used in the heating, air conditioning, and refrigeration industry. The electrical characteristics, not the working conditions, are emphasized. Some older motors are still in operation, but they are not discussed in this text.

17.8 SINGLE-PHASE OPEN MOTORS

The power supply for most *single-phase* motors is either 120 V or 208 V to 230 V. A home furnace has a power supply of 120 V, whereas the air conditioner outside uses a power supply of 230 V, Figure 17–9. A commercial building may have either 230 V or 208 V, depending on the power company. Some single-phase motors are dual voltage. The motor has two run windings and one start winding. The two run windings have the same resistance and the start winding has a high resistance. The motor will operate with the two run windings in parallel in the low-voltage mode. When it is required to run in the high-voltage mode, the technician changes the numbered motor leads according to the manufacturer's instructions. This wires the run windings in series with each other and delivers an effective voltage of 120 V to each winding. It can be said that the motor windings are actually only 120 V because they only operate on

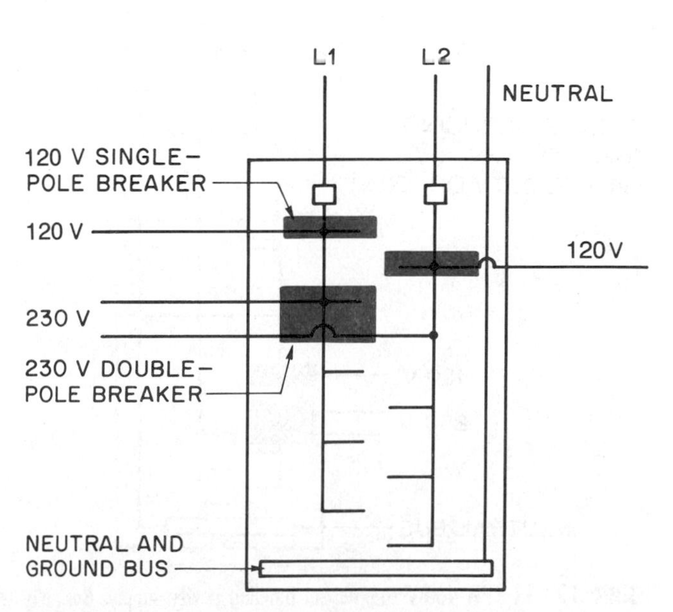

FIGURE 17–9 Main breaker panel for a typical residence.

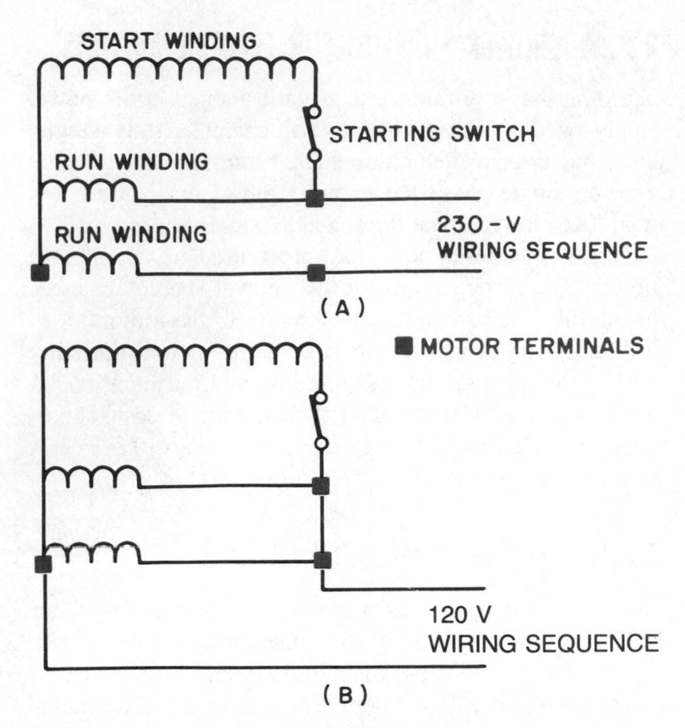

FIGURE 17–10 Wiring diagram of dual-voltage motor. It is made to operate using 120 V or 230 V, depending on how the motor is wired. (A) 230-V wiring sequence. (B) 120-V wiring sequence.

120 V, no matter which mode they are in. The technician can change the voltage at the motor terminal box, Figure 17–10.

Some commercial and industrial installations may use a 460-V power supply for large motors. The 460 V may be reduced to a lower voltage to operate the small motors. The smaller motors may be single phase and must operate from the same power supply, Figure 17–11.

A motor can rotate clockwise or counterclockwise. Some motors are reversible from the motor terminal box, Figure 17–12.

17.9 SPLIT-PHASE MOTORS

Split-phase motors have two distinctly different windings, Figure 17–13. They have a medium amount of starting torque and a good operating efficiency. The split-phase motor is normally used for operating fans in the fractional horsepower range. Its normal operating ranges are 1800 rpm and 3600 rpm. An 1800-rpm motor will normally operate at 1725 rpm to 1750 rpm under a load. The difference in the rated rpm and the actual rpm is called *slip*. If the motor is loaded to the point where the speed falls below 1725 rpm, the current draw will climb above the rated amperage. Motors rated at 3600 rpm will normally slip in speed to about

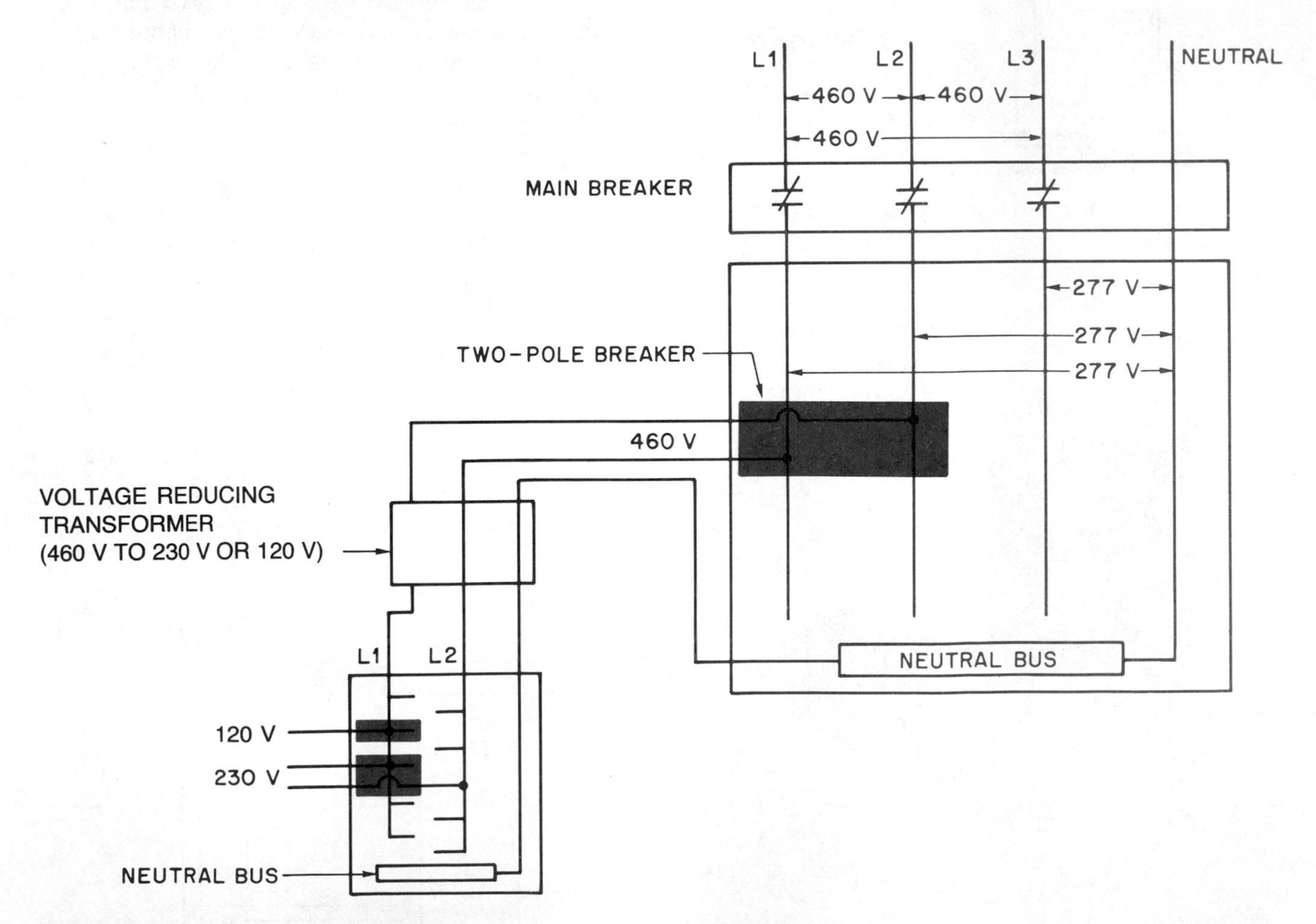

FIGURE 17–11 A 460-V commercial building power supply. Normally when a building has a 460-V power supply, it will have a step-down transformer to 120 V for office machines and small appliances.

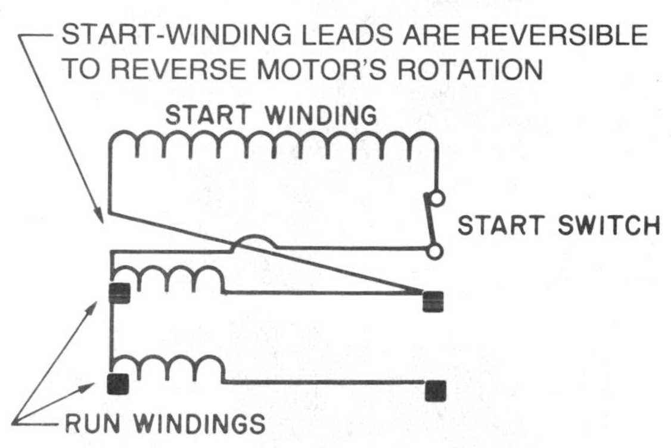

FIGURE 17–12 Single-phase motor can be reversed by changing the connections in the motor terminal box. The direction the motor turns is determined by the start winding. This can be shown by disconnecting the start-winding leads and applying power to the motor. It will hum and will not start. The shaft can be turned in either direction and the motor will run in that direction.

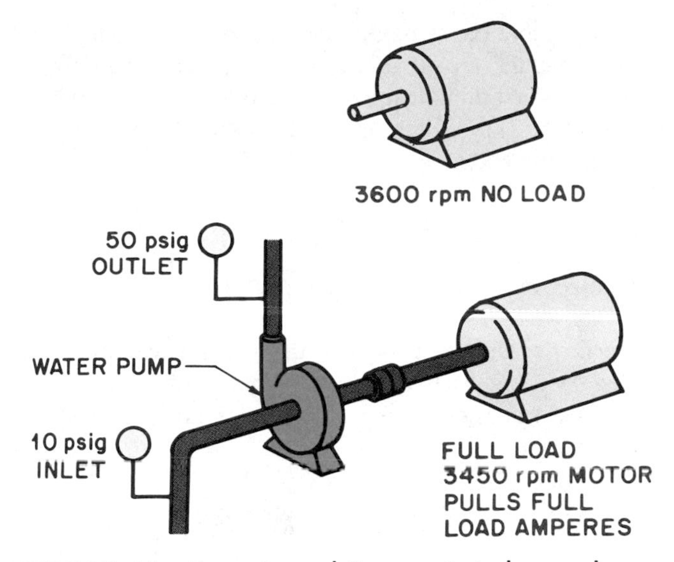

FIGURE 17–14 How motor revolutions per minute change under motor load.

3450 rpm to 3500 rpm, Figure 17–14. Some of these motors are designed to operate at either speed, 1750 rpm or 3450 rpm. The speed of the motor is determined by the number of motor poles and by the method of wiring the motor poles. The technician can change the speed of a two-speed motor at the motor terminal box.

17.10 THE CENTRIFUGAL SWITCH

The *centrifugal switch* is used to disconnect the start winding from the circuit when the motor reaches 75% of the rated speed. Motors described here are those that run in the atmosphere. (Hermetic motors that run in the refrigerant environment are discussed later.) When a motor is started in the air, the arc from the centrifugal switch will not harm the atmosphere. (It will harm the refrigerant, so there must be no arc in a refrigerant atmosphere.)

The centrifugal switch is a mechanical device attached to the end of the shaft with weights that will sling outward when the motor reaches 75% speed. For example, if the motor has a rated speed of 1725 rpm, at 1294 rpm (1725 × 0.75) the centrifugal weights will change position and open a switch to remove the start winding from the circuit. This switch is under a fairly large current load, so a spark will occur. If the switch fails to open its contacts and remove the start winding, the motor will draw too much current and the overload device will cause it to stop.

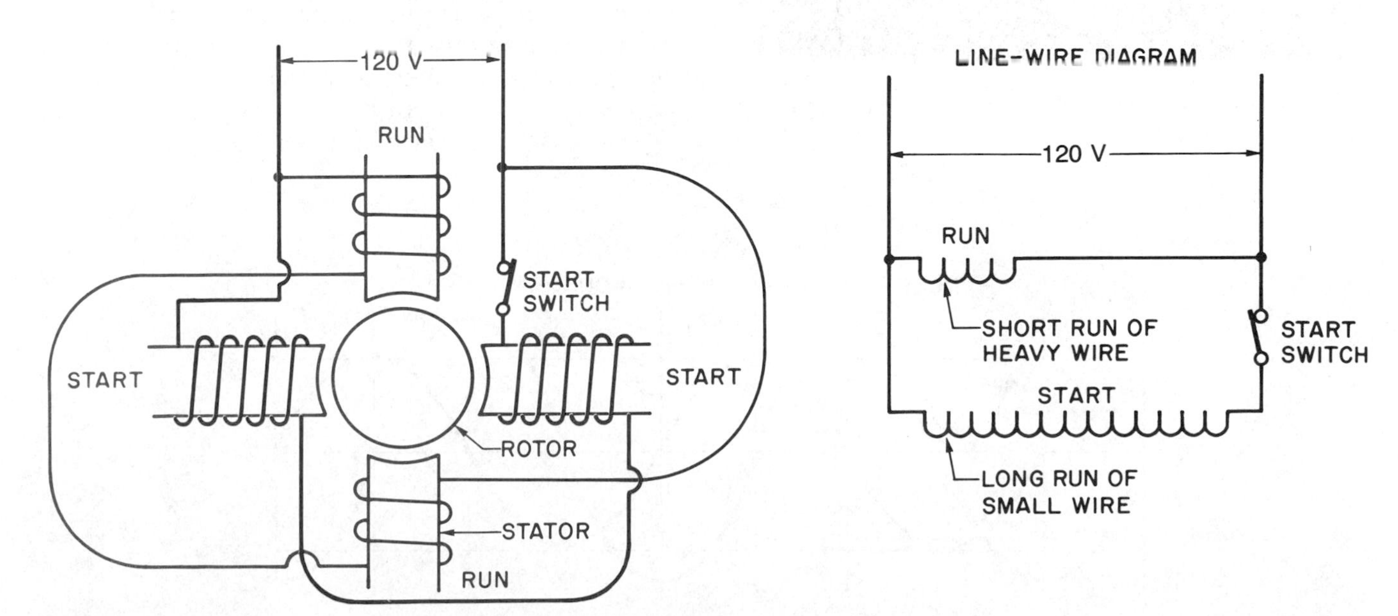

FIGURE 17–13 Diagram shows difference in the resistance in start and run windings.

The more the switch is used, the more its contacts will burn from the arc. If this type of motor is started many times, the first thing that will likely fail will be the centrifugal switch. This switch makes an audible sound when the motor starts and stops, Figure 17–15.

17.11 THE ELECTRONIC RELAY

The *electronic relay* is used with some motors to open the starting windings after the motor has started. This is a solid-state device designed to open the starting winding circuit when the design speed has been obtained. Other devices are also used to perform this function; they are described with hermetic motors.

17.12 CAPACITOR-START MOTORS

The capacitor-start motor is the same basic motor as the split-phase motor, Figure 17–16. It has the two distinctly different windings for starting and running. The previously mentioned methods may be used to interrupt the power to the starting windings while the motor is running. A start capacitor is wired in series with the starting windings to give the motor more starting torque. Figure 17–17 shows voltage and current cycles in an induction motor. In an inductive circuit the current *lags* the voltage. In a capacitive circuit the current *leads* the voltage. The amount by which the current leads or lags the voltage is the *phase angle.* A capacitor is chosen to make the phase angle such that it is most efficient for starting the motor, Figure 17–18. This capacitor is not designed to be used while the motor is running, and it must be switched out of the circuit soon after the motor starts. This is done at the same time the starting windings are taken out of the circuit.

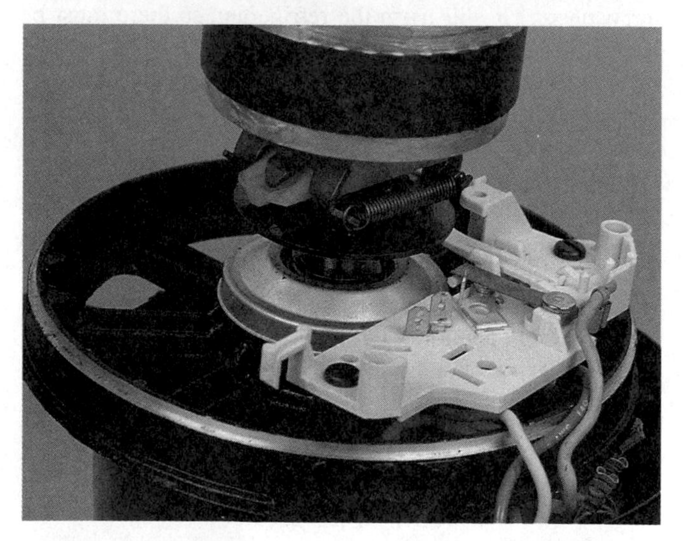

FIGURE 17–15 Centrifugal switch located at the end of the motor. *Photo by Bill Johnson*

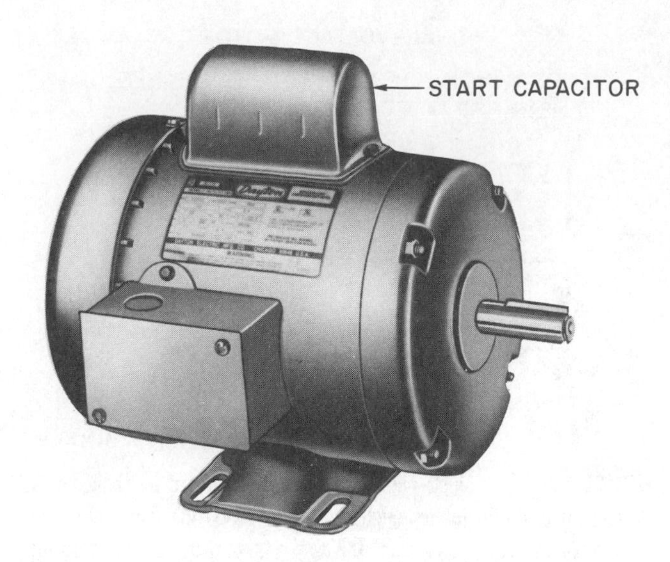

FIGURE 17–16 Capacitor-start motor. *Courtesy W. W. Gainger, Inc.*

17.13 CAPACITOR-START, CAPACITOR-RUN MOTORS

Capacitor-start, capacitor-run motors are much the same as the split-phase motors. The run capacitor is wired into the circuit to provide the most efficient phase angle between the current and voltage when the motor is running. The run capacitor is in the circuit at any time the motor is running. If

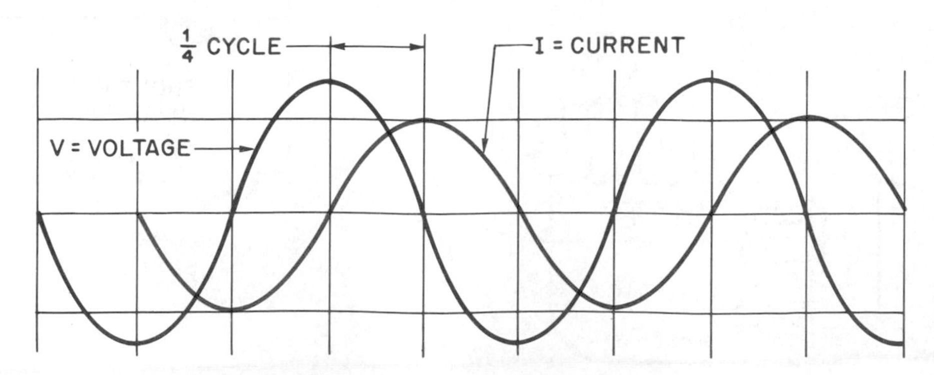

FIGURE 17–17 AC cycle, voltage, and current in an induction circuit. The current lags the voltage.

FIGURE 17–18 A start capacitor. *Photo by Bill Johnson*

FIGURE 17–20 Permanent split-capacitor motor. *Courtesy Universal Electric Company*

a run capacitor fails because of an open circuit within the capacitor, the motor may start, but the running amperage will be about 10% too high and the motor will get hot if operated at full load, Figure 17–19. The capacitor-start, capacitor-run motor is one of the most efficient motors used in refrigeration and air conditioning equipment. It is normally used with belt-drive fans and compressors.

17.14 PERMANENT SPLIT-CAPACITOR MOTORS

The permanent split-capacitor (PSC) motor has windings very similar to the split-phase motor, Figure 17–20, but it does not have a start capacitor. Instead it uses one run capacitor wired into the circuit in a similar way to the run capacitor in the capacitor-start, capacitor-run motor. This is the simplest split-phase motor. It is very efficient and has no moving parts for the starting of the motor; however, the starting torque is very low so the motor can only be used in low starting torque applications, Figure 17–21.

The PSC motor may have several speeds. A multispeed motor can be identified by the many wires at the motor

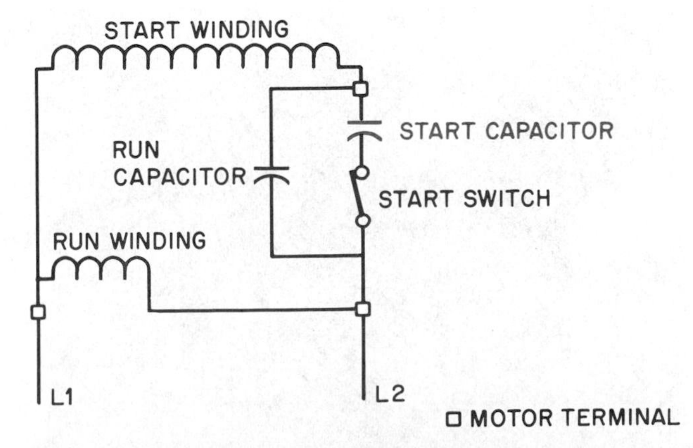

FIGURE 17–19 Wiring diagram of capacitor-start, capacitor-run motor. The start capacitor is only in the circuit during starting, and the run capacitor is in the circuit during starting and running of the motor.

FIGURE 17–21 Open PSC motor that may be used to turn a fan. *Courtesy Universal Electric Company*

electrical connections, Figures 17-22 and 17–23. As the resistance of the motor winding decreases, the speed of the motor increases. When more resistance is wired into the circuit, the motor speed decreases. Most manufacturers use this motor in the fan section in air conditioning and heating systems. Motor speed can be changed by switching the wires. Earlier systems used a capacitor-start, capacitor-run motor and a belt drive, and air volumes were adjusted by varying the drive pulley diameter.

The PSC motor may be used to obtain slow fan speeds during the winter heating season for higher leaving-air temperatures with gas, oil, and electric furnaces. The fan speed can be increased by switching to a different resistance in the winding using a relay. This will provide more airflow in summer to satisfy cooling requirements, Figure 17–24.

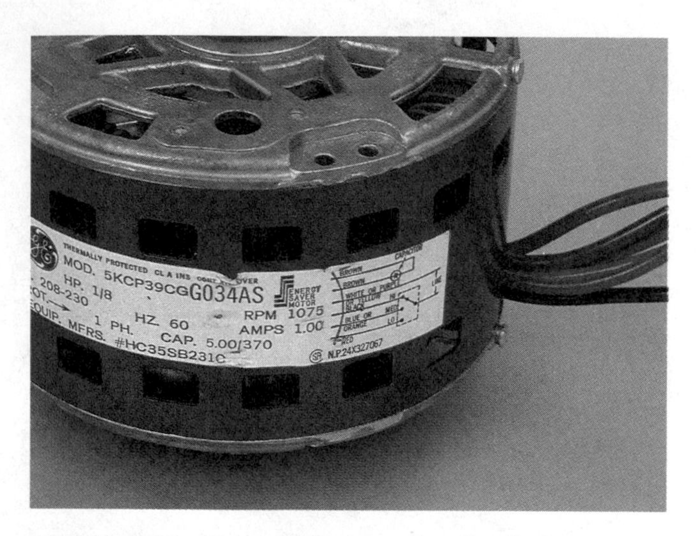

FIGURE 17–22 Multispeed PSC motor. *Photo by Bill Johnson*

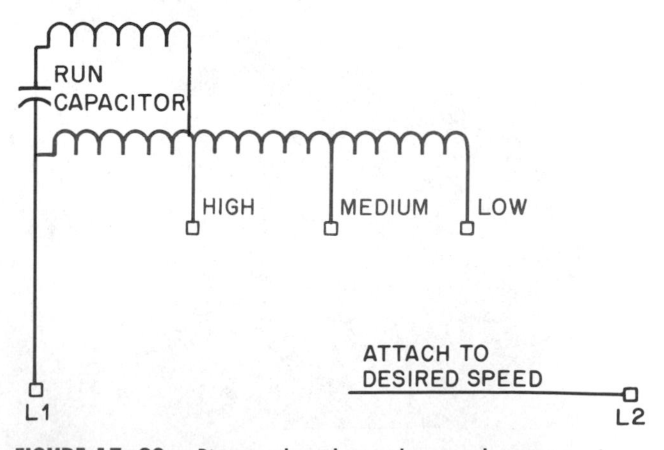

FIGURE 17–23 Diagram shows how a three-speed motor may be wired to run at a slow speed in the winter and high speed in the summer.

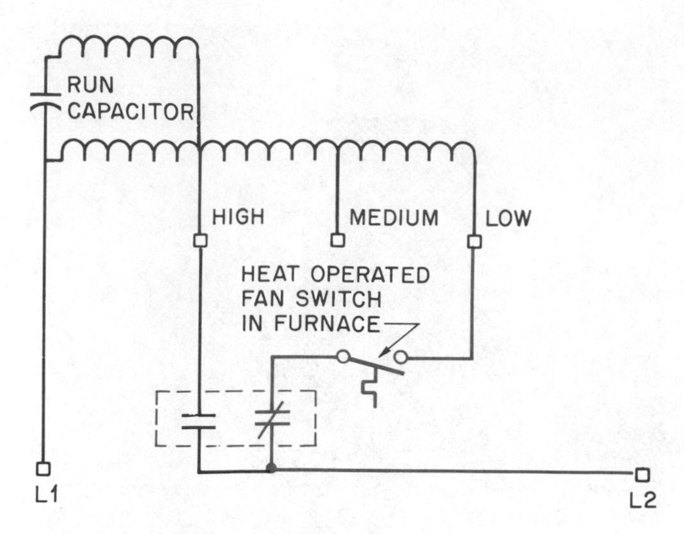

FAN RELAY: WHEN ENERGIZED, SUCH AS IN COOLING, THE FAN CANNOT RUN IN THE LOW-SPEED MODE. WHEN DEENERGIZED THE FAN CAN START IN THE LOW-SPEED MODE THROUGH THE CONTACTS IN THE HEAT OPERATED FAN SWITCH.

IF THE FAN SWITCH AT THE THERMOSTAT IS ENERGIZED WHILE THE FURNACE IS HEATING, THE FAN WILL MERELY SWITCH FROM LOW TO HIGH. THIS RELAY PROTECTS THE MOTOR FROM TRYING TO OPERATE AT 2 SPEEDS AT ONCE.

FIGURE 17–24 Diagram of a PSC motor shows how the motor can be applied for high air volume in the summer and low air volume in the winter.

17.15 THE SHADED-POLE MOTORS

The shaded-pole motor has very little starting torque and is not as efficient as the PSC motor, so it is only used for light-duty applications. These motors have small starting windings at the corner of each pole that help the motor start by providing an induced current and a rotating field, Figure 17–25. It is an economical motor from the standpoint of

FIGURE 17–25 Shaded-pole motor. *Photo by Bill Johnson*

initial cost. The shaded-pole motor is normally manufactured in the fractional horsepower range. For years it has been used in air-cooled condensers to turn the fans, Figure 17–26.

17.16 THREE-PHASE MOTORS

Three-phase motors are used on commercial equipment. The building power supply must have three-phase power available. (Three-phase power is seldom found in a home.) Three-phase motors have no starting windings or capacitors. They can be thought of as having three single-phase power supplies, Figure 17–27. Each of the phases can have either two or four poles. A 3600-rpm motor will have three sets, each with two poles (total of six), and an 1800-rpm motor will have three sets, each with four poles (total of 12). Each phase changes direction of current flow at different times but always in the same order. A three-phase motor has a high starting torque because of the three phases of current that operate the motor. At any given part of the rotation of the motor, one of the windings is in position for high torque. This makes starting large fans and compressors very easy, Figure 17–28.

The three-phase motor rpm also slips to about 1750 rpm and 3450 rpm when under full load. The motor is not normally available with dual speed; it is either an 1800-rpm or a 3600-rpm motor.

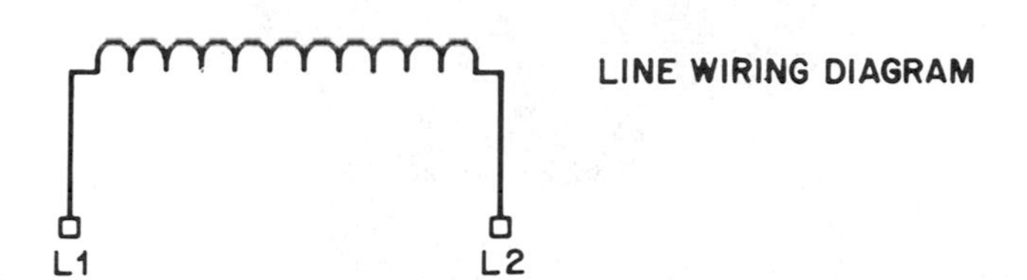

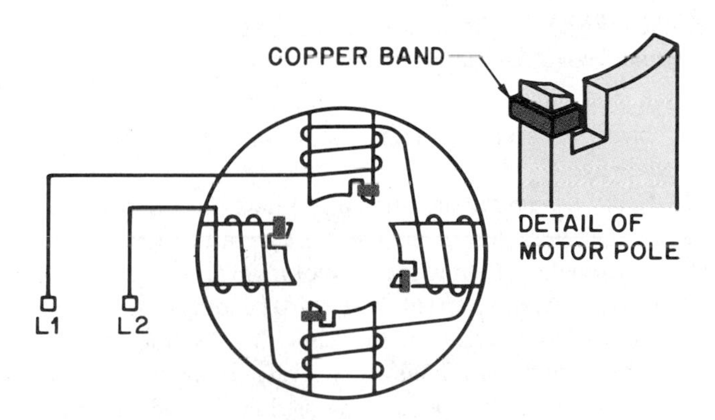

FIGURE 17–26 Wiring diagram of a shaded-pole motor.

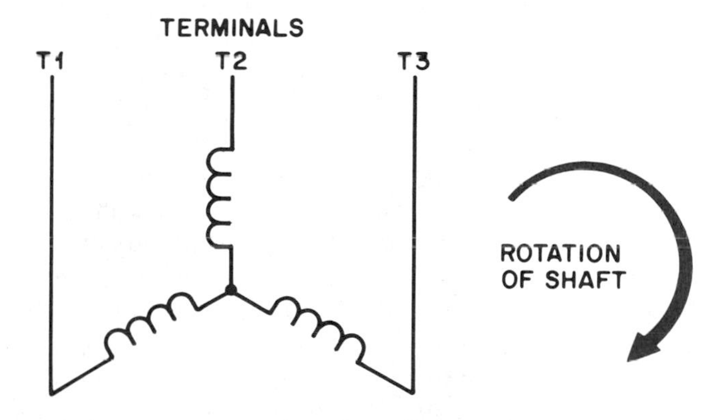

FIGURE 17–28 Diagram of a three-phase motor.

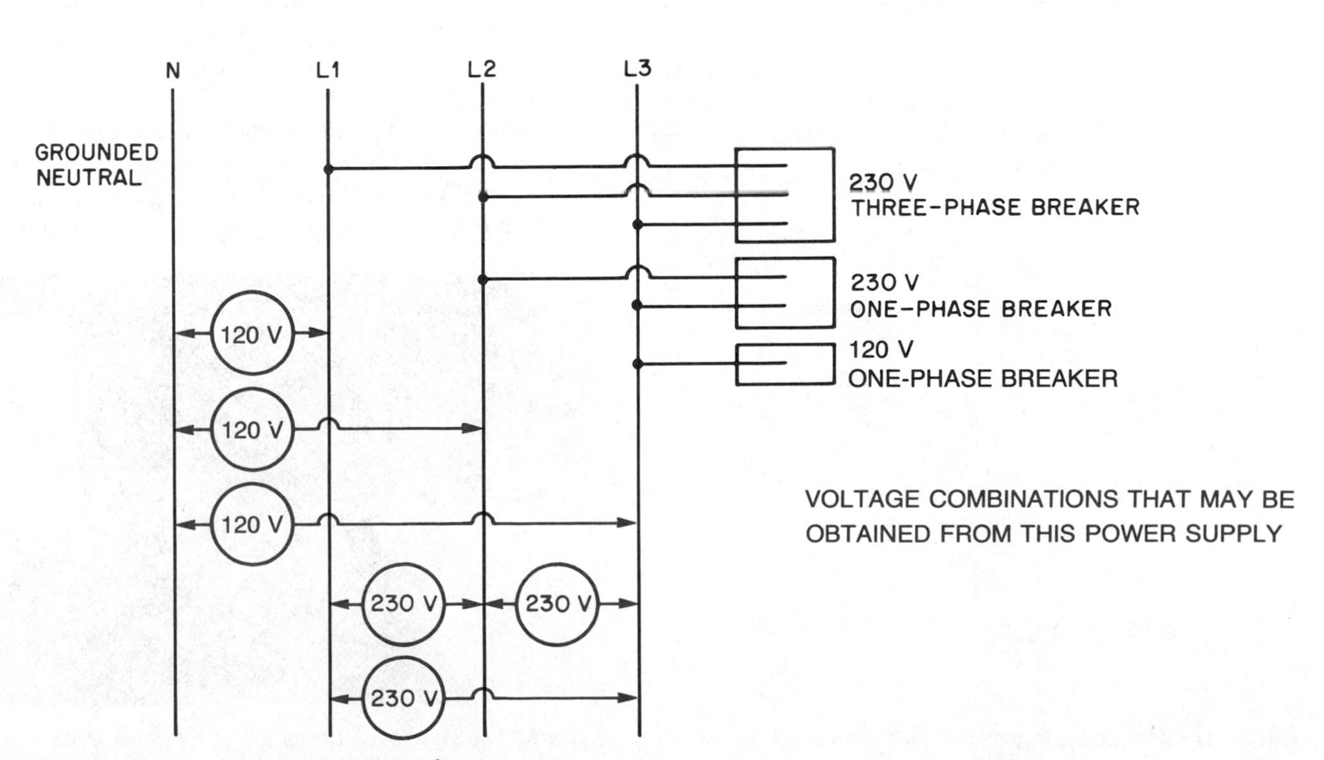

FIGURE 17–27 Diagram of three-phase power supply.

The rotation of a three-phase motor may be changed by switching any two motor leads, Figure 17–29. This rotation must be carefully observed when three-phase fans are used. If a fan rotates in the wrong direction, it will move only about half as much air. If this occurs, reverse the motor leads and the fan will turn in the correct direction.

All of the motors we have described are considered to be *open* motors and are used for fans and pumps. These motors have other characteristics that must be considered when selecting a motor for a particular job: for example, the motor mounting. Is the motor solidly mounted to a base, or is there a flexible mount to minimize noise?

The sound level is another factor. Will the motor be used where ball bearings would make too much noise? If so, sleeve bearings should be used.

Still another factor is the operating temperature of the motor surroundings. A condenser fan motor that pulls the air over the condenser coil and then over the motor requires a motor that will operate in a warmer atmosphere. The best advice is to replace any motor with an exact replacement.

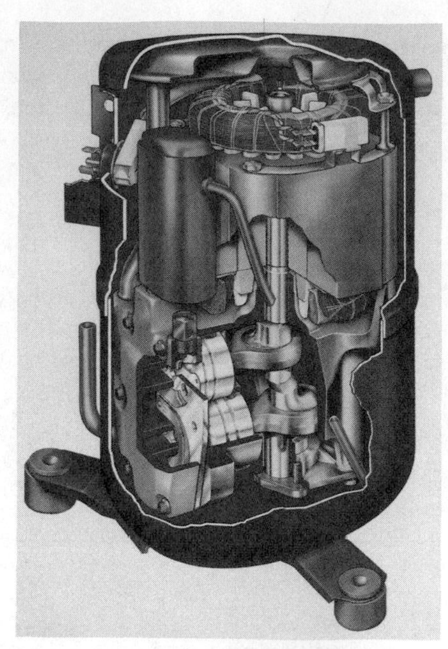

FIGURE 17–30 Typical motor for a hermetic compressor. *Courtesy Tecumseh Products Company*

17.17 SINGLE-PHASE HERMETIC MOTORS

The wiring in a single-phase hermetic motor is similar to that in a split-phase motor. It has start and run windings, each with a different resistance. The motor runs with the run winding and uses a potential relay to open the circuit to the start windings. A run capacitor is often used to improve running efficiency. A hermetic motor is designed to operate in a refrigerant, usually vapor, atmosphere. It is undesirable for liquid refrigerant to enter the shell as by an overcharge of refrigerant. Single-phase hermetic compressors usually are manufactured up to 5 hp, Figure 17–30. If more capacity is needed, multiple systems or larger three-phase units are used.

Hermetic compressor motor materials must be compatible with the refrigerant and oil circulating in the system. The coatings on the windings, the materials used to tie the motor windings, and the papers used as wedges must be of the correct material. The motor is assembled in a dry, clean atmosphere.

Hermetic motors are started in much the same way as the other motors described. The start windings must be removed from the circuit when the motor gets to about 75% of its normal operating speed. The start windings are not removed from the circuit in the same way as for an open motor because the windings are in a refrigerant atmosphere. Open single-phase motors are operated in air, and a spark is allowed when the start winding is disconnected. This cannot be allowed in a hermetic motor because the spark will deteriorate the refrigerant. Special devices determine when the compressor motor is running at the correct speed to disconnect the start winding.

Because the hermetic motor is enclosed in refrigerant, the motor leads must pass through the compressor shell to the outside. A terminal box on the outside houses the three motor terminals, Figure 17–31, one for the run winding, one

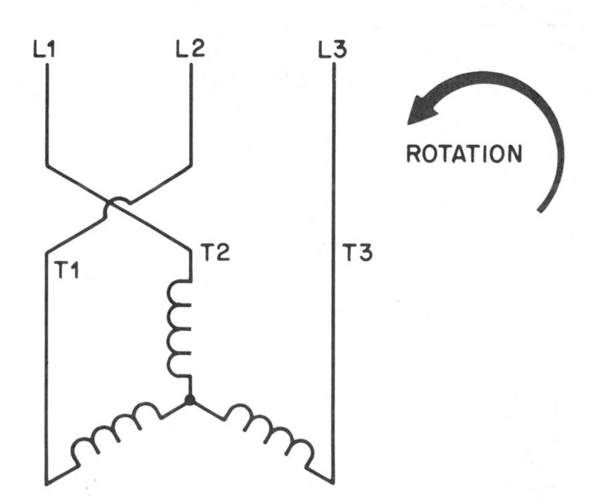

FIGURE 17–29 Wiring diagram of a three-phase motor. The rotation of this motor can be reversed by changing any two motor leads.

FIGURE 17–31 Motor terminal box on the outside of the compressor. *Photo by Bill Johnson*

for the start winding, and one for the line common to the run and start windings. See Figure 17–32 for a wiring diagram of a three-terminal compressor. The start winding has more resistance than the run winding.

The motor leads are insulated from the steel compressor shell. Neoprene was the most popular insulating material for years. However, if the motor terminal becomes too hot, due to a loose connection, the neoprene may eventually become brittle and possibly leak, Figure 17–33. Many compressors now use a ceramic material to insulate the motor leads.

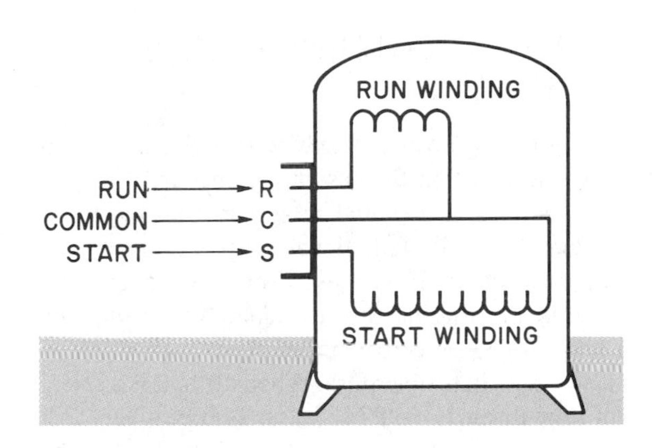

FIGURE 17–32 Wiring diagram of what is behind the three terminals on a single-phase compressor.

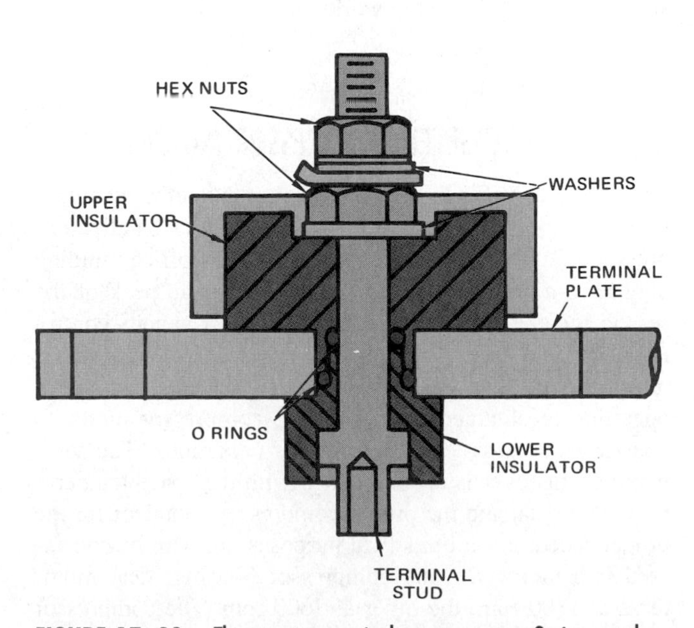

FIGURE 17–33 These motor terminals use neoprene O rings as the insulator between the terminal and the compressor housing. *Courtesy of Trane Company*

17.18 THE POTENTIAL RELAY

The *potential relay* is often used to break the circuit to the start winding when the motor reaches approximately 75% of its normal speed. This relay has a normally closed set of contacts. The coil is designed to operate at a slightly higher voltage than the applied (line) voltage. When the rotor begins to turn, a transformer action (called back electromotive force [emf]) takes place at the start winding; as the rotor approaches 75% of its design speed, the voltage exceeds the applied voltage and is sufficient to energize the coil. This opens the contacts, which open the start winding circuit, Figure 17–34.

17.19 THE CURRENT RELAY

The *current relay* also breaks the circuit to the start winding. It uses the inrush current of the motor to determine when the motor is running up to speed. A motor draws locked-rotor current during the time the power is applied to the windings and the motor has not started turning. As the motor starts turning, the current peaks; it begins to reduce as the motor turns faster. The current relay has a set of normally open contacts that close when the inrush current flows through its coil, energizing the start windings. When the motor speed

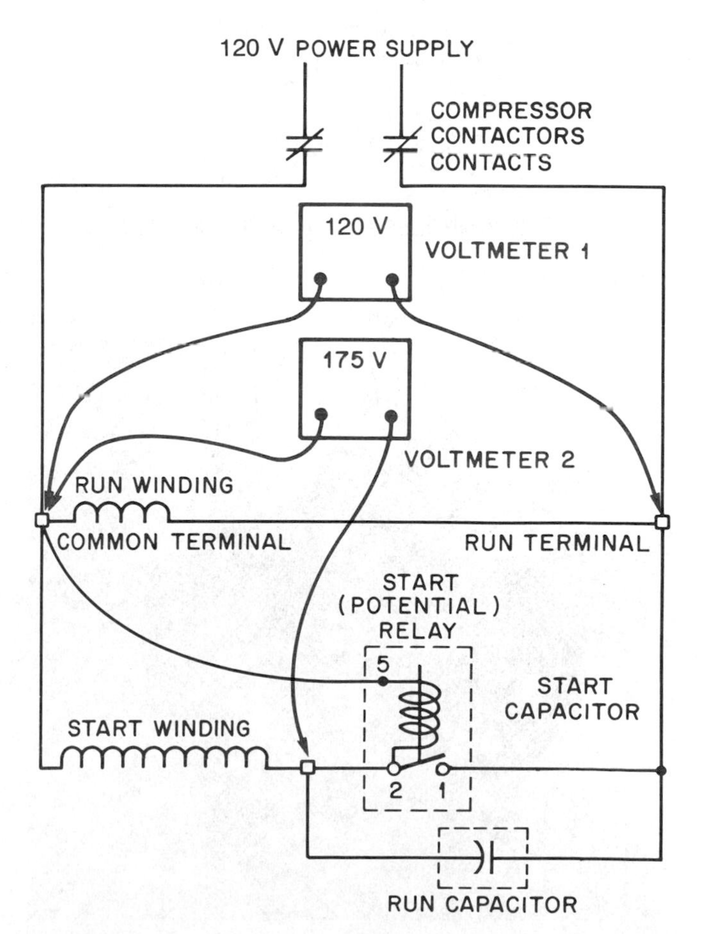

FIGURE 17–34 Wiring diagram illustrates the higher voltage of the start winding in a typical motor.

reaches about 75% of the rated rpm, the current relay opens its contacts, either by gravity or by a spring, Figure 17–35. The coil is wired in series with the run winding of the motor. The full current of the motor must flow through the coil of the current relay. **The current relay may always be identified by the size of the wire in the relay coil. This wire is unusually large because it must carry the full-load current of the motor,** Figure 17–36.

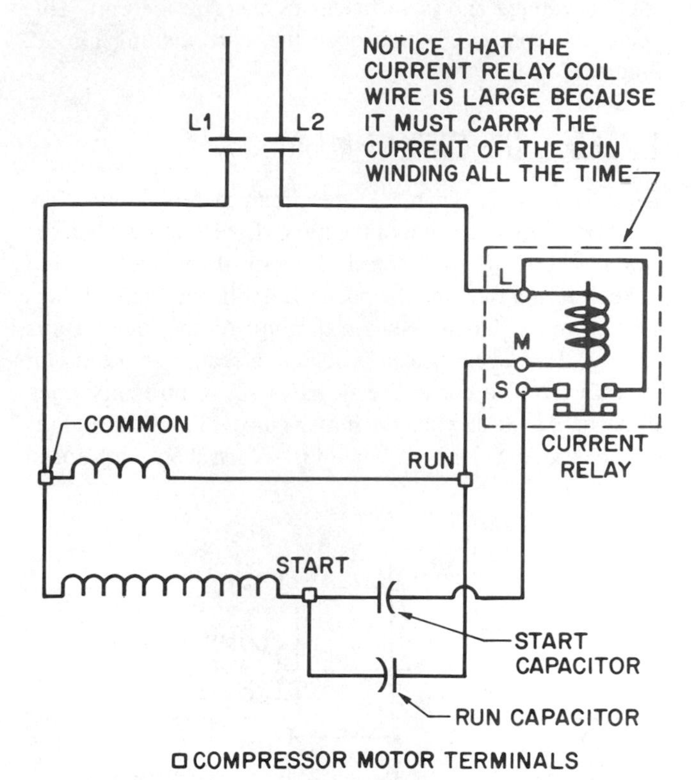

FIGURE 17–35 Wiring diagram of a current relay.

FIGURE 17–36 Current relay is identified by the size of the wire in the holding coil. *Photo by Bill Johnson*

These starting methods are used on many compressors with split-phase motors that need high starting torque. If a system has a capillary tube metering device or a fixed-bore orifice metering device, the pressures will equalize during the off cycle and a high starting torque compressor may not be necessary. A PSC motor may be used for this application. PSC motors are often used in residential air conditioning and heat pumps with the pressure equalizing type refrigeration cycle using fixed-bore metering devices. In the PSC motor the starting and running windings are both energized whenever there is power to the motor. The motor does use a running capacitor wired between the run and start terminals, so line voltage is not applied directly to the start windings.

17.20 POSITIVE TEMPERATURE COEFFICIENT START DEVICE

The PSC motor may not need any start assistance when conditions are well within the design parameters. If it does need start assistance, a potential relay and start capacitor may be added to provide additional torque or a positive temperature coefficient (PTC) device may be added. The PTC is a thermistor that has no resistance to current flow when the unit is off. Remember, a thermistor changes resistance with a change in temperature. When the unit is started, the current flow through the PTC causes it to heat very fast and create a high resistance in its circuit. This changes the phase angle of the start windings. It will not give a motor the starting torque that a start capacitor will, but it is advantageous because it has no moving parts. The PTC is wired in parallel with the run capacitor and acts like a short across the run capacitor during starting. This provides full-line voltage to the start windings during starting. Figure 17–37 shows how a PTC device works in a circuit.

17.21 TWO-SPEED COMPRESSOR MOTORS

Two-speed compressor motors are used by some manufacturers to control the capacity required from small compressors. For example, a residence or small office building may have a 5-ton air conditioning load at the peak of the season and a 2 1/2-ton load as a minimum. Capacity control is desirable for this application. A two-speed compressor may be used to accomplish capacity control. Two-speed operation is obtained by wiring the compressor motor to operate as a two-pole motor or a four-pole motor. The automatic changeover is accomplished with the space temperature thermostat and the proper compressor contactor for the proper speed. For all practical purposes, this can be considered two motors in one compressor housing. One motor turns at 1800 rpm, the other at 3600 rpm. The compressor uses either motor, based on the capacity needs. This compressor has more than three motor terminals to operate the two motors in the compressor.

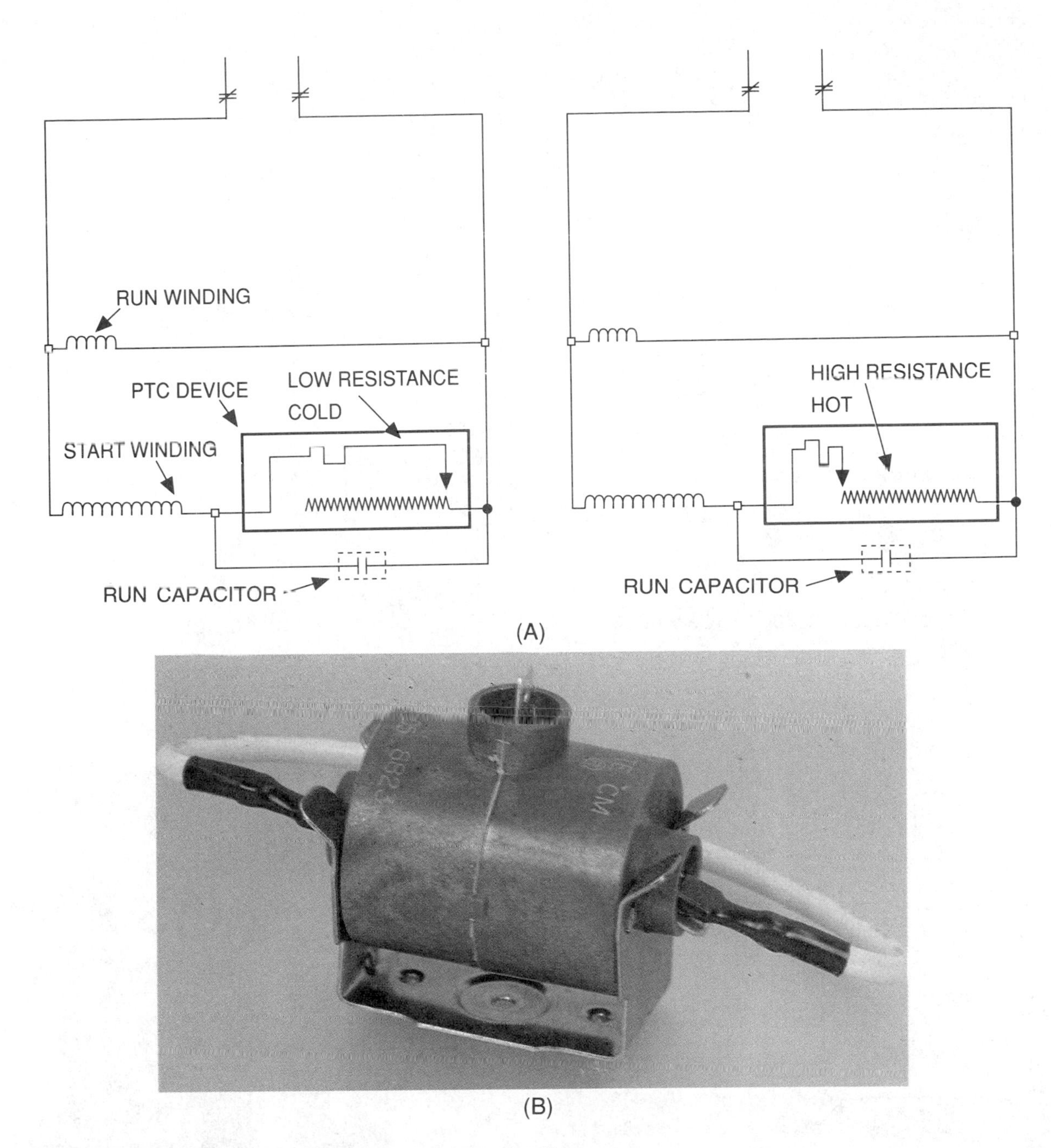

FIGURE 17–37 A positive temperature coefficient device (PTC). *Photo by Bill Johnson*

17.22 SPECIAL APPLICATION MOTORS

There are some special application single-phase motors, which may have more than three motor terminals, that are not two-speed motors. Some manufacturers design an auxiliary winding in the compressor to give the motor more efficiency. These motors are normally in the 5-hp and smaller range. Other special motors may have the winding thermostat wired through the shell with two extra motor terminals. The winding thermostat can be wired out of the circuit if it should fail with its circuit open, Figure 17–38.

17.23 THREE-PHASE MOTOR COMPRESSORS

Large commercial and industrial installations will have three-phase power for the air conditioning and refrigeration equipment. Three-phase compressor motors normally have three motor terminals, but the resistance across each winding is the same, Figure 17–39. As explained earlier, three-phase motors have a high starting torque and, consequently, should experience no starting problems.

Welded hermetic compressors were limited to 7 1/2 tons for many years but are now being manufactured in

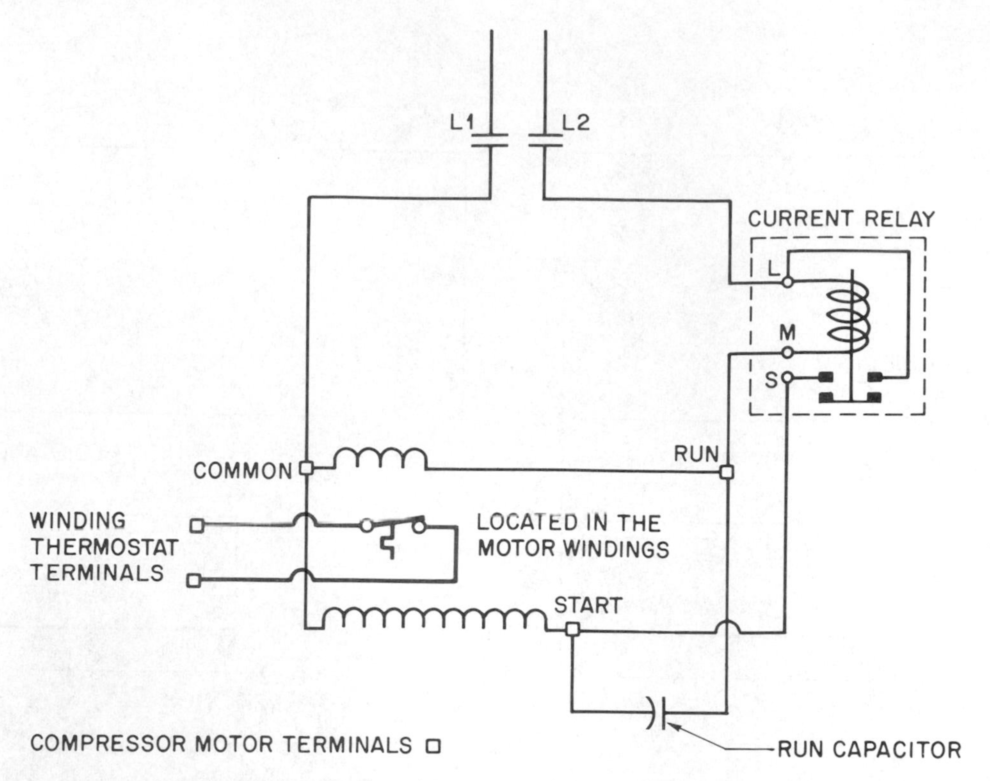

FIGURE 17–38 Compressor with five terminals in the terminal box, two of which are wired to the winding thermostat.

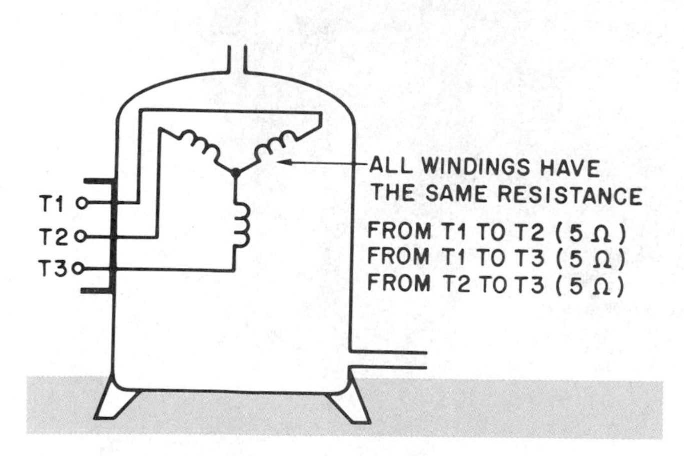

FIGURE 17–39 Three-phase compressor with three leads for the three windings. The windings all have the same resistance.

FIGURE 17–40 Large welded hermetic compressor. *Courtesy of Trane Company*

sizes up to about 50 tons. The larger welded hermetic compressors are traded to the manufacturer when they fail and are remanufactured. These must be cut open for service, Figure 17–40.

Serviceable hermetic compressors of the reciprocating type are manufactured in sizes up to about 125 tons, Figure 17–41. These compressors may have dual-voltage motors for 208 V to 230 V or 460-V operation, Figure 17–42. These compressors are normally rebuilt or remanufactured when the motor fails, so an overhaul is considered when a motor fails. The compressors may be rebuilt in the field or traded for remanufactured compressors. Most large metropolitan areas will have companies that can rebuild the compressor to the proper specifications. The difference in rebuilding and remanufacturing is that one is done by an independent rebuilder and the other by the original manufacturer or an authorized rebuilder.

FIGURE 17–41 Serviceable hermetic compressors. *Courtesy of Trane Company*

17.24 VARIABLE SPEED MOTORS

The desire to control motors to provide a greater efficiency for the fans, pumps, and compressors has led the industry to explore development of and use of variable speed motors. Most motors do not need to operate at full speed and load except during the peak temperature of the season and could easily satisfy the heating or air conditioning load at other times by operating at a slower speed. When the motor speed is reduced, the power to operate the motor reduces proportionately. For example, if a home or building needs only 50% of the capacity of the air conditioning unit to satisfy the space temperature, it would be advantageous to reduce the capacity of the unit rather than stop and restart the unit. When the power consumption can be reduced in this manner, the unit becomes more efficient.

The voltage and frequency (cycles per second) of the power supply determine the speed of a conventional motor. New motors are being used that can operate at different speeds by the use of electronic circuits. Several methods are used to vary the frequency of the power supply and the

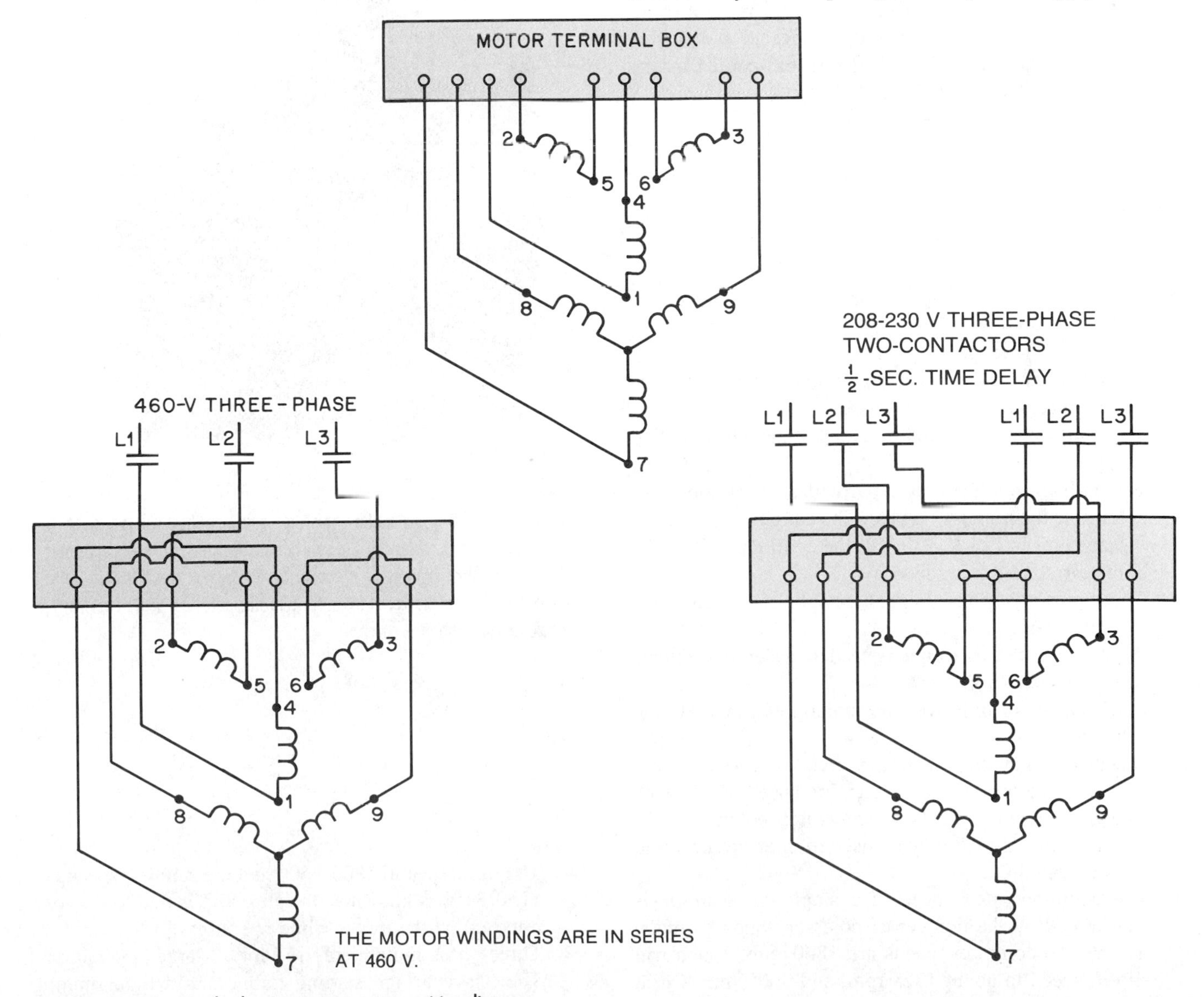

FIGURE 17–42 Dual-voltage compressor motor wiring diagram.

voltage depending on the type of motor. The compressor motor and the fan motors may be controlled through any number of speed combinations based on the needs.

Some benefits of variable speed technology are:

- power savings
- load reduction based on demand
- soft starting of the motor (no locked rotor amperage)
- better space temperature and humidity control
- solid state motor starters do not have open contacts
- a unit may be oversized for future expansion and run at part load until the expansion

17.25 COOLING ELECTRIC MOTORS

All motors must be cooled because part of the electrical energy input to the motor is given off as heat. Most open motors are air cooled. Hermetic motors may be cooled by air or refrigerant gas, Figure 17–43. Small and medium-sized motors are normally cooled by refrigerant gas. Only very large motors are water cooled. An air-cooled motor has fins on the surface to help give the surface more area for dissipating heat. These motors must be located in a moving airstream. To cool properly, refrigerant gas-cooled motors must have an adequate refrigerant charge.

(A)

(B)

FIGURE 17–43 All motors must be cooled or they will overheat. (A) This compressor is cooled by the refrigerant gas passing over the motor winding. (B) This compressor is cooled by air from a fan. The air-cooled motor has fins on the compressor to help dissipate the heat. *Courtesy Copeland Corporation*

SUMMARY

- Motors turn fans, compressors, and pumps.
- Some of these applications need high starting torque and good running efficiencies; some need low starting torque with average or good running efficiencies.
- Compressors applied to refrigeration normally require motors with high starting torque.
- Small fans normally need motors that have low starting torque.
- The voltage supplied to a particular installation will determine the motor's voltage. The common voltage for furnace fans is 120 V; 230 V is the common voltage for home air conditioning systems.
- Common single-phase motors are split-phase, PSC, and shaded-pole.
- When more starting torque is needed, a start capacitor is added to the motor.
- A run capacitor improves the running efficiency of the split-phase motor.
- A centrifugal switch breaks the circuit to the start winding when the motor is up to running speed. The switch changes position with the speed of the motor.
- An electronic switch may be used to interrupt power to the start winding.
- The common rated speed of a single-phase motor is determined by the number of poles or windings in the motor. The common speeds are 1800 rpm, which will slip in speed to about 1725 rpm, and 3600 rpm, which will slip to about 3450 rpm.
- The difference in 1800/3600 and the running speeds of 1750/3450 is known as the slip. Slip is due to the load imposed on the motor while operating.
- Three-phase motors are used for all large applications. They have a high starting torque and a high running efficiency. Three-phase power is not available at most

residences, so these motors are limited to commercial and industrial installations.

- The power to operate hermetic motors must be conducted through the shell of the compressor by way of insulated motor terminals.
- Since the winding of a hermetic compressor is in the refrigerant atmosphere, a centrifugal switch may not be used to interrupt the power to the start winding.
- A potential relay takes the start winding out of the circuit using back emf.
- A current relay breaks the circuit to the start winding using the motor's run current.
- The PSC motor is used when high starting torque is not required. It needs no starting device other than the run capacitor.
- The PTC device is used with some PSC motors to give small amounts of starting torque. It has no moving parts.
- When compressors are larger than 5 tons, they are normally three phase.
- Dual-voltage three-phase compressors are built with two motors wired into the housing.
- Three-phase reciprocating compressors come in sizes up to about 125 tons.
- Variable speed motors operate at higher efficiencies with varying loads.

REVIEW QUESTIONS

1. Name the two popular operating voltages in residences.
2. What device takes the start winding out of the circuit when an open motor gets up to running speed?
3. Describe the difference in resistances of the starting and running windings of a split-phase motor.
4. What device may be wired into the starting circuit to improve the starting torque of a compressor?
5. What device may be wired into the circuit to improve the running efficiency of a compressor?
6. What is back emf?
7. Why is a hermetic compressor manufactured from special materials?
8. Name the two types of motors used for single-phase hermetic compressors.
9. How does the power pass through the compressor shell to the motor windings?
10. Name the two types of relays used to start the single-phase hermetic compressor.
11. What is a PTC device?
12. How are some small compressors operated at two different speeds?
13. Why are two speeds desirable?
14. How are some compressor motors operated at different voltages?
15. Name two methods to cool most hermetic compressors.

U N I T 18

Application of Motors

OBJECTIVES

After studying this unit, you should be able to

- identify the proper power supply for a motor.
- describe the application of three-phase versus single-phase motors.
- describe other motor applications.
- explain how the noise level in a motor can be isolated from the conditioned space.
- describe the different types of motor mounts.
- identify the various types of motor drive mechanisms.

SAFETY CHECKLIST

* Ensure that electric motor ground straps are connected properly when appropriate so that the motor will be grounded.
* Never touch a motor drive belt when it is moving. Be sure that your fingers do not get between the belt and the pulley.

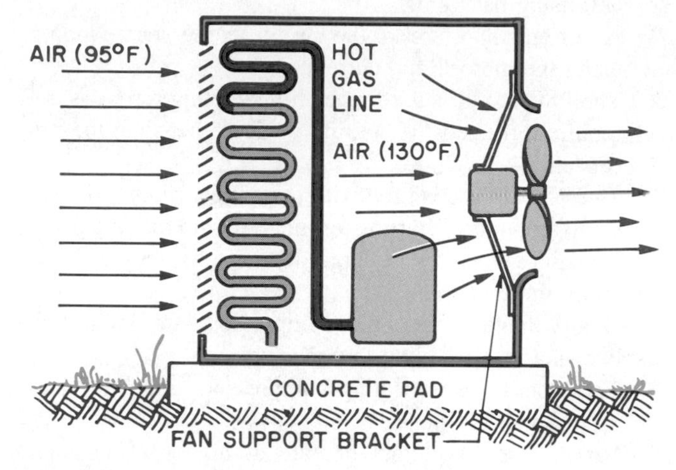

FIGURE 18–1 Motor operating in the hot airstream after the air has been pulled through the condenser.

18.1 MOTOR APPLICATIONS

Because electric motors perform so many different functions, choosing the proper motor is necessary for safe and effective performance. Usually the manufacturer or design engineer for a particular job chooses the motor for each piece of equipment. However, as a technician you often will need to substitute a motor when an exact replacement is not available, so you should understand the reasons for choosing a particular motor for a job. For example, when a fan motor burns out in the air conditioning condensing unit, the correct motor must be obtained or another failure may occur. ***In an air-cooled condenser the air is normally pulled through the hot condenser coil and passed over the fan motor. This hot air is used to cool the motor. You must be aware of this, or you will install the wrong motor.*** The motor must be able to withstand the operating temperatures of the condenser air, which may be as high as 130°F, Figure 18–1.

Open motors are discussed in this unit because they are the only ones from which a technician has a choice of selection. Some design differences that influence the application are:

- the power supply
- the work requirements
- the motor insulation type or class
- the bearing types
- the mounting characteristics

18.2 THE POWER SUPPLY

The power supply must provide the correct voltage and sufficient current. For example, the power supply in a small shop building may be capable of operating a 5-hp air compressor. But suppose air conditioning is desired. If the air conditioning contractor prices the job expecting the electrician to use the existing power supply, the customer may be in for a surprise. The electrical service for the whole building may have to be changed. The motor equipment nameplate information and the manufacturer's catalogs provide the needed information for the additional service, but someone must put the whole project together. The installing air conditioning contractor may have that responsibility. See Figure 18–2 for a typical motor nameplate and Figure 18–3 for a part of a page from manufacturer's catalog with the electrical data included. Figure 18–4 is an example of a typical electrical panel rating.

The power supply data contains:

1. The voltage (120 V, 208 V, 230 V, 460 V)
2. The current capacity in amperes
3. The frequency in hertz or cycles per second (60 cps in the United States and 50 cps in many foreign countries)
4. The phase (single or three phase)

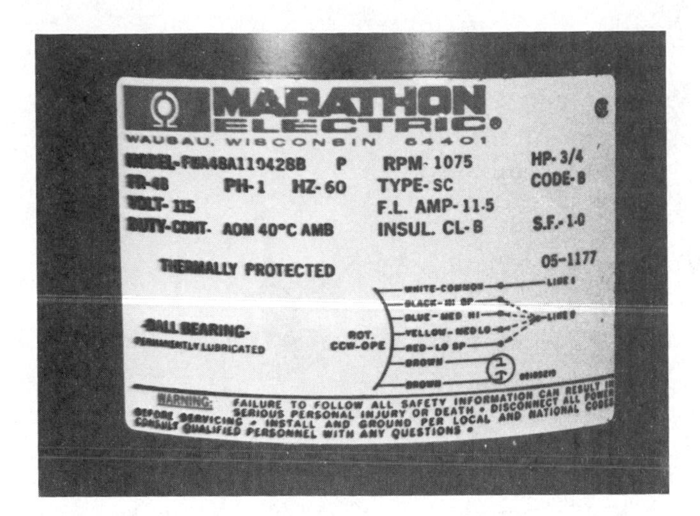

FIGURE 18–2 Motor nameplate. *Photo by Bill Johnson*

Motors and electrical equipment must fit within the system's total electrical capacity, or failures may occur. Voltage and current are most often the inadequate characteristic encountered. The technician can check them, but it is usually preferable to have a licensed electrician make the final calculations.

Voltage

The voltage of an installation is important because every motor operates within a specified voltage range, usually within ±10%. Figure 18–5 gives the upper and lower limits of common voltages. If the voltage is too low, the motor will draw a high current. For example, if a motor is designed to operate on 230 V but the supply voltage is really 200 V, the motor's current draw will go up. The motor is trying to do its job, but it lacks the power and it will overheat, Figure 18–6.

If the applied voltage is too high, the motor may develop local hot spots within its windings, but it will not always experience high amperage. The high voltage will actually give the motor more power than it can use. A 1-hp motor

TYPE CS • NEMA SERVICE FACTOR • 60 HERTZ • CUSHION BASE

HP	RPM	Bearings	Overload Protector	Full Load (12) Amps	Frame
1/6	1800	Sleeve	Auto	4.4	K48
1/4	1800	Sleeve	Auto	2.7	K48
		Ball	Auto	2.7	K48
1/3	3600	Sleeve	Auto	3.1	L48
	1800	Sleeve	Auto	2.9	L48
		Sleeve	Auto	2.6	J56
		Ball	No	2.9	L48
		Ball	No	2.6	J56
		Ball	Auto	2.6	J56
	1200	Sleeve	No	3.0	K56
1/2	3600	Sleeve	Auto	4.0	L48
	1800	Sleeve	No	3.6	J56
		Sleeve	Auto	3.6	J56
		Ball	No	3.6	J56
		Ball	Auto	3.6	J56
3/4	3600	Sleeve	Auto	4.6	J56
	1800	Sleeve	No	5.2	K56
		Sleeve	Auto	5.2	K56
		Ball	No	5.2	K56
		Ball	Auto	5.2	K56
1	3600	Sleeve	Auto	6.0	K56
		Ball	No	6.0	K56
	1800	Sleeve	Auto	6.5	L56
		Ball	No	6.5	L56
		Ball	Auto	6.5	L56
1 1/2	3600	Ball	Auto	8.0	L56
	1800	Ball	Auto	7.5	M56
2	3600	Ball	No	9.5	L56

FIGURE 18–3 Part of a page from a manufacturer's catalog. *Courtesy Century Electric Inc.*

FIGURE 18–4 Nameplate from an electrical panel used for a power supply. *Photo by Bill Johnson*

208 VOLT-RATED MOTOR	+10%	228.8 VOLTS
	−10%	187.2 VOLTS
230 VOLT-RATED MOTOR	+10%	253 VOLTS
	−10%	207 VOLTS
208–230 VOLT-RATED MOTOR	+10%	253 VOLTS
	−10%	187.2 VOLTS

FIGURE 18–5 Table shows the maximum and minimum operating voltages of typical motors.

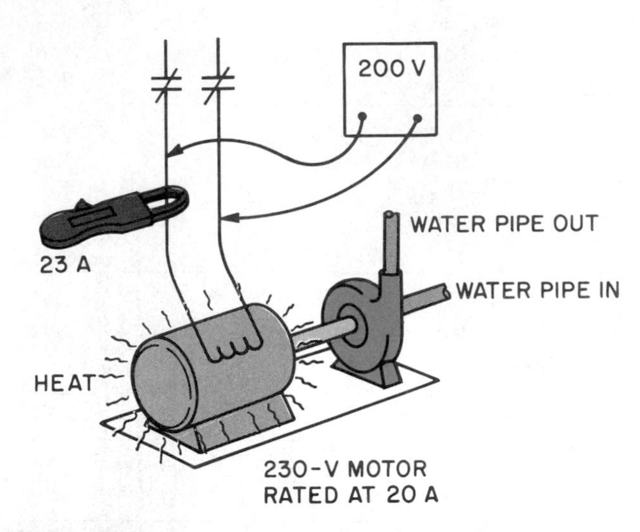

FIGURE 18–6 Motor operating at low-voltage condition.

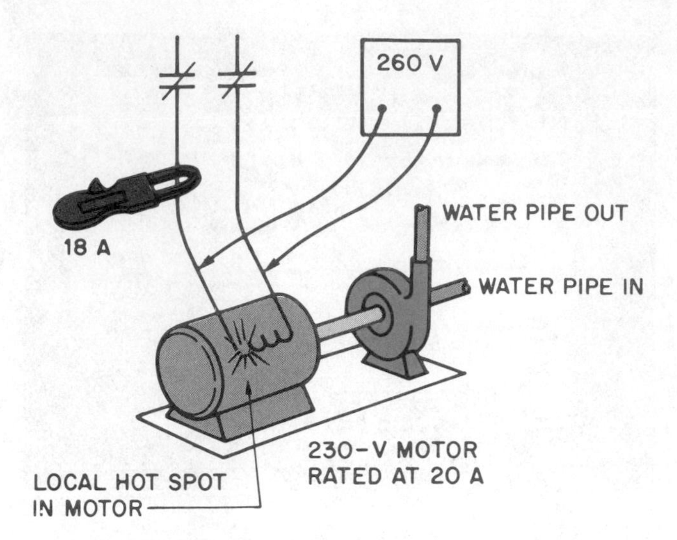

FIGURE 18–7 Motor rated as 230 V and operating at 260 V.

APPROXIMATE FULL LOAD AMPERAGE VALUES FOR ALTERNATING CURRENT MOTORS

Motor	Single Phase		3-Phase-Squirrel Cage Induction		
HP	120V	230V	230V	460V	575V
$\frac{1}{6}$	4.4	2.2			
$\frac{1}{4}$	5.8	2.9			
$\frac{1}{3}$	7.2	3.6			
$\frac{1}{2}$	9.8	4.9	2	1.0	0.8
$\frac{3}{4}$	13.8	6.9	2.8	1.4	1.1
1	16	8	3.6	1.8	1.4
$1\frac{1}{2}$	20	10	5.2	2.6	2.1
2	24	12	6.8	3.4	2.7
3	34	17	9.6	4.8	3.9
5	56	28	15.2	7.6	6.1
$7\frac{1}{2}$			22	11.0	9.0
10			28	14.0	11.0

Does not include shaded pole.

FIGURE 18–8 Chart shows approximate full-load amperage values. *Courtesy BDP Company*

with a voltage rating of 230 V that is operating at 260 V is running above its 10% maximum. This motor may be able to develop 1 1/4 hp at this higher-than-rated voltage, but the windings are not designed to operate at that level. The motor can overheat and eventually burn out if it continually runs overloaded. This can happen *without* drawing excessive current, Figure 18–7.

Current Capacity

There are two current ratings for a motor. The full-load amperage (FLA) is the current the motor draws while operating at a full-load condition at the rated voltage. This is also called the run-load amperage (RLA). For example, a 1-hp motor will draw approximately 16 A at 120 V or 8 A at 230 V in a single-phase circuit. Figure 18–8 is a chart that shows approximate amperages for some typical motors.

The other amperage rating that may be given for a motor is the locked-rotor amperage (LRA). These two current ratings are available for every motor and are stamped on the motor nameplate for an open motor. Some compressors do not have both ratings printed. The LRA or the FLA (RLA) tells the technician if the motor is operating outside its design parameters. Normally the LRA is about five times the FLA. For example, a motor that has an FLA of 5 A will normally have an LRA of about 25 A. If the LRA is given on the nameplate and the FLA is not given, divide the LRA by 5 to get an approximate FLA or RLA. For example, if a compressor nameplate shows an LRA of 80 A the approximate FLA is 80/5 = 16 A.

Every motor has a service factor that may be listed in the manufacturer's literature. This service factor is actually reserve horsepower. A service factor of 1.15 applied to a motor means that the motor can operate at 15% over the nameplate horsepower before it is out of its design parameters. A motor operating with a variable load and above normal conditions for short periods of time should have a larger service factor. If the voltage varies at a particular installation, a motor with a high service factor may be chosen. The service factor is standardized by the National Electrical Manufacturer's Association (NEMA). Figure 18–9 is a typical manufacturer's chart showing service factors.

Frequency

The frequency in cycles per second (cps) is the frequency of the electrical current the power company supplies. The technician has no control over this. Most motors are 60 cps in the United States but could be 50 cps in a foreign country. Most 60-cps motors will run on 50 cps, but they will devel

THREE PHASE • DRIPPROOF

Type SC Squirrel Cage • Fractional HP

- 60 Hertz
- Ball Bearing
- 40° C Ambient
  Class B Insulation

- NEMA Service Factor
  1/20 thru 1/8 HP 1.40
  1/6 thru 1/3 HP—1.32
  1/2 thru 3/4 HP—1.25
  1 thru 200 HP—1.15

- Versatile 208-430/460 volt motors available in many ratings.

HP	RPM	Volts	Full Load (5) Amps	Frame
Rigid Base				
1/4	1800	200-230/460	0.8	K48
	1200	230/460	0.6	H56
1/3	3600	200-230/460	0.7	B56
	1800	200-230/460	0.8	K48
		208-230/460	0.8	B56
	1200	200-208	1.7	J56
		230/460	0.8	J56
1/2	3600	208-230/460	0.9	B56
	1800	200-208	2.4	B56
		230/460	1.1	B56
		208-230/460	1.1	B56
	1200	200-208	2.0	J56
		230/460	1.0	J56
3/4	3600	208-230/460	1.2	J56
	1800	200-208	3.2	H56
		230/460	1.3	H56
		200-230/460	1.3	H56
	1200	200-208	3.3	J56
		200-208	3.3	M143T
		230/460	1.6	J56
		230/460	1.6	M143T
1	3600	200-208	3.2	J56
		230/460	1.5	J56
	1800	200-208	3.8	J56
		200-230/460	1.7	L143T
		200-230/460	1.7	J56
		575	1.4	L143T
	1200	200-208	3.8	N145T
		230/460	1.9	K56
		230/460	1.9	N145T

FIGURE 18–9 Chart shows service factors for motors. *Courtesy Century Electric, Inc.*

op only five sixths of their rated speed (50/60). If you believe that the supply voltage is not 60 cps, contact the local power company. When motors are operated with local generators as the power supply, the generator's speed will determine the frequency. A cps meter is normally mounted on the generator and can be checked to determine the frequency.

Phase

The number of phases of power supplied to a particular installation is determined by the power company. They make this determination based on the amount of electrical load and the types of equipment that must operate from their power supply. Normally, single-phase power is supplied to residences and three-phase power is supplied to commercial and industrial installations. Single-phase motors will operate on two phases of three-phase power, Figure 18–10. Three-phase motors will not operate on single-phase power, Figure 18–11. The technician must match the motors to the number of phases of the power supply, Figure 18–12.

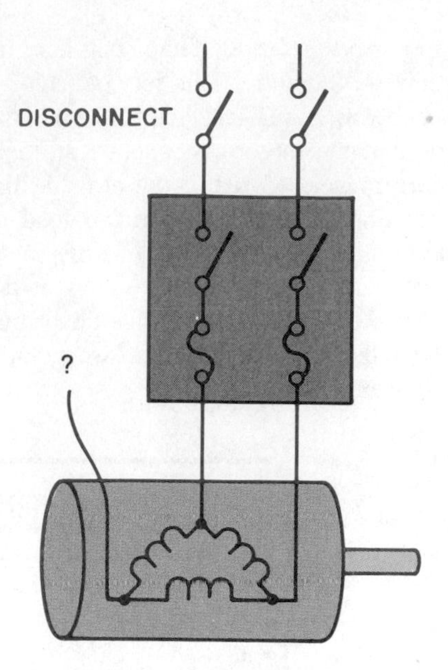

FIGURE 18–11 Three-phase motor with three leads that cannot be wired into single-phase circuit.

18.3 ELECTRICAL MOTOR WORKING CONDITIONS

The motor's working conditions determine which motor is the most economical for the particular job. An open motor with a centrifugal starting switch (single phase) for the air conditioning fan may not be used in a room with explosive gases. When the motor's centrifugal switch opens to interrupt the power to the start winding, the gas may ignite. An explosion-proof motor enclosed in a housing must be used, Figure 18–13. Local codes should be checked and adhered to. The explosion-proof motor is too expensive to be installed in a standard office building, so a proper choice of motors should be made. A motor operated in a very dirty area may need to be enclosed, giving no ventilation for the motor windings. This motor must have some method to dissipate the heat from the windings, Figure 18–14.

A drip-proof motor should be used where water can fall on it. It is designed to shed water, Figure 18–15.

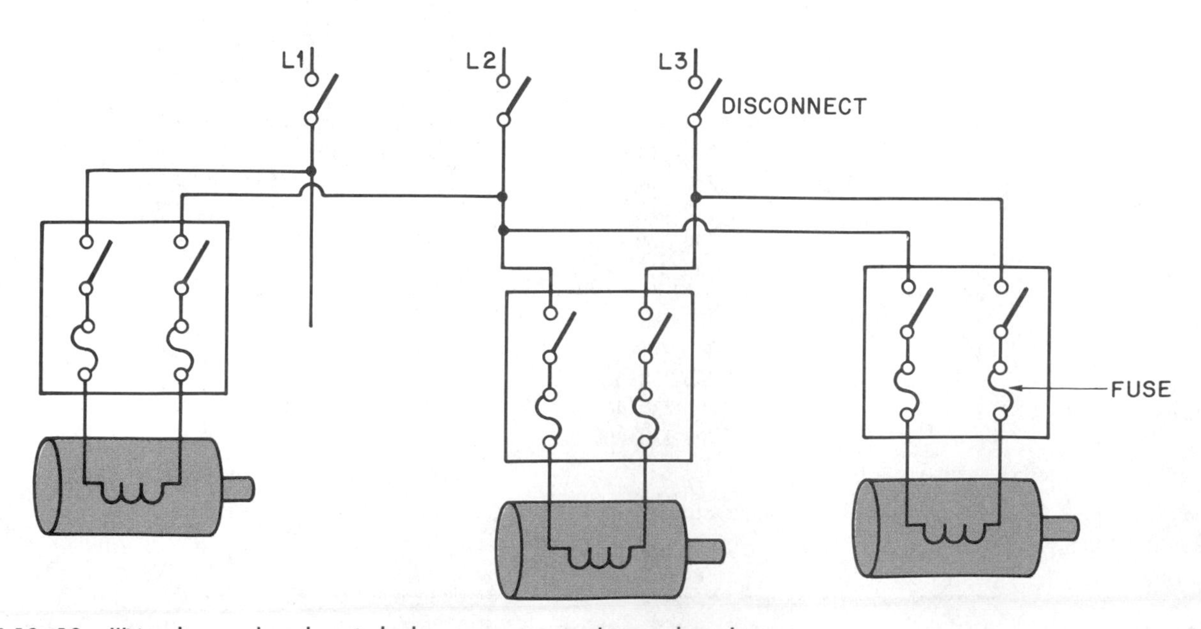

FIGURE 18–10 Wiring diagram shows how single-phase motors are wired into a three-phase circuit.

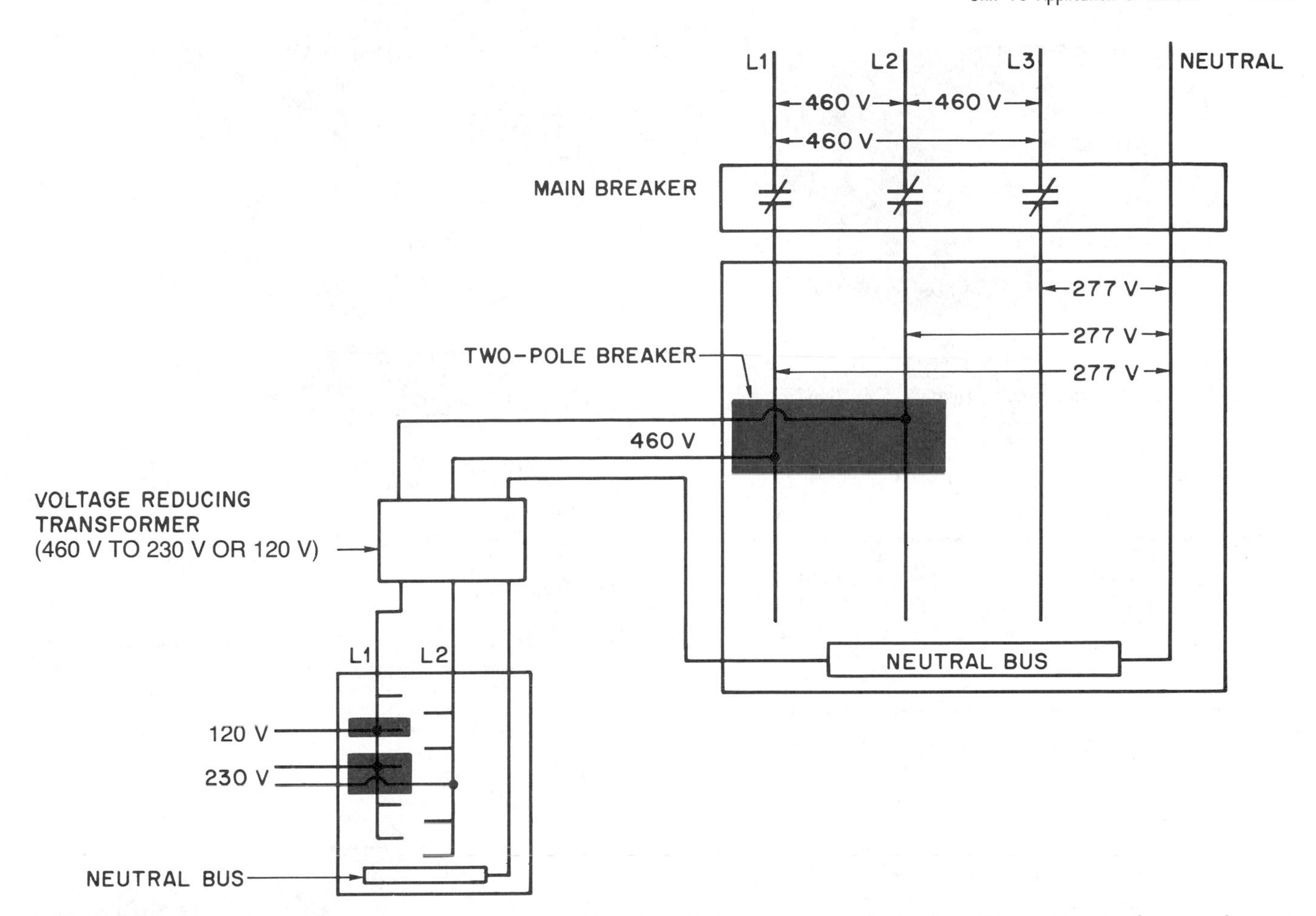

FIGURE 18–12 460-V panel power supply for a commercial building shows distribution of various loads. The 460-V to 120-V step-down transformer is used to operate appliances and office machines.

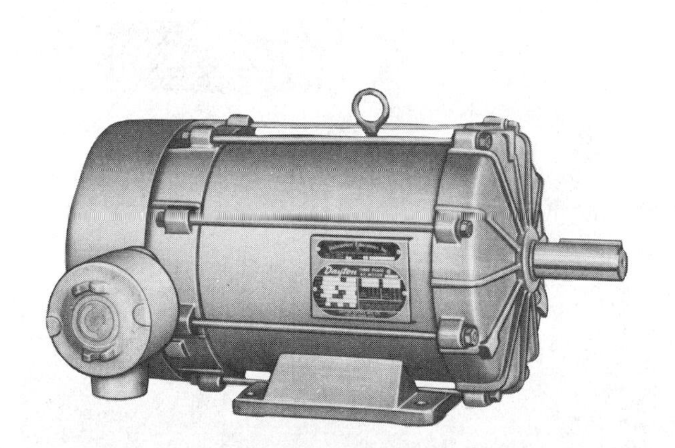

FIGURE 18–13 Explosion-proof motor. *Courtesy W. W. Grainger, Inc.*

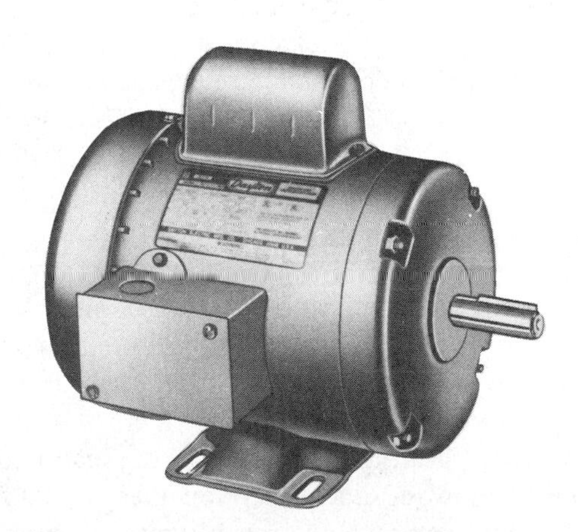

FIGURE 18–14 Totally enclosed motor. *Courtesy W. W. Grainger, Inc.*

18.4 INSULATION TYPE OR CLASS

The insulation type or class describes how hot the motor can safely operate in a particular ambient temperature condition. The example of the motor used earlier in this unit in an air conditioning condensing unit is typical. This motor must be designed to operate in a high ambient temperature. Motors are classified by the maximum allowable operating temperatures of the motor windings, Figure 18–16.

For many years motors were rated by the allowable temperature rise of the motor above the ambient temperature. Many motors still in service are rated this way. A typical motor has a temperature rise of 40°C. If the maximum ambient temperature is 40°C (104°F), the motor

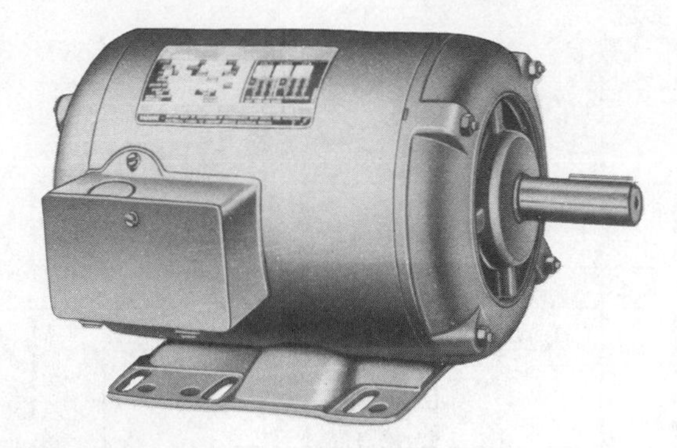

FIGURE 18–15 Drip-proof motor. *Courtesy W. W. Grainger, Inc.*

Class A	221° F	(105° C)
Class B	266° F	(130° C)
Class F	311° F	(155° C)
Class H	356° F	(180° C)

FIGURE 18–16 Temperature classification of typical motors.

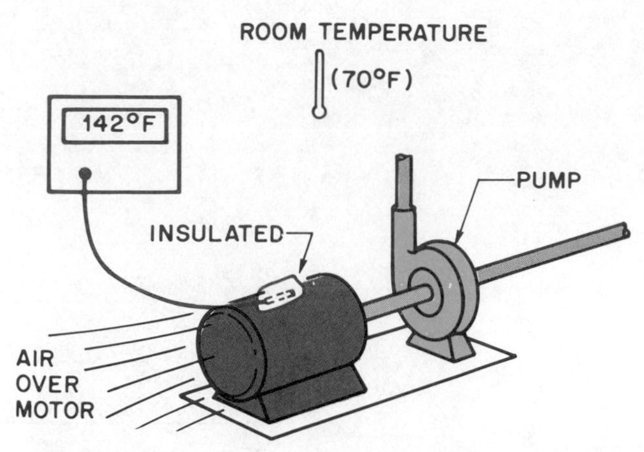

NOTE: CARE MUST BE TAKEN THAT TEMPERATURE TESTER LEAD IS TIGHT ON MOTOR AND NEXT TO MOTOR WINDINGS. THE LEAD MUST BE INSULATED FROM SURROUNDINGS. THE TEMPERATURE CHECK POINT MUST BE AS CLOSE TO THE WINDING AS POSSIBLE.

FIGURE 18–17 This motor is being operated in an ambient that is less than the maximum allowable for its insulation.

should have a winding temperature of 40°C + 40°C = 80°C (176°F). When troubleshooting these motors, the technician may have to convert from Celsius to Fahrenheit or may need a conversion table if the temperature rating is given only in Celsius. If the temperature of the motor winding can be determined, the technician can tell if a motor is running too hot for the conditions. For example, a motor in a 70°F room is allowed a 40°C rise on the motor. The maximum winding temperature is 142°F (70°F = 21°C; 21°C + 40°C rise = 61°C). This is 142°F, Figure 18–17.

18.5 TYPES OF BEARINGS

Load characteristics and noise level determine the type of bearings that should be selected for the motor. Two common types are the *sleeve* bearing and the *ball* bearing, Figure 18–18.

The sleeve bearing is used where the load is light and noise must be low (eg, a fan motor on a residential furnace). A ball-bearing motor would probably make excessive noise in the conditioned space. Metal duct work is an excellent sound carrier. Any noise in the system is carried throughout the entire system. Sleeve bearings are normally used in smaller applications, such as in residential and light commercial air conditioning systems for this reason. They are quiet and dependable, but they cannot stand great pressures (eg, if fan belts are too tight). These motors have either vertical or horizontal shaft applications. The typical air-cooled condenser has a vertical motor shaft and is pushing the air out the top of the unit. This results in a downward thrust on the motor bearings, Figure 18–19. A furnace fan

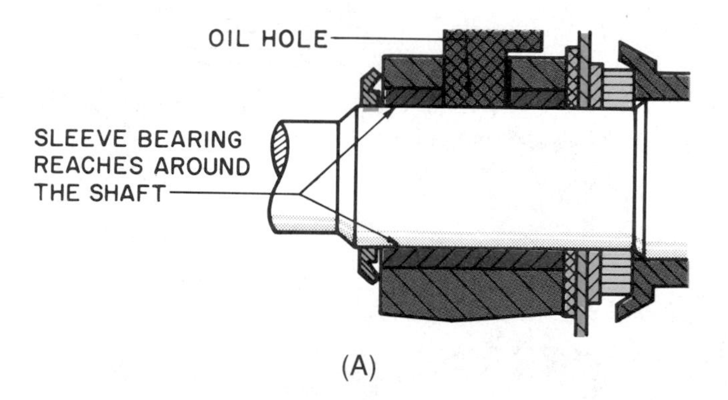

(A)

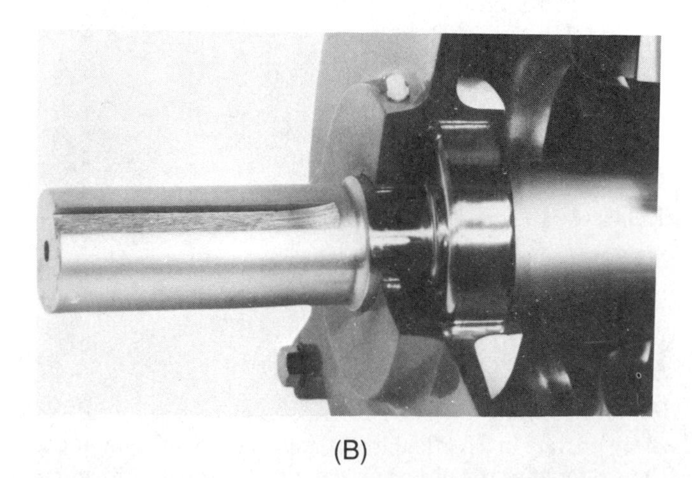

(B)

FIGURE 18–18 (A) Sleeve bearings and (B) ball bearings. (B) *Courtesy Century Electric, Inc.*

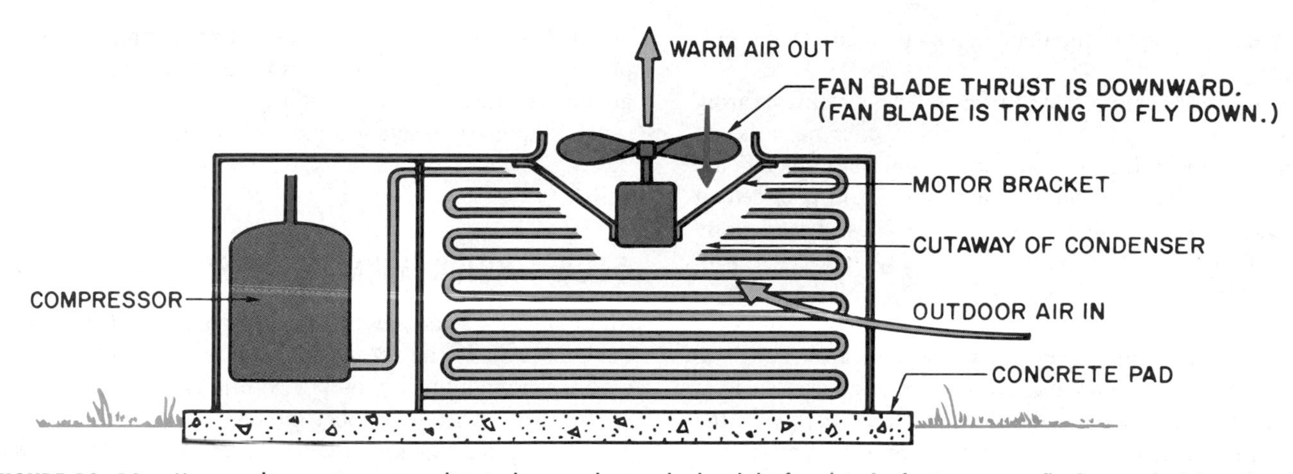

FIGURE 18–19 Motor working against two conditions: the normal motor load and the fact that the fan is trying to fly downward while pushing air upward.

has a horizontal motor shaft, Figure 18–20. These two types may not look very different, but they are. The vertical condenser fan is trying to fly downward into the unit to push air out the top. This puts a real load on the end of the bearing (called the *thrust surface*), Figure 18–21.

Sleeve bearings are made from material that is softer than the motor shaft. The bearing must have lubrication—an oil film between the shaft and the bearing surface. The shaft actually floats on this oil film and never should touch the bearing surface. The oil film is supplied by the lubrication system. Two types of lubrication systems for sleeve bearings are the *oil port* and the *permanently lubricated* bearing.

The oil port bearing has an oil reservoir that is filled from the outside by means of an access port. This bearing must be lubricated at regular intervals with the correct type of oil, which is usually 20-weight nondetergent motor oil or an oil specially formulated for electric motors. If the oil is too thin, it will allow the shaft to run against the bearing surface. If the oil is too thick, it will not run into the clearance between the shaft and the bearing surface. The correct interval for lubricating a sleeve bearing depends on the design and use of the motor. Manufacturer's instructions will indicate the recommended interval. Some motors have large reservoirs and do not need lubricating for years. This is good if there is limited access to the motor.

The permanently lubricated sleeve bearing is constructed with a large reservoir and a wick to gradually feed the oil to the bearing. This bearing truly does not need lubrication until the oil deteriorates. If the motor has been running hot for many hours, the oil will deteriorate and fail. Shaded-pole motors operating in the heat and weather have

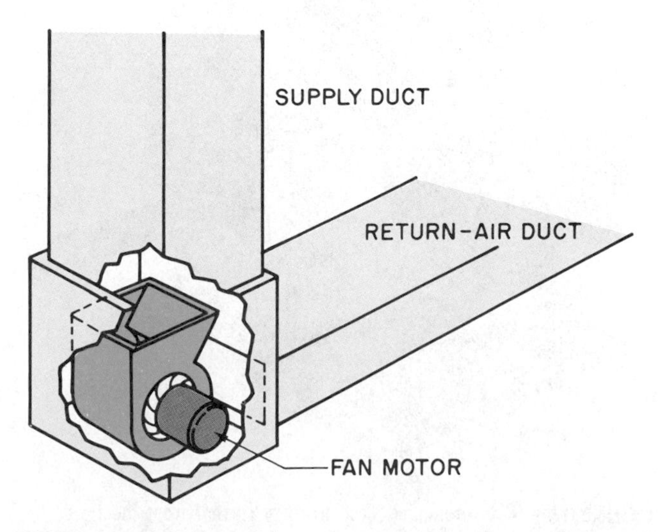

FIGURE 18–20 Furnace fan motor mounted in a horizontal position.

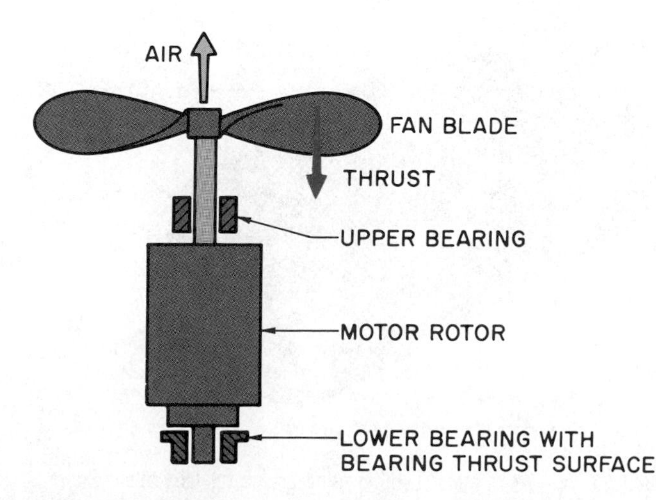

FIGURE 18–21 Thrust surface on a fan motor bearing.

these bearing systems and many have operated without failure for years.

Ball-bearing motors are not as quiet as sleeve-bearing motors and are used in locations where their noise levels will not be noticed. Large fan motors and pump motors are normally located far enough from the conditioned space that the bearing noises will not be noticed. These bearings are made of very hard material and usually lubricated with grease rather than oil. Motors with ball bearings generally have permanently lubricated bearings or grease fittings.

Permanently lubricated ball bearings are similar to sleeve bearings, but they have reservoirs of grease sealed in the bearing. They are designed to last for years with the lubrication furnished by the manufacturer, unless the conditions in which they operate are worse than those for which the motor was designed.

Bearings needing lubrication have grease fittings, so a grease gun can force grease into the bearing. This is often done by hand, Figure 18–22. Only an approved grease must be used. In Figure 18–23 the slotted screw at the bottom of the bearing housing is a *relief* screw. When grease is pumped into the bearing, this screw must be removed, or the pressure of the grease may push the grease seal out and grease will leak down the motor shaft.

Large motors use a type of ball bearing called a *roller* bearing, which has cylindrically shaped rollers instead of balls.

18.6 MOTOR MOUNTING CHARACTERISTICS

Mounting characteristics of a motor determine how it will be secured during its operation. Noise level must be considered when mounting a motor. Two primary means are *rigid* mount and *resilient* or rubber mount.

Rigid-mount motors are bolted, metal to metal, to the frame of the fan or pump and will transmit any motor noise into the piping or duct work. Motor hum is an electrical noise, which is different from bearing noise, and must also be isolated in some installations.

Resilient-mount motors use different methods of isolating the motor noise and bearing noise from the metal framework of the system. Notice the ground strap on the resilient-mount motor, Figure 18–24. This motor is electrically and mechanically isolated from the metal frame. If the motor were to have a ground (circuit from the hot line to the frame of the motor), the motor frame would be electrically hot without the ground strap. ***When replacing a motor *always* connect the ground strap properly or the motor could become dangerous, Figure 18–25.***

The four basic mounting styles are *cradle mount, rigid-base mount, end mount* (with tabs or studs or flange mount), and *belly-band-strap mount,* all of which fit standard di-

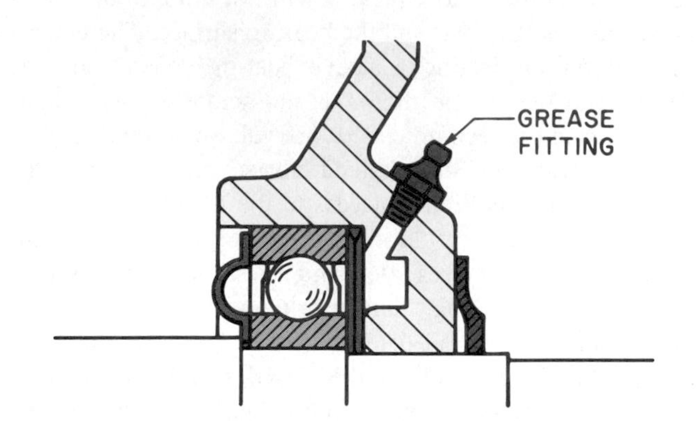

FIGURE 18–22 Fitting through which the bearing is greased.

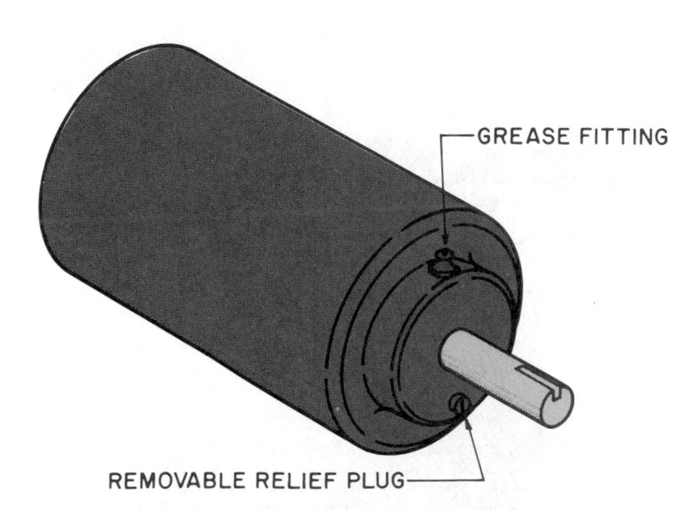

FIGURE 18–23 Motor bearing using grease for lubrication. Notice the relief plug.

FIGURE 18–24 Grounding strap to carry current from the frame if the motor has a grounded winding. *Courtesy Universal Electric Company*

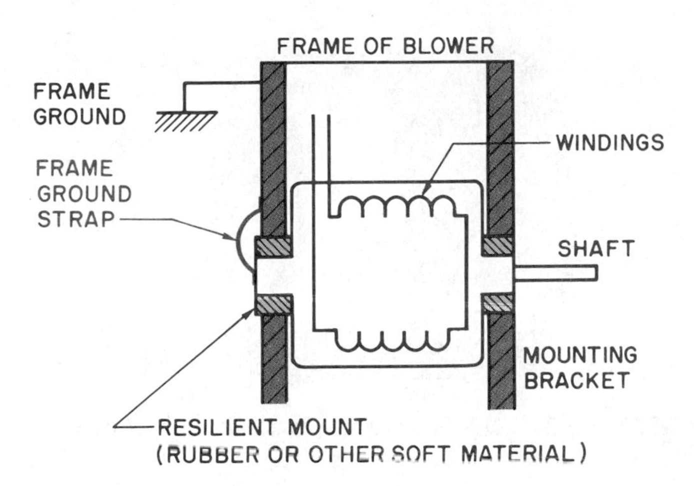

FIGURE 18–25 How the grounding strap works.

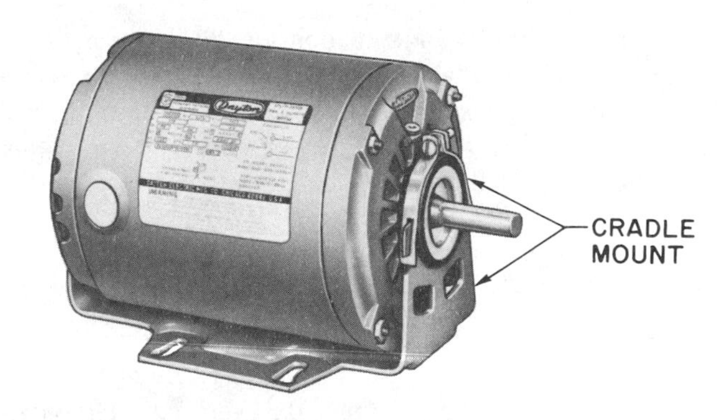

FIGURE 18–27 Cradle-mount motor. *Courtesy W. W. Grainger, Inc.*

mensions established by NEMA and which are distinguished from each other by *frame numbers*. Figure 18–26 shows some typical examples.

Cradle-Mount Motors

Cradle-mount motors are used for either direct-drive or belt-drive applications. They have a cradle that fits the motor end housing on each end. The end housing is held down with a bracket, Figure 18–27. The cradle is fastened to the equipment or pump base with machine screws, Figure 18–28. Cradle-mount motors are available only in the small horsepower range. A handy service feature is that the motor can easily be removed.

Rigid-Mount Motors

Rigid-base-mount motors are similar to cradle-mount motors except that the base is fastened to the motor body,

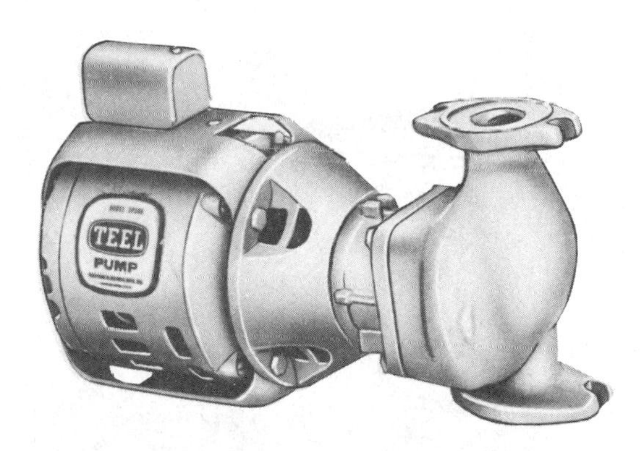

FIGURE 18–28 Cradle fastened to base of pump. *Courtesy W. W. Grainger, Inc.*

MOTOR DIMENSIONS FOR NEMA FRAMES

Standardized motor dimensions as established by the National Electrical Manufacturers Association (NEMA) are tabulated below and apply to all base-mounted motors listed herein which carry a NEMA frame designation.

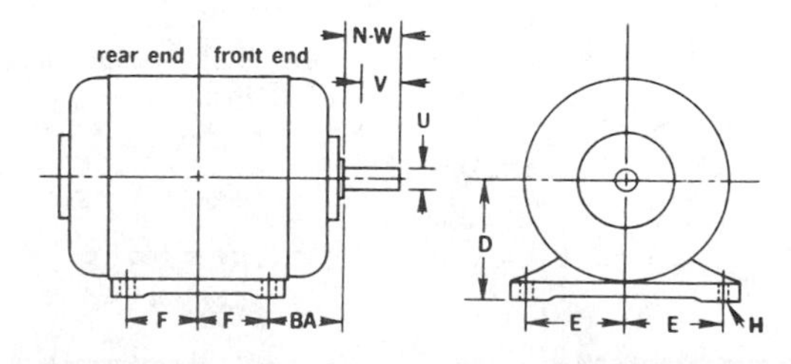

NEMA FRAME	D(*)	2E	2F	BA	H	N-W	U	V(§) Min.	Key Wide	Key Thick	Key Long	NEMA FRAME
	All Dimensions in Inches											
42	2 5/8	3 1/2	1 11/16	2 1/16	9/32 slot	1 1/8	3/8	—	—	21/64 flat	—	42
48	3	4 1/4	2 3/4	2 1/2	11/32 slot	1 1/2	1/2	—	—	29/64 flat	—	48
56 56H	3 1/2	4 7/8	3 3&5(‡)	2 3/4	11/32 slot	1 7/8(†)	5/8(†)	—	3/16(†)	3/16(†)	1 3/8(†)	56 56H
56HZ	3 1/2	**	**	**	**	2 1/4	7/8	2	3/16	3/16	1 3/8	56HZ

FIGURE 18–26 Dimensions of typical motor frames. *Courtesy W. W. Grainger, Inc.*

Figure 18–29. The sound isolation for this motor is in the belt, if one is used, that drives the prime mover. The belt is flexible and dampens motor noise. This motor is often used as a direct drive to turn a compressor or pump. A flexible coupling is used between the motor and prime mover in a direct-drive installation, Figure 18–30.

End-Mount Motors

End-mount motors are very small motors mounted to the prime mover with tabs or studs fastened to the motor housing, Figure 18–31. Flange-mounted motors have a flange as a part of the motor housing, for example, an oil burner motor, Figure 18–32.

Belly-Band-Mount Motors

Belly-band-mount motors have a strap that wraps around the motor to secure it with brackets mounted to the strap. These motors are often used in air conditioning air handlers. Several universal types of motor kits are belly-band mounted and will fit many different applications. These motors are all direct drive, Figure 18–33.

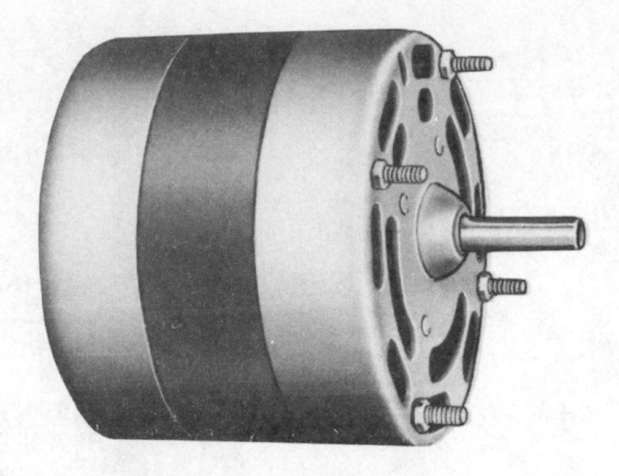

FIGURE 18–31 Motor end mounted with tabs and studs. *Courtesy W. W. Grainger, Inc.*

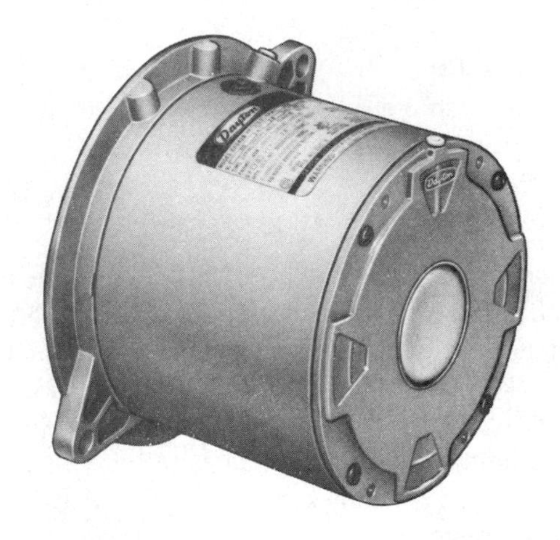

FIGURE 18–32 Motor for an oil burner with a flange on the end to hold the motor to the equipment that is being turned. *Courtesy W. W. Grainger, Inc.*

FIGURE 18–29 Rigid-mount motor. *Courtesy W. W. Grainger, Inc.*

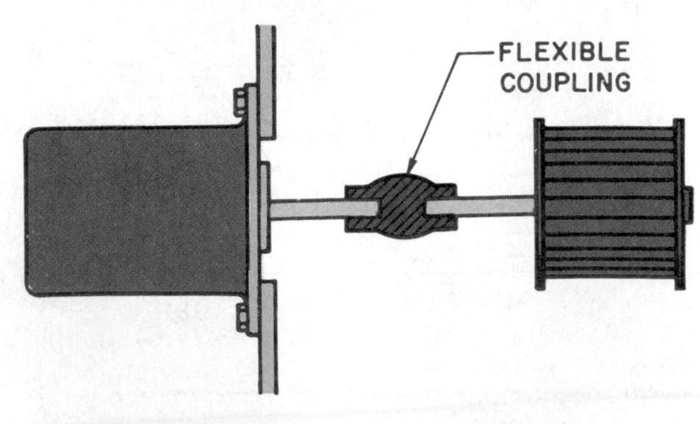

FIGURE 18–30 Direct-drive motor with flexible coupling.

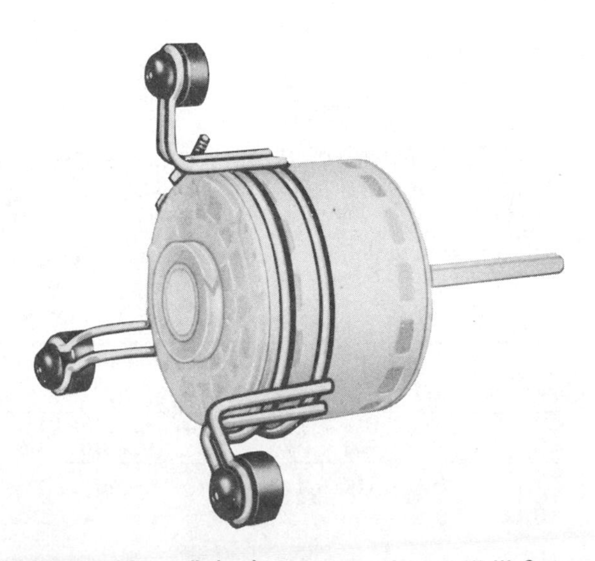

FIGURE 18–33 Belly-band-mount motor. *Courtesy W. W. Grainger, Inc.*

18.7 MOTOR DRIVES

Motor drives are devices or systems that connect a motor to the driven load. For instance, the motor is a driving device, and a fan is a driven component. All motors drive their loads by belts, direct drives through couplings, or the driven component may be mounted on the motor shaft. Gear drives are a form of direct drive and will not be covered in this text because they are used mainly in large industrial applications.

The drive mechanism is intended to transfer the motor's rotating power or energy to the driven device. For example, a compressor motor is designed to transfer the motor's power to the compressor, which compresses the refrigerant vapor and pumps refrigerant from the low side to the high side of the system. Efficiency, speed of the driven device, and noise level are some factors involved in this transfer. It takes energy to turn the belts and pulleys on a belt-drive system in addition to the compressor load. Therefore a direct-drive motor may be better suited for this application. Figure 18–34 is an example of both direct and belt drives.

Belt-drive applications have been used for years to drive both fans and compressors. Pulley sizes can be changed, and the speed of the driven device may be changed. This is a versatility of the belt-drive type of system, Figure 18–35. This can be a great advantage if the capacity of a compressor or a fan speed needs to be changed. However, the changes must be made within the capacity of the drive motor.

Belts are manufactured in different types and sizes. Some have different fibers inside to prevent stretching. ***Handle belts carefully during installation. A belt designed for minimum stretch must not be installed by forcing it over the side of the pulley because it may not stretch enough. Fibers will break and weaken the belt, Figure 18–36. Do not get your fingers between the belt and the pulley. Never touch the belt when it is moving.***

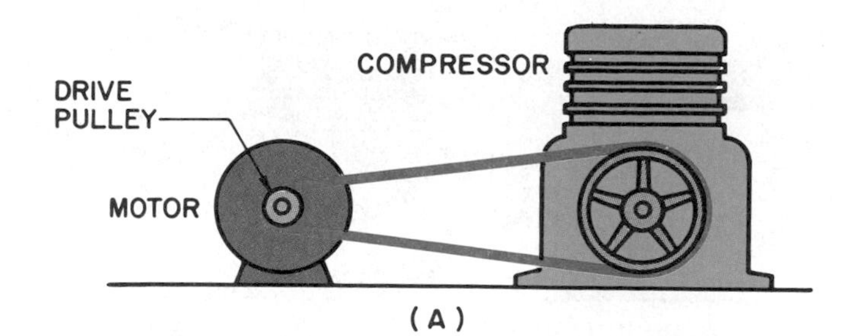

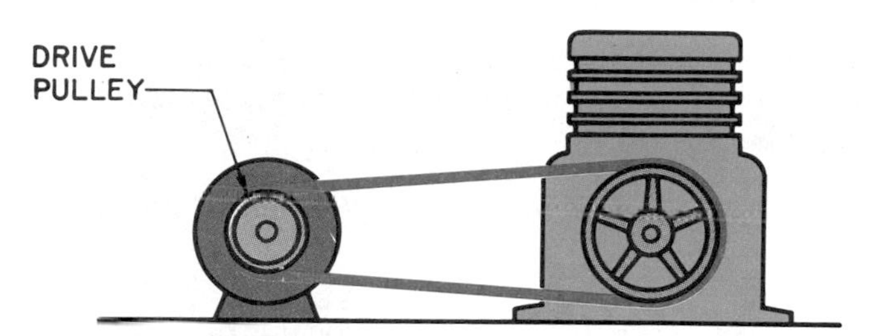

FIGURE 18–35 Belt drive.

Belt widths are denoted by "A" and "B." An A width belt must not be used with a B width pulley nor vice versa,

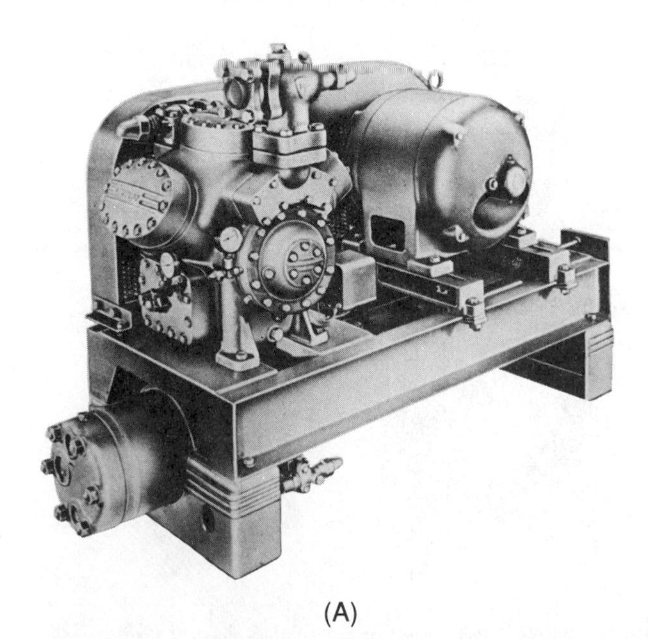
(A)

(B)

FIGURE 18–34 Motor drive mechanisms. *Reproduced courtesy of Carrier Corporation*

MOTOR IS ADJUSTED TOWARD COMPRESSOR
FOR BELTS TO BE INSTALLED.

FIGURE 18–36 Correct method for installing belts over a pulley. The adjustment is loosened to the point that the belts may be passed over the pulley side.

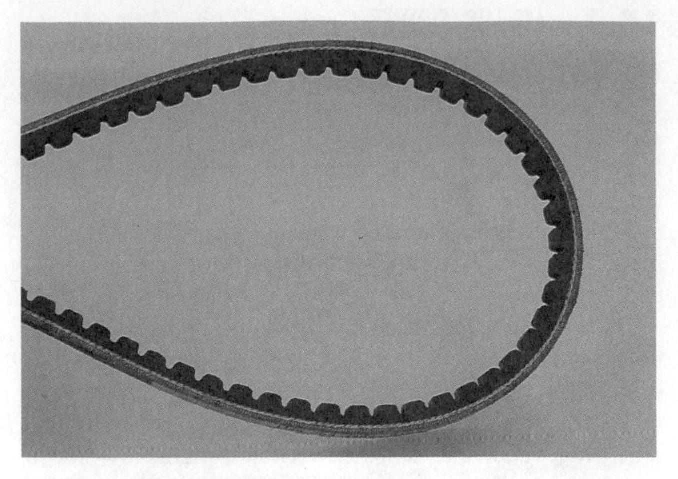

FIGURE 18–38 Belt with grooves has a tractor type grip. *Photo by Bill Johnson*

Figure 18–37. Belts can also have different grips, Figure 18–38.

When a drive has more than one belt, the belts must be matched. Two belts with the same length marked on the belt are not necessarily matched. They may not be exactly the same length. A *matched* set of belts means the belts are *exactly* the same length. A set of 42-in. belts *marked* as a matched set means each belt is exactly 42 in. If the belts are not marked as a matched set, one may be 42 1/2 in., and the other may be 41 3/4 in. Thus the belts will not pull evenly—one belt will take most of the load and will wear out first.

Belts and pulleys wear like any moving or sliding surface. When a pulley begins to wear, the surface roughens and wears out the belts. Normal pulley wear is caused by use or running time. Belt slippage will cause premature wear. Pulleys must be inspected occasionally, Figure 18–39.

Belts must have the correct tension, or they will cause the motor to operate in an overloaded condition. A belt tension gage should be used to correctly adjust belts to the proper tension. This gage is used in conjunction with a chart that gives the correct tension for different types of belts of various lengths. Figure 18–40 shows two types of belt tension gages.

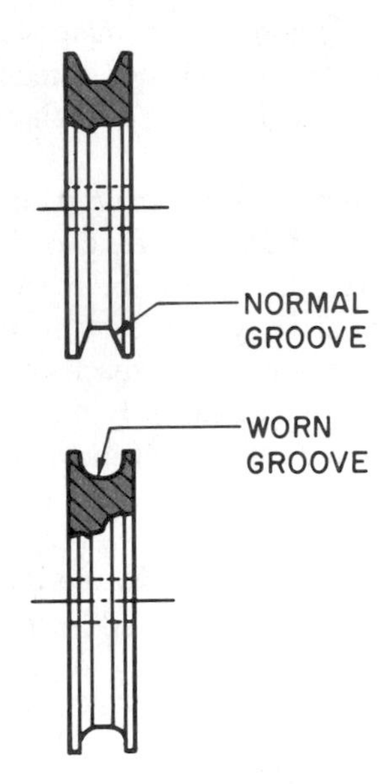

FIGURE 18–39 Normal and worn pulley comparison.

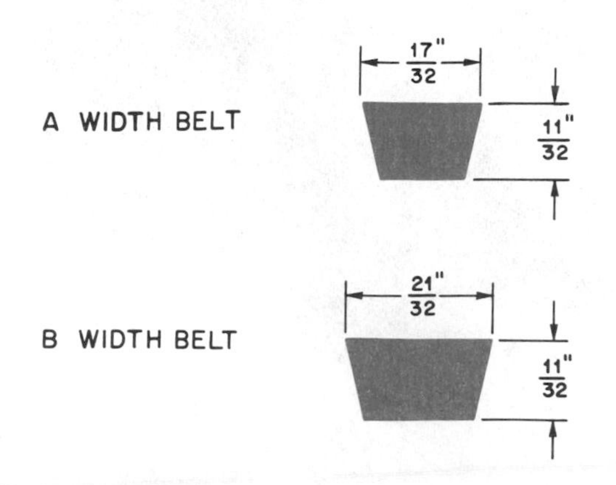

FIGURE 18–37 A and B width belts.

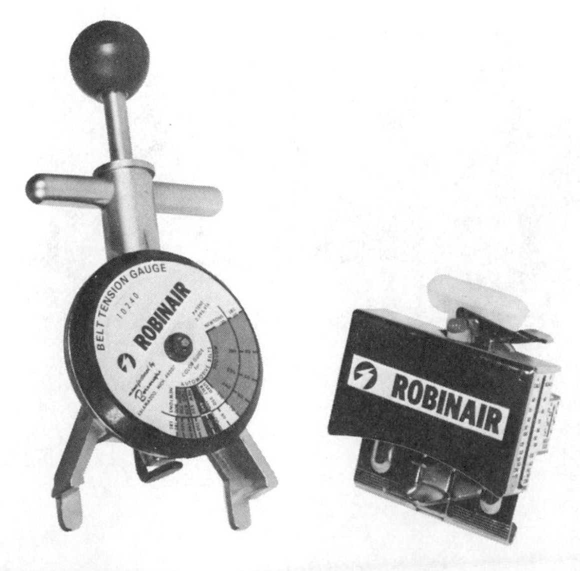

FIGURE 18–40 Belt tension gages. *Courtesy Robinair Division, SPX Corp.*

Direct-drive motor applications are normally used with drive motors for fans, pumps, and compressors. Small fans and hermetic compressors are direct drive, but the motor shaft is actually an extended shaft with the fan or compressor on the end, Figure 18–41. The technician can do nothing to alter these. When this type is used in an open-drive application, some sort of coupling must be installed between the motor and the driven device, Figure 18–42. Some couplings have springs that connect the two coupling halves together to absorb small amounts of vibration from the motor or pump.

A more complicated coupling is used between the motor and a larger pump or a compressor, Figure 18–43. This coupling and shaft must be in very close alignment, or vibration will occur. The alignment must be checked to see that the motor shaft is parallel with the compressor or pump shaft. Alignment is a very precise operation and is done by experienced technicians. If two shafts are aligned to within tolerance while the motor and driven mechanism is at room temperature, the alignment must be checked again after the system is run long enough to get the system up to operating temperature. The motor may not expand and move the same distance as the driven mechanism and the alignment may need to be adjusted to the warm value, Figure 18–44.

When a new motor must be installed to replace an old one, try to use an exact replacement. An exact replacement

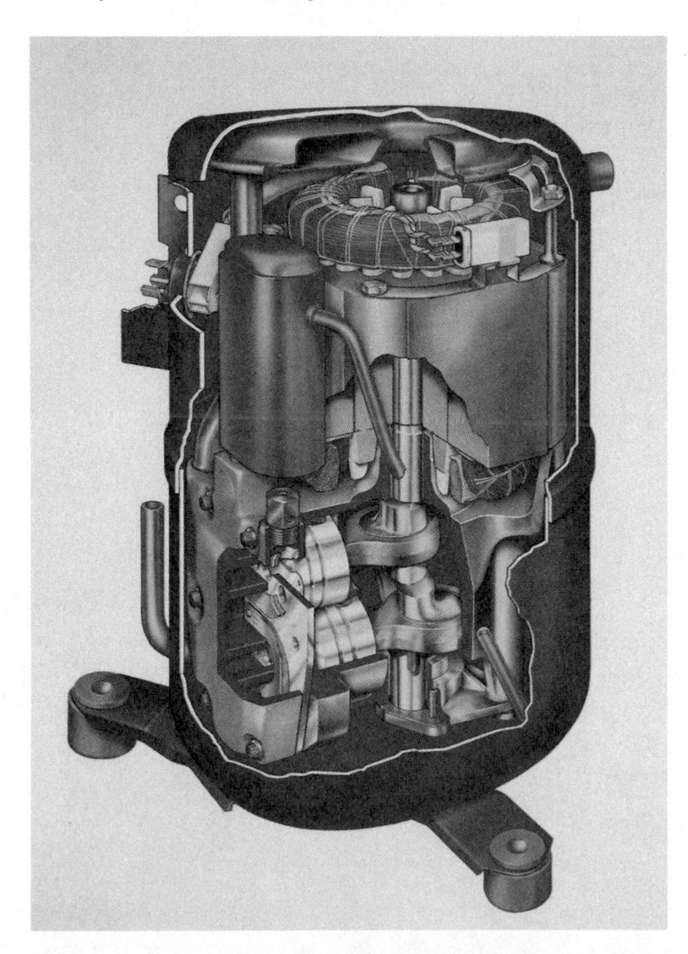

FIGURE 18–41 Direct-drive compressor. *Courtesy Tecumseh Products Company*

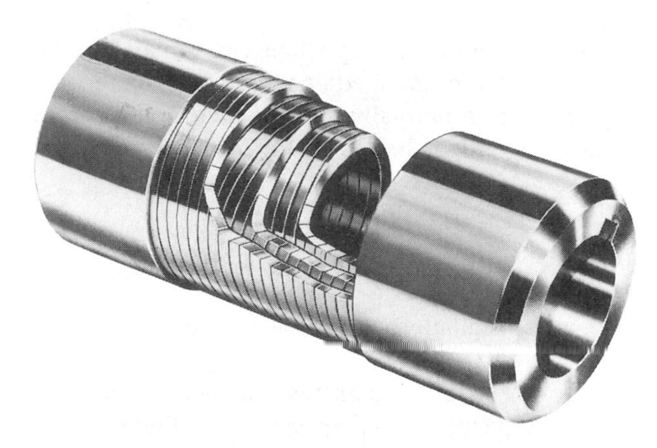

FIGURE 18–42 Small flexible coupling. *Courtesy Lovejoy, Inc.*

FIGURE 18–43 More complicated coupling used to connect larger motors to large compressors and pumps. *Courtesy Lovejoy, Inc.*

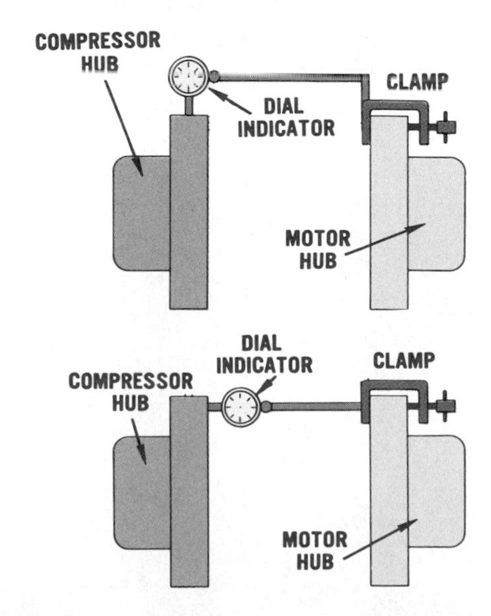

FIGURE 18–44 The alignment of the two shafts must be very close for the system to operate correctly. *Courtesy Trane Company*

may be found at a motor supply house or it may have to be obtained from the original equipment manufacturer. When the motor is not a normally stocked motor, you can save much time by taking the old motor to the distributor and asking for one "just like this."

SUMMARY

- In many installations only one type of motor can be used.
- The power supply determines the applied voltage, the current capacity, the frequency, and the number of phases.
- The working conditions (duty) for a motor deal with the atmosphere in which the motor must operate (ie, wet, explosive, or dirty).
- Motors are also classified according to the insulation of the motor windings and the motor temperature under which they operate.
- Each motor has sleeve, ball, or roller bearings. Sleeve bearings are the quietest but will not stand heavy loads.
- ***Motors must be mounted in the fashion designed for the installation.***
- ***An exact motor replacement should be obtained whenever possible.***
- The drive mechanism transfers the motor's energy to the driven device (the fan, pump, or compressor).

REVIEW QUESTIONS

1. Name four items to consider in a power supply for a system and its motors.
2. What is the allowable voltage variation for typical motors?
3. What are the two main power-supply characteristics over which the technician has some control?
4. What is meant by the service factor of a motor?
5. What two categories of electric motors concern the power supply?
6. Name some of the typical conditions in which a motor must operate.
7. How does the insulation class of a motor affect its use?
8. Name two types of bearings commonly used on small motors.
9. Name four types of motor mounts.
10. What is the best replacement motor to use for a special application?
11. What does the drive mechanism do?
12. What is a matched set of belts?
13. Name the different types of belts.
14. Why must direct-drive couplings be aligned?
15. Why are springs used in a small coupling?

UNIT 19

Motor Starting

OBJECTIVES

After studying this unit, you should be able to

- **describe the differences between a relay, a contactor, and a starter.**
- **state how the locked-rotor current of a motor affects the choice of a motor starter.**
- **list the basic components of a contactor and starter.**
- **compare two types of external motor overload protection.**
- **describe conditions that must be considered when resetting safety devices to restart electric motors.**

SAFETY CHECKLIST

* When a motor is stopped for safety reasons don't restart it immediately. If possible, determine the reason for the overload condition before restarting.
* Conductor wiring should not be allowed to pass too much current or it will overheat causing conductor failure or fire.
* When replacing or servicing line voltage components, ensure that power is off, panel is locked and tagged, and that you have the only key.

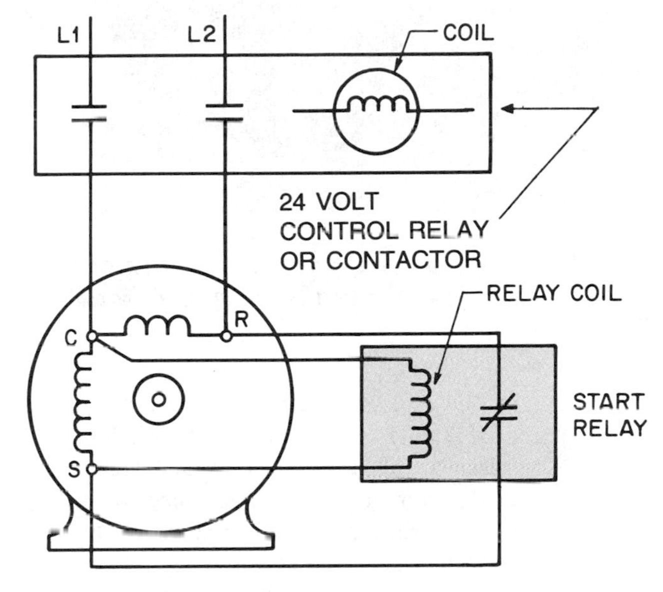

FIGURE 19–1 The starting relay is actually part of the motor circuit. The control relay or contactor passes power to the starting relay.

19.1 INTRODUCTION TO MOTOR CONTROL DEVICES

This unit concerns those components used to close or to open the power-supply circuit to the motor. These devices are called relays, contactors, and starters.

For example, a compressor in a residential air conditioner is controlled in the following manner. With a temperature rise in the space to be conditioned, the thermostat contacts close and pass low voltage to energize the coil on the compressor relay. This closes the contacts in the compressor relay, allowing the applied or line voltage to pass to the compressor motor windings. Relays, contactors, and starters are names given to these motor controls even though they perform the same function. Figure 19–1 is a diagram of a motor starting relay. They all have coils that, when energized, close contacts and pass the line current to the motor. Notice the relay is a major part of the motor circuit even though it is external to the motor windings. Power is fed to the motor windings and starting relay from the control relay.

The size of the motor and the application usually determine the type of switching device (relay, contactor, or starter). For example, a small manual switch can start and stop a handheld drier, and a person operates the swtich. A large 100-hp motor that drives an air conditioning compressor must start, run, and stop unattended. It will also consume much more current than the hair drier. The components that start and stop large motors must be more elaborate than those that start and stop small motors.

19.2 RUN-LOAD AND LOCKED-ROTOR CURRENT

Electric motors have two current (amperage) ratings: the *run-load amperage* (RLA), sometimes referred to as the *full-load amperage* (FLA), and the *locked-rotor amperage* (LRA). The RLA or FLA is the current drawn while the motor is running. The LRA is the current drawn by the motor just as it begins to start. Both currents must be considered when choosing the component (relay, contactor, starter) that passes the line voltage to the motor.

19.3 THE RELAY

The *relay* has a magnetic coil that closes one or more sets of contacts, Figure 19–2. It is considered a throw-away device because parts are not available for rebuilding it when it no longer functions properly.

FIGURE 19–2 Relay for starting a motor. It has a magnetic coil that closes the contacts when the coil is energized. *Photo by Bill Johnson*

Relays are designed for light duty. *Pilot* relays can switch (on and off) larger contactors or starters. Pilot relays for switching circuits are very light duty and are not designed to directly start motors. Relays designed for starting motors are not really suitable as switching relays because they have more resistance in the contacts.

The pilot-duty relay contacts are often made of a fine silver alloy and designed for low-level current switching. Use on a higher load would melt the contacts. Heavier-duty motor switching relays are often made of silver cadmium oxide with a higher surface resistance and are physically larger than pilot-duty relays.

If a relay starts the indoor fan in the cooling mode in a central air conditioning system, it must be able to withstand the inrush current of the fan motor on startup, Figure 19–3.

FIGURE 19–3 Fan relay that may be used to start an evaporator fan on a central air conditioning system. This is a throw-away relay. *Photo by Bill Johnson*

(A motor normally has a starting current of five times the running current.) Relays are often rated in horsepower: If a relay is rated for a 3-hp motor, it will be able to stand the inrush or locked-rotor current of a 3-hp motor.

A relay may have more than one type of contact configuration. It could have two sets of contacts that close when the magnetic coil is energized, Figure 19–4, or it may have two sets of contacts that close and one set that opens when the coil is energized, Figure 19–5. A relay with a single set of contacts that close when the coil is energized is called a *single-pole–single-throw, normally open relay* (spst, NO). A relay with two contacts that close and one that opens is called a *triple-pole–single-throw,* with two *normally open* contacts, and one *normally closed contact, tpst.* Figure 19–6 shows some different relay contact arrangements.

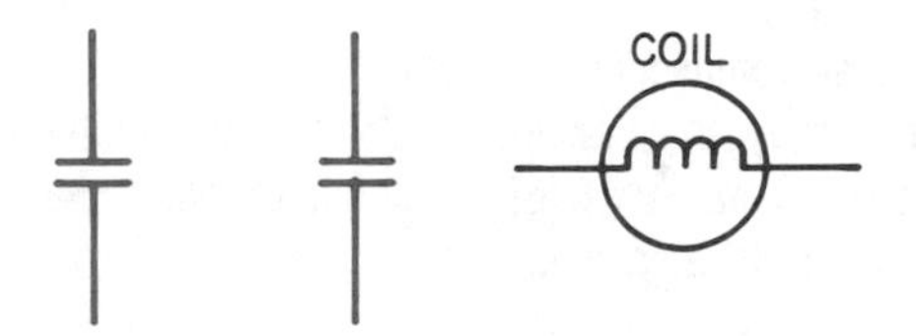

FIGURE 19–4 Double-pole–single-throw relay with two sets of contacts that close when the coil is energized.

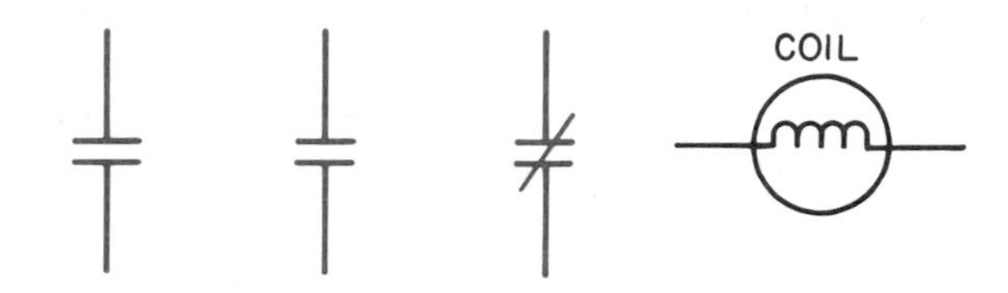

FIGURE 19–5 Relay with two sets of contacts that close and one set that opens when the coil is energized. It has two normally open contacts (NO) and one normally closed (NC).

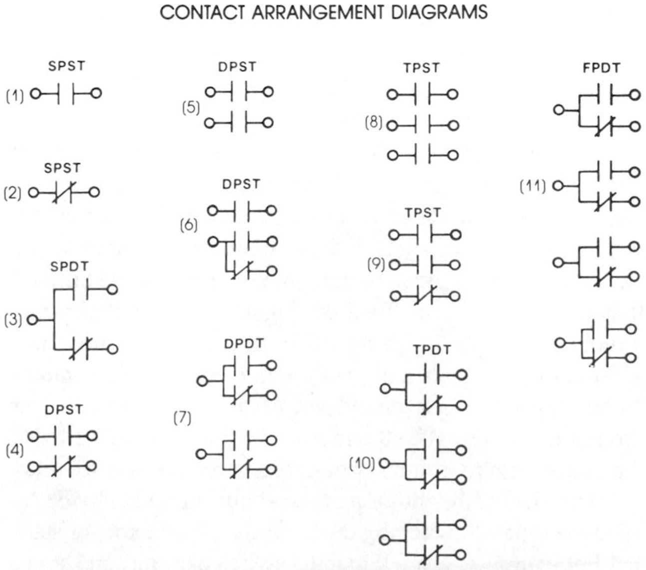

CONTACT ARRANGEMENT DIAGRAMS

SPST (1) SPST (2) SPDT (3) DPST (4) DPST (5) DPST (6) DPDT (7) TPST (8) TPST (9) TPDT (10) FPDT (11)

FIGURE 19–6 Some of the more common combinations of contacts supplied with relays. *Reproduced courtesy of Carrier Corporation*

19.4 THE CONTACTOR

The *contactor* is a larger version of a relay, Figure 19–7. It can be rebuilt if it fails, Figure 19–8. A contactor has movable and stationary contacts. The holding coil can be designed for various operating voltages: 24-V, 120-V, 208-V to 230-V, or 460-V operation.

Contactors can be very small or very large, but all have similar parts. A contactor may have many configurations of contacts, from the most basic, which has just one contact, to more elaborate types with several. The single-contact type contactor is used on many residential air conditioning units. It interrupts power to one leg of the compressor, which is all that is necessary to stop a single-phase compressor. Sometimes a single-contact contactor is needed to provide crankcase heat to the compressor. A trickle charge of electricity passes through line 2 to the motor windings when the contacts are open, during the off cycle, Figure 19–9. If you use a contactor with two contacts for a replacement, the compressor would not have crankcase heat. Once again, an exact replacement is the best choice. Figure 19–10 shows another method of supplying crankcase (oil sump) heat only during the off cycle.

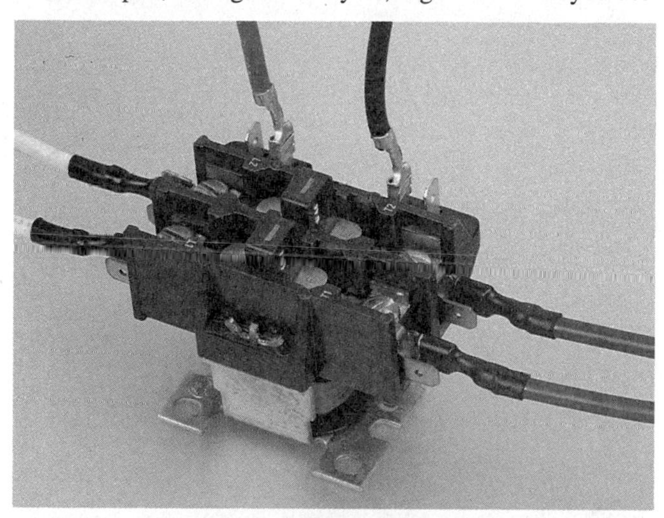

FIGURE 19–7 Contactor. *Photo by Bill Johnson*

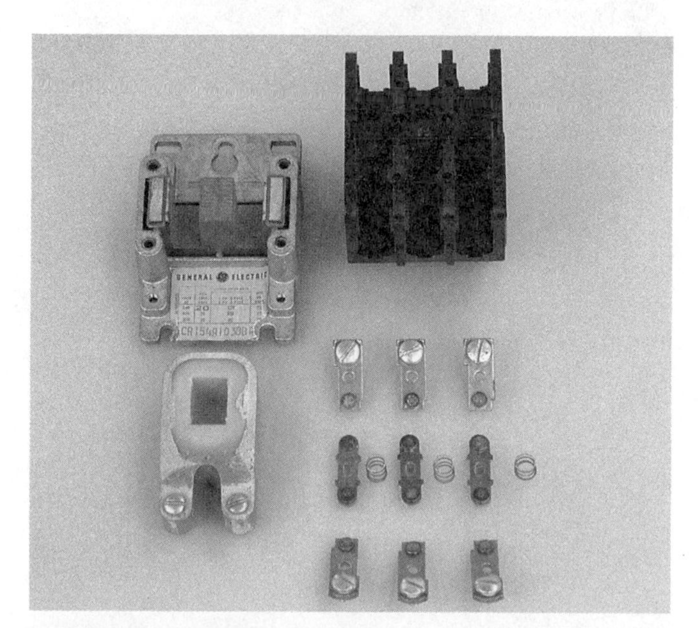

FIGURE 19–8 Parts that may be replaced in a contactor: the contacts, both movable and stationary; the springs that hold the contacts; and the holding coil. *Photo by Bill Johnson*

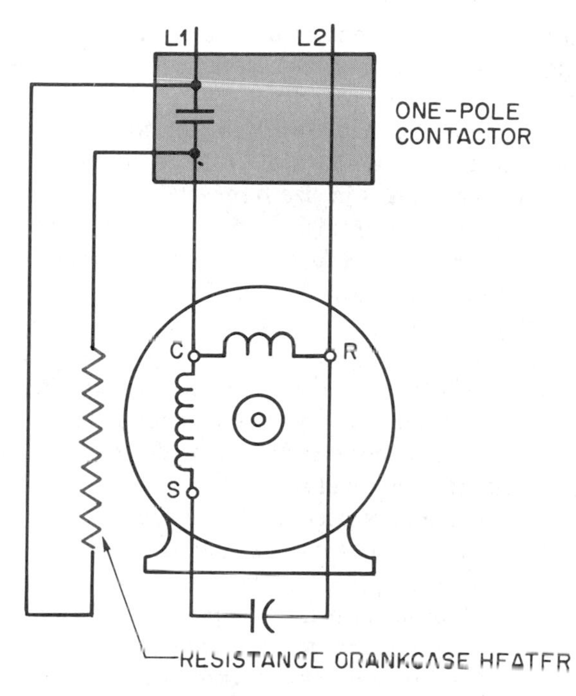

FIGURE 19–9 This contactor has only one set of contacts. Only one line of the power needs to be broken to stop a motor like the one in the diagram. This is a method of supplying crankcase heat to the compressor only during the off cycle. The current during the running cycle (with the contacts closed) bypasses the heater and goes to the run winding.

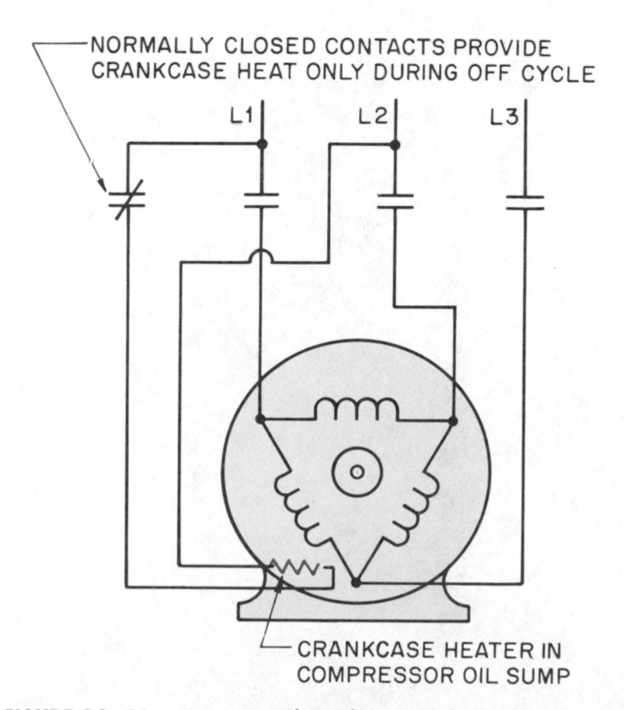

FIGURE 19–10 Contactor with auxiliary contacts for switching other circuits.

Some contactors have as many as five or six sets of contacts. Generally for larger motors there will be three heavy-duty sets to start and stop the motor; the rest (*auxiliary contacts*) can be used to switch auxiliary circuits.

19.5 MOTOR STARTERS

The *starter,* or motor starter as it is sometimes called, is similar to the contactor. In some cases, a contactor may be converted to a motor starter. The motor starter differs from a contactor because it has motor overload protection built into its framework, Figures 19–11 and 19–12. A motor starter may be rebuilt and ranges in size from small to very large.

Motor protection is discussed more fully later in this unit, but we should know that overload protection protects that particular motor. See Figure 19–13 for a melting alloy-type overload heater. The fuse or circuit breaker cannot be wholly relied on for protection because it protects the entire circuit, which may have many components. In some cases the motor starter protection is a better indication of a motor problem than the motor winding thermostat protection will provide.

The contact surfaces become dirtier and more pitted with each motor starting sequence, Figure 19–14. Some technicians believe that these contacts may be buffed or cleaned with a file or sandpaper. ***Contacts may be cleaned, but this should be done as a temporary measure only. Filing or sanding exposes base metal under the silver**

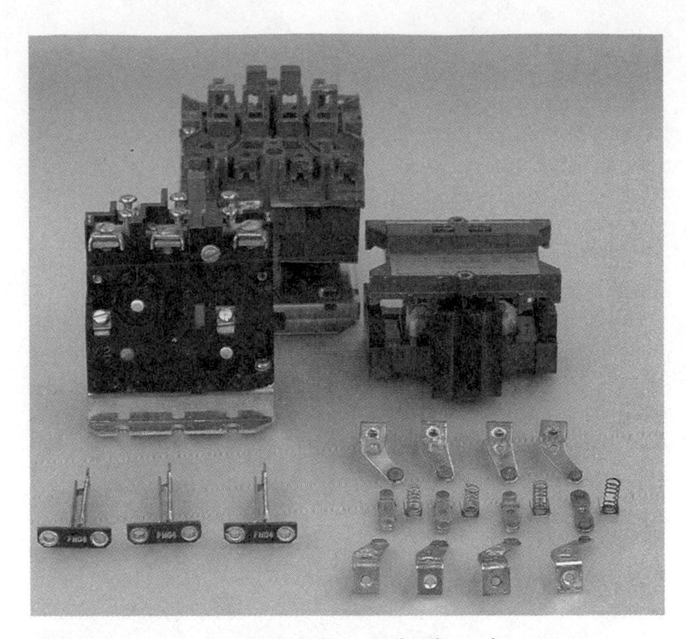

FIGURE 19–12 Components that may be changed on a starter: contacts, both movable and stationary; the springs; the coil; and the overload protection devices. *Photo by Bill Johnson*

FIGURE 19–11 Motor starter. *Photo by Bill Johnson*

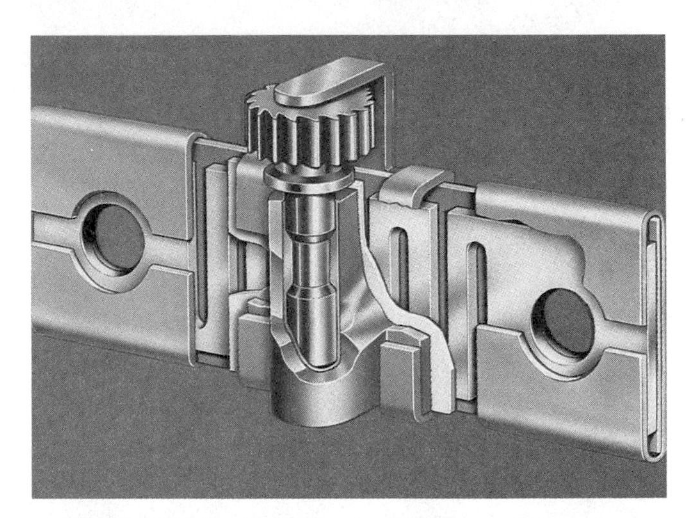

FIGURE 19–13 Melting alloy-type overload heater. *Courtesy Square D Company*

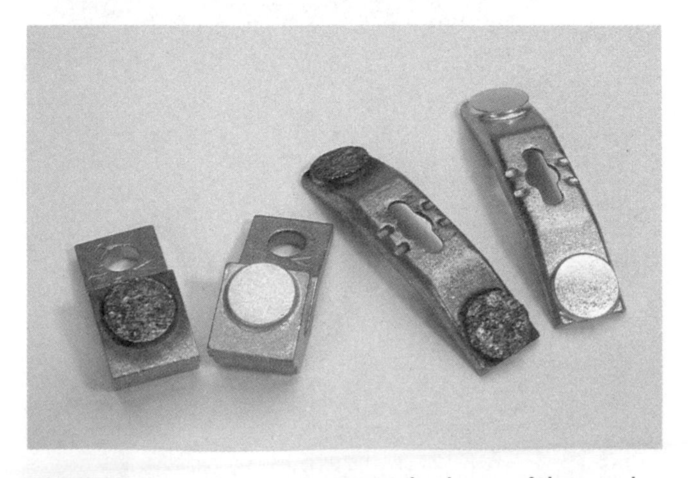

FIGURE 19–14 Clean contacts contrasted with a set of dirty pitted contacts. *Courtesy Square D Company*

plating and speeds its deterioration. Replacing the contacts is the only recommended repair.* If this device is a relay, the complete relay must be replaced. If the device is a contactor or a starter, the contacts may be replaced with new ones, Figure 19–14. There are both movable and stationary contacts and springs that hold tension on the contacts.

You can use a voltmeter to check resistance across a set of contacts under full load. When the meter leads are placed on each side of a set of contacts and there is a measurable voltage, there is resistance in the contacts. When the contacts are new and have no resistance, no voltage should be read on the meter. An old set of contacts will have some slight voltage drop and will produce heat, due to resistance in the contact's surface, where the voltage drop occurs. Figure 19–15 shows an example of how contacts may be checked with an ohmmeter.

Each time a motor starts, the contacts will be exposed to the inrush current (the same as LRA). These contacts are under a tremendous load for this moment. When the contacts open, there is an arc caused by the breaking of the electrical circuit. The contacts must make and break as fast as possible. The contacts have a magnet that pulls them together with a mechanism to take up any slack. For example, there may be three large movable contacts in a row that must all be held equally tight against the stationary contacts. The springs behind the movable contacts keep this pressure even, Figure 19–16. If there is a resistance in the contact surface, the contacts may get hot enough to take the tension out of the springs. This will make the contact-to-contact pressure even less and make the heat greater. The tension in these springs must be maintained.

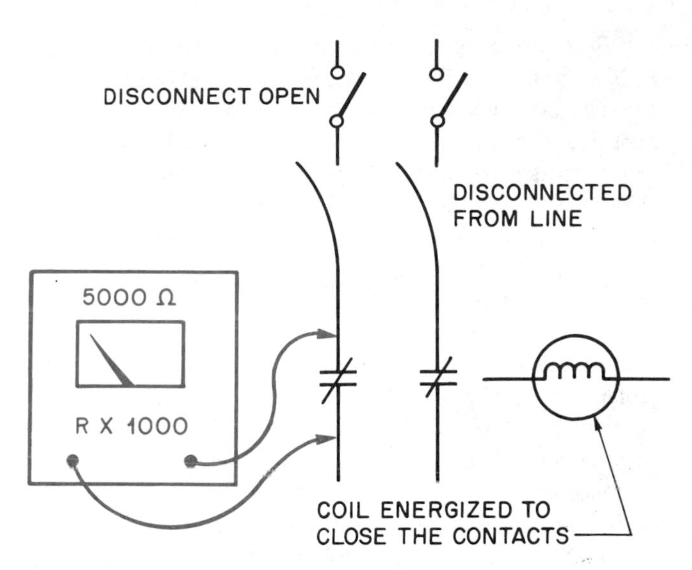

FIGURE 19–15 Checking contacts with ohmmeter shows resistance in contacts caused by dirty surface.

19.6 MOTOR PROTECTION

The electric motors used in air conditioning, heating, and refrigeration equipment are the most expensive single components in the system. These motors consume large amounts of electrical energy, and considerable stress is placed on the motor windings. Therefore, they deserve the best protection possible within the economic boundaries of a well-designed system. The more expensive the motor, the more expensive and elaborate the protection should be.

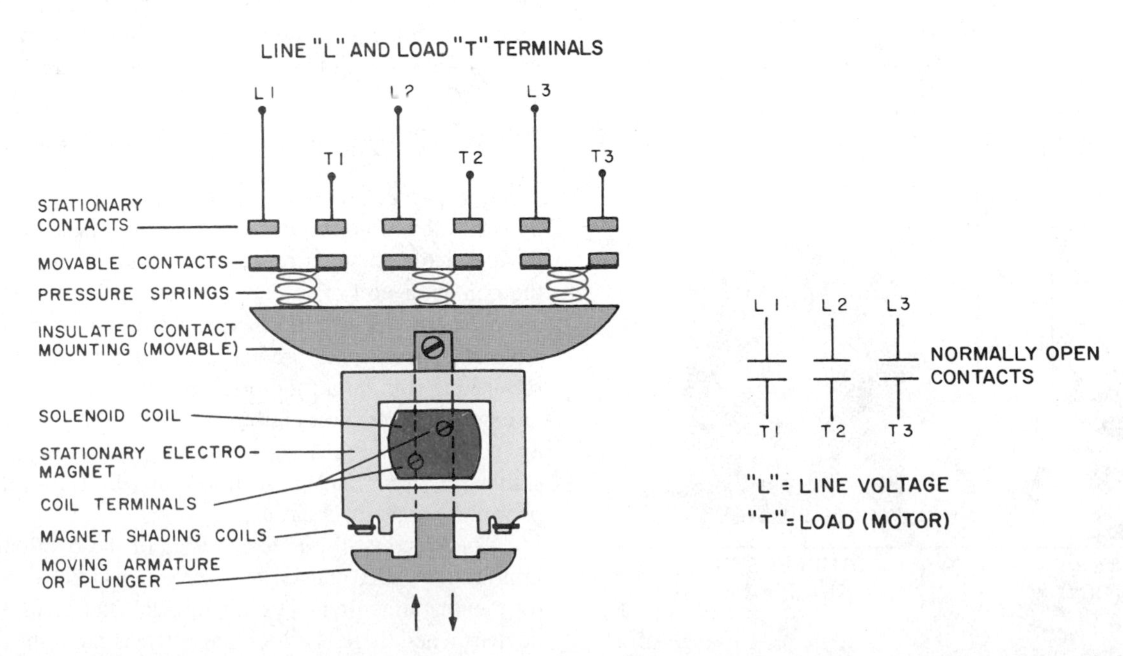

FIGURE 19–16 When energized, springs will hold the three movable contacts tightly against the stationary contacts. From Alerich, *Electric Motor Control,* © 1993 by Delmar Publishers Inc.

Fuses are usually used as circuit (not motor) protectors. ***The conductor wiring in the circuit must not be allowed to pass too much current or it will overheat and cause conductor failures or fires.*** A motor may be operating at an overloaded condition that would not cause the conductor to be overloaded; hence, the fuse will not open the circuit. Let's use a central air conditioning system as an example. There are two motors in the condensing unit: the compressor (the largest) and the condenser fan motor (the smallest). In a typical unit the compressor may have a current draw of 23 A, and the fan motor may use 3 A. The fuse protects the total circuit. If one motor is overloaded, the fuse may not open the circuit, Figure 19–17. Each motor should be protected within its own operating range.

Motors can operate without harm for short periods at a slight overcurrent condition. The overload protection is designed to disconnect the motor at some current draw value that is slightly more than the FLA value, so the motor can be operated at its full-load design condition. Time is involved in this value in such a manner that the higher the current value above FLA, the more quickly the overload should react. The amount of the overload and the time are both figured into the design of the particular overload device.

Overload current protection is applied to motors in different ways. Overload protection, for example, is not needed for some small motors that will not cause circuit overheating or will not damage themselves. Some small motors do not have overload protection because they will not consume enough power to damage the motor unless shorted from winding to winding or from the winding to the frame (to ground). See Figure 19–18 for an example of a small condenser fan motor that does not draw enough amperage at the LRA condition to overheat. This motor does not have overload protection. This motor is described as "impedance protected." It may only generate 10 W of power, the same as a 10-W light bulb. It is not an expensive motor. If the motor fails because of a burnout, the current draw will be interrupted by the circuit protector.

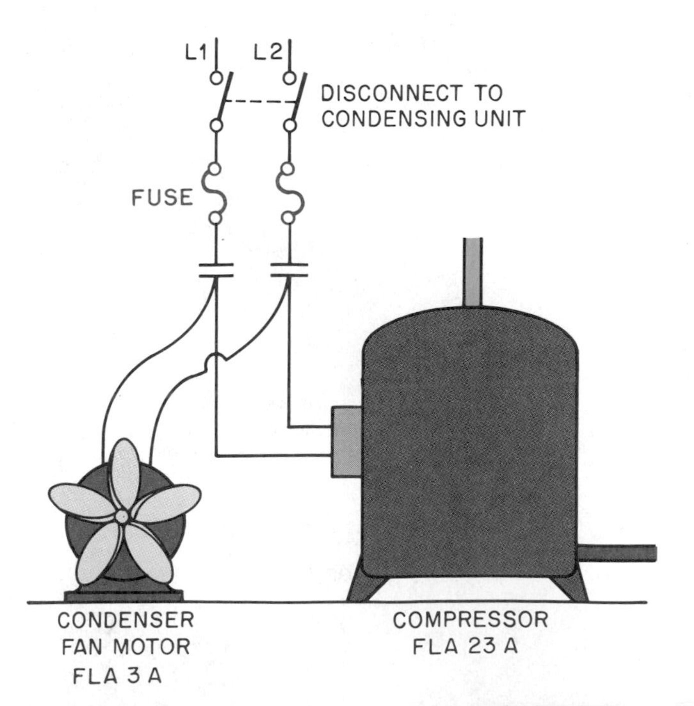

FIGURE 19–17 Two motors in the same circuit are served by the same conductor and have different overload requirements.

FIGURE 19–18 Small condenser fan motor does not pull enough amperage at locked rotor to create enough heat to be a problem. This is known as impedance motor protection. *Photo by Bill Johnson*

Overload protection is divided into inherent (internal) protection and external protection.

19.7 INHERENT MOTOR PROTECTION

Inherent protection is that provided by internal thermal overloads in the motor windings or the thermally activated snap-disc (bimetal), Figure 19–19. The same types of devices are used with open and hermetic motors.

19.8 EXTERNAL MOTOR PROTECTION

External protection is often applied to the device passing power to the motor contactor or starter. These are normally devices actuated by current overload and break the circuit to the contactor coil. The contactor stops the motor. When a motor is started with a relay, the motor is normally small and only has internal protection, Figure 19–20. Contactors are used to start larger motors, and either inherent protection or external protection is used. Large motors (above 5 hp in air conditioning, heating, and refrigeration systems) use starters and overload protection either built into the starter or in the contactor's circuit.

The value (trip point) and type of the overload protection are normally chosen by the system design engineer or by the manufacturer. The technician checks the overload devices when there is a problem, such as random shutdowns because of an overload tripping. The technician must be able to understand the designer's intent with regard to the

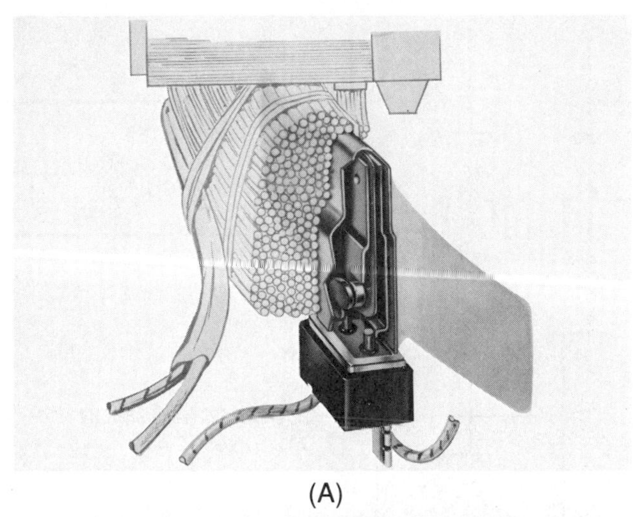

(A)

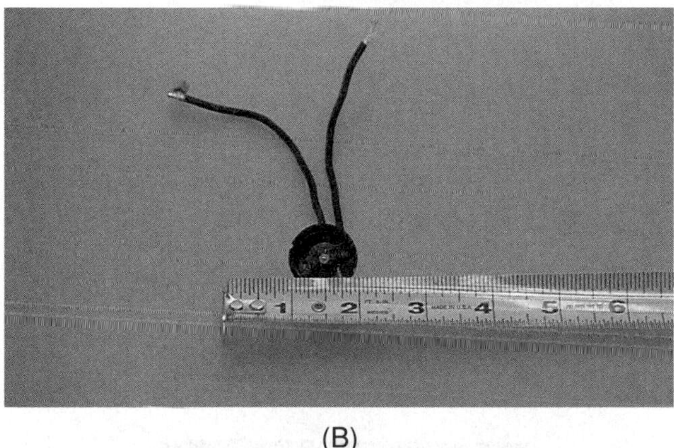

(B)

FIGURE 19–19 **(A) Inherent overload protector. (B) External overload protector.** ***(A) Courtesy Tecumseh Products Company, (B) Photo by Bill Johnson***

FIGURE 19–20 **Fan motor that is normally started with a relay. Motor protection is internal.** ***Courtesy W. W. Grainger, Inc.***

motor's operation and the overload device operation because they are closely related in a working system.

Many motors have a service factor, which is the reserve capacity of the motor. The motor can operate above the FLA and within the service factor without harm. Typical service factors are 1.15 to 1.40. The smaller the motor, the larger the service factor. For example, a motor with an FLA of 10 A and a service factor of 1.25 can operate at 12.5 A without damaging the motor. The overload protection for a particular motor takes the service factor into account.

19.9 NATIONAL ELECTRICAL CODE STANDARDS

The National Electrical Code (NEC) sets the standard for all electrical installations, including motor overload protection. The code book published by the National Electrical Manufacturer's Association should be consulted for any overload problems or misunderstandings that may occur regarding correct selection of the overload device.

The purpose of the overload protection device is to disconnect the motor from the circuit when an overload condition occurs. Detecting the overload condition and opening the circuit to the motor can be done in several ways: by an overload device mounted on the motor starter, or by a separate overload relay applied to a system with a contactor. Figure 19–21 shows an example of a thermal overload relay.

19.10 TEMPERATURE-SENSING DEVICES

Various sensing devices are used for overcurrent situations. The most popular ones are those sensitive to temperature changes.

The bimetal element is an example. The line current of the motor passes through a heater (that can be changed to suit a particular motor amperage) that heats a bimetal strip. When the current is excessive, the heater warps the bimetal, opening a set of contacts that interrupt power to the contactor's coil circuit. All bimetal overload devices are designed with snap action to avoid excessive arcing. These thermal-type overloads are sensitive to any temperature and condi-

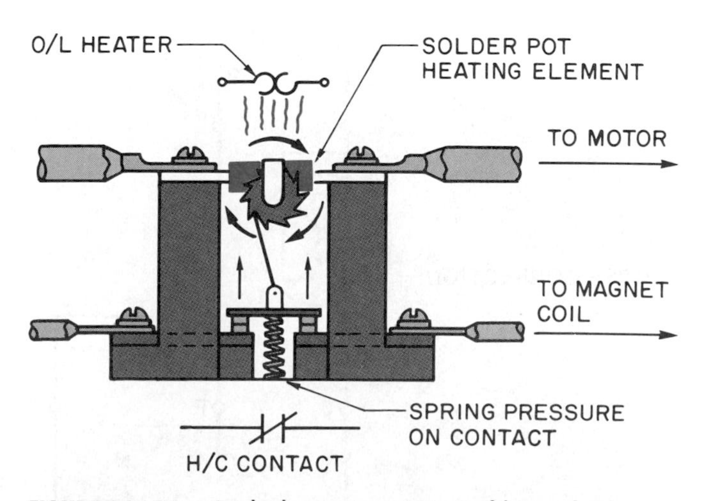

FIGURE 19–21 **Overload using a resistor-type of heater that heats a low-temperature solder. From Alerich, *Electric Motor Control,* © 1993 by Delmar Publishers Inc.**

tions around them. High ambient temperature and loose connections are frequent problems encountered. Figure 19–22 shows an example of a thermal overload with a loose connection.

A low-melting solder may be used in place of the bimetal. This is called a *solder pot*. The solder will melt from the heat caused by an overcurrent condition. The overload heater is sized for the particular amperage draw of the motor it is protecting. The overload control circuit will interrupt the power to the motor contactor coil and stop the motor in case of overload. The solder melts and the overload mechanism turns because it is spring loaded. It can be reset when it cools, Figure 19–21.

Both of these overload protection devices are sensitive to temperature. The temperature of the heater causes them to function. Heat from any source, even if it has nothing to do with motor overload, makes the protection devices more sensitive. For example, if the overload device is located in a metal control panel in the sun, the heat from the sun may affect the performance of the overload protection device. A loose connection on one of the overload device leads will cause local heat and may cause it to open the circuit to the motor even though there is actually no overload, Figure 19–23.

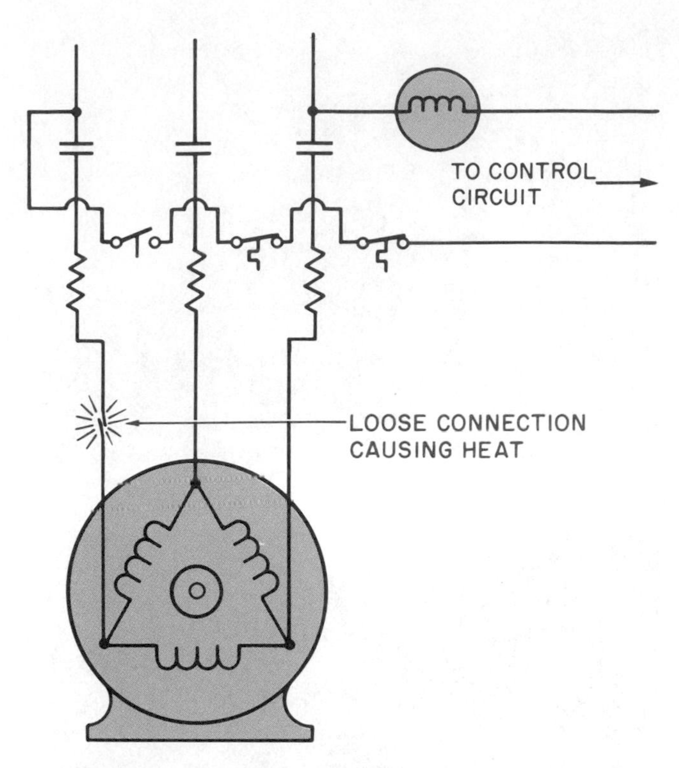

FIGURE 19–23 Local heat, such as from a loose connection, will influence the thermal type of overload.

19.11 MAGNETIC OVERLOAD DEVICES

Magnetic overload devices are separate components and are not attached to the motor starter. This component is very accurate and not affected by ambient temperature, Figure 19–24. The advantage of this overload device is that it can be located in a hot cabinet on the roof and the temperature will not affect it. It will shut the motor off at an accurate ampere rating regardless of the temperature.

FIGURE 19–24 Magnetic overload. *Photo by Bill Johnson*

19.12 RESTARTING THE MOTOR

When a motor is found stopped for safety reasons, such as an overload, don't restart it immediately. Look around for the possible problem before a restart. When a motor stops because it is overloaded, the overload condition at the instant the motor is stopped, is reduced to 0 A. This does *not* mean that the motor should be restarted immediately. The cause of the overload may still exist, and the motor could be too hot and may need to cool.

There are various ways of restarting a motor after an overload condition has occurred. Some manufacturers design their control circuits with a manual reset to keep the motor from restarting, and some use a time delay to keep the motor off for a predetermined time. Others use a relay that will keep the motor off until the thermostat is reset. The

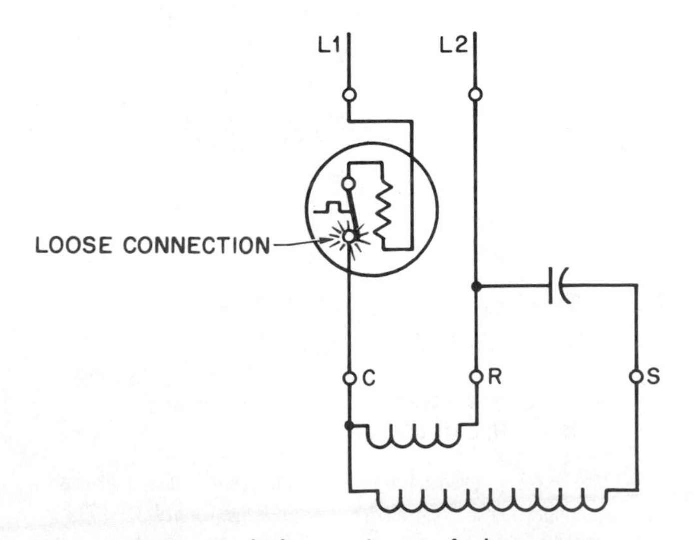

FIGURE 19–22 Overload tripping because of a loose connection.

units that have a manual reset at the overload device may require someone to go to the roof to reset the overload if the unit is located on the roof. See Figure 19–25 for an example of a manual reset. When the reset is in the thermostat circuit, the protection devices may be reset from the room thermostat. This is convenient, but several controls may be reset at the same time and the technician may not know which control is being reset. When the manual reset button is pushed and a restart occurs, there is no doubt which control has been reset to start the motor.

Time-delay reset devices keep the unit from short cycling but may reset themselves with a problem condition still existing.

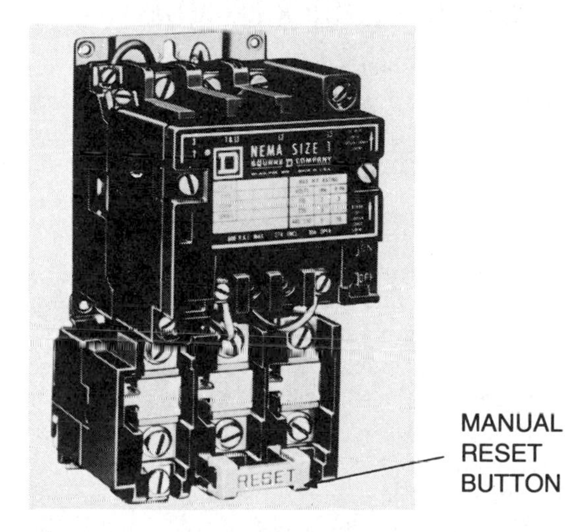

FIGURE 19–25 Manually reset overload. *Courtesy Square D Company*

SUMMARY

- The relay, the contactor, and the motor starter are three types of motor starting and stopping devices.
- The relay is used for switching circuits and motor starting.
- Motor starting relays are used for heavier-duty jobs than are switching relays.
- Contactors are large relays that may be rebuilt.
- Starters are contactors with motor overload protection built into the framework of the contactor.
- The contacts on relays, contactors, and starters should not be filed or sanded.
- Large motors should be protected from overload conditions other than by normal circuit overload protection devices.
- Inherent motor overload protection is provided by sensing devices within the motor.
- External motor overload protection is applied to the current-passing device: the relay, contactor, or starter.
- The technician must have an understanding of the motor operation and the overload device operation.
- The service factor is the reserve capacity of the motor.
- Bimetal and solder-pot devices are thermally operated.
- Magnetic overload devices are very accurate and not affected by ambient temperature.
- ***Most motors should not be restarted immediately after shutdown from an overload condition because they may need time to cool. When possible, determine the reason for the overload condition before restarting the motor.***

REVIEW QUESTIONS

1. What is the recommended repair for a defective relay?
2. What components can be changed on a contactor and a starter for rebuilding purposes?
3. What are the two types of relays?
4. What two amperages influence the choice for replacing a motor starting component?
5. What is the difference between a contactor and a starter?
6. Can a contactor ever be converted to a starter?
7. What are the contact surfaces of relays, contactors, and starters made of?
8. What causes an overload protection device to function?
9. What are typical coil voltages used for relays, contactors, and starters?
10. Why is it not a good idea to file or sand the contactor contacts?
11. Why is it not a good idea to use circuit protection devices to protect large motors from overload conditions?
12. Under what conditions are motors allowed to operate with slightly higher than design loads?
13. Describe the difference between inherent and external overload protection.
14. What is the purpose of overload protection at the motor?
15. State reasons why a motor should not be restarted immediately.

UNIT 20

Troubleshooting Electric Motors

OBJECTIVES

After studying this unit, you should be able to

- **describe different types of electric motor problems.**
- **list common electrical problems in electric motors.**
- **identify various mechanical problems in electric motors.**
- **describe a capacitor checkout procedure.**
- **explain the difference in troubleshooting a hermetic motor problem and an open motor problem.**

SAFETY CHECKLIST

* If it is suspected that a motor has electrical problems, pull the motor disconnect to prevent further damage or an unsafe condition, lock and tag the panel, and keep the only key on your person.
* Before checking a motor capacitor, short from one terminal to the other with a 20,000-Ω 5-W resistor to discharge the capacitor. This practice is recommended even if the capacitor has its own bleed resistor. Use insulated pliers.
* When wiring a motor run capacitor be sure to connect the lead that feeds power to the capacitor to the terminal identified for this purpose.
* Turn the power off before trying to turn the open drive compressor over using a wrench.

20.1 ELECTRIC MOTOR TROUBLESHOOTING

Electric motor problems are either mechanical or electrical. Mechanical problems may appear to be electrical. For example, a bearing dragging in a small permanent split-capacitor (PSC) fan motor may not make any noise. The motor may not start, and it appears to be an electrical problem. The technician must know how to diagnose the problem correctly. This is particularly true with open motors because if the driven component is stuck, a motor may be changed unnecessarily. If the stuck component is a hermetic compressor, the whole compressor must be changed; if it is a serviceable hermetic compressor, the motor can be replaced or the compressor running gear can be rebuilt.

20.2 MECHANICAL MOTOR PROBLEMS

Mechanical motor problems normally occur in the bearings or the shaft where the drive is attached. The bearings can be tight or worn due to lack of lubrication. Grit can easily get into the bearings of some open motors and cause them to wear.

Motor problems are not usually repaired by heating, air conditioning, and refrigeration technicians. They are handled by technicians trained in rebuilding motor and rotating equipment. A motor vibration may require you to seek help from a qualified balancing technician. Explore every possibility to ensure that the vibration is not caused by a field problem, such as a fan loaded with dirt or liquid flooding into a compressor.

Motor bearing failure with roller and ball bearings can often be determined by the bearing noise. When sleeve bearings fail, they normally lock up (will not turn) or sag to the point that the motor is out of its magnetic center. At this point the motor will not start. .

When motor bearings fail, they can be replaced. If the motor is small, the motor is normally replaced because it would cost more to change the bearings than to purchase and install a new motor. The labor involved in obtaining bearings and disassembling the motor can take too much time to make a profit. This is particularly true for fractional horsepower fan motors. These small motors almost always have sleeve bearings pressed into the end bells of the motor, and special tools may be needed to remove and to install new bearings, Figure 20–1.

20.3 REMOVING DRIVE ASSEMBLIES

To remove the motor, you need to remove the pulley, coupling, or fan wheel from the motor shaft. The fit between the shaft and whatever assembly it is fastened to may be very tight. ***Removing the assembly from the motor shaft must be done with care.*** The assembly may have been running on this shaft for years, and it may have rust between the shaft and the assembly. You must remove the assembly without damaging it. Special pulley pullers, Figure 20–2, will help, but other tools or procedures may be required.

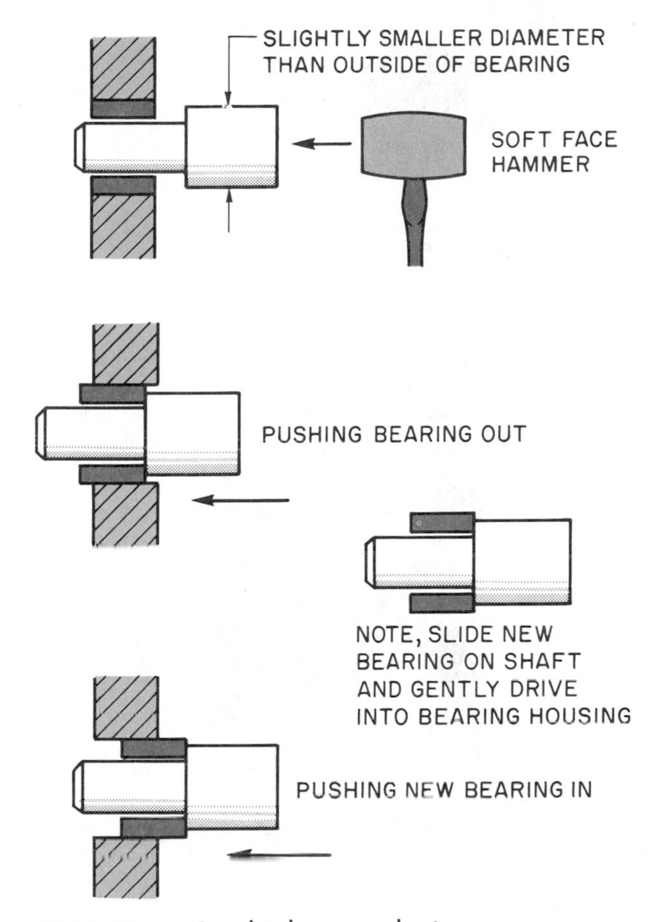

FIGURE 20–1 Special tool to remove bearings.

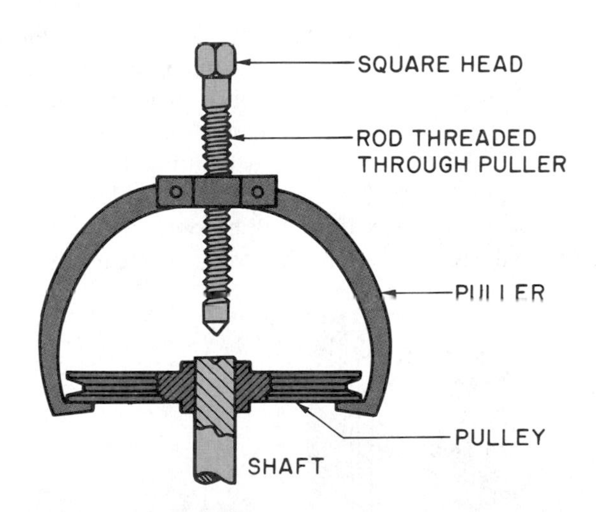

FIGURE 20–2 Pulley puller.

Most assemblies are held to the motor shaft with set screws threaded through the assembly and tightened against the shaft. A flat spot is usually provided on the shaft for the set screw to be seated and to keep it from damaging the shaft surface, Figure 20–3. The set screw is made of very hard steel, much harder than the motor shaft. When larger motors with more torque are used, a matching keyway is normally machined in the shaft and assembly. This keyway provides a better bond between the motor and assembly, Figure 20–4. A set screw is then often tightened down on the top of the key to secure the assembly to the motor shaft.

FIGURE 20–3 Flat spot on motor shaft where pulley set screw is tightened. *Photo by Bill Johnson*

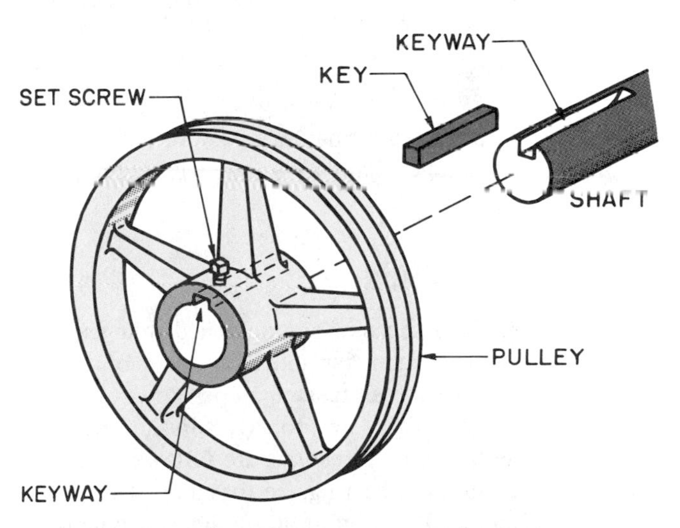

FIGURE 20–4 Pulley with a groove cut in it that matches a groove in the shaft. A key is placed in these grooves and a set screw is often tightened down on top of the key.

Many technicians make the mistake of trying to drive a motor shaft out of the assembly fastened to the shaft. In doing so, they blunt or distort the end of the motor shaft. When it is distorted, the motor shaft will never go through the assembly without damaging it, Figure 20–5. The shaft is made from mild steel and can be easily damaged. If the

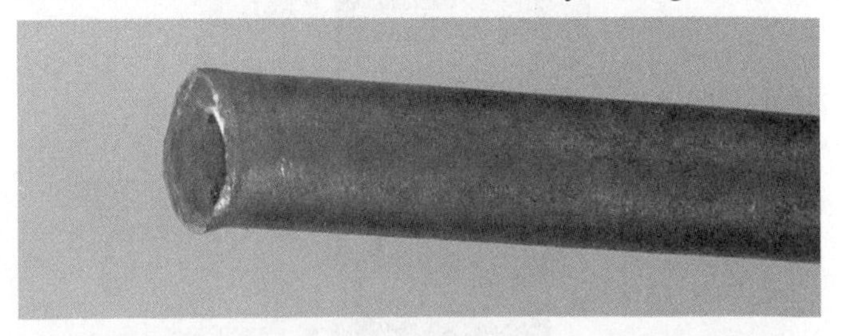

FIGURE 20–5 Motor shaft damaged from trying to drive it through the pulley. *Photo by Bill Johnson*

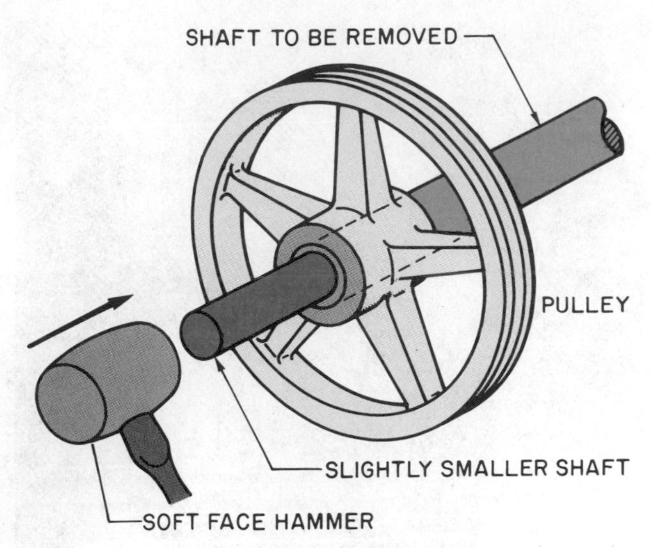

FIGURE 20–6 Shaft driven through the pulley with another shaft as the contact surface. The shaft that is used as the contact surface is smaller in diameter than the original shaft.

shaft must be driven, you may need to use a similar shaft with a slightly smaller diameter as the driving tool, Figure 20–6.

20.4 BELT TENSION

Many motors fail because of overtightened belts and incorrect alignment. The technician should be aware of the specifications for the motor belt tension on belt-drive systems. A belt tension gage will ensure properly adjusted belts when the gage manufacturer's directions are followed. Belts that are too tight strain the bearings so that they wear out prematurely, Figure 20–7. Figure 20–8 shows a type of belt tension gage.

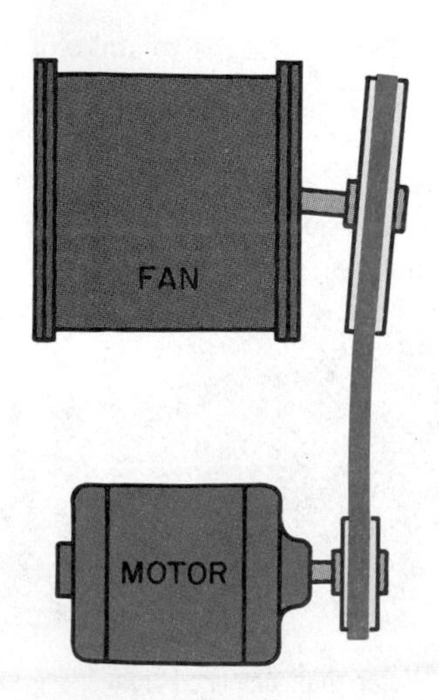

FIGURE 20–7 Belt that is too tight.

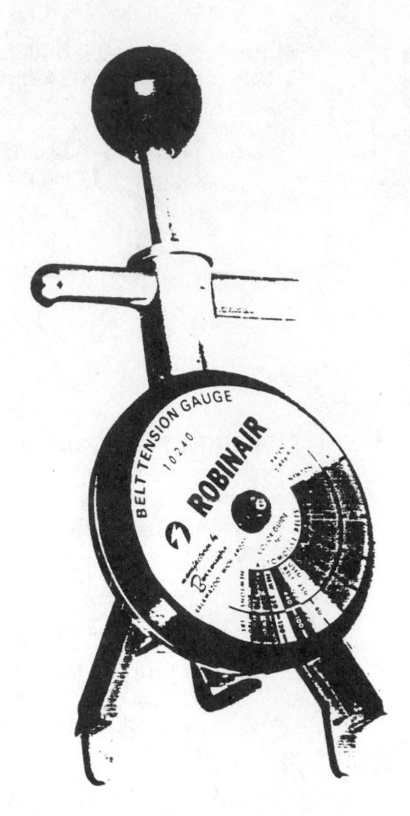

FIGURE 20–8 Belt tension gage. *Courtesy Robinair Division, SPX Corp.*

20.5 PULLEY ALIGNMENT

Pulley alignment is very important. If the drive pulley and driven pulley are not in line, a strain is imposed on the shafts' drive mechanisms. The pulleys may be made in line with the help of a straightedge, Figure 20–9. A certain amount of adjustment tolerance is built into the motor base on small motors, and this may be enough to allow the motor to be out of alignment if the belt becomes loose. Aligning shafts on the furnace may not be easy to do under a house by

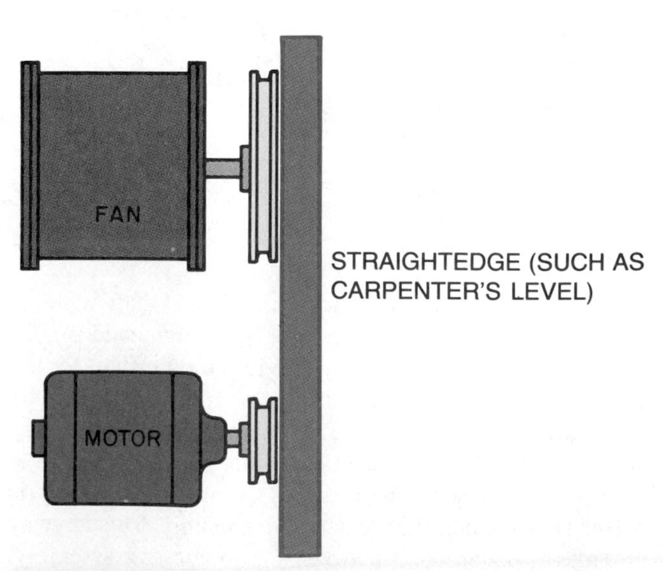

FIGURE 20–9 Pulleys must be in proper alignment or belt and bearing wear will occur. The pulleys may be aligned using a straightedge.

flashlight, but it must be done or the motor or drive mechanism will not last.

When mechanical problems occur with a motor, the motor is normally either replaced or taken to a motor repair shop. Bearings can be replaced in the field by a competent technician, but it is generally better to leave this type of repair to motor experts. When the problem is the pulley or drive mechanism, the air conditioning, heating, and refrigeration technician may be responsible for the repair. ***Proper tools must be used for motor repair, or shaft and motor damage may occur.***

20.6 ELECTRICAL PROBLEMS

Electrical motor problems are the same for hermetic and open motors. Open motor problems are a little easier to understand or diagnose because they can often be seen. When an open motor burns up, this can often be easily diagnosed because the winding may be seen through the end bells. With a hermetic motor, instruments must be used because they are the only means of diagnosing problems that are not visible inside the compressor. There are three common electrical motor problems: (1) an open winding, (2) a short circuit from the winding to ground, and (3) a short circuit from winding to winding.

20.7 OPEN WINDINGS

Open windings in a motor can be found with an ohmmeter. There should be a known measurable resistance from terminal to terminal on every motor for it to run when power is applied to the windings. Single-phase motors must have the applied system voltage at the run winding to run and at the start winding during starting, Figure 20–10. Figure 20–11 is an illustration of a motor with an open start winding.

20.8 SHORTED MOTOR WINDINGS

Short circuits in windings occur when the conductors in the winding touch each other where the insulation is worn or in some way defective. This creates a short path through which the electrical energy flows. This path has a lower resistance

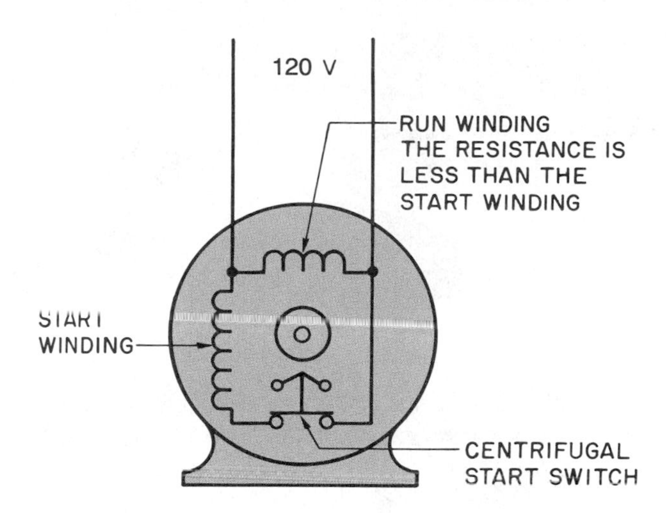

FIGURE 20–10 Wiring diagram with run and start windings.

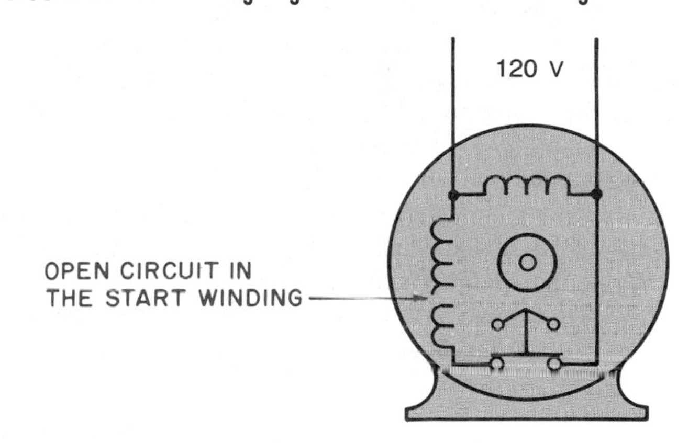

FIGURE 20–11 Motor with open winding.

and increases the current flow in the winding. Although motor windings appear to be made from bare copper wire, they are coated with an insulator to keep the copper wires from touching each other. The measurable resistance mentioned in the previous paragraph is known for all motors. Some motors have a published resistance for their windings. The best way to check a motor for electrical soundness is to *know what the measurable resistance should be for a particular winding and verify it with a good ohmmeter,* Figure 20–12. This measurable resistance will normally be less

Compressor Model	Voltage	MOTOR AMPS				FUSE SIZE		Winding Resistance in Ohms
		Full Winding		1/2 Winding		Recommended Max		
		Rated Load	Locked Rotor	Rated Load	Locked Rotor	Fusetron	Std.	
9RA - 0500 - CFB	230/1/60	27.5	125.0			FRN-40	50	Start 1.5 Run 0.40
9RB TFC	208-230/3/60	22.0	115.0			FRN-25	40	0.51-0.61
9RJ TFD	460/3/60	12.1	53.0			FRS-15	15	2.22-2.78
9TK TFE	575/3/60	7.8	42.0			FRS-10	15	3.40-3.96
MRA FSR	200-240/3/50	17.0	90.0	8.5	58.0	FRN-25	35	0.58-0.69
MRB FSM	380-420/3/50	9.5	50.0	4.8	32.5	FRS-15	20	1.80-2.15
MRF								

FIGURE 20–12 Resistances for some typical hermetic compressors. *Courtesy Copeland Corporation*

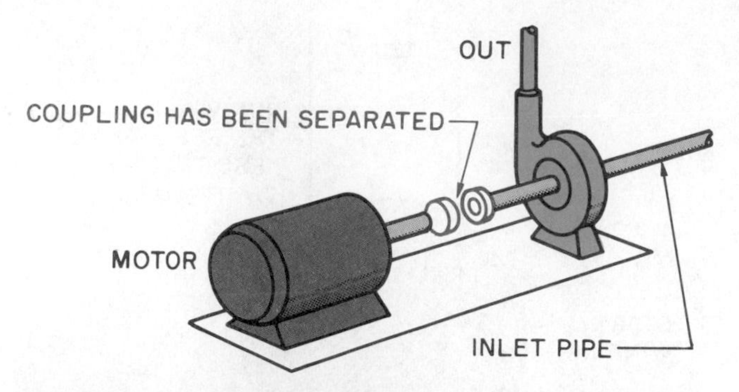

FIGURE 20–13 The coupling was disconnected between this motor and pump because it was suspected that the motor or pump was locked up and would not turn.

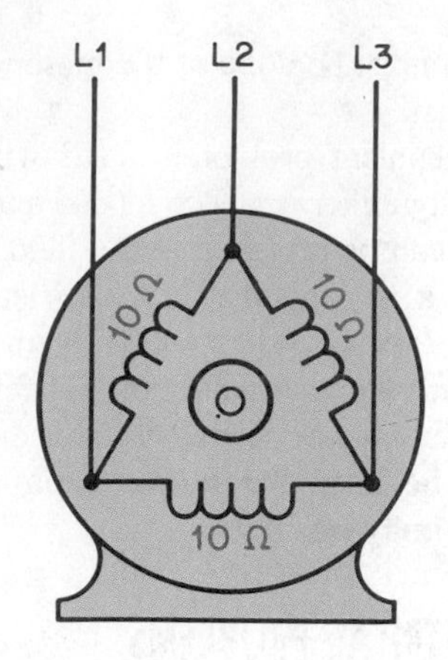

FIGURE 20–14 Wiring diagram of three-phase motor. The resistance is the same across all three windings.

than the rated value when a motor has short-circuit problems. The decrease in resistance causes the current to rise, which causes motor overload devices to open the circuit and possibly even the circuit overload protection may trip. If the resistance does not read within these tolerances, there is a problem with the winding. This table is helpful when you troubleshoot a hermetic compressor. Tables may not be easy to obtain for open motors, and the windings are not as easy to check because the individual windings do not all come out to terminals as they do on a hermetic compressor.

If the decrease in resistance in the windings is in the start winding, the motor may not start. If it is in the run winding, the motor may start and draw too much current while running. If the motor winding resistance cannot be determined, then it is hard to know whether a motor is overloaded or whether it has a defective winding when only a few of its coils are shortened.

When the motor is an open motor, the load can be removed. For example, the belts can be removed or the

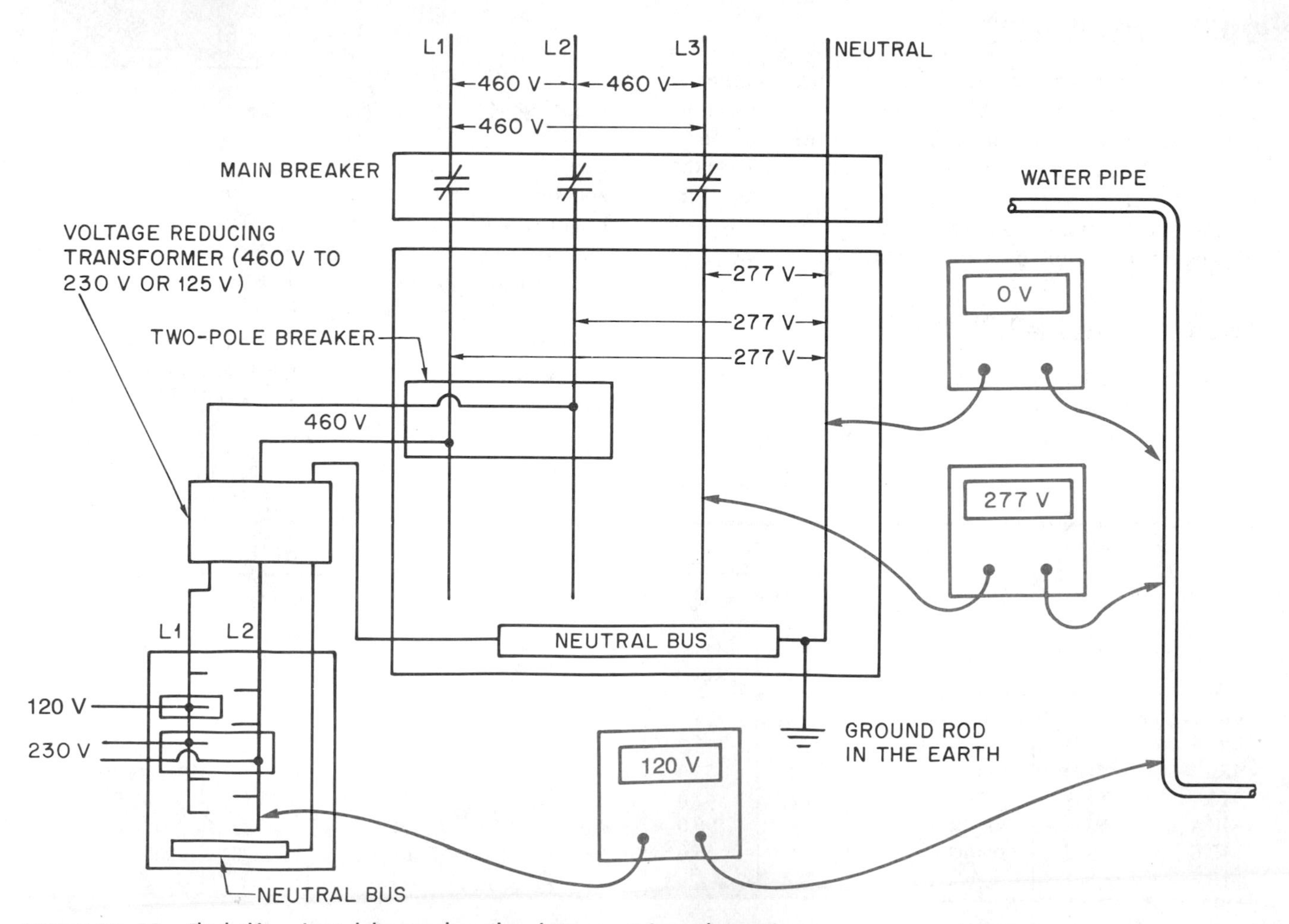

FIGURE 20–15 This building electrical diagram shows the relationship of the earth ground system to the system's piping.

coupling can be taken apart, and the motor can be started without the load, Figure 20–13. If the motor starts and runs correctly without the load, the load may be too great.

Three-phase motors must have the same resistance for each winding. (There are three identical windings.) Otherwise there is a problem. An ohmmeter check will quickly reveal an incorrect winding resistance, Figure 20–14.

20.9 SHORT CIRCUIT TO GROUND (FRAME)

A short circuit from winding to ground or the frame of the motor may be detected with a good ohmmeter. No circuit should be detectable from the winding to ground. The copper suction line on a compressor is a good source for checking to the ground. ***"Ground" or "frame" are interchangeable terms because the frame should be grounded to the earth ground through the building electrical system, Figure 20–15.***

To check a motor for a ground, use a good ohmmeter with an R × 10,000-Ω scale. Special instruments for finding very high resistances to ground are used for larger, more sophisticated motors. Most technicians use ohmmeters. Top-quality instruments can detect a ground in the 10,000,000 Ω and higher range. A *megger* even has an internal high-voltage direct current (DC) supply to help create conditions to detect the ground, Figure 20–16. The term "megger" means megohmmeter and will measure resistances in the millions of ohms. It has to do with the capacity of the meter to detect very high resistances.

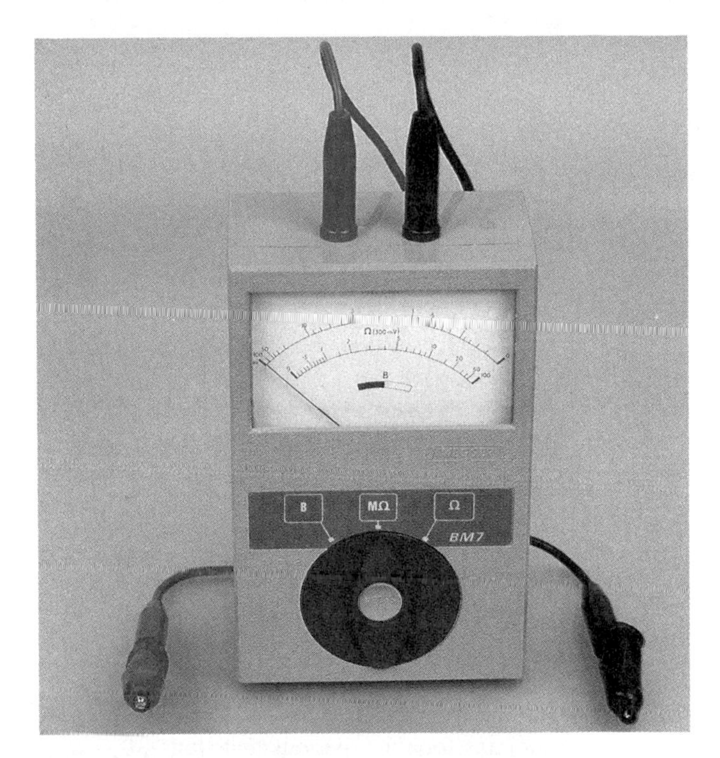

FIGURE 20–16 A megger. *Photo by Bill Johnson*

A typical ohmmeter will detect a ground of about 1,000,000 Ω or less. The rule of thumb is that if an ohmmeter set to the R × 10,000 scale will even move the needle, with one lead touching the motor terminal and the other touching a ground (such as a copper suction line), the motor should be started with care. ***If the meter needle moves to the midscale area, do *not* start the motor, Figure 20–17.*** When a meter reads a very slight resistance to ground, the windings may be dirty and damp if it is an open motor. Clean the motor and the ground will probably be eliminated. Some motors operating in a dirty, damp atmosphere may indicate a slight circuit to ground in damp weather. Air-cooled condenser fan motors are an example. When the motor is started and allowed to run long enough to get warm and dry, the ground circuit may disappear.

Hermetic compressors may occasionally have a slight ground due to the oil and liquid refrigerant in the motor splashing on the windings. The oil may have dirt suspended in it and show a slight ground. Liquid refrigerant causes this condition to be worse. If the ohmmeter shows a slight ground but the motor starts, run the motor for a little while and check again. If the ground persists, the motor is probably going to fail soon if the system is not cleaned. A suction line filter-drier may help remove particles that are circulating in the system and causing the slight ground.

For troubleshooting electric motors, the ammeter and the voltmeter are the main instruments used. If the resistance is correct, the motor is electrically sound. Other problems may be found using the ammeter.

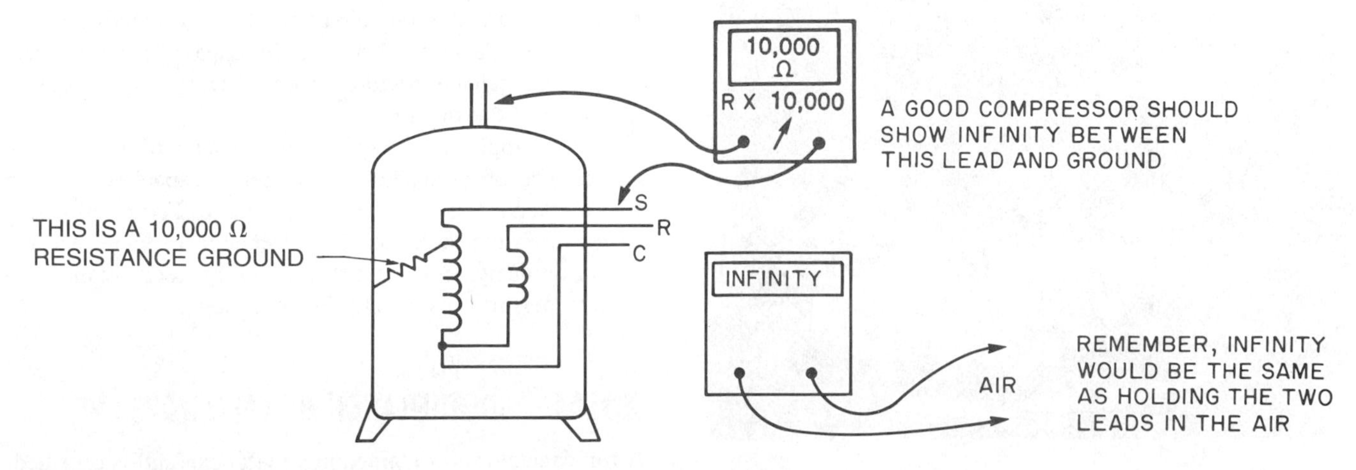

FIGURE 20–17 Volt-ohmmeter detecting a circuit to ground in a compressor winding.

FIGURE 20–18 Example of how a pump or compressor coupling can be disconnected and the component's shaft turned by hand to check for a hard-to-turn shaft or a stuck component. ***Turn off power to motor and lock out.***

FIGURE 20–20 Start capacitor with bleeder resistor across the terminals to bleed off the charge during the off cycle. *Photo by Bill Johnson*

20.10 MOTOR STARTING PROBLEMS

Symptoms of electric motor starting problems are:

1. The motor hums and then shuts off.
2. The motor runs for short periods and shuts off.
3. The motor will not try to start at all.

The technician must decide whether there is a motor mechanical problem, a motor electrical problem, a circuit problem, or a load problem. ***If the motor is an open motor, turn the power off. Lock and tag panel. Try to rotate the motor by hand. If it is a fan or a pump, it should be easy to turn, Figure 20–18. If it is a compressor, the shaft may be hard to turn. Use a wrench to grip the coupling when trying to turn a compressor. Be sure the power is off.***

If the motor and the load turn freely, examine the motor windings and components. If the motor is humming and not starting, the starting switch may need replacing or the windings may be burned. If the motor is open, you may be able to visually check them; if you can't, remove the motor end bell, Figure 20–19.

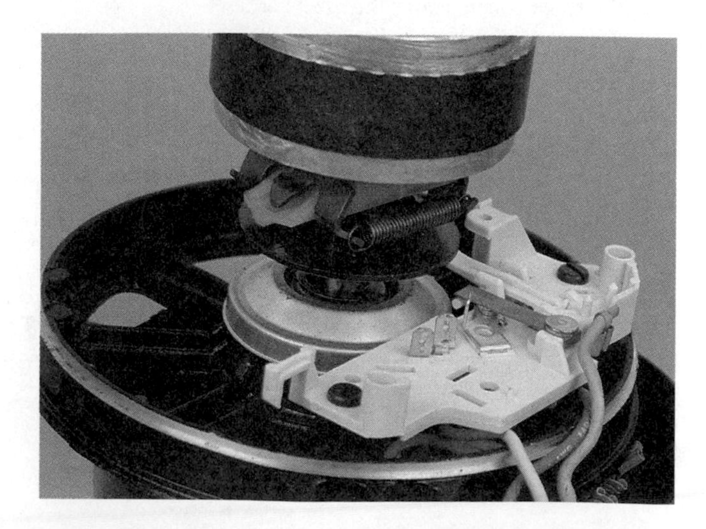

FIGURE 20–19 The end bell has been removed from this motor to examine the start switch and the windings. *Photo by Bill Johnson*

20.11 CHECKING CAPACITORS

Motor capacitors may be checked to some extent with an ohmmeter in the following manner. Turn off the power to the motor and remove one lead of the capacitor. ***Short from one terminal to the other with a 20,000-Ω 5-W resistor to discharge the capacitor in case the capacitor has a charge stored in it. Some start capacitors have a resistor across these terminals to bleed off the charge during the off cycle, Figure 20–20. Short across the capacitor anyway. Use insulated pliers. The resistor may be open and won't bleed the charge from the capacitor. If you place ohmmeter leads across a charged capacitor, the meter movement may be damaged.***

Set the ohmmeter scale to the R × 10 scale and touch the leads to the capacitor terminals, Figure 20–21. If the capacitor is good, the meter needle will go to 0 and begin to fall back toward infinite resistance. If the leads are left on the capacitor for a long time, the needle will fall back to infinite resistance. If the needle falls part of the way back and will drop no more, the capacitor has an internal short. The electrolyte is touching the other pole of electrolyte. If the needle will not rise at all, try the R × 100 scale. Try reversing the leads. The ohmmeter is charging the capacitor with its internal battery. This is DC voltage. When the capacitor is charged in one direction, the meter leads must be reversed for the next check. If the capacitor has a bleed resistor, the capacitor will charge to 0 Ω, then drop back to the value of the resistor.

This simple check with an ohmmeter will not give the capacitance of a capacitor. You need a capacitor tester to find the actual capacitance of a capacitor, Figure 20–22. It is not often that a capacitor will change in value, so this capacitor analyzer is used primarily by technicians who need to perform many capacitor checkouts.

20.12 IDENTIFICATION OF CAPACITORS

A run capacitor is contained in a metal can and is oil filled. If this capacitor gets hot due to overcurrent, it will often swell, Figure 20–23. The capacitor should then be changed.

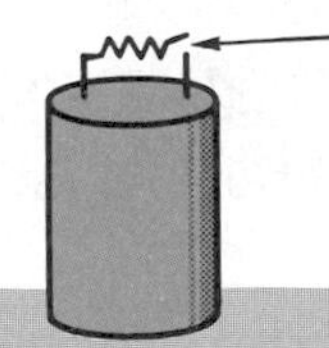

DISCONNECT BLEED RESISTOR FOR TEST IF DESIRED. WHEN BLEED RESISTOR IS LEFT IN THE CIRCUIT, THE METER NEEDLE WILL RISE FAST AND FALL BACK TO THE BLEED RESISTOR VALUE.
FOR BOTH RUN AND START CAPACITORS

1. FIRST SHORT THE CAPACITOR FROM POLE TO POLE USING A 20 000 OHM 5-WATT RESISTOR.
2. USING THE RX 100 OR RX 1000 SCALE TOUCH THE METER'S LEADS TO THE CAPACITOR'S TERMINALS. METER NEEDLE SHOULD RISE FAST AND FALL BACK SLOWLY. IT WILL EVENTUALLY FALL BACK TO INFINITY IF THE CAPACITOR IS GOOD. (PROVIDED THERE IS NO BLEED RESISTOR).

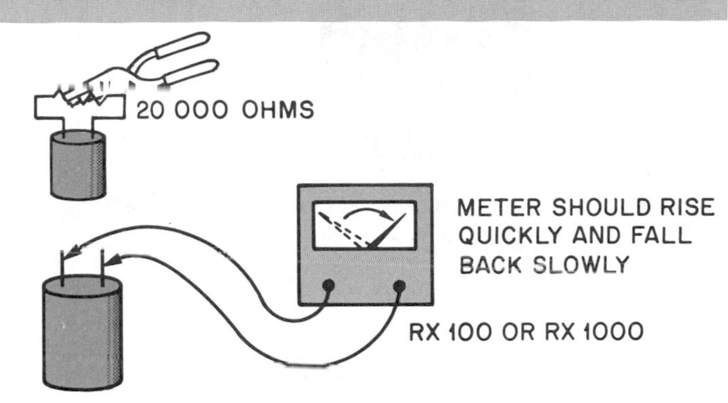

3. YOU CAN REVERSE THE LEADS FOR A REPEAT TEST, OR SHORT THE CAPACITOR TERMINALS AGAIN. IF YOU REVERSE THE LEADS, THE METER NEEDLE MAY RISE EXCESSIVELY HIGH AS THERE IS STILL A SMALL CHARGE LEFT IN THE CAPACITOR.
4. FOR RUN CAPACITORS THAT ARE IN A METAL CAN: WHEN ONE LEAD IS PLACED ON THE CAN AND THE OTHER LEAD ON A TERMINAL INFINITY SHOULD BE INDICATED ON THE METER USING THE RX 10 000 OR RX 1000 SCALE

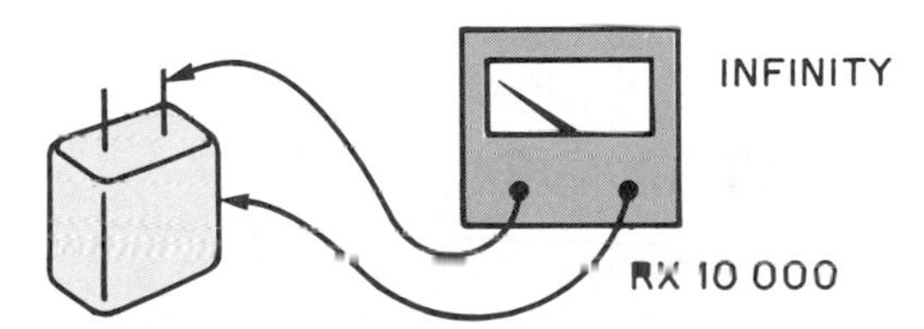

FIGURE 20–21 Procedure to field check a capacitor.

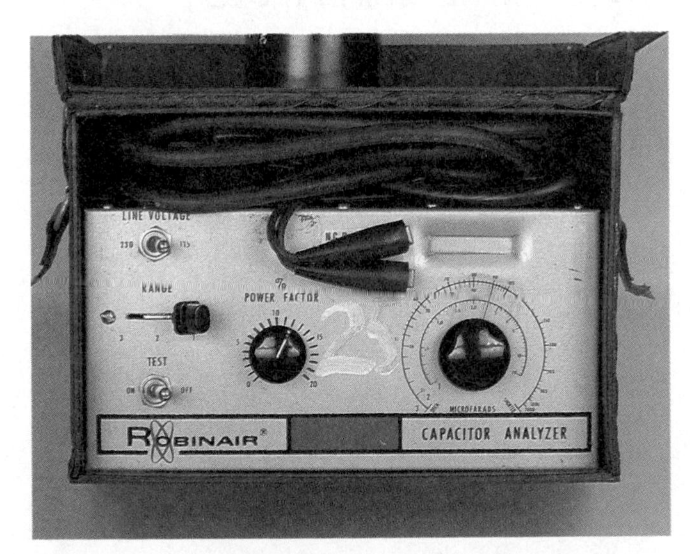

FIGURE 20–22 Capacitor analyzer. *Photo by Bill Johnson*

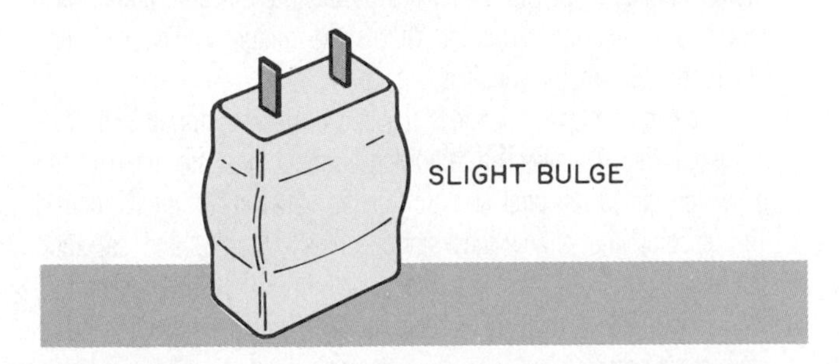

FIGURE 20–23 Capacitor swelled because of internal pressure.

Run capacitors have an identified terminal to which the lead that feeds power to the capacitor should be connected. When the capacitor is wired in this manner, a fuse will blow if the capacitor is shorted to the container. If the capacitor is not wired in this manner and a short occurs, current can flow through the motor winding to ground during the off cycle and overheat the motor, Figure 20–24.

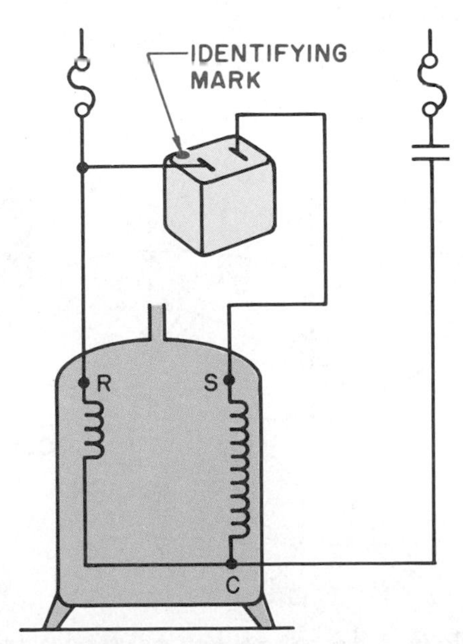

FIGURE 20–24 The wiring to a run capacitor for proper fuse protection of the circuit.

FIGURE 20–25 When a start capacitor has overheated, the small rubber diaphragm in the top will bulge.

The start capacitor is a dry type and may be contained in a shell of paper or plastic. Paper containers are no longer used, but you may find one in an older motor. If this capacitor has been exposed to overcurrent, it may have a bulge at the vent at the top of the container, Figure 20–25. The capacitor should then be changed.

20.13 WIRING AND CONNECTORS

The wiring and connectors that carry the power to a motor must be in good condition. When a connection becomes loose, oxidation of the copper wire occurs. The oxidation acts as an electrical resistance and causes the connection to heat more, which in turn causes more oxidation. This condition will only get worse. Loose connections will result in low voltages at the motor and in overcurrent conditions. Loose connections will appear the same as a set of dirty contacts and may be located with a voltmeter, Figure 20–26. If a connection is loose enough to create an overcurrent condition, it can often be located by a temperature rise at the loose connection.

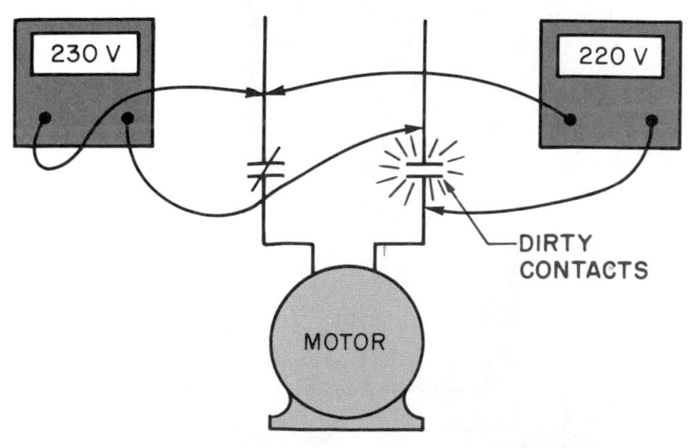

FIGURE 20–26 When voltmeter leads are applied to both sides of a dirty contact, the meter will read voltage drop across the contact.

20.14 TROUBLESHOOTING HERMETIC MOTORS

Diagnosing hermetic compressor motor problems differs from diagnosing open motor problems because the motor is enclosed in a shell and cannot be seen; the motor sound level may be dampened by the compressor shell. Motor noises that are obvious in an open motor may become hard to hear in a hermetic compressor.

The motor inside a hermetic compressor can only be checked electrically from the outside. It will have the same problems as an open motor, that is, open circuit, short circuit, or grounded circuit. The technician must give the motor a complete electrical checkout from the outside of the shell; if it cannot be made to operate, the compressor must be changed. A motor checkout includes the starting and running components for a single-phase compressor, such as the run and start capacitors and the start relay.

As mentioned earlier, compressors operate in remote locations and are not normally attended. A compressor can throw a rod, tearing it up, and the motor may keep running until damaged. The compressor parts damage the motor, which may then burn out. The compressor damage is not detected because the damage cannot be seen. When a compressor is changed because of a bad motor, you should suspect mechanical damage.

20.15 SERVICE TECHNICIAN CALLS

SERVICE CALL 1

The manager of a retail store calls. *They have no air conditioning and the piping going into the unit is frozen solid. The air handler is in the stock room. The evaporator fan motor winding is open, and the outdoor unit has continued to run without air passing over the evaporator, freezing it solid.*

On arriving at the store and noticing that the fan is not running the technician turns the cooling thermostat to OFF to stop the condensing unit, then turns the fan switch to ON. ***The technician then turns off the power, locks and tags the panel, and removes the fan compartment cover.*** The fan motor body is cool to the touch, indicating the motor has been off for some time. The technician then turns on the power and checks power at the motor terminal block; the motor has power. The power is turned off, locked out, and the technician checks the motor winding with an ohmmeter; the motor windings are open.

The fan motor is changed. The unit cannot be started until the coil is allowed to thaw. The technician instructs the store manager to allow the fan to run until air is felt at the outlet in the men's room. This ensures the airflow has started. Wait 15 min, and set the thermostat to COOL and the fan switch to AUTO. The technician leaves. A call back later in the day indicates that the system is now cooling correctly.

SERVICE CALL 2

A residential customer calls. *There is no heat. The fan motor on the gas furnace has a defective run capacitor. The furnace burner is cycling on and off with the high-limit control. The fan will run for a few minutes each cycle before shutting off from over temperature. The defective capacitor causes the fan motor current to run about 15% too high.*

The technician arrives and goes under the house where the horizontal furnace is located. The furnace is hot. The technician hears the burner ignite and watches it for a few minutes. The burner shuts off, and about that time the fan comes on. The technician removes the fan compartment door and sees the fan turning very slowly. ***The power is shut off. The panel is locked and tagged.*** The motor is very hot. It seems to turn freely, so the bearings must be normal. The technician removes the run capacitor and uses an ohmmeter for a capacitor check, Figure 20–21. The capacitor is open—no circuit or continuity. The capacitor is replaced. The fan compartment door is shut and power is restored. It is evident from the sound of the motor that it is turning faster than before the capacitor was changed.

Before leaving the job, the technician turns off the power, oils the motor, and replaces the air filters.

SERVICE CALL 3

A retail store customer calls. *There is no air conditioning. The compressor, located outside, starts up, then shuts off. The fan motor is not running. The dispatcher tells the customer to shut the unit off until the technician arrives. The PSC fan motor for the condenser has bad bearings. The motor feels free to turn the shaft, but if power is applied to the motor when it is turning, the motor will stop.*

The technician arrives and talks to the store manager. The technician goes to the outdoor unit and disconnects the power so that the system can be controlled from outside. The room thermostat is set to call for cooling.

The technician goes outside and turns the disconnect on and observes. The compressor starts, but the fan motor does not. ***The power is turned off. The panel is locked and tagged. The compressor leads are disconnected at the load side of the contactor so that the compressor will not be in a strain.*** The technician makes the disconnect switch again and, using the clamp-on ammeter, checks the current going to the fan motor. It is drawing current and trying to turn. It is not known at this point if the motor or the capacitor is defective. The power is shut off. The technician spins the motor and turns the power on while the motor is turning. The motor acts like it has brakes and stops. This indicates bad bearings or an internal electrical motor problem.

The motor is changed and the new motor performs normally even with the old capacitor. The compressor is reconnected, and power is resumed. The system now cools normally. The technician changes the air filter and oils the indoor fan motor before leaving.

The following service calls do not have the solution described. The solution can be found in the Instructor's Guide.

SERVICE CALL 4

An insurance office customer calls. There is no air conditioning. This system has an electric furnace with the air handler (fan section) mounted above the suspended ceiling in the stock room. The condensing unit is on the roof. The low-voltage power supply is at the air handler.

The technician arrives and goes to the room thermostat. It is set on cooling, and the indoor fan is running. The technician goes to the roof and discovers the breaker is tripped. Before resetting the breaker, the technician decides to find out why the breaker tripped. The cover to the electrical panel to the unit is removed. A voltage check shows that the breaker has all the voltage off. The contactor's 24-V coil is the only thing energized because its power comes from downstairs at the air handler.

The technician checks the motors for a ground circuit by placing one lead on the load side of the contactor and the other on the ground with the meter set on R $\times$ 10,000. The meter shows a short circuit to ground (no resistance to ground). See Figure 20–17 for an example of a motor ground check. From this test, the technician does not know if the problem is the compressor or the fan motor. The wires are disconnected from the line side of the contactor to isolate the two motors. The compressor shows normal, no reading. The fan motor shows 0 Ω resistance to ground.

What is the problem and the recommended solution?

SERVICE CALL 5

A commercial customer calls. There is no air conditioning in the upstairs office. This is a three-story building with an air handler on each floor and a chiller in the basement.

The technician arrives and goes to the fan room on the second floor where the complaint is. There is no need to go to the chiller or the other floors because there would be complaints if the chiller was not furnishing cooling to them.

The chilled water coil piping feels cold; the chiller is definitely running and furnishing cold water to the coil. The fan motor is not running. ***Since the motor may have electrical problems, the technician proceeds with caution by opening the electrical disconnect. Lock and tag the disconnect.*** The technician pushes the reset button on the fan motor starter and hears the ratchet mechanism reset. The motor will not try to start while pushing the reset button because the disconnect switch is open. The unit must have been pulling too much current for the overload to trip.

Using an ohmmeter, the technician checks for a ground by touching one lead to a ground terminal and the other to one of the motor leads on the load side of the starter. The meter is set on R $\times$ 10,000 and will detect a fairly high resistance to ground. See Figure 20–17 for an example of a motor ground check. ***Any movement of the meter needle when touched to the motor lead**

would indicate a ground so caution is necessary. If the meter needle moves as much as one fourth of its scale, the circuit should not be energized until the ground is cleared up, or physical damage may occur.* The motor is not grounded. The resistance between each winding is the same, so the motor appears to be normal. The technician then turns the motor over by hand to see if the bearings are too tight; the motor turns normally.

The technician shuts the fan compartment door and then fastens a clamp-on ammeter to one of the motor leads at the load side of the starter. The electrical disconnect is then closed to start the motor. When it is closed the motor tries to start. It will not start and pulls a high amperage. The motor seems to be normal from an electrical standpoint and turns freely, so the power supply is now suspected.

The technician quickly pulls the disconnect to the off position and gets an ohmmeter. Each fuse is checked with the ohmmeter. The fuse in L2 is open. The fuse is replaced and the motor is started again. The motor starts and runs normally with normal amperage on all three phases. The question is, why did the fuse blow?

What is the problem and the recommended solution?

SERVICE CALL 6

A customer who owns a truck stop calls. There is no cooling in the main dining room. The customer does some of the maintenance, so the technician can expect some problem related to this maintenance program.

The technician arrives and can hear the condensing unit on the roof running; it shuts off about the time the technician gets out of the truck. The customer had just serviced the system by changing the filters, oiling the motors, and checking the belts. Since then, the fan motor has been shutting off. It can be reset by pressing the reset button at the fan contactor, but it will not run long.

The technician and the customer go to the stock room where the air handler is located. The fan is not running, and the condensing unit is shut down from low pressure. The technician fastens the clamp-on ammeter on one of the motor leads and restarts the motor. The motor is pulling too much current on all three phases and the motor seems to be making too much noise.

What is the problem and the recommended solution?

SUMMARY

- Motor problems can be divided into mechanical problems and electrical problems.
- Mechanical problems are bearing or shaft problems.
- Bearing problems are often caused by belt tension.
- Shaft problems may be caused by the technician while removing pulleys or couplings.
- Motor balancing problems are normally not handled by the heating, air conditioning, and refrigeration technician.
- ***Make sure that vibration is not caused by the system.***
- Most electrical problems are open windings, short-circuited windings, or grounded windings.
- The laws of electrical current flow must be used while troubleshooting motors.
- If the motor is receiving the correct voltage and is electrically sound, check the motor components.
- Troubleshooting hermetic compressor motors is different from troubleshooting open motors because hermetic motors are enclosed.

REVIEW QUESTIONS

1. What are the two main categories of motor problems?
2. In what category is a motor shaft problem?
3. Who normally works with motor balancing problems?
4. How is a stuck hermetic compressor repaired?
5. What can be done to check an open motor to see if it is locked up or if the load component is locked?
6. What three electrical conditions must be met to have current flow?
7. What are the two electrical test instruments most used for motor problems?
8. Where is a convenient place to check a motor for an electrical ground?
9. If the motor is electrically sound, what should then be checked?
10. How does checking a hermetic compressor motor differ from checking an open motor?

UNIT 20 DIAGNOSTIC CHART FOR OPEN-TYPE ELECTRIC MOTORS

Electric motors are used to turn the prime movers in all heating and cooling air-conditioning systems as well as refrigeration systems. The prime movers are the fans, pumps, and compressors in these systems. Compressors are usually hermetically sealed in the system and are discussd in the units covering compressors. This discussion will cover open-type electric motors single phase and three phase. When possible, the technician should listen to the customer for possible causes of the problem. Often the customer's comments will result in quickly resolving the problem in the system.

PROBLEM	POSSIBLE CAUSE	POSSIBLE REPAIR	PARAGRAPH NUMBER
Motor does *not* attempt to start—makes no sound	Open disconnect switch	Close disconnect switch	15.1, 15.2
	Open fuse or breaker	Replace fuse or reset breaker and determine why it opened	15.1, 15.2
	Tripped overload	Reset overload and determine why it tripped	19.6
	Faulty fan switch	Repair or replace switch mechanism	15.4
	Faulty wiring	Repair or replace faulty wiring or connectors	20.13
Motor will not start—hums and trips on overload	Incorrect power supply	Correct the power supply	15.1
Single Phase A. Shaded pole and split phase motors	Tight or dragging bearings	Change motor	20.10
	Tight belt or overloaded	Adjust belt	20.2
		Close fan compartment door	20.15
		Reduce load	20.15
		Change to larger motor	18.1
	Srart circuit contacts on split phase	Change motor	17.10
	Defective start capacitor	Change capacitor	17.12, 20.11
B. Permanent split capacitor motor	Defective run capacitor	Change capacitor	17.14, 20.11
Three Phase	Incorrect power supply—voltage unbalance	Correct voltage balance	17.7
	One phase out—single phase condition	Correct before starting motor	17.7
	Tight or dragging bearings	Change bearings or motor	20.10
	Tight belt	Adjust belt	20.2
	Overloaded	Close fan compartment door	20.15
		Reduce load	20.15
		Change to larger motor	17.10
Motor starts and runs for short time then shuts off due to overload	Defective overload	Check actual load on overload and replace if it trips below specifications	19.6
	Excess current in overload circuit—check for added load, such as fans or pumps	Correct load to match motor and overload	20.15
	Low voltage	Determine reason and correct	17.7
Three Phase	One phase out—single phase condition	Correct before starting motor	17.7
	Voltage unbalance	Correct voltage balance	17.7
	Excess load on motor	Fan moving too much air—correct	20.15
		Motor undersized—change to larger motor	18.1
	Fan or pump bearings too tight	Disconnect load and check bearings—repair as needed	18.5

SECTION

5 Commercial Refrigeration

UNIT 21

Evaporators and the Refrigeration System

OBJECTIVES

After studying this unit, you should be able to

- define high-, medium-, and low-temperature refrigeration.
- determine the boiling temperature in an evaporator.
- identify different types of evaporators.
- describe multiple- and single-circuit evaporators.

SAFETY CHECKLIST

* Wear goggles and gloves when attaching or removing gages to transfer refrigerant or to check pressures.
* Wear warm clothing when working in a walk-in cooler or freezer.

21.1 REFRIGERATION

Refrigeration is the process of removing heat from a place where it is not wanted and transferring that heat to a place where it makes little or no difference. Commercial refrigeration is similar to the refrigeration that occurs in your household refrigerator. The food that you keep in the refrigerator is stored at a temperature lower than the room temperature. Typically, the fresh food compartment temperature is about 35°F. Heat from the room moves through the walls of the refrigerator from the warm room (typically 75°F) to the cooler temperature in the refrigerator. Heat travels normally and naturally from a warm to a cool medium.

If the heat that is transferred into the refrigerator remains in the refrigerator, it will warm the food products and spoilage will occur. This heat may be removed from the refrigerator by mechanical means using the refrigeration equipment furnished with the refrigerator. This mechanical means requires energy, or work. Figure 21–1 shows how this heat is removed with the compression cycle. Because it is 35°F in the box and 75°F in the room, the mechanical

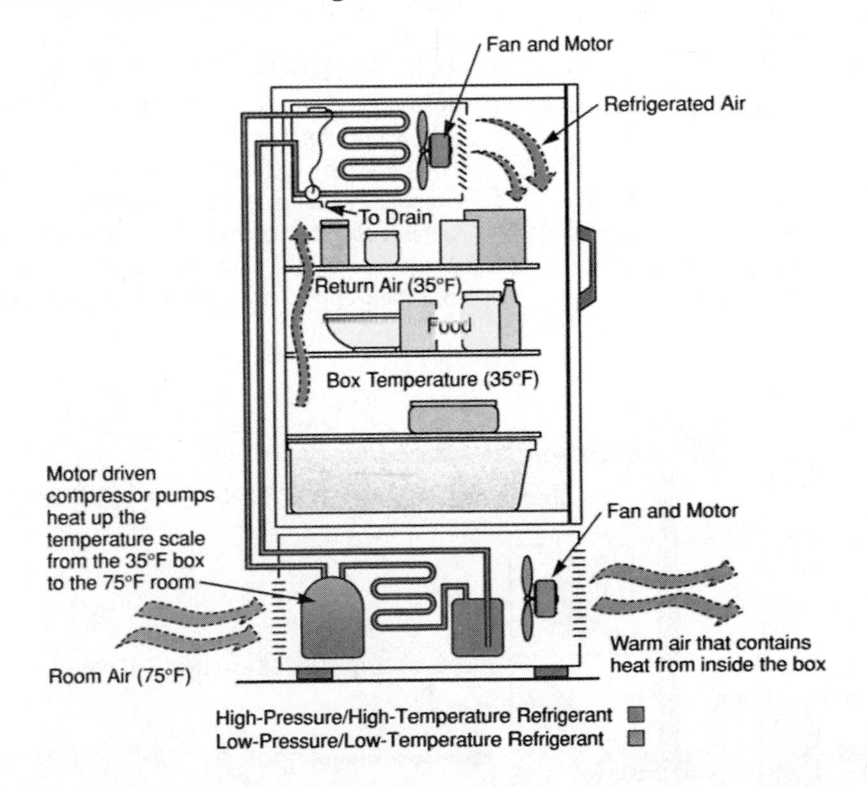

FIGURE 21–1 Heat normally flows from a warm place to a colder place. When it is desirable for heat to be moved from a colder to a warmer place, the heat has to be moved by force. The compressor in the refrigeration system is the pump that pumps the heat up the temperature scale by force. These compressors are normally driven by electric motors.

energy in the compression cycle actually pumps the heat to a warmer environment from the box temperature to the room temperature.

The heat is transferred into a cold refrigerant coil and pumped with the system compressor to the condenser where it is released into the room. This is much like using a sponge to move water from one place to another. When a dry sponge is allowed to absorb water in a puddle and you take the wet sponge to a container and squeeze it, you exert energy, much like a compressor in the refrigeration system, Figure 21–2.

Another example of refrigeration is a central air conditioning system in a residence. It absorbs heat from the home by passing indoor air at about 75°F over a coil that is cooled to about 40°F. Heat will transfer from the room air to the coil cooling the air. This cooled air may be mixed with the room air, lowering its temperature, Figure 21–3. This process is called air conditioning, but it is also refrigeration at a higher temperature level than the household refrigerator. It is frequently called high-temperature refrigeration.

FIGURE 21–2 A sponge absorbs water. The water can then be carried in the sponge to another place. When the sponge is squeezed, the water is rejected to another place. The squeezing of the sponge may be considered the energy that it takes to pump the water.

Commercial refrigeration differs from domestic (household) refrigeration because it is located in commercial business locations. The food store, fast food restaurant, drug store, flower shop, and food processing plant are only a few of the applications for commercial refrigeration. Some of the systems are plug-in appliances, such as a small reach-in ice storage bin at the local convenience store. The refrigeration system is all located within the one unit. Some systems consist of individual boxes with single remote condensing units and some are complex systems with several compressors in a rack serving several reach-in display cases in a supermarket. Most commercial refrigeration is installed and serviced by a special group of technicians who only work with commercial refrigeration and the food service business.

21.2 TEMPERATURE RANGES OF REFRIGERATION

The temperature ranges for commercial refrigeration may refer to the temperature of the refrigerated box, or the boiling temperature of the refrigerant in the coil. When discussing box temperatures, the following temperatures will illustrate some of the guidelines used in industry.

HIGH-TEMPERATURE APPLICATIONS. High-temperature refrigeration applications will normally provide box temperatures of 40°F to 60°F. The storage of such products as flowers and candy may require these temperatures.

MEDIUM-TEMPERATURE APPLICATIONS. The household refrigerator fresh food compartment is a good example of medium-temperature refrigeration, which typically ranges from 28°F to 40°F. Many different products are stored at the medium-temperature range. The medium-temperature refrigeration range is above freezing for most products. Few products are stored below 32°F. Items such as eggs, lettuce, and tomatoes lose their appeal if they freeze in a refrigerator.

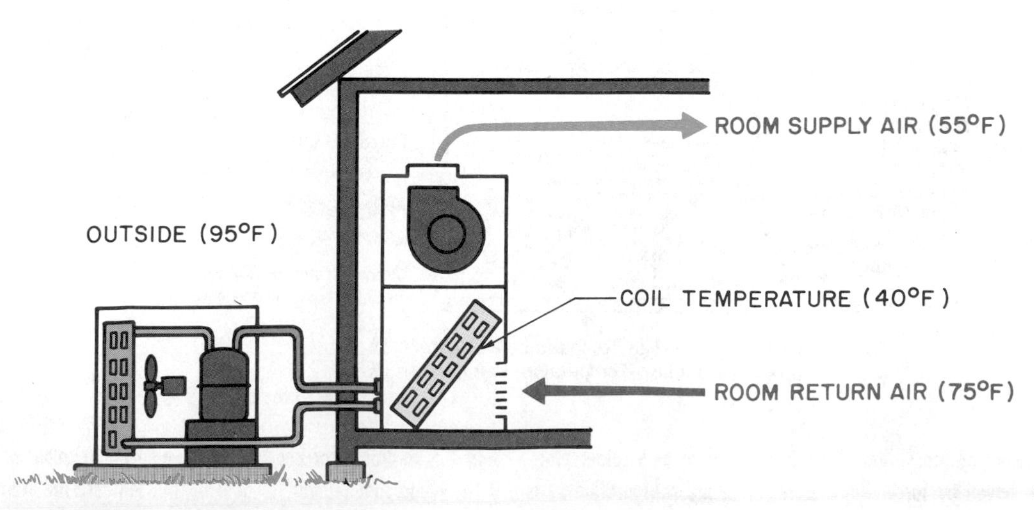

FIGURE 21–3 Air conditioning example of refrigeration.

LOW-TEMPERATURE APPLICATION. Low-temperature refrigeration produces temperatures below the freezing point of water, 32°F. One of the higher low-temperature applications is the making of ice, which usually occurs at about 32°F.

Low-temperature food storage applications generally start at 0°F and go as low as -20°F. At this temperature ice cream would be frozen hard. Frozen meats, vegetables, and dairy products are only a few of the foods preserved by freezing. Some foods may be kept for long periods of time and are appetizing when thawed for cooking, provided they are frozen correctly and kept frozen.

21.3 THE EVAPORATOR

The evaporator in a refrigeration system is responsible for absorbing heat into the system from whatever medium is to be cooled. This heat-absorbing process is accomplished by maintaining the evaporator coil at a lower temperature than the medium to be cooled. For example, if a walk-in cooler is to be maintained at 35°F to preserve food products, the coil in the cooler must be maintained at a lower temperature than the 35°F air that will be passing over it. The refrigerant boiling in Figure 21–4 shows the refrigerant in the evaporator boiling at 20°F, which is 15°F lower than the entering air.

The evaporator operating at these low temperatures removes latent and sensible heat from the cooler. Operating at 20°F, as in the example above, the evaporator will collect moisture from the air in the cooler, latent heat. The removal of sensible heat reduces the food temperature.

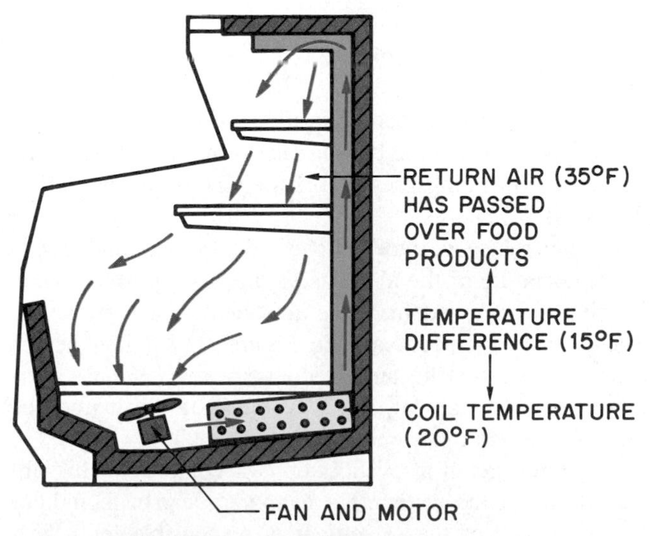

FIGURE 21–4 Relationship of the coil's boiling temperature to the air passing over the coil while operating in the design range.

21.4 BOILING AND CONDENSING

Two important factors in understanding refrigeration are (1) boiling temperature and (2) condensing temperature. The boiling temperature and its relationship to the system exist in the evaporator. The condensing temperature exists in the condenser and will be discussed in the next unit.

These temperatures can be followed by using the pressure/temperature chart in conjunction with a set of refrigeration pressure gages, Figures 21–5 and 21–6.

21.5 THE EVAPORATOR AND BOILING TEMPERATURE

The *boiling temperature* of the liquid refrigerant determines the coil operating temperature. In an air conditioning system a 40°F evaporator coil with 75°F air passing over it produces conditions used for air conditioning or high-temperature refrigeration. Boiling is normally associated with high temperatures and water. Unit 3 discussed the fact that water boils at 212°F at atmospheric pressure. It also discussed the fact that water boils at other temperatures, depending on the pressure. When the pressure is reduced, water will boil at 40°F. This is still boiling—changing a liquid to a vapor. In a refrigeration system, the refrigerant may boil at 20°F by absorbing heat from the 35°F food.

The service technician must be able to determine what operating pressures and temperatures are correct for the various systems being serviced under different load conditions. Much of this knowledge comes from experience. When the thermometers and gages are observed, the readings must be evaluated. There can be as many different readings as there are changing conditions.

Guidelines can help the technician know the pressure and temperature ranges at which the equipment should operate. Relationships exist between the entering air temperature and the evaporator for each system. These relationships are similar from installation to installation.

21.6 REMOVING MOISTURE

Dehumidifying the air means to remove the moisture, and this is frequently desirable in refrigeration systems. Moisture removal is similar from one refrigeration system to another. Knowing this relationship can help the technician know what conditions to look for. The load on the coil would rise or fall accordingly as the return air temperature rises or falls. Warmer return air in the box will also have more moisture content, which imposes further load on the coil. If the cooler is warm due to food added to it, the coil would have more heat to remove because it has more load on it. It would be much like boiling water in an open pan on the stove. The water boils at one rate with the burner on medium and at an increased rate with the burner on high. The boiling pressure stays the same in the boiling water in a pan because the pan is open to the atmosphere. When this

TEMPERATURE	REFRIGERANT				TEMPERATURE	REFRIGERANT				TEMPERATURE	REFRIGERANT			
°F	12	22	134a	502	°F	12	22	134a	502	°F	12	22	134a	502
−60	19.0	12.0		7.2	12	15.8	34.7	13.2	43.2	42	38.8	71.4	37.0	83.8
−55	17.3	9.2		3.8	13	16.4	35.7	13.8	44.3	43	39.8	73.0	38.0	85.4
−50	15.4	6.2		0.2	14	17.1	36.7	14.4	45.4	44	40.7	74.5	39.0	87.0
−45	13.3	2.7		1.9	15	17.7	37.7	15.1	46.5	45	41.7	76.0	40.1	88.7
−40	11.0	0.5	14.7	4.1	16	18.4	38.7	15.7	47.7	46	42.6	77.6	41.1	90.4
−35	8.4	2.6	12.4	6.5	17	19.0	39.8	16.4	48.8	47	43.6	79.2	42.2	92.1
−30	5.5	4.9	9.7	9.2	18	19.7	40.8	17.1	50.0	48	44.6	80.8	43.3	93.9
−25	2.3	7.4	6.8	12.1	19	20.4	41.9	17.7	51.2	49	45.7	82.4	44.4	95.6
−20	0.6	10.1	3.6	15.3	20	21.0	43.0	18.4	52.4	50	46.7	84.0	45.5	97.4
−18	1.3	11.3	2.2	16.7	21	21.7	44.1	19.2	53.7	55	52.0	92.6	51.3	106.6
−16	2.0	12.5	0.7	18.1	22	22.4	45.3	19.9	54.9	60	57.7	101.6	57.3	116.4
−14	2.8	13.8	0.3	19.5	23	23.2	46.4	20.6	56.2	65	63.8	111.2	64.1	126.7
−12	3.6	15.1	1.2	21.0	24	23.9	47.6	21.4	57.5	70	70.2	121.4	71.2	137.6
−10	4.5	16.5	2.0	22.6	25	24.6	48.8	22.0	58.8	75	77.0	132.2	78.7	149.1
−8	5.4	17.9	2.8	24.2	26	25.4	49.9	22.9	60.1	80	84.2	143.6	86.8	161.2
−6	6.3	19.3	3.7	25.8	27	26.1	51.2	23.7	61.5	85	91.8	155.7	95.3	174.0
−4	7.2	20.8	4.6	27.5	28	26.9	52.4	24.5	62.8	90	99.8	168.4	104.4	187.4
−2	8.2	22.4	5.5	29.3	29	27.7	53.6	25.3	64.2	95	108.2	181.8	114.0	201.4
0	9.2	24.0	6.5	31.1	30	28.4	54.9	26.1	65.6	100	117.2	195.9	124.2	216.2
1	9.7	24.8	7.0	32.0	31	29.2	56.2	26.9	67.0	105	126.6	210.8	135.0	231.7
2	10.2	25.6	7.5	32.9	32	30.1	57.5	27.8	68.4	110	136.4	226.4	146.4	247.9
3	10.7	26.4	8.0	33.9	33	30.9	58.8	28.7	69.9	115	146.8	242.7	158.5	264.9
4	11.2	27.3	8.6	34.9	34	31.7	60.1	29.5	71.3	120	157.6	259.9	171.2	282.7
5	11.8	28.2	9.1	35.8	35	32.6	61.5	30.4	72.8	125	169.1	277.9	184.6	301.4
6	12.3	29.1	9.7	36.8	36	33.4	62.8	31.3	74.3	130	181.0	296.8	198.7	320.8
7	12.9	30.0	10.2	37.9	37	34.3	64.2	32.2	75.8	135	193.5	316.6	213.5	341.2
8	13.5	30.9	10.8	38.9	38	35.2	65.6	33.2	77.4	140	206.6	337.2	229.1	362.6
9	14.0	31.8	11.4	39.9	39	36.1	67.1	34.1	79.0	145	220.3	358.9	245.5	385.0
10	14.6	32.8	11.9	41.0	40	37.0	68.5	35.1	80.5	150	234.6	381.5	262.7	408.4
11	15.2	33.7	12.5	42.1	41	37.9	70.0	36.0	82.1	155	249.5	405.1	280.7	432.9

Vacuum—Red Figures
Gage Pressure—Bold Figures

FIGURE 21–5 Pressure/temperature chart in inches mercury vacuum or psig.

FIGURE 21–6 Pressure gages have pressure/temperature relationship printed on the gage. *Photo by Bill Johnson*

same boiling process occurs in an enclosed coil, the pressures will rise when the boiling occurs at a faster rate. This causes the operating pressure of the whole system to rise, Figure 21–7.

When the evaporator removes heat from air and lowers the temperature of the air, sensible heat is removed. When moisture is removed from the air, latent heat is removed. The moisture is piped to a drain, Figure 21–8. Latent heat is called hidden heat because it does not register on a thermometer, but it is heat, like sensible heat and it must be removed, which takes energy.

The refrigeration evaporator is a component that absorbs heat into the system. The evaporator can be thought of as the *sponge* of the system. It is responsible for a heat exchange between the conditioned space or product and the refrigerant inside the system. Some evaporators absorb heat more efficiently than others. Figure 21–9 illustrates this heat exchange between air and refrigerant.

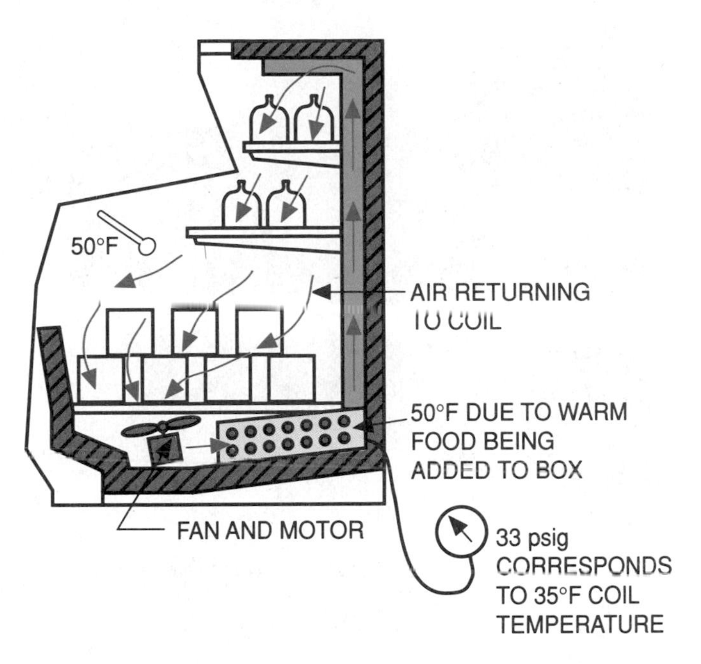

FIGURE 21–7 Coil-to-air temperature relationship under increased coil load.

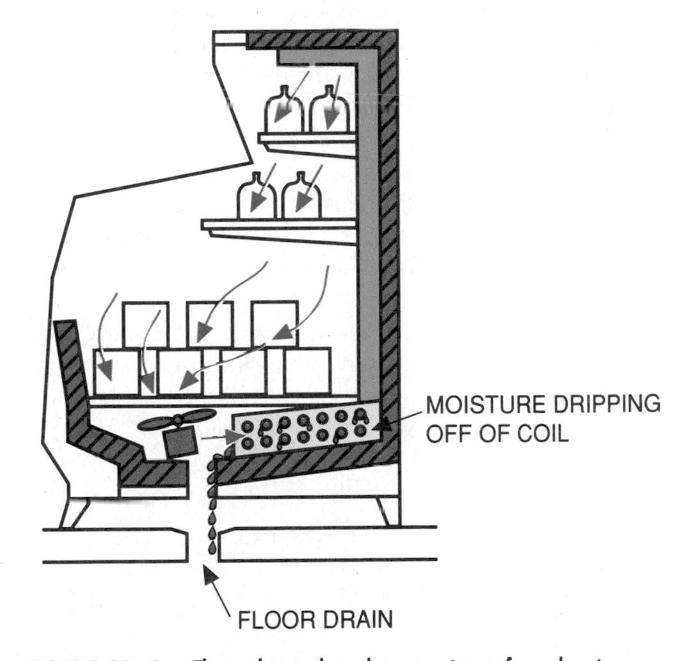

FIGURE 21–8 The cooling coil condenses moisture from the air.

21.7 HEAT EXCHANGE CHARACTERISTICS OF THE EVAPORATOR

Conditions that govern the rate of heat exchange are:

1. The evaporator *material* through which the heat has to be exchanged. Evaporators may be manufactured from copper, steel, brass, stainless steel, or aluminum. Corrosion is one factor that determines what material is used. For instance, when acidic materials need to be cooled,

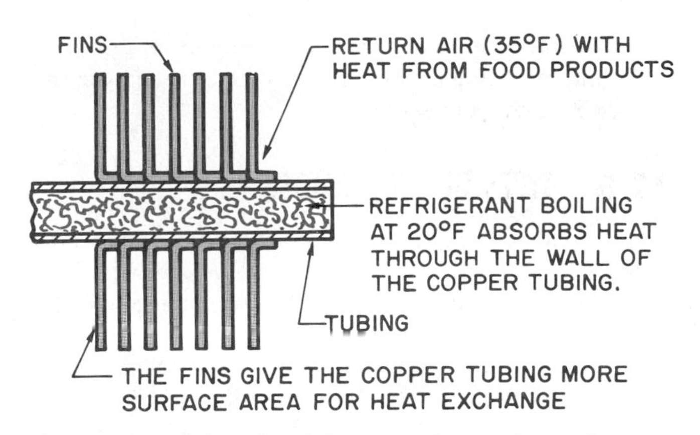

FIGURE 21–9 Relationship of the air to refrigerant heat exchange.

copper or aluminum coils would be eaten away. Stainless steel may be used instead, but stainless steel does not conduct heat as well as copper.
2. The *medium* to which the heat is exchanged. Giving heat up from air to refrigerant is an example. The best heat exchange occurs between two liquids, such as water to liquid refrigerant. However, this is not always practical because heat frequently has to be exchanged between air and vapor refrigerant. The vapor-to-vapor exchange is slower than the liquid-to-liquid exchange, Figure 21–10.
3. The *film factor*. This is a relationship between the medium giving up heat or absorbing heat and the heat-exchange surface. The film factor relates to the velocity of the medium passing over the exchange surface. When the velocity is too slow, the film between the medium and the surface becomes an insulator and

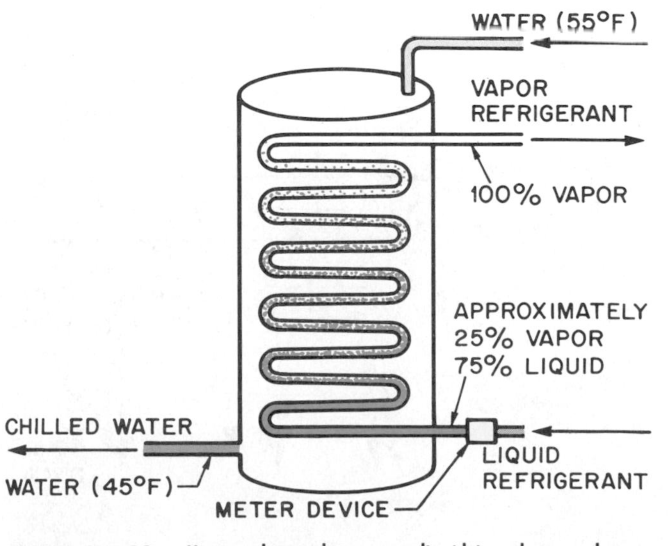

FIGURE 21–10 Heat exchange between a liquid in a heat exchanger and the refrigerant inside the coil.

slows the heat exchange. The velocity keeps the film to a minimum, Figure 21–11. The correct velocity is chosen by the manufacturer.

21.8 TYPES OF EVAPORATORS

Numerous types of evaporators are available, and each has its purpose. The first evaporators for cooling air were natural convection type. They were actually bare pipe evaporators with refrigerant circulating through them, Figure 21–12. This evaporator was used in early walk-in coolers and was mounted high in the ceiling. It relied on the air being cooled, falling to the floor, and setting up a natural air current. The evaporator had to be quite large for the particular application because the velocity of the air passing over the coil was so slow. It is still occasionally used today.

The use of a blower to force or induce air over the coil improved the efficiency of the heat exchange. This meant that smaller evaporators could be used to do the same job. Design trends in industry have always been to smaller, more efficient equipment, Figure 21–13.

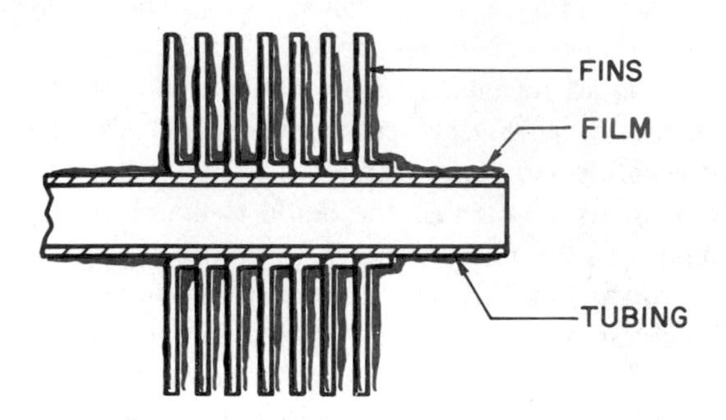

FIGURE 21–11 One of the deterring factors in a normal heat exchange. The film factor is the film of air or liquid next to the tube in the heat exchange.

FIGURE 21–13 Forced-draft evaporator. *Courtesy Bally Case and Cooler, Inc.*

The expansion of the evaporator surface to a surface larger than the pipe itself gives a more efficient heat exchange. The *stamped evaporator* is a result of the first designs to create a large pipe surface. The stamped evaporator is two pieces of metal stamped with the impression of a pipe passage through it, Figure 21–14.

Pipe with fins attached, called a *finned tube evaporator,* is used today more than any other type of heat exchange between air and refrigerant. This heat exchanger is efficient because the fins are in good contact with the pipe carrying the refrigerant. Figure 21–15 shows an example of a finned tube evaporator.

Multiple circuits improve evaporator performance and efficiency by reducing pressure drop inside the evaporator. The pipes inside the evaporator can be polished smooth, but

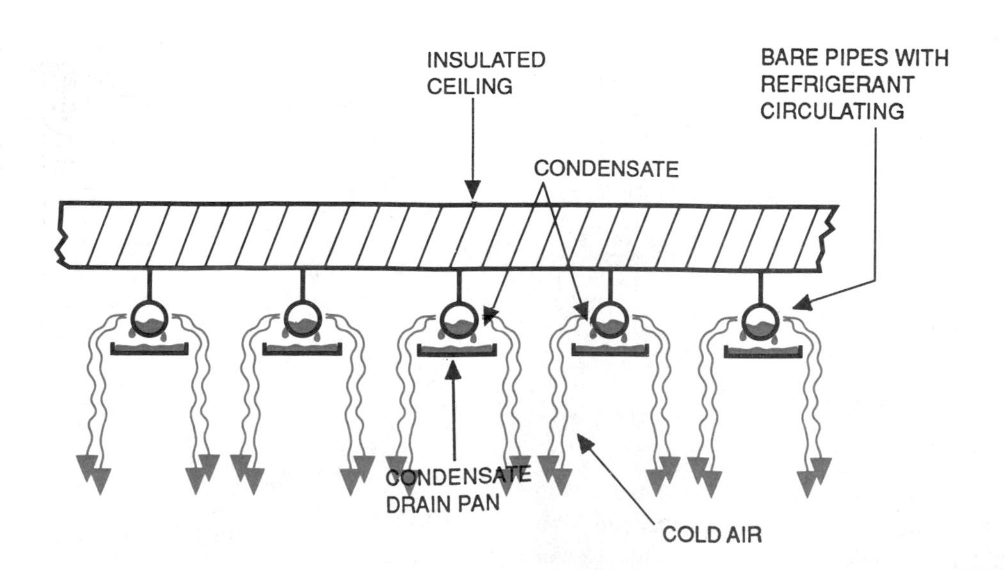

FIGURE 21–12 Bare pipe evaporator.

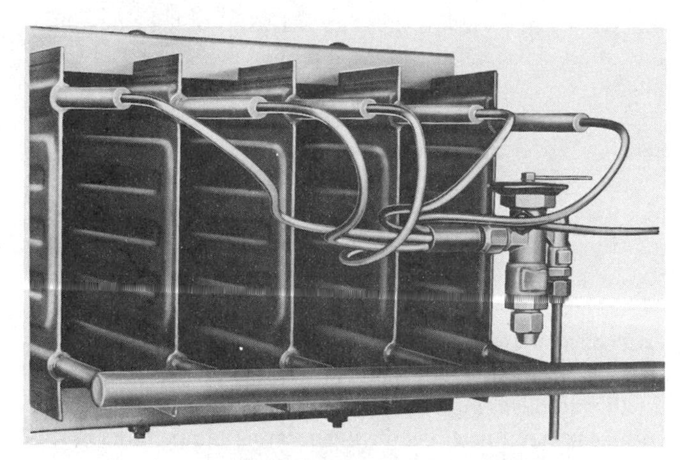

FIGURE 21–14 Stamped evaporator. *Courtesy Sporlan Valve Company*

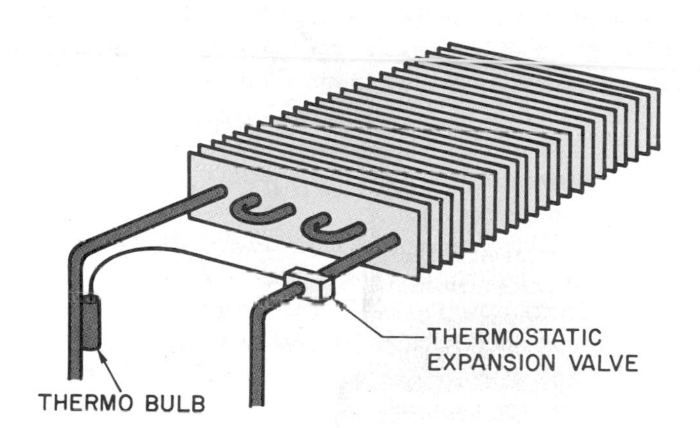

FIGURE 21–15 Finned evaporator.

they still offer resistance to the flow of both liquid and vapor refrigerants. The shorter the evaporator is, the less resistance there is to this flow. When an evaporator becomes quite long, it is common to cut it off and run another circuit in parallel next to it, Figure 21–16.

The evaporator for cooling liquids or making ice operates under the same principles as one for cooling air but is designed differently. It may be strapped on the side of a cylinder with liquid inside, submerged inside the liquid container, or be a double-pipe system with the refrigerant inside one pipe and the liquid to be cooled circulated inside an outer pipe, Figure 21–17.

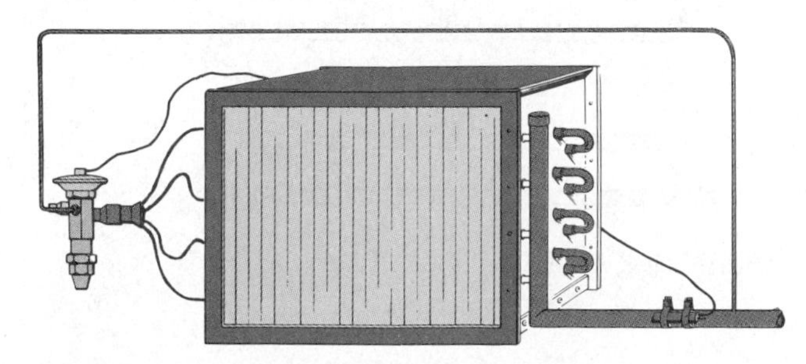

FIGURE 21–16 Multicircuit evaporator. *Courtesy Sporlan Valve Company.*

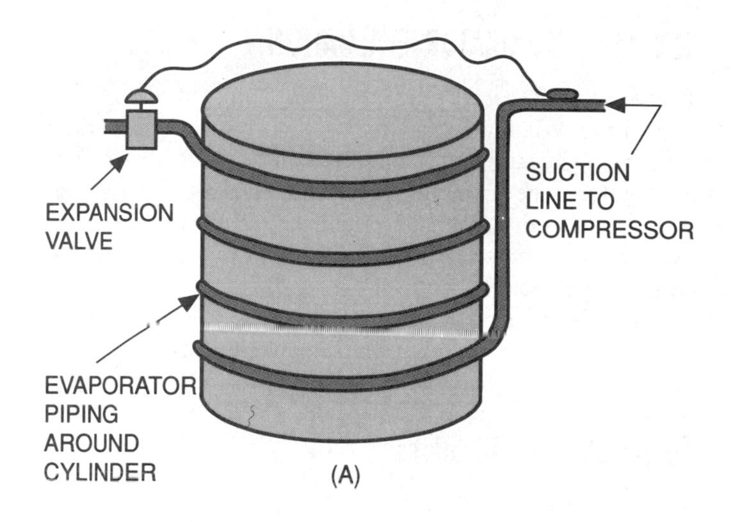

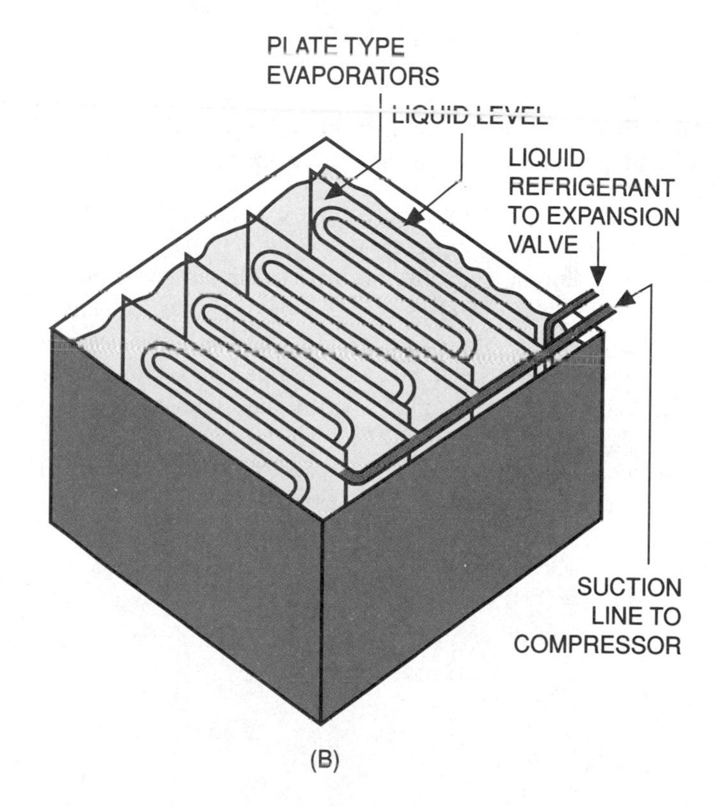

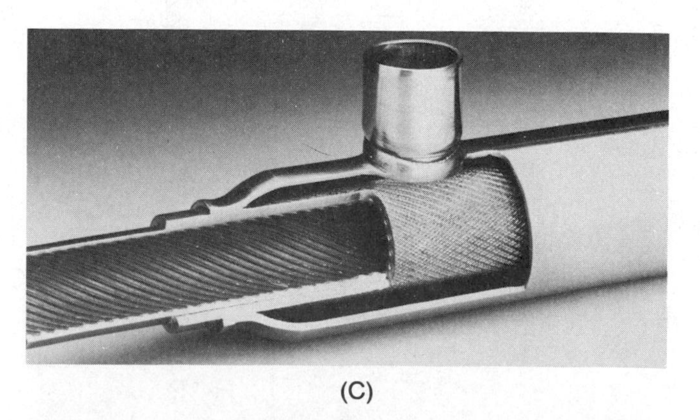

FIGURE 21–17 Liquid heat exchangers. (A) Drum-type evaporator. (B) Plate-type evaporators in a tank. (C) Pipe in a pipe evaporator. *(C) Courtesy Noranda Metal Industries Inc.*

21.9 EVAPORATOR EVALUATION

Knowing the design considerations helps in *evaporator evaluation*. When the service technician arrives at the job, it may be necessary to evaluate whether or not a particular evaporator is performing properly. This can be considered one of the starting points in organized troubleshooting. The evaporator absorbs heat, the compressor pumps it, and the condenser rejects it. The following example pertains to a medium-temperature walk-in box. However, the evaporator evaluation would be about the same for any typical application.

Evaporator Specifications

1. Copper pipe coil
2. Aluminum fins attached to the copper pipe coil
3. Forced draft with a prop-type fan
4. One continual refrigerant circuit
5. R-12
6. Evaporator to maintain space temperature at 35°F
7. Evaporator clean and in good working condition

First is a description of how the evaporator functions when it is working correctly.

Entering the evaporator is a partial liquid-partial vapor mixture at 20°F and 21 psig; it is approximately 75% liquid and 25% vapor. Approximately 25% of the liquid entering the expansion device at the evaporator is changed to a vapor and cools the remaining 75% of the liquid to the evaporator's boiling temperature (20°F). This is accomplished with the pressure drop across the expansion device. When the warm liquid passes through the small opening in the expansion device into the low pressure (21 psig) of the evaporator side of the device, some of the liquid flashes to a gas, Figure 21–18.

As the partial liquid-partial vapor mixture moves through the evaporator, more of the liquid changes to a vapor. This is called *boiling* and is a result of heat absorbed into the coil from whatever medium the evaporator is cooling. Finally, near the end of the evaporator the liquid is all boiled away to a vapor. At this point the refrigerant is known as *saturated vapor.* This means that the refrigerant vapor is saturated with heat. If any more heat is added to it, it will rise in temperature. If any heat is taken away from it, it will start changing back to a liquid. This vapor is saturated with heat, but it is still at the evaporating temperature corresponding to the boiling point, 20°F. **This is a most important point in the function of an evaporator because all of the liquid must be boiled away as close to the end of the coil as possible. This is necessary to (1) keep the coil efficiency up and (2) ensure that liquid refrigerant does not leave the evaporator and move into the compressor.** For the evaporator to run efficiently, it must operate as full of liquid as possible without boiling over because the best heat exchange is between the liquid refrigerant and the air passing over the coil.

The pressure/enthalpy chart in Figure 21–19 shows graphically what happens inside the evaporator for the walk-in cooler example above. The refrigerant enters the evaporator at point A (after leaving the expansion valve). The liquid pressure is 20 psig and contains 30 Btu/lb of heat at this point. Approximately 25% of the liquid was flashed to a vapor passing through the expansion valve. As the liquid proceeds through the evaporator, it is changing to a

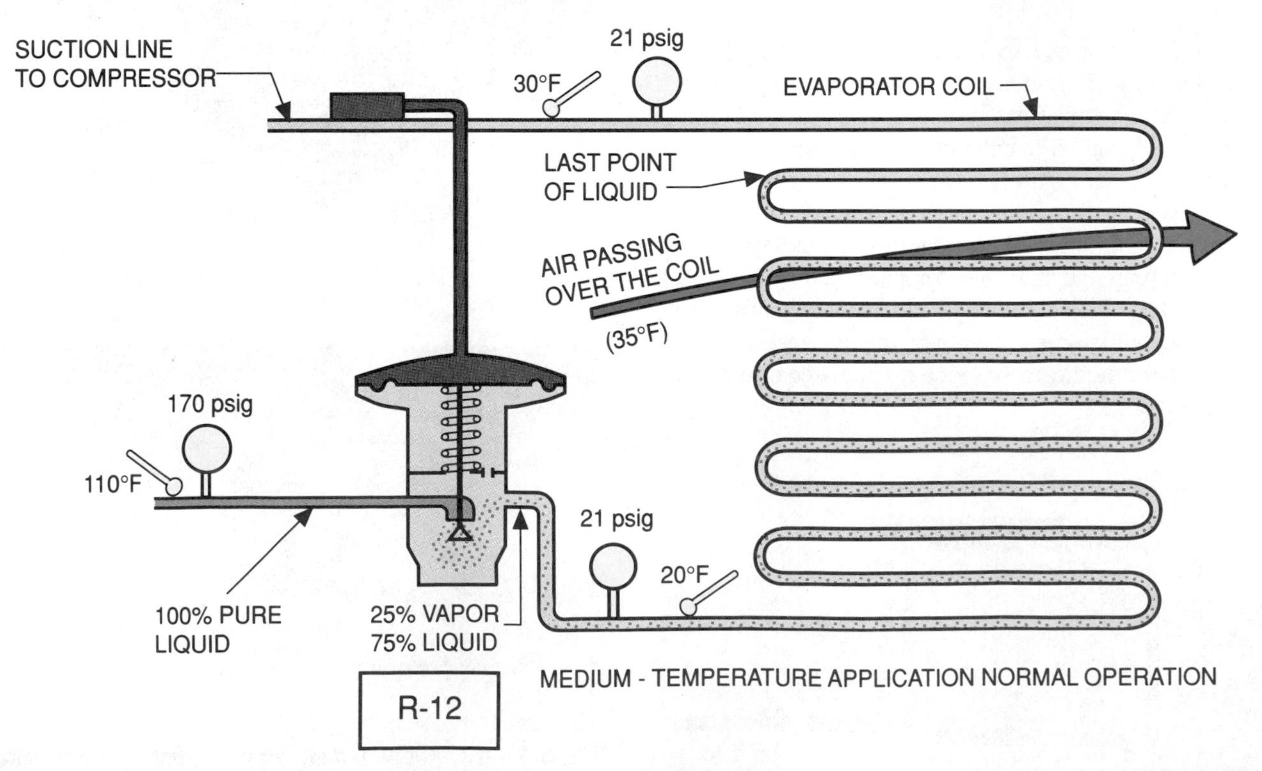

FIGURE 21–18 When the 110°F liquid passes through the expansion valve orifice, some of the liquid flashes to a vapor and cools the remaining liquid to the evaporating temperature of 20°F.

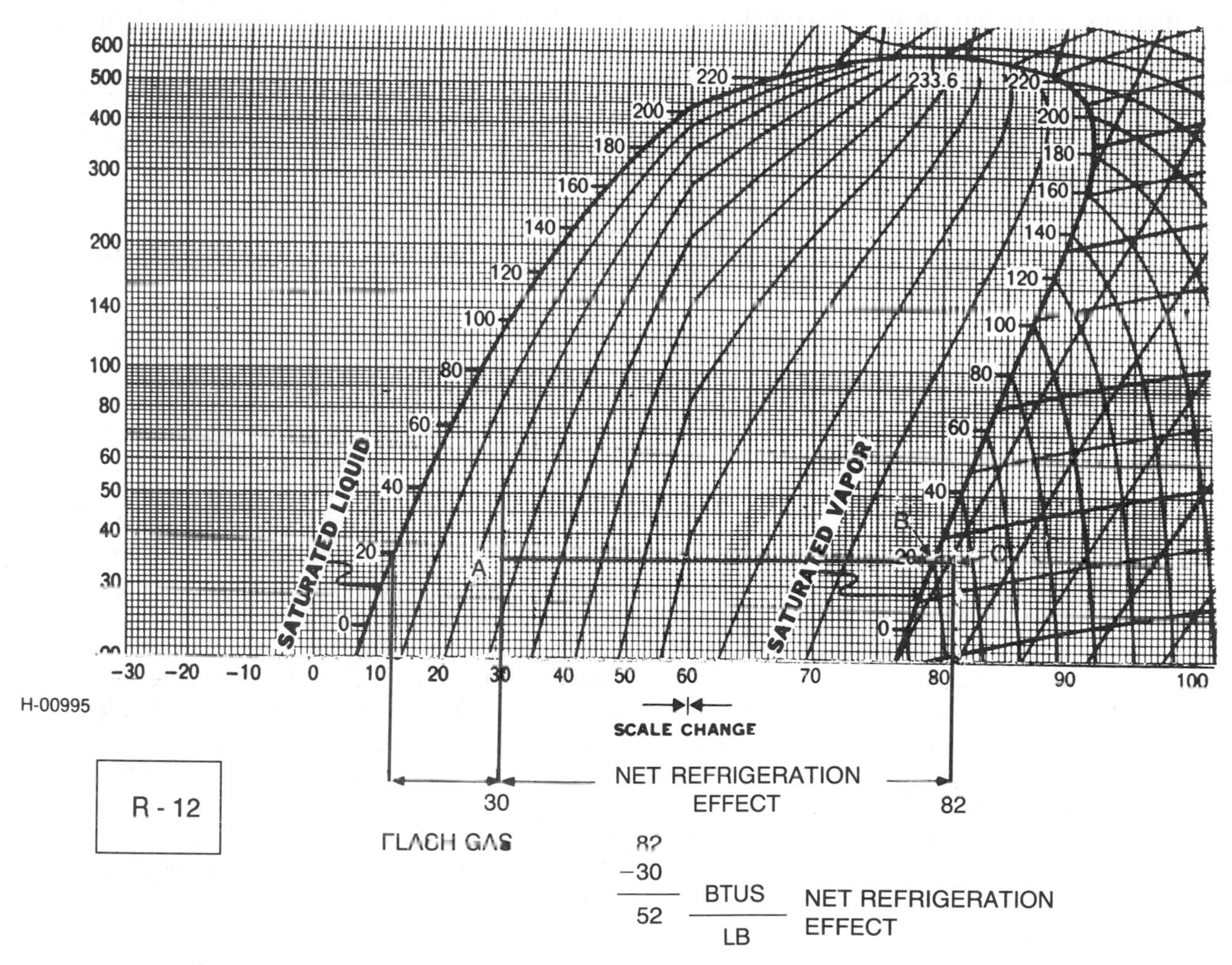

FIGURE 21–19 The refrigeration effect in the evaporator.

vapor. It is all changed to a vapor at point B, but the vapor temperature is still 20°F and capable of absorbing heat, in the form of superheat. The vapor temperature starts to rise while it is still in the evaporator until the temperature is 30°F (containing 10° of superheat). The vapor leaves the evaporator at point C with a heat content of 82 Btu/lb. The usable refrigeration in the evaporator is from point A to C where the refrigerant absorbed 52 Btu/lb of heat energy. You only need to know how many Btu/h capacity is needed to determine the amount of refrigerant that needs to be circulated. For example, if the evaporator needs to have a capacity of 35,000 Btu/h, it must have 673 lb of refrigerant circulate through it per hour (35,000 Btu/h ÷ 52 Btu/lb = 673 lb/h). This sounds like a lot of refrigerant, but it is only 11.2 lb/min (673 lb/h ÷ 60 min/h = 11.2 lb/min). The size of the compressor determines how much refrigerant can be pumped.

21.10 LATENT HEAT IN THE EVAPORATOR

The **latent heat absorbed during the change of state is much more concentrated than the sensible heat that would be added to the vapor leaving the coil.** Refer to the example of heat in Unit 1 that showed how it takes 1 Btu to change the temperature of 1 lb of 211°F water to 212°F water. It also showed that it takes 970 Btu to change 1 lb of 212°F water to 212°F steam. This is true for water or refrigerant. The change of state is where the great amount of heat is absorbed into the system. The example above showed 52 Btu/lb of refrigerant circulated was absorbed at a boiling temperature of 20°F without a change in pressure.

21.11 THE FLOODED EVAPORATOR

To get the maximum efficiency from the evaporator heat exchange, some evaporators are operated full of liquid or flooded, with a device to keep the liquid refrigerant from passing to the compressor. These *flooded evaporators* are specially made and normally use a float metering device to keep the liquid level as high as possible in the evaporator. This text will not go into detail about this system because it is not a device encountered often. Manufacturer's literature should be consulted for any special application.

When an evaporator is flooded, it would be much like water boiling in a pot with a compressor taking the vapor off the top of the liquid. There would always be a liquid level. If the evaporator is not flooded, that is, when the refrigerant starts out as a partial liquid and boils away to a vapor in the heat exchange pipes, it is known as a *dry-type* or direct expansion evaporator.

21.12 DRY-TYPE EVAPORATOR PERFORMANCE

To check the performance of a dry-type evaporator, the service technician would first make sure that the refrigerant coil is operating with enough liquid inside the coil. This is generally done by comparing the boiling temperature inside the coil to the line temperature leaving the coil. The difference in temperatures is usually 8°F to 12°F. For example, in the coil pictured in Figure 21–20, the following temperature in the coil was arrived at by converting the coil pressure (suction pressure) to temperature. In this example, the pressure was 21 psig, which corresponds to 20°F. This suction pressure reading is important to the technician because the boiling temperature must be known to arrive at the superheat reading for the coil.

21.13 SUPERHEAT

The difference in temperature between the boiling refrigerant and the suction line temperature is known as superheat. Superheat is the sensible heat added to the vapor refrigerant after the change of state has occurred. Superheat is the best method of checking to see when a refrigerant coil has a proper level of refrigerant. When a metering device is not feeding enough refrigerant to the coil, the coil is said to be a *starved coil,* and the superheat is greater, Figure 21–21. It can be seen from the example that all of the refrigeration takes place at the beginning of the coil. The suction pressure is very low, below freezing, but only a portion of the coil is being used. This coil would quickly freeze solid and no air would pass through it. The freeze line would creep upward until the whole coil is a block of ice and the refrigeration would do no good. The refrigerated box temperature would rise.

21.14 HOT PULL DOWN (EXCESSIVELY LOADED EVAPORATOR)

When the refrigerated space has been allowed to warm up considerably, the system must go through a hot pull down. On a hot pull down the evaporator and metering device are not expected to act exactly as they would in a typical design condition. For instance, if a walk-in cooler were supposed to maintain 35°F and it were allowed to warm up to 60°F and have some food or beverages inside, it would take an extended time to pull the air and product temperature down. The coil may be boiling the refrigerant so fast that the superheat may not come down to 8°F to 12°F until the box has cooled down closer to the design temperature. **A superheat reading on a hot pull down should be interpreted with caution,** Figure 21–22. The superheat reading will only be correct when the coil is at or near normal conditions.

When a dry-type coil is fed too much refrigerant, the refrigerant does not all change to a vapor. This coil is thought of as a *flooded* coil, flooded with liquid refrigerants, Figure 21–23. **Do not confuse this with a coil flooded by design.**

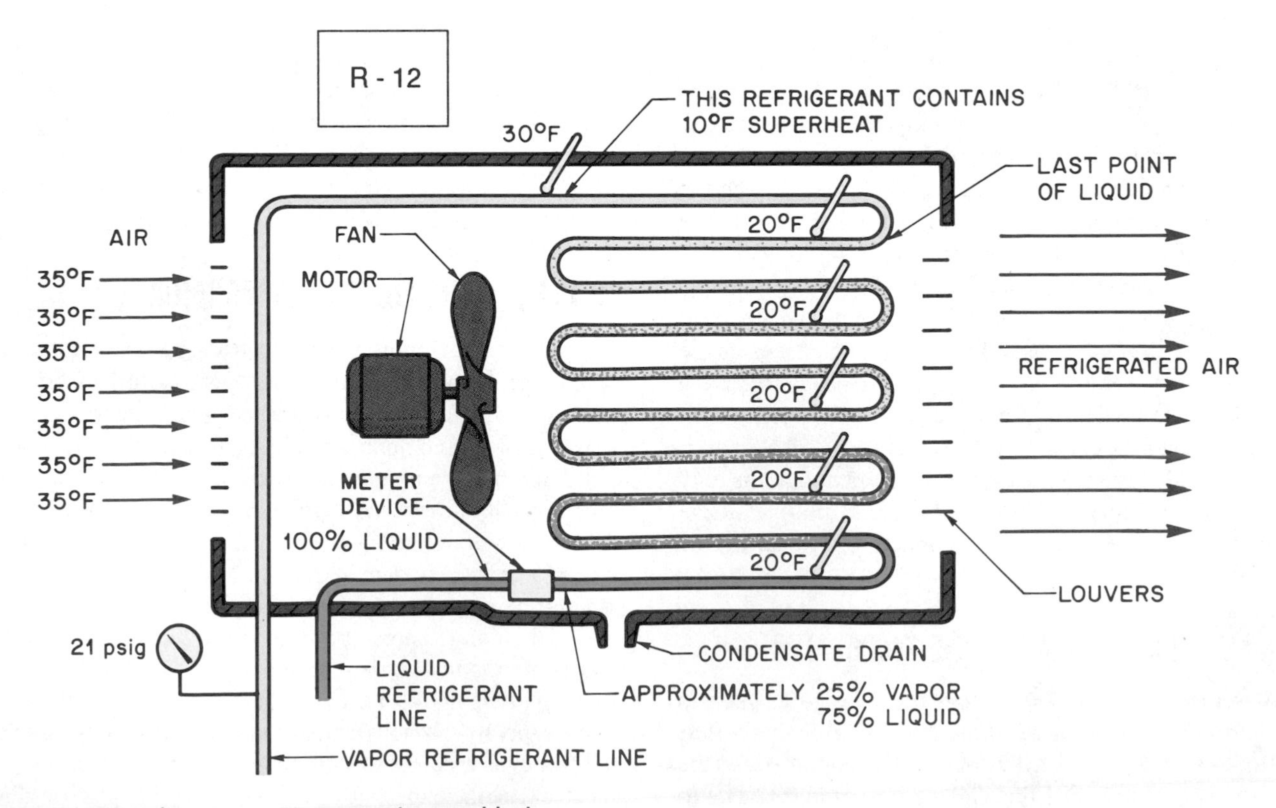

FIGURE 21–20 The evaporator operating under normal load.

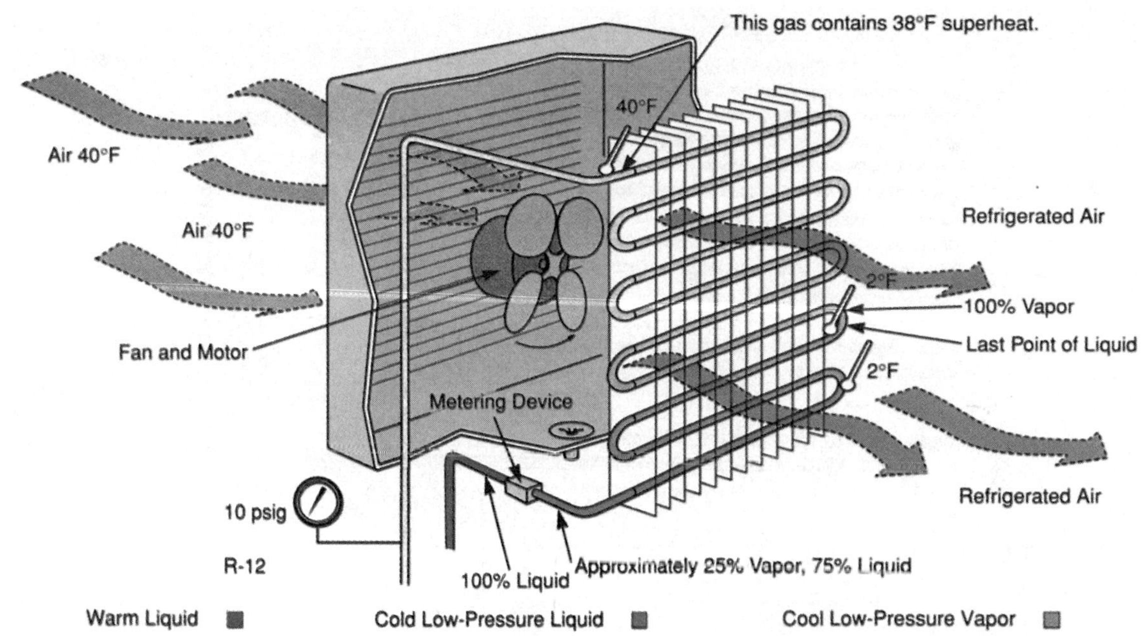

FIGURE 21–21

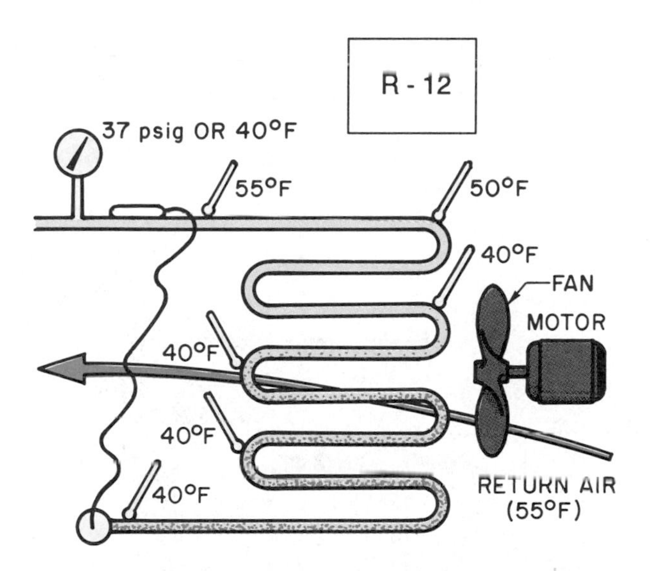

FIGURE 21–22 Hot pull down with a coil. This is a medium-temperature evaporator that should be operating at 21 psig (R-12 20°F). The return air is 55°F instead of 35°F. This causes the pressure in the coil to be higher. The warm box boils the refrigerant at a faster rate. The expansion valve is not able to feed the evaporator fast enough to keep the superheat at 10°F. The evaporator has 15°F of superheat.

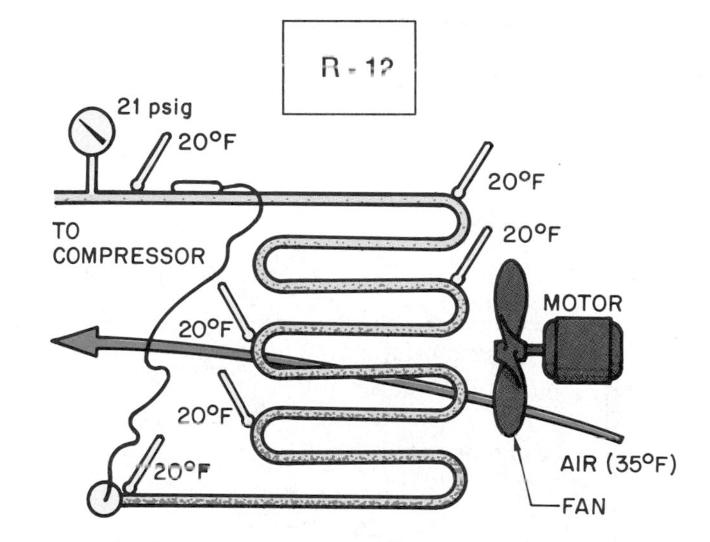

FIGURE 21–23 Evaporator flooding because expansion device is not controlling refrigerant flow.

This is a symptom that can cause real trouble because unless this liquid in the suction line boils to a vapor before it reaches the compressor, compressor damage may occur. **Remember, the evaporator is supposed to boil all of the liquid to a vapor.**

21.15 PRESSURE DROP IN EVAPORATORS

Multicircuit evaporators are used when the coil would become too long for a single circuit, Figure 21–24. The same evaluating procedures hold true for a multicircuit evaporator as for a single-circuit evaporator.

A dry-type evaporator has to be as full as possible to be efficient. Each circuit should be feeding the same amount of

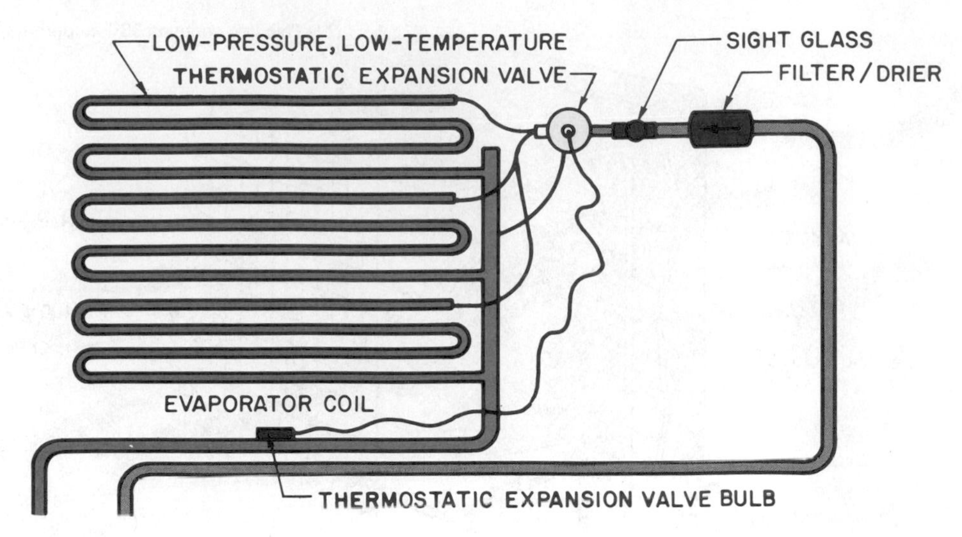

FIGURE 21–24 Multicircuit evaporator.

refrigerant. If this needs to be checked, the service technician can check the common pressure tap for the boiling pressure. This pressure can be converted to temperature. Then the temperature will have to be checked at the outlet of each circuit to see if any circuit is overfeeding or starving, Figures 21–25 and 21–26.

Some reasons for uneven feeding for a multicircuit evaporator are:

1. Stopped-up distribution system
2. Dirty coil
3. Uneven air distribution
4. Coil circuits of different lengths

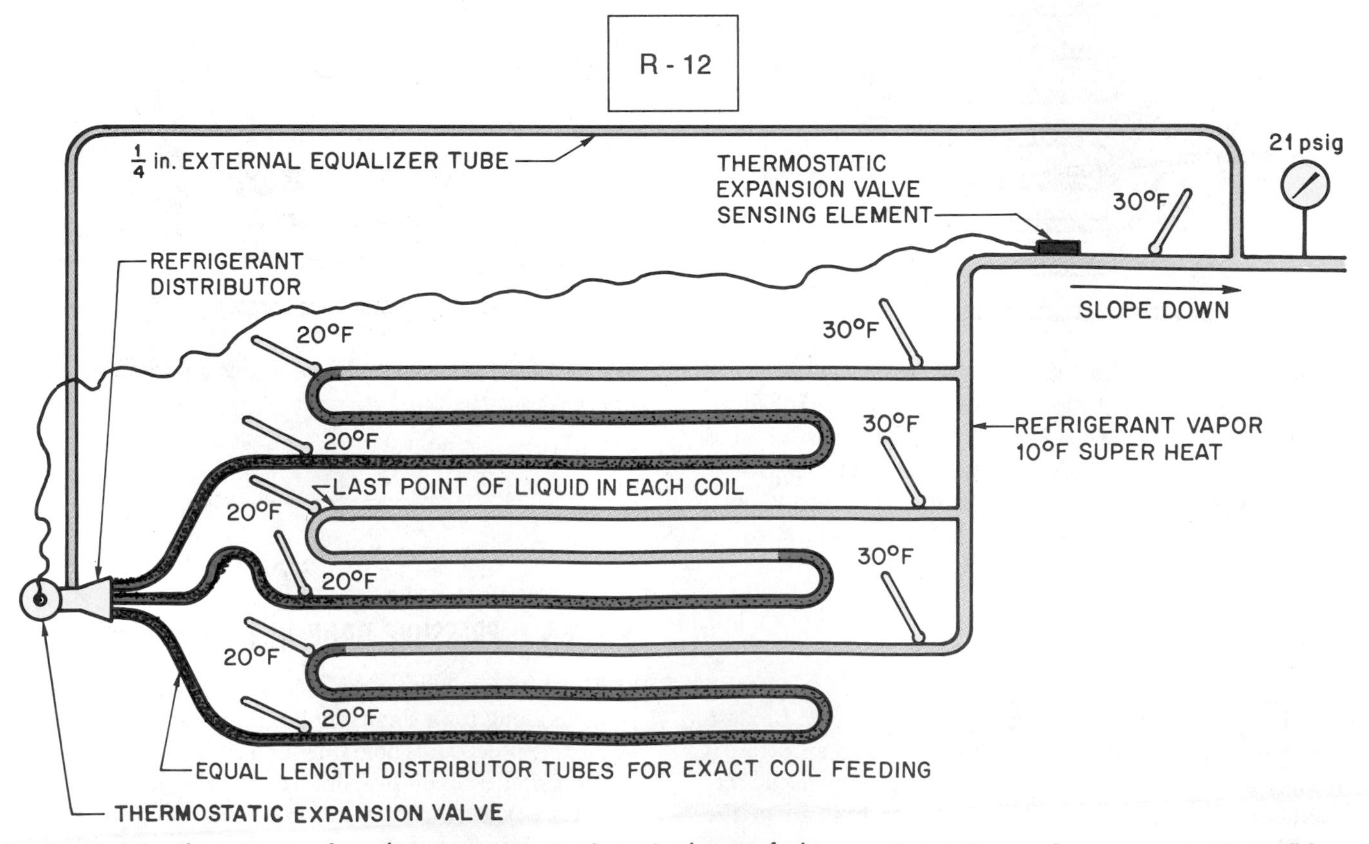

FIGURE 21–25 The appearance of a multicircuit evaporator on the inside when it is feeding correctly. It is like several evaporators piped in parallel.

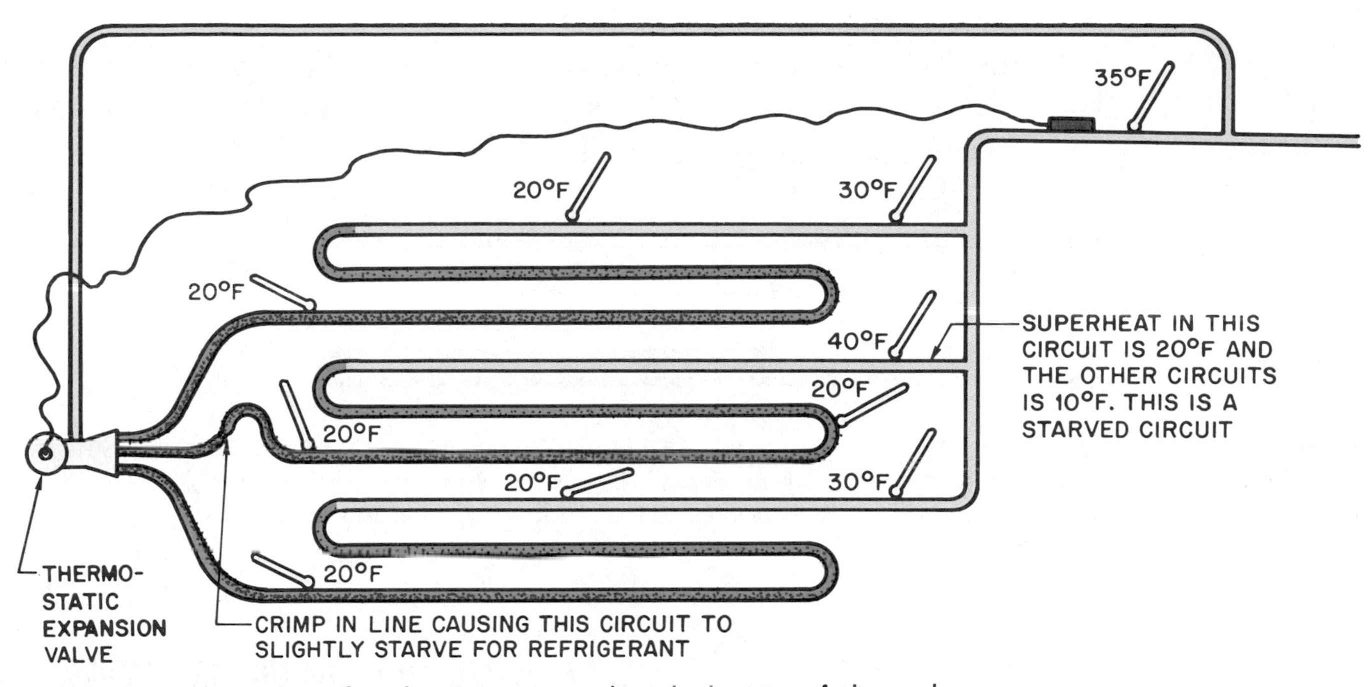

FIGURE 21–26 The appearance of a multicircuit evaporator on the inside when it is not feeding evenly.

21.16 LIQUID COOLING EVAPORATORS (CHILLERS)

A different type of evaporator is required for liquid cooling. It functions much the same as that for cooling air. They are normally of the dry type expansion evaporator in the smaller systems, Figure 21–27. These evaporators have more than one refrigerant circuit to prevent pressure drop in the evaporator. The use of refrigeration gages and some accurate method of checking temperature of the suction line are very important. These evaporators sometimes have to be checked for performance to see if they are absorbing heat like they should. They have a normal superheat range similar to air-type evaporators (8°F to 12°F). When the superheat is within this range and all circuits in a multicircuit evaporator are performing alike, the evaporator is doing its job on the refrigerant side. However, this does not mean that it will cool properly. The liquid side of the evaporator must be clean so that the liquid will come in proper contact with the evaporator.

Typical problems on the liquid side of the evaporator are:

1. Mineral deposits that may build up on the liquid side and cause a poor heat exchange. They would act like an insulator.
2. Poor circulation of the liquid to be cooled where a circulating pump is concerned.

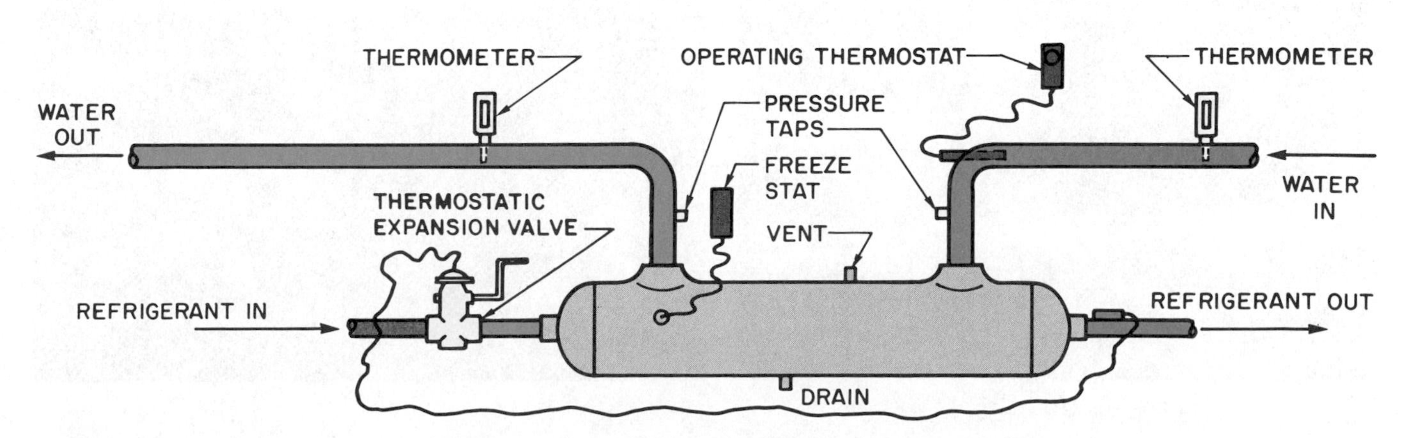

FIGURE 21–27 Evaporators used for exchanging heat between liquids and refrigerant. Most of these evaporators are of the direct expansion type.

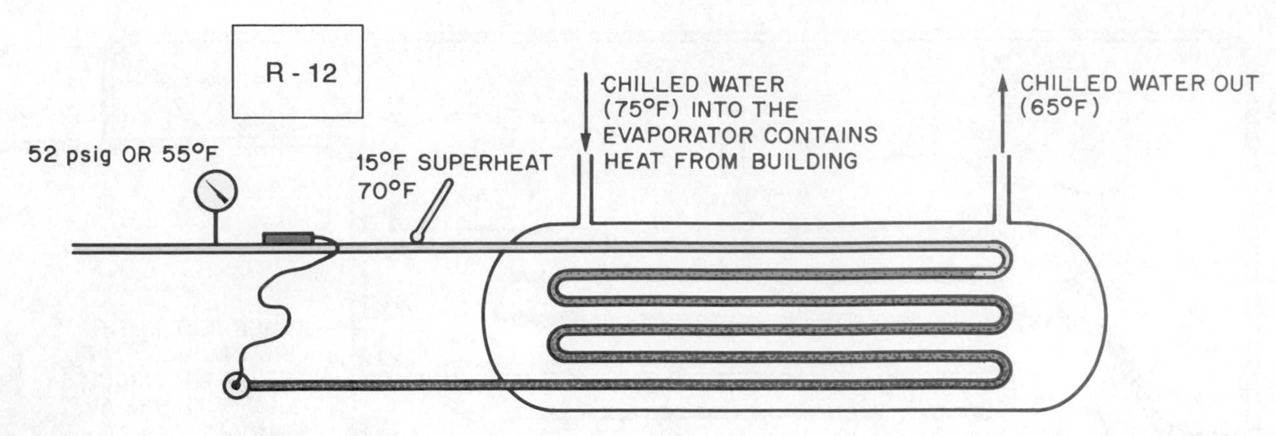

FIGURE 21–28 A hot pull down on a liquid evaporator giving up its heat to refrigerant. This evaporator normally has 55°F water in and 45°F water out. The hot pull down with 75°F water instead of 55°F water boils the refrigerant at a faster rate. The expansion valve may not be able to feed the evaporator fast enough to maintain 10°F superheat. No conclusions should be made until the system approaches design conditions.

When the superheat is found to be correct and the coil is feeding correctly on a multicircuit system, the technician should consider the temperature of the liquid. The superheat may not be within the prescribed limits if the liquid to be cooled is not cooled down close to the design temperature. On a hot pull down of a liquid product, the heat exchange can be such that the coil appears to be starved for refrigerant because the coil is so loaded up that it is boiling the refrigerant faster than normal. The technician must be patient because a pull down cannot be rushed, Figure 21–28. Air-to-refrigerant evaporators do not have quite the pronounced difference in pull down that liquid heat exchange evaporators do because of the excellent heat exchange properties of the liquid to the refrigerant.

21.17 EVAPORATORS FOR LOW-TEMPERATURE APPLICATION

Low-temperature evaporators used for cooling space or product to below freezing are designed differently because they require the coil to operate below freezing.

In an airflow application, the water that accumulates on the coil will freeze and will have to be removed. The design of the fin spacing must be carefully chosen. A very small amount of ice accumulated on the fins will restrict the airflow. Low-temperature coils have fin spacings that are wider than medium-temperature coils, Figure 21–29. Other than the airflow blockage due to ice buildup, these low-temperature evaporators perform much the same as medium-temperature evaporators. They are normally dry-

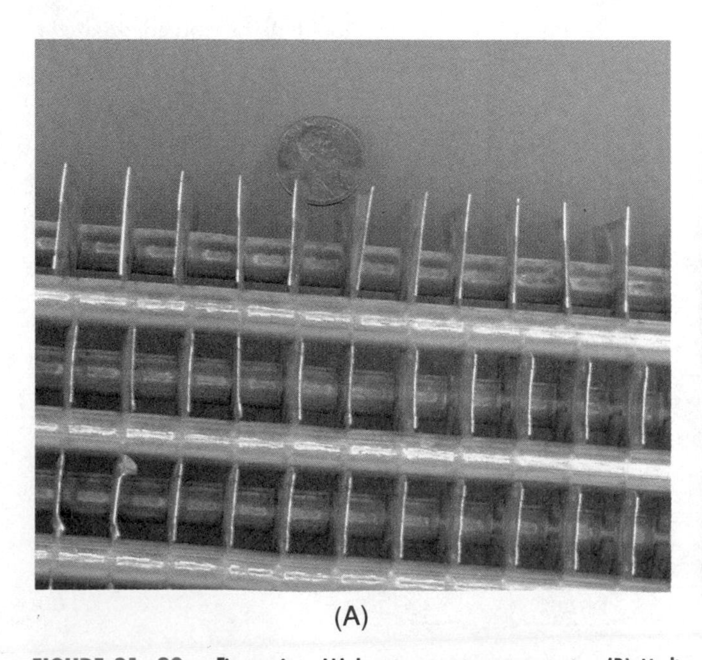

(A)

(B)

FIGURE 21–29 Fin spacing. (A) Low-temperature evaporator. (B) Medium-temperature evaporator. *Photos by Bill Johnson*

type evaporators and have one or more fans to circulate the air across the coil. The defrosting of the coil has to be done by raising the coil temperature above freezing to melt the ice. Then the condensate water has to be drained off and kept from freezing.

Defrost is sometimes accomplished with heat from outside the system. Electric heat can be added to the evaporator to melt the ice, but this heat adds to the load of the system and needs to be pumped out after defrost.

21.18 DEFROST OF ACCUMULATED MOISTURE

Defrost can be accomplished with heat from inside the system using the hot gas from the discharge line of the compressor. This is accomplished by routing a hot gas line from the compressor discharge line to the outlet of the expansion valve and installing a solenoid valve to control the flow. When defrost is needed, hot gas is released inside the evaporator, which will quickly melt any ice, Figure 21–30.

When the hot gas enters the evaporator, it is likely that liquid refrigerant will be pushed out the suction line toward the compressor. To prevent this liquid from entering the compressor, often a suction line accumulator will be added to the suction piping. Accumulators are discussed later.

The hot gas defrost system is economical because power does not have to be purchased for defrost using external heat, such as electric heaters that will heat the evaporator. The heat is already in the system.

Electric defrost is accomplished using electric heating elements located at the evaporator. The compressor is stopped and these heaters are energized on a call for defrost and allowed to operate until the frost is melted from the coil, Figure 21–31. These heaters are often embedded in the actual evaporator fins and cannot be removed if they were to burn out. Frequently, in the event that the heaters do burn out, hot gas defrost can be added to the system and the electric heat defrost procedures discontinued.

When either system is used for defrost, the evaporator fan must be turned off during defrost, or two things will happen:

(1) The heat from defrost will be transferred directly to the conditioned space.
(2) The cold conditioned air would slow down the defrost process.

Evaporators applied to some ice-making processes have similar defrost methods. They must have some method of applying heat to the evaporator to melt the ice. Sometimes the heat is electric or hot gas. When the evaporator is

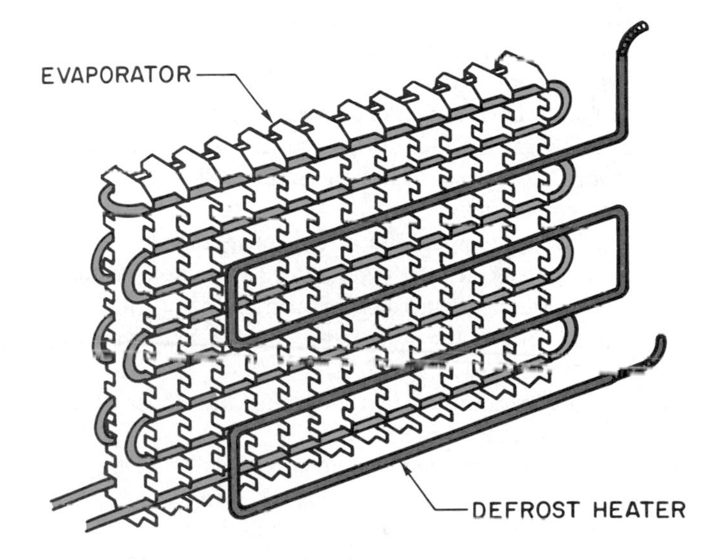

FIGURE 21–31 Heater used for electric defrost of low-temperature evaporators.

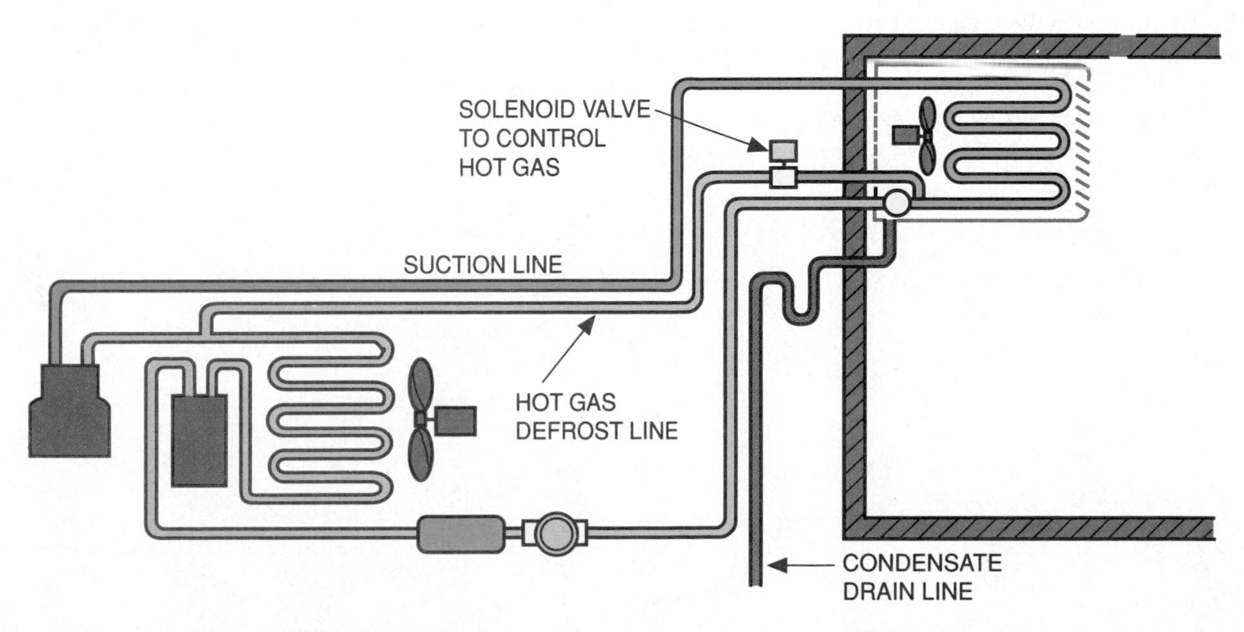

FIGURE 21–30 Using hot gas to defrost an evaporator.

being used to make ice, the makeup water for the ice maker is sometimes used for defrost.

In summation, when checking an evaporator remember that its job is to absorb heat into the refrigeration system.

SUMMARY

- Refrigeration is defined as the process of "removing heat from a place where it is not wanted and depositing it in a place where it makes little or no difference."
- Heat travels normally from a warm substance to a cool substance.
- For heat to travel from a cool substance to a warm substance, work must be performed. The motor that drives the compressor in the refrigeration cycle does this work.
- The evaporator is the component that absorbs the heat into the refrigeration system.
- The evaporator must be cooler than the medium to be cooled to have a heat exchange.
- The refrigerant boils to a vapor in the evaporator and absorbs heat because it is boiling at a low pressure and low temperature.
- The boiling temperature of the refrigerant in the evaporator determines the evaporator (low-side) pressure.
- Medium-temperature systems can use off-cycle defrost. The product is above freezing, and the heat from it can be used to cause the defrost.
- Low-temperature refrigeration must have heat added to the evaporator to melt the ice.
- Evaporators have the same characteristics for the same types of installations regardless of location.
- The first evaporator coils were bare pipe.
- Stamped plate steel evaporators were among the first attempts to extend the surface area of a bare pipe.
- Fins were later added to further extend the bare tube and give more efficiency.
- Most refrigeration coils are copper with aluminum fins.
- The starting point in organized troubleshooting is to determine if the evaporator is operating efficiently.
- Checking the superheat is the best method the service technician has for evaluating evaporator performance.
- Some evaporators are called dry-type because they use a minimum of refrigerant.
- Dry-type evaporators are also called direct expansion evaporators and maintain a constant superheat when operating correctly within their design range.
- Some evaporators are flooded and use a float to meter the refrigerant. ***Superheat checks on these evaporators should be interpreted with caution.***
- Some evaporators have a single circuit, and some have multiple circuits.
- Multicircuit evaporators keep excessive pressure drop from occurring in the evaporator.
- There is a relationship between the boiling temperature of the refrigerant in the evaporator and the temperature of the medium being cooled.
- The coil normally operates from 10°F to 20°F colder than the air passing over it.

REVIEW QUESTIONS

1. What is the function of the evaporator in the refrigeration system?
2. What happens to the refrigerant in the evaporator?
3. What is the heat called that is added to the vapor after the liquid is boiled away?
4. What determines the pressure on the low-pressure side of the system?
5. What is considered a typical superheat for a refrigeration system evaporator?
6. What does a high superheat indicate?
7. What does a low superheat indicate?
8. Why is a multicircuit evaporator used?
9. What expansion device do flooded evaporators use?
10. An evaporator that is not flooded is thought of as what type of evaporator?
11. When the evaporator experiences a load increase, what happens to the suction pressure?
12. What is used to defrost the ice from a medium-temperature evaporator?
13. What is commonly used to defrost the ice from a low-temperature evaporator?
14. A medium-temperature refrigeration box operates within what temperature range?

UNIT 22

Condensers

OBJECTIVES

After studying this unit, you should be able to

- explain the purpose of the condenser in a refrigeration system.
- describe the differences in operating characteristics between water-cooled and air-cooled systems.
- describe the basics of exchanging heat in a condenser.
- explain the difference between a tube within a tube-coil type condenser and a tube within a tube-serviceable condenser.
- describe the difference between a shell and coil condenser and a shell and tube condenser.
- describe a wastewater system.
- describe a recirculated water system.
- describe a cooling tower.
- explain the relationship between the condensing refrigerant and the condensing medium.
- compare an air-cooled, high-efficiency condenser to a standard condenser.

SAFETY CHECKLIST

* Wear goggles and gloves when attaching or removing gages to transfer refrigerant or to check pressures.
* Wear warm clothing when working in a walk-in cooler or freezer. A technician does not think properly when chilled.
* Keep hands well away from moving fans. Do not try to stop a fan blade when waiting for it to stop turning after the power has been turned off.
* Do not touch the hot gas line.

22.1 THE CONDENSER

The *condenser* is a heat exchange device similar to the evaporator that rejects the heat from the system absorbed by the evaporator. This heat is in the form of hot gas that must be cooled down to the point where it will condense. When heat was being absorbed into the system, we pointed out that it is at the point of change of state (liquid to a vapor) of the refrigerant that the greatest amount of heat is absorbed. The same thing, in reverse, is true in the condenser. The point where the change of state (vapor to a liquid) occurs is where the greatest amount of heat is rejected.

The condenser is operated at higher pressures and temperatures than the evaporator and is often located outside. The same principles apply to heat exchange in the condenser as in the evaporator. The materials a condenser is made of and the medium used to transfer heat into make a difference in the efficiency of the heat exchanger.

22.2 WATER-COOLED CONDENSERS

The first commercial refrigeration condensers were water cooled. These condensers were crude compared to modern water-cooled devices, Figure 22–1. Water-cooled condensers are efficient compared to air-cooled condensers and operate at much lower condensing temperatures. Water-cooled equipment comes in several styles. The tube within a tube, the shell and coil, and the shell and tube are the most common.

FIGURE 22–1 An early water-cooled condensing unit. *Courtesy Tecumseh Products Company*

22.3 TUBE WITHIN A TUBE CONDENSERS

The *tube within a tube* condenser comes in two styles: the coil type and the cleanable type with flanged ends, Figure 22–2.

The tube within a tube that is fabricated into a coil is manufactured by slipping one pipe inside another and sealing the ends in such a manner that the outer tube becomes one container and the inner tube becomes another container, Figure 22–3. The two pipes are then formed into a coil to save space. The heat exchange occurs between the fluid inside the outer pipe and the fluid inside the inner pipe, Figure 22–4.

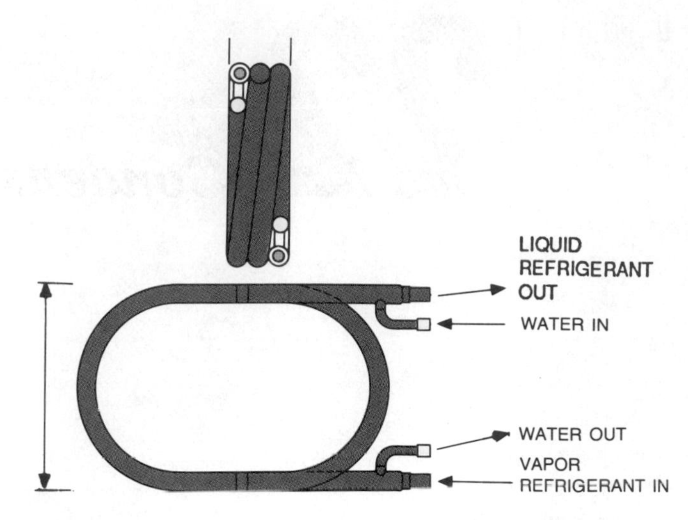

FIGURE 22–3 Tube within a tube condenser constructed by sliding one tube through the other tube. The tubes are sealed in such a manner that the inside tube is separate from the outside tube. *Courtesy Noranda Metal Industries Inc.*

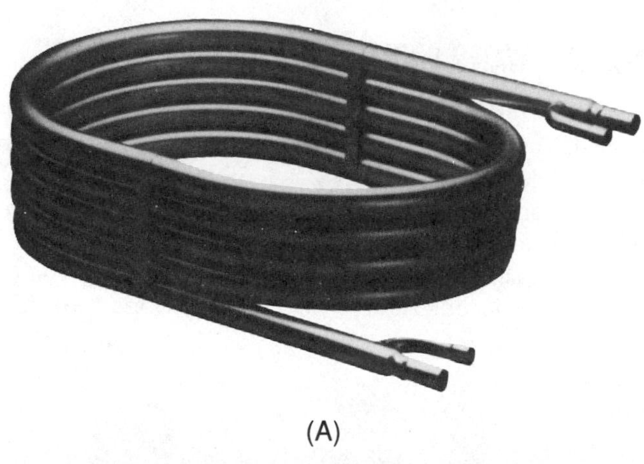

(A)

(B)

FIGURE 22–2 Two types of tube within a tube condensers. (A) A pipe within a pipe. (B) A flanged type of condenser. The flanged condenser can be cleaned by removing the flanges. Removal of the flanges opens only the water circuit not the refrigerant circuit. *(A) Courtesy Noranda Metal Industries Inc., (B) Photo by Bill Johnson*

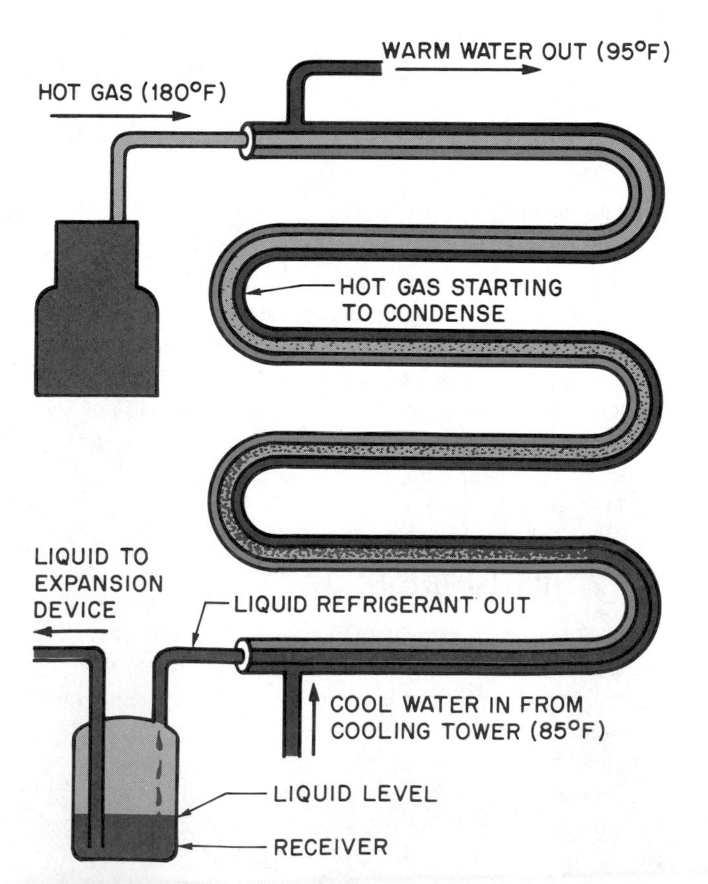

FIGURE 22–4 Fluid flow through the condenser. The refrigerant is flowing in one direction, water in the other.

22.4 MINERAL DEPOSITS

Because water flows through one of the tubes, mineral deposits and scale will form even in the best water. The heat in the vicinity of the discharge gas has a tendency to cause any minerals in the water to deposit onto the tube surface. This is a slow process, but it will happen in time to any water-type condenser. These mineral deposits act as an insulator between the tube and the water and must be kept to a minimum. Water treatment can be furnished to help prevent this buildup of mineral scale. This treatment is normally added at the tower or injected into the water by chemical feed pumps. Figure 22–5 shows an example of treatment being added to a tower. Figure 22–6 is an example of water treatment being pumped into the water piping.

In some mild cases of scale buildup more water circulation to improve the heat exchange will occur. Later in this unit, variable water-flow controls are introduced. These controls will step up the water flow on an increase in head pressure. This type of control causes more water to flow through the condenser automatically when the mineral deposits cause an increase in head pressure because of a poor heat exchange. A dirty water-cooled condenser causes high head pressure and increased energy cost. If the water is wasted instead of cooled and used again, the water bill would go up before the operator would notice there was a condenser problem.

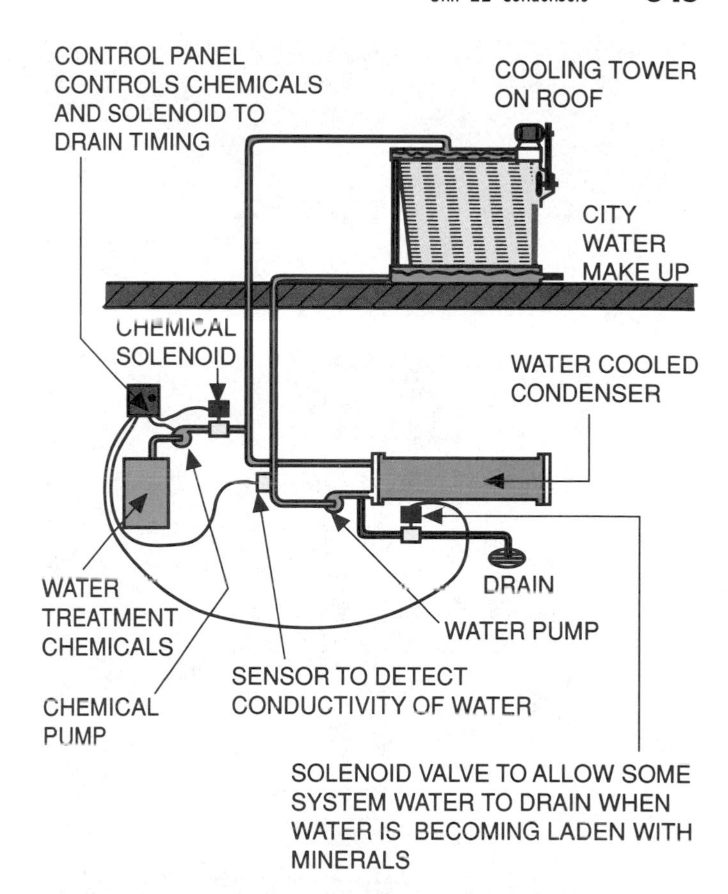

FIGURE 22–6 Automatic system of feeding the treatment chemicals to the water. It includes an automatic monitoring system that determines when the system actually needs chemicals added. This type is normally used on larger systems because of the economics of the total system cost.

FIGURE 22–5 Method of adding water treatment to a cooling tower. Treatment is being metered at a rate that will last about a month so that the operator will not have to be in attendance at the tower all of the time. There is a constant bleed of the tower water to the drain to keep the water from being overconcentrated with minerals. *Courtesy Calgon Corporation*

The tube within a tube condenser that is made into a coil cannot be cleaned mechanically with brushes. This type must be cleaned with chemicals designed not to harm the metal in the condenser. Professional help from a chemical company that specializes in water treatment is recommended when a condenser must be cleaned with chemicals. Condensers of this type are normally made from copper or steel; some special condensers are made of stainless steel or copper and nickel.

22.5 CLEANABLE CONDENSERS

The tube within a tube condenser that is fabricated with flanges on the end can be mechanically cleaned. The flanges can be removed, and the tubes can be examined and brushed with an approved brush, Figure 22–7. The flanges and gaskets on this type of condenser are in the water circuit with the refrigerant flowing around the tubes. The refrigerant circuit is not opened to clean the tubes. Figure 22–8 shows how tubes are cleaned. Consult the manufacturer for the correct brush. Fiber is usually preferable. This is a more expensive type of condenser, but it is serviceable.

FIGURE 22–7 Condenser is flanged for service. When the flanges are removed, the refrigerant circuit is not disturbed. *Photo by Bill Johnson*

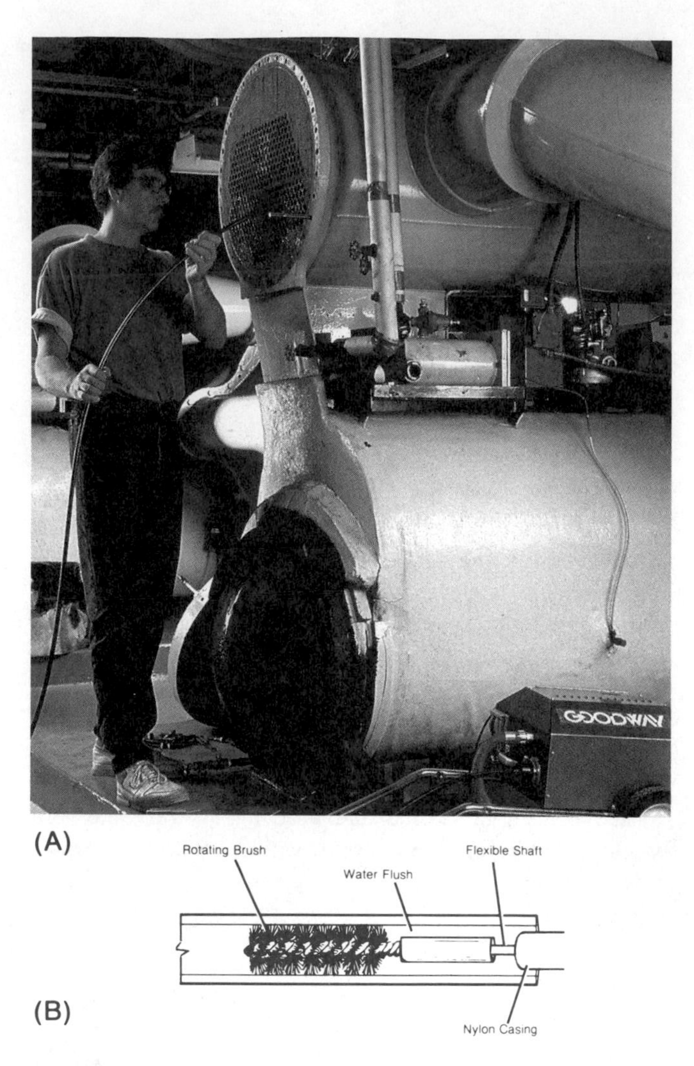

FIGURE 22–8 Brushes actually being pushed through the water side of the condenser. Use only approved brushes. *Courtesy Goodway Tools Corporation*

22.6 SHELL AND COIL CONDENSERS

The shell and coil condenser is similar to the tube within a tube coil. It is a coil of tubing packed into a shell that is then closed and welded. Normally the refrigerant gas is discharged into the shell, and the water is circulated in the tube located in the shell. The shell of the condenser serves as a receiver storage tank for the extra refrigerant in the system. This condenser is not mechanically serviceable because the coil is not straight, Figure 22–9. It must be cleaned chemically.

22.7 SHELL AND TUBE CONDENSERS

Shell and tube condensers are more expensive than shell and coil condensers, but they can be cleaned mechanically with brushes. They are constructed with the tubes fastened into an end sheet in the shell. The refrigerant is discharged into the shell, and the water is circulated through the tubes. The ends of the shell are like end caps (known as water boxes)

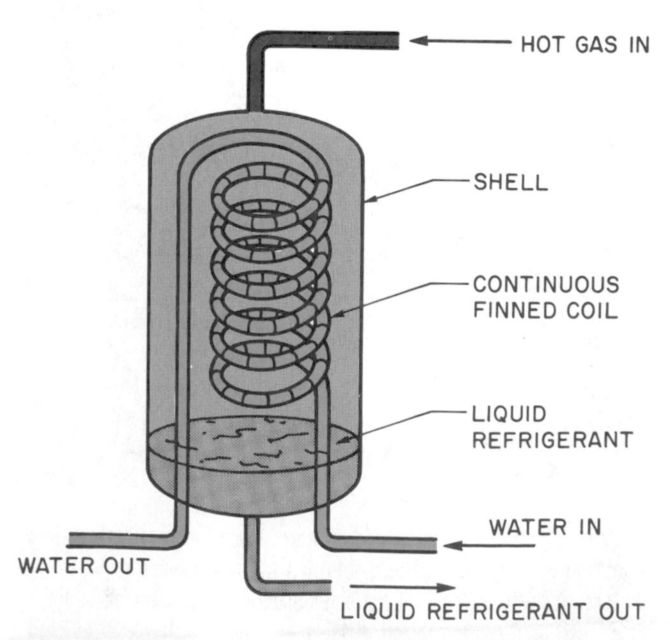

FIGURE 22–9 Shell and coil condenser. The hot refrigerant gas is piped into the shell, and the water is contained inside the tubes.

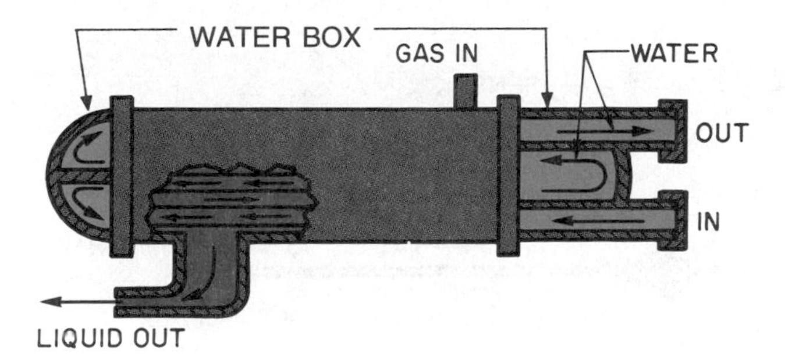

FIGURE 22–10 Water can be circulated back and forth through the condenser by using the end caps to give the water the proper direction.

with the water circulating in them, Figure 22–10. These end caps can be removed, so the tubes can be inspected and brushed out if needed. The shell acts as a receiver storage tank for extra refrigerant. This is the most expensive condenser and is normally used in larger applications.

The water-cooled condenser is used to remove the heat from the refrigerant. When it is removed, the heat is in the water. Two things can be done at this point: (1) waste the water, or (2) pump the water to a remote place, remove the heat, and reuse the water.

22.8 WASTEWATER SYSTEMS

Wastewater systems are just what the name implies. The water is used once, then wasted down the drain, Figure 22–11. This is worthwhile if water is free or if only a small amount is used. Where large amounts of water are used, it is probably more economical to save the water, cool it in an outside water tower, and reuse it.

The water supplied to systems that use the water only once and waste it has a broad temperature range. For instance, the water in summer may be 75°F out of the city mains and as low as 40°F in the winter, Figure 22–12. Water that runs through a building with long pipe runs may have warm water in the beginning from standing in the pipes; however, the main water temperature may be quite low. This change in water temperature has an effect on the refrigerant head pressure (condensing temperature). The head pressure must be about 100 psig higher than the suction pressure for the expansion device to function properly. A typical wastewater system with 75°F entering water uses 1.5 gal/min per ton of refrigeration.

22.9 REFRIGERANT-TO-WATER TEMPERATURE RELATIONSHIP

There is a relationship between the condensing refrigerant temperature and the temperature of the water used to condense the refrigerant. Normally the condensing temperature is about 10°F higher than the leaving water. For example, in summer the water from the mains may be 80°F. This would step up the water flow to the point that the leaving water may be 90°F. This 90°F can be used as a reference point to find out what the head pressure should be (90°F plus 10°F = 100°F); this would be the approximate condensing temperature. The head pressure for R-12 corresponding to this 100°F condensing temperature is 117 psig. In winter when the water temperature in the main is down to 40°F, the leaving water will still be about 90°F due to the automatic water valve that throttles the water to maintain a constant head pressure, Figure 22–12. Because the entering water is much cooler, less water is needed to condense the refrigerant.

In a wastewater system, the water flow can be varied to suit the need by means of a regulating valve for the water. This valve has a pressure tap that fastens the bellows in the

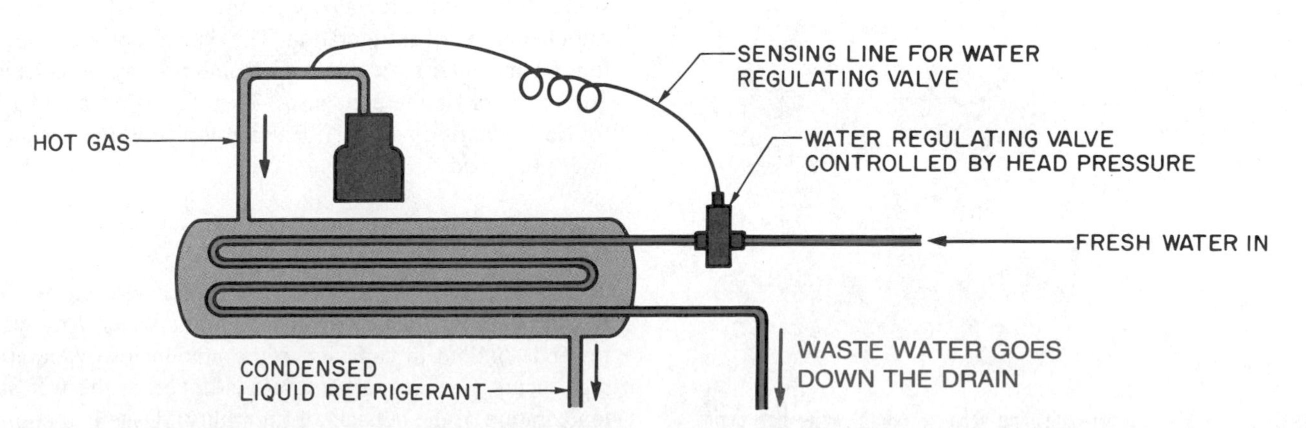

FIGURE 22–11 Wastewater system used when water is plentiful at a low cost, such as from a well or lake.

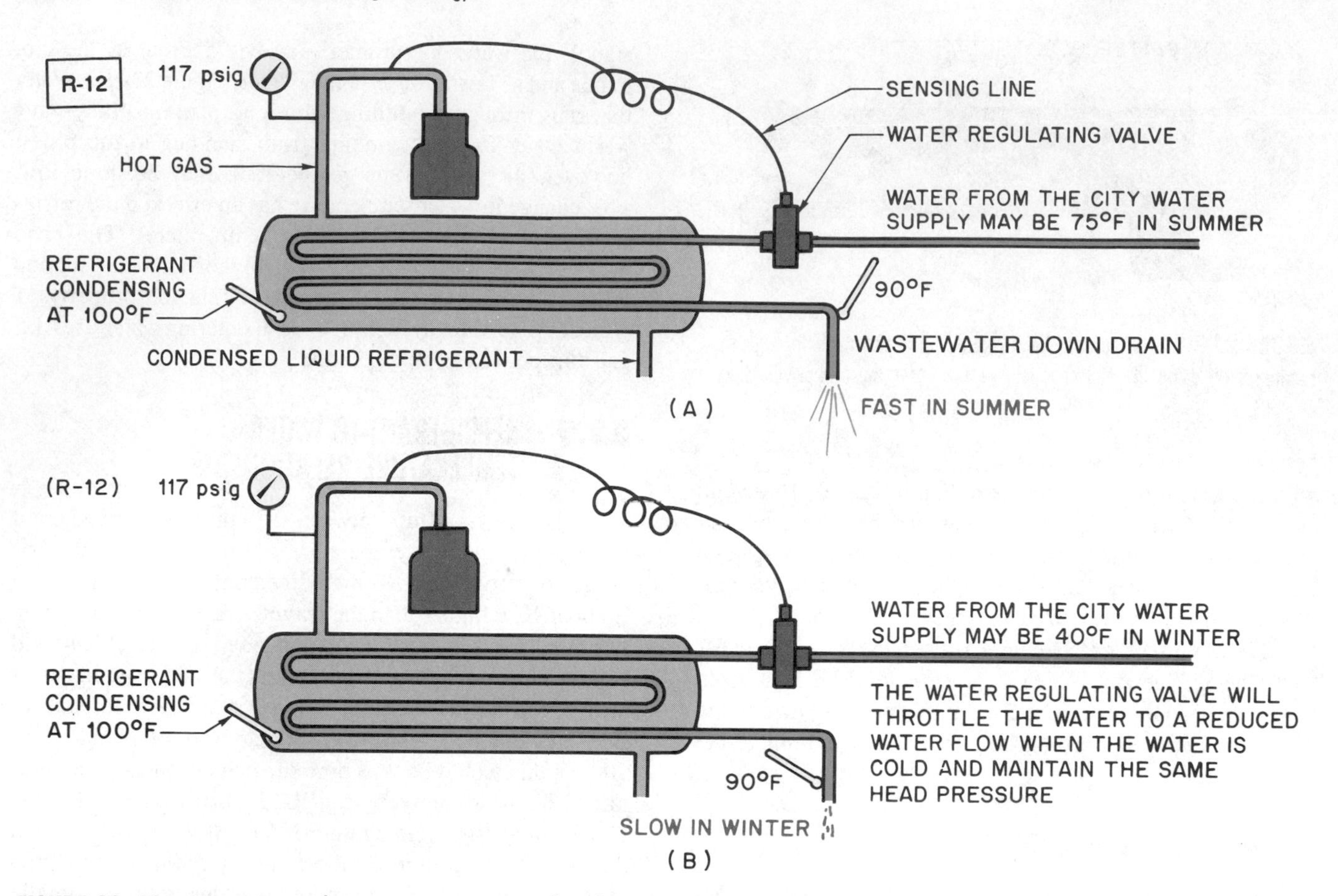

FIGURE 22–12 Wastewater condenser system at two sets of conditions. (A) Summer with warm water entering the system. (B) Winter when the water in the city mains is colder.

control to the high-pressure side of the system. When the head pressure goes up, the valve opens and allows more water to flow through the condenser to keep the head pressure correct, Figure 22–13.

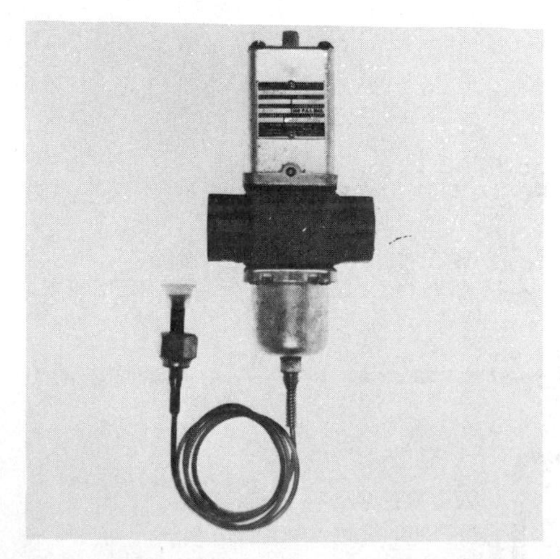

FIGURE 22–13 Water-regulating valve to control water flow during different demand periods. *Photo by Bill Johnson*

22.10 RECIRCULATED WATER SYSTEMS

When the system gets large enough that saving water is a concern, then a system that will recirculate the water is considered. This system uses the condenser to absorb the heat into the water just as in the wastewater system. The water is then pumped to an area away from the condenser where the heat is removed from the water, Figure 22–14. There is a known temperature relationship between the water and the refrigerant. The refrigerant will normally condense at a temperature of about 10°F higher than the leaving-water temperature, Figure 22–15. A recirculating water system will circulate approximately 3 gal of water per minute per ton of refrigeration. The typical design temperature for the entering water is 85°F and most systems have a 10°F rise across the condenser. Therefore, you could look for 85°F entering and 95°F leaving at a time the system is under full load.

22.11 COOLING TOWERS

The *cooling tower* is a device that passes outside air over the water to remove the system heat from the water. Any water tower is limited in capacity to the amount of evaporation that occurs. The evaporation rate is linked to the wet-bulb temperature of the outside air (humidity). Usually a cooling tower can cool the water that returns back to the condenser

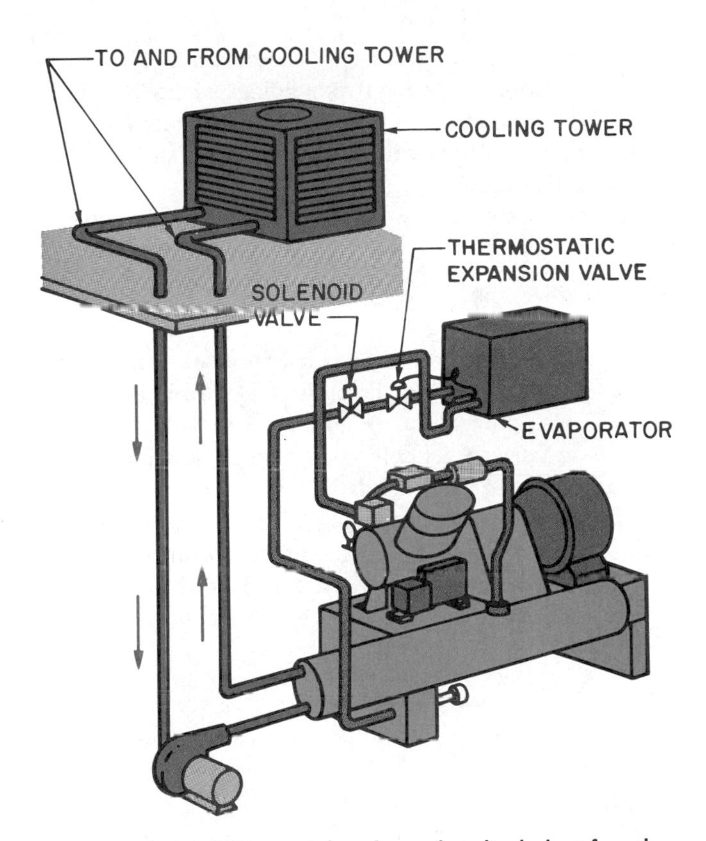

FIGURE 22–14 Water-cooled condenser that absorbs heat from the refrigerant and pumps the water to a cooling tower at a remote location. The condenser is located close to the compressor, and the tower is on the roof outside the structure.

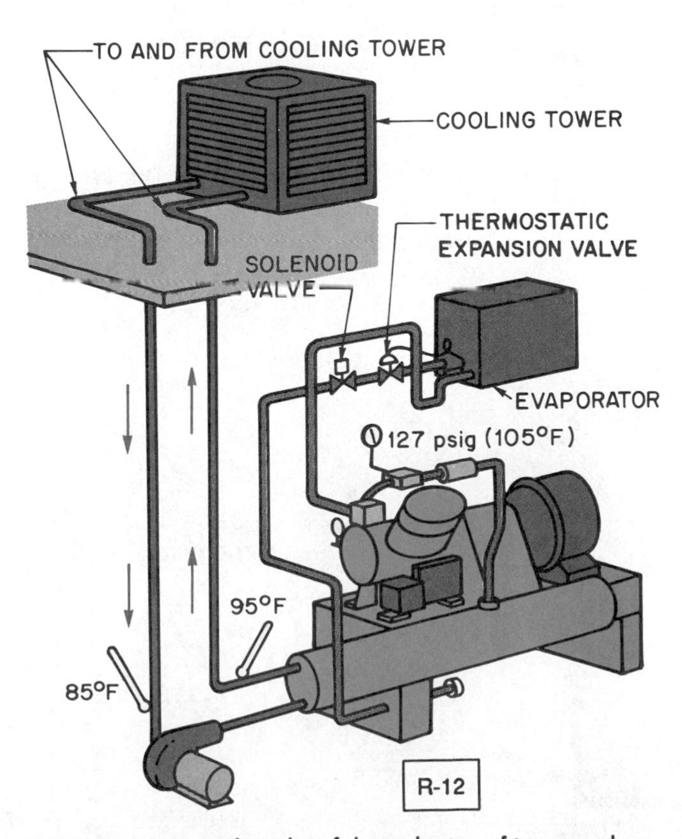

FIGURE 22–15 Relationship of the condensing refrigerant to the leaving-water temperature.

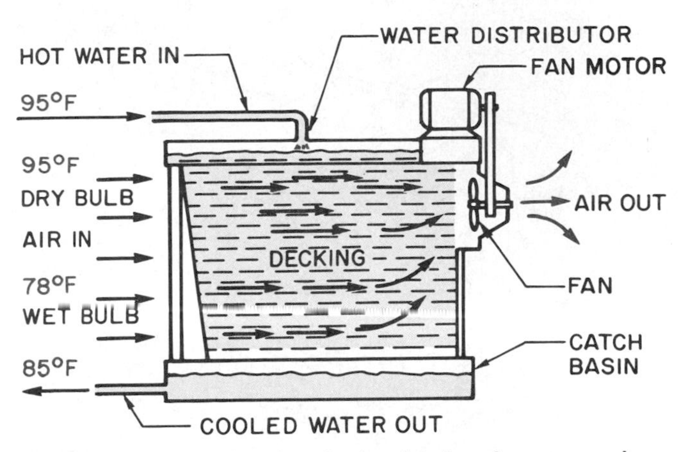

FIGURE 22–16 Relationship of a forced-draft cooling tower to the ambient air. Cooling tower performance depends on the wet-bulb temperature of the air. This relates to the humidity and the ability of the air to absorb moisture.

to within 7°F of the wet-bulb temperature of the outside air, Figure 22–16. This tower arrangement comes in sizes of about 2 tons of refrigeration and up. Towers can either be (1) natural draft, (2) forced draft, or (3) evaporative.

22.12 NATURAL-DRAFT TOWERS

The *natural draft tower* does not have a blower to move air through the tower. It is customarily made of some material that the weather will not deteriorate, such as redwood, fiberglass, or galvanized sheet metal.

Because natural-draft towers rely on the natural prevailing breezes to blow through them, they need to be located in the prevailing wind. The water is sprayed into the top of the tower through spray heads, and some of the water evaporates as it falls to the bottom of the tower where the water is collected in a basin. This evaporation takes heat from the remaining water and adds to the capacity of the tower. Water must be made up for the evaporated water. A makeup system using a float assembly connected to the water supply makes up for evaporated water automatically by adding fresh water, Figure 22–17.

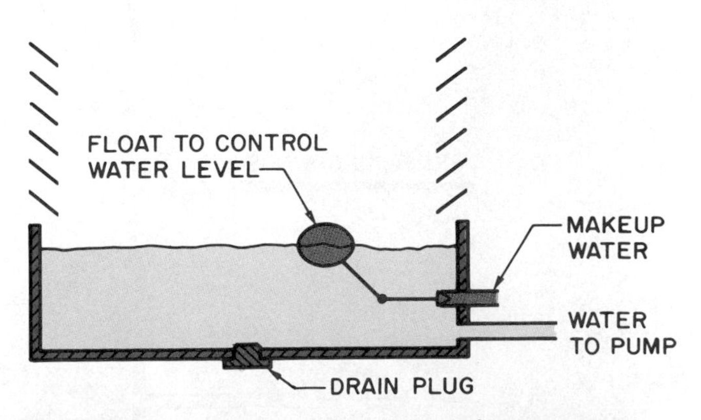

FIGURE 22–17 Makeup water system in a cooling tower. Because the cooling tower performance depends partly on evaporation of water from the tower, this makeup is necessary.

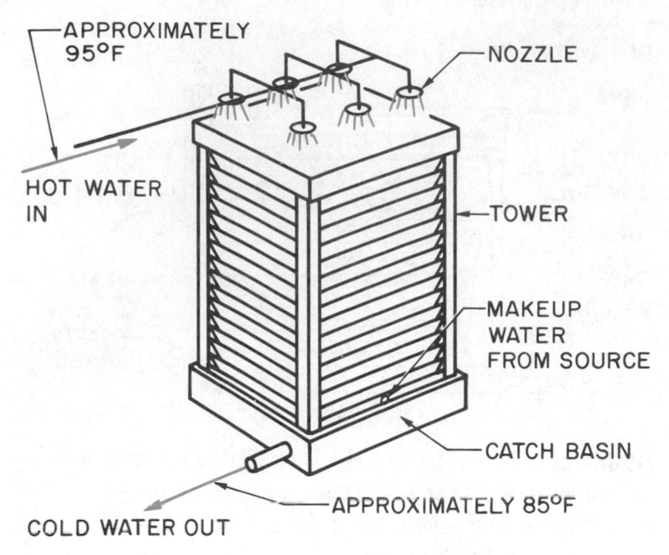

FIGURE 22–18 Natural-draft cooling tower. (Must be located in prevailing winds.)

The tower location must be chosen carefully. If it is located in a corner between two buildings where the breeze cannot blow through it, higher than normal water temperatures will occur, which will cause higher than normal head pressures, Figure 22–18.

These towers have two weather-related conditions that must be considered.

(1) The tower must be operated in the winter on refrigeration systems, and the water will freeze in some climates if freeze protection is not provided. Heat can be added to the water in the basin of the tower; antifreeze will also prevent this, Figure 22–19.

(2) The water can get cold enough to cause a head pressure drop. A water-regulating valve can be installed to prevent this from happening. Natural-draft towers can be seen on top of buildings as structures that look like they are made of slats. These slats help keep the water from blowing out of the tower, Figure 22–20.

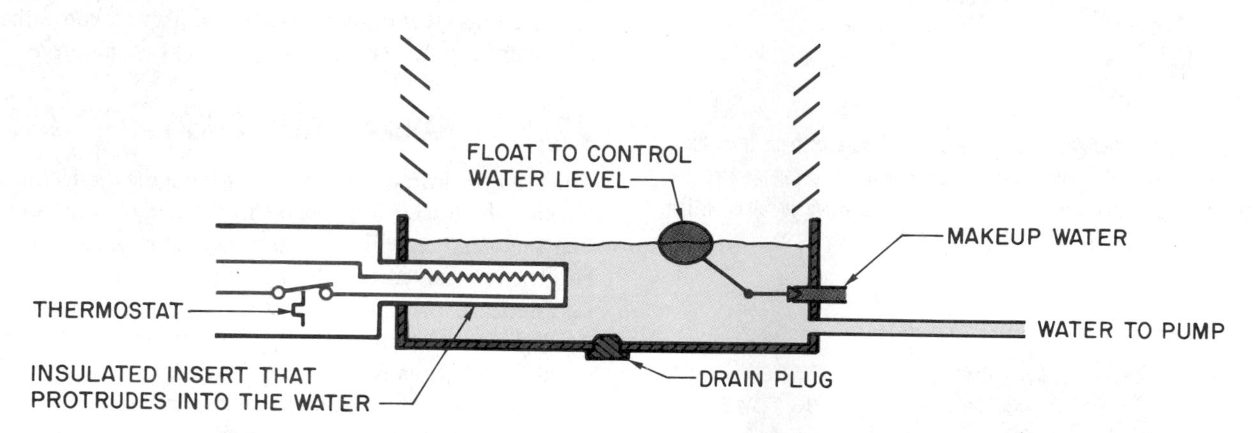

FIGURE 22–19 Type of heat that may be applied to keep the water in the basin from freezing in winter. This heat can be controlled thermostatically to prevent it from being left on when not needed.

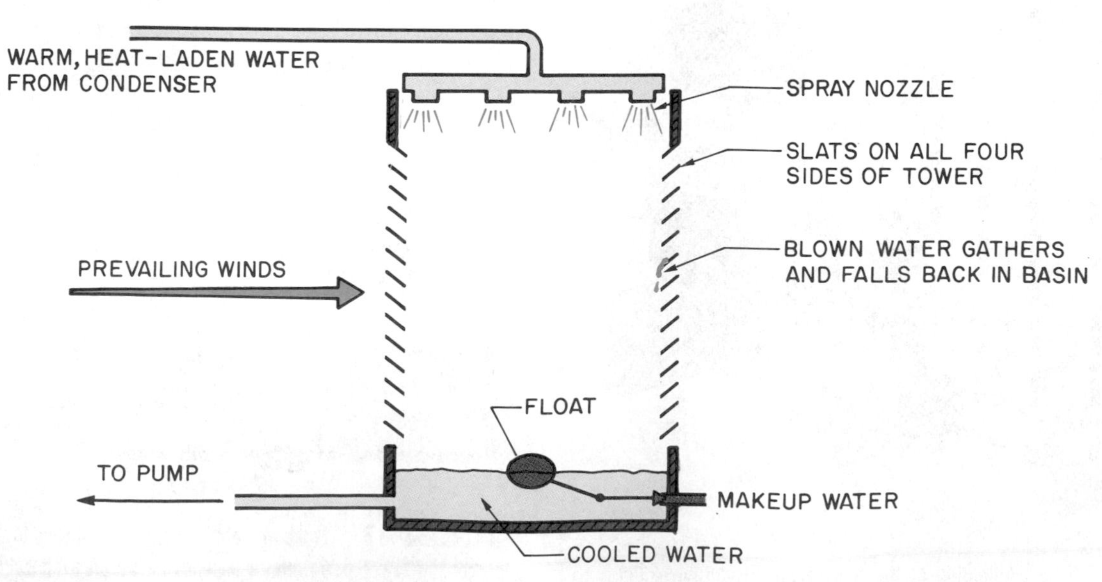

FIGURE 22–20 Slats on the sides of the natural-draft cooling tower keep the water inside the tower when the wind is blowing.

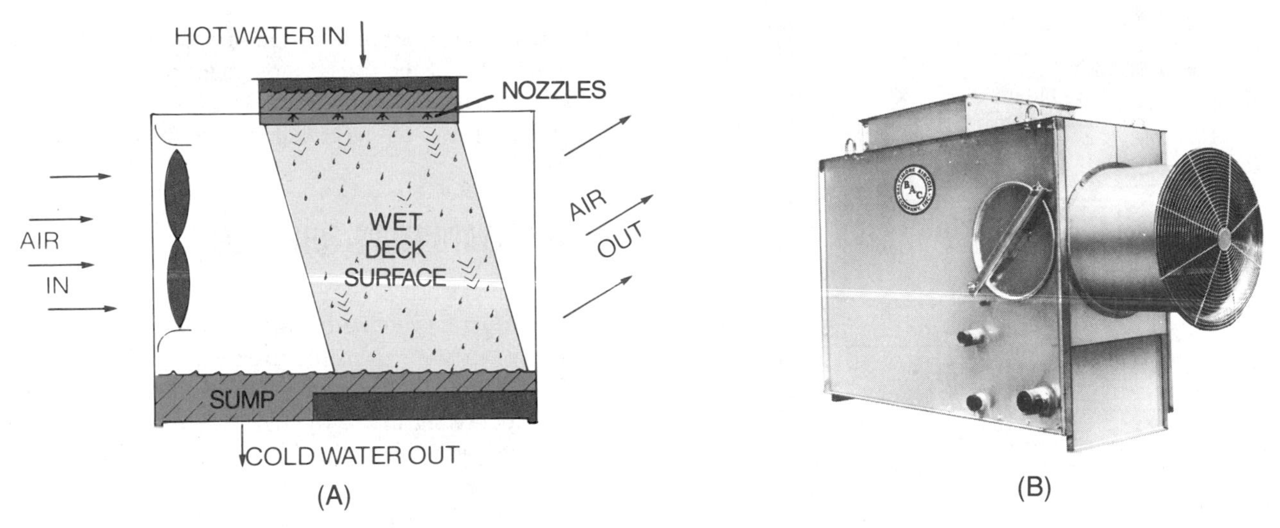

FIGURE 22–21 (A) Forced-draft tower. (B) Induced-draft tower. *Courtesy of Baltimore Aircoil Company Inc.*

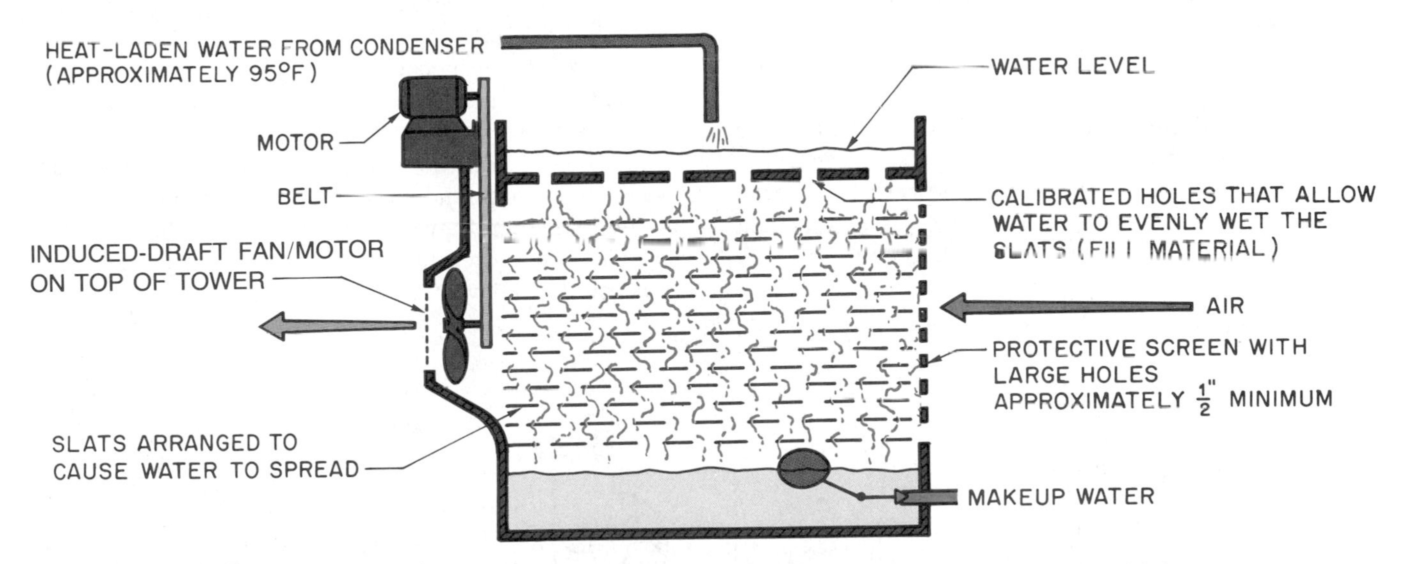

FIGURE 22–22 Calibrated holes at the top of an induced-draft tower. The holes distribute the water over the fill material below.

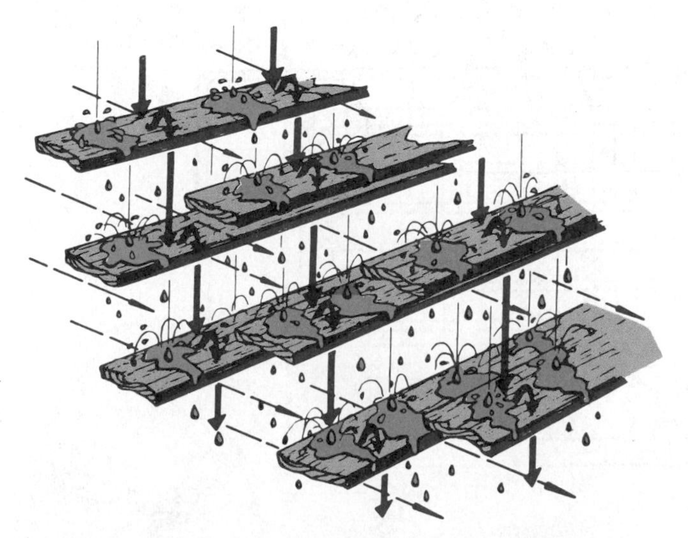

FIGURE 22–23 The water trickles down through the fill material. *Courtesy Marley Cooling Tower Company*

22.13 FORCED- OR INDUCED-DRAFT TOWERS

Forced- or *induced-draft* cooling towers differ from natural-draft towers because they have a fan to move air over a wetted surface, Figure 22–21. They are customarily designed with the warm water from the condenser pumped into a flat basin at the top of the tower. This basin has calibrated holes drilled in it to allow a measured amount of water to pass downward through the fill material, Figure 22–22. The fill material is usually redwood or manmade fiber and gives the water surface area for the fan to blow air over to evaporate and cool the water, Figure 22–23. As the water is evaporated, it is replaced with a water makeup system using a float, similar to the natural-draft tower.

Forced- or induced-draft towers can be located almost anywhere because the fan can move the air. They can even be located inside buildings, where the air is brought in and out through ducts, Figure 22–24. The tower is fairly en-

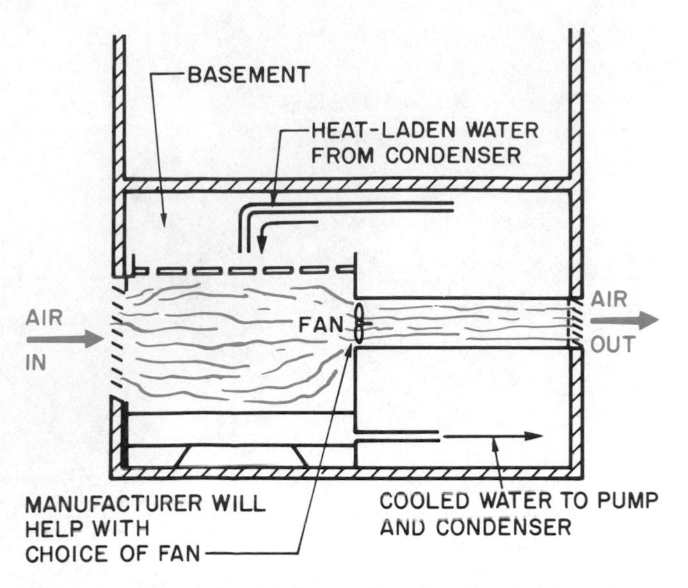

FIGURE 22–24 Induced-draft tower located inside a building with air ducted to the outside.

closed to the prevailing winds, so no water-regulating valve is normally necessary. The fan can be cycled off and on to control the water temperature and thus control the head pressure. The mass of the water in the tower provides a long cycle between the time the fan starts and stops. Forced-draft towers are small compared to natural-draft towers. They are versatile because of the forced movement of air.

NOTE: An induced-draft tower is similar to a forced-draft tower, but the air is pulled, not pushed, across the wetted surface.

22.14 EVAPORATIVE CONDENSERS

Evaporative condensers are a different type altogether because the refrigerant condenser is actually located inside the tower. These types are often confused with cooling towers, Figure 22–25. In the water towers discussed previously, the condenser containing the refrigerant was remote from the tower, and the water was piped through the condenser to the tower. The evaporative condenser uses the same water over and over with a pump located at the tower. As the water is evaporated, it is replaced with a makeup system using a float, as with the other towers. When the evaporative condenser is used in cold climates, freeze protection must be provided in winter.

As water is evaporated from any cooling tower system, the minerals in the water will become more concentrated in the remaining water. If these minerals are allowed to become overconcentrated, they will begin to deposit on the condenser surface and cause head pressure problems. To prevent this from happening, water must be allowed to escape the system on a continuous basis. This escaping water is called "blow down." It is actually water that is allowed to

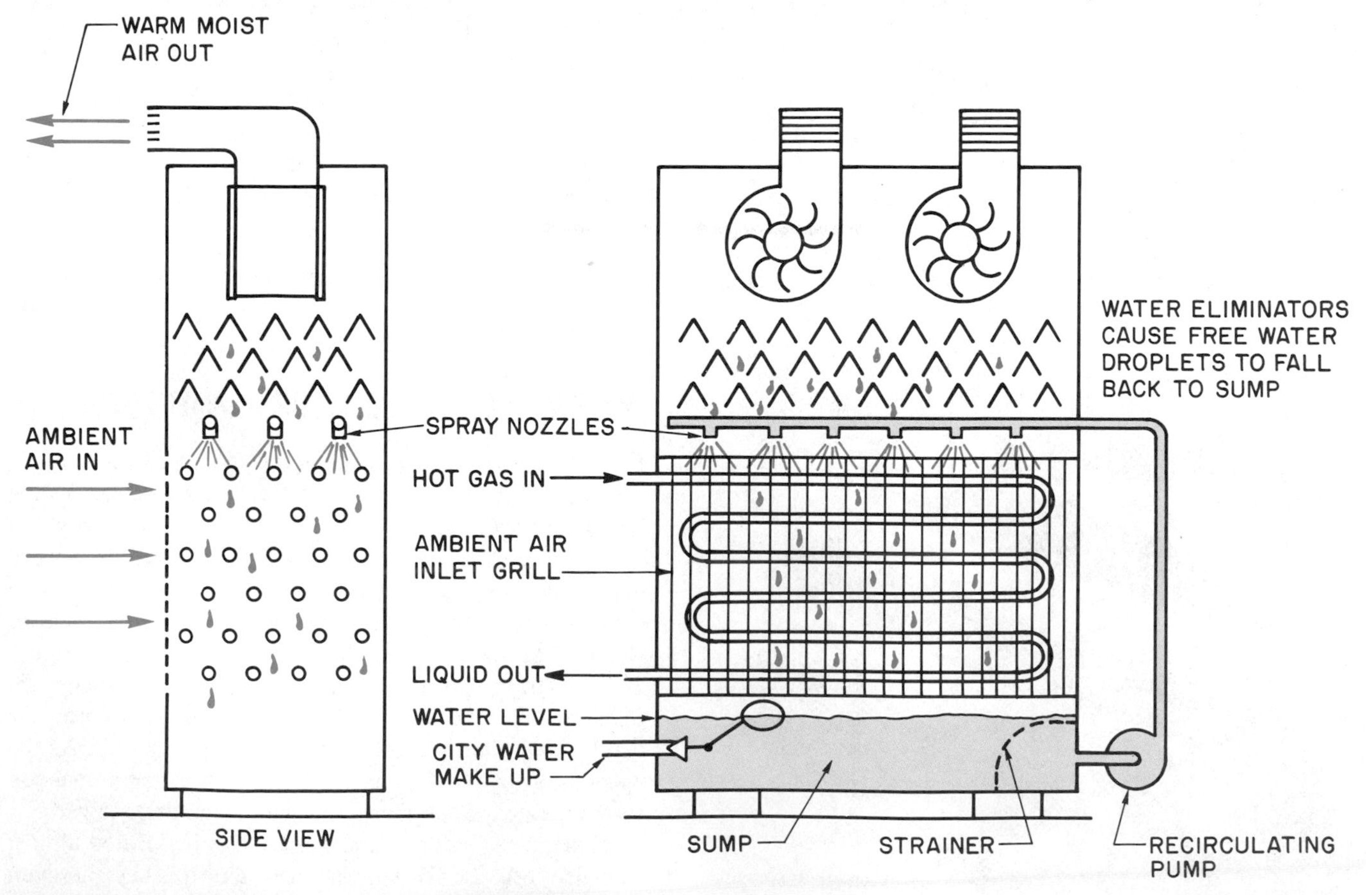

FIGURE 22–25 Water recirculates in the evaporative condenser. The condenser tubes are in the tower, rather than in a condenser shell located in a building.

escape down the drain to be made up with fresh water and the float system. It is not unusual for people who do not know the purpose of "blow down" to shut off the drain line of what appears to be clean water wasted down the drain. Problems will be the result. Technicians must establish a regular procedure for checking the rate of "blow down."

22.15 AIR-COOLED CONDENSERS

Air-cooled condensers use air as the medium to reject the heat into. This can be advantageous where it is difficult to use water. The first air-cooled condensers were bare pipe with air from the compressor flywheel blowing over the condenser. The compressors were open drive at this time. To improve the efficiency of the condenser and to make it smaller, the surface area was then extended with fins. The condensers at this time were normally steel with steel fins, Figure 22–26. These condensers resembled radiators and were sometimes referred to as radiators.

Steel air-cooled condensers are still used in many small installations on refrigeration units. Figure 22–27 is a photo of a condenser for a large refrigeration system.

Air-cooled condensers come in a variety of styles. In some the air blows horizontally through them and are subject to prevailing winds, Figure 22–28. In some air-cooled condensers the airflow pattern is vertical. They take air into the bottom and discharge it out the top. The prevailing winds do not affect these condensers to any extent, Figure 22–29. Another style of air-cooled condenser takes the air in the sides and discharges it out the top. This condenser can be affected by prevailing winds, Figure 22–30.

The smaller refrigeration systems are often located in the conditioned space, such as a restaurant or store. These air-cooled condensers normally have widely spaced steel fins on a steel coil, which allow more time before the coil will stop up with dust and other airborne material.

The hot gas normally enters the air-cooled condenser at the top. The beginning tubes of the condenser will be re-

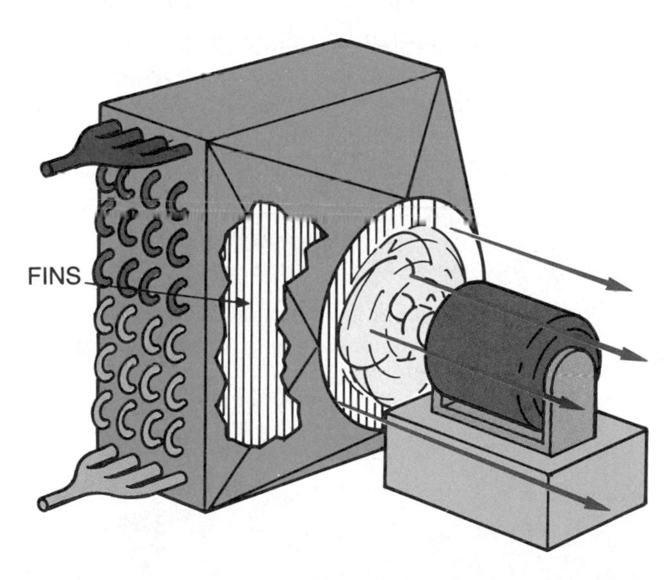

FIGURE 22–26 Fins designed to give the coil more surface area.

FIGURE 22–28 Horizontal air-cooled condenser subject to prevailing winds blowing through it. *Courtesy Copeland Corporation*

FIGURE 22–27 Larger refrigeration air-cooled condenser. It resembles an air conditioning condenser. *Courtesy Heatcraft Inc., Refrigeration Products Division*

FIGURE 22–29 Condenser with vertical airflow pattern. Air enters the bottom and is blown out the top. It is unaffected by prevailing winds. *Courtesy Heatcraft Inc., Refrigeration Products Division*

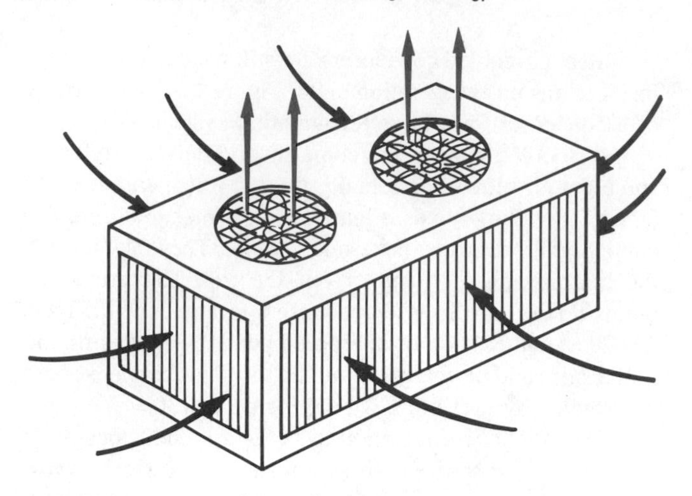

FIGURE 22–30 Condenser takes the air in from the sides and discharges it out the top. The prevailing winds could affect it.

ceiving the hot gas straight from the compressor. This gas will be highly superheated. (Remember that superheat is heat that is added to the refrigerant after the change of state in the evaporator.) When the superheated refrigerant from the evaporator reaches the compressor and is compressed, more heat is added to the gas. Part of the energy applied to the compressor transfers into the refrigerant in the form of heat energy. This additional heat added by the compressor causes the refrigerant leaving the compressor to be heavily heat laden. On a hot day (95°F) the hot gas leaving the compressor could easily reach 210°F. ***DO NOT TOUCH the hot gas discharge line from the compressor. It will burn your fingers.*** The condenser must remove this heat down to the condensing temperature before any condensation can occur.

Air-cooled condensers have a relationship to the temperature of the air passing over them as the evaporator did in the previous unit. For instance, the refrigerant inside the coil will normally condense at a 30°F higher temperature than the air passing over it (also known as the ambient air). This statement is true for most standard-efficiency condensers that have been in service long enough to have a typical dirt deposit on the fins and tubing. The relationship can be improved by adding condenser surface. With an outside air temperature of 95°F, the condensing temperature will be about 125°F. With the refrigerant R-12 condensing at 125°F, the head pressure or high-pressure gage should read 169 psig. (See the pressure/temperature chart for R-12.) This is important because it helps the service technician establish what the head pressure should be.

See Figure 22–31 for an illustration of the following description of an air-cooled, R-12 condenser located outside a store. It is responsible for rejecting the heat absorbed inside a medium-temperature walk-in cooler. This cooler has reach-in doors typical of those in a convenience store. In this box the beverages are on shelves, and store personnel can stock the shelves from the walk-in portion behind the shelves. The cooler is maintained at 35°F. The outside air temperature is 95°F. The refrigeration system must absorb heat at 35°F and reject the same heat to the outside where the condensing medium is 95°F.

1. The hot gas is entering the condenser at 200°F. The condensing temperature is going to be 30°F warmer than the outside (ambient) air.

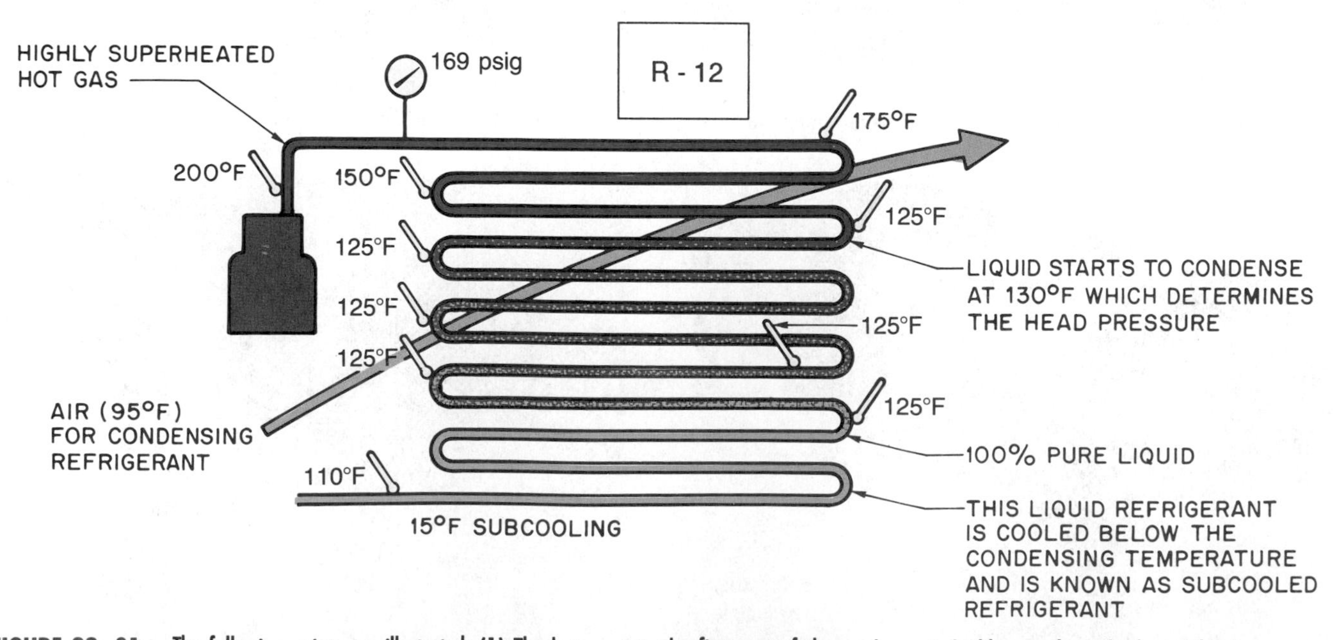

FIGURE 22–31 The following points are illustrated: (1) The hot gas into the first part of the condenser is highly superheated. The condensing temperature is 125°F, and the discharge gas must be cooled from 200° to 125°F before any condensing will occur. (2) The best heat exchange is between the liquid and the air on the outside of the coil. More heat is removed during the change of state than while de-superheating the vapor. (3) When the liquid is all condensed and pure liquid is in the coil, the liquid can be subcooled to below the condensing temperature.

2. The outside air temperature is 95°F. The condensing temperature is 95°F plus 30°F = 125°F. The refrigerant must be cooled to 125°F before any actual condensing occurs. Thus the condenser has to lower the hot-gas temperature 75°F in the first part of the coil. This is called *de-superheating*. It is the first job of the condenser.
3. Part way through the coil the superheat is removed down to the actual condensing temperature of 125°F, and liquid begins to form in the coil.
4. When the refrigerant gets to the end of the coil, the condenser tubes will be full of liquid and then drain into the receiver. If the condenser is long enough, the liquid may even cool below the condensing temperature of 125°F. This is called *subcooling*.
5. The liquid in the bottom of the condenser draining into the receiver may cool to 110°F.

All air-cooled condensers don't have the same 30°F relationship with the ambient air. It is good practice to determine the temperature relationship between the air and refrigerant at startup time and record it. Then if the relationship changes, you would suspect trouble, such as a dirty condenser or an overcharge of refrigerant that has been added without your knowledge.

22.16 HIGH-EFFICIENCY CONDENSERS

Condensing temperatures can be reduced by increasing the condenser surface area. The larger the condenser surface area, the closer the condensing temperature is to the ambient temperature. A perfect condenser would condense the refrigerant at the same temperature as the ambient, but would be so large that it would not be practical. In the previous examples, we have used a temperature difference of 30°F between the ambient and the condensing temperature. Figure 22–32 shows an example of a condenser condensing at 110°F, 15°F above the ambient temperature. The head pressure for R-12 condensing at 110°F would be 136.4 psig. (R-502 would be 247.9 psig.) This reduction of head pressure provides for a more efficient system—less power is consumed for the same amount of usable refrigeration. Condensing temperatures to within 10°F of the ambient have been used on some installations, particularly in extra–low-temperature systems.

Compressors are affected by a decrease in head pressure and will be discussed in the next unit.

22.17 THE CONDENSER AND LOW AMBIENT CONDITIONS

The foregoing example describes how an air-cooled condenser operates on a hot day. An example of a condenser operating under different conditions might be in a supermarket with a small package-display case located inside the store. Because this is a package-display case, the condenser is also located inside the store, Figure 22–33. The inlet air to the case is in the store itself and may be quite cool if the air conditioner is operating. The store temperature may be 70°F or even cooler at times. This reduces the operating pressure on the high-pressure side of the system. When the condenser relationship temperature rule is applied, we see that the new condensing temperature would be 70°F plus 30°F = 100°F. The head pressure would be 117 psig. This may be enough to effect the performance of the expansion device. This will be covered in more detail later, but for now

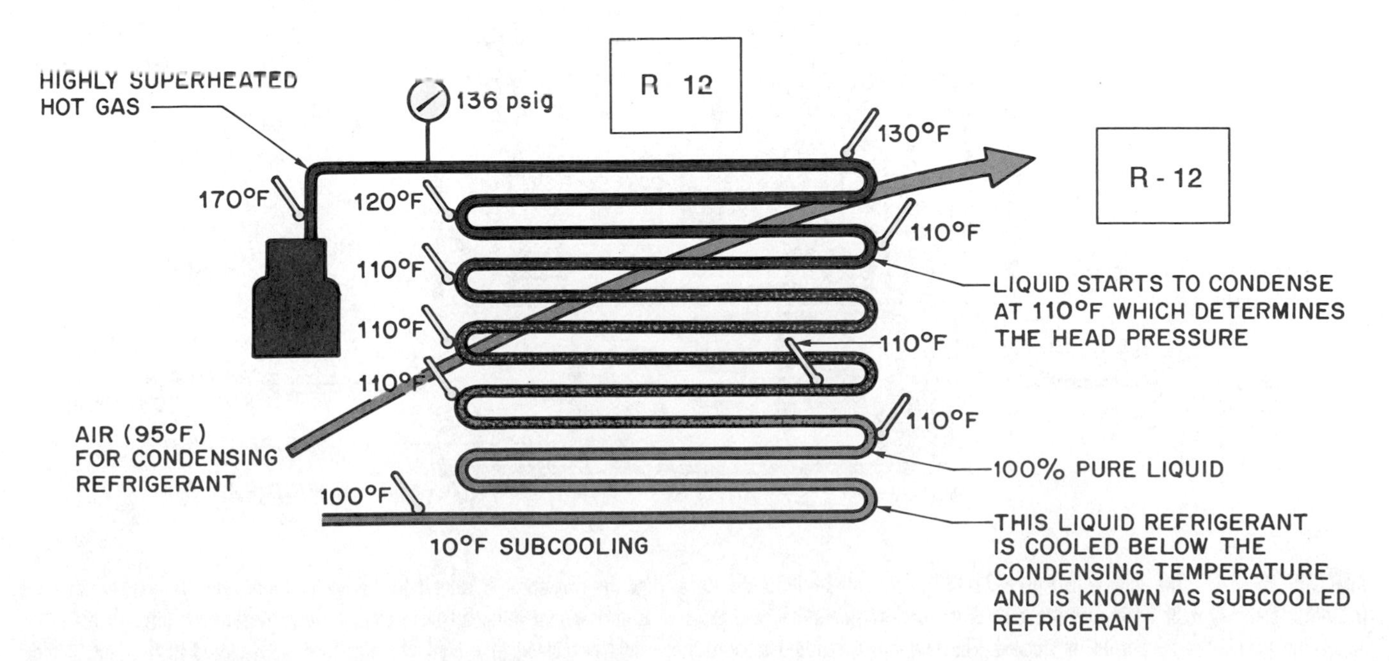

FIGURE 22–32 Operating conditions for a high-efficiency condenser.

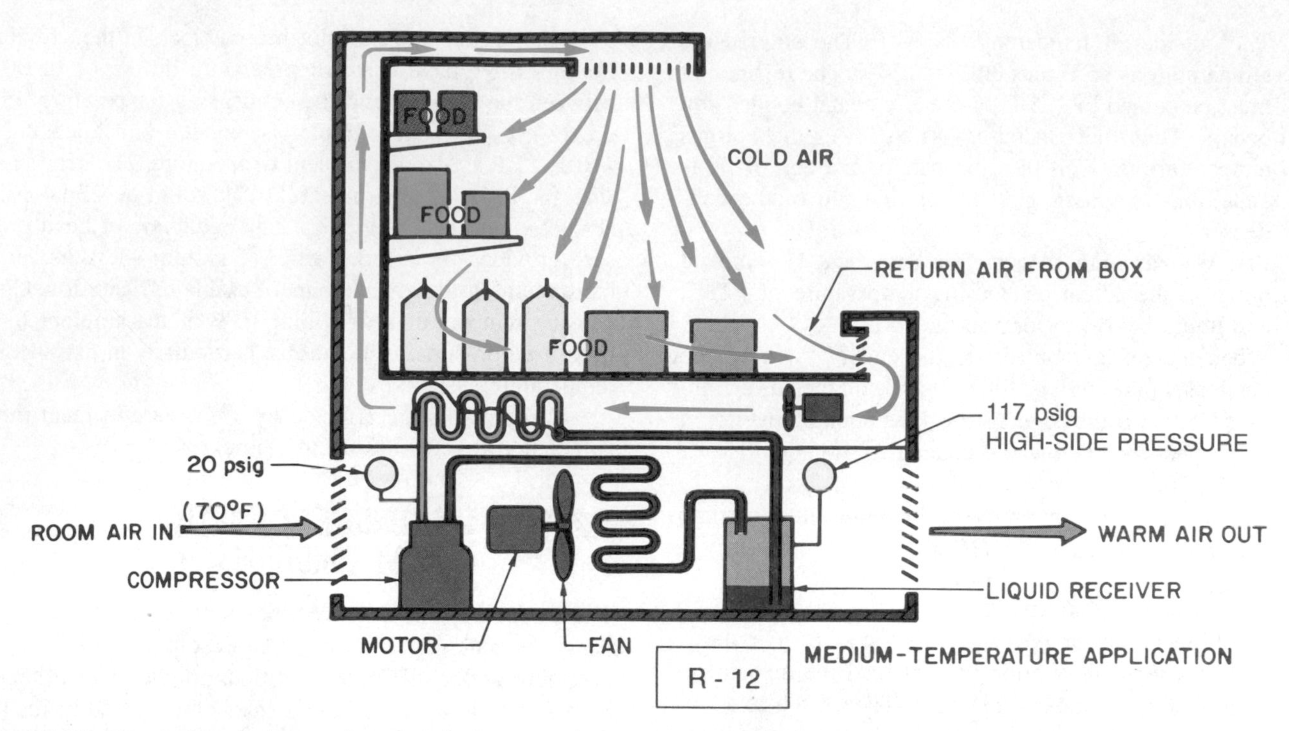

FIGURE 22–33 Package-display case located inside the store with the compressor and the condenser located inside the cabinet. This is a plug-in device; no piping is required.

it is necessary to know that the head pressure must be about 75 to 100 psig higher than the suction or low-side pressure for the expansion device to operate correctly. The low-side pressure for a medium-temperature fixture will be in the neighborhood of 20 psig at the lowest point using R-12. When the relationship of the low-side pressure of 20 psig is subtracted from the high-side pressure of 117 psig for a condenser with 70°F air passing over it, the difference is 97 psig. This is close to the 75- to 100-psig minimum difference. Should this fixture be exposed to temperatures of

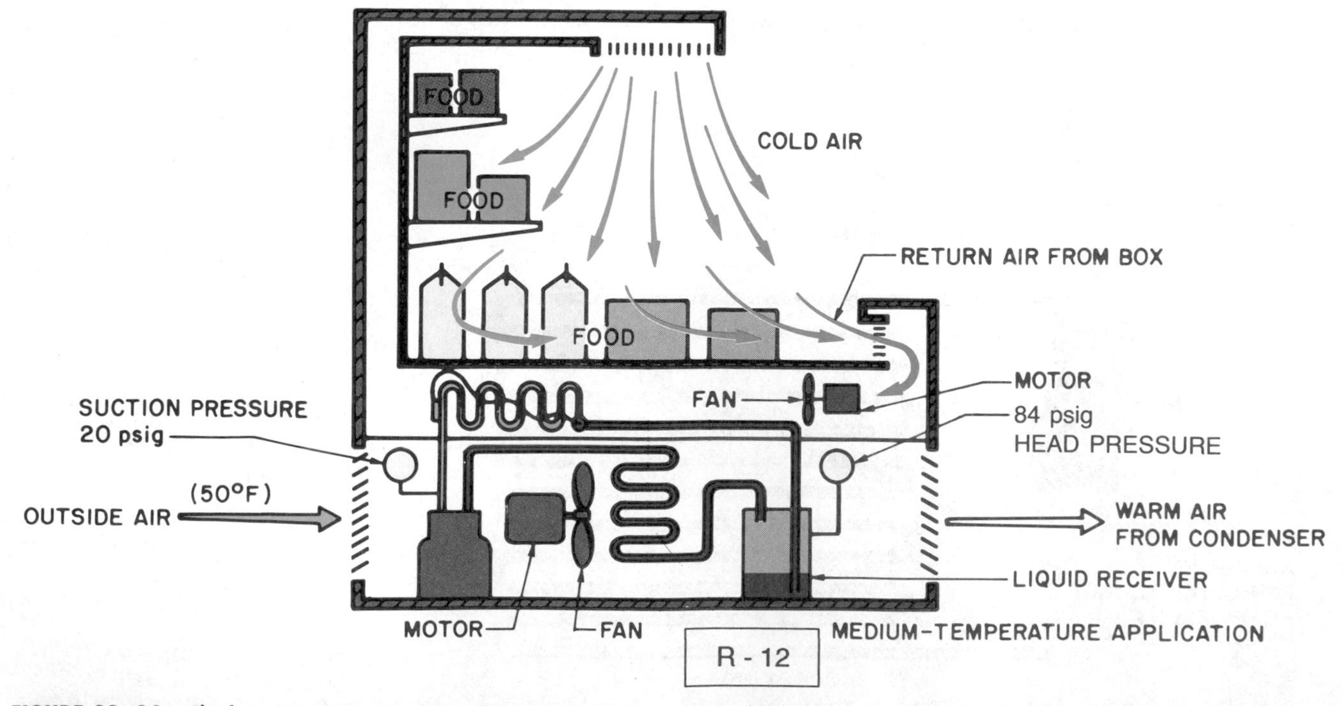

FIGURE 22–34 The fixture in Figure 22–33 is relocated outside the store. Now the condenser is subject to the winter conditions. The performance will fall off if some type of head pressure control is not furnished. The head pressure is so low that it cannot push enough liquid refrigerant through the expansion valve. The evaporator is starved. The unit has a reduced capacity that would be evident if a load of warm food were placed in the box. It may not pull the food temperature down. The coil would ice up because there would not be any off-cycle defrost.

lower than 70°F, difficulties with a starved evaporator coil may result.

If the medium-temperature box were moved outside the store, the fixture may quit working to capacity during winter. If the air temperature over the coil drops to 50°F outside, the head pressure is going to drop to 50°F plus 30°F = 80°F condensing temperature, which corresponds to 84 psig, Figure 22–34. The pressure difference of 84 psig − 20 psig = 64 psig isn't sufficient to feed enough liquid refrigerant through the expansion device to properly feed the coil, Figure 22–34. Notice in the figure the pressure and temperature difficulties encountered. The head pressure must be regulated.

Another example of a condenser operating outside the design parameter would be an ice-holding box like those found at service stations and convenience stores. These fixtures hold ice made at another location. They must keep the ice hard in all types of weather and often operate at about 0°F to 20°F inside the box.

It may be 30°F outside the box where the small air-cooled condenser is rejecting the heat. The condenser would be operating at about 30°F plus 30°F = 60°F, or at a head pressure of 58 psig, Figure 22–35. The evaporator should be operating at about −15°F, or at a suction pressure of 2.5 psig to maintain 0°F. This gives a pressure difference of 58 psig − 2.5 psig = 55.5 psig, which will starve the evaporator. This unit has to run to keep the ice frozen, so something has to be done to get the head pressure up to a value at least 75 to 100 psig higher than the suction pressure of 2.5 psig.

One thing that helps prevent problems for some equipment in low ambient conditions is that the load is reduced. The ice-holding box, for example, does not have to run as much at 30°F weather to keep the ice frozen at 0°F. It may perform poorly, but you will not know it unless it must run for a long period of time for some reason, such as if a load of ice that is barely frozen is loaded into it. The unit may have a long running cycle pulling the load temperature down and the evaporator may freeze solid.

22.18 HEAD PRESSURE CONTROL

Practical methods to maintain the correct workable head pressure automatically and not cause equipment wear are fan cycle control, dampers, and condenser flooding. Each type has different characteristics and features. Different companies recommend different head pressure control types for their own reasons.

Fan Cycling Devices

The air-cooled condenser normally has a small fan that passes the air over the condenser. When this fan is cycled off, the head pressure will go up in any kind of weather, provided the prevailing winds do not take over and do the fan's job. To cycle off the fan, a pressure control can be piped into the high-pressure side of the system that will close a set of electrical contacts on a preset rise in pressure, Figure 22–36. The electrical contacts will stop and start the fan motor on pressure changes.

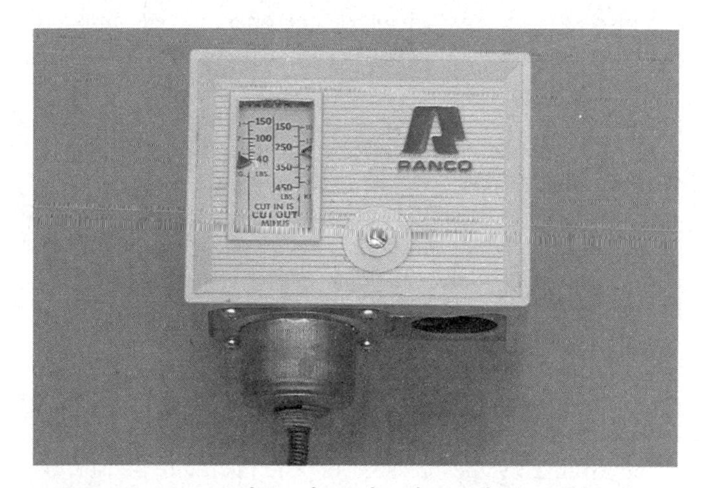

FIGURE 22–36 Condenser fan cycling device. This control has the same action (make on a rise in pressure) as the low-pressure control except that it operates at a much higher range. *Photo by Bill Johnson*

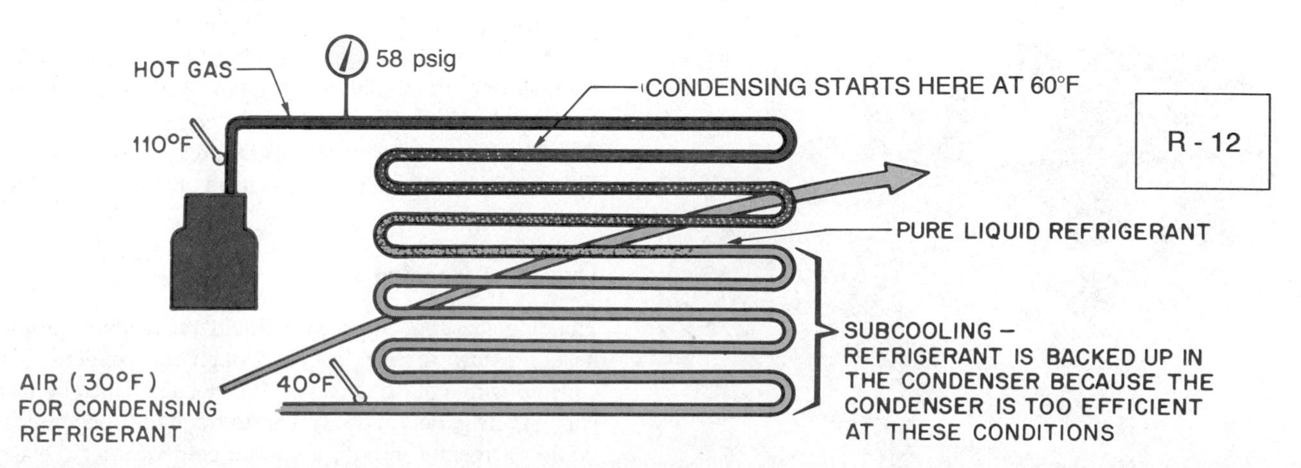

FIGURE 22–35 An illustration of how the condenser acts in a low ambient condition. Notice that this condenser has much more liquid refrigerant in it and the head pressure is very low. It may be so low that it may starve the expansion device. It relies on this pressure to force liquid refrigerant through its metering device.

A common setup for R-12 may call for the fan to cut off when the head pressure falls to 125 psig and to restart the fan when the head pressure reaches 175 psig. This setting will not interfere with the summer operation of the system and will give good performance in the winter. The settings are far enough apart to keep the fan from short cycling any more than necessary. When the thermostat calls for the compressor to come on and the ambient air is cold, the head pressure would be so low that the system would never get up to good running capacity because the condenser fan would be moving more cold air over the condenser. The condenser fan cycling control would keep the condenser fan off until the head pressure is within the correct operating range.

The fan cycling device is one method of maintaining a correct operating head pressure. It can be added to the system without much expense for the control, and the piping normally does not have to be altered. One problem with this device is that it has a tendency to cause the head pressure to swing up and down as the fan is stopped and started. This can also affect the expansion device operation because the pressure may be 175 psig for part of the cycle and then, when the fan comes on, drop rapidly to 125 psig. This pressure swing causes the expansion device to operate erratically. There are other means to control the head pressure at a steady state on air-cooled condensers.

When a condenser has more than one fan, one fan can be put in the lead, and the other fans can be cycled off by temperature, Figure 22–37. The lead fan can be cycled off by pressure like a single fan. This can help prevent the pressure swings from being so close together like the single-fan application. For example, when three fans are used, the first will cycle off at approximately 70°F, and the second will cycle off at approximately 60°F. The remaining fan may be controlled by the pressure or a temperature sensor located on the liquid line.

Air Shutters or Dampers

Shutters may be located either at the inlet to the condenser or at the outlet. The air shutter has a pressure-operated motor that pushes a shaft to open the shutters when the head pressure rises to a predetermined pressure, Figure 22–38. This pressure-operated motor is actually a piston in a cylinder. When the pressure rises, the piston arm extends to open the shutters, Figure 22–39. Shutters open and close slowly and provide even head pressure control.

FIGURE 22–38 Condenser with air shutter. With one fan it is the only control. With multiple fans the other fans can be cycled by temperature with the shutter controlling the final fan. *Courtesy of Trane Company*

FIGURE 22–39 Piston-type shutter operator. With this piston the high-pressure discharge gas is on one side of the bellows and the atmosphere is on the other side. When the head pressure rises to a predetermined point, the shutters begin to open. In summer the shutters will remain wide open during the running cycle. *Courtesy Robertshaw Controls Company*

FIGURE 22–37 Multitple-fan condenser. *Courtesy Heatcraft Inc., Refrigeration Products Division*

When a single fan is used, the shutter is installed over the inlet or over the outlet to the fan. When there are multiple fans, the shutter-covered fan is operated all the time, and the other fans can be cycled off by pressure or temperature. This arrangement can give good head pressure control down to low temperatures.

Condenser Flooding

Flooding the condenser with liquid refrigerant causes the head pressure to rise just as though the condenser were covered with a plastic blanket. It is accomplished by having enough refrigerant in the system to flood the condenser with liquid refrigerant in both mild and cold weather. This calls for a large refrigerant charge and a place to keep it. In addition to the charge there must be a valve arrangement to allow the refrigerant liquid to fill the condenser during both

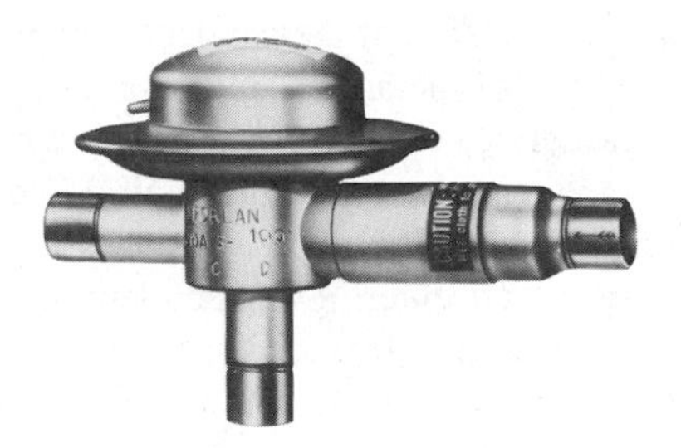

FIGURE 22–40 Head pressure control for condenser flooding. This valve allows the refrigerant to flood the condenser during both mild and cold weather. This method requires enough refrigerant to flood the condenser and has a large receiver to hold the refrigerant during the warm season when it is not needed to keep the head pressure up. *Courtesy Sporlan Valve Company*

mild and cold weather. This condenser flooding method is designed to maintain the correct head pressure in the coldest weather during start up and while operating, Figure 22–40.

22.19 USING THE CONDENSER SUPERHEAT

Air-cooled condensers have the characteristic of high discharge line temperatures even in winter. This can be used to advantage because the heat can be captured in winter and redistributed as heat for the structure or to heat water. The refrigeration system is rejecting heat out of the refrigerated box, heat that leaked into the box from the store itself. This heat has to be rejected to a place that is unobjectionable. In summer the heat should be rejected outside the store or possibly into a domestic hot water system, Figure 22–41. Water temperatures of 140°F are obtainable for use around the store. This can be a big energy saving.

22.20 HEAT RECLAIM

In winter the store needs heat. If heat could be rejected inside the store, it could reduce the heating cost. Any heat that is recovered from the system is heat that does not have to be purchased, Figure 22–42.

With air-cooled equipment heat recovery can be accomplished easily. The discharge gas can be passed to the rooftop condenser or to a coil mounted in the ductwork that supplies heat to the store. The condensing temperature of air-cooled equipment is high enough to be used as heat, and the quantity of heat is sizable enough to be important. Some stores in moderate climates will be able to extract enough heat from the refrigeration system to supply the full amount of heat to the store.

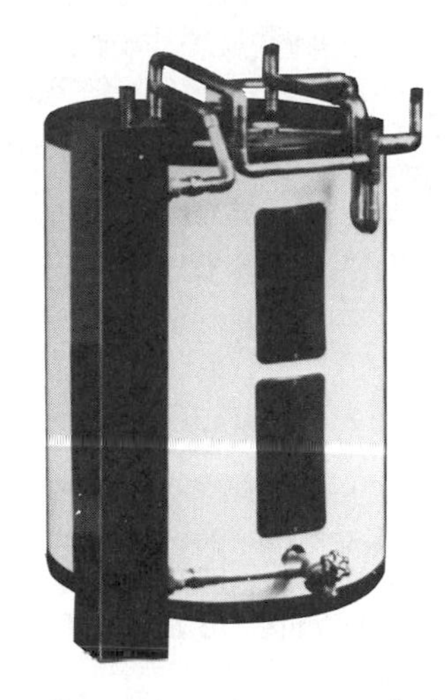

FIGURE 22–41 Heat exchanger to capture the heat from the highly superheated discharge line and use it to heat domestic hot water. This line can easily be 210°F and can furnish 140°F water in a supply limited by the size of the refrigeration system. The larger the system, the more heat available. *Courtesy Noranda Metal Industries Inc.*

22.21 CONDENSER EVALUATION

A final note: Don't get lost in the details of the equipment.

All of the system's components interact with each other, for example, the condenser operating pressure affecting

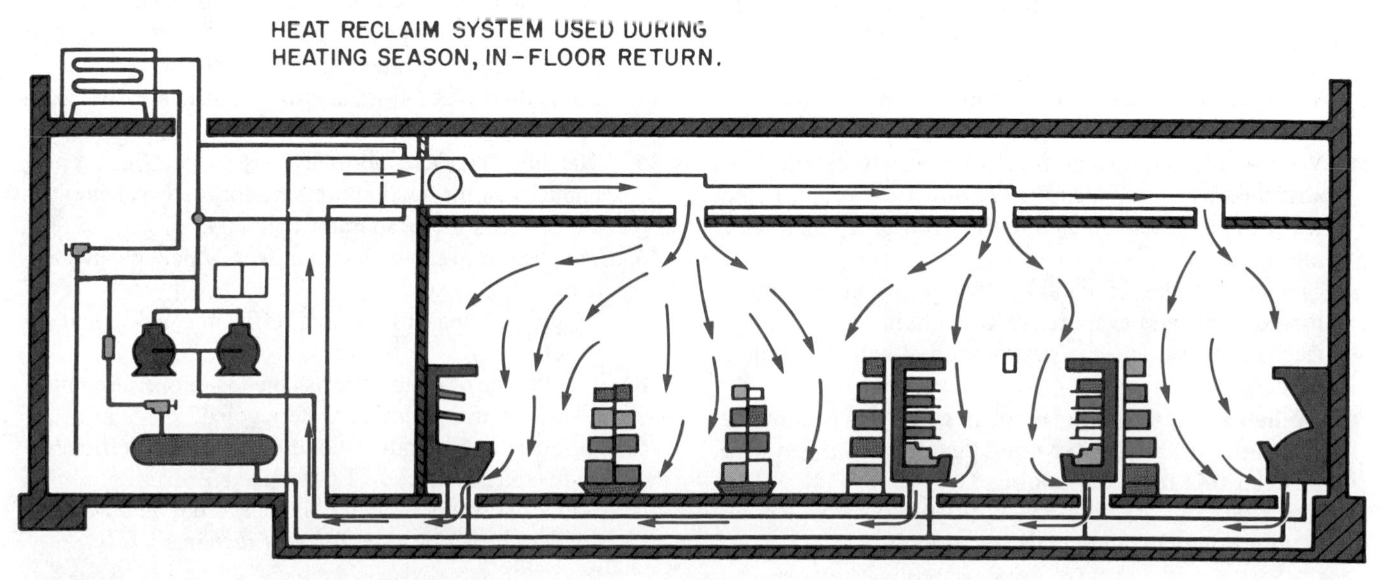
HEAT RECLAIM SYSTEM USED DURING
HEATING SEASON, IN-FLOOR RETURN.

FIGURE 22–42 System that can supply heat to the store.

the evaporator. You will be able to draw the correct conclusions with a little experience. Every compression system has a condenser to reject the heat from the system. Examination of the equipment will disclose the condenser, whatever type it may be. The condenser will be hot on air-cooled equipment and warm on water-cooled equipment.

SUMMARY

- The condenser is the component that rejects the heat from the refrigeration system.
- The refrigerant condenses to a liquid in the condenser and gives up heat.
- Water was the first medium that heat was rejected into through the water-cooled condenser.
- There are three types of water-cooled condensers: the tube within a tube, the shell and coil, and the shell and tube.
- The tube within a tube with flanges and the shell and tube condensers can be cleaned with brushes.
- The tube within a tube and the shell and coil condensers must be cleaned with chemicals.
- ***When condenser cleaning is desirable, consult the manufacturer.***
- The greatest amount of heat is given up from the refrigerant while the condensing process is taking place.
- The refrigerant normally condenses about 10°F higher than the leaving condensing medium in a water-cooled condenser.
- The first job of the condenser is to de-superheat the hot gas flowing from the compressor.
- After the refrigerant is condensed to a liquid, the liquid can be further cooled below the condensing temperature. This is called subcooling.
- When the condensing medium is cold enough to reduce the head pressure to the point that the expansion device will starve the evaporator, head pressure control must be used, whether air or water is used as the condensing medium.
- Water-cooled equipment is more efficient than air-cooled equipment.
- Water-cooled equipment has two places to deposit heat absorbed by the condenser: (1) down the drain, in the wastewater system or (2) into the atmosphere, in a cooling tower.
- The three types of cooling towers are natural draft, forced draft, and evaporative condenser.
- Recirculated water uses evaporation to help the cooling process.
- ***When water is evaporated, it will overconcentrate the minerals, and water must be added to the system to keep this from happening.***
- Three types of common head pressure controls are fan cycling, shutters, and condenser flooding.
- There is a relationship of the condensing temperature to the temperature of the air passing over a condenser that can help the service technician determine what the high-pressure gage reading should be on air-cooled condensers. Normally the refrigerant condenses at about 30°F higher than the air entering the condenser.
- The heat from a refrigeration system can be captured and used to heat water or to add heat to the conditioned space.

REVIEW QUESTIONS

1. What is the responsibility of the condenser in the refrigeration system?
2. Name three types of water-cooled condensers.
3. Why do some condensers have to be cleaned with brushes and others with chemicals?
4. Name three materials that condensers are normally made of.
5. Who should be consulted when condenser cleaning is needed?
6. When a water-cooled condenser is operating, the refrigerant normally condenses at ______°F higher than the leaving water.
7. The first job of the condenser is to ______ the gas before condensing can occur.
8. When is the most heat removed from the refrigerant in the condensing process?
9. After the refrigerant is condensed to a liquid, the remaining liquid can be further cooled down. This further cooling is called ______.
10. After heat is absorbed into a condenser medium in a water-cooled condenser, the heat can be deposited in one of two places. What are they?
11. Why does a water-cooling tower overconcentrate in mineral content?
12. A water-cooling tower capacity is governed by what aspect of the ambient air?
13. When an air-cooled condenser is used, the condensing refrigerant will normally be ______°F higher in temperature than the entering air temperature.
14. Name three methods of controlling head pressure in an air-cooled condenser.
15. The high pressure normally has to be ______ psig higher than the suction pressure for the expansion device to feed the evaporator correctly.
16. The prevailing winds can affect which of the air-cooled condensers?
17. Which condenser is more efficient: air-cooled or water-cooled?
18. Which type of condenser has the lower operating head pressures: air-cooled or water-cooled?
19. Compare the standard conditions of high-efficiency and standard air-cooled condensers.
20. Name two ways in which the heat from an air-cooled condenser can be used for heat reclaim.

UNIT 23 *Compressors*

OBJECTIVES

After studying this unit, you should be able to

- explain the function of the compressor in a refrigeration system.
- discuss compression ratio.
- describe four different methods of compression.
- state specific conditions under which a compressor is expected to operate.
- explain the difference between a hermetic compressor and a semi-hermetic compressor.
- describe the various working parts of reciprocating and rotary compressors.

SAFETY CHECKLIST

* Wear goggles and gloves when attaching or removing gages to transfer refrigerant or to check pressures.
* Wear warm clothing when working in a walk-in cooler or freezer.
* Do not touch the compressor discharge line with your bare hands.
* Be careful not to get your hands or clothing caught in moving parts such as pulleys, belts, or fan blades.
* Wear an approved back brace belt when lifting and use your legs keeping your back straight and upright.
* Observe all electrical safety precautions. Be careful at all times and use common sense.

23.1 THE FUNCTION OF THE COMPRESSOR

The *compressor* is considered the heart of the refrigeration system. The term that best describes a compressor is a *vapor pump.* The compressor actually increases the pressure from the suction pressure level to the discharge pressure level. For example, in a low-temperature system the suction pressure for a system that has R-12 as the refrigerant may have a suction pressure of 3 psig and a discharge pressure of 169 psig. The compressor increases the pressure 166 psig (169 − 3 = 166), Figure 23–1. A system next to the low-temperature system may have a different pressure increase. It may be a medium-temperature system and have a suction pressure of 21 psig with a discharge pressure of 169 psig.

R - 12
LOW-TEMPERATURE APPLICATION
REFRIGERANT ON THE LOW-PRESSURE SIDE OF THE SYSTEM IS EVAPORATING AT −13°F AND 3 psig
DIFFERENCE OF 166 psig
THE REFRIGERANT ON THE HIGH-PRESSURE SIDE OF THE SYSTEM IS CONDENSING AT 130°F AND 181 psig
169 psig
3 psig
COLD, HEAT-LADEN GAS
HOT SUPERHEATED GAS
OIL LEVEL

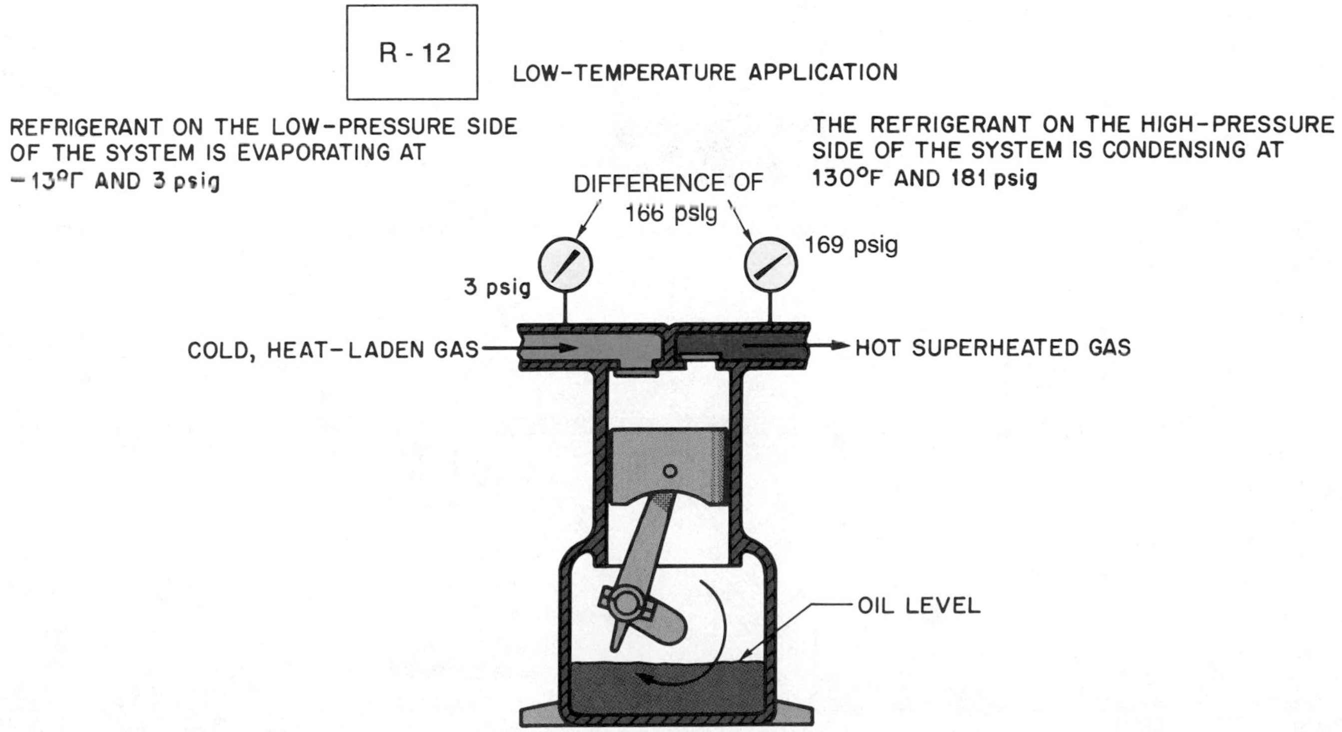

FIGURE 23–1 Pressure difference between the suction and discharge side of the compressor.

This system has an increase of 148 psig (169 − 21 = 148), Figure 23–2.

Compression ratio is the technical expression of pressure difference; it is the high-side absolute pressure divided by the low-side absolute pressure. Compression ratio is expressed in absolute pressures. For example, when a compressor is operating with R-12, a head pressure of 169 psig (125°F) and a suction pressure of 2 psig (−16°F), the compression ratio would be:

$$\text{Compression Ratio} = \frac{\text{Absolute Discharge}}{\text{Absolute Suction}}$$

$$\text{CR} = \frac{169 \text{ psig} + 14.7 \text{ atmosphere}}{2 \text{ psig} + 14.7 \text{ atmosphere}}$$

$$\text{CR} = \frac{183.7}{16.7}$$

$$\text{CR} = 11$$

Compression ratio is used to compare pumping conditions for a compressor. When compression ratios become too high, above approximately 12:1 for a hermetic reciprocating compressor, the refrigerant gas temperature leaving the compressor rises to the point that oil for lubrication may become overheated. Overheated oil may turn to carbon and create acid in the system. Compression ratios can be reduced by two-stage compression. One compressor discharges into the suction side of the second compressor, Figure 23–3.

The compressor has cool refrigerant entering the suction valve to fill the cylinders. This cool vapor contains the heat absorbed in the evaporator. The compressor pumps this heat-laden vapor to the condenser so that it can be rejected from the system.

The vapor leaving the compressor can be very warm. With a discharge pressure of 181 psig, the discharge line at

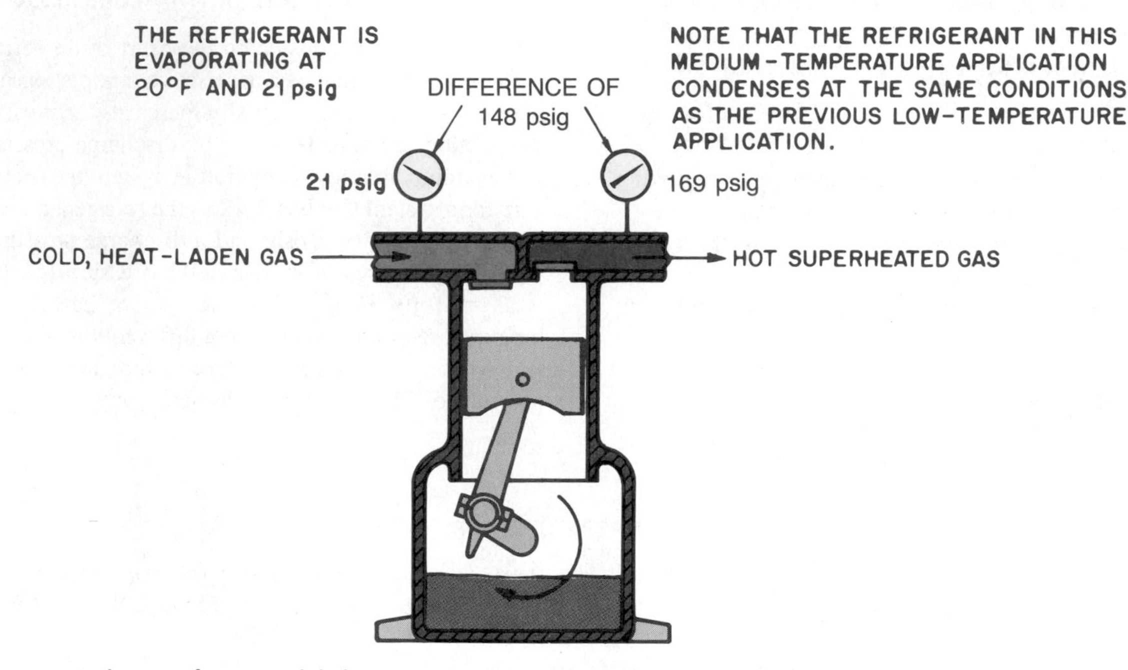
THE REFRIGERANT IS EVAPORATING AT 20°F AND 21 psig
DIFFERENCE OF 148 psig
NOTE THAT THE REFRIGERANT IN THIS MEDIUM-TEMPERATURE APPLICATION CONDENSES AT THE SAME CONDITIONS AS THE PREVIOUS LOW-TEMPERATURE APPLICATION.
21 psig
169 psig
COLD, HEAT-LADEN GAS
HOT SUPERHEATED GAS

FIGURE 23–2 Combination of suction and discharge pressure. This is R-12 applied to a medium-temperature system. The compressor only has to lift the suction gas 148 psi.

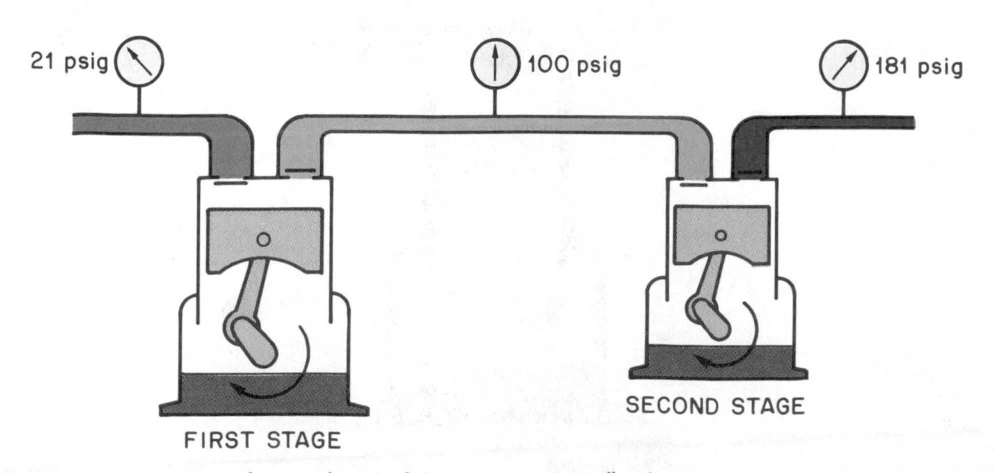
21 psig
100 psig
181 psig
FIRST STAGE
SECOND STAGE

FIGURE 23–3 Two-stage compressor. Notice the second stage of the compressor is smaller than the first stage.

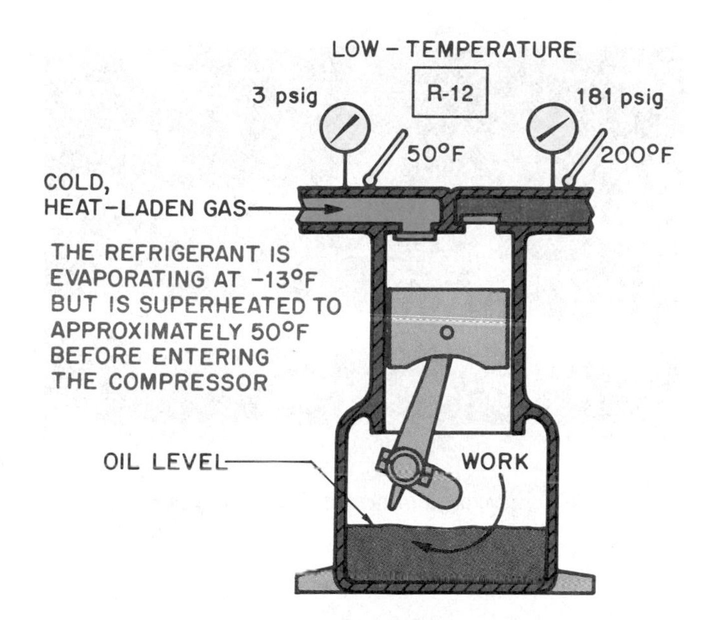

FIGURE 23–4 The refrigerant entering the compressor is called heat laden because it contains the heat that was picked up in the evaporator from the boiling process. The gas is cool but full of heat that was absorbed at a low pressure and temperature level. When this gas is compressed in the compressor, the heat concentrates. In addition to the heat that was absorbed in the evaporator, the act of compression converts some energy to heat.

the compressor could easily be 200°F or higher. ***Do not touch the compressor discharge line because you can burn your fingers.*** The vapor is compressed with the heat from the suction gas concentrated in the gas leaving the compressor, Figure 23–4.

23.2 TYPES OF COMPRESSORS

Five major types of compressors are used in the refrigeration and air conditioning industry. These are the reciprocating, screw, rotary, scroll, and centrifugal. The reciprocating, Figure 23–5, is the compressor used most frequently in small and medium-sized commercial refrigeration systems and will be described in detail in this unit. The screw compressor, Figure 23–6, is used in large commercial and industrial systems and will be described only briefly here because it is used in larger systems described later in this text. The rotary, Figure 23–7, and the scroll, Figure 23–8, along with

FIGURE 23–5 Reciprocating compressor. *Courtesy Copeland Corporation*

FIGURE 23–6 Screw compressor. *Courtesy Frick Company*

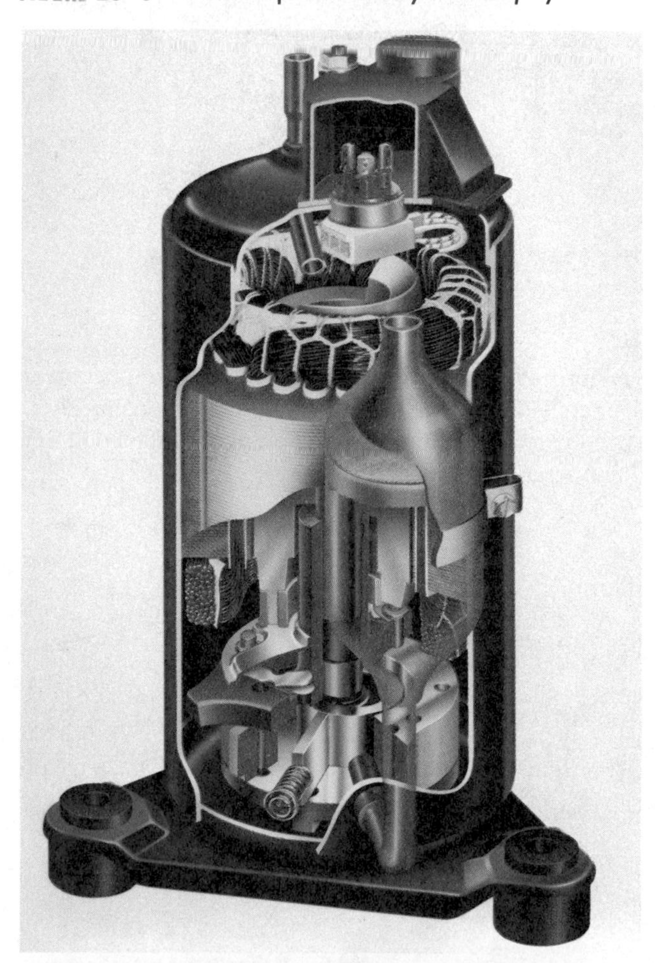

FIGURE 23–7 Rotary compressor. *Reprinted with permission of Motors and Armatures, Inc.*

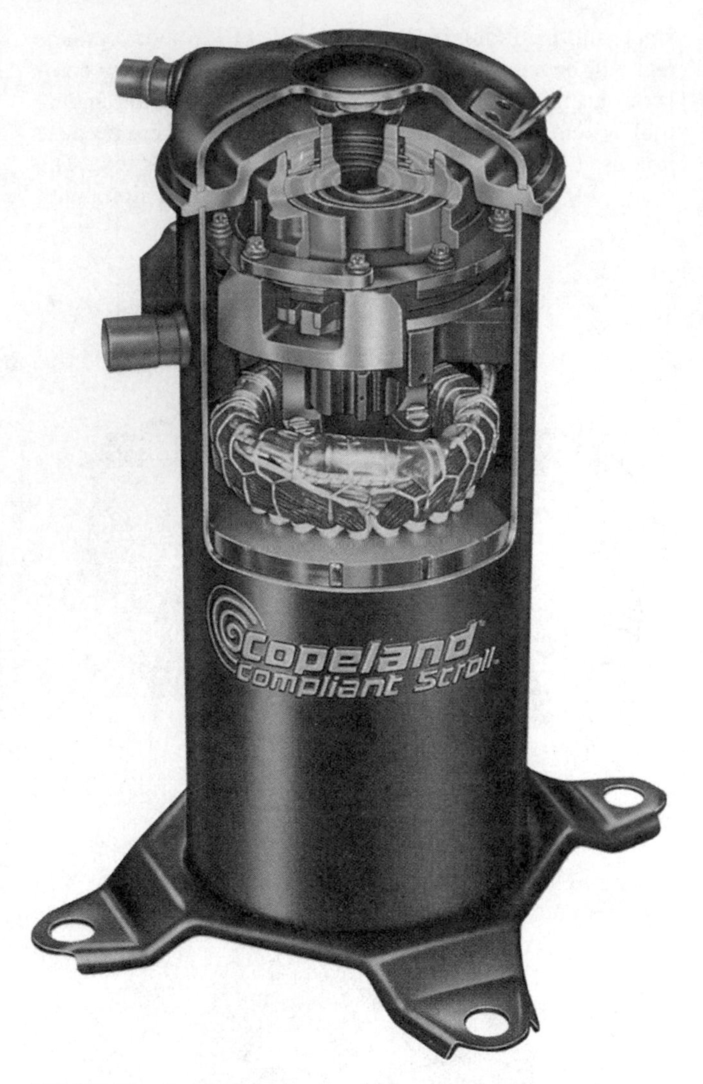

FIGURE 23–8 Scroll compressor. *Courtesy Copeland Corporation.*

the reciprocating, are used in residential and light commercial air conditioning. Centrifugal compressors, Figure 23–9, are used extensively for air conditioning in large buildings and are described later in this text.

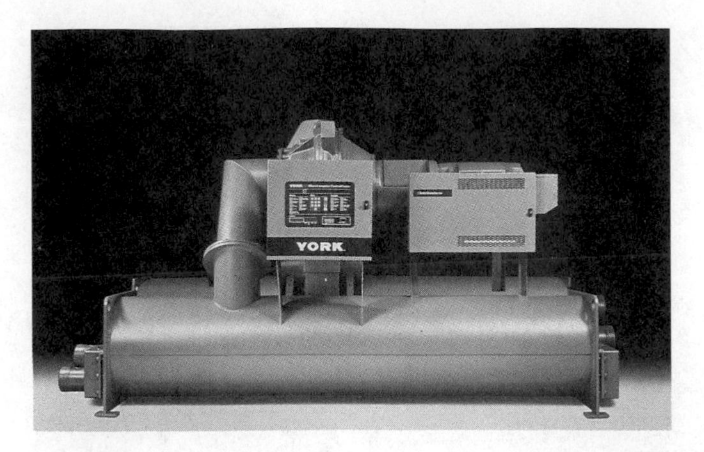

FIGURE 23–9 Centrifugal compressor. *Courtesy York International Corp.*

The Reciprocating Compressor

Reciprocating compressors are categorized by the compressor housing and by the drive mechanisms. The two housing categories are *open* and *hermetic* compressors, Figure 23–10. Hermetic refers to the type of housing the compressor is contained in and is divided into two types: fully welded, Figure 23–11, and serviceable, Figure 23–10(B). The drive mechanisms may either be enclosed inside the shell or outside the shell. When the compressor is hermetic, the drive mechanism is direct. The compressor and motor shaft are the same shaft.

FULLY WELDED HERMETIC COMPRESSORS. The motor and compressor are contained inside a single shell that is welded closed when a welded hermetic compressor is manufactured, Figure 23–11. This unit is sometimes called the *tin can* compressor because it cannot be serviced without cut-

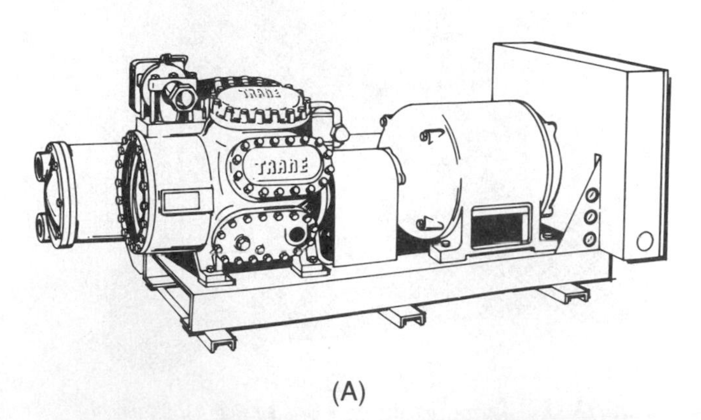

(A)

(B)

FIGURE 23–10 (A) Open-drive compressor. (B) Serviceable hermetic compressor. *Courtesy (A) Trane Company, (B) Copeland Corporation*

FIGURE 23–11 The welded hermetic compressor is used in the smaller compressor sizes, from 1 ton to 24 tons. Most welded hermetic compressors have a few things in common. The suction line usually is piped directly into the shell and is open to the crankcase. The discharge line normally is piped from the compressor inside the shell to the outside of the shell. The compressor shell is typically thought of as a low-side component. *Courtesy Bristol Compressors Inc.*

ting open the shell. Characteristics of the fully welded hermetic compressor are:

1. There is no access to the inside of the shell except by cutting the shell open.
2. They are only opened by a very few companies that specialize in this type of work. Otherwise, unless the manufacturer wants the compressor back for examination, it is a throw-away compressor.
3. The motor shaft and the compressor crankshaft are one shaft.
4. It is usually considered a low-side device because the suction gas is vented to the whole inside of the shell, which includes the crankcase. The discharge (high-pressure) line is normally piped to the outside of the shell so that the shell only has to be rated at the low-side working-pressure value.
5. Generally, they are cooled with suction gas.
6. They usually have a pressure lubrication system.
7. The combination motor and crankshaft are customarily in a vertical position with a bearing at the bottom of the shaft next to the oil pump. The second bearing is located about halfway on the shaft between the compressor and the motor.
8. The pistons and rods work outward from the crankshaft, so they are working at a 90° angle in relation to the crankshaft, Figure 23–12.

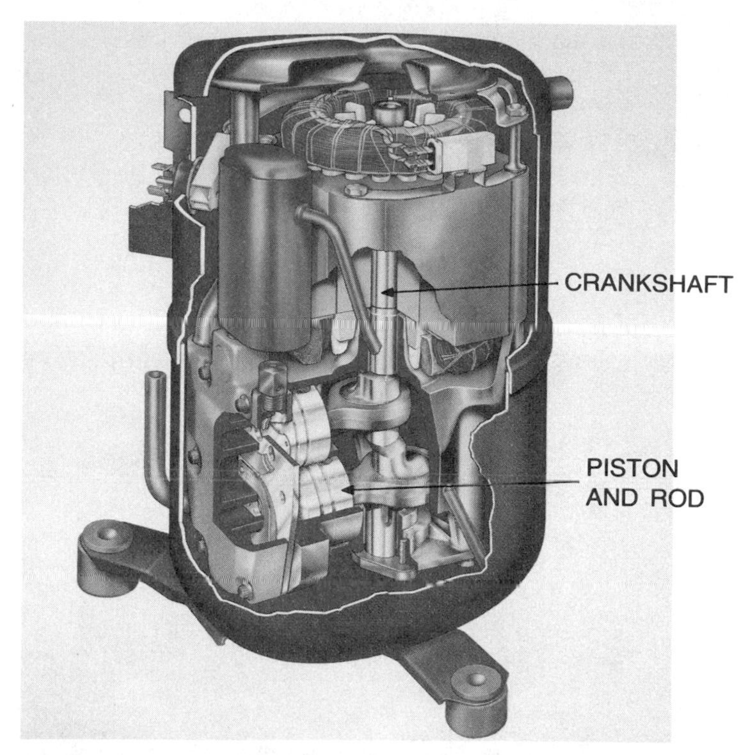

FIGURE 23–12 Internal workings of a welded hermetic compressor. *Courtesy Techumseh Products Company*

SERVICEABLE HERMETIC COMPRESSORS. When a serviceable hermetic (semihermetic) compressor is manufactured, the motor and compressor are contained inside a single shell that is bolted together. This unit can be serviced by removing the bolts and opening the shell at the appropriate place, Figure 23–13. Following are the characteristics of the serviceable hermetic compressor:

1. The unit is bolted together at locations that facilitate service and repair.

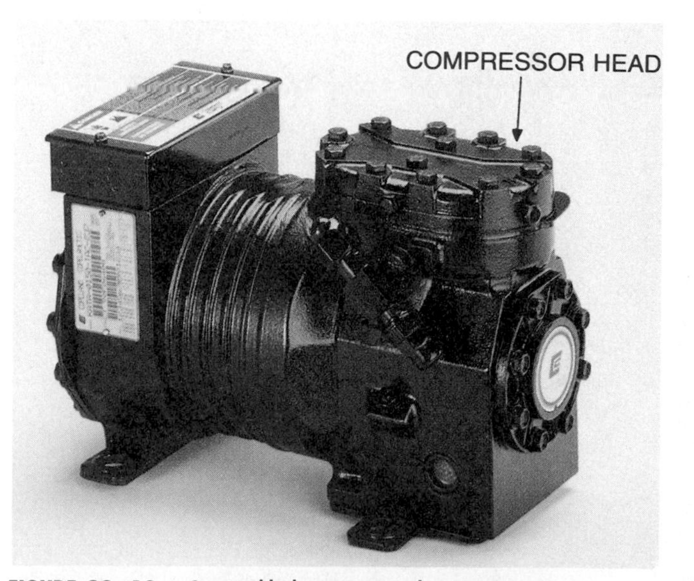

FIGURE 23–13 Serviceable hermetic (semihermetic) compressor designed in such a manner that it can be serviced in the field. *Courtesy Copeland Corporation*

2. The housing is normally cast iron and may have a steel housing fastened to the cast iron compressor. They are normally heavier than the fully welded type.
3. The motor and crankshaft combination are similar to that in the fully welded type except that the crankshaft is usually horizontal.
4. They generally use a splash-type lubrication system in the smaller compressors and a pressure lubrication system in the larger compressors.
5. They are often air cooled and can be recognized by the fins in the casting or extra sheet metal on the outside of the housing to give the shell more surface area.
6. The piston heads are normally at the top or near the top of the compressor and work up and down from the center of the crankshaft, Figure 23–14.

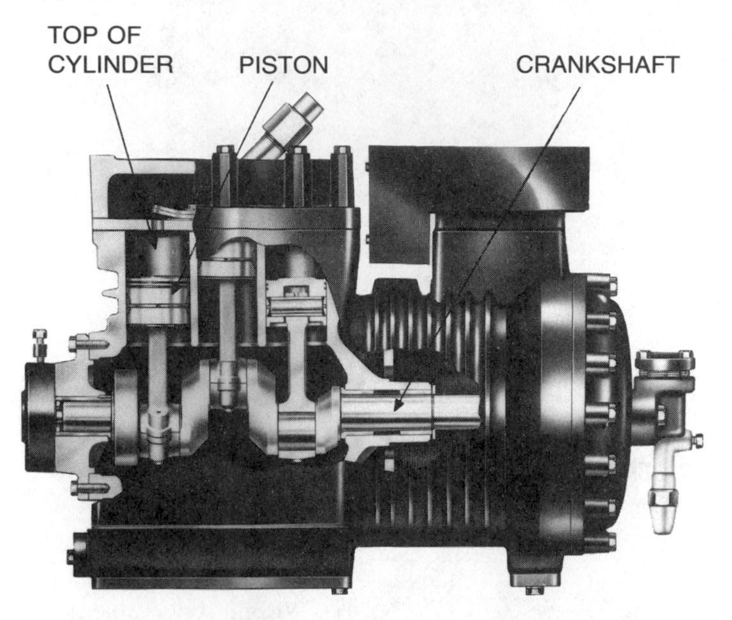

FIGURE 23–14 Working parts of the serviceable hermetic compressor. The crankshaft is in the horizontal position, and the rods and pistons move in and out from the center of the shaft. The oil pump is on the end of the shaft and draws the oil from the crankcase at the bottom of the compressor. *Courtesy Copeland Corporation*

OPEN-DRIVE COMPRESSORS. Open-drive compressors are manufactured in two styles: belt drive and direct drive, Figure 23–15. Any compressor with the drive on the outside of the casing must have a shaft seal to keep the refrigerant from escaping to the atmosphere. This seal arrangement has not changed much in many years. Open-drive compressors are bolted together and may be disassembled for service to the internal parts.

BELT-DRIVE COMPRESSORS. The belt-drive compressor was the first type of compressor and is still used to some extent. With the belt-drive unit the motor and its shaft are parallel with the compressor's shaft. The motor is beside the compressor. Notice that since the compressor and motor shaft are parallel, there is a sideways pull on both shafts to tighten the belts. This strains both shafts and requires the manufacturer to compensate for this in the shaft bearings. Figure 23–16 indicates correct and incorrect alignments of belt-drive compressors and motors.

DIRECT-DRIVE COMPRESSORS. The direct-drive compressor differs from the belt-drive in that the compressor shaft is end to end with the motor shaft. These shafts have a coupling

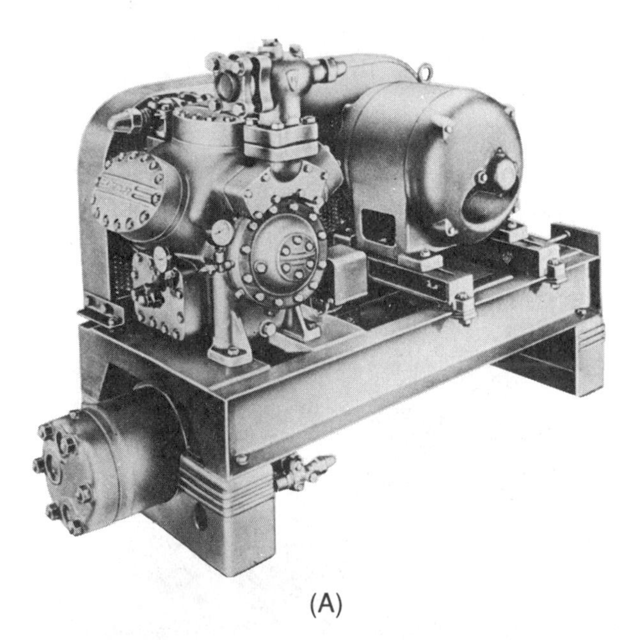

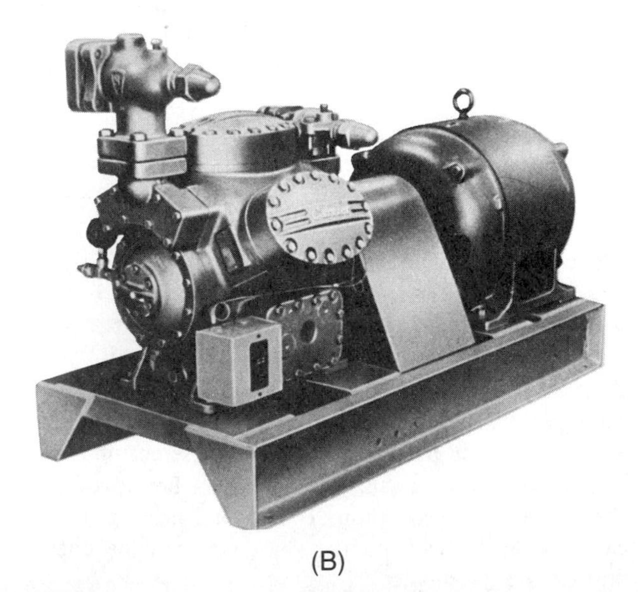

FIGURE 23–15 (A) Belt-drive compressor. (B) Direct-drive compressor. The belt-drive compressor may have different speeds, depending on the pulley sizes. The direct-drive compressor turns at the speed of the motor because it is attached directly to the motor. Common speeds are 1750 rpm and 3450 rpm. The coupling between the motor and the compressor is slightly flexible. *Reproduced courtesy of Carrier Corporation*

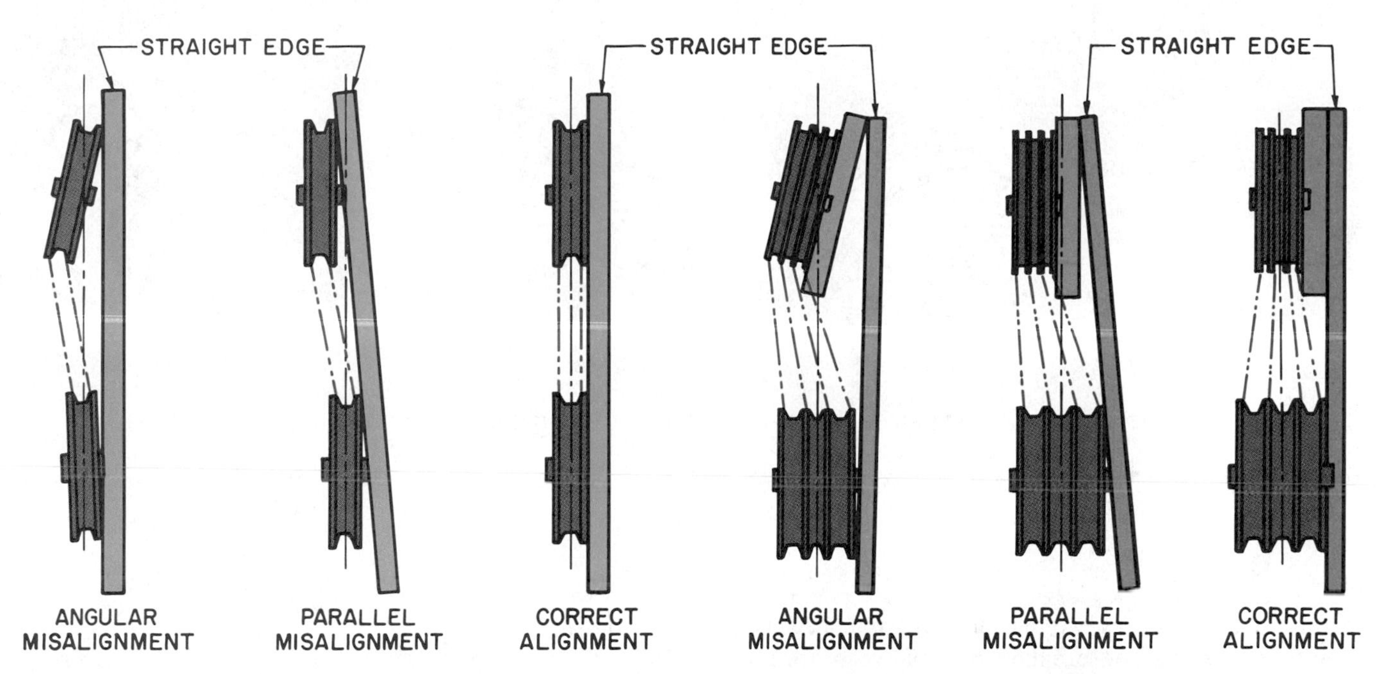

FIGURE 23–16 Correct and incorrect alignments of belt-drive compressor with its motor. The belts used on a multiple-belt installation are matched at the factory for the correct length.

between them with a small amount of flexibility. The two shafts have to be in very good alignment to run correctly, Figure 23–17.

THE ROTARY SCREW COMPRESSOR. The rotary screw compressor is another mechanical method of compressing refrigerant gas used in larger installations. Instead of a piston and cylin-

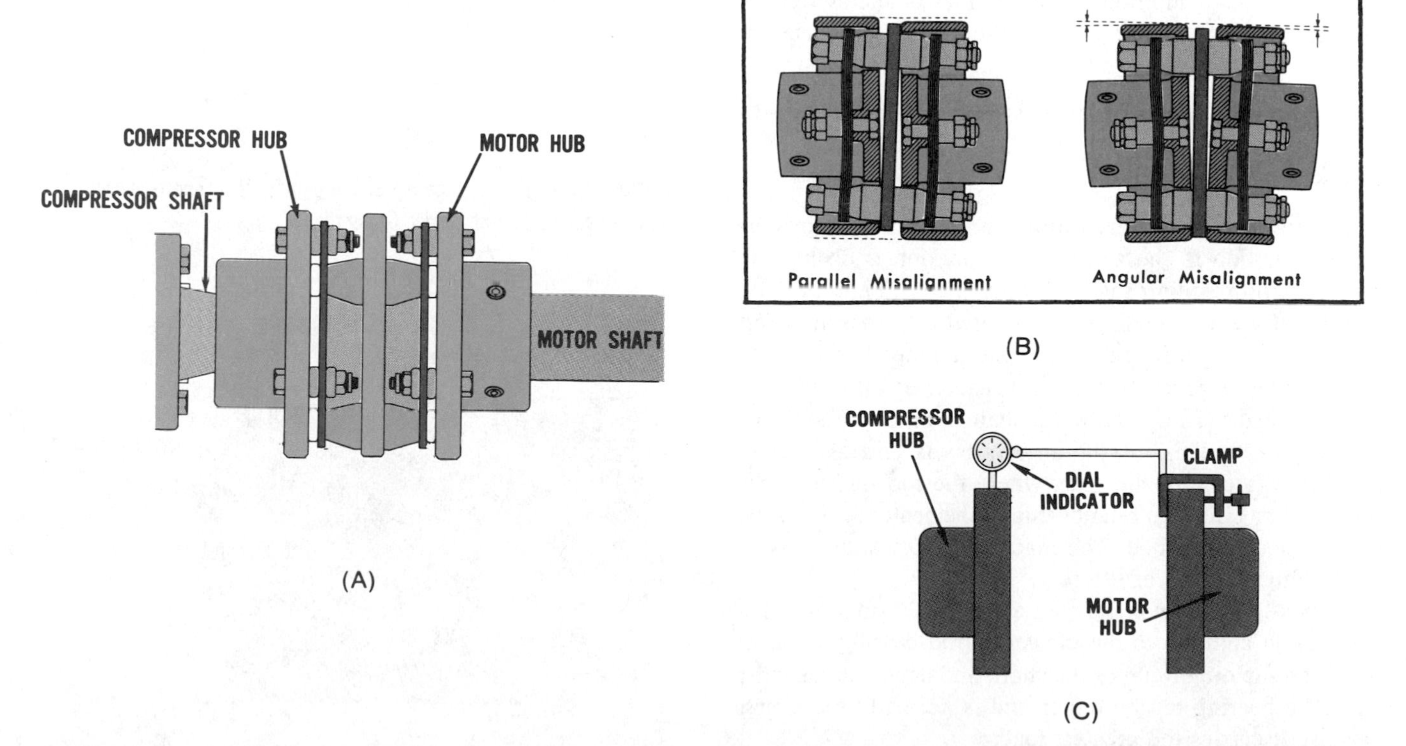

FIGURE 23–17 Alignment of a direct-drive compressor and motor. The alignment must be within the compressor manufacturer's specifications. These compressors turn so fast that if the alignment is not correct, a seal or bearing will soon fail. When the correct alignment is attained, the compressor and motor are normally fastened permanently to the base they are mounted on. The motor and the compressor can be rebuilt in place in larger installations. *Courtesy Trane Company*

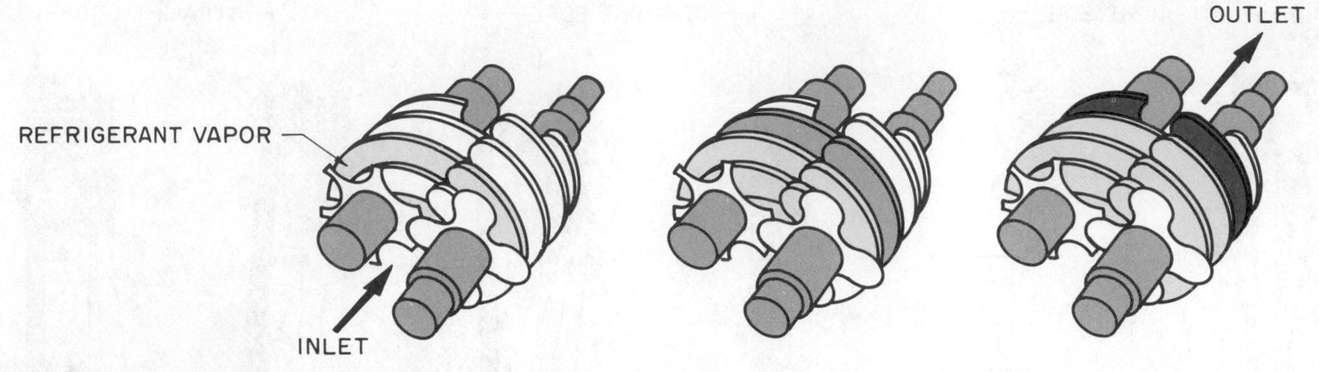

FIGURE 23–18 Tapered machined screw-type gears in a screw compressor.

der, this compressor uses two matching, tapered, machined, screw-type gears that squeeze the refrigerant vapor from the inlet to the outlet, Figure 23–18.

The rotary screw compressor uses an open motor instead of the hermetic design. A shaft seal traps the refrigerant in the compressor housing where the rotating shaft leaves the compressor housing. A flexible coupling is used to connect the motor shaft to the compressor shaft to prevent minor misalignment from causing seal or bearing damage.

Screw compressor applications are for large systems (see Figure 23–6). The refrigerant may be any of the common refrigerants, R-12, R-22 or R-502. The operating pressure on the low- and high-pressure sides of the system are the same as for a reciprocating system of like application.

23.3 RECIPROCATING COMPRESSOR COMPONENTS

Crankshafts

The *crankshaft* of a reciprocating compressor transmits the circular motion to the rods, and the motion is changed to back and forth (reciprocating) for the pistons, Figure 23–19. Crankshafts are normally manufactured of cast iron or soft steel. The crankshaft can be cast (molten metal poured into a mold) into a general shape and machined into the exact size and shape. In this case the shaft is made of cast iron, Figure 23–20. This machining process is critical because the throw (the off-center part where the rod fastens) does not turn in a circle (in relationship to the center of the shaft) when placed in a lathe. The machinist must know how to work with this type of setup.

These off-center shafts normally have two main-bearing surfaces in addition to the off-center rod-bearing surfaces; one is on the motor end of the shaft, and one is on the other end. The bearing on the motor end is normally the largest because it carries the greatest load.

Some shafts are straight and have a cam-type arrangement called an *eccentric*. This allows the shaft to be manufactured straight and from steel. The shaft may not be any more durable, but it is easier to machine. The eccentrics can

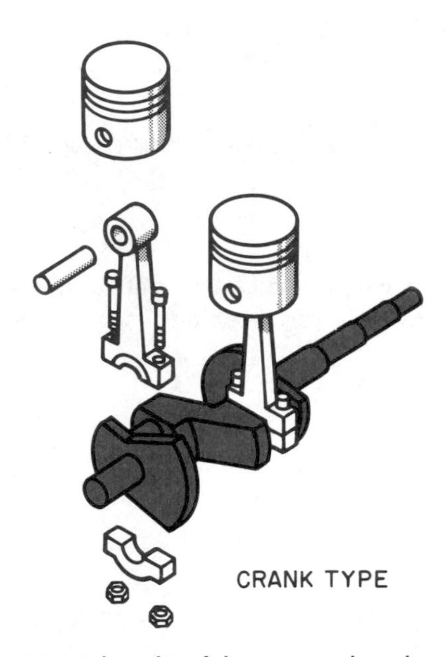

FIGURE 23–19 Relationship of the pistons, rods, and crankshaft. *Reproduced courtesy of Carrier Corporation*

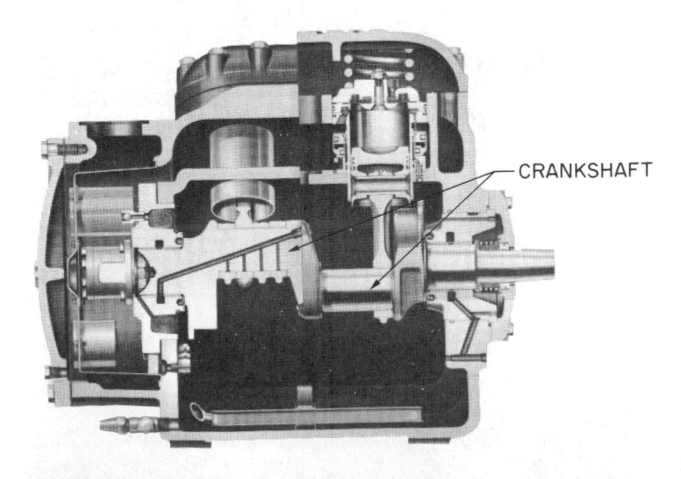

FIGURE 23–20 Crankshafts are cast into the general shape and machined to the correct shape in a machine shop. This crankshaft is in an open-drive compressor but is typical also for hermetic compressors. *Courtesy Trane Company*

be machined off-center to the shaft to accomplish the reciprocating action, Figure 23–21. Notice that the rod has to be different for the eccentric shaft because the end of the rod has to fit over the large eccentric on the crankshaft.

All of these shafts must be lubricated. The smaller compressors using the splash system may have a catch basin to catch oil and cause it to flow down the center of the shaft, Figure 23–22. It is then slung to the outside of the crankshaft surface when the compressor runs. This causes the oil to move to the other parts, such as the rods on both ends.

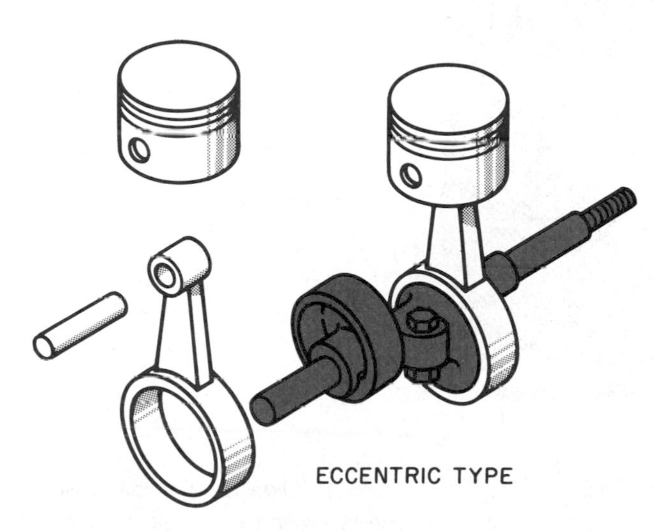

FIGURE 23–21 This crankshaft obtains the off-center action with a straight shaft and an eccentric. The eccentric is much like a cam lobe. The rods on this shaft have large bottom throws and slide off the end of the shaft. This means that to remove the rods, the shaft must be taken out of the compressor. *Reproduced courtesy of Carrier Corporation*

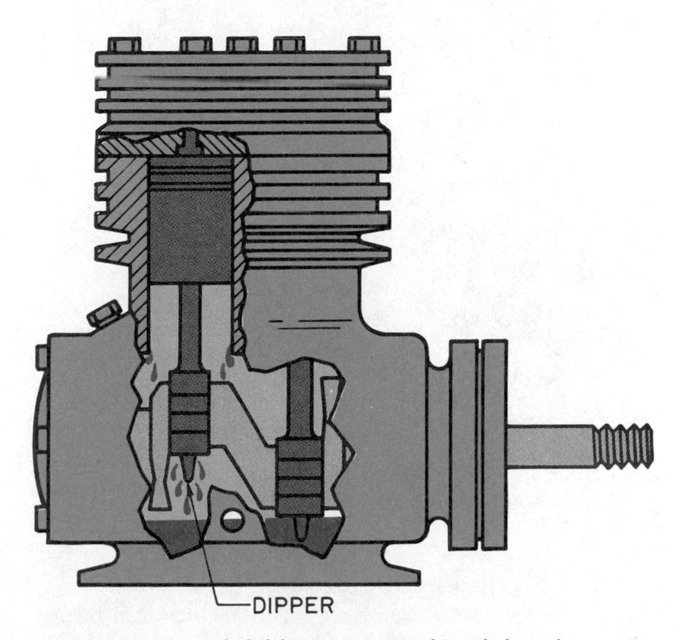

FIGURE 23–22 Splash lubrication system that splashes oil up onto the parts. *Reproduced courtesy of Carrier Corporation*

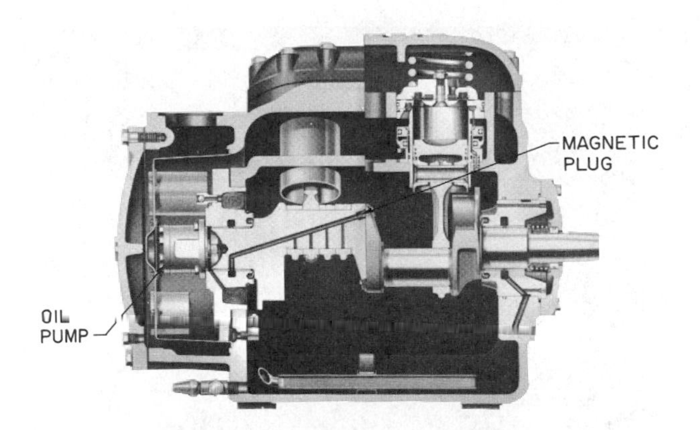

FIGURE 23–23 Crankshaft drilled for the oil pump to force the oil up the shaft to the rods, then up the rods to the wrist pins. Magnetic elements are sometimes placed along the passage to capture iron filings. *Courtesy Trane Company*

Some of the shafts are drilled and lubricated with a pressure lubrication system. These compressors have an oil pump mounted on the end of the crankshaft that turns with the crankshaft, Figure 23–23.

NOTE: There is no pressure lubrication when the compressor first starts. The compressor must be running up to speed before the lubrication system is fully effective.

Because some compressors have vertical shafts and some have horizontal, manufacturers have been challenged to provide proper lubrication where needed. You may have to consult the compressor manufacturer for questions about a specific application.

Connecting Rods

Connecting rods connect the crankshaft to the piston. These rods are normally made in two styles: the type that fit the crankshaft with off-center throws, and the type to fit the eccentric crankshaft, Figures 23–19 and 23-21. Rods can be made of several different metals such as iron, brass, and aluminum. The rod design is important because it takes a lot of the load in the compressor. If the crankshaft is connected directly to the motor and the motor is running at 3450 rpm the piston at the top of the rod is changing directions 6900 times per minute. The rod is the connection between this piston and the crankshaft and is the link between this changing of direction.

The rods with the large holes in the shaft end are for eccentric shafts. They cannot be taken off with the shaft in place. The shaft has to be removed to take the piston out of the cylinder. The rods with the small holes are for the off-center shafts, are split, and have rod bolts, Figure 23–24. These rods can be separated at the crankshaft, and the rod and piston can be removed with the crankshaft in place.

The rod is small on the piston end and fastens to the piston by a different method. The rod normally has a connector called a *wrist pin* that slips through the piston and the upper end of the rod. This almost always has a *snap ring* to

FIGURE 23–24 Service technician removing the rods from a compressor. The rods have the split bottoms. *Courtesy Trane Company*

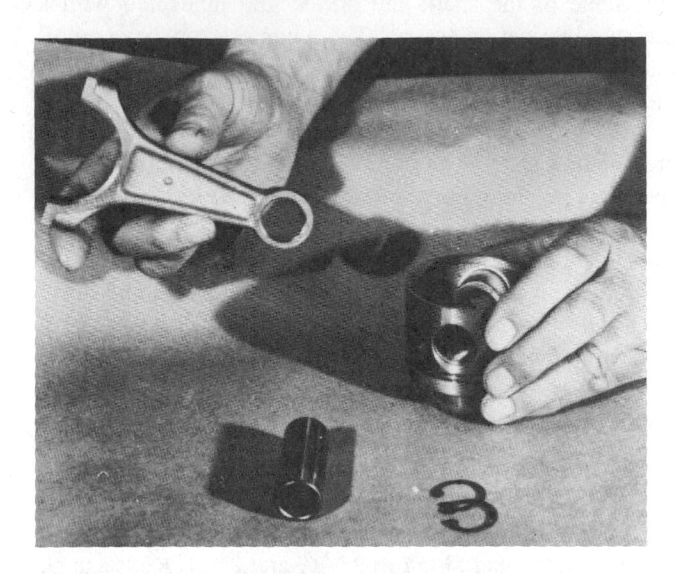

FIGURE 23–25 The end of the rod that fits into the piston. The wrist pin holds the rod to the piston while allowing the pivot action that takes place at the top of the stroke. The wrist pin is held secure in the piston with snap rings. *Courtesy Trane Company*

keep the wrist pin from sliding against the cylinder wall, Figure 23–25.

The Piston

The *piston* is the part of the cylinder assembly exposed to the high-pressure gas during compression. Pistons have high-pressure gas on top and suction or low-pressure gas on the bottom during the upstroke. They have to slide up and down in the cylinder in order to pump. They must have some method of preventing the high-pressure gas from slipping by to the crankcase. Piston rings like those used in automobile engines are used on the larger pistons. These rings are of two types: compression and oil. The smaller compressors use the oil on the cylinder walls as the seal. A cross-sectional view of these rings can be seen in Figure 23–26.

Refrigerant Cylinder Valves

The valves in the top of the compressor determine the direction in which the gas entering the compressor will flow. A cutaway of a compressor cylinder is shown in Figure 23–27. These valves are made of very hard steel. The two

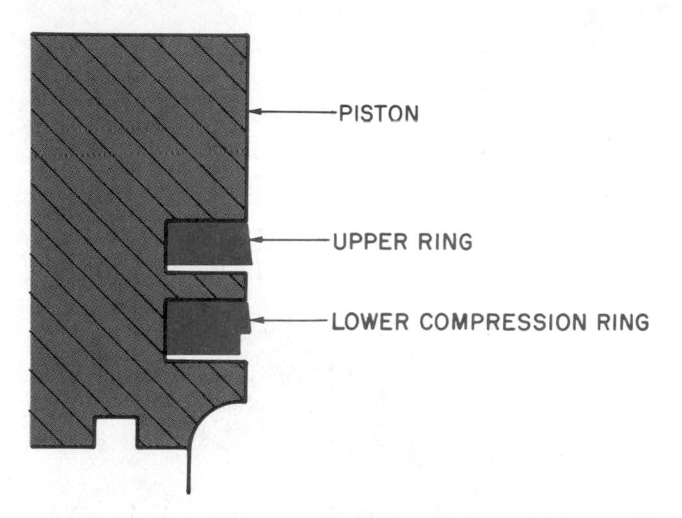

FIGURE 23–26 Piston rings for a refrigeration compressor resemble the rings used on automobile pistons. *Courtesy Trane Company*

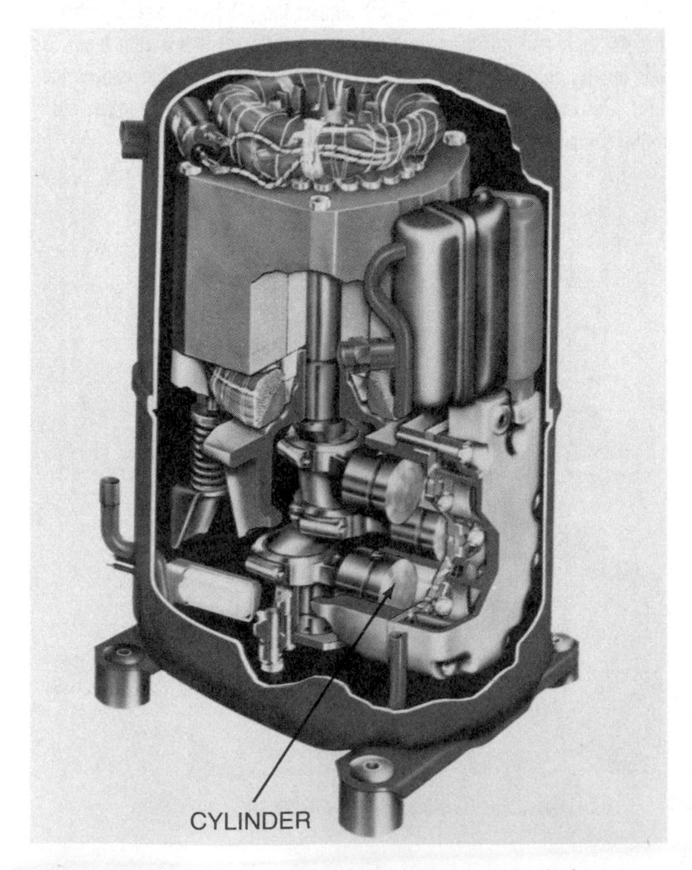

FIGURE 23–27 Cutaway of compressor shows typical cylinder. *Courtesy Tecumseh Products Company*

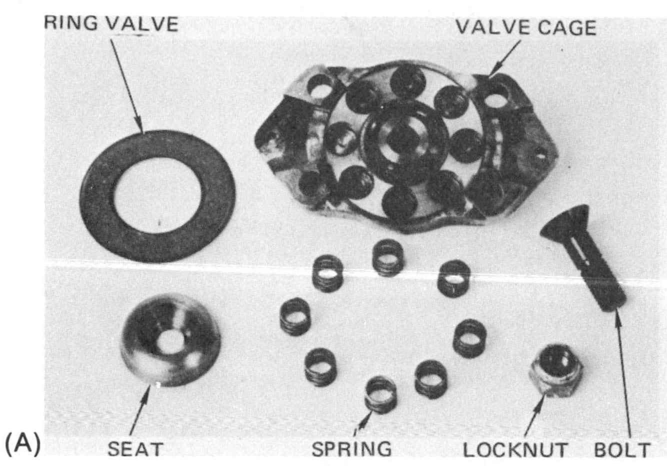

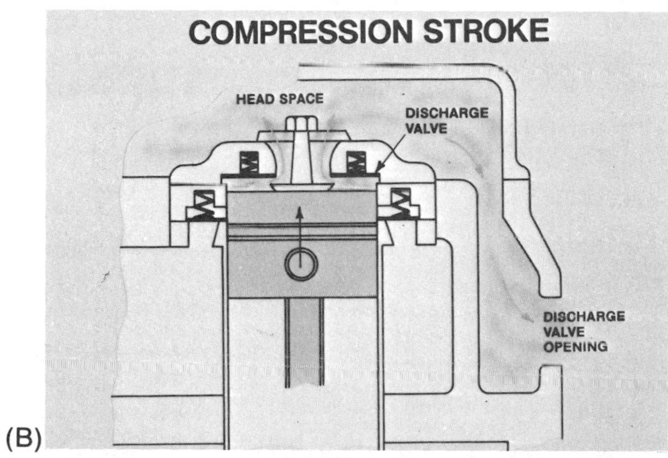

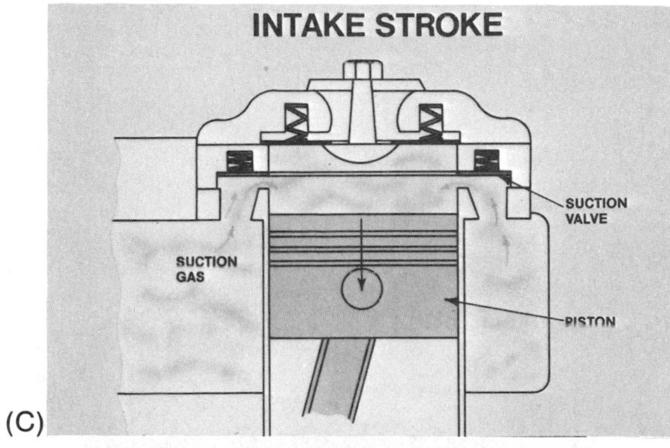

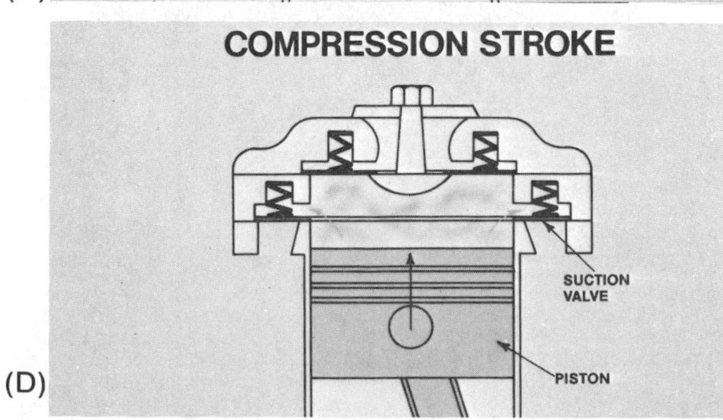

FIGURE 23–28 Ring valves. They normally have a set of small springs to close them. *Courtesy Trane Company*

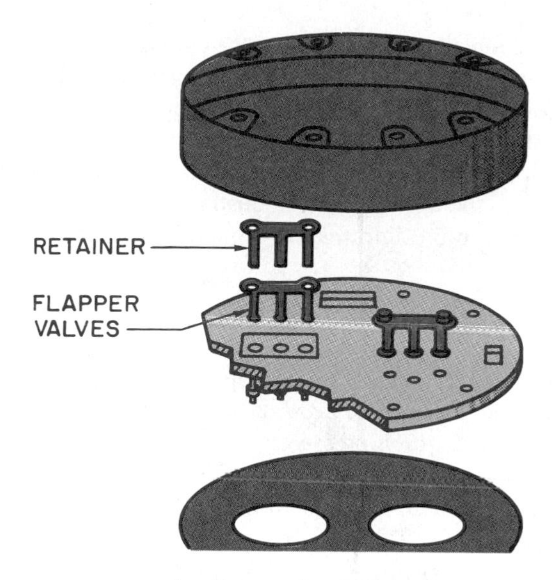

FIGURE 23–29 Reed or flapper valves held down on one end. This provides enough spring action to close the valve when reverse flow occurs. *Reproduced courtesy of Carrier Corporation*

styles that make up the majority of the valves on the market are the *ring valve* and the *flapper (reed) valve*. They serve both the suction and the discharge ports of the compressor.

The ring valve is made in a circle with springs under it. If ring valves are used for the suction and the discharge, the larger one will be the suction valve, Figure 23–28.

Flapper valves have been made in many different shapes. Each manufacturer has its own design, Figure 23–29.

The Valve Plate

The *valve plate* holds the suction and discharge flapper valves. It is located between the head of the compressor and the top of the cylinder wall, Figure 23–30. Many different methods have been used to hold the valves in place without taking up any more space than necessary. The bottom of the plate actually protrudes into the cylinder. Any volume of

FIGURE 23–30 Valve plate typical of those used to hold the valves. They can be replaced or rebuilt if not badly damaged. There is a gasket on both sides. *Reproduced courtesy of Carrier Corporation*

gas that cannot be pumped out of the cylinder because of the valve design will reexpand on the downstroke of the piston. This makes the compressor less efficient.

There are other versions of the crankshaft and valve arrangements than those listed here. They are not used enough in the refrigeration industry to justify coverage. If you need more information, contact the specific manufacturer.

The Head of the Compressor

The component that holds the top of the cylinder and its assembly together is the *head*. It sets on top of the cylinder and contains the high-pressure gas from the cylinder until it moves into the discharge line. It often contains the suction chamber, separated from the discharge chamber by a partition and gaskets. These heads have many different design configurations and need to accomplish two things. They hold the pressure in and hold the valve plate on the cylinder. They are made of steel in some welded hermetic compressors and of cast iron in a serviceable hermetic type. The cast iron heads may be in the moving airstream and have fins on them to help dissipate the heat from the top of the cylinder, Figure 23–31.

Mufflers

Mufflers are used in many fully hermetic compressors to muffle compressor pulsation noise. There are audible suction and discharge pulsations that can be transmitted into the piping if they are not muffled. Mufflers must be designed to have a low pressure drop and still muffle the discharge pulsations, Figure 23–32.

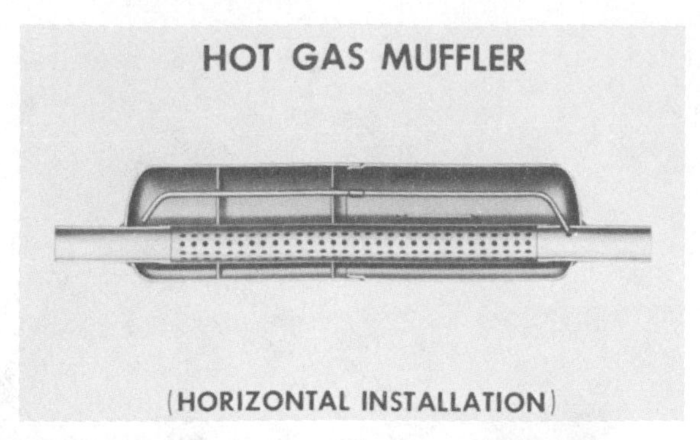

FIGURE 23–32 Compressor muffler. *Courtesy Trane Company*

The Compressor Housing

The *housing* holds the compressor and sometimes the motor. It is made of stamped steel for the welded hermetic and of cast iron for the serviceable hermetic.

The welded hermetic compressor is designed so the compressor shell is under low-side pressure and will normally have a working pressure of 150 psig. The compressor is mounted inside the shell, and the discharge line is normally piped to the outside of the shell. This means the shell does not need to have a test pressure as high as the high-side pressure. A cutaway of a hermetic compressor inside a welded shell and the method used to weld the shell together are shown in Figure 23–27.

Two methods are used to mount the compressor inside the shell: rigid mount and spring mount.

FIGURE 23–31 Typical compressor heads. (A) Suction-cooled compressor. (B) Air-cooled compressor. The compressors that have air-cooled motors are located in a moving airstream, or overheating will occur. *Courtesy Copeland Corporation*

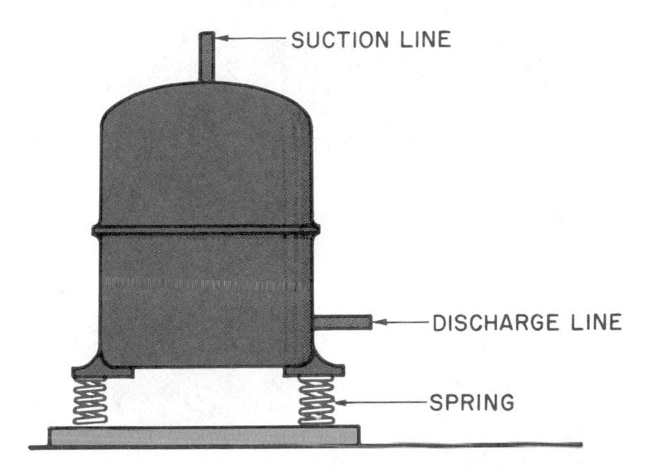

FIGURE 23–33 Compressor motor pressed into its steel shell. It requires experience to remove the motor. The compressor has springs under the mounting feet to help eliminate vibration. This compressor is shipped with a bolt tightened down through the springs. This bolt must be loosened by the installing contractor.

Rigid-mounted compressors were used for many years. The compressor shell was mounted on external springs that had to be bolted tightly for shipment. The springs were supposed to be loosened when installed, Figure 23–33. Occasionally they often were not, and the compressor vibrated because, without the springs, it was mounted rigidly to the condenser casing. External springs can also rust, especially where there is a lot of salt in the air.

The internal spring-mounted compressors actually suspend the compressor from springs inside the shell. These compressors have methods of keeping the compressor from moving too much during shipment. Sometimes a compressor will come loose from one or two of the internal springs. When this happens, the compressor will normally run and pump just like it is supposed to but will make a noise on startup or shutdown or both. If the compressor comes off the springs and they are internal, there is nothing that can be done to repair it in the field, Figure 23–34.

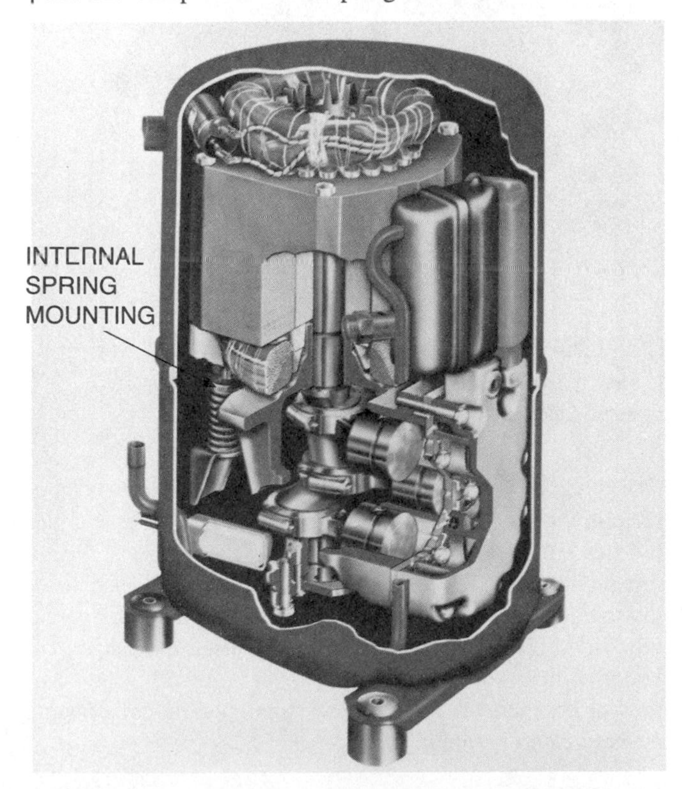

FIGURE 23–34 Compressor mounted inside the welded hermetic shell on springs. The springs have guides that only allow them to move a certain amount during shipment. *Courtesy Tecumseh Products Company*

Compressor Motors in a Refrigerant Atmosphere

The compressor motor operating inside the refrigerant atmosphere must have special consideration. Motors for hermetic compressors differ from standard electric motors. The materials used in a hermetic motor are not the same materials that would be used in a fan or pump motor that would run in air. Hermetic motors must be manufactured of materials compatible with the system refrigerants. For instance, rubber cannot be used because the refrigerant would dissolve it. The motors are assembled in a very clean atmosphere and kept dry. When a hermetic motor malfunctions, it cannot be repaired in the field.

MOTOR ELECTRICAL TERMINALS. There must be some conductor to carry the power from the external power supply to the internal motor. The power to operate the compressor must be carried through the compressor housing without the refrigerant leaking. The connection also has to insulate the electric current from the compressor shell. These terminals are sometimes fused glass with a terminal stud through the middle on the smaller compressors. When large terminals are required, the terminals are sometimes placed in a fiber block with an O ring-type seal, Figure 23–35.

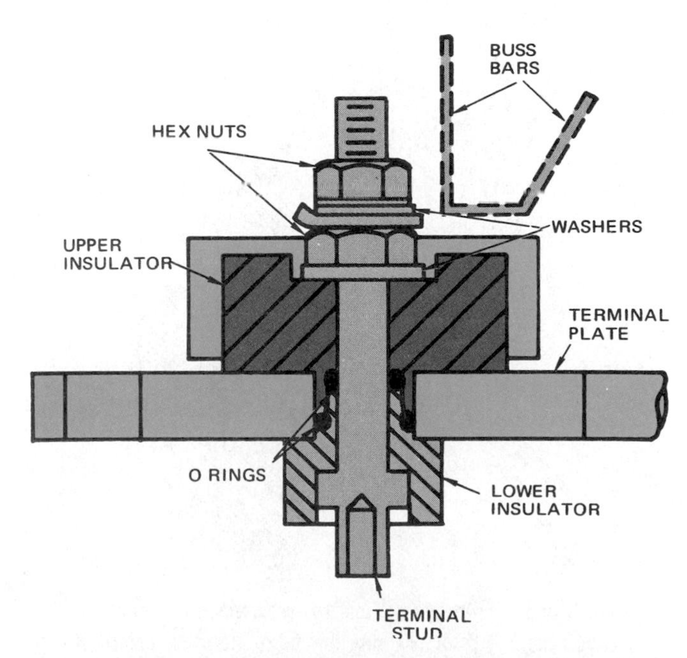

FIGURE 23–35 Motor terminals. The power to operate the compressor is carried through the compressor shell but must be insulated from the shell. This is a fiber block used as the insulator. O rings keep the refrigerant in. *Courtesy Trane Company*

Care must be taken with these terminals (due to loose electrical connections) to prevent overheating. Should the terminal overheat, a leak could occur. If the terminal block is a fused-glass type, it would be hard to repair. The fused-glass type will stand more heat, but there is a limit to how much heat it can take. Less heat can be tolerated with the O ring and fiber type of terminal board. When the O ring and fiber board are damaged, they can be replaced with new parts. However, refrigerant loss can result before the problem is discovered.

INTERNAL MOTOR PROTECTION DEVICES. Internal overload protection devices in hermetic motors protect the motor from overheating. These devices are embedded in the windings and are wired in two different ways. One style breaks the line circuit inside the compressor. Because it is internal and carries the line current, it is limited to smaller compressors. It has to be enclosed to prevent the electrical arc from affecting the refrigerant, Figure 23–36. If contact of this line type remained open, the compressor cannot be restarted. The compressor would have to be replaced.

Another type of motor overload protection device breaks the control circuit. This is wired to the outside of the compressor to the control circuit. If the pilot-duty type wired to the outside of the compressor were to remain open, an external overload device could be substituted.

THE SERVICEABLE HERMETIC COMPRESSOR. The serviceable hermetic normally has a cast iron shell and is considered a low-side device, Figure 23–37. Because of the piping arrangement in the head, the discharge gas is contained either under the head or out the discharge line. The motor is rigidly mounted to the shell, and the compressor must be externally mounted on springs or other flexible mounts to prevent vibration. The serviceable hermetic is used exclusively in larger compressor sizes because it can be rebuilt. The compressor components are much the same as the components in the welded hermetic type.

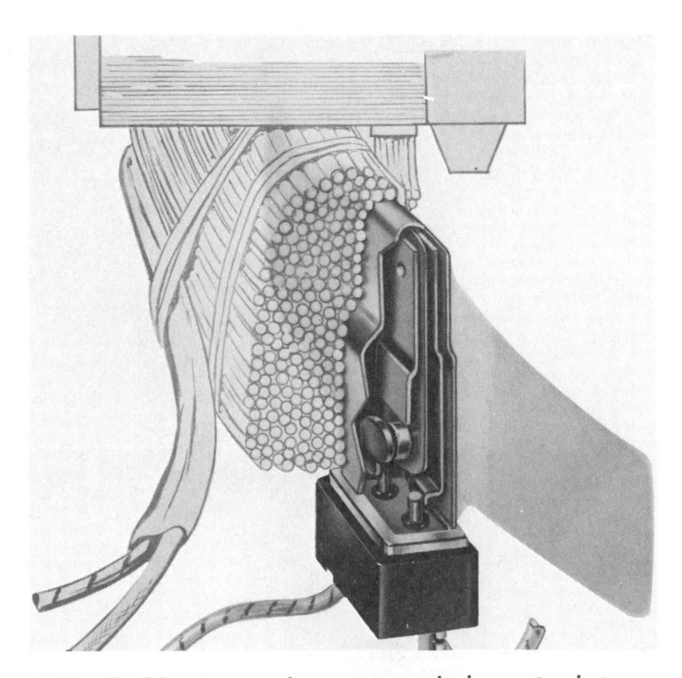

FIGURE 23–36 An internal compressor overload protection device that breaks the line circuit. Because this set of electrical contacts is inside the refrigerant atmosphere, they are contained inside a hermetic container of their own. If the electrical arc were allowed inside the refrigerant atmosphere, the refrigerant would deteriorate in the vicinity of the arc. *Courtesy Tecumseh Products Company*

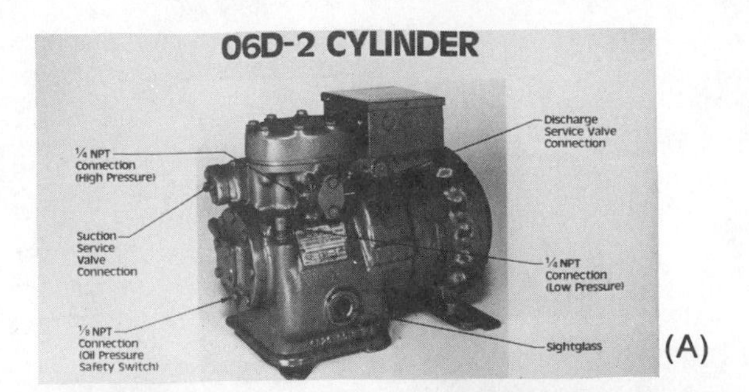

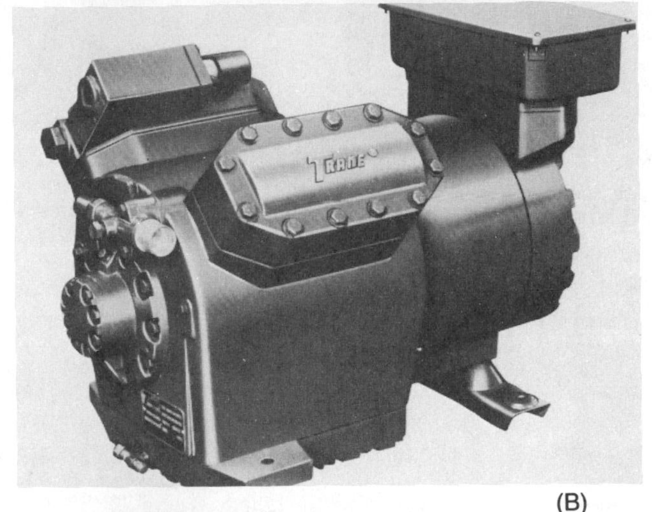

FIGURE 23–37 Serviceable hermetic compressors. *(A) Reproduced courtesy of Carrier Corporation, Courtesy (B) Trane Company, (C) Copeland Corporation*

OPEN-DRIVE COMPRESSOR. Open-drive compressors are manufactured with the motor external to the compressor shell. The shaft protrudes through the casing to the outside where

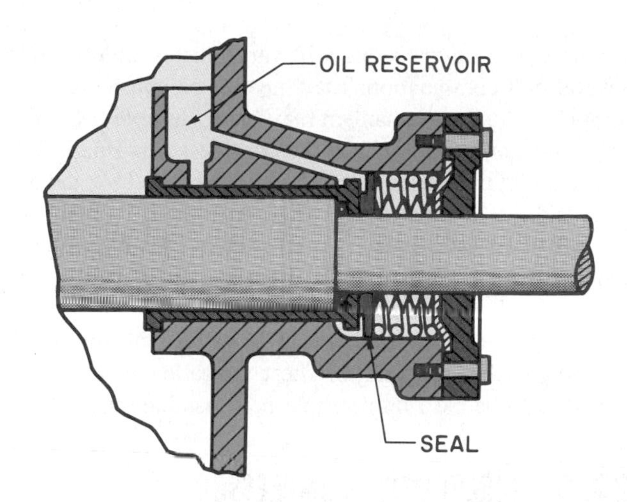

FIGURE 23–38 Shaft seal is responsible for keeping the refrigerant inside the crankcase and allowing the shaft to turn at high speed. This seal must be installed correctly. If the seal is installed on a belt-drive compressor, the belt tension is important. If it is installed on a direct-drive compressor, the shaft alignment is important. *Reproduced courtesy of Carrier Corporation*

either a pulley or a coupling is attached. This compressor is normally heavy duty in nature. It must be mounted tightly to a foundation. The motor is either mounted end to end with the compressor shaft or beside the compressor and belts used to turn the compressor.

The Shaft Seal

The pressure inside the compressor crankcase can be either in a vacuum (below atmosphere) or a positive pressure. If the unit were an extra–low-temperature unit using R-12 as the refrigerant, the crankcase pressure could easily be in a vacuum. If the shaft seal were to leak, the atmosphere would enter the crankcase. When the compressor is setting still, it could have a high positive pressure on it. For example, when R-502 is used and the system is off for extended periods (the whole system may get up to 100°F in a hot climate), the crankcase pressure may go over 200 psig. The crankcase shaft seal must be able to hold refrigerant inside the compressor under all of these conditions and while the shaft is turning at high speed, Figure 23–38.

The shaft seal has a rubbing surface to keep the refrigerant and the atmosphere separated. This surface is normally a carbon material rubbing against a steel surface. If assembled correctly, these two surfaces can rub together for years and not wear out. This correct assembly normally consists of the shafts being aligned correctly on a system where the bearings are in good working order. The belts must have the correct tension if the unit is a belt drive. If the unit is direct drive, the shafts have to be aligned according to manufacturer's instructions.

23.4 BELT-DRIVE MECHANISM CHARACTERISTICS

With the belt-drive compressor, the motor is mounted at the side of the compressor and has a pulley on the motor as well as on the compressor. The pulley on the motor is called the *drive pulley,* and the pulley on the compressor is called the *driven pulley*. The drive pulley is sometimes adjustable, which allows the compressor speed to be adjustable. The drive pulley can also be changed to a different size to vary the compressor speed. This can be advantageous when a compressor is too large for a job (too much capacity) and needs to be slowed down to compensate. Different pulleys are shown in Figure 23–39.

The compressor can also be speeded up for more capacity if the motor has enough horsepower in reserve (if the motor is not already running at maximum horsepower). If the pulley size change is needed, consult the compressor manufacturer to be sure that the design limits of the compressor are not exceeded.

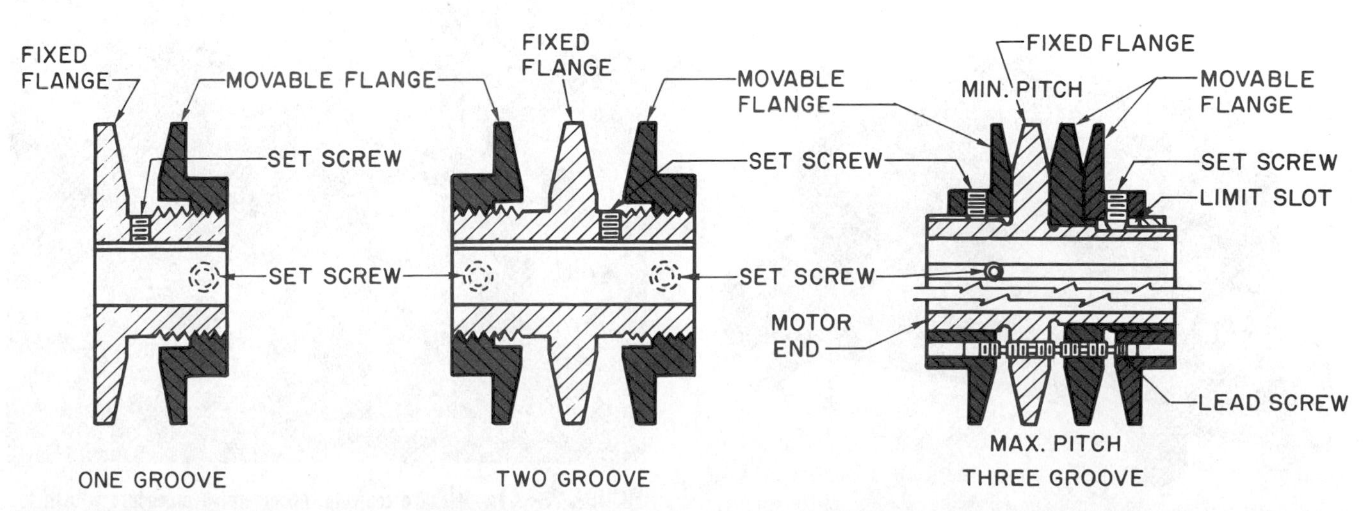

FIGURE 23–39 Different pulleys used on belt-drive compressors. They are single-groove and multiple-groove pulleys with different belt widths. Some pulleys are adjustable for changing the compressor speed.

The formula for determining the size of the drive pulley is determined from information gathered from the existing system. The motor speed is normally fixed, 1725 or 3450 rpm and the pulley (flywheel) on the compressor is normally a fixed size. The only variable is the drive pulley. A typical problem may be: a motor with an rpm of 1725 and a drive pulley of 4 in. is used to drive a compressor at 575 rpm with a pulley size of 12 in. It is desirable to reduce the compressor speed to 500 rpm to reduce the capacity. What would the new drive pulley size be? The formula is

Drive Pulley Size × Drive Pulley rpm =
Driven Pulley Size × Driven Pulley rpm, or
Pulley 1 × rpm 1 = Pulley 2 × rpm 2
(the desired new rpm)

Solve the problem for pulley 1, restated as

$$\text{Pulley } 1 = \frac{\text{Pulley } 2 \times \text{rpm } 2}{\text{rpm } 1}$$

$$\text{Pulley } 1 = \frac{12 \text{ in.} \times 500 \text{ rpm}}{1725 \text{ rpm}}$$

$$\text{Pulley } 1 = 3.48 \text{ or } 3\tfrac{1}{2} \text{ in.}$$

Most compressors are not designed to operate over a certain rpm, and the compressor manufacturer has this information. When a particular pulley size is needed, the pulley supplier will help you choose the correct size. Belt sizes also must be calculated. Choosing the correct belts and pulleys is important and should be done by an experienced person.

The compressor and motor shafts must be in correct alignment with the proper tension applied to the belts. The motor base and the compressor base must be tightened rigidly so that no dimensions will vary during the operation. Several belt combinations may have to be considered. The compressor drive mechanism may have multiple belts, Figure 23–40, or a single belt if the compressor is small. Belts also come in different types. The width of the belt, the grip type, and the material have to be considered.

When multiple belts are used, they have to be bought as matched sets. For example, if a particular compressor and motor drive has four V belts of B width and 88 in. long, you should order four 88-in. belts of B width that are factory matched for the exact length. These are called *matched belts* and have to be used on multiple-belt installations.

23.5 DIRECT-DRIVE COMPRESSOR CHARACTERISTICS

The direct-drive compressor is limited to the motor speed that the drive motor is turning. In this type of installation the motor and compressor shafts are end to end. These shafts have a slightly flexible coupling between them but must be in very close alignment, or the bearings and seal will fail prematurely, Figure 23–41. This compressor and motor combination is mounted on a common rigid base.

Both the motor and compressor are customarily manufactured so that they can be rebuilt in place. Thus, once the shafts are aligned, the shell of the motor and the shell of the compressor can be fastened down and always remain in place. If the motor or compressor has to be rebuilt, the internal parts may be removed for rebuilding, then the shafts will automatically line back up when reassembled.

The reciprocating compressor has not changed appreciably for many years. Manufacturers continuously try to improve the motor and the pumping efficiencies. The valve arrangements can make a difference in the pumping efficiency and are being studied for improvement.

FIGURE 23–40 Multiple-belt compressor. *Courtesy Tecumseh Products Company*

FIGURE 23–41 Flexible coupling. An extensive procedure is used to obtain the correct shaft alignment. This must be done or bearings and seal will fail prematurely. *Courtesy Lovejoy, Inc.*

23.6 RECIPROCATING COMPRESSOR EFFICIENCY

A compressor's efficiency is determined by the design of the compressor. The efficiency of a compressor starts with the filling of the cylinder. The following sequence of events takes place inside a reciprocating compressor during the pumping action.

A medium-temperature application will be used as an example for the pumping sequence. The suction pressure is 20 psig, and the discharge pressure is 180 psig.

1. **Piston at the top of the stroke and starting down.** When the piston has moved down far enough to create less pressure in the cylinder than is in the suction line, the intake flapper valve will open and the cylinder will start to fill with gas, Figure 23–42.
2. **Piston continues to the bottom of the stroke.** At this point the cylinder is nearly as full as it is going to get. There is a very slight time lag at the bottom of the stroke as the crankshaft carries the rod around the bottom of the stroke, Figure 23–43.
3. **Piston is starting up.** The rod throw is past dead-center, and the piston starts up. When the cylinder is as full as it is going to get, the suction flapper valve closes.
4. **The piston proceeds to the top of the stroke.** When the piston reaches a point that is nearly at the top, the pressure in the cylinder becomes greater than the pressure in the discharge line. If the discharge pressure is 180 psig, the pressure inside the cylinder may have to reach 190 psig to overcome the discharge valves and spring tension, Figure 23–44.

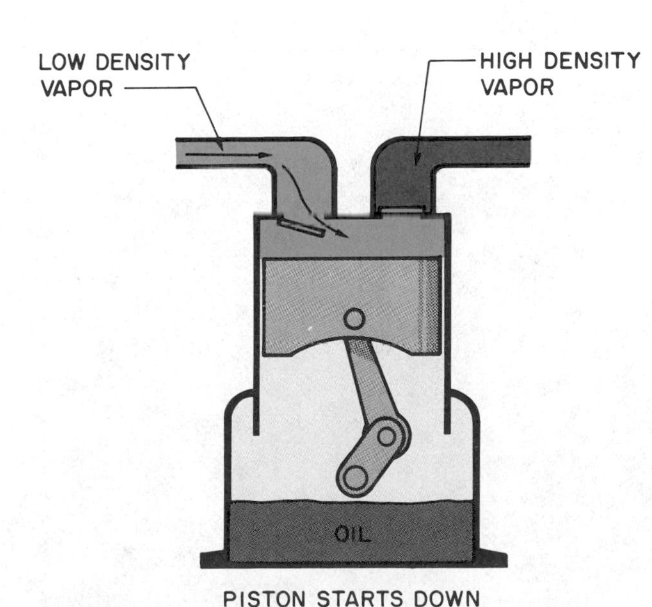

FIGURE 23–42 An illustration of what happens inside the reciprocating compressor while it is pumping. When the piston starts down, a low pressure is formed under the suction reed valve. When this pressure becomes less than the suction pressure and the valve spring tension, the cylinder will begin to fill. Gas will rush into the cylinder through the suction reed valve.

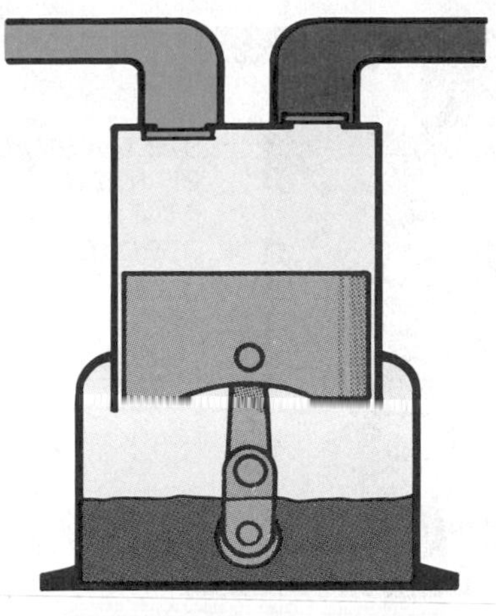

FIGURE 23–43 When the piston gets near the bottom of the stroke, the cylinder is nearly as full as it is going to get. There is a short time lag as the crankshaft circles through bottom dead-center, during which a small amount of gas can still flow into the cylinder.

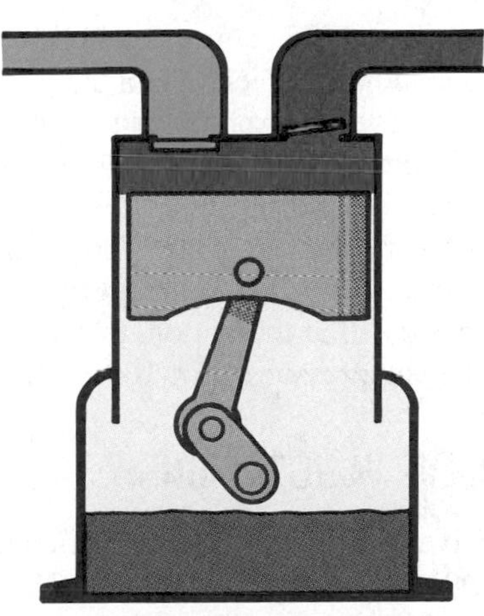

FIGURE 23–44 When the piston starts back up and gets just off the bottom of the cylinder, the suction valve will have closed and pressure will begin to build in the cylinder. When the piston gets close to the top of the cylinder, the pressure will start to approach the pressure in the discharge line. When the pressure inside the cylinder is greater than the pressure on the top side of the discharge reed valve, the valve will open, and the discharge gas will empty out into the high side of the system.

5. **The piston is at exactly top dead-center.** This is as close to the top of the head as it can go. There has to be a certain amount of clearance in the valve assemblies and between the piston and the head or they would touch. This clearance is known as *clearance volume*. The piston is going to push as much gas out of the cylinder as time and clearance volume will allow. There will be a small amount of gas left in the clearance

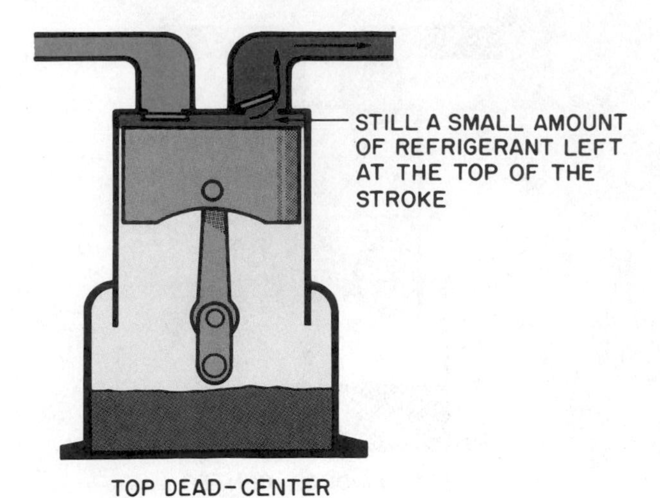

FIGURE 23–45 A reciprocating compressor cylinder cannot completely empty because of the clearance volume at the top of the cylinder. The manufacturers try to keep this clearance volume to a minimum but cannot completely do away with it.

volume, Figure 23–45. This gas will be at the discharge pressure mentioned earlier. When the piston starts back down, this gas will reexpand, and the cylinder will not start to fill until the cylinder pressure is lower than the suction pressure of 20 psig. This reexpanded refrigerant is part of the reason that the compressor is not 100% efficient. Valve design and the short period of time the cylinder has to fill at the bottom of the stroke are other reasons the compressor is not 100% efficient.

23.7 DISCUS VALVE DESIGN

The Discus valve design allows a closer tolerance inside the compressor cylinder at top-dead-center. This closer tolerance gives the compressor more efficiency because of less clearance volume. The discus valve also has a larger bore and allows more gas through the port in a short period of time. Figure 23–46 shows a Discus compressor, a conventional valve and plate, and a Discus valve and plate.

23.8 LIQUID IN THE COMPRESSOR CYLINDER

The piston-type reciprocating compressor is known as a positive displacement device. This means that when the cylinder starts on the upstroke, it is going to empty itself or stall. ***If the cylinder is filled with liquid refrigerant that does not compress (liquids are not compressible), something is going to break. Piston breakage, valve breakage, and rod breakage can all occur if a large amount of liquid reaches the cylinder, Figure 23–47.***

Liquid in compressors can be a problem from more than one standpoint. Large amounts of liquid, called a *slug* of liquid or *liquid slugging*, usually cause immediate damage. Small amounts of liquid floodback can be just as detrimental but with a slower action. When small amounts of

(A)

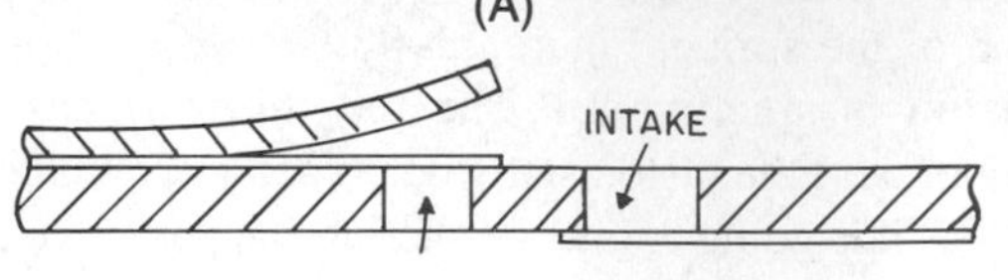

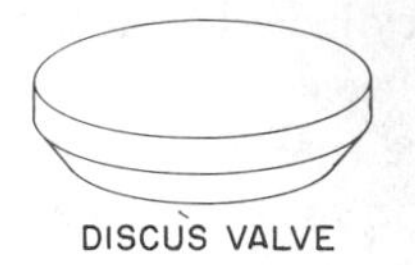

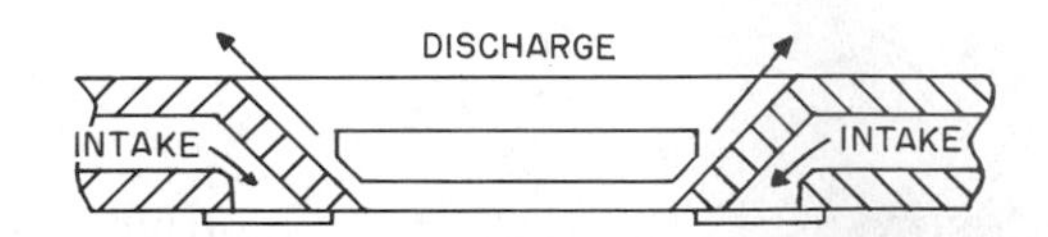

(B)

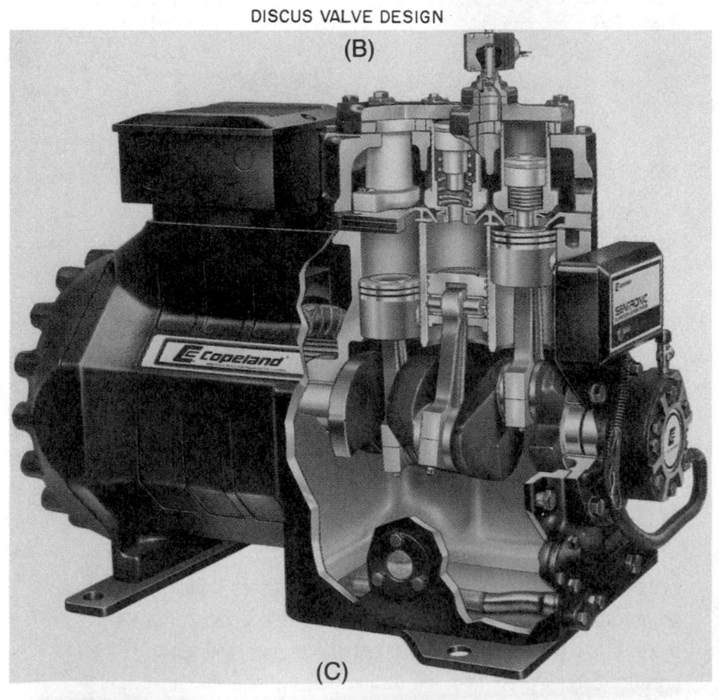

(C)

FIGURE 23–46 (A) Discus compressor. (B) Discus valve design. (C) Cutaway view of Discus compressor. *Courtesy Copeland Corporation*

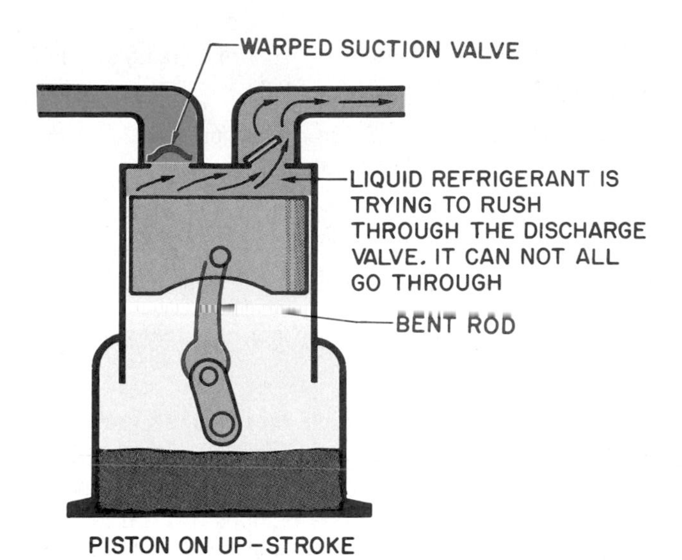

FIGURE 23–47 Cylinder trying to compress liquid. Something has to give.

liquid enter the compressor, they can cause oil dilution. If the compressor has no oil lubrication protection, this may not be noticed until the compressor fails. One of the aspects of this failure is that marginal oil pressure may cause the compressor to throw a rod. Throwing a rod would be called mechanical failure. This may cause motor damage and burn out the motor. The technician may erroneously diagnose this as an electrical problem. If the compressor is a welded hermetic, the technician may have difficulty diagnosing the problem until another failure occurs. ***Never take a compressor failure for granted. Give the system a thorough checkout on startup and actively look for a problem.***

23.9 SYSTEM MAINTENANCE AND COMPRESSOR EFFICIENCY

The compressor's overall efficiency can be improved by maintaining the correct working conditions. This involves keeping the suction pressure as high as practical and the head pressure as low as practical within the design parameters.

A dirty evaporator will cause the suction pressure to drop. When the suction pressure goes below normal, the vapor that the compressor is pumping becomes less dense and gets thin, sometimes called rarified vapor. The compressor performance decreases.

In the past, not much attention has been given to dirty evaporators. Technicians are now beginning to realize that low evaporator pressures cause high compression ratios. For example, if an ice cream storage walk-in cooler is designed to operate at −5°F room temperature it would result in a coil temperature of about −20°F (using a coil to return air temperature of 15°F). If the refrigerant is R-12, the suction pressure would be 0.6 psig, slightly above atmospheric pressure. If the condenser operates at a 25°F temperature difference with the outside air and it is 95°F the following calculation would show the compression ratio to be 11:1, Figure 23–48A.

Compression Ratio = Absolute Discharge Pressure ÷ Absolute Suction Pressure

CR = 157.6 psig + 15 (atmosphere) ÷ .6 psig + 15 (atmosphere)

CR = 11.1:1

157.6 psig = condensing temperature for R-12 at 120°F (95°F + 25 = 120°)

0.6 psig = evaporator or boiling temperature at −20°F

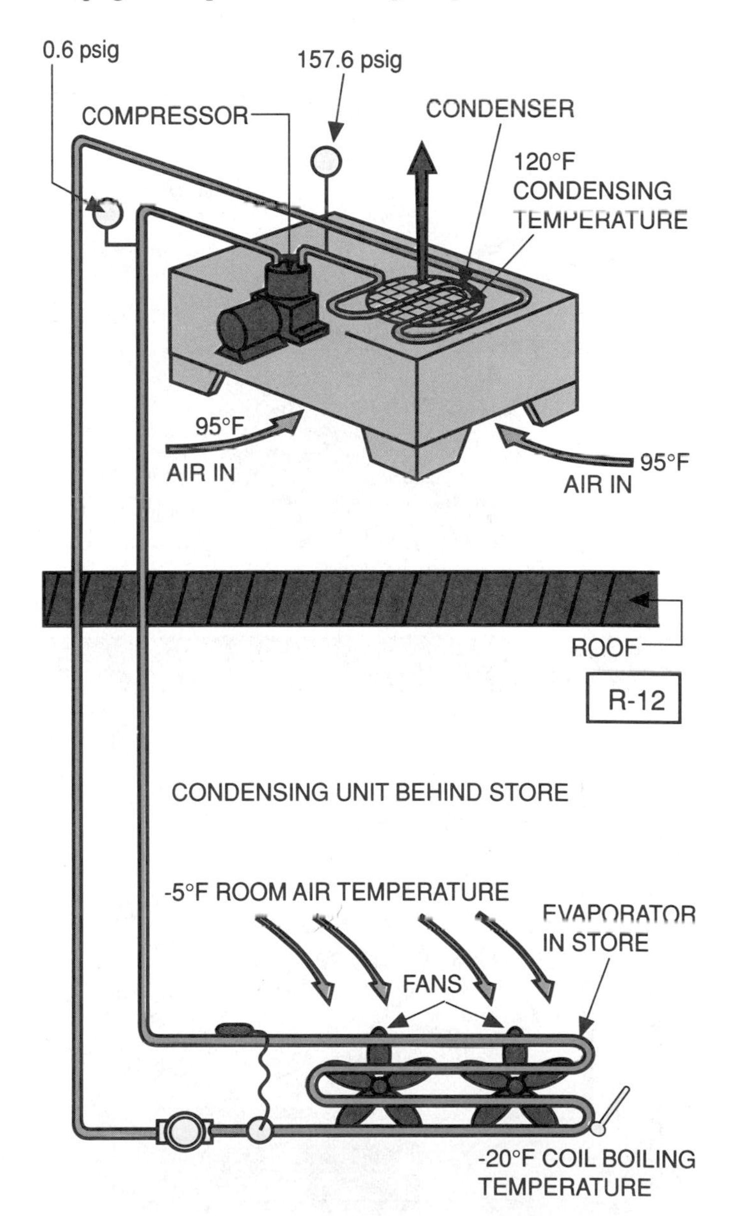

FIGURE 23–48A System operating with normal compression ratio. Notice the condenser is large enough to keep the head pressure low.

If the coil were old and dirty, Figure 23–48B, and the coil temperature reduced to −30°F, the head pressure would remain the same, but a new refrigerant boiling temperature of −30°F would produce a suction pressure of 5.5 in. Hg vacuum. This is below atmospheric pressure. It is an absolute pressure of 11.99 psia. (This was arrived at by converting inches of mercury vacuum to inches of mercury absolute then converting this to psia.)

29.92 in. Hg absolute − 5.5 in. Hg vacuum = 24.42 in. Hg absolute

24.42 in. Hg absolute × 2.036 in. Hg per psi = 11.99 psia

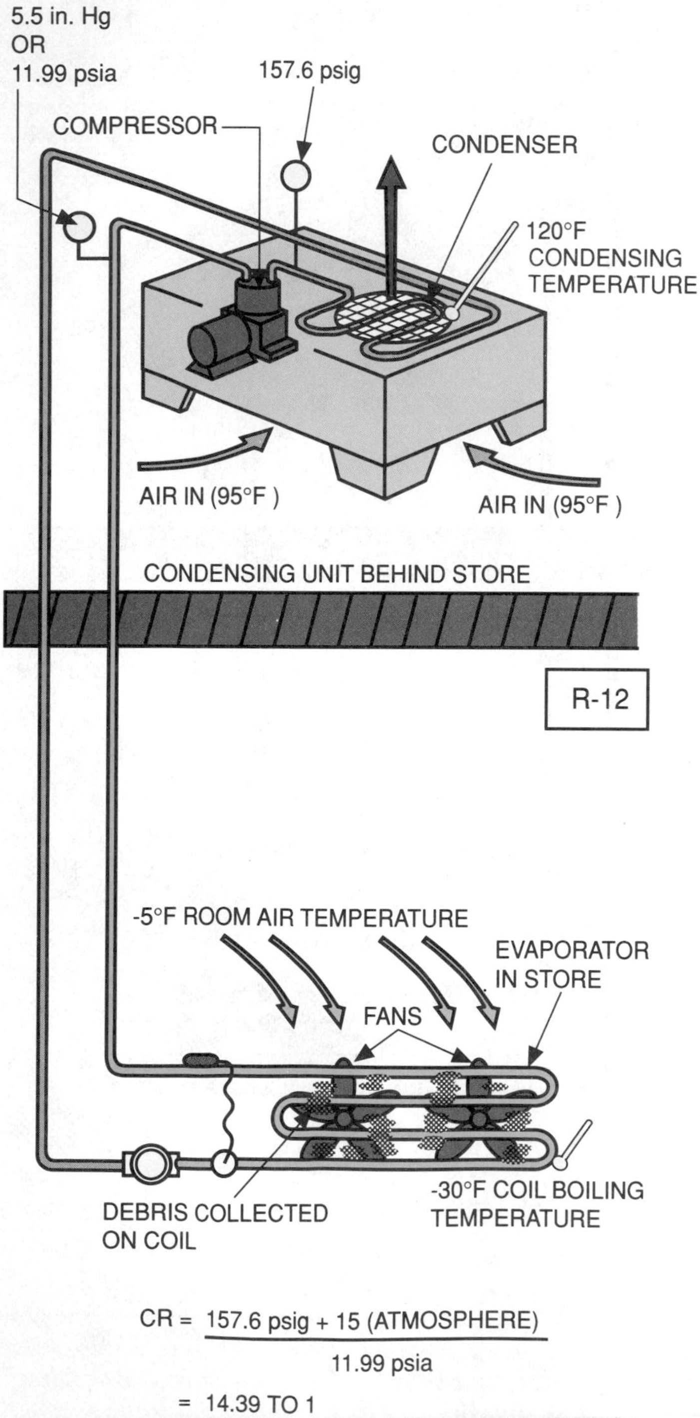

The new compression ratio is 14.39.

CR = 157.6 + 15 (atmosphere) ÷ 11.99 psia

CR = 14.39:1

This compression ratio is too high. Most manufacturers recommend that the compression ratio not be over 12:1.

The same situation could occur if the store manager were to turn the refrigerated box thermostat down to -15°. The coil temperature would go down to -30°F and the compression ratio would be 14.39:1. This is too high. Many low-temperature coolers are operated at temperatures far below what is necessary. This condition often causes compressor problems.

Dirty condensers also cause compression ratios to rise, but they do not rise as fast as with a dirty evaporator. For example, if the above ice cream storage box were operated at 0.6 psig evaporator and the condenser became dirty to the point that the head pressure rises to 175 psig, the compression ratio would be 12.18:1, Figure 23–48C. This is not good, but not as bad as the 14.39:1 ratio in the previous example.

CR = 175 psig + 15 (atmosphere) ÷ .6 psig + 15 (atmosphere)

CR = 12.18:1

If both situations were to occur at the same time, as in Figure 23–48D, the compressor would have a compression ratio of 15.85:1. The compressor would probably not last very long under these conditions.

CR = 175 psig + 15 (atmosphere) ÷ 11.99 psia

CR = 15.85:1

The technician has a responsibility to maintain the equipment to obtain the greatest efficiency and to otherwise protect the equipment with a maintenance program. Clean coils are part of this program. The owner usually does not know the difference.

A dirty condenser makes the head pressure rise. This causes the amount of refrigerant in the clearance volume (at the top of the compressor cylinder) to be greater than the design conditions allow. This makes the compressor efficiency drop. If there is a dirty condenser (high head pressure) and a dirty evaporator (low suction pressure), the compressor will run longer to keep the refrigerated space at the design temperature. The overall efficiency drops. A customer with a lot of equipment may not be aware of this if only part of the equipment is not efficient.

When the efficiency of a compressor drops, the owner is paying more money for less refrigeration. A good maintenance program is economical in the long term.

5.5 in. Hg OR 11.99 psia
157.6 psig
COMPRESSOR
CONDENSER
120°F CONDENSING TEMPERATURE
AIR IN (95°F)
AIR IN (95°F)
CONDENSING UNIT BEHIND STORE
R-12
-5°F ROOM AIR TEMPERATURE
EVAPORATOR IN STORE
FANS
-30°F COIL BOILING TEMPERATURE
DEBRIS COLLECTED ON COIL

$$CR = \frac{157.6 \text{ psig} + 15 \text{ (ATMOSPHERE)}}{11.99 \text{ psia}} = 14.39 \text{ TO } 1$$

FIGURE 23–48B This system has a dirty evaporator.

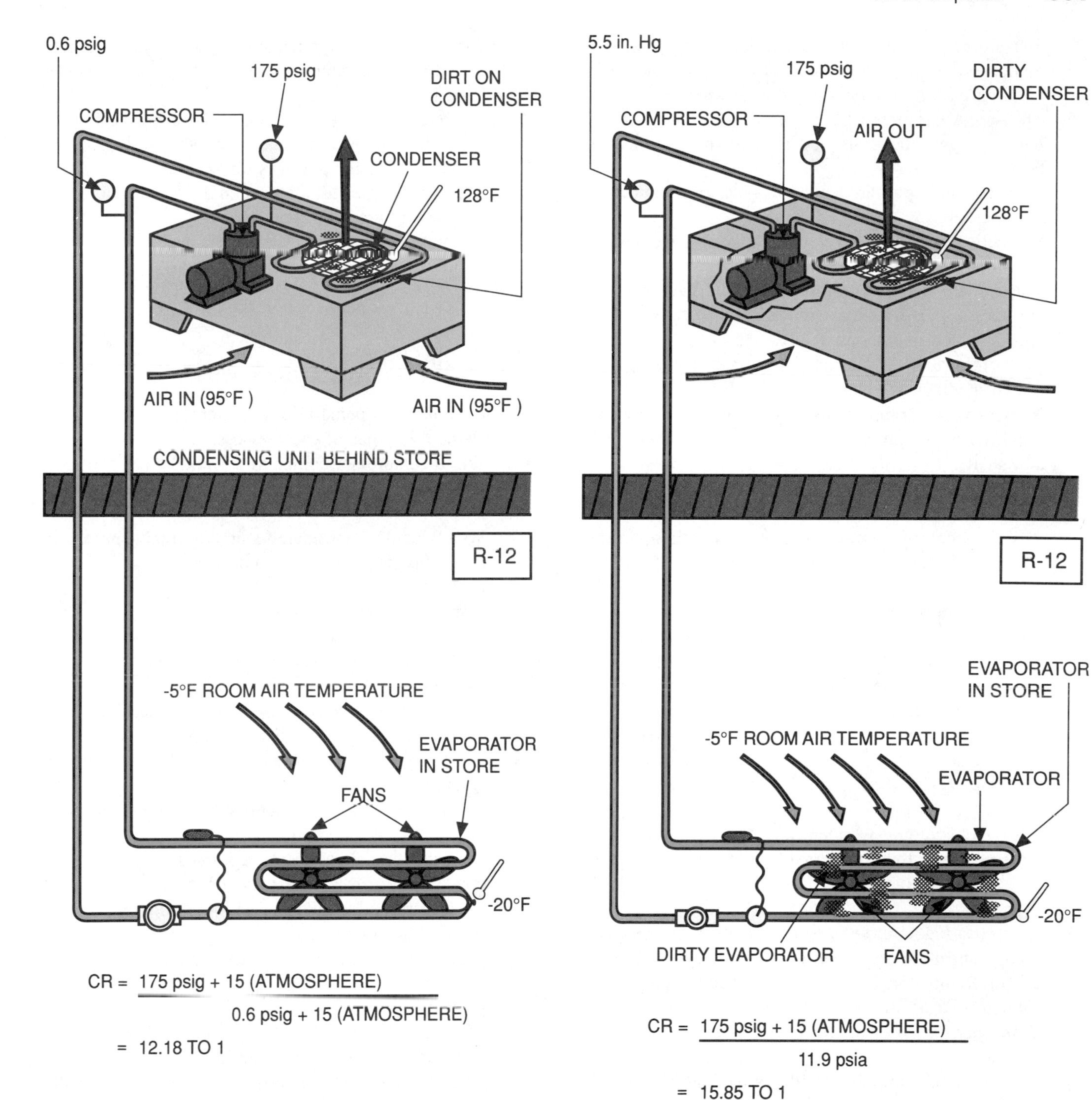

FIGURE 23–48C System with high head pressure because the condenser is dirty.

FIGURE 23–48D This system has a dirty condenser and a dirty evaporator.

SUMMARY

- The compressor is a vapor pump.
- The compressor cannot compress liquid refrigerant.
- The compressor lifts the low-pressure gas from the suction side of the system to the discharge side of the system.
- The discharge gas can be quite hot because the heat contained in the cool suction gas is concentrated when compressed in the compressor.
- Additional heat is added to the gas as it passes through the compressor because some of the work energy does not convert directly to compression but converts to heat.
- A discharge line on a compressor on a hot day can be as hot as 200°F and still be normal.

- Two types of compressors are usually used to achieve compression in commercial refrigeration: the reciprocating and the rotary screw compressor.
- The screw compressor compresses vapor between tapered, screw-type gears and is used in large installations.
- Hermetic and open-drive are two types of reciprocating compressors.
- Hermetic compressors are manufactured as welded hermetic and serviceable hermetic.
- The welded hermetic compressor must have the shell cut open for it to be serviced, and this work is done by special rebuilding shops only.
- In the serviceable hermetic compressor the shell is bolted together and can be disassembled in the field.
- Reciprocating compressors have similar components: crankshafts, oil pump rods, pistons, valve plates, heads, and shells.
- Most reciprocating compressor motors are cooled by suction gas. Some are air cooled.
- ***In all hermetic compressors the motor is operating in the refrigerant atmosphere and special precautions must be taken in the manufacture and service of these motors.***
- Hermetic compressors have a shaft with the motor on one end and the compressor on the other end.
- Special internal overload devices are used on hermetic motors that operate inside the refrigerant atmosphere.
- One type of internal overload protection device interrupts the actual line current. If this overload device does not close back when it should, the compressor is defective.
- The other internal overload device is a pilot-duty type that interrupts the control voltage. If something were to happen to this device, an external type could be installed.
- Reciprocating compressors are positive displacement pumps, meaning that when they have a cylinder full of gas or liquid, the cylinder is going to be emptied or damage will occur.
- With open compressors the motor is on the outside of the shell.
- The motor can either be found mounted beside the compressor, with the compressor and motor shafts side by side, or the motor may be mounted at the end of the compressor shaft with a flexible coupling between them.
- In either case, shaft-to-motor alignment is very important.
- Belts for belt-drive applications come in many different types. The manufacturer's supplier should be consulted for advice.
- Reciprocating compressor efficiencies depend primarily on the clearance volume and the motor efficiency.
- Discus valve design provides for smaller clearance volume, a greater area through which the gas can flow, and consequently greater efficiency.
- Continued compressor efficiencies depend to a great extent on a good maintenance program.

REVIEW QUESTIONS

1. Describe the operation of a compressor.
2. Name five types of compressors.
3. Will a compressor compress a liquid?
4. Why is the discharge gas leaving the compressor so hot?
5. What would be considered a high normal temperature for a discharge gas line on a reciprocating compressor?
6. How is the compressor motor normally cooled in a welded hermetic compressor?
7. Which compressor type uses pistons to compress the gas?
8. Which compressor type uses tapered, machined gear components to trap the gas for compression?
9. What type of compressor uses belts to turn the compressor?
10. What would be done to increase the compressor speed on a belt-driven compressor?
11. Describe the reciprocating compressor piston, rod, crankshaft, valves, valve plate, head, shaft seal, internal motor overload device, pilot-duty motor overload device, and coupling.
12. Name two things in the design of the reciprocating compressor that control the efficiency of the compressor.
13. Why is a Discus valve design more efficient than a conventional flapper valve design?
14. At what speeds does a hermetic compressor normally turn?
15. What effect does a slight amount of liquid refrigerant have on a compressor over a long period of time?
16. What lubricates the refrigeration compressor?
17. What can the service technician do to keep the refrigeration system operating at peak efficiency?

UNIT 24

Expansion Devices

OBJECTIVES

After studying this unit, you should be able to

- describe the three most popular types of expansion devices.
- describe the operating characteristics of the three most popular expansion devices.
- describe how the three expansion devices respond to load changes.

SAFETY CHECKLIST

* Wear warm clothing when working in a walk-in cooler or freezer.

24.1 EXPANSION DEVICES

The *expansion device,* often called the *metering device,* is the fourth component necessary for the compression refrigeration cycle to function. The expansion device is not as visible as the evaporator, the condenser, or the compressor. Generally, the device is concealed inside the cabinet and not obvious to the causal observer. It can either be a valve or a fixed-bore device. Figure 24–1 illustrates an expansion valve installed inside the evaporator cabinet.

The expansion device is one of the division lines between the high side of the system and the low side of the system (the compressor is the other). Figure 24–2 shows the

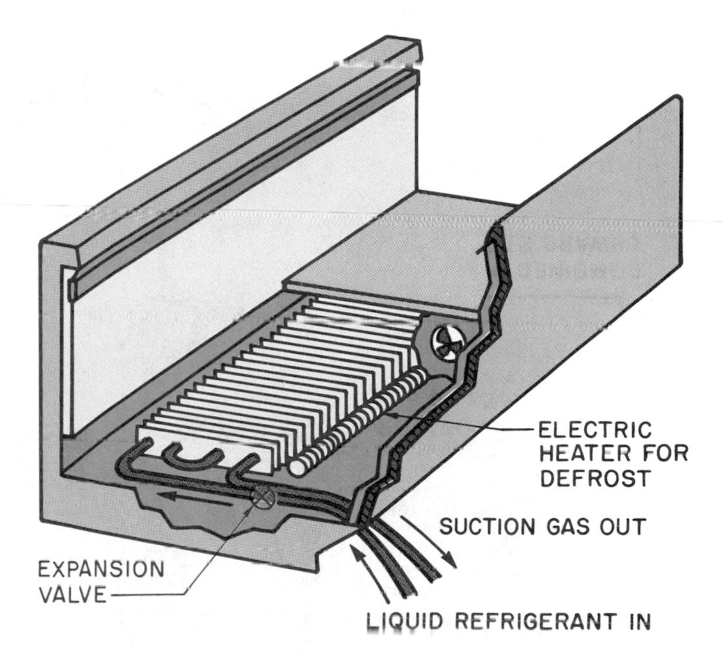

FIGURE 24–1 Metering device installed on a refrigerated case. The valve is not out in the open and is not as visible as the compressor, condenser, or evaporator.

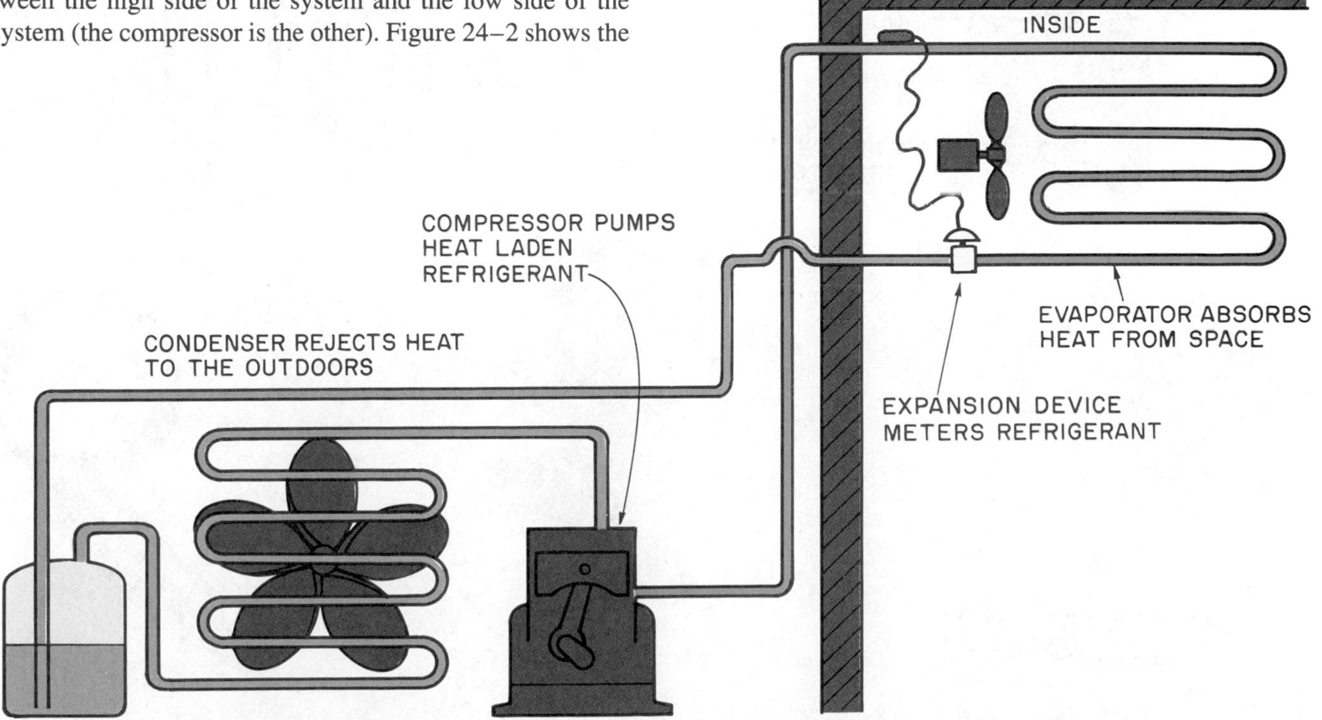

FIGURE 24–2 The complete refrigeration cycle with the four basic components: compressor, condenser, evaporator, and expansion device.

location of the device. The expansion device is responsible for metering the correct amount of refrigerant to the evaporator. The evaporator performs best when it is as full of liquid refrigerant as possible without any leaving in the suction line. Any liquid refrigerant that enters the suction line may reach the compressor because there should not be any appreciable heat added to the refrigerant in the suction line to boil the liquid to a vapor. Later in this section a suction line heat exchanger for special applications will be discussed, which is used to boil away liquid that may be in the suction line. Usually, liquid in the suction line is a problem.

The expansion device is normally installed in the liquid line between the condenser and the evaporator. The liquid line may be warm to the touch on a hot day and can be followed quite easily to the expansion device where there is

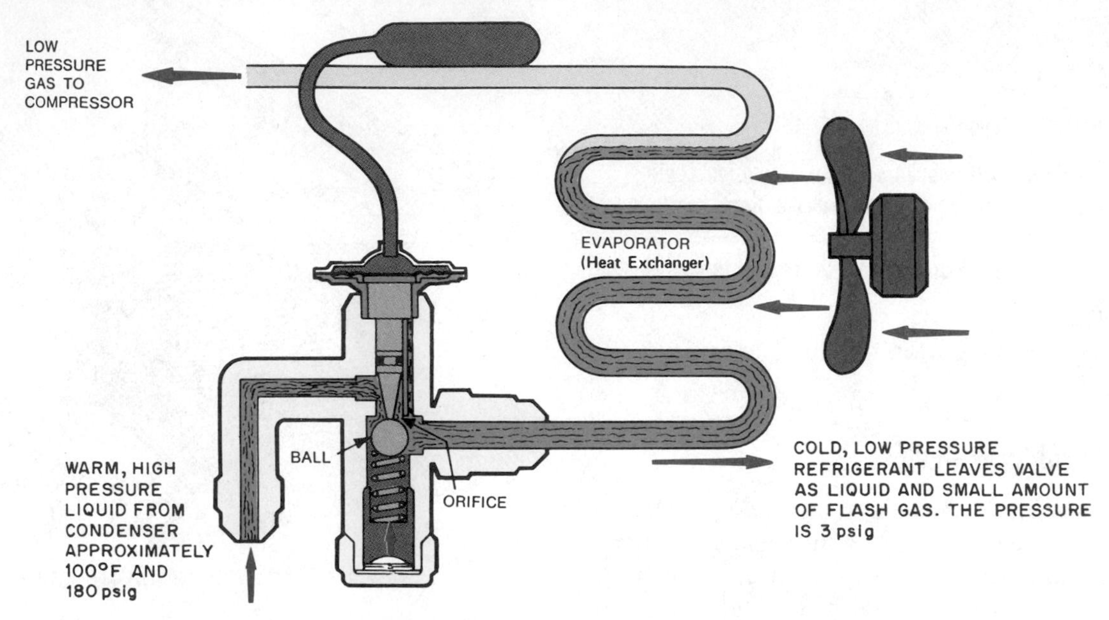

FIGURE 24–3 The expansion device has a dramatic temperature change from one side to the other. *Courtesy Parker Hannifin Corp.*

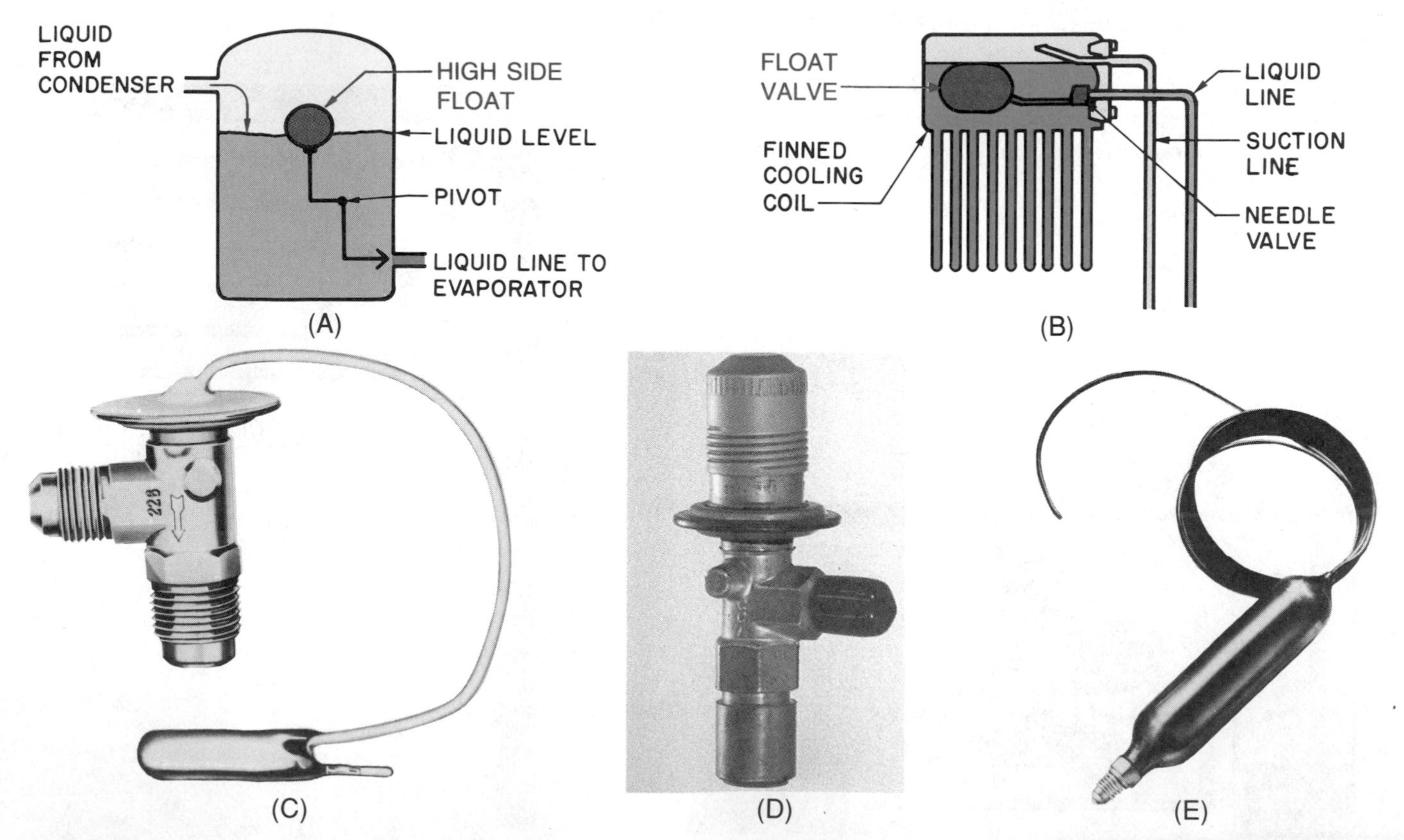

FIGURE 24–4 Five metering devices. (A) High-side float. (B) Low-side float. (C) Thermostatic expansion valve. (D) Automatic expansion valve.(E) Capillary tube. *Photos (C) Courtesy Singer Controls, (D) by Bill Johnson, (E) Courtesy Parker Hannifin Corp.*

a pressure drop and an accompanying temperature drop. For example, on a hot day the liquid line entering the expansion device may be 110°F. If this is a low-temperature cooler using R-12, the low-side pressure on the evaporator side may be 3 psig at a temperature of −15°F. This is a dramatic temperature drop and can be easily detected when found. The device may be warm on one side and frosted on the other, Figure 24–3. Because some expansion devices are valves and some are fixed-bore devices, this change can occur in a very short space—less than an inch on a valve, or a more gradual change on some fixed-bore devices.

Expansion devices come in five different types: (1) high-side float, (2) low-side float, (3) thermostatic expansion valve, (4) automatic expansion valve, and (5) fixed bore, such as the capillary tube, Figure 24–4. However, only three are currently being furnished with refrigeration equipment. The high-side float and the low-side float are not currently being used on typical refrigeration equipment and should not be encountered in the field.

24.2 THERMOSTATIC EXPANSION VALVE

The *thermostatic expansion valve* (TXV) meters the refrigerant to the evaporator using a thermal sensing element to monitor the superheat. This valve opens or closes in response to a thermal element. The TXV maintains a constant superheat in the evaporator. Remember, when there is superheat, there is no liquid refrigerant. Excess superheat is *not* desirable, but a small amount is necessary with this valve to ensure that no liquid refrigerant leaves the evaporator.

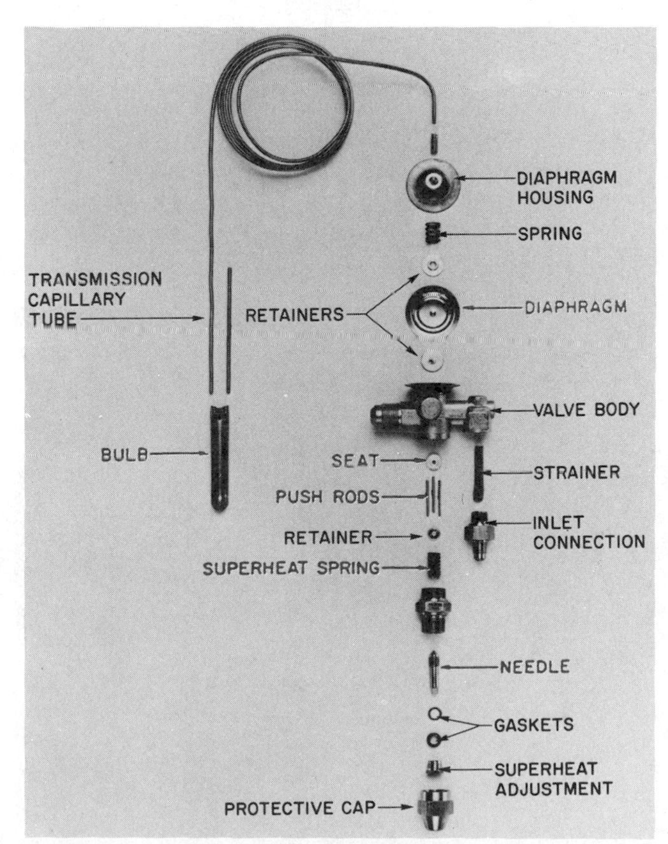

FIGURE 24–5 Exploded view of TXV. All parts are visible in the order in which they go together in the valve. *Courtesy Singer Controls Division, Schiller Park, Illinois*

24.3 TXV COMPONENTS

The TXV consists of the (1) valve body, (2) diaphragm (3) needle and seat, (4) spring, (5) adjustment and packing gland, and (6) the sensing bulb and transmission tube, Figure 24–5.

24.4 THE VALVE BODY

On common refrigerant systems the *valve body* is an accurately machined piece of solid brass or stainless steel that holds the rest of the components and fastens the valve to the refrigerant piping circuit, Figure 24–6. Notice that the valves have different configurations. Some of them are one

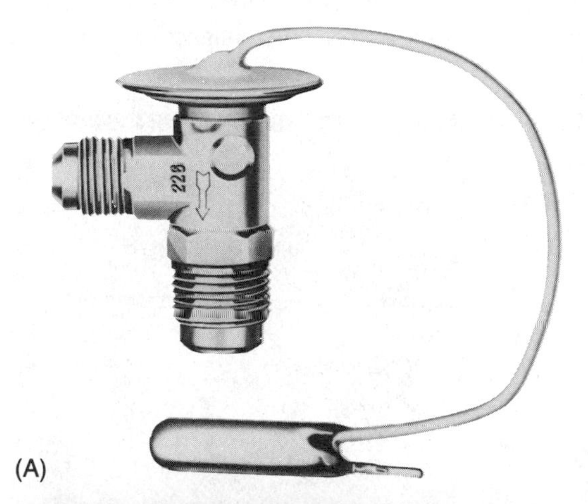

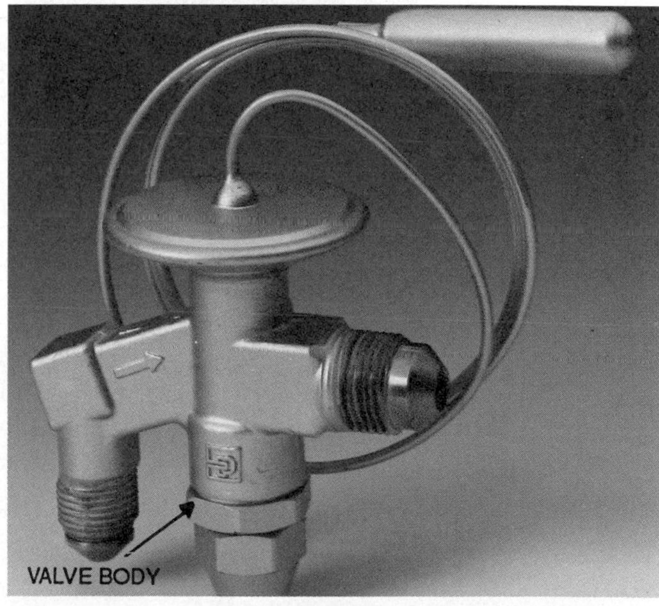

FIGURE 24–6 TXVs. All have remote sensing element. (A) Some of these valves are one-piece valves that are thrown away when defective. (B) Some can be rebuilt in place. This feature is particularly good when the valve is soldered into the system. *Courtesy (A) Singer Controls Division, Schiller Park, Illinois, (B) Parker Hannifin Corp.*

piece and cannot be disassembled, and some are made so that they can be taken apart.

These valves may be fastened to the system by three methods: flare, solder, or flange. When a valve is installed in a refrigeration system, future service should be considered, so a flare connection or a flange-type valve should be used, Figure 24–7. If a solder connection is used, a valve that can be disassembled and rebuilt in place is desirable, Figure 24–8. The valve often has an inlet screen with a very fine mesh to strain out any small particles that may stop up the needle and seat, Figure 24–9.

Some valves have a third connection called an *external equalizer.* This connection is normally a 1/4-in. flare or 1/4-in. solder and is on the side of the valve close to the diaphragm, Figure 24–10. As will be discussed later, the evaporator pressure has to be represented under the diaphragm of the expansion valve. When an evaporator has a very long circuit, a pressure drop in the evaporator may occur, and an external equalizer is used. A pressure connection is made at

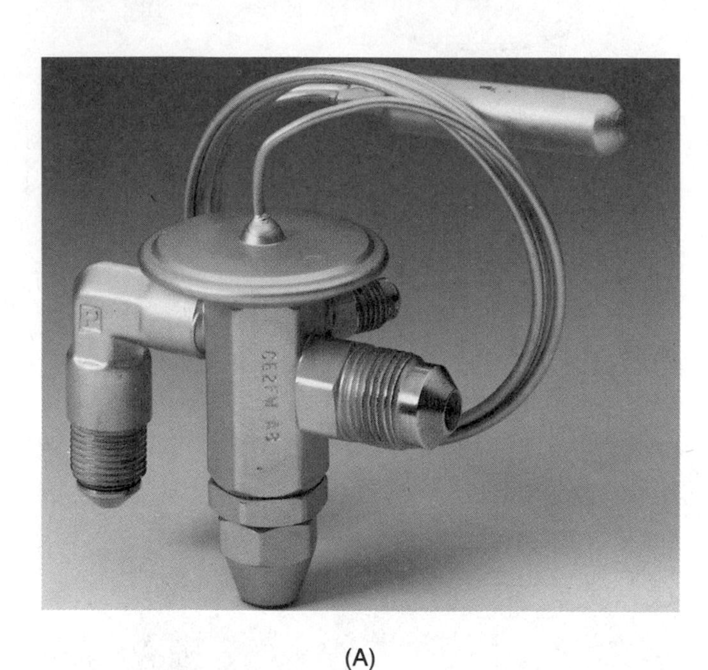

(A)

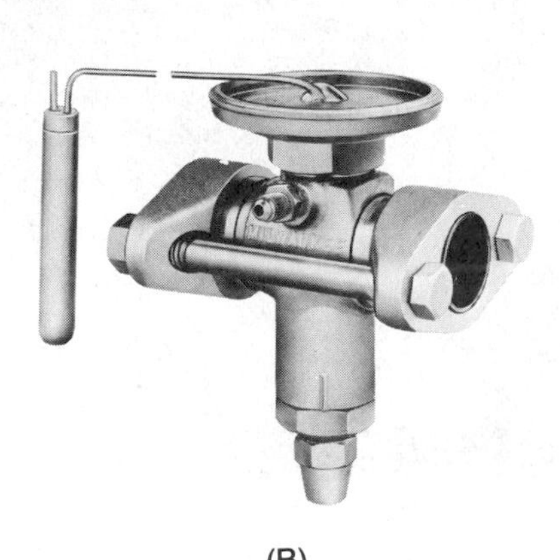

(B)

FIGURE 24–7 (A) Flare and the (B) flange-type valve. It can be removed from the system and replaced easily when it is installed where it can be reached with wrenches. *Courtesy (A) Parker Hannifin Corp., (B) Singer Controls Division, Schiller Park, Illinois*

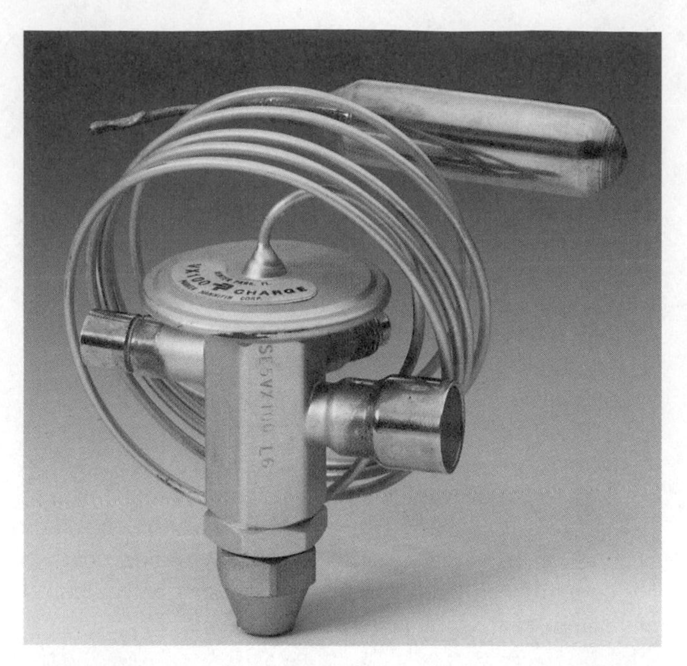

FIGURE 24–8 Solder-type valve that can be disassembled and rebuilt without taking it out of the system. This valve can be quite serviceable in some situations. *Courtesy Parker Hannifin Corp.*

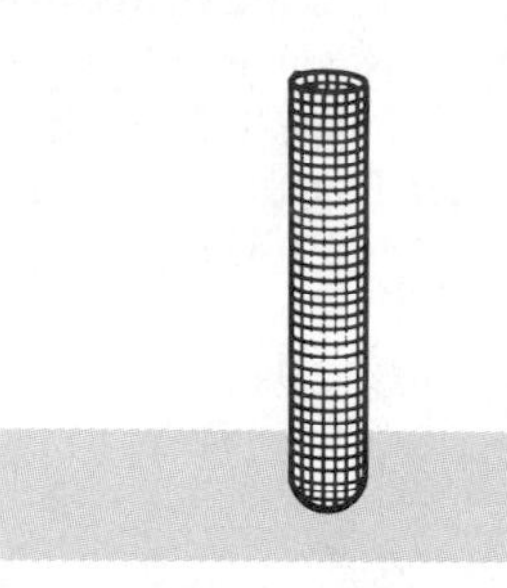

FIGURE 24–9 Most valves have some sort of inlet screen to strain any small particles out of the liquid refrigerant before they reach the very small opening in the expansion valve.

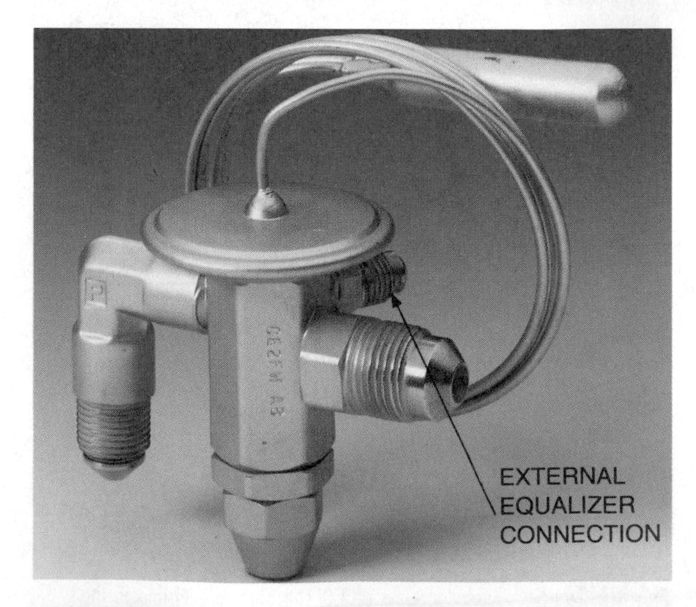

FIGURE 24–10 The third connection on this expansion valve is called the external equalizer. *Courtesy Parker Hannifin Corp.*

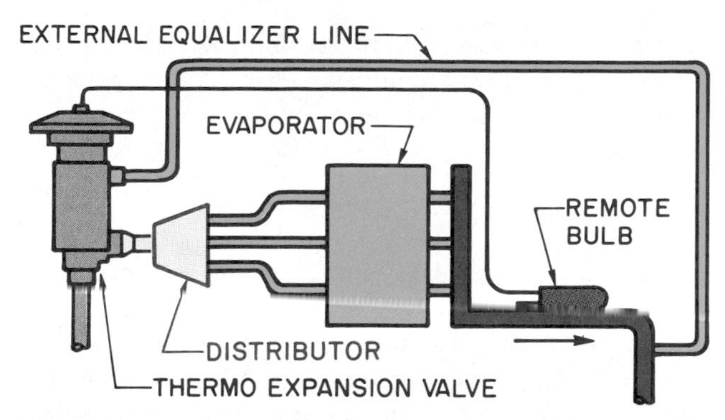

FIGURE 24–11 Evaporator with multiple circuits and an external equalizer line to keep pressure drop to a minimum. When an evaporator becomes so large that the length would create pressure drop, the evaporator is divided into circuits, and each circuit must have the correct amount of refrigerant.

the end of the evaporator that supplies the evaporator pressure under the diaphragm. Some evaporators have several circuits and a method of distributing the refrigerant that will cause pressure drop between the expansion valve outlet and the evaporator inlet. This installation must have an external equalizer for the expansion valve to have correct control of the refrigerant, Figure 24–11. This distributor will be discussed in more detail under refrigerant components.

24.5 THE DIAPHRAGM

The *diaphragm* is located inside the valve body and moves the needle in and out of the seat in response to system load changes. The diaphragm is made of thin metal and is under the round domelike top of the valve, Figure 24–12.

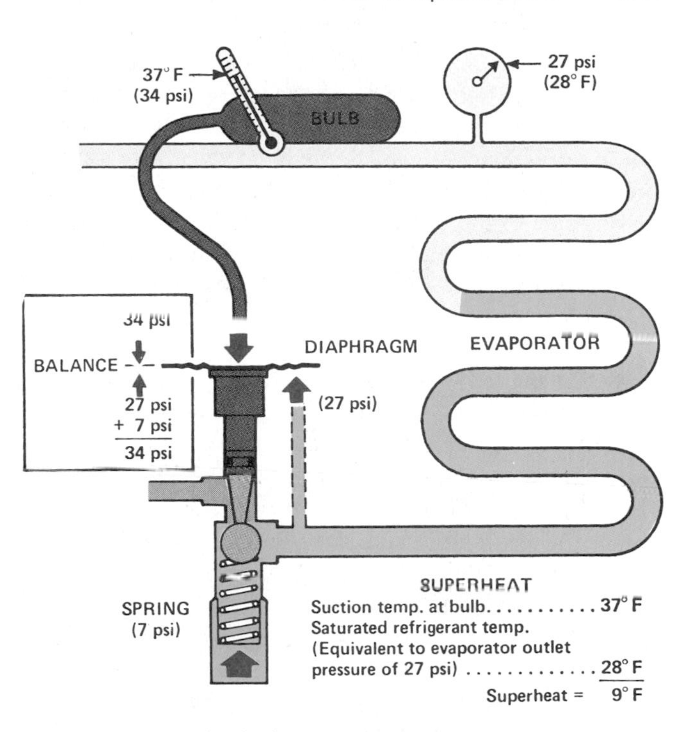

FIGURE 24–12 The diaphragm in the expansion valve is a thin membrane that has a certain amount of flexibility. It is normally made of a hard metal such as stainless steel. *Courtesy Parker Hannifin Corp.*

24.6 NEEDLE AND SEAT

The *needle and seat* control the flow of refrigerant through the valve. They are normally made of some type of very hard metal, such as stainless steel, to prevent the refrigerant passing through from eroding the seat. The needle and seat are used in a metering device so that close control of the refrigerant can be obtained, Figure 24–13. Some valve man-

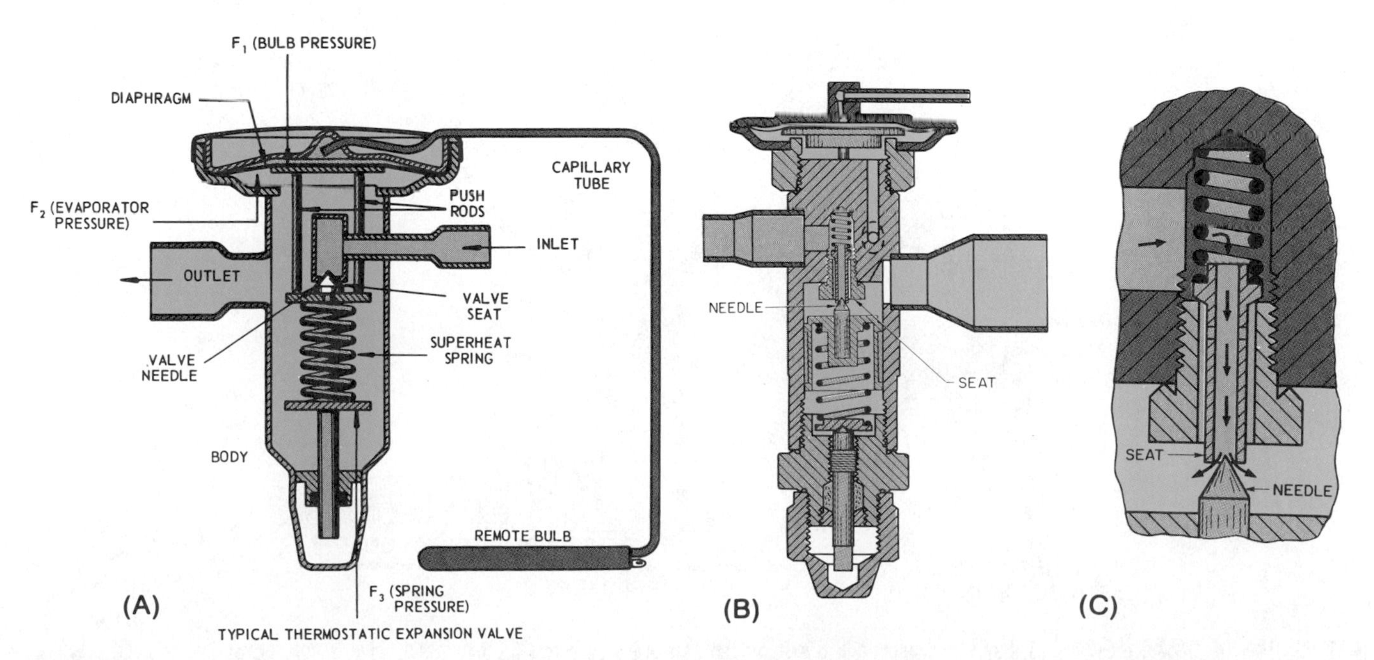

FIGURE 24–13 Needle and seat devices used in expansion valves. *Courtesy (A) Singer Controls Division, Schiller Park, Illinois, (B) and (C) Sporlan Valve Company*

ufacturers have needle and seat mechanisms that can be changed for different capacities or to correct a problem.

The size of the needle and seat determines how much liquid refrigerant will pass through the valve with a specific pressure drop. For example, when the pressure is 170 psig on one side of the valve and 2 psig on the other side of the valve, a measured and predictable amount of liquid refrigerant will pass through the valve. If this same valve were used when the pressure is 100 psig and 3 psig, the valve would not be able to pass as much refrigerant. The conditions that the valve will operate under must be considered when selecting the valve. The manufacturer's manual for a specific valve is the best place to get the proper information to make these decisions. The pressure difference from one side of the valve to the other is not necessarily the discharge versus the suction pressure. Pressure drop in the condenser and interconnecting piping may be enough to cause a problem if not considered, Figure 24–14. Notice in the figure that the actual discharge and suction pressure are not the same as the pressure drop across the expansion valve.

Thermostatic expansion valves are rated in tons of refrigeration at a particular pressure drop condition. The capacity of the system and the working conditions of the system must be known. For example, using the manufacturer's catalog in Figure 24–15, we see that a 1-ton valve is needed in a medium-temperature cooler with a 20°F evaporator and a 1-ton capacity, which is going to operate inside the store. The inside of the store is expected to stay at 70°F year around. The head pressure is expected to remain at a constant 125 psig. The 1-ton valve has a capacity of 1.3 ton at these conditions. If the same cooler is moved outside where the temperature is warmer and a head pressure control is used, the same valve will have a capacity of 1.6 ton. This is not enough to create a problem. The outside ambient temperature may vary from 0°F to 95°F. The head pressure control will vary the head pressure from 126 psig at the low end to 170 psig at the high end.

24.7 THE SPRING

The *spring* is one of the three forces that act on the diaphragm. It raises the diaphragm and closes the valve by pushing the needle into the seat. When a valve has an adjustment, the adjustment applies more or less pressure to the spring to change the tension for different superheat settings. The spring tension is factory set for a predetermined superheat of 8°F to 12°F, Figure 24–16.

The adjustment part of this valve can either be a screw slot or a square-headed shaft. Either type is normally covered with a cap to prevent water, ice, or other foreign matter from collecting on the stem. The cap also serves as a backup leak prevention. Most adjustment stems on expansion valves have a packing gland that can be tightened to prevent refrigerant from leaking. The cap would cover the stem and the gland, Figure 24–17. Normally one complete turn of the stem can change the superheat reading from 0.5°F to 1°F of superheat, depending on the manufacturer.

24.8 THE SENSING BULB AND TRANSMISSION TUBE

The *sensing bulb* and *transmission tube* are extensions of the valve diaphragm. The bulb detects the temperature at the end of the evaporator on the suction line and transmits this temperature, converted to pressure, to the top of the diaphragm. The bulb contains a fluid, such as refrigerant, that responds to a pressure/ temperature relationship chart just

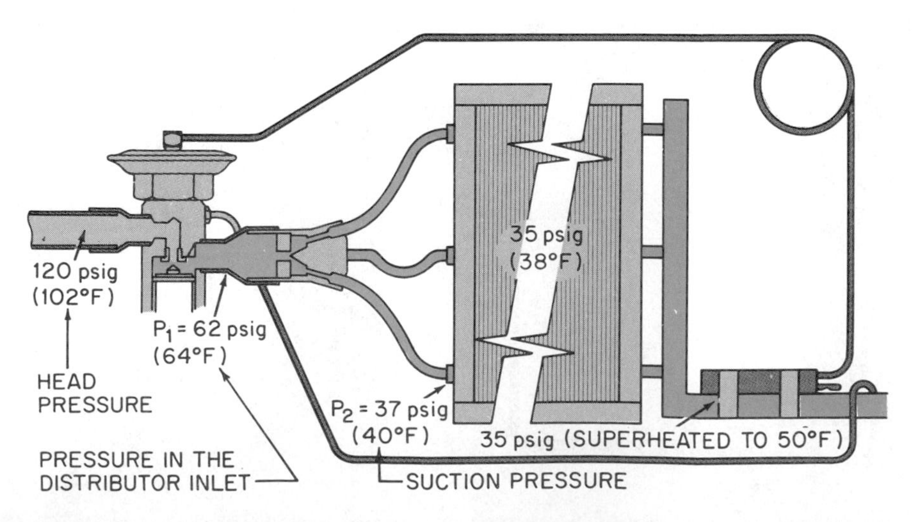

FIGURE 24–14 Real pressures as they would appear in a system. The pressure drop across the expansion valve is not necessarily the head pressure versus the suction pressure. The pressure drop through the refrigerant distributor is one of the big pressure drops that has to be considered. The refrigerant distributor is between the expansion valve and the evaporator coil. *Courtesy Sporlan Valve Company*

		EVAPORATOR TEMPERATURE (°F)															
		+50								+40							
	Nominal Capacity	PRESSURE DROP ACROSS VALVE (psi)															
Model	(tons)	40	60	80	100	125	150	175	200	40	60	80	100	125	150	175	200
128	1/4	0.32	0.38	0.43	0.48	0.55	0.60	0.65	0.72	0.30	0.34	0.38	0.43	0.49	0.53	0.55	0.57
223	1/2	0.50	0.67	0.79	0.86	0.93	1.2	1.3	1.4	0.45	0.65	0.75	0.84	0.90	1.0	1.1	1.3
226	1	0.95	1.1	1.2	1.5	1.7	1.8	2.1	2.3	0.90	1.0	1.2	1.4	1.6	1.7	2.0	2.2
228	1-1/2	1.5	1.6	1.9	2.1	2.4	2.5	2.6	2.7	1.4	1.5	1.7	1.9	2.1	2.2	2.3	2.4
326	2	2.0	2.3	2.6	2.9	3.3	3.4	3.5	3.6	1.9	2.2	2.4	2.6	3.0	3.1	3.2	3.3
328	3	3.1	3.5	4.0	4.5	4.9	5.0	5.1	5.2	3.0	3.4	3.7	3.9	4.0	4.2	4.3	4.4
	4	4.0	5.0	5.9	6.8	8.0	8.5	9.4	10.0	3.5	4.5	5.2	6.0	6.6	7.4	8.0	8.5
426/428	5	4.6	5.7	6.5	7.3	8.5	9.3	10.1	10.8	4.4	5.0	5.3	6.5	7.1	7.9	8.6	9.5
407	7-1/2	7.4	9.1	10.5	11.7	13.0	13.5	14.0	14.4	7.0	8.0	9.3	10.3	11.5	11.7	12.0	12.4
	10	9.4	11.5	13.3	14.8	16.6	17.0	17.6	17.9	9.0	10.2	11.8	13.2	14.6	14.9	15.3	15.6
419	12-1/2	10.0	12.3	14.1	15.8	17.6	18.4	19.0	19.5	10.0	10.8	12.5	14.0	15.6	17.1	18.2	19.3
420	16	13.8	15.7	18.1	20.2	22.6	23.1	24.2	25.0	13.0	13.9	16.0	17.9	20.0	21.9	22.9	24.3
	19	16.2	18.6	21.5	23.9	26.6	27.3	28.4	29.3	15.9	16.5	19.0	21.2	23.6	24.5	25.6	27.5
	25	22.0	24.5	28.3	31.5	34.8	35.5	36.7	37.4	21.0	21.7	25.0	28.0	30.9	31.9	33.7	34.9

Condition 1

Condition 2

		EVAPORATOR TEMPERATURE (°F)															
		+20								0							
	Nominal Capacity	PRESSURE DROP ACROSS VALVE (psi)															
Model	(tons)	60	80	100	125	150	175	200	225	60	80	100	125	150	175	200	225
128	1/4	0.24	0.28	0.31	0.34	0.39	0.44	0.50	0.53	0.22	0.25	0.28	0.31	0.34	0.36	0.39	0.40
223	1/2	0.50	0.63	0.70	0.79	0.92	1.0	1.2	1.3	0.48	0.55	0.63	0.70	0.88	0.97	1.1	1.2
226	1	1.0	1.1	1.2	1.3	1.5	1.6	1.7	1.9	0.89	1.0	1.1	1.2	1.4	1.5	1.6	1.8
228	1-1/2	1.5	1.6	1.7	1.8	2.0	2.1	2.2	2.4	1.3	1.4	1.5	1.7	1.9	2.0	2.1	2.2
326	2	1.9	2.1	2.3	2.5	2.7	2.9	3.0	3.2	1.7	2.0	2.1	2.3	2.5	2.6	2.7	3.1
328	3	3.0	3.3	3.5	3.7	3.9	4.0	4.2	4.4	2.8	2.9	3.0	3.2	3.3	3.4	3.5	3.6
	4	3.2	4.0	4.5	5.6	6.3	7.0	7.6	8.2	3.0	3.5	4.0	4.8	5.9	6.6	7.0	7.8
426/428	5	3.7	4.2	4.7	6.0	7.1	7.6	8.2	8.6	3.5	4.0	4.3	5.2	6.3	7.0	7.5	8.0
407	7-1/2	5.9	6.8	7.6	8.4	9.3	9.6	9.9	10.1	4.6	5.0	5.4	6.0	6.6	7.2	7.8	8.4
	10	7.5	8.6	9.6	10.7	11.6	12.1	12.4	12.7	5.8	6.3	6.9	7.7	8.4	9.2	9.9	10.3
419	12-1/2	7.9	9.2	10.2	11.6	12.4	13.1	14.0	14.9	6.0	6.6	7.3	8.2	9.0	9.7	10.5	11.2
420	16	10.1	11.7	13.1	14.7	16.0	17.2	18.1	19.3	7.9	8.4	9.4	10.5	11.5	12.4	13.1	14.2
	19	12.0	13.9	15.6	17.6	19.0	20.5	21.8	23.0	9.4	10.0	11.1	12.4	13.6	14.7	15.3	16.4
	25	15.9	18.3	20.5	23.0	25.0	26.3	27.7	29.1	12.2	13.0	14.6	16.4	17.9	19.4	20.6	22.0

FIGURE 24–15 Manufacturer's table shows the capacity of valves at different pressure drops. Notice that the same valve has different capacities at different pressure drops. The more the pressure drop, the more capacity a valve has. This is a partial table and not intended for use in design of a system. *Courtesy Singer Controls Division, Schiller Park, Illinois*

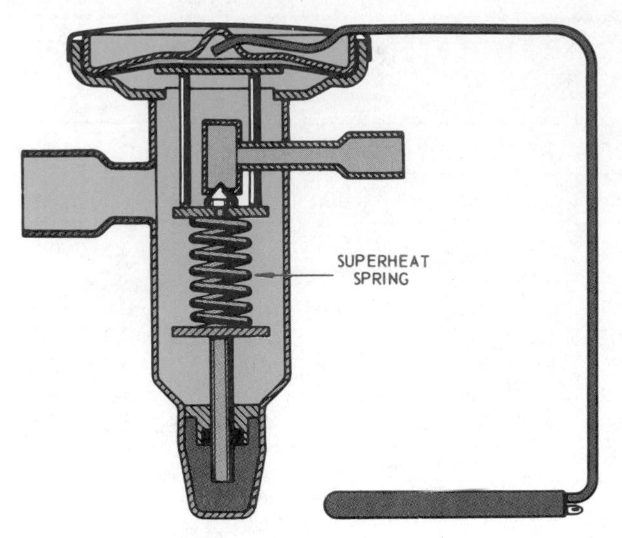

FIGURE 24–16 Spring used in the TXV. *Courtesy Singer Controls Division, Schiller Park, Illinois*

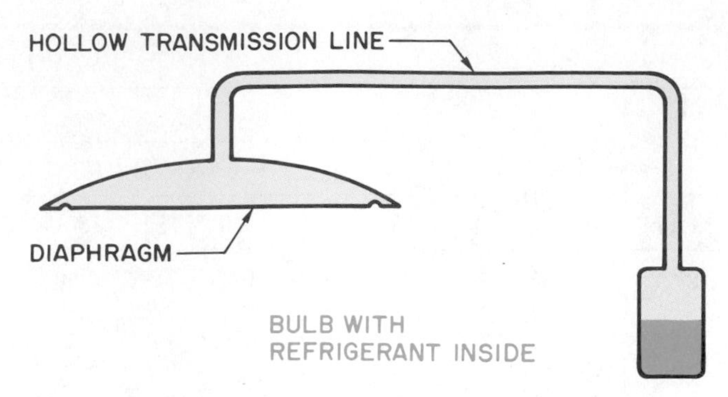

FIGURE 24–18 Illustration of the diaphragm, the bulb, and the transmission tube.

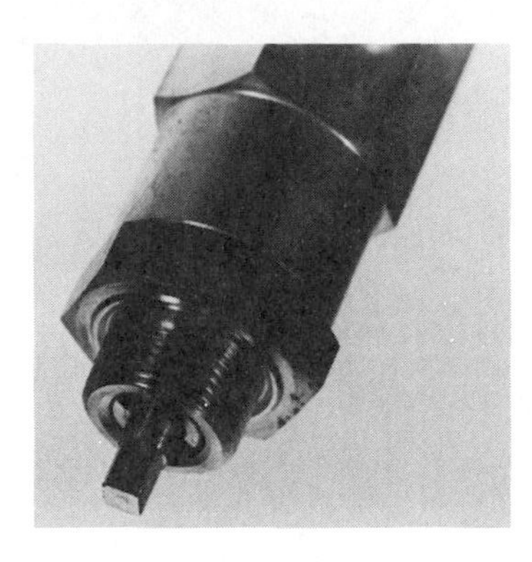

(A)

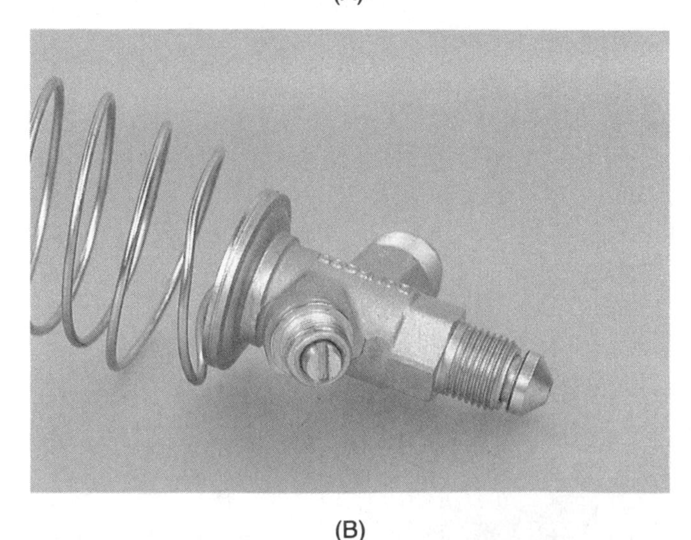

(B)

FIGURE 24–17 Adjustment stems on expansion valves. Some of them are adjusted with a screwdriver and some with a valve wrench. One full turn of the stem will normally change the superheat 0.5°F to 1°F. *Photos by Bill Johnson*

like R-12 or R-22 would. When the suction line temperature goes up, this temperature change occurs inside the bulb. When there is a pressure change, the transmission line (which is nothing more than a small-diameter hollow tube) allows the pressure between the bulb and diaphragm to equalize back and forth, Figure 24–18.

The seat in the valve is stationary in the valve body, and the needle is moved by the diaphragm. One side of the diaphragm gets its pressure from the bulb, and the other side gets its pressure from the evaporator. The diaphragm moves up and down in response to three different pressures. These three different pressures act at the proper time to open, close, or modulate the valve needle between open and closed. These three pressures are the *bulb pressure*, the *evaporator pressure*, and the *spring pressure*. They all work as a team to position the valve needle at the correct position for the load conditions at any particular time. Figure 24–19 is a series of illustrations that show the TXV functions under different load conditions.

24.9 TYPES OF BULB CHARGE

The fluid inside the expansion valve bulb is known as the *charge* for the valve. Four types of charge can be obtained with the TXV: liquid charge, cross liquid charge, vapor charge, and cross vapor charge.

24.10 THE LIQUID CHARGE BULB

The *liquid charge bulb* is a valve and bulb charged with a fluid characteristic of the refrigerant in the system. The diaphragm and bulb are not actually full of liquid. They have enough liquid, however, that they always have some liquid inside them. The liquid will not all boil away. The pressure/temperature relationship is almost a straight line on a graph. When the temperature goes up a degree, the pressure goes up a specific amount and can be followed on a pressure/temperature chart. When this pressure/temperature concept is carried far enough, it can easily be seen that when

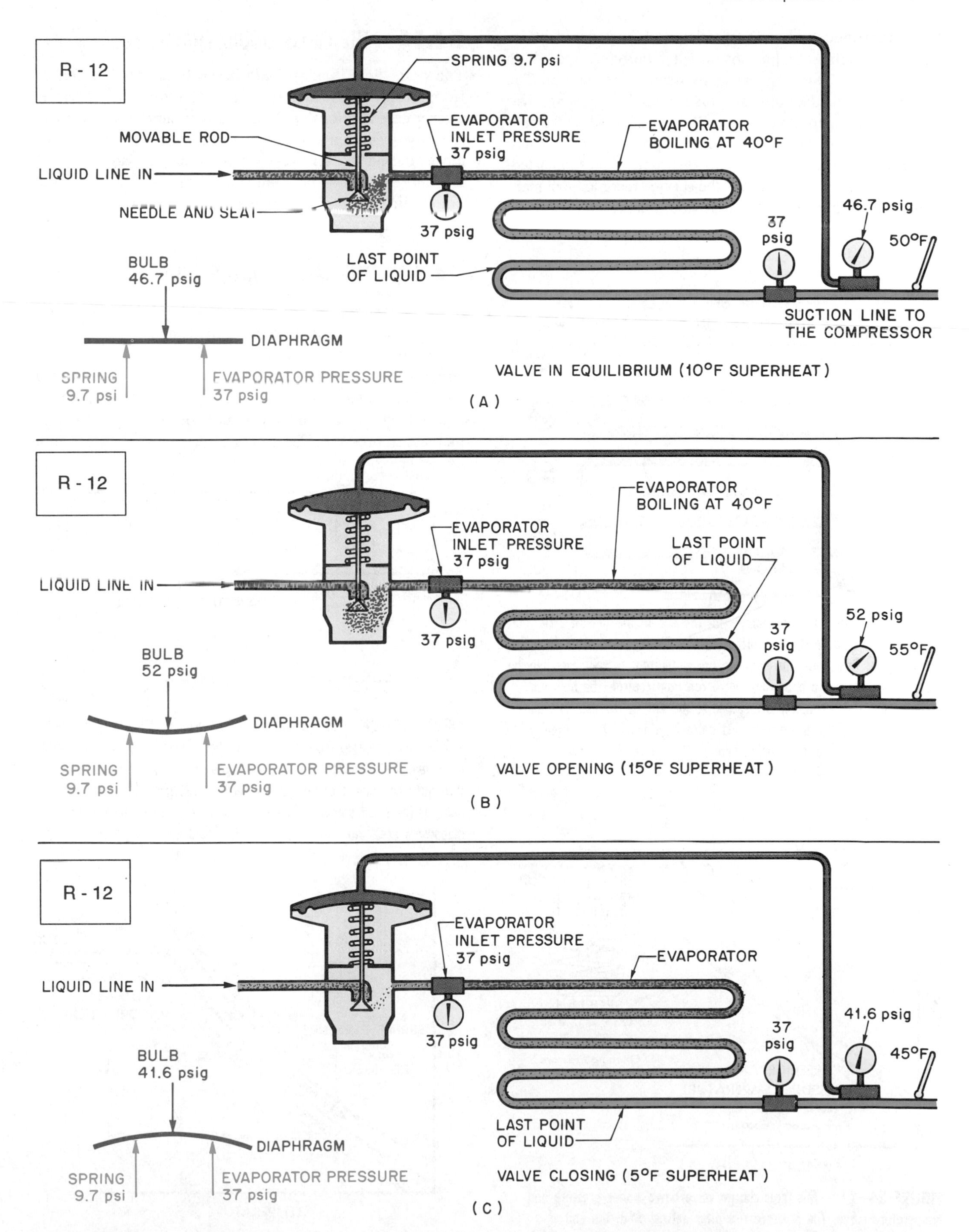

FIGURE 24–19 All components work together to hold the needle and seat in the correct position to maintain stable operation with the correct superheat. (A) Valve in equilibrium. (B) Valve opening. (C) Valve closing.

high temperatures are encountered, high pressures will exist. During defrost the expansion valve bulb may reach high temperatures. This can cause two things to happen. The pressures inside the bulb can cause excessive pressures over the diaphragm, and the valve will open wide when the bulb gets warm. This will overfeed the evaporator and can cause liquid to flood the compressor when defrost is terminated, Figure 24–20. This can cause the service technician or manufacturer to change to the cross liquid charge bulb.

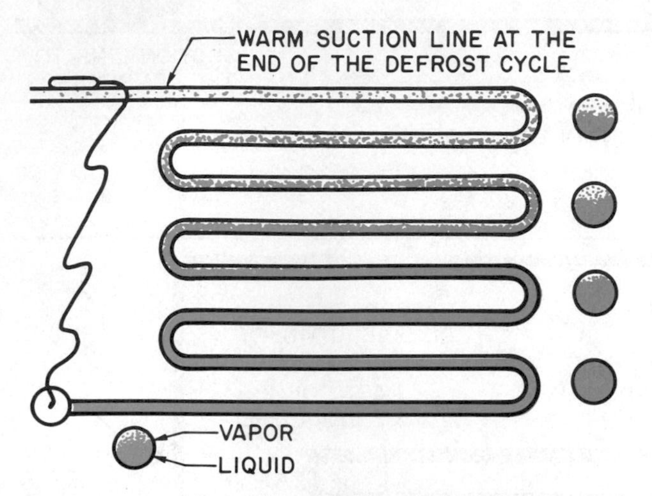

FIGURE 24–20 This can happen at the end of the defrost cycle. The valve is open because the evaporator suction line is warm after the defrost cycle. This allows the liquid in the evaporator to spill over into the suction line before the expansion valve can gain control. The bulb can become very hot because it is fastened to the end of the coil. The coil can become quite hot in an extended defrost cycle, and the remaining liquid can move on to the suction line.

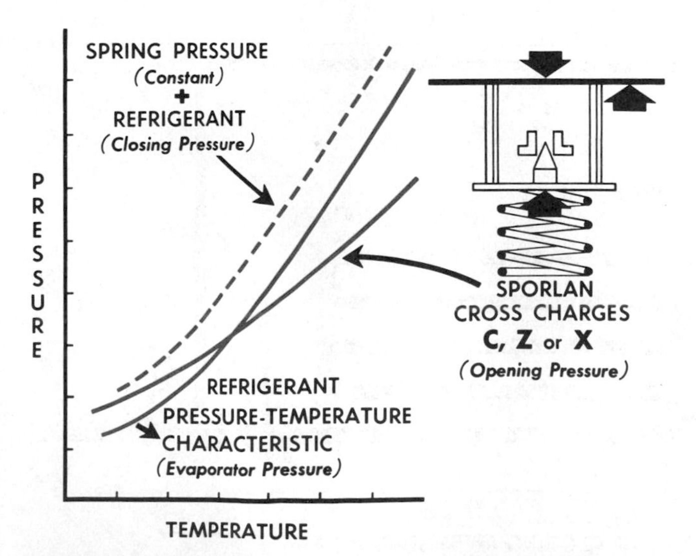

FIGURE 24–21 This cross charge valve has a flatter pressure and temperature curve. The pressure rise after defrost or at the end of a cycle will have a tendency to close the valve and prevent liquid problems. *Courtesy Sporlan Valve Company*

24.11 THE CROSS LIQUID CHARGE BULB

The *cross liquid charge bulb* has a fluid that is different from the system fluid. It does not follow the pressure/temperature relationship. It has a flatter curve and will close the valve faster on a rise in evaporator pressure. This valve closes during the off cycle when the compressor shuts off and the evaporator pressure rises. This will help prevent liquid refrigerant from flooding over into the compressor at start-up, Figure 24–21.

24.12 THE VAPOR CHARGE BULB

The *vapor charge bulb* is actually a valve that has only a small amount of liquid refrigerant in the bulb, Figure 24–22. It is sometimes called a *critical charge bulb*. When the bulb temperature rises, more and more of the liquid will boil to a vapor until there is no more liquid. When this point is reached, an increase in temperature will no longer bring an increase in pressure. The pressure curve will be flat, Figure 24–23.

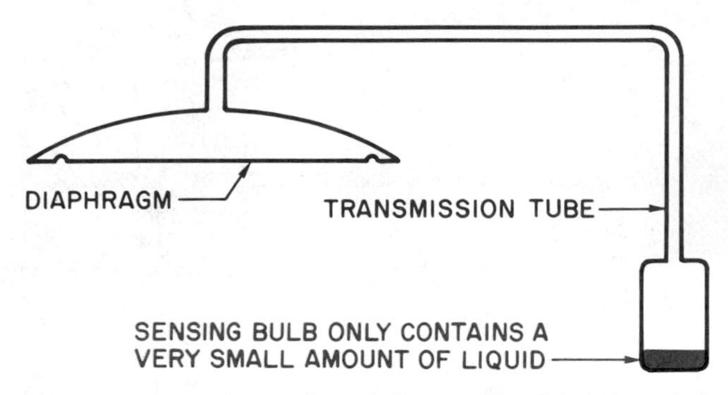

FIGURE 24–22 This gas charge bulb is actually a liquid charge bulb with a critical charge. When the bulb reaches a point, the liquid is boiled away and the pressure won't rise any more. When this valve is used, care must be taken that the valve body does not get colder than the bulb, or the small amount of liquid in the valve will condense in the diaphragm area. The control is where the liquid is.

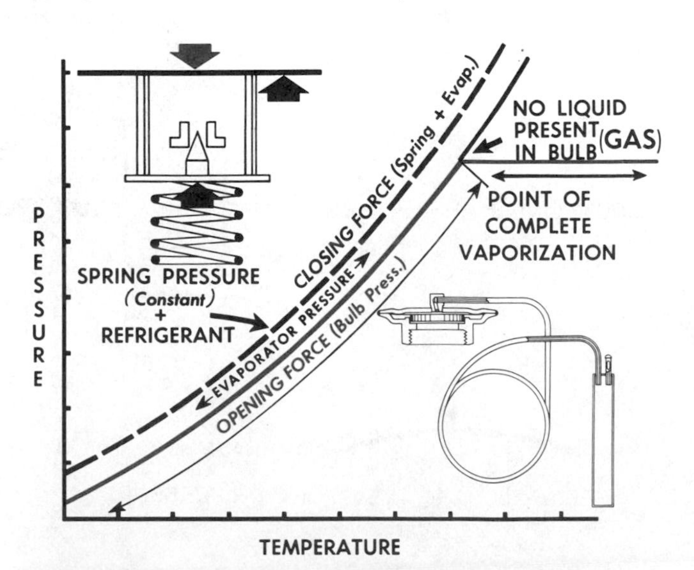

FIGURE 24–23 Graph shows pressures inside the valve when the sensing bulb is in a hot area. *Courtesy Sporlan Valve Company*

When this valve is applied, care must be taken that the valve body does not get colder than the bulb, or the liquid that is in the bulb will condense above the diaphragm. When this happens, the control at the end of the suction line by the bulb will be lost to the valve diaphragm area, Figure 24–24. The valve will be controlled by the temperature of the liquid that is at the diaphragm and the valve will lose control. Small heaters have been installed at the valve body to keep this from happening.

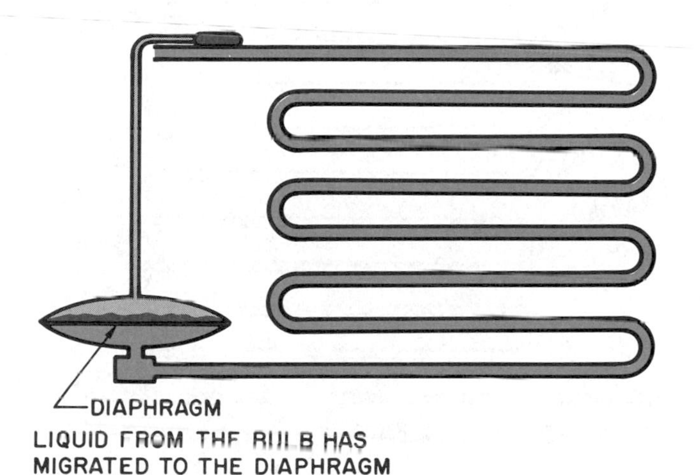

FIGURE 24–24 The liquid can migrate to the valve body when the body becomes cooler than the bulb. When this happens, the valve body can be heated with a warm cloth, and the liquid will move back to the bulb.

24.13 THE CROSS VAPOR CHARGE BULB

The *cross vapor charge bulb* has a similar characteristic to the vapor charged valve, but it has a fluid different from the refrigerant in the system. This gives a different pressure and temperature relationship under different conditions. These special valves are applied to special systems. Manufacturers or suppliers should be consulted when questions are encountered.

24.14 FUNCTIONING EXAMPLE OF A TXV WITH INTERNAL EQUALIZER

When all of these components are assembled, the expansion valve will function as follows. Note: This is a liquid-filled bulb.

1. **Normal load conditions.** The valve is operating in equilibrium (no change, stable), Figure 24–25. The evaporator is operating at a medium-temperature application and just before the cut-off point. The suction pressure is 21 psig, and the refrigerant, R-12, is boiling (evaporating) in the evaporator at 20°F. The expansion valve is maintaining 10°F of superheat, so the suction line temperature is 30°F at the bulb location. The bulb has been on the line long enough that it is the same temperature as the line, 30°F. For now, suppose that the liquid in the bulb will be 28.5 psig, corresponding to 30°F. The spring is exerting a pressure equal to the difference in pressure to hold the needle at the correct position in the seat to maintain this condition. The spring pressure in this example is 7.5 psi.

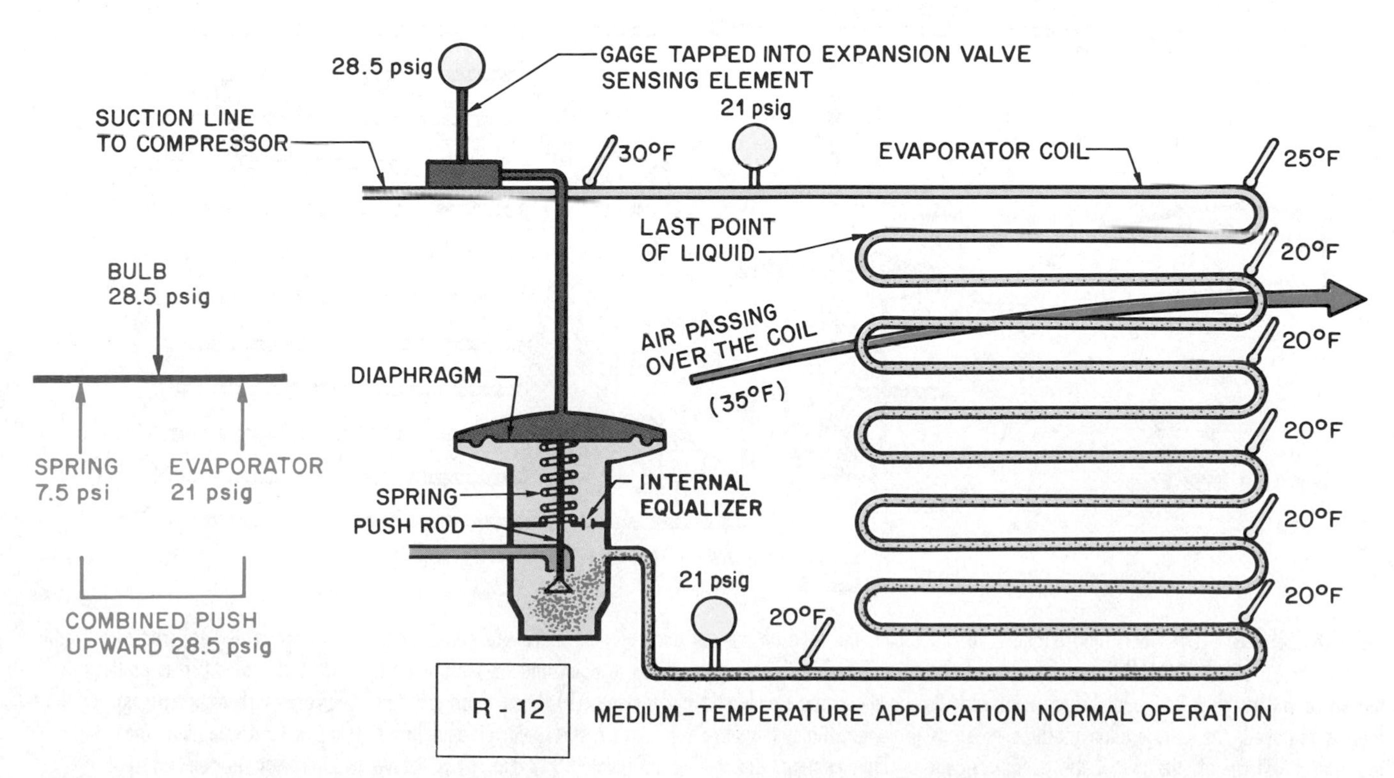

FIGURE 24–25 TXV under a normal load condition. The valve is said to be in equilibrium. The needle is stationary.

2. **Load changes with food added to the cooler.** See Figure 24–26(A). When food that is warmer than the inside of the cooler is added to the cooler, the load of the evaporator changes. The warmer food warms the air inside the cooler, and the load is added to the refrigeration coil by the air. This warmer air passing over the coil causes the liquid refrigerant inside the coil to boil faster. The suction pressure will also rise. The net effect

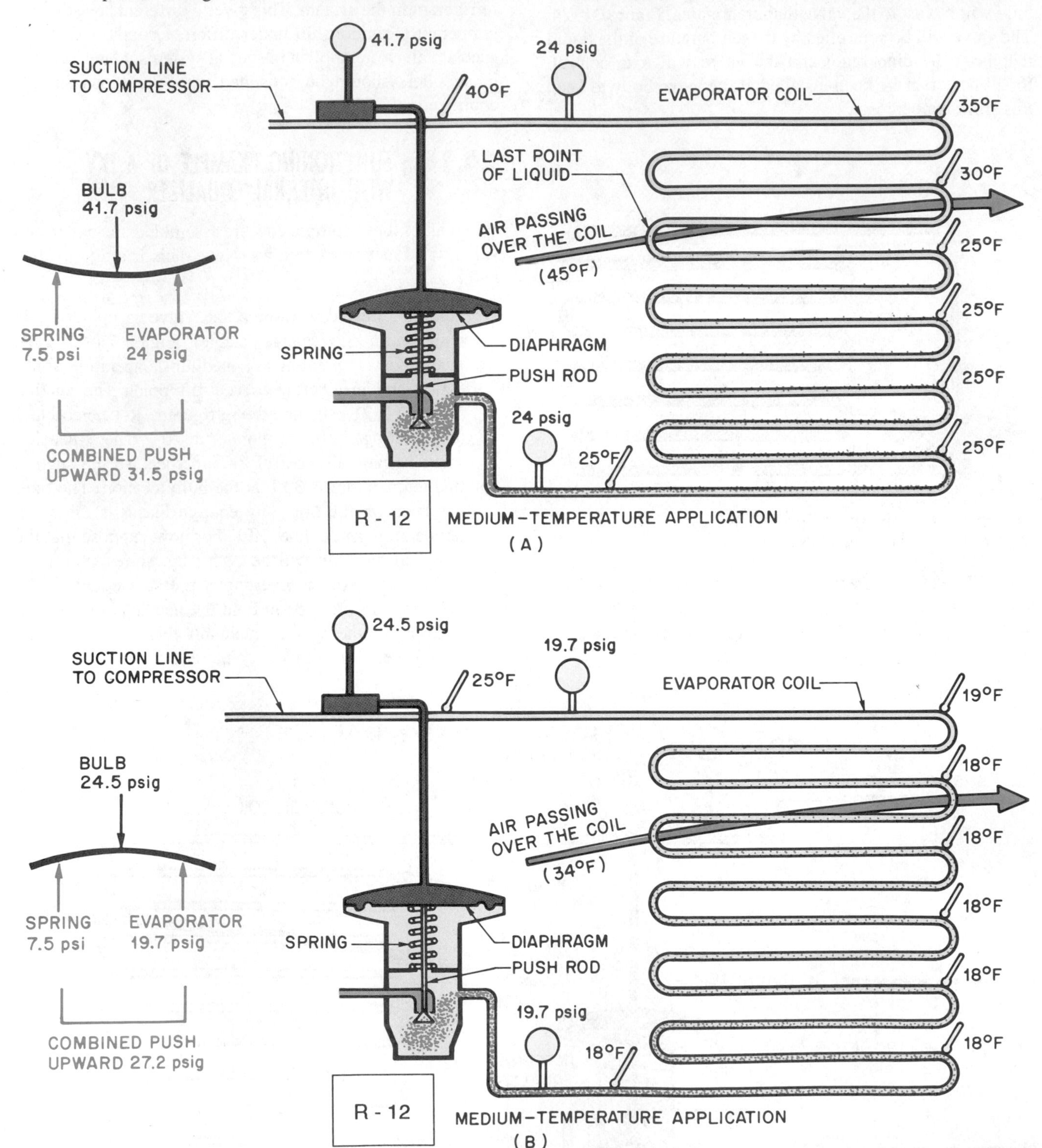

FIGURE 24–26 (A) When food is added to the cooler, the load on the coil goes up. The extra Btu added to the air in the cooler are transferred into the coil, which causes a temperature increase of the suction line. This causes the valve to open and feed more refrigerant. If this condition is prolonged, the valve will reach a new equilibrium point with the needle being steady and not moving. (B) When there is a load decrease, such as when some of the food is removed, the refrigeration machine now has prolonged time, the valve will reach a new point of equilibrium at a low load condition. The thermostat will cut off the unit if this condition goes on long enough. The equilibrium point in (A) and (B) is the valve superheat set point of 10°F superheat. The valve should always return to this 10°F set point with prolonged steady-state operation.

of this condition will be that the last point of liquid in the coil will be farther from the end of the coil than when the coil was under normal working conditions. The coil will start to starve for liquid refrigerant. The TXV will start to feed more refrigerant to compensate for this shortage. When this condition of increased load has gone on for an extended time, the TXV will stabilize in the feeding of refrigerant and reach a new point of equilibrium, where no adjustment occurs.

3. **Load changes with food removed from the cooler.** See Figure 24–26(B). When a large portion of the food is removed from the cooler, the load will decrease on the evaporator coil. There will no longer be enough load to boil the amount of refrigerant that the expansion valve is feeding into the coil, so the expansion valve will start overfeeding the coil. The coil is beginning to flood with liquid refrigerant. The TXV needs to throttle the refrigerant flow. When the TXV has operated at this

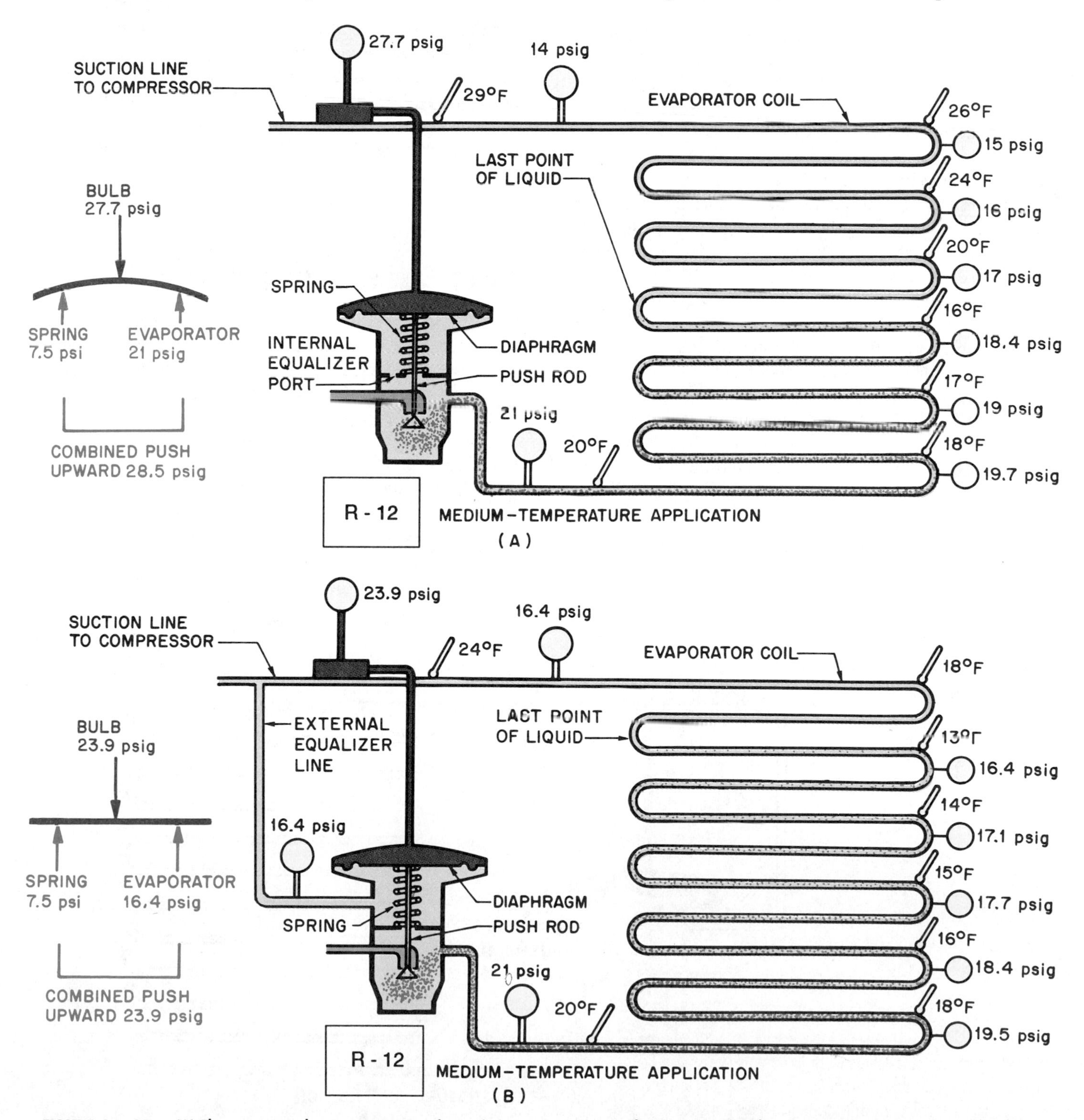

FIGURE 24–27 (A) This evaporator has 7 psig pressure drop, which is causing a starved evaporator. (B) When a TXV with an external equalizer is added, the coil has the correct amount of refrigerant for best coil efficiency.

condition for a period of time, it will stabilize and reach another point of equilibrium. When the condition exists for a long enough time, the thermostat will stop the compressor because the air in the cooler is reduced to the thermostat 'cut-out' point.

24.15 TXV WITH EXTERNAL EQUALIZERS

Evaporators with pressure drop from the inlet to the outlet are often designed and applied. This pressure drop may be from a distributor located after the expansion valve or because the evaporator has a long piping circuit. If for any reason there is a pressure drop in excess of about 5 psig, a TXV with an external equalizer should be used so the valve will properly feed refrigerant to the coil. Excess pressure drop in a coil where a TXV is used will cause the valve to starve the coil of refrigerant, Figure 24–27. Remember, the evaporator is more efficient when it has the maximum amount of refrigerant without flooding back to the compressor. An expansion valve with an external equalizer was shown in Figure 24–10.

The external equalizer line should always be piped to the suction line after the expansion valve sensing bulb to prevent superheat problems from internal leaks in the valve, Figure 24–28. If a TXV with an external equalizer line were to have an internal leak, very small amounts of liquid would vent to the suction line through the external equalizer line. If the sensing bulb for the valve were to have this small amount of liquid touching it, it would throttle toward closed because this would simulate a flooded coil, Figure 24–29. Sometimes this condition can be sensed by feeling the line leading to the suction line. It should not become cold as it leaves the expansion valve.

24.16 TXV RESPONSE TO LOAD CHANGES

The TXV responds to a change in load in the following manner. When the load is increased, for example when a load of food warmer than the cooler (refrigerated box) is placed inside the cooler, the TXV opens, allowing more refrigerant into the coil. The evaporator needs more refrigerant at this time because the increased load is evaporating the refrigerant in the evaporator faster. The suction pressure increases, Figure 24–26. When there is a load decrease (eg, when some of the food is removed from the cooler), the liquid in the evaporator evaporates slower and the suction pressure decreases. At this time the TXV will throttle back by slightly closing the needle and seat to maintain the correct superheat. **The TXV responds to a load increase by metering more refrigerant into the coil. This causes an increase in suction pressure.**

24.17 TXV VALVE SELECTION

The TXV must be carefully chosen for a particular application. **Each TXV is designed for a particular refrigerant.** One valve on the market uses a spring changeout to change refrigerant, but the refrigerant still has to be considered in the selection of the spring.

The capacity of the system is very important. If the system calls for a 1/2-ton expansion valve and a 1-ton valve is used, the valve will not control correctly because the needle and seat are too large. If a 1/4-ton valve were used, it would not pass enough liquid refrigerant to keep the coil full, and a starving coil condition would occur.

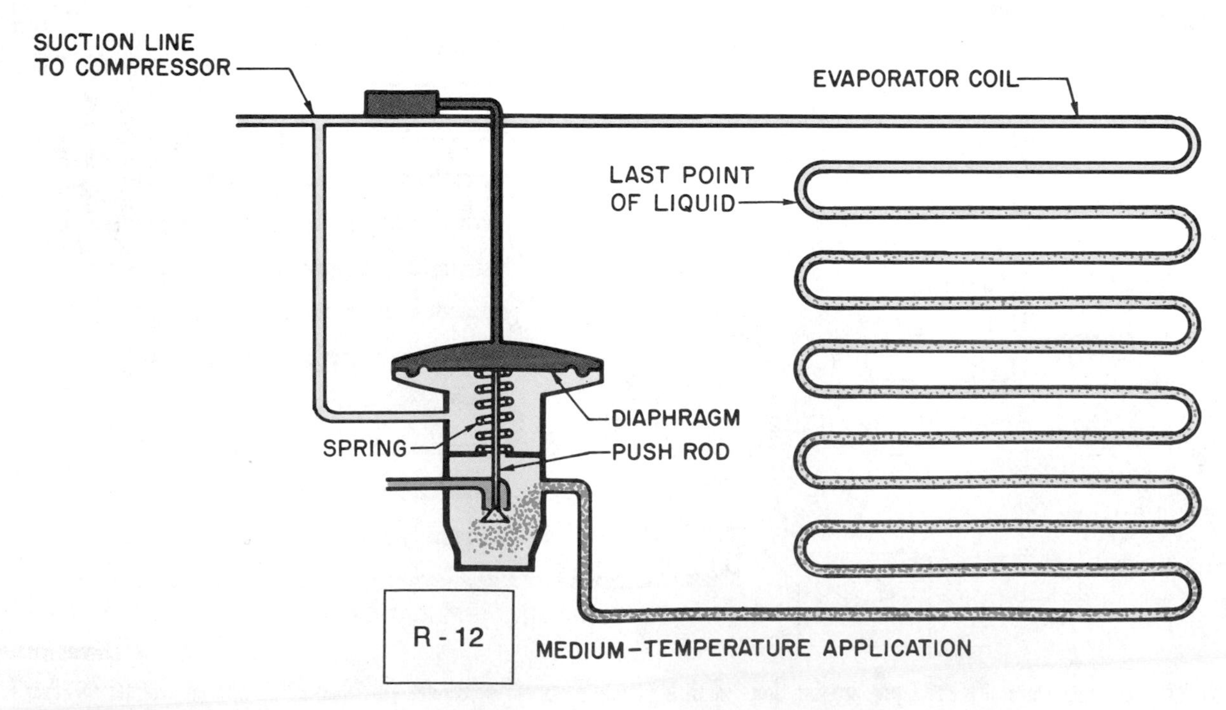

FIGURE 24–28 This illustration shows the correct method for connecting an external equalizer line.

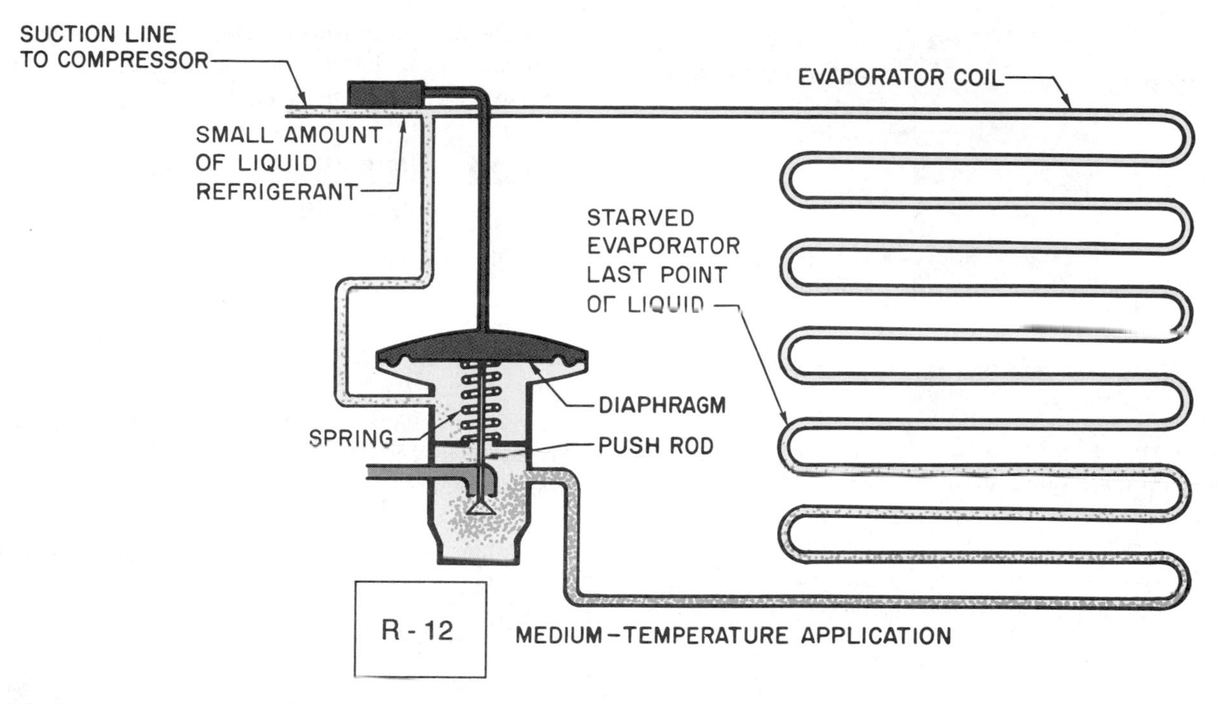

FIGURE 24–29 When the equalizer line is connected before the sensing bulb, a small leak in the internal parts of the valve will simulate a flooded evaporator.

24.18 BALANCED PORT TXV

Low ambient temperatures have led valve manufacturers to develop TXVs that will feed at the same rate when the ambient temperatures are low. These valves allow refrigerant flows that do not reduce when the head pressure is low in mild weather. The evaporator will then have the correct amount of refrigerant and be able to operate at design conditions at lower outdoor temperatures.

You cannot tell a balanced port expansion valve from a regular expansion valve by its appearance. It will have a different model number and you can look it up in a manufacturer's catalog to determine the exact type of valve.

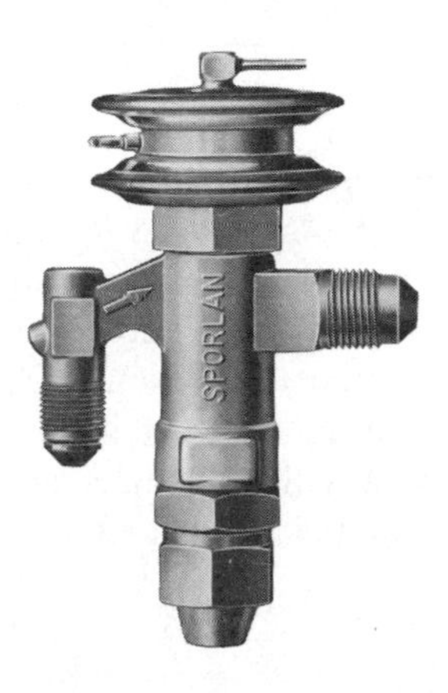

FIGURE 24–30A Pressure-limiting expansion valve. When the pressure is high in the evaporator, such as during a hot pull down, this valve will override the thermostatic element with a pressure element and throttle the refrigerant. *Courtesy Sporlan Valve Company*

24.19 PRESSURE-LIMITING TXV

The pressure-limiting TXV has another bellows that will only allow the evaporator to build to a predetermined pressure, and then the valve will shut off the flow of liquid. This valve is desirable on low-temperature applications because it keeps the suction pressure to the compressor down during a hot pull down that could overload the compressor. For example, when a low-temperature cooler is started with the inside of the box hot, the compressor will operate under an overloaded condition until the box cools down. The pressure-limiting TXV valve will prevent this from happening, Figure 24–30A.

24.20 SERVICING THE TXV

When any TXV is chosen, care should be taken that the valve is serviceable and will perform correctly. Several things should be considered: (1)type of fastener (flare, solder, or flange), (2) location of the valve for service and performance, and (3) expansion valve bulb location. This valve has moving parts that are subject to wear. When a valve must be replaced, an exact replacement is usually best. When this is not possible, a supplier can furnish you with the information needed to change to another valve.

Many technicians (and owners) will adjust the TXV. This may cause problems because it will change pressures and temperatures in a coil. If the valve sensing element is mounted properly and in the correct location where it can sense refrigerant vapor, the valve should work correctly as

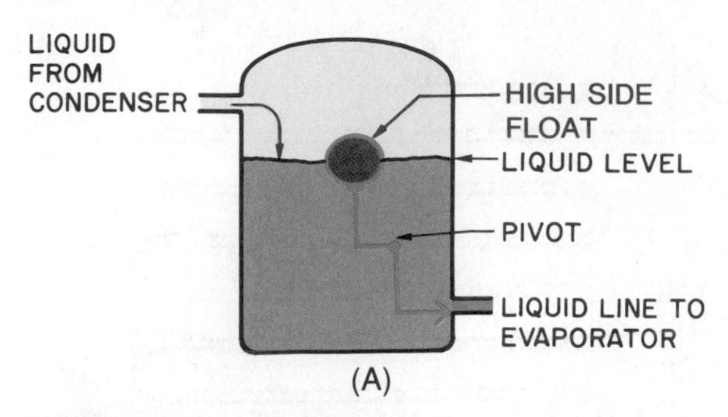

FIGURE 24–30B This expansion valve is not adjustable.

shipped from the factory. These valves are very reliable and normally do not require any adjustment. If there are signs that the valve has been adjusted, look for other problems, because the valve probably did not need adjustment to begin with. Many manufacturers are producing nonadjustable valves to prevent people from adjusting the valve when not needed, Figure 24–30B.

24.21 SENSING ELEMENT INSTALLATION

Particular care should be taken when installing the expansion valve sensing element. Each manufacturer has a recommended method for this installation, but they are all similar. The valve sensing bulb has to be mounted at the end of the evaporator on the suction line. The best location is near the bottom of the line on a horizontal run where the bulb can be mounted flat and not be raised by a fitting, Figure 24–31. The bulb should not be located on the bottom of the line because there will be oil returning that will act as an insulator to the sensing element. The object of the sensing element is to sense the temperature of the suction line. To do this, the line should be very clean and the bulb fastened to this line very securely. Normally the manufactuer suggests that the bulb be insulated from the ambient temperature if the ambient is much warmer than the suction line temperature because the bulb will be influenced by the ambient temperature as well as by the line temperature.

24.22 THE SOLID-STATE CONTROLLED EXPANSION VALVE

The solid-state controlled expansion valve uses a thermistor as a sensing element to vary the voltage to a heat motor-operated valve (a valve with a bimetal element). This valve normally uses 24 V as the control voltage to operate the valve, Figure 24–32A.

When voltage is applied to the coil in the valve, the valve opens. Modulation is accomplished by varying the voltage, Figure 24–32B. The valve is versatile and can be used to accomplish different functions in the system. When the voltage is cut off at the end of the cycle, the valve will

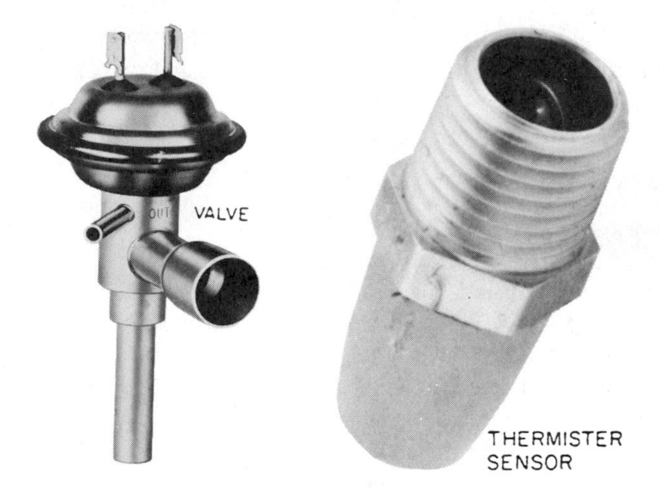

FIGURE 24–32A Expansion valve controlled with a thermistor and a heat motor. *Courtesy Singer Controls Division, Schiller Park, Illinois*

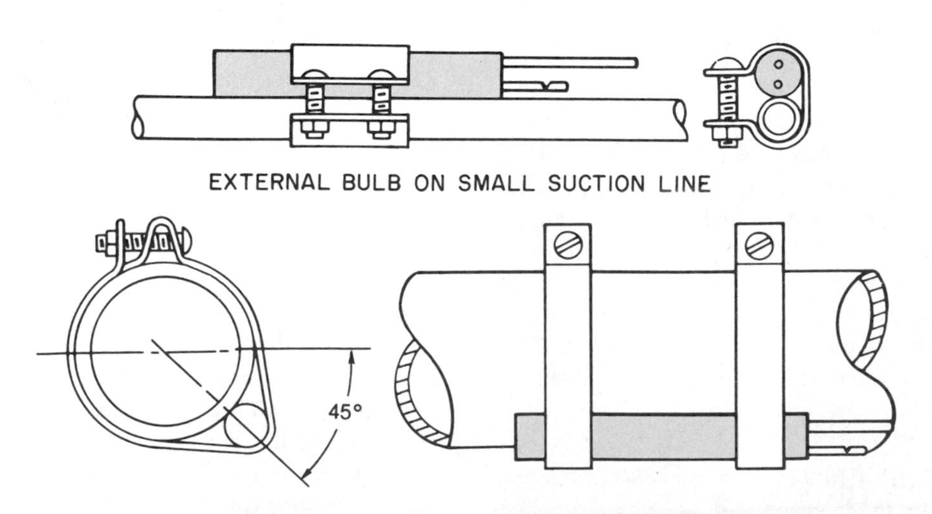

FIGURE 24–31 Best positions for mounting the expansion valve sensing bulb. *Courtesy ALCO Controls Division, Emerson Electric Company*

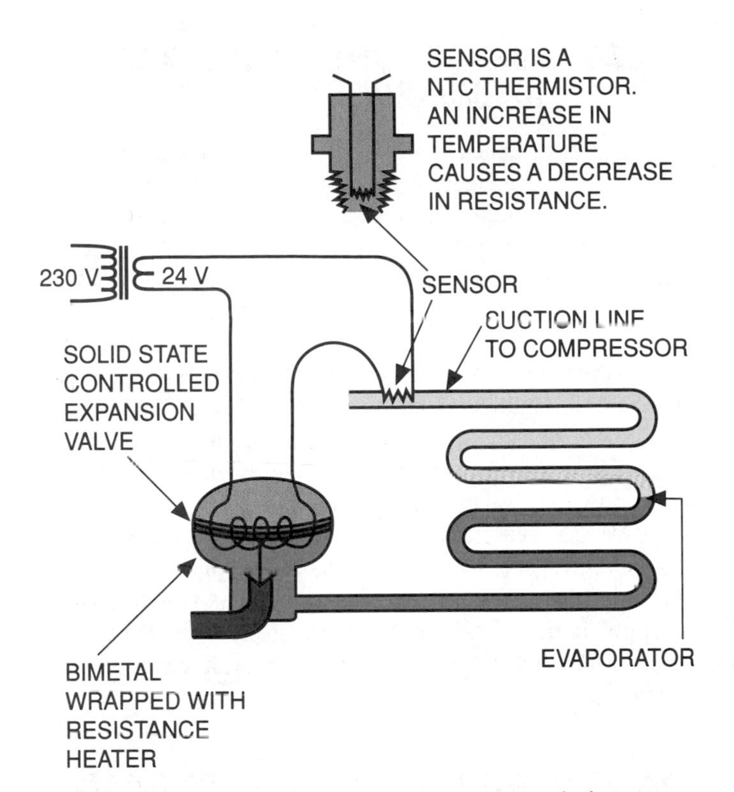

FIGURE 24–32B The temperature of the sensor controls the current flow to the bimetal in the valve. More current flow causes the bimetal to warp and open the valve. Less current flow cools the bimetal and throttles the valve closed.

shut off and the system can be pumped down. If the voltage is allowed to remain on the element, the valve will remain open during the off cycle, and the pressures will equalize.

The thermistor is inserted into the vapor stream at the end of the evaporator. It is very small in mass and will respond very quickly to temperature changes.

The solid-state controlled expansion valve responds to the change in temperature of the sensing element like the typical TXV except that it does not have a spring. When the thermistor is suspended in dry vapor, it is heated by the current passing through it. This creates a faster response than does merely measuring the vapor temperature, Figure 24–32C. When the valve opens and saturated vapor reaches the element, the valve begins to close slightly. This valve controls to a very low superheat, which allows the evaporator to use maximum surface area.

The solid-state controlled expansion valve is unique because refrigerant can flow in either direction through the valve body. It is suitable for heat pump applications because of this and is used on packaged heat pumps where the manufacturer can build a system with only one expansion valve allowing refrigerant to flow in either direction.

24.23 THE AUTOMATIC EXPANSION VALVE

The *automatic expansion valve* is an expansion device that meters the refrigerant to the evaporator by using a pressure-sensing device. This device is also a valve that changes in inside dimension in response to its sensing element. **The**

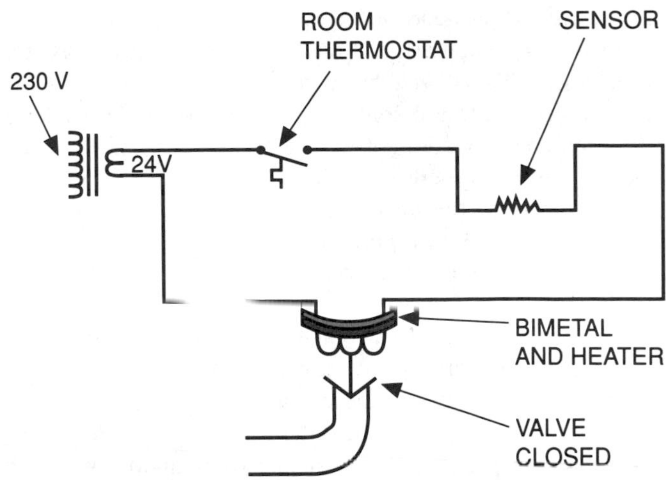

THE ROOM THERMOSTAT MAY BE USED TO CONTROL REFRIGERANT FLOW. IN THIS APPLICATION, THE ROOM THERMOSTAT CONTACTS ARE OPENED, NO CURRENT FLOW AND THE VALVE CLOSES. NO MORE REFRIGERANT WILL BE FED INTO THE EVAPORATOR DURING THE OFF CYCLE.

(1)

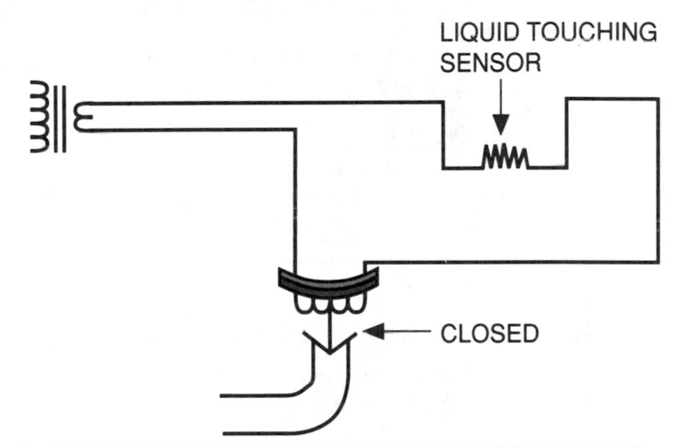

LIQUID REFRIGERANT TOUCHING THE SENSOR WILL COOL IT AND INCREASE THE RESISTANCE, DECREASING THE CURRENT FLOW AND THROTTLE THE VALVE TOWARD CLOSED.

(2)

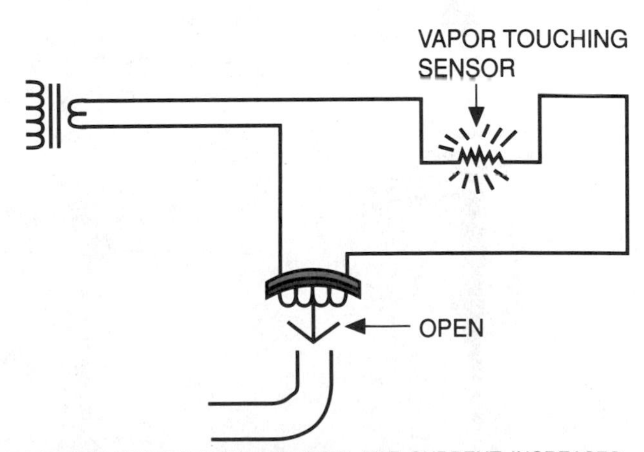

WHEN VAPOR TOUCHES THE SENSOR, THE CURRENT INCREASES, HEATING THE BIMETAL AND OPENING THE VALVE. WITHOUT THE THERMOSTAT SHOWN IN (A), THE ENSOR WOULD BE WARM DURING THE OFF CYCLE WHICH WOULD OPEN THE VALVE AND EQUALIZE THE PRESSURE.

(3)

FIGURE 24–32C Different applications and actions for the solid-state (thermistor) controlled valve.

automatic expansion valve maintains a constant pressure in the evaporator. Notice that superheat was not mentioned. This device has a needle and seat like the TXV that is fastened to a diaphragm, Figure 24–33. One side of the diaphragm is common to the evaporator and the other side to the atmosphere. When the evaporator pressure drops for any reason, the valve begins to open and feed more refrigerant into the evaporator.

The automatic expansion valve is built much the same as the TXV except that it does not have a sensing bulb. The body is normally made of machined brass. The adjustment of this valve is normally at the top of the valve. There may be a cap to remove, or there may be a cap to turn. This adjustment changes the spring tension that supports the atmosphere in pushing down on the diaphragm. When the tension is increased, the valve will feed more refrigerant and increase the suction pressure, Figure 24–34.

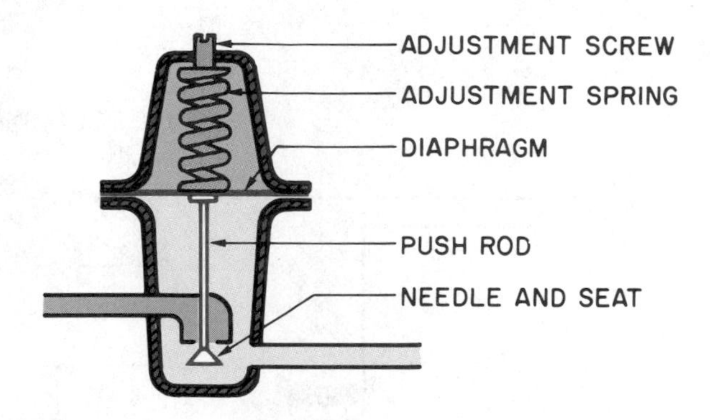

FIGURE 24–33 The automatic expansion valve uses the diaphragm as the sensing element and maintains a constant pressure in the evaporator but does not control superheat.

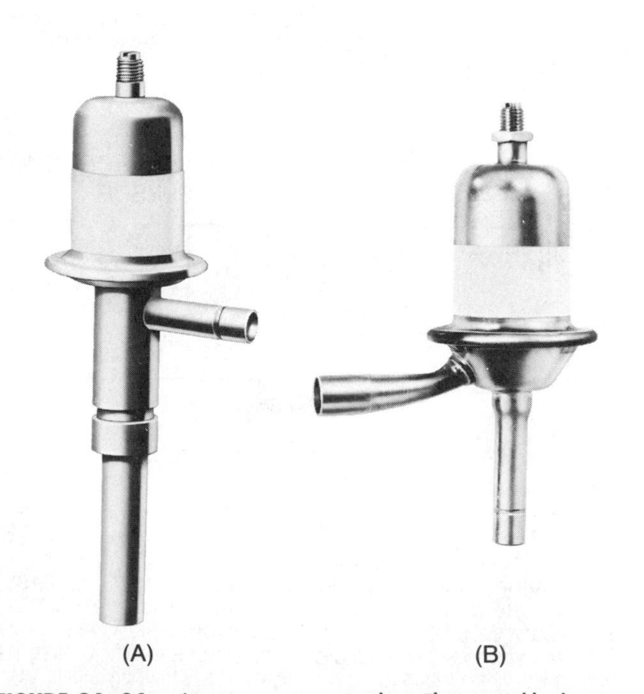

FIGURE 24–34 Automatic expansion valves. They resemble the TXV, but they do not have the bulb for sensing temperature at the end of the suction line. *Courtesy Singer Controls Division, Schiller Park, Illinois*

24.24 AUTOMATIC EXPANSION VALVE RESPONSE TO LOAD CHANGES

The automatic expansion valve responds differently than the TXV to load changes. It actually acts in reverse. When a load is added to the coil, the suction pressure starts to rise. The automatic expansion valve will start to throttle the refrigerant by closing enough to maintain the suction pressure at the set point. This has the effect of starving the coil slightly. A large increase in load will cause more starving. When the load is decreased and the suction pressure starts to fall, the automatic expansion valve will start to open and feed more refrigerant into the coil. If the load reduces too much, liquid could actually leave the evaporator and proceed down the suction line, Figure 24–35.

We can see from these examples that this valve responds in reverse to the load. The best application for this valve is where there is a fairly constant load. One of its best features is that it can hold a constant pressure. When this valve is applied to a water-type evaporator, freezing will not occur.

24.25 SPECIAL CONSIDERATIONS FOR THE TXV AND AUTOMATIC EXPANSION VALVE

The TXV and the automatic expansion valve both are expansion devices that allow more or less refrigerant flow, depending on the load. Both need a storage device (receiver) for refrigerant when it is not needed. The receiver is a small tank located between the condenser and the expansion device. Normally the condenser is close to the receiver. It has a *king* valve that functions as a service valve. This valve stops the refrigerant from leaving the receiver when the low side of the system is serviced. This receiver can serve both as a storage tank for different load conditions and as a tank into which the refrigerant can be pumped when servicing the system, Figure 24–36.

24.26 THE CAPILLARY TUBE METERING DEVICE

The *capillary tube* metering device controls refrigerant flow by pressure drop. It is a copper tube with a very small calibrated inside diameter, Figure 24–37A. The diameter and the length of the tube determine how much liquid will pass through the tube at any given pressure drop, Figure 24–37B. It is much like a garden hose; the larger the hose, the more gallons of water will pass at 100 psig pressure at the hose inlet, Figure 24–37C. The water hose has pressure drop all along the length, so a longer hose will have less pressure at the outlet even though the pressure at the inlet is still 100 psig.

This length-to-pressure drop relationship is used by manufacturers to arrive at the correct pressure drop that will allow the correct amount of refrigerant to pass through the capillary tube to correctly fill the evaporator. The capillary tube can be quite long on some installations and may be wound in a coil to store the extra tubing length.

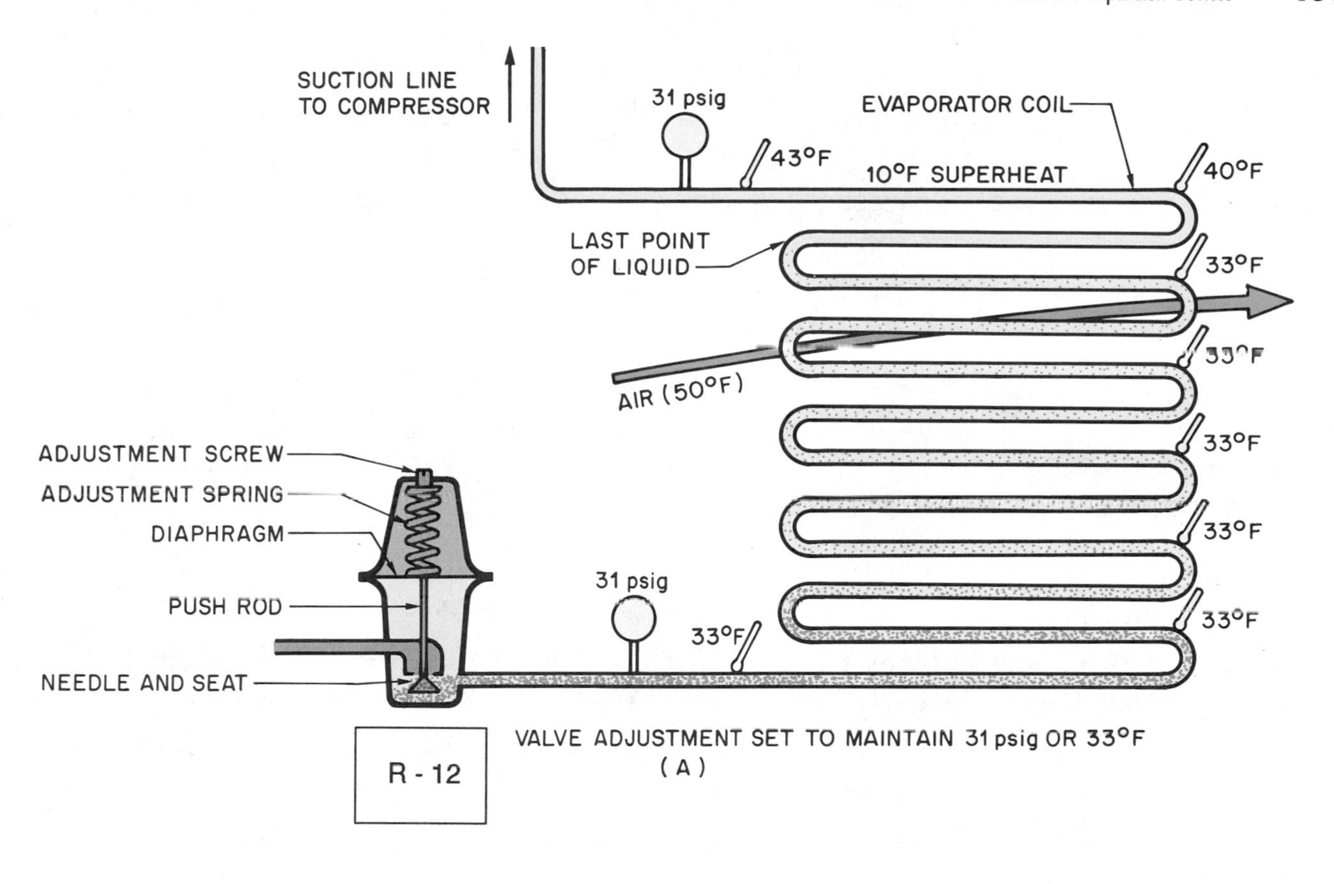

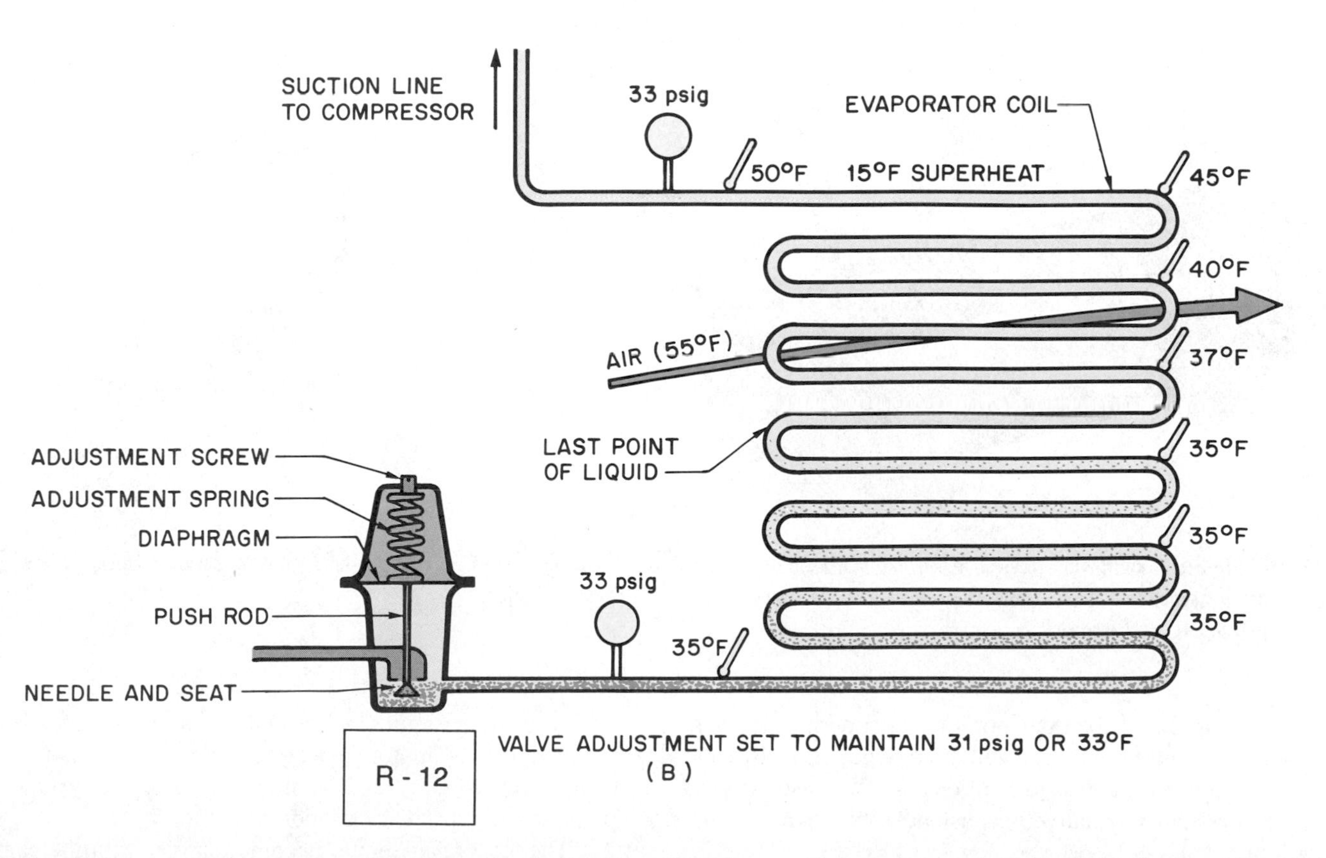

FIGURE 24–35 Automatic expansion valve under varying load conditions. This valve responds in reverse to a load change. (A) Normal operation. (B) When the load goes up the valve closes down and starts to starve the coil slightly to keep the evaporator pressure from rising. (C) When the load goes down, the valve opens up to keep the evaporator pressure up. This valve is best applied where the load is relatively constant. When applied to a water-type evaporator, freeze protection can be a big advantage with this valve.

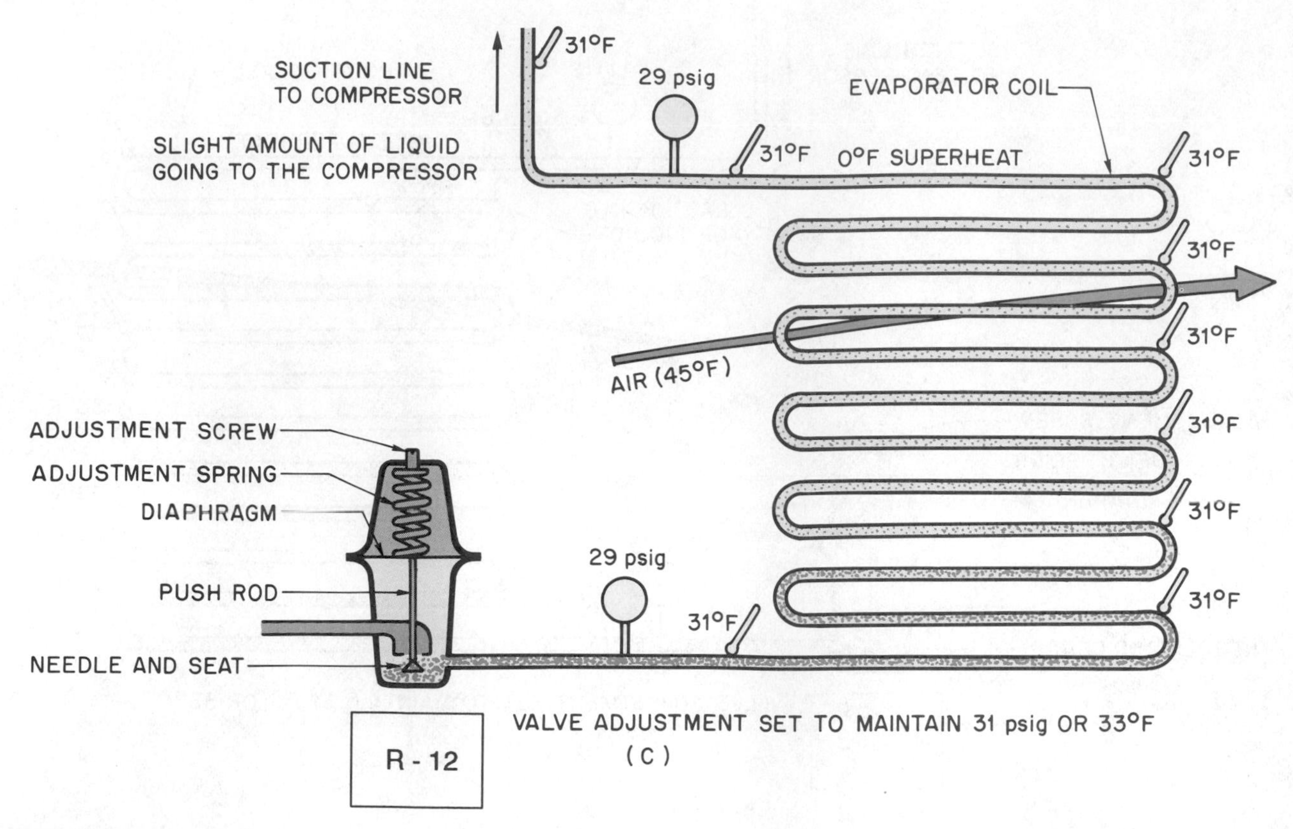

FIGURE 24–35 *(Continued)*

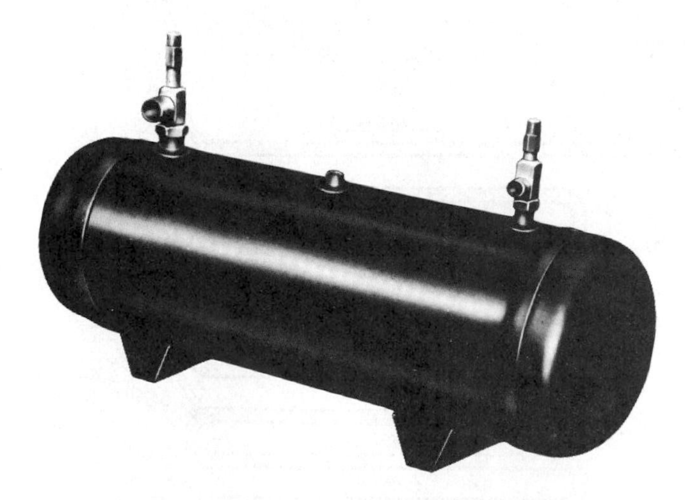

FIGURE 24–36 Refrigerant receiver. When the load increases, more refrigerant is needed and moves from the receiver into the system. *Courtesy Refrigeration Research*

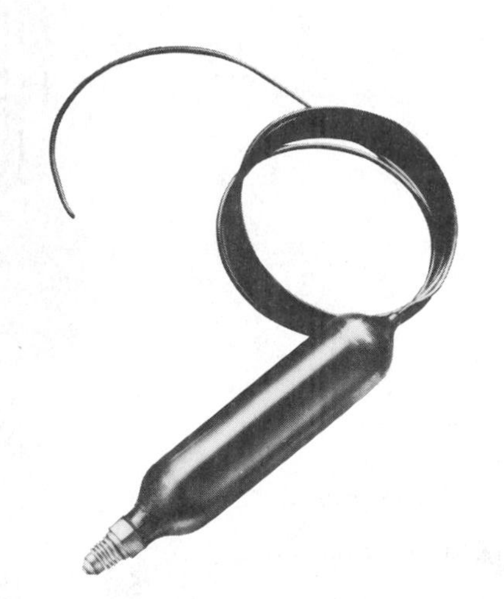

FIGURE 24-37A Capillary tube metering device. *Courtesy Parker Hannifin Corp.*

The capillary tube does not control superheat or pressure. It is a fixed-bore device with no moving parts. Because this device cannot adjust to load change, it is usually used where the load is relatively constant with no large fluctuations.

The capillary tube is an inexpensive device for the control of refrigerant and is used often in small equipment. This device does not have a valve and does not stop the liquid from moving to the low side of the system during the off cycle, so the pressures will equalize during the off cycle. This reduces the motor starting torque requirements for the compressor. Figure 24–38A illustrates a capillary tube in place at the inlet of the evaporator.

The technician should become familiar with the capillary tube metering device because it is probably the most used device for metering refrigerant. It has no moving parts, so it will not wear out. About the only problems it may have would be small particles that may block or partially block

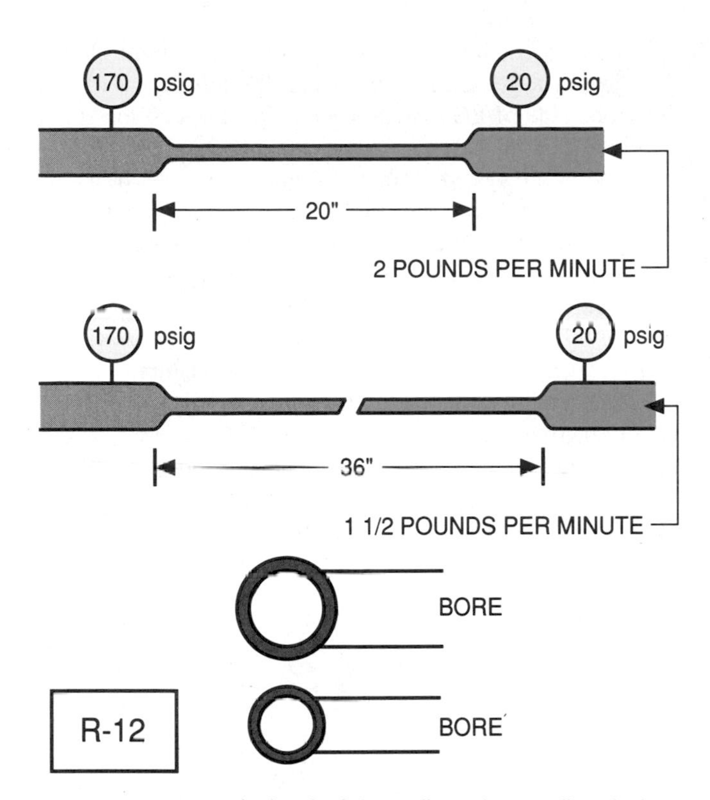

FIGURE 24–37B The length of the capillary tube as well as the bore determines the flow rate of the refrigerant, which is rated in pounds per minute.

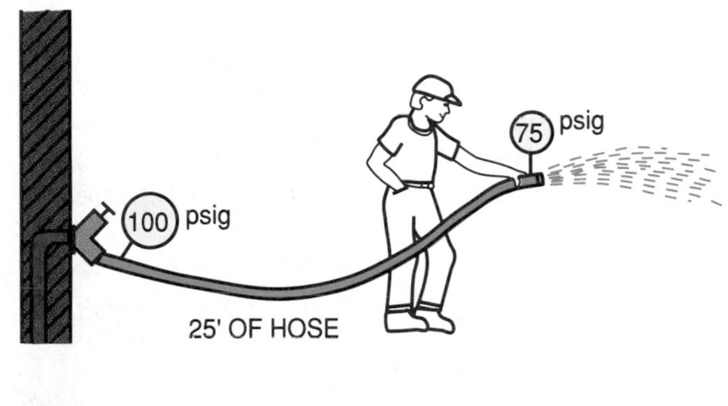

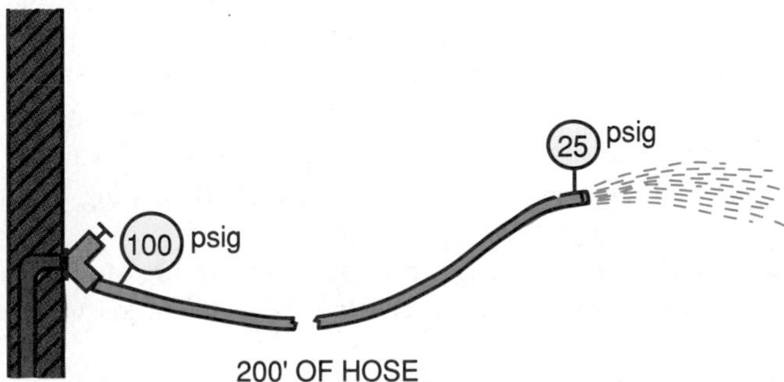

FIGURE 24–37C The longer the hose, the less water pressure at the end. A capillary tube functions the same way.

the tube. The bore is so small that a small piece of flux, carbon, or solder would cause a problem if it were to reach the tube inlet. Manufacturers always place a strainer or strainer-drier just before the capillary tube to prevent this from happening, Figure 24–38B.

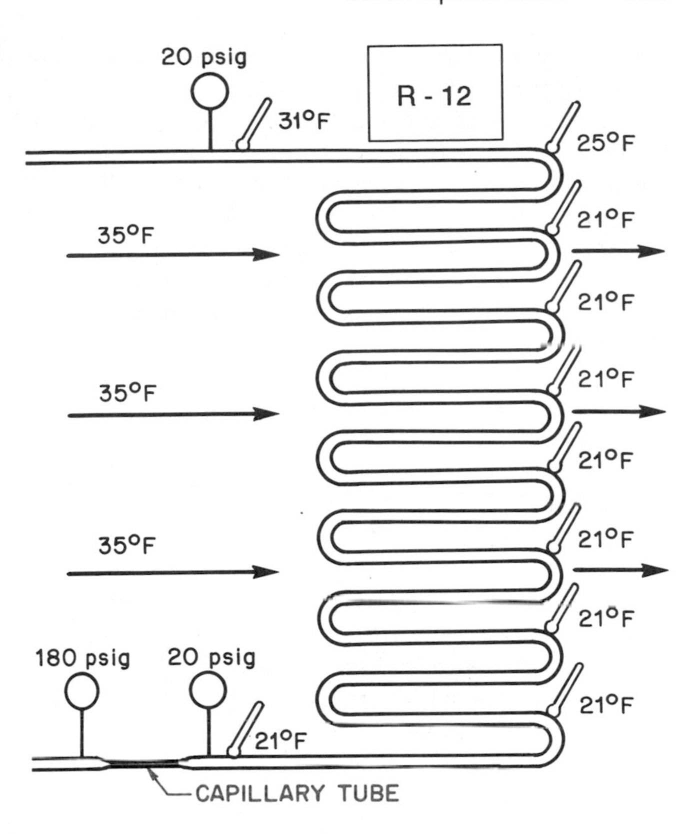

FIGURE 24–38A Typical operating conditions for a medium-temperature application.

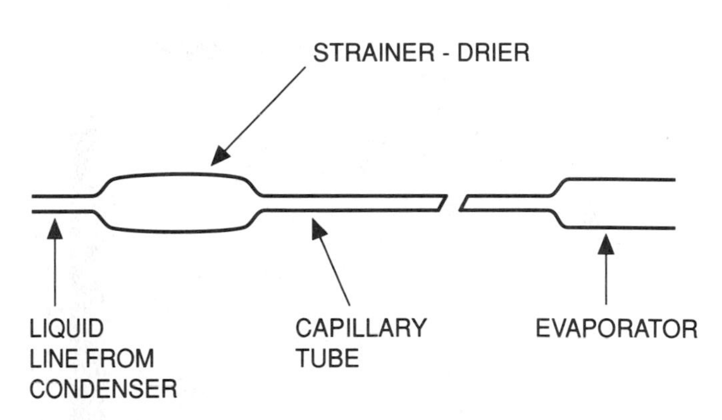

FIGURE 24–38B A strainer-drier protects the capillary tube from circulating particles.

24.27 OPERATING CHARGE FOR THE CAPILLARY TUBE SYSTEM

The capillary tube system requires only a small amount of refrigerant because it does not modulate (feed more or less) refrigerant according to the load. Capillary tube systems are known as critically charged systems. When the refrigerant charge is analyzed, when the unit is operating at the design conditions there is a specific amount of refrigerant in the evaporator and a specific amount of refrigerant in the condenser. This is the amount of refrigerant required for proper refrigeration. Any other refrigerant that is in the system is in the pipes for circulating purposes only.

The amount of refrigerant in the system is critical in capillary tube systems. It is easy for technicians to overcharge the system if they are not careful and familiar with the system. In most capillary tube systems, the charge is printed on the nameplate of the equipment. The manufacturer always recommends measuring the refrigerant into these systems either by scales or by liquid-charging cylinders. A properly charged system would maintain about 10°F superheat at the end of the coil, Figure 24–38A.

In many capillary tube systems, the capillary tube is fastened to the suction line between the condenser and the evaporator to exchange heat between the capillary tube and the suction line. To troubleshoot a possible capillary tube problem the correct superheat reading must be taken. The capillary tube should hold the refrigerant to about 10°F superheat at the end of the evaporator. *To obtain the correct superheat reading it must be taken before the capillary tube/suction line heat exchanger, Figure 24–39.*

The capillary tube is very slow to respond to load changes or charge modifications. For example, if a technician were to add a small amount of refrigerant to the capillary tube system, it would take at least 15 min for the charge to adjust. The reason for this is that the refrigerant moves from one side of the system to the other through the small bore and this takes time. When you add refrigerant to the low side of the system, it moves into the compressor and is pumped into the condenser. It now must move to the evaporator before the charge is in balance. Many technicians get in a hurry and overcharge this type of system. Manufacturers do not recommend adding refrigerant to top off the charge; they recommend starting over from a deep vacuum and measuring a complete charge into the system.

The capillary tube metering device is used primarily on small fractional horsepower refrigeration systems. These systems are hermetically sealed and have no bolted gasketed connections and provide a leak-free system. They are factory assembled in a very clean environment and should run troublefree for many years.

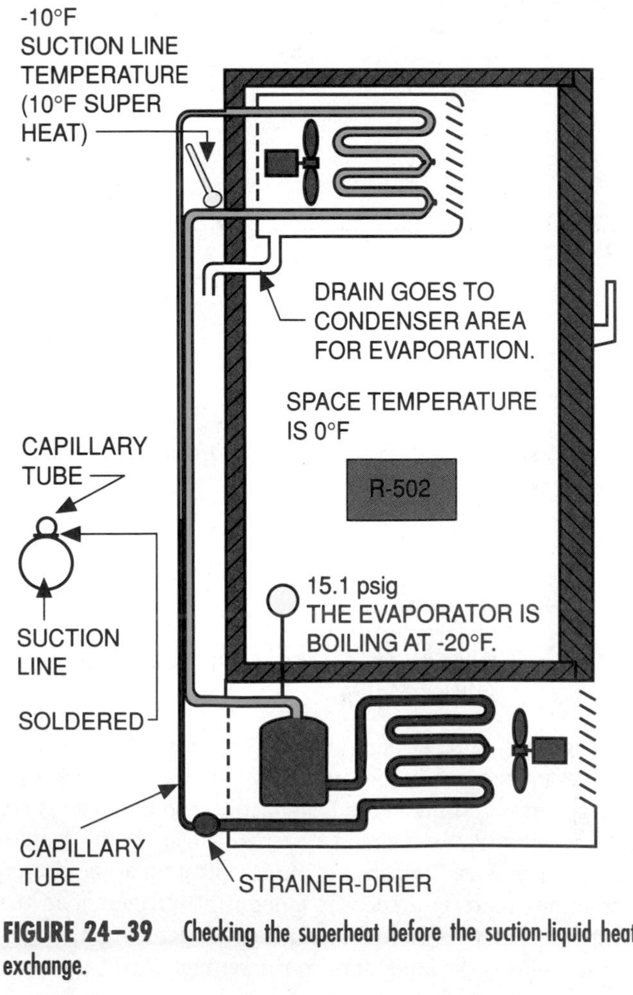

FIGURE 24–39 Checking the superheat before the suction-liquid heat exchange.

SUMMARY

- The expansion device is one of the dividing points between the high and low sides of the system.
- The TXV valve maintains a constant *superheat* in the evaporator.
- The TXV is composed of a body, a diaphragm, a needle and seat, a spring, an adjustment, and a bulb and transmission tube.
- The TXV has three forces acting on its needle and seat: the bulb, the evaporator, and the spring pressure.
- The bulb pressure is the only force that acts to open the valve.
- The forces inside the expansion valve all work together to hold the needle and seat in the correct position so that the evaporator will have the correct amount of refrigerant under all load conditions.
- The bulb and diaphragm are charged with one of four different charges: the liquid charge, the cross liquid charge, the vapor charge, or the cross vapor charge.
- The TXV bulb must be mounted securely to the suction line to accurately sense the suction line temperature.
- The TXV responds to a load increase by feeding more refrigerant into the evaporator.
- An external equalizer prevents excessive pressure drop in the evaporator from causing the TXV to starve the evaporator.
- Balanced port TXVs are used where low ambient temperatures exist.
- The solid-state controlled expansion valve uses a thermistor to monitor the suction line temperature to control refrigerant flow to the evaporator.
- The automatic expansion valve responds in reverse to a load change; when the load increases, the automatic expansion valve throttles the refrigerant instead of feeding more refrigerant as the TXV does.

- The capillary tube expansion device is a fixed-bore metering device usually made of copper, with a very small inside diameter, and no moving parts.
- The capillary tube system uses a very limited amount of refrigerant compared to other metering devices and is popular for small systems.

REVIEW QUESTIONS

1. What are the three forces acting on the TXV diaphragm?
2. What material are the needle and seat normally made of?
3. The TXV maintains what condition in the evaporator?
4. Name the four types of charge that the TXV uses to control refrigerant flow.
5. Where is the bulb of the TXV mounted?
6. When do some TXVs require an external equalizer?
7. Draw a diagram showing how an external equalizer line is connected into a system.
8. How does a TXV respond to an increase in load?
9. What condition does the automatic expansion valve maintain in the evaporator?
10. How does the automatic expansion valve respond to a load increase?
11. What determines the amount of refrigerant that flows through a capillary tube metering device?

UNIT

25 Special Refrigeration System Components

OBJECTIVES

After studying this unit, you should be able to

- distinguish between mechanical and electrical controls.
- explain how and why mechanical controls function.
- define low ambient operation.
- describe electrical controls that apply to refrigeration.
- describe off-cycle defrost.
- describe random and planned defrost.
- explain temperature-terminated defrost.
- describe the various refrigeration accessories.
- describe the low-side components.
- describe the high-side components.

SAFETY CHECKLIST

* Wear goggles and gloves when attaching or removing gages to transfer refrigerant or to check pressures.
* Wear warm clothing when working in a walk-in cooler or freezer. A technician does not think properly when chilled.
* Be careful not to get your hands or clothing caught in moving parts such as pulleys, belts, or fan blades.
* Wear an approved back brace belt when lifting and use your legs, keeping your back straight and upright.
* Observe all electrical safety precautions. Be careful at all times and use common sense.
* Never work near live electrical circuits unless absolutely necessary. Turn the power off at the nearest disconnect, lock and tag the panel, and keep the only key in your possession.

25.1 THE FOUR BASIC COMPONENTS

The compression refrigeration cycle must have the four basic components to function: the compressor, the condenser, the evaporator, and the expansion device. However, many more devices and components can enhance the performance and the reliability of the refrigeration system. Some of these protect the components, and some improve the reliability under various conditions.

Special refrigeration system components can be divided into two broad categories: controls and accessories. Control components are divided into mechanical, electrical, and electromechanical.

25.2 MECHANICAL CONTROLS

Mechanical controls generally stop, start, or modulate fluid flow and can be operated by pressure, temperature, or electricity. These controls can usually be identified because they are almost always found in the piping.

25.3 TWO-TEMPERATURE CONTROLS

Two-temperature operation is desirable when more than one evaporator is used with one compressor. This occurs if two or more evaporators are designed to operate in different temperature ranges, such as when an evaporator operating at 30°F is used on the same compressor with an evaporator operating at 20°F, Figure 25–1. Two-temperature applica-

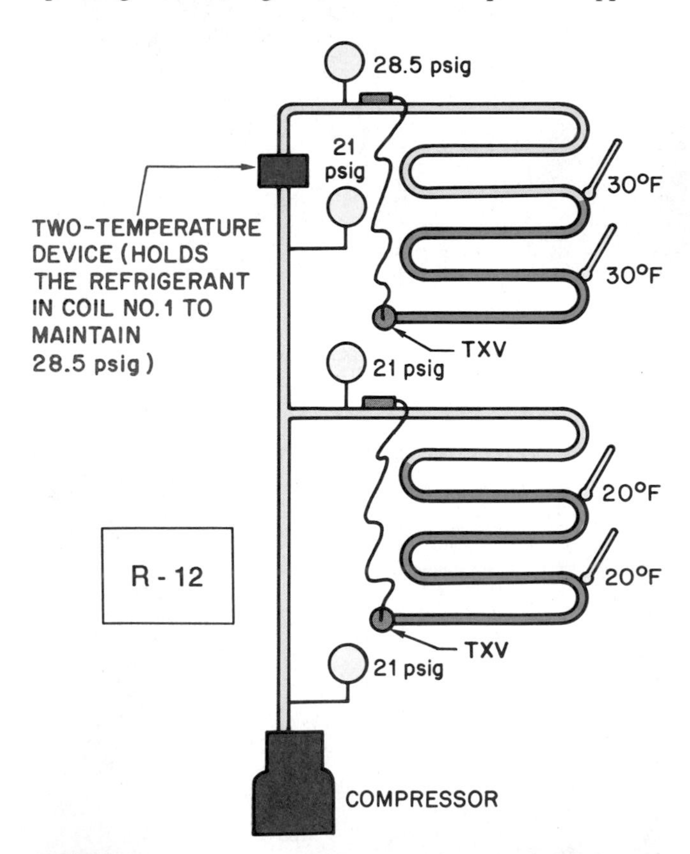

FIGURE 25–1 Two-temperature operation. Two evaporators are operating on one compressor. One evaporator operates at 20°F (21 psig), the other at 30°F (28.5 psig).

tion is normally accomplished with a purely mechanical valve or a temperature-controlled valve.

25.4 EVAPORATOR PRESSURE CONTROL

The *evaporator pressure regulator valve (EPR valve)* is a mechanical control that keeps the refrigerant in the evaporator from going below a predetermined point. The EPR valve is installed in the suction line at the evaporator outlet. The bellows in the EPR valve senses evaporator pressure and throttles (modulates) the suction gas to the compressor. This will then allow the evaporator pressure to go as low as the pressure setting on the valve. When the EPR valve is used with the thermostatic expansion valve (TXV), the system now has the characteristics of maintaining a constant superheat by keeping the pressure from going too low, Figure 25–2.

The EPR valve can be applied to a system that cools water. The evaporator will not go below the predetermined point, which could be freezing. For example, when the system is started, with a load on it, the EPR valve would be wide open. The TXV would be throttling the liquid refrigerant into the evaporator. When the refrigerant is cooled to the point at which the EPR valve is set, it will begin to throttle a slow flow of refrigerant. If this setting is just above freezing, the valve will throttle off enough to keep the evaporator from freezing until the thermostat responds and shuts the system off, Figure 25–3. This valve is often called a "hold back valve" in industry because it holds the refrigerant back in the evaporator, preventing the suction pressure in the evaporator from going below a set point.

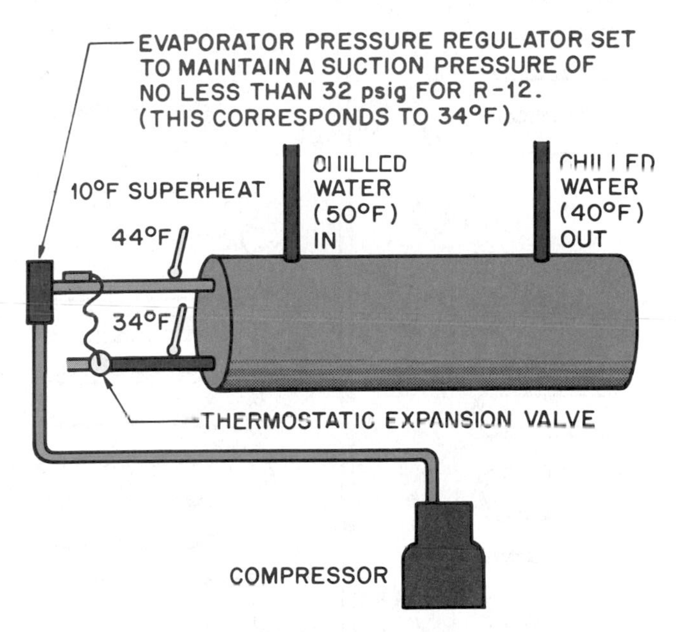
R - 12
EVAPORATOR PRESSURE REGULATOR SET TO MAINTAIN A SUCTION PRESSURE OF NO LESS THAN 32 psig FOR R-12. (THIS CORRESPONDS TO 34°F)
10°F SUPERHEAT
44°F
34°F
CHILLED WATER (50°F) IN
CHILLED WATER (40°F) OUT
THERMOSTATIC EXPANSION VALVE
COMPRESSOR

FIGURE 25–3 TXV keeps the evaporator at the proper level of refrigerant while the EPR valve keeps the evaporator from getting too cold. This combination of controls has all of the advantages of the TXV and the automatic expansion valve in one system. The EPR valve may be used on a water-type evaporator to keep it from operating below freezing. The thermostat should cut off the compressor soon after the EPR valve starts to throttle.

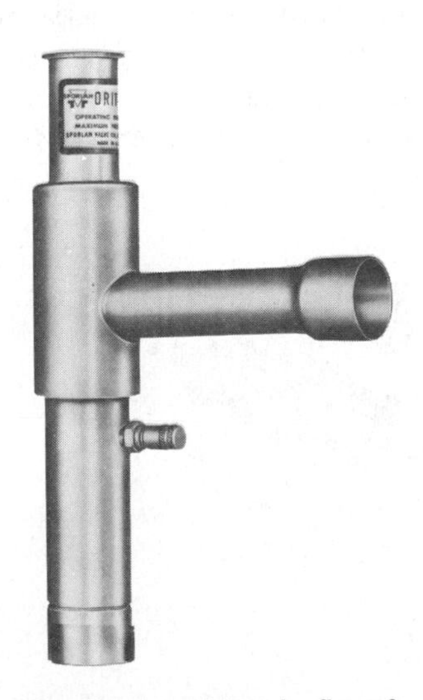

FIGURE 25–2 EPR valve to modulate the flow of vapor refrigerant leaving the evaporator. It limits the pressure in the evaporator and keeps it from dropping below the set point. *Courtesy Sporlan Valve Company*

25.5 MULTIPLE EVAPORATORS

When more than two evaporators are used with one compressor, one of them may sometimes be at a different temperature and pressure range. For example, when an evaporator that needs to operate at 20 psig (15°F for R-12) is piped with another evaporator of 25 psig (25°F for R-12) to the same compressor, an EPR valve is needed in the highest temperature evaporator's suction line. The true suction pressure for the system will be the low value of 20 psig, but the other evaporator will operate at the correct pressure of 25 psig. Several evaporators of different pressure requirements can be piped together using this method, Figure 25–4.

It is desirable to know the actual pressure in the evaporator because the evaporator pressure is not the same as the true suction pressure. Normally there is a gage port, known as a Schrader valve port, permanently installed in the EPR valve body that allows the service technician to take gage readings on the evaporator side of the valve. The true suction pressure can be obtained at the compressor service valve, Figure 25–2.

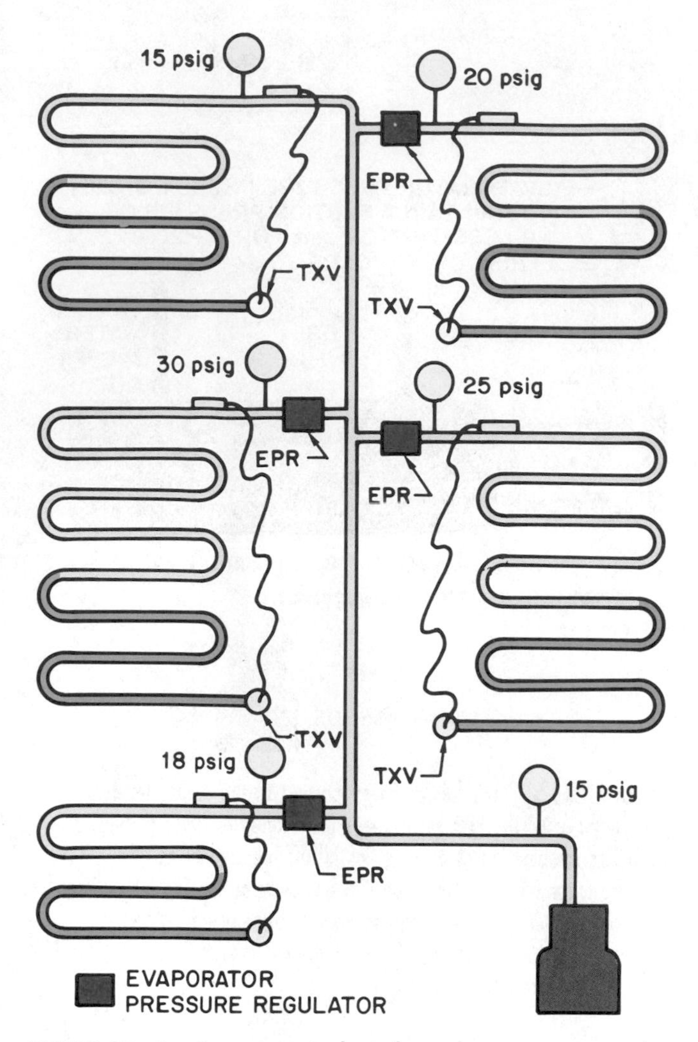

FIGURE 25–4 Evaporators piped together and one compressor used to maintain a suction pressure for the lowest pressure evaporator. Evaporators that have pressures higher than the true suction pressure of the lowest evaporator all have EPR valves set to their individual needs.

25.6 TWO-TEMPERATURE VALVE

The two-temperature valve operates in much the same way as the EPR valve except it is more definitive in its action. It generally has some method to open or close the valve instead of modulating it. This can be accomplished either with an overcenter movement inside the valve or with a thermostat and an electric valve.

The mechanical type of two-temperature valve uses the overcenter device and gives a distinct temperature spread in the evaporator. It has low- and high-temperature settings that can be seen. The high-temperature setting can be set above freezing for the coil to ensure defrost before the valve will open and allow refrigerant to flow out of the coil, Figure 25–5.

The electric type of device could be a solenoid valve placed either at the evaporator outlet or the evaporator inlet before the expansion valve. If it is placed at the outlet, it would act much as the mechanical type except it is electrically controlled with a thermostat in the box. If the valve is placed in the liquid line before the expansion valve, when the valve closes the liquid would be pumped out of the coil into the receiver. Each type has its advantage. An example of a solenoid valve can be seen in Figure 25–16.

The thermostatic type of the two-temperature valve acts much like the TXV except it controls the vapor leaving the evaporator. The sensing bulb can be located either in the air inlet or in the air outlet to the evaporator. This would be a modulating type of valve, Figure 25–6.

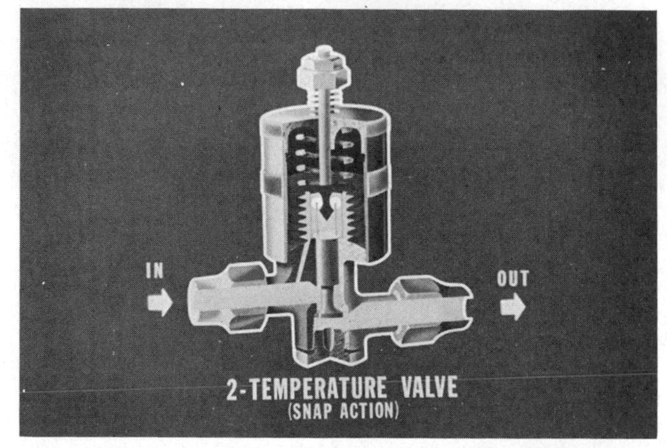

FIGURE 25–5 This two-temperature valve uses an overcenter device that actually gives two different temperatures inside the evaporator. This control has a low-temperature event and a high-temperature event. The high-temperature event can be set above freezing and used as defrost. This is not a modulating type of control. *Reproduced courtesy of Carrier Corporation*

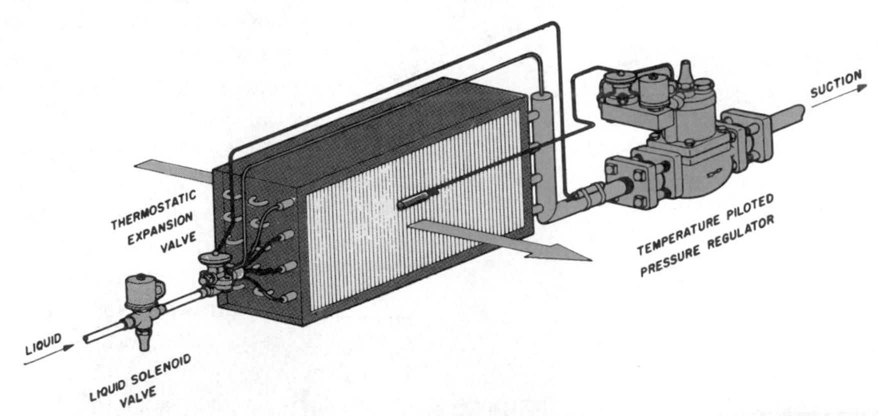

FIGURE 25–6 Thermostatic two-temperature device. This valve may have its sensing bulb in the entering airstream or the leaving airstream of the evaporator. *Courtesy ALCO Controls Division, Emerson Electric Company*

25.7 CRANKCASE PRESSURE REGULATOR

The crankcase pressure regulator (*CPR valve*) looks much the same as the EPR valve, but it has a different function. The CPR valve is in the suction line also, but it is usually located close to the compressor rather than at the evaporator outlet. The CPR valve sensing bellows is on the true compressor suction side of the valve and would normally have a gage port on the evaporator side of the valve, Figure 25–7.

This valve is used to keep the low-temperature compressor from overloading on a hot pull down. A hot pull down would occur (1) when the compressor has been off for a long enough time and the foodstuff has a rise in temperature, or (2) on start-up with a warm box. In either case the temperature in the refrigerated box influences the suction pressure. When the temperature is high, the suction pressure is high. When the suction pressure goes up, the density of the suction gas goes up. The compressor is a constant-volume pump and does not know when the gas it is pumping is dense enough to create a motor overload.

When a compressor is started and the refrigerated cooler is warmer than the design range of the compressor, an overload occurs. The refrigeration compressor motor can be operated at about 10% over its rated capacity without harm during this pull down. When a motor has a full-load amperage rating of 20 A, it could run at 22 A for an extended pull down, and there would be no harm to the motor. If the cooler temperature were to be 75°F or 80°F, the motor would be overloaded to the point that the motor overcurrent protection would shut it off. This protection is automatically reset, so the compressor would try to restart immediately, causing it to short cycle. This would continue until the manual reset motor control, the fuse, or breaker stopped the compressor, or the system would slowly pull down during the brief running times.

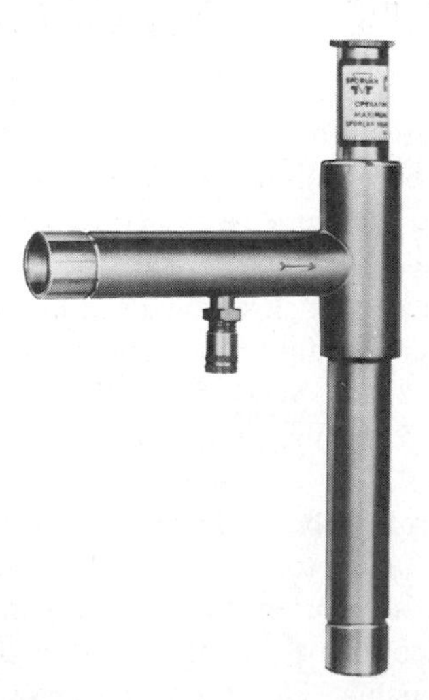

FIGURE 25–7 CPR valve to keep the compressor from running in an overloaded condition during a hot pull down. *Courtesy Sporlan Valve Company*

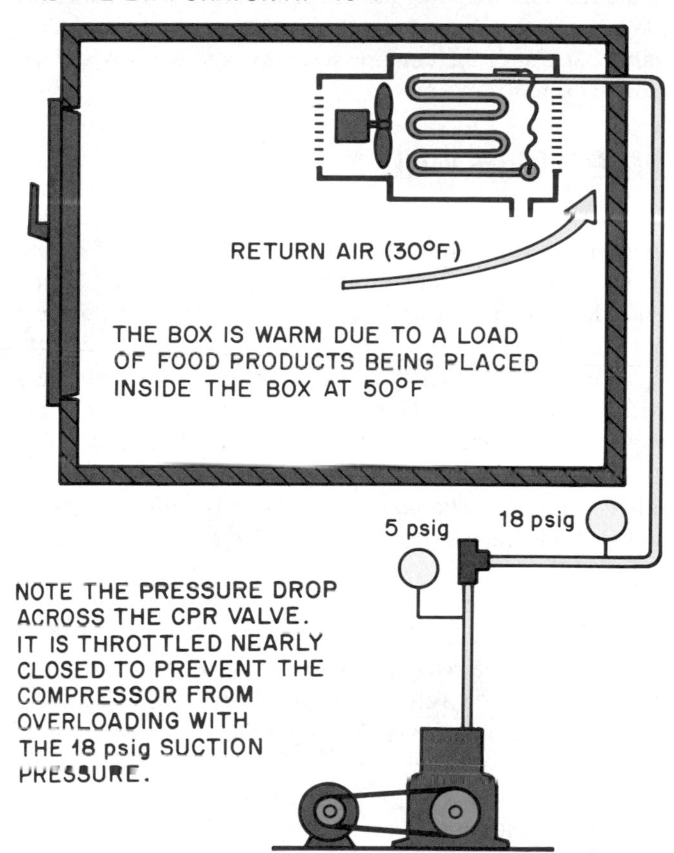
THIS IS A LOW-TEMPERATURE FREEZER WITH THE COMPRESSOR DESIGNED TO OPERATE AT 2 psig AND THE EVAPORATOR AT -16°F
RETURN AIR (30°F)
THE BOX IS WARM DUE TO A LOAD OF FOOD PRODUCTS BEING PLACED INSIDE THE BOX AT 50°F
5 psig
18 psig
NOTE THE PRESSURE DROP ACROSS THE CPR VALVE. IT IS THROTTLED NEARLY CLOSED TO PREVENT THE COMPRESSOR FROM OVERLOADING WITH THE 18 psig SUCTION PRESSURE.
COMPRESSOR MOTOR FULL-LOAD AMPERAGE IS 20 A WHEN THE SUCTION PRESSURE IS 5 psig

FIGURE 25–8 The CPR valve throttles the vapor refrigerant entering the compressor. There will be a pressure drop across the valve when the evaporator is too warm. This valve is adjusted and set up using the compressor full-load amperage on a hot pull down. The CPR valve can be set to throttle the suction gas to the compressor and maintain a suction pressure on the compressor side of the valve that will not allow the compressor to overload. When the refrigerated box temperature is pulled down to the design range, the suction gas in the evaporator will be the same as the pressure at the compressor because the valve is wide open.

The CPR valve would throttle the suction gas into the compressor to keep the compressor current at no more than the rated value, Figure 25–8. Before CPR valves were used, the service technician had to start the system manually by throttling the suction service valve on a hot start-up.

25.8 ADJUSTING THE CPR VALVE

The setting of the CPR valve should be accomplished on a hot pull down or at least when the compressor has enough load that it is trying to run overloaded. This can be done by shutting off the unit until the box or cooler is warm enough to create a load on the evaporator. This loads up the compressor enough to cause it to run at a high current. Use the ammeter on the compressor while adjusting the CPR valve to the full-load amperage of the compressor. For example, if the compressor is supposed to draw 20 A, throttle the CPR valve back until the compressor amperage is 20 A and the valve is set.

25.9 RELIEF VALVES

Relief valves are designed to release refrigerant from a system when predetermined high pressures exist. Refrigerant relief valves come in two different types: the spring-loaded type that will reseat and the one-time type that does not close back.

The *spring-loaded* type of relief valve is normally brass with a neoprene seat. This valve is piped so that it is in the vapor space of the condenser or receiver. The relief valve must be in the vapor space, not in the liquid space, for it to relieve pressure. The object is to let vapor off to vaporize some of the remaining liquid and lower its temperature, Figure 25–9.

The top of the valve normally has threads, so the refrigerant can be piped to the outside of the building. ***When relief valves are used to protect vessels in a fire, it is desirable for the refrigerant to be removed from the area. Refrigerant gives off a noxious gas when it is burned.***

The *one-time* relief valve is caused to relieve by temperature. These valves, often called fusible plugs, are designed with one of the following methods: a fitting filled with a low melting-temperature solder; a patch of copper soldered on a drilled hole in the copper line using low-temperature solder; or a fitting that has been drilled out at the end with a spot of low-temperature solder over the hole, Figure 25–10. Sometimes the melting temperature is printed on the fitting. The melting temperature will be very low, about 220°F.

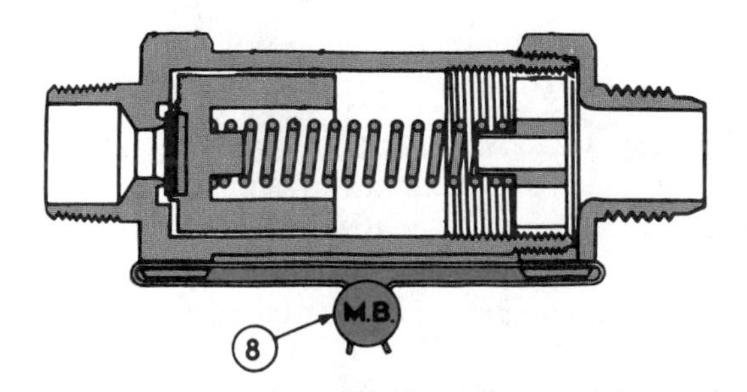

FIGURE 25–9 Spring-loaded type of relief valve used to protect the system from very high pressures that might occur if the condenser fan failed or if a water supply failed on a water-cooled system. This relief valve is designed to reseat after the pressure is reduced. There are threads on the valve outlet, so the valve can be piped to the outside if needed. *Courtesy Superior Valve Company*

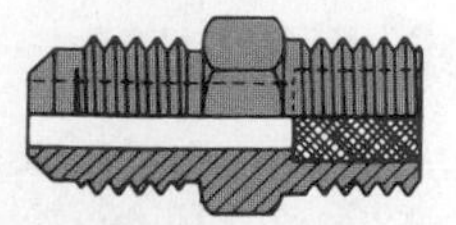

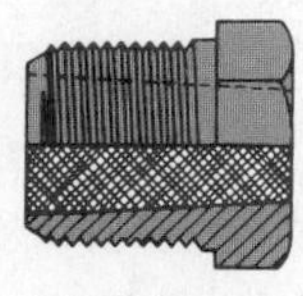

FIGURE 25–10 Fusible relief. It may be a low-temperature solder patch that covers a hole drilled in a fitting or pipe. Care should be taken when soldering in the vicinity of this relief device, or it will melt. *Courtesy Mueller Brass Company*

When soldering around the compressor make sure that the solder in the fusible plug is not melted away. This type of device will normally never relieve unless there is a fire. It can often be found on the suction side of the system close to the compressor to protect the system and the public. This will keep the compressor shell from experiencing high pressures and rupturing during a fire. Remember, the compressor shell has a working pressure of 150 psig.

25.10 LOW AMBIENT CONTROLS

Low ambient controls are important in refrigeration systems because refrigeration is needed all year. When the condenser is located outside, the head pressure will go down in the winter to the point that the expansion valve will not have enough pressure drop across it to feed refrigerant correctly. When this happens, some method must be used to keep the head pressure up to an acceptable level. The most common methods are:

1. Fan cycling using a pressure control
2. Fan speed control
3. Air volume control using shutters and fan cycling
4. Condenser flooding devices

25.11 FAN CYCLING HEAD PRESSURE CONTROL

Fan cycling has been used for years because it is simple. When a unit has one small fan, this is a simple and reliable method because only one fan is cycled. When there is more than one fan, the extra fans may be cycled by temperature, and the last fan can be cycled by head pressure. The control used is a pressure control that closes on a rise in pressure to start the fan and opens on a fall in pressure to stop the fan when the head pressure falls. When fan cycling is used, the fan will cycle in the winter but will run constantly in hot weather, Figure 25–11.

This type of control can be hard on a fan motor because of all the cycling. The fan motor needs to be a type that does not have a high starting current. The control must have enough contact surface area to be able to start the motor many times to be reliable. The motor current should be carefully compared to the control capabilities.

Fan cycling can vary the pressure to the expansion device a great deal. The technician must choose whether to

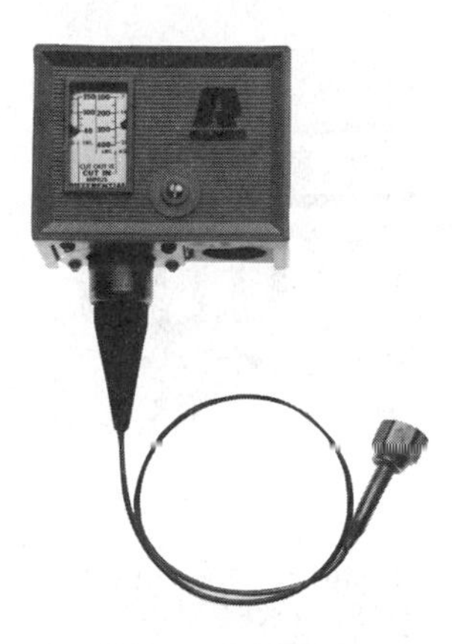

FIGURE 25–11 Pressure control with the same action as a low-pressure control; it makes on a rise in head pressure to start the condenser fan. This pressure control operates at a higher pressure range than the low-pressure control normally used on the low side of the system. Fan cycle devices will cause the head pressure to fluctuate up and down. This can affect the operation of the expansion device. *Courtesy Ranco*

have the control set points close together for the best expansion device performance or far apart to keep the fan from short cycling. The best application for fan cycling devices is in the use of multiple fans, where the last fan has a whole condenser to absorb the fluctuation. Vertical condensers may be affected by the prevailing winds. If these winds are directed into the coil, they can provide airflow similar to a fan. Sometimes a shield is installed to prevent this, Figure 25–12.

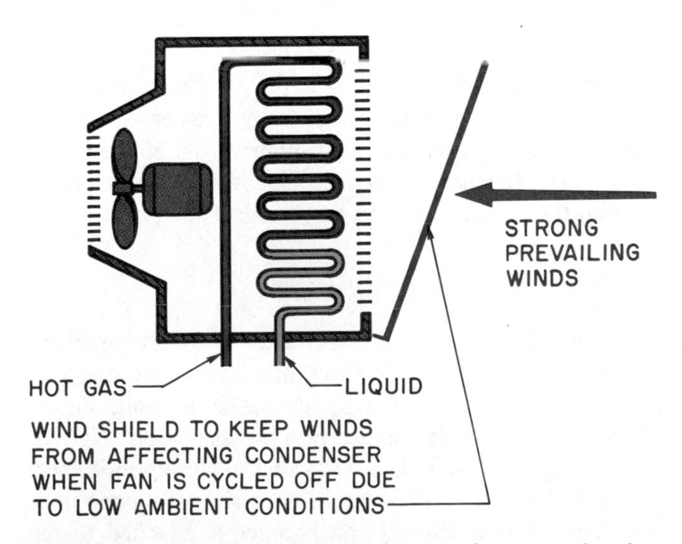

FIGURE 25–12 If fan cycling controls are used, care must be taken if the condenser is vertical. The prevailing winds can cause the head pressures to be too low even with the fan off. A shield can be used to prevent the wind from affecting the condenser.

25.12 FAN SPEED CONTROL FOR CONTROLLING HEAD PRESSURE

Fan speed control devices have been used successfully in many installations. This device can be used with multiple fans, where the first fans are cycled off by temperature and the last fan is controlled by head pressure or condensing temperature. The device that controls this fan is normally a transducer that converts pressure to motor speed control. A temperature sensor may also be used as a transducer. As the temperature drops on a cool day, the motor speed is reduced. As the temperature rises, the motor speed increases. At some predetermined point the additional fans are started. On a hot day all fans will be running, with the variable fan running at maximum speed. Some fan speed controls use a temperature sensor to monitor the condenser's temperature, Figure 25–13.

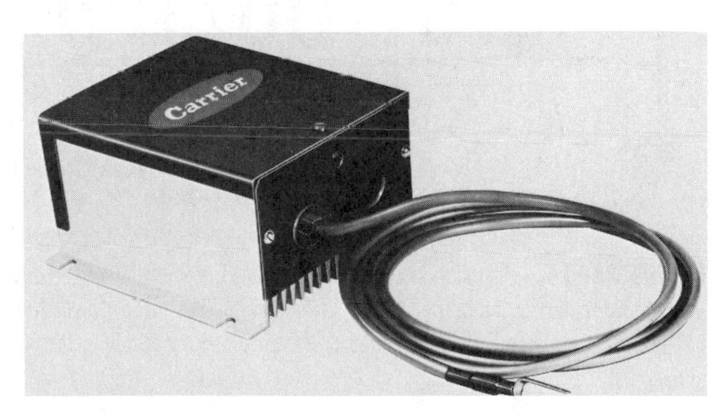

FIGURE 25–13 Control used to vary the fan speed on a special motor based on condenser temperature. *Reproduced courtesy of Carrier Corporation*

25.13 AIR VOLUME CONTROL FOR CONTROLLING HEAD PRESSURE

Air volume control using shutters is accomplished with a damper motor driven with the high-pressure refrigerant. When there are multiple fans, the first fans can be cycled off using temperature, and the shutter can be used on the last fan, as in the previous two systems. This system results in a steady head pressure, as with fan speed control. The expansion valve inlet pressure does not go up and down like it does when one fan is cycled on and off. The shutter can be located either on the fan inlet or on the fan outlet, Figure 25–14. Be careful that the fan motor is not overloaded with a damper.

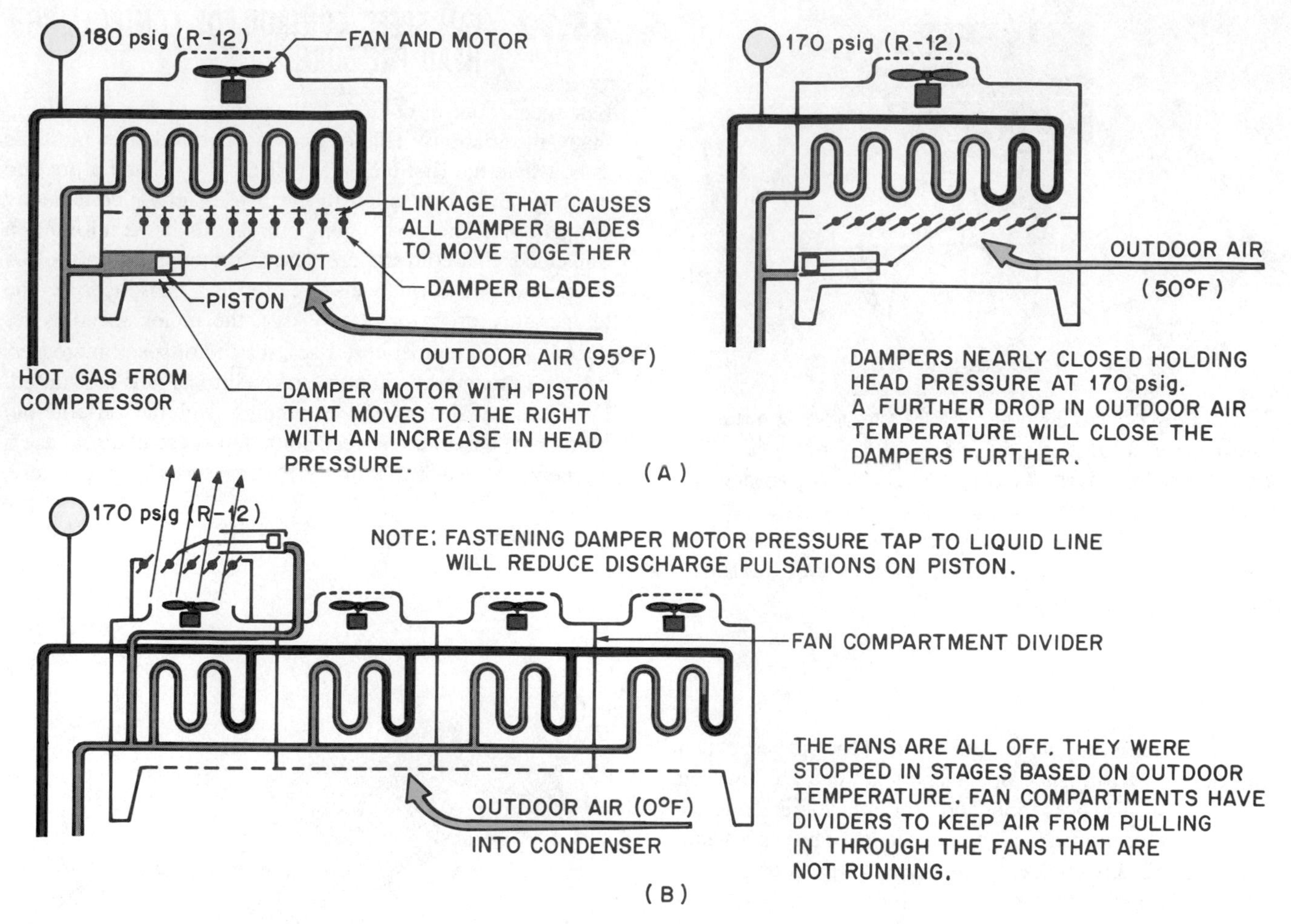

FIGURE 25–14 Installation using dampers to vary the air volume instead of fan speed. This is accomplished with a refrigerant-operated damper motor that operates the shutters. This application modulates the airflow and gives steady head pressure control. If more than one fan is used, the first fans can be cycled off based on ambient temperature. Care should be taken that the condenser fan will not be overloaded when the shutters are placed at the fan outlet.

25.14 CONDENSER FLOODING FOR CONTROLLING HEAD PRESSURE

Condenser flooding devices are used in both mild and cold weather to cause the refrigerant to move from an oversized receiver tank into the condenser, Figure 25–15. This excess refrigerant in the condenser acts like an overcharge of refrigerant and causes the head pressure to be much higher than it normally would be on a mild or cold day. The head pressure will remain the same as it would on a warm day. This method gives very steady control with no fluctuations while running. This system requires a large amount of refrigerant because in addition to the normal operating charge, there must be enough refrigerant to flood the condenser in the winter. A large receiver is used to store the refrigerant in the summer when the valves will divert the excess liquid into the receiver.

Condenser flooding has one added benefit in the winter. Because the condenser is nearly full of refrigerant, the liquid refrigerant that is furnished to the expansion devices is well below the condensing temperature. Remember, this is called subcooling and will help improve the efficiency of the system. The liquid may be subcooled to well below freezing for systems operating in a cold climate.

25.15 THE SOLENOID VALVE

The *solenoid valve* is the most frequently used component to control fluid flow. This valve has a magnetic coil that when energized will lift a plunger into the coil, Figure 25–16. This valve can either be normally open (NO) or normally closed (NC). The NC valve is closed until energized; then it opens. The NO valve is normally open until energized; then it closes. The plunger is attached to the valve so the plunger action moves the valve.

Solenoid valves are snap-acting valves that open and close very fast with the electrical energy to the coil. They can be used to control either liquid or vapor flow. The snap

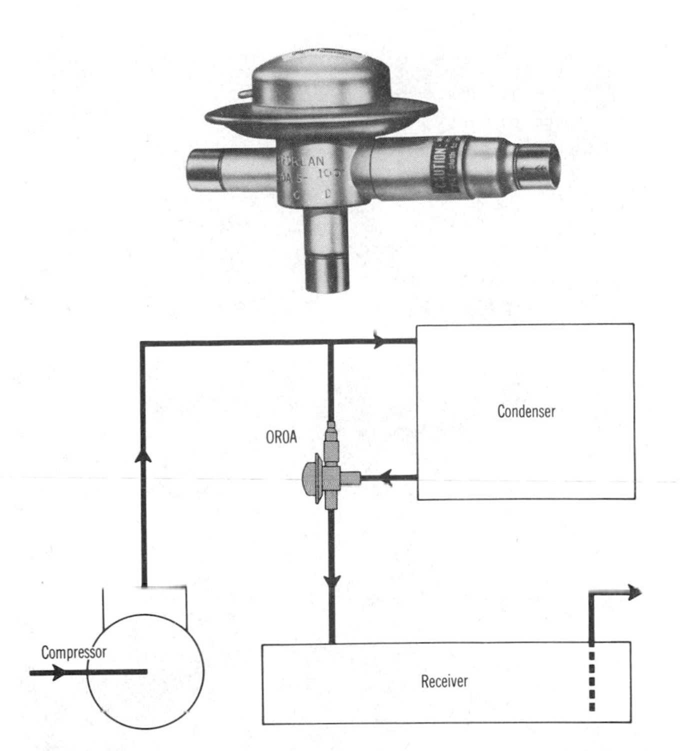
OROA
Condenser
Compressor
Receiver

FIGURE 25–15 Low ambient head pressure control accomplished with refrigerant. A large receiver stores the refrigerant when it is not needed (the summer cycle). The valve in the system diverts the hot gas to the condenser during summer operation, and the system acts like a typical system with a large receiver. In the winter cycle part of the gas goes to the condenser, depending on the outdoor temperature, and part of the gas to the top of the receiver. The gas going to the receiver keeps the pressure up for correct expansion valve operation. The gas going to the condenser is changed to a liquid and subcooled in the condenser. It then returns to the receiver to subcool the remaining liquid. *Courtesy Sporlan Valve Company*

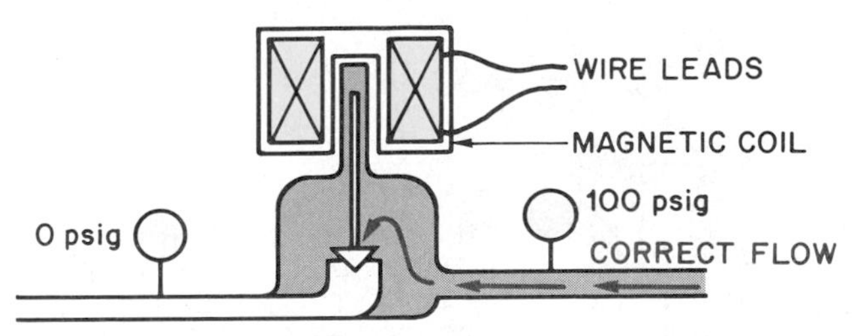
WIRE LEADS
MAGNETIC COIL
0 psig
100 psig
CORRECT FLOW
FLUID PRESSURE WILL HELP SEAT THE VALVE
WHEN THE VALVE IS INSTALLED CORRECTLY.

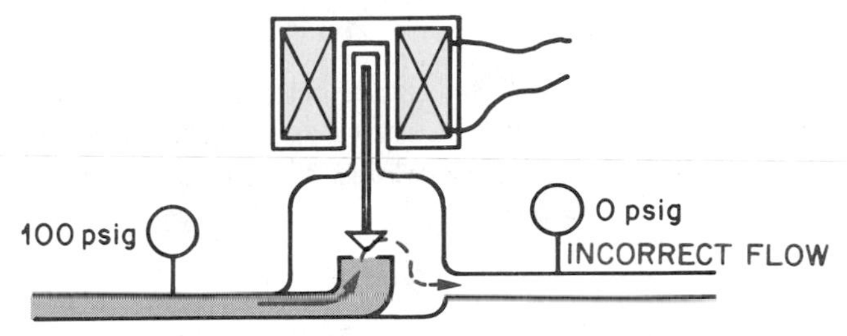
100 psig
0 psig
INCORRECT FLOW
THE PRESSURE IS RAISING THE VALVE NEEDLE
UP AND LEAKING BECAUSE THE HIGH PRESSURE
IS UNDER THE SEAT.

FIGURE 25–17 The fluid flow in a solenoid valve must be in the correct direction, or the valve will not close tight. If the high-pressure fluid is under the seat, it will have a tendency to raise the valve off the seat because the fluid helps to hold the valve on the seat.

action can cause liquid hammer when installed in a liquid line, so be careful when locating the valve. Follow the manufacturer's instructions as to the location and placement of solenoid valves. Liquid hammer occurs when the fast-moving liquid is shut off abruptly by the solenoid valve, causing the liquid to stop abruptly.

The solenoid valve is responsible for stopping and starting fluid flow. Two common mistakes in installation can prevent the solenoid valve from functioning correctly: the direction in which the valve is mounted and the position in which the valve is installed. The fluid flow has to be in the correct direction with a solenoid valve, or the valve may not close tightly, Figure 25–17. The valve is mounted in the

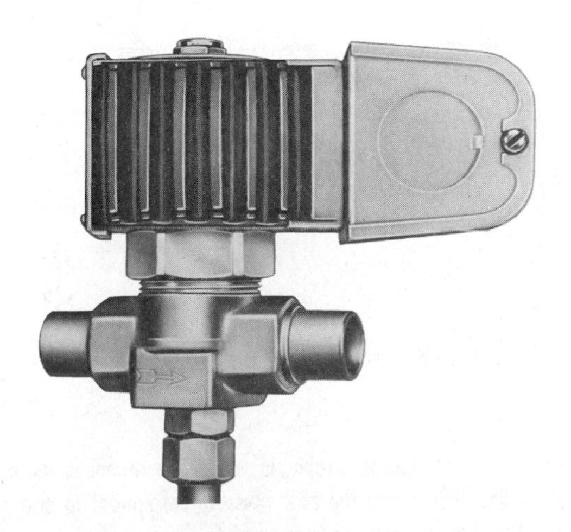
(A)

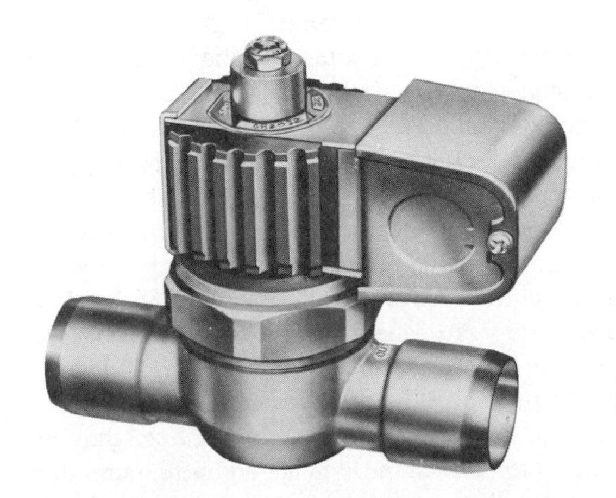
(B)

FIGURE 25–16 Electrically operated solenoid valves. The valve is moved by the plunger attached to the seat. The plunger moves into the magnetic coil when the coil is energized to either open or close the valve. *Courtesy Sporlan Valve Company*

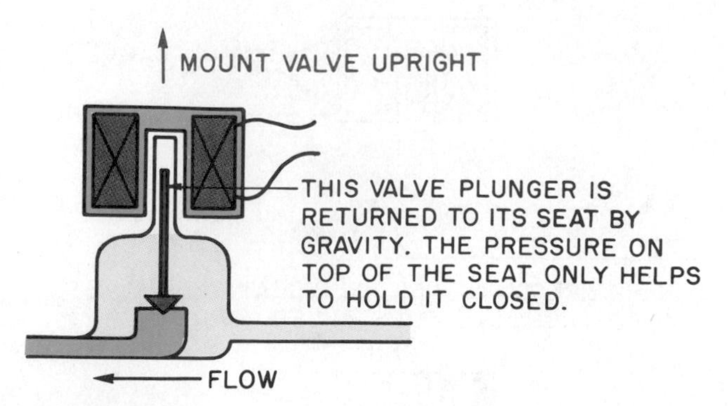

FIGURE 25–18 Solenoid valve installed in the correct position. Most valves use gravity to seat the valve, and the magnetic force raises the valve off its seat. If the valve is installed on its side or upside down, the valve will not seat when deenergized.

correct direction when the fluid helps to close the valve. If the high pressure is under the valve seat, the valve may have a tendency to lift off its seat. The valve will have an arrow to indicate the direction of flow. When placing the solenoid valve in the correct direction, the position of the valve must be considered. Most solenoid valves have a heavy plunger that is lifted to open the valve. When the plunger is not energized, the weight of this plunger holds the valve on its sealing seat. If this type of valve is installed on its side or upside down, the valve will remain in the energized position when it should be closed, Figure 25–18.

The solenoid valve must be fastened to the refrigerant line so that it will not leak refrigerant. It can be fastened by flare, flange, or solder connections. Most valves have to be serviced at some time. The valves that are soldered in the line can be serviced easily if they can be disassembled.

A special solenoid valve known as *pilot acting* is available for larger applications for controlling fluid flow. This valve uses a very small valve seat to divert high-pressure gas that causes a larger valve to change position. This type of solenoid valve uses this difference in pressure to cause the large movement while the solenoid's magnetic coil only has to lift a small seat. These valves are used when large vapor or liquid lines must be switched; they can have more than one inlet and outlet. Some are known as *four-way* valves, others as *three-way* valves. These have special functions. If an electrical coil had to be designed to cause the switching action of a large valve, it would be very large and draw too much current for practical use, Figure 25–19. Pilot action reduces the size of the electrical coil and the overall size of the valve.

Solenoid valves must be sized correctly for their particular application. For the refrigerant liquid line, they are sized according to refrigerant tonnage and a pressure drop through the valve that is acceptable. If a valve needs to be replaced, the manufacturer's data will help you choose the correct valve.

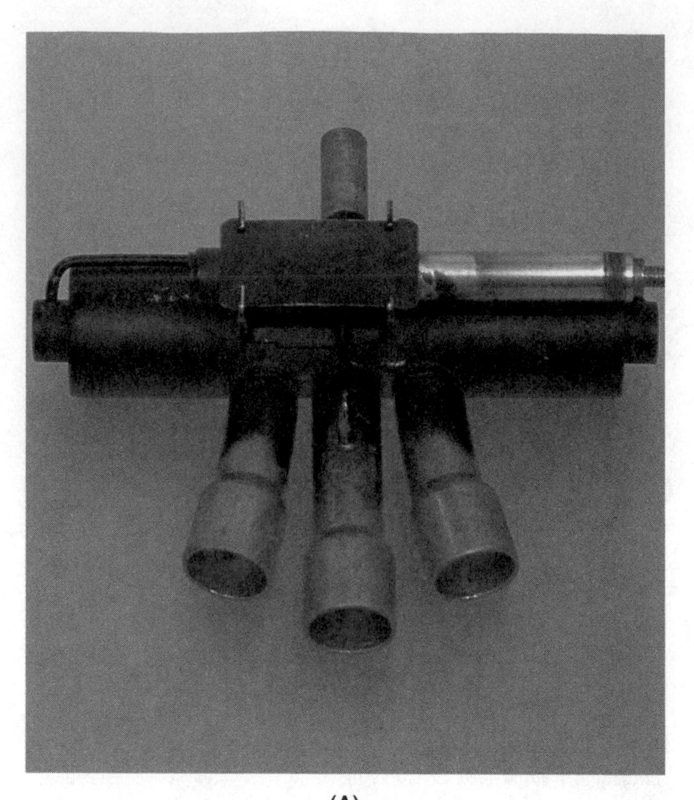

(A)

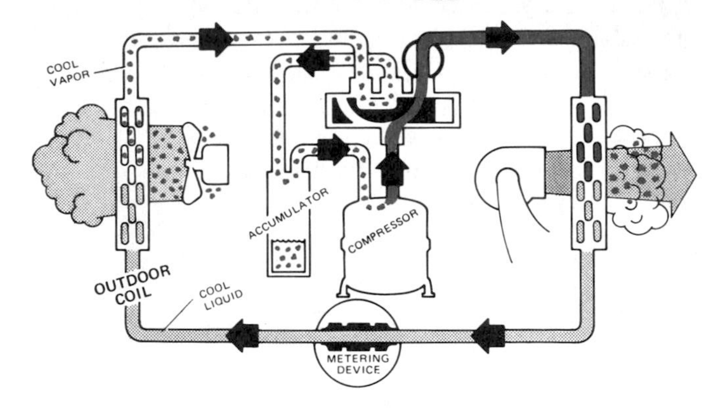

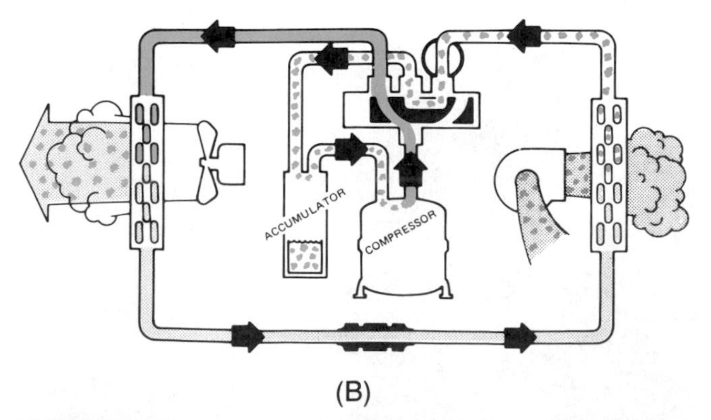

(B)

FIGURE 25–19 Pilot-operated solenoid valve. The magnetic coil only controls a small line that directs the high-pressure refrigerant to one end of a sliding piston. The magnetic coil only has to do a small amount of work, and the difference in pressure does the rest. *Photo by Bill Johnson, Line Drawing—Courtesy Carrier Corporation*

25.16 PRESSURE SWITCHES

Pressure switches are used to stop and start electrical current flow to refrigeration components. They play a very important part in the function of refrigeration equipment. The typical pressure switch can be a:

1. Low-pressure switch—closes on a rise in pressure
2. High-pressure switch—opens on a rise in pressure
3. Low ambient control—closes on a rise in pressure
4. Oil safety switch—has a time delay; closes on a rise in pressure

25.17 LOW-PRESSURE SWITCH

The low-pressure switch is used in two applications in refrigeration: low-charge protection and control of space temperature.

The low-pressure control can be used as a low-charge protection by setting the control to cut out at a value that is below the typical evaporator operating pressure. For example, in a medium-temperature cooler the air temperature may be expected to be no lower than 34°F. When this cooler uses R-12 for the refrigerant, the lowest pressure at which the evaporator would be expected to operate would be around 20 psig because the coil would normally operate at about 15°F colder than the air temperature in the coil (34°F − 15°F = 19°F). The refrigerant in the coil should not operate below this temperature of 19°F, which converts to 20 psig.

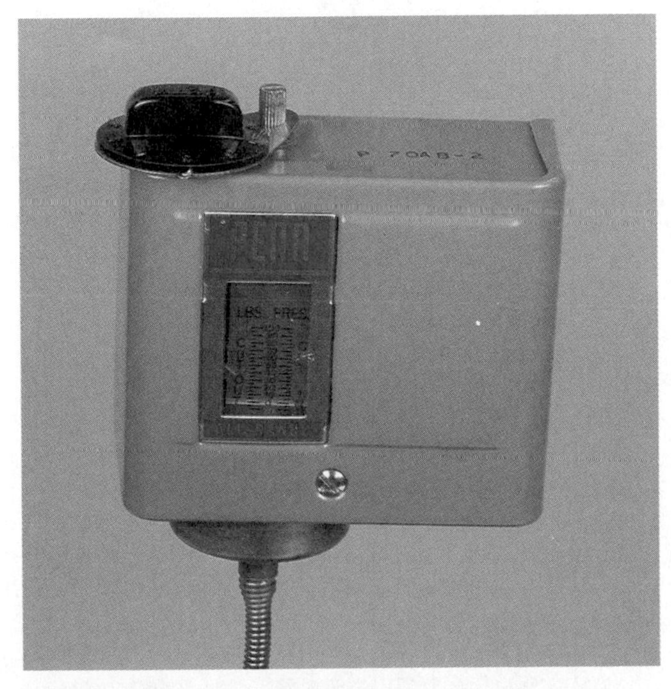

FIGURE 25–20 Low-pressure control. The control contacts make on a rise in pressure (open on a fall). If the low-side pressure goes down below the control set point for any reason, the control stops the compressor. *Photo by Bill Johnson*

The low-pressure cutout should be set below the expected operating condition of 20 psig and above the atmospheric pressure of 0 psig. A setting that would cut off the compressor at 5 psig would keep the low-pressure side of the system from going into a vacuum in the event of refrigerant loss. This setting would be well below the typical operating condition. This control would normally be automatically reset. When a low-charge condition exists, the compressor would be cut off and on with this control and maintain some refrigeration. The store owner may call with a complaint that the unit is cutting off and on but not lowering the cooler temperature properly. The cut-in setting for this application should be a pressure that is below the highest temperature the cooler is expected to experience in a typical cycle and just lower than the thermostat cut-in point. For example, the thermostat may be set at 45°F, which would mean that the pressure in the evaporator may rise as high as 42 psig. The low-pressure control should be set to cut in at about 40 psig. The previous example of a low-pressure control set to cut out at 5 psig and to cut in at 40 psig means that the control differential is 35 psig. The wide differential helps to keep the compressor from coming back on too soon or from short cycling, Figure 25–20.

25.18 LOW-PRESSURE CONTROL APPLIED AS A THERMOSTAT

The low-pressure control setup described in the previous example is for low-charge protection only, to keep the system from going into a vacuum. The same control can be set to operate the compressor to maintain the space temperature in the cooler and to serve as a low-charge protection. Using the same temperatures as in the preceding paragraph, 34°F and 45°F, as the operating conditions, the low-pressure control can be set to cut out when the low pressure reaches 21 psig. This corresponds to a coil temperature of 20°F. When the air in the cooler reaches 34°F, the coil temperature should be 19°F, with a corresponding pressure of 20 psig. This system has a room-to-coil temperature difference of 15°F (34°F room temperature − 19°F coil temperature = 15°F temperature difference). When the compressor cuts off, the air in the cooler is going to raise the temperature on the coil to 34°F and a corresponding pressure of 32 psig. As the cooler temperature goes up, it will raise the temperature of the refrigerant. When the temperature of the air increases to 45°F, the coil temperature should be 45°F and have a corresponding pressure of 42 psig. This could be the cut-in point of the low-pressure control. The settings would be cut out at 20 psig and cut in at 42 psig. This is a differential of 22 psig and would maintain a cooler temperature of 34°F to 45°F. You could say that you are using the refrigerant in the evaporator coil as the thermostat.

One of the advantages of this type of control arrangement is that there are no interconnecting wires between the inside of the cooler and the condensing unit. If a thermostat is used to control the air temperature in the cooler, a pair of wires must be run between the condensing unit and the inside of the cooler. In some installations the condensing unit is a considerable distance from the cooler, which makes this impractical. With the temperature being controlled at the condensing unit, the owner is less likely to turn the control and cause problems.

There are as many low-pressure control settings as there are applications. Different situations call for different settings. Figure 25–21 is a chart of recommended settings by one company.

Low-pressure controls are rated by their pressure range and the current draw of the contacts. A low-pressure control that is suitable for R-12 may not be suitable for R-502 because of pressure range. For the same application in the previous paragraph using R-502, the cutout would be 53 psig and the cut-in would be 88 psig. A control for the correct pressure range must be chosen. Some of these controls are single-pole–double-throw. They can make or break on a rise. This control can serve as one component for two different functions.

The contact rating for a pressure control has to do with the size of electrical load the control can carry. If the pressure control is expected to start a small compressor, the inrush current should also be considered. Normally a pressure control used for refrigeration is rated so it can directly start up to a 3-hp single-phase compressor. If the compressor is any larger, or three phase, a contactor is normally used. The pressure control can then control the contactor's coil, Figure 25–22.

25.19 HIGH-PRESSURE CONTROL

The high-pressure control is normally not as complicated as the low-pressure control (switch). It is used to keep the compressor from operating with a high head pressure. This control is necessary on water-cooled equipment because an interruption of water is more likely than an interruption of air. This control opens on a rise in pressure and should be set above the typical high pressure that the machine would normally encounter. The high-pressure control may be either automatic or manual reset.

When an air-cooled condenser is placed outside, the condenser can be expected to operate at no more than 30°F warmer than the ambient air. This condition is true after the condenser has run long enough to have a coat of dirt built up on the coil. It is true that a clean condenser is important, but it is more often slightly dirty than clean. If the ambient air is 95°F, the condenser would be operating at about 170 psig if the system used R-12 (95°F + 30°F = 125°F condensing temperature, which corresponds to 169 psig). The high-pressure control should be set well above 170 psig for the R-12 system. If the control were set to cut out at 250 psig, there should be no interference with normal operation, and it still would give good protection, Figure 25–23.

Most condensers operate at 30°F temperature difference as shipped from the factory and will continue to operate at this condition when properly maintained. Many condensers for low-temperature applications may operate at

APPROXIMATE PRESSURE/CONTROL SETTINGS

Application	Refrigerant					
	12		22		502	
	Out	In	Out	In	Out	In
Ice Cube Maker—Dry Type Coil	4	17	16	37	22	45
Sweet Water Bath—Soda Fountain	21	29	43	56	52	66
Beer, Water, Milk Cooler, Wet Type	19	29	40	56	48	66
Ice Cream Trucks, Hardening Rooms	2	15	13	34	18	41
Eutectic Plates, Ice Cream Truck	1	4	11	16	16	22
Walk In, Defrost Cycle	14	34	32	64	40	75
Reach In, Defrost Cycle	19	36	40	68	48	78
Vegetable Display, Defrost Cycle	13	35	30	66	38	77
Vegetable Display Case—Open Type	16	42	35	77	44	89
Beverage Cooler, Blower Dry Type	15	34	34	64	42	75
Retail Florist—Blower Coil	28	42	55	77	65	89
Meat Display Case, Defrost Cycle	17	35	37	66	45	77
Meat Display Case—Open Type	11	27	27	53	35	63
Dairy Case—Open Type	10	35	26	66	33	77
Frozen Food—Open Type	7	5	4	17	8	24
Frozen Food—Open Type—Thermostat	2°F	10°F	—	—	—	—
Frozen Food—Closed Type	1	8	11	22	16	29

FIGURE 25–21 Table to be used as a guide for setting low-pressure controls for the different applications. *Courtesy C. C. Dickson*

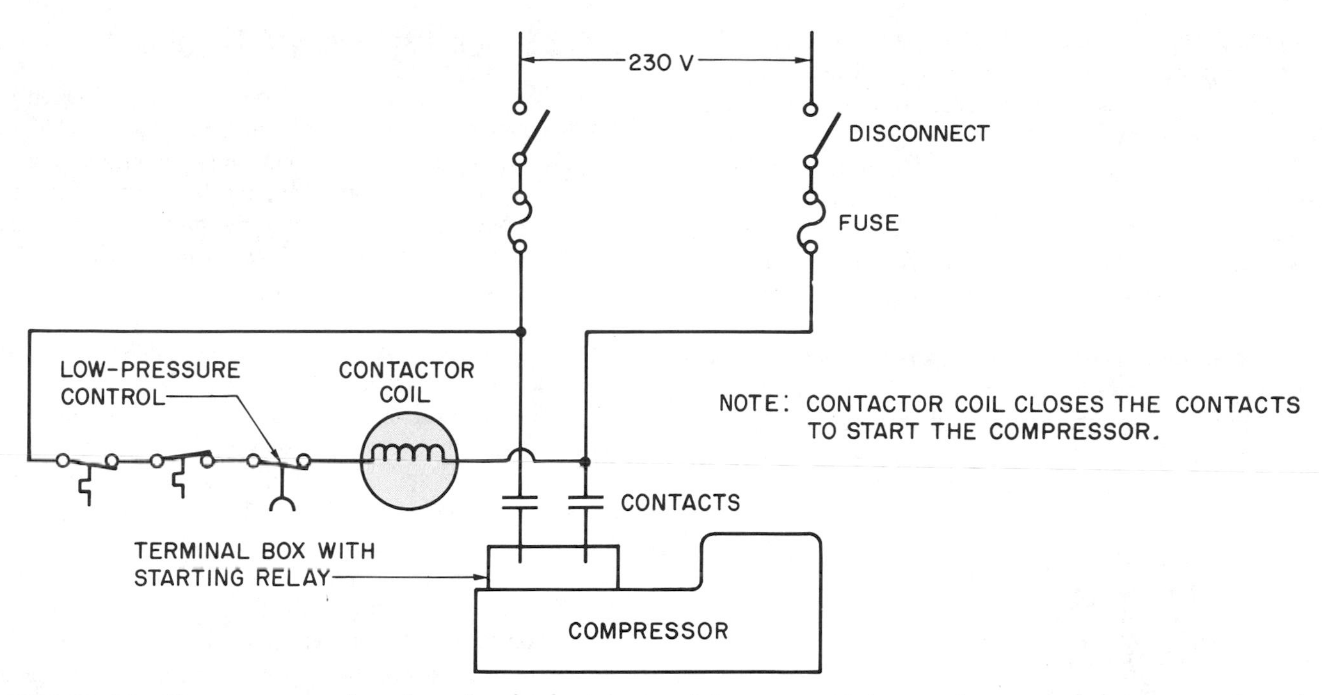

FIGURE 25–22 When a compressor that is larger than the electrical rating of the pressure control contacts is encountered, the compressor has to be started with a contactor. The wiring diagram shows how this is accomplished.

FIGURE 25–23 High-pressure switch to keep the compressor from running when high system pressures occur. *Courtesy Ranco*

temperature differences as low as 15°F to maintain low compression ratios.

The control cut-in point must be above the pressure corresponding to the ambient temperature of 95°F. If the compressor cuts off and the outdoor fan continues to run, the temperature inside the condenser will quickly reach the ambient temperature. For example, if the ambient is 95°F, the pressure will quickly fall to 108 psig for an R-12 system. If the high-pressure control were set to cut in at 125 psig, the compressor could come back on with a safe differential of 125 psig (cutout 250 − cut-in 125 = 125 psig differential).

Some manufacturers specify a manual reset high-pressure control. When this control cuts out, someone must press the reset button to start the compressor. This calls attention to the fact that a problem exists. The manual reset control provides better equipment protection, but the automatic reset control may save the food by allowing the compressor to run at short intervals. An observant owner or operator should notice the short cycle of the automatic control if the compressor is near the workspace.

25.20 LOW AMBIENT FAN CONTROL

The low ambient fan control has the same switch action as the low-pressure control but operates at a higher pressure range. This control stops and starts the condenser fan in response to head pressure. This control must be coordinated with the high-pressure control to keep them from working against each other. The high-pressure control stops the compressor when the head pressure gets too high, and the low ambient control starts the fan when the pressure gets to a predetermined point before the high-pressure control stops the compressor.

When a low ambient control is used, the high-pressure cutout should be checked to make sure that it is higher than the cut-in point of the low ambient control. For example, if the low ambient control is set to maintain the head pressure between 125 psig and 175 psig, a high-pressure control setting of 250 psig should not interfere with the low ambient control setting. This can easily be verified by installing a gage and stopping the condenser fan to make sure that the high-pressure control is cutting out where it is supposed to. With the gage on the high side, the fan action can be observed also to see that the fan is cutting off and on as it should. The fan will operate all of the time on high ambient days, and the fan control will not stop the fan. The low ambient pressure control can be identified by the terminology on the control's action. It is described as a "close on a rise" in pressure, Figure 25–24. This is the same terminology as used on a low-pressure control. The difference is the low ambient fan control has a higher operating pressure range.

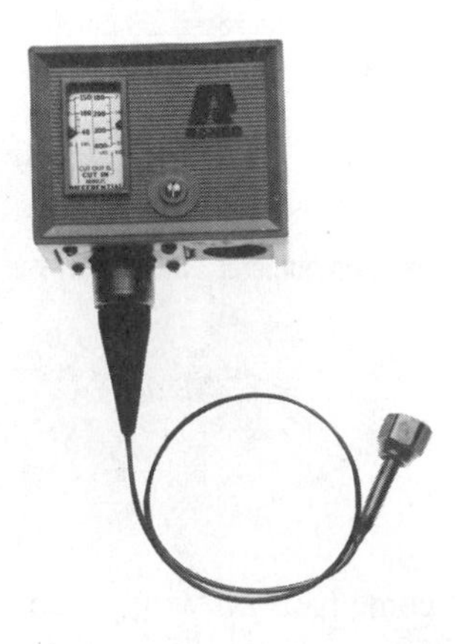

FIGURE 25–24 Low ambient control used to open the contacts on a drop in head pressure to stop the condenser fan. *Courtesy Ranco*

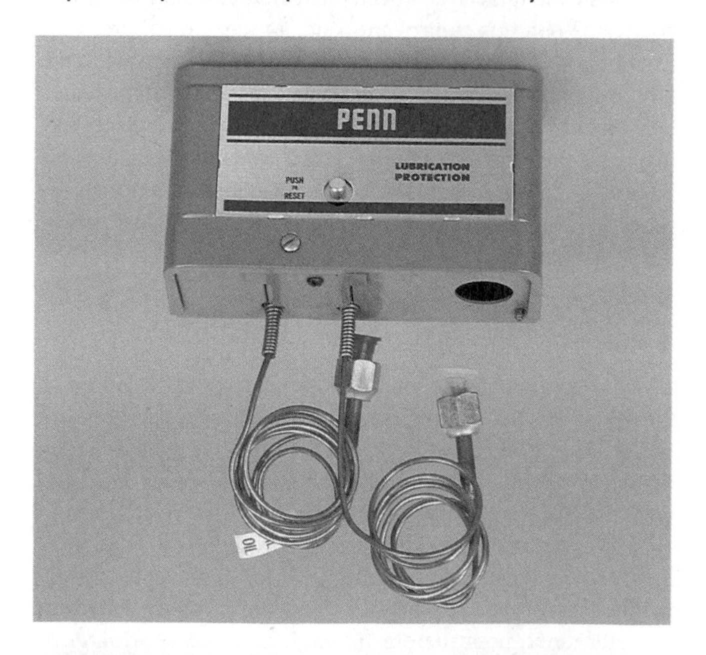

FIGURE 25–25A Oil safety controls have a bellows on each side of the control. They are opposed in their forces and measure the net oil pressure: the difference in the suction pressure (oil pump inlet) and the oil pump outlet pressure. This control has a 90-sec time delay to allow the compressor to get up to speed and establish oil pressure before it shuts down. *Photo by Bill Johnson*

25.21 OIL PRESSURE SAFETY CONTROL

The oil pressure safety control (switch) is used to ensure that the compressor has oil pressure when operating, Figure 25–25A. This control is used on larger compressors and has a different sensing arrangement than the high- and low-pressure controls. The high- and low-pressure controls are single diaphragm or single bellows controls because they are comparing atmospheric pressure to the pressures inside the system. Atmospheric pressure can be considered a constant for any particular locality because it does not vary more than a small amount.

The oil pressure safety control is a pressure differential control. This control actually measures a difference in pressure to establish that positive oil pressure is present. The pressure in the compressor crankcase (this is the oil pump suction inlet) is the same as the compressor suction pressure, Figure 25–25B. The suction pressure will vary from the off or standing reading to the actual running reading, not to mention the reading that would occur when a low charge is experienced. For example, when a system is using R-22 as the refrigerant, the pressures may be similar to the fol-

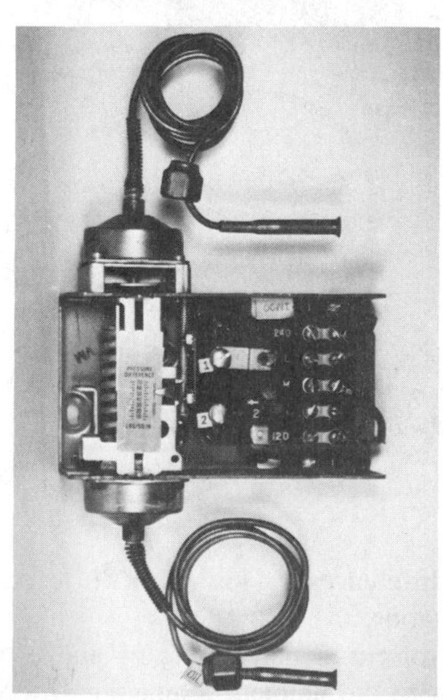

FIGURE 25–25B Two views of an oil pressure safety control. This control satisfies two requirements: how to measure net oil pressure effectively and how to get the compressor started to build oil pressure *Photos by Bill Johnson*

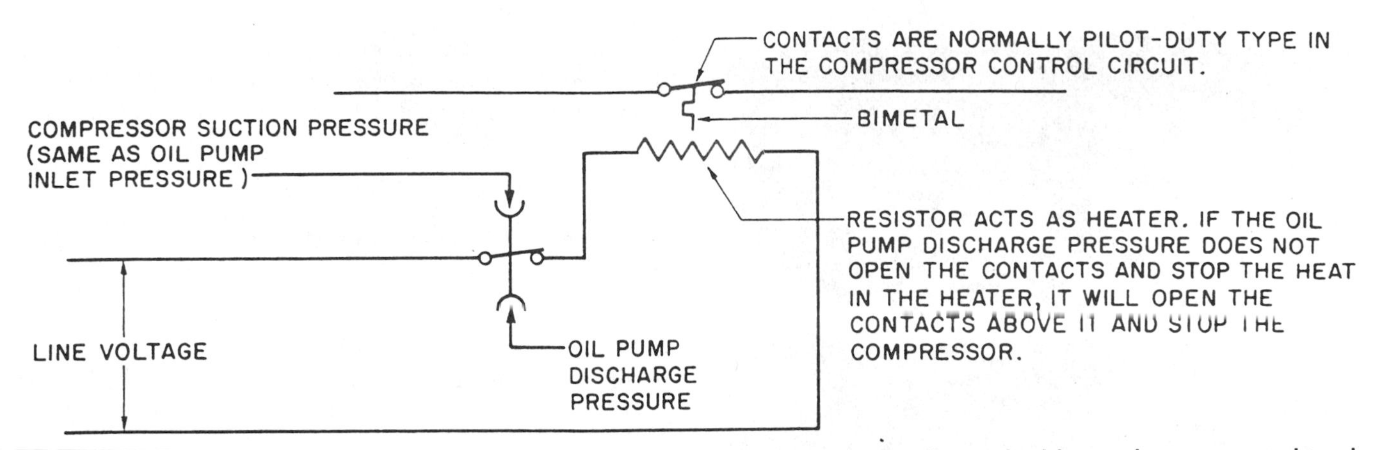

FIGURE 25–25C The oil pressure control is used for lubrication protection for the compressor. The oil pump that lubricates the compressor is driven by the compressor crankshaft. Therefore, there has to be a time delay to allow the compressor to start and build oil pressure. This time delay is normally 90 sec. The time delay is accomplished either with a heater circuit heating a bimetal or an electronic circuit.

lowing: 125 psig while standing, 70 psig while operating, and 20 psig during a low-charge situation.

A plain low-pressure cut-out control would not function at all of these levels, so a control had to be devised that would.

Most compressors need at least 30 psig of actual oil pressure for proper lubrication. This means that whatever the suction pressure is, the oil pressure has to be at least 30 lb greater because the oil pump inlet pressure is the same as the suction pressure. For example, if the suction pressure is 70 psig, the oil pump outlet pressure must be 100 psig for the bearings to have a net oil pressure of 30 psig. This difference in the suction pressure and the oil pump outlet pressure is called the *net oil pressure*.

In the basic low-pressure control the pressure is under the diaphragm or bellows and the atmospheric pressure on the other side. The oil pressure control uses a double bellows—one bellows opposing the other—to detect the net or actual oil pressure. The pump inlet pressure is under one bellows, and the pump outlet pressure is under the other bellows. These bellows are opposite each other either physically or by linkage. The bellows with the most pressure is the oil pump outlet, and it overrides the bellows with the least amount of pressure. This override reads out in net pressure and is attached to a linkage that can stop the compressor when the net pressure drops for a predetermined time.

Because the control needs a differential in pressure to allow power to pass to the compressor, it must have some means to allow the compressor to get started. There is no pressure differential until the compressor starts to turn because the oil pump is attached to the compressor crankshaft. A time delay is built into the control to allow the compressor to get started and to prevent unneeded cutouts when oil pressure may vary for only a moment. This time delay is normally about 90 sec. It is accomplished with a heater circuit and a bimetal device or electronically, Figure 25–25C. ***The manufacturer's instructions should be consulted when working with any compressor that has an oil safety control.***

25.22 DEFROST CYCLE

The defrost cycle in refrigeration is divided into medium-temperature and low-temperature ranges, and the components that serve the defrost cycle are different.

25.23 MEDIUM-TEMPERATURE REFRIGERATION

The medium-temperature refrigeration coil normally operates below freezing and rises above freezing during the off cycle. A typical temperature range would be from 34°F to 45°F space temperature inside the cooler. The coil temperature is normally 10°F to 15°F cooler than the space temperature in the refrigerated box. This means that the coil temperatures would normally operate as low as 19°F (34°F – 15°F = 19°F). The air temperature inside the box will always rise above the freezing point during the off cycle and can be used for the defrost. This is called *off-cycle* defrost and can either be random or planned.

25.24 RANDOM OR OFF-CYCLE DEFROST

Random defrost will occur when the refrigeration system has enough reserve capacity to cool more than the load requirement. When the system has reserve capacity, it will be shut down from time to time by the thermostat and the air in the cooler can defrost the ice from the coil. When the compressor is off, the evaporator fans will continue to run, and the air in the cooler will defrost the ice from the coil.

FIGURE 25–26 **Timer to program off-cycle defrost. This is a 24-h timer that can have several defrost times programmed for the convenience of the installation.** ***Photo by Bill Johnson***

When the refrigeration system does not have enough capacity or the refrigerated box has a constant load, there may not be enough off time to accomplish defrost. This is when it has to be planned.

25.25 PLANNED DEFROST

Planned defrost is accomplished by forcing the compressor to shut down for short periods of time so that the air in the cooler can defrost the ice from the coil. This is accomplished with a timer that can be programmed. Normally the timer stops the compressor during times that the refrigerated box is under the least amount of load. For example, a restaurant unit may defrost at 2 AM and 2 PM to avoid the rush hours, Figure 25–26.

25.26 LOW-TEMPERATURE EVAPORATOR DESIGN

Low-temperature evaporators all operate below freezing and must have planned defrost. Because the air inside the refrigerated box is well below freezing, heat must be added to the evaporator for defrost. This defrost is normally accomplished with *internal heat* or *external heat.*

25.27 DEFROST USING INTERNAL HEAT (HOT GAS DEFROST)

The internal heat method of defrost normally uses the hot gas from the compressor. This hot gas can be introduced into the evaporator from the compressor discharge line to the inlet of the evaporator and allowed to flow until the evaporator is defrosted, Figure 25–27. A portion of the energy used for hot gas defrost is available in the system. This makes it attractive from an energy-saving standpoint.

Injecting hot gas into the evaporator is rather simple if the evaporator is a single-circuit type because a T in the expansion valve outlet is all that is necessary. When a multicircuit evaporator is used, the hot gas must be injected between the expansion valve and the refrigerant distributor. This gives an equal distribution of hot gas to defrost all of the coils equally.

The defrost cycle is normally started with a timer in space temperature applications, where forced air is used to cool the product. When the defrost cycle is started, some

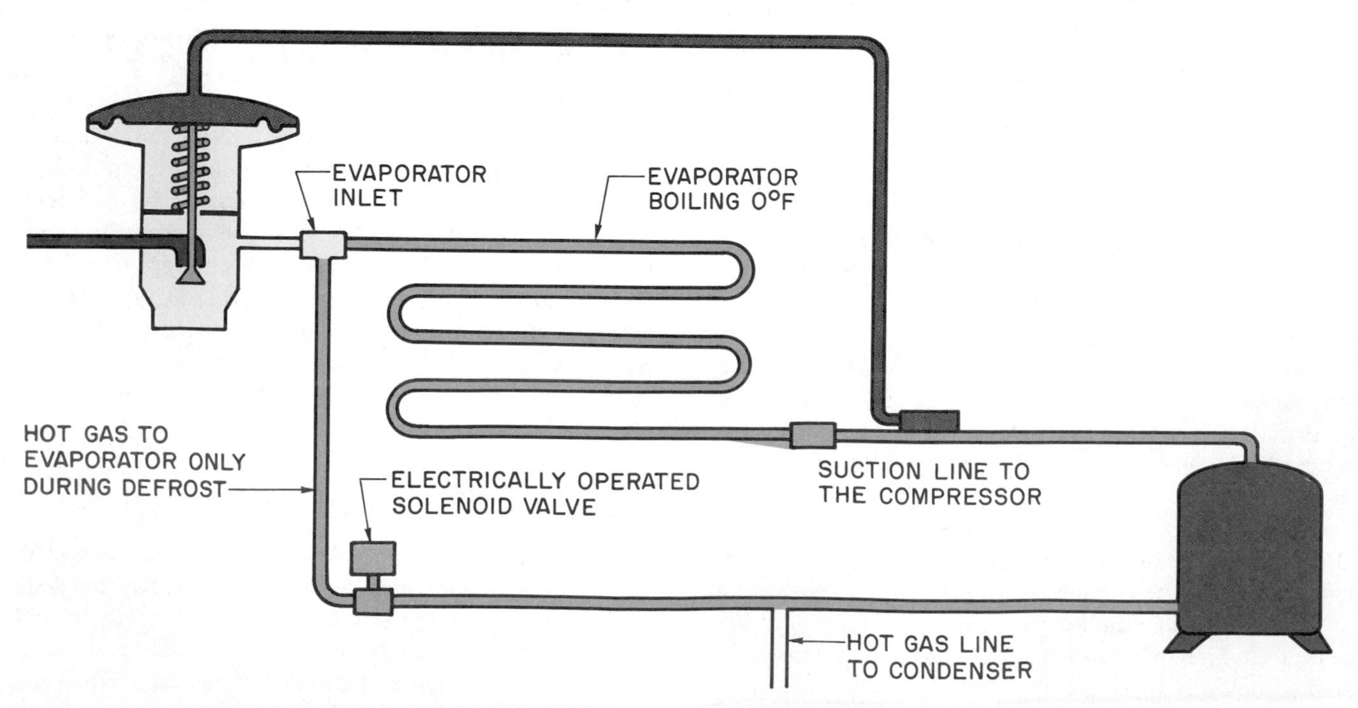

FIGURE 25–27 Hot gas defrost.

method must be used to terminate it. Defrost can be terminated by time or temperature. The amount of time it takes to defrost the coil must be known before time alone can be used efficiently to terminate defrost. Because this time can vary from one situation to another, the timer could be set for too long a time, and the unit would run in defrost when it is not desirable, causing energy loss.

Defrost can be started with time and terminated with temperature. When this is done, a temperature-sensing device is used to determine that the coil is above freezing. The hot gas entering the evaporator is stopped, and the system goes back to normal operation, Figure 25–28.

During the hot gas defrost cycle, several things must happen at one time. The timer is used to coordinate the following functions:

1. The hot gas solenoid must open.
2. The evaporator fans must stop, or cold air will keep defrost from occurring.
3. The compressor must continue to run.
4. A maximum defrost time must be determined and programmed into the timer in the event the defrost termination switch failed to terminate defrost.
5. Drain pan heaters may be energized.

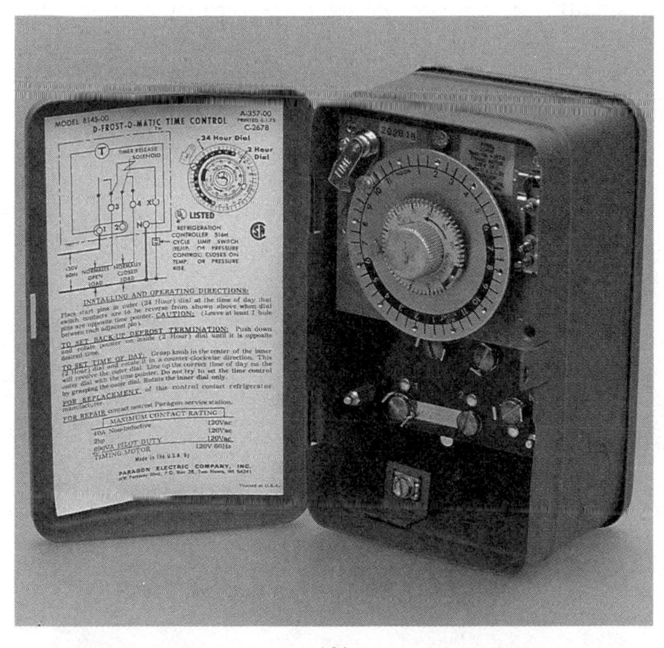

(A)

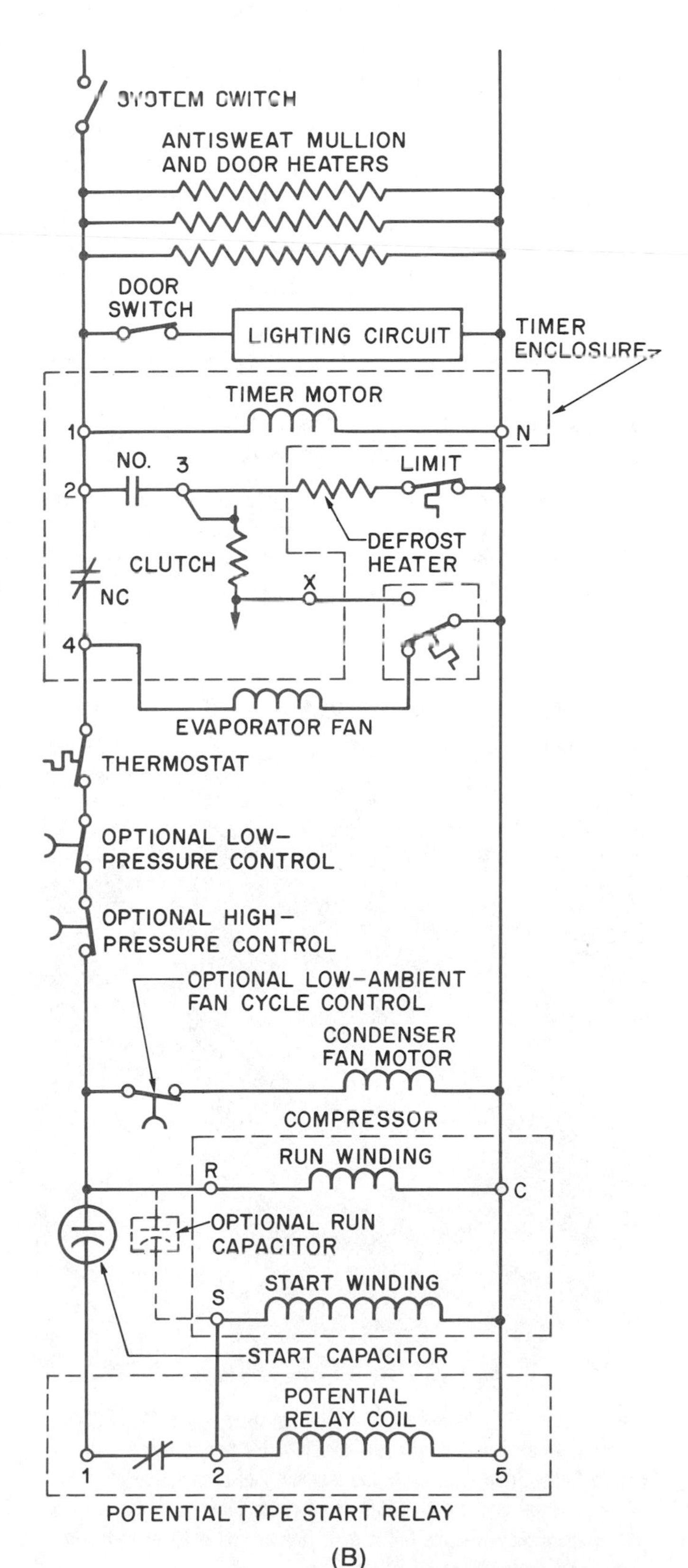

(B)

FIGURE 25–28 (A) Timer with a mechanism that can stop defrost with an electrical signal from a temperature-sensing element. When defrost is over, the coil temperature will rise above freezing. There is no reason for the defrost to continue after the ice has melted. (B) Wiring diagram of a circuit to control the defrost cycle. The events happen like this: When there is a defrost call (the timer's contacts close), the solenoid valve opens and the compressor continues to run and pump the hot gas into the evaporator. The coil gets warm enough to cause the thermostat to change from the cold contacts to the hot contacts; the defrost will be terminated by the × terminal on the timer. When the coil cools off enough for the thermostat to change back to the cold contacts, the fan will restart. This is another method of keeping the compressor from running overloaded on a hot pull down. *Photo by Bill Johnson*

25.28 EXTERNAL HEAT TYPE OF DEFROST

The external heat method of defrost is often accomplished with electric heating elements that are factory mounted next to the evaporator coil. This type of defrost is also a planned defrost that is controlled by a timer. The external heat method is not as efficient as the internal heat method because energy has to be purchased for defrost. Electric defrost may be more efficient if long runs are required for the hot gas lines. Refrigerant can condense in these long runs and cause slow defrost and liquid refrigerant can reach the compressor causing slugging and compressor damage. When electric defrost is used, it is more critical that the defrost be terminated at the earliest possible time. The timer controls the following events for electric defrost:

1. The evaporator fan stops.
2. The compressor stops. (There may be a pump-down cycle to pump the refrigerant out of the evaporator to the condenser and receiver.)
3. The electric heaters are energized.
4. Drain pan heaters may be energized.

NOTE: A temperature sensor may be used to terminate defrost when the coil is above freezing. A maximum defrost time should be programmed into the timer in the event the defrost termination switch failed to terminate defrost, Figure 25–29.

25.29 REFRIGERATION ACCESSORIES

Accessories in the refrigeration cycle are devices that improve the system performance and service functions. This text will start at the condenser, where the liquid refrigerant leaves the coil, and add various accessories as they are encountered in systems. Each system does not have all the accessories.

25.30 RECEIVERS

The *receiver* is located in the liquid line and is used to store the liquid refrigerant after it leaves the condenser. The receiver should be lower than the condenser so the refrigerant has an incentive to flow into it naturally. This is not always possible. The receiver is a tanklike device that can either be upright or horizontal, depending on the installation, Figure 25–30. Receivers can be quite large on systems that need to store large amounts of refrigerant. Figure 25–31 is a photo of a large receiver that will hold several hundred pounds of refrigerant.

The receiver inlet and outlet connection can be at almost any location on the outside of the tank body. On the inside of the receiver, however, the refrigerant must enter the receiver at the top in some manner. The refrigerant that is leaving the receiver must be taken from the bottom. This is accomplished with a dip tube if the line is at the top.

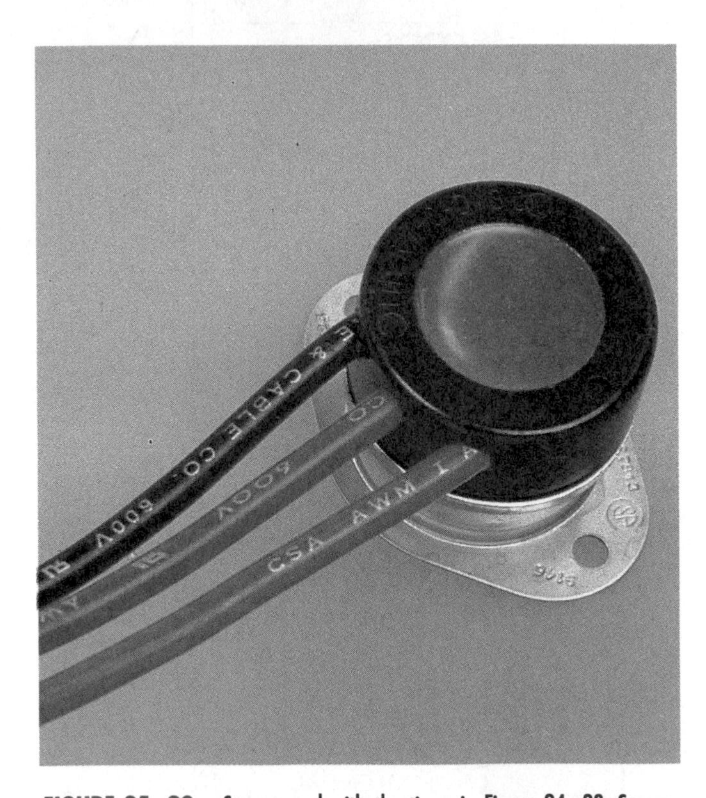

FIGURE 25–29 Sensor used with the timer in Figure 24–28. Sensor has three wires; it is a single-pole–double-throw device. It has a hot contact (made from common to the terminal that is energized on a rise in temperature) and a cold contact (made from common to the terminal that is energized when the coil is cold). This control is either in the hot or the cold mode. *Photo by Bill Johnson*

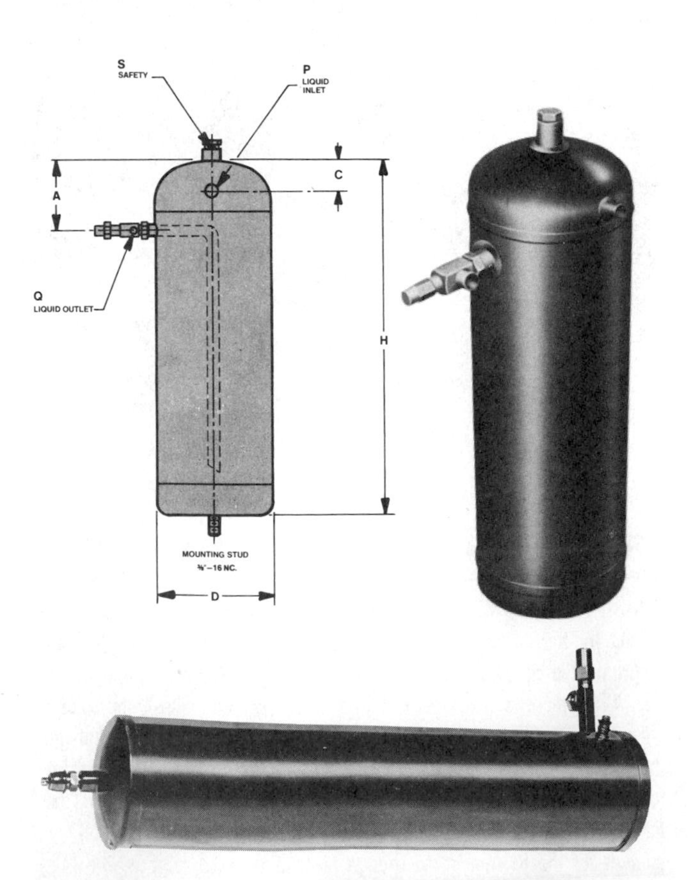

FIGURE 25–30 Vertical and horizontal receivers. *Courtesy Refrigeration Research*

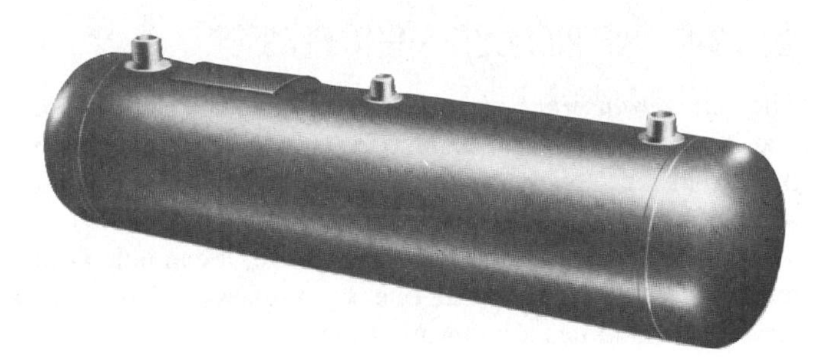

FIGURE 25–31 Large receivers store the charge for a condenser flooding system. It may hold more than 100 lb of refrigerant. *Courtesy Refrigeration Research*

25.31 THE KING VALVE ON THE RECEIVER

The *king* valve is located in the liquid line between the receiver and the expansion valve. It is often fastened to the receiver tank at the outlet, Figure 25–32. The king valve is important in service work because when it is front seated, no refrigerant can leave the receiver. If the compressor is operated with this valve closed, the refrigerant will all be pumped into the condenser and receiver. Most of it will flow into the receiver. The other valves in the system can then be closed and the low-pressure side of the system can be opened for service. When service is complete and the low side of the system is ready for operation, the king valve can be opened, and the system can be put back into operation.

The king valve has a pressure service port that a gage manifold can be fastened to. When the valve stem is turned away from the back seat, this port will give a pressure reading in the liquid line between the expansion valve and the compressor high-side gage port. When the valve is front seated and the system is being pumped down, this gage port is on the low-pressure side of the system common to the liquid line (this becomes the low side because the king valve is trapping the high-pressure refrigerant in the receiver). Gage line removal may be difficult until the king valve is back seated after the repair is completed if there is any pressure in the liquid line, Figure 25–32.

Another component that can be found in many systems is a solenoid valve. This was covered earlier. It is a valve that stops and starts the liquid flow.

25.32 FILTER-DRIERS

The *refrigerant filter-drier* can be found at any point on the liquid line after the king valve. The filter-drier is a device that removes foreign matter from the refrigerant. This foreign matter can be dirt, flux from soldering, solder beads, filings, moisture, parts, and acid caused by moisture. Filter-driers can remove construction dirt (filter only), moisture, and acid.

These filter-drying operations are accomplished with a variety of materials that are packed inside. Some of the manufacturers furnish beads of chemicals and some use a porous block made from the drying agent. The most common agents found in the filter-driers are activated alumina, molecular sieve, or silica gel. The component has a fine screen at the outlet to catch any fine particles that may be moving in the system, Figure 25–33A.

The filter-drier comes in two styles: permanent and replaceable core. Both types of driers can be used in the suction line when chosen for that application. This is discussed in more detail later.

The filter-drier can be fastened to the liquid line by a flare connection in the smaller sizes up to 5/8-in. Solder connections can be used for a full range of sizes from 1/4 to 1 5/8 in. The larger solder filter-driers are all replaceable core types. The replaceable core filter-drier can be useful for future service, Figure 25–33B. The king valve may be front seated to pump the refrigerant into the condenser and receiver so the drier cores may be replaced.

The filter-drier can be installed anywhere in the liquid line and is found in many locations. The closer it is to the metering device, the better it cleans the refrigerant before it

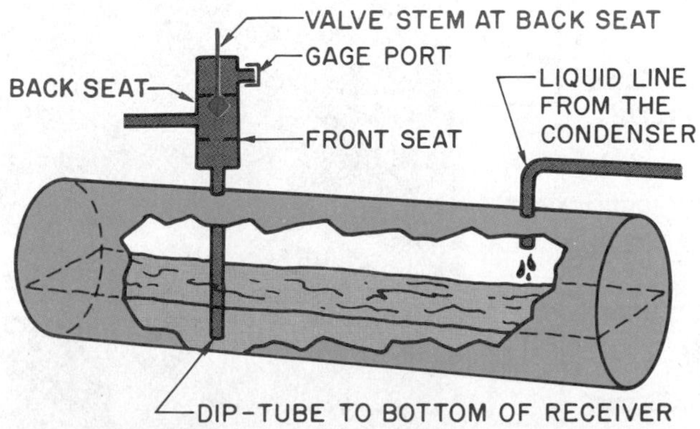

FIGURE 25–32 King valve piped in the circuit. The back seat is open to the gage port when the valve is front seated. This valve has to be back seated for the gage to be removed if there is refrigerant in the system.

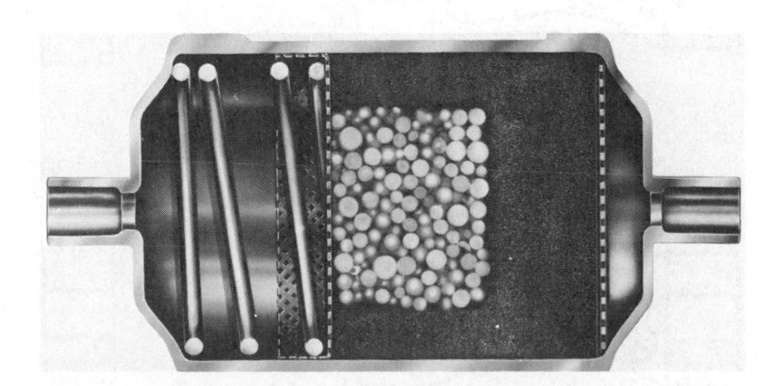

FIGURE 25–33A Filter-drier. This device is for a liquid line application and removes particles and moisture. The moisture removal is accomplished with a desiccant inside the drier shell. This desiccant material will be in bead or block form. *Courtesy Parker Hannifin Corporation*

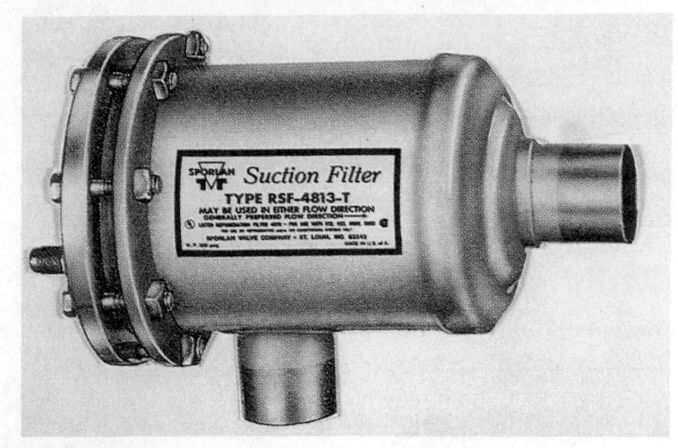

FIGURE 25–33B This filter-drier casing has removable cores. It can be left in the system if needed. *Courtesy Sporlan Valve Comany*

enters the tiny orifice in the metering device. The closer it is to the king valve, the easier it is to service. Service technicians and engineers choose their own placement, based on experience and judgment.

25.33 REFRIGERANT CHECK VALVES

A check valve is a device that will allow fluid flow in only one direction. It is used to prevent refrigerant from backing up in a line.

Several types of mechanisms are used to allow flow in one direction. Two common types of check valves are the ball check and magnetic check, Figure 25–34 (A). Both of these check valves cause some pressure drop, depending on the flow rate. The technician must consult the manufacturer's catalog for flow rate and application. The check valve can be identified by an arrow on the outside that indicates the direction of flow, Figure 25–34 (B).

25.34 REFRIGERANT SIGHT GLASSES

The *refrigerant sight glass* is normally located anywhere that it can serve a purpose. When it is installed just prior to the expansion device, the technician can be assured that a solid column of liquid is reaching the expansion device. When it is installed at the condensing unit, it can help with troubleshooting. More than one sight glass, one at each place, is sometimes a good investment.

Sight glasses come in two basic styles: plain glass and glass with a moisture indicator. The plain glass type is used to observe the refrigerant as it moves along the line. The sight glass with a moisture indicator in it can tell the technician what the moisture content is in the system. It has a small element in it that changes color when moisture is present. An example can be seen in Figure 25–35.

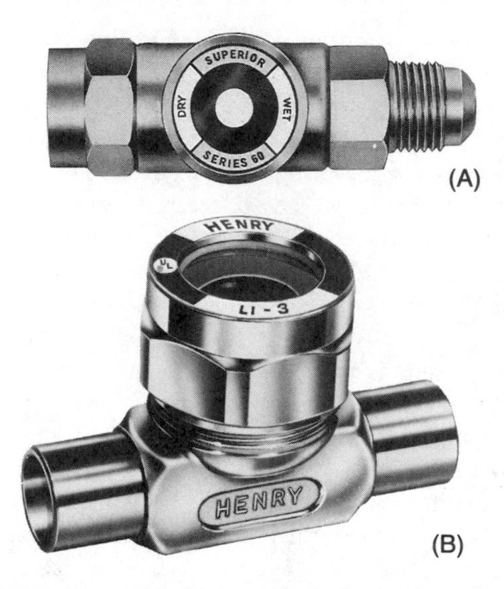

FIGURE 25–35 (A) Sight glass with an element that indicates the presence of moisture in the system. (B) Sight glass used only to view the liquid refrigerant to be sure there are no vapor bubbles in the liquid line. *Courtesy (A) Superior Valve Company, (B) Henry Valve Company*

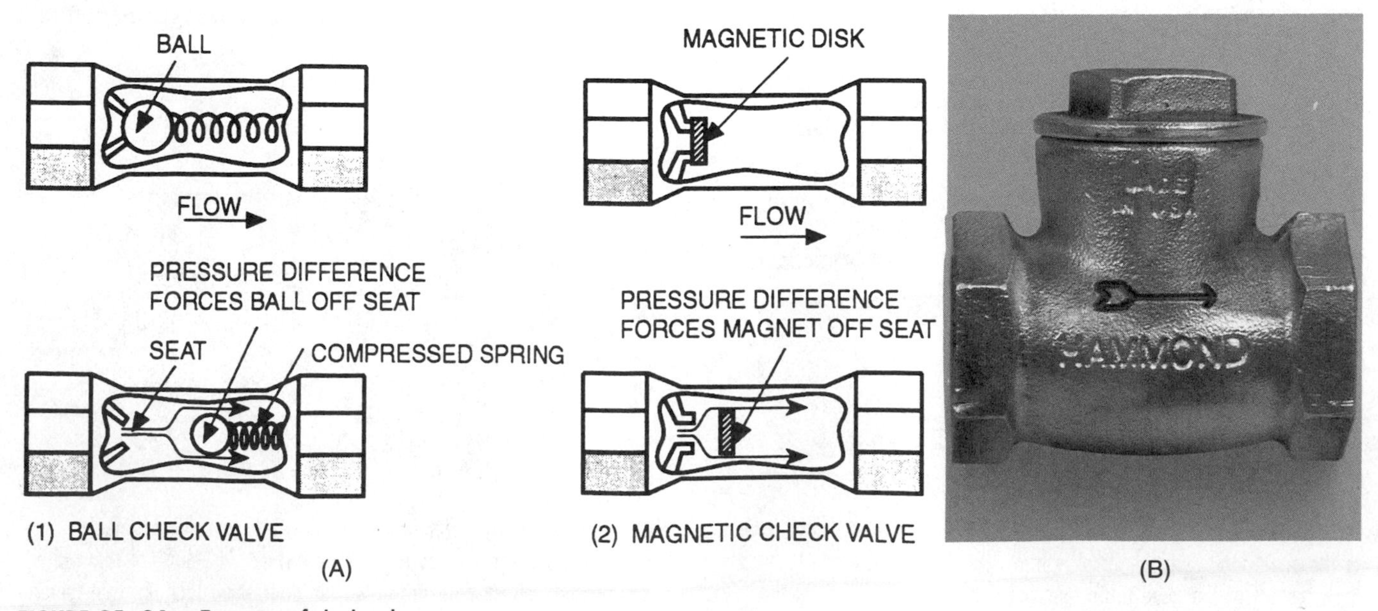

FIGURE 25–34 Two types of check valves.

25.35 LIQUID REFRIGERANT DISTRIBUTORS

The *refrigerant distributor* is fastened to the outlet of the expansion valve. It distributes the refrigerant to each individual evaporator circuit. This is a precision-machined device that ensures that the refrigerant is divided equally to each circuit, Figure 25–36. This is not a simple task because the refrigerant is not all liquid or all vapor. It is a mixture of liquid and vapor. A mixture has a tendency to stratify and feed more liquid to the bottom and the vapor to the top. Sizing a distributor to a system can be a complex decision because it must have exactly the correct pressure drop to perform correctly. The expansion valve and the distributor must be considered together because of the pressure reduction through the distributor. This sizing is determined by the manufacturer for a new installation. ⊕**In the future, as refrigerants are changed from the chlorofluorocarbon refrigerants to the new refrigerants, the refrigerant distributor will need to be considered. A close relationship with the manufacturers will be necessary or mismatched distributors to expansion valves may occur.**⊕

Some distributors have a side inlet that hot gas can be injected into for hot gas defrost, Figure 25–37.

FIGURE 25–36 Multicircuit refrigerant distributor. Distributor to feed equal amounts of liquid refrigerant to the different refrigerant circuits. The combination of liquid and vapor refrigerant has a tendency to stratify with the liquid moving to the bottom. This is a precision-machined device to separate the mixture evenly. *Photo by Bill Johnson*

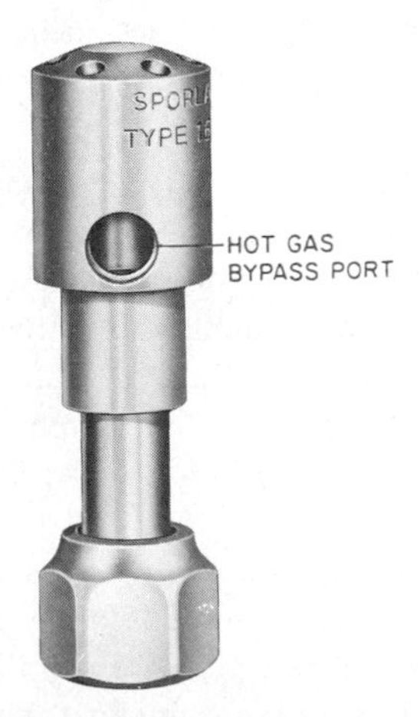

FIGURE 25–37 Same type of distributor as in Figure 25–36 except that it has a side inlet for allowing hot gas to enter the evaporator evenly during defrost. *Courtesy Sporlan Valve Company*

25.36 HEAT EXCHANGERS

The *heat exchanger* is often placed in or at the suction line leaving the evaporator. This heat exchange is between the suction and the liquid line, Figure 25–38A.

1. It improves the capacity of the evaporator by subcooling the liquid refrigerant entering it, which can allow a smaller evaporator in the refrigerated box. The cooler the refrigerant entering the expansion device, the more net refrigerant effect in the evaporator. This is not necessarily a system efficiency improvement because the heat is given up to the suction gas, still in the system.
2. It prevents liquid refrigerant from moving out of the evaporator to the suction line and into the compressor. Most of these heat exchangers are simple and straightforward.

The heat exchanger has no electrical circuit or wires. It can be recognized by the suction line and the liquid line piped to the same device. In some small refrigeration devices the capillary tube is soldered to the suction line. This accomplishes the same thing that a larger heat exchanger does, Figure 25–38B.

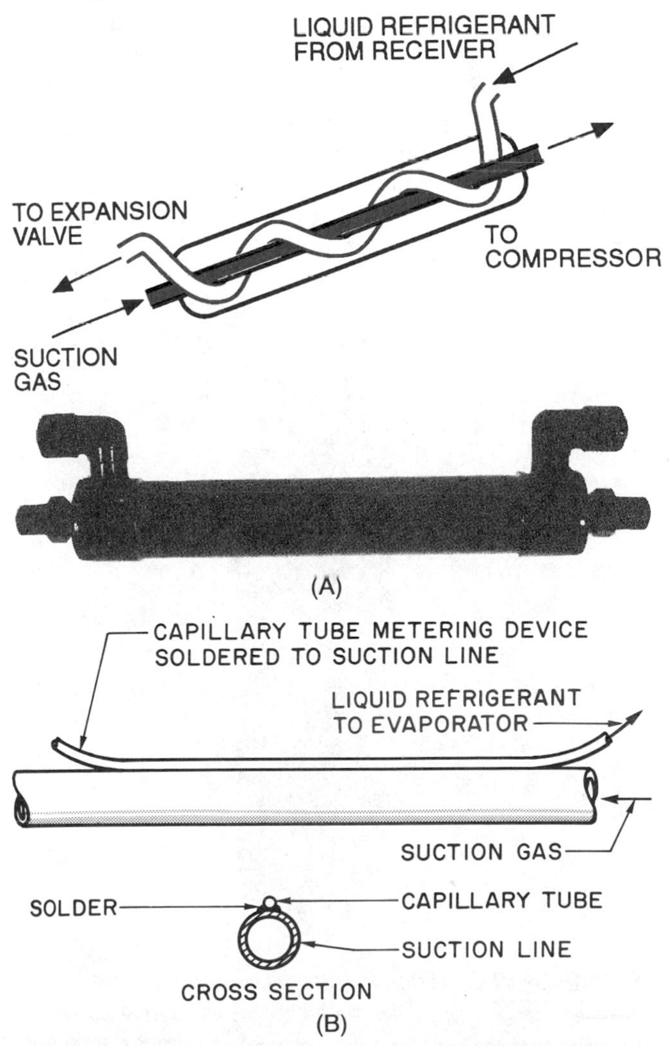

FIGURE 25–38 Heat exchangers. (A) Plain liquid-to-suction type. (B) Capillary tube fastened to the suction line that accomplishes the same thing. *Photo by Bill Johnson*

The next two components that could be installed in the suction line are the EPR valve and the CPR valve. These were covered earlier in this unit as control components.

25.37 SUCTION LINE ACCUMULATORS

The *suction line accumulator* can be located in the suction line to prevent liquid refrigerant from passing into the compressor. Under certain conditions, usually during defrost, liquid refrigerant may leave the evaporator. The suction line is insulated and will not evaporate much refrigerant. The suction line accumulator collects this liquid and gives it a place to boil off (evaporate) to a gas before continuing on to the compressor.

The suction line accumulator will also collect oil from the suction line. After it is collected, the oil is moved on to the compressor by velocity or by a small hole in the piping close to the bottom of the accumulator, Figure 25–39A. The accumulator is usually located close to the compressor. It should not be insulated so that any liquid that is in it may have a chance to vaporize.

The suction line accumulator may have a small coil in the bottom where liquid from the liquid line may be routed to boil off the refrigerant in the bottom of the accumulator, Figure 25–39B. Otherwise, part of the refrigerant charge for the system will stay in the accumulator until the heat from the vapor passing through boils it to a vapor. Actually, not much heat will exchange unless liquid is in the accumulator so it does not have much effect on the system while vapor only is in the accumulator.

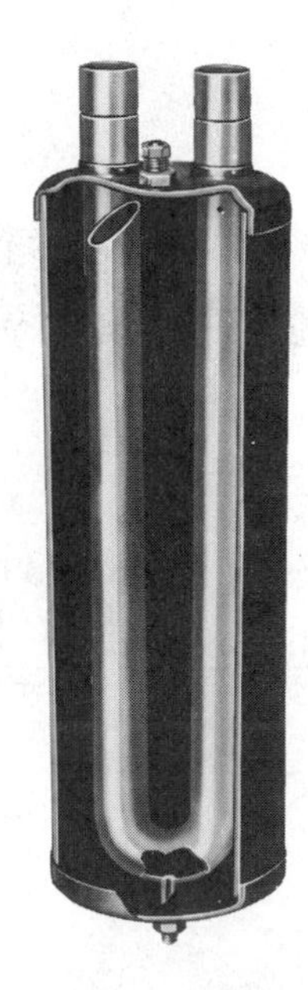

FIGURE 25–39A Suction line accumulator. Liquid refrigerant returning down the suction line can be trapped and allowed to boil away before entering the compressor. The coil bleed hole allows a small amount of any liquid in the accumulator to be returned to the compressor. If the liquid is refrigerant, there will not be enough of it getting back through the bleed hole to cause damage. *Courtesy AC and R Components, Inc.*

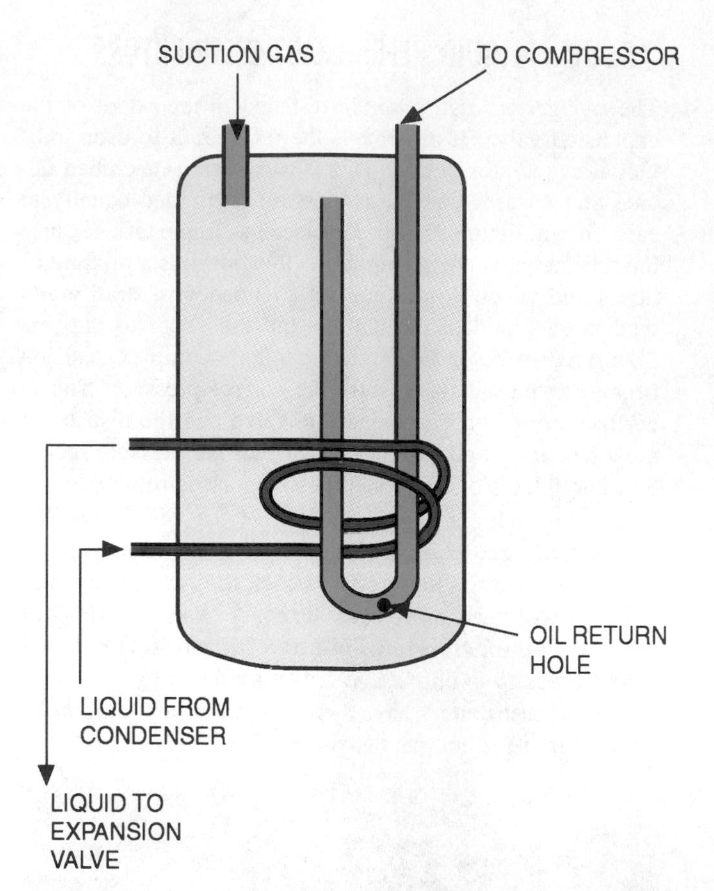

FIGURE 25–39B Suction line accumulator with heat exchanger.

Suction line accumulators are typically made of steel. They also sweat, which causes rust. Older accumulators should be observed from time to time for refrigerant leaks. Cleaning and painting the solder connection will help to prevent rust.

25.38 SUCTION LINE FILTER-DRIERS

The *suction line filter-drier* is similar to the drier in the liquid line. It is rated for suction line use. A filter-drier is placed in the suction line to protect the compressor and it is good insurance in any installation. It is essential to install a filter-drier after any failure that contaminates a system. A motor burnout in a compressor shell usually moves acid and contamination into the whole system. When a new compressor is installed, a suction line filter-drier can be installed to clean the refrigerant and oil before it reaches the compressor, Figure 25–40A. *Driers that have the capability to remove high levels of acid are typically used after motor burnout problems.*

Most suction line filter driers have some means to check the pressure drop across the cores. This is important because only a small pressure drop can cause the suction pressure to drop to the point that the compression ratio is increased. Compressor efficiencies will be reduced. Figure 25–40B shows the pressure fittings on a typical suction line drier.

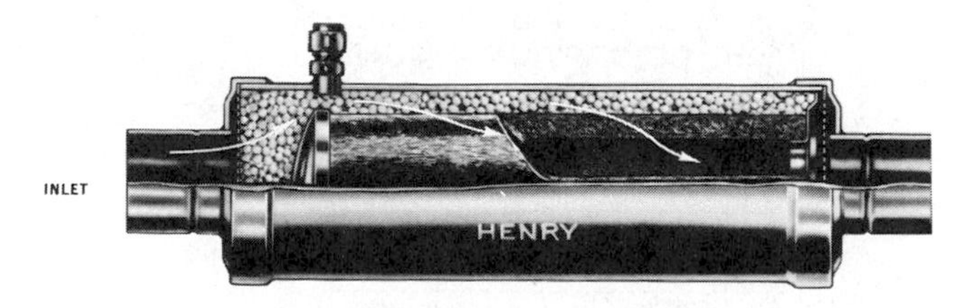

FIGURE 25–40A Suction line filter-drier. This device cleans any vapor that is moving toward the compressor. *Courtesy Henry Valve Company*

FIGURE 25–40B Suction line filter-drier with access ports to check pressure drop. *Photo by Bill Johnson*

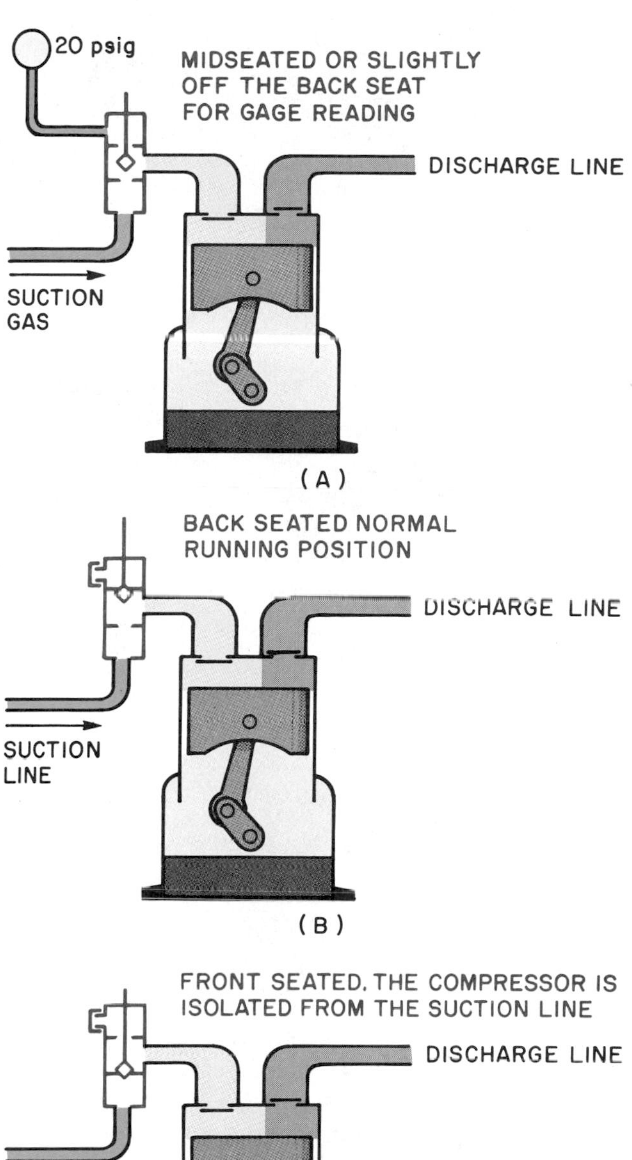

FIGURE 25–41 Three positions for the suction service valve.

25.39 SUCTION SERVICE VALVES

The *suction service valve* is normally attached to the compressor. Equipment used for refrigeration installations usually has service valves. This is not always the case on air conditioning equipment. The suction service valve can never be totally closed because of the valve's seat design. The service valve is either back seated, front seated, or midseated. An example of these functions is shown in Figure 25–41.

The suction service valve is used:

1. As a gage port
2. To throttle the gas flow to the compressor
3. To valve off the compressor from the evaporator for service

The service valve consists of the valve cap, valve stem, packing gland, inlet, outlet, and valve body, Figure 25–42.

The *valve cap* is used as a backup to the packing gland to prevent refrigerant from leaking around the stem. It is normally made of steel and should be kept dry on the inside. An oil coating on the inside will help prevent rust. *Valve covers should always be in place except during service operations.*

The *valve stem* has a square head to accomodate the valve wrench. This stem is normally steel and should be kept rust free. If rust does build up on the stem, a light sanding and a coat of oil will help. The valve turns in and out through the valve packing gland. Rust on the stem will destroy the packing in the gland.

The *packing gland* can either be a permanent type or an adjustable type. If the gland is not adjustable and a leak is started, it can only be stopped with the valve cap. If the gland is adjustable, the packing can be replaced. Normally it

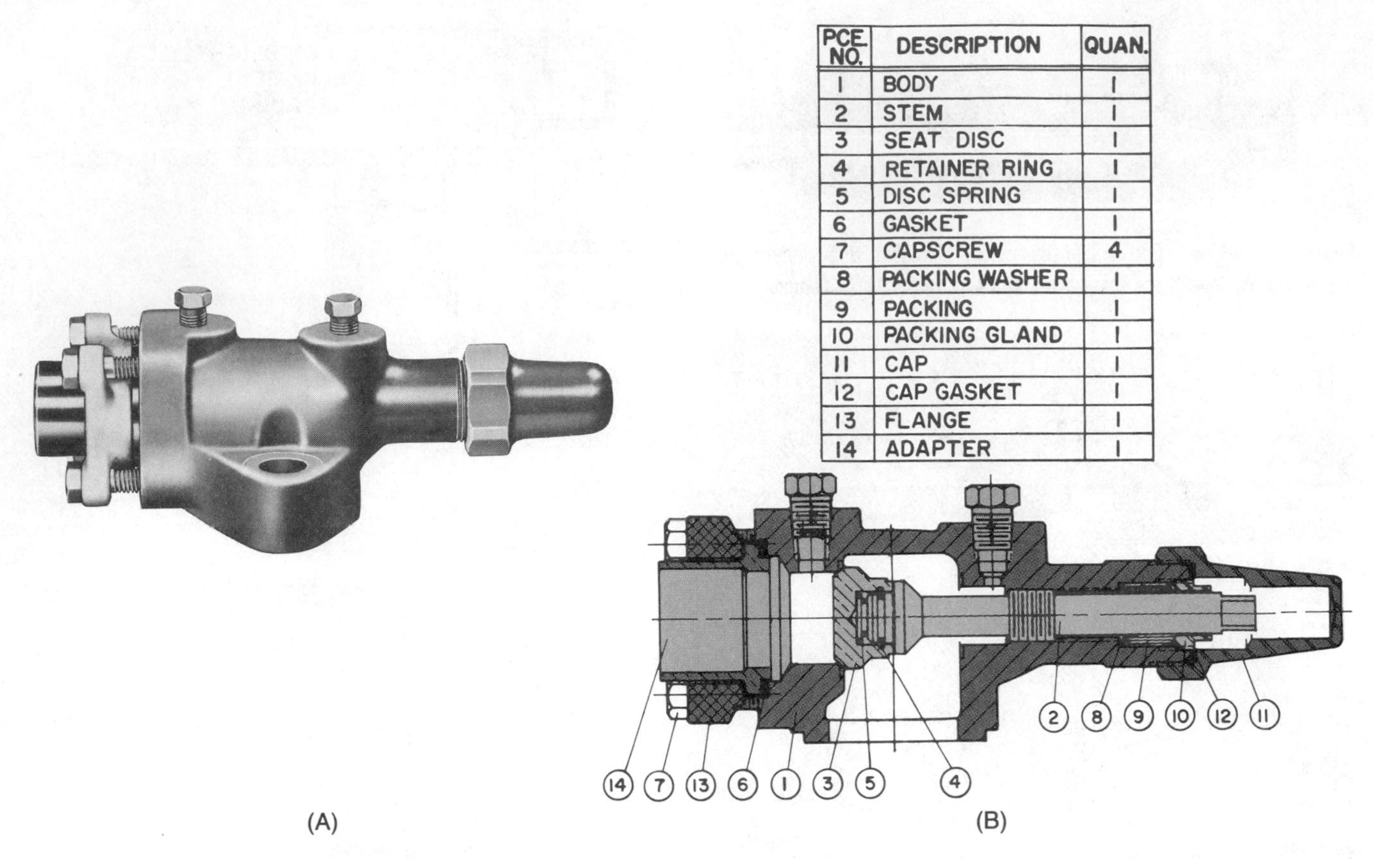

PCE. NO.	DESCRIPTION	QUAN.
1	BODY	1
2	STEM	1
3	SEAT DISC	1
4	RETAINER RING	1
5	DISC SPRING	1
6	GASKET	1
7	CAPSCREW	4
8	PACKING WASHER	1
9	PACKING	1
10	PACKING GLAND	1
11	CAP	1
12	CAP GASKET	1
13	FLANGE	1
14	ADAPTER	1

FIGURE 25–42 Components of a service valve. The suction and discharge valves are made the same except the suction valve is often larger. *Courtesy Henry Valve Company*

is graphite rope, Figure 25–43. Any time the valve stem is turned, the gland should be first loosened if it is adjustable. This will keep the valve stem from wearing the packing when the valve stem is turned.

Because the suction valve sweats, it is not uncommon for the valve stem to rust and cause pits in the stem. This only occurs when poor service practices have been followed. When this occurs, the valve stem should be carefully sanded smooth, then refrigerant oil applied to the stem before turning. *If the stem is not smoothed, damage to the packing will occur. The technician should always dry the valve stem and the inside of the protective cap and apply a light coat of oil before placing the cap on the valve stem.*

The suction service valve attaches to the compressor on one side. The refrigerant piping from the evaporator fastens to the other side of the valve. The piping can either be flared or soldered where fastened to the service valve.

PCE. NO.	DESCRIPTION	QUAN.
1	BODY	1
2	STEM	1
3	SEAT DISC ASS'Y	1
4	DISC SPRING	1
5	DISC PIN	4
6	RETAINER RING	1
7	PACKING WASHER	1
8	PACKING	2 *
9	PACKING GLAND	1
10	CAP	1
11	CAP GASKET	1
12	FLANGE	1
13	ADAPTER	1
14	GASKET	1
15	CAPSCREW	4
16	PIPE PLUG	2

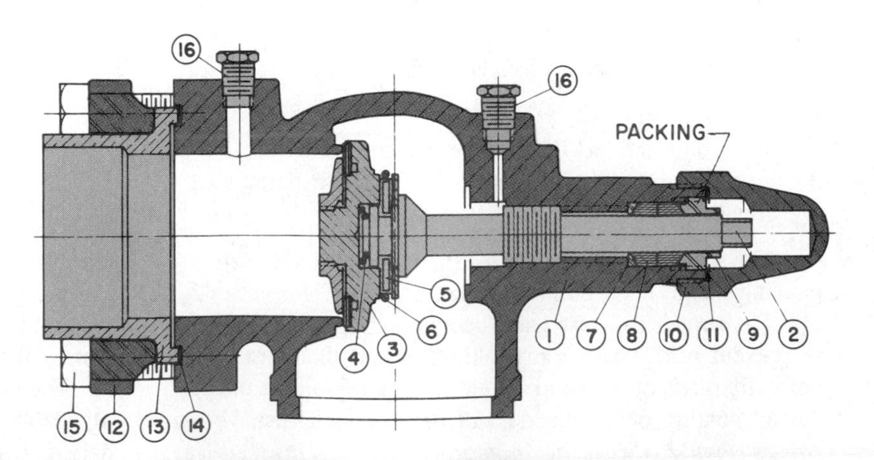

FIGURE 25–43 Service valve showing packing material. This can be replaced if it leaks. *Courtesy Henry Valve Company*

25.40 DISCHARGE SERVICE VALVES

The *discharge service valve* is the same as the suction service valve except that it is located in the discharge line. This valve can be used as a gage port and to valve off the compressor for service. ***The compressor cannot be operated with this valve front seated except for closed-loop capacity checks under experienced supervision. Extremely high pressures will result if the test equipment is not properly applied.***

Compressor service valves are used for many service functions. One of the most important is when changing out the compressor. ⊛**When both service valves are front seated, the refrigerant may be recovered from between the valves where it is trapped inside the compressor, Figure 25–44.**⊛ The compressor is now isolated and can be removed. When a new one is installed, the only part of the system that must be evacuated when the new compressor is installed is the new compressor. A compressor can be changed in this fashion with no loss of refrigerant.

25.41 REFRIGERATION SERVICE VALVES

Refrigeration service valves are normally hand-operated specialty valves used for service purposes. These valves can be used in any line that may have to be valved off for any reason. They come in two types: the diaphragm valve and the ball valve.

25.42 DIAPHRAGM VALVES

The *diaphragm valve* has the same internal flow pattern that a "globe" valve does. The fluid has to rise up and over a seat, Figure 25–45. There is a measurable pressure drop through this type of valve. The valve can be tightened by

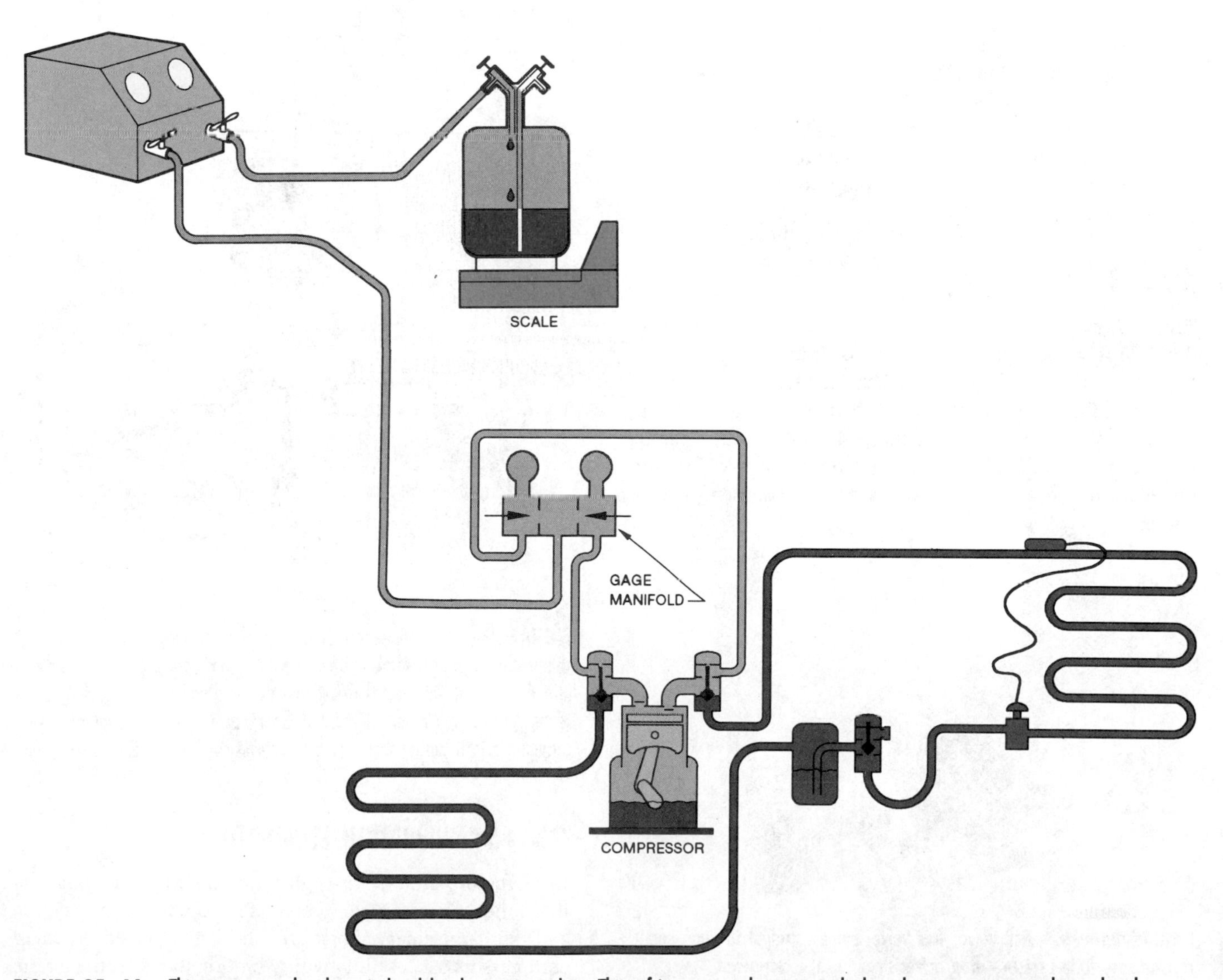

FIGURE 25–44 This compressor has been isolated by the service valves. The refrigerant can be recovered, then the compressor can be replaced.

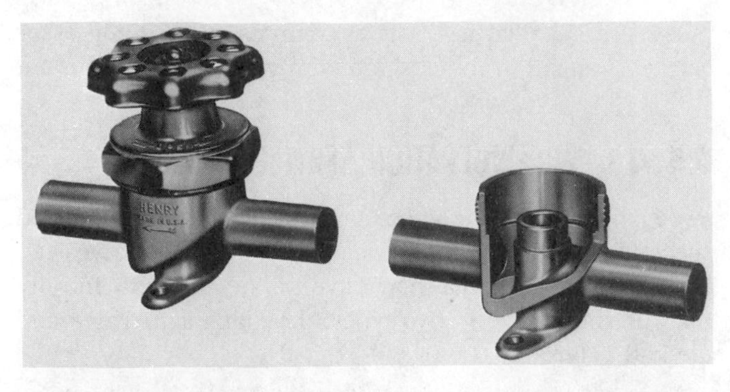

FIGURE 25–45 Diaphragm-type hand valve used when servicing a system. This valve is either open or closed, unlike the suction and discharge valve with a gage port. The valve has some resistance to fluid flow, called pressure drop. It can be used anywhere a valve is needed. The larger sizes of this valve are soldered in the line. Care should be used that the valve is not overheated when soldering to the valve. *Courtesy Henry Valve Company*

hand enough to hold back high pressures. This valve can be installed into the system with either a flare or soldered connection. When it is soldered, care should be taken that it is not overheated. Most of these valves have seats made of materials that would melt when the valve was being soldered into a line.

25.43 BALL VALVES

The *ball valve* is a straight-through valve with little pressure drop. This valve can also be soldered into the line, but the temperature has to be considered. All manufacturers furnish directions that show how to install their valve, Figure 25–46.

One of the advantages of the ball valve is that it can be opened or closed easily because it only takes a 90° turn to either open or close it. It is known as a quick open or close valve.

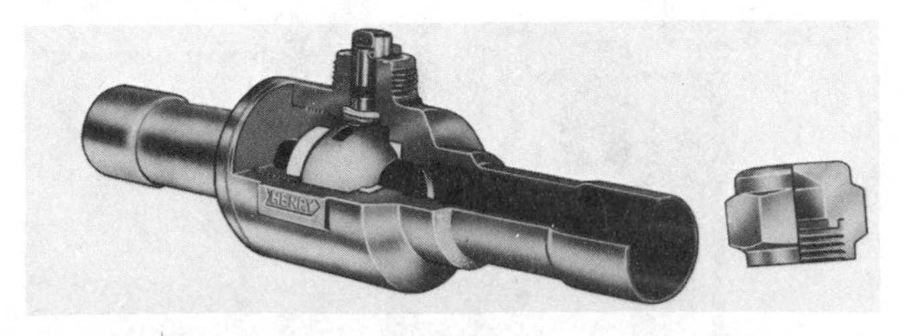

FIGURE 25–46 Ball valve. This valve is open straight through and creates very little pressure drop or resistance to the refrigerant flow. *Courtesy Henry Valve Company*

25.44 OIL SEPARATORS

The *oil separator* is installed in the discharge line to separate the oil from the refrigerant and return the oil to the compressor crankcase. All reciprocating and rotary compressors allow a small amount of oil to pass through into the discharge line. Once the oil leaves the compressor, it would have to go through the complete system to get back to the compressor crankcase by way of the system piping and coils. The oil separator has a float to allow this oil to short-cut and return to the crankcase, Figure 25–47. The oil separator should be kept warm to keep liquid refrigerant from condensing in it during the off cycle. The float does not distinguish between oil and liquid refrigerant. If liquid refrigerant were in the separator, it would return the refrigerant to the compressor crankcase. This would dilute the oil and cause marginal lubrication.

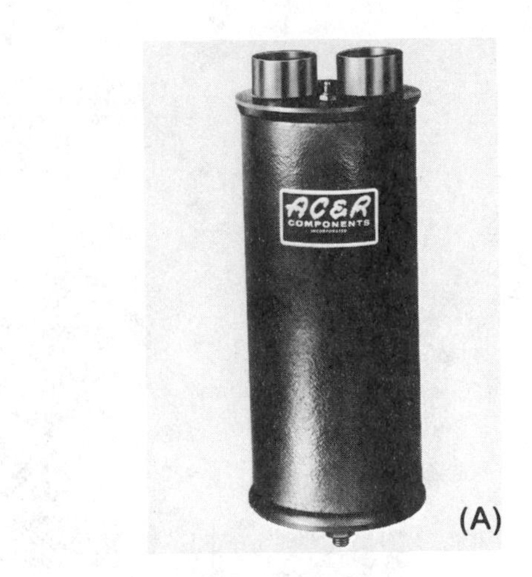

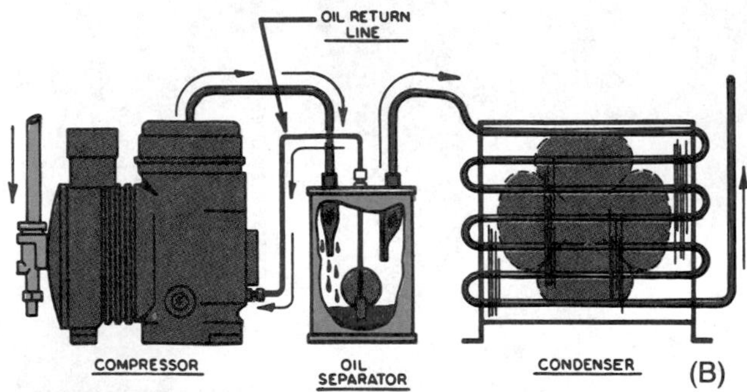

FIGURE 25–47 Oil separator used in the discharge line of a compressor to return some of the oil to the compressor before it gets out into the system. This is a float-action valve and will return liquid refrigerant as well as oil. It must be kept warm to keep refrigerant from condensing in it during the off cycle. *Courtesy AC and R Components, Inc.*

25.45 VIBRATION ELIMINATORS

Compressors produce enough vibration while running that it is often necessary to protect the tubing at the suction and discharge lines. Vibration can be eliminated on small compressors successfully with vibration loops. Large tubing can not be routed in loops, so special vibration eliminators are often installed. These are constructed with a bellows-type lining and a flexible outer protective support, Figure 25–48A.

These devices must be installed correctly or other problems may occur, such as extra vibration. Typically, it is recommended that they be installed close to the compressor and routed in the same direction as the compressor crankshaft, Figure 25–48B. If mounted cross ways of the crankshaft, excess vibration may occur. Follow the manufacturer's directions.

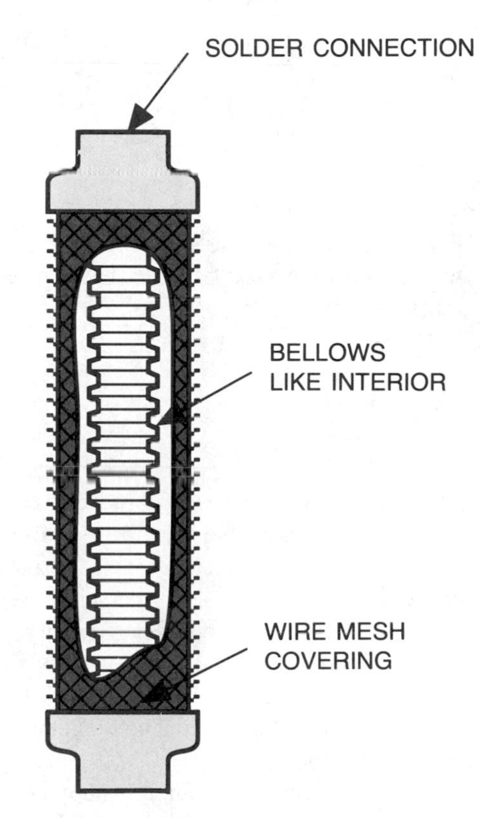

FIGURE 25–48A Construction of vibration eliminator.

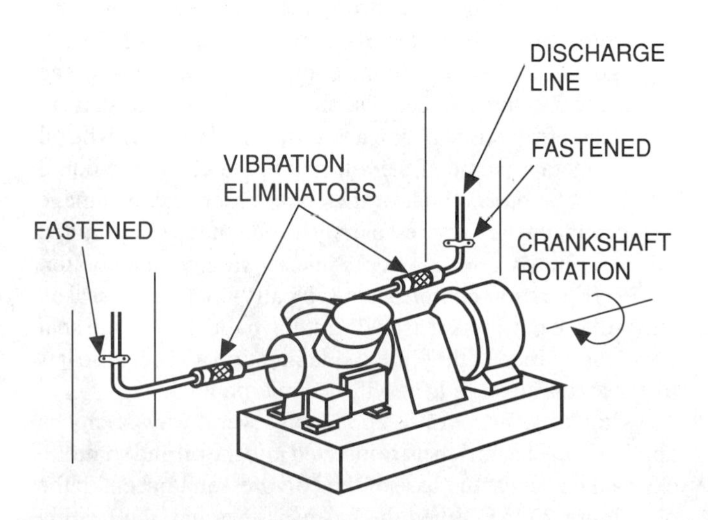

FIGURE 25–48B Vibration eliminators installed on compressor.

25.46 PRESSURE ACCESS PORTS

Pressure access ports are a method of taking pressure readings at places that do not have service ports. Several types can be used effectively. Some can be attached to a line while the unit is operating. This can be helpful when a pressure reading is needed in a hurry and the system needs to keep on running. There are two types that can be fastened while running. One type is bolted on the line and has a gasket. When this valve is bolted in place, a pointed plunger is forced through the pipe. A very small hole is pierced in the line, just enough to take pressure readings and transfer small amounts of refrigerant if needed, Figure 25–49.

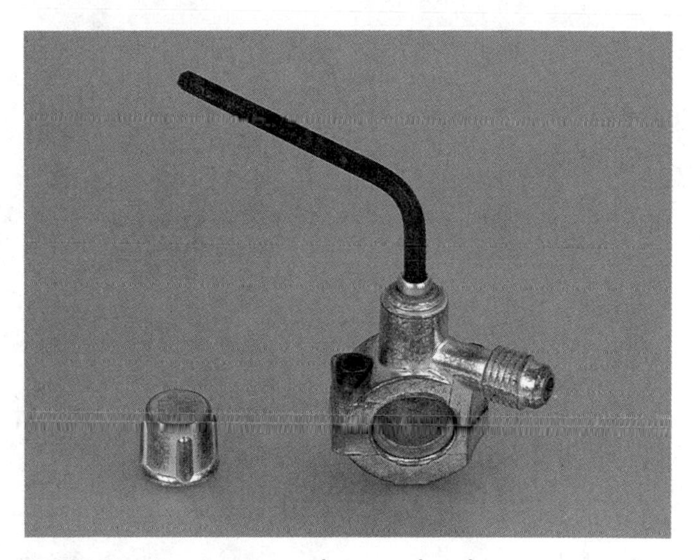

FIGURE 25–49 Pressure tap devices used to obtain pressure readings when there are no gage ports. *Photo by Bill Johnson*

Line-piercing valves should be installed using the manufacturer's directions or a leak is likely to occur. It is good practice not to locate a line tap valve on the hot gas line if a gasket is involved because, in time, the heat will deteriorate the gasket and a leak will occur. If a high-side pressure reading is required, use the liquid line; it is much cooler and less likely to leak.

The other type of valve that can be installed while running requires the valve to be soldered on the line with a low-temperature solder. This can only be done on a vapor line because a liquid line will not heat up as long as there is liquid in it. Manufacturers claim that there is no damage to the refrigerant in the line for this type of soldering application. This valve may be more leak free because it is soldered. After the valve is soldered on the line, a puncture is made similar to the valve previously described, Figure 25–50.

Other valves used as ports must have a hole drilled in the refrigerant line. The valve stem is inserted into the line and soldered. This must be done when there is no pressure in the line, Figure 25–51.

Attaching a gage hose to the valves can be done in two ways. Some of the valves have handles that shut off the

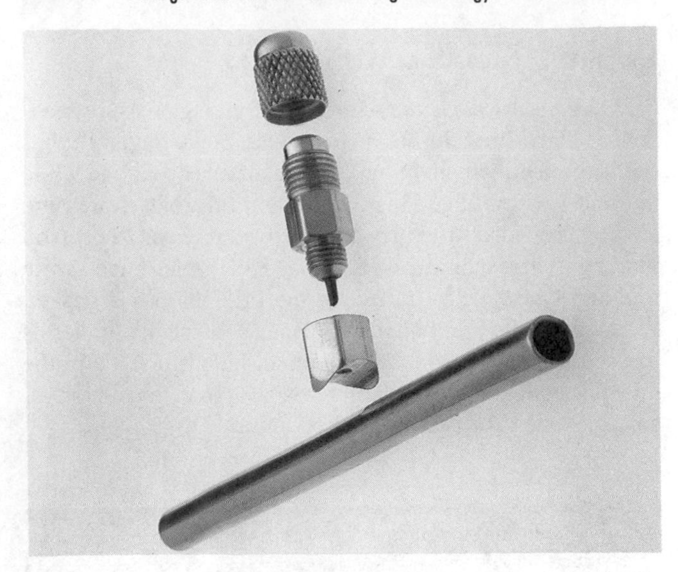

FIGURE 25–50 This valve can be soldered on a vapor line while the system has pressure in it. Low-temperature solder is used. It cannot be soldered to a line with liquid in it because the liquid will not allow the line to get hot enough. *Courtesy J/B Industries*

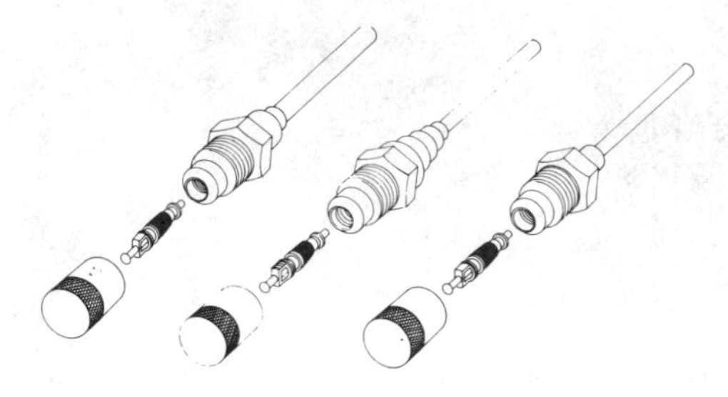

FIGURE 25–51 This valve port must have a hole drilled in the line and can only be installed with the system at atmospheric pressure. *Courtesy J/B Industries*

valve to the atmosphere. The others normally use a *Schrader* connection, which is like a tire valve on a car or bicycle except that it has threads that accept the 1/4-in. gage hose connector from a gage manifold, Figure 25–52.

25.47 CRANKCASE HEAT

Some refrigerants require that heat be applied to the crankcase of the compressor to keep the refrigerant from migrating to the oil during the off cycle or during any long periods of off time. Some refrigerants have an affinity for oil, almost like a magnetic attraction. ⊛The refrigerant in common use today that migrates the most is R-22. It has not been used for very many applications in refrigeration, but it is a hydrochlorofluorocarbon refrigerant, and it is being used for a replacement for many R-12 and R-502 systems because both of them are on the list to be phased out first.

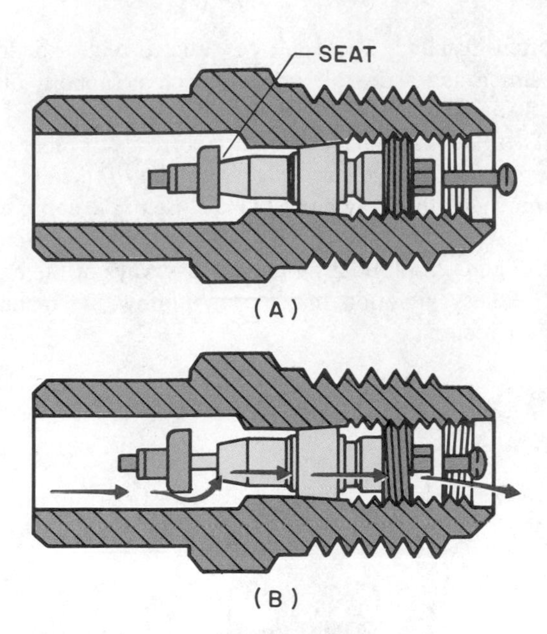

FIGURE 25–52 The Schrader valve is similar to the valve on a car tire. It has threads to accept the service technician's gage line threads. *Courtesy J/B Industries*

We are likely to see more and more R-22 in refrigeration systems until a suitable substitute is found.⊛

If you take a container of R-22 and a container of refrigerant oil and pipe them together at the vapor space, the R-22 will move to the oil container in a very short time, Figure 25–53. This migration can be slowed by adding heat to the container of oil, Figure 25–54. Think of the refrigeration system as two containers attached together with piping, like the example above.

Crankcase heat is common in air conditioning (cooling) systems where R-22 has been used for many years. The homeowner often shuts off the disconnect to the outdoor condensing unit for the winter, leaving the compressor without crankcase heat. If the homeowner then starts the unit without some time for the heat to boil the refrigerant out of the oil, damage will likely occur to the compressor. On start-up, the crankcase pressure reduces as soon as the compressor starts to turn. The refrigerant will boil and turn the oil to foam. The oil and refrigerant (some of the refrigerant may be in the liquid state) will be pumped out of the compressor. Valve damage may occur and the compressor may be operated with a limited oil charge until it returns to the crankcase from the evaporator.

Refrigeration compressors typically operate all season and will not have this seasonal shutdown, but if they are shut down for a period of time, the above situation will occur so it is likely that crankcase heat will become popular.

Crankcase heat can be applied in several ways. Some is applied using heaters that are inserted into the oil and some are external heaters at the base of the compressor adjacent to the oil, Figure 25–55. When the heater is external, the heating element must be in good contact with the compressor housing

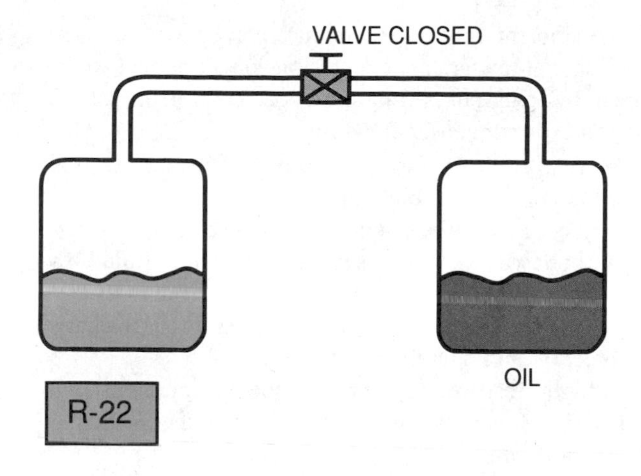

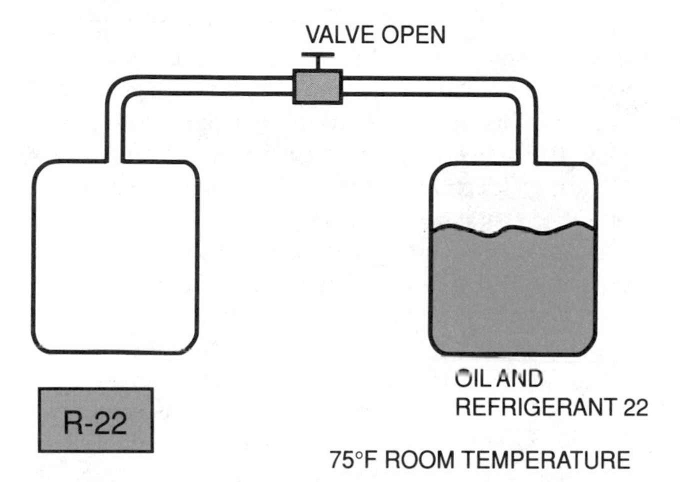

FIGURE 25–53 When the valve is opened, the refrigerant moves to the oil container.

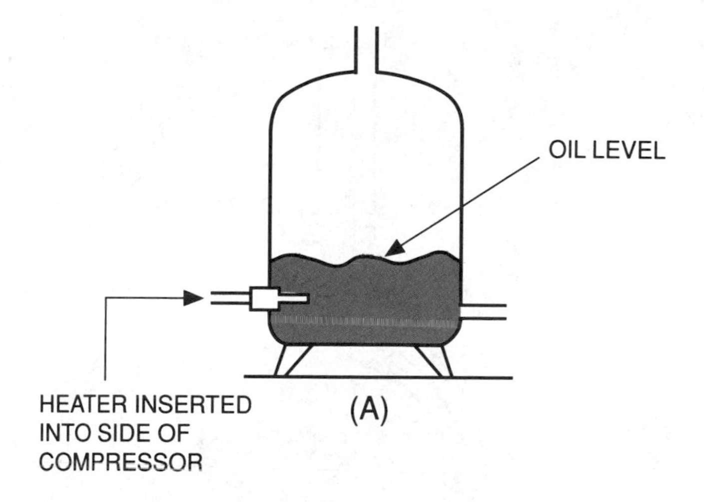

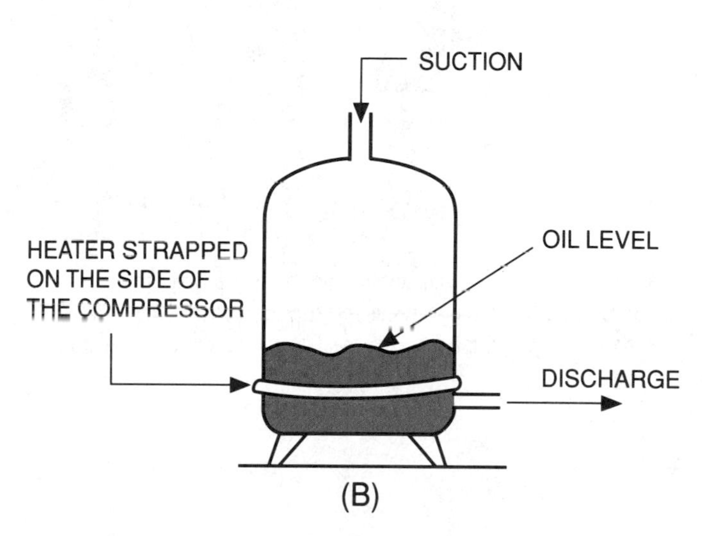

FIGURE 25–55 Two types of crankcase heat.

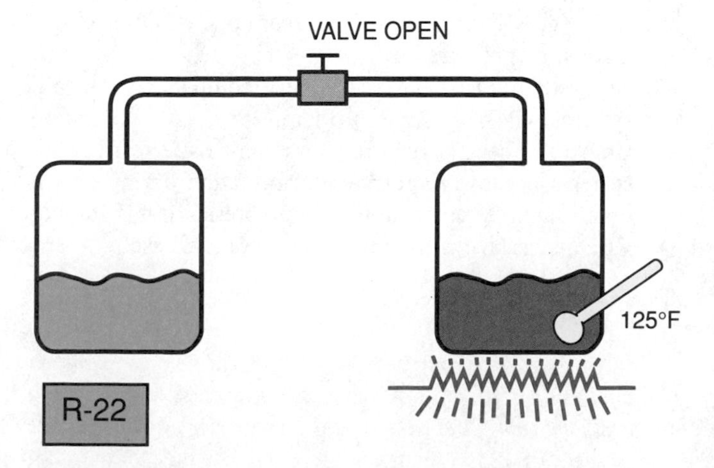

FIGURE 25–54 Heat keeps the refrigerant out of the oil.

or it will overheat the element and not transfer heat into the compressor. These heaters can be installed incorrectly and left loose unless directions are followed.

Because the crankcase heat is only needed (and desired) during the off cycle, manufacturers are likely to use relays to shut the heat off during the running cycle. (Heat added during the running cycle only reduces system efficiency.) This can also be accomplished with a set of NC contacts in the compressor contactor, contacts that are closed when the contactor is deenergized, Figure 25–56.

The technician should be well aware of crankcase heat. If you approach a compressor and it has been off for a long time, the housing should be warm to the touch if it has crankcase heat. If the manufacturer has installed crankcase heat, don't start the compressor up cold or damage may occur.

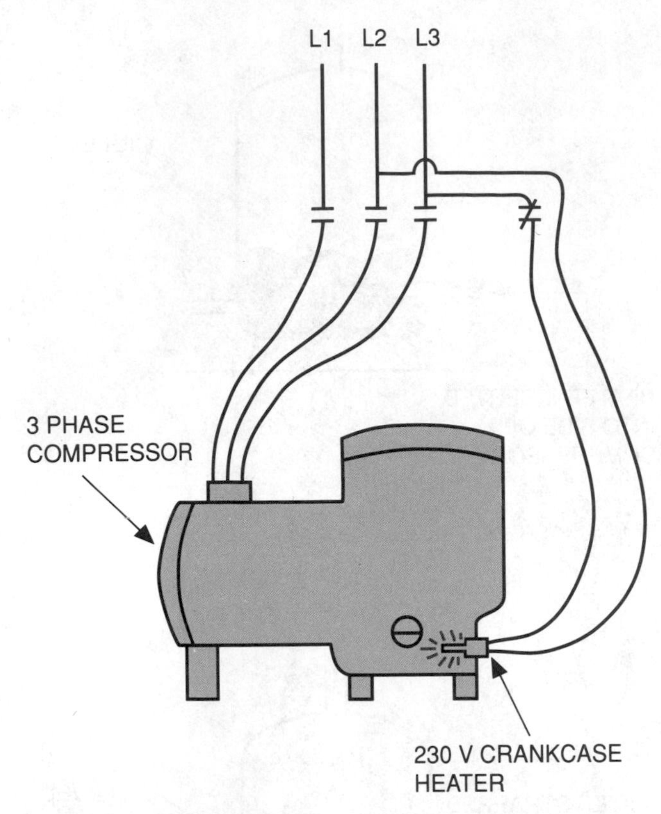

FIGURE 25–56 The compressor is not running but the crankcase heater is hot. When the compressor contactor is energized, the compressor contacts will close and the crankcase heater contacts will open, stopping the heat.

SUMMARY

- Four basic components of the compression cycle are the compressor, condenser, evaporator, and expansion device.
- Two other types of components that enhance the refrigeration cycle are control components and accessories.
- Control components can be classified as mechanical, electrical, and electromechanical.
- Mechanical components are used to stop, start, or modulate either fluid flow or electrical energy.
- Two-temperature operation may utilize an evaporator pressure regulator, two-temperature-valve, or solenoid valve for multiple evaporators at different temperatures.
- Crankcase pressure regulators limit the amount of gas that a compressor can pump under an overloaded condition, such as during a hot pull down.
- Relief valves prevent high pressures from occurring in the system.
- Low ambient controls maintain a proper working head pressure at low ambient conditions on air-cooled equipment by using fan cycling, fan speed control, air volume control, or condenser flooding.
- Electrical controls stop, start, or modulate electron flow for the control of motors and fluid flow.
- When energized, the solenoid valve, a valve with a magnetic coil, will open or close a valve to control fluid flow.
- Pressure switches stop and start system components.
- The low-pressure switch can be used for low-charge protection and as a thermostat.
- The high-pressure switch protects the system against high operating pressures. This control can be either manual or automatic reset.
- The low ambient control (switch) maintains the correct operating head pressures on air-cooled equipment in both mild and cold weather by cycling the condenser fan.
- The oil safety switch ensures that the correct oil pressure is available 90 sec after start-up on larger compressors (normally above 5 hp). This control is manual reset.
- Defrost cycle controls are either for medium-temperature or low-temperature applications.
- Medium-temperature applications either use random off-cycle defrost, where the air during the off cycle is used for defrost when the compressor cuts off with the thermostat, or planned off-cycle defrost which is accomplished with a timer.
- Low-temperature defrost must be accomplished with heat from either within or without the system. The evaporator fan must be stopped in both cases because the air in the cooler is so cold that defrost cannot occur with the fans on.
- Defrost with internal heat can be accomplished with hot gas from the compressor.
- Defrost with external heat is normally done with electric heaters in the vicinity of the evaporator. The compressor must be stopped during this defrost.
- Refrigeration accessories normally *do not* automatically change the flow of refrigerant but enhance the operation of the system.
- Refrigeration system accessories can be service valves, filter-driers, sight glasses, refrigerant distributors, heat exchangers, storage tanks, oil separators, vibration eliminators, or pressure taps.
- No system must have all of these components, but all systems will have some of them.
- Crankcase heat is required for many R-22 refrigerant systems because of refrigerant migration during down time. Damage can occur to the compressor if it is started while excessive refrigerant is in the crankcase.

REVIEW QUESTIONS

1. Name the four basic components necessary for the compression cycle.
2. Describe modulating fluid flow.
3. What is the purpose of the two-temperature valve?
4. What is the purpose of a crankcase pressure regulator?
5. What does a relief valve protect the system from?
6. Name two types of relief valves.
7. Why is low ambient control necessary?
8. Name four methods used for low ambient control on air-cooled equipment.

Describe the devices in Questions 9–12.

9. Low-pressure switch
10. High-pressure switch
11. Low ambient control (switch)
12. Oil safety switch
13. What is off-cycle defrost?
14. Name two types of off-cycle defrost.
15. Name two methods for accomplishing defrost in low-temperature refrigeration systems.

Describe how the following components enhance the refrigeration cycle:

16. Filter-drier
17. Heat exchanger
18. Suction accumulator
19. Receiver
20. Pressure taps

UNIT 26

Application of Refrigeration Systems

OBJECTIVES

After studying this unit, you should be able to

- **describe the different types of display equipment.**
- **discuss heat reclaim.**
- **describe package versus remote-condensing applications.**
- **describe mullion heat.**
- **describe the various defrost methods.**
- **discuss walk-in refrigeration applications.**
- **describe ice-making equipment.**
- **describe basic vending machine refrigeration operation.**
- **explain basic refrigerated air-dry unit operation.**

SAFETY CHECKLIST

* Wear goggles and gloves when attaching or removing gages and when using gages to transfer refrigerant or check pressures.
* Be careful not to get your hands or clothing caught in moving parts such as pulleys, belts, or fan blades.
* Observe all electrical safety precautions. Be careful at all times and use common sense.
* Follow manufacturers' instructions when using any cleaning or other types of chemicals. Many of these chemicals are hazardous so ensure that they are kept away from ice, drinking water, and other edible products.

26.1 APPLICATION DECISIONS

When refrigeration equipment is needed, a decision must be made as to the specific equipment to be installed. The factors that enter into this decision process are the first cost of the equipment to be installed, the conditions to be maintained, the operating cost of the equipment, and the long-term intent of the installation. When first cost alone is considered, the equipment may not perform acceptably. For example, if the wrong equipment choice is made for storing meat, the humidity may be too low in the cooler, and dehydration may occur. When too much moisture is taken out of the meat, there will be a weight loss. This weight loss is actually part of the condensate that goes down the drain during defrost.

This unit provides information regarding some of the different options to consider when equipment is purchased or installed. The service technician should know these options to help make service decisions.

26.2 REACH-IN REFRIGERATION MERCHANDISING

Retail stores use reach-in refrigeration units for merchandising their products. Customers can go from one section of the store to another and choose items to purchase. These reach-in display cases are available in high-, medium-, and low-temperature ranges. Each type of display case has open display and closed display types of cases to choose from. These open or closed styles may be chest type, upright type with display shelves, or upright type with doors. The display boxes can be placed end to end for a continuous length of display. When this is done, the frozen food is kept together, and fresh foods (medium-temperature applications) are grouped, Figure 26–1.

All of these combinations of reach-in display add to the customer's convenience and enhance the sales appeal of the products. The two broad categories of open and closed displays are purely for merchandising the product. A closed case is more efficient than an open case. The open case is more appealing to the customer because the food is more visible and easier to reach. Open display cases maintain conditions because the refrigerated air is heavy and settles to the bottom. These display cases can even be upright and store low-temperature products. This is accomplished by designing and maintaining air patterns to keep the refriger-

FIGURE 26–1 Display cases used to safely store food displayed for sale. Some display cases are open to allow the customer to reach in without opening a door; some have doors. Closed cases are the most energy efficient. *Courtesy Hill Refrigeration*

ated air from leaving the case, Figure 26–2. When reach-in refrigerated cases are used as storage, such as in restaurants, they do not display products, although they have the same components, and doors are used to keep the room air out.

FIGURE 26–2 Open display cases accomplish even, low-temperature refrigeration because cold air can be controlled. It falls naturally. When these cases have shelves that are up high, air patterns are formed to create air curtains. These air curtains must not be disturbed by the air conditioning systems's air discharge. All open cases have explicit directions as to how to load them with load lines conspicuously marked. *Courtesy Tyler Refrigeration Corporation*

26.3 SELF-CONTAINED REACH-IN FIXTURES

Reach-in refrigeration fixtures can be either self-contained with the condensing unit built inside the box or they may have a remote condensing unit. To make the best decision as to which type of fixture to purchase depends on several factors.

Self-contained equipment rejects the heat at the actual case. The compressor and condenser are located at the fixture, and the condenser rejects its heat back into the store, Figure 26–3. This is good in winter. In summer it is desirable in warmer climates to reject this heat outside. Self-contained equipment can be moved around to new locations without much difficulty. It normally plugs into either a 120-V or 230-V electrical outlet. Only the outlet may have to be moved. Figure 26–4 shows examples of wiring diagrams of medium- and low-temperature fixtures.

When service problems arise, only one fixture is affected, and the food can be moved to another cooler. Self-contained equipment is located in the conditioned space, and the condenser is subject to any airborne particles in the conditioned space. Keeping the several condensers clean at a large installation may be difficult. For example, in a restaurant kitchen large amounts of grease will deposit on the condenser fins and coil, which will then collect dust. The combination of grease and dust is not easy to clean in the kitchen area without contaminating other areas. The versatility of self-contained equipment is good for a kitchen,

FIGURE 26–3 Self-contained refrigerated box has its condensing unit built in. Refrigeration tubing does not have to be run. This box rejects the heat it absorbs back into the store. It is movable because it only needs an electrical outlet and, sometimes, a drain, depending on the box. *Courtesy Hill Refrigeration*

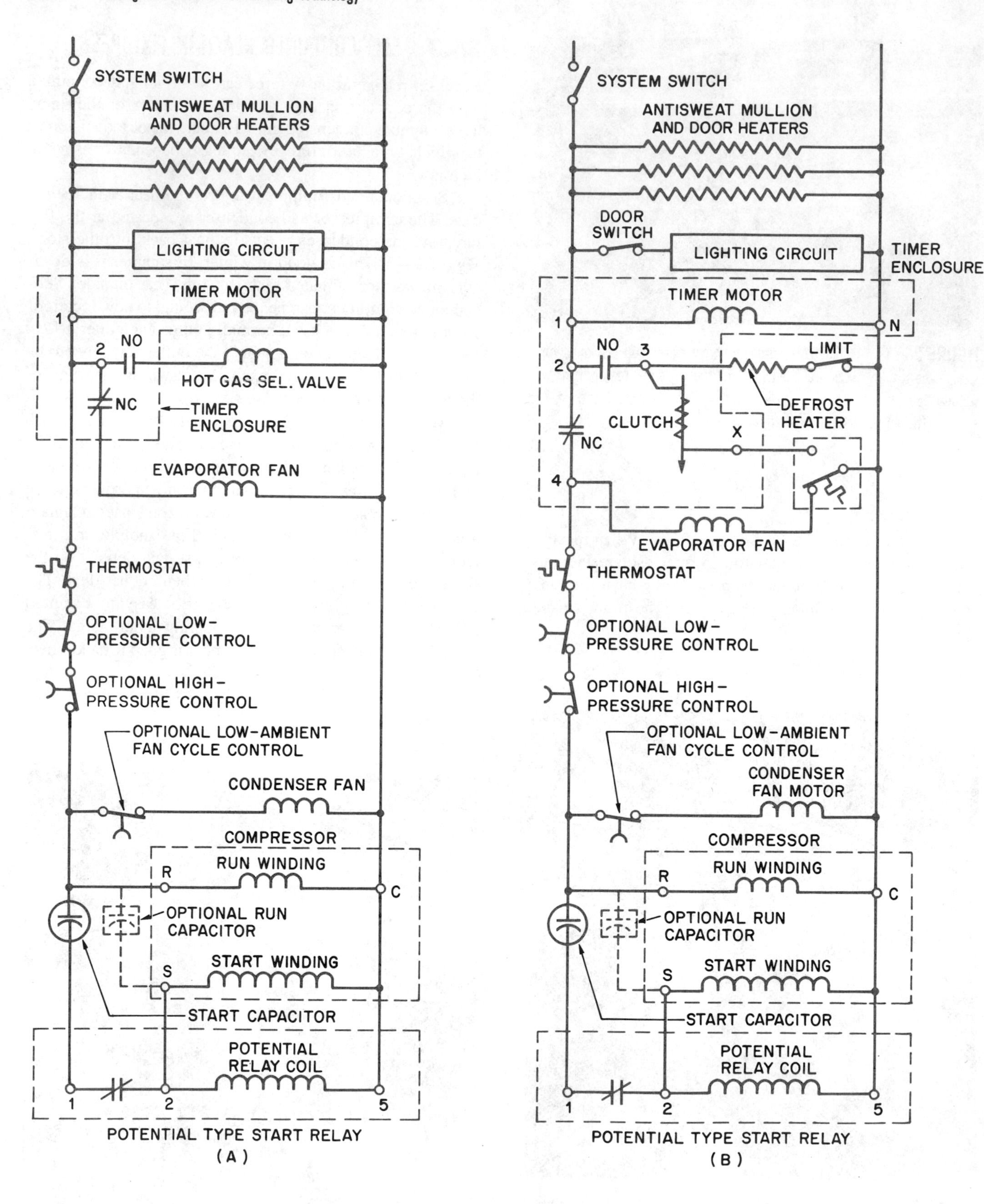

FIGURE 26–4 Wiring diagrams for self-contained reach-in boxes. (A) Low-temperature with hot gas defrost. (B) Low-temperature with electric defrost. (C) Medium-temperature with planned off-cycle defrost. (D) Medium- or high-temperature with random defrost.

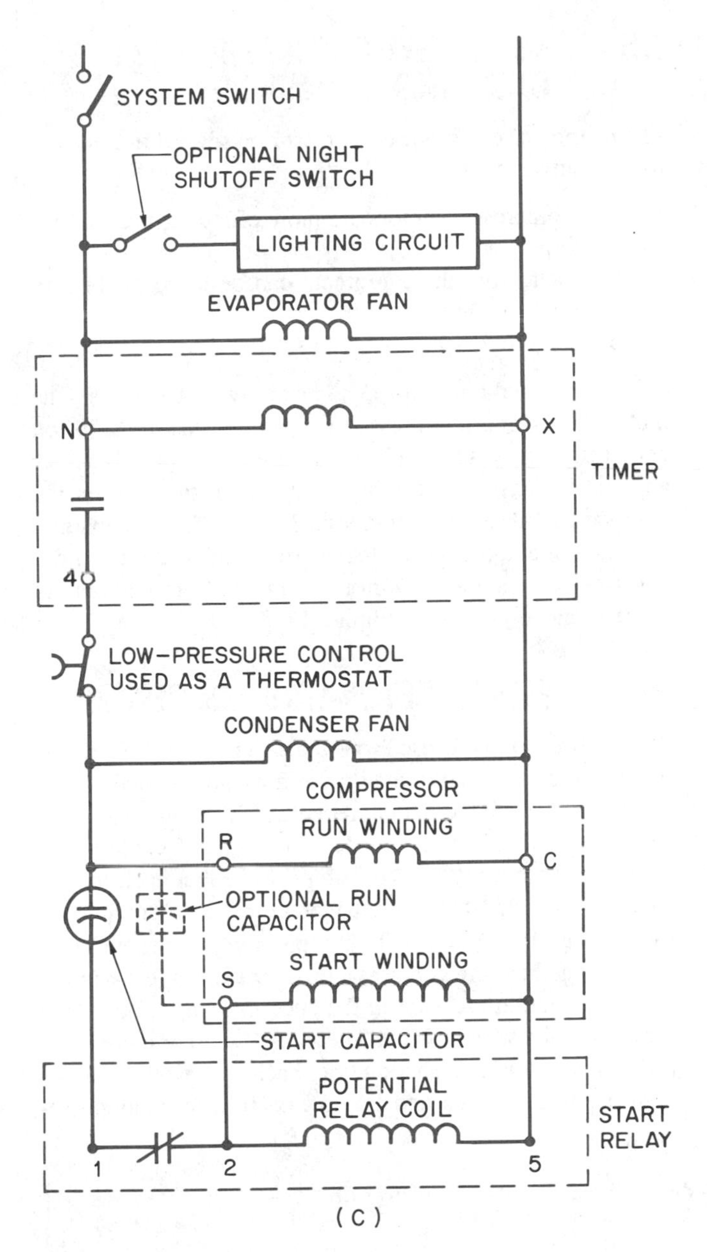

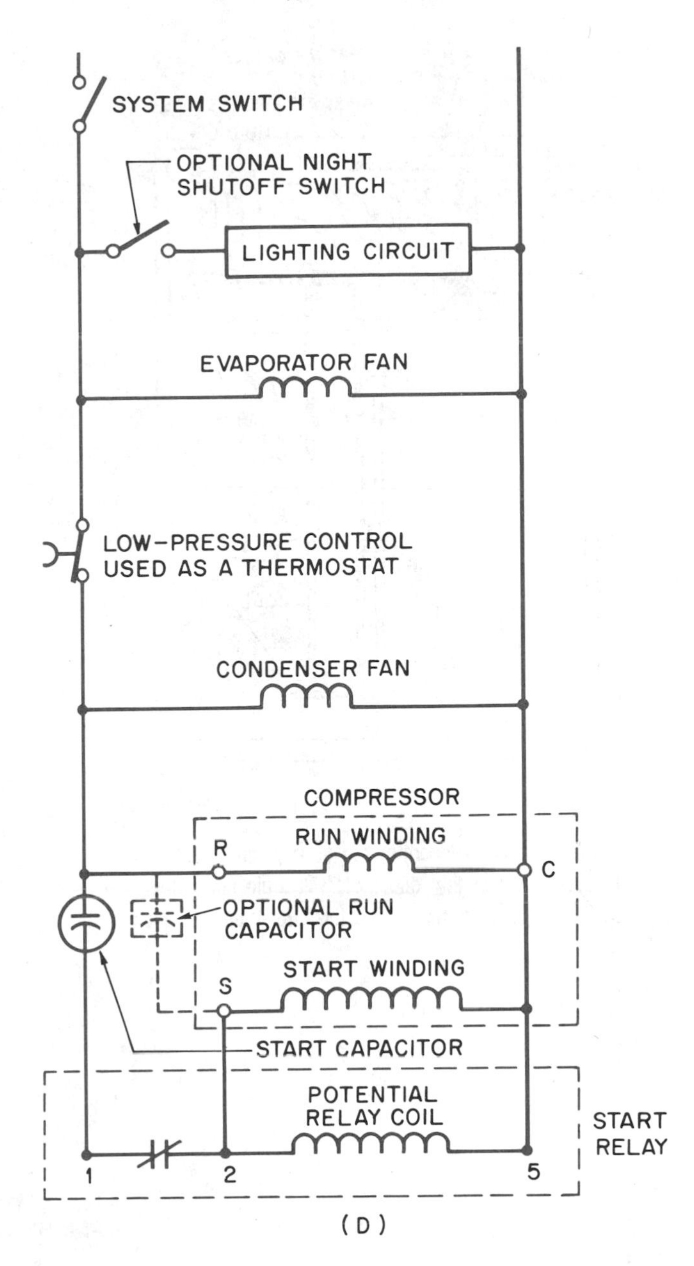

FIGURE 26–4 *(Continued)*

but remote condensing units may be easier to maintain. The warm condensers are also a natural place for pests to locate.

All refrigeration equipment must have some means to dispose of the condensate that is gathered on the evaporator. This condensate must either be piped to a drain or be evaporated. Self-contained equipment can use the heat that the compressor gives off to evaporate this condensate, provided the condensate can be drained to the area of the condenser. To do this the condenser must be lower than the evaporator. This is not easy to do when the condenser is on top of the unit, Figure 26–5.

26.4 REMOTE CONDENSING UNIT EQUIPMENT

Remote equipment can be designed in two ways: an individual condensing unit for each fixture, or one compressor or multiple compressors manifolded together to serve several cases.

26.5 INDIVIDUAL CONDENSING UNITS

When individual condensing units are used and trouble is encountered with the compressor or a refrigerant leak occurs, only one system is affected. The condensing unit can be located outside, with proper weather protection, or in a common equipment room, where all equipment can be observed at a glance for routine service. The air in the room can be controlled with dampers for proper head pressure control. The equipment room can be arranged so that it recirculates store air in the winter and reuses the heat that it took out of the store through the refrigeration system. This saves buying heat from a local utility, Figure 26–6.

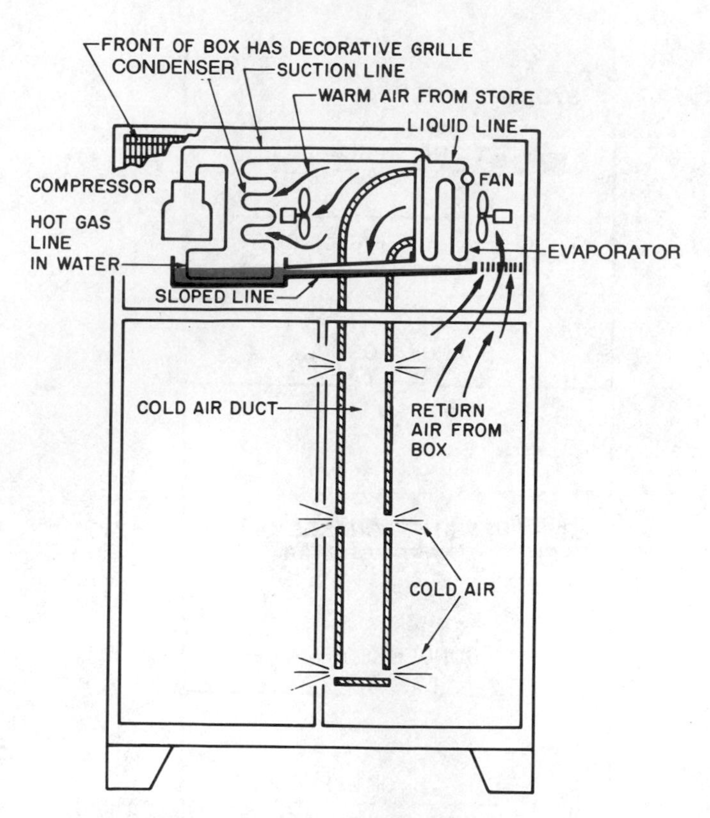

FIGURE 26–5 The evaporator can be high enough on a self-contained cooler so that the condensate will drain into a pan and be evaporated by heat from the hot gas line. No drain is needed in this application.

26.6 MULTIPLE EVAPORATORS AND SINGLE-COMPRESSOR APPLICATIONS

The method of using one compressor on several fixtures has its own advantages:

1. The compressor motors are more efficient because they are larger.
2. The heat from the equipment can be more easily captured for use in heating the store or hot water for store use, Figure 26–7.

These installations are designed in two basic ways. One method uses one compressor for several cases. A 30-ton compressor may serve 10 or more cases. This particular application may have cycling problems at times when the load varies unless the compressor has a cylinder unloading design, a method to vary the capacity of a multicylinder compressor by stopping various cylinders from pumping at predetermined pressures, Figure 26–8.

26.7 MULTIPLE MEDIUM-SIZED COMPRESSORS

Another method is to use more than one compressor manifolded on a rack. They may have a common suction and discharge line and a common receiver. The compressors can be cycled on and off as needed for capacity control. For example, there may be four compressors that each have a capacity of 7 1/2 tons of refrigeration. Some compressor racks have different capacity compressors, Figure 26–9. In this arrangement the compressors are responsible for maintaining a suction pressure in the suction manifold that will serve the coldest evaporator. There may be 10 or more cases piped to this common suction line. Each case would have its own expansion valve and liquid line from a common re-

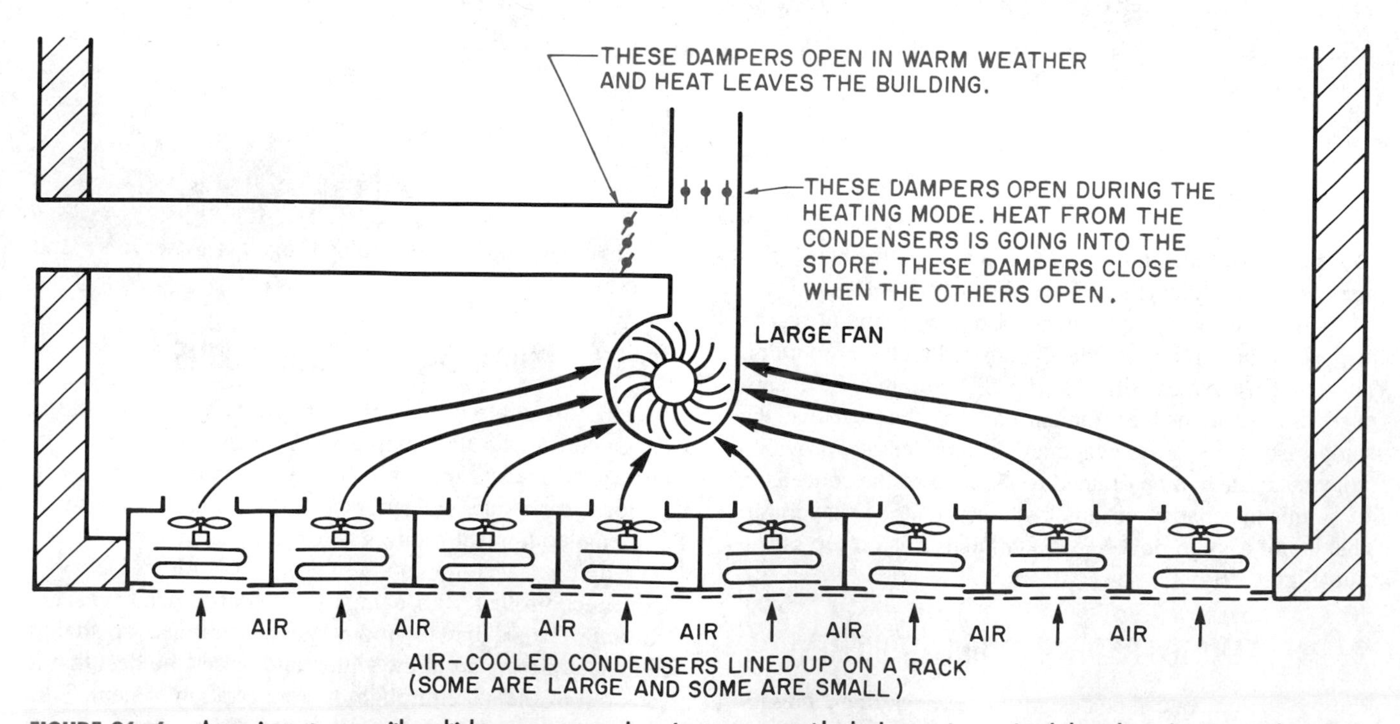

FIGURE 26–6 An equipment room with multiple compressors and condensers can provide the heat to be circulated throughout the store in the winter.

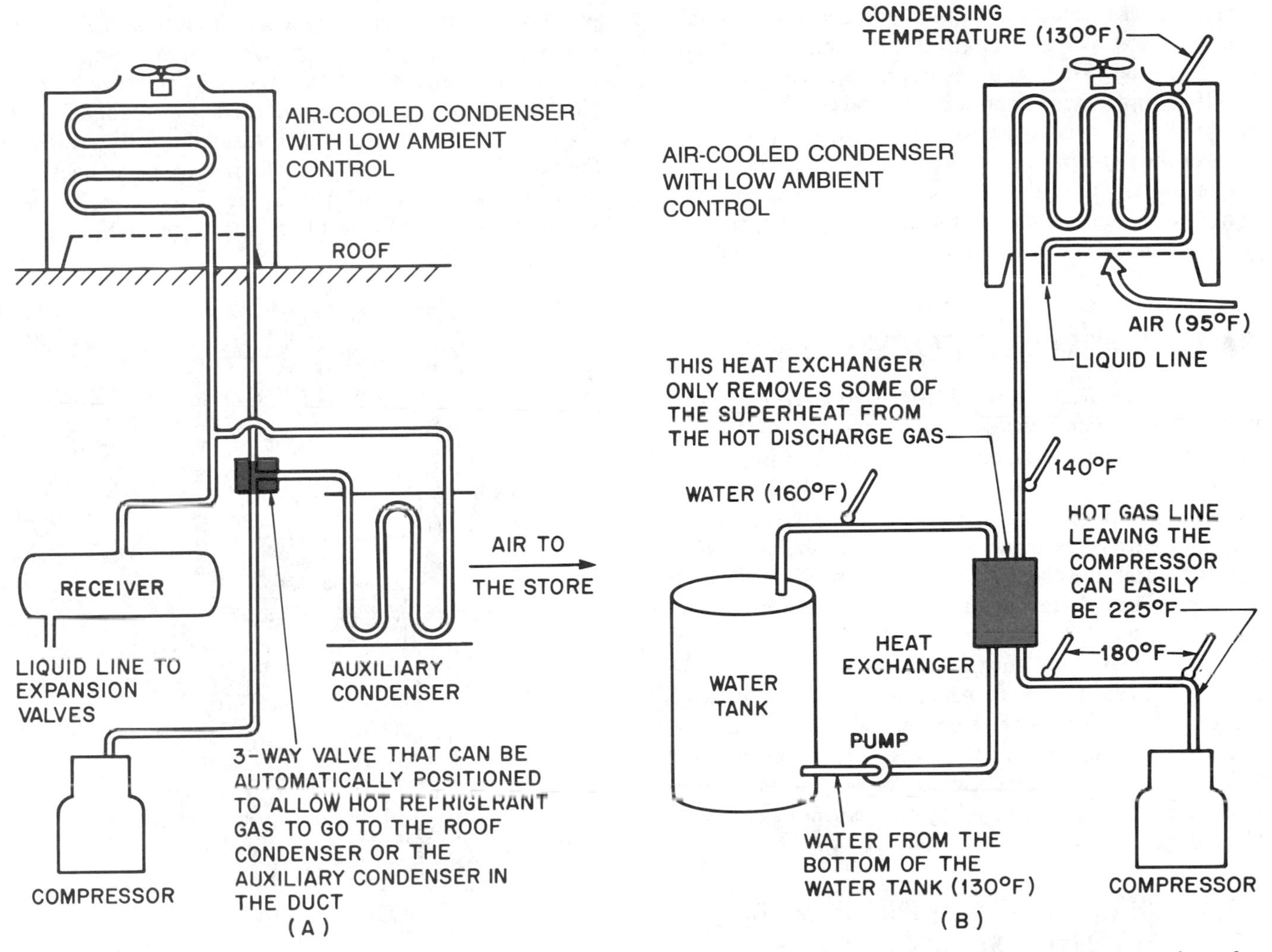

FIGURE 26–7 (A) The heat from a common discharge line can be captured and reused. This heat can be discharged from a condenser on the roof or from a coil mounted in the duct work. (B) This device is a hot water heat-reclaim device. It uses the heat from the hot gas line to heat water.

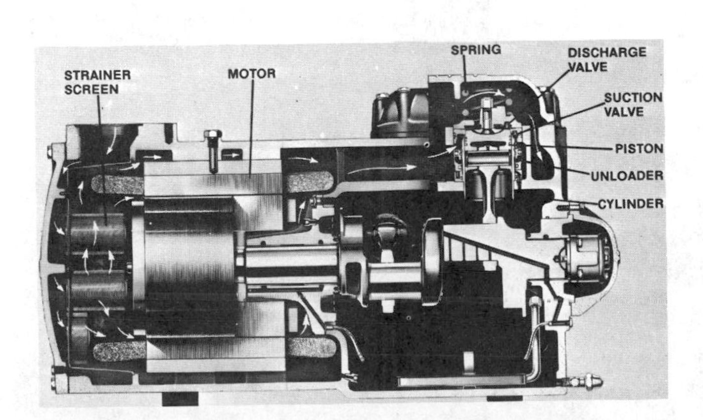

FIGURE 26–8 Compressor with cylinder unloading. This system is furnished with certain compressors to vary the capacity at reduced loads. This is accomplished by preventing various cylinders from pumping on demand. For example, a 4-cylinder compressor with 40 tons capacity has 10 tons capacity per cylinder. With the proper controls this compressor can produce 10, 20, 30, or 40 tons. *Courtesy Trane Company*

FIGURE 26–9 These compressors have varying capacities but are manifolded together with a common suction and common discharge line. These compressors can be cycled on and off individually on demand. *Courtesy Tyler Refrigeration Corporation*

ceiver. Either method can be used and has advantages and disadvantages. For example, when a single compressor is used, the total system depends on this one component.

If multiple compressors are used and one fails, the other compressors can continue to run. However, a bad motor burn on one compressor will contaminate the other three compressors. The system may be cleaned using filter-driers and may prevent damage to the remaining compressors if accomplished quickly. In either case a refrigerant leak will affect the whole system.

26.8 EVAPORATOR TEMPERATURE CONTROL

When multiple evaporators are used, they may not all have the same temperature rating. For example, the coldest evaporator may require a suction pressure of 20 psig, and the warmest evaporator may require a suction pressure of 28 psig. Evaporator pressure regulator (EPR) valves can be located in each of the higher temperature evaporators. When the load varies, the compressors can be cycled on and off to maintain a suction pressure of 20 psig. The compressors can be cycled using low-pressure control devices.

This type of installation is used in some larger installations such as supermarkets. An advantage of this system is that the heat being removed from several fixtures is concentrated into one discharge line. This heat can be piped in such a manner that it can be rejected to the atmosphere or back into the store. This can be a big supplement to the store's heating system in the winter, Figure 26–10.

26.9 INTERCONNECTING PIPING IN MULTIPLE-EVAPORATOR INSTALLATIONS

When the fixtures are located in the store and the compressors are in a common equipment room, the liquid line must be piped to the fixture, and the suction line must be piped back to the compressor area. The suction line should be insulated to prevent it from picking up heat on the way back to the compressor. This can be accomplished by preplanning and by providing a pipe chase in the floor. Preferably the pipe chase should be accessible in case of a leak, Figure 26–11. In some installations refrigerant lines are routed through individual plastic pipes in the floor. If a leak should occur, the individual lines may be replaced by pull-

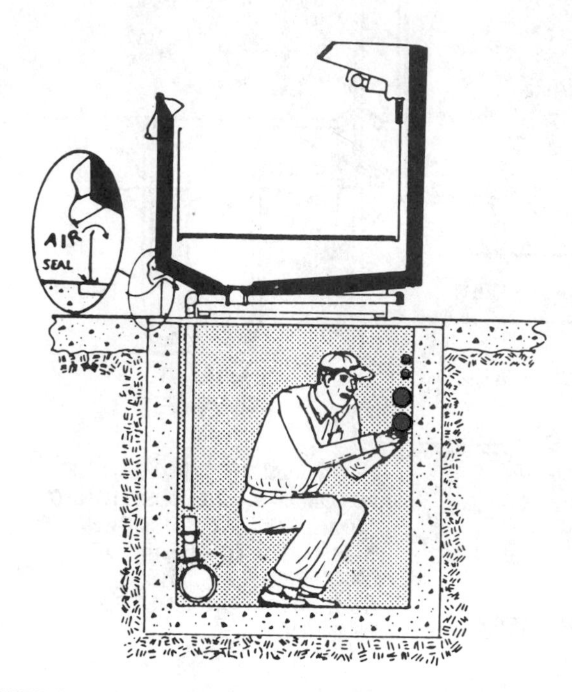

FIGURE 26–11 A pipe chase is one method of running piping from the fixture to the equipment room and is effective. When a chase like this is used, the piping can be serviced if needed. *Courtesy Tyler Refrigeration Corporation*

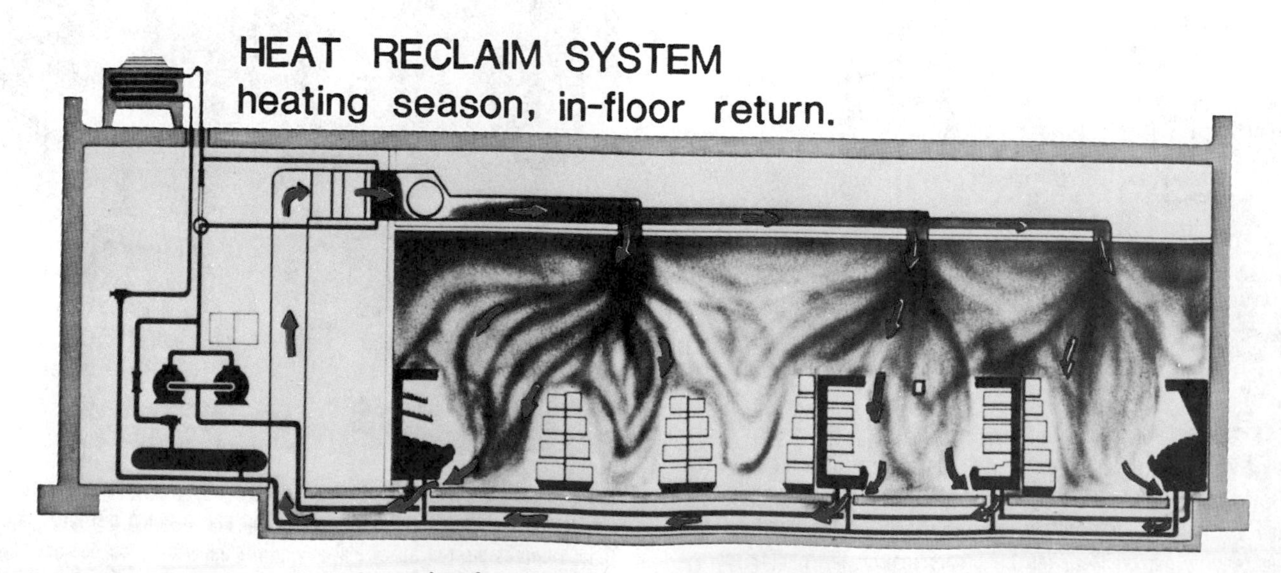

FIGURE 26–10 Piping arrangement for heat reclaim for space-heating purposes. *Courtesy Hill Refrigeration*

ing the old ones and replacing them with new ones. This may be more popular than the trench method in modern stores because it gives the designer more versatility for locating equipment. It is hard to plan a large trench to every fixture. Long refrigerant lines can cause oil return problems. Individual runs can reduce the line length. Careful design factors must be followed for correct oil return.

26.10 TEMPERATURE CONTROL OF THE FIXTURE

Control of this type of remote medium-temperature application can be accomplished without interconnecting wiring being installed between the fixture and the equipment room. A power supply for the fixture has to be located at the fixture. Where the application is medium temperature, planned off-cycle defrost can be accomplished at the equipment room with a time clock and a liquid line solenoid valve. The clock and solenoid valve can be located at the case, but the equipment room gives a more central location for all controls. The clock can activate the solenoid, closing the valve for a predetermined time. The refrigerant will be pumped out of the individual fixture. The evaporator fan will continue to run, and the air in the fixture will defrost the coil. When the proper amount of time has passed, the solenoid-activated valve will open and the coil will begin to operate normally. The compressor will continue to run during the defrost of the individual cases. The defrost times can be staggered by offsetting them. This is sometimes accomplished with a master time clock with many circuits. Use of electronic timing devices allows many different events to be controlled at one time.

Low-temperature installations must have a more extensive method of defrost because heat must be furnished to the coil in the fixture, and the fan must be stopped. This can be accomplished at the fixture with a time clock and heating elements. The power supply for the fixture is in the vicinity of the fixture. It can also be accomplished with hot gas, but a hot gas line must then be run from the compressor to the evaporator. This is a third line to be run to each case. It must also be insulated to keep the gas hot until it reaches the evaporator. The defrost of each case can be staggered with different time clock settings, Figure 26–12. This type of defrost is the most efficient because the heat from the defrost is coming from the other cases; however, electric defrost probably has fewer problems and is easier to troubleshoot.

These methods of defrost are typical, but by no means the only methods. Different manufacturers devise their own methods of defrost to suit their equipment.

26.11 THE EVAPORATOR AND MERCHANDISING

The evaporator is the device that absorbs the heat into the refrigeration system. It is located at the point where the public is choosing the product. It can be built in several ways. At best, an evaporator is bulky. A certain amount of planning must be done by the manufacturers and their engineering staff to provide attractive fixtures that are also functional, Figure 26–13.

Customer appeal must be considered in the choice of equipment. The service technician may not understand why some equipment is installed the way it is because customer convenience or merchandising may have played a major part in the decision. Each supermarket chain is involved in staying ahead of the competition in the marketplace and customer convenience is part of it.

Display fixtures are available as (1) chest type (open, open with refrigerated shelves, closed) and (2) upright (open with shelves, with doors).

26.12 CHEST-TYPE DISPLAY FIXTURES

Chest-type reach-in equipment can be designed with an open top or lids. Vegetables, for example, can easily be displayed in the open type. The vegetables can be stocked, rotated, and kept damp with a sprinkler hose, Figure 26–14. The product can be covered at night with plastic lids or film. The customer can see the product because the fixture may have its own lights. Meat that is stored in the open is normally packaged in clear plastic. In this type of fixture the evaporator may be at the bottom of the box. Fans blow the cold air through grilles to give good air circulation. Service for the coil components and fan is usually through removable panels under the vegetable storage or on the front side of the fixture. The appliance is normally placed with a wall on the back side, Figure 26–15.

The chest-type fixture can be furnished with the condensing unit built in or designed so that it can be located at a remote location, such as in an equipment room. If the condensing unit is furnished with the cabinet, it is usually located underneath the front and can be serviced by removing a front panel. When the condensing unit is furnished with a fixture, it is called self-contained.

26.13 REFRIGERATED SHELVES

In some chest-type fixtures refrigerated shelves are located at the top. These shelves must have correct air flow around them or they must have plate-type evaporators to maintain the correct food conditions. When there are evaporators at the top, they are normally piped in series with the evaporators at the bottom, Figure 26–16.

26.14 CLOSED-TYPE CHEST FIXTURES

The closed-type chest is normally a low-temperature fixture and can store ice cream or frozen foods. The lids may be lifted off, raised, or slide from side to side. These boxes are not as popular now as in the past because the lids are a barrier to the customer.

The upright closed display normally has doors through which the customer can look to see the product. This type of cooler may have a self-contained or remote condensing unit. It can also be piped into a system with one compressor. This

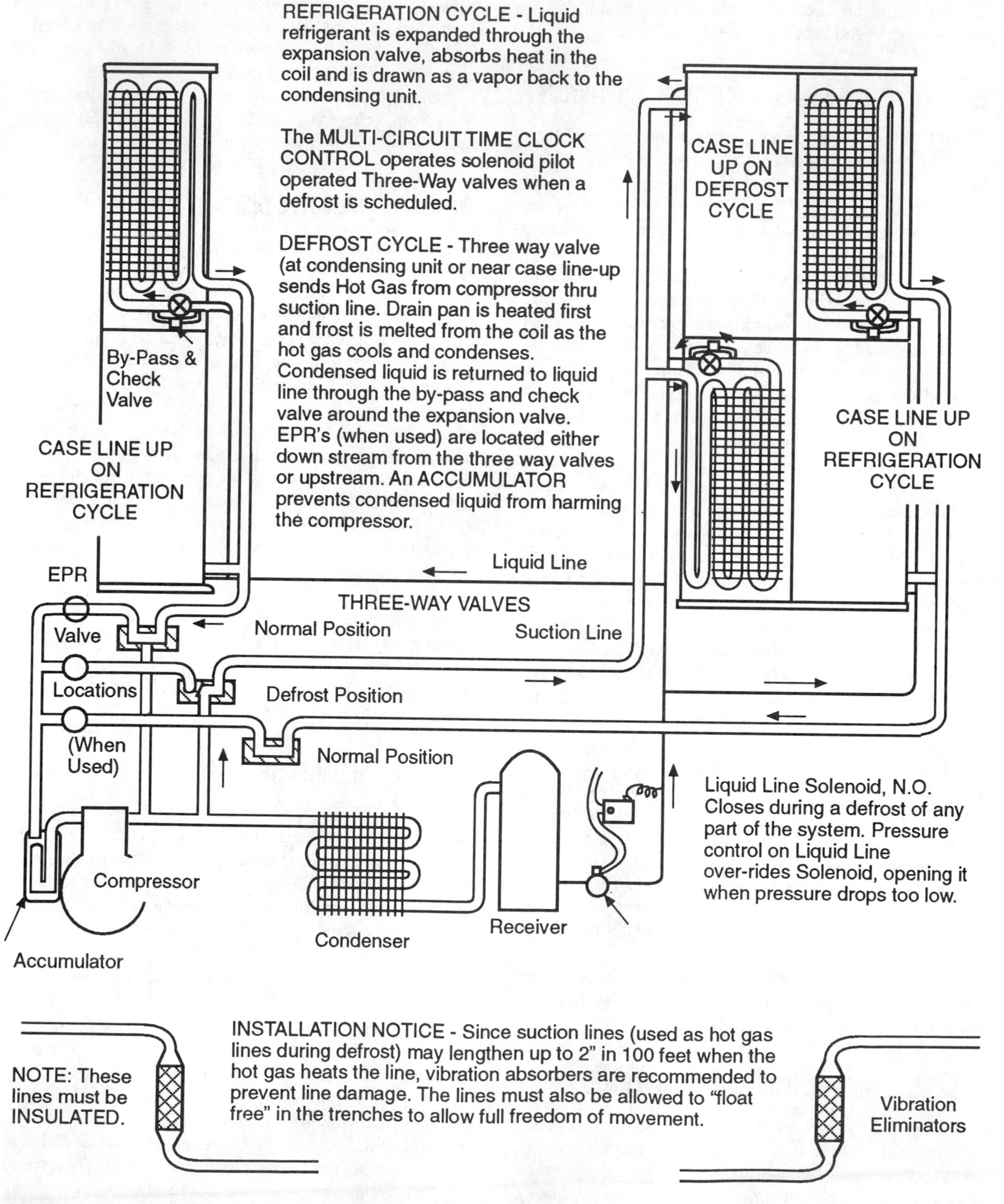
LATENT HEAT DEFROST AS USED ON A COMBINED SYSTEM
(Do not confuse this with Reverse Cycle Defrost with all cases being defrosted at once.)
Only one-third to one-fourth of the total Btuh load can be defrosted at any one time.
REFRIGERATION CYCLE - Liquid refrigerant is expanded through the expansion valve, absorbs heat in the coil and is drawn as a vapor back to the condensing unit.
The MULTI-CIRCUIT TIME CLOCK CONTROL operates solenoid pilot operated Three-Way valves when a defrost is scheduled.
DEFROST CYCLE - Three way valve (at condensing unit or near case line-up sends Hot Gas from compressor thru suction line. Drain pan is heated first and frost is melted from the coil as the hot gas cools and condenses. Condensed liquid is returned to liquid line through the by-pass and check valve around the expansion valve. EPR's (when used) are located either down stream from the three way valves or upstream. An ACCUMULATOR prevents condensed liquid from harming the compressor.
CASE LINE UP ON DEFROST CYCLE
By-Pass & Check Valve
CASE LINE UP ON REFRIGERATION CYCLE
CASE LINE UP ON REFRIGERATION CYCLE
Liquid Line
EPR
THREE-WAY VALVES
Valve
Normal Position
Suction Line
Locations
Defrost Position
(When Used)
Normal Position
Liquid Line Solenoid, N.O. Closes during a defrost of any part of the system. Pressure control on Liquid Line over-rides Solenoid, opening it when pressure drops too low.
Compressor
Receiver
Condenser
Accumulator
INSTALLATION NOTICE - Since suction lines (used as hot gas lines during defrost) may lengthen up to 2" in 100 feet when the hot gas heats the line, vibration absorbers are recommended to prevent line damage. The lines must also be allowed to "float free" in the trenches to allow full freedom of movement.
NOTE: These lines must be INSULATED.
Vibration Eliminators

FIGURE 26–12 Hot gas defrost with multiple cases. *Courtesy Tyler Refrigeration Corporation*

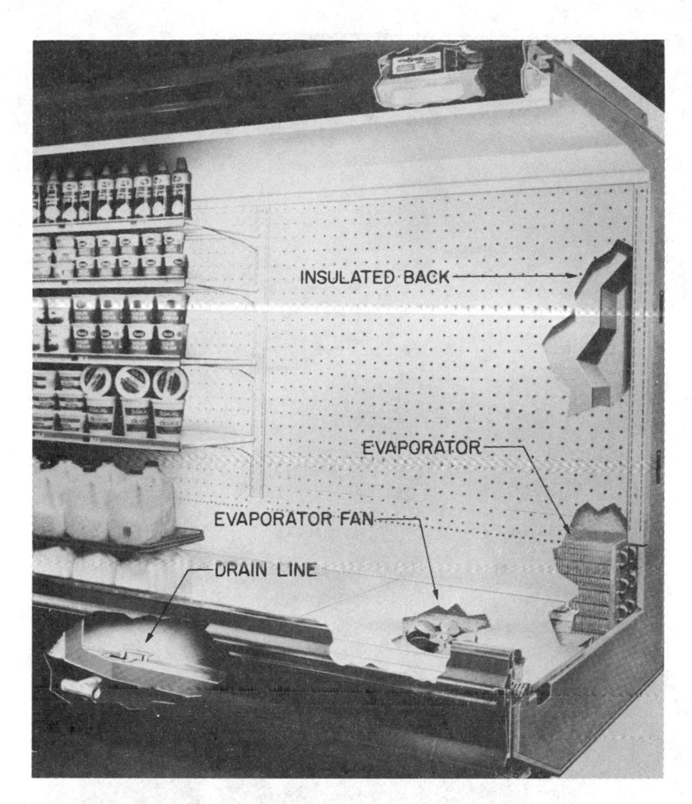

FIGURE 26–13 Dairy case showing the location of the evaporator. *Courtesy Hill Refrigeration*

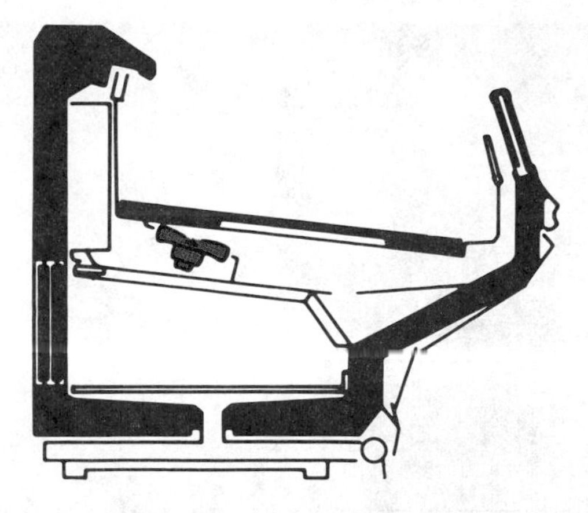

FIGURE 26–15 Cutaway side view of a chest-type display case. The fans can be serviced from inside the case. The case is normally located with its back to a wall or another fixture. *Courtesy Hill Refrigeration*

FIGURE 26–14 Open display case used to store fresh vegetables. The case has a drain, so the vegetables can be moistened with a hand-operated sprinkler. *Courtesy Tyler Refrigeration Corporation*

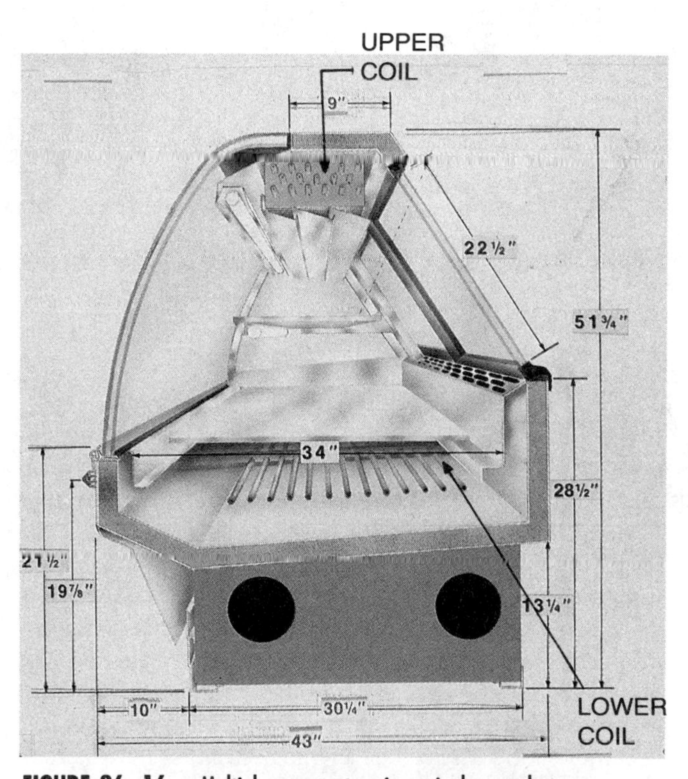

FIGURE 26–16 Multiple evaporators in a single case because refrigeration is needed in more than one place. *Courtesy Tyler Refrigeration Corporation*

display is sometimes used as one side of the wall for a walk-in cooler. The display shelves can then be loaded from inside the walk-in cooler, Figure 26–17. These fixtures may be found in many convenience stores.

26.15 CONTROLLING SWEATING ON THE CABINET OF FIXTURES

Display cases that have any cabinet surfaces that may operate below the dew point temperature of the room (the tem-

FIGURE 26–17 Display case with walk-in cooler as its back wall. The display case can be supplied from inside the cooler. *Courtesy Hill Refrigeration*

perature at which moisture will form) must have some means (usually small heaters) to keep this moisture from forming. This cabinet heating normally occurs around doors on closed-type equipment. The colder the refrigerated fixture, the more need there is for the protection. The heaters are made with a resistance-type wire that is run just under the surface of the cabinet. Some of these are called *mullion* heaters. Mullion means division between panels. In refrigeration equipment, it is the panel between the doors that becomes cold. There can be a large network of these heaters in some equipment, Figure 26–18. Any surface that may collect moisture from sweating needs a heater. *Some of them can be thermostatically controlled or controlled by a humidistat based on the humidity in the store.* Air conditioning (cooling) systems are used for controlling the humidity in the stores more than ever. You may have noticed how cold some supermarkets are in the shopping areas. This is designed to keep some of the load off the refrigeration equipment. Any heat and moisture that is removed with the air conditioning system is removed at a higher efficiency level. The air conditioning system has a much lower compression ratio than any of the refrigeration systems and will remove more moisture. Owners realize that with a low humidity in the store, they do not have to use as much mullion heat and coils operating below freezing do not need as much defrost time.

26.16 MAINTAINING STORE AMBIENT CONDITIONS

The humidity is taken out of the store in the warm weather with the air conditioning system and the display cases. The more humidity removed by the air conditioning system, the less is needed to be removed by the refrigeration fixtures. This means less defrost time. Some stores keep a positive air pressure in the store with makeup air that is conditioned through the air conditioning system. This method is different from the random method of taking in outside air when the doors are opened by customers. It is a more carefully planned approach to the infiltration of the outside air, Figure 26–19. You'll notice this system when the front doors are opened and a slight volume of air blows in your face.

The doors on display cases are usually constructed of double-pane glass and sealed around the edges to keep

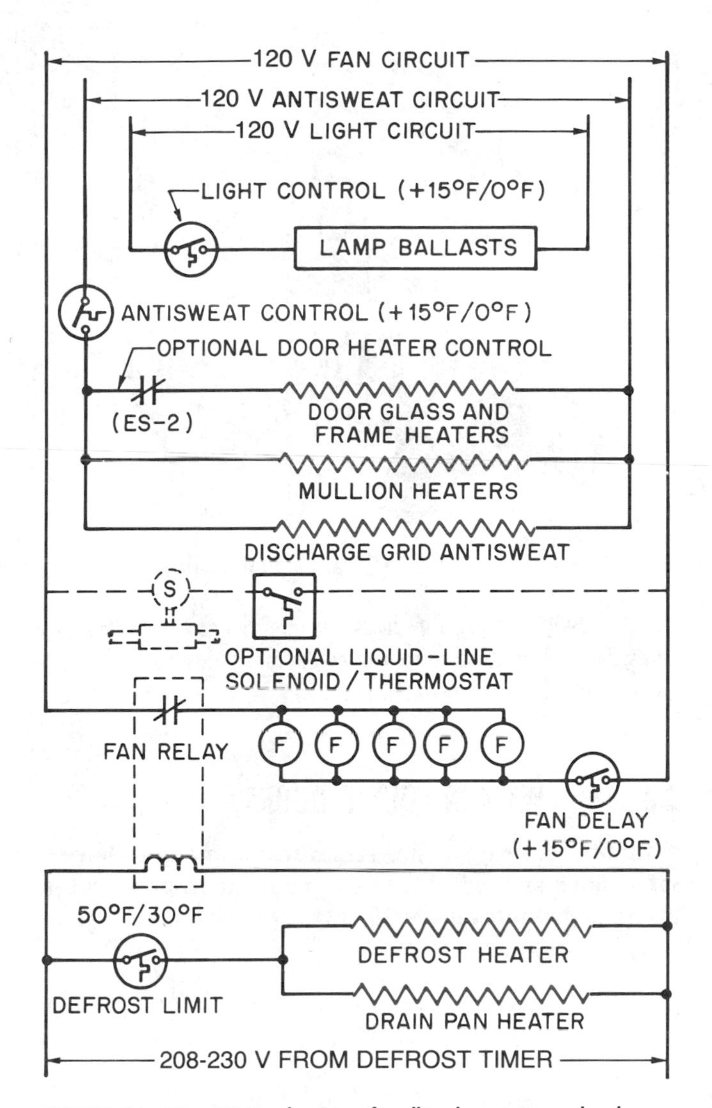

FIGURE 26–18 Wiring diagram of mullion heaters on a closed display case. There can be several circuits that apply to the heaters. This unit has two power supplies, 120 V and 208 V or 230 V.

moisture from entering between the panes. These doors must be rugged because any fixture that the public uses is subject to abuse, Figure 26–20.

The cabinets of these fixtures must be made of a strong material that is easy to clean. Stainless steel, aluminum, porcelain, and vinyl are commonly used. The most expensive cabinets are made of stainless steel, and they are the longest lasting.

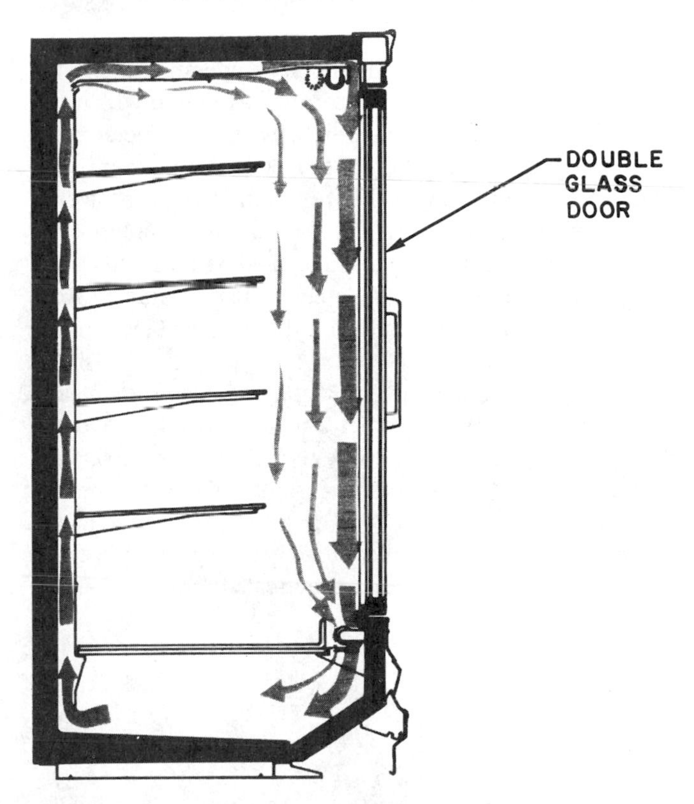

FIGURE 26–20 Doors are rugged and reliable. They are constructed of double glass sealed on the edges to keep them from sweating between the glass. *Courtesy Hill Refrigeration*

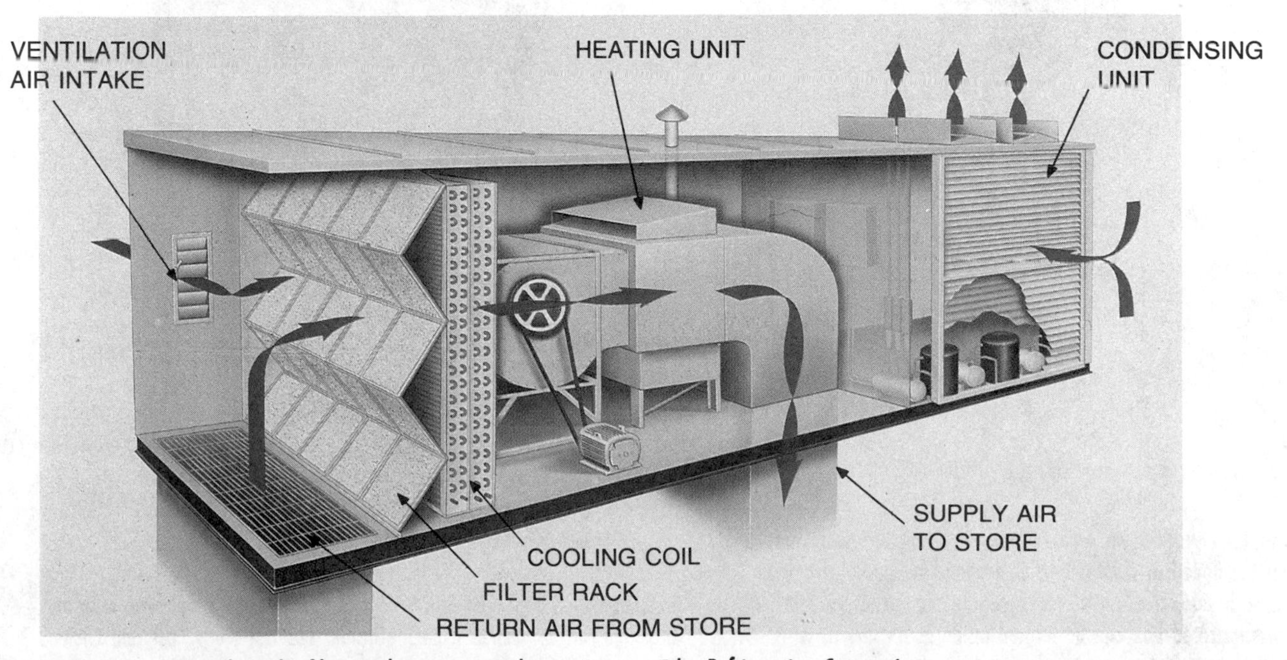

FIGURE 26–19 Planned infiltration known as ventilation. *Courtesy Tyler Refrigeration Corporation*

26.17 WALK-IN REFRIGERATION

Walk-in refrigeration equipment is either permanently erected or of the knock-down type. The permanently erected refrigerated boxes cannot be moved. Very large installations are permanent.

26.18 KNOCK-DOWN WALK-IN COOLERS

Knock-down walk-in coolers are constructed of panels from 1 in. to 4 in. thick, depending on the temperature inside the cooler. They are a sandwich type of construction with metal on each side and foam insulation between. The metal in the panels may be galvanized sheet metal or aluminum. The panels are strong enough that no internal support steel is needed for small coolers. The panels can be interlocked together. They are shipped disassembled on flats and can be assembled at the job site. This cooler can be moved from one location to another and can be reassembled, Figure 26–21.

Walk-in coolers come in a variety of sizes and applications. One wall of the cooler can be a display, and the shelves can be filled from inside the cooler. The coolers are normally waterproof and can be installed outdoors. The outside finish may be aluminum or galvanized sheet metal. When the panels are pulled together with their locking mechanism, they become a prefabricated structure, Figure 26–22.

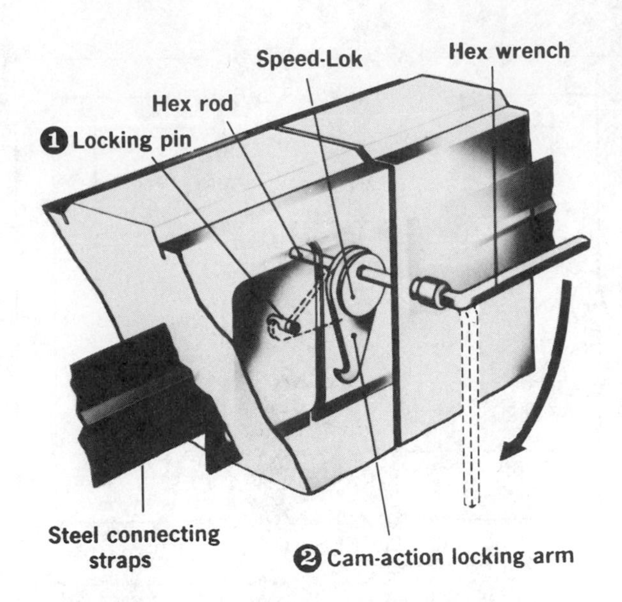

FIGURE 26–22 Locking mechanism for prefabricated walk-in cooler panels. They can be unlocked for moving the cooler when needed. *Courtesy Bally Case and Cooler, Inc.*

26.19 WALK-IN COOLER DOORS

***Walk-in cooler doors are very durable and must have a safety latch on the inside to allow anyone trapped on the inside to get out, Figure 26–23*.**

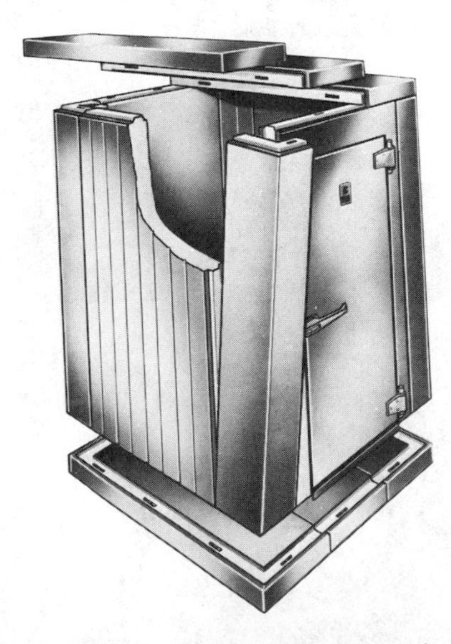

FIGURE 26–21 Knock-down walk-in cooler that can be assembled on the job. It can be moved at a later time if needed. The panels are foam with metal on each side, creating its own structure that needs no internal braces. This prefabricated box can be located inside or outside. It is weatherproof. *Courtesy Bally Case and Cooler, Inc.*

FIGURE 26–23 Walk-in cooler doors are rugged. They have a safety latch and can be opened from the inside. *Courtesy Bally Case and Cooler, Inc.*

26.20 EVAPORATOR LOCATION IN A WALK-IN COOLER

Refrigerating a walk-in cooler is much like cooling any large space. The methods that are used today take advantage of evaporators that have fans to improve the air circulation and make them compact. Following is a list of types of systems used to refrigerate these coolers:

1. Evaporators piped to condensers using field-assembled pipe
2. Evaporators with precharged piping
3. Package units, wall-hung or top-mounted with condensing units built in

The evaporators should be mounted in such a manner that the air currents blowing out of them do not blow all of the air out the door when the door is opened. They can be located on a side wall or in a corner. These evaporators are normally in aluminum cabinets in which the expansion device and electrical connections are accessed at the end panel or through the bottom of the cabinet, Figure 26–24. Some installations have a fan switch that shuts the fan off when the door is opened to prevent cold air from being pushed out of the cooler. This works well if the door is not propped open for long periods of time. The compressor is operating without the fans and this can cause liquid flooding to the compressor. Some coolers deenergize the liquid line solenoid valve and pump the refrigerant from the evaporator to the condenser and receiver while the fans are off. This works also but is an added step in the control sequence.

(A)

(B)

FIGURE 26–24 Fan coil evaporators are used in walk-in coolers. *Courtesy Bally Case and Cooler, Inc.*

26.21 CONDENSATE REMOVAL

The bottom of the cabinet on the evaporator contains the condensate drain pan that must be piped to the outside of the cooler. When the inside is below freezing, heat has to be provided to keep the condensate in the line from freezing, Figure 26–25. This heat is normally provided by an electrical resistance heater that can be field installed. The line is piped to a drain and must have a trap to prevent the atmosphere from being pulled into the cooler. The line and trap must be heated if the line is run through below-freezing surroundings. These drain line heaters sometimes have their own thermostats to keep them from using energy during warm weather.

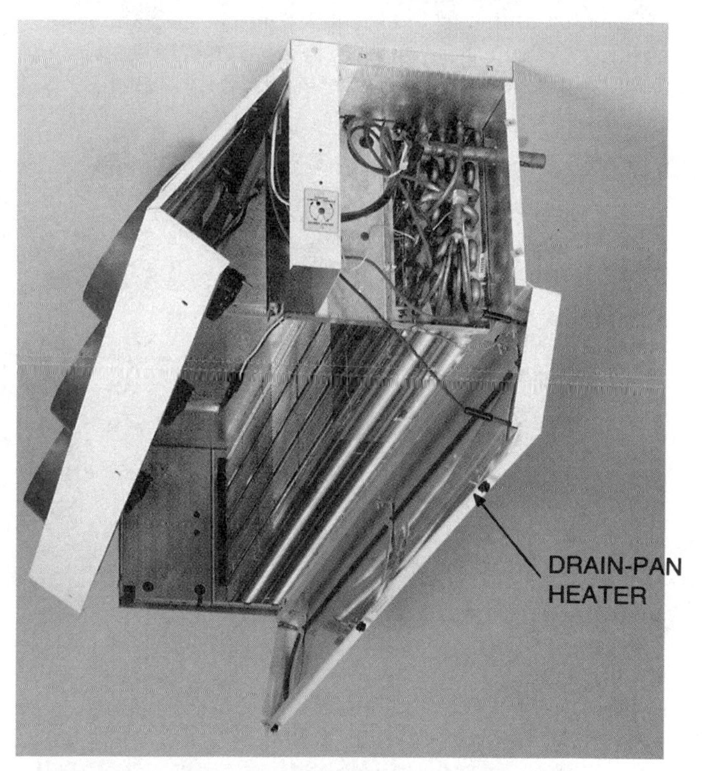

FIGURE 26–25 Drain pan heaters to keep drain pans from freezing. *Courtesy Larkin Refrigeration Products*

26.22 REFRIGERATION PIPING

One of two methods may be used to install refrigeration piping. One method is to pipe the cooler in the conventional manner. The other is to use precharged tubing, called a *line set*. When the conventional method is used, the installing contractor usually furnishes the copper pipe and fittings as part of the agreement. This piping can have straight runs and factory elbows for the corners. When the piping is completed, the installing contractor has to leak check, evacuate, and charge the system. This requires the service of an experienced technician.

When precharged tubing is used, the tubing is furnished by the equipment manufacturer. It is sealed on both ends and has quick-connect fittings. This piping has no fittings for corners and must be handled with care when bends are made. The condenser has its operating charge in it, and the evaporator has its operating charge in it. The tubing has the correct operating charge for the particular length chosen for the job. This has some advantage because the installation crew does not have to balance the operating charge in the system. The system is a factory-sealed system with field connections, Figure 26–26. No soldering or flare connections have to be made. The system can be installed by someone with limited experience.

If, because of miscalculation, the piping is too long for the installation, you can obtain a new line set from the supplier, or you can cut the existing line set to fit. ⊛**If the existing line set is to be altered, the refrigerant can be leak checked, evacuated, and charged to specifications.**⊛ If the line set is too long, the extra tubing should be coiled and placed in a horizontal position to ensure good oil return. Figure 26–27 shows a step-by-step illustration of how the line set can be altered.

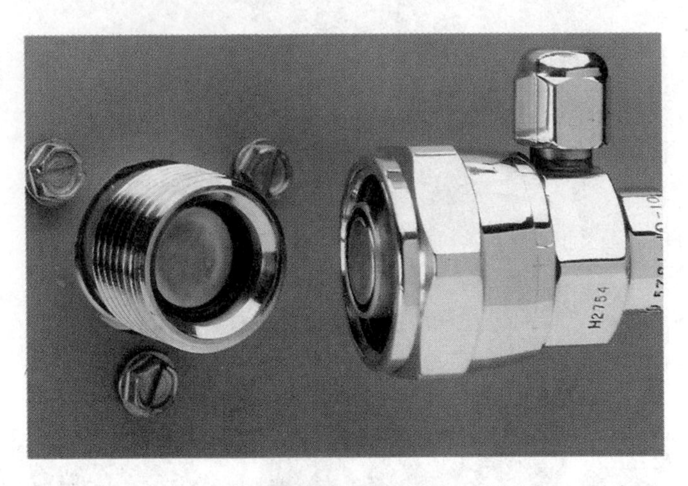

FIGURE 26–26 Quick-connect fittings providing the customer with a system that is factory sealed and charged. *Courtesy Aeroquip Corporation*

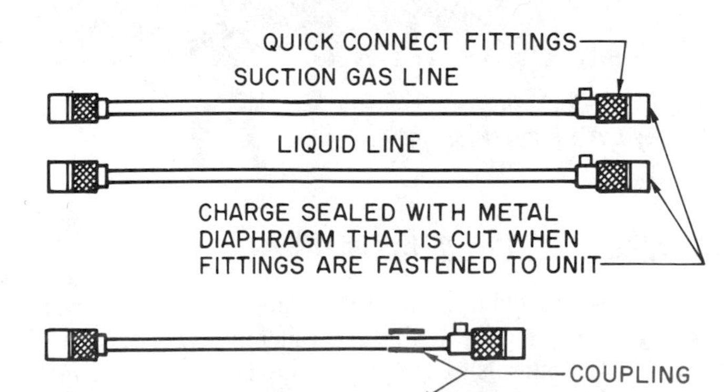

FIGURE 26–27 Altering quick-connect lines. Note: The suction line is charged with vapor. The liquid line has a liquid charge. It contains about as much liquid as may be pulled into it under a deep vacuum. ⊛**Recover the charge in each tube and cut to the desired length.**⊛ Fasten together with couplings. ***Leak check, evacuate, and charge: (1) the suction line with vapor (2) the liquid line with liquid. You now have a short line set with the correct charge.*** See Unit 8, System Evacuation, and Unit 10, System Charging.

26.23 PACKAGE REFRIGERATION FOR WALK-IN COOLERS

Wall-hung or ceiling-mount units are package-type units, Figure 26–28. They can be installed by personnel who do not understand refrigeration. The installation may be compared to installing a window air conditioner. This makes them very popular for some applications because the owner can keep the cost down. Care must be used as to where the condenser air is discharged outside. It must have room to discharge without recirculating to the air inlet or head pressure problems will occur. These units are factory assembled and require no field evacuation or charging. They come in high-, medium-, and low-temperature ranges. Only the electrical connections need to be made in the field.

FIGURE 26–28 This wall-hung unit, "saddle unit," actually hangs on the wall of a cooler. The weight is distributed on both sides. This is a package unit that only has to be connected to the power supply to be operable. *Courtesy Bally Case and Cooler, Inc.*

26.24 ICE-MAKING EQUIPMENT, PACKAGED TYPE

Ice making is accomplished in a temperature range that is somewhat different from the low-, medium-, or high-temperature refrigeration. The ice is made with evaporator temperatures between medium- and low-temperature ranges of about 10°F with the ice at 32°F. Most refrigeration applications use tube and fin evaporators and a defrost cycle to clear the ice buildup from the evaporator. Ice making is accomplished by accumulating the ice on some type of evaporator surface and then catching and saving it after a defrost cycle, commonly called the harvest cycle.

Large commercial block ice makers use cans and freeze the ice in the cans. We will only discuss small ice-making equipment, such as found in commercial kitchens, motels, and hotels. These ice makers are usually of the package type. The power is supplied by a power cord, or in some cases wired directly to the electrical circuit, called hard wired. Some ice makers may be of the split system type with the condenser on the outside.

Package ice machines store their own ice at 32°F in a bin below the ice maker. This bin is refrigerated by the melting ice in the bin (which explains the storage temperature of 32°F), so there is some melting; the hotter the day, the more melting.

A drain must be provided for the melting ice and any water that may overflow from the ice-making process. Do not confuse an ice machine with an ice-holding machine such as those seen on the outside of a convenience store or service station. Ice-holding machines hold ice, usually in bags, at a temperature well below freezing. This ice is made and bagged at one location, then stored and dispensed at the retail location.

Most package ice makers are air cooled and must be located where the correct airflow across the condenser is maintained. Some package ice makers are water cooled and use wastewater systems whereby the water is used to cool the condenser and then passes down the drain.

The installation of a package machine is usually kept simple by the following steps:

- Set the machine, much like a refrigerator
- Supply power, plug, or hard wire
- Provide a water supply
- Provide a drain for the ice bin

Package ice machines make two types of ice, flake ice and several forms of cube (solid) ice.

26.25 MAKING FLAKE ICE

Flake ice is normally made in a vertical refrigerated cylinder surrounded by the evaporator on the outside. Flake ice is the very thin pieces of ice often seen in restaurants and vending machine cups. Some people prefer flake ice to solid forms of ice when served in drinks. It melts fast because there is little weight to flake ice. It has a good bit of air contained in the flake. Solid forms of ice last longer because there is more weight; remember it takes 144 Btu/lb to make or melt ice. It takes less ice by the pound to fill a cup. Solid ice costs more and some of it is likely to be thrown out.

Flake ice is formed on the inside of the cylinder by maintaining a water level inside the cylinder. An auger is constantly turning inside the cylinder and scraping the formed ice off the evaporator surface. The auger forces the ice upward and out a chute to the side where it moves by gravity to the ice storage bin. This type of ice-making machine has a constant harvest cycle, as long as the compressor and auger are operating.

A geared motor is used to turn the auger. The geared motor may be located on the top or bottom of the auger. Figure 26–29 shows a bottom gear motor machine. The auger is in a vertical position and must have a shaft seal where the auger leaves the gear box. If the gear motor is on top of the auger, the shaft seal prevents gear oil from seeping into the ice, Figure 26–30. When the gear box is on the bottom of the auger, the shaft seal prevents water from entering the gear box, Figure 26–31.

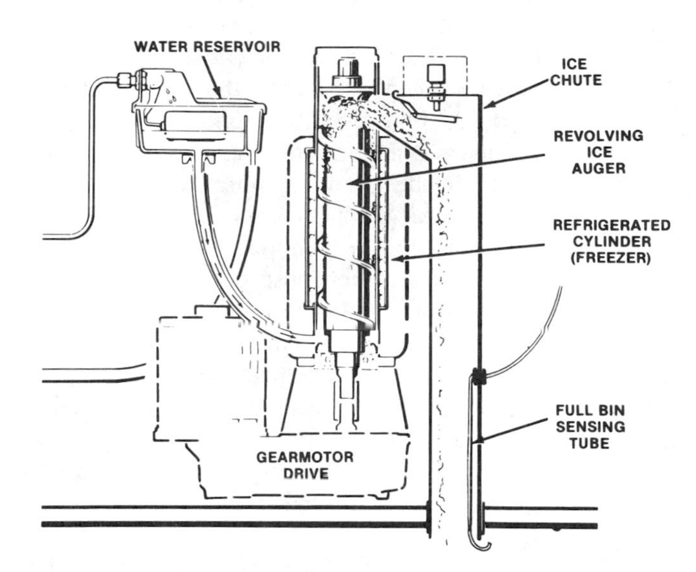

FIGURE 26–29 A flake ice maker, its evaporator, and auger. The auger turns and shaves the ice off the evaporator. The tolerances are very close. *Courtesy Scotsman*

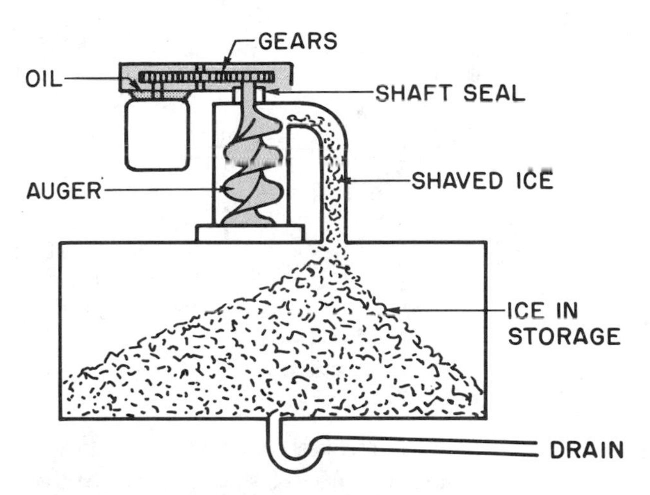

FIGURE 26–30 The auger is on top of the ice maker. The shaft seal prevents soil from seeping into the ice.

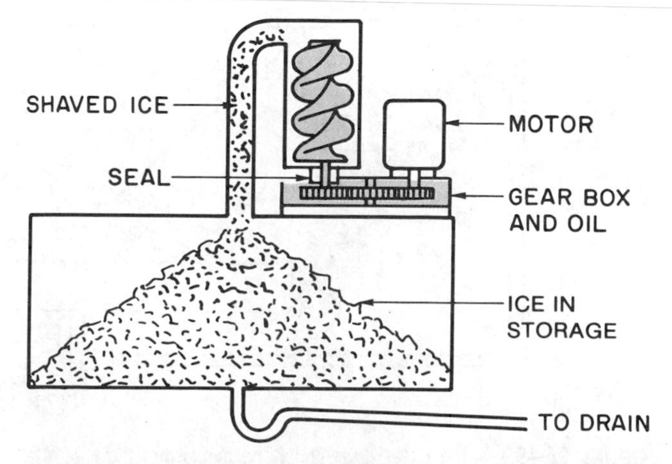

FIGURE 26–31 The gear box is on the bottom of the auger. A seal leak would allow water to enter the gear box.

The level of the water in the evaporator is determined by a float chamber located at the top of the evaporator. This float level is critical. If the level is too high, it may impose an extra load on the gear motor. If the water level is too high in the evaporator, the auger is doing more work. The motor driving the auger may have a manual reset overload that will shut the motor off and stop the compressor. When the motor reset button must be reset often, suspect the water level as the problem. Check the seat in the float chamber. A float that will not control the water level may allow water to pour over the top of the chamber during the off cycle. This would impose an extra load on the refrigeration capacity because it must refrigerate this extra water to the freezing point to make ice. Water overflowing the evaporator will drip into the ice bin and melt ice that is already made causing a loss of ice capacity, Figure 26–32.

If the level is too low, the ice quality may be too hard and the ice maker may not be efficient. In extreme cases, the auger may freeze to the evaporator. In this case the power of the motor and gears may turn the evaporator out of alignment.

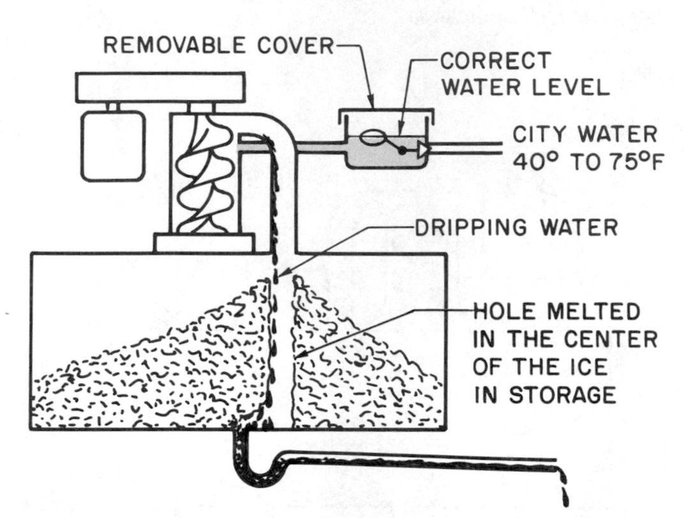

FIGURE 26–32 When the float level is too high, water will run over and melt ice that has been made. The fresh water must be refrigerated to the freezing condition after start-up.

26.26 MAKING CUBE ICE

Several types of cube ice are made in package ice makers using several methods. Cube ice should be clear, not cloudy. Clear ice is usually preferred for beverage cooling. Ice with a high mineral content or that is aerated (has minute air bubbles inside) will appear cloudy and is not as desirable to some people because of its appearance. Some of the design features of the cube ice maker results in ice that is very clear.

Flat Ice Cut to Cubes

Flat cube ice is made on a flat evaporator to a predetermined thickness. When harvest occurs, the ice sheet is slightly defrosted and melted loose from the evaporator. The ice sheet then falls or slides to a grid of cutter wires. These wires have a very low voltage, approximately 5 V depending on the manufacturer, and will provide enough heat to cut the ice into squares or triangular shapes. The ice can be 1/4 in. to 1/2 in. thick, depending on the length of time ice is made before defrost occurs.

The evaporator may be designed for water to flow on top of the plate or under the plate by means of a water distribution manifold. When the water flows under the plate, the surface tension of the water tends to hold the water to the plate as it flows under it, Figure 26–33. This evaporator must be kept clean or the water will not cling to the plate, falling into the ice bin below and melting ice.

Some plate-type ice machines are designed so that the water flow is over the plate. These use a water distribution system designed for this purpose. Some of these use a small motor and cam to determine the thickness of the ice and when defrost should occur, Figure 26–34.

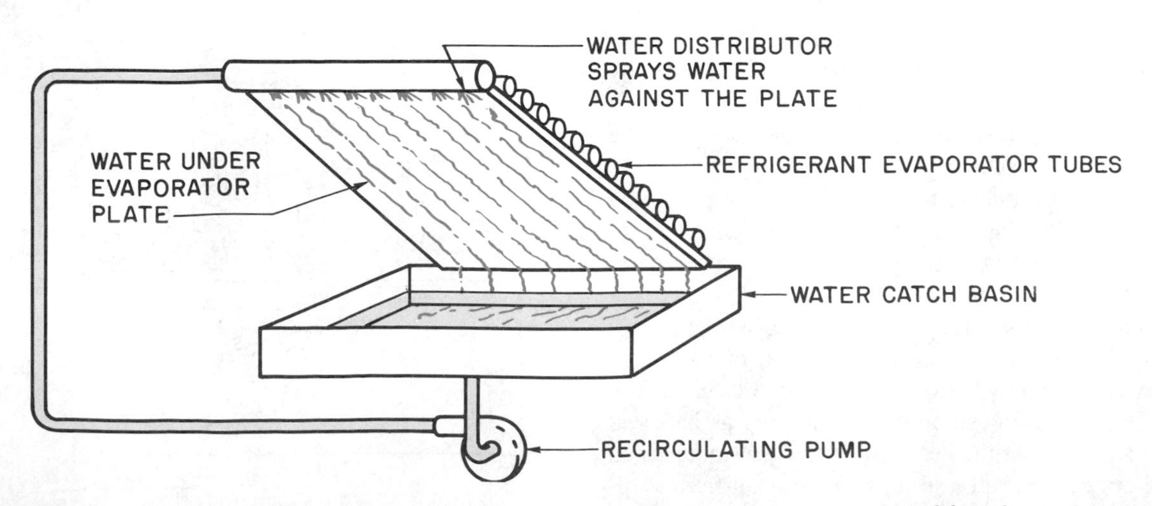

FIGURE 26–33 The water flows under the evaporator plate with this ice maker. This evaporator must be clean and free of mineral deposits or the water will not follow the plate. In this case it will fall into the ice bin below and melt ice already made. Then the makeup water will have to be refrigerated and the capacity to make ice will be reduced.

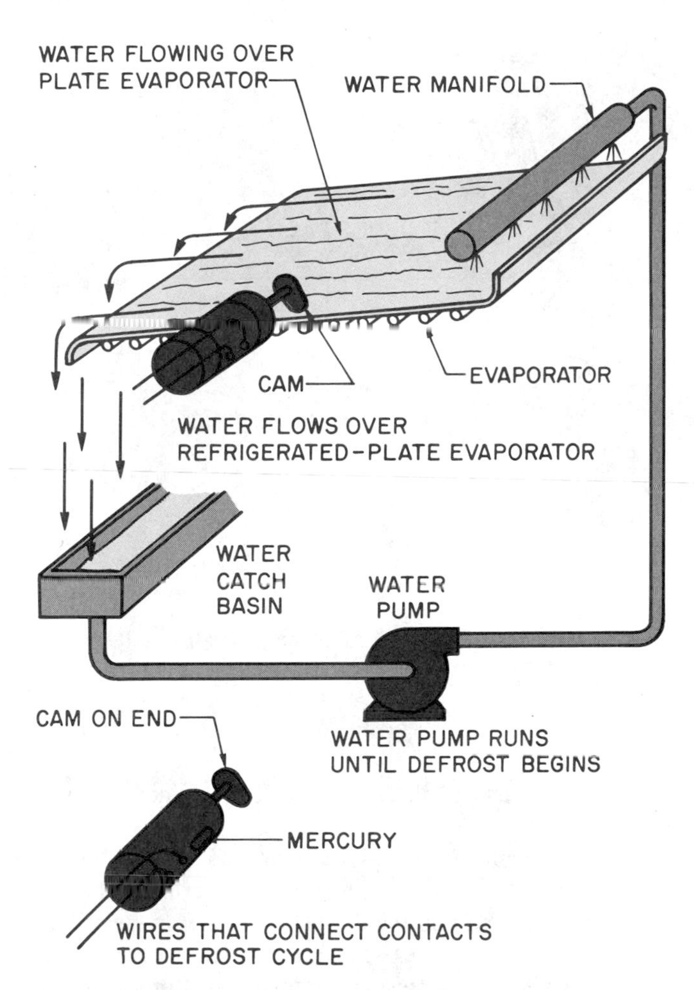

FIGURE 26–34 Sensors that control ice thickness in this machine. The ice thickness switch rotates when ice begins to form on the evaporator plate. The cam touches the ice during rotation (approximately 1 rpm). The switch has mercury contacts in the rear. As the cam touches the ice it causes the mercury to roll to the back. The contacts are made and defrost begins.

Water level is maintained in the water reservoir for recirculation by means of a float.

Cube Ice

Many methods are used to make cube ice. Some common ones are discussed in this text. Cube ice is made by flooding water over an evaporator with cups shaped like the cube desired. When the ice reaches the desired thickness, defrost or harvest occurs. The evaporator may be horizontal or vertical.

When the evaporator is horizontal, water is sprayed up into the cups where the desired cube is formed. Some of the water falls to the catch basin and is recirculated. When defrost occurs, the water is shut off and defrost allows the cubes to fall to the first level of the catch basin. The water wand with the spray heads wipes the cubes to a chute leading to the catch basin, Figure 26–35.

Vertical evaporators have ice cups designed to cause the ice to fall out of the cups during defrost. Water flows

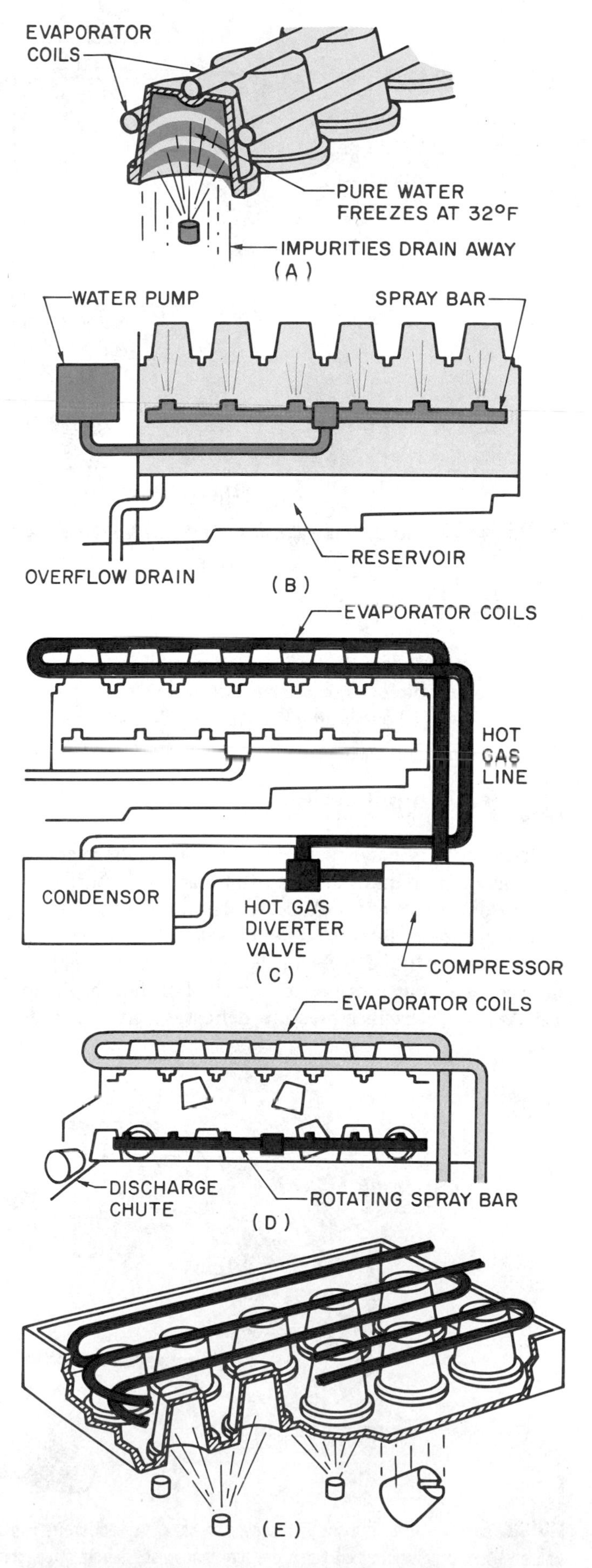

FIGURE 26–35 Ice cubes and the evaporator they are made on.

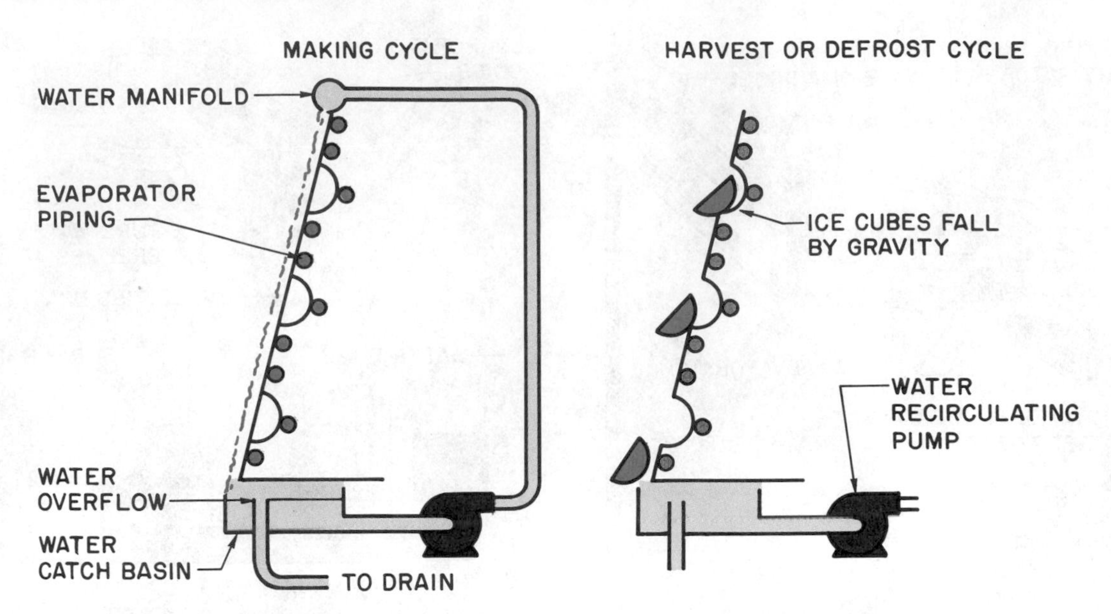

FIGURE 26–36 Vertical evaporator. Water flows over the cups during the make cycle. During harvest, cubes fall out of the cups to the bin below.

over these evaporators by means of gravity, like a waterfall. The evaporator piping is on the back of the evaporator plate, Figure 26–36. The cubes are caught during defrost in the ice bin.

26.27 MAKING CYLINDRICAL ICE

Cylindrical ice is made inside a tube within a tube evaporator. Water flows into the center tube and the refrigerant evaporator is in the outside tube. As the water flows through the center tube, it freezes on the outside walls of the tube. Toward the end of the cycle, the hole in the center begins to close and the pump pressure at the inlet of the tube begins to rise. At a predetermined pressure, defrost is started. The ice shoots out the end of the tube as a long cylinder, Figure 26–37.

The evaporator is normally wound into a perfect circle and the ice coming from it has a slight curve. As the ice leaves the evaporator during defrost, a breaker at the end of the tube will break the ice to length. Winding the evaporator in a circle allows a large evaporator to be formed into a small machine.

26.28 DEFROST

When solid-type ice is manufactured there must be some form of defrost, regardless of the type of ice maker. This is normally accomplished with hot refrigerant gas. The system

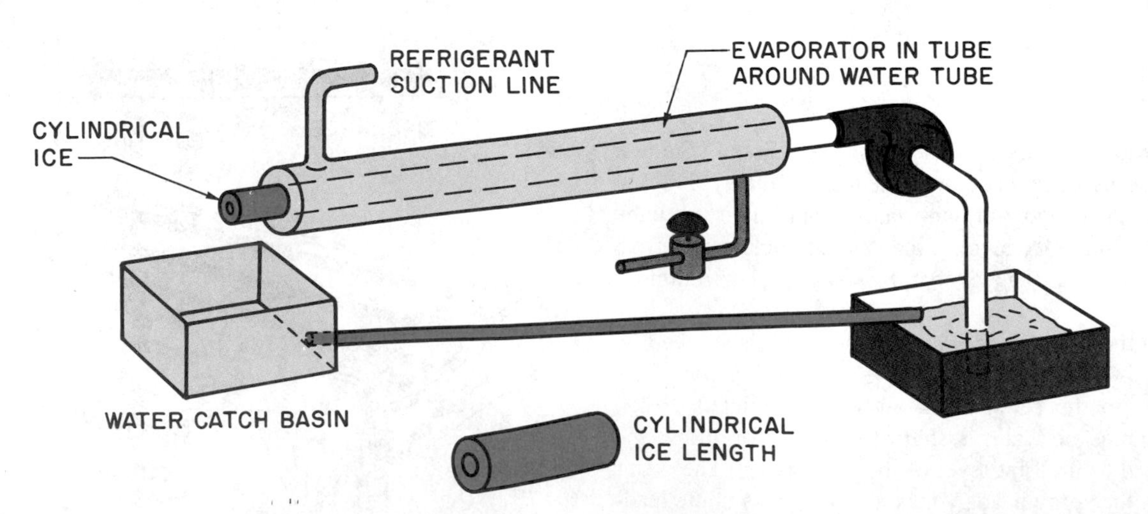

FIGURE 26–37 This drawing is a simulation of how cylindrical ice is made. As ice is formed on the inside of the tube, the hole in the ice becomes smaller. When a predetermined pump pressure is reached, defrost occurs and the ice shoots out the end, as from a gun. The ice is caught and broken to length, then moved to a bin. In the actual machine, the evaporator would normally be wound in a coil and the ice would have a slight curvature.

has a natural built-in supply of heat that may be distributed evenly over the evaporator. Hot gas is usually piped to the expansion valve outlet so that the evaporator circuit carries the hot gas, Figure 26–38.

As the defrost cycle begins, the water pump is turned off, the hot gas solenoid valve is energized (opened) and the condenser fan is deenergized (turned off).

NOTE: The water pump is not turned off during defrost in a cylindrical ice maker. The water pressure is used to force the ice out of the tube within a tube evaporator.

The water is stopped to keep from transferring any heat to the water sump from the hot gas. When the water pump is turned off, the excess water from the pumping system will normally cause an overflow in the sump. This is planned so that some of the mineral buildup in the sump water will move down the drain. Entering supply water will replace a portion of the sump water. If there is no overflow, mineral deposits will concentrate in the water and some will deposit on the evaporator.

As the hot gas solenoid is energized, hot gas will flow to the low-pressure areas in the evaporator and begin defrost. Prolonged defrost is not economical and will reduce the ice-making capacity of the machine. Different manufacturers use different methods for the duration of the defrost. Time, temperature, and pressure are some common methods. These may be accomplished with mechanical-electrical or even electronic controls.

The condenser fan is stopped during defrost to help keep the head pressure up so there will be enough heat to complete the defrost cycle.

26.29 WATER QUALITY FOR ICE MAKERS

The water used to make ice must be of the best quality or it will cause problems in the future. Mineral deposits will form on the evaporator and water piping. When the deposits form on the evaporator, water may not flow over the evaporator plates uniformly when solid or cube ice is made. The longer the unit runs with these deposits, the harder they will be to remove.

Many of the cube makers have a clean cycle to circulate water with chemicals added.

SAFETY PRECAUTION: *All of these chemicals are hazardous to humans. Follow the manufacturer's instructions. The best time to circulate chemicals is when there is no ice in the bin, so no contamination will seep to the stored ice. Because it will be rare to find an ice maker empty, you may need to move the ice from the bin to storage. Then clean and rinse the bin according to the manufacturer's directions.*

Many ice makers require filtration for the entering supply water. The manufacturer's directions must be followed. When a used ice maker is purchased, the directions may be lost. You should write to the manufacturer for copies of the operating and maintenance instructions and follow them.

There are several grades of filtration. The most basic will remove any sand, rust particles, or other large particles. The more sophisticated types will even remove minerals suspended in the water. Some localities have a very high mineral content and if this is the case, these must be removed or they will deposit on the evaporator and cause mineral build-up problems. During defrost, the ice may not turn loose from the evaporator until much of it has melted. Some ice makers use the conduction characteristics of the ice as a means for signaling for defrost. When the ice builds up to a point, feeler contacts touch the ice. The ice must have some electrical conductivity for defrost to start. If the water is too pure because too many minerals have been removed, conductivity will not be correct and defrost will

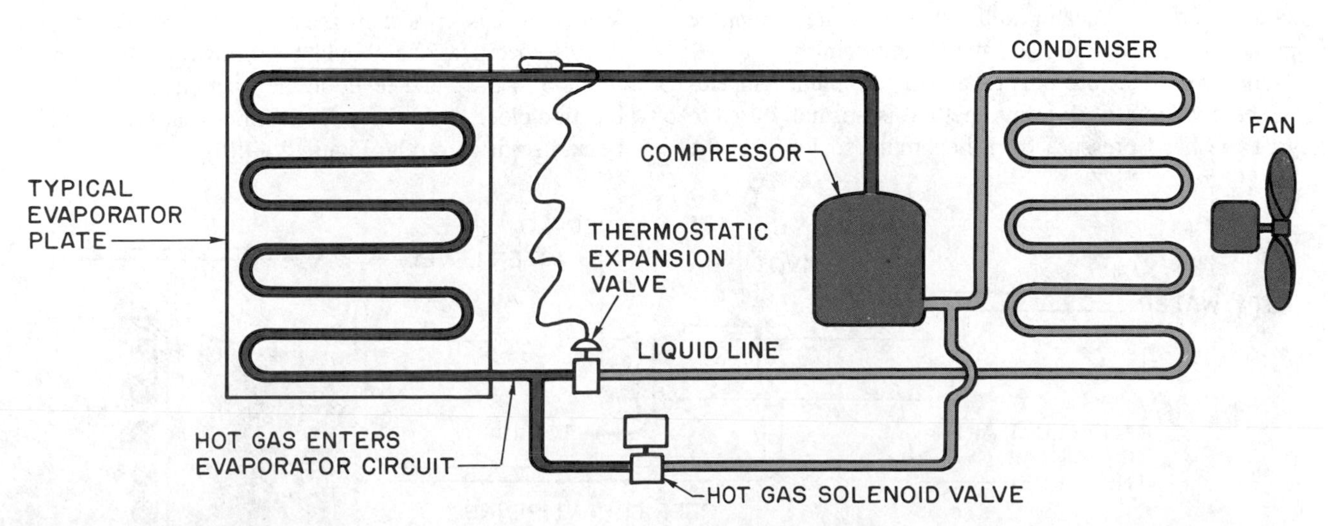

FIGURE 26–38 Hot gas will move to the cold low-pressure evaporator and defrost the ice. The defrost solenoid is energized and the condenser fan deenergized during a typical defrost.

not occur. These units may be recognized because they have electronic circuit boards for defrost. A pinch of salt may be added to the circulating water to provide conductivity in the water and start defrost when the evaporator is frozen solid. If a defrost cycle occurs after adding salt, it indicates that the water is too pure. Special controls can be obtained from the ice-maker manufacturer so that defrost will occur under pure water conditions.

The temperature of the entering water will vary from the water main (or well) temperature in the winter to the main temperature in the summer. Well water temperature should be constant all year. Well water temperature may be as low as 40°F. Water in city mains may be as low as 40°F in the winter to 80°F in the summer. The ice maker capacity will vary from summer to winter because of this temperature variable, resulting in less ice in the summer when you may need it the most.

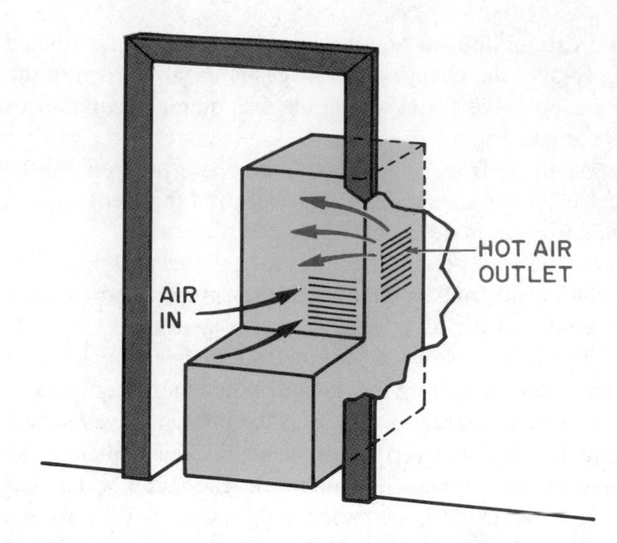

FIGURE 26–39 Ice machine located in an alcove with condenser air recirculating.

26.30 PACKAGE ICE MACHINE LOCATION

Most package ice machines are designed to be located in ambient temperatures of 40°F to 115°F. Follow the manufacturer's recommendations. If the ambient is too cold for the ice maker, freezing may occur and damage the float or other components where water is left standing. If it is too hot, compressor failures and low ice capacities may occur. The ambient conditions under which the ice maker must operate are important.

Winter Operation

A machine located outdoors at a motel may not make ice below about 40°F ambient temperature because the thermostat in the bin is set at about 32°F. With this setting when ice touches the thermostat sensor, it will shut off the machine. The combination of 40°F outside and some ice in the bin will normally satisfy the thermostat. When the machine is located outside, the water should be shut off when the outside temperature approaches freezing. The water should be drained to prevent freezing and damage to the pump, the evaporator on some units, and the float assembly.

When machines are outside and the ambient temperature is below about 65°F, some method is normally used to prevent the head pressure from becoming so low that the expansion device will not feed enough refrigerant to the evaporator. Low ambient control causing fan cycling is a common method used.

The correct location of an ice machine indoors is critical because of the air flow across the condenser. Air cannot be allowed to recirculate across the condenser or high head pressures will occur. This will reduce the capacity and may in some cases harm the compressor, Figure 26–39.

TROUBLESHOOTING ICE MAKERS

When an ice maker is not functioning properly, the problem can usually be found in the water circuit, the refrigeration circuit, or the electrical circuit.

Water Circuit Problems

Water level is critical in most ice makers. The manufacturer's specifications should be followed. Most manufacturers post the water level on the float chamber or at some location close to the float. Determine what the level should be and set it correctly, Figure 26–40.

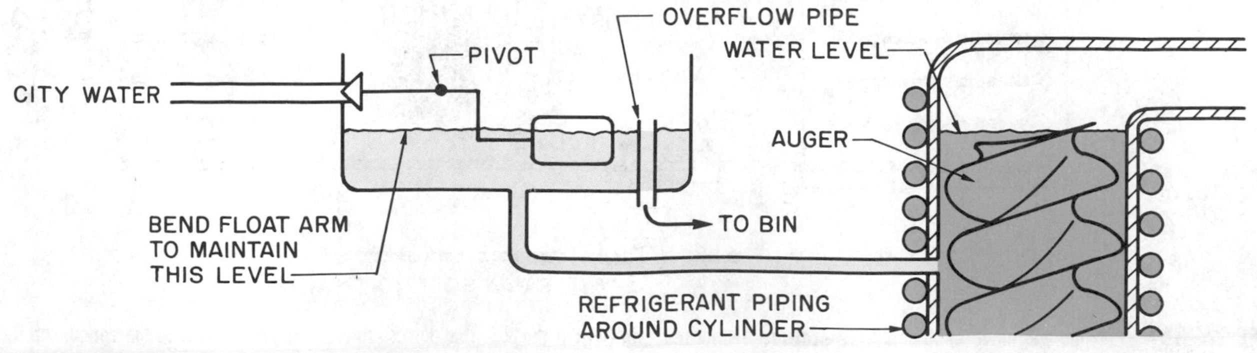

FIGURE 26–40 Maintain the correct float level.

The float valve has a soft seat, usually made of neoprene. This seat may become worn and a new one may be required. A float seat is inexpensive compared to an ice maker operating inefficiently for any period of time. When a float does not seal the incoming water, water at the supply temperature flows through the system. This water must be refrigerated to the freezing temperature and it may also melt ice that has already been made. A temporary repair may be to turn the float seat over if it can be removed from its holder, Figure 26–41.

The best time to determine whether or not a float is sealing correctly is when the ice maker has no ice. Remember, when there is ice in the bin, it is constantly melting. You cannot tell if the water in the drain is from melting ice or a leaking float. When there is no ice in the bin, the only water in the drain would be coming from a leaking float, Figure 26–42.

Water circulation over evaporators that make cube ice is accomplished by a water circulating system that is carefully designed and tested by the manufacturer. You should make every effort to understand what the manufacturer has intended before you modify anything. It is not unusual for a technician to try various methods to obtain the correct water flow over the evaporator when the manufacturer's manual would explain the correct procedure in detail. Don't overlook something as simple as a dirty evaporator. If it must be cleaned, check the manufacturer's suggestions.

The quality of the water entering the machine must be at least as good as the manufacturer recommends. Check the filter system if the water seems to have too many minerals. A water treatment company may be contacted to help you determine the water condition and the possible correction procedures.

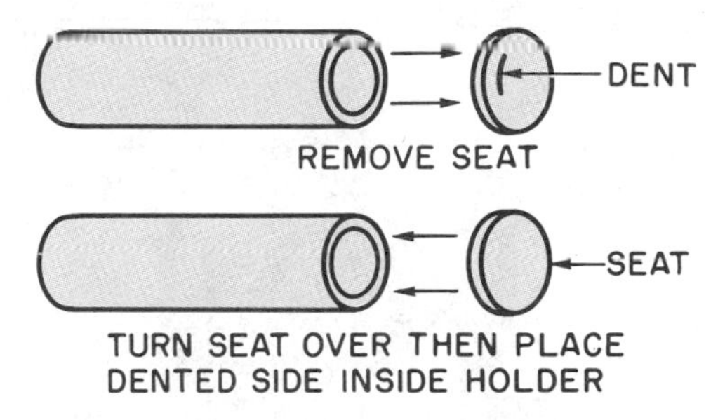

FIGURE 26–41 Sometimes a float seat may be reversed or turned over with the good side out for a temporary repair.

FIGURE 26–42 The ice machine storage bin is dry. There is no ice, but water is dripping out the drain. The float valve is leaking.

Refrigeration Circuit Problems

Refrigeration circuit problems are either high-pressure or low-pressure side problems.

HIGH-PRESSURE SIDE PROBLEMS. High-pressure side problems can be caused by:

- poor air circulation over the condenser
- recirculated condenser air
- a dirty condenser
- a defective fan motor
- overcharge of refrigerant

Another high-pressure side problem is a result of trying to operate a system when the ambient temperature is below the design temperature for the unit. Low head pressure can cause low suction pressure, particularly in systems with capillary tube metering devices. Owners tend to try to place the machines outside and expect them to perform all year. The design conditions the manufacturer recommends should be followed.

LOW-PRESSURE SIDE PROBLEMS. Low-pressure side problems can be caused by:

- low refrigerant charge
- incorrect water flow
- mineral deposits on the water side of the evaporator
- restricted liquid line or drier
- defective metering device, thermostatic expansion valve
- moisture in the refrigerant
- inefficient compressor

Low-side problems will show a pressure that is too high or too low as indicated on the refrigeration gages. When the pressure is too low, you should suspect a starved evaporator. When the ice maker makes solid-type ice, a starved evaporator will often cause irregular ice patterns on the evaporator. The ice will become thicker where the evaporator surface is colder.

When the pressure is too high, you should suspect an expansion device that is feeding too much refrigerant, a defective thermostatic expansion valve, or a compressor that is not pumping to capacity. When the expansion valve is overfeeding, the compressor will normally be cold from the flooding refrigerant, Figure 26–43.

A stuck water float can cause excess load if the entering water is warmer than normal, Figure 26–44. As mentioned earlier, it is hard to detect a leaking float. Look for the correct level in the float chamber. If adjusting the float does not help, the seat is probably defective.

Electrical Problems

A defrost problem may look like a low-pressure problem if the unit fails to defrost at the correct time. Heavy amounts of ice may build up. Remember, defrost is only used on machines that make cube ice. When heavy amounts of ice are found on a machine, look for the method of defrost. The machine may have a forced defrost cycle to allow you to clear the evaporator and start again. Some manufacturers may suggest that you not cause the machine to defrost or shut the machine off, that you should wait for the normal defrost cycle. This is so you can determine what type of problem caused the condition. If you arrive and find the evaporator frozen solid and force a defrost, you may not find the problem because it may not recur in the next cycle.

Line voltage problems may be detected with a voltmeter. The unit nameplate will tell you what the operating voltage for the machine should be. Make sure the machine is within the correct voltage range. The plug on the end of the power cord may become damaged or overheated due to a poor connection. Examine it while the machine is running by touching it. If it is warmer than body temperature, it is not properly conducting power to the machine, Figure 26–45.

Once the power is wired to the junction box in the ice maker, it goes to the control circuit that operates the machine. Normally, the control voltage is the same as the line voltage. Larger machines may use relays to energize contactors for starting the compressor. Figure 26–46 shows typical diagrams. It is suggested that you study these diagrams. You may wish to copy these diagrams and use colored pencils to trace the various circuits.

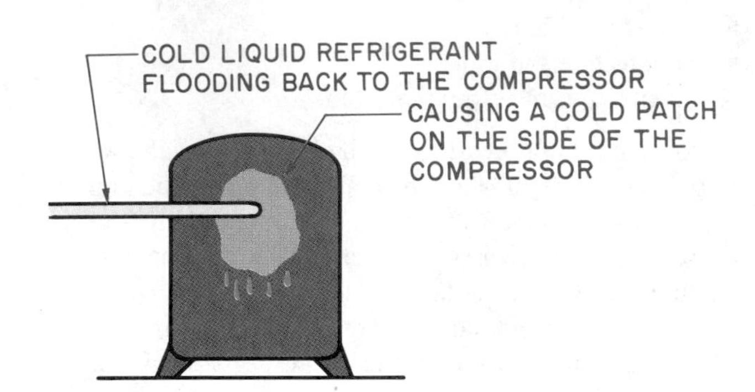

FIGURE 26–43 The expansion valve is allowing refrigerant to flood back to the compressor.

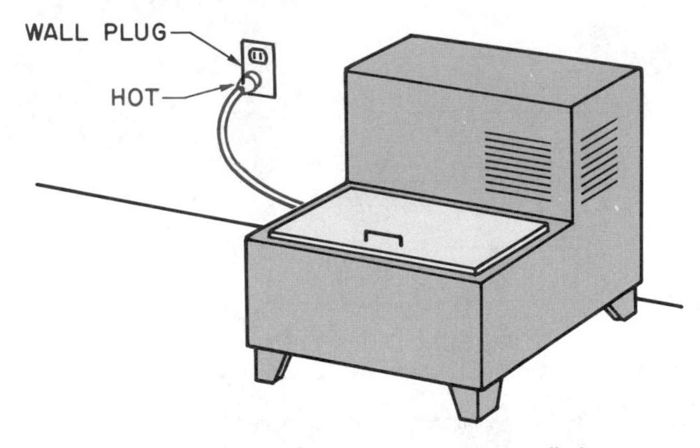

FIGURE 26–45 This unit has a poor connection at wall plug.

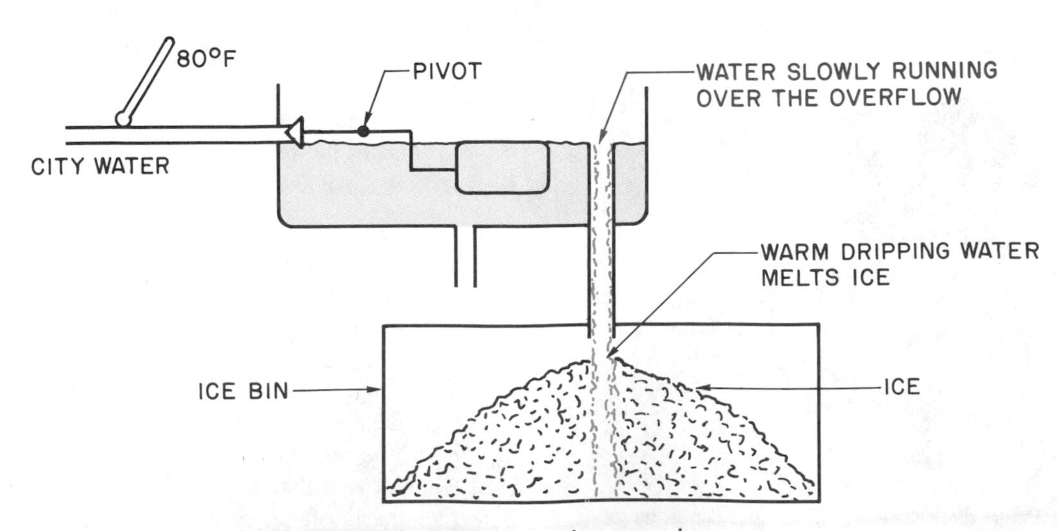

FIGURE 26–44 A float that does not seal will reduce the usable ice a machine can make.

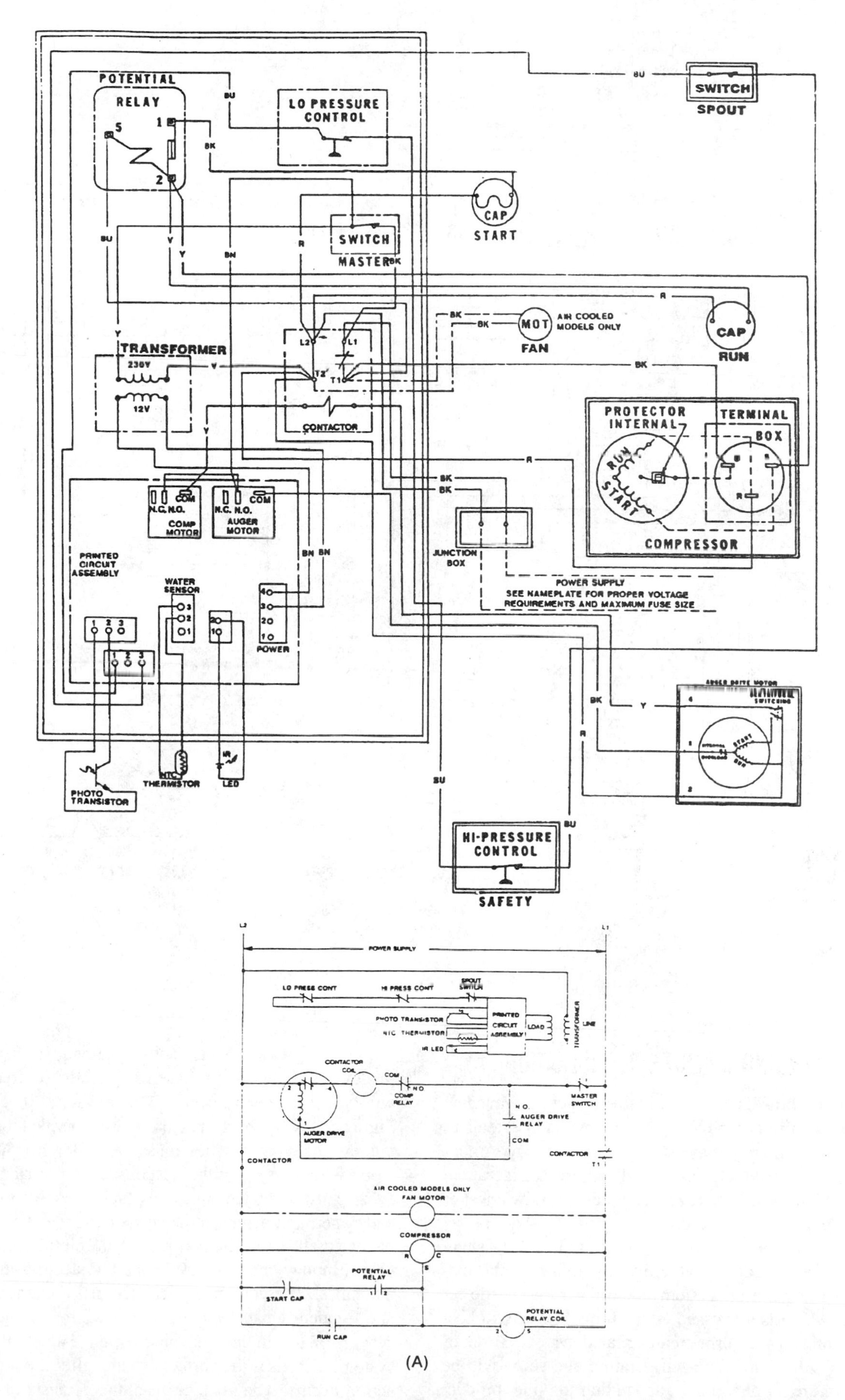

(A)

FIGURE 26–46 Typical wiring diagrams for ice makers. (A) Flaker. (B) Cuber. *Courtesy Scotsman*

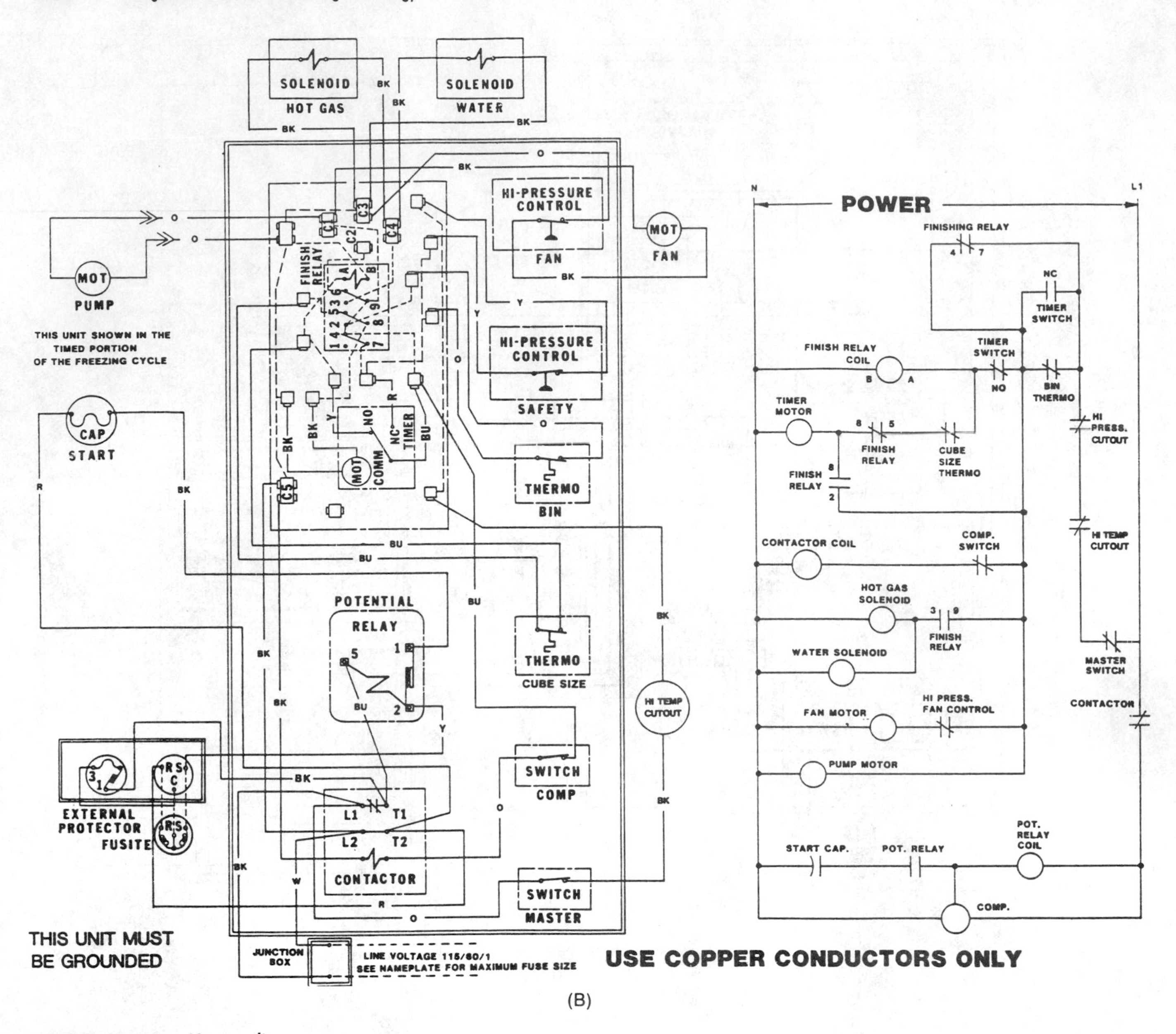

(B)

FIGURE 26–46 *(Continued)*

26.32 VENDING MACHINE REFRIGERATION

A vending machine is a self-contained package machine that dispenses various products when money is inserted into the pay slot. The money may be either coins or paper money. Some of the products dispensed require refrigeration. These may be beverages (frozen or liquid), sandwiches, or frozen products such as ice cream, Figure 26–47. The refrigeration portion of the vending machine is only a small part of the mechanical and electrical operation of the machine. This refrigeration system is similar to domestic refrigeration, which is discussed later in this text. It will be a fractional horsepower hermetically sealed system. Some of the systems will be medium-temperature and some will be low-temperature applications, depending on the product dispensed.

Vending machines are self-contained, so they are plug-in devices, like a household refrigerator or freezer. They will typically operate from a 20-A electrical plug-in circuit, Figure 26–48. These machines are complex because the money changing devices must be built into the machine along with the system that dispenses the product. The money dispensing devices must be able to receive money, coin and paper, and dispense the correct change. This is accomplished with sophisticated electronic circuits and sensors. Often, money changers are located adjacent to the vending machines. These money changers make change for coins and frequently paper money to coins. The product handling system of a vending machine, often called the conveyer system, dispenses the correct product after the money transaction occurs. The vending machine consists of the refrigeration system, the money changing system, and the product

FIGURE 26–47 (A) Cold drink dispenser. (B) Sandwich and snack dispenser. (C) Ice cream vending machine. *Courtesy Rowe International Inc.*

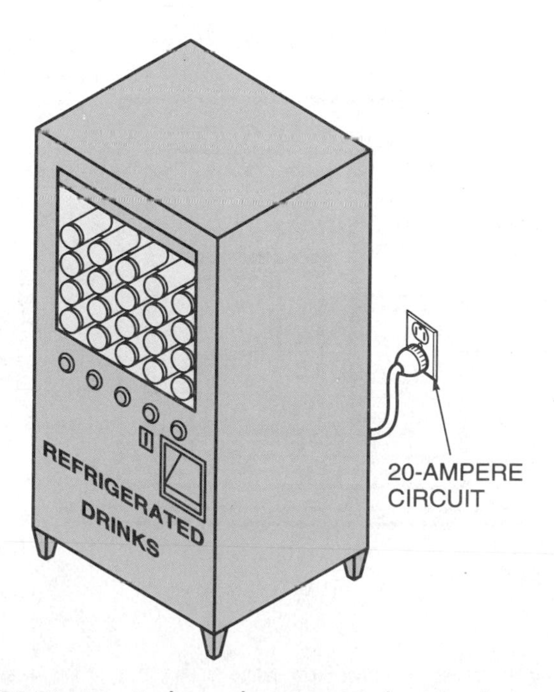

FIGURE 26–48 Vending machines use 20-A electrical outlets.

dispensing system. Each system in the machine can be intricate. This text discusses some of the basic refrigeration systems used in vending. Further training in the other systems of the machine may be obtained in factory schools and from manufacturers' literature.

BEVERAGE COOLING. Beverage coolers are used to cool either canned or bulk beverages. They are usually designed to dispense the product when money is fed into the machine. Medium-temperature refrigeration systems are used because the beverages must be maintained above freezing. When the beverages are in cans or bottles, the evaporators can be sized small and operate at low temperatures because evaporation of the product does not have to be considered.

The evaporators in vending machines operate at below freezing so they must have a defrost cycle. Typically, the air in the vending machine is warm enough for off-cycle defrost. This may have to be timed off-cycle defrost in some cases because the machine may not have enough off time. The timed off-cycle defrost can be controlled using the electronic circuit board that is typically furnished with the vending machine. When defrost occurs, the moisture from the coil usually runs through a liquid trap to prevent atmosphere from entering the refrigerated area and then to a pan where it is evaporated using either the hot gas line or air from the condenser, Figure 26–49. The condensate pan can be a very dirty place where rodents and insects can be attracted because any drink can or bottle that may break will drain to the condensate pan and be evaporated, leaving behind sugar and flavoring from the drink.

In a hot climate, the drinks may be hot enough to keep the compressor running all the time if loaded off a truck in the sun to the machine. It is a good idea to precool the drinks before adding them to the machine. For example, a vending machine located close to a medium-temperature walk-in cooler is a good idea. The drinks can be brought into the area and cooled from the temperature on the back of the delivery truck to room temperature or to walk-in cooler temperature, then moved to the vending machine, Figure 26–50. This reduces the load on the vending machine refrigeration system and ensures cold drinks when needed. Some vending machines have a holding area in the bottom of the machine where drinks may be stored and chilled before depositing them in the dispensing racks. This again helps to ensure the customer of cold drinks when desired.

In vending machines the drinks are usually stacked on top of each other and are dispensed from the bottom. When the money is placed in the slot, the drink drops out by gravity. Change for any additional money may be dispensed also. When the machine is loaded, the drinks are loaded at the top and if they are warmer than the machine's refrigerated set point, they have time to cool before dropping out the vending slot, Figure 26–51. This also serves as a method for rotating the stock because the first cans or bottles dropped in the slot will be the first out.

Because space is important in vending machines, they use air-cooled condensers and forced-draft evaporators with capillary tube metering devices, Figure 26–52. The evaporator fan blows the cold air over the top of the drinks and it returns back to the evaporator at the bottom of the box.

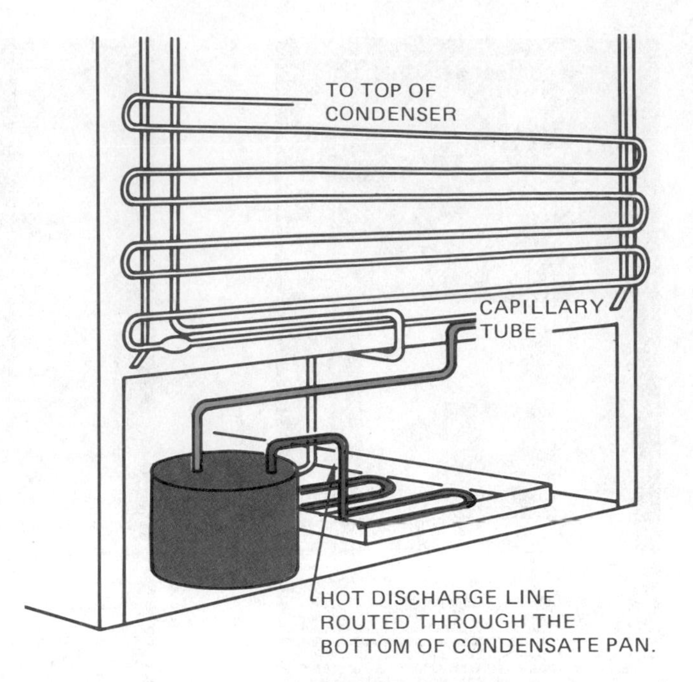

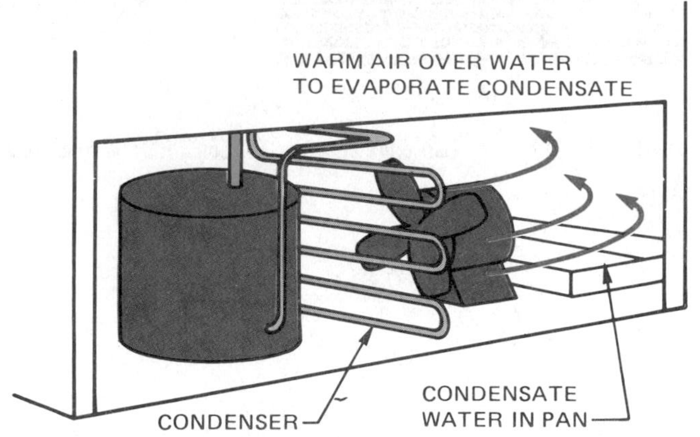

FIGURE 26–49 Evaporating condensate.

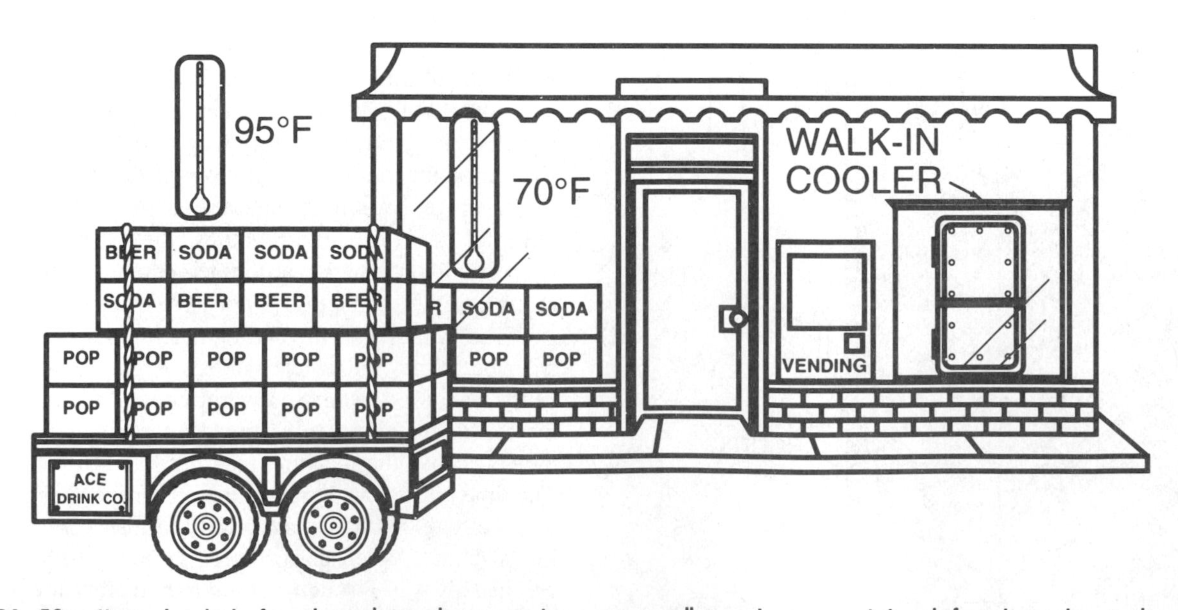

FIGURE 26–50 Moving hot drinks from the truck into the store and even into a walk-in cooler to precool them before placing them in the vending machine will take some of the load off the refrigeration system in the vending machine.

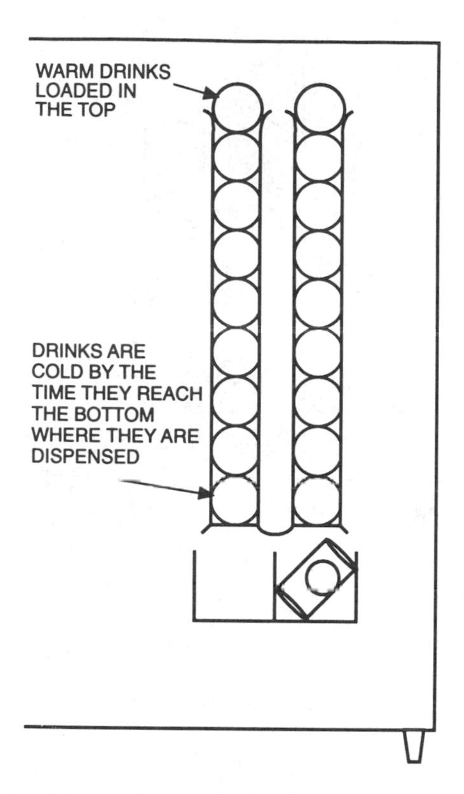

FIGURE 26–51 Loading a canned drink dispensing machine.

Because a vending machine may be located outside or in a cool location, the condenser will typically have a low ambient control to cycle the condenser fan to keep the head pressure high enough so the capillary tube will feed refrigerant correctly to the evaporator. This can be accomplished with either pressure control or liquid line temperature control, Figure 26–53.

These systems are much like a household refrigerator using a very small hermetic compressor. Service for the refrigeration system is basically the same as with a household refrigerator. You may refer to the unit on domestic refrigerators for more information on the service of small systems.

Other beverage coolers may dispense refrigerated liquids and ice, such as soft drinks with a cup full of ice. The dispensed liquid should be refrigerated and dispensed with

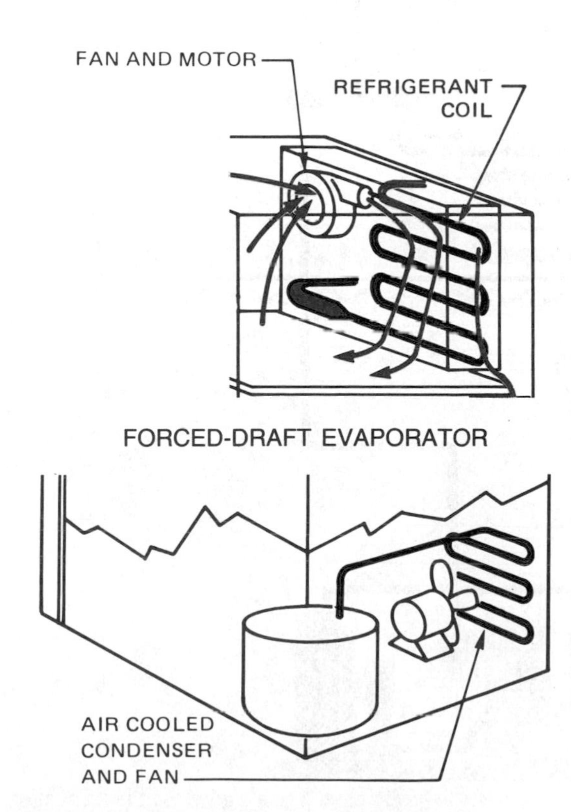

FIGURE 26–52 Forced-draft evaporator and condenser.

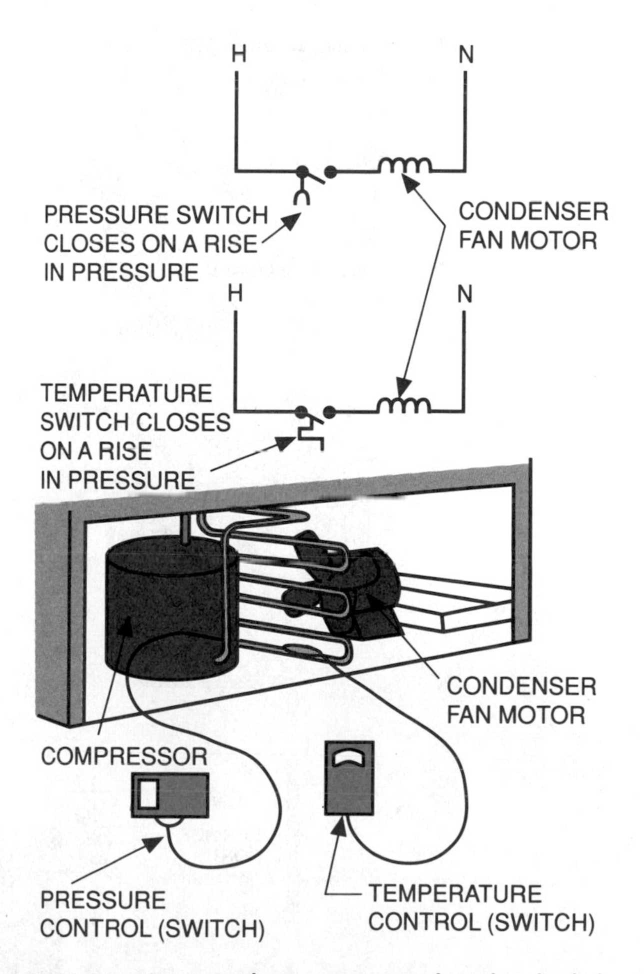

FIGURE 26–53 Use either a pressure control piped to the discharge line or a temperature sensor strapped on the liquid line to cycle the condenser fan motor for low ambient control.

the ice, so the drink is ready to drink when dispensed. If the liquid is dispensed warm and into ice, it would take a few minutes to become cool. Liquids at various temperatures would also require varying amounts of ice to reach the correct temperature and the quality of the drink would be hard to maintain.

The refrigeration system of some machines has a dual purpose: to make ice and to prechill the entering water. This is accomplished with two evaporators and one condensing unit built into the vending machine. A three-way valve in the liquid line can be used to direct the liquid to either of the evaporators, Figure 26–54.

The ice maker used in the typical vending machine may be a flake ice maker such as explained earlier in this unit. It would have an evaporator around a chamber with an auger inside. As the ice is made, the auger scrapes it off the evaporator surface and pushes it over into a bin, Figure 26–55. A bin level switch will shut the compressor off or switch the compressor operation to the ice bank evaporator, depending on the demand on the machine. Because water is the big percentage of the drink's content, the water is prechilled before mixing in the consumer's container (usually a

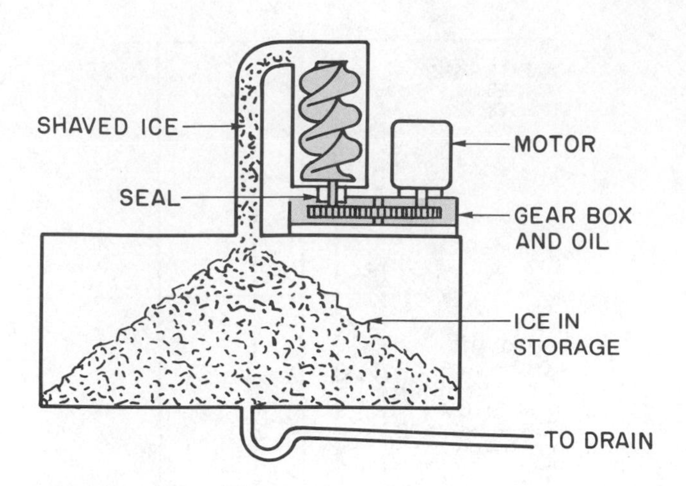

FIGURE 26–55 Flake ice maker.

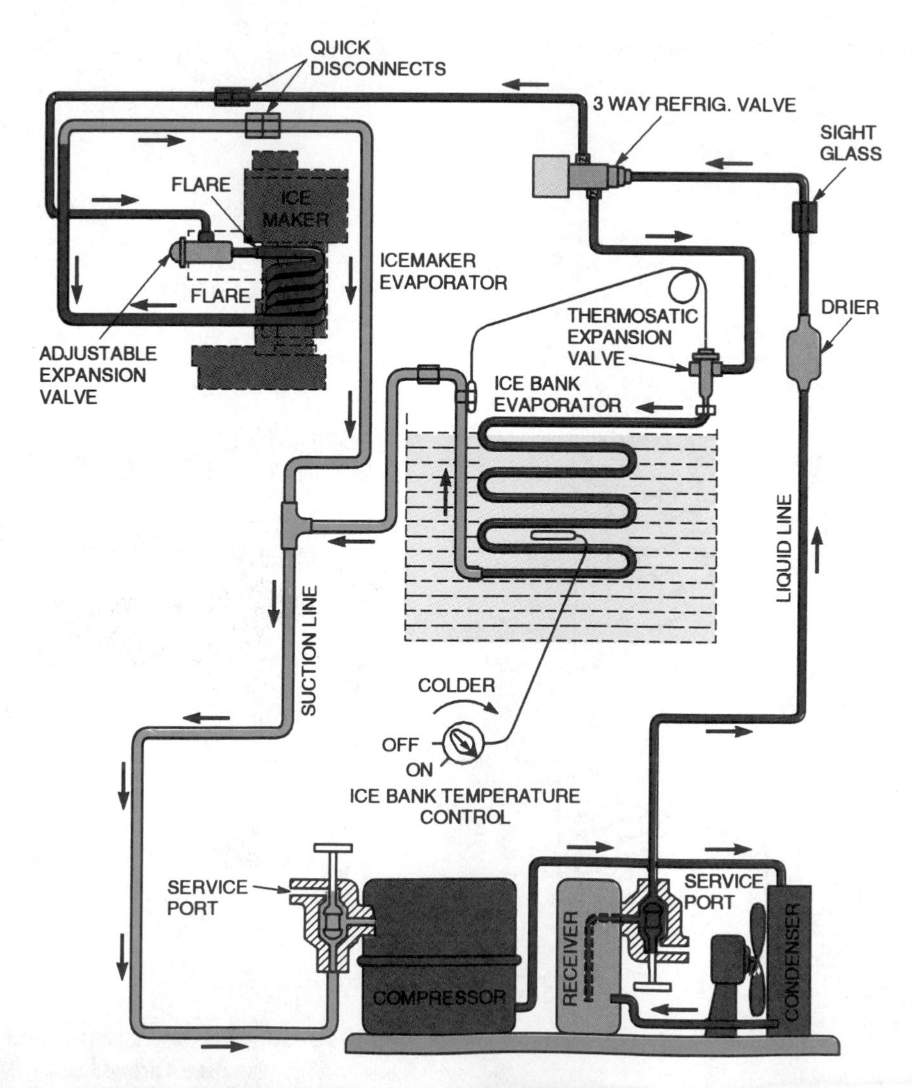

FIGURE 26–54 Ice maker and ice bank evaporators. *Courtesy Rowe International Inc.*

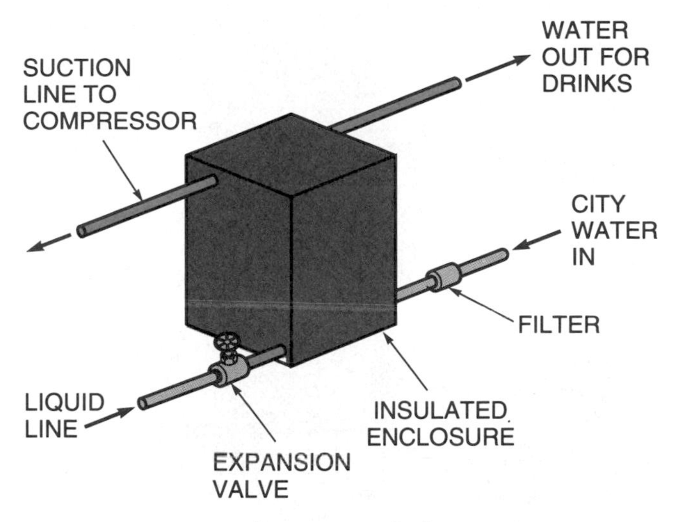

FIGURE 26–56 Precooling evaporator for beverage dispenser.

paper or plastic cup). The water may be prechilled in two ways. One is a direct heat exchange with the refrigerant, Figure 26–56. Another method is to use a water bath evaporator with an ice bank. The entering water piping passes through the water bath heat exchange, which has circulating water in a tank. The evaporator is fastened to the outside wall of the tank and ice can accumulate on this evaporator plate, which gives the system reserve capacity for times of heavy use, Figure 26–57. When there is heavy use, the compressor can be used to make ice, which will always ensure a cold drink. The ice-making portion of the machine has first choice for use of the compressor. The ice bank has second choice.

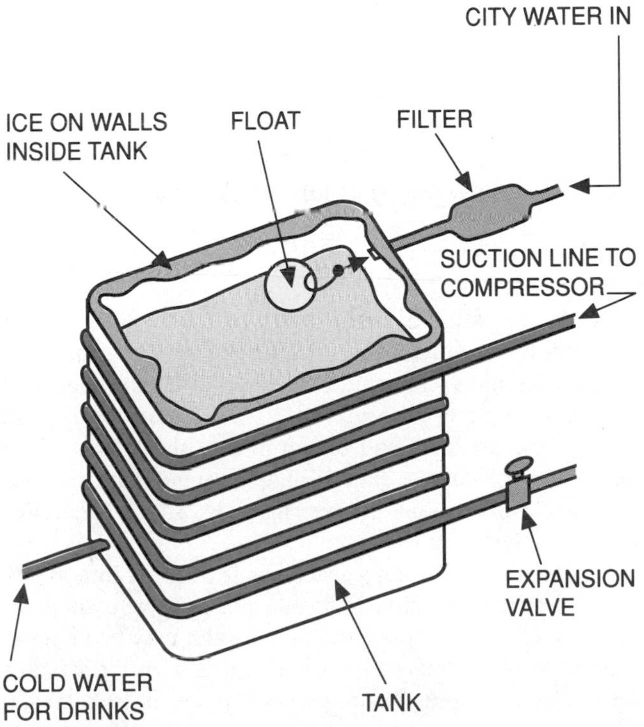

CITY WATER IN
ICE ON WALLS
INSIDE TANK
FLOAT
FILTER
SUCTION LINE TO
COMPRESSOR
EXPANSION
VALVE
COLD WATER
FOR DRINKS
TANK

FIGURE 26–57 Ice bank precooler has some reserve capacity in the frozen ice.

Carbonation of cold drinks is accomplished by injecting $C0_2$ into the drink from a small cylinder inside the vending machine. The $C0_2$ cylinder has a regulator to reduce the pressure of the cylinder contents as it is injected into the water while filling the cup.

SANDWICH AND OTHER PRODUCTS VENDING MACHINES. There are other products that must be refrigerated before being dispensed at medium temperature. Sandwiches and other items may be located in slots in the vending machine and dispensed like canned drinks. These machines typically have a transparent front so the customer can see the product before purchase, Figure 26–58.

Medium-temperature vending machines may have a totally self-contained refrigeration system that may be removed from the machine, Figure 26–59. This can be handy from a service standpoint. If there is a problem, the complete system can be changed out and taken to the workbench for service.

FIGURE 26–58 Food vending machine with glass front for viewing the product.

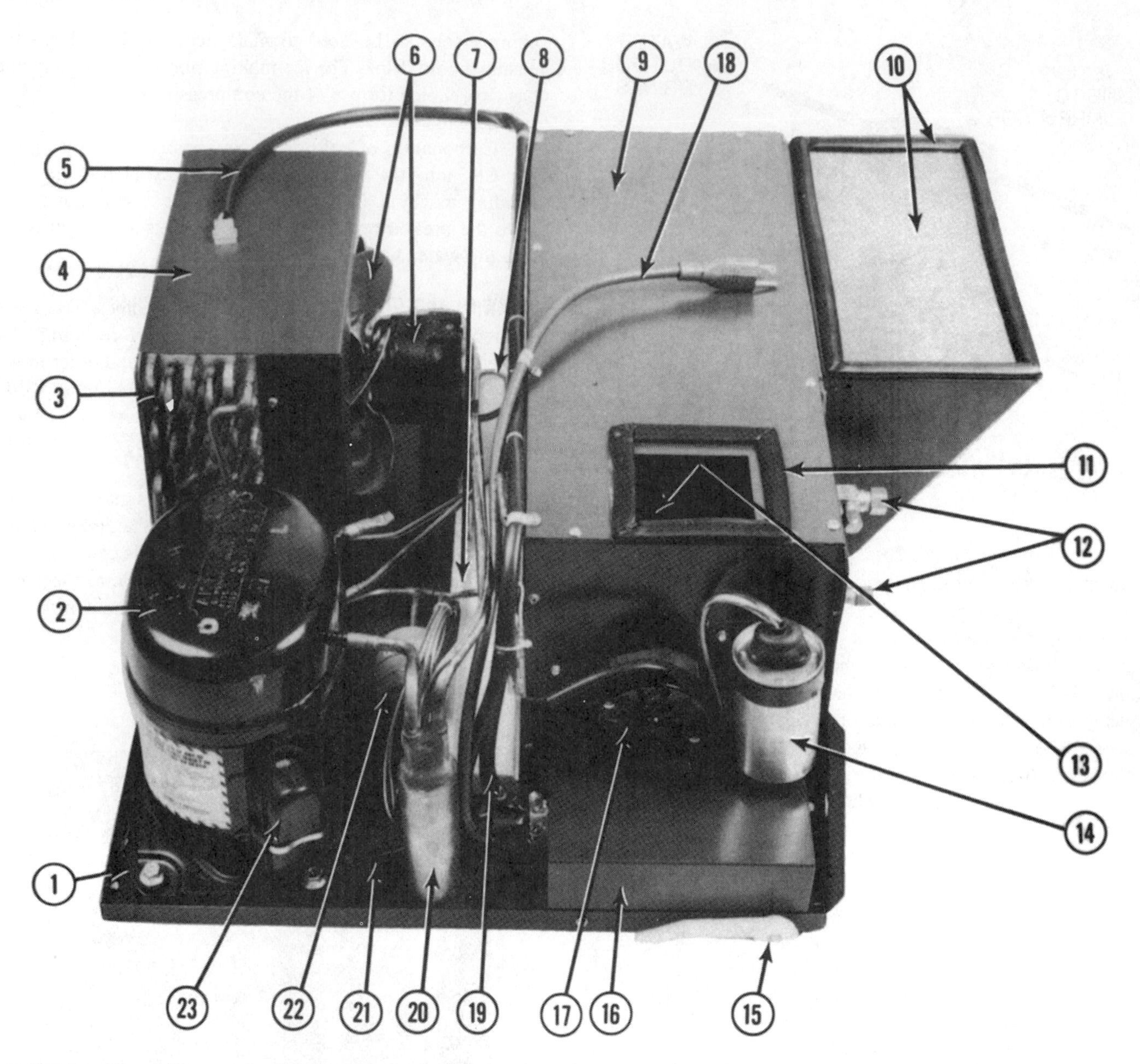

FIGURE 26–59 Self-contained refrigeration portion of vending machine. This unit may be easily replaced in the field and taken to a shop for repair. *Courtesy Rowe International Inc.*

Other machines may dispense frozen products, such as ice cream. These machines must operate at about 0°F to maintain the food frozen hard and are considered low-temperature applications. These machines may have plate-type evaporators that cover the inside of the product section of the machine, Figure 26–60. When a large evaporator is used, automatic defrost is not practical, so defrost is handled when the machine is serviced for routine preventive maintenance.

Low-temperature machines will often have a shield inside that keeps the ambient air from loading the evaporator with moisture while the machine is being filled with product. This prevents excessive frost buildup on the evaporator.

Most vending machines have what is known as a health switch. This is a thermostat that alerts the operator that the machine is operating above the safe operating range for a long enough period of time that the product quality may be affected. For example, if a vending machine is dispensing sandwiches and a temperature of above 45°F is reached for a particular length of time, usually 30 min, bacteria may grow in the food products and be harmful to the health of the customers. Electronic circuits and warning lights alert the operator. The machine lighting may also shut off, alerting the public not to buy.

Sanitation of vending machines is of great importance for all vending operations. The machine must be maintained in a clean fashion or the customers' health may be affected. The health department may have some say in the vending operation if these machines are not kept clean. Regular service of all moving parts, defrost of the refrigeration evaporator, and machine cleanup must be done on a regular basis.

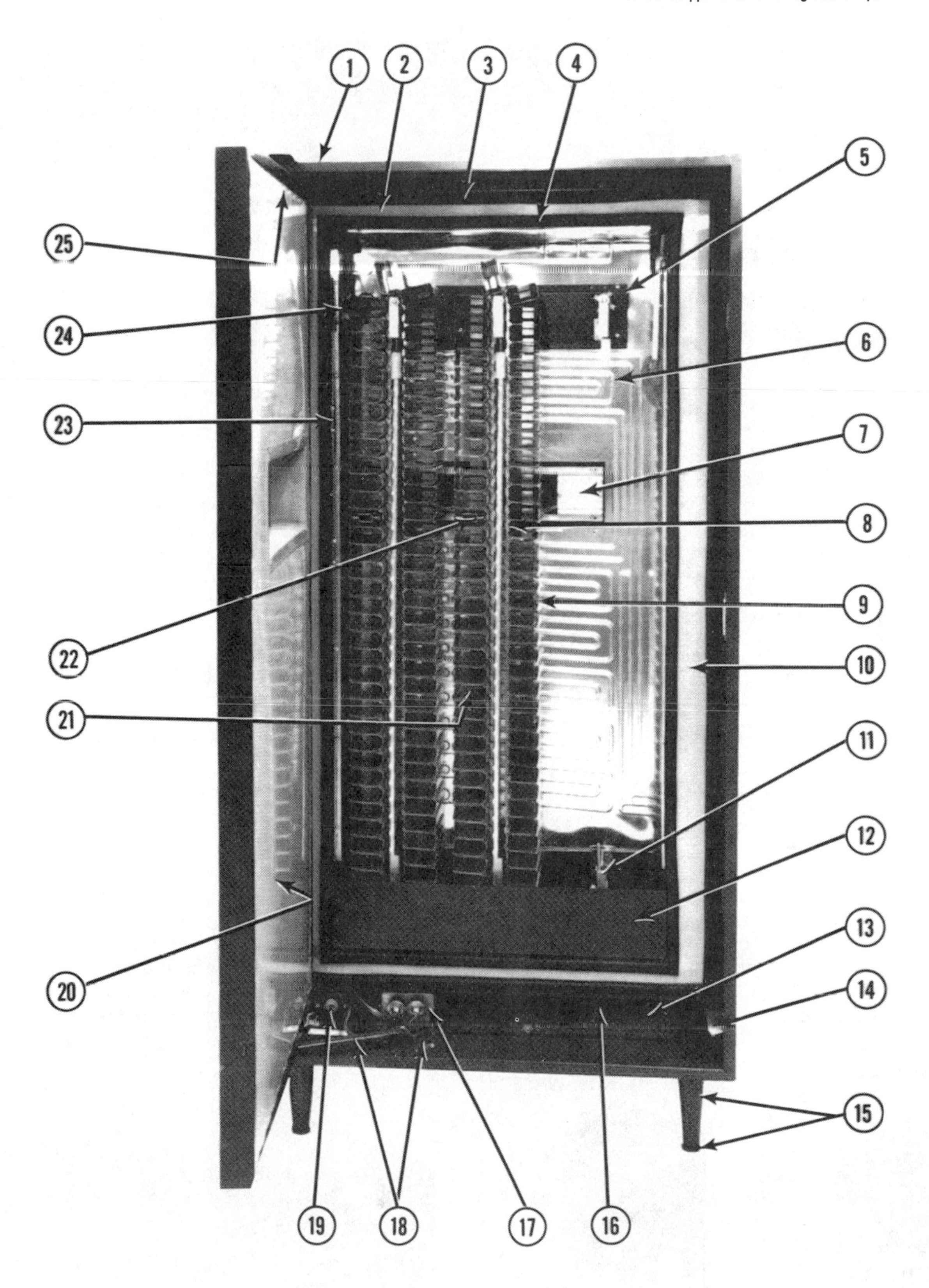

FIGURE 26–60 Low-temperature vending machine. Notice the plate evaporator. *Courtesy Rowe International Inc.*

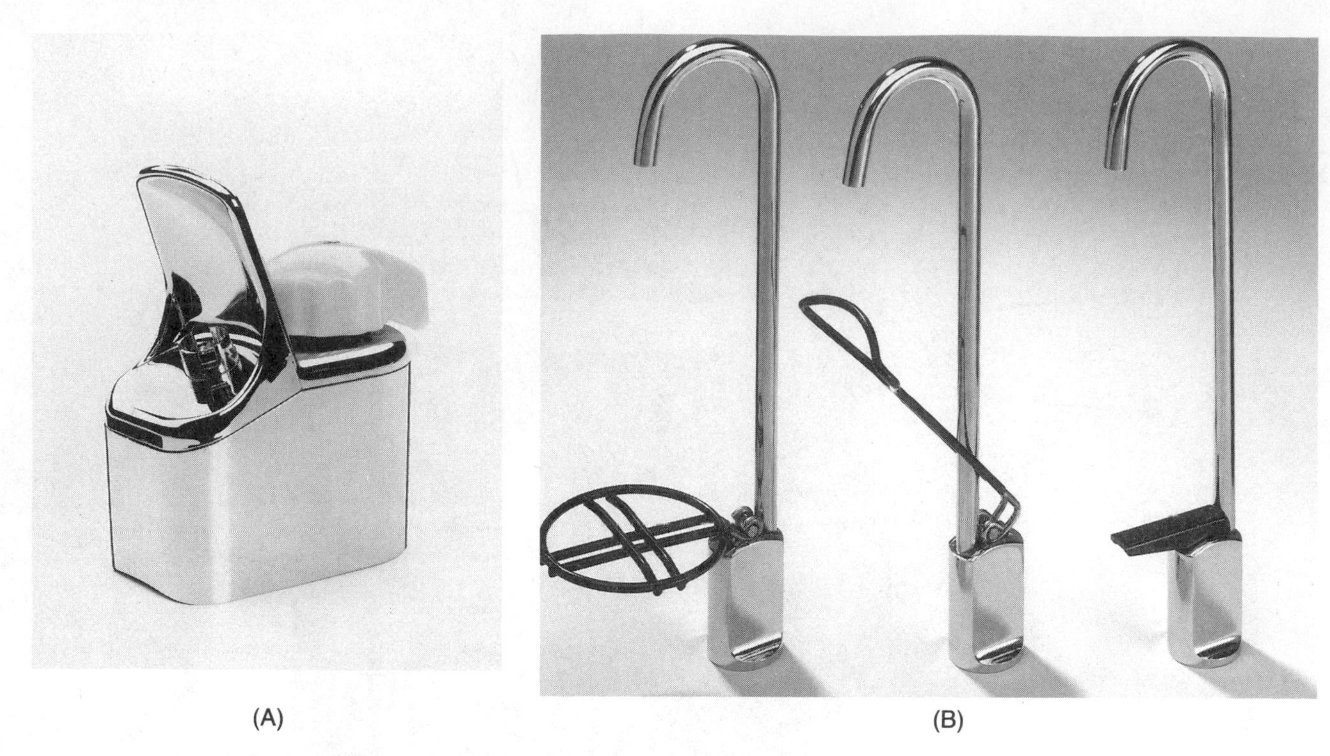

FIGURE 26–61 Two methods of dispensing cold water. *Courtesy EBCO Manufacturing Company*

26.33 WATER COOLERS

Water coolers, often called water fountains, are used in many public buildings to cool drinking water to a temperature similar to ice water. The water is dispensed either through a fixture called a bubbler, Figure 26–61 (A), or through a snout used to fill a cup, Figure 26–61 (B). There are typically two types of package water coolers, one that has a large bottle of water located above the cooler, Figure 26–62 (A), and one that is under regular water main pressure, Figure 26–62 (B).

Pressure-type systems require less maintenance because they are automatic in their dispensing of water. The water may be under regular water main pressure entering the cooler and it must be reduced to a pressure that will develop an arc for people to drink from, Figure 26–63. The

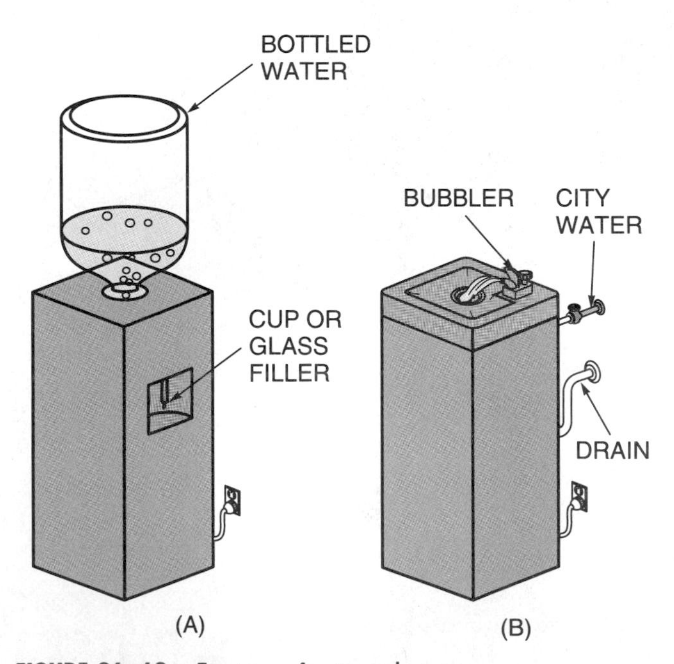

FIGURE 26–62 Two types of water cooler.

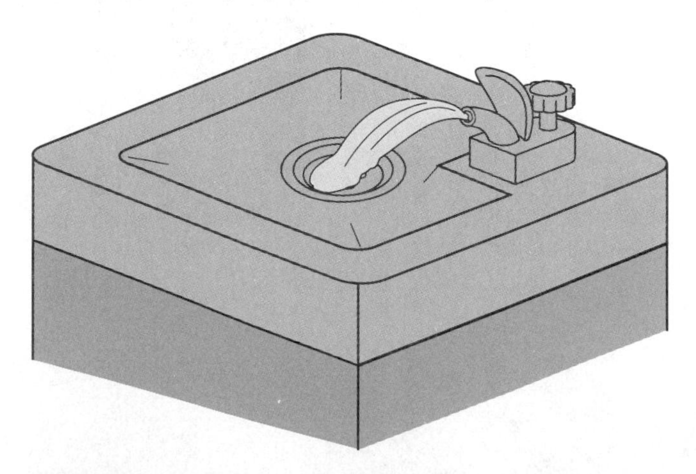

FIGURE 26–63 Bubbler creates arc for drinking water.

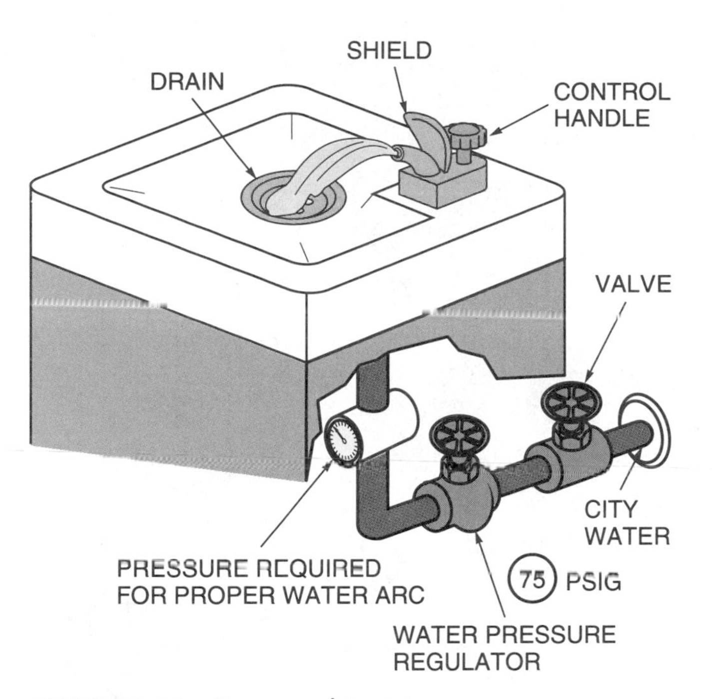

FIGURE 26–64 Pressure regulation is important.

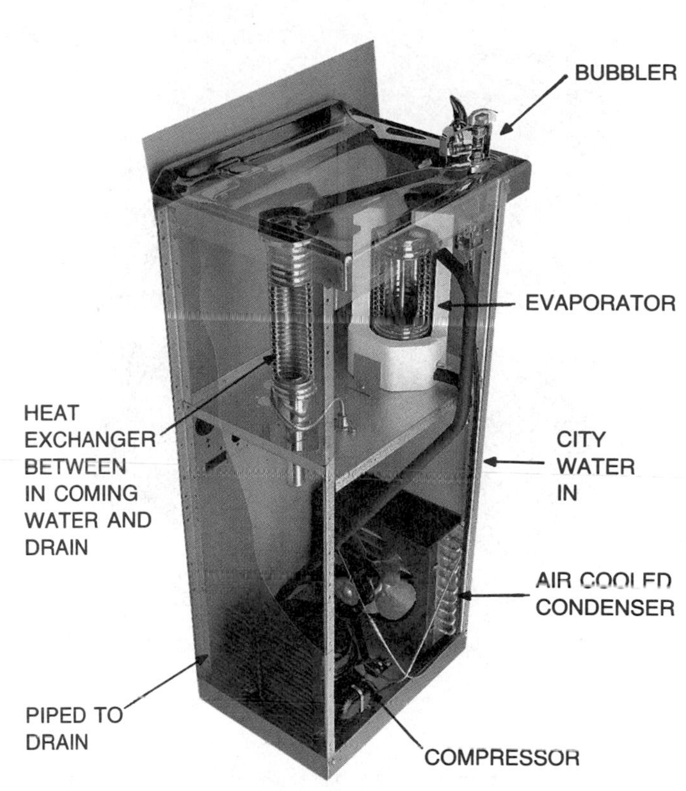

FIGURE 26–65 Water cooler. *Courtesy EBCO Manufacturing Company*

water pressure regulator for reducing the pressure is part of the system that must be serviced by the technician. The regulator typically has a screwdriver slot for adjusting the arc of the water at the bubbler, Figure 26–64.

The refrigeration section of the appliance is a small self-contained hermetic system consisting of a compressor (fractional horsepower), a typical air-cooled condenser with a small fan and an evaporator. The evaporator is the only special part of the refrigeration system. It is typically a small tank made of nonferrous metal such as copper, brass, or stainless steel. The refrigerant piping is wrapped around the tank. Water coolers dispense drinking water, called "potable water," and they must be sanitary, so the choice of tank is important.

It is estimated that 60% of the water dispensed in a bubbler for drinking goes down the drain. This is cold refrigerated water. Most manufacturers have some form of heat exchange to take advantage of this loss. The heat may either be exchanged between the incoming water and the drain or the liquid refrigerant line and the drain, Figure 26–65.

When water is dispensed from a bottle, the refrigeration system is below the bottle and enclosed in the housing, Figure 26–66. Water is typically dispensed by gravity into a cup when this type of cooler is used so there is no loss of water down the drain.

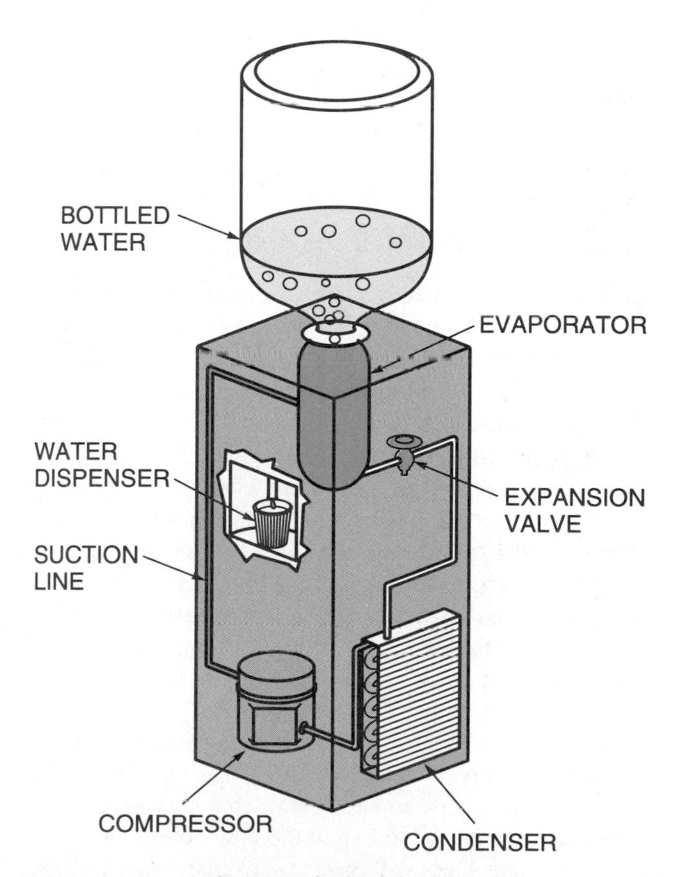

FIGURE 26–66 Refrigeration for bottled water cooler.

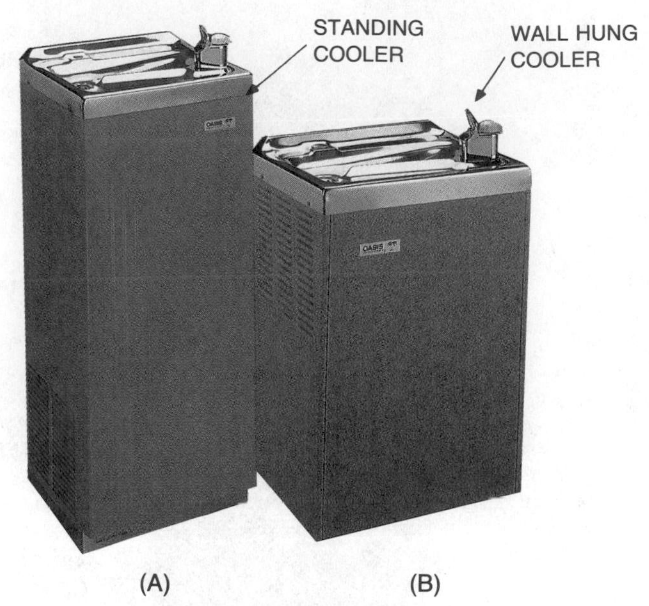

FIGURE 26–67 Two types of water cooler. *Courtesy EBCO Manufacturing Company*

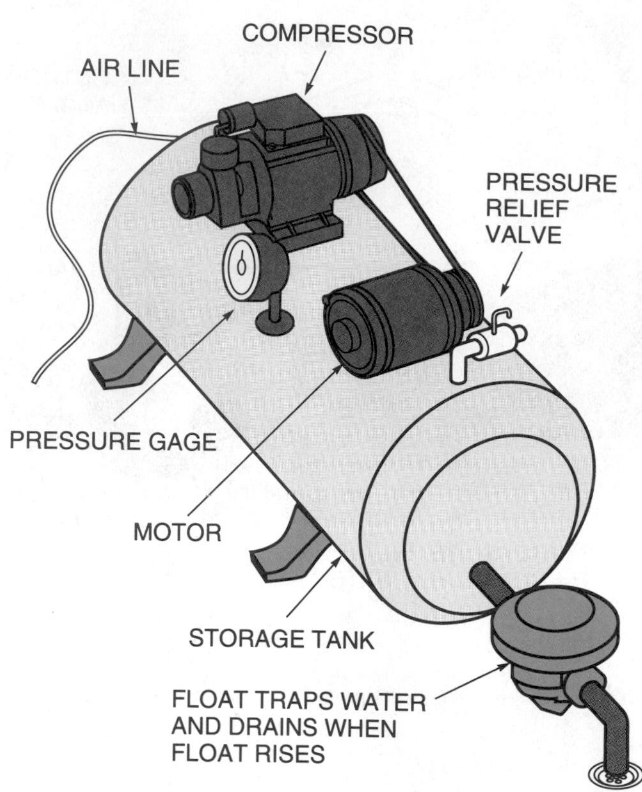

FIGURE 26–68 Air compressor showing moisture-trapping system.

Some water coolers are free standing, Figure 26–67 (A), and some are wall mounted, Figure 26–67 (B). When major repairs are required, most water coolers are removed from their location and taken to a central shop for repair.

Central water cooling systems are used for some large buildings. These systems use a central water cooler and circulate chilled water to the bubblers and cup-type dispensers throughout the building. These are popular in large buildings because they are more efficient and place all of the refrigeration service at a central location. Each bubbler must have a drain back to the central drain system.

25.34 REFRIGERATED AIR DRIERS

There are many applications where air from the atmosphere is compressed for use and the air must be dehydrated. When air is compressed, it leaves the compressor saturated with moisture. Some of this moisture condenses in the air storage tank and is exhausted through a float. The air is still very close to saturated as it leaves the storage tank. Some applications where the air must be dry are when the air is used for special manufacturing processes, or when air is used for controls, called "pneumatic control." This air may be dehydrated using refrigeration.

Refrigerated air driers are basically refrigeration systems located in the air supply, after the storage tank. The air may be cooled in a heat exchanger, then moved to the storage tank where much of the water will separate from the air. This water may be drained using a float that rises and provides for the drainage at a predetermined level, Figure 26–68. The air then passes through another heat exchange where the air temperature is reduced to below the dew point temperature (the point where water will condense from the air). This heat exchanger consists of an evaporator that normally operates at a temperature just above freezing, Figure 26–69.

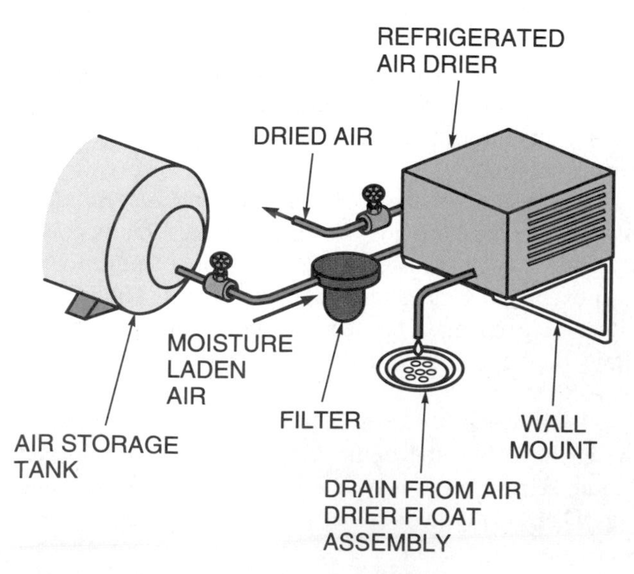

FIGURE 26–69 Refrigerated air drier installation.

This system typically runs 100% of the time because it does not know when there is a demand for air flowing through the system. Because there are times when there is no air flow, there are times when there is virtually no load on the refrigeration system. The only time there is a load on the system is when air is passing through that needs to be dried. Because the refrigeration system is running, in case there is a demand, a method must be used to apply a false load on the evaporator when there is no load. This is usually accomplished with a special refrigeration valve called a hot gas bypass valve, Figure 26–70. This valve is situated between the hot gas line and the inlet to the evaporator. It

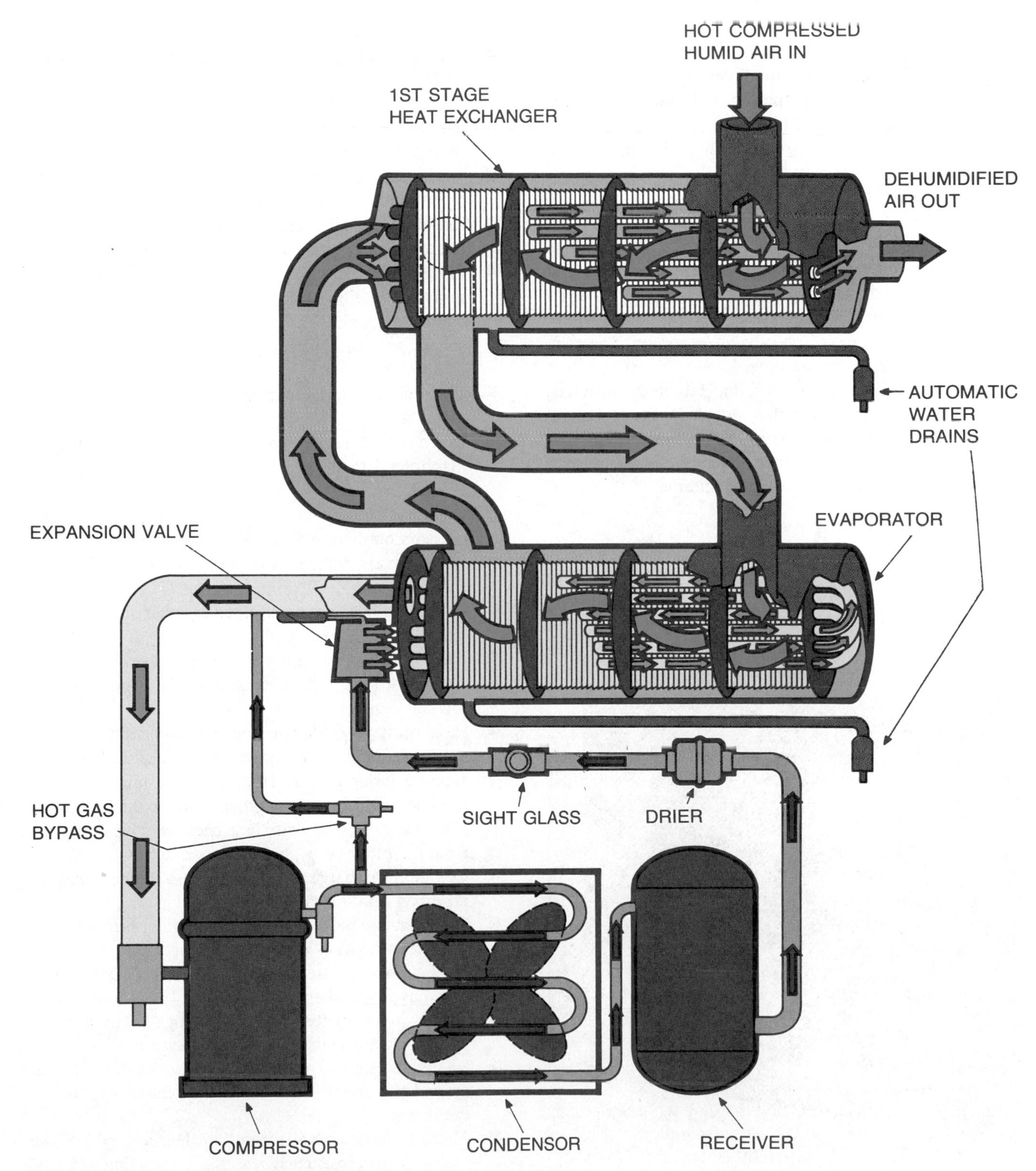

FIGURE 26–70 Flow schematic for refrigerated air drier. *Courtesy Van Air Systems Inc.*

monitors the pressure in the evaporator and allows hot gas to enter the evaporator at any time there is no or reduced load on the evaporator. This is accomplished by monitoring the suction pressure with a pressure-sensitive valve. It may be called a pressure regulator because it does not allow the pressure in the evaporator to drop below the set point. The difference in this pressure regulator and others discussed in this text is that it is located between the high- and low-pressure side of the system. The set point is usually just above the freezing point of water. For example, for R-22, the set point of the hot gas bypass valve may be 61.5 psig (35°F). The valve will not allow the evaporator pressure to reduce below 35°F, so the evaporator will not freeze during any no- or reduced-load situations.

Refrigerated air driers are normally self-contained refrigeration systems that may be air cooled or water cooled, Figure 26–71. The compressor can either be a hermetic unit for the smaller sizes or a semihermetic for larger sizes. The air line is piped to and through the evaporator of the air drier and then moves on to the system where it is used. The water is collected and exhausted to a convenient drain.

The evaporator may be one of several types; a pipe within a pipe or a honeycomb type are common. The compressor is usually a hermetic type.

The service technician should understand the manufacturer's intent for each application. At times, the technician must improvise when replacing components that are not available. As equipment becomes outdated, the components for exact replacemnent may not be available.

FIGURE 26–71 Refrigerated air drier. *Courtesy Van Air Systems Inc.*

SUMMARY

- Items to consider when choosing refrigeration equipment for a job are (1) first cost, (2) conditions to be maintained, (3) operating cost of the equipment, and (4) long-range intent of the equipment.
- Product dehydration can be a factor in the choice of equipment.
- Display equipment may be package, or it may have remote condensers.
- With package or self-contained equipment, the condenser rejects the heat back into the store when located inside.
- The condensing unit in packaged equipment is normally on top or underneath the fixture.
- All fixtures have condensate that must be drained away or evaporated.
- Individual small compressors are not as efficient as larger compressors because of motor efficiency.
- Fixtures may also be piped to a common equipment room with either individual compressors or single-compressor units. The single-compressor unit may consist of two or more large compressors or one large compressor.
- When a single large compressor is used, capacity control is desirable.
- When two or more large compressors are used, they are manifolded together with a common suction and discharge line.
- Capacity control is accomplished by cycling compressors on the multiple-compressor racks.
- Several evaporators may be piped together to a common suction line.
- The liquid lines are manifolded together after the liquid leaves the receiver.
- Each fixture has its own expansion valve.
- There must be a defrost cycle on both medium- and low-temperature fixtures.
- Heat must be added to the coil for defrost in low-temperature refrigeration systems. The fan must be stopped during defrost. Hot gas defrost must have a hot gas line run from the equipment room to the fixture. Hot gas defrost is more efficient because it uses the heat from the other cases to accomplish defrost.
- Common discharge lines concentrate the heat that was absorbed into the system.
- This heat can be used to heat the store or heat hot water with heat-recovery devices.
- Air patterns are used to cool open display boxes when they have high shelves.
- When doors are used, they are rugged, usually two panes of glass with an air space between.
- Heaters are used around the doors to keep the cabinet around the doors above the dew point temperature of the room.
- These heaters are called mullion heaters and are normally electric resistance heat that is sometimes thermostatically controlled.

- Some stores are controlling humidity by introducing fresh air from the outside through the air conditioning system.
- Humidity, usually in the summer, leaves the store by two means, the refrigeration equipment and the air conditioning equipment.
- It is less expensive to remove humidity with the air conditioning equipment than with the refrigeration equipment defrost cycle.
- Some walk-in coolers are permanent structures, and some are knock-down units that can be assembled in the field and moved.
- All coolers must have drains for the condensate if the piping is located where it is below freezing; heat must be applied to the drain lines.
- Drain heaters are normally electric resistance heat and may have thermostats to keep them from operating when they are not needed.
- Some coolers have remote condensing units, and some have wall-hung or roof-mount package equipment.
- Some of the remote units are field piped, and some are piped with quick-connect tubing.
- Most small ice-making applications use package equipment that needs only a water supply, electrical supply, and drain.
- Ice-holding machines may operate at 0°F.
- Ice-making machines normally make ice at above 0°F and hold the ice in a bin at about 32°F.
- Cube ice is normally made in inverted or vertical evaporators with water spraying up into the evaporator or flowing on flat plate evaporators.
- Flake ice is normally made in a circular evaporator, and the ice is scraped off with a turning auger.
- Most package ice makers are air cooled; however, when water cooled, they are normally wastewater systems.
- The refrigeration vending machine systems consist of the money receiving and changing mechanism, the food handling section (often called the conveyer), and the refrigeration system.
- Beverage cooling is one application for refrigerated vending and uses medium-temperature refrigeration. Off-cycle defrost is typically used to keep ice off the evaporator.
- Defrosted condensate flows to a sealed drain to be evaporated using moving air or the hot gas line.
- Canned drinks are typically stacked on top of each other when loading, the first loaded is the first used because they are dispensed out the bottom.
- Vending refrigeration uses air-cooled condensers with fractional horsepower compressors and often forced-draft evaporators.
- Some beverage coolers dispense liquids, such as soft drink machines that dispense a cup filled with ice and a chilled liquid. These machines have one compressor and two evaporators, one for the ice maker and one for chilling the incoming water.
- Sandwiches and other food products may be dispensed from machines that may be medium-temperature applications used for fresh foods or low-temperature applications where frozen foods may be dispensed.
- Sanitation is extremely important in vending machines.
- Vending machines often have a health switch that tells the operator if the machine has been operating above the safe temperature for some period of time.
- Water coolers dispense cold water, are self-contained, use a fractional horsepower compressor and an air cooled condenser. The evaporator is a container with the evaporator piping wrapped around it.
- Water coolers are package refrigeration machines that may be either free standing or hung on the wall. A drain must be connected back to the building drain system.
- Refrigerated air driers are refrigeration systems that are used to dehydrate air that is compressed with an air compressor. Some manufacturing processes and all pneumatic air control systems must have air with a low moisture content.
- Refrigerated air driers typically operate all of the time, whether there is a load or not. They provide cooling to about 35°F.
- The typical refrigerated air drier uses a hermetic compressor and either an air- or water-cooled condenser.
- ***There are many applications for refrigeration. There is no substitute for knowing the manufacturer's intent. Don't improvise without this knowledge.***

REVIEW QUESTIONS

1. Name three things that should be considered when choosing equipment for a refrigeration installation.
2. What are the two broad categories of display cases?
3. How are conditions maintained in open display cases when there are high shelves?
4. Describe the doors used on closed display cases.
5. What are mullion heaters?
6. What are the three temperature ranges of refrigeration systems?
7. Name two methods of rejecting heat from refrigerated cases.
8. Describe how multiple cases operate with the same compressor?
9. Name two advantages of having multiple cases applied to one compressor.
10. Where is the piping routed when the compressors are in the equipment room?
11. When one large compressor is used, what desirable feature should this compressor have?
12. How is defrost accomplished when the equipment room is in the back and the fixtures are in the front on medium-temperature fixtures?
13. Name two types of defrost used with low-temperature fixtures.
14. What materials are normally used in the construction of display fixtures?

15. What materials are used in the construction of walk-in coolers?
16. What two special precautions should be taken with drain lines in walk-in coolers?
17. What holds up the structure of a walk-in cooler?
18. Name three types of ice made by package ice makers.
19. Which type of ice is made in an ice maker using a geared motor and auger?
20. Which type of ice maker has a continuous harvest cycle?
21. What is used to refrigerate the ice in the storage bin in a package ice maker?
22. What would be the symptoms of a defective float on an ice maker?
23. What electrical service size must be provided for a typical vending machine?
24. What type of defrost is normally used for vending machine medium-temperature applications?
25. Name two methods for prechilling entering water in a liquid dispensing vending machine.
26. Describe two methods for dispensing water in a water cooler?
27. What is the special valve used in a refrigerated air drier?

UNIT 27

Special Refrigeration Applications

OBJECTIVES

After studying this unit, you should be able to

- describe methods used for refrigerating trucks.
- discuss phase change plates used in truck refrigeration.
- describe two methods used to haul refrigerated freight using the railroad.
- discuss the basics of blast cooling for refrigeration.
- describe cascade refrigeration.
- describe basic methods used for ship refrigeration.

SAFETY CHECKLIST

* Ensure that nitrogen and carbon dioxide systems are off before entering an enclosed area where they have been in use. The area should also be ventilated so that oxygen is available for breathing and has not been replaced by these gases.
* Liquid nitrogen and liquid carbon dioxide are very cold when released to the atmosphere. Take precautions that they do not get on your skin. Wear gloves and goggles if you are working in an area where one of these gases may escape.
* Ammonia is toxic to humans. Wear protective clothing, goggles, gloves, and breathing apparatus when servicing ammonia refrigeration systems.

27.1 SPECIAL APPLICATION REFRIGERATION

There are many special applications of refrigeration. Some technicians may never work with these special applications but others will. Keep in mind the theory involved with refrigeration as you study this unit.

27.2 TRANSPORT REFRIGERATION

Transport refrigeration is the process of refrigerating products while they are being transported from one place to another. This can be by truck, rail, air, or water. For example, vegetables may be harvested in California and shipped to New York City for consumption. These products may need refrigeration along the way.

There are many different products that must be shipped in many different ways. Some of the products are shipped in the fresh state, such as lettuce and celery, which are shipped at about 34°F. Some products are shipped in the frozen state, such as ice cream, which is shipped at about −10°F. The fresh vegetables may be shipped through climates that are very hot, such as passing through Arizona and New Mexico. They may also be shipped through very cold climates, such as in the winter when fresh vegetables are shipped through Minnesota. Different situations require different types of systems. For example, refrigeration may be needed for a load of lettuce when leaving California but when it gets to the northern states, it may need to be heated to prevent it from freezing, Figure 27–1.

FIGURE 27–1 This load of lettuce needs refrigeration when crossing the desert and heat in the northern states.

Products that fall into the above categories are carried on trucks, trains, airplanes, and ships. The refrigeration system is responsible for getting the load to the destination in good condition.

27.3 TRUCK REFRIGERATION SYSTEMS

Several methods are used to refrigerate trucks. One of the most basic systems that is still being used is to merely pack the product in ice that is made at an ice plant. This works well for a well-insulated truck for short periods of time. The melting ice produces water that must be handled. For years, you could spot these trucks because of the trail of water left behind, Figure 27–2. This method can only hold the load temperature to above freezing.

Dry ice, which is solidified and compressed carbon dioxide, is also used for some short-haul loads. It is somewhat harder to hold the load temperature at the correct level using dry ice because it is so cold. Dry ice changes state from a solid to a vapor (called sublimation) without going through the liquid state at −109°F. By using these temperatures, the food products could be reduced to very low temperatures. Dehydration of food products that are not in airtight containers can become a problem using dry ice because the very low temperature of −109°F is a great attraction for moisture in the food.

Liquid nitrogen or liquid carbon dioxide (CO_2) may also be used for refrigeration of products. This is accomplished by using nitrogen that has been refrigerated and condensed to a liquid and stored in a cylinder on the truck at a low pressure. The cylinder has a relief valve that releases some of the vapor to the atmosphere if the pressure in the cylinder rises above the set point of about 25 psig in the cylinder. This boils some of the remaining liquid to drop the cylinder temperature and pressure, Figure 27–3. Liquid is piped to a manifold where the refrigerated product is located and released to cool the air in the truck and the product. A thermostat located in the food space controls a solenoid valve in the liquid nitrogen line, stopping and starting the flow of liquid, Figure 27–4. Fans may be used to disperse the cold vapor evenly over the product.

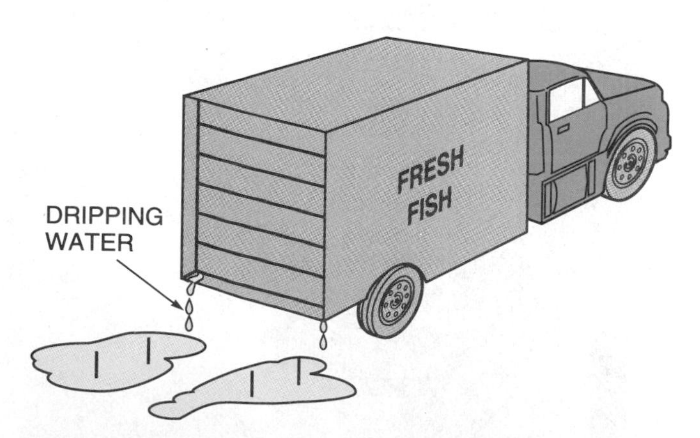

FIGURE 27–2 When ice is used for preserving products, it melts and often leaves a trail of water on the highway or a puddle in a parking area.

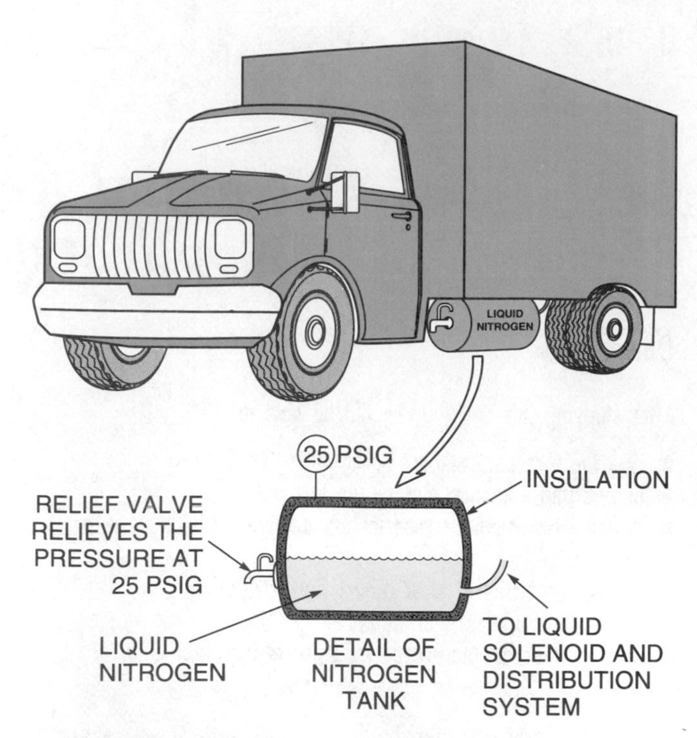

FIGURE 27–3 Nitrogen cylinder maintained at about 25 psig.

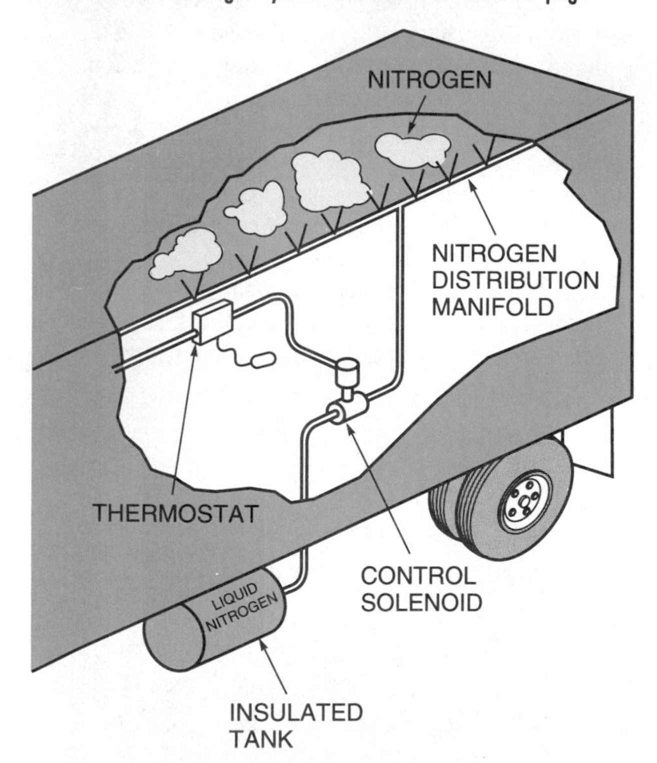

FIGURE 27–4 Control for a liquid nitrogen cooling system.

Air in the atmosphere is 78% nitrogen and is the source for the nitrogen used for refrigeration. It is not toxic; it can be released to the atmosphere with confidence that no harm will occur to the environment. ***Nitrogen has no oxygen in it so it is not suitable for breathing. Special safety interlocks shut the refrigeration off if the door to the refrigerated truck is opened to prevent personnel from being smothered from lack of oxygen. Nitrogen is also very cold; it evaporates at −320°F at atmospheric pressure.**

Any contact with the skin would cause instant freezing of flesh. Do not ever allow liquid nitrogen to get on your skin.*

Liquid CO_2 can be used in the same manner as liquid nitrogen to refrigerate product. The system would work much the same as a liquid nitrogen system only CO_2 boils at −109°F.

Even though liquid nitrogen and CO_2 are both very cold, they can be used to refrigerate medium-temperature loads at 35°F with proper controls. These controls would include a distribution system for the refrigerant that does not impinge on the product and a space temperature thermostat that accurately senses the refrigerated space temperature.

Both of the liquid injection methods have been used for many years as a replacement for mechanical refrigeration for permanent installations and for emergency use for refrigeration. They are probably more expensive in the long run than a permanent installation but they also have some good points. For example, when used to refrigerate fresh foods, the presence of nitrogen or CO_2 vapor displaces any oxygen and helps preserve the food. The system is simple and easy to control because there is only a solenoid and a distribution manifold to maintain.

Many truck bodies are equipped with refrigerated plates (called cold plates) that have a phase change solution inside called "Eutectic solution." Phase change means to change the state of matter. This is much like the change of state from ice to water, except at a different temperature. When the change of state occurs, much more heat is absorbed per pound of material. This "Eutectic solution" has the capability of changing the state at many different temperatures, depending on the composition of the product.

The "Eutectic solution" in the cold plate is brine, a form of salt water using either sodium chloride, a form of table salt, or calcium chloride. These brines are very corrosive and must be handled with care.

Various strengths of solution of the brine may be used to arrive at the desirable melting temperature below freezing, called the "Eutectic" temperature. Actually, the brine does not freeze solid; salt crystals form and the remaining solution is liquid. Different strengths of brine will cause the crystals to form at different temperatures.

The brine solution is contained in plates that are from 1 in. to 3 in. thick. They are mounted on the wall or ceiling of the truck in such a manner that air can flow all around them, Figure 27–5. The room air transfers heat to the plates. The brine solution changes from a solid to a liquid while absorbing heat. To recharge the plates, they must be cooled to the point that the change of state occurs back to solid crystals. These plates will hold a refrigerated load at the correct temperature (for medium- or low-temperature application) for an all-day delivery route to be run. For example, a truck loaded with ice cream may be loaded at night for delivery the next day. The ice cream must be at its prescribed temperature and the truck cold plates must be at the design temperature. For ice cream, this may be −10°F or lower. When the truck comes back at night, it may still be −10°F, but most of the brine solution will be changed back to a

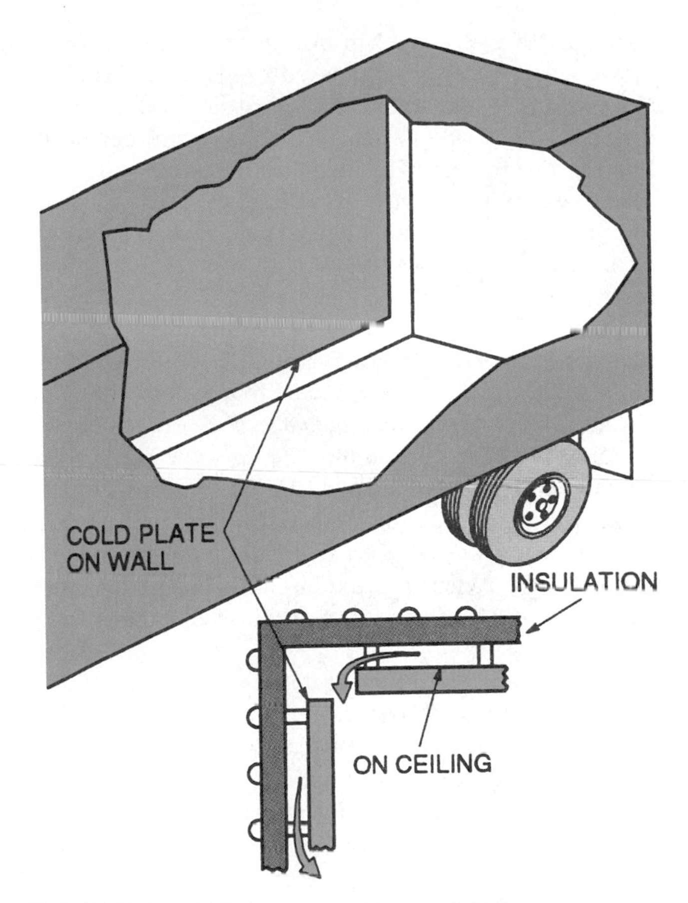

FIGURE 27–5 Cold plates mounted in a truck body.

liquid and will need recharging to the solid crystal state. The heat absorbed during the day is in the brine solution and must be removed.

The plates are recharged in several ways. Some systems have a direct expansion coil located in the cold plate. This coil is connected to the refrigeration system at the loading dock, Figure 27–6.

The refrigeration system may use one of the typical refrigerants, R-12, R-500, or R-502, or it may be R-717

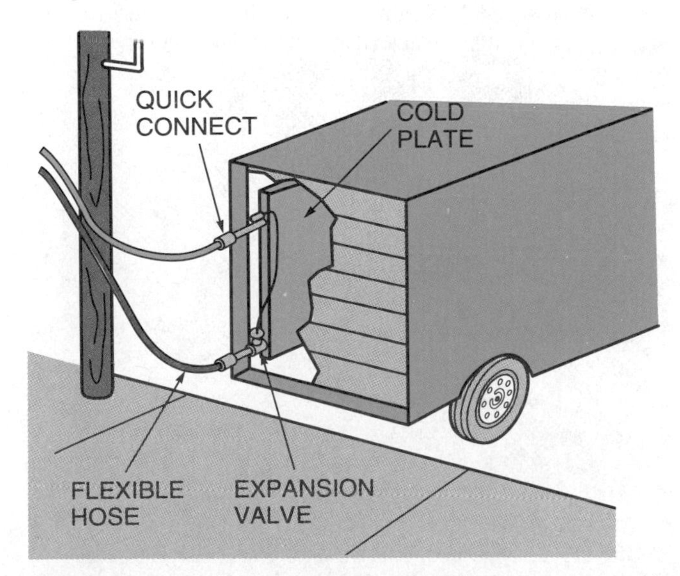

FIGURE 27–6 Cold plate connected to the refrigeration system at the dock.

ammonia. When the truck backs up to the loading dock, quick-connect hoses are connected from the central refrigeration plant to the truck. The expansion valve may be located on the loading dock or at the actual coil in the "Euctectic" cold plate. The central refrigeration system will reduce the pressure in the evaporator in the cold plate to the point that the brine solution will change back to solid crystals in a few hours. The truck may then be loaded with product.

Another method of recharging the cold plate is to circulate refrigerated brine solution through a coil located in the cold plate. Brine that is refrigerated to a lower temperature than the cold plate will then recharge or freeze the brine in the cold plate. The circulating brine must be much colder than the solution in the cold plate. Connecting this system to the truck also involves quick-connect fittings at the loading dock. There is always the possibility that some brine will be lost during the connecting and disconnecting of the quick-connect fittings. This brine must be cleaned up or corrosion will occur.

Trucks may also be refrigerated using a refrigeration system on the truck. Whatever the refrigeration source is, it must have a power supply. This power supply may be from the truck power supply, from a diesel or gas engine-driven compressor, from a land line (electric power to a building), or from an electric compressor operated by a motor generator.

Van-type delivery trucks often use a refrigeration compressor mounted under the hood that is turned by the truck engine. The refrigeration may be controlled by cycling the compressor using an electric clutch that is thermostatically controlled, Figure 27–7. This system works well as long as the truck engine is running. Because the truck is usually moving all day during delivery, refrigeration can easily be maintained. Some of these trucks have an auxiliary electric motor-driven compressor that may be plugged in at the loading dock while the truck engine is not running.

Trucks with small condensing units under the truck body are also used to refrigerate the load. These often are used to recharge cold plates at night when the truck is at the dock or other power supply and the cold plates hold the load during the day, Figure 27–8. The advantage of this system is that it is self-contained and does not require a connection to a central refrigeration system at the loading dock. These condensing units may be powered while on the road using a small motor generator unit mounted under the chassis. In this case, there would be no need for cold plates because the generator would operate on the road and the compressor could be plugged in at night.

The motor generators have either a gas or diesel engine that turns the generator to generate 230-V, 60-cycle current to operate the compressor. Diesel is the preferred fuel because diesel engines will operate longer with less fuel and with less maintenance. Diesel engines have a greater first cost but seem to be less expensive in the long run.

A generator producing 230 V from the truck engine has been used in some cases. It must have a rather sophisticated control because the truck engine turns at so many different speeds. The stationary generator has only one speed. Different speeds would affect the power supplied to the compressor. As the engine turns faster, the voltage and the cycles per second vary; 60 cps is normal for alternating current electric motors. Electronic circuits may be used to maintain a constant voltage.

Larger trucks use either nose-mount or under-belly units that may either be operated using a compressor-driven diesel engine or a motor generator driven by a diesel engine, Figure 27–9. When the engine turns the compressor, it has a governor to regulate the speed of the compressor. The system may be two-speed, high and low. Control of the compressor may be accomplished using cylinder unloading on the compressor. For example, suppose the compressor is a 4-cylinder compressor and has a capacity of 1 ton per cylinder. The compressor may be operated from 1 ton to 4 tons by unloading cylinders, Figure 27–10. There should be a minimum load of 1 ton at all times or the diesel engine will

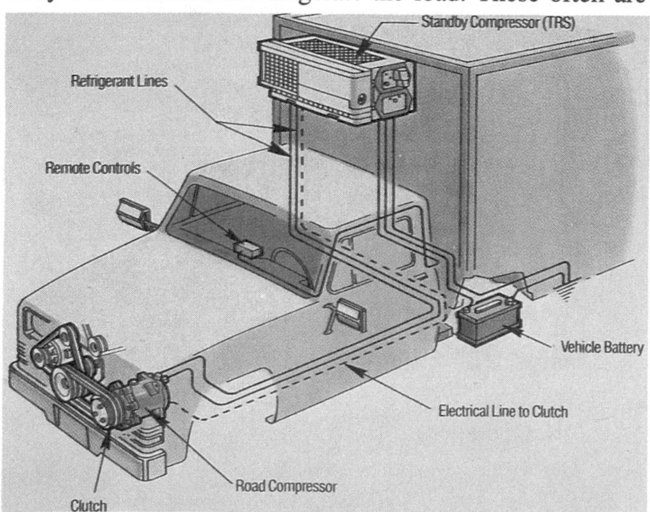

FIGURE 27–7 This truck refrigeration is driven from the truck's engine and may be cycled by means of an electric clutch. *Reproduced courtesy of Carrier Corporation*

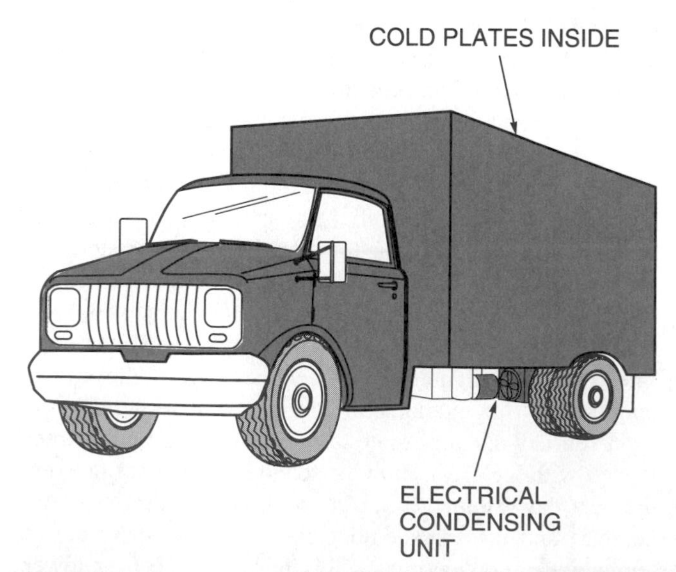

FIGURE 27–8 The truck uses a small refrigeration system on the truck to recharge the cold plates while parked at night.

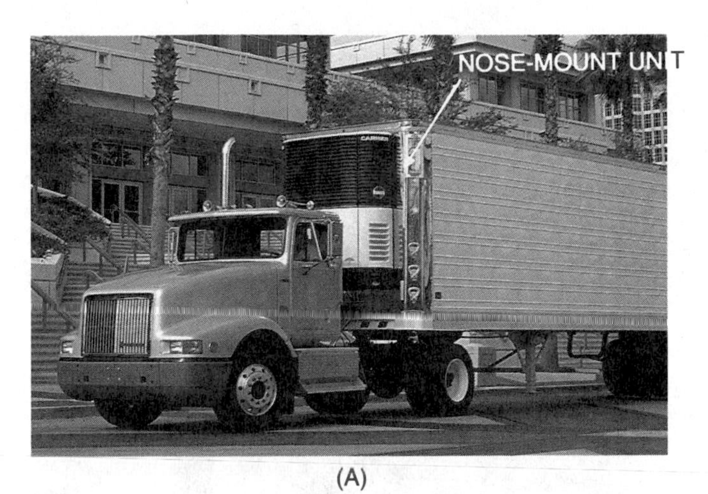

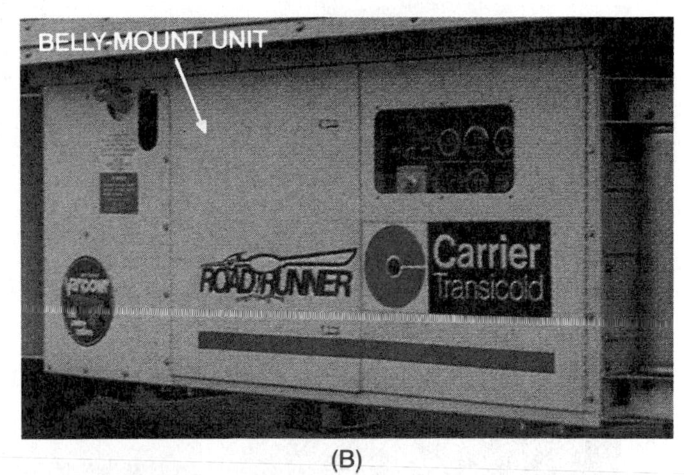

FIGURE 27–9 Nose-mount and belly-mount refrigerated truck. *Reproduced courtesy of Carrier Corporation*

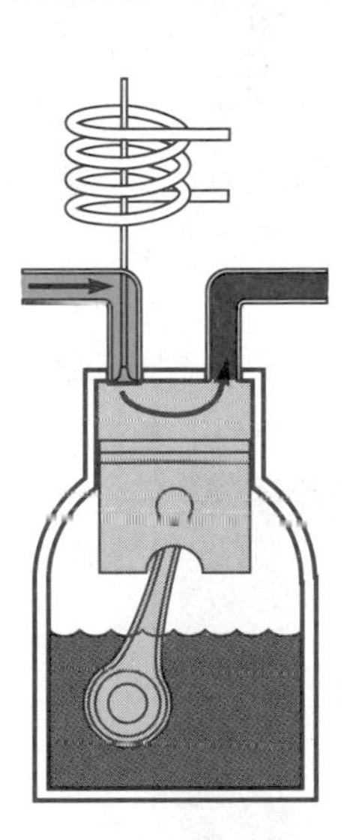

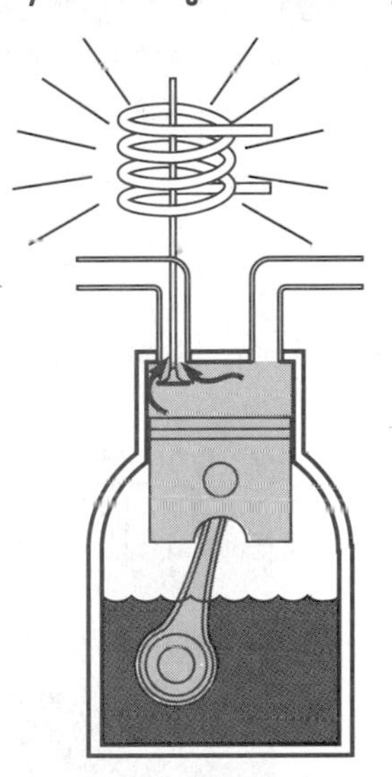

THIS CYLINDER IS NOT PUMPING BECAUSE THE SUCTION VALVE IS BEING HELD OPEN BY THE SOLENOID. GAS ENTERS THE CYLINDER AND IS PUSHED BACK INTO THE SUCTION LINE WITHOUT BEING COMPRESSED. COMPRESSION CAN BE CONTROLLED.

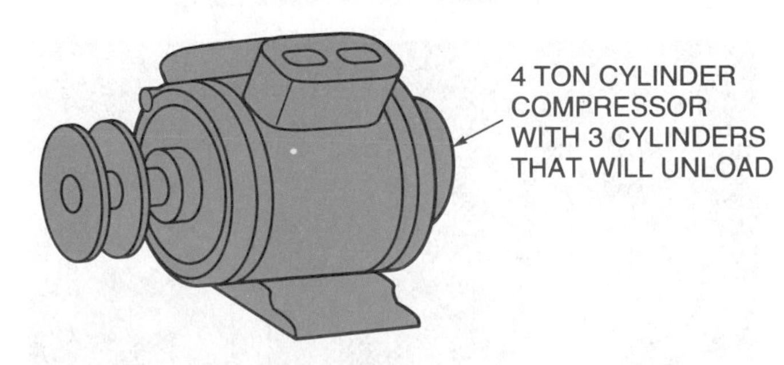

FIGURE 27–10 Capacity control for a compressor.

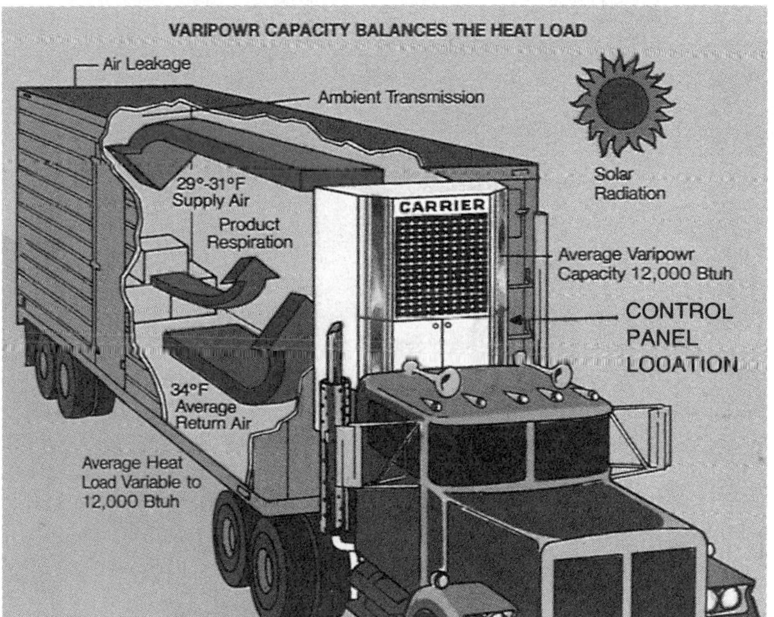

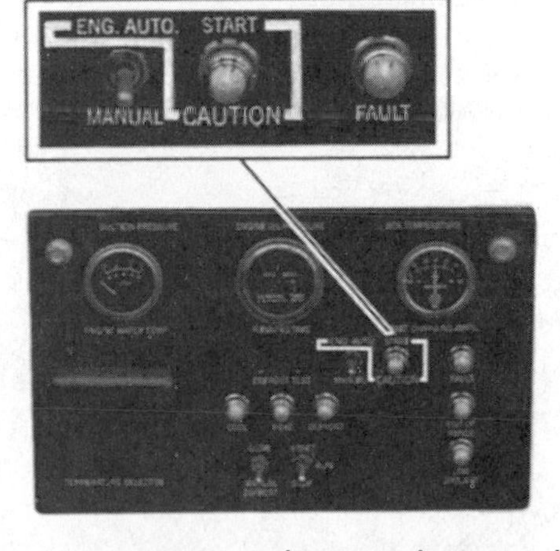

FIGURE 27–11 Automatic control to stop and start a truck refrigeration system. *Reproduced courtesy of Carrier Corporation*

have to be cycled off and on. It is most desirable to keep the engine running; however, controls for diesel engines have been developed that will automatically start and stop the engine, Figure 27–11.

The evaporator for these truck refrigeration units is typically located at the front of the truck and blows air toward the back of the truck, Figure 27–12. The fan is usually a form of centrifugal fan that distributes a high volume of air at a high velocity so the air will reach the back

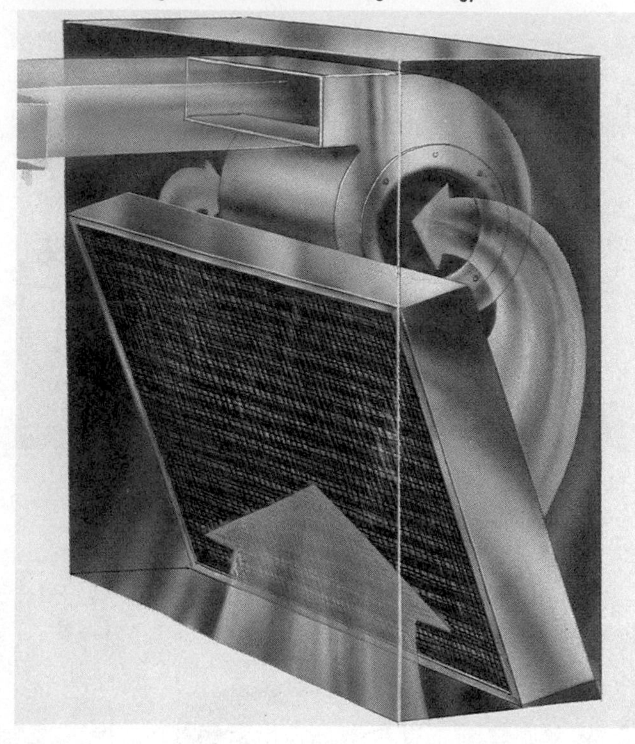

FIGURE 27–12 Evaporator for a nose-mount refrigeration unit. *Courtesy Thermo King Corp.*

of the truck, Figure 27–13. The air is then returned to the inlet of the air handler over the product load. The fan is driven from the engine that drives the compressor and is driven with belts or gear boxes. These components are normally located just inside the access door to the unit where they can be checked and serviced easily from outside the refrigerated space, Figure 27–14. The evaporator fan is usually a centrifugal-type fan. The condenser fan is often driven from the same drive mechanism and is typically a prop-type fan, Figure 27–15.

The evaporator and condenser coils are usually made of copper with aluminum fins.

These truck systems must be designed and built to be both rugged and lightweight because each truck has a maxi-

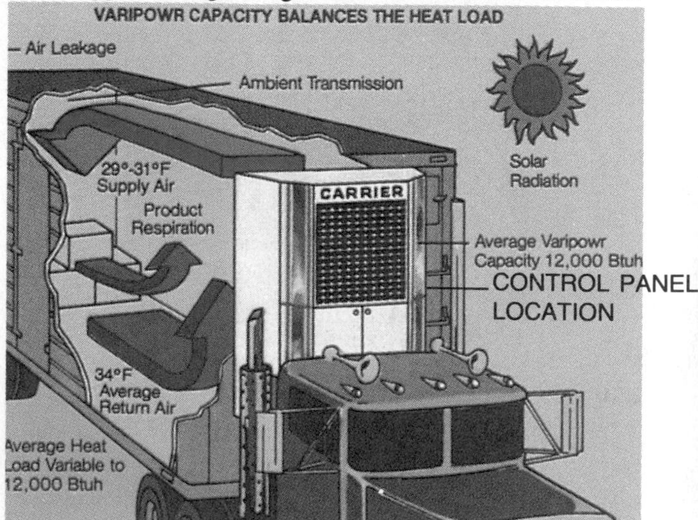

FIGURE 27–13 Air distribution system for truck. *Reproduced courtesy of Carrier Corporation*

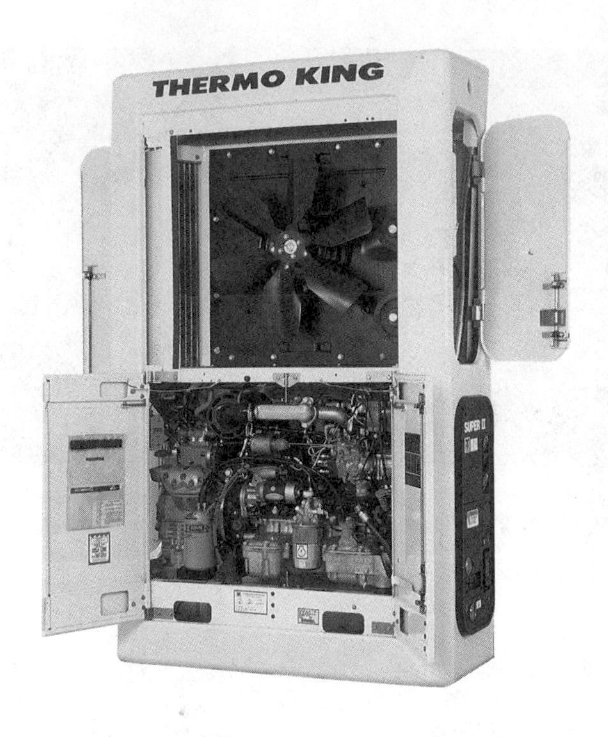

FIGURE 27–14 Access for service for a nose-mount unit. *Courtesy Thermo King Corp.*

(A)

(B)

FIGURE 27–15 Fan drive mechanisms. *(A) Courtesy Thermo King Corp. and (B) Reproduced courtesy of Carrier Corporation*

mum load limit. Any weight added to the refrigeration unit is weight that cannot be hauled as a paying load. Many of the components are aluminum, such as the compressor and any other component that can be successfully manufactured and maintained with this lightweight material.

Truck refrigeration systems must be designed to haul either low- or medium-temperature loads. Part of the time, they may be hauling fresh vegetables and part of the time the load may be frozen foods. The space temperature is controlled with a thermostat that starts out at full load and when the space temperature is within approximately 2°F of the set point, the system starts to unload the compressor. If the space temperature continues to drop, the unit will shut down.

The air distribution for truck refrigeration is designed to hold the load at the specific temperature, not to pull the load temperature down. As mentioned earlier, the food should be at the desired storage temperature before loading it onto the truck. The truck does not have much reserve capacity to pull the load temperature down, especially if the interior of the truck warmed very much. If the refrigeration system for a medium-temperature load of food has to reduce the load temperature, the air distribution system may overcool some of the highest parts of the load because the truck body is so small that the cold air from the evaporator may impinge directly on the top product, Figure 27–16. If the load is frozen food and has warmed, it may be partially thawed. The truck unit may not have the capacity to refreeze the product and if it does, it may be so slow that product damage may occur. Various alarm methods are used to alert the driver of problems with the load, Figure 27–17. The driver should stop and check the refrigeration at the first sign of problems. It may not be easy to locate a technician when the truck is away from metropolitan areas where service can be found.

FIGURE 27–17 An alarm may alert the driver if the load's condition is not being maintained. *Courtesy Thermo King Corp.*

Truck bodies are insulated to keep the heat out. This is normally accomplished using foam insulation sprayed onto the walls with wallboard fastened to the foam. This makes a sandwich type of wall that is very strong, Figure 27–18. The walls must be rugged because of the loading and unloading of the product. Aluminum- and fiber-reinforced plastic products are often used for the walls. The floor must be strong enough to support the weight of the load while traveling on a

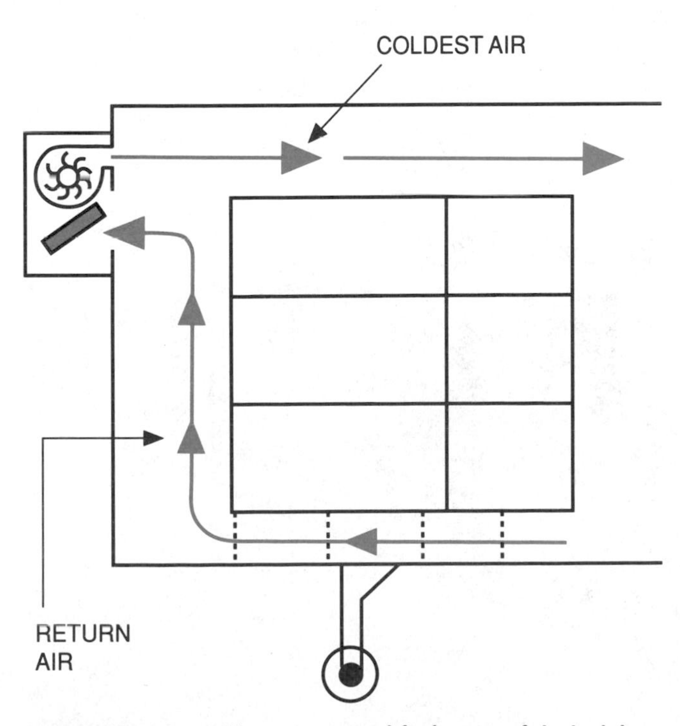

FIGURE 27–16 Cold air may overcool food on top of the load due to air distribution.

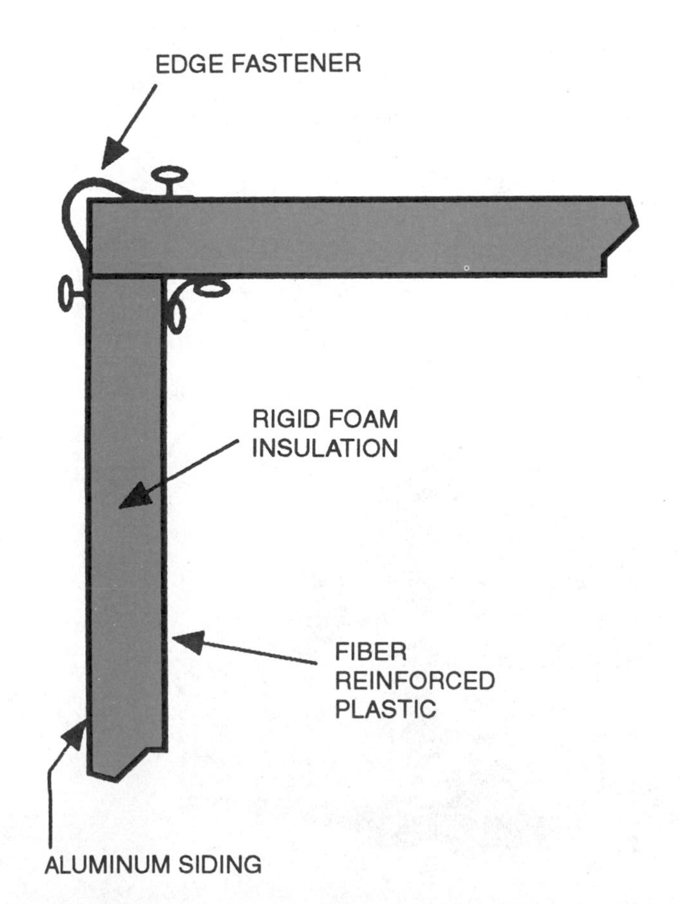

FIGURE 27–18 Sandwich construction for truck walls.

rough highway; it must also be able to support a forklift that may be used for loading.

The inside construction of a refrigerated truck must be manufactured in such a manner that it may be kept clean because food is being handled. The doors are normally in the back so that the truck may be loaded with a forklift; however, some trucks have side doors and compartments for different types of product. For example, a refrigerated truck may need to carry both low- and medium-temperature loads at the same time. This is accomplished using one refrigeration condensing unit with more than one evaporator.

The doors of a refrigerated truck must have a good gasket seal to prevent infiltration of the outside air to inside the truck. The doors must also have a secure lock or the load may be stolen.

The truck transport refrigeration technician should be a diesel technician as well as a refrigeration technician. These technicians are trained in both fields and they usually do not work on other types of refrigeration. Many times the technician is factory trained to service a particular type of refrigeration system and engine combination. A good knowledge of basic refrigeration would be good preparation for truck transportation and knowledge regarding diesel engines can be added.

27.4 RAILWAY REFRIGERATION

Products shipped and refrigerated by rail are mechanically refrigerated by either a self-contained unit located at one end of the refrigerated car or by an axle-driven unit. The axle unit is an older type and not being used to any extent now. The rail car would only have refrigeration while the train is in motion using axle-drive units.

The self-contained unit is the most popular. It is generally powered by a diesel-driven motor generator, similar to truck refrigeration, Figure 27–19. The compressor is a standard voltage and may be powered with a power cord while the car is in a station or on a rail siding for long periods. The evaporator blower is mounted at the end of the rail car with the motor generator. All of the serviceable components can be serviced from this compartment, Figure 27–20.

Two-speed engines are often used where the generator has an output frequency of 60 cps at full speed and 40 cps at reduced load. The compressor electric motor must be able to operate at the different frequencies. The diesel engine speed is adjusted by the space temperature thermostat. The diesel

FIGURE 27–19 Motor generator used to power a conventional compressor. *Courtesy Fruit Growers Express*

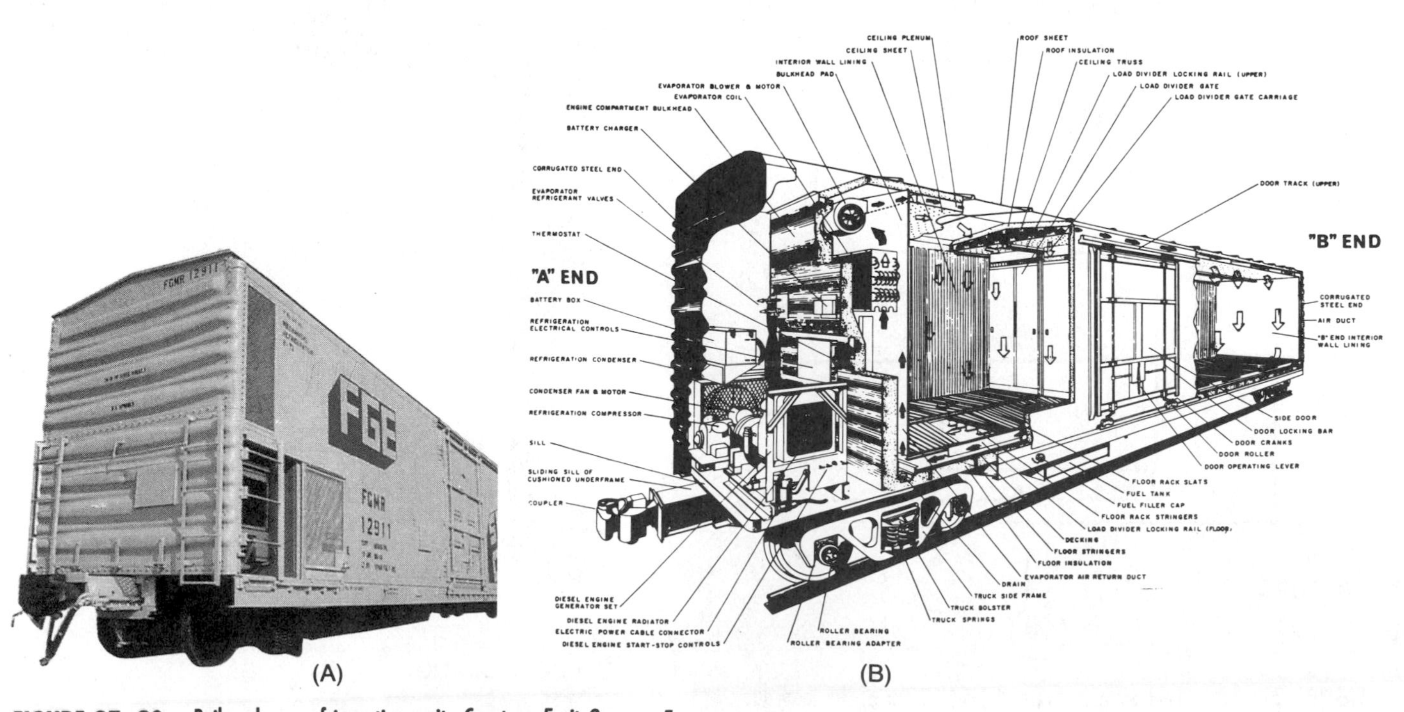

FIGURE 27–20 Railroad car refrigeration unit. *Courtesy Fruit Growers Express*

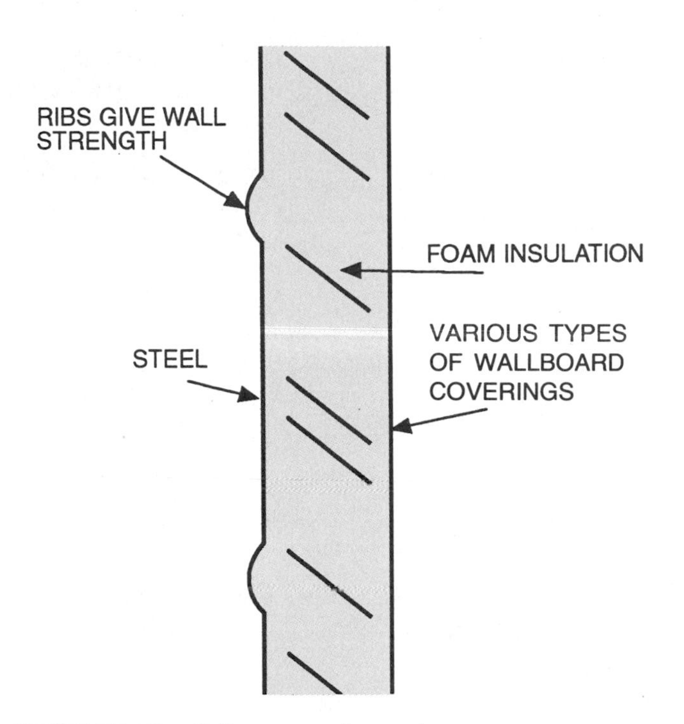

FIGURE 27–21 Wall construction for a rail car.

engine operates at a much lower fuel cost at low speed. When low speed is reached, the compressor may be unloaded using cylinder unloading. These systems have a rather wide range of capacities to suit the application. Like trucks, they are capable of reducing the space temperature to 0°F or below for holding frozen foods or they may be operated at medium temperature for fresh foods.

Rail cars have an air distribution system much like trucks and are designed to hold the refrigerated load, not reduce the temperature. They do not have much reserve capacity.

Rail cars are made of steel on the outside with typically a rigid foam insulation in the walls, Figure 27–21. The walls are covered with various types of wallboard that can be cleaned easily. Railroad cars would ride rough, with metal wheels on metal rails, and have a lot of impact vibration when being connected and disconnected. However, they have a sophisticated suspension system for en route travel and to protect the load from the rough coupling of cars when the train is connected together. They have wide doors for forklift access. These doors must have a very good gasket to prevent air leakage. Typically, the rail car is operated under a positive air pressure and must have good door seals.

The railroad also ships refrigerated trucks in a system known as "piggy-back" cars. These are cars with trucks loaded on the rail cars and shipped across country by rail, Figure 27–22. This is popular because the refrigerated load often would have to be off loaded to a refrigerated truck in the locality where the food is consumed or processed. With the "piggy-back" cars, the refrigerated load arrives on the truck for final destination hauling.

27.5 EXTRA–LOW-TEMPERATURE REFRIGERATION

Extra–low-temperature refrigeration may be described as refrigeration systems that will hold the product temperature at temperatures below −10°F. There are probably many different opinions as to what constitutes extra–low-temperature refrigeration, but this text will define it as below −10°F. Temperatures below −10°F are not used to any great extent for storing food; however, these temperatures are used to fast freeze foods on a commercial basis. These temperatures may be in the vicinity of −50°F.

When the product temperature needs to be reduced using temperatures lower than −50°F, a different type of system may be used. The compression ratio for the compressor becomes too great for single-stage refrigeration at extremely low evaporator temperatures. Another method, however, is to use two stages of compression. It is accom-

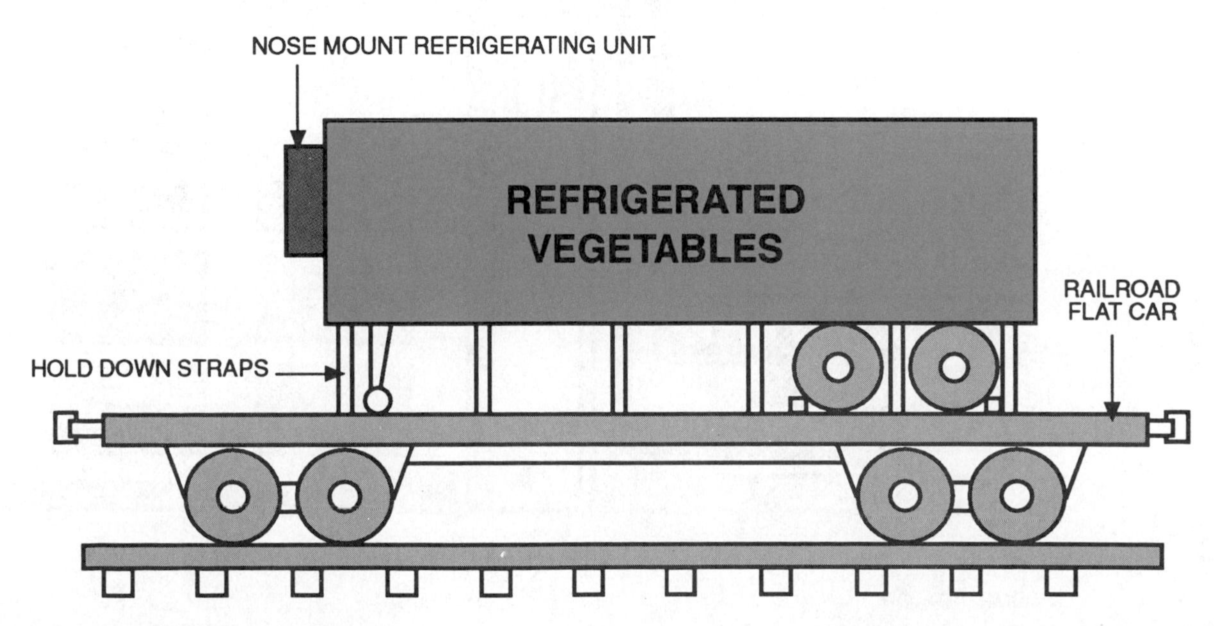

FIGURE 27–22 "Piggy-back" rail car refrigeration.

plished with a first-stage compressor (or cylinder) pumping into a second-stage compressor (or cylinder), Figure 27–23. The compression ratio is split over two stages of compression. If the above single-stage system were operating at a room temperature of −50°F, it would have a boiling refrigerant temperature of about −60°F, using a coil to air temperature difference of 10°F, Figure 27–24. This would make the evaporator pressure 7.2 in. Hg (11.2 psia) for R-502. If

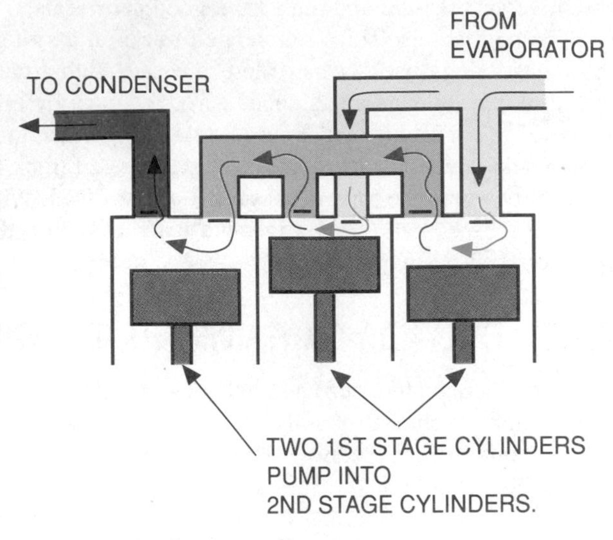

FIGURE 27–23 Two-stage refrigeration.

the condensing temperature were to be 115°F, the head pressure would be 265 psig (280 psia). The compression ratio would be:

CR = Absolute Discharge Pressure ÷ Absolute Suction Pressure

CR = 280 ÷ 11.2

CR = 25:1

This is not a usable compression ratio. It can be reduced by having a two-stage compressor, Figure 27–25. The first stage of compression has a suction pressure of 11.2 psia and a head pressure of 40 psig (55 psia). The compression ratio would be:

CR = 55 ÷ 11.2

CR = 4.9:1

The second-stage compressor would have a suction pressure of 40 psig (55 psia) and a discharge pressure of 265 psig (280 psia).

CR = 280 ÷ 55

CR = 5.09:1

This will allow the first stage of compression to operate at a much better efficiency. The amount of refrigerant gas trapped in the clearance volume at the top of the stroke greatly affects the compressor efficiency.

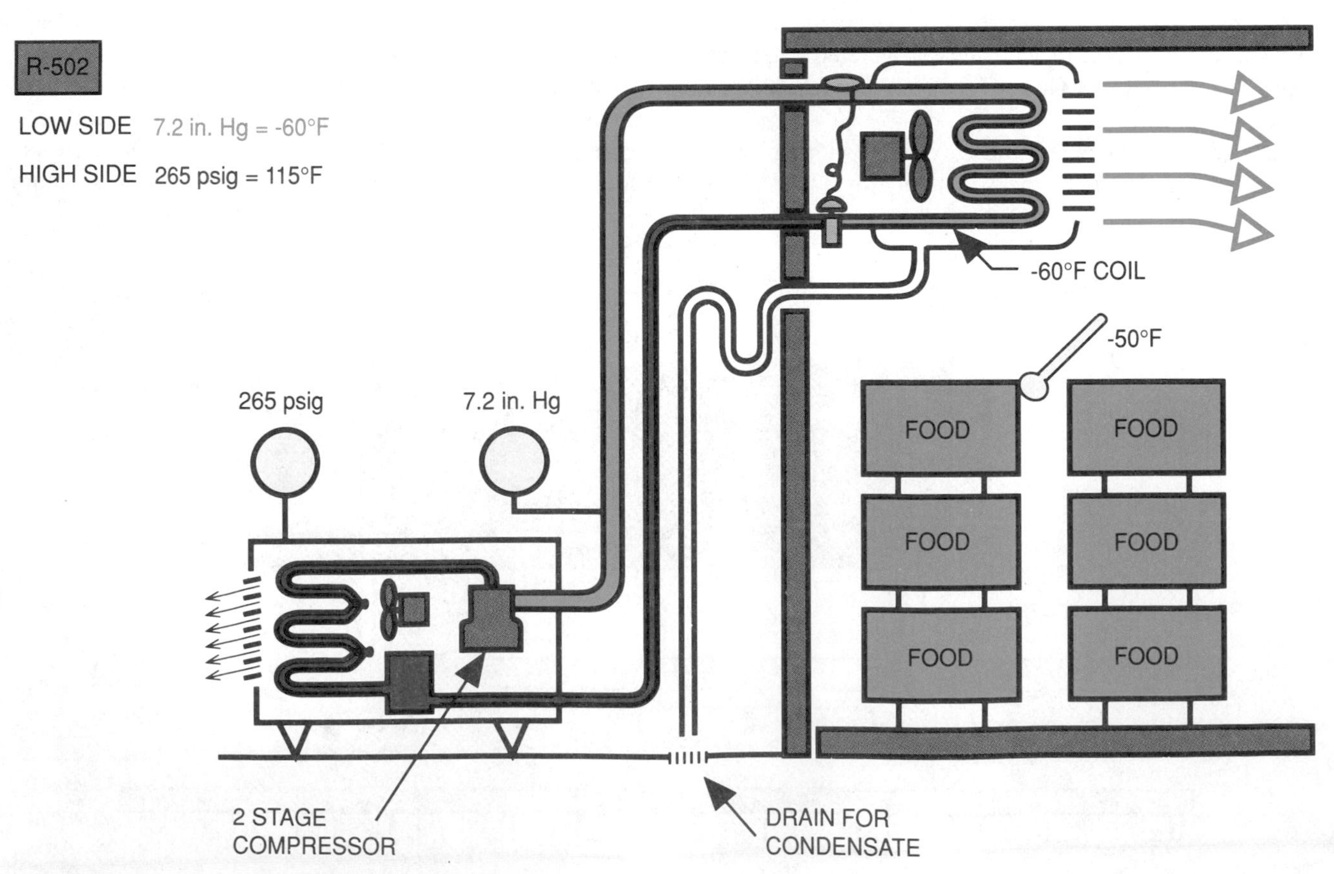

FIGURE 27–24 Two-stage refrigeration system in operation.

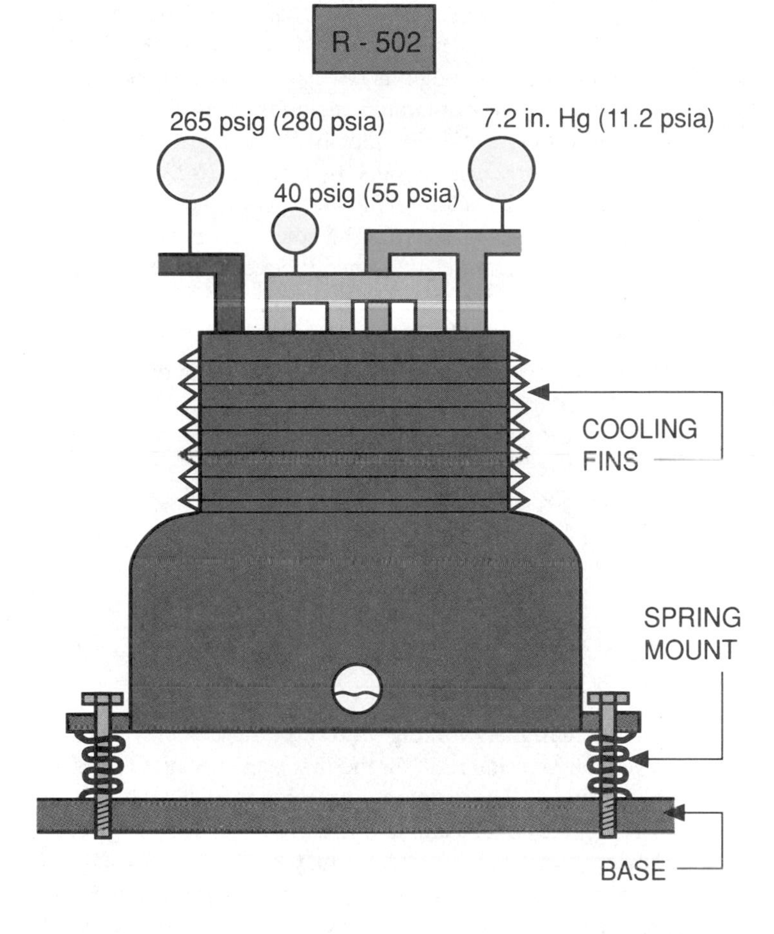

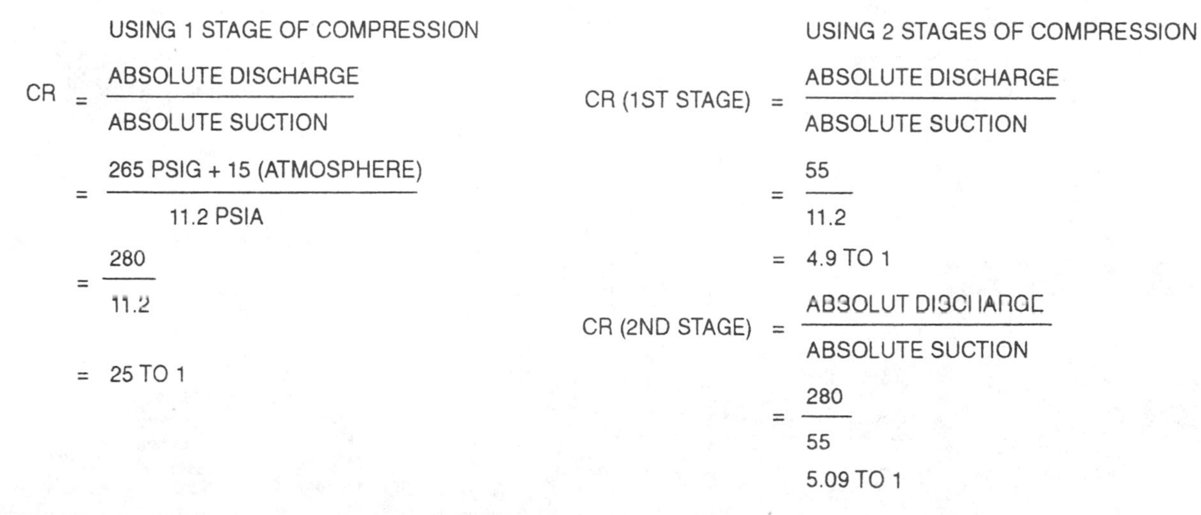

FIGURE 27–25 Compression ratio and two-stage refrigeration.

Manufacturers would always like to have the system operate in a positive pressure. The system above is operating in a vacuum and if a leak occurs on the low-pressure side of the system, atmosphere would enter the system. This system may not be very practical, but it is very possible. The lower the compression ratio the better the compressor efficiency. This is particularly true for the first-stage compressor. Designers will pick the best operating conditions for the system, selecting several compressor combinations until the best efficiency is found. Some three-stage systems may even use three different refrigerants, one for each stage.

Systems with operating temperatures down to about −160°F using the compression cycle may use a system called cascade refrigeration. Cascade refrigeration uses two or three stages of refrigeration, depending on how low the lower temperature range may need to be. This is accomplished when the condenser of one stage exchanges heat with the evaporator in another stage. The condenser rejects heat from the system and the evaporator absorbs heat into the system. This involves more than one refrigeration system, working at the same time and exchanging heat between them.

The first stage of the system uses a refrigerant like R-13 that boils at a very low temperature before it goes into a vacuum. R-13 boils at -114.6°F at atmospheric pressure and would have a vapor pressure of 7.58 psig at an evaporator boiling temperature of -100°F, a pressure considerably above vacuum. The problem with this refrigerant is the high-side pressure. It has a critical temperature of 83.9°F and a critical pressure of 561 psia (546 psig). The critical temperature and pressure is the highest point (temperature and pressure) at which this refrigerant will condense. For example, R-13 will not condense at pressures above 546 psig. Figure 27–26 shows a cascade system with three stages of compression. Note the following conditions in the system.

1. The evaporator is in the first stage of refrigeration and is operating at −100°F with a suction pressure of 7.58 psig using R-13. The discharge pressure is 101.68 because condensing is occurring at −25°F. This is possible because the condenser heat is being removed with an evaporator operating at −35°F. This is the second-stage evaporator.
2. The second-stage evaporator is operating at −35°F and using R-22 and has a suction pressure of 2.6 psig.
3. The second-stage compressor has a discharge pressure of 75.5 psig because it is condensing at 44°F. The reason it is condensing this low is because the condenser is cooled by an evaporator operating at 34°F.
4. The third-stage evaporator is operating at 34°F with a suction pressure of 60.1 psig using R-22. The compressor is discharging into an air-cooled condenser with 90°F air passing over it. The condenser is operating at a 20°F temperature difference (between the air and condensing temperature) and is condensing at 110°F with a head pressure of 226.4 psig.

The compression ratios for the various stages are:

1. Stage 1
   CR = Absolute Suction ÷ Absolute Suction
   CR = 116.68 ÷ 22.58
   CR = 5.17:1
2. Stage 2
   CR = 90 ÷ 17.6
   CR = 5.14:1
3. Stage 3
   CR = 241.4 ÷ 75.1
   CR = 3.21:1

The designer working with these figures may try for a lower compression ratio for the first stage because this is the most critical. The compression ratio may be reduced by lowering the condensing temperature of the first-stage compressor. A different refrigerant may be used in this stage. If the evaporator temperature is dropped much lower in the second stage to lower the condensing temperature in the first stage using R-22, the second-stage evaporator will be operating in a vacuum. R-502 could be used for an evapora-

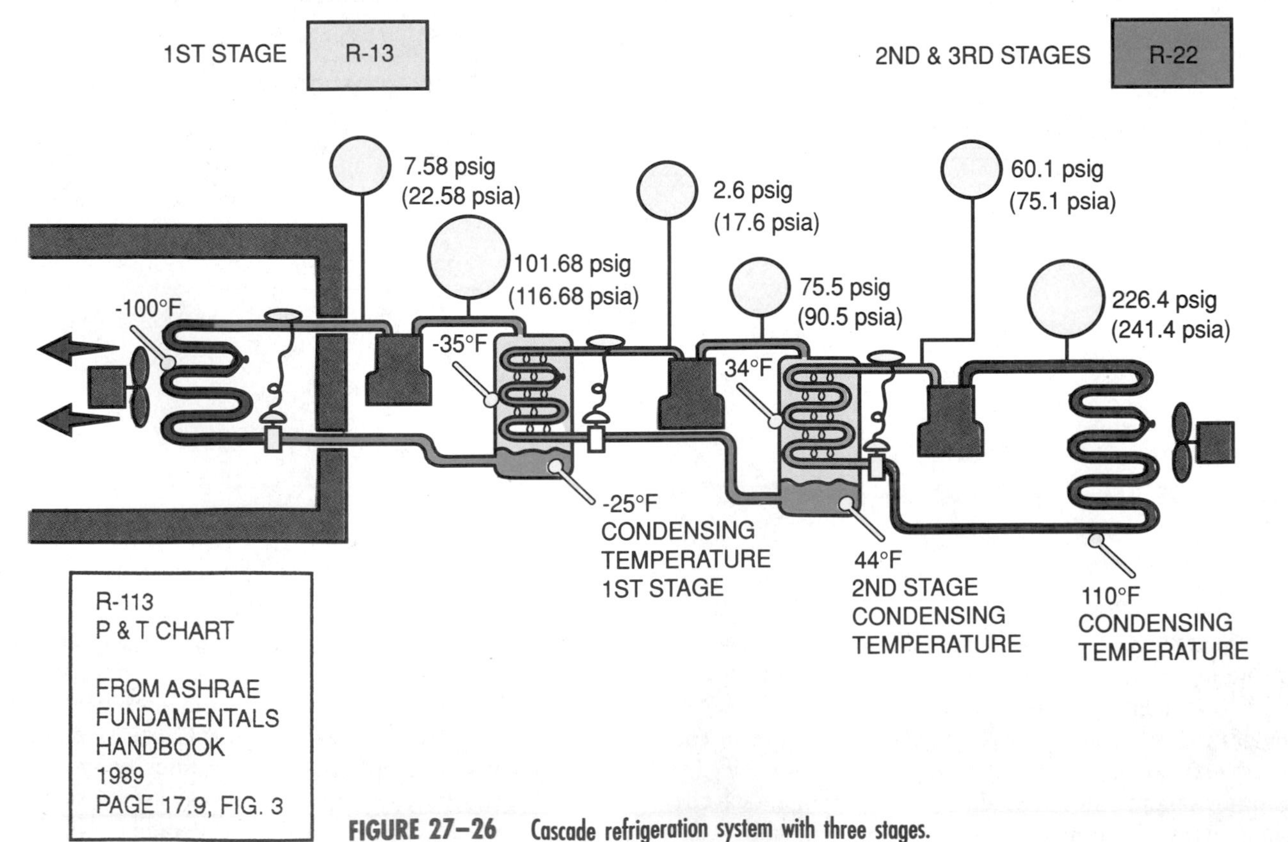

FIGURE 27–26 Cascade refrigeration system with three stages.

tor in the second stage down to −49.8°F before it would operate in a vacuum. The system efficiency can also be improved by operating the condenser in the conditioned space if the system does not have too much capacity. Then room temperature air could be passed over the third-stage condenser.

Several refrigerants may be used for cascade refrigeration. The experienced designer can choose the most efficient refrigerant for the application.

27.6 QUICK FREEZING METHODS

Food that is frozen for later use must be frozen in the correct manner or it will deteriorate, the quality will be reduced, or the color may be affected. When this food is frozen, the faster the temperature is lowered, the better the quality of the food. This can be done in several ways using a refrigeration system at very low temperatures. This method is often called "quick freezing."

Generally, the meats that you buy at the grocery store and freeze in your home freezer do not have the same quality as the meats you buy already frozen. The reason for this is the home freezer does not freeze the meat fast enough. When meat is frozen slowly, ice crystals grow in the cells and often puncture them. This is why you often see a puddle of liquid when meat is thawed. This is food value and flavor escaping. When foods are frozen commercially, they are frozen quickly. This is accomplished using very cold air moving across the product at a high velocity to remove the heat quickly. When most fresh food is frozen very quickly and thawed, it should still have the quality of fresh food.

Different foods have different requirements for freezing. For example, some poultry must be frozen very quickly to preserve the color. To accomplish this, one method would be to seal the meat in an airtight plastic bag; it is then dipped in a very cold solution of brine causing the surface to freeze. The bag is then washed and the poultry removed to another area for freezing the core of the meat at a slower rate.

Foods are often frozen in bulk in packages to be thawed later for processing or cooking. When foods are frozen in this manner, they may stick together, but it makes no difference because they are going to be used in bulk.

Many foods are frozen in individual pieces. For example, shrimp may be frozen individually so they can be sold by the pound while frozen. This process may be accomplished by a process called blast freezing, which uses trays in a blast tunnel or on a conveyer system in a blast freezer room, Figure 27–27. The blast freeze method is accomplished at very low temperatures and a high air velocity. If the conveyer is placed in a room where the temperature is

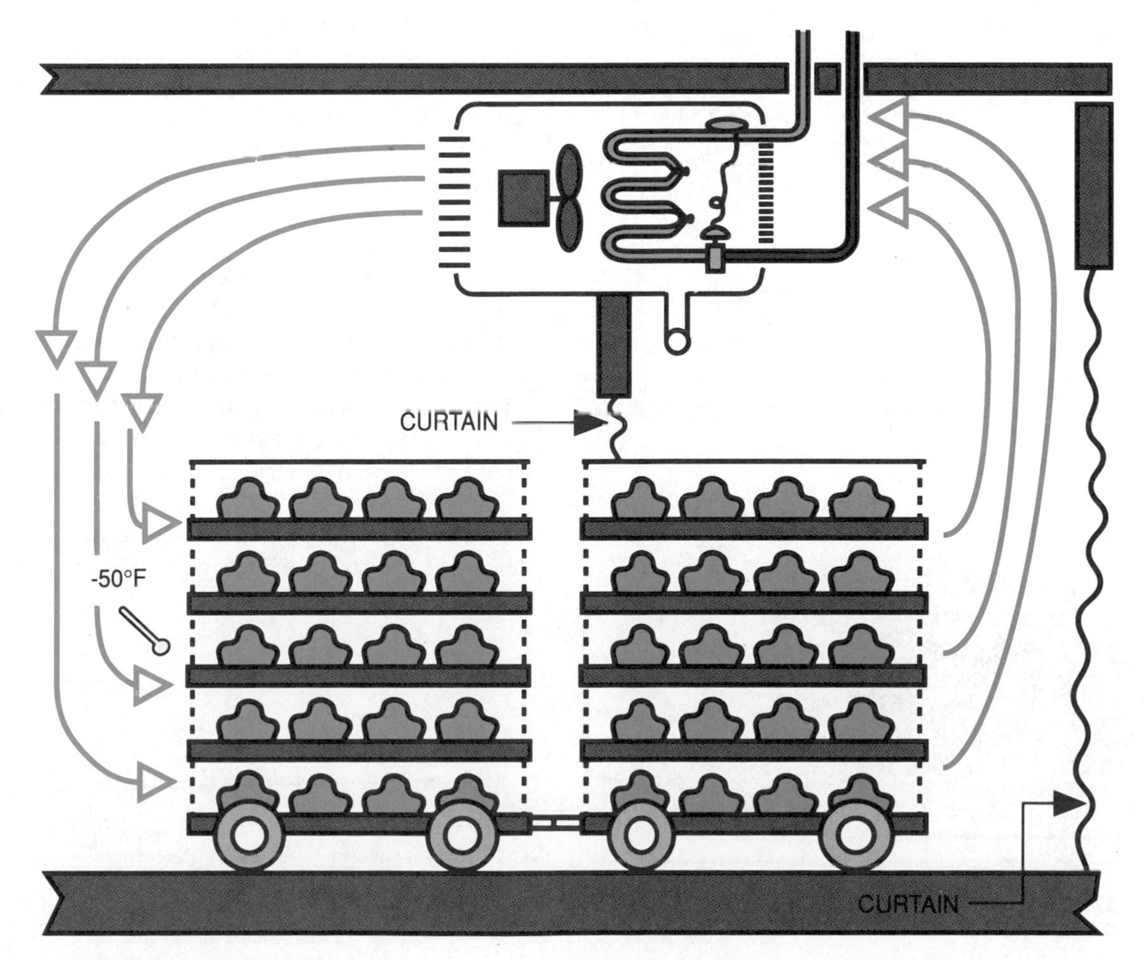

FIGURE 27–27 Blast freezing tunnel.

-40°F or lower and the air is moving fast, small size food pieces may travel the length of the conveyer and leave frozen hard. When the food is placed on the conveyer in a separated fashion, it is not touching other pieces and will be frozen individually, Figure 27–28. This food may then be packaged and dispensed individually. Different weights of food may be placed on longer or shorter conveyer belts or the belt speed may be adjusted to the correct time for freezing the food.

Food trays and tunnels may be used for fast freezing food. The food is placed on trays with plenty of space around the food. The trays are loaded on carts and moved into a tunnel, Figure 27–29. When the tunnel is full and a cart is pushed in, a cart is pushed out the other end with frozen food, ready to package. This method requires more labor, but is less expensive to build from a first-cost standpoint than a conveyer system, which is more automated.

Larger parcels of food, such as a carcass of beef, would take longer to freeze. These would be precooled to very close to freezing and then moved to the blast freeze room for quick freezing.

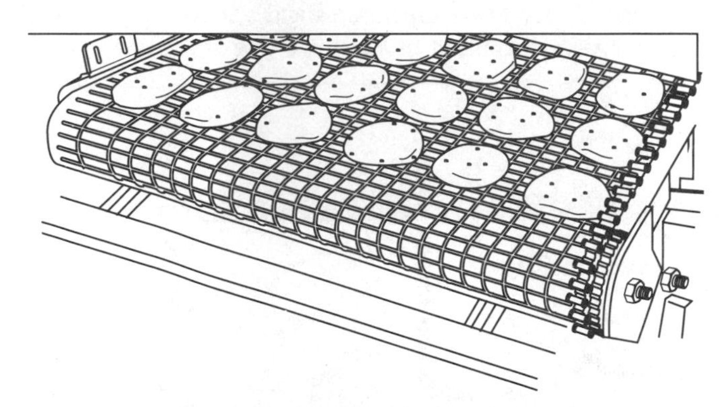

FIGURE 27–28 Conveyer system for a blast cooler.

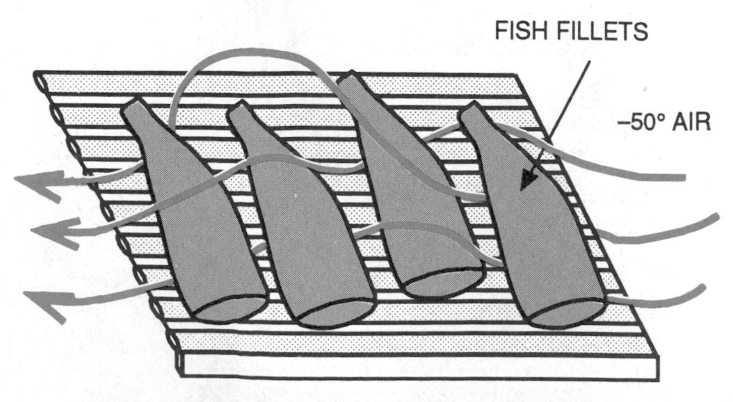
THE FISH FILLETS ARE PRE-COOLED TO JUST ABOVE FREEZING BEFORE ENTERING THE QUICK FREEZE PROCESS
FISH FILLETS
-50° AIR
THE COLD AIR COMES IN CONTACT WITH ALL SIDES OF THE FISH

FIGURE 27–29 Spacing of food for tunnel quick cooling.

This quick freezing method may be accomplished using standard refrigeration equipment. The larger refrigeration systems may use ammonia as the refrigerant with reciprocating compressors. Other systems may use R-22 and screw compressors.

⊛The chlorofluorocarbon (CFC)/ozone depletion issue is causing ammonia systems to make a comeback. They are efficient and do not pollute.⊛

27.7 MARINE REFRIGERATION

Ship refrigeration may be used for several different applications. For example, a ship may pick up a perishable product requiring medium-temperature refrigeration (34°F) in South America and deliver it to New York City. The ship may then pick up a load of frozen food to be delivered to Spain at −15°F, a low-temperature application. The ship's refrigeration system must be able to operate at either temperature range.

Another application for ship refrigeration may be for preserving fish caught at sea. If the ship is only out to sea for a few days, ice carried from shore is often used. This would be the case of a shrimp boat operating off the coast. Auxiliary refrigeration is also used to aid the ice in refrigerating the load, to prevent excess melting. If the ship is to be at sea for long periods, the load must be frozen for preservation. It is not unusual for a ship to stay at sea for several months gathering a load of fish. This is often accomplished with a fleet of small boats that catch the fish and a large refrigerated processing ship called a mother ship, Figure 27–30. In this case, the refrigeration system may have to be very large, possibly several hundred tons. These ships must have a method for quick freezing fish to preserve the flavor and color.

Another method used for refrigerated freight is to haul it in containerized compartments that are loaded onto the ship. These containers may have their own electric refrigeration system that may be operated at the dock and can be plugged into the ship's electrical system, Figure 27–31. These units must be located in such a manner that air can circulate over the condensers or they will have head pressure problems. It is typical for them to be located on the deck of an open ship.

The ships that have on-board large refrigeration systems may use any of the refrigerants listed in this text, such

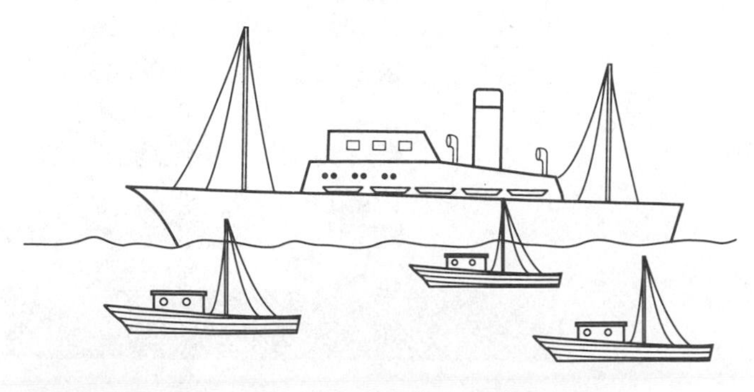

FIGURE 27–30 Mother ship and a fleet of fishing boats.

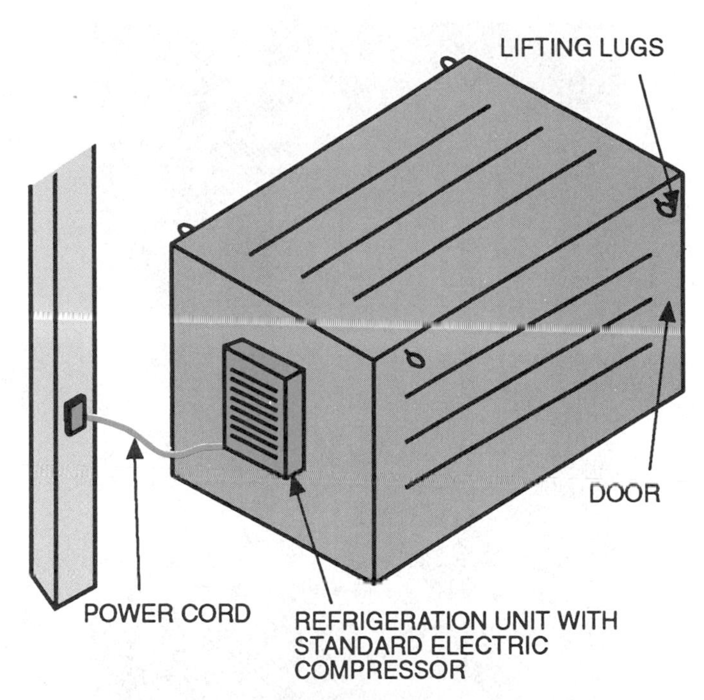

FIGURE 27–31 Containerized freight for ships.

as R-12, R-22, R-502, or R-11, depending on the age and the design of the refrigeration system. **The CFC refrigerants, R-11, R-12, and R-502, will be phased out and it is not clear at this time what the replacement refrigerants will be for these applications.** Ammonia refrigeration systems are also used for ships. This refrigerant is not considered dangerous to the environment so these systems will be ongoing. It is, however, extremely toxic to humans and must be used with care. ***It is necessary for technicians to wear breathing apparatus as well as protective clothing, gloves, and goggles when working with ammonia systems.***

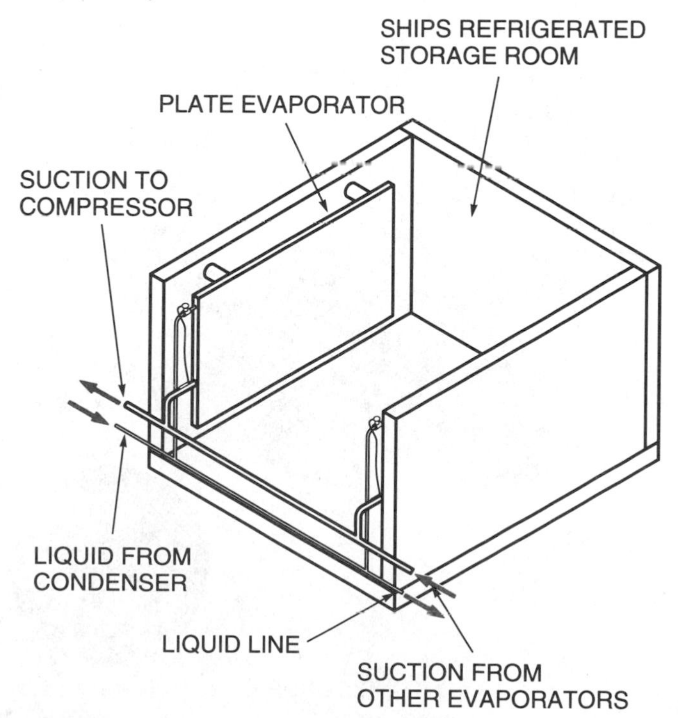

FIGURE 27–32 Plate evaporators for a ship.

Large system compressors may be reciprocating, screw, or centrifugal when R-11 is used. They may be driven using the ship's electrical system if it has a large enough generating plant using belts or direct drive. When the system becomes large, it may be steam turbine driven using the ship's steam system.

The evaporators may be large plates in the individual refrigerated rooms or bare pipes for some systems, Figure 27–32. Systems using forced air are also used, Figure 27–33. These systems are direct expansion, using thermostatic expansion valves. The ship may have several refrigerated compartments so all the refrigeration is not in one large area.

These evaporators must have some means for defrost because they will normally be operating below freezing at the coils. This may be accomplished with hot gas if the evaporator is close enough to the compressor. If not, sea water may be used. If it is too cold, it may be heated and sprayed over the evaporator. This must be done without the fan running on a fan coil unit and the coil must not be refrigerating during defrost time. This can be done by stopping the fan and shutting off the liquid line solenoid coil during defrost. The compressor will pump the refrigerant out of the evaporator during defrost. Defrost can be terminated using coil temperature or time. The refrigerant that is pumped out of the coil will be in the receiver during defrost and the compressor can continue to run to other refrigerated compartments on the ship. Defrost can be staggered so all coils will not call for defrost at the same time.

The condensers will use sea water, which is readily available everywhere the ship goes. Sea water contains debris and must be filtered with filters that are easy to access and clean. This is commonly done with valves that allow the pumps to reverse the flow through the filters and piping to backwash the system, Figure 27–34. The refrigerant condensers for the typical refrigerants, R-11, R-12, R-22 and R-502, must be designed and built for easy cleaning because they will become dirty. These condensers have removable water box covers, called marine water boxes, where the

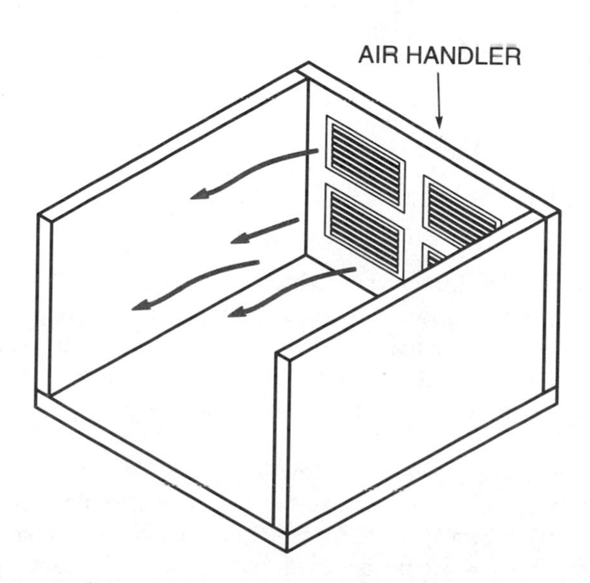

FIGURE 27–33 Forced-air refrigeration for ship.

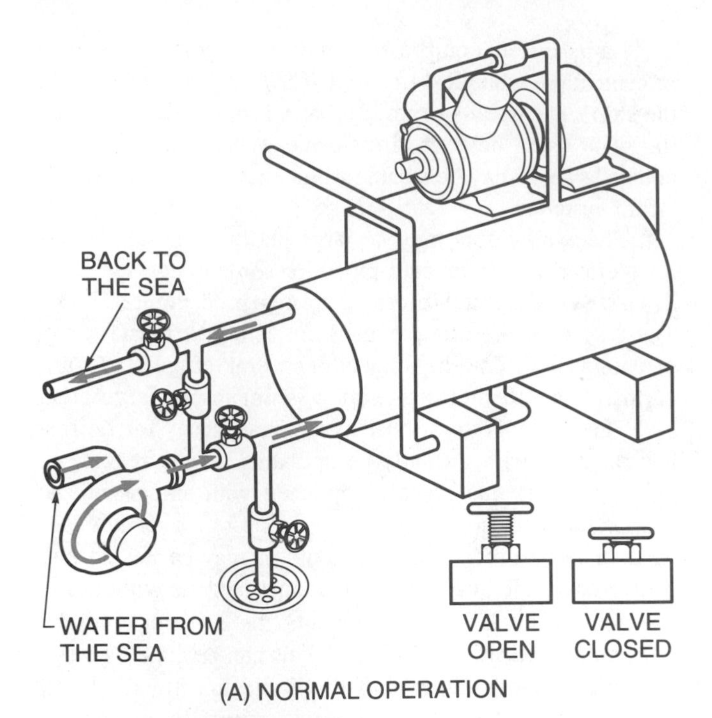

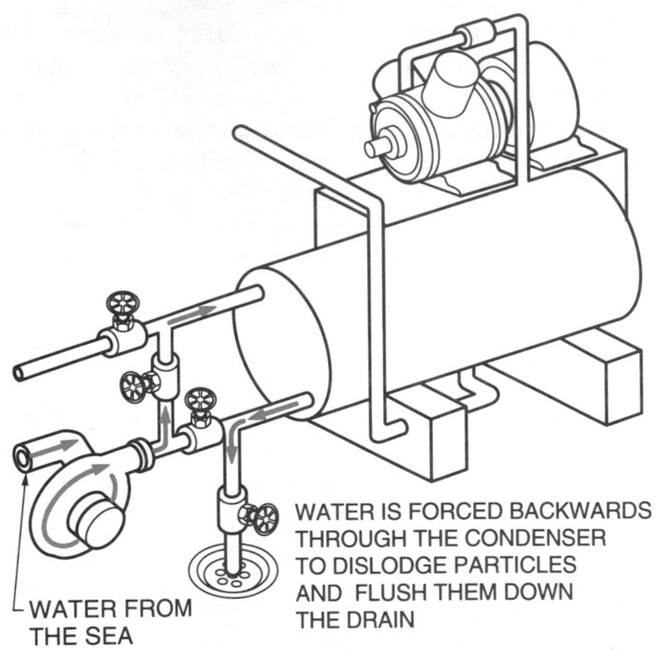

FIGURE 27–34 Flushing the ship's sea water-cooled condensers.

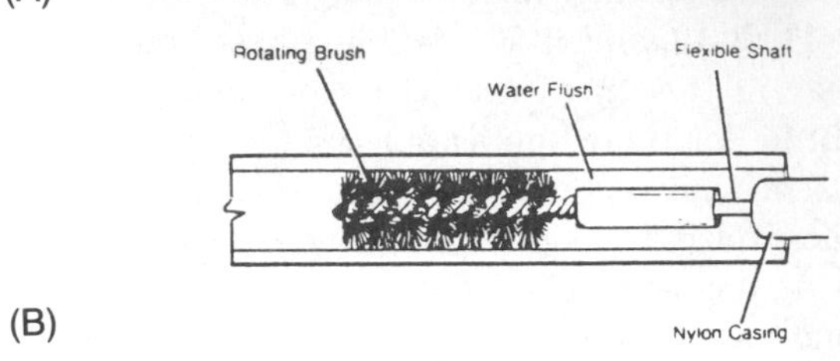

FIGURE 27–35 Cleaning a condenser using a brush.

tubes can be cleaned using a brush, Figure 27–35. The tubes for these refrigerants are made of cupronickel (copper and nickel), a metal that does not corrode easily and that is tough. The tube sheet that holds the tubes is typically made of steel and the water box of brass where possible, Figure 27–36.

Reserve capacity refrigeration and a spare parts inventory for making repairs at sea are recommended by the American Bureau of Shipping. They recommend that two compressor systems be used, either one of which with enough capacity to maintain the load, running continuously

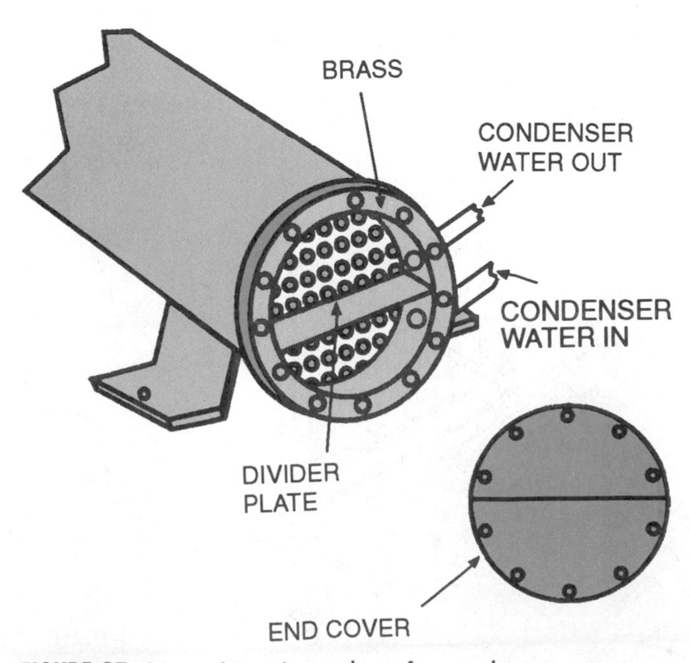

FIGURE 27–36 Tubes and water boxes for a condenser.

while in tropical waters. This is 100% standby capacity of the compressor, which is the most likely part of the system to cause a problem. These systems can become quite large. There must be someone aboard with experience who is familiar with the system and can make any necessary repairs. Repairs at sea are not uncommon. It can take days to reach a port with a repair facility.

The individual evaporators can be located at great distances from the ship's condensers and compressors and long refrigerant lines will have to be run. This brings on the problems of returning lubricating oil that leaves the compressor. Oil separators may be used at the compressor to keep most of the oil in the compressor, but some will get into the system and will have to be returned.

Circulating brine systems have some advantages over individual evaporators in the refrigerated spaces. The refrigeration and the refrigerant can all be contained in the equipment room with a central compressor or compressors. The brine may be chilled and circulated to the various applications, Figure 27–37.

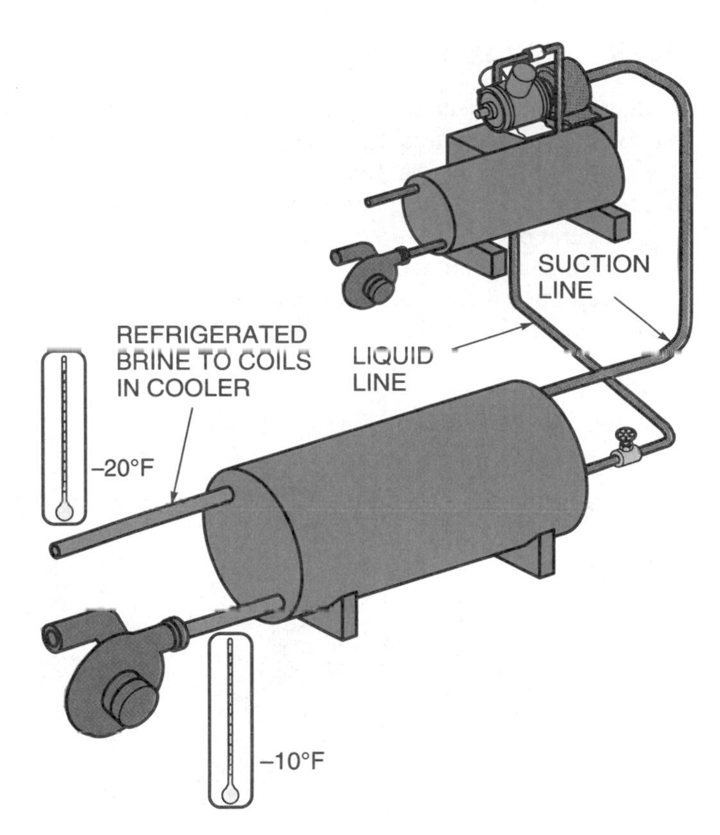

FIGURE 27–37 Brine circulation system for a central refrigeration system.

As with truck refrigeration, cargo ship refrigeration is designed to hold the refrigerated load temperature. It usually does not have much reserve capacity to reduce the load temperature. Even if there is a reserve refrigeration system, it will only be the heavy running gear, such as the compressor, fans, and motors, and not a complete system. The ship storage rooms are not designed and built for the type of air circulation to reduce the temperature of the load even if reserve capacity is available.

When the ship is loaded, every square foot of storage counts. The ship must be designed to accept the load and store it for the trip even if the seas are rough. The placement of the load will have much to do with the success of the refrigerated load. Correct circulation of air over the load is important when the load is fresh food, such as fruit or vegetables. These foods give off heat called heat of respiration as they are stored, Figure 27–38. This is not like storing a load of frozen food that only has to be protected from heat that may enter from the outside, like a walk-in cooler. It is heat generated by the load and is calculated in the size of the refrigeration equipment.

All ships must have refrigeration for the ship storage of foods for the crew and or passengers. Refrigeration for the ship stores may be drawn from the main refrigeration system as long as it is running when the ship hauls refrigerated cargo. When the ship is not hauling refrigerated cargo, auxiliary systems may be used or the ship may have a separate system for the crew. Many of the same appliances may be used on the ship as in a supermarket. Some of these units may be individual package units with the compressor condenser built into the unit or central refrigeration rooms may be used.

Thermal Properties of Food

	Heat of Respiration, Btu/ton · d						
Commodity	**32°F**	**41°F**	**50°F**	**59°F**	**68°F**	**77°F**	**Reference**
Beans, Lima Unshelled	2306-6628	4323-7925	—	22,046-27,449	29,250-39-480	—	Lutz and Hardenburg (1968) Tewfik and Scott (1954)
Shelled	3890-7709	6412-13,436	—	—	46,577-59,509	—	Lutz and Hardenburg (1968) Tewfik and Scott (1954)
Beans, Snap	*b	7529-7709	12,032-12,824	18,731-20533	26,044-28,673	—	Ryall and Lipton (1972) Watada and Morris (1966)
Beets, Red, Roots	1189-1585	2017-2089	2594-2990	3711-5115	—	—	Ryall and Lipton (1972)
Sweet	901-1189	2089-3098	—	5512-9907	6196-7025	—	Lutz and Hardenburg (1968), Micke *et al.* (1965), Gerhardt *et al.* (1942)
Corn, Sweet with Husk, Texas	9366	17,111	24,676	35,878	63,543	89,695	Scholz *et al.* (1963)
Cucumbers, Calif.	*b	*b	5079-6376	5295-7313	6844-10,591	—	Eaks and Morris (1956)

FIGURE 27–38 Heat of respiration for some fresh vegetables. *Reprinted with permission of the American Society of Heating, Refrigeration and Air Conditioning Engineers from the 1990 ASHRAE Handbook–Refrigeration*

27.8 AIR CARGO HAULING

Products are often hauled by air to hasten the product trip to market. Flowers are frequently shipped from places like Hawaii to the rest of the United States by air. On-board refrigeration is out of the question because of the weight of the refrigeration systems. When the product needs to be cooled during flight, ice or dry ice is used.

Specially designed containers are built to slide into the cargo compartment of the plane, Figure 27–39.

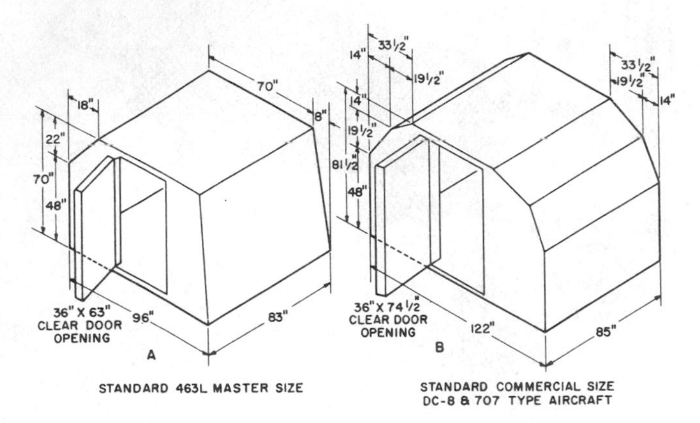

FIGURE 27–39 Containers for shipping refrigerated products for airlines. *Reprinted with permission of the American Society of Heating, Refrigeration and Air Conditioning Engineers from the 1990 ASHRAE Handbook–Refrigeration*

SUMMARY

- Transport refrigeration is the process of transporting various refrigerated products by truck, rail, air, or water.
- Vegetables shipped from California to New York will need refrigeration while crossing the hot desert and may even need heat when traveling in the cold northern states.
- Some products are shipped at medium temperatures and some at low temperatures in the frozen state.
- Products are shipped by truck, train, plane, and ship.
- The simplest transport refrigeration is to use ice made at a central plant.
- Dry ice, made from compressed carbon dioxide (CO_2), may be used to refrigerate loads. It changes state from a solid to a vapor without going through the liquid state, called sublimation, at -109°F. It is particularly good for low-temperature loads.
- Liquid nitrogen or CO_2 may also be used for refrigeration. The liquid is stored in low-pressure cylinders with a relief valve system that maintains the cylinder pressure at about 25°F by relieving excess pressure from the cylinder. Nitrogen evaporates at −320°F and CO_2 at −109°F.
- These systems are simple, with only a thermostat controlling a solenoid valve. A distribution system distributes the cold nitrogen vapor.
- ***Nitrogen and CO_2 must be handled with care because they are so cold. They can freeze flesh immediately on contact. Be sure to wear protective clothing, gloves, and goggles when working with either of these systems. Both of these vapors also displace oxygen so the system cannot be operating while people are in the refrigerated space. Special controls shut the system off if the door is opened to the refrigerated space.***
- Trucks are often equipped with refrigerated plates that contain a solution called a "Eutectic solution" that goes through a phase change when cooled and warmed. The solution is typically a brine solution of either sodium chloride (table salt) or calcium chloride.
- These brine solutions are corrosive to ferrous metals.
- The brine solution does not actually turn solid; it turns to crystals when frozen and the crystals turn to liquid when warmed.
- Trucks may also be refrigerated using mechanical refrigeration located on the truck. Some of the common systems for powering truck refrigeration systems are using the truck's engine and a compressor with an electric clutch; an electric motor drive may be an option for this system, or a small refrigeration system either under the truck or on the front of the truck that is driven with a small gas or diesel engine.
- When at the dock, the compressor may be plugged into a power supply.
- Larger trucks use either nose-mount or under-belly refrigeration units.
- The evaporator and fan are located in the front of the truck on a typical unit.
- Truck systems are not designed to have the capacity to refrigerate a load that is not down to temperature. They are meant to hold the load temperature while in transit. There is not much reserve capacity.
- Truck bodies are insulated and the doors have gaskets to prevent infiltration.
- The inside of the truck body must be rugged and able to stand the weight of a forklift. It must also be made of materials that can be cleaned easily for sanitary food storage.
- Railway cars have been refrigerated by axle-driven units and compression systems located in the front of the car. These compression system units generally use electric motor-driven compressors with a motor generator as the power supply. They can be plugged into local power while at the station for long periods.

- Two-speed engines and compressors that unload are used for capacity control.
- Extra–low-temperature refrigerated applications at temperatures below -10°F often use the compression cycle by either using two-stage compressors or cascade refrigeration systems.
- Cascade refrigeration systems use more than one stage of refrigeration to reach very low temperatures.
- Fresh food that is frozen quickly retains its fresh color and taste. The quicker the food is frozen, the better these qualities will be maintained.
- When meats are frozen slowly, the ice crystals form slowly and puncture the cell walls of the meat. This allows the meat to drain when thawed.
- Food may be frozen quickly by dipping it in a cold solution, such as brine (salt water), or it may be frozen quickly by using blast freezers, very cold air blowing directly on the food. The cold air and the velocity help to remove heat from the food quickly.
- Food that is frozen in bulk may be contained in packages for defrost and processing later. Food that needs to be used individually must be individually frozen.
- Conventional refrigeration systems may be used for blast freezing.
- Ship refrigeration may have several applications, from medium temperature to low temperature.
- A fish processing ship must have the capability of quick freezing and holding the fish after they are processed.
- Refrigerated cargo ships may have large refrigerated holds where plate-type or forced-draft evaporators are used to maintain the temperature. Circulated brine systems may have an advantage over individual evaporators in each hold because of long refrigerant lines and possible long oil return lines.
- The condensers for ship refrigeration usually use sea water and are made of cupronickel.
- Experienced technicians and reserve refrigeration capacity is necessary for a ship refrigeration system.
- Another method for shipping refrigerated cargo is to use self-contained air-cooled refrigeration on containerized freight.
- Ships must also have refrigeration for the ship's stores.
- Many products are shipped by air to speed them to market (eg, flowers). They are typically kept cool using ice or dry ice.

REVIEW QUESTIONS

1. Name three different types of transport refrigeration.
2. What is the evaporating temperature for liquid nitrogen, CO_2?
3. What is the working pressure of a typical liquid nitrogen storage tank and how is this pressure maintained?
4. What are the two different solutions used in "Eutectic" plate refrigeration?
5. Describe phase change refrigeration in the "Eutectic" cold plates?
6. What is the advantage in cold plate refrigeration?
7. What is the method of refrigerating cold plates?
8. Name two locations for truck refrigeration systems.
9. Why are truck refrigeration systems primarily made of aluminum?
10. Name three different power supplies used for truck refrigeration.
11. What type of reserve capacity does truck refrigeration have?
12. Why is truck refrigeration designed to operate on low- and medium-temperature refrigeration?
13. How is the air distributed over the cargo in truck refrigeration?
14. Name two types of engines used for truck refrigeration systems.
15. How are rail cars refrigerated?
16. What is "piggy-back" refrigeration?
17. Why is two-stage compression popular for extra–low-temperature refrigeration systems?
18. An R-502 system is operating with a −50°F evaporator and the condenser is operating at 115°F. What is the compression ratio of this system?
19. Describe cascade refrigeration and tell why more than one refrigerant is used.
20. Why is a circulating brine system more attractive for ship refrigeration than individual evaporators?
21. Name two methods used to power the compressors on a ship.
22. What is a mother ship and what is the application?
23. Why do ships have so many reserve parts for the refrigeration system?
24. What are the ship's stores?
25. How are flowers kept cool on an airplane?

UNIT 28

Troubleshooting and Typical Operating Conditions for Commercial Refrigeration

OBJECTIVES

After studying this unit, you should be able to

- list the typical operating temperatures and pressures for the low-pressure side of a refrigeration system for high, medium, and low temperatures.
- list the typical operating pressures and temperatures for the high-pressure side of the systems.
- state how R-12 and R-502 compare on the high-pressure and low-pressure sides of the system.
- diagnose an inefficient evaporator.
- diagnose an inefficient condenser.
- diagnose an inefficient compressor.

SAFETY CHECKLIST

* Wear goggles and gloves when attaching and removing gages and when transferring refrigerant or checking pressures.

* Wear warm clothing when working in a cooler or freezer. A cold technician does not make good decisions.

* Observe all electrical safety precautions. Be careful at all times and use common sense.

* Disconnect electrical power whenever possible when working on a system. Lock and tag the panel and keep the key on your person.

* Wear rubber gloves when using chemicals to clean an evaporator or condenser. Use only approved chemicals. This is particularly important when cleaning an evaporator in a food storage unit.

* Do not attempt to perform a compressor vacuum or closed-loop test without the close supervision of an instructor or other experienced person.

28.1 ORGANIZED TROUBLESHOOTING

To begin troubleshooting any area of a refrigeration system, you need some idea of what the typical conditions should be. In commercial refrigeration we may be dealing with air temperatures inside a refrigerated box or outside at the condenser. We may need to know the current draw of the compressor or fan motors. The pressures inside the system may be important for troubleshooting and diagnosis. There are many conditions both outside and within the system that are important to know and understand.

It is helpful to keep in mind that when a piece of equipment has been running well for some period of time without noticeable problems that normally only one problem will trigger any sequence of events that may be encountered. For instance, it is common for the technician to approach a piece of equipment and think that two or three components have failed at one time. The chance of two parts failing at once is remote, unless one part causes the other to fail. These failures can almost always be traced to one original cause.

Knowing how the equipment is supposed to be functioning helps. You should know how it is supposed to sound, where it should be cool or hot, and when a particular fan is supposed to be operating. Knowing the correct operating pressures for a typical system can help you get started. **Before you do anything, look over the whole system for obvious problems.**

These troubleshooting procedures are divided into high-, medium-, and low-temperature ranges with some typical pressures for each. Each system has its own temperature range, depending on what it is supposed to be refrigerating. The table in Figure 28–1 shows the temperature ranges for some refrigerated storage applications. Notice that there is a different temperature requirement for almost every food under more than one situation. There may be a different temperature requirement for long-term storage than for short-term storage. The temperatures and pressures on the low-pressure side of the system are a result of the product load. **An evaporator acts the same whether there is a single compressor close to the evaporator or a large compressor in the equipment room.** The examples given relate to evaporators. The function of the compressor is to lower the pressure in the evaporator to the correct boiling point to achieve a desired condition in the evaporator.

COMMODITY	STORAGE TEMP. °F	RELATIVE HUMIDITY %	APPROXIMATE STORAGE LIFE
Carrots			
Prepackaged	32	80–90	3–4 weeks
Topped	32	90–95	4–5 months
Celery	31–32	90–95	2–4 months
Cucumbers	45–50	90–95	10–14 days
Dairy products			
Cheese	30–45	65–70	
Butter	32–40	80–85	2 months
Butter	0 to −10	80–85	1 year
Cream (sweetened)	−15	—	several months
Ice cream	−15	—	several months
Milk, fluid whole			
Pasteurized Grade A	33	—	7 days
Condensed, sweetened	40	—	several months
Evaporated	Room Temp	—	1 year, plus
Milk, dried			
Whole milk	45–55	low	few months
Non-fat	45–55	low	several months
Eggs			
Shell	29–31	80–85	6–9 months
Shell, farm cooler	50–55	70–75	
Frozen, whole	0 or below	—	1 year, plus
Frozen, yolk	0 or below	—	1 year, plus
Frozen, white	0 or below	—	1 year, plus
Oranges	32–34	85–90	8–12 weeks
Potatoes			
Early crop	50–55	85–90	—
Late crop	38–50	85–90	—

(A)

COIL-TO-AIR TEMPERATURE DIFFERENCES TO MAINTAIN PROPER BOX HUMIDITY

TEMPERATURE RANGE	DESIRED RELATIVE HUMIDITY	TD (REFRIGERANT TO AIR)
25° F to 45° F	90%	8° F to 12° F
25° F to 45° F	85%	10° F to 14° F
25° F to 45° F	80%	12° F to 16° F
25° F to 45° F	75%	16° F to 22° F
10° F and Below	—	15° F or less

(B)

FIGURE 28–1 (A) Chart shows some storage requirements for typical products. (B) Coil-to-air temperature difference chart to select desired relative humidity. *Courtesy (A) Used with permission from ASHRAE, Inc., 1791 Tullie Circle NE, Atlanta, GA 30329, (B) Copeland Corporation*

28.2 TROUBLESHOOTING HIGH-TEMPERATURE APPLICATIONS

High-temperature refrigerated box temperatures start at about 45°F and go up to about 60°F. Normally a product temperature will be at one end of the range or the other depending on the product (flowers may be at 60°F and candy at 45°F). The intent is to provide the low- and high-temperature and pressure conditions that may be encountered for high-temperature refrigeration systems. The coil temperature is normally 10°F to 20°F cooler than the box temperature. This difference is referred to as temperature difference or TD. This means that the coil will normally be operating at 26 psig (45°F – 20°F = 25°F coil temperature) at the lowest temperature for an R-12 system. The pressure would be 58 psig (60°F coil temperature) at the end of the cycle when the compressor is off for the highest temperature for a high-temperature application. Figure 28–2 is a chart of some typical operating conditions for high-, medium-, and low-temperature applications. These figures are for R-12 and R-502 for their typical applications. If any other refrigerant or application is used, the temperature readings and pressure readings may be converted to the different refrigerant.

When the technician suspects problems and installs the gages for pressure readings, the following readings should be found on a 20°F TD high-temperature system using R-12:

1. With the compressor running just before the cut-out point, a reading below 25 psig would be considered too low.
2. With the compressor running just after it started up with a normal box temperature, 60°F (60°F − 20°F = 40°F coil temperature) a reading above 37 psig would be considered too high.
3. When the compressor is off and the box temperature is up to the highest point, the coil pressure would correspond to the air temperature in the box.

If a 60°F box is the application, the suction pressure could be 58 psig. This would be just before the compressor starts in a normal operation.

These pressures are fairly far apart but serve as reference points for most high-temperature applications. The preceding illustration is for a 20°F TD. There will be times when a 10°F TD is considered normal for a particular application. The best procedure for the technician is to consult the chart in Figure 28–1 for the correct application and determine what the pressures should be by converting the box temperatures to pressure. This chart shows what the storage temperatures are.

The correct box humidity conditions may also be found by studying Figure 28–1. The box humidity conditions will determine the coil-to-air temperature difference for a particular application. For example, you may want to store cucumbers in a cooler. The chart shows that cucumbers should be stored at a temperature of 45°F to 50°F and a humidity of 90% to 95% to prevent dehydration. Examining the second part of the chart indicates that to maintain these conditions the coil-to-air temperature difference should be 8°F to 12°F; probably 10°F should be used.

The pressures from one piece of equipment to another may vary slightly. *Do not forget that any time the compressor is running, the coil will be 10°F to 20°F cooler than the fixture air temperature.*

28.3 TROUBLESHOOTING MEDIUM-TEMPERATURE APPLICATIONS

Medium-temperature refrigerated box temperatures range from about 30°F to 45°F in the refrigerated space. Many products will not freeze when the box temperature is as low as 30°F. Using the same methods that were used in the high-temperature examples to find the low- and high-pressure readings that would be considered normal, we find the following. The lowest pressures that would normally be encountered would be at the lowest box temperatures of 30°F. If the coil were to be boiling the refrigerant at 20°F below the inlet air temperature, the refrigerant would be boiling at 10°F (30°F – 20°F = 10°F). This corresponds to a pressure for R-12 of 15 psig. If the service technician encounters pressures below 15 psig for an R-12 medium-temperature system, there is most likely a problem. The highest pressures that would normally be encountered while the system is running would be at the boiling temperature just after the compressor starts at 45°F. This is 45°F – 20°F = 25°F, or a corresponding pressure of 25 psig, Figure 28–2. The pressure in the low side while the compressor is off would correspond to the air temperature inside the cooler. When 45°F is considered the highest temperature before the compressor comes back on, the pressure would be 42 psig. Consult the chart in Figure 28–1 for any specific application because each is different. The chart is used as a guideline for the product type. The equipment operating conditions may vary to some extent from one manufacturer to another.

28.4 TROUBLESHOOTING LOW-TEMPERATURE APPLICATIONS

Low-temperature applications start at freezing and go down in temperature. There is not much application at just below freezing. The first usable application is the making of ice, which normally occurs from 20°F evaporator temperature to about 5°F evaporator temperature. Ice making has many different variables. These variables are associated with the type of ice being made. Flake ice is a continuous process so the pressures are about the same regardless of the manufacturer. The manufacturer of the specific piece of equipment should be consulted for exact temperatures and pressures for flake ice machines.

Normally for cube ice makers you will find that a suction pressure of 21 psig (20°F) at the beginning of the cycle will begin to make ice using R-12. The ice will normally be harvested before the suction pressure reaches the pressure corresponding to 5°F or 11 psig, Figure 28–2. If you discover that the suction pressure would not go below 21 psig for an R-12 machine, suspect a problem. If the suction pressure goes below 11 psig, suspect a problem. These pressures are for the normal running cycle.

The low-temperature application for frozen foods is generally considered to start at 5°F space temperature and goes down from there. Foods require different temperatures to be frozen hard. Some are hard at 5°F, others at −10°F. The designer or operator will use the most economical design or operation. For instance, ice cream may be frozen hard at −10°F, but it may be maintained at −30°F without harm. The economies of cooling it to a cooler temperature may not be wise. As the temperature goes down, the compressor has less capacity because the suction gas becomes

This table is intended to show the typical operating pressures for commercial refrigeration systems. Column 1 is with the compressor off and the box temperature just at the cut-in point. Column 2 is just after the compressor comes on. Column 3 is just before the compressor cuts off. This is not intended to be the only operating pressures, but the upper and lower limits as applied to high-, medium- and low-temperature typical applications. A coil-to-inlet air temperature of 20° F (TD) was used as an average. Many systems will use 10° F or 20° F TD. See Figure 25–1 (A) for different TD applications.

R-12

	Column 1 Compressor Off	Column 2 Compressor On	Column 3 Compressor On
HIGH-TEMPERATURE			
Box Temperature	60° F	60° F	45° F
Coil Temperature	60° F	40° F	25° F
Temperature Difference	0° F	20° F	20° F
Suction Pressure	58 psig	37 psig	25 psig
MED-TEMPERATURE			
Box Temperature	45° F	45° F	30° F
Coil Temperature	45° F	25° F	10° F
Temperature Difference	0° F	20° F	20° F
Suction Pressure	42 psig	25 psig	15 psig
LOW-TEMPERATURE			
Box Temperature	5° F	5° F	−20° F
Coil Temperature	5° F	−15° F	−40° F
Temperature Difference	0° F	20° F	20° F
Suction Pressure	12 psig	2.5 psig	11 in. Hg

R-502

	Column 1 Compressor Off	Column 2 Compressor On	Column 3 Compressor On
LOW-TEMPERATURE			
Box Temperature	5° F	5° F	−20° F
Coil Temperature	5° F	−15° F	−40° F
Temperature Difference	0° F	20° F	20° F
Suction Pressure	36 psig	19 psig	4.1 psig

R-12

ICE MAKING

Ice begins to form at 20° F coil temperature
Suction pressure 21#
End of Cycle, Ice about to Harvest 5° F coil temperature
Suction Pressure 11#

When a service technician installs the gage manifold, the readings for a typical installation should not exceed or go below these readings. For instance, if the gages were applied to a medium-temperature vegetable box, the highest suction pressure that should be encountered on a typical box should be 25 psig with the compressor running and a low of 15 psig at the end of the cycle. Any readings above or below these will be out of range for a box that has a 20° F temperature difference between the air temperature entering the coil and the coil's refrigerant boiling temperature.

(A)

FIGURE 28–2 These charts can be used for typical operating temperatures and pressures for refrigerating systems using forced-air evaporators.

TEMPERATURE AND PRESSURE RANGES FOR THE LOW PRESSURE SIDE OF THE SYSTEM

TEMPERATURE °F	REFRIGERANT 12	22	134a	502	
−60	19.0	12.0		7.2	
−55	17.3	9.2		3.8	
−50	15.4	6.2		0.2	
−45	13.3	2.7		**1.9**	
−40	11.0	**0.5**	14.7	**4.1**	
−35	8.4	**2.6**	12.4	**6.5**	
−30	5.5	**4.9**	9.7	**9.2**	
−25	2.3	**7.4**	6.8	**12.1**	← LOW TEMPERATURE
−20	**0.6**	**10.1**	3.6	**15.3**	
−18	**1.3**	**11.3**	2.2	**16.7**	
−16	**2.0**	**12.5**	0.7	**18.1**	
−14	**2.8**	**13.8**	**0.3**	**19.5**	
−12	**3.6**	**15.1**	**1.2**	**21.0**	
−10	**4.5**	**16.5**	**2.0**	**22.6**	
−8	**5.4**	**17.9**	**2.8**	**24.2**	
−6	**6.3**	**19.3**	**3.7**	**25.8**	
−4	**7.2**	**20.8**	**4.6**	**27.5**	
−2	**8.2**	**22.4**	**5.5**	**29.3**	
0	**9.2**	**24.0**	**6.5**	**31.1**	
1	**9.7**	**24.8**	**7.0**	**32.0**	
2	**10.2**	**25.6**	**7.5**	**32.9**	
3	**10.7**	**26.4**	**8.0**	**33.9**	
4	**11.2**	**27.3**	**8.6**	**34.9**	
5	**11.8**	**28.2**	**9.1**	**35.8**	
6	**12.3**	**29.1**	**9.7**	**36.8**	
7	**12.9**	**30.0**	**10.2**	**37.9**	
8	**13.5**	**30.9**	**10.8**	**38.9**	
9	**14.0**	**31.8**	**11.4**	**39.9**	
10	**14.6**	**32.8**	**11.9**	**41.0**	
11	**15.2**	**33.7**	**12.5**	**42.1**	
12	**15.8**	**34.7**	**13.2**	**43.2**	
13	**16.4**	**35.7**	**13.8**	**44.3**	
14	**17.1**	**36.7**	**14.4**	**45.4**	
15	**17.7**	**37.7**	**15.1**	**46.5**	
16	**18.4**	**38.7**	**15.7**	**47.7**	
17	**19.0**	**39.8**	**16.4**	**48.8**	← MEDIUM TEMPERATURE
18	**19.7**	**40.8**	**17.1**	**50.0**	
19	**20.4**	**41.9**	**17.7**	**51.2**	
20	**21.0**	**43.0**	**18.4**	**52.4**	
21	**21.7**	**44.1**	**19.2**	**53.7**	
22	**22.4**	**45.3**	**19.9**	**54.9**	
23	**23.2**	**46.4**	**20.6**	**56.2**	
24	**23.9**	**47.6**	**21.4**	**57.5**	
25	**24.6**	**48.8**	**22.0**	**58.8**	
26	**25.4**	**49.9**	**22.9**	**60.1**	
27	**26.1**	**51.2**	**23.7**	**61.5**	
28	**26.9**	**52.4**	**24.5**	**62.8**	
29	**27.7**	**53.6**	**25.3**	**64.2**	
30	**28.4**	**54.9**	**26.1**	**65.6**	
31	**29.2**	**56.2**	**26.9**	**67.0**	
32	**30.1**	**57.5**	**27.8**	**68.4**	← HIGH TEMPERATURE
33	**30.9**	**58.8**	**28.7**	**69.9**	
34	**31.7**	**60.1**	**29.5**	**71.3**	
35	**32.6**	**61.5**	**30.4**	**72.8**	
36	**33.4**	**62.8**	**31.3**	**74.3**	
37	**34.3**	**64.2**	**32.2**	**75.8**	
38	**35.2**	**65.6**	**33.2**	**77.4**	
39	**36.1**	**67.1**	**34.1**	**79.0**	
40	**37.0**	**68.5**	**35.1**	**80.5**	
41	**37.9**	**70.0**	**36.0**	**82.1**	

Vacuum—Red Figures
Gage Pressure—Bold Figures

(B)

FIGURE 28–2B

thinner. The compressor may not even cut off at the very low temperatures if the box thermostat is set this low.

Using the same guidelines as in high- and medium-temperature applications, we find the highest normal refrigerant boiling temperature for low-temperature refrigeration applications to be 5°F − 20°F = −15°F for R-12. The refrigerant boils at 20°F below the space temperature because the compressor is running and creating a suction pressure of 2.5 psig. The lowest suction pressure may be anything down to the pressure corresponding to −20°F space temperature. This would be −20°F minus −20°F = −40°F or 11 in. Hg vacuum for an R-12 system, Figure 28–2. With the compressor off just at the time the thermostat would call for cooling for the highest temperature application, the evaporator would be 5°F, the same as the space temperature or 12 psig.

A different refrigerant for low-temperature applications may be used to keep the pressures positive, above atmospheric. If we used R-502 for this application, we would find a new set of figures. 5°F corresponds to 36 psig when the space temperature is 5°F with the compressor off. When the space temperature is 5°F and the compressor is running, the refrigerant will be boiling at about 20°F below the space temperature or 5°F − 20°F = −15°F. This −15°F corresponds to a pressure of 10 psig. The lowest temperature normally encountered in low-temperature refrigeration is a space temperature of about −20°F. This would make the suction pressure about −20°F minus −20°F = −40°F or 4.1 psig, Figure 28–2. We can easily see that by using R-502 in this application the system is going to operate well above a vacuum. R-502 boils at −50°F at 0 psig, so the cooler would get down to −30°F air temperature before the suction would get down to atmospheric pressure of 0 psig with a 20°F TD.

Temperatures below −20°F space or product temperature are obtainable and used.

The chart in Figure 28–2 can be used as a guideline for typical low-side operating conditions for the foregoing systems. Its intent is to illustrate the high and low temperatures and pressures for each application. The machine you may be working on should fall within these limits.

28.5 TYPICAL HIGH-PRESSURE CONDITIONS

Typical high-side pressure operating conditions can either be applied to air-cooled equipment or water-cooled equipment. Troubleshooting is very different for each.

28.6 TYPICAL AIR-COOLED CONDENSER OPERATING CONDITIONS

Air-cooled condensers have an operating temperature range from cold to hot as they are exposed to outside temperatures. The equipment may even be located inside a conditioned building. The condenser has to maintain a pressure that will create enough pressure drop across the expansion device for it to feed correctly. This pressure drop across the expansion device will push enough refrigerant through the device for proper evaporator refrigerant levels in the evaporator. Expansion valves for R-12 are normally sized at a 75- to 100-psig pressure drop. For example, when a valve manufacturer indicates that a valve has a capacity of 1 ton at an evaporator temperature of 20°F, this is for a pressure drop of 80 psig. The same valve will have less capacity if the pressure drop goes to 60 psig.

Because any piece of equipment located outside must have head pressure control, the minimum values at which the head pressure control is set would be the minimum expected head pressure. For R-12 this is normally about 127 psig for air-cooled equipment. This corresponds to 105°F condensing temperature. R-502 has a head pressure of 232 psig at 105°F, Figure 28–3. When the discharge pressure is less than 127 psig for an R-12 machine, the condenser is operating at the lowest pressure to maintain 80 psig drop across the expansion device and a lower pressure may not be feeding the expansion valve correctly.

28.7 CALCULATING THE CORRECT HEAD PRESSURE FOR AIR-COOLED EQUIPMENT

The maximum normal high-side pressure would correspond to the maximum ambient temperature in which the condens-

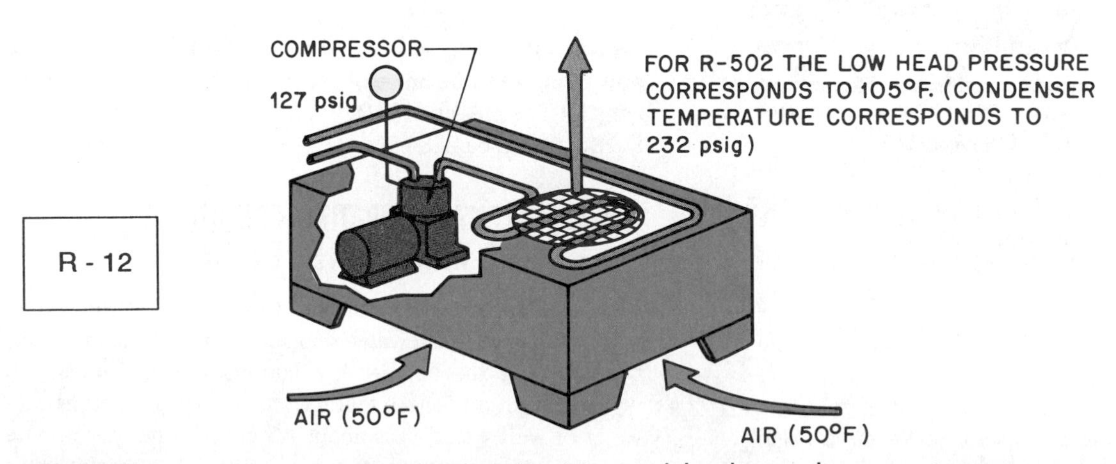

FIGURE 28–3 Refrigeration system shows the relationship of the air-cooled condenser to the entering air temperature on a cool day. This unit has head pressure control for mild weather. The outside air entering the condenser is 50°F. This would normally create a head pressure of about 85 psig. This unit has a fan cycle device to hold the head pressure up to 125 psig minimum.

er is operated. Most air-cooled condensers will condense the refrigerant at a temperature of about 30°F above the ambient temperature when the ambient is above 70°F. Figure 28–4 presents an example of a refrigeration system operating under low ambient conditions without head pressure control. When the ambient temperature drops below 70°F, the relationship changes to a lower value.

If the condenser is a higher-efficiency condenser or oversized, the condenser may condense at 25°F above the ambient. Using the 30°F figure, we see that if a unit were located outside and the temperature were 95°F, the condensing temperature would be 95°F + 30°F = 125°F. This corresponds to 169 psig for R-12 and 301 psig for R-502, Figure 28–5. ⊛**These are the two most common refrigerants for**

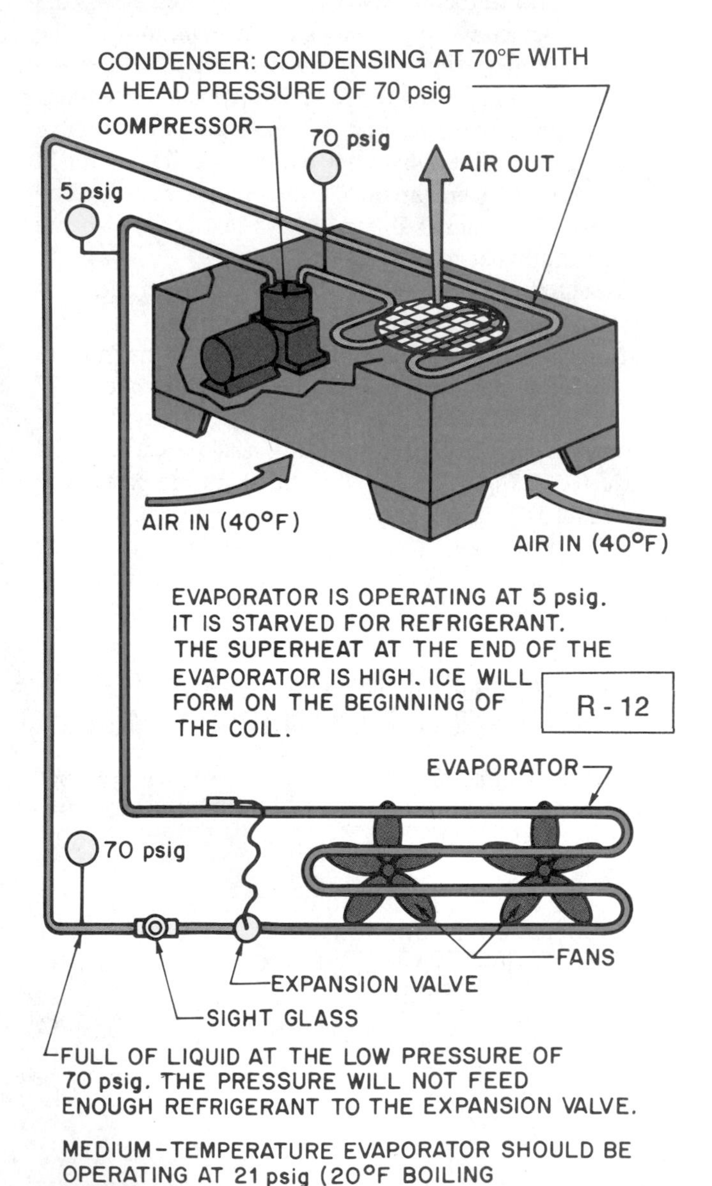
CONDENSER: CONDENSING AT 70°F WITH
A HEAD PRESSURE OF 70 psig
COMPRESSOR
70 psig
AIR OUT
5 psig
AIR IN (40°F)
AIR IN (40°F)
EVAPORATOR IS OPERATING AT 5 psig.
IT IS STARVED FOR REFRIGERANT.
THE SUPERHEAT AT THE END OF THE
EVAPORATOR IS HIGH. ICE WILL
FORM ON THE BEGINNING OF
THE COIL.
R - 12
EVAPORATOR
70 psig
FANS
EXPANSION VALVE
SIGHT GLASS
FULL OF LIQUID AT THE LOW PRESSURE OF
70 psig. THE PRESSURE WILL NOT FEED
ENOUGH REFRIGERANT TO THE EXPANSION VALVE.
MEDIUM-TEMPERATURE EVAPORATOR SHOULD BE
OPERATING AT 21 psig (20°F BOILING
TEMPERATURE)

FIGURE 28–4 Refrigeration system operating at an ambient temperature that caused the head pressure to be too low for normal operation. The expansion valve is not feeding correctly because there is not enough head pressure to push the liquid refrigerant through the valve.

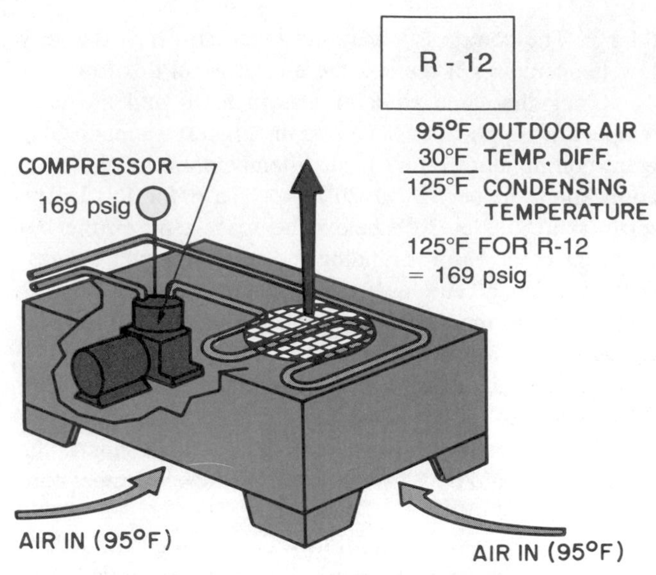
R - 12
95°F OUTDOOR AIR
30°F TEMP. DIFF.
125°F CONDENSING
TEMPERATURE
COMPRESSOR
169 psig
125°F FOR R-12
= 169 psig
AIR IN (95°F)
AIR IN (95°F)
STANDARD EFFICIENCY AIR-COOLED CONDENSER
CONDITIONS WITH A TYPICAL LIGHT FILM OF
AIRBORNE DIRT CONDENSING AT 35°F ABOVE
THE AMBIENT ENTERING AIR TEMPERATURE.

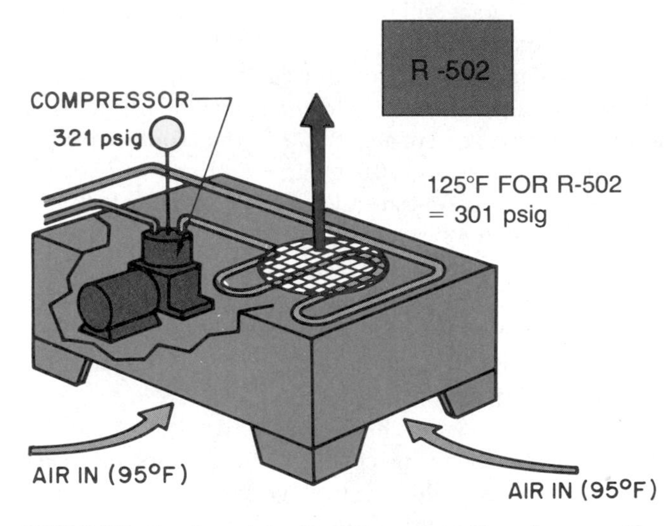
R -502
COMPRESSOR
321 psig
125°F FOR R-502
= 301 psig
AIR IN (95°F)
AIR IN (95°F)

FIGURE 28–5 Air-cooled refrigeration system with condenser outside. Typical pressures are shown.

commercial refrigeration; however, R-134A and R-22 are now being used because of the chlorofluorocarbon problems.⊛ The pressure for R-134A would be 185 psig and for R-22 it would be 278 psig at 125°F condensing temperature.

28.8 TYPICAL OPERATING CONDITIONS FOR WATER-COOLED EQUIPMENT

Water-cooled condensers are used in many systems in two ways. Some are wastewater, and some reuse the same water by extracting the heat with a cooling tower. Typically a water-cooled condenser that uses fresh water, such as city water or well water, uses about 1.5 gal of water per minute per ton while a system using a cooling tower circulates about 3 gal/ton. These two applications have different operating conditions and will be discussed separately.

28.9 TYPICAL OPERATING CONDITIONS FOR WASTEWATER CONDENSER SYSTEMS

Wastewater systems use the same condensers as the cooling tower applications, but the water is wasted down the drain. Normally a water-regulating valve is used to regulate the water flow for economy and to regulate the head pressure. The condensers are either cleanable or a coil type that cannot be cleaned. With either type condenser, it is advantageous to know from the outside how the condenser is performing on the inside.

When a water-regulating valve is used to control the water flow, the water flow will be greater if the condenser is not performing correctly. When the head pressure goes up, the water will start to flow faster to compensate. This will take place until the capacity of the valve opening is reached. Then the head pressure will increase with maximum water flow. The head pressure has a more consistent relationship with the outlet water temperature because the outlet water temperature is more of a constant than the inlet water. Sometimes the inlet water is colder, such as in winter. It would not be unusual for the inlet water to be 45°F in the winter. 90°F may be the high value if the water travels through a hot ceiling in the summer, Figure 28–6.

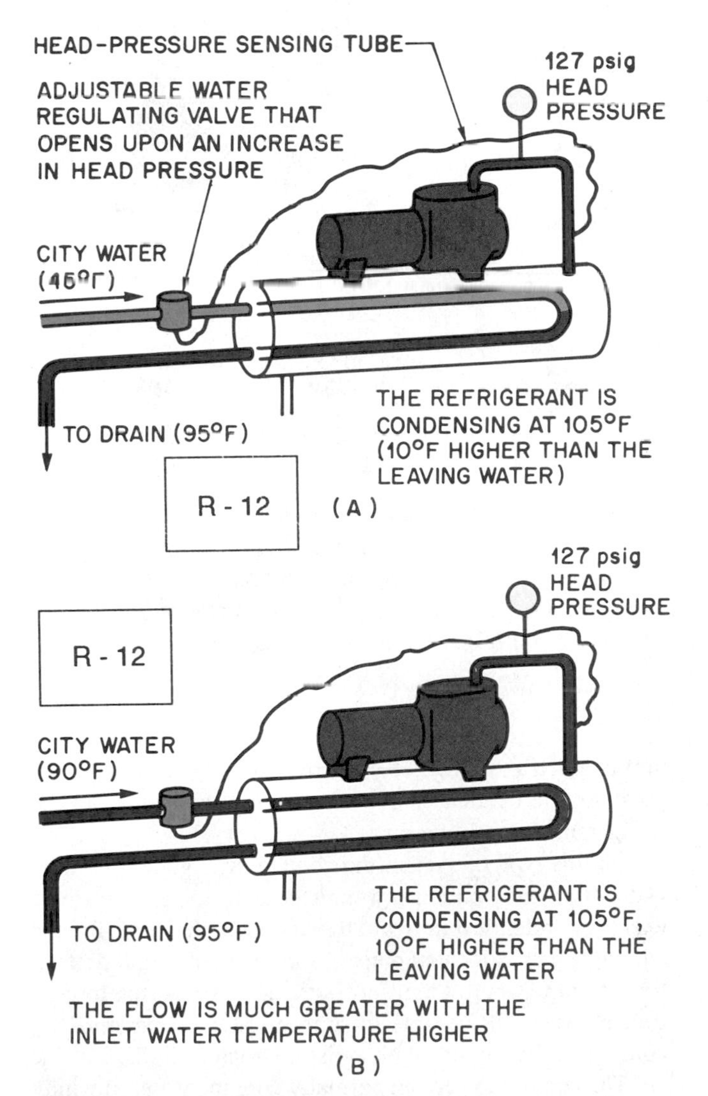

FIGURE 28–6 Water-cooled wastewater refrigeration system. The water flow is adjusted by the water-regulating valve, which keeps too much water from flowing, maintains a constant head pressure, and shuts the water off at the end of the cycle. There is a relationship between the condensing refrigerant and the leaving-water temperature. This is normally a 10°F TD.

The condensing temperature is normally 10°F higher than the water temperature leaving the condenser. If the system water valve were set to maintain 127 psig head pressure for an R-12 system, the condensing temperature of the refrigerant would be 105°F, and the leaving water should be about 95°F, Figure 28–6. If the condensing temperature were much above this, you would suspect a dirty condenser. If the gages indicated 220 psig (145°F condensing temperature) and the leaving water temperature were to be 95°F, the difference in temperature of the condensing refrigerant is 45°F – 95°F = 50°F. This indicates that the condenser is not removing the heat; the coil is dirty, Figure 28–7. It can be cleaned chemically or with brushes if it is a cleanable condenser. Whenever a noticeable increase in water flow occurs, a dirty condenser is suspected.

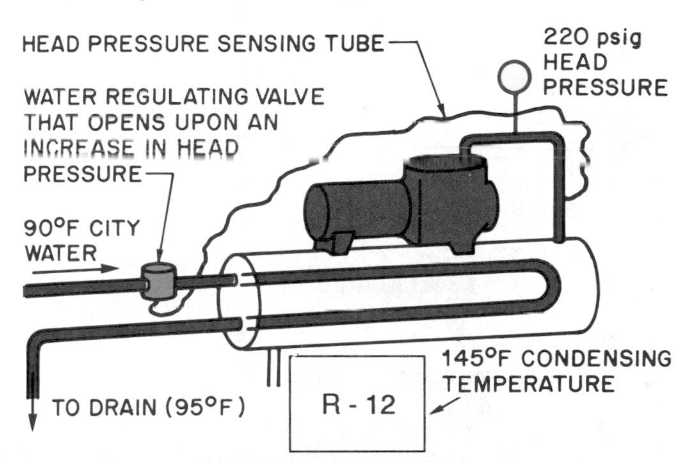

FIGURE 28–7 Water-cooled refrigeration system has dirty condenser tubes, and the water is not taking the heat out of the system.

28.10 TYPICAL OPERATING CONDITIONS FOR RECIRCULATED WATER SYSTEMS

Water-cooled condensers that use a cooling tower to remove the heat from the water normally do not use water-regulating valves to control the water flow. It is normally a constant volume of water that is pumped by a pump. The volume is customarily designed into the system in the beginning and can be verified by checking the pressure drop across the water circuit at the condenser inlet to outlet. There has to be some pressure drop for there to be water flow. The original specifications of the system should include the engineer's intent with regard to the water flow, but these may not be obtainable on an old installation.

Most systems that reuse the same water with a cooling tower have a standard 10°F temperature rise across the condenser. For example, if the water from the tower were to be 85°F entering the condenser, the water leaving the condenser should be 95°F, Figure 28–8. If the difference were to be 15°F or 20°F, you might think that the condenser is doing its job of removing the heat from the refrigerant, but the water flow is insufficient, Figure 28–9. If the water temperature entering the condenser were to be 85°F and the leaving water were to be 90°F, it may be that there is too much water flow. If the head pressure is not high, the condenser is removing the heat, and there is too much water flow. If the head pressure is high and there is the right amount of water flow, the condenser is dirty. See Figure 28–10 for an example of a dirty condenser.

The condenser has a relationship of heat exchange with the condenser water like the wastewater system. The refrigerant normally condenses about 10°F warmer than the leaving water. If the leaving water is 95°F, a properly operating condenser should be condensing at 105°F or a discharge pressure of 127 psig for R-12 and 232 psig for R-502, Figure 28–11. A pressure of 135 psig for R-134A and 211 psig for R-22 could be expected. If the condensing temperature is much more than 10°F above the leaving water temperature, the condenser is not removing the heat from the refrigerant. It could be dirty.

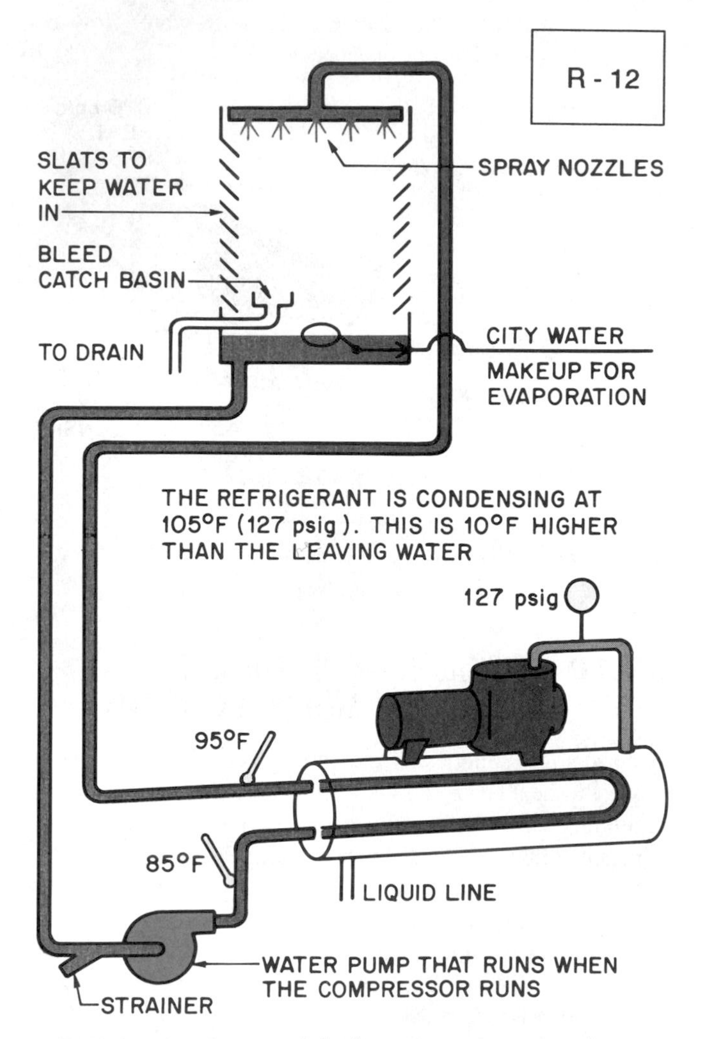

FIGURE 28–8 A water-cooled refrigeration system reusing the water after the heat is rejected to the atmosphere. There is a constant bleed of water to keep the system from overconcentrating with the minerals left behind when the water is evaporated. The difference in the incoming water and the leaving water is 10°F. This is the typical temperature rise across a water tower system.

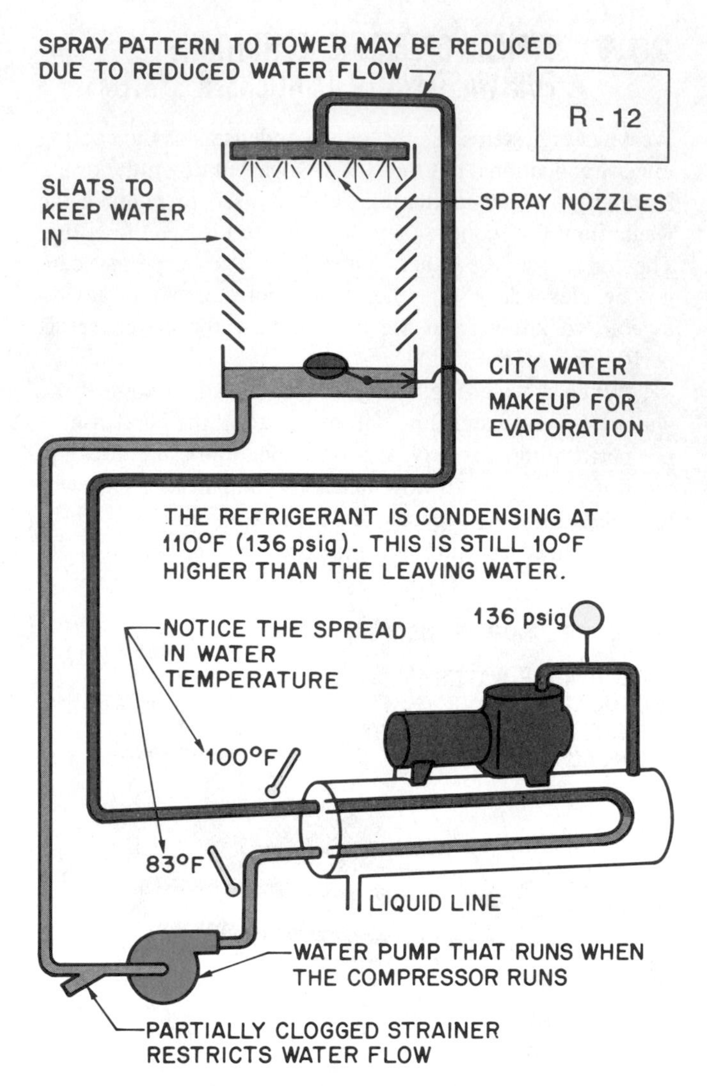

FIGURE 28–9 This system has too much temperature rise, indicating there is not enough water flow. The water strainers may be stopped up. The condensing temperature is higher than normal, which causes the head pressure to rise. A decrease in water flow may be detected by water pressure drop if it is known what it is supposed to be.

In this type of system the condenser is getting its inlet water from the cooling tower. A cooling tower has a heat exchange relationship with the ambient air. Usually the cooling tower is located outside and may be natural draft or forced draft (where a fan forces the air through the tower). Either tower will be able to supply water temperature according to the humidity or moisture content in the outside air. The cooling tower can normally cool the water to within 7°F of the wet-bulb temperature of the outside air, Figure 28–12. If the outside wet-bulb temperature (taken with a psychrometer) is 78°F, the leaving water will be about 85°F if the tower is performing correctly. Wet-bulb temperature concerns the moisture content in air and is discussed in more detail in Unit 35.

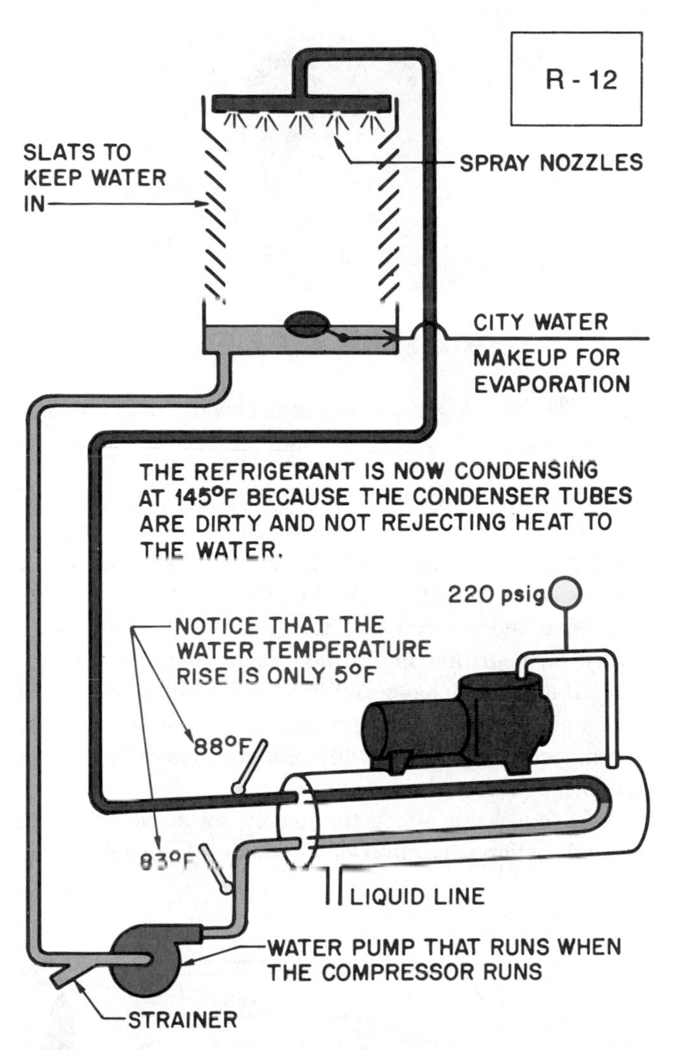

FIGURE 28–10 System has dirty condenser tubes.

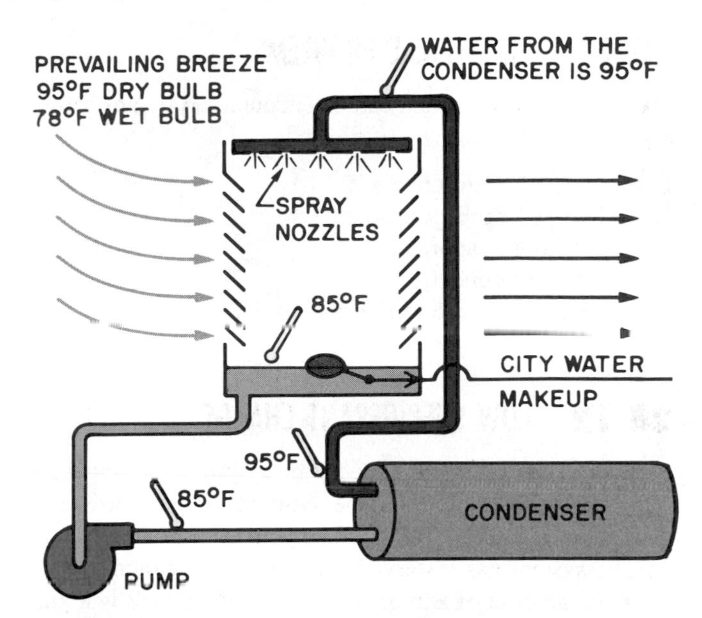

COOLED WATER IN BASIN OF TOWER IS 85°F. IT CAN NORMALLY BE COOLED TO WITHIN 7°F OF THE OUTDOOR WET-BULB TEMPERATURE DUE TO EVAPORATION.
NOTICE THAT THE FINAL WATER TEMPERATURE IS MUCH COOLER THAN THE OUTDOOR DRY-BULB TEMPERATURE.

FIGURE 28–12 Cooling tower has a temperature relationship with the air that is cooling the water. Most cooling towers can cool the water that goes back to the condenser to within 7°F of the wet-bulb temperature of the ambient air. For example, if the wet-bulb temperature is 78°F, the tower should be able to cool the water to 85°F. If it does not, a tower problem should be suspected.

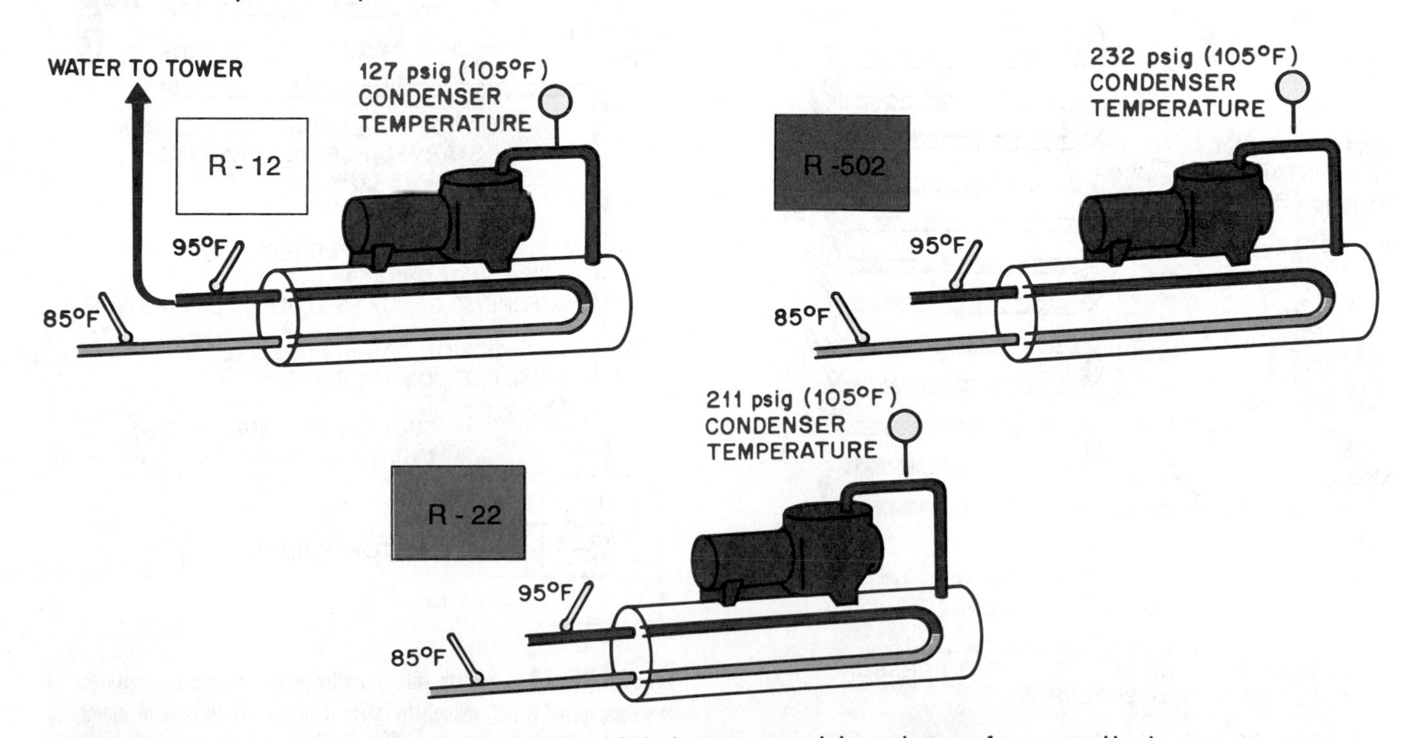

FIGURE 28–11 The water tower system has a relationship with the leaving water and the condensing refrigerant just like the wastewater system. Normally the refrigerant condenses at 10°F warmer than the water leaving the condenser.

28.11 SIX TYPICAL PROBLEMS

Six typical problems that can be encountered by any refrigeration system are:

1. Low refrigerant charge
2. Excess refrigerant charge
3. Inefficient evaporator
4. Inefficient condenser
5. Restriction in the refrigerant circuit
6. Inefficient compressor

28.12 LOW REFRIGERANT CHARGE

A low refrigerant charge affects most systems in about the same way, depending on the amount of the refrigerant needed to be correct. The normal symptoms are low capacity. The system has a starved evaporator and cannot absorb the rated amount of Btu or heat. The suction gage will read low, and the discharge gage will read low. The exception to this is a system with an automatic expansion valve, Figure 28–13. It will be discussed later in this unit.

If the system has a sight glass, it will have bubbles in it that look like air but which are actually vapor refrigerant. Figure 28–14 is a typical liquid line sight glass. Remember, a sight glass that is full of vapor or liquid may look the same. If there is only vapor in the glass, a slight film of oil may be present. This is a good indicator of vapor only.

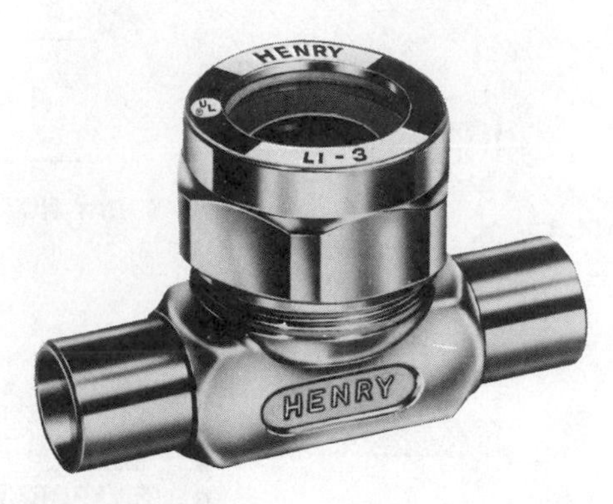

FIGURE 28–14 A sight glass to indicate when pure liquid is in the liquid line. *Courtesy Henry Valve Company*

When a system has a sight glass, it will generally have a thermostatic expansion valve (TXV) or automatic expansion valve and a receiver. These valves will hiss when a partial vapor–partial liquid mixture is going through the valve. If the system has a capillary tube, it will probably not have a sight glass. The technician needs to know how the system feels to the touch at different points to determine the gas charge level without using gages.

The low charge affects the compressor by not supplying the cool suction vapor to cool the motor. Most compressors

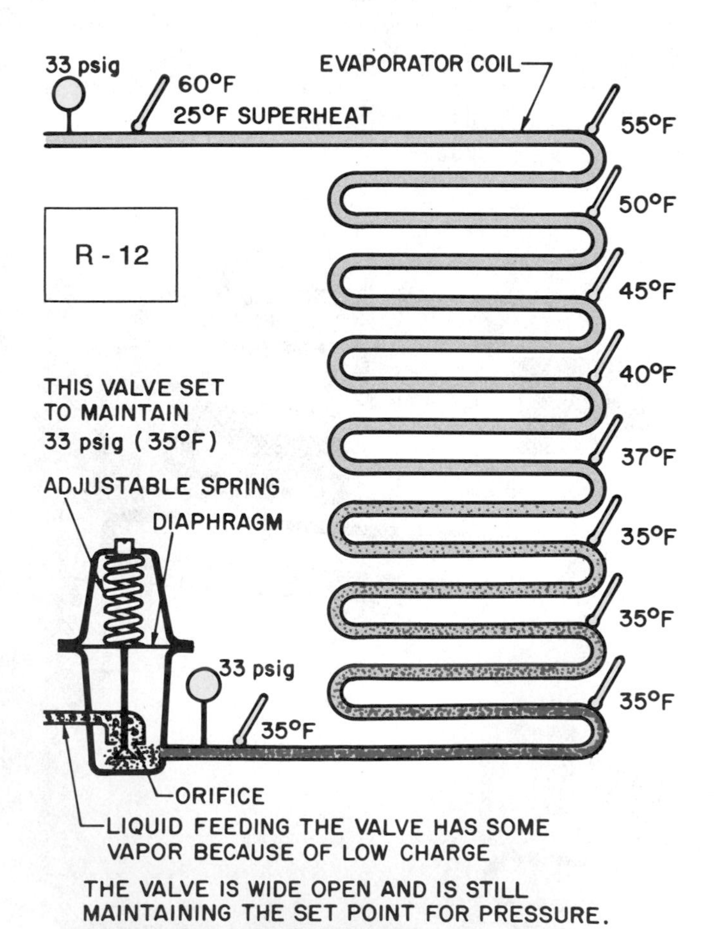

FIGURE 28–13 Low-refrigerant charge system characteristics when the metering device is an automatic expansion valve.

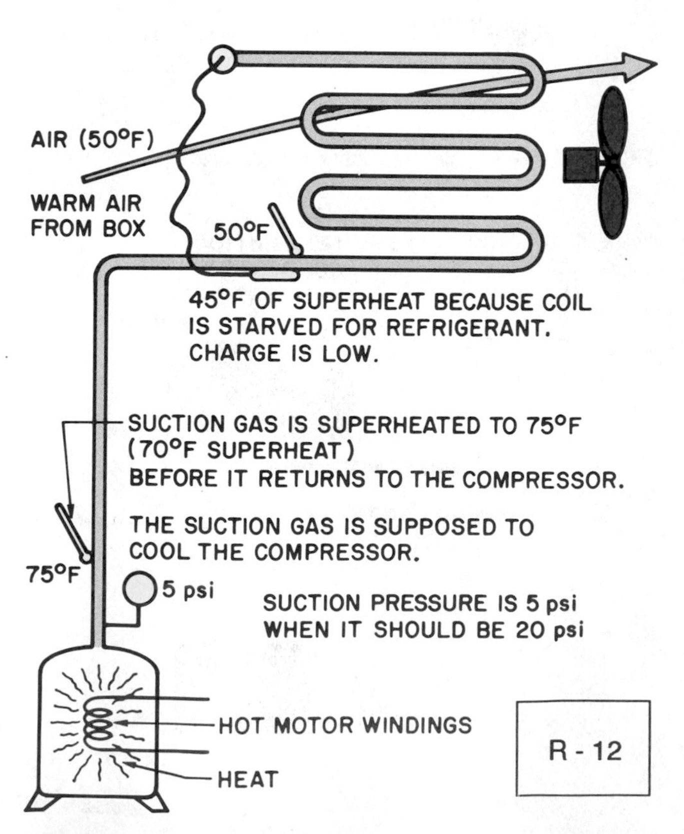

FIGURE 28–15 System with a suction-cooled hermetic compressor. The compressor is hot enough to cause it to cut off because of motor temperature. It is hard to cool off a hot hermetic compressor from the outside because the motor is suspended inside a vapor atmosphere.

FIGURE 28–16 An air-cooled compressor. The compressor motor will not normally get hot as a result of a low refrigerant charge, but the discharge gas leaving the compressor may get too warm because the refrigerant entering the compressor is too warm. Most compressor manufacturers require the suction gas not to be over 65°F for continuous operation. *Courtesy Tyler Refrigeration Company*

are suction cooled, so the result is a hot compressor motor. It may even be off due to the motor-winding thermostat, Figure 28–15. If the compressor is air cooled, the suction line coming back to the compressor will not need to be as cool as a suction-cooled compressor. It may be warm by comparison, Figure 28–16.

28.13 REFRIGERANT OVERCHARGE

A refrigerant overcharge also acts much the same way from system to system. The discharge pressure is high, and the suction pressure may be high. The automatic expansion valve system will not have a high suction pressure because it maintains a constant suction pressure, Figure 28–17. The TXV may have a slightly higher suction pressure if the head pressure is excessively high because the system capacity may be down, Figure 28–18.

The capillary tube will have a high suction pressure because the amount of refrigerant flowing through it depends on the difference in pressure across it. The more head pressure, the more liquid it will pass. The capillary tube will allow enough refrigerant to pass so that it will allow liquid into the compressor. When the compressor is sweating down the side or all over, it is a sign of liquid refrigerant in the compressor, Figures 28–19 and 28–20.

The compressor should only have vapor entering it. Vapor will rise in temperature as soon as it touches the compressor shell. When liquid is present, it will not rise in temperature and will cool the compressor shell. **Liquid refrigerant still has its latent heat absorption capability**

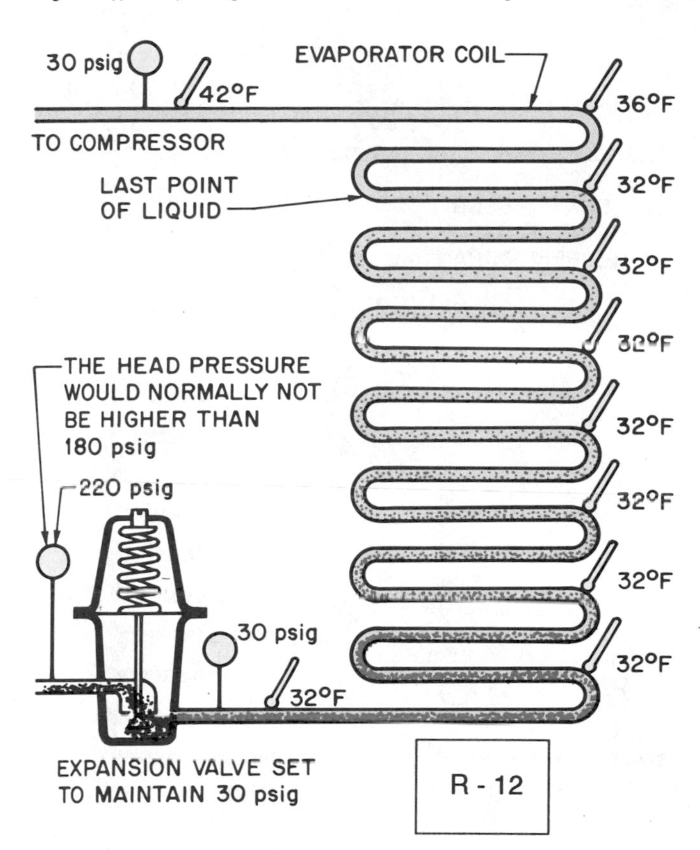

FIGURE 28–17 System has an automatic expansion valve for a metering device. This device maintains a constant suction pressure. The head pressure is higher than normal, but the suction pressure remains the same.

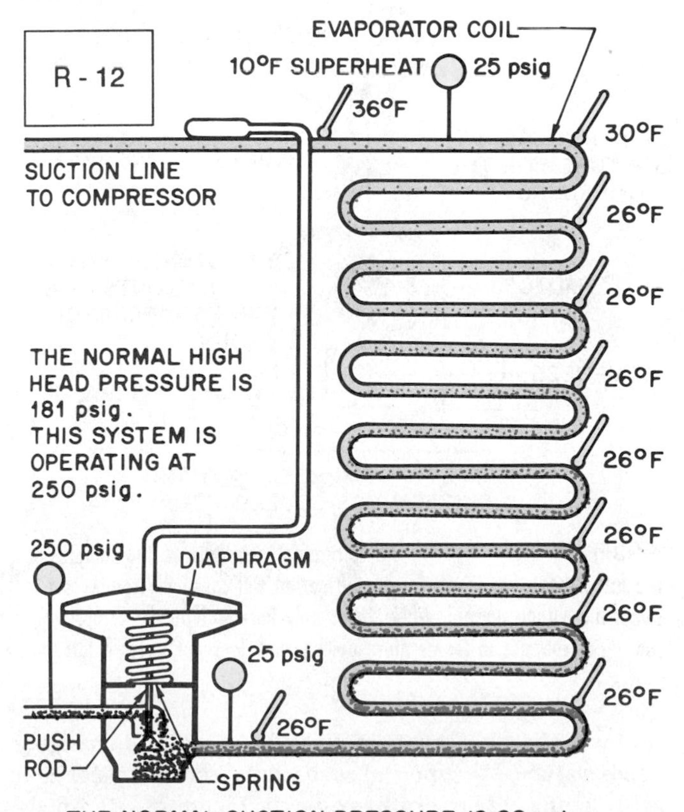

FIGURE 28–18 A TXV system using R-12 that has a higher than normal head pressure.

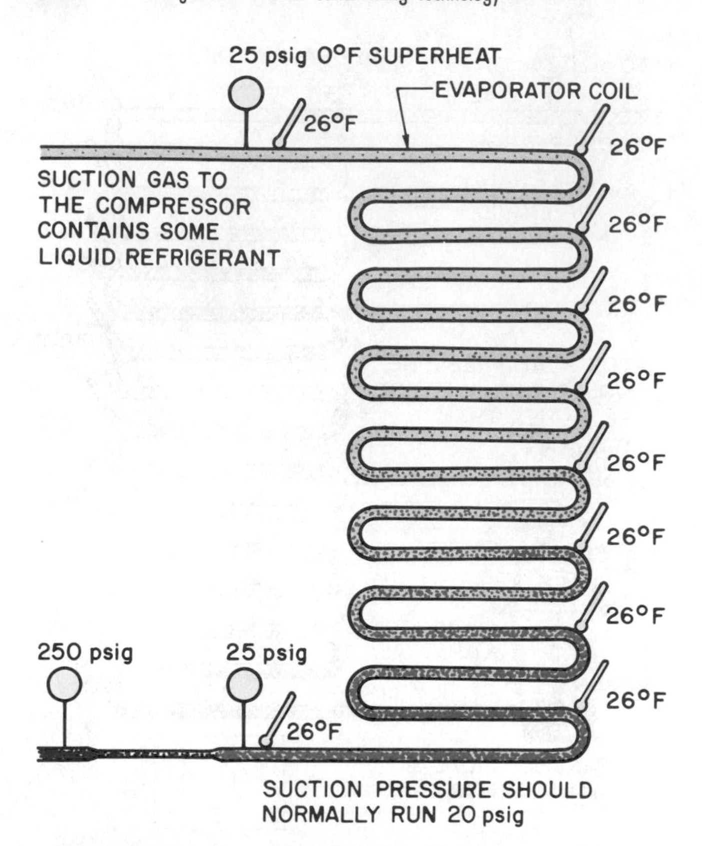

FIGURE 28–19 The capillary tube system using R-12 has an overcharge of refrigerant, and the head pressure is higher than normal. This has a tendency to push more refrigerant than normal through the metering device.

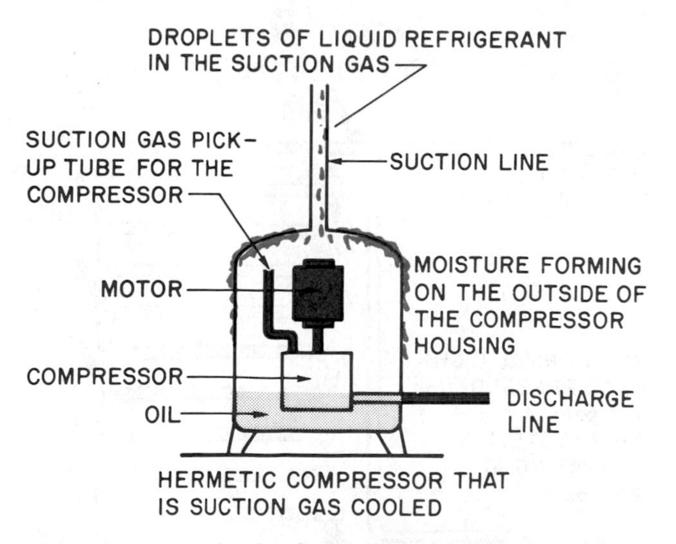

FIGURE 28–20 When liquid refrigerant gets back to the compressor, the latent heat that is left in the refrigerant will cause the compressor to sweat more than normal. When vapor only gets back to the compressor, the vapor changes in temperature quickly and does not sweat a lot.

and will absorb a great amount of heat without changing temperature. A vapor absorbs only sensible heat and will change in temperature quickly, Figure 28–20. Another reason that liquid may get back to the compressor is poor heat exchange in the evaporator in a capillary tube system. If liquid is getting back to the compressor, the evapora-

Canned goods and other non-refrigerated items are great for imposing piles and eye-catching displays. Don't try the same gimmicks with perishables. The case will fail to refrigerate any of the merchandise when the air ducts are blocked.

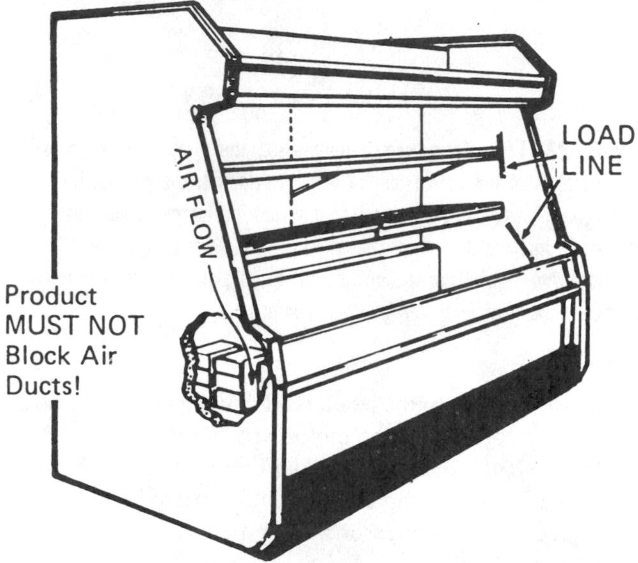

Low temperature multi-shelf cases are particularly sensitive to air pattern changes as well as extremes of humidity and temperature in the store.

Jumble displays may have some sales benefits, but they really foul up the protective layer of cold air in the case. Observe the LOAD LINE stickers!

FIGURE 28–21 Evaporator that cannot absorb the required Btu because of product interference. There is a load line on the inside where the product is stored. *Courtesy Tyler Refrigeration Company*

tor heat exchange should be checked before removing refrigerant.

28.14 INEFFICIENT EVAPORATOR

An inefficient evaporator does not absorb the heat into the system and will have a low suction pressure. The suction line may be sweating or frosting back to the compressor. An inefficient evaporator can be caused by a dirty coil, a fan running too slowly, an expansion valve starving the coil, recirculated air, ice buildup, or product interference causing blocked airflow, Figure 28–21.

All of these can be checked with an evaporator performance check. This check can be performed by making sure that the evaporator has the correct amount of refrigerant with a superheat check, Figure 28–22. The heat exchange surface should be clean. The fans should be blowing enough air and not recirculating it from the discharge to the inlet of the coil.

The refrigerant boiling temperature should not be more than 20°F colder than the entering air on an air evaporator coil, Figure 28–2. A water coil should have no more than a 10°F TD in the boiling refrigerant and the leaving water, Figure 28–23. When the boiling refrigerant relationship to the medium being cooled starts increasing, the heat exchange is decreasing.

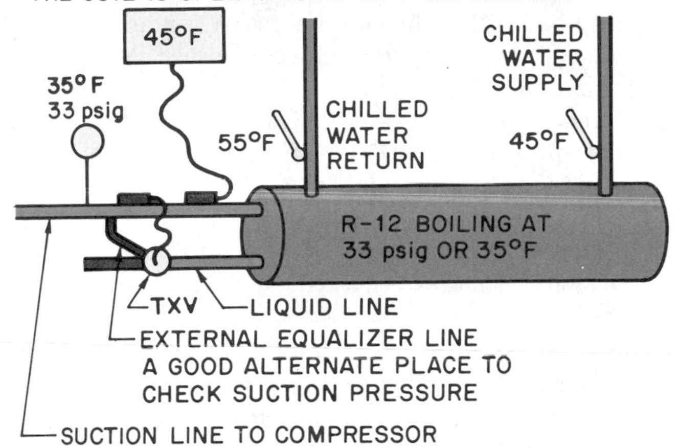
ELECTRONIC TEMPERATURE TESTER SHOWS THE SUCTION LINE TO BE 10°F HIGHER THAN THE BOILING REFRIGERANT TEMPERATURE OF 35°F. THE COIL IS OPERATING AT 10°F SUPERHEAT.
45°F
35° F
33 psig
55°F
CHILLED WATER RETURN
CHILLED WATER SUPPLY
45°F
R-12 BOILING AT 33 psig OR 35°F
TXV
LIQUID LINE
EXTERNAL EQUALIZER LINE A GOOD ALTERNATE PLACE TO CHECK SUCTION PRESSURE
SUCTION LINE TO COMPRESSOR

FIGURE 28–23 Water coils can be analyzed in much the same way as air coils. The refrigerant temperature and pressure at the end of the evaporator will indicate the refrigerant level. The refrigerant boiling temperature should not be more than 10°F cooler than the leaving water.

28.15 INEFFICIENT CONDENSER

An inefficient condenser acts the same whether it is water cooled or air cooled. If the condenser cannot remove the heat from the refrigerant, the head pressure will go up. The

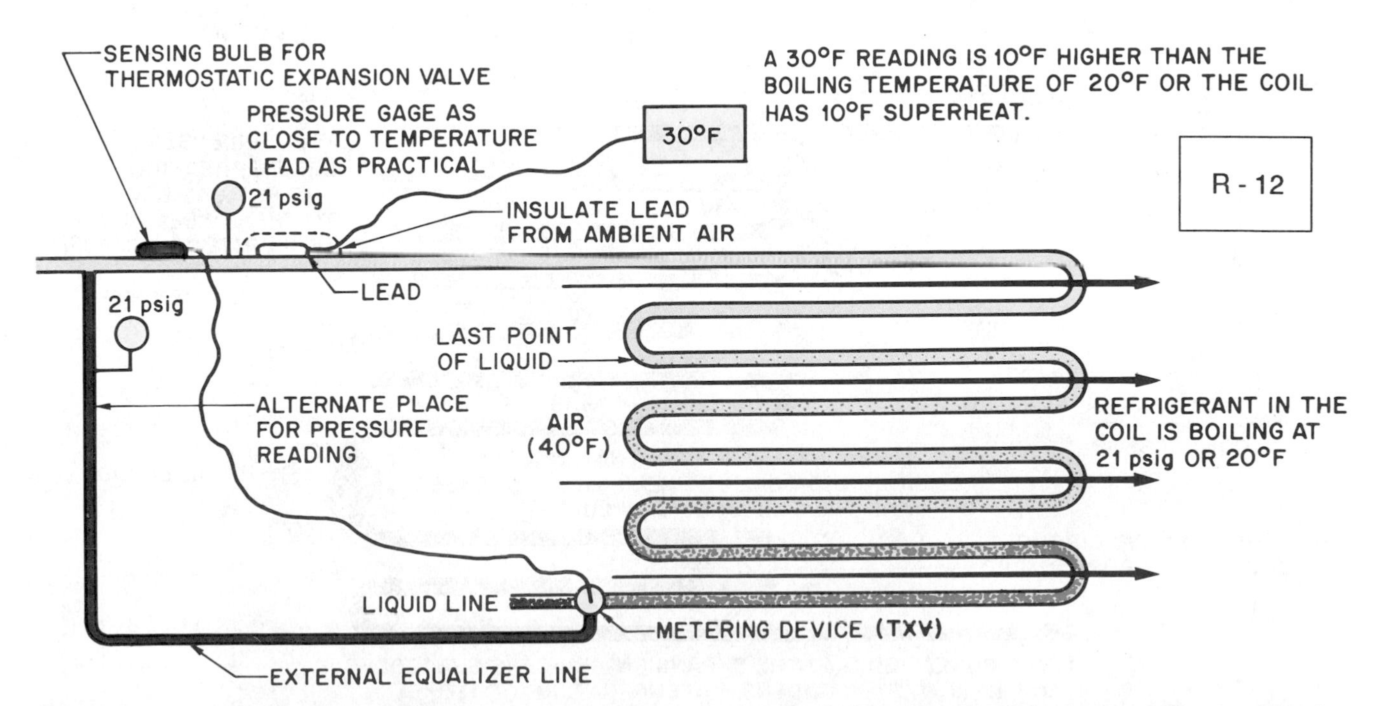
SENSING BULB FOR THERMOSTATIC EXPANSION VALVE
PRESSURE GAGE AS CLOSE TO TEMPERATURE LEAD AS PRACTICAL
30°F
A 30°F READING IS 10°F HIGHER THAN THE BOILING TEMPERATURE OF 20°F OR THE COIL HAS 10°F SUPERHEAT.
R - 12
21 psig
INSULATE LEAD FROM AMBIENT AIR
LEAD
21 psig
LAST POINT OF LIQUID
ALTERNATE PLACE FOR PRESSURE READING
AIR (40°F)
REFRIGERANT IN THE COIL IS BOILING AT 21 psig OR 20°F
LIQUID LINE
METERING DEVICE (TXV)
EXTERNAL EQUALIZER LINE

FIGURE 28–22 Analyzing a coil for efficiency. This requires temperature and pressure checks at the outlet of the evaporator. When a coil has the correct refrigerant level, checked by superheat and the correct air-to-refrigerant heat exchange, the coil will absorb the correct amount of heat. The correct heat exchange is taking place if the refrigerant is boiling at 10°F to 20°F cooler than the entering air. Note, this occurs when the cooler temperature is within the design range, 30°F to 45°F for medium-temperature for example. Every circuit on coils with multiple circuits should be checked for even distribution.

condenser does three things and has to be able to do them correctly, or excessive pressures will occur.

1. De-superheat the hot gas from the compressor. This gas may be 200°F or hotter on a hot day on an air-cooled system. De-superheating is accomplished in the beginning of the coil.
2. Condense the refrigerant. This is done in the middle of the coil. The middle of the coil where the condensing is occurring is the only place that the coil temperature will correspond to the head pressure. You could check the temperature against the head pressure if a correct temperature reading can be taken, but the fins are usually in the way.
3. Subcool the refrigerant before it leaves the coil. This subcooling is cooling the refrigerant to a point below the actual condensing temperature. A subcooling of 5°F to 20°F is typical. The subcooling can be checked just like the superheat, only the temperature is checked at the liquid line and compared to the high-side pressure converted to condensing temperature, Figure 28–24.

The condenser must have the correct amount of cooling medium (air or water). This medium must **not** be recirculated (mixed with the incoming medium) without being cooled. Ensure that air-cooled equipment is not located so that the air leaving the condenser circulates back into the inlet. This air is hot and will cause the head pressure to go up in proportion to the amount of recirculation, Figure 28–25. An air-cooled condenser should not be located down low, close to the roof, even though the air comes in the side. The temperature is higher at the roof level than it is a few

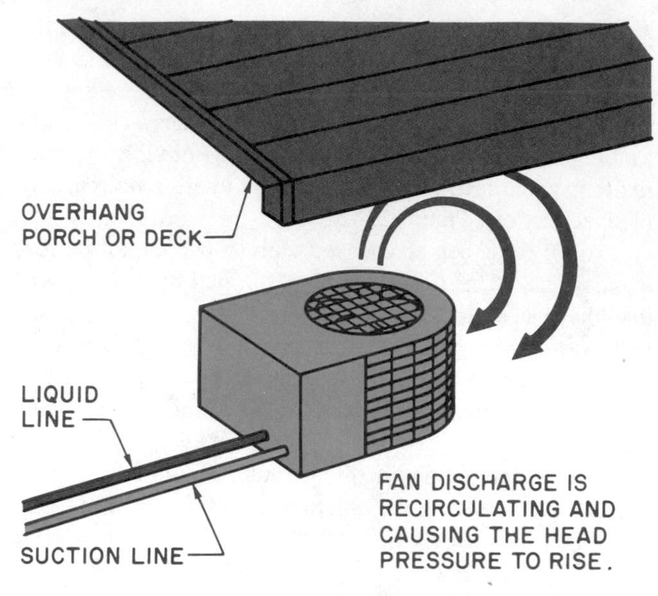
OVERHANG
PORCH OR DECK
LIQUID
LINE
SUCTION LINE
FAN DISCHARGE IS
RECIRCULATING AND
CAUSING THE HEAD
PRESSURE TO RISE.
THE CONDENSER SHOULD NOT
HAVE ANY AIR OBSTRUCTIONS

FIGURE 28–25 The air-cooled condenser is located too close to an obstacle, and the hot air leaving the condenser is recirculated back into the inlet of the coil.

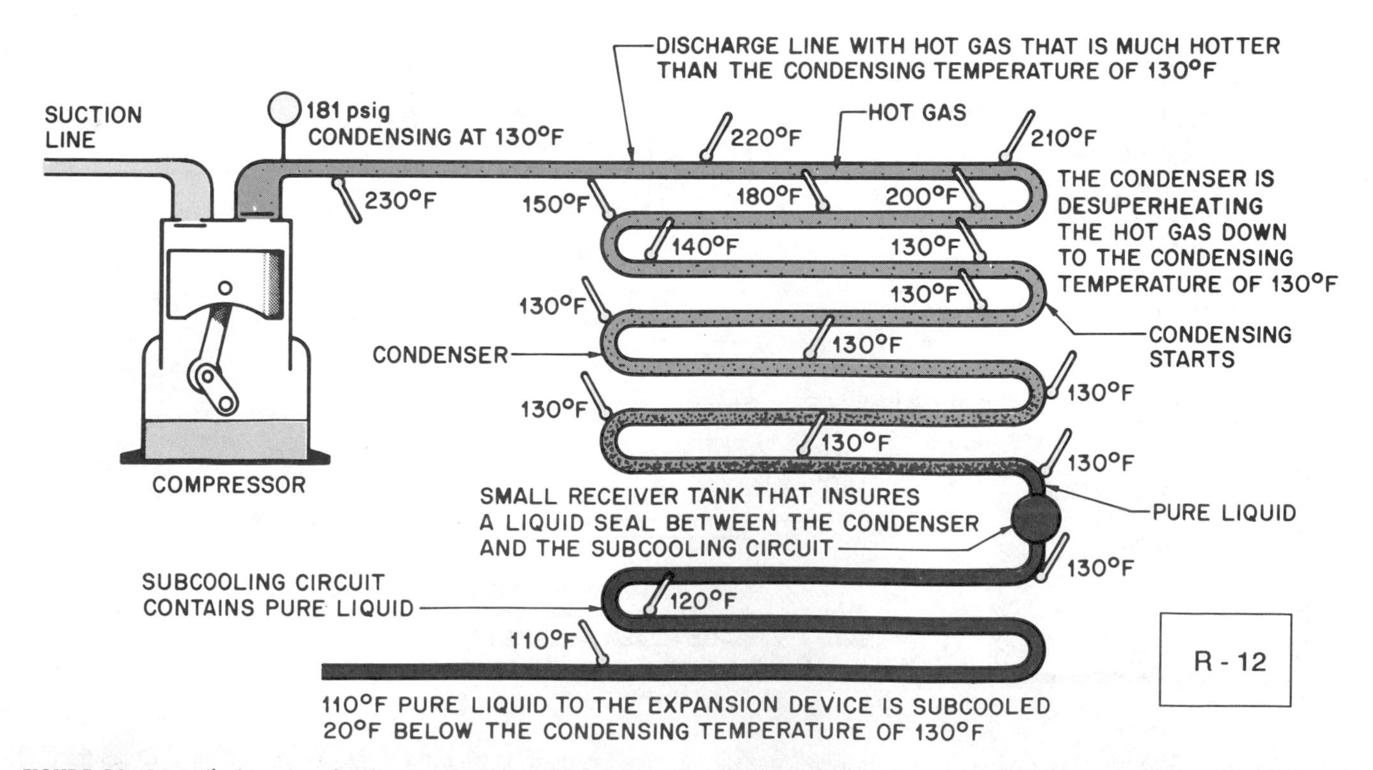
DISCHARGE LINE WITH HOT GAS THAT IS MUCH HOTTER
THAN THE CONDENSING TEMPERATURE OF 130°F
SUCTION
LINE
181 psig
CONDENSING AT 130°F
220°F
HOT GAS
210°F
230°F
150°F
180°F
200°F
THE CONDENSER IS
DESUPERHEATING
THE HOT GAS DOWN
TO THE CONDENSING
TEMPERATURE OF 130°F
140°F
130°F
130°F
130°F
CONDENSING
STARTS
CONDENSER
130°F
130°F
130°F
130°F
130°F
COMPRESSOR
SMALL RECEIVER TANK THAT INSURES
A LIQUID SEAL BETWEEN THE CONDENSER
AND THE SUBCOOLING CIRCUIT
PURE LIQUID
130°F
SUBCOOLING CIRCUIT
CONTAINS PURE LIQUID
120°F
110°F
R - 12
110°F PURE LIQUID TO THE EXPANSION DEVICE IS SUBCOOLED
20°F BELOW THE CONDENSING TEMPERATURE OF 130°F

FIGURE 28–24 Checking the subcooling on a condenser. The condenser does three jobs: (1) De-superheats the hot gas, in the first part of the condenser. (2) Condenses the vapor refrigerant to a liquid, in the middle of the condenser. (3) Subcools the refrigerant at the end of the condenser. The condensing temperature corresponds to the head pressure. Subcooling is the temperature of the liquid line subtracted from the condensing temperature. A typical condenser can subcool the liquid refrigerant 5°F to 20°F cooler than the condensing temperature.

inches higher. A clearance of about 18 in. will give better condenser performance, Figure 28–26. An air-cooled condenser that has a vertical coil may be influenced by prevailing winds. If the fan is trying to discharge its air into a 20-mph wind, it may not move the correct amount of air, and a high head pressure may occur, Figure 28–27.

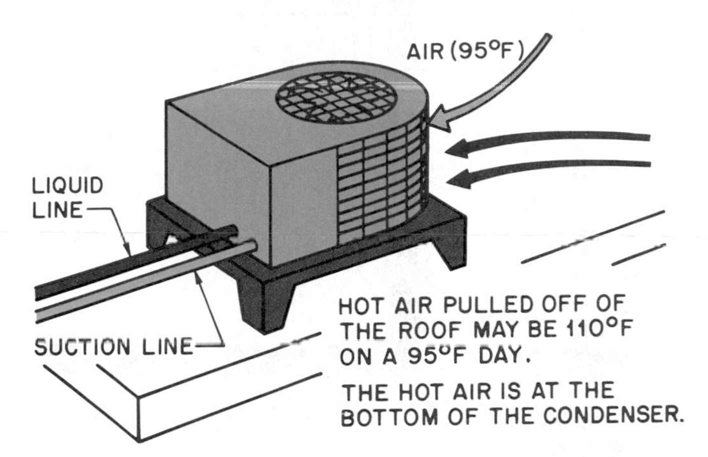

FIGURE 28–26 An air-cooled condenser should not be located close to a roof. The temperature of the air coming directly off the roof is much warmer than the ambient air because the roof acts like a solar collector.

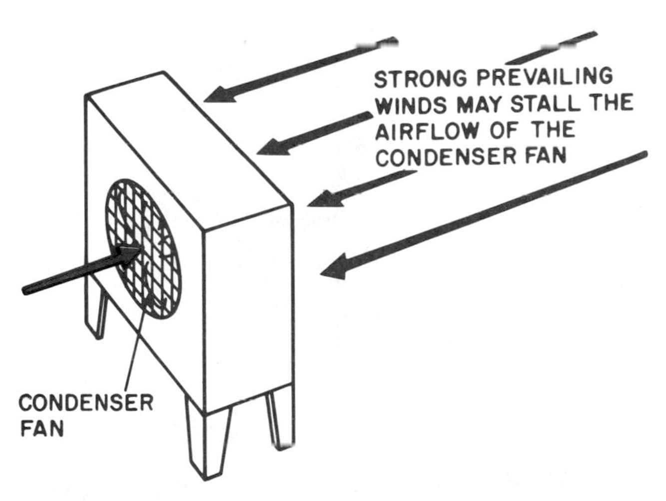

FIGURE 28–27 A condenser that is located in such a manner that it is discharging its air into a strong prevailing wind may not get enough air across the coil.

28.16 REFRIGERANT FLOW RESTRICTIONS

Restrictions that occur in the refrigeration circuit are either partial or full. A partial restriction may be in the vapor or the liquid line. A restriction always causes pressure drop at the point of the restriction. Different conditions will occur, depending on where the restriction is. Pressure drop can always be detected with gages, but the gages cannot always indicate the correct location of the restriction. Gage ports may need to be installed for pressure testing.

If a restriction occurs due to something outside the system, it is usually physical damage, such as flattened or bent tubing. These can be hard to find if they are in hidden places such as under the insulation or behind a fixture.

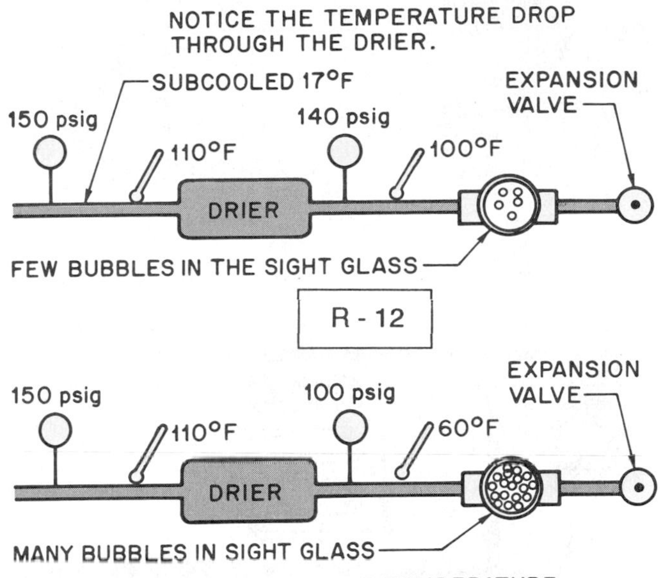

FIGURE 28–28 The driers each have a restriction. One of them is very slight. Where there is pressure drop in a liquid line, there is temperature drop. If the temperature drop is very slight, it can be detected with a thermometer. Sometimes gages are not easy to install on each side of a drier to check for pressure drop.

If a partial restriction occurs in a liquid line, it will be evident because the refrigerant will have a pressure drop and will start to act like an expansion device at the point of the restriction. **When there is a pressure drop in a liquid line, there is always a temperature change.** A temperature check on each side of a restriction will locate the place. Sometimes when the drop is across a drier, the temperature difference from one side to the other may not be enough to feel with bare hands, but a thermometer will detect it, Figure 28–28.

If a system has been running for a long time and a restriction occurs, physical damage may have occurred. If the restriction occurs soon after start, a filter or drier may be stopping up. When this occurs in the liquid line drier, bubbles will appear in the sight glass when the drier is located before the sight glass, Figure 28–29.

Another occurrence that may create a partial restriction could be valves that do not open all the way. Normally the TXV either works or it doesn't. It will function correctly, but if it loses its charge in the thermal bulb, it will close and cause a full restriction, Figure 28–30.

There is a strainer at the inlet to most expansion devices that can trap particles and stop up slowly. If the device is a valve that can be removed, it can be inspected and cleaned if necessary. If it is a capillary tube, it will be soldered into the line and not easy to inspect, Figure 28–31.

Water circulating in any system that operates below freezing will freeze at the first cold place it passes through. This would be in the expansion device. One drop of water

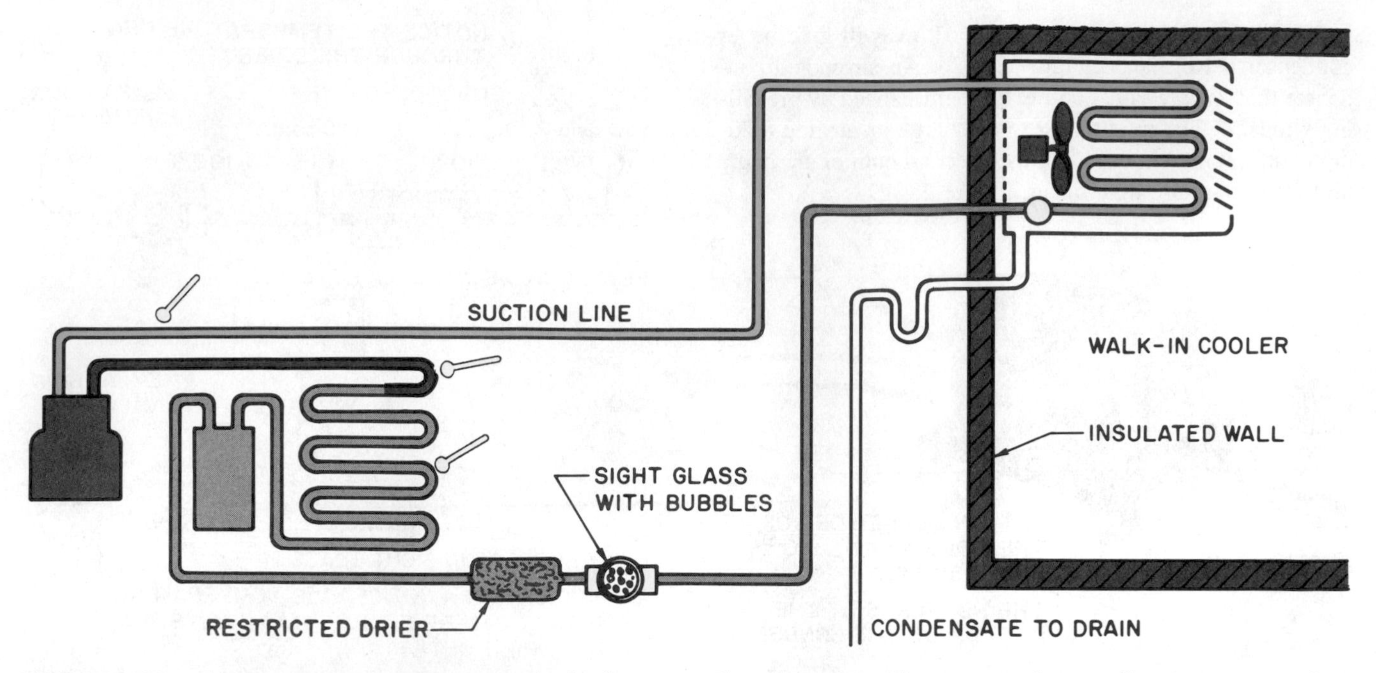

FIGURE 28–29 The restriction in a system may occur shortly after startup. This indicates that solid contaminants from installing the system must be in the drier. If the restriction occurs after the system has been running for a long time, the restriction may be physical damage, such as a bent pipe. Normally, loose contamination will make its way through the system to the drier in a matter of hours.

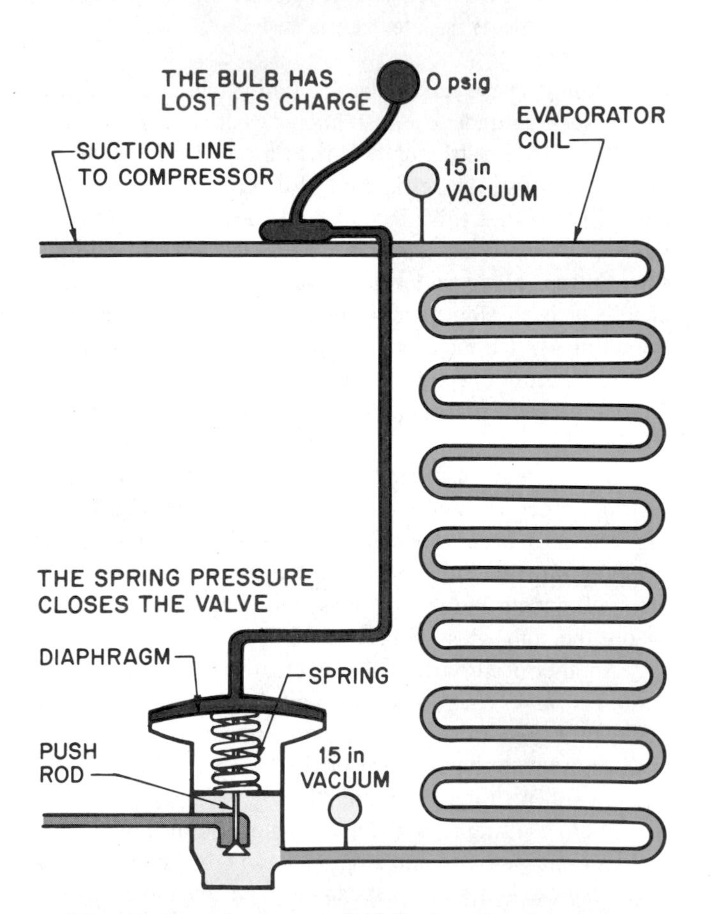

FIGURE 28–30 The charge in the bulb of a TXV is the only force that opens the valve, so when it loses its pressure, the valve closes. The system can go into a vacuum if there is no low-pressure control to stop the compressor. A partial restriction can occur if the bulb loses part of the charge or if the inlet strainer stops up.

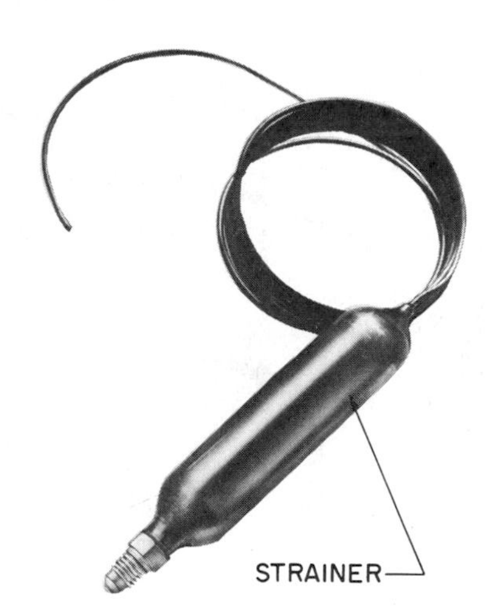

FIGURE 28–31 Capillary tube metering device with the strainer at the inlet. This strainer is soldered into the line and is not easy to service. *Courtesy Parker Hannifin Corporation*

can stop a refrigeration system. Sometimes a piece of equipment that has just been serviced will show signs of moisture on the first hot day, Figure 28–32. This is because the drier in the liquid line will have more capacity to hold moisture when it is cool. The first hot day, the drier may turn a drop or two of water loose, and it will freeze in the expansion device, Figure 28–33. When you suspect this, apply heat to thaw the water to a liquid. ***Care must be used when applying heat. A hot wet cloth is a good source.*** If applying a hot cloth to the metering device causes the system to start functioning properly, the problem is free water in the system. ⊕**Recover the refrigerant, change the drier, and evacuate the system.**⊕

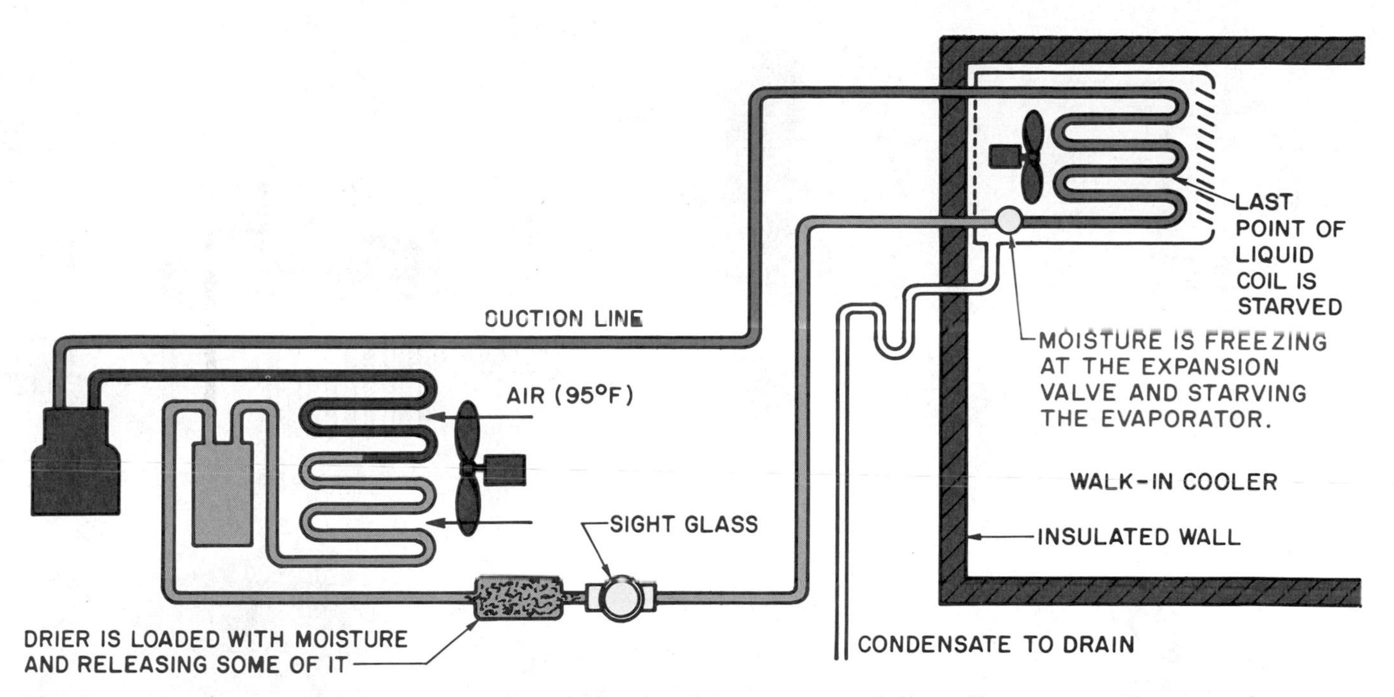

FIGURE 28–32 The system has had some moisture in it. The drier has all the moisture it can hold at mild temperatures. When the weather gets warm, the drier cannot hold all of the moisture and turns some of it loose. The moisture will freeze at the expansion device, where the first refrigeration is experienced if the system operates below freezing.

Other components that may cause restrictions are automatic valves in the lines, such as the liquid line solenoid, crankcase pressure regulator, or the evaporator pressure regulator. These valves may easily be checked with gages applied to both sides of them where pressure taps are provided, Figure 28–34.

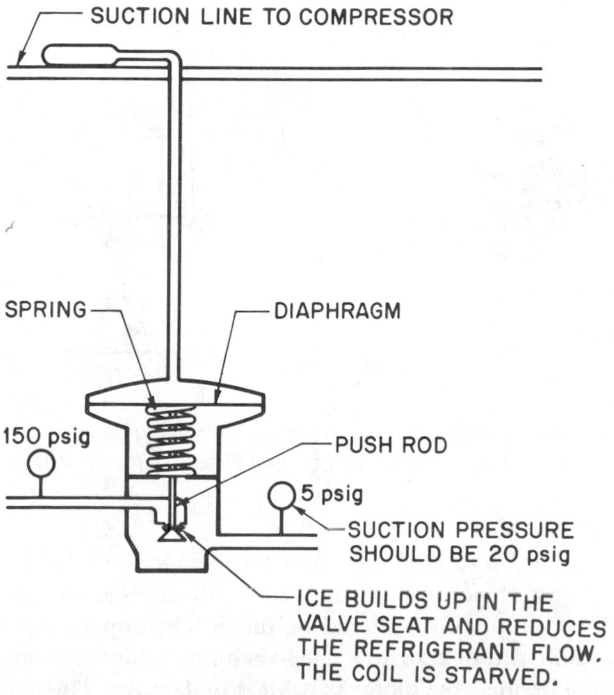

SUCTION LINE TO COMPRESSOR
SPRING
DIAPHRAGM
150 psig
PUSH ROD
5 psig
SUCTION PRESSURE SHOULD BE 20 psig
ICE BUILDS UP IN THE VALVE SEAT AND REDUCES THE REFRIGERANT FLOW. THE COIL IS STARVED.

FIGURE 28–33 If it is suspected that moisture is frozen in the metering device, mild heat can be added. Heat from a hot wet cloth will normally thaw the moisture out, and the system will start refrigerating again.

28.17 INEFFICIENT COMPRESSOR

Inefficient compressor operation can be one of the most difficult problems to find. When a compressor will not run, it is evident where the problem is. Motor troubleshooting procedures are covered in a separate unit. When a compressor is pumping at slightly less than capacity, it is hard to determine the problem. It helps at this point to remember that a compressor is a vapor pump. It should be able to create a pressure from the low side of the system to the high side under design conditions.

The following methods are all used by service technicians to discover the compressor problems.

28.18 COMPRESSOR VACUUM TEST

The compressor vacuum test is usually performed on a test bench with the compressor out of the system. This test may be performed in the system when the system has service valves. ***Care should be taken not to pull air into the system while in a vacuum.*** All reciprocating compressors should immediately go into a vacuum if the suction line is valved off when the compressor is running. This test proves that the suction valves are seating correctly on at least one cylinder. This test is *not satisfactory* on a multicylinder compressor. If one cylinder will pump correctly, a vacuum will be pulled. A reciprocating compressor should pull 26 in. to 28 in. Hg vacuum with the atmosphere as the dis-

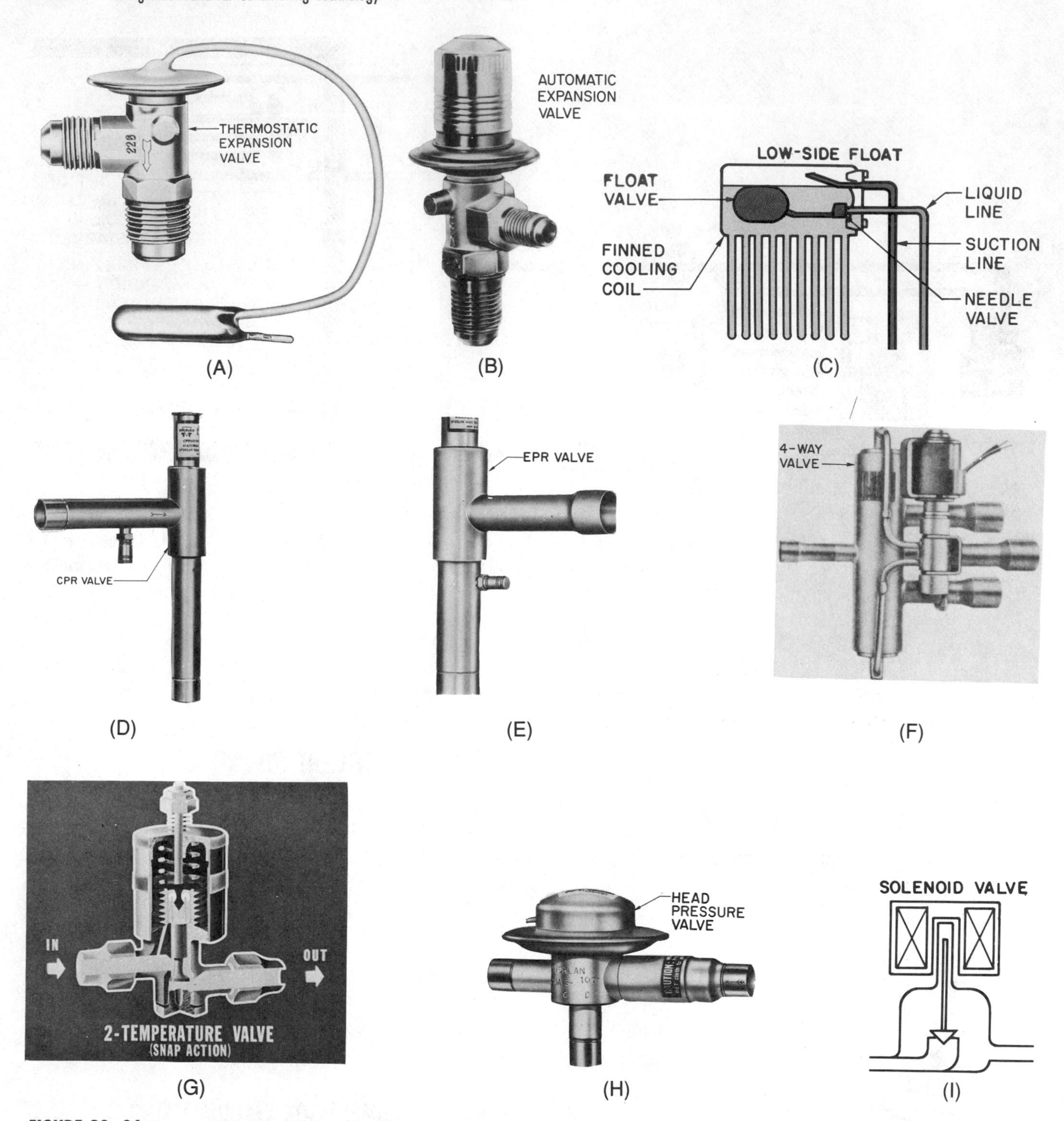

FIGURE 28–34 Valve components that can close are all subject to closing and causing a restriction. *Courtesy (A), (B) and (D) Singer Controls Division, Schiller Park, Illinois, (E) and (H) Sporlan Valve Company, (F) ALCO Controls Division, Emerson Electric Company, (G) Carrier Corporation*

charge pressure, Figure 28–35. The compressor should pull about 24 in. Hg vacuum, against 100 psig discharge pressure, Figure 28–36. When the compressor has pumped a differential pressure and is stopped, the pressures should not equalize. For example, a compressor has been operated until the suction pressure is 24 in. Hg vacuum and the head pressure is 100 psig, then it is stopped. These pressures should stay the same while the compressor is off. When refrigerant is used for this pumping test, the 100 psig will drop some because of the condensing refrigerant. Nitrogen is a better choice to pump in the test because the pressure will not drop. ***Care should be taken when operating a hermetic compressor in a deep vacuum (below 1000 microns) because the motor is subject to damage. This low vacuum cannot be obtained with a reciprocating compressor. Also, most compressor motors are cooled with suction gas and will get hot if operated for any length of time performing these tests. This vacuum test should not**

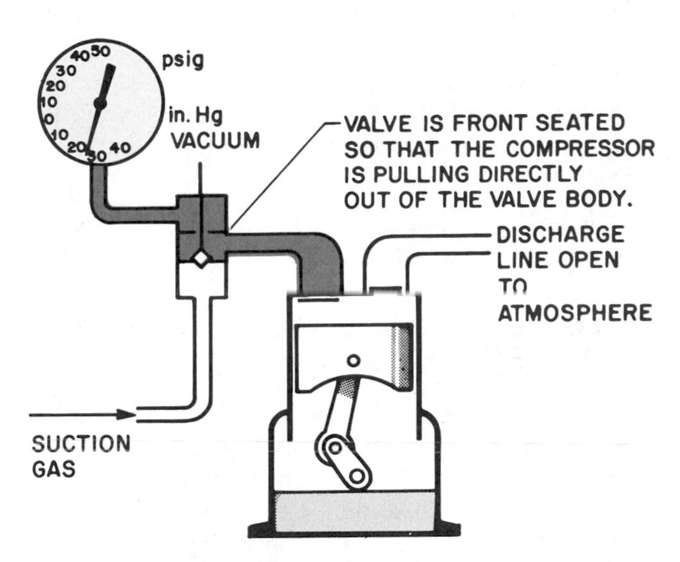

FIGURE 28–35 Compressor pulling a vacuum with the atmosphere as the head pressure. Most reciprocating compressors can pull 26 in. to 28 in. Hg vacuum with the atmosphere as the discharge pressure.

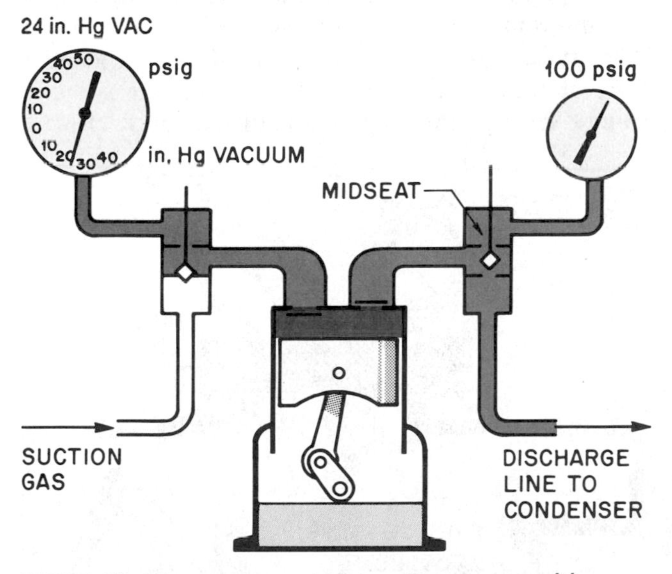

FIGURE 28–36 Compressor check by pulling a vacuum while pumping against a head. A reciprocating compressor can normally pull 24 in. Hg vacuum against 100 psig head pressure.

take more than 3 to 5 min. The motor will not overheat in this period of time. The test should only be performed by experienced technicians.*

28.19 CLOSED-LOOP COMPRESSOR RUNNING BENCH TEST

Running bench testing the compressor can be accomplished by connecting a line from the discharge to the suction of the compressor and operating the compressor in a closed loop. A difference in pressure can be obtained with a valve arrangement or gage manifold. This will prove the compressor will pump. ***When the compressor is hermetic, it should operate at close to full-load current in the closed loop when design pressures are duplicated.*** Nitrogen or refrigerant can be used as the gas to compress. Typical operating pressures will have to be duplicated for the compressor to operate at near the full-load current rating. For example, for a medium-temperature compressor, a suction pressure of 20 psig and a head pressure of 170 psig will duplicate a typical condition for R-12 on a hot day. The compressor should operate at near to nameplate full-load current when the design voltage is supplied to the motor.

The following is a step-by-step procedure for performing this test. Use this procedure with the information in Figure 28–37. ***This test should be performed only under the close supervision of an instructor or by a qualified person and on equipment under 3 hp. Safety goggles must be worn.***

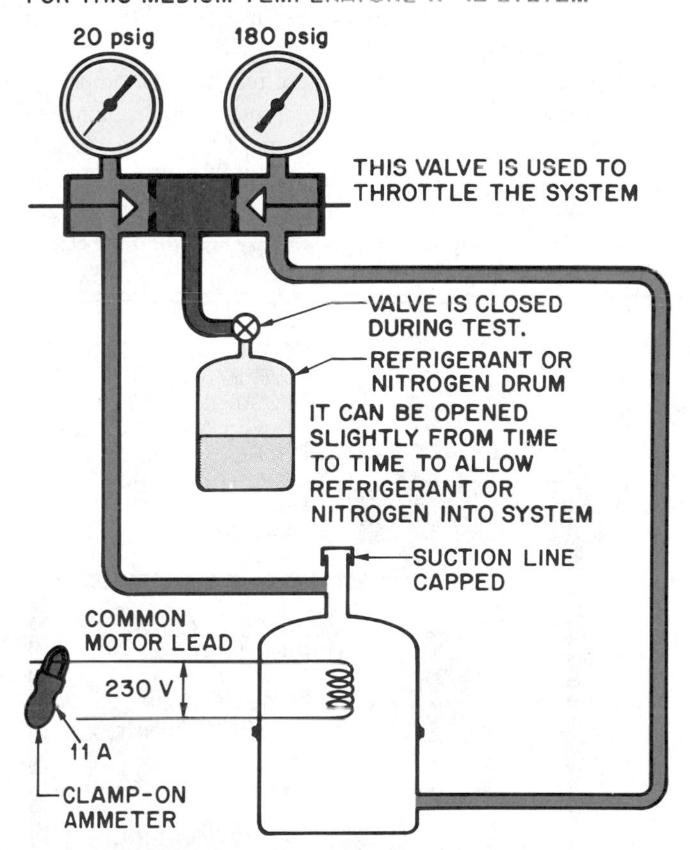

FIGURE 28–37 Checking a compressor's pumping capacity by using a closed loop. The test is accomplished by routing the discharge gas back into the suction with a piping loop. The discharge gage manifold valve is gradually throttled towards closed ***(do not close entirely)*** *Note:* It can never be closed, or tremendous pressure will occur. When the design suction and discharge pressures are reached, the compressor should be pulling close to nameplate full-load amps. *Note: A suction-cooled compressor cannot be run for long in this manner, or the motor will get hot.* The refrigerant characteristic to the compressor or nitrogen can be used to circulate for pumping. Nitrogen will not produce the correct amperage, but it will be close enough. ***This test should only be performed by experienced technicians***

Steps to perform a closed-loop test with the compressor out of the system:

1. Use the gage manifold and fasten the suction line to the gage line in such a manner that the compressor is pumping only from the gage line.
2. Fasten the discharge gage line to the discharge valve port in such a manner that the compressor is pumping only into the gage manifold.
3. Plug the center line of the gage manifold.
4. ***Open both gage manifold valves wide open, counterclockwise.***
5. ***Start the compressor, keep your hand on the off switch.***
6. The compressor should now be pumping out the discharge line and back into the suction line. The discharge gage manifold valve may be slowly throttled ***(do not close entirely)*** toward closed until the discharge pressure rises to the design level and the suction pressure drops to the design level.
7. When the desired pressures are reached, the amperage reading on the compressor motor should compare closely to the full-load amperage of the compressor. If the correct pressures cannot be obtained with the amount of gas in the compressor, a small amount of gas may be added to the loop system by attaching the center line to a refrigerant cylinder and slightly opening the cylinder valve.
8. The amperage may vary slightly from full load because of the input voltage. For example, a voltage above the nameplate will cause an amperage below full load and vice versa.

Figure 28–38 illustrates a situation where a technician accidently closed the wrong valve.

28.20 CLOSED-LOOP COMPRESSOR RUNNING FIELD TEST

When a compressor has service valves, this test can be performed in place in the system using a gage manifold as the loop, Figure 28–39. The compressor is started with the compressor service valves turned all the way to the front seat and the gage manifold valves open. The center line on the gage manifold must be plugged. ***This can be a dangerous start-up and should only be performed under supervision of an experienced person. The compressor has to have a place to pump the discharge gas because reciprocating compressors are positive displacement pumps. When the compressor cylinder is full of gas, it is going to pump it somewhere or stall. In this test the gas goes around through the gage manifold and back into the**

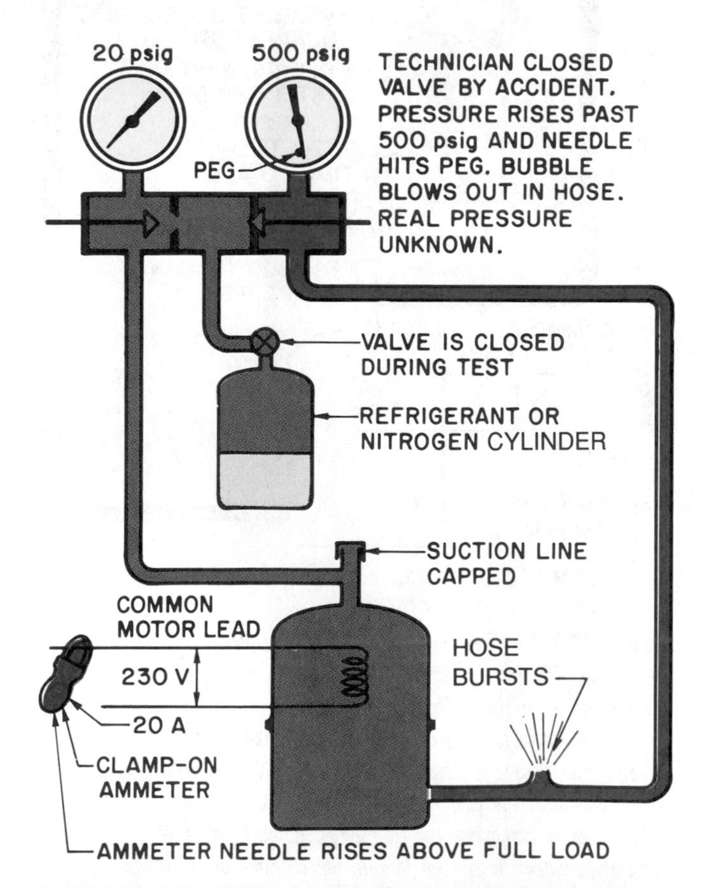

FIGURE 28–38 *What happens if a compressor is started up in a closed loop with no place for the discharge gas to go (discharge gage manifold closed by accident). One cylinder full of gas could be enough to build tremendous pressures.*

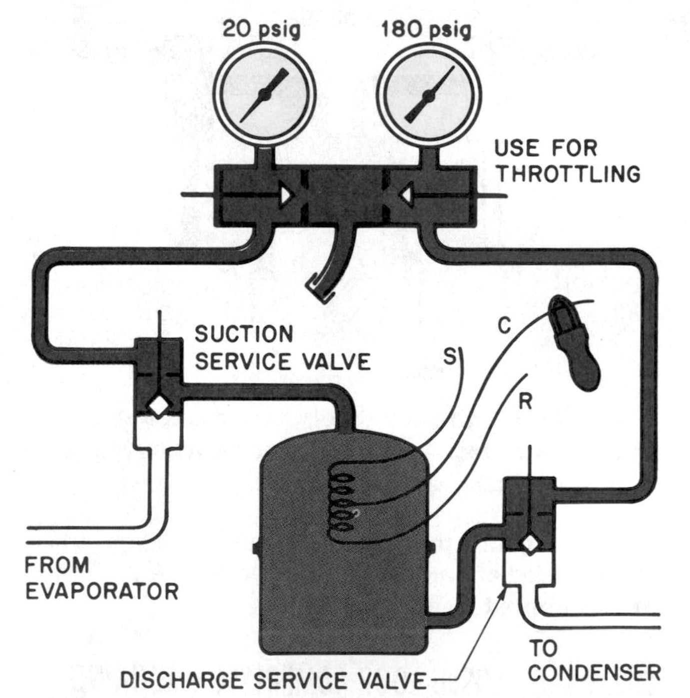

FIGURE 28–39 Compressor test being performed in a system using a gage manifold as a closed loop. Notice that this test can be performed in the system because the compressor has service valves.

suction. Should the compressor be started before the gage manifold is open for the escape route through the gage manifold, tremendous pressures will result before you can stop the compressor. It only takes one cylinder full of vapor to fill the gage manifold to more than capacity. This test should only be performed on compressors under 3 hp. Safety goggles must be worn.*

Steps to perform the closed-loop compressor test with the compressor in the system:

1. Turn the unit OFF and fasten the suction line of the gage manifold to the suction service valve. Plug the center line of the gage manifold.
2. Turn the suction service valve stem to the front seat.
3. Fasten the discharge line to the discharge valve and turn the discharge service valve stem to the front seat.
4. ***Open both gage manifold valves all the way, counterclockwise.***
5. ***Start the compressor and keep your hand on the switch.***
6. The compressor should now be pumping out through the discharge line, through the gage manifold, and back into the suction line.
7. The discharge gage manifold valve may be throttled toward the seat to restrict the flow of refrigerant and create a differential in pressure. Throttle ***(do not close entirely)*** the valve until the design head and suction pressure for the system is attained and the compressor should then be pulling near to full-load amperage. As in the bench test the amperage may vary slightly because of the line voltage.

28.21 COMPRESSOR RUNNING TEST IN THE SYSTEM

Running test in the system can be performed by creating typical design conditions in the system. Typically a compressor will operate at a *high suction* and a *low head pressure* when it is not pumping to capacity. This will cause the compressor to operate at a low current. When the technician gets to the job, the conditions are not usually at the design level. The fixture is usually not refrigerating correctly—this is what instigated the call to begin with. The technician may not be able to create design conditions, but the following approach should be tried if the compressor capacity is suspected.

1. Install the high- and low-side gages.
2. Make sure that the charge is correct (not over nor under), using manufacturer's recommendations.
3. Check the compressor current and compare to full load.
4. Block the condenser airflow and build up the head pressure.

If the compressor will not pump the head pressure up to the equivalent of a 95°F (95°F + 30°F = 125°F condensing temperature or 170 psig for R-12, 185 psig for R-134A, 278 psig for R-22 and 301 psig for R-502) day and draw close to nameplate full-load current, the compressor is not pumping. When the compressor is a sealed hermetic type, it may whistle when it is shut down. This whistle is evidence of an internal leak from the high side to the low side.

If the compressor has service valves a closed-loop test can be performed on the compressor using the methods we have explained, while it is in the system. ***Make sure no air is drawn into the system while it is in a vacuum.***

The compressor temperature should be monitored at any time these tests are being conducted. If the compressor gets too warm, it should be stopped and allowed to cool. If the compressor has internal motor-temperature safety controls, don't operate it when the control is trying to stop the compressor.

HVAC GOLDEN RULES

When making a service call to a business:

- Never park your truck or van in a space reserved for customers.
- Look professional and be professional.
- Before starting troubleshooting procedures, get all the information you can regarding the problem.
- Be extremely careful not to scratch tile floors or to soil carpeting with your tools or by moving equipment.
- Be sure to practice good sanitary and hygiene habits when working in a food preparation area.
- Keep your tools and equipment out of the customers' and employees' way if the equipment you are servicing is located in a normal traffic pattern.
- Be prepared with the correct tools and ensure that they are in good condition.
- Always clean up after you have finished. Try to provide a little extra service by cleaning filters, oiling motors or by providing some other service that will impress the customer.
- Always discuss the results of your service call with the owner or representative of the company. Try to persuade the owner to call if there are any questions as a result of the service call.

PREVENTIVE MAINTENANCE FOR REFRIGERATION

PACKAGED EQUIPMENT. Packaged equipment is built and designed for minimum maintenance because the owner may be the person that takes care of it until a breakdown occurs. Most of the fan motors are permanently lubricated and will run until they quit, at which time they are replaced with new ones.

The owners should be educated to keep the condensers clean and not to stack inventory so close as to block the condenser air flow. When the unit is a reach-in cooler, the owner should be cautioned to follow the manufacturer's directions in loading the box. The load line on the inside should be observed for proper air distribution.

The owner or manager should examine each refrigerated unit frequently and be aware of any peculiar noises or actions from the box. Each box should have a thermometer that should be monitored each day by management. Any rise in temperature that does not reverse itself should alert the manager that a problem is occurring. This can prevent unnecessary loss of perishable foods.

The electrical service for all package equipment should be visually inspected for frayed wires and overheating. Power cord connections may become loose and start to build up heat. If the end that plugs into the wall receptacle becomes hot, the machine should be shut down until the problem is found and corrected. ***If only the plug is replaced, the wall receptacle may still be a problem. The whole connection should be repaired, the wall receptacle and the cord plug.***

Ice machines require special attention if they are to be reliable. Management should know what the quality of the ice should be when the machine is operating correctly. When the quality begins to deteriorate, the reason should be found. This can be as simple as looking for the correct water flow on a cube maker, or the correct ice cutting pattern on another type of cube maker. The ice falling into the bin of a flake maker should have the correct quality, neither soft and mushy nor brittle and hard. The level of the ice in the bin the morning after the machine has been left to run all night can reveal if the machine is making enough ice. For example, if it is observed that the bin is full every morning and the bin thermostat is satisfied but then one morning the bin is only half full, trouble can be suspected. Keep a close watch on the machine.

Drains and drain lines for all refrigeration equipment should be maintained and kept clean and free. This can normally be accomplished when the unit is cleaned and sanitized on the inside. When the cleaning water is flushed down the drain, the speed that the drain moves the water should indicate if the drain is partially plugged or draining freely.

SPLIT SYSTEM REFRIGERATION EQUIPMENT, EVAPORATOR SECTION, CONDENSER SECTION, AND INTERCONNECTING PIPING. In refrigeration split systems the evaporator section is located in one place and the condensing unit in another. The interconnecting piping makes the system complete. The evaporator section contains the evaporator, motor, defrost heaters, drain pan, and drain line. Refrigeration evaporators do not need cleaning often, but need to be inspected and cleaned when needed. The technician cannot always tell when a coil is dirty by looking at the evaporator. Grease or dirt may be in the core of the coil. Routine cleaning once a year for the evaporator will usually keep the coil clean. ***Use only approved cleaning compounds for use where food is present. Turn off the power before cleaning any system. Cover the fan motors and all electrical connections when cleaning to prevent water and detergent from getting in them.***

The motors in the evaporator unit are normally sealed motors and permanently lubricated. If not, they should be lubricated at recommended intervals. These are normally marked on the motor. Observe the fan blade for alignment and look for bearing wear. This may be found by lifting the motor shaft. Most small motors have considerable end play. Don't mistake this for bearing wear. You must lift the shaft to discover bearing wear. The fan blade may also gather weight from dirt and become out of balance.

All wiring in the evaporator section should be visually inspected. If it is cracked or frayed, shut off the power and replace it. Don't forget the defrost heaters and any heater tapes that may be used to keep the drain line warm during cold weather.

The complete case should be cleaned and sanitized at regular intervals. You may need to be the judge of these intervals. Don't let the unit become dirty, including the floor and storage racks in walk-in coolers. ***Keep all ice off the floor of walk-in coolers, or an accident may occur.***

The condensing unit may be located inside an equipment room in the store or it may be outside. When the condenser is located inside an equipment room, it must have proper ventilation. Most equipment rooms have automatic exhaust systems that turn fans on when the equipment room reaches a certain temperature. In large equipment rooms, the temperature will stay warm enough for the exhaust fan to run all the time.

Condensers in equipment rooms become dirty like they do outdoors and must be cleaned. Most system condensers should be cleaned at least once a year to ensure the best efficiency and to prevent problems. The condensers may be cleaned with an approved cleaner for condensers and then washed. ***Turn off the power before cleaning any system. All motors and wiring must be covered to prevent them from getting wet. Watch the equipment room floor, if oil is present and water gets on it, it will become slippery.***

The technician can find many future problems by close inspection of the equipment room. Leaks may be found by observing oil spots on piping. Touch testing the various components will often tell the technician that a compressor is operating too hot or too cold. Tape-on-type temperature indicators may be fastened to discharge lines that indicate the highest temperature the discharge line has ever reached. For example, most manufacturers would consider that oil inside the compressor will begin to break down at discharge temperatures above 250°F. A tape-on temperature indicator will tell if high temperatures are occurring while no one is around, such as a defrost problem in the middle of the night. This can lead to a search for a problem that is not apparent.

The crankcase of a compressor should not be cold to the touch below the oil level. This is a sure sign that liquid refrigerant is in the refrigerant oil on some systems, not necessarily in great amounts, but enough to cause diluted refrigerant oil. Diluted refrigerant oil causes marginal lubrication and bearing wear.

Fan motors may be inspected for bearing wear. Again, do not mistake end play for bearing wear. Some fan motors require lubrication at the correct intervals.

A general cleaning of the equipment room at regular intervals will ensure the technician good working conditions when a failure occurs.

The interconnecting piping should be inspected for loose insulation, oil spots (indicating a leak) and to ensure the pipe is secure. Some piping is in trenches in the floor. A trench that is full of water because of a plugged floor drain will hurt capacity of the equipment because the insulation value of the suction line insulation is not as great when wet. A heat exchange between the liquid lines and suction lines or the ground and suction lines will occur. The pipe trenches should be kept as clean as practical.

28.22 SERVICE TECHNICIAN CALLS

In addition to the six typical problems encountered in the refrigeration system, there are many more not so typical problems. The following service situations will help you understand troubleshooting. Most of these service situations have already been described, although not as an actual troubleshooting procedure. Sometimes the symptoms do not describe the problem and a wrong diagnosis is made. Do not draw any conclusion until the whole system has been examined. Become a system doctor. Examine the system and say, "This needs further examination," then do it.

Refrigeration systems often cool large amounts of food so they are slow to respond. The temperature may drop very slowly on a hot or warm pull down.

SERVICE CALL 1

A customer calls and complains that a medium-temperature walk-in cooler with a remote condensing unit has a compressor that is short cycling and not cooling correctly. *The evaporator has two fans, and one is burned. The unit is short cycling on the low-pressure control because there is not enough load on the coil.*

On the way to the job the technician goes over the possibilities of the problem. This is where it helps to have some familiarity with the job. The technician remembers that the unit has a low-pressure control, a high-pressure control, a thermostat, an overload, and an oil safety control. Defrost is accomplished with an off cycle using the refrigerated space air with the fans running. Process of elimination helps make the decision that the thermostat and the oil safety control are not at fault. The thermostat has a 10°F differential, and the cooler temperature should not vary 10°F in a short cycle. The oil safety control is manual reset and will not short cycle. This narrows the possibilities down to the motor protection device and the high- or low-pressure controls.

On arrival, the technician looks over the whole system before doing anything and notices that one evaporator fan is not running and the coil is iced so the suction is too low, Figure 28–40. This causes the low-pressure control to shut off the compressor. When the examination is complete, the fan motor is changed, and the system is put back in operation. The compressor is shut off, and the fans allowed to run long enough to defrost the coil. The temperature inside the cooler is 50°F, and because of the amount of food it will take a long while for the cooler to pull down to the cutout point of 35°F. The technician cautions the owner to watch the thermometer in the cooler to make sure that the temperature is going down. A call later in the day will confirm that the unit is working properly.

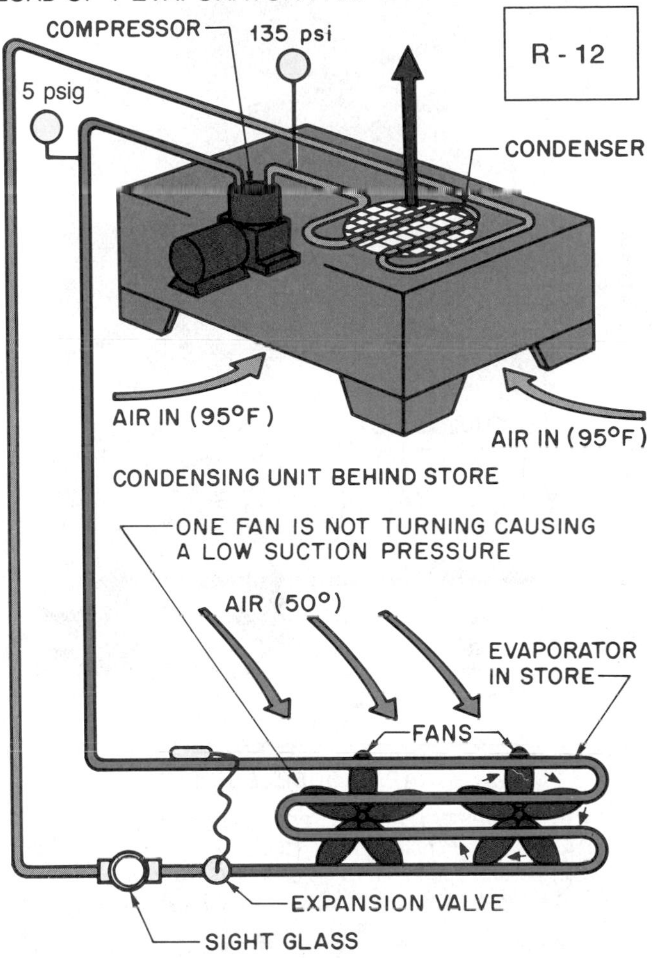
NOTICE THAT THE HEAD PRESSURE WOULD NORMALLY BE 169 psig (95°F + 30°F = 125°F) BUT IT IS 135° psig BECAUSE OF THE REDUCED LOAD OF 1 EVAPORATOR FAN.
COMPRESSOR
135 psi
R - 12
5 psig
CONDENSER
AIR IN (95°F)
AIR IN (95°F)
CONDENSING UNIT BEHIND STORE
ONE FAN IS NOT TURNING CAUSING A LOW SUCTION PRESSURE
AIR (50°)
EVAPORATOR IN STORE
FANS
EXPANSION VALVE
SIGHT GLASS
MEDIUM-TEMPERATURE EVAPORATOR SHOULD BE OPERATING AT 21 psig (20°F BOILING TEMPERATURE) BUT IT IS OPERATING ALL OF THE TIME AND THE COOLER TEMPERATURE IS 50°F.

FIGURE 28–40 Symptoms of a medium-temperature system with one of the two evaporator fans burned out. This system has a TXV.

SERVICE CALL 2

A medium-temperature reach-in cooler is not cooling, and the compressor is cutting on and off. In this system the condensing unit is on top. *The unit has a TXV and a low charge. Two tubes rubbed together and created a leak.*

The technician arrives at the job and finds the unit is short cycling on the low-pressure control. The sight glass has bubbles in it, indicating a low charge, Figure 28–41. The technician discovers that the small tube leading to the low-pressure control has rubbed a hole in the suction line. There is oil around the point of the leak. The system is pumped down by closing the king valve, and the leak is repaired. The system has not operated in a vacuum because of the low-pressure control, so an evacuation is not necessary.

After the repair a leak check is performed. The system is then started and charged to the correct charge. A call later in the day verifies that the system is functioning correctly.

SERVICE CALL 3

A factory cafeteria manager reports that the reach-in cooler in the lunch room is not cooling and is running all the time. This is a

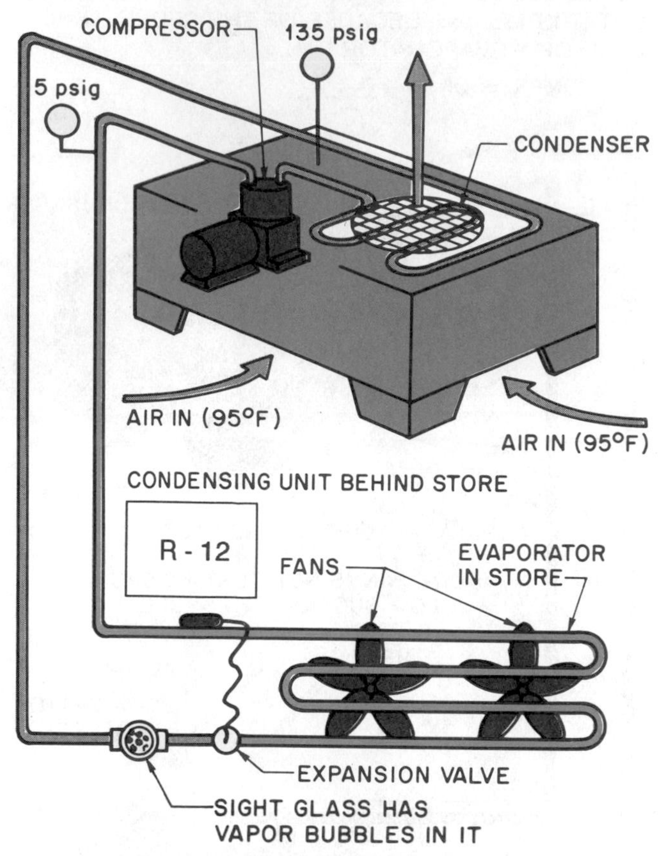
NOTICE THAT THE HEAD PRESSURE IS DOWN DUE TO REDUCED LOAD.
COMPRESSOR
135 psig
5 psig
CONDENSER
AIR IN (95°F)
AIR IN (95°F)
CONDENSING UNIT BEHIND STORE
R - 12
FANS
EVAPORATOR IN STORE
EXPANSION VALVE
SIGHT GLASS HAS VAPOR BUBBLES IN IT
MEDIUM-TEMPERATURE EVAPORATOR SHOULD BE OPERATING AT 21 psig (20°F BOILING TEMPERATURE) BUT IT IS OPERATING AT 5 psig BECAUSE OF A LOW CHARGE CAUSED BY A LEAK.

FIGURE 28–41 Symptoms of a medium-temperature system operating in a low-charge situation; TXV.

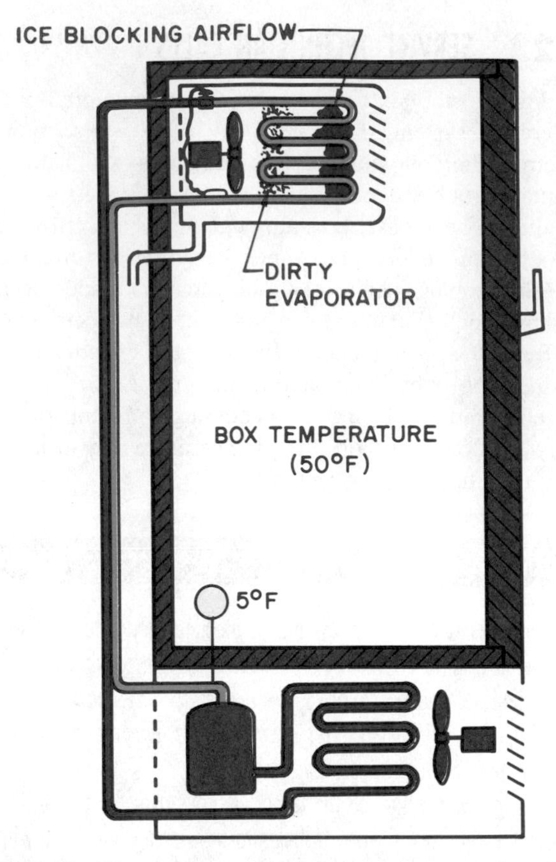
ICE BLOCKING AIRFLOW
DIRTY EVAPORATOR
BOX TEMPERATURE (50°F)
5°F

FIGURE 28–42 Symptoms of a medium-temperature cooler with a dirty evaporator coil; TXV.

medium-temperature cooler with the condensing unit at the bottom of the fixture. It has not had a service call in a long time. *The evaporator is dirty and icing over, Figure 28–42. The unit has a TXV and there is no off time for defrost because the box does not get cool enough.*

The technician arrives and sees that the system is iced. The compressor is sweating down the side; liquid is slowly coming back to the compressor—not enough to cause a noise from slugging. The TXV should maintain a constant superheat, but it may lose control if the pressure drops too low. The first thing that has to be done is to defrost the evaporator. This is done by stopping the compressor and using a heat gun (like a high-powered hair drier). The evaporator has a lot of dirt on it. ***The evaporator is cleaned with a coil cleaner that is approved for food-handling areas.*** When the evaporator is cleaned, the system is started. The unit is now operating with a full sight glass, and the suction line leaving the evaporator is cold. From this point it will take time for the unit to pull the cooler temperature down. The service technician leaves and calls later in the day to confirm that the repair has solved the problem.

SERVICE CALL 4

A convenience store manager reports that the reach-in cooler that stores the dairy products is not cooling properly. *The temperature is 55°F. It has been cooling well until early this morning. There is a leak due to a stress crack (caused by age and vibration) in the suction line near the compressor. The compressor is vibrating because the customer has moved the unit, and the condensing unit has fallen down in the frame. This system has a capillary tube metering device,* Figure 28–43.

The technician's examination discloses that the compressor is vibrating because the condensing unit is not setting straight in the frame. While securing the condensing unit, the technician notices an oil spot on the bottom of the suction line and that the compressor shell is hot. Gages are installed, and it is discovered that the suction pressure is operating in a vacuum. ***Care must be used when installing gages if a vacuum is suspected or air may be drawn into the system.*** The compressor is stopped, and the low-side pressure is allowed to rise. There is not enough pressure in the system to accomplish a good leak test, so refrigerant is added. A leak check in the vicinity of the suction line reveals a leak. This appears to be a stress crack due to the vibration. **The refrigerant is recovered from the system to repair the leak.** ***Air must have been pulled into the system while it was operating in a vacuum, so the refrigerant charge must be***

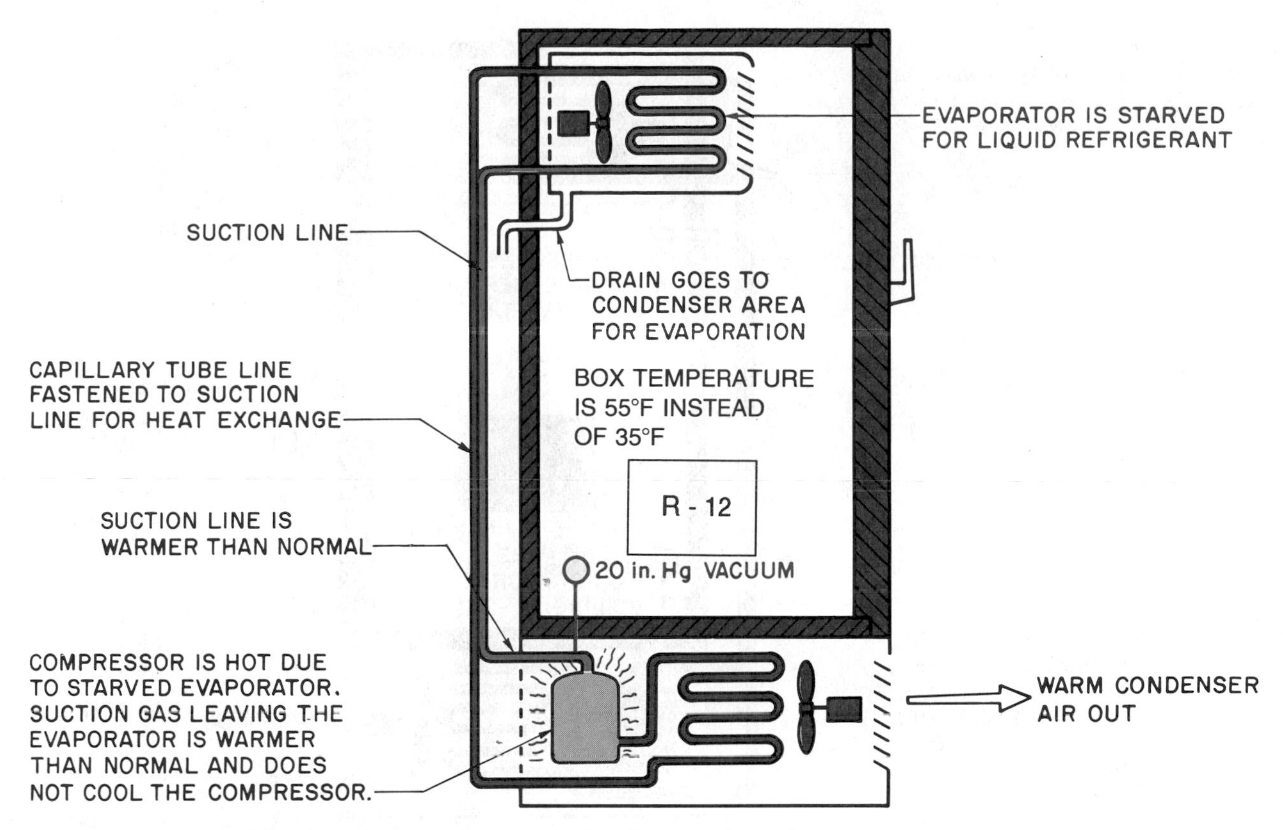
EVAPORATOR IS STARVED FOR LIQUID REFRIGERANT
SUCTION LINE
DRAIN GOES TO CONDENSER AREA FOR EVAPORATION
CAPILLARY TUBE LINE FASTENED TO SUCTION LINE FOR HEAT EXCHANGE
BOX TEMPERATURE IS 55°F INSTEAD OF 35°F
R - 12
SUCTION LINE IS WARMER THAN NORMAL
20 in. Hg VACUUM
COMPRESSOR IS HOT DUE TO STARVED EVAPORATOR. SUCTION GAS LEAVING THE EVAPORATOR IS WARMER THAN NORMAL AND DOES NOT COOL THE COMPRESSOR.
WARM CONDENSER AIR OUT

FIGURE 28–43 Symptoms of a medium-temperature system with a low charge; capillary tube.

removed and recovered. **This system does not have a low-pressure control.** A short length of pipe is installed where the stress crack was found. A new liquid line drier is installed because the old one may not have any capacity left. The system is leak checked, triple evacuated, and charged, using a measured charge that the manufacturer recommended. The technician calls the manager later in the day and learns that the unit is functioning correctly.

SERVICE CALL 5

A restaurant manager calls to indicate that the reach-in freezer used for ice storage is running all the time. This system was worked on by a competitor in the early spring. A leak was found, repaired, and the unit was recharged. Hot weather is here, and the unit is running all the time. *This system has a capillary tube metering device and the system has an overcharge of refrigerant. The other service technician did not measure the charge into the system,* Figure 28–44.

The service technician examines the fixture and notices that the compressor is sweating down the side. The condenser feels hot for the first few rows and then warm. This appears to be an overcharge of refrigerant. The condenser should be warm near the bottom where the condensing is occurring. The evaporator fan is running, and the evaporator looks clean, so the evaporator must be doing its job. Gages are installed, and the head pressure is 400 psig with an outside temperature of 95°F; the refrigerant is R-502. The head pressure should be no more than 301 psig on a 95°F day (95°F + 30°F = 125°F condensing temperature or 301 psig). This system calls for a measured charge of 2 lb 8 oz. **There are two approaches that can be taken: (1) alter the existing charge; or (2) recover the charge and measure a new charge into the unit while the unit is in a deep vacuum. It is a time-consuming process to recover the charge and evacuate the system.**

The technician chooses to alter the existing charge. This is a plain capillary tube system with a heat exchanger (the capillary tube is soldered to the suction line after it leaves the evaporator). A thermometer lead is fastened to the suction line **after the evaporator** but **before the heat exchanger.** The suction pressure is checked for the boiling point of the refrigerant and compared with the suction line temperature. The superheat is 0°F with the existing charge of refrigerant. **Refrigerant is removed to an approved recovery cylinder until the superheat is 5°F at this point.**

NOTE: The heat exchanger will allow a lower superheat than normal. When the system charge is balanced, the technician leaves the job and will call the manager later in the day for a report on how the system is functioning.

SERVICE CALL 6

An office manager calls indicating that the reach-in medium-temperature beverage cooler in the employee cafeteria is running all the time. The suction line is covered with frost back to the compressor. *A small drier is*

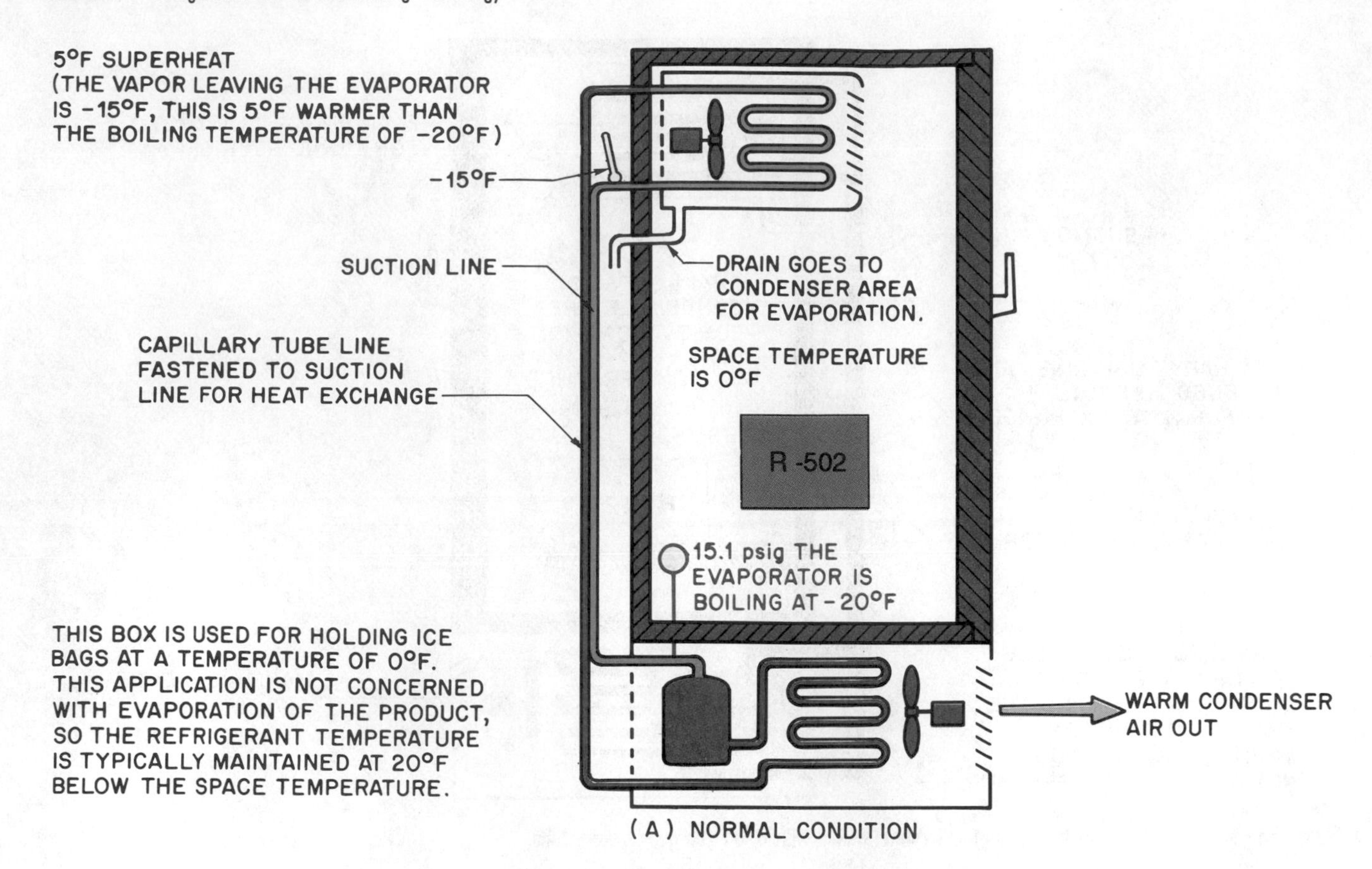
5°F SUPERHEAT
(THE VAPOR LEAVING THE EVAPORATOR IS −15°F, THIS IS 5°F WARMER THAN THE BOILING TEMPERATURE OF −20°F)
−15°F
SUCTION LINE
DRAIN GOES TO CONDENSER AREA FOR EVAPORATION.
CAPILLARY TUBE LINE FASTENED TO SUCTION LINE FOR HEAT EXCHANGE
SPACE TEMPERATURE IS 0°F
R -502
15.1 psig THE EVAPORATOR IS BOILING AT −20°F
THIS BOX IS USED FOR HOLDING ICE BAGS AT A TEMPERATURE OF 0°F. THIS APPLICATION IS NOT CONCERNED WITH EVAPORATION OF THE PRODUCT, SO THE REFRIGERANT TEMPERATURE IS TYPICALLY MAINTAINED AT 20°F BELOW THE SPACE TEMPERATURE.
WARM CONDENSER AIR OUT
(A) NORMAL CONDITION

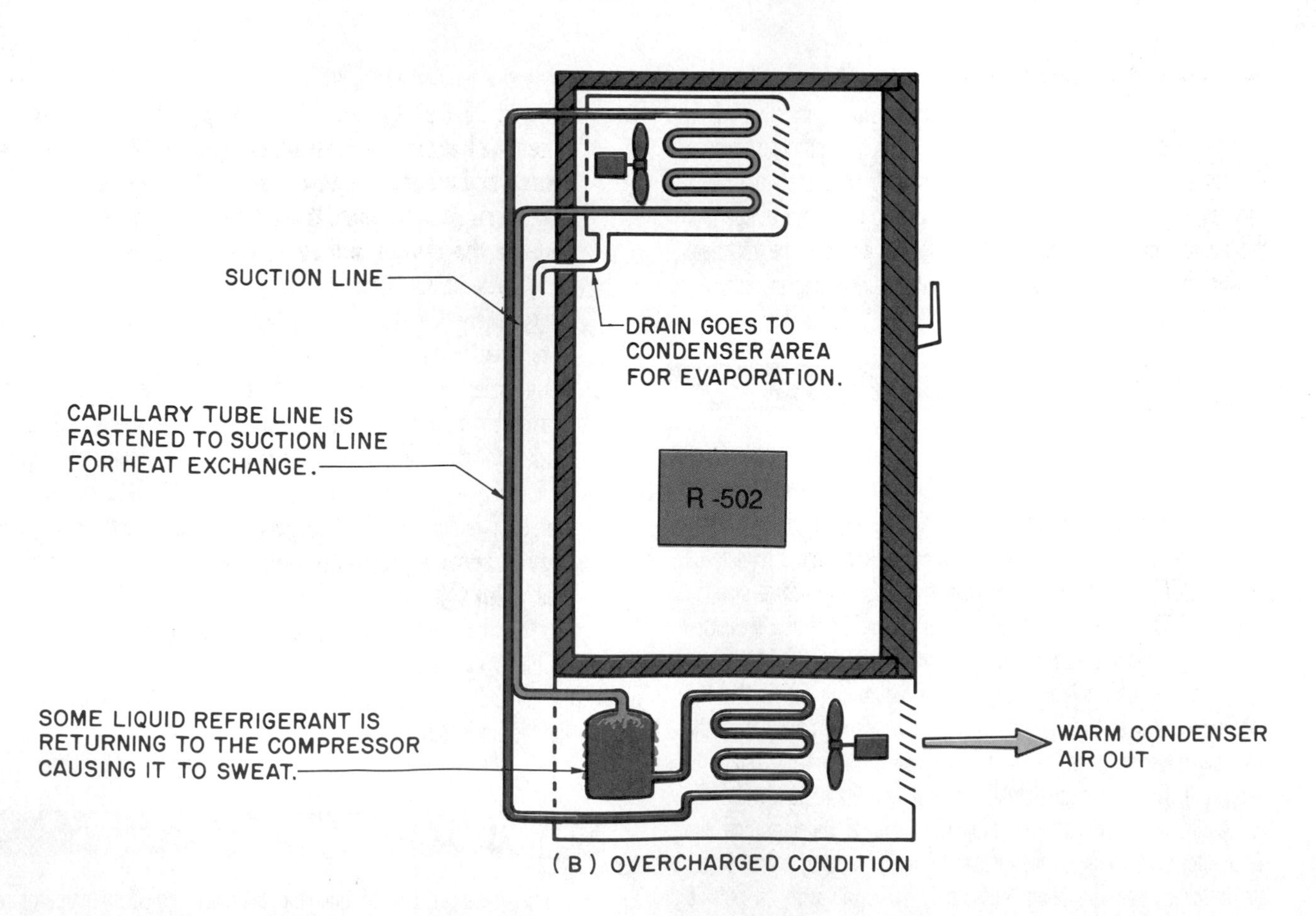
SUCTION LINE
DRAIN GOES TO CONDENSER AREA FOR EVAPORATION.
CAPILLARY TUBE LINE IS FASTENED TO SUCTION LINE FOR HEAT EXCHANGE.
R -502
SOME LIQUID REFRIGERANT IS RETURNING TO THE COMPRESSOR CAUSING IT TO SWEAT.
WARM CONDENSER AIR OUT
(B) OVERCHARGED CONDITION

FIGURE 28–44 Symptoms of a reach-in freezer with an overcharge of refrigerant; capillary tube.

stopped up with sludge from a compressor changeout after a motor burnout. A suction line drier should have been installed, Figure 28–45.

The technician looks the system over closely and sees that the suction line is frosting and that the compressor has a frost patch on it. A first glance indicates that the fan at the evaporator is off and that the coil is dirty. Further examination, however, shows that the liquid line is frosting starting at the outlet of the drier. This means the drier is partially stopped up. The pressure drop across the drier makes the drier act like an expansion device. This effectively means there are two expansion devices in series because the drier is feeding the capillary tube. **The refrigerant charge must be recovered, and the drier replaced.** After the refrigerant has been recovered, the drier is sweated into the liquid line. There are no service valves. While the system is open, the technician solders a suction line drier in the suction line close to the compressor.

The reason the unit was frosting instead of sweating is that the evaporator was starved for liquid refrigerant. The suction pressure went down to a point below freezing. The unit's capacity was reduced to the point where it was running constantly and had no defrost. The frost will become more dense, blocking the air through the coil and act as an insulator. This will cause the coil to get even colder with more frost. This condition will continue with the frost line moving on to the compressor. The ice or frost acts as an insulator and an air blockage. Air has to circulate across the coil for the unit to produce at its rated capacity.

SERVICE CALL 7

A restaurant owner calls indicating that the reach-in freezer for ice cream is running but the temperature is rising. *The system has a capillary tube metering device and the evaporator fan is defective. The customer hears the compressor running and thinks the whole unit is running,* Figure 28–46.

The technician has never been to this installation and has to examine the system thoroughly. It is discovered that the frost on the suction line goes all the way to and down the side of the compressor. The first thought is that the system is not going through the proper defrost cycle. The coil is iced and has to be defrosted before anything can be done. The defrost timer is advanced until the unit goes through defrost. The defrost cycle clears the coil of ice. When defrost is over, the fan does not restart as it should. The technician takes off the panel to the fan compartment and checks the voltage. There is voltage, but the fan will not run. When the motor is checked for continuity through the windings, it is discovered that the fan motor winding is open. The fan motor is changed, and the system is started again. It will take several hours for the fixture to pull back down to the normal running temperature of −10°F. A call later in the day verifies that the freezer temperature is going down.

SERVICE CALL 8

A call is received indicating that the customer's reach-in dairy case used for milk storage is running constantly, and the temperature is

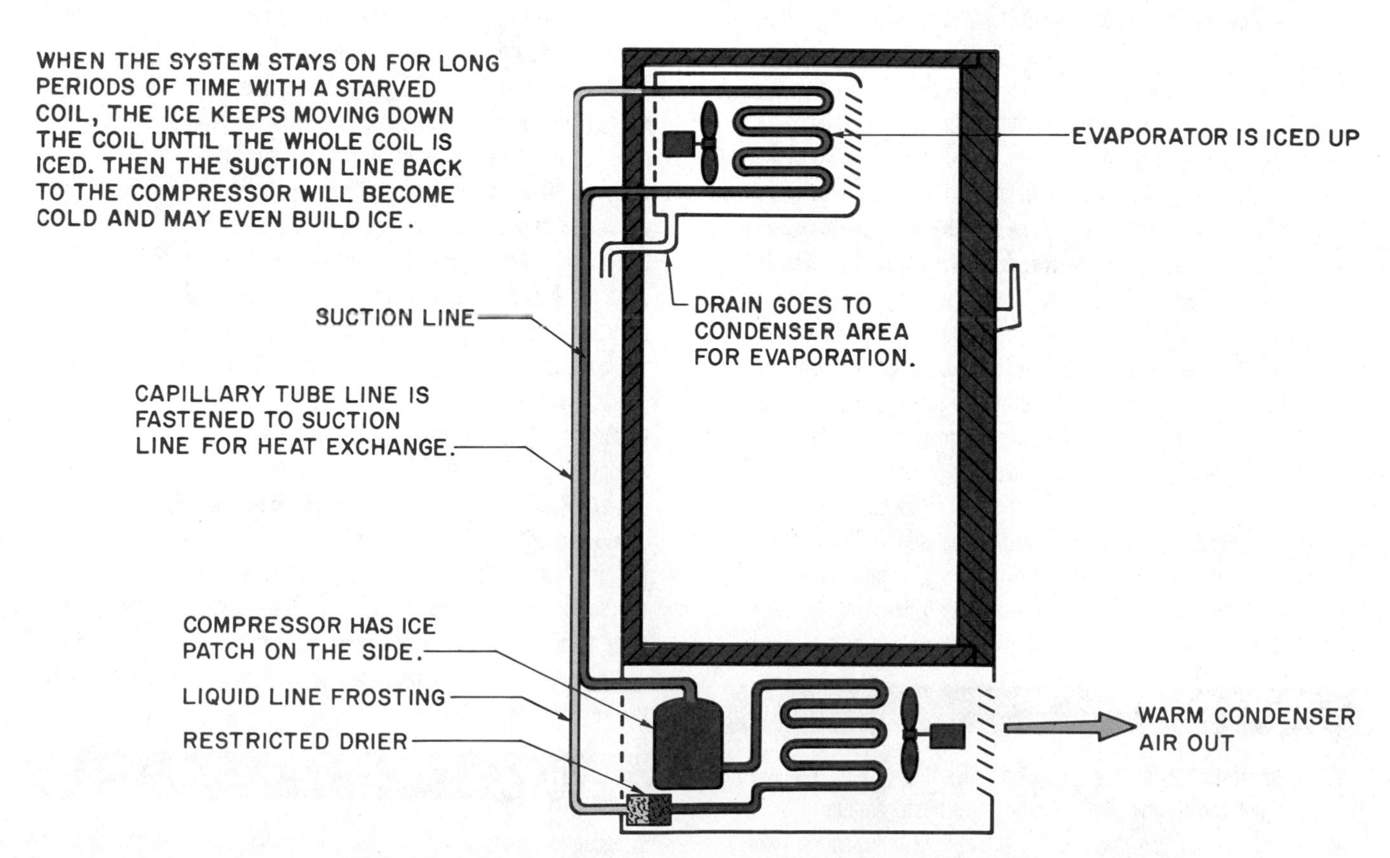
WHEN THE SYSTEM STAYS ON FOR LONG PERIODS OF TIME WITH A STARVED COIL, THE ICE KEEPS MOVING DOWN THE COIL UNTIL THE WHOLE COIL IS ICED. THEN THE SUCTION LINE BACK TO THE COMPRESSOR WILL BECOME COLD AND MAY EVEN BUILD ICE.
EVAPORATOR IS ICED UP
SUCTION LINE
DRAIN GOES TO CONDENSER AREA FOR EVAPORATION.
CAPILLARY TUBE LINE IS FASTENED TO SUCTION LINE FOR HEAT EXCHANGE.
COMPRESSOR HAS ICE PATCH ON THE SIDE.
LIQUID LINE FROSTING
RESTRICTED DRIER
WARM CONDENSER AIR OUT

FIGURE 28–45 Symptoms of a medium-temperature reach-in cooler with a partially stopped-up liquid line drier.

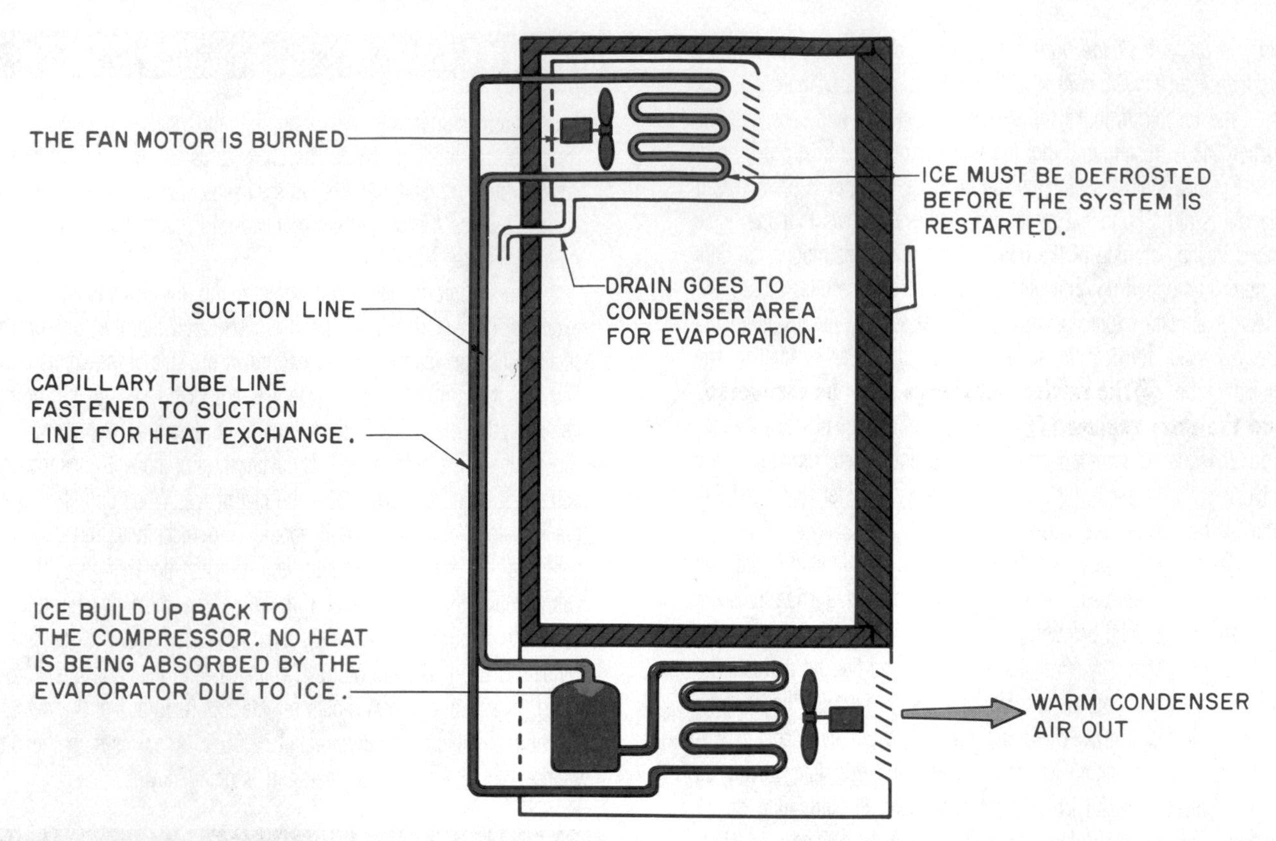
THE FAN MOTOR IS BURNED
ICE MUST BE DEFROSTED BEFORE THE SYSTEM IS RESTARTED.
DRAIN GOES TO CONDENSER AREA FOR EVAPORATION.
SUCTION LINE
CAPILLARY TUBE LINE FASTENED TO SUCTION LINE FOR HEAT EXCHANGE.
ICE BUILD UP BACK TO THE COMPRESSOR. NO HEAT IS BEING ABSORBED BY THE EVAPORATOR DUE TO ICE.
WARM CONDENSER AIR OUT

FIGURE 28–46 Symptoms of a reach-in freezer with a defective evaporator fan motor; capillary tube.

48°F. This unit has not had a service call in 10 years of service. *The evaporator is dirty. This unit has an automatic expansion valve. These coils never have filters, and years of dust will accumulate on the coils,* Figure 28–47.

The technician finds the suction line very cold. The compressor is sweating. This is evidence of an overcharge, or that the refrigerant is not boiling to a vapor in the evaporator. A close examination of the evaporator indicates that it is not exchanging heat with the air in the cooler. ***The evaporator is cleaned with a special detergent approved for evaporator cleaning and for use in food-handling areas.*** Areas that have dairy products are particularly difficult because dairy products absorb odors easily. The system is started; the sweat gradually moves off the compressor, and the suction line feels normally cool. The technician leaves the job and will call back later to see how the unit is performing.

An automatic expansion valve maintains a constant pressure. It responds in reverse to a load change. If a load of additional product is added to the cooler, the rise in suction pressure will cause the automatic expansion valve to throttle back and slightly starve the evaporator. If the load is reduced, such as with a dirty coil, the valve will overfeed to keep the refrigerant pressure up.

SERVICE CALL 9

A golf club restaurant manager calls and says that a small beverage cooler is not cooling the drinks to the correct temperature. *A small leak at the flare nut on the outlet to the automatic expansion valve has caused a partial loss of refrigerant,* Figure 28–48.

The technician arrives and hears the expansion valve hissing. This means the expansion valve is passing vapor along with the liquid it is supposed to pass. The sight glass shows some bubbles. Gages are installed; the suction pressure is normal, and the head pressure is low. The suction pressure is 17.5 psig; this corresponds to 15°F boiling temperature for R-12. The boiling refrigerant normally is about 20°F cooler than the beverages to be cooled. The liquid beverage in the cooler is 50°F and should be 35°F. The head pressure should be about 126 psig, the pressure corresponding to 105°F. The ambient is 75°F, and the condensing temperature should be 105°F (75°F + 30°F = 105°F). The head pressure is 100 psig. All of these signs point to an undercharge. The technician turns the cooler off and allows the low-side pressure to rise so that there will be a better chance to detect a leak. A leak is found at the flare nut leaving the expansion valve. The nut is tightened, but it doesn't stop the refrigerant from leaking. The flare connection must be defective.

⊛**The charge is recovered, and the flare nut is removed.**⊛ There is a crack in the tubing at the base of the flare nut. The flare is repaired, and the system is leak checked and triple evacuated. A new charge of refrigerant is measured into the system from a vacuum. This gives the most accurate operating charge. The system is started. The technician will call back later to ensure that the unit is operating properly.

SERVICE CALL 10

A restaurant manager reports that the compressor in the pie case is cutting off and on and sounds like it is straining while it is running,

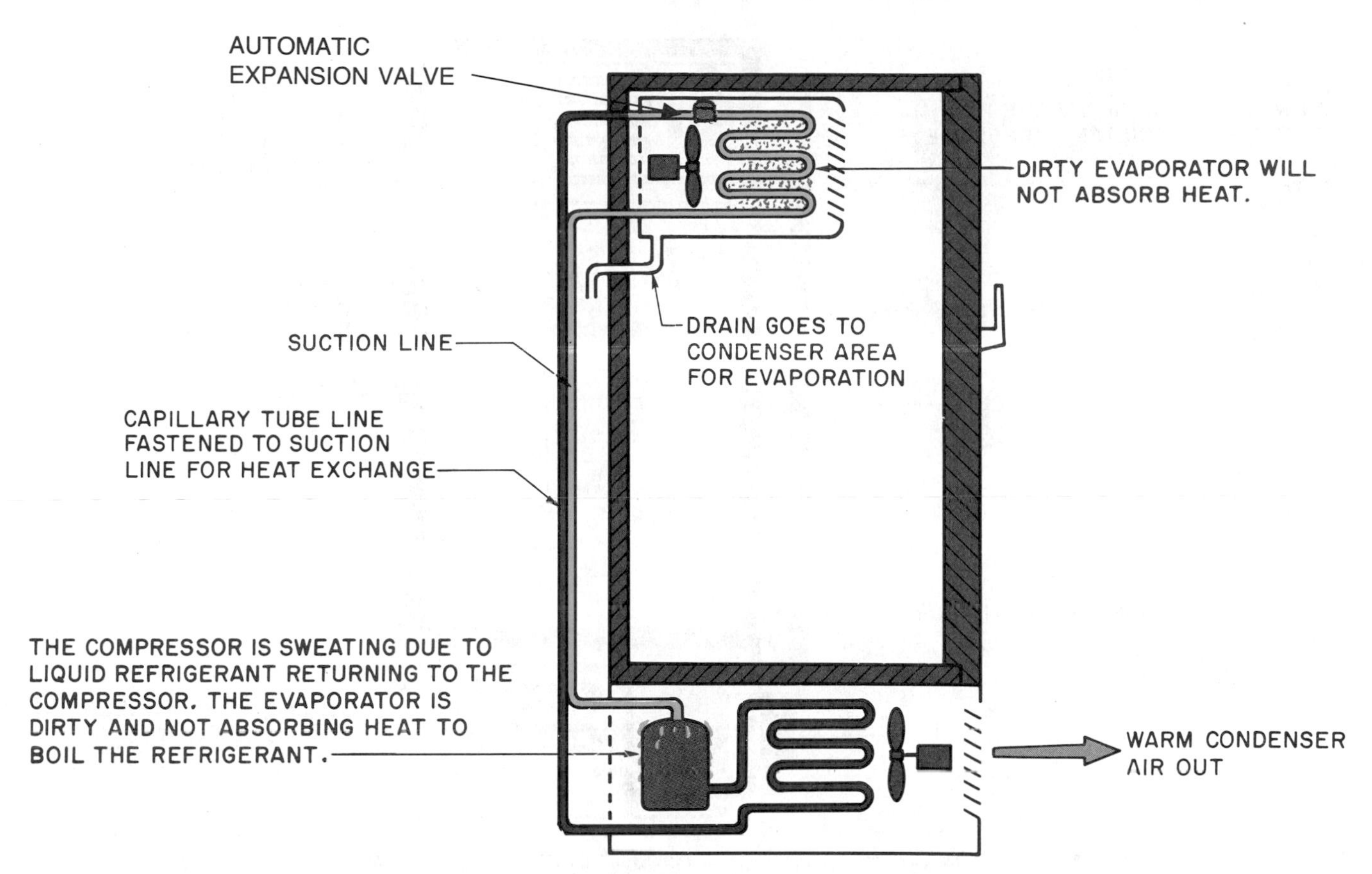
AUTOMATIC
EXPANSION VALVE
DIRTY EVAPORATOR WILL
NOT ABSORB HEAT.
SUCTION LINE
DRAIN GOES TO
CONDENSER AREA
FOR EVAPORATION
CAPILLARY TUBE LINE
FASTENED TO SUCTION
LINE FOR HEAT EXCHANGE
THE COMPRESSOR IS SWEATING DUE TO
LIQUID REFRIGERANT RETURNING TO THE
COMPRESSOR. THE EVAPORATOR IS
DIRTY AND NOT ABSORBING HEAT TO
BOIL THE REFRIGERANT.
WARM CONDENSER
AIR OUT

FIGURE 28–47 Symptoms of a reach-in dairy case with a dirty evaporator coil; automatic expansion valve.

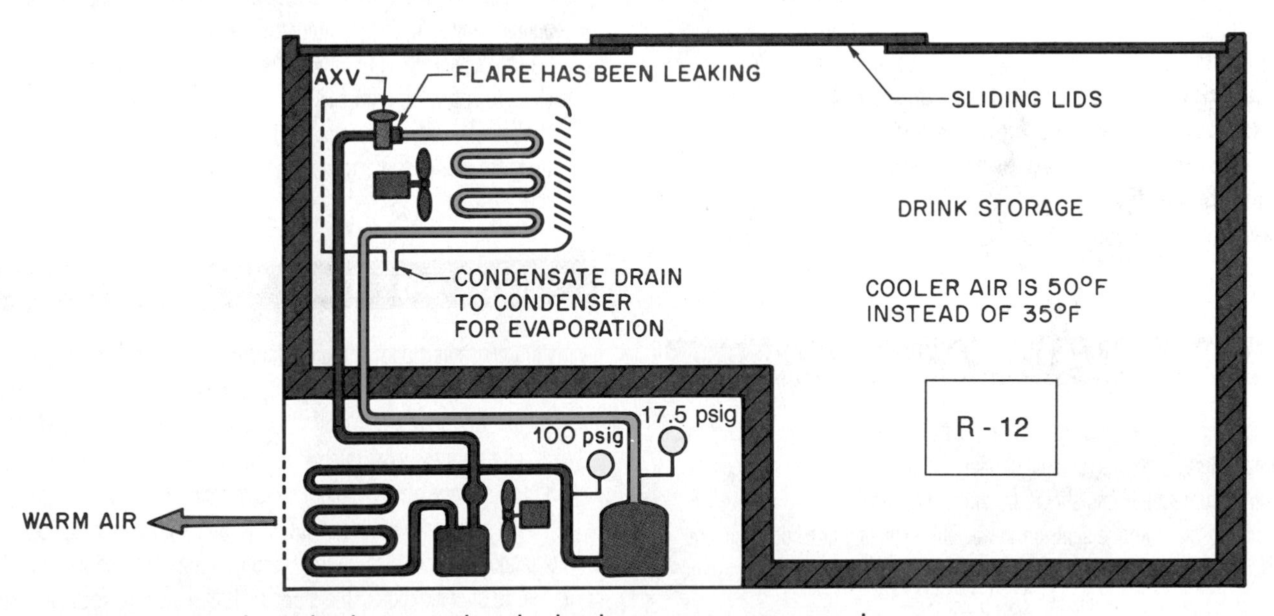
AXV
FLARE HAS BEEN LEAKING
SLIDING LIDS
DRINK STORAGE
CONDENSATE DRAIN
TO CONDENSER
FOR EVAPORATION
COOLER AIR IS 50°F
INSTEAD OF 35°F
17.5 psig
100 psig
R - 12
WARM AIR

FIGURE 28–48 Symptoms of a reach-in beverage cooler with a low charge; automatic expansion valve.

Figure 28–49. This unit was charged after a leak, then shut down and put in storage. It has not been operated in several months. *The unit has an automatic expansion valve and has an overcharge of refrigerant. When the unit was charged it was in the back room in the winter, and it was cold in the room. The unit was charged to a full sight glass with a cold condenser.*

The service technician remembers that the unit was started and charged in a cold ambient and assumes that it may have an overcharge of refrigerant. When the technician arrives, the unit is started in the location where it is going to stay; the ambient is warm. Gages are installed, and the head pressure is 250 psig. The pressure should be 136 psig at the highest because the ambient temperature

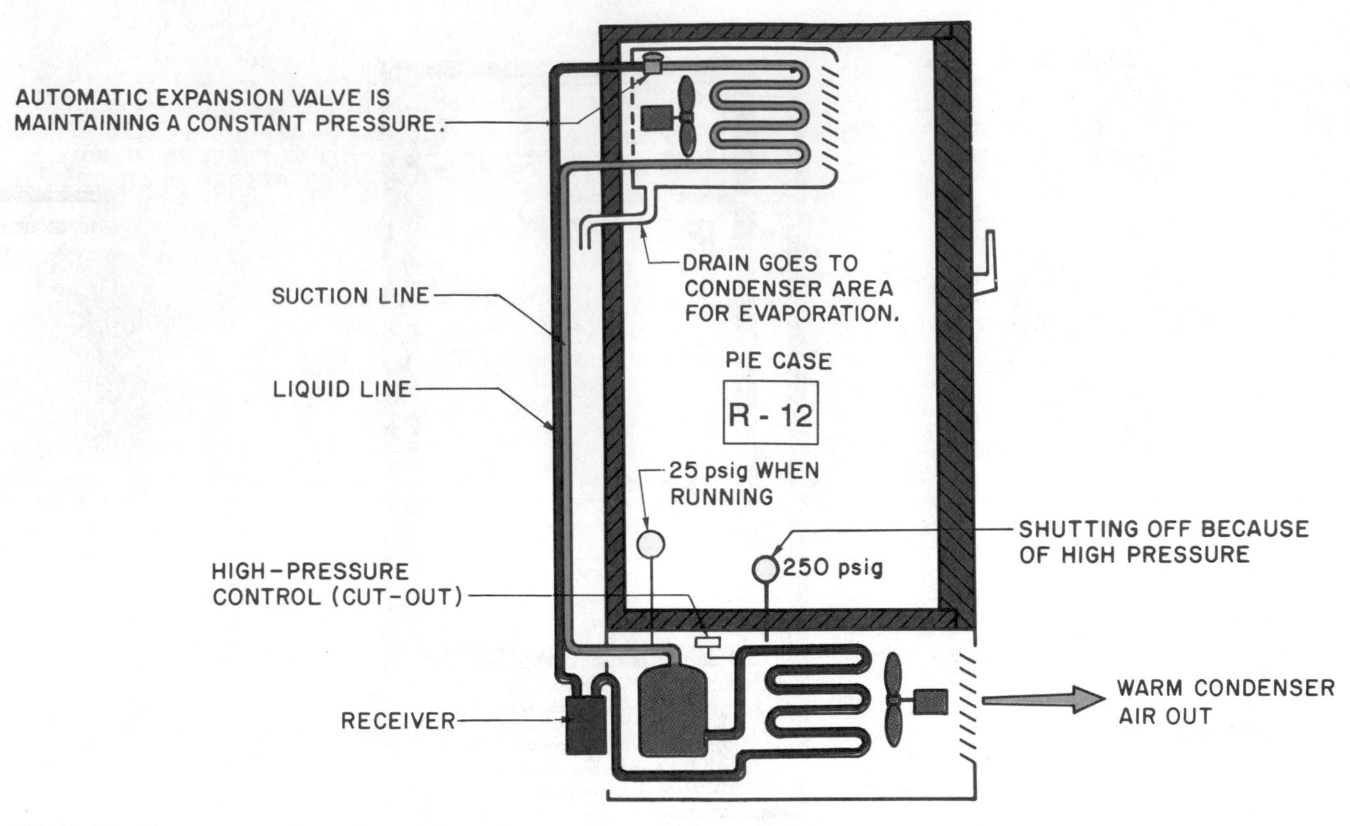

FIGURE 28–49 Symptoms of a reach-in pie case with an overcharge of refrigerant; automatic expansion valve.

is 80°F. This should create a condensing temperature of no more than 110°F (80°F + 30°F = 110°F) or 136 psig for R-12. The compressor is cutting off because of high pressure at 260 psig. **Refrigerant is removed to an approved recovery cylinder from the machine until the head pressure is down to 136 psig, the correct head pressure for the ambient temperature.** The sight glass is still full. A call back to the owner later in the day verifies that the system is working correctly.

SERVICE CALL 11

A store manager calls to say that a reach-in cooler is rising in temperature, and the unit is running all the time. It is a medium-temperature cooler and should be operating between 35°F and 45°F. It has an automatic expansion valve. This unit has not had a service call in 3 years. *The evaporator fan motor is not running. This system has off-cycle defrost,* Figure 28–50.

The service technician arrives at the site and examines the system. The compressor is on top of the fixture and is sweating down the side. Because this uses an automatic expansion valve, the sweating is a sign that the evaporator is not boiling the refrigerant to a vapor. The evaporator could be dirty or the fan may not be moving enough air. After removing the fan panel, the technician sees that the evaporator fan motor is not running. The blades are hard to turn, indicating the bearings are tight. The bearings are lubricated, the fan is started, and seems to run like it should. This is a nonstandard motor and cannot be purchased locally. The system is left in running condition until a new fan motor can be obtained.

The technician returns in a week with the correct motor and exchanges it for the old one. The system has been working correctly all week. The old motor would fail again because the bearings are scored, the reason for exchanging a working motor for a new one.

SERVICE CALL 12

A store assistant manager reports that a reach-in medium-temperature cooler is rising in temperature. *This unit has been performing satisfactorily for several years. A new stock clerk has loaded the product too high, and the product is interfering with the airflow of the evaporator fan. The system has an automatic expansion valve,* Figure 28–51.

The technician arrives at the store and immediately notices that the product is too high in the product area. The extra product is moved to another cooler. The stock clerk is shown the load level lines in the cooler. A call back later in the day indicates that the cooler is now working correctly.

SERVICE CALL 13

A customer reports that the condensing unit on a medium-temperature walk-in cooler is running all the time, and the cooler is rising in temperature. *This is the first hot day of the season, and some boxes*

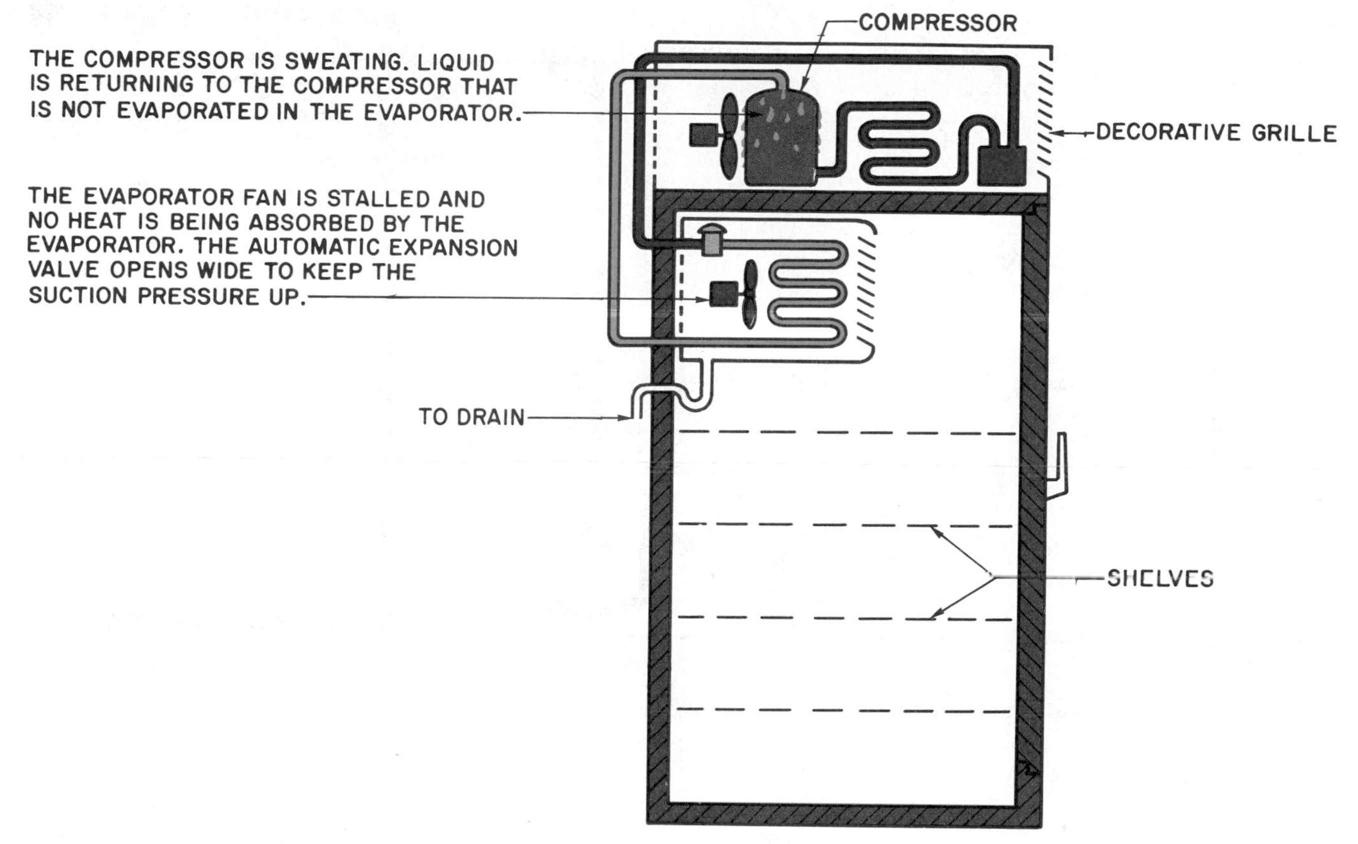
COMPRESSOR
THE COMPRESSOR IS SWEATING. LIQUID IS RETURNING TO THE COMPRESSOR THAT IS NOT EVAPORATED IN THE EVAPORATOR.
DECORATIVE GRILLE
THE EVAPORATOR FAN IS STALLED AND NO HEAT IS BEING ABSORBED BY THE EVAPORATOR. THE AUTOMATIC EXPANSION VALVE OPENS WIDE TO KEEP THE SUCTION PRESSURE UP.
TO DRAIN
SHELVES

FIGURE 28–50 Symptoms of a reach-in cooler with a defective fan motor; automatic expansion valve.

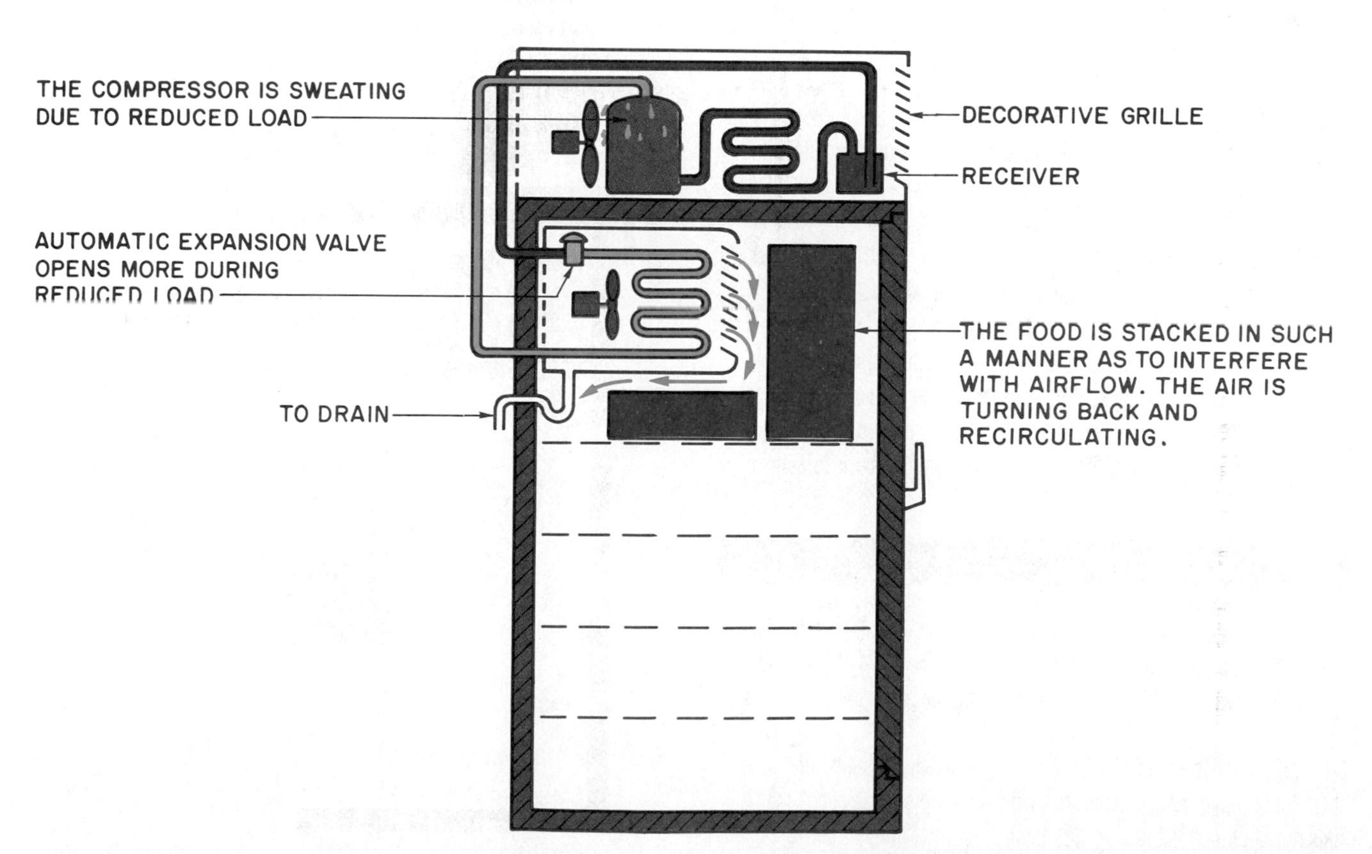
THE COMPRESSOR IS SWEATING DUE TO REDUCED LOAD
DECORATIVE GRILLE
RECEIVER
AUTOMATIC EXPANSION VALVE OPENS MORE DURING REDUCED LOAD
THE FOOD IS STACKED IN SUCH A MANNER AS TO INTERFERE WITH AIRFLOW. THE AIR IS TURNING BACK AND RECIRCULATING.
TO DRAIN

FIGURE 28–51 Symptoms of a reach-in medium-temperature cooler where the product is interfering with the air pattern; automatic expansion valve.

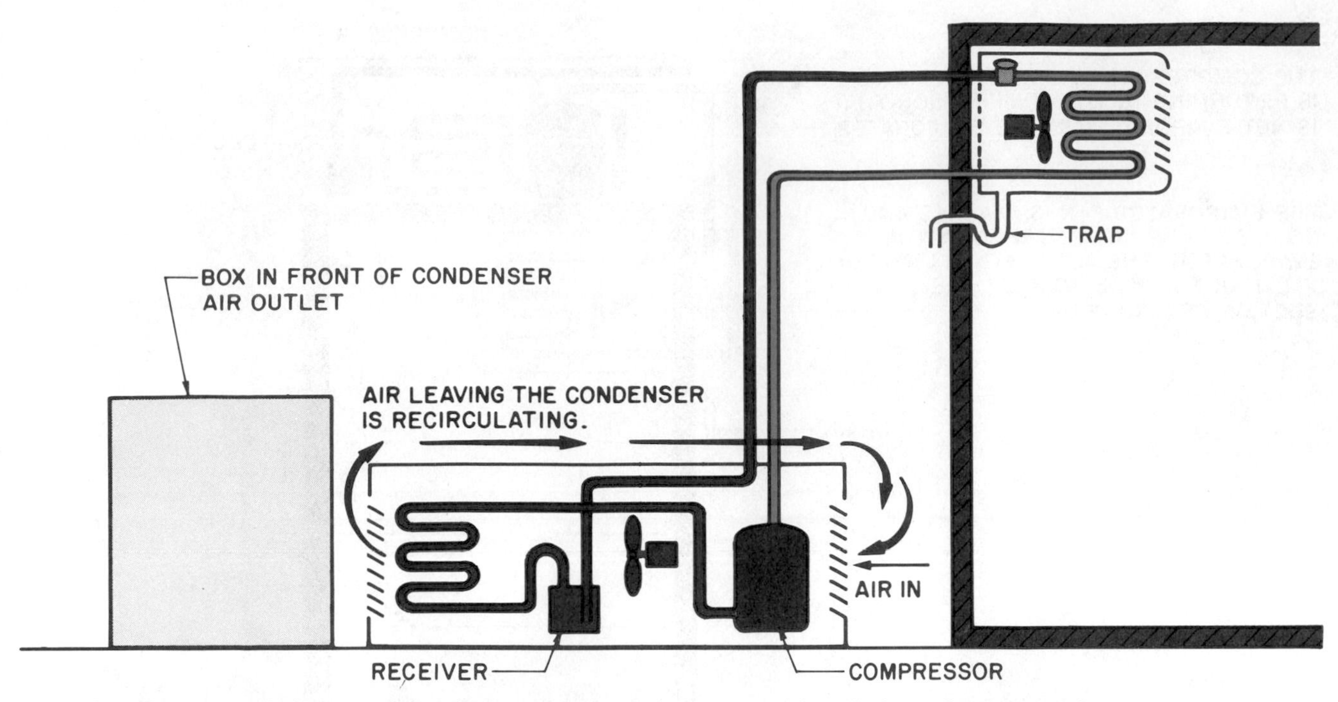

TRAP
BOX IN FRONT OF CONDENSER AIR OUTLET
AIR LEAVING THE CONDENSER IS RECIRCULATING.
AIR IN
RECEIVER
COMPRESSOR

FIGURE 28–52 Symptoms of an air-cooled condenser with the hot discharge air recirculating back into the condenser inlet.

have been stacked too close to the condenser outlet. The hot air is leaving the condenser and recirculating back into the fan inlet, Figure 28–52.

The service technician arrives and looks over the system. The temperature is 52°F inside the cooler. The liquid line temperature is hot instead of warm. This is a TXV system, and the sight glass is clear. The unit appears to have a full charge with a high head pressure. Gages are installed, and the head pressure is 275 psig. The system is using R-12 as the refrigerant. The head pressure should be no more than 158 psig, which is a condensing temperature of 120°F; it is 90°F (90°F + 30°F = 120°F or 158 psig). Further examination indicates that air is leaving the condenser and recirculating back to the condenser inlet. Several boxes have been stored in front of the condenser. These are moved, and the head pressure drops to 158 psig. The reason there has not been a complaint up to now is that the air has been cold enough in the previous mild weather to keep the head pressure down even with the recirculation problem. A call back later in the day verifies that the cooler temperature is back to normal.

SERVICE CALL 14

A store frozen food manager calls to say that the temperature is going up in the low-temperature walk in freezer. *The defrost time clock motor is burned, and the system will not go into defrost,* Figure 28–53.

The service technician arrives, examines the system, and sees that the evaporator coil is coated with thick ice. It is evident that it has not been defrosting. The first thing to do is to force a defrost. The technician goes to the timer and when the clock dial is examined, it is found that the time indicator says 4:00 AM, but it is 2:00 PM. Either

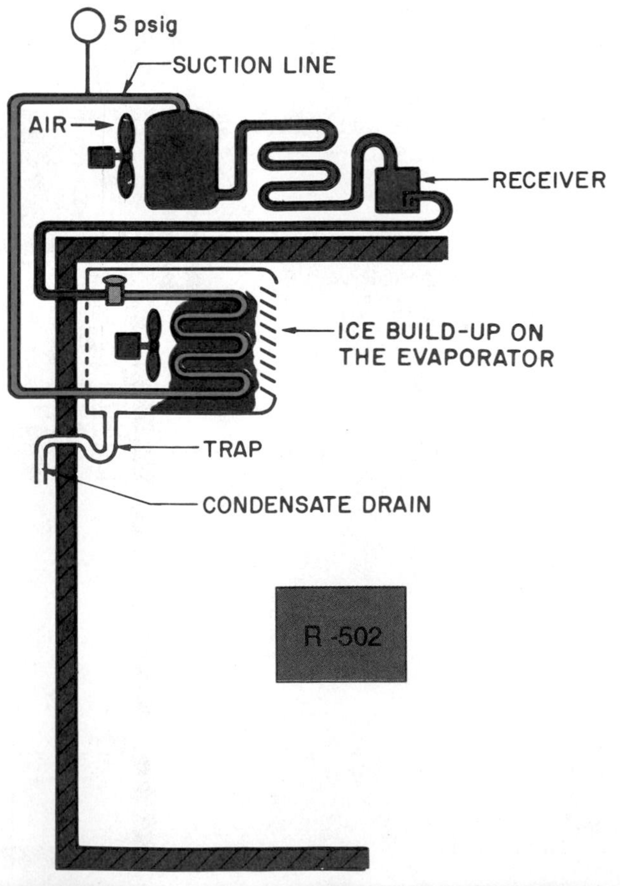

THE SUCTION PRESSURE SHOULD BE ABOUT 18 psig (-16°F) FOR A 0°F FREEZER TEMPERATURE. THIS FREEZER IS WARMER THAN 0°F AND THE SUCTION PRESSURE IS LOW.
5 psig
SUCTION LINE
AIR
RECEIVER
ICE BUILD-UP ON THE EVAPORATOR
TRAP
CONDENSATE DRAIN
R-502

FIGURE 28–53 Symptoms of a defrost problem with a defective time clock.

the power has been off, or the timer motor is not advancing the timer. The technician advances the timer by hand until the defrost cycle starts and then marks the time. After the defrost is terminated by the temperature-sensing device, the system goes into normal operation. The technician gives the clock about one-half hour and sees that it has not moved. A new time clock is installed. The customer is cautioned to look out for heavy ice buildup. A call the next day verifies that the system is working correctly.

SERVICE CALL 15

A customer states that the defrost seems to be lasting too long in the low-temperature walk-in cooler. It used to defrost, and the fans would start back up in about 10 min. It is now taking 30 min. *The defrost termination switch is not terminating defrost with the temperature setting. This is an electric defrost system. The system is staying in defrost until the time override in the timer takes it out of defrost. This also causes the compressor to operate at too high a current because the fans should not start until the coil cools to below 30°F. With the defrost termination switch stuck in the cold position, the fans will come on when the timer terminates defrost. The compressor will be overloaded by the heat left in the coil from the defrost heaters,* Figure 28–54.

The technician arrives at the job and examines the system. The coil is free of ice so defrost has been working. The cooler is cold, −5°F. The technician advances the time clock to the point that defrost starts. The termination setting on the defrost timer is 30 min. This may be a little long unless the cooler is only defrosted once a day. The timer settings are to defrost twice a day, at 2:00 PM and 2:00 AM. The coils are defrosted in approximately 5 min, but the defrost continues. The technician allows defrost to continue for 15 min to allow the coil to get to maximum temperature, so the defrost termination switch will have a chance to make. After 15 min the system is still in defrost so the timer is advanced to the end of the timed cycle. When the system goes out of defrost, the fans start. They should not start until the coil gets down to below 30°F. The defrost termination switch is removed. It has three terminals, and the wires are marked for easy replacement of a new control. A new control is acquired and installed. The system is started. The owner called the next day, reported a normal defrost, and that the system was acting normally. The old defrost termination switch is checked with an ohmmeter after having been room temperature for several hours, and it is found that the cold contacts are still closed. The control is definitely defective.

SERVICE CALL 16

The frozen food manager in a supermarket calls indicating that the walls between the doors are sweating on a reach-in freezer. *This has never happened. The freezer is about 15 years old. The mullion heater in the wall of the cooler that keeps the panel above the dew point temperature of the room is not functioning,* Figure 28–55.

The technician knows what the problem is before going to the site. After arriving, the technician examines the fixture for the reason the mullion heater is not getting hot. If the heater is burned out, the panel needs to be removed. The circuit is traced to the back of the cooler, and the wires to the heater are found. An ohm check proves the heater still has continuity. A voltage check shows there is no

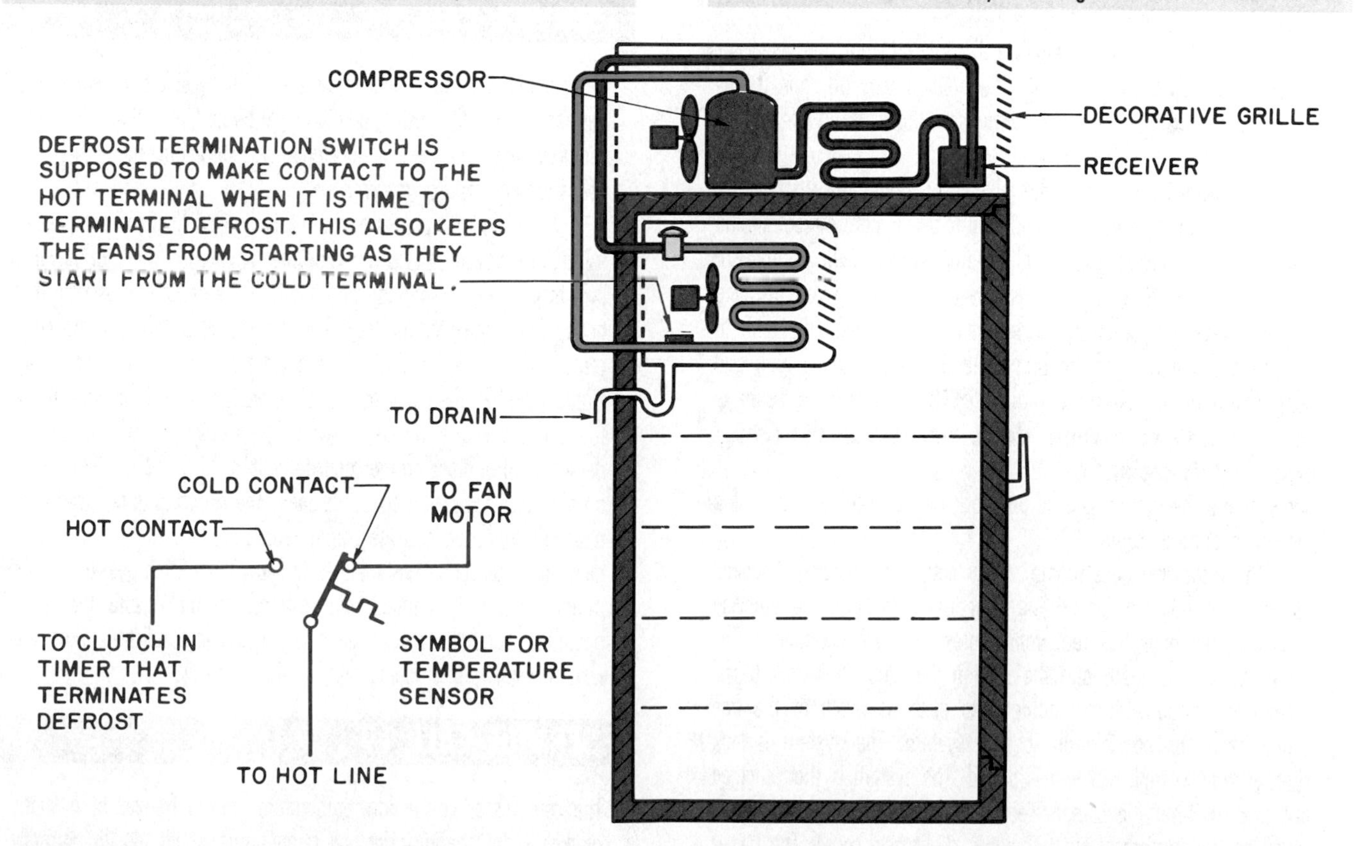

FIGURE 28–54 Symptoms of a defective defrost termination thermostat (stuck with cold contacts closed) in a low-temperature freezer.

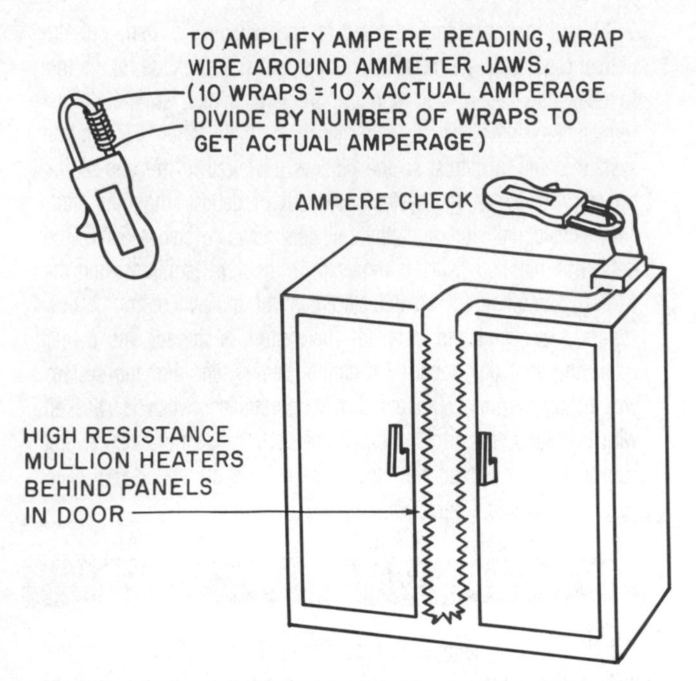
TO AMPLIFY AMPERE READING, WRAP WIRE AROUND AMMETER JAWS. (10 WRAPS = 10 X ACTUAL AMPERAGE DIVIDE BY NUMBER OF WRAPS TO GET ACTUAL AMPERAGE)
AMPERE CHECK
HIGH RESISTANCE MULLION HEATERS BEHIND PANELS IN DOOR

FIGURE 28–55 Mullion heater on a low-temperature freezer.

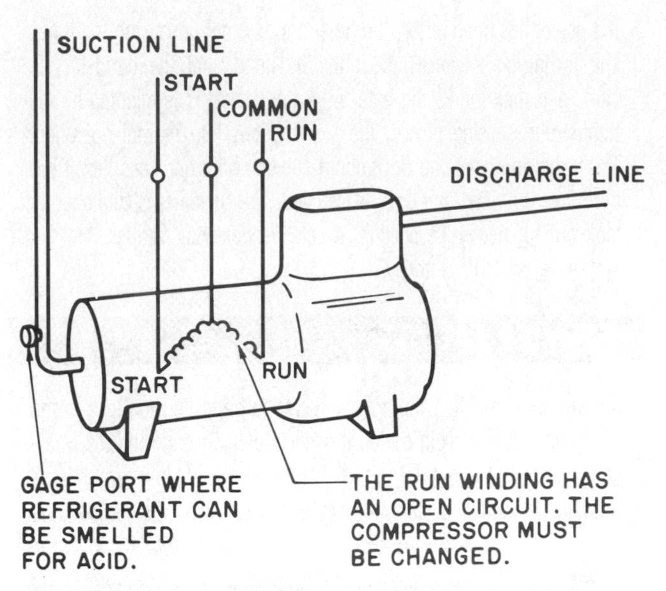
SUCTION LINE
START
COMMON
RUN
DISCHARGE LINE
START
RUN
GAGE PORT WHERE REFRIGERANT CAN BE SMELLED FOR ACID.
THE RUN WINDING HAS AN OPEN CIRCUIT. THE COMPRESSOR MUST BE CHANGED.

FIGURE 28–56 A system with a burned compressor motor.

voltage going to the heater. After further tracing by the technician, a loose connection is located in a junction box. The wires are connected properly, and a current check is performed to prove the heater is working.

SERVICE CALL 17

A restaurant maintenance person calls to say that the medium-temperature walk-in cooler is off. The breaker was off, and it was reset, but it tripped again. *The compressor motor is burned,* Figure 28–56.

The technician arrives at the job and examines the system. The food is warming up, and the cooler is up to 55°F. Before resetting the breaker, the technician uses an ohmmeter and finds the compressor has an open circuit through the run winding. Nothing can be done except change the compressor. A refrigerant line is opened slightly to determine the extend of the burn. The refrigerant has a high acid odor, indicating a bad motor burn. **⊛The technician uses a recovery unit to capture, clean, and recycle the refrigerant in this cooler.⊛**
hours before the system can be put back into service, so the food is moved to another cooler.

The technician goes to the supply house and gets the required materials, including a suction line filter-drier with high acid-removing qualities. This drier has removable cores for each changeout. The compressor is changed, and the suction line drier is installed. The system is purged with dry nitrogen to push as much of the free contaminants as possible out of the system. The system is leak checked, and a triple vacuum is pulled. The system is then charged and started. After the compressor runs for an hour, an acid test is taken on the crankcase oil. It shows a slight acid count. The unit is run for another 4 hr, and an acid check shows even less acid. The system is pumped down, and the cores are changed in the suction line drier. The liquid line drier is changed, and the system is allowed to operate overnight.

The technician returns the next day and takes another acid check. It shows no sign of acid. The system is left to run in this condition with the drier cores still in place in case some acid becomes loose at a later date. *The motor burn is attributed to a random motor failure.*

SERVICE CALL 18

A medium-temperature reach-in cooler is not cooling properly. The temperature is 55°F inside, and it should be no higher than 45°F. The compressor sounds like it is trying to start but then cuts off. *The starting capacitor is defective.*

The technician arrives at the job in time to hear the compressor try to start. Several things can keep the compressor from starting. It is best to give the compressor the benefit of the doubt and assume it is good. Check the starting components first and then check the compressor. The starting capacitor is removed from the circuit for checking, Figure 28–57. The capacitor has discoloration around the vent at the top, and the vent is pushed upward. These are signs the capacitor is defective. After bleeding the capacitor with a 20,000-Ω, 5-W resistor, an ohm test shows the capacitor open. The capacitor is replaced with a similar capacitor, and the compressor is started. The compressor is allowed to run for several minutes to allow the suction gas to cool the motor; then it is stopped and restarted to make sure that it is operating correctly. A call back the next day indicates that the compressor is still stopping and starting correctly.

SERVICE CALL 19

The compressor in the low-temperature walk-in freezer of a supermarket is not starting. The box temperature is 0°F, and it normally

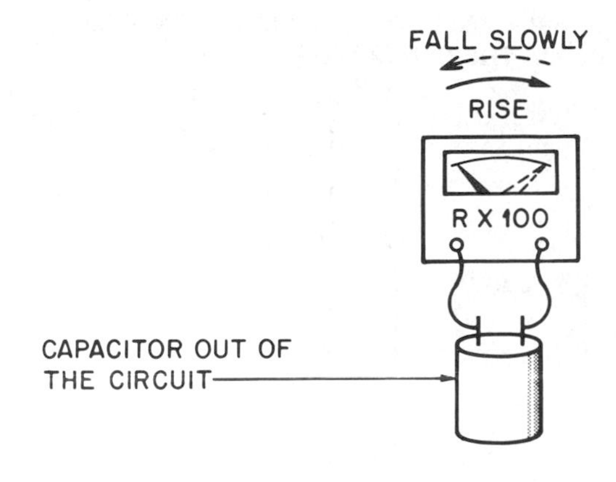

GOOD CAPACITOR

THE CAPACITOR HAS BEEN CHECKED TO SEE THAT IT IS NOT CHARGED BY PLACING A 20,000 OHM 5 WATT RESISTOR FROM TERMINAL TO TERMINAL.

1. TOUCH ONE LEAD, THEN THE OTHER TO THE CAPACITOR TERMINALS. THE NEEDLE SHOULD RISE TO THE O RESISTANCE SIDE THEN FALL BACK.
2. IF THERE IS A RESISTOR BETWEEN THE TERMINALS THE NEEDLE WILL FALL BACK TO THE VALUE OF THE RESISTOR (KNOWN AS A BLEED RESISTOR).
3. TO PERFORM THE TEST AGAIN, THE LEADS MUST BE REVERSED. THE METER'S BATTERY IS DIRECT CURRENT. YOU WILL NOTICE AN EVEN FASTER RISE IN THE NEEDLE THE SECOND TIME BECAUSE THE CAPACITOR HAS THE METER'S BATTERY CHARGE.

10 000 Ω

BLEED RESISTOR

DEFECTIVE CAPACITOR

BLEED THE CAPACITOR AS DESCRIBED ABOVE

1. TOUCH THE METER LEADS TO THE CAPACITOR TERMINALS. IF IT IS DEFECTIVE, IT WILL NOT RISE ANY HIGHER THAN THE VALUE OF THE BLEED RESISTOR.

FIGURE 28–57 Symptoms of a system with a defective starting capacitor.

operates at −10°F. *The compressor is locked. This is a multiple-evaporator installation with four evaporators piped into one suction line. An expansion valve on one of the evaporators has been allowing a small amount of liquid refrigerant to get back to the compressor. This has caused marginal lubrication, and the compressor has scored bearings that are bad enough to lock the compressor.*

The service technician arrives and examines the system. The compressor is a three-phase compressor and does not have a starting relay or starting capacitor. It is important that the compressor be started within 5 hours, or the food must be moved. The technician turns off the power to the compressor starter. The leads are then removed from the load side of the starter. The motor is checked for continuity with an ohmmeter, Figure 28–58. The motor windings appear to be normal. **NOTE: Some compressor manufacturers furnish data that tell what the motor-winding resistances should be.** ■ The meter is then turned to the voltage selection for a 230-V circuit. The starter is then energized to test the load side of the starter to make sure that each of the three phases has the correct power. This is a no-load test. The leads have to be connected to the motor terminals and the voltage applied to the leads while the motor is trying to start before the technician knows for sure there is a full 230-V under the starting load.

The technician connects the motor leads back up and gets two more meters for checking voltage. This allows the voltage to be checked from phase 1 to phase 2, phase 1 to phase 3, and phase 2 to phase 3 at the same time while the motor is being started. When the power is applied to the motor, there is 230-V from each phase to the other, and the motor will not start. **This is conclusive. The motor is locked.** The motor can be reversed (by changing any two leads, such as L1 and L2) and it may start, but it is not likely that it will run for long.

The technician now has 4 hr to either change the compressor or move the food products. A call to the local compressor supplier shows that a compressor is in stock in town. The technician calls the shop for help. Another technician is dispatched to pick up the new compressor and a crane is ordered to set the new compressor in and take the old one out. The compressor is 30 hp and weighs about 800 lb.

When the crane gets to the job, the old compressor is disconnected and ready to set out. This was accomplished by front seating the suction and discharge service valves to isolate the compressor. When the new compressor is installed, all that will need to be done after evacuating the compressor is to open the service valves and start it. The original charge is still in the system. By the time the old compressor is removed, the new compressor arrives. The new compressor is set in place and the service valves are connected. While a vacuum is being pulled, the motor leads are connected. In an hour the new compressor has been triple evacuated and is ready to start.

When the new compressor is started, it is noticed that the suction line is frosting on the side of the compressor. Liquid is getting back to the compressor. A thermometer lead is fastened to each of the evaporators at the evaporator outlet. It is discovered that one of the evaporator suction lines is much colder than the others. The expansion valve bulb is examined and is found to be loose and not sensing the suction line temperature. Someone has taken the screws out of the mounting strap.

The expansion valve bulb is then secured to the suction line, and the system is allowed to operate. The frost line moves back from the side of the compressor, and the system begins to function normally.

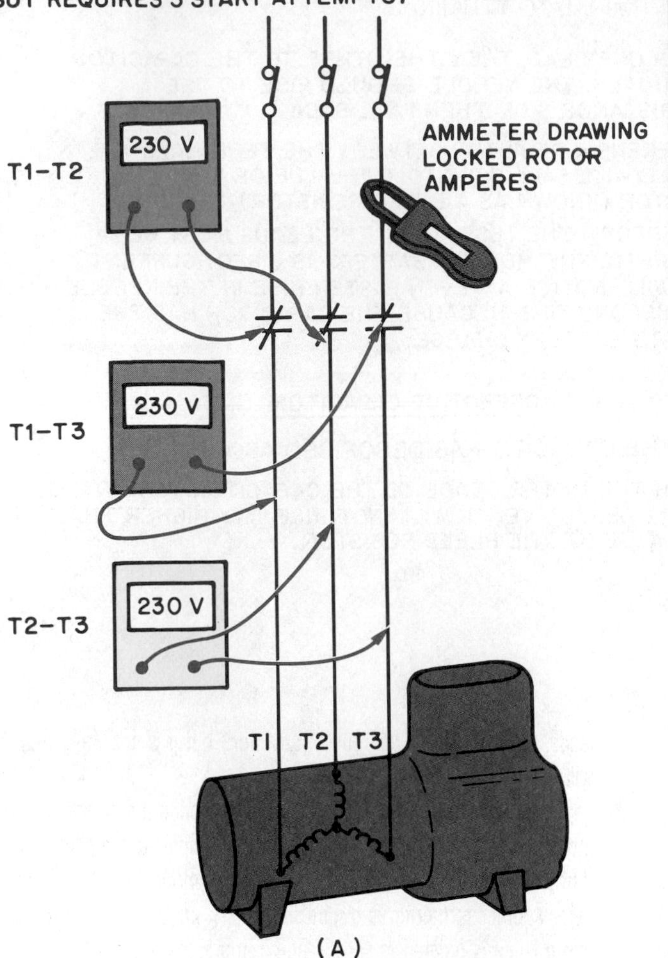

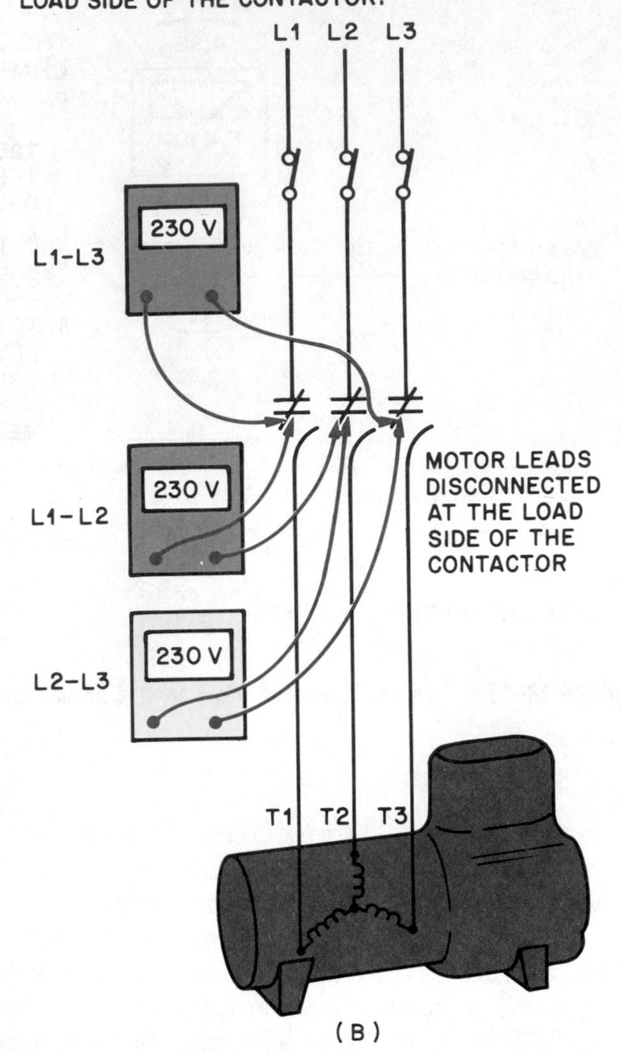

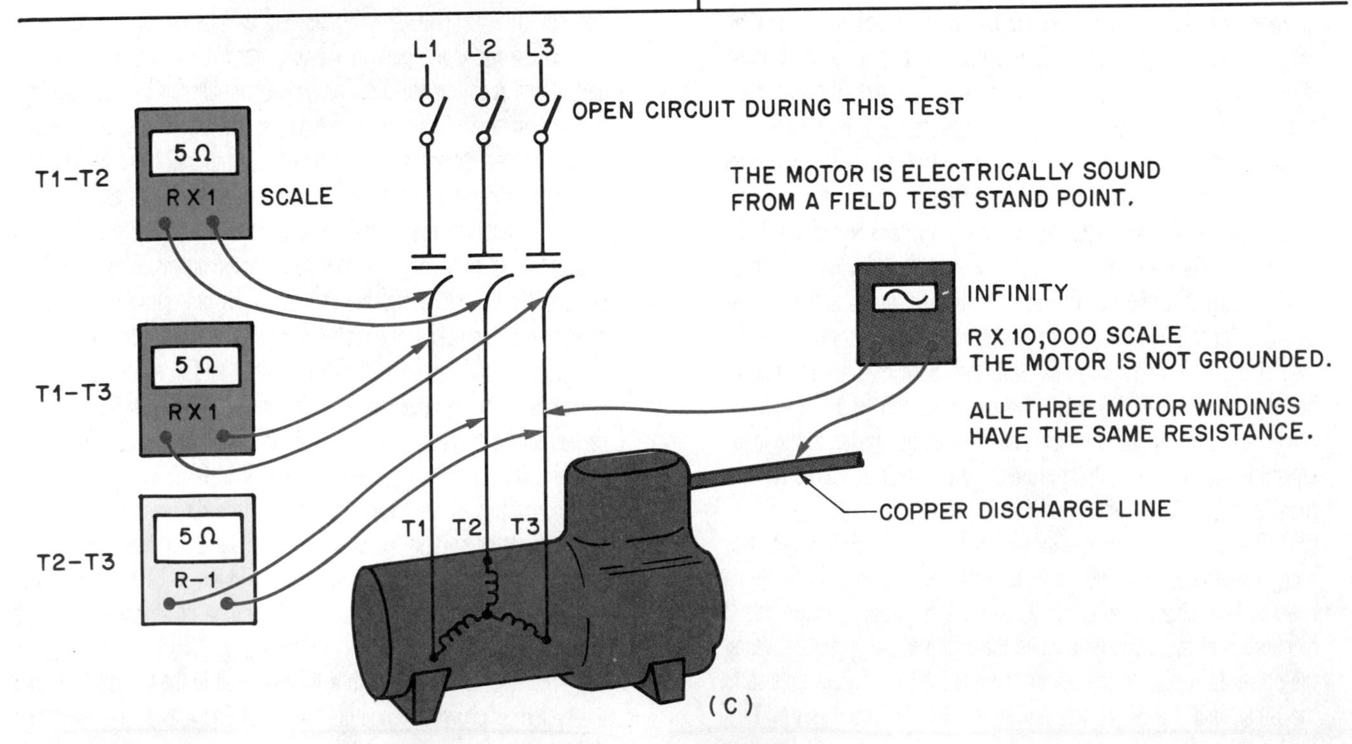

FIGURE 28–58 Symptoms of a system with a three-phase compressor that is locked.

The technician stops by the store the next day and determines that the frost line to the compressor is correct, not frosting the compressor.

SERVICE CALL 20

A customer with an old cube ice maker calls and reports that his ice maker is not making good quality ice. It is the type that makes plate ice and cuts it into squares. *The problem is the unit has a low charge and the ice pattern is not correct. It is making thick ice on one side and thin ice on the other.*

The technician arrives and looks at the ice. It is evident that something is wrong because of the difference in the thickness of the cubes. The technician takes the top off the unit and can see the refrigerant pattern on the back of the evaporator. It is evident from the irregular ice pattern that the evaporator is starved for refrigerant. This unit has a sight glass and bubbles are showing in the glass. The ambient temperature is high (90°F), so the head pressure cannot be too low because of the ambient conditions.

The technician connects gages to the unit. The suction pressure is 5 psig and the unit uses R-12. The suction pressure should be no lower than 11 psig, just before harvest. The technician shuts the machine off and decides to raise the suction pressure for a leak check. The gage manifold used for this job has only been used for R-12 and the center line is plugged. The technician opens the gage manifold valves so the high pressure refrigerant will move to the low side and boost the low side pressure.

A leak check shows a leaking flare nut on the suction side of the compressor. The flare nut is tightened and this stops the leak. The gage manifold valves are closed and the machine is started. Refrigerant is added until the sight glass bubbles clear. The machine makes a uniform plate of ice and goes into defrost. The ice falls to the grid wires. The technician then replaces all panels and leaves, telling the owner to look in on the machine later in the day and to give the service company a call if all is not normal.

SERVICE CALL 21

A customer with a plate-type cube ice maker called in to state that the ice had stacked up on the cutting grid. *The problem is the cutting grid has a loose connection and no current is reaching the low-voltage terminals.*

The technician arrives, raises the bin door, and sees the ice laying on top of the grid. The machine is shut off and the power is turned off. The top is removed to access the wiring to the grid. It looks to be intact. The machine is turned on and an ampere check of the grid is performed. The amperage the grid should draw is penciled in the top of the machine where the last technician had worked on the machine and grid. It should draw 30 A at 3.2 V. The ammeter showed 0 A. The voltage was checked and found to be 3.5 V. The problem must be the grid, so the machine is shut off again and the grid wires are disconnected.

The technician opens the bin door and removes the grid. It has so much ice on it that nothing can be seen. The grid is moved to a nearby sink and the ice removed. The grid is then visually checked. All of the wires seem to be in place in the middle of the grid, so the connections on the end where power enters the grid are checked. A very loose connection is found, the terminal screw is missing, and the wire is just hanging.

The technician replaces the screw and fastens the grid back in place. The machine is started. Amperage is now checked and found to be correct. The technician determines that the machine was making ice before and only needed grid repair. One batch of ice is made, then the technician leaves the job telling the owner to watch the machine from time to time the rest of the day and call if there are problems.

SERVICE CALL 22

A restaurant owner with a cube ice making machine calls and indicates that the evaporator is frozen solid and no ice is dropping into the bin. *The problem is the defrost probes are dirty. These probes sense the ice thickness by making contact with the ice and creating a circuit by conducting through the ice to start defrost when the ice reaches a predetermined thickness. The probe is monitored by an electronic circuit to determine when the correct thickness has been reached by the conductivity of the ice.*

The technician arrives and examines the entire system carefully. This technician has worked on this type of ice maker before and knows that the manufacturer recommends that the machine not be shut off at this time until the problem is found. If the technician were to shut the machine off and defrost the evaporator, the condition may not happen again and the trouble might not be found.

The technician knows this machine is older and suspects dirty defrost sensor probes. The manufacturer has furnished the technician with a special jumper cord with a fixed resistor of the value of the resistance through ice. This jumper can be used to jump the ice probes and determine if the sensor is defective. The technician jumps the probe connections and the unit starts a defrost cycle. This is a sure sign that the sensor is dirty where it contacts the ice.

The technician waits until the unit completes the defrost cycle and the ice drops to the storage bin, then shuts the machine off and turns the power off. The probe is removed and cleaned. It is reinstalled and the machine power is restored. The machine is started. About 25 min later, the machine starts another defrost cycle on its own. The technician leaves the job, telling the owner to watch the machine for the rest of the day for correct operation.

SERVICE CALL 23

The owner of a restaurant calls and states that one of the ice machines is not dropping new ice to the bin. This is a new installation in a new restaurant. *The problem is the water has been filtered to the point that so many minerals have been removed the water does not have enough conductivity for the electronic circuit to determine when the ice is touching the probes.*

The technician arrives and surveys the situation. The evaporator is frozen solid. This is a new machine, with the best water filtration system. There is a chance the filtration is so effective that the conductivity of the water has been reduced to where the defrost sensor probe will not sense the conductivity. The technician remembers the last service school attended where these symptoms were described and goes to the shelf for a box of salt. A pinch of salt (about 1/8 teaspoon) is placed in the water sump where the pump will pick it up and circulate the slightly salted water over the probes. Within a very few minutes, the machine starts to defrost.

The technician knows the owner cannot put salt in the filtered water every day so the manual is consulted. A resistor is available from the manufacturer that can be placed across certain terminals on the circuit board and allow the probe and electronic circuit board to sense the conductivity of water that has been filtered. Normal defrost can then be expected.

The technician obtains the resistor from the supplier and installs it across the appropriate terminals. The machine is started; the technician watches it through two defrost cycles and is then satisfied the machine will operate satisfactorily.

SERVICE CALL 24

A restaurant owner calls to request a checkup for the cube type ice maker in the kitchen. This maker has been in the kitchen for several years and the owner can't remember when the last service was performed. *There is no particular problem.*

The technician arrives and examines the machine thoroughly. The condenser appears to be dirty, and the evaporator section needs cleaning. The machine is full of ice. The technician decides to remove the ice and clean the bin at the same time.

The machine is turned off and the power supply shut off locked and tagged. The technician starts transferring the ice to trays to be moved to the walk-in cooler. As the technician gets toward the bottom of the bin, it is found to be dirty. So cleaning the bin is a good idea.

The panels to the condenser are removed. A vacuum cleaner is used to remove lint and light dust. The condenser fan motor and terminal box are covered with a plastic bag to prevent moisture from entering. An approved coil cleaner is applied to the condenser and allowed to set on the coil so it can soak into the dirty surface. While the condenser coil cleaner is working, the bin is cleaned and sanitized. When the coil cleaner has been on the condenser for about 30 min, the technician takes a water hose with a nozzle to flush the coil cleaner off the condenser and down a nearby drain. The restaurant owner is surprised at the dirt washed out of the condenser.

The water is allowed to drain off the machine and mopped from the floor around the machine. The plastic is removed from the motor and the unit is dried with towels.

This unit has a clean cycle for cleaning the evaporator. The clean cycle is a cycle where only the water pump circulates and chemicals may be circulated with the water for the prescribed period of time. The technician pours the recommended amount of chemicals in the sump and starts the pump. The chemicals circulate for the prescribed time and the technician dumps the chemicals down the drain and flushes fresh water through the system.

The evaporator section is then sanitized using the clean cycle and the water dumped. A hose is used to wash the bin area one more time to be sure no chemicals are still present and the machine is started.

The technician allows the machine to make two harvests of ice and dumps them down the drain. This ensures there are no residual chemicals and that the machine is making proper ice.

The machine is then put back into service.

The following service calls do not have the solution described. The solution can be found in the Instructor's Guide.

SERVICE CALL 25

A store manager reports that the food is thawing out in the low-temperature walk-in freezer. This freezer lost its charge in the winter, the leak was repaired, and the unit recharged. The service technician remembers that this unit was serviced by a new technician when the loss-of-charge incident happened. The technician suspects that the unit may be off because of the manual reset high-pressure control. This is the only control on the unit that will keep it off, unless there was a power problem. The technician arrives at the job and finds that the unit is off because of the high-pressure control. Before resetting the control, the technician installs a gage manifold so that the pressures can be observed. When the high-pressure control is reset, the compressor starts, but the head pressure rises to 375 psig and shuts the compressor off. The system uses R-502. The ambient temperature is 90°F, so the head pressure should not be more than the pressure corresponding to 120°F. This was arrived at by adding 30°F to the ambient temperature: 30°F + 90°F = 120°F. The head pressure corresponding to 120°F is 282 psig. It is obvious the head pressure is too high. The condenser seems clean enough, and air is not recirculating back to the condenser inlet.

What is the likely problem and recommended solution?

SERVICE CALL 26

A customer calls to indicate that the medium-temperature walk-in cooler unit that was just installed is running all the time and not cooling the box.

The technician goes to the job and looks it over. This is a new installation of a walk-in medium-temperature cooler and was just started up yesterday. The hermetic compressor is sweating down the side of the housing. The evaporator fan is operating, and the box temperature is not down to the cut-off point, so there is enough load to boil the refrigerant to a vapor.

What is the likely problem and the recommended solution?

SERVICE CALL 27

A customer reports that the low-temperature reach-in cooler that had a burned compressor last week is not cooling properly. The food is beginning to thaw.

The technician arrives and examines the whole system. The evaporator seems to be starving for refrigerant. When the technician approaches the condensing unit in the back of the store, it is noticed that the liquid line is sweating where it leaves the drier. A suction line drier was not installed at the time of the motor burn.

What is the likely problem and the recommended solution?

SERVICE CALL 28

A customer calls on the first hot day in the spring and says the walk-in freezer is rising in temperature. The food will soon start to thaw if something is not done.

The service technician arrives, examines the whole system, and notices that the liquid line is hot, not warm. This is a TXV system and has a sight glass. It is full, indicating a full charge of refrigerant. The compressor is cutting off because of high pressure and then restarting periodically. This system uses R-502, and the head of the compressor is painted purple to signify the refrigerant type. Gages are installed, and the head pressure is starting out at 345 psig and rising to the cutout point or 400 psig.

What is the likely problem the recommended solution?

SERVICE CALL 29

A restaurant manager indicates that the compressor on a water-cooled unit is cutting off from time to time, *and the walk-in cooler is losing temperature. This is a medium-temperature cooler.*

The service technician arrives to find a wastewater condenser (the water is regulated by a water-regulating valve to control the head pressure and then goes down the drain). The water is coming out of the condenser and going down the drain at a rapid rate. The liquid line is hot, not warm. The water is not taking the heat out of the refrigerant like it should. Gages are installed on the compressor, and the head pressure is 200 psig. The compressor is cutting off and on because of the high-pressure control. The head pressure should be about 125 psig with a condensing temperature of 105°F.

What is the likely problem and the recommended solution?

SERVICE CALL 30

A customer calls and says the temperature is rising in a frozen food walk-in freezer. Although it tries, the compressor will not start.

The service technician listens to the compressor try to start and hears the overload cut it off, making a clicking sound. The compressor is hot, too hot to start many more times before it will overheat to the point that it will be off for a long time. This is because every start the motor goes through makes it hotter. The technician needs to make sure that it starts and stays on the line the next time it is started. This is a single-phase compressor and has a starting relay and starting capacitor. The problem can be in several places: (1) the starting capacitor, (2) the starting relay, or (3) the compressor. The technician uses an ohmmeter to check for continuity in the compressor, from the common terminal to the run terminal, and from the common terminal to the start terminal. The compressor has continuity and should start, provided it is not locked.

An ohm check of the starting capacitor shows that it will charge and discharge. (This test is accomplished by first shorting the capacitor terminals with a 20,000-Ω, 5-W resistor and then using the R × 100 scale on the ohmmeter.) The capacitor should cause the meter to rise and then fall. Reversing the leads will cause it to rise and fall again.)

NOTE: This check does not indicate the actual capacity of a capacitor. It indicates that the capacitor will charge and discharge.

The starting relay is checked for continuity from terminal 1 to terminal 2. This is the circuit the starting capacitor is energized through. The circuit shows no resistance; the contacts are good.

What is the likely problem and the recommended solution?

SERVICE CALL 31

A motel maintenance person calls indicating that the ice maker on the second floor of the motel is not making ice. This ice maker is a flake ice machine. It sits outside and the temperature is hot.

The technician arrives and notices right away there is no ice in the bin. The compressor is running and hot air is coming from the condenser. Therefore, it must be refrigerating. The suction line leaving the evaporator is cold.

Gages are fastened to the compressor. The suction pressure is high, but so is the discharge pressure. The machine uses R-502 and the suction pressure is 55 psig (30°F boiling temperature). No ice can be made under these conditions. The discharge pressure is 341 psig (135°F condensing temperature). It is a hot day, 97°F, but not hot enough to cause a head pressure this high. The technician wonders for a minute if the unit is overcharged or if the condenser is dirty. This could cause high head pressure, which would then cause high suction pressure and low or no ice production.

After sitting down and thinking for a minute, the technician thinks of the water circuit then feels the water line. It is cooler than hand temperature (which is about 90°F). This is not the answer. The technician then looks at the float chamber and discovers that water is running into the chamber.

What is the likely problem and the recommended solution?

SUMMARY

- To begin to troubleshoot you need to know what the equipment performance should be.
- Usually only one component fails at a time. If there is more than one component defective, it is because one component caused the others to fail. There is normally a sequence of events.

- The technician should know how the equipment should sound, where it should be cool, and where it should be warm. This comes with experience.
- ***The technician should look the whole system over before doing anything.***
- All evaporators that cool air have a relationship with the air they are cooling. The coil will generally boil the refrigerant between 10°F and 20°F colder than the air entering the evaporator.
- When higher humidities are desired inside the cooler to prevent product dehydration, a 10°F coil-to-air relationship is customarily used.
- High-temperature evaporators normally operate between 25°F and 40°F boiling temperature of the liquid refrigerant.
- Medium-temperature evaporators normally operate between 10°F and 25°F boiling refrigerant temperature of the liquid refrigerant.
- Low-temperature applications are in two areas: ice making and food storage.
- Ice making normally starts at a coil temperature of about 20°F.
- Low-temperature food storage systems normally operate from −15°F down to about −49°F refrigerant boiling temperature.
- Tables show the various storage temperatures for different products for both short- and long-term storage.
- Three temperatures should be considered when any evaporator is evaluated: (1) When the compressor is off, just before the thermostat calls for the compressor to start back up. The coil temperature is the same as the box temperature at this point. (2) Just after the compressor starts back up before the box begins to cool down. The compressor is running, so the coil will be cooler than the box temperature. (3) While the compressor is running, just before the thermostat calls for the compressor to stop. The coil temperature will be cooler than the box temperature.
- Typical conditions for the high-pressure side of the system are divided into two categories: air cooled and water cooled.
- There is a relationship between the entering air and the condensing refrigerant. Usually the refrigerant should not condense at more than 30°F higher than the entering air.
- High-efficiency condensers may have lower relationships, down to about 25°F higher than the entering air.
- When the refrigerant is condensing at a temperature greater than 30°F higher than the entering air, something is wrong.
- Water-cooled condensers have a relationship with the leaving water. The condensing temperature is usually 10°F warmer than the leaving water.
- A low refrigerant charge is indicated by a low suction pressure and a low head pressure when the metering device is a TXV or a capillary tube.
- When a low charge is encountered and the metering device is an automatic expansion valve, low head pressure and bubbles in the sight glass will occur.
- Hissing can be heard during low charge operation at the expansion valve when a TXV or automatic expansion valve is the metering device.
- The automatic expansion valve maintains a constant low-side pressure. This is the reason that the low-side pressure does not go down during low charge operation as with a capillary tube and a TXV.
- An overcharge of refrigerant always causes an increase in head pressure.
- An increase in head pressure causes an increase in suction pressure when the TXV or the capillary tube is the metering device.
- ***An increase in head pressure causes more refrigerant to flow through a capillary tube. Liquid refrigerant may flood into the compressor with an overcharge.***
- The automatic expansion valve responds in reverse to load changes. When the load is increased, the valve throttles back and will slightly starve the evaporator. When the load is decreased excessively, the valve is subject to overfeed the coil and may allow small amounts of liquid refrigerant to enter the compressor.
- When an inefficient evaporator is encountered, the refrigerant boiling temperature will be too low for the entering air temperature.
- When an inefficient condenser is encountered, the condensing refrigerant temperature will be too high for the heat rejection medium. For air-cooled equipment the condensing refrigerant should be no more than 30°F higher than the entering air. For water-cooled equipment the condensing refrigerant temperature should be no more than 10°F higher than the leaving water.
- The best check for an inefficient compressor is to see if it will pump to capacity.
- Design load conditions can almost always be duplicated by building the head pressure up and duplicating the design suction pressure.

REVIEW QUESTIONS

1. What is the first thing a technician should know before beginning troubleshooting?
2. Name the refrigerated space temperatures that apply to the three temperature ranges.
3. What is the coil-to-air temperature relationship for a refrigeration system designed for minimum food dehydration?

4. When food dehydration is not a factor, what is the coil-to-air temperature relationship?
5. How is minimum food dehydration accomplished?
6. Why is minimum dehydration not used on every job?
7. What is the coil-to-air temperature relationship for an average coil where dehydration is not a factor?
8. At what temperature does an ice maker usually begin to make ice?
9. What is the evaporator coil-to-air temperature relationship when the compressor is not running?
10. What is the coil-to-air temperature relationship between the entering air and the condensing temperature for an air-cooled condenser?
11. What is the leaving-water temperature to the condensing temperature relationship for a water-cooled condenser?
12. How can an evaporator be tested for efficiency?
13. How can a condenser be tested for efficiency?
14. How can a hermetic compressor be tested for efficiency?

How do the following systems respond to an overcharge?

15. TXV
16. Automatic expansion valve
17. Capillary tube

How do the following systems respond to an undercharge?

18. TXV
19. Automatic expansion valve
20. Capillary tube

UNIT 28 DIAGNOSTIC CHART FOR COMMERCIAL REFRIGERATION

Commercial refrigeration equipment which controls the storage temperatures for perishable products must be kept operating properly to prevent spoilage. The technician must restore the system back to good working order in the least amount of time. Orderly troubleshooting is the most effective way to arrive at sound conclusions in a minimum of time. Following are some recommendations to follow for commercial refrigeration systems.

1. Listen to the customer's description of the system's symptoms. Most customers of commercial refrigeration equipment monitor the system and the temperatures carefully. Question the customer about sounds, cycle times and temperatures on normal days and what is different now.
2. Observation may be one of the most important steps you can follow. Inspect the complete system before making any adjustments or repairs. Check the condenser for dirt, leaves or trash in the coil or fan. Look for blocked airflow. Inspect the wiring for discoloration of connections. Look for loose or missing insulation on the suction line or kinks or flattened tubing. Check the evaporator for air blockage due to frost, ice, or dirt. Make sure the fan is circulating air.
3. Once the customer has been consulted and the system has been observed, proceed and determine what is functioning and what is not functioning. The following chart will help to determine possible causes of the problem.

PROBLEM	POSSIBLE CAUSE	POSSIBLE REPAIR	PARAGRAPH NUMBER
Compressor will not start or attempt to start, makes no sound	Open disconnect switch	Close disconnect switch	15.1, 15.2
	Open fuse or breaker	Replace fuse or reset breaker and determine why it opened	15.1, 15.2
	Tripped overload	Reset and determine why it tripped	19.6
	Pressure control stuck in open position	Reset or replace and determine why it opened	25.16–25.19
	Faulty wiring	Repair or replace faulty wiring connectors	20.13
Compressor will not start, hums and trips on overload	Incorrect wiring, **Single Phase**	Check common, run and start connections	17.17–17.19
	Incorrect wiring, **Three Phase**	Check all three phases, phase-to-phase	17.7
	Start relay wired wrong or defective, **Single Phase**	Rewire relay if needed, or replace if defective	17.17–17.19
	Start capacitor defective	Replace capacitor	17.12, 20.11
	Compressor winding open or shorted.	Replace compressor	20.7–20.10
	Watch for open internal overload due to hot compressor	Allow wiring to cool down	
	Internal compressor problems, stuck or tight	Replace compressor	20.14
Compressor starts but stays in start, **Single Phase**	Wired incorrectly	Check wiring compared to diagram	17.17–17.19
	Low voltage supply	Correct low voltage	17.8
	Defective start capacitor	Replace capacitor	17.12, 20.11
	Defective start relay	Replace relay	17.17–17.19
	Compressor has open winding	Replace compressor	20.6, 20.7
	Defective run capacitor	Replace capacitor	17.12, 20.11
	High discharge pressure	Overcharge of refrigerant, not enough condenser cooling, closed discharge service valve	22.9, 28.15
	Tight or binding compressor	Replace compressor	25.22

(*continued*)

UNIT 28 DIAGNOSTIC CHART FOR COMMERCIAL REFRIGERATION (Continued)

PROBLEM	POSSIBLE CAUSE	POSSIBLE REPAIR	PARAGRAPH NUMBER
Compressor starts and runs for a short time then shuts off due to overload	Defective overload	Check acutal load on overload and replace if it trips below specifications	19.6
	Excess current in overload circuit	Check circuit for added load, such as fans or pumps	20.15
	Low voltage	Determine reason and correct	18.2
	Unbalanced voltage, **Three Phase**	Correct unbalance, possible redistribution of loads	18.2
	Defective run capacitor	Replace	17.14, 20.11
	Excessive load on compressor		
	A. HIGH SUCTION: Pressure, defrost heaters on all the time	Check and repair defrost control circuit	25.28
	B. Hot food placed in cooler	Operator education	21.14
	C. HIGH DISCHARGE: Pressure recirculating condenser air	Move object causing recirculation	28.15
	Reduced airflow	Remove blockage, make sure all fans are on	28.15
	Dirty condenser	Clean condenser	28.15
	Compressor windings shorted	Replace compressor	20.7–20.10
Compressor starts but short cycles	Overload protector shutting compressor off	See above category	19.6
	Thermostat	Differential too close, thermostat setting; bulb in cold air stream, move location	14.4
	High pressure condition	Check airflow or water flow	28.15
		Air blockage or recirculating, correct	28.15
		Overcharge, remove refrigerant	28.6
		Air in system, remove	28.7
	Low pressure condition with pump down control		
	A. Solenoid valve leaking	Replace solenoid	25.15
	B. Compressor discharge valve leaking during off cycle (will also leak while running)	Repair valve or replace compressor	28.17–28.21
	With or without pump down control		
	A. Low refrigerant charge	Repair leak and add refrigerant	28.12
	B. Restriction in system such as expansion valve, drier or crimped pipe	Repair or replace	28.16
Compressor runs continuously or space temperature is too high	Low refrigerant charge	Repair leak, add refrigerant	28.12
	Operating control (thermostat or low pressure control) contacts stuck closed	Replace control	28.1–28.4
	Excessive load		
	A. Unit undersized	Reduce load or replace unit	28.1–28.4
	B. Door of cooler open too often	Train operator	20.15
	C. Product too hot when placed in cooler	Precool food in air conditioned space	26.32
	D. Poor door gaskets, or infiltration	Seal air leaks	26.17–26.19
	E. Cooler setting in too hot a location	Move or reduce heat	26.1
	Evaporator coil not defrosting	Check defrost method and repair	25.28
	Restriction in refrigerant circuit, drier, metering device, crimped pipe	Find and repair	28.16
	Poor condenser performance		
	A. Restricted air flow	Restore airflow	25.14
	B. Recirculated air	Move objects causing recirculation	25.14
	C. Dirty condenser	Clean condenser	28.15

SECTION

6 Air Conditioning (Heating and Humidification)

UNIT 29

Electric Heat

OBJECTIVES

After studying this unit, you should be able to

- discuss the efficiency, relative purchase and installation costs, and operating cost of electric heat.
- list types of electric heaters and state their uses.
- describe how sequencers operate in electric forced-air furnaces.
- trace the circuitry in a diagram of an electric forced-air furnace.
- perform basic tests in troubleshooting electrical problems in an electric forced-air furnace.
- describe typical preventive maintenance procedures used in electric heating units and systems.

SAFETY CHECKLIST

* Be careful that nails or other objects driven into or mounted on radiant heating panels do not damage the electrical circuits.
* Any heater designed to have air forced across the element should not be operated without the fan.
* Always observe all electrical safety precautions.

29.1 INTRODUCTION

Electric heat is produced by converting electrical energy to heat. This is done by placing a known resistance of a particular material in an electrical circuit. The resistance has relatively few free electrons and does not conduct electricity easily. The resistance to electron flow produces heat at the point of resistance. One type of material used for this resistance is a special wire called nichrome, for nickel chromium.

Electric heat is efficient but is more expensive to operate compared to many sources of heat. It is efficient because there is very little loss of electrical energy from the meter to the heating element. It is expensive because it takes large amounts of electrical energy to produce the heat, and the cost of electrical energy in most areas of the country can be expensive compared to fossil fuels (coal, oil, and gas).

The purchase price of electrical heating systems is usually less than other systems. The installation and maintenance is also usually less expensive. This makes electric heating systems attractive to many purchasers.

When electric heat is used as the primary heat source, a high value of insulation is normally used throughout the structure to lower the amount of heat required.

This unit briefly describes several types of electric heating devices. Emphasis, however, is placed on central forced-air electric heat because it is frequently serviced by technicians in this industry in some areas of the country.

29.2 PORTABLE ELECTRIC HEATING DEVICES

Portable or small space heaters are sold in many retail stores and by many industrial distributors and manufacturers, Figure 29–1. Some have glowing coils (due to the resistance of the wire to electron flow). These transfer heat by radiation (infrared rays) to the solid objects in front of the heater. The radiant heat travels in a straight line and is absorbed by solid objects, which warm the space around them. Radiant heat also provides heating comfort to individuals. The heat concentration decreases by the square of the distance and is soon dissipated into the space. Quartz and glass panel heaters are also used to heat small spaces by radiation, Figure 29–2.

Other space heaters use fans to move air over the heating elements and into the space. This is called forced-convection heat because the heat-laden air is moved me-

FIGURE 29–1 Portable electric space heater. *Courtesy Fostoria Industries, Inc.*

FIGURE 29–2 Quartz heater. *Courtesy Fostoria Industries, Inc.*

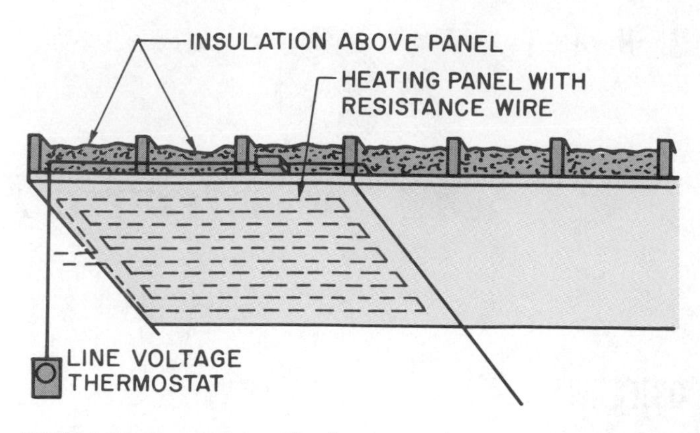

FIGURE 29–3 Radiant ceiling heating panel.

chanically. The units may be designed to move enough air across the heating elements so that they do not glow.

Radiant spot heating can be effectively used at doorways, warehouses, work areas, and even outdoors. The effect is much like a sunlamp pointing at the heated area. Distance has a great bearing on the effectiveness of radiant heating.

29.3 RADIANT HEATING PANELS

Radiant electric heating panels are used in residential and light commercial buildings. The panel is made of gypsum board with wire heating circuits running throughout. The panels are usually installed in ceilings and controlled with individual room thermostats, Figure 29–3. This provides room or zone control. ***Be careful that nails or other objects driven into or mounted on the panels do not damage the electrical circuits.***

Heating panels must have good insulation behind (or above) to keep heat from escaping. The electrical connections are also on the back, so the junctions are easily accessible from an attic.

The heat produced is even, easy to control, and tends to keep the mass of the room warm by its radiation. This makes items in the room pleasingly warm to the touch.

29.4 ELECTRIC BASEBOARD HEATING

Baseboard heaters are popular convection heaters used for whole house, spot, or individual room heating, Figure 29–4. They are economical to install and can be controlled by individual room thermostats. These thermostats are normally line voltage thermostats. Unused rooms can be closed off and the heat turned down, which makes baseboard heaters economical in certain applications.

Baseboard units are mounted on the wall just above the floor or carpet. Outside walls are usually used. The heater is a natural-draft unit. Air enters near the bottom, passes over the electric element, is heated, and rises in the room. As

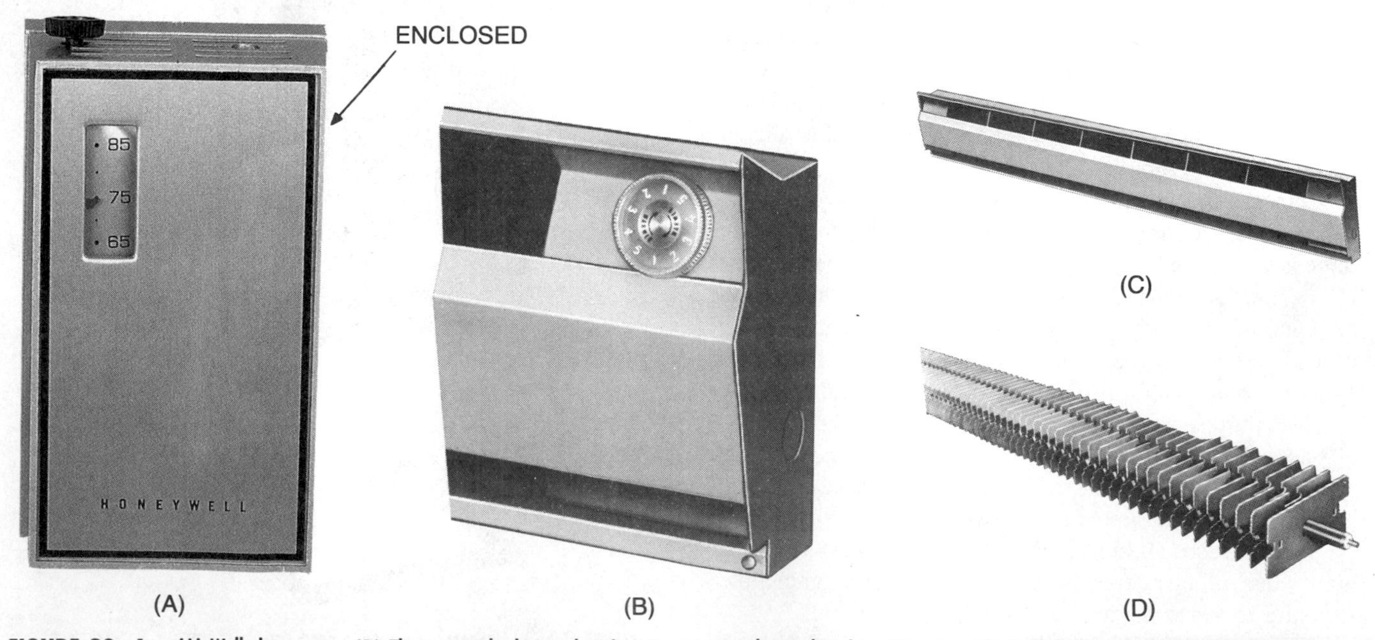

FIGURE 29–4 (A) Wall thermostat. (B) Thermostat built into baseboard unit. (C) Electric baseboard heater. (D) Finned element used in electric baseboard heater. *(A) Photo by Bill Johnson, (B), (C), and (D) Courtesy W. W. Grainger, Inc.*

the air cools, it settles, setting up a natural-convection air current.

Baseboard heat is easy to control, safe, quiet, and evenly distributed throughout the house.

29.5 UNIT HEATERS

Unit heaters are suspended from the ceiling and use a fan to force the air across the elements into the space to be heated, Figure 29–5. These heaters are controlled with line voltage thermostats. ***Any heater designed to have air forced across the element should not be operated without the fan.***

FIGURE 29–5 Electric unit heater. *Courtesy International Telephone and Telegraph Corporation—-Reznor Division*

29.6 ELECTRIC HYDRONIC BOILERS

The electric hydronic (hot water) boiler system is used for some residential and light commercial applications. Except for the control arrangement and safety devices the boiler is somewhat like an electric domestic water heater. It is also connected to a closed loop of piping and requires a pump to move water through the loop consisting of the boiler and the terminal heating units, Figure 29–6.

The electric boiler is small and compact for easy location and installation. The first cost of a small electric boiler is comparable to that of an oil or gas boiler. It is relatively easy to troubleshoot and repair.

Any boiler handling water is subject to all the problems of a water-circulating system. For instance, if the boiler shuts down and the room temperature drops below freezing, pipes will burst and the boiler may be damaged; water treatment against scale and corrosion is also needed in a boiler installation.

The boiler itself is efficient because it converts virtually all of its electrical energy input into heat energy and transfers it to the water. When the boiler is located in the conditioned space, any heat loss through the boiler's walls or fittings is not lost from the heated structure.

FIGURE 29–6 Electric hydronic boiler. *Courtesy Burnham Corporation*

This system is quiet and reliable, but cooling and humidification systems cannot easily be added.

29.7 CENTRAL FORCED-AIR ELECTRIC FURNACES

Central forced-air furnaces are used with ductwork to distribute the heated air to rooms or spaces away from the furnace. The heating elements are factory installed in the furnace unit with the air-handling equipment (fan), or they can be purchased as duct heaters and installed within the ductwork, Figures 29–7 and 29–8.

Central heaters are usually controlled by a single thermostat, resulting in one control point—there is no individual room or zone control. An advantage with duct heaters is that temperature in individual rooms or zones can be controlled with multiple heaters placed in the ductwork system. However, an interlock system must be incorporated to ensure that the fan is running to pass air over the heater to prevent it from burning up.

The heating elements are made of nichrome resistance wire mounted on ceramic or mica insulation. They are enclosed in the ductwork or the furnace housing in an electric furnace.

Air conditioning (cooling and humidification) can usually be added because of the air-handling feature.

Be careful when servicing these systems because many exposed electrical connections are behind the inspection panels. Electric shock can be fatal.

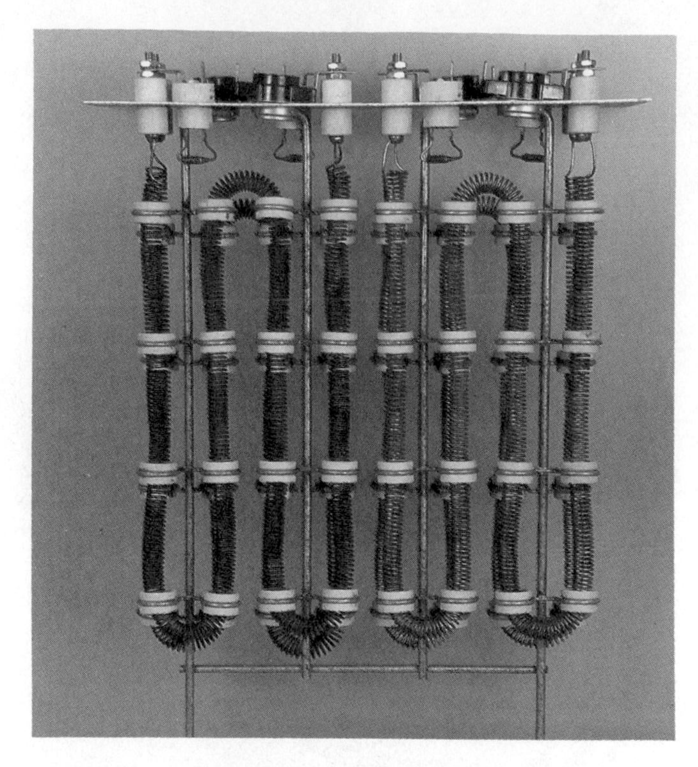

FIGURE 29–7 Electric duct heater. *Photo by Bill Johnson*

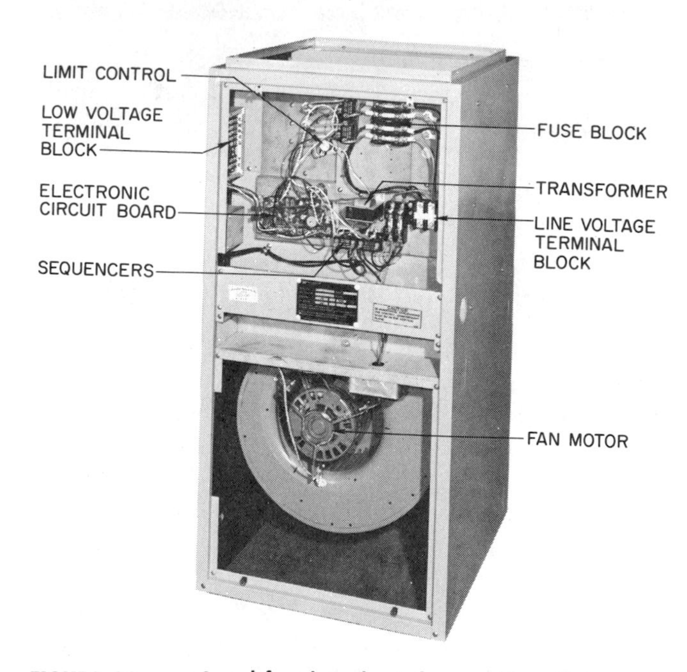

FIGURE 29–8 Central forced-air electric furnace with multiple sequencers. *Courtesy The Williamson Company*

29.8 AUTOMATIC CONTROLS FOR FORCED-AIR ELECTRIC FURNACES

Automatic controls are used to maintain temperature at desired levels in given spaces and to protect the equipment and occupant. Three common controls used in electric heat applications are *thermostats, sequencers* (discussed later in this unit), and *contactors (or relays)*. Figure 29–9 shows a typical thermostat.

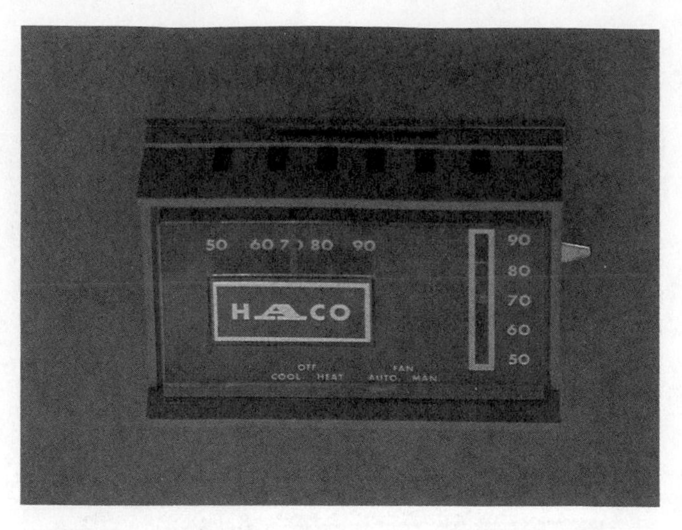

FIGURE 29–9 Low-voltage thermostat. *Courtesy Robertshaw Controls Company*

29.9 THE LOW-VOLTAGE THERMOSTAT

The low-voltage thermostat is used for sequencers and contactors because it is compact, very responsive, safe, and easy to install and troubleshoot. The low-voltage wiring may be run without an electrical license in many localities. Figure 29–10 shows a wiring diagram of a low-voltage thermostat typical of those used with an electric forced-air heating furnace.

The thermostat has an isolated subbase that allows two power supplies to be run to the thermostat. This isolated subbase may be needed when air conditioning is added after the furnace is installed. The furnace will have its low-voltage power supply, and the air conditioning unit may have its own low-voltage power supply. If one power supply is used for both, a jumper is required from the R terminal to the 4 terminal. This is shown in Figure 29–10 by a dotted line. The R terminal is the cooling terminal, and the 4 terminal is assigned to heating if the subbase has two power supplies.

The *heat anticipator* used in a thermostat with electric heat must be set at the time of the installation. The setting is determined by adding all the current draw in the 24 V circuit that passes through the thermostat control bulb. When this current is determined, it is then set with the indicator on the heat anticipator. For example, suppose that the current passing through terminal 4 is 0.75 A. This number is used to set the heat anticipator in the low-voltage thermostat subbase. Most sequencers will have the ampere load of the low-voltage circuit printed on the sequencer. If there is more than one, they can be added. For example, for three sequencers, each with a heater load of 0.3 A, the heat anticipator is set on 0.3 + 0.3 + 0.3 = 0.9 A.

29.10 CONTROLLING MULTIPLE STAGES

Most electric heating furnaces have several heating elements activated in stages to avoid putting a high-power load

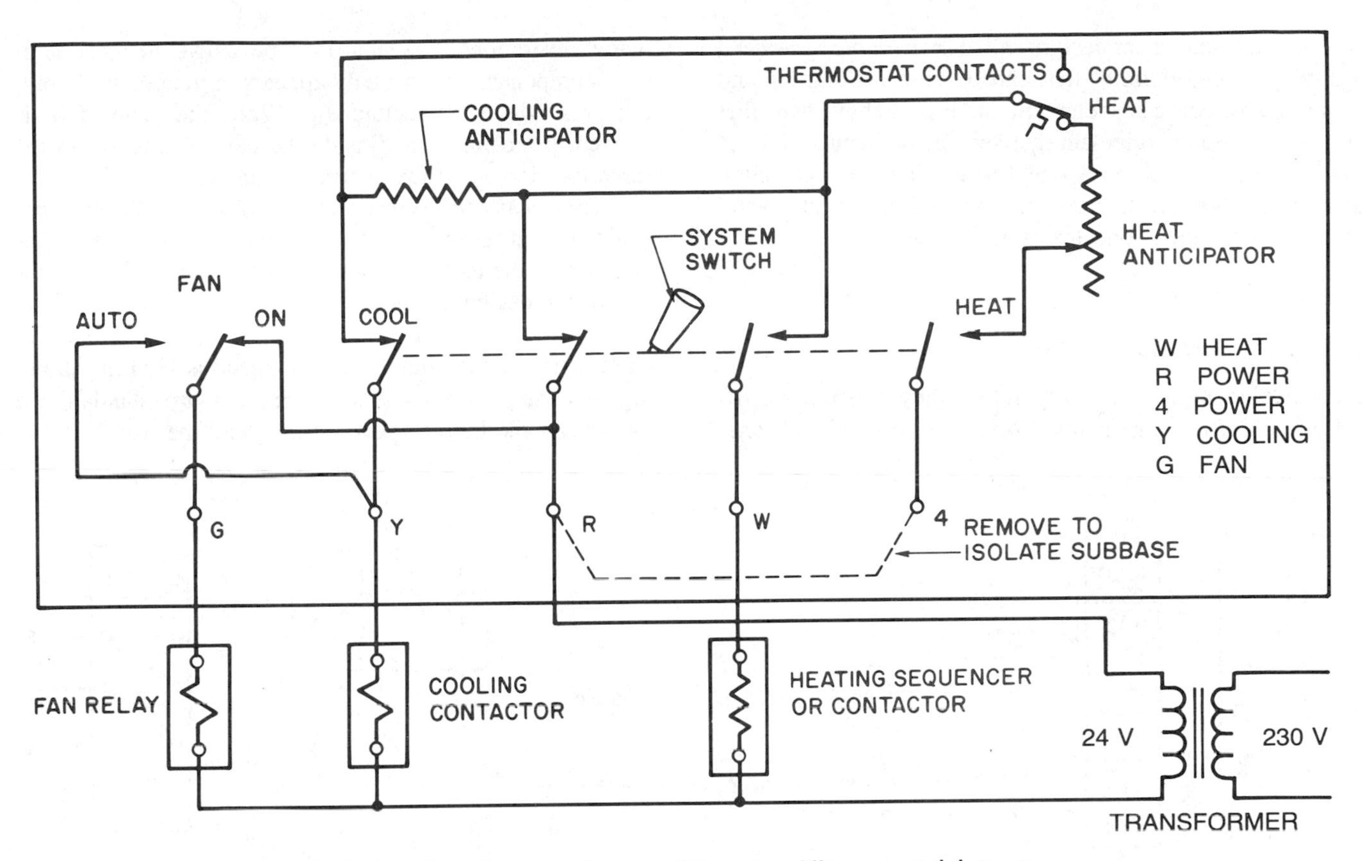

FIGURE 29–10 Wiring diagram of a low-voltage thermostat. Some manufacturers use different terminal designations.

in service all at once. Some furnaces may have as many as six heaters to be connected to the electrical load at the proper time. The *sequencer* is used to do this. It uses low-voltage control power to start and stop the electric heaters. A sequencer can be described as a heat motor type of device. It uses a bimetal strip with a low-voltage wire wrapped around it. When the thermostat calls for heat, the low-voltage wire heats the strip and warps it out of shape in a known direction for a known distance. As it warps or bends, it closes electrical contacts to the electric heat circuit. This bending takes time, and each set of contacts closes quietly and in a certain order with a time delay between the closings.

See Figure 29–11 for a diagram of a package sequencer. This sequencer can start or stop three stages of strip heat. Some sequencers have five circuits, Figure 29–12: three heat circuits, a fan circuit, and a circuit to pass low-voltage power to another sequencer for three more stages of heat. There are five sets of contacts, and none is in the same circuit.

NOTE
DIFFERENT SPACING
BETWEEN CONTACTS
(A)
LINE
POWER
TO HEATER
ELEMENTS
BIMETAL STRIP WITH
24-V HEATER COIL
TO THERMOSTAT
TO
TRANSFORMER
(B)

FIGURE 29–11 Sequencer with three contacts. (A) OFF position. (B) ON position.

FIGURE 29–12 Multiple-type sequencer. *Photo by Bill Johnson*

Another sequencer design, called *individual sequencers,* has only a single circuit that can be used for starting and stopping an electric heat element, but it could have two other circuits: one to energize another sequencer through a set of low-voltage contacts and one for starting the fan motor. Several stages of electric heat may be controlled with several of these sequencers, one for each heat strip.

29.11 WIRING DIAGRAMS

Individual manufacturers vary in how they illustrate electrical circuits and components. Some use pictorial and schematic illustrations. The *pictorial* type shows the location of each component as it actually appears to the person installing or servicing the equipment. When the panel door is opened, the components inside the control box are in the same location as on the pictorial diagram.

The *schematic* (sometimes called *line* or *ladder type*) shows the current path to the components. Schematic diagrams help the technician to understand and follow the intent of the design engineer.

Figure 29–13 contains a legend of electrical symbols in an electric heating circuit. Such a legend is vital in following a wiring diagram. Many symbols are standardized throughout the industry, even throughout the world.

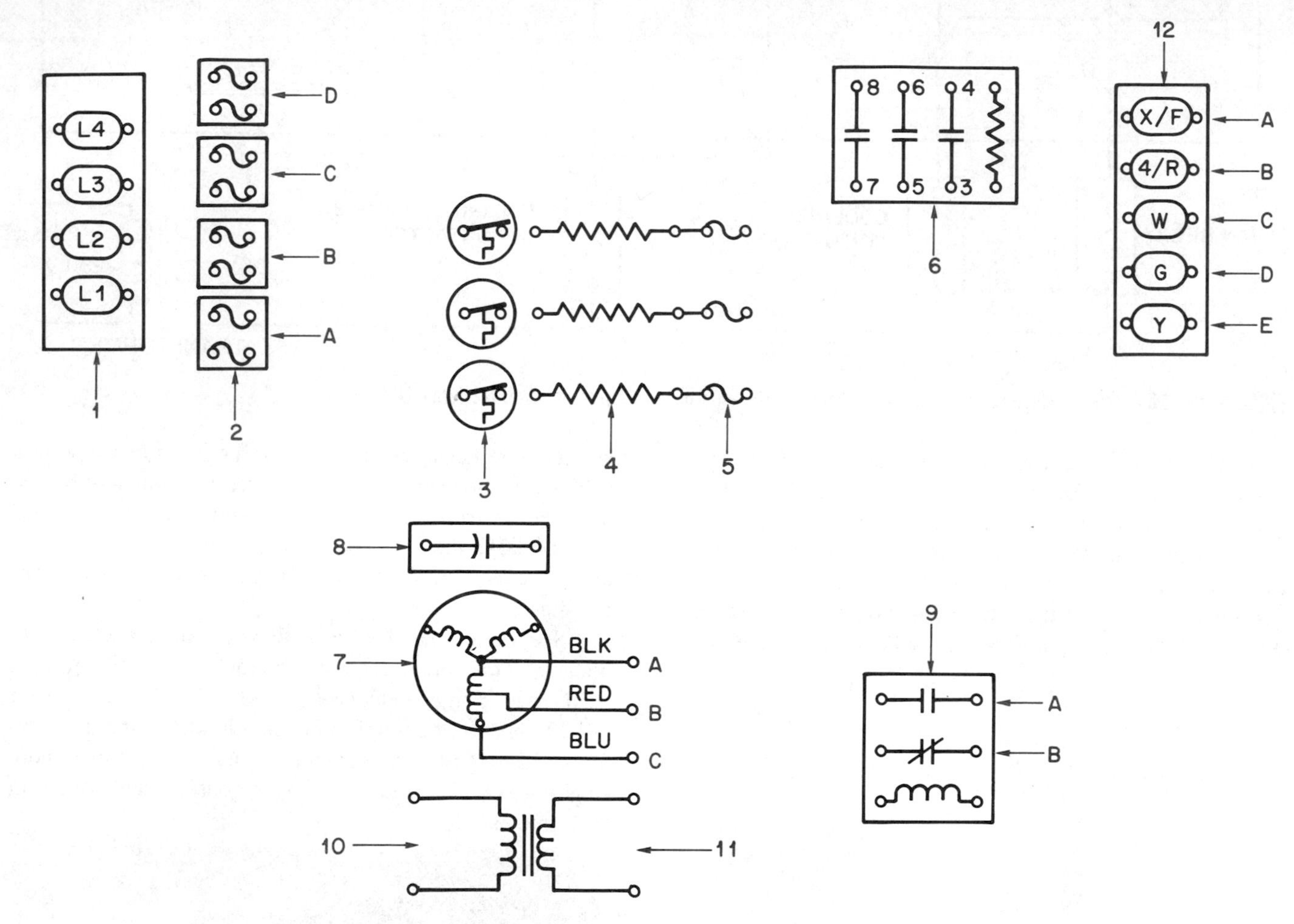

1. Line voltage terminal block from disconnect
2. Fuse block
   A. L1 Fuse block
   B. L2 Fuse block
   C. L3 Fuse block
   D. L4 Fuse block
3. Automatic reset limit switch
4. Electric heat element
5. Fusible link
6. Sequencer
7. Fan motor
   A. High speed motor lead
   B. Medium speed motor lead
   C. Low speed motor lead
8. Fan motor capacitor
9. Fan relay (double pole single throw)
   A. Normally open relay
   B. Normally closed relay
10. Primary side of transformer (line voltage)
11. Secondary side of transformer (low voltage—24 V)
12. Low voltage terminal block
   A. Common terminal from transformer
   B. Terminal for hot lead from transformer
   C. Heating
   D. Fan relay for cooling
   E. Cooling

FIGURE 29–13 Components of pictorial diagram.

29.12 CONTROL CIRCUITS FOR FORCED-AIR ELECTRIC FURNACES

The low-voltage control circuit safely and effectively controls the heating elements that do the work and pull the most current. The circuit contains devices for safety and for control. For example, the limit switch, a safety device, shuts off the unit if high temperature occurs; the room thermostat, control, or operating device stops and starts the heat based on room temperature.

Safety and operating devices can consume power and do work or pass power to a power-consuming device. For example, the sequencer heater coil that operates the bimetal and a magnetic solenoid that moves the armature in a contactor are power-consuming devices in the low-voltage circuit. The contacts of a contactor or a limit control pass power to the power-consuming devices and are in the high-voltage circuit. **Power-passing devices are wired in series with the power-consuming devices. Power-consuming devices are wired in parallel with each other.**

Figure 29–14 is a diagram of a low-voltage control circuit. By tracing this circuit, you can see that when the thermostat contacts are closed, a circuit is completed. This in turn activates the contacts in the sequencer, which activate the heating elements in the high-voltage or line voltage circuit. The line to the G terminal is used for the cooling circuit and has no other purpose. The heating elements and the fan circuit have been omitted to simplify the circuit.

Single heating element control is illustrated in Figure 29–15. The electrical current from the L1 fuse block goes

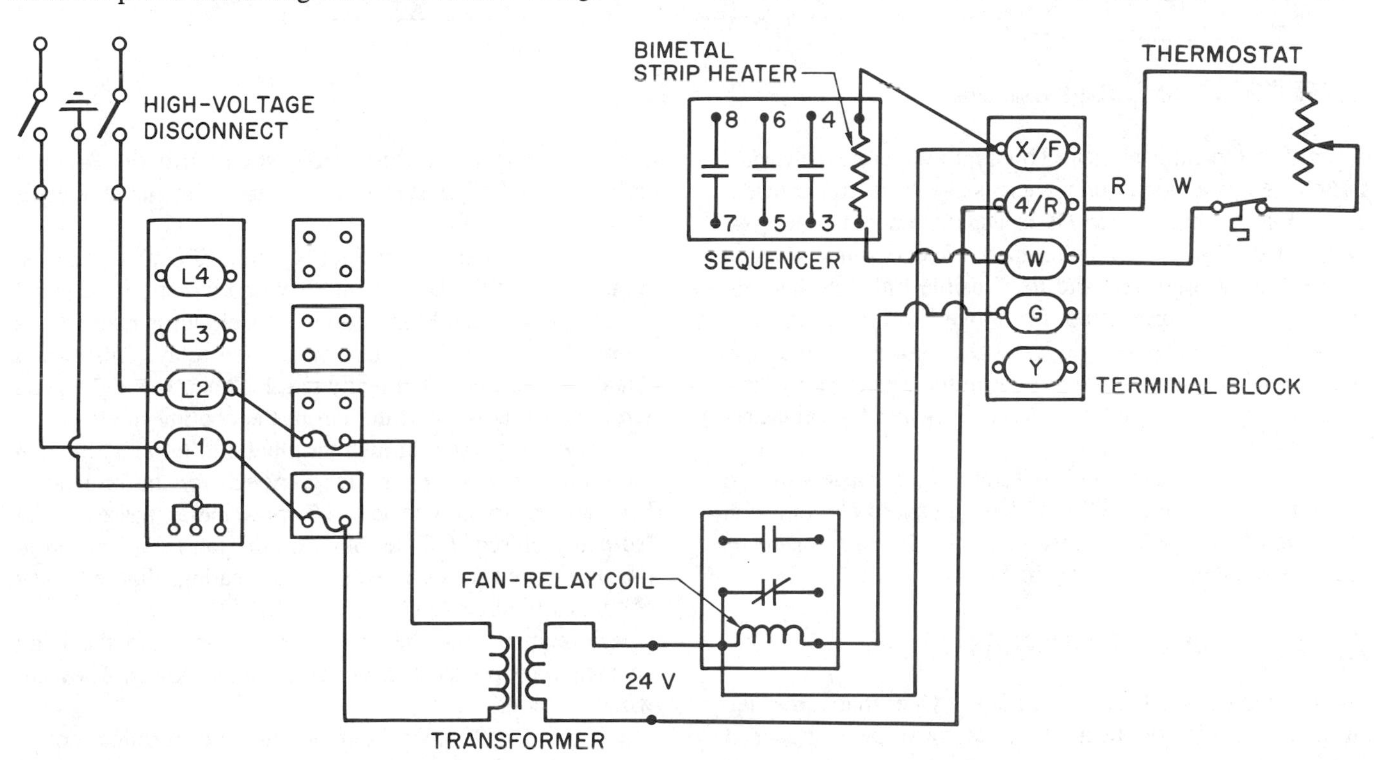

FIGURE 29–14 Diagram of low voltage control circuit.

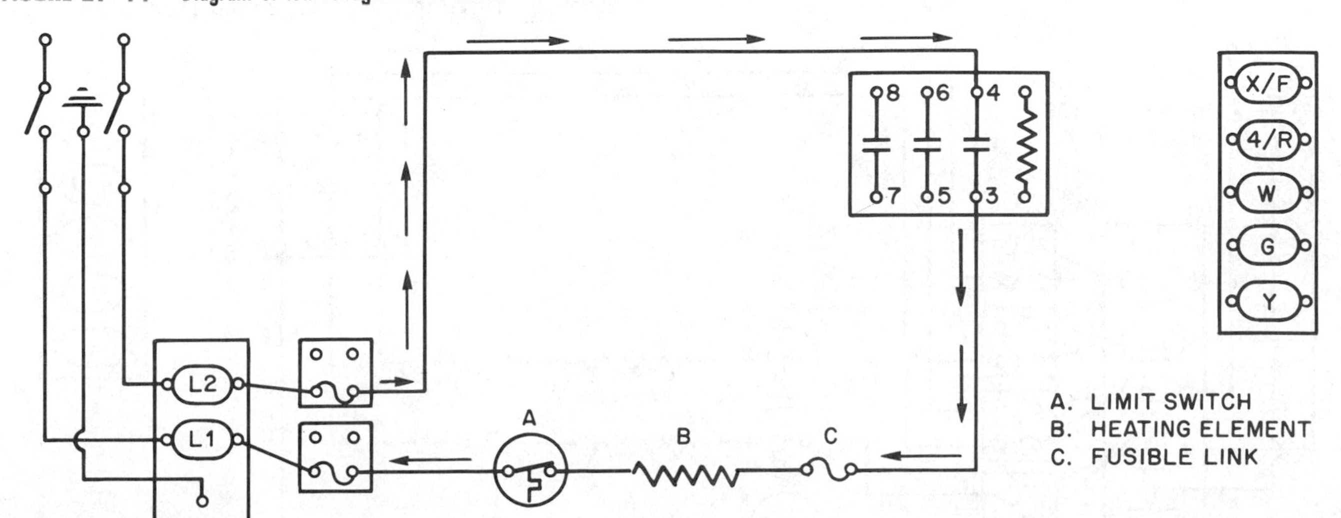

FIGURE 29–15 Single-heating element circuit with sequencer (high voltage).

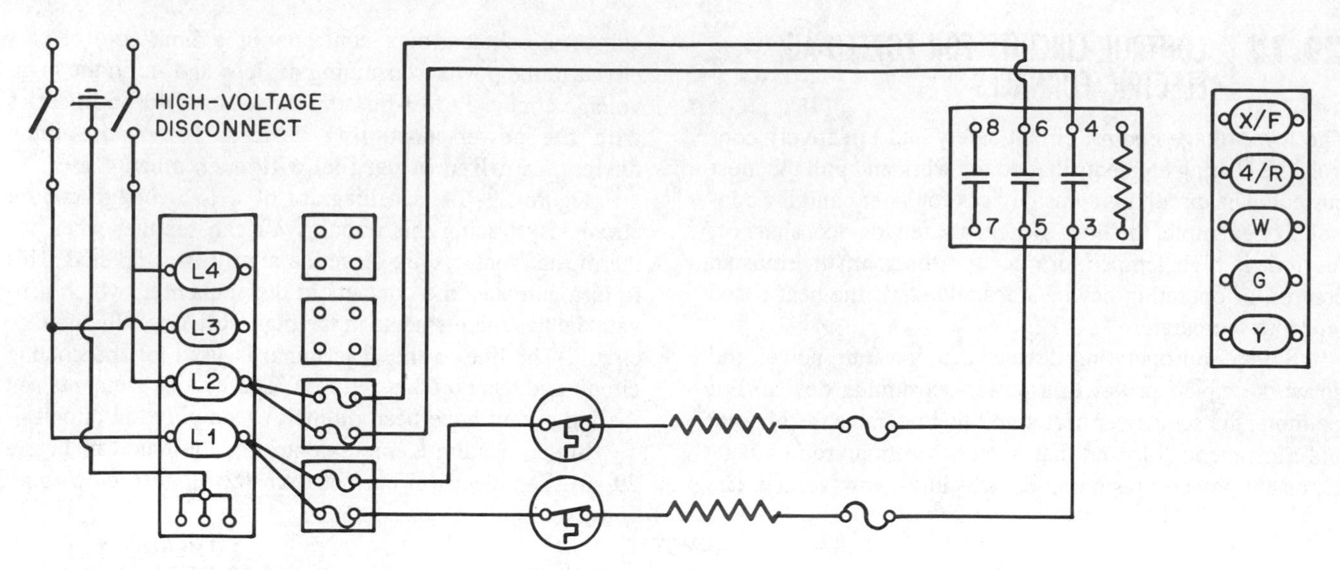

FIGURE 29–16 Two-heating element circuits.

directly to the limit switch. This is a temperature-actuated switch (A) that opens under excessive heat and provides protection to the furnace. It is usually an automatic rest switch that closes when the temperature cools. It is wired to the heating element (B) and to a fusible link (C) that provides additional protection from overheating of the elements. Under higher temperatures this link melts and must be replaced. The link is wired to terminal 3 of the sequencer with the circuit completed through L2 when the sequencer contact is closed.

Figure 29–16 presents a diagram of a furnace with two heating elements. Figure 29–17 has an example of one with three elements. Each of these examples is using the same package sequencer as shown in Figure 29–12.

29.13 FAN MOTOR CIRCUITS

The fan motor, a power-consuming device, that forces the air over the electric heat elements must be started and stopped at the correct time. Figure 29–18 is an example of the fan wiring circuit. ***It must run before the furnace gets too hot and continue to run until the furnace cools down.***

Note that the L1 terminal is wired directly to the fan motor and that the L2 terminal is wired directly to terminal 4 on the sequencer. From terminal 4 a circuit is made to the normally open (NO) contact on the fan relay. This circuit could have been made directly to L2. This is the high-speed fan circuit used to start the fan in the cooling mode.

Power is passed through terminal 4 to terminal 3 when the sequencer is energized long enough for the bimetal to bend and close the contacts. The power then passes to the Normally closed (NC) terminal on the fan relay and on to the slow-speed winding required for heating. Figure 29–19 combines Figures 29–17 and 29–18.

Remember that the sequencer contacts and the relay contacts *pass* power and do not consume power. They are wired in series.

Figure 29–20 is an example legend of terminal designations used on electric forced-air furnaces that use multiple sequencers instead of one package sequencer.

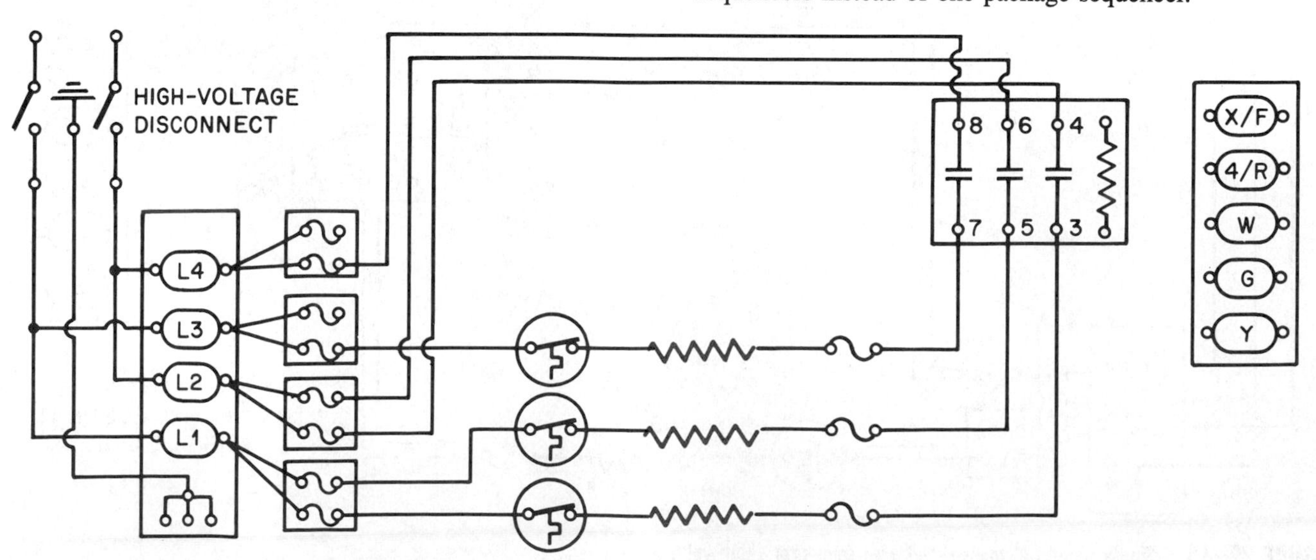

FIGURE 29–17 Three-heating element circuits.

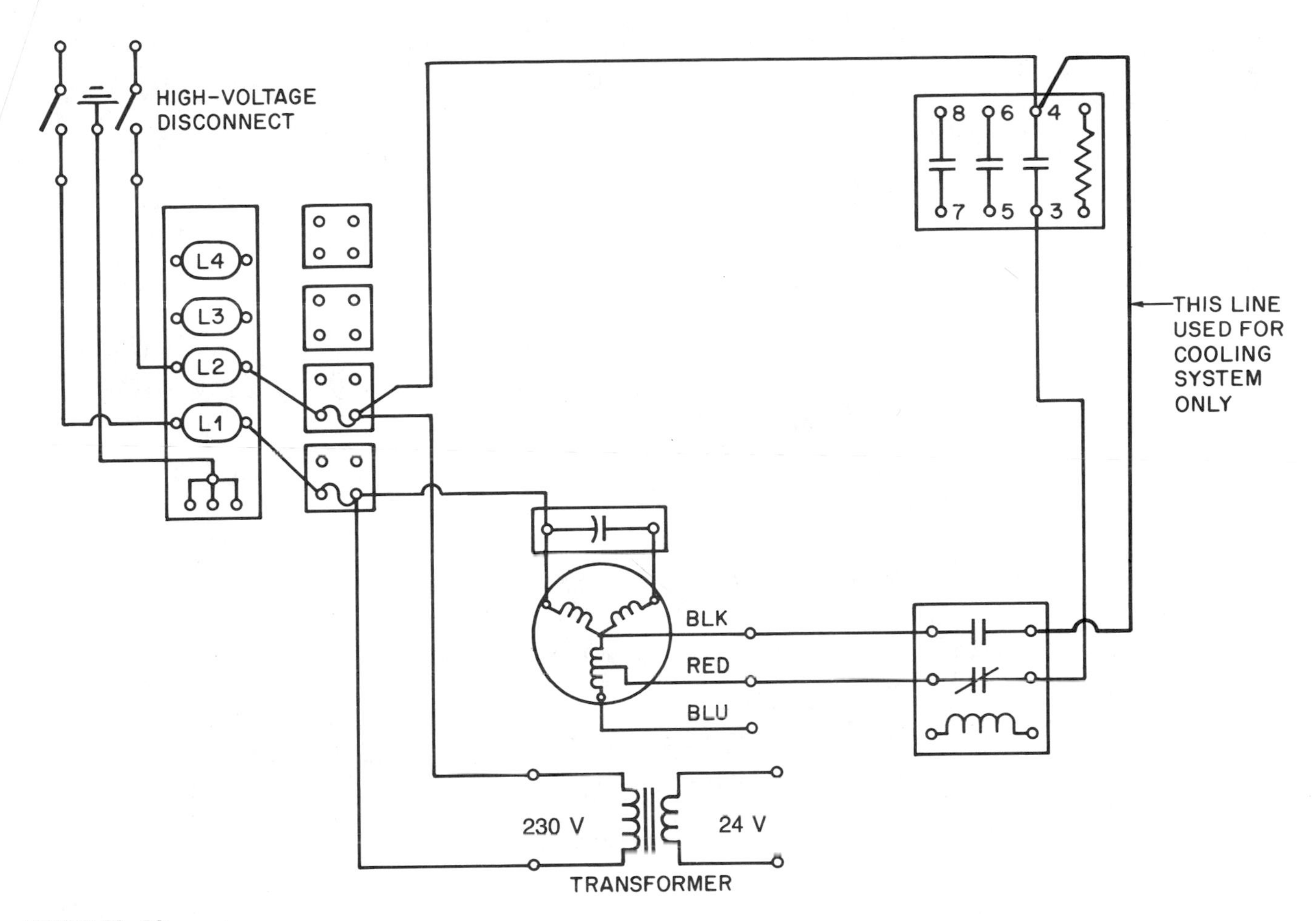

FIGURE 29–18

Figure 29–21 is a wiring diagram for a forced-air furnace with individual sequencers. Note the low-voltage circuit wiring sequence:

1. Power is wired directly from the transformer to terminal C and on to all power-consuming devices. There are no power-passing devices in this circuit. This can be called the *common* circuit.
2. Power is wired directly from the other side of the transformer to terminal R and then on to the R terminal on the room thermostat. The room thermostat passes (or distributes) power to the respective power-consuming device circuits on a call for HEAT, COOL, or FAN ON.
3. On a call for cooling, the Y and G terminals are energized and start the air conditioning and the indoor fan in high speed.
4. On a call for heat, the thermostat energizes the W terminal and passes power to the timed fan control and the first sequencer. The first sequencer passes power to the next sequencer heater coil, then the next.

See Figure 29–21 for the high-voltage sequence.

1. Although there are six fuses, they are labeled either L1 or L2, meaning that there are only two lines supplying power to the unit. The six fuses break the power-consuming heaters and fan motor down into smaller amperage increments.
2. On a call for heat, the contacts in sequencer 1 close and pass power from L2 to the limit switch and on to heater 1. At the same time the timed fan control is energized and its contacts close, starting the fan in low speed.
3. When the sequencer contacts close for the high-voltage load, the low-voltage contacts also close and pass power in the low-voltage circuit to the next sequencer's operating coil, which sets up the same sequence for the second heater. This continues down the line until all sequencer contacts are closed and all the heat is on.

NOTE: This sequence may take 20 sec or more per sequencer, and there are five heat sequencers for the heat to be completely on. When the room thermostat is satisfied, it only takes about 20 sec to turn off the heat because the power to all sequencers is interrupted by the room thermostat. The temperature-operated fan switch takes command of the fan at this time and keeps the fan on until the heaters cool.

The fan circuit is a circuit by itself. The fan will start in low speed if the fan relay is not energized. If the fan is running in slow speed and someone turns the fan switch to ON, the fan will switch to high speed. It cannot run in both speeds at the same time due to the relay with NO and NC contacts.

29.14 CONTACTORS TO CONTROL ELECTRIC FURNACES

Using contactors (or relays) to control electric heat means that all of the heat will be started at one time unless separate

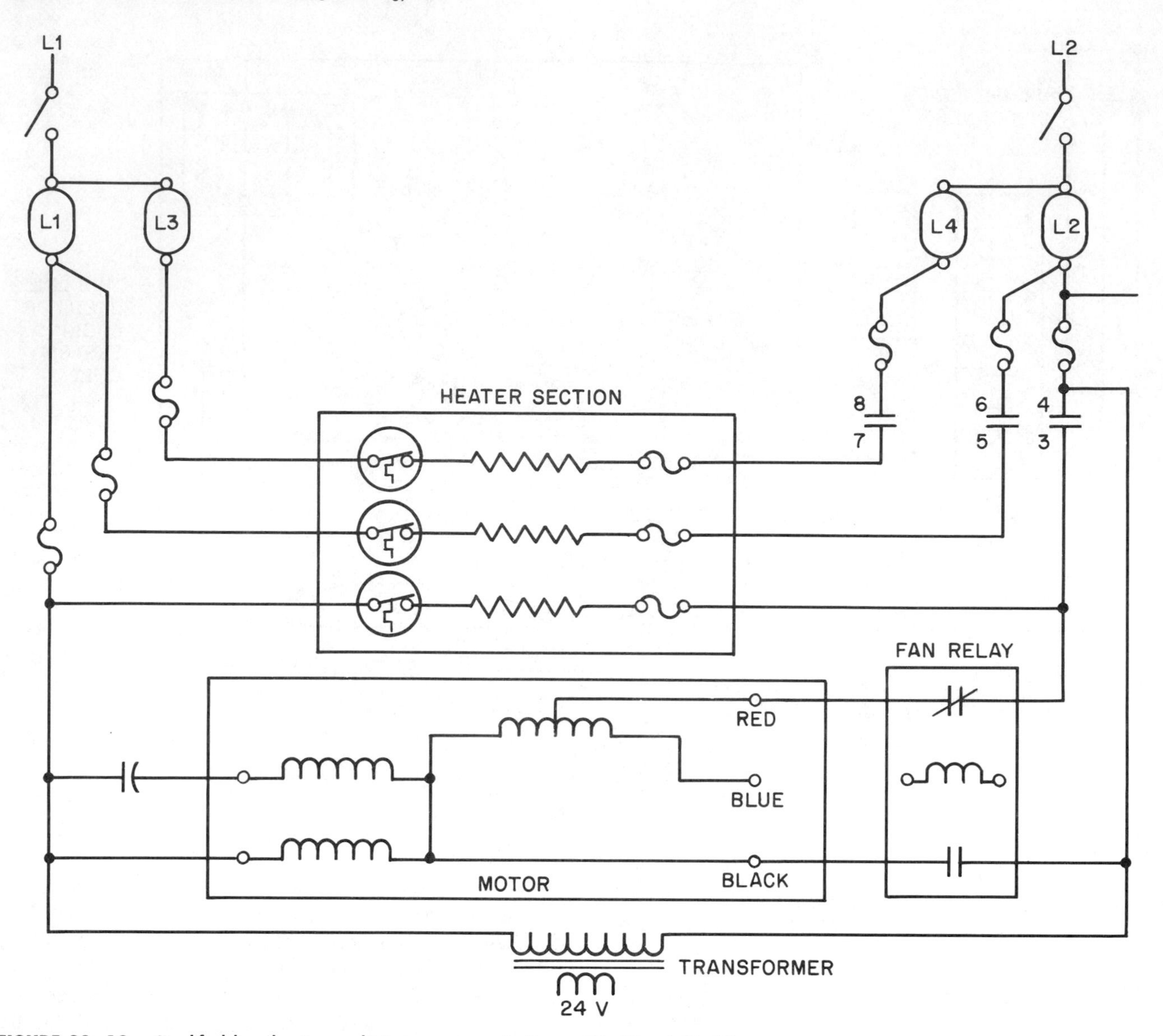

FIGURE 29–19 Simplified line drawing combining components in Figures 29–17 and 29–18.

time delay relays are used. Contactors are magnetic and make noise. They snap in and may hum. Contactors are normally used on commercial systems where they may not be directly attached to the ductwork. This prevents the noise from traveling through the system, Figure 29–22. The contactor magnetic coils are 24 V and 230 V. The 24-V coils are controlled by the room thermostat and they energize the 230-V coil. There are four contactors and their coils in this diagram for the electric heat and one for the fan. The 24-V coils are energized at one time and pull in together.

29.15 AIRFLOW IN ELECTRIC FURNACES

The airflow in cubic feet per minute (cfm) may be verified in an electric heating system using the following formula and example. This formula is often called the sensible heat formula and is restated as it is seen in many other text calculations as

$$Q_s = 1.08 \times \text{cfm} \times \text{TD}$$

Q_s = Sensible heat in Btuh
1.08 = A constant
cfm = Cubic feet per minute
TD = Temperature difference across furnace

The formula is restated and solved for cfm:

$$\text{cfm} = \frac{Q_s}{1.08 \times TD}$$

The total heat (in watts) added to the airstream may be calculated by taking the total amperage to the electric furnace and multiplying it times the applied voltage. For example, a furnace ampere draw is 85 A. At the same time the applied voltage is 208 V and the temperature rise across the furnace is 50°F.

$$\text{Watts} = \text{Amperes} \times \text{Volts}$$
$$\text{Watts} = 85 \times 208$$
$$\text{Watts} = 17{,}680$$

NOTE: This calculation includes the fan motor, which is treated like a resistance load. This will give a slight error on the motor calculation. The motor would actually consume

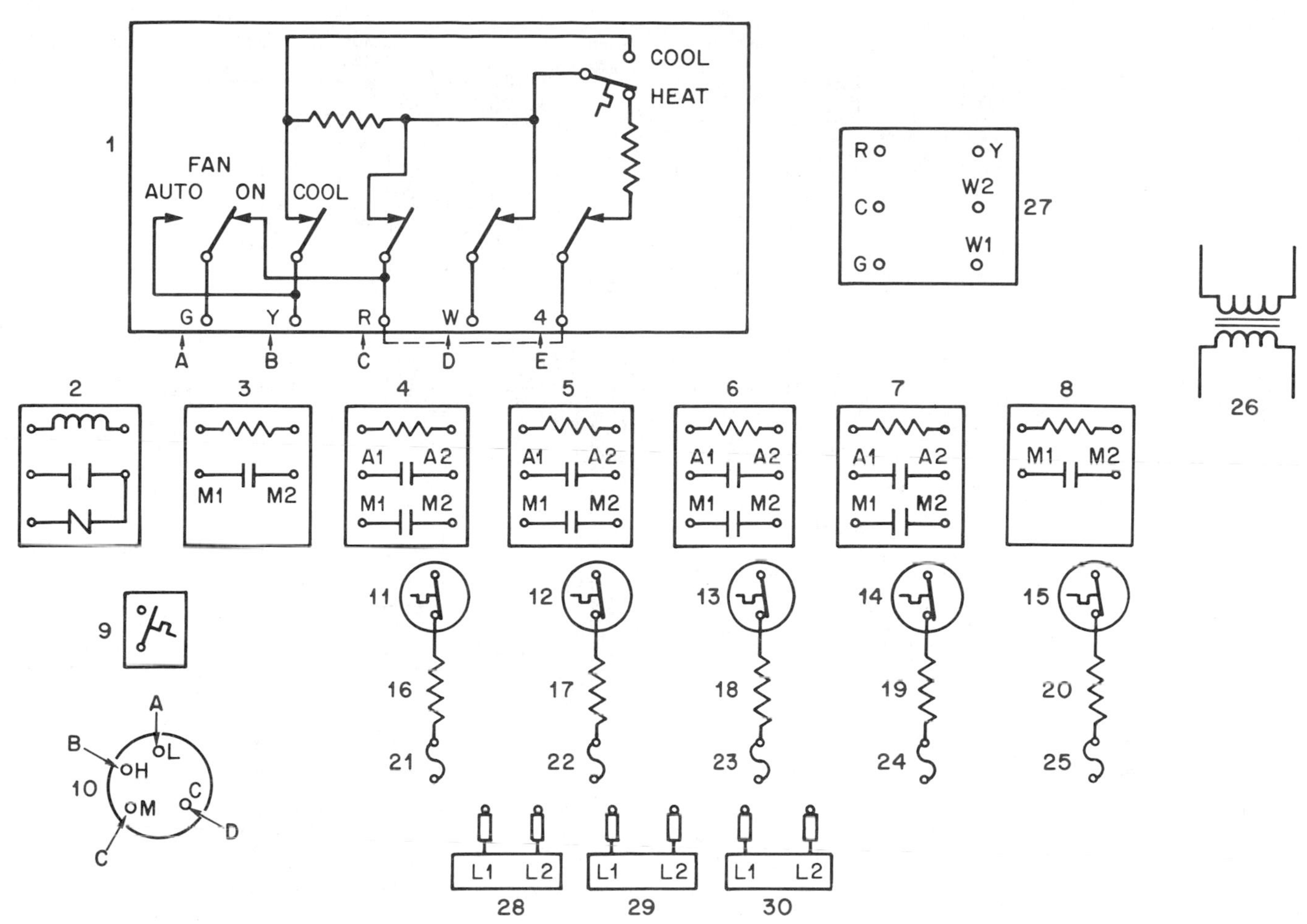

1. Thermostat
   A. G terminal (to fan relay)
   B. Y terminal (to cooling contactor)
   C. R terminal (from transformer secondary)
   D. W terminal (to heating sequencer)
   E. 4 terminal (red or jumped)
2. Fan relay (for cooling)
3. Timed fan control
4. Heat Sequencer I
5. Heat Sequencer II
6. Heat Sequencer III
7. Heat Sequencer IV
8. Heat Sequencer V
   (A1 and A2 terminals for 24 volts only. These are for auxiliary sequencer heater strips which energize next sequencer. M1 and M2 are terminals for heating elements.)
9. Temperature actuated fan control
10. Fan motor
    A. Low speed
    B. High speed
    C. Medium speed
    D. Common

11–15. Temperature actuated limit switches (automatic reset)
16–20. Electric heating elements
21–25. Fusible links
26. Primary side of transformer
27. Low voltage terminal board
28–30. Line terminal blocks

FIGURE 29–20 Legend for Figure 29–21.

about 200 W less power if measured with a wattmeter. Because this is a fraction of a percent error, it is ignored.

The watts must be converted to Btuh by multiplying it by 3.413 (There are 3.413 Btu heat energy in each watt.)

$$\text{Btuh} = \text{Watts} \times 3.413$$

$$\text{Btuh} = 17{,}680 \times 3.413$$

$$\text{Btuh} = 60{,}341.8 \text{ (Figure 29–23)}$$

We can now use the sensible heat formula to find the cfm.

$$\text{cfm} = \frac{Q_s}{1.08 \times TD}$$

$$\text{cfm} = \frac{60,\ 341.8}{1.08 \times 50}$$

$$\text{cfm} = 1117.4$$

The technician may use the electric heat section of a system to establish the correct cfm for the cooling system. Control the fan speed by operating the fan in the FAN ON position. This will ensure that the fan is running at the same speed for cooling. Reduced airflow causes real problems with cooling systems and is covered in a later unit.

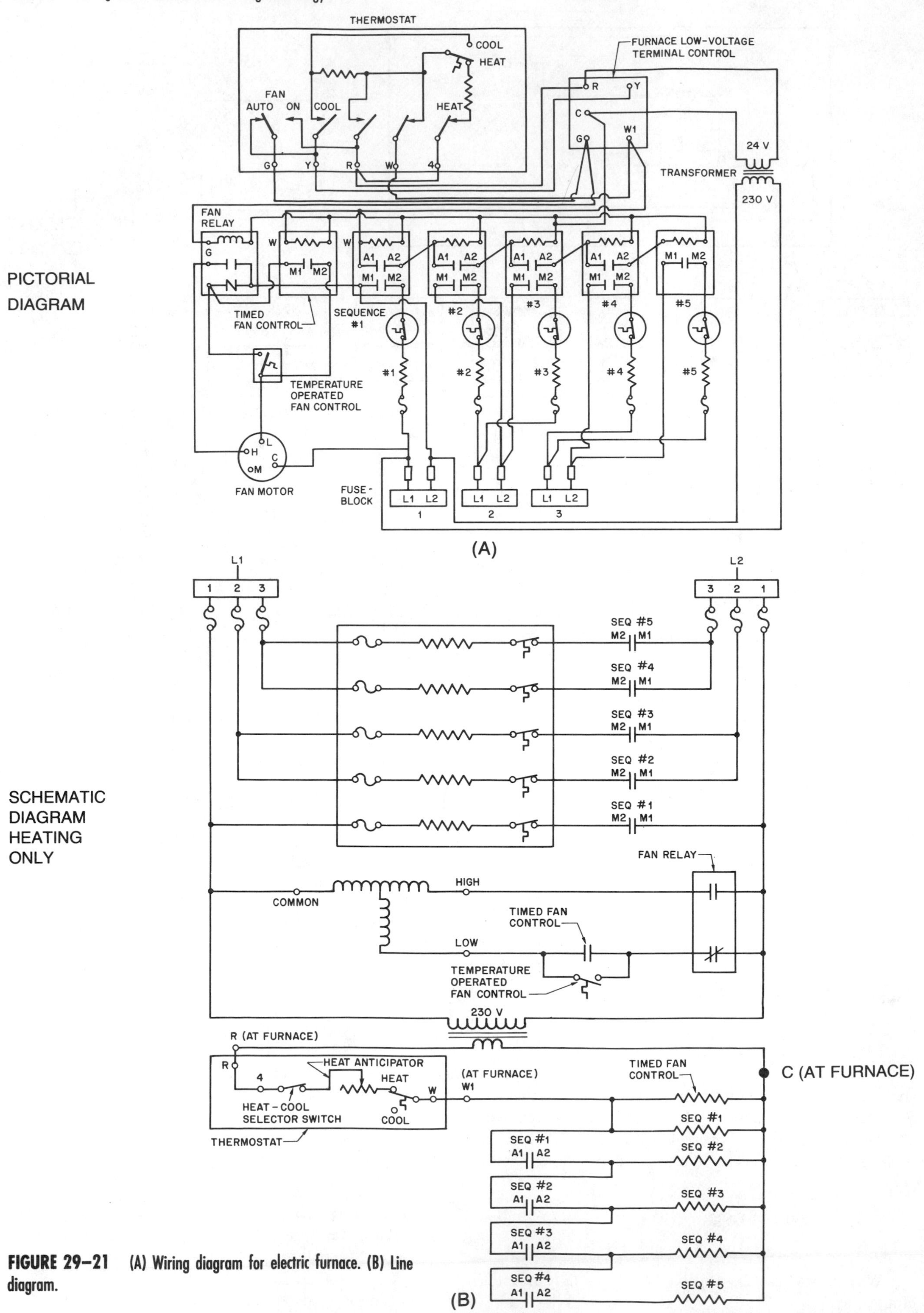

FIGURE 29–21 (A) Wiring diagram for electric furnace. (B) Line diagram.

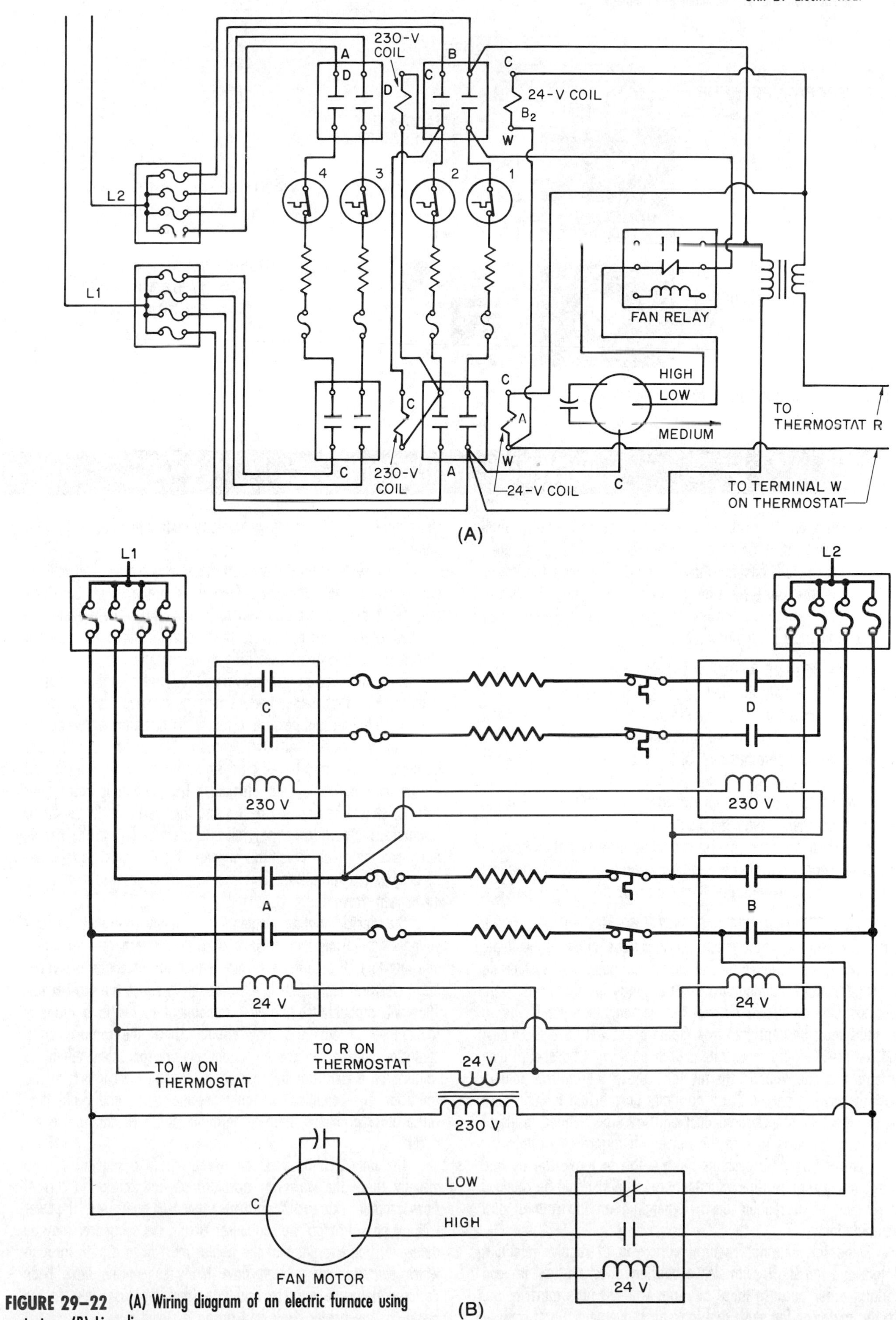

FIGURE 29–22 (A) Wiring diagram of an electric furnace using contactors. (B) Line diagram.

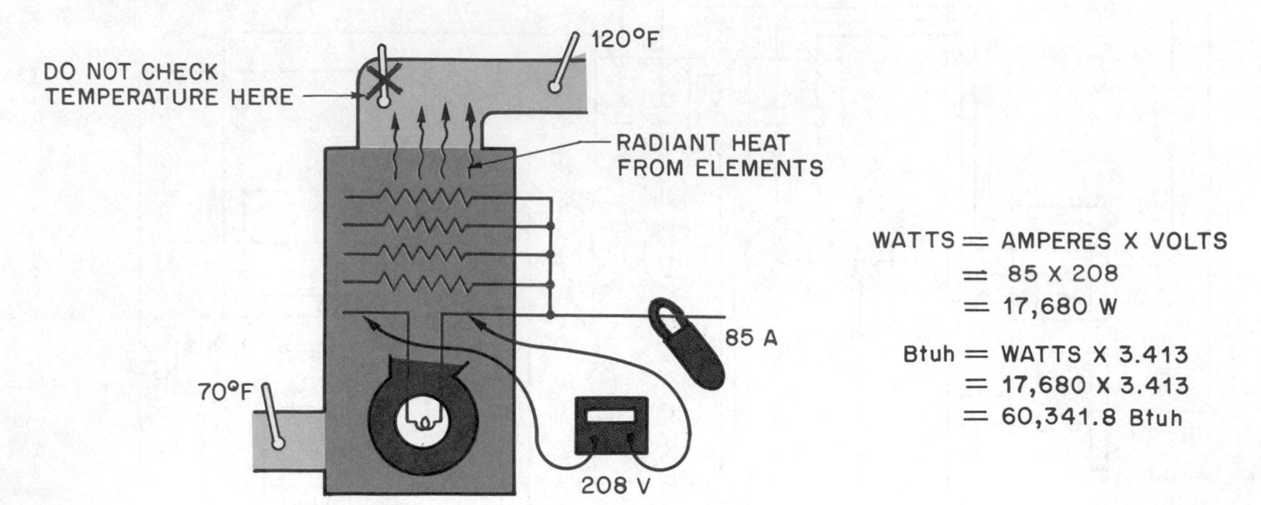

FIGURE 29–23 Electric furnace airflow calculation.

PREVENTIVE MAINTENANCE

Maintenance of electrical heating equipment may involve small appliances or central heating systems. Electric heat appliances, units, or systems all consume large amounts of power. This power is routed to the units by way of wires, wall plugs (appliance heaters), and connectors. These wires, plugs, and connectors carry the load and must be maintained.

ELECTRICAL HEATING APPLIANCES. The power cord at the plug-in connector is usually the first place that trouble may occur on a heating appliance. The heater should be taken out of service at the first sign of excess heat at the wall plug. It may become warm to the touch, but if it becomes too hot to hold comfortably, shut it off and check the plug and the wall receptacle. Fire may start in the vicinity of the plug if this is not corrected. It is a situation that will become worse every time the heater is operated.

An electric appliance heater should be observed at all times to ensure safe operation. Don't locate the heater too close to combustible materials. Only operate the heater in the upright position.

CENTRAL ELECTRICAL HEATING SYSTEMS. The maintenance for large electrical heating systems may include either panel type, baseboard type, or forced air. The panel and baseboard systems do not require much maintenance when properly installed. Baseboard heat has more air circulation over the elements than panel type, so the units must be kept dust free. Some areas will have more dust and require cleaning more often. Both panel and baseboard type systems use line voltage thermostats. These thermostats contain the only moving part and will normally be the first to fail. These thermostats will collect some dust and must be cleaned inside at some time. Be sure to turn the power off before servicing.

Central forced-air systems have a fan to move the air and consequently require filter maintenance. Filters should be changed or cleaned on a regular basis, depending on the rate of dust accumulation.

Some fan bearings require lubrication at regular intervals. Motors and fan shafts with sleeve bearings must be kept in good working order because these bearings will not take much abuse. Lift the motor or fan shaft and look for movement. Don't confuse shaft end-play with bearing wear. Many motors have up to 1/8 in. end-play.

Some systems may have belt drives and require belt tension adjustment or belt changing. Frayed or broken belts must be replaced to ensure safe and trouble-free operation. Belts should be checked at least once a year. It is very dangerous to operate electric heating systems with little or no airflow.

Filters should be checked every 30 days of operation unless it is determined that longer periods may be allowed. Use the filters that fit the holder and be sure to follow the direction arrows on the filter for the correct airflow direction. When the filter system becomes clogged on a forced-air system, the airflow is reduced and causes overheating of the elements. The automatic reset limit switches should be the first to trip and then reset. If the condition continues, these switches may fail and the fuse link in the heater may open the circuit. When this happens, the link must be replaced by a qualified technician. This requires a service call that could have been prevented.

The electric heating elements are usually controlled by low-voltage thermostats that open or close the contacts on sequencers or contactors. The contacts in the sequencers or contactors carry large electrical loads, which will probably produce the most stress. They will probably be the first component to fail in a properly maintained system. You may visually check the contacts on a contactor, but the contacts on sequencers are concealed. When the contacts on a contactor become pitted, change the contacts or the contactor. The condition will only become worse and could then cause other problems, namely, terminal and wire damage to the circuit.

The only method you can use to check a sequencer is to visually check the wires for discoloration and check the current flow through each circuit with the sequencer energized. If power will not pass through the sequencer circuit, the sequencer may be defective. Don't forget that the heater must have a path through which electrical energy can flow. Many sequencers have been declared defective only to find later that the fuse links in the heater or the heater itself had an open circuit.

Line voltage electrical connections in the heater junction box should be examined closely. If there is any wire discoloration or connectors that appear to have been hot, these components must be changed. If the wire is discolored and it is long enough, it may be cut back until the conductor is its normal color.

Fuse holders may lose their tension and not hold the fuses tightly. This is caused from heating and will be apparent because the fuses will blow when there is no overload and for no other apparent reason. These fuses and holders must be changed for the unit to perform safely and correctly.

HVAC GOLDEN RULES

When making a service call to a residence:

- Try to park your vehicle so that you do not block the customer's driveway.
- Look professional and be professional. Clean clothes and shoes must be maintained. Wear coveralls for under the house work and remove them before entering the house.
- Ask the customer about the problem. The customer can often help you solve the problem and save you time.
- Check humidifiers where applicable. Annual service is highly recommended for health reasons.

Added Value to the Customer

Here are some simple, inexpensive procedures that may be included with the basic service call.

- Repair all frayed wires.
- Clean or replace dirty filters.
- Lubricate fan bearings when needed.
- Fasten all cabinet doors securely, replace any missing screws.

29.16 SERVICE TECHNICIAN CALLS

SERVICE CALL 1

A customer with a 20-kW electric furnace calls. *There is no heat. The customer says that the fan comes on, but the heat does not. This is a system that has individual sequencers with a fan-starting sequencer. The first-stage sequencer has a burned-out coil and will not close its contacts. The first stage starts the rest of the heat, so there is none.*

The technician is familiar with this system and had started it after installation. The technician realizes that because the fan starts the 24-V power supply is working. On arriving at the job, the thermostat is set to call for heat. The technician goes under the house with a spare sequencer, a volt-ohmmeter, and an ammeter.

When approaching the electric furnace, the fan can be heard running. The panel covering the electric heat elements and sequencers is removed. The ammeter is used to check to see if any of the heaters are using current; none are. ***The technician observes electrical safety precautions.*** The voltage is checked at the coils of all sequencers. See Figure 29–21 for a diagram. It is found that 24 V is present at the timed fan control and the first-stage sequencer, but the contacts are not closed. The electrical disconnect is opened, and the continuity is checked across the first-stage sequencer coil. The circuit is open. When the sequencer is changed and power is turned on, the electric heat comes on.

SERVICE CALL 2

A residential customer calls. *There is no heat, but the customer smells smoke. The company dispatcher advises that the customer should turn the system off until the technician arrives. The fan motor has an open circuit and will not run. The smoke smell is coming from the heating unit cycling on the limit control.*

The technician arrives and goes to the room thermostat and turns the system to heat. This is a system that does not have cooling or the thermostat would be turned to FAN ON to see if the fan will start. If it does not start, this helps to solve the problem. When the technician goes to the electric furnace in the hall closet, it is noticed that a smoke smell is present. When an ammeter is used to check the current at the electric heater, it is found that it is pulling 40 A, but the fan is not running. ***The technician observes electrical safety precautions.***

The technician looks at the wiring diagram and discovers that the fan should start with the first-stage heat from a set of contacts on the sequencer. The voltage is checked to the fan motor; it is getting voltage but is not running. The unit power is turned off, locked and tagged, and a continuity check of the motor proves that the motor winding is open. The technician changes the motor, and the system operates normally.

SERVICE CALL 3

The service technician is on a routine service contract call and inspection. *The terminals on the electric heat units have been hot. The insulation on the wire is burned at the terminals.*

When removing the panels on the electric heat panel, the technician sees the burned insulation on the wires and shows this to the store manager. It is suggested that those wires and connectors be replaced. The store manager asks what the consequences of waiting for them to fail would be. The technician explains that they may fail on a cold weekend and allow the building water to freeze. If the overhead sprinkler freezes and thaws, all of the merchandise will get wet; it is not worth the chance. ***The technician turns off the power and completes the job, using the correct wire size, high-temperature insulation, and connectors.***

The following service call does not have the solution described. The solution can be found in the Instructor's Guide.

SERVICE CALL 4

A small business calls. Their heating system is not putting out enough heat. This is the first day of very cold weather, and the space temperature is only getting up to 65°F with the thermostat set at 72°F. The system is a package air conditioning unit with 30 kW (six stages of 5 kW each) of strip heat located on the roof. The fuse links in two of the heaters are open due to previously dirty air filters.

The technician arrives, goes to the room thermostat, checks the setting of the thermostat, and finds that the space temperature is 5°F lower than the setting. The technician goes to the roof with a volt-ohmmeter and an ammeter ***After removing the panels on the side of the unit where the strip heat is located, the amperage is carefully checked.*** Two of the six stages are not pulling any current. It looks like a sequencer problem at first. A voltage check of the individual sequencer coils shows that all of the sequencers should have their contacts closed; there is 24 V at each coil.

A voltage check at each heater terminal shows that all stages have voltage but are not pulling any current.

What is the problem and the recommended solution?

SUMMARY

- Electric heat is a convenient way to heat individual rooms and small spaces.
- Central electric heating systems are usually less expensive to install and maintain than other types.
- Operational costs may be more than with other types of fuels.
- The low-voltage thermostat, sequencer, and relays are control mechanisms used in central electric forced-air systems.
- The sequencer is used to activate the heating elements in stages. This avoids putting heavy kilowatt loads in service at one time. If this were done, it could cause fluctuations in power, resulting in voltage drop, flickering lights, and other disturbances.
- Sequencers have bimetal strips with 24-V heaters. These heaters cause the bimetal strip to bend, closing contacts, and activating the heating elements.
- ***The fan in a central system must operate while the heating elements are on. The systems must be wired to ensure that this occurs.***
- Systems are protected with limit switches and fusible links (temperature controlled).
- Preventive maintenance inspections should be made periodically on electric heating units and systems.

REVIEW QUESTIONS

1. What are two economic advantages of electric heat?
2. What is an economic disadvantage of electric heat?
3. List four types of electric heating devices or systems.
4. What is a disadvantage in the control of central electric forced-air heating systems?
5. List three types of controls used in central electric forced-air heating systems.
6. Of what material are the heating elements made in these systems?
7. Briefly describe how a sequencer operates.
8. Why are sequencers used in electric forced-air furnaces?
9. Why is a limit switch used in the heating element circuit?
10. Power-consuming devices are wired in ______. Power-passing devices are wired in ______.
11. List five components or parts that should be inspected during a preventive maintenance check.

UNIT 29 DIAGNOSTIC CHART FOR ELECTRIC HEAT

Electric heat may involve either boiler systems or forced air systems. The discussion in this text will be for forced air systems as they are the most common. These forced air systems may be found as the primary heating system, or the secondary heating system when used with heat pumps. The technician should always listen to the customer for symptoms. The customer can often lead you to the cause of the problem with a simple description of how the system is performing.

PROBLEM	POSSIBLE CAUSE	POSSIBLE REPAIR	PARAGRAPH NUMBER
No heat—thermostat calling for heat	Open disconnect switch	Close disconnect switch	29.11
	Open fuse or breaker	Replace fuse or reset breaker and determine why it opened	29.11
	High temperature fuse link (open circuit)	Loosen connection at fuse link causing heat	29.12
	Faulty high voltage wiring or connections	Repair or replace faulty wiring or connections	29.15
	Control voltage power supply off	Check control voltage fuses and safety devices	15.1–15.3
	Faulty control voltage wiring or connections	Repair or replace faulty wiring or connections	20.13
	Heating element burned (open circuit)	Replace heating element—check airflow	29.15
Insufficient heat	Portion of heaters or limits open circuit	See above	29.15
	Low voltage	Correct voltage	29.11

UNIT 30 Gas Heat

OBJECTIVES

After studying this unit, you should be able to

- describe each of the major components of a gas furnace.
- list three fuels burned in gas furnaces and describe characteristics of each.
- discuss gas pressure measurement in inches of water column and describe how a manometer is used to make this measurement.
- discuss gas combustion.
- describe a solenoid, diaphragm, and heat motor gas valve.
- list the functions of an automatic combination gas valve.
- describe the standing-pilot, electric spark-to-pilot, direct-spark, and hot surface ignition systems.
- state conditions under which glow-coil reignition systems might be used.
- list three flame-proving devices and describe the operation of each.
- discuss reasons and the systems used for the delay in starting and stopping the furnace fan.
- state the purpose of a limit switch.
- describe flue-gas venting systems.
- discuss the types of vent dampers.
- sketch a gas piping system as it should be installed immediately upstream from the gas valve.
- describe procedures used to check leaks in gas piping systems.
- sketch a basic wiring diagram for a gas furnace.
- sketch a basic wiring diagram of a glow-coil reignition circuit.
- describe flame rectification when used with electric spark-to-pilot ignition systems.
- compare the designs of a high-efficiency gas furnace and a conventional furnace.
- describe procedures for taking flue-gas carbon dioxide and temperature readings.
- describe typical preventive maintenance procedures.

SAFETY CHECKLIST

* Fuel gases are dangerous in poorly vented areas. They can replace oxygen in the air and cause suffocation. They are also explosive. If a gas fuel leak is suspected, prevent ignition by preventing any open flame or spark from occurring. ***A spark from a flashlight switch can ignite gas***. Vent all enclosed areas adequately before working in them.
* Always shut off gas supply when installing or servicing a gas furnace.
* A gas furnace that is not functioning properly can produce carbon monoxide, a poisonous gas. It is absolutely necessary that the production of this gas be avoided.
* Yellow tips on the flame indicate an air-starved flame emitting poisonous carbon monoxide.
* It is essential that the furnace be vented properly so that the flue gases will all be dissipated into the atmosphere.
* When taking flue-gas samples do not touch the hot vent pipe.
* A furnace with a defective heat exchanger (one with a hole or crack) must not be allowed to operate because the flue gases can mix with the air being distributed throughout the building.
* Wear goggles when using compressed air for cleaning parts.
* The limit switch must operate correctly to keep the furnace from overheating due to a restriction in the airflow or other furnace malfunction.
* Proper replacement air must be available in the furnace area.
* All new gas piping assemblies should be tested for leaks.
* Gas piping systems should be purged only in well-ventilated areas.
* Follow all safety procedures when troubleshooting electrical systems. Electric shock can be fatal.
* Turn off the power before attempting to replace any electrical component. Lock and tag the disconnect panel. There should be only one key. Keep this key on your person while making any repairs.
* Perform a gas leak check after completing an installation or repair.

30.1 INTRODUCTION TO GAS-FIRED FORCED-HOT AIR FURNACES

Gas-fired forced-hot air furnaces have a heat-producing system and a heated air distribution system. The heat-producing system includes the manifold and controls, burners, heat exchanger, and the venting system. The heated air distribution system consists of the blower that moves the air throughout the ductwork and the ductwork assembly. See

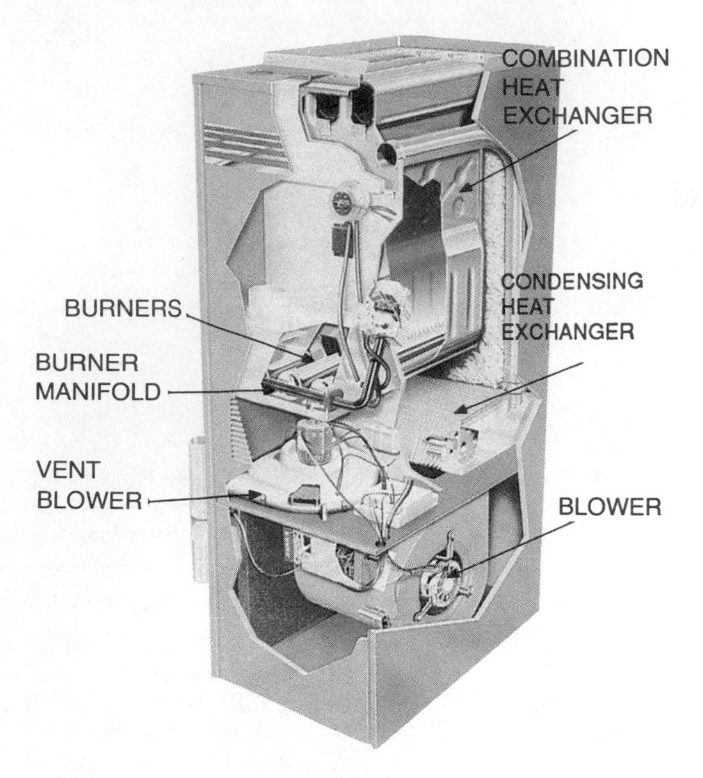

FIGURE 30–1 A modern condensing gas furnace. *Courtesy Heil-Quaker Corporation*

Figure 30–1 for a photo of a modern high-efficiency condensing gas furnace.

The manifold and controls meter the gas to the burners where the gas is burned, which creates flue gases in the heat exchanger and heats the air in and surrounding the heat exchanger. The venting system allows the flue gases to be exhausted into the atmosphere. The blower distributes the heated air through ductwork to the areas where the heat is wanted.

30.2 TYPES OF FURNACES

Upflow

The *upflow* furnace stands vertically and needs headroom. It is designed for the first floor installation with the ductwork in the attic or for basement installation with the ductwork between or under the first floor joists. The furnace takes in cool air from the rear, bottom, or sides near the bottom. It discharges hot air out the top, Figure 30–2.

"Low-boy"

The *low-boy* furnace is approximately 4 ft high. It is used primarily in basement installations with low headroom where the ductwork is located under the first floor. Air intake and discharge are both at the top, Figure 30–3.

Downflow

The *downflow* furnace, sometimes referred to as a *counterflow* furnace, looks like the upflow furnace. The ductwork may be in a concrete slab floor or in a crawl space under the

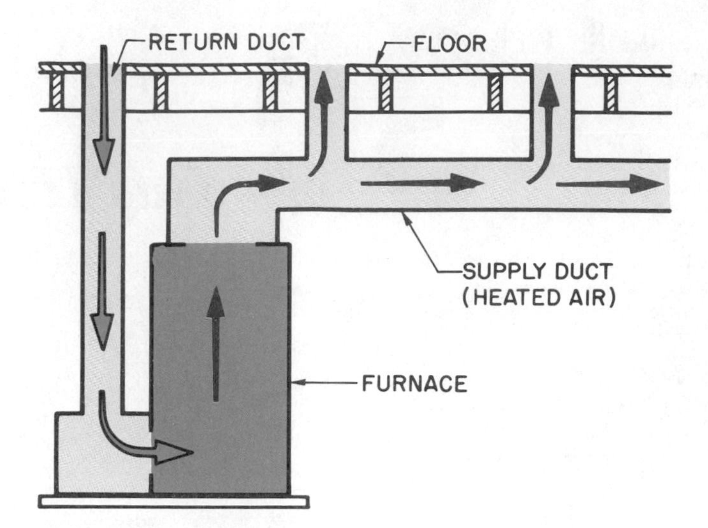

FIGURE 30–2 Upflow gas furnace airflow.

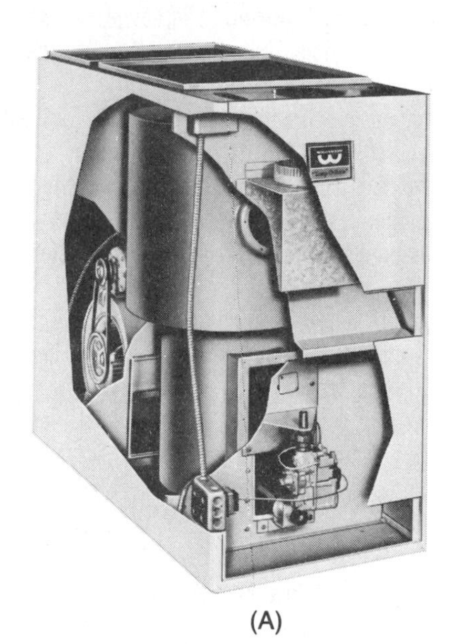

(A)

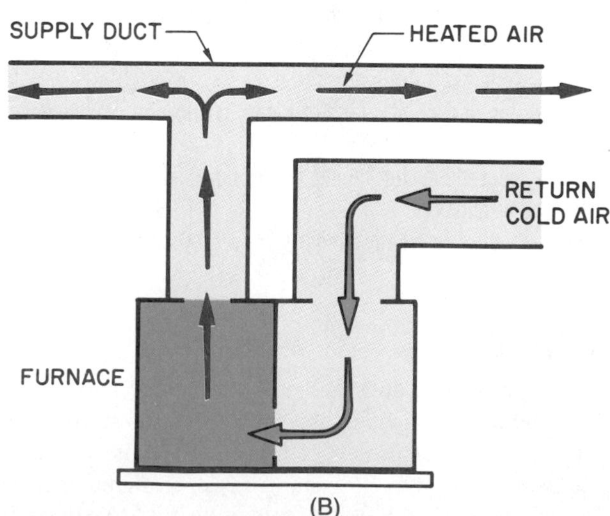

(B)

FIGURE 30–3 (A) Low-boy gas furnace. (B) Airflow for low-boy furnace. *(A) Courtesy The Williamson Company*

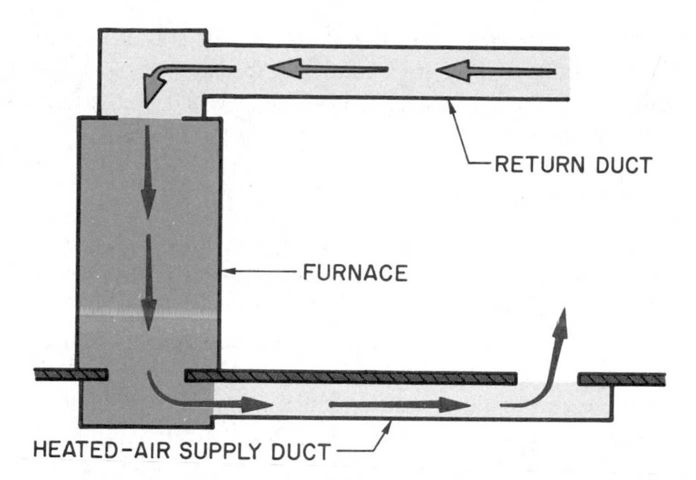

FIGURE 30–4 Airflow for a downflow or counterflow gas furnace.

house. The air intake is at the top, and the discharge is at the bottom, Figure 30–4.

Horizontal

The *horizontal* furnace is positioned on its side. It is installed in crawl spaces, in attics, or suspended from floor joists in basements. In these installations it takes no floor space. The air intake is at one end, the discharge at the other, Figure 30–5.

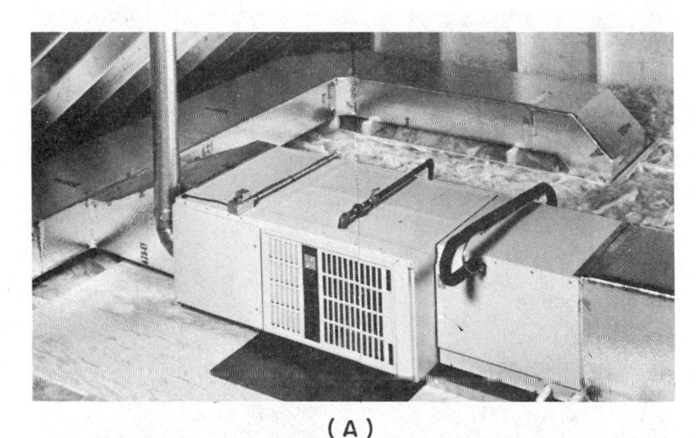

(A)

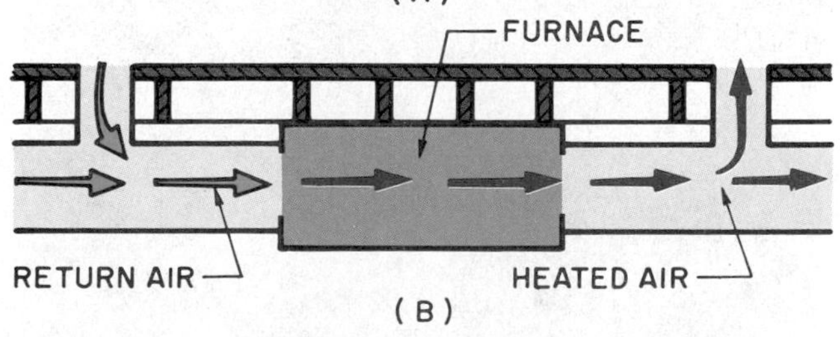

(B)

FIGURE 30–5 (A) Horizontal gas furnace with air conditioning. (B) Airflow for horizontal gas furnace. *(A) Courtesy BPD Company*

30.3 GAS FUELS

Natural gas, manufactured gas, and liquified petroleum (LP) are commonly used in gas furnaces.

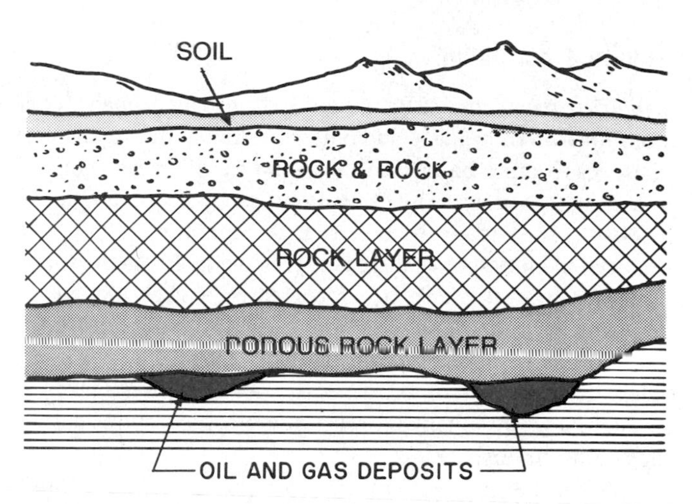

FIGURE 30–6 Gas and oil deposits deep in the earth.

Natural Gas

Natural gas has been forming along with oil for millions of years from dead plants and animals. This organic material accumulated for years and years and gradually was washed or deposited into hollow spots in the earth, Figure 30–6. This material accumulated to great depths, causing tremendous pressure and high temperature from its own weight. This caused a chemical reaction, changing this organic material to oil and gas, which accumulated in pockets and porous rocks deep in the earth.

Natural gas is composed of 90% to 95% methane and other hydrocarbons almost all of which are combustible, which makes this gas efficient and clean burning. Natural gas has a specific gravity of 0.60, and dry air has a specific gravity of 1.0, so natural gas weighs 60% as much as dry air. Therefore, natural gas rises when discharged into the air. The characteristics of natural gas will vary somewhat from one location to another. When burned with air, 1 ft^3 of natural gas will produce about 1050 Btu of heat energy. Contact the utility company providing the gas in your area for its specific characteristics.

Gas from these pockets deep in the earth (more generally called wells) may contain moisture and other gases that must be removed before distributing the gas.

Natural gas by itself has neither odor nor color and is not poisonous. ***But it is dangerous because it can displace oxygen in the air, causing suffocation, and when it accumulates it can explode.***

Sulfur compounds, called *odorants,* which have a garlic smell, are added to the gas to make leak detection easier. No odor is left, however, when the gas has been burned.

Manufactured Gas

Manufactured gas is produced in many ways, often as a by-product of burning coal to make coke. The average cubic foot of manufactured gas produces 530+ Btu/ft^3. Manufactured gas is used primarily for special applications in particular localities. Its specific gravity is also 0.60, so it is also lighter than air.

Liquified Petroleum

Liquified petroleum (LP) is liquified propane, butane, or a combination of propane and butane. When in the vapor state these gases may be burned as a mixture of one or both of these gases with air. The gas is liquified by keeping it under pressure until ready to be used. A regulator at the tank reduces the pressure as the gas leaves the tank. As the vapor is drawn from the tank, it is replaced by a slight boiling or vaporizing of the remaining liquid. LP gas is obtained from natural gas or as a by-product of the oil-refining process. The boiling point of propane is −44°F, which makes it feasible to store in tanks for use in low temperatures during northern winters.

When burned, 1 ft^3 of *propane* produces 2500 Btu of heat. However, it requires 24 ft^3 of air to support this combustion. The specific gravity of propane gas is 1.52, which means it is 1 1/2 times heavier than air, so it sinks when released in air. ***This is dangerous because it will replace oxygen and cause suffocation; it will also collect in low places to make pockets of highly explosive gas.***

Butane produces approximately 3200 Btu/ft^3 of gas when burned. LP gas contains more atoms of hydrogen and carbon than natural gas does, which causes the Btu output to be greater. Occasionally, butane and propane are mixed with air to alter its characteristics. It is then called propane or butane air gas. This mixture is accomplished with blending equipment under pressure. This reduces the Btu/ft^3 and the specific gravity. The characteristics may be changed with air to closely duplicate natural gas. This allows it to be substituted and burned in natural gas burners without adjustment for standby purposes. LP gas is used primarily where natural gas is not available.

The specific gravity of a gas is important because it affects the gas flow through the piping and through the orifice of the furnace. The orifice is a small hole in a fitting (called a *spud*) through which the gas must flow to the burners, Figure 30–7. The gas flow rate also depends on the pressure and on the size of the orifice. At the same pressure more light gas than heavy gas will flow through a given orifice.

SAFETY PRECAUTION: *LP gas alone must not be used in a furnace that is set up for natural gas because the orifice will be too large. This results in overfiring.*

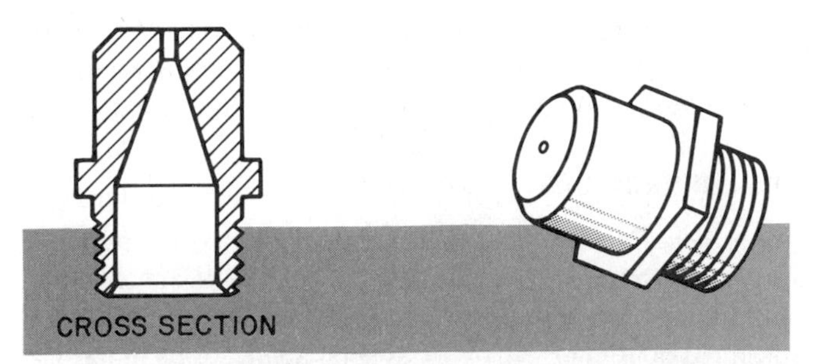

FIGURE 30–7 Spud showing the orifice through which the gas enters the burner.

When the proper mixture of LP gas and air is used, it will operate satisfactorily in natural gas furnaces because the proper mixture will match the orifice size and the burning rate will be the same.

Manifold Pressures

The manifold pressure at the furnace should be set according to the manufacturer's specifications because they relate to the characteristics of the gas to be burned. The manifold gas pressure is much lower than 1 psi and is expressed in inches of water column (in. W.C.). *One inch of water column* pressure is the pressure required to push a column of water up 1 in.

SOME COMMON MANIFOLD PRESSURES	
Natural gas and propane/air mixture	3–3 1/2 in. W.C.
Manufactured gas (below 800 Btu quality)	2 1/2 in. W.C.
LP gas	11 in. W.C.

A *water manometer* is used to measure gas pressure. It is a tube of glass or plastic formed into a U. The tube is about half full of plain water, Figure 30–8. The standard manometer used in domestic and light commercial installations is graduated in. W.C. The pressure is determined by the difference in levels in the two columns. Gas pressure is piped to one side of the tube, and the other side is open to the

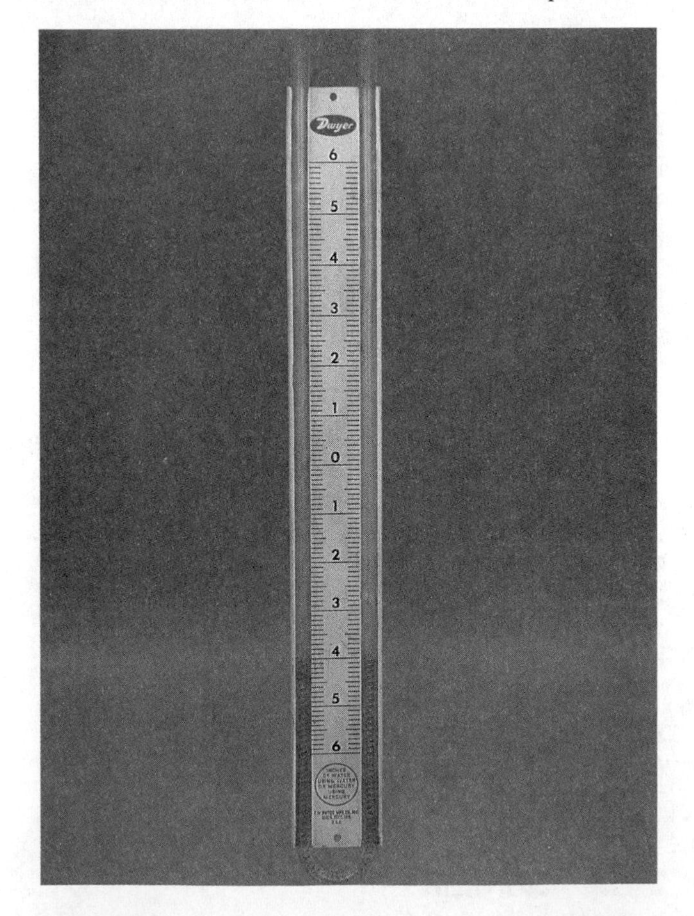

FIGURE 30–8 A manometer used for measuring pressure in inches of water column. *Photo by Bill Johnson*

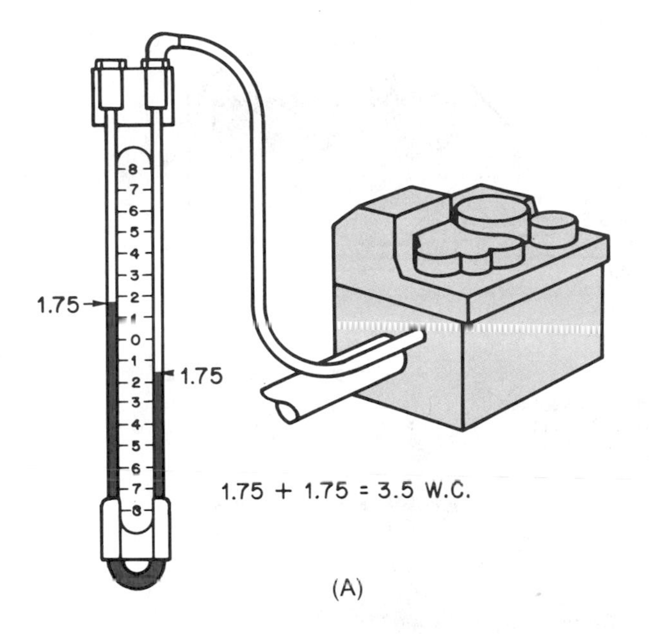

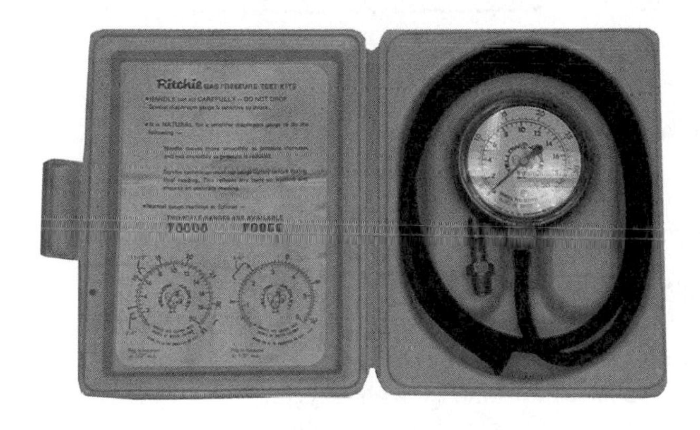

FIGURE 30–9 (A) Manometer connected to a gas valve to check the manifold pressure. (B) Gage calibrated in inches of water column. *Photo by Bill Johnson*

atmosphere to allow the water to rise, Figure 30–9. The instrument often has a sliding scale, so it can be adjusted for easier reading. Some common pressure conversions follow.

PRESSURES PER SQUARE INCH
1 lb/in^2 = 27.71 in. W.C.
1 oz/in^2 = 1.732 in. W.C.
2.02 oz/in^2 = 3.5 in. W.C. (standard pressure for natural gas)
6.35 oz/in^2 = 11 in. W.C. (standard pressure for LP gas)

SAFETY PRECAUTION: *Turn the gas off while connecting the water manometer.*

30.4 GAS COMBUSTION

To properly install and service gas heating systems, the heating technician must know the fundamentals of combustion. These systems must be installed to operate safely and efficiently.

Combustion needs fuel, oxygen, and heat. A reaction between the fuel, oxygen, and heat is known as *rapid oxidation* or the process of burning.

The fuel in this case is the natural gas (methane), propane, or butane; oxygen comes from the air, and the heat comes from a pilot flame or other type of ignition system. The fuels contain hydrocarbons that unite with the oxygen when heated to the ignition temperature. The air contains approximately 21% oxygen. Enough air must be supplied to furnish the proper amount of oxygen for the combustion process. The ignition temperature for natural gas is 1100°F to 1200°F. The pilot flame or other type of ignition must provide this heat, which is the minimum temperature for burning. When the burning process occurs, the chemical formula is

$$CH_4 \text{ (methane)} + 2O_2 \text{ (oxygen)} \rightarrow CO_2 \text{ (carbon dioxide)} + 2H_2O \text{ (water vapor)} + \text{heat}$$

This formula represents perfect combustion. This process produces carbon dioxide, water vapor, and heat. Although perfect combustion seldom occurs, slight variations create no danger. ***The technician must see to it that the combustion is as close to perfect as possible. By-products of poor combustion are carbon monoxide—a poisonous gas—soot, and minute amounts of other products. It should be obvious that carbon monoxide production must be avoided.*** Soot lowers furnace efficiency because it collects on the heat exchanger and acts as an insulator, so it also must be kept at a minimum.

Most gas furnaces and appliances use an atmospheric burner because the gas and air mixtures are at atmospheric pressure during burning. The gas is metered to the burner through the orifice. The velocity of the gas pulls in the primary air around the orifice, Figure 30–10(A).

The burner tube diameter is reduced where the gas is passing through it to induce the air. This burner diameter reduction is called the *venturi,* Figure 30–10(A). The gas-air mixture moves on through the venturi into the mixing tube where the gas and air are mixed for better combustion. The mixture is forced by its own velocity through the burner ports or slots where it is ignited as it leaves the port, Figure 30–10(B).

When the gas is ignited at the port, secondary air is drawn in around the burner ports to support combustion. The flame should be a well-defined blue with slightly orange, *not yellow,* tips, Figure 30–10(C). ***Yellow tips indicate an air-starved flame emitting poisonous carbon monoxide.*** Orange streaks in the flame are not to be confused with yellow. Orange streaks are dust particles burning.

Other than pressure to the gas orifice, the primary air is the only adjustment that can be made to the flame. Modern furnaces have very little adjustability; this is intentional design to maintain minimum standards of flame quality. The only thing that will change the furnace combustion characteristics to any extent, except gas pressure, is dirt or lint

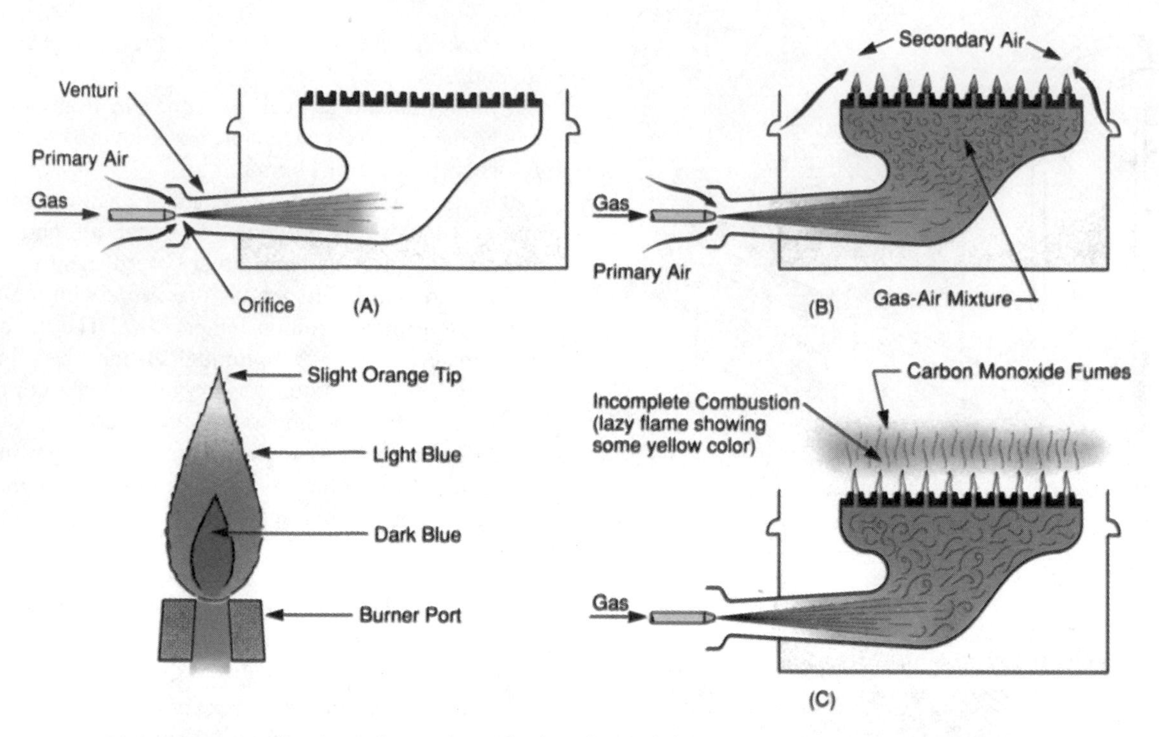

FIGURE 30–10 (A) Primary air is induced into the air shutter by the velocity of the gas stream from the orifice. (B) Ignition of the gas is on top of the burner. (C) Incomplete combustion yields yellow "lazy" flame. Any orange color indicates dust particles drawn in with primary air.

drawn in with the primary air. If the flame begins to burn *yellow,* primary air restriction should be suspected.

For a burner to operate efficiently, the gas flow rate must be correct and the proper quantity of air must be supplied. Modern furnaces use a primary and a secondary air supply, and the gas supply or flow rate is determined by the orifice size and the gas pressure. A normal pressure at the manifold for natural gas is 3 1/2 in. W.C. It is important, however, to use the manufacturer's specifications when adjusting the gas pressure.

A gas-air mixture that is too lean (not enough gas) or too rich (too much gas) will not burn. If the mixture for natural gas contains 0% to 4% natural gas, it will not burn. If the mixture contains from 4% to 15% gas, it will burn. ***However, if the mixture is allowed to accumulate, it will explode when ignited.*** If the gas in the mixture is 15% to 100%, the mixture will not burn or explode. The burning mixtures are known as the *limits of flammability* and are different for different gases.

The rate at which the flame travels through the gas and air mixture is known as the *ignition or flame velocity* and is determined by the type of gas and the air-to-gas ratio. Natural gas (CH_4) has a lower burning rate than butane or propane because it has fewer hydrogen atoms per molecule. Butane (C_4H_{10}) has 10 hydrogen atoms, whereas natural gas has 4. The speed is also increased in higher gas-air mixtures within the limits of flammability; therefore, the burning speed can be changed by adjusting the air flow.

Perfect combustion requires two parts oxygen to one part methane. The atmosphere consists approximately of one-fifth oxygen and four-fifths nitrogen with very small quantities of other gases. Approximately 10 ft^3 of air is necessary to obtain 2 ft^3 of oxygen to mix with 1 ft^3 of methane. This produces 1050 Btu and approximately 11 ft^3 of flue gases, Figure 30–11.

The air and gas never mix completely in the combustion chamber. Extra primary air is always supplied so that the methane will find enough oxygen to burn completely. This excess air is normally supplied at 50% over what would be needed if it were mixed thoroughly. This means that to burn 1 ft^3 of methane, 15 ft^3 of air is supplied. There will be 16 ft^3 of flue gases to be vented from the combustion area, Figure 30–11. There will also be additional air added at the draft hood. This is discussed later.

Flue gases contain approximately 1 ft^3 of oxygen (O_2), 12 ft^3 of nitrogen (N_2), 1 ft^3 of carbon dioxide (CO_2), and 2 ft^3 of water vapor (H_2O). ***It is essential that the furnace be vented properly so that these gases will all be dissipated into the atmosphere.***

Cooling the flame will cause inefficient combustion. This happens when the flame strikes the sides of the combustion chamber (due to burner misalignment) and is called *flame impingement.* When the flame strikes the cooler metal of the chamber, the temperature of that part of the flame is lowered below the ignition temperature. This results in poor combustion and produces carbon monoxide and soot.

The percentages and quantities indicated above are for the burning of natural gas only. These figures vary considerably when propane, butane, propane-air, or butane-air are used. Always use the manufacturer's specifications for each furnace when making adjustments.

SAFETY PRECAUTION: *Care should be taken while taking flue-gas samples not to touch the hot vent pipe.*

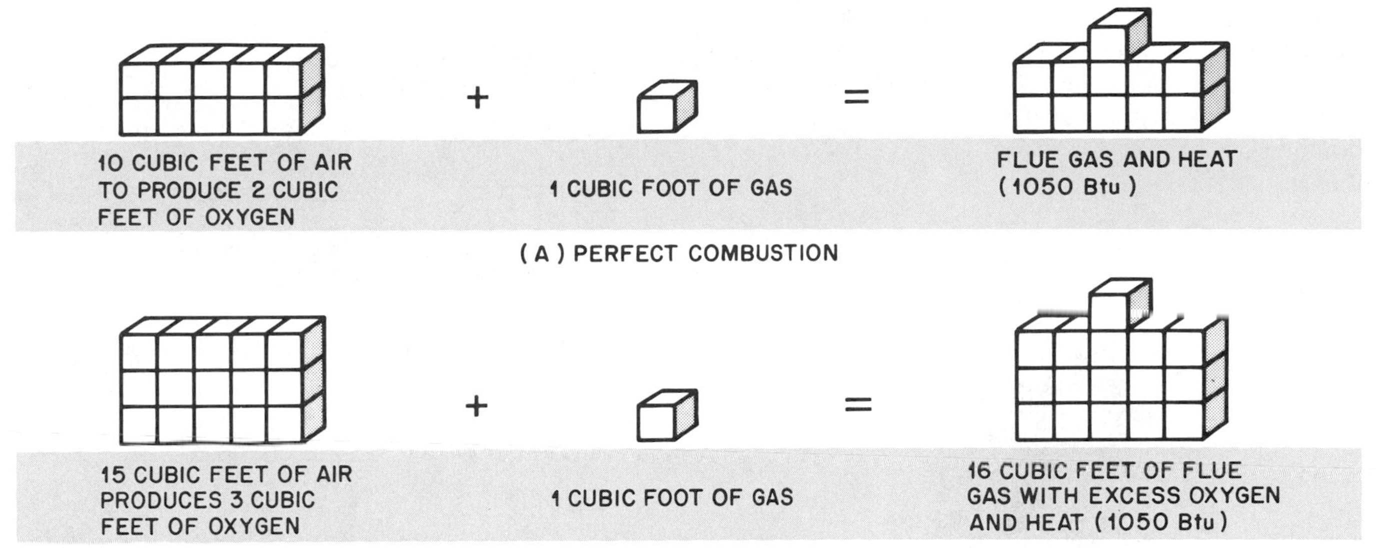

FIGURE 30–11 Quantity of air for combustion.

30.5 GAS REGULATORS

Natural gas pressure in the supply line does not remain constant and is always at a much higher pressure than required at the manifold. The *gas regulator* drops the pressure to the proper level (in. W.C.) and maintains this constant pressure at the outlet where the gas is fed to the gas valve. Many regulators can be adjusted over a pressure range of several inches of water column, Figure 30–12. The pressure is increased when the adjusting screw is turned clockwise; it is decreased when the screw is turned counterclockwise. Some regulators have limited adjustment capabilities; others have no adjustment. Such regulators are either fixed permanently or sealed so that an adjustment cannot be made in the field. The natural gas utility company should be contacted to determine the proper setting of the regulator. In most modern furnaces the regulator is built into the gas valve, and the manifold pressure is factory set at the most common pressure: 3 1/2 in. W.C. for natural gas.

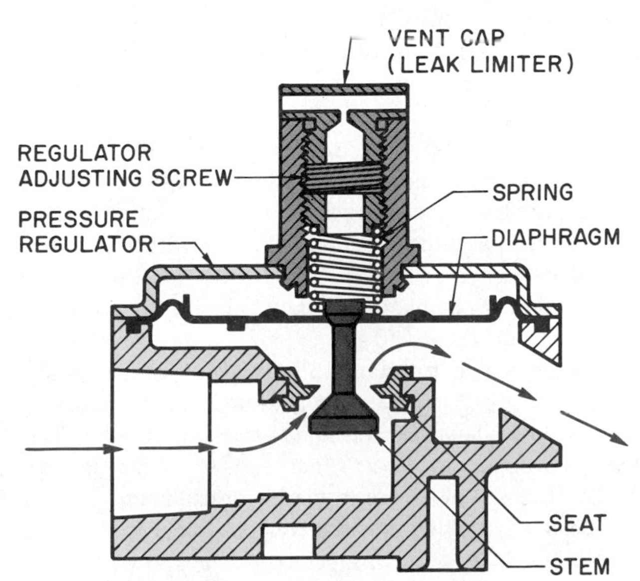

FIGURE 30–12 Diagram of a standard gas pressure regulator.

LP gas regulators are located at the supply tank. These regulators are furnished by the gas supplier. Check with the supplier to determine the proper pressure at the outlet from the regulator. The outlet pressure range is normally 10 to 11 in. W.C. The gas distributor would normally adjust this regulator. In some localities a higher pressure is supplied by the distributor. The installer then provides a regulator at the appliance and sets the water column pressure to the manufacturer's specifications, usually 11 in. W.C.. LP gas installations normally *do not* use gas valves with built-in regulators.

SAFETY PRECAUTION: *Only experienced technicians should adjust gas pressures.*

30.6 GAS VALVE

From the regulator the gas is piped to the *gas valve* at the manifold, Figure 30–13. There are several types of gas

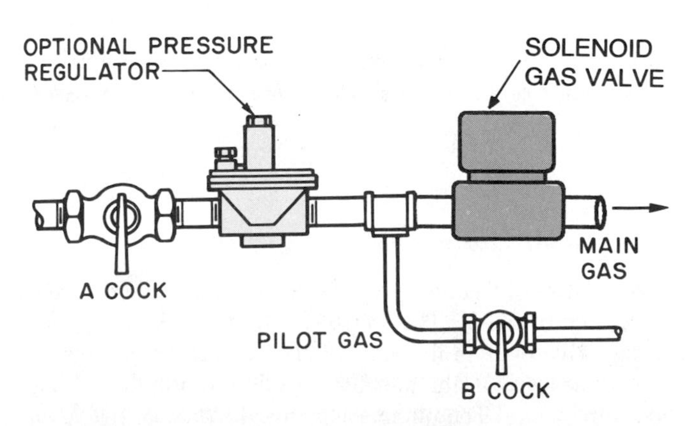

FIGURE 30–13 Natural gas installation where the gas passes through a separate regulator to the gas valve and then to the manifold.

valves. Many are combined with pilot valves and then called *combination* gas valves, Figure 30–14. We will first consider the gas valve separately and then in combination. Valves are generally classified as solenoid, diaphragm, and heat motor.

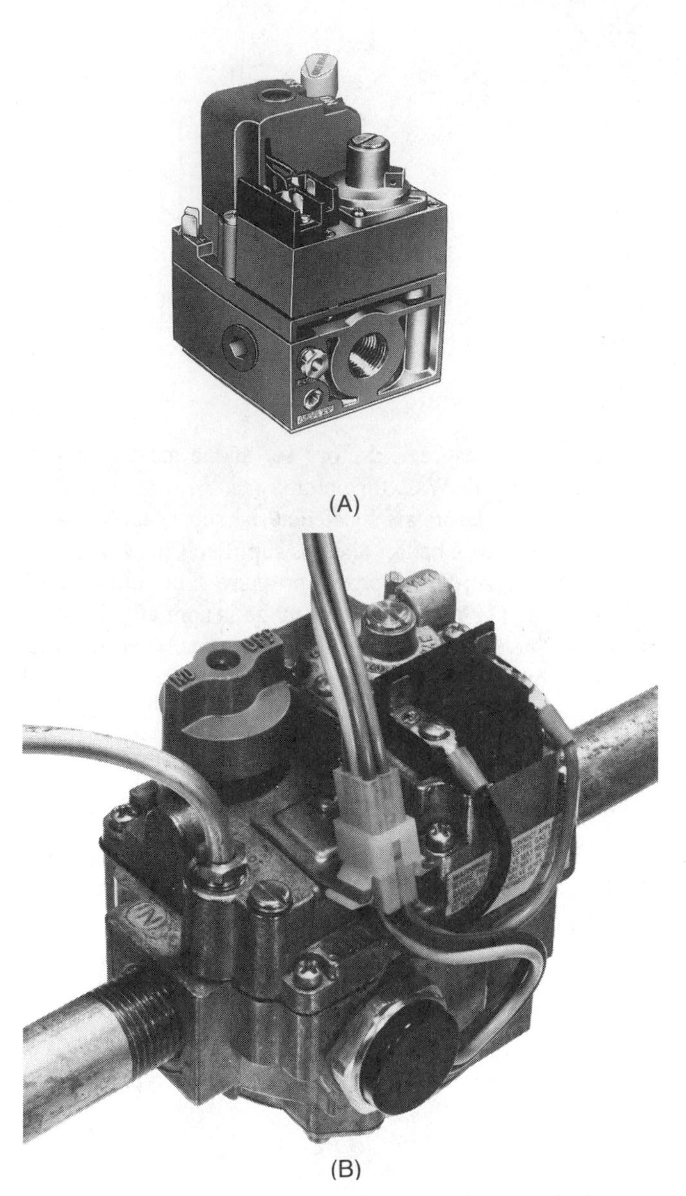
(A)
(B)

FIGURE 30–14 Two gas valves with pressure regulator combined. *Courtesy (A) Honeywell, Inc., Residential Division, (B) Robertshaw Controls Company*

30.7 SOLENOID VALVE

The gas-type *solenoid* valve is a normally closed (NC) valve, Figure 30–15. The plunger in the solenoid is attached to the valve or is in the valve. When an electric current is applied to the coil, the plunger is pulled into the coil. This opens the valve. The plunger is spring loaded so that when the current is turned off the spring forces the plunger to its NC position, shutting off the gas, Figure 30–16.

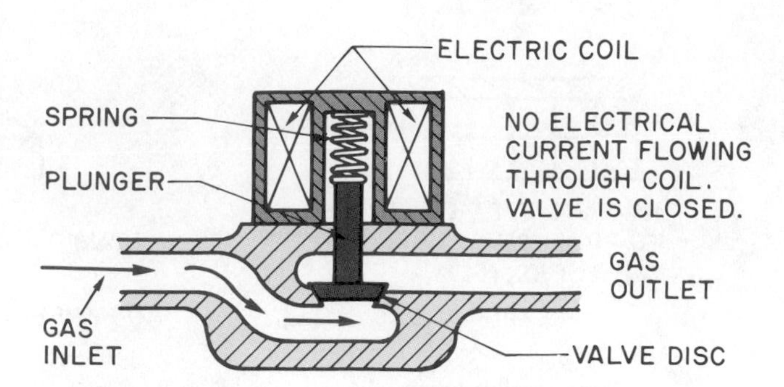
ELECTRIC COIL
SPRING
PLUNGER
NO ELECTRICAL CURRENT FLOWING THROUGH COIL. VALVE IS CLOSED.
GAS OUTLET
GAS INLET
VALVE DISC

FIGURE 30–15 Solenoid gas valve in normally closed position.

ELECTRIC CURRENT APPLIED TO COIL PULLS PLUNGER INTO COIL, OPENING VALVE AND ALLOWING GAS TO FLOW. THE SPRING MAKES THE VALVE MORE QUIET AND HELPS START THE PLUNGER DOWN WHEN DEENERGIZED AND HOLDS IT DOWN.
SPRING COMPRESSED
GAS INLET
GAS OUTLET

FIGURE 30–16 Solenoid valve in open position.

30.8 THE DIAPHRAGM VALVE

The *diaphragm* valve uses gas pressure on one side of the diaphragm to open the valve. When there is gas pressure above the diaphragm and atmospheric pressure below it, the diaphragm will be pushed down and the main valve port will be closed, Figure 30–17. When the gas is removed from above the diaphragm, the pressure from below will push the diaphragm up and open the main valve port, Figure 30–17. This is done by a very small valve, called a *pilot-operated* valve because of its small size. It has two ports—one open while the other is closed. When the port to the upper chamber is closed and not allowing gas into the chamber above the diaphragm, the port to the atmosphere is opened. The gas already in this chamber is vented or bled to the pilot where it is burned. The valve controlling the gas into this upper chamber is operated electrically by a small magnetic coil, Figure 30–18.

When the thermostat calls for heat in the thermally operated valve, a bimetal strip is heated, which causes it to warp. A small heater is attached to the strip, or a resistance wire is wound around it, Figure 30–19. When the strip warps, it closes the valve to the upper chamber and opens the bleed valve. The gas in the upper chamber is bled to the pilot where it is burned, reducing the pressure above the diaphragm. The gas pressure below the diaphragm pushes the valve open, Figure 30–20.

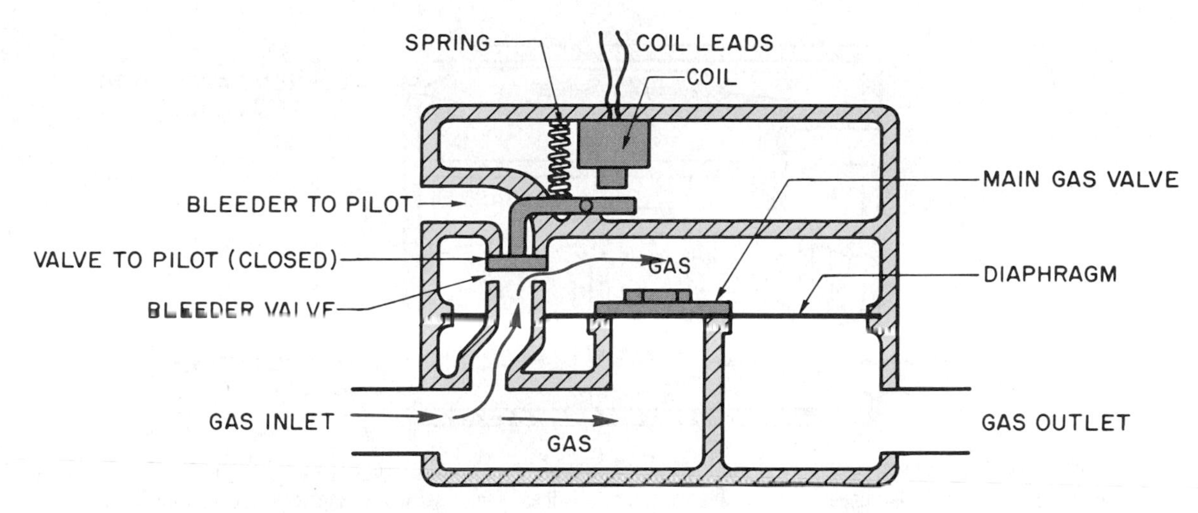

FIGURE 30–17 Electrically operated magnetic diaphragm valve.

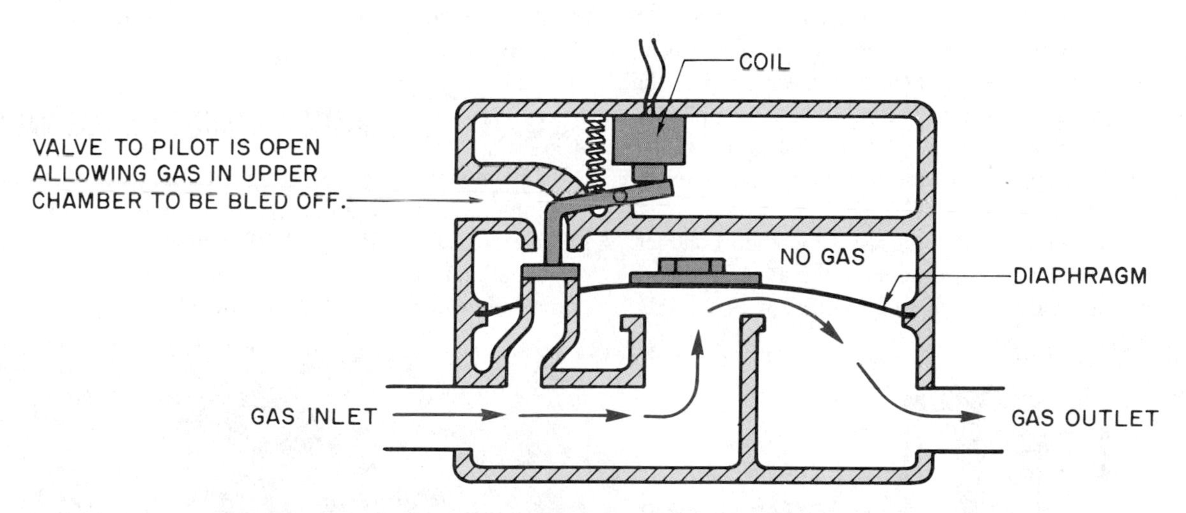

FIGURE 30–18 When an electric current is applied to the coil, the valve to the upper chamber is closed as the lever is attracted to the coil. The gas in the upper chamber bleeds off to the pilot, reducing the pressure in this chamber. The gas pressure from below the diaphragm pushes the valve open.

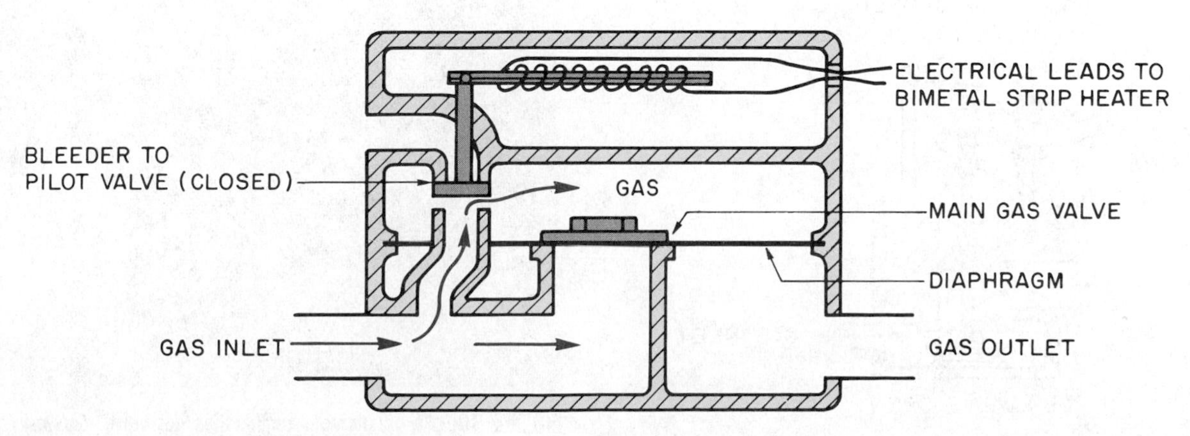

FIGURE 30–19 Thermally operated diaphragm gas valve.

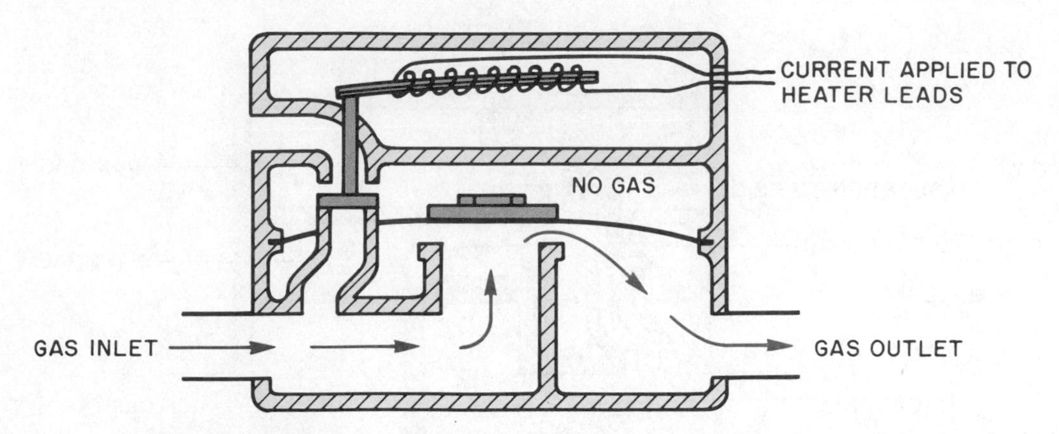

FIGURE 30–20 When an electric current is applied to the leads of the bimetal strip heater, the bimetal warps, closing the valve to the upper chamber, opening the valve to bleed the gas from the upper chamber. The gas pressure is then greater below the diaphragm, pushing the valve open.

30.9 HEAT MOTOR-CONTROLLED VALVE

In a heat motor-controlled valve an electric heating element or resistance wire is wound around a rod attached to the valve, Figure 30–21. When the thermostat calls for heat, this heating coil or wire is energized and produces heat, which expands the rod. When expanded, the rod opens the valve, allowing the gas to flow. As long as heat is applied to the rod, the valve remains open. When the heating coil is deenergized by the thermostat, the rod contracts. A spring will close the valve.

It takes time for the rod to expand and then contract. This varies with the particular model but the average time is 20 sec to open the valve and 40 sec to close it.

SAFETY PRECAUTION: *Be careful while working with heat motor gas valves because of the time delay. Because there is no audible click, you cannot determine the valve's position. Gas may be escaping without your knowledge.*

30.10 AUTOMATIC COMBINATION GAS VALVE

Many modern furnaces designed for residential and light commercial installations use an automatic combination gas valve (ACGV), Figure 30–22. These valves incorporate a manual control, the gas supply for the pilot, the adjustment and safety shutoff features for the pilot, the pressure regula-

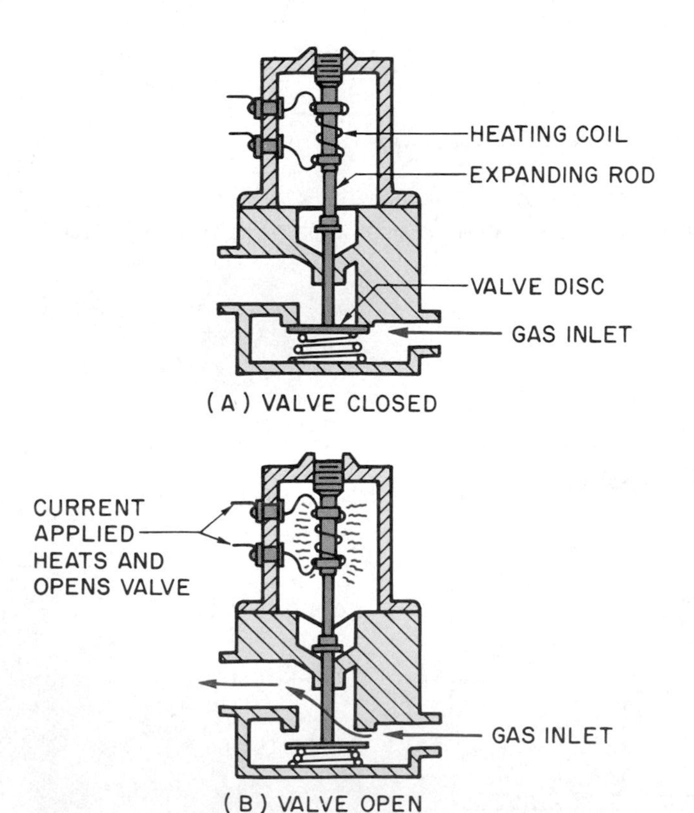

FIGURE 30–21 Heat motor-operated valve.

FIGURE 30–22 Automatic combination gas valve. *Courtesy Honeywell, Inc., Residential Division*

tor, and the controls to operate the main gas valve. They often have dual shutoff seats for extra safety protection. This is also called the redundant gas valve. These valves also combine the features described earlier relating to the control and safety shutoff of the gas.

30.11 MANIFOLD

The *manifold* in the gas furnace is a pipe through which the gas flows to the burners and on which the burners are mounted. The gas orifices are threaded into the manifold and direct the gas into the venturi in the burner. The manifold is attached to the outlet of the gas valve, Figure 30–23.

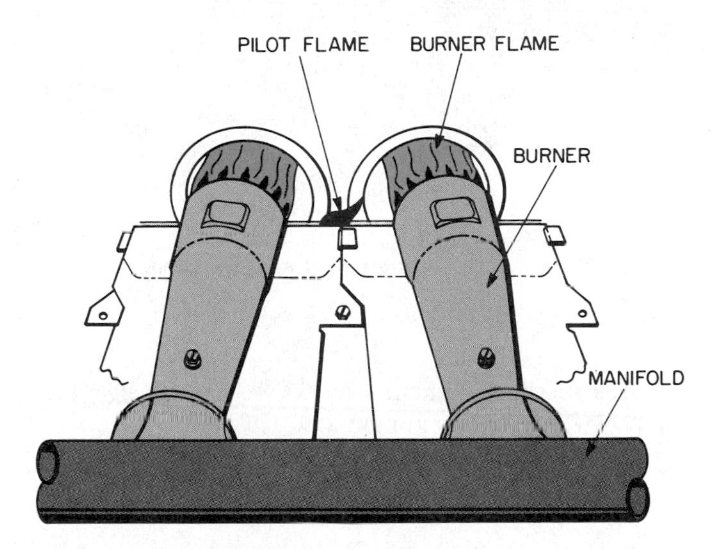

FIGURE 30–23 Manifold. *Courtesy BDP Company*

30.12 ORIFICE

The *orifice* is a precisely sized hole through which the gas flows from the manifold to the burners. The orifice is located in the spud, Figure 30–7. The spud is screwed into the manifold. The orifice allows the correct amount of gas into the burner.

30.13 BURNERS

Gas combustion takes place at the burners. Combustion uses primary and secondary air. Primary air enters the burner from near the orifice, Figure 30–10. The gas leaves the orifice with enough velocity to create a low-pressure area around it. The primary air is forced into this low-pressure area and enters the burner with the gas. The procedure for adjusting the amount of primary air entering the burner is explained later in this unit.

The primary air is not sufficient for proper combustion. Additional air, called secondary air, is available in the combustion area. Secondary air is vented into this area through ventilated panels in the furnace. Both primary and secondary air must be available in the correct quantities for proper combustion. The gas is ignited at the burner by the pilot flame.

The drilled port burner is generally made of cast iron with the ports drilled. The slotted port burner is similar to the drilled port except the ports are slots, Figures 30–24 and 30–25. The ribbon burner produces a solid flame down the top of the burner, Figure 30–26. The single port burner is the simplest and has, as the name implies, one port. This is often called the *inshot* or *upshot* burner, Figure 30–27. All of these burners are known as *atmospheric* burners because the air for the burning process is at atmospheric pressure. In some larger forced-draft gas burners, air is forced in with blowers.

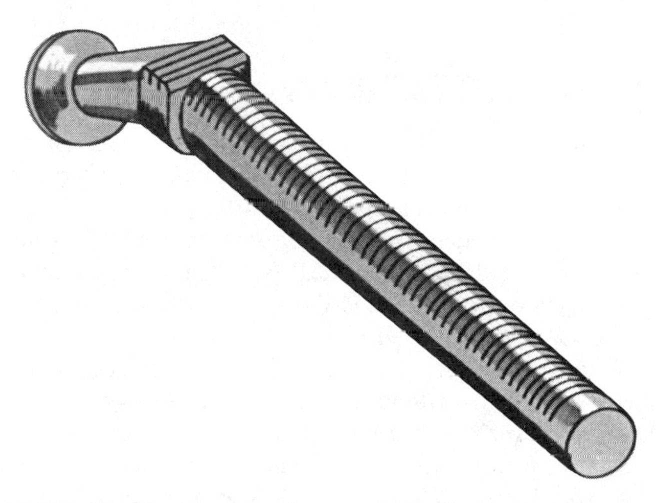

FIGURE 30–24 A cast iron burner with slotted ports. *Reproduced courtesy of Carrier Corporation*

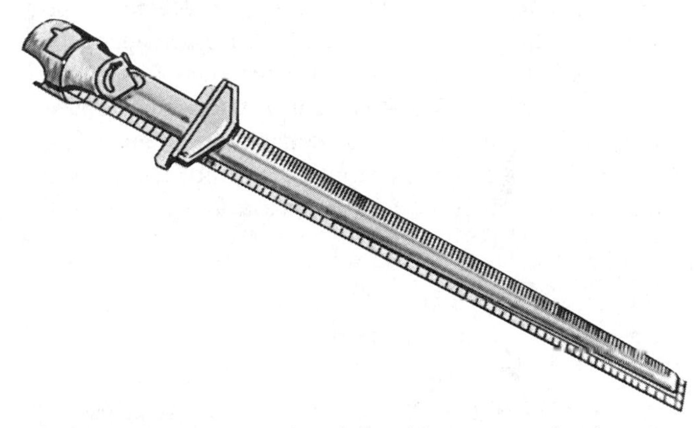

FIGURE 30–25 A stamped steel slotted burner. *Reproduced courtesy of Carrier Corporation*

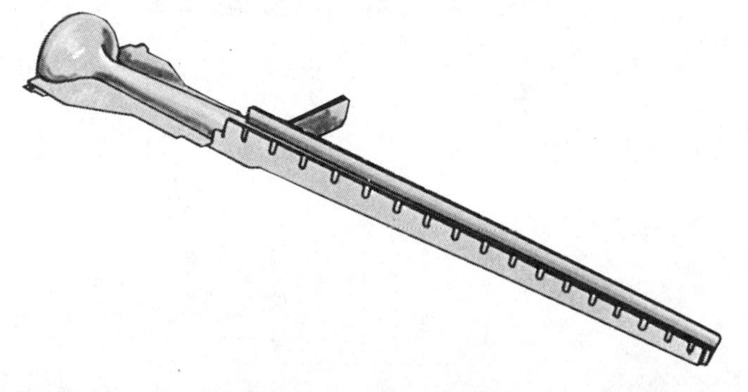

FIGURE 30–26 A ribbon burner. *Reproduced courtesy of Carrier Corporation*

FIGURE 30–27 An inshot burner. *Reproduced courtesy of Carrier Corporation*

30.14 HEAT EXCHANGERS

The burners are located at the bottom of the heat exchanger, Figure 30–28. The *heat exchanger* is divided into sections with a burner in each section.

Heat exchangers are made of sheet steel and designed to provide rapid transfer of heat from the hot combustion products through the steel to the air that will be distributed to the space to be heated.

Heat exchangers must have the correct airflow across them or problems will occur. If too much air flows across a heat exchanger, the flue gas will become too cool and the products of combustion may condense and run down the flue pipe. These products are slightly acid and will deteriorate the flue system. If there is not enough airflow, the combustion chamber will overheat and stress will occur. Furnace manufacturers print the recommended air temperature rise for most furnaces on the furnace nameplate. Normally it is between 40°F and 70°F for a gas furnace. This can be verified by taking the return air temperature and subtracting it from the leaving-air temperature, Figure 30–29. Be sure to use the correct temperature probe locations. Radiant heat can effect the temperature readings.

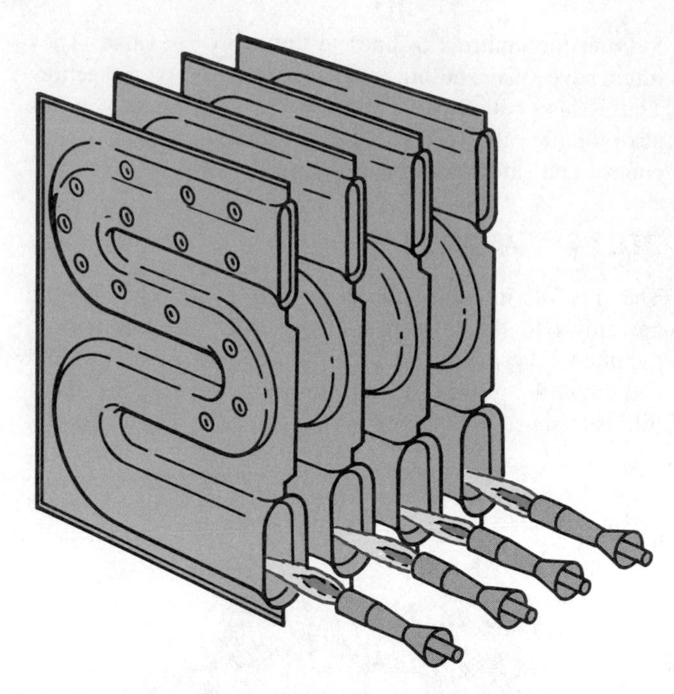

FIGURE 30–28 Modern heat exchanger with four sections.

The following formula may be used to calculate the exact air flow across a gas furnace:

$$\text{cfm} = \frac{Qs}{1.08 \times TD}$$

where Q_s = Sensible heat in Btuh
cfm = Cubic feet of air per minute
1.08 = Constant used to change cfm to pounds of air per minute for the formula
TD = Temperature difference between supply and return air

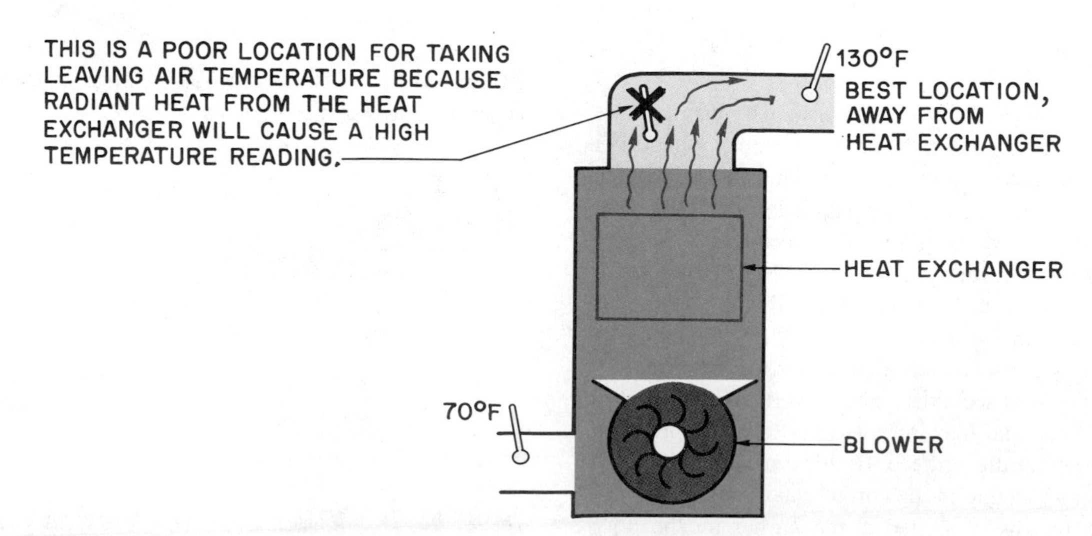
THIS IS A POOR LOCATION FOR TAKING LEAVING AIR TEMPERATURE BECAUSE RADIANT HEAT FROM THE HEAT EXCHANGER WILL CAUSE A HIGH TEMPERATURE READING.
130°F
BEST LOCATION, AWAY FROM HEAT EXCHANGER
HEAT EXCHANGER
70°F
BLOWER

FIGURE 30–29 Checking the air temperature rise across a gas furnace.

This requires some additional calculations because gas furnaces are rated by the input. Because most gas furnaces are 80% efficient during steady-state operation, you may multiply the input times the efficiency to find the output. For example, if a gas furnace has an input rating of 80,000 Btuh and a temperature rise of 55°F, how much air is moving across the heat exchanger in cfm? The first thing we must do is find out what the actual heat output to the air stream is. 80,000 × 0.80 (80% estimated furnace efficiency) = 64,000 Btuh. Use this formula:

$$\text{cfm} = \frac{Q_s}{1.08 \times TD}$$

$$\text{cfm} = \frac{64{,}000}{1.08 \times 55}$$

$$\text{cfm} = 1077.4$$

There is some room for error in the calculation unless you take several temperature readings at both the supply and return duct and average them. The more readings you take, the more accurate your answer. Also, the more accurate your thermometer is, the more accurate the results. Glass stem thermometers graduated in 1/4 degree increments are preferred. There is also the heat added by the fan motor. This will be about 300 W for a furnace this size. If you will multiply 300 W × 3.413 Btu/W, you will find some extra heat added to the air. 300 × 3.413 = 1023.9 Btu. For an even more accurate reading you may take a watt reading of the motor or multiply the amperage reading times the applied voltage for a watt reading approximation.

$$\text{Watts} = \text{Amperage} \times \text{Applied Voltage}$$

Poor combustion can corrode the heat exchanger so good combustion is preferred. The steel in the exchangers may be coated or bonded with aluminum, glass, or ceramic material. These materials are more corrosion resistant, and the unit will last longer. Some exchangers are made of stainless steel, which is more expensive but resists corrosion extremely well. ***The exchanger must not be corroded or pitted enough to leak since one of its functions is to keep combustion gases separated from the air to be heated and circulated throughout the building.***

30.15 FAN SWITCH

The fan switch automatically turns the blower on and off. The blower circulates the heated air to the conditioned space. The switch can be temperature controlled or time delay. In either instance, the heat exchanger is given a chance to heat up before the fan is turned on. This delay keeps cold air from being circulated through the duct system before the heat exchanger gets hot at the beginning of the cycle. The heat exchanger must be hot in a conventional furnace because the heat provides a good draft to properly vent the combustion gases.

There is also a delay in shutting off the fan. This allows the heat exchanger time to cool off and dissipate the furnace heat at the end of the cycle.

The temperature-sensing element of the switch, usually a bimetal helix, is located in the airstream near the heat exchanger, Figure 30–30. When the furnace comes on, the air is heated, which expands the bimetal and closes the contacts, thus activating the blower motor. This is called a *temperature on–temperature off* fan switch.

The fan switch could activate the blower with a time delay and shut it off with a temperature-sensing device. When the thermostat calls for heat and the furnace starts, a small resistance heating device is activated. This heats a bimetal strip that will close electrical contacts to the blower when heated. This provides a time delay and allows the furnace to heat the air before the blower comes on. The bimetal helix in the airstream keeps the contacts closed even after the room thermostat is satisfied and the burner flame goes out. When the furnace shuts down, the bimetal cools at the same time as the heat exchanger and turns off the blower. This fan switch provides a positive starting of the fan with temperature stopping it and is called *time on–temperature off.*

A third type, called a *time on–time off,* switch uses a small heating device, such as the one used in the time on-temperature off switch. The difference is that this switch is not mounted in such a way that the heat exchanger heat will influence it. The time delay is designed into the switch and is not adjustable. Most models of the other two switches are adjustable. Procedures to make these adjustments will be discussed later in this unit.

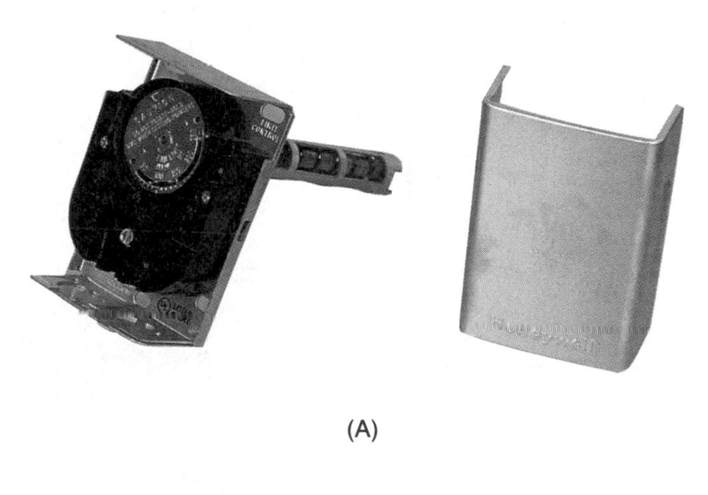

(A)

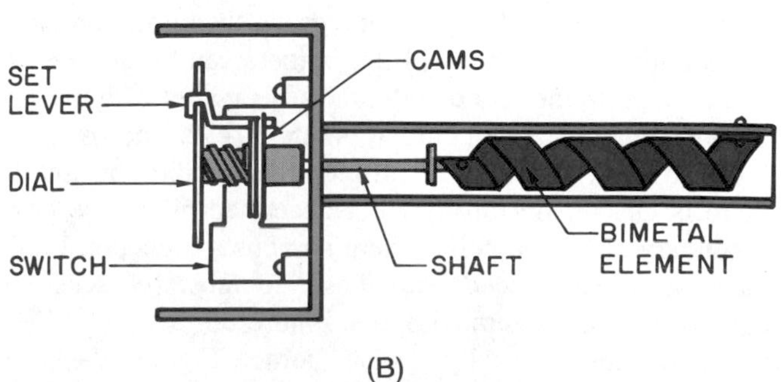

(B)

FIGURE 30–30 Temperature on-temperature off fan switch. *Photo by Bill Johnson*

30.16 LIMIT SWITCH

The *limit switch* is a safety device. If the fan does not come on or if there is another problem causing the heat exchanger to overheat, the limit switch will open its contacts, which closes the gas valve. ***Almost any circumstance causing a restriction in the airflow to the conditioned space can make the furnace overheat, for example, dirty filters, a blocked duct, dampers closed, fan malfunctioning, or a loose or broken fan belt. The furnace may also be overfired due to an improper setting or malfunctioning of the gas valve. It is extremely important that the limit switch operate as it is designed to do.***

The limit switch has a heat-sensing element. When the furnace overheats this element opens contacts closing the main gas valve. This switch can be combined with a fan switch, Figure 30–31. Different state codes have different high-limit cut-out requirements, often around 250°F.

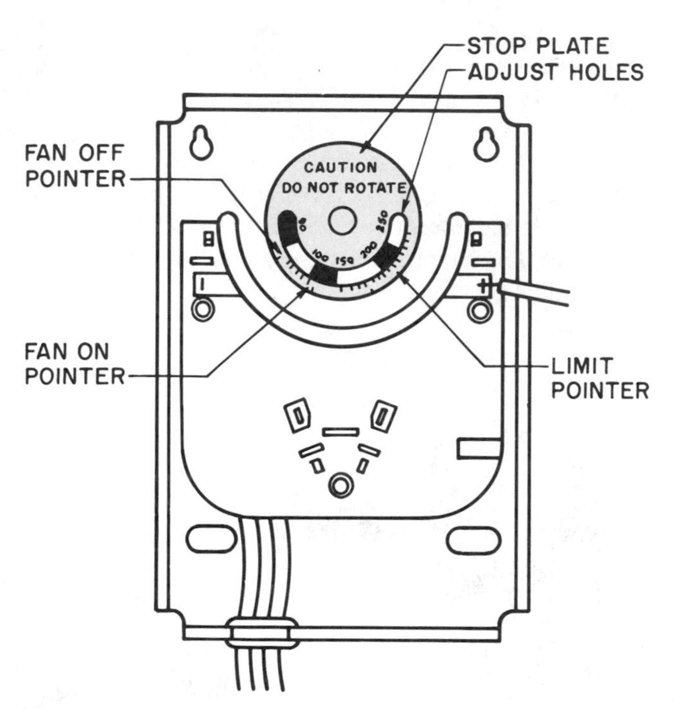

FIGURE 30–31 Fan on-off and limit control.

30.17 PILOTS

Pilot flames are used to ignite the gas at the burner on most conventional gas furnaces. Pilot burners can be *aerated* or *nonaerated*. In the aerated pilot the air is mixed with the gas before it enters the pilot burner, Figure 30–32. The air openings, however, often clog and require periodic cleaning if there is dust or lint in the air. Nonaerated pilots use only secondary air at the point where combustion occurs. Little maintenance is needed with these, so most furnaces are equipped with nonaerated pilots, Figure 30–33.

The pilot is actually a small burner, Figure 30–34. It has an orifice, similar to the main burner, through which the gas passes. If the pilot goes out or does not perform properly, a safety device will stop the gas flow.

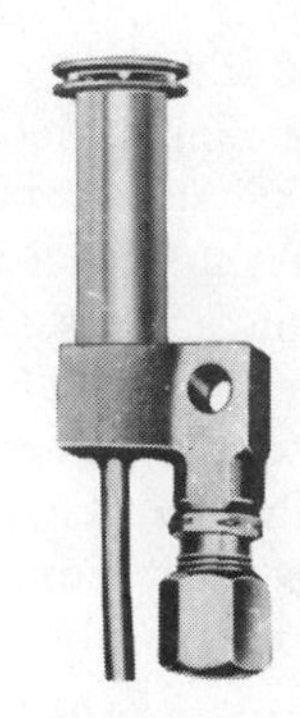

FIGURE 30–32 Aerated pilot. *Courtesy Robertshaw Controls Company*

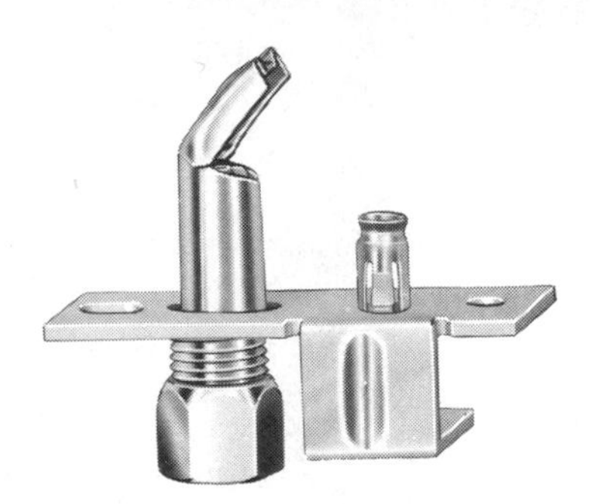

FIGURE 30–33 Nonaerated pilot. *Courtesy Robertshaw Controls Company*

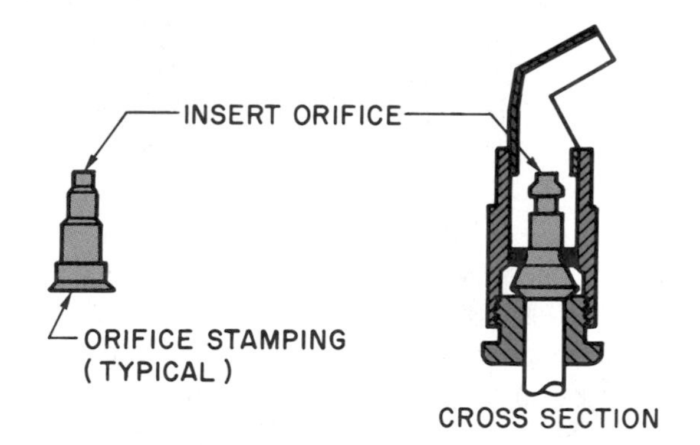

FIGURE 30–34 Nonaerated pilot burner.

Standing pilots burn continuously; other pilots are ignited by an electric spark or other ignition device when the thermostat calls for heat. In furnaces without pilots, the ignition system ignites the gas at the burner. In furnaces with pilots, the pilot must be ignited and burning, and this must be proved before the gas valve to the main burner will open.

The pilot burner must direct the flame so there will be ignition at the main burners, Figure 30–35. The pilot flame also provides heat for the safety device that shuts off the gas flow if the pilot flame goes out.

30.18 SAFETY DEVICES AT THE STANDING PILOT

Three main types of safety devices, called *flame-proving* devices, keep the gas from flowing through the main valve if the pilot flame goes out: the thermocouple or thermopile, the bimetallic strip, and the liquid-filled remote bulb.

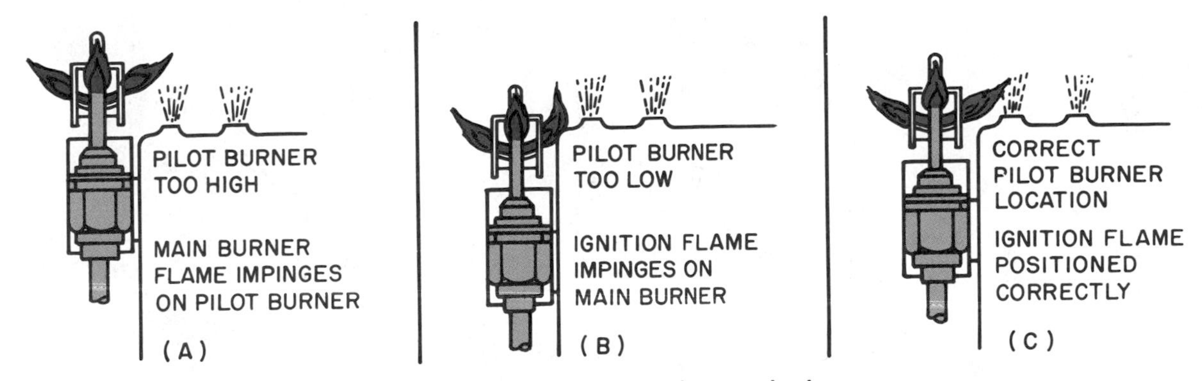

FIGURE 30–35 The pilot flame must be directed at the burners and adjusted to the proper height.

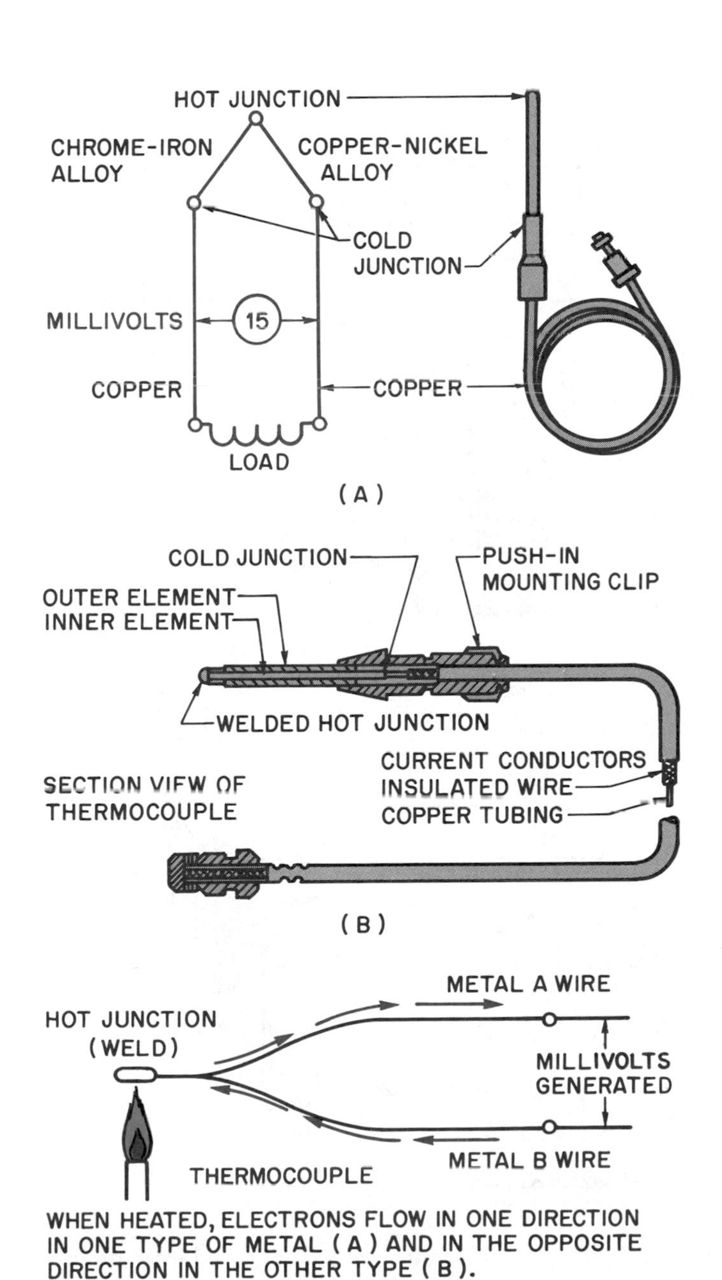

FIGURE 30–36 Thermocouple. *Courtesy Robertshaw Controls Company*

30.19 THERMOCOUPLES AND THERMOPILES

The *thermocouple* consists of two dissimilar metals welded together at one end, Figure 30–36, called the "hot junction." When this junction is heated, it generates a small voltage (approximately 15 mV with load; 30 mV without load) across the two wires or metals at the other end. The other end is called the "cold junction." The thermocouple is connected to a shutoff valve, Figure 30–37. As long as the

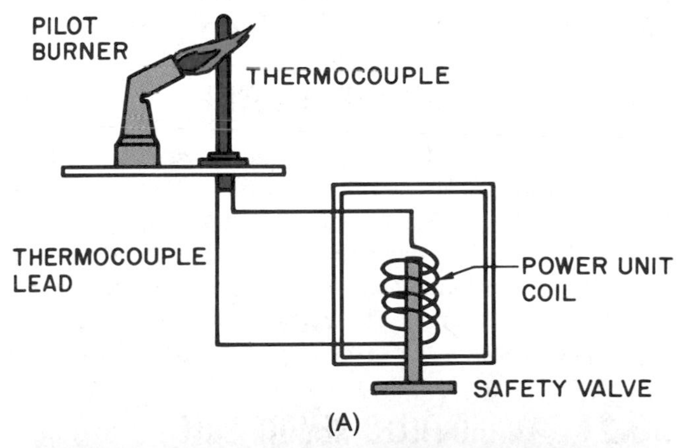

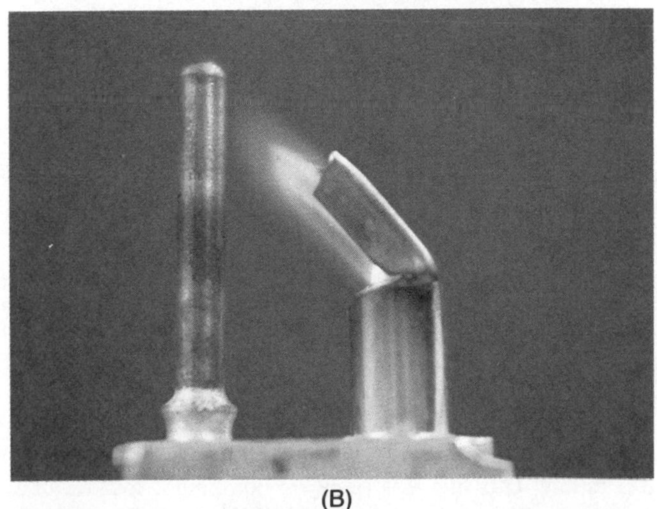

FIGURE 30–37 The thermocouple generates electrical current when heated by the pilot flame. This induces a magnetic field in the coil of the safety valve holding it open. If flame is not present, the coil will deactivate, closing the valve, and the gas will not flow. *Photo by Bill Johnson*

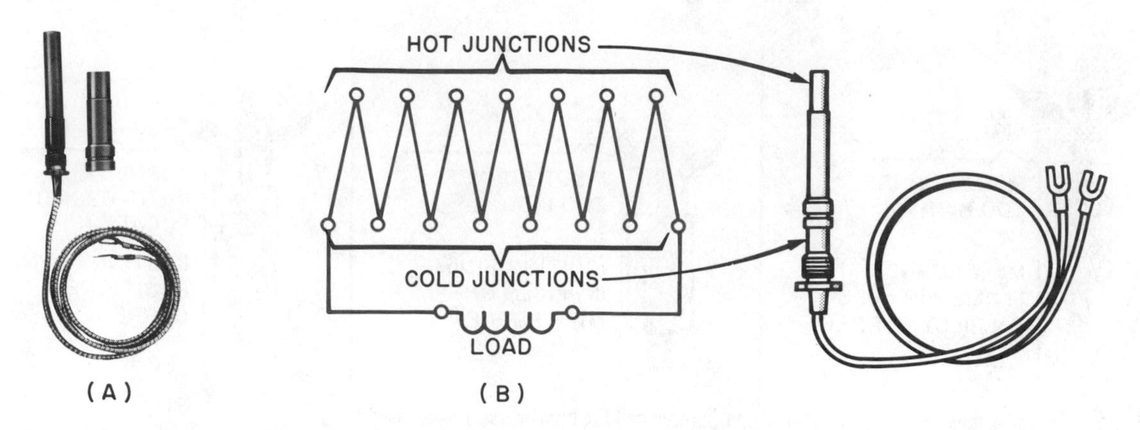

FIGURE 30–38 A thermopile consists of a series of thermocouples in one housing. *Photo courtesy Honeywell, Inc., Residential Division*

electrical current in the thermocouple energizes a coil, the gas can flow. If the flame goes out, the thermocouple will cool off in about 30 sec, and no current will flow and the gas valve will close. A *thermopile* consists of several thermocouples wired in series to increase the voltage. If a thermopile is used, it performs the same function as the thermocouple, Figure 30–38.

30.20 BIMETALLIC SAFETY DEVICE

In the *bimetallic* safety device the pilot heats the bimetal strip, which closes electrical contacts wired to the gas safety valve, Figure 30–39(A). As long as the pilot flame heats the bimetal, the gas safety valve remains energized and gas will flow when called for. When the pilot goes out, the bimetal strip cools within about 30 sec and straightens, opening contacts and causing the valve to close, Figure 30–39(B).

30.21 LIQUID-FILLED REMOTE BULB

The *liquid-filled remote bulb* includes a diaphragm, a tube, and a bulb, all filled with a liquid, usually mercury. The remote bulb is positioned to be heated by the pilot flame, Figure 30–40. The pilot flame heats the liquid at the remote bulb. The liquid expands, causing the diaphragm to expand, which closes contacts wired to the gas safety valve. As long as the pilot flame is on, the liquid is heated and the valve is open, allowing gas to flow. If the pilot flame goes out, the liquid cools in about 30 sec and contracts, opening the electrical contacts and closing the gas safety valve.

30.22 GLOW-COIL IGNITION CIRCUIT

The glow coil is used to automatically reignite a pilot light if it goes out. This happens often on equipment located in a drafty area. The glow coil is normally applied to equipment with a standing pilot that is *not* 100% shut off. A thermal type of main gas safety valve is used to allow the glow coil enough time to heat up and light the pilot. The pilot has gas going to it at all times; if the pilot goes out, the circuit to the glow coil will be energized at the same time that the

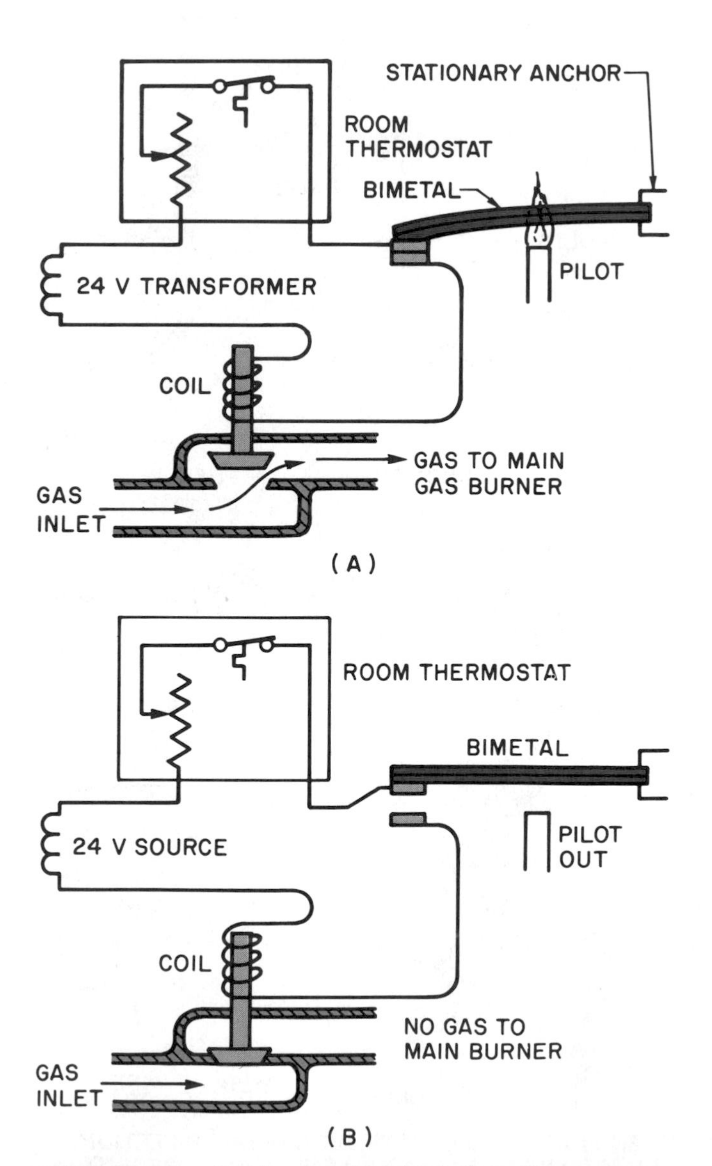

FIGURE 30–39 (A) When the pilot is lit, the bimetal warps, causing the contacts to close. The coil is energized, pulling the plunger into the coil opening the valve. This allows the gas to flow to the main gas valve. (B) When the pilot is out, the bimetal straightens, opening the contacts. The safety valve closes. No gas flows to the furnace burners.

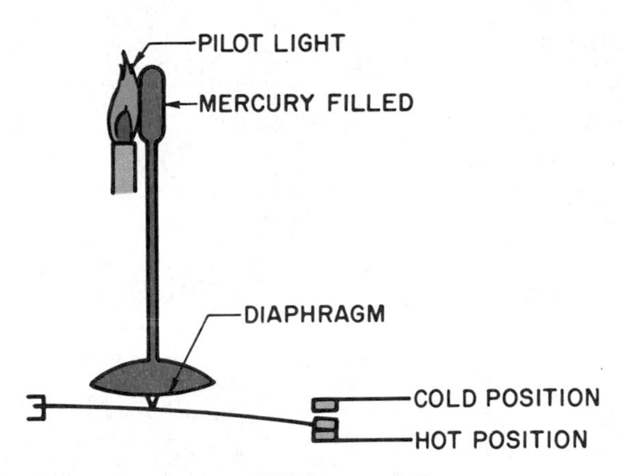

FIGURE 30–40 Liquid-filled remote bulb.

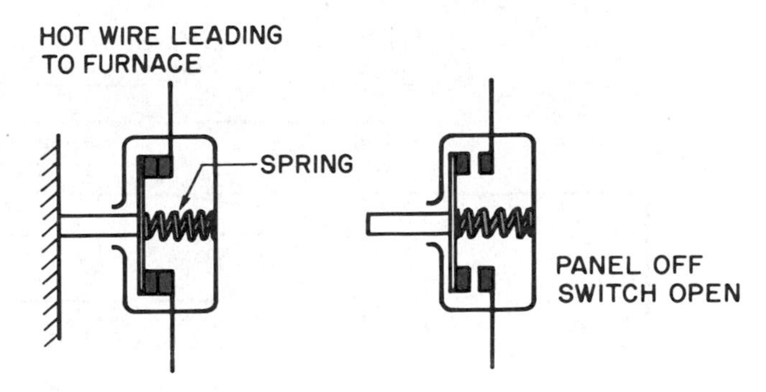

FIGURE 30–42 Panel switch used to interrupt power to the furnace circuit for safety purposes.

circuit to the gas valve is deenergized, Figure 30–41. If the glow coil fails to ignite the gas, the small amount of gas from the pilot will go up the flue. Notice the door switch in the circuit. This switch will not allow power to pass when the front panel is removed. It will have to be held shut to check voltage inside the furnace. It will appear like a light switch in a refrigerator door, Figure 30–42.

The power supply in this glow-coil circuit is from the low-voltage transformer; however, it is not 24 V but 12 V. It may be taken from a third tap or wire from the middle of the control transformer's coil. This is not the only method used to obtain the low voltage for a glow coil. Sometimes, the full transformer 24 V may go to the glow-coil component. When this is the case, the glow coil will more than likely still be a 12-V glow coil and have another method for reducing the voltage (such as using a resistor), Figure 30–43.

The glow coil as a component may also have a high-temperature device to keep the coil from glowing in the event that the pilot does not light. The glow coil itself is normally made of platinum and would burn up if left on

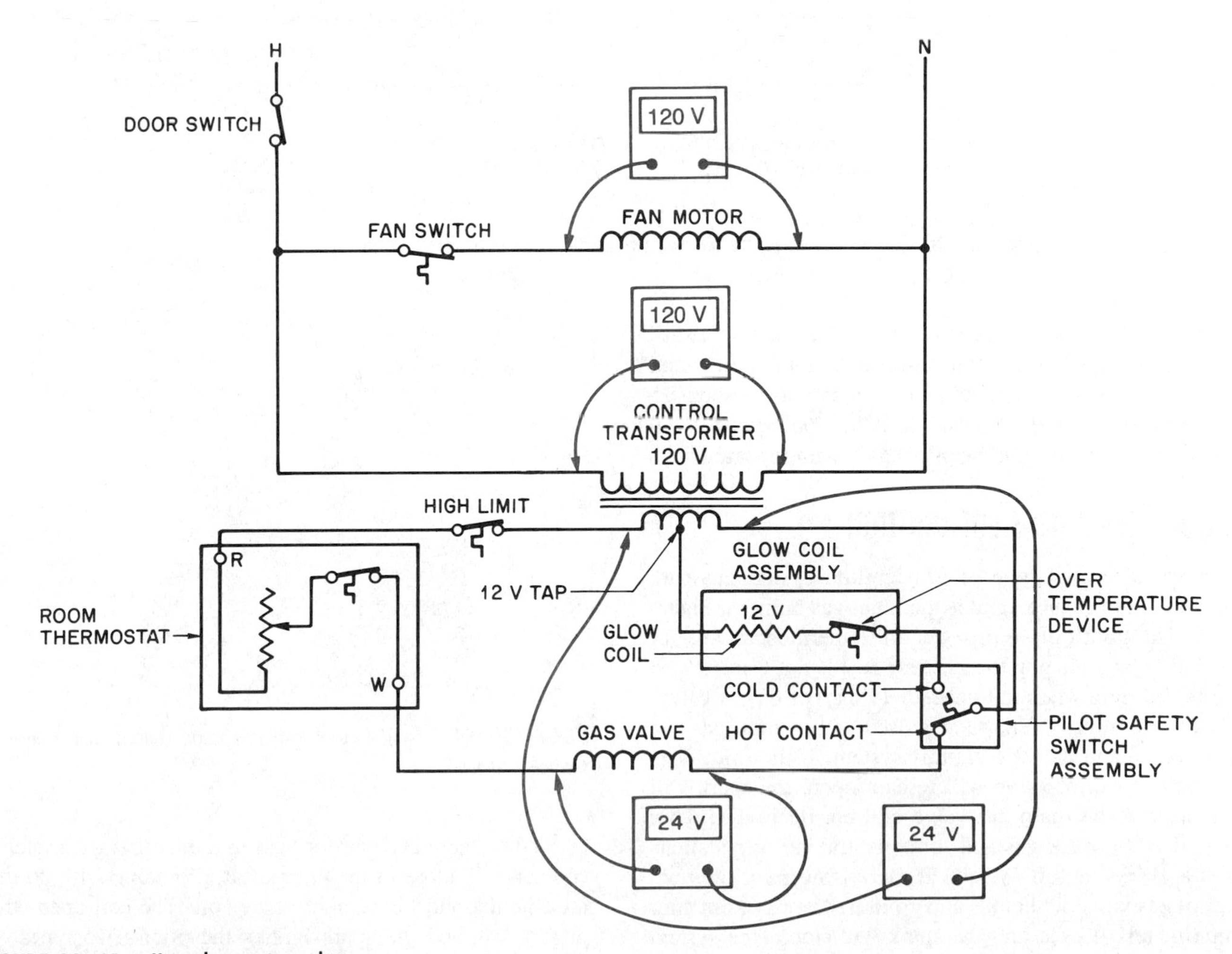

FIGURE 30–41 Normal operating voltages.

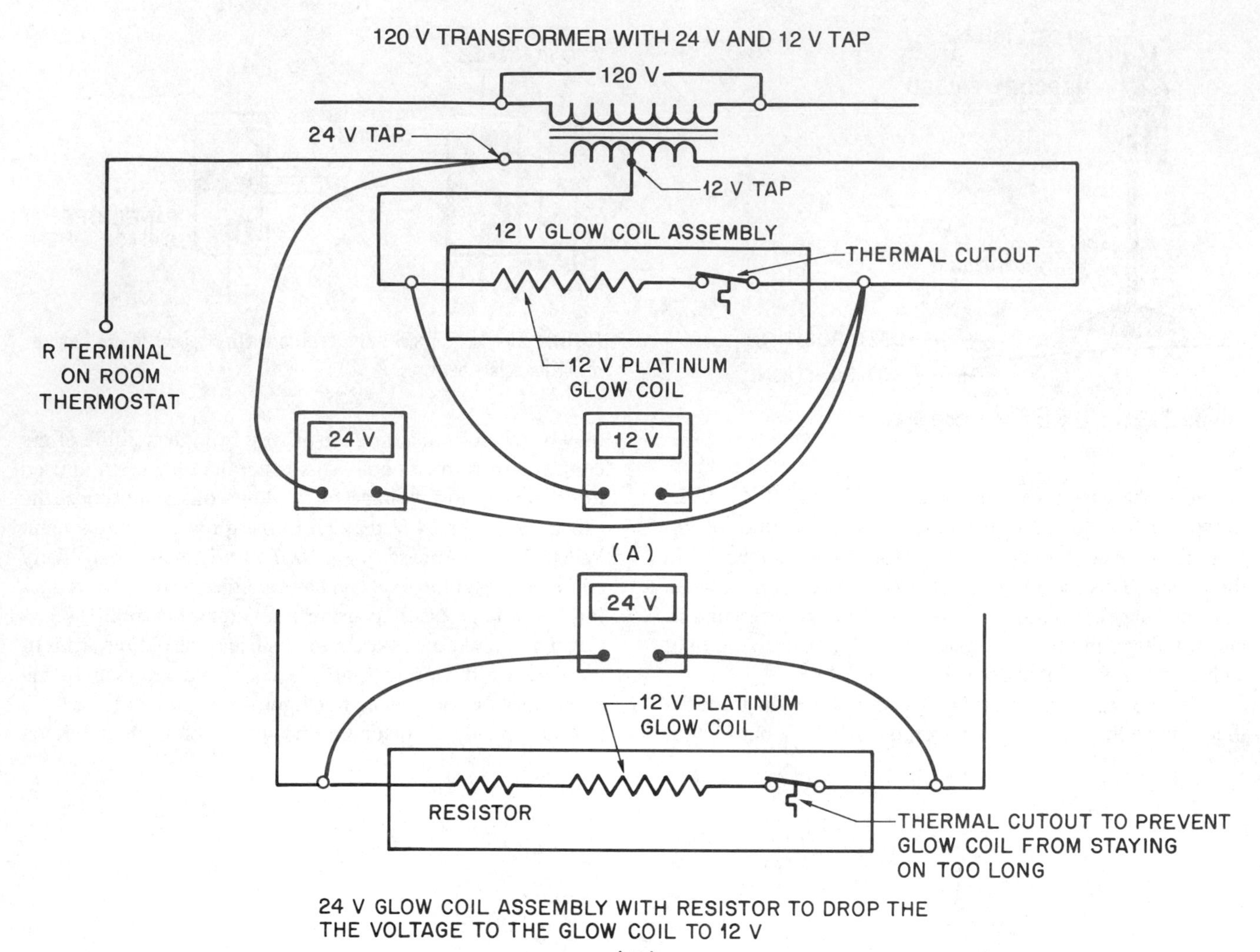

FIGURE 30–43 **(A) Voltage reduced using third wire on transformer. (B) Resistor used in glow-coil circuit.**

indefinitely. Stopping the glow coil is often accomplished with a high-temperature cutout furnished with the glow-coil assembly. The cutout shuts off the glow coil after about 15 or 20 sec if the pilot does not light. After cooling off it will reenergize and try to light the pilot again with repeated tries.

30.23 SPARK-TO-PILOT IGNITION

In the spark-to-pilot-type of gas ignition system, a spark ignites the pilot, which ignites the main gas burners, Figure 30–44. The pilot only burns when the thermostat calls for heat. This system is popular because fuel is not wasted with the pilot burning when not needed. Two types of gas valves are used with this system. One is used with natural gas and is not considered a 100% shutoff system. If the pilot does not ignite, the pilot valve will remain open, the spark will continue, and the main gas valve will not open. The other type is used with LP gas and some natural gas applications and is a 100% shutoff system. If there is no pilot ignition, the pilot gas valve will close and go into safety lockout after approximately 45 sec and the spark will stop. This system must be manually reset, usually at the thermostat.

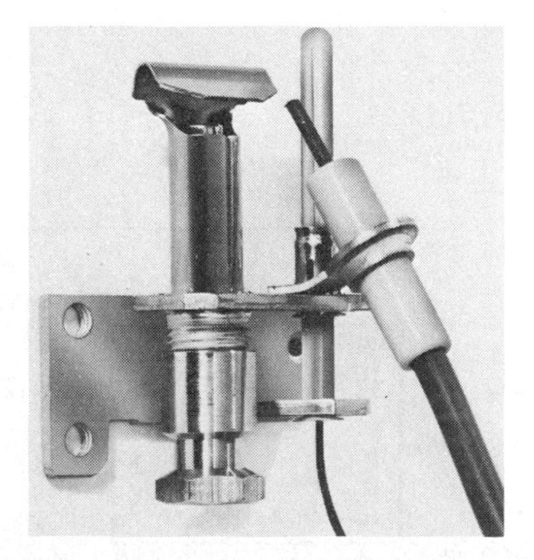

FIGURE 30–44 **Spark-to-pilot ignition system.** ***Courtesy Robertshaw Controls Company***

When there is a call for heat in the natural gas system, contacts will close in the thermostat, providing 24 V to the pilot igniter and to the pilot valve coil. The coil opens the pilot valve, and the spark ignites the pilot. Two types of systems are used to open the main gas valve: a mercury vapor tube or a flame rectification system. In the *mercury vapor tube* system the remote bulb is located at the pilot. When this bulb is heated by the pilot flame, it expands a bellows or diaphragm that makes a switch and opens the main gas valve. The pilot flame then ignites the main burners.

In the *flame rectification* system the heat from the pilot flame changes the normal alternating current to direct current. The electronic components in the system will only energize and open the main gas valve with a direct current. It is important that the pilot flame quality be correct to ensure proper operation. Consequently, the main gas valve will open only when pilot flame is present.

The spark is intermittent and arcs approximately 100 times per minute. It must be a high-quality arc or the pilot will not ignite. The main gas valve should open 50 to 70 sec after the thermostat contacts close. This delay may vary depending on the manufacturer of the furnace controls.

30.24 DIRECT-SPARK IGNITION (DSI)

Many modern furnaces are designed with a spark ignition direct to the main burner. No pilot is used in this system. Components in the system are the igniter/sensor assembly and the DSI module, Figure 30–45. The sensor rod verifies that the furnace has fired and sends a microamp signal to the DSI module confirming this. The furnace will then continue to operate. This system goes into a "safety lockout" if the flame is not established within the "trial for ignition" period (approximately 6 sec). Gas is being furnished to the main burner so there cannot be much time delay. The system can then only be reset by turning the power off to the system control and waiting 1 min before reapplying the power. This is a "typical" system. The technician should follow the wiring diagram and manufacturer's instructions for the specific furnace being installed or serviced.

Most ignition problems are caused from improperly adjusted spark gap, igniter positioning, and bad grounding, Figure 30–46A. The igniter is centered over the left port. Most manufacturers also provide specific troubleshooting instructions. Intermittent sparking after the furnace is lit is normal in most systems. Continued sparking may be due to excessive resistance in the igniter wire or to a defective DSI module. The continual sparking may not be harmful to the system, but it is noisy and often can be heard in the living or working area. Read and follow the manufacture's instructions to repair the system.

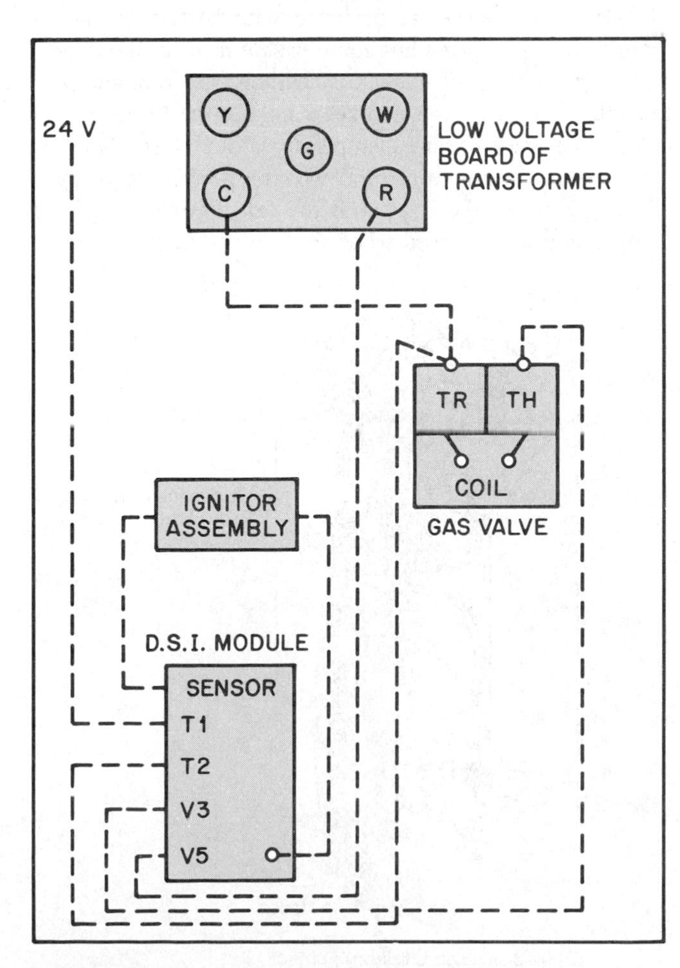

FIGURE 30–45 Diagram of components in a DSI system. *Courtesy Heil-Quaker Corporation*

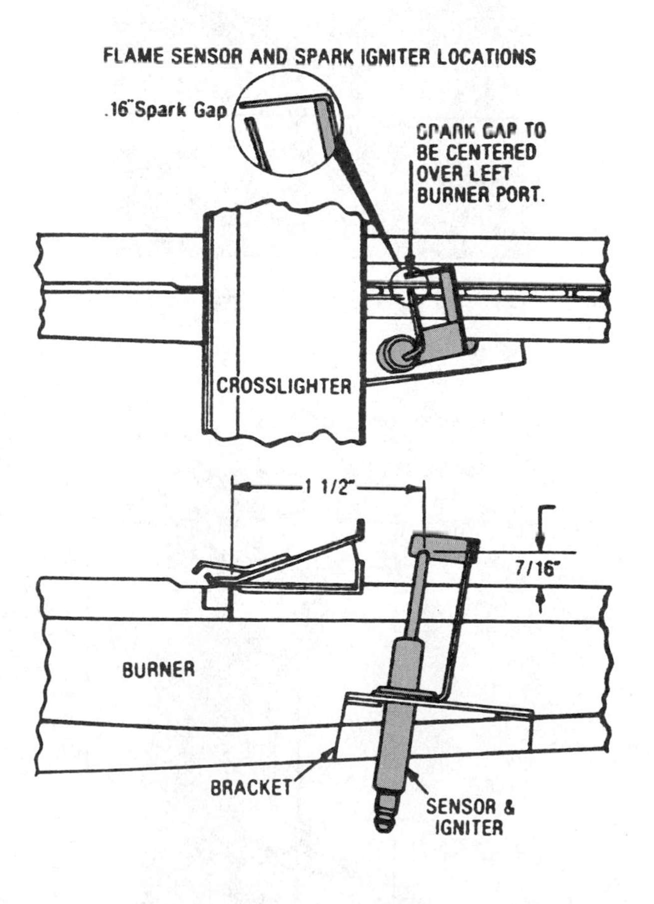

FIGURE 30–46A Spark gap and igniter position for a DSI system. *Courtesy Heil-Quaker Corporation*

30.25 HOT SURFACE IGNITION

The hot surface ignition system uses a special product called silicon carbide that offers a high resistance to electrical current flow, but is very tough and will not burn up, like a glow coil. This substance is placed in the gas stream and is allowed to get very hot before the gas is allowed to impinge on the glowing hot surface. Immediate ignition should occur when the gas valve opens.

The hot surface igniter is usually operated off 120 V, the line voltage to the furnace, and draws considerable current when energized. It is only energized for a short period of time during start-up of the furnace so this current is not present for more than a very few minutes per day. Figure 30–46B shows the hot surface igniter while hot.

FIGURE 30–46B Hot surface igniter. *Reproduced courtesy of Carrier Corporation*

The hot surface igniter may be used for lighting the pilot or for direct ignition of the main furnace burner. When it is used as the direct igniter for the burner, it will have a very few seconds to ignite the burner, then a safety lockout will occur to prevent too much gas from escaping.

30.26 HIGH-EFFICIENCY GAS FURNACES

Gas furnaces with high-efficiency ratings have been developed and are being installed in many homes and businesses. The U.S. Federal Trade Commission requires manufacturers to provide an annual fuel utilization efficiency rating (AFUE). This rating allows the consumer to compare furnace performances before buying. Annual efficiency ratings have been increased in some instances from 65% to 97+%. A part of this increase in efficiency is accomplished by keeping excessive heat from being vented to the atmosphere. Stack temperatures in conventional furnaces are kept high to provide a good draft for proper venting. They also prevent condensation and therefore corrosion in the vent.

High-efficiency furnaces are now designed with a forced-vent draft so that flue temperatures can be reduced. Figure 30–47 shows a furnace with an efficiency of about 85%. These furnaces are known as 80+ furnaces because their efficiencies are 80% or higher.

Heat exchangers are designed with baffles to absorb more of the heat, providing more usable heat for the conditioned space. Some installations have bimetal vent dampers to control the flue gases and keep them in the furnace until heated to a predetermined temperature. Other vent dampers are opened during running time and closed when the furnace stops holding in the heat until the fan dissipates it. This damper is always closed when the furnace is off. In older

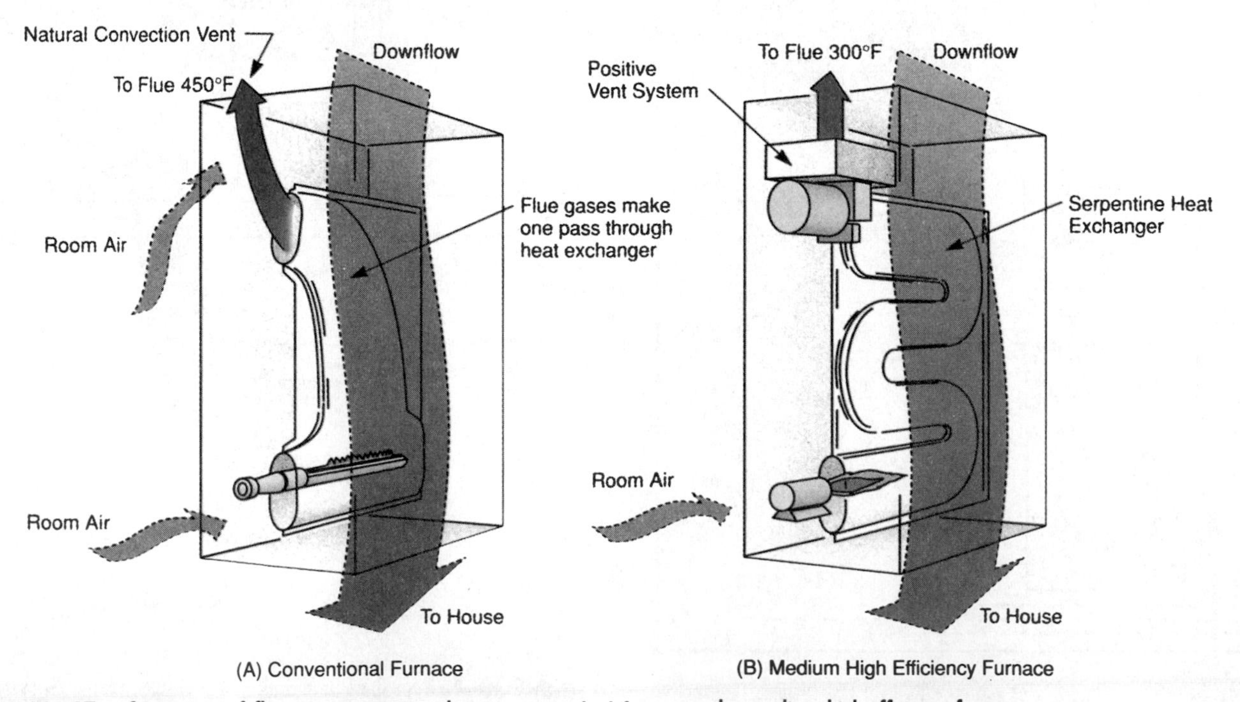

FIGURE 30–47 Comparison of flue-gas temperatures between a standard furnace and a medium high-efficiency furnace.

systems some heated room air is vented up the flue 24 h a day whether or not the furnace is operating.

The system shown in Figure 30–47 reduces the flue temperature by as much as 150°F. Efficiencies of 97+% have been attained with some modern furnaces. The *pulse* and the *condensing furnaces* are examples of high-efficiency furnaces.

Pulse Furnace

The pulse furnace, Figure 30–48A, ignites minute quantities of gas 60 to 70 times per second in a closed combustion chamber. The process begins when small amounts of natural gas and air enter the combustion chamber. This mixture is ignited with a spark igniter. This ignition forces the combus-

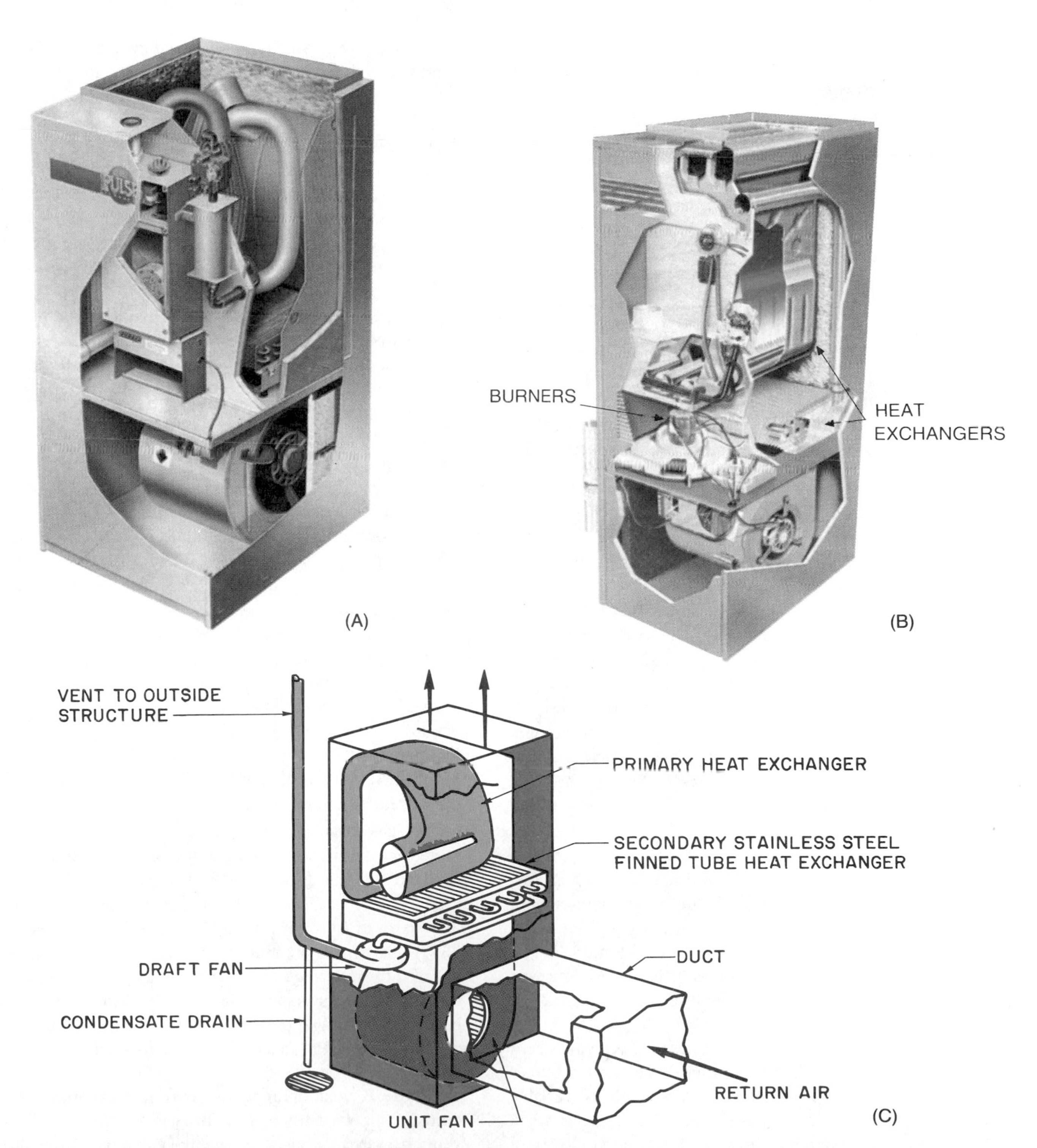

FIGURE 30–48 (A) High-efficiency pulse furnace. (B) High-efficiency condensing furnace. (C) Condensing gas furnace. Notice that the cool return air passes over the flue gas exchanger just before it vents to the outside. This condenses the moisture from the flue gas. This condensing process provides about 1000 Btu for each pint (or pound) of water condensed. *Courtesy (A) Lennox Industries, Inc., (B) Heil-Quaker Corporation*

tion materials down a tailpipe to an exhaust decoupler. The pulse is reflected back to the combustion chamber, igniting another gas and air mixture, and the process is repeated. Once this process is started, the spark igniter can be turned off because the pulsing will continue on its own 60 to 70 times per second. This furnace also uses heat exchangers to absorb much of the heat produced. Air is circulated over the exchanger and then to the space to be heated. This furnace also uses a condensing component described in the next paragraph.

Condensing Furnace

The condensing furnace, Figure 30–48B, pipes the flue gas through a heat exchanger in the return air and cools the flue gas to the point that the moisture actually condenses out of it. This changing from a gas to a liquid is a change of state and produces heat. It produces approximately 1000 Btuh for each pint of liquid condensed. This is latent heat and adds significantly to the efficiency of the furnace. This moisture is very corrosive and at a low temperature. The flue gas can be piped away from the furnace with polyvinyl chloride pipe, and the moisture can be piped down the drain. These condensing furnaces may reach efficiencies of 97%.

The manufacturer's literature or other sources must be consulted for installation and troubleshooting procedures because the furnaces are highly specialized.

30.27 VARIABLE SPEED GAS FURNACES

The industry has always needed a furnace that follows the heat loss of the structure. For example, a house that requires a 100,000-Btuh furnace on the coldest day of winter, which may be 0°F may only require 60,000 Btuh of heat when it is 20°F. The furnace begins to cycle off and back on as the weather warms. The warmer the weather, the more cycle times per day the furnace has. These cycle times cause the furnace to be less efficient. It is estimated that a furnace that is oversized may have a seasonal efficiency as low as 50%, while having a steady-state efficiency of 80%. When this cycling can be reduced or eliminated, the efficiency of the furnace becomes greater.

For many years, two-stage gas furnaces have been used in northern climates where more cycling occurs. These furnaces have a low-fire function and a high-fire function with one fan speed. This is accomplished with a gas valve and a furnace that is designed for this type of operation. It is a satisfactory operation but still not as efficient as a furnace that could track the heat loss of the house and vary the fan speeds on demand.

The variable speed gas furnace would typically be a condensing furnace with a main air blower and a burner combustion blower. These blower motors would be variable speed and would respond to the space temperature; when the space temperature begins to drop, the fans would speed up. When the space temperature rises, instead of the furnace shutting off, the fans would slow down. The fan operation along with high and low fire give these furnaces a very good efficiency. The gas valve may be two staged, with high and low fire. These furnaces have electronic circuit boards to control the furnaces.

30.28 VENTING

Conventional gas furnaces use a hot flue gas and natural convection to vent the products of combustion (flue gases). The hot gas is vented quickly, primarily to prevent cooling of the flue gas, which produces condensation and other corrosive actions. However, these furnaces lose some efficiency because considerable heat is lost up the flue.

High-efficiency furnaces (90% and higher) recirculate the flue gases through a special extra heat exchanger to keep more of the heat available for space heating in the building. The gases are then pushed out the flue by a small fan. This causes condensation and thus more corrosion in the flue. Plastic pipe is used because it is not damaged by the corrosive materials. These furnaces may have efficiencies of up to 97%.

Regardless of the type, a gas furnace must be properly vented. The venting must provide a safe and effective means of moving the flue gases to the outside air. In a conventional furnace, it is important for the flue gases to be vented as quickly as possible. Conventional furnaces are equipped with a draft hood that blends some room air with the rising flue gases, Figure 30–49(A). The products of combustion enter the draft hood and are mixed with air from the area around the furnace. Approximately 100% additional air (called *dilution* air) enters the draft hood at a lower temperature than the flue gases. The heated gases rise rapidly and create a draft, bringing the dilution air in to mix and move up the vent.

If wind conditions produce a downdraft, the opening in the draft hood provides a place for the gases and air to go, diverting it away from the pilot and main burner flame. This *draft diverter* helps reduce the chance of the pilot flame being blown out and the main burner flame being altered, Figure 30–49(B).

Make sure replacement air is available in the furnace area. Remember it takes 10 ft^3 of air for each 1 ft^3 of natural gas to support combustion. To this is added 5 ft^3 more (excess air) to ensure enough oxygen. Another 15 ft^3 is added at the draft hood, making a total of 30 ft^3 of air for each 1 ft^3 of gas. A 100,000 Btuh furnace would require 2857 ft^3/h of fresh air ($100{,}000 \div 1050$ Btu $ft^3 \times 30\ ft^3 = 2857$). All of this air must be replaced in the furnace area and also must be vented as flue gases and dilution air rise up the vent. If the air is not replaced in the furnace area, it will become a negative pressure area, and air will be pulled down the flue. Products of combustion will then fill the area.

Type B vent or approved masonry materials are required for conventional gas furnace installations. *Type B* venting systems consist of metal vent pipe of the proper thickness, which is approved by a recognized testing laboratory, Figure 30–50. The venting system must be continuous from the furnace to the proper height above the roof. Type B

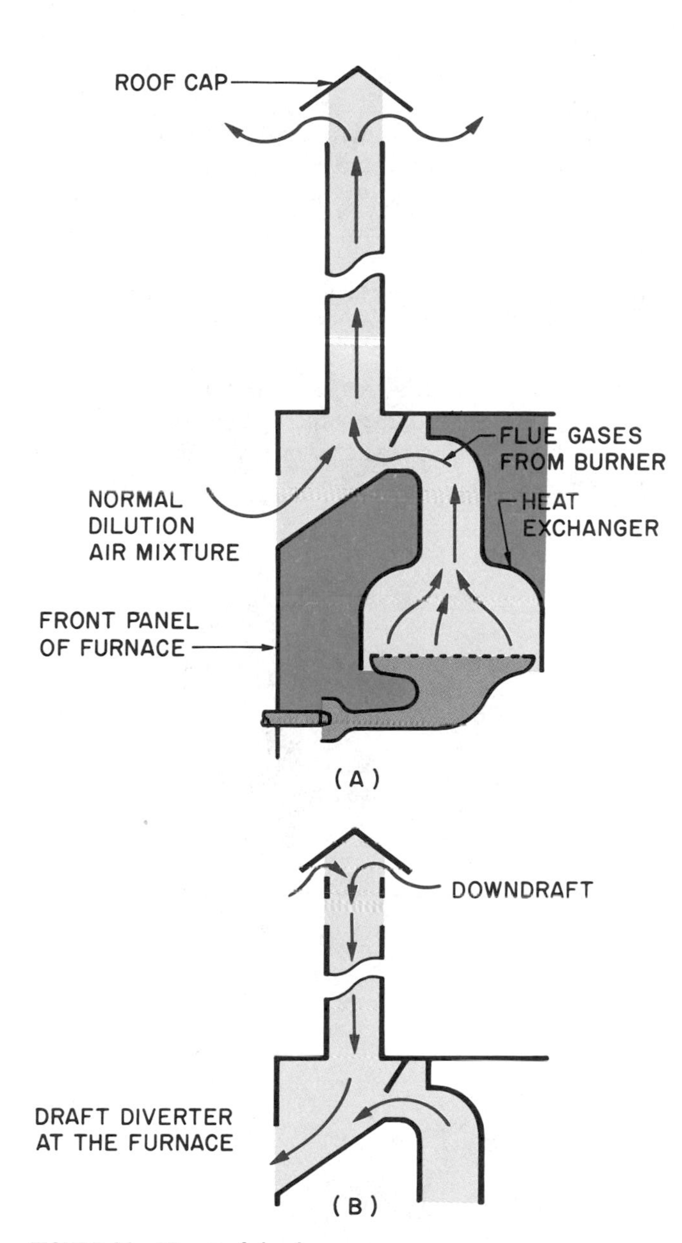

FIGURE 30–49 Draft hood.

FIGURE 30–50 Section of type B gas vent.

vent pipe is usually of a double-wall construction with air space between. The inner wall may be made of aluminum and the outer wall constructed of steel or aluminum.

The vent pipe should be at least the same diameter as the vent opening at the furnace. The horizontal run should be as short as possible. Horizontal runs should always be sloped upward as the vent pipe leaves the furnace. A slope of 1/4 in. per foot of run is the minimum recommended, Figure 30–51. Use as few elbows as possible. Long runs lower the temperature of the flue gases before they reach the

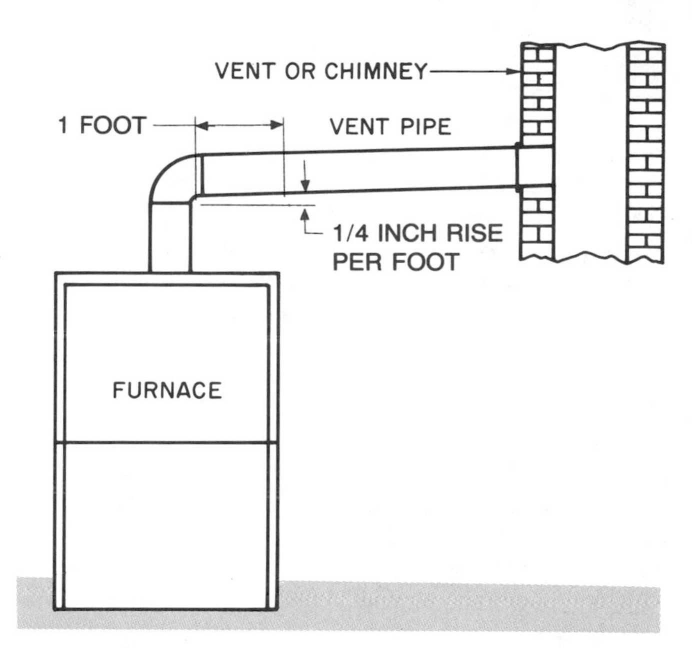

FIGURE 30–51 The minimum rise should be 1/4 in. per foot of horizontal run.

vertical vent or chimney and they reduce the draft. Do not insert the vent connector beyond the inside of the wall of the chimney when the gases are vented into a masonry-type chimney, Figure 30–52.

If two or more gas appliances are vented into a common flue, the common flue size should be calculated based on recommendations from the National Fire Protection Association (NFPA) or local codes.

The top of each gas furnace vent should have an approved vent cap to prevent the entrance of rain and debris and to help prevent wind from blowing the gases back into the building through the draft hood, Figure 30–53.

Metal vents reach operating temperatures quicker than masonry vents do. Masonry chimneys tend to cool the flue

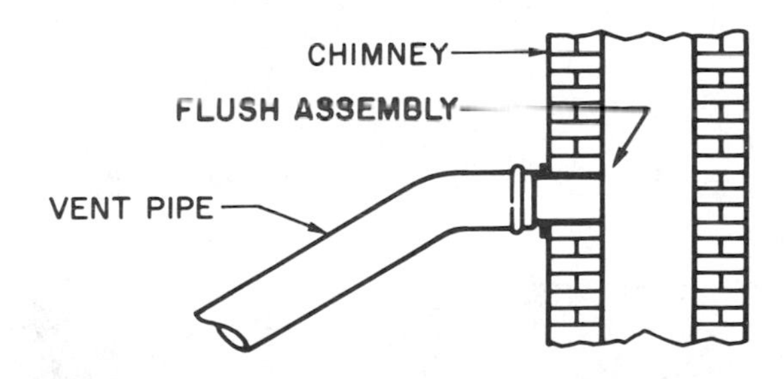

FIGURE 30–52 The vent pipe should not extend beyond the inside of the flue lining.

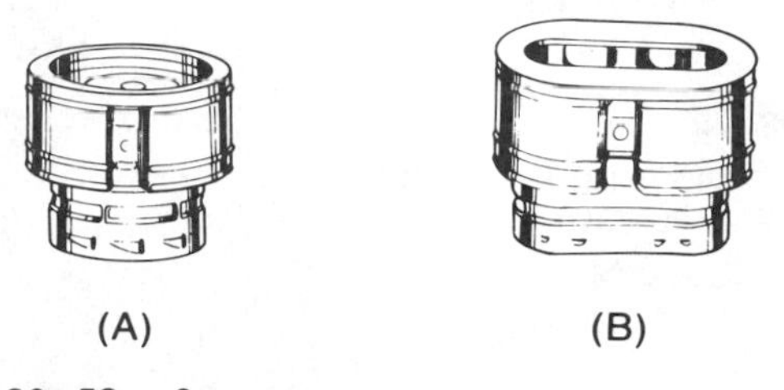

FIGURE 30–53 Gas vent caps.

gases. They must be lined with a glazed-type tile or vent pipe installed in the chimney. If an unlined chimney is used, the corrosive materials from the flue gas will destroy the mortar joints. Condensation occurs at start-up and short running times in mild weather. It takes a long running cycle to heat a heavy chimney. The warmer the gases, the faster they rise and the less damage there is from condensation and other corrosive actions. Vertical vents can be masonry with lining and glazing or prefabricated metal chimneys approved by a recognized testing laboratory.

When the furnace is off, heated air can leave the structure through the draft hood. Automatic vent dampers that close when the furnace is off and open when the furnace is started can be placed in the vent to prevent air loss.

Several types of dampers are available. One type uses a helical bimetal with a linkage to the damper. When the bimetal expands, it turns the shaft fastened to the helical strip and opens the damper, Figure 30–54. Another type uses damper blades constructed of bimetal. When the system is off, the damper is closed. When the vent is heated by the flue gases, the bimetal action on the damper blades opens them, Figure 30–55.

In high-efficiency systems a small blower is installed in the vent system near the heat exchanger. This blower operates only when the furnace is on so that heated room air is not being vented 24 h a day. This blower will vent gases at a lower temperature than is possible with natural draft, allowing the heat exchanger to absorb more of the heat from the burner for distribution to the conditioned space, Figure 30–56. When blowers are used for furnace venting, the room thermostat normally energizes the blower. Then when blower air is established, the burner ignites. Blower air can be established by airflow switches or a centrifugal switch on the motor.

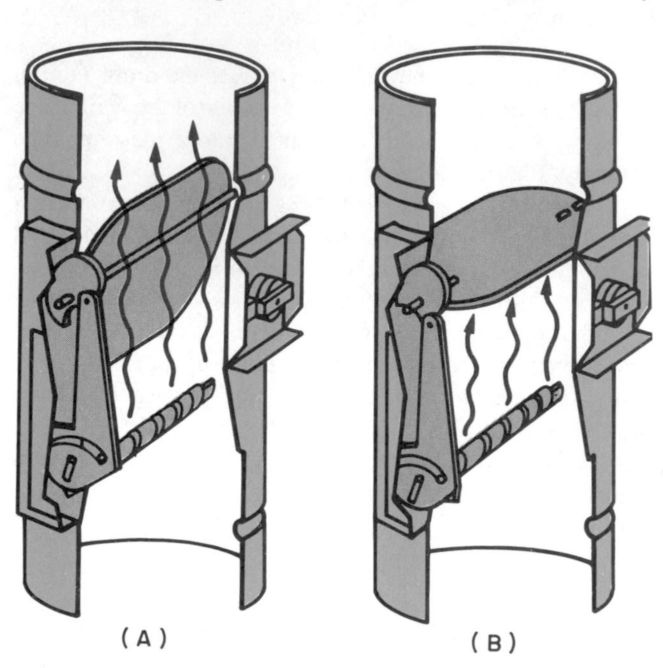

FIGURE 30–54 Vent damper operated with a bimetal helix.

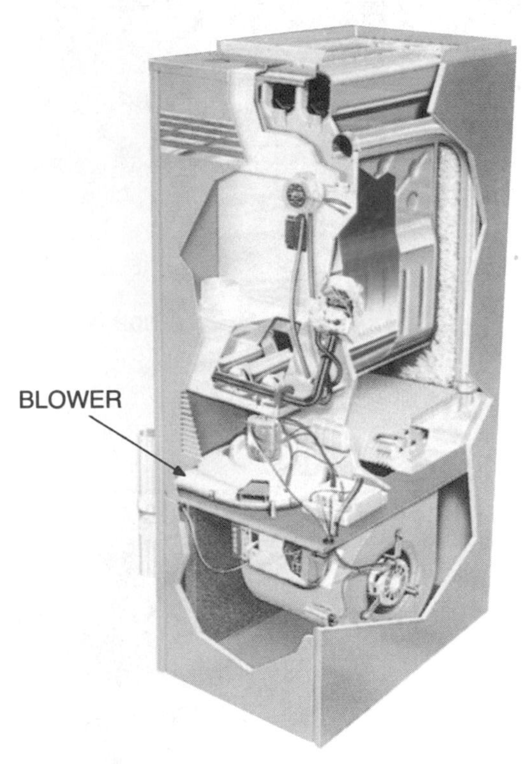

FIGURE 30–56 Forced venting with a vent blower. *Courtesy Heil-Quaker Corporation*

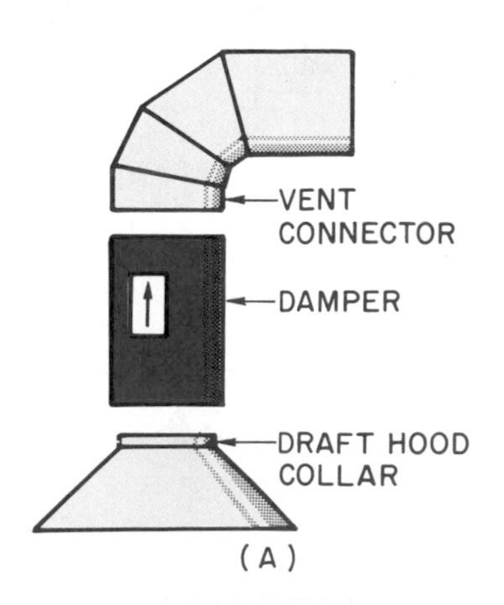

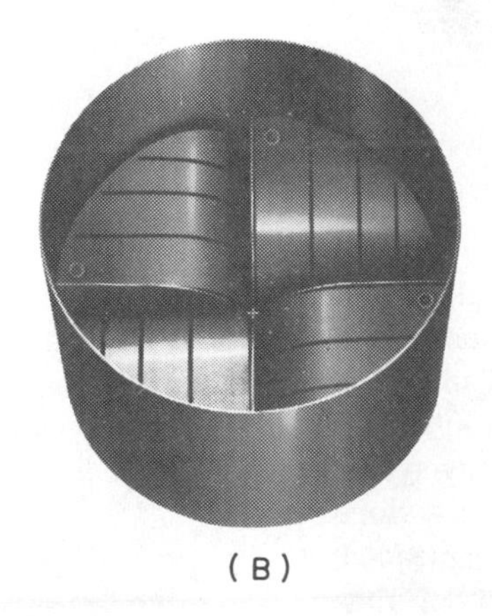

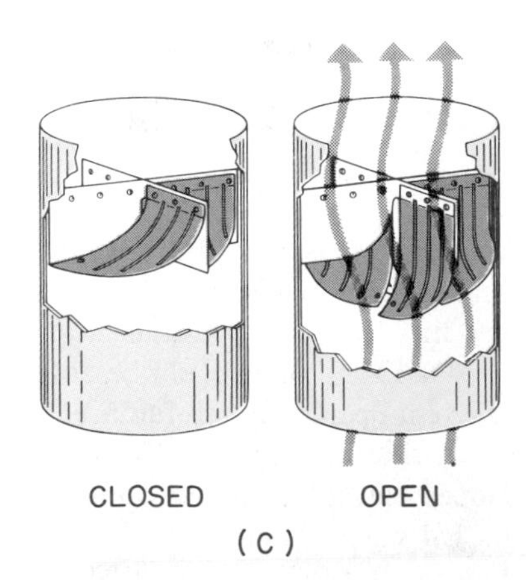

FIGURE 30–55 Automatic vent damper with bimetal blades.

30.29 GAS PIPING

Installing or replacing gas piping can be an important part of a technician's job. The first thing a technician should do is to become familiar with the national and local codes governing gas piping. Local codes may vary from national codes. A technician should also be familiar with the characteristics of natural gas and LP gas in the particular area in which he or she works. Pipe sizing and furnace Btu ratings will vary from area to area due to varying gas characteristics.

The piping should be kept as simple as possible with the pipe run as direct as possible. Distributing or piping gas is similar to distributing electricity. The piping must be large enough, and there should be as few fittings as possible in the system because each of these fittings creates a resistance. Pipe that is too small with other resistances in the system will cause a pressure drop, and the proper amount of gas will not reach the furnace. The specific gravity of the gas must also be taken into consideration when designing the piping. Systems should be designed for a maximum pressure drop of 0.35 in. W.C. In designing the piping system the amount of gas to be consumed by the furnace must be determined. The gas company should be contacted to determine the heating value of the gas in that area. They can also furnish you with pipe sizing tables and helpful suggestions. Most natural gas will supply 1050 Btu/ft^3 of gas. To determine the gas to be consumed in 1 h for a typical natural gas, use the following formula:

$$\frac{\text{Furnace Btu input per hour rating}}{\text{Gas heating rating}} = \text{Cubic feet of gas needed per hour}$$

Suppose the Btu rating of the furnace were 100,000 and the gas heating rating were 1050. Then

$$\frac{100{,}000}{1050} = 95.2 \text{ ft}^3 \text{ of natural gas needed per hour}$$

Tables give the size (diameter) of pipe needed for the length of the pipe to provide the proper amount of gas needed for this furnace. The designer of the piping system would have to know whether natural gas or LP gas were going to be used because the pipe sizing would be different due to the differences in specific gravities of the gases.

Steel or wrought iron pipe should be used. Aluminum or copper tubing may be used, but each requires treatment to prevent corrosion. They are used only in special circumstances.

SAFETY PRECAUTION: *Ensure that all piping is free of burrs and that threads are not damaged, Figure 30–57. All scale, dirt, or other loose material should be cleaned from pipe threads and from the inside of the pipe. Any loose particles can move through the pipe to the gas valve and keep it from closing properly. It may also stop up the small pilot light orifice.*

When assembling threaded pipe and fittings, use a joint compound, commonly called *pipe dope*. Do not apply this compound on the last two threads at the end of the pipe because it could get into the pipe and plug the orifice or prevent the gas valve from closing, Figure 30–58. Also be careful when using Teflon tape to seal the pipe threads because it can also get into the pipe and cause similar problems.

At the furnace the piping should provide for a drip trap, a shutoff valve, and a union, Figure 30–59. A manual shutoff valve within 2 ft of the furnace is required by most local codes. The drip trap is installed to catch dirt, scale, or condensate (moisture) from the supply line. A union between

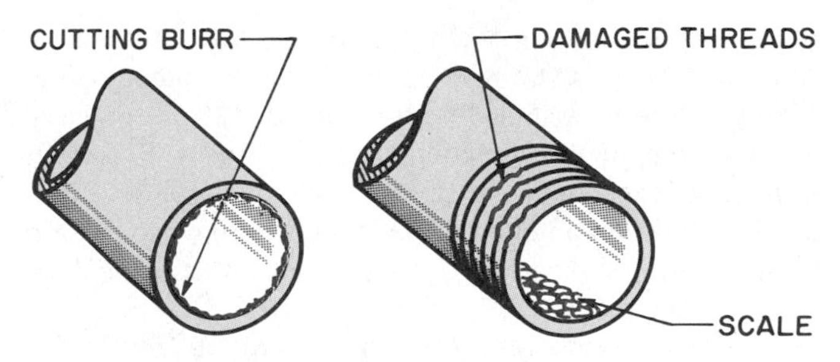

FIGURE 30–57 Ensure that threads are not damaged. Deburr pipe. Clean all scale, dirt, and other loose material from pipe threads and inside of pipe.

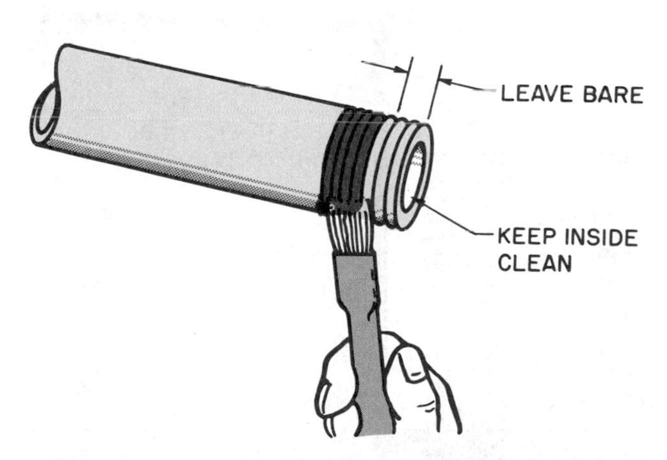

FIGURE 30–58 Pipe dope should be applied to male threads but do not apply to last two threads at end of pipe.

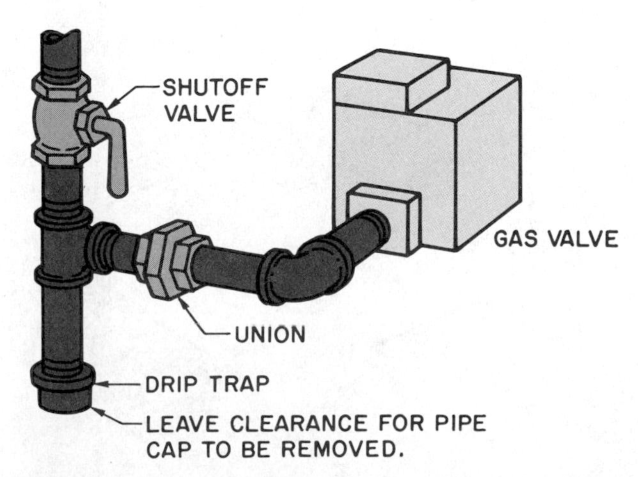

FIGURE 30–59 A shutoff valve, drip trap, and union should be installed ahead of the gas valve.

the tee and the gas valve allows the gas valve or the entire furnace to be removed without disassembling other piping. Piping should be installed with a pitch of 1/4 in. for every 15 ft of run in the direction of flow. This will prevent trapping of the moisture that could block the gas flow, Figure 30–60. Small amounts of moisture will move to the drip leg and slowly evaporate. Use pipe hooks or straps to support the piping adequately, Figure 30–61.

When completed, test the piping assembly for leaks. There are several methods, one is to use the gas in the system. Do not use other gases, and especially *don't* use oxygen. Turn off the manual shutoff to the furnace. Make sure all joints are secure. Turn the gas on in the system. Watch the gas meter dial to see if it moves, indicating that gas is passing through the system. A check for 5 to 10 min will indicate a large leak if the meter dial continues to show gas flow. An overnight standing check is better. If the dial moves, indicating a leak, check each joint with soap and water, Figure 30–62. A leak will make the soap bubble. When you find the leak and repair it, repeat the same procedure to ensure that the leak has been repaired and that there are no other leaks.

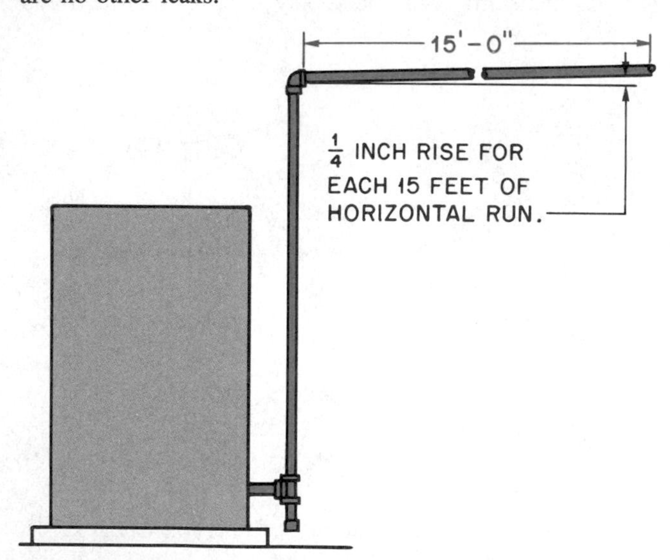

FIGURE 30–60 Piping from furnace should have a rise of 1/4 in. for each 15 ft of horizontal run.

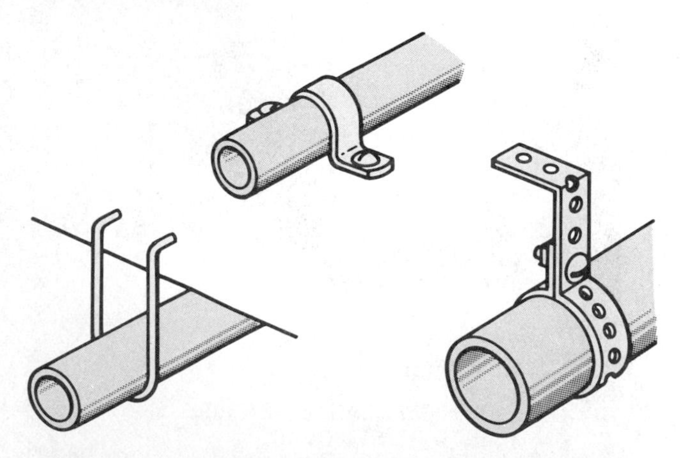

FIGURE 30–61 Piping should be supported adequately with pipe hooks or straps.

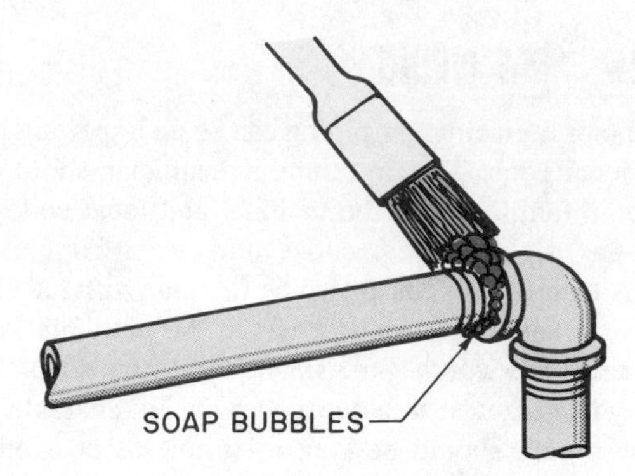

FIGURE 30–62 Apply liquid soap to joints. Bubbling indicates a leak.

Leaks can also be detected with a manometer. Install the manometer in the system to measure the gas pressure. Turn the gas on and read the pressure in. W.C. on the instrument. Now turn the gas off. There should be no loss of pressure. Some technicians allow an overnight standing period with no pressure drop as their standard. Use soap and water to locate the leak as before.

NOTE: If a standing pilot system that is not 100% shut off is being checked, the pilot light valve *must be closed* because it will indicate a leak.

A high-pressure test is more efficient. For a high-pressure leak check, use air pressure from a bicycle tire pump. Ten pounds of pressure for 10 min with no leakdown will prove there are no leaks. ***Don't let this pressure reach the automatic gas valve. The manual valve must be closed during the high-pressure test.***

If there are no gas leaks, the system must be *purged,* that is, the air must be bled off to rid the system of air or other gases. ***Purging should take place in a well-ventilated area. It can often be done by disconnecting the pilot tubing or loosening the union in the piping between the manual shutoff and the gas valve. Do not purge where the gas will collect in the combustion area.*** After the system has been purged, allow it to set for at least 15 min to allow any accumulated gas to dissipate. If you are concerned about gas collecting in the area, wait for a longer period. Moving the air with a hand-operated fan can speed this up. When it is evident that no gas has accumulated, the pilot can be lit.

SAFETY PRECAUTION: *Be extremely careful because gas is explosive. The standing leak test must be performed and all conditions satisfied before a gas system is put in operation. Be aware of and practice any local code requirements for gas piping.*

30.30 GAS FURNACE WIRING DIAGRAMS

Figure 30–63 is a wiring diagram for a gas furnace. It does not show the safety features necessary in a furnace for proving that the pilot flame is lit or for a glow-coil or intermittent-spark ignition. These are discussed later.

The path for the 120-V electrical current (white wire) goes from L1 to the limit switch (which is normally closed),

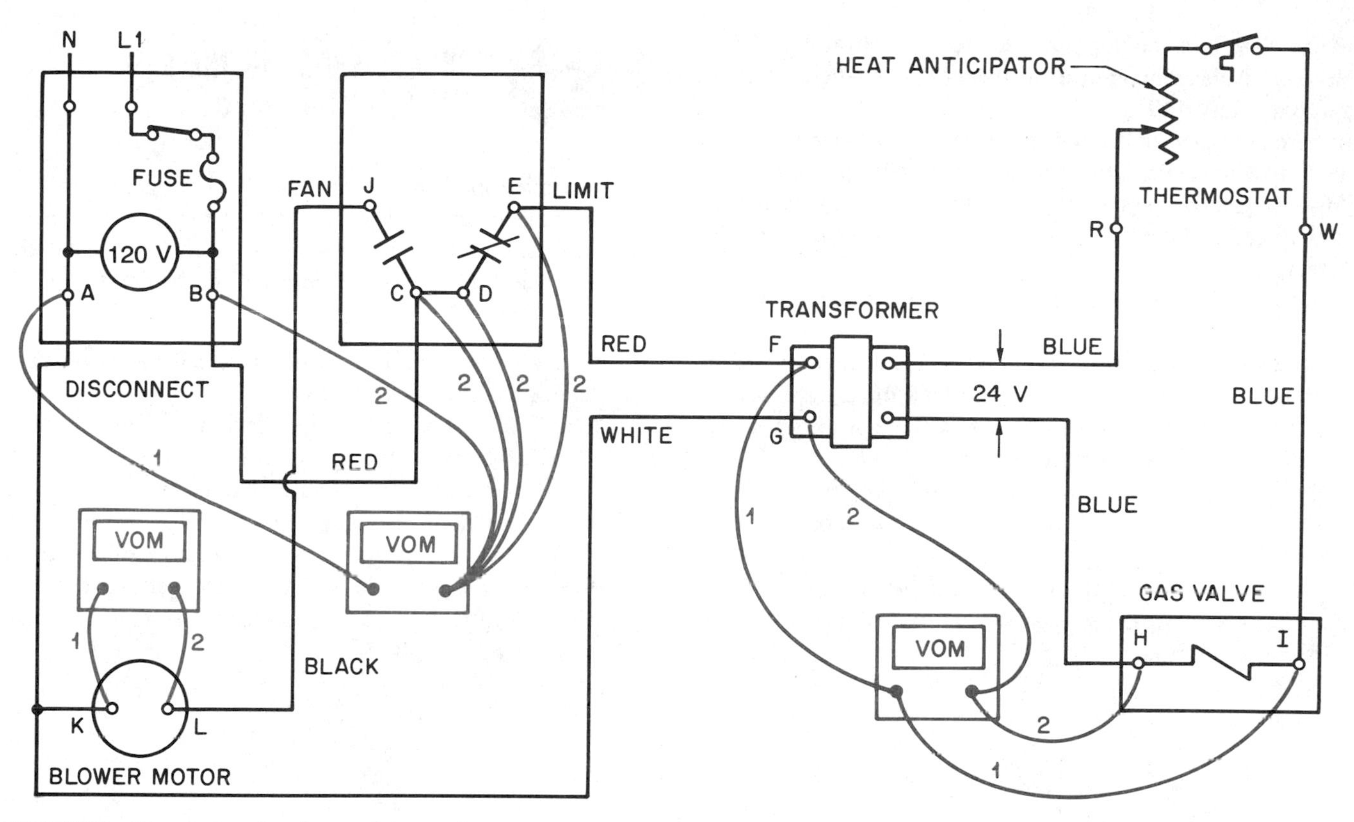

FIGURE 30–63 Basic wiring diagram for a gas furnace.

through the transformer, and back to the N terminal on the white wire.

The transformer reduces the voltage to 24 V. The thermostat and the gas valve are in this circuit (blue wire). It goes from the transformer secondary to the thermostat, to the gas valve, and back to the other terminal on the transformer secondary making a complete circuit.

The fan circuit is shown with the black wire with part of it in red. It goes from L1 through the fan switch to the blower motor and back to the disconnect to N making a complete circuit.

The transformer is usually energized all the time. The current path can be traced through the red wire and the NC limit switch. When there is a call for heat, the contacts in the thermostat will close. This will cause a current to flow in the secondary (blue wire). A coil in the gas valve will be energized, opening the gas valve. The pilot safety and flame-proving device are in another circuit. When enough heat is produced, a bimetal control will warp, closing the contacts to the fan. This will cause the fan to operate, distributing the heat to the space where it is needed.

The limit switch is a safety device. If the furnace overheats, a bimetal will warp, opening this switch in the 120-V circuit. This will stop the current flow to the transformer, closing the gas valve. This will not interrupt the power to the fan, which will continue to operate and distribute the heat and cool the heat exchanger.

SAFETY PRECAUTION: *Treat electrical troubleshooting with great respect. Electric shock can be fatal*.

30.31 TROUBLESHOOTING TECHNIQUES

To troubleshoot these circuits, first set the selector switch on a VOM (volt-ohm-milliammeter) to 250 V to be sure that the VOM will not be damaged by overvoltage. To check the 120-V circuit, place probe 1 on terminal A or the neutral wire connection and probe 2 at B. A reading of 120 V indicates that power is being supplied to the unit. No reading would indicate that the fuse, circuit breaker, or switch was defective.

Leave probe 1 at a neutral terminal and move probe 2 to C: no power indicates that the conductor is defective; D: no reading indicates that the jumper in the limit switch is open; E: no reading indicates that the limit switch is open. (Note that the limit switch is heat operated. It should not open until the temperature in the heat exchanger is greater than 200°F).

To check power to the transformer, place probe 1 at F and probe 2 at G. 120 V should be read. This shows that the correct voltage is supplied to the transformer. If there is no reading here, there must be a break in the conductor.

On a service call check the voltage at the transformer first. If there is no voltage, then follow the other steps to determine where the interruption is.

To check the fan circuit, do the following: Place probe 1 at A and probe 2 at B. If you read 120 V, place probe 2 at C. If there is no power, a conductor must be open. If there is power, move probe 2 to J.

NOTE: This is a temperature-actuated switch and will not close until the temperature in the heat exchanger reaches approximately 140°F.

If there is no power at J after proper heat has been achieved, the fan switch is defective and should be replaced. If there is power at J, move probe 1 to K and probe 2 to L. If there is a 120-V reading here but the motor does not run, the motor is defective.

To troubleshoot the output voltage from the transformer, place probe 1 on I and probe 2 on H. If the meter reads 24 V, the gas valve should open. If the valve does not open, turn the power off. Take one lead off the gas valve, adjust the meter to measure ohms, place one probe on H and the other at I. If there is no resistance, the coil in the valve is defective and the valve should be replaced. If there is a resistance, the valve itself is stuck and needs to be replaced. If there is no voltage at H and I, place a jumper across R and W at the thermostat. If there is voltage at H and I, the thermostat should be replaced. If there is still no voltage, a wire must be broken.

30.32 TROUBLESHOOTING THE GAS VALVE CIRCUIT

This is much the same as the troubleshooting sequence in the previous example. The pilot light is burning. The problem is that the pilot safety switch is not making to the hot contact. The following procedure will determine the problem. See Figure 30–64 for the voltmeter readings.

1. Check to see that the thermostat is calling for heat.
2. Check for 120 V at the unit.
3. Check for 24 V at the secondary of the transformer. *Note:* If there is voltage to the unit and not to the transformer, the limit control may be open. If a limit control suddenly becomes open, be sure to check the airflow. The limit control normally *never* moves from the made position unless the furnace heat exchanger is above 200°F. See Figure 30–65 for voltage readings with an open limit.

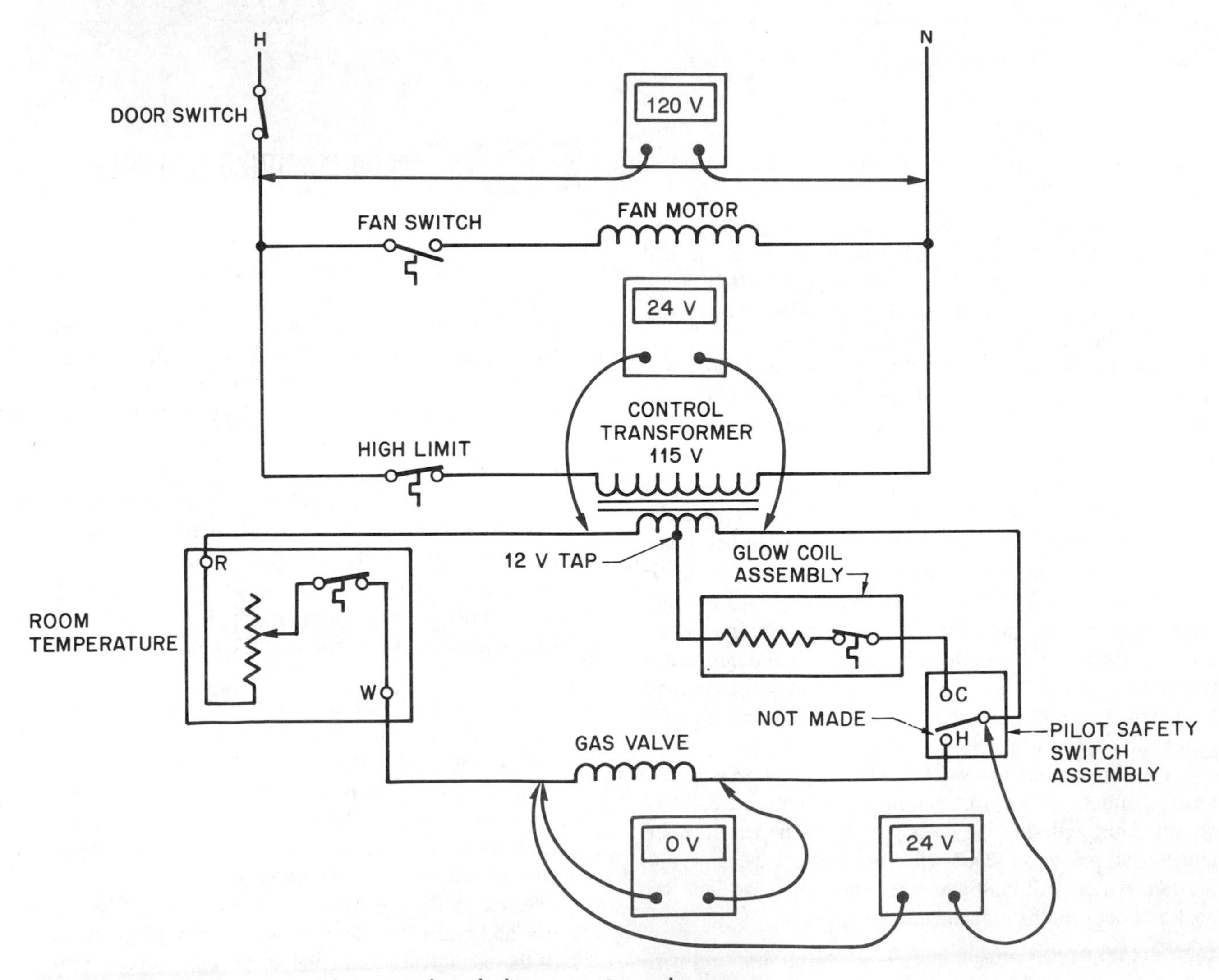

FIGURE 30–64 The pilot safety switch is not made to the hot contact after a pilot outage.

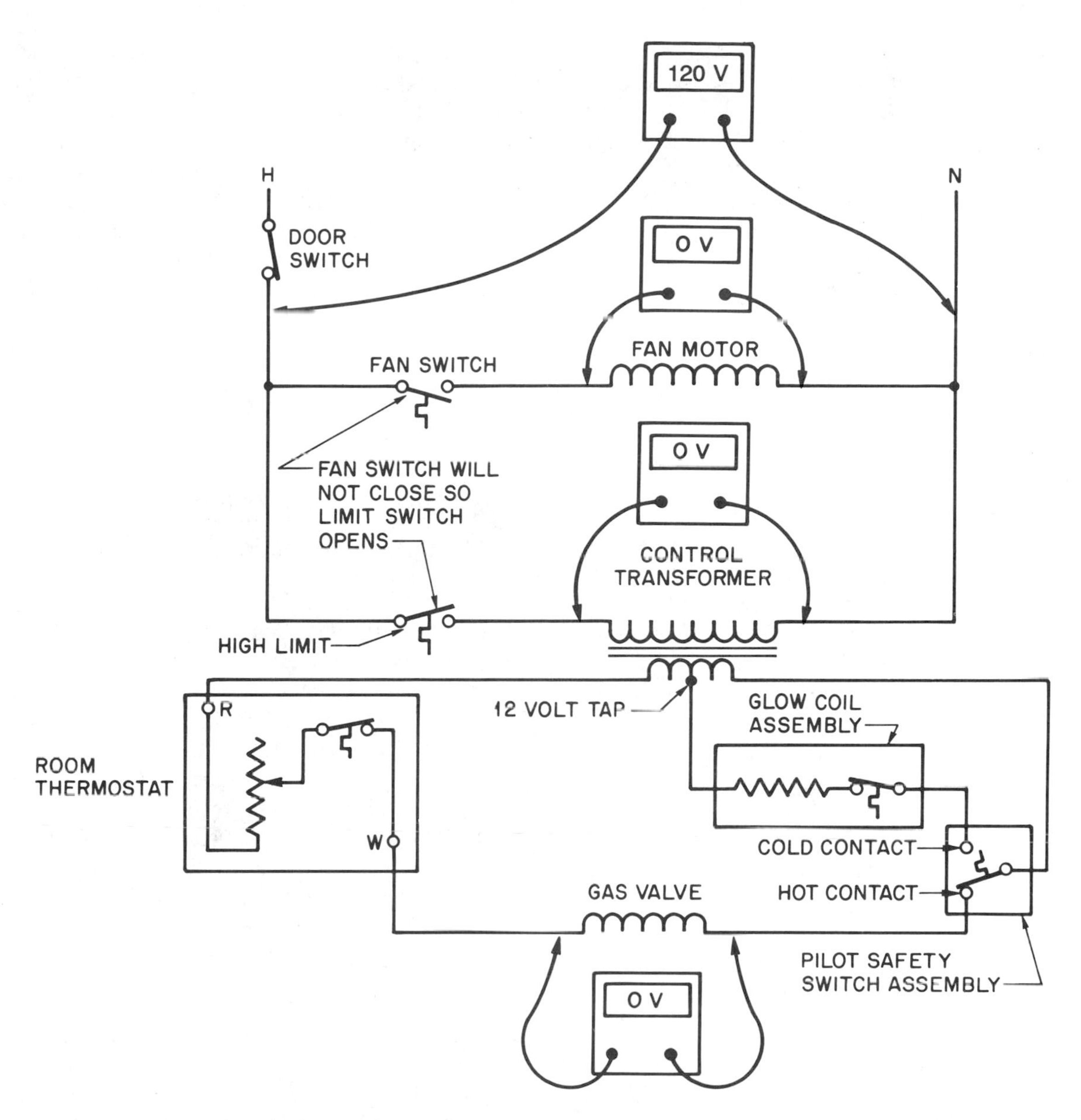

FIGURE 30–65 The pilot light is lit—the limit switch is open.

4. Check for voltage at the gas valve. There is no voltage. There are three possibilities: (1) The thermostat contacts, or heat anticipator, are open; (2) the pilot safety switch is defective; (3) the conductors leading to the gas valve from the 24-V power supply are open.

Figure 30–65 shows still another problem that can occur. The fan switch is not making. The fan will not come on to dissipate the heat in the furnace, so the limit control shuts the gas off. Notice that voltage is absent everywhere except at the line coming in. This will be true when the limit switch is open. When the limit switch closes, the furnace will light for short periods of time.

NOTE: When approaching a furnace in this condition, see if it is hot to the touch. If the furnace is hot and the fan is not running, there is a fan or fan circuit problem.

30.33 TROUBLESHOOTING THE SAFETY PILOT-PROVING DEVICE—THE THERMOCOUPLE

The thermocouple generates a small electrical current when one end is heated. When the pilot flame is lit and heating the thermocouple, it generates a current that energizes a coil holding a safety valve open. This valve is manually opened when lighting the pilot, the coil only holds it open. If the pilot flame goes out, the current would no longer be generated and the safety valve would close, shutting off the supply of gas.

To light the pilot, turn the gas valve control to PILOT position. Depress the control knob and light the pilot. Hold the knob down for approximately 45 sec, release, and turn the valve to the ON position. The thermocouple may be defective if the pilot goes out when the knob is released. Figure 30–66 describes the no-load test for a thermocouple.

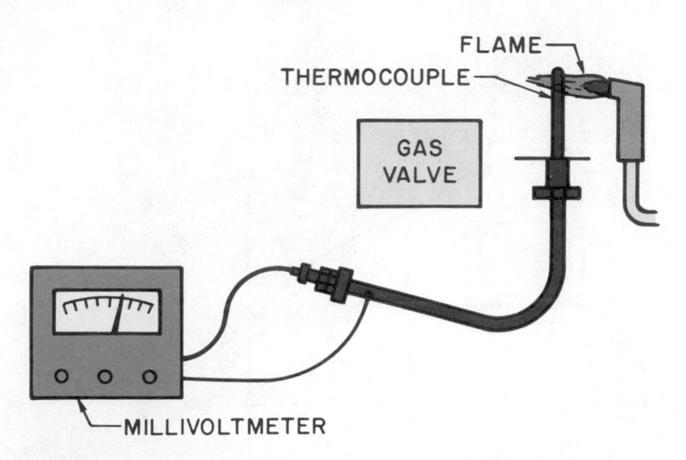

FIGURE 30–66 Thermocouple no load test. Operate the pilot flame for at least 5 min with the main burner off. (Hold the gas cock knob in the PILOT position and depress.) Disconnect the thermocouple from the gas valve while still holding the knob down to maintain the flame on the thermocouple. Attach the leads from the millivoltmeter to the thermocouple. Use direct current voltage. Any reading below 20 mV indicates a defective thermocouple or poor pilot flame. If pilot flame is adjusted correctly, test thermocouple under load conditions.

It is important to remember that the thermocouple only generates enough current to *hold* the coil armature in. The armature is pushed in when the valve handle is depressed.

To check under-load conditions, unscrew the thermocouple from the gas valve. Insert the thermocouple testing adapter into the gas valve. The testing adapter allows you to take voltage readings while the thermocouple is operating. Screw the thermocouple into the top of this adapter, Figure 30–67. Relight the pilot and check the voltage produced

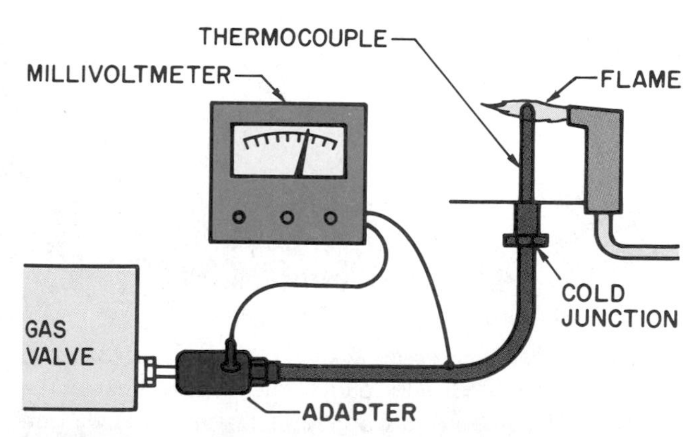

FIGURE 30–67 Thermocouple test under load conditions. Disconnect thermocouple from gas valve. Screw thermocouple test adapter into gas valve. Screw thermocouple into test adapter. Light the pilot and main burner and allow to operate for 5 min. Attach one lead from the millivoltmeter to either connecting post of the adapter and the other lead to the thermocouple tubing. Any reading under 9 mV would indicate a defective thermocouple or insufficient pilot flame. Adjust pilot flame. Replace thermocouple if necessary. Ensure that pilot, pilot shield, and thermocouple are positioned correctly. Too much heat at the cold junction would cause a satisfactory voltage under no-load conditions but unsatisfactory under a load condition.

with a millivoltmeter, using the terminals on the adapter. The pilot flame must cover the entire top of the thermocouple rod. If the thermocouple produces at least 9 mV under load (while connected to its coil), it is good. Otherwise replace it. Manufacturer's specifications must be checked to determine the acceptable voltage for holding the different valve coils. If the thermocouple functions properly but the flame will not continue to burn when the knob is released, the coil in the safety valve must be replaced. This is normally done by replacing the entire gas valve. Thermopiles produce a much greater voltage than thermocouples.

30.34 TROUBLESHOOTING THE GLOW-COIL CIRCUIT

Troubleshooting this circuit is very much like troubleshooting any other electrical circuit. There has to be a power supply, the transformer, interconnecting wiring, and a load —the glow coil. To get these in proper perspective, let's use a customer complaint. The customer calls and says there is no heat. The problem is that the pilot has gone out and the pilot safety valve did not make to the cold contact to provide power to the glow coil. See Figure 30–68 for the voltmeter readings that should appear at the various points.

Procedures to use:

1. Make sure that the thermostat is calling for heat.
2. Check to see that there is power to the unit (approximately 120 V).
3. Check for 24 V at the control transformer on the secondary side.
4. Check for 24 V at the pilot safety terminal on the line (inlet side) with the other probe at the glow coil terminal.
5. Check for voltage at the glow coil's other terminal. Suppose you get 0 V. This indicates that the pilot safety switch is not making contact on the cold contact. It needs to be replaced.

   NOTE: The glow-coil component should have power to it whenever there is no pilot. This power is supplied through the pilot safety switch.

If you had 12 V to the glow coil and it would not glow, wait for a few minutes to allow the thermal element to cool. If it did not cool and come on, then the glow-coil component needs to be replaced. When you get the pilot component out of the unit, check it with an ohmmeter. Sometimes the thermal contacts will close back while removing the glow coil assembly. Because this is a standing pilot type of device, the pilot can be lit with a match until a replacement can be obtained.

The glow coil has to be positioned in such a fashion that the gas from the pilot is directed toward the glow coil. If the pilot light is dirty or the gas is not hitting the glow

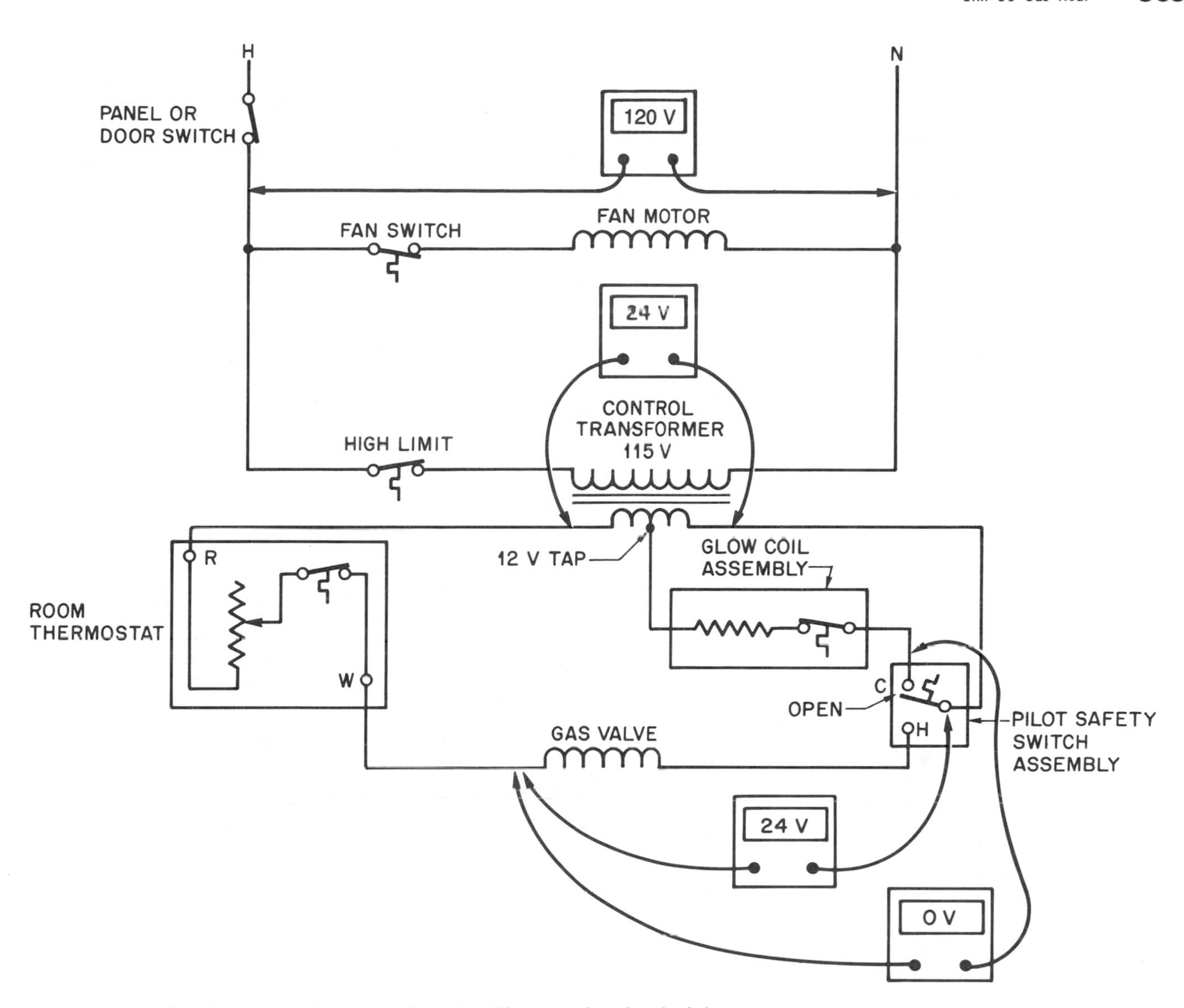

FIGURE 30–68 The pilot safety switch is not made to the cold contact when the pilot light is out.

coil, it may not light. If the glow coil glows and will not light the pilot, then check:

1. The pilot gas stream hitting the glow coil
2. If the glow coil is hot enough (check for full voltage)

The glow coil can be used for applications other than just the standing pilot. For instance, it can be used for intermittent pilot. This is a system where the pilot is extinguished at the end of each cycle and relit on a call for heat. This is discussed later in the unit. The glow coil can be used to ignite gas in applications that are designed for it. The same troubleshooting procedures would be used. Voltage must be at the coil for it to glow.

30.35 TROUBLESHOOTING SPARK-IGNITION PILOT LIGHTS

Most spark-ignition pilot light assemblies have internal circuits or printed circuit boards. The technician can only troubleshoot the circuit to the board, not circuits within the board. If the trouble is in the circuit board, the board must be changed. Figure 30–69 shows how one manufacturer accomplishes spark ignition using the flame rectification method. Flame rectification uses the flame to change alternating current to direct current. This current is picked up by an electronic circuit to prove that a pilot light is present before the main gas valve is opened. There are two diagrams, one is a pictorial and the other is a line type. Notice that the manufacturer has placed the terminal board on the bottom of the pictorial and on the side of the line diagram. Components are not placed in the same position from one diagram to another.

The spark-ignition board in the previous example has terminals provided for adding other components, such as an electronic air cleaner and a vent damper shutoff motor. This is handy because the installing technician only has to follow the directions to wire these components when they are used in conjunction with this furnace. The electronic air cleaner is properly interlocked to operate only when the fan is run-

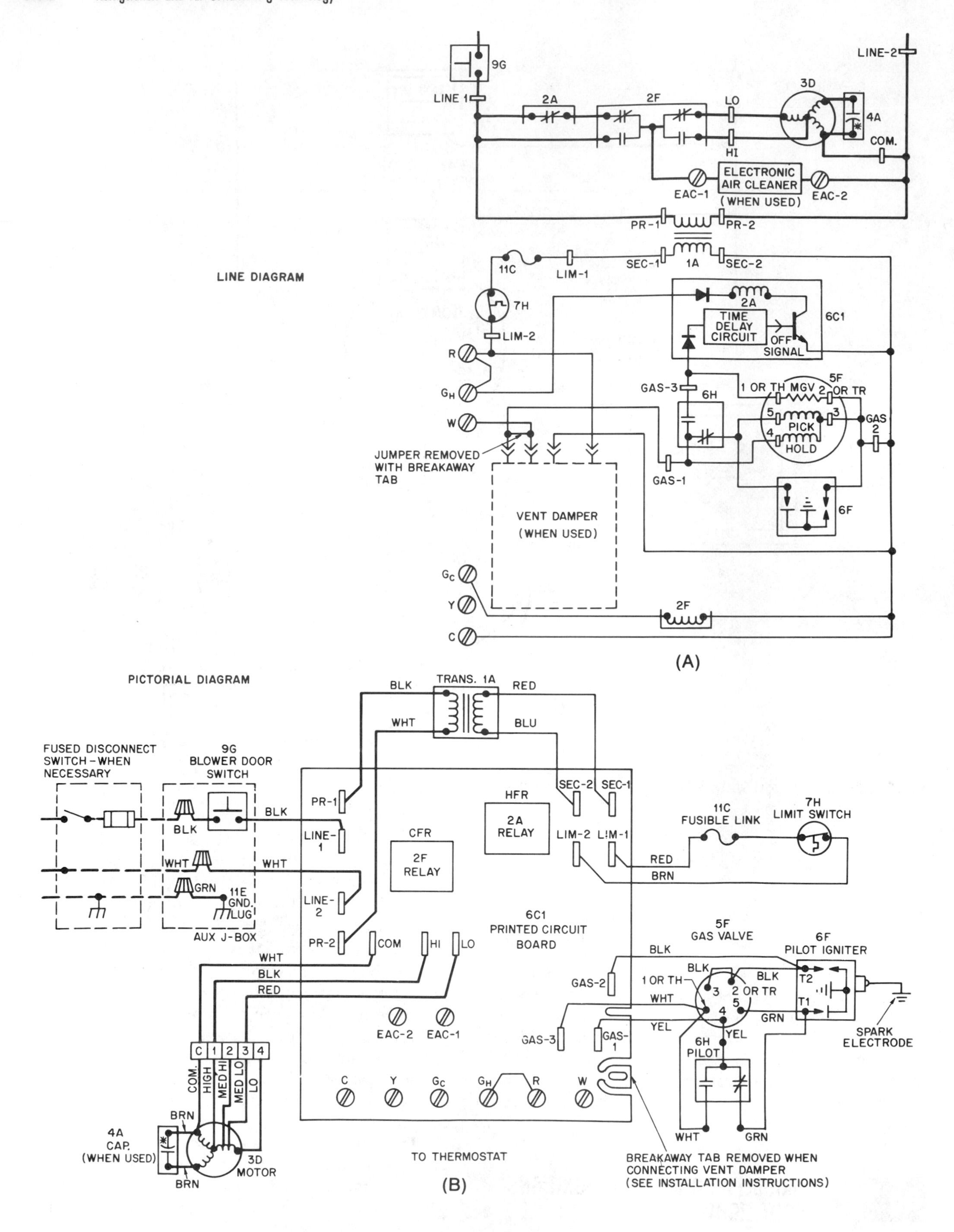

FIGURE 30–69 (A) Line diagram. (B) Pictorial diagram. *Courtesy BDP Company*

ning. This is not always easy to wire into the circuit to accomplish the proper sequence, particularly with two-speed fan applications.

The following reference points can be used for this circuit board:

1. ***When the front panel is removed, everything stops until the switch is blocked closed. This should only be done by a qualified service technician.***
2. The fan motor is started through the 2A contacts and in single speed.
3. The 2A contacts are held open during the off cycle by energizing the 2A coil. This coil is deenergized during the on cycle.
4. If the control transformer is not functioning, the fan will run all the time. If the fan runs constantly but nothing else operates, suspect a bad control transformer.
5. The fan starts and stops through the time delay in the heating cycle.

The following description of the electrical circuit can be used for an orderly troubleshooting procedure. See Figures 30–70 and 30–71 for the voltages in the pictorial and line diagrams.

1. Line power should be established at the primary of the control transformer.
2. 24 V should be detected from the C terminal (common to all power-consuming devices) to the LIM-1 terminal. See the meter lead on the 24-V meter labeled 1.
3. Leaving the probe on the C terminal, move the other probe to positions 2, 3, and 4. 24 V should be detected at each terminal, or there is no need to proceed.
4. The key to this circuit is to have 24 V between the C terminal and the GAS-1 terminal. At this time the pilot should be trying to light. The spark at the pilot should be noticeable. It makes a ticking sound with about one-half second between sounds.
5. If the pilot light is lit and the ticking has stopped, the 6H relay should have changed over. The NC contacts in the 6H relay should be open and the NO contacts should be closed. This means that you should have 24 V between the C terminal, which is the same as the GAS-2 terminal, and the GAS-3 terminal. The 6H relay is actually a pilot-proving relay and heat from the pilot changes its position. If it will not pass power, it must be replaced.
6. When the circuit board will pass power to the GAS-3 terminal (probe position 5) and the burner will not light, the gas valve is the problem. Before changing it, be sure that its valve handle is turned to the correct direction and the interconnecting wire is good. The procedure of leaving one probe on the C terminal and moving the other probe can be done quickly when a furnace has a circuit board with terminals. This makes troubleshooting easier.
7. If the R terminal has 24 V and the W terminal does not, the thermostat and the interconnecting wiring are the problem. You can place a jumper from R to W for a moment; if the spark starts to arc, the problem is the wiring or the thermostat.
8. The fan is started through the circuit board and is part of the sequence of operation also. The fan is started when the 2A relay is deenergized through the time-delay circuit. This is accomplished when the 6H relay in the circuit board changes position. When 24 V is detected from terminal C to terminal GAS-3, the time-delay circuit starts timing. When the time is up, the contacts open and deenergize the fan relay coil; the contacts close, and the fan starts.
9. The voltmeter can be relocated to the COM terminal and the LO terminal. When relay 2A is passing power, 120 V should be detected here. If there is power to the fan motor and it will not start, open up the fan section and check the motor for continuity. Check the capacitor. See if the motor is hot. If the motor cannot be started with a good capacitor, replace the motor.

See Figure 30–72 for an example of the same furnace circuit board that has been expanded to include the cooling thermostat and two-speed fan operation. The high-speed fan is desirable in some applications where the quantity of air for cooling is more than that for heating. Notice how the electronic air cleaner is started in either cooling or heating through a different relay application. This board has two relays: the same 2A relay as before, and the double-pole–double-throw (dpdt) relay 2F. The 2F relay is energized by the thermostat in the cooling mode through the Gc terminal.

The key to troubleshooting these circuits is to keep in mind which components are controlling other components. A study of the previous diagrams will help you decide which component is actually defective—the board or the component that the board is controlling.

30.36 COMBUSTION EFFICIENCY

With the high and ever-increasing cost of fuel, it is essential that adjustments be made to produce the most efficient but safe combustion possible. It is also necessary to ensure that an overcorrection is not made that would produce carbon monoxide.

Atmospheric gas burners use primary air, which is sucked in with the gas, and secondary air, which is pulled into the combustion area by the draft produced when the ignition takes place. ***Incomplete combustion produces carbon monoxide. Enough secondary air must be supplied so that carbon monoxide will not be produced.***

Primary air intake can be adjusted with shutters near the orifice. This adjustment is made so that the flame has the characteristics specified by the manufacturer. Generally, however, the flame is blue with a small orange tip when it is burning efficiently, Figure 30–73.

Modern furnaces often do not have an air adjustment because of the types of burners used. With the end shot burner, air adjustment is not as critical.

Conversion burners are used to convert coal or oil furnaces or boilers to gas fired. They normally have secondary

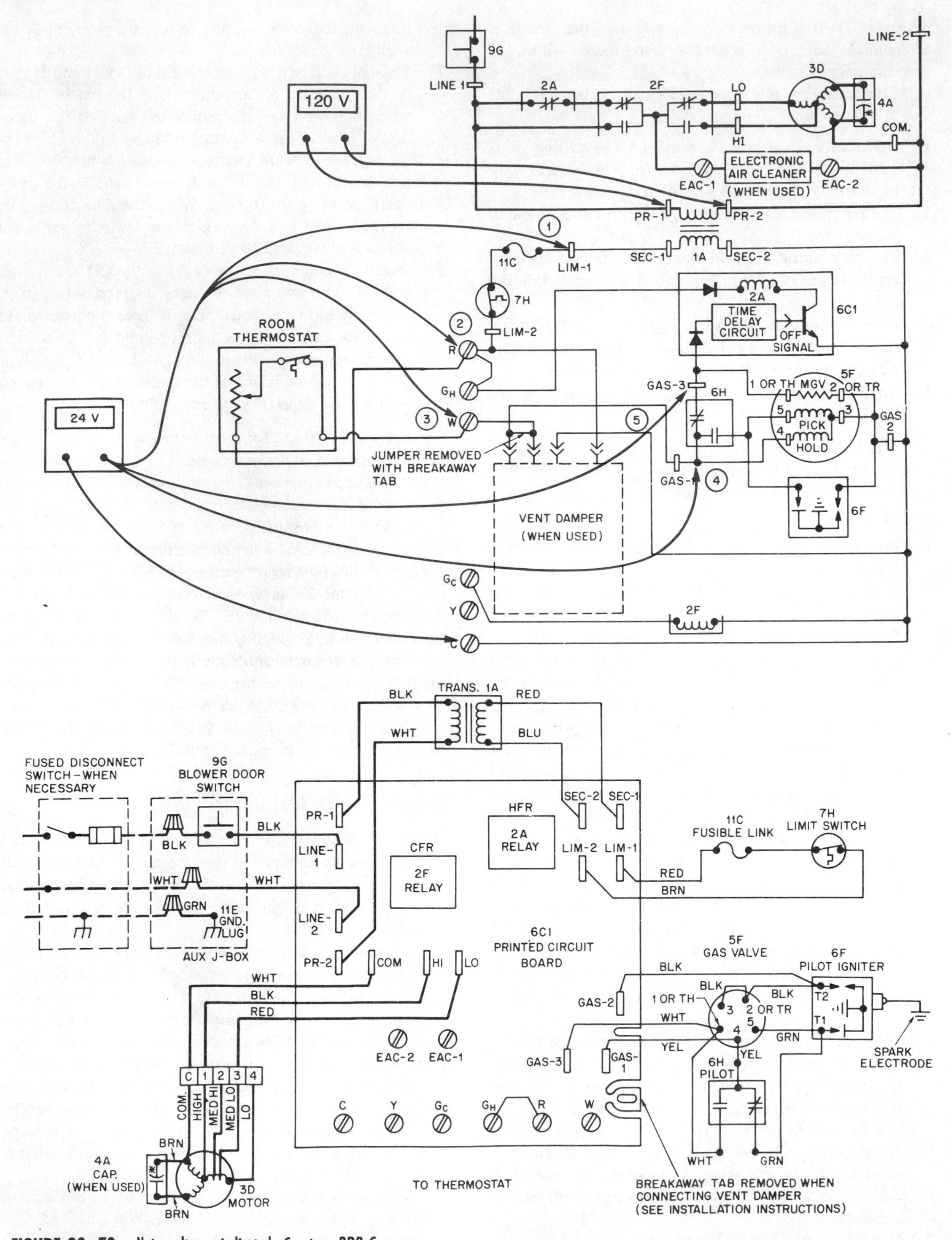

FIGURE 30–70 Note voltages indicated. *Courtesy BDP Company*

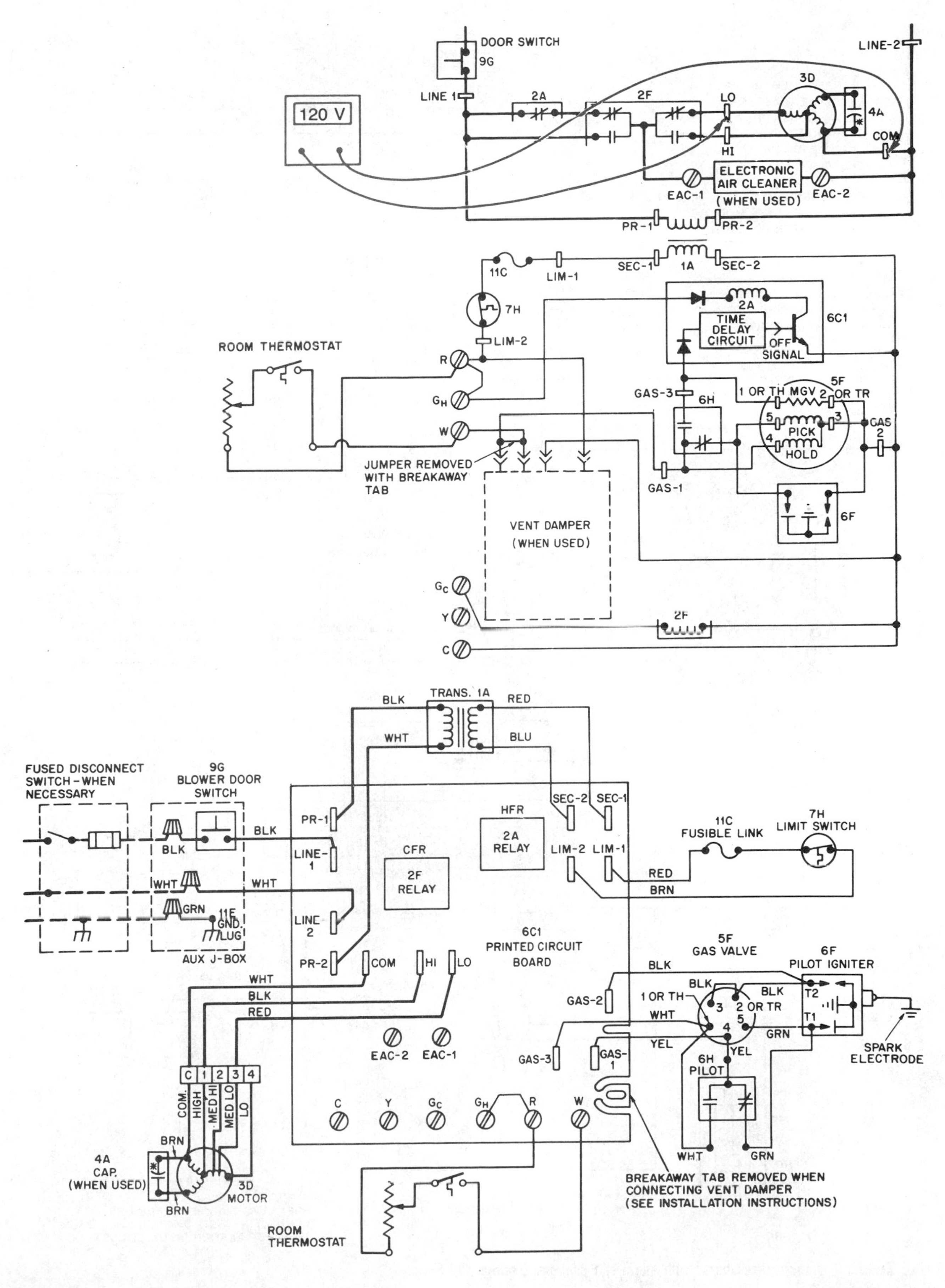

FIGURE 30–71 Note voltage indicated. *Courtesy BDP Company*

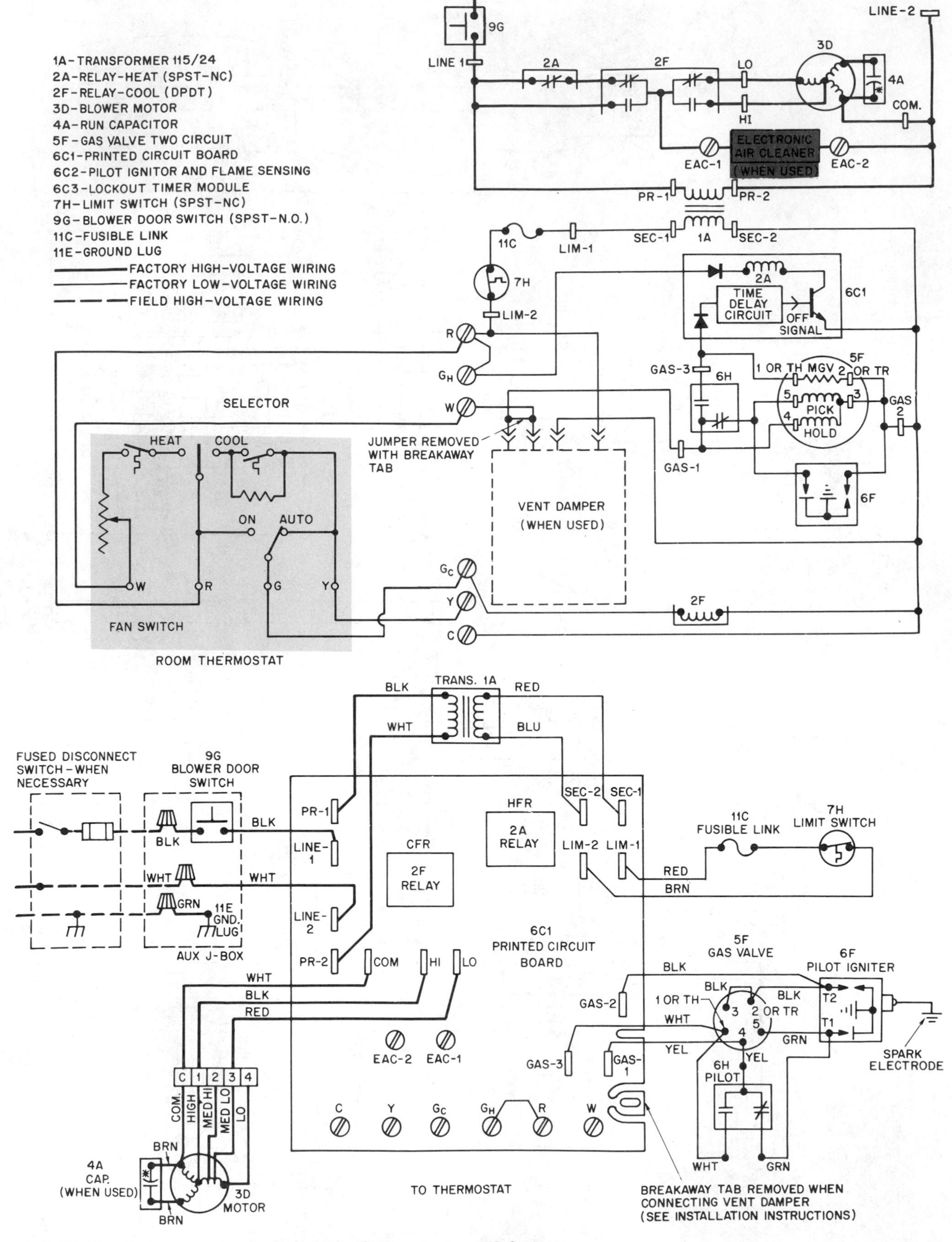

FIGURE 30–72 Furnace circuit board with thermostat diagram. *Courtesy BDP Company*

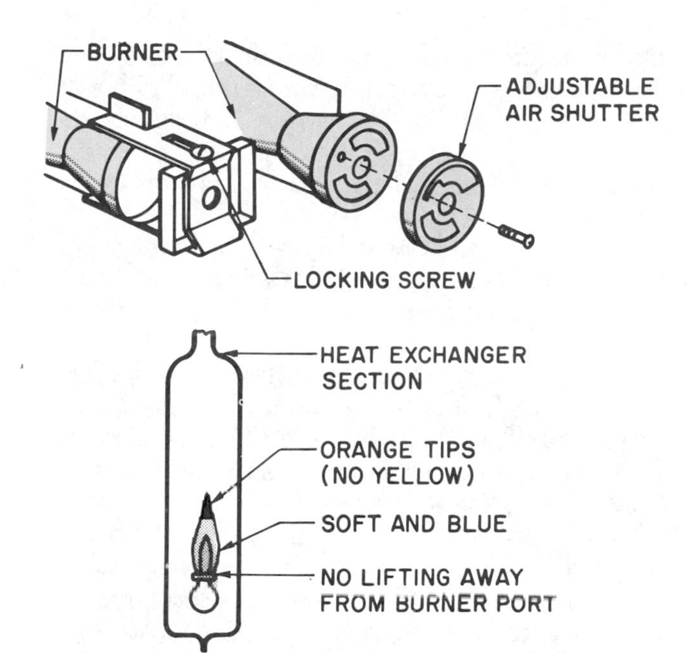

FIGURE 30–73 Main burner flame adjustment. The air shutter is turned to produce the proper flame. To check the flame, turn the furnace on at the thermostat. Wait a few minutes to ensure that any dust or other particles have been burned and no longer have an effect on the flame. The flames should be stable, quiet, soft and blue with slightly orange tips. They should not be yellow. They should extend directly upward from the burner ports without curling downward, floating or lifting off the ports and they should not touch the side of the heat exchanger.

air adjustments, and these must be adjusted in the field for maximum efficiency. The carbon dioxide test is used to get the necessary information from the flue gases to make the secondary air adjustment. Carbon dioxide in the flue gases is actually a measurement of the excess secondary air supplied. As the secondary air supplied is decreased, the percentage of carbon dioxide in the flue gases increases, and the chance of producing carbon monoxide becomes greater.

If the exact amount of air is supplied and perfect combustion takes place, a specific amount of carbon dioxide by volume will be present in the flue gases. This is called the ultimate carbon dioxide content. The following table indicates the ultimate percentage of carbon dioxide content for various gases: (***These figures are for perfect combustion and are not used as the carbon dioxide reading for a gas-burning appliance.***)

FOR PERFECT COMBUSTION	
Natural gas	11.7—12.2
Butane gas	14.0
Propane gas	13.7

A gas furnace is never adjusted to produce the ultimate carbon dioxide because there is too much danger of producing carbon monoxide. Therefore, the accepted practice is to adjust the secondary air to produce excess air resulting in less carbon dioxide by volume in the flue gases.

HVAC GOLDEN RULES

When making a service call to a residence:

- Always carry the proper tools and avoid extra trips through the customer's house.
- Make and keep firm appointments. Call ahead if you are delayed. The customer's time is valuable also.
- Keep customers informed if you must leave the job for parts or other reasons. Customers should not be upset if you inform them of your return schedule if it is reasonable.

Here are some simple, inexpensive procedures that may be included in the basic service call.

- Leave the furnace and area clean.
- If standing pilot is used, perform service check.
- Clean or replace dirty filters.
- Lubricate bearings when needed.
- Check burner compartment to ensure burners are clean and burning correctly.
- Replace all panel covers.

PREVENTIVE MAINTENANCE

Gas equipment preventive maintenance consists of servicing the air-side components: filter, fan, belt and drive, and the burner section. Air-side components are the same for gas as for electric forced-air heat, so only the burner section will be discussed here.

The burner section consists of the actual burner, the heat exchanger, and the venting system for the products of combustion.

The burner section of the modern gas furnace does not need adjustment from season to season. It will burn efficiently and be reliable if it is kept clean. Rust, dust, or scale must be kept from accumulating on top of the burner where the secondary air supports combustion or in the actual burner tube where primary air is induced into the burner. When a burner is located in a clean atmosphere (no heavy particles in the air as in a manufacturing area) most of the dust particles in the air will burn and exit through the venting system. The burner may not need cleaning for many years. A vacuum cleaner with a crevice tool may be used to remove small amounts of scale or rust deposits.

The burner and manifold may be removed for more extensive cleaning inside the burner tube where deposits may be located. Each burner may be removed and tapped lightly to break the scale loose. An air hose should be used to blow down through the burner ports while the burner is out of the system. The particles will blow back out the primary air shutter and out of the system. ***Wear goggles when using compressed air for cleaning.*** This cleaning function cannot be accomplished while the burners are in the furnace.

Burner alignment should be checked to make sure that no flame impingement to the heat exchanger is occurring. The burners should set straight and feel secure while in place. Large deposits of rust scale may cause flame impingement and must be removed.

A combustion analysis is not necessary as routine maintenance on the modern gas furnace. The primary air is the only adjustment on new furnaces, and it cannot be out of adjustment enough to create much inefficiency when natural gas is the fuel. Liquified petroleum fuel burners may need adjustment for best performance. All gas burners should burn with a clear blue flame, with only orange streaks in the flame. **There should never be any yellow or yellow tips on the flame. If yellow tips cannot be adjusted from the flame, the burner contains dirt or trash and it should be removed for cleaning.**

Observe the burner for correct ignition as it is lit. Often, the crossover tubes that carry the burner flame from one burner to the next get out of alignment and the burner lights with an irregular pattern. Sometimes it will "puff" when slow ignition of one of the burners occurs.

The pilot light flame should be observed for proper characteristics. It should not have any yellow tips and it should not impinge on the burner. It should impinge on the thermocouple only. If the pilot flame characteristics are not correct, the pilot assembly may have to be removed and cleaned. Compressed air will usually clear any obstructions from the pilot parts. ***Do not use a needle to make the orifice holes larger. If they are too small, obtain the correct size from a supply house.***

The venting system should be examined for obstructions. Birds may nest in the flue pipe, or the vent cap may become damaged. Visual inspection is the first procedure. Then, with the furnace operating, the technician should strike a match in the vicinity of the draft diverter and look for the flame to pull into the diverter. This is a sign that the furnace is drafting correctly.

Observe the burner flame with the furnace fan running. A flame that blows with the fan running indicates a crack in the heat exchanger. This would warrant further examination of the heat exchanger. It is difficult to find a small crack in a heat exchanger and one may only be seen with the heat exchanger out of the furnace housing. Extensive tests have been developed for proving whether or not a gas furnace heat exchanger has a crack. These tests may be beyond what a typical technician would use because the recommended tests require expensive equipment and experience.

High-efficiency condensing furnaces require maintenance of the forced-draft blower motor if it is designed to be lubricated. The condensate drain system should be checked to be sure it is free of obstructions and that it drains correctly. Manufacturers of condensing furnaces sometimes locate the condensing portion in the return air section where all air must move through the coil. This is a finned coil and will collect dirt if the filters are not cleaned. Routine inspection of this condensing coil for obstruction is a good idea.

All installations should be observed by setting the room thermostat to call for heating and ensure the proper sequence of events occurs. For example, when checking a standard furnace, the pilot may be lit all the time and the burner should ignite. Then the fan should start. Some furnaces may have intermittent pilot lights. The pilot should light first, the burner next, and then the fan should start. With a condensing furnace, the power vent motor may start first, the pilot light next, then the burner, and the fan motor last.

Some technicians disconnect the circulating fan motor and allow the furnace to operate until the high limit control shuts it off. This proves the safety device will shut off the furnace. Technicians will often perform a combustion analysis test to check furnace efficiency and then make air shutter adjustment. The adjustment should be made to produce from 65% to 80% of the ultimate carbon dioxide. This would produce 8% to 9.5% carbon dioxide in the flue. It is important that manufacturer's specifications be followed when making any adjustments.

The flue-gas temperature should also be taken if the percent of combustion efficiency is to be determined. The percent combustion efficiency is an index of the useful heat obtained. It is not possible to obtain 100% efficiency because heated air does leave the building through the flue.

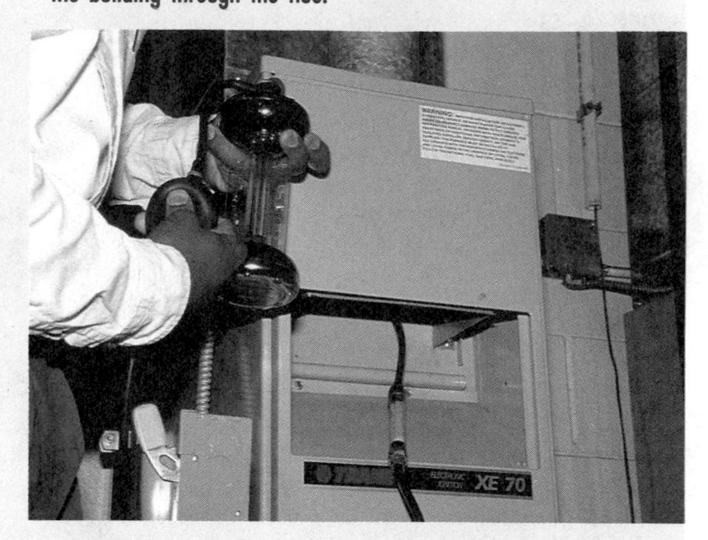

FIGURE 30–74 Sampling tube in draft diverter. *Photo by Bill Johnson*

The carbon dioxide and flue-gas temperatures are taken in the flue between the draft diverter and the furnace. A sampling tube can be inserted through the draft diverter projecting into the inlet side of a gas-burning furnace, Figure 30–74. An oxygen and carbon dioxide analyzer and a stack thermometer are used for making these tests, Figure 30–75. You can refer to a chart provided by the manufacturer of the test equipment to determine the best efficiency. Average gas-burning furnaces manufactured before 1982 should produce 75% to 80% efficiency.

Test equipment is also available to check carbon monoxide content in the flue gases. Carbon monoxide-free combustion is defined as less than 0.04% carbon monoxide in an air-free sample of the flue gas. Carbon monoxide can be produced by flame impingement on a cool surface, insufficient primary air (yellow flame) as well as insufficient secondary air.

Other adjustments that can be checked and compared with manufacturer's specifications are suction at the draft hood, draft at the chimney, gas pressure, and gas input.

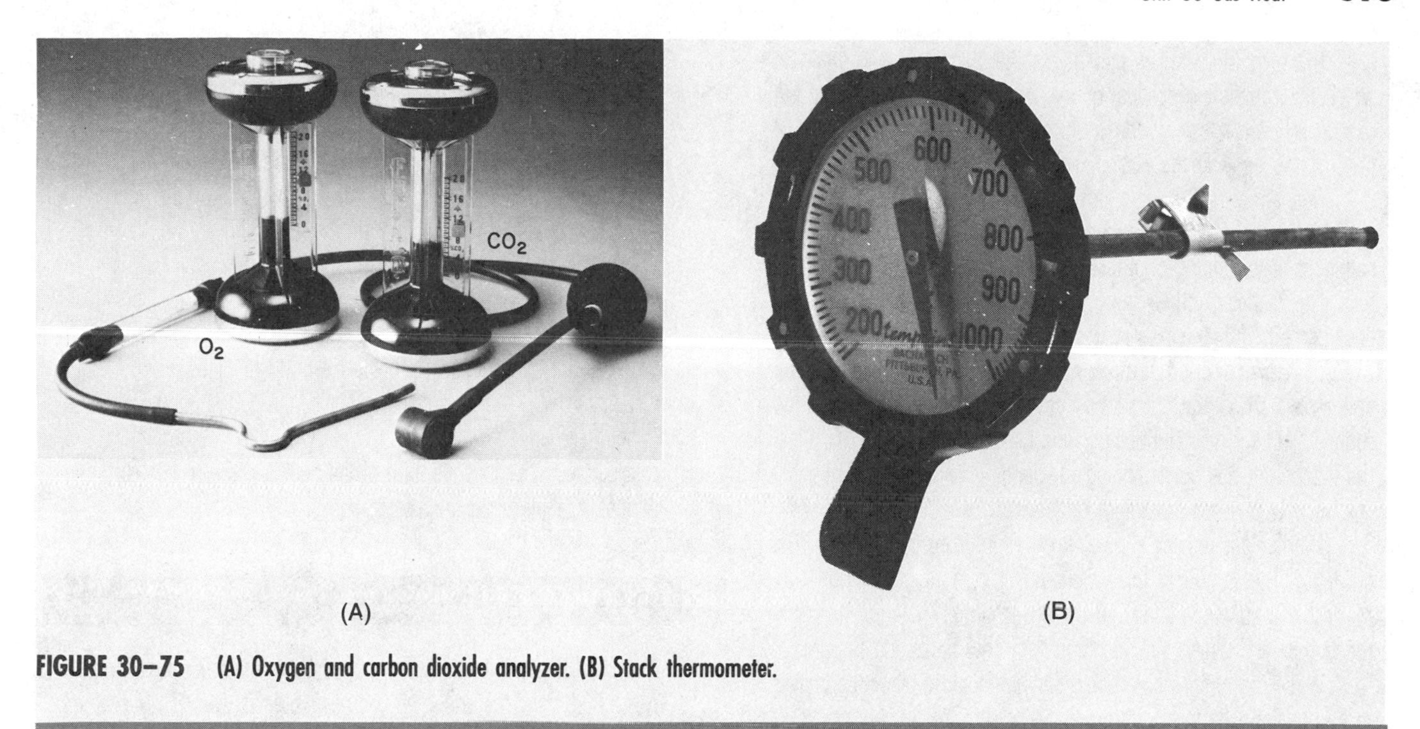

FIGURE 30–75 **(A) Oxygen and carbon dioxide analyzer. (B) Stack thermometer.**

30.37 SERVICE TECHNICIAN CALLS

SERVICE CALL 1

The manager of a retail store calls. *There is no heat. This is an upflow gas furnace with a standing pilot and air conditioning. The low-voltage transformer is burned out because the gas valve coil is shorted. This causes excess current for the transformer. The furnace is located in the stockroom.*

The technician arrives at the job, goes to the room thermostat and places the fan switch to ON to see if the indoor fan will start. It will not, so the technician suspects that the low-voltage power supply is not working. The thermostat is set to call for heat. The technician goes to the stockroom where the furnace is located. The voltage is checked at the transformer secondary, and it indicates 0 V. The power is turned off and the ohmmeter is used to check continuity of the transformer. The secondary (low-voltage) coil has an open circuit, so the transformer is changed. Before connecting the secondary wires, the continuity of the low-voltage circuit is checked. There is only 2 Ω of resistance in the gas valve coil. This is so low that it will cause a high current flow in the secondary circuit. (The resistance should be at least 20 Ω). The technician goes to the truck and checks the continuity of another gas valve and finds it to be 50 Ω, so the gas valve is also changed.

The system is started with the new gas valve and transformer. A current check of the gas valve circuit shows that it is only pulling 0.5 A, which is not overloaded. If the technician had just installed the transformer and turned the power on, it is likely that another transformer failure would have occurred quickly.

The technician changes the air filters and calls the store manager back to the stockroom for a conference before leaving. The store manager is informed that the boxes of inventory must be kept away from the furnace because they may present a fire hazard.

SERVICE CALL 2

A customer in a residence reports that *fumes can be smelled and are probably coming from the furnace in the hall closet. The furnace has not been operated for the past 2 weeks because the weather has been mild. Because the fumes might be harmful, the dispatcher tells the customer to shut the furnace off until a technician arrives. The flue is stopped up with a shingle laid on top of the brick chimney by a roofer who had been making repairs.*

The technician arrives and starts the furnace. A match is held at the draft diverter to see if the flue has a negative pressure. The flue-gas fumes are not rising up the flue for some reason; the match flame blows away from the flue pipe and draft diverter. The technician turns off the burner and examines it and the heat exchanger area with a flashlight. There is no soot that would indicate the burner had been burning incorrectly. The technician goes to the roof to check the flue and notices a shingle on top of the chimney. The shingle is lifted off and heat then rises out of the chimney.

The technician goes back to the basement and starts the furnace again. A match is held at the draft diverter, and the flame is drawn toward the flue. The furnace is now operating correctly. The furnace filter is changed, and the fan motor is oiled before the technician leaves the job.

SERVICE CALL 3

A customer requests that an efficiency check be performed on a furnace. *This customer thinks that the gas bill is too high for the conditions and wants the system completely checked out.*

The technician arrives and meets the customer, who wants to watch the complete procedure of the service call. The technician will run an efficiency test on the furnace burners at the beginning and then at the end, so the customer can see the difference.

The furnace is in the basement and easily accessible. The technician turns the thermostat to 10°F above the room temperature setting to ensure that the furnace will not shut off in the middle of the test. The technician then goes to the truck and gets the flue-gas analyzer kit. ***Being careful not to touch the stack, the stack temperature is taken at the heat exchanger side of the draft diverter.*** A sample of the flue gas is drawn into the sample chamber after the flue-gas temperature has stopped rising and the indoor fan has been running for about 5 min. This ensures that the furnace is up to maximum temperature.

The flue-gas reading shows that the furnace is operating at 80% efficiency. This is normal for a standard efficiency-type furnace. After the test is completed and recorded, the technician shuts off the gas to the burner and allows the fan to run to cool down the furnace.

When the furnace has cooled down enough to allow the burners to be handled, the technician takes them out. The burners are easily removed in this furnace by removing the burner shield and pushing the burners forward one at a time while raising up the back. They will then clear the gas manifold and can be removed.

There is a small amount of rust, which is normal for a furnace in a basement. The draft diverter and the flue pipe are removed so that the technician and customer can see the top of the heat exchanger. All is normal—no rust or scale.

With a vacuum cleaner the technician removes the small amount of dirt and loose rust from the heat exchanger and burner area. The burners are taken outside where they are blown out with a compressed air tank at the truck ***The technician wears eye protection.*** After the cleaning is complete, the technician assembles the furnace and tells the customer that the furnace was in good condition and that no difference in efficiency will be seen. Modern gas burners do not stop up as badly as the older ones, and the air adjustments will not allow the burner to get out of adjustment more than 2% or 3% at the most.

The fan motor is oiled, and the filter is changed. The system is started and allowed to get up to normal operating temperature. While the furnace is heating to that temperature, the technician checks the fan current and finds it to be running at full load. The fan is doing all the work that it can. The efficiency check is run again, and the furnace is still operating at about 80%. It is now clean, and the customer has peace of mind. The thermostat is set before leaving.

SERVICE CALL 4

A customer indicates that *the pilot light will not stay lit. The pilot light goes out after it has been lit a few minutes. The heat exchanger has a hole in it very close to the pilot light. The pilot light will light, but when the fan starts it blows out the pilot light.*

The technician arrives and checks the room thermostat to see that it is set above the room temperature. The technician then goes to the basement. The standing pilot light is not lit, so the technician lights it and holds the button down until the thermocouple will keep the pilot light lit. When the gas valve is turned to ON, the main burner lights. Everything is normal until the fan starts, then the pilot light flame starts to wave around and after a short period it goes out. The thermocouple cools, and the gas to the main burner shuts off. The technician shows the customer the hole in the heat exchanger. The gas valve is turned off, and it is explained to the customer that the furnace cannot be operated in this condition because of the potential danger of gas fumes.

The technician explains to the customer that this furnace is 18 years old, and the customer really should consider getting a new one. A heat exchanger can be changed in this furnace, but it requires considerable labor plus the price of the heat exchanger. The customer decides to replace the furnace.

SERVICE CALL 5

A residential customer calls. *There is no heat. The main gas valve coil is shorted, and the low-voltage transformer is burned. This is an electronic intermittent ignition system, Figure 30–69. This symptom can lead the technician to believe that the printed circuit board is defective.*

The technician arrives and notices the indoor fan is running. The technician cannot stop the fan from running by turning the room thermostat to OFF. (The low-voltage circuit holds the fan motor off in this system.) A low-voltage problem is suspected. The technician goes to the furnace and checks the output of the low-voltage transformer; the meter reads 0 V at the transformer. The transformer is burned. ***The technician turns off the power and replaces the transformer.*** When the power is restored, the electronic ignition circuit lights the pilot light. Everything seems normal, but the transformer becomes extra hot and smells like it is burning.

The technician turns off the power and changes the electronic circuit board and 1-A fuse. When power is restored, the pilot lights and everything is normal for a moment, but the transformer is still getting hot.

The technician turns off the power and fastens a clamp-on ammeter in the low-voltage system, using 10 wraps of wire to amplify the ampere reading, Figure 12–52. The amperage is normal. When power is restored, the pilot lights again and the amperage is normal (about 0.5 A). When the pilot proves and it is time for the main gas valve to be energized, the current goes up to 3 A and the transformer begins to heat up.

The technician turns off the power and uses the ohmmeter to check the resistance of all power-consuming components in the circuit. When the main gas valve coil in the gas valve is checked, it shows 2 Ω of resistance.

The gas valve is replaced. Power is restored. The pilot light is lit and proved. The main gas valve opens and the burner ignites. The system operates normally.

The technician turns off the power and changes the circuit board back to the original board and then restores the power. The furnace goes through a normal startup.

The following service calls do not have the solutions described. The solutions can be found in the Instructor's Guide.

SERVICE CALL 6

A residential customer called to indicate that the furnace stopped in the middle of the night. The furnace is old and has a thermocouple for the safety pilot.

The technician arrives and goes directly to the thermostat. The thermostat is set correctly for heating. The thermometer in the house shows a full 10°F below the thermostat setting. This unit has air conditioning, so the technician turns the fan switch to ON to see if the indoor fan will start; it does.

The furnace is in an upstairs closet. The technician sees that the pilot light is not burning. This system has 100% shutoff, so there is no gas to the pilot unless the thermocouple holds the pilot-valve solenoid open. The technician positions the main gas valve to the PILOT position and presses the red button to allow gas to the pilot light. The pilot light burns when a lit match on an extender is placed next to it. The technician then holds the red button down for 30 seconds and slowly releases it. The pilot light goes out.

What is the likely problem and the recommended solution?

SERVICE CALL 7

A residential customer calls. The furnace is not heating the house. The furnace is located in the basement and is very hot, but no heat is moving into the house. The dispatcher tells the customer to turn the furnace off until a service technician can get to the job.

The technician has some idea what the problem is from the symptoms. The technician knows what kind of furnace it is from previous service calls, so some parts are brought along. When the technician arrives, the room thermostat is set to call for heat. From the service request it is obvious that the low-voltage circuit is working because the customer says the burner will come on. The technician then goes to the basement and hears the burner operating. When the furnace has had enough time to get warm and the fan has not started, the technician takes the front control panel off the furnace and notices that the temperature-operated fan switch dial (circular dial type, Figure 30–31) has rotated as if it were sensing heat. ***The technician carefully checks the voltage entering and leaving the fan switch.***

What is the likely problem and the recommended solution?

SERVICE CALL 8

A residential customer calls. There is no heat. The furnace is in the basement, and there is a sound like a clock ticking.

The service technician arrives and goes to the basement. The furnace door is removed. The technician hears the arcing sound. The shield in front of the burner is removed. The technician sees the arc. A match on an extender is lit and placed near the pilot to see if there is gas at the pilot and to see if it will light. It does. The arcing stops, as it should. The burner lights after the proper time delay.

What is the likely problem and the recommended solution?

SUMMARY

- Forced-hot air furnaces are normally classified as upflow, downflow (counterflow), horizontal, or low-boys.
- Furnace components consist of the cabinet, gas valve, manifold, pilot, burners, heat exchangers, blower, electrical components, and venting system.
- Gas fuels are natural gas, manufactured gas (used in few applications), and LP gas (propane, butane, or a mixture of the two).
- Inches of water column is the term used when determining or setting gas fuel pressures.
- A water manometer measures gas pressure in inches of water column.
- For combustion to occur there must be fuel, oxygen, and heat. The fuel is the gas, the oxygen comes from the air, and the heat comes from the pilot flame or other igniter.
- Gas burners use primary and secondary air. Excess air is always supplied to ensure as complete combustion as possible.
- Gas valves control the gas flowing to the burners. The valves are controlled automatically and allow gas to flow only when the pilot is lit or when the ignition device is operable.
- Some common gas valves are classified as solenoid, diaphragm, or heat motor valves.
- Ignition at the main burners is caused by heat from the pilot or from an electric spark. There are standing pilots that burn continuously, and intermittent pilots that are ignited by a spark when the thermostat calls for heat. There is also a direct-spark ignition, in which the spark ignites the gas at the burners.
- The thermocouple, the bimetal, and the liquid-filled remote bulb are three types of safety devices (flame-proving devices) to ensure that gas does not flow unless the pilot is lit.
- The manifold is a pipe through which the gas flows to the burners and on which the burners are mounted.
- The orifice is a precisely sized hole in a spud through which the gas flows from the manifold to the burners.
- The burners have holes or various designs of slots through which the gas flows. The gas burns immediately on the outside of the burners at the top or end depending on the type.
- The gas burns in an opening in the heat exchanger. Air passing over the heat exchanger is heated and circulated to the conditioned space.
- A blower circulates this heated air. The blower is turned on and off by a fan switch, which is controlled either by time or by temperature.

- The limit switch is a safety device. If the fan does not operate or if the furnace overheats for another reason, the limit switch causes the gas valve to close.
- ***Venting systems must provide a safe and effective means of moving the flue gases to the outside atmosphere. Flue gases are mixed with other air through the draft hood. Venting may be by natural draft or by forced draft. Flue gases are corrosive.***
- Gas piping should be kept simple with as few turns and fittings as possible. It is important to use the correct size pipe.
- ***All piping systems should be tested carefully for leaks.***
- A glow coil is a reignition device that ignites the pilot if it goes out.
- High-efficiency gas furnaces have been developed. More heat is exchanged in the furnace for distribution to the conditioned space.
- The combustion efficiency of furnaces should be checked and adjustments made when needed.
- Preventive maintenance calls should be made periodically on gas furnaces.

REVIEW QUESTIONS

1. List the four types of gas furnaces and describe each.
2. What two gas fuels are most commonly used in residential furnaces?
3. What is the specific gravity of each type of gas fuel? How does the specific gravity affect each type?
4. Describe a manometer.
5. What is an inch of water column?
6. What percentage of oxygen is contained in air?
7. Why is the percentage of oxygen in air a factor when determining proper air supply to support gas combustion?
8. What is a typical pressure in inches of water column for natural gas at the manifold of the furnace?
9. Why is excess air supplied to support combustion in a gas furnace?
10. Why is a gas regulator necessary before the gas reaches the valve?
11. List three types of gas valves.
12. Describe an automatic combination gas valve.
13. Describe two types of pilots.
14. When is a glow-coil reignition system used?
15. Describe how a thermocouple flame-proving safety device works.
16. Describe the bimetallic flame-proving safety device.
17. Describe the liquid-filled remote-bulb flame-proving safety device.
18. Why are the devices in Questions 15–17 considered safety devices?
19. What does the manifold do?
20. What is an orifice?
21. List four types of gas burners.
22. Describe the function of a heat exchanger.
23. Why is it important that a heat exchanger not be corroded?
24. How does a temperature on-temperature off fan switch operate?
25. How does a time-delay fan switch differ from a temperature on-temperature off switch?
26. What is a limit switch? How does it work?
27. Sketch a vent hood and describe how it functions. What is dilution air?
28. How does a vent hood or draft diverter operate in downdraft conditions?
29. What two types of venting systems are used for gas-fuel installations?
30. Why is a metal vent often preferred over a masonry chimney?
31. Describe how at least one type of automatic vent damper operates.
32. What kind of metal is most gas piping made from?
33. Why does the size (diameter) of gas piping have to be considered when designing a piping system?
34. Describe the proper procedures for making a threaded joint in gas piping.
35. What three types of fittings are needed in a piping system just before the gas valve?
36. Describe the procedure to test for leaks in a piping system.
37. Sketch a wiring diagram for a basic gas furnace.
38. Describe the load and no-load thermocouple test.
39. Why is a spark-to-pilot ignition system often preferred over a standing pilot system?
40. How does a flame rectification system operate?
41. Describe a direct-spark ignition system.
42. List differences between modern high-efficiency gas furnaces and older models.
43. Why is a carbon dioxide test made on a gas furnace?
44. Describe five preventive maintenance procedures performed on a gas furnace.

UNIT 30 DIAGNOSTIC CHART FOR GAS HEAT

The gas-heat appliances discussed here are the forced-air type. (Many of these troubleshooting procedures may be used for gas boilers as well.) Forced-air systems may be found in homes and businesses and are used as primary heating systems. The gas-heat appliances may either be the standing pilot or the electronic-ignition type. The electronic-ignition type of furnace discussion covers basic procedure because each manufacturer has its own sophisticated troubleshooting procedure. Always listen to customers for their observations as to what is happening. They can often supply the clue for finding the problem. Remember, they are in contact with the system daily.

PROBLEM	POSSIBLE CAUSE	POSSIBLE REPAIR	PARAGRAPH NUMBER
STANDING PILOT-LIGHT SYSTEMS			
No heat—burner does not light	Open disconnect switch	Close disconnect switch	30.30
	Open fuse or breaker	Replace fuse or reset breaker and determine why it opened	30.30
	Faulty wiring	Repair or replace faulty wiring or connections	30.30
	Defective low voltage transformer	Replace transformer and look for possible overload condition	30.30, 30.31
	Open pilot safety (bimetal) contact	Change pilot assembly	30.20
	Pilot light not lit	Light pilot	30.17
	Open limit control circuit	Change limit control if defective	30.16
Pilot light will not stay lit	**A.** Thermocouple systems		
	Defective thermocouple	Check and replace when defective	30.33
	Pilot light flame directed wrong	Redirect pilot flame	30.17
	Defective gas Valve for thermocouple	Change coil of gas valve	30.19
	B. Liquid filled remote bulb systems		
	Pilot-light flame directed wrong	Redirect pilot flame	30.17
	Defective sensor	Change sensor or valve containing sensor assembly	30.21
Burner lights—fan does not run	Faulty wiring or connectors in fan circuit	Repair or replace faulty wiring or connectors	30.30
	Defective fan switch	Replace fan switch	30.15
	Defective fan motor	Change fan motor	20.6–20.9
	Defective fan relay	Change fan relay	19.3
Insufficient heat	Furnace undersized	Add insulation to structure to reduce load or change to larger furnace	—
	Insufficient air across furnace causing burner to cycle due to high temperature	Increase airflow	42.5
	Leaking supply duct to unconditioned area	Repair duct	37.16–37.20
	Leaking return air pulling in unconditioned air	Repair duct	37.16–37.20
	Underfired due to low gas pressure	Set gas pressure to correct setting using manometer	30.5
ELECTRONIC IGNITION SYSTEMS			
No heat—burner does not light	See no heat for standing pilot		
No heat. Fan runs	Gas valve may be in "off" position	Turn gas valve to "on" position. Follow manufacturer's check-out procedures	30.23–30.25
	Defective circuit board	If defective, change circuit board	30.35

UNIT 31

Oil Heat

OBJECTIVES

After studying this unit, you should be able to

- **describe how fuel oil and air are prepared and mixed in the oil burner unit for combustion.**
- **list products produced as a result of combustion of the fuel oil.**
- **list the components of gun-type oil burners.**
- **describe basic service procedures for oil burner components.**
- **sketch wiring diagrams of the oil burner primary control system and the fan circuit.**
- **state tests used to determine oil burner efficiency.**
- **explain corrective actions that may be taken to improve burner efficiency, as indicated from the results of each test.**
- **describe preventive maintenance procedures.**

SAFETY CHECKLIST

* Do not reset any primary control too many times because unburned oil may accumulate after each reset. If a puddle of oil is ignited, it will burn intensely. If this should happen, stop the burner motor but allow the furnace fan to run. The air shutter should be shut off to reduce the air to the burner. Notify the fire department. Do not try to open the inspection door to put the fire out; let it burn itself out with reduced air.
* Do not start a burner if the heat exchanger is cracked or otherwise defective. The flue gases from combustion should never mix with the circulating air.
* Do not start the burner with the fuel pump bypass plug in place unless the return line to the tank is in place. Without this line, there is no place for the excess oil to go except to possibly rupture the shaft seal.
* When taking flue-gas efficiency tests avoid burns by being careful to avoid touching the hot flue-gas pipe.
* Do not apply more than 10 psig to the oil line filter housing.
* The ignition is produced by an arc of 10 000 V. Keep your distance from this arc.
* Observe all electrical safety precautions.

31.1 INTRODUCTION TO OIL-FIRED FORCED-WARM AIR FURNACES

The oil-fired forced-warm air furnace has two main systems: a heat-producing system and a heat-distributing system. The *heat-producing system* consists of the oil burner, the fuel supply components, combustion chamber, and heat exchanger. The *heat-distributing system* is composed of the blower fan, which moves the air through the ductwork, and other related components, Figure 31–1.

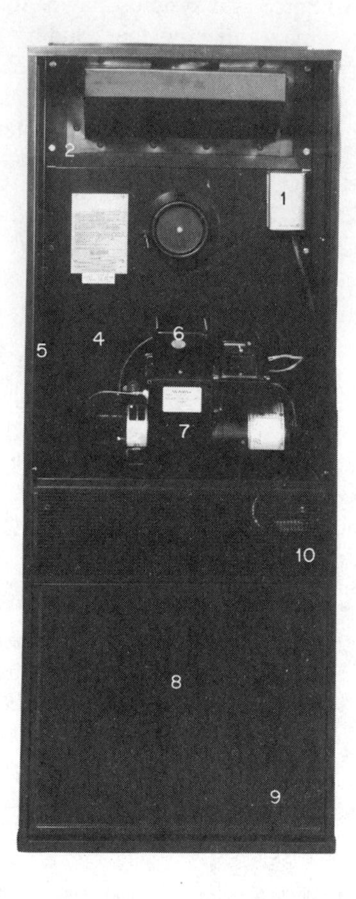

FIGURE 31–1 **Forced-warm air oil furnace.** ***Courtesy Ducane Corporation***

Following is a brief description of how such a furnace operates. When the thermostat calls for heat, the oil burner is started. The oil-air mixture is ignited. After the mixture has heated the combustion chamber and the heat exchanger reaches a set temperature, the fan comes on and distributes the heated air through the ductwork to the space to be heated. When this space has reached a predetermined temperature, the thermostat will cause the burner to shut down. The fan will continue to run until the heat exchanger cools to a set temperature.

31.2 PHYSICAL CHARACTERISTICS

The physical appearance and characteristics of forced-air furnaces vary to some extent. The *low-boy* is often used when there is not much headroom, Figure 31–2. Low-boys may have a cooling coil on top to provide air conditioning.

An *upflow* furnace is a vertical furnace in which the air is taken in at the bottom and is forced across the heat exchanger and out the top, Figure 31–1. This furnace is installed with the ductwork above it such as in the attic, or it can be installed in a basement with the ductwork above it under the floor.

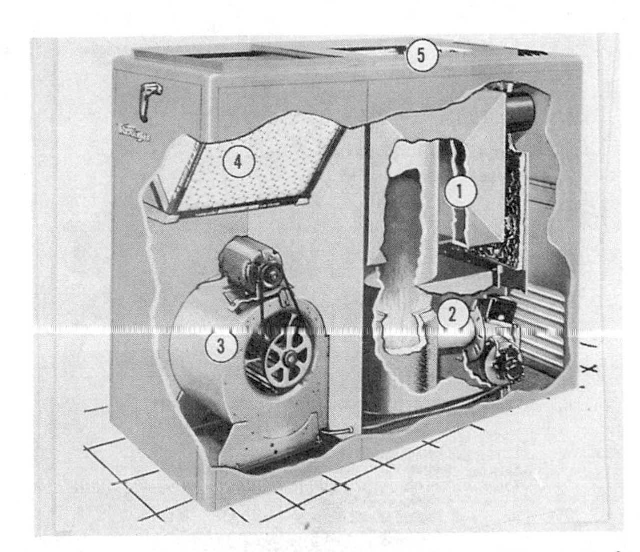

FIGURE 31–2 Low-boy forced-air furnace. *Courtesy Metzger Machine Corp.*

A *downflow* furnace looks similar to an upflow except that the air is drawn in from the top and forced out the bottom, Figure 31–3. The ductwork in this case would be just below or in the floor below the furnace.

A *horizontal* furnace is usually installed in a crawl space under a house or in an attic, Figure 31–4. These units are available with right-to-left or left-to-right airflow.

Oil-fired forced-air heating is a popular method of heating residences and light commercial buildings. Several million furnaces of this type are being used in the country at the

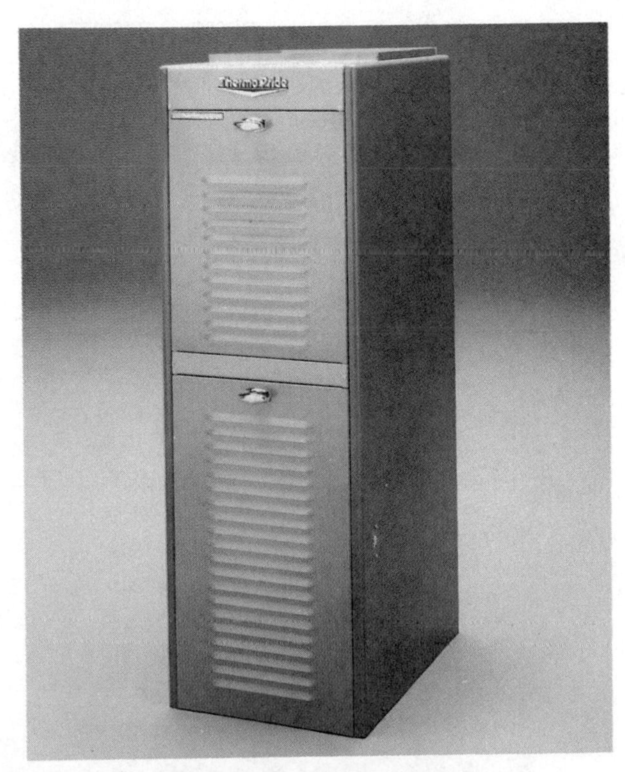

FIGURE 31–3 A downflow or counterflow furnace. *Courtesy Thermopride Williamson Company*

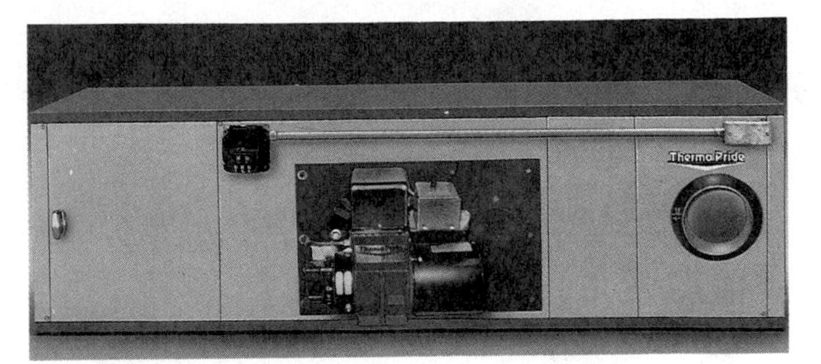

FIGURE 31–4 Horizontal forced-air furnace. *Courtesy ThermoPride Williamson Company*

present time. It is essential that the air conditioning and heating technician understand proper installation and servicing techniques involved with oil burners. Safety is always a concern of technicians but with the high cost of fuel oils, efficiency in the combustion process is also extremely important.

31.3 FUEL OILS

Six grades of fuel oil are used as heating oils, with the No. 2 grade the one most commonly used to heat residences and light commercial buildings. Fuel oil is obtained from petroleum pumped from the ground. The lighter fuel oils are products of a distillation process at the oil refinery during which the petroleum is vaporized, condensed, and the different grades separated.

Fuel oil is a combination of liquid hydrocarbons containing hydrogen and carbon in chemical combination. Some of these hydrocarbons in the fuel oil are light, and others are heavy. Combustion occurs when the hydrocarbons unite rapidly with oxygen (O_2). Heat, carbon dioxide (CO_2), and water vapor are produced when the combustion occurs.

31.4 PREPARATION OF FUEL OIL FOR COMBUSTION

Fuel oil must be prepared for combustion. It must first be converted to a gaseous state by forcing the fuel oil, under pressure, through a nozzle. The nozzle breaks the fuel oil up to form tiny droplets. This process is called *atomization*, Figure 31–5. The oil droplets are then mixed with air, which contains oxygen. The lighter hydrocarbons form gas covers around the droplets. Heat is introduced at this point with a spark. The vapor ignites (combustion takes place), and the temperature rises, causing the droplets to vaporize and burn.

Perfect combustion exists in theory only. In reality, products other than carbon dioxide and water vapor are produced in this burning process. These other products are carbon monoxide, soot, and unburned fuel.

High-pressure gun-type oil burners are generally used to achieve this combustion. Oil is fed under a pressure of 100 psi to a nozzle. Air is forced through a tube that surrounds this nozzle. Usually the air is swirled in one direc-

FIGURE 31–5 Atomized fuel oil droplets and air leaving burner nozzle. *Courtesy Delavan Inc.*

tion and the oil in the opposite direction. The oil forms into tiny droplets combining with the air. The ignition transformer furnishes a high-voltage spark between two electrodes located near the front of the nozzle, and combustion occurs.

31.5 GUN-TYPE OIL BURNERS

The main parts of a gun-type oil burner are the burner motor, blower or fan wheel, pump, nozzle, air tube, electrodes, transformer, and primary controls, Figures 31–6 and 31–7.

FIGURE 31–6 An oil burner. *Courtesy American Burner Corporation*

Burner Motor

The oil burner *motor* is usually a split-phase fractional horsepower motor that provides power for both the fan and the fuel pump. A flexible coupling is used to connect the shaft of the motor to the shaft of the pump, Figure 31–7. The motor speeds may be either 1750 or 3450 rpm. The pump should always match the motor rpm.

Burner Fan or Blower

The burner *fan* or blower is a squirrel cage type with adjustable air inlet openings in a collar attached to the blower housing. This provides a means for regulating the volume of air being drawn into the blower. The fan forces air through the air tube to the combustion chamber where it is mixed

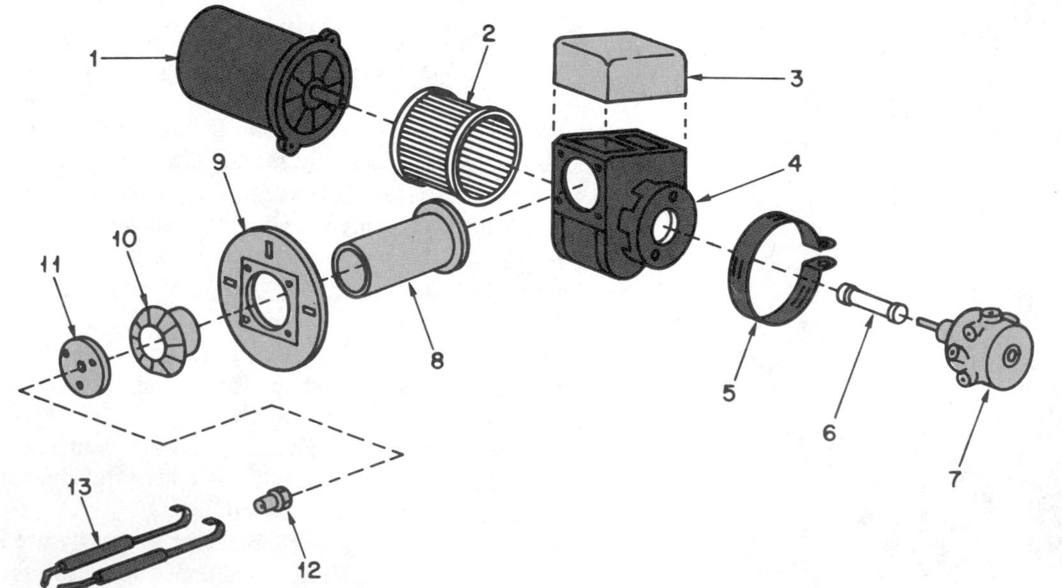

FIGURE 31–7 Exploded view of an oil burner assembly.

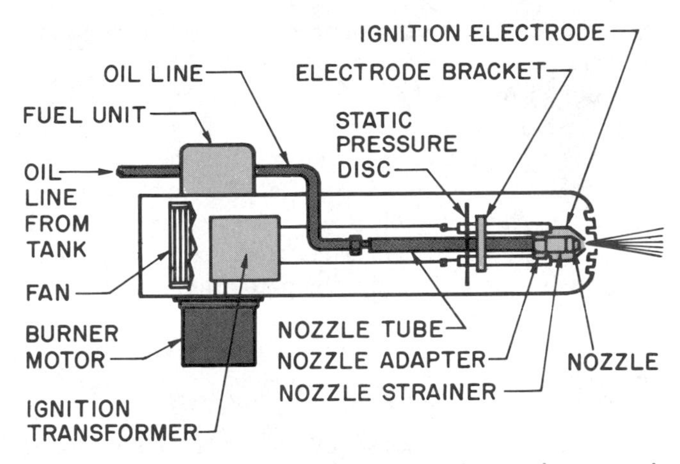

FIGURE 31–8 Schematic drawing (top view) of a typical gun-type oil burner. *Courtesy Honeywell, Inc., Residential Division*

with the atomized fuel oil to provide the necessary oxygen to support the combustion, Figure 31–8.

Fuel Oil Pumps

Several types of fuel oil pumps are used in gun-type oil burners, Figure 31–9. A *single-stage pump* is used when the fuel oil storage tank is above the burner. The fuel oil flows to the burner by gravity, and the pump provides oil pressure to the nozzle, Figure 31–10.

A *one-pipe* supply system may be used with a single-stage pump. This means that there is one pipe from the tank to the burner. In a normal operation a surplus of oil is pumped to the nozzle. The nozzle cannot handle this sur-

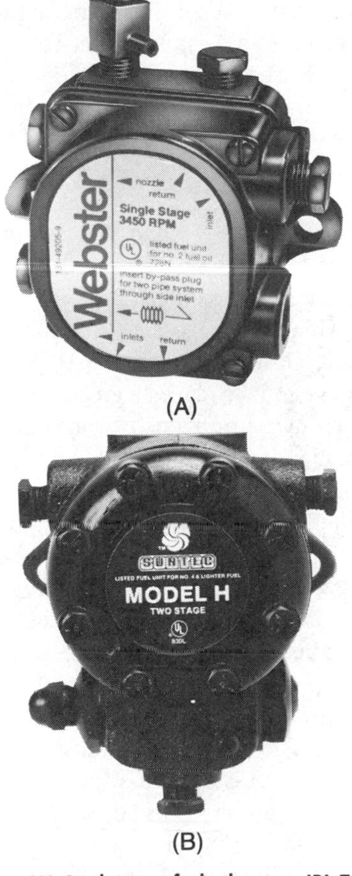

FIGURE 31–9 (A) Single-stage fuel oil pump. (B) Two-stage fuel oil pump. *Courtesy (A) Webster Electric Company, (B) Suntec Industries Incorporated, Rockford, Illinois*

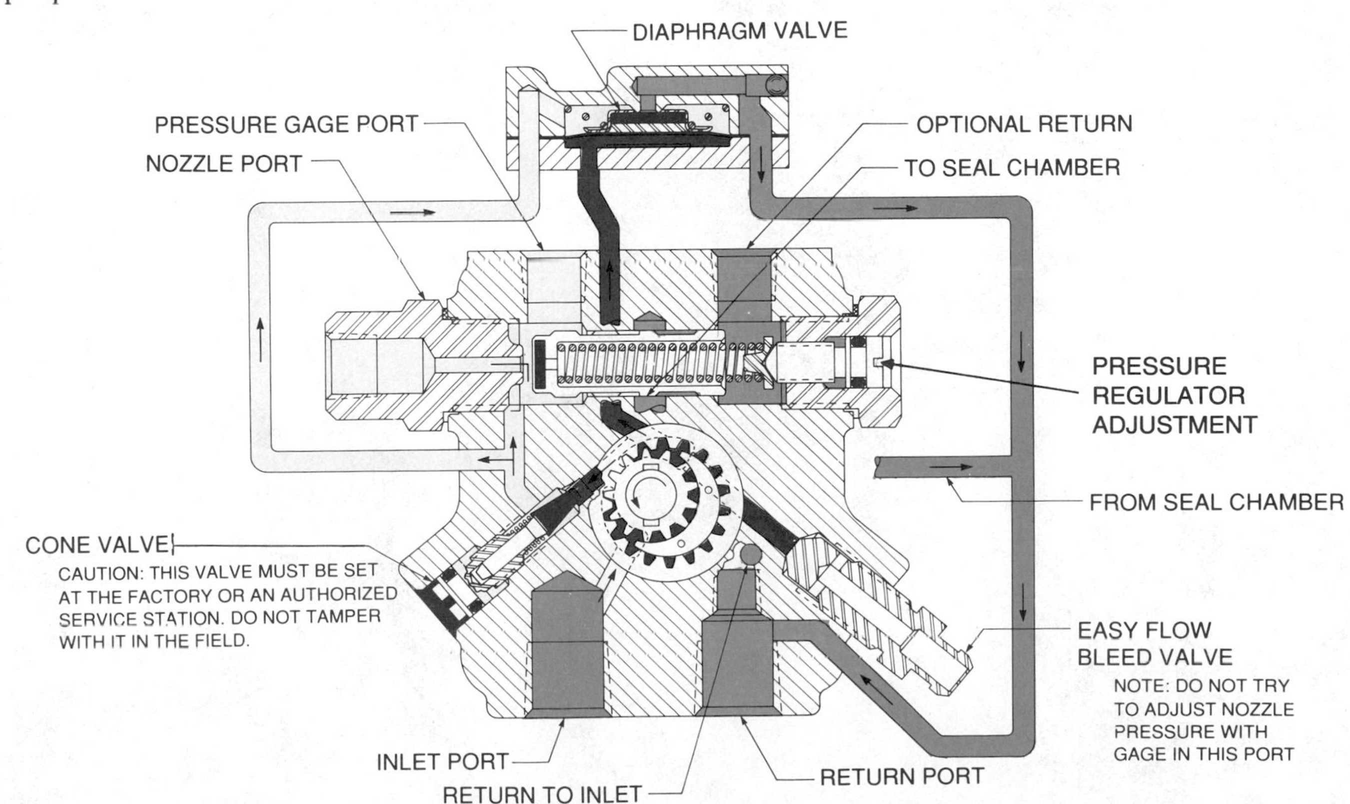

FIGURE 31–10 Single-stage fuel oil pump showing flow of oil. *Courtesy Suntec Industries, Incorporated, Rockford, Illinois*

plus, and the excess fuel is returned to the low-pressure or inlet side of the pump. When installing a one-pipe system, make sure that the bypass plug is not in place so that the surplus oil can return to the inlet side of the pump, Figure 31–11. If the fuel oil gets too low in the storage tank or if the piping is opened for any reason in a one-pipe system, the air must be bled from the system at the pump. This is discussed in more detail later.

A *dual- or two-stage pump* is used when the oil is stored below the burner. One stage of the pump lifts the oil to the pump inlet, and the other stage provides oil pressure to the nozzle. This dual-stage unit does the same job as the single-stage unit. However, it has an extra pump gear, called a suction pump gear. It produces a greater vacuum for lifting the fuel oil than can be obtained from a single-stage unit. The suction pump supplies the fuel oil needed by the pressure pump, Figure 31–12.

If the burner has a dual- or two-stage pump, the supply system should have two pipes. Many single-stage pumps are also installed with a two-pipe system. Two pipes are required with a dual-stage pump when the supply tank is below the pump because there must be a vacuum on the tank side of the pump to lift the oil. The surplus oil pumped to the nozzle in this case is returned to the storage tank through the second pipe. The bypass plug must be inserted in the pump. Two-pipe systems are self-venting (they do not have to be bled) and are preferred in all installations.

The pumps used in high-pressure gun-type burners are rotary pumps using either a cam system or gears or a combination of the two to provide the pressure, Figure 31–12. The pump itself should not be repaired by the technician. Pumps are normally replaced if defective, so it is not really important to the technician whether cams or gears are used in the pump.

Built into each pump is a pressure-regulating valve, Figure 31–10. The pump provides excessive pressure. The pressure-regulating valve can be set so that the fuel oil being delivered to the nozzle is under a set pressure of 100 psi. As mentioned previously, oil is also delivered in greater quantities than the nozzle can handle. This excess oil is diverted back to the inlet side of the pump in one pipe systems and back to the storage tank in two pipe systems.

Nozzle

The *nozzle* prepares the fuel oil for combustion by atomizing, metering, and patterning the fuel oil. It does this by separating the fuel into tiny droplets. The smallest of these droplets will be ignited first. The larger droplets (there are more of these) provide more heat transfer to the heat exchanger when they are ignited. The atomization of the fuel oil is a complex process. The straight lateral movement of the fuel oil must be changed to a circular motion. This circular movement opposes a similar airflow. The fuel enters the orifice of the nozzle through the swirl chamber. Figure 31–13 shows a typical nozzle.

The bore size of the nozzle is designed to allow a certain amount of fuel through at a given pressure to produce the Btu heat desired. Each nozzle is marked as to the amount of fuel it will deliver. A nozzle marked 1.00 will deliver 1 gallon of fuel oil per hour (gph) if the input pressure is 100 psi. With No. 2 fuel oil at a temperature of 60°F approximately 140,000 Btu would be produced each hour. A 0.8 nozzle would deliver 0.8 gph, and so on.

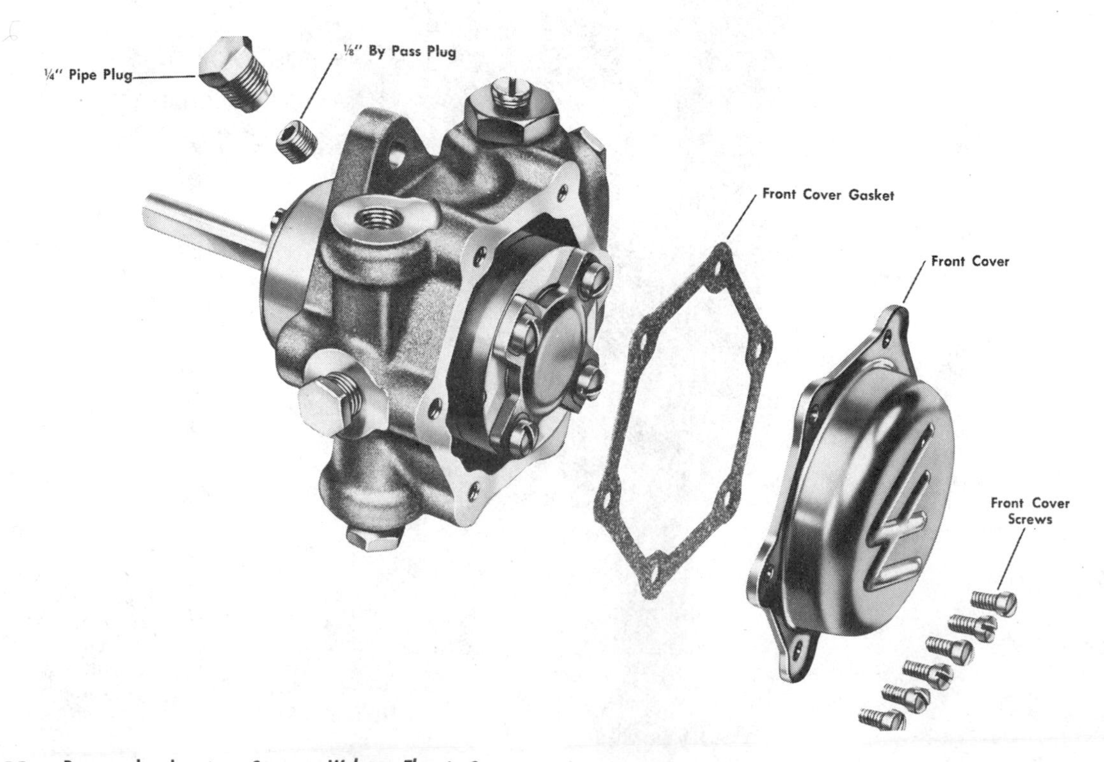

FIGURE 31–11 Bypass plug location. *Courtesy Webster Electric Company*

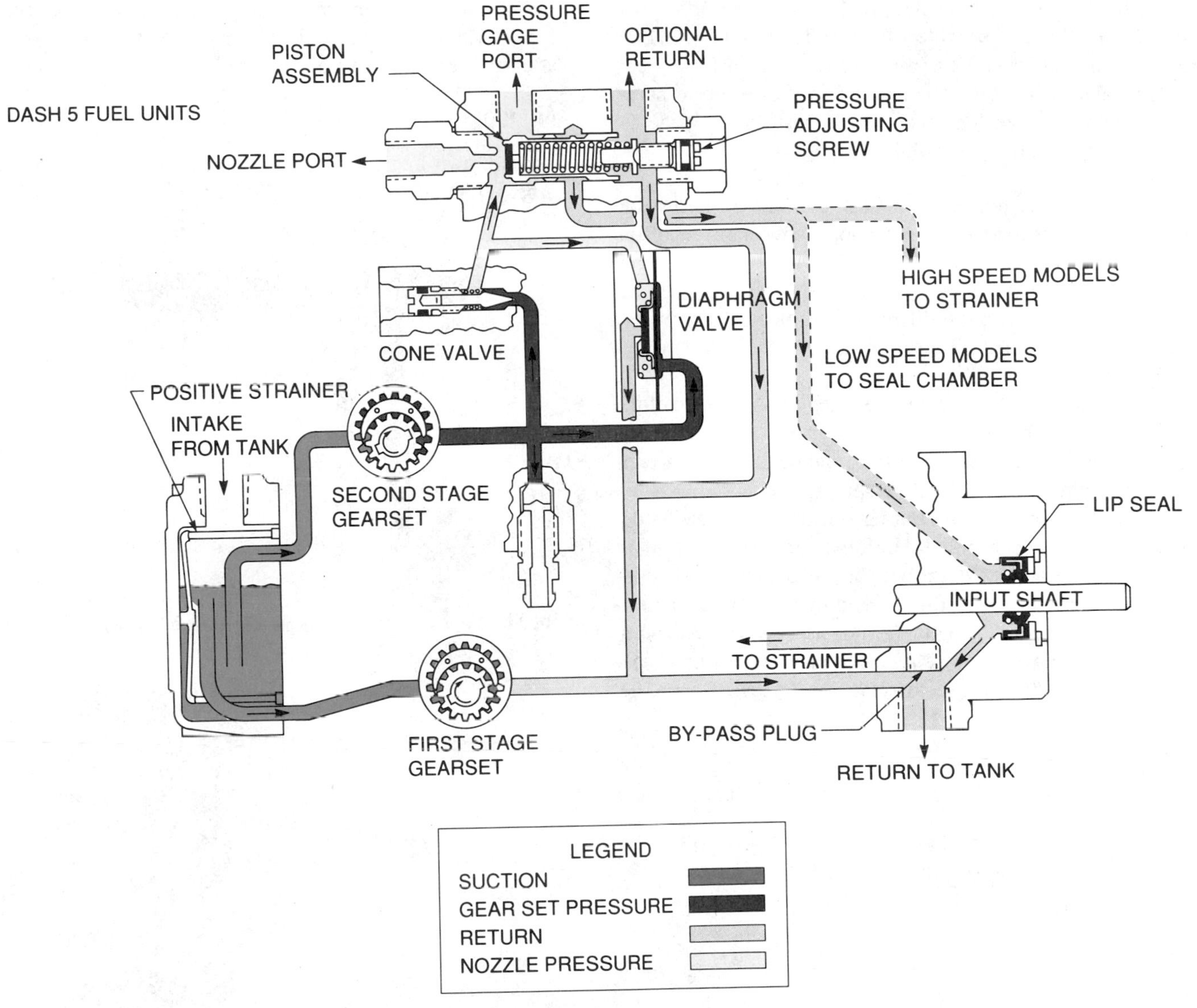

FIGURE 31–12 Two-stage fuel oil pump. *Courtesy Suntec Industries, Incorporated, Rockford, Illinois*

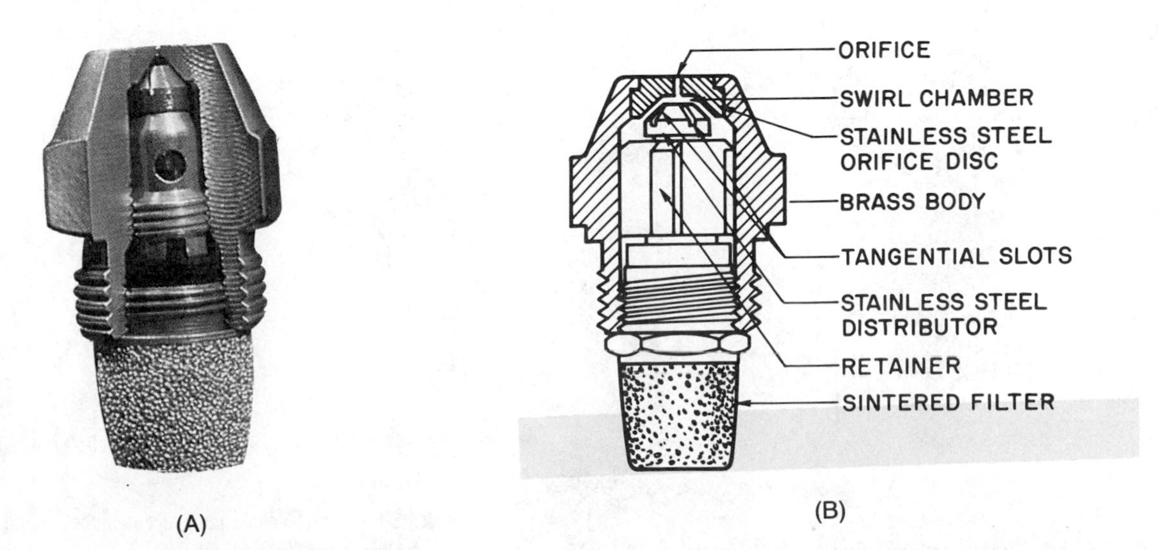

FIGURE 31–13 Oil burner nozzle. *Courtesy Delavan Inc.*

The nozzle must be designed so that the spray will ignite smoothly and provide a steady quiet fire that will burn cleanly and efficiently. It must provide a uniform spray pattern and angle that is best suited to the requirements of the specific burner. There are three basic spray patterns: hollow, semihollow, and solid with angles from 30° to 90°, Figure 31–14.

Hollow cone nozzles generally produce a more stable spray angle and pattern than do solid cone nozzles of the same flow rate. These nozzles are often used where the flow rate is under 1 gph.

Solid cone nozzles distribute droplets fairly evenly throughout the pattern. These nozzles are often used in larger burners.

Semihollow cone nozzles are often used in place of the hollow or solid cone nozzles. The higher flow rates tend to produce a more solid spray pattern, and the lower flow rates tend to produce a more hollow spray pattern.

The high-pressure oil burner nozzle is a precision device that must be handled carefully. ***Do not attempt to clean these nozzles. Metal brushes or other cleaning devices will distort or otherwise damage the precision-machined surfaces. When the nozzle is not performing properly, replace it.*** Use a nozzle changer designed specifically for this purpose to remove and install a nozzle, Figure 31–15.

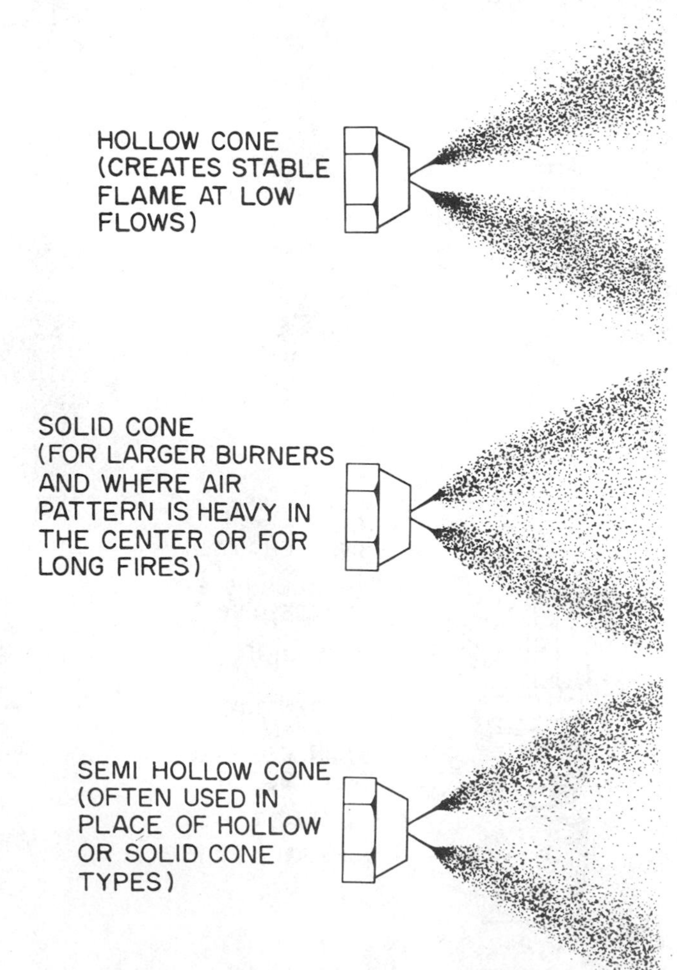

FIGURE 31–14 Spray patterns. *Courtesy Delavan Inc.*

To avoid an after-fire drip at the nozzle, a solenoid-type cut-off valve can be installed. This valve reduces smoking after the burner is shut off, Figure 31–16.

Air Tube

Air is blown into the combustion chamber through the *air tube*. Within this tube the straight or lateral movement of the

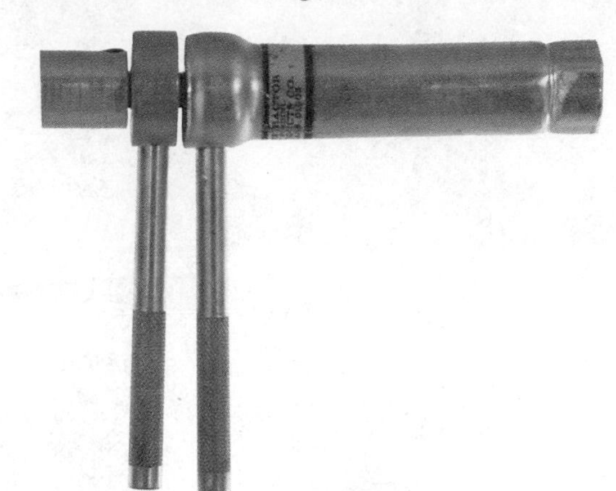

FIGURE 31–15 A nozzle changer. *Photo by Bill Johnson*

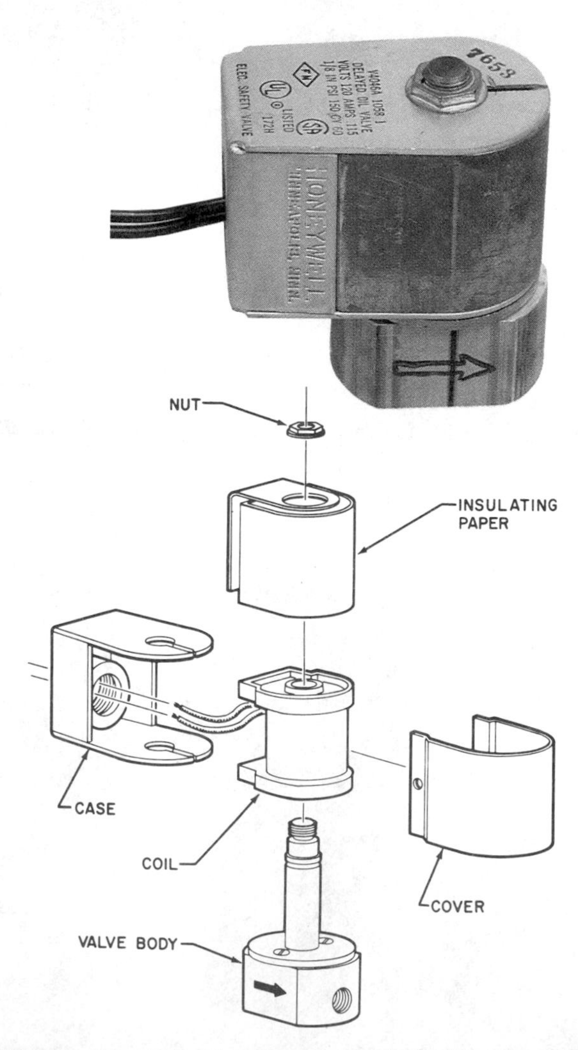

FIGURE 31–16 Solenoid positive automatic fuel oil cut-off control to avoid afterfire drip. *Photo by Bill Johnson*

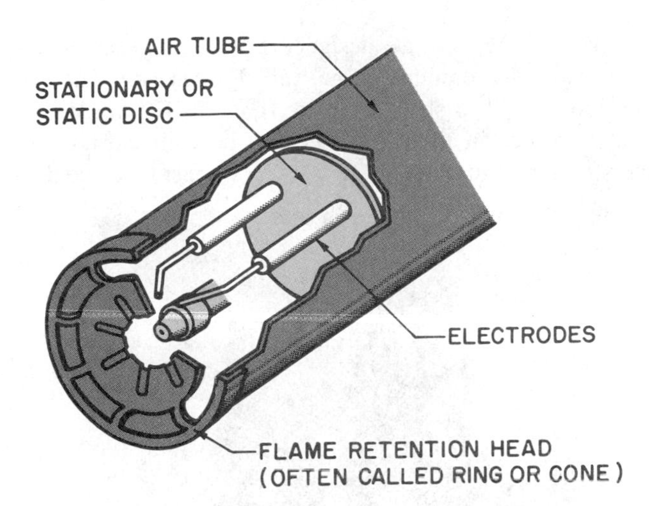

FIGURE 31–17 Oil burner air tube.

FIGURE 31–18 Flame retention device. *Courtesy Ducane Corporation*

air is changed to a circular motion. This circular motion is opposite that of the circular motion of the fuel oil. The air moving in a circular motion in one direction mixes with the fuel oil moving in a circular motion in the opposite direction in the combustion chamber. The air tube of some burners contains a stationary disc, Figure 31–17. This disc increases the static air pressure and reduces the air volume. The increase in static pressure causes an increase in air velocity for mixing with the oil droplets. The circular air motion is achieved by vanes located in the air tube after the nozzle. The head of the air tube chokes down the air and the fuel oil mixture, causing a higher velocity.

The amount of air required for the fuel oil preignition treatment is greater than the amount required for combustion. Prior to ignition approximately 2000 ft^3 of air is needed per gallon of fuel oil. Combustion requires 1540 ft^3 of air per gallon of fuel oil. The air blown into the combustion chamber is greater than what would be needed for the theoretical perfect combustion. Should the amount of air exceed the 2000 ft^3, the burner flame will be long and narrow and possibly impinge on or strike the rear of the combustion chamber. Airflow adjustments should be made only after analyzing the product of combustion, the flue gases. This is discussed in detail later in this unit.

Some burners are equipped with a *flame retention ring* or *cone* located at the front of the air tube. These rings or cones are designed to provide greater burner efficiency by creating more turbulence within the air-oil mixture, providing a stable flame front. The flame is locked to the burner cone, Figures 31–17 and 31–18.

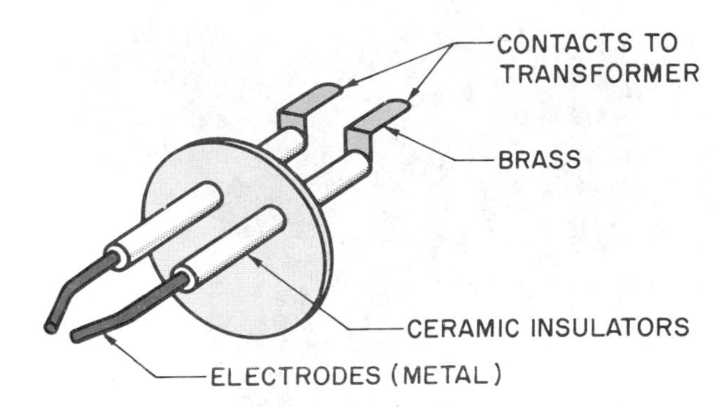

FIGURE 31–19 Electrode assembly.

Electrodes

Two *electrodes* are located within the air tube of each oil burner unit, Figure 31–19. The electrodes are metal rods insulated with a ceramic material to prevent an electrical ground. The rear portions of the electrodes are made of a flat brass alloy that must make firm contact with the transformer terminals. The position of the electrodes is adjustable. This is discussed later in this unit.

There are two types of ignition. Modern furnaces are designed so that the electrodes are providing the high-voltage spark during the entire burning cycle. This is *continuous* or *constant* ignition. Some older furnaces were designed so that the electrodes provide the spark only for a short period of time during the beginning of the burning cycle. This is called *intermittent* ignition. This system will be rarely found and only in older furnaces.

Ignition Transformer

Oil burners use a *step-up transformer* to provide high voltage to the electrodes, which produce the spark for ignition. The voltage is increased from 120 V to 10,000 V. The transformer is located on the burner assembly and is normally hinged for service purposes, Figure 31–20. The transformer terminals are often held in firm contact with the electrodes with springs.

FIGURE 31–20 Transformer mounted on oil burner assembly. *Courtesy American Burner Corporation*

FIGURE 31–21 High-voltage ignition transformer. *Photo by Bill Johnson*

Transformers cannot be serviced in the field. When a weak or defective transformer is detected, replace it with one that meets the manufacturer's specifications, Figure 31–21.

Primary Control Unit

The oil burner *primary controls* provide a means for operating the burner and a safety function whereby the burner is shut down in the event that combustion does not occur.

A primary control must turn the burner on and off in response to the low-voltage operating controls (thermostat). When the thermostat closes its contacts, low voltage is supplied to a coil that energizes and closes a switch by magnetic force. This switch transfers line voltage to the burner, and it begins to operate.

If the burner does not ignite or if it flames out during the combustion cycle, the safety function of the primary control must shut down the burner to prevent large quantities of unburned oil from accumulating in the combustion chamber.

Older installations may use a bimetal-actuated switch, called a *stack switch* or *stack relay,* for this safety feature. This switch is installed in the stack (flue) between the heat exchanger and the draft damper, Figure 31–22. This device is a heat-sensing component and is wired to shut down the burner if heat is not detected in the stack.

FIGURE 31–22 Stack switch. *Photo by Bill Johnson*

Modern furnaces are designed with a light-sensing device made of cadmium sulfide, called a *cad cell.* The cad cell works with a special primary control, Figure 31–23, and is located in the oil burner unit beneath the ignition transformer, Figures 31-24 and 31-25. It must be located so

FIGURE 31–23 This primary control works in conjunction with the cad cell for flame detection. *Photo by Bill Johnson*

FIGURE 31–24 Cad cell. *Photo by Bill Johnson*

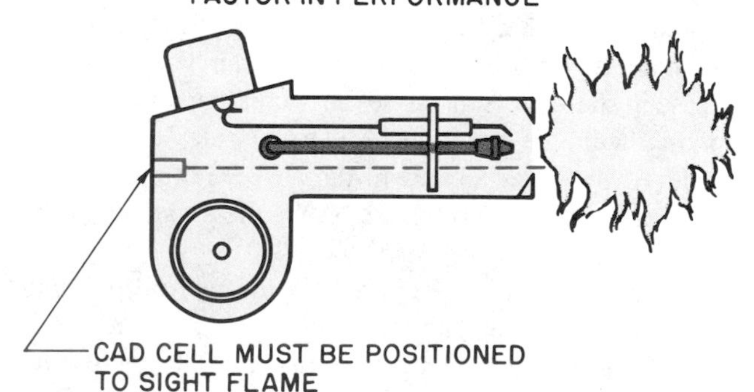

1. CELL REQUIRES A DIRECT VIEW OF FLAME.
2. ADEQUATE LIGHT FROM THE FLAME MUST REACH THE CELL TO LOWER ITS RESISTANCE SUFFICIENTLY.
3. CELL MUST BE PROTECTED FROM EXTERNAL LIGHT.
4. AMBIENT TEMPERATURE MUST BE UNDER 140°F.
5. LOCATION MUST PROVIDE ADEQUATE CLEARANCE. METAL SURFACES MUST NOT AFFECT CELL BY MOVEMENT, SHIELDING OR RADIATION.

FIGURE 31–25 Cad cell location.

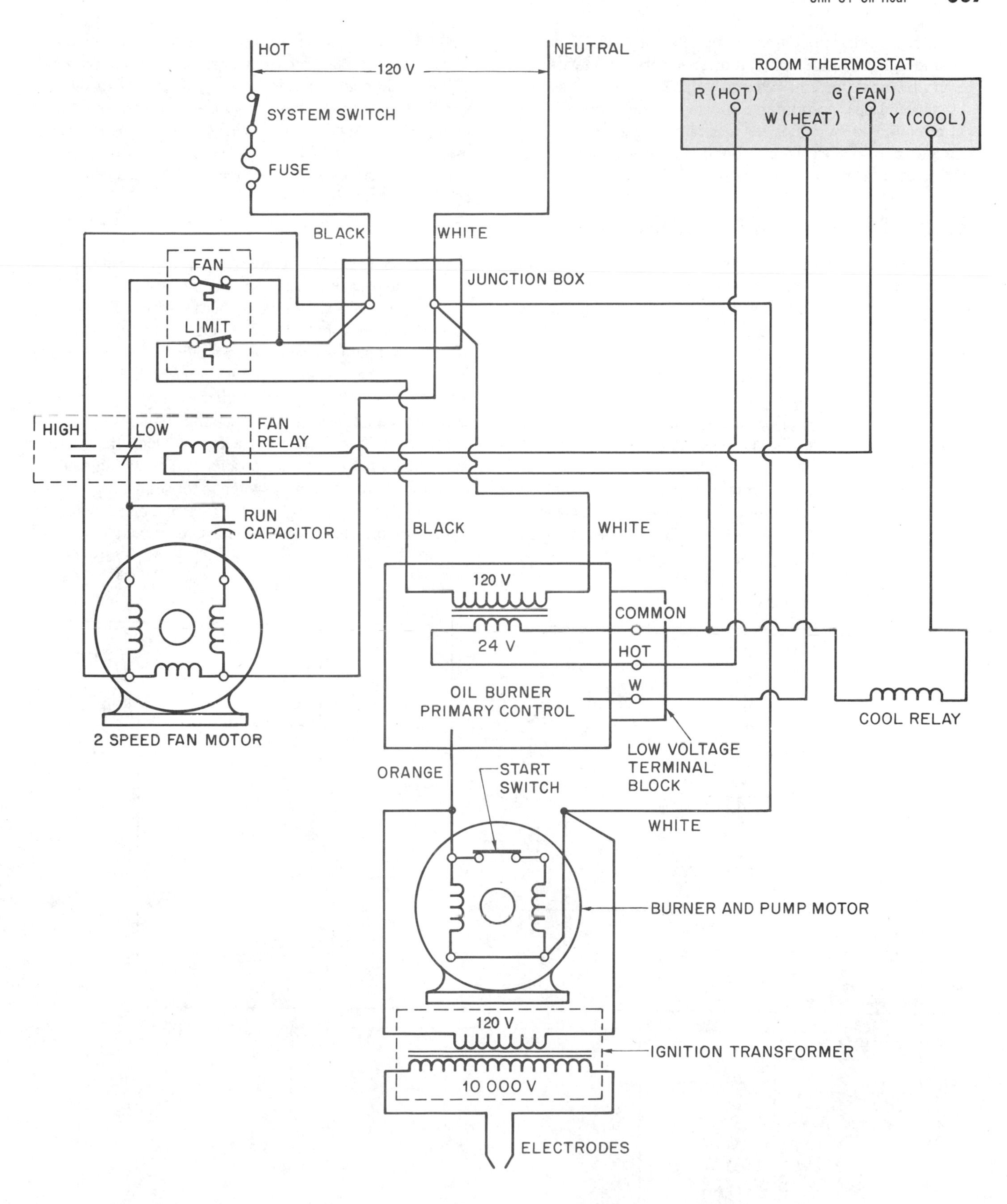

FIGURE 31–26 The primary control has a hot wire (black) and a neutral wire (white) feeding in. When a call from the thermostat for heat occurs, the orange wire is energized and the burner motor starts at the same time as the arc at the electrode.

that it can "sight" the flame through the air tube. When not sensing light from the flame, it offers a very high resistance to an electrical current. When the flame is on, the cad cell resistance drops, permitting a greater current flow. Electrical circuits and switching devices are designed so that this change in resistance (flame on or flame off) allows the burner to remain on or shut down.

31.6 OIL FURNACE WIRING DIAGRAMS

Figure 31–26 is a wiring diagram for a typical forced-warm air oil furnace. This diagram is designed to be used with an air conditioning system as well. It includes wiring for the fan circuit for the air distribution, the oil burner primary control circuit, and the 24-V control circuit. Note that this diagram illustrates the circuitry for a multispeed blower motor.

Figure 31–27 illustrates the wiring for the blower fan motor operation only. The fan relay Ⓐ protects the blower motor from current flowing in the high- and low-speed circuits at the same time. If the normally open (NO) contacts on the high-speed circuit close, the NC contacts on the low-speed circuit will open. The NO fan limit switch Ⓑ is operated by a temperature-sensing bimetal device. When the temperature at the heat exchanger reaches approximately 170°F, these contacts close, providing current to the low-speed terminal on the fan motor. The high-speed circuit can be activated only from the thermostat. The high speed on the fan motor is *not* used for heating. It will operate only when the thermostat is set to FAN ON manually or for cooling (air conditioning).

The fan circuit for heating would then be from the power source (hot leg) through the temperature-actuated fan limit switch, to the NC fan relay, to the low-speed fan motor terminal, and back through the neutral wire.

Figure 31–28 shows the oil burner wiring diagram. The burner is wired through a NC limit switch Ⓐ. This limit switch is temperature actuated to protect the furnace and the building. If there is an excessive temperature buildup from the furnace overheating, this switch will open, causing the burner to shut down. This overheating could result from the burner being overfired, from a fan problem, from airflow restriction, or from similar problems. The wiring can then be followed from the limit switch to the primary control Ⓑ.

The black wire and orange wire at the primary pass the same current, but the orange is wired through the 24-V

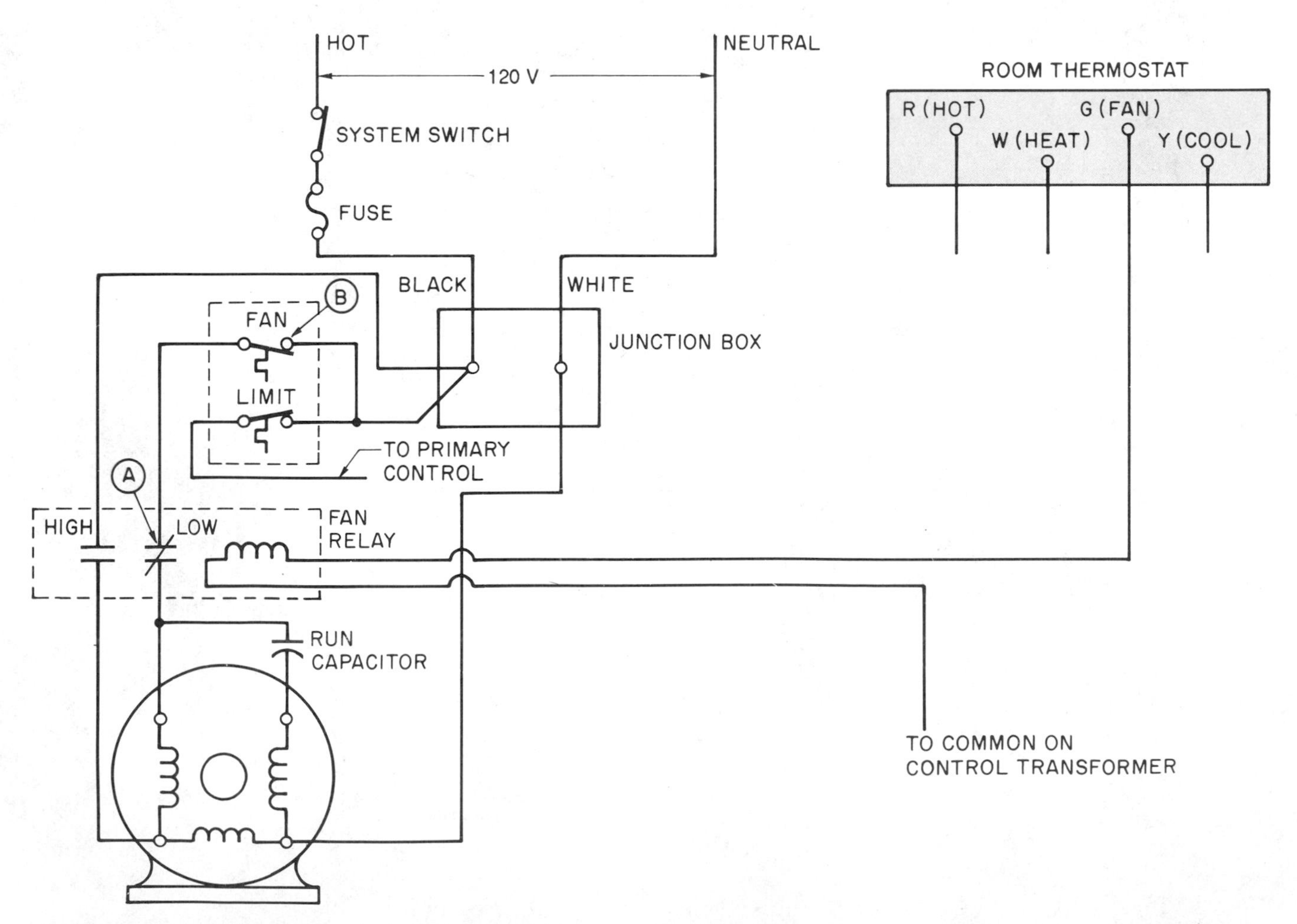

FIGURE 31–27 The fan runs using the thermally activated fan switch in the winter cycle. If the fan relay were to be energized while in the HEAT mode, the relay would switch from low to high speed.

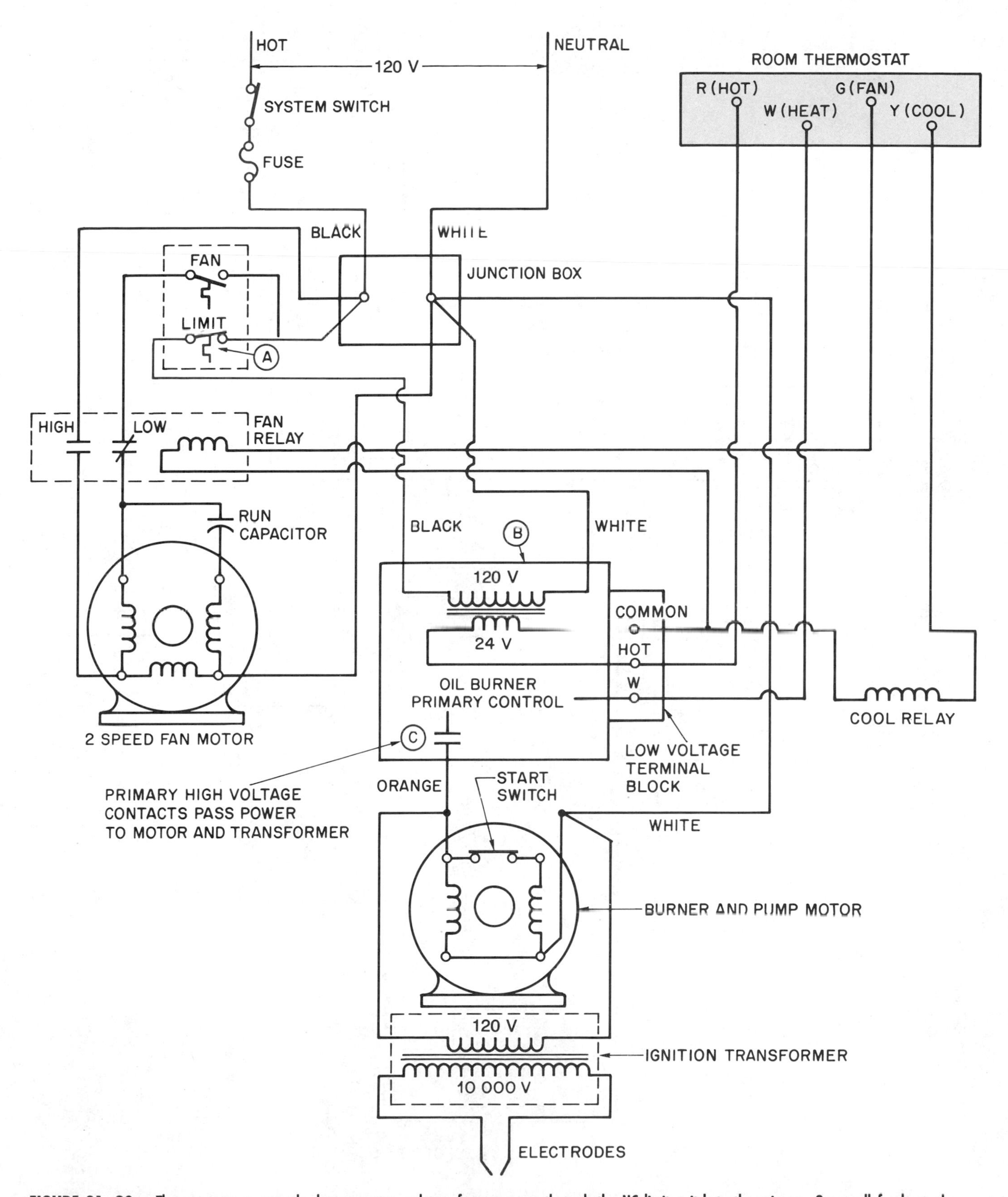

FIGURE 31–28 The power to operate the burner motor and transformer passes through the NC limit switch to the primary. On a call for heat, the thermostat energizes an internal 24-V relay in the primary. If the primary safety circuit is satisfied, power will pass through the orange wire to the burner motor and ignition transformer.

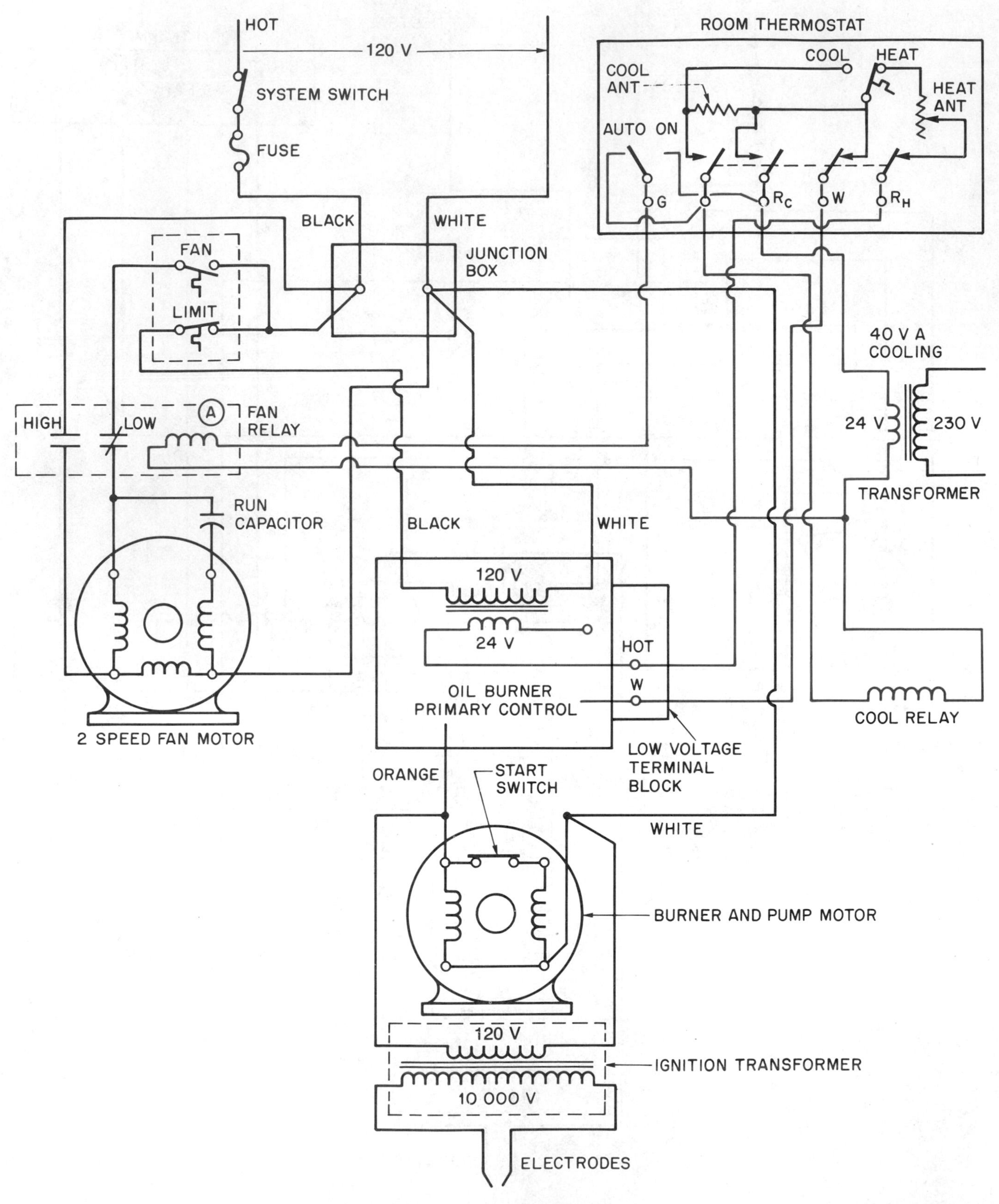

FIGURE 31–29 The common wire in the primary transformer does not extend into the field circuit. The room thermostat has a split subbase with the hot wire from two different transformers feeding it. These two hot circuits do not come in contact because of the split subbase.

primary high-voltage contacts ©. When the thermostat calls for heat, these contacts close, providing current to the ignition transformer, burner motor, and fuel valve, if one is used. The return is through the neutral (white wire) to the power source. The transformer in the primary control reduces the 120-V line voltage to 24 V for the thermostat control circuit.

If there is a problem in the start-up of the burner, the safety device (stack switch or cad cell) will shut down the burner. Figure 31–29 illustrates the wiring for the 24-V thermostat control circuit. The letter designations on the diagram indicate what the wire colors would normally be: R (red), "hot" leg from the transformer; W (white), heat; Y (yellow), cooling; G (green), manual fan relay.

It will occasionally be necessary to convert a system from one designed for heating only, to one that will also accommodate air conditioning. A heating-cooling system requires a larger 24-V control transformer than one normally supplied with a heating only system. To make this conversion, a 40-VA transformer is required in addition to the one supplied in the normal primary control circuit. The thermostat in this conversion should also have an isolating subbase so that one transformer is not connected electrically with the other. A fan relay Ⓐ is also needed in the fan circuit to activate the high-speed fan.

Power-consuming devices must be connected to both legs (ie, connected in parallel). Power-passing devices will be wired in series through the hot leg.

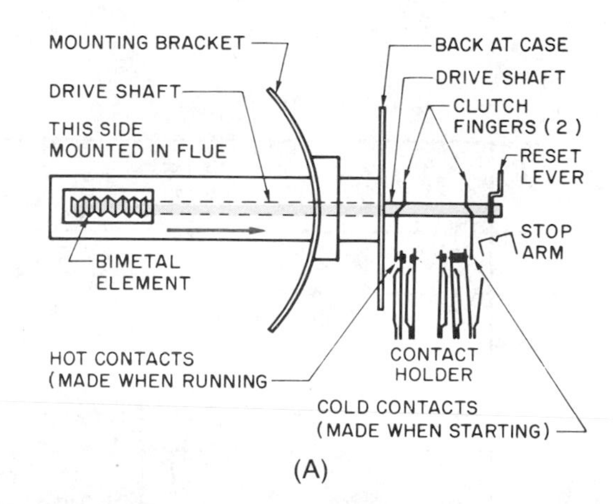

(A)

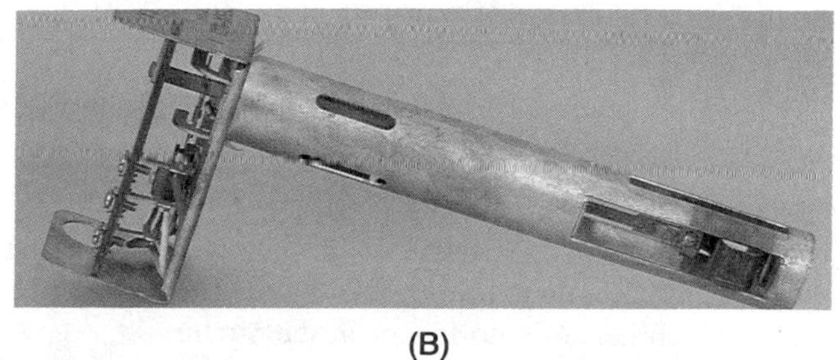

(B)

FIGURE 31–30 **(A) Illustration of a stack switch. (B) Photo of a stack switch.** ***(A) Courtesy Honeywell, Inc., Residential Division, (B) Photo by Bill Johnson***

31.7 STACK SWITCH SAFETY CONTROL

The bimetal element of the stack switch is positioned in the flue pipe. When the bimetal element is heated, proving that there is ignition, it expands and pushes the drive shaft in the direction of the arrow, Figure 31–30. This closes the hot contacts and opens the cold contacts.

Figure 31–31 is a wiring diagram showing the current flow during the initial start-up of the oil burner. The 24-V room thermostat calls for heat. This will energize the 1K coil, closing the 1K1 and 1K2 contacts. Current will flow through the safety switch heater and cold contacts. The hot contacts remain open. The closing of the 1K1 contacts provides current to the oil burner motor, oil valve, and ignition transformer. Under normal conditions there will be ignition and heat produced from the combustion of the oil-air mixture. This will provide heat to the stack switch in the flue pipe, causing the hot contacts to close and the cold contacts to open. The current flow then is shown in Figure 31–32. The circuit is completed through the 1K2 contact, the hot

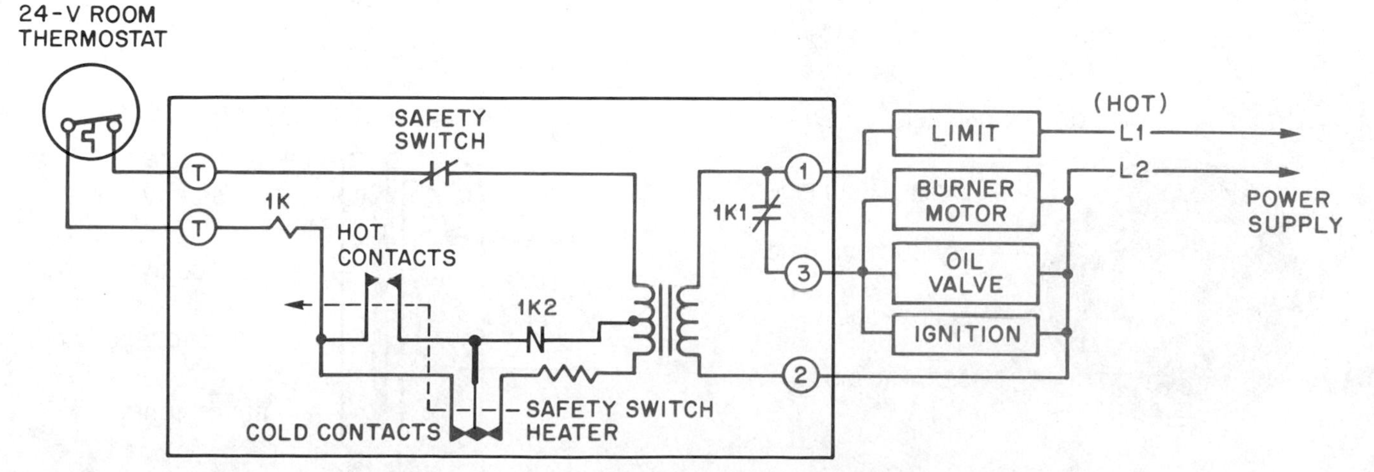

FIGURE 31–31 Stack switch circuit when burner first starts. Note current flow through safety switch heater.

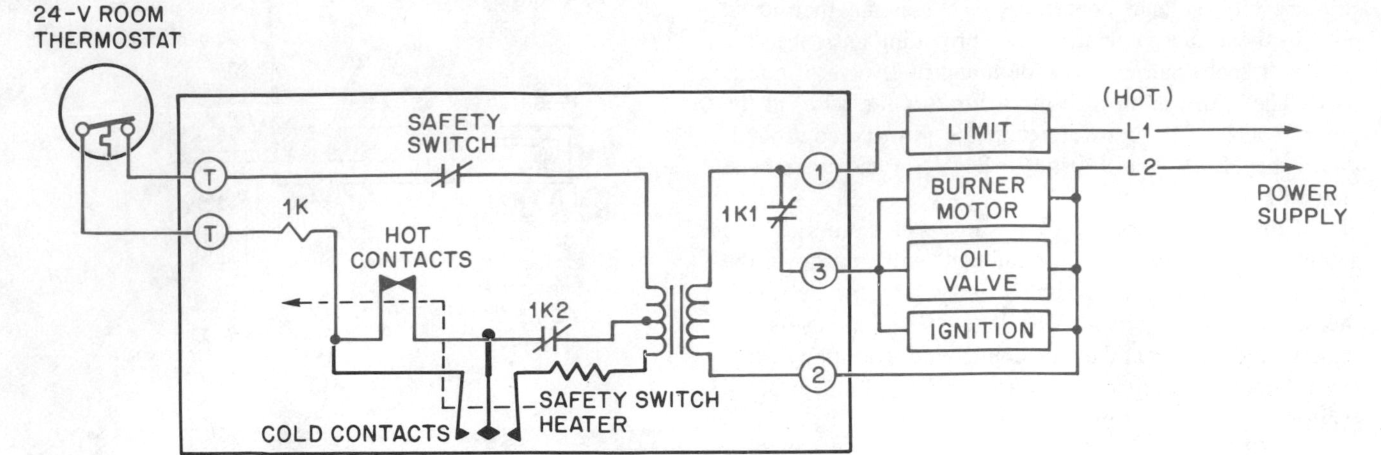

FIGURE 31–32 Typical stack switch circuitry with flame on, hot contacts closed, cold contacts open. Safety switch heater no longer is in the circuit.

contacts to the 1K coil. The 1K1 coil remains closed, and the furnace continues to run as in a normal safe start-up. The safety switch heater is no longer in the circuit.

If there is no ignition and consequently no heat in the stack, the safety switch heater remains in the circuit. In approximately 90 sec the safety switch heater opens the safety switch, which causes the burner to shutdown. To start the cycle again, depress the manual reset button. Allow approximately 2 min for the safety switch heater to cool, before attempting to restart.

31.8 CAD CELL SAFETY CONTROL

The cad cell is designed to offer a low resistance when it senses light. When there is ignition (flame), the cad cell has a low resistance. It has a very high resistance when there is no light (no flame), Figure 31–33. The cad cell must be positioned properly to sense sufficient light for the burner to operate efficiently.

In a standard modern primary circuit, the cad cell may be coupled with a triac, which is a form of a solid-state device designed to conduct current when the cad cell circuit resistance is high (no flame).

In Figure 31–34 the circuit shows the current flow in a normal start-up. There is low resistance in the cad cell. The triac does not conduct readily; therefore, the current bypasses the safety switch heater, and the burner continues to run.

In Figure 31–35 there is no flame and a high resistance in the cad cell. By design the triac will then conduct current that will pass through the safety switch heater. If this current continues through the heater, it will open the safety switch and cause the burner to shut down. To recycle, depress the reset button.

In most modern installations the cad cell is preferred over the thermal stack switch. It acts faster, has no mechanical moving parts, and consequently is considered more dependable.

SAFETY PRECAUTION: *Care should be taken not to reset any primary control too many times because unburned oil may accumulate after each reset. If a puddle of oil is ignited, it will burn intensely. If this should happen, the technician should stop the burner motor but should allow the furnace fan to run. The air shutter should be shut off to reduce the air to the burner. Notify the fire

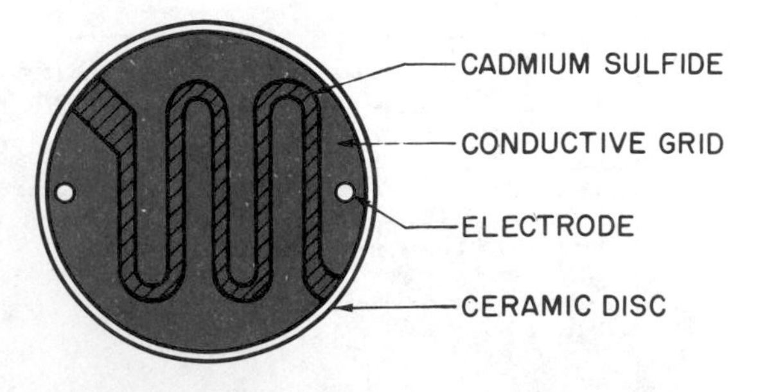

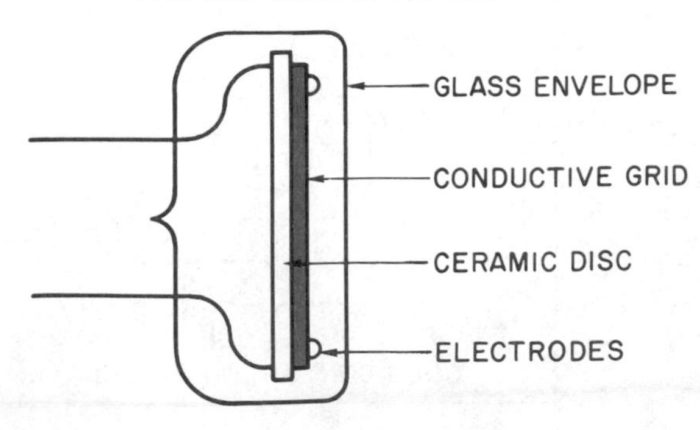

FIGURE 31–33 Diagram of the face and side of a cad cell.

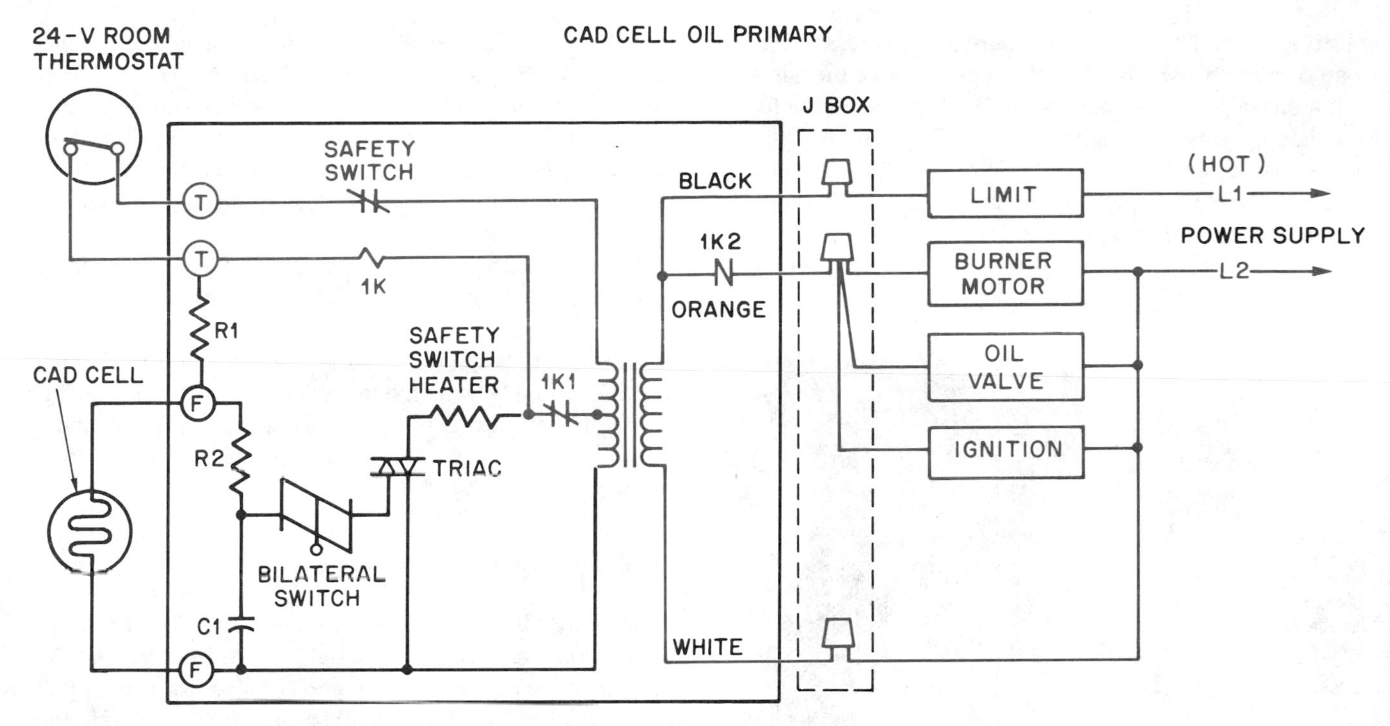

FIGURE 31–34 Cad cell circuitry when cell "sights" flame. Safety switch heater is not in the circuit.

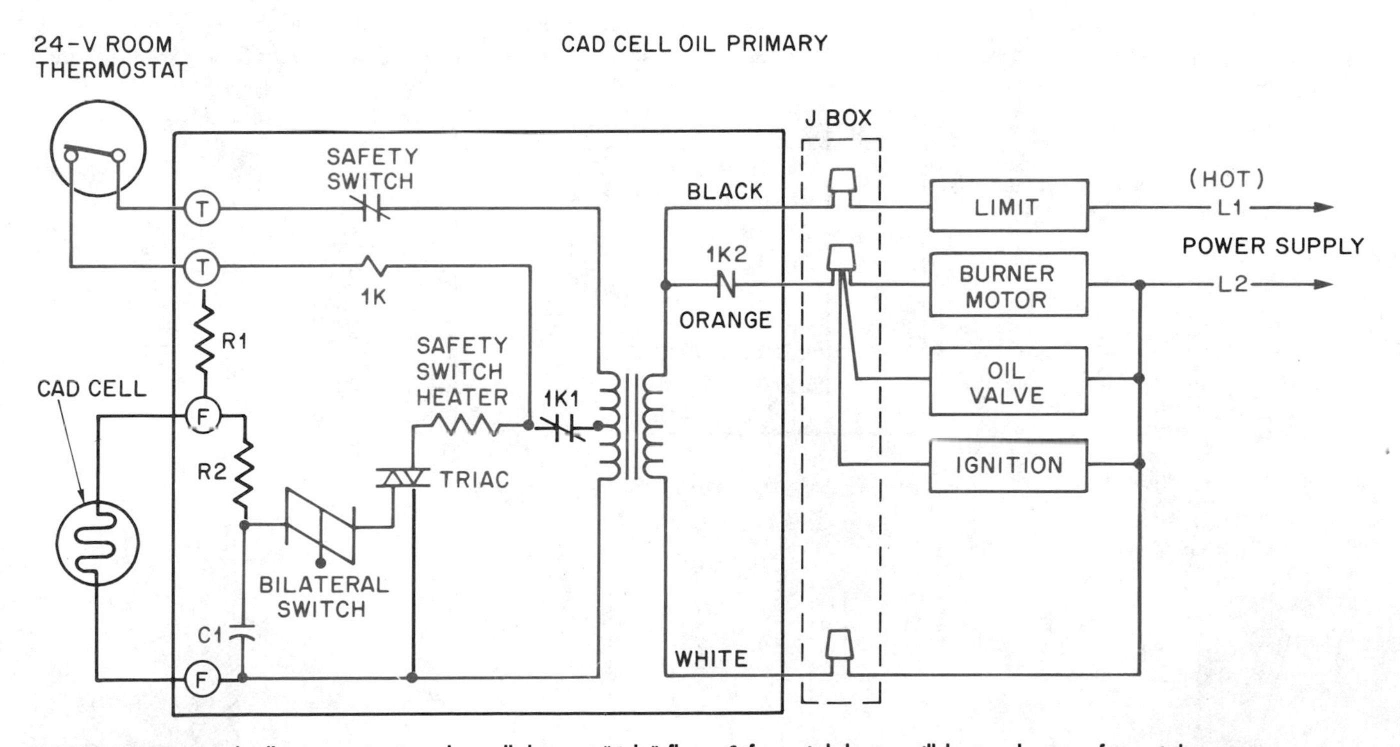

FIGURE 31–35 Cad cell primary circuitry when cell does not "sight" flame. Safety switch heater will heat and open safety switch.

department. Do *not* try to open the inspection door to put the fire out. Let the fire burn itself out with reduced air.*

31.9 FUEL OIL SUPPLY SYSTEMS

The fuel oil supply system for a residential or light commercial system uses one or two fuel lines. The location of the fuel storage tank in relation to the burner determines whether one or two lines are used. If the tank is above the burner level, a one-pipe system may be used. The flow of the fuel oil is then by gravity through the fuel filter, Figure 31–36, then to the burner, Figure 31–37. When the tank is below the burner level, a two-pipe system is used. Fuel oil is drawn or pumped from the tank, Figure 31–38.

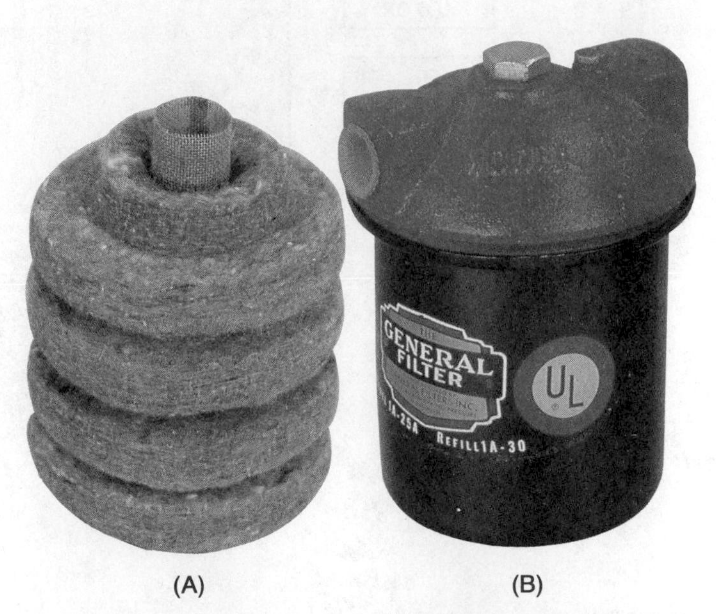

FIGURE 31–36 Filter cartridge and housing. *Photos by Bill Johnson*

Whenever the fuel oil line filter or pump is serviced in a one-pipe system, all air in the fuel line and the oil burner unit must be bled off to ensure proper fuel oil flow. Bleeding the system may also be necessary if the fuel oil tank is allowed to become empty. Air bleeding is automatic in the two-pipe system. Because of this, a two-pipe system is preferred.

The *viscosity* or thickness of the fuel oil is determined to a great extent by its temperature. The viscosity controls the rate at which the fuel oil flows. The lower the temperature of the oil, the slower it flows. In cold climates the fuel oil tank should be installed in the basement or buried underground to ensure that the oil does not thicken to the extent that its flow rate would be reduced below that needed for a normal operation.

Auxiliary Fuel Supply Systems

In some installations, oil burner units will be located above the fuel oil supply tank at a height that will be beyond the lift capabilities of the burner pump. This occurs primarily in commercial applications. Heights exceeding 15 ft from the bottom of the tank are beyond the capabilities of even the two-stage systems. In such installations an auxiliary fuel system must be used to get the fuel supply to the burner.

In this instance, a fuel oil booster pump is used to pump the fuel to an accumulator or reservoir tank, Figure 31–39. This booster pump is wired separately from the burner unit

ITEM	DESCRIPTION
1	OIL BURNER
2	ANTIPULSATION LOOP
3	OIL FILTER
4	SHUTOFF VALVE (FUSIBLE)
5	INSTRUCTION CARD
6	OIL GAGE
7	VENT ALARM
8	VENT
9	FILL LINE COVER

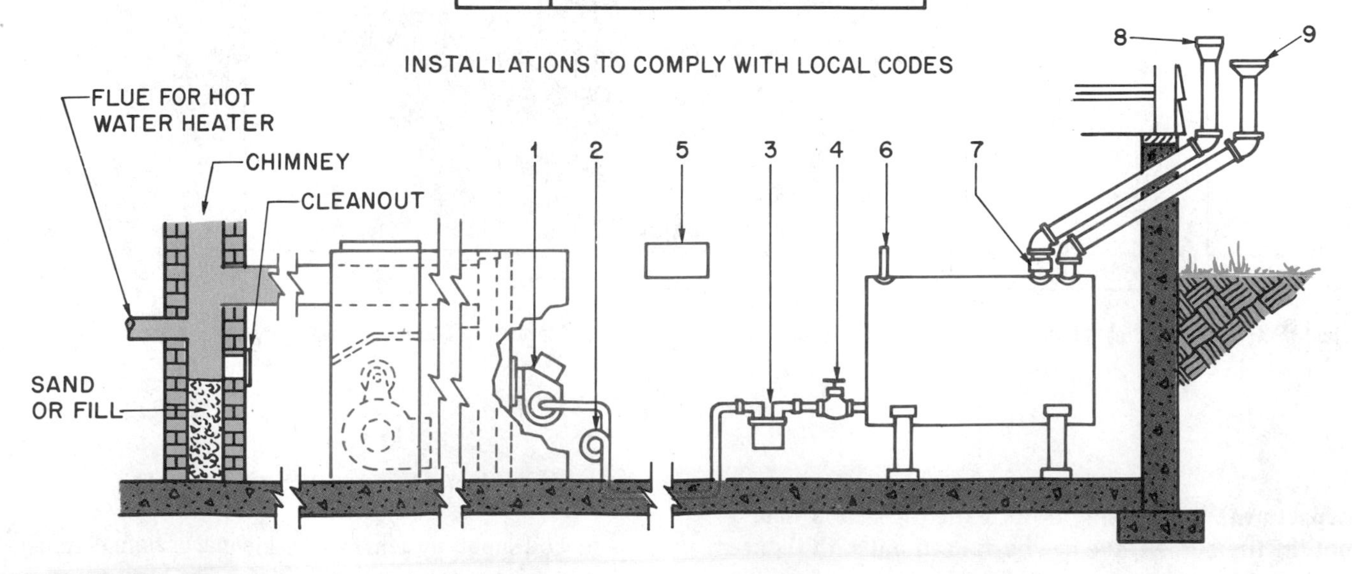

FIGURE 31–37 One-pipe system from tank to burner.

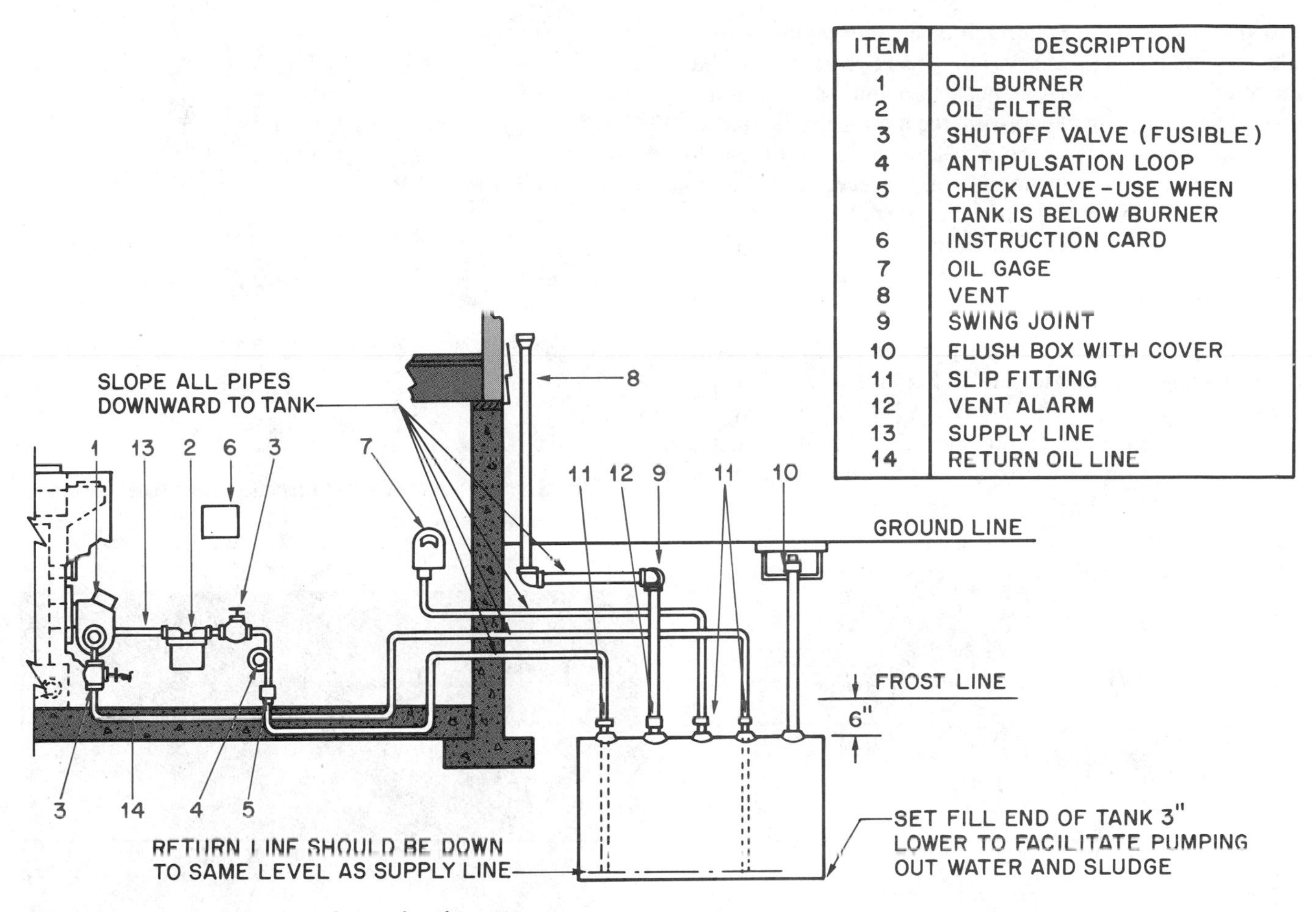

ITEM	DESCRIPTION
1	OIL BURNER
2	OIL FILTER
3	SHUTOFF VALVE (FUSIBLE)
4	ANTIPULSATION LOOP
5	CHECK VALVE-USE WHEN TANK IS BELOW BURNER
6	INSTRUCTION CARD
7	OIL GAGE
8	VENT
9	SWING JOINT
10	FLUSH BOX WITH COVER
11	SLIP FITTING
12	VENT ALARM
13	SUPPLY LINE
14	RETURN OIL LINE

FIGURE 31–38 Two-pipe system from tank to burner.

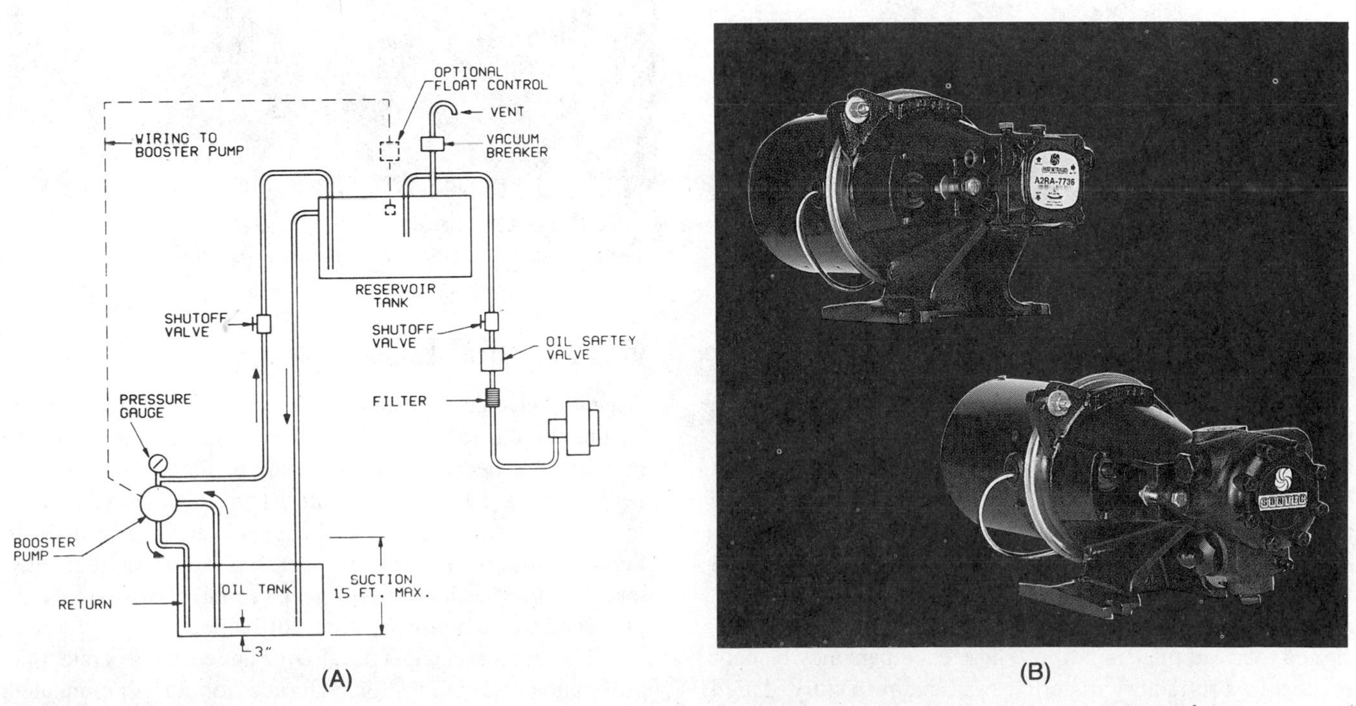

FIGURE 31–39 (A) Diagram of system using auxiliary booster pump and reservoir tank. (B) Booster pump. *Courtesy (B) Suntec Industries Incorporated, Rockford, Illinois*

and pumps the fuel oil through a separate piping system. The accumulator tank is kept full. Check valves are used to maintain the prime of the booster pump when it is not operating. An adjustable pressure regulator valve is a part of the fuel oil booster pump, and this regulator must be set to maintain a fuel oil pressure of 5 psi measured at the accumulator tank. Failure to properly adjust this regulator could result in seal damage to the fuel unit.

Fuel Line Filters

A filter should be located between the tank and the pump to remove many fine solid impurities from the fuel oil before it reaches the pump, Figures 31-36 and 31-38. Filters may also be located in the fuel line at the pump outlet to further reduce the impurities that may reach tiny oil passages in the nozzle, Figure 31–40.

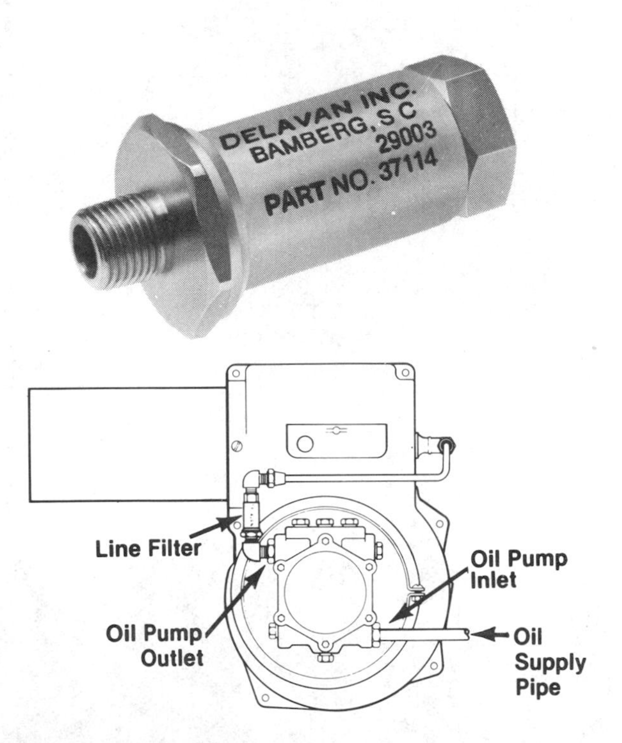

FIGURE 31–40 Fuel-line filter at pump outlet. *Courtesy Delavan Inc.*

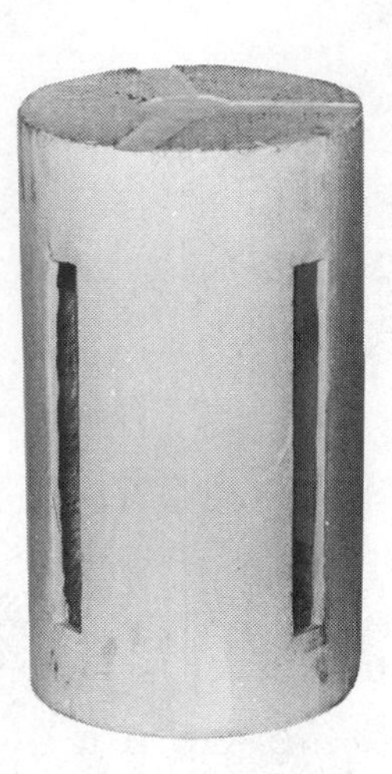

FIGURE 31–41 An alumna silicon refractory combustion chamber. *Courtesy Ducane Corporation*

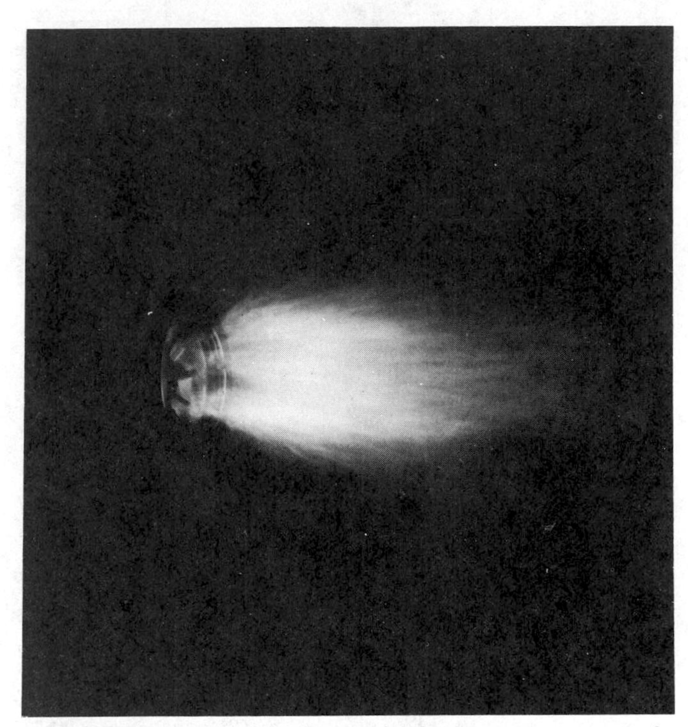

FIGURE 31–42 Inside combustion chamber. Note retention ring with flame locked to it. *Courtesy R. W. Beckett Corporation*

31.10 COMBUSTION CHAMBER

The atomized oil and air mixture is blown from the air tube into the combustion chamber where it is ignited by a spark from the electrodes. The atomized oil must be burned in suspension; that is, it must be burned while in the air in the combustion chamber. If the flame hits the wall of the combustion chamber before it is totally ignited, the cooler wall will cause the oil vapor to condense, and efficient combustion will not occur. Combustion chambers should be designed to avoid this condition. These chambers may be built of steel or a refractory material. A silicon refractory is used in many modern furnaces, Figure 31–41. Figure 31–42 shows a flame in a combustion chamber.

31.11 HEAT EXCHANGER

A heat exchanger in a forced-air system takes heat caused by the combustion in the furnace and transfers it to the air that is circulated to heat the building, Figure 31–43. The heat exchanger is made of material that will cause a rapid transfer of heat. Modern exchangers, particularly in residential furnaces, are made of sheet steel, which is frequently coated with special substances to resist corrosion. Acids produced by combustion cause corrosion.

The heat exchanger is also designed to separate flue gases and other combustion materials from the air circulated through the building. ***The flue gases from combustion should never mix with the circulating air.***

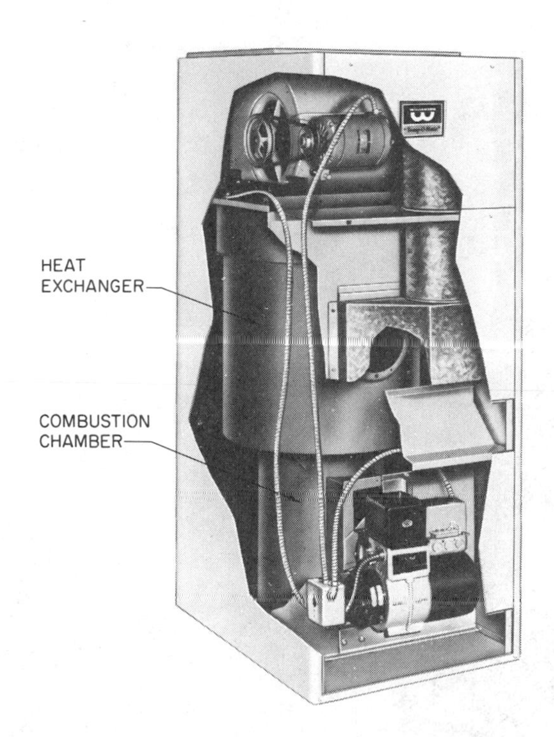

FIGURE 31–43 Heat exchanger in oil-fired forced-air furnace. *Courtesy The Williamson Company*

Heat exchangers and combustion chambers are often inspected by the technician. A mirror made of chrome-plated steel is an ideal tool for this job, Figure 31–44.

Oil furnaces should have the correct airflow across the heat exchanger. An air temperature rise check may be performed if the installation is new or if it is suspected that the airflow is not correct. Often, the quantity of airflow must be known for air conditioning equipment. The unit capacity may be found on the unit nameplate or calculated from the oil burner nozzle size. For example, the unit may have a 0.75 gph nozzle and a 65°F temperature rise. What should the cfm be? A gallon of fuel oil has a Btu capacity of 140,000 Btu.

$$140{,}000 \times 0.75 = 105{,}000$$

Btuh input for the furnace

$$105{,}000 \times 0.70 \text{ (70\% efficient)} =$$

73,500 Btuh output

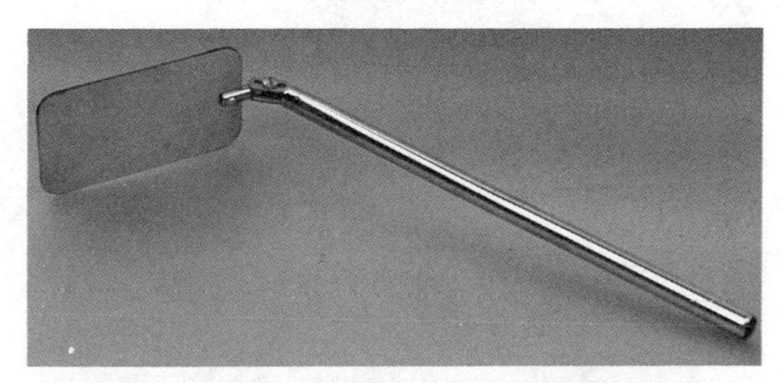

FIGURE 31–44 Mirror used by the heating technician to view the heat exchanger, the inside of the combustion chamber, and the flame. The handles on some models telescope and the mirror is adjustable. *Photo by Bill Johnson*

This is the amount of heat transferring to the air.

$$\text{cfm} = \frac{Qs}{1.08 \times TD}$$

$$\text{cfm} = \frac{105{,}000}{1.08 \times 65}$$

$$\text{cfm} = 1495.7$$

This does not account for the fan motor which would add approximately 1365 Btuh (400 W × 3413 = 1365).

31.12 CONDENSING OIL FURNACE

Some manufacturers have worked to design more efficient oil furnaces since energy prices have increased and since the need for fossil-fuel conservation has become so evident. The *condensing oil furnace* is one of the latest product designs for increased efficiency.

The condensing oil furnace, like the more conventional furnace, has two systems: the combustion or heat-producing system and the heated air circulation system. The combustion system includes the burner and its related components: the combustion chamber, as many as three heat exchangers, and a vent fan and pipe. The heated air circulation system includes the blower fan, housing, motor, the plenum, and duct system.

The following describes the operation of a condensing oil furnace manufactured by Yukon Energy Corporation, Figure 31–45. The burner forces air and fuel oil into the combustion chamber where they are mixed and ignited. Much of this combustion heat is transferred to the main heat exchanger, which surrounds the combustion chamber. Combustion gases still containing heat are forced through a second heat exchanger where additional heat is transferred. These gases still containing some heat are forced through a third heat exchanger where nearly all the remaining heat is removed. This third heat exchanger is a coil-type where the temperature is reduced below the dew point, low enough for a change of state to occur. This results in the moisture condensing from the flue gases. This is a latent-heat high-efficiency exchange. About 1000 Btu are transferred to the airstream for each pound of moisture condensed (1 lb water = about 1 pint). This can result in considerable savings. The remaining exhaust gases are vented to the outside. These gases are normally forced out by a blower through polyvinyl chloride (PVC) pipe. The condensing coil must be made of stainless steel to resist the acid in the flue gases. A drain is located at the condensing coil to remove the condensate to a container or to a suitable area or drain outside.

In the heated air circulation system the blower draws air through the return air duct and moves it across the condensing heat exchanger where this air is preheated by removing the sensible and latent heat from the coil. The air then passes around the second heat exchanger where additional heat is added.

The circulating air then moves over the main heat exchanger, after which it is circulated to the rooms to be heated through the duct system. This leaving air is approx-

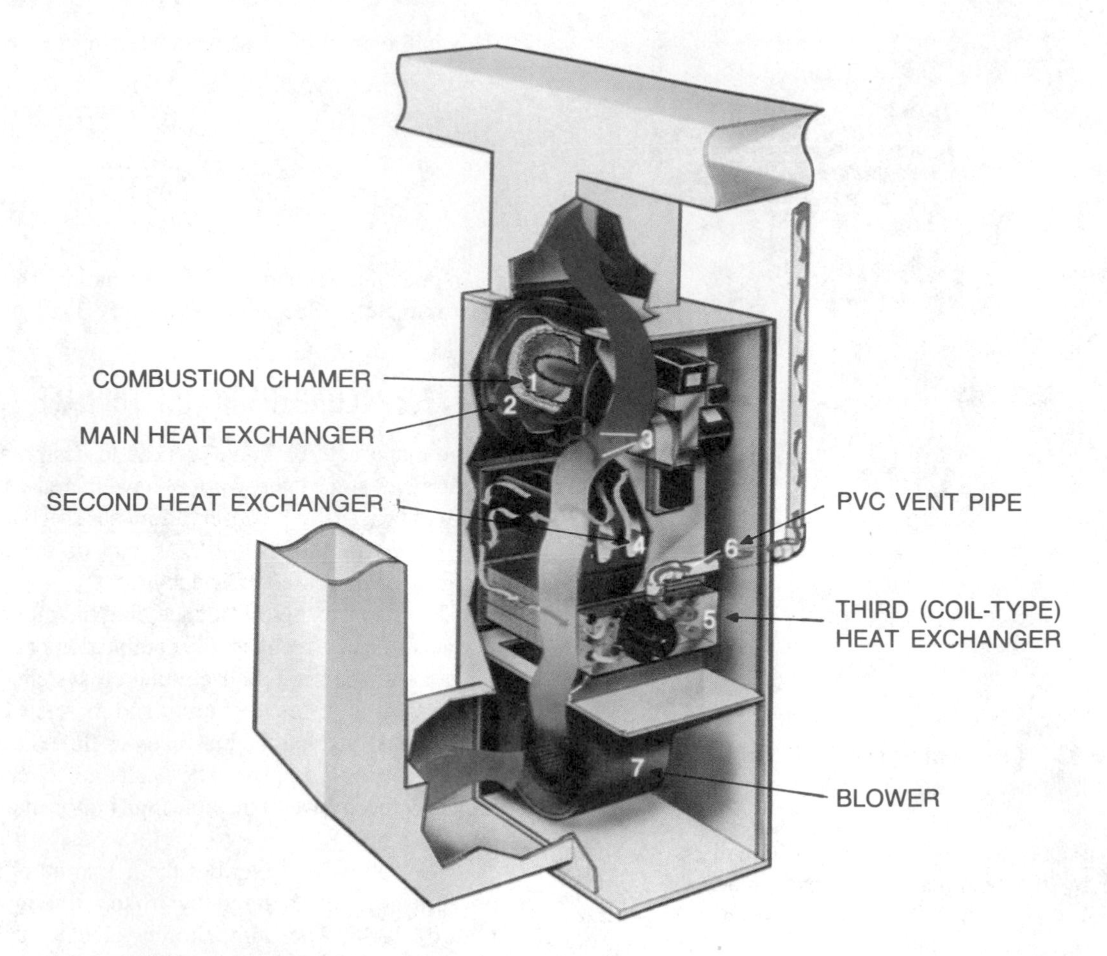

FIGURE 31–45 Condensing oil furnace. *Courtesy Yukon Energy Corporation*

imately 60°F warmer than when it entered the furnace through the return air duct.

These furnaces have a 90+ annual fuel utilization efficiency (AFUE). They also can be installed without a chimney as they are vented through a PVC pipe.

31.13 SERVICE PROCEDURES

Pumps

The performance of the fuel oil system from the tank to the nozzle in residential and commercial units without booster pumps can be determined with a vacuum gage and a pressure gage. Before the vacuum and pressure checks are made, however, the following should be determined:

1. The tank has sufficient fuel oil.
2. The tank shutoff valve is open.
3. The tank location is noted (above or below burner).
4. The type of system is noted (one or two pipe).

Connect the vacuum and pressure gages to the fuel oil unit as shown in Figure 31–46. If the tank is above the burner level, the supply system may be either a one- or two-pipe system. If the tank is above the burner and with the unit in operation, the vacuum gage should read 0 in. Hg. If the

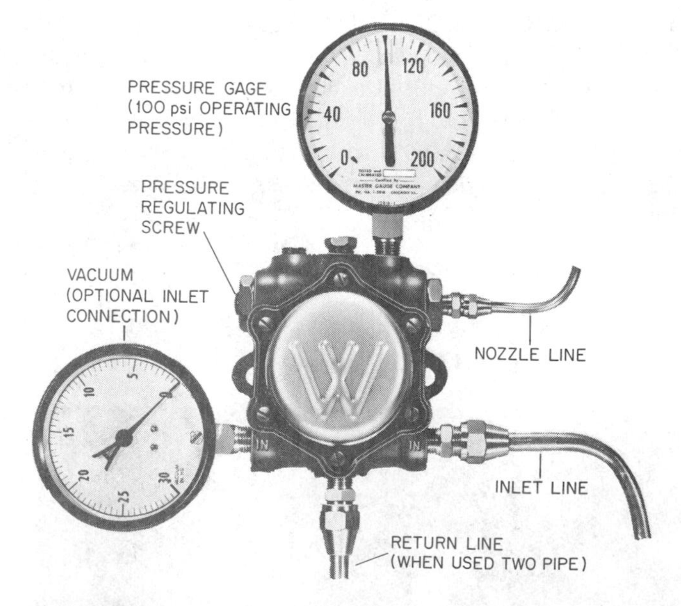

FIGURE 31–46 Pump with pressure and vacuum gage in place for tests and servicing. *Courtesy Webster Electric Company*

vacuum gage indicates a vacuum, one or more of the following may be the problem:

- The fuel line may be kinked.
- The line filter may be clogged or blocked.
- The tank shutoff valve may be partially closed.

If the supply tank is below the burner level, the vacuum gage should indicate a reading. Generally, the gage should read 1 in. Hg for every foot of vertical lift and 1 in. Hg for every 10 ft of horizontal run. The lift and horizontal run combination should not total more than 17 in. Hg. (Remember, the lift capability of a dual-stage pump is approximately 15 ft.) Vacuum readings in excess of this formula may indicate a tubing size too small for the run. This may cause the oil to separate into a partial vapor and become milky.

The pressure gage indicates the performance of the pump and its ability to supply a steady even pressure to the nozzle. The pressure regulator should be adjusted to 100 psi. See Figure 31–47 for an illustration of a pressure adjusting valve. To adjust the regulator, with the pressure gage in place, turn the adjusting screw until the gage indicates 100 psi. With the pump shown in Figure 31–47, the valve-screw cover screw should be removed and the valve screw turned with a 1/8-in. Allen wrench.

With the regulator set and the unit in operation, the pressure gage should indicate a steady reading. If the gage pulsates, one or more of the following may be the problem:

- Partially clogged supply filter element
- Partially clogged unit filter or screen
- Air leak in fuel oil supply line
- Air leak in fuel oil pump cover

A pressure check of the valve differential is necessary when oil enters the combustion chamber after the unit has been shut down. To make this check, insert the pressure gage in the outlet for the nozzle line, Figure 31–48. With the gage in place and the unit running, record the pressure. With the unit shutdown, the pressure gage should drop about 15 psi and hold. If the gage drops to 0 in. psi, the fuel oil cutoff inside the pump is defective and the pump may need to be replaced. If the pressure gage indicates a hold of 15 psi but there is still an indication of improper fuel oil shut off, the problem may be:

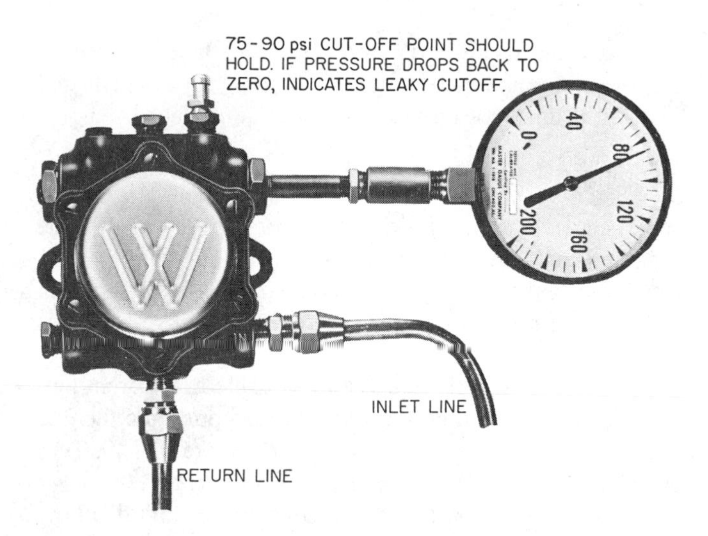

FIGURE 31–48 **Pressure gage in place to check valve differential.** *Courtesy Webster Electric Company*

- Possible air leak in fuel unit (pump)
- Possible air leak in fuel oil supply system
- Nozzle strainer may be clogged

Burner Motor

If the burner motor does not operate:

1. Press the reset button
   NOTE: Reset may not function if motor is too hot. Wait for motor to cool down.
2. Check the electrical power to the motor. This can be done with a voltmeter across the orange and white or

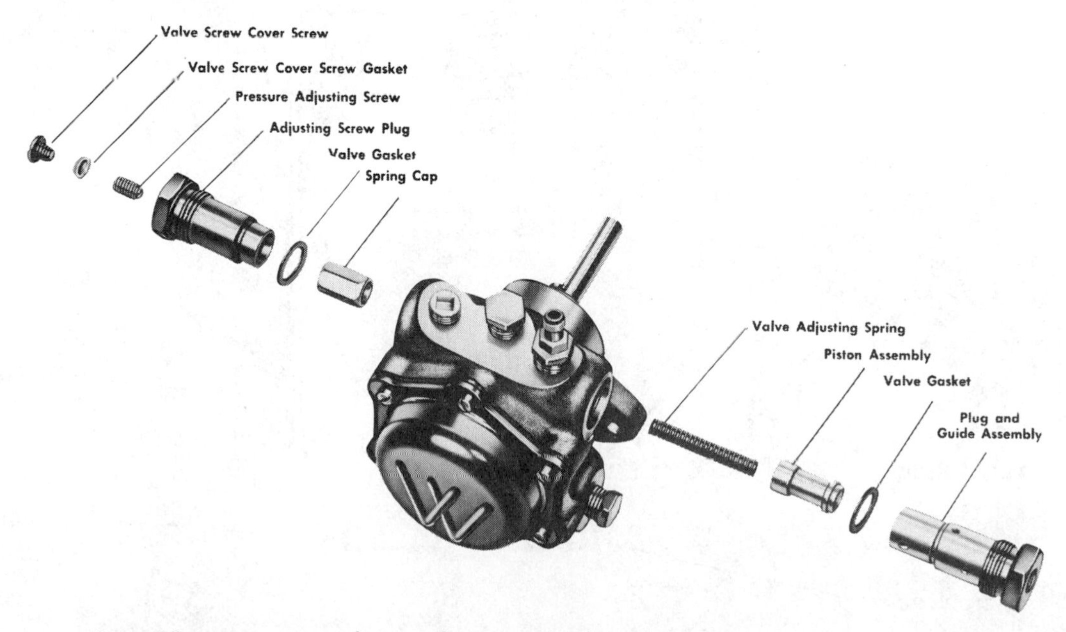

FIGURE 31–47 **Fuel pump illustrating pressure adjusting valve.** *Courtesy Webster Electric Company*

black and white leads to the motor. If there is power to the motor and there is nothing binding it physically and it still does not operate, it should be replaced.

Bleeding a One-Pipe System

When starting an oil burner with a one-pipe system for the first time or whenever the fuel-line filter or pump is serviced, the system piping will have air in it. The system must be bled to remove this air. The system should also be bled if the supply tank is allowed to become empty.

To bleed the system, a 1/4-in. flexible transparent tubing should be placed on the vent or bleed port, and the free end of the tube placed in a container, Figure 31–49. Turn the bleed port counterclockwise. Usually from 1/8 to 1/4 turn is sufficient. The burner should then be started, allowing the fuel oil to flow into the container. Continue until the fuel flow is steady, and there is no further evidence of air in the system; then shut off the valve.

Converting a One-Pipe System to a Two-Pipe System

Most residential burners are shipped to be used with a one-pipe system. To convert these units to a two-pipe system, do the following:

1. Shut down all electrical power to the unit.
2. Close the fuel oil supply valve at the tank.
3. Remove inlet port plug from the unit, Figure 31–50.
4. Insert bypass plug shipped with unit into the deep seat of inlet port with an Allen wrench.

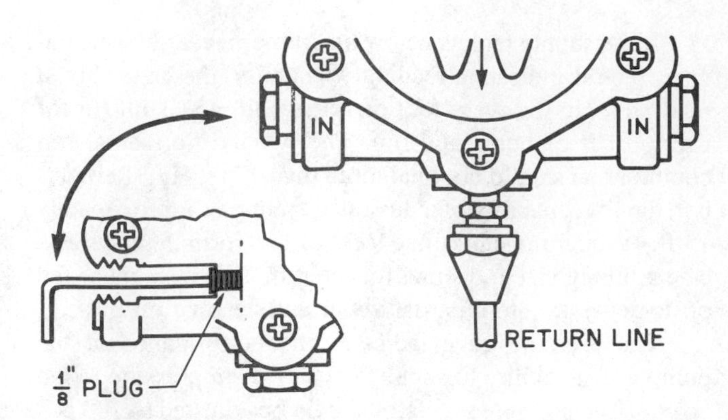

FIGURE 31–50 Installation of bypass plug for conversion to two-pipe system. *Courtesy Webster Electric Company*

5. Replace inlet port plug and install flare fitting and copper line returning to the tank.

SAFETY PRECAUTION: *Do not start the pump with the bypass plug in place unless the return line to the tank is in place. Without this line, there is no place for the excess oil to go except to possibly rupture the shaft seal.*

Nozzles

Nozzle problems may be discovered by observing the condition of the flame in the combustion chamber, taking readings on a pressure gage, and analyzing the flue gases. (Flue-gas analysis is discussed later.)

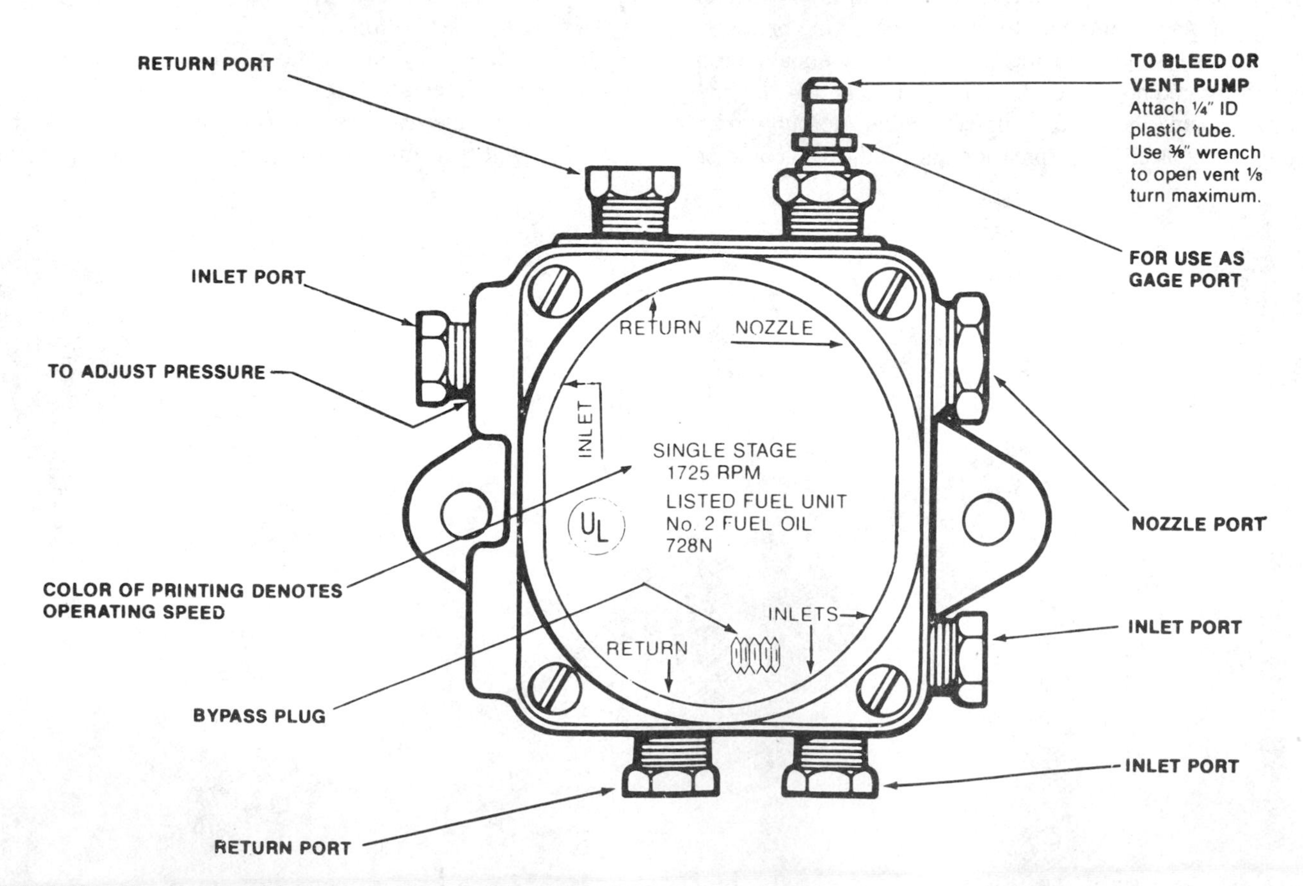

FIGURE 31–49 Diagram of fuel oil pump. *Courtesy Webster Electric Company*

Common conditions that may relate to the nozzle are:

- Pulsating pressure gage
- Flame changing in size and shape
- Flame impinging (striking) on sides of combustion chamber
- Sparks in flame
- Low carbon dioxide reading in flue gases (less than 8%)
- Delayed ignition
- Odors present

When nozzle problems are apparent, the nozzle should be replaced. The nozzle is a delicate finely machined component, and wire brushes and other cleaning tools should not be used on it. Generally, a nozzle should be replaced annually as a normal servicing procedure. ***Do *not* attempt to clean the nozzle strainer.*** If the nozzle strainer is clogged, the cause should be determined. Clean the fuel oil nozzle line and the fuel unit (pump) strainer. Change the supply filter element. There may also be water in the supply tank, and the fuel oil may otherwise be contaminated.

Carbon formation at the nozzle or fuel oil burning at the nozzle may be caused by bent or distorted nozzle features. ***Nozzles should be removed and replaced with a special nozzle wrench. Never use adjustable pliers or a pipe wrench.***

When the nozzle has been removed, keep the oil from running from the tube by plugging the end of the nozzle assembly with a small cork. If oil leaves this tube, air will enter and cause an erratic flame when the burner is put back into operation. It may also cause an afterdrip or afterfire when the burner is shut down.

It is important that nozzles do not overheat. Overheating causes the oil within the swirl chamber to break down and cause a varnish-like substance to build up. Major causes of overheating and varnish buildup are:

- Fire burning too close to the nozzle
- Firebox too small
- Nozzle too far forward
- Inadequate air-handling components on burner
- Burner overfired

If the fire burns too close to the nozzle, the blast tube opening may be narrowed. This will increase the air velocity as it leaves the tube and move the flame away from the nozzle. Increasing the pressure of the pump will increase the velocity of the oil droplets leaving the nozzle and will help alleviate this problem.

A nozzle that is too far forward will clog from reflected heat from the firebox from the gum formation caused by the cracking (overheated) oil. The nozzle may be moved back into the air tube by using a short adapter or by shortening the oil line.

Do not allow an afterdrip or afterfire to go unchecked. Any leakage or afterfire will result in carbon formation and clogging of the nozzle. Check for an oil leak, air in the line, or a defective cut-off valve if there is one. An air bubble behind the nozzle will not always be pushed out during burning. It may expand when the pump is shut off and cause an afterdrip.

Ignition System

Of all phases of oil burner servicing, ignition problems are among the easiest to recognize and solve. Much ignition service consists of making the proper spark gap adjustment, cleaning the electrodes, and making all connections secure.

To check the ignition transformer:

1. Turn off power to the oil burner unit.
2. Swing back the ignition transformer.
3. Shut off fuel supply or disconnect burner motor lead.
4. Turn power back on to the oil burner. With a proper voltmeter and high-voltage leads, measure the voltage of the transformer output. This reading should be 10,000 V. If a lesser voltage is read, the transformer is defective and should be replaced. ***Follow the manufacturer's instructions and keep your distance from the 10,000 V leads.***

To check the electrodes:

1. Ensure that the three spark gap settings of the electrodes are set properly. Check Figure 31–51 for the settings of the gap, height above the center of the nozzle, and distance of the electrode tips forward from the nozzle center.
2. Make sure the tips of the electrodes are in back of the oil spray. This can be checked by using a flame mirror, Figure 31–44. If this is not possible, remove the elec-

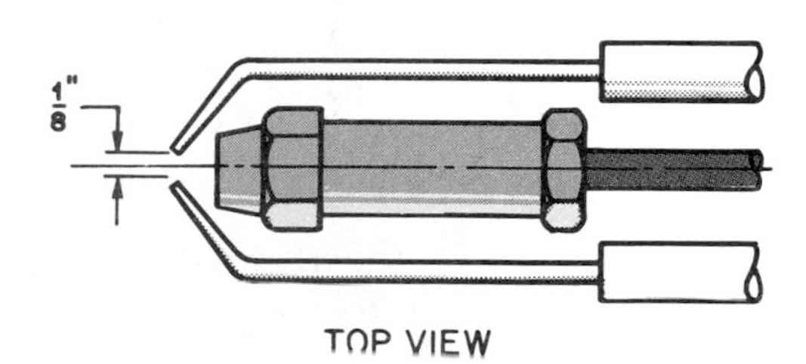

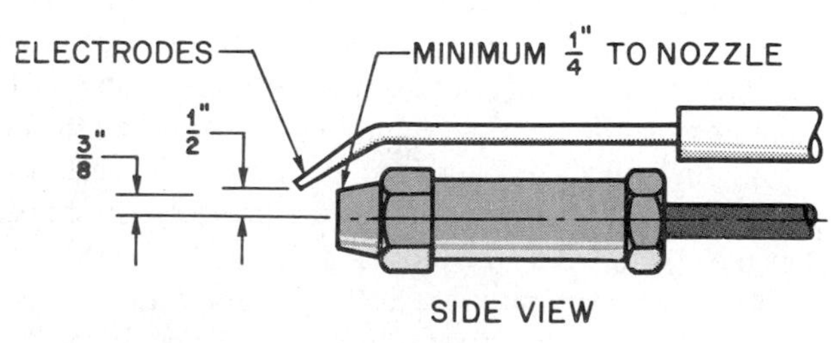

HEIGHT ADJUSTMENT

$\frac{1}{2}$" RESIDENTIAL INSTALLATION

$\frac{3}{8}$" COMMERCIAL INSTALLATION

POSITION OF ELECTRODES IN FRONT OF NOZZLE IS DETERMINED BY SPRAY ANGLE OF NOZZLE

FIGURE 31–51 Electrode adjustments. Electrodes cannot be closer than 1/4 in. to any metal part.

trode assembly and spray the fuel oil into an open container, making sure the electrodes are not in the fuel oil path.

ELECTRODE INSULATORS. Wipe the insulators clean with a cloth moistened with a solvent. If they remain discolored after a good cleaning, they are filled with carbon throughout their porous surfaces and should be replaced.

To check the insulator:

1. Disconnect burner motor lead.
2. Remove one side of the high-tension lead and position it close to the discolored portion of the other insulator, which is still connected to its lead.
3. Turn on the burner switch.
4. If any spark jumps to the insulator from the lead, the electrode should be replaced. Check both sides, using this procedure.

SAFETY PRECAUTION: *This is 10,000 V and can cause electric shock. There is actually very little current available, so it is not likely to be fatal. This spark is much like the spark on a lawn mower engine. It will make you jump and hurt yourself.*

31.14 COMBUSTION EFFICIENCY

Until a few years ago the heating technician was primarily concerned with ensuring that the heating equipment operated cleanly and safely. Technicians made adjustments by using their eyes and ears. High costs of fuel, however, now make it necessary for the technician to make adjustments using test equipment to ensure efficient combustion.

Fuels consist mainly of hydrocarbons in various amounts. In the combustion process new compounds are formed and heat is released. The following are simplified formulas showing what happens during a perfect combustion process:

$$C + O_2 \rightarrow CO_2 + \text{heat}$$
$$H_2 + 1/2O_2 \rightarrow H_2O + \text{heat}$$

During a normal combustion process, carbon monoxide, soot, smoke, and other impurities are produced along with heat. Excess air is supplied to ensure that the oil has enough oxygen for complete combustion, but even then the air and fuel may not be mixed perfectly. The technician must make adjustments so that near perfect combustion takes place, producing the most heat and reducing the quantity of unwanted impurities. On the other hand, too much excess air will absorb heat, which will be lost in the stack (flue gas) and reduce efficiency. Air contains only about 21% oxygen, so the remaining air does not contribute to the heating process.

The following tests can be made for proper combustion:

1. Draft
2. Smoke
3. Net temperature (flue stack)
4. Carbon dioxide

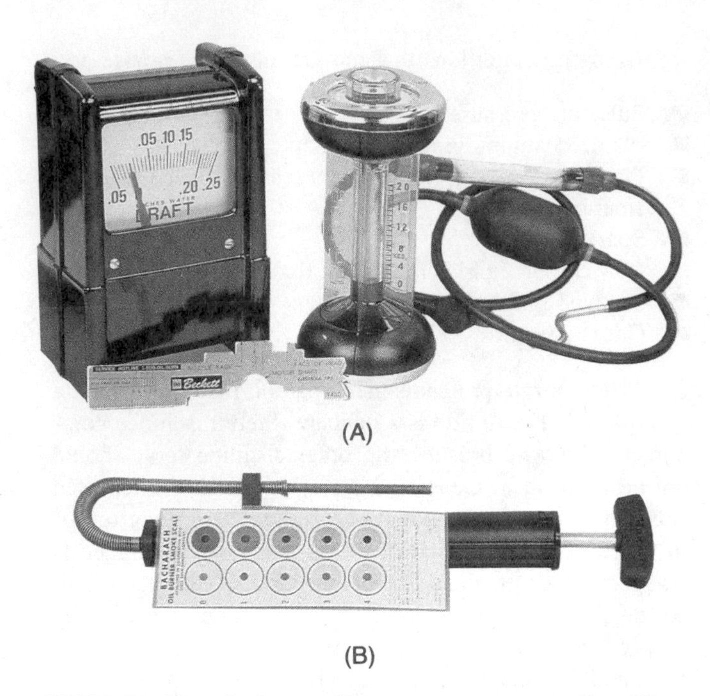

FIGURE 31–52 Combustion efficiency testing equipment. *Photos by Bill Johnson*

Technicians develop their own procedures in combustion testing. Making an adjustment to help correct one problem will often correct or help to correct others. Compromises have to be made also. When correcting one problem, another may be created. In some instances problems may be caused by the furnace design.

FIGURE 31–53 Combustion efficiency analyzer. *Courtesy United Technologies Bacharach*

Some technicians use individual testing devices for each test, Figure 31–52. Others use electronic combustion analyzers with digital readouts, Figure 31–53.

To make these tests with the individual testing devices, a hole must be drilled or punched in the flue pipe 12 in. from the furnace breeching, on the furnace side of the draft regulator, and at least 6 in. away from it. The hole should be of the correct size so that the stem or sampling tube of the instrument can be inserted into it. ***The manufacturer's instructions furnished with the instruments should be followed carefully. Procedures indicated here are very general and should not take the place of the manufacturer's instructions.***

Draft Test

Correct draft is essential for efficient burner operation. The draft determines the rate at which combustion gases pass through the furnace, and it governs the amount of air supplied for combustion. The draft is created by the difference in temperatures of the hot flue gases and is negative pressure in relation to the atmosphere. Excessive draft can increase the stack temperature and reduce the amount of carbon dioxide in the flue gases. Insufficient draft may cause pressure in the combustion chamber, resulting in smoke and odor around the furnace. Adjust the draft before you make other adjustments to obtain maximum efficiency.

To make the test:

1. Drill a hole into the combustion area for the draft tube. (A bolt may be removed on some furnaces and the bolt hole used for this access.) This is necessary to determine the overfire draft, Figure 31–54.
2. Place the draft gage on a level surface near the furnace and adjust to 0 in.
3. Turn the burner on and let run for at least 5 min.
4. Insert the draft tube into the combustion area to check the overfire draft, Figure 31–54.
5. Insert the draft tube into the flue pipe to check the flue draft.

The overfire draft should be at least −0.02 in. W.C. The flue draft should be adjusted to maintain the proper overfire draft. Most residential oil burners require a flue draft of −0.04 in. to −0.06 in. W.C. to maintain the proper overfire draft. These drafts are updrafts and are negative in relationship to atmospheric pressure. Longer flue passages require a higher flue draft than shorter flue passages.

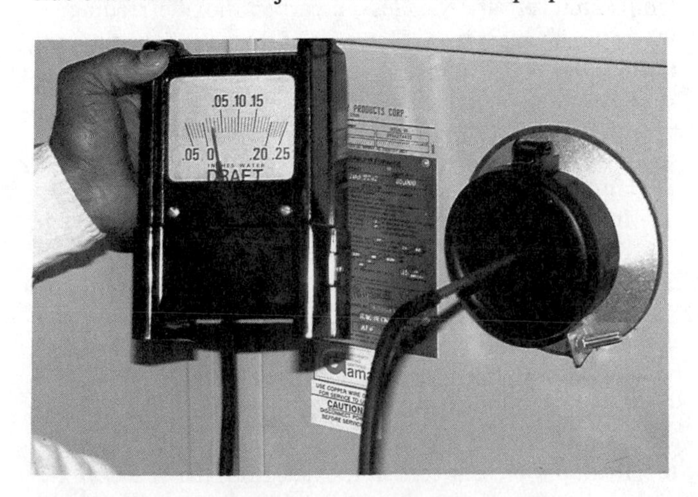

FIGURE 31–54 Checking overfire draft. *Photo by Bill Johnson*

Smoke Test

Excessive smoke is evidence of incomplete combustion. This incomplete combustion can result in a fuel waste of up to 15%. A 5% fuel waste is not unusual. Excessive smoke also results in a soot buildup on the heat exchanger and other heat-absorbing areas of the furnace. Soot is an insulator. This results in less heat being absorbed by the heat exchanger and increased heat loss to the flue. A 1/16-in. layer of soot can cause a 4.5% increase in fuel consumption, Figure 31–55.

The smoke test is accomplished by drawing a prescribed number of cubic inches of smoke-laden flue products through a specific area of filter paper. The residue on this filter paper is then compared with a scale furnished with the testing device. The degree of sooting can be read off the scale. A smoke tester such as the one illustrated in Figure 31–56 may be used.

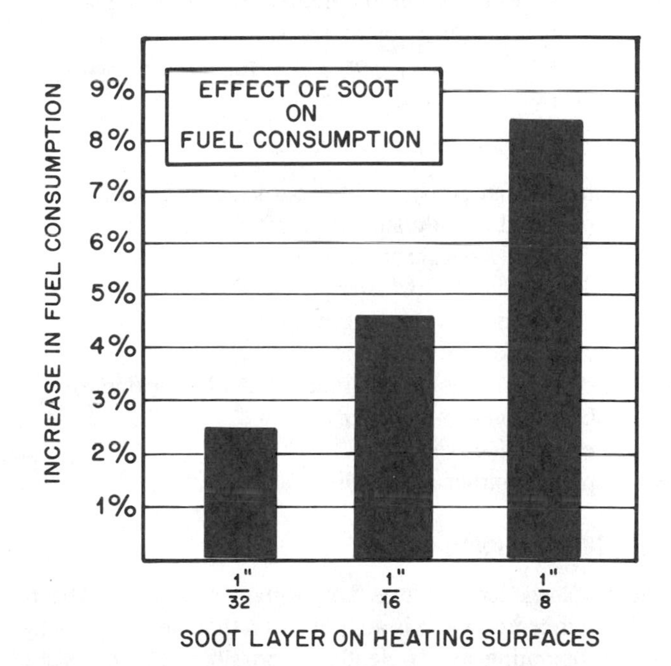

FIGURE 31–55 Effect of soot on fuel consumption. *Courtesy United Technologies Bacharach*

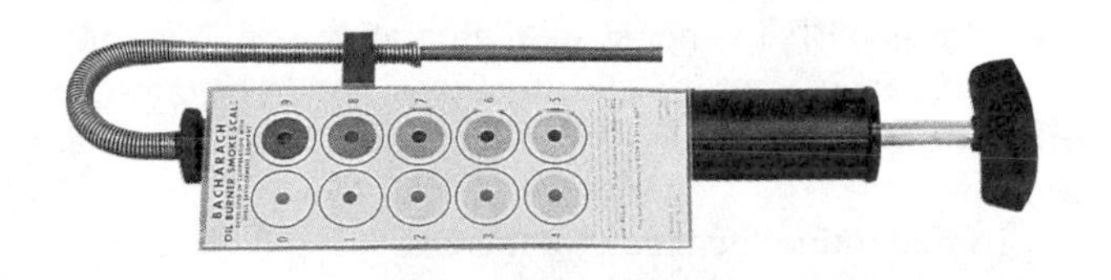

FIGURE 31–56 Smoke tester. *Photo by Bill Johnson*

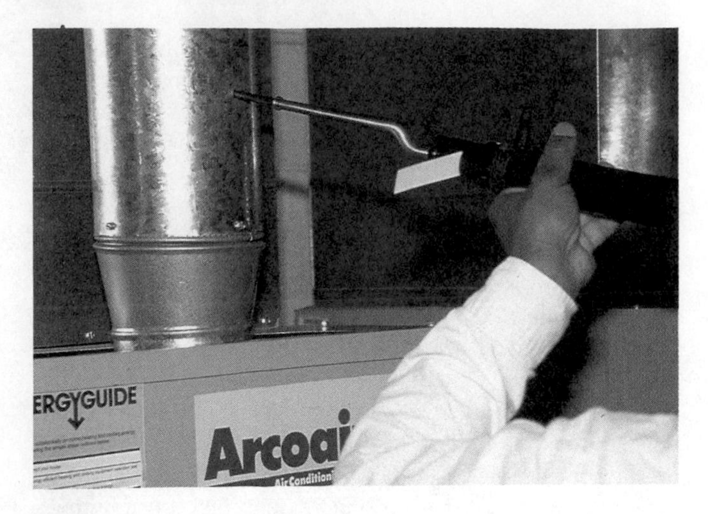

FIGURE 31–57 Making smoke test. *Photo by Bill Johnson*

FIGURE 31–58 Making stack gas temperature test. *Photo by Bill Johnson*

To make the smoke test, Figure 31–57:

1. Turn on the burner and let run for at least 5 min or until the stack thermometer stops rising.
2. Ensure that filter paper has been inserted in tester.
3. Insert sampling tube of test instrument into hole in flue.
4. Pull the tester handle the number of times indicated by the manufacturer's instructions.
5. Remove the filter paper and compare with the scale furnished with the instrument.

Excessive smoke can be caused by:

- Improper fan collar setting (burner air adjustment)
- Improper draft adjustment (draft regulator may be required or need adjustment)
- Poor fuel supply (pressure)
- Oil pump not functioning properly
- Nozzle defective or of incorrect type
- Excessive air leaks in furnace (air diluting flame)
- Improper fuel-to-air ratio
- Firebox defective
- Improper burner air-handling parts

Net Stack Temperature

The net stack temperature is important because an abnormally high temperature is an indication that the furnace may not be operating as efficiently as possible. The net stack temperature is determined by subtracting the air temperature around the furnace from the measured stack or flue temperature. For instance, if the flue temperature reading was 650°F and the basement air temperature where the furnace was located was 60°F, the net stack temperature is 650°F − 60°F = 590°F. Manufacturer's specifications should be consulted to determine normal net stack temperatures for particular furnaces.

To determine the stack temperature, Figure 31–58:

1. Insert the thermometer stem into the hole in the flue.
2. Turn on the burner and allow to run for at least 5 min or until the stack thermometer stops rising in temperature.
3. Subtract the basement or ambient temperature from the stack temperature reading.
4. Compare with manufacturer's specifications.

A high stack temperature may be caused by one or more of the following:

- Excessive draft through the combustion chamber
- Dirty or soot-covered heat exchanger
- Lack of baffling
- Undersized furnace
- Incorrect or defective combustion chamber
- Overfiring

Carbon Dioxide Test

The carbon dioxide test is an important combustion efficiency test. A high carbon dioxide reading is good. If the test reading is low, it indicates that the fuel oil has not burned efficiently or completely. The reading should be considered with all of the test readings. Under most normal conditions a 10+% carbon dioxide reading should be obtained. If problems exist that would be very difficult to correct but the furnace was considered safe to operate and had a net stack temperature of 400°F or less, a carbon dioxide reading of 8% could be acceptable. However, if the net stack temperature is over 500°F, a reading of at least 9% carbon dioxide should be obtained.

To make the carbon dioxide test, Figure 31–59:

1. Turn on the burner and operate for at least 5 min.
2. Insert thermometer and wait for the temperature to stop rising.
3. Insert the sampling tube into the hole previously made in the flue pipe.
4. Remove a test sampling, using the procedure provided by the manufacturer of the test instrument.
5. Mix the fluid in the test instrument with the sample gases from the flue according to instructions.
6. Read the present carbon dioxide from the scale on the instrument.

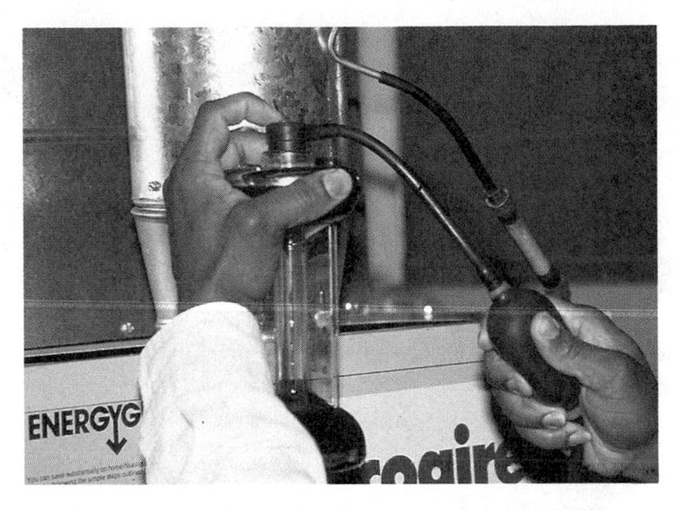

FIGURE 31–59 Making carbon dioxide test. *Photo by Bill Johnson*

A low percent carbon dioxide reading may be caused by one or more of the following:

- High draft or draft regulator not working properly
- Excess combustion air
- Air leakage into combustion chamber
- Poor oil atomization
- Worn, clogged, or incorrect nozzle
- Oil-pressure regulator set incorrectly

Electronic Combustion Analyzers

The technician may have an electronic combustion analyzer. This equipment provides information similar to that provided by the test instruments already described. These instruments may indicate carbon monoxide and oxygen concentrations rather than carbon dioxide percentages, but the technician can determine acceptable levels and necessary corrections to be made on the furnace from the manufacturer's instructions.

Generally, the tests are made from samplings from the same position in the flue, as with the other instruments. These electronic instruments are easier to use and save time. The readings are provided by a digital readout system that is convenient to use, Figure 31–53.

PREVENTIVE MAINTENANCE

Oil heat, like gas heat has air-side and burner-side components. The air-side has been discussed in the electric heat unit. Only the burner section will be discussed here.

Oil heat requires more consistent regular service than any heat system discussed in this book. The reason is the unit uses fuel oil that must be properly metered and burned for the best efficiency. The burning efficiency is accomplished with the proper fuel supply and properly adjusted and maintained burner section. This includes the combustion chamber, heat exchanger, and the flue system. They must all be working perfectly for proper burning efficiency. Improper burning efficiency will cause soot to form, which will slow down the heat exchange between the combustion products and the room air. This condition will deteriorate and cause the combustion efficiency to worsen. It will pay the customer to have the system serviced each year before the season starts, using the appropriate sections of the following procedures. This may look like a long service call for each furnace every year, but a competent service technician should be able to accomplish the maintenance portions in an hour for a typical furnace. Corrective actions are not considered as part of the maintenance. Discussion of the preventive maintenance procedures will start at the tank and move to the flue.

The tank will either be above or below ground and can be either above or below the burner. Below-ground tanks are exposed to the various soils that may corrode the tank. Leaks in the tank may allow oil to seep out or water to seep in. Oil leaks to the soil may not be discovered, but water leaks into the oil will cause fuel burning problems. They will either be discovered at the time of routine preventive maintenance or when an emergency service call is made.

Every year, the service technician should check the tank for water accumulation in the bottom of the tank. A commercial paste that changes color when in contact with water may be spread on the oil tank measuring stick and inserted into the tank. If there is water in the bottom of the tank when the stick is pulled out of the oil, the paste will show the level of the water. If the tank is leaking, it should be replaced. Often, an above-ground tank may be substituted.

The oil lines should be inspected for bends, dents, and rust where they leave the ground and enter the house. A small pinhole in a supply line will cause air to be pulled into the system if the tank is below the burner. This will cause an erratic flame at the burner. The burner may also have an afterburn when the unit is stopped. This is caused by the expanding air in the system. Any flame problems, problems priming the pump, or afterburn problems should prompt you to check the supply line for leaks.

A vacuum test may be performed by disconnecting the oil supply line from the tank and letting the system pump the oil supply line into a vacuum and shutting the pump off. The line should stay in a vacuum if it is leak free. ***Oil and air may leak backward through the pump. This may be prevented by disconnecting the oil return line and holding your finger over it once the pump is stopped.*** A pressure test on the oil supply line will also prove whether or not there is a leak in the supply line. The line must be disconnected and plugged on each end. Apply about 10 psig of pressure (the filter housing is the weakest point and its working pressure is approximately 10 psig) to the line and it should hold indefinitely. No pressure drop in 15 min should satisfy you that no leak is present. If there is a pressure drop, the leak must be found and repaired. If the line is

full of oil, you should see it seep out. Don't forget to check the filter housing gasket.

The gun burner should be removed from the unit and the combustion chamber inspected to be sure the refractory is in good condition. Look for signs of soot, and vacuum out as needed. While the burner is out, change the oil nozzle; be sure to replace the old one with the correct nozzle and use only a nozzle wrench. Check the burner head for overheating and cracks. Examine the static tube for signs of oil or soot, and clean as needed. Set the electrodes using a gage, and examine the insulators for cracks and soot deposits. Clean as needed and change if cracked. Examine the flexible coupling for signs of wear. Make sure the cad cell is clean and aligned correctly and that the bracket is tight. Lubricate the burner motor. Replace the gun burner back into the furnace. Be sure to install the gasket and tighten all bolts.

Change the in-line oil filter. Be sure to fill the cartridge with fresh oil. Change the gasket and tighten it correctly. This will help ensure a good start.

Start the furnace and while it is heating up to operating temperature, prepare to perform a combustion analysis. Check the draft. If the draft is not correct (see text for where to check and approximate draft readings), clean the flue and heat exchanger if needed.

Insert the thermometer in the flue and when it reaches the correct temperature, perform a smoke test. If the fire is smoking, adjust the air until the correct smoke spot is obtained. If you suspect any problem, check the oil pressure to be sure that you have 100 psig at the nozzle. When the smoke is correct, perform a combustion analysis and adjust to the correct carbon dioxide reading.

When this is accomplished, the furnace should operate correctly for the next year.

HVAC GOLDEN RULES

When making a service call to a residence:

- Ask the customer about the problem. The customer can often help you solve the problem and save you time.
- Check humidifiers where applicable. Annual service is highly recommended for health reasons.
- Be prepared and do not let any fuel oil escape on the ground or floor around the furnace. The customer will notice the odor.

Added Value to the Customer

Here are some simple, inexpensive procedures that may be included in the basic service call.

- Check flame characteristics; a smoke test is good insurance and may lead to extra paying service work.
- Replace all panels with the correct fasteners.
- Inspect heat exchanger for soot buildup.

31.15 SERVICE TECHNICIAN CALLS

SERVICE CALL 1

A new customer calls requesting *a complete checkup of the oil furnace, including an efficiency test. This customer will stay with the service technician and watch the complete procedure.*

The technician arrives and explains to the customer that the first thing to do is run the furnace and perform an efficiency test. By running a test before and after adjustment, the technician can report results. The thermostat is set to about 10°F above the room temperature to allow time for it to warm up while the technician sets up. This will also ensure that the furnace will not shut off during the test. The technician gets the proper tools and goes to the basement where the furnace is located. The customer is already there and asks if the technician would mind an observer. The technician explains that a good technician should not mind being watched.

The technician inserts the stack thermometer in the flue and observes the temperature; it is no longer rising. The indoor fan is operating. A sample of the combustion gas is taken and checked for efficiency. A smoke test is also performed. The unit is operating with a slight amount of smoke, and the test shows the efficiency to be 65%. This is about 5% to 10% lower than normal.

The technician then removes a low-voltage wire from the primary, which shuts off the oil burner and allows the fan to continue to run and cool the furnace. The technician then removes the burner nozzle assembly and replaces the nozzle with an exact replacement. Before returning the nozzle assembly to its place, the technician also sets the electrode spacing and then changes the oil filter, the air filter, and oils the furnace motor. The furnace is ready to start again.

The technician starts the furnace and allows it to heat up. When the stack temperature has stopped rising, the technician pulls another sample of flue gas. The furnace is now operating at about 67% efficiency and still a little smokey in the smoke test. The technician checks the furnace nameplate and notices that this furnace needs a 0.75-gph nozzle but has a 1-gph nozzle. Evidently the last service technician put in the wrong nozzle. It is common (but poor practice) for some technicians to use what they have, even if it is not exactly correct.

This technician removes the nozzle just installed as an exact replacement and then gets more serious about the furnace because it has been mishandled. The technician installs the correct nozzle, starts the unit, and checks the oil pressure. The pressure has been reduced to 75 psig to correct for the oversized nozzle. The technician changes the pressure to 100 psig (the correct pressure), standard for residential gun-type burners.

The air to the burner is adjusted; meanwhile the stack temperature is reading 660°F (this is a net temperature of 600°F because the room temperature is 60°F). The technician runs another smoke test, which now shows minimum smoke. The efficiency is now 73%. This is much better.

The customer has been kept informed all through the process and is surprised to learn that oil burner service is so exact. The technician sets the room thermostat back to normal and leaves a satisfied customer.

SERVICE CALL 2

A residential customer calls and reports that *the oil furnace in the basement under the family room is making a noise when it shuts off. The oil pump is not shutting off the oil fast enough.*

The technician arrives, goes to the room thermostat, and turns it up above the room temperature to keep the furnace running. This furnace has been serviced in the last 60 days, so a nozzle and oil filter change is not needed. The technician goes to the furnace and removes a low-voltage wire from the primary control to stop the burner. The fire does not extinguish immediately but shuts down slowly with a rumble.

The technician installs a solenoid in the small oil line that runs from the pump to the nozzle and wires the solenoid coil in parallel with the burner motor so that it will be energized only when the burner is operating. The technician disconnects the line where it goes into the burner housing and places the end into a bottle to catch any oil that may escape. The burner is then turned on for a few seconds. This clears any air that may be trapped in the solenoid out of the line leading to the nozzle.

The technician reconnects the line to the housing and starts the burner. After it has been running for a few minutes, the technician shuts down the burner. It has a normal shutdown. The start-up and shutdown is repeated several times. The furnace operates correctly. The technician then sets the room thermostat to the correct setting before leaving.

SERVICE CALL 3

A customer from a duplex apartment calls. There is no heat. The customer is out of fuel. The customer had fuel delivered last week, but the driver filled the wrong tank. This system has an underground tank and it is sometimes hard to get the oil pump to prime (pull fuel to the pump). The technician is misled for some time, thinking that there is fuel and that the pump is defective.

The technician arrives and goes to the room thermostat. It is set at 10°F above the room temperature setting. The technician goes to the furnace, which is in the garage at the end of the apartment. The furnace is off because the primary control has tripped. The technician takes a flashlight and examines the combustion chamber for any oil buildup that may have accumulated if the customer had been resetting the primary. ***If a customer had repeatedly reset the primary trying to get the furnace to fire and the technician then starts the furnace, this excess oil would be dangerous.*** There is no excess oil in the combustion chamber, so the reset button is pushed. The burner does not fire.

The technician suspects that the electrodes are not firing correctly or that the pump is not pumping. The first thing to do is to fasten a piece of flexible tubing to the bleed port of the side of the oil pump. The hose is placed in a bottle to catch any fuel that may escape, then the bleed port is opened and the primary control is reset. This is a two-pipe system and when the correct oil quality is found at the bleed port the oil supply is verified. The burner and pump motor start. No oil is coming out of the tubing, so the pump must be defective. Before changing the pump, the technician decides to check to make sure there is oil in the line. The line entering the pump is removed. There is no oil in the line. The technician opens the filter housing and finds very little oil in the filter. The technician now decides to check the tank and borrows the stick the customer uses to check the oil level. The stick is pushed through the fill hole in the tank. There is no oil.

The technician tells the customer, who calls the oil company. The driver that delivered the oil is close by and dispatched to the job. The driver points to the tank that was filled, the wrong one, and then fills the correct tank.

The technician starts the furnace and bleeds the pump until a full line of liquid oil flows, then closes the bleed port. The burner ignites and goes through a normal cycle. This furnace is not under contract for maintenance, so the technician suggests to the occupant that a complete service call with nozzle change, electrode adjustment, and filter change be done. The customer agrees. The technician completes the service and turns the thermostat to the normal setting before leaving.

The following service calls do not have the solutions given. The solutions may be found in the Instructor's Guide.

SERVICE CALL 4

A customer calls indicating a smell of smoke when the furnace starts up. This customer does not have the furnace serviced each year.

The technician goes to the furnace in the basement, and the customer follows. The technician turns the system switch off and goes back upstairs and sets the thermostat 5°F above the room temperature so the furnace can be started from the basement. When the technician gets back to the basement, the furnace is started and a puff of smoke is observed. The technician inserts a draft gage in the burner door port; the draft is +0.01 in. W.C. positive pressure.

What is the likely problem and the recommended solution?

SERVICE CALL 5

A retail store manager calls stating that there is no heat in the small retail store.

The technician arrives, goes to the room thermostat, and discovers it set at 10°F higher than the room temperature. The thermostat is calling for heat. The technician goes to the furnace in the basement and discovers it needs resetting for it to run. The technician examines the combustion chamber with a flashlight and finds no oil accumulation. The primary control is then reset. The burner motor starts, and the fuel ignites. It runs for 90 sec and shuts down.

What is the likely problem and the recommended solution.

SERVICE CALL 6

A customer reports that there is no heat and the furnace will not start when the reset button is pushed.

The technician arrives and checks to see that the thermostat is calling for heat. The set point is much higher than the room temperature. The technician goes to the garage where the furnace is located and presses the reset; nothing happens. The primary is carefully checked to see if there is power to the primary control. It shows 120 V. Next the circuit leaving the primary, the orange wire, is checked. (The technician realizes that power comes into the primary on the white and black wires, the white being neutral, and leaves on the orange wire.) There is power on the white to black but none on white to orange, Figure 31–29.

What is the likely problem and the recommended solution?

SUMMARY

- Number 2 grade fuel oil is most commonly used in heating residences and light commercial buildings.
- Fuel oil is composed primarily of hydrogen and carbon in chemical combination.
- Fuel oil is prepared for combustion by converting it to a gaseous state. This is done by atomization—breaking it up into tiny droplets and mixing with air.
- Gun-type oil burner parts are the burner motor, blower or fan wheel, pump, nozzle, air tube, electrodes, transformer, and primary controls.
- The motor is a split-phase fractional horsepower motor.
- The fan is a squirrel cage type.
- The pump can be single or dual stage. Single stage or dual stage is used when the fuel supply is above the burner. Dual stage is used when the fuel supply is below the burner, and the fuel must be lifted to the pump. Pumps have pressure regulating valves and must be set at 100 psi.
- The nozzle atomizes, meters, and patterns the fuel oil.
- Air is blown through the air tube into the combustion chamber and mixed with the atomized fuel oil.
- The electrodes provide the spark for the ignition of the fuel oil and air mixture.
- The transformer provides the high voltage, producing the spark across the electrodes.
- The primary controls provide the means for operating the furnace and the means for shutting down the furnace if there is no ignition.
- Fuel storage and supply systems can be one-pipe or two-pipe design. A two-pipe system is preferred in all installations. A one-pipe system may be used in installations where the tank is above the burner. A two-pipe system is used in all installations where the tank is below the burner.
- An auxiliary or booster supply system must be used where the burner is more than 15 ft above the storage tank.
- The atomized oil and air mixture is ignited in the combustion chamber. Modern chambers are often built of or lined with a refractory material.
- The heat exchanger takes heat caused by the combustion and transfers it to the air that is circulated to heat the building.
- The performance of the pump can be checked with a vacuum gage and a pressure gage.
- One-pipe systems must be bled before the burner is started for the first time or whenever the fuel supply lines have been opened.
- To convert a pump from a one-pipe to a two-pipe system, insert a bypass plug in the return or inlet port. The manufacturer's specifications must be checked to determine this procedure.
- ***Nozzles should not be cleaned or unplugged. They should be replaced.***
- Electrodes should be clean, connections should all be secure, and the spark gap should be adjusted accurately.
- Combustion efficiency tests should be made when servicing oil burners, and corrective action taken to obtain maximum efficiency.
- Preventive maintenance procedures should be performed annually.

REVIEW QUESTIONS

1. How many grades of fuel oil are normally considered as heating oils?
2. What grade is most commonly used in heating residences and light commercial buildings?
3. Describe how the fuel oil is prepared for combustion.
4. What products would be produced if pure combustion took place?
5. List other products that may be produced in the burning process in an oil burner.
6. What is the standard pressure under which oil is fed to the nozzle of an oil burner?
7. List the main components of a gun-type oil burner.
8. Describe the two types of pumps used in gun-type oil burners.
9. Describe the three functions of the nozzle.
10. Sketch and label the three basic spray patterns.

11. When should air adjustments be made?
12. What is the purpose of the electrodes?
13. What is the purpose of the ignition transformer?
14. What is the normal output voltage of the transformer?
15. What are the two functions of the primary control unit?
16. Where is the stack switch located? How does it operate?
17. Where is the cad cell located? How does it operate?
18. When can a one-pipe supply system be used?
19. What is the disadvantage in a one-pipe supply system?
20. When are two-pipe systems used?
21. Describe an auxiliary fuel supply system.
22. When are auxiliary fuel supply systems used?
23. What will happen in the combustion process if the flame hits the combustion chamber wall before it is fully ignited?
24. Describe the purpose of the heat exchanger.
25. What are the two gages used when servicing oil burner pumps?
26. What should the vacuum gage reading be with the pump in operation in a gravity fed system?
27. List the conditions considered in determining the vacuum gage reading if the supply tank is located below the burner.
28. List possible problems if the pressure gage installed on the pump pulsates.
29. List five possible problems that may relate to a defective nozzle.
30. Describe the purpose of the flame retention ring
31. Sketch a wiring diagram of a primary control circuit using a cad cell.
32. Sketch a wiring diagram for an air-handling system.
33. List four tests to determine oil burner efficiency.
34. List two corrective actions that might be taken as a result of readings on each of the four tests in Question 33.
35. List six preventive maintenance procedures.

UNIT 31 DIAGNOSTIC CHART FOR OIL HEAT

The oil burning appliances discussed here are the forced-air type. (Many of the same rules apply to oil boilers as far as their burner operation is concerned.) These oil burning systems may be found in homes and businesses as the primary heating systems. There are two basic types of oil safety controls, the cad cell and the stack switch. Both are discussed in the text. Only the basic procedures are discussed here. For more specific information about any particular step of the procedure, consult the oil heat unit in the text or the manufacturer. Always listen to customers for their input to locate the problem. Remember, they are present most of the time and often pay attention to how the equipment functions.

PROBLEM	POSSIBLE CAUSE	POSSIBLE REPAIR	PARAGRAPH NUMBER
Furnace will not start—no heat	Open disconnect switch	Close disconnect switch	31.6
	Open fuse or breaker	Replace fuse or reset breaker and determine why it opened	31.6
	Faulty wiring	Repair or replace faulty wiring or connections	31.6
	Defective low voltage transformer	Replace transformer and look for possible overload condition	31.6
	Primary safety control off—needs reset	Check for oil accumulation in combustion chamber, if none, reset control and observe fire and flame characteristics	31.6–31.8
	Tripped Burner motor reset	Press reset button, check amperage, if too much, check motor or pump for binding	—
Furnace starts after reset but shuts off after 90 seconds	Cad cell may be out of alignment or dirty	Check cad cell for alignment and smoke on lens—if smoke deposits on lens adjust burner for correct fire	31.8
	Defective cad cell	Replace cad cell and reset	31.8
	Defective primary control	Change primary control	—
Furnace starts but no ignition occurs	No fuel	Fill fuel tank	31.7, 31.8
	Electrodes out of alignment	Align electrodes	31.13
	Defective ignition transformer	Replace transformer	31.13
	Defective oil pump	Replace oil pump	31.13
	Restricted fuel filter	Replace filter	31.13
	Defective coupling between pump and motor	Replace coupling	31.5
Burner runs and ignition occurs, but fan does not start	Faulty wiring or connections in fan circuit	Repair or replace faulty wiring or connectors	31.6
	Defective fan switch	Replace fan switch	31.6
	Defective fan motor	Replace fan motor	31.6
Burner has delayed ignition, makes noise on start up	Electrode out of alignment	Align electrodes	31.13
	Clogged nozzle	Change nozzle and line filter	31.13
	Transformer weak	Change transformer	31.13
	Too much or too little air	Adjust air	31.13
	Nozzle position in burner	Adjust nozzle position	31.13
Burner makes noise on shut down	Fuel cut-off at fuel pump	Check fuel cut-off using gages. If not correct, change fuel pump or add solenoid in oil supply line to burner	31.13

UNIT 31 DIAGNOSTIC CHART FOR OIL HEAT (Continued)

PROBLEM	POSSIBLE CAUSE	POSSIBLE REPAIR	PARAGRAPH NUMBER
High stack temperature	Too much air to burner Too much draft Over fired	Correct air by adjustment Adjust or add draft regulator Change to correct nozzle size with correct oil pressure	31.14 31.14 31.14
Smoking	Too little air Not enough draft	Correct air by adjustment Clean flue and clean furnace heat exchanger on oil side	31.14 31.14
Smoke in conditioned space	Cracked heat exchanger Smoke puffing out around burner or inspection door and pulling in through fan compartment	Replace heat exchanger See smoking problem above	31.2, 31.11
Oil smell	Oil leak or spill during service	Clean up oil spill or leak	31.14

UNIT 32

Hydronic Heat

OBJECTIVES

After studying this unit, you should be able to

- describe a basic hydronic heating system.
- describe reasons for a hydronic heating system to have more than one zone.
- list the three heat sources commonly used in hydronic systems.
- state the reason a boiler is constructed in sections or tubes.
- discuss the purposes for eliminating air from the system.
- describe the purpose of limit controls and low-water cutoff devices.
- state the purpose of a pressure relief valve.
- describe the two purposes of an air cushion or expansion tank.
- state the purpose of a zone control valve.
- explain centrifugal force as it applies to hydronic circulating pumps.
- sketch a finned-tube baseboard unit.
- sketch a one-pipe hydronic heating system.
- describe a two-pipe reverse return system.
- describe a tankless domestic hot water heater used with a hydronic space-heating system.
- list preventive maintenance procedures for hydronic heating systems.

SAFETY CHECKLIST

* When checking a motor pump coupling, turn off the power and lock and tag the disconnect before removing the cover to the coupling.
* If you are checking the electrical service to a pump that has a contactor, shut off the power and remove the cover to the disconnect. Do not touch any electrical connections with meter leads until you have placed one voltmeter lead on a ground source such as a conduit. Then touch the other meter lead to every electrical connection in the contactor box. The reason for this is that some pumps are interlocked electrically with other disconnects and there may be a circuit from another source in the box.
* Ensure that all controls are working properly in a hot water system because an overheated boiler has great explosion potential.

32.1 INTRODUCTION TO HYDRONIC HEATING

Hydronic heating systems are systems where water or steam carries the heat through pipes to the areas to be heated. This text will cover only hot water systems because they are generally used in residential and light commercial installations.

In hydronic systems water is heated in a boiler and circulated through pipes to a heat transfer component called a *terminal unit,* such as a radiator or finned-tube baseboard unit. Here heat is given off to the air in the room. The water is then returned to the boiler. There is no forced moving air in these systems in most residential installations, and if properly installed there will be no hot or cold spots in the conditioned space. Hot water stays in the tubing and heating units even when the boiler is not running, so there are no sensations of rapid cooling or heating, which might occur with forced-warm air systems.

These systems are designed to include more than one zone when necessary. If the system is heating in a small home or area, it may have one zone. If the house is a long ranch type or multilevel, there may be several zones. Separate zones are often installed in bedrooms so that these temperatures can be kept lower than those in the rest of the house, Figure 32–1.

The water is heated at the boiler, using an oil, gas, or electrical heat source. These burners or heating elements are similar to those discussed in other units in this section of the book. Sensing elements in the boiler start and stop the heat source according to the boiler temperature. The water is circulated with a centrifugal pump. A thermostatically controlled zone control valve allows the heated water into the

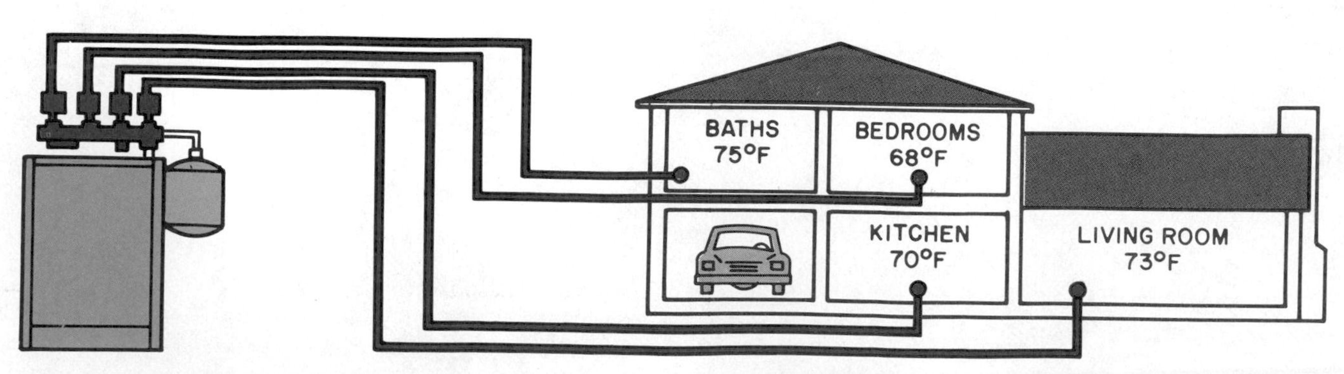

FIGURE 32–1 Four-zone hydronic heating system.

zone needing the heat. Most residential installations use finned-tube baseboard units to transfer heat from the water to the air.

The design process of sizing of the system will not be covered in detail in this unit, but it is necessary that the boiler, piping, and terminal units be the proper size. All components including the pump and valves should also be sized properly for the correct water flow.

32.2 BOILER

A *boiler* is, in its simplest form, a furnace that heats water, using oil, gas, or electricity as the heat source. Some larger commercial boilers use a combination of two fuels. When one fuel is more available than another, it can be used, or the boiler can easily be changed over or converted to use the other fuel. The part of the boiler containing the water is usually constructed in sections or in tubes when oil or gas heat sources are used, Figure 32–2. This provides more surface area for the burner to heat and is more efficient than heating a tank of water. See Figure 32–3 for an example of an electric boiler.

It is important to eliminate air from the water in a boiler for several reasons. Air in water at normal atmospheric temperatures causes a significant amount of corrosion. As the temperature of the water is increased, the corrosive factor of this air in the water is increased many times. Air pockets can also form in the system, causing a blockage of the water circulation. Air in the system can also cause undesirable noise. Many boilers are designed to speed up elimination of air by trapping it and forcing it to be vented to the outside air or into an expansion tank, Figure 32–4. Once the system has been operating and most air has been eliminated, the only new air to be concerned with is that entering with

FIGURE 32–3 An electrically heated boiler. *Courtesy Weil-McLain/A United Dominion Company*

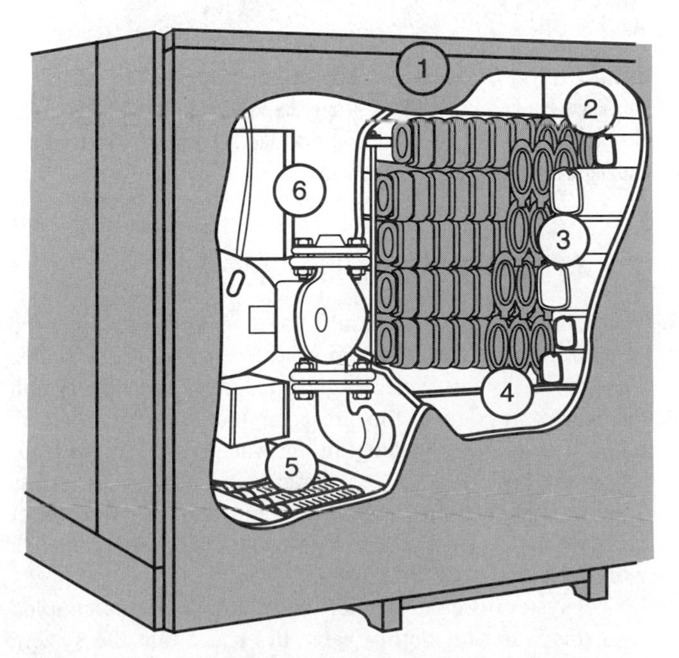

FIGURE 32–2 A gas-heated hydronic boiler. *Courtesy The Peerless Heater Company*

the makeup water. This is usually a small amount of water and a small amount of air that with proper venting will not be a problem. Figure 32–5 is a photo of a manual type and an automatic air vent. These vents are placed on top of the boiler, at high points in the piping system, and at the finned-tube baseboard units. Boilers are also equipped with several safety controls.

32.3 LIMIT CONTROL

A *limit control* in a hot water boiler shuts down the heating source if the water temperature gets too high, Figure 32–6. This control is emersed into the water in the boiler or into a dry well in the boiler water for the best temperature sensing results. Most water systems for homes and small commercial buildings are low-pressure types that operate below 30 psig and 200°F. The limit control would be a temperature control set at some point less than 200°F. For example, if the system water temperature is supposed to operate at 190°F, the limit control would shut the boiler off at 200°F. These controls are normally not adjustable. They are set from the factory.

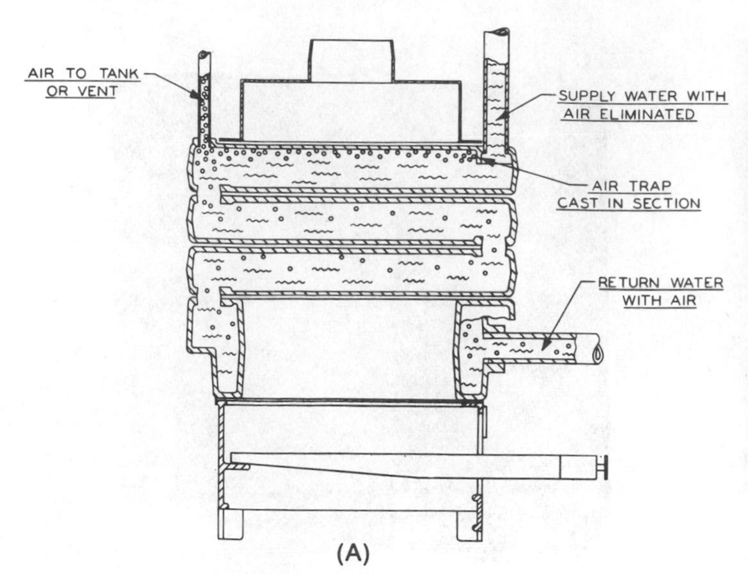

(A)

TOP OUTLET BOILER

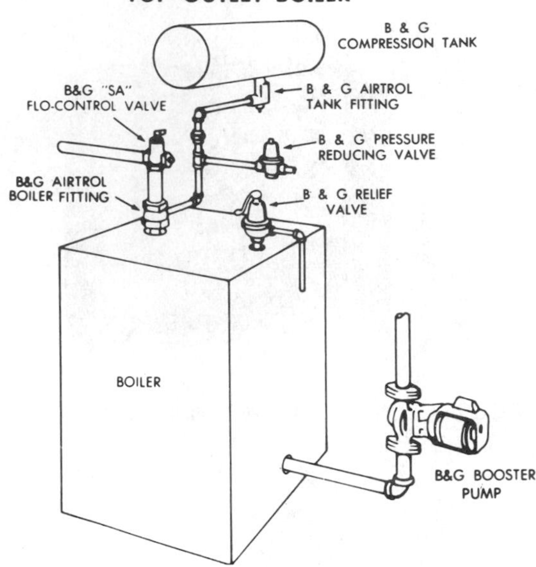

Hot water heating boilers having large internal passages are often excellent low velocity points for air separation. The dip tube which should be pushed down into the boiler (always install ABF with a short nipple) permits only bubble free water from a point well below the top of the boiler to circulate out to the system.

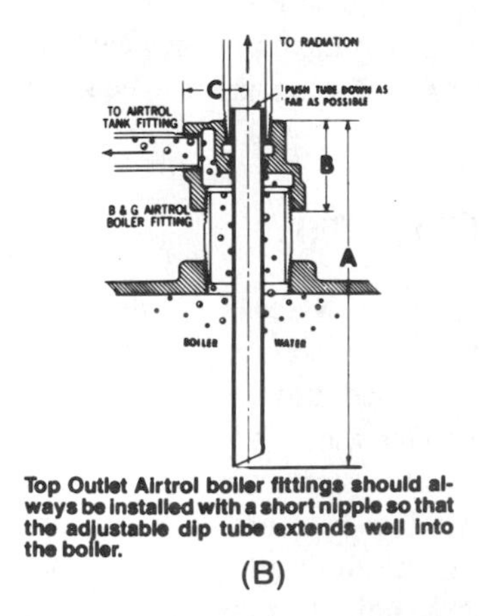

Top Outlet Airtrol boiler fittings should always be installed with a short nipple so that the adjustable dip tube extends well into the boiler.

(B)

FIGURE 32–4 (A) An air-eliminating device that traps the air at the top, venting it to the expansion tank. (B) Another type air eliminator. *Courtesy (A) The Peerless Heater Company, (B) ITT Fluid Handling Division*

FIGURE 32–5 (A) Manual air vent. (B) Automatic air vent. *Photo by Bill Johnson*

FIGURE 32–6 High limit control. This is set so that under normal conditions the boiler shuts down before the relief valve opens. This avoids the problem of water being released from the furnace under pressure, causing damage. *Photo by Bill Johnson*

32.4 WATER-REGULATING VALVE

Water heating systems should have an automatic method of adding back to the system if water is lost due to leaks. The source for this water might be a city water supply and the water pressure is too great for the system. A *water-regulating valve* is installed in the water makeup line leading to the boiler and is set to maintain the pressure on its leaving side (entering the boiler) and less than the relief valve on the boiler. These low-pressure water boilers have a working pressure of 30 psig.

A system pressure relief valve is sometimes coupled with this water-regulating valve to ensure that the system pressure does not exceed the boiler's pressure relief setting, which would be less than 30 psig, Figure 32–7.

FIGURE 32–7 An automatic water-regulating valve with a system pressure relief valve. *Courtesy ITT Fluid Handling Division*

FIGURE 32–8 Safety relief valve. *Courtesy ITT Fluid Handling Division*

32.5 PRESSURE RELIEF VALVE

The American Society of Mechanical Engineers (ASME) Boiler and Pressure Vessel Codes require that each hot water heating boiler have at least one officially rated *pressure relief valve* set to relieve at or below the maximum allowable working pressure of the low-pressure boiler. This would be 30 psig. This valve discharges excessive water when pressure is created by expansion. It also releases excessive pressure if there is a runaway overfiring emergency, Figure 32–8.

32.6 AIR CUSHION TANK OR EXPANSION TANK

When water is heated, it expands. A hot water heating system operates with all components and piping full of water. When water is heated from its source temperature to over 200°F, it will expand considerably, and some provision must be made for this expansion. Thus an *air cushion tank* or *expansion tank* is used, Figure 32–9. This is an airtight tank located above the boiler that provides space for air initially trapped in the system and for the expanded water when it is heated. This tank should provide the only air space within the system. Often when initially filling the system too much air is trapped in the tank. Figure 32–10 illustrates a system using a vent tube through which air can be released one time. Another expansion tank type has a flexible diaphragm to keep the air and water separated. The air above the diaphragm is compressed when water from expansion is pushed up into the tank.

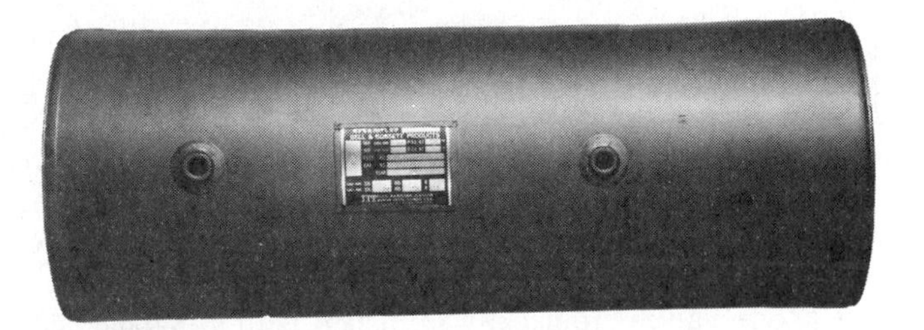

FIGURE 32–9 Air cushion or expansion tank. *Courtesy ITT Fluid Handling Division*

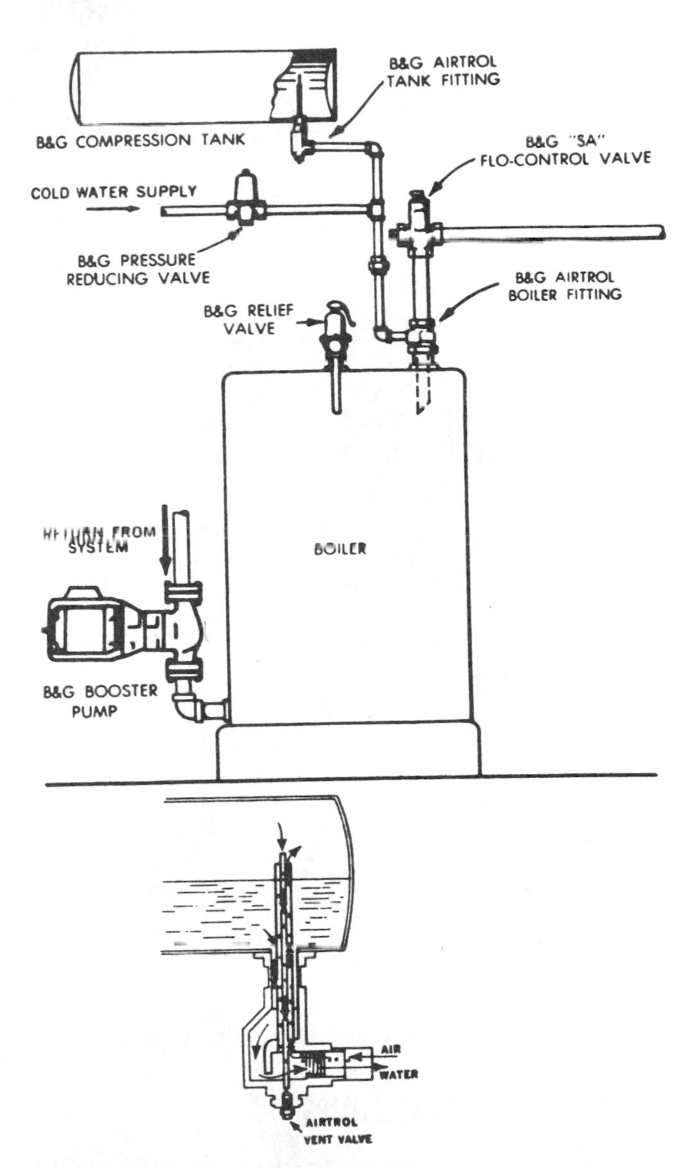

FIGURE 32–10 A system with a vent tube that will release air one time after initially starting up the boiler. This air is vented out the valve at the bottom. *Courtesy ITT Fluid Handling Division*

32.7 ZONE CONTROL VALVES

Zone control valves are thermostatically controlled valves that control water flow to the various zones in the system.

Many are heat motor operated, Figure 32–11. When the thermostat calls for heat, a resistance wire around the valve heats and causes the valve's bimetal element to expand and open slowly. Other types are electric motor operated, which allows slow opening and closing of the valve. Slow opening and closing reduces expansion noise and prevents water hammer.

Zone valves are usually furnished with a manual open feature to allow the technician to open the valve by hand. For example, you may have a question as to whether a valve is stuck or if water is actually circulating to the coil. You may open the valve by hand and feel the coil for hot water circulation. You may also be called to a residence on a "no heat" call and find the valve coil or motor burned. You could manually open the valve to allow hot water to circulate until you return. However, this may cause overheating of the space but that may be more desirable than getting too cold.

FIGURE 32–12 When the impeller of the pump is rotated, it "throws" the water away from the center of the rotation and out through the opening.

FIGURE 32–11 A heat motor zone control valve. *Courtesy Taco, Inc.*

32.8 CENTRIFUGAL PUMPS

Centrifugal pumps, also called *circulators,* force the hot water from the boiler through the piping to the heat transfer units and back to the boiler. These pumps use centrifugal force to circulate the water through the system. Centrifugal force is generated whenever an object is rotated around a central axis. The object or matter being rotated tends to fly away from the center due to its velocity. This force increases proportionately with the speed of the rotation, Figure 32–12.

The *impeller* is that part of the pump that spins and forces water through the system. The proper direction of rotation of the impeller is essential. The vanes or blades in the impeller must "slap" and then throw the water, Figure 32–13. Impellers used on circulating pumps in hot water heating systems usually have sides enclosing the vanes. Such types are called *closed* impellers, Figure 32–13. Many pumps, Figure 32–14, used in closed systems where some makeup water is used (which will cause corrosion) are called *bronze-fitted* pumps. They generally have a cast iron body with the impeller and other moving parts made of bronze or nonferrous metals. Others have stainless steel or all bronze parts.

Centrifugal water pumps are not positive displacement pumps as are compressors. The term circulating pump is used because the centrifugal pumps used in small systems do not add much pressure to the water from the inlet to the outlet. It is important to understand how a centrifugal pump responds to pressure and load changes.

A centrifugal pump may resemble a vegetable can with holes drilled around the bottom as in Figure 32–15. If water is allowed to flow into the can, it will flow out the bottom holes by gravity. If the can is rotated, the water will be thrown outward by the centrifugal force. The water flow may then be captured with a pump housing and directed to

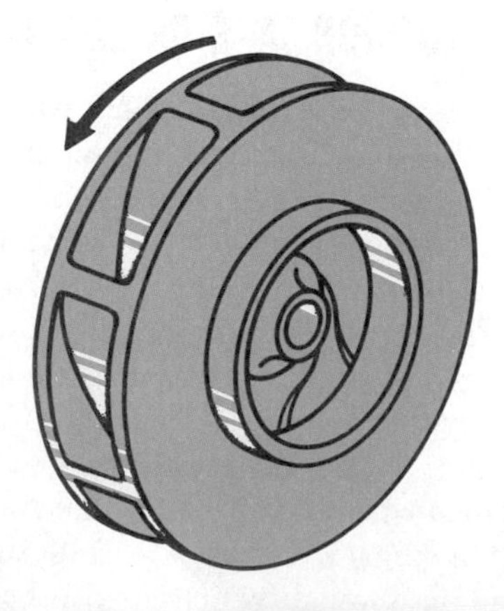

FIGURE 32–13 Pump impeller. Note the direction of the rotation indicated by the arrow. This is an example of a closed impeller.

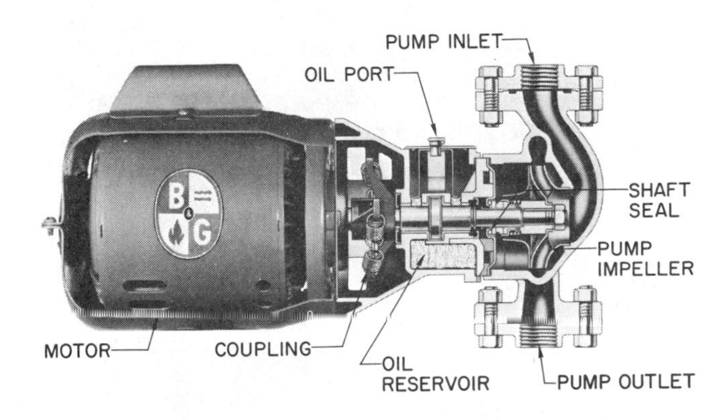

FIGURE 32–14 Centrifugal pump. *Courtesy of ITT Fluid Handling Division*

the pump discharge. The faster the can is rotated, the more water the pump will move from the inlet to the outlet. The larger the can, the more water the pump will pump and the more pressure the rotation will impart to the water. The larger the holes in the can, the more water the pump will circulate.

The following example will demonstrate how a centrifugal pump performance curve is developed. This is a progressive example. See Figure 32–16 (A) through (J).

(A) A pump is connected to a reservoir that maintains the water level just above the pump inlet. The pump

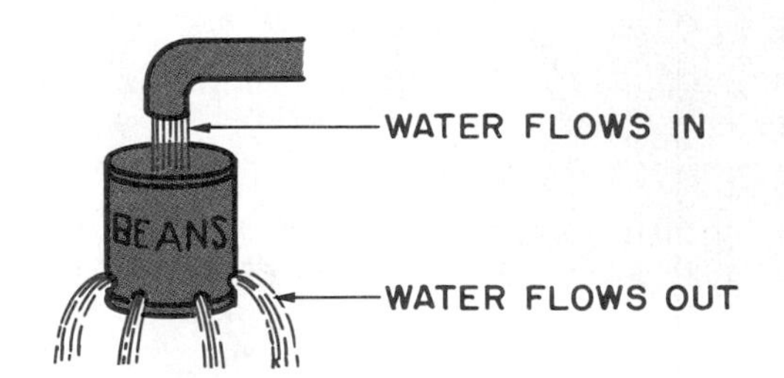

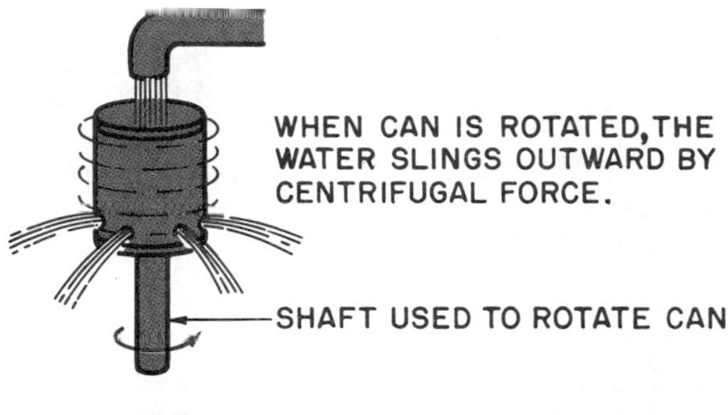

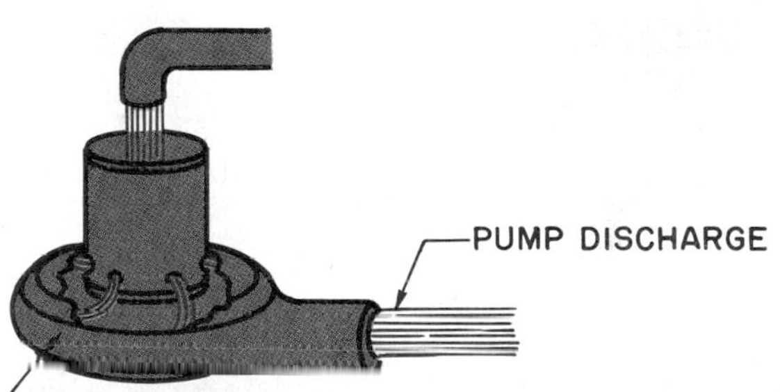

FIGURE 32–15 A fundamental centrifugal water pump.

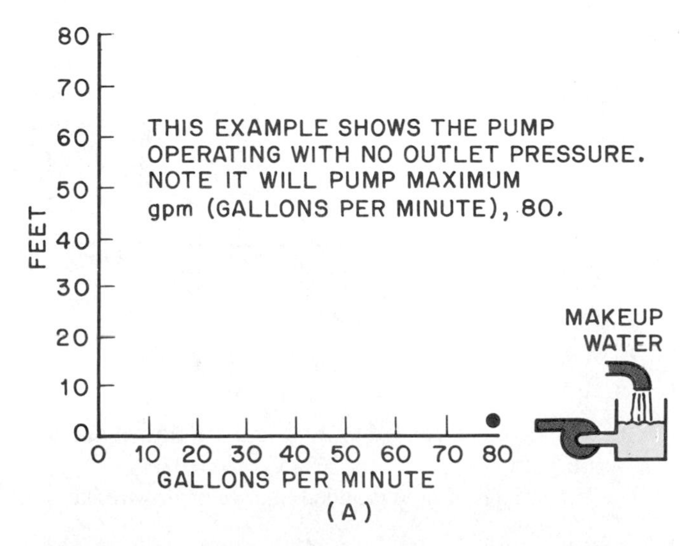

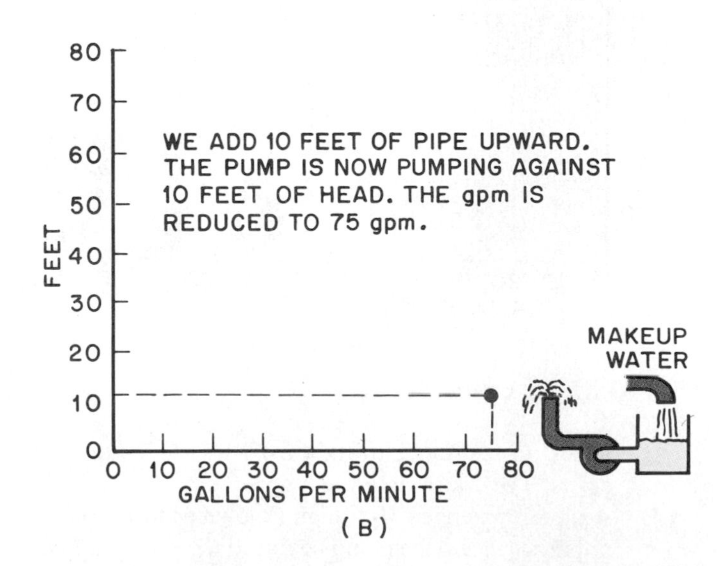

FIGURE 32–16 Developing a centrifugal pump performance curve.

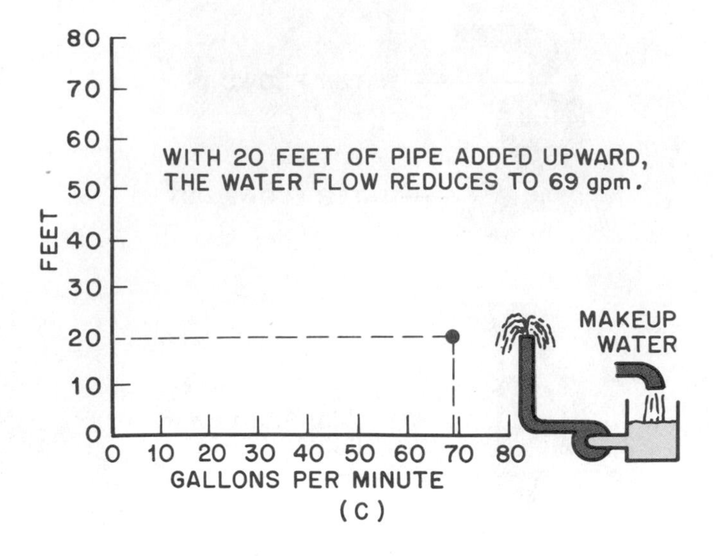

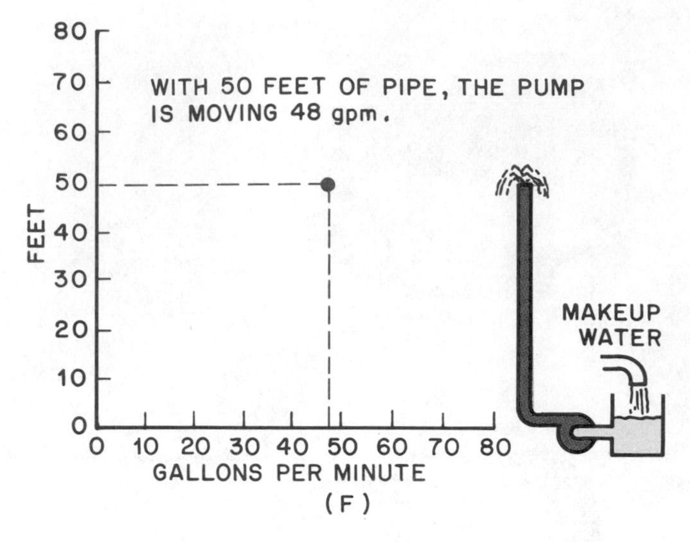

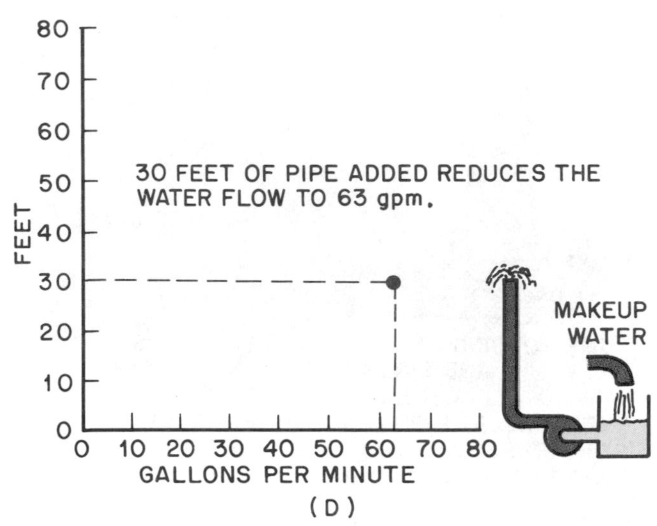

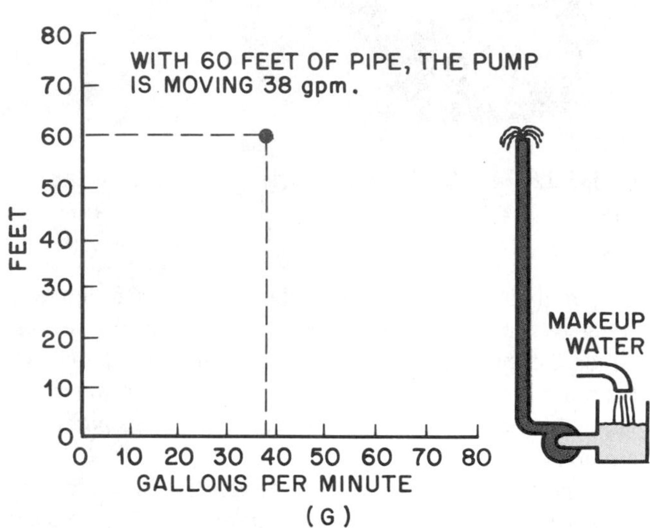

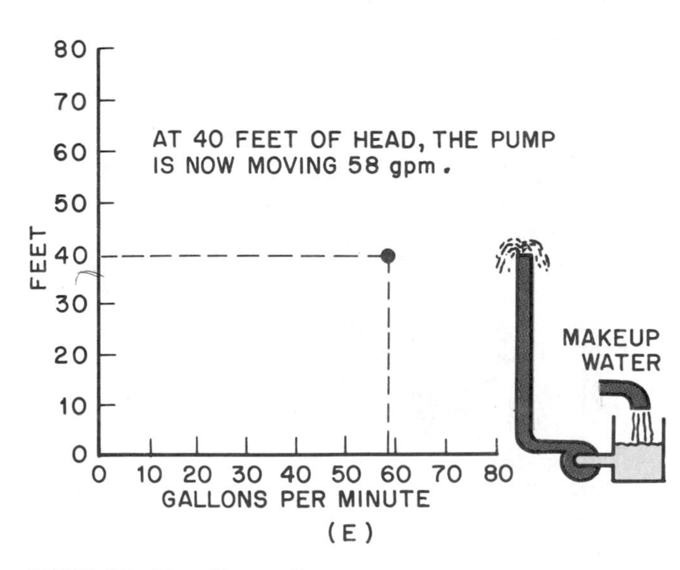

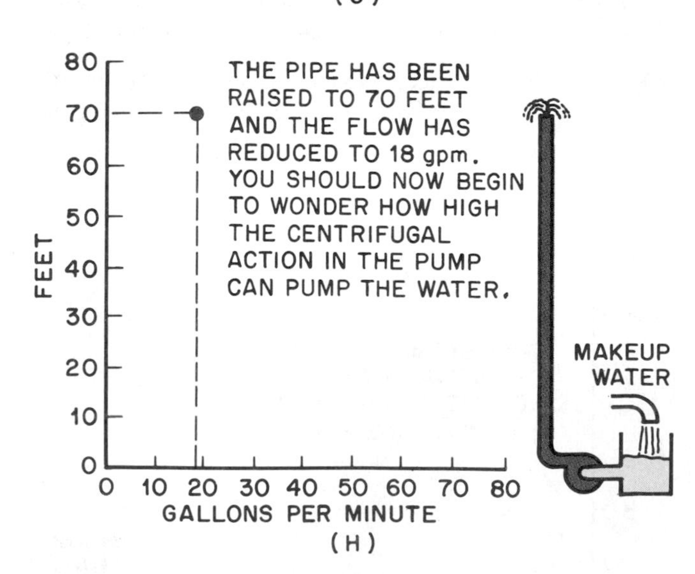

FIGURE 32–16 *(Continued)*

meets no resistance to the flow and maximum flow exists. The water flow is 80 gpm.

(B) A pipe is extended 10 ft high (known as feet of head, or pressure) and the pump begins to meet some pumping resistance. The water flow reduces to 75 gpm.

(C) The pipe is extended to 20 ft and the flow reduces to 68 gpm.

(D) When 30 ft of pipe is added, the water flow reduces to 63 gpm.

(E) The pipe is extended to 40 ft. The pump is meeting

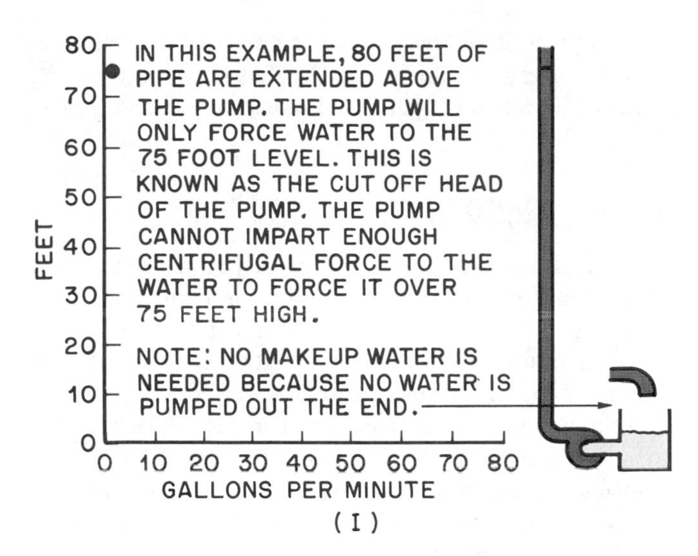

IN THIS EXAMPLE, 80 FEET OF PIPE ARE EXTENDED ABOVE THE PUMP. THE PUMP WILL ONLY FORCE WATER TO THE 75 FOOT LEVEL. THIS IS KNOWN AS THE CUT OFF HEAD OF THE PUMP. THE PUMP CANNOT IMPART ENOUGH CENTRIFUGAL FORCE TO THE WATER TO FORCE IT OVER 75 FEET HIGH.
NOTE: NO MAKEUP WATER IS NEEDED BECAUSE NO WATER IS PUMPED OUT THE END.
FEET
0
10
20
30
40
50
60
70
80
0 10 20 30 40 50 60 70 80
GALLONS PER MINUTE
(I)

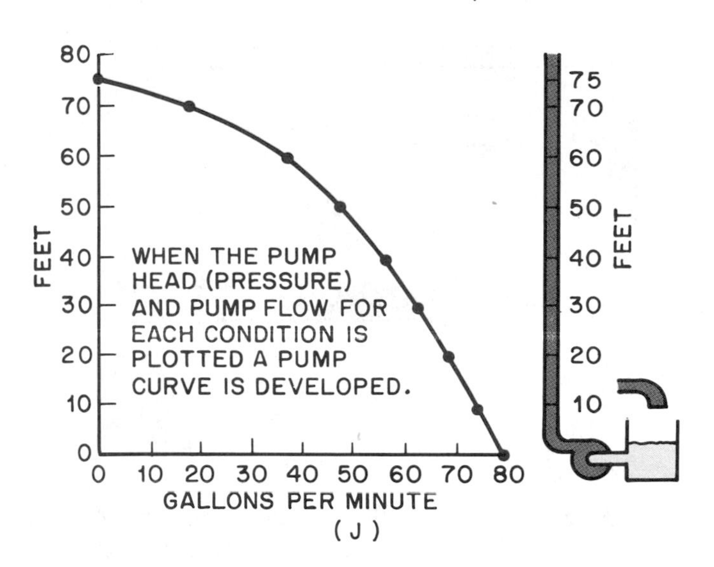

FIGURE 32–16 *(Continued)*

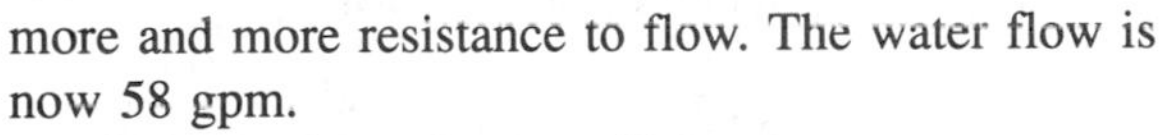

more and more resistance to flow. The water flow is now 58 gpm.

(F) At 50 ft, the flow slows to 48 gpm.

(G) At 60 ft, the flow slows to 38 gpm.

(H) At 70 ft, the water flow has been reduced to 18 gpm. The pump has almost reached its pumping limit. The pump can only impart a fixed amount of centrifugal force.

(I) When 80 ft of pipe is added to the pump outlet, the water will rise to the 75 ft level. If you could look down into the pipe, you would be able to see the water gently moving up and down. The pump has reached its pumping pressure head capacity. This is known as the shut off point of the pump.

(J) This is a manufacturer's pump curve and the preceding examples show how the manufacturer arrives at the pump curve.

It would be awkward for the manufacturer to add vertical pipe to all size pumps for testing purposes so they use valves to simulate vertical head. Before we can use valves, we must be able to convert liquid water vertical head to gage readings. Gages read in psig. When a gage reads 1 psig, it is the same pressure that a column of water 2.309 ft (27.7 in.) high exerts at the bottom, Figure 32–17. This would be the same as saying that a column of water 1 ft high (feet of head) is equal to 0.433 psig because 1 psig divided by 2.3093 ft = 0.433. Some examples of water columns and pressure are shown in Figure 32–18.

A pump manufacturer may use gages to establish pump head (pressure) by placing a gage at the inlet and a gage at the outlet and reading the difference. This is known as psi difference. Psi difference is then converted to feet of head, Figure 32–19.

The power consumed by the motor driving a centrifugal pump is in proportion to the quantity of water the pump circulates. For example, in the first pump example in Figure 32–16(A), the pump would require more horsepower than

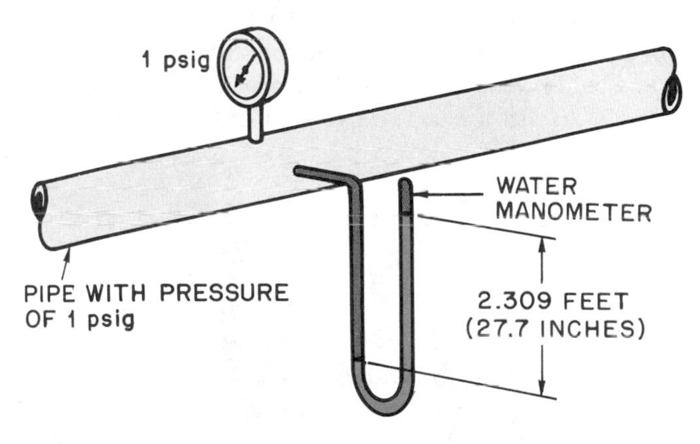

FIGURE 32–17 A pressure reading on a gage of 1 psig equals 2.309 ft of water.

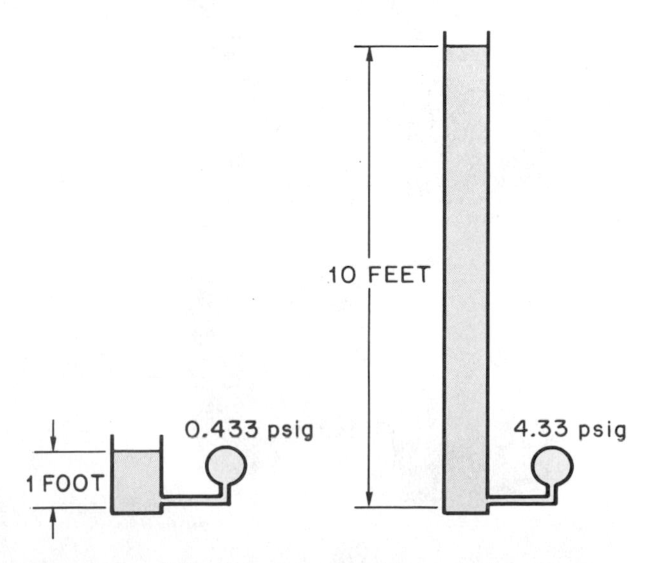

FIGURE 32–18 This figure shows that different heights of water columns exert different pressures. Each time you move the column 1 ft, the pressure rises 0.433 psig.

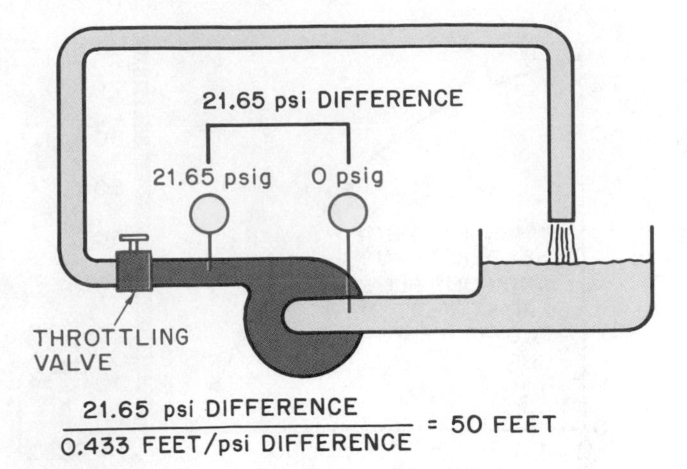

FIGURE 32–19 This figure illustrates how a pump manufacturer may test a pump for pumping head. More elaborate systems that measure gallons per minute at the same time are used to establish a pump curve.

in example (I) where the pump was pumping against the pump shutoff pressure. When a valve is installed in a piping circuit and closed, it would seem that the power consumption would rise because of an increase in pump head, but it doesn't. As the discharge valve is closed, the water begins to recirculate in the pump and the power consumption reduces. This can be demonstrated by shutting the valve at the outlet of a centrifugal pump and monitoring the amperage, Figure 32–20.

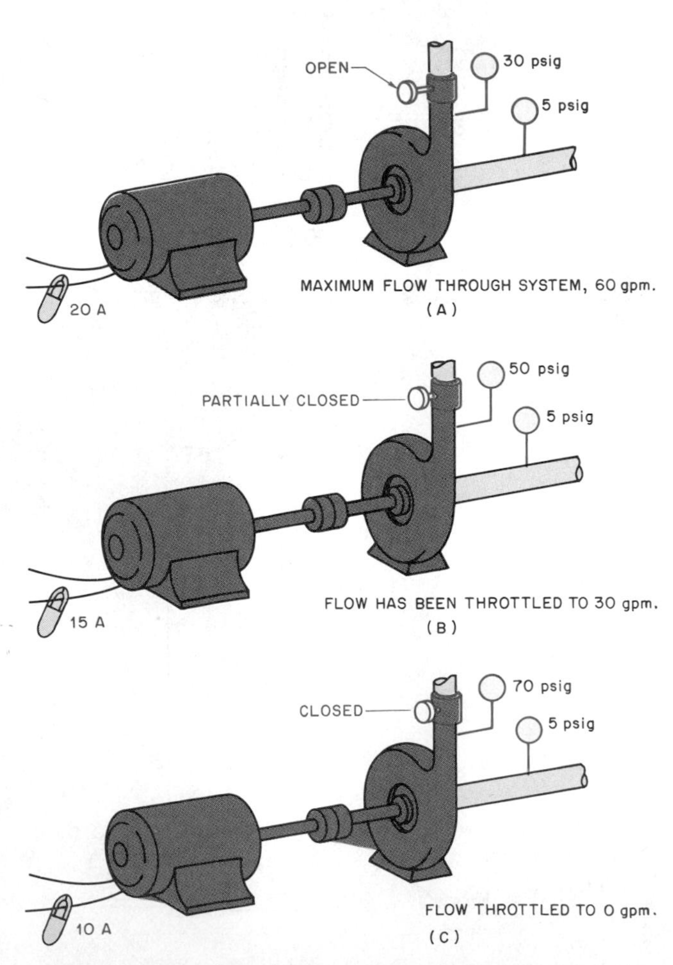

FIGURE 32–20 As the flow is reduced by throttling the outlet valve, the power to drive the motor reduces the current required to turn the pump impeller in the water.

32.9 FINNED-TUBE BASEBOARD UNITS

Most residences with modern hydronic heating systems use finned-tube baseboard heat transfer for terminal heating units, Figure 32–21. Air enters the bottom of these units and passes over the hot fins. Heat is given off to the cooler air, causing it to rise by convection. The heated air leaves the unit through the damper area. The damper can be adjusted to regulate the heat flow. Baseboard units are generally available in lengths from 2 ft to 8 ft and are relatively easy to install following the manufacturer's instructions. Figure 32–22 shows a radiator and fan coil terminal units.

Finned tube heating units are rated in Btuh/ft of pipe at two different flow rates—500 lb/h and 2000 lb/h. Five hundred lb/h is the equivalent to 1 gal of water flow per minute, and 4 gal/min equals 2000 lb/h.

$$1 \text{ gal/min} = 8.33 \text{ lb/min}$$
$$8.33 \text{ lb/min} \times 60 \text{ min/h} = 499.8 \text{ lb/h Rounded to } 500 \text{ lb/h}$$

$$8.33 \text{ lb/min} \times 4 \text{ gal/min} = 33.32 \text{ lb/min}$$
$$33.32 \text{ lb/min} \times 60 \text{ min/h} = 1999.2 \text{ lb/h Rounded to } 2000 \text{ lb/h}$$

The manufacturer will rate the finned-tube radiation at so many Btuh/ft at different temperatures.

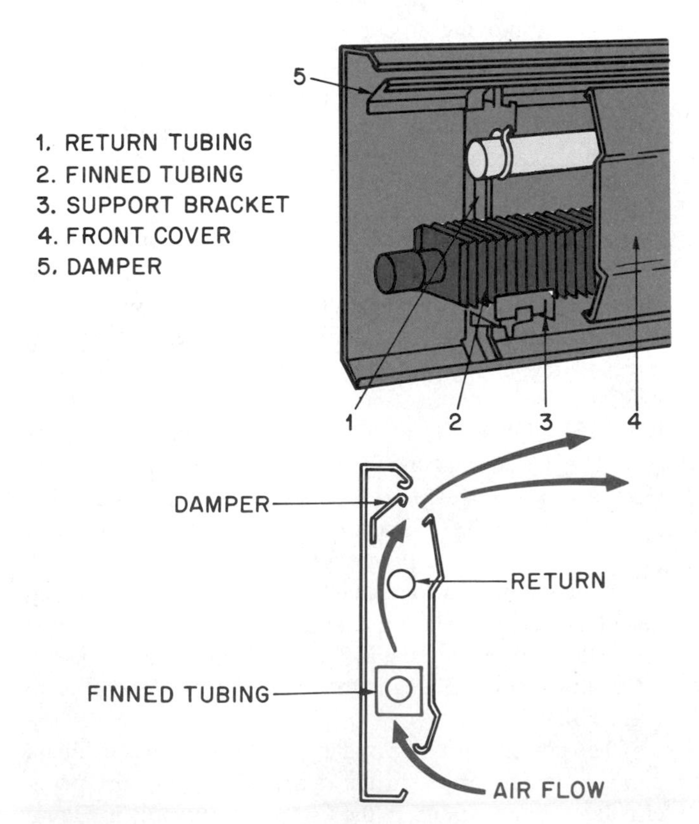

FIGURE 32–21 Two-pipe finned-tube baseboard unit.

(A)

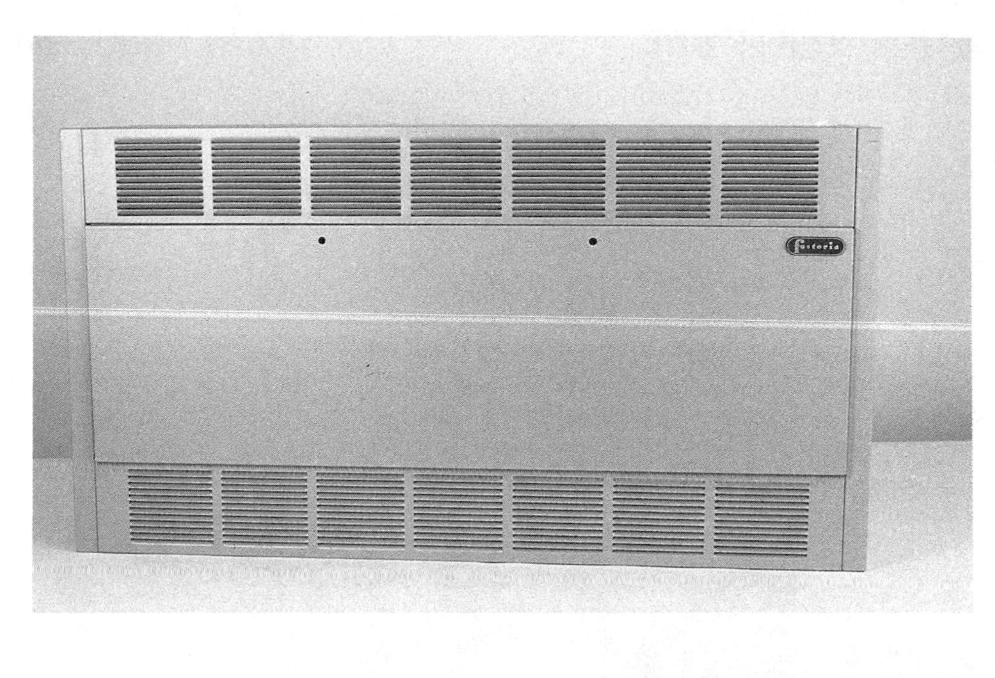

(B)

FIGURE 32–22 Terminal hydronic heating units. (A) Water in the radiator radiates heat out into the room. (B) The fan coil units are similar to the baseboard units but have a fan that blows the hot air into the room. *(A) Photo by Bill Johnson, (B) Courtesy Fostoria Industries, Inc.*

FROM A MANUFACTURER'S TABLE (Btuh/ft)

1/2 in. ID FINNED-TUBE CONVECTOR AT 500 lb/h

170°F	180°F	190°F	200°F	210°F	220°F
520	580	650	710	780	850

1/2 in. ID FINNED-TUBE CONVECTOR AT 2000 lb/h

170°F	180°F	190°F	200°F	210°F	220°F
550	610	690	750	820	900

These same convectors are available in 3/4-in. ID tubing. This will give you some idea of how the convectors are chosen. If a room needs 16,000 Btuh and the flow rate has been determined to be 1 gal/min and the system's average water temperature is 180°F, the convector length would be 27.6 ft.

$$16{,}000 \text{ Btuh}/580 \text{ Btu/ft} = 27.6 \text{ ft}$$

Other types of terminal heating units are selected from manufacturer's literature in much the same manner.

When installing baseboard terminal units, provide for expansion. The hot water passing through the unit makes it expand. Expansion occurs toward both ends of the unit so the ends should not be restricted. It is a good practice to install expansion joints in longer units, Figure 32–23A. If one expansion joint is used, place it in the center. If two are used, space them evenly at intervals one third of the length of the unit. When piping drops below the floor level, the vertical risers should each be at least 1 ft long. The table in Figure 32–23B shows the diameter of the holes needed through the floor for the various pipe sizes. The table in Figure 32–24 lists the maximum lengths that can be installed when the water temperature is raised from 70°F to the temperature shown.

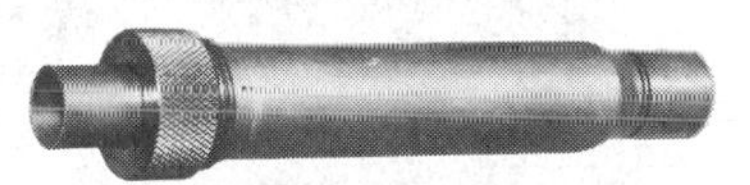

FIGURE 32–23A An expansion joint. *Courtesy Edwards Engineering Corporation*

TUBE SIZE		
NOMINAL (INCHES)	O.D. (INCHES)	RECOMMENDED MINIMUM HOLE (INCHES)
$\frac{1}{2}$	$\frac{5}{8}$	1
$\frac{3}{4}$	$\frac{7}{8}$	$1\frac{1}{4}$

FIGURE 32–23B This table indicates the diameter of the holes required for each pipe size to allow for expansion.

AVERAGE WATER TEMPERATURE °F	MAXIMUM LENGTH OF STRAIGHT RUN (FEET)
220	26
210	28
200	30
190	33
180	35
170	39
160	42
150	47

FIGURE 32–24 Maximum lengths of baseboard unit that can be installed when the temperature is raised from 70°F to the temperature indicated in the table.

32.10 BALANCING VALVES

When designing a hot water heating system, consideration must be given to the flow rate of the water and the friction in the system. Friction is caused by the resistance of the water flowing through the piping, valves, and fittings in the system. The flow rate is the number of gallons of water flowing each minute through the system. A system is considered to be in balance when the resistance to the water flow is the same in each flow path. A means for balancing the system should be provided in all installations. One method is to install a *balancing valve,* Figure 32–25, on the return side of each heating unit. This valve is adjustable. ***The valve adjustment is generally determined by the system designer, and the installer sets them accordingly. Be sure to read the manufacturer's instructions for the correct procedure to set the valve.***

FIGURE 32–25 Balancing valve. *Photo by Bill Johnson*

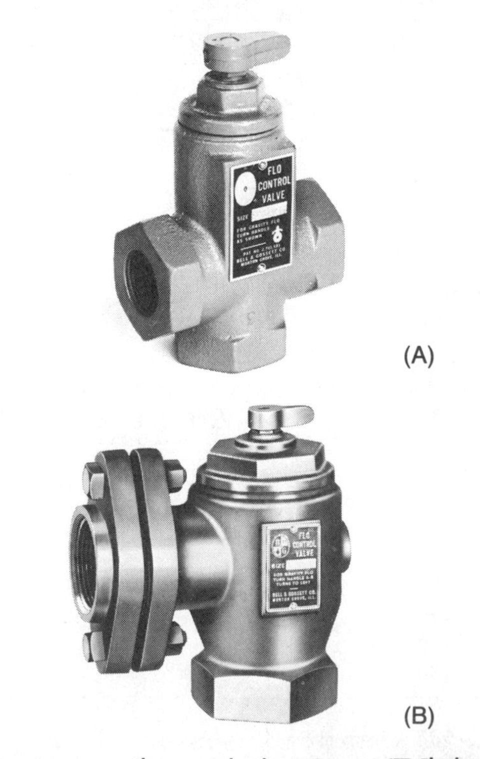

FIGURE 32–26 Flow control valves. *Courtesy ITT Fluid Handling Division*

32.11 FLOW CONTROL VALVES

Flow control valves or *check valves* are necessary so that the water will not flow by gravity when the circulator is not operating, Figure 32–26.

32.12 HORIZONTAL AND VERTICAL (DOWNFLOW) FORCED-AIR DISCHARGE UNIT HEATERS

Horizontal and vertical forced-air discharge unit heaters (used generally in commercial and industrial applications) have fans to blow the air across the heat transfer elements. The heating elements are normally made of heavy copper tubing through which the hot water is circulated. The tubing is surrounded with aluminum fins to more efficiently give off heat to the air blown across them, Figure 32–27.

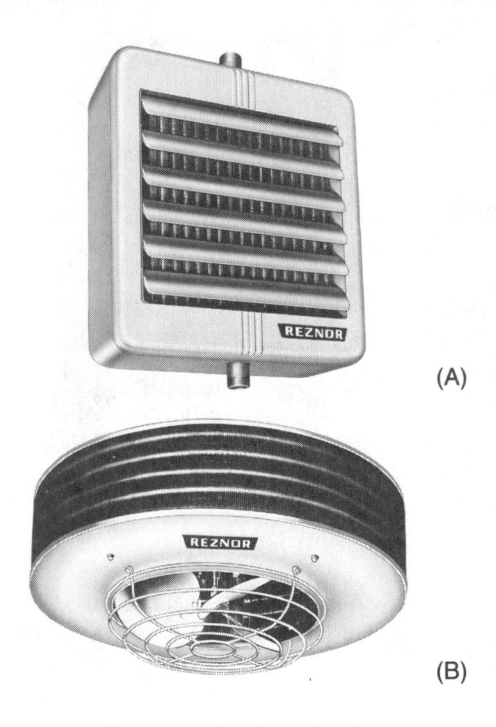

FIGURE 32–27 (A) Horizontal and (B) downflow air discharge unit heaters. *Courtesy of ITT Reznor Division*

32.13 HYDRONIC HEATING PIPING SYSTEMS

Most residential systems use a *series loop* or a *one-pipe system* layout for each zone. Figure 32–28 illustrates a series loop system. In this system, all of the hot water flows through all the heating units. Neither the temperature nor the amount of flow can be varied from one unit to the next without affecting the entire system. The water temperature in the last unit will be lower than in the first because heat is given off as the water moves through the system. This system is simple and economical but without much flexibility.

A basic one-pipe system is illustrated in Figure 32–29. It has a one-pipe main supply with branches to each of the heating units. The piping in the branches is smaller in diameter than the main pipe. One of the tees to each of the units is a special tee called a *one-pipe fitting,* Figure 32–30. This

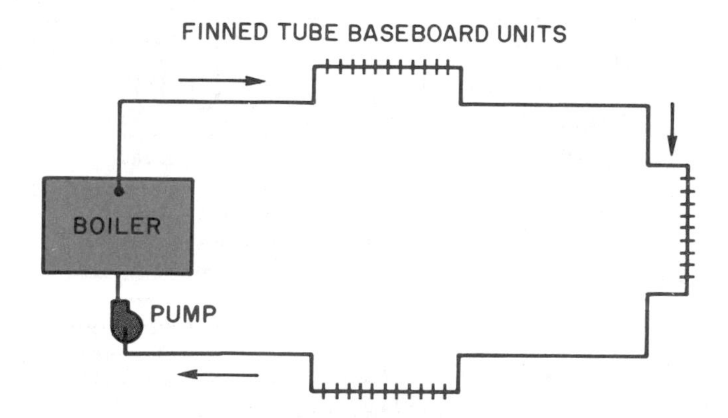

FIGURE 32–28 Series loop hydronic heating system.

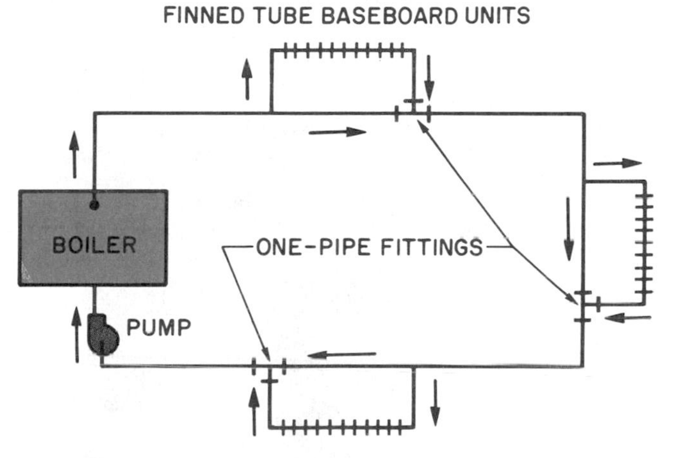

FIGURE 32–29 One-pipe hydronic heating system.

fitting forces some of the water in the line to go through the heating unit and the rest to continue in the main supply line.

The *two-pipe reverse return system,* Figure 32–31, has two pipes running parallel with each other and the water flowing in the same direction in each pipe. Notice that the water going into the first heating unit is the last to be returned to the boiler. The water going into the last heating unit is the first to be returned to the boiler. This system equalizes the distance of the water flow in the system.

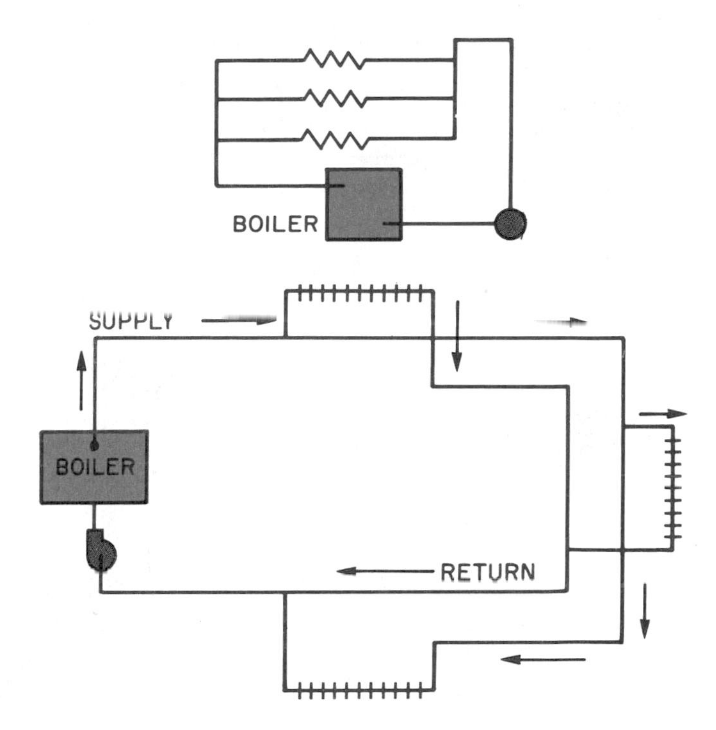

FIGURE 32–31 Two-pipe reverse return system.

In the *two-pipe direct return system,* Figure 32–32, the water flowing through the nearest heating unit to the boiler is the first back to the boiler. It therefore has the shortest run. The water flowing through the unit farthest away has the longest run.

Hot water heating systems installed in floors and ceilings are known as *panel systems.* They are normally designed with each coil as an individual zone or circuit and are

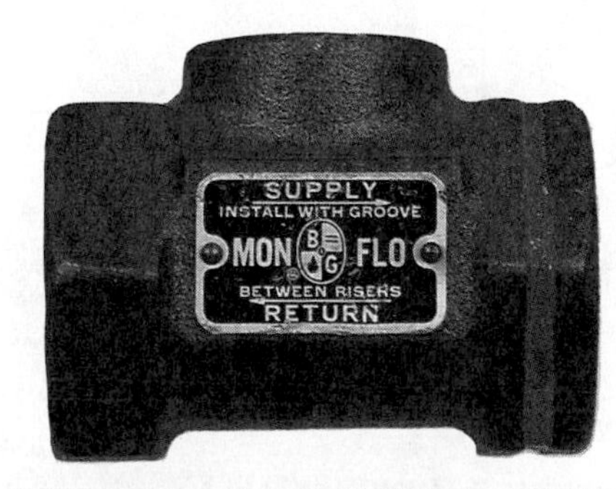

HOW TO USE B & G MONOFLO FITTINGS

- Be sure the RING is between the risers OR
- Be sure the SUPPLY arrow on the supply riser and the RETURN arrow on the return riser point in the direction of flow.

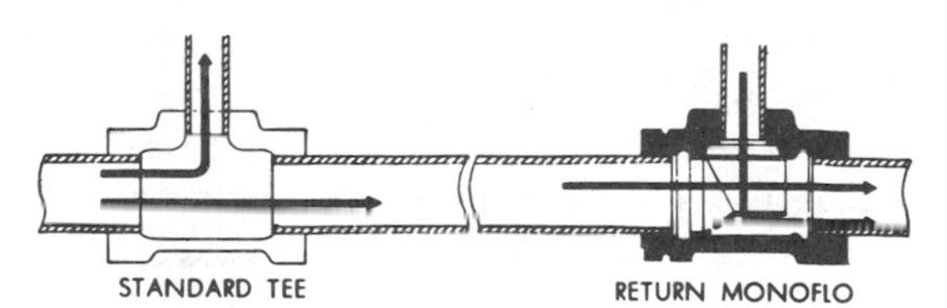

For radiators above the main—normal resistance
For most installations where radiators are above the main, only one Monoflo Fitting need be used for each radiator.

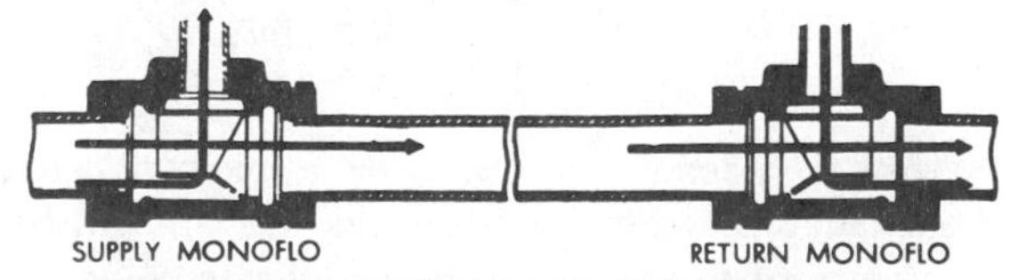

For radiators above the main—high resistance
Where characteristics of the installation are such that, resistance to circulation is high, two Fittings will supply the diversion capacity necessary.

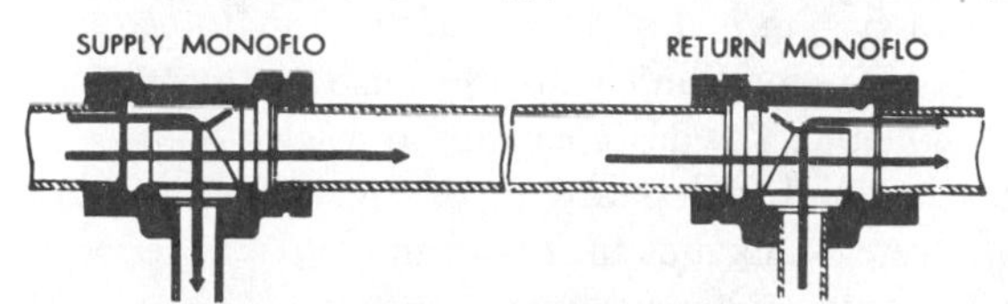

For radiators below the main
Radiators below the main require the use of both a Supply and Return Monoflo Fitting, except on a ¾" main use a single return fitting.

FIGURE 32–30 A tee for one-pipe system. Partial water flow is diverted through the tee to the terminal unit. *Courtesy ITT Fluid Handling Division*

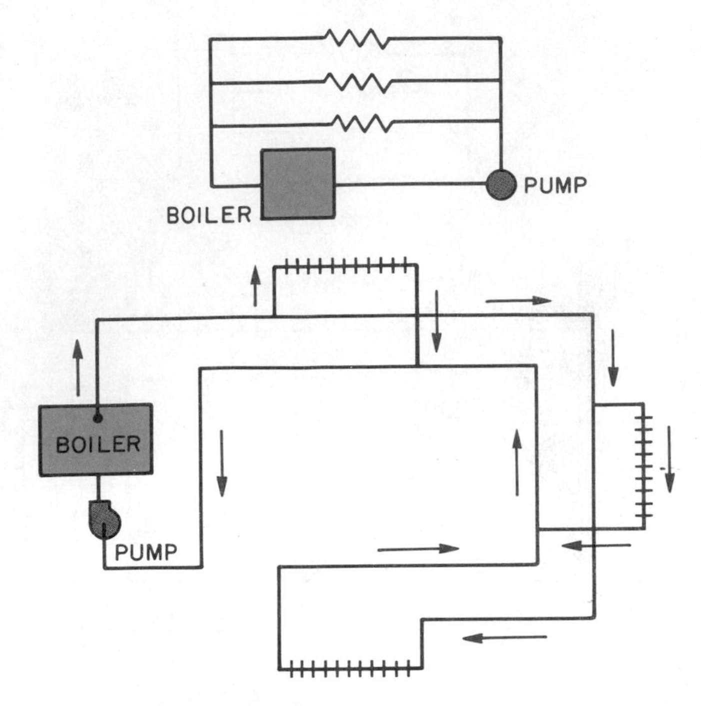

FIGURE 32–32 Two-pipe direct return system.

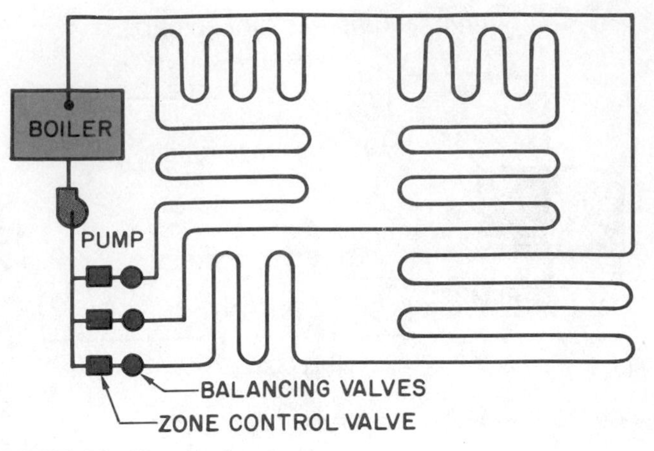

FIGURE 32–33 A radiant panel system.

each balanced individually, usually at a place where they come together, Figure 32–33.

Most installations will be multiple zone applications. Each zone is treated as a separate system, Figure 32–34. The zone control valve controls water circulated through that zone. Each unit is balanced at the terminal unit.

Regardless of the type of piping layout, the estimator or engineer will pay close attention to the sizing of the pipe for any system. Pipe sizing and pump size determines how much water will flow in each circuit. If the pipe is too large, it will be expensive and it may not meet the low-limit flow requirements for convectors. For example, the minimum design flow rate for different diameter convector pipes is:

1/2 in. ID	0.3 gpm
3/4 in. ID	0.5 gpm
1 in. ID	0.9 gpm
1 1/4 in. ID	1.6 gpm

If the flow rates below these are used at the convectors, the heat exchange between the water and the air will be less. The velocity of the water in the convector helps the heat exchange. Flow that is too slow is called laminar flow.

If a system is designed with the flow too fast, the pipe will be too small and velocity noise of the water will be objectionable. The designer tries to reach a balance between oversized and undersized pipe. The following table lists some water velocities in ft per minute (fpm) recommended for hydronic piping systems.

Pump discharge	8 to 12 fpm
Pump suction	4 to 7 fpm

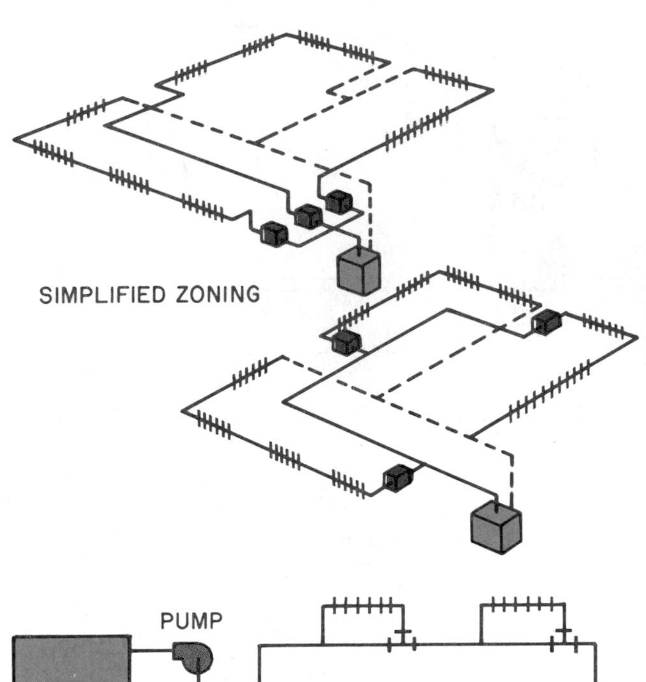

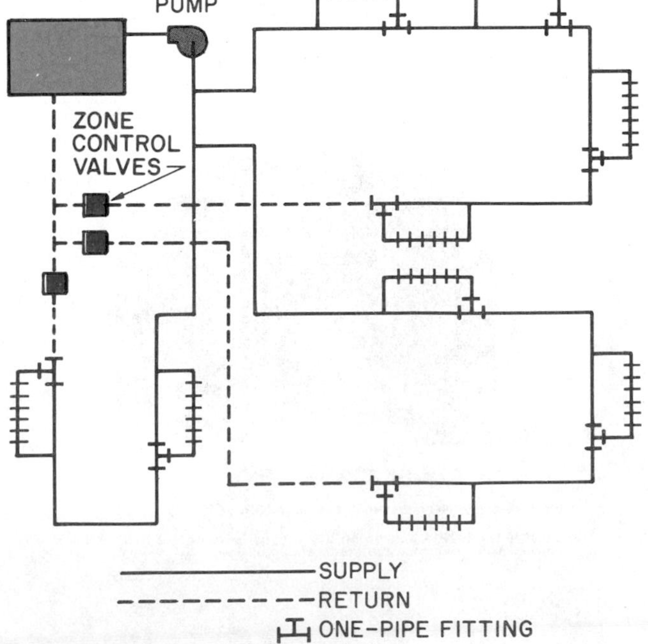

FIGURE 32–34 Multizone installation.

Drain lines	4 to 7 fpm
Header	5 to 15 fpm
Riser	3 to 10 fpm
General service	5 to 10 fpm
City water lines	3 to 7 fpm

32.14 TANKLESS DOMESTIC HOT WATER HEATERS

Most hot water heating boilers (oil and gas fired) can be furnished with a domestic hot water heater consisting of a coil inserted into the boiler. The domestic hot water is contained within the coil and heated by the boiler quickly, which eliminates the need for a storage tank. This system can meet most domestic hot water needs without using a separate tank, Figure 32–35. This is generally an efficient way to produce hot water. This heater is installed with a flow control valve to allow only the proper amount of cold water into the system to be heated.

FIGURE 32–35 Tankless domestic water heater installed in a boiler. *Courtesy Weil-McLain/A United Dominion Company*

PREVENTIVE MAINTENANCE

The air side of hydronic systems can be either natural draft or forced draft. If the system is natural draft, the convectors will be next to the floor and will be subjected to dust on the bottom side where the air enters. There are no filters on these natural draft systems and consequently there is no filter maintenance. The coil becomes the filter but the air moves so slowly that not much dust is picked up. The cover should be removed after several years and the coil vacuumed clean.

If the system is forced air, routine filter maintenance will be required as described in Unit 29. Coils may become dirty after many years of good filter maintenance because all particles of dust are not filtered and some will deposit on the coil. When the coil becomes dirty, capacity will be reduced. Most coils are oversized and the capacity reduction may not be noticed for many years.

Cleaning a coil is accomplished with approved liquid detergents mixed with water and applied to the coil. If the coil is in an equipment room, it may be cleaned in place. If the coil is in the attic or above a ceiling, it may have to be removed from the air handler for cleaning.

The fan blades may also have dust deposits built up on them and need cleaning. The fan wheel may be cleaned in the fan if the motor is not in the way. This may be accomplished by scraping the dust buildup with a screwdriver blade, using a vacuum cleaner to remove it. If the fan wheel is very dirty, it may be best to remove the fan section and clean it outdoors. ***Do not allow water to enter the bearings or motor. Plastic bags may be used to protect them.*** If the bearings get soaked, purge them with oil or grease to remove the water. If the motor gets wet, set it in the sun and use an ohmmeter to verify that the motor is not grounded. Setting the motor on an operating air conditioning condensing unit fan outlet will quickly dry it as the warm, dry air passes over it.

The water side of a hydronic system requires some maintenance for the system to give long, trouble-free performance. Fresh water contains oxygen, which causes corrosion in the pipes. Corrosion is a form of rust. The proper procedure for the start-up of a hydronic system calls for the system to be washed inside with an approved solution to remove any oils and construction debris. This may or may not have been performed on any system you may encounter. When the system is cleaned initially, it is supposed to be filled with fresh water and chemicals that will treat the initial charge of water from corrosion. Again, this may or may not have been done. If the system is several years old and no water treatment has been used, the system will surely need maintenance. It should be drained and flushed. Water treatment should be added when the system is filled with fresh water. Be sure to follow the water treatment company's instructions.

Systems with leaks have a continuous water makeup of fresh water, causing continuing corrosion because the fresh water contains oxygen. Water leaks must be repaired. It is common for water from leaks in coils to be carried away in drain lines and not be discovered. When a system has an automatic air bleed, air will not build up.

Water pumps must be maintained. They have lubrication ports for oil or grease. Keep the oil reservoir at the correct level on the pump with the correct oil. The motor will not have an oil reservoir but will require a certain amount of oil. Do not exceed this recommended amount.

When a motor has grease fittings, proper procedures must be used when applying grease. You cannot continue to pump grease into the bearings without some relief. The seal will rupture. ***Unscrew the relief fitting on the bottom of the bearing and allow grease to escape through the relief hole. Grease**

may be added until fresh grease is forced out the relief opening.*

Pumps also have couplings that should be inspected. ***Turn off the power and lock and tag the disconnect before removing the cover to the coupling.***
filings caused by wear that may be in the housing. Check for loose bolts and play in the coupling if it is a rubber type. Play in the coupling is natural for the spring type.

The electrical service should be checked if the pump has a contactor. ***Shut off the power and remove the cover to the disconnect. Do not touch any electrical connections with meter leads until you have placed one voltmeter lead on a ground source, a conduit pipe will do, and touch the other meter lead to every electrical connection in the contactor box. The reason for this is some pumps are interlocked electrically with other disconnects and there may be a circuit from another source in the box. Use all cautions.***

When contactor contacts are found burned, replace them; if wire is found to be frayed or burned, repair or otherwise correct the situation.

HVAC GOLDEN RULES

When making a service call to a residence:

- Try to park your vehicle so that you do not block the customer's driveway.
- Look professional and be professional. Clean clothes and shoes must be maintained. Wear coveralls for under the house work and remove them before entering the house.
- Check humidifiers where applicable. Annual service is highly recommended for health reasons.

Added Value to the Customer

Here are some simple, inexpensive procedures that may be added to the basic service call.

- Don't leave water leaks.
- Check the boiler for service needs, this may lead to more paying service work, depending on the type. Oil-heated boilers require the most extra work.
- Check for the correct water treatment.
- Lubricate all bearings on pumps and fans as needed.

32.15 SERVICE TECHNICIAN CALLS

SERVICE CALL 1

A motel manager calls indicating that there is no heat at the motel. *The problem is on the top two floors of a four-story building. The system has just been started for the first time this season, and it has air in the circulating hot water. The automatic vent valves have rust and scale in them due to lack of water treatment and proper maintenance.*

The technician arrives and goes to the boiler room. The boiler is hot, so the heat source is working. The water pump is running, because the technician can hear water circulating. The technician goes to the top floor to one of the room units and removes the cover; the coil is at room temperature. The technician listens closely to the coil and can hear nothing circulating. There may be air in the system and the air vents sometimes called bleed ports must be located. The technician knows they must be at the high point in the system and normally would be on the top of the water risers from the basement.

The technician goes to the basement where the water lines start up through the building and finds a reference point. The technician then goes to the top floor and finds the top portions of these pipes. The pipes rise through the building next to the elevator shaft. The technician goes to the top floor to the approximate spot and finds a service panel in the hall in the ceiling. Using a ladder, the automatic bleed port for the water supply is found over the panel. This is the pump discharge line. See Figure 32–5 for an example of an automatic air vent or bleed port.

The technician removes the rubber line from the top of the automatic vent. The rubber line carries any water that bleeds to a drain. There is a valve stem in the automatic vent (as in an automobile tire). The technician presses the stem, but nothing escapes. There is a hand valve under the automatic vent, so the valve is closed and the automatic vent is removed. The technician then carefully opens the hand valve and air begins to flow out. Air is allowed to bleed out until water starts to run. The water is very dirty. ***Care should be used around hot water.***

The technician then bleeds the return pipe in the same manner until all the air is out, and water only is at the top of both pipes. The technician then goes to the heating coil where the cover was previously removed. The coil now has hot water circulating.

The technician then takes the motel manager to the basement and drains some water from the system and shows the manager the dirty water. The water should be clear or slightly colored with water treatment. The technician suggests that the manager call a water treatment company that specializes in boiler water treatment, or the system will have some major troubles in the future.

The technician then replaces the two automatic vents with new ones and opens the hand valves so that they can operate correctly.

SERVICE CALL 2

A customer in a small building calls. *There is no heat in one section of the building though the system is heating well in most of the building. The building has four hot water circulating pumps, and one of them is locked up—the bearings are seized. These pumps are each 230 V, three phase, and 2 hp.*

The technician arrives at the building and consults the customer. They go to the part of the building where there is no heat. The thermostat is calling for heat. They go to the basement where the boiler and pumps are located. They can tell from examination that the number 3 pump, which serves the part of the building that is cool, is not turning. The technician carefully touches the pump motor; it is cool and has not been trying to run. Either the motor or the pump has problems.

The technician chooses to check the voltage to the motor first, then goes to the disconnect switch on the wall and measures 230 V across phases 1 to 2, 2 to 3, and 1 to 3. This is a 230-V three-phase motor with a motor starter. The motor overload protection is in the starter, and it is tripped.

The technician still does not know if the motor or the pump has problems. All three windings have equal resistance, and there is no ground circuit in the motor. The power is turned on, an ammeter is clamped on one of the motor leads, and the overload reset is pressed. The technician can hear the motor try to start, and it is pulling locked-rotor current. The disconnect is pulled to stop the power from continuing to the motor. It is evident that either the pump or the motor is stuck.

The technician turns off the power and locks it out, then returns to the pump. It is on the floor next to the others. The guard over the pump shaft coupling is removed. The technician tries to turn the pump over by hand—remember, the power is off. The pump is very hard to turn. The question now is whether it is the pump or the motor that is tight.

The technician disassembles the pump coupling and tries to turn the motor by hand. It is free. The technician then tries the pump; it is too tight. The pump bearings must be defective. The technician gets permission from the owner to disassemble the pump. The technician shuts off the valves at the pump inlet and outlet, removes the bolts around the pump housing, then removes the impeller and housing from the main pump body. The technician takes the pump impeller housing and bearings to the shop and replaces them, installs a new shaft seal, and returns to the job.

While the technician is assembling the pump, the customer asks why the whole pump was not taken instead of just the impeller and housing assembly. The technician shows the customer where the pump housing is fastened to the pump with bolts and dowel pins. The relationship of the pump shaft and motor shaft have been maintained by not removing the complete pump. They will not have to be aligned after they are reassembled. The pump is manufactured to be rebuilt in place to avoid this.

After completing the pump assembly, the technician turns it over by hand, assembles the pump coupling, and affirms that all fasteners are right. Then the pump is started. It runs and the amperage is normal. The technician turns off the power, locks it out, and assembles the pump coupling guard. The power is turned on and the technician leaves.

SERVICE CALL 3

An apartment house manager reports that *there is no heat in one of the apartments. This apartment house has 25 fan coil units with a zone control valve on each and a central boiler. One of the zone control valves is defective.*

The technician arrives and goes to the apartment with no heat. The room thermostat is set above the room setting. The technician goes to the fan coil unit located in the hallway ceiling and opens the fan coil compartment door. It drops down on hinges. The technician stands on a short ladder to reach the controls. There is no heat in the coil, and there is no water flowing.

This unit has a zone control valve with a small heat motor. The technician checks for power to the valve's coil; it has 24 V, as it should. See Figure 32–11 for an example of this type of valve. This valve has a manual open feature, so the valve is opened by hand. The technician can hear the water start to flow and can feel the coil get hot. The valve heat motor or valve assembly must be changed. The technician chooses to change the assembly because the valve and its valve seat are old.

The technician obtains a valve assembly from the stock of parts at the apartment and proceeds to change it. ***The technician shuts off the power and the water to the valve and water coil. A plastic drop cloth is then spread to catch any water that may fall. Then a bucket is placed under the valve before the top assembly is removed.*** The old top is removed, and the new top installed. The electrical connections are made, and the system valves are opened allowing water back into the coil.

The technician turns on the power and can see the valve begin to move after a few seconds (this is a heat motor valve, and it responds slowly). The technician can now feel heat in the coil. The compartment door is closed. The plastic drop cloth is removed, the room thermostat is set, and the technician leaves.

The following service calls do not have the solutions described. The solutions can be found in the Instructor's Guide.

SERVICE CALL 4

A customer calls and indicates that one of the pumps beside the boiler is making a noise. This is a large home with three pumps, one for each zone.

The technician arrives and goes to the basement, hears the coupling, and notices there are metal filings around the pump shaft.

What is the problem and the recommended solution?

SERVICE CALL 5

A homeowner calls and states that there is water on the floor around the boiler in the basement. The relief valve is relieving water.

The technician goes to the basement and sees the boiler relief valve seeping. The boiler gage reads 30 psig, which is the rating of the relief valve. The system normally operates at 20 psig. The technician looks at the flame in the burner section and sees the burner burning at a low fire. A voltmeter check shows there is no voltage to the gas valve operating coil.

What is the problem and the recommended solution?

SUMMARY

- Hydronic heating systems use hot water or steam to carry heat to the areas to be heated.
- Water is heated by gas, oil, or electricity and circulated through pipes to terminal units where the heat is exchanged to the air of the conditioned space.
- A boiler is a furnace that heats water. In gas and oil systems the part of the boiler containing the water is constructed in sections or tubes.
- Air should be eliminated from the water to prevent corrosion, noise, and water flow blockage.
- A limit control is used to shut the heat source off if the water temperature gets too high.
- A low-water cutoff shuts down the heat source if the water level gets too low. Some of these devices have an automatic water feeder that allows water to flow into the system. If the proper water level still cannot be maintained, the low-water cutoff activates a switch that cuts off electrical power to the heat source.
- A pressure relief valve is required to discharge water when excessive pressure is created by expansion due to overheating.
- An air cushion tank or expansion tank is necessary to provide space for trapped air and to allow for expansion of the heated water.
- Zone control valves are thermostatically operated and control the flow of hot water into individual zones.
- A centrifugal pump circulates the water through the system.
- The quantity of water pumped determines the power consumed by the pump motor.
- Finned-tube baseboard terminal units are commonly used in residential hydronic heating systems.
- A balancing valve is used to equalize the flow rate throughout the system.
- One-pipe systems and two-pipe reverse return piping systems are most commonly used in residential and light commercial installations.
- A one-pipe system requires a specially designed tee to divert some water to the baseboard unit while allowing the rest to flow through the main pipe.
- Tankless domestic hot-water heaters are commonly used with hot-water heating systems. The water is contained within a coil and heated by the boiler.
- Hydronic systems should be serviced with proper preventive maintenance procedures annually.

REVIEW QUESTIONS

1. Describe a basic hydronic heating system.
2. Why do hot water heating systems often have more than one zone?
3. List the three heat sources commonly used with hydronic systems.
4. Why are gas- or oil-heated boilers constructed in sections or with tubes?
5. List three reasons for eliminating air from the system.
6. Where is the air vented?
7. What is the purpose of a limit control on a boiler?
8. Describe the purpose of a low-water cutoff used with an automatic water feeder.
9. What is the purpose of a pressure relief valve?
10. What is the purpose of an air cushion tank or expansion tank?
11. Describe two types of zone control valves.
12. What is the purpose of zone control valves?
13. Describe centrifugal force as it applies to hydronic circulating pumps.
14. What is the purpose of the impeller on a circulating pump?
15. Make a sketch of an impeller and show the direction of rotation.
16. What is another term for pumping pressure?
17. What determines the amount of power consumed by the pump motor?
18. What is the purpose of the fins on a finned-tube baseboard unit?
19. Make a sketch of a finned-tube baseboard unit and show the circulation of the air.
20. How are finned-tube baseboard units rated?
21. Why must expansion be considered when installing baseboard units?
22. Why are balancing valves necessary?
23. Sketch a one-pipe hydronic heating system.
24. What is the purpose of the one-pipe fitting (tee) used in these systems?
25. How does a two-pipe reverse return system operate?
26. Describe a radiant panel heating system.
27. How does a tankless domestic hot water system operate?

UNIT 32 DIAGNOSTIC CHART FOR HYDRONIC HEAT

The hydronic heat appliances discussed here are limited to the hot-water type. These systems are found in many homes and businesses as the primary heating system. These systems are typically assembled by an installing contractor who may use specific controls for specific reasons. Usually a plan is developed and left with the customer. If this plan is not available, contact the original contractor for any specifics. Always listen to customers for their input. They can often supply a clue for finding the problem.

PROBLEM	POSSIBLE CAUSE	POSSIBLE REPAIR	PARAGRAPH NUMBER
Boiler will not fire	See, electric, gas, or oil heat diagnostic charts for possible causes		Units 29–31
HOT WATER SYSTEMS Boiler is hot but conditioned space is cold	Pump not running—see electric motor diagnostic chart	Check fuses and breakers—change or reset, look for problem	Unit 20
	Tripped pump overload	Reset and check amperage	32.8
	Pump running, but water not circulating	Open proper valves	32.15
		Check system water level	32.15
		Check and clean strainers	32.15
		Bleed air from system	32.15
	Fans at terminal units not running	See electric motor diagnostic chart	Unit 20
	Terminal unit valves not opening	Repair or replace valves if needed	32.15
Boiler off because of low water cut-off	City water supply not operating	Re-establish city water supply **SAFETY PRECAUTION: *Do not put cold water into hot dry boiler***	

UNIT 33 Alternative Heating (Wood and Solar)

Codes, regulations, and laws regarding wood-heating appliances and components change frequently. It is essential that you be familiar with the appropriate current National Fire Protection Association information and local codes, regulations, and laws.

OBJECTIVES

After studying this unit, you should be able to

- list reasons for the formation of creosote.
- describe methods to help prevent the formation of creosote.
- explain how stovepipe should be assembled and installed.
- describe the operation of a catalytic combustor.
- list safety hazards that may be encountered with wood stoves.
- explain how a fireplace insert can improve on the heating efficiency of a fireplace.
- describe procedures for installing a stove to be vented through a fireplace chimney.
- describe a dual-fuel furnace.
- describe the difference between passive and active solar systems.
- describe the declination angle and the effect it has on the sun's radiation during winter and summer.
- describe a refrigerant-charged collector system.
- list the typical components in an air-based and in a liquid- based solar system and describe the function of each.
- describe the difference between sensible-heat storage and latent-heat storage.
- describe the operation of various solar space-heating systems.

SAFETY CHECKLIST

* There are many safety practices that must be adhered to when installing or using a wood-burning appliance. Many of these safety practices are stated within this unit, printed in red. In addition it is also necessary to use good judgment and common sense at all times.
* If an antifreeze solution is used in a solar domestic hot water system, a double-walled heat exchanger must be used.

Many people are burning wood as an alternative fuel for heating their homes. Some homes are heated entirely by wood; others use wood as supplemental heating. Wood must be readily available to make this type of heating practical. Heating technicians should have an understanding of this type of heat if they live and work in areas where wood is available.

33.1 ORGANIC MAKEUP AND CHARACTERISTICS OF WOOD

Wood plants manufacture glucose. Some of this glucose or sugar turns into cellulose. Approximately 88% of wood is composed of cellulose and lignin in about equal parts. Cellulose is an inert substance and forms the solid part of wood plants. It forms the main or supporting structure of each cell in a tree. Lignin is a fibrous material that forms an essential part of woody tissue. The remaining 12% of the wood is composed of resins, gums, and a small quantity of other organic material.

Water in green or freshly cut wood constitutes from one third to two thirds of its weight. Thoroughly air-dried wood may have as little as 15% moisture content by weight.

Wood is classified as hardwood or softwood. Hickory, oak, maple, and ash are examples of hardwood. Pine and cedar are examples of softwood. Figure 33–1 lists some common types of wood and the heat value in each type in millions of Btu/cord. A cord of wood can be split or unsplit or mixed. It is important to know whether the wood is split because there is more wood in a stack that is split. Wood is sold by the cord, which is a stack 4 ft × 4 ft × 8 ft or 128 ft^3.

Wood should be dry before burning. Approximately 20% more heat is available in dry wood than in green wood.

TYPE	WEIGHT CORD	BTU PER CORD AIR DRIED WOOD	EQUIVALENT VALUE #2 FUEL OIL, GALS.
White Pine	1800#	17,000,000	120
Aspen	1900	17,500,000	125
Spruce	2100	18,000,000	130
Ash	2900	22,500,000	160
Tamarack	2500	24,000,000	170
Soft Maple	2500	24,000,000	170
Yellow Birch	3000	26,000,000	185
Red Oak	3250	27,000,000	195
Hard Maple	3000	29,000,000	200
Hickory	3600	30,500,000	215

FIGURE 33–1 The table indicates the weight per cord, the Btu per cord of air dried wood, and the equivalent value of No. 2 fuel oil in gallons. *Courtesy Yukon Energy Corporation*

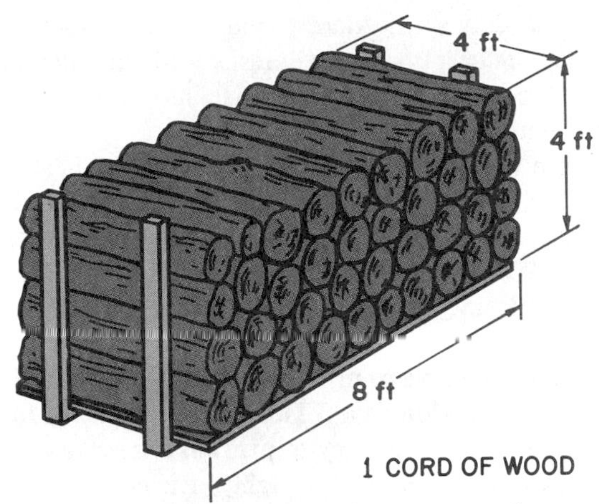

FIGURE 33–2 Wood should be stacked on runners so that air can circulate through it.

Wood should be stacked off the ground on runners, well ventilated, and exposed to the sun when it is drying, Figure 33–2. When possible it should be covered, but air should be allowed to circulate through it. Wood splits more easily when it is green or freshly cut and dries better when it has been split. When green wood is burned, combustion will be incomplete, resulting in unburned carbon, oils, and resins, which leave the fire as smoke.

During oxidation or burning, oxygen is added to the chemical process. This actually turns wood back into the products that helped it grow as a plant: primarily carbon dioxide (CO_2), water (H_2O), other miscellaneous materials, and, of course, heat. Some woods produce more heat than others, Figure 33–1. Generally, dry hardwoods are the most efficient.

33.2 CREOSOTE

The combustion in a wood-burning appliance is never complete. The smoke contains the gases and other products of the incomplete combustion including a substance called *creosote*. Chemically, creosote is a mixture of unburned organic material. When it is hot, it is a thick dark-brown liquid. When it cools, it forms into a residue like tar. It then often turns into a black flaky substance inside the chimney or stovepipe. Some of the primary causes of excessive creosote are:

- Smoldering low-heat fires
- Smoke in contact with cool surfaces in the stovepipe or chimney
- Burning green wood
- Burning softwood

Creosote buildup in the stovepipe and chimney is dangerous. It can ignite and burn with enough force to cause a fire in the building. Stovepipes have been known to be blown apart, and chimney fires are common with this excessive buildup.

The best prevention for the formation of excessive creosote is to use a stove with an insulated secondary combustion chamber where the temperature is normally higher than 1000°F or to use a stove with a catalytic combustor. To help prevent the formation of creosote in conventional stoves, burn dry hardwood. Fires should burn with some intensity. When the stovepipe or chimney flue temperature drops below 250°F, creosote will condense on the surfaces. Use well-insulated stovepipe with as little run to the chimney as possible. Use as few elbows or bends as possible and ensure that the stovepipe has a rise at all points from the stove to the chimney. The minimum rise should be 1/4 in. per foot of horizontal run. A rise of 30° is recommended, Figure 33–3. Anything that slows the movement of the gases allows them to cool, which will cause more creosote condensation. Stovepipes and chimneys should be cleaned regularly.

Assemble the stovepipe with the crimped end down, Figure 33–4. This will keep the creosote inside the pipe,

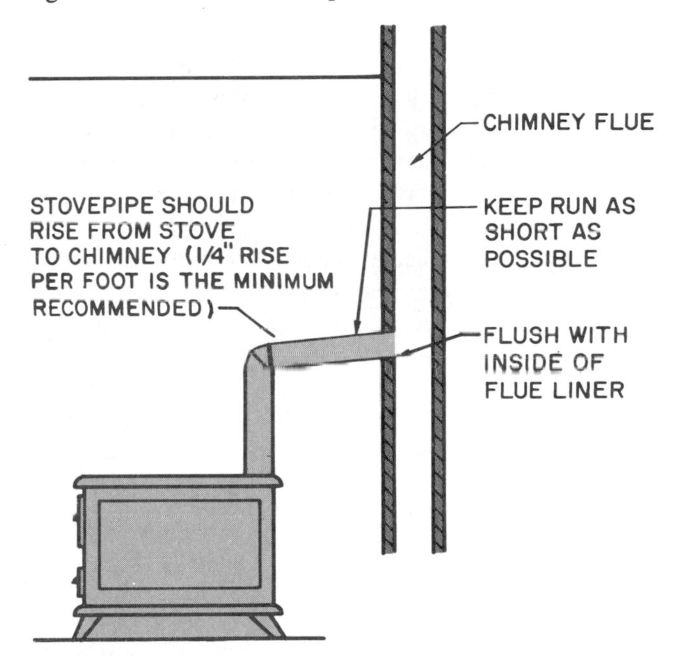

FIGURE 33–3 Stovepipe should rise from stove to chimney.

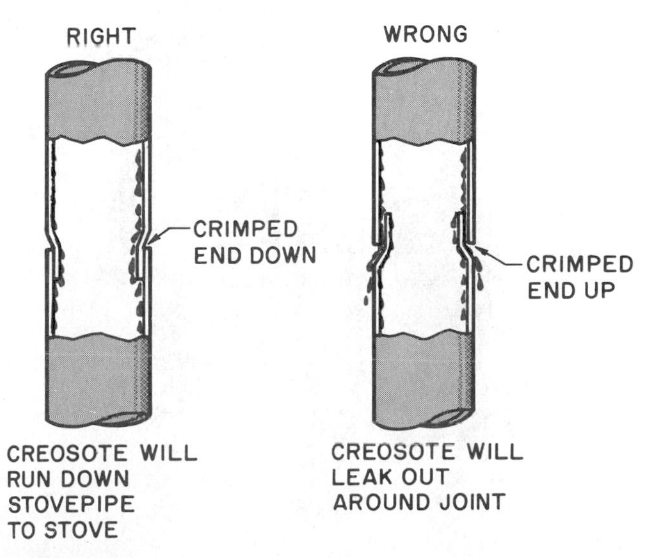

FIGURE 33–4 Stovepipe should be assembled with crimped end down so that creosote will not leak out of the joint.

allow it to drain to the stove, and be burned. Regardless of the preventive measures, proper design, and installation, creosote will form. Never let the creosote build up. Even a buildup of 1/8 to 1/4 in. can burn with dangerous intensity.

33.3 TYPES OF WOOD-BURNING APPLIANCES

Wood can be burned in wood-burning stoves, fireplace inserts, wood-burning furnaces, and dual-fuel furnaces.

33.4 WOOD-BURNING STOVES

Heat from wood-burning stoves radiates into the room until it is absorbed by some object(s), the walls, and floor of the room. After these have been heated, the stove heats the air in the room. The combustion also heats the stove, which in turn heats the air around it. Some stoves are classified as radiant stoves. Others, however, have an outer jacket, usually with openings or grill work. This is to allow for air circulation around the stove. Some of these are designed for natural convection, others for forced-air movement with a blower, Figure 33–5. Wood-burning stoves heat by both radiation and convection. Whether one is considered a radiant-heating or convection-heating stove depends on the design. Good-quality stoves are built of a heavy-duty steel plate, 1/4 in. or thicker, or of cast iron. Doors are often of cast iron to prevent warping. Warped doors cause air leaks and uncontrolled combustion.

Better-quality stoves are airtight. No air can get to the fire except through the draft dampers. This allows good control of the combustion process. The more air that reaches the fire through cracks and other openings, the less control there is of the combustion process and the lower the efficiency of the stove. The primary source of air leaks is around the doors. The doors should be constructed of a heavy steel or cast iron that will resist warping. They should also have ceramic rope or other insulation where the door makes contact with the stove.

When wood stoves are used as the primary heat source in a home, they should be of the best quality, which will normally result in the greatest efficiency. It is necessary to keep the heat in the stove as long as possible to achieve the greatest efficiency. To do this, many stoves are designed with baffles or heat chambers, Figure 33–6. ***Enough heat must leave the stove to provide a good draft condition to vent the combustion gases. This is also necessary to heat the stovepipe and chimney to prevent excessive creosote condensation.*** Good-quality highly efficient stoves have a firebrick lining to help retain heat and increase the life of the stove by providing added protection to the metal from the heat. Some stoves are designed with thermostatic controls. These controls open and close the damper. Closing the dampers when the space is heated will result in better use of the fuel, thus providing more efficiency.

Environmental Protection Agency (EPA) Regulations

On July 1, 1988, the Environmental Protection Agency's first national woodstove emissions standards went into effect. Emissions from wood-burning appliances were adding to the pollutants in the air along with all the other sources of air pollutants. ***Wood smoke contains both POM (polycyclic organic matter) and non-POM, which are considered to be health hazards and are of concern to many people.*** All wood-burning stoves manufactured after July 1, 1989, must be EPA certified. There are two EPA standards —one for catalytic stoves and one for noncatalytic stoves. Catalytic stoves contain a catalytic combustor and may not emit more than 5.5 g/h of particulates per the July 1989

(A)

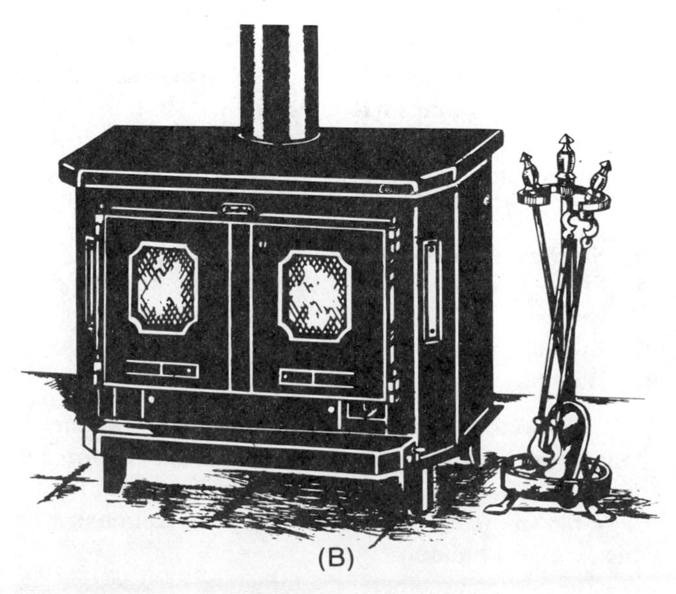
(B)

FIGURE 33–5 Two types of wood-burning stoves. (A) Radiant. (B) Forced convection. *(B) Courtesy Blue Ridge Mountain Stove Works*

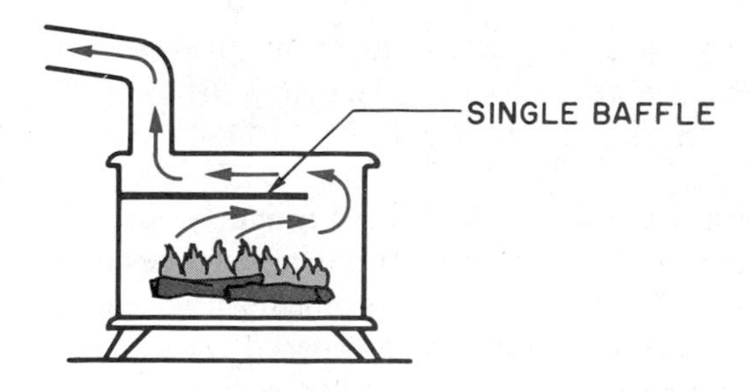

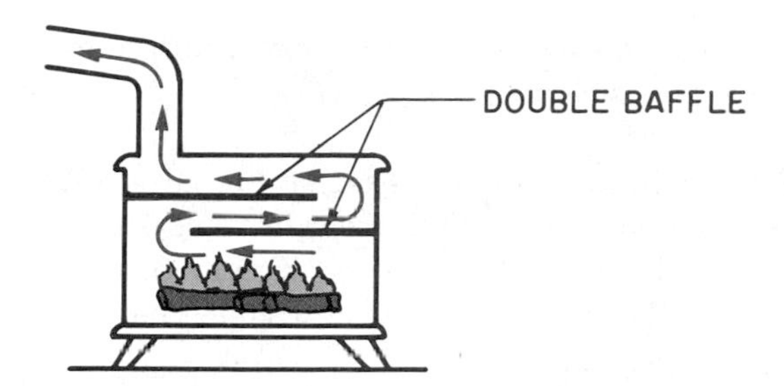

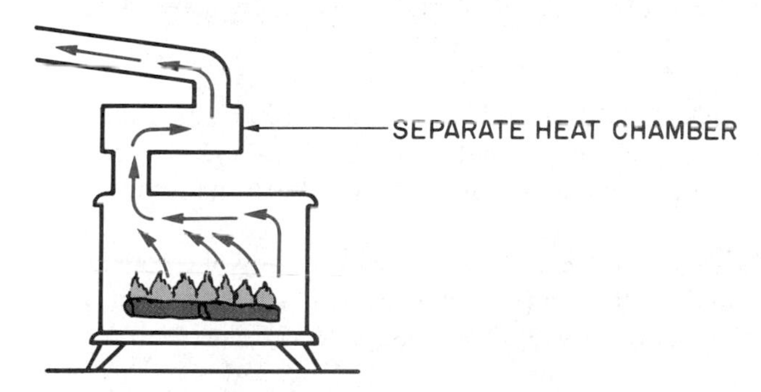

FIGURE 33–6 To help retain the heat within the stove for better efficiency, baffles or heat chambers are used.

standards. The noncatalytic stoves may not emit more than 8.5 g/h. The particulate emission of the catalytic stove is approximately 10% of the emissions of the older stoves. These limits are reduced further for those stoves manufactured after July 1, 1990, to 4.1 for the catalytic models and 7.5 for the noncatalytics. As new stoves are purchased and installed, particularly when replacing older stoves, the air pollutants from wood-burning stoves will be reduced drastically.

Noncatalytic Certified Stoves

These stoves force the unburned gases to pass through an insulated secondary combustion chamber. Temperatures higher than 1000°F are maintained in this chamber, burning the gases. This burns the pollutants, including most of the creosote, and produces more heat, providing a better stove operating efficiency.

Catalytic Combustor Stoves

The development of stoves with catalytic combustors is the second way manufacturers have been able to meet EPA certification requirements. Catalytic stoves reduce pollutants, including creosote, and improve the stove efficiency significantly.

The catalytic combustor actually causes most of the products that make up the flue gases or smoke to burn. Smoke from a wood-burning stove contains products that will burn and consequently are a potential source for more heat. For this smoke to burn, it must be heated to a temperature much higher than is possible or desirable in most stoves, even those with secondary combustion. In noncatalytic stoves these temperatures must reach 1000°F to 1500°F; additional oxygen is needed, and the smoke must be exposed to these temperatures and the oxygen for a longer time.

The catalytic combustor is a form of an afterburner. Afterburners increase the burning of all by-products of wood, but an additional fuel is needed to do this. Afterburners are not practical devices. Catalytic combustors use a catalyst to cause combustion without the additional fuel. The catalyst causes a chemical reaction, which causes the flue gases to burn at about 500°F rather than the 1000°F to 1500°F otherwise required.

Catalytic combustors come in various styles. Most are similar and use the same technology. They are cell-like structures, Figure 33–7, and consist of a substrate, washcoat, and catalyst. The *substrate* is a ceramic material formed into a honeycomb shape. Ceramic material is used because of its stability in extreme cold and hot conditions. The *washcoat,* usually made of an aluminum-based substance called *alumina,* covers the ceramic material and helps disperse the catalyst across the combustor surface. The *catalyst* is made of a noble metal, usually platinum,

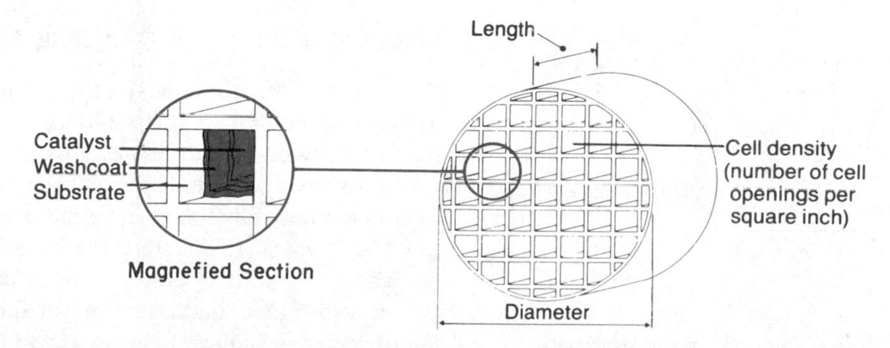

FIGURE 33–7 Catalytic combustor element. *Courtesy Corning Glassworks*

palladium, or rhodium, which are chemically stable in extreme temperatures.

Combustion in the Catalytic Combustor

For combustion to occur in the catalytic combustor, the fuel is the unburned flue gases or smoke, the oxygen is either excess from the main combustion chamber or supplied directly to the combustor, and the heat needed is 500°F to 600°F furnished from the combustion of the wood. Once the combustor starts to function, however, the stove firebox temperature can be as low as 250°F to 400°F because the combustor will generate enough of its own heat to maintain the combustion.

SAFETY PRECAUTION: *Very high temperatures can be generated by the burning of the flue gases in the combustor. Iron and steel stove parts can be subject to metal oxidation from this heat. These materials should be spaced at least 2 in. from the combustor; otherwise stainless steel should be used.*

Typical Stove Design with a Catalytic Element

Figure 33–8 shows a typical stove design with a catalytic element (F). Following are brief descriptions of each component of the stove:

WOOD COMBUSTION CHAMBER (A) (FIREBOX).

HEAT EXCHANGER AREA (B). In the heat exchanger area, much of the heat from the firebox and combustor is transferred to the surface area and surrounding space before the gases are exhausted through the flue.

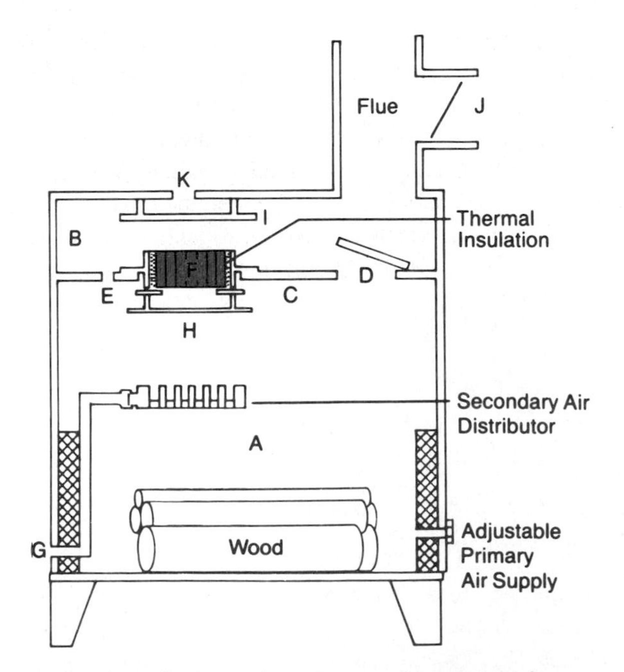

FIGURE 33–8 Illustration of wood stove with catalytic combustor. *Courtesy Corning Glassworks*

BAFFLE PLATE (C). The baffle plate separates the firebox from the heat exchanger area. It contains the bypass damper (D), safety bypass (E), and combustor (F).

BYPASS DAMPER (D). The bypass damper should be opened before opening the firebox door. The combustor restricts smoke exhaust. If this damper were not opened, smoke would come out the firebox door when opened. Most manufacturers recommend that this or a similar damper arrangement be opened when a hot fire exists. This allows the gases to escape and keeps the combustor from being damaged by the high fire and possible flame impingement.

SAFETY BYPASS (E). The safety bypass is a relatively small hole that allows some smoke to escape by the combustor. This is necessary in case the combustor becomes plugged. Much of this smoke will still be burned as it passes over the combustor on its way to the flue.

COMBUSTOR (F). The combustor will normally be wrapped or enclosed in a stainless steel "can."

SECONDARY AIR INLET (G). Some stoves provide a preheated secondary air supply to the combustor through the secondary air distributor, which distributes the air uniformly.

FLAME GUARD (H). The flame guard protects the combustor from damage from flame impingement and from physically being damaged while the wood is loaded into the firebox.

RADIATION SHIELD (I). The radiation shield is provided in many stove designs to protect the metal directly above the combustor from extreme heat. It also reflects radiant heat back to the combustor, causing it to operate hotter and thus more efficiently.

BAROMETRIC DAMPER (J). The barometric damper is used in controlling airflow and in keeping the combustor from overfiring or operating at too high a temperature.

VIEW PORT (K). Many manufacturers provide a view port to actually see how the combustor is operating. It will glow when above 1000°F. Much of the time it will not be operating at a temperature this high, however. It does not have to glow to be operating properly.

Operating a Stove with a Catalytic Element

When starting the stove, the gas temperatures must be raised to 500°F to 600°F (260°C to 320°C). This is a medium to high firing rate, and this temperature should be maintained for approximately 20 min. This is called achieving *catalytic light-off.* When refueling a stove that has been burning below 500°F, you should let it be fired at a higher temperature for about 10 min to ensure proper catalytic operation. Stoves with older combustors may require slightly higher temperatures to achieve light-off. The catalytic activity decreases as the combustor ages. The catalyst itself does not burn but promotes reactions in the hydrocarbons in the smoke. There are two separate reactions: a hydrogen reaction and a carbon monoxide reaction. Both cause burning or oxidation to occur more easily and at a lower temperature.

Causes of Catalytic Failure

Failure of the ceramic structure or substrate can be caused by severe temperature cycling (extreme high temperatures at frequent intervals) and uneven air distribution, producing hot and cool zones, Figure 33–9. Flame impingement is another cause of failure.

SAFETY PRECAUTION: *Catalyst failure can be caused by extremely high temperatures (over 1800°F) and by what is known as *catalytic poisoning*. Only natural wood should be burned in a catalytic stove. Materials such as lead, zinc, and sulfur will form alloys with the noble metal used as the catalyst, or compounds will be formed to cover the catalyst.*

Either of the above will result in the combustor being ineffective.

Combustor Life

Most combustors will last from 2 to 5 years, depending on the amount of use. Most stoves are designed for easy replacement of the combustor. An ineffective catalytic combustor will cause a stove to act sluggishly. It will be difficult to start and hard to maintain a good fire. When this happens, the catalytic element should be replaced. Figure 33–10 shows the catalytic element. It is placed in a stainless steel container.

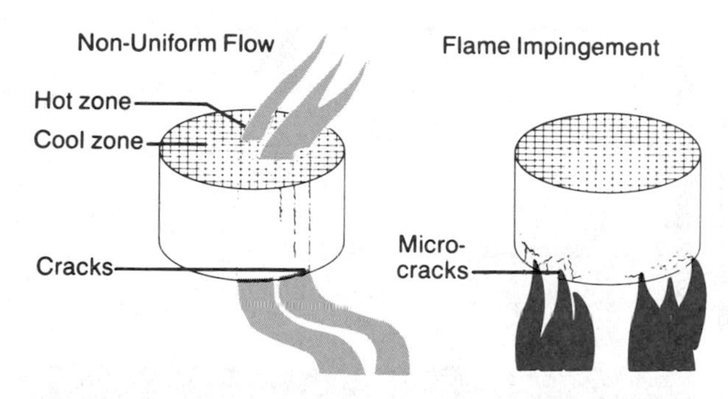

FIGURE 33–9 Uneven air distribution or flame impingement can cause catalytic failure. *Courtesy Corning Glassworks*

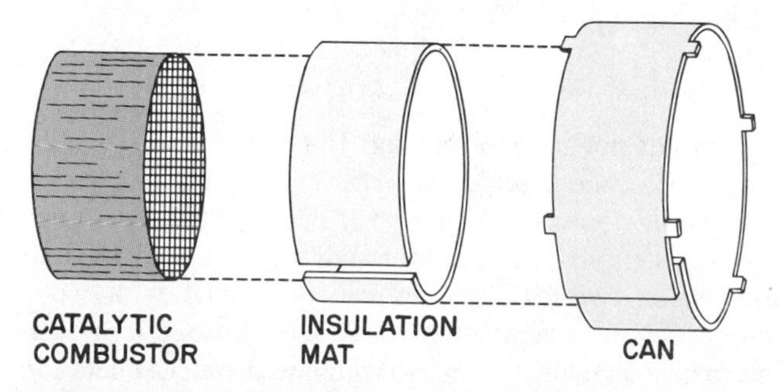

FIGURE 33–10 Catalytic element is placed in a container ("can") for protection. Elements are very fragile particularly after they have been used. *Courtesy Corning Glassworks*

General Operating and Safety Procedures

***1. Check the combustor regularly to ensure that it is not plugged and that it is otherwise operating properly.**
2. Check the chimney and flue regularly for creosote buildup. When operating properly, the combustor will not eliminate all creosote, and normal safe practices should be followed.
3. Ensure that safety bypasses are open. A blockage can result in smoke and flue-gas leakage.
4. Ensure that all clearances from combustible materials are maintained when installing the stove.
5. Never burn materials other than natural wood. Never burn paper, plastic, coal, or painted wood. These materials will poison the catalyst.
6. Do not allow the flame to impinge on the combustor.*

Location of the Stove

Because heat rises, many people think the most efficient location for a stove is in the basement. This is true only if the basement area is the primary space heated. Remember, many stoves are radiant stoves. The heat will radiate through the air, heating solid objects. Most of the heat will be absorbed by the basement walls and solid objects, which in turn will heat the basement space. The heated basement air will then move up through registers in the first floor. However, too much of the heat will be used in heating the basement.

The stove should be placed as near as practical to the space where the heat is desired. This is often near the center of the first floor area. This is not often practical, however, because there may not be a chimney located where the stove should be placed. The stove must be in the same room as the chimney inlet. An approved prefabricated chimney may be installed where it would be most appropriate for the stove and the room. Avoid long stovepipe runs connecting the stove to the chimney. Do not connect the stovepipe to a chimney flue used for another purpose. Therefore, it may be the location of the chimney that determines where the stove will be placed.

Heat Distribution

The radiant heat from the stove warms the objects and the walls in the room, which heat the air and make it rise. The heat reaches the ceiling area, cools, and descends to the floor area to be reheated and recirculated, Figure 33–11. To heat rooms above the stove, place registers in the ceiling to allow the heated air to pass through the registers and heat the room above. There should also be a register for the cooler air to descend to the lower floor, Figure 33–12.

A stairway near the stove is a good way for the heated air to rise to the upper floor. If the stairway is open to the

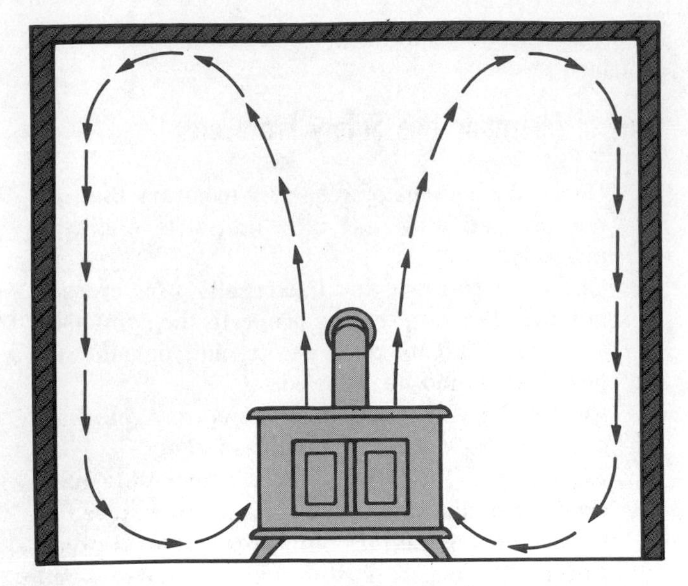

FIGURE 33–11 Convection circulation pattern for one room.

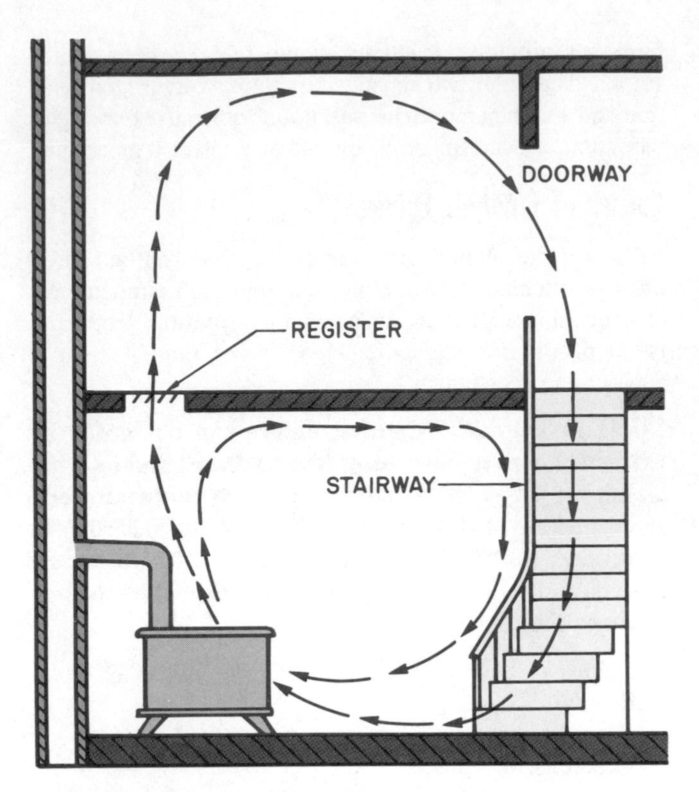

FIGURE 33–13 Convection circulation pattern using stairway to return cooled air.

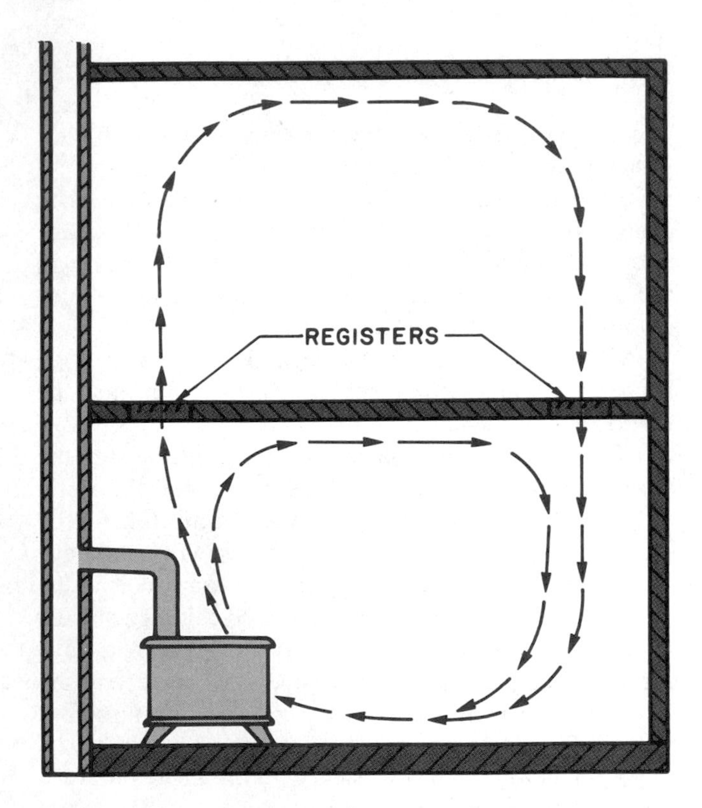

FIGURE 33–12 Convection circulation pattern for two rooms, one above the other.

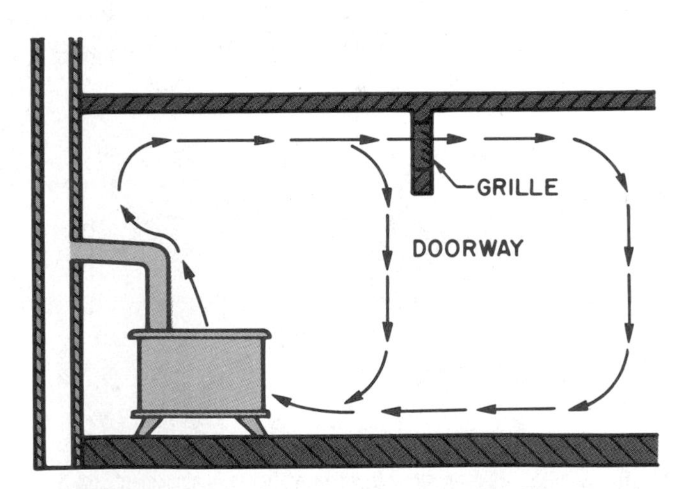

FIGURE 33–14 Convection circulation pattern with register near ceiling. Cooled air returns to the stove area through the doorway. Grilles may have electric fans for more efficient circulation.

room where the stove is located, the heated air can rise through the registers over the stove and the cooled air can return down the stairwell, Figure 33–13. If heat is desired in a room beside the one where the stove is located, registers should be cut through the wall next to the ceiling. The heated air rises to the ceiling and passes through these registers near the ceiling, Figure 33–14. These registers may be equipped with fans. It is difficult for the heat to pass from one room to another through doorways because the door openings do not go to the ceiling. The partial wall above the door blocks the circulation. The cooled air will return through the doorways, however. If there are no door openings, registers should be cut through the wall near the floor also. Small two-level houses can be heated with wood stoves because heat will rise to the second floor. It is difficult to heat a ranch style house with one stove. The heat can be ducted from the basement with forced air, but more than one stove is usually required for most satisfactory installations.

Makeup Air

Air used in combustion must be made up or resupplied from the outside, for example, with an inlet from outside directly to the stove, Figure 33–15. Often an older home has many air leaks. The hot air may leak out near the ceiling or through the attic. Cool air can leak in through cracks around windows, under doors, and elsewhere. Although this may disturb the normal heating cycle, it does help to make up air for that air leaving the chimney as a result of the combustion and venting. ***In modern homes that are sealed and insulated well, some provision may have to be made to supply the makeup air. A door may have to be opened a crack to provide a proper draft. A stove could actually burn enough oxygen in a small home to make it difficult to get enough oxygen to breathe. Always make sure there is enough makeup air.***

Safety Hazards

***Live coals in the stove in the living area of a house along with creosote in stovepipes and chimneys make it absolutely necessary to install, maintain, and operate stoves safely. This cannot be emphasized enough.**

Following are some of the safety hazards that may be encountered.

- **A hot fire can ignite a buildup of creosote, resulting in a stovepipe or chimney fire.**
- **Radiation from the stove or stovepipe may overheat walls, ceilings, or other combustible materials in the house and start a fire.**
- **Sparks may get out of the stove, land on combustible materials, and ignite them. This could happen through a defect in the stove, while the door is left ajar, while the firebox is being filled, or while ashes are removed.**
- **Flames could leak out through faulty chimneys, or heat could be conducted through cracks to a combustible material.**
- **Burning materials coming out of the top of the chimney can also start a fire at the outside of the house. These sparks or glowing materials can ignite roofing materials, leaves, brush, or other matter outside the house.***

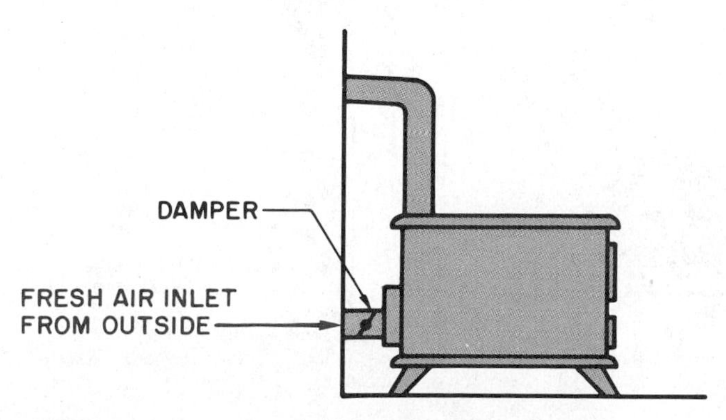

FIGURE 33–15 Stove with fresh air tube from outside.

Installation Procedures

***A wood-burning stove or appliance should be approved by a national testing laboratory. Before installing a stove, stovepipe, or prefabricated chimney, be sure that all building and fire marshal's codes are followed as well as the instructions of the testing laboratory and manufacturer. If one code or set of instructions is more restrictive than another, it is absolutely necessary that the most restrictive instructions be followed.**

These instructions should include the distance the stove should be located from any combustible material, such as a wall or the floor. They should indicate the minimum required protective material between the stove and the wall and between the stove and the floor. Excessive heat from the stove can heat the walls or floor to the point where a fire can be started.

The stove must be connected to the chimney with an approved stovepipe, often called a *connector* pipe. Single-wall or double-wall stovepipe may be used for this, Figure 33–16. The double-wall pipe has an air space between the two walls. Normally this type of stovepipe installation would require less spacing to the nearest combustible material than a single-wall pipe would. Codes and instructions must be followed.

A stove collar adapter should be used to install the stovepipe to the stove. The stovepipe should be placed on the outside of the stove collar and yet it must also be fitted inside the collar so that the creosote will run into, not on top of, the stove, Figure 33–17.

Figure 33–18 shows three different types of installations. The stovepipe must not run through any combustible material such as a ceiling or wall. Approved chimney sections with necessary fittings should be used. Figure 33–19 illustrates details for adapting the stovepipe to the "through-the-wall" chimney fittings. Remember to

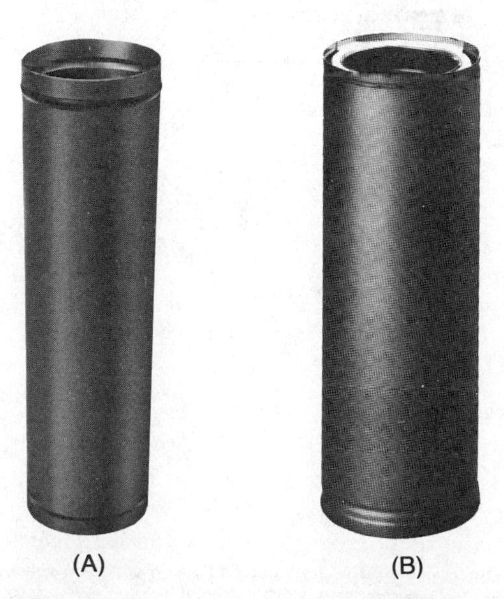

FIGURE 33–16 (A) Single-wall stovepipe section. (B) Double-wall stovepipe section.

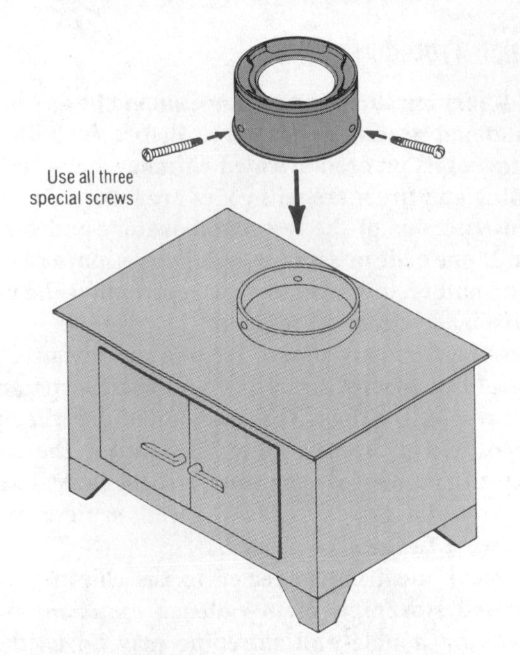

FIGURE 33–17 Stove collar adapter for double-wall stovepipe.

keep horizontal stovepipe runs to a minimum. A rise of 1/4 in./ft should be considered a minimum, Figure 33–3. Local codes or manufacturers may require a greater rise. Figure 33–20 shows various prefabricated chimney fittings and sections. Be sure that approved thimbles, joist shields, insulation shields, and other necessary fittings are used where required. Be sure that all materials are approved by a recognized national testing laboratory.

Chimneys may be double walled with insulation between the walls or triple walled with air space between the walls. This insulating material may be a ceramic fiber refractory blanket, Figure 33–21.

Only one stove can be connected to a chimney. For masonry chimneys with more than one flue, no more than one stove can be connected to each flue. A factory-built chimney used for wood stoves should be rated as residential and building heating appliance chimney. Check local codes.

Many stoves are vented through masonry chimneys. If a new chimney is being used, it should have been constructed according to applicable building codes and

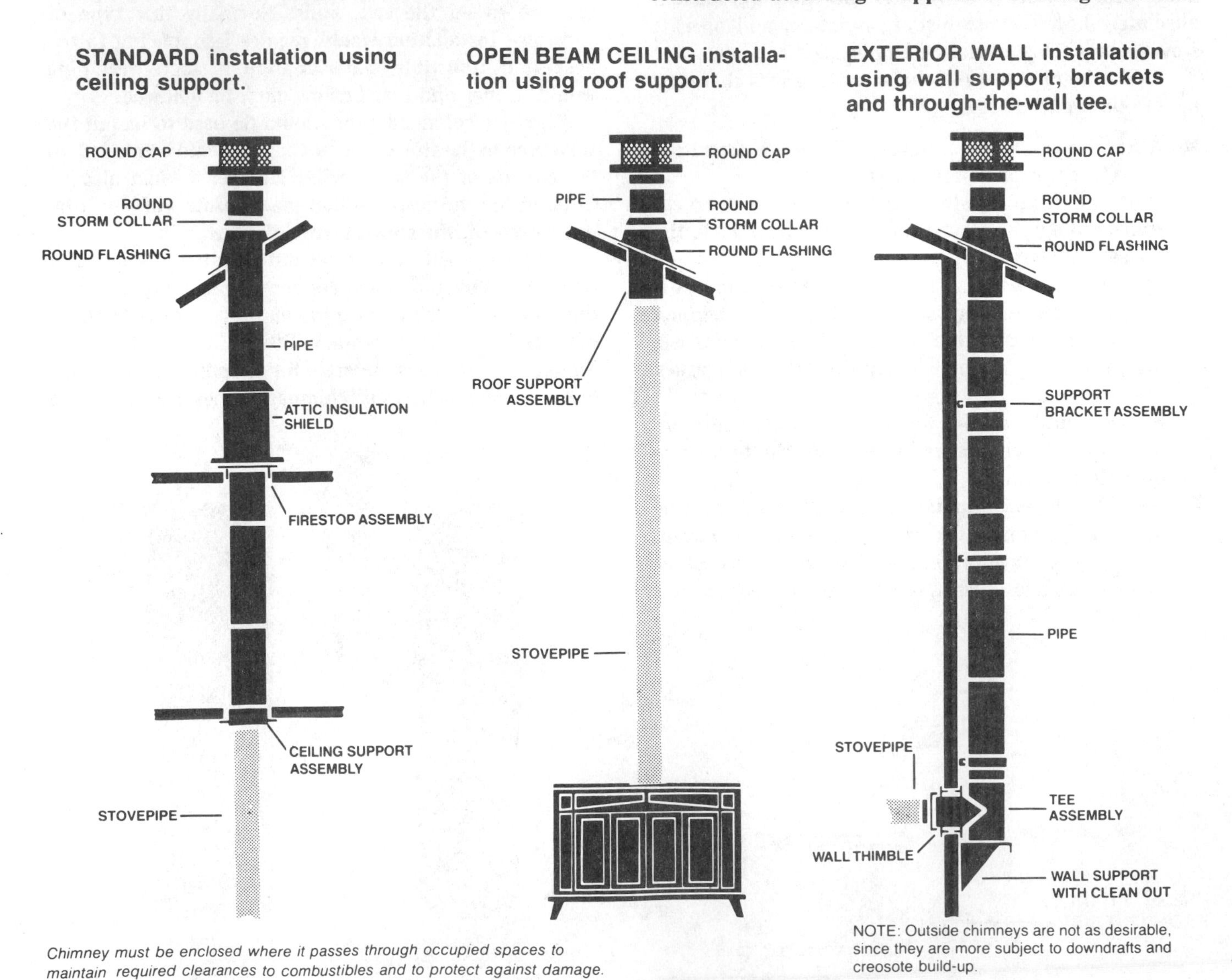

FIGURE 33–18 Three different types of stove installations.

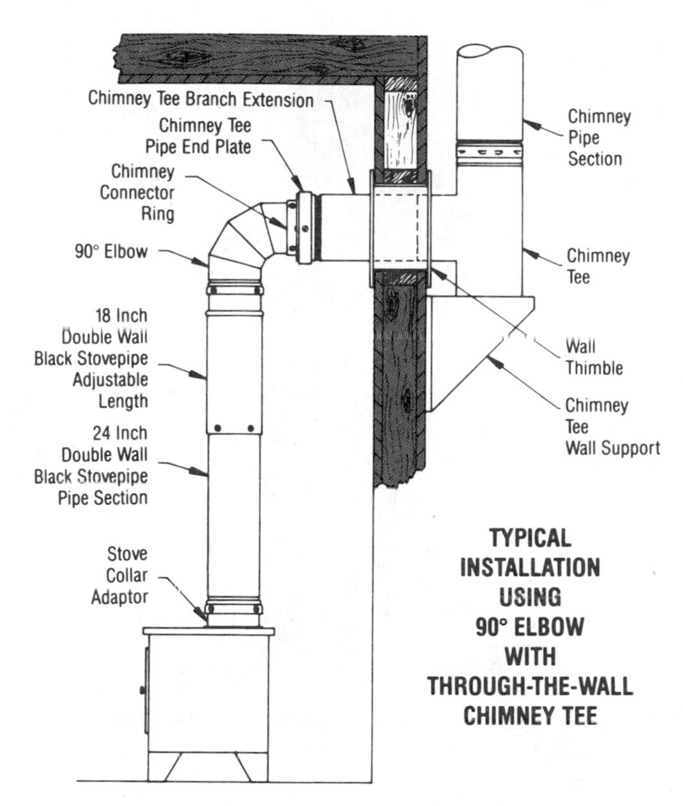

FIGURE 33–19 Details of a "through-the-wall" installation.

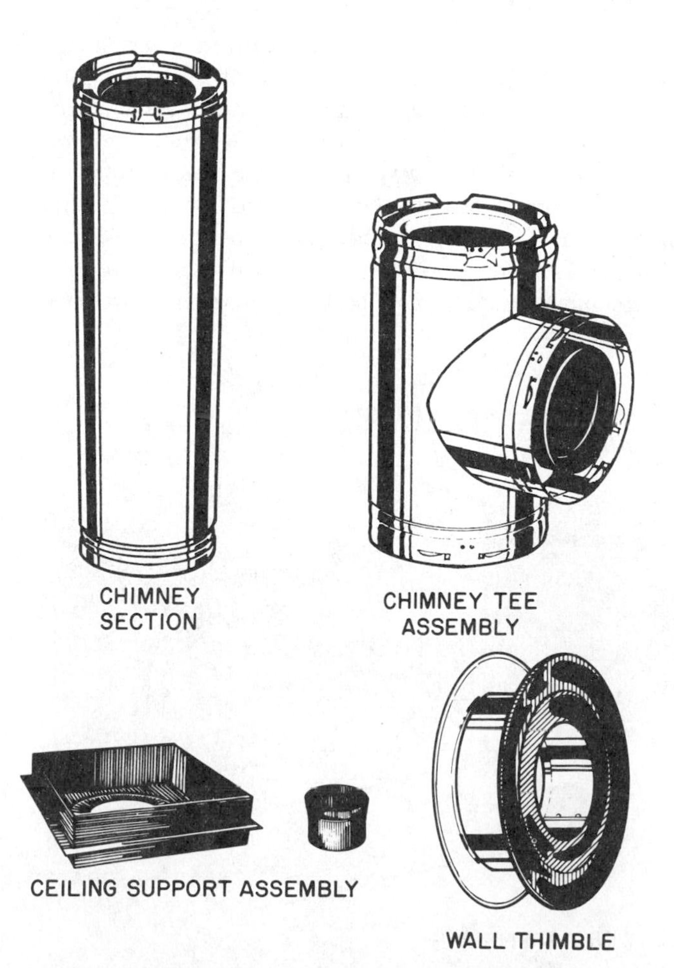

FIGURE 33–20 Prefabricated chimney fittings and sections.

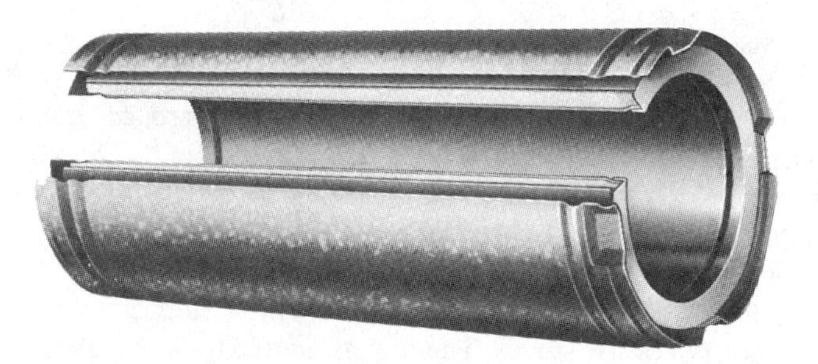

FIGURE 33–21 Double-walled chimney section with ceramic fiber refractory blanket.

carefully inspected. If an old chimney is being used, it should be inspected by an experienced person to ensure that it is safe to use. The mortar in the joints may be deteriorated and loose or the chimney may have been cracked by a serious chimney fire. All necessary repairs should be made to the chimney by a competent mason before connecting the stove to it.*

33.5 SMOKE DETECTORS

Smoke detectors should be used in all homes regardless of the type of heating appliance used. It is even more important to install smoke detectors when burning wood fuel. Many types of detectors are available. Either an AC-powered photoelectric type or an AC- or battery-powered ionization chamber detector are usually satisfactory.

33.6 PELLET STOVES

Pellet stoves are a more recent design, are energy efficient, and use a renewable energy source, Figure 33–22. The fuel

FIGURE 33–22 Pellet stove. *Courtesy Vermont Castings*

consists of small compressed pellets made from waste wood, primarily sawdust from lumber saw mills. However, the pellets may be made from waste cardboard or even agricultural wastes such as sunflower seed hulls. Stove owners should be sure that only the type of pellets recommended by the manufacturer are used. Pellets are normally provided in 40-lb bags.

Most pellet stoves are designed with a hopper at the back or top of the stoves. The pellets are fed to the combustion chamber with an auger powered by an electric motor often controlled with a thermostat.

The room air is generally circulated through the heat exchanger and into the conditioned space by a multispeed blower.

Very little air pollution is produced with as little as 1% of the pellet material remaining as ash. Most stoves do not require a chimney because exhaust gases are forced outside through a vent pipe by a fan. Combustion air may be drawn in from outdoors so that heated room air is not used for combustion. This provides additional efficiency. Venting of exhaust gases may be through a horizontally positioned vent pipe with an end cap to prevent wind from blowing air and the exhaust gases into the stove, Figure 33–23. A lined masonry chimney that meets all appropriate codes may be used. However, most masonry chimneys will be large enough so that they will affect the operation of pellet stoves adversely. These chimneys may be relined with a stainless steel liner, Figure 33–24. Consult the stove manufacturer's literature for the recommended diameter for these liners.

It is essential that all manufacturers' instructions and all national, state, and local codes are followed when installing and venting these stoves.

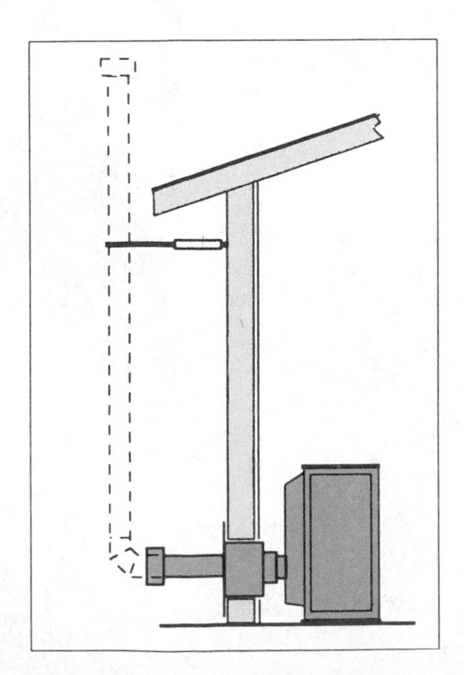

FIGURE 33–23 Horizontal vent pipe must have end cap to prevent air or exhaust gases from being blown into stove. Vent pipe may be extended vertically above eaves. *Courtesy Vermont Castings*

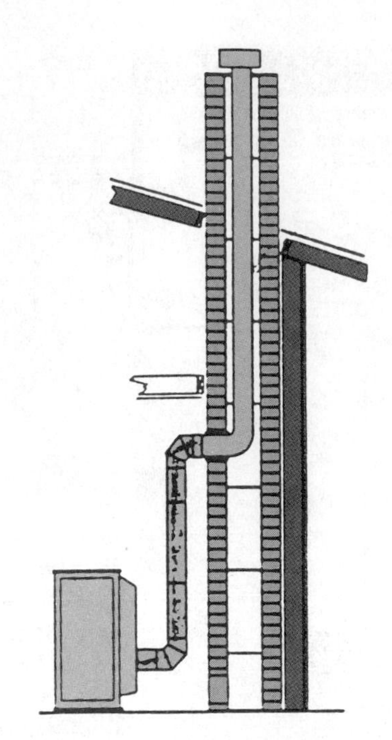

FIGURE 33–24 Pellet stoves may operate more efficiently if masonry chimneys are relined with a stainless steel liner. *Courtesy Vermont Castings*

33.7 FIREPLACE INSERTS

Fireplace inserts can convert a fireplace from a very inefficient heating source to one that is fairly efficient. Very little heat can be obtained from a fireplace without some device to get the heat from the fireplace into the room. The fireplace insert provides a way to retain the heat and blow it into the room. When fireplace fires are burned regularly without such a device, the heat is almost totally wasted, Figures 33–25 and 33–26. Many fireplace inserts have catalytic combustors. ***For an insert to be used, the fireplace**

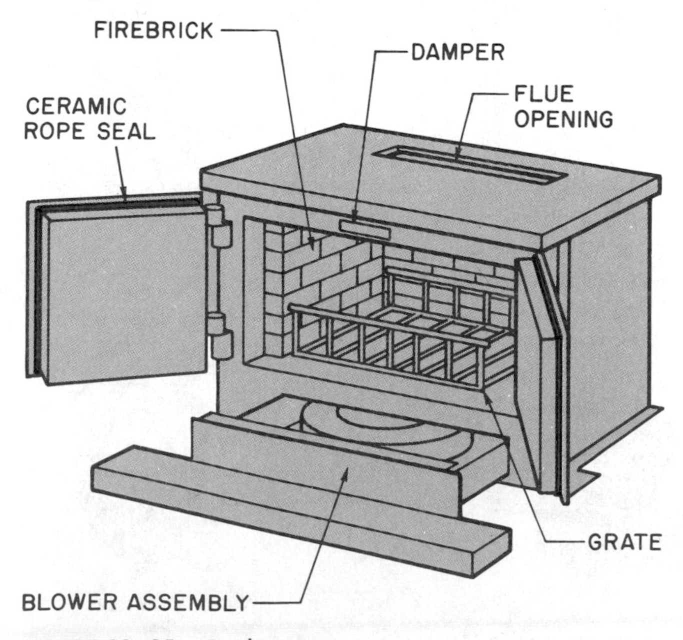

FIGURE 33–25 Fireplace insert.

FIGURE 33–26 Fireplace insert installed. *Courtesy Blue Ridge Mountain Stove Works*

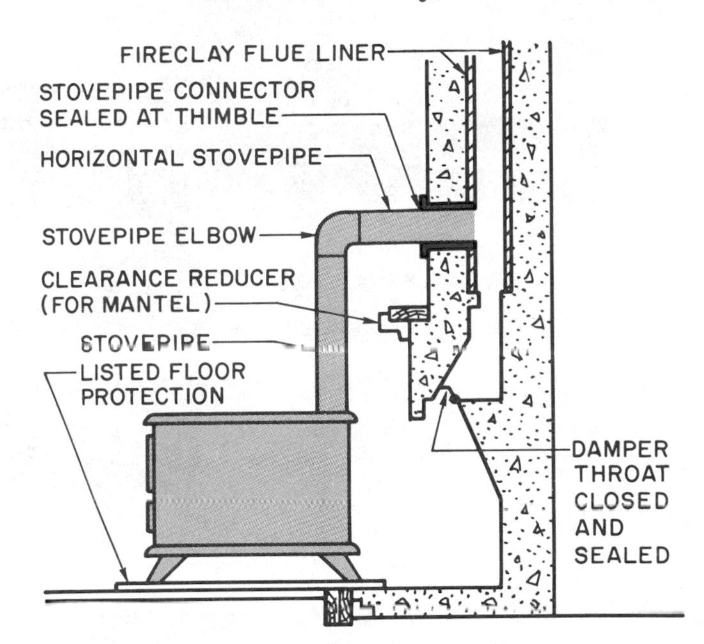

FIGURE 33–27 Flue opening above fireplace opening.

and chimney must generally be of masonry construction. Local building codes and the manufacturer's specifications for the particular insert should be consulted to ensure that the size and construction of the fireplace will accommodate the insert. Most factory-built or prefabricated fireplaces are not approved for inserts. Before an insert is installed, the chimney should be inspected. If there is creosote buildup, the chimney must be cleaned. Check for cracks or flaws in the chimney that could be a fire hazard. Inserts should never be used in a chimney flue used for another purpose.*

Instead of an insert many people wish to install a wood stove and exhaust the flue gases through the fireplace. There are two methods that may be used to make this conversion. An opening can be cut into the flue liner of the chimney above the fireplace opening, Figure 33–27, or a flue adapter can be extended from the stove connector pipe into the chimney flue as in Figure 33–28. Connecting the pipe directly into the flue liner generally yields a better draft. ***A mason should be employed to cut the opening and install the thimble. Be sure that there is adequate clearance between the pipe and the ceiling, any wooden mantel, or other combustible material. The stovepipe should extend into the thimble as far as possible without protruding into the flue. The fireplace damper should be securely shut. An approved sealer should be used between the thimble and the masonry.***

To make the installation through the fireplace opening, first remove the damper. If it cannot be removed, fasten it in the open position. A prefabricated adapter should be purchased to provide the proper fitting and correct size to accept the connector pipe.

33.8 ADD-ON WOOD-BURNING FURNACES

Wood-burning furnaces may be installed to operate in conjunction with an existing furnace, Figure 33–29. ***Manu-**

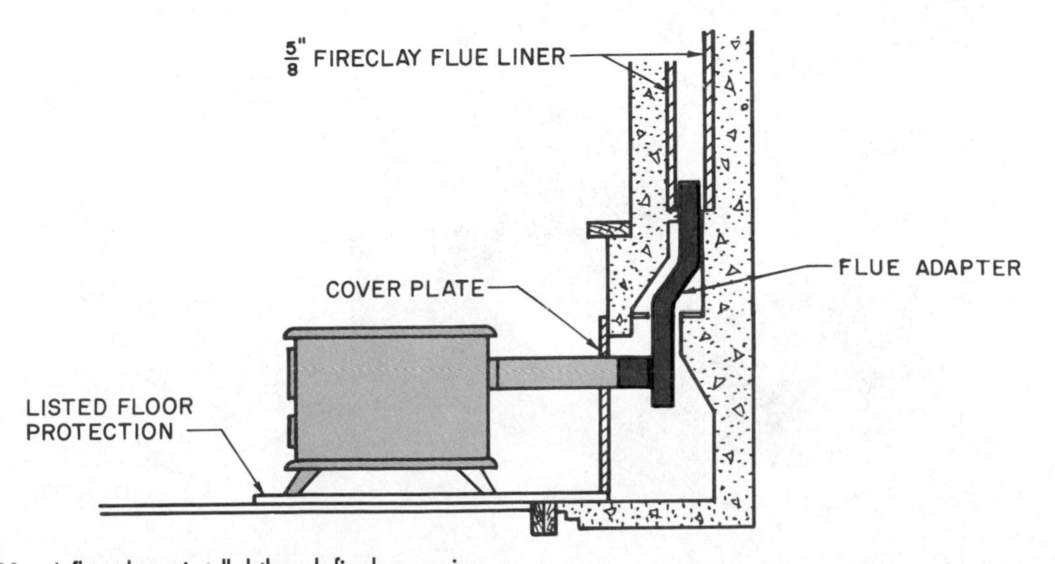

FIGURE 33–28 A flue adapter installed through fireplace opening.

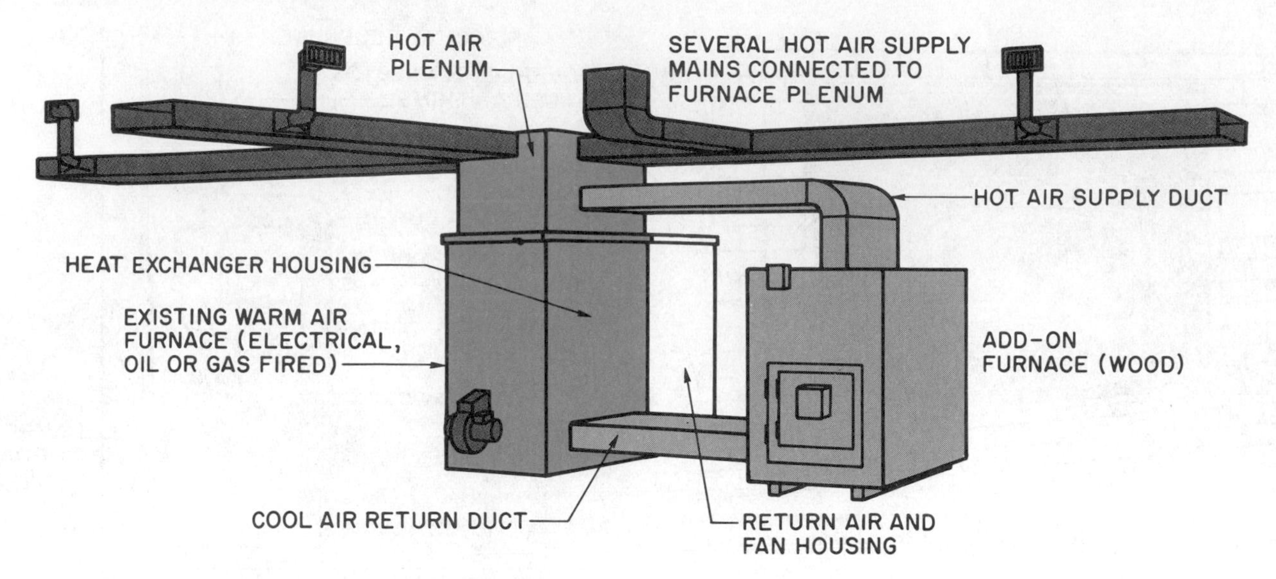

FIGURE 33–29 Add-on furnace. *Courtesy National Stove Works*

facturer's installation instructions and specifications must be followed carefully.* This add-on furnace may be used as the primary source of heat with the existing furnace supplementing it in extreme cold weather, or the wood add-on furnace may be used as the supplementary source of heat with the original furnace providing the primary heat.

33.9 DUAL-FUEL FURNACES

Dual-fuel furnaces provide the same service as the add-on and existing furnaces do except they are two furnaces in one unit. They are combination wood-burning furnaces that operate in conjunction with another type of fuel such as oil, gas, or electricity. Some of these furnaces are designed so that the wood is ignited by the combustion from the oil or gas side. Wood can be used as the primary fuel with the oil, gas, or electricity, or vice versa. Figure 33–30 is an illustration of a typical wood/oil dual furnace.

Thermostats for dual-fuel units are two-stage units. One stage controls the wood fire and the second the other source of heat. If the stage controlling the wood furnace is set higher than the other, the thermostat will call for heat from

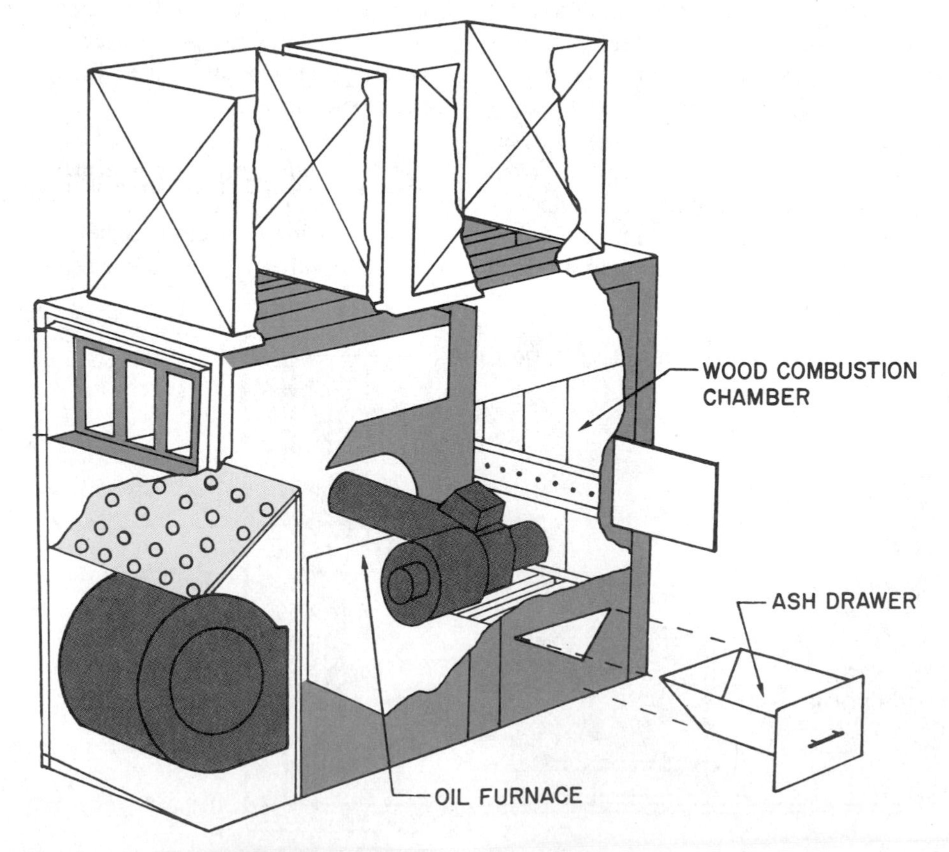

FIGURE 33–30 Dual-fuel furnace. *Courtesy Yukon Energy Corporation*

the wood furnace first. When it calls for heat, the damper will open, allowing primary air to the wood fire. The wood will then increase its burning rate, and when the thermostat is satisfied the damper will close. The primary air will be shut way down, and the fire will burn at a very slow rate. When the wood burns down to the point where it cannot produce the heat needed, the backup furnace will come on. The settings on the thermostat may be reversed if another fuel is the primary heat source.

Again, when installing the furnace be sure that proper clearances to combustible materials are maintained. Proper clearances are also required for the ductwork. The duct should also be large enough to handle natural convection airflow in the event there is an electrical power failure or a furnace fan failure. This guarantees heat to the house when power is off or the fan is not functioning properly and allows the heated air to pass through without overheating the duct.

It is very important to locate the draft regulator properly. It should be located in the connector pipe as close to the furnace as possible, Figure 33–31. It must be in the same room as the furnace.

Outside air to ensure proper combustion must be available or provided. The fire in the furnace uses oxygen and must have a continuous supply. As with a wood-burning stove, the air in the house contains only enough oxygen to support the combustion for a short time. This makeup air should be provided through a fresh air duct similar to that shown in Figure 33–32.

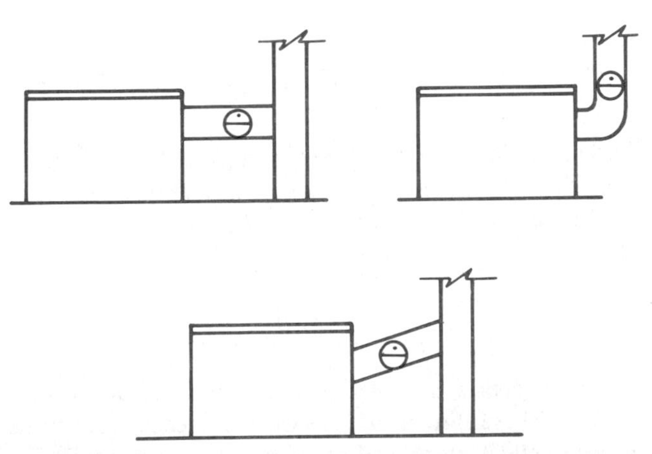

FIGURE 33–31 Suggested draft regulator locations. *Courtesy Yukon Energy Corporation*

33.10 FACTORY-BUILT CHIMNEYS

Factory-built chimneys (prefabricated) must be installed in accordance with the conditions stated by the listing laboratory. These conditions will be found in the manufacturer's instructions. Factory-built chimneys used for wood furnaces must be listed as residential and building heating appliance chimneys. Masonry chimneys must be constructed according to local codes.

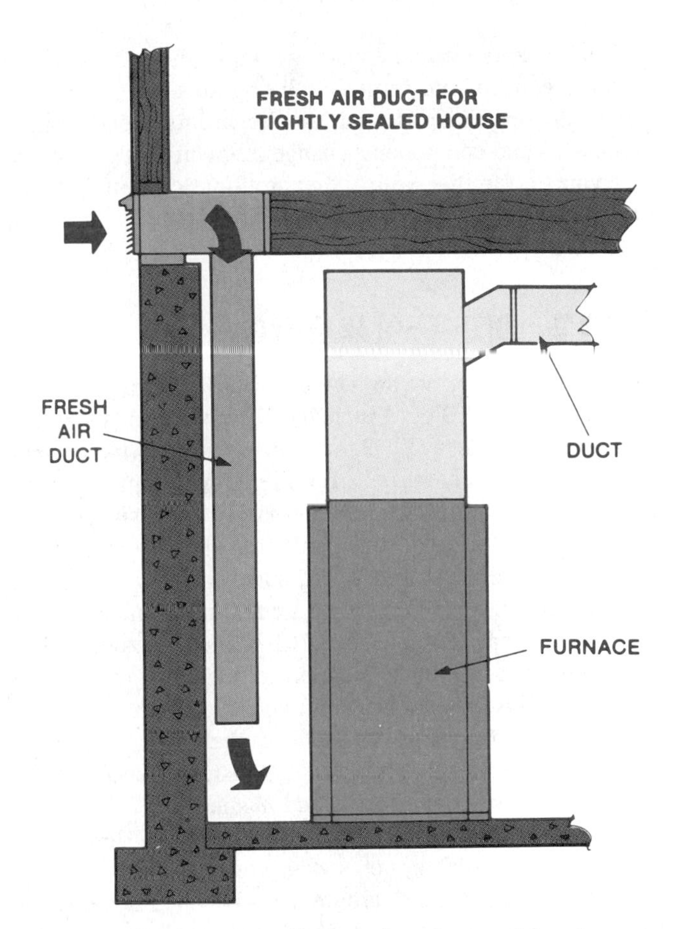

FIGURE 33–32 Typical makeup air duct. *Courtesy Yukon Energy Corporation*

33.11 CLEANING THE CHIMNEY, CONNECTOR PIPE, AND HEAT EXCHANGER

Chimney fires should be avoided at all costs. A check for creosote and soot buildup should be made twice a month. Creosote cannot only cause fire, but soot buildup on the heat exchanger in a furnace or stove can result in a great loss in efficiency.

Round steel brushes are the best for cleaning metal connector pipe and metal and clay flue chimneys. It is generally better to clean chimneys from the top. If the chimney is from a fireplace, it may be cleaned through the fireplace. An extendable handle should be used so that the brush can be inserted into the chimney, brushing and sweeping downward until the entire length of the chimney is cleaned. A weight may be placed under the brush, and the chimney can be cleaned by lowering the brush into the chimney. The brush will sweep the sides of the flue lining as it is raised.

The debris will be at the bottom of the chimney at the cleanout opening. Sweep the debris into a container. The connector pipe can be cleaned in a similar manner with a brush of the proper size. A smaller flat steel brush can be used for cleaning the heat exchanger in the furnace. Be sure to provide a covering for flooring, carpeting, or furniture where you may be working. You should have an industrial vacuum cleaner available when working where the soot and creosote can fall and settle in a living area.

Codes, regulations, and laws regarding wood heating appliances and components change frequently. It is essential that you be familiar with the appropriate current National Fire Protection Association information and local codes and laws.

33.12 DIRECT SOLAR ENERGY

Stored solar energy is the energy from fossil fuels such as coal, gas, and oil. Fossil fuels have formed over thousands of years from decayed plants and animals, which, when alive depended on the sun for life. This supply is being rapidly used and becoming increasingly expensive.

The sun furnishes the earth tremendous amounts of direct energy each day. It is estimated that 2 weeks of the sun's energy reaching the earth is equal to all of the known deposits of coal, gas, and oil. The challenge facing scientists, engineers, and technicians is to better use this energy. We know the sun heats the earth, which heats the air immediately above the earth. One of the challenges is to learn how to collect this heat, store it, and distribute it to provide heat and hot water for homes and business.

Many advances have been made, but the design and installation of solar systems has progressed very slowly. However, many feel that progress has been made compared to the progress in the design of other technologies over the years. It is assumed, though, that as the resources are further depleted and as economics or political actions cause energy crises, it will be only a matter of time before direct energy of the sun is used extensively.

33.13 PASSIVE AND ACTIVE SOLAR DESIGN

Many structures being constructed presently are using *passive solar* designs. These designs use nonmoving parts of a building or structure to provide heat or cooling, or they eliminate certain parts of a building that help cause inefficient heating or cooling. Some examples follow.

- Place more windows on the east, south, or west sides of homes and fewer on the north side. This allows warming from the morning sun in the east and from the sun throughout the rest of the day from the south and west. By eliminating windows on the north side, this coldest side of the house can be better insulated.
- Place greenhouses, usually on the south side, to collect heat from the sun to help heat the house, Figure 33–33.
- Design roof overhangs to shade windows from the sun in the summer but to allow sun to shine through the windows in the winter, Figure 33–34.
- Provide a large mass such as a concrete or brick wall to absorb heat from the sun and temper the inside environment naturally.
- Place latent-heat storage tubes containing phase-change materials in windows where they can collect heat to be released at a later time, Figure 33–35.

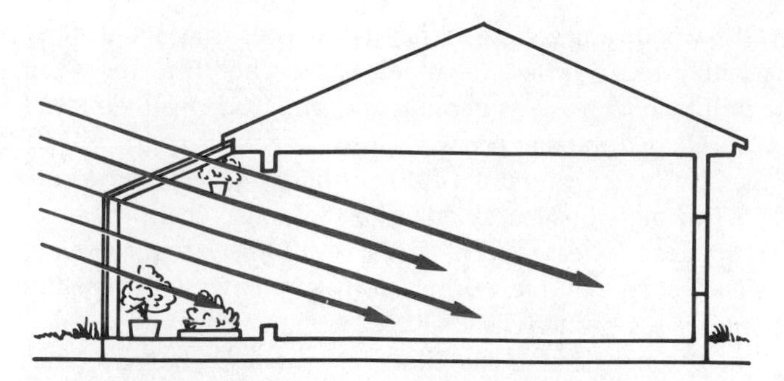

FIGURE 33–33 A greenhouse will allow the sun to shine into the house. It may also have a masonry floor or barrels of water may be located in the greenhouse to help store the heat until evening when the sun goes down. This heat then may be circulated throughout the house.

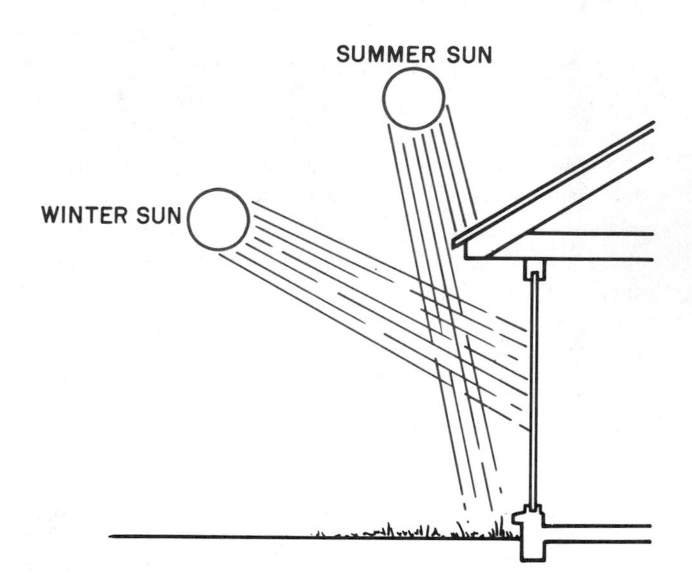

FIGURE 33–34 An overhang may be constructed to allow the sun to shine into the house in the winter when it is lower in the sky and yet shade the window in the summer.

This section is limited to a discussion of *active solar systems*. These systems use electrical or mechanical devices to help collect, store, and distribute the sun's energy. The distribution of this heat to the conditioned space is by means of the same type of equipment used in fossil-fuel furnaces.

33.14 DIRECT AND DIFFUSE RADIATION

The sun is a star often called the "daystar." A very small amount of the sun's energy reaches the earth. Much of the energy that does reach the earth's atmosphere is reflected into space or absorbed by moisture and pollutants before reaching the earth. The energy reaching the earth directly is *direct* radiation. The reflected or scattered energy is *diffuse* radiation, Figure 33–36.

FIGURE 33–35 Latent-heat storage tubes may be placed in windows where heat is stored when the sun is shining. *Courtesy of Calortherm Associates*

33.15 SOLAR CONSTANT AND DECLINATION ANGLE

The rate of solar energy reaching the outer limits of the earth's atmosphere is the same at all times. It has been determined that the radiation from the sun at these outer limits produces 429 Btu/ft²/h on a surface perpendicular (90°) to the direction of the sun's rays. This is known as the *solar constant*. The energy from the sun is often called *insolation*.

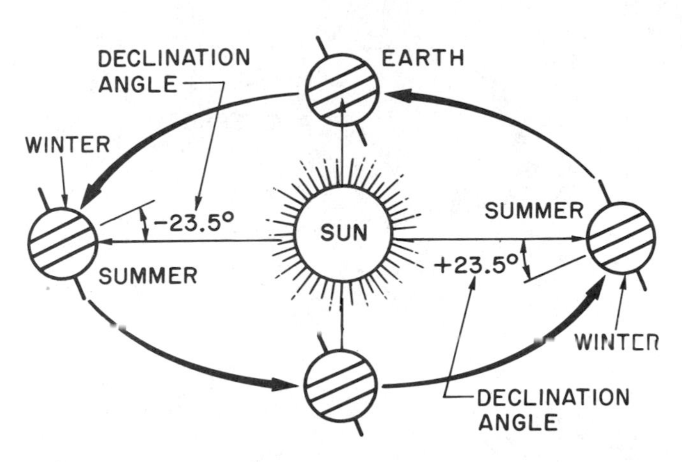

FIGURE 33–37 The angle of declination (23.5°) tilts the earth so that the angle from the sun north of the equator is greater in the winter. The sun's rays are not as direct, and the normal temperatures are colder than during the summer.

The earth revolves once each day around an axis that passes through the north and south poles. This axis is tilted, so the intensity of the sun's energy reaching the northern and southern atmospheres varies as the earth orbits around the sun. This tilt or angle is called the *declination angle* and is responsible for differences during the year in the distribution of the intensity of the solar radiation, Figure 33–37.

The amount of radiation reaching the earth also varies according to the distance it travels through the atmosphere. The shortest distance is when the sun is perpendicular (90°) to a particular surface. This is when the greatest energy reaches that section of the earth, Figure 33–38. The angle of the sun's rays with regard to a particular place on the earth plays an important part in the collection of the sun's energy.

33.16 SOLAR COLLECTORS

Solar collectors are used in active solar systems to collect this energy from the sun. Most solar collectors are flat and are positioned and tilted to catch as many of the sun's rays as close to 90° as possible in winter, Figure 33–39. For the greatest efficiency these collectors could be mounted on a

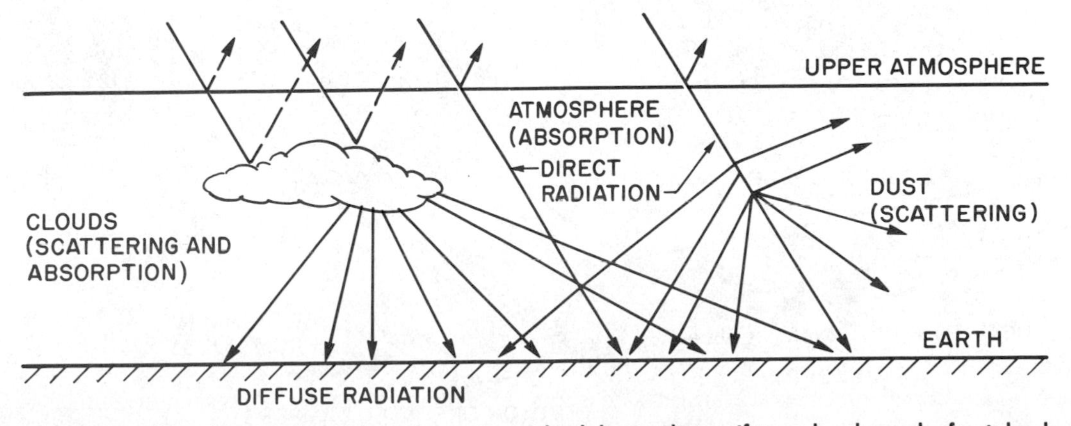

FIGURE 33–36 Radiation striking the earth directly from the sun is considered direct radiation. If it reaches the earth after it has been deflected by clouds or dust particles it is called diffuse radiation.

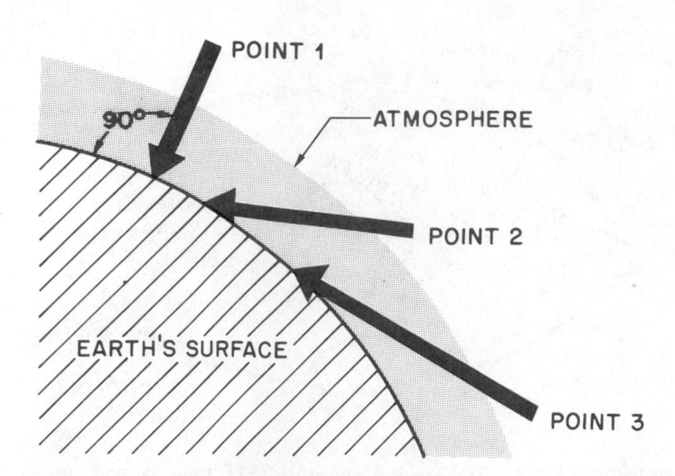

FIGURE 33–38 Radiation from the sun will be warmer at point 1 as it travels the shortest distance through the atmosphere.

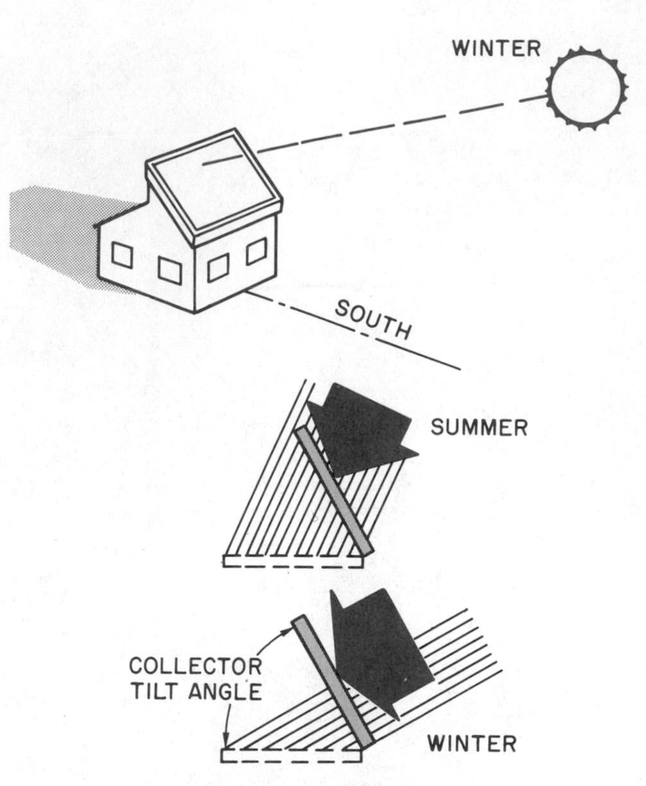

FIGURE 33–39 Collectors should face south. The collector tilt angle should be so that the sun's rays are 90° to the collector in the winter when the heat is needed the most.

device that follows the sun's path throughout the day and throughout the season. At the present time this type of device is too expensive, so the positioning of the collectors is done to collect the most energy possible. These collectors will collect direct radiation and, to a lesser extent, diffuse radiation. This means that some heat can be generated on hazy days and, to some extent, through clouds.

In an *active solar air space-heating system,* air is circulated through the collector where it is heated by the sun's radiation. The blower forces this heated air into a storage unit or to the space to be heated. Automatic dampers control whether this air is blown into storage or into the conditioned space, Figure 33–40.

Air Solar Collectors

Air solar collectors take in cold air from a manifold duct. The air is circulated through the collector, often through baffles to contain it in the collector longer so that it will be heated to the desired or maximum temperature before being blown to the storage unit or space to be heated, Figure 33–41. Air collectors are usually constructed with one or two tempered glass panels, an absorber plate, insulation, and a metal frame, Figure 33–42.

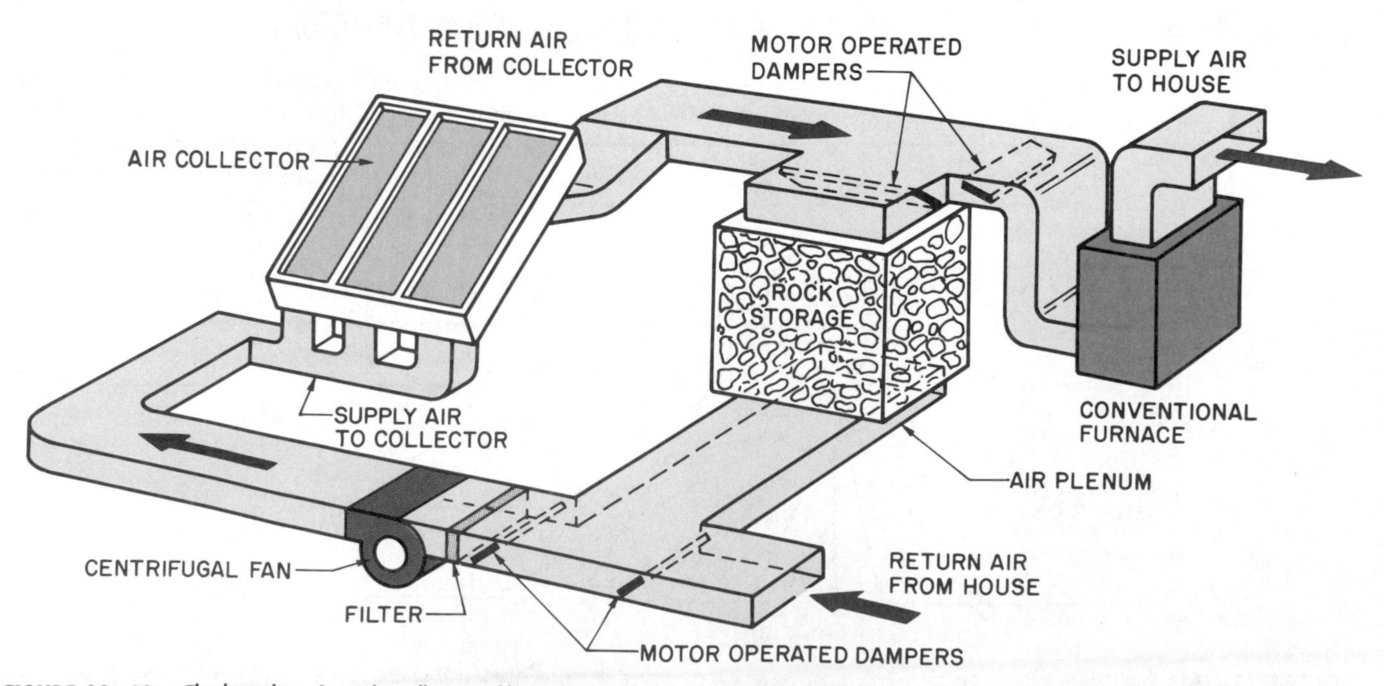

FIGURE 33–40 The heated air from the collector is blown through the duct to the conventional furnace where it is distributed throughout the house.

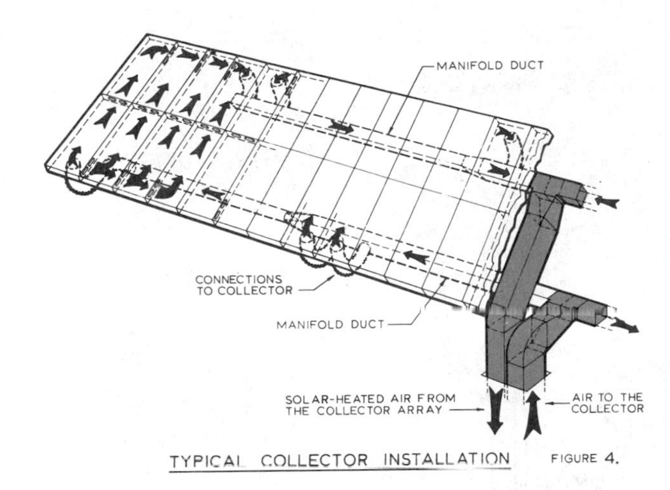

FIGURE 33–41 Solar air collector showing manifold ducts. *Courtesy Solaron Corporation*

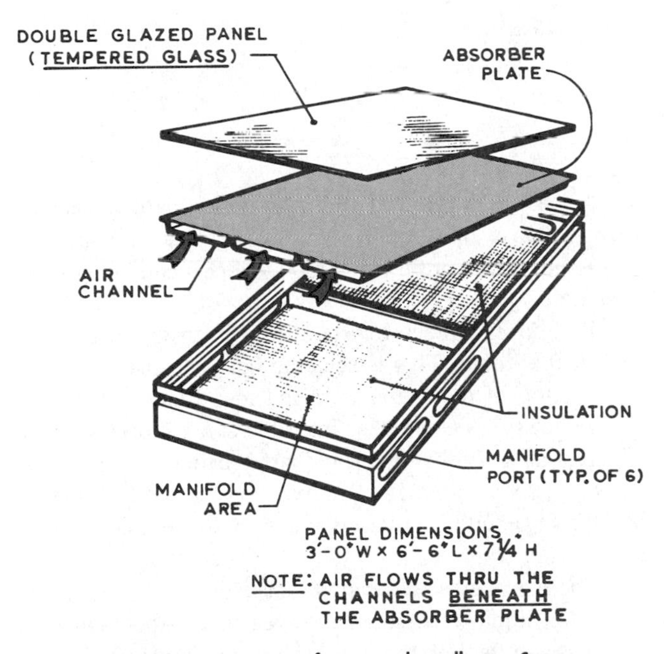

FIGURE 33–42 Construction of an air solar collector. *Courtesy Solaron Corporation*

The sun radiates through the one or two panels of glass. There may be some reflection and loss of heat, but the glass is designed to allow the maximum radiation to pass through. These glass panels, usually constructed of a low-iron tempered glass, are also designed to keep the heat from being radiated from the absorber plate out of the collector. Low-iron glass is used because it allows more of the sun's radiation to pass through. The absorber plate is painted or treated with a special coating to absorb as much energy as possible. Often a selective coating designed to absorb certain wavelengths of the sun's radiation is used. It may not absorb as much of the heat as a normal black-painted coating would, but it absorbs those longer wavelengths that do not escape back through the glass. Thus, more heat is trapped in the collector. The coated surface is often rough, allowing more heat to be absorbed.

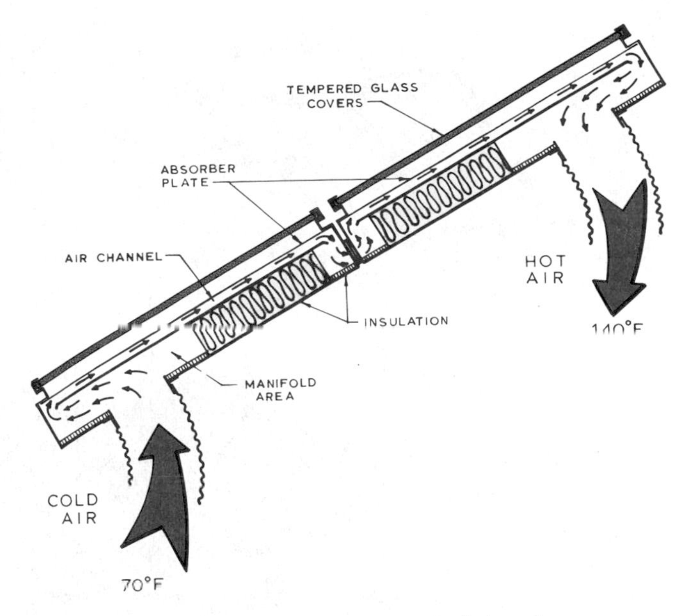

FIGURE 33–43 Two collectors are mounted side by side. Arrows show the air flow. *Courtesy Solaron Corporation*

The air to be heated is circulated underneath the absorber plate. This cool air absorbs heat from the underside of the absorber plate as it moves through the collector and is returned either to storage or to the heated space. It is important that the panels be insulated properly to keep unwanted heat loss from the collector to a minimum. Remember that the collector is located outside where the air is often very cold. This air could absorb significant amounts of the heat from the collector if it were not properly insulated.

As many collectors as needed can be mounted side by side to obtain sufficient surface area to give the desired heating results, Figure 33–43. Note, in the figure, that the air passes under the absorber plate.

In most areas these solar heating systems are either auxiliary systems to a conventional heating system, or the conventional system is a backup for the solar system.

Liquid Solar Collectors

Liquid solar collectors are flat-plate collectors similar to air collectors except that a liquid (either water or antifreeze solution) is passed through it in copper tubing. The liquid is heated as it passes through the collector. There are many designs for a system using liquid collectors. A basic one is illustrated in Figure 33–44. The water in the collector's tubing is heated and pumped to the storage tank. Water from this storage tank is pumped to a coil in a conventional furnace where air is blown across it. This air absorbs heat from the coil, and this heated air is distributed to the space to be heated.

Liquid collectors also use one or two panels of glass. Again, this glass may vary in quality and design, but a low-iron tempered glass is recommended.

The absorber plate is usually made of copper and formed tightly around the copper tubing, Figure 33–45. For this type of design the tubing should be soldered the entire

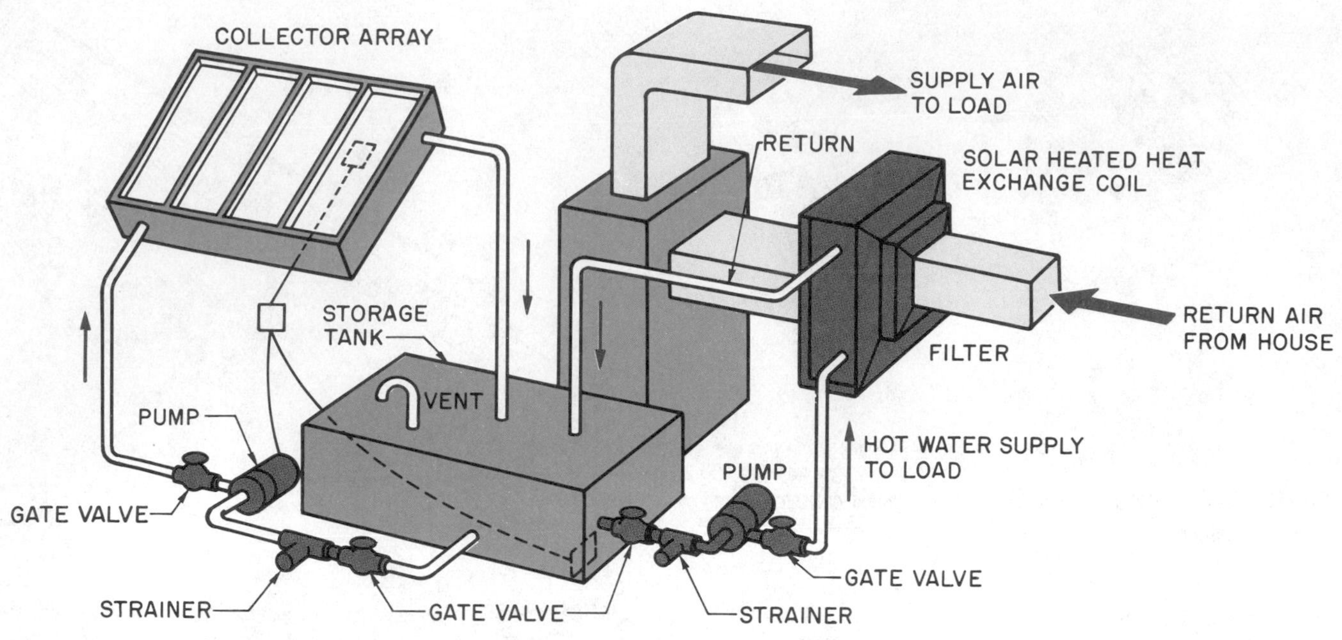

FIGURE 33–44 A liquid-based solar space-heating system.

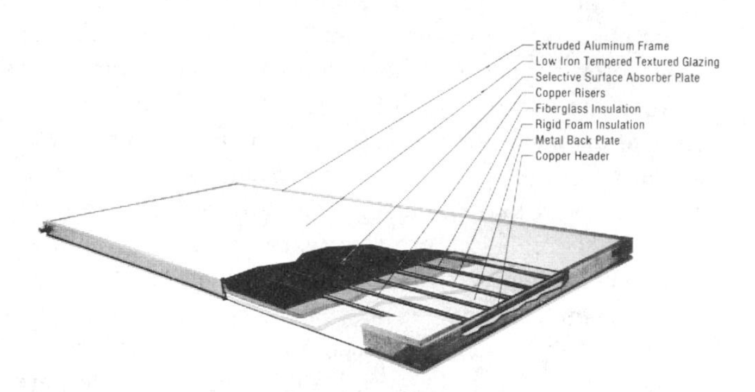

FIGURE 33–45 A liquid solar collector. *Courtesy Solaron Corporation*

length to the absorber plate. In some designs the tubes and plates are mechanically pressed or fastened together to achieve maximum contact. The absorber plate is painted or coated, generally with a selective coating similar to that described for the air collectors. Copper is generally used for the absorber plate and tubing because it has very good heat transfer characteristics. The absorber plate absorbs the radiation from the sun, and this heat is transferred to the tubing and then to the liquid by conduction. The continuous soldered joints, if used, help with this conduction transfer. The fluid is pumped through the collector, removing heat as it passes through. These collectors must also be insulated very well to keep heat loss from the collector to a minimum.

Refrigerant-Charged Collectors

Refrigerant-charged collector systems can be either passive or active systems. These systems operate in a similar manner to those discussed previously. Heat is absorbed into the refrigerant in the collector and transferred to a heat exchanger where the heat can be used for space heating, domestic hot water, or put in storage where it can be used at a later time.

In a liquid collector system, heat is absorbed from the sun by the liquid. This is sensible heat because the actual temperature of the liquid is raised to a higher level. In a refrigerant system, the heat from the sun vaporizes the refrigerant as it absorbs the heat. This is called *heat of vaporization* or *latent heat.* Heat is absorbed but the temperature of the refrigerant is not raised to a higher level unless superheat occurs.

These systems are called phase-change systems because the liquid refrigerant changes to a vapor and then, after passing through the heat exchanger, condenses back to a liquid. Typically, the refrigerant vapor moves by its own increased pressure to a heat exchanger. This heat exchanger is normally a coil filled with water surrounded by the refrigerant vapor. At this point the refrigerant cools and condenses. The efficiency of these systems is thought to be greater than liquid and air systems because latent rather than sensible heat is absorbed into the system. Due to the use of latent heat rather than sensible heat, the temperature at the collector may be from 20°F to 40°F lower than with a liquid or air system.

A **passive refrigerant-charged system** is a simple and often trouble-free system. There is one requirement and that is that the heat exchanger must be located above the collector, Figure 33–46. This allows the liquid refrigerant to return to the collector by gravity. The heat can be transferred to a coil for space heating, to a coil for heating domestic hot water, or to storage where it can be used at a later time. This system requires a pump to circulate the water in the secondary system.

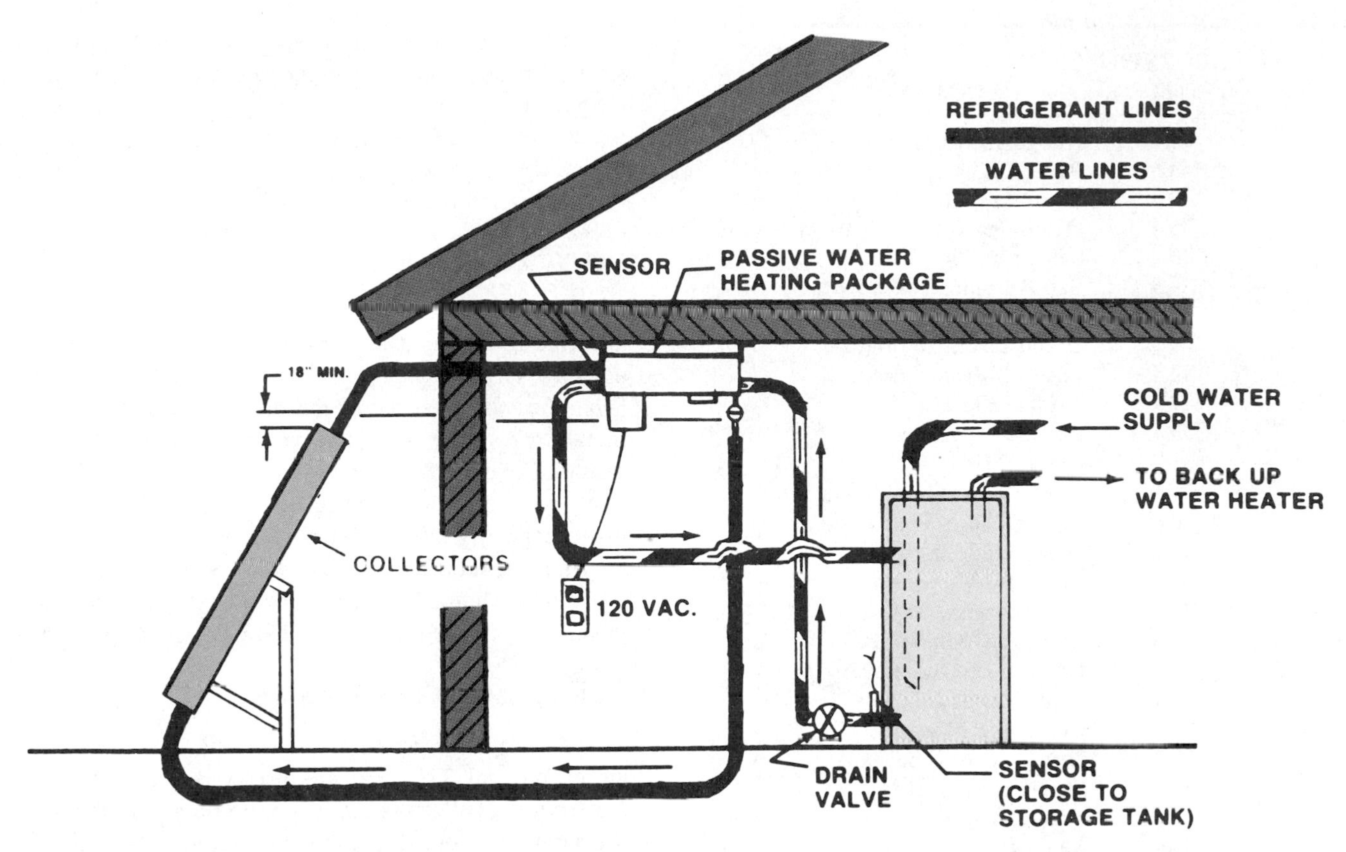

FIGURE 33–46 Passive refrigerant–charged system. Liquid refrigerant returns to collector by gravity. *Courtesy of Solar Research, Division of Refrigeration Research, Inc.*

An **active refrigerant-charged system** operates similar to the passive system except that the heat exchanger can be located below the collector. This is often more convenient, Figure 33–47. A pump is required to circulate the condensed refrigerant back to the collector.

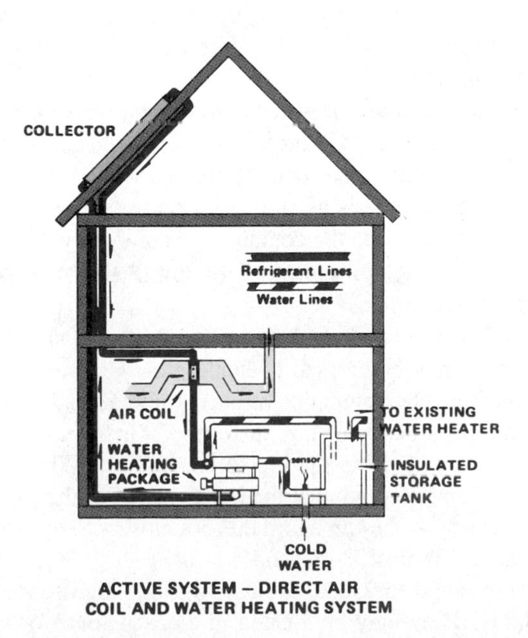

FIGURE 33–47 Active refrigerant–charged system. Liquid refrigerant must be pumped back to collector. *Courtesy of Solar Research, Division of Refrigeration Research, Inc.*

33.17 COLLECTOR LOCATION AND POSITIONING

Flat-plate collectors are usually positioned permanently to receive the maximum energy from the sun. This location and positioning is determined according to the position of the sun at midday during the winter when the most heat is needed. The collectors should face south. The tilt will vary from one location to another because of the location's latitude. The *latitude* is the distance from the equator and is expressed in degrees. The collector should be tilted so that it will be at a 90° angle to the sun's rays at midday during the coldest season. Figures 33–48 and 33–49 indicate the approximate latitude of various cities. The rule of thumb for the angle of the collector is to add 15° to the degrees of latitude for that particular location. In other words, for the latitude at Albuquerque, New Mexico of 35.1°, the collector tilt for a system in this city would be 35° + 15° = 50°. This angle is not critical so the 0.1° can be disregarded, Figure 33–50. An *inclinometer* is often used to position the collector at the correct angle, Figure 33–51.

Collectors can be mounted on roofs or on the ground. They should be mounted close to the rest of the system, Figure 33–52. This is particularly essential for air systems. Many times the roof may not face the south, or the roof may be shaded. In these cases the collectors may have to be located on the ground. The collector should not be in the shade from 9 AM to 3 PM during the heating season.

DEGREES LATITUDE	LOCATION
48.2	Glasgow, Montana
43.6	Boise, Idaho
40.0	Columbus, Ohio
35.4	Oklahoma City, Oklahoma
40	Salt Lake City, Utah
29.5	San Antonio, Texas
32.8	Fort Worth, Texas
40.3	Grand Lake, Colorado
42.4	Boston, Massachusetts
27.9	Tampa, Florida
33.4	Phoenix, Arizona
33.7	Atlanta, Georgia
35.1	Albuquerque, New Mexico
40.8	State College, Pennsylvania
42.8	Schenectady, New York
43.1	Madison, Wisconsin
33.9	Los Angeles, California
45.6	St. Cloud, Minnesota
36.1	Greensboro, North Carolina
36.1	Nashville, Tennessee
39.0	Columbia, Missouri
30.0	New Orleans, Louisiana
32.5	Shreveport, Louisiana
42.0	Ames, Iowa
42.4	Medford, Oregon
44.2	Rapid City, South Dakota
38.6	Davis, California
38.0	Lexington, Kentucky
42.7	East Lansing, Michigan
40.5	New York, New York
41.7	Lemont, Illinois
46.8	Bismark, North Dakota
39.3	Ely, Nevada
31.9	Midland, Texas
34.7	Little Rock, Arkansas
39.7	Indianapolis, Indiana

FIGURE 33–48 U.S. cities with their approximate degrees latitude north.

33.18 SOLAR HEAT STORAGE SYSTEMS

The collectors will absorb heat only during the day when the sun is shining. They will collect some heat on cloudy days but probably not enough to heat the space to the temperature desired in cold weather. On bright sunny days they may collect more than enough to heat the conditioned space. It is therefore necessary to store the heat generated on sunny days to help heat the structure at night and on cloudy days. *Sensible-heat storage* and *latent-heat storage* are the two general types of storage systems.

Sensible-Heat Storage

In sensible-heat storage the heat flowing into the storage raises the temperature of the storage material. When heating the space in a home or business, heat is removed from the storage material and its temperature drops. Most sensible-heat storage systems use either an air system with a rock bed or a liquid-based system. Regardless of the design or type of the system, a storage material, a container, and provisions for adding and removing heat will always be needed.

Latent-Heat Storage

In latent-heat storage systems the material used for the storage changes state. It absorbs heat without changing temperature. Heat flowing into the storage melts the material. When heat is taken out of the material, it solidifies or refreezes. (You may wish to review Unit 1 on latent heat.) This type of material is called phase-change material (PCM). Many types of PCM have been investigated. There are types that change by melting-freezing, boiling-condensing, changing from one solid crystalline structure to another, and changing from one liquid structure to another. The only PCM that is practical to use at this time is the melting-freezing type.

Air-Based Sensible-Heat Storage

Air-based systems generally use rocks as the storage medium or material. The rocks are placed in a container or bin. Air from the collectors that is not used directly for heating the building is forced into the bin and heats the rocks. Heat is stored in the rocks until needed and then air is blown through the rock bin and distributed to the space to be heated.

Rocks should be of medium size (1- to 3-in. diameter). They should be rounded if possible and should all be approximately the same size. A rule of thumb is that they should not be smaller than three-fourths the average size and should not be larger than 1 1/2 times the average size. If rocks are too large, only the outside will be heated, and the space taken up by the rest of the rocks will be wasted. If they are too small, they will fill up all space in the bin and reduce air circulation.

Some rocks are very soft; they break and cause dust. Others, particularly of volcanic origin, may have an odor. Avoid all of these. The rocks should be washed thoroughly to ensure that there is no dust. Mix them (small and large) and shovel them into the container or bin evenly. ***The bins must be strong enough to withstand the tremendous weight of the rocks.***

The design of the rock bed depends on the space available. Bins may be vertical or horizontal. Vertical beds provide the best characteristics because they take advantage of the natural tendency of hot air to rise. Air from the collectors is forced in from the top to the bottom. When heat is removed, air is forced in at the bottom and out the top. Most of the heat is at the top and therefore closest to the space to be heated, Figures 33–53 and 33–54.

Horizontal rock beds are used when a high space is not available. They may be located in a crawl space or similar area. Horizontal bins do have some disadvantages: (1) The airflow is from side to side rather than from bottom to top or top to bottom (when being charged with heat). The rocks

FIGURE 33–49 Map to determine the approximate latitude north.

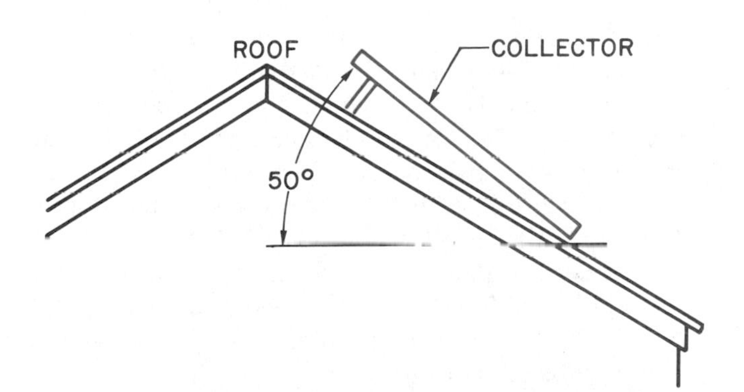

FIGURE 33–50 Collector tilt angle for Albuquerque, New Mexico.

FIGURE 33–51 An inclinometer. *Photo by Bill Johnson*

FIGURE 33–52 Typical collector installations.

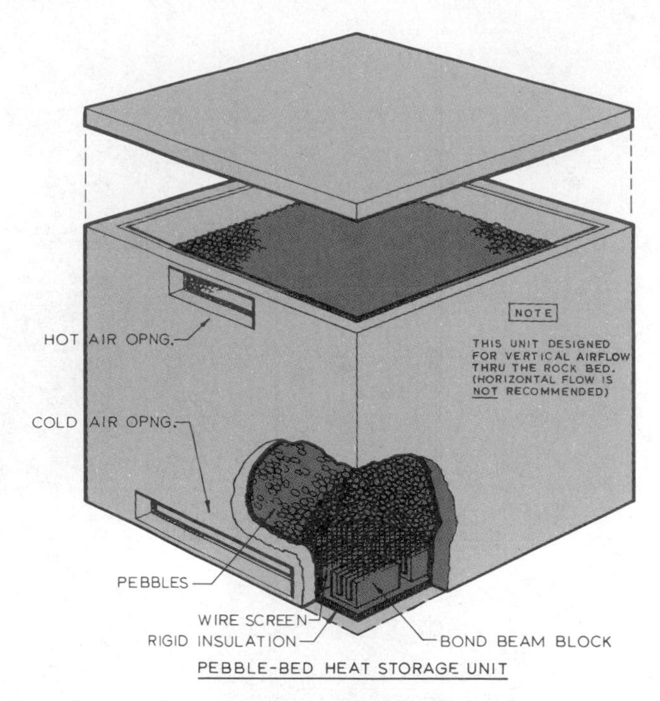

FIGURE 33–53 A rock bed air-based storage unit. Note the air space at the bottom between the beam blocks and at the top. These air spaces are called plenums and are necessary to allow the proper air flow through the unit. *Courtesy Solaron Corporation*

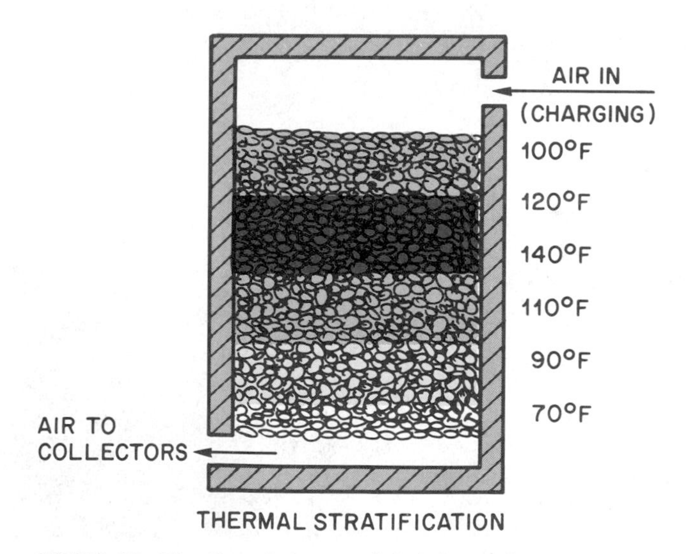

FIGURE 33–54 Vertical storage rock bed showing heat stratification.

will settle, particularly with the movement from expansion and contraction from heat and cooling, and leave an air space at the top. The air will flow horizontally through this space, bypassing the rocks and not picking up sufficient heat. A floating top will eliminate this air space. (2) As air from the collectors flows into the bin, it will rise and heat the rocks at the top more than the rocks at the bottom. When air is forced through for space heating, it will blow through the cool and warm rocks. When the air is mixed, its overall temperature will be lower than air taken out the top, Figure 33–55.

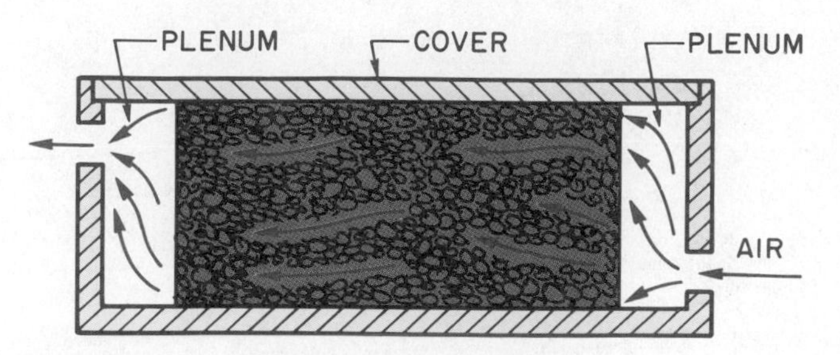

FIGURE 33–55 Horizontal rock storage bin.

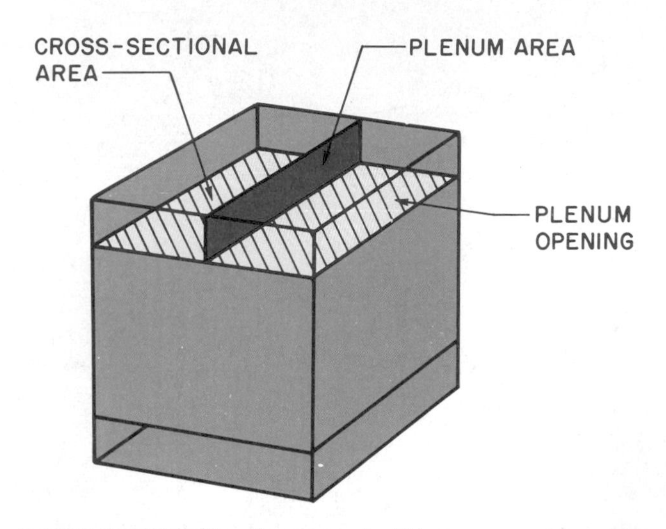

FIGURE 33–56 The plenum area should be approximately 8% of the bin cross-section area.

The rock bin should have *plenums* (air spaces) at the top and at the bottom of a vertical bed or at the ends of a horizontal bed. The plenum allows the air to circulate across the entire top and bottom, permitting air to circulate by all the rocks. The efficiency would drop significantly if there were obstructions in the plenum. Each plenum cross section should be approximately 8% of the bin cross section, Figure 33–56.

The airflow from the collector to the bed and back to the collector and from the bed to the house and back to the bed is controlled by dampers. Dampers are discussed in more detail later in this unit.

Liquid-Based Storage Systems

These storage systems usually use water as the storage medium. Many things must be considered when designing this type of system. The designer must determine the most appropriate size of the storage container or tank, its shape, material, and location. In addition, leak protection methods, temperature limits, insulation, and pressurization methods (if any) must be determined.

Storage tanks may be steel, fiberglass, concrete, or wood with a plastic lining. Normally a single tank is used for space-heating applications, but if the proper size is not available or if there are space limitations more than one tank may be used.

Direct solar space heating usually requires storage temperatures of 160°F or less. Space heating with solar-assisted heat pumps usually requires a storage temperature of 100°F or less—the lower the required storage temperature, the less insulation needed for the storage container.

Steel Tanks

Steel tanks have been used for years in similar applications and considerable information concerning them is available. They are easy to manufacture, and piping and fittings can be attached easily. However, they are expensive, subject to corrosion, and difficult to install indoors.

Steel tanks should be purchased with a baked-on phenolic epoxy on the inside. If at all possible, limit the temperature to 160°F because the corrosion rate doubles with every 20°F rise in temperature. The outside of buried tanks should be protected with a coal-tar epoxy, above-ground tanks with primer and enamel.

If a steel tank is used, steps may have to be taken to prevent excessive electrochemical corrosion, oxidation (rusting), galvanic corrosion, and pitting. Manufacturers of these tanks should be consulted for information as to how to prevent these conditions.

Fiberglass Tanks

These tanks should be designed specifically for solar storage. They should be factory insulated. Fiberglass tanks usually have two shells, an inner and outer shell, with insulation between. These tanks do not rust or corrode and may be installed above ground or in the ground. ***The maximum temperature is limited with these tanks, and they normally cannot be pressurized. They are easily damaged, so inspect them before installing to ensure that there will be no leaks.***

Concrete Tanks

Concrete tanks can be purchased precast, or they may be cast in place and constructed to fit many different space limitations. The concrete itself becomes part of the storage, providing some additional capacity. They do not corrode or rust. The weight of this type of tank must be considered when locating it, and enough support must be provided. The weight can be an advantage for buried tanks because they will not be easily floated up or out of the ground by groundwater. They should be insulated. Because heating and cooling will cause expansion and possibly result in leaks, some provision, such as a spray-on butyl rubber coating or a plastic liner, should be used to help prevent leaks.

Insulation Requirements

Indoor tanks located in heated areas require the least insulation. Tanks in unheated areas and outdoor installations require more. Simple covers are sufficient to protect the insulation in indoor installations, but underground tanks require waterproof insulation.

Latent-Heat Storage Systems

Latent-heat storage material melts when it absorbs heat, and it refreezes or solidifies when it gives up heat. The material does this without changing temperature. One feature of this system is that it can store more heat in less space than either of the other systems. Therefore it requires less space, a smaller container, and less insulation. If the storage material is chosen so that the melting point is slightly above the temperature required by the system, the collector output would need to be only a few degrees warmer than this melting point. The material will increase in temperature only until it reaches its melting point. It then changes state until it is completely melted. It has more even temperatures than rock or liquid storage systems. The collector output for a rock or liquid storage system would need to operate at a much higher temperature when the storage system is partially or nearly heated to its maximum temperature.

The melting point of the PCM is one of the most important factors in choosing the material. There is a temperature drop in heat exchangers, and forced-air heating systems should distribute heated air at a considerably warmer temperature than the desired room temperature. Therefore, in an air-based system the minimum melting point of the PCM should be approximately 90°F. Because liquid-based systems go through an additional heat exchange process, the minimum melting point in these systems should be about 100°F. Solar-assisted heat pumps can use materials with a melting point between 50°F and 90°F.

Some latent-heat systems can store a small amount of sensible heat by heating the material above the melting point. The amount of sensible heat stored is generally very small compared to the latent heat stored.

Figure 33–57 illustrates a storage system using tubes filled with the PCM spaced apart in a container. Air is

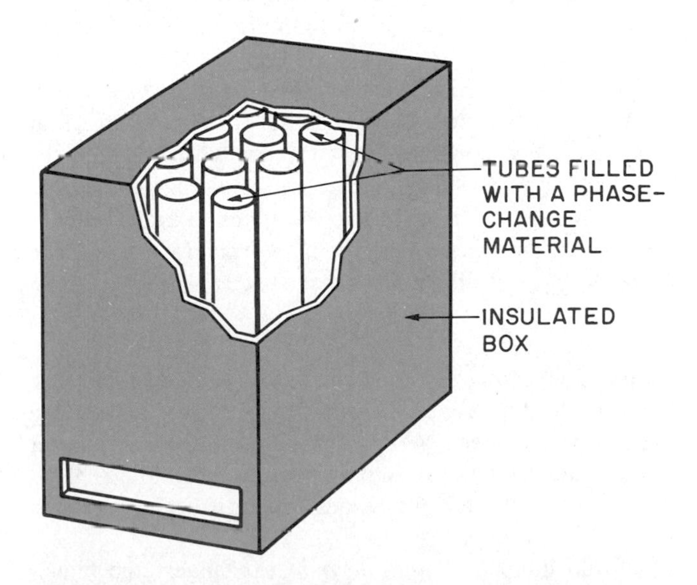

FIGURE 33–57 Tubes filled with a PCM can be placed in an insulated container. Air can be forced into the top of the container from the collectors, charging the storage, or it can be circulated from the bottom and out the top, extracting heat from the PCM and heating the conditioned space.

blown around the tubes to either charge the system from the collectors or to remove heat for circulation to the area to be heated. Figure 33–58 shows a system using flat trays containing the PCM. Again, air is circulated through the container to charge the system with heat from the collectors or to remove the heat for space heating.

Salts and waxes can be used in these systems. They have melting points ranging from approximately 80°F to 120°F.

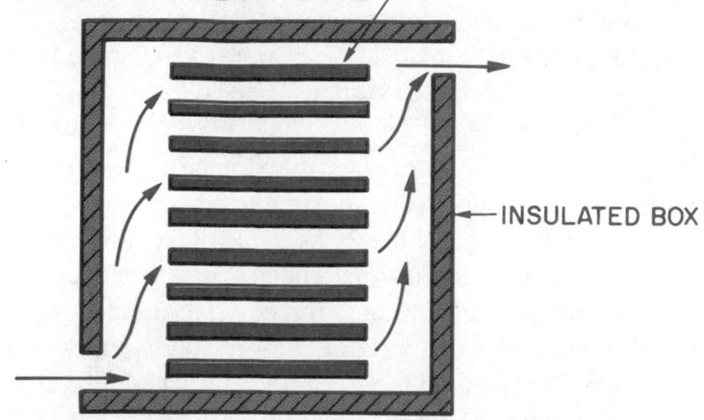

FIGURE 33–58 Trays with a PCM can be placed in an insulated storage container. This functions similar to the PCM tube storage in Figure 33–57.

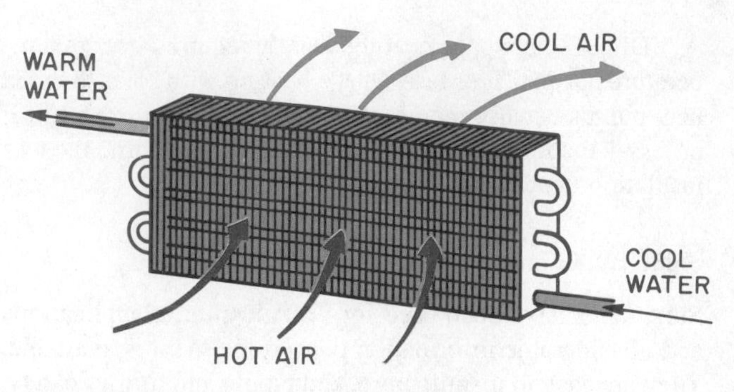

FIGURE 33–59 An air-to-water heat exchanger.

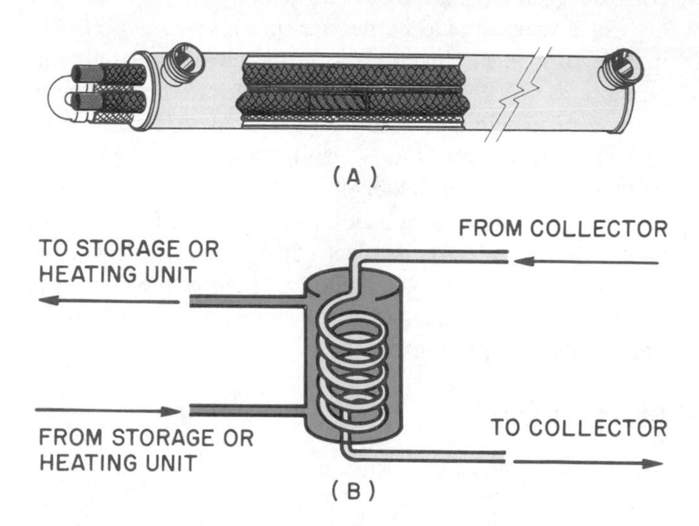

FIGURE 33–60 Liquid-to-liquid heat exchangers. *Courtesy Noranda Metal Industries, Inc.*

33.19 OTHER SYSTEM COMPONENTS

Heat Exchangers

Heat exchangers transfer heat from one medium to another, usually without mixing either of the materials. They may transfer heat from air to air, air to liquid, liquid to liquid, or liquid to air.

In the air-based space-heating system, a separate heat exchanger is not needed because the rock bed is the heat exchanger. The collectors heat the air, which charges the rock bed with heat. When the system is discharged or air is forced through the rock bed to the space to be heated, the rock bed is the place where the heat exchange occurs. The air being blown to the conditioned space absorbs heat from the rocks.

AIR-TO-LIQUID. *Air-to-liquid* exchangers are used in domestic hot water heating. An exchanger, Figure 33–59, is placed in the heated air duct near the domestic hot water heater. It absorbs heat from the air into the heat exchange fluid, which is pumped to the hot water tank where it heats the water.

LIQUID-TO-LIQUID. In *liquid-to-liquid* exchangers the liquid-based solar collectors heat the fluid in the collectors. It is then piped into a coil in the storage heat exchanger, Figure 33–60. Heated fluid from the outlet of the heat exchanger is then piped either to the storage tank or to the water heating element, such as a hydronic baseboard heating unit.

LIQUID-TO-AIR. *Liquid-to-air* exchangers may use a heat exchanger like that used in the air-to-liquid system. These exchangers usually are constructed with finned water tubing. The component is placed in the air distribution ductwork, where air forced through it absorbs heat from the fins and tubing. The tubing may contain hot water directly from the solar collectors or from the storage tank.

Any system loses efficiency each time a heat exchanger is used. This is because of the difference in temperature between the heat exchange mediums. It is therefore necessary to limit the number of heat exchangers.

Filters

Filters should be installed at the rock bed inlet and outlet. It is essential to filter as much dust as possible from the rock bed and from the conditioned space.

Automatically Controlled Dampers

Dampers must be installed in solar air systems to control the air from the collectors to the storage or to the conditioned space and return to the storage bin and collector. Motor-driven, thermostatically controlled dampers are used, Figure 33–61. Back-draft dampers should also be installed between the rock bed and the collectors to prevent heat loss

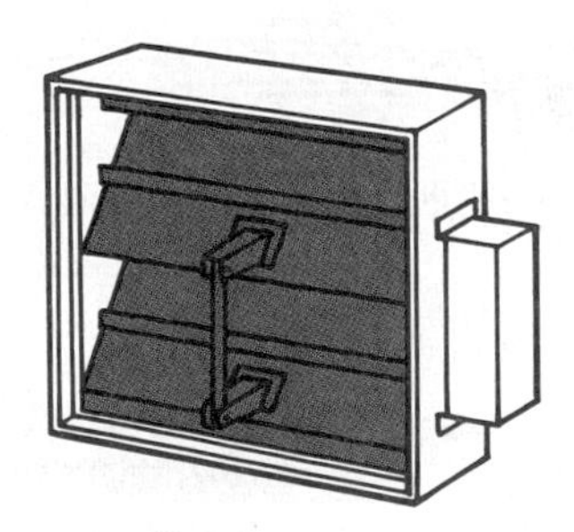

FIGURE 33–61 A motorized damper.

from the storage bed to the collectors when they are not absorbing heat from the sun.

Package Air Handlers

Air handlers that include a motor-driven fan and dampers are often used from the air collectors to the furnace or to the rock storage bin, Figure 33–62.

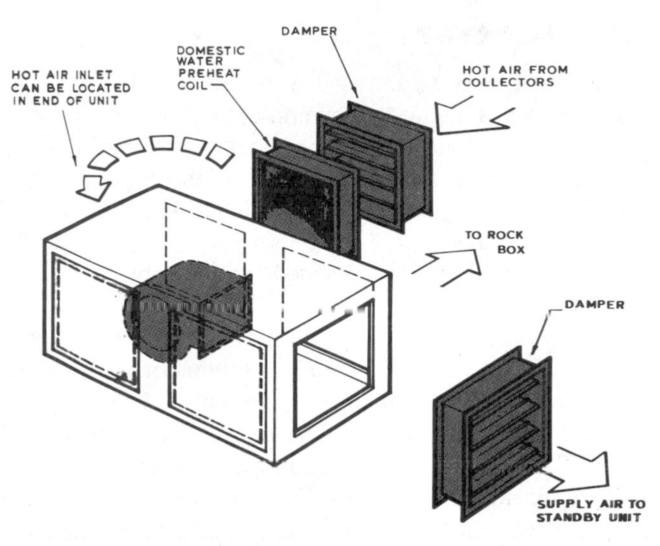

FIGURE 33–62 Diagram of air handler, showing possible locations of heating coil and dampers. The fan is shown with the dotted line inside the unit. *Courtesy Solaron Corporation*

Controllers

Controllers turn pumps on and off, control dampers, and open and close valves, Figure 33–63. The differential thermostat is the main component in a controller. In a house heated by a conventional furnace, only one temperature setting is needed on the thermostat to turn the furnace on and off. In a solar-heated house the temperature at the collector, in the storage, and in the house must all be sensed so that the controller will operate the pumps, valves, and dampers for the system to function efficiently.

When the liquid temperature at the collectors is hotter than the temperature at the storage, the collector pump will be activated to circulate liquid from the collector to the storage tank. When the room thermostat calls for heat and the storage tank is hot enough, the pump to the space heating coil is activated; when the room thermostat calls for heat and the storage is not warm enough, the furnace is activated. The controller must "know" the temperature at all these points to "determine" which circuits to activate, Figure 33–64.

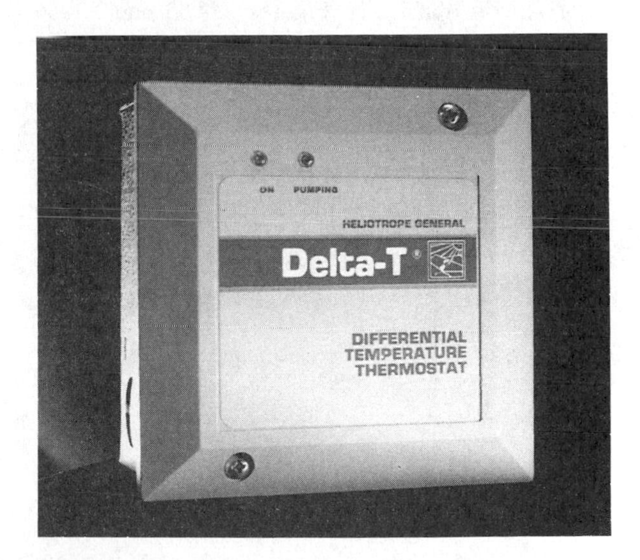

FIGURE 33–63 A typical controller for a space-heating solar system. *Courtesy Heliotrope General*

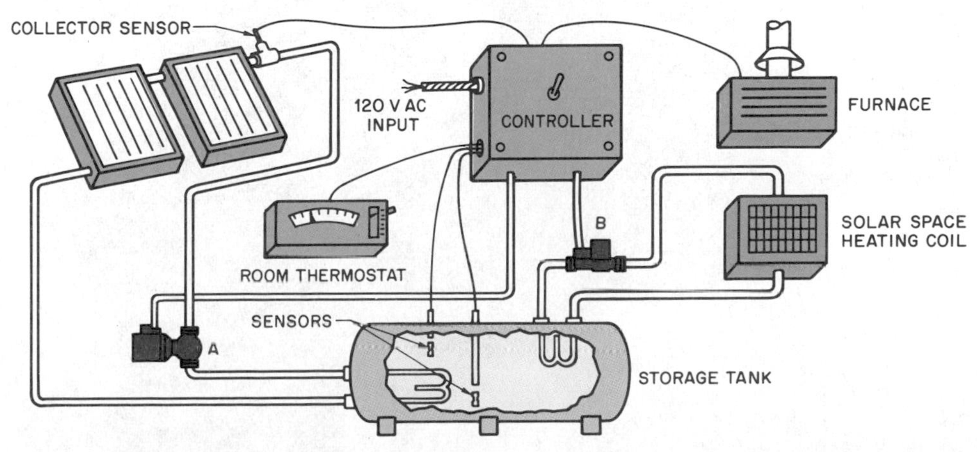

FIGURE 33–64 When the collectors are hotter by a predetermined amount than the storage temperature, pump A is activated, circulating hot liquid from the collectors to the storage. When the room thermostat calls for heat and the water in the storage tank is hot enough, pump B is activated, circulating heated water to the solar space-heating coil. When the room thermostat calls for heat and the storage liquid is not hot enough, the furnace is activated.

Temperature Sensors

Temperature sensors are used in the rock bed to send the temperature reading at the top and bottom of the bed to the controller. Usually one sensor is installed 6 in. from the top and another 6 in. from the bottom of the rock bed. The sensors are connected to controllers that activate the auxiliary heat or dampers.

Temperature sensors, located in the storage tank, are also used in liquid-based systems. The sensors tell the controller when the water at the top is hot enough for space heating or when the temperature at the bottom is such that the collector pump should be turned on or when the temperature is so hot that it is unsafe and should be cooled down. Sensors may be located in the tank, in wells in the tank, or outside the tank if it is metallic. The sensors should be protected from the water and, if installed on the tank, should be in good contact with the metal and insulated from the ambient. Sensors may be thermocouples, bimetal controls, liquid or vapor expansion tubes and bulbs, thermistors, or silicon transistors. Figures 33–65 and 33–66 are examples of typical thermistor sensors. They should always be matched with the control. Figure 33–67 illustrates how a bulb-type thermistor sensor may be installed in tanks.

Thermometers should also be installed at the top and bottom of the water storage tank. They are used to adjust the system's operation and to troubleshoot the system if it doesn't function properly.

FIGURE 33–66 Thermistor sensor in brass plug.

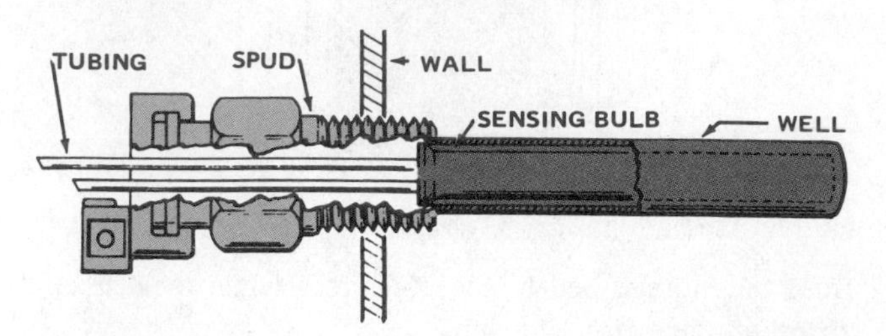

FIGURE 33–67 Thermistor sensor installed through wall of tank. *Courtesy Heliotrope General*

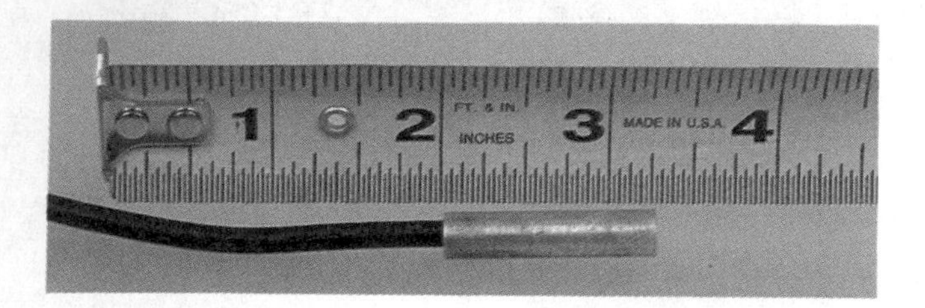

FIGURE 33–65 Thermistor sensor. *Photo by Bill Johnson*

Pumps

Liquid-based solar heating systems require *pumps* to move the liquid through the collectors to the storage and back to the collectors. The fluid may at times bypass the storage under certain conditions and be fed directly to the heat exchanger serving the heat distribution system. In this instance another pump is often used. This may be the same

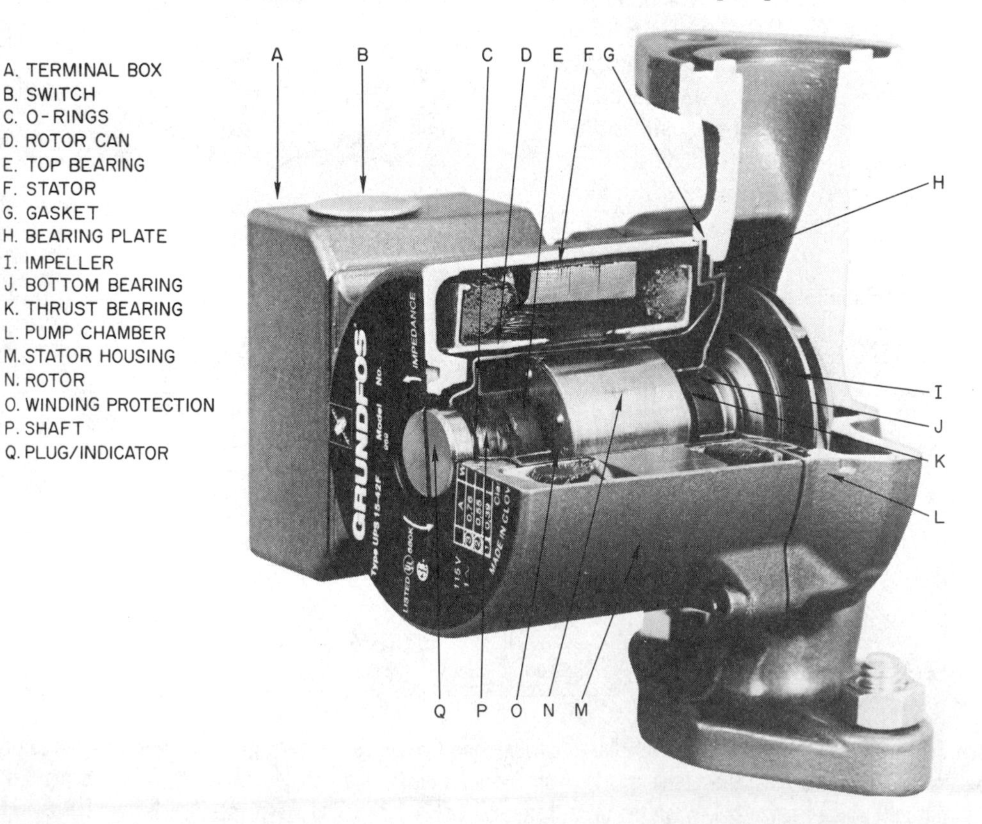

FIGURE 33–68 A cutaway of a typical circulating pump used in a solar space-heating system. *Courtesy Grundfos Pumps Corporation*

pump that pumps the water from storage to the heat exchanger.

Centrifugal pumps are generally used in solar systems. An impeller turns in a pump housing and forces the fluid through the system. These pumps produce a low pressure but can move a large volume of liquid. You may read in some manufacturer's specifications or other materials that pumps have a low or high head. This simply means a low- or high-pressure difference that the pump can develop from the inlet to the outlet, Figure 33–68.

Valves

The following is a brief description of some valves in a liquid-based solar system.

HAND-OPERATED VALVES. *Hand-operated valves* are used to isolate a particular section of the system. This may be necessary when the system is shut down for a period of time or when it is being serviced. The three common types of hand-operated valves are the *globe valve*, the *gate valve*, and the *ball valve*, Figure 33–69. The globe valve is used to shut off the system or to control the amount of flow. It offers resistance to the flow even when open, and it does not fully drain when installed horizontally. Gate valves are used as shutoff valves; they will drain and offer little resistance. Gate valves are generally used in solar systems. Ball valves offer virtually no resistance. They may be used as shut-off valves but may be more expensive.

CHECK VALVES. *Check valves* are installed between the pump outlet and the collector inlet to prevent siphoning back to the collector on cold nights when the collectors have been drained. Collectors using only water must be drained in cold weather at night and at other times when the water might freeze. Check valves must be installed properly. ***Gravity operated check valves must be installed in the correct direction. If they are installed upside down or at any angle other than their intended design, they will not operate properly; therefore siphoning, and possible freezing, will occur.***

SPRING-OPERATED CHECK VALVES. *Spring-operated check valves* may be installed at any angle, but they must be installed in the line with the correct direction of flow. Although they are more dependable than gravity operated check valves, they offer more resistance, which causes a greater pressure drop.

PRESSURE RELIEF VALVES *Pressure relief valves* are necessary to relieve a closed system of excessive pressure to prevent damage to a system and to eliminate a hazardous condition. These valves are normally selected for the pressure and temperature at which they will open, Figure 33–70. ***Always use exact replacement relief valves.***

Expansion Tanks

All closed systems must have space into which liquids can expand when heated to keep the system pressure under control. Therefore, an *expansion tank* must be included in the system. The expansion tank must be sized properly to avoid damage to the system. The kind of liquid in the system must

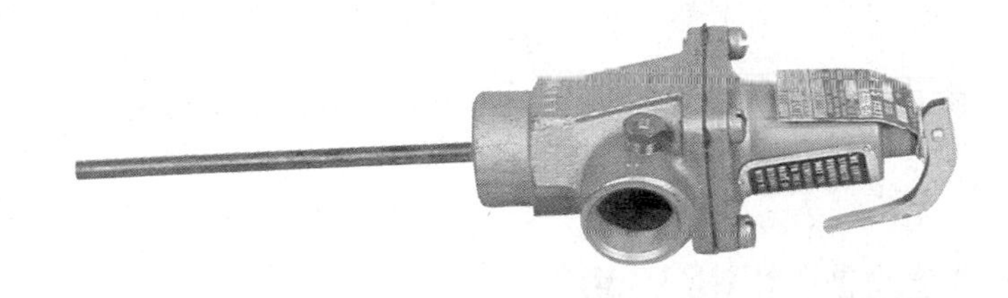

FIGURE 33–70 A pressure-temperature relief valve. *Photo by Bill Johnson*

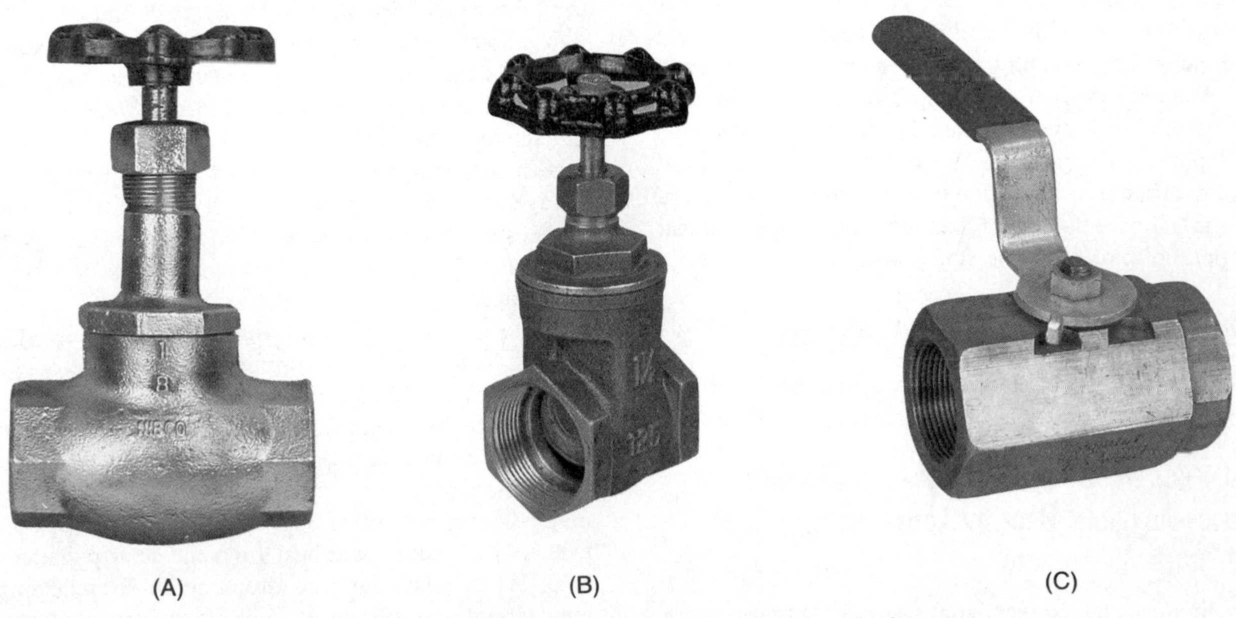

FIGURE 33–69 Hand-operated valves. (A) Globe valve. (B) Gate valve. (C) Ball valve. *Photos by Bill Johnson*

be considered when determining the size of the tank. For instance, antifreeze solutions will expand more than water, so more space is needed for the antifreeze solutions.

Expansion tanks may be designed for the water or fluid to mix with the air in the tank, or they may be designed so that the liquid is kept separate from the air, Figure 33–71. This latter type of tank has a flexible diaphragm separating the liquid from the air, which prevents the liquid from absorbing the air.

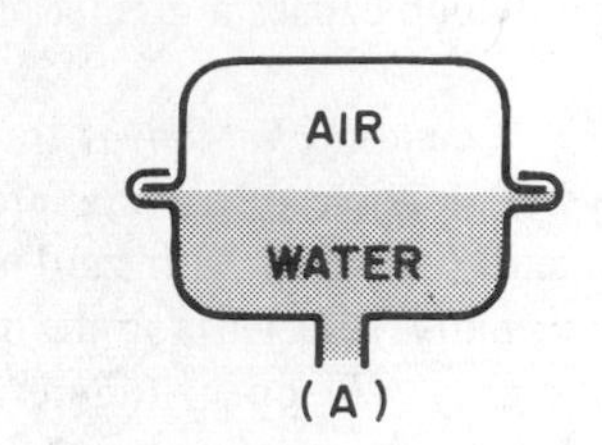

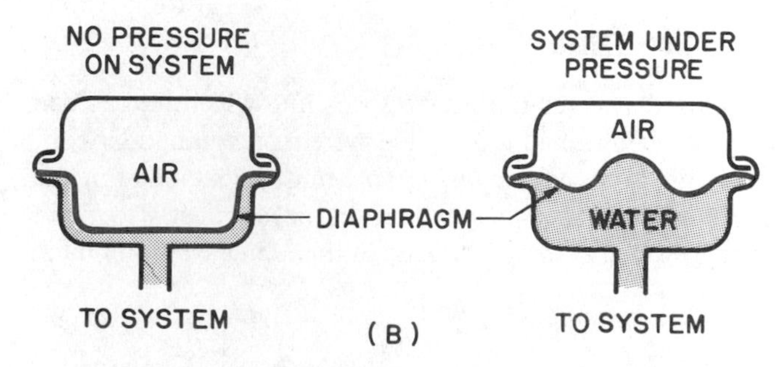

FIGURE 33–71 (A) No diaphragm in this type. (B) Expansion tank with a diaphragm to separate the water and air.

33.20 INSULATION

Insulation is a key factor in efficient solar systems. Insulation should be properly installed wherever there is a chance for the heat to escape. This includes the sides and bottom of collectors, the storage container, and all heat transfer components such as piping, tubing, and ductwork. Insulation designed specifically for duct and piping applications should be used rather than wrapping insulation around the duct and piping. Adapting insulation for applications for which it is not intended is inefficient because insulation should not be crushed, packed, or squeezed. This reduces the air space and lowers the efficiency. Insulation on exposed piping, such as on the roof, should be treated, painted, or of a material that resists deterioration from the sun's radiation.

33.21 SOLAR SPACE-HEATING SYSTEMS

The following includes a few of the combinations possible in using solar radiation for space heating applications.

Air Collection–Rock Bed Storage–Air Distribution to House–Auxiliary Heat by Conventional Forced-Warm Air Furnace

This system uses the conventional warm air furnace to distribute the heat from the collectors and the storage bin to the house. The furnace is also used as the auxiliary heat supply when heat from the solar system is not available, Figure 33–72. When the collectors are absorbing heat and have reached a predetermined temperature and the house thermostat is not calling for heat, dampers A and B will open to the rock storage bin. The collector fan will start, and the heat will be blown into the top of the rock bed. This is called *charging* the storage bin. The rocks near the top will be heated first and will absorb the most heat.

When heat is not being collected and there is a call for heat from the house, dampers A and B will close to the collector, C will open to the house duct, and the furnace fan will activate. The airflow will be reversed; that is, it will now be from the bottom of the rock bed to the top, forcing out the hottest air, which will be distributed throughout the house by the furnace fan.

When heat is being collected and there is a simultaneous demand for heat in the house, dampers A and B will close to the storage, and the collector fan and furnace fan will start blowing heated air from the collectors directly to the furnace distribution system, returning through damper C and the solar duct system to the collector fan and back to the collectors. If the solar collectors and storage cannot produce enough heat, the conventional furnace will activate to satisfy the demand.

Liquid Collection–Water Storage–Air Distribution to House–Auxiliary Heat by Conventional Forced-Warm Air Furnace

This system has collectors, a water storage tank, a liquid-to-air heat exchanger, and centrifugal pumps and uses a conventional forced-warm air furnace to distribute the heat and for auxiliary heat. It also has a domestic hot water tank with a heat exchanger in the storage tank, Figure 33–73.

When sensors indicate that heat is ready for collection at the collector and if the house thermostat is not calling for heat, the controller will activate pump A and circulate the water directly into the storage tank from the collectors. All liquids in this tank would be water because the collector liquid mixes with the storage liquid. Water is pumped through the collectors as long as the collectors are absorbing heat at a higher temperature than the temperature of the storage water. Should there be a call for heat from the house thermostat, the controller will activate circulator pump B and pump the heated storage water through the heat exchanger.

The liquid-to-air heat exchangers for these systems may be located in the return air duct, just as it enters the furnace, or above the heat source as shown in Figure 33–73. A two-stage thermostat is often used in these systems. If the solar heat cannot supply the heat needed, the second stage of the thermostat will activate the conventional furnace. It may be designed so that both systems are operating simultaneously because some heat from the solar radiation will help. When this solar heat drops below a predetermined temperature, this system will cut off and the auxiliary heat will take over.

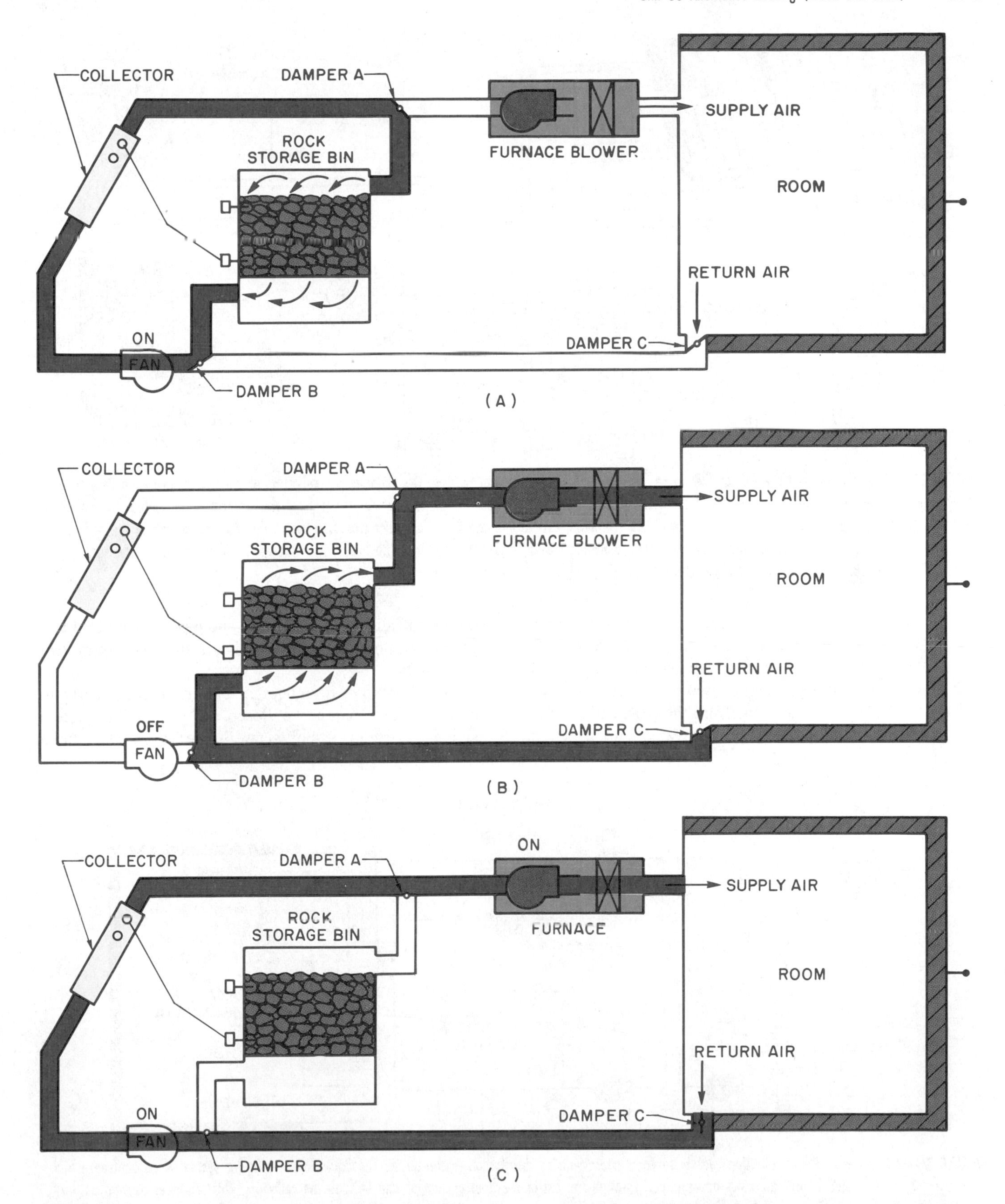

FIGURE 33–72 (A) When collectors are charging the storage bin, damper A closes airflow to the furnace and damper B allows the air to flow through the fan back to the collector. (B) When the room thermostat calls for heat and the collector temperature is not sufficient to supply the heat, the furnace blower will come on; damper A closes airflow from the collector allowing air to flow from the rock storage to the room and return to the storage. (C) When the house thermostat is calling for heat and the collectors are producing sufficient heat, dampers A and B will close to the storage, providing a path directly from the collectors to the house. Return air damper C will be open, providing a path for the air back to the collectors.

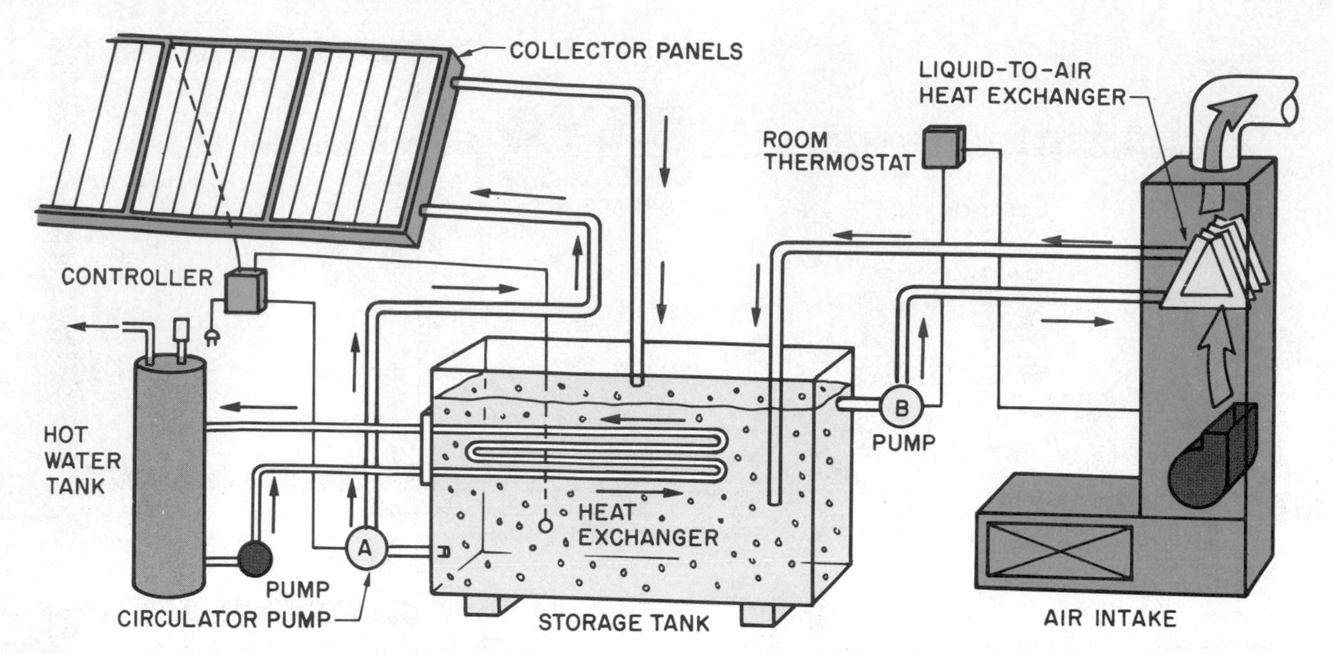

FIGURE 33–73 Collectors use water as the liquid. This is a drain-down system because the water in the collectors would freeze on cold nights. When collectors are producing heat at a higher temperature than the storage water, circulator pump A pumps water from the storage to be heated and from the collectors back to storage. When the house thermostat calls for heat, circulator pump B pumps water from storage to the heat exchanger in the furnace duct. The furnace fan blows air across the heat exchanger, warming the air, and distributes it to the house. Also shown is a domestic hot water tank, heated through a heat exchanger in the water storage.

This solar system is called a *drain-down* system. When circulator pump A shuts down, the collector water will all drain to a heated area to keep the water from freezing in cold weather. Any freezing can cause considerable damage to the system. Note that the pipe from the collector into the storage tank does not extend into the water. This provides venting and allows the collectors and piping to drain. If this space is not provided, other venting in the collector system is necessary.

An alternative design often preferred, particularly in colder climates, requires a closed collector piping system using antifreeze instead of water, Figure 33–74. Note that a

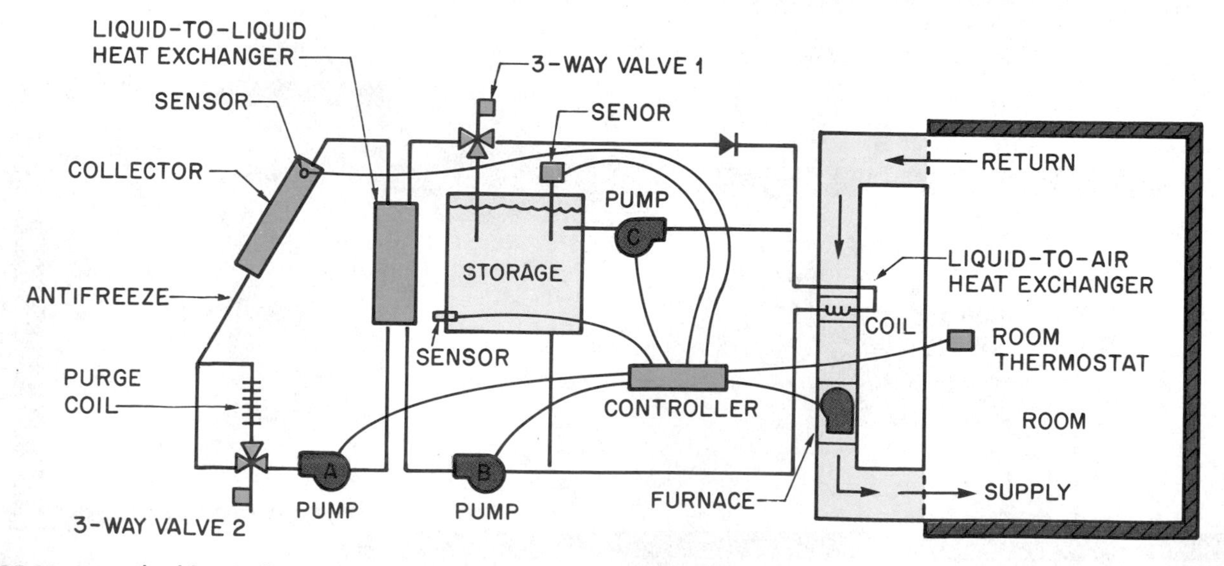

FIGURE 33–74 Closed liquid collection, water storage system with air distribution to the house. The liquid in the collector system is an antifreeze and water solution. It is heated at the collectors and pumped through the coil at the heat exchanger and back to the collectors. This continues as long as the collectors are absorbing heat at a predetermined temperature. When the room thermostat is calling for heat, three-way valve 1 allows the water to be circulated by pump B through the liquid to air heat exchanger and back to the liquid-to-liquid heat exchanger where it absorbs more heat. If the room thermostat is not calling for heat, three-way valve 1 diverts the water to storage. If the storage temperature has reached a predetermined temperature, the collector solution is circulated through the purge coil. The solution is diverted to this purge coil through three-way valve 2. (The heat is dissipated into the air outside through this coil.) This is a necessary safety precaution because it is possible for the collector systems to overheat and damage the equipment.

heat exchanger is used to heat the storage water. The collector antifreeze fluid must be kept separated from the storage water. This system is less efficient than the drain-down system, which does not require the extra heat exchanger.

Many closed collector systems such as this are designed with a *purge coil.* On a hot day when the storage is fully charged with heat and the room thermostat is not calling for heat, the solution in the collector system may overheat. This could cause the safety valve to open, resulting in loss of solution and overheating, which will damage the equipment. When a predetermined temperature is reached, the controller diverts the solution through the three-way valve 2 to the purge coil. This is normally a finned-tube coil and dissipates the heat from the system to the air.

Liquid Collection—Water Storage—Water Distribution—Auxiliary Heat by Conventional Hot Water Boiler

A liquid-based solar collector system may be paired with a hot water finned-tube convector heating furnace. The collector system can be of the same design as an air distribution system. It can be a drain-down water design or one that uses antifreeze with a heat exchanger in the storage unit. Solar collector and storage temperatures are lower than those produced by a hot water boiler. This can be compensated for to some extent by using more fin tubing than would be used with a conventional hot water system.

Figure 33–75(A) illustrates a design using a liquid-based solar collector with a hot water boiler system. This

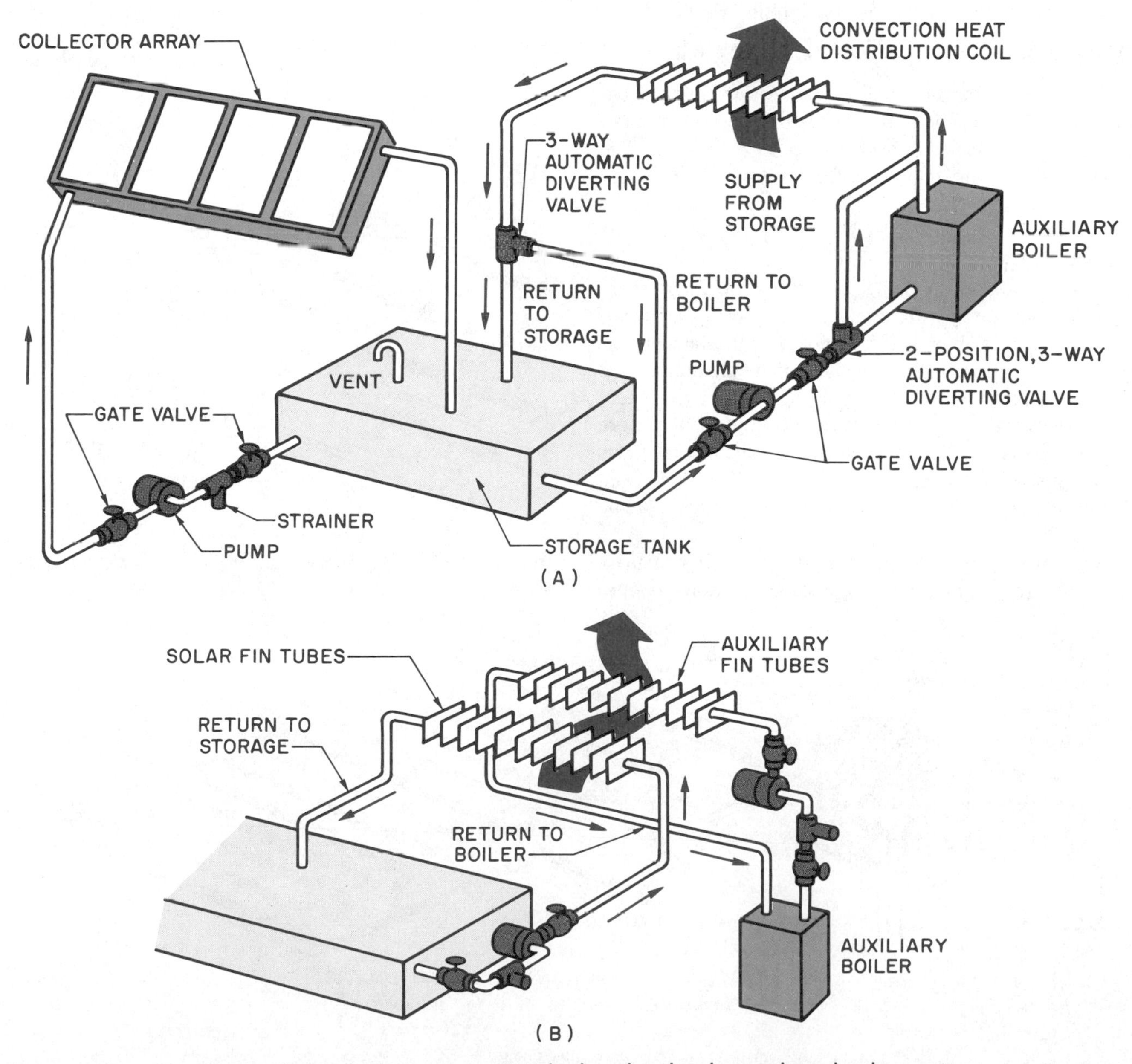

FIGURE 33–75 (A) Liquid-to-liquid solar space-heating system. (B) Both solar and auxiliary heat can be used at the same time.

design uses the auxiliary boiler when the storage water is not hot enough. If the solar storage is hot enough, the boiler does not operate. Water is pumped from the storage tank and bypasses the boiler. It goes directly to the baseboard fin tubing. The three-way *diverting valve* automatically causes this boiler bypass. When auxiliary heat is called for, the two three-way diverting valves cause the water to bypass the storage. It would be very inefficient to pump heated water through the storage tank.

This system can be designed as two separate systems in parallel as in Figure 33–75(B). These separate systems can both be operated at the same time. With this design the solar system can help the auxiliary system when the temperature of the storage water is below that which would be satisfactory for it to operate alone. Sensors are used in either system to tell the controller when to start and stop the pumps and when to control the diverting valves.

Water Collection–Water Storage-Radiant Heat–Conventional Furnace Auxiliary Heat (any type)

This system is appropriate for use in a solar heating application, Figure 33–76. The heating coils may be imbedded in concrete in the floor or in plaster in ceilings or walls. The normal surface temperature for floor heating is 85°F, for wall or ceiling panels 120°F. The same type controls, pumps, and valves are used with this system as with the baseboard hot water system.

Floor installations are most common. The coils are imbedded in concrete approximately 1 in. below the surface. If this is a concrete slab on grade, it should be insulated underneath. In ceiling and wall applications there should be coils for each room to be heated. Centering the coils in walls, ceiling joists, or floor joists between rooms to heat more than one area is usually not satisfactory. When installations are on outside walls they should be insulated very well between the coils and outside surfaces to prevent extreme heat loss.

Radiant heating installed in the floor is very comfortable and there is little temperature variation from room to room. Figure 33–77 illustrates a polybutylene pipe being installed in a floor. The pipe is tied to a wire mesh and propped on blocks to keep it near the top of the concrete slab and the concrete is poured around it, covering it by approximately 1 in. There should be no joints in this pipe in the floor. Polybutylene pipe (plastic) is popular for this installation. Any joints necessary should be manifolded outside the slab. Almost any type of flooring material may be used over the slab. Common materials are tile, wood, and carpeting. A controller is installed with sensors at the collectors, the water storage, and outdoors. It is also connected to the room thermostat. The controller modulates the temperature of the water being circulated in the floor according to the outside temperature and the room temperature.

Water Collection–Water Storage–Water-to-Air Heat Pump Space Heating

The solar heating systems described previously are only effective when the storage is sufficiently charged with heat

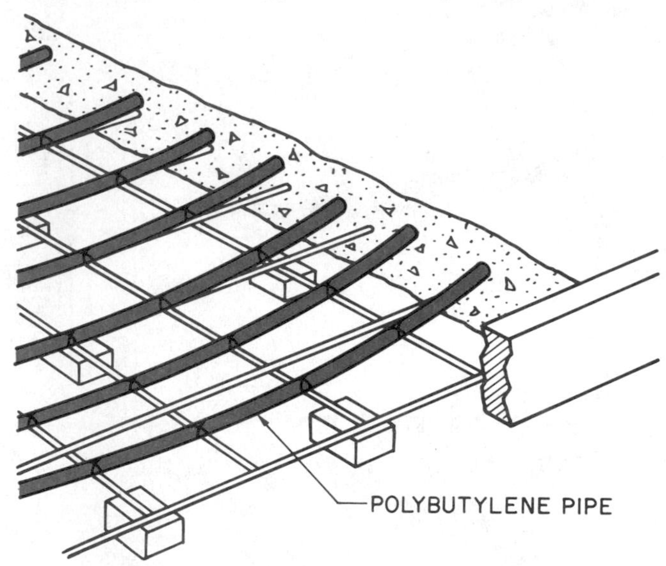

FIGURE 33–77 Polybutylene pipe being installed in a concrete slab to be used in a solar radiant heating system.

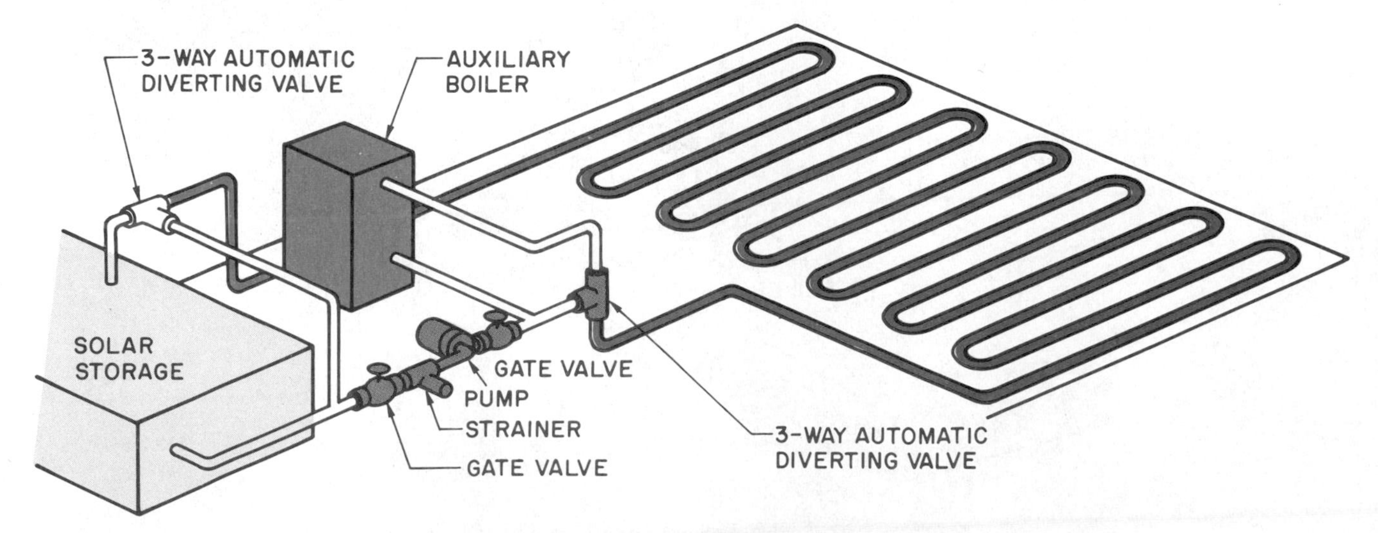

FIGURE 33–76 A solar space-heating radiant system. The radiant heating coils are normally imbedded in a concrete floor.

to provide the space-heating temperature desired. The heat required of the storage medium to be effective varies somewhat with the type system and equipment used. However, most systems are not effective when the storage temperature is below 90°F. In cold climates with cloudy weather it will not take long for the storage temperature to drop below 90°F.

A heat pump, however, can extract heat from temperatures far below 90°F. This makes it practical to connect a heat pump to a coil supplying heated storage water. If the storage water temperature is high enough, this will operate in a manner similar to the water-to-air system. If the temperature of the storage material drops below that which can effectively be used in this manner, the storage water will be diverted to the heat pump where it uses the heat available to help it provide the necessary heat for the conditioned space, Figure 33–78.

Solar Heated Domestic Hot Water

Many solar systems are designed and installed specifically to heat or to assist in heating domestic hot water. These use similar components to those used in space heating systems, Figure 33–79. The heated collector water or antifreeze is

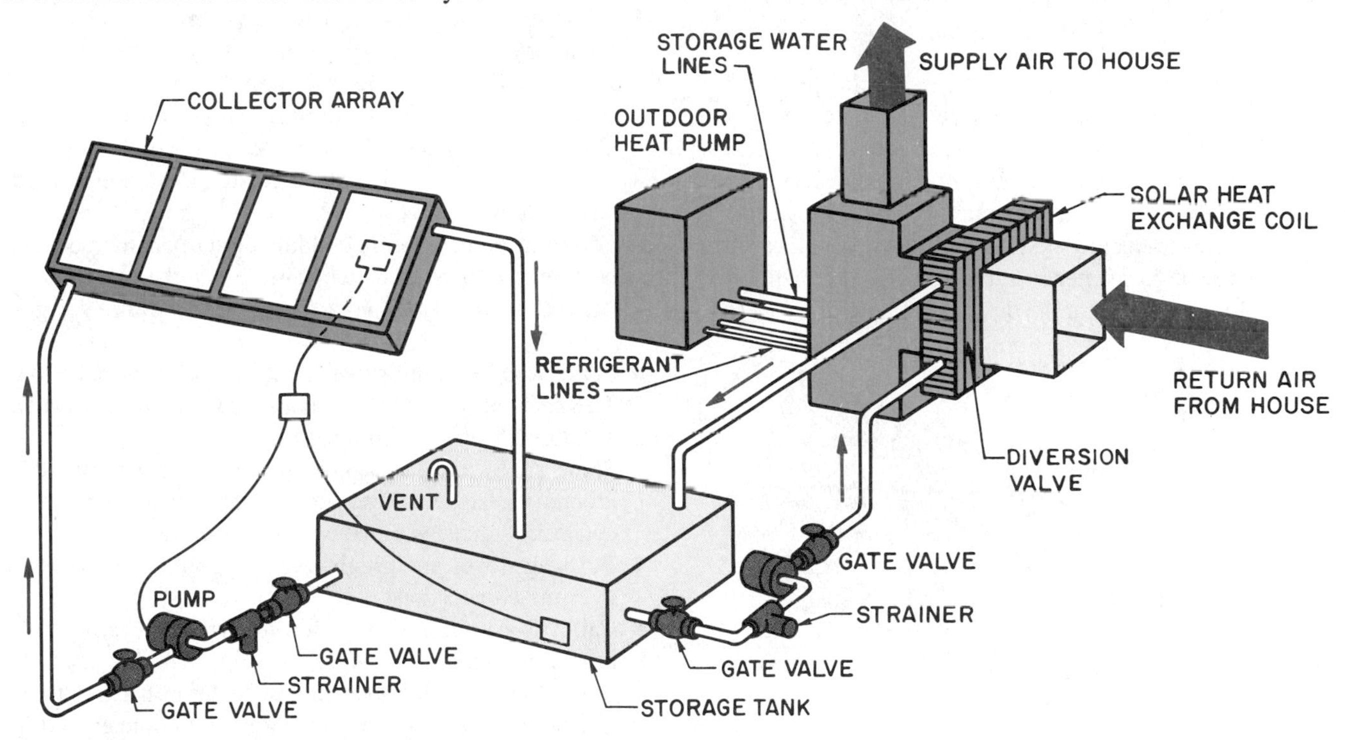

FIGURE 33–78 A liquid-based solar space-heating system.

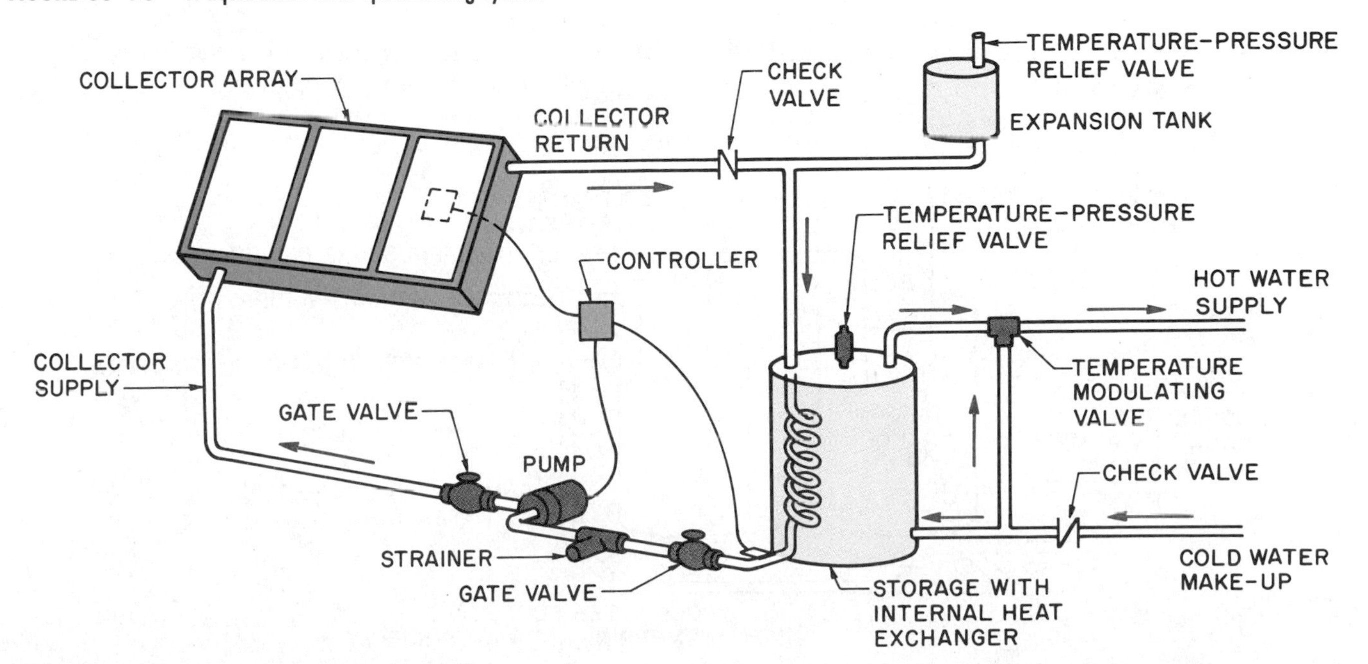

FIGURE 33–79 Solar domestic hot water system.

pumped through the heat exchanger where it provides heat to the water in the tank. This tank must have conventional heating components also because the solar heating will probably not meet the demand at all times. ***If an antifreeze solution is used, the heat exchanger must have a double wall to prevent the toxic solution from mixing with the water used in the home or business, Figure 33–80.***

SAFETY PRECAUTION: *Ethylene glycol may be used as an antifreeze if allowed by local or state codes. It is toxic and must be used with a double-wall heat exchanger. Propylene glycol is a "food grade" antifreeze and is recommended for use in a domestic hot water system.*

Many systems are designed to use a solar preheat or storage tank in conjunction with a conventional hot water tank. This gives additional solar-heated water to the system. Conventional heating (electric, gas, oil) will not have to be used as much with this extra storage, Figure 33–81.

When a space-heating system is installed, it will usually include domestic hot water heating as well. There are several ways in which this can be designed, Figure 33–73. The cold makeup water is passed through the heat exchanger in the storage tank. It absorbs whatever heat is available before entering the hot water tank, thus reducing the amount of heat that must be supplied by the conventional means.

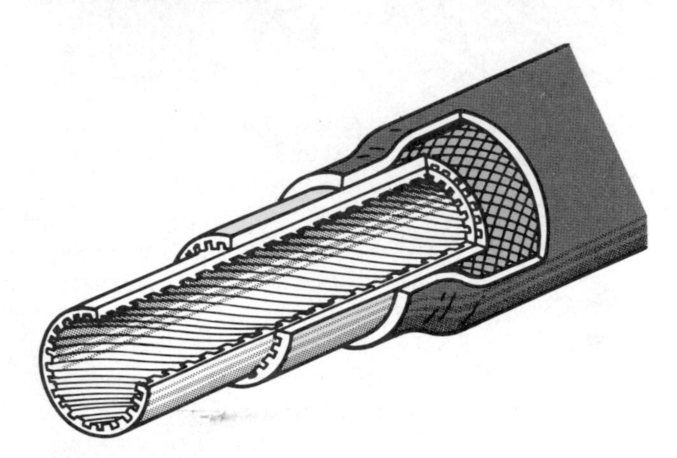

FIGURE 33–80 Heat exchanger piping has a double wall when an antifreeze solution is used in the collectors.

SUMMARY

- Wood is classified as hardwood or softwood.
- Generally, hardwoods provide more heat than softwoods.
- Wood is sold by the cord which is a stack 4 ft × 4 ft × 8 ft or 128 ft^3.
- ***Creosote is a product of incomplete combustion. It condenses in the stovepipe and chimney and can be hazardous. It can ignite and burn with intensity. The following conditions contribute to creosote buildup: smoldering low-heat fires, smoke in contact with cool surfaces in the stovepipe and chimney, burning green wood, burning softwood.***
- To help prevent creosote buildup, burn dry hardwood and be sure that there is an adequate rise in the stovepipe.
- ***Inspect and clean the stovepipe and chimney regularly.***
- Some wood-burning stoves heat primarily by radiation. The heat radiates into the room, heating objects, walls, and floor. These in turn heat the air.
- Some stoves have a jacket that allows air to circulate around the stove. These are available in both natural-convection and forced-convection designs.
- Efficient stoves are airtight, which permits more control over the combustion.
- Catalytic combustors are used to improve heating efficiency and reduce creosote.
- Stoves should be located near the center of the area to be heated, but they must be located near a chimney and in the same room as the inlet to the chimney. Avoid long runs of stovepipe.
- ***Makeup air for air used in combustion should be provided through a fresh air duct.**

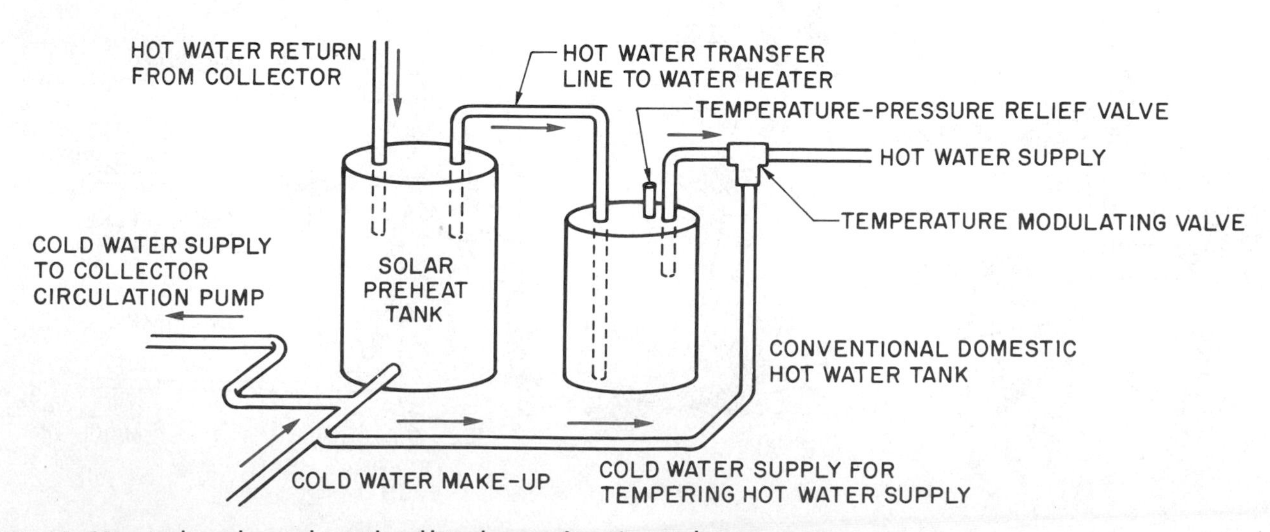

FIGURE 33–81 A solar preheat tank provides additional storage for a domestic hot water system.

- **Possible safety hazards that may be encountered with stoves:**
  - **Stovepipe and chimney fires may result from excessive creosote buildup.**
  - **Radiation may overheat walls, ceilings, or other combustible materials.**
  - **Sparks or glowing coals may land on combustible materials and ignite them.**
  - **Flames could leak out through cracks in a faulty chimney.**
  - **Burning materials coming out of the top of a chimney can ignite roofing materials or combustibles on the ground.**
- **Minimum clearances between stove and combustibles should be adhered to.**
- **Use only approved stovepipe, chimneys, and other components. Follow all manufacturer's instructions and specifications.**
- **Use smoke detectors.**
- **Install fireplace inserts only in approved fireplaces.**
- **If a stove is vented through the fireplace chimney, the damper opening must be completely sealed.***
- Typical thermostatic damper controls for stoves use a bimetal device to open and close the damper.
- Dual-fuel furnaces are two furnaces in one unit. They are normally wood furnaces with either an oil, gas, or electric furnace operating in conjunction with it.
- Two-stage thermostats are used for these furnaces. The stage for the primary heat (usually the wood) is set a few degrees above the other.
- ***Adequate combustion makeup air should be provided.***
- ***When installing the furnace and ductwork, clearances to combustibles must be maintained.***
- Passive solar designs are nonmoving parts of a building to provide or enhance heating or cooling.
- Active solar systems use electrical or mechanical devices to help collect, store, and distribute the sun's energy.
- An active system includes collectors, a means for storing heat, and a system for distributing it.
- The tilt or angle of the earth on its axis is called the angle of declination. This angle is responsible for the differences in solar radiation during the seasons.
- The amount of radiation reaching the earth also varies according to the distance it travels through the atmosphere. The shortest distance is when the sun is perpendicular (90°) to a particular surface.
- Air solar collectors heat the air, which is then circulated either to a storage component or directly to the conditioned space. The cooled air is then returned to the collector where the heating and circulation process is continued.
- Liquid solar collectors circulate water or an antifreeze solution through coils where the liquid is heated by the sun's radiation. It is then pumped to a storage tank or to a heat exchanger where the heat is distributed to the space to be heated.
- Refrigerant-charged collection systems use the latent heat absorbed while boiling, which is released while condensing in the indoor coil.
- Collectors are positioned to face south. They are tilted to collect the sun's rays at a 90° angle in winter.
- Sensible heat is normally stored in rocks or in water.
- Latent heat is stored in a salt or wax material.
- Latent-heat storage uses phase-change materials. These materials do not require as high a temperature as water or air and use less space, but they form a more expensive system.
- Other solar system components include heat exchangers, dampers, package air handlers, sensors, valves, pumps, and expansion tanks.
- Many combinations of components may be used for a complete solar space-heating system.

REVIEW QUESTIONS

1. When is it practical to burn wood for home heating?
2. Describe the organic makeup of wood.
3. List three hardwoods.
4. List two softwoods.
5. What are the primary causes of creosote buildup?
6. Why is creosote buildup dangerous?
7. What procedures can help prevent excessive creosote buildup?
8. Why should stovepipe be assembled with the crimped end down?
9. Why are airtight stoves preferred over those with air leaks?
10. Why are some stoves designed with one or more internal baffles?
11. What factors should be considered when locating a stove in a house?
12. Sketch the paths of air heated by (a) a stove in a room; (b) in a room with a room above it; (c) in a room with a room beside it.
13. What is the difference between a catalytic and a noncatalytic stove?
14. Describe the operation of a catalytic combustor.
15. How can air used for combustion in a room be made up?
16. List safety hazards that may be encountered with stoves.
17. Where is a thimble used?
18. How can a fireplace insert improve on the heating efficiency of a fireplace?
19. What are two methods for venting a stove through a fireplace or fireplace chimney?
20. What is an add-on furnace?
21. What is a dual-fuel furnace?
22. Sketch a design for getting combustion air makeup through a fresh air duct.
23. How do active solar systems differ from passive designs?
24. What is the solar constant?

25. What is the declination angle?
26. How does the declination angle affect the sun's radiation during summer and winter?
27. Describe the difference between an active air-based solar heating system and an active liquid-based solar heating system.
28. Describe how air-based collectors function.
29. Sketch a diagram of a liquid-based collector.
30. Describe how liquid-based collectors function.
31. Describe how refrigerant-charged collector systems function.
32. Why is a special selective coating often used on absorber plates?
33. How is the angle of the collector tilt determined?
34. Describe the difference between sensible-heat and latent-heat storage systems.
35. Describe the types of tanks that may be used in liquid storage systems.
36. What are the advantages of latent-heat storage?
37. Describe one use of a heat exchanger in a solar space-heating system.
38. In what kind of system are air dampers used?
39. What is the purpose of using temperature sensors?
40. Why is an expansion tank necessary in a closed liquid collection space-heating system?

UNIT 34

Air Humidification and Filtration

OBJECTIVES

After studying this unit, you should be able to

- explain relative humidity.
- list reasons for providing humidification in winter.
- discuss the differences between evaporative and atomizing humidifiers.
- describe bypass and under-duct-mount humidifiers.
- describe disc, plate, pad, and drum humidifier designs and the media used.
- explain the operation of the infrared humidifier.
- explain why a humidifier used with a heat pump or electric furnace may have its own independent heat source.
- describe the spray-nozzle and centrifugal atomizing humidifiers.
- state the reasons for installing self-contained humidifiers.
- list general factors used when sizing humidifiers.
- describe general procedures for installing humidifiers.
- explain why cleaning air in buildings is necessary.
- list five types of air-filtering or purification materials or devices.

SAFETY CHECKLIST

* Use all electrical safety precautions when servicing or troubleshooting electrical/electronic circuits. Use common sense.
* When replacing or servicing line voltage components ensure that power is off, that the panel is locked and tagged, and that you have the only key.

34.1 RELATIVE HUMIDITY

In fall and winter, homes often are dry because cold air from outside infiltrates the conditioned space. The infiltration air in the home is artificially dried out when it is heated because it expands, spreading out the moisture. The amount of moisture in the air is measured or stated by a term called *relative humidity.* It is the percentage of moisture in the air compared to the capacity of the air to hold moisture. In other words, if the relative humidity is 50%, each cubic foot of air is holding one half the moisture it is capable of holding. The relative humidity of the air decreases as the temperature increases because air with higher temperatures can hold more moisture. When a cubic foot of 20°F outside air at 50% relative humidity is heated to room temperature (75°F), the relative humidity of that air drops.

For comfort, the dried-out air should have its moisture replenished. The recommended relative humidity for a home is between 40% and 60%. Indoor air quality (IAQ) has become increasingly important. When the relative humidity varies above or below these limits, studies have shown that bacteria, viruses, fungi, and other organisms become more active. The dry warm air draws moisture from everything in the conditioned space, including carpets, furniture, woodwork, plants, and people. Furniture joints loosen, nasal and throat passages dry out, and skin becomes dry. Dry air causes more energy consumption than necessary because the air gets moisture from the human body through evaporation from the skin. The person then feels cold and sets the thermostat a few degrees higher to become comfortable. With more humidity in the air, a person is more comfortable at a lower temperature.

Static electricity is also much greater in dry air. It produces discomfort because of the small electrical shock a person receives when touching something after having walked across the room.

34.2 HUMIDIFICATION

Years ago people placed pans of water on radiators or on stoves. They even boiled water on the stove to make moisture available to the air. The water evaporated into the air and raised the relative humidity. Although this may still be done in some homes, efficient and effective equipment called *humidifiers* produce this moisture and make it available to the air by *evaporation.* The evaporation process is speeded up by using power or heat or by passing air over large areas of water. The area of the water can be increased by spreading it over pads or by atomizing it.

34.3 HUMIDIFIERS

Evaporative humidifiers work on the principle of providing moisture on a surface called a *media* and exposing it to the dry air. This is normally done by forcing the air through or around the media and picking up the moisture from the media as a vapor or a gas. There are several types of evaporative humidifiers.

BYPASS HUMIDIFIER. The bypass humidifier relies on the difference in pressure between the supply (warm) side of the furnace and the return (cool) side. It may be mounted on either the supply plenum or duct or the cold air return plenum or duct. Piping must be run from the plenum or duct where it is mounted to the other plenum or duct. If mounted in the supply duct, it must be piped to the cold air return,

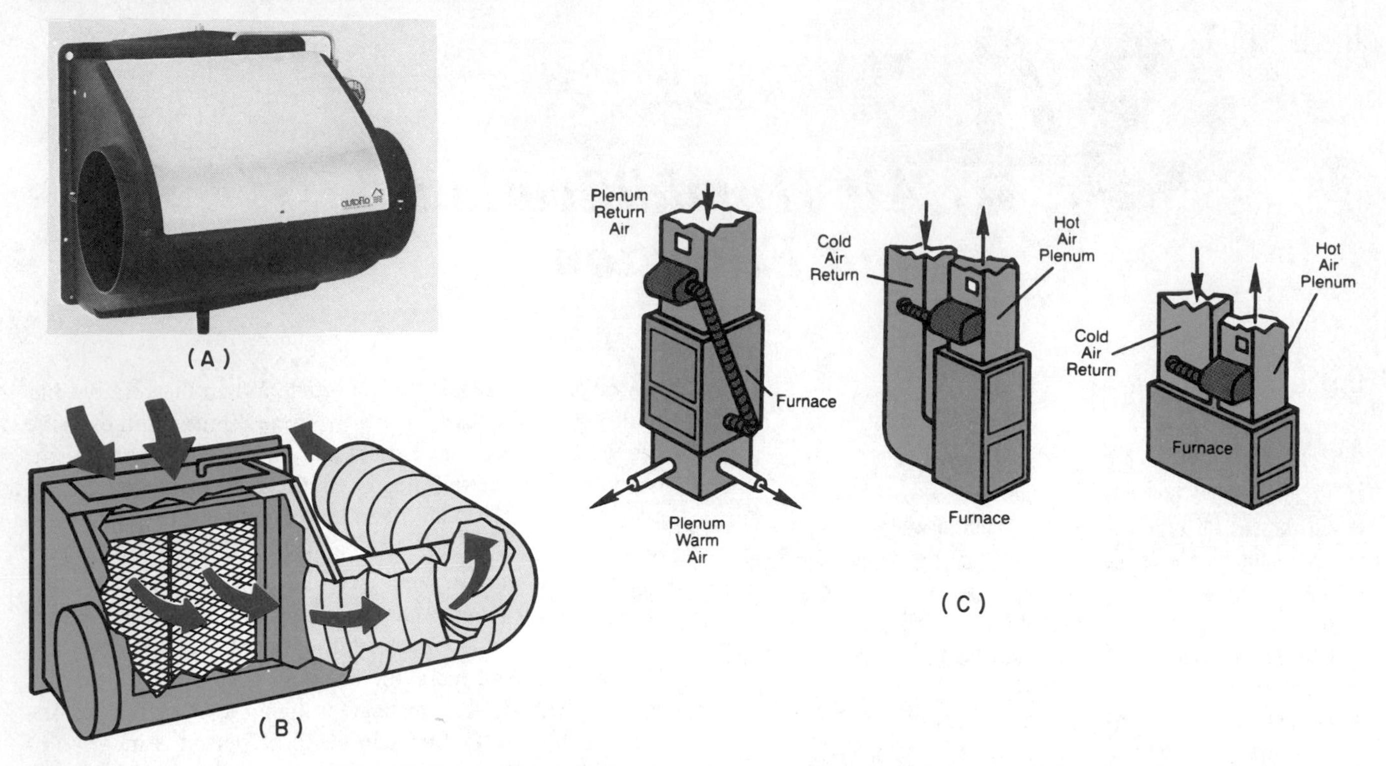

FIGURE 34–1 (A) Bypass humidifier. (B) Cutaway showing airflow from plenum through media to the return. (C) Typical installations. *Courtesy AutoFlo Company*

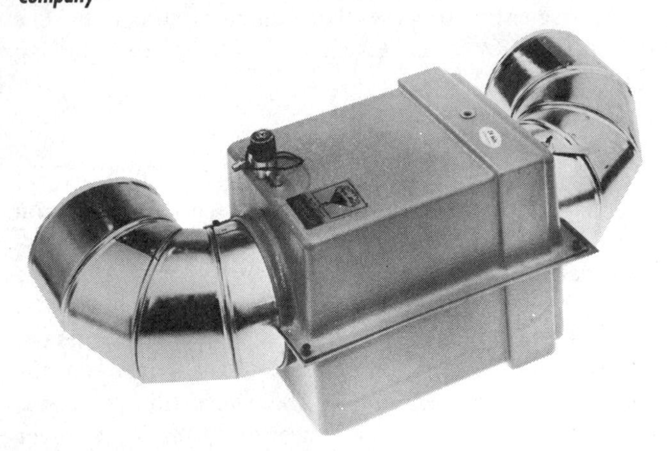

FIGURE 34–2 Bypass humidifier used between plenums with a pipe to each. *Courtesy Aqua-Mist, Inc.*

FIGURE 34–3 A plenum-mounted humidifier with a plate-type media that absorbs water from the reservoir and evaporates into the air in the plenum. *Courtesy AutoFlo Company*

Figures 34–1 and 34–2. The difference in pressure between the two plenums draws some air through the humidifier to the supply duct and is distributed throughout the house.

PLENUM-MOUNT HUMIDIFIER. The plenum-mount humidifier is mounted in the supply plenum or the return air plenum. The furnace fan forces air through the media where it picks up moisture. The air and moisture are then distributed throughout the conditioned space, Figure 34–3.

UNDER-DUCT-MOUNT HUMIDIFIER. The under-duct-mount humidifier is mounted on the underside of the supply duct so that the media is extending into the airflow where moisture is picked up in the airstream. Figure 34–4 illustrates an under-duct humidifier.

34.4 HUMIDIFIER MEDIA

Humidifiers are available in several designs with various kinds of media. Figure 34–4 is a photo of a type using disc

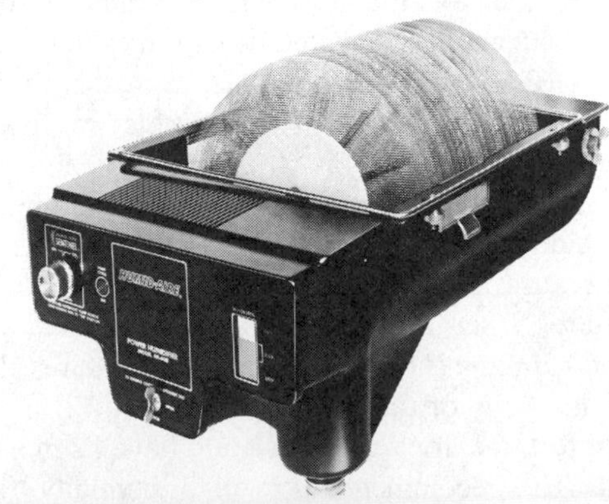

FIGURE 34–4 Under-duct humidifier using disc screens as the media. *Courtesy Humid Aire Division, Adams Manufacturing Company*

FIGURE 34–5 Under-duct humidifier using drum-style media. *Courtesy Herrmidifier Company, Inc.*

screens mounted at an angle. These discs are mounted on a rotating shaft causing the slanted discs to pick up moisture from the reservoir. The moisture is then evaporated into the moving airstream. The discs are separated to prevent electrolysis, which causes the minerals in the water to form on the media. The wobble from the discs mounted at an angle washes the minerals off and into the reservoir. The minerals can then be drained from the bottom of the reservoir.

Figure 34–5 illustrates a type of media in a drum design. A motor turns the drum, which picks up moisture from the reservoir. The moisture is then evaporated from the drum into the moving airstream. The drums can be screen or sponge types.

A plate- or pad-type media is shown in Figure 34–3. The plates form a wick that absorbs water from the reservoir. The airstream in the duct or plenum causes the water to evaporate from the wicks or plates.

Figure 34–6 shows an electrically heated water humidifier. In electric furnace and heat pump installations, the temperature in the duct is not as high as in other types of

FIGURE 34–6 Humidifier with electric heating elements. *Courtesy AutoFlo Company*

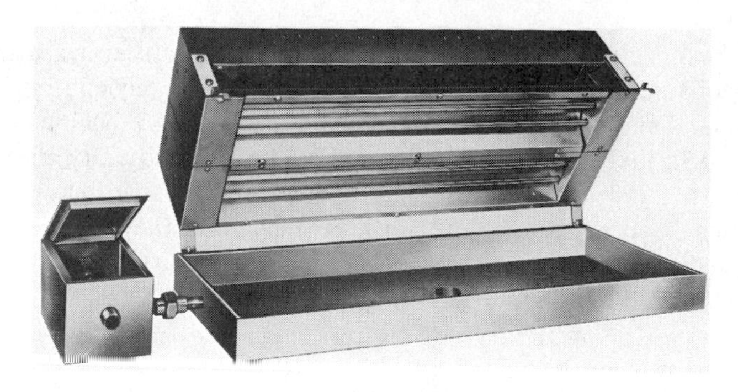

FIGURE 34–7 Infrared humidifier. *Courtesy Humid Aire Division, Adams Manufacturing Company*

hot-air furnaces. Media evaporation is not as easy with lower temperatures. The electrically heated humidifier heats the water with an electric element, causing it to evaporate and be carried into the conditioned space by the airstream in the duct.

Another type is the infrared humidifier, Figure 34–7, which is mounted in the duct and has infrared lamps with reflectors to reflect the infrared energy onto the water. The water thus evaporates rapidly into the duct airstream and is carried throughout the conditioned space. This action is similar to the sun's rays shining on a large lake and evaporating the water into the air.

Humidifiers are generally controlled by a *humidistat*, Figure 34–8. The humidistat controls the motor and the

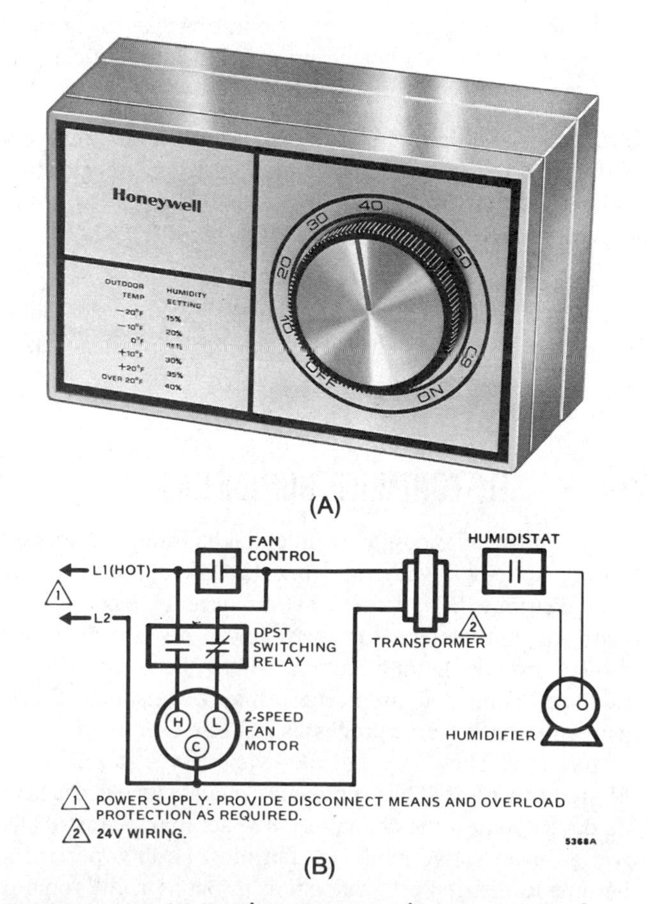

FIGURE 34–8 (A) Humidistat. (B) Wiring diagram. *Courtesy of Honeywell, Inc., Residential Division*

heating elements in the humidifier. The humidistat has a moisture-sensitive element, often made of hair or nylon ribbon. This material is wound around two or more bobbins and shrinks or expands, depending on the humidity. Dry air causes the element to shrink, which activates a snap-action switch and starts the humidifier. Many other devices are used, including electronic components that vary in resistance with the humidity.

34.5 ATOMIZING HUMIDIFIERS

Atomizing humidifiers discharge tiny water droplets (mist) into the air, which evaporate very rapidly into the duct airstream or directly into the conditioned space. These humidifiers can be *spray-nozzle* or *centrifugal* types, but they should not be used with hard water because it contains minerals (lime, iron, etc) that leave the water vapor as dust and will be distributed throughout the house or building. Eight to 10 grains of water hardness is the maximum recommended for atomizing humidifiers.

The spray-nozzle type sprays water through a metered bore of a nozzle into the duct airstream where it is distributed to the occupied space. Another type sprays the water onto an evaporative media where it is absorbed by the airstream as a vapor. They can be mounted in the plenum, under the duct, or on the side of the duct. It is generally recommended that atomizing humidifiers be mounted on the hot-air or supply side of the furnace. Figures 34–9 and 34–10 illustrate two types.

The centrifugal atomizing humidifier uses an impeller or slinger to throw the water and break it into particles that are evaporated in the airstream, Figure 34–11.

SAFETY PRECAUTIONS: *Atomizing humidifiers should operate only when the furnace is operating, or moisture will accumulate and cause corrosion, mildew, and a major moisture problem where it is located.*

Some models operate with a thermostat that controls a solenoid valve turning the unit on and off. The furnace must be on and heating before this type will operate. Others, wired in parallel with the blower motor, operate with it. Most are also controlled with a humidistat.

34.6 SELF-CONTAINED HUMIDIFIERS

Many residences and light commercial buildings do not have heating equipment with ductwork through which the heated air is distributed. Hydronic heating systems, electric baseboard, or unit heaters, for example, do not use ductwork.

To provide humidification where these systems are used, self-contained humidifiers may be installed. These generally use the same processes as those used with forced-air furnaces. They may use the evaporative, atomizing, or infrared processes. These units may include an electric heating device to heat the water, or the water may be distributed over an evaporative media. A fan must be incorporated in the unit to distribute the moisture throughout the room or area. Figure 34–12 illustrates a drum type. A design using

FIGURE 34–9 Combination spray-nozzle and evaporative pad humidifier. *Courtesy Aqua-Mist, Inc.*

FIGURE 34–10 Atomizing humidifier. *Courtesy AutoFlo Company*

FIGURE 34–11 Centrifugal humidifier. *Courtesy Herrmidifier Company, Inc.*

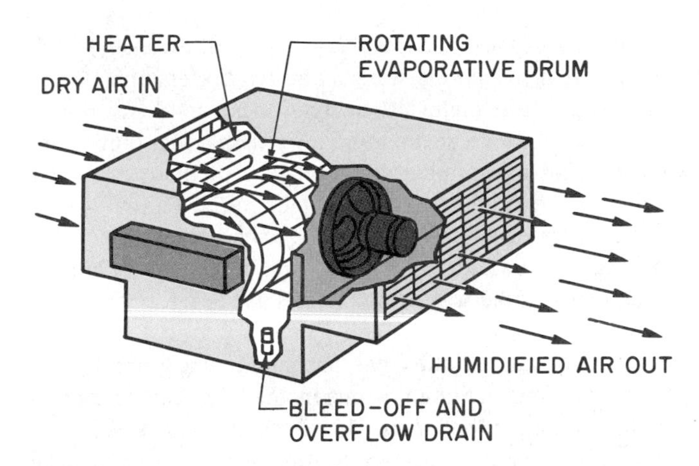

FIGURE 34–12 Drum-type self-contained humidifier.

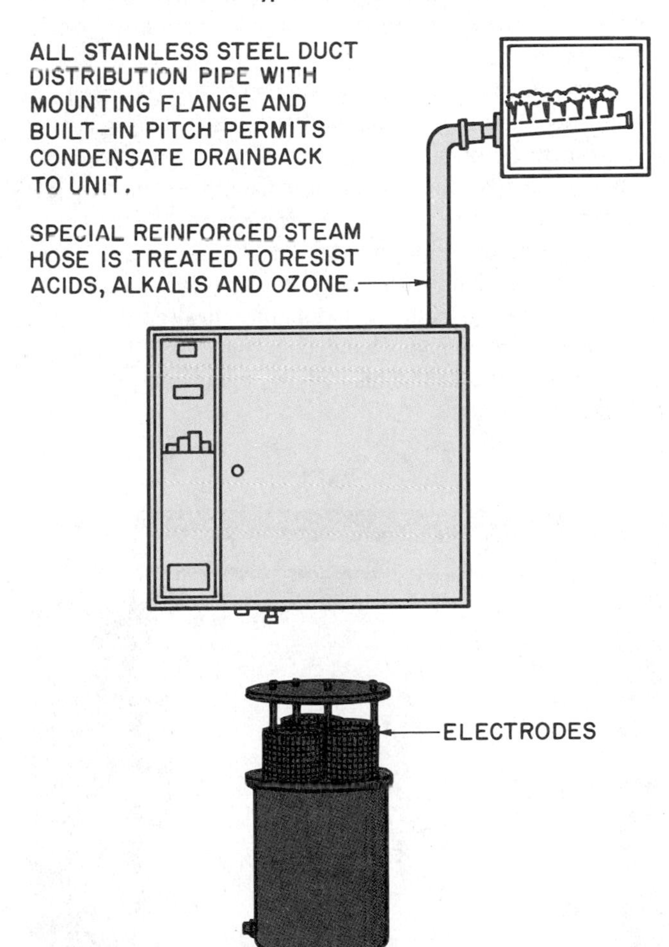

FIGURE 34–13 Self-contained steam humidifier.

steam is shown in Figure 34–13. In this system the electrodes heat the water, converting it to steam. The steam passes through a hose to a stainless steel duct. Steam humidification is also used in large industrial applications where steam boilers are available. The steam is distributed through a duct system or directly into the air.

34.7 PNEUMATIC ATOMIZING SYSTEMS

Pneumatic atomizing systems use air pressure to break up the water into a mist of tiny droplets and disperse them.

SAFETY PRECAUTION: *These systems as well as other atomizing systems should only be applied where the atmosphere does not have to be kept clean or where the water has a very low mineral content because the minerals in the water are also dispersed in the mist throughout the air. The minerals fall out and accumulate on surfaces in the area. These are often used in manufacturing areas, such as textile mills.*

34.8 SIZING HUMIDIFIERS

The proper size humidifier should be installed. This text emphasizes installation and service, so the details for determining the size or capacity of humidifiers will not be covered. However, the technician should be aware of some general factors involved in the sizing process:

1. The number of cubic feet of space to be humidified. This is determined by taking the number of heated square feet of the house and multiplying it by the ceiling height. A 1500-ft^2 house with an 8-ft ceiling height would have 1500 $ft^2 \times 8$ ft $= 12{,}000$ ft^3.
2. The construction of the building. This includes quality of insulation, storm windows, fireplaces, building "tightness," and so on.
3. The amount of air change per hour and the approximate lowest outdoor temperature.
4. The level of relative humidity desired.

34.9 INSTALLATION

The most important factor regarding installation of humidifiers is to follow the manufacturer's instructions. Evaporative humidifiers are often operated independent of the furnace. It is normally recommended that they be controlled by a humidistat, but it does no real harm for them to operate continuously, even when the furnace is not operating. Atomizing humidifiers, however, should not operate when the furnace and blower are not operating. Moisture will accumulate in the duct if allowed to do so.

Particular attention should be given to clearances within the duct or plenum. The humidifier should not exhaust directly onto air conditioning coils, air filters, electronic air cleaners, blowers, or turns in the duct.

If mounting on a supply duct, choose one that serves the largest space in the house. The humid air will spread throughout the house, but the process will be more efficient when given the best distribution possible.

Plan the installation carefully, including locating the humidifier, as already discussed, and providing the wiring and plumbing (with drain). A licensed electrician or plumber must provide the service where required by code or law.

SERVICE, TROUBLESHOOTING, AND PREVENTIVE MAINTENANCE

Proper service, troubleshooting, and preventive maintenance play a big part in keeping humidifying equipment operating efficiently. Cleaning the components that are in contact with the water is the most important factor. The frequency of cleaning depends on the hardness of the water: the harder the water, the more minerals in the water. In evaporative systems these minerals collect on the media, on other moving parts, and in the reservoir. In addition, algae, bacteria, and virus growth can cause problems, even to the extent of blocking the output of the humidifier. Algicides can be used to help neutralize algae growth. The reservoir should be drained regularly if possible, and components, particularly the media, should be cleaned periodically.

HUMIDIFIER NOT RUNNING. When the humidifier doesn't run, the problem is usually electrical, or a component is bound tight or locked due to a mineral buildup. A locked condition may cause a thermal overload protector to open. Using troubleshooting techniques described in other units in this text, check overload protection, circuit breakers, humidistat, and low-voltage controls if there are any. Check the motor to see if it is burned out. Clean all components and disinfect.

EXCESSIVE DUST. If excessive dust is caused by the humidifier, the dust will be white due to mineral buildup on the media. Clean or replace the media. If excessive dust occurs in an atomizing humidifier, the wrong equipment has been installed.

WATER OVERFLOW. Water overflow indicates a defective float-valve assembly. It may need cleaning, adjusting, or replacing.

MOISTURE IN OR AROUND DUCTS. Moisture in ducts is found only in atomizing humidifiers. ***Remember, this equipment should operate only when the furnace operates.*** Check the control to see if it operates at other times, such as COOL or FAN ON modes. A restricted airflow may also cause this problem.

LOW OR HIGH LEVELS OF HUMIDITY. If the humidity level is too high or too low, check the calibration of the humidistat by using a sling psychrometer. If it is out of calibration, it may be possible to adjust it. Ensure that the humidifier is clean and operating properly.

34.11 INDOOR AIR QUALITY

Energy conservation, air pollution, and other concerns have resulted in newer buildings being constructed with as little air infiltration from the outside as possible. This means that air within a building carrying dust, dirt, smoke, and other contaminants is recirculated many times. These impurities may include calcium carbonate, sodium chloride, and copper oxide. These contaminants must be "caught" or the air will get progressively dirty and impure, resulting in an unhealthy indoor environment. If reservoir-type humidifiers are not maintained properly, fungi, bacteria, algae, and viruses can collect and be discharged into the air. This is especially true if there are lengthy periods of time when the system is not used and the water stagnates. Steam-type humidifiers, in which the water is heated hot enough to produce steam, often alleviate many of these problems. Figure 34–14 illustrates a steam humidifier with a microprocessor control and information source. Many of these types use water as the conductor between electrodes, Figure 34–13. As the current passes through the water, it heats it to the boiling temperature, producing steam. In this type of humidifier, many of the water-borne minerals are left behind; and the heat kills the bacteria and viruses.

Indoor air pollution has become a major problem in many buildings. Dilution is one method that can be used to control indoor air pollution. This air dilution can be achieved by allowing more natural infiltration of outside air into the building or by using mechanical equipment to force outside air inside. However, these methods reduce the energy efficiency of the heating and cooling equipment.

One design of mechanical equipment that has been developed to provide more dilution consists of two insulated ducts run to an outside wall. One duct provides for fresh outside air to enter the building and the other exhausts stale polluted air to the outside. A desiccant-coated heat-transfer disc rotates between the two airstreams. In winter this disc recovers heat and moisture from the exhaust air, transferring it to the incoming fresh air. In summer, heat and moisture are removed from incoming air and transferred to the exhaust air. Figure 34–15 illustrates how this equipment can be installed.

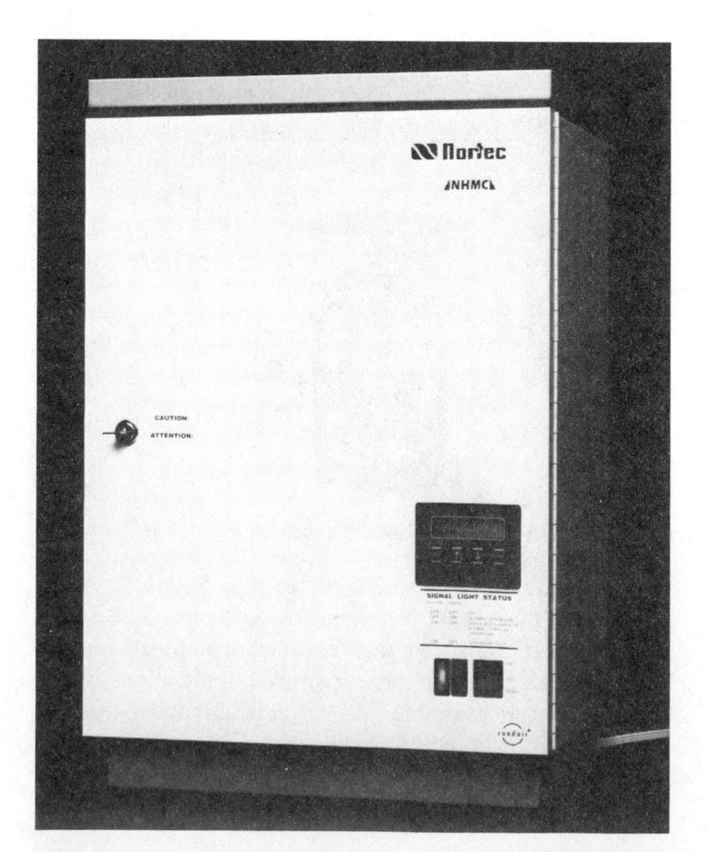

FIGURE 34–14 Steam humidifier with microprocessor control.
Courtesy Nortec

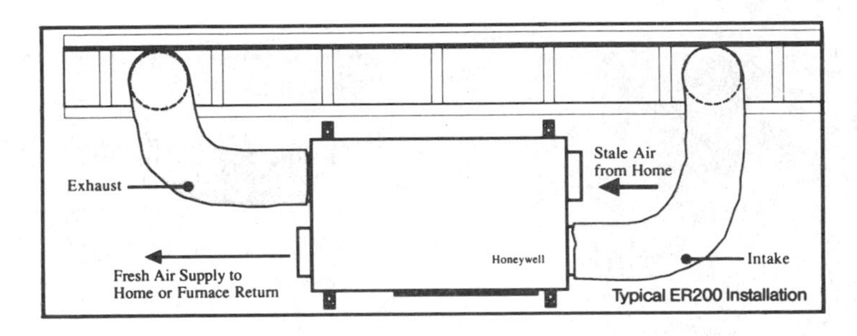

FIGURE 34–15 Air dilution system with energy recovery. *Courtesy Honeywell, Inc.*

FIGURE 34–16 Fiberglass filter media purchased in bulk. *Courtesy W. W. Grainger, Inc.*

FIGURE 34–17 Fiberglass filter media purchased in a frame. *Courtesy W. W. Grainger, Inc.*

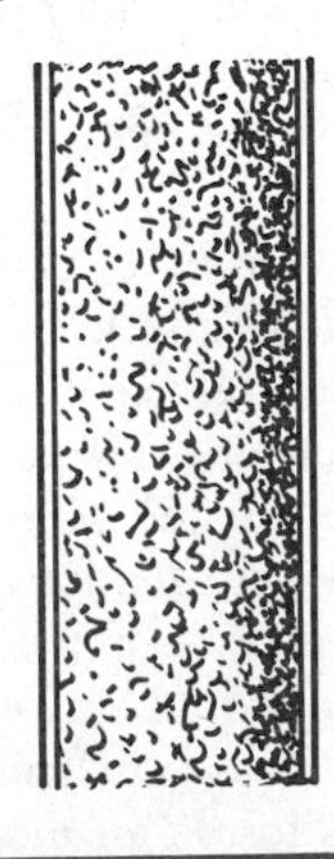

FIGURE 34–18 Section of fiberglass media.

Other methods used to help control indoor air pollution are to filter the air with one or more types of filters, by using electrostatic precipitators, often called electronic air cleaners, or by using an electronic air purification system.

Fiberglass Filter Media

This media can be purchased in bulk, Figure 34–16, or in frames, Figure 34–17. This material is usually 1 in. thick and coated with a special nondrying, nontoxic adhesive on each fiber. Many high-density types are designed to remove up to 90% of dust and pollen in the air. This filter material is designed so that it gets progressively dense as the air passes through it, Figure 34–18. These filters must be placed in the filter rack correctly, with the arrow pointing in the direction of the airflow. This allows the larger particles to be caught first, providing better efficiency. These filters must be changed regularly and be disposed of.

Extended Surface Air Filters

Some applications do not permit the use of fiberglass as a filter media or require a higher air velocity than fiberglass allows. Extended surface filters are often made of nonwoven cotton, producing air cleaning efficiencies of up to three times greater and longer wear than fiberglass, Figure 34–19. This type of filter is often used in computer and electronic equipment rooms.

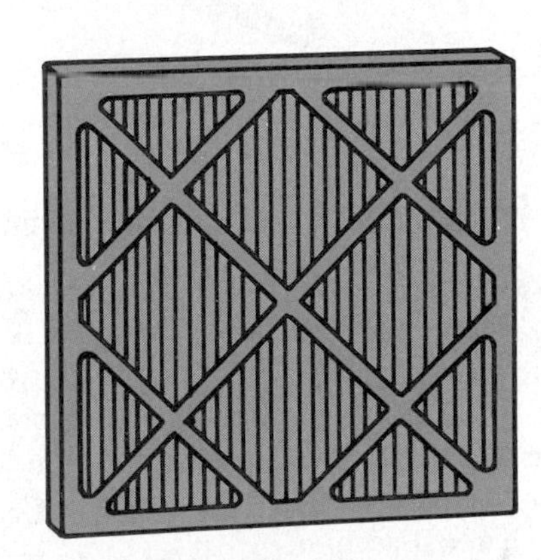

FIGURE 34–19 Extended surface air filters.

Steel Washable Air Filters

These filters are permanent and are washed rather than replaced, Figure 34–20. They are usually used in commercial applications such as in restaurants, hotels, and schools.

FIGURE 34–20 A steel washable filter. *Courtesy W. W. Grainger, Inc.*

Bag-Type Air Filters

Figure 34–21 illustrates another type of filter that produces greater filtering efficiencies. These filters use fine fiberglass media within the bags and remove microscopic particles. These may be used in hospital operating rooms, electronic equipment assembly rooms, and computer equipment rooms.

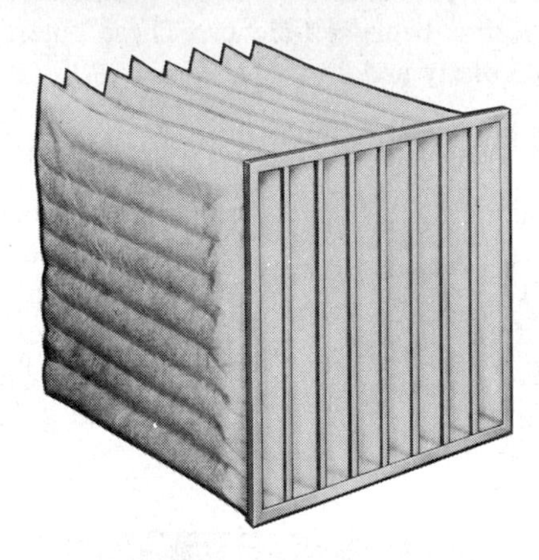

FIGURE 34–21 Bag-type filter. *Courtesy W. W. Grainger, Inc.*

Electrostatic Precipitators (Electronic Air Cleaners)

There are several types of electrostatic precipitators. These cleaners are (1) designed to be mounted at the furnace, (2) within the throw-away filter type frame, or (3) within duct systems; and (4) there are stand-alone portable systems. These systems generally have a prefilter section that filters out the larger airborne particles, an ionizing section, and some have a charcoal section.

The prefilter traps larger particles and airborne contaminants. In the ionizing section particles are charged with a positive charge. These particles then pass through a series of negative- and positive-charged plates. The charged contaminants are repelled by the positive plates and attracted to the negative plates. The air then passes through a charcoal filter, on those systems that have them, to remove many of the odors. Figure 34–22 shows some of these electrostatic filtering units.

Electronic Air Purification Devices

Other devices have been developed to help clean the air and reduce pollution. One type uses an electronic purification process, Figure 34–23. This device produces trivalent oxygen, which helps to eliminate airborne particulates and odors. Trivalent oxygen is an oxidant that is unstable and readily decomposes to ordinary oxygen.

32.12 DUCT AND EQUIPMENT CLEANING

A building may be found to be polluted to the extent that a thorough cleaning of the air-handling equipment, coils, plenums, and duct is necessary. The duct system should be disassembled and cleaned with a heavy-duty vacuum system. Coils and fans may be cleaned with an approved nonacid cleaner. The duct and air handling equipment may then be treated with a disinfectant approved by the Environmental Protection Agency for use in air systems.

Condensate pans should be installed so that they drain completely, otherwise they can provide conditions for growth of algae, bacteria, and many kinds of pollutants.

SUMMARY

- In cool or cold weather, homes are dry because colder air infiltrates into the home and is heated. When heated, the air is artificially dried out by expansion.
- The relative humidity is the percentage of moisture in the air compared to the capacity of the air to hold the moisture.
- Dry air draws moisture from carpets, furniture, woodwork, plants, and people and frequently has a detrimental effect.
- Humidifiers put moisture back into the air.
- Evaporative humidifiers provide moisture to a media and force air through it, evaporating the moisture.
- Evaporative humidifiers may be a bypass type mounted outside the ductwork on a forced-air furnace with piping from the hot air through the humidifier to the cold air side of the furnace. They may be mounted in the plenum or under the duct.
- The media may be disc, drum, or plate types.
- For heat pump or electric heat installations the humidifier may have an independent electric heater to increase the temperature of the water for evaporation.
- Infrared humidifiers use infrared lamps and a reflector to cause the water to be evaporated into the airstream.
- Atomizing humidifiers may be spray-nozzle or centrifugal types. They discharge a mist (tiny water droplets) into the air. They should not be used in hard water conditions, and only where there is forced air movement.
- Self-contained humidifiers are used with hydronic, electric baseboard, or unit heaters, where forced air is not available. They have their own fan and often a heater.

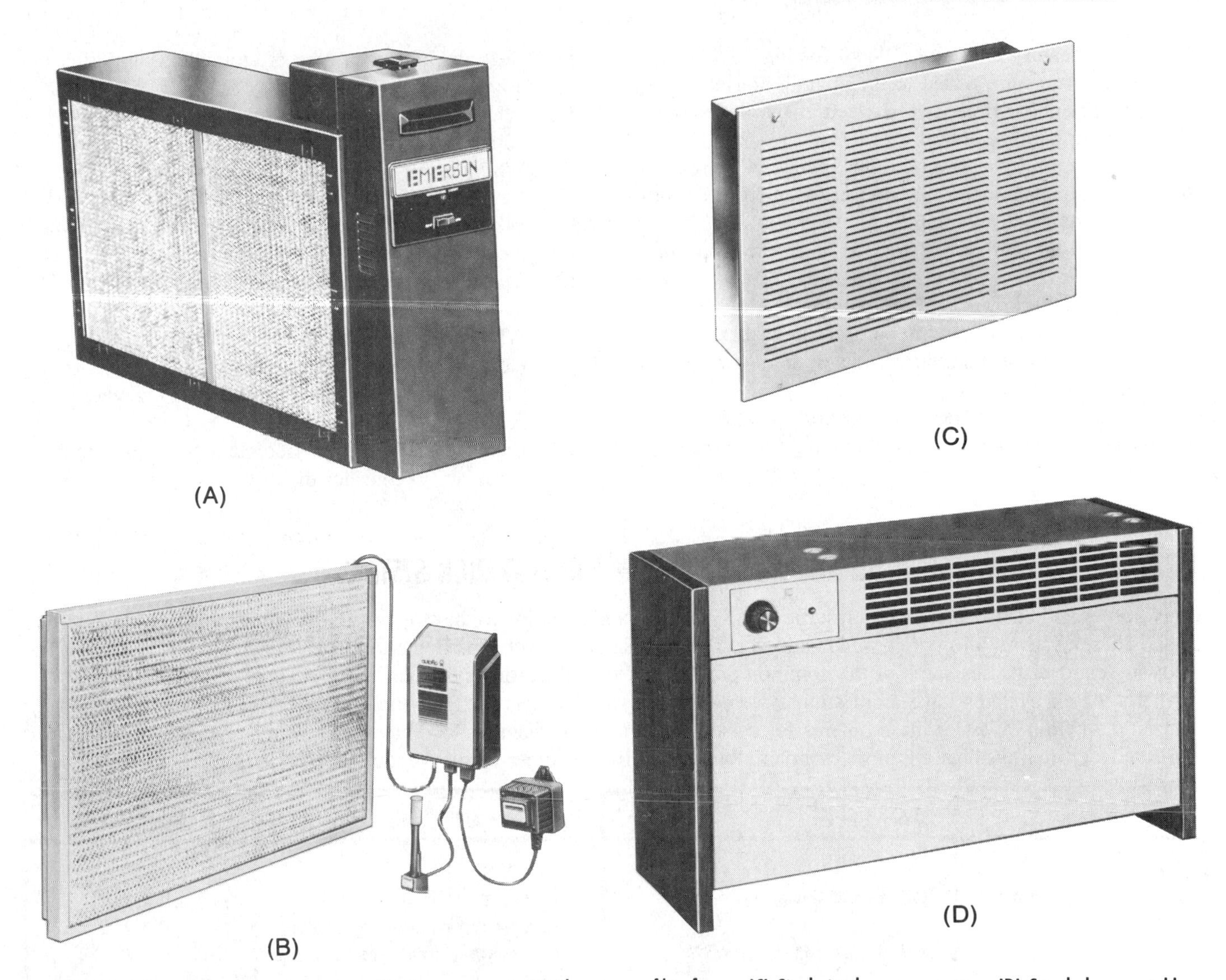

FIGURE 34–22 Electrostatic precipitators. (A) Furnace mount. (B) Throw-away filter frame. (C) Single-intake return system. (D) Stand-alone portable system. *Courtesy W. W. Grainger, Inc.*

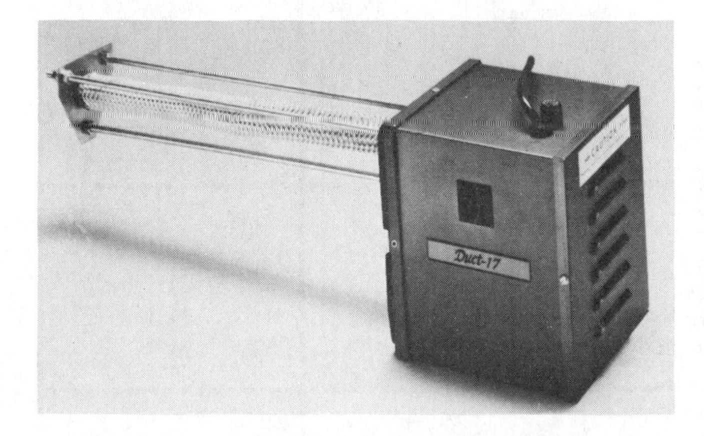

FIGURE 34–23 An electronic air purification system. *Courtesy Aqua-Mist, Inc.*

- The number of cubic feet in the house, the type of construction, the amount of air change per hour, the lowest outdoor temperatures, and the relative humidity level desired determine the size of the humidifier.
- ***Manufacturer's instructions should be followed carefully when installing humidifiers.***
- Modern buildings are being constructed to allow little air infiltration. Filtering and purification devices are used to clean the air.
- Indoor air quality has become a major factor in many buildings.
- Steam humidifiers may be used to kill bacteria, algae, and viruses.
- Filtration devices include fiberglass filter media, extended surface air filters, steel washable air filters, bag-type air filters, electronic air cleaners, and electronic air purification devices.

REVIEW QUESTIONS

1. Why are homes drier in winter than in summer?
2. Explain relative humidity.
3. Why does relative humidity decrease as the temperature increases?
4. How does dry warm air dry out household furnishings?

5. Why does dry air cause a "cool feeling"?
6. How is moisture added to the air in a house?
7. Describe the differences between evaporative and atomizing humidifiers.
8. How does a bypass humidifier operate?
9. Is a plenum-mounted humidifier installed in the supply or return air plenum?
10. What is the purpose of the media in an evaporative humidifier?
11. Describe disc, plate, and drum humidifier designs. Describe the media that can be used on each.
12. Why do some humidifiers have their own water-heating device?
13. Explain how an infrared humidifier operates.
14. What are the two types of atomizing humidifiers?
15. Describe how each type of atomizing humidifier operates.
16. Why is it essential for the furnace to be running when an atomizing humidifier is operating?
17. Why are self-contained humidifiers used?
18. Describe a self-contained humidifier.
19. What general factors are considered when determining the size or capacity of a humidifier?
20. Describe, in general terms, installation procedures for humidifiers used with forced-air furnaces.
21. Why are systems of cleaning air more necessary in modern buildings than in older ones?
22. List five types of air filtering and cleaning materials or devices.
23. Describe the design of fiberglass filters that make it necessary to consider the direction of the airflow.

UNIT 34 DIAGNOSTIC CHART FOR HUMIDIFICATION AND FILTRATION SYSTEMS

Many homes and businesses have humidification systems to add humidity to the structure in the winter when inside humidity drops. These systems all contain water and water can cause some problems if the system is not properly cared for. The following chart discusses some of the common problems of humidification systems.

All forced-air systems have a filtration system of some sort. These vary from the basic air filter to the more complex electronic air filter. Most of the problems for basic air filters are reduced airflow due to lack of maintenance. Problems for electronic air filter systems are more complex. Remember, the customer may have the clue for finding the problem.

PROBLEM	POSSIBLE CAUSE	POSSIBLE REPAIR	PARAGRAPH NUMBER
HUMIDIFIERS			
Unit not operating—depends on electricity	Faulty electrical circuits	Change fuse or reset breaker	15.2, 15.3
		Repair faulty electrical circuits or connections	15.2, 15.3
	Humidistat not calling for humidity	Check humidistat calibration; adjust or change humidistat	34.4
	Interlock with fan circuit	Determine what type of interlock and correct	34.4
Unit operating, but humidity is low	No water supply	Re-establish water supply	34.10
	Defective float	Change float	34.10
	Evaporation media saturated with minerals and will not absorb water	Change media	34.10
	Power assembly to turn evaporation media not turning	Make sure power is available, if it is change unit	34.10
FILTERS			
Media type			
Restricted	Filters are not changed often enough	Change on a more regular schedule	34.11
Dust entering conditioned space, possibly causing dirty evaporator coil and fan blades	Filter media too course	Change to finer filter media	34.11
	Air by passing filter	Install correct size filters	34.11
Electronic Filter			
No power to filter	Filter interlock with fan	Establish correct fan interlock circuit	34.11
Power to filter but filter not operating	Filter grounded in element	Clean filter	34.11
	Contacts in filter element not making contact	Clean contacts and reassemble	34.11
	Defective power assembly	Change power assembly	34.11
Filter makes cracking sound too often	High dust content in air	Change to finer pre-filter	34.11

SECTION

7 Air Conditioning (Cooling)

U N I T

35 Comfort and Psychrometrics

OBJECTIVES

After studying this unit, you should be able to

- recognize the four factors involved in comfort.
- explain the relationship of body temperature to room temperature.
- describe why one person is comfortable when another is not.
- define psychrometrics.
- define wet-bulb and dry-bulb temperature.
- define dew point temperature.
- explain vapor pressure of water in air.
- describe humidity.
- plot air conditions using a psychrometric chart.

35.1 COMFORT

Comfort describes a delicate balance of pleasant feeling in the body produced by its surroundings. A comfortable atmosphere describes our surroundings when we are not aware of discomfort. Providing a comfortable atmosphere for people becomes the job of the heating and air conditioning profession. Comfort involves four things: (1) temperature, (2) humidity, (3) air movement, and (4) air cleanliness.

The human body has a sophisticated control system for both protection and comfort. The human body can move from a warm house to 0°F outside, and it starts to compensate for the surroundings. It can move from a cool house to 95°F outside, and it will start to adjust to keep the body comfortable and from overheating. Body adjustments are accomplished by the circulatory and respiratory systems. When the body is exposed to a climate that is too cold, it starts to shiver, an involuntary reaction, to warm the body. When the body gets too warm, the vessels next to the skin dilate to get the blood closer to the surrounding air in an effort to increase the heat exchange with the air. If this does not cool the body, it will perspire. When this perspiration is evaporated, it takes heat from the body and cools it.

35.2 FOOD ENERGY AND THE BODY

The human body may be compared to a coal hot water boiler. The coal is burned in the boiler to create heat. Heat is energy. Food to the human body is like coal to a hot water boiler. The coal in a boiler is converted to heat for space heating. Some heat goes up the flue, some escapes to the surroundings, and some is carried away in the ashes. If fuel is added to the fire and the heat cannot be dissipated, the boiler will overheat, Figure 35–1.

The body uses food to produce energy. Some energy is stored as fatty tissue, some leaves as waste, some leaves as heat, and some is used as energy to keep the body functioning. If the body needs to dissipate some of its heat to the surroundings and cannot, it will overheat, Figure 35–2.

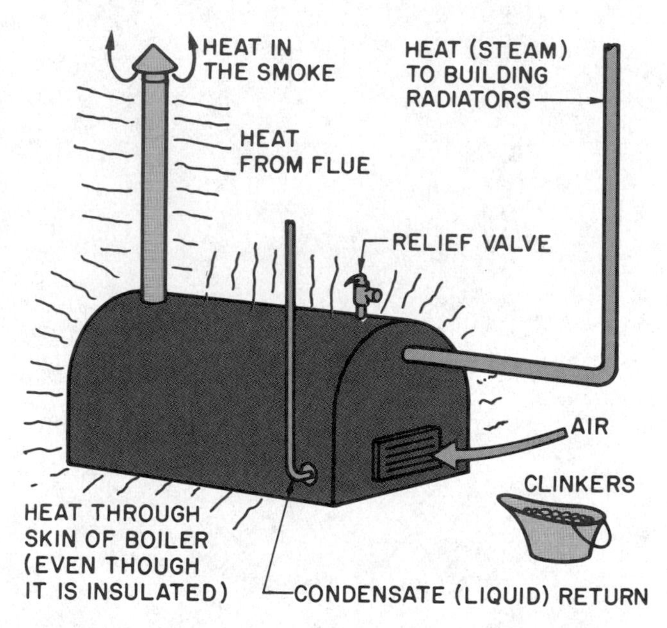

FIGURE 35–1 The boiler can be compared to the human body in that it uses fuel for energy.

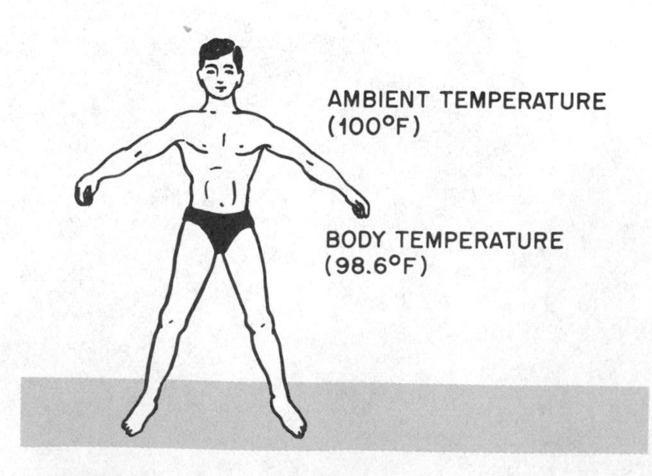

FIGURE 35–2 The human body must give off some of its generated heat to the surroundings or it will overheat.

35.3 BODY TEMPERATURE

Human body temperature is normally 98.6°F, Figure 35–3. We are comfortable when the heat level in our body due to food intake is transferring to the surroundings at the correct rate. But certain conditions must be met for this comfortable, or balanced, condition to exist.

The body gives off and absorbs heat by the three methods of heat transfer: conduction, convection, and radiation, Figure 35–4. The evaporation of perspiration could be considered a fourth way, Figure 35–5. When the surroundings are at a particular comfort condition, the body is giving up heat at a steady rate that is comfortable. The surroundings must be cooler than the body for the body to be comfortable. Typically, when the body is at rest (sitting) and in surroundings of 75°F and 50% humidity with a slight air movement, the body is close to being comfortable. Notice that the room air at this condition is 23.6°F cooler than the human body, Figure 35–6. In cooler weather a different set of conditions applies (eg, we wear more clothing). The following statements can be used as guidelines for comfort.

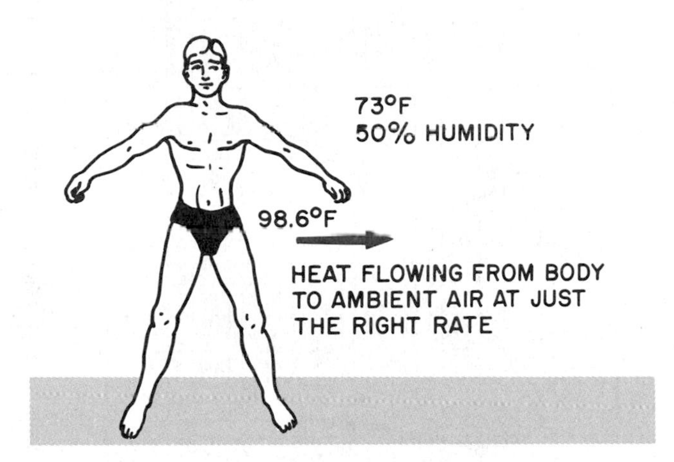

FIGURE 35–3 The body's normal temperature is 98.6°F.

1. In winter:
   A. Lower temperature can be offset with higher humidity.
   B. The lower the humidity is, the higher the temperature must be.
   C. Air movement is more noticeable.
2. In summer:
   A. When the humidity is high, air movement helps.
   B. Higher temperatures can be offset with lower humidity.
3. The comfort conditions in winter and in summer are different.
4. Styles of clothes in different parts of the country make a slight difference in the conditioned space-temperature requirements for comfort. For example, in Maine the

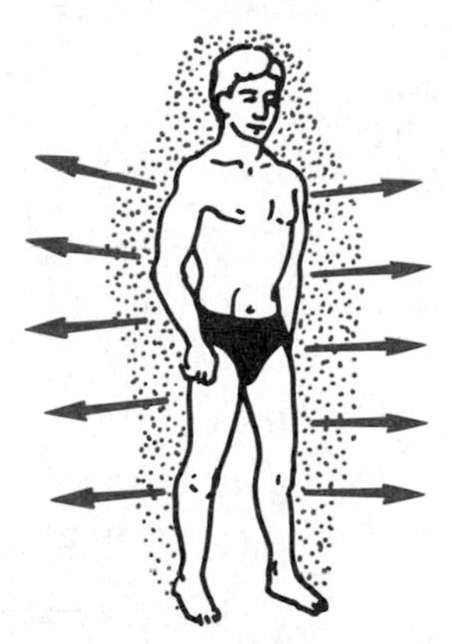

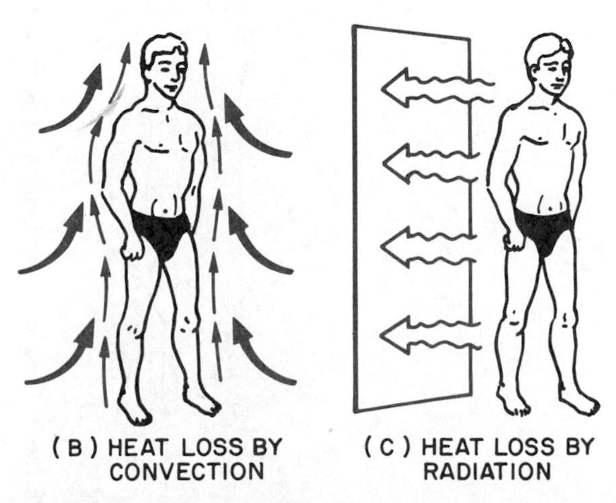

FIGURE 35–4 The three direct ways the human body gives off heat. (A) Conduction. (B) Convection. (C) Radiation.

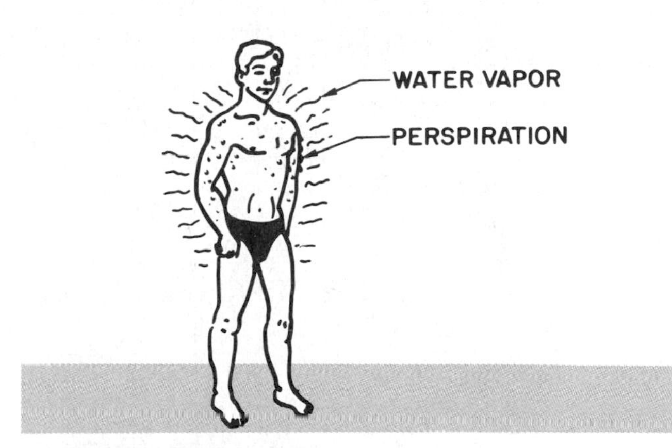

FIGURE 35–5 This could be called a fourth way the body gives off heat—the evaporation of perspiration.

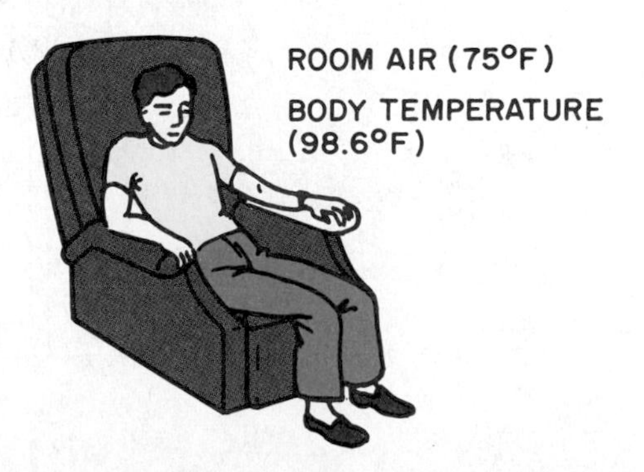

FIGURE 35–6 Relationship of the human body at rest with the atmosphere surrounding it.

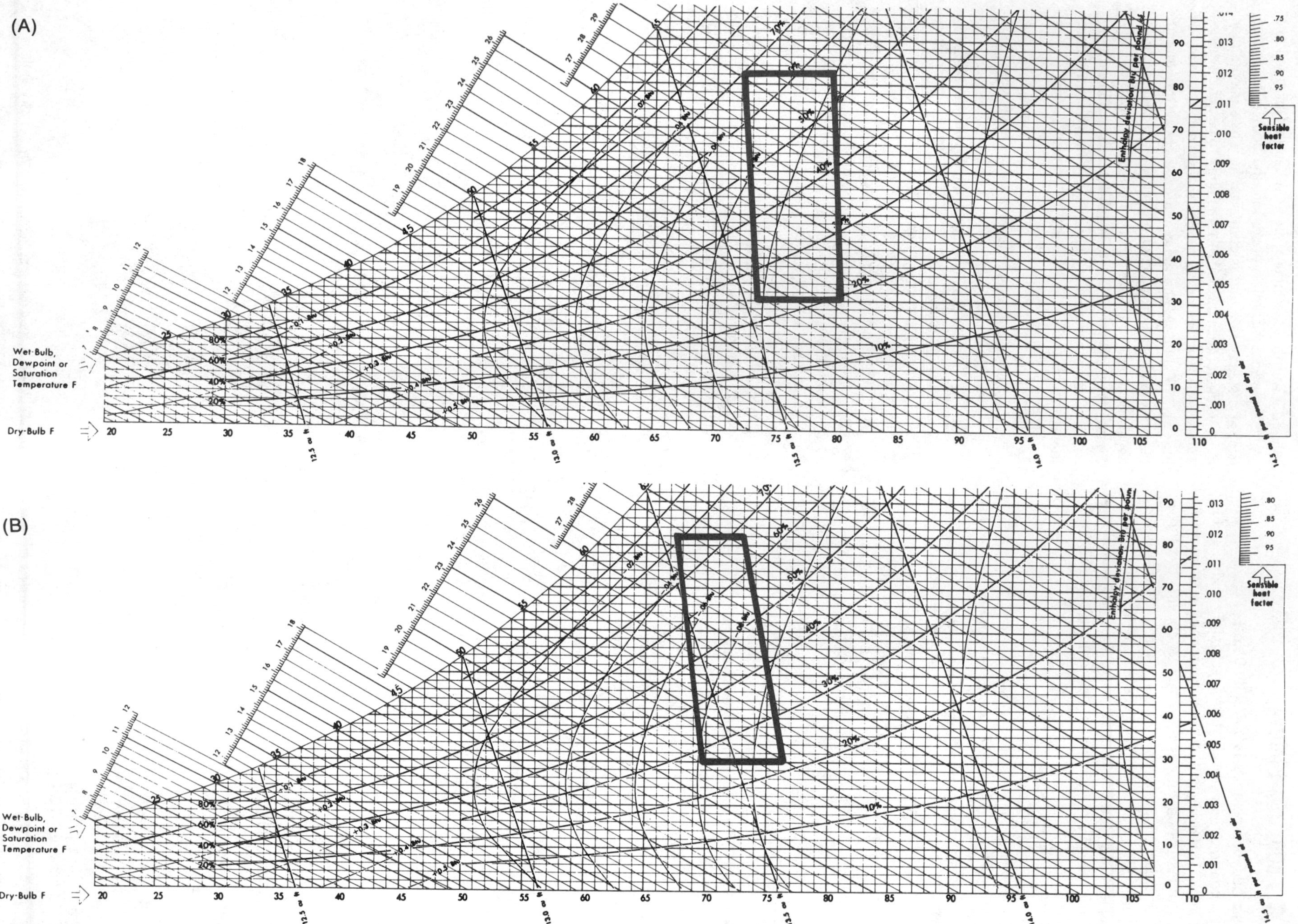

FIGURE 35–7 These are generalized comfort charts for different temperature and humidity conditions. (A) Summer. (B) Winter. *Adapted from Carrier Corporation Psychrometric Chart*

styles would be warmer in the winter than in Georgia, so the inside temperature of a home or office will not have the same comfort level.

5. Body metabolism varies from person to person. Women, for example, are not as warm natured as men. The circulatory system generally does not work in older people as well as in younger people.

35.4 THE COMFORT CHART

The chart in Figure 35–7, often called a *generalized comfort chart,* can be used to compare one situation with another. It shows the different combinations of temperatures and humidity for summer and winter. This table is for one air movement condition. A technician may use this chart to plot and compare comfort conditions in an occupied space. In general, the closer the plot falls to the middle of the chart, the more people would be comfortable at that condition. Notice that there is a different chart for summer and winter conditions.

Cooling, heating, humidifying, and cleaning our air describes the air conditioning profession. Air consists of approximately 78% nitrogen, 21% oxygen, and 1% other gases. Water, in the form of low-pressure vapor, is suspended in the air and called *humidity,* Figure 35–8. The moisture in the air is part of the conditions plotted on the comfort chart to determine comfort level.

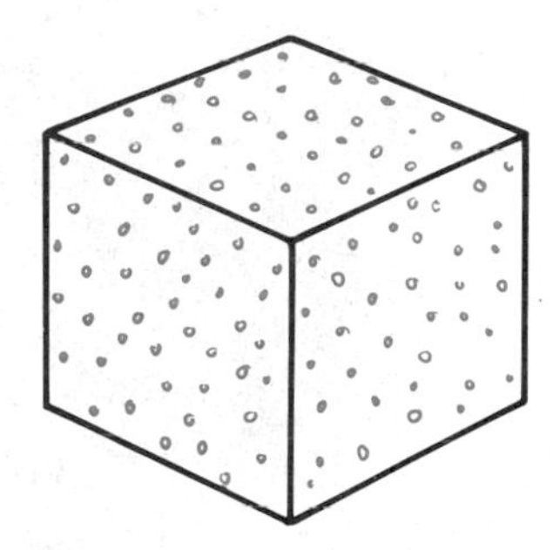

FIGURE 35–8 Moisture suspended in air.

35.5 PSYCHROMETRICS

The study of air and its properties is called *psychrometrics.* When we move through a room, we are not aware of the air inside the room, but the air has weight and occupies space like water in a swimming pool. The water in a swimming pool is more dense than the air in the room; it weighs more per unit of volume. Air weighs 0.075 lb/ft^3, Figure 35–9. The density of water is 62.4 lb/ft^3.

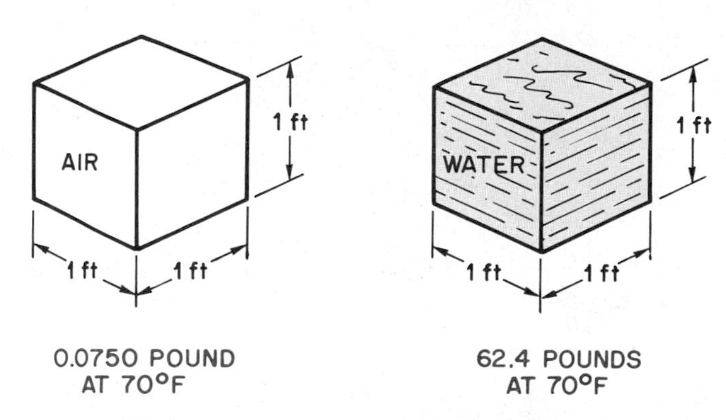

FIGURE 35–9 Relative difference between the weight of air and the weight of water.

Air offers resistance to movement. To prove this, take a large piece of cardboard and try to swing it around with the flat side moving through the air, Figure 35–10. It is hard to do because of the resistance of the air. The larger the area of the cardboard, the more resistance there will be. For example, if you took a large piece of cardboard outside on a windy day, the wind will try to take the cardboard from you. The cardboard acts like a sail on a boat, Figure 35–11. For another example, Figure 35–12, invert an empty glass and push it down in water. The air in the glass resists the water going up into the glass.

FIGURE 35–10 This man is having a hard time swinging a piece of cardboard around because the air around him is taking up space and causing a resistance to the movement of the cardboard.

FIGURE 35–11 This man walks out of a building and into a breeze. He is pushed along by the breeze against the cardboard because the air has weight and takes up space.

The weight of air in a room can be calculated by multiplying the room volume by the weight of a cubic foot of air.

FIGURE 35–12 This illustration proves in another way that air takes up space.

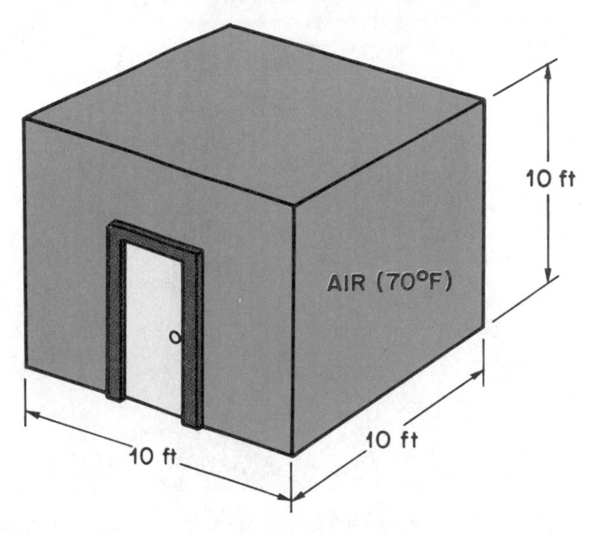

FIGURE 35–13 This room is 10 ft × 10 ft × 10 ft. The room contains 1000 ft³ of air that weighs 0.075 lb/ft³. 1000 ft³ × 0.075 lb/ft³ = 75.0 lb of air in the room.

$1 \div 0.075\ \text{lb/ft}^3 = 13.33\ \text{ft}^3/\text{lb}$
1 pound of air occupies 13.33 cubic feet.

FIGURE 35–14 This example shows how to find the number of cubic feet a pound of air occupies.

In a room 10 ft × 10 ft × 10 ft, the volume is 1000 ft³, so the room air weighs 1000 ft³ × 0.075 lb/ft³ = 75 lb, Figure 35–13. The number of cubic feet of air to make a pound of air can be obtained by taking the reciprocal of the density. The reciprocal of a number is 1 divided by that number. The reciprocal of the density of air at 70°F is 1 divided by 0.075 or 13.33 ft/lb of air, Figure 35–14.

35.6 MOISTURE IN AIR

Air is not totally dry. Surface water and rain keep moisture in the atmosphere everywhere (even in a desert) at all times. (The earth's surface is approximately 65% water.) Moisture in the air is called *humidity*.

35.7 SUPERHEATED GASES IN AIR

Because air is made of several different gases, it is not a pure element or gas. Air is made up of nitrogen (78%), oxygen (21%), and approximately 1% other gases, Figure 35–15. These gases in the air are highly superheated. Nitrogen, for instance, boils at −319°F, and oxygen boils at −297°F at atmospheric pressure, Figure 35–16. Hence, nitrogen and oxygen in the atmosphere are *superheated gases*—they are superheated several hundred degrees above absolute (0° Rankine). Each gas exerts pressure according to Dalton's Law of Partial Pressures. This law states that each gas in a mixture of gases acts independently of the other gases and the total pressure of a gas mixture is the sum of the pressures of each gas in the mixture. More than one gas can occupy a space at the same time.

Water vapor suspended in air is a gas that exerts its own pressure and occupies space with the other gases. Water at 70°F in a dish in the atmosphere exerts a pressure of 0.7392 in. Hg, Figure 35–17. If the water vapor pressure in the air is less than the water vapor pressure in the dish, the water in the dish will evaporate slowly to the lower pressure area of the water vapor in the air. For example, the room may be at a dry-bulb temperature of 70°F with a humidity of 30%. The vapor pressure for the moisture suspended in the air is 0.101 psia × 2.036 = 0.206 in. Hg., Figure 35–18. Vapor pressure

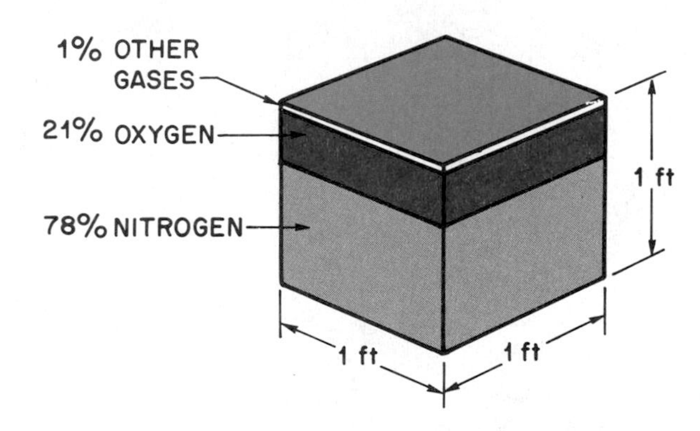

FIGURE 35–15 Relationship of nitrogen to oxygen in air.

ROOM TEMPERATURE (70°F)

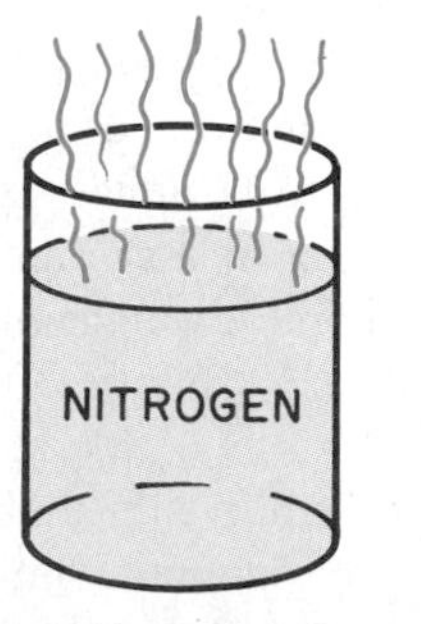

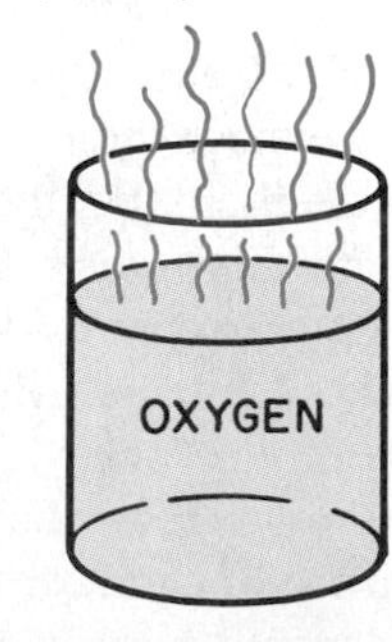

LIQUID NITROGEN WILL EVAPORATE (BOIL) AT −319°F.

LIQUID OXYGEN WILL EVAPORATE (BOIL) AT −297°F.

FIGURE 35–16 If you place a container of liquid nitrogen and a container of liquid oxygen in a room, they will start to evaporate or boil.

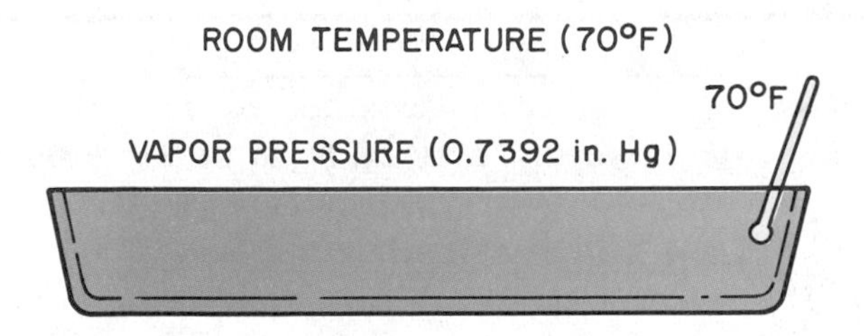

FIGURE 35–17 Vapor pressure at 70°F of water in an open dish in a room.

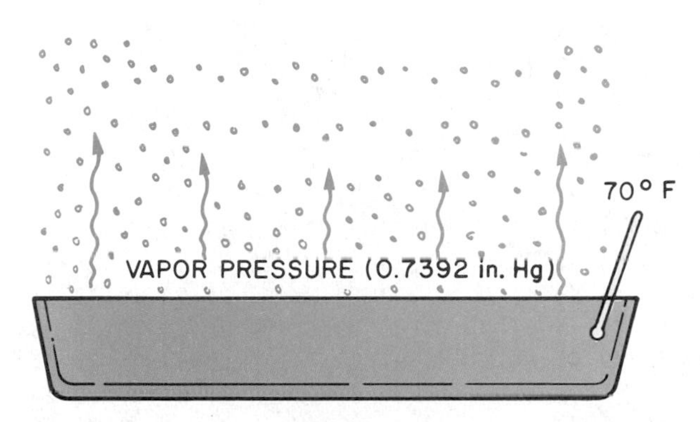

FIGURE 35–18 Moisture suspended in the air has a pressure controlled by the humidity in the room.

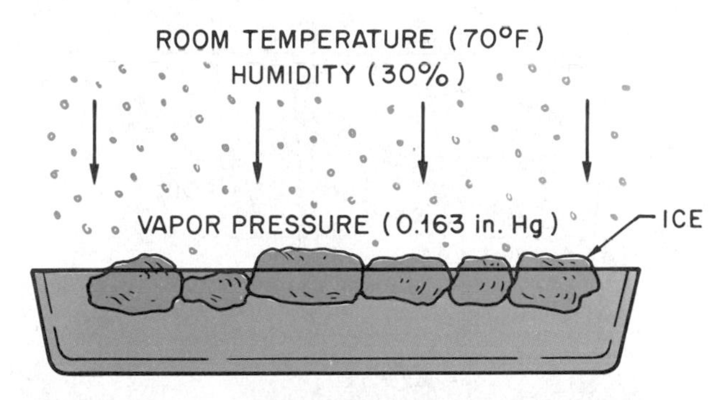

FIGURE 35–19 The moisture in the dish has ice in it, which lowers the vapor pressure to 0.163 (0.08 psia × 2.036 = 0.163 in. Hg). The room temperature is still 70°F with a humidity of 30%, which has a vapor pressure of 0.206 in. Hg.

for moisture in air can be found in some psychrometric charts and in saturated water tables. When reverse pressures occur, the action of the water vapor reverses. For example, if the water vapor pressure in the dish is less than the pressure of the vapor in the air, water from the air will condense into the water in the dish, Figure 35–19.

When water vapor is suspended in the air, the air is sometimes called "wet air." If the air has a large amount of moisture, the moisture can be seen (eg, fog or a cloud). Actually, the air is not wet because the moisture is suspended in the air. This could more accurately be called a nitrogen, oxygen, and water vapor mixture.

35.8 HUMIDITY

The moisture content in air (humidity) is measured by weight, expressed in pounds or grains (7000 gr/lb). Air can hold very little water vapor. 100% humid air is 29.92 in. Hg and 70°F can hold 110.5 gr of moisture (0.01578 lb) per pound of air. Several methods are used to calculate the percentage of moisture content in the air. *Relative humidity* is the most practical and most used for field measurements. It is based on the weight of water vapor in a pound of air compared to the weight of water vapor that a pound of air could hold if it were 100% saturated, Figure 35–20.

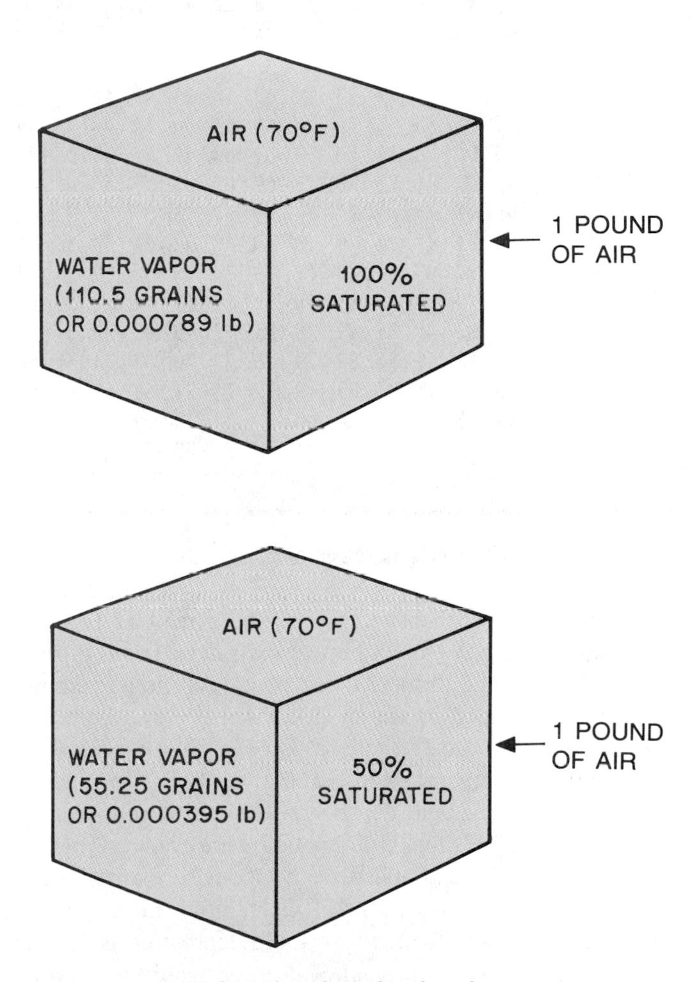

FIGURE 35–20 Relative humidity is based on the water vapor suspended in a pound of air compared to the weight of water vapor a pound of air could hold if 100% saturated.

35.9 DRY-BULB AND WET-BULB TEMPERATURES

The moisture content of air can be checked by using a combination of dry-bulb and wet-bulb temperatures. *Dry-bulb* temperature is the sensible-heat level of air and is taken with an ordinary thermometer. Wet-bulb temperature is taken with a thermometer with a wick on the end that is soaked with distilled water. The reading from a wet-bulb thermometer takes into account the moisture content of the air. It reflects the total heat content of air. The wet-bulb thermometer will get cooler than the dry-bulb thermometer because of the evaporation of the distilled water. Distilled water is used because some water has undesirable mineral deposits. Some minerals will change the boiling temperature.

DB	WB DEPRESSION																													
TEMP.	1	2	3	4	5	6	7	8	9	10	11	12	13	14	15	16	17	18	19	20	21	22	23	24	25	26	27	28	29	30
32	90	79	69	60	50	41	31	22	13	4																				
36	91	82	73	65	56	48	39	31	23	14	6																			
40	92	84	76	68	61	53	46	38	31	23	16	9	2																	
44	93	85	78	71	64	57	51	44	37	31	24	18	12	5																
48	93	87	80	73	67	60	54	48	42	36	34	25	19	14	8															
52	94	88	81	75	69	63	58	52	46	41	36	30	25	20	15	10	6	0												
56	94	88	82	77	71	66	61	55	50	45	40	35	34	26	24	17	12	8	4											
60	94	89	84	78	73	68	63	58	53	49	44	40	35	31	27	22	18	14	6	2										
64	95	90	85	79	75	70	66	61	56	52	48	43	39	35	34	27	23	20	16	12	9									
68	95	90	85	81	76	72	67	63	59	55	51	47	43	39	35	31	28	24	21	17	14									
72	95	91	86	82	78	73	69	65	61	57	53	49	46	42	39	35	32	28	25	22	19									
76	96	91	87	83	78	74	70	67	63	59	55	52	48	45	42	38	35	32	29	26	23									
80	96	91	87	83	79	76	72	68	64	61	57	54	54	47	44	41	38	35	32	29	27	24	21	18	16	13	11	8	6	1
84	96	92	88	84	80	77	73	70	66	63	59	56	53	50	47	44	41	38	35	32	30	27	25	22	20	17	15	12	10	8
88	96	92	88	85	81	78	74	71	57	64	61	58	55	52	49	46	43	41	38	35	33	30	28	25	23	21	18	16	14	12
92	96	92	89	85	82	78	75	72	69	65	62	59	57	54	51	48	45	43	40	38	35	33	30	28	26	24	22	19	17	15
96	96	93	89	86	82	79	76	73	70	67	74	61	58	55	53	50	47	45	42	40	37	35	33	31	29	26	24	22	20	18
100	96	93	90	86	83	80	77	74	71	68	65	62	59	57	54	52	49	47	44	42	40	37	35	33	31	29	27	25	23	21
104	97	93	90	87	84	80	77	74	72	69	66	63	61	58	56	53	51	48	46	44	41	39	37	35	33	31	29	27	25	24
108	97	93	90	87	84	81	78	75	72	70	67	64	62	59	57	54	52	50	47	45	43	41	39	37	35	33	31	29	28	26

FIGURE 35–21 Wet-bulb depression chart.

The difference between the dry-bulb reading and the wet-bulb reading is called the *wet-bulb depression*. Figure 35–21 has a wet-bulb depression chart. As the amount of moisture suspended in the air decreases, the wet-bulb depression increases and vice versa. For example, a room with a dry-bulb temperature of 76°F and a wet-bulb temperature of 64°F has a wet bulb depression of 12°F and a relative humidity of 52%. If the 76°F dry-bulb temperature is maintained and moisture is added to the room so that the wet-bulb temperature rises to 74°F, the relative humidity increases to 91% and the new wet-bulb depression is 2°F. If the wet-bulb depression is allowed to go to 0°F, (eg, 76°F dry bulb and 76°F wet bulb), the relative humidity will be 100%. The air is holding all of the moisture it can—it is *saturated* with moisture.

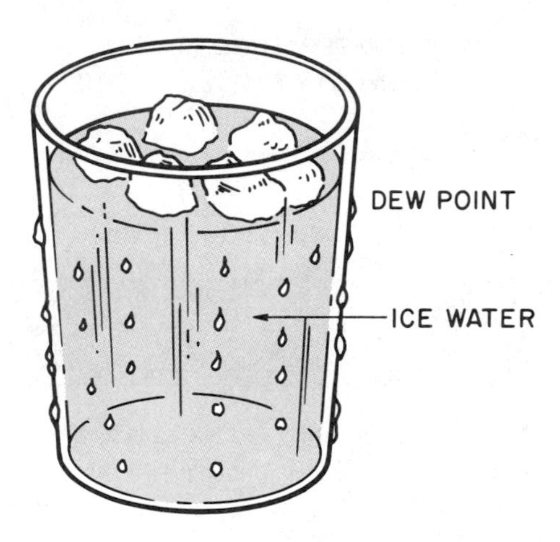

FIGURE 35–22 The glass was gradually cooled until beads of water began to form on the outside of the glass.

35.10 DEW POINT TEMPERATURE

The *dew point temperature* is the temperature at which moisture begins to condense out of the air. For example, if you were to set a glass of warm water in a room with a temperature of 75°F and 50% relative humidity, the water in the glass would evaporate slowly to the room. If you gradually cool the glass with ice, when the glass surface temperature becomes 55.5°F, water will begin to form on the surface of the glass, Figure 35–22. Moisture from the room will also collect in the water in the glass and the level will begin to rise. This temperature at which water forms is called the dew point temperature of the air. Air can be dehumidified by passing it over a surface that is below the dew point temperature of the air; moisture will collect on the cold surface, for example, an air conditioning coil, Figure 35–23. The condensed moisture is drained. This is the moisture that you see running out of the condensate line of an air conditioner, Figure 35–24.

35.11 THE PSYCHROMETRIC CHART

The foregoing description can all be plotted on a psychrometric chart, Figure 35–25A. The chart looks complicated, but a clear plastic straightedge and a pencil will help you understand it. See Figures 35–25B through 35–25G for some examples of plottings of the different conditions on a psychrometric chart.

If you know any two conditions previously mentioned, you can plot any of the other conditions. The easiest condi-

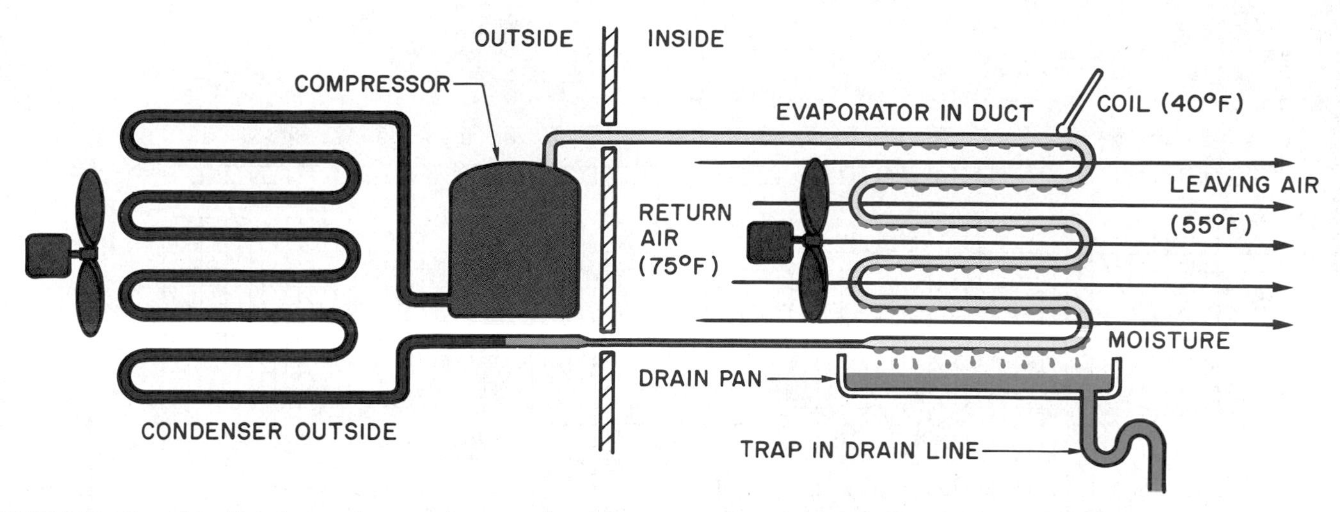

FIGURE 35–23 The cold surface on the air conditioning coil condenses moisture from the air passed over it.

FIGURE 35–24 The water dripping out of the back of this window air conditioner is moisture that was collected from the room onto the air conditioner's evaporator coil.

tions to determine from room air are the wet-bulb and the dry-bulb temperatures. For example, you can take an electronic thermometer and make a wet-bulb thermometer if you do not have one. Take two leads and tape them together with one lead about 2 in. below the other one, Figure 35–26. A simple wick can be made from a piece of white cotton. Make sure that it does not have any perspiration on it. Wet the lower bulb (with the wick on it) with distilled water that is warmer than the room air. Water from a clean condensate drain line may be used if other distilled water is not available. Water from the city system can be used but may give slightly incorrect results bcause of the impurities in the water. Hold the leads about 3 ft back from the element on the end and slowly spin them in the air. The wet lead will drop to a colder temperature than the dry lead. Keep spinning them until the lower lead stops dropping in temperature but is still damp. Quickly read wet-bulb and dry-bulb temperatures without touching the bulbs. Suppose the reading is 75°F DB (dry bulb) and 62.5°F WB (wet bulb). Put your pencil point at this place on the psychrometric chart, Figure 35–25A, and make a dot. Draw a light circle around it so you can find the dot again. The following information can be concluded from this plot:

1.	Dry-bulb temperature	**75°F**
2.	Wet-bulb temperature	**62.5°F**
3.	Dew point temperature	**55.5°F**
4.	Total heat content of 1 lb of air	**27.2 Btu**
5.	Moisture content of 1 lb of air	**65 gr**
6.	Relative humidity	**50%**
7.	Specific volume of air	**13.6 ft³/lb**

Figure 35–27 illustrates a technician using a sling psychrometer to obtain the wet-bulb dry-bulb reading.

35.12 PLOTTING ON THE PSYCHROMETRIC CHART

The condition of air can be plotted on the psychrometric chart as it is being conditioned. The following examples show how different applications of air conditioning are plotted.

- Air is heated. Movement through the heating equipment can be followed as a sensible-heat direction on the chart, Figure 35–28A.
- Air is cooled. There is no moisture removal. This shows a sensible-heat direction on the chart, Figure 35–28B.
- Air is humidified. No heat is added or removed. An increase in moisture content and dew point temperature shows, Figure 35–28C.
- Air is dehumidified. No heat is added or removed. A decrease in moisture content and dew point temperature shows, Figure 35–28D.
- Air is cooled and humidified using an evaporative cooler. These are popular in hot, dry climates, Figure 35–28E.

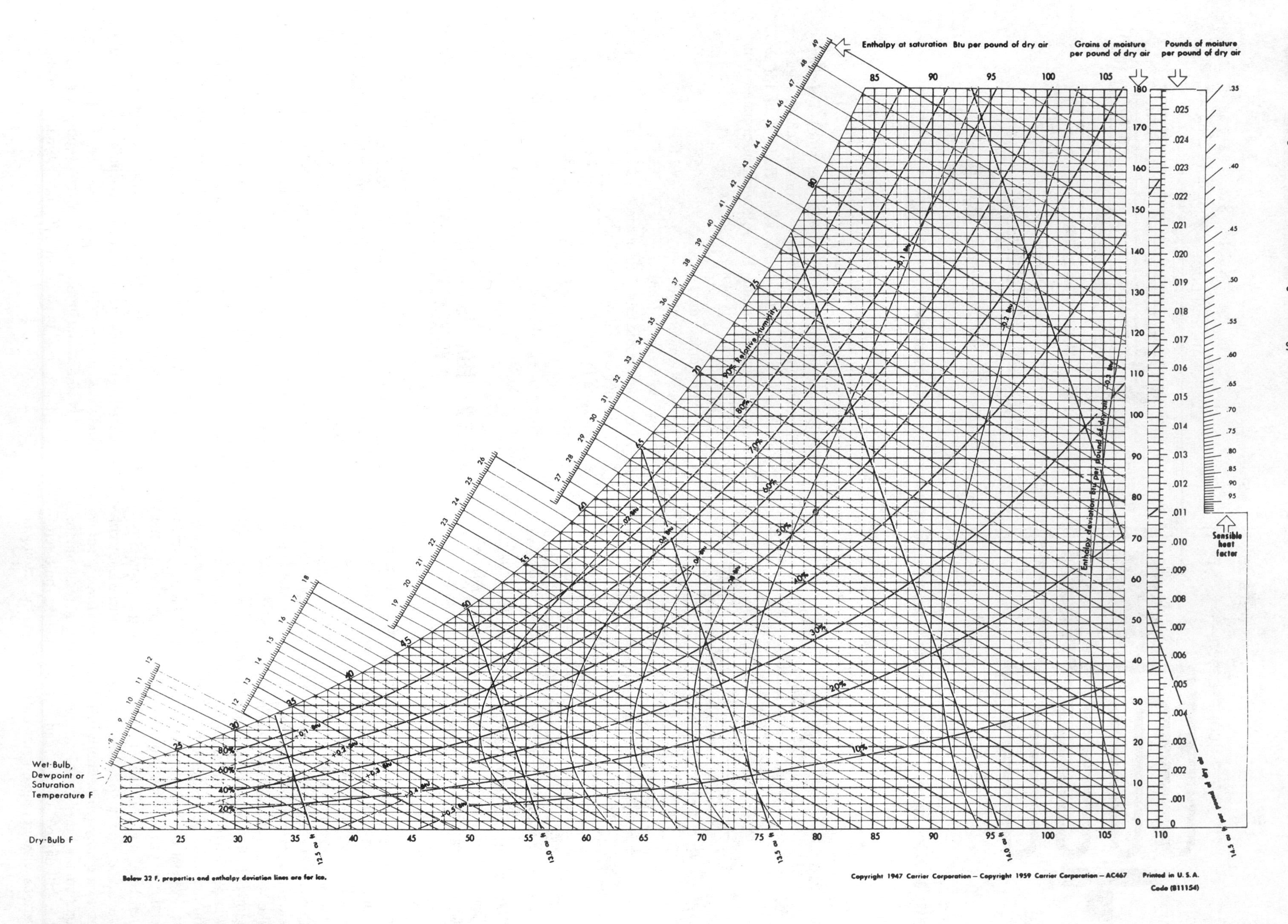

FIGURE 35–25A Psychrometric chart. *Reproduced courtesy of Carrier Corporation*

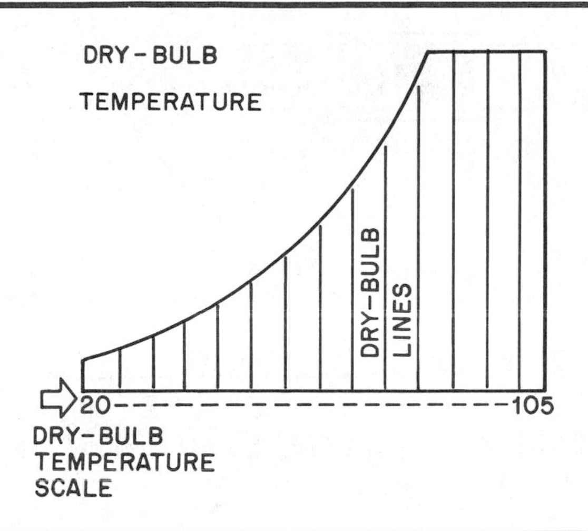

FIGURE 35–25B Skeleton chart showing the dry-bulb temperature lines. From Lang, *Principles of Air Conditioning* © 1987 by Delmar Publishers Inc.

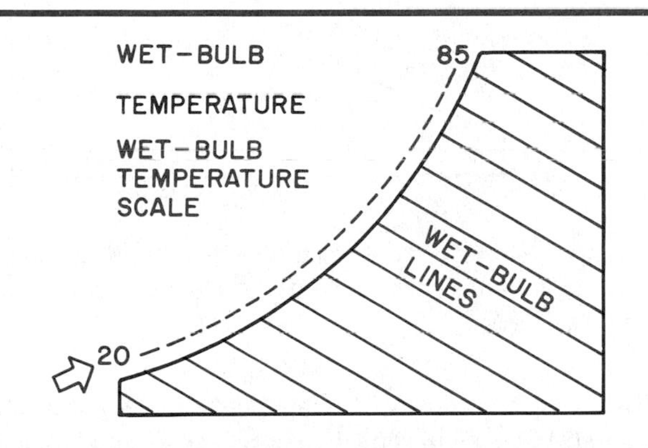

FIGURE 35–25C Skeleton chart showing the wet-bulb lines. From Lang, *Principles of Air Conditioning* © 1987 by Delmar Publishers Inc.

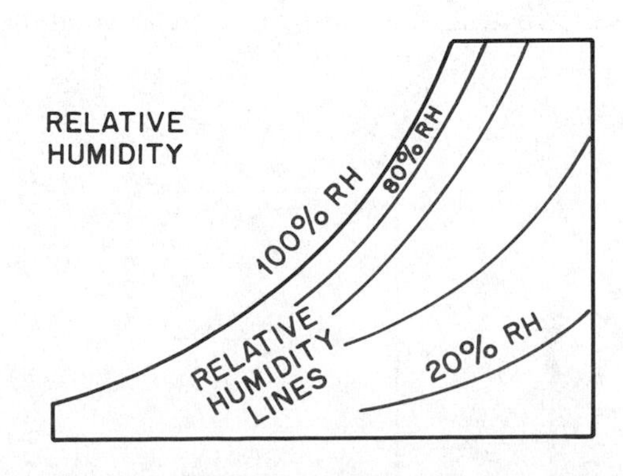

FIGURE 35–25D Skeleton chart showing the relative humidity lines. From Lang, *Principles of Air Conditioning* © 1987 by Delmar Publishers Inc.

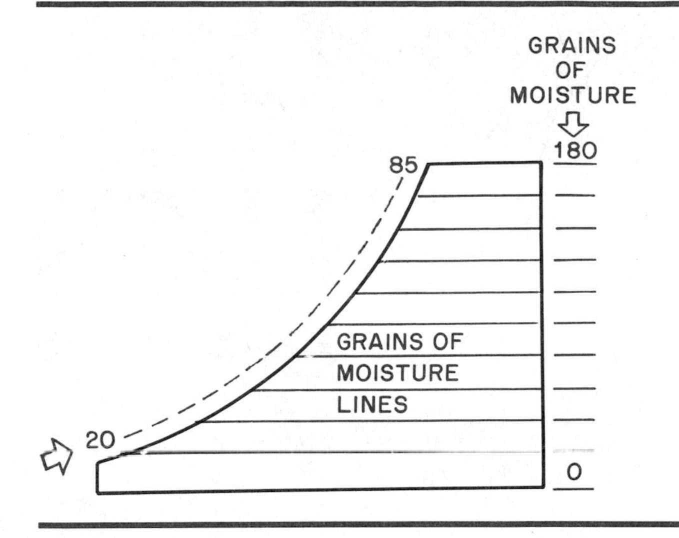

FIGURE 35–25E Skeleton chart showing the moisture content of air expressed in grains per pound of air. From Lang, *Principles of Air Conditioning* © 1987 Delmar Publishers Inc.

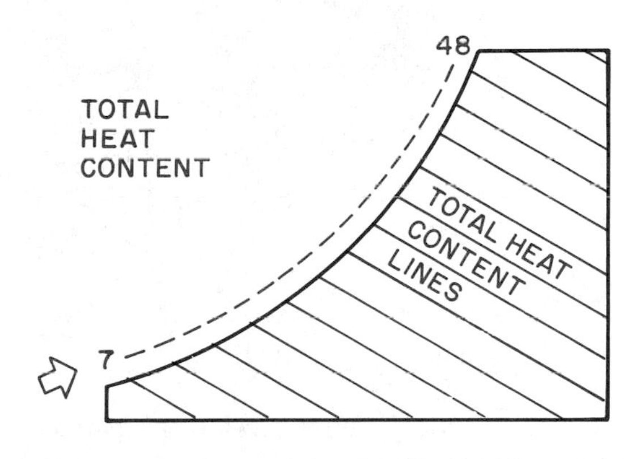

FIGURE 35–25F Skeleton chart showing the total heat content of air in Btu/lb. These lines are almost parallel to the wet-bulb lines. From Lang, *Principles of Air Conditioning* © 1987 by Delmar Publishers Inc.

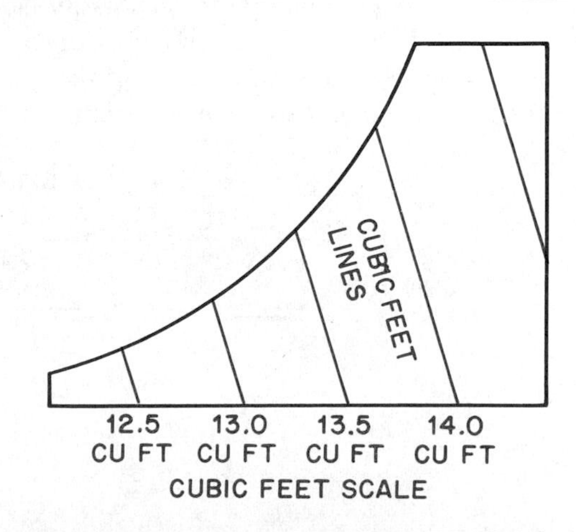

FIGURE 35–25G Skeleton chart showing the specific volume of air at different conditions. From Lang, *Principles of Air Conditioning* © 1987 by Delmar Publishers Inc.

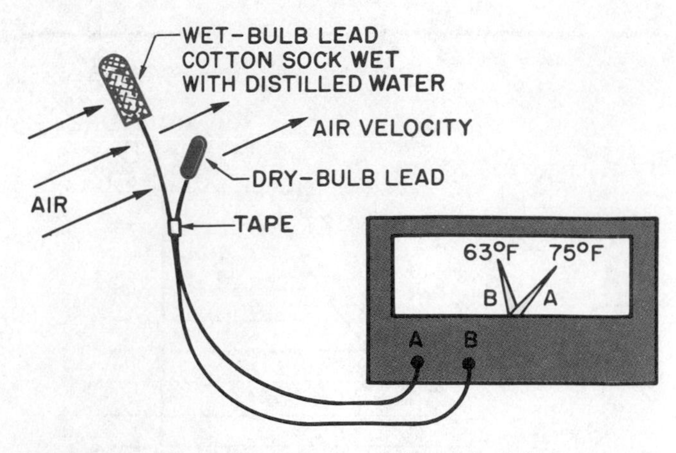

FIGURE 35–26 How to make a wet-bulb and dry-bulb thermometer from an electronic thermometer.

FIGURE 35–27 Technician spinning a wet-bulb and dry-bulb thermometer, known as a sling psychrometer.

When air enters air conditioning equipment it can be plotted on a chart. Figure 35–29 shows a chart indicating that from the reference point in the middle, the air may be conditioned to heat, cool, humidify, or dehumidify. Some apparatus will both add heat and moisture, or cool and remove moisture. The following examples will show what happens in the most common heating and cooling systems.

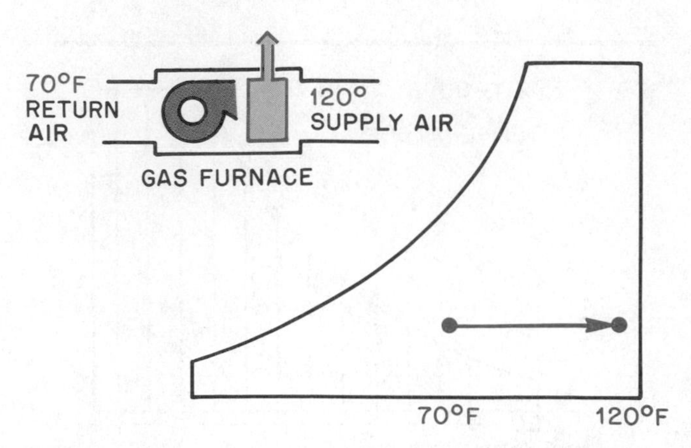

FIGURE 35–28A Air passing through a sensible-heat exchange furnace.

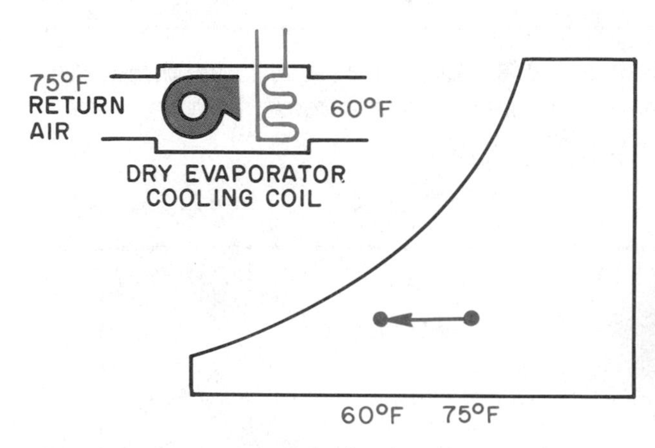

FIGURE 35–28B Air is cooled with a dry evaporator coil, operating above the dew point temperature of the air. No moisture is removed. This is not a typical situation.

- The most common winter application is to heat and humidify. This will show both a rise in temperature and an increase in moisture and dew point temperature, Figure 35–30.
- The most common summer application is to cool and dehumidify air. A decrease in temperature, moisture content, and dew point will occur, Figure 35–31.

It is important to notice that any change in heat content or moisture content of air will cause a change in the wet-

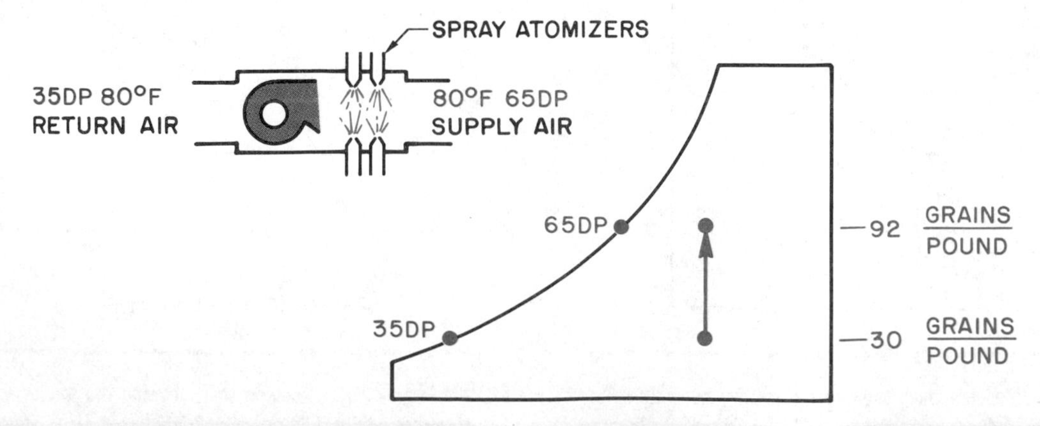

FIGURE 35–28C Spray atomizers are used to add moisture to the air. The dew point temperature and moisture content both increase.

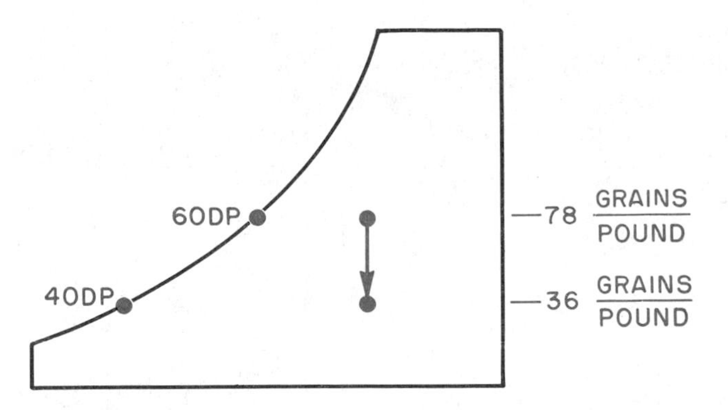

FIGURE 35–28D Moisture is removed from the air. This is not a typical application and is shown as an example only.

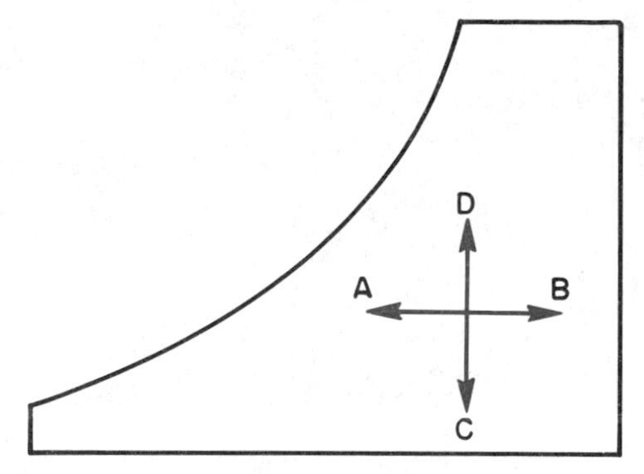

D
A
B
C
WHEN AIR IS CONDITIONED AND THE PLOT MOVES IN THE DIRECTION OF:
(A) SENSIBLE HEAT IS REMOVED.
(B) SENSIBLE HEAT IS ADDED.
(C) LATENT HEAT IS REMOVED, MOISTURE REMOVED.
(D) LATENT HEAT IS ADDED, MOISTURE ADDED.

FIGURE 35–29 A summation of sensible and latent heat.

bulb reading and, therefore, a change in the total heat content.

35.13 TOTAL HEAT

The capacity of a heating or cooling unit may be field checked with the total heat feature of the psychrometric chart. If the amount of air passing over a heat exchanger is known, the total heat can be checked where it enters and leaves the heat exchanger. This will give a fairly accurate account of the performance of the heat exchanger, Figure 35–32.

The air that surrounds us has to be maintained at the correct conditions for us to be comfortable. The air in our

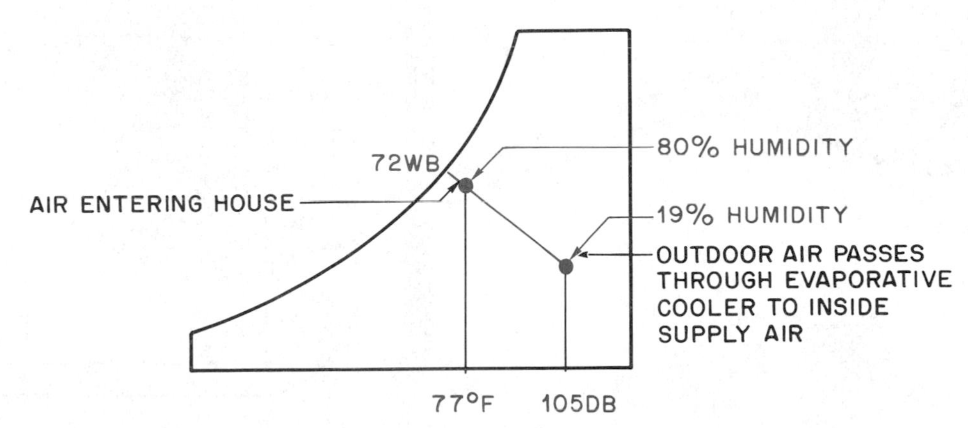

FIGURE 35–28E Hot, dry air passes through the water circuit of the evaporative cooler. Heat is given up to the cooler water and the air entering the house is cooled and humidified; water evaporates to cool the water.

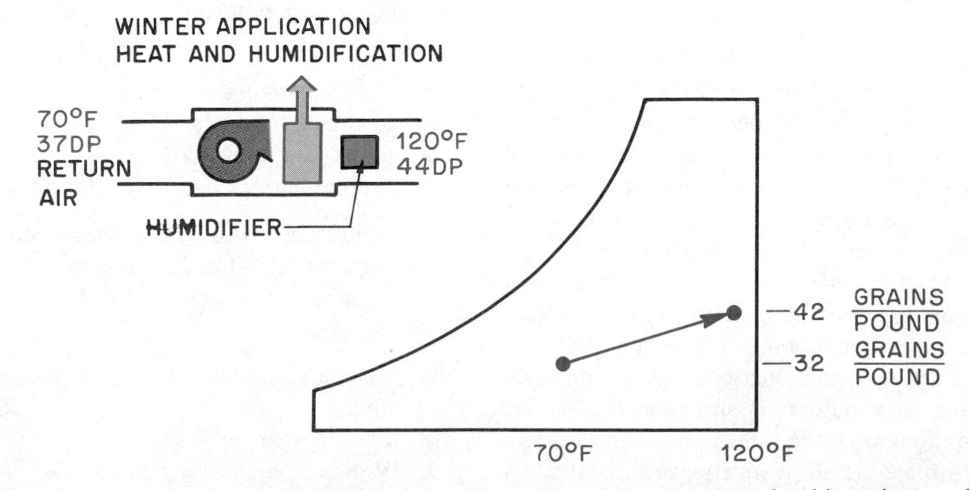

FIGURE 35–30 Sensible heat raises the temperature of the air from 70°F to 120°F. Moisture is evaporated and latent heat is added to the air.

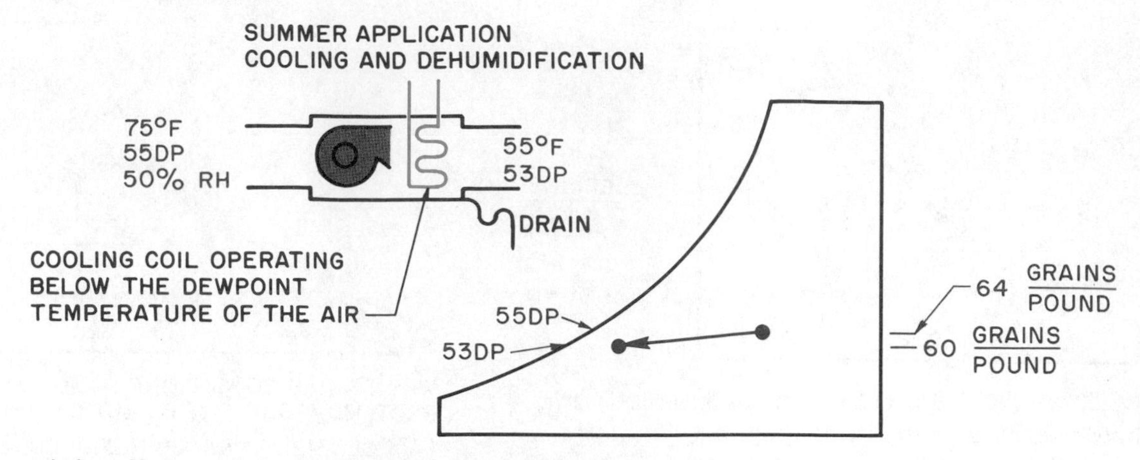

FIGURE 35–31 Removal of sensible heat cools the air. Removal of latent heat removes moisture from the air.

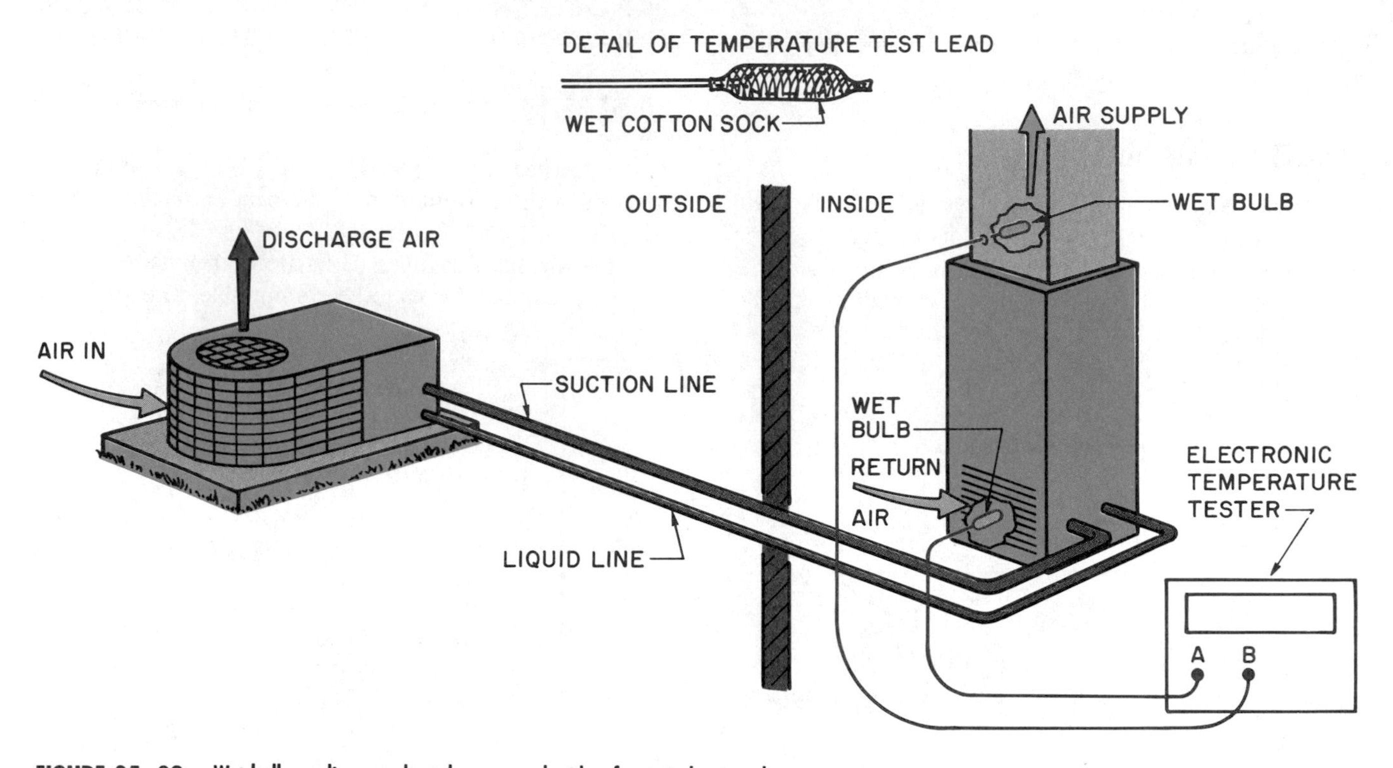

FIGURE 35–32 Wet-bulb reading can be taken on each side of an air heat exchanger.

homes is treated by heating it, cooling it, dehumidifying it, humidifying it, and cleaning it so that our bodies will give off the correct amount of heat for comfort. Air is first picked up from the room air, conditioned, and blended with the room air. A small amount of air is induced from the outside into the conditioner to keep the air from becoming oxygen starved and stagnant. This is called fresh air intake or *ventilation.* If a system has no ventilation, it is relying on air infiltrating the structure around doors and windows.

Modern energy-efficient homes can be built so tight that infiltration does not provide enough fresh air. Recent studies show that indoor pollution in homes and buildings has increased as a result of an increase in energy conservation in heating and air conditioning structures. People are more energy conscious now than in the past, and modern buildings are constructed to allow much less air infiltration from the outside. A typical homeowner may take the following measures to prevent outside air from entering the structure:

- Install storm windows
- Install storm doors
- Caulk around windows and doors
- Install dampers on exhaust fans and dryer vents

All of these will reduce the amount of outside air that leaks into the structure and improve energy costs. This may not be all good because of the following indoor pollution sources:

- Chemicals in new carpets, drapes, and upholstered furniture
- Cooking odors
- Vapors from cleaning chemicals
- Bathroom odors

- Vapors from freshly painted rooms
- Vapors from aerosol cans, hair sprays, and room deodorizers
- Vapors from particle building board epoxy resins
- Pets and their upkeep
- Radon gas leaking into the structure from the soil

These indoor pollutants may be diluted with outdoor air in the form of ventilation. Infiltration is the term used for random air that leaks into a structure. Ventilation is planned, fresh air added to the structure. When air is introduced into the system before the heating or air conditioning system, it is called ventilation. This may be accomplished with a duct from the outside to the return air side of the equipment, Figure 35–33.

There is some discussion as to how much air should be introduced, but it is generally agreed that at least a 0.25 air change per hour for the entire structure is desirable. This means that 25% of the indoor air is pushed out by inducing air into the system. For example, suppose a 2000-ft² home with an 8-ft ceiling needs ventilation because the home is very tight. How many cubic feet of air per minute must be introduced to change 25% of the air per hour?

$$2000 \text{ ft}^2 \times 8\text{-ft ceiling} \times 0.25 = 4000 \text{ ft}^3/\text{h}$$

$$\frac{4000 \text{ ft}^3/\text{h}}{60 \text{ min/h}} = 67 \text{ cfm}$$

This adds a considerable load to the equipment. For example, suppose the house is located in Atlanta, Georgia, where the outdoor design temperature is 17°F in the winter. (See Figure 36–13, Design Dry-Bulb column 99%.) If the home is to be maintained at 70°F indoors when the outdoor temperature is 17°F, there is a 53°F temperature difference (TD). The load on the heating equipment due to ventilation alone will be:

$$Q_s = 1.08 \times \text{cfm} \times \text{TD}$$
$$Q_s = 1.08 \times 67 \times 53$$
$$Q_s = 3835 \text{ Btuh}$$

Outside temperature of 17°F is only for 1% of the year. The fresh air will be warmer than 17°F the other 99% of the year. This problem depicts the worst possible case. Many system designers will use the 97.5% column for design of the system. Smaller equipment may be selected because of the warmer design temperatures. This may be taking a chance in the event of a cold winter.

The summertime calculation is made in much the same manner, except a different formula is used. Sensible and latent heat must be considered. In the summer, the design temperatures are 94°F dry-bulb and 74°F wet-bulb. The total heat formula may be used for this calculation.

$$Q_t = 4.5 \times \text{cfm} \times \text{total heat difference}$$

where Q_t = total heat
4.5 = a constant used to change pounds of air to cfm
total heat difference = difference in total heat indoors and outdoors

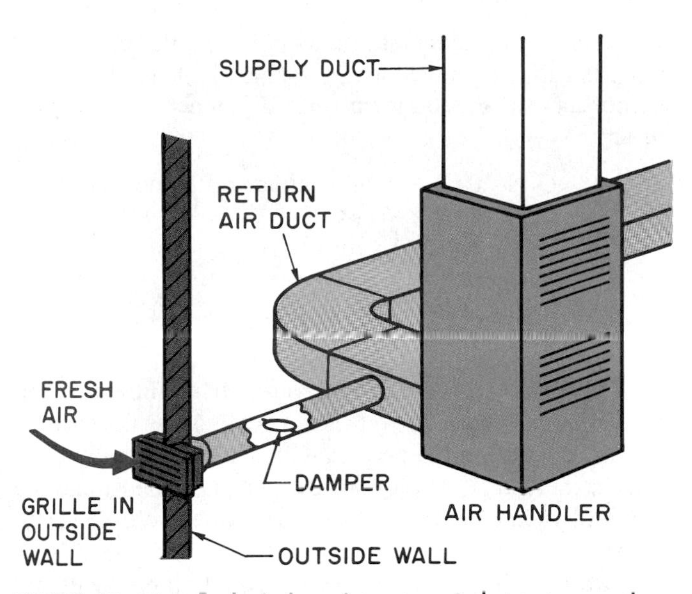

FIGURE 35–33 Fresh air drawn into return air duct to improve the air quality inside the house.

To solve the problem of fresh air, plot the indoor air and the outdoor air on the psychrometric chart.

Outdoors: 95°F dry-bulb and 74°F wet-bulb = 37.68 Btu/lb

Indoors: 85°F dry-bulb and 50% relative humidity indoors = 28.10 Btu/lb

Total heat difference = 9.58 Btu/lb difference

$$Q_t = 4.5 \times \text{cfm} \times \text{total heat difference}$$
$$Q_t = 4.5 \times 67 \times 9.58$$
$$Q_t = 2888 \text{ Btu/h total heat added due to ventilation}$$

System designers would require this total heat calculation to be broken down into the sensible-heat gain and the latent-heat gain. Equipment must be selected for the correct sensible- and latent-heat capacities or space humidity will not be correct. This calculation is done by using the sensible-heat and the latent-heat formulas as separate calculations. You will not arrive at the exact same total heat, because you cannot see the lines on the psychrometric chart close enough for total accuracy.

$$Q_s = 1.08 \times \text{cfm} \times \text{TD}$$

plus

$$Q_L = 0.68 \times \text{cfm} \times \text{grain difference}$$

where Q_L = latent heat
0.68 = a constant used to change cfm to pounds of air and grains per pound
grains difference = the difference in the grains per pound of air for the indoor air and the outdoor air

Example:

$$Q_s = 1.08 \times \text{cfm} \times \text{TD}$$
$$Q_s = 1.08 \times 67 \times 19$$
$$Q_s = 1375 \text{ Btuh sensible heat}$$

From the psychrometric chart points plotted earlier, you will find that the outdoor air contains 92.8 gr/lb and the indoor air contains 64 gr/lb for a difference of 28.8 gr/lb of air.

$$Q_L = 0.68 \times \text{cfm} \times \text{grains difference}$$
$$Q_L = 0.68 \times 67 \times 28.8$$
$$Q_L = 1312$$
$$\text{total heat} = Q_s + Q_L$$
$$\text{total heat} = 1375 + 1312$$
$$\text{total heat} = 2687 \text{ Btuh}$$

The total heat answer using wet-bulb is a little different. It is 2888 Btuh because you cannot see and line the chart up perfectly. The printing process will also cause small distortion in the charts. A very small error multiplied times the cfm will yield a rather large error because the cfm is a large number.

Office buildings have the same pollution problems, except there are more sources of pollution because there are more people and different and varied activities. Office buildings have a tendency to be remodeled more often leading to more construction-type pollution. Copying machines that use liquid copy methods give off vapors. Also, there are more people per square foot in office buildings.

The national code requirements for fresh air in buildings where the public works is based on the number of people in the building and the type of use. Different buildings may have different indoor pollution rates because of the activity in the building. For example, buildings that have a lot of copying machines or blueprint machines may have more indoor pollution than ones with only electric processors, such as computers. Department stores have different requirements than restaurants. Figure 35–34 shows some of the fresh air requirements for different applications recommended by the American Society of Heating, Refrigeration, and Air Conditioning Engineers (ASHRAE).

The design engineer is responsible for choosing the correct fresh air makeup for a building, and the service technician is responsible for regulating the airflow in the field. The technician may be given a set of specifications for a building and told to regulate the outdoor air dampers to induce the correct amount of outdoor air into the building.

The mixed air condition may be determined by calculation for the purpose of setting the outside air dampers. The condition may be plotted on the psychrometric chart for ease of understanding. The dampers may then be adjusted for the correct conditions. For example, suppose a building with a return air condition of 73°F dry-bulb and 60°F wet-bulb is taking in and mixing air from the outside of 90°F dry-bulb and 75°F wet-bulb. When these two conditions are plotted on the psychrometric chart and a line is drawn between them, the condition of the mixture air will fall on the line, Figure 35–35. If the mixture were half outdoor air and half indoor air, the condition of the mixture air would be midway between points A and B.

The point on line A to B may be calculated for any air mixture if the percentage of either air, outdoor or indoor, is known. For example, when the point for 25% outdoor air

RECOMMENDED OUTDOOR AIR VENTILATING RATES
Abbreviated from ASHRAE Standard 62-1989

	cfm or see footnote
Dining Room	20
Bars & cocktail lounges	30
Hotel conference rooms	20
Office spaces	20
Office conference rooms	35
Retail stores	.02 to .03(a)
Beauty shops	25
Ballrooms & discos	25
Spectator areas	15
Theater auditoriums	15
Transportation waiting rooms	15
Classrooms	15
Hospital patient rooms	25
Residences	.35(b)
Smoking lounges	60

a cfm per square feet of space
b air changes per hour

FIGURE 35–34 Fresh air requirements for some typical applications. *Used with permission from ASHRAE, Inc.*

and 75% indoor air mixture was calculated, the following calculation was used to find the dry-bulb point on the line. When the dry-bulb point is known and it falls on the line, you have a coordinate to find any other information for this air condition.

$$\text{25\% outdoor air: } 0.25 \times 90°\text{F} = 22.50$$
$$\text{75\% indoor air: } 0.75 \times 72°\text{F} = 54.75$$
$$\text{The mixture dry-bulb} = 77.25$$

When you look on the chart in Figure 35–36, you will find this point (C). Note, the point always falls closest to the air temperature with the most percentage of air.

A real problem may be something like this:

Building Air Conditioning Specifications

- Total building air = 50,000 cfm
- Recirculated air = 37,500 cfm, 75% return air
- Makeup air = 12,500 cfm, 25% outdoor air
- Indoor design temperature = 75°F dry-bulb, 50% relative humidity (62.5 wet-bulb)
- Outdoor design temperature (Atlanta, Georgia) = 94°F dry-bulb, 74°F wet-bulb

A technician arrives on the job to check the outdoor air percentage to find that the outdoor temperature is not at the design temperature. It is 93°F dry-bulb 75°F wet-bulb and the indoor conditions are 73°F dry-bulb 59°F wet-bulb. What should the mixture air temperature be?

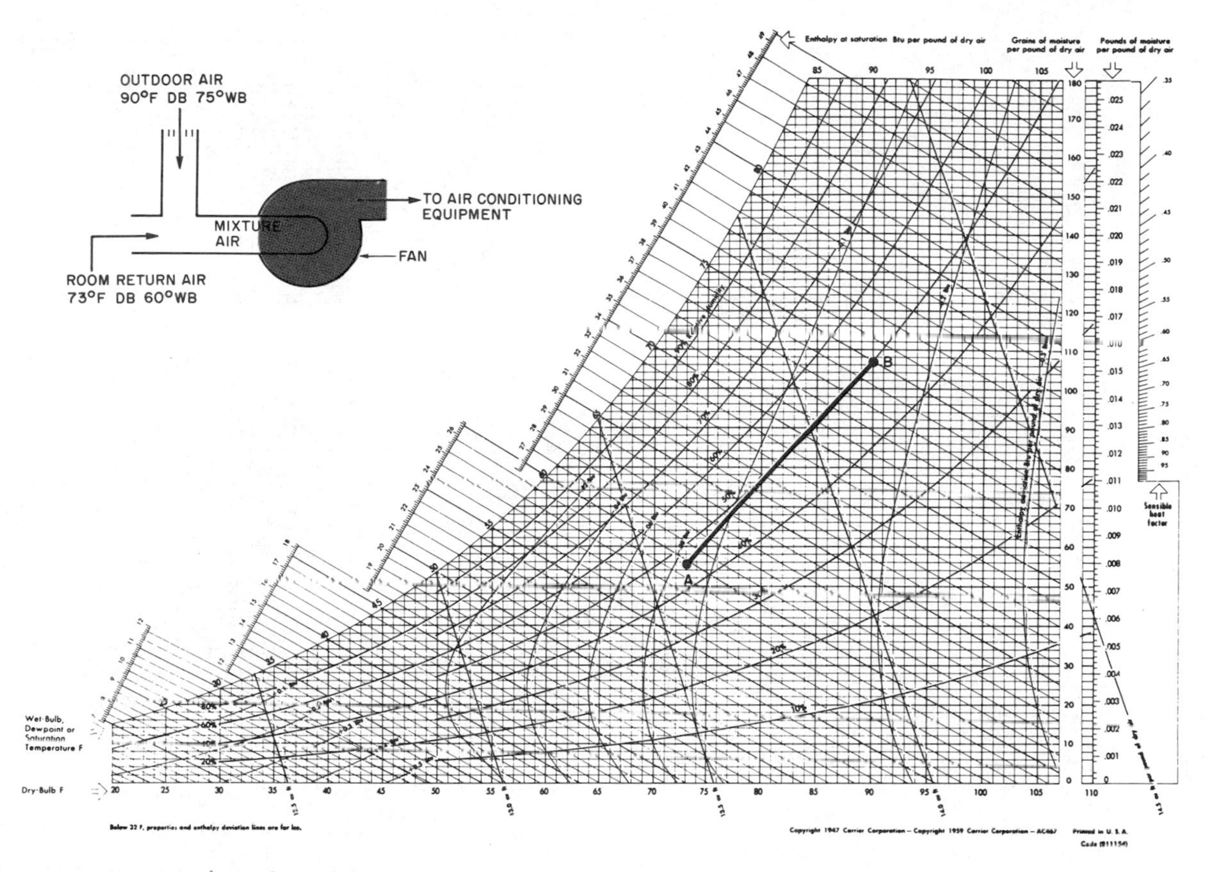

FIGURE 35–35 Point A is the condition of the indoor air and point B is the condition of the outdoor air. When the two are mixed, the mixed air condition will fall on the line drawn from A to B. *Adapted from Carrier Corporation Psychrometric Chart*

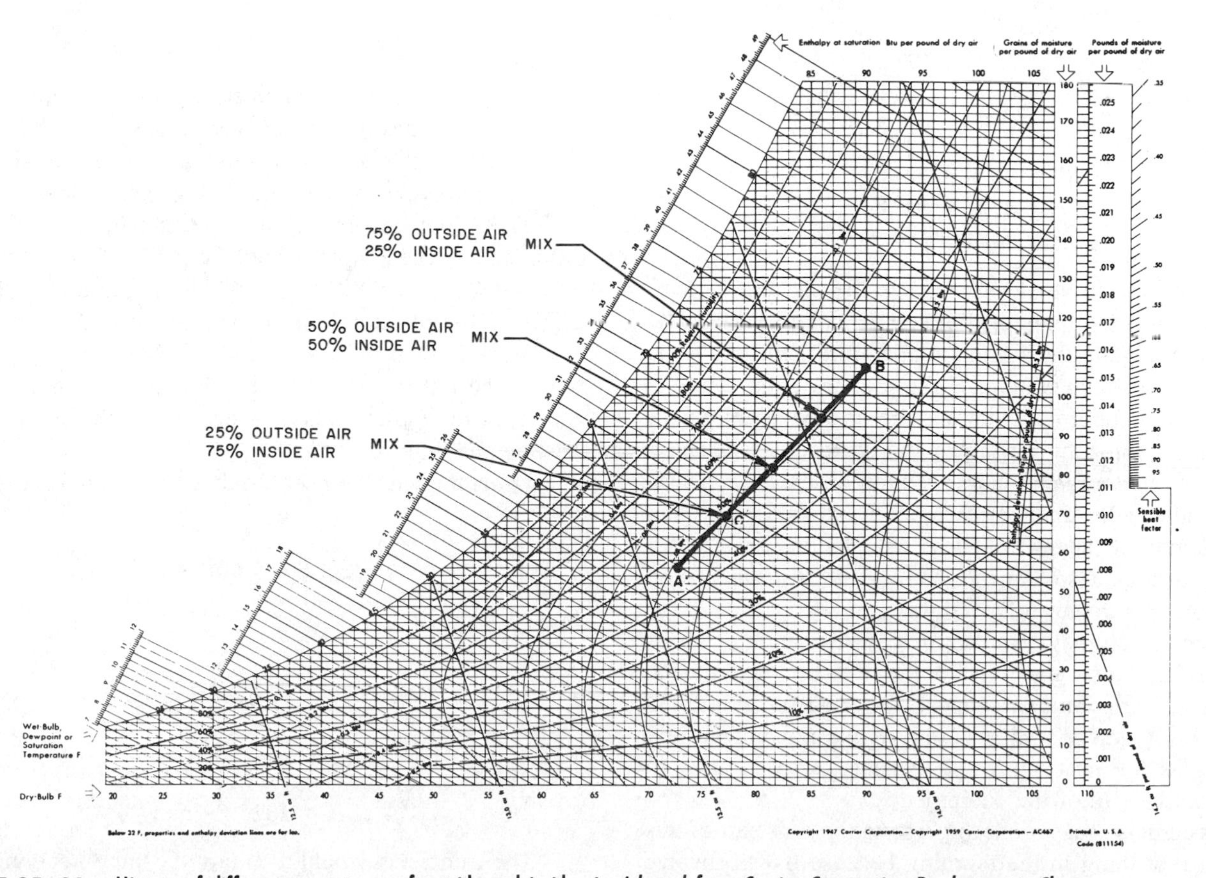

FIGURE 35–36 Mixture of different percentages of outside and inside air. *Adapted from Carrier Corporation Psychrometric Chart*

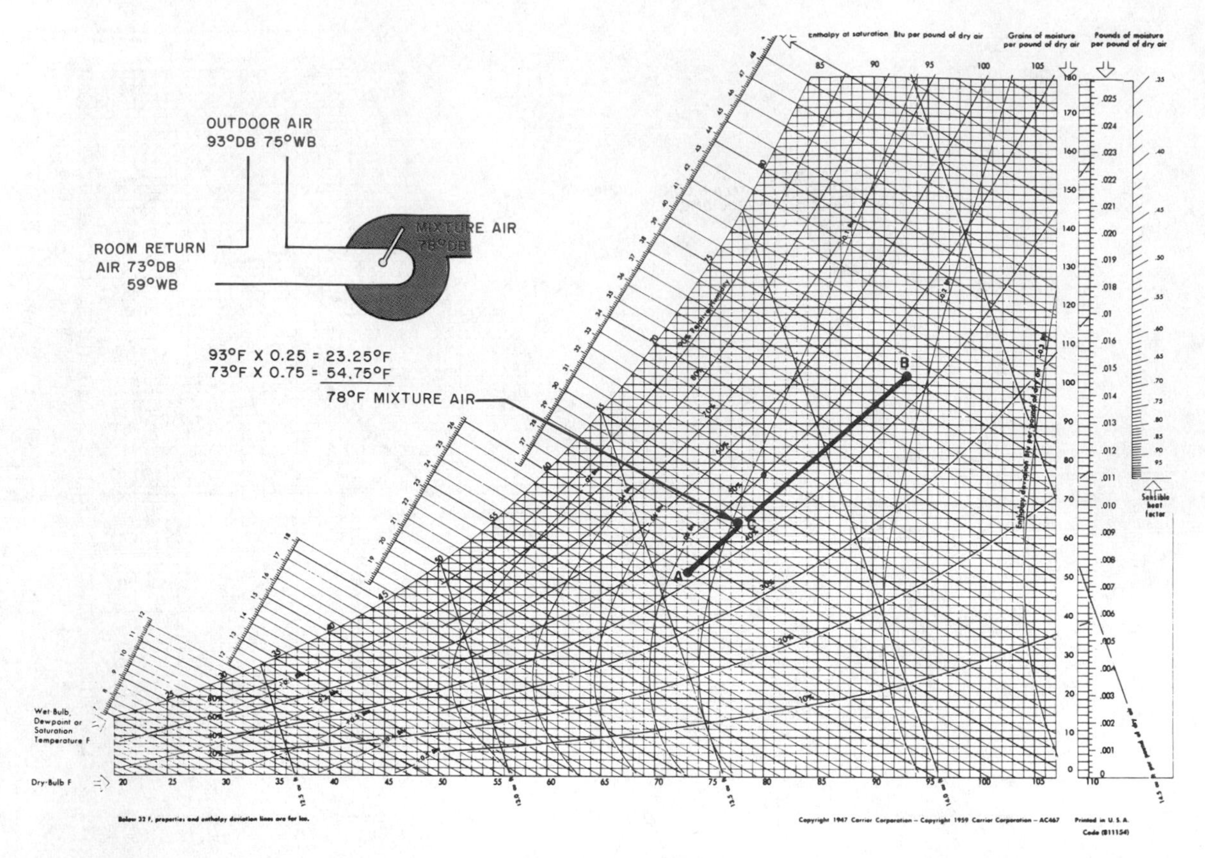

FIGURE 35–37 This practical problem of finding the mixture air condition is shown plotted on the chart. Point C is the mixture air. *Adapted from Carrier Corporation Psychrometric Chart*

A calculation for the mixture air is made. The actual conditions are plotted in Figure 35–37.

$$93°F \times 0.25 = 23.25$$
$$73°F \times 0.75 = 54.75$$
$$\text{Mix air dry-bulb} = 78°F$$

The technician would now check the mixture air temperature. If it is too high, the outdoor air dampers would be slightly closed until the air mix temperature stabilizes at the correct mixture temperature. If the mixture temperature is too low, the outdoor dampers may be opened slightly until the air mix temperature is stable and correct.

NOTE: The outdoor and indoor conditions should be rechecked after adjustment because of weather changes and indoor condition changes. These calculations and proper damper settings also depend on correct instrumentation while checking the air temperatures.

A technician may also use the psychrometric chart for field checking the capacity of a piece of air conditioning equipment. For example, suppose the capacity of a 5-ton (60,000 Btuh) unit on a shoe store is in question and the unit uses gas as the heat source. The heat source is used to determine the cfm of the equipment.

A technician may want to perform the cfm part of this problem first thing in the morning because it will involve operating the heating system long enough to arrive at the cfm. The unit heat is turned on and an accurate temperature rise across the gas heat exchanger is taken at 69.5°F.

NOTE: The technician was careful not to let radiant heat from the heat exchanger influence the thermometer lead.

The unit nameplate shows the unit to have 187,500 Btuh input. The output would be 80% of the input, or 187,500 Btuh × 0.80 = 150,000 Btuh output to the airstream. The fan motor would add about 600 W × 3431 Btu/W = 2048 Btu for a total of 152,048 Btuh.

NOTE: The fan is operated using the FAN ON position on the room thermostat. This ensures the same fan speed in heating and cooling.

Using the sensible-heat formula, the technician can find the cfm.

$$Q_s = 1.08 \times \text{cfm} \times \text{TD}$$

Solved for cfm:

$$\text{cfm} = \frac{Q_s}{1.08 \times TD}$$

$$\text{cfm} = \frac{152{,}048}{1.08 \times 69.6}$$

$$\text{cfm} = 2026 \text{ ft}^3/\text{min}$$

The technician would then turn the unit to cooling and check the wet-bulb temperature in the entering and leaving

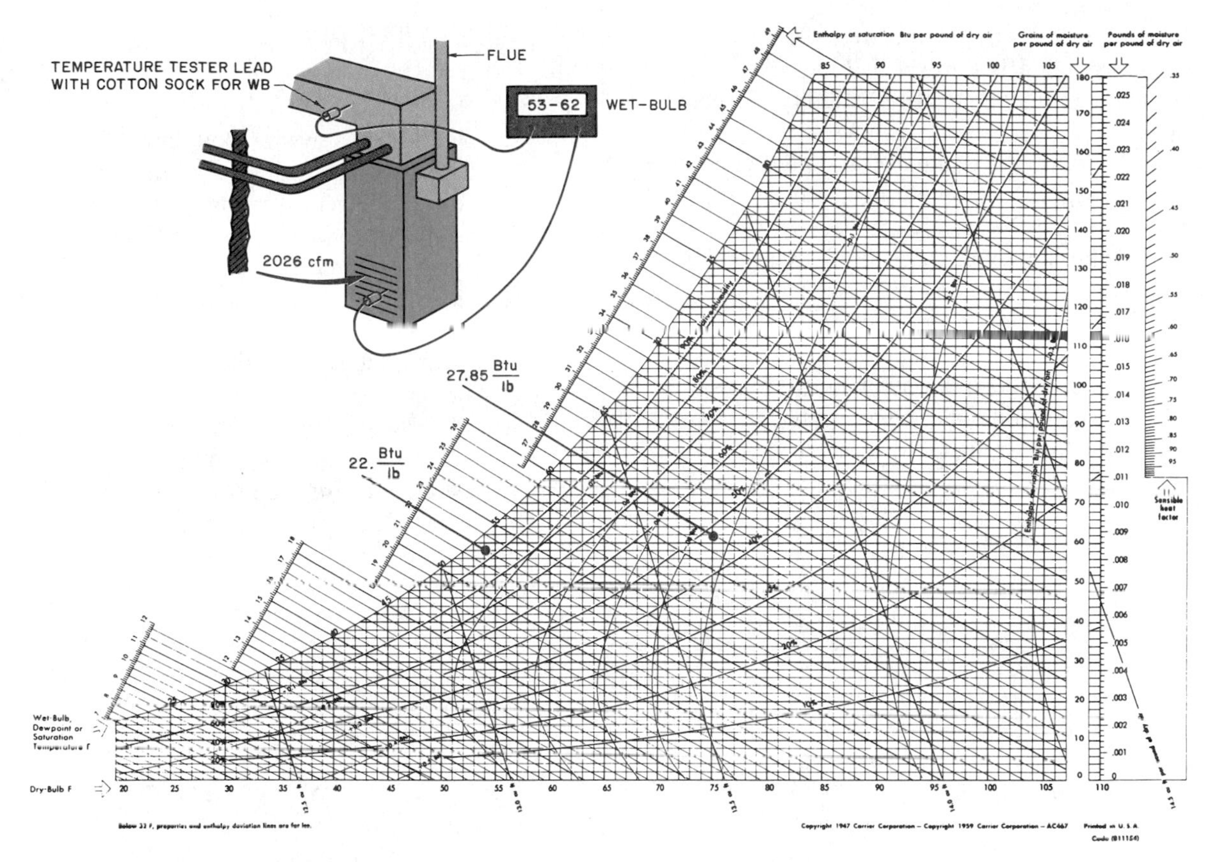

FIGURE 35–38 Capacity check for cooling mode. *Adapted from Carrier Corporation Psychrometric Chart*

airstream. Suppose the entering wet-bulb is 62°F and the leaving wet-bulb is 53°F, Figure 35–38:

Total heat at 62°F wet-bulb is 27.85

Total heat at 53°F wet-bulb is 22.00.

Total heat difference = 5.85 Btu/lb of air

Using the total heat formula:

$$Q_t = 4.5 \times \text{cfm} \times \text{total heat difference}$$

$$Q_t = 4.5 \times 2026 \times 5.85$$

$$Q_t = 53{,}334 \text{ Btuh total heat}$$

The unit has a capacity of 60,000 Btuh so it is operating very close to capacity. This is about as close as a technician can expect to obtain.

SUMMARY

- *Comfort* describes the delicate balance of feeling in relationship to our surroundings.
- The body burns food and turns it to energy.
- The body stores energy, wastes it, consumes it in work, or gives the heat off to the surroundings.
- For the body to be comfortable, it has to be warmer than the surroundings, so it can give up excess heat to the surroundings.
- The body gives off heat in three conventional ways: conduction, convection, and radiation. Evaporation may be considered a fourth way.
- Air contains 78% nitrogen 21% oxygen, 1% other gases, and suspended water vapor.
- The density of air at 70°F and 29.92 in. Hg is 0.075 lb.ft^3.
- The specific volume of air is the reciprocal of the density: 1/0.075 = 13.33 ft^3/lb.
- The moisture content of air can vary the transfer of heat from the human body; therefore, different temperatures and moisture content can give the same relative comfort level.
- Slight air movement can help to offset higher temperatures in the summer.
- Dry-bulb temperature is registered with a regular thermometer.
- Wet-bulb temperature is registered with a thermometer that has a wet wick. The wet bulb thermometer lead gets colder than the dry-bulb thermometer lead because the moisture on the wick evaporates.

- The difference between the wet-bulb reading and the dry-bulb reading is the wet-bulb depression. It can be used to determine the relative humidity of a conditioned space.
- The dew point temperature of air is the point at which moisture begins to drop out of the air.
- Water vapor in the air creates its own vapor pressure.
- The psychrometric chart is used to plot various air conditions.
- The wet-bulb reading on a psychrometric chart shows the total heat content of a pound of air.
- When the cubic feet of air per minute is known, the wet-bulb reading in and out of an air exchanger can give the total heat being exchanged. This can be used for field calculating the capacity of a unit.

REVIEW QUESTIONS

1. Name four comfort factors.
2. Name three ways the body gives off heat.
3. How does perspiration cool the body?
4. How can lower room temperatures be offset in winter?
5. What two conditions can offset higher room temperatures in summer?
6. How is the relative humidity of a conditioned space measured?
7. What is the name of the chart used to plot the various air conditions?
8. What two unknowns are the easiest to obtain in the field to plot air conditions?
9. Describe the dew point temperature.
10. What is the density of air at 70°F?

U N I T 36

Refrigeration Applied to Air Conditioning

OBJECTIVES

After studying this unit, you should be able to

- explain three ways in which heat transfers into a structure.
- state two ways that air is conditioned for cooling.
- explain refrigeration as applied to air conditioning.
- describe an air conditioning evaporator.
- describe three types of air conditioning compressors.
- describe an air conditioning condenser.
- describe an air conditioning metering device.
- list different types of evaporator coils.
- identify different types of condensers.
- explain how "high efficiency" is accomplished.
- describe package air conditioning equipment.
- describe split-system equipment.

SAFETY CHECKLIST

* Use precaution when loading or unloading equipment from the dock or truck. Wear an approved back brace belt when lifting and use your legs, keeping your back straight and upright. It may be necessary to use a small crane. Trucks with lift gates may be used to lower the equipment to the ground.
* When installing equipment in attics and under buildings be alert for stinging insects such as spiders. Avoid being scratched or cut by exposed nails and other sharp objects in attics and in crawl spaces under buildings.
* Wear goggles and gloves when connecting charged line sets or gages because escaping refrigerant can cause serious frostbite.

36.1 REFRIGERATION

Air conditioning (cooling) is refrigeration applied to keeping the space temperature of a building cool during the hot summer months. The air conditioning system (refrigeration) removes the heat that leaks into the structure from the outside and deposits it outside the structure where it came from. Some people living in the warmer climate sections of the country may never have air conditioning, but there are times when they are probably uncomfortable. When the nights are warm (above 75°F) and the humidity is high, it is hard to be comfortable enough to rest well.

The basics of refrigeration and its components were discussed earlier in this book. Some of the material is covered again here so that it will be readily available. As you study this unit, you will find that many of the components used for refrigeration applied to air conditioning are different from the components applied to commercial refrigeration.

36.2 STRUCTURAL HEAT GAIN

Heat leaks into a structure by conduction, infiltration, and radiation (the sun's rays or solar load). The summer *solar load* on a structure is greater on the east and west sides because the sun shines for longer periods of time on these parts of the structure, Figure 36–1. If a building has an attic, the air space can be ventilated to help relieve the solar load on the ceiling, Figure 36–2. If the structure has no attic, it is at the mercy of the sun unless it is well insulated, Figure 36–3.

Conduction heat enters through walls, windows, and doors. The rate depends on the temperature difference be-

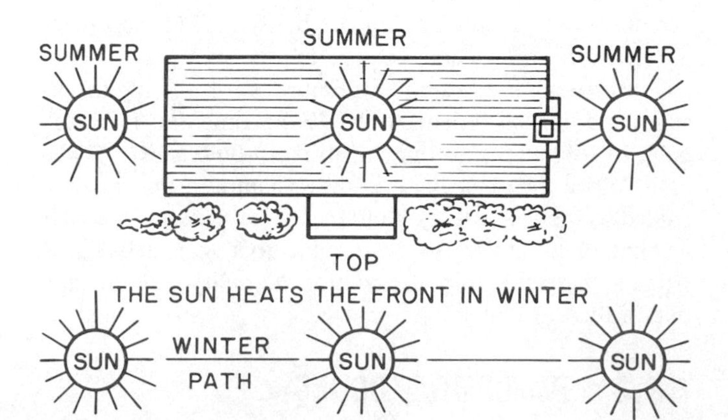

FIGURE 36–1 Solar load on a home.

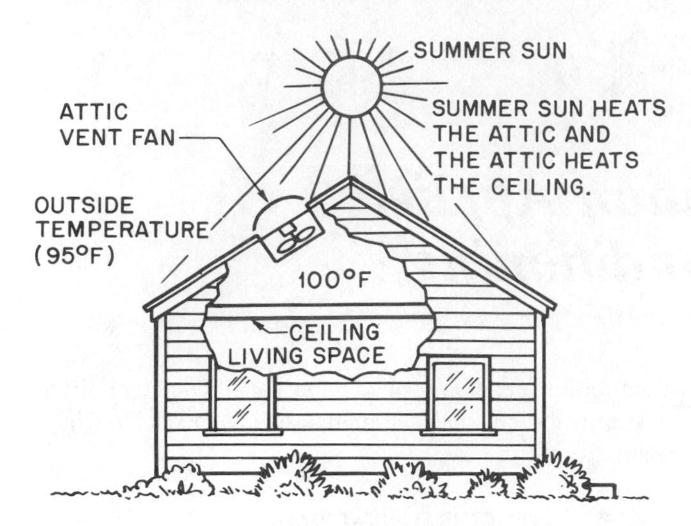

FIGURE 36–2 Ventilated attic to keep the solar heat from the ceiling of the house.

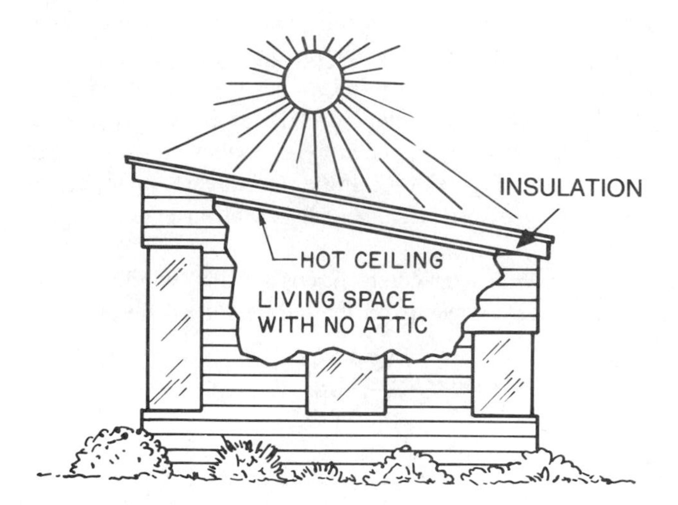

FIGURE 36–3 House has no attic. The sun shines directly on the ceiling of the living space.

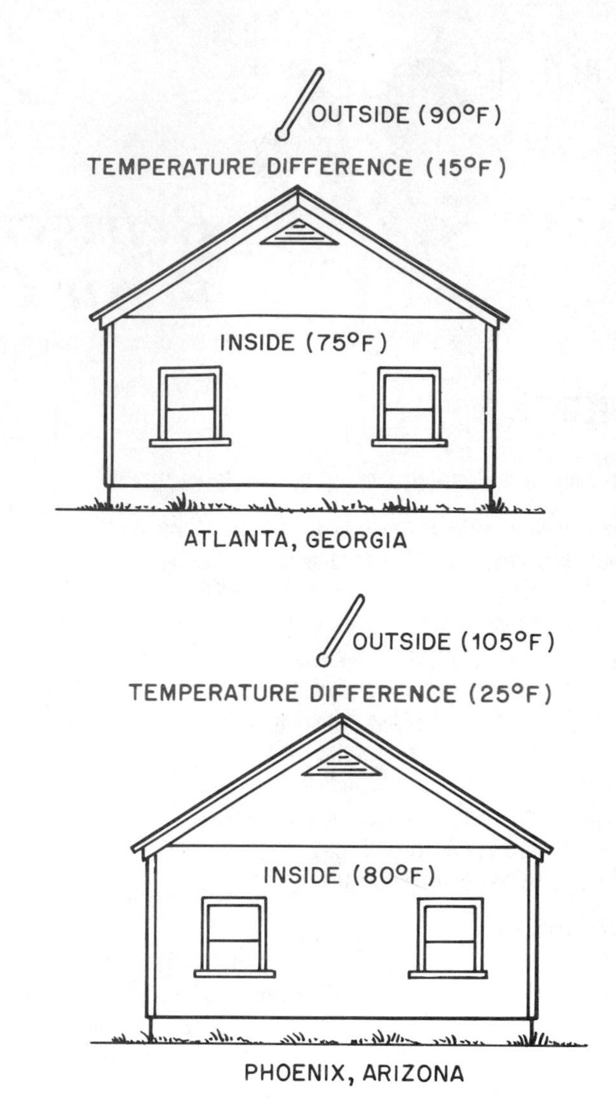

FIGURE 36–4 Difference in the inside and outside temperatures of a home in Atlanta, Georgia and a home in Phoenix, Arizona.

tween the inside and the outside of the house, Figure 36–4.

Some of the warm air that gets into the structure is infiltrated through the cracks around the windows and doors. Air also leaks in when the doors are opened for persons entering and leaving the building. This infiltrated air has different characteristics in different parts of the country. Using the example in Figure 36–4, the typical design condition in Phoenix is 105°F dry-bulb and 71°F wet-bulb. In Atlanta the air may be 90°F dry-bulb and 73°F wet-bulb. When the air leaks into the structure in Phoenix it is cooled to the space temperature. This air contains a certain amount of humidity for each cubic foot that leaks in. In Atlanta, there will normally be more humidity with the infiltration than in Phoenix.

36.3 EVAPORATIVE COOLING

Air has been conditioned in more than one way to achieve comfort. In the climates where the humidity is low, a device called an *evaporative cooler,* Figure 36–5, has been used for years. This device uses fiber mounted in a frame with water slowly running down the fiber as the cooling media. Fresh air is drawn through the water-soaked fiber and cooled by evaporation to a point close to the wet-bulb temperature of the air. The air entering the structure is very humid but cooler than the dry-bulb temperature. For example, in Phoenix, Arizona, the design dry-bulb temperature in summer is 105°F. At the same time the dry-bulb is 105°F, the wet-bulb temperature may be 70°F. An evaporative cooler may lower the air temperature entering the room to 80°F dry-bulb. 80°F air is cool compared to 105°F, even if the humidity is high.

36.4 REFRIGERATED COOLING OR AIR CONDITIONING

Refrigerated air conditioning is similar to commercial refrigeration because the same components are used to cool the air: (1) the evaporator, (2) the compressor, (3) the condenser, and (4) the metering device. These components are assembled in several ways to accomplish the same goal,

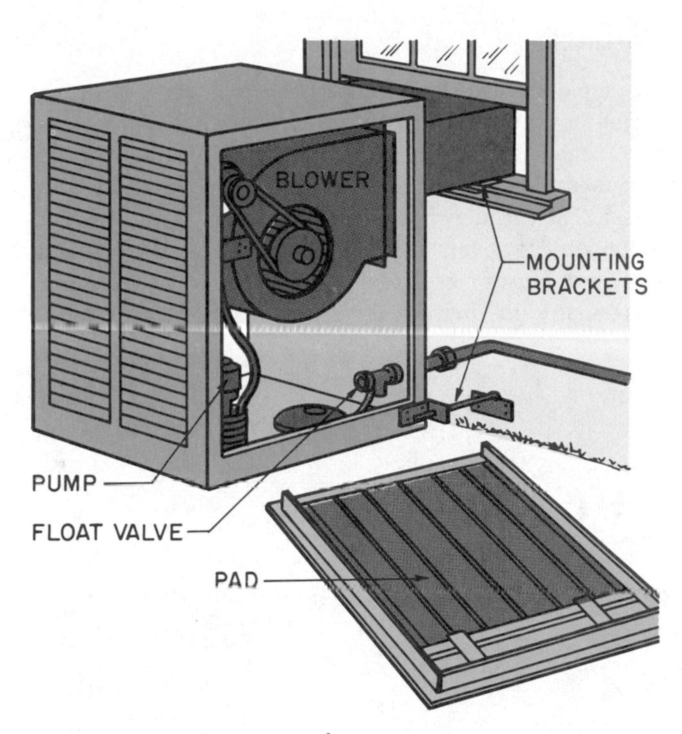

FIGURE 36–5 Evaporative cooler.

refrigerated air to cool space. Review Unit 3 if you are not familiar with the basics of refrigeration.

Package Air Conditioning

The four components are assembled into two basic types of equipment for air conditioning purposes: package equipment and split-system equipment. With *package equipment* all of the components are built into one cabinet. It is also called *self-contained* equipment, Figure 36–6. Air is ducted to and from the equipment. Package equipment may be located beside the structure or on top of it. In some instances the heating equipment is built into the same cabinet.

Split-System Air Conditioning

In *split system air conditioning* the condenser is located outside, remote from the evaporator, and uses interconnecting refrigerant lines. The evaporator may be located in the attic, a crawl space, or a closet for upflow or downflow applications. The fan to blow the air across the evaporator may be included in the heating equipment, or a separate fan may be used for the air conditioning system, Figure 36–7.

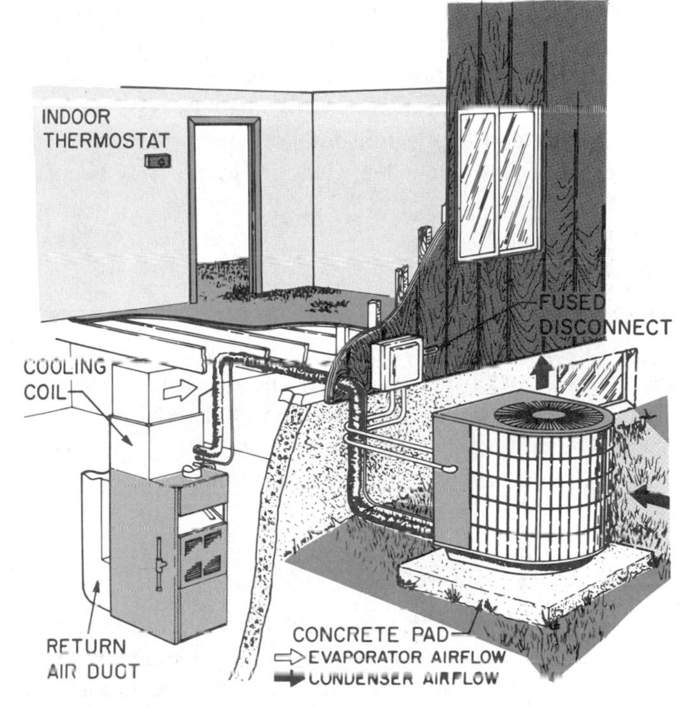

FIGURE 36–7 Split air conditioning system. *Courtesy Climate Control*

36.5 THE EVAPORATOR

The *evaporator* is the component that absorbs heat into the refrigeration system. It is a refrigeration coil made of aluminum or copper with aluminum fins on either type attached to the coil to give it more surface area for better heat exchange. The evaporator coil has several designs for airflow through the coil and draining the condensate water from the coil, depending on the installation. The different designs are known as the *A* coil, the *slant* coil, and the *H* coil.

FIGURE 36–6 Package air conditioner. *Courtesy Heil Heating and Cooling Products*

The A Coil

The *A coil* is used for upflow, downflow, and horizontal flow applications. It consists of two coils with their circuits side by side and spread apart at the bottom in the shape of the letter A, Figure 36–8. When used for upflow or downflow, the condensate pan is at the bottom of the A pattern. When used for horizontal flow, a pan is placed at the bottom of the coil and the coil is turned on its side. The airflow through an A coil is through the core of the coil. It cannot be from side to side with the two coils in series. The A coil is not the best coil application for horizontal airflow. When horizontal airflow is needed, slant or H coils may be more desirable.

The Slant Coil

The *slant* coil is a one-piece coil mounted in the duct on an angle (usually 60°) or slant to give the coil more surface area. The slant of the coil causes the condensate water to drain to the condensate pan located at the bottom of the slant. The coil can be used for upflow, downflow, or horizontal flow when designed for these applications, Figure 36–9.

The H Coil

The *H coil* is normally applied to horizontal applications although it can be adapted to vertical applications by using special drain pan configurations. The drain is normally at the bottom of the H pattern, Figure 36–10.

FIGURE 36–8 An A coil. *Courtesy Carrier Corporation*

FIGURE 36–9 A slant coil. *Courtesy BDP Company*

Coil Circuits

All of the aforementioned coils may have more than one circuit for the refrigerant. As was indicated in Unit 21, when a coil becomes too long and excessive pressure drop occurs, it is advisable to have more than one coil in parallel, Figure 36–11. The coil may have as many circuits as necessary to do the job. However, when more than one circuit is used, a distributor must be used to distribute the correct amount of refrigerant to the individual circuits, Figure 36–12.

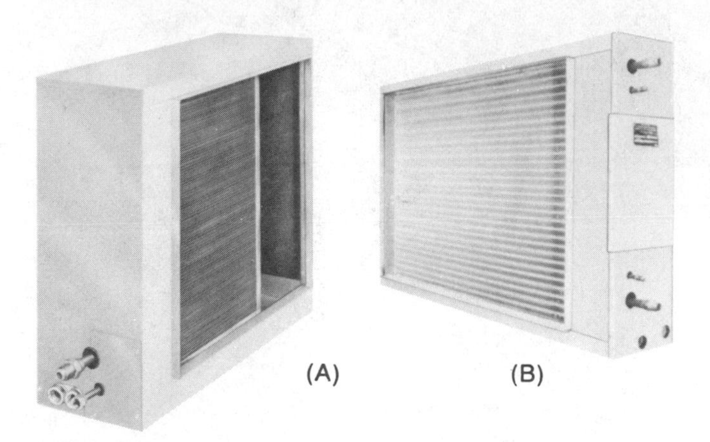

FIGURE 36–10 An H coil. *Courtesy BDP Company*

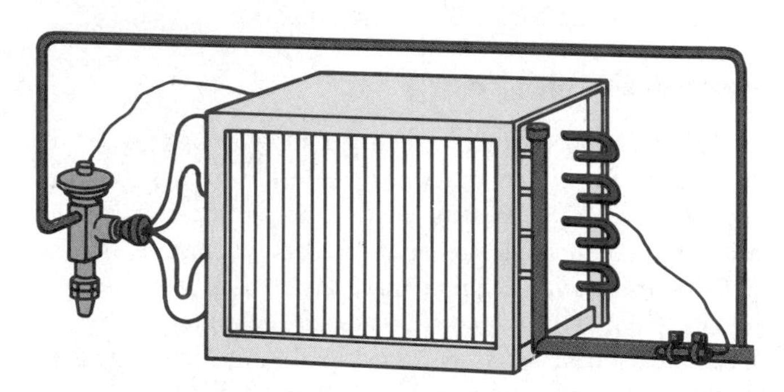

FIGURE 36–11 Multicircuit coil. *Courtesy Sporlon Valve Company*

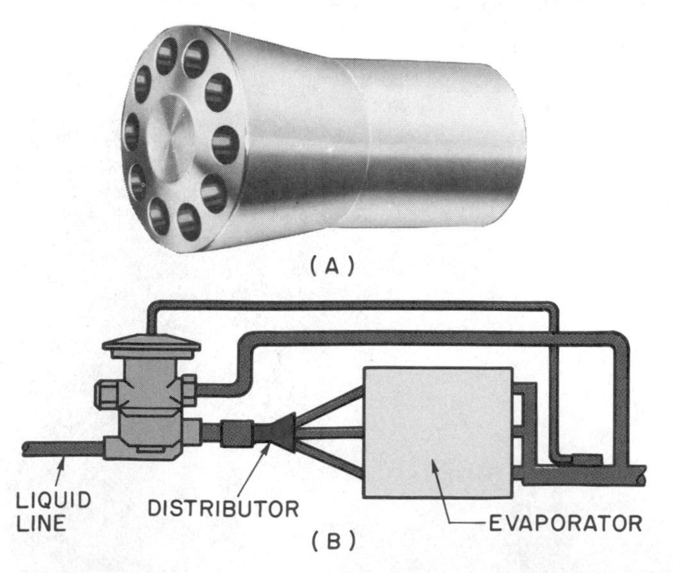

FIGURE 36–12 Refrigerant distributor. *Courtesy Sporlon Valve Company*

CLIMATIC CONDITIONS FOR THE UNITED STATES[a]

						Winter[d]		Summer[e]						
Col. 1	Col. 2		Col. 3		Col. 4	Col. 5		Col. 6			Col. 7	Col. 8		
State and Station	Latitude[b]		Longitude[b]		Elevation[c]	Design Dry-Bulb		Design Dry-Bulb and Mean Coincident Wet-Bulb			Mean Daily Range	Design Wet-Bulb		
	°	′	°	′	Ft	99%	97.5%	1%	2.5%	5%		1%	2.5%	5%
ARIZONA														
Douglas AP	31	3	109	3	4098	27	31	98/63	95/63	93/63	31	70	69	68
Flagstaff AP	35	1	111	4	6973	−2	4	84/55	82/55	80/54	31	61	60	59
Fort Huachuca AP (S)	31	3	110	2	4664	24	28	95/62	92/62	90/62	27	69	68	67
Kingman AP	35	2	114	0	3446	18	25	103/65	100/64	97/64	30	70	69	69
Nogales	31	2	111	0	3800	28	32	99/64	96/64	94/64	31	71	70	69
Phoenix AP (S)	33	3	112	0	1117	31	34	109/71	107/71	105/71	27	76	75	75
Prescott AP	34	4	112	3	5014	4	9	96/61	94/60	92/60	30	66	65	64
Tuscon AP (S)	32	1	111	0	2584	28	32	104/66	102/66	100/66	26	72	71	71
Winslow AP	35	0	110	4	4880	5	10	97/61	95/60	93/60	32	66	65	64
Yuma AP	32	4	114	4	199	36	39	111/72	109/72	107/71	27	79	78	77
CALIFORNIA														
Bakersfield AP	35	2	119	0	495	30	32	104/70	101/69	98/68	32	73	71	70
Barstow AP	34	5	116	5	2142	26	29	106/68	104/68	102/67	37	73	71	70
Blythe AP	33	4	114	3	390	30	33	112/71	110/71	108/70	28	75	75	74
Burbank AP	34	1	118	2	699	37	39	95/68	91/68	88/67	25	71	70	69
Chico	39	5	121	5	205	28	30	103/69	101/68	98/67	36	71	70	68
Los Angeles AP (S)	34	0	118	2	99	41	43	83/68	80/68	77/67	15	70	69	68
Los Angeles CO (S)	34	0	118	1	312	37	40	93/70	89/70	86/69	20	72	71	70
Merced-Castle AFB	37	2	120	3	178	29	31	102/70	99/69	96/68	36	72	71	70
Modesto	37	4	121	0	91	28	30	101/69	98/68	95/67	36	71	70	69
Monterey	36	4	121	5	38	35	38	75/63	71/61	68/61	20	64	62	61
Napa	38	2	122	2	16	30	32	100/69	96/68	92/67	30	71	69	68
Needles AP	34	5	114	4	913	30	33	112/71	110/71	108/70	27	75	75	74
Oakland AP	37	4	122	1	3	34	36	85/64	80/63	75/62	19	66	64	63
Oceanside	33	1	117	2	30	41	43	83/68	80/68	77/67	13	70	69	68
Ontario	34	0	117	36	995	31	33	102/70	99/69	96/67	36	74	72	71
GEORGIA														
Albany, Turner AFB	31	3	84	1	224	25	29	97/77	95/76	93/76	20	80	79	78
Americus	32	0	84	2	476	21	25	97/77	94/76	92/75	20	79	78	77
Athens	34	0	83	2	700	18	22	94/74	92/74	90/74	21	78	77	76
Atlanta AP (S)	33	4	84	3	1005	17	22	94/74	92/74	90/73	19	77	76	75
Augusta AP	33	2	82	0	143	20	23	97/77	95/76	93/76	19	80	79	78
Brunswick	31	1	81	3	14	29	32	92/78	89/78	87/78	18	80	79	79
Columbus, Lawson AFB	32	3	85	0	242	21	24	95/76	93/76	91/75	21	79	78	77
Dalton	34	5	85	0	720	17	22	94/76	93/76	91/76	22	79	78	77
Dublin	32	3	83	0	215	21	25	96/77	93/76	91/75	20	79	78	77
Gainesville	34	2	83	5	1254	16	21	93/74	91/74	89/73	21	77	76	75
NEW YORK														
Albany AP (S)	42	5	73	5	277	−6	−1	91/73	88/72	85/70	23	75	74	72
Albany CO	42	5	73	5	19	−4	1	91/73	88/72	85/70	20	75	74	72
Auburn	43	0	76	3	715	−3	2	90/73	87/71	84/70	22	75	73	72
Batavia	43	0	78	1	900	1	5	90/72	87/71	84/70	22	75	73	72
Binghamton AP	42	1	76	0	1590	−2	1	86/71	83/69	81/68	20	73	72	70
Buffalo AP	43	0	78	4	705r	2	6	88/71	85/70	83/69	21	74	73	72
Cortland	42	4	76	1	1129	−5	0	88/71	85/71	82/70	23	74	73	71
Dunkirk	42	3	79	2	590	4	9	88/73	85/72	83/71	18	75	74	72
Elmira AP	42	1	76	5	860	−4	1	89/71	86/71	83/70	24	74	73	71
Geneva (S)	42	5	77	0	590	−3	2	90/73	87/71	84/70	22	75	73	72

[a]Table 1 was prepared by ASHRAE Technical Committee 4.2, Weather Data, from data compiled from official weather stations where hourly weather observations are made by trained observers, See also Ref 1, 2, 3, 5 and 6.

[b]Latitude, for use in calculating solar loads, and longitude are given to the nearest 10 minutes. For example, the latitude and longitude for Anniston, Alabama are given as 33 34 and 85 55 respectively, or 33° 40, and 85° 50.

[c]Elevations are ground elevations for each station. Temperature readings are generally made at an elevation of 5 ft above ground, except for locations marked r, indicating roof exposure of thermometer.

[d]Percentage of winter design data shows the percent of the 3-month period, December through February.

[e]Percentage of summer design data shows the percent of 4-month period, June through September.

FIGURE 36–13 Excerpt from a table shows different design conditions for various parts of the United States. *Used with permission from ASHRAE, Inc., 1791 Tullie Circle, NE, Atlanta, GA 30329*

36.6 THE FUNCTION OF THE EVAPORATOR

The evaporator is a heat exchanger that takes the heat from the room air and transfers it to the refrigerant. Two kinds of heat must be transferred: sensible heat and latent heat. *Sensible heat* lowers the air temperature, and *latent heat* changes the water vapor in the air to condensate. The condensate collects on the coil and runs through the drain pan to a trap (to stop air from pulling into the drain) and then it is normally piped to a drain.

Typically, room air may be 75°F dry-bulb and have a humidity of 50%, which is 62.5°F wet-bulb. The coil generally operates at a refrigerant temperature of 40°F to remove the required amount of sensible heat (to lower the air temperature) and latent heat (to remove the correct amount of moisture). The air leaving the coil is approximately 55°F dry-bulb with a humidity about 95%, which is 54°F wet-bulb. Notice the high humidity leaving the coil. This is because the air has been cooled. When it mixes with room air, it is heated and expands where it can absorb moisture from the room air. The result is dehumidification. These conditions are average for a climate with high humidity. If the humidity is very high, such as in coastal locations, the coil temperature may be a little lower to remove more humidity. If the system is in a locality where the humidity is very low, such as in the desert areas of the country, the coil temperature may be a little higher than 40°F. The coil temperature is controlled with the airflow across the evaporator coil. More airflow will cause higher coil temperature and less airflow lower coil temperatures. The same equipment can be sold in all parts of the country and the airflow can be varied to accomplish the proper evaporator temperature for the proper humidity.

36.7 DESIGN CONDITIONS

A house built in Atlanta has less sensible-heat load and more latent-heat load than one constructed in Phoenix. The designer or engineer must be familiar with local design practices, Figure 36–13. The airflow across the same coil in the two different parts of the country may be varied to accomplish different air conditions.

36.8 EVAPORATOR APPLICATION

The evaporator may be installed in the airstream in several different ways. It may have a coil case that encloses the coil, Figure 36–14, or it may be located in the ductwork. The coil will normally operate below the dew point temperature. The coil enclosure should be insulated to keep it from absorbing heat from the surroundings. An insulated coil cabinet will not sweat on the outside, Figure 36–15. Many evaporators

FIGURE 36–14 Evaporator with a coil case. *Courtesy BDP Company*

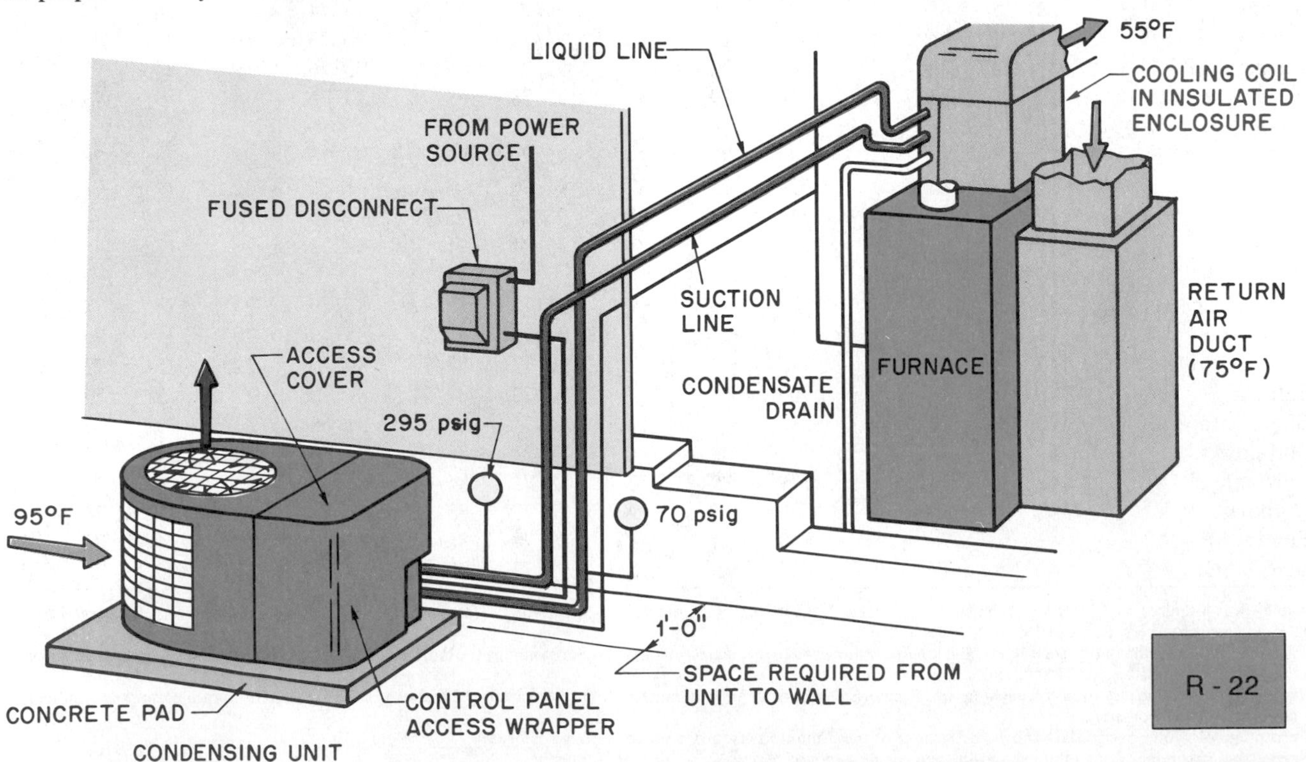

FIGURE 36–15 Operating conditions of an evaporator and a condenser. Notice the cooling coil is in an insulated enclosure to prevent sweating on the outside.

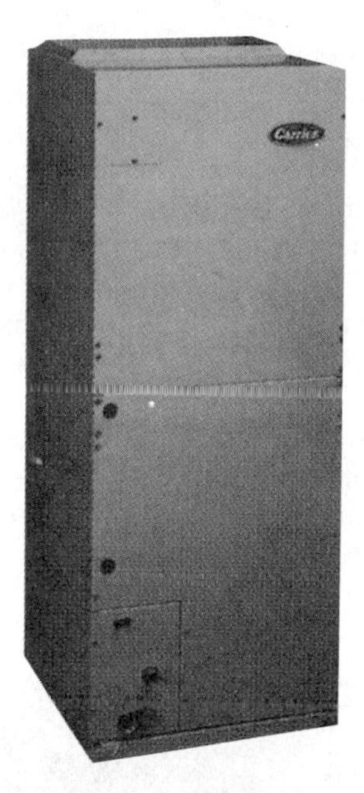

FIGURE 36–16 Evaporator mounted in the air handler. *Courtesy Carrier Corporation*

are built into the air handler by the manufacturer, Figure 36–16.

36.9 THE COMPRESSOR

The following types of compressors are used in air conditioning systems: reciprocating, rotary, scroll, centrifugal, and screw. These are of the same design as similar types used in commercial refrigeration. The centrifugal and screw compressors, described briefly in Unit 23, are used primarily in large commercial and industrial applications and will be discussed further later in this text.

Compressors are vapor pumps that pump heat-laden refrigerant vapor from the low-pressure side of the system, the evaporator side, to the high-pressure or condenser side. To do this they compress the vapor from the low side, increasing the pressure and raising the temperature.

36.10 THE RECIPROCATING COMPRESSOR

The reciprocating compressor used in residential and light commercial air conditioning applications is similar to those discussed in Unit 23.

These compressors may be either the fully hermetic or the serviceable hermetic type, Figure 36–17. Modern residential systems with reciprocating compressors usually are equipped with the fully hermetic type. These are suction gas cooled, Figure 36–18. These compressors are positive-displacement compressors and use R-22 refrigerant. Some units built before 1965 may use R-12 or R-500.

Serviceable hermetic compressors are often used in larger systems found in commercial applications. These compressors may be either gas or air cooled. If suction gas cooled, the suction line is piped to the motor end of the compressor, Figure 36–17. The maximum high-suction gas temperature is 70°F.

(A)

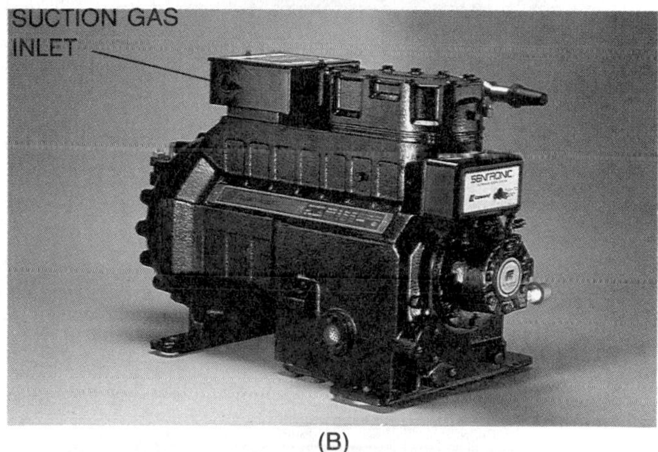

(B)

FIGURE 36–17 (A) Hermetic compressor. (B) Serviceable hermetic compressor. *Courtesy (A) Bristol Compressors, Inc., (B) Copeland Corporation*

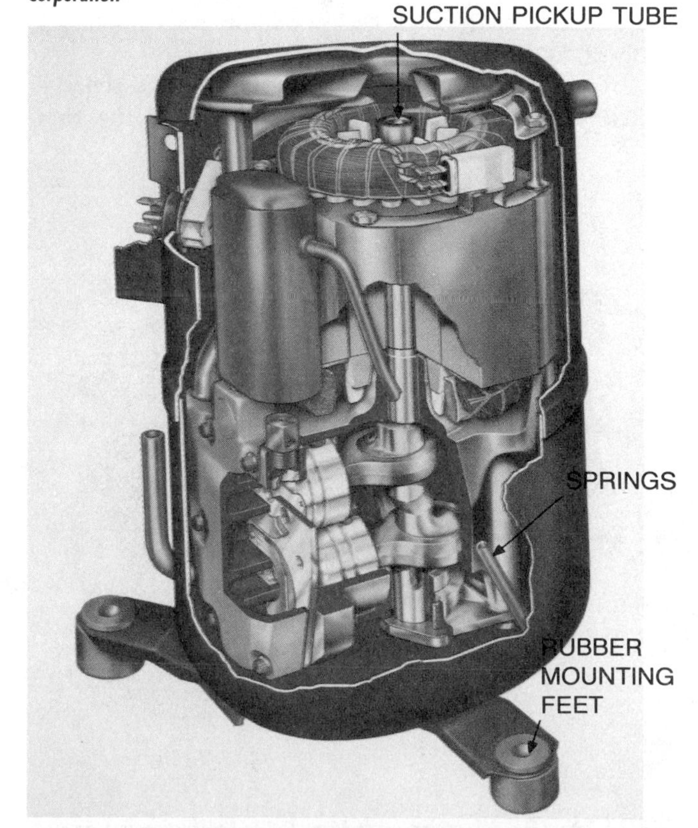

FIGURE 36–18 Suction gas-cooled compressor. *Courtesy Tecumseh Products Company*

36.11 COMPRESSOR SPEEDS (rpm)

Modern air conditioning compressors used in the small and medium size ranges must turn standard motor speeds of 3450 rpm or 1750 rpm. Early compressors were 1750 rpm, used R-12, and were large and heavy. Present compressors use the faster motor and the more efficient refrigerant R-22, so equipment can be smaller and lighter.

36.12 COOLING THE COMPRESSOR AND MOTOR

All compressors must be cooled because some of the mechanical energy in the compressor converts to heat energy. Without cooling, the compressor motor would burn and the oil would become hot enough to break down and form carbon. Hermetic compressors have always been considered suction gas cooled. The large ones are directly suction gas cooled. Small compressors, up to 7 1/2 horsepower may not be directly cooled by the suction gas. Some compressors, such as the rotary and scroll compressors, are cooled by the discharge gas leaving the compressor. This discharge gas temperature is influenced by the suction gas temperature. If the correct superheat is not maintained at the compressor, the compressor will overheat because the discharge gas temperature will rise. Because of this, the compressor motor temperature is still controlled by the suction gas temperature.

The newer compressor motors are able to run at higher temperatures because of better materials in the motor windings and higher quality lubricating oils. These are materials that will still insulate the winding wires and not break down while running under high heat conditions and oils that do not break down at reasonably high temperatures.

The technician should be aware of the discharge-cooled compressor and not determine that the compressor is running too hot. The case of the compressor is hotter than a suction-cooled compressor. A check of the discharge line temperature would be the sure test. Most manufacturers agree that the discharge line leaving the compressor should not exceed 225°F.

FIGURE 36–19 Air-cooled compressor. Notice the suction line enters at the side of the cylinder. *Courtesy Copeland Corporation*

Some serviceable hermetic compressors are air cooled, Figure 36–19. The suction line enters the compressor at the side of the cylinder rather than near the motor. These compressors have ribs that help dissipate the heat into the air passing over it. It is essential that these compressors be located in a moving airstream.

In some compressors used on water-cooled equipment, the inlet water line is wrapped around the compressor motor body, Figure 36–20.

FIGURE 36–20 Compressor similar to that in Figure 36–19, but it is applied to a water-cooled condenser. *Photo by Bill Johnson*

36.13 COMPRESSOR MOUNTINGS

Welded hermetic reciprocating compressors all have rubber mounting feet on the outside, and the compressor is mounted on springs inside the shell, Figure 36–18. Older compressors were mounted on springs outside the shell and the compressor was pressed into the shell. New compressors have a vapor space between the motor and the shell, so the motor-temperature sensor must be on the inside to sense the motor temperature quickly.

The suction gas dumps out into the shell, usually in the vicinity of the motor. Some compressors dump the suction gas directly into the rotor. The turning rotor tends to dissipate any liquid drops in the return suction gas. The suction pickup tube for the compressor, which is inside the shell, is normally located in a high position so that liquid refrigerant or foaming oil cannot enter the compressor cylinders, Figure 36–18.

The compressor for air conditioning equipment is normally located outside with the condenser. These hermetic compressors cannot be field serviced and often are not factory serviced. When one becomes defective, the manufacturer may authorize the technician to discard it or may require that it be returned to the factory to determine what made it fail.

36.14 REBUILDING THE HERMETIC COMPRESSOR

Some manufacturers are remanufacturing hermetic compressors by opening the shell and repairing the compressor inside. The manufacturer will be the one to decide if this is economical or not. The larger the compressor, the more advantage there would be to repairing it.

The standard serviceable hermetic compressor is cast iron, and the manufacturer will want this compressor returned for remanufacturing. Small cast iron compressors are not used widely in small air conditioning equipment because of the initial cost.

38.15 THE ROTARY COMPRESSOR

The rotary compressor is small and light. It is sometimes cooled with compressor discharge gas, which makes the compressor appear to be running too hot. A warning may be posted in the compressor compartment that the housing will appear to be too hot. The compressor and motor are pressed into the shell of a rotary compressor and there is no vapor space between the compressor and the shell. This also reduces the size of the rotary compressor compared to a reciprocating compressor.

The rotary compressor is more efficient than the reciprocating compressor and is used in small- to medium-sized systems, Figure 36–21. These compressors are manufactured in two basic design types: the stationary vane and the rotary vane.

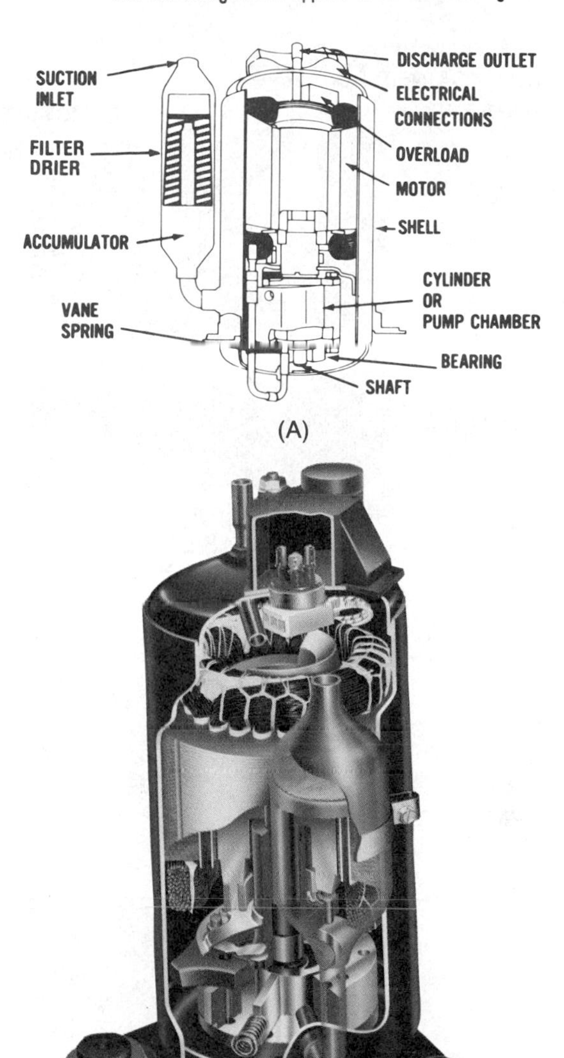

FIGURE 36–21 Rotary compressor. *Courtesy Motors and Armatures, Inc.*

Stationary Vane Rotary Compressor

The components for this compressor are the housing, a blade or vane, a shaft with an off-center (eccentric) rotor, and a discharge valve. The shaft turns the off-center rotor so that it "rolls" around the cylinder, Figure 36–22. The blade or vane keeps the intake and compression chambers of the cylinder separate. The tolerances between the rotor and cylinder must be very close, and the vane must be machined so that gas does not escape from the discharge side to the intake. As the rotor turns, the vane slides in and out to remain tight against the rotor. The valve at the discharge keeps the compressed gas from leaking back into the chamber and into the suction side during the off cycle.

As the shaft turns, the rotor rolls around the cylinder, allowing suction gas to enter through the intake and compresses the gas on the compression side. This is a continuous process as long as the compressor is running. The results are similar to those with a reciprocating compressor. Low-pressure suction gas enters the cylinder, is compressed (which also causes its temperature to rise), and leaves through the discharge opening.

Rotary Vane Rotary Compressor

The rotary vane compressor has a rotor fitted to the center of the shaft. The rotor and the shaft have the same center. The rotor has two or more vanes that slide in and out and trap and compress the gas, Figure 36–23. The shaft and rotor are positioned off center in the cylinder. As a vane passes by the suction intake opening, low-pressure gas follows it. This gas continues to enter and is trapped by the next vane, which compresses it and pushes it out the discharge opening. Notice that the intake opening is much larger than the discharge opening. This is to allow the low-pressure gas to enter more readily. The intake and compression is a continuous process as long as the compressor is running.

All the refrigerant that enters the intake port is discharged through the exhaust or discharge port. There is no clearance volume as in the reciprocating compressor. This is the primary reason why the rotary compressor is highly efficient.

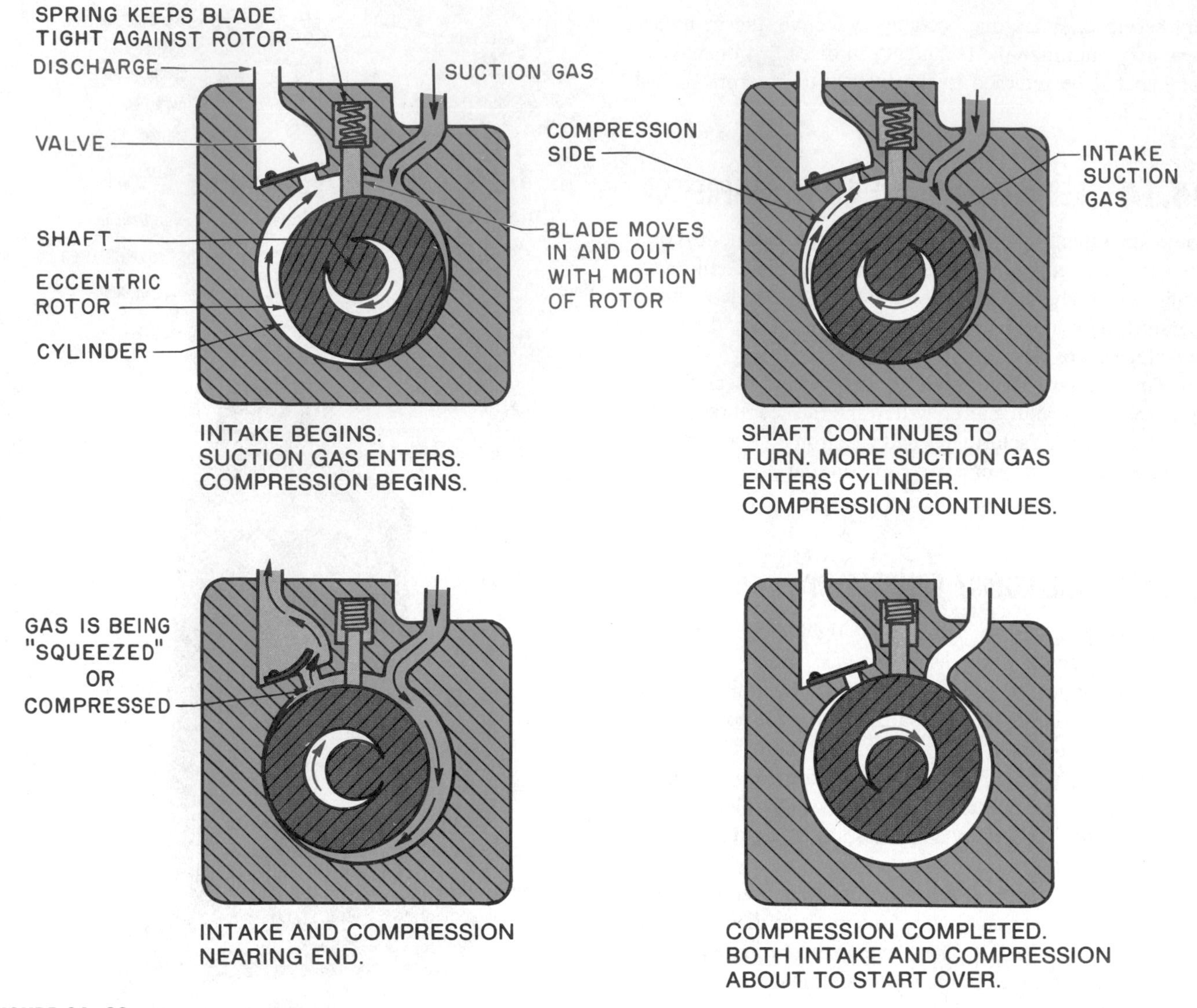
SPRING KEEPS BLADE TIGHT AGAINST ROTOR
DISCHARGE
SUCTION GAS
VALVE
SHAFT
ECCENTRIC ROTOR
CYLINDER
BLADE MOVES IN AND OUT WITH MOTION OF ROTOR
INTAKE BEGINS.
SUCTION GAS ENTERS.
COMPRESSION BEGINS.
COMPRESSION SIDE
INTAKE SUCTION GAS
SHAFT CONTINUES TO TURN. MORE SUCTION GAS ENTERS CYLINDER.
COMPRESSION CONTINUES.
GAS IS BEING "SQUEEZED" OR COMPRESSED
INTAKE AND COMPRESSION NEARING END.
COMPRESSION COMPLETED.
BOTH INTAKE AND COMPRESSION ABOUT TO START OVER.

FIGURE 36–22 Operation of stationary vane rotary compressor.

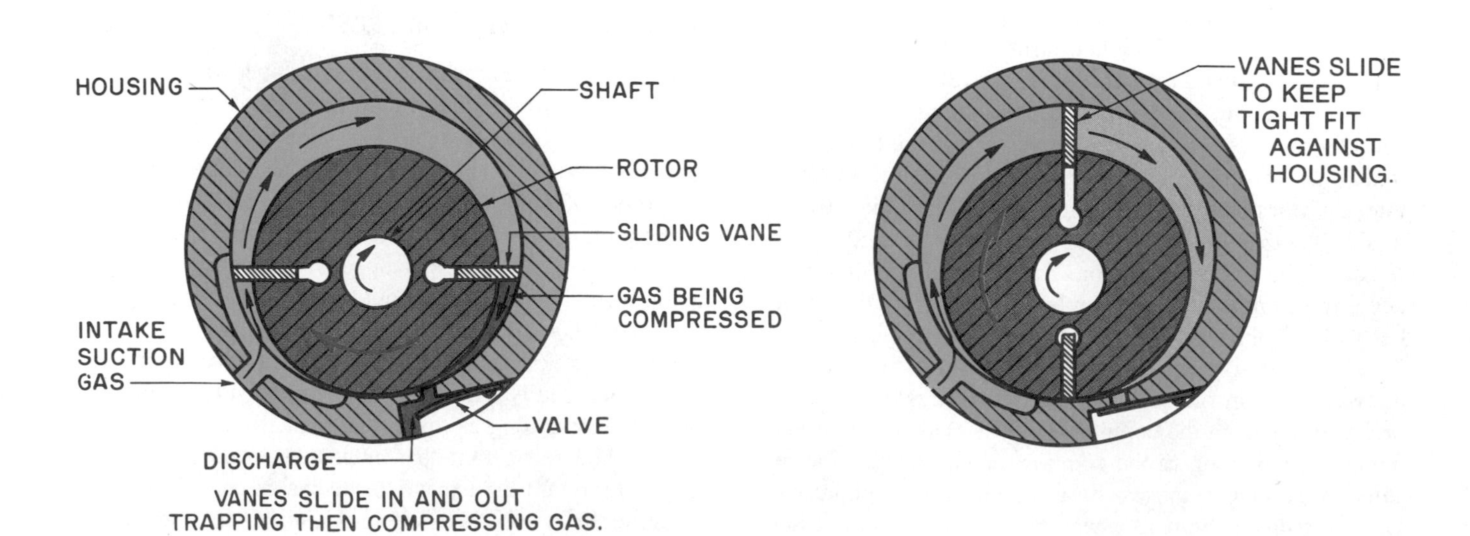
HOUSING
SHAFT
ROTOR
SLIDING VANE
GAS BEING COMPRESSED
INTAKE SUCTION GAS
VALVE
DISCHARGE
VANES SLIDE IN AND OUT TRAPPING THEN COMPRESSING GAS.
VANES SLIDE TO KEEP TIGHT FIT AGAINST HOUSING.

FIGURE 36–23 Operation of rotary vane rotary compressor.

36.16 THE SCROLL COMPRESSOR

In a scroll compressor, Figure 36–24, the compression takes place between two spiral-shaped forms. One of these is stationary and the other operates with an orbiting action within the other, Figure 36–25. This orbiting movement draws gas into a pocket between the two spirals. As this action continues, the gas opening is sealed off and the gas is forced into a smaller pocket at the center, Figure 36–26. This figure illustrates only one pocket of gas. Actually, most of the scroll or spiral is filled with gas and compression or "squeezing" occurs in all the pockets. Several stages of compression are occurring at the same time, which acts to balance the compression action. The scroll compressor is quiet compared to the same size reciprocating compressor.

The contact between the spirals sealing off the pockets is achieved using centrifugal force. This minimizes gas leakage. However, should liquid refrigerant or debris be present in the system, the scrolls will separate without damage or stress to the compressor. Because of its design, the scroll tips remain in contact without seals. Therefore, there is little wear.

The scroll compressor provides high efficiency because of the following:

- It requires no valves; consequently, there are no valve compression losses.
- The suction and discharge locations are separate, which reduces the heat transfer between the suction and discharge gas. There is less temperature difference between the various stages of compression occuring in the compressor.
- There is no clearance volume providing reexpansion gas as in a reciprocating compressor.

FIGURE 36–24 The scroll compressor. *Courtesy Copeland Corporation*

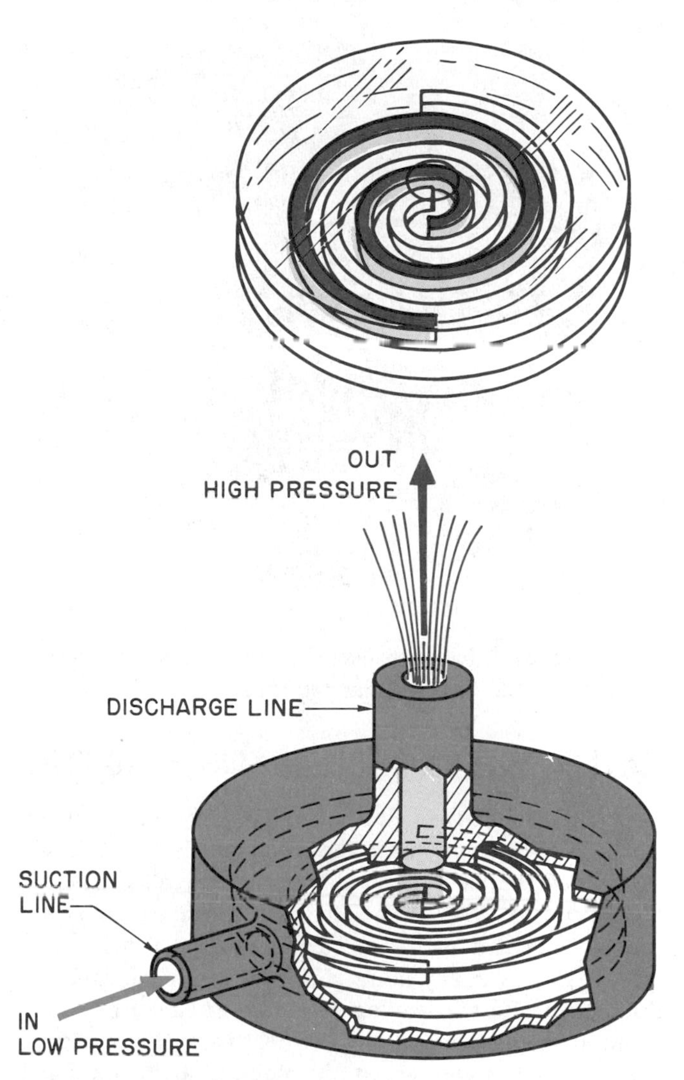

FIGURE 36–25 Orbiting action between scrolls in scroll compressor.

FIGURE 36–26 Compression of one pocket of refrigerant in scroll compressor. *Courtesy Copeland Corporation*

36.17 THE CONDENSER

The condenser for air conditioning equipment is the component that rejects the heat from the system. Most equipment is air cooled and rejects heat to the air. The coils are copper or aluminum and both types have aluminum fins to add to the heat exchange surface area, Figure 36–27.

FIGURE 36–27 Condensers have either copper or aluminum tubes with aluminum fins. *Courtesy Carrier Corporation*

36.18 SIDE-AIR-DISCHARGE CONDENSING UNITS

Air-cooled condensers all discharge hot air loaded with the heat absorbed from within the structure. Early condensers discharged the air out the side and were called side discharge, Figure 36–28. The advantage of this equipment is that the fan and motor are under the top panel, Figure 36–29. However, any noise generated inside the cabinet is discharged into the leaving airstream and may be clearly heard in a neighbor's yard. ***The heat from the condenser coil can be hot enough to kill plants that it blows on.*** These condensers are still being used.

36.19 TOP-AIR-DISCHARGE CONDENSERS

The modern trend in residential equipment is for the condenser to be a top-discharge type, Figure 36–30. In this type of unit the hot air and noise are discharged from the top of the unit into the air. This is advantageous as far as air and noise are concerned, but the fan and motor are on top of the equipment and rain, snow, and leaves can fall directly into the unit, so the fan motor should be protected with a rain shield, Figure 36–31. The fan motor bearings are in a vertical position, which means there is more thrust on the end of the bearing, so the bearing needs a thrust surface for this type of application, Figure 36–32.

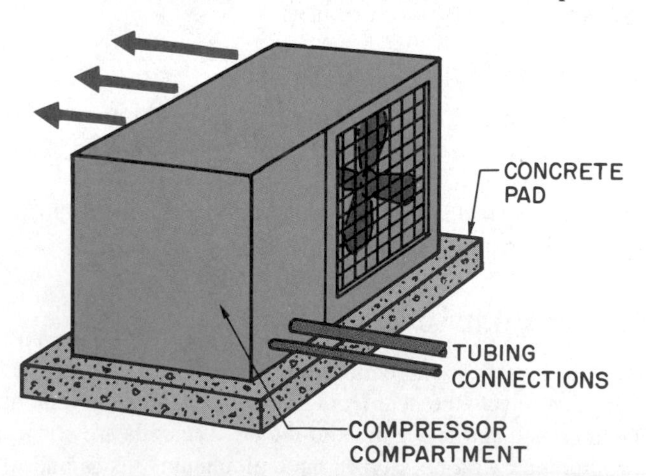

FIGURE 36–28 Condensing unit discharges hot air out the side.

FIGURE 36–29 Fan motor and all components are located under the top panel. *Courtesy Carrier Corporation*

FIGURE 36–30 Equipment with air discharged out the top of the cabinet. *Courtesy York International Corporation*

36.20 CONDENSER COIL DESIGN

Some coil surfaces are positioned vertically, and grass and dirt can easily get into the bottom of the coil. The coils must be clean for the condenser to operate efficiently. Some equipment uses the bottom few rows for a subcooling circuit to lower the condensed refrigerant temperature below the condensing temperature. It is common for the liquid line to be 10°F to 15°F cooler than the condensing temperature. Each degree of subcooling will add approximately 1% to the efficiency of the system. If the subcoooling circuit is dirty, it could affect the capacity by 10% to 15%. This could mean the difference in a piece of equipment being able to cool or

FIGURE 36–31 In the top-air-discharge units the fan is usually on top.

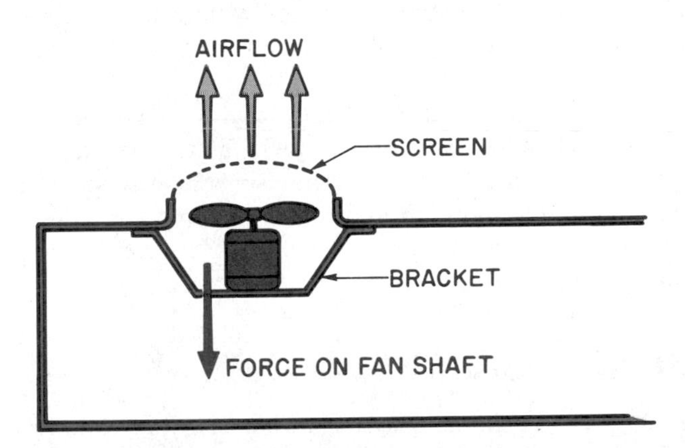

FIGURE 36–32 Top-air-discharge units have an additional load placed on the bearings. The fan is trying to fly down while it is running and pushing air out the top. During the starting and stopping of the fan, the bottom bearing has the fan blade and shaft resting on it. This bearing must have a thrust surface.

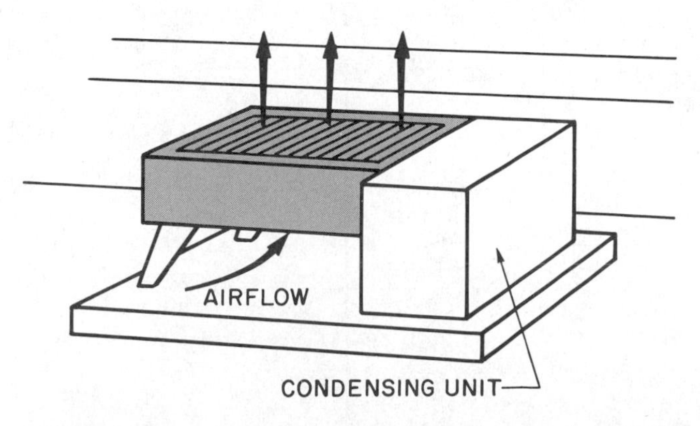

FIGURE 36–33 Some condensers are designed so that the coil is off the ground.

not being able to cool a structure if the equipment were sized too close to the design cooling load.

Some manufacturers use horizontal or slant-type condensers to position the coil off the ground. These condensers are less likely to pick up leaves and grass at the ground level, Figure 36–33.

36.21 HIGH-EFFICIENCY CONDENSERS

Modern times and the federal government have demanded that air conditioning equipment become more efficient. Probably the best way to improve efficiency is to lower the head pressure so the compressor is not working as hard. More surface area in the condenser reduces the compressor head pressure even in the hottest weather. This means lower compressor current and less power consumed for the same amount of air conditioning. Some manufacturers with oversized condensers use two-speed condenser fans—one speed for mild weather, one for hot. Without two-speed fans the condensers would be too efficient in mild weather, causing the head pressure to be too low, starving the expansion device and resulting in less capacity.

36.22 CABINET DESIGN

The condenser cabinet is usually located outside, so it needs weatherproofing. Most cabinets are galvanized and painted to give them more years of life without rusting. Some cabinets are made of aluminum, which is lighter but may not last as long if in a salty environment in coastal areas.

Most small equipment is assembled with self-tapping sheet metal screws. These screws are held by a drill screw holder during manufacturing and are threaded into the cabinet when turned with an electric drill. See Figure 36–34 for an example of a portable electric drill and holder. These

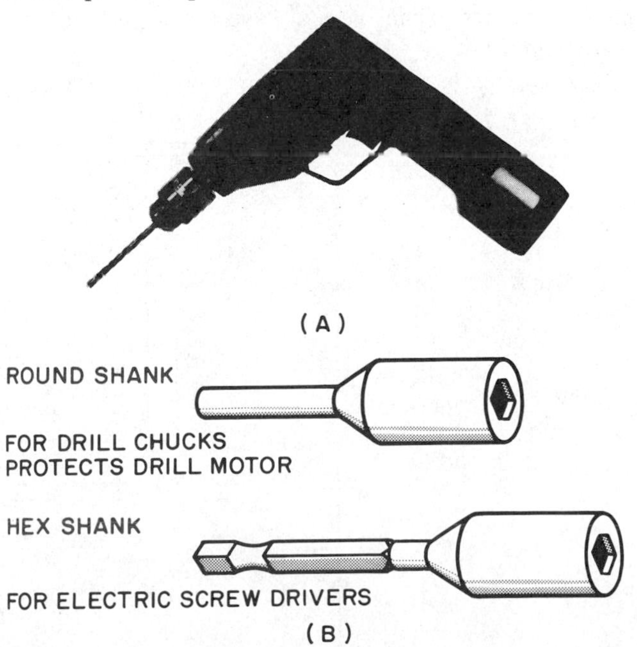

FIGURE 36–34 Portable electric drill and screw holder. *Photo by Bill Johnson*

screws should be made of weather-resistant material that will last for years out in the weather. The weather may be salt air as in the coastal areas. In these locations, stainless steel is a good choice of metal for sheet metal screws. When equipment that is assembled with drill screws is installed in the field, all of the screws should be fastened back into the cabinet tightly or the unit may rattle. After being threaded many times, the screw holes may become oversized; if so, use the next size screw to tighten the cabinet panels.

36.23 EXPANSION DEVICES

The expansion device meters the refrigerant to the evaporator. The thermostatic expansion valve (TXV) and the fixed-bore metering device (either a capillary tube or an orifice) are the types most often used.

The TXVs are the same types as those described in Unit 24 except they have a different temperature range. Air conditioning expansion devices are in the high-temperature range. TXVs are more efficient than fixed-bore devices because they allow the evaporator to reach peak performance faster. They allow more refrigerant into the evaporator coil during a hot pull down—when the conditioned space is allowed to get too warm before the unit is started.

When a TXV is used, the refrigerant pressures do not equalize during the off cycle unless the valve is made to equalize them with a planned bleed port. When the pressures do not equalize during the off cycle, a high-starting torque motor must be used for the compressor. This means that the compressor must have a start capacitor for start-up after the system has been off. Some valve manufacturers have a bleed port that always allows a small amount of refrigerant to bleed through. During the off cycle this valve will allow pressures to equalize and a compressor with a low-starting torque motor can be used. Low-starting torque motors are less expensive and have fewer parts that may malfunction.

36.24 AIR-SIDE COMPONENTS

The air side of an air conditioning system consists of the supply air and the return air systems. The airflow in an air conditioning system is normally 400 cfm/ton in the average humidity climates. For other climates, a different airflow may be used. In humid coastal areas 350 cfm/ton may be used, whereas 450 cfm/ton may be used in desert areas. The air leaving the air handler in a typical application could be expected to be about 55°F in the average system. The ductwork carrying this air must be insulated when it runs through an unconditioned space, or it will sweat and gain unwanted heat from the surroundings. The insulation should have a vapor barrier to keep the moisture from penetrating the insulation and collecting on the metal duct. All connections of the insulation must be fastened tight, and a vapor barrier must be used at all seams.

The return air will normally be about 75°F. If the return air duct is run through the unconditioned space, it may not need insulation. If the unconditioned space is a crawl space or basement, the temperature may be 75°F, and no heat will exchange. Even if a small amount of heat did exchange, a cost evaluation should be made to see if it is more economical to insulate the return duct or allow the small heat exchange. If the duct is run through a hot attic, the duct must be insulated.

Cool air will distribute better from a supply register located high in the room because it will fall after leaving the register. The final distribution point is where the cool air is mixed with the room air to arrive at a comfort condition. Figure 36–35 shows some examples of air conditioning diffusers.

36.25 INSTALLATION PROCEDURES

Package air conditioning systems were described earlier as systems where the whole air conditioner is built into one

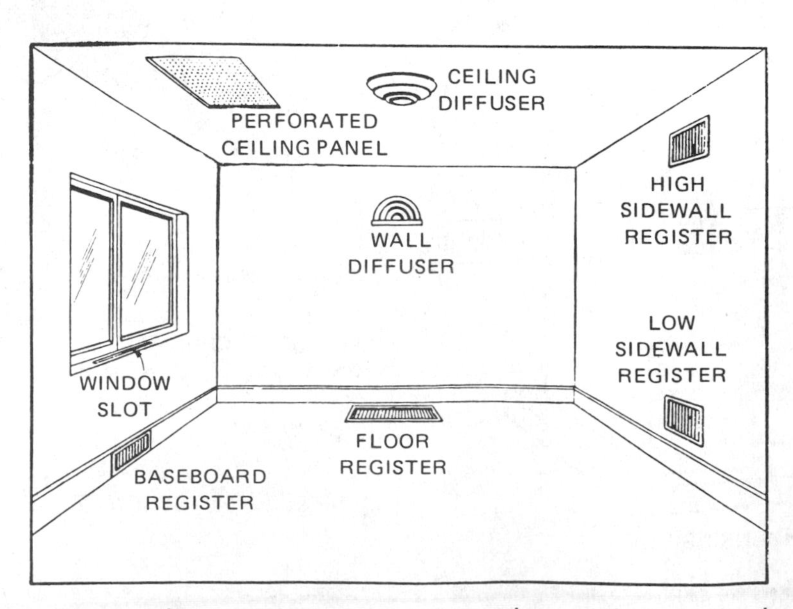

FIGURE 36–35 The final task of the air conditioning system is to get the refrigerated air properly distributed in the conditioned space. From Lang, *Principles of Air Conditioning,* © 1979 by Delmar Publishers Inc.

cabinet. Actually, a window unit is a package air conditioner designed to blow freely into the conditioned space. Most package air conditioners described in this text have two fan motors, one for the evaporator and one for the condenser, Figure 36–36. A window unit, however, uses one fan motor with a double shaft that drives both fans. The indoor fan is not designed to push air down a duct.

The package air conditioner has the advantage that all equipment is located outside the structure, so all service can be performed on the outside. This equipment is probably more efficient than the split system because it is completely factory assembled and charged with refrigerant. The refrigerant lines are short, so there is less efficency loss in the refrigeration lines.

The installation of a package system consists of mounting the unit on a firm foundation, fastening the package unit to the ductwork, and connecting the electrical service to the

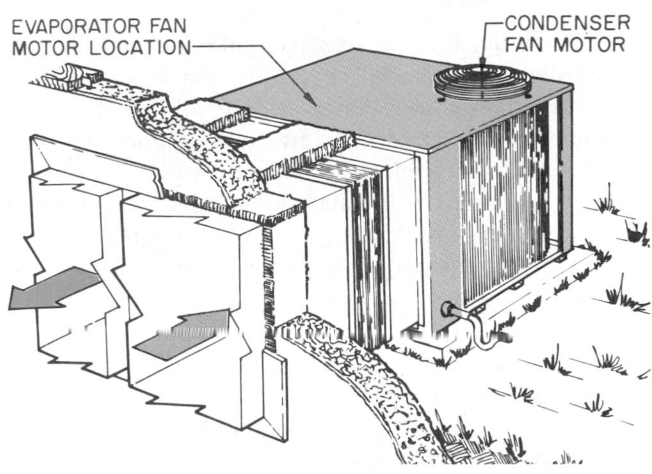

FIGURE 36–36 Package air condenser with two fan motors—one for the evaporator coil and one for the condenser coil. *Courtesy Climate Control*

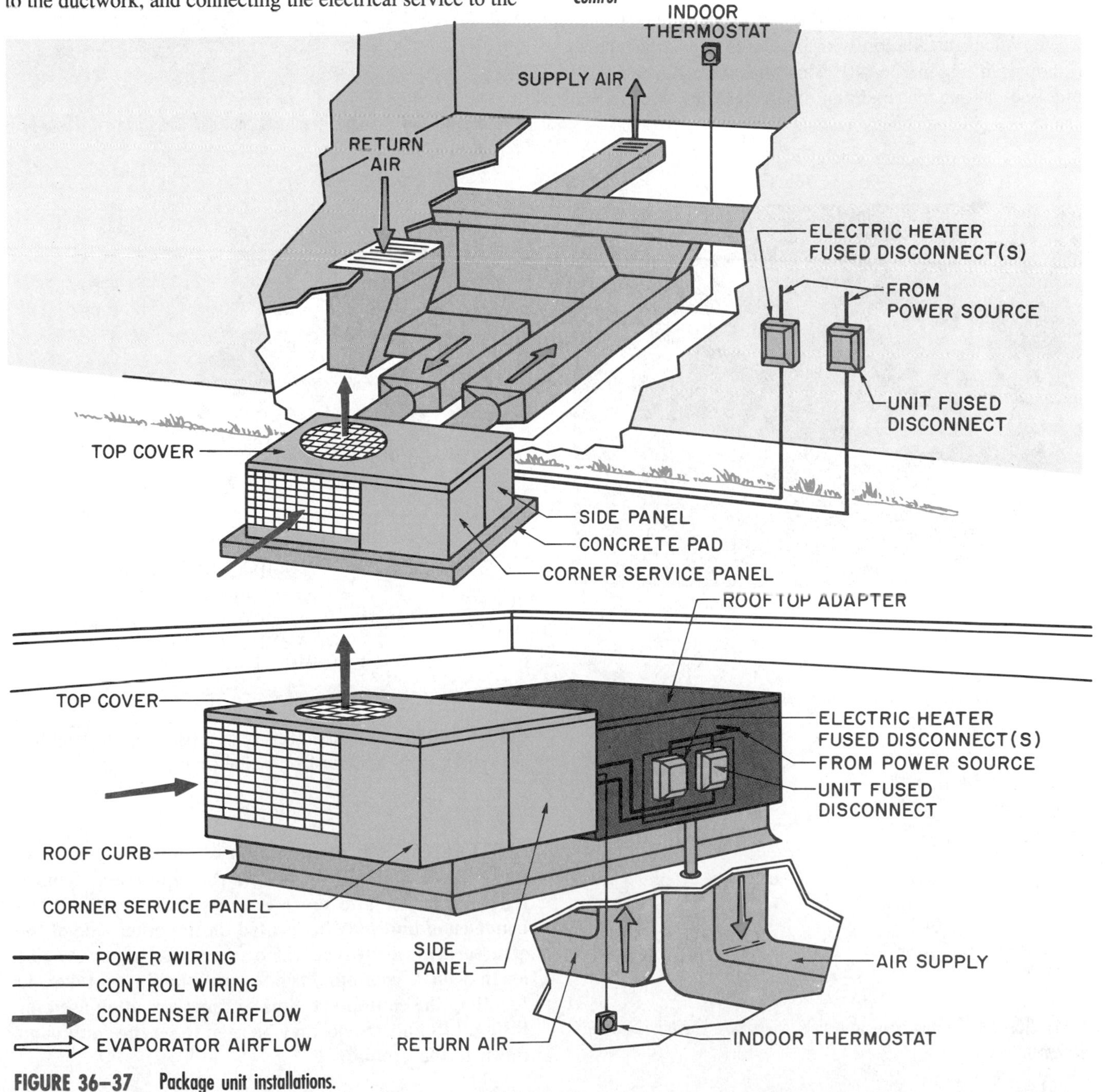

FIGURE 36–37 Package unit installations.

unit. These units can be easily installed in some locations where the ductwork can be readily attached to the unit. Some common installation types are rooftop, beside the structure, and in the eaves of structures, Figure 36–37.

The split air conditioning system is used when the condensing unit must be located away from the evaporator. There must be interconnecting piping furnished by the installing contractor or the equipment supplier. When furnished by the equipment supplier, the tubing may be charged with its own operating charge or with dry nitrogen. Figure 36–38 shows an example of three types of tubing connections.

The condensing unit should be located as close as possible to the evaporator to keep the interconnecting tubing short. The tubing consists of a cool gas line and a liquid line. The cool gas line is insulated and is the larger of the two lines. The insulation keeps unwanted heat from conducting into the line and keeps the line from sweating and dripping. Figure 36–39 is an example of a precharged line set.

A typical installation with pressures and temperatures is shown in Figure 36–40. This illustration gives some guidelines about the operating characteristics of a typical system.

FIGURE 36–39 Suction (cool gas line) and liquid line set. *Photo by Bill Johnson*

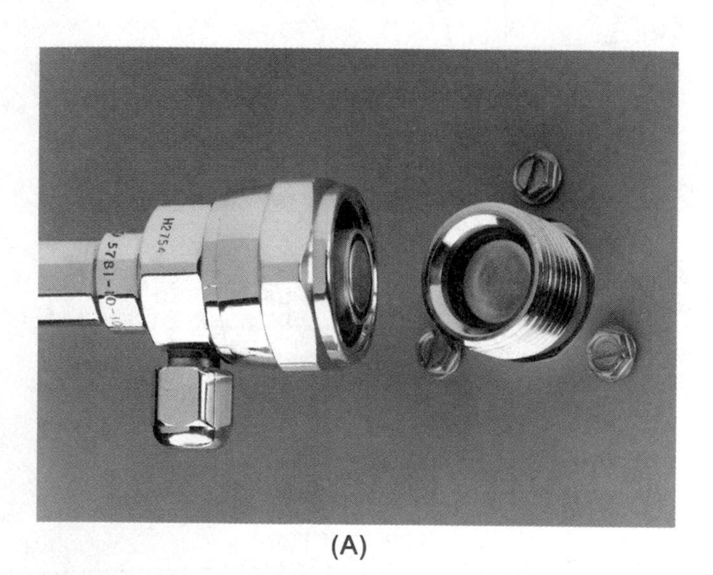

(A)

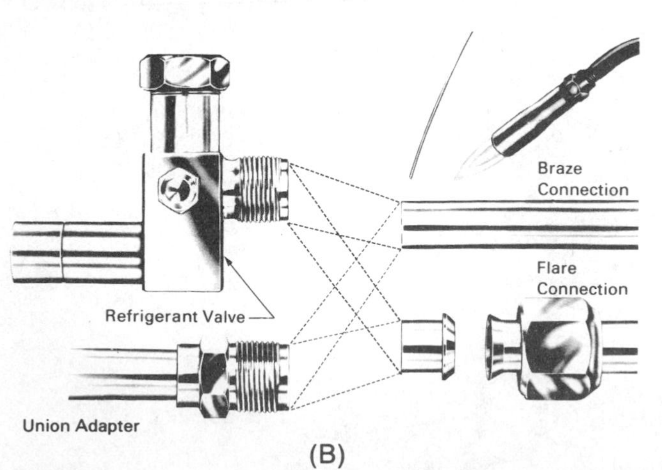

(B)

FIGURE 36–38 Three types of tubing connections. *Courtesy Aeroquip Corporation*

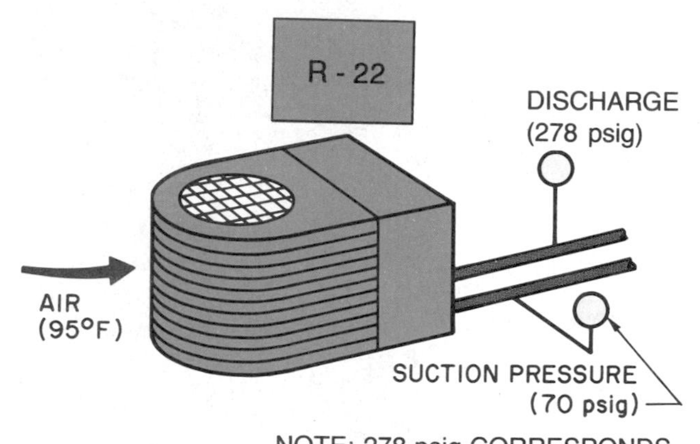

FIGURE 36–40 Installation showing the pressures and temperatures for a system in the humid southern part of the United States.

SAFETY PRECAUTION: *Installation of equipment may require the technician to unload the equipment from a truck or trailer and move it to various locations. The condensing unit may be located on the other side of the structure, far away from the driveway, or it may even be located on a rooftop. Proper care should be taken in handling the equipment. Small cranes are often used for lifting. Lift gate trucks may be used to set the equipment down to the ground.

When installing the equipment, care should be taken while working under structures and in attics. Spiders and other types of stinging insects are often found in these places and sharp objects such as nails are often left uncovered. When working in an attic be careful not to step through a ceiling.

Care should be taken while handling the line sets during connection if they have refrigerant in them. Liquid R-22 boils at −41°F in the atmosphere and can inflict serious frostbite to the hands and eyes. Goggles and gloves must be worn when connecting line sets.*

SUMMARY

- Evaporative cooling and refrigerated air conditioning are two methods used for comfort air conditioning.
- Evaporative cooling may be used in areas where the temperature is high and the humidity is low.
- Refrigerated air conditioning cools the air and removes moisture.
- Refrigerated air conditioning is used in hot temperatures with high or low humidity.
- Refrigerated air conditioning systems have four major components: the evaporator, the compressor, the condenser, and the metering device.
- Evaporators are made in three types: the A coil, slant coil, and H coil. The type of coil used depends on its position in the system and the particular manufacturer's design, upflow, downflow, or horizontal flow.
- Air conditioning evaporators operate at about 40°F and remove sensible heat and latent heat.
- Removal of sensible heat lowers the air temperature; removal of latent heat removes moisture.
- The compressor is the positive displacement pump that pumps the heat-laden vapor from the evaporator to the condenser.
- Compressors are cooled by suction gas, air, or discharge gas.
- Currently R-22 is the most commonly used refrigerant for air conditioning; R-500 and R-12 were often used in the past.
- Reciprocating, rotary, and scroll compressors are the types most commonly used in residential and light commercial air conditioning.
- Condensers are located outside to reject the heat to the outside.
- High efficiency in a condenser is achieved by increasing the condenser surface area. Two-speed fans may be used, one speed for mild weather and the other for hot weather.
- Package air conditioners are installed through the roof or through the wall at the end of the structure, wherever the duct can be fastened.
- The package unit has been charged at the factory and is factory assembled. Under similar conditions, it is probably more efficient than the split system.
- The split system has two interconnecting pipes. The large line is the insulated cool gas line; the small line is the liquid refrigerant line.

REVIEW QUESTIONS

1. Name two methods for cooling air for air conditioning.
2. What are the advantages of refrigerated air conditioning?
3. Name the four major components of a refrigerated air conditioning system.
4. Name two types of refrigerated air conditioning equipment.
5. Name three methods used to cool the compressor motor.
6. List two types of metering devices used in air conditioning.
7. Which type of metering device requires a compressor motor with a high starting torque?
8. Why do the other metering devices not require a high-starting torque motor?
9. What is the most popular refrigerant used in air conditioning?
10. Name three refrigerants that may be found in air conditioning equipment?
11. What two types of heat must be removed with the air conditioning equipment?
12. Which component in the air conditioning system absorbs heat into the system?
13. Which component pumps heat?
14. What three types of compressors are most commonly used in residential and light commercial air conditioning installations?
15. What component rejects heat from the system?
16. What is the compressor normally mounted on?
17. Name the large and small interconnecting lines on an air conditioning system.

UNIT 37 Air Distribution and Balance

OBJECTIVES

After studying this unit, you should be able to

- describe the prime mover of air for an air conditioning system.
- describe characteristics of the propeller and the centrifugal blowers.
- take basic air pressure measurements.
- measure air quantities.
- list the different types of air-measuring devices.
- describe the common types of motors and drives.
- describe duct systems.
- explain what constitutes good airflow through a duct system.
- describe a return air system.
- plot air flow conditions on the air friction chart.

SAFETY CHECKLIST

- Use a grounded, double-insulated, or cordless portable electric drill when installing or drilling into metal duct.
- Sheet metal can cause serious cuts. Wear gloves when installing or working with sheet metal.
- Wear goggles whenever working around duct that has been opened with the fan on.

37.1 CONDITIONING EQUIPMENT

As indicated previously, in many climates air has to be conditioned for us to be comfortable. One way to condition air is to use a fan to move the air over the conditioning equipment. This equipment may consist of a cooling coil, a heating device, a device to add humidity, or a device to clean the air. The forced-air system uses the same room air over and over again. Air from the room enters the system, is conditioned, and returned to the room. Fresh air enters the structure either by infiltration around the windows and doors or by ventilation from a fresh air inlet connected to the outside, Figure 37–1.

The forced-air system is different from a natural-draft system, where the air passes naturally over the conditioning equipment. Baseboard hot water heat is an example of natural-draft heat. The warmer water in the pipe heats the air in the vicinity of the pipe. The warmer air expands and rises. New air from the floor at a cooler temperature takes the place of the heated air, Figure 37–2. There is very little concern for the amount of air moving in a natural convection system.

37.2 CORRECT AIR QUANTITY

The object of the forced air system is to *deliver the correct quantity of conditioned air to the occupied space.* When this occurs, the air mixes with the room air and creates a comfortable atmosphere in that space. Different spaces have different air quantity requirements; the same structure may have several different cooling requirements. For example, a house has rooms of different sizes with different requirements. A bedroom requires less heat and cooling than a large living room. Different amounts of air need to be delivered to these rooms to maintain comfort conditions, Figures 37–3 and 37–4. Another example is a small office building with a high cooling requirement in the lobby and a low requirement in the individual offices. The correct amount of

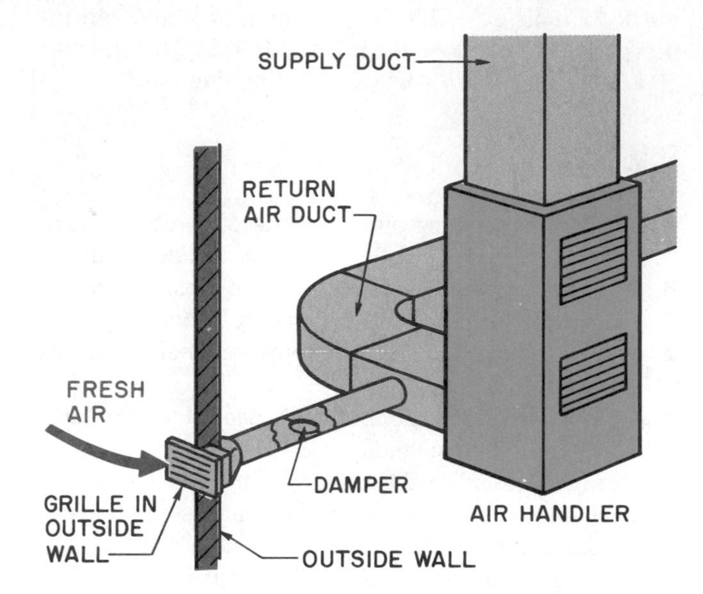

FIGURE 37–1 Ventilation.

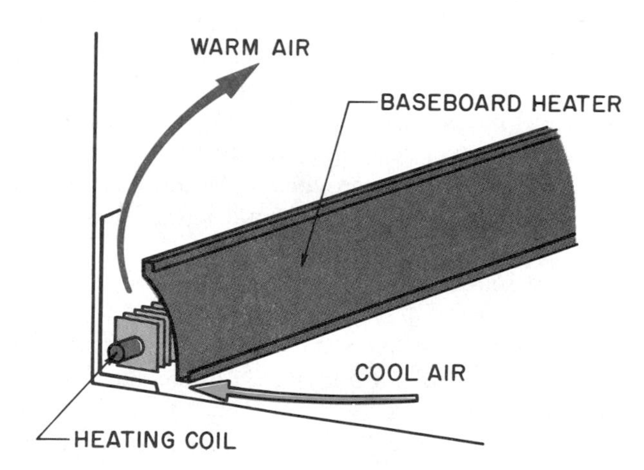

FIGURE 37–2 Air near heat will become heated, then expand and rise.

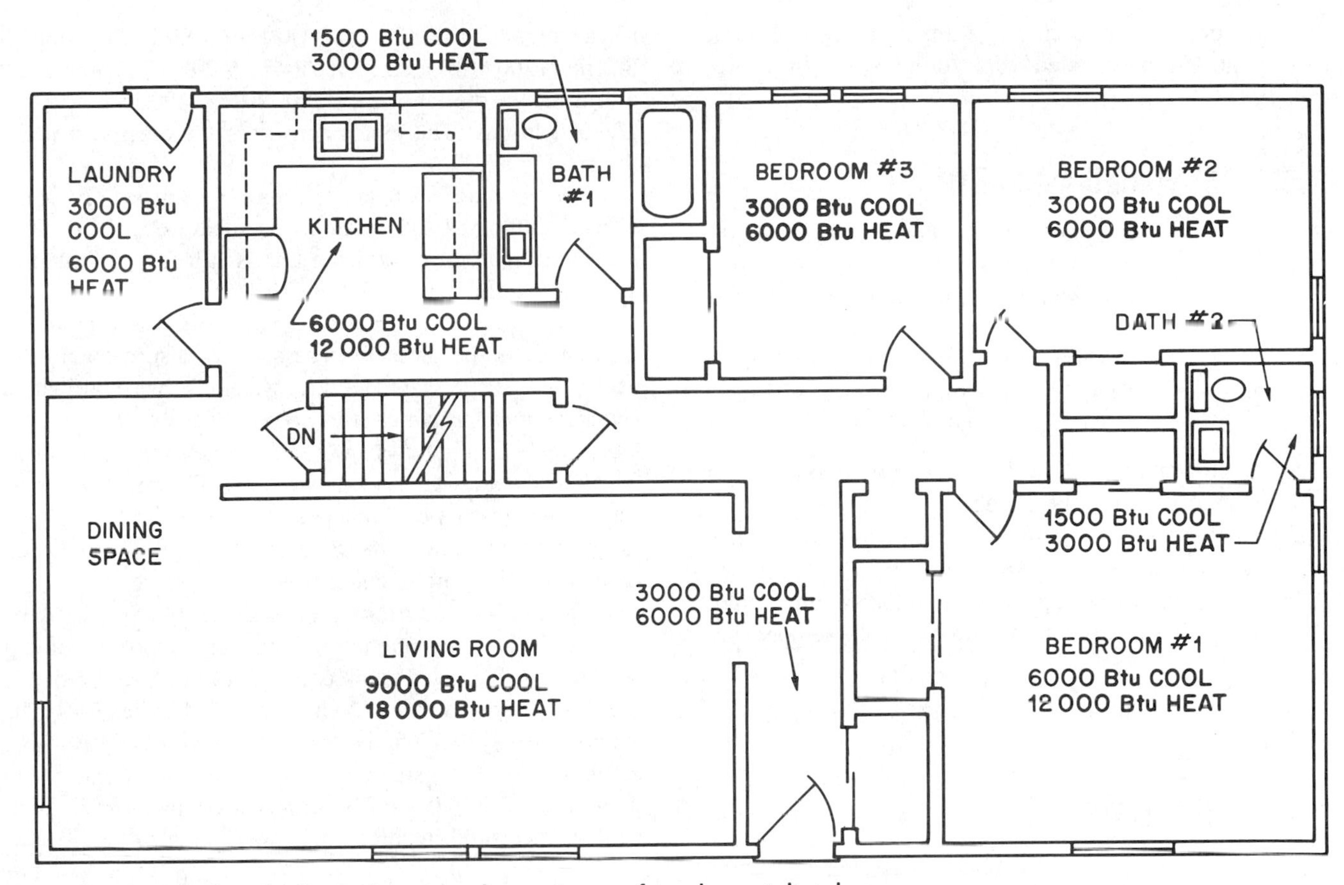

FIGURE 37–3 The floor plan has the heat and cooling requirements for each room indicated.

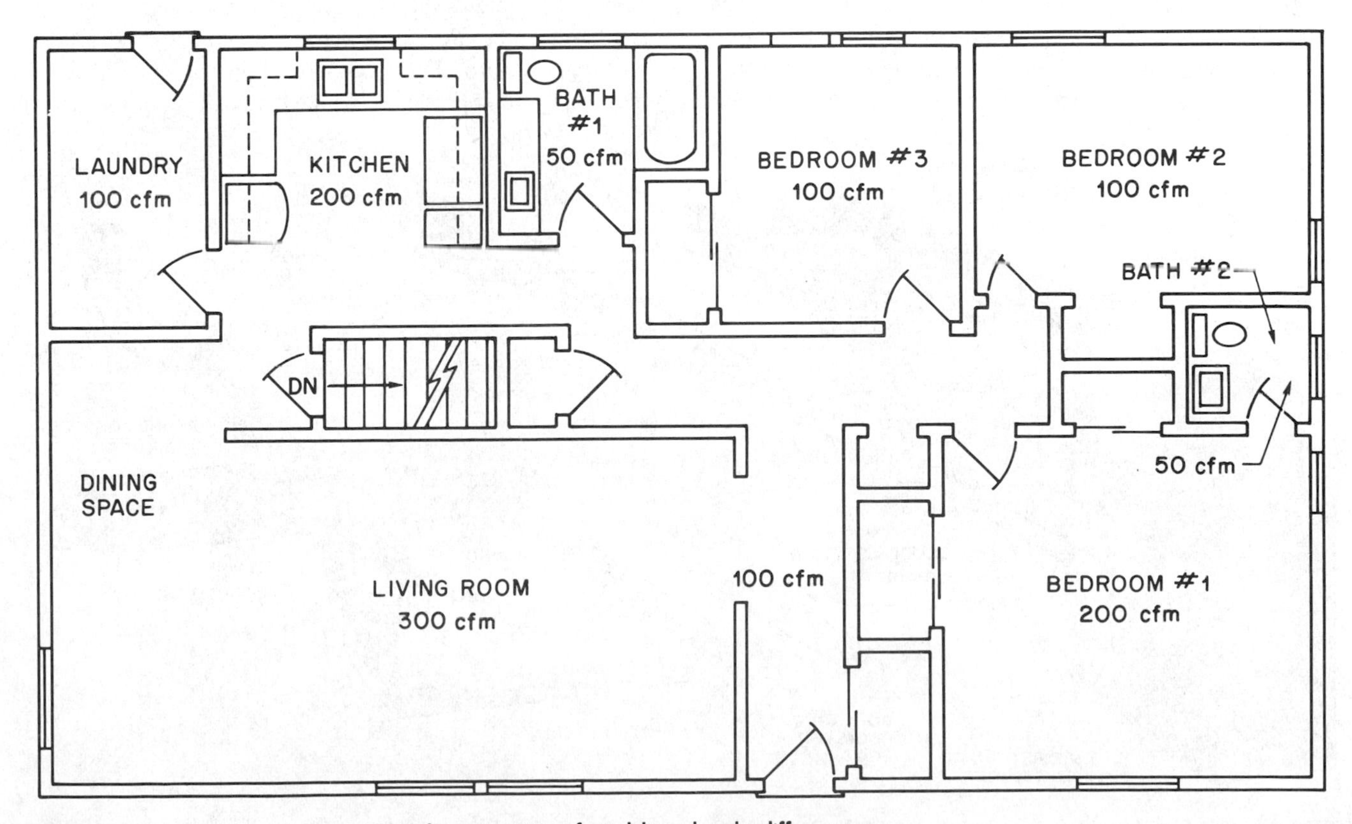

FIGURE 37–4 Floor plan of Figure 37–3 shows quantities of air delivered to the different rooms.

air must be delivered to each part of the building so that one area will not be overcooled while other areas are cooled correctly.

37.3 THE FORCED-AIR SYSTEM

The components that make up the *forced-air system* are the *blower* (fan), the *air supply system*, the *return air system*, and the *grilles* and *registers* where the circulated air enters the room and returns to the conditioning equipment. See Figure 37–5 for an example of duct fittings. When these components are correctly chosen, they work together as a system with the following characteristics:

1. No air movement will be felt in the conditioned space that would normally be occupied.
2. No air noise will be noticed in the conditioned space.
3. No temperature swings will be felt by the occupants.
4. The occupants will not be aware that the system is on or off unless it stops for a long time and the temperature changes.

The lack of awareness of the conditioning system is important.

37.4 THE BLOWER

The *blower* or fan provides the pressure difference to force the air into the duct system, through the grilles and registers, and into the room. Air has weight and has a resistance to movement. This means that it takes energy to move the air to the conditioned space. The fan may be required to push enough air through the evaporator and ductwork for 3 tons of air conditioning. Typically 400 ft^3 of air must be moved per minute per ton of air conditioning. This system would move 400 cfm $\times$ 3 tons or 1200 ft^3 of air per minute. Air has a weight of 1 lb/13.35 ft^3. This fan would be moving 90 lb/min (1200 cfm $\div$ 13.35 ft^3/lb = 89.99). 90 lb/min $\times$ 60 min/h means the fan moves 5400 lb/h. 5400 lb/h $\times$ 24 = 129,600 lb/day. The motor consumes the energy to move the air, Figure 37–6.

The pressure in a duct system for a residence or a small office building is too small to be measured in psi. It is measured in a unit of pressure that is still force per unit of area but in a smaller graduation. The pressure in ductwork is measured in *inches of water column* (in W.C.). A pressure of 1 in. W.C. is the pressure necessary to raise a column of water 1 in. Air pressure in a duct system is measured with a *water manometer,* which uses colored water that rises up a tube. Figure 37–7 shows a water manometer that is inclined for more accuracy at very low pressures. Figure 37–8 shows some other instruments that may be used to measure very low air pressures; they are graduated in inches of water even though they may not contain water.

The atmosphere exerts a pressure of 14.696 psi at sea level at 70°F. Atmospheric pressure will support a column of water 34 ft high. This is equal to 14.696 psi. 1 psi will raise a column of water 27.7 in. or 2.31 ft, Figure 37–9. The average duct system will not exceed a pressure of 1 in. W.C. A pressure of 0.05 psig will support a column of water 1.39 in. high (27.7 $\times$ 0.05 = 1.39 in.). Airflow pressure in ductwork is measured in some very low figures.

37.5 SYSTEM PRESSURES

A duct system is pressurized by three pressures—*static pressure, velocity pressure,* and *total pressure*. Static pressure is the same as the pressure on an enclosed vessel. This is like the pressure of the refrigerant in a cylinder that is pushing outward. Figure 37–10 shows a manometer for measuring static pressure. Notice the position of the sensing

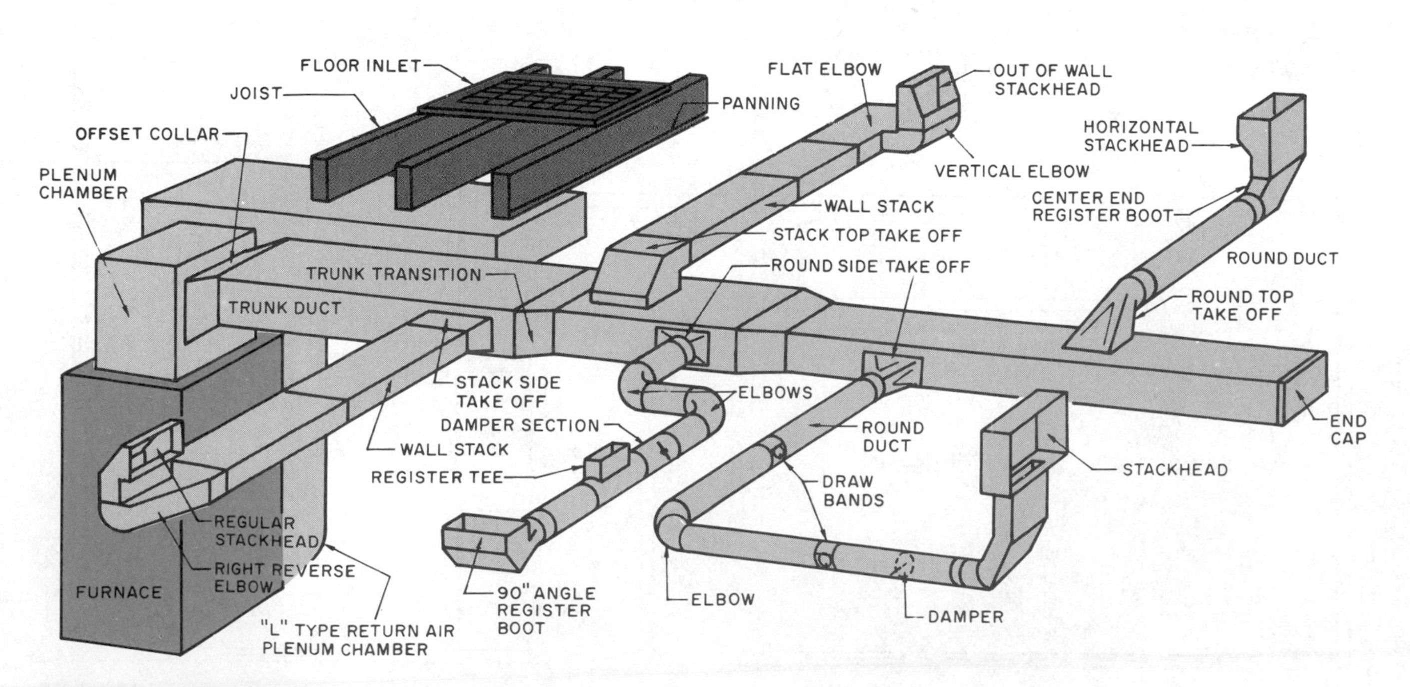

FIGURE 37–5 Duct fittings.

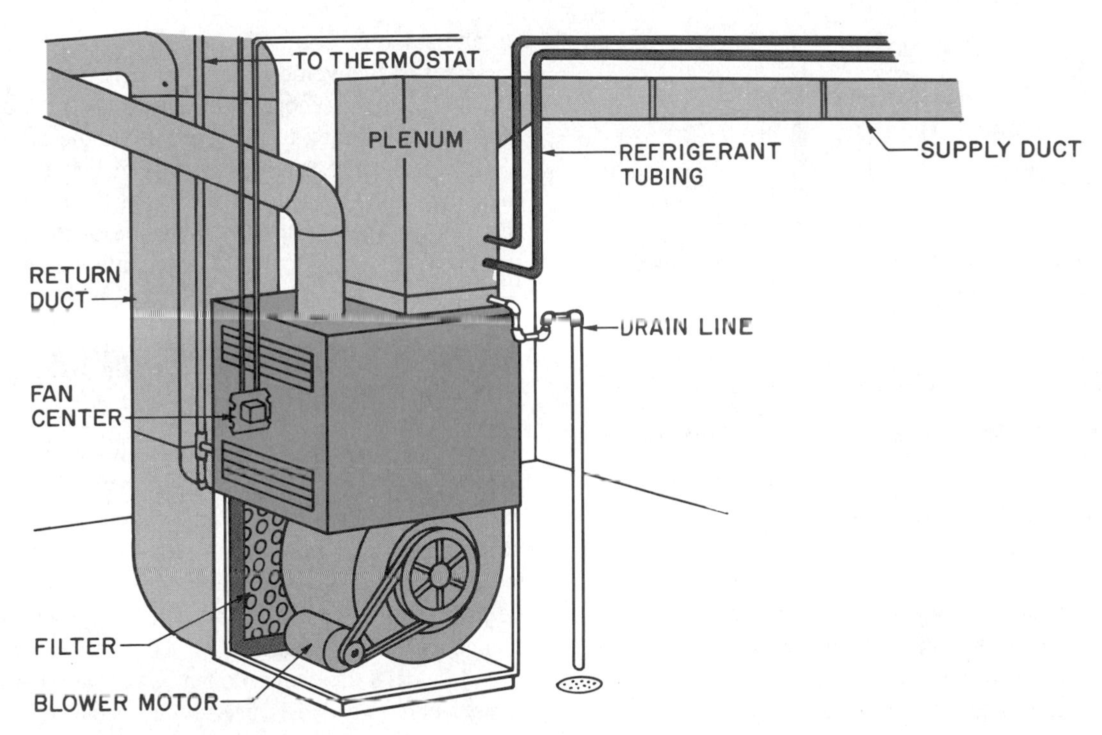

FIGURE 37–6 Fan and motor to move air.

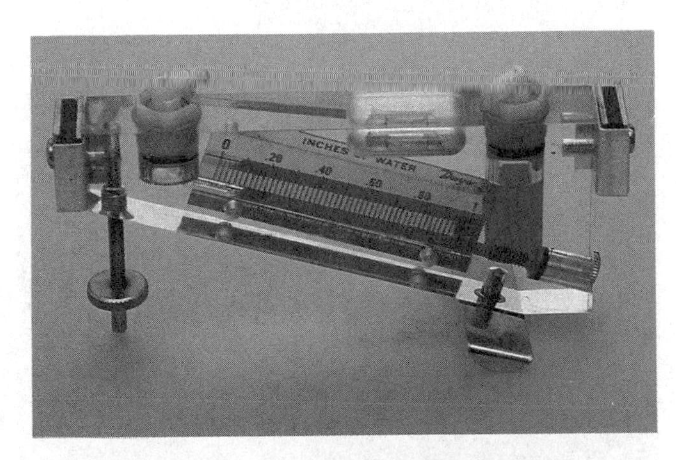

FIGURE 37–7 Water manometer is inclined to allow the scale to be extended for more graduations. *Photo by Bill Johnson*

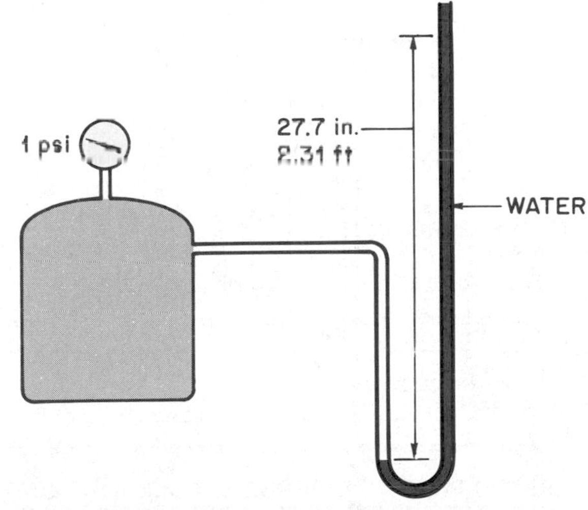

FIGURE 37–9 Vessel with a pressure of 1 psi inside. The water manometer has a column 27.7 in. high (2.31 ft).

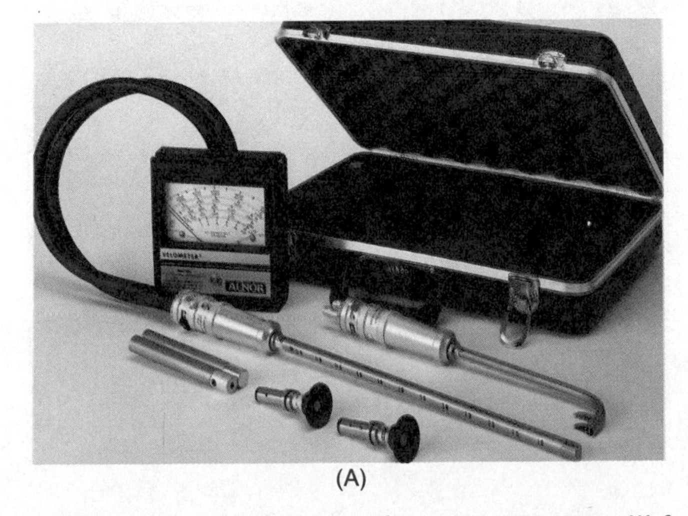

(A)

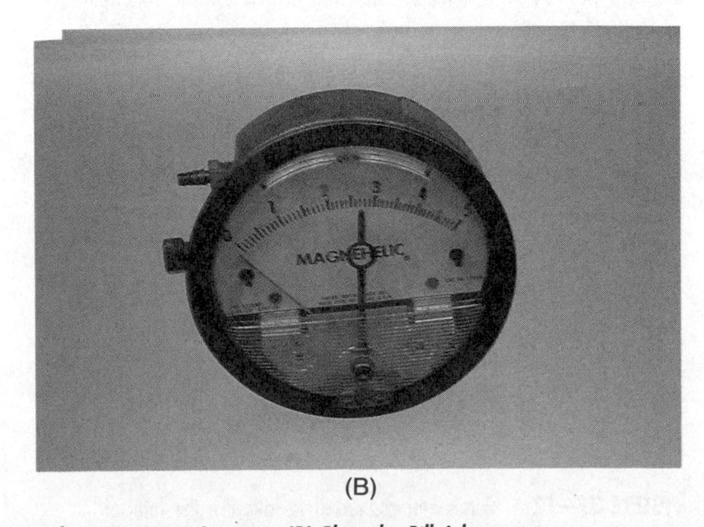

(B)

FIGURE 37–8 Other instruments used to measure air pressures. *(A) Courtesy Alnor Instrument Company, (B) Photo by Bill Johnson*

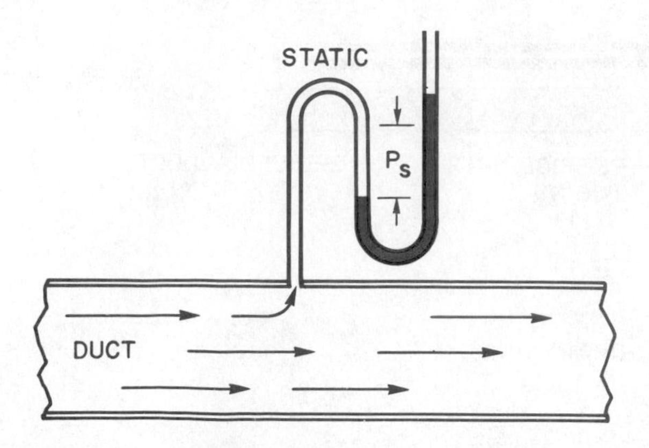

FIGURE 37–10 Manometer connected to measure the static pressure.

tube. The probe has a very small hole in the end, so the air rushing by the probe opening will not cause incorrect readings.

The air in a duct system is moving along the duct and therefore has velocity. The velocity of the air and the weight of the air create velocity pressure. Figure 37–11 shows a manometer for measuring velocity pressure in an air duct. Notice the position of the sensing tube. The air velocity goes straight into the tube inlet, which registers both velocity and static pressure. The probe and manometer arrangement reads the velocity pressure by canceling the static pressure with the second probe. The static pressures balance each other and the velocity pressure is the difference.

The total pressure of a duct can be measured with a manometer applied a little differently, Figure 37–12. Notice that the velocity component or probe of the manometer is positioned so that the air is directed into the end of the tube. This will register static and velocity pressures.

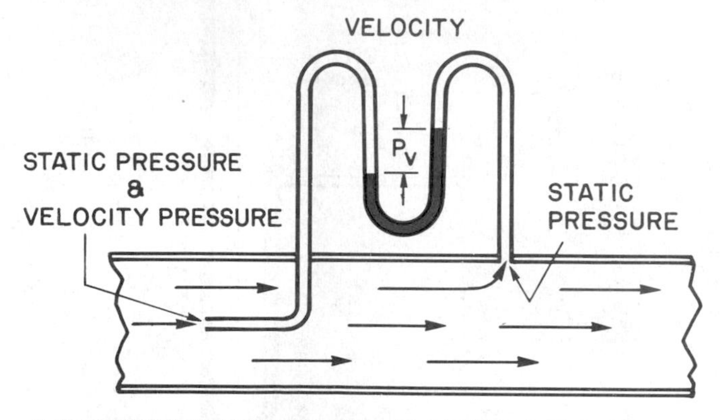

FIGURE 37–11 Manometer connected to measure the velocity pressure of the air moving in the duct.

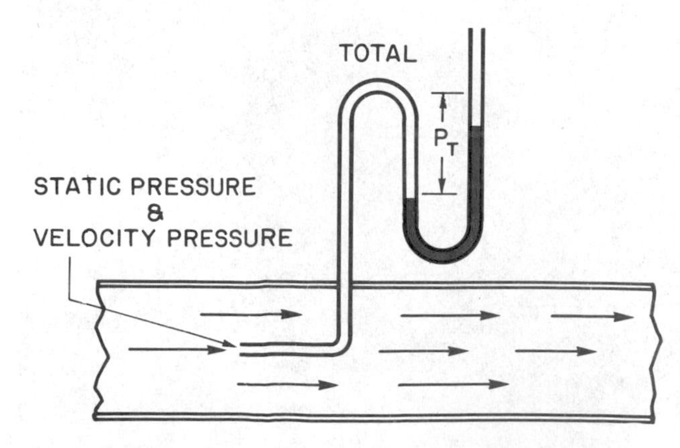

FIGURE 37–12 Manometer connected to measure the total air pressure in the duct.

37.6 AIR-MEASURING INSTRUMENTS FOR DUCT SYSTEMS

The water manometer has been mentioned as an air pressure-measuring instrument. An instrument used to measure the actual air velocity is the *velometer.* This instrument actually measures how fast the air is moving past a particular point in the system. Figure 37–13 shows two different velometers. These instruments should be used as the particular manufacturer suggests.

A special device called a *pitot tube* was developed many years ago and is used with special manometers for checking duct air pressure at most pressure levels, Figure 37–14.

(A)

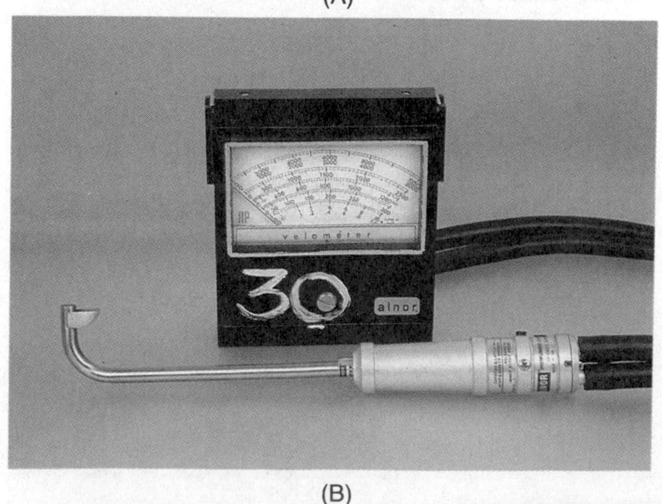

(B)

FIGURE 37–13 Two types of velometers. *Photos by Bill Johnson*

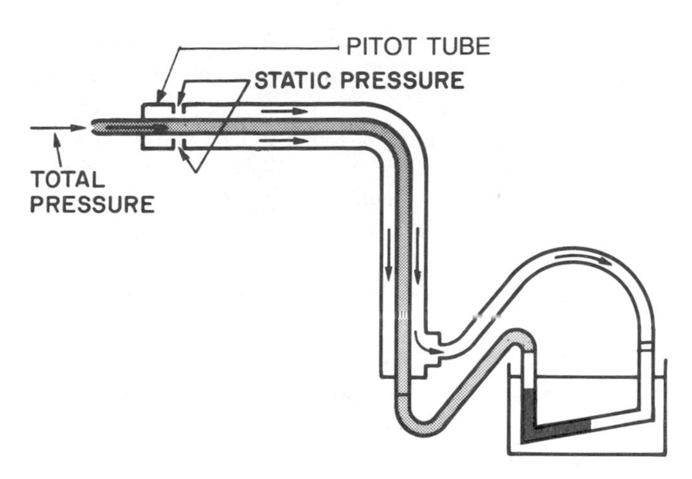
PITOT TUBE
STATIC PRESSURE
TOTAL PRESSURE

FIGURE 37–14 Pitot tube set up to measure velocity pressure.

37.7 TYPES OF FANS

The blower, or fan as it is sometimes called, can be described as a device that produces airflow or movement. Several different types of blowers produce this movement, but all can be described as non–positive-displacement air movers. Remember, the compressor is a positive-displacement pump. When the cylinder is full of refrigerant (or air for an air compressor), the compressor is going to empty that cylinder or break something. The fan is not positive displacement—it cannot build the kind of pressure that a compressor can. The fan has other characteristics that have to be dealt with.

The two fans that we discuss in this text are the *propeller fan* and the *forward curve centrifugal fan*, also called the *squirrel cage fan wheel*.

The propeller fan is used in exhaust fan and condenser fan applications. It will handle large volumes of air at low-pressure differentials. The propeller can be cast iron, aluminum, or stamped steel and is set into a housing called a *venturi* to encourage airflow in a straight line from one side of the fan to the other, Figure 37–15. The propeller fan makes more noise than the centrifugal fan so it is normally used where noise is not a factor.

The squirrel cage or centrifugal fan has characteristics that make it desirable for ductwork. It builds more pressure from the inlet to the outlet and moves more air against more pressure. This fan has a forward curved blade and a cutoff to

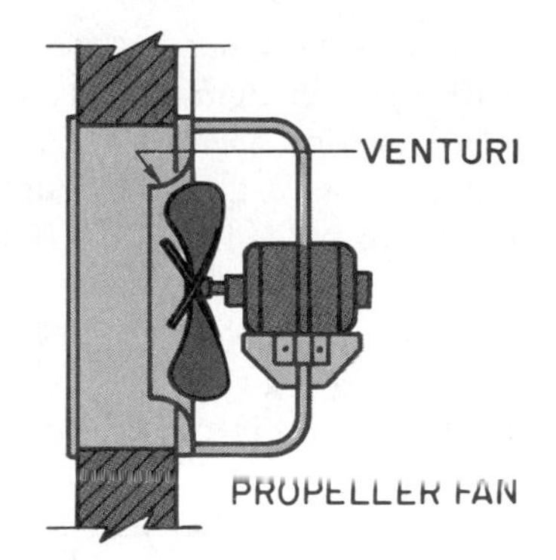
VENTURI
PROPELLER FAN

FIGURE 37–15 Propeller-type fan.

shear the air spinning around the fan wheel. This air is thrown by centrifugal action to the outer perimeter of the fan wheel. Some of it would keep going around with the fan wheel if it were not for the shear that cuts off the air and sends it out the fan outlet, Figure 37–16. The centrifugal fan is very quiet when properly applied. It meets all requirements of duct systems up to very large systems that are considered high-pressure systems. High-pressure systems have pressures of 1 in. W.C. and more. Different types of fans, some of them similar to the forward curve centrifugal fan, are used in larger systems.

One characteristic that makes troubleshooting the centrifugal fan easier is the volume of air-to-horsepower requirement. This fan uses energy at the rate at which it moves air through the ductwork. The current draw of the fan motor is in proportion to the pounds of air it moves or pumps. For example, a fan motor that pulls full-load amperage at the rated fan capacity will pull less than full-load

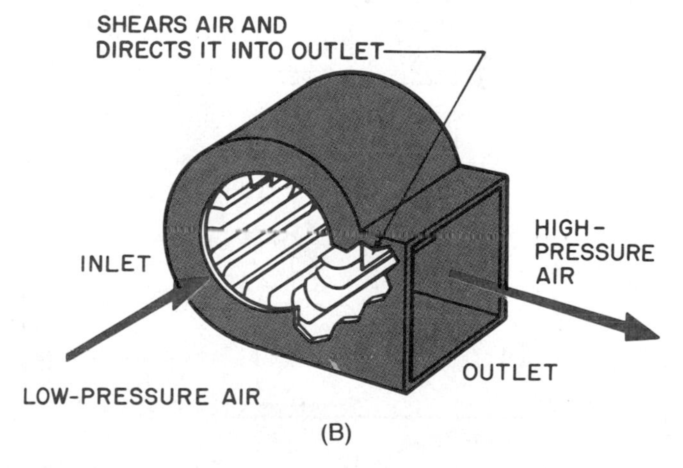
SHEARS AIR AND DIRECTS IT INTO OUTLET
INLET
HIGH-PRESSURE AIR
OUTLET
LOW-PRESSURE AIR
(B)

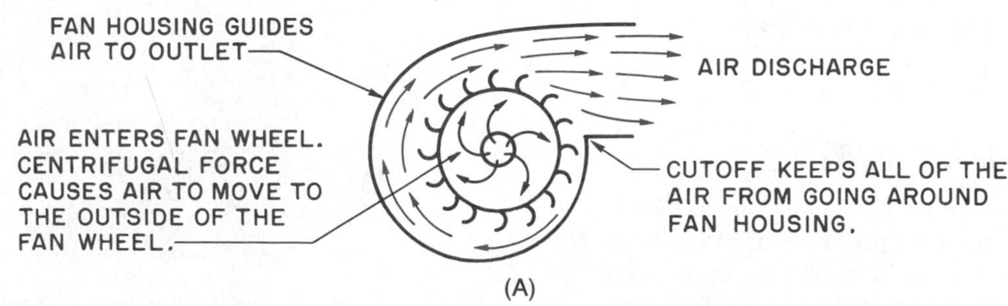
FAN HOUSING GUIDES AIR TO OUTLET
AIR DISCHARGE
AIR ENTERS FAN WHEEL. CENTRIFUGAL FORCE CAUSES AIR TO MOVE TO THE OUTSIDE OF THE FAN WHEEL.
CUTOFF KEEPS ALL OF THE AIR FROM GOING AROUND FAN HOUSING.
(A)

FIGURE 37–16 Centrifugal fan.

amperage at any value less than the fan capacity. If the fan is supposed to pull 10 A at maximum capacity of 3000 cfm, it will pull 10 A while moving only this amount of air. The weight of this volume of air can be calculated by dividing ($3000 \text{ cfm} \div 13.35 \text{ ft}^3/\text{lb} = 224.7 \text{ lb/min}$). If the fan inlet is blocked, the suction side of the fan will be starved for air and the current will go down. If the discharge side of the fan is blocked, the pressure will go up in the discharge of the fan and the current will go down because the fan is not handling as many pounds of air, Figure 37–17. The air is merely spinning around in the fan and housing and not being forced into the ductwork. This particular type of fan can be checked for airflow with an ammeter when you make simple field measurements. If the current is down, the airflow is down. If the airflow is increased, the current will go up. For example, if the door on the blower compartment is opened, the fan current will go up because the fan will have access to the large opening of the fan compartment.

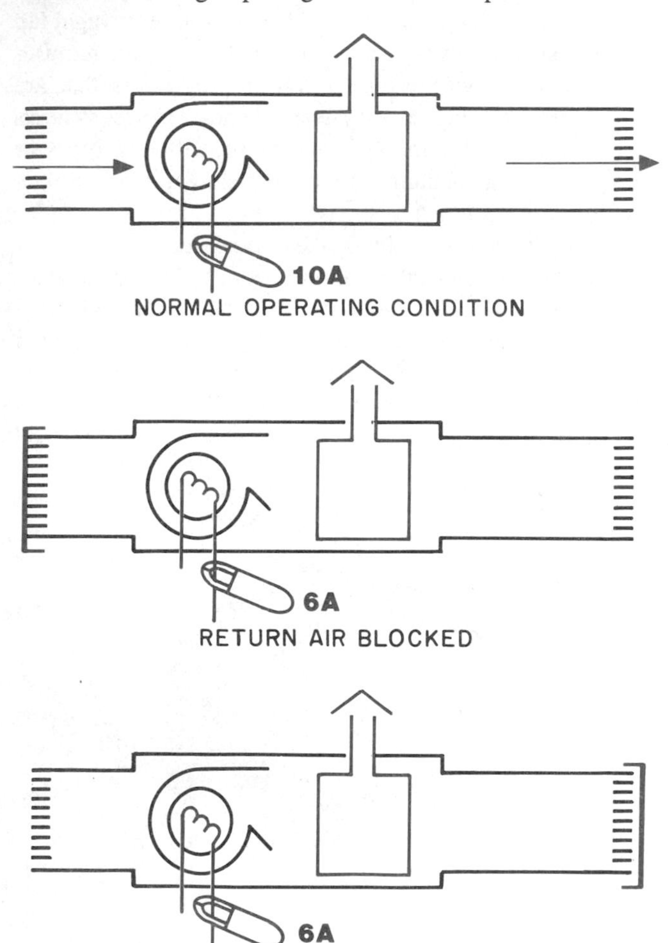

FIGURE 37–17 Airflow situations.

37.8 TYPES OF FAN DRIVES

The centrifugal blower must be turned by a motor. Two drive mechanisms are used: belt drive or direct drive. The belt-drive blower was used exclusively for many years. The motors were usually 1800 rpm. They actually run at 1750 rpm under load and operate very quietly. The motor normally has a capacitor and will go from a stopped position to 1750 rpm in about 1 sec. The motor may make more noise starting than running.

Later, manufacturers began making equipment more compact and began using smaller blowers with 3600-rpm motors. These motors actually turn at 3450 rpm under load. They had to turn from 0 rpm to 3450 rpm in about 1 sec and could make quite a noise on start-up.

Sleeve bearings and resilient (rubber) mountings are used to keep bearing noise out of the blower section. Belt-drive blowers have two bearings on the fan shaft and two bearings on the motor. Sometimes these bearings are permanently lubricated by the manufacturer, so a technician cannot oil them.

The drive pulley on the motor, the driven pulley on the fan shaft, and the belt must be maintained. This motor and blower combination has many uses because the pulleys can be adjusted or changed to change fan speeds, Figure 37–18.

Recently, most manufacturers have been using a direct-drive blower. The motor is mounted on the blower housing, usually with rubber mounts, with the fan shaft extending into the fan wheel. The motor is a permanent split-capacitor (PSC) motor that starts up very slowly, taking several seconds to get up to speed, Figure 37–19. It is very quiet and does not have a belt and pulleys to wear out or to adjust.

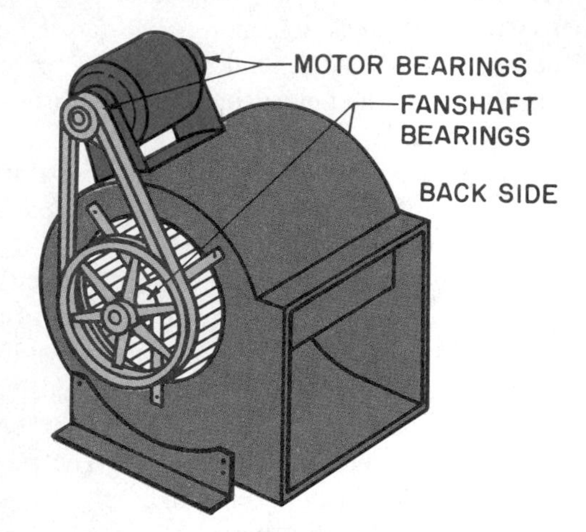

FIGURE 37–18 Blower driven with a motor using a belt and two pulleys to transfer the motor energy to the fan wheel.

FIGURE 37–19 Fan motor that is normally started with a relay. Motor protection is internal. *Courtesy Universal Electric Company*

Shaded-pole motors are used on some direct-drive blowers; however, they are not as efficient as PSC motors.

With PSC motors the fan wheel bearing is located in the motor, which reduces the bearing surfaces from four to two. The bearings may be permanently lubricated at the factory. The front bearing in the fan wheel may be hard to lubricate if a special oil port is not furnished. There are no belts or pulleys to maintain or adjust. The fan turns at the same speed as the motor, so multispeed motors are common. The air volume may be adjusted with the different fan motor speeds instead of a pulley. The motor may have up to four different speeds that can be changed by switching wires at the motor terminal box. Common speeds are from about 1500 rpm down to about 800 rpm. The motor can be operated at a faster speed in the summer (as two-speed motors) for more airflow for cooling, Figure 37–20.

37.9 THE SUPPLY DUCT SYSTEM

The supply duct system distributes air to the terminal units, registers, or diffusers into the conditioned space. Starting at the fan outlet, the duct can be fastened to the blower or blower housing directly or have a vibration eliminator between the blower and the ductwork. The vibration eliminator is recommended on all installations but is not always used. If the blower is quiet it may not be necessary, Figure 37–21.

The duct system must be designed to allow air moving toward the conditioned space to move as freely as possible, but the duct must not be oversized. Oversized duct is not economical and can cause airflow problems. Duct systems can be *plenum, extended plenum, reducing plenum,* or *perimeter loop,* Figure 37–22. Each system has its advantages and disadvantages.

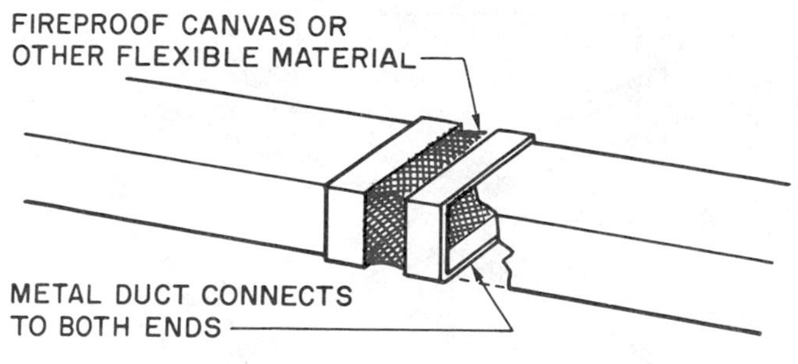

FIGURE 37–21 Vibration eliminator.

37.10 THE PLENUM SYSTEM

The plenum system has an individual supply system that makes it well suited for a job where the room outlets are all close to the unit. This system is economical from a first-cost standpoint and can easily be installed by an installer with a minimum of training and experience. The supply diffusers (where the air is diffused and blown into the room) are normally located on the inside walls and are used for heating systems that have very warm or hot air as the heating source. Plenum systems work better on fossil-fuel (coal, oil, or gas) systems than with heat pumps because the leaving-air temperatures are much warmer in fossil-fuel systems. They are often applied to inside walls for short runs, for example, in apartment houses.

When the supply diffusers are located on the inside walls, a warmer air is more desirable. The supply air temperature on a heat pump without strip heat is rarely more than 100°F, whereas on a fossil-fuel system it could easily reach 130°F.

The return air system can be a single return located at the air handler, which makes materials economical. The

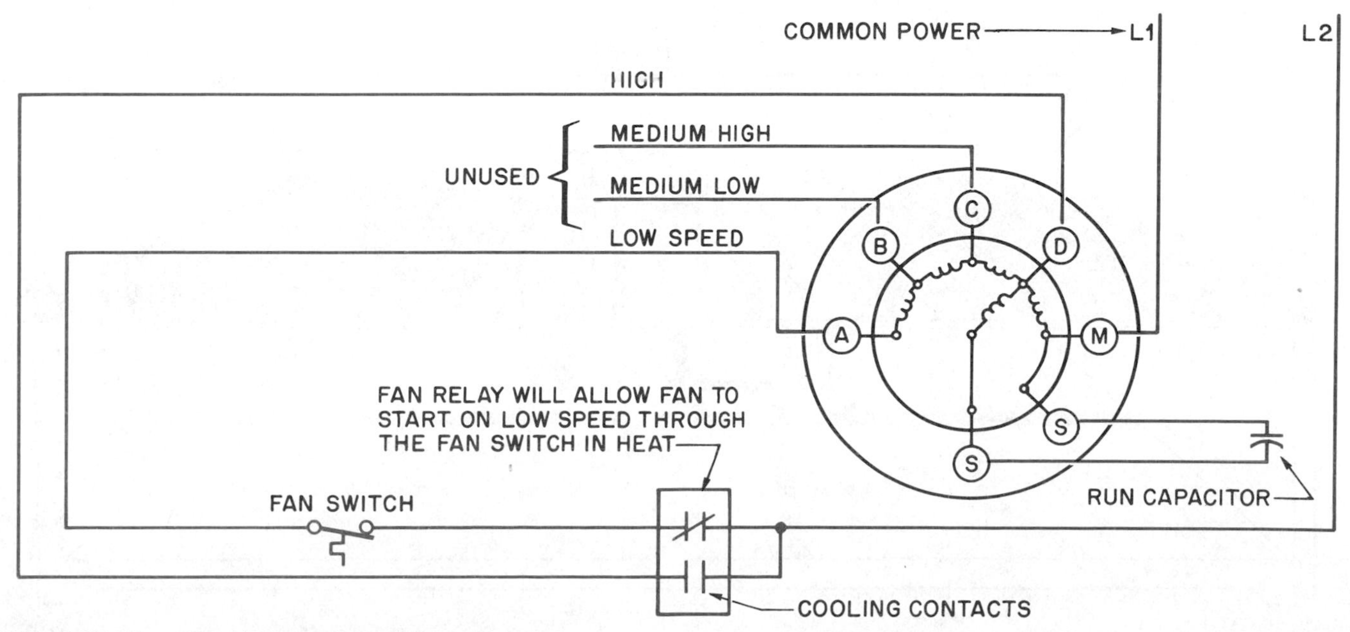

FIGURE 37–20 Wiring diagram of a multiple-speed motor.

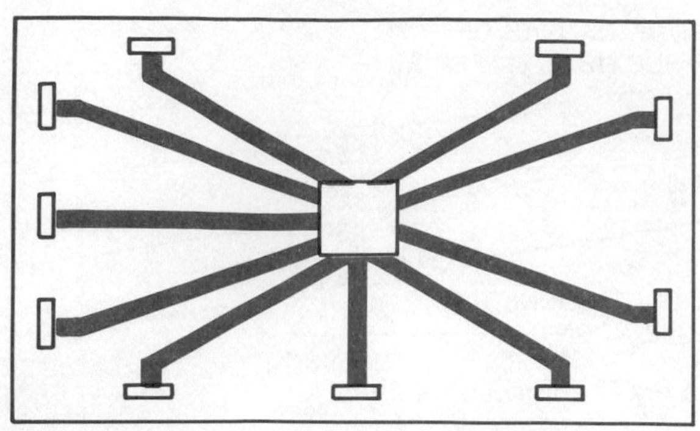

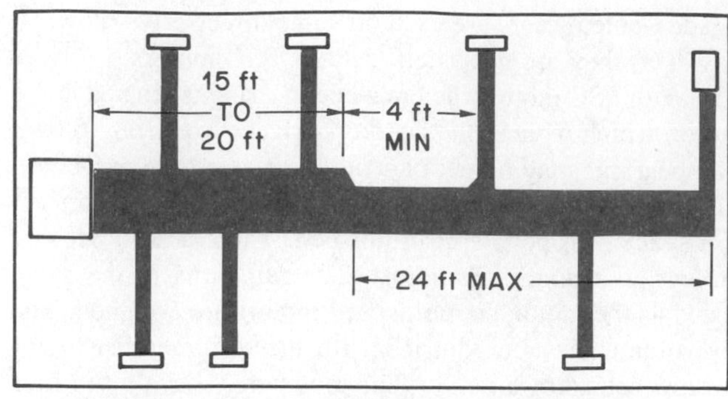

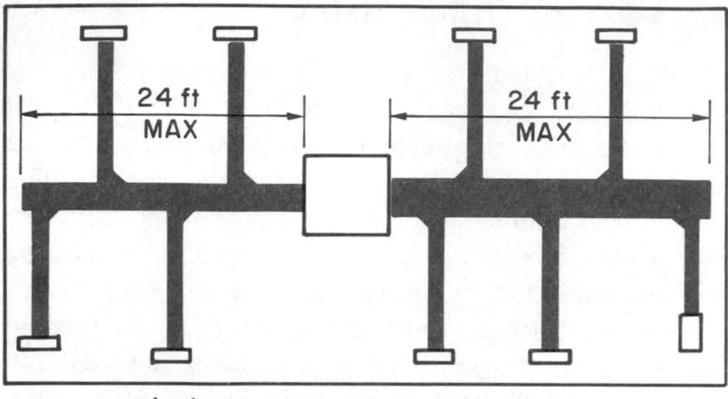

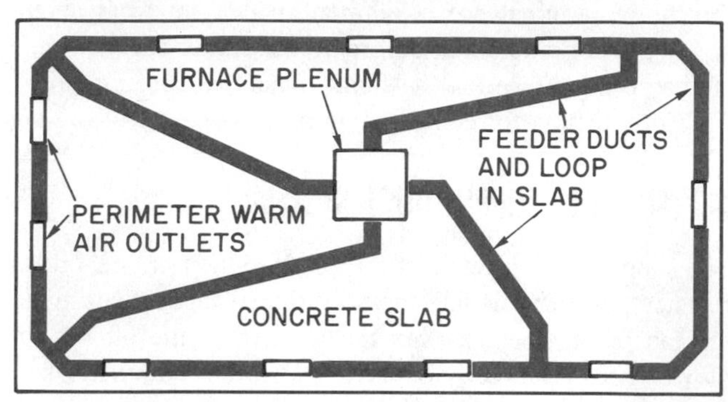

FIGURE 37–22 (A) Plenum system. (B) Extended plenum system. (C) Reducing plenum system. (D) Perimeter loop system.

single-return system will be discussed in more detail later. Figure 37–23 shows an example of a plenum system with registers in the ceiling.

37.11 THE EXTENDED PLENUM SYSTEM

The extended plenum system can be applied to a long structure such as the ranch-style house. This system takes the plenum closer to the farthest point. The extended plenum is called the *trunk duct* and can be round, square, or rectangular, Figure 37–24. The system uses small ducts called *branches* to complete the connection to the terminal units. These small ducts can be round, square, or rectangular. In small sizes they are usually round because it is less expensive to manufacture and assemble round duct in the smaller sizes. An average home probably has 6-in. round duct for the branches.

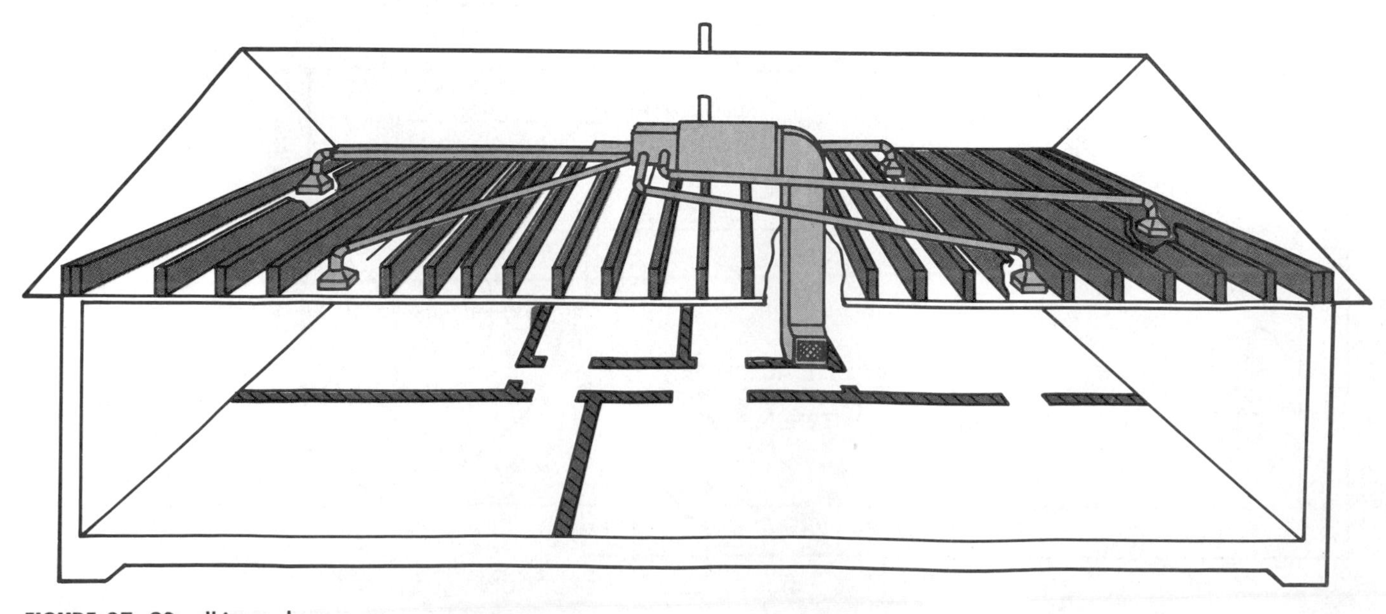

FIGURE 37–23 Using a plenum system.

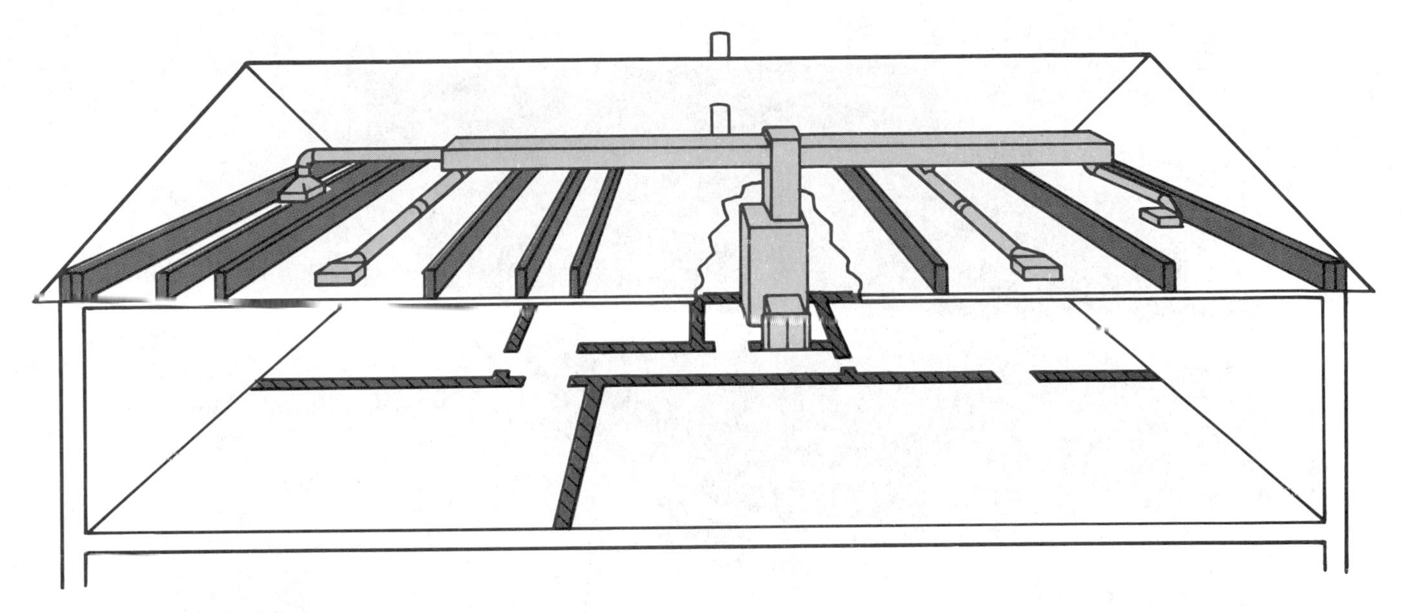

FIGURE 37–24 Extended plenum system.

37.12 THE REDUCING PLENUM SYSTEM

The reducing plenum system reduces the trunk duct size as branch ducts are added. This system has the advantage of saving materials and keeping the same pressure from one end of the duct system to the other, when properly sized. This ensures that each branch duct has approximately the same pressure pushing air into its takeoff from the trunk duct, Figure 37–25.

37.13 THE PERIMETER LOOP SYSTEM

The perimeter loop duct system is particularly well suited for installation in a concrete floor in a colder climate. The loop can be run under the slab close to the outer walls with the outlets next to the wall. There is warm air in the whole loop when the furnace fan is running, and this keeps the slab at a more even temperature. The loop has a constant pressure around the system and provides the same pressure to all outlets, Figure 37–26.

37.14 DUCT SYSTEM STANDARDS

All localities should have some minimum standard for air conditioning and heating systems. The regulating body might use a state standard, or it might have a local standard. These are called codes. Many state and local standards adopt other known standards, such as BOCA (Building Officials and Code Administrators International). All technicians should become familiar with the code requirements and follow them. These code requirements are not standard across the nation.

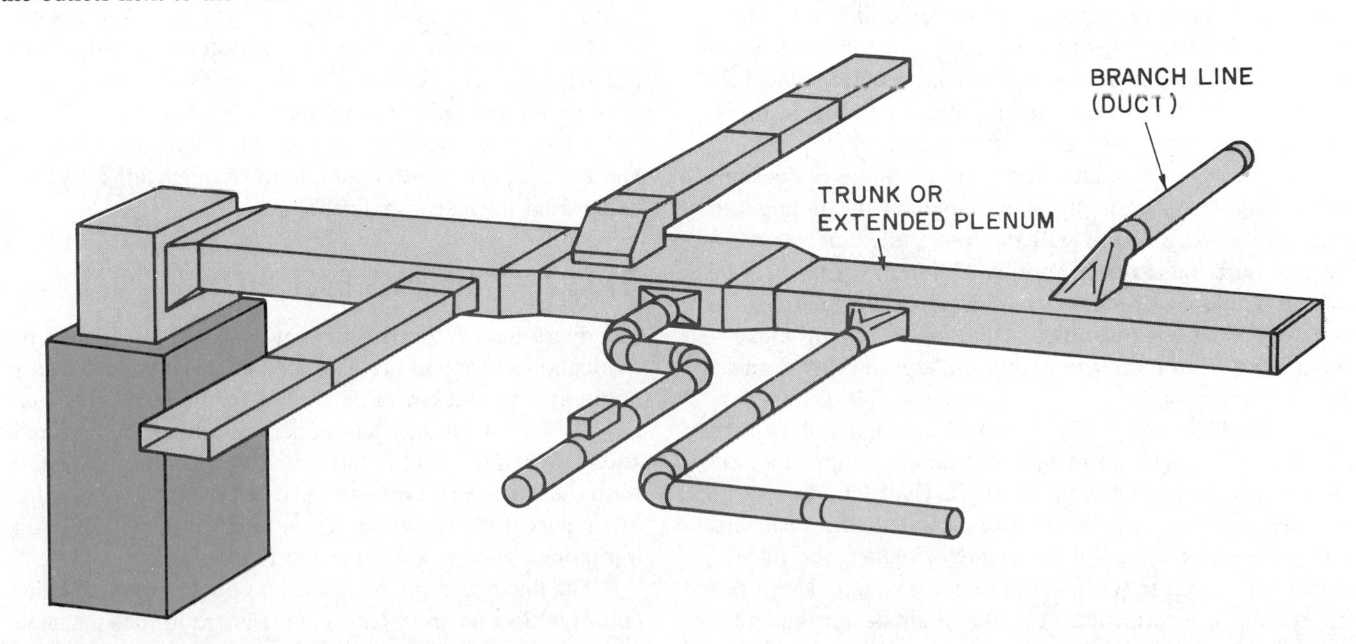

FIGURE 37–25 Reducing plenum system. *Courtesy Climate Control*

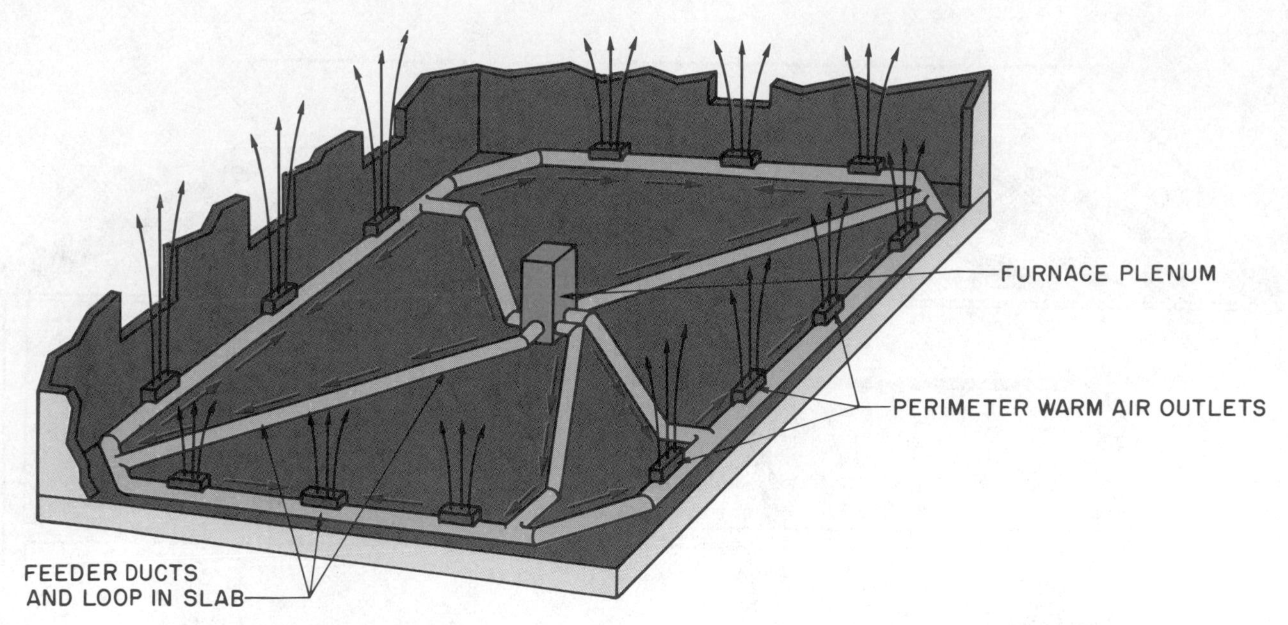

FIGURE 37–26 Perimeter loop system.

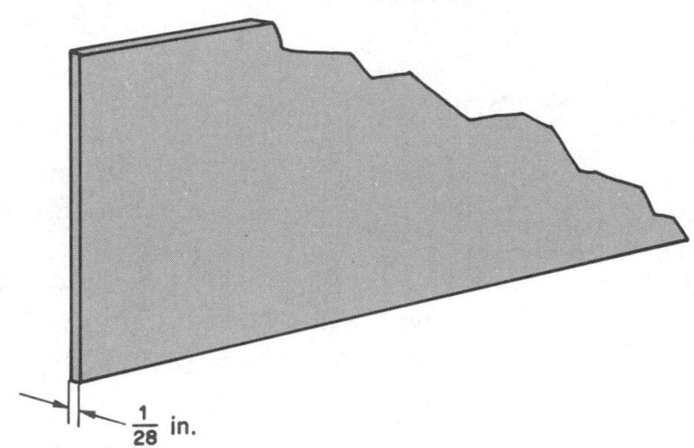

FIGURE 37–27 Duct thickness, 28 gauge.

37.15 DUCT MATERIALS

The ductwork for carrying the air from the fan to the conditioned space can be made of different materials. For many years galvanized sheet metal was used exclusively, but it is expensive to manufacture and assemble at the job. Galvanized metal is by far the most durable material. It can be used in walls where easy access for servicing is not available. Aluminum, fiberglass ductboard, spiral metal duct, and flexible duct have all been used successfully. ***The duct material must meet the local codes for fire protection.***

37.16 GALVANIZED STEEL DUCT

Galvanized steel duct comes in several different thicknesses, called the *gauge* of the metal. When a metal thickness is 28 gauge, it means that it takes approximately 28 thicknesses of the metal to make a piece 1-in. thick, or the metal is approximately 1/28 of an inch thick, Figure 37–27. The thickness of the duct can be less when the dimensions of the ductwork are small. When the ductwork is larger, it must be more rigid or it will swell and make noises when the fan starts or stops. Figure 37–28 shows a table to be used as a guideline for choosing duct metal thickness. Quite often the duct manufacturer will cross-break or make a slight bend from corner to corner on large fittings to make the duct more rigid.

Metal duct is normally furnished in lengths of 4 ft and can be round, square, or rectangular. Smaller round duct can be purchased in lengths up to 10 ft. Duct lengths can be fastened together with special fasteners, called S fasteners, and drive cleats if the duct is square or rectangular or self-tapping sheet metal screws if the duct is round. These fasteners make a secure connection that is almost airtight at the low pressures at which the duct is normally operated. See Figure 37–29 for an example of these fasteners. If there is any question of the air leaking out, special tape can be applied to the connections to ensure that no air will leak out. Figure 37–30 shows a duct that is fastened together with self-tapping sheet metal screws. Figure 37–31 shows a connection that has been taped after being fastened.

Aluminum duct follows the same guidelines as galvanized duct. The cost of aluminum prevents this duct from being used for many applications.

37.17 FIBERGLASS DUCT

Fiberglass duct is furnished in two styles: flat sheets for fabrication and round prefabricated duct. Fiberglass duct is normally 1 in. thick with an aluminum foil backing, Figure 37–32. The foil backing has a fiber reinforcement to make it strong. When the duct is fabricated, the fiberglass is cut to form the edges, and the reinforced foil backing is left intact to support the connection. These duct systems are easily transported and assembled in the field.

The ductboard can be made into duct in several different ways. Special knives that cut the board in such a manner as to produce overlapping connections can be used in the

GAGES OF METAL DUCTS AND PLENUMS USED FOR
COMFORT HEATING OR COOLING FOR A SINGLE DWELLING UNIT

	COMFORT HEATING OR COOLING			Comfort Heating Only
	Galvanized Steel			
	Nominal Thickness (In Inches)	Equivalent Galvanized Sheet Gage No.	Approximate Aluminum B & S Gage	Minimum Weight Tin-Plate Pounds Per Base Box
Round Ducts and Enclosed Rectangular Ducts				
14″ or less	0.016	30	26	135
Over 14″	0.019	28	24	—
Exposed Rectangular Ducts				
14″ or less	0.019	28	24	—
Over 14″	0.022	26	23	—

FIGURE 37–28 Table of recommended metal thickness for different sizes of duct.

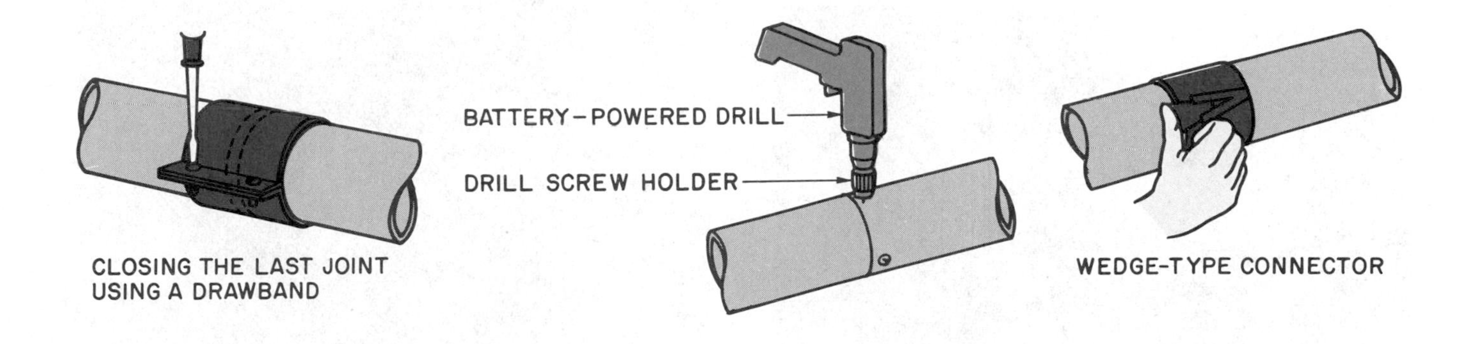

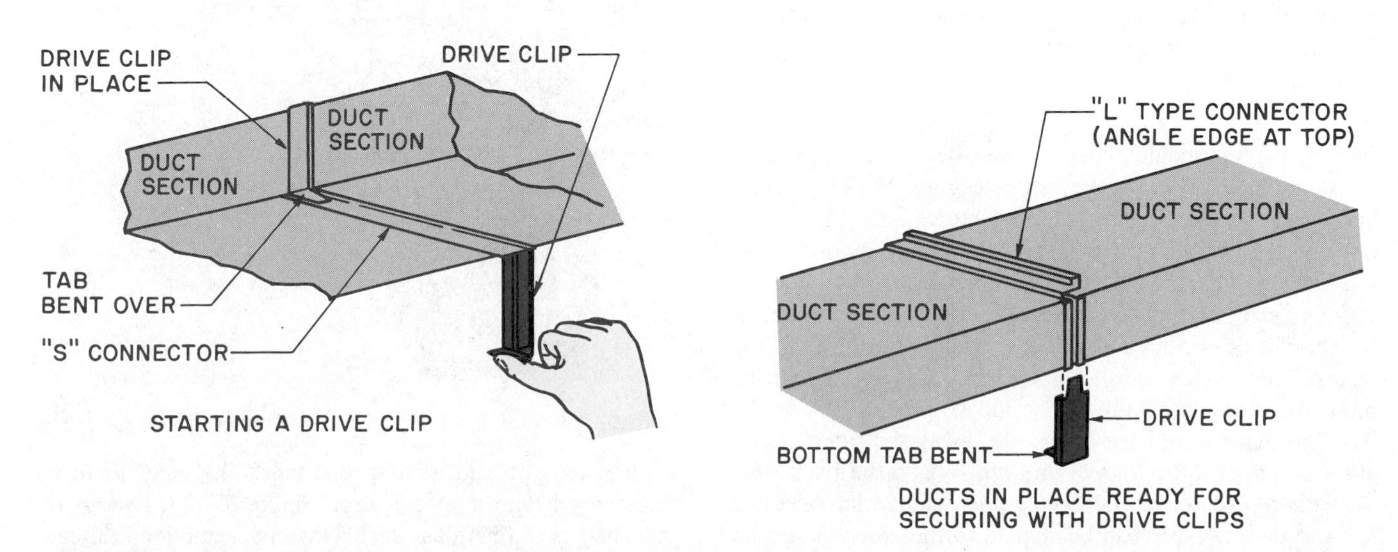

FIGURE 37–29 Fasteners for square and round duct for low-pressure systems only.

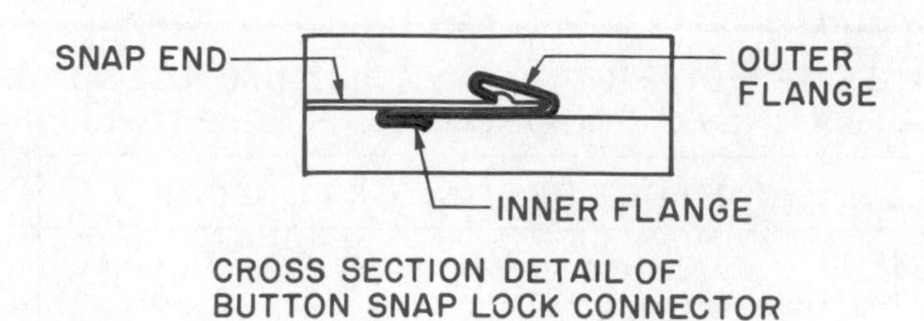

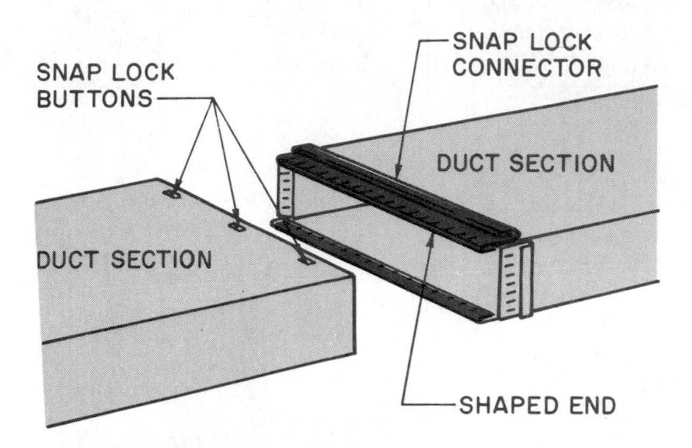

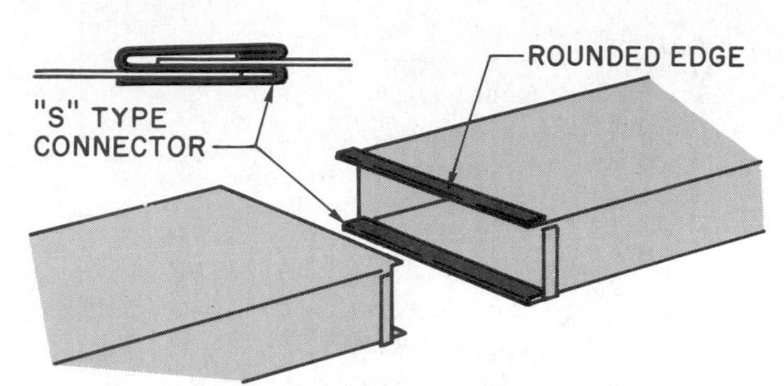

FIGURE 37–29 *(Continued)*

FIGURE 37–30 Fastening duct using a portable electric drill and sheet metal screws. *Photo by Bill Johnson*

FIGURE 37–31 Taped duct connection. *Photo by Bill Johnson*

field by placing the ductboard on any flat surface and using a straightedge to guide the knife, Figure 37–33. Special ductboard machines can be used to fabricate the duct in the field, at the job sight, or in the shop to be transported to the job. An operator has to be able to set up the machine for different sizes of duct and fittings. When the duct is made in the shop, it can be cut and left flat and stored in the original boxes. This makes transportation easy. The pieces can be marked for easy assembly at the job site.

The machines or the knives cut away the fiberglass to allow the duct to be folded with the foil on the back side. When two pieces are fastened together, an overlap of foil is left so that one piece can be stapled to the other. A special staple is used that turns out and up on the ends. Then the connection is taped with special tape. The tape should be pressed on with a special iron, Figure 37–34. One of the advantages of fiberglass duct is that the insulation is already on the duct when it is assembled.

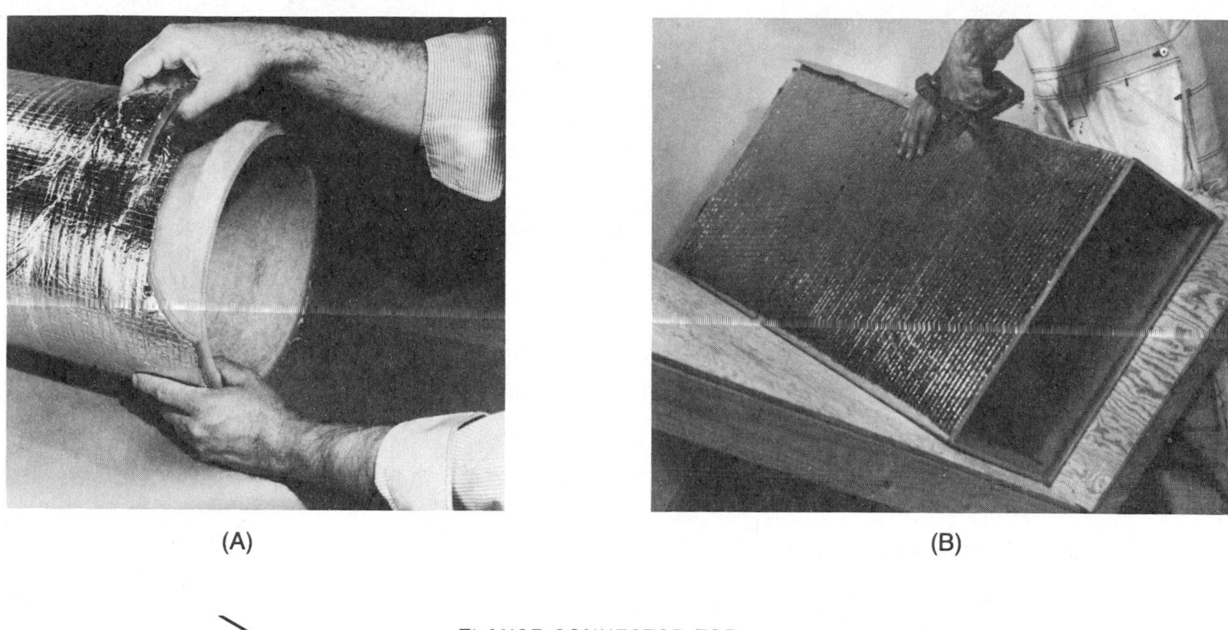

(A) (B)

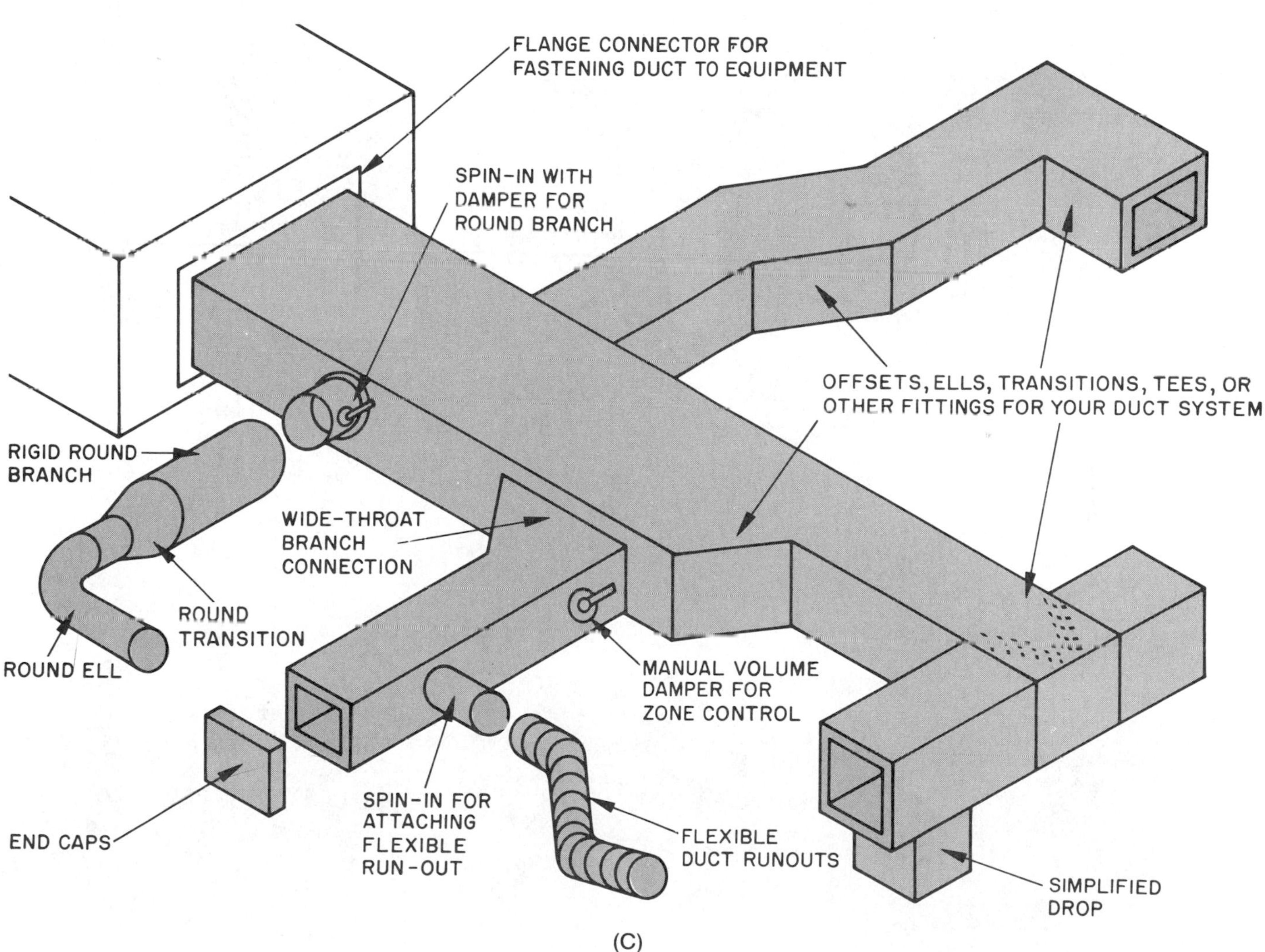

(C)

FIGURE 37–32 Ducts manufactured out of compressed fiberglass with a foil backing. *Courtesy Manville Corporation*

37.18 SPIRAL METAL DUCT

Spiral metal duct is used more on large systems. It is normally manufactured at the job sight with a special machine. The duct comes in rolls of flat narrow metal. The machine winds the metal off the spool and folds a seam in it. The length of a run of duct can be made very long with this method.

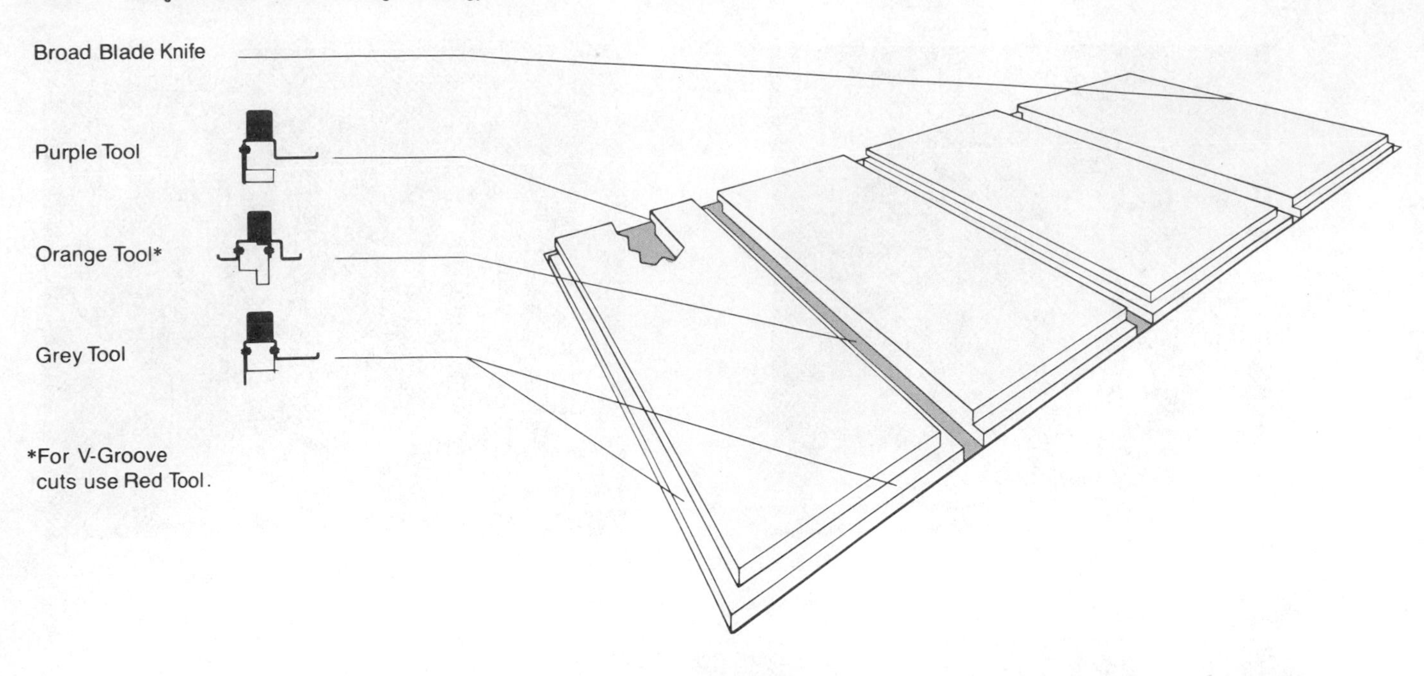

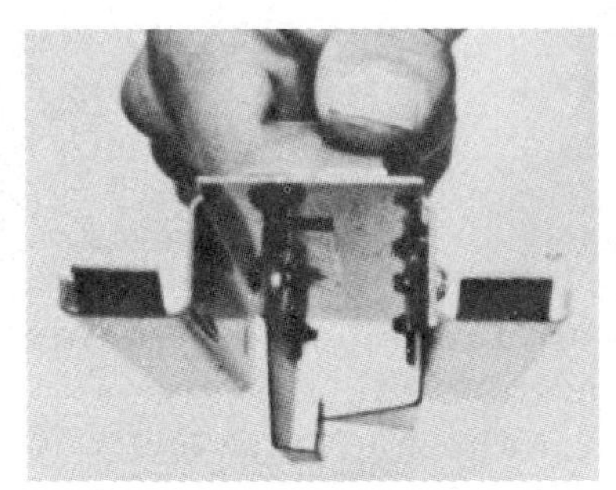
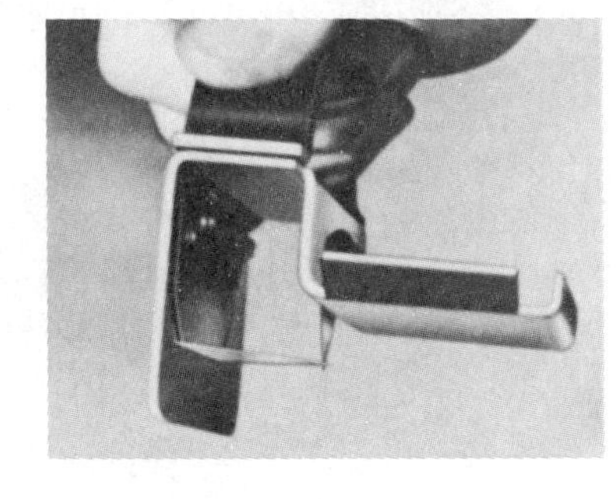
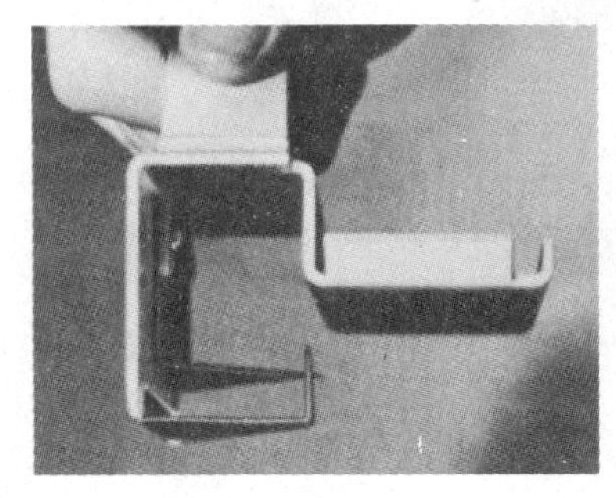
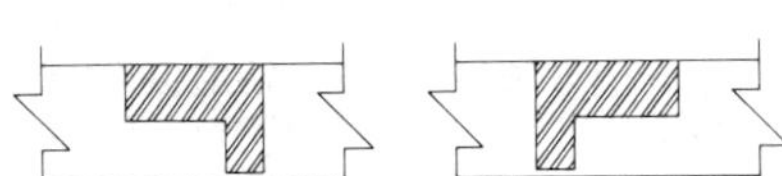
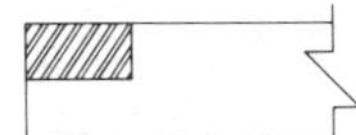
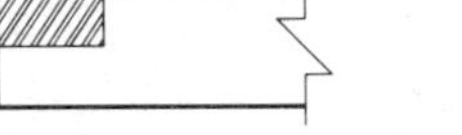
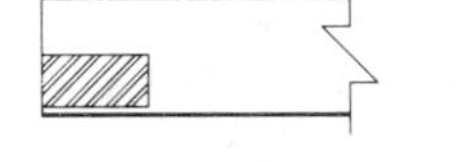

Orange Tool:
Cuts modified shiplap for forming corners.

Grey Tool:
The Grey Tool can also be used to make the closing corner joint. And, when making short sections, or when using square-edge board which does not have the premolded M/F edges, the Grey Tool cuts the female slip-joint used to join sections.

Purple Tool:
Cuts the male slip-joint used to join sections.

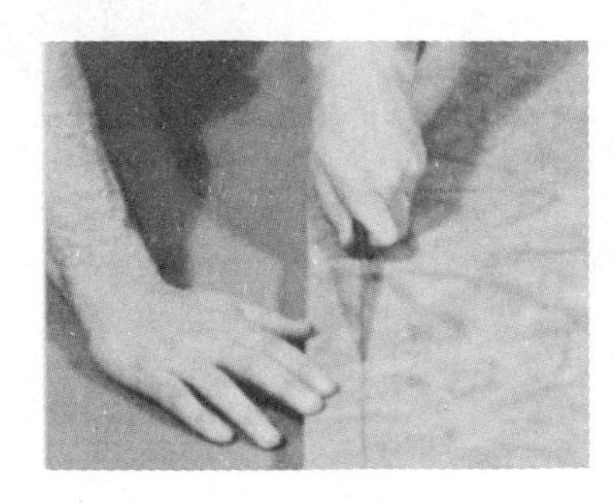
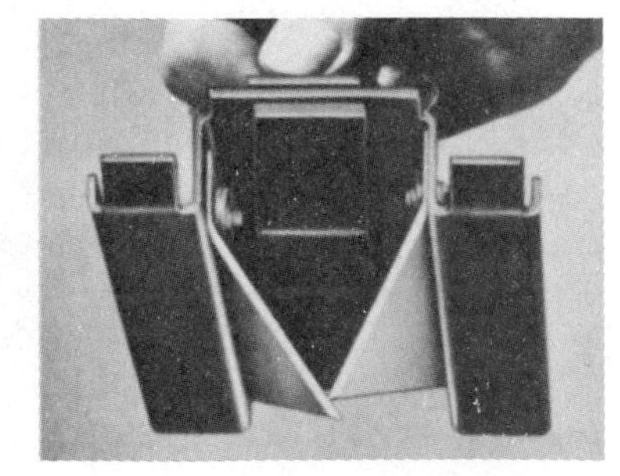
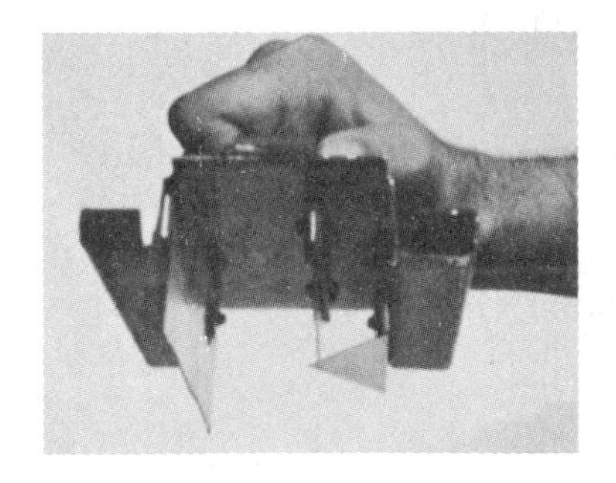
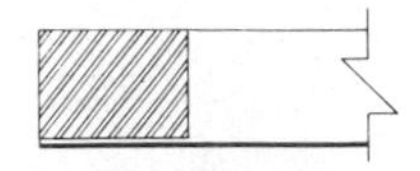
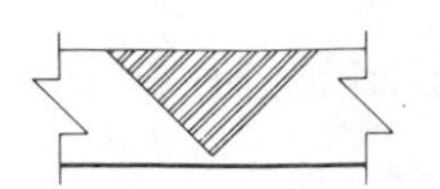
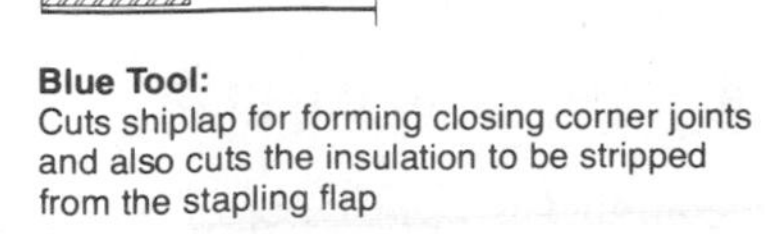

Broad 3lade Knife:
Used in cutting the board to size and in cutting insulation from the staple flap.

Red Tool:
Cuts "V" or miter for forming corners.

Blue Tool:
Cuts shiplap for forming closing corner joints and also cuts the insulation to be stripped from the stapling flap

FIGURE 37–33 Knives for fabricating fiberglass duct. *Courtesy Manville Corporation*

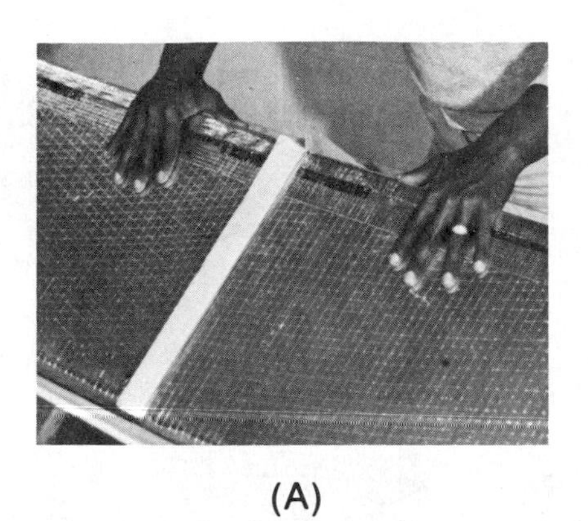

(A)

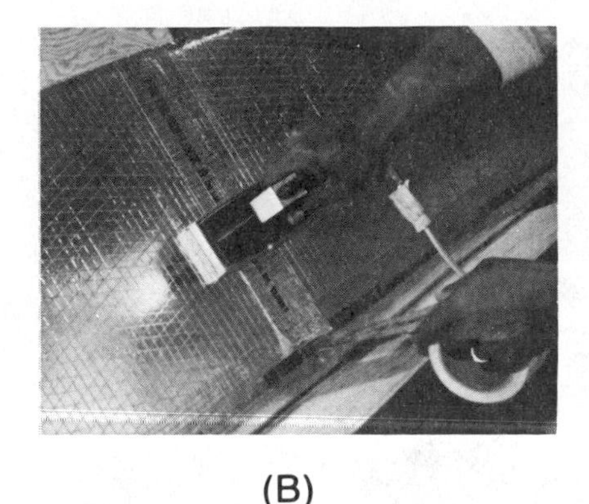

(B)

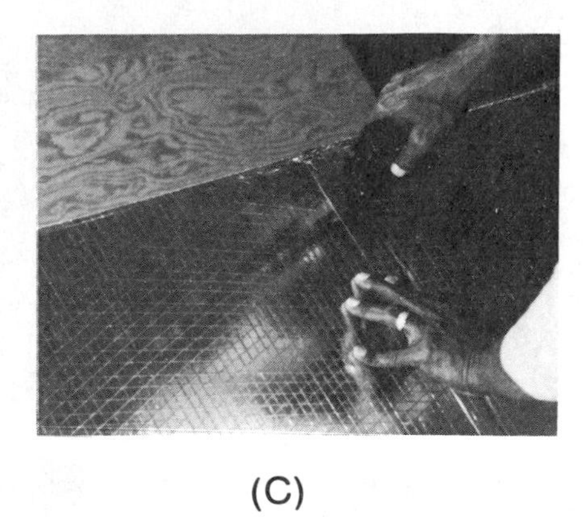

(C)

FIGURE 37–34 Fiberglass duct assembly. *Courtesy Manville Corporation*

37.19 FLEXIBLE DUCT

Flexible round duct comes in sizes up to about 24 in. in diameter. Some of it has a reinforced aluminum foil backing and comes in a short box with the duct material compressed. Without the insulation on it the duct looks like a coil spring with a flexible foil backing.

Some flexible duct comes with vinyl or foil backing and insulation on it, in lengths of 25 ft to a box, Figure 37–35. It is compressed into the box. Flexible duct is easy to route around corners. Keep the duct as short as practical and don't allow tight turns that may cause the duct to collapse. This duct has more friction loss inside it than metal duct does, but it also serves as a sound attenuator to reduce blower noise down the duct. For best airflow, flexible duct should be stretched as tight as is practical.

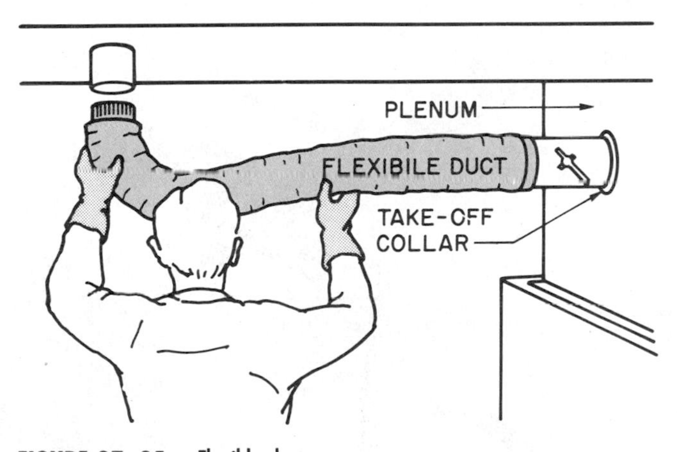

FIGURE 37–35 Flexible duct.

37.20 COMBINATION DUCT SYSTEMS

Duct systems can be combined in various ways. For example:

1. All square or rectangular metal duct, the trunk line, and the branches are the same shape, Figure 37–36.

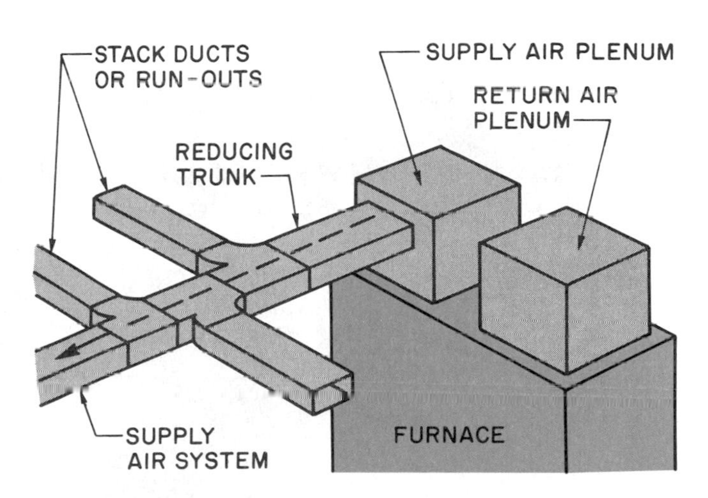

FIGURE 37–36 Square or rectangular duct.

2. Metal trunk lines with round metal branch duct, Figure 37–37.
3. Metal trunk lines with round fiberglass branch duct, Figure 37–38.
4. Metal trunk lines with flexible branch duct, Figure 37–39.
5. Ductboard trunk lines with round fiberglass branches, Figure 37–40.
6. Ductboard trunk lines with round metal branches, Figure 37–41.
7. Ductboard trunk lines with flexible branches, Figure 37–42.
8. All round metal duct with round metal branch ducts, Figure 37–43.
9. All round metal trunk lines with flexible branch ducts, Figure 37–44.

37.21 DUCT AIR MOVEMENT

Special attention should be given to the point where the branch duct leaves the main trunk duct to get the correct amount of air into the branch duct. The branch duct must be fastened to the main trunk line with a *takeoff fitting*. Several fittings have been designed for *takeoff fitting*. The takeoff that has a larger throat area than the runout duct will allow

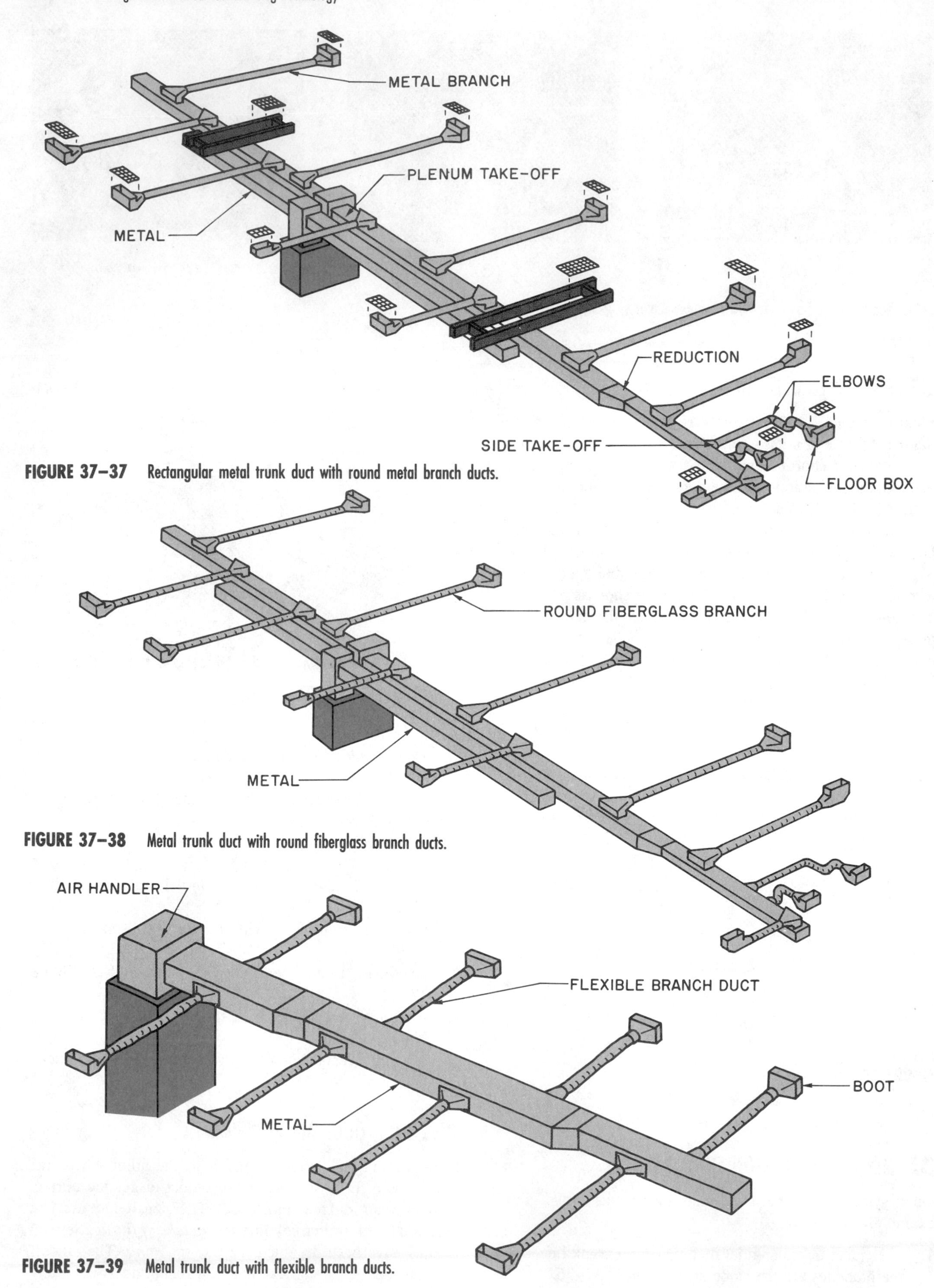

FIGURE 37–37 Rectangular metal trunk duct with round metal branch ducts.

FIGURE 37–38 Metal trunk duct with round fiberglass branch ducts.

FIGURE 37–39 Metal trunk duct with flexible branch ducts.

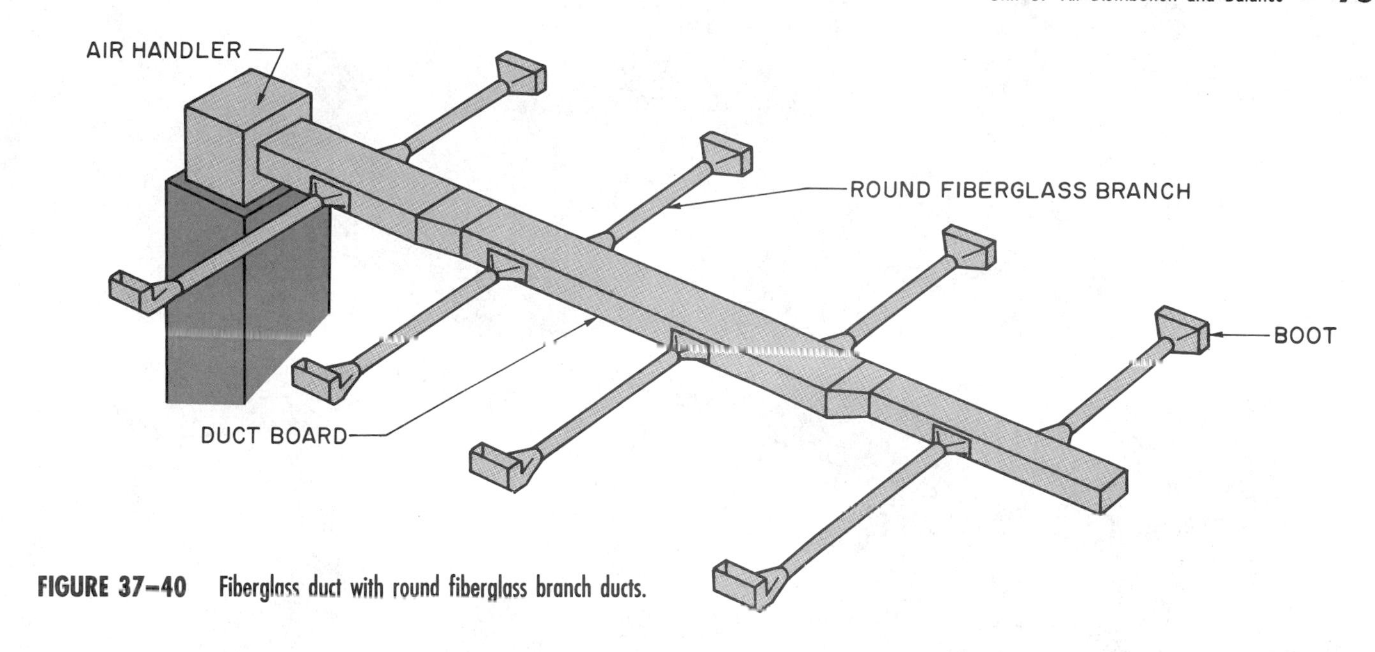

FIGURE 37–40 Fiberglass duct with round fiberglass branch ducts.

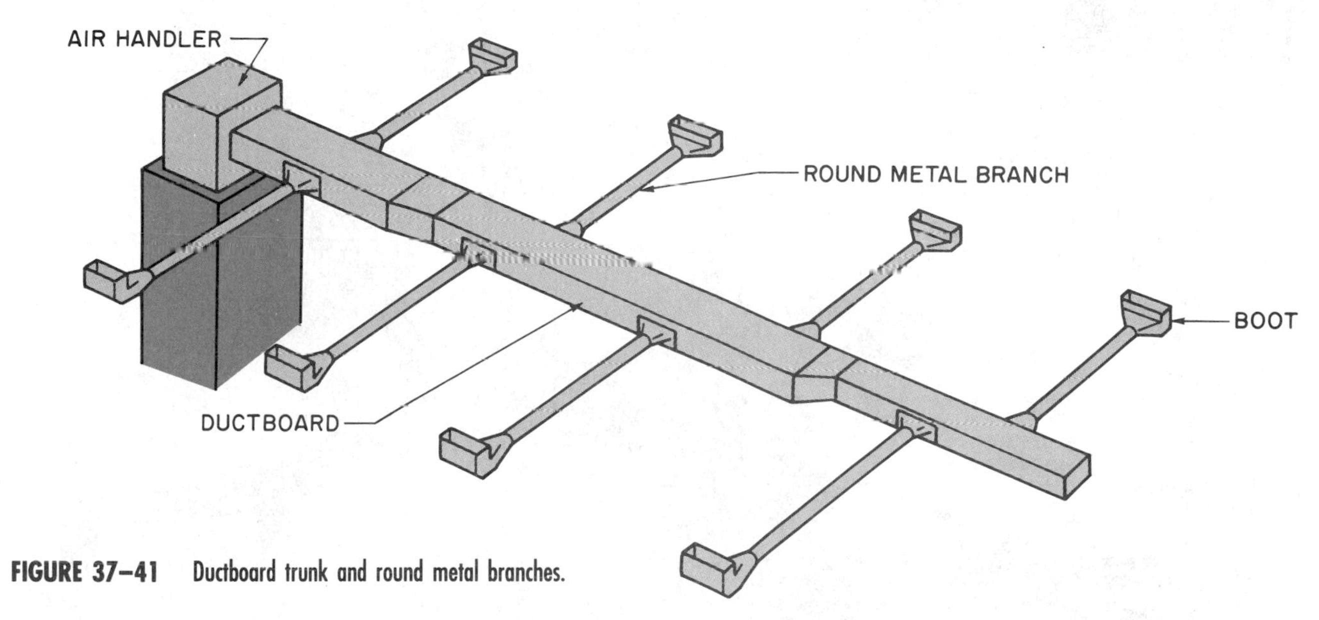

FIGURE 37–41 Ductboard trunk and round metal branches.

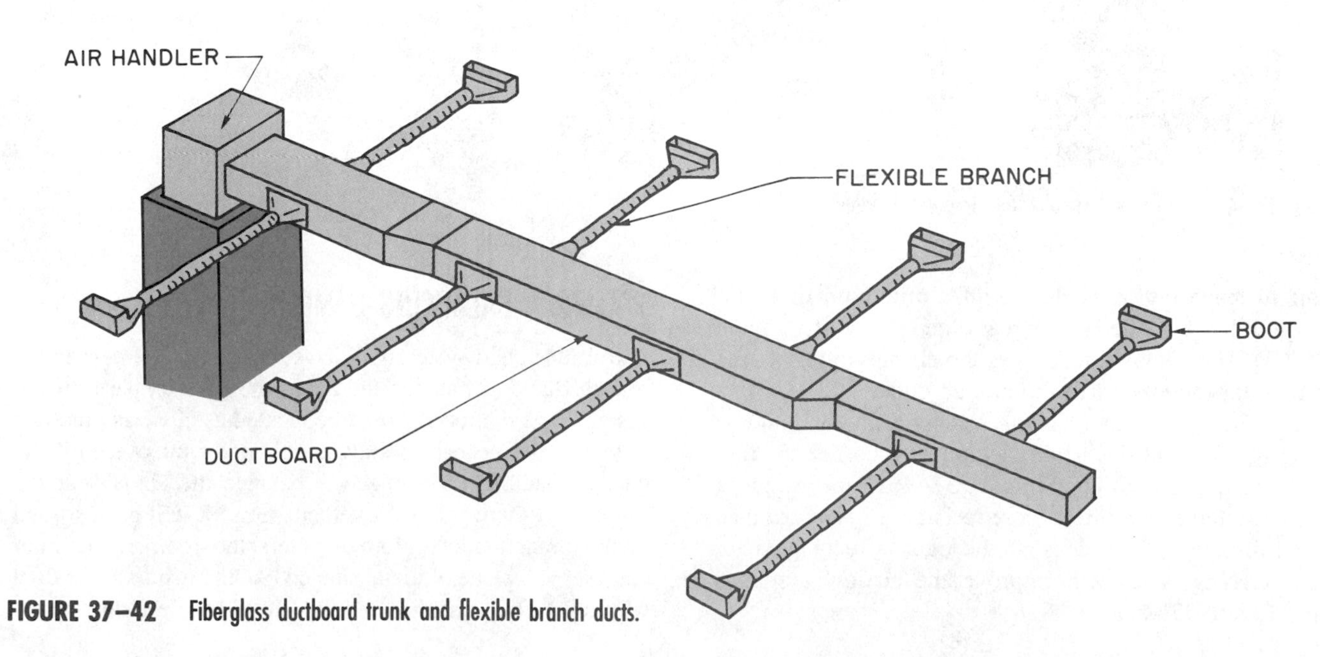

FIGURE 37–42 Fiberglass ductboard trunk and flexible branch ducts.

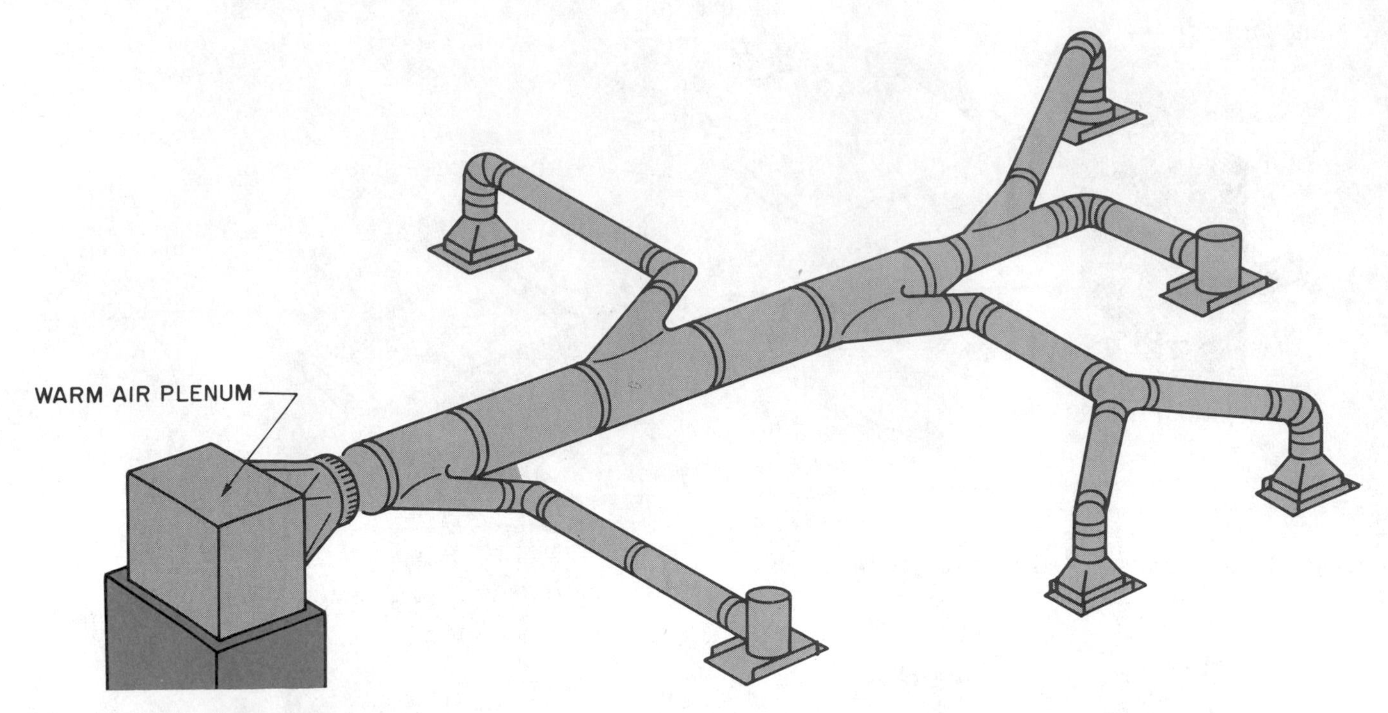

FIGURE 37–43 Metal round duct.

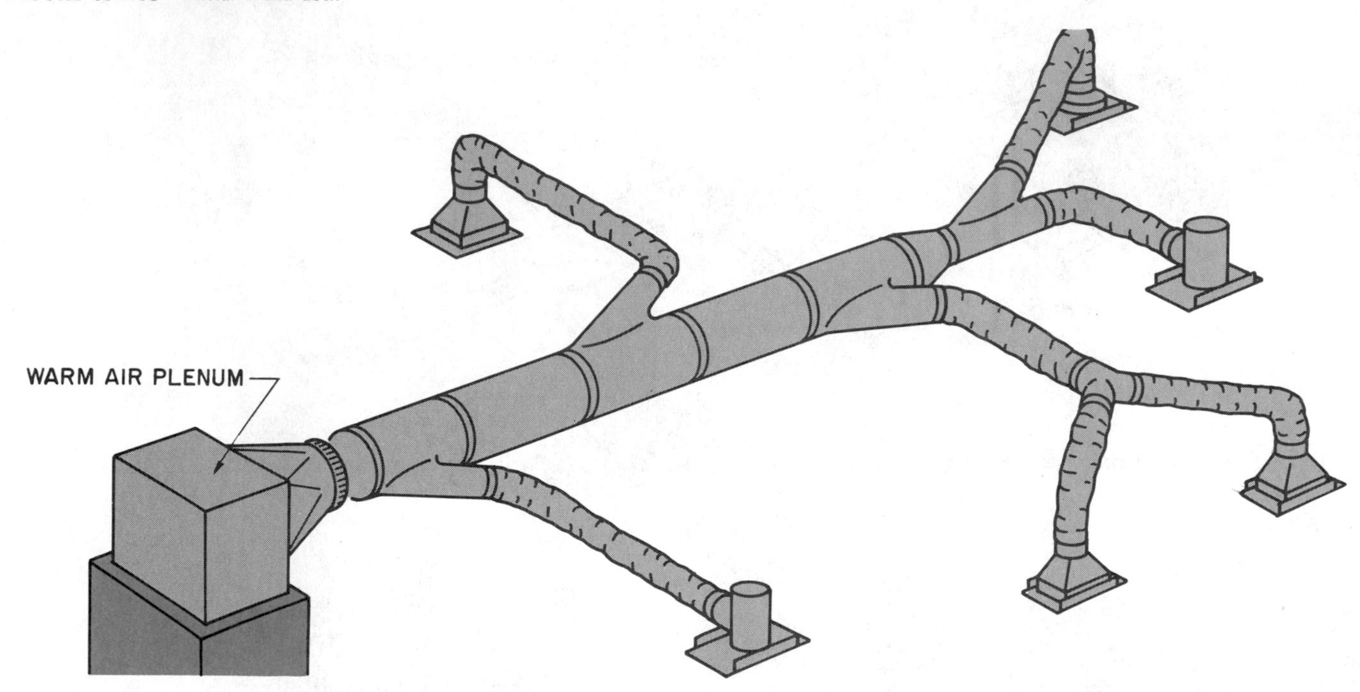

FIGURE 37–44 All round metal trunk line with flexible branch lines.

the air to leave the trunk duct with a minimum of effort. This could be called a streamlined takeoff, Figure 37–45(B). The takeoff encourages the air moving down the duct to enter the takeoff to the branch duct.

Air moving in a duct has *inertia*—it wants to continue moving in a straight line. If air has to turn a corner, the turn should be carefully designed, Figure 37–46. For example, a square-throated elbow offers more resistance to airflow than a round-throated elbow does. If the duct is rectangular or square, turning vanes will improve the airflow around a corner, Figure 37–47.

37.22 BALANCING DAMPERS

A well-designed system will have *balancing dampers* in the branch ducts to balance the air in the various parts of the system. The dampers should be located as close as practical to the trunk line, with the damper handles uncovered if the duct is insulated. The place to balance the air is near the trunk, so if there is any air velocity noise it will be absorbed in the branch duct before it enters the room. A damper consists of a piece of metal shaped like the inside of the duct with a handle protruding through the side of the duct to the

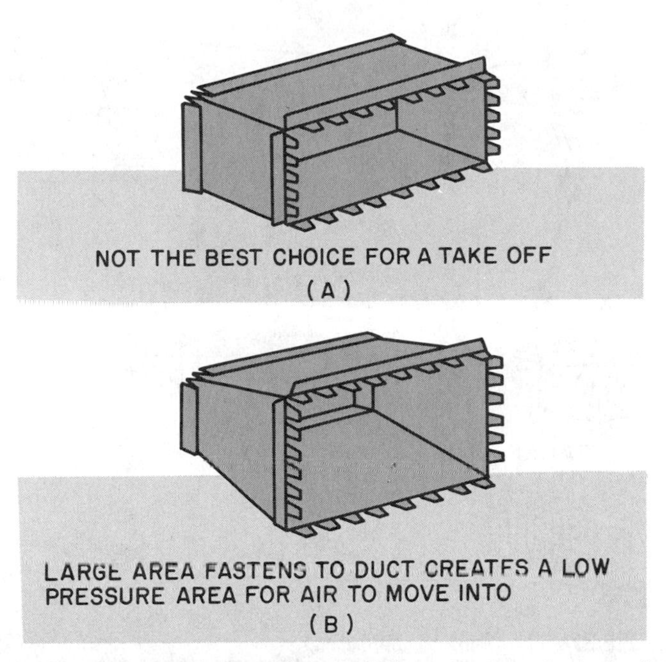

FIGURE 37–45 (A) Standard takeoff fitting. (B) Streamlined takeoff fitting.

outside. The handle allows the damper to be turned at an angle to the airstream to slow the air down, Figure 37–48.

37.23 DUCT INSULATION

When ductwork passes through an unconditioned space, heat transfer may take place between the air in the duct and the air in the unconditioned space. If the heat exchange adds or removes very much heat from the conditioned air, insulation should be applied to the ductwork. A 15°F temperature difference from inside the duct to outside the duct is considered the maximum difference allowed before insulation is necessary.

The insulation is built into a fiberglass duct by the manufacturer. Metal duct can be insulated in two ways: on the outside or on the inside. When applied to the outside, the insulation is usually a foil- or vinyl-backed fiberglass. It comes in several thicknesses, with 2-in. thickness the most common. The backing creates a moisture vapor barrier. This

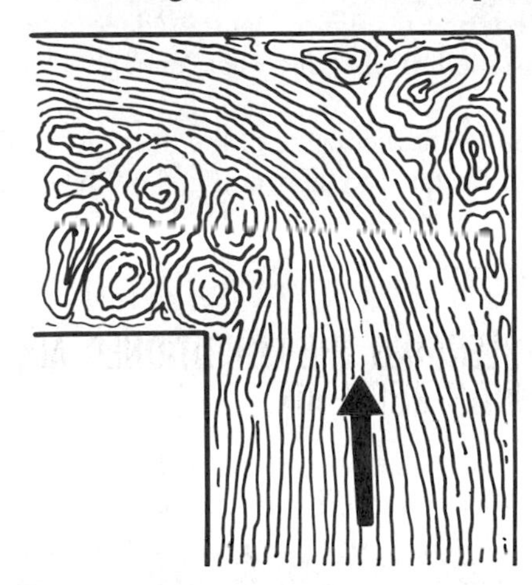

FIGURE 37–46 This illustration shows what happens when air tries to go around a corner.

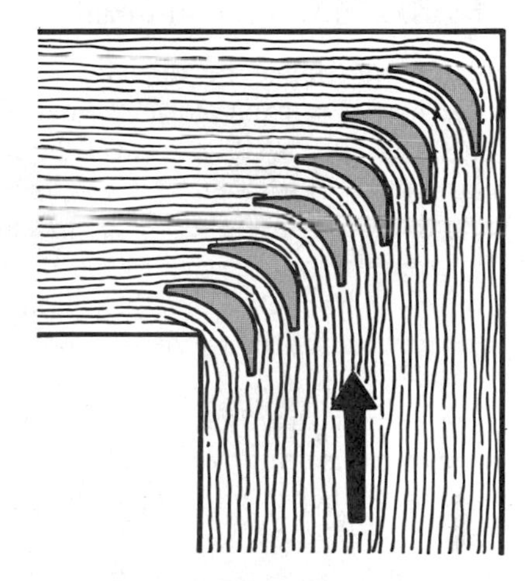

FIGURE 37–47 Square elbow with turning vanes.

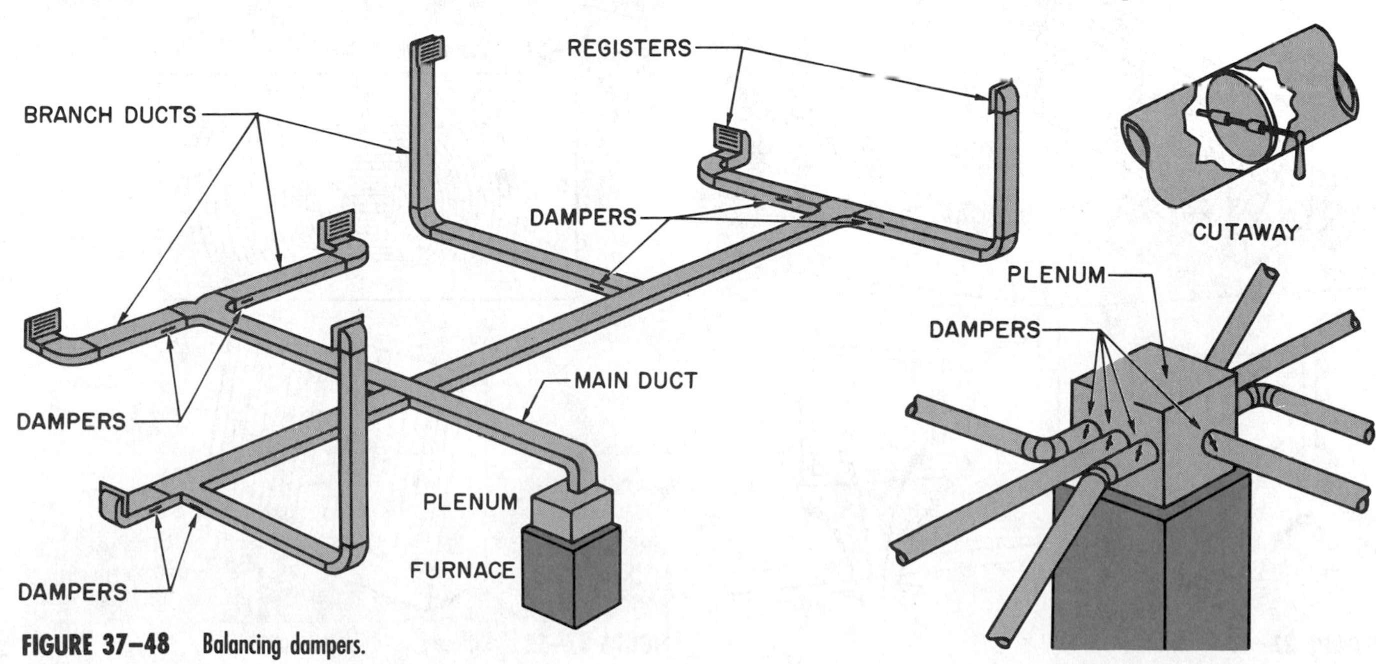

FIGURE 37–48 Balancing dampers.

is important where the duct may operate below the dew point temperature of the surroundings and moisture would form on the duct. The insulation is joined by lapping it and stapling it. It is then taped to prevent moisture from entering the seams. External insulation can be added after the duct has been installed if the duct has enough clearance all around.

When applied to the inside of the duct, the insulation is either glued or fastened to tabs mounted on the duct by spot weld or glue. This insulation must be applied when the duct is being manufactured.

37.24 BLENDING THE CONDITIONED AIR WITH ROOM AIR

When the air reaches the conditioned space, it must be properly distributed into the room so that the room will be comfortable without anyone being aware that a conditioning system is operating. This means the final components in the system must place the air in the proper area of the conditioned space for proper air blending. Following are guidelines that can be used for room air distribution.

1. When possible, air should be directed on the walls. They are the load. For example, in winter air can be directed on the outside walls to cancel the load (cold wall) and keep the wall warmer. This will keep the wall from absorbing heat from the room air. In summer the same distribution will work; it will keep the wall cool and keep room air from absorbing heat from the wall. The *diffuser* spreads the air to the desired air pattern, Figure 37–49.
2. Warm air for heating distributes better from the floor because it tends to rise, Figure 37–50.
3. Cool air distributes better from the ceiling because it tends to fall, Figure 37–51.

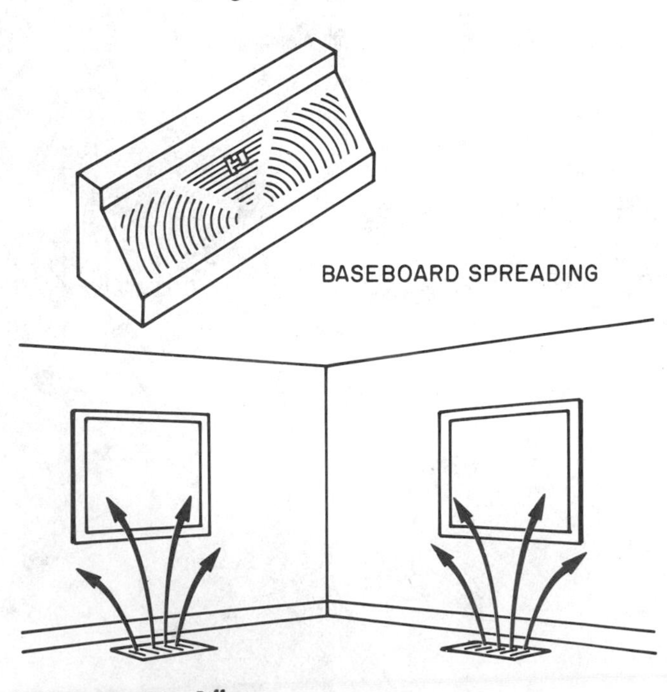

FIGURE 37–49 Diffuser.

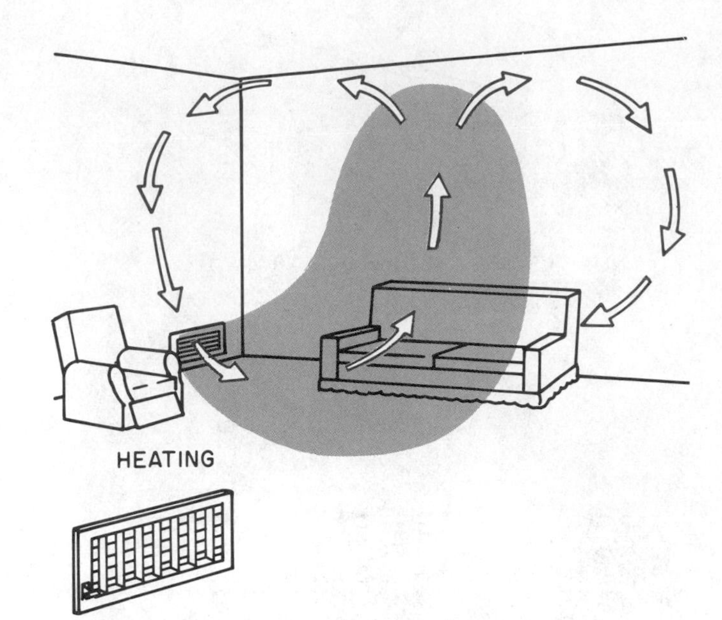

FIGURE 37–50 Warm air distribution.

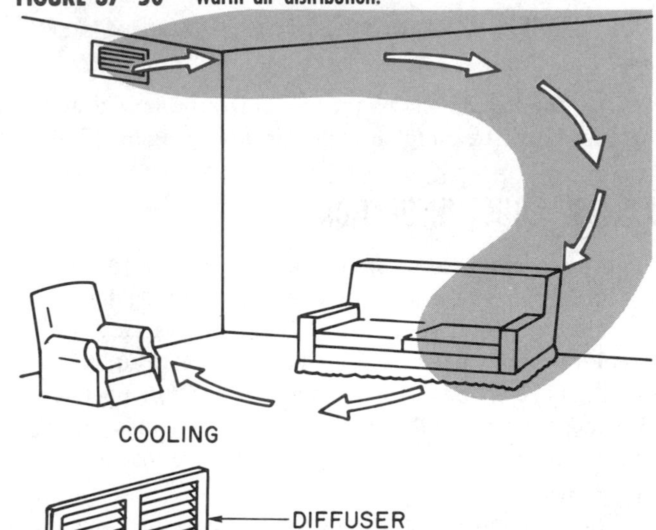

FIGURE 37–51 Cool air distribution.

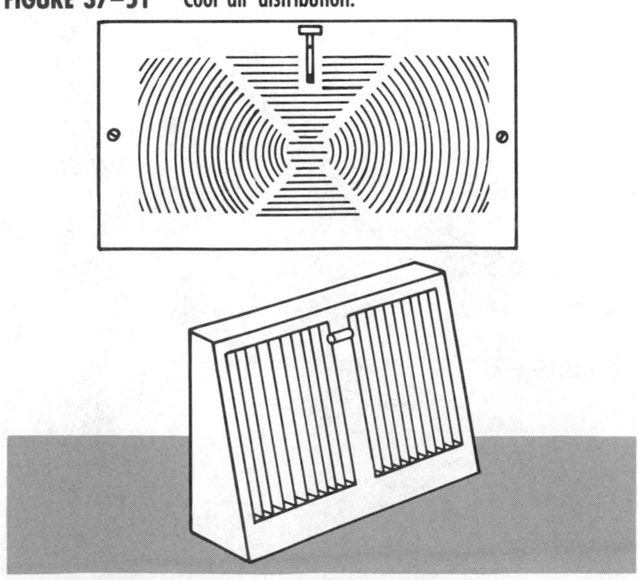

FIGURE 37–52 Diffusers used on the outside wall.

4. The most modern concept for both heating and cooling is to place the diffusers next to the outside walls to accomplish this load-canceling effect, Figure 37–52.
5. The amount of throw (how far the air from the diffuser will blow into the room from the diffuser) depends on the air pressure behind the diffuser and the style of the diffuser blades. Air pressure in the duct behind the diffuser creates the velocity of the air leaving the diffuser.

Figure 37–53 shows some air registers and diffusers. The various types can be used for low side wall, high side wall, floor, ceiling, or baseboard.

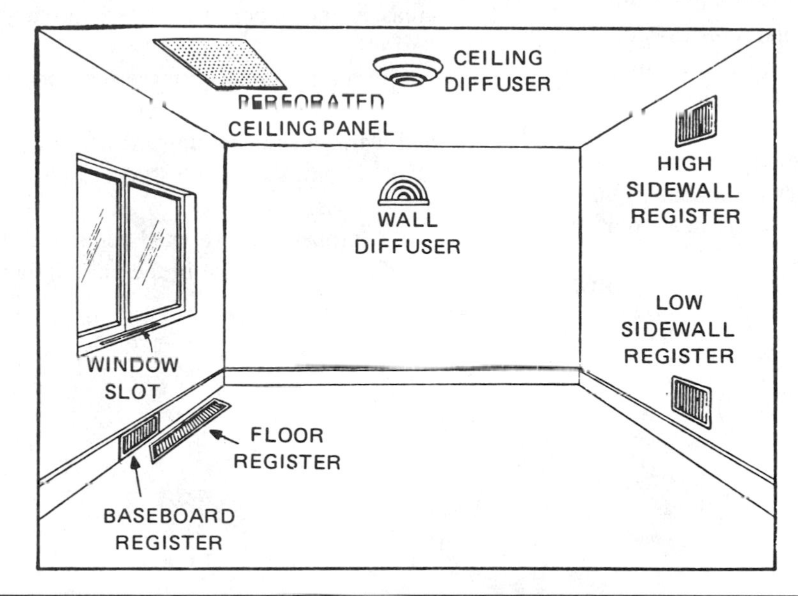

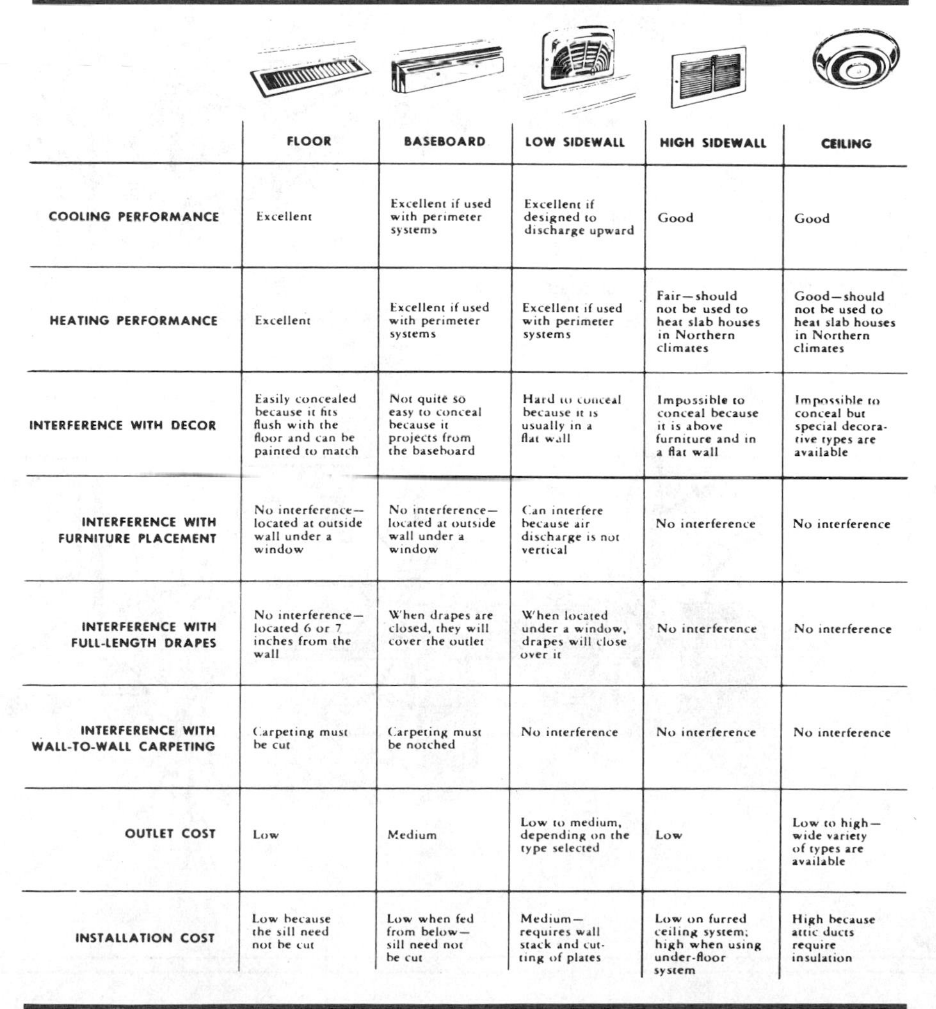

	FLOOR	BASEBOARD	LOW SIDEWALL	HIGH SIDEWALL	CEILING
COOLING PERFORMANCE	Excellent	Excellent if used with perimeter systems	Excellent if designed to discharge upward	Good	Good
HEATING PERFORMANCE	Excellent	Excellent if used with perimeter systems	Excellent if used with perimeter systems	Fair—should not be used to heat slab houses in Northern climates	Good—should not be used to heat slab houses in Northern climates
INTERFERENCE WITH DECOR	Easily concealed because it fits flush with the floor and can be painted to match	Not quite so easy to conceal because it projects from the baseboard	Hard to conceal because it is usually in a flat wall	Impossible to conceal because it is above furniture and in a flat wall	Impossible to conceal but special decorative types are available
INTERFERENCE WITH FURNITURE PLACEMENT	No interference—located at outside wall under a window	No interference—located at outside wall under a window	Can interfere because air discharge is not vertical	No interference	No interference
INTERFERENCE WITH FULL-LENGTH DRAPES	No interference—located 6 or 7 inches from the wall	When drapes are closed, they will cover the outlet	When located under a window, drapes will close over it	No interference	No interference
INTERFERENCE WITH WALL-TO-WALL CARPETING	Carpeting must be cut	Carpeting must be notched	No interference	No interference	No interference
OUTLET COST	Low	Medium	Low to medium, depending on the type selected	Low	Low to high—wide variety of types are available
INSTALLATION COST	Low because the sill need not be cut	Low when fed from below—sill need not be cut	Medium—requires wall stack and cutting of plates	Low on furred ceiling system; high when using under-floor system	High because attic ducts require insulation

FIGURE 37–53 Air registers and diffusers. From Lang, *Principals of Air Conditioning,* © 1979 by Delmar Publishers Inc.

37.25 THE RETURN AIR DUCT SYSTEM

The return air duct is constructed in much the same manner as the supply duct except that some installations are built with central returns instead of individual room returns. Individual return air systems have a return air grille in each room that has a supply diffuser (with the exception of rest rooms and kitchens). The individual return system will give the most positive return air system, but they are expensive. The return air duct is normally sized at least slightly larger than the supply duct, so there is less resistance to the airflow in the return system than in the supply system. There will be more details when we discuss duct sizing. See Figure 37–54 for an example of a system with individual room returns.

The central return system is usually satisfactory for a one-level residence. Larger return air grilles are located so that air from common rooms can easily move back to the common returns. For air to return to central returns, there must be a path, such as doors with grille work, open doorways, and undercut doors in common hallways. These open areas in the doors can prevent privacy that some people desire.

In a structure with more than one floor level, install a return at each level. Remember, cold air moves downward and warm air moves upward naturally without encouragement. Figure 37–55 illustrates air stratification in a two-level house.

A properly constructed central return air system helps to eliminate fan noise in the conditioned space. The return

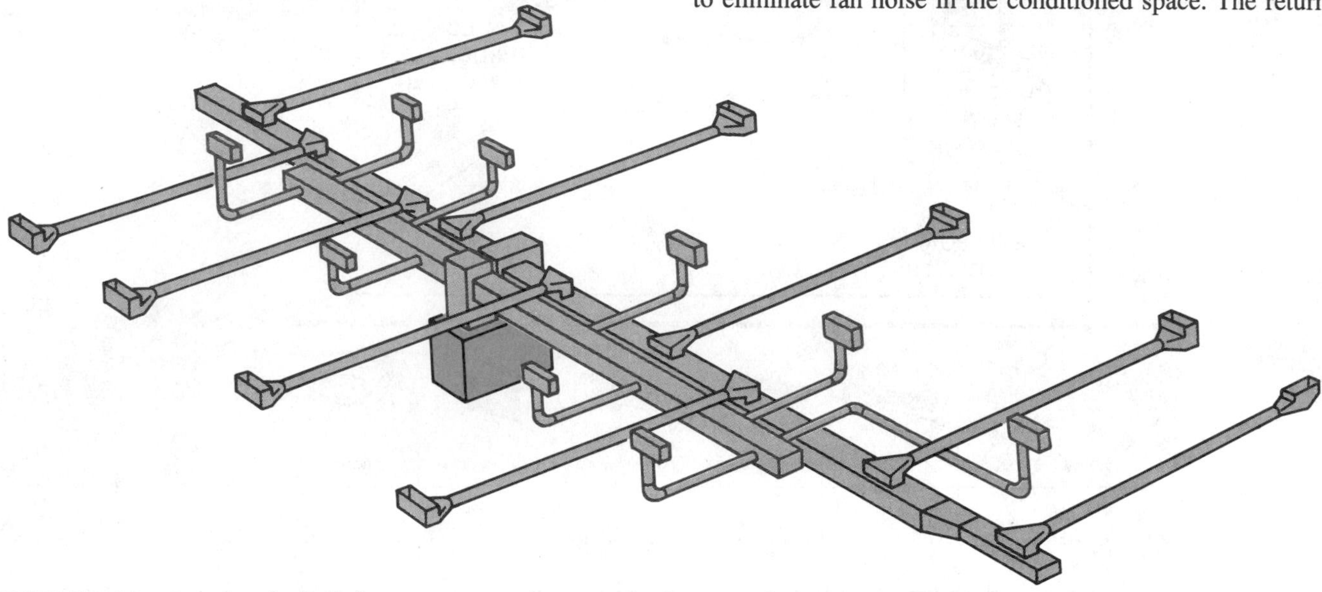

FIGURE 37–54 Duct plan of individual room return air inlets. *Reproduced courtesy of Carrier Corporation*

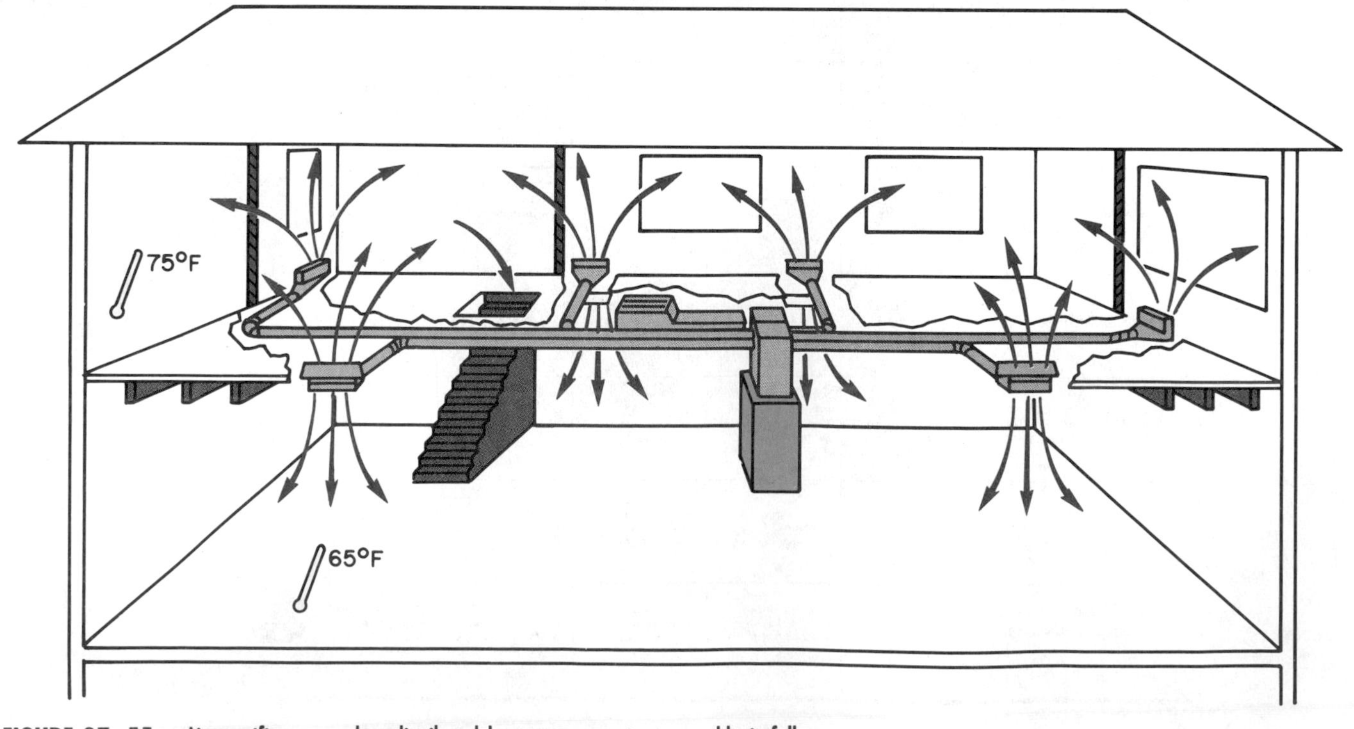

FIGURE 37–55 Air stratifies even when distributed because warm air rises, cold air falls.

air plenum should not be located on the furnace because fan running noise will be noticeable several feet away. The return air grille should be around an elbow from the furnace. If this cannot be done, the return air plenum can be insulated on the inside to help deaden the fan noise.

Return air grilles are normally large and meant to be decorative. They do not have another function unless they house a filter. They are usually made of stamped metal or have a metal frame with grille work, Figure 37–56.

37.26 SIZING DUCT FOR MOVING AIR

To move air takes energy because (1) air has weight, (2) the air tumbles down the duct, rubbing against itself and the ductwork, and (3) fittings create resistance to the airflow.

Friction loss in ductwork is due to the actual rubbing action of the air against the side of the duct and the turbulence of the air rubbing against itself while moving down the duct.

Friction due to rubbing the walls of the ductwork cannot be eliminated but can be minimized with good design practices. Proper duct sizing for the amount of airflow helps the system performance. The smoother the duct surface is, the less friction there is. The slower the air is moving, the less friction there will be. It is beyond the scope of this text to go into details of duct design. However, the following information can be used as basic guidelines for a typical residential installation.

Each foot of duct offers a known resistance to airflow. This is called *friction loss*. It can be determined from tables and special slide calculators designed for this purpose. The following example is used to explain friction loss in a duct system, Figure 37–57.

1. Ranch-style home requires 3 tons of cooling.
2. Cooling provided by a 3-ton cooling coil in the ductwork.
3. The heat and fan are provided by a 100,000 Btu/h furnace input, 80,000 Btu/h output.
4. The fan has a capacity of 1360 cfm of air while operating against 0.40 in. W.C. static pressure with the system fan operating at medium to high speed. The system only needs 1200 cfm of air in the cooling mode. The system fan will easily be able to achieve this with a small amount of reserve capacity using a 1/2-hp motor, Figure 37–58. The cooling mode usually requires more air than the heating mode. Cooling normally requires 400 cfm of air per ton; 3 tons × 400 = 1200 cfm.
5. The system has 11 outlets, each requiring 100 cfm, in the main part of the house and 2 outlets, each requiring 50 cfm, located in the bathrooms. Most of these outlets are on the exterior walls of the house and distribute the conditioned air on the outside walls.
6. The return air is taken into the system from a common hallway, one return at each end of the hall.
7. While reviewing this system, think of the entire house as the system. The supply air must leave the supply registers and sweep the walls. It then makes its way across the rooms to the door adjacent to the hall. The air is at room temperature at this time and goes under the hall door to make its way to the return air grille.
8. The return air grille is where the duct system starts. There is a slight negative pressure (in relation to the room pressure) at the grille to give the air the incentive to enter the system. The filters are located in the return air grilles. The pressure on the fan side of the filter will be −0.03 in. W.C., which is less than the pressure in the room, so the room pressure pushes the air through the filter into the return duct.
9. As the air proceeds down the duct toward the fan, the pressure continues to decrease. The lowest pressure in the system is in the fan inlet, −0.20 in. W.C. below the room pressure.
10. The air is forced through the fan, and the pressure increases. The greatest pressure in the system is at the fan outlet, 0.20 in. W.C. above the room pressure. The pressure difference in the inlet and the outlet of the fan is 0.40 in. W.C.
11. The air is then pushed through the heat exchanger in the furnace where it drops to a new pressure that is not useful to the service technician.
12. The air then moves through the cooling coil where it enters the supply duct system at a pressure of 0.10 in. W.C.

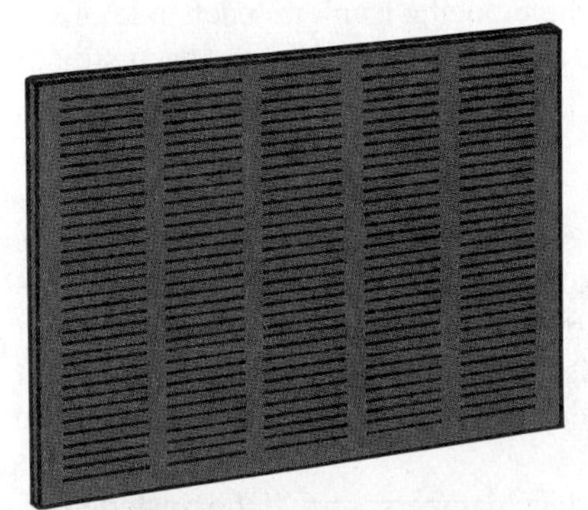

(A) STAMPED LARGE-VOLUME AIR INLET

(B) FLOOR AIR INLET

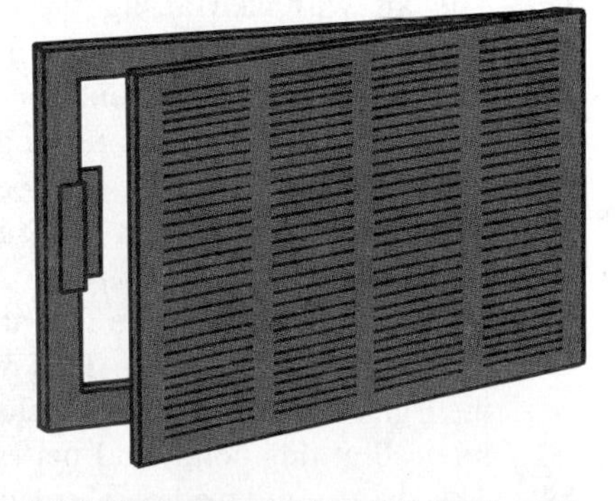

(C) FILTER AIR INLET GRILLES

FIGURE 37–56 Return air grilles.

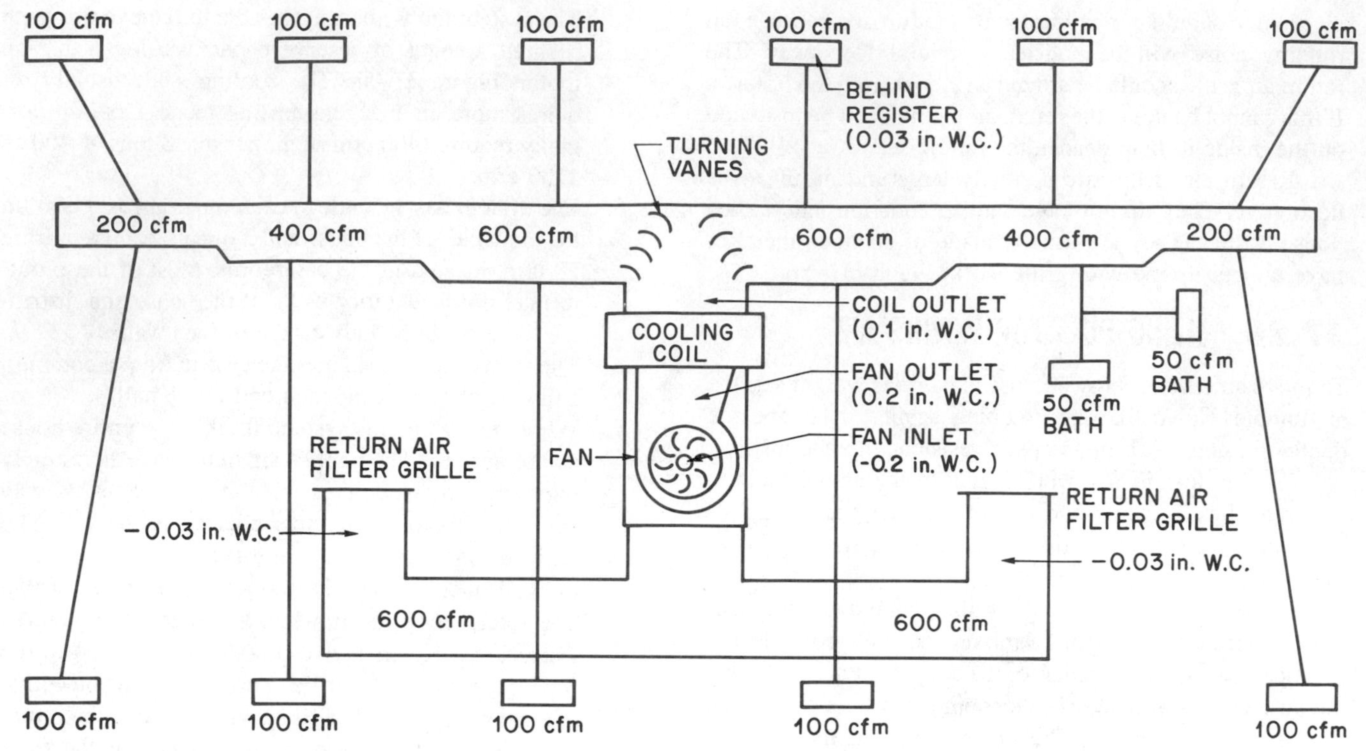

FIGURE 37–57 Duct system.

SIZE	Blower Motor HP	Speed	External Static Pressure in. W.C.							
			0.1	0.2	0.3	0.4	0.5	0.6	0.7	0.8
048100	1/2 PSC	High	1750	1750	1720	1685	1610	1530	1430	—
		Med-High	1360	1370	1370	1360	1340	1315	—	—
		Med-Low	1090	1120	1140	1130	1100	—	—	—
		Low	930	960	980	980	965	945	—	—

FIGURE 37–58 Manufacturer's table for furnace airflow characteristics. *Courtesy BDP Company*

13. The air will take a slight pressure drop as it goes around the corner of the tee that splits the duct into two reducing plenums, one for each end of the house. This tee in the duct has turning vanes to help reduce the pressure drop as the air goes around the corner.
14. The first section of each reducing trunk has to handle an equal amount of air, 600 cfm each. Two branch ducts are supplied in the first trunk run, each with an air quantity of 100 cfm. This reduces the capacity of the trunk to 400 cfm on each side. A smaller trunk can be used at this point, and materials can be saved.
15. The duct is reduced to a smaller size to handle 400 cfm on each side. Because another 200 cfm of air is distributed to the conditioned space, another reduction can be made.
16. The last part of the reducing trunk on each side of the system needs only to handle 200 cfm for each side of the system.

This supply duct system will distribute the air for this house with minimal noise and maximum comfort. The pressure in the duct will be about the same all along the duct because as air was distributed off the trunk line, the duct size was reduced to keep the pressure inside the duct at the prescribed value.

At each branch duct dampers should be installed to balance the system air supply to each room. This system will furnish 100 cfm to each outlet, but if a room did not need that much air, the dampers could be adjusted. The branch to each bathroom will need to be adjusted to 50 cfm each.

The return air system is the same size on each side of the system. It returns 600 cfm per side with the filters located in the return air grilles in the halls. The furnace fan is located far enough from the grilles so that it won't be heard.

Complete books have been written on duct sizing. Manufacturer's representatives may also help you with specific applications. They also offer schools in duct sizing that use their methods and techniques.

37.27 MEASURING AIR MOVEMENT FOR BALANCING

Air balancing is sometimes accomplished by measuring the air leaving each supply register. When one outlet has too much air, the damper in that run is throttled to slow down the air. This, of course, redistributes air to the other outlets and will increase their flow. See Figure 5–49 for an air instrument to determine airflow at an outlet.

The air quantity of an individual duct can be measured in the field to some degree of accuracy by using instruments to determine the velocity of the air in the duct. A velometer can be inserted into the duct to do this. The velocity must be measured in a cross section of the duct, and an average of the readings taken is used for the calculation. This is called *traversing* the duct. For example, if the air in a 1-ft² duct (12 in. × 12 in. = 144 in²) is traveling at a velocity of 1 ft/min, the volume of air passing a point in the duct is 1 cfm. If the velocity is 100 ft/min, the volume of air passing the same point is 100 cfm. The cross-sectional area of the duct is multiplied by the average velocity of the air to determine the volume of the moving air, Figure 37–59. Figure 37–60 shows the patterns used while traversing the duct.

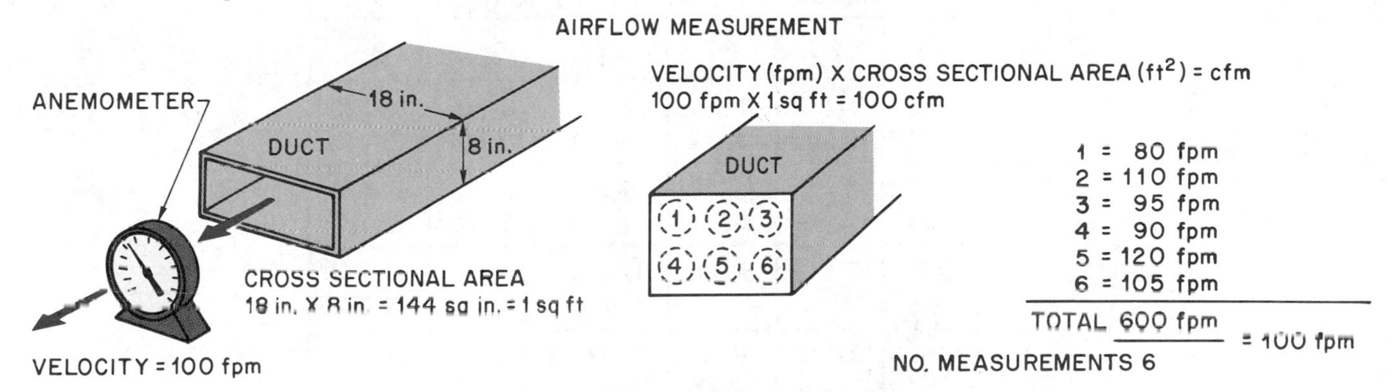

FIGURE 37–59 Cross section of duct with airflow shows how to measure duct area and air velocity.

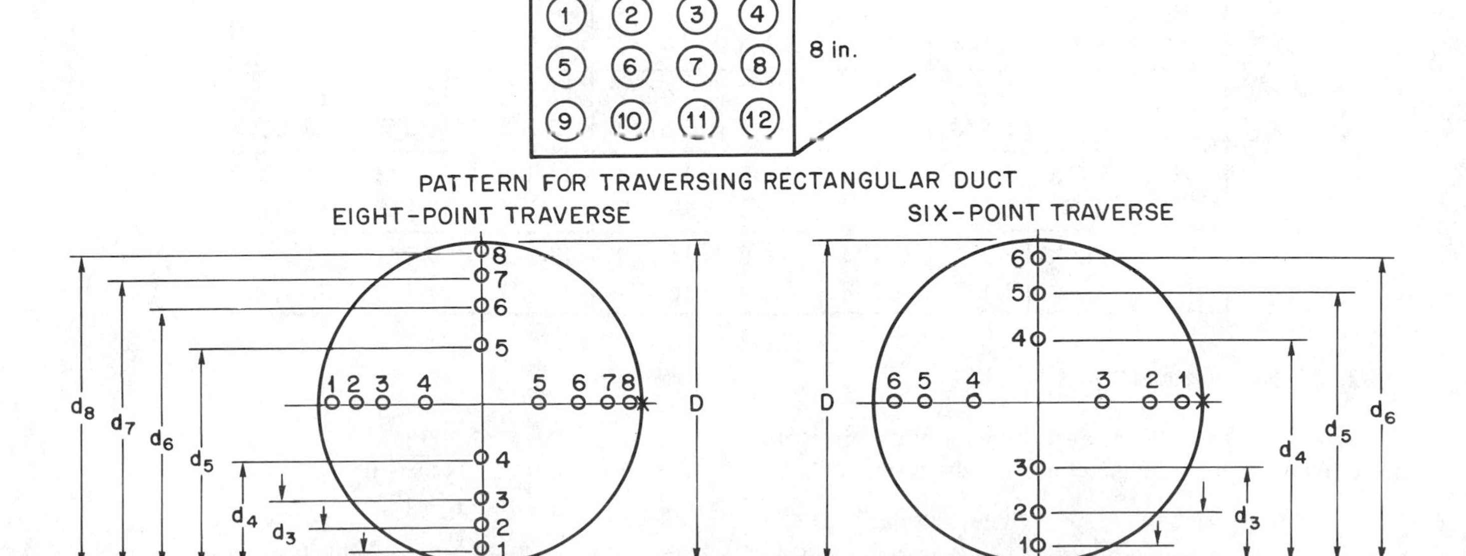

FIGURE 37–60 This figure shows the pattern for square, rectangular, and round duct that should be used for traversing the duct for the most accurate average reading.

A traverse should be made across two perpendicular diameters.
For the 6, 8 and 10 point traverse method, the distance from the reference point for a variety of duct diameters have been calculated. Figure B shows the distance of the individual measuring points form the reference point expressed in pipe diameters. Figures C and D show the distance of the individual measuring points from the reference point expressed in inches. These dimensions are equivalent to the probe immersion distance.

TRAVERSE METHOD	PROBE IMMERSION IN DUCT DIAMETERS									
	d_1	d_2	d_3	d_4	d_5	d_6	d_7	d_8	d_9	d_{10}
6 POINT	0.043	0.147	0.296	0.704	0.853	0.957	–	–	–	–
8 POINT	0.032	0.105	0.194	0.323	0.677	0.806	0.895	0.968	–	–
10 POINT	0.025	0.082	0.146	0.226	0.342	0.658	0.774	0.854	0.918	0.975

(B)

DUCT. DIA. (IN.)	PROBE IMMERSION FOR 6 PT. TRAVERSE					
	d_1	d_2	d_3	d_4	d_5	d_6
10	3/3	1-1/2	3	7	8-1/2	9-5/8
12	1/2	1-3/4	3-1/2	8-1/2	10-1/4	11-1/2
14	5/8	2	4-1/8	9-7/8	12	13-3/8
16	3/4	2-3/8	4-3/4	11-1/4	13-5/8	15-1/4
18	3/4	2-5/8	5-3/8	12-5/8	15-3/8	17-1/4
20	7/8	3	6	14	17	19-1/8
22	1	3-1/4	6-1/2	15-1/2	18-3/4	21
24	1	3-1/2	7-1/8	16-7/8	20-1/2	23

(C)

DUCT. DIA. (IN.)	PROBE IMMERSION FOR 8 PT. TRAVERSE							
	d_1	d_2	d_3	d_4	d_5	d_6	d_7	d_8
10	5/16	1	2	3-1/4	6-3/4	8	9	9-5/8
12	3/8	1-1/4	2-3/8	3-7/8	8-1/8	9-5/8	10-3/4	11-1/2
14	7/16	1-1/2	2-3/4	4-1/2	9-1/2	11-1/4	12-1/2	13-1/2
16	1/2	1-5/8	3-1/8	5-1/8	10-7/8	12-7/8	14-3/8	15-1/2
18	9/16	1-7/8	3-1/2	5-7/8	12-1/4	13-1/2	16-1/8	17-1/2
20	5/8	2-1/8	3-7/8	6-1/2	18-1/2	16-1/8	17-7/8	19-3/8
22	11/16	2-3/8	4-1/4	7-1/8	14-7/8	17-3/4	19-3/4	21-1/4
24	3/4	2-1/2	4-5/8	7-3/4	16-1/4	19-1/2	21-1/2	23-1/4

(D)

FIGURE 37–60 *(Continued)*

When you know the average velocity of the air in any duct, you can determine the cfm using the formula:

$$\text{cfm} = \text{Area (in ft}^2\text{)} \times \text{Velocity (in fpm)}$$

For example, suppose a duct is 20 in. × 30 in. and the average velocity is 850 fpm. The cfm can be found by first finding the area in square feet:

$$\text{Area} = \text{Width} \times \text{Height}$$

$$\text{Area} = 20 \text{ in.} \times 30 \text{ in.}$$

$$\text{Area} = \frac{600 \text{ in}^2}{144 \text{ in}^2/\text{ft}^2}$$

$$\text{Area} = 4.2 \text{ ft}^2$$

$$\text{cfm} = \text{Area} \times \text{Velocity}$$

$$\text{cfm} = 4.2 \text{ ft}^2 \times 850 \text{ fpm}$$

$$\text{cfm} = 3750 \text{ ft}^3 \text{ of air per minute}$$

Suppose the duct were round and had a diameter of 12 in. with an average velocity of 900 fpm.

$$\text{Area} = \mu \times \text{Radius squared}$$
$$\text{Area} = 3.14 \times 6 \times 6$$
$$\text{Area} = \frac{113 \text{ in}^2}{144 \text{ in}^2/\text{ft}^2}$$
$$\text{Area} = 0.78 \text{ ft}^2$$
$$\text{cfm} = \text{Area} \times \text{Velocity}$$
$$\text{cfm} = 0.79 \times 900$$
$$\text{cfm} = 711 \text{ ft}^3 \text{ of air per minute}$$

Special techniques and good instrumentation must be used to find the correct average velocity in a duct. For example, readings should not be taken within 10 duct diameters of the nearest fitting. This means that long straight runs must be used for taking accurate readings. This is not always possible because of the fittings and takeoffs in a typical system, particularly residential systems.

The proper duct traverse must be used and it is different for round duct versus square or rectangular duct. Figure 37–60 shows the pattern for square, rectangular, and round duct that should be used for traversing the duct for the most accurate average reading.

37.28 THE AIR FRICTION CHART

The previous system can be plotted on the friction chart in Figure 37–61. This chart is for volumes of air up to 2000

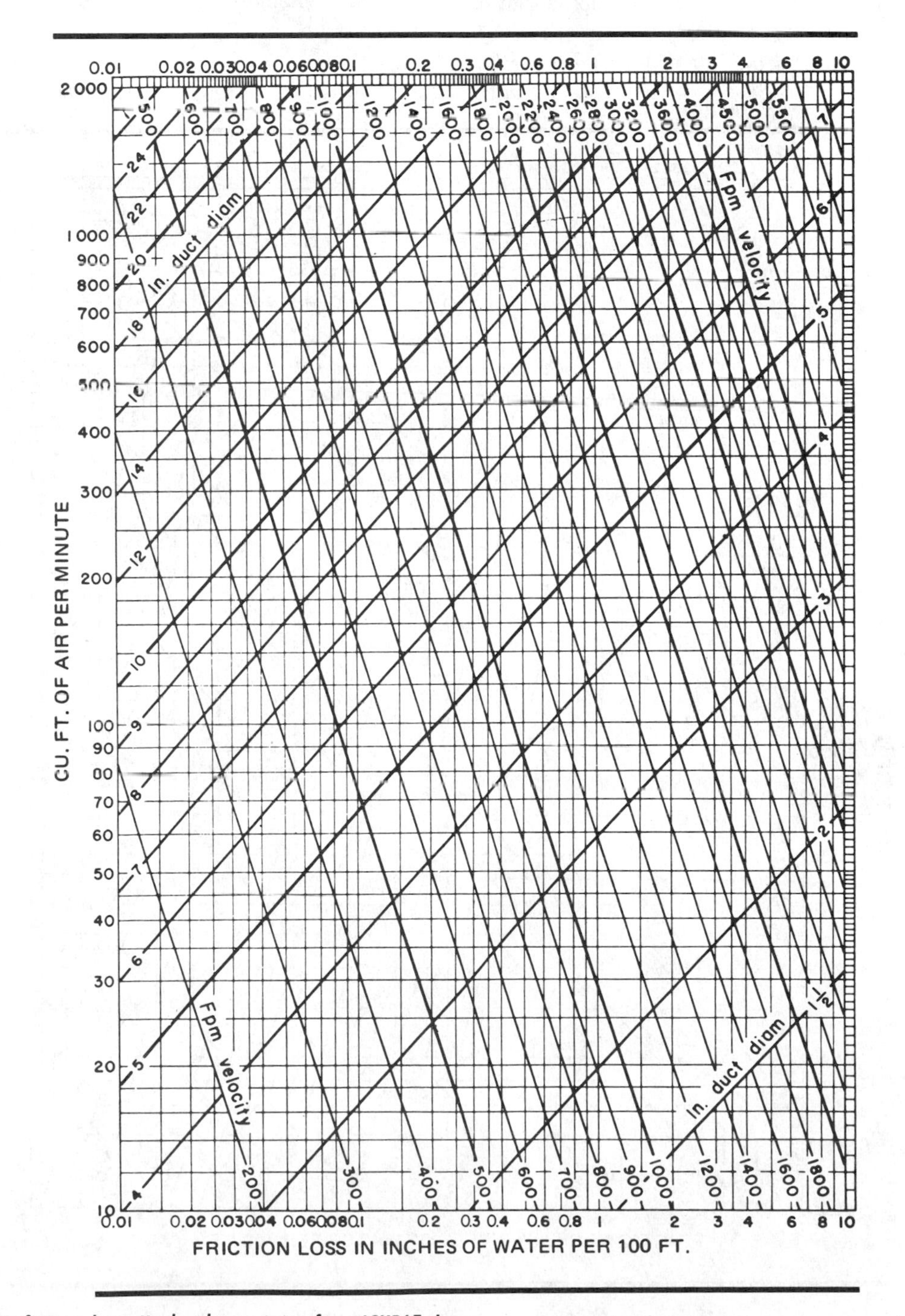

FIGURE 37–61 Air friction chart. *Used with permission from ASHRAE, Inc.*

cfm. Using the 400 cfm/ton mentioned earlier for air conditioning, we can use this chart for systems up to 5 tons of cooling. Figure 37–62 shows a chart for larger systems, up to 100,000 cfm.

The friction chart has cubic feet at the left and round-pipe sizes angle from left to right toward the top of the page. These duct sizes are rated in round-pipe size on the chart and can be converted to square or rectangular duct by using the table in Figure 37–63. The round-pipe sizes are for air with a density of 0.075 lb/ft^3 using galvanized pipe. Other charts are available for ductboard and flexible duct. The air velocity in the pipe is shown on the diagonal lines that run from left to right toward the bottom of the page. The friction loss in inches of water column per 100 ft of duct is shown along the top and bottom of the chart. For example, a run of pipe that carries 100 cfm (on the left of the chart) can be plotted over the intersection of the 6-in. pipe line. When this pipe is carrying 100 cfm of air, it has a velocity of just over 500 ft/min and a pressure drop of 0.085 in. W.C./100 feet. A 50-ft run would have half the pressure drop, 0.0425 in. W.C.

The friction chart can be used by the designer to size the duct system before the job price is quoted. This duct sizing provides the sizes that will be used for figuring the duct materials. The duct should be sized using the chart in Figure 37–61, and the recommended velocities from the table in Figure 37–64. In the previous example a 4-in. pipe could have been used, but the velocity would have been nearly 1200 ft/min. This would be noisy, and the fan may not have enough capacity to push sufficient air through the duct.

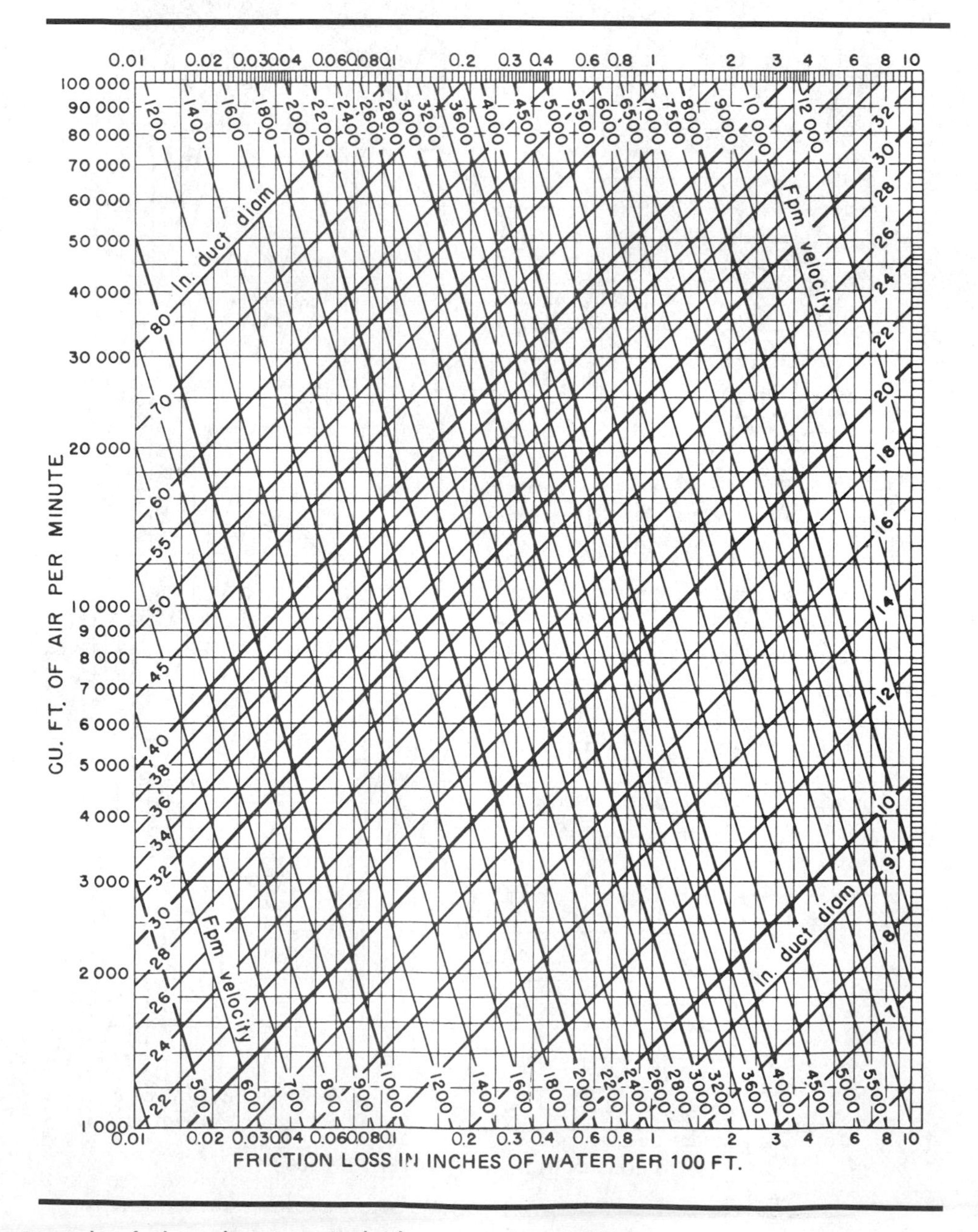

FIGURE 37–62 Friction chart for larger duct system. *Used with permission from ASHRAE, Inc.*

Lgth Adj.[b]	6	7	8	9	10	11	12	13	14	15	16	17	18	19	20	22	24	26	28	30	Lgth Adj.[b]
	Length of One Side of Rectangular Duct (a), in.																				
6	6.6																				6
7	7.1	7.7																			7
8	7.6	8.2	8.7																		8
9	8.0	8.7	9.3	9.8																	9
10	8.4	9.1	9.8	10.4	10.9																10
11	8.8	9.5	10.2	10.9	11.5	12.0															11
12	9.1	9.9	10.7	11.3	12.0	12.6	13.1														12
13	9.5	10.3	11.1	11.8	12.4	13.1	13.7	14.2													13
14	9.8	10.7	11.5	12.2	12.9	13.5	14.2	14.7	15.3												14
15	10.1	11.0	11.8	12.6	13.3	14.0	14.6	15.3	15.8	16.4											15
16	10.4	11.3	12.2	13.0	13.7	14.4	15.1	15.7	16.4	16.9	17.5										16
17	10.7	11.6	12.5	13.4	14.1	14.9	15.6	16.2	16.8	17.4	18.0	18.6									17
18	11.0	11.9	12.9	13.7	14.5	15.3	16.0	16.7	17.3	17.9	18.5	19.1	19.7								18
19	11.2	12.2	13.2	14.1	14.9	15.7	16.4	17.1	17.8	18.4	19.0	19.6	20.2	20.8							19
20	11.5	12.5	13.5	14.4	15.2	16.0	16.8	17.5	18.2	18.9	19.5	20.1	20.7	21.3	21.9						20
22	12.0	13.0	14.1	15.0	15.9	16.8	17.6	18.3	19.1	19.8	20.4	21.1	21.7	22.3	22.9	24.0					22
24	12.4	13.5	14.6	15.6	16.5	17.4	18.3	19.1	19.9	20.6	21.3	22.0	22.7	23.3	23.9	25.1	26.2				24
26	12.8	14.0	15.1	16.2	17.1	18.1	19.0	19.8	20.6	21.4	22.1	22.9	23.5	24.2	24.9	26.1	27.3	28.4			26
28	13.2	14.5	15.6	16.7	17.7	18.7	19.6	20.5	21.3	22.1	22.9	23.7	24.4	25.1	25.8	27.1	28.3	29.5	30.6		28
30	13.6	14.9	16.1	17.2	18.3	19.3	20.2	21.1	22.0	22.9	23.7	24.4	25.2	25.9	26.6	28.0	29.3	30.5	31.7	32.8	30
32	14.0	15.3	16.5	17.7	18.8	19.8	20.8	21.8	22.7	23.5	24.4	25.2	26.0	26.7	27.5	28.9	30.2	31.5	32.7	33.9	32
34	14.4	15.7	17.0	18.2	19.3	20.4	21.4	22.4	23.3	24.2	25.1	25.9	26.7	27.5	28.3	29.7	31.0	32.4	33.7	34.9	34
36	14.7	16.1	17.4	18.6	19.8	20.9	21.9	22.9	23.9	24.8	25.7	26.6	27.4	28.2	29.0	30.5	32.0	33.3	34.6	35.9	36
38	15.0	16.5	17.8	19.0	20.2	21.4	22.4	23.5	24.5	25.4	26.4	27.2	28.1	28.9	29.8	31.3	32.8	34.2	35.6	36.8	38
40	15.3	16.8	18.2	19.5	20.7	21.8	22.9	24.0	25.0	26.0	27.0	27.9	28.8	29.6	30.5	32.1	33.6	35.1	36.4	37.8	40
42	15.6	17.1	18.5	19.9	21.1	22.3	23.4	24.5	25.6	26.6	27.6	28.5	29.4	30.3	31.2	32.8	34.4	35.9	37.3	38.7	42
44	15.9	17.5	18.9	20.3	21.5	22.7	23.9	25.0	26.1	27.1	28.1	29.1	30.0	30.9	31.8	33.5	35.1	36.7	38.1	39.5	44
46	16.2	17.8	19.3	20.6	21.9	23.2	24.4	25.5	26.6	27.7	28.7	29.7	30.6	31.6	32.5	34.2	35.9	37.4	38.9	40.4	46
48	16.5	18.1	19.6	21.0	22.3	23.6	24.8	26.0	27.1	28.2	29.2	30.2	31.2	32.2	33.1	34.9	36.6	38.2	39.7	41.2	48
50	16.8	18.4	19.9	21.4	22.7	24.0	25.2	26.4	27.6	28.7	29.8	30.8	31.8	32.8	33.7	35.5	37.2	38.9	40.5	42.0	50
52	17.1	18.7	20.2	21.7	23.1	24.4	25.7	26.9	28.0	29.2	30.3	31.3	32.3	33.3	34.3	36.2	37.9	39.6	41.2	42.8	52
54	17.3	19.0	20.6	22.0	23.5	24.8	26.1	27.3	28.5	29.7	30.8	31.8	32.9	33.9	34.9	36.8	38.6	40.3	41.9	43.5	54
56	17.6	19.3	20.9	22.4	23.8	25.2	26.5	27.7	28.9	30.1	31.2	32.3	33.4	34.4	35.4	37.4	39.2	41.0	42.7	44.3	56
58	17.8	19.5	21.2	22.7	24.2	25.5	26.9	28.2	29.4	30.6	31.7	32.8	33.9	35.0	36.0	38.0	39.8	41.6	43.3	45.0	58
60	18.1	19.8	21.5	23.0	24.5	25.9	27.3	28.6	29.8	31.0	32.2	33.3	34.4	35.5	36.5	38.5	40.4	42.3	44.0	45.7	60
62		20.1	21.7	23.3	24.8	26.3	27.6	28.9	30.2	31.5	32.6	33.8	34.9	36.0	37.1	39.1	41.0	42.9	44.7	46.4	62
64		20.3	22.0	23.6	25.1	26.6	28.0	29.3	30.6	31.9	33.1	34.3	35.4	36.5	37.6	39.6	41.6	43.5	45.3	47.1	64
66		20.6	22.3	23.9	25.5	26.9	28.4	29.7	31.0	32.3	33.5	34.7	35.9	37.0	38.1	40.2	42.2	44.1	46.0	47.7	66
68		20.8	22.6	24.2	25.8	27.3	28.7	30.1	31.4	32.7	33.9	35.2	36.3	37.5	38.6	40.7	42.8	44.7	46.6	48.4	68
70		21.1	22.8	24.5	26.1	27.6	29.1	30.4	31.8	33.1	34.4	35.6	36.8	37.9	39.1	41.2	43.3	45.3	47.2	49.0	70
72			23.1	24.8	26.4	27.9	29.4	30.8	32.2	33.5	34.8	36.0	37.2	38.4	39.5	41.7	43.8	45.8	47.8	49.6	72
74			23.3	25.1	26.7	28.2	29.7	31.2	32.5	33.9	35.2	36.4	37.7	38.8	40.0	42.2	44.4	46.4	48.4	50.3	74
76			23.6	25.3	27.0	28.5	30.0	31.5	32.9	34.3	35.6	36.8	38.1	39.3	40.5	42.7	44.9	47.0	48.9	50.9	76
78			23.8	25.6	27.3	28.8	30.4	31.8	33.3	34.6	36.0	37.2	38.5	39.7	40.9	43.2	45.4	47.5	49.5	51.4	78
80			24.1	25.8	27.5	29.1	30.7	32.2	33.6	35.0	36.3	37.6	38.9	40.2	41.4	43.7	45.9	48.0	50.1	52.0	80
82				26.1	27.8	29.4	31.0	32.5	34.0	35.4	36.7	38.0	39.3	40.6	41.8	44.1	46.4	48.5	50.6	52.6	82
84				26.4	28.1	29.7	31.3	32.8	34.3	35.7	37.1	38.4	39.7	41.0	42.2	44.6	46.9	49.0	51.1	53.2	84
86				26.6	28.3	30.0	31.6	33.1	34.6	36.1	37.4	38.8	40.1	41.4	42.6	45.0	47.3	49.6	51.7	53.7	86
88				26.9	28.6	30.3	31.9	33.4	34.9	36.4	37.8	39.2	40.5	41.8	43.1	45.5	47.8	50.0	52.2	54.3	88
90				27.1	28.9	30.6	32.2	33.8	35.3	36.7	38.2	39.5	40.9	42.2	43.5	45.9	48.3	50.5	52.7	54.8	90
92					29.1	30.8	32.5	34.1	35.6	37.1	38.5	39.9	41.3	42.6	43.9	46.4	48.7	51.0	53.2	55.3	92
96					29.6	31.4	33.0	34.7	36.2	37.7	39.2	40.6	42.0	43.3	44.7	47.2	49.6	52.0	54.2	56.4	96

FIGURE 37–63 Chart to convert from round to square or rectangular duct. *1985 chart used with permission from ASHRAE, Inc.*

High-velocity systems have been designed and used successfully in small applications. Such systems normally have a high air velocity in the trunk and branch ducts; the velocity is then reduced at the register to avoid drafts from the high-velocity air. If the air velocity is not reduced at the register, the register has a streamlined effect and is normally located in the corners of the room where someone is not likely to walk under it.

The friction chart can also be used by the field technician to troubleshoot airflow problems. Airflow problems come from system design, installation, and owner problems.

System design problems can be a result of a poor choice of fan or incorrect duct sizes. Design problems may result from not understanding the fact that the air friction chart is for 100 ft of duct only. The technician should have

Structure	Supply Outlet	Return Openings	Main Supply	Branch Supply	Main Return	Branch Return
Residential	500–750	500	1,000	600	800	600
Apartments, Hotel Bedrooms, Hospital Bedrooms	500–750	500	1,200	800	1,000	800
Private Offices, Churches, Libraries, Schools	500–1,000	600	1,500	1,200	1,200	1,000
General Offices, Deluxe Restaurants, Deluxe Stores, Banks	1,200–1,500	700	1,700	1,600	1,500	1,200
Average Stores, Cafeterias	1,500	800	2,000	1,600	1,500	1,200

FIGURE 37–64 Chart of recommended velocities for different duct designs. From Lang, *Principles of Air Conditioning,* © 1987 by Delmar Publishers Inc.

an understanding of how to size duct to determine whether or not the duct is sized correctly.

To size duct properly the technician must understand the calculation of pressure drop in fittings and pipe. The pressure drop in duct is calculated and read on the friction chart in feet of duct, from the air handler to the terminal outlet. This is a straightforward measurement. The fittings are calculated in a different manner. Pressure drop across different types of fittings in inches of water column has been determined by laboratory experiment and is known. This pressure drop has been converted to equivalent feet of duct for the convenience of the designer and technician. The technician must know what type of fittings are used in the system and then a chart may be consulted for the equivalent feet of duct, Figure 37–65. The equivalent feet of duct for all of the fittings for a particular run must be added together and then added to the actual duct length for the technician to determine the proper duct size for a particular run.

When a duct run is under 100 ft, a corrected friction factor must be used or the duct will not have enough resistance to airflow and too much flow will occur. The corrected friction factor may be determined by multiplying the design friction loss in the duct system times 100 (the friction loss used on the chart) and dividing by the actual length of run. For example, a system may have a duct run handling 300 cfm of air to an inside room. This duct is very close to the main duct and air handler. There are 60 effective feet, including the friction loss in the fittings and the duct length, Figure 37–66.

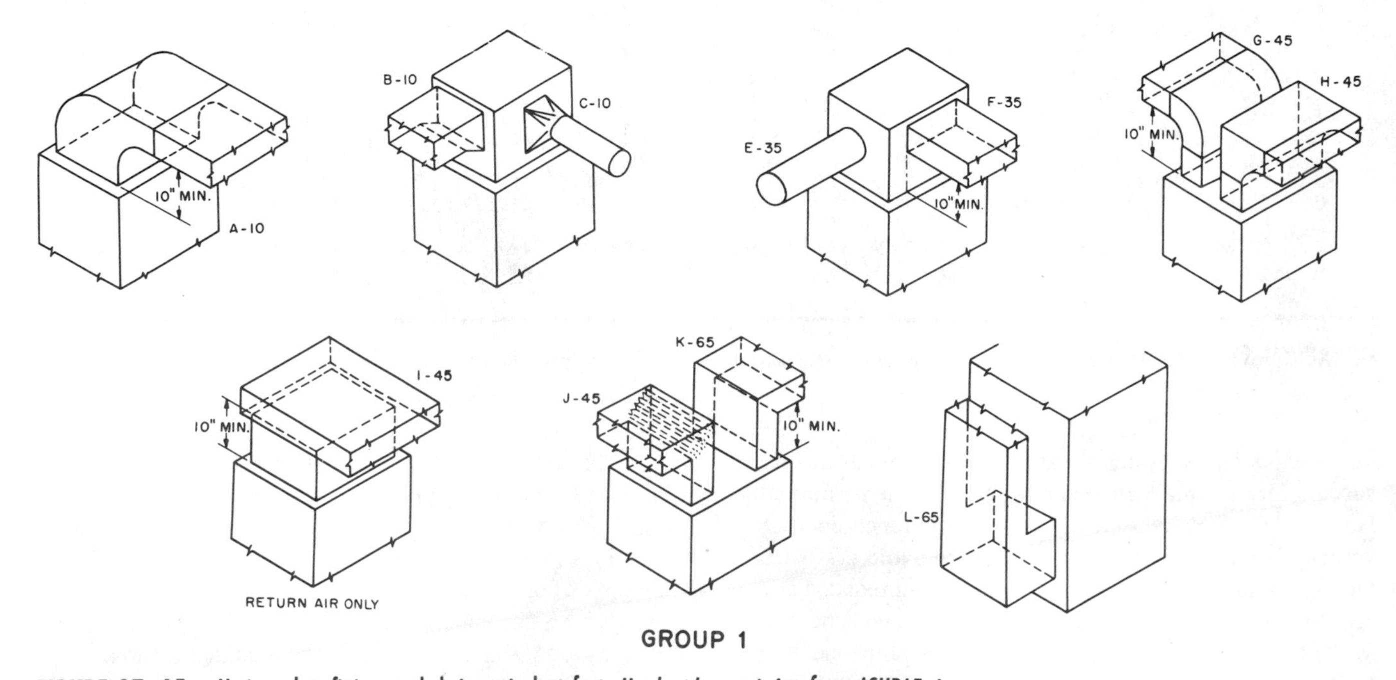

FIGURE 37–65 Various duct fittings and their equivalent feet. *Used with permission from ASHRAE, Inc.*

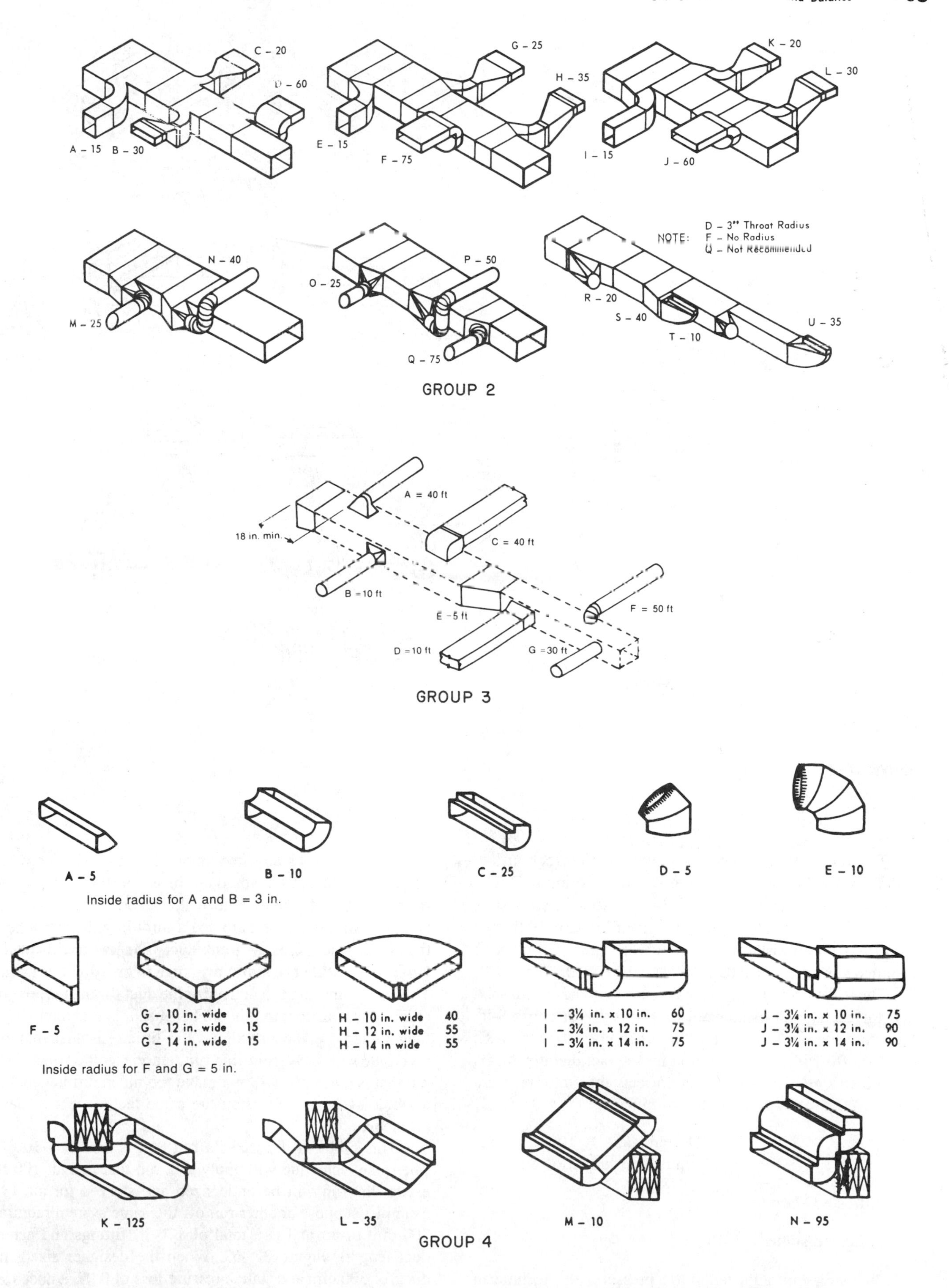
C – 20
D – 60
A – 15 B – 30
G – 25
H – 35
E – 15
F – 75
K – 20
L – 30
I – 15
J – 60
NOTE: D – 3" Throat Radius
F – No Radius
Q – Not Recommended
N – 40
M – 25
O – 25
P – 50
Q – 75
R – 20
S – 40
T – 10
U – 35
GROUP 2
A = 40 ft
18 in. min.
C = 40 ft
B = 10 ft
E – 5 ft
F = 50 ft
D = 10 ft
G = 30 ft
GROUP 3
A – 5
B – 10
C – 25
D – 5
E – 10
Inside radius for A and B = 3 in.
F – 5
G – 10 in. wide 10
G – 12 in. wide 15
G – 14 in. wide 15
H – 10 in. wide 40
H – 12 in. wide 55
H – 14 in wide 55
I – 3¼ in. x 10 in. 60
I – 3¼ in. x 12 in. 75
I – 3¼ in. x 14 in. 75
J – 3¼ in. x 10 in. 75
J – 3¼ in. x 12 in. 90
J – 3¼ in. x 14 in. 90
Inside radius for F and G = 5 in.
K – 125
L – 35
M – 10
N – 95
GROUP 4

FIGURE 37–65 *(Continued)*

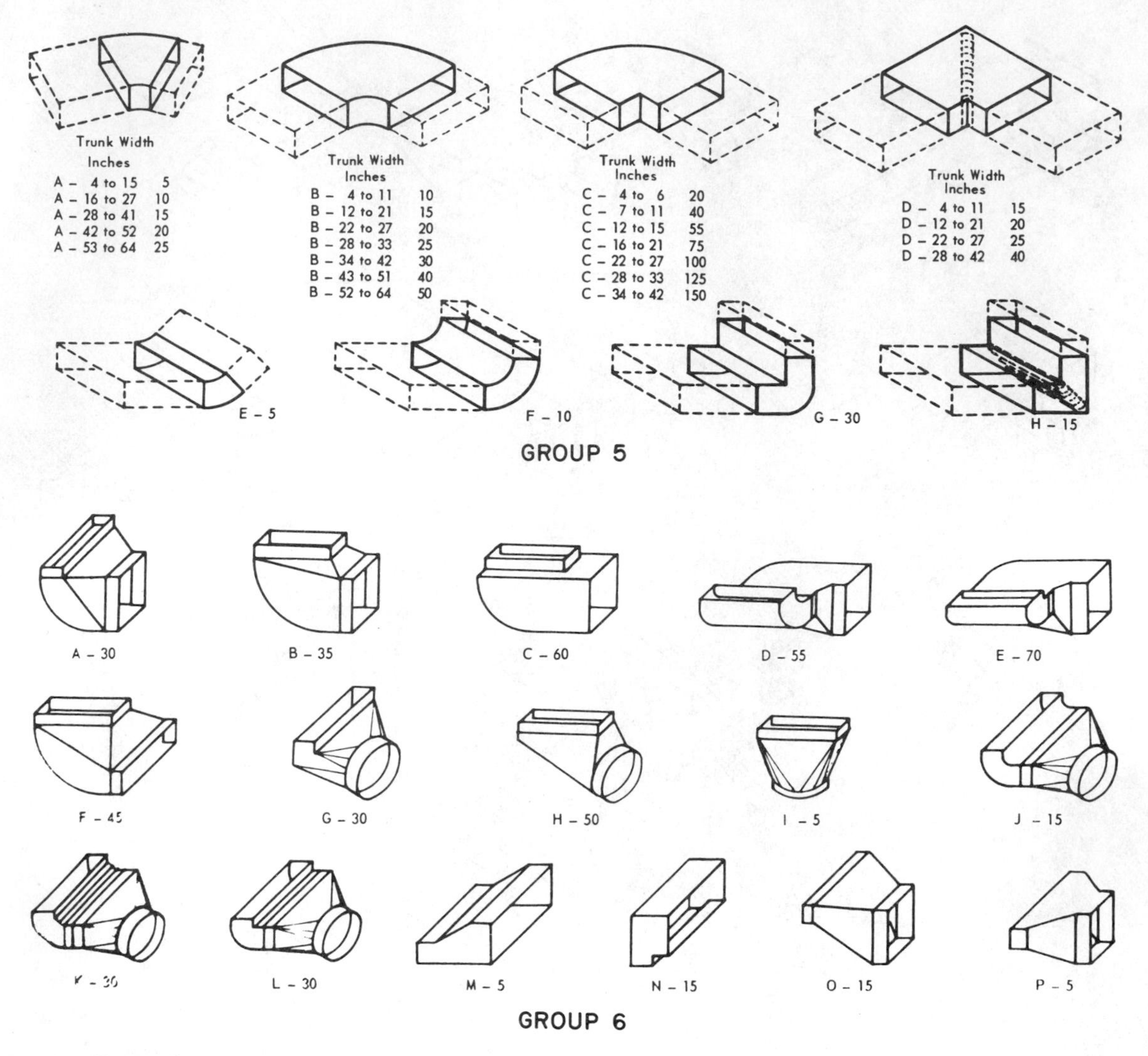

FIGURE 37–65 *(Continued)*

The fan may be capable of moving the correct airflow in the supply duct with a friction loss of 0.36 in. W.C. The cooling coil may have a loss of 0.24 and the supply registers a loss of 0.03. When the coil and register loss are subtracted from the friction loss for the supply duct, the duct must be designed for a loss of 0.09 in. W.C. (0.36 — 0.27 = 0.09).

The following is the mistake the designer made and how the technician discovered it. The designer chose 9-in. duct from the friction chart at the junction of 0.09 in. W.C. and 300 cfm without a correction for the fact that the duct is only 60 ft long. The technician corrects the static pressure loss for 60 ft of duct using the formula:

$$\text{Adjusted static} = \frac{\text{Design static} \times 100}{\text{Total equivalent length of duct}}$$

$$\text{Adjusted static} = \frac{0.09 \times 100}{60}$$

$$\text{Adjusted static} = 0.15$$

By using this adjusted static pressure, the technician refers to the friction chart and plots at 300 cfm, then moves to the right to 0.15 adjusted static. The duct size now becomes 8-in. duct. This would not be a significant savings in materials, but it would reduce the airflow to the correct amount. An alternative is to use a throttling damper where the branch line leaves the trunk duct. **The key to this problem is correcting the static pressure to an adjusted value.** This causes the chart to undersize the duct for the purpose of reducing the airflow to this run because it is less than 100 ft long. Otherwise, the air will have so little resistance that too much air will flow from this run, starving other runs. One run that is oversized will not make too much difference, but a whole system sized using the same technique will have problems.

This problem illustrates a run that is less than 100 ft. The same technique will apply to a run that is over 100 ft, except the run will be undersized and starved for air. For example, suppose another run off the same system requires 300 cfm of air and is a total of 170 ft (fittings and actual duct length), Figure 37–67. When the designer sized the duct for 300 cfm at a static pressure loss of 0.09, a duct size of 9 in. round is chosen. The technician finds the airflow is

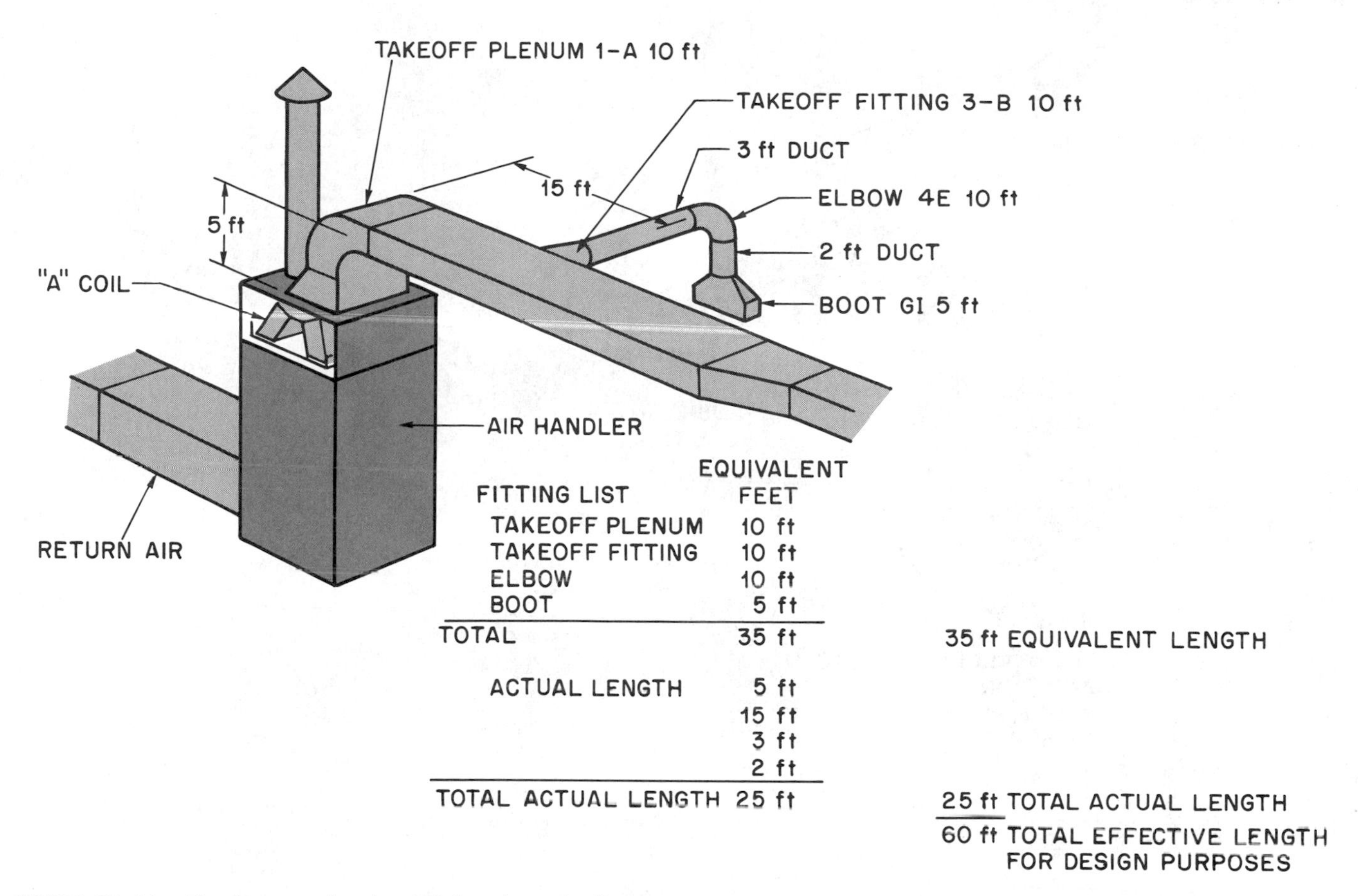

FIGURE 37–66 This duct is run less than 100 ft to the air handler.

too low and checks the figures using the formula:

$$\text{Adjusted static} = \frac{\text{Design static} \times 100}{\text{Total equivalent length of duct}}$$

$$\text{Adjusted static} = \frac{0.09 \times 100}{170}$$

$$\text{Adjusted static} = 0.053$$

When the duct is sized using the adjusted static, the adjustment causes the chart to oversize the duct and it moves the correct amount of air. In this case, the technician enters the chart at 300 cfm, moves to the right to 0.053, and finds the correct duct size to be 10 in. instead of 9 in. This is the reason this system is starved for air. All duct is first sized in round, then converted to square or rectangular.

NOTE: All runs of duct are measured from the air handler to the end of the run for sizing.

As shown from the previous examples, when some duct is undersized and some oversized the result will probably produce problems. The technician should pay close attention to any job with airflow problems and try to find the original duct sizing calculations. This would enable the technician to fully analyze a problem duct installation. Any equipment supplier will help you choose the correct fan and equipment. They may even help with the duct design.

The installation of the duct system can make a difference. ***The installers should protect the duct during the construction of the job. Stray material can make its way into the duct and block the air, Figure 37–68. The installers must not collapse the duct and then insulate over the collapsed part. Ductwork that is run below a concrete slab can easily be damaged to the point that it will not pass the correct amount of air. When the insulation is applied to the inside of the duct, it must be fastened correctly or it may come loose and fall into the airstream.***

Experience is a valuable asset when designing duct systems.

SAFETY PRECAUTION: *The technician is responsible for obtaining air pressure and velocity readings in ductwork. This involves drilling and punching holes in metal duct. Use a grounded drill cord or a cordless drill and be careful with drill bits and the rotating drill. Sheet metal can cause serious cuts and air blowing from the duct can blow chips in your eyes. Protective eye covering is necessary.*

SUMMARY

- Air is passed through conditioning equipment and then circulated into the room to condition it.
- The conditioning equipment may heat, cool, humidify, dehumidify, clean, or a combination of these to make air comfortable.
- Infiltration is air leaking into a structure.

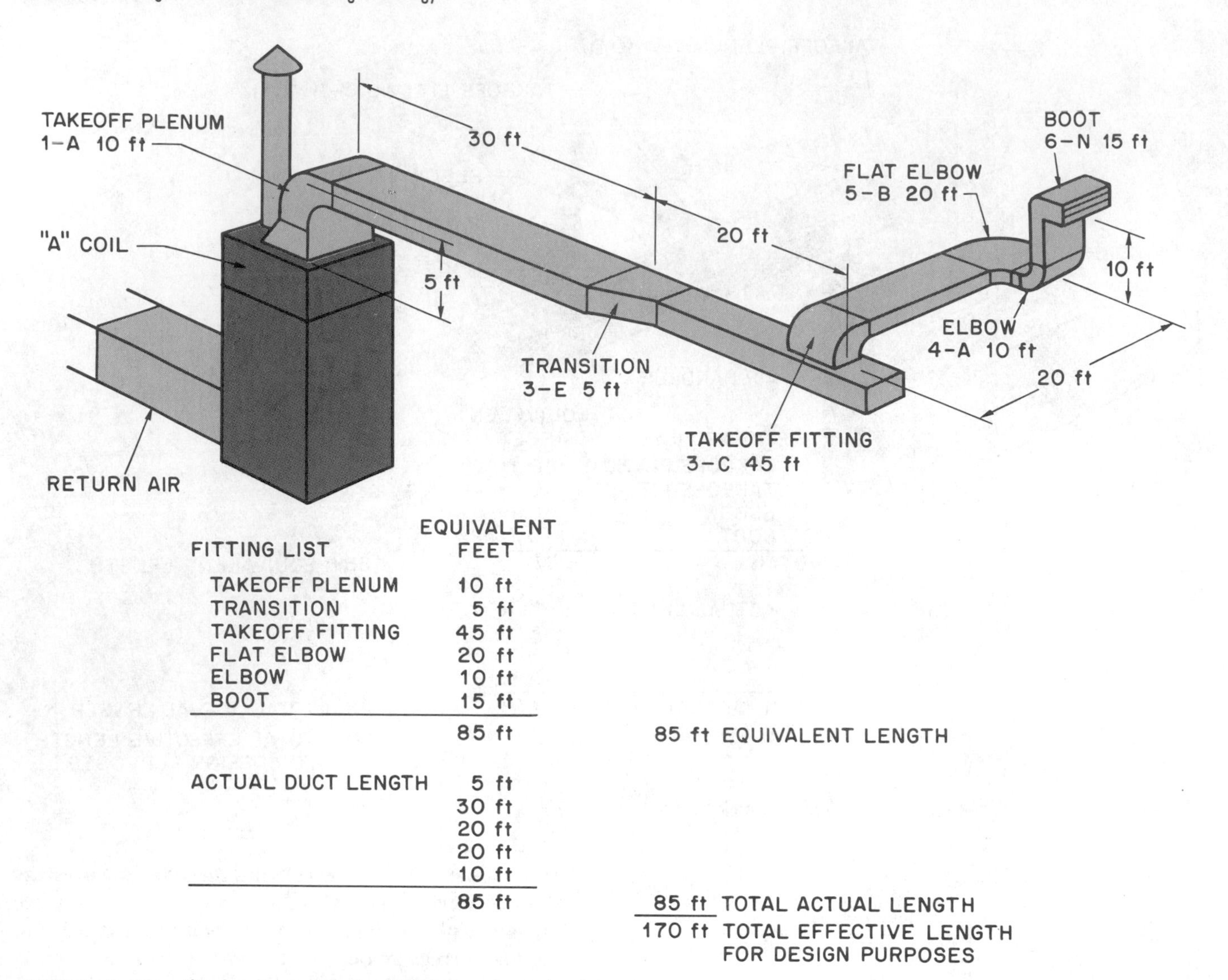

FIGURE 37–67 This duct is run more than 100 ft from the air handler.

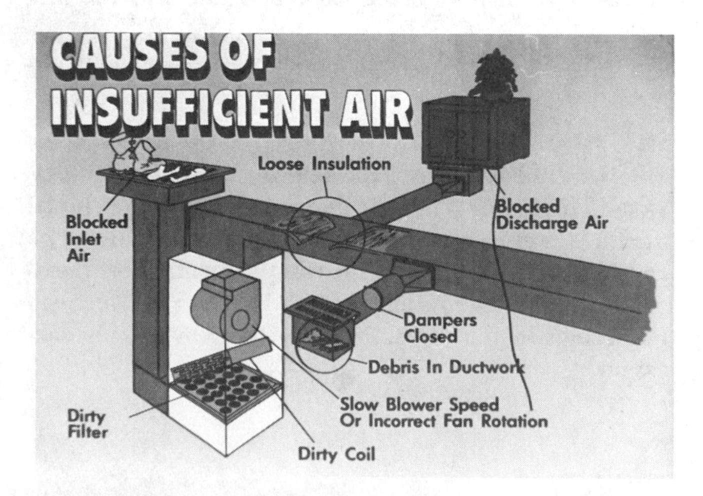

FIGURE 37–68 Obstructed airflow. *Reproduced courtesy of Carrier Corporation*

- Ventilation is air being induced into the conditioning equipment and conditioned before it is allowed to enter the conditioned space.
- The duct system distributes air to the conditioned space. It consists of the blower (or fan), the supply duct, and the return duct.
- The blower or fan uses energy to move the air.
- Propeller and centrifugal fans are commonly used in residential and small commercial systems.
- The propeller type is used to move a lot of air against a small pressure. It can make noise.
- The centrifugal type is used to move large amounts of air in ductwork, which offers resistance to the movement of air.
- Fans are turned either by belt-drive or direct-drive motors.
- A fan is not a positive-displacement device.
- Small centrifugal fans use energy in proportion to the amount of air they move.
- Duct systems, both supply and return, are large pipes or tunnels that the air flows through.
- Duct can be made from aluminum, galvanized steel, flexible tubes, and fiberglass board.
- The pressure the fan creates in the duct is very small and is measured in inches of water column.

- 1 in. W.C. is the amount of pressure needed to raise a column of water 1 in.
- The atmospheric pressure of 14.696 psia will support a column of water 34 ft high.
- 1 psi will support a column of water 27.1 in. or 2.31 ft high.
- Moving air in a duct system creates static pressure, velocity pressure, and total pressure.
- Static pressure is the pressure pushing outward on the duct.
- Velocity pressure is moving pressure created by the velocity of the air in the duct.
- Total pressure is the velocity pressure plus the static pressure.
- The pitot tube is a probe device used to measure the air pressures.
- Air velocity (fpm) in a duct can be multiplied by the cross-sectional area (ft^2) of the duct to obtain the amount of air passing that particular point in cubic feet per minute.
- Typical supply duct systems used are plenum, extended plenum, reducing plenum, and perimeter loop.
- The plenum system is economical and can easily be installed.
- The extended plenum system takes the trunk duct closer to the farthest outlets but is more expensive.
- The reducing plenum system reduces the trunk duct size when some of the air volume has been reduced.
- The perimeter loop system supplies an equal pressure to all of the outlets. It is useful in a concrete slab floor.
- Branch ducts should always have balancing dampers to balance the air to the individual areas.
- When the air is distributed in the conditioned space, it is common practice to distribute the air on the outside wall to cancel the load.
- Warm air distributes better from the floor because it rises.
- Cold air distributes better from the ceiling because it falls.
- The amount of throw tells how far the air from a diffuser will reach into the conditioned space.
- Return air systems normally used are individual room return and common or central return.
- Each foot of duct, supply or return, has a friction loss that can be plotted on a friction chart for round duct.
- Round duct sizes can be converted to square or rectangular equivalents for sizing and friction readings.

REVIEW QUESTIONS

1. Name the five changes that are made in air to condition it.
2. Name the two ways that a structure obtains fresh air.
3. Name two types of blowers that move the air in a forced-air system.
4. Which type of blower is used to move large amounts of air against low pressures?
5. Which type of blower is used to move large amounts of air in ductwork?
6. Name two reasons why ductwork resists airflow.
7. Name four types of duct distribution systems.
8. Name two types of blower drives.
9. What is the pressure in ductwork expressed in?
10. What is a common instrument used to measure pressure in ductwork?
11. Why is pounds per square inch not used to measure the pressure in ductwork?
12. Name the three types of pressures created in moving air in ductwork.
13. Name two types of return air systems.
14. What component distributes the air in the conditioned space?
15. Where is the best place to distribute warm air? Why?
16. Where is the best place to distribute cold air? Why?
17. What chart is used to size ductwork?
18. Name four materials used to manufacture duct.
19. What fitting leaves the trunk duct and directs the air into the branch duct?
20. The duct sizing chart is expressed in round duct. How can this be converted to square or rectangular duct?

UNIT 38 *Installation*

OBJECTIVES

After completing this unit, you should be able to

- list three crafts involved in air conditioning installation.
- identify types of duct system installations.
- describe the installation of metal duct.
- describe the installation of ductboard systems.
- describe the installation of flexible duct.
- recognize good installation practices for package air conditioning equipment.
- discuss different connections for package air conditioning equipment.
- describe the split air conditioning system installation.
- recognize correct refrigerant piping practices.
- state start-up procedures for air conditioning equipment.

SAFETY CHECKLIST

- * Be careful when handling sharp tools, fasteners, and metal duct. Wear gloves whenever practicable.
- * Use only grounded, double-insulated, or cordless electrical tools.
- * Wear gloves, goggles, and clothing that will protect your skin when working with or around fiberglass insulation.
- * Extreme care should be used when working with any electrical circuit. All safety rules must be adhered to. Particular care should be taken while working with any primary power supply to a building because the fuse that protects that circuit may be on the power pole outside. Keep the power off whenever possible when installing or servicing equipment. Lock and tag the panel. There should be only one key. Keep this on your person at all times.
- * Although low-voltage circuits are generally harmless, you can experience shock if you are wet and touch live wires. Use precautions for all electrical circuits.
- * Before starting a system for the first time examine all electrical connections and moving parts. Connections, pulleys, or other moving parts may be loose.
- * Follow all manufacturer's instructions before and during initial start-up.

38.1 INTRODUCTION TO EQUIPMENT INSTALLATION

Installing air conditioning equipment requires three crafts: duct, electrical, and mechanical, which includes refrigeration. Some contractors use separate crews to carry out the different tasks. Others may perform the duties of two of the crafts within their own company and subcontract the third to a more specialized contractor. Some small contractors do all three jobs with a few highly skilled people. The three job disciplines are often licensed at local and state levels; they may be licensed by different departments.

Included in the following are duct installation procedures for (1) all-metal square or rectangular, (2) all-metal round, (3) fiberglass ductboard, and (4) flexible duct. ***Local codes must be followed while performing all work.***

38.2 SQUARE AND RECTANGULAR DUCT

Square metal duct is fabricated in a sheet metal shop by qualified sheet metal layout and fabrication personnel. The duct is then moved to the job site and assembled to make a system. Because the all-metal duct system is rigid, all dimensions have to be precise or the job will not go together. The duct must sometimes rise over objects or go beneath them and still measure out correctly so the takeoff will reach the correct branch location. The branch duct must be the correct dimension for it to reach the terminal point, the boot for the room register. Figure 38–1 illustrates an example of a duct system layout.

Square duct or rectangular duct is assembled with S fasteners and drive clips, which make a duct connection nearly airtight. If further sealing is needed, the connection can be caulked and taped. While the duct is being assembled, it has to be fastened to the structure for support. This can be accomplished in several ways. It may lay flat in an attic installation or be hung by hanger straps. The duct should be supported so that it will be steady when the fan starts and will not transmit noise to the structure. The rush of air (which has weight) down the duct will move the duct if it is not fastened. Vibration eliminators next to the fan section will prevent the transmission of fan vibrations down the duct. See Figure 38–2 for a flexible duct connector that can be installed between the fan and the metal duct. These are always recommended, but not always used. Figure 38–3 illustrates an example of a flexible connector that can be installed at the end of each run used to dampen sound.

Metal duct can be purchased in the popular sizes from some supply houses for small and medium systems. This makes metal duct systems available to the small contractor who may not have a sheet metal shop. The assortment of standard duct sizes can be assembled with an assortment of standard fittings to build a system that appears to be custom made for the job, Figure 38–4.

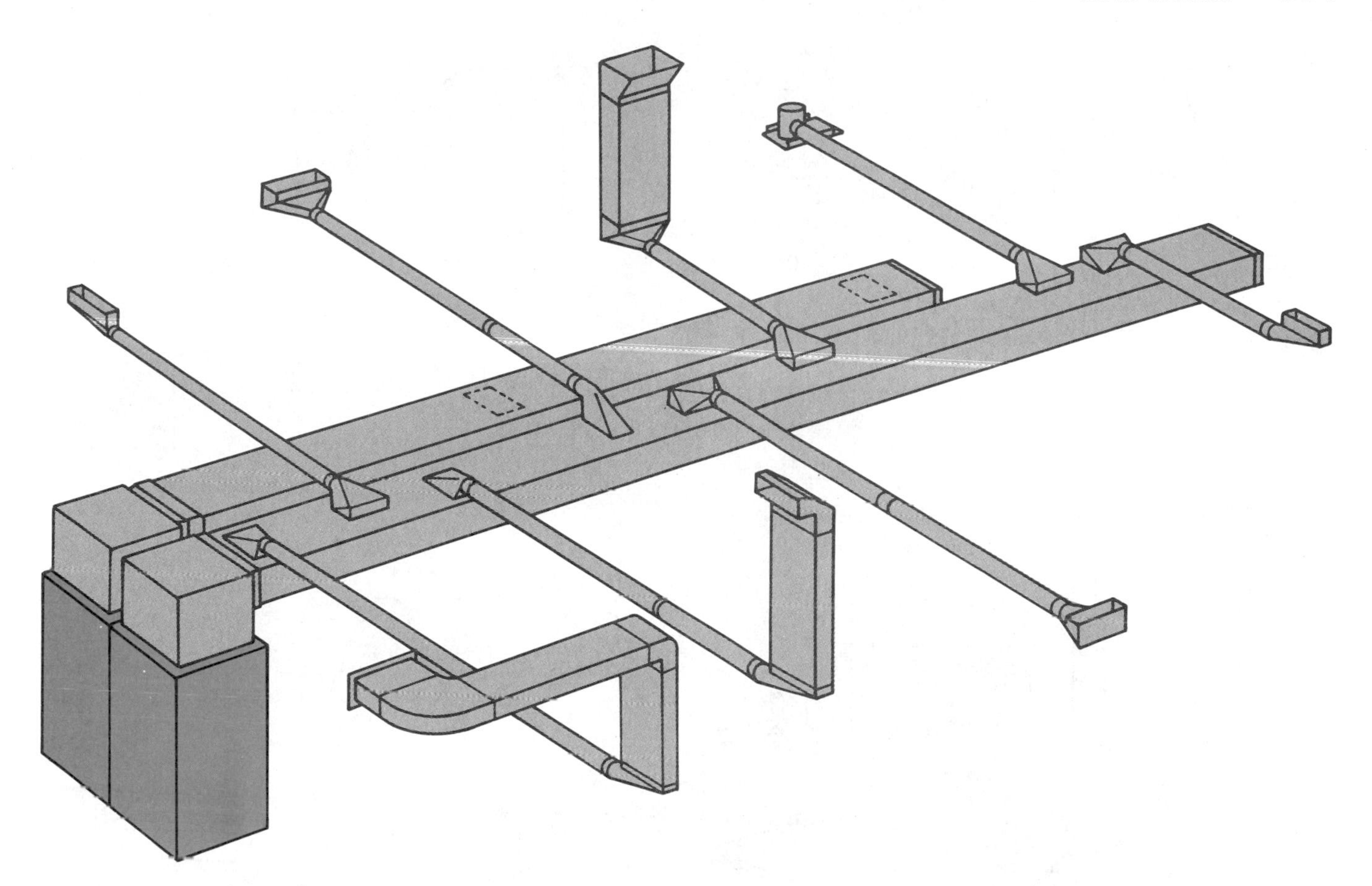

FIGURE 38–1 Duct system layout.

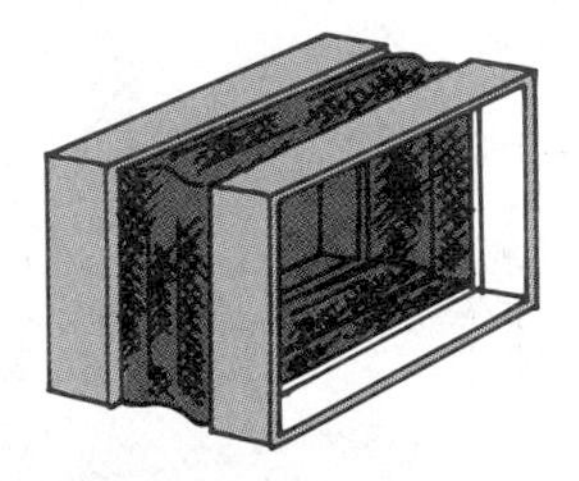

FIGURE 38–2 Flexible duct connector.

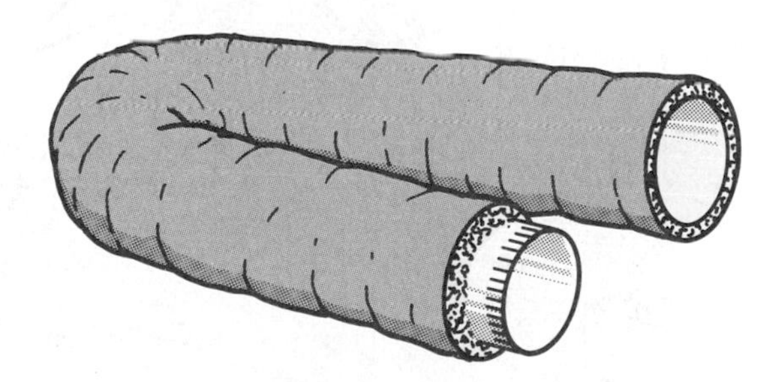

FIGURE 38–3 Round flexible duct connector.

38.3 ROUND METAL DUCT SYSTEMS

Round metal duct systems are easy to install and are available from some supply houses in standard sizes for small and medium systems. These systems use reducing fittings from a main trunk line and may be assembled in the field. This type of system must be fastened together at all connections. Self-tapping sheet metal screws are popular fasteners. The screws are held by a magnetized screwholder while an electric drill turns and starts the screw, Figure 38–5. Each connection should have a minimum of three screws spaced evenly to keep the duct steady. A good installer can fasten the connections as fast as the screws can be placed in the screwholder, Figure 38–6. A reversible, cordless, variable-speed drill is the ideal tool to use.

Round metal duct takes more clearance space than square or rectangular duct. It must be supported and mounted at the correct intervals to keep it straight. Exposed, round metal duct does not look as good as square duct and rectangular duct, so it is often used in places that are out of sight.

SAFETY PRECAUTION: *Be careful when handling the sharp tools and fasteners. Use grounded, double-insulated, or cordless electrical tools.*

38.4 INSULATION FOR METAL DUCT

Insulation for metal duct can be applied to the inside or the outside of the duct. When insulation is applied to the inside, it is usually done in the fabrication shop. The insulation can be fastened with tabs, glue, or both. The tabs are fastened to the inside of the duct and have a shaft that looks like a nail protruding from the duct wall. The liner and a washer are pushed over the tab shaft. The tabs have a base that the shaft is fastened to, Figure 38–7. This base is fastened to the inside of the duct by glue or spot welding. Spot welding is the most permanent method but is difficult to do and is

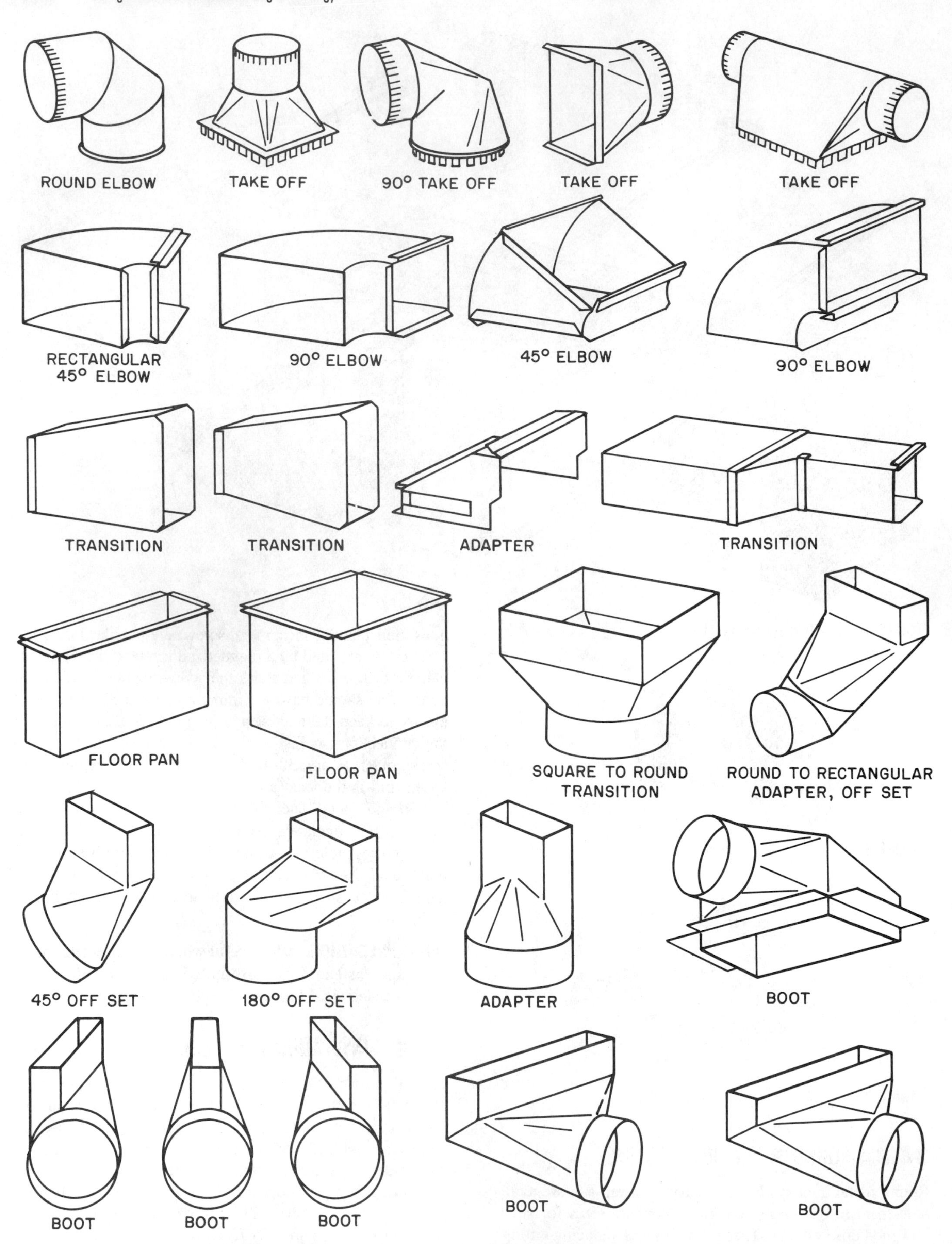

FIGURE 38–4 Standard duct fittings.

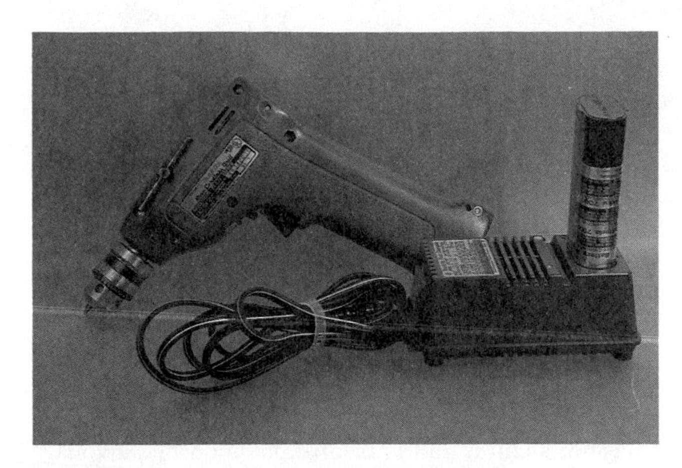

FIGURE 38–5 Electric hand drill. *Photo by Bill Johnson*

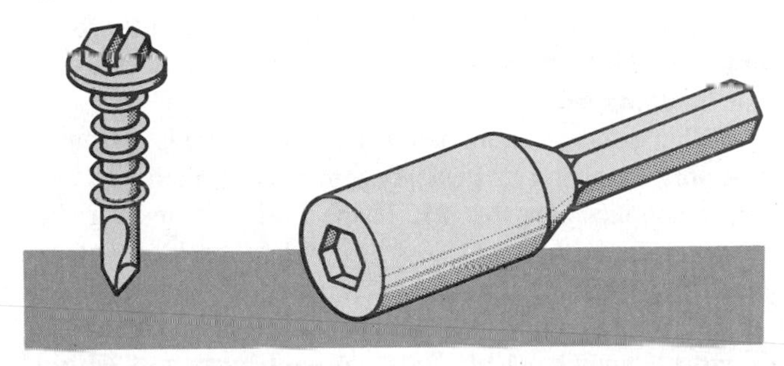

FIGURE 38–6 Self-tapping sheet metal screw and a magnetic screw holder.

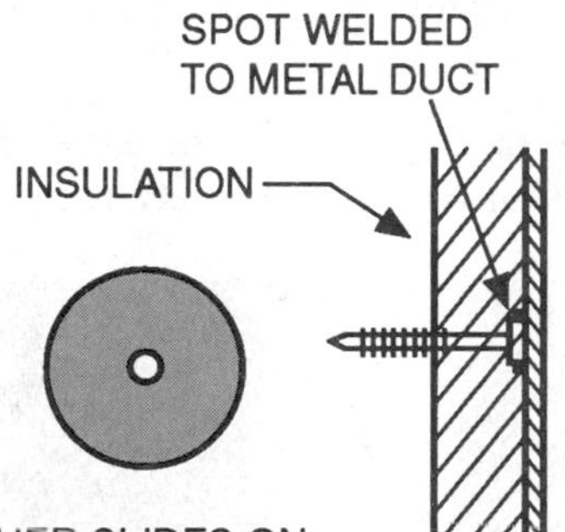
SPOT WELDED
TO METAL DUCT
INSULATION
METAL WASHER SLIDES ON
SPOT WELDED STUD TO HOLD
INSULATION ON THE SIDE OF THE DUCT.
THESE FASTENERS ARE EVENLY SPACED
DOWN THE DUCT.

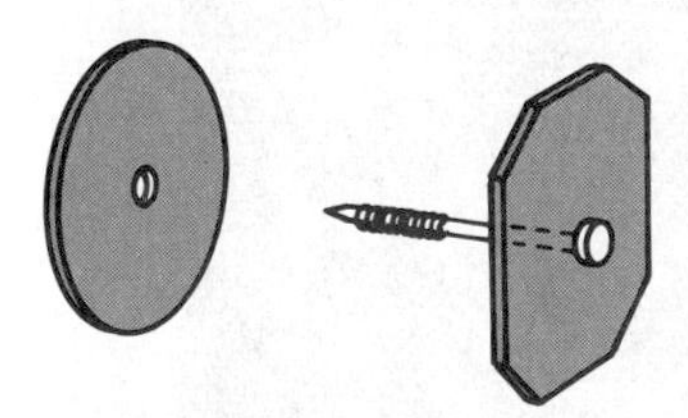
THIS TAB MAY BE GLUED
ON THE DUCT WORK.
CONTACT GLUE IS
FURNISHED ON THE
BACK OF TAB.

FIGURE 38–7 Tab to hold fiberglass duct liner to the inside of the duct.

expensive. An electrical spot welder must be used to weld the tab.

The liner can be glued to the duct, but it may come loose and block airflow, perhaps years from the installation date. This is difficult to find and repair, particularly if the duct is in a wall or framed in by the building structure. The glue *must* be applied correctly and even then may not hold forever. Using tabs with glue is a more permanent method.

The liner is normally fiberglass and is coated on the air side to keep the airflow from eroding the fibers. Many pounds of air pass through the duct each season. An average air conditioning system handles 400 cfm/ton of air. A 3-ton system handles 1200 cfm or 72,000 cfh (400 cfm × 60 min/h = 72,000 cfh). Then 72,000 cfh ÷ 13.35 ft^3/lb = 5393 lb of air per hour. This is more than 2 1/2 tons of air per hour.

SAFETY PRECAUTION: *Fiberglass insulation can irritate the skin. Gloves and goggles should be worn while handling it.*

38.5 DUCTBOARD SYSTEMS

Fiberglass ductboard is popular among many contractors because it requires little special training to construct a system. Special knives can be used to fabricate the duct in the field. When the knives are not available, simple cuts and joints can be made with a utility or contractor's knife. With this duct the insulation is already attached to the outside skin. The skin is made of foil with a fiber running through it to give it strength. When assembling this duct, it is important to cut some of the insulation away so that the outer skin can lap over the surface that it is joining. The two skins are strengthened by stapling them together and taping to make them airtight, Figure 38–8.

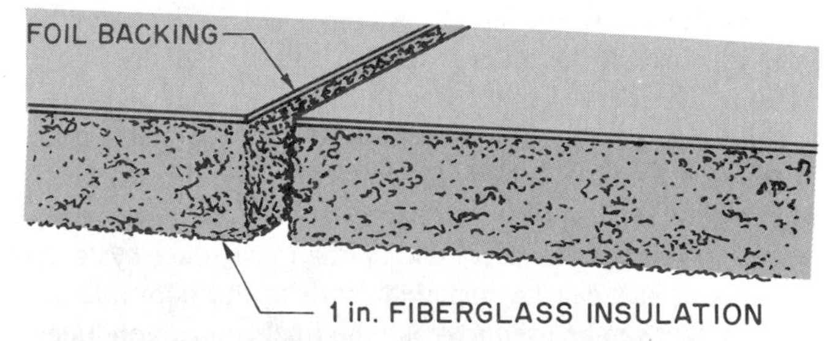
FOIL BACKING
1 in. FIBERGLASS INSULATION

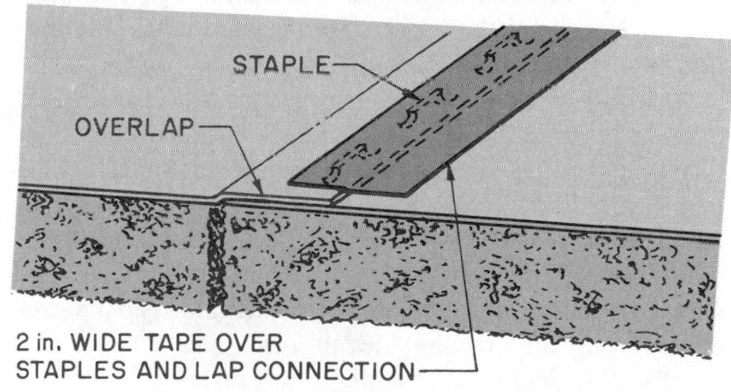
STAPLE
OVERLAP
2 in. WIDE TAPE OVER
STAPLES AND LAP CONNECTION

CROSS SECTION OF STAPLE

FIGURE 38–8 Outer skin is lapped over, and the backing of one piece is stapled to the backing of the other piece.

Fiberglass ductboard can be made into almost any configuration that metal duct can be made into. It is lightweight and easy to transport because the duct can be cut, laid out flat, and assembled at the job site. Metal duct fittings, on the other hand, are large and take up a lot of space in a truck. The original shipping boxes can be used to transport the ductboard and to keep it dry.

Round fiberglass duct is as easy to install as the ductboard because it can also be cut with a knife.

Fiberglass duct must be supported like metal duct to keep it straight. The weight of the ductboard itself will cause it to sag over long spans. A broad type of hanger that will not cut the ductboard cover is necessary.

Fiberglass duct deadens sound because the inside of the duct has a coating (like the coating in metal duct liner to keep the duct fibers from eroding) that helps to deaden any air or fan noise that may be transmitted into the duct. The duct itself is not rigid and does not carry sound.

38.6 FLEXIBLE DUCT

Flexible duct has a flexible liner and may have a fiberglass outer jacket for insulation if needed, for example, where a heat exchange takes place between the air in the duct and the space where the duct is routed. The outer jacket is held by a moisture-resistant cover made of fiber-reinforced foil or vinyl. The duct may be used for the supply or return and should be run in a direct path to keep bends from closing the duct. Sharp bends can greatly reduce the airflow and should be avoided.

When used in the supply system, flexible duct may be used to connect the main trunk to the boot at the room diffuser. The boot is the fitting that goes through the floor. In this case, it has a round connection on one end for the flexible duct and a rectangular connection on the other end where the floor register fastens to it. The duct flexibility makes it valuable as a connector for metal duct systems. The metal duct can be installed close to the boot and the flexible duct can be used to make the final connection. Long runs of flexible duct are not recommended unless the friction loss is taken into account.

Flexible duct must be properly supported. A band 1 in. or more wide is the best method to keep the duct from collapsing and reducing the inside dimension. Some flexible duct has built-in eyelet holes for hanging the duct. Tight turns should be avoided to keep the duct from collapsing on the inside and reducing the inside dimension. Flexible duct should be stretched to a comfortable length to keep the liner from closing and creating friction loss.

Flexible duct used at the end of a metal duct run will help reduce any noise that may be traveling through the duct. This can help reduce the noise in the conditioned space if the fan or heater is noisy.

38.7 ELECTRICAL INSTALLATION

The air conditioning technician should be familiar with certain guidelines regarding electrical installation to make sure that the unit has the correct power supply and that the power supply is safe for the equipment, the service technician, and the owner. The control voltage for the space thermostat is often installed by the air conditioning contractor even if the line-voltage power supply is installed by a licensed electrical technician.

The power supply must include the correct voltage and wiring practices, including wire size. The law requires the manufacturer to provide a nameplate with each electrical device that gives the voltage requirements and the current the unit will draw, Figure 38–9. The applied voltage (the voltage that the unit will actually be using) should be within ±10% of the rated voltage of the unit. That is, if the rated voltage of a unit is 230 V, the maximum operating voltage that the unit should be allowed to operate at would be 253 volts ($230 \times 1.10 = 253$). The minimum voltage would be 207 volts ($230 \times 9.90 = 207$). ***If the unit operates for a long time beyond these limits, the motors and controls will be damaged.***

When package equipment is installed, there is one power supply for the unit. If the system is split, there are two power supplies, one for the inside unit and one for the outside unit. Both power supplies will go back to a main panel, but there will be a separate fuse or breaker at the main panel, Figure 38–10. ***There should be a disconnect or cut-off switch within 25 ft of each unit and within sight of the unit.*** This is a safety precaution for the technician working on the unit. If the disconnect is around a corner, someone may turn on the unit while the technician is working on the electrical system.

SAFETY PRECAUTION: *Care should be used while working with any electrical circuit. All safety rules *must* be ad-

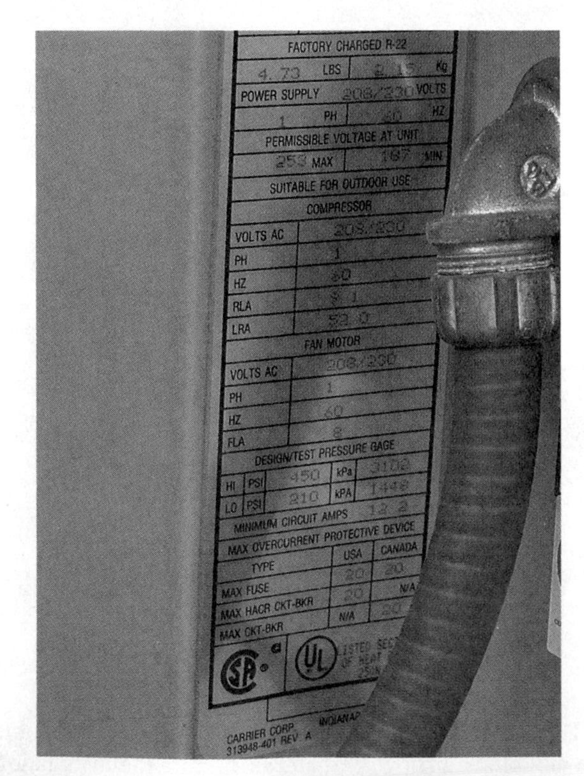

FIGURE 38–9 Air conditioning unit nameplate. *Photo by Bill Johnson*

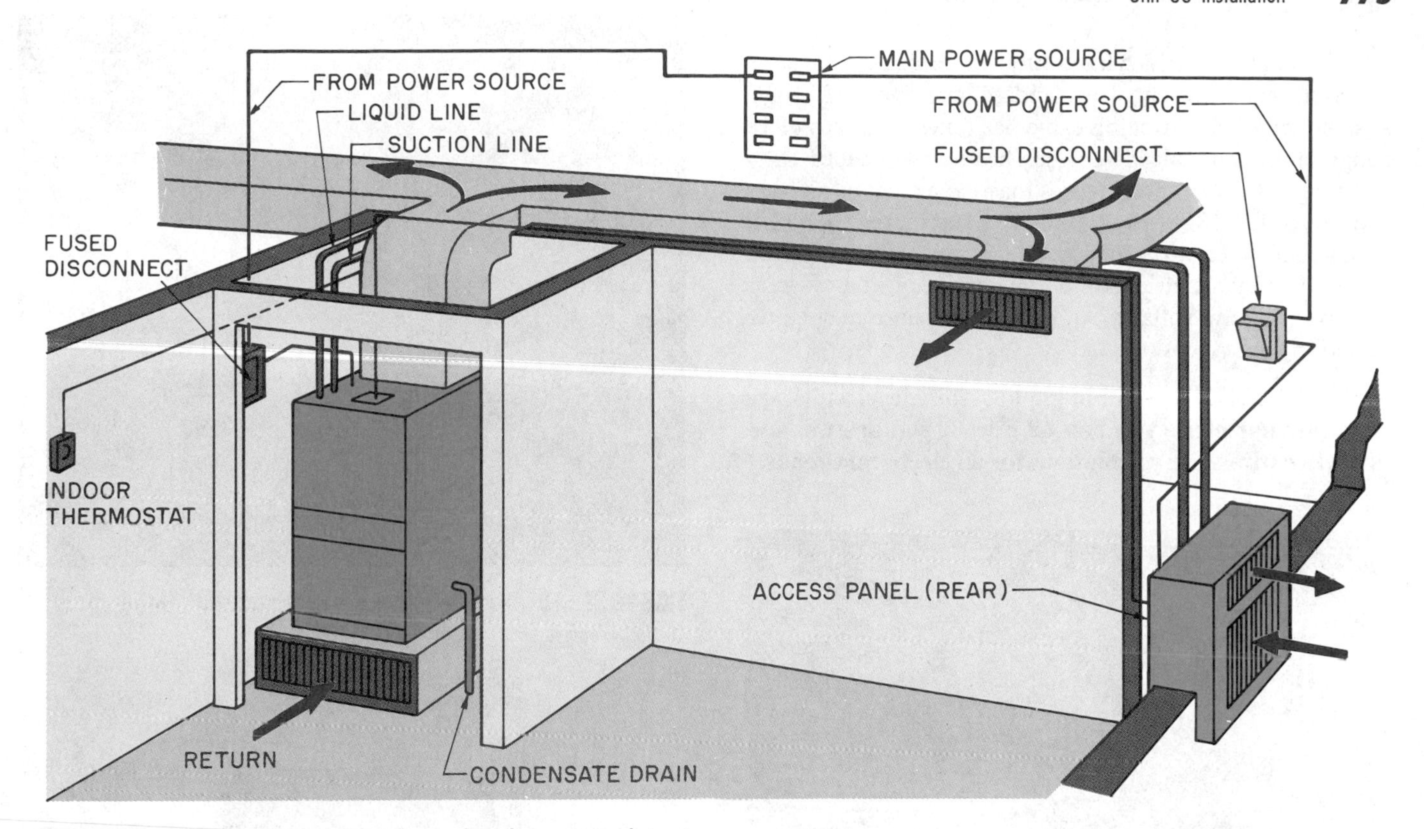

FIGURE 38–10 Wiring connecting indoor and outdoor units to the main power service.

hered to. Particular care should be taken while working in the primary power supply to any building because the fuse that protects that circuit may be on the power pole outside. A screwdriver touched across phases may throw off pieces of hot metal before the outside fuse responds.*

Wire sizing tables specify the wire size that each component will need. The National Electrical Code (NEC) provides installation standards, including workmanship, wire sizes, methods of routing wires, and types of enclosures for wiring and disconnects. Use it for all electrical wiring installation unless a local code prevails.

The control-voltage wiring in air conditioning and heating equipment is the line voltage reduced through a step-down transformer. It is installed with color-coded or numbered wires so that the circuit can be followed through the various components. For example, the 230-V power supply is often at the air handler, which may be in a closet, an attic, a basement, or a crawl space. The interconnecting wiring may have to leave the air handler and go to the room thermostat and the condensing unit at the back of the house. The air handler may be used as the junction for these connections, Figure 38–11.

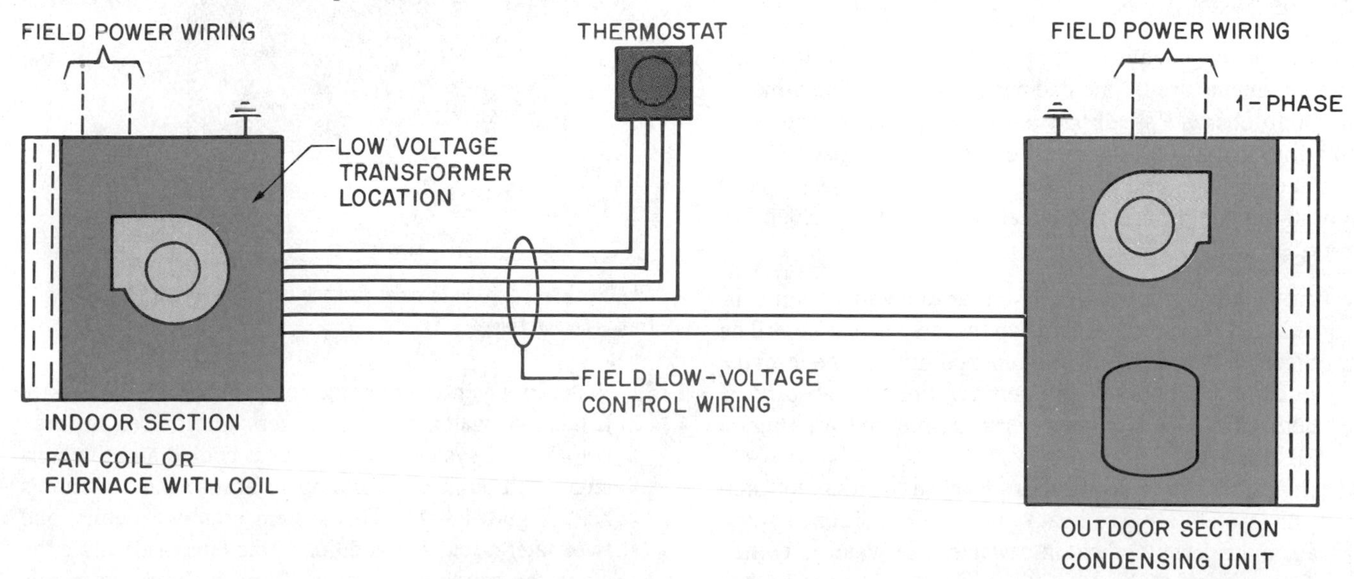

FIGURE 38–11 Pictorial diagram shows the relative position of the wiring as it is routed to the room thermostat, the air handler, and the condensing unit.

The control wire is a light-duty wire because it carries a low voltage and current. The standard wire size is 18 gauge. A standard air conditioning cable has four wires each of 18 gauge in the same plastic-coated sheath and is called 18–4 (18 gauge, 4 wires). Some wires may have eight conductors and are called 18–8, Figure 38–12. There are red, white, yellow, and green wires in the cable. The cable or sheath can be installed by the air conditioning contractor in most areas because it is low voltage. An electrical license may be required in some areas.

SAFETY PRECAUTION: *Although low-voltage circuits are generally harmless, you can be hurt if you are wet and touch live wires. Use precautions for all electrical circuits.*

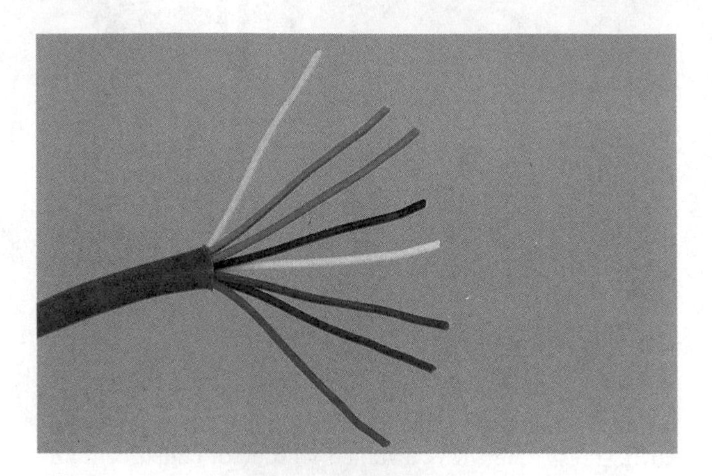

FIGURE 38–12 18-8 thermostat wire. *Photo by Bill Johnson*

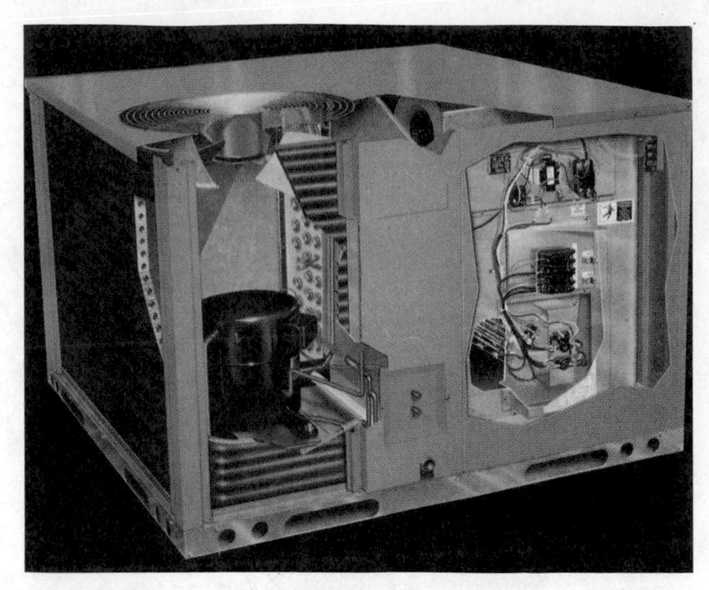

FIGURE 38–13 Air-to-air package unit. *Courtesy Heil Heating and Cooling Products*

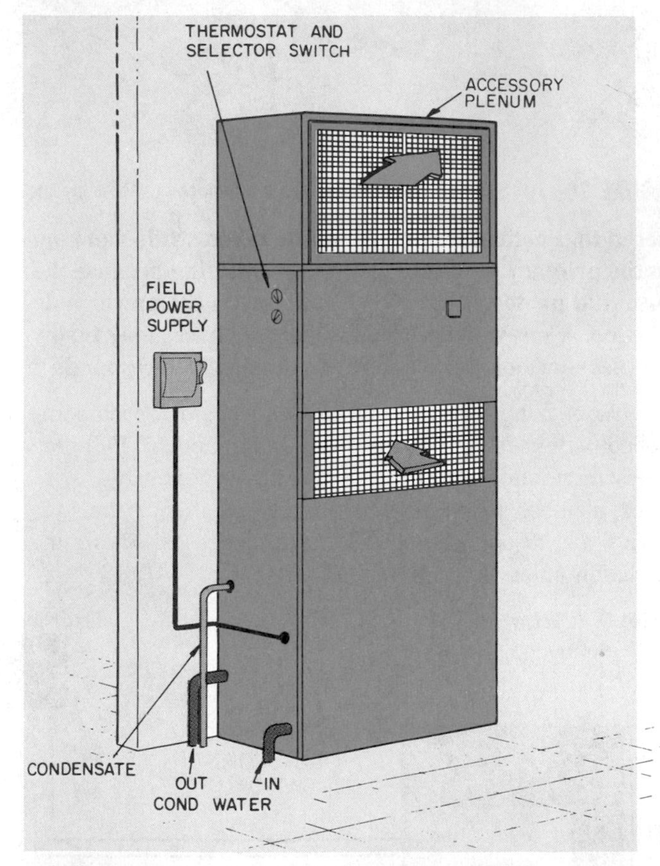

FIGURE 38–14 Air-to-water package unit. *Reproduced courtesy of Carrier Corporation*

38.8 INSTALLING THE REFRIGERATION SYSTEM

The mechanical or refrigeration part of the air conditioning installation is included in either a package or split system.

Package Systems

Package or self-contained equipment is equipment in which all components are in one cabinet or housing. The window air conditioner is a small package unit. Larger package systems may have 100 tons of air conditioning capacity.

The package unit is available in several different configurations for different applications, some of which are described below.

1. An air-to-air application is similar to a window air conditioner except that it has two motors. This unit will be discussed because it is the most common. The term *air to air* is used because the refrigeration unit absorbs its heat from the air and rejects it into the air, Figure 38–13.
2. Figure 38–14 is an illustration of an air-to-water unit. This system absorbs heat out of the conditioned space air and rejects the heat into water. The water is wasted or passed through a cooling tower to reject the heat to the atmosphere. This system is sometimes called a water-cooled package unit.
3. Water-to-water describes the equipment in Figure 38–15. It has two water heat exchangers and is used in large commercial systems. The water is cooled and then circulated through the building to absorb the structure's heat, Figure 38–16. This system uses two pumps and two water circuits in addition to the fans to circulate the air in the conditioned space. To properly maintain this system, you need to be able to service pumps and water circuits.

FIGURE 38–15 Water-to-water package unit. *Reproduced courtesy of Carrier Corporation*

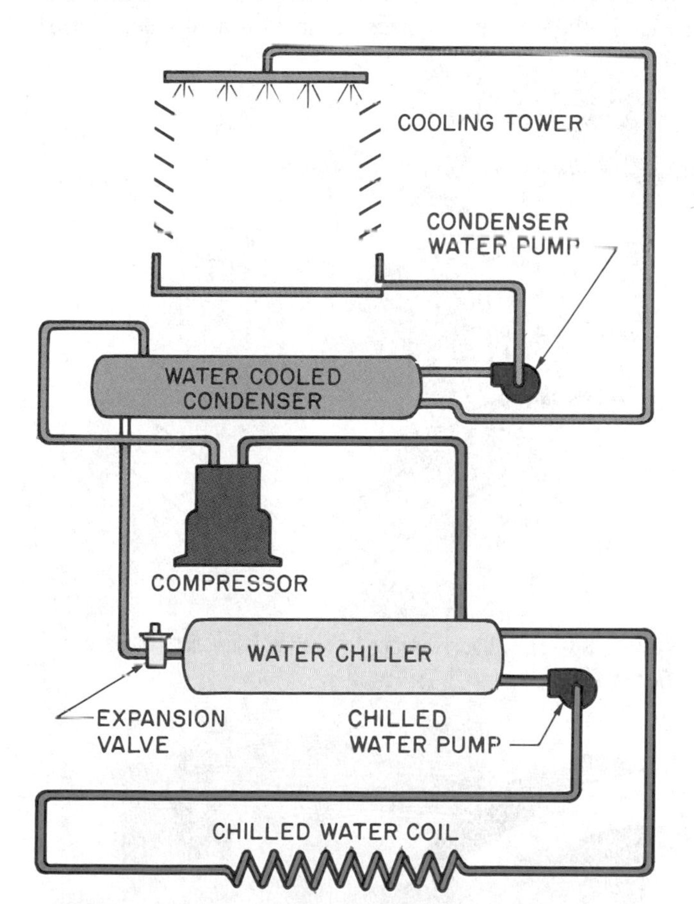

FIGURE 38–16 The total water-to-water system must have two pumps to move the water in addition to the fans.

4. Water-to-air describes the equipment in Figure 38–17. The unit absorbs heat from the water circuit and rejects the heat directly into the atmosphere. This equipment is used for larger commercial systems.

Air-cooled air conditioning systems are by far the most common systems in residential and light commercial instal-

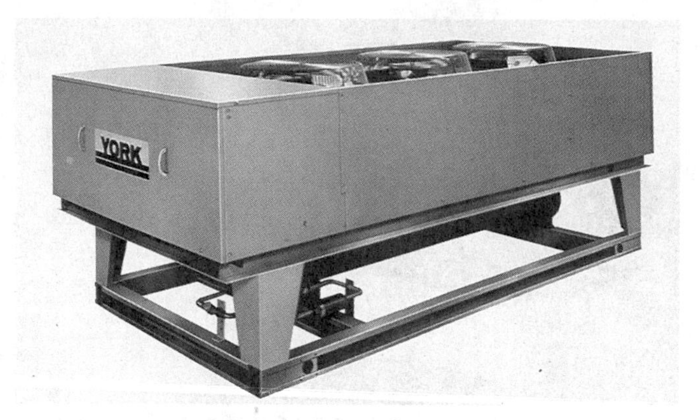

FIGURE 38–17 Water-to-air package unit. *Courtesy York International Corporation*

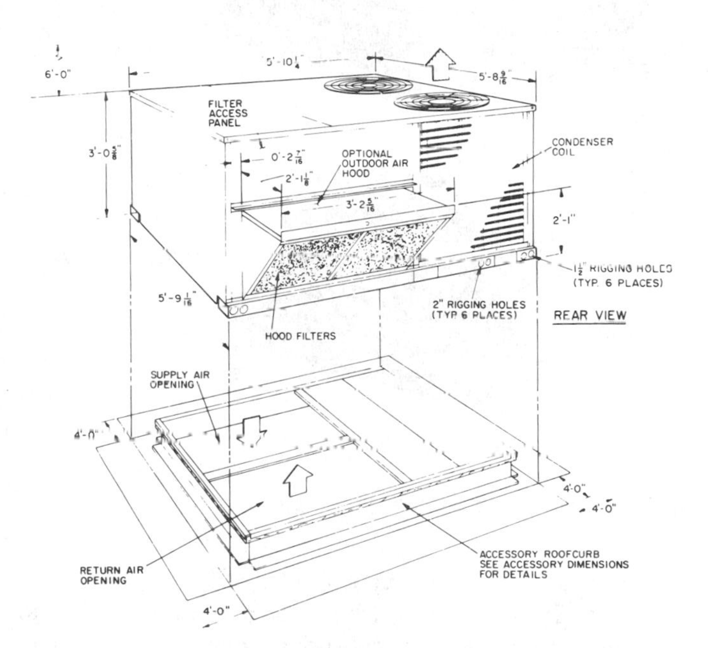

FIGURE 38–18 Unit will set on the roof curb. *Courtesy Heil Heating and Cooling Products*

lations. The air-to-air system installation requires that the unit be set on a firm foundation. The unit may be furnished with a roof curb for rooftop installations. The roof curb will raise the unit off the roof and provide waterproof duct connections to the conditioned space below. When a unit is to be placed on a roof in new construction, the roof curb can be shipped separately, and a roofer can install it, Figure 38–18. The air conditioning contractor can then set the package unit on the roof curb for a watertight installation. The foundation for another type of installation may be located beside the conditioned space and may be a high-impact plastic pad, a concrete pad, or a metal frame, Figure 38–19. The unit should be placed where the water level cannot reach the unit.

Vibration Isolation

The foundation of the unit should be placed in such a manner that the unit vibration is not transmitted into the build-

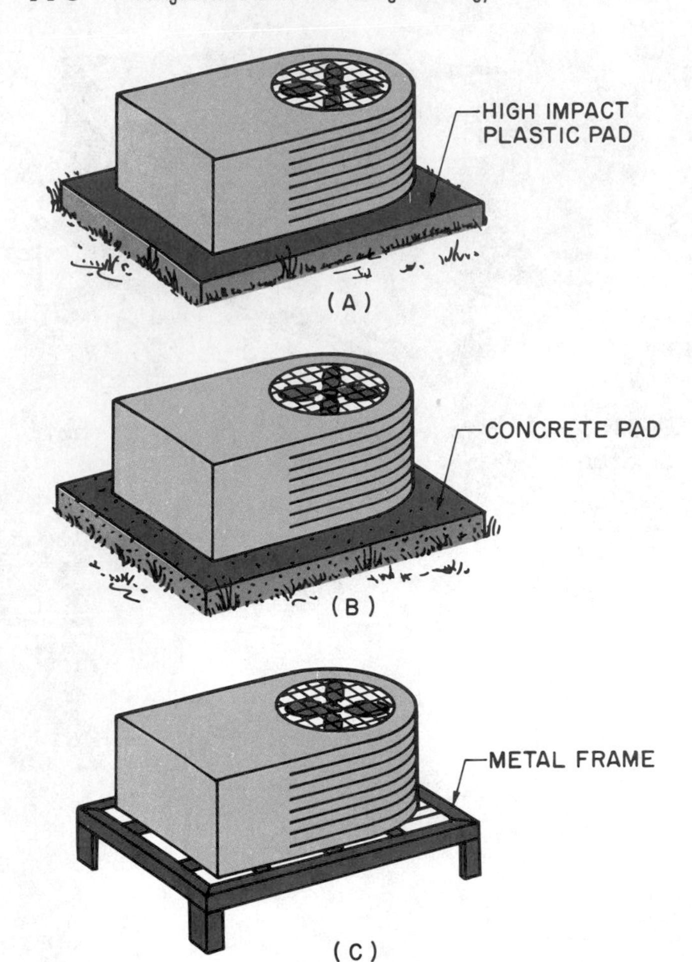

FIGURE 38–19 Various types of unit pads.

ing. The unit may need vibration isolation from the building structure. Two common methods for preventing vibration are rubber and cork pads and spring isolators. Rubber and cork pads are the simplest and least expensive method for simple installations, Figure 38–20. The pads come in sheets that can be cut to the desired size. They are placed under the unit at the point where the unit rests on its foundation. If

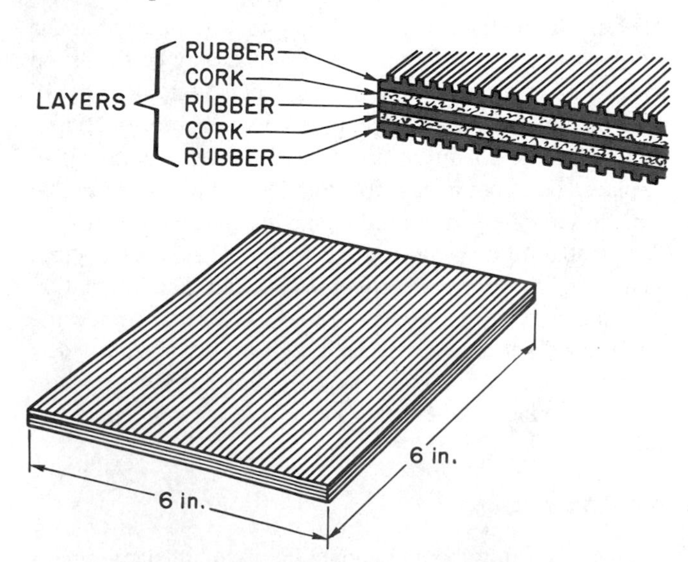

FIGURE 38–20 Rubber and cork pads may be placed under a unit to reduce the vibration.

there are raised areas on the bottom of the unit for this contact, the pads may be placed there.

Spring isolators may also be used and need to be chosen correctly for the particular weight of the unit. The springs will compress and lose their effectiveness under too heavy a load. They also may be too stiff for a light load if not chosen correctly.

Duct Connections for Package Equipment

The ductwork entering the structure may be furnished by the manufacturer or made by the contractor. Manufacturers furnish a variety of connections for rooftop installations but do not normally have many factory-made fittings for going through a wall. See Figure 38–21 for a rooftop installation.

Package equipment comes in two different duct connection configurations relative to the placement of the return and supply ducts, either side by side or over and under. In the *side-by-side* pattern the return duct connection is beside the supply connection, Figure 38–22. The connections are almost the same size. There will be a large difference in the connection sizes on an over-and-under unit. If

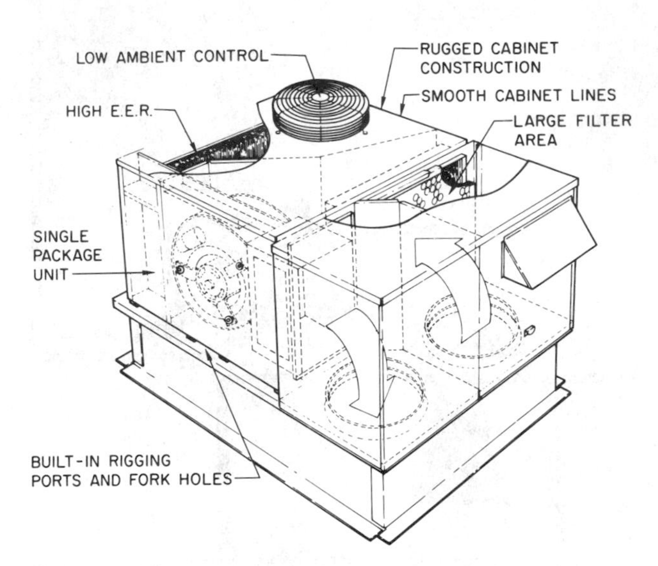

FIGURE 38–21 Duct connector for rooftop installation. *Courtesy Climate Control*

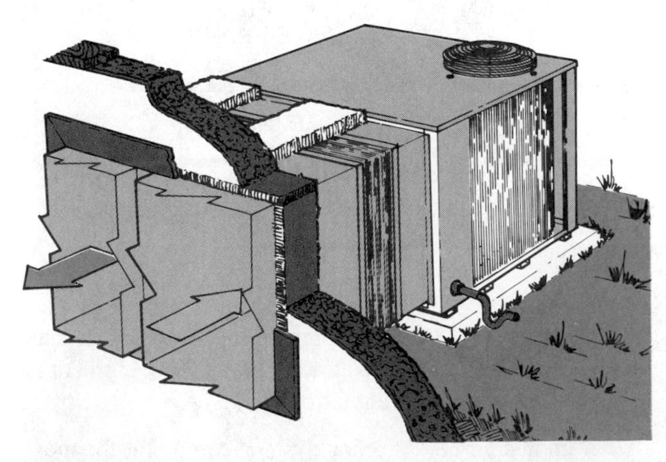

FIGURE 38–22 This air-to-air package unit has side-by-side duct connections. *Courtesy Climate Control*

the unit is installed in a crawl space, the return duct must run on the correct side of the supply duct if the crawl space is low. It is difficult to cross a return duct over or under a supply duct if there is not enough room. The side-by-side duct design makes it easier to connect the duct to the unit because of the sizes of the connections on the unit. They are closer to square in dimension than the connections on an over-and-under duct connection. If the duct has to rise or fall going into the crawl space, a standard duct size can be used.

The *over-and-under* duct connection is more difficult to fasten ductwork to because the connections are more oblong, Figure 38–23. They are wide because they extend from one side of the unit to the other and shallow because one is on top of the other. A duct transition fitting is almost always necessary to connect this unit to the ductwork. The transition is normally from wide shallow duct to almost square. When side-by-side systems are used, the duct system may be designed to be shallow and wide to keep the duct transition from being so complicated.

The duct connection must be watertight and insulated, Figure 38–24. Some contractors cover the duct with a weatherproof hood.

At some time the package equipment will need replacing. The duct system usually outlasts two or three units.

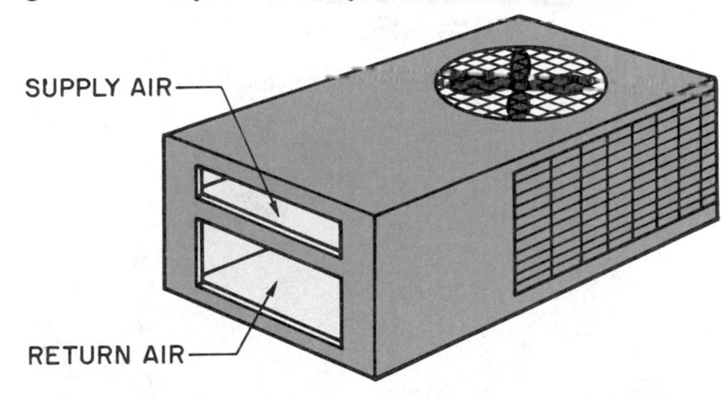

FIGURE 38–23 Over-and-under duct connections.

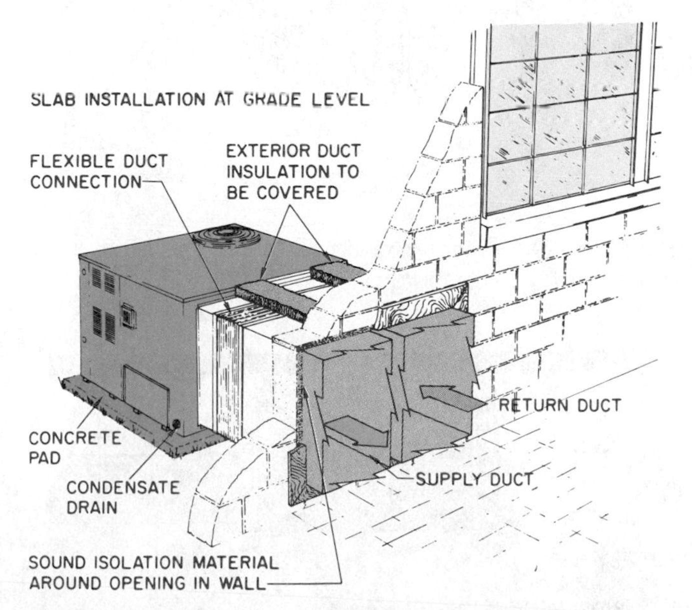

FIGURE 38–24 Air-to-air package unit installed through a wall. *Courtesy Climate Control*

Manufacturers may change their equipment style from over and under to side by side. This complicates the new choice of equipment because an over-and-under unit may not be as easy to find as a side-by-side unit, and the duct already exists for an over-and-under connection, Figure 38–25.

Package air conditioners have no field run refrigerant piping. The refrigerant piping is assembled within the unit at the factory. The refrigerant charge is included in the price of the equipment. The equipment is ready to start except for electrical and duct connections. One precaution must be observed, however. Most of these systems use R-22 and the oil in the system has an affinity for R-22. The R-22 in the unit will all be in the compressor if it is not forced out by a crankcase heater. This was not mentioned in the unit on refrigeration because the refrigerants used in refrigeration equipment are not attracted to the oil like R-22. ***All manufacturers that use crankcase heaters on their compressors supply a warning to leave the crankcase heater energized for some time, perhaps as long as 12 hours, before starting the compressor.*** The installing contractor must coordinate the electrical connection and start-up times closely because the unit should not be started immediately when the power is connected.

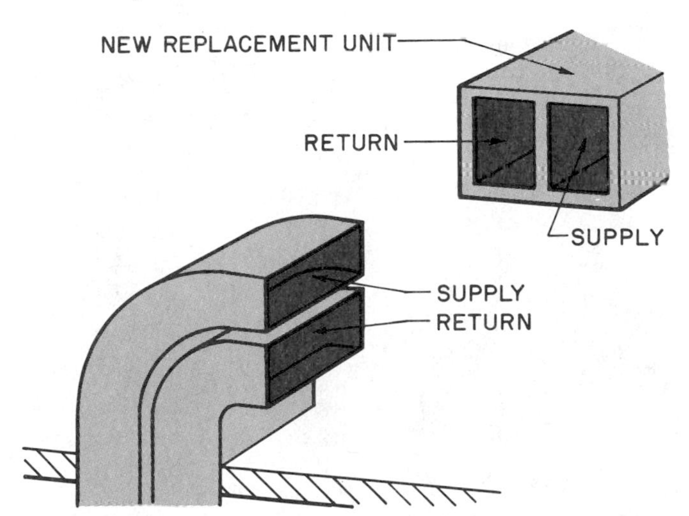

FIGURE 38–25 Air-cooled package unit with over-and-under duct connections. The manufacturer has changed design to a side-by-side duct connection and the installing contractor has a problem.

38.9 INSTALLING THE SPLIT SYSTEM

In split-system air conditioning equipment installations, the condenser is in a different location from the evaporator.

38.10 THE EVAPORATOR SECTION

The evaporator is normally located close to the fan section regardless of whether the fan is in a furnace or in a special air handler. The air handler (fan section) and coil must be located on a solid base or suspended from a strong support. Upflow and downflow equipment often has a base that may be fireproof and rigid for the air handler to rest on. Some vertical-mount air-handler installations have a wall-hanging support for the unit.

When the unit is installed horizontally, it may rest on the ceiling joists in an attic or on a foundation of blocks or concrete in a crawl space, Figure 38–26. The air handler may be hung from the floor or ceiling joists in different ways, Figure 38–27. If the air handler is hung from above, vibration isolators are often located under it to keep fan noise or vibration from being transmitted into the structure. Figure 38–28 is a trapeze hanger using vibration isolation pads.

The air handler (fan section) should always be installed so that it will be easily accessible for future service. The air handler will always contain the blower and sometimes the controls and heat exchanger.

Most manufacturers have designed their air handlers or furnaces so that, when installed vertically, they are totally accessible from the front. This works well for closet installations where there is insufficient room between the closet walls and the side walls of the furnace or air handler. Figure 38–29 is an electric furnace used as an air conditioning air handler.

When a furnace or air handler is located in a crawl space, side access is important, so the air handler should be located off the ground next to the floor joists. The ductwork will also be installed off the ground, Figure 38–30. If the air handler is top access, it will have to be located lower than the duct and the duct must make a transition downward to the air handler or the supply and return ends, Figure 38–31.

The best location for a top-access air handler is in an attic crawl space where the air handler can be set on the joists. The technician can work on the unit from above, Figure 38–32.

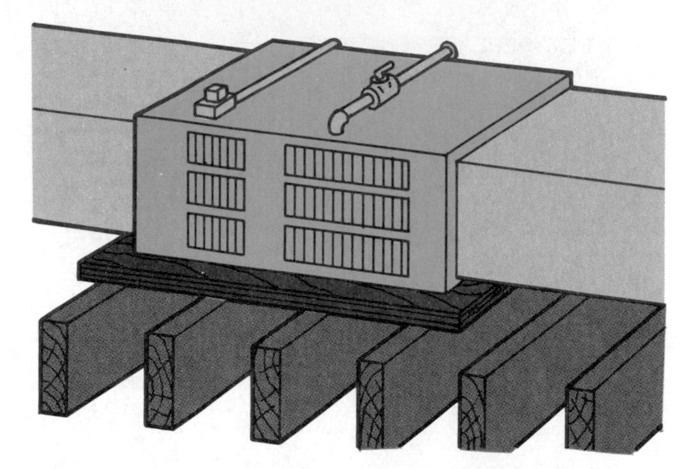

FIGURE 38–26 Furnace located in an attic crawl space. See manufacturer's recommendations for suggested base.

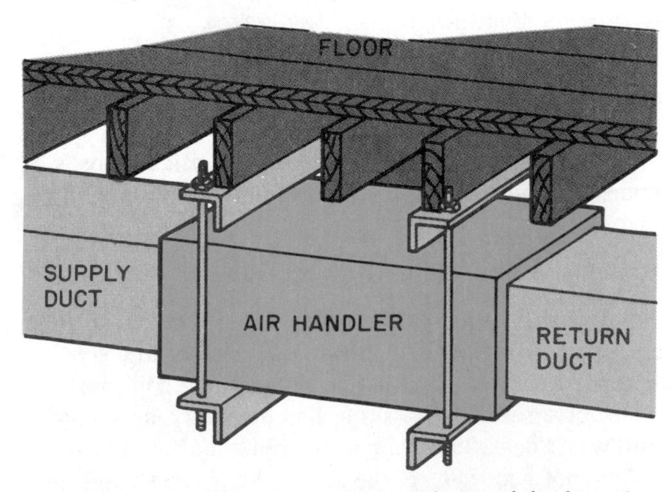

FIGURE 38–27 Air handler is hung from above, and the ductwork is connected and hung at the same height.

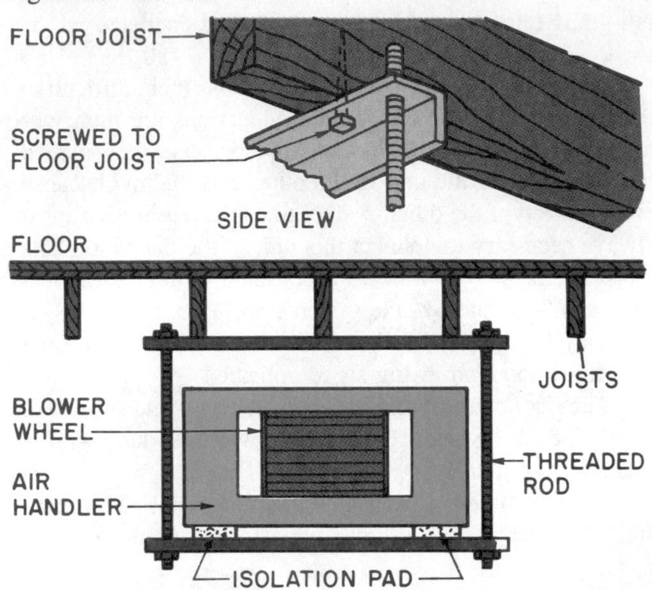

FIGURE 38–28 Trapeze hanger with vibration isolation pads.

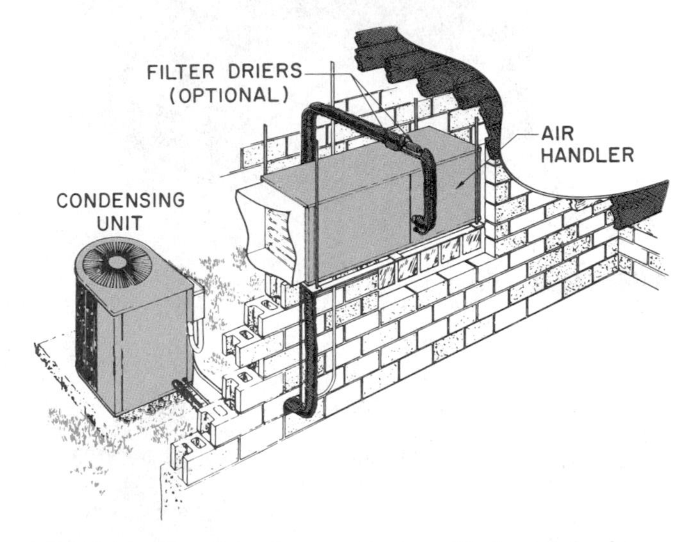

FIGURE 38–29 Electric furnace with air conditioning coil placed in the duct. *Courtesy Climate Control*

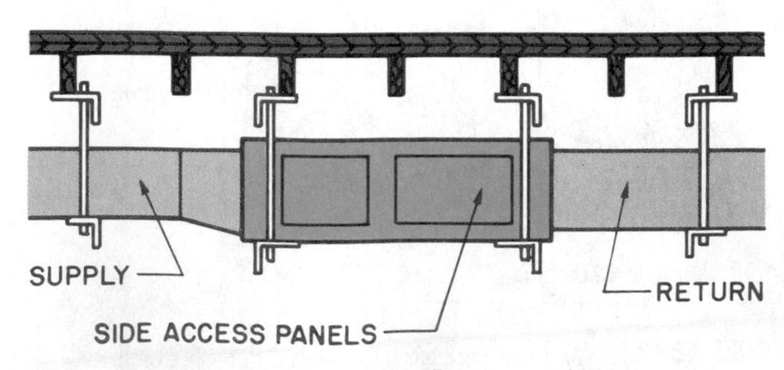

FIGURE 38–30 Side-access air handler in a crawl space.

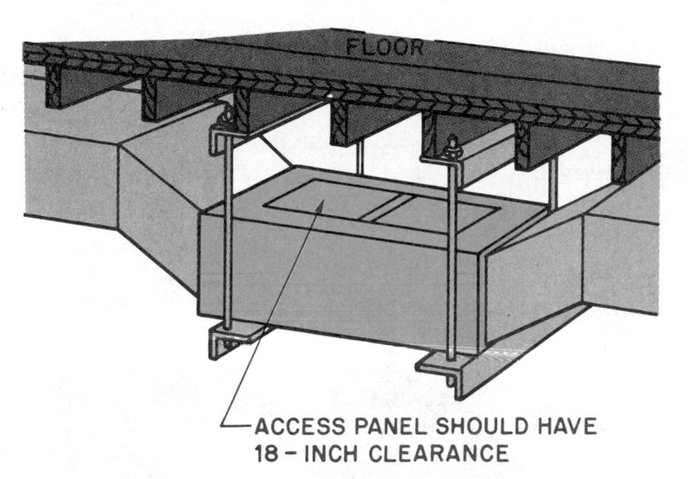

FIGURE 38–31 Top-access air handler.

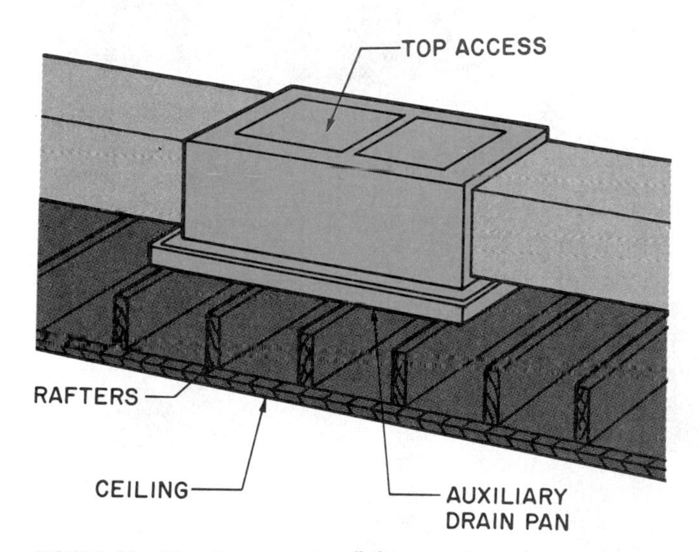

FIGURE 38–32 Evaporator installed in an attic crawl space.

Condensate Drain Piping

When the evaporator is installed, provisions must be made for the condensate that will be collected in the air conditioning cycle. An air conditioner in a climate with average humidity will collect about 3 pints (pt) of condensate per hour of operation for each ton of air conditioning. A 3-ton system would condense about 9 pt per hour of operation. This is more than a gallon of condensate per hour or more than 24 gal in a 24-h operating period. This can add up to a great deal of water over a period of time. If the unit is near a drain that is below the drain pan, simply pipe the condensate to the drain, Figure 38–33. A trap in the drain line will hold some water and keep air from pulling into the unit from the termination point of the drain. The drain may terminate in an area where foreign particles may be pulled into the drain pan. The trap will prevent this, Figure 38–34. If there is no

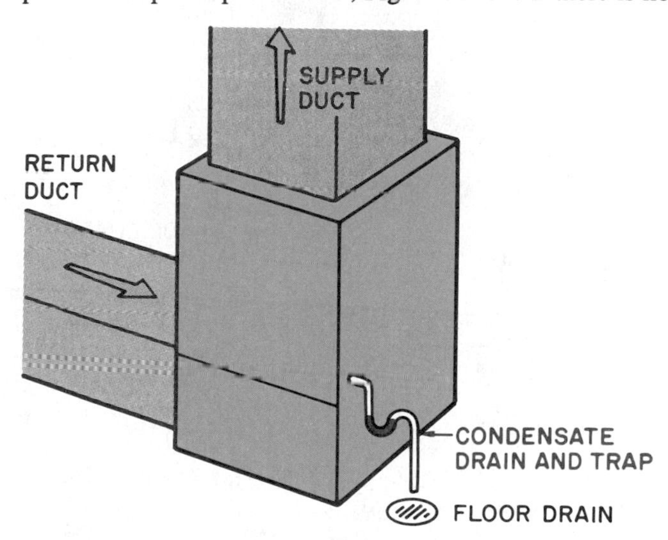

FIGURE 38–33 Condensate piped to a drain below the evaporator drain pan.

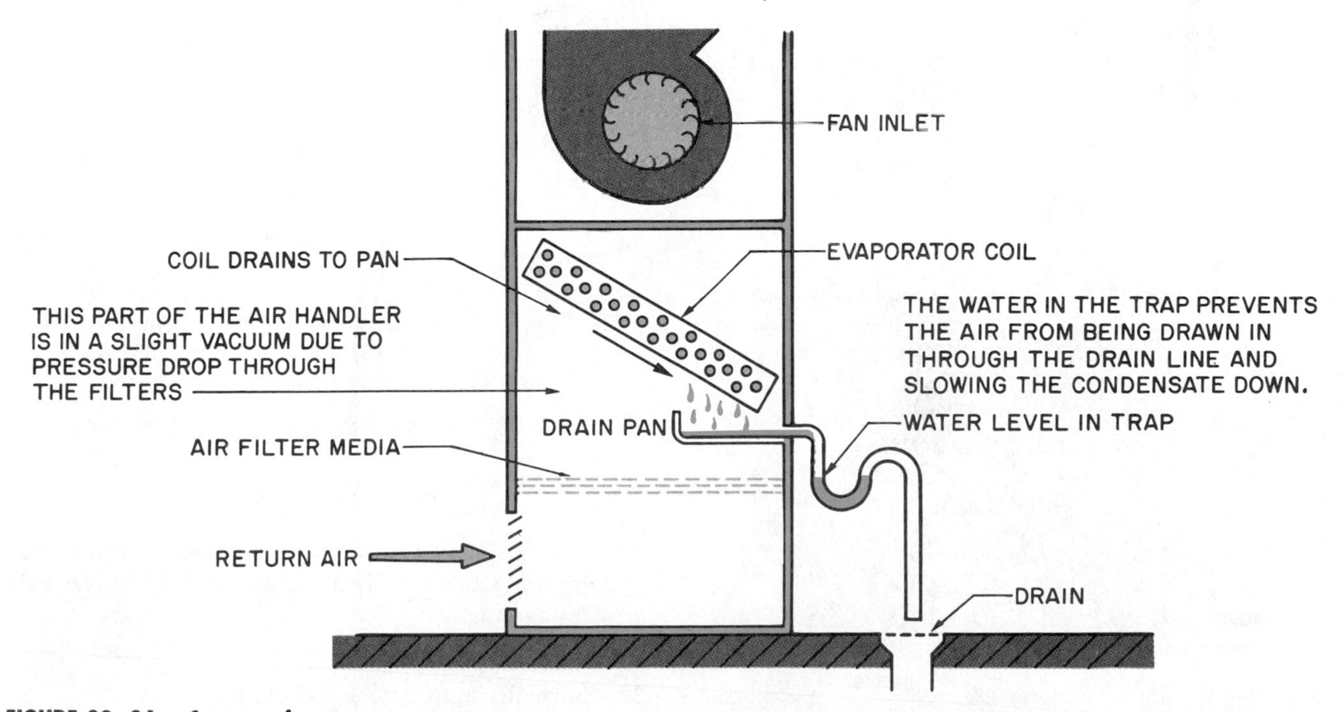

FIGURE 38–34 Cut-away of a trap.

drain close to the unit, the condensate must be drained or pumped to another location, Figure 38–35.

Some locations call for the condensate to be piped to a dry well. A dry well is a hole in the ground filled with stones and gravel. The condensate is drained into the well and absorbed into the ground, Figure 38–36. For this to be successful, the soil must be able to absorb the amount of water that the unit will collect.

When the evaporator and drain are located above the conditioned space, an auxiliary drain pan under the unit is recommended and often required, Figure 38–37. Airborne particles, such as dust and pollen, can get into the drains. Algae will also grow in the water in the lines, traps, and pans and may eventually plug the drain. If the drain system is plugged, the auxiliary drain pan will catch the overflow and keep the water from damaging whatever is below it.

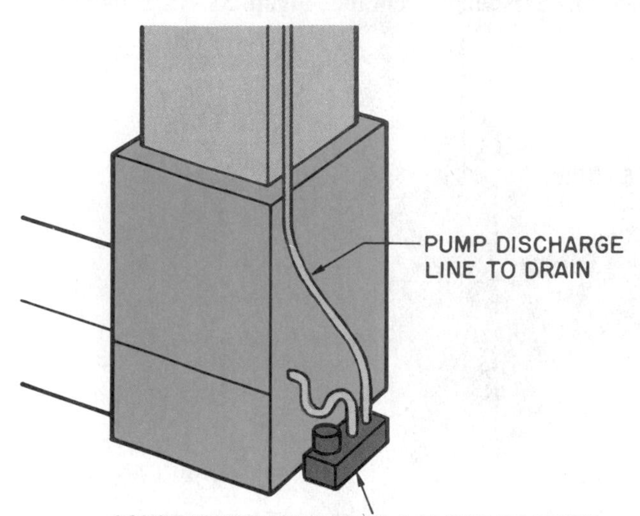

FIGURE 38–35 The drain in this installation is above the drain connection on the evaporator, and the condensate must be pumped to a drain at a higher level.

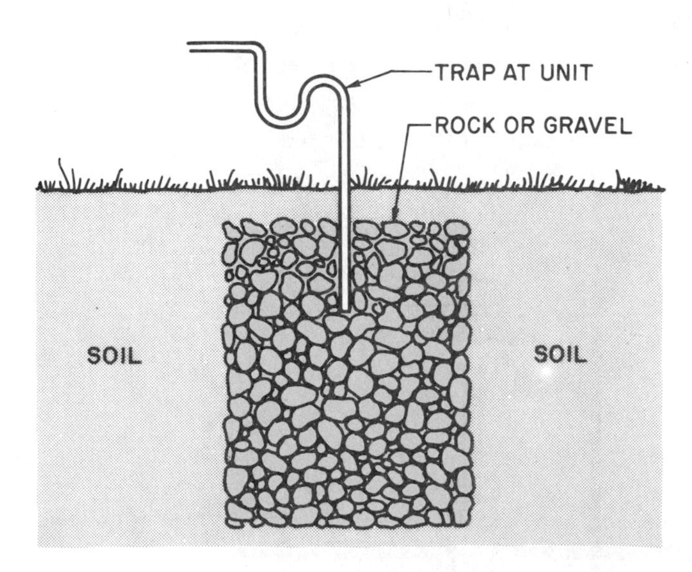

FIGURE 38–36 Dry well for condensate.

This auxiliary drain should be piped to a conspicuous place. The owner should be warned that if water ever comes from this drain line, a service call is necessary. Some contractors pipe this drain to the end of the house and out next to a driveway or patio so that if water ever drains from this point it would be readily noticed, Figure 38–38.

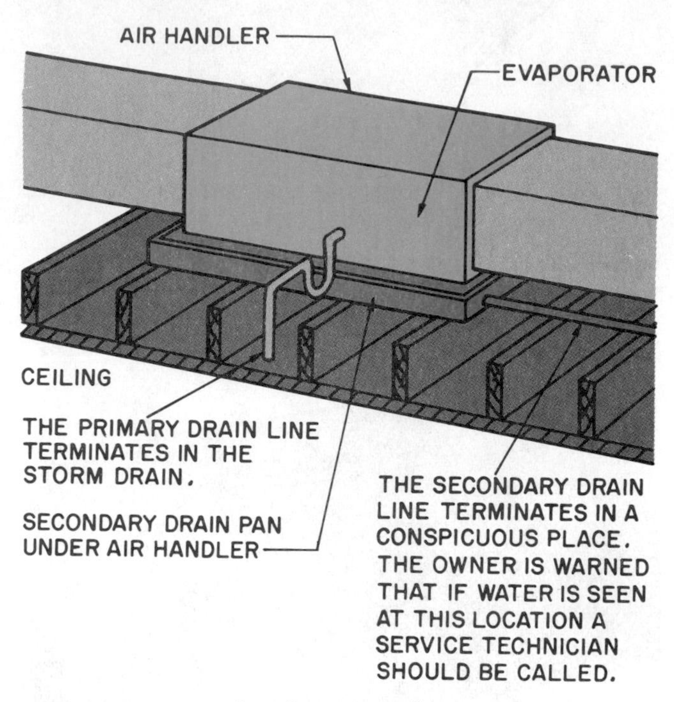

FIGURE 38–37 Auxiliary drain pan installation.

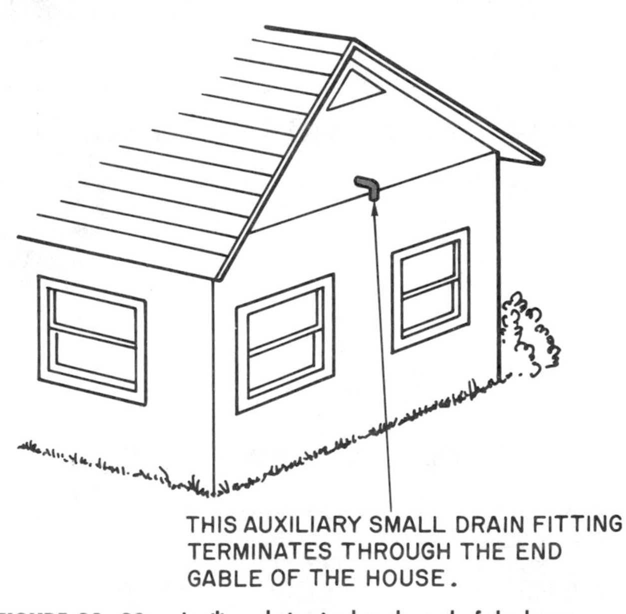

FIGURE 38–38 Auxiliary drain piped to the end of the house.

38.11 THE CONDENSING UNIT

The condensing unit location is remote from the evaporator. The following must be considered carefully when placing a condensing unit:

1. Proper air circulation
2. Convenience for piping and electrical service

3. Future service
4. Natural water and roof drainage
5. Solar influence
6. Appearance

Air Circulation and Installation

The unit must have adequate air circulation. The air discharge from the condenser may be from the side or from the top. Discharged air must not hit an object and circulate ba through the condenser. The air leaving the condenser warm or hot and will create high head pressure and po operating efficiency if it passes through the condenser again. Follow manufacturer's literature for minimum clearances for the unit, Figure 38–39.

Electrical and Piping Considerations

The refrigerant piping and the electrical service must be connected to the condensing unit. Piping is discussed later in this unit, but, for now, realize that the piping will be routed between the evaporator and the condenser. When the condenser is located next to a house, the piping must be routed between the house and the unit, usually behind or beside the house with the piping routed next to the ground. If the unit is placed too far from the house, the piping and electrical service become natural obstructions between the house and the unit. Children may jump on the piping and electrical conduit. If the unit is placed too close to the house, it may be difficult to remove the service panels. Before placing the unit, study the electrical and the refrigeration line connections and make them as short as practical, leaving adequate room for service.

Service Accessibility

Unit placement may determine whether a service technician gives good service or barely adequate service. The technician must be able to see what is being worked on. A unit is often placed so that the technician can touch a particular component but is unable to see it, or, by shifting positions, can see the component but not touch it. The technician should be able to both see and touch the work at the same time.

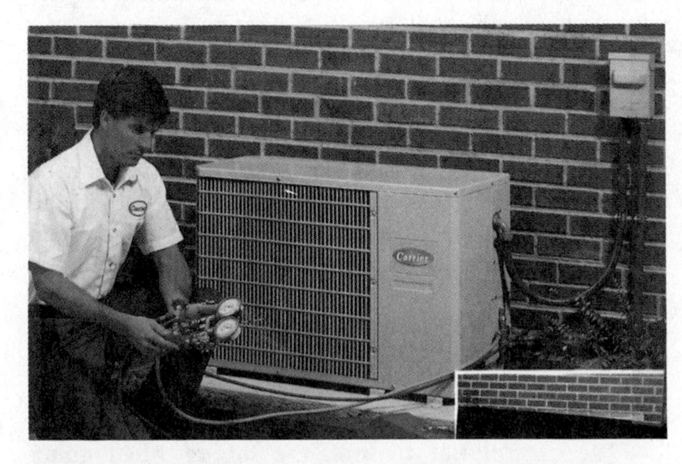

FIGURE 38–39 Condensing unit for a split system located so that it has adequate airflow and service room. *Courtesy Carrier Corporation*

Water Drainage From Natural Sources

The natural drainage of the ground water and roof water should be considered in unit placement. ***The unit should not be located in a low place where ground water will rise in the unit. If this happens, the controls may short to ground, and the wiring will be ruined.*** All units should be placed on a base pad of some sort. Concrete and high-impact plastic are commonly used. Metal frames can be used to raise the unit when needed.

The roof drainage on a structure may run off in gutters. The condensing unit should not be located where the gutter drain or roof drainage will pour down onto the unit. Condensing units are made to withstand rain water but not large volumes of drainage. If the unit is a top-discharge unit, the drainage from above will be directly into the fan motor, Figure 38–40.

Solar Influence

If possible, put the condensing unit on the shady side of the house because the sun shining on the panels and coils will lower the efficiency. However, the difference is not crucial, and it would not pay to pipe the refrigerant tubing and electrical lines long distances just to keep the sun from shining on the unit.

Shade helps cool the unit, but it may also cause problems. Some trees have small leaves, sap, berries, or flowers that may harm the finish of the unit. Pine needles may fall into the unit, and the pine pitch that falls on the unit's cabinet may harm the finish and outweigh the benefit of placing the unit in the shade of many types of trees.

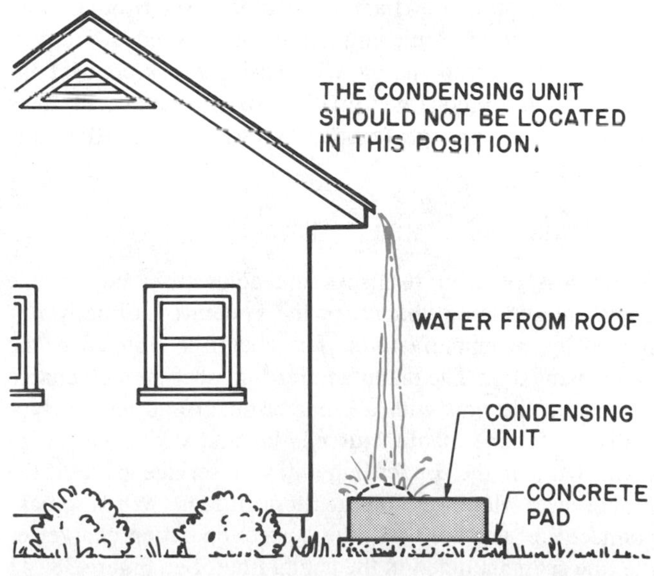

FIGURE 38–40 This unit is located improperly and allows the roof drain to pour down into the top of the unit.

Placing a Condensing Unit for Best Appearance

The condensing unit should be located where it will not be noticeable or will not make objectionable noise. When located on the side of a house, the unit may be hidden from the street with a low shrub. If the unit is a side-discharge type, the fan discharge must be away from the shrub or the shrub may not live. If the unit is top discharge, the shrub is unaffected but the unit's sound will rise. This noise may be objectionable in a bedroom located above the unit.

Locating the condensing unit at the back of the house may place the unit closer to the evaporator and would mean shorter piping, but the back of the house is the usual location of patios and porches, and the homeowner, while sitting outside, may not want to hear the unit. In such a case, a side location at the end of the house where there are no bedrooms may be the best choice.

Each location has its considerations. The salesperson and the technician should consult with the owner about locating the various components. The salesperson should be familiar with all local code requirements. A floor plan of the structure may help. Some companies use large graph paper to draw a rough floor plan to scale when estimating the job. The equipment can be located on the rough floor plan to help solve these problems. The floor plan can be shown to the homeowners to help them understand the contractor's suggestions.

38.12 INSTALLING REFRIGERANT PIPING

The refrigerant piping is always a big consideration when installing a split-system air conditioner. The choice of the piping system may make a difference in the start-up time for the system. The piping should always be kept as short as practical. For an air conditioning installation three methods are used to connect the evaporator and the condensing unit on almost all equipment under 5 tons (refrigeration systems for commercial refrigeration are not the same): (1) contractor-furnished piping, (2) flare or compression fittings, with the manufacturer furnishing the tubing (called a line set), and (3) precharged tubing with sealed quick-connect fittings, called a precharged line set. **In all of these systems the operating charge for the system is shipped in the equipment.**

The Refrigerant Charge

Regardless of who furnishes the connecting tubing, the complete operating charge for the system is normally furnished by the manufacturer. The charge is shipped in the condensing unit. The manufacturer furnishes enough charge to operate the unit with a predetermined line length, typically 30 ft. The manufacturer holds and stores the refrigerant charge in the condensing unit with service valves if the tubing uses flare or compression fittings. When quick-connect line sets are used, the correct operating charge for the line set is included in the actual lines. See Figure 38–41 for an example of both service valves and quick-connect fittings.

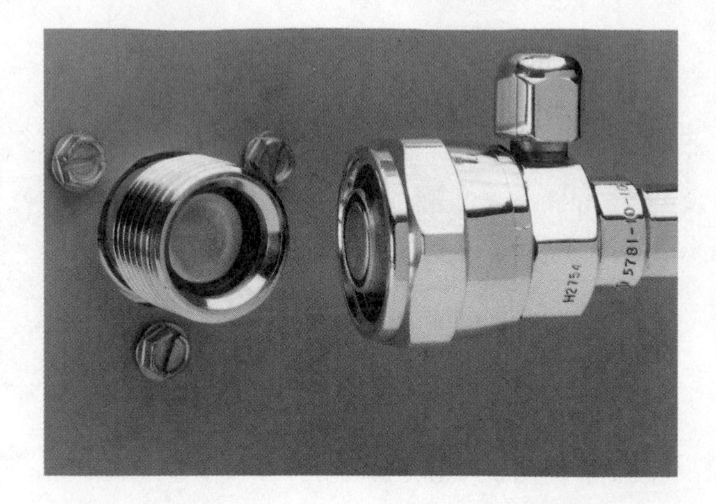

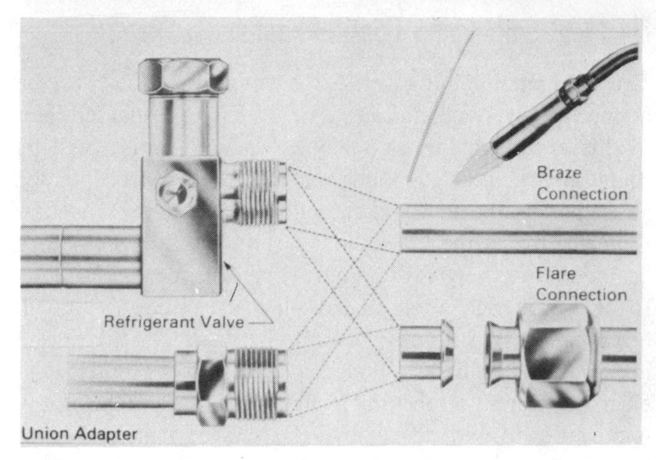

FIGURE 38–41 Two basic piping connections for split-system air conditioning. *Courtesy Aeroquip Corporation*

When service valves are used, the piping is fastened to the service valves by flare or compression fittings. Some manufacturers provide the option of soldering to the service valve connection. The piping is always hard-drawn or soft copper. In installations where the piping is exposed, straight pipe may look better. Hard-drawn tubing may be used for this with factory elbows used for the turns. When the piping is not exposed, soft copper is easily formed around corners where a long radius will be satisfactory.

The Line Set

When the manufacturer furnishes the tubing, it is called a *line set.* The suction line is insulated, and the tubing may be charged with nitrogen, contained with rubber plugs in the tube ends. When the rubber plug is removed, the nitrogen rushes out with a loud hiss. This indicates that the tubing is not leaking.

When the tubing is used, uncoil it from the end of the coil. Place one end of the coil on the ground and roll it while keeping your foot on the tubing end on the ground, Figure 38–42.

Be careful not to kink the tubing when going around corners, or it may collapse. Because of the insulation on the pipe, you may not even see the kink.

FIGURE 38–42 Uncoiling tubing. *Photo by Bill Johnson*

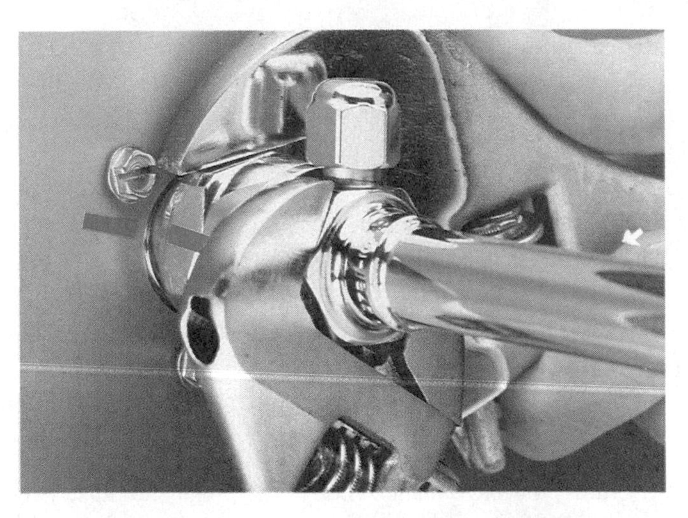

FIGURE 38–43 How to tighten fittings. *Courtesy Aeroquip Corporation*

Tubing Leak Test and Evacuation

When the piping is routed and in place, the following procedure is commonly used to make the final connections:

1. The tubing is fastened at the evaporator end. ***When the tubing has flare nuts, a drop of oil applied to the back of the flare will help prevent the flare nut from turning the tubing when the flare connection is tightened.*** The suction line may be as large as 1 1/8-in. OD tubing. Large tubing will have to be made very tight. Two adjustable wrenches are recommended, Figure 38–43. The liquid line will not be any smaller than 1/4-in. OD tubing, and it may be as large as 1/2-in. OD on larger systems.
2. The smaller (liquid) line is fastened to the condensing unit service valve.
3. The larger (suction) line is fastened hand tight.
4. The liquid-line service valve is slightly opened, and refrigerant is allowed to purge through the liquid line, the evaporator, and the suction line. This pushes the nitrogen out of the piping and the evaporator, and it escapes at the hand-tightened suction-line flare nut, Figure 38–44.

   NOTE: ⊕EPA rules allow this process under deminimus loss of refrigerant.⊕

Line sets come in standard lengths of 10, 20, 30, 40, or 50 ft. If lengths other than these are needed, the manufacturer should be consulted. Some manufacturers have a maximum allowable line length, normally in the vicinity of 50 ft. If the line length has to be changed from the standard length, the manufacturer will also recommend how to adjust the unit charge for the new line length. Most units are

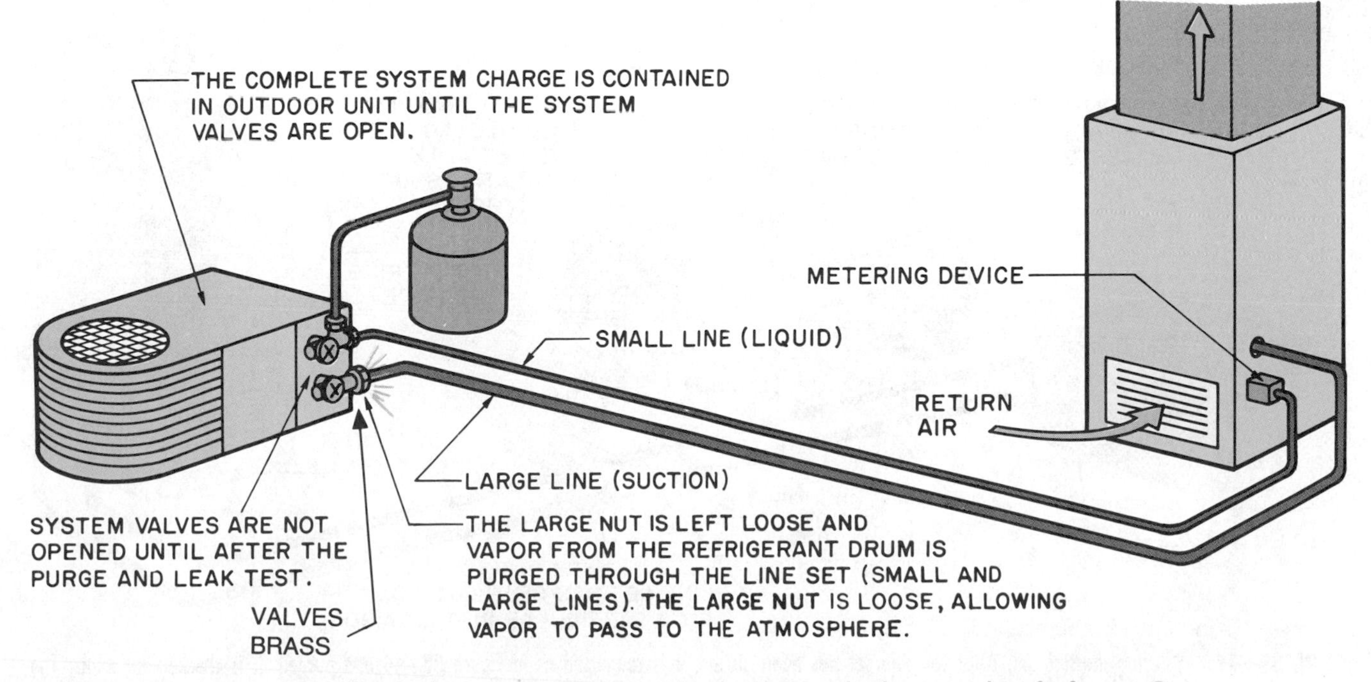
THE COMPLETE SYSTEM CHARGE IS CONTAINED IN OUTDOOR UNIT UNTIL THE SYSTEM VALVES ARE OPEN.
METERING DEVICE
SMALL LINE (LIQUID)
RETURN AIR
LARGE LINE (SUCTION)
SYSTEM VALVES ARE NOT OPENED UNTIL AFTER THE PURGE AND LEAK TEST.
THE LARGE NUT IS LEFT LOOSE AND VAPOR FROM THE REFRIGERANT DRUM IS PURGED THROUGH THE LINE SET (SMALL AND LARGE LINES). THE LARGE NUT IS LOOSE, ALLOWING VAPOR TO PASS TO THE ATMOSPHERE.
VALVES
BRASS

FIGURE 38–44 Purging nitrogen shipping charge. NOTE: ⊕EPA rules allow this process under de minimus loss of refrigerant.⊕

shipped with a charge for 30 ft of line. If the line is shortened, refrigerant must be removed. If the line is longer than 30 ft, refrigerant must be added.

Altered Line-Set Lengths

When line sets must be altered, they may be treated as a self-contained system of their own. The following procedures should be followed:

1. Alter the line sets as needed for the proper length. The nitrogen charge will escape during this alteration. **Do not remove the rubber plugs.**
2. When the alterations are complete, the lines may be pressured to about 25 psig to leak check any connections you have made. The rubber plugs should stay in place. If you desire to pressure test with higher pressures, hook up the evaporator and condenser ends and you may safely pressurize to 150 psig with nitrogen. The line's service port is common to the line side of the system. If the valves are not opened at this time, the line set and evaporator may be thought of as a sealed system of their own for pressure testing and evacuation, Figure 38–45.
3. After the leak test has been completed, evacuate the line set (and evaporator if connected). Break the vacuum to about 10 psig and go through the line-set purge just as though it is a factory installation. **After starting the unit, you will have to alter the charge to meet the manufacturer's guidelines.**

Precharged Line Sets (Quick-Connect Line Sets)

Precharged line sets with quick-connect fittings are shipped in most of the standard lengths. The difference in line sets and precharged line sets is that the correct refrigerant operating charge is shipped in the precharged line set. Refrigerant does not have to be added or taken out unless the line set is altered. The following procedures are recommended for connecting precharged line sets with quick-connect fittings.

1. Roll the tubing out straight.
2. Determine the routing of the tubing from the evaporator to the condenser and put it in place.
3. Remove the protective plastic caps on the evaporator fittings. Place a drop of refrigerant oil (this is sometimes furnished with the tubing set) on the neoprene O rings of each line fitting. The O ring is used to prevent refrigerant from leaking out while the fitting is being connected, and it serves no purpose after the connection is made tight. It may tear while making the connection if it is not lubricated.
4. Start the threaded fitting and tighten hand tight. Making sure that several threads can be tightened by hand will ensure that the fitting is not cross-threaded.
5. When the fitting is tightened hand tight, finish tightening the connection with a wrench. You may hear a slight escaping of vapor while making the fitting tight; this purges the fitting of any air that may have been in the fitting. The O ring should be seating at this time. Once you have started tightening the fitting, do not stop until you are finished. If you stop in the middle, some of the system charge may be lost.
6. Tighten all fittings as indicated.
7. After all connections have been tightened, leak check all connections that you have made.

Altered Precharged Line Sets

When the lengths of quick-connect line sets are altered, the charge will have to be altered. The line set may be treated as a self-contained system for alterations. The following is the recommended procedure.

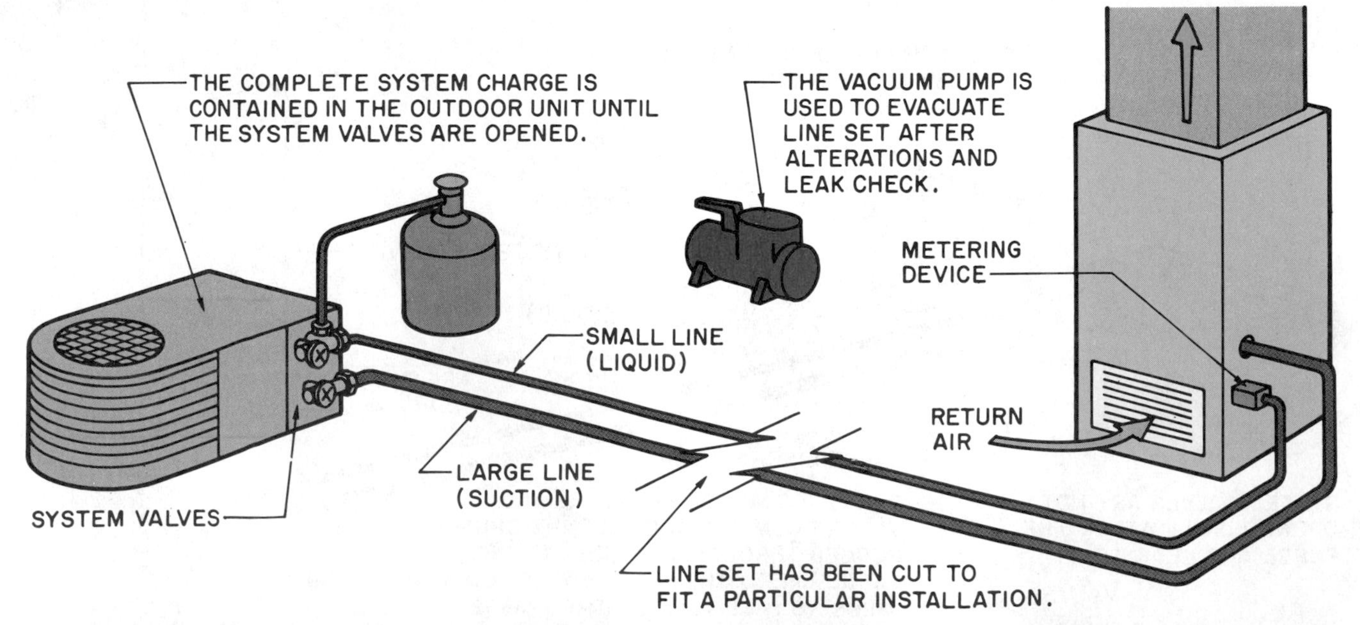

THE COMPLETE SYSTEM CHARGE IS CONTAINED IN THE OUTDOOR UNIT UNTIL THE SYSTEM VALVES ARE OPENED.
THE VACUUM PUMP IS USED TO EVACUATE LINE SET AFTER ALTERATIONS AND LEAK CHECK.
METERING DEVICE
SMALL LINE (LIQUID)
RETURN AIR
LARGE LINE (SUCTION)
SYSTEM VALVES
LINE SET HAS BEEN CUT TO FIT A PARTICULAR INSTALLATION.

FIGURE 38–45 Altering the line set. After the line set has been altered, it is connected as in Figure 38–44 and purged. It is checked for leaks. The line set and evaporator are evacuated. After evacuation, a vapor holding charge is added, about 20 psig of the refrigerant, characteristic to the system.

LIQUID LINE DIAMETER INCHES	OUNCES OF R-22 PER FOOT OF LENGTH OF LIQUID LINE
$\frac{3}{8}$	0.58
$\frac{5}{16}$	0.36
$\frac{1}{4}$	0.21

If 3 feet of liquid line must be added to a ($\frac{3}{8}$ in.) 30-foot line, an additional 1.74 ounces must be added to the system upon starting the unit. ($3 \times 0.58 = 1.74$)
If the unit is shipped with a precharged line set, the complete charge for that particular line set is contained in the lines. If this set is altered, the complete liquid line length must be used. For example, if a 50-foot set ($\frac{1}{4}$ in.) is cut to 25 feet of length, the charge is exhausted from the lines, and 5.25 ounces must be added to the charge contained in the condenser when started. ($25 \times 0.21 = 5.25$)

FIGURE 38–46 Table of liquid-line capacities.

1. ⊕**Before connecting the line set, recover the refrigerant from the line set. Do not just cut it because the liquid and suction lines contain some liquid refrigerant. Recovering the refrigerant is required.**⊕
2. Cut the line set and alter the length as needed.
3. Pressure test the line set.
4. Evacuate the line set to a low vacuum.
5. Valve off the vacuum pump and pressurize to about 10 psig on just the line set and connect it to the system using the procedures already given.
6. Read the manufacturer's recommendation as to the amount of charge for the new line lengths and add this much refrigerant to the system. If there are no recommendations, see Figure 38–46 for a table of liquid-line capacities. The liquid line should contain the most refrigerant, and the proper charge should be added to the liquid line to make up for what was lost when the line was cut. No extra refrigerant charge needs to be added for the suction line if it is filled with vapor before assembly.

Piping Advice

The piping practices given for air conditioning equipment are typical of most manufacturers' recommendations. **The manufacturer's recommendation should always be followed.** Each manufacturer ships installation and start-up literature inside the shipping crate. If it is not there, request it from the manufacturer and use it.

38.13 EQUIPMENT START-UP

The final step in the installation is starting the equipment. The manufacturer will furnish start-up instructions. After the unit or units are in place, leak checked, have the correct factory charge furnished by the manufacturer, and are wired for line and control voltage, follow these guidelines:

1. The line voltage must be connected to the unit disconnect panel. This should be done by a licensed electrician. Disconnect a low-voltage wire, such as the Y wire at the condensing unit, to prevent the compressor from starting. Turn on the line voltage. The line voltage allows the crankcase heater to heat the compressor crankcase. ***This applies to any unit that has crankcase heaters. Heat must be applied to the compressor crankcase for the amount of time the manufacturer recommends (usually not more than 12 h). Heating the crankcase boils any refrigerant out of the crankcase before the compressor is started. If the compressor is started with liquid in the crankcase, some of the liquid will reach the compressor cylinders and may cause damage. The oil will also foam and provide only marginal lubrication until the system has run for some time.***
2. It is normally a good idea to plan on energizing the crankcase heater in the afternoon and starting the unit the next day. Before you leave, make sure that the crankcase heater is hot.
3. If the system has service valves, open them.
   NOTE: Some units have valves that *do not* have back seats.
   Do not try to back seat these valves that do not have back seats or you may damage the valve. Open the valve until resistance is felt, then stop, Figure 38–47.
4. Check the line voltage at the installation site and make sure it is within the recommended limits.
5. Check all electrical connections including those made at the factory and ensure that they are tight and secure. ***Turn the power off when checking electrical connections.***
6. Set the fan switch on the room thermostat to FAN ON and check the indoor fan for proper operation, rotation, and current draw. You should feel air at all registers, normally about 2 to 3 ft above a floor register. Make sure there are no air blockages at the supply and return openings.
7. Turn the fan switch to FAN AUTO and with the HEAT-OFF-COOL selector switch at the off position, replace

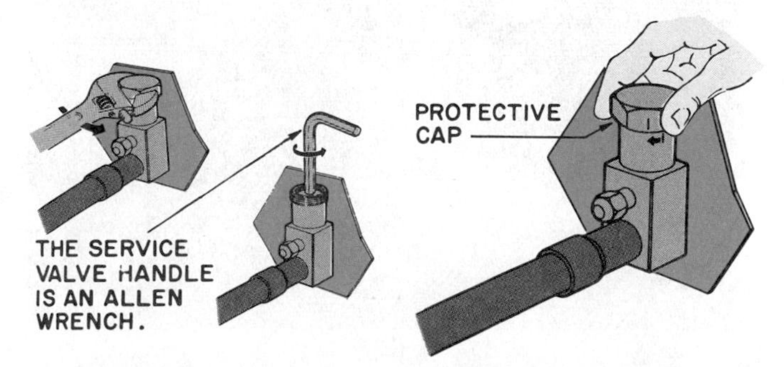

FIGURE 38–47 Service valves furnished with equipment may not have back seats. *Courtesy Aeroquip Corporation*

the Y wire at the condensing unit. ***The power at the unit is turned off and the panel is locked and tagged while making this connection.***

8. Place your ammeter on the common wire to the compressor, have someone move the HEAT-OFF-COOL selector switch to COOL, and slide the temperature setting to call for cooling. The compressor should start. ***Some manufacturer's literature will recommend that you have a set of gages on the system at this time. Be careful of the line length on the gage you install on the high-pressure side of the system. If the system has a critical charge and a 6-ft gage line is installed, the line will fill up with liquid refrigerant and will alter the charge, possibly enough to affect performance. A short gage line is recommended, Figure 38–48.*** Leak check the gage port when the gages are removed. Replace the gage port cover.

    If the manufacturer recommends that gages be installed for start-up, install them before you start the system. If they do not recommend them, follow these recommendations:

9. When gages are not installed, there are certain signs that can indicate correct performance. The suction line coming back to the compressor should be cool, although the "coolness" can vary. Two things will cause the suction-line temperature to vary and still be correct: the metering device and the ambient temperature. Most modern systems are fixed-bore metering devices (capillary tube or orifice). When the outside temperature is down to 75°F or 80°F, for example, the suction line will not be as cool as on a hot day because the condenser becomes more efficient and liquid refrigerant is retained in the condenser, partially starving the evaporator. If the day is cool, 65°F or 70°F, some of the air to the condenser may be blocked, which causes the head pressure to rise and the suction line to become cooler. The ambient temperature in the conditioned space causes the evaporator to have a large load, which makes the suction line not as cool until the inside temperature is reduced to near the design temperature, about 75°F.

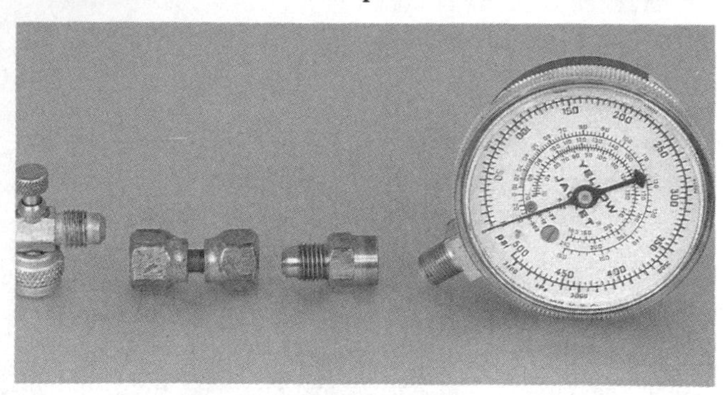

(B)

FIGURE 38–48 Short gauge line for high-pressure side of the system. *Photo by Bill Johnson*

The amperage of the compressor is a good indicator of system performance. If the outside temperature is hot and there is a good load on the evaporator, the compressor will be pumping at near capacity. The motor current will be near nameplate amperage. It is rare for the compressor amperage to be more than the nameplate rating due to only the system load. It is also normal for the compressor amperage to be slightly below the nameplate rating.

When you are satisfied that the unit is running satisfactorily, inspect the installation. Check that:

1. All air registers are open.
2. There are no air restrictions.
3. The duct is hung correctly, and all connections are taped.
4. All panels are in place with all screws.
5. The customer knows how to operate the system.
6. All warranty information is completed.
7. The customer has the operation manual.
8. The customer knows how to contact you.

SAFETY PRECAUTION: *Before starting the system examine all electrical connections and moving parts. The equipment may have faults and defects that can be harmful (eg, a loose pulley may fly off when the motor starts to turn). Remember that vessels and hoses are under pressure. Always be mindful of potential electrical shock.*

SUMMARY

- The installation of air conditioning equipment normally involves three crafts: ductwork, electrical, and mechanical or refrigeration.
- Duct systems are normally constructed of square, rectangular, or round metal, ductboard, or flexible material.
- Square and rectangular metal duct systems are assembled with S fasteners and drive clips.
- Round metal duct comes in many standard sizes with fittings.
- Round duct must be fastened with sheet metal screws at each connection.

- The first fitting in a duct system may be a vibration eliminator to keep any fan noise or vibration from being transmitted into the duct.
- Insulation may be applied to the inside or outside of any metal duct system that may exchange heat with the ambient.
- When applied to the inside, insulation can be glued or glued and fastened with a tab.
- When the duct is insulated on the outside, a vapor barrier must be used to keep moisture from forming on the duct surface if the duct surface is below the dew point temperature of the ambient.
- Fiberglass ductboard is compressed fiberglass with a reinforced foil backing that helps support it and create a vapor barrier.
- Flexible duct is a flexible liner that may have a cover of fiberglass that is held in place with vinyl or reinforced foil.
- ***The electrical installation includes choosing the correct enclosures, wire sizes, and fuses or breakers.***
- The electrical contractor will normally install the line-voltage wiring, and the air conditioning contractor will usually install the low-voltage control wiring. ***Local codes should be consulted before any wiring is done.***
- The low-voltage control wiring is normally color coded.
- Air conditioning equipment is manufactured in package systems and split systems.
- Package equipment is completely assembled in one cabinet.
- Package equipment has two types of duct connections: over and under and side by side.
- Package units come in small to very large sizes.
- Air-to-air package equipment installation consists of placing the unit on a foundation, connecting the ductwork, and connecting the electrical service and control wiring.
- The duct connections in package equipment may be made through a roof or through a wall at the end of a structure.
- Roof installations have waterproof roof curbs and factory-made duct systems.
- Isolation pads or springs placed under the equipment prevent equipment noise from travelling into the structure.
- Outside units may be installed on pads or frames; the pads are normally made of high-impact plastic or concrete.
- In split-system air conditioning equipment the evaporator is at an inside air handler (blower). The air handler may be an existing furnace or a separate blower package. The compressor is located outside with the condenser.
- Two refrigerant lines connect the evaporator and the condensing unit; the large line is the insulated suction line, and the small line is the liquid line.
- The air handler and condenser should be installed so that they are accessible for service.
- A condensate drain provision must be made for the evaporator.
- A secondary drain pan should be provided if the evaporator is located above the conditioned space.
- The condensing unit should be located as close as practical to the evaporator and the electrical service. It should not be located where natural water or roof water will drain directly into the top of the unit.
- The condensing unit should not be located where its noise will be bothersome.
- Line sets come in standard lengths of 10, 20, 30, 40, and 50 ft. The system charge is normally for 30 ft when the whole charge is stored in the condensing unit.
- Line sets may be altered in length. The refrigerant charge must be adjusted when the lines are altered.
- The start-up procedure for the equipment is in the manufacturer's literature. Before start-up, check the electrical connections, the fans, the airflow, and the refrigerant charge.

REVIEW QUESTIONS

1. Name the three crafts normally involved in an air conditioning installation.
2. Which type of duct system is the most expensive and longest lasting?
3. What fasteners fasten square metal duct at the joints?
4. What is normally used to fasten round metal duct at the joints?
5. Name two methods for fastening insulation to the duct when it is insulated on the inside?
6. Why does insulation installed on the outside of duct have a vapor barrier when used for air conditioning?
7. What happens if insulation on the inside of duct comes loose?
8. What material is most duct insulation made from?
9. Why is a flexible connector installed in a metal duct system?
10. What happens if flexible duct is turned too sharply around a corner?
11. Why must flexible duct be pulled as tight as practical?
12. How should flexible duct be hung from above?
13. What is the difference between a package unit and a split system?
14. What are some of the accessories that a manufacturer may supply to help the installing contractor in a rooftop installation with package equipment?
15. What are some advantages of a split air conditioning system?
16. What are three advantages of package equipment?
17. Name the two duct connection configurations on a package unit.
18. Name the two refrigerant lines connecting the condensing unit to the evaporator.
19. Name three criteria of a good location for an air-cooled condensing unit.

UNIT 39 Controls

OBJECTIVES

After studying this unit, you should be able to

- **describe the control sequence for an air conditioning system.**
- **explain the function of the 24-V control voltage.**
- **describe the space thermostat.**
- **describe the compressor contactor.**
- **explain the operation of the high- and low-pressure controls.**
- **discuss the function of the overloads and motor-winding thermostat.**
- **discuss the winding thermostat and the internal relief valve.**
- **identify operating and safety controls.**
- **compare modern and older control concepts.**
- **describe how crankcase heat is applied in some modern equipment.**

39.1 CONTROLS FOR AIR CONDITIONING

Equipment control for maintaining correct air conditions involves the control of three components: the indoor fan, the compressor, and the outdoor fan. These components are used in air-cooled air conditioning equipment. Water-cooled equipment for small air conditioning applications is seldom used. Some of these controls were covered in earlier units. They are discussed again here for the convenience of the reader.

The indoor fan, compressor, and outdoor fan must be started and stopped automatically at the correct times. The normal sequence of operation is:

1. The indoor fan must operate when the compressor operates.
2. The outdoor fan must operate when the compressor operates (except for units that may have a fan cycle device that allows the fan to be cycled off for short periods of time during cool weather for head pressure control).
3. The indoor fan may have a continuous-operation switch at the thermostat. In this position the indoor fan will run continuously, and the compressor and outdoor fan will cycle on and off on demand.

Operating and safety controls are the two main types of functional controls. A room thermostat, for example, is an operating control used to sense the space temperature to stop and start the compression system (compressor and outdoor fan), Figure 39–1. The high-pressure control is a safety control that keeps the unit from operating when the head pressure is too high, Figure 39–2. Both of these controls have something in common: neither consumes power. They pass power to other devices, such as the compressor's magnetic contactor.

FIGURE 39–1 **Room thermostat for controlling space temperature.** ***Photo by Bill Johnson***

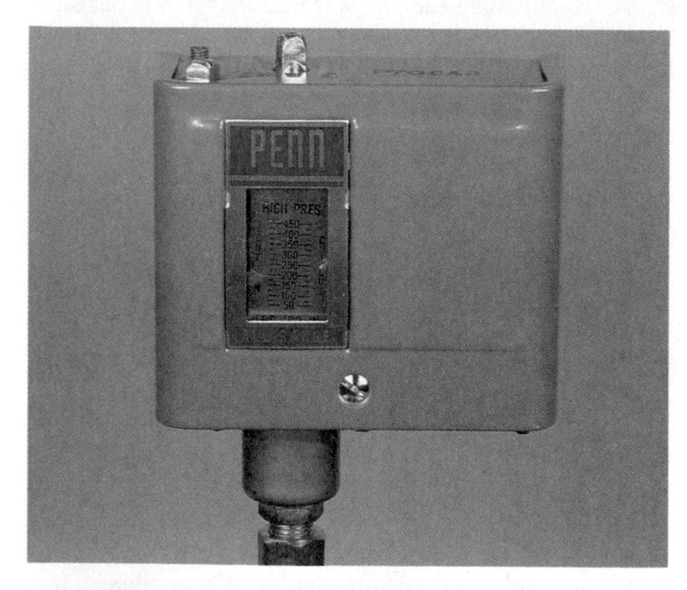

FIGURE 39–2 **Commercial type of high-pressure control used in earlier residential equipment.** ***Photo by Bill Johnson***

39.2 PRIME MOVERS—COMPRESSORS AND FANS

The prime movers of the system are the fans (indoor and outdoor) and the compressor. These devices consume most of the power in the air conditioning process and are operated with high voltage, usually 230 V in a residence. The controls are operated with 24 V for safety and convenience. The voltage reduction is accomplished by a transformer located in the condensing unit or air handler, Figure 39–3.

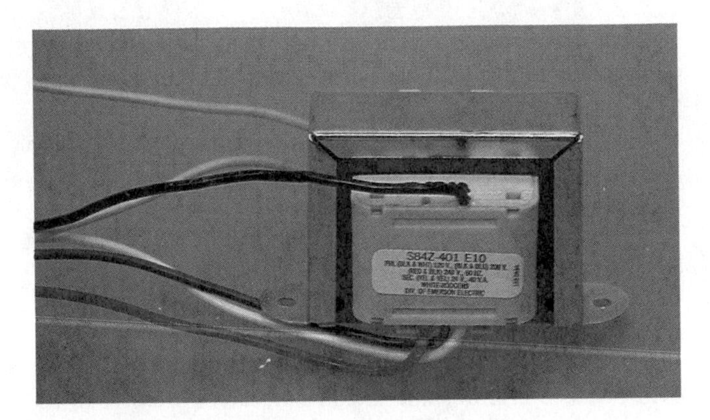

FIGURE 39–3 Low-voltage transformer. *Photo by Bill Johnson*

39.3 LOW-VOLTAGE CONTROLS

The low-voltage circuit operates the various power-consuming devices in the control system that start and stop the compressor and fan motors. These low-voltage devices do not use much current. The control transformer is usually a 40-VA transformer. This means that the highest amperage the control transformer will be able to produce on the secondary side is 1.7 A. This was arrived at by dividing the VA rating of 40 by the rated voltage (40 VA ÷ 24 V = 1.666 A rounded to 1.7). If the circuit current is greater than 1.666 A, the voltage will begin to drop and the transformer will get hot. In this situation a larger transformer is needed.

The contactor that actually starts and stops the compressor is considered one of the controls. It consumes power in its 24-V magnetic holding coil, Figure 39–4. The contacts of the contactor pass power when the magnetic coil is energized, Figure 39–5. The contact circuit to the compressor is 230 V.

39.4 SOME HISTORY OF RESIDENTIAL CENTRAL AIR CONDITIONING

Residential central air conditioning became popular in the late 1950s, and its popularity has continued to grow. This popularity has caused air conditioning equipment to become competitive in price. In the warm climates in the United

FIGURE 39–4 Contactor. *Photo by Bill Johnson*

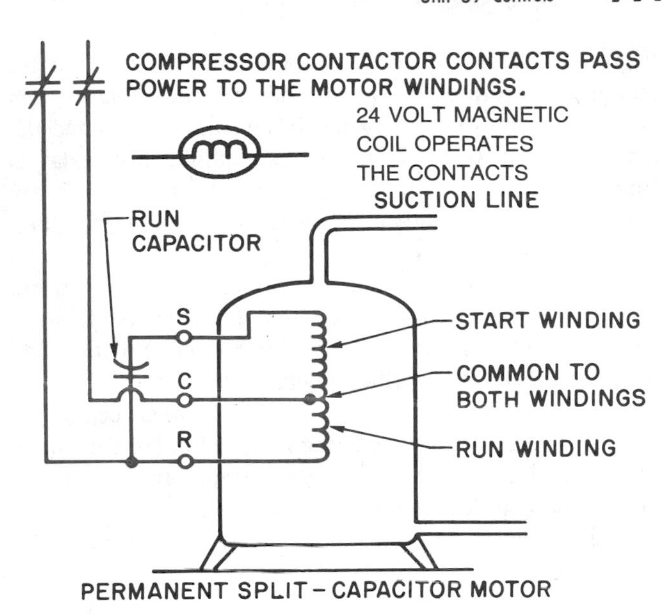

FIGURE 39–5 Diagram of circuit that passes power to the motor windings.

States, central air conditioning is desirable in all new construction because of the resale value of the structure, even if the owner or builder does not desire air conditioning. Lending institutions may even require that a new home have air conditioning before they will finance the loan. The reason is that the lender may have to resell the house, and central air conditioning is a strong selling feature, which may not be easy to add in the future.

The first central air conditioning systems installed in residences were small commercial systems applied to homes. They were all water cooled and very efficient and reliable. Air-cooled equipment became popular later. Air-cooled systems do not pump and handle water. Freezing water and mineral deposits are not problems.

39.5 ECONOMICS OF EQUIPMENT DESIGN

The first air-cooled systems were heavy duty, bulky, and hard to handle. Slow-speed hermetic compressors (1800 rpm) or belt-drive open compressors were used with R-12 as the refrigerant. The residential and light commercial equipment market began to grow to the point that price-conscious buyers began to look for less-expensive equipment. Speculating home builders began to build large developments and wanted to save on the air conditioning equipment. Manufacturers were thus forced to find more economical ways to build equipment. A more efficient refrigerant was also needed. When R-22 became available, the evaporator and the condenser tubing could be made smaller, making equipment more compact and lighter, which made storage and shipping much easier. In addition, the suction and liquid lines that connect the condensing unit and the evaporator are one size smaller and reduce the price of the installation. The physical size of the installation was scaled down and the manufacturing and installation costs were reduced.

The compressors were then manufactured to turn at a higher speed, so a small compressor could pump more refrigerant and do the job of a larger compressor. Present compressors turn at 3600 rpm and are much smaller than earlier models. Because of the faster compressors and the more efficient R-22 refrigerant, the equipment of today is smaller and more efficient than the earlier equipment. Every manufacturer constantly seeks to make equipment lighter, smaller, more efficient, and less expensive.

Air conditioning equipment manufactured to compete economically with more expensive equipment will have only the essential controls yet may be very reliable. The following sections describe in detail typical controls for an air-cooled system built in the mid-1960s when central air conditioning systems were being installed in large numbers for the first time. Many of these systems are still in use today. The detailed description is important because many of these controls are no longer used in recently manufactured equipment.

39.6 OPERATING CONTROLS FOR OLDER AIR-COOLED SYSTEMS

The room thermostat senses and controls the space temperature. The thermostat is not a power-consuming device, but it passes power to the power-consuming devices. It is sensitive to temperature changes because of the bimetal element under the cover. Early thermostats were larger than modern ones.

The fan relay starts and stops the indoor fan. The relay is a power-consuming device that receives power through the room thermostat to energize the magnetic coil and to close the contacts that start the fan.

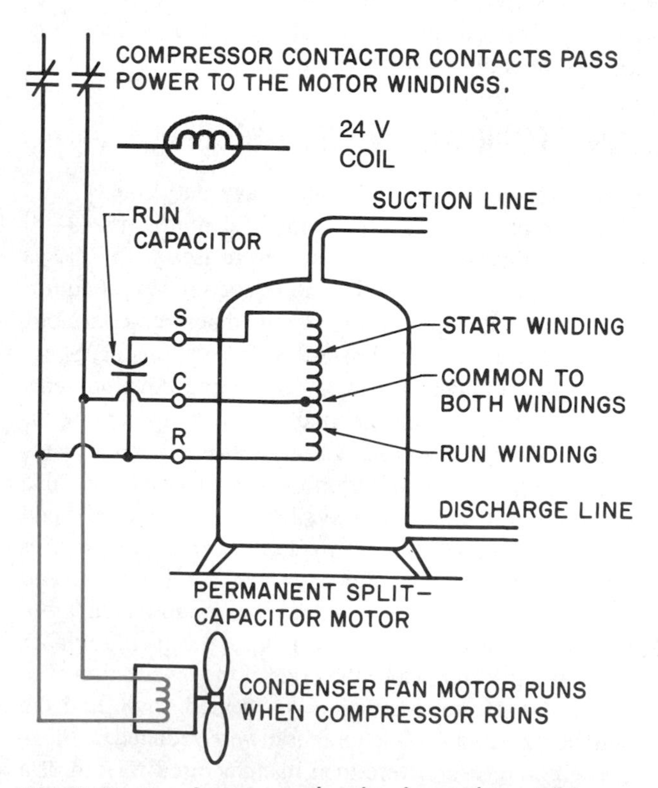

FIGURE 39–6 **Compressor and outdoor fan wired in parallel.**

The compressor contactor starts and stops the compressor and outdoor fan, which are normally wired in parallel, Figure 39–6. This control is a power-consuming device because of the magnetic coil. When the coil is energized, it closes the contacts and starts the compressor and the outdoor fan motor, Figure 39–7.

The compressor starting and running circuits are actually not controls in the strictest sense, but most service technicians treat them as such. The compressors that were used for many years had capacitor-start, capacitor-run motors. These motors have a high starting torque and were used because thermostatic expansion valves were used as the metering devices. They do not equalize the high-and low-side pressures during the off cycle, and a high-torque compressor motor is required. The following components were part of the starting system: (1) a potential starting relay, (2) a start capacitor, and (3) a run capacitor, Figure 39–8. See Figure 39–9 for an example of the starting circuit diagram.

FIGURE 39–7 **Typical compressor contactor.** ***Photo by Bill Johnson***

FIGURE 39–8 **Components to start and run a capacitor-start, capacitor-run motor.** ***Photo by Bill Johnson***

39.7 SAFETY CONTROLS FOR OLDER AIR-COOLED SYSTEMS

The high-pressure control stops the compressor when a high-pressure condition exists, for example, if the condenser

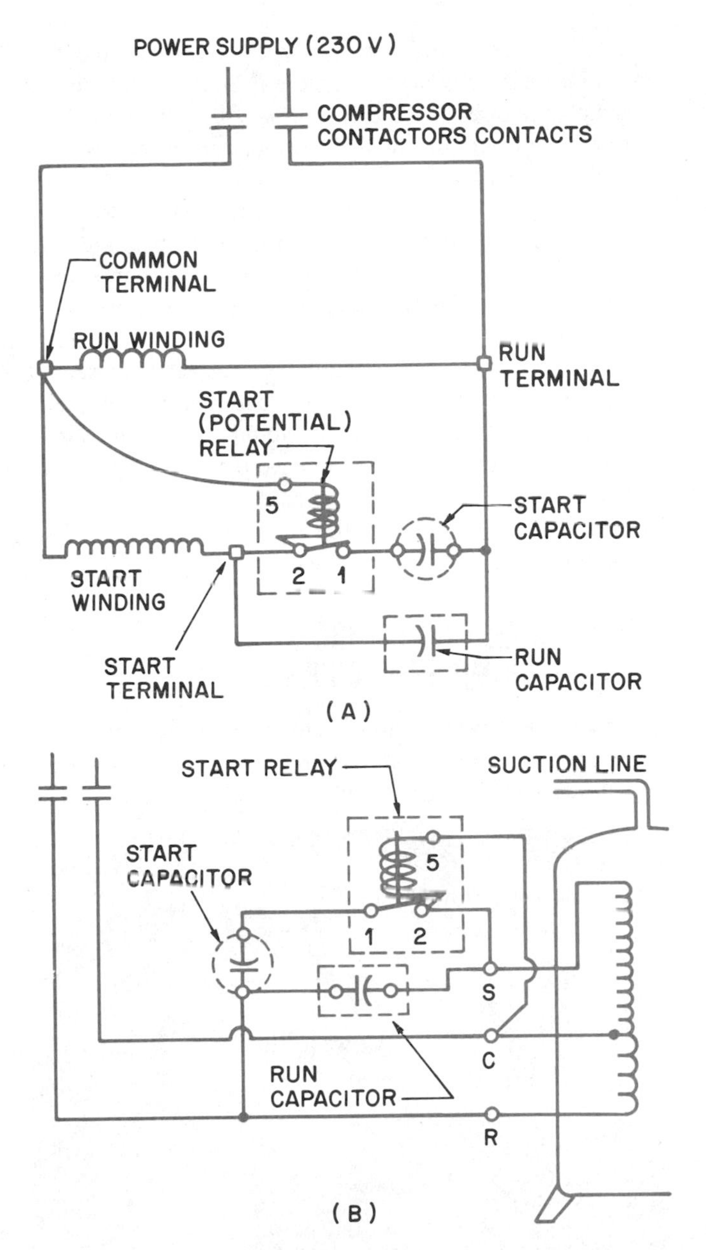

FIGURE 39–9 Starting circuit for a capacitor-start, capacitor-run motor. (A) Schematic. (B) Pictorial.

fan stops when it should be running. The compressor has no way of knowing that the fan has stopped, and it keeps pumping refrigerant into the condenser. ***The pressures can quickly rise to a dangerous level, so the compressor must be stopped. Figure 39–10 illustrates a commercial high-pressure control.***

For protection, the low-pressure control stops the compressor from pulling the suction pressure below a predetermined point. If the system loses all or part of its refrigerant charge, the low side of the system may pump into a vacuum without a low-pressure control. The low-pressure control can be set to shut off the compressor before a vacuum is reached. Figure 39–11 is a photo of a commercial low-pressure control. If the leak is on the low side of the system and the compressor pumps into a vacuum, air will be drawn into the system. Then there would be two problems: a leak and air in the system. The low-pressure control also pro-

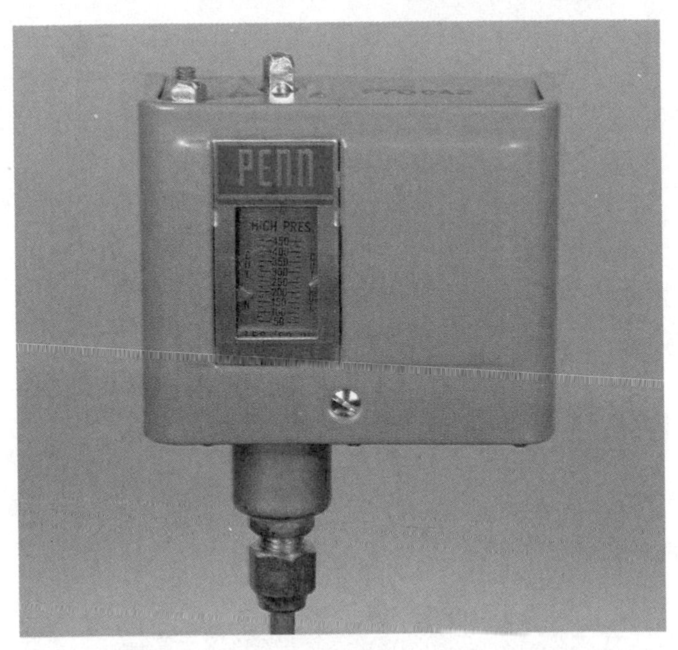

FIGURE 39–10 This commercial high-pressure control is much larger than currently used residential high-pressure controls. *Photo by Bill Johnson*

FIGURE 39–11 Type of low-pressure control used on commercial equipment and on early residential equipment. *Photo by Bill Johnson*

vides freeze protection. When the air across the indoor coil is reduced, the suction pressure will be reduced to the point that a freeze condition may result. The condensate on the evaporator coil may turn to ice and block the airflow even more. The evaporator coil may turn to a solid block of ice. Reduced airflow may also be caused from dirty filters, closed air registers, a blocked return air grille, or a dirty evaporator coil.

The early systems had overload protection for the common and the run circuit of the compressor. This overload protection was usually a heat-sensitive bimetal device with a current rating. If the motor current went above the rating on the overload device, the bimetal strip was heated and opened the circuit to the compressor contactor coil to stop the compressor before damage occurred, Figure 39–12. This overload protection was typically located in the motor terminal box outside the refrigerant atmosphere.

Internal motor protection senses the actual motor temperature and stops the motor when it is too hot. Compressor internal protection may be the type that breaks the line power inside the compressor, or it may be the pilot type that interrupts the circuit to the contactor coil. This motor protection is located inside the compressor inside its own sealed container.

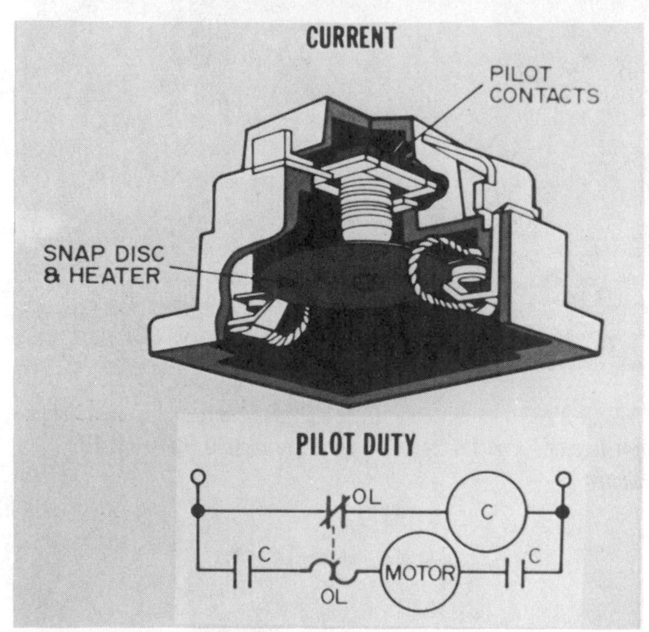

FIGURE 39–12 This overload device is a heat-sensitive bimetal strip. *Reproduced courtesy of Carrier Corporation*

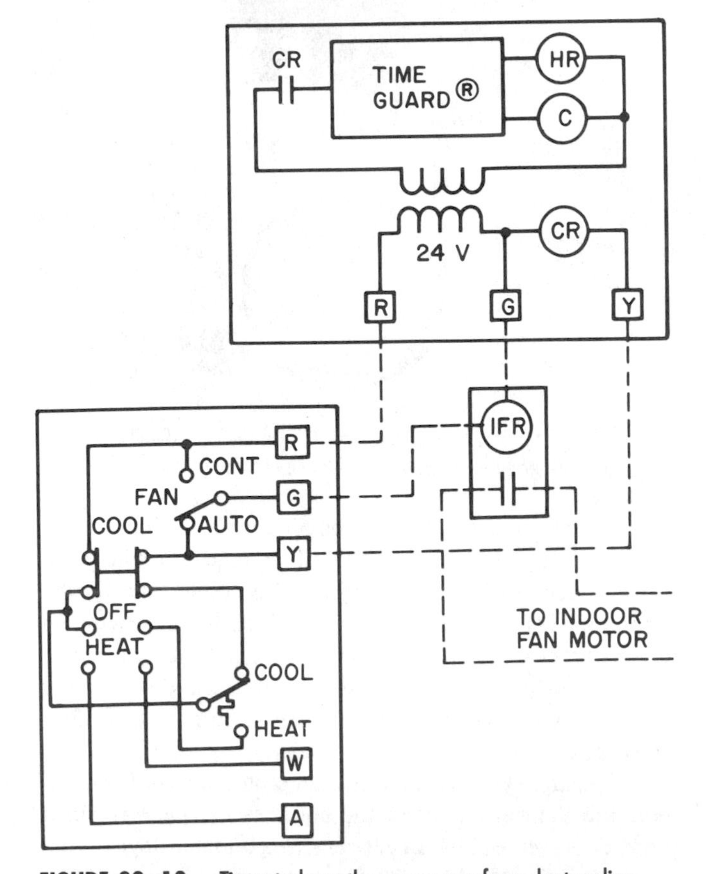

FIGURE 39–13 Timer to keep the compressor from short cycling.

Short-cycle protection was used to prevent the compressor from short cycling when a safety or operating control would open the circuit and then make back in a short cycle. The overload is an example. Some types of overloads are current sensitive. These controls will open the circuit when the current is too high but will make back the instant it is reduced. If a compressor had a bad starting relay and stopped due to high current, it would try to start again immediately after stopping if there were no short-cycle control, Figure 39–13. This same control protects the compressor from short cycling if the homeowner turns the thermostat up and down several times while adjusting it.

39.8 OPERATING CONTROLS FOR MODERN EQUIPMENT

The room thermostat is now much smaller, Figure 39–14. Some room thermostats are electronic and use a thermistor as the sensing element, Figure 39–15. These thermostats

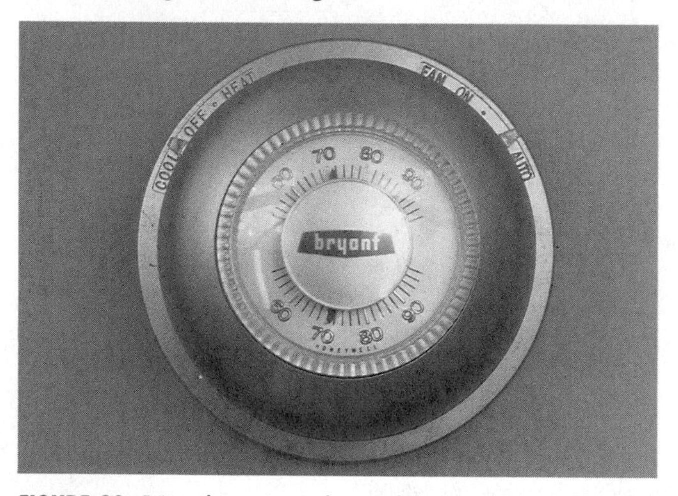

FIGURE 39–14 This is a typical room thermostat. It is a combination heating and cooling thermostat with a bimetal element for sensing room temperature. *Photo by Bill Johnson*

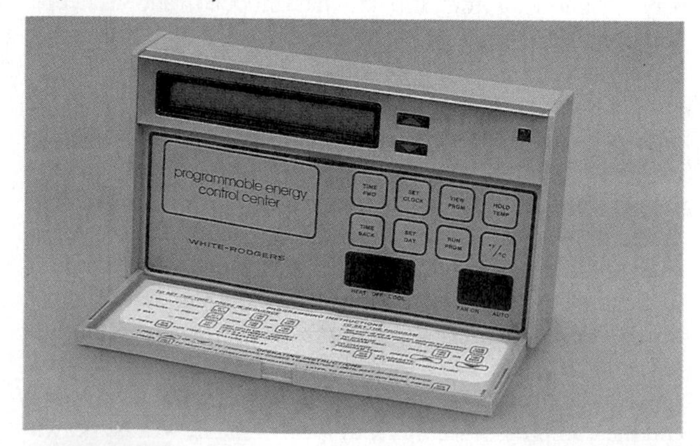

FIGURE 39–15 Electronic thermostat with a thermistor for the sensing element. *Photo by Bill Johnson*

may have programs for night and daytime work schedule adjustments for both the heating and the cooling settings. They can be thought of as power-consuming because they must have a circuit with a timer. They also pass power to various components (the fan relay and compressor contactor). The circuit that passes power is the contact circuit.

The fan relay for the modern equipment may be smaller than the older versions, but it performs the same function, Figure 39–16.

The modern compressor contactor has one big difference on some manufacturers' equipment. Some contactors have only one set of contacts, Figure 39–17. Older contactors always had two sets of contacts that interrupted the power to the common and run circuits to the compressor, both line 1 and line 2. The modern contactor may only interrupt the power to one circuit. *Note:* The circuit with continuous power may feed power through a run capacitor to provide a small amount of current through the compressor windings to keep the compressor warm during the off cycle. This is a substitute for crankcase heat. If a contactor with two contacts is substituted, there will not be heat for the crankcase.

The starting circuit does not have as many components. If a retired service technician suddenly opened the panel of a modern piece of equipment, the conclusion would quickly be drawn that the equipment could not work correctly. It would seem unprotected because there are not as many components as before. The components used to start and run modern equipment are a run capacitor and possibly a start-assist unit such as a positive temperature coefficient (PTC) device. The PTC device has no moving parts as a start relay does. It makes a circuit from the start terminal to the run terminal for a short time to get the compressor started. These devices are used on permanent split-capacitor (PSC) motors as a start assist. The PSC motor has little starting torque and is used with metering devices that equalize during the off cycle, Figure 39–18.

FIGURE 39–16 Fan relay. *Photo by Bill Johnson*

FIGURE 39–17 ompressor contactor with one set of contacts. *Photo by Bill Johnson*

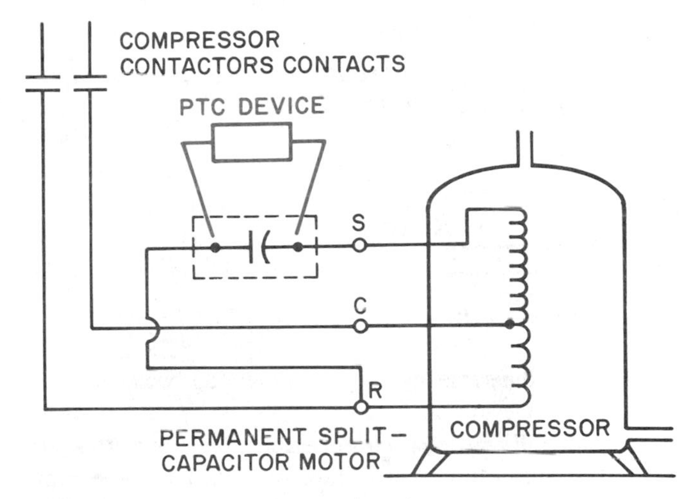

FIGURE 39–18 Wiring diagram shows how a PSC motor is wired when a PTC device is used as a start assist for the compressor.

39.9 SAFETY CONTROLS FOR MODERN EQUIPMENT

Modern equipment may not have any visible safety controls. It may have only a motor-temperature control that senses the temperature of the motor windings. This control may be mounted in the windings and have no external terminals. A note on the compressor may read "this compressor is internally protected with a winding thermostat," Figure 39–19. The intent of the manufacturer is to let this control function cut off the compressor during the following situations:

1. When a system is operating with a low refrigerant charge, the motor windings will overheat and the motor-winding thermostat will stop the compressor. The suction gas is supposed to cool the compressor windings. It is functioning as low-charge protection. When the motor gets warm or hot, the heat in the mass of the motor prevents the control from closing its contacts until the motor cools. This is a built-in short-cycle protection.
2. When the condenser is dirty, the compressor head pressure rises. This makes the current increase and the motor gets hot and cuts off because of the motor temperature cutout. The motor-winding thermostat is functioning as a high-pressure control in this situation. It

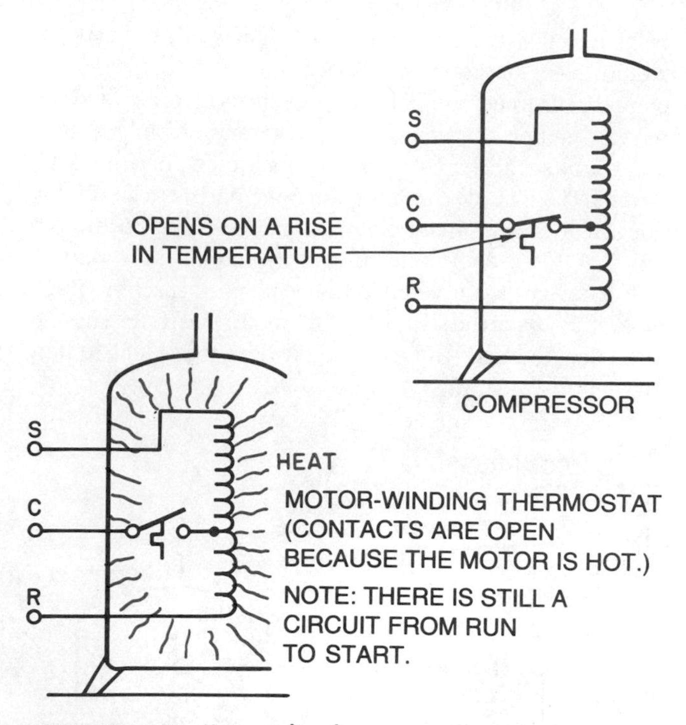
S
C
R
OPENS ON A RISE IN TEMPERATURE
COMPRESSOR
HEAT
MOTOR-WINDING THERMOSTAT (CONTACTS ARE OPEN BECAUSE THE MOTOR IS HOT.)
NOTE: THERE IS STILL A CIRCUIT FROM RUN TO START.

FIGURE 39–19 Motor-winding thermostat.

still has the same short-cycle protection because of the mass of the motor.

3. ***Internally in most compressors there is a pressure relief valve that will relieve hot gas from the compressor discharge into the motor-winding thermostat if a great pressure differential is experienced, Figure 39–20.*** Suppose the condenser fan motor burned and stopped. High head pressure would occur immediately. When these pressures exceed the relief valve setting, the valve relieves and blows hot gas on the winding thermostat to cut the compressor off. Internal relief valve settings may have a differential of 450 psig. This means the head pressure has to be 450 psig higher than the suction pressure for the relief valve to open. If a condenser fan stopped, the head pressure could go to 540 psig, and the suction pressure could go to 90 psig before the internal relief valve opened (540 − 90 = 450 psig). ***A reading of the high-pressure gage is not enough to determine that the relief valve should function.*** The winding thermostat in this application serves as the high-pressure cutout when fast action is needed.
4. The run capacitor may become defective, and the compressor will pull too much current trying to start. The motor will heat up and cut out on the winding thermostat. The winding thermostat is functioning as an overload in this situation.
5. The homeowner or children may fool with the space temperature thermostat and turn it up and down several times in quick succession. The compressor will try to start each time, and the motor will get hot enough to cause the motor-winding thermostat to cut it off. The motor-winding thermostat is preventing the compressor from short cycling in this situation.

When an overcharge of refrigerant is great enough to cause the compressor windings to be cool, the winding thermostat may not give adequate protection, Figure 39–21. In this situation the only high-pressure protection the compressor has is the internal relief valve.

Some modern systems have a loss-of-charge protection in the form of a low-pressure cut-out control located on the liquid line. It may have a setting as low as 5 psig and will stop the system only if there is complete loss of charge,

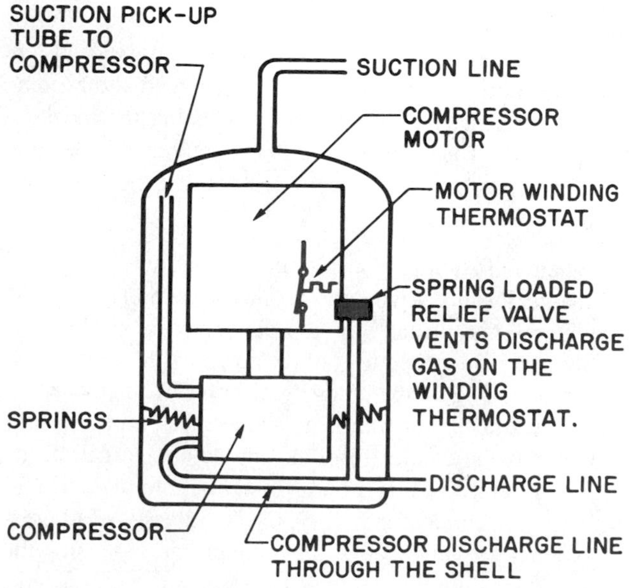

SUCTION PICK-UP TUBE TO COMPRESSOR
SUCTION LINE
COMPRESSOR MOTOR
MOTOR WINDING THERMOSTAT
SPRING LOADED RELIEF VALVE VENTS DISCHARGE GAS ON THE WINDING THERMOSTAT.
SPRINGS
DISCHARGE LINE
COMPRESSOR
COMPRESSOR DISCHARGE LINE THROUGH THE SHELL

FIGURE 39–20 The internal relief valve in a welded hermetic compressor.

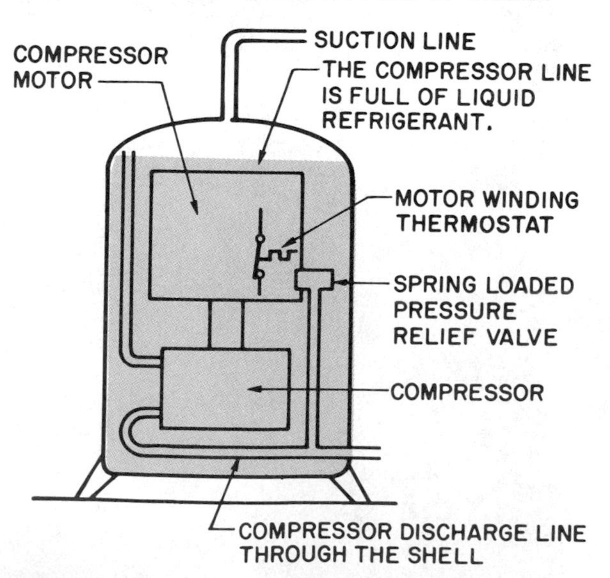
THE MOTOR CAN BE OPERATING UNDER TREMENDOUS LOADS AND REMAIN COOL WHEN THE COMPRESSOR IS FULL OF LIQUID.
COMPRESSOR MOTOR
SUCTION LINE
THE COMPRESSOR LINE IS FULL OF LIQUID REFRIGERANT.
MOTOR WINDING THERMOSTAT
SPRING LOADED PRESSURE RELIEF VALVE
COMPRESSOR
COMPRESSOR DISCHARGE LINE THROUGH THE SHELL

FIGURE 39–21 This is an example of one time when the relief valve and motor temperature combination may not give high-pressure protection to stop the compressor.

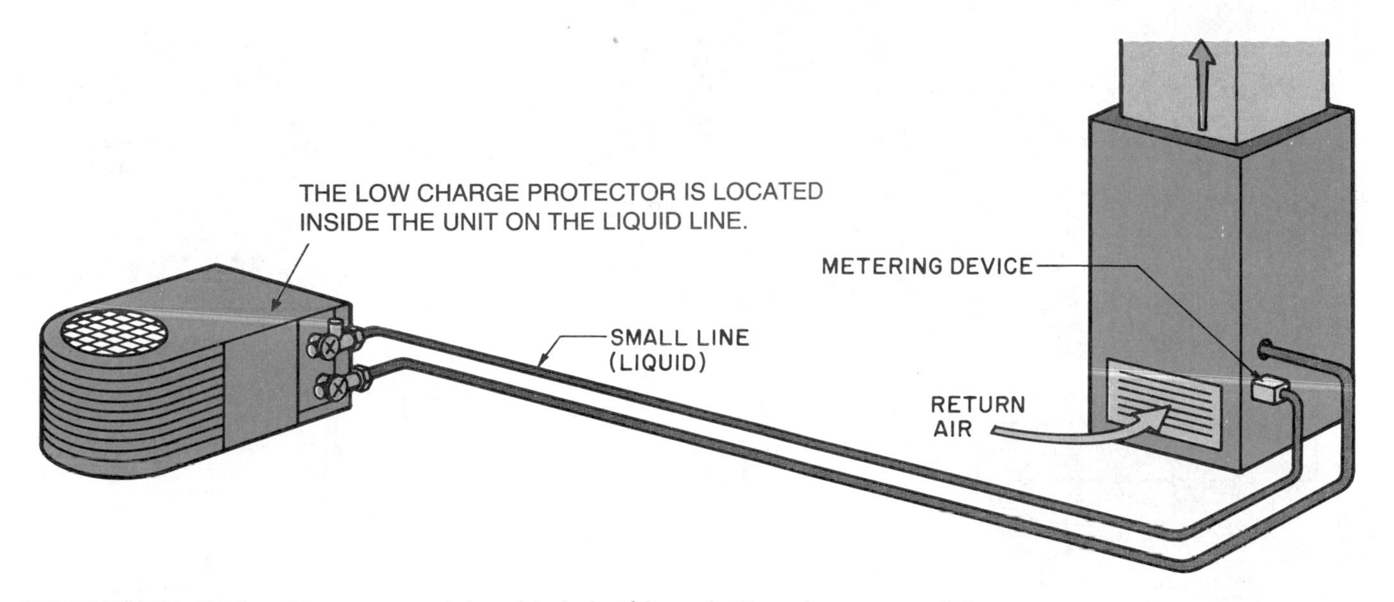
THE LOW CHARGE PROTECTOR IS LOCATED
INSIDE THE UNIT ON THE LIQUID LINE.
METERING DEVICE
SMALL LINE
(LIQUID)
RETURN
AIR

FIGURE 39–22 This loss-of-charge protector is located in the liquid line and will stop the compressor if the pressure goes as low as 5 psig.

Figure 39–22. This may not seem like much protection, but it will shut the compressor off when a complete loss of charge occurs. The compressor will not try to restart until the charge is put back into the system. One thing to keep in mind for this type of system is when a low-side leak occurs, air will enter the system when the low side goes into a vacuum because this control is on the high-pressure side of the system.

39.10 THE WORKING CONTROL PACKAGE

All controls must be assembled into a working assembly. Manufacturers must be competitive, so they are always trying to provide a simple, effective control package that will protect the equipment and give it a long life span. Figure 39–23 illustrates some typical diagrams that apply to simple residential equipment.

Figure 39–24 summarizes the controls used on earlier equipment and newer equipment. The technician should be familiar with both because some manufacturers still use the older types of controls and older style controls may still be in use on older equipment.

In some modern high-efficiency equipment, part of the efficiency package is the method of controlling the equipment. Solid-state circuit boards are used by some manufacturers for some of the listed functions. For example, the short-cycle timer can easily be built into the electronic circuit, Figure 39–25.

39.11 ELECTRONIC CONTROLS AND AIR CONDITIONING EQUIPMENT

An electronic circuit board can monitor high voltage, whereas electromechanical controls cannot easily do this. Power companies may supply voltage that is higher than the equipment specification. For example, when a system is rated as 230 V, the equipment may be operated at ±10% of this voltage. If the voltage is too low, the motor will pull more than normal current and the overload device (or winding thermostat) should stop the motor. If the motor operates at a slightly high-voltage condition, the overload devices will not help because the motor current will be below normal. The electronic circuit board may have a voltage monitor that will stop the compressor at a high voltage even if the current is low. The electronic circuit board can also react quickly to low voltage (known as "brownout"). It can monitor and cut off the compressor before an overload has time to heat up and react.

Each manufacturer that uses a circuit board has a recommended checkout procedure. For practical troubleshooting, remember that a circuit board usually looks like a single control on the wiring diagram. The control circuit goes into and comes out of the board. However, sometimes the circuits on the board may be checked by using a jumper wire from one circuit to another to determine whether the board is defective.

SUMMARY

- The control of air conditioning equipment consists of controlling the various components to start and stop them in the correct sequence to maintain space temperature for comfort.
- The three main components controlled are the indoor fan, the compressor, and the outdoor fan.
- The control circuit uses low voltage (24 V) for safety.
- Low voltage is obtained from high voltage by a transformer located in the condensing unit or in the air handler.
- Some electronic thermostats use a thermistor as the sensing device and may incorporate other features, such as temperature set-back, for night and day.
- Air-cooled equipment is used more widely today than water-cooled equipment.

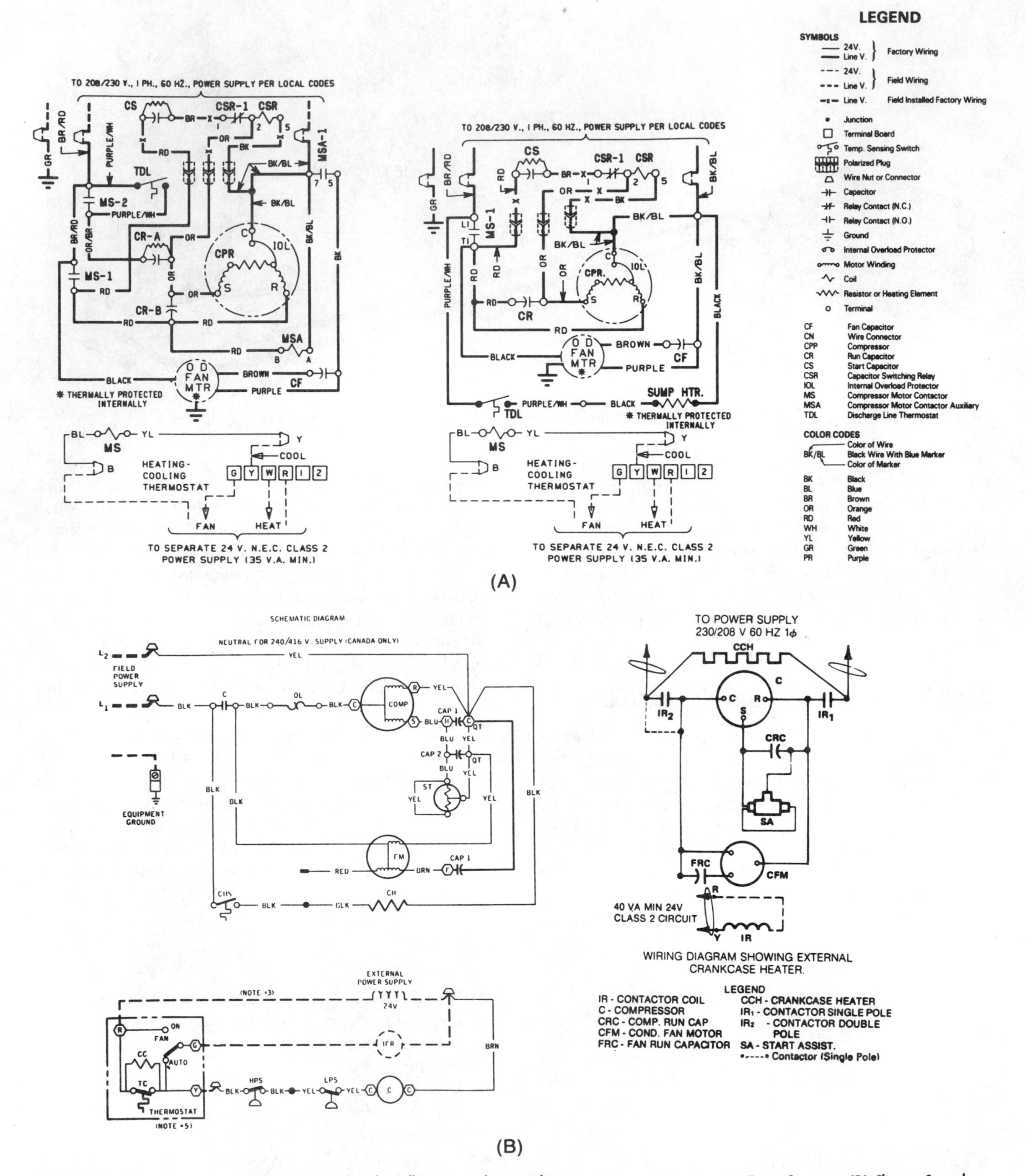

FIGURE 39–23 Wiring diagrams for residential and small commercial air conditioning equipment. *Courtesy (A) Trane Company, (B) Climate Control*

OLDER EQUIPMENT	NEWER EQUIPMENT
Room thermostat	Room thermostat
Fan relay	Fan relay
Compressor contactor	Compressor contactor
Winding thermostat (maybe)	Winding thermostat
Run capacitor	Run capacitor
Internal relief valve (maybe)	Internal relief valve
Low-pressure control	No charge protection
High-pressure control	Winding thermostat
Short-cycle protection	Winding thermostat
Overloads (usually two)	Winding thermostat
Crankcase heater	Through the capacitor

FIGURE 39–24 Controls used on older and newer equipment.

COMPLETE CIRCUIT BOARD

FIGURE 39–25 Electronic circuit board with voltage monitors built in. Also included is a timer circuit that is not evident. *Photo by Bill Johnson*

- Controls are either operating controls or safety controls.
- Some controls, such as contactors and electronic thermostats have more than one circuit. Some circuits may pass power and some may consume power.
- The compressor overload protectors are used to keep the compressor from drawing too much current and overworking.
- The motor winding thermostat shuts off the compressor when the compressor motor winding gets too hot.
- The compressor internal relief valve vents high-pressure gas from the high-pressure side to the low-pressure side of the compressor internally. It can be used as a high-pressure indicator by directing this high-pressure, high-temperature gas onto the winding thermostat.
- Modern equipment may have only a few controls and components. It is common for the unit to have a single-pole contactor and a run capacitor as the only visible components.
- Electronic circuit boards are furnished with some equipment. The circuit boards may include short-cycle protection, as well as low- and high-voltage protection.

REVIEW QUESTIONS

1. Name two types of controls.
2. Some controls ____________ power, and some controls ____________ power.
3. What type of control is the room thermostat?

What protection do the controls in Questions 4–8 offer?

4. High-pressure control
5. Low-pressure control
6. Winding thermostat
7. Internal relief valve
8. Electronic circuit board
9. Why can't you substitute a two-pole contactor for a single-pole contactor on some units?
10. What is the standard low-voltage control voltage for residential air conditioning?
11. What changes have manufacturers made in residential air conditioning equipment?
12. What component starts the compressor in a residential air conditioner?
13. What starts the indoor fan?
14. What starts the outdoor fan?
15. What stops the compressor in a low-charge condition if there is no low-pressure control?

UNIT 40

Typical Operating Conditions

OBJECTIVES

After studying this unit, you should be able to

- explain what conditions will vary the evaporator pressures and temperatures.
- define how the various conditions in the evaporator and ambient air affect condenser performance.
- state the relationship of the evaporator to the rest of the system.
- describe the relationship of the condenser to the total system performance.
- compare high-efficiency equipment and standard-efficiency equipment.
- establish reference points when working on unfamiliar equipment to know what the typical conditions should be.
- describe how humidity affects equipment suction and discharge pressure.
- explain three methods that manufacturers use to make air conditioning equipment more efficient.

SAFETY CHECKLIST

- The technician must use caution while observing equipment to obtain operating conditions. Electrical, pressure, and temperature readings must be taken while the technician observes the equipment, and many times the readings must be taken while the equipment is in operation.
- Wear goggles and gloves when attaching gages to a system.
- Observe all electrical safety precautions. Be careful at all times and use common sense.

Air conditioning technicians must be able to evaluate both mechanical and electrical systems. The mechanical operating conditions are determined or evaluated with gages and thermometers; electrical conditions are determined with electrical instruments.

40.1 MECHANICAL OPERATING CONDITIONS

Air conditioning equipment is designed to operate at its rated capacity and efficiency at one set of design conditions. This design condition is generally considered to be at an outside temperature of 95°F and at an inside temperature of 80°F with a humidity of 50%. This rating is established by the Air Conditioning and Refrigeration Institute (ARI). Equipment must have a rating as a standard from which to work. Equipment is also rated so that the buyer will have a common basis with which to compare one piece of equipment with another. All equipment in the ARI directory is rated under the same conditions. When an estimator or buyer finds that a piece of equipment is rated at 3 tons, it will perform at a 3-ton capacity or 36,000 Btu/h under the stated conditions. When the conditions are different, the equipment will perform differently. For example, most homeowners will not be comfortable at 80°F and a relative humidity of 50%. They will normally operate their system at about 75°F, and the relative humidity in the conditioned space will be close to 50%. The equipment will not have quite the capacity at 75°F that it had at 80°F. If the designer wants the system to have a capacity of 3 tons at 75°F and 50% relative humidity, the manufacturer's literature may be consulted to make the change in equipment choice. **This 75°F 50% humidity condition will be used as the design condition for this unit because it is a common operating condition of the equipment in the field.** ARI rates condensers with 95°F air passing over them. A new standard-efficiency condenser will condense refrigerant at about 125°F with 95°F air passing over it. As the condenser ages, dirt accumulates on the outdoor coil and the efficiency decreases. The refrigerant will then condense at a higher temperature. This higher temperature can easily approach 130°F. This is the value often found in the field. The text will use 125°F condensing temperature, assuming that equipment is properly maintained in the field. It is not easy to see the change in load conditions on your gages and instruments used in the field. An increase in humidity is not followed with a proportional rise in suction pressure and amperage. Perhaps the most noticeable aspect of an increase in humidity is an increase in the condensate accumulated in the condensate drain system.

40.2 RELATIVE HUMIDITY AND THE LOAD

The inside relative humidity adds a significant load to the evaporator coil and has to be considered as part of the load. When conditions vary from the design conditions, the equipment will vary in capacity. The pressures and temperatures will also change.

40.3 SYSTEM COMPONENT RELATIONSHIPS UNDER LOAD CHANGES

If the outside temperature increases from 95°F to 100°F, the equipment will be operating at a higher head pressure and will not have as much capacity. The capacity also varies when the space temperature goes up or down or when the humidity varies. There is a relationship among the various components in the system. The evaporator absorbs heat.

When anything happens to increase the amount of heat that is absorbed into the system, the system pressures will rise. The condenser rejects heat. If anything happens to prevent the condenser from rejecting heat from the system, the system pressures will rise. The compressor pumps heat-laden vapor. Vapor at different pressure levels and saturation points (in reference to the amount of superheat) will hold different amounts of heat.

40.4 EVAPORATOR OPERATING CONDITIONS

The evaporator will normally operate at a 40°F boiling temperature when operating at the 75°F 50% humidity condition. This will cause the suction pressure to be 70 psig for R-22. (The actual pressure corresponding to 40°F is 68.5 psig; for our purposes we will round this off to 70 psig.) This example is at design conditions and a steady-state load. At this condition the evaporator is boiling the refrigerant exactly as fast as the expansion device is metering it into the evaporator. As an example, suppose the evaporator has a return air temperature of 75°F and the air has a relative humidity of 50%. The liquid refrigerant goes nearly to the end of the coil, and the coil has a superheat of 10°F. This coil is operating as intended at this typical condition, Figure 40–1.

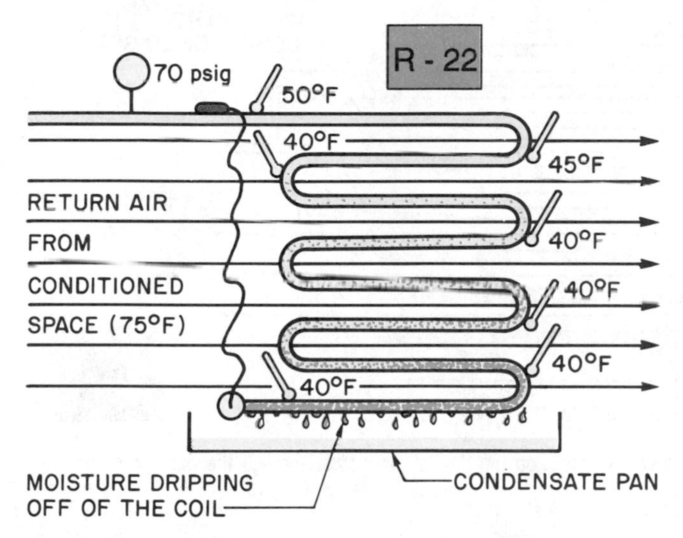

FIGURE 40–1 Evaporator operating at typical conditions. The refrigerant is boiling at 40°F in the coil. This corresponds to 68.5 psig typically rounded off to 70 psig for R-22.

Late in the day, after the sun has been shining on the house, the heat load inside becomes greater. A new condition is going to be established. The example evaporator in Figure 40–2 has a fixed-bore metering device that will only feed a certain amount of liquid refrigerant. The space temperature in the house has climbed to 77°F and is causing the liquid refrigerant in the evaporator to boil faster, Figure 40–3. This causes the suction pressure and the superheat to go up slightly. The new suction pressure is 73 psig, and the new superheat is 13°F. This is well within the range of typical operating conditions for an evaporator. The system actually has a little more capacity at this point if the outside temperature is not too much above design. If the head pressure goes up because the outside temperature is 100°F, for example, the suction pressure will even go higher than 73 psig because head pressure will influence it, Figure 40–4.

Many different conditions will affect the operating pressures and air temperatures to the conditioned space. There can actually be as many different pressures as there are different inside and outside temperature and humidity variations. This can be confusing, particularly to the new service technician. However, the service technician can use some common conditions in troubleshooting. This is necessary because there are very few times when the technician has a chance to work on a piece of equipment when the conditions are perfect. Most of the time when the technician is assigned the job, the system has been off for some time or not operating correctly for long enough that the conditioned space temperature and humidity are higher than normal, Figure 40–5. After all, a rise in space temperature is what often prompts a customer to call for service.

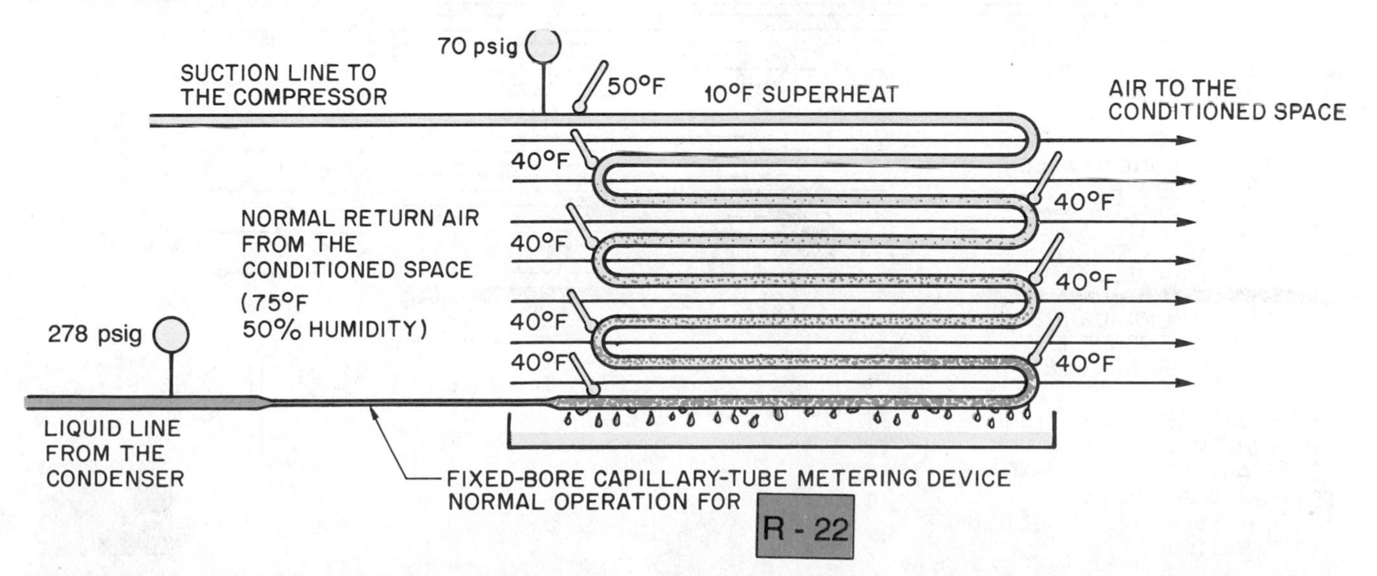

FIGURE 40–2 This fixed-bore metering device will not vary the amount of refrigerant feeding the evaporator as much as the thermostatic expansion valve. It is operating at an efficient state of 10°F superheat.

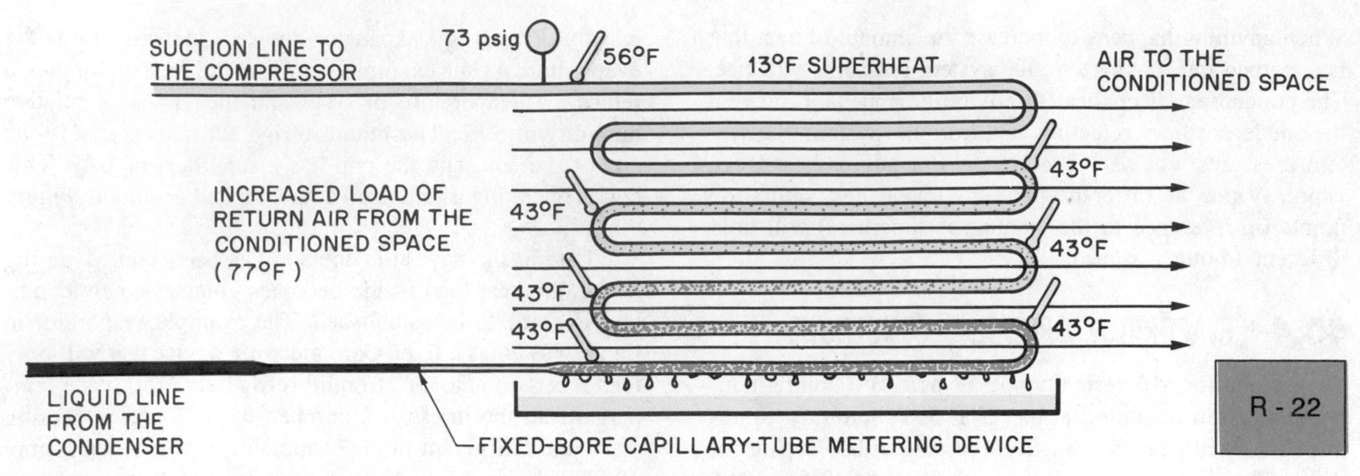

FIGURE 40–3 This is a fixed-bore metering device when an increase in load has caused the suction pressure to rise.

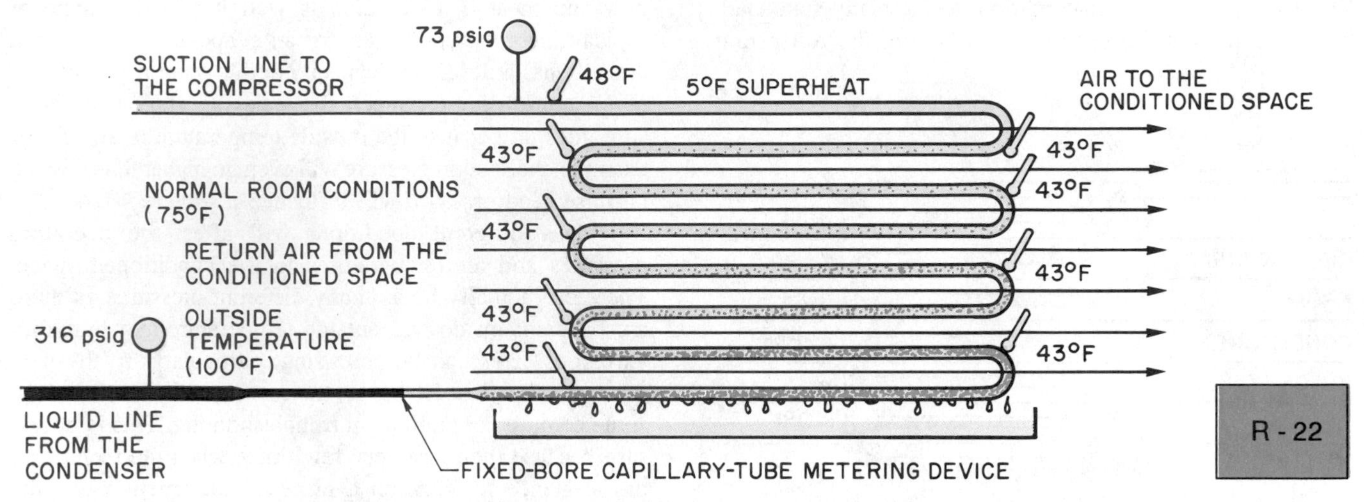

FIGURE 40–4 System with a high head pressure. The increase in head pressure has increased the flow of refrigerant through the fixed-bore metering device and the superheat is decreased.

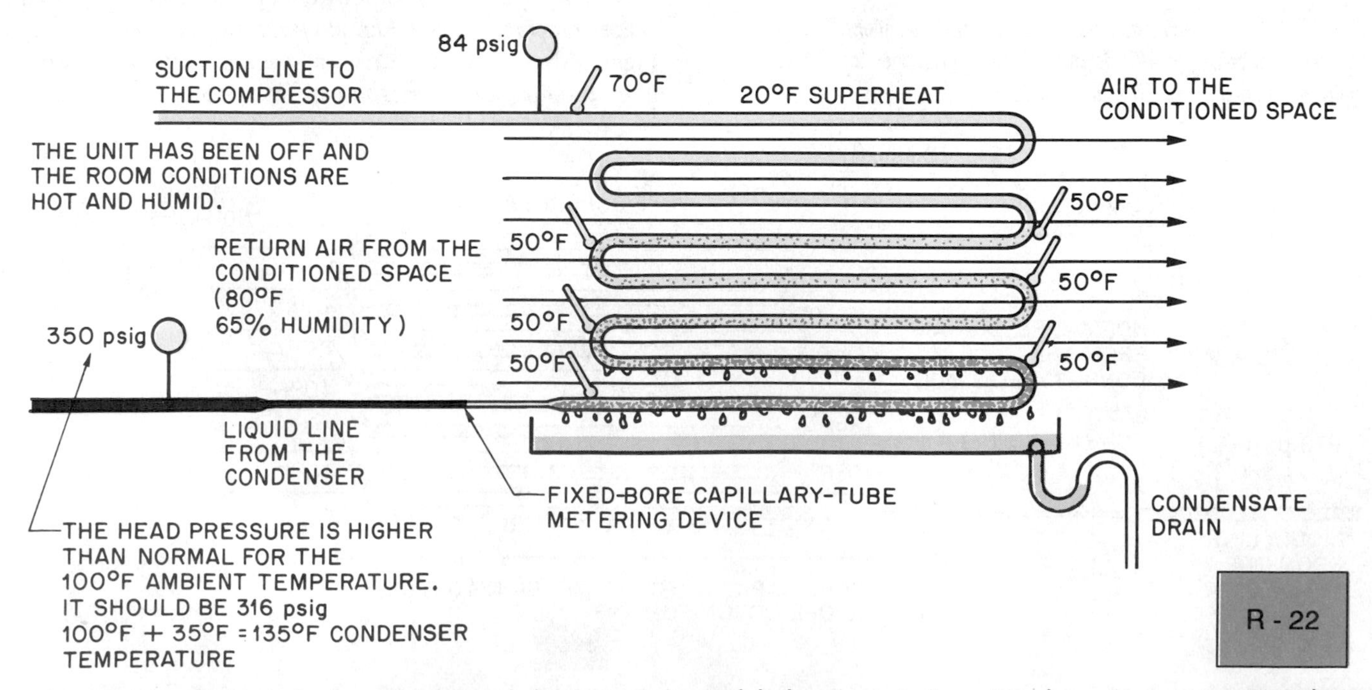

FIGURE 40–5 This system has been off long enough that the temperature and the humidity inside the conditioned space have gone up. Notice the excess moisture forming on the coil and going down the drain.

40.5 HIGH EVAPORATOR LOAD AND A COOL CONDENSER

High temperature in the conditioned space is not the only thing that will cause the system to have different pressures and capacity. The reverse can happen if the inside temperature is warm and the outside temperature is cooler than normal, Figure 40–6. For example, before going to work, a couple may turn the air conditioner off to save electricity. They may not get home until after dark to turn the air conditioner back on. By this time it may be 75°F outside but still be 85°F inside the structure. The air passing over the condenser is now cooler than the air passing over the evaporator. The evaporator may also have a large humidity load. The condenser becomes so efficient that it will hold some of the charge in the condenser because the condenser starts to condense refrigerant in the first part of the condenser. Thus more of the refrigerant charge is in the condenser tubes, and this will slightly starve the evaporator. The system may not have enough capacity to cool the home for several hours. The condenser may hold back enough refrigerant to cause the evaporator to operate below freezing and freeze up before it can cool the house and satisfy the thermostat. This condition has been improved by some manufacturers by means of two-speed fan operation for the condenser fan. The fan is controlled by a single-pole–double-throw thermostat that operates the fan at high speed when the outdoor temperature is high and low speed in mild temperatures. Typically, the fan will run on high speed when the outdoor temperature is above 85°F and at low speed below 85°F. When the fan is operating at low speed in mild weather, the head pressure is increased, reducing the chances of the evaporator condensate freezing.

40.6 GRADES OF EQUIPMENT

Manufacturers have been constantly working on the design of air conditioning equipment to make it more efficient. There are normally three grades of equipment: economy grade, standard-efficiency grade, and high-efficiency grade. Some companies manufacture all three grades and offer them to the supplier. Some manufacturers only offer one grade and may take offense at someone calling their equipment the lower grade. Economy grade and standard-efficiency grade are about equal in efficiency, but the materials they are made of and their appearances are different. High-efficiency equipment may be much more efficient and will not have the same operating characteristics, Figure 40–7. A condenser will normally condense the refrigerant at a temperature of about 30°F to 35°F higher than the ambient temperature. For example, when the outside temperature is 95°F, the average condenser will condense the refrigerant at 125°F to 130°F, and the head pressure for these condensing temperatures, would be 278 to 297 psig for R-22. High-efficiency air conditioning equipment may have a much lower operating head pressure. The high efficiency is gained by using a larger condenser surface and the same compressor or a more efficient compressor. The condensing temperature may be as low as 20°F greater than the ambient temperature. This would bring the head pressure down to a temperature corresponding to 115°F or 243 psig. The compressor will not use as much power at this condition, Figure 40–8. When large condensers are used, the head pressure is reduced. Lower power requirements are a result. More efficient condensers require some means of head pressure control, such as that mentioned above.

40.7 DOCUMENTATION WITH THE UNIT

The technician needs to know what the typical operating pressures should be at different conditions. Some manufacturers furnish a chart with the unit to tell what the suction and discharge pressures should be at different conditions, Figures 40–9 and 40–10. Some publish a bulletin that lists all of their equipment along with the typical operating pressures and temperatures. Others furnish this information with

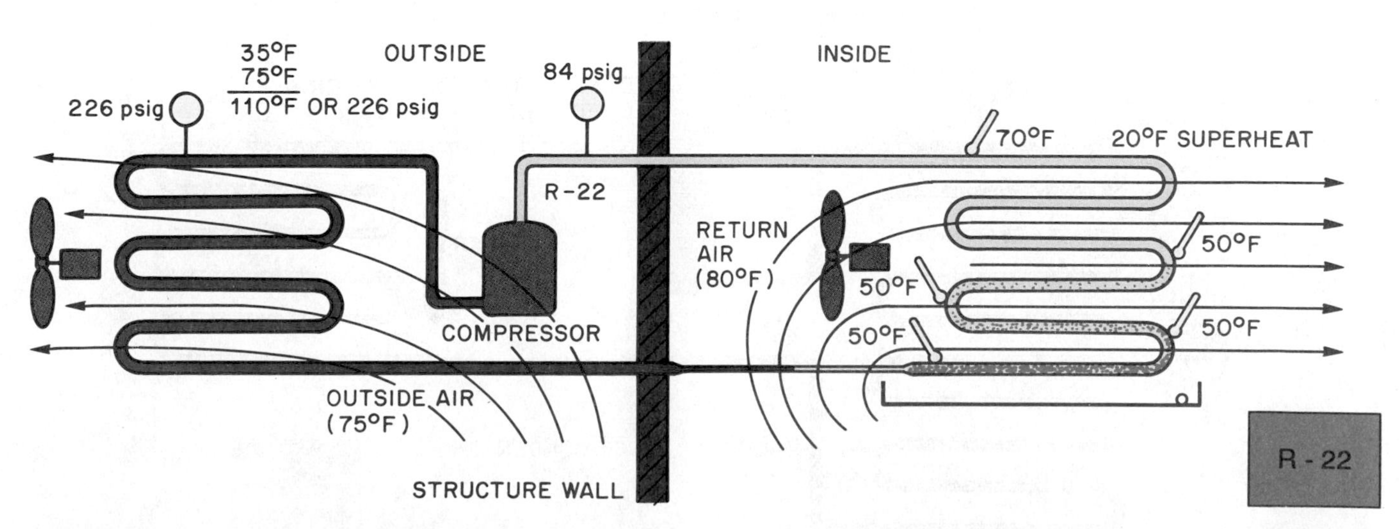

FIGURE 40–6 System operating with the outside ambient air cooler than the inside space temperature air. As the return air cools down, the suction pressure and boiling temperature will decrease.

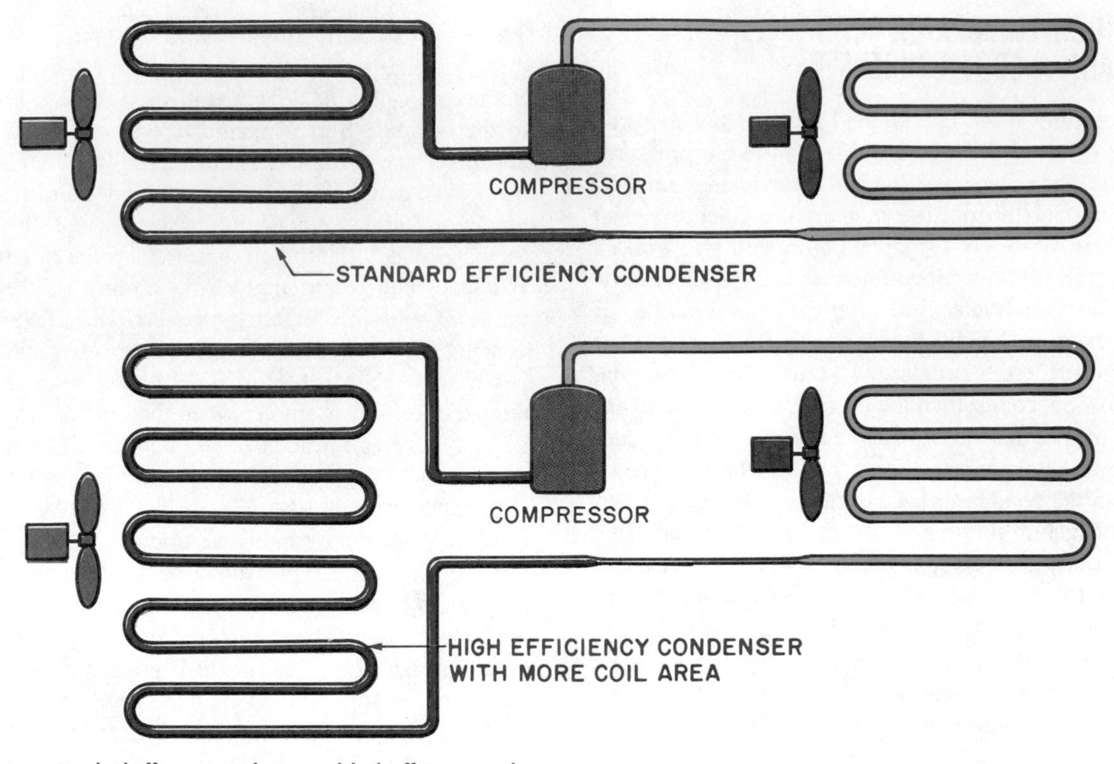

FIGURE 40–7 Standard-efficiency condenser and high-efficiency condenser.

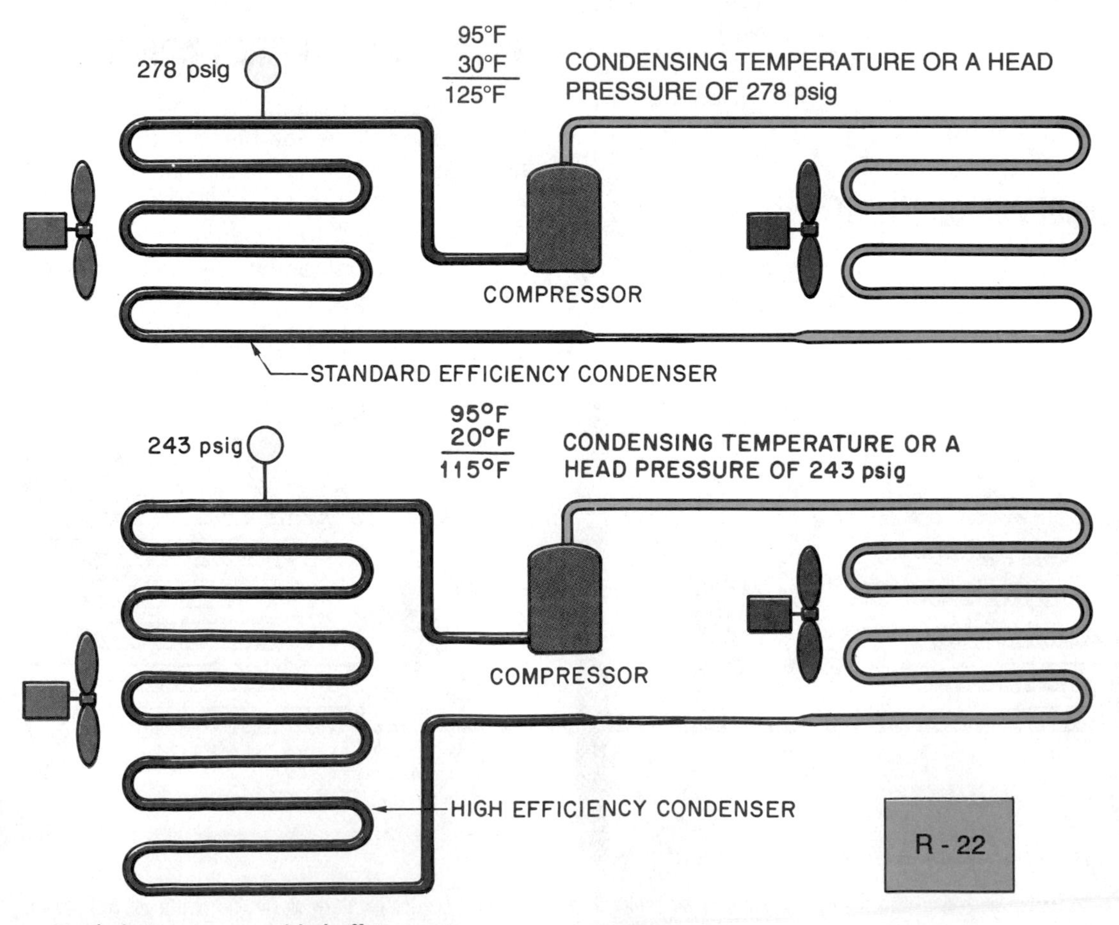

FIGURE 40–8 Standard-efficiency unit and high-efficiency unit.

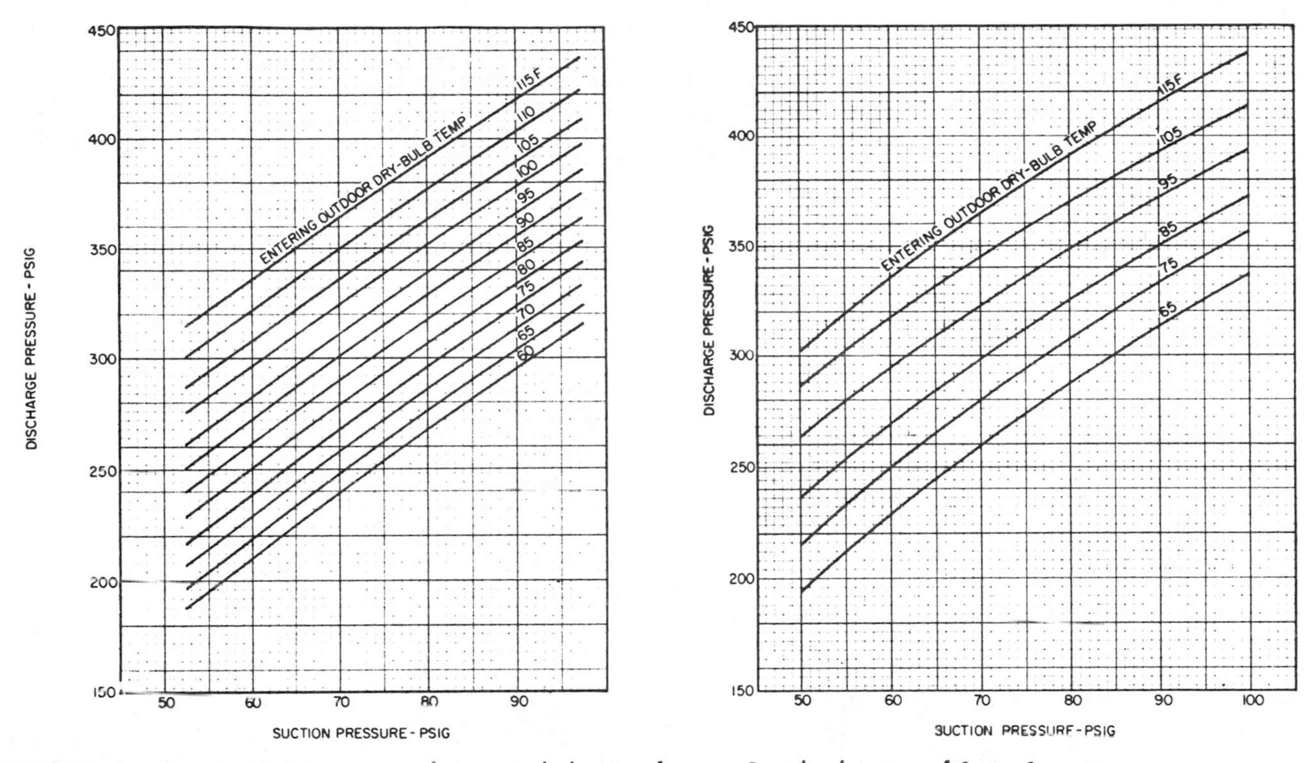

FIGURE 40–9 Charts furnished by some manufacturers to check unit performance. *Reproduced courtesy of Carrier Corporation*

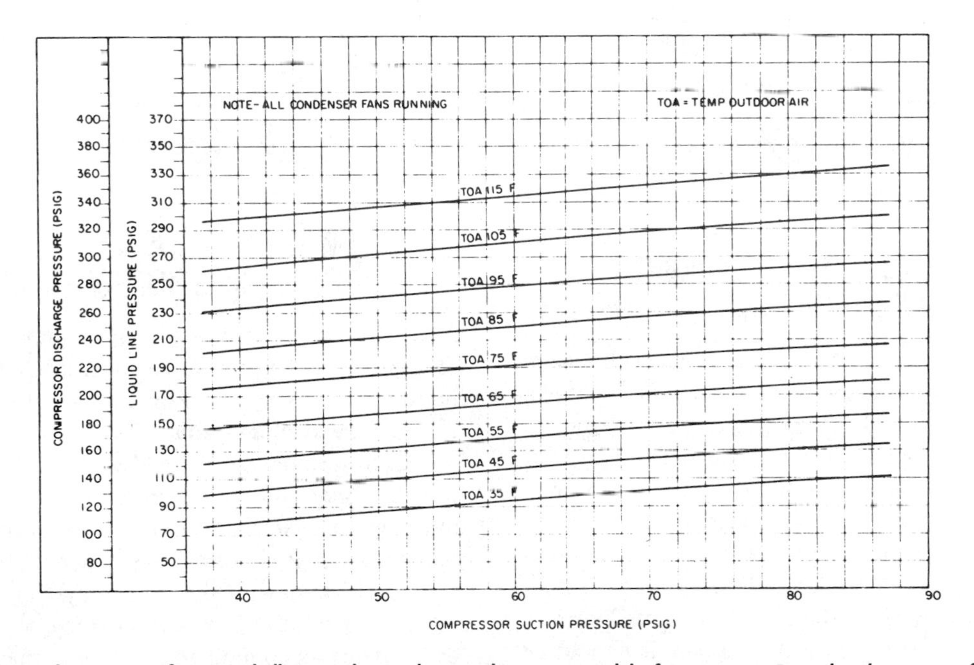

FIGURE 40–10 Page from a manufacturer's bulletin explaining how to charge one model of equipment. *Reproduced courtesy of Carrier Corporation*

the unit in the installation and start-up manual, Figure 40–11. The homeowner may have this booklet, which may be helpful to the technician.

Three things must be considered when the manufacturer publishes typical operating conditions:

1. The load on the outdoor coil, which is influenced by the outdoor temperature
2. The sensible-heat load on the indoor coil, which is influenced by the indoor dry-bulb temperature
3. The latent-heat load on the indoor coil, which is influenced by the humidity. The humidity is determined by taking the wet-bulb temperature of the indoor air.

The manufacturer would require the technician to record these temperatures and plot them against a graph or table to determine the performance of the equipment.

REFRIGERANT CHARGING

SAFETY PRECAUTION:** **To prevent personal injury, wear safty glasses and gloves when handling refrigerant. Do not overcharge system. This can cause compressor flooding.

1. Operate unit a minimum of 15 minutes before checking charge.
2. Measure suction pressure by attaching a gage to suction valve service port.
3. Measure suction line temperature by attaching a service thermometer to unit suction line near suction valve. Insulate thermometer for accurate readings.
4. Measure outdoor coil inlet air dry-bulb temperature with a second thermometer.
5. Measure indoor coil inlet air wet-bulb temperature with a sling psychrometer.
6. Refer to table. Find air temperature entering outdoor coil and wet-bulb temperature entering indoor coil. At this intersection note the superheat.
7. If unit has higher suction line temperature than charted temperature, add refrigerant until charted temperature is reached.
8. If unit has lower suction line temperature than charted temperature, bleed refrigerant until charted temperature is reached.
9. If air temperature entering outdoor coil or pressure at suction valve changes, charge to new suction line temperature indicated on chart.
10. This procedure is valid, independent of indoor air quantity.

SUPERHEAT CHARGING TABLE
(SUPERHEAT ENTERING SUCTION SERVICE VALVE)

Outdoor	INDOOR COIL ENTERING AIR °F WB													
Temp (°F)	50	52	54	56	58	60	62	64	66	68	70	72	74	76
55	9	12	14	17	20	23	26	29	32	35	37	40	42	45
60	7	10	12	15	18	21	24	27	30	33	35	38	40	43
65	—	6	10	13	16	19	21	24	27	30	33	36	38	41
70	—	—	7	10	13	16	19	21	24	27	30	33	36	39
75	—	—	—	6	9	12	15	18	21	24	28	31	34	37
80	—	—	—	—	5	8	12	15	18	21	25	28	31	35
85	—	—	—	—	—	—	8	11	15	19	22	26	30	33
90	—	—	—	—	—	—	5	9	13	16	20	24	27	31
95	—	—	—	—	—	—	—	6	10	14	18	22	25	29
100	—	—	—	—	—	—	—	—	8	12	15	20	23	27
105	—	—	—	—	—	—	—	—	5	9	13	17	22	26
110	—	—	—	—	—	—	—	—	—	6	11	15	20	25
115	—	—	—	—	—	—	—	—	—	—	8	14	18	23

FIGURE 40–11 Chart furnished by manufacturer in the installation and start-up literature. *Reproduced courtesy of Carrier Corporation*

40.8 ESTABLISHING A REFERENCE POINT ON UNKNOWN EQUIPMENT

When a technician arrives at the job and finds no literature and cannot obtain any, what should be done? The first thing is to try to establish some known condition as a reference point. For example, is the equipment standard efficiency or high efficiency? This will help establish a reference point for the suction and head pressures mentioned earlier. The high-efficiency equipment is often larger than normal. The equipment is not always marked as high efficiency, and the technician may have to compare the size of the condenser to another one to determine the head pressure. It should be obvious that a larger or oversized condenser would have a lower head pressure. For example, a 3-ton compressor will have a full-load amperage (FLA) rating of about 17 A. The amperage rating of the compressor may help to determine the rating of the equipment, Figure 40–12. Although a 3-ton high-efficiency piece of equipment will have a lesser amperage rating than a standard piece of equipment, the ratings will be close enough to compare and to determine the capacity of the equipment. If the condenser is very large for an amperage rating that should be 3 tons, the equipment is probably high efficiency, and **the head pressure will not be as high as in a standard piece of equipment.**

40.9 METERING DEVICES FOR HIGH-EFFICIENCY EQUIPMENT

High-efficiency air conditioning equipment often uses a thermostatic expansion valve rather than a fixed-bore metering device because a certain amount of efficiency may be gained. The evaporator may also be larger than normal. An oversized evaporator will add to the efficiency of the system. It is more difficult to make a determination regarding the evaporator size because, being enclosed in a casing or the ductwork, it is not as easy to see.

The operating conditions of a system can vary so much that all possibilities cannot be covered; however, a few general statements committed to memory may help. Standard-efficiency equipment and high-efficiency equipment will follow the general conditions listed in the following subsections.

Operating Conditions Near Design Space Conditions for Standard-Efficiency Equipment

1. The suction temperature is 40°F (70 psig for R-22 and 37 psig for R-12), Figure 40–13.
2. The head pressure should correspond to a temperature of no more than 35°F above the outside ambient temperature. This would be 296 psig for R-22 and 181 psig for R-12 when the outside temperature is 95°F (95°F +

APPROXIMATE FULL LOAD AMPERAGE VALUES
FOR ALTERNATING CURRENT MOTORS

Motor	Single Phase		3-Phase-Squirrel Cage Induction		
HP	115 V	230 V	230 V	460 V	575 V
1/6	4.4	2.2			
1/4	5.8	2.9			
1/3	7.2	3.6			
1/2	9.8	4.9	2	1.0	0.8
3/4	13.8	6.9	2.8	1.4	1.1
1	16	8	3.6	1.8	1.4
1 1/2	20	10	5.2	2.6	2.1
2	24	12	6.8	3.4	2.7
3	34	17	9.6	4.8	3.9
5	56	28	15.2	7.6	6.1
7 1/2			22	11.0	9.0
10			28	14.0	11.0

Does not include shaded pole.

FIGURE 40–12 Table of current ratings for different sized motors at different voltages. *Courtesy BDP Company*

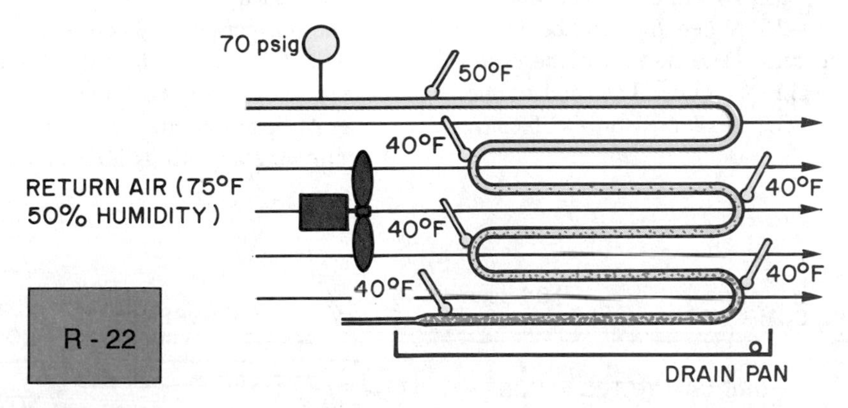

FIGURE 40–13 Evaporator operating at close to design conditions.

35°F = 130°F condensing temperature). A typical condition would be condensing at 30°F above the ambient, or 95°F + 30°F = 125°F, Figure 40–14. The head pressure would be 278 psig for R-22 and 169 psig for R-12.

Space Temperature Higher Than Normal for Standard-Efficiency Equipment

1. The suction pressure will be higher than normal. Normally, the refrigerant boiling temperature is about 35°F cooler than the entering air temperature. (Recall the relationship of the evaporator to the entering air temperature described in the refrigeration section.) When the conditions are normal, the refrigerant boiling temperature would be 40°F when the return air temperature is 75°F. This is true when the humidity is normal. When the space temperature is higher than normal because the equipment has been off for a long time, the return air temperature may go up to 85°F and the humidity may be high also. The suction temperature may then go up to 50°F, with a corresponding pressure of 84 to 93 psig for

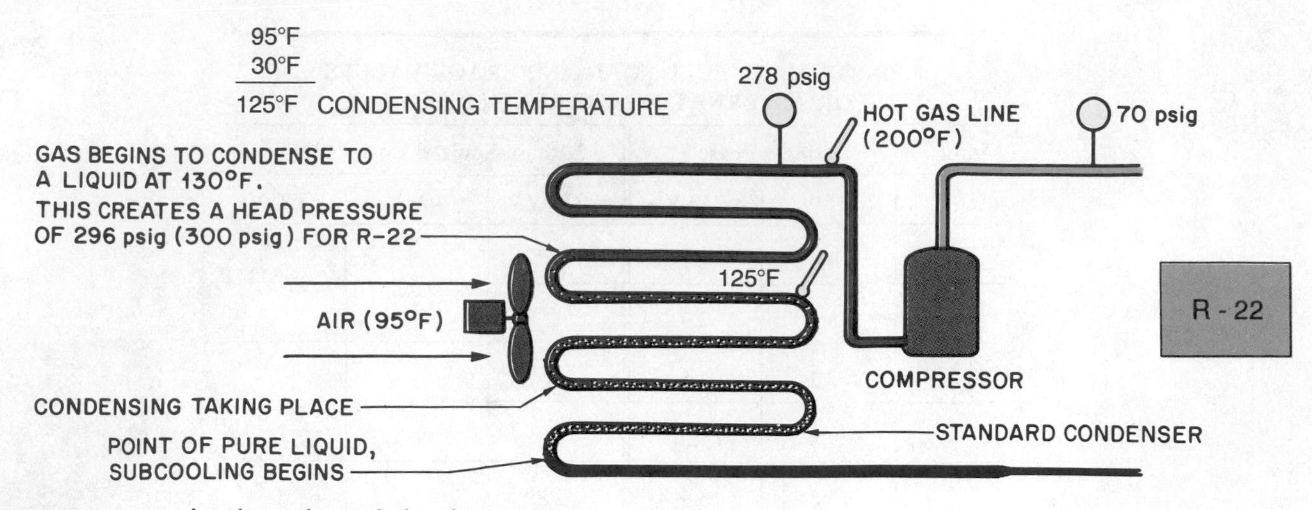

FIGURE 40–14 Normal conditions of a standard condenser operating on a 95°F day.

R-22, Figure 40–15. This higher-than-normal suction pressure may cause the discharge pressure to rise also.

2. The discharge pressure is influenced by the outside temperature and the suction pressure. For example, the discharge pressure is supposed to correspond to a temperature of no more than 30°F higher than the ambient temperature under normal conditions. When the suction pressure is 80 psig, the discharge pressure is going to rise accordingly. It may move up to a new condensing temperature 10°F higher than normal. This would mean that the discharge pressure for R-22 could be 317 psig (95°F + 30°F + 10°F = 135°F) while the suction pressure is high, Figure 40–16. When the unit begins to reduce the space temperature and humidity, the evaporator pressure will begin to reduce. The load on the evaporator is reduced. The head pressure will come down also.

Operating Conditions Near Design Conditions for High-Efficiency Equipment

1. The evaporator design temperature may in some cases operate at a slightly higher pressure and temperature on high-efficiency equipment because the evaporator is larger. The boiling refrigerant temperature may be 45°F with the larger evaporator at design conditions, and this would create a suction pressure of 76 psig for R-22 and be normal, Figure 40–17. R-12 is not mentioned in this coverage of high-efficiency equipment because it is not being used.
2. The refrigerant may condense at a temperature as low as 20°F more than the outside ambient temperature. For a 95°F day with R-22, the head pressure may be as low as 243 psig, Figure 40–17. If the condensing temperature were as high as 30°F above the ambient tempera-

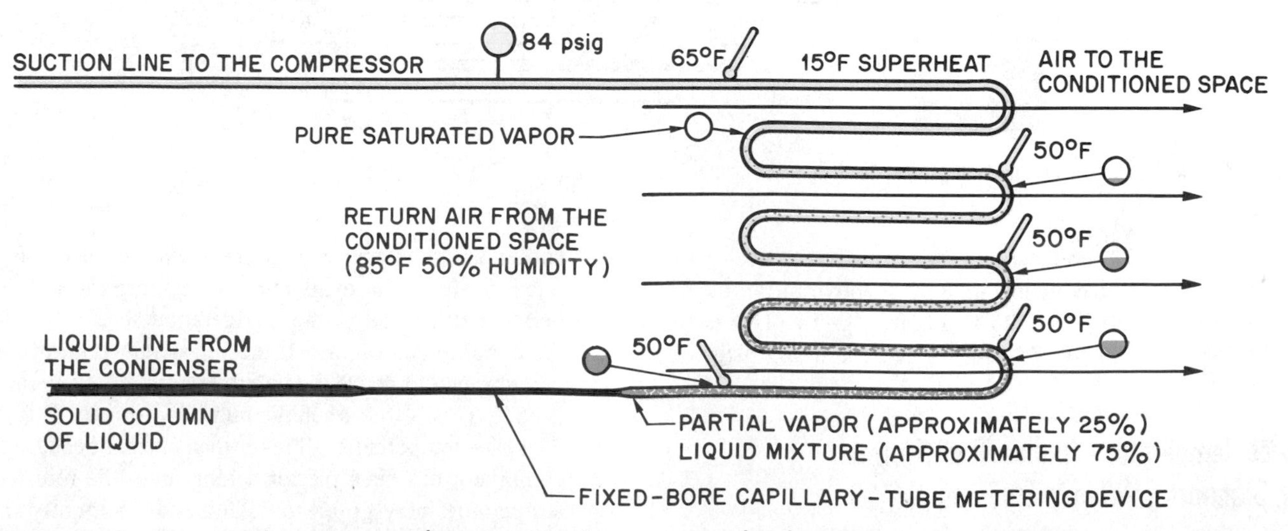

FIGURE 40–15 Pressures and temperatures as they may occur with an evaporator when the space temperature and humidity are above design conditions.

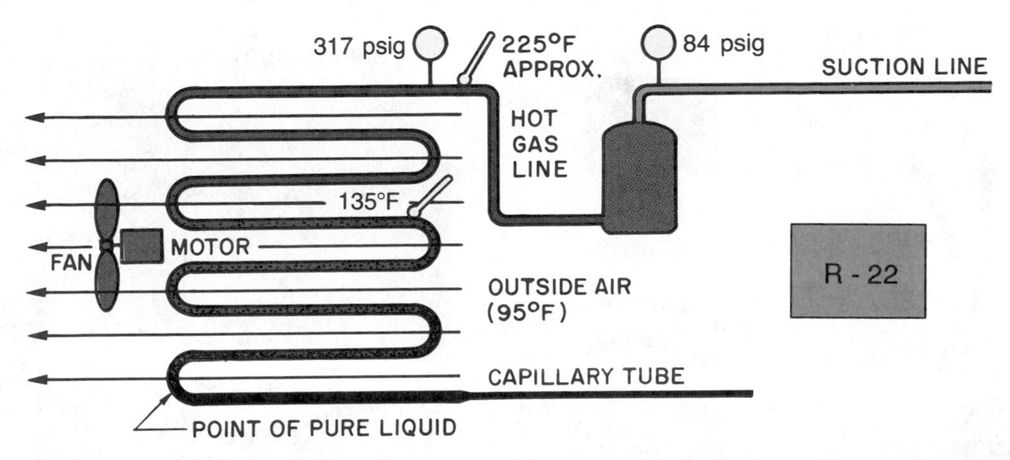

FIGURE 40–16 Condenser operating in design condition as far as the outdoor ambient air is concerned, but the pressure is high because the evaporator is under a load that is above design.

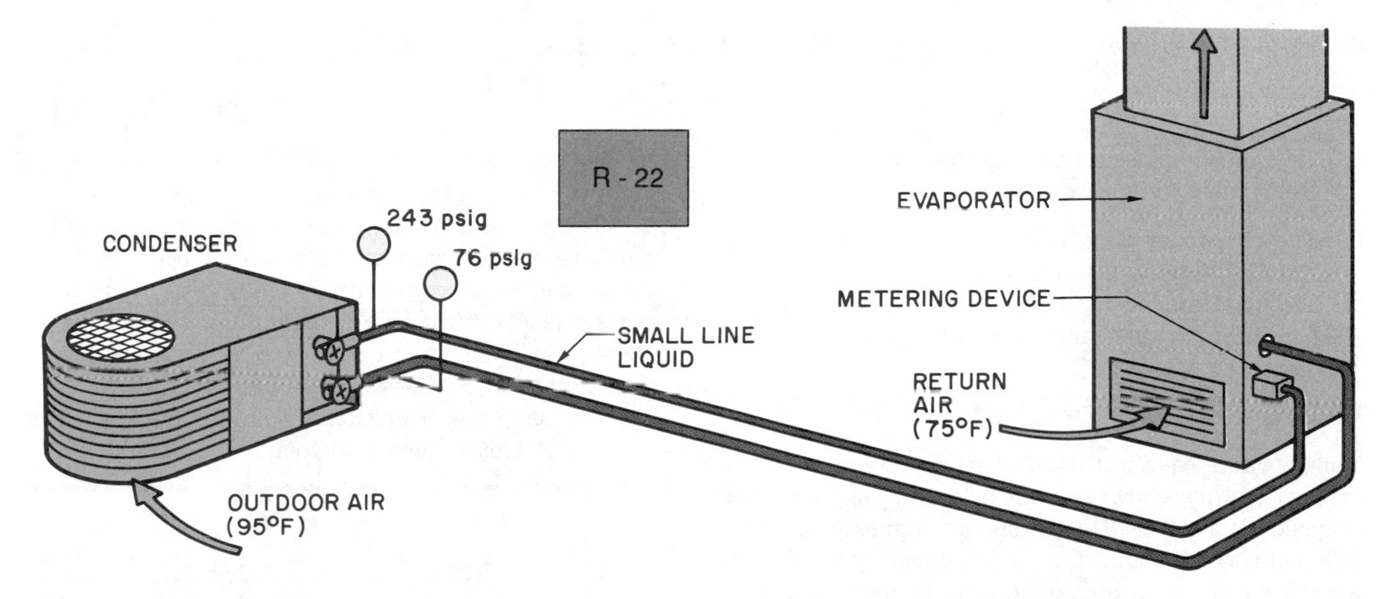

FIGURE 40–17 High-efficiency system operating conditions, 45°F evaporator temperature and 115°F condensing temperature.

ture, you should suspect a problem. For example, the head pressure should not be more than 277 psig on a 95°F day for R-22. This is discussed later.

Other Than Design Conditions for High-Efficiency Equipment

1. When the unit has been off long enough for the load to build up, the space temperature and humidity are above design conditions and the high-efficiency system pressures will be higher than normal, as they would in the standard-efficiency system. With standard equipment, when the return air temperature is 75°F and the humidity is approximately 50%, the refrigerant boils at about 40°F. This is a temperature difference of 35°F. A high-efficiency evaporator is larger, and the refrigerant may boil at a temperature difference of 30°F, or around 55°F when the space temperature is 85°F, Figure 40–18. The exact boiling temperature relationship depends on the manufacturer and how much coil surface area was selected.
2. The high-efficiency condenser, like the evaporator, will operate at a higher pressure when the load is increased. The head pressure will not be as high as it would with a standard-efficiency condenser because the condenser has extra surface area.

The capacity of high-efficiency systems is not up to the rated capacity when the outdoor temperature is much below design. The earlier condition of the family that shut off the air conditioning system before going to work and then turned it back on when they came home from work will be much worse with a high-efficiency system. The fact that the condenser became too efficient at night when the air was cooler with a standard condenser will be much more evident with the larger high-efficiency condenser. Most manufacturers produce two-speed condenser fans so that a lower fan speed can be used in mild weather to help compensate for this temperature difference. The service technician who tries to analyze a component of high-efficiency equipment on a mild day will find the head pressure low. This will cause the suction pressure to also be low. Using the coil-to-air relationships for the condenser and the evaporator will

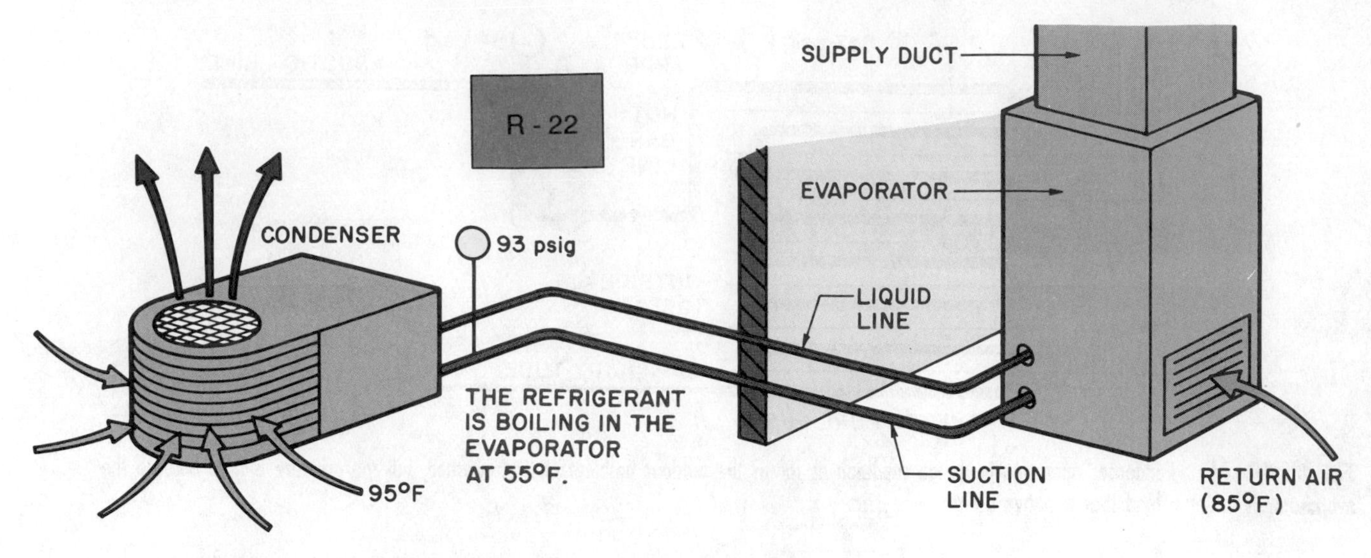

FIGURE 40–18 Evaporator in a high-efficiency system with a load above design conditions.

help determine the correct pressures and temperatures.

Remember these two statements:

1. The evaporator absorbs heat, which is related to its operating pressures and temperatures.
2. The condenser rejects heat and has a predictable relationship to the load and ambient temperature.

40.10 EQUIPMENT EFFICIENCY RATING

Manufacturers have a method of rating equipment so the designer and the owner can tell high efficiency from low efficiency at a glance. This rating was originally called the EER rating. EER stands for energy efficiency ratio and is actually the output in Btuh divided by the input in watts of power used to produce the output. For example, a system may have an output of 36,000 Btuh with an input of 4000 W.

$$\frac{36{,}000 \text{ Btuh}}{4000 \text{ W}} = \text{an EER of } 9$$

The larger the EER rating, the more efficient the equipment. For example, suppose the 36,000 Btuh air conditioner only required 3600 W input.

$$\frac{36{,}000 \text{ Btuh}}{3600 \text{ Btuh}} = \text{an EER of } 10$$

The customer is getting the same capacity using less power. The equipment is more efficient. **The larger the EER rating, the more efficient the equipment.**

The EER rating is a steady-state rating and does not account for the time the unit operates before reaching peak efficiency. This operating time has an unknown efficiency. It also does not account for shutting the system down at the end of the cycle (when the thermostat satisfies) leaving a cold coil in the duct. The cold coil continues to absorb heat from the surroundings, not the conditioned space. Refrigerant equalizes to the cold coil that must be pumped out at the beginning of the next cycle. This accounts for some of the inefficiency at the beginning of the cycle.

The picture is not complete using the EER rating system, so a rating of seasonal efficiency has been developed, called seasonal energy efficiency ratio (SEER). This rating is tested and verified by a rating agency and includes the start-up and shut-down cycles. The rating agency is the ARI. Ratings of all manufacturers' equipment are published by ARI, and these manufacturers list the ratings in their catalogs. A typical rating may look like Figure 40–19.

MODEL	CAPACITY (Btuh)	SEER
A	24,000	9.00
B	24,000	10.00
C	24,000	10.50
D	24,000	11.00
E	24,000	11.50
F	24,000	12.00

FIGURE 40–19 Examples of SEER ratings.

40.11 TYPICAL ELECTRICAL OPERATING CONDITIONS

The electrical operating conditions are measured with a volt-ohmmeter and an ammeter. Three major power-consuming devices may have to be analyzed from time to time: the indoor fan motor, the outdoor fan motor, and the compressor. The control circuit is considered a separate function.

The starting point for considering electrical operating conditions is to know what the system supply voltage is supposed to be. For residential units, 230 V is the typical voltage. Light commercial equipment will nearly always use 208 V or 230 V single phase or three phase. Single and three phase may be obtained from a three-phase power supply. The equipment rating may be 208/230 V. The reason for the two different ratings is that 208 V is the supply voltage that some power companies provide, and 230 V is the supply voltage provided by other power companies. Some light commercial equipment may have a supply voltage of 460 V three phase if the equipment is at a large commercial installation. For example, an office may have a 3- or 4-ton air conditioning unit that operates separately from the main central system. If the supply voltage is 460 V, three phase, the small unit may operate from the same power supply. When 208/230-V equipment is used at a commercial installation, the compressor may be three phase, and the fan motors single phase. The number of phases that the power company furnishes makes a difference in the method of starting the compressor. Single-phase compressors may have a start assist, such as the positive temperature coefficient device or a start relay and start capacitor. Three-phase compressors will have no start-assist accessories.

40.12 MATCHING THE UNIT TO THE CORRECT POWER SUPPLY

The typical operating voltages for any air conditioning system must be within the manufacturer's specifications. This is $\pm 10\%$ of the rated voltage. For the 208/230-V motor the minimum allowable operating voltage would be $208 \times 0.90 = 187.2$ V. The maximum allowable operating voltage would be $230 \times 1.10 = 253$ V. Notice that the calculations used the 208 V for the base for figuring the low voltage and the 230 V rating for calculating the highest voltage. This is because this application is 208/230 V. If the motor is rated at 208 V or 230 V alone, that value (208 V or 230 V) is used for evaluating the voltage. The equipment may be started under some conditions that are beyond the rated conditions. The technician must use some judgment. For example, if 180 V is measured, the motor should not be started because the voltage will drop further with the current draw of the motor. If the voltage reads 260 V, the motor may be started because the voltage may drop slightly when the motor is started. If the voltage drops to within the limits, the motor is allowed to run.

40.13 STARTING THE EQUIPMENT WITH THE CORRECT DATA

When the correct rated voltage is known and the minimum and maximum voltages are determined, the equipment may be started if the voltages are within the limits. The three motors—indoor fan, outdoor fan, and the compressor—can be checked for the correct current draw.

The indoor fan is building the air pressure to move the air through the ductwork, filters, and grilles to the conditioned space. By law, the voltage characteristics must be printed on the motor in such a manner that they will not come off. This information may be printed on the motor, but the motor might be mounted so that the data cannot be easily seen. In some cases the motor may be inside the squirrel cage blower. If so, removing the motor is the only way of determining the fan current. When the supplier can be easily contacted, you may obtain the information there. If the motor electrical characteristics cannot be obtained from the nameplate, you might be able to get it from the unit nameplate. However, the fan motor may have been changed to a larger motor for more fan capacity.

40.14 FINDING A POINT OF REFERENCE FOR AN UNKNOWN MOTOR RATING

When a motor is mounted so that the electrical characteristics cannot be determined, you must improvise. We know that air conditioning systems normally move about 400 cfm of air per ton. This can help you determine the amperage of the indoor fan motor by comparing the fan amperage on an unknown system to the amperage of a known system. All you need is the approximate system capacity. As discussed earlier, you can find this by comparing compressor amperages of the unit in question and a known unit. For example, the compressor amperage of a 3-ton system is about 17 A when operating on 230V. If you notice that the amperage of the compressor on the system you're checking is 17 A, you can assume that the system is close to 3 tons. The fan motor for a 3-ton system should be about 1/3 hp for a typical duct system. The fan motor amperage for a 1/3-hp permanent split-capacitor motor is 3.6 A at 230 V. If the fan in question were pulling 5 A, suspect a problem.

Fan motors may be shipped in a warm-air furnace and may have been changed if air conditioning was added at a later date. In this case the furnace nameplate may not give the correct fan motor data. The condensing unit will have a nameplate for the condenser fan motor. This motor should be sized fairly close to its actual load and should pull close to nameplate amperage.

40.15 DETERMINING THE COMPRESSOR RUNNING AMPERAGE

The compressor current draw may not be as easy to determine as the fan motor current draw because compressor manufacturers do not all stamp the compressor run-load amperage (RLA) rating on the compressor nameplate. There are so many different compressor sizes it is hard to state the correct full load amperage. For example, motors normally come in the following increments: 1, 1 1/2, 2, 3, and 5 hp. A unit rated at 34,000 Btu/h is called a 3-ton unit, although it actually takes 36,000 Btu/h to be a 3-ton unit. Ratings that are not completely accurate are known as *nominal ratings*. They are rounded off to the closest rating. A typical 3-ton unit would have a 3-hp motor. A unit with a rating of 34,000 Btu/h does not need a full 3-hp motor, but it is supplied with one because there is no standard horsepower motor to meet its needs. If the motor amperage for a 3-hp motor were stamped on the compressor, it could cause confusion because the motor may never operate at that amperage. A unit nameplate lists electrical information, Figure 40–20. ***The manufacturer may stamp the compressor RLA on the unit nameplate. This amperage should not be exceeded.***

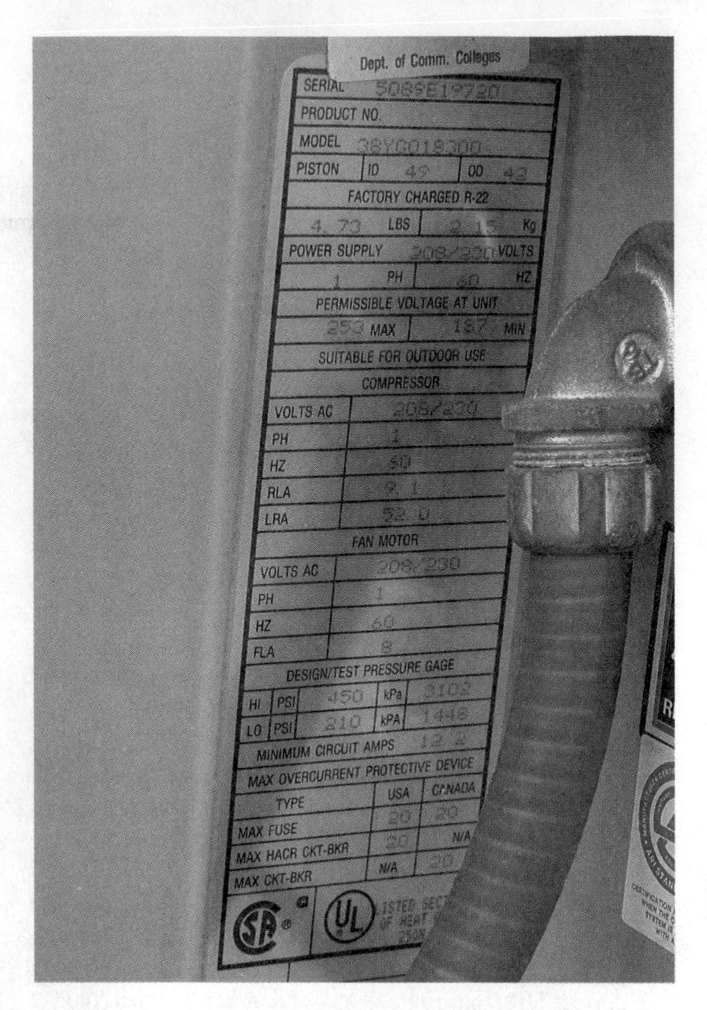

FIGURE 40–20 Condensing unit nameplate. *Photo by Bill Johnson*

40.16 COMPRESSORS OPERATING AT FULL-LOAD CURRENT

It is rare for a compressor motor to operate at its RLA rating. If design or above-design conditions were in effect, the compressor would operate at close to full load. When the unit is operating at a condition greater than design, such as when the unit has been off for some time in very hot weather, it might appear that the compressor is operating at more than RLA. However, there are usually other conditions that keep the compressor from drawing too much current. The compressor is pumping vapor and vapor is very light. It takes a substantial increase in pressure difference to create a significantly greater work load.

40.17 HIGH VOLTAGE, THE COMPRESSOR, AND CURRENT DRAW

A motor operating at a voltage higher than the voltage rating of the motor is a condition that will prevent the motor from drawing too much current. A motor rated at 208/230 V has an ampere rating at some value between 208 and 230 V. Therefore, if the voltage is 230 V, the amperage may be lower than the nameplate amperage even during overload. The compressor motor may be larger than needed and may not reach its rated horsepower until it gets to the maximum rating for the system. It may be designed to operate at 105°F or 115°F outdoor ambient for very hot regions, but when the unit is rated at 3 tons, it would be rated down at the higher temperatures. The unit nameplate may contain the compressor amperage at the highest operating condition that the unit is rated for, 115°F ambient temperature.

40.18 CURRENT DRAW AND THE TWO-SPEED COMPRESSOR

Some air conditioning manufacturers use two-speed compressors to achieve better seasonal efficiencies. These compressors may use a motor capable of operating as a two-pole or a four-pole motor. A four-pole motor runs at 1800 rpm, and a two-pole motor runs at 3600 rpm. A motor control circuit will slow the motor to reduced speed for mild weather.

Variable speed may be obtained with electronic circuits. The system's efficiency is greatly improved because there is no need to stop the compressor and restart it at the beginning of the next cycle. However, this means more factors for a technician to consider. Manufacturer's literature must be consulted to establish typical running conditions under other-than-design conditions. The equipment should perform just as typical high-efficiency equipment would when operating at design conditions.

SAFETY PRECAUTION: *The technician must use caution while observing equipment to obtain the operating conditions. Electrical, pressure, and temperature readings must be taken while the technician observes the equipment. Many times the readings must be made while the equipment is in operation.*

SUMMARY

- The technician must be familiar with mechanical and electrical operating conditions.
- Mechanical conditions deal with pressures and temperatures.
- The inside conditions that vary the load on the system are the space temperature and humidity.
- When the discharge pressure rises, the system capacity decreases.
- When suction pressure rises, head pressure also rises.
- When head pressure rises, suction pressure also rises.
- High-efficiency systems often use larger or oversized evaporators, and the refrigerant will boil at a temperature of about 45°F at typical operating conditions. This is a temperature difference of 30°F compared to the return air.
- High-efficiency equipment has a lower operating head pressure than standard-efficiency equipment partly because the condenser is larger. It has a different relation-

ship to the outside ambient temperature, normally 20°F to 25°F.

- Electrical instruments used to check working conditions are the ammeter and the voltmeter.
- The first thing the technician needs to know for electrical troubleshooting is what the operating voltage for the unit is supposed to be.
- Typical voltages are 230 V single phase for residential and 208/230 V single or three phase, or 460 V three phase for commercial installations.
- Equipment manufacturers require that operating voltages of ±10% of the rated voltage of the equipment be maintained.
- ***The unit may have the compressor's RLA printed on it. It should *not* be exceeded.***
- Some manufacturers are now using two-speed motors and variable-speed motors for variable capacity.

REVIEW QUESTIONS

1. What are the standard conditions at which air conditioning equipment is designed to operate?
2. What is the typical temperature relationship between a standard air-cooled condenser and the ambient temperature?
3. What is the temperature relationship between a high-efficiency condenser and the ambient temperature?
4. How is high efficiency obtained with a condenser?
5. How does the temperature relationship between the condenser and the ambient temperature change as the ambient temperature changes?
6. What is the temperature relationship between the boiling refrigerant temperature and the entering air temperature with a standard-efficiency evaporator?
7. What does the head pressure do if the suction pressure rises?
8. What will cause the suction pressure to rise?
9. What happens to the suction pressure if the head pressure rises?
10. What can cause the head pressure to rise?
11. What two instruments are used to troubleshoot the system when an electrical problem is suspected?
12. What is the common voltage supplied to a residence?
13. How can an unknown current be found for a fan motor?
14. What are the typical motor horsepower sizes?
15. Name two methods that manufacturers use to vary the capacity of residential and small commercial compressors.

UNIT

41 Troubleshooting

OBJECTIVES

After studying this unit, you should be able to

- select the correct instruments for checking an air conditioning unit with a mechanical problem.
- determine the standard operating suction pressures for both standard- and high-efficiency equipment.
- calculate the correct operating suction pressures for both standard- and high-efficiency air conditioning equipment with other-than-design conditions.
- calculate the standard operating discharge pressures at various ambient conditions.
- select the correct instruments to troubleshoot electrical problems in an air conditioning system.
- check the line- and low-voltage power supplies.
- troubleshoot basic electrical problems in an air conditioning system.
- use an ohmmeter to check the various components of the electrical system.

SAFETY CHECKLIST

* Be extremely careful when installing or removing gages. Escaping refrigerant can injure your skin and eyes. Wear gloves and goggles. If possible turn off the system to install a gage on the high-pressure side. Most air conditioning systems are designed so that the pressures will equalize when turned off.
* ⊛**Wear goggles and gloves when transferring refrigerant from a cylinder to a system or when recovering refrigerant from a system. Do not recover refrigerant into a disposable cylinder. Use only tanks or cylinders approved by the Department of Transportation (DOT).**⊛
* Be very careful when working around a compressor discharge line. The temperature may be as high as 220°F.
* All safety practices must be observed when troubleshooting electrical systems. Often, readings must be taken while the power is on. Only let the insulated meter leads touch the hot terminals. Never use a screwdriver or other tools around terminals that are energized. Never use tools in a hot electrical panel.
* Turn the power off whenever possible when working on a system. Lock and tag the disconnect panel and keep the only key on your person.
* Do not stand in water when making any electrical measurements.
* Short capacitor terminals with a 20,000-Ω resistor before checking with an ohmmeter.
* Students and other inexperienced persons should perform troubleshooting tasks only under the supervision of an experienced person.

41.1 INTRODUCTION

Troubleshooting air conditioning equipment involves troubleshooting mechanical and electrical systems and they may have symptoms that overlap. For example, if an evaporator fan motor capacitor fails, the motor will slow down and begin to get hot. It may even get hot enough for the internal overload protector to stop it. While it is running slowly, the suction pressure will go down and give symptoms of a restriction or low charge. If the technician diagnoses the problem based on suction pressure readings only, a wrong decision may be made.

41.2 MECHANICAL TROUBLESHOOTING

Gages and temperature-testing equipment are used when performing mechanical troubleshooting. The gages used are those on the gage manifold as shown in Figure 41–1. The suction or low-side gage is on the left side of the manifold and the discharge or the high-side gage is on the right side. The most common refrigerant used in air conditioning is R-22. An R-22 temperature chart is printed on each gage for determining the saturation temperature for the low- and high-pressure sides of the system. Because these same gages are used for refrigeration, a temperature scale for R-12 and R-502 is printed on the gage also, Figure 41–2. The R-12 scale is not used for residential air conditioning unless the

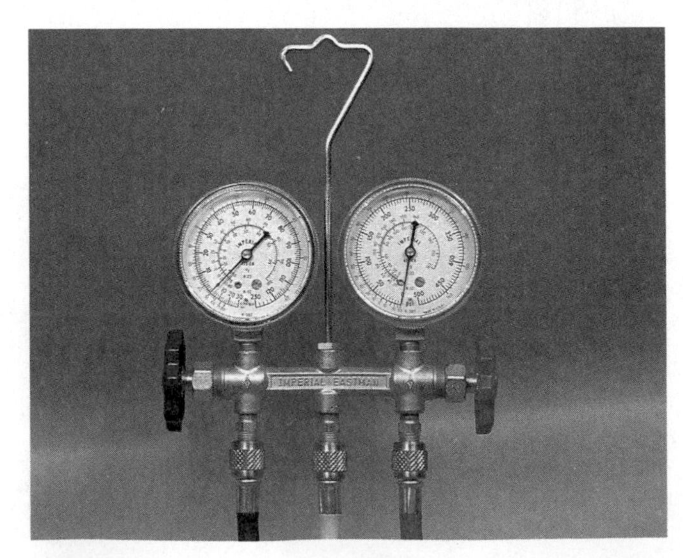

FIGURE 41–1 Low-side gage on the left and high-side gage on the right. *Photo by Bill Johnson*

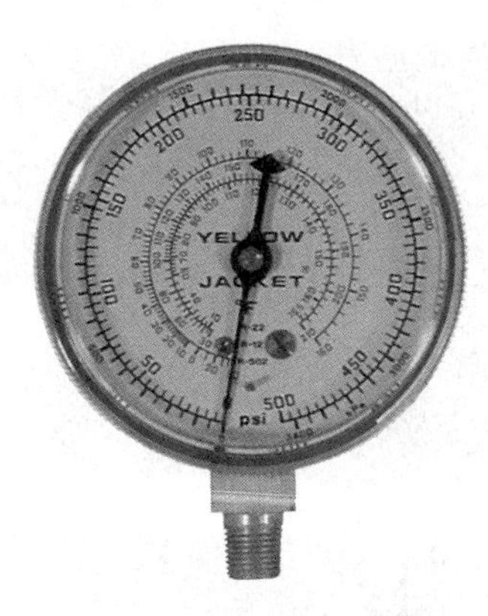

FIGURE 41–2 Gages have common refrigerant temperature relationships for R-12, R-22, and R-502 printed on them. *Photo by Bill Johnson*

equipment is old enough that R-12 is used. ⊛**The common refrigerant for automobile air conditioning starting in 1992 is R-134a.**⊛ Older automobiles would normally use R-12.

41.3 GAGE MANIFOLD USAGE

The gage manifold displays the low- and high-side pressures while the unit is operating. These pressures can be converted to the saturation temperatures for the evaporating (boiling) refrigerant and the condensing refrigerant by using the pressure and temperature relationship. The low-side pressure can be converted to the boiling temperature. If the boiling temperature for a system should be close to 40°F, it can be converted to 70 psig for R-22. The superheat at the evaporator should be close to 10°F at this time. It is difficult to read the suction pressure at the evaporator because it normally has no gage port. Therefore, the technician takes the pressure and temperature readings at the condensing unit suction line to determine the system performance. Guidelines for checking the superheat at the condensing unit will be discussed later in this unit. If the suction pressure were 48 psig, the refrigerant would be boiling at about 24°F, which is cold enough to freeze the condensate on the evaporator coil and too low for continuous operation. The probable causes of low boiling temperatures are low charge or restricted airflow, Figure 41–3.

SAFETY PRECAUTION: *You must be extremely careful while installing manifold gages. High-pressure refrigerant will injure your skin and eyes. Wearing goggles and gloves are necessary. The danger from attaching the high-pressure gages can be reduced by shutting off the unit and allowing the pressures to equalize to a lower pressure. Liquid R-22 can cause serious frostbite because it boils at −44°F at atmospheric pressure.*

The high-side gage may be used to convert pressures to condensing temperatures. For example, if the high-side gage reads 278 psig and the outside ambient temperature is 80°F, the head pressure may seem too high. The gage manifold chart shows that the condenser is condensing at 125°F. However, a condenser should not condense at a temperature more than 30°F higher than the ambient temperature. The ambient temperature is 80°F + 30°F = 110°F, so the condensing temperature is actually 15°F too high. Probable causes are a dirty condenser or an overcharge of refrigerant.

The gage manifold is used whenever the pressures need to be known. Two types of pressure connections are used with air conditioning equipment: the Schrader valve and the service valve, Figure 41–4. The *Schrader valve* is a pressure connection only. The *service valve* can be used to isolate the system for service. It may have a Schrader connection for the gage port and a service valve for isolation, Figure 41–5.

41.4 WHEN TO CONNECT THE GAGES

When servicing small systems, a gage manifold should not be connected every time the system is serviced. A small amount of refrigerant escapes each time the gage is connected. Some residential and small commercial systems have a critical refrigerant charge. When the high-side line is connected, high-pressure refrigerant will condense in the gage line. The refrigerant will escape when the gage line is disconnected from the Schrader connector. A gage line full of liquid refrigerant lost while checking pressures may be enough to affect the system charge. A short gage line connector for the high side will help prevent refrigerant loss, Figure 41–6. This can be used only to check pressure. You cannot use it to transfer refrigerant out of the system because it is not a manifold.

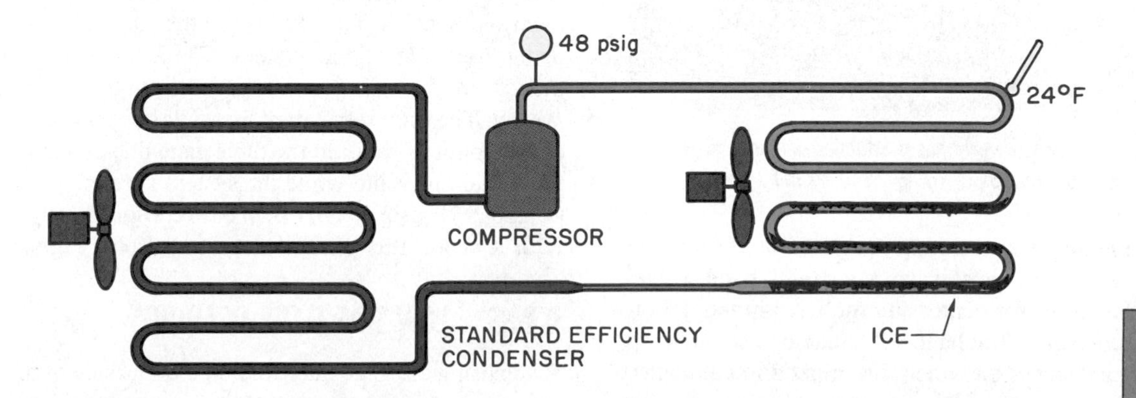

FIGURE 41–3 This coil was operated below freezing until the condensate on the coil froze.

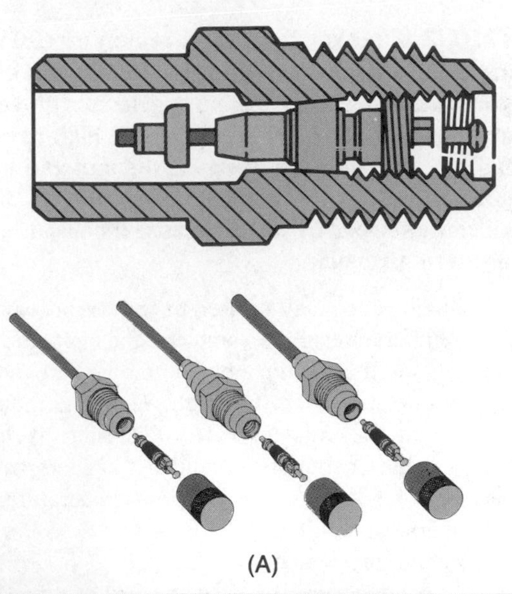

(A)

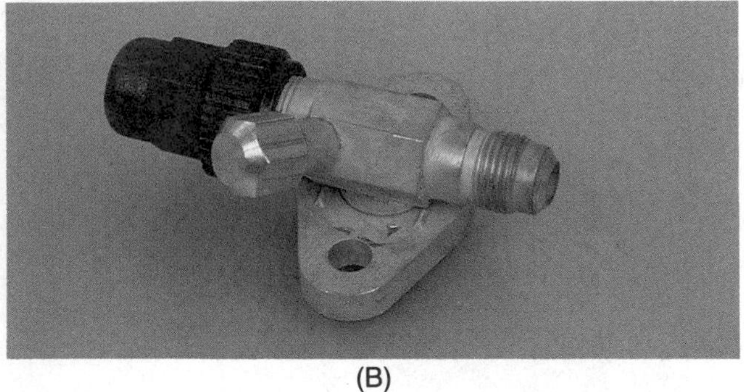

(B)

FIGURE 41–4 Two valves commonly used when taking pressure readings on modern air conditioning equipment. *(A) Courtesy J/B Industries, (B) Photo by Bill Johnson*

FIGURE 41–5 This service valve has a Schrader port instead of a back seat like a refrigeration service valve. *Photo by Bill Johnson*

Another method of connecting the gage manifold to the Schrader valve service port is with a small hand valve, which has a depressor for depressing the stem in the Schrader valve, Figure 41–7. The hand valve can be used to keep the refrigerant in the system when disconnecting the gages. Follow these steps: (1) Turn the stem on the valve out; this allows the Schrader valve to close. (2) Make sure that a plug

FIGURE 41–6 This short service connection is used on the high-side gage port to keep too much refrigerant from condensing in the gage line. *Photo by Bill Johnson*

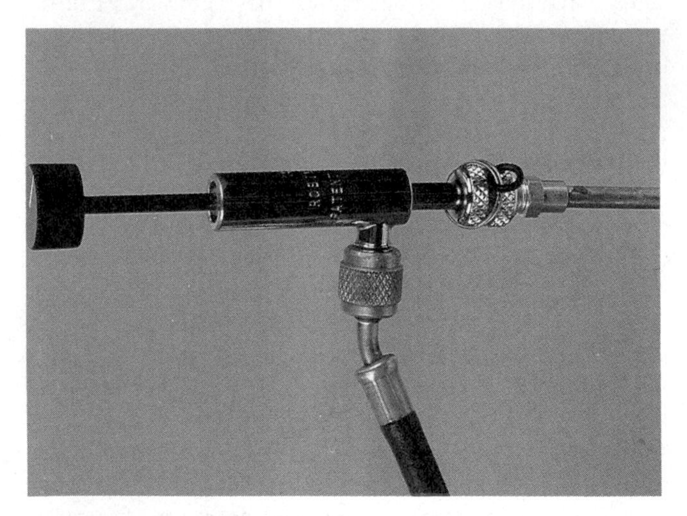

FIGURE 41–7 Hand valve used to press the Schrader valve stem and control the pressure. It can also be used to back the valve stem out. *Photo by Bill Johnson*

is in the center line of the gage manifold. (3) Open the gage manifold handles. This will cause the gage lines to equalize to the low side of the system. The liquid refrigerant will move from the high-pressure gage line to the low side of the system. The only refrigerant that will be lost is an insignificant amount of vapor in the three gage lines. This vapor is at the suction pressure while the system is running and should be about 70 psig. ***This procedure should only be done with a clean and purged gage manifold, Figure 41–8.***

41.5 LOW-SIDE GAGE READINGS

When using the gage manifold on the low side of the system you can compare the actual evaporating pressure to the normal evaporating pressure. This verifies that the refrig-

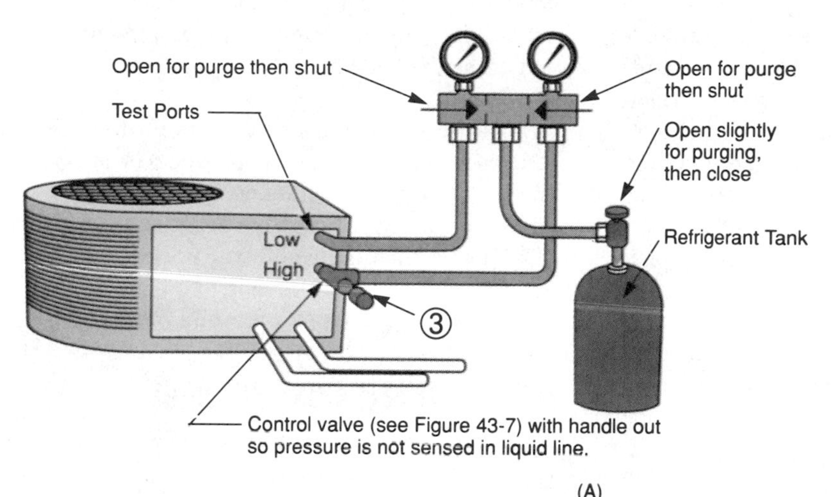

Open for purge then shut
Test Ports
Open for purge then shut
Open slightly for purging, then close
Refrigerant Tank
Low
High
③
Control valve (see Figure 43-7) with handle out so pressure is not sensed in liquid line.
(A)
① Before fastening the line to a Schrader valve, allow vapor to purge through the refrigerant lines from the refrigerant tank to push any contaminates out of the lines.
② Turn the tank off and shut the manifold valves.
③ Tighten fittings down onto the Schrader ports and obtain a gage reading as needed. The reading on the high-side port will be obtained by tightening the control valve handle against the gage port.
Room Temperature Vapor

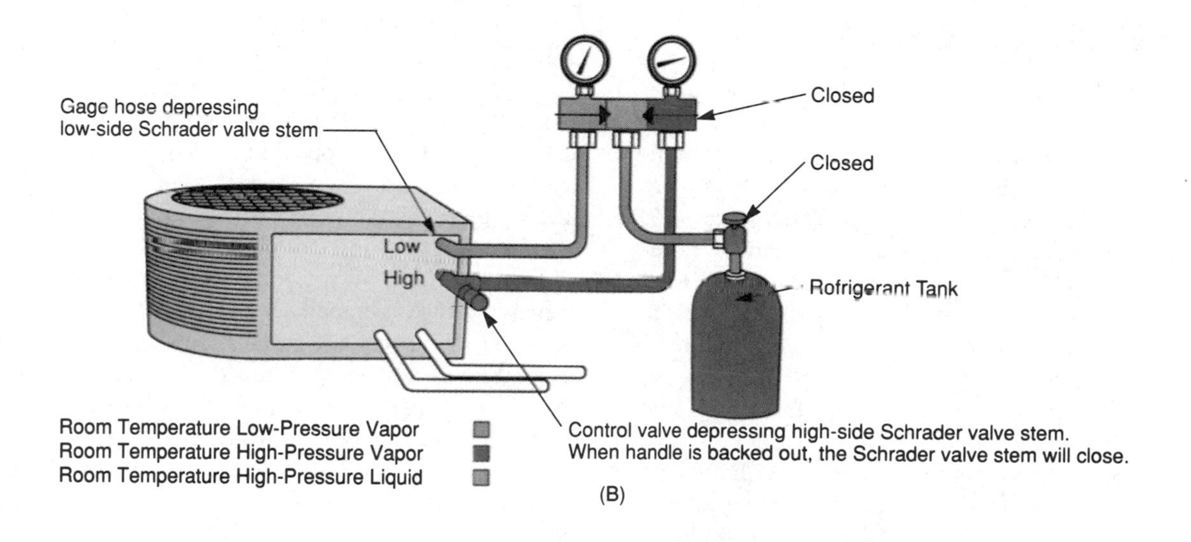

Gage hose depressing low-side Schrader valve stem
Closed
Closed
Low
High
Rofrigerant Tank
Room Temperature Low-Pressure Vapor
Room Temperature High-Pressure Vapor
Room Temperature High-Pressure Liquid
Control valve depressing high-side Schrader valve stem. When handle is backed out, the Schrader valve stem will close.
(B)

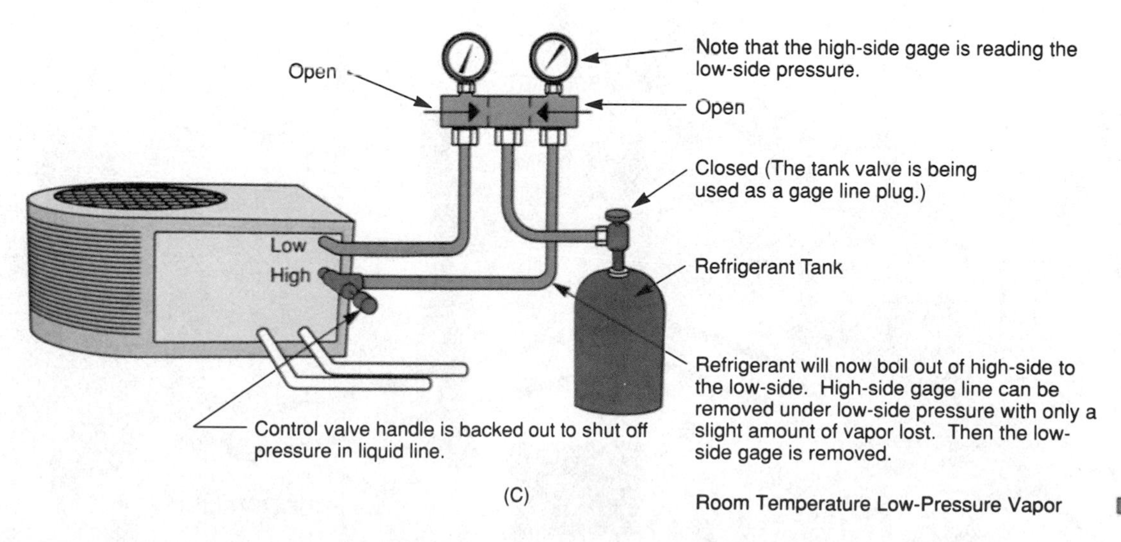

Open
Note that the high-side gage is reading the low-side pressure.
Open
Closed (The tank valve is being used as a gage line plug.)
Low
High
Refrigerant Tank
Refrigerant will now boil out of high-side to the low-side. High-side gage line can be removed under low-side pressure with only a slight amount of vapor lost. Then the low-side gage is removed.
Control valve handle is backed out to shut off pressure in liquid line.
(C)
Room Temperature Low-Pressure Vapor

FIGURE 41–8 Method of pumping refrigerant condensed in the high-pressure line over into the suction line. (A) Purging manifold. (B) Taking pressure readings. (C) Removing liquid refrigerant from high-side gage line.

erant is boiling at the correct temperature for the low side of the system at some load condition. It has been indicated previously that there are high-efficiency systems and standard-efficiency systems and that high-efficiency systems often have oversized evaporators. This makes the suction pressure slightly higher than normal. A standard-efficiency system usually has a refrigerant boiling temperature of about 35°F cooler than the entering air temperature at the standard operating condition of 75°F return air with a 50% humidity.

If the space temperature is 85°F and the humidity is 70%, the evaporator has an oversized load. It is absorbing an extra heat load, both sensible heat and latent heat, from the moisture in the air. You need to wait a sufficient time for the system to reduce the load before you can determine if the equipment is functioning correctly. Gage readings at this time will not reveal the kind of information that will verify the system performance unless there is a manufacturer's performance chart available.

41.6 HIGH-SIDE GAGE READINGS

Gage readings obtained on the high-pressure side of the system are used to check the relationship of the condensing refrigerant to the ambient air temperature. Standard air conditioning equipment condenses the refrigerant at no more than 30°F higher than the ambient temperature. For a 95°F entering air temperature, the head pressure should correspond to 95°F + 30°F = 125°F for a corresponding head pressure of 278 psig for R-22 and 169 psig for R-12. If the head pressure shows that the condensing temperature is higher than this, something is wrong.

When checking the condenser entering air temperature, be sure to check the actual temperature. Don't take the weather report as the ambient temperature. Air conditioning equipment located on a black roof has solar influence from the air being pulled across the roof, Figure 41–9. If the condenser is located close to any obstacles, such as below a sundeck, air may be circulated back through the condenser and cause the head pressure to be higher than normal, Figure 41–10.

High-efficiency condensers perform the same as standard-efficiency condensers except they operate at lower pressures and condensing temperatures. High-efficiency condensers normally condense the refrigerant at a temperature as low as 20°F higher than the ambient temperature. On a 95°F day the head pressure corresponds to a temperature of 95°F + 20°F = 115°F, which is a pressure of 243 psig for R-22.

41.7 TEMPERATURE READINGS

Temperature readings also can be useful. The four-lead electronic thermometer performs very well as a temperature reading instrument, Figure 41–11. It has small temperature leads that respond quickly to temperature changes. The leads can be easily attached to the refrigerant piping with electrical tape and insulated from the ambient temperature with short pieces of foam line insulation, Figure 41–12.

It is important that the lead be insulated from the ambient if the ambient temperature is different from the line temperature. This temperature lead is better and easier to use than the glass thermometers used in the past. It was almost impossible to get a true line reading by strapping a glass thermometer to a copper line, Figure 41–13.

Temperatures vary from system to system. The technician must be prepared to record accurately these temperatures to evaluate the various types of equipment. Some technicians record temperature readings of various equipment under different conditions for future reference. The common temperatures used would be inlet air wet- and dry-bulb, outdoor air dry-bulb, and the suction-line temperature. Sometimes, the compressor discharge temperature needs to be known. A thermometer with a range from -50°F to +250°F is a common instrument to be used for all of these tests.

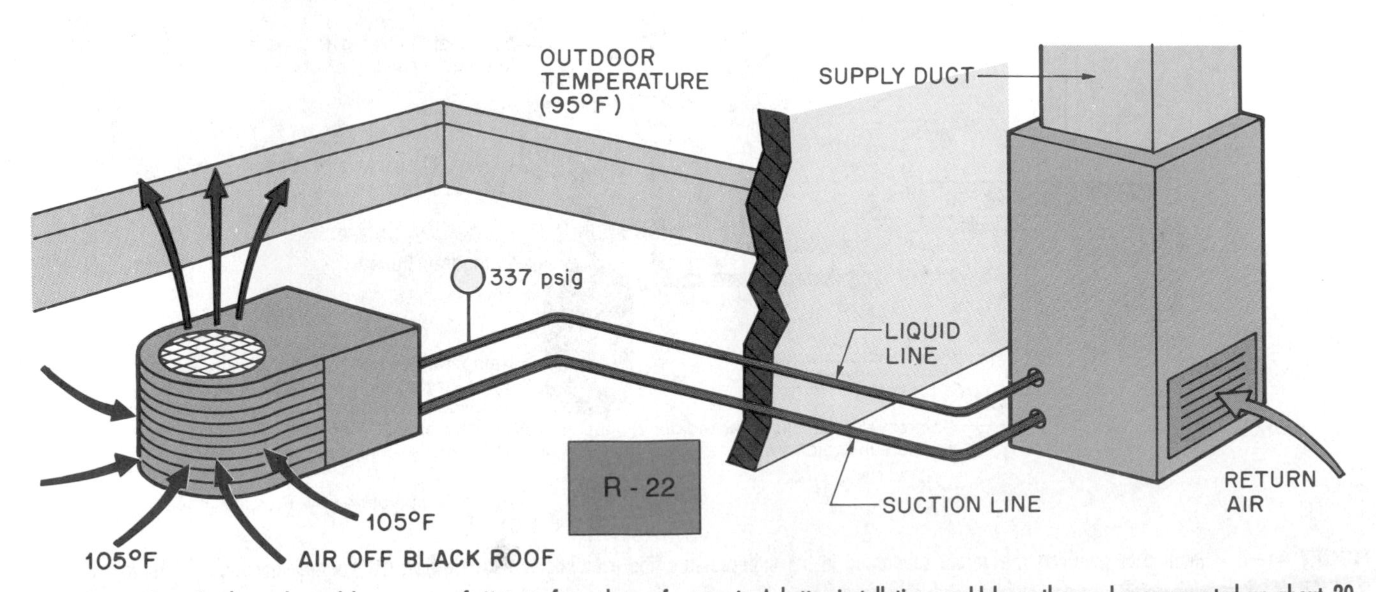

FIGURE 41–9 Condenser located low on a roof. Hot air from the roof enters it. A better installation would have the condenser mounted up about 20 in. so that the air could at least mix with the ambient air. Ambient air is 95°F but air off the roof is 105°F.

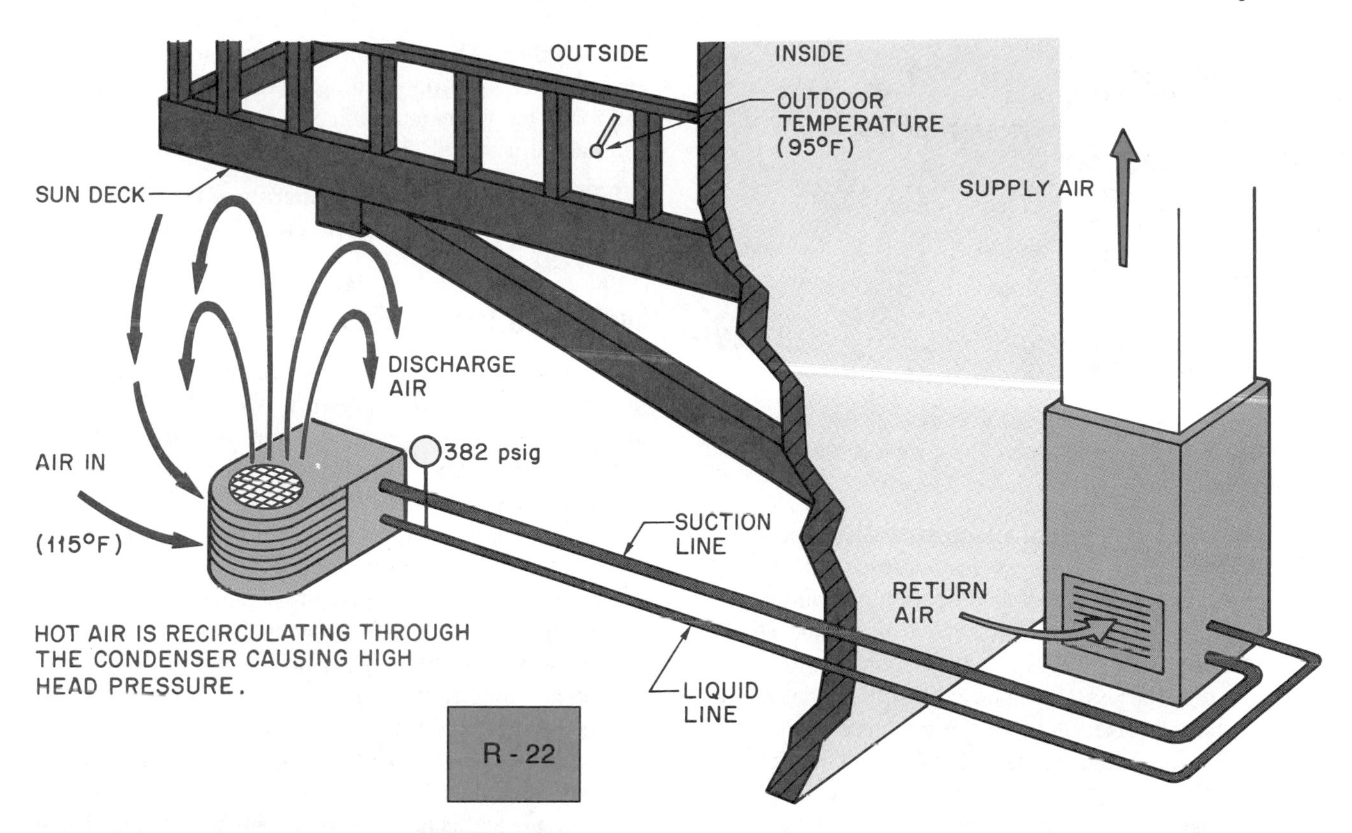

FIGURE 41–10 Condenser installed so that the outlet air is hitting a barrier in front of it.

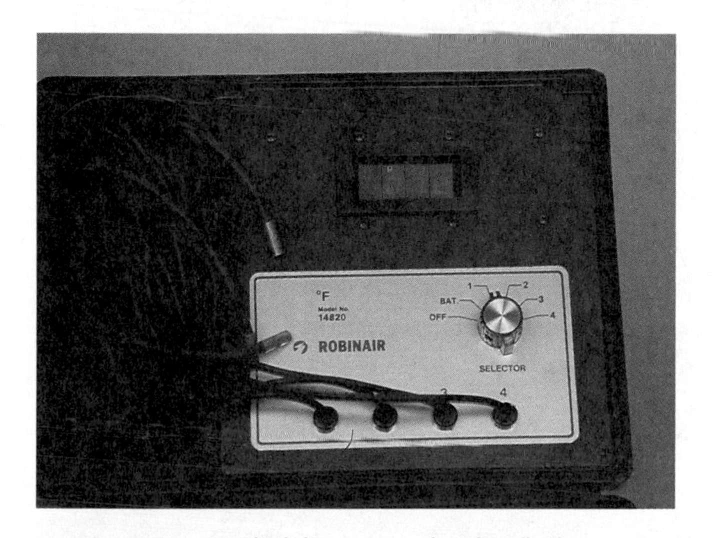

FIGURE 41–11 Four-lead thermometer. *Photo by Bill Johnson*

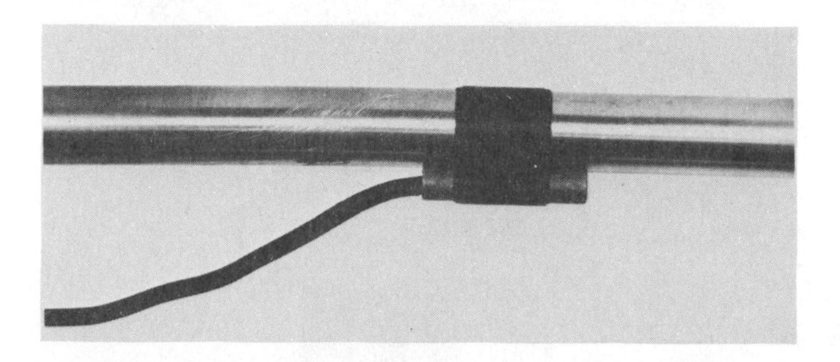

FIGURE 41–12 Temperature lead attached to a refrigerant line in the correct manner. It must also be insulated. *Photo by Bill Johnson*

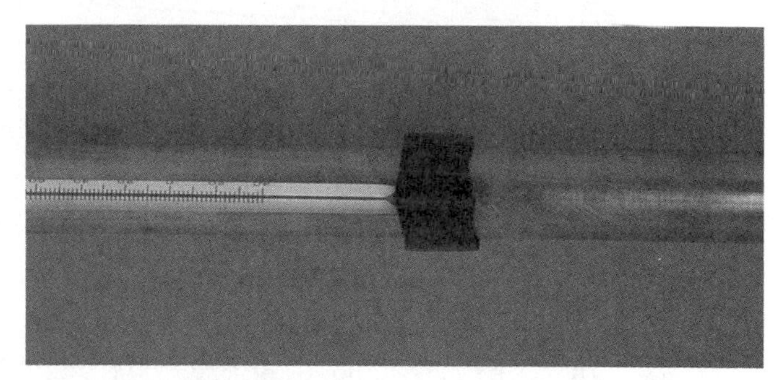

FIGURE 41–13 Glass thermometer strapped to a refrigerant line. *Photo by Bill Johnson*

Inlet Air Temperatures

It may be necessary to know the inlet air temperature to the evaporator for a complete analysis of a system. A wet-bulb reading for determining the humidity may be necessary. Such a reading can be obtained by using one of the temperature leads with a cotton sock that is saturated with pure water. A wet-bulb and a dry-bulb reading may be obtained by placing a dry-bulb temperature lead next to a wet-bulb temperature lead in the return airstream, Figure 41–14. The velocity of the return air will be enough to accomplish the evaporation for the wet-bulb reading.

Evaporator Outlet Temperatures

The evaporator outlet air temperature is seldom important. It may be obtained, however, in the same manner as the inlet air temperature. The outlet air dry-bulb temperature will

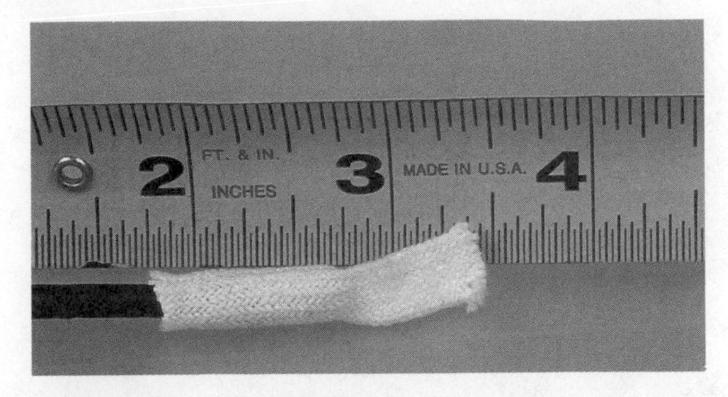

FIGURE 41–14 Electronic thermometer temperature lead has a damp cotton sock wrapped around it to convert it to a wet-bulb lead. *Photo by Bill Johnson*

normally be about 20°F less than the inlet air temperature. The temperature drop across an evaporator coil is about 20°F when it is operating at typical operating conditions of 75°F and 50% relative humidity return air. If the conditioned space temperature is high with a high humidity, the temperature drop across the same coil will be much less because of the latent-heat load of the moisture in the air.

If a wet-bulb reading is taken, there will be approximately a 10°F wet-bulb drop from the inlet to the outlet during standard operating conditions. The outlet humidity will be almost 90%. This air is going to mix with the room air and will soon drop in humidity because it will expand to the room air temperature. It is very humid because it is contracted or shrunk from the cooling process.

SAFETY PRECAUTION: *The temperature lead must not be allowed to touch the moving fan while taking air temperature readings.*

Suction-Line Temperatures

The temperature of the suction line returning to the compressor and the suction pressure will help the technician understand the characteristics of the suction gas. The suction gas may be part liquid if the filters are stopped up or if the evaporator coil is dirty, Figure 41–15. The suction gas may have a high superheat if the unit has a low charge or if there is a refrigerant restriction, Figure 41–16. The combination of suction-line temperature and pressure will help the service technician decide whether the system has a low charge or a stopped-up air filter in the air handler. For example, if the suction pressure is too low and the suction line is warm, the system has a starved evaporator, Figure 41–17. If the suction-line temperature is cold and the pressure indicates that the refrigerant is boiling at a low temperature, the coil is not absorbing heat as it should. The coil

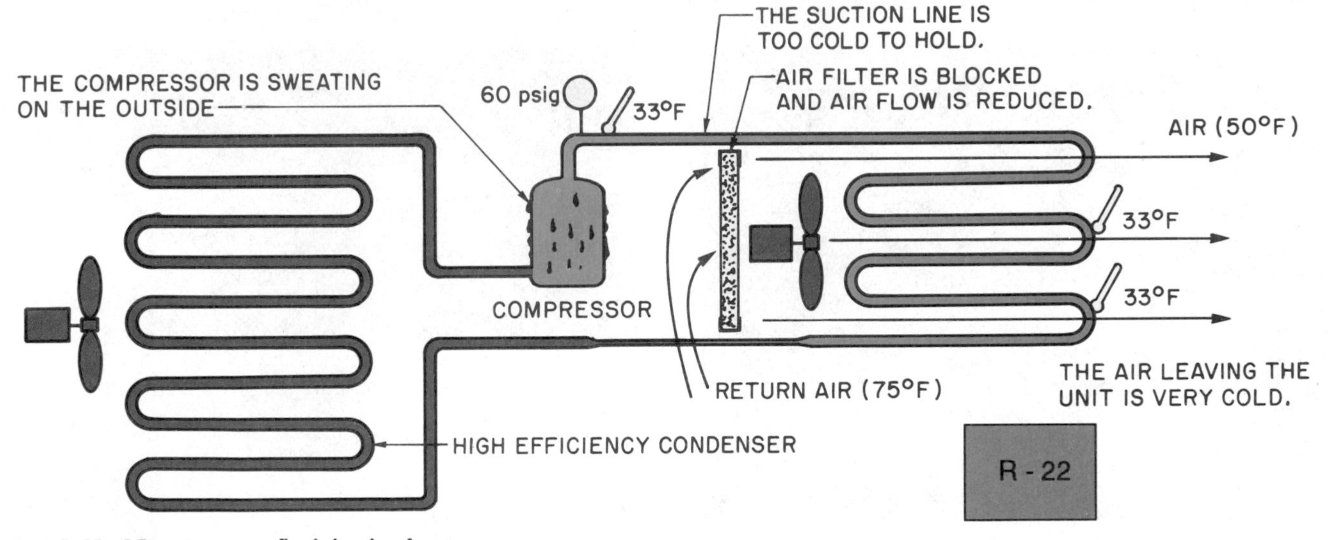

FIGURE 41–15 Evaporator flooded with refrigerant.

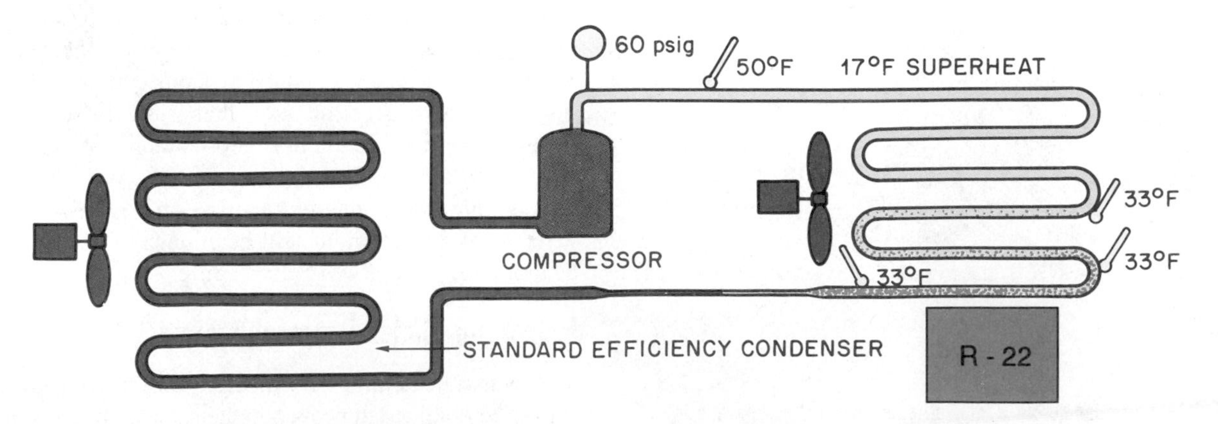

FIGURE 41–16 Evaporator is starved for refrigerant because of a low refrigerant charge.

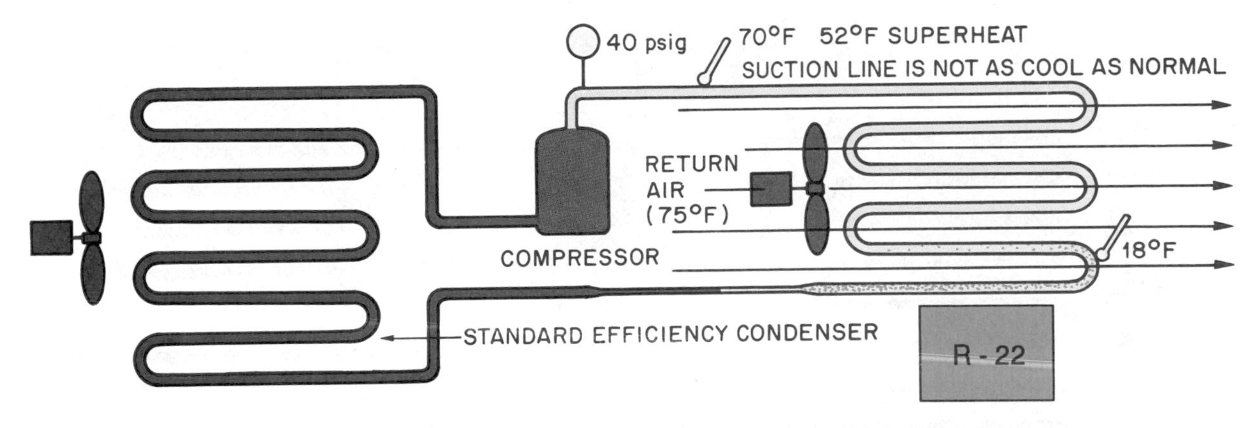

FIGURE 41–17 If the suction pressure is too low, and the suction line is not as cool as normal, the evaporator is starved for refrigerant. The unit may be low in refrigerant charge.

may be dirty, or the airflow may be insufficient, Figure 41–18. The cold suction line indicates that the unit has enough charge because the evaporator must be full for the refrigerant to get back to the suction line, Figure 41–19.

Discharge-Line Temperatures

The temperature of the discharge line may tell the technician that something is wrong inside the compressor. If there is an internal refrigerant leak from the high-pressure side to the low-pressure side, the discharge gas temperature will go up, Figure 41–20. Normally the discharge-line temperature at the compressor would not exceed 220°F for an air conditioning application even in very hot weather. When a high discharge-line temperature is discovered, the probable cause is an internal leak. The technician can prove this by building up the head pressure as high as 300 psig and then shutting off the unit. If there is an internal leak, this pressure difference between the high and the low sides can often be heard (as a whistle) equalizing through the compressor. If the suction line at the compressor shell starts to warm up immediately, the heat is coming from the discharge of the compressor.

SAFETY PRECAUTION: *The discharge line of a compressor may be as hot as 220°F under normal conditions, so be careful while attaching a temperature lead to this line.*

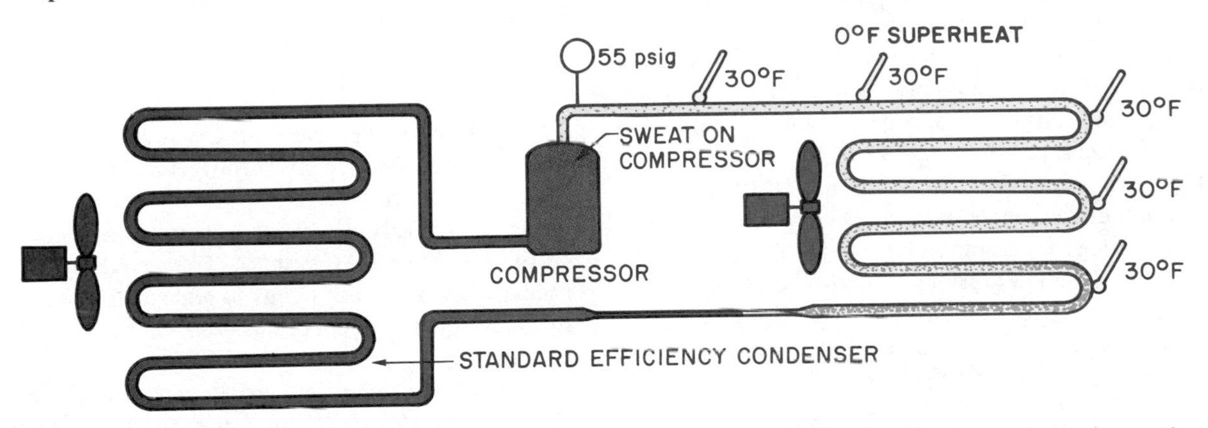

FIGURE 41–18 When the suction pressure is too low and the superheat is low, the unit is not boiling the refrigerant in the evaporator. The coil is flooded with liquid refrigerant.

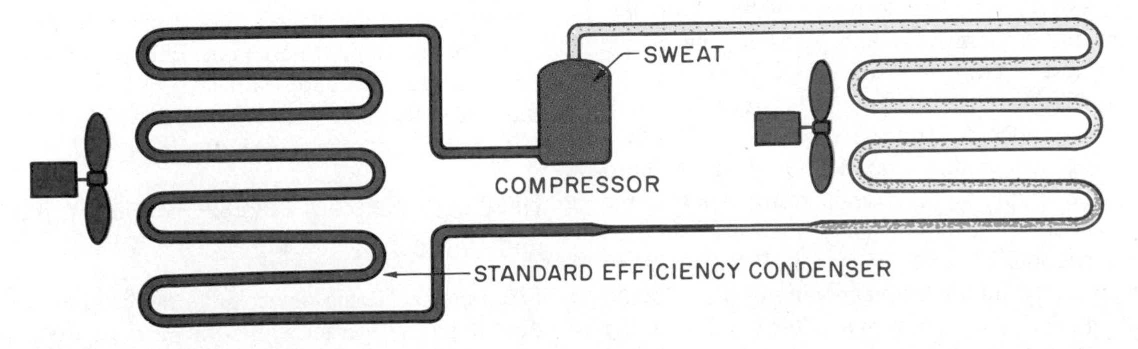

FIGURE 41–19 Evaporator full of refrigerant.

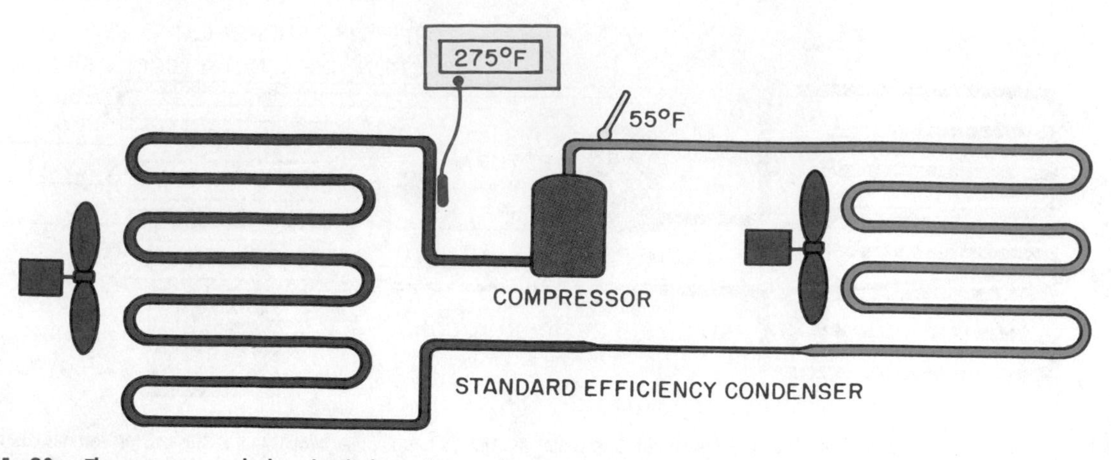

FIGURE 41–20 Thermometer attached to the discharge line on this compressor. If the compressor has an internal leak, the hot discharge gas will circulate back through the compressor. The discharge gas will be abnormally hot. When a compressor is cooled by suction gas, a high superheat will cause the compressor discharge gas to be extra hot.

Liquid-Line Temperatures

Liquid-line temperature may be used to check the subcooling efficiency of a condenser. Most condensers will subcool the refrigerant to between 10°F and 20°F below the condensing temperature of the refrigerant. If the condensing temperature is 125°F on a 95°F day, the liquid line leaving the condenser may be 105°F to 115°F when the system is operating normally. If there is a slight low charge, there might not be as much subcooling and the system efficiency therefore will not be as good. The condenser performs three functions: (1) removes the superheat from the discharge gas, (2) condenses the refrigerant to a liquid, and (3) subcools the liquid refrigerant below the condensing temperature. All three of these functions must be successfully accomplished for the condenser to operate at its rated capacity.

It's a good idea for a new technician to take the time to completely check out a working system operating at the correct pressures. Apply the temperature probes and gages to all points to actually verify the readings. This will provide reference points to remember.

41.8 CHARGING PROCEDURES IN THE FIELD

While establishing field charging procedures the technician should keep in mind what the designers of the equipment intended, how the equipment should perform. The charge consists of the correct amount of refrigerant in the evaporator, the liquid line, the discharge line between the compressor and the condenser, and the suction line. The discharge line and the suction line do not hold as much refrigerant as the liquid line because the refrigerant is in the liquid state. Actually, the liquid line is the only interconnecting line that contains much refrigerant. When a system is operating correctly, under design conditions, there should be a prescribed amount of refrigerant in the condenser, the evaporator, and the liquid line.

Understanding the following statements is important to the technician in the field:

1. The amount of refrigerant in the evaporator can be measured by the superheat method.
2. The amount of refrigerant in the condenser can be measured by the subcooling method.
3. The amount of refrigerant in the liquid line may be determined by measuring the length and calculating the refrigerant charge. However, in field service work, if the evaporator is performing correctly, the liquid line has the correct charge.

When the above statements are understood, the technician can check for the correct refrigerant level to determine the correct charge.

A field charging procedure may be used to check the charge of some typical systems. The technician sometimes needs typical reference points to add small amounts of gas for adjusting the amount of refrigerant in equipment that has no charging directions.

Often the technician arrives and finds the system charge needs adjusting. This can occur due to an over- or undercharge from the factory or from a previous technician's work. It can also occur from system leaks. *System leaks should always be repaired to prevent further loss of charge to the atmosphere.* The technician must establish charging procedures to use in the field for all types of equipment. These procedures will help get the system back on line under emergency situations. Following are some methods used for different types of equipment.

Fixed-Bore Metering Devices—Capillary Tube and Orifice Type

Fixed-bore metering devices like the capillary tube do not throttle the refrigerant as the thermostatic expansion valve (TXV) does. They allow refrigerant flow based on the difference in the inlet and the outlet pressures. The one time when the system can be checked for the correct charge and everything will read normal is at the typical operating condition of 75°F and 50% humidity return air and 95°F outside ambient air. If other conditions exist, different pressures and different superheat readings will occur. The item that most affects the readings is the outside ambient temperature. When it is lower than normal, the condenser will become more efficient and will condense the refrigerant sooner in the coil. This will have the effect of partially starving the evaporator for refrigerant. Refrigerant that is in the condenser that should be in the evaporator starves the evaporator.

When you need to check the system for correct charge or to add refrigerant, the best method is to follow the manufacturer's instructions. If they are not available, the typical operating condition may be simulated by reducing the airflow across the condenser to cause the head pressure to rise. On a 95°F day the highest condenser head pressure is usually 278 psig for R-22 (95°F ambient + 30°F added condensing temperature difference = 125°F condensing temperature or 278 psig). Because the high pressure pushes the refrigerant through the metering device, when the head pressure is up to the high normal end of the operating conditions there is no refrigerant held back in the condenser.

When the condenser is pushing the refrigerant through the metering device at the correct rate, the remainder of the charge must be in the evaporator. A superheat check at the evaporator is not always easy with a split air conditioning system, so a superheat check at the condensing unit for a split system can be made. The suction line from the evaporator to the condensing unit may be long or short. Let's use two different lengths: up to 30 ft and from 30 to 50 ft for a test comparison. When the system is correctly charged, the superheat should be 10°F to 15°F at the condensing unit with a line length of 10 to 30 ft. The superheat should be 15°F to 18°F when the line is 30 to 50 ft long. Both of these conditions are with a head pressure of 278 psig ±10 psig. At these conditions the actual superheat at the evaporator will be close to the correct superheat of 10°F. When using this method, be sure that you allow enough time for the system to settle down after adding refrigerant, before you draw any conclusions, Figure 41–21. Orifice-type metering devices have a tendency to hunt while reaching steady-state operation. This hunting can be observed by watching the suction pressure rise and fall accompanied by the superheat as the suction-line temperature changes. When this occurs, the technician will have to use averaging to arrive at the proper superheat for the coil. For example, if it is varying between 6°F and 14°F, the average value of 10°F superheat may be used.

Field Charging the TXV System

The TXV system can be charged in much the same way as the fixed-bore system, with some modifications. The condenser on a TXV system will also hold refrigerant back in mild ambient conditions. This system always has a refrigerant reservoir or receiver to store refrigerant and will not be affected as much as the capillary tube by lower ambient conditions. To check the charge, restrict the airflow across the condenser until the head pressure simulates a 95°F ambient, 278 psig head pressure for R-22. Using the superheat method will not work for this valve because if there is an overcharge, the superheat will remain the same. Superheats of 15°F to 18°F are not unusual when measured at the con-

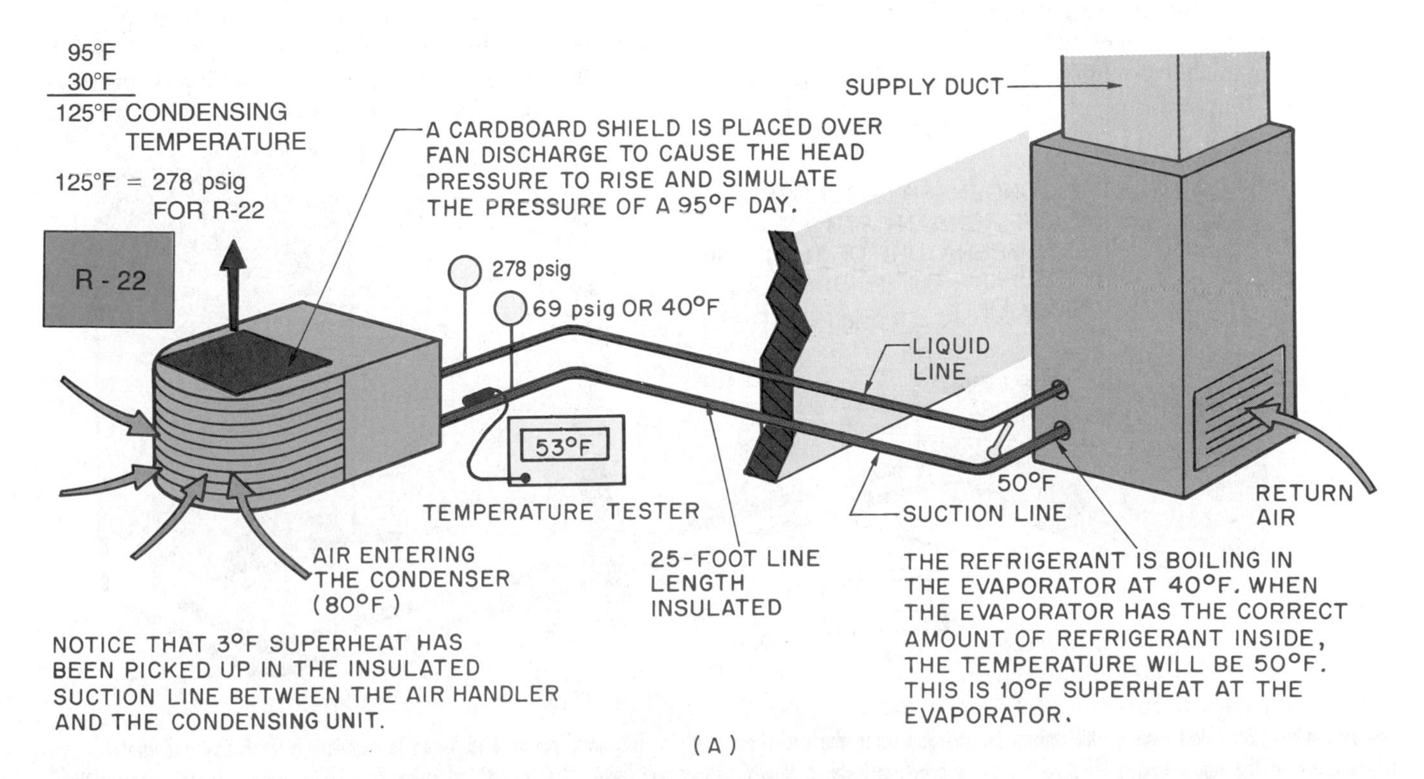
95°F
30°F
125°F CONDENSING TEMPERATURE
125°F = 278 psig FOR R-22
R - 22
SUPPLY DUCT
A CARDBOARD SHIELD IS PLACED OVER FAN DISCHARGE TO CAUSE THE HEAD PRESSURE TO RISE AND SIMULATE THE PRESSURE OF A 95°F DAY.
278 psig
69 psig OR 40°F
LIQUID LINE
53°F
TEMPERATURE TESTER
50°F
SUCTION LINE
RETURN AIR
AIR ENTERING THE CONDENSER (80°F)
25-FOOT LINE LENGTH INSULATED
THE REFRIGERANT IS BOILING IN THE EVAPORATOR AT 40°F. WHEN THE EVAPORATOR HAS THE CORRECT AMOUNT OF REFRIGERANT INSIDE, THE TEMPERATURE WILL BE 50°F. THIS IS 10°F SUPERHEAT AT THE EVAPORATOR.
NOTICE THAT 3°F SUPERHEAT HAS BEEN PICKED UP IN THE INSULATED SUCTION LINE BETWEEN THE AIR HANDLER AND THE CONDENSING UNIT.
(A)

FIGURE 41–21 System charged by raising the discharge pressure to simulate a 95°F day.

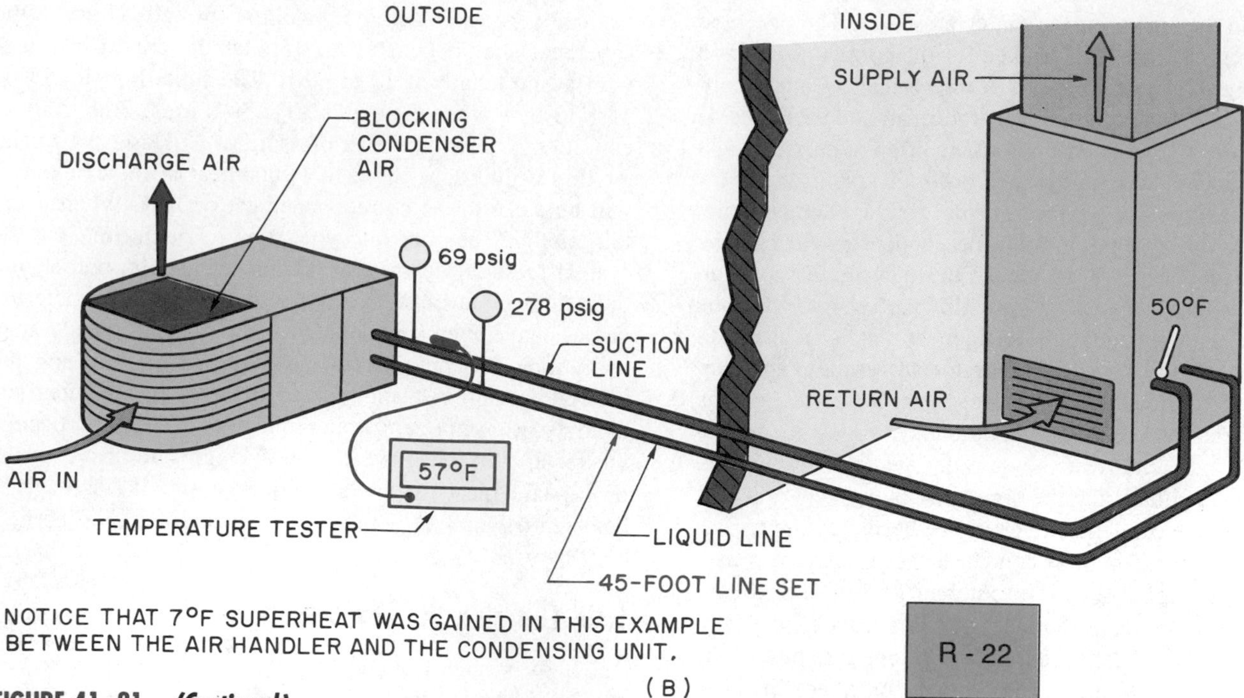

FIGURE 41–21 ***(Continued)***

densing unit for TXVs. If the sight glass is full, the unit has at least enough refrigerant, but it may have an overcharge. If the unit does not have a sight glass, a measure of the subcooling of the condenser may tell you what you want to know. For example, a typical subcooling circuit will subcool the liquid refrigerant from 10°F to 20°F cooler than the condensing temperature. A temperature lead attached to the liquid line should read 115°F to 105°F, or 10°F to 20°F cooler than the condensing temperature of 130°F, Figure 41–22. If the subcooling temperature is 20°F to 25°F cooler than the condensing temperature, the unit has an overcharge of refrigerant and the bottom of the condenser is acting as a large subcooling surface.

The charging procedures just described will also work for high-efficiency equipment. The head pressure does not need to be operated quite as high. A head pressure of 250 psig will be sufficient for an R-22 system when charging.

SAFETY PRECAUTION: *Refrigerant in cylinders and in the system is under great pressure (as high as 300 psi for R-22).* ⊛Use proper safety precautions when transferring refrigerant, and be careful not to overfill tanks or cylinders when recovering refrigerant. Do not use disposable cylinders by refilling them. Use only DOT-approved tanks or cylinders.⊛ *The high-side pressure may be reduced for attaching and removing gage lines

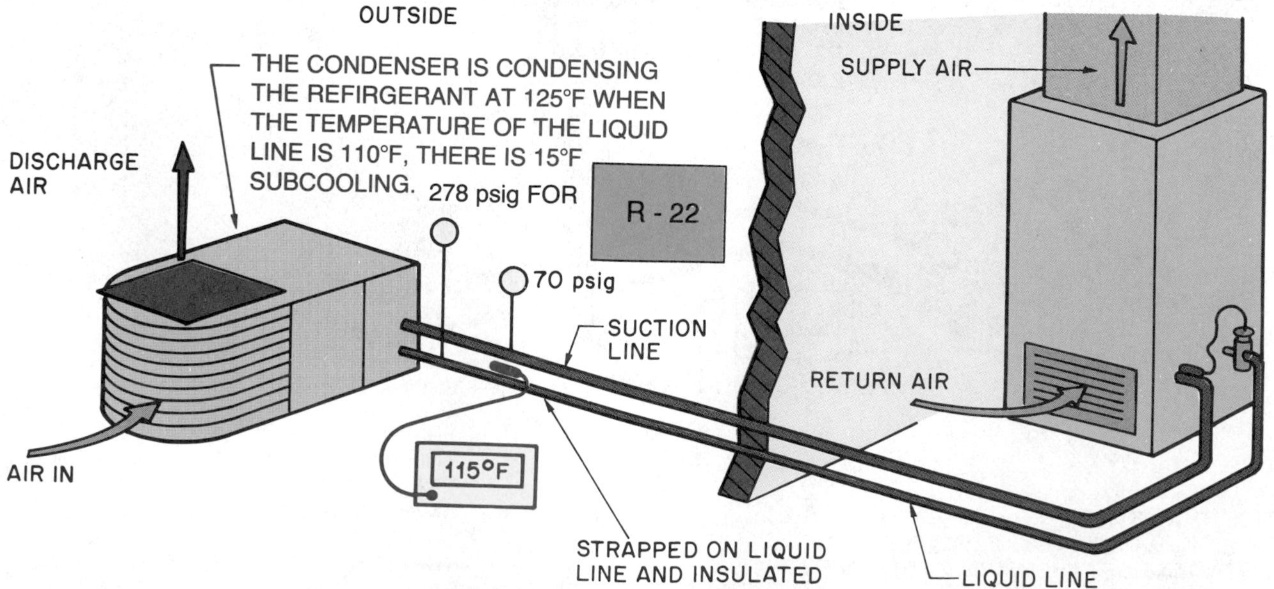

FIGURE 41–22 Unit with a TXV cannot be charged using the superheat method. The head pressure is raised to simulate a 95°F day and the temperature of the liquid line is checked for the subcooling level. A typical system will have 10°F to 20°F of subcooling when the condenser contains the correct charge.

by shutting off the unit and allowing the unit pressures to equalize.*

41.9 ELECTRICAL TROUBLESHOOTING

Electrical troubleshooting is often required at the same time as mechanical troubleshooting. The volt-ohm-milliammeter (VOM) and the clamp-on ammeter are the primary instruments used, Figure 41–23.

You need to know what the readings should be to know whether the existing readings on a particular unit are correct. This is often not easy to determine because the desired reading may not be furnished. Figure 18–8 shows typical horsepower-to-amperage ratings. It is a valuable tool for determining the correct amperage for a particular motor.

For a residence or small commercial building one main power panel will normally serve the building. This panel is divided into many circuits. For a split-system type of cooling system there are usually separate breakers (or fuses) in the main panel for the air handler or furnace for the indoor unit and for the outdoor unit. For a package or self-contained system, usually one breaker (or fuse) serves the unit, Figure 41–24. The power supply voltage is stepped down by the control transformer to the control voltage of 24 V.

Begin any electrical troubleshooting by verifying that the power supply is energized and that the voltage is correct. One way to do this is to go to the room thermostat and see if the indoor fan will start with the FAN ON switch. See Figure 41–25 for a wiring diagram of a typical split-system air conditioner.

The air handler or furnace, where the low-voltage transformer is located, is frequently under the house or in the attic. This quick check with the FAN ON switch can save you a trip under the house or to the attic. If the fan will start with the fan relay, several things are apparent: (1) the indoor fan operates; (2) there is control voltage; and (3) there is line voltage to the unit because the fan will run. When taking a service call over the phone, ask the homeowner if the indoor fan will run. If it doesn't, take a transformer. This could save a trip to the supply house or the shop.

SAFETY PRECAUTION: *All safety practices must be observed while troubleshooting electrical systems. Many times the system must be inspected while power is on. Only let the insulated meter leads touch the hot terminals. Special care should be taken while troubleshooting the main power supply because the fuses may be correctly sized large enough to allow great amounts of current to flow before they blow (eg, when a screwdriver slips in the panel and shorts across hot terminals). ***Never* use a screwdriver in a hot panel.***

If the power supply voltages are correct, move on to the various components. The path to the load may be the next item to check. If you're trying to get the compressor to run, remember that the compressor motor is operated by the compressor contactor. Is the contactor energized? Are the contacts closed? See Figure 41–26 for a diagram. Note that in the diagram the only thing that will keep the contactor coil from being energized is the thermostat, the path, or the low-pressure control. If the outdoor fan is operating and the

(A)

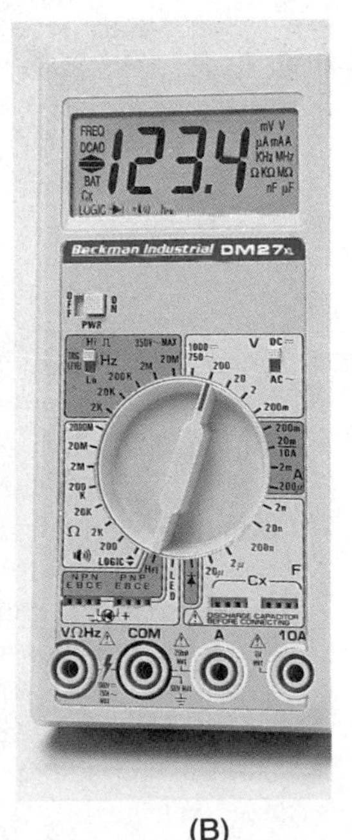

(B)

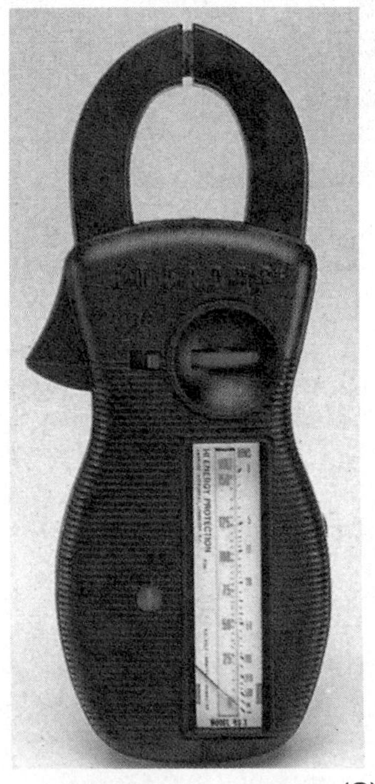

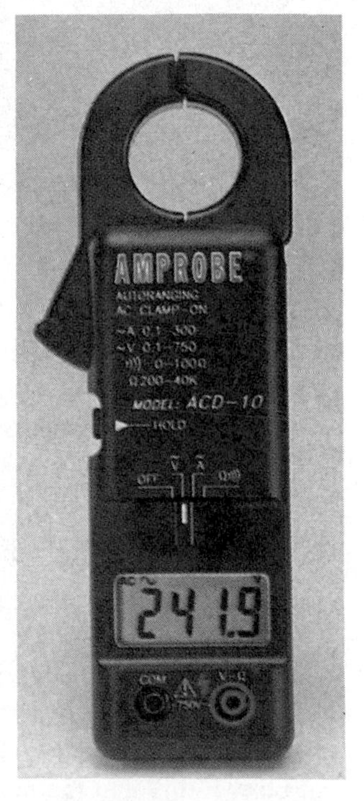

(C)

FIGURE 41–23 Instruments used to troubleshoot the electrical part of an air conditioning system. (A) Analog VOM. (B) Digital VOM. (C) Clamp-on ammeter. *(A) Photo by Bill Johnson, Courtesy (B) Beckman Industrial Corporation, (C) Amprobe Instrument Division of Core Industries Inc.*

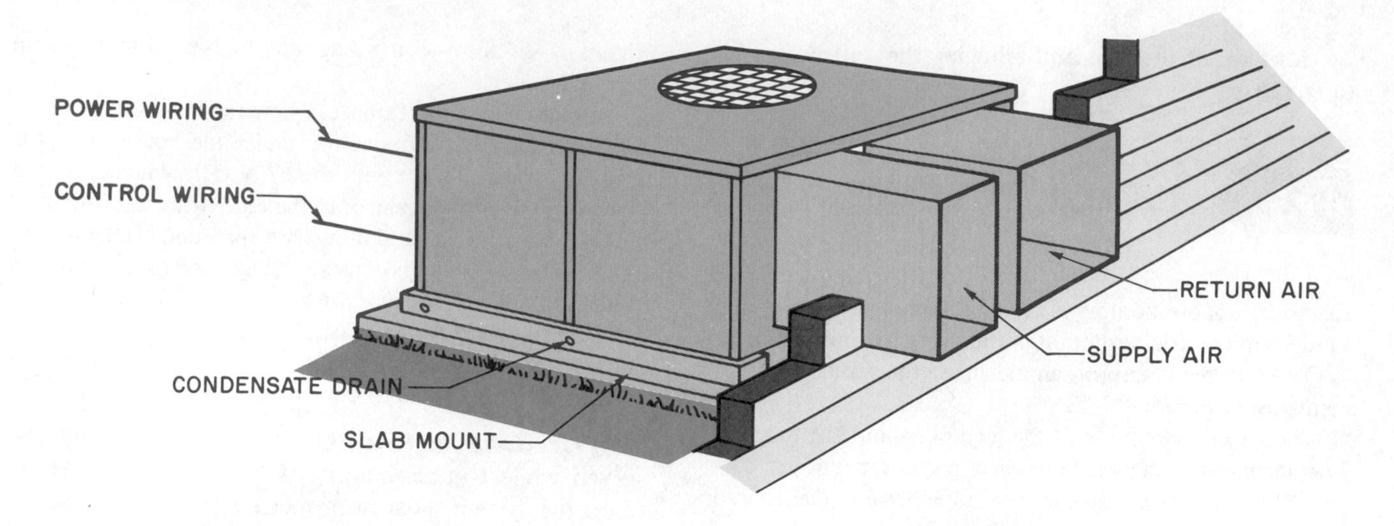

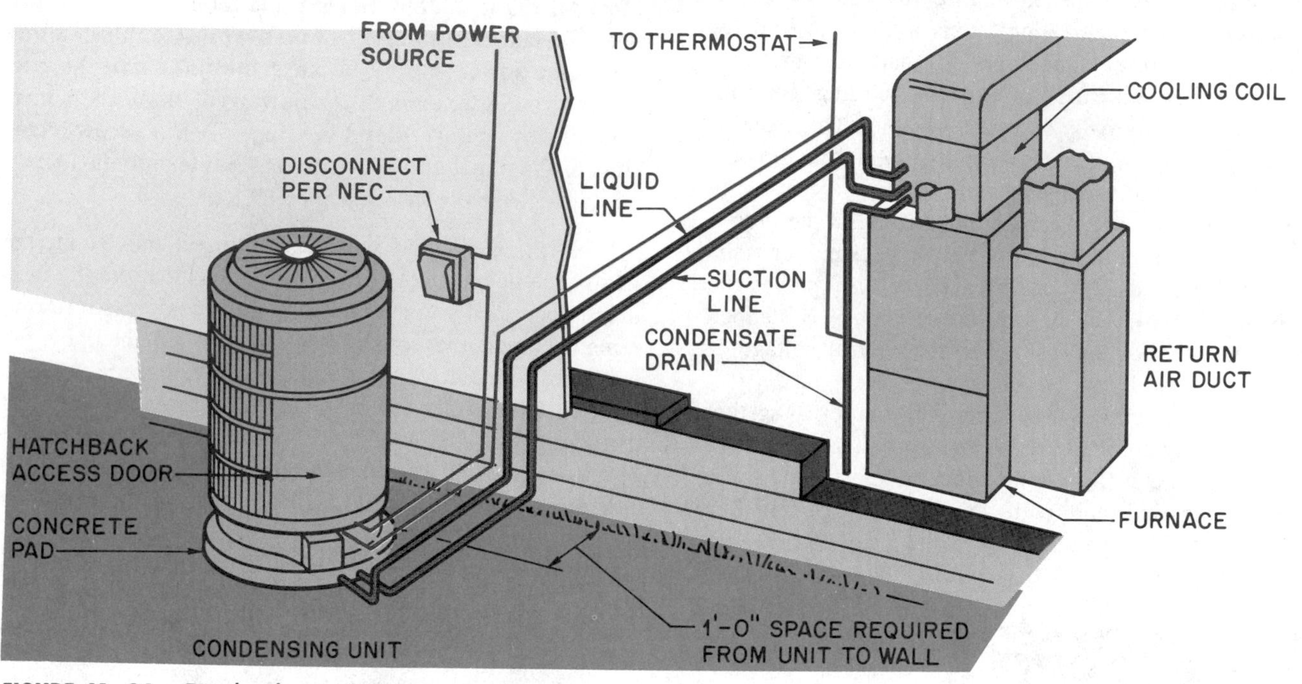

FIGURE 41–24 Typical package and split-system installation.

compressor is not, the contactor is energized because it also starts the fan. If the fan is running and the compressor is not, either the path (wiring or terminals) is not making good contact or the compressor internal overload protector is open.

41.10 COMPRESSOR OVERLOAD PROBLEMS

When the compressor overload protector is open, touch the motor housing to see if it is hot. If you cannot hold your hand on the compressor shell, the motor is too hot. Ask yourself these questions: Can the charge be low (this compressor is suction gas cooled)? Can the start-assist circuit not be working and the compressor not starting?

Allow the compressor to cool before restarting it. It is best that the unit be fixed so that it will not come back on for several hours. The best way to do this is to remove a low-voltage wire. If you pull the disconnect switch and come back the next day, the refrigerant charge may be in the crankcase because there was no crankcase heat. If you want to start the unit within the hour rather than waiting, pull the disconnect switch and run a small amount of water through a hose over the compressor. ***The standing water poses a potential electrical hazard. Be sure that all electrical components are protected with plastic or other waterproof covering. Don't come in contact with the water or electrical current when working around live electricity.*** It will take about 30 min to cool. Have the gages on the unit and a cylinder of refrigerant connected because when the compressor is started up by closing the disconnect, it may need refrigerant. If the system has a low charge and you have to get set up to charge after starting the system, the compressor may cut off again from overheating before you have a chance to get the gages connected.

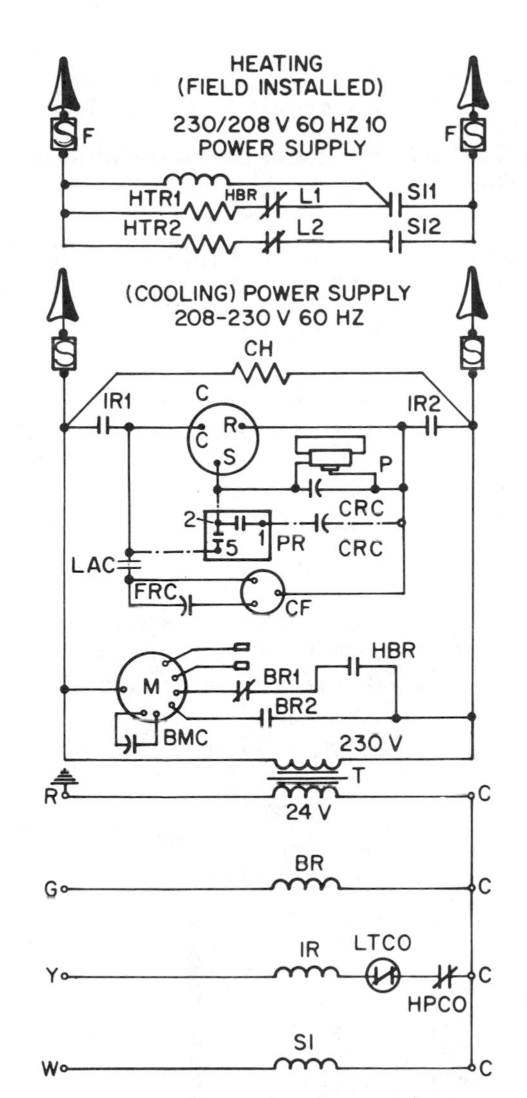

COMPONENT PART IDENTIFICATION

BMC	BLOWER MOTOR CAPACITOR
BR	BLOWER RELAY
C	COMPRESSOR
CF	CONDENSER FAN MOTOR
CH	CHRANKCASE HEATER
CRC	COMPRESSOR RUN CAPACITOR
CS	EXHAUSTER CENTRIFUGAL SWITCH
CSC	COMPRESSOR START CAPACITOR
EM	EXHAUSTER MOTOR
F	FUSE
FC	FAN CONTROL KLIXON
FRC	FAN RUN CAPACITOR
FT	FAN TIMER
GV	GAS VALVE
HBR	BLOWER RELAY (HEATING)
HPCO	HIGH-PRESSURE CUT-OUT
HR	HEAT RELAY
HTR	ELECTRIC HEATER
IR	COMPRESSOR CONTACTOR
L	LIMIT
LAC	LOW AMBIENT CONTROL (0°F) (ALL RD/RG-D UNITS & ALL R-H RG-H3 PHASE UNITS)
LTCO	LOW TEMPERATURE CUT-OUT (50°F) (ALL R-H/RG-H 1 PHASE UNITS)
M	BLOWER MOTOR
P	PTC STANDARD START ASSIST
PPK	POST PURGE KLIXON
PR	POTENTIAL RELAY
PR+ CDC	OPTIONAL START ASSIST (REPLACES "P")
S1	SEQUENCER
T	TRANSFORMER

FIGURE 41–25 Wiring diagram of a split-system summer air conditioner. *Courtesy Climate Control*

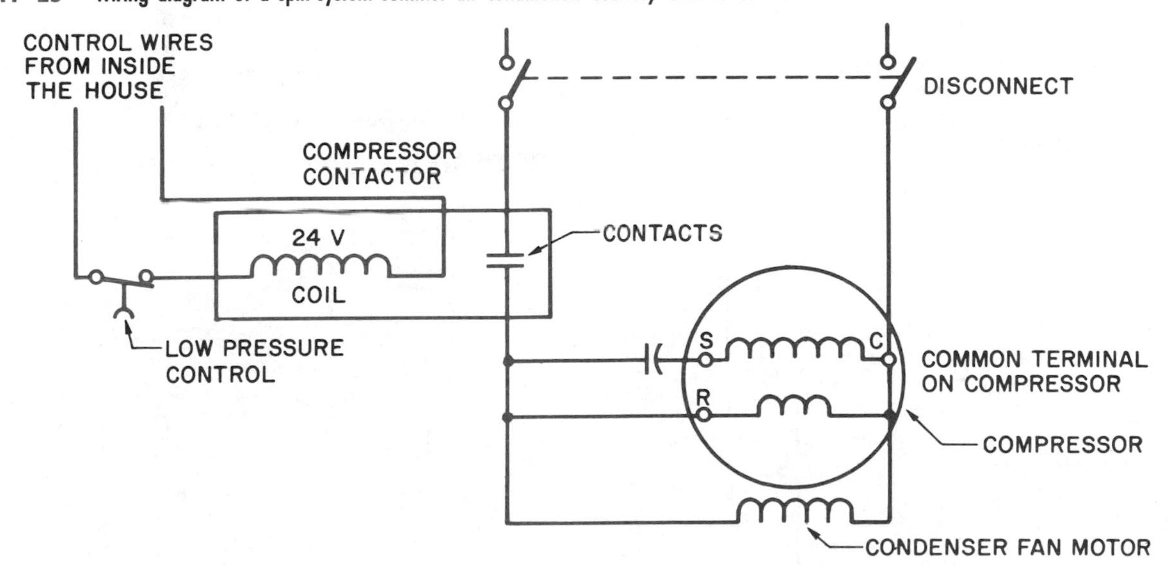

FIGURE 41–26 Wiring diagram of basic components that appear in a control and compressor circuit.

41.11 COMPRESSOR ELECTRICAL CHECKUP

You may need to perform an electrical check of the compressor if the compressor will not start or if a circuit protector has opened. For example, suppose that the compressor can be heard trying to start. It will make a humming noise but will not turn. Check the compressor with an ohmmeter to see if all of the windings are correct. Remember, a load has to be present before current flows. The load must have the correct resistance. Let's say a compressor specification calls for the run winding to have a resistance of 4 Ω and the start winding a resistance of 15 Ω. If the ohmmeter indicates that the start winding has only 10 Ω, then the winding has a short circuit (sometimes called a *shunt*). This will change the winding characteristics, and the compressor will not start. It is defective and must be changed, Figure 41–27. It is important to have quality instruments for making these checks.

The ohmmeter check may show the compressor to have an open circuit in the start or run windings. Suppose that the same symptom of a hot compressor is discovered. The compressor is allowed to cool and it still will not start. An ohm check shows that the start or run winding is open. This compressor is also defective and must be changed, Figure 41–28.

When the continuity check indicates that the common circuit in the compressor is open, it could be that the internal overload protector is open because the motor may not have cooled enough. The motor is suspended in a vapor space inside the shell, and it takes time to cool, Figure 41–29.

SAFETY PRECAUTION: *Never use an ohmmeter to check a live circuit.*

41.12 TROUBLESHOOTING THE CIRCUIT ELECTRICAL PROTECTORS—FUSES AND BREAKERS

One service call that must be treated cautiously is when a circuit protector such as a fuse or breaker opens the circuit. The compressor and fan motors have protection that will normally guard them from minor problems. The breaker or fuse is for large current surges in the circuit. When one is tripped, don't simply reset it. Perform a resistance check of the compressor section, including the fan motor. ***The compressor might be grounded (has a circuit to the case of the compressor), and it will be harmful to try to start it.* Be sure to isolate the compressor circuit before condemning the compressor.** Take the motor leads off the compressor to check for a ground circuit in the compressor, Figure 41–30.

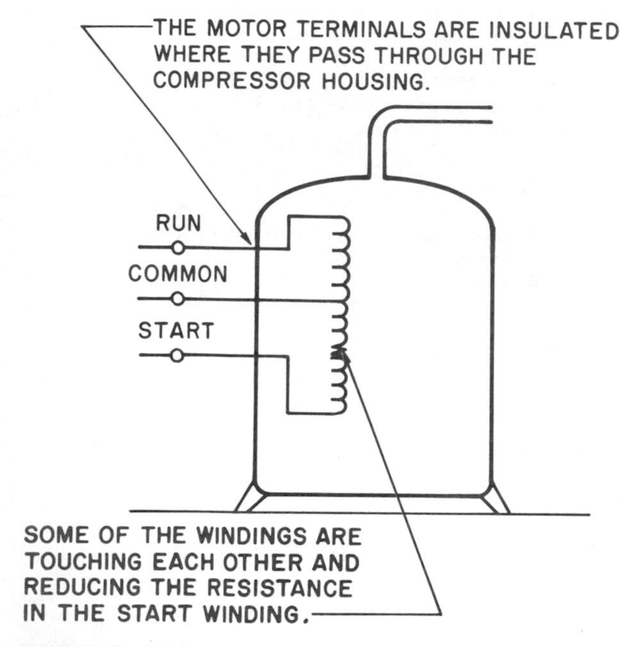

FIGURE 41–27 Compressor with shorted winding.

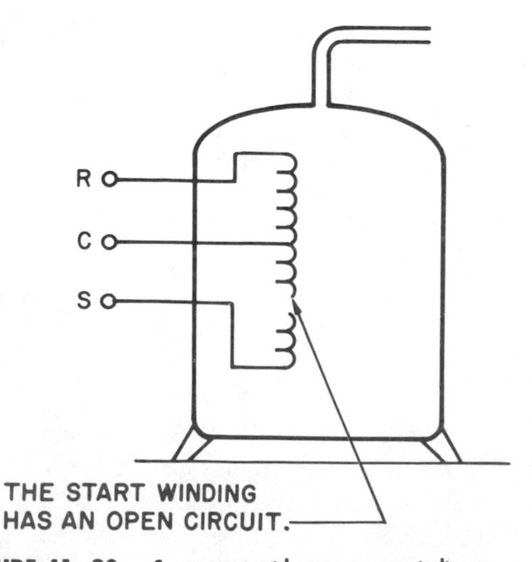

FIGURE 41–28 Compressor with open start winding.

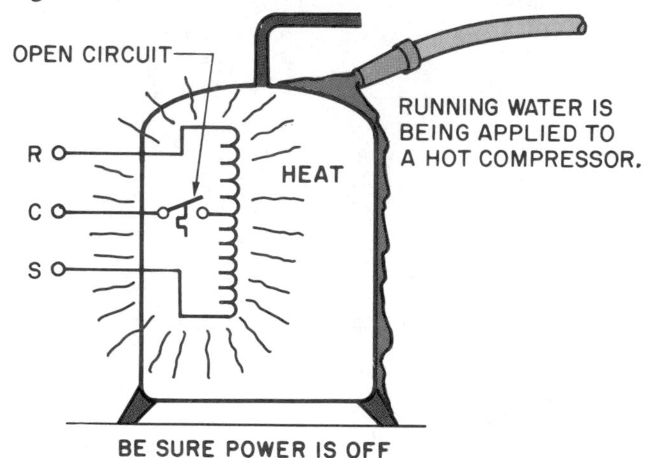

FIGURE 41–29 Cooling a compressor with water.

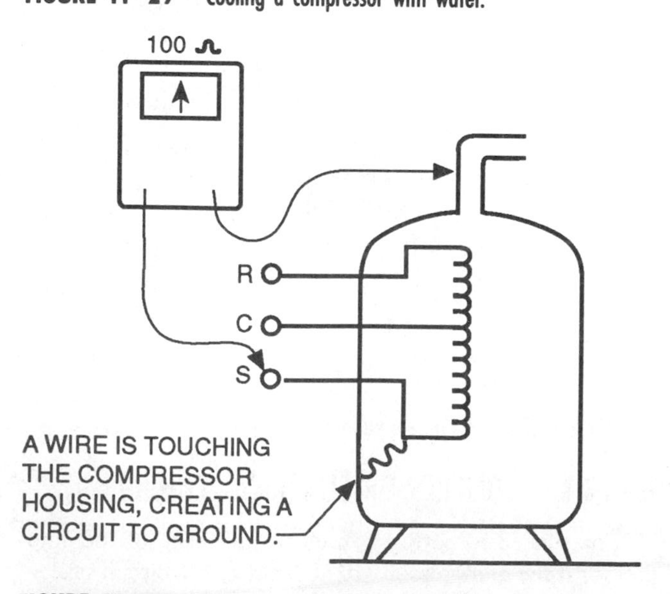

FIGURE 41–30 Compressor with winding shorted to the casing.

PREVENTIVE MAINTENANCE

Air conditioning equipment preventive maintenance involves the indoor airside, the outdoor airside (air cooled and water cooled), and electrical circuits.

The indoor airside maintenance is much the same for air conditioning as for electric heat, where motor and filter maintenance is discussed. The only difference is that the evaporator coil operates below the dew point temperature of the air and is wet. It will become a superfilter for any dust particles that may pass through the filter or leak in around loose panel compartment doors. Most air handlers have draw-through coil-fan combinations. The air passes through the coil before the fan. When the fan blades become dirty and loaded with dirt, it is a sure sign that the coil is dirty. The dirt has to pass through the coil first, and it is wet much of the time. When the fan is dirty, the coil must be cleaned.

The coil may be cleaned in place in the unit by two methods. One is by the use of a special detergent manufactured to work while the coil is wet. This detergent is sprayed on the coil and into the core of the coil with a hand pump type sprayer, similar to a garden sprayer. When the unit is started, the condensate will carry the dirt down the coil and down the condensate drain line. This type of cleaner is for light-duty cleaning. Care must be taken that the condensate drain line does not become clogged with the dirt from the coil and pan.

The coil may also be cleaned by shutting the unit down and applying a more powerful detergent to the coil, forcing it into the coil core. After the detergent has had time to work, the coil is then sprayed with a water source, such as a water hose. Care must be taken not to force too much water into the coil, or the drain pan will overflow. The water must not be applied faster than the drain can accept it.

Special pressure cleaners that have a nozzle pressure of 500 to 1000 psig may be used. These units have a low water flow, about 5 gal/min, and will not overflow an adequate drain system. The high nozzle pressure allows the water to clean the core of the coil.

It is always best to "back-wash" a coil when cleaning with water. The water is forced through the coil in the opposite direction to the airflow for back-washing. The reason for this is most of the dirt will accumulate on the inlet of the coil and be progressively less as the air moves through the coil. If the coil is not back-washed, you may only drive dirt accumulation to the center of the coil. ***Never use hot water or a steam cleaner on refrigeration equipment if there is refrigerant in the unit because pressures will rise high enough to burst the unit at the weakest point. This may be the compressor shell. A refrigerant system must be open to the atmosphere before hot water or steam is used for cleaning.***

In some cases, the coil cannot be cleaned in the unit and must be removed for cleaning. If the unit has service valves, the refrigerant may be pumped into the condenser-receiver and the coil removed from the unit. The coil will not have any refrigerant in it and may be cleaned with approved detergent and hot water or with a steam cleaner. Approved cleaners may be purchased at any air conditioning supply house. ***Follow the directions. Make sure no water can enter the coil through the piping connections.***

The outdoor unit may be either air cooled or water cooled. Air-cooled units have fan motors that must be lubricated. Some motors only require lubrication after several years of operation. Then a recommended amount of approved oil is added to the oil cup. Some motors require more frequent lubrication.

The fan blades should be checked to make sure they are secure on the shaft. At the same time, the rain shield on top of the fan motor should be checked if there is one. It should not allow water to enter the motor on upflow units where the motor is out in the open.

Coils may become dirty from pulling dust through the coils. It is hard to look at any coil and determine how much dirt is in the core. The coil should be cleaned at the first sign of dirt buildup or high operating head pressure.

Condenser coils may be easier to clean because they are on the outside. ***Make sure all power is turned off, locked, and tagged where someone can't accidentally turn it on. The fan motor must be covered and care should be taken that water does not enter the controls. They may not all be in the control cabinet.*** Apply an approved detergent on the coil using a handpump sprayer. Soak the coil to the middle. Let the detergent soak for 15 to 30 min, then back-wash the coil using a garden hose or spray cleaner. ***Never use hot water or steam if the unit has any refrigerant in it. Heat from hot water or steam would cause internal pressures that far exceed the system working pressure and may burst the system at the weakest point, which may be the compressor housing. The system should be open to the atmosphere before adding heat of any kind.***

Water-cooled equipment must be maintained like air-cooled equipment. Two things are done as part of the maintenance program to minimize the mineral concentrations in the water. One is to make sure the water tower has an adequate water bleed system. This is a measured amount of water that goes down the drain. This water is added back to the system with the supply water makeup line. Owners may suggest that the bleed system be shut off because all they see is water going down the drain. They may not realize that the purpose of this is to dilute the minerals in the water.

The other maintenance procedure is to use the correct water treatment. A water treatment specialist should supervise the setup of the treatment procedures. This will ensure the best quality water in the system.

Electrical preventive maintenance consists of examining any contactors and relays for frayed wire and pitted contacts. When these occur, they should be taken care of before you leave the job. There can be no prediction as to how long they will last. Single-phase compressors will just stop running if a wire burns in two or a contact burns away. A three-phase compressor will try to run on two phases and this is very hard on the motor. Wires should be examined to make sure they are not rubbing on the frame and wearing the insulation off the wire. Wires resting on copper lines may rub a hole in the line, resulting in a refrigerant leak.

HVAC GOLDEN RULES

- Always carry the proper tools and avoid extra trips through the customer's house.
- Make and keep firm appointments. Call ahead if you are delayed. The customer's time is valuable too.

Added Value to the Customer. Here are some simple, inexpensive procedures that may be included in the basic service call.

- Touch test the suction and liquid lines to determine possible over- or undercharge.
- Clean or change filters as needed.
- Lubricate all bearings as needed.
- Replace all panels with the correct fasteners.
- Inspect all contactors and wiring. This may lead to more paying service.
- Make sure the condensate line is draining properly. Clean if needed. Algicide tablets may be necessary.
- Check evaporator and condenser coils for blockage or dirt.

41.13 SERVICE TECHNICIAN CALLS

Troubleshooting can take many forms and cover many situations. The following actual troubleshooting situations will help you understand what the service technician does while solving actual problems.

SERVICE CALL 1

A residential customer calls to report that the central air conditioner at a residence is not cooling enough and runs continuously. *The problem is low refrigerant charge.*

The technician arrives and finds the unit running. The temperature indicator on the room thermostat shows the thermostat is set at 72°F, and the thermometer on the thermostat indicates that the space temperature is 80°F. The air feels very humid. The technician notices that the indoor fan motor is running. The velocity of the air coming out of the registers seems adequate, so the filters are not stopped up.

The technician goes to the condensing unit and hears the fan running. The air coming out of the fan is not warm. This indicates that the compressor is not running. The door to the compressor compartment is removed and it is noticed that the compressor is hot to the touch. This is an indication that the compressor has been trying to run. The gages are installed on the service ports, and they both read the same because the system has been off. The gages read 144 psig, which corresponds to 80°F. The residence is 80°F inside, and the ambient is 85°F, so it can be assumed that the unit has some liquid refrigerant in the system because the pressure corresponds so closely to the chart. If the system pressure were 100 psig, it would be obvious that little liquid refrigerant is left in the system. A large leak should be suspected.

The technician decides that the unit must have a low charge and that the compressor is off because of the internal overload protector. ***The technician pulls the electrical disconnect and takes a resistance reading across the compressor terminals with the motor leads removed.*** The meter shows an open circuit from common to run and common to start. It shows a measurable resistance between run and start. This indicates that the motor winding thermostat must be open. This indication is verified by the hot compressor.

All electrical components and terminals are covered with plastic. The electrical disconnect is locked out. A water hose is connected, and a small amount of water is allowed to run over the top of the compressor shell to cool it. A cylinder of refrigerant is connected to the gage manifold so that when it is time to start the compressor, the technician will be ready to add gas and keep the compressor running. The gage lines are purged of any air that may be in them. While the compressor is cooling, the technician changes the air filters and lubricates the condenser and the indoor fan motors. After about 30 min the compressor seems cool. The water hose is removed, and the water around the unit is allowed a few minutes to run off. Do not stand in the water. When the disconnect switch is closed, the compressor starts. The suction pressure drops to 40 psig. The normal suction pressure is 70 psig for the system because the refrigerant is R-22. Refrigerant is added to the system to bring the charge up to normal. The space temperature is 80°F, so the suction pressure will be higher than normal until the space temperature becomes normal. The technician must have a reference point to get the correct charge. The following reference points are used in this situation because there is no factory chart.

1. The outside temperature is 90°F; the normal operating head pressure should correspond to a temperature of 90°F + 30°F = 120°F or 260 psig. It should not exceed this when the space temperature is down to a normal 75°F. This is a standard-efficiency unit. The technician restricts the airflow to the condenser and causes the head pressure to rise to 275 psig as refrigerant is added.
2. The suction pressure should correspond to a temperature of about 35°F cooler than the space temperature of 80°F, or 45°F, which corresponds to a pressure of 76 psig for R-22.
3. The system has a capillary tube metering device, so some conclusions can be drawn from the temperature of the refrigerant coming back to the compressor. A thermometer lead is attached to the suction line at the condensing unit. As refrigerant is added, the technician notices that the refrigerant returning from the evaporator is getting cooler. The evaporator is about 30 ft from the condensing unit, and some heat will be absorbed into the suction line returning to the condensing unit. The technician uses a guideline of 15°F of superheat with a suction line this length. This assumes that the refrigerant leaving the evaporator has about 10°F of superheat and that another 5°F of superheat is absorbed along the line. When these conditions are reached, the charge is very close to correct and no more refrigerant is added.

A leak check is performed and a flare nut is found to be leaking. It is tightened and the leak is stopped. The technician loads the truck and leaves. A call later in the day shows that the system is working correctly.

SERVICE CALL 2

A residential customer calls the air conditioning service company. The air conditioning unit has been cooling correctly until afternoon, when it quit cooling. This is a residential unit that has been in operation for several years. *The problem is a complete loss of charge.*

The technician arrives at the residence and finds the thermostat to be set at 75°F and the space temperature to be 80°F. The air coming out of the registers feels the same as the return air temperature. There is plenty of air velocity at the registers, so the filters appear to be clean.

The technician goes to the back of the house and finds that the fan and compressor in the condensing unit are not running. The breaker at the electrical box is in the ON position, so power must be available. A voltage check shows that the voltage is 235 V. A look at the wiring diagram shows that there should be 24 V between the C and Y terminals to energize the contactor. The voltage actually reads 25 V, slightly above normal, but so is the line voltage of 235 V. The conclusion is that the thermostat is calling for cooling, but the contactor is not energized. The only safety control in the contactor coil circuit is the no-charge protector, so the unit must be out of refrigerant.

Gages are fastened to the gage ports, and the pressure is 0 psig. The technician connects a cylinder of refrigerant and starts adding refrigerant for the purpose of leak checking. When the pressure is up to about 2 psig, the refrigerant is stopped. Nitrogen is added to push the pressure up to 50 psig and gas can be heard leaking from the vicinity of the compressor suction line. A hole is found in the suction line where the cabinet had been rubbing against it. This accounts for the fact that the unit worked well up to a point and quit working almost immediately.

The trace refrigerant and nitrogen is allowed to escape from the system, and the hole is patched with silver solder. A liquid-line drier is installed in the liquid line, and a triple evacuation is performed to remove any contaminants that may have been pulled into the system. The system is charged and started. The technician follows the manufacturer's charging chart to ensure the correct charge.

SERVICE CALL 3

A commercial customer called and reported that the air conditioning unit at a small office building was not cooling. The unit was operating and cooling yesterday afternoon when the office closed. *The problem is that someone turned the thermostat down to 55°F late yesterday afternoon, and the condensate on the evaporator froze solid overnight trying to pull the space temperature down to 55°F.*

The technician arrives at the site and notices that the thermostat is set at 55°F. An inquiry shows that one of the employees was too warm in an office at the end of the building and turned down the thermostat to cool that office. The technician notices there is no air coming out of the registers. The air handler is located in a closet at the front of the building. Examination shows that the fan is running, and the suction line is frozen solid at the air handler.

The technician stops the compressor and leaves the evaporator fan operating by turning the heat-cool selector switch to OFF and the fan switch to ON. It is going to take a long time to thaw the evaporator, probably an hour. The technician leaves the following directions with the office manager.

1. Let the fan run until air comes out of the registers, which will verify that air flow has started through the ice.
2. When air is felt at the registers, wait 30 min to ensure that all ice is melted and turn the thermostat back to COOL to start the compressor.

The technician then looks in the ceiling to check the air damper on the duct run serving the back office of the employee who was not cool enough and turned down the thermostat. The damper is nearly closed; it probably was brushed against by a telephone technician who had been working in the ceiling. The damper is reopened.

A call later in the day indicated that the system was working again.

SERVICE CALL 4

A residential customer reports that the air conditioning unit is not cooling correctly. *The problem is a dirty condenser.* The unit is in a residence and next to the side yard. The homeowner mows the grass, and the lawnmower throws grass on the condenser. This is the first hot day of summer, and the unit had been cooling the house until the weather became hot.

The technician arrives at the job and notices that the thermostat is set for 75°F but the space temperature is 80°F. There is plenty of slightly cool air coming out of the registers. The technician goes to the side of the house where the unit is located. The suction line feels cool, but the liquid line is very hot. An examination of the condenser coil shows that the coil is clogged with grass and is dirty.

The technician shuts off the unit with the breaker at the unit and locks it out, then takes enough panels off to be able to spray coil cleaner on the coil. A high-detergent coil cleaner is applied to the coil and allowed to stand for about 15 min to soak into the coil dirt. While this is occurring, the motors are oiled and the filters are changed. ***To keep the condenser fan motor from getting wet when the coil cleaner is washed off, it is covered with a plastic bag.***

A water hose with a nozzle to concentrate the water stream is used to wash the coil in the opposite direction of the airflow. One washing is not enough, so the coil cleaner is applied to the coil again and allowed to set for another 15 min. The coil is washed again and is now clean.

The unit is assembled and started. The suction line is cool and the liquid line is warm to the touch. The technician decides not to put gages on the system and leaves. A call back later in the day to the homeowner indicated that the system was operating correctly.

SERVICE CALL 5

A homeowner indicates that the air conditioning unit is not cooling. This is a residential high-efficiency unit. *The problem is the TXV is defective.*

The service technician arrives and finds the house warm. The thermostat is set at 74°F, and the house temperature is 82°F. The indoor fan is running, and plenty of air is coming out of the registers.

The technician goes to the condensing unit and finds that the fan and compressor are running, but they quickly stop. The suction line is not cool. Gages are attached to the gage ports. The low-side pressure is 25 psig, and the head pressure is 170 psig, this corresponds closely to the ambient air temperature. The unit is off because of the low-pressure control. When the pressure rises, the compressor restarts but stops in about 15 sec. The liquid-line sight glass is full of refrigerant. The technician concludes that there is a restriction on the low side of the system. The TXV is a good place to start when there is an almost complete blockage. ⊛**The system does not have service valves, so the charge has to be removed and recovered in an approved cylinder.**⊛

The TXV is soldered into the system. It takes an hour to complete the TXV change. When the valve is changed, the system is leak checked and evacuated three times. A charge is measured into the system, and the system is started. It is evident from the beginning that the valve change has repaired the unit.

SERVICE CALL 6

An office manager calls to report that the air conditioning unit in a small office building is not cooling. The unit was cooling correctly yesterday afternoon when the office closed. *The problem is a night electrical storm tripped a breaker on the air conditioner air handler.* The power supply is at the air handler.

The technician arrives and goes to the space thermostat. It is set at 73°F, and the temperature is 78°F. There is no air coming out of the registers. The fan switch is turned to ON, and the fan still does not start. It is decided to check the low-voltage power supply, which is in the attic at the air handler. The technician finds the tripped breaker. ***Before resetting it, the technician decides to check the unit electrically.***

A resistance check of the fan circuit and the low-voltage control transformer proves there is a measurable resistance. The circuit seems to be safe. The circuit breaker is reset and stays in. The thermostat is set at COOL, and the system starts. A call later shows that the system is operating properly.

SERVICE CALL 7

A residential customer calls. The unit is not running. The homeowner found the breaker at the condensing unit tripped and reset it several times, but it did not stay set. *The problem is the compressor motor winding is grounded to the compressor shell and tripping the breaker.* ***The homeowner should be warned to reset a breaker no more than once.***

The technician arrives at the job and goes straight to the condensing unit with electrical test equipment. The breaker is in the tripped position and is moved to the OFF position. ***The voltage is checked at the load side of the breaker to ensure that there is no voltage. The breaker has been reset several times and is not to be trusted.*** When it is determined that the power is definitely off, the ohmmeter is connected to the load side of the breaker. The ohmmeter reads infinity, meaning no circuit. The compressor contactor is not energized, so the fan and compressor are not included in the reading. The technician pushes in the armature of the compressor contactor to make the contacts close. The ohmmeter now reads 0, or no resistance indicating that a short exists. It cannot be determined whether the fan or the compressor is the problem so the compressor wires are disconnected from the bottom of the contactor. The short does not exist when the compressor is disconnected. This verifies that the short is in the compressor circuit, possibly the wiring. The meter is moved to the compressor terminal box and the wiring is disconnected. The ohmmeter is attached to the motor terminals at the compressor. The short is still there, so the compressor motor is condemned.

The technician must return the next day to change the compressor. The technician disconnects the control wiring and insulates the disconnected compressor wiring so that power can be restored. This must be done to keep crankcase heat on the unit until the following day. If it is not done, most of the refrigerant charge will be in the compressor crankcase and too much oil will move with the refrigerant when it is recovered.

The technician performs one more task before leaving the job. The Schrader valve fitting at the compressor is slightly depressed to determine the smell of the refrigerant. The refrigerant has a strong acid odor. A suction-line filter-drier with high acid removal capacity needs to be installed along with the normal liquid-line drier.

⊛**The technician returns on the following day, recovers the refrigerant from the system and then changes the compressor, adding the suction and liquid-line driers.**⊛ The unit is leak checked, evacuated, and the charge measured into the system with a set of accurate scales. The system is started, and the technician asks the customer to leave the air conditioner running, even though the weather is mild, to keep refrigerant circulating through the driers to clean it in case any acid is left in the system.

The technician returns on the fourth day and measures the pressure drop across the suction-line drier to make sure that it is not stopped up with acid from the burned-out compressor. It is well within the manufacturer's specification.

SERVICE CALL 8

A residential customer calls. The customer can hear the indoor fan motor running for a short time, then it stops. This happens repeatedly. *The problem is the indoor fan motor is cycling on and off on its internal thermal overload protector because the fan capacitor is defective.* The fan will start and run slowly then stop.

The technician knows from the work order to go straight to the indoor fan section. It is in the crawl space under the house, so electrical instruments are carried on the first trip. The fan has been off long enough to cause the suction pressure to go so low that the evaporator coil is frozen. The breaker is turned off during the check.

The technician suspects the fan capacitor or the bearings, so a fan capacitor is carried along. After bleeding the charge from the fan capacitor it is checked with the ohmmeter to see if it will charge and discharge. ***It is important to ensure that there is no electrical charge in the capacitor before checking it with an ohmmeter. A 20,000-Ω resistor should be used between the two terminals to bleed off any charge.*** The capacitor will not charge and discharge, so it is changed.

The technician oils the fan motor and starts it from under the house with the breaker. The fan motor is drawing the correct amperage. The coil is still frozen, and the technician must set the space thermostat to operate only the fan until the homeowner feels air coming out of the registers. Then only the fan should be operated for one-half hour (to melt the rest of the ice from the coil) before the compressor is started.

The following service calls do not have the solutions described. The solution is in the Instructor's Guide.

SERVICE CALL 9

A commercial customer reports that an air conditioning unit in a small office building is not cooling on the first warm day.

The technician first goes to the thermostat and finds that it is set at 75°F, and the space temperature is 82°F. The air feels very humid. There is no air coming from the registers. The air handler is in the attic. An examination of the condensing unit shows that the suction line has ice all the way back to the compressor. The compressor is still running.

What is the likely problem and the recommended solution?

SERVICE CALL 10

A residential customer calls. The customer reports the unit in their residence was worked on in the early spring under the service contract. The earlier work order shows that refrigerant was added by a technician new to the company.

The technician finds that the thermostat is set at 73°F, and the house is 77°F. The unit is running, and cold air is coming out of the registers. Everything seems normal until the condensing unit is examined. The suction line is cold and sweating, but the liquid line is only warm.

Gages are fastened to the gage ports, and the suction pressure is 85 psig, far from the correct pressure of 74 psig (77°F − 35°F = 42°F, which corresponds to about 74 psig). The head pressure is supposed to be 260 psig to correspond to a condensing temperature of 120°F (85°F + 35°F = 120°F or 260 psig). The head pressure is 350 psig. The unit shuts off after about 10 min running time.

What is the likely problem and the recommended solution?

SERVICE CALL 11

A residential customer calls to report that the air conditioning unit is not cooling enough to reduce the space temperature to the thermostat setting. It is running continuously on this first hot day.

The technician arrives and finds the thermostat set at 73°F; the house temperature is 78°F. Air is coming out of the registers, so the filters must be clean. The air is cool but not as cold as it should be. The air temperature should be about 55°F and it is 63°F.

At the condensing unit the technician finds that the suction line is cool but not cold. The liquid line seems extra cool. It is 90°F outside, and the condensing unit should be condensing at about 120°F. If the unit had 15°F of subcooling, the liquid line should be warmer than hand temperature, yet it is not.

Gages are fastened to the service ports to check the suction pressure. The suction pressure is 95 psig, and the discharge pressure is 225 psig. The airflow is restricted to the condenser, and the head pressure gradually climbs to 250 psig; the suction pressure goes up to 110 psig. The compressor should have a current draw of 27 A, but it only draws 15 A.

What is the likely problem and the recommended solution?

SERVICE CALL 12

A commercial customer calls to indicate that the air conditioning unit in a small office building was running but suddenly shut off.

The technician arrives, goes to the space thermostat and finds it set at 74°F; and the space temperature is 77°F. The fan is not running. The fan switch is moved to ON, but the fan motor does not start. It is decided that the control voltage power supply should be checked first. The power supply is in the roof condensing unit on this particular installation. ***The ladder is placed against the building away from the power line entrance.*** Electrical test instruments are taken to the roof along with tools to remove the panels. The breakers are checked and seem to be in the correct ON position. The panel is removed where the low-voltage terminal block is mounted. A voltmeter check shows there is line voltage but no control voltage.

What is the likely problem and the recommended solution?

SAFETY PRECAUTION: *All troubleshooting by students or inexperienced people should be performed under the supervision of an experienced person. If you do not know whether a situation is safe, consider it unsafe.*

SUMMARY

- Troubleshooting air conditioning equipment involves both mechanical and electrical problems.
- Mechanical troubleshooting uses gages and thermometers for the instruments.
- The gage manifold is used along with the pressure/temperature chart to determine the boiling temperature for the low side and the condensing temperature for the high side of the system.
- The superheat for an operating evaporator coil is used to prove coil performance.
- Standard air conditioning conditions for rating equipment are 80°F return air with a humidity of 50% when the outside temperature is 95°F.

- The typical customer operates the equipment at 75°F return air with a humidity of 50%. This is the condition that the normal pressures and temperatures in this unit are based on.
- There is a relationship between the evaporator and the return air at this typical condition. For a standard-efficiency unit, it is 35°F. The refrigerant in the evaporator normally boils at 35°F lower than the 75°F return air or 40°F.
- A high-efficiency evaporator normally has a boiling temperature of 45°F.
- Gages fastened to the high side of the system are used to check the head pressure. The head pressure is a result of the refrigerant condensing temperature.
- The condensing refrigerant temperature has a relationship to the medium to which it is giving up heat.
- A standard-efficiency unit normally condenses the refrigerant at no more than 30°F higher than the air to which the heat is rejected.
- A high-efficiency condensing unit normally condenses the refrigerant at a temperature as low as 20°F warmer than the air used as a condensing medium.
- High efficiency is obtained with more condenser surface area.
- Superheat is normally checked at the condensing unit with a thermometer.
- The electronic thermometer may be used to check wet-bulb (using a wet wick on one bulb) and dry-bulb temperatures.
- The condenser has three functions: to take the superheat out of the discharge gas, to condense the hot gas to a liquid, and to subcool the refrigerant.
- A typical condenser may subcool the refrigerant 10°F to 20°F lower than the condensing temperature.
- Two types of metering devices are normally used on air conditioning equipment: the fixed bore (orifice or capillary tube) and the TXV.
- The fixed-bore metering device uses the pressure difference between the inlet and outlet of the device for refrigerant flow. It does not vary in size.
- The TXV modulates or throttles the refrigerant to maintain a constant superheat.
- To correctly charge an air conditioning unit, the manufacturer's recommendations must be followed.
- The TXV system normally has a sight glass in the liquid line to aid in charging. A subcooling temperature check may be used when there is no sight glass.
- The tools for electrical troubleshooting are the ammeter and the volt-ohmmeter.
- ***Before checking a unit electrically, the proper voltage and the current draw of the unit should be known.***
- The main electrical power panel may be divided into many circuits. The air conditioning system is normally on two separate circuits for a split system and one circuit for a package system.
- When a compressor is hot, such as when it has been running at low charge, the compressor internal overload protector may stop the compressor.
- ***When a hot compressor is started, assume that the system is low in refrigerant. A cylinder of refrigerant should be connected, so refrigerant may be added before the compressor shuts off again.***
- A compressor can be checked electrically with an ohmmeter. The run and start windings should have a known resistance, and there should be no circuit to ground.

REVIEW QUESTIONS

1. What are the two main troubleshooting areas in the air conditioning service field?
2. What information does the low-side pressure reading give the service technician?
3. What information does the high-side pressure reading give the service technician?
4. What is the correct operating superheat for an air conditioning evaporator coil?
5. What is the typical indoor air condition at which a homeowner operates a system?
6. What is the temperature at which the typical air conditioning evaporator coil operates?
7. What is the typical temperature difference between the entering air and the boiling refrigerant temperatures on a standard air conditioning evaporator?
8. What is the best method of charging an air conditioning system that is low in refrigerant?
9. When the outside ambient air temperature is 90°F, at what temperature should the refrigerant in the condenser be condensing, and what is the head pressure for an R-22 system?
10. What numbers are printed on a refrigerant gage besides pressure?
11. How can a service technician measure the superheat at the condensing unit if the system is a split system?
12. When the ambient air temperature is 95°F and the unit is a high-efficiency unit, what is the lowest condensing refrigerant temperature when the unit is operating properly, and what would the head pressure be for R-22?
13. How is high efficiency accomplished in an air conditioning condensing unit?
14. Name two methods for fastening gage manifolds to an air conditioning system.
15. Which control in the compressor stops the compressor when the motor gets hot?
16. When the compressor is off due to the control in Question 15, which terminals on the compressor will have continuity?
17. What instrument is used to measure the current draw of a compressor?
18. What instrument is used to measure continuity in a compressor winding?
19. What normally happens when a compressor that is electrically grounded tries to start?
20. What should a technician do if the circuit breaker is tripped?

UNIT 41 DIAGNOSTIC CHART FOR SUMMER AIR CONDITIONING SYSTEMS

Many homes and businesses have air conditioning (summer cooling). These systems operate in the spring, summer, and fall for the purpose of comfort for the occupants. These systems are often connected to the heating system. Air conditioning may also be combined with any heat pump system. The air conditioning system controls will be discussed here. Always listen to the owner to help to identify the potential problem.

PROBLEM	POSSIBLE CAUSE	POSSIBLE REPAIR	PARAGRAPH NUMBER
No cooling, outdoor unit not running indoor fan running	Open outdoor disconnect switch	Close disconnect switch	41.12
	Open fuse or breaker	Replace fuse, reset breaker and determine problem	41.12
	Faulty wiring	Repair or replace faulty wiring or connections	41.12
No cooling, indoor fan and outdoor unit will not run	Low voltage control problem		
	A. Thermostat	Repair loose connections or replace thermostat and or subbase	41.9
	B. Interconnecting, wiring or connections	Repair or replace wiring or connections	41.9
	C. Transformer	Replace if defective	41.9
		SAFETY PRECAUTION: Look out for too much current draw due to ground circuit or shorted coil	
No cooling indoor and outdoor fan running, but compressor not running	Tripped compressor internal overload		
	A. Low line voltage	Correct low voltage, power company or loose connections	41.9
	B. High head pressure		41.6
	1. Dirty outdoor (condenser)	Clean condenser	41.6
	2. Condenser fan not running all the time	Check condenser fan motor and capacitor	41.6
	3. Condenser air recirculating	Correct recirculation problem	41.6
	4. Overcharge of refrigerant	Correct charge	41.8
	C. Low charge, motor not being properly cooled	Correct charge, if due to leak, repair leak	41.8
Indoor coil freezing	Restricted airflow	Change filters	41.5
		Open all supply register dampers	41.5
		Clear return air blockage	41.5
		Clean fan blades	41.5
		Speed fan up to higher speed	41.5
	Low charge	Adjust unit charge—repair leak if refrigerant has been lost	41.8
	Metering device	Change or clean metering device	41.8
	Restricted filter drier	Change filter drier	41.8
	Operating unit during low ambient conditions without proper head pressure control	Add proper low ambient head pressure control	22.17

SECTION

8 All-Weather Systems

UNIT 42

Electric, Gas, and Oil Heat With Electric Air Conditioning

OBJECTIVES

After studying this unit, you should be able to

- describe year-round air conditioning.
- discuss the three typical year-round air conditioning systems.
- list the five ways to condition air.
- state how much air is normally required to be distributed per ton for a cooling system.
- describe why a heating system normally uses less air than a cooling system.
- explain two methods used to vary the airflow in the heating season from that in the cooling season.
- describe two types of control voltage power supplies used in add-on air conditioning.
- describe add-on air conditioning.
- explain package all weather systems

SAFETY CHECKLIST

* Care should be used around any power supply because electrical shock hazard is possible.
* If you suspect a gas leak, ensure that ventilation around a gas furnace is adequate before troubleshooting.

42.1 COMFORT ALL YEAR

Year-round air conditioning describes a system that conditions the living space for heating and cooling throughout the year. This is done in several ways. The most common are electric air conditioning with electric resistance heat, electric air conditioning with gas heat, and electric air conditioning with oil heat. This unit describes how these systems work together. Each system has been covered individually in other units. Another common method is the heat pump, which is discussed in Unit 43. A less frequently used method is gas heat with gas air conditioning. This is a special system and will not be covered in this text.

42.2 FIVE PROCESSES FOR CONDITIONING AIR

Air is *conditioned* when it is heated, cooled, humidified, dehumidified, or cleaned. This unit describes how air is heated and cooled with the same system. The systems discussed in this unit are called *forced-air* systems. Air is distributed through ductwork to the conditioned space. The fan is normally a component of the heating system and provides the force to move the air through the duct system. A typical system may have an electric, gas, or oil furnace with an evaporator in the airstream for cooling, Figure 42–1. The evaporator, located in the indoor airstream, is used in conjunction with a condensing unit outside the conditioned space. Interconnecting piping connects the evaporator to the condensing unit, Figure 42–2.

42.3 ADD-ON AIR CONDITIONING

Many systems are installed in stages. The furnace may be installed when the structure is built. The air conditioning may be added at the same time or later (*add-on air conditioning*). When the heating system is installed first, some considerations must be made for air conditioning if it is to be added later.

Air conditioning systems must have the correct air circulation. Typically, they require an airflow of 400 cfm/ton, which is more than needed for an average forced-air heating system. A 3-ton cooling system, for example, requires 1200 cfm. The furnace on an existing heating system where 3 tons of cooling are added must be able to furnish the required amount of air, and the ductwork must be sized for the airflow.

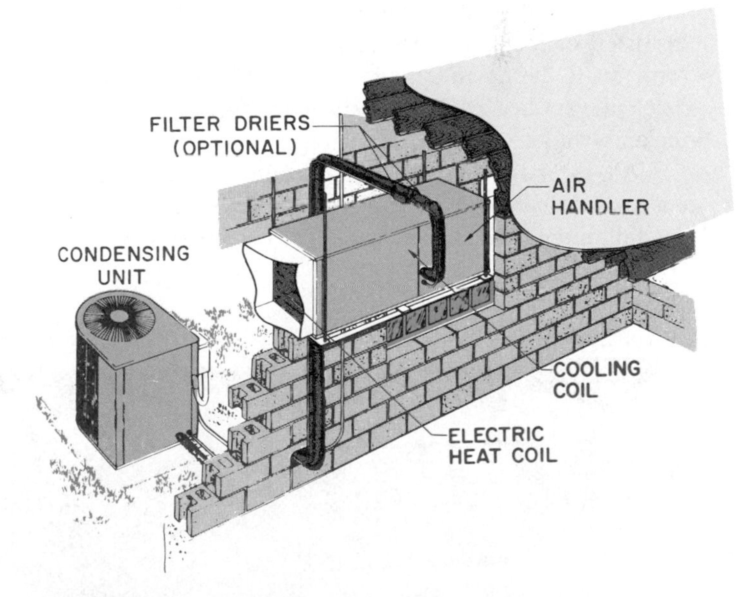

FIGURE 42–1 This is a typical electric furnace that has an air conditioning coil placed in the duct. The fan in the furnace is used to circulate air for both air conditioning and heating. *Courtesy Climate Control*

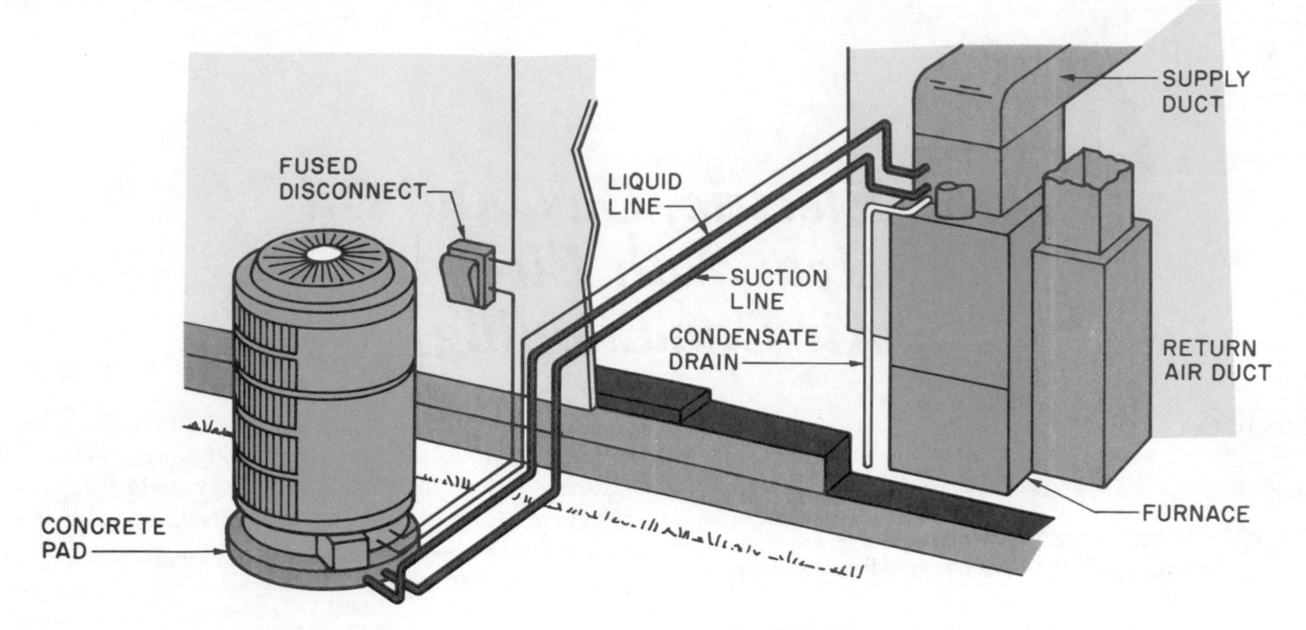

FIGURE 42–2 Complete installation with the piping shown. The evaporator is piped to the condensing unit on the outside of the structure.

42.4 INSULATION FOR EXISTING DUCTWORK

It is popular in warm climates to install heating systems in the crawl space below the structure. The heat from the warm duct rises and warms the floor, Figure 42–3. Some people believe that heat lost under the house is not all lost because some of it rises through the floor. This makes a significant difference only when the space under the house is sealed, and duct systems are not insulated. If air conditioning (cooling) is added to these systems, the duct will sweat. The system would be more efficient if it were insulated from the beginning.

Therefore, if the ductwork for a system with air conditioning is installed outside the conditioned space, it must be insulated or moisture ("sweat") from the ambient air will form on it during the cooling cycle. The insulation also helps prevent heat exchange between the air in the duct and the ambient air.

A typical system installed for heating only may have an undersized duct system and a blower that is too small if air conditioning is added. For example, a customer may decide to add 3 tons of air conditioning that requires 1200 cfm of air to an existing furnace system. A typical gas furnace must have a minimum temperature drop of 45°F for proper venting. This is a minimum furnace size of 58,320 Btuh for the airflow of 1200 cfm. The smallest standard furnace has a 60,000 Btuh output.

$$Q = 1.08 \times \text{cfm} \times TD$$
$$Q = 1.08 \times 1200 \times 45$$
$$Q = 58{,}320 \text{ Btuh output}$$

Using a 60,000 Btuh output

$$\frac{60{,}000}{0.08} = 75{,}000 \text{ Btuh}$$

or a 75,000-Btuh input gas furnace for a standard 80% efficiency furnace.

In a hot climate location, the furnace requirements may only be 60,000 Btuh. This may be the furnace already installed at the job site. Adding a 3-ton unit requiring 1200 cfm may cause airflow problems if the duct system is sized for the airflow for heating only. This may be a 70°F rise, or

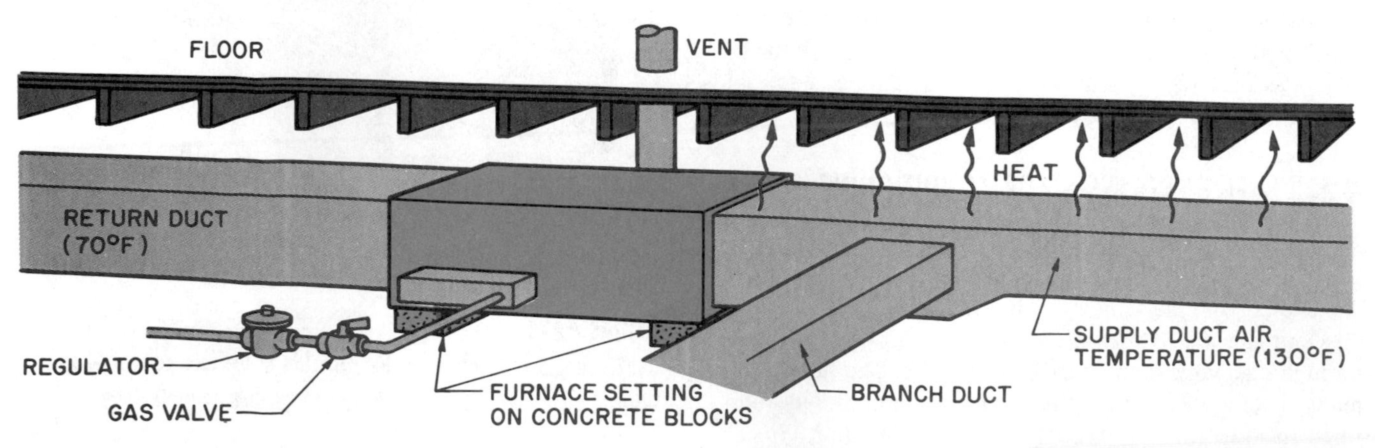

FIGURE 42–3 Heat rising off the duct in a crawl space under a house.

$$\text{cfm} = \frac{Q}{1.08 \times TD} = \frac{60{,}000}{1.08 \times 70} = 794 \text{ (round to 800)}$$

The furnace fan may not be capable of moving 1200 cfm of air, so some installation technicians will install a larger fan motor (for a belt-drive system) or a larger fan wheel and motor (for a direct-drive system). This is not what the furnace manufacturer intended and may cause venting problems. If the technician succeeds in moving 1200 cfm through duct sized for 800 cfm, the air will make noise because of the increased velocity. The alternative would be to reduce the air conditioning load, increase the furnace size, or install a system producing two different fan volumes, one for summer and one for winter. Two-speed fan operation is discussed later in this unit.

The air conditioning load may be reduced by attic ventilation, insulation, storm windows, shades, or awnings. For example, if the air conditioning load in the preceding problem could be reduced to 2 tons, the airflow could be reduced to 800 cfm and the current furnace would be adequate for the airflow. The duct would probably be adequate also. The customer would save money on the installation and the technician would gain a satisfied customer.

Computer load calculation programs can be used to work "what if" situations for a typical structure. For example, when a computer load calculation is entered, storm windows may be added with only a few key strokes to determine the Btuh savings. Then attic ventilation may be tried. Different combinations may be recalculated until the most economical combination of reductions is found. The customer benefits from monthly power reduction in addition to a less expensive air conditioning installation.

42.5 EVALUATION OF AN EXISTING DUCT SYSTEM

When an air conditioning system is to be added after the original furnace installation, the following considerations are the most important: the airflow, the ductwork, and the registers and grilles.

Air distribution was discussed in detail in Unit 37. A heating system that is already installed needs to be evaluated to see if air distribution changes need to be made before air conditioning is installed. The blower size and the motor horsepower may be guides as to the amount of air the fan section is capable of moving. The manufacturer of the air handler or furnace is the best source of this information. However, it is not always readily available. Figure 42–4 is a chart used by estimators and service technicians to help determine what the blower capabilities of a typical system may be. The blower wheel dimensions are an important factor.

The duct system may be evaluated with an evaluation chart also, Figure 42–4. ***This chart should not be used when designing a system from the beginning.*** It should only be used as a reference when estimating or troubleshooting. The estimator or service technician who suspects an airflow problem may consult the chart and compare the duct system to the chart. If the system does not meet the chart's minimum requirements, further investigation should be made.

For example, the system described above (a system with a 75,000-Btuh furnace, 80% efficiency, an output of 60,000 Btuh, and an airflow of 800 cfm) can be evaluated using Figure 42–4. The output capacity falls between 60,000 and 70,000 Btuh. The airflow requirement is 700 cfm for this range of furnace. If 2 tons of air conditioning are added, Figure 42–4, the required airflow is 800 cfm. Looking to the right, the blower horsepower and blower wheel size can be found. The existing furnace may be compared to the data from the chart. The fan motor may be too small but may be changed to the larger one required for air conditioning if the fan wheel dimensions are correct. Various combinations of duct size may be found. A survey of the existing "heat only" duct system may show that the return air is undersized and an extra return duct may be required. Further to the right you will find the required supply duct information. It can be compared to the system and changed to suit the application. This type of chart is valuable for job surveys.

The grilles and registers used for the final air distribution are important also. Figure 42–4 will help you choose the correct return air grille. It is important that the air volume meet the minimum before the air conditioning system is added. The registers are responsible for distributing an air pattern into the conditioned space. If the original purpose of the installation was heating only, the registers may not provide the correct air pattern. This is particularly true for floor baseboard registers. They may be designed to keep the air down low on the floor if they are for heating only. Heated air will rise and warm the room. These registers will not work well in the air conditioning season because the cool air will stay on the floor, Figure 42–5. Contact a supplier for the correct supply register replacements.

When a distribution system does not meet the minimum standards of the charts discussed previously, it should be changed. If it is not changed, there will probably not be enough air for the air conditioning system, and the air might not be properly distributed. The changes may involve adding new duct or return air runs or changing the blower or motor. The installation may require an airflow that is too great for the existing furnace because it will not allow enough air to be circulated. Because the furnace heat exchanger is not large enough, a new furnace may be the only solution. The air distribution registers are usually easy to change to accommodate air conditioning.

42.6 COOLING VERSUS HEATING AIR QUANTITY

It is usually desirable for less airflow in the winter, so the air that is entering the room through the registers will be hot. If the airflow of 400 cfm/ton of air conditioning is used in the winter, the air may create drafts of slightly warm air instead of hot air. Correct airflow is necessary for gas and oil furnaces for correct venting. Too much air will cause products of combustion to be too cool and condensation may occur. Changing the air volume is accomplished with dampers or a multispeed fan.

Dampers are sometimes installed by the original contractor. They have summer and winter positions. Someone

QUICK-SIZING TABLE for HEATING (FURNACE) ONLY DUCT SYSTEM (Without a Cooling Coil in Place)

OUTPUT CAPACITY (See Notes)	Min. Air Flow Req'd.	Supply Duct or Extended Plenum @ 800 FPM	Min. Sq. Inch Needed for Spec. CFM (Total Area of All Supply Duct)	Min. Number Supply Runs @ 600 FPM			MINIMUM SIZE	
				5" RUNS 80 CFM	6" RUNS 115 CFM	7" RUNS 165 CFM	Return Duct Furnace or Air Handler @ 800 FPM	Return Air Grille (or equivalent) @ Face Velocity of 500 FPM
45,000 to 55,000	500 CFM	14" x 8" or 12" round	100	7	5	4	14" x 8" or 12" round	12" x 12"
60,000 to 70,000	700 CFM	18" x 8" or 14" round	140	10	6	5	18" x 8" or 14" round	24" x 10"
75,000 to 85,000	800 CFM	22" x 8" or 14" round	170	10	7	5	22" x 8" or 14" round	24" x 12"
95,000 to 105,000	900 CFM	24" x 8" or 15" round	190	12	8	6	24" x 8" or 14" round	24" x 12"
105,000 to 115,000	1100 CFM	22" x 10" or 16" round	220	—	10	7	22" x 10" or 16" round	30" x 12"
125,000 to 150,000	1400 CFM	24" x 12" or 18" round	280	—	12	9	24" x 12" or 18" round	30" x 14"
155,000 to 160,000	1600 CFM	1-35" x 10" or 20" round or 2-22" x 8"	360	—	14	10	32" x 10" or 20" round	30" x 18"

Notes:
1. BTUH with **maximum** temperature rise.
2. Gas furnaces are rated in input capacity. Rated output capacity is 80% of input.
3. Oil and electric furnace are rated in output capacity.

QUICK-SIZING TABLE for HEATING and COOLING DUCT SYSTEM

Air conditioning systems should never be sized on the basis of floor area only, but knowledge of the approximate floor area (sq. ft.) that can be cooled with a ton of air conditioning will be of invaluable assistance to you in avoiding serious mathematical errors.

Size of O.D. Unit	Normal Air Flow Req'd @ 400 CFM per Ton	Furnace		Supply Duct or Extended Plenum @ 800 FPM	Min. Number Supply Runs @ 600 FPM				Min. Return Duct. Size at Furnace or Air Handler @ 800 FPM	Min. Return Air Grille Size (or equivalent) @ Face Velocity of 500 FPM
		Blower Motor H.P.	Blower Wheel Dia. X Width		5" RUNS 80 CFM	6" RUNS 115 CFM	7" RUNS 165 CFM	3½ X 14" 170 CFM		
1½ ton 18,000 BTUH	600 CFM	1/4 H.P.	9" X 8" 10" X 8"	16" X 8" or 12" round	8	5	4	4	16" X 8" or 12" round	24" X 8"
2 ton 24,000 BTUH	800 CFM	1/4 H.P.	9" X 9" 10" X 8"	22" X 8" or 14" round	10	7	5	5	22" X 8" or 14" round	22" X 12"
2½ ton 30,000 BTUH	1000 CFM	1/3 H.P.	10" X 8" 10" X 10" 12" X 9"	20" X 10" or 18" round	13	9	7	6	20" X 10" or 16" round	30" X 12"
3 ton 36,000 BTUH	1200 CFM	1/3 H.P.	10" X 8" 10" X 10" 12" X 9"	24" X 10" or 18" round	—	11	8	7	24" X 10" or 18" round	30" X 12"
3½ ton 42,000 BTUH	1400 CFM	1/2 H.P. 3/4 H.P.	10" X 8" 10" X 10" 12" X 9" 12" X 10"	24" X 12" or 18" round	—	12	9	8	24" X 12" or 18" round	30" X 14"
4 ton 48,000 BTUH	1600 CFM	1/2 H.P. 3/4 H.P.	10" X 10" 12" X 9" 12" X 10" 12" X 12"	32" X 10" or 20" round	—	14	11	10	28" X 12" or 20" round	30" X 18"

FIGURE 42–4 (A) Duct estimating table. (B) This chart shows many system characteristics including the return air characteristics of typical systems used for cooling and heating. *Copyright American Standard, Inc. 1985*

QUICK-SIZING TABLE FOR HEATING AND COOLING DUCT SYSTEM

Air conditioning systems should never be sized on the basis of floor area only, but knowledge of the approximate floor area (sq. ft.) that can be cooled with a ton of air conditioning will be of invaluable assistance to you in avoiding serious mathematical errors.

Size of O.D. Unit	Normal Air Flow Req'd @ 400 cfm per Ton	Furnace		Supply Duct or Extended Plenum @ 800 fpm	Min. Number Supply Runs @ 600 fpm				Min. Return Duct Size at Furnace or Air Handler @ 800 fpm	Min. Return Air Grille Size (or equivalent) @ Face Velocity of 500 fpm
		Blower Motor hp	Blower Wheel Dia. X Width		5″ Runs 80 cfm	6″ Runs 115 cfm	7″ Runs 155 cfm	$3\frac{1}{2}$ x 14″ 170 cfm		
$1\frac{1}{2}$ ton 18,000 Btuh	600 cfm	$\frac{1}{4}$ hp	9″ x 8″ 10″ x 8″	16″ x 8″ or 12″ round	8	5	4	4	16″ x 8″ or 12″ round	24″ x 8″
2 ton 24,000 Btuh	800 cfm	$\frac{1}{4}$ hp	9″ x 9″ 10″ x 8″	22″ x 8″ or 14″ round	10	7	5	5	22″ x 8″ or 14″ round	22″ x 12″
$2\frac{1}{2}$ ton 30,000 Btuh	1000 cfm	$\frac{1}{3}$ hp	10″ x 8″ 10″ x 10″ 12″ x 9″	20″ x 10″ or 18″ round	13	9	7	6	20″ x 10″ or 16″ round	30″ x 12″
3 ton 36,000 Btuh	1200 cfm	$\frac{1}{3}$ hp	10″ x 8″ 10″ x 10″ 12″ x 9″	24″ x 10″ or 18″ round	—	11	8	7	24″ x 10″ or 18″ round	30″ x 12″
$3\frac{1}{2}$ ton 42,000 Btuh	1400 cfm	$\frac{1}{2}$ hp $\frac{3}{4}$ hp	10″ x 8″ 10″ x 10″ 12″ x 9″ 12″ x 10″	24″ x 12″ or 18″ round	—	12	9	8	24″ x 12″ or 18″ round	30″ x 14″
4 ton 48,000 Btuh	1600 cfm	$\frac{1}{2}$ hp $\frac{3}{4}$ hp	10″ x 10″ 12″ x 9″ 12″ x 10″ 12″ x 12″	32″ x 10″ or 20″ round	—	14	11	10	28″ x 12″ or 20″ round	30″ x 18″

FIGURE 42–4 ***(Continued)***

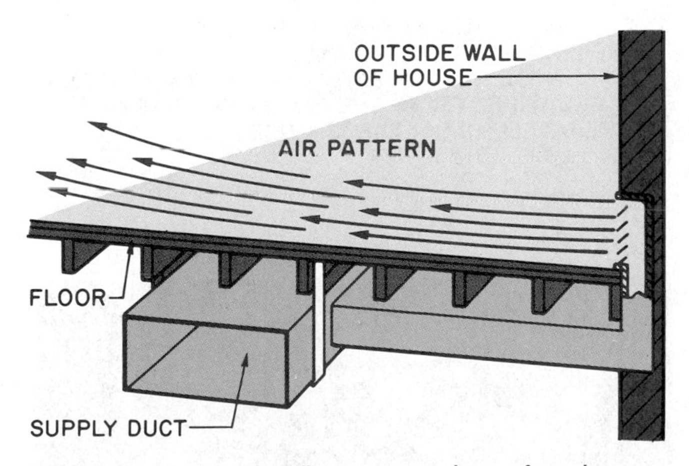

FIGURE 42–5 **Low side wall registers, not adequate for cooling.**

must change the damper position for each change of season. Because this is often overlooked, some contractors install a low-voltage end switch that will not allow the cooling contactor to be energized unless the damper is in the summer position, Figure 42–6.

Multispeeds from the fan may be achieved by adjusting the fan motor pulley or using multispeed motors. When the pulley must be adjusted to change fan speeds for a new season, a service technician must visit the location twice a year. The pulley may be adjusted at the spring start-up for air conditioning and at the fall furnace checkout. The filters may be changed and the motors oiled at the same time. This is a routine service for some installations each year. Changing the pulley is not a job that homeowners should perform unless they have experience with this type of service. Multispeed motors have a winding for each motor speed, Figure 42–7.

42.7 CONTROL WIRING FOR COOLING AND HEATING

Another consideration for year-round air conditioning is the control wiring. The control system must be capable of operating heating and air conditioning equipment at the proper times. ***The heating must not be operated at the same time as the cooling.*** The thermostat is the control that accomplishes this. Figure 42–8 shows an example of a typical HEAT-COOL thermostat with manual changeover from season to season. Figure 42–9 illustrates a HEAT-COOL thermostat with automatic changeover.

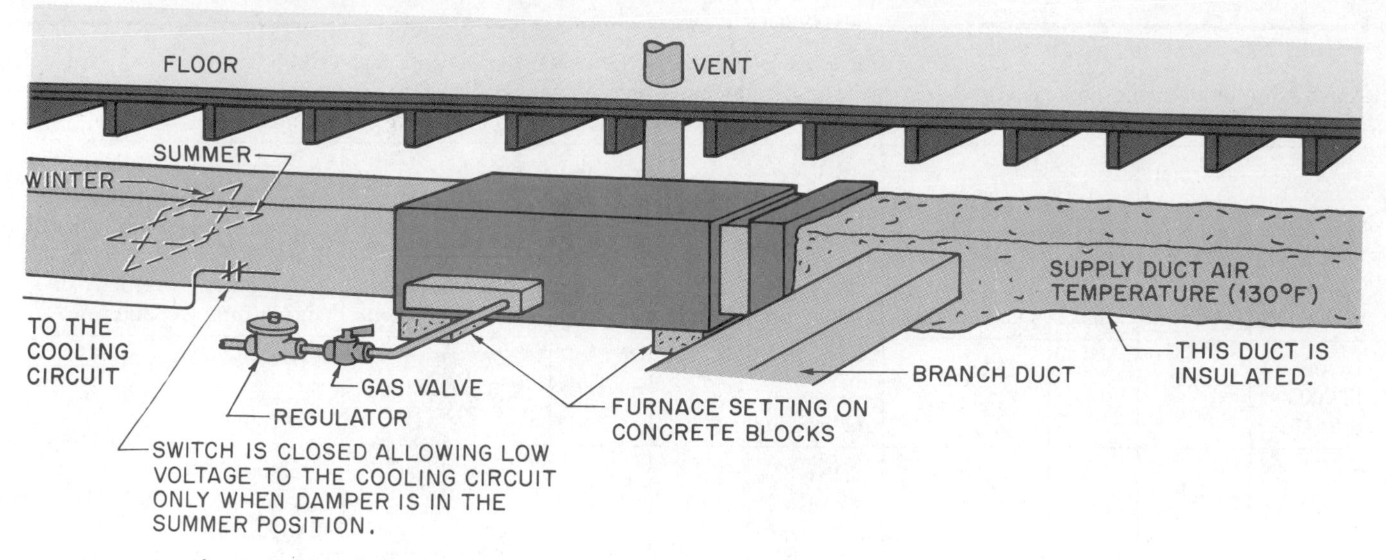

FIGURE 42–6 This system has a damper to allow more airflow in the summer than in the winter.

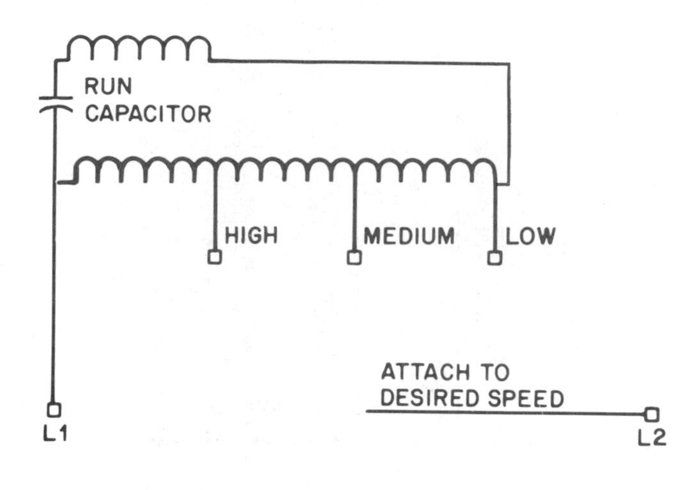

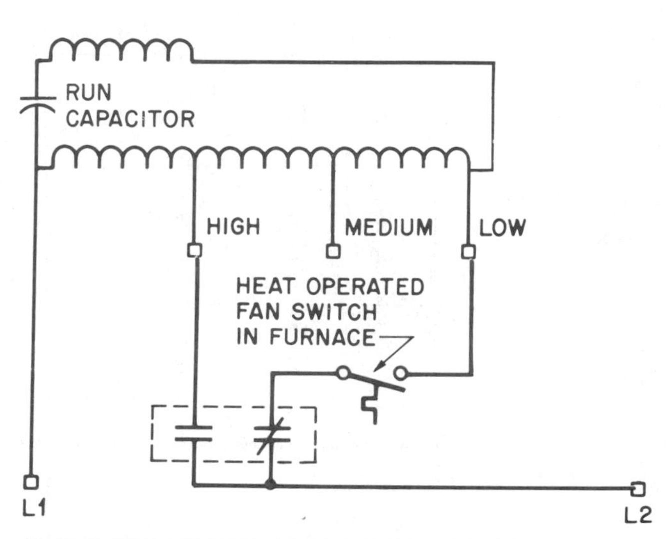

FAN RELAY: WHEN ENERGIZED, SUCH AS IN COOLING, THE FAN CANNOT RUN IN THE LOW-SPEED MODE. WHEN DEENERGIZED THE FAN CAN START IN THE LOW-SPEED MODE THROUGH THE CONTACTS IN THE HEAT OPERATED FAN SWITCH.

IF THE FAN SWITCH AT THE THERMOSTAT IS ENERGIZED WHILE THE FURNACE IS HEATING, THE FAN WILL MERELY SWITCH FROM LOW TO HIGH. THIS RELAY PROTECTS THE MOTOR FROM TRYING TO OPERATE AT 2 SPEEDS AT ONCE.

FIGURE 42–7 Three-speed fan motor.

42.8 TWO LOW-VOLTAGE POWER SUPPLIES

The control circuit may have more than one low-voltage power supply, which can cause much confusion for the technician trying to troubleshoot a problem. If the heating system was installed first and the air conditioning added later, there may be two power supplies. The furnace must have a low-voltage transformer to operate in the heating mode. When the air conditioning was added, a transformer may have been furnished with it because the air conditioning manufacturer did not know if the low-voltage power supply furnished with the furnace was adequate. Air conditioning systems usually require a 40-VA transformer to supply enough current to energize the compressor contactor and the fan relay. Most basic furnaces can operate on a 25-VA transformer and may be furnished with one.

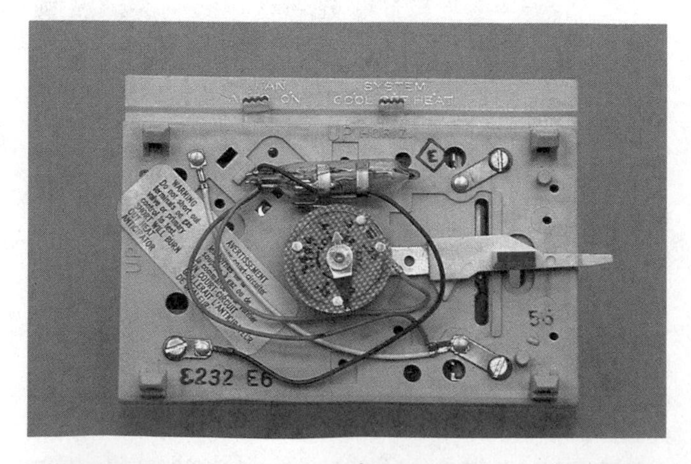

FIGURE 42–8 Typical HEAT-COOL thermostat with cover off. *Photo by Bill Johnson*

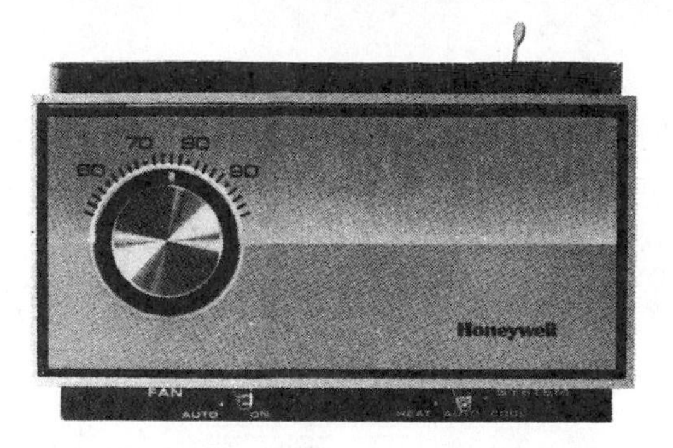

FIGURE 42–9 Thermostat with automatic changeover from cooling to heating. *Photo by Bill Johnson*

When two power supplies are used, a special arrangement must be made in the thermostat and subbase. In one arrangement the two circuits are separated; in the other the two transformers are wired in parallel by phasing them. ***If the transformers are to be separated, two hot wires are wired into the subbase. The two circuits must be kept apart or damage may occur, Figure 42–10. Care should be used around any power supply because electrical shock hazard is possible. Notice that one leg of power from the heating transformer is in the cooling circuit. No problem should occur unless the other leg is involved.***

42.9 PHASING TWO LOW-VOLTAGE TRANSFORMERS

When the two transformers are wired in parallel, they must be kept in phase or the transformer will have opposing phases, Figure 42–11.

The preferred method to determine whether or not they are in phase is to use a voltmeter and apply it to the two hot connections. In the diagram in Figure 42–12 they are terminals RC and 4. High voltage, about 50 V, indicates the transformers are out of phase. The meter will read 0 V from hot lead to hot lead if the transformers are in phase. If out of phase, it can be corrected with the primary or secondary wiring, Figures 42–11 and 42–12.

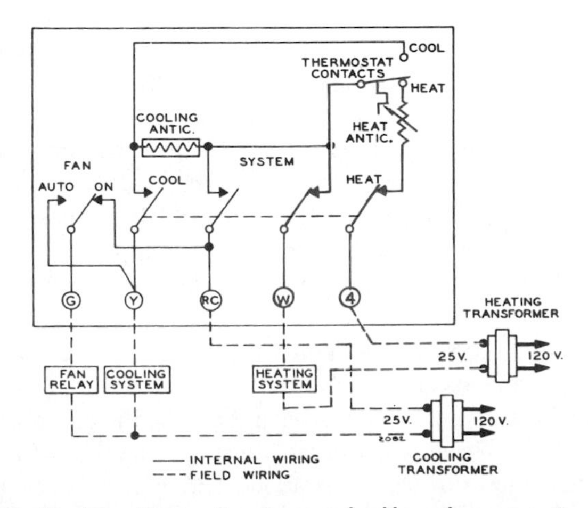

FIGURE 42–10 Wiring of an integrated subbase thermostat. *Courtesy White-Rodgers Division, Emerson Electric Co.*

42.10 ADDING A FAN RELAY

When air conditioning is added to an existing system, the furnace may not have a fan relay to start the fan. Most fossil-fuel furnaces (oil or gas) will start the fan with a thermal type of fan switch in the heating mode and will have no provision to start the fan in the cooling mode. Some electric furnaces have a fan relay already built in. When there is none, a separate fan relay must be added when the air conditioning system is installed. This relay is sometimes furnished in a package with the control transformer called a *transformer relay package* or *fan center,* Figure 42–13.

SAFETY PRECAUTION: *The fan relay is part of the low-voltage and high-voltage circuits. Use proper caution whenever you work with electricity.*

42.11 NEW ALL-WEATHER SYSTEMS

When an all-weather system is installed as an original system, the foregoing considerations can be correctly designed into the initial installation. The ductwork is always designed around the cooling system because it requires more airflow.

New all-weather system installations are split systems or package systems. If split, a gas, oil, or electric furnace is used for heating, and electric air conditioning is used for cooling. A package system will be gas or electric heat with electric cooling. Heat pumps also come in package systems; they are covered in the next unit.

42.12 ALL-WEATHER SPLIT SYSTEMS

Split systems are installed in the same way as a "furnace only" system with the duct sized to handle the air for the cooling. Remember, a cooling system must have more cubic feet per minute of air than a heating system alone. Some consideration may be given to the air distribution at the grilles for proper air distribution for cooling. Most furnace manufacturers have matching coil packages that will fit their furnaces. Using the matching coil package for a furnace is a good idea because the coil will come in its own insulated enclosure and will properly fit the furnace for the best air flow through the coil. Using the matching coil package may work a hardship at installation time because the existing ductwork will need to be cut back for the coil package to be fitted.

42.13 PACKAGE OR SELF-CONTAINED ALL-WEATHER SYSTEMS

Package systems are not normally made for oil with air conditioning. These systems are gas and electric or electric and electric, Figure 42–14, and are installed like electric air conditioning package systems. A review of the unit on air conditioning (cooling) will show how this is done. The units that use gas must have a gas line installed. The flue installed with a gas package unit is a fixed component and part of the unit. These units are sometimes called *gas packs,* Figure 42–15.

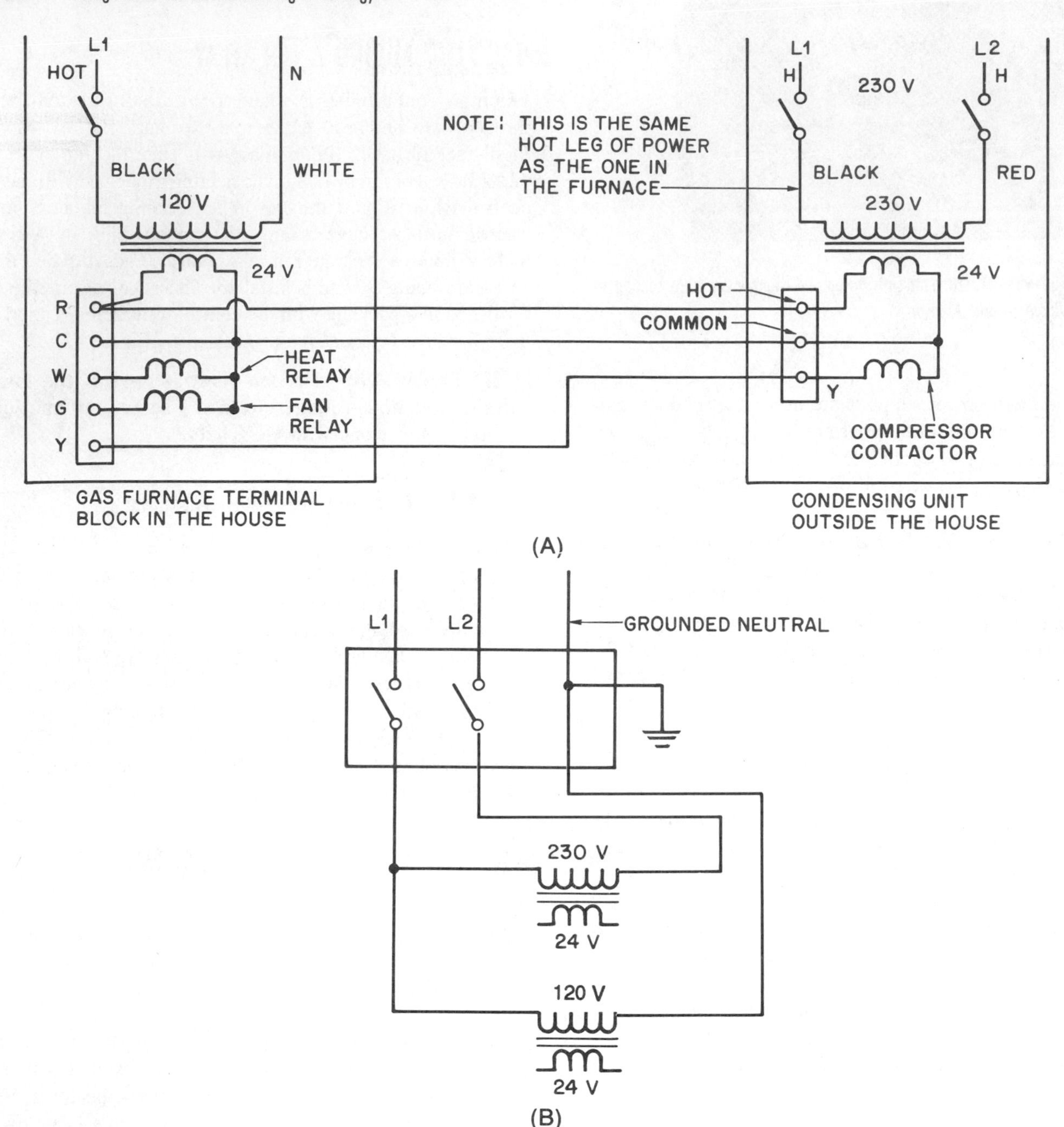

FIGURE 42–11 (A) Transformers wired in parallel. (B) Line diagram.

42.14 WIRING THE ALL-WEATHER SYSTEM

The wiring of all-weather systems is similar to that of air conditioning systems except for the extra power that may be required to operate the electric heat, Figure 42–16. The control wiring is much the same as a gas furnace and electric air conditioning except that the wiring is all done between the thermostat and the package unit. The wires that normally run to a remote furnace are not needed.

42.15 SERVICING THE ALL-WEATHER SYSTEM

Package equipment installations have the advantage that the whole system is outside the house. ***Any gas hazard is virtually eliminated because gas leaks are dissipated outside.*** A technician can service the unit without crawling to a furnace in an attic or under a house. All of the control wiring is at the unit and is easily accessible. Figure 42–17 shows an example of a typical installation.

When major repairs are needed on the compressor or expansion device, for example, the repair is simplified because they are all together and accessible. The proper panels may be removed to service the various components.

When troubleshooting an all-weather system, the technician usually is called to deal with a heating problem in winter or a cooling problem in summer. Seldom does a heating problem carry over to the cooling season, or a cooling problem to the heating season, because the controls are separated sufficiently. Dirty air filters are one type of problem that may not show up at the end of the cooling or heating season but will cause problems at the peak of next season. For example, when a homeowner does not change

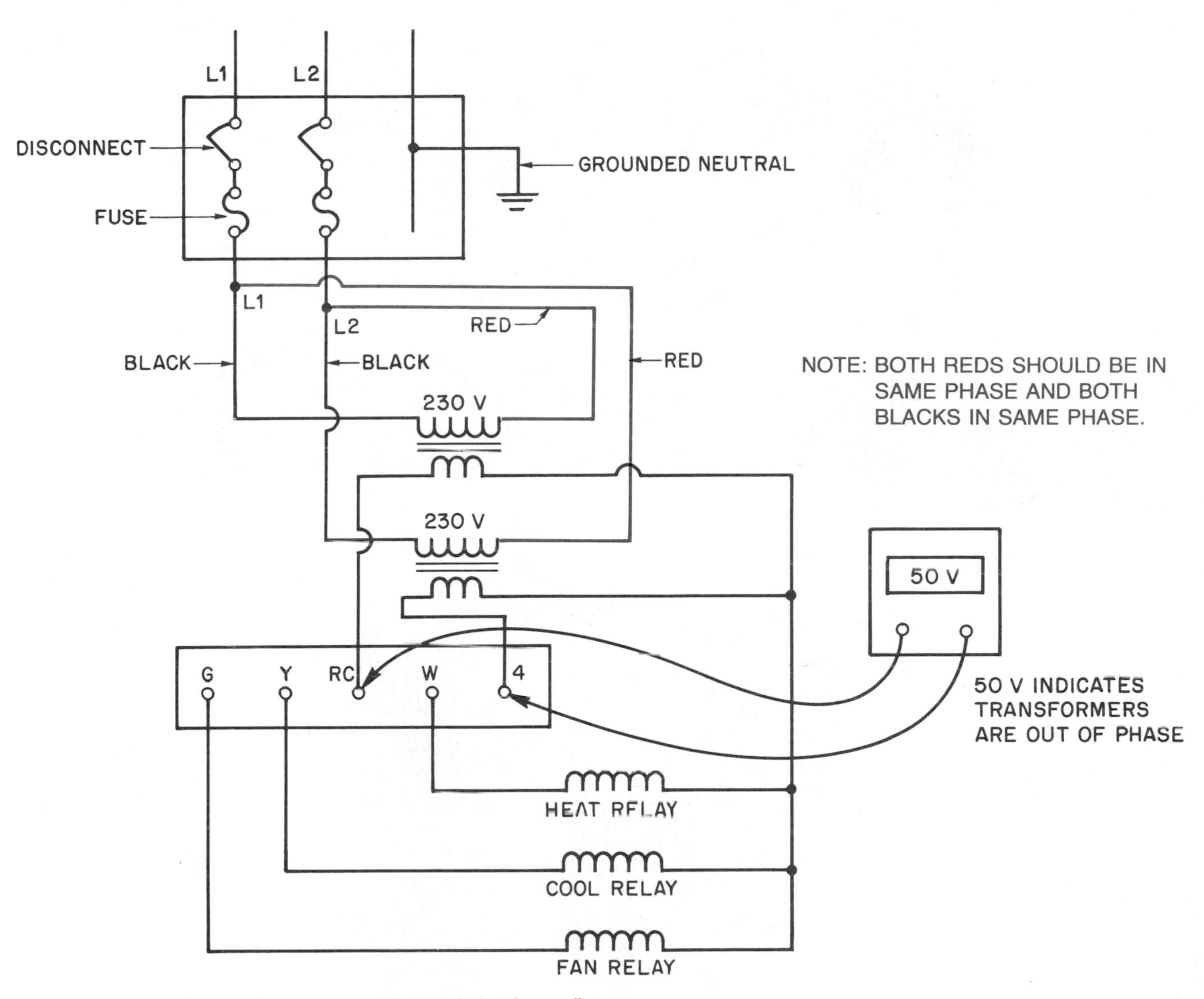

FIGURE 42–12 When transformers are out of phase, high voltage will occur.

FIGURE 42–13 Transformer and fan relay package. *Photo by Bill Johnson*

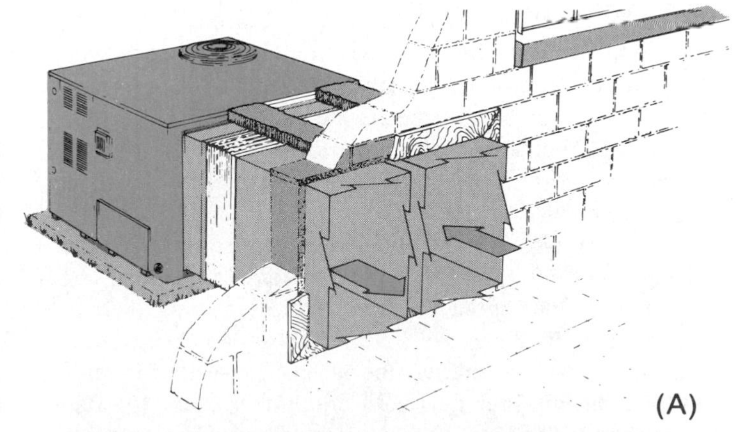

FIGURE 42–14 These two package units look alike. (A) Gas used as the heat source and electricity used for cooling. (B) Electric heat used with electric air conditioning. *(A) Courtesy Climate Control, (B) Reproduced courtesy of Carrier Corporation*

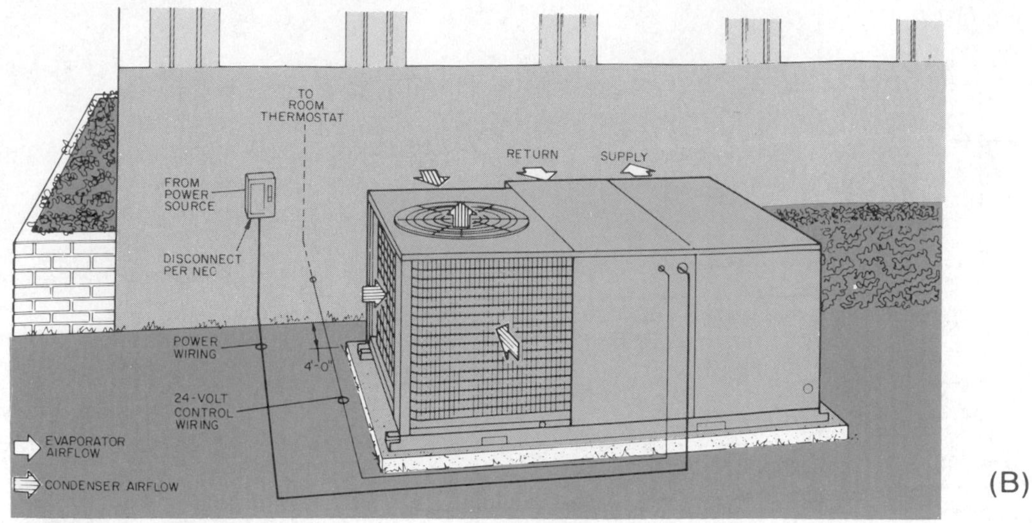

FIGURE 42–14 ***(Continued)***

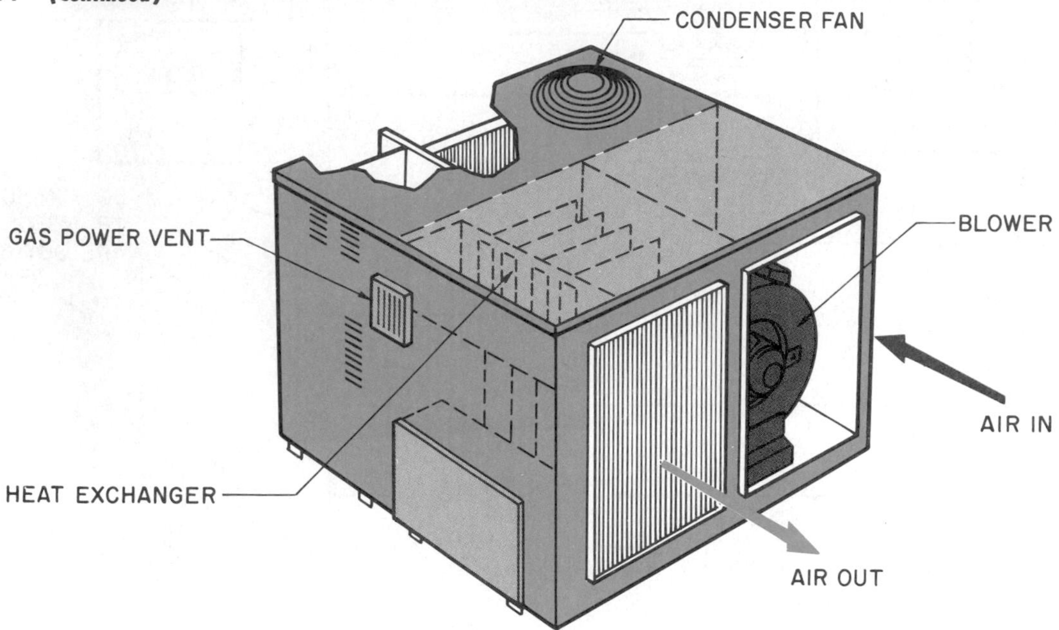

FIGURE 42–15 Flue vent on a gas package unit. *Courtesy Climate Control*

the air filters at the end of the heating season, the cooling will normally work fine until the first hot day that the unit has long running times. When the running times are short, the unit may run below freezing due to the dirty filters but will cut off before the coil freezes up. When the running time is longer, the coil may freeze solid before the room thermostat shuts off the unit.

The reverse is true with filters when they are not changed at the end of the cooling season. The cooling unit may not run long enough to freeze solid before the thermostat shuts it off. When the heating season starts, the furnace will appear to be working fine when the weather is mild. ***When the weather gets cold enough to cause the furnace to run for long periods, it may overheat with the high limit control shutting it off.***

The other components of year-round air conditioning, such as the humidifier and the air cleaner, are discussed in Section Six in this text.

SUMMARY

- All-weather systems heat in the winter and cool in the summer.
- This is accomplished with combinations of equipment such as gas heat and electric cooling, oil heat and electric cooling, and electric heat and electric cooling.
- Air is conditioned by heating, cooling, humidifying, dehumidifying, and cleaning.
- Summer air conditioning is sometimes added to an existing heating system. This is called add-on cooling.
- When summer air conditioning is added to an existing heating system, the ductwork, the terminal air distribution system, the fan on the existing furnace, and the control wiring must be considered.
- Different air volumes are sometimes desirable for the heating and cooling seasons. The air in the heating season is warmer at the terminal units when the air volume is reduced in the heating season.
- Different air volumes are accomplished with dampers and variable fan speeds.

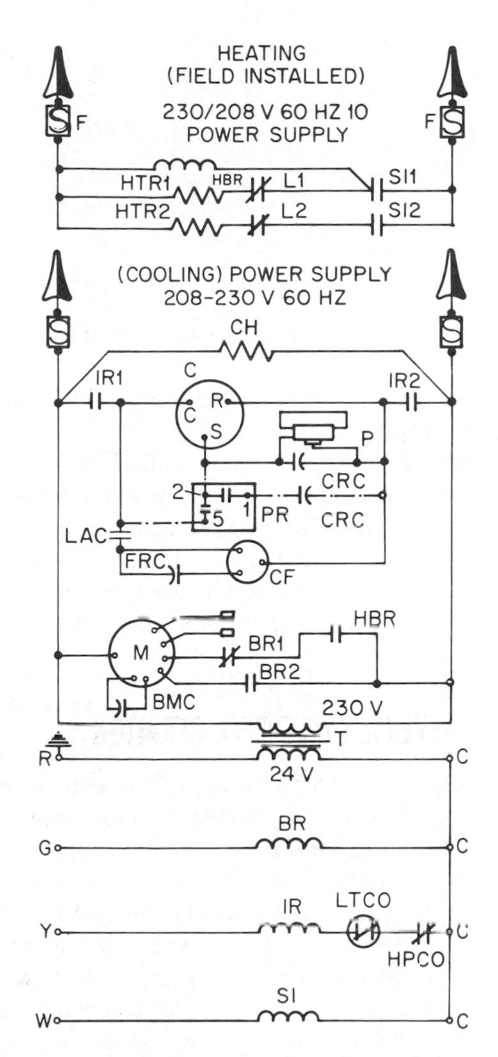

COMPONENT PART IDENTIFICATION

BMC	BLOWER MOTOR CAPACITOR
BR	BLOWER RELAY
C	COMPRESSOR
CF	CONDENSER FAN MOTOR
CH	CHRANKCASE HEATER
CRC	COMPRESSOR RUN CAPACITOR
CS	EXHAUSTER CENTRIFUGAL SWITCH
CSC	COMPRESSOR START CAPACITOR
EM	EXHAUSTER MOTOR
F	FUSE
FC	FAN CONTROL KLIXON
FRC	FAN RUN CAPACITOR
FT	FAN TIMER
GV	GAS VALVE
HBR	BLOWER RELAY (HEATING)
HPCO	HIGH-PRESSURE CUT-OUT
HR	HEAT RELAY
HTR	ELECTRIC HEATER
IR	COMPRESSOR CONTACTOR
L	LIMIT
LAC	LOW AMBIENT CONTROL (0°F) (ALL RD/RG-D UNITS & ALL R-H RG-H3 PHASE UNITS)
LTCO	LOW TEMPERATURE CUT-OUT (50°F) (ALL R-H/RG-H 1 PHASE UNITS)
M	BLOWER MOTOR
P	PTC STANDARD START ASSIST
PPK	POST PURGE KLIXON
PR	POTENTIAL RELAY
PR+ CDC	OPTIONAL START ASSIST (REPLACES "P")
S1	SEQUENCER
T	TRANSFORMER

FIGURE 42–16 Wiring diagram for electric heat and electric air conditioning package unit. *Courtesy Climate Control*

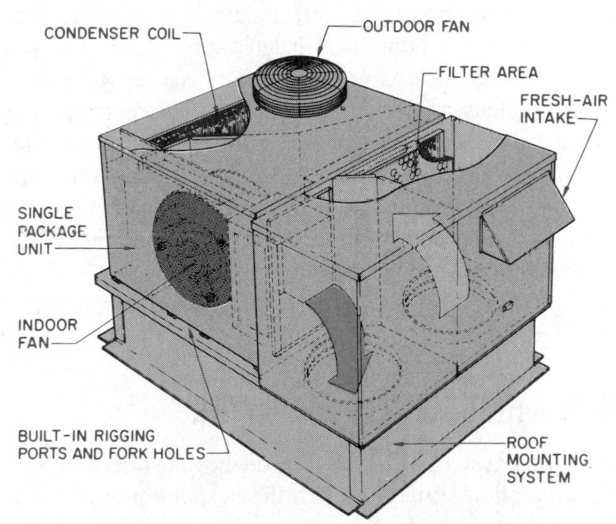

FIGURE 42–17 Installation of a package unit shows how the ductwork is attached. *Courtesy Climate Control*

- The control circuit may have two transformers—one furnished with the furnace and one with the air conditioning unit.
- When there are two power supplies (transformers), the two may be kept separated in the thermostat or they may be wired in parallel.
- Package all-weather systems normally consist of gas heat and electric air conditioning or electric heat and electric air conditioning.
- An advantage of package systems is that the whole system is located outside the structure.

REVIEW QUESTIONS

1. Name the five things that may be done to condition air.
2. What is an all-weather system?
3. What are the three common fuels used for heating in all-weather systems?
4. What is add-on summer air conditioning?
5. What are some of the things that must be considered with add-on air conditioning?
6. What are two methods to vary the airflow in an all-weather system?
7. Why is it desirable to have less airflow in the heating season?
8. What is the recommended airflow for summer air conditioning?
9. Name two ways to wire a system with two control transformers.
10. What is one advantage of a package all-weather system?

UNIT 43 Heat Pumps

OBJECTIVES

After studying this unit, you should be able to

- describe a reverse-cycle heat pump.
- list the components of a reverse-cycle heat pump.
- explain a four-way valve.
- state the various heat sources for heat pumps.
- compare electric heat to heat with a heat pump.
- state how heat pump efficiency is rated.
- determine by the line temperatures if a heat pump is in cooling or heating.
- discuss the terminology of heat pump components.
- define coefficient of performance.
- explain auxiliary heat.
- describe the control sequence on an air-to-air heat pump.
- describe techniques being used to improve the efficiency of heat pump systems.
- discuss recommended preventive maintenance procedures for heat pump systems.

SAFETY CHECKLIST

* When installing insulated ductwork or otherwise working around fiberglass, wear gloves, goggles, and clothing that will cover your skin. If fiberglass particles are in the air, wear a mask that covers your nose and mouth.
* Be careful not to cut your hands when working with metal duct, panels, and fasteners.
* Exercise all precautions when installing units on rooftops or in other hazardous locations.
* Observe all safety precautions when troubleshooting electrical components of a heat pump. When units are located in crawl spaces and you are in contact with the ground, be careful not to establish a path for current to flow through your body to ground.
* Troubleshoot with the electrical power on only when it is absolutely necessary. At all other times ensure that the power is off. When the power is off, lock and tag the disconnect panel. There should be only one key. Keep it with you so that another person will not be able to turn the power on while you are working on the equipment.
* Be extra cautious if conditions are wet when you are troubleshooting an outside unit.
* When connecting gage lines or transferring refrigerant, wear gloves, goggles, and clothing to protect your skin because refrigerant can freeze your skin and eye tissue. High-pressure refrigerant can pierce your skin and blow particles into your eyes.
* The hot gas line can cause serious burns and should be avoided.
* If water is used to melt the ice from the coil of the outdoor unit, turn the power off and protect all electrical components. If troubleshooting later with the power on, do not stand in this water.
* Use common sense and caution at all times while installing, maintaining, or troubleshooting any piece of equipment. As a technician you are constantly exposed to *potential danger*. Refrigerant, electrical shock hazard, rotating equipment, hot metal, sharp metal, and lifting heavy objects are among the most common hazards you will face.

43.1 REVERSE-CYCLE REFRIGERATION

Heat pumps, like refrigerators, are refrigeration machines. Refrigeration involves the removal of heat from a place where it is not wanted and depositing it in a place where it makes little or no difference. The heat can actually be deposited in a place where it is wanted as heat reclaim. This is the difference between a heat pump and a cooling air conditioner. The air conditioner can only pump heat one way. The heat pump is a refrigeration system that can pump heat two ways. It is normally used for space conditioning: heating and cooling.

All compression cycle refrigeration systems are heat pumps in that they pump heat-laden vapor. The evaporator of a heat pump absorbs heat into the refrigeration system, and the condenser rejects the heat. The compressor pumps the heat-laden vapor. The metering device controls the refrigerant flow. These four components—the evaporator, the condenser, the compressor, and the metering device—are essential to compression cycle refrigeration equipment. The same components are in a heat pump system along with the *four-way valve,* which is used to control the direction of heat flow.

43.2 HEAT SOURCES FOR WINTER

The cooling system in a typical residence absorbs heat into the refrigeration system through the evaporator and rejects this heat to the outside of the house through the condenser. The house might be 75°F inside while the outside temperature might be 95°F or higher. The summer air conditioner pumps heat from a low temperature inside the house to a higher temperature outside the house; that is, it pumps heat up the temperature scale. A freezer in a supermarket takes the heat out of ice cream at 0°F to cool it to −10°F so that it will be frozen hard. The heat removed can be felt at the condenser as hot air. This example shows that there is usable heat in a substance even at 0°F, Figure 43–1. There is heat in any substance until it is cooled down to −460°F. Review Unit 1 for examples of heat level.

If heat can be removed from 0°F ice cream, it can be removed from 0°F outside air. The typical heat pump does just that. It removes heat from the outside air in the winter and deposits it in the conditioned space to heat the house. (Actually, about 85% of the usable heat is still in the air at 0°F.) Hence, it is called an *air-to-air* heat pump, Figure 43–2. In summer the heat pump acts like a conventional air

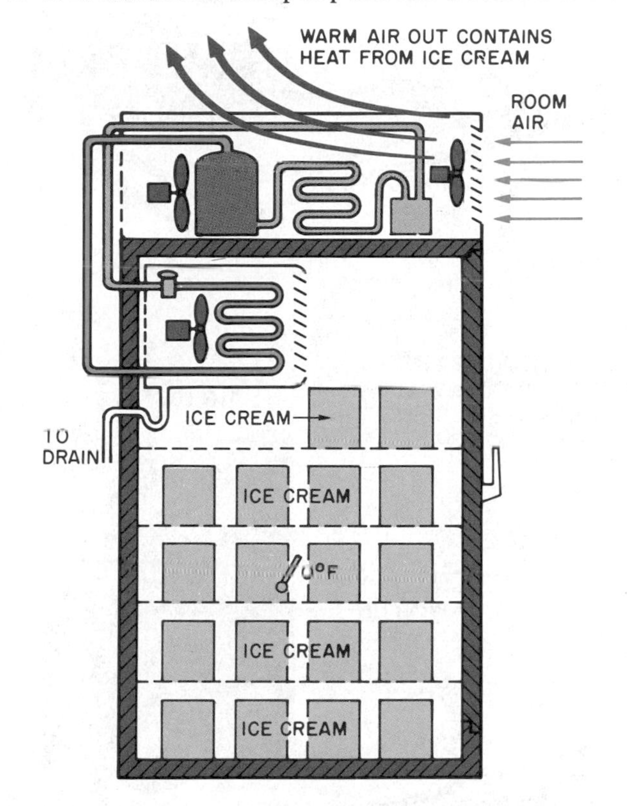

FIGURE 43–1 Low-temperature refrigerated box is removing heat from ice cream at 0°F. Part of the heat coming out of the back of the box is coming from the ice cream.

FIGURE 43–3 Outdoor portion of an air-to-air heat pump. It absorbs heat from the outside air for use inside the structure. *Reproduced courtesy of Carrier Corporation*

conditioner and removes heat from the house and deposits it outside. From the outside an air-to-air heat pump looks like a central cooling air conditioner, Figure 43–3.

43.3 THE FOUR-WAY VALVE

The refrigeration principles that a heat pump uses are the same as those stated previously. However, a new component is added to allow the refrigeration equipment to pump heat in either direction. The air-to-air heat pump in Figure 43–4 shows the heat pump moving heat from inside the conditioned space in summer to the outside. Then in winter the heat is moved from the outside to the inside, Figure 43–5. This change of direction is accomplished with a special

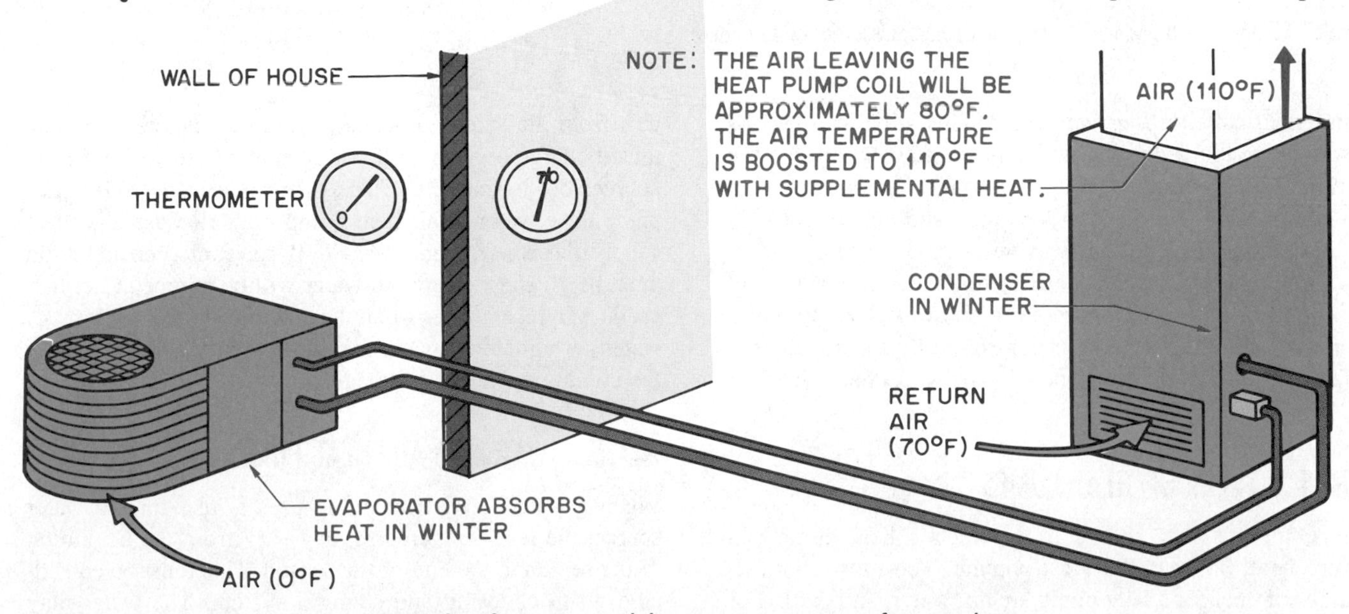

FIGURE 43–2 Air-to-air heat pump removing heat from 0°F air and depositing it in a structure for winter heat.

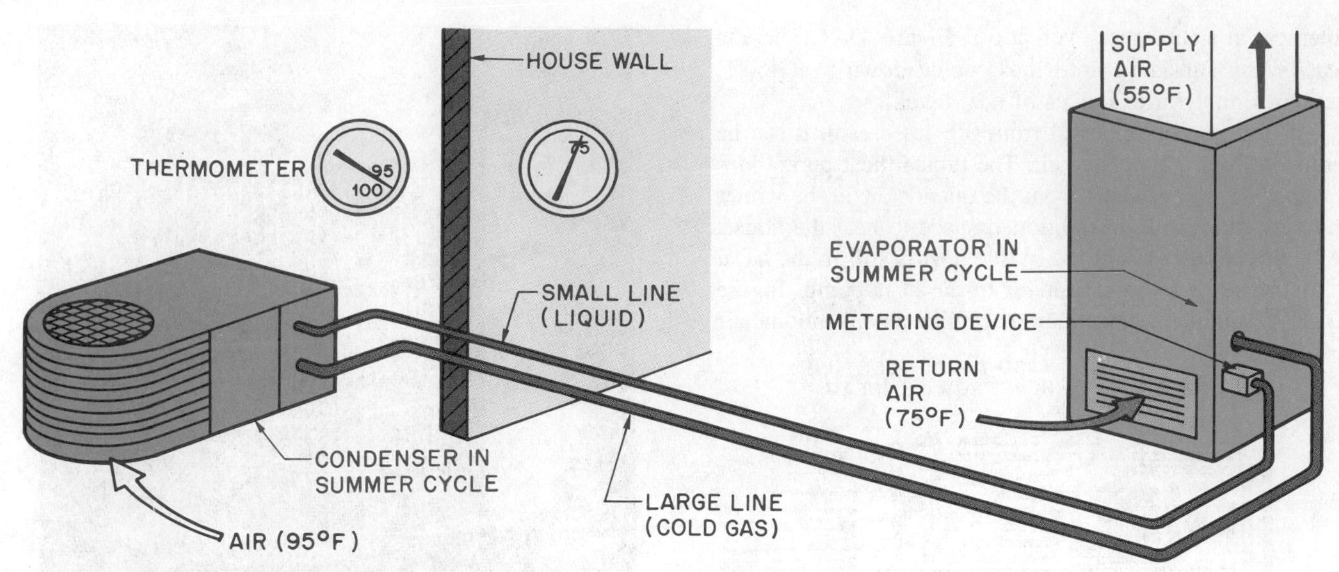

FIGURE 43–4 Air-to-air heat pump moving heat from the inside of a structure to the outside.

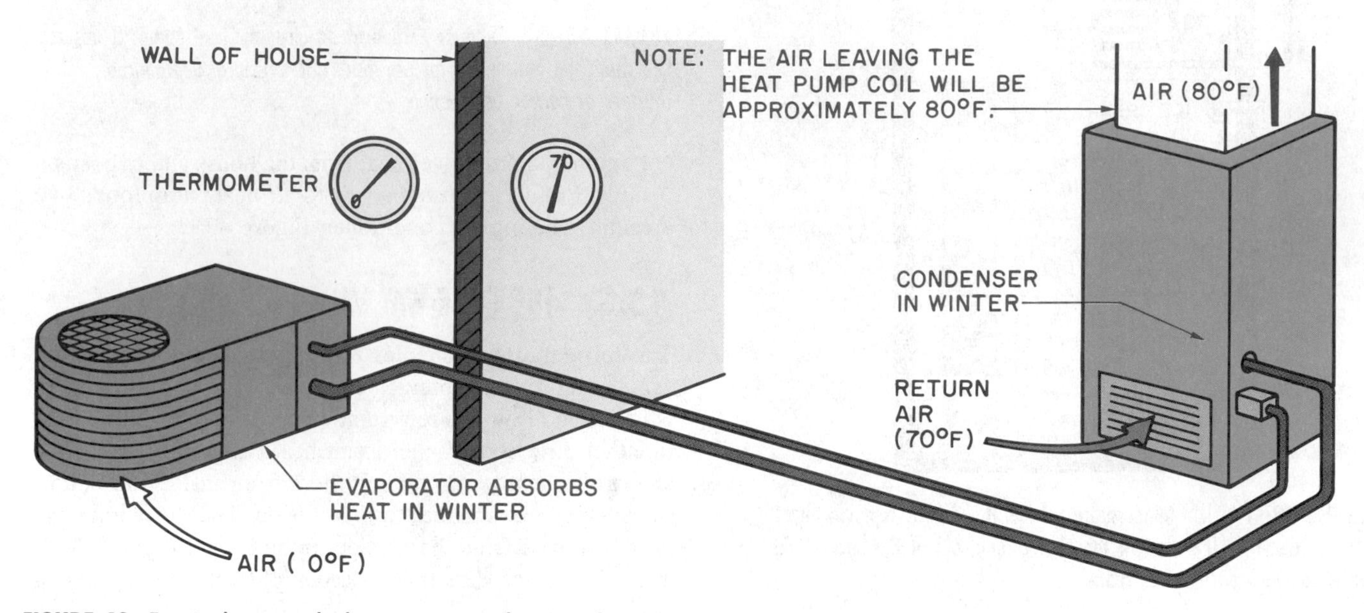

FIGURE 43–5 In the winter the heat pump pumps heat into the structure.

component called a *four-way valve*. This valve can best be described in the following way. The heat absorbed into the refrigeration system is pumped through the system with the compressor. The heat is contained and concentrated in the discharge gas. The four-way valve diverts the discharge gas and the heat in the proper direction to either heat or cool the conditioned space. Figure 43–6 shows an example of a four-way valve. This valve is controlled by the space temperature thermostat, which positions it to either HEAT or COOL.

43.4 TYPES OF HEAT PUMPS

Air is not the only source from which a heat pump can absorb heat, but it is the most popular. The most common sources of heat for a heat pump are air, water, and earth. For example, a structure located next to a large lake can remove heat from the lake and deposit it in the structure. The lake must be large enough so that the temperature will not drop appreciably in the lake. This system is a *water-to-air* heat pump. The water-to-air heat pump may also use a well to supply the water, Figure 43–7. If so, some consideration must be given to where the water will be pumped after it is used. A typical water-to-air heat pump uses 3 gallons of water per minute (gpm) in the heating cycle and 1.5 gpm in the cooling cycle per ton of refrigeration (12,000 Btu/h).

43.5 WATER-TO-AIR HEAT PUMPS

When a water-to-air heat pump uses a lake for the water source, the lake temperature may vary from the beginning of the season to the end of the season. This must be considered when choosing and sizing a system. The water may usually be pumped back into the lake after being used, so

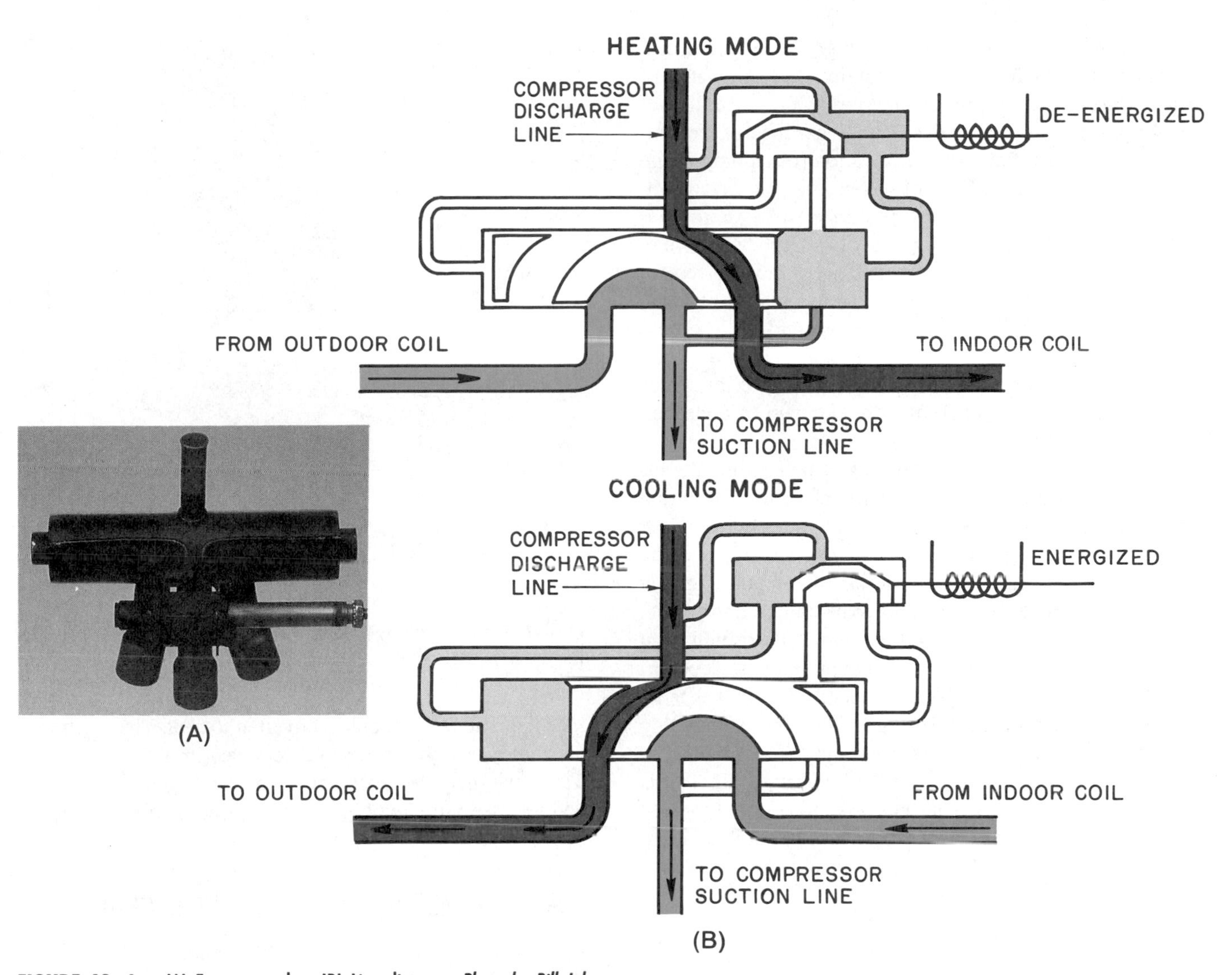

FIGURE 43–6 (A) Four-way valve. (B) Line diagram. *Photo by Bill Johnson*

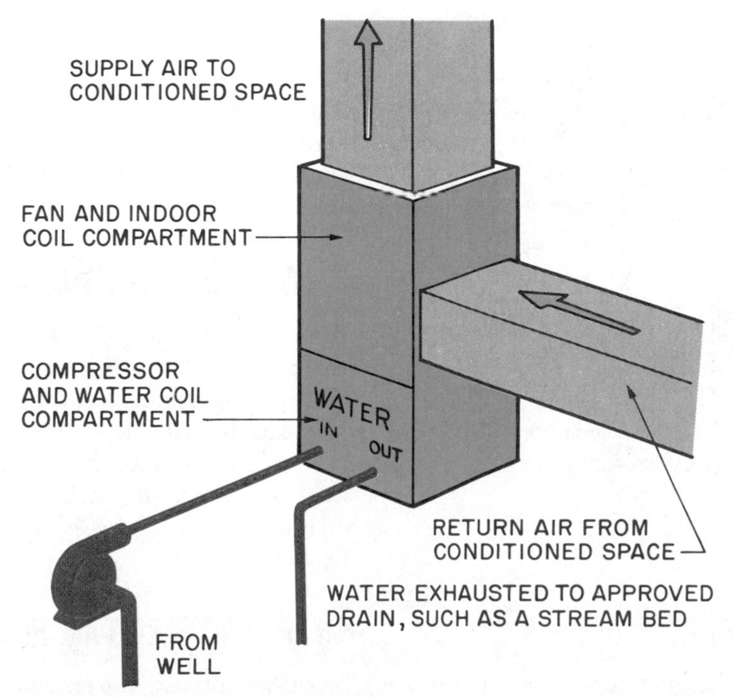

FIGURE 43–7 Water-to-air heat pump is absorbing heat from well water.

this is normally no problem. **However, local codes and authority must be followed for any installation.** Water-to-air heat pumps may get more attention from the local governing body than air-to-air heat pumps because of the handling of the water. For example, a 3-ton heat pump in the heating mode will use 9 gpm of water. This is a rate of 540 gallons per hour (gph) that must be drained off (9 gpm × 60 min/h = 540 gph) or 12,960 gal/day (540 gph × 24 h/day = 12,960 gal/24 h). Some installations have diversion valves to heat pools in the summer and to water the lawn, but at the rate of 12,960 gal/day the ground will become saturated very soon if there is not plenty of land. This amount of water cannot normally be just piped to the corner of the property to run off. It must be pumped to a planned location, such as a stream.

Industry and commercial buildings often use heat pumps to move heat from one part of the building or manufacturing process to another part. For example, a large office building may require cooling for some areas even in winter. The offices that are located in the interior of the building may have heat buildup due to lights, office machines, and people. It is economical in some office buildings to absorb this heat into a refrigeration system and to pump it to the outer perimeter of the building where heat is needed. This makes more sense than wasting this heat to the atmosphere and buying energy to heat the outside of the building. This moving of heat from one part of a commercial building to another is sometimes accomplished with many small residential-sized water-to-air heat pumps. The advantage is that it is economical to operate and each zone has its own unit for zone control, Figure 43–8.

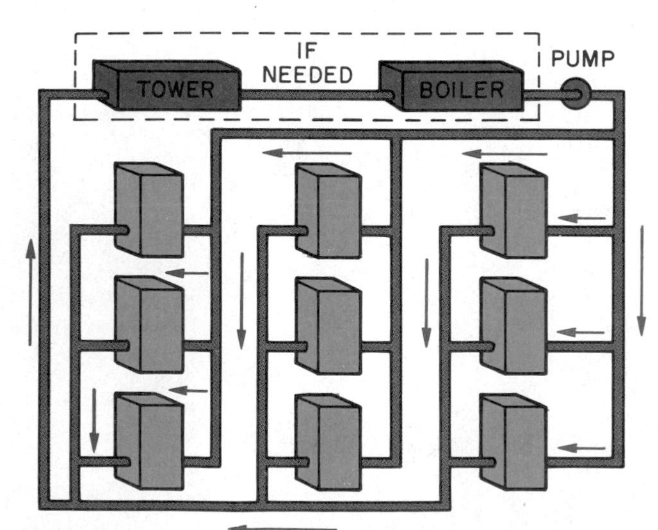

FIGURE 43–8 Water-to-air system that can absorb heat from one part of a building and deposit it in another. *Reproduced courtesy of Carrier Corporation*

43.6 REMOVING HEAT FROM MANUFACTURING PROCESSES

Some manufacturing processes, such as metalworking, generate heat. Some of the energy that is used to form or cut

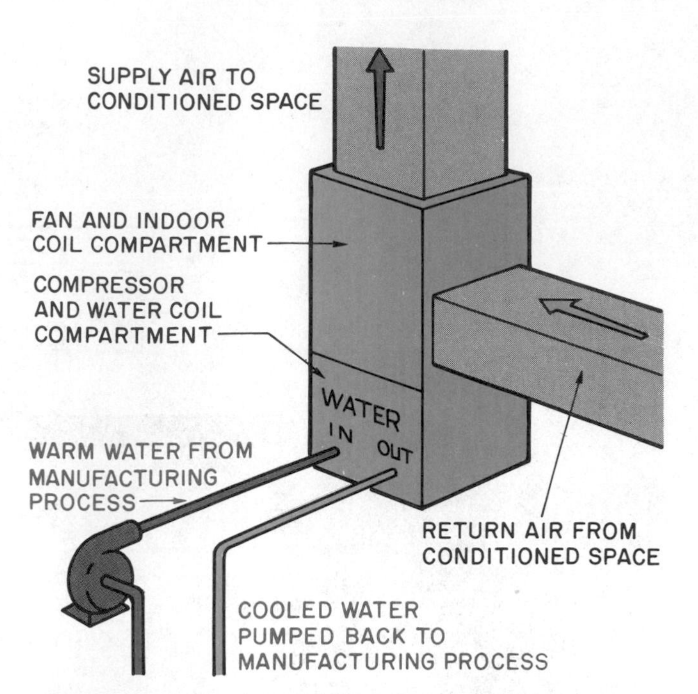

FIGURE 43–9 Heat from cooled manufacturing process used with heat pump to heat office.

metal in a machine shop is converted to heat. The machines are cooled, and the excess heat removed from the machines. This heat may be absorbed into a refrigeration system and released into the office space to heat the offices, Figure 43–9.

43.7 REMOVING HEAT FROM THE GROUND

Groundwater heat pumps must have some method of removing heat from the ground. One way is to use a long ditch and buried pipe, Figure 43–10. Another method is to use well water as a heat source in the winter by removing the heat from it and pumping it to another well, Figure 43–11. A deep well with the pipe loop going to the bottom of the well may also be used. Be sure to check the local code for either type of installation. When one of these systems is considered, someone with experience in this field should be consulted.

43.8 SOLAR-ASSISTED HEAT PUMPS

Many ways are being explored for using the reverse-cycle heat pump. Solar-assisted heat pumps are one example. These pumps capture heat from the sun and boost the level of the heat to usable levels for heating homes. Some manufacturers are designing systems especially for this purpose. This text describes the basic heat pump cycle and the air-to-air heat pump in detail because it is currently the most popular.

43.9 THE AIR-TO-AIR HEAT PUMP

The air-to-air heat pump resembles the central air conditioning system. There are indoor and outdoor system compo-

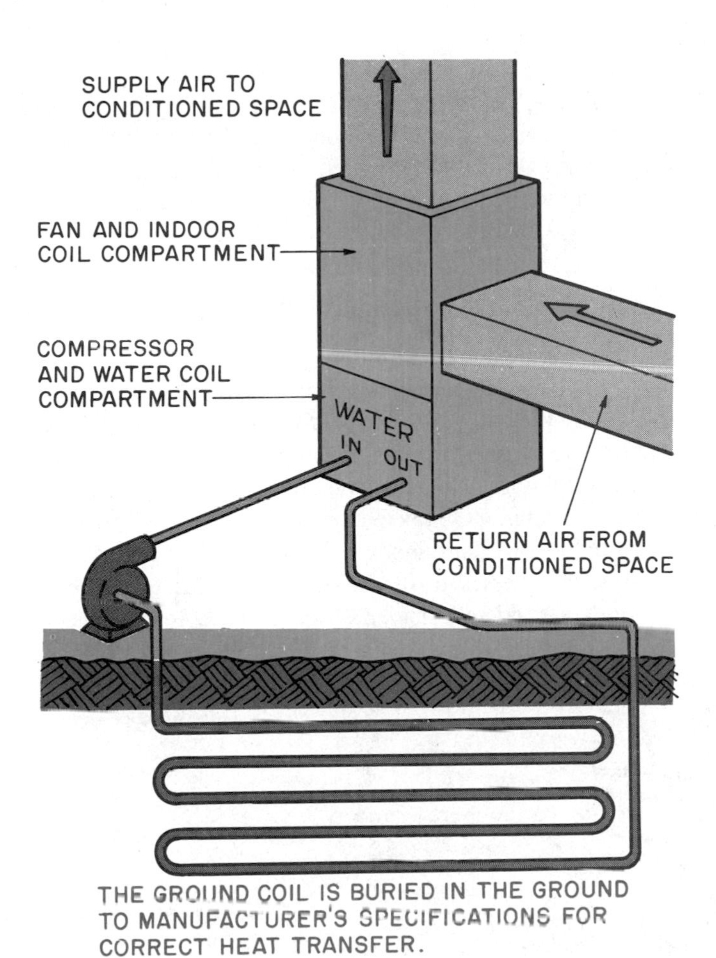

FIGURE 43–10 Water-to-air heat pump using a long underground loop.

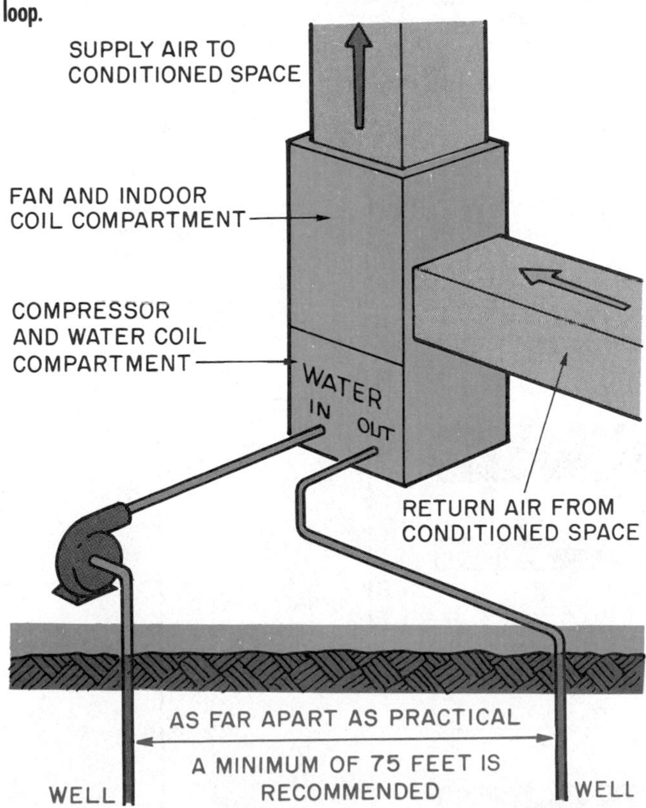

FIGURE 43–11 Water-to-air heat pump using well water and pumping the water into another well.

nents. When discussing typical air conditioning systems, these components are often called the *evaporator* (indoor unit) and the *condenser* (outdoor unit). This terminology will work for air conditioning but not for a heat pump.

The system's coils have new names when applied to a heat pump. The coil that serves the inside of the house is called the *indoor coil*. The unit outside the house is called the *outdoor unit* and contains the *outdoor coil*. The reason is that the indoor coil is a condenser in the heating mode and an evaporator in the cooling mode. The outdoor coil is a condenser in the cooling mode and an evaporator in the heating mode. This is all determined by which way the hot gas is flowing. In winter the hot gas is flowing toward the indoor unit and will give up heat to the conditioned space. The heat must come from the outdoor unit, which is the evaporator. See Figure 43–12 for an example of the direction of the gas flow. The system mode can easily be determined by gently touching the gas line to the indoor unit. ***If it is hot (it can be 200°F, so be careful), the unit is in the heating mode.***

Like cooling equipment, heat pumps are manufactured in split systems and package systems.

43.10 REFRIGERANT LINE IDENTIFICATION

When an air-to-air heat pump is a split system, the same lines are connected between the indoor unit and the outdoor unit except that they have a new name. The large line is called a *gas line* because it is always a gas line. In previous

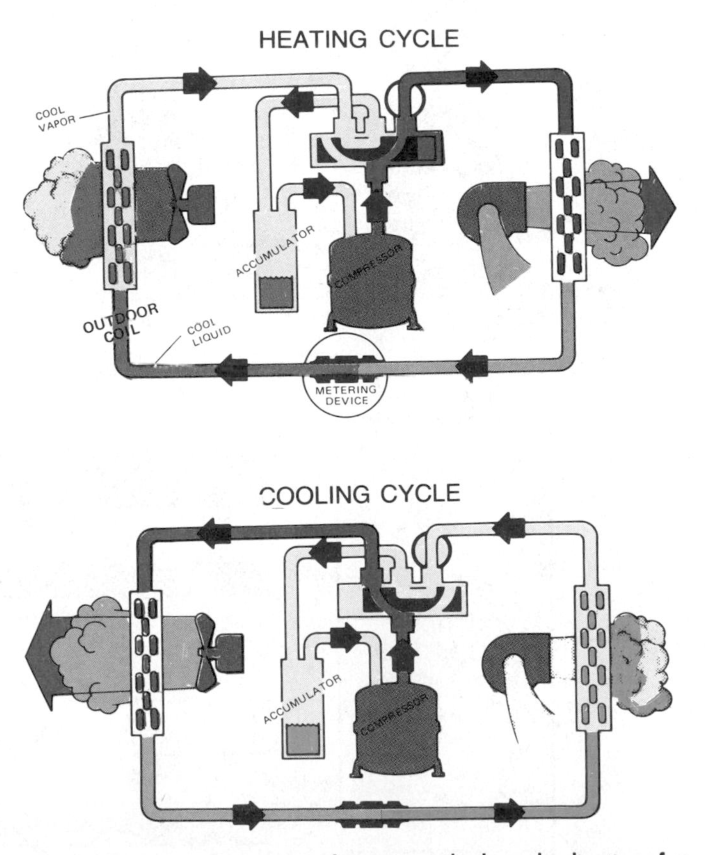

FIGURE 43–12 Heat pump refrigeration cycle shows the direction of the refrigerant gas flow. *Reproduced courtesy of Carrier Corporation*

units it was called a cold gas line or a suction line. In a heat pump it is always a gas line because it is a suction line or cold gas line in summer and a hot gas line in winter, Figure 43–13.

The small line is a *liquid line* in summer and winter, so it keeps the same name. Some changes do occur in the liquid line between summer operation and winter operation. The liquid flows toward the inside unit in summer and toward the outside unit in winter, Figure 43–14. The line is the same size as for cooling; the liquid direction is just reversed.

43.11 METERING DEVICES

Because the direction of the liquid flow is reversed from one season to another, some of the refrigeration components are slightly different. The metering devices used on heat pumps are different because there must be a metering device at the indoor unit as well as at the outdoor unit at the proper time. For example, when the unit is in the cooling mode, the metering device is at the indoor unit. When the system changes over to heating, a metering device must then meter refrigerant to the outdoor unit. This is accomplished in various ways with several combinations of metering devices.

43.12 THERMOSTATIC EXPANSION VALVES

The thermostatic expansion valve (TXV) was the first metering device in common use with heat pumps. Because this device will only allow liquid to flow in one direction, it had to have a check valve piped in parallel with it to allow flow around it in the other mode of operation. For example, in the cooling mode the valve needs to meter the flow with the liquid refrigerant moving toward the indoor unit. When the system reverses to the heating mode, the indoor unit becomes a condenser and the liquid needs to be able to

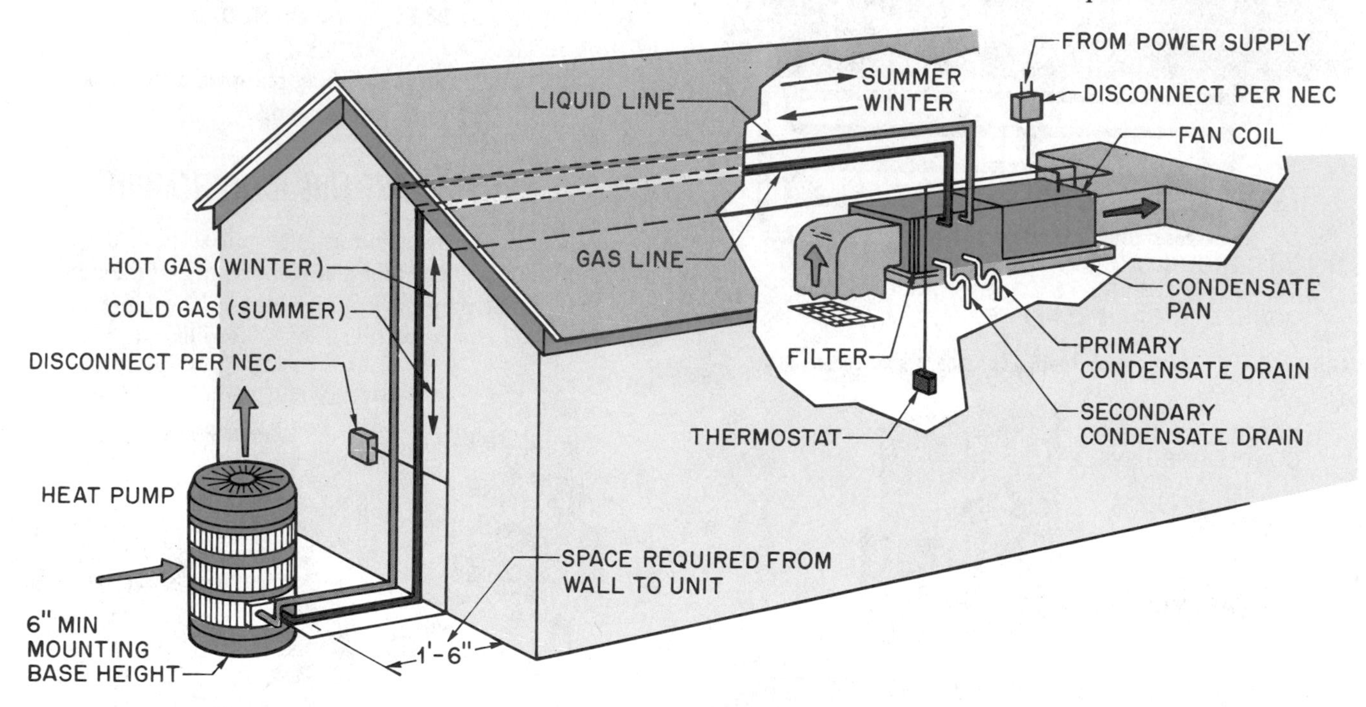

FIGURE 43–13 Split-system heat pump shows the interconnecting refrigerant lines. The large line is the gas line.

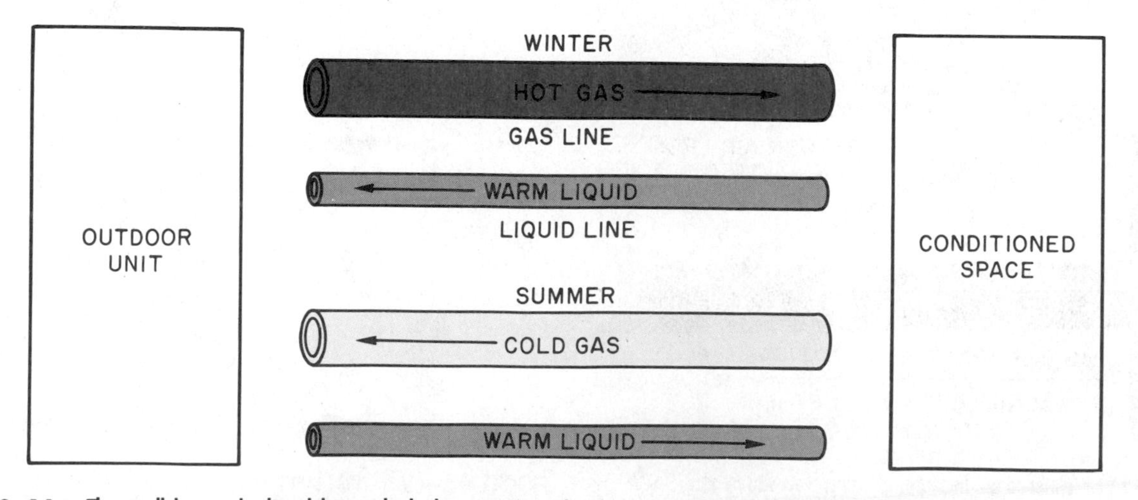

FIGURE 43–14 The small line is the liquid line in both the summer and winter operation.

move freely toward the outdoor unit. The liquid flows through the check valve in winter and is metered through the TXV in the summer cycle, Figure 43–15.

The TXV applied to a heat pump must be chosen carefully. The sensing bulb for the valve is on the gas line, which may become too hot for a typical TXV bulb. It is not unusual for the hot gas line to reach temperatures of 200°F. If the sensing bulb for a typical TXV is exposed to these temperatures, it is subject to rupture, Figure 43–16.

A TXV with a limited charge in the bulb is usually used for this application. Remember, from Unit 24, a limited charge bulb does not continue to build pressure with added heat. It only builds a certain amount of pressure then additional heat causes no appreciable increase in pressure.

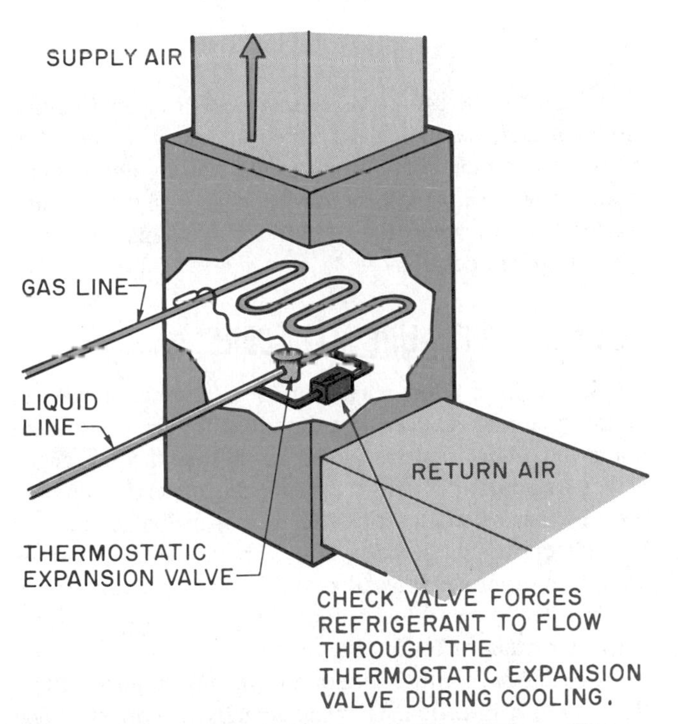

FIGURE 43–15 Piping diagram that must be used when a TXV is used for a metering device on a heat pump. Notice the check valve.

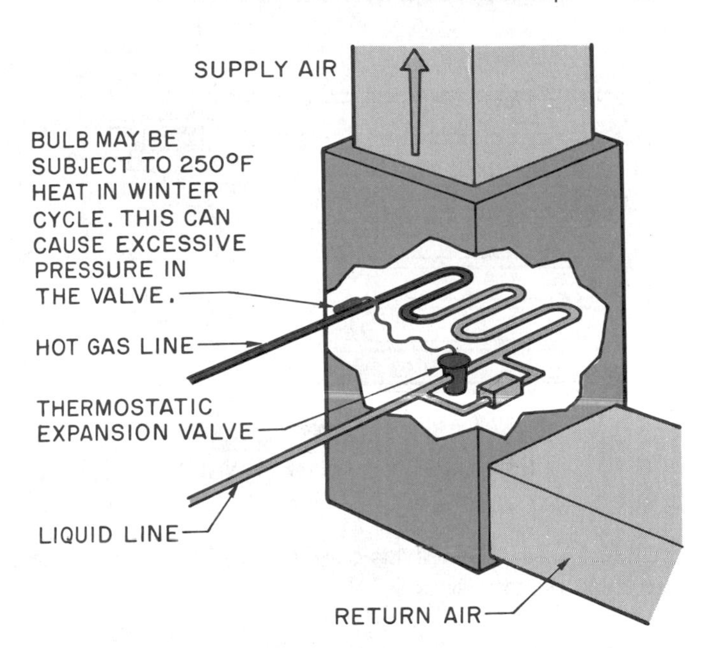

FIGURE 43–16 This illustration shows that care must be taken in the choice of TXVs. If an ordinary TXV sensing bulb were to be mounted on this gas line, it would get hot enough to rupture the valve diaphragm.

43.13 THE CAPILLARY TUBE

The capillary tube metering device is also used on heat pumps. This device allows refrigerant flow in either direction, but it is not normally used in this manner because capillary tubes of different sizes are required in summer and winter. The capillary tube may be used in several ways. One is to use a check valve to reverse the flow. When this is done, two capillary tubes are usually used, Figure 43–17.

43.14 COMBINATIONS OF METERING DEVICES

Sometimes a combination of two metering devices is used. One popular combination is to use the capillary tube for the indoor metering device for summer operation, with a check valve piped parallel to allow flow in the other direction for winter operation. Then the outside unit may have a TXV

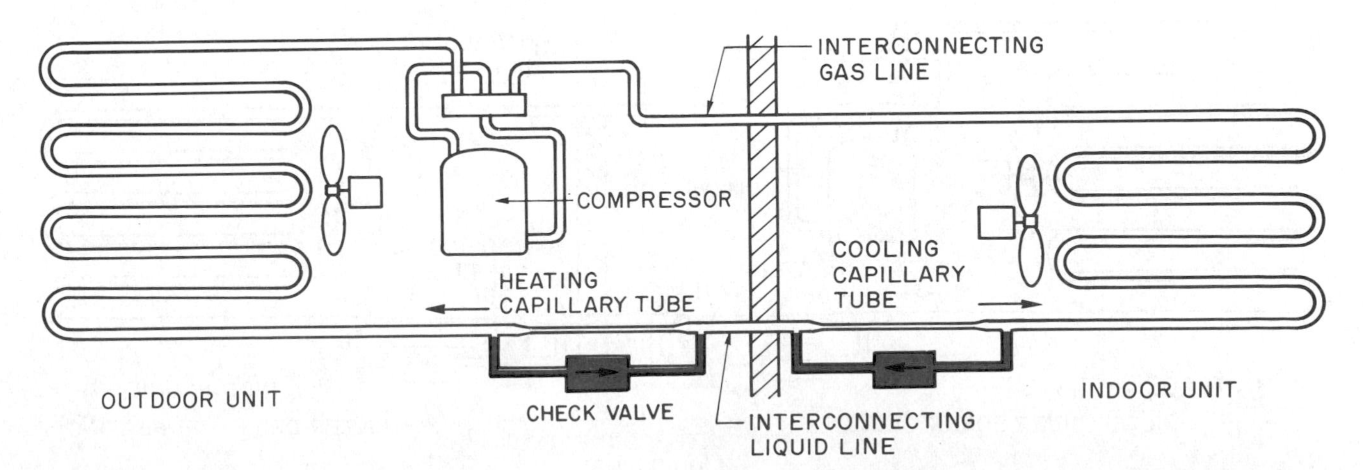

FIGURE 43–17 Capillary tube metering device used on a heat pump. Notice that check valves are used and there are two different capillary tubes.

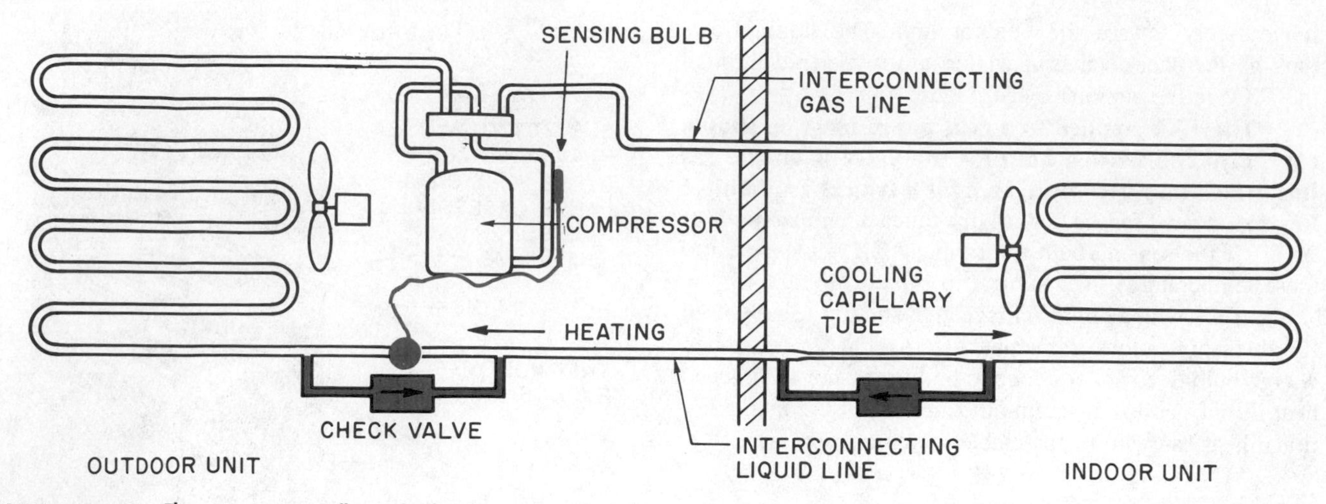

FIGURE 43–18 This unit uses a capillary tube for the indoor cycle and a TXV for the outdoor unit. This TXV has the sensing element mounted to the compressor permanent suction line.

with a check valve. This system is efficient because it uses the capillary tube in the summer mode only. It is an efficient metering device for summer operation because the load conditions are nearly constant. The TXV is used in the winter on the outdoor coil. This is efficient for the winter cycle because winter conditions are not constant. The TXV will reach maximum efficiency sooner than the capillary tube in the same application because it can open its metering port when needed. This allows the evaporator to become full of refrigerant earlier in the running cycle. This dual system uses each device at its best, Figure 43–18.

43.15 ELECTRONIC EXPANSION VALVES

The electronic expansion valve was discussed in the unit on refrigeration and is sometimes applied to heat pumps. If the heat pump is a close-coupled unit with the indoor coil close to the outdoor unit, such as a package system (discussed later), a single valve can be used. The reason the valve is best applied to a close-coupled unit is that the valve can meter in both directions, and if the liquid line were long it would need to be insulated. The valve will meter in either direction and maintain the correct superheat at the compressor's common suction line in the heating and cooling modes. The sensing element may be located on the common suction line just before the compressor to maintain this correct refrigerant control.

43.16 LIQUID-LINE ACCESSORIES

Liquid-line filter-driers must be used in conjunction with all of the metering devices just described. When a standard liquid-line filter-drier is installed in a system that has check valves to control the flow through the metering devices, the drier is installed in series with the expansion device. The flow direction is the same as with the metering device. If the same filter-drier were installed in the common liquid line, it would filter in one direction and the particles would wash out in the other direction, Figure 43–19.

Special *biflow* filter-driers are manufactured to allow flow in either direction. They are actually two driers in one shell with check valves inside the drier shell to cause the

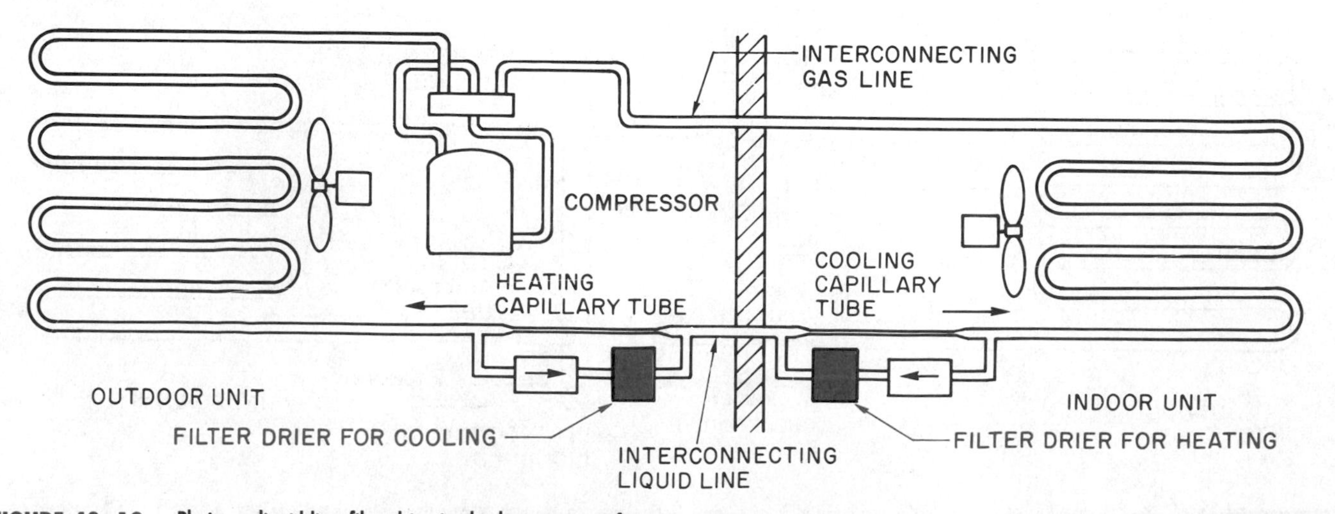

FIGURE 43–19 Placing a liquid-line filter-drier in the heat pump refrigerant piping.

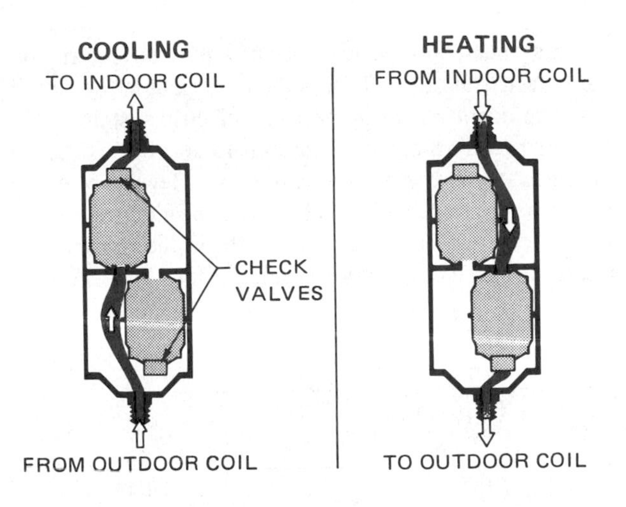

FIGURE 43–20 Biflow drier. *Reproduced courtesy of Carrier Corporation*

liquid to flow in the proper direction at the proper time, Figure 43–20.

43.17 ORIFICE METERING DEVICES

Another common metering device used by some manufacturers is a combination flow device and check valve. This device allows full flow in one direction and restricted flow in the other direction, Figure 43–21. Two of these devices are necessary with a split system—one at the indoor coil and one at the outdoor coil. The metering device at the indoor coil has a larger bore than the one at the outdoor coil because the two coils have different flow characteristics. The summer cycle uses more refrigerant in normal operation. When this device is used, a biflow filter-drier is normally used in the liquid line when field repairs are done.

All of these components may be found on water-to-air heat pumps or air-to-air heat pumps. The components that control and operate the systems are much the same.

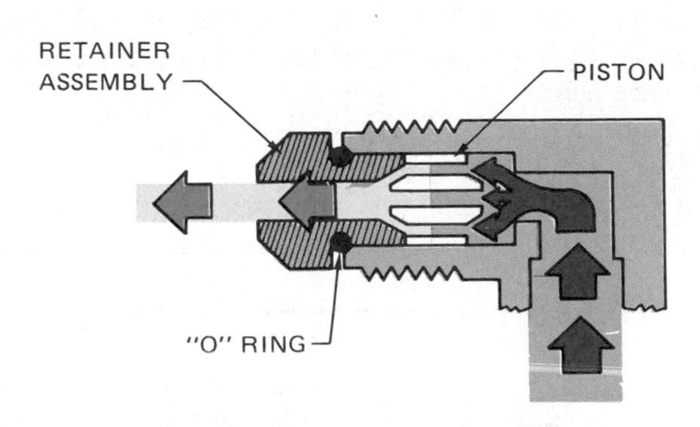

FIGURE 43–21 Combination fixed-bore (orifice) and check valve metering device. *Reproduced courtesy of Carrier Corporation*

43.18 APPLICATION OF THE AIR-TO-AIR HEAT PUMP

Air-to-air systems are normally installed in milder climates —in the "heat pump belt," which is basically those parts of the United States where winter temperatures can be as low as 10°F.

The reason for this geographical line is the characteristic of the air-to-air heat pump. It absorbs heat from the outside air; as the outside air temperature drops, it is more difficult to absorb heat from it. For example, the evaporator must be cooler than the outside air for the air to transfer heat into the evaporator. Normally, in cold weather, the heat pump evaporator will be about 20°F to 25°F cooler than the air from which it is absorbing heat. We will use 25°F temperature difference (TD) in this text as the example, Figure 43–22. On a 10°F day the heat pump evaporator will be boiling (evaporating) the liquid refrigerant at 25°F lower than the 10°F air. This means that the boiling refrigerant temperature will be −15°F. As the evaporator temperature goes down, the compressor loses capacity. The compressor is a fixed-size pump. It will have more capacity on a 30°F day than on a 10°F day. **The heat pump loses capacity as**

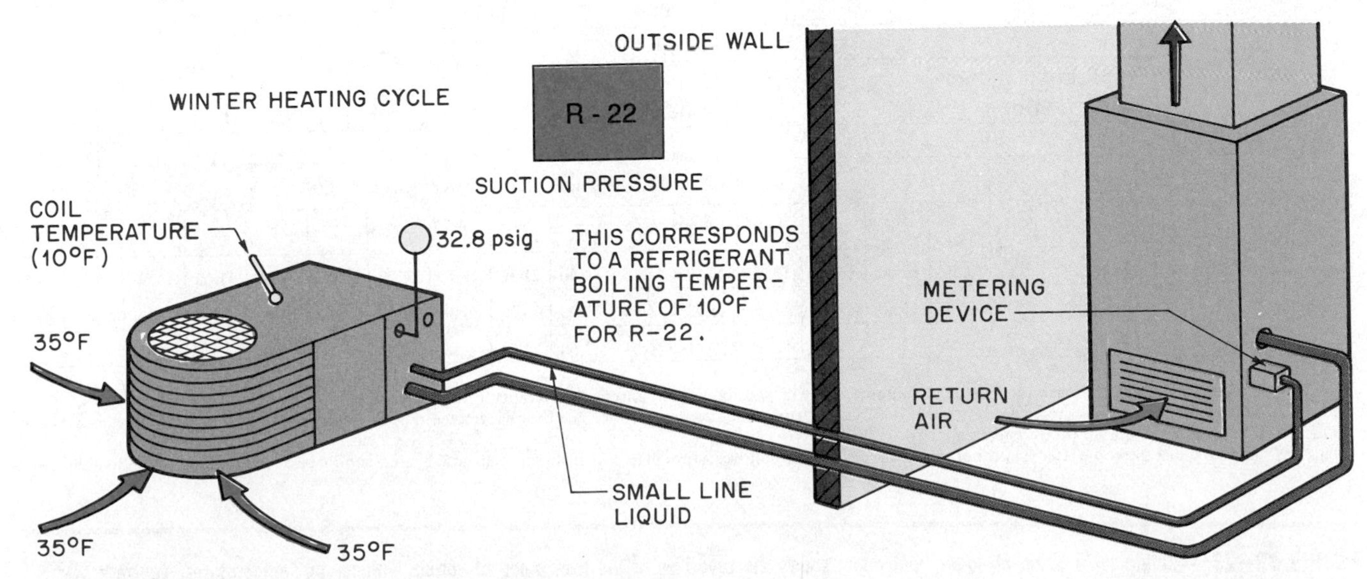

FIGURE 43–22 Heat pump outdoor coil operating in the winter cycle.

the capacity need of the structure increases. On a 10°F day the structure needs more capacity than on a 30°F day, but the heat pump's capacity is less. Thus, the heat pump must have help.

43.19 AUXILIARY HEAT

In an air-to-air heat pump system, the help that the heat pump gets is called *auxiliary heat*. The heat pump itself is the primary heat, and the auxiliary heat may be electric, oil, or gas. Electric auxiliary heat is the most popular because it is easier to adapt a heat pump to an electric system.

The structure that a heat pump is to heat has a different requirement for every outside temperature level. For example, a house might require 30,000 Btu of heat during the day when the outside temperature is 30°F. As the outside temperature drops, the structure requires more heat. It might

SPECIFICATIONS

MODEL	541B034SHP					
SERIES	A					
ELECTRICAL						
Unit Volts—Hertz—Phase	208-230—60—1					
Operating Voltage Range	197-253					
Unit Ampacity for Wire Sizing	28.6					
Min Wire Size (60° Copper) (AWG)	10					
Max Branch Circuit Fuse Size (Amps)	45					
Total Unit Amps	23.1					
Compressor Rated Load Amps	21.8					
Locked Rotor Amps	88					
Fan Motor	1/6 HP, PSC					
Full Load Amps	1.3					
PERFORMANCE DATA						
ARI Sound Rating Number	19					
517B/HPFC (D)	030	036	—	—	—	—
519A/MCC	—	—	036	036	042	042
520B/BP	—	—	042	—	042	—
Rated Heating Capacity—47° F	33,000	34,500	34,500	34,000	35,000	34,500
Watts	4100	3950	3900	3850	3950	3850
COP	2.4	2.6	2.6	2.6	2.6	2.6
Rated Heating Capacity—17° F	18,000	19,000	19,000	18,500	19,000	18,500
Watts	3300	3200	3150	3100	3200	3100
COP	1.6	1.7	1.8	1.8	1.7	1.8
Rated Cooling Capacity (Btuh)	30,000	32,000	32,000	32,500	32,500	32,500
Watts	4500	4500	4450	4400	4600	4500
EER	6.7	7.1	7.2	7.4	7.1	7.2
Min Application Indoor Airflow (Ft^3/Min)	1000	1050	1050	1050	1100	1100

541DJ030/ 517E030	Indoor Coil Airflow Ft^3/Min* 1100	EDB* 70° F	OUTDOOR COIL ENTERING AIR TEMPERATURE °F															
			−13	−8	−3	2	7	12	17	22	27	32	37	42	47	52	57	62
Instantaneous Capacity (MBtuh)			10.50	12.00	13.50	15.00	16.50	18.10	19.60	21.30	22.90	24.70	26.80	28.00	31.00	33.60	36.50	39.50
Integrated Capacity (MBtuh)†			9.68	11.00	12.40	13.80	15.20	16.60	17.90	19.20	20.30	21.40	24.40	28.60	31.00	33.60	36.50	39.50
Total Power Input (KW)‡			2.04	2.13	2.22	2.31	2.41	2.51	2.60	2.71	2.80	2.91	3.04	3.14	3.28	3.44	3.61	3.80

*See the Heating Performance Correction Factors Table for Ft^3/Min and indoor coil entering air temperature adjustments.
† The Btuh heating capacity values shown are net "integrated" values from which the defrost effect has been subtracted. The Btuh heating from supplement heaters should be added to those values to obtain total system capacity.
‡ The KW values include the compressor, outdoor fan motor, and indoor blower motor. The KW from supplemental heaters should be added to these values to obtain total system KW.

FIGURE 43–23 Rating table of an air-to-air heat pump shows the capacities of the heat pump at various outdoor air temperatures. ***Courtesy BDP Company***

need 60,000 Btu of heat in the middle of the night when the temperature has dropped to 0°F. The heat pump could have a capacity of 30,000 Btu/h at 30°F and 20,000 Btu/h at 0°F. The difference of 40,000 Btu/h must be made up with auxiliary heat.

43.20 BALANCE POINT

The *balance point* occurs when the heat pump can pump in exactly as much heat as the structure is leaking out. At this point the heat pump will completely heat the structure by running continuously. Above this point the heat pump will cycle off and on. Below this point the heat pump will run continuously but will not be able to heat the structure by itself.

43.21 COEFFICIENT OF PERFORMANCE

If the heat pump will not heat the structure all winter, why is it so popular? This can be answered in one word, *efficiency*. To understand the efficiency of an air-to-air heat pump, you need an understanding of electric heat. A customer receives 1 W of usable heat for each watt of energy purchased from the power company while using electric resistance heat. This is called 100% efficient or a *coefficient of performance* (COP) of 1:1. The output is the same as the input. With a heat pump the efficiency may be improved as much as 3:1 for an air-to-air system with typical equipment. When the 1 W of electrical energy is used in the compression cycle to absorb heat from the outside air and pump this heat into the structure, the unit could furnish 3 W of usable heat. Thus its COP is 3:1, or it can be thought of as 300% efficient. See Figure 43–23 for a manufacturer's rating table for an air-to-air heat pump.

A high COP only occurs during higher outdoor winter temperatures. As the temperature falls, the COP also falls. A typical air-to-air heat pump will have a COP of 1.5:1 at 0°F, so it is still economical to operate the compression system at these temperatures along with the auxiliary heating system. Some manufacturers have controls to shut off the compressor at low temperatures. When the temperature rises, the compressor will come back on. Some temperatures that manufacturers use to stop the compressor are 0°F to 10°F. The long running times of the compressor will not hurt it or wear it out to any extent, so some manufacturers do not shut off the compressor at all. They would rather have the compressor run all the time than to restart it with a cold crankcase when the temperature rises.

The *water-to-air* heat pump might not need auxiliary heat because the heat source (the water) can be a constant temperature all winter. The heat loss (the heat requirement) and the heat gain (the cooling requirement) of the structure might almost be equal. Consequently, these heat pumps have COP ratings as high as 3.5 or even 4:1. Just remember that the higher COP ratings of water-to-air heat pumps are due not to the components of the refrigeration equipment but to the temperatures of the heat source. Earth and lakes have a more constant temperature than air. Such pumps are very efficient for winter heating because of the COP and for summer cooling because they operate at water-cooled air conditioning temperatures and head pressures.

43.22 SPLIT-SYSTEM AIR-TO-AIR HEAT PUMP

Like cooling air conditioning equipment, air-to-air heat pumps come in two styles: split systems and package (self-contained) systems.

The split-system air-to-air heat pump resembles the split-system cooling air conditioning system. The components look exactly alike. An expert normally cannot tell if the equipment is air conditioning or heat pump equipment from the outside.

43.23 THE INDOOR UNIT

The indoor unit of an air-to-air system may be an electric furnace with a heat pump indoor coil where the summer air conditioning evaporator would be placed, Figure 43–24. The outdoor unit may resemble a cooling condensing unit.

The indoor unit is the part of the system that circulates the air for the structure. It contains the fan and coil. The airflow pattern may be upflow, downflow, or horizontal to serve different applications. Some manufacturers have cleverly designed their units so that one unit may be adapted to all of these flow patterns. This is done by correctly placing the pan for catching and containing the summer condensate, Figure 43–25.

Coil placement in the airstream is important in the indoor unit. The electric auxiliary heat and the primary heat (the heat pump) may need to operate at the same time in weather below the balance point of the house. **The refrigerant coil must be located in the airstream before the auxiliary heating coil.** Otherwise, heat from the auxiliary

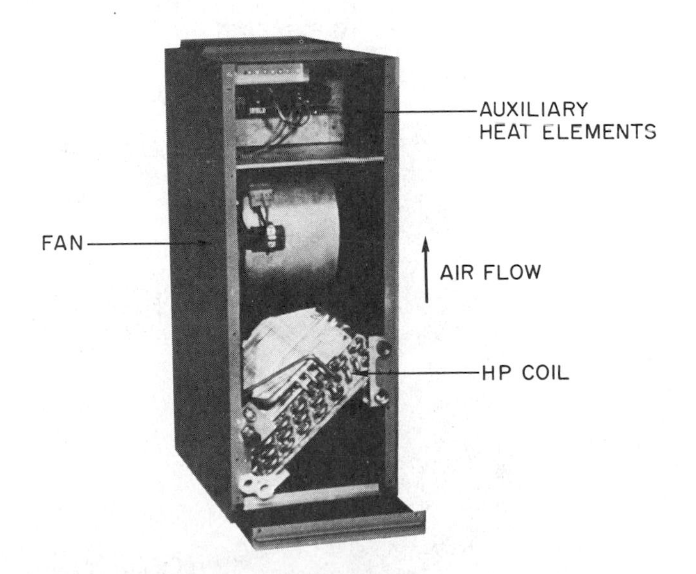

FIGURE 43–24 Air-to-air heat pump indoor unit. It is an electric furnace with a heat pump coil in it. Notice that the air flows through the heat pump coil and then through the heating elements. *Reproduced courtesy of Carrier Corporation*

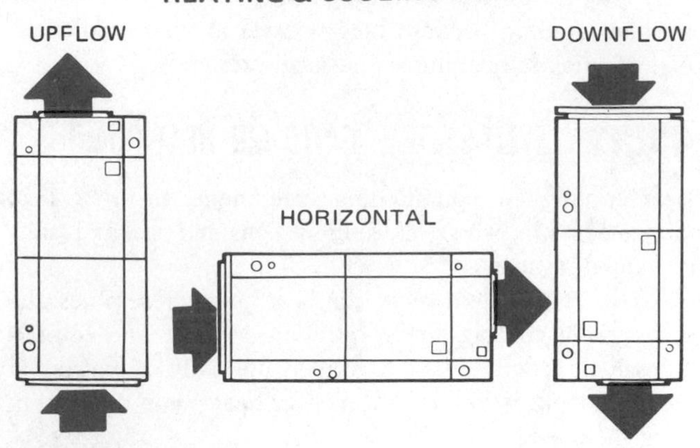

FIGURE 43–25 This heat pump indoor unit can be applied either in the vertical (upflow, downflow) or horizontal mode by placing the condensate pan under the coil in the respective position. *Reproduced courtesy of Carrier Corporation*

unit will pass through the refrigerant coil when both are operating in winter. If the auxiliary heat is operating and is located before the heat pump coil, the head pressure will be too high and could rupture the coil or burn the compressor motor. Remember, the coil is operating as a *condenser* in the heating mode. It rejects heat from the refrigeration system, so any heat added to it will cause the head pressure to rise, Figure 43–26.

The indoor unit may be a gas or oil furnace. If this is the case, the indoor coil must be located in the outlet airstream of the furnace. When the auxiliary heat is gas or oil, the heat pump does not run while the auxiliary heat is operating. If it isn't, the furnace heat exchanger would sweat in summer. The indoor coil then would be operating as an evaporator in summer, and the outlet air temperature could be lower than the dew point temperature of the air surrounding the heat exchanger. Most local codes will not allow a gas or oil heat

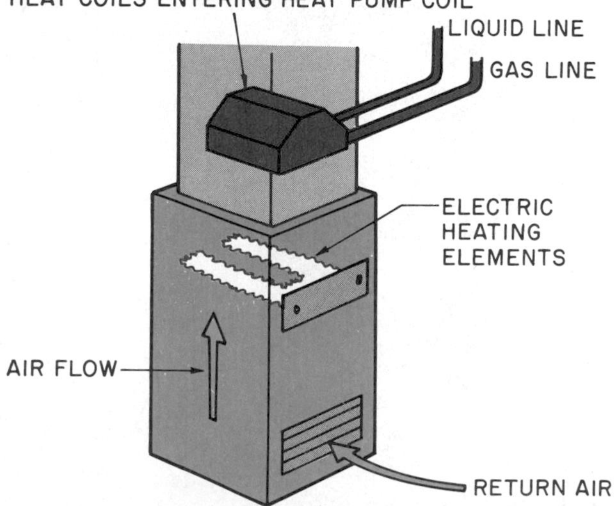

FIGURE 43–26 This illustration shows what will happen if a heat pump coil is applied to an electric furnace (oil or gas are the same) with the indoor coil after the electric heating coil.

exchanger to be located in a cold airstream for this reason. When the auxiliary heat is gas or oil, special control arrangements must be made so that the heat pump will not operate while the gas or oil heat is operating. When a heat pump is added onto an electric furnace and the coil must be located after the heat strips, follow the same rules as for gas or oil. See Figure 43–27 for an example of this installation.

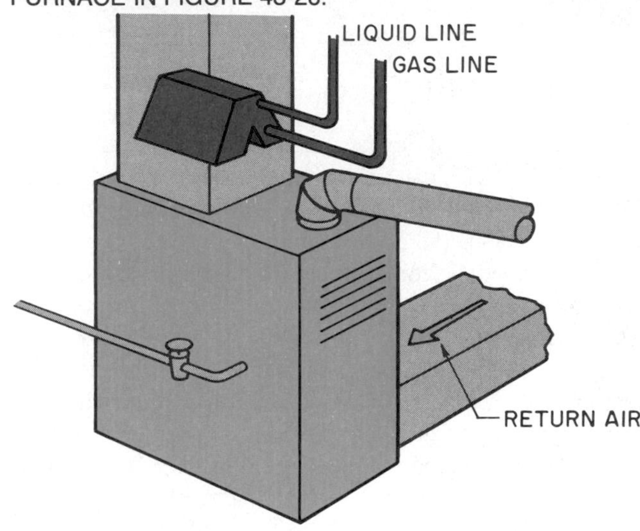

THE HEAT PUMP COIL IS INSTALLED IN THE
HOT AIR STREAM OF THE GAS FURNACE. THIS
SYSTEM HAS A CONTROL ARRANGEMENT THAT
WILL NOT ALLOW THE FURNACE TO RUN AT THE
SAME TIME AS THE HEAT PUMP. THIS CONTROL
ARRANGEMENT CAN BE USED FOR THE ELECTRIC
FURNACE IN FIGURE 43-26.
LIQUID LINE
GAS LINE
RETURN AIR

FIGURE 43–27 Gas furnace with a heat pump indoor coil instead of an air conditioner evaporator.

43.24 AIR TEMPERATURE OF THE CONDITIONED AIR

The heat pump indoor unit is installed in much the same way as a split-system cooling unit. The air distribution system must be designed more precisely due to the air temperatures leaving the air handler during the heating mode. The air temperature is not as hot as with gas and oil systems that are normally the heating system with electric cooling. The heat pump usually has leaving-air temperatures of 100°F or lower when just the heat pump is operating. If the airflow is restricted, the air temperature will go up slightly but the unit COP will not be as good. The efficiency will be reduced. Most heat pumps require a minimum of 400 cfm/ton of capacity in the cooling mode. This will equate to approximately 400 cfm/ton in heating.

When 100°F air is distributed in the conditioned space, it must be distributed carefully or drafts will occur. Normally, the air distribution system is on the outside walls. The air registers are either in the ceiling or the floor, depending on the structure. It is common in two-story houses for the first-story registers to be in the floor with the air handler in the crawl space below the house. The upstairs unit is commonly located in the attic crawl space with the registers in the ceiling near the outside walls. The outside walls are where the heat leaks out of a house in the winter and leaks into the house in the summer. If these walls are slightly heated in the winter with the airstream, less heat is taken out of the air in the conditioned space. The room air stays warmer, Figure 43–28. 100°F air mixed with room air does not feel like 130°F air mixed with room air from a gas or oil furnace. It may feel like a draft.

When 100°F air is distributed from the inside walls, such as high side wall registers, the system may not heat satisfactorily. The air will mix with the room air and might feel drafty. The outside walls will also be lower in temperature than when the air is directed on them, and a "cold wall" effect could be noticed in cold weather, Figure 43–29. This is generally considered a poor application for a heat pump.

43.25 THE OUTDOOR UNIT INSTALLATION

The outdoor installation for a heat pump is much like a central air conditioning system from an airflow standpoint. The unit must have a good air circulation around it, and the discharge air must not be allowed to recirculate.

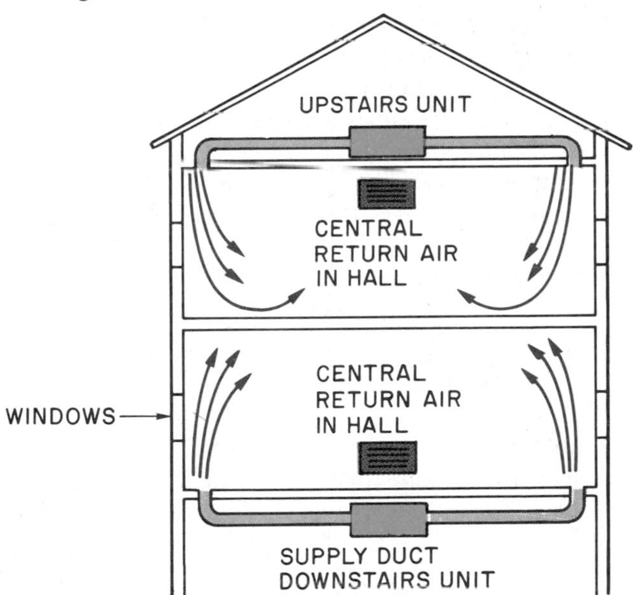

FIGURE 43–28 This heat pump installation shows a good method of distributing the warm air in the heating mode.

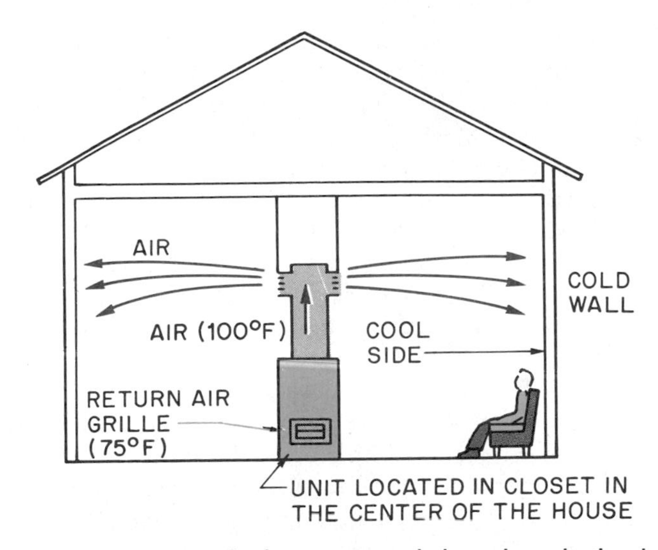

FIGURE 43–29 Air distribution system with the air being distributed from the inside using high side wall registers shows the heat pump 100°F air mixing with the room air of 75°F.

Some more serious aspects should be considered. The direction of the prevailing wind in the winter could lower the heat pump performance. If the unit is located in a prevailing north wind or a prevailing wind from a lake, the performance may not be up to standard. A prevailing north wind might cause the evaporator to operate at a lower than normal temperature. A wind blowing inland off a lake will be very humid and might cause freezing problems in the winter.

The outdoor unit must *not* be located where roof water will pour into it. The outdoor unit will be operating at below freezing much of the time, and any moisture or water that is not in the air itself should be kept from the unit's coil. If not, excess freezing will occur.

The outdoor unit is an evaporator in winter and will attract moisture from the outside air. If the coil is operating below freezing, the moisture will freeze on the coil. If the coil is above freezing, the moisture will run off the coil as it does in an air conditioning evaporator. This moisture must have a place to go. If the unit is in a yard, the moisture will soak into the ground. ***If the unit is on a porch or walk, the moisture could freeze and create slippery conditions, Figure 43–30.***

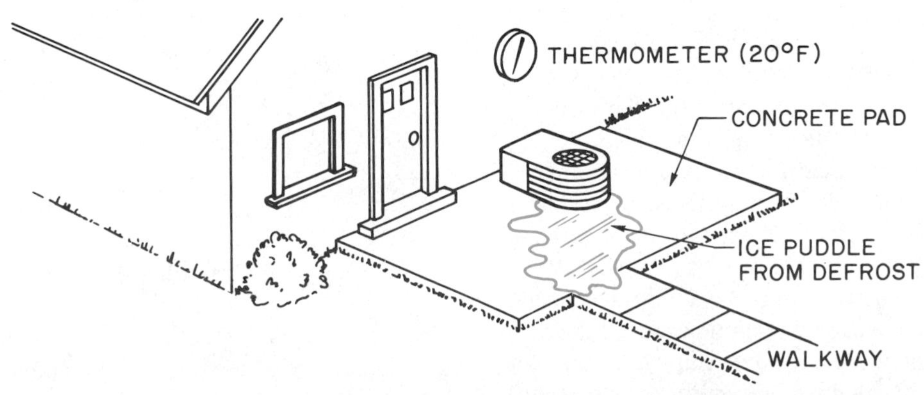

FIGURE 43–30 This heat pump was placed in the wrong place because the water that forms on the outside coil will run onto the walkway next to the coil and freeze.

The outdoor unit is designed with drain holes or pans in the bottom of the unit to allow free movement of the water away from the coil. If they are inadequate, the coil will become a solid block of ice in cold weather. When the coil is frozen solid, it is a poor heat exchanger with the outside air, and the COP will be reduced. Defrosting methods are discussed later.

Manufacturer's have recommended installation procedures for their particular units that will show the installer how to locate the unit for the best efficiency and water drainage.

The refrigerant lines that connect the indoor unit to the outdoor unit are much the same as for air conditioning. They come in line sets with the large line insulated. Quick-connect fittings with precharged lines or flare connectors with nitrogen-charged lines are typical. See Unit 38 for line installation. The only difference that should be considered is the large line, the gas line. In winter it may be 200°F and should be treated as a hot line in which heat must be contained. Therefore, many manufacturers use a thicker insulation on the gas line for heat pumps. The gas line should not be located next to any object that will be affected by its warm outside temperature. If the line set must be run underground, it must be kept dry because moisture will transfer heat to the ground. Underground piping should be routed through plastic sleeves for waterproofing and water should not be allowed to flood the sleeves.

SAFETY PRECAUTION: *Installing a heat pump involves observing the same safety precautions as installing a "cooling only" air conditioning system. The ductwork must be insulated, which requires working with fiberglass. Be careful when working with metal duct and fasteners and especially when installing units on rooftops and other hazardous locations.*

43.26 PACKAGE AIR-TO-AIR HEAT PUMPS

Package air-to-air heat pumps are much like package air conditioners. They look alike and are installed in the same way. Therefore give the same considerations to the prevailing wind and water conditions. The heat pump outdoor coil must have drainage in summer and winter. The package heat pump has all of the components in one housing and is easy to service, Figure 43–31. They have optional electric heat compartments that usually accept different electric heat sizes from 5000 W (5 kW) to 25,000 W (25 kW). When a system needs only 10 kW of auxiliary heat, the installing contractor need only install 10 kW of heat.

The metering devices used for package air-to-air heat pumps are much like those for split systems. Some manufacturers use a common metering device for the indoor and outdoor coils because the two are so close together. When one metering device is used, it must be able to meter both ways at a different rate in each direction because a different amount of refrigerant is used in the summer evaporator (indoor coil) than in the winter evaporator (outdoor coil).

Package air-to-air heat pumps must be installed correctly. For example, the duct must extend to the unit. Because the unit contains the outdoor coil also, the duct will need to extend all the way to an outside wall of the structure for a crawl space installation, Figure 43–32. Supply and return ducts must be insulated to prevent heat exchange between the duct and the ambient air.

FIGURE 43–31 Package heat pump. All components are contained in one housing. *Reproduced courtesy of Carrier Corporation*

Package air-to-air heat pumps have the same advantage from a service standpoint as package cooling air conditioners or package gas-burning equipment. All of the controls and components may be serviced from the outside. They also have the advantage of being factory assembled and charged. They are leak checked and may even be operated before leaving the factory. This reduces the likelihood of having a noisy fan motor or other defect.

43.27 CONTROLS FOR THE AIR-TO-AIR HEAT PUMP

The air-to-air heat pump is different from any other combination of heating and cooling equipment. The following sequences of operation must be controlled at the same time

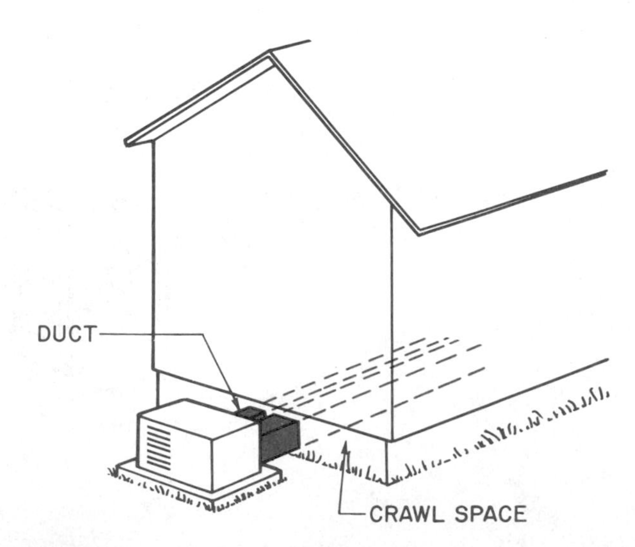

FIGURE 43–32 The self-contained unit contains the outdoor coil and the indoor coil. The ductwork must be routed to the unit.

for the heat pump to be efficient: space temperature, defrost cycle, indoor fan, the compressor, the outdoor fan, auxiliary heat, and emergency heat. Each manufacturer has its own method of controlling the heating and cooling sequence.

43.28 SPACE TEMPERATURE CONTROL

The space temperature for an air-to-air heat pump is not controlled in the same way as a typical heating and cooling system. There are actually two complete heating systems and one cooling system. The two heating systems are the refrigerated heating cycle from the heat pump and the auxiliary heat from the supplementary heating system—electric, oil, or gas. The auxiliary heating system may be operated as a system by itself if the heat pump fails. When the auxiliary heat becomes the primary heating system because of heat pump failure, it is called *emergency heat* and is normally only operated long enough to get the heat pump repaired. The reason is that the COP of the auxiliary heating system is not as good as the COP of the heat pump.

The space temperature thermostat is the key to controlling the system. It is normally, a two-stage heating and two-stage cooling type of thermostat used exclusively for heat pump applications. Other variations in the thermostat concern the number of stages of heating or cooling. Figure 43–33 shows a typical heat pump thermostat.

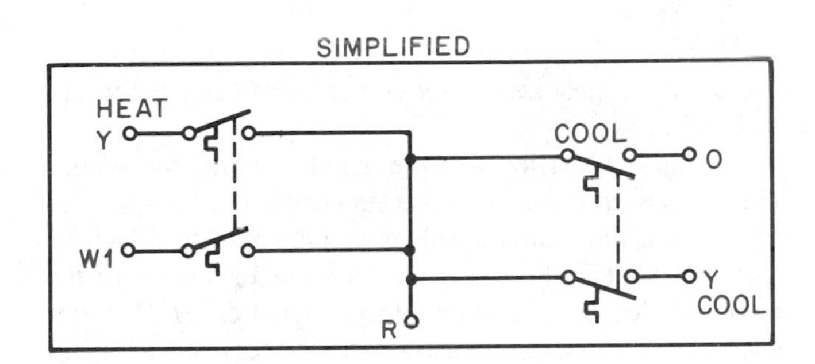

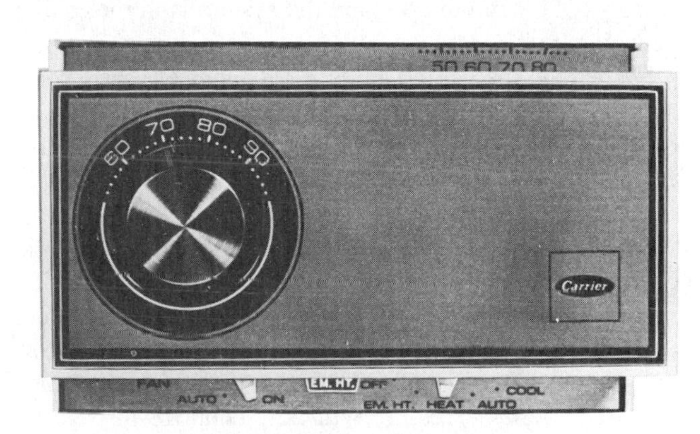

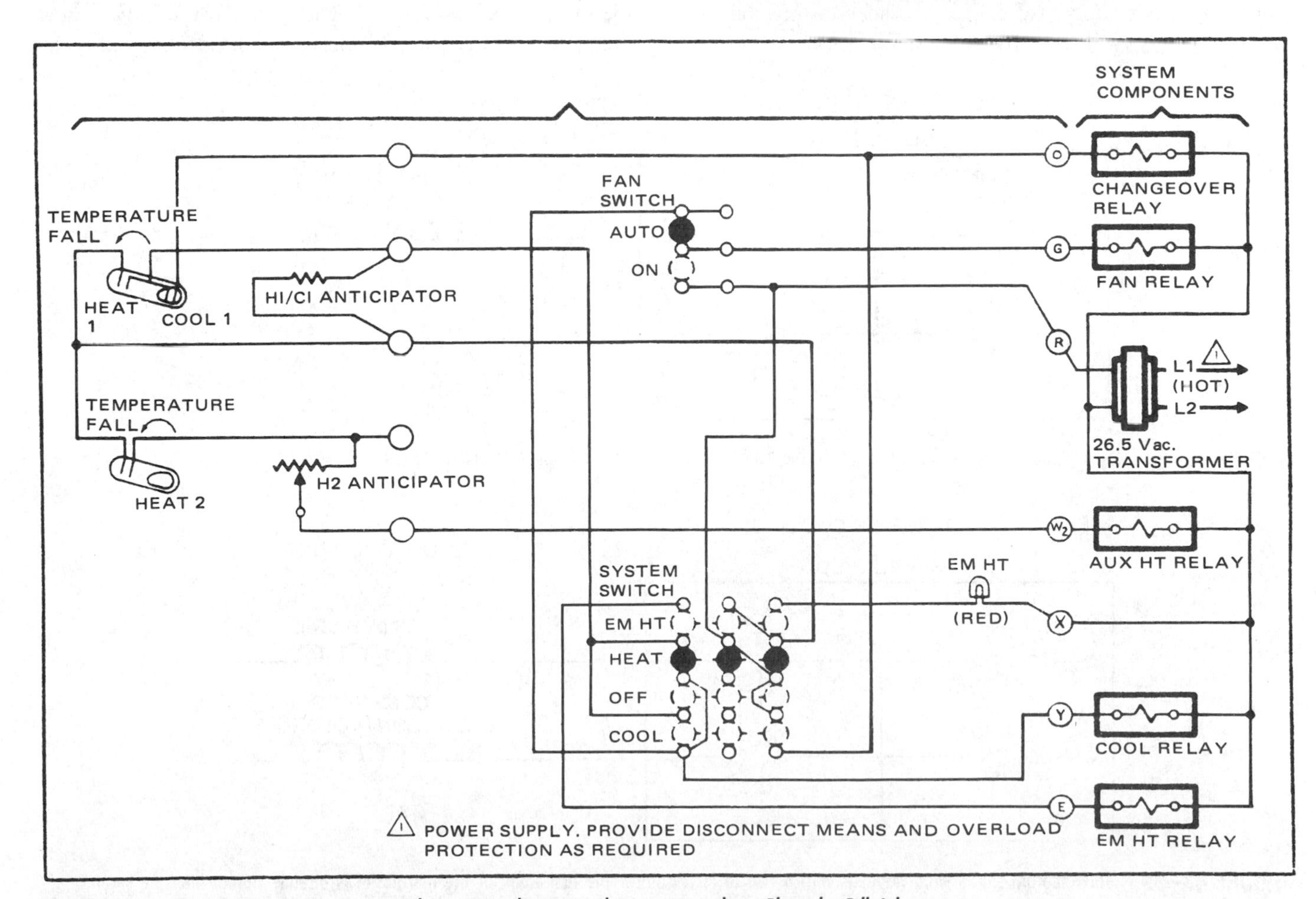

FIGURE 43–33 Heat pump thermostat with two-stage heating and two-stage cooling. *Photo by Bill Johnson.*

The following is the sequence of cooling and heating control for an automatic changeover thermostat—one that automatically changes from cooling to heating and back. **Note: This control sequence is only one of many. Each manufacturer has its own. The temperature-sensing element is a bimetal that controls mercury bulb contacts. The auxiliary heat is electric. The thermostat fan switch is in AUTO.**

Cooling Cycle Control

When the first-stage bulb of the thermostat closes its contacts (on a rise in space temperature), the four-way valve magnetic coil is energized. When the compressor starts the second stage of operation, it diverts the hot gas from the compressor to the outdoor coil, and the system is in the cooling cycle, Figure 43–34.

When the space temperature rises about 1°F, the second-stage contacts of the thermostat close. The second-stage contacts energize the compressor contactor and the indoor fan relay to start the compressor and outdoor fan and the indoor fan, Figure 43–35.

When the space temperature begins to fall (because the compressor is running and removing heat), the first contacts to open are the second-stage contacts. The compressor (and outdoor fan) and the indoor fan stop. The first-stage contacts remain closed, and the four-way valve remains energized. The system pressure will now equalize through the metering device for ease of compressor starting, but the four-way valve will not change position.

When the space temperature rises again, the compressor (and outdoor fan) and indoor fan will start again. If the outdoor temperature is getting cooler (eg, in autumn), the space temperature will continue to fall, and the second-stage contacts will open. This deenergizes the four-way valve magnetic coil, and the unit can change over to heat when it starts up. **The four-way valve is a pilot-operated valve. It will not change over until the compressor starts and a pressure differential is built up.**

Space Heating Control

When the space temperature continues to fall due to outdoor conditions, the first-stage heating contacts close. This starts the compressor (and outdoor fan) and the indoor fan. The four-way valve is not energized, so the compressor hot gas is directed to the indoor coil. The unit is now in the heating mode, Figure 43–36.

When the space temperature warms, the first-stage heating contacts open, and the compressor (and outdoor fan) and indoor fan stop.

If the outdoor temperature is cold, below the balance point of the structure, the space temperature will continue to fall because the heat pump alone will not heat the structure. The second-stage heating contacts will close and start the auxiliary electric heat to help the compressor. The second-stage heat was the last component to be energized, so it will be the first component off. This is the key to how a heat pump can get the best efficiency from the refrigerated cycle.

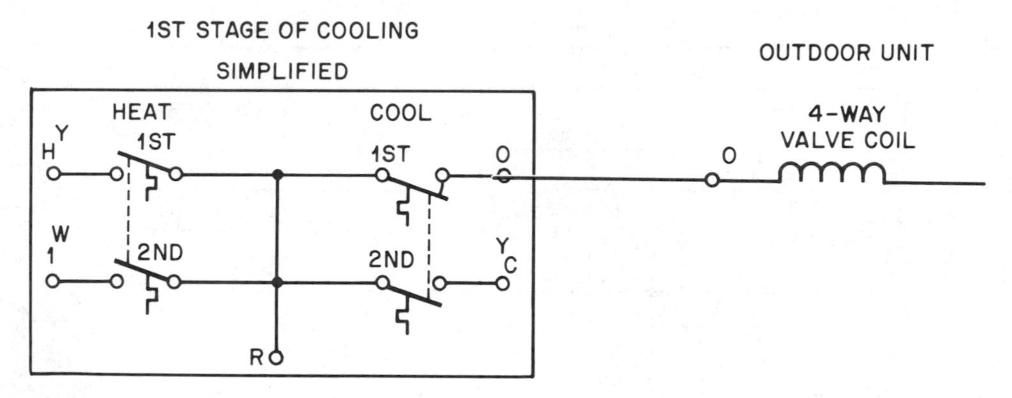

FIGURE 43–34 First stage of the cooling cycle where the four-way valve magnetic holding coil is energized.

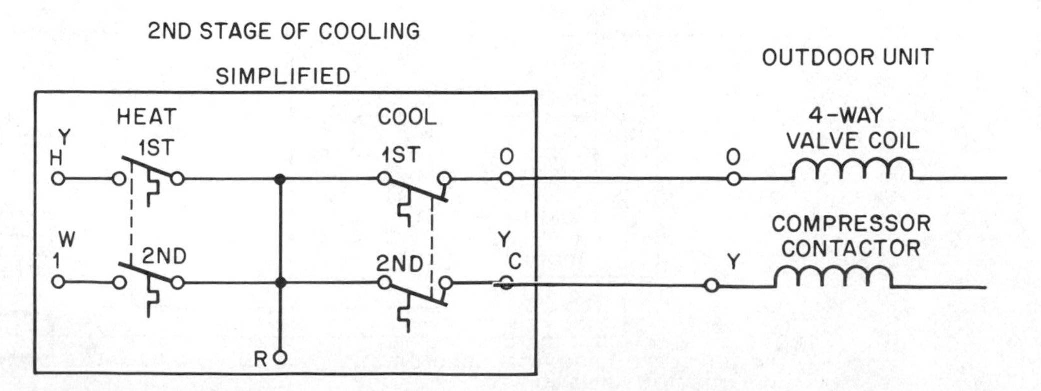

FIGURE 43–35 This diagram shows what happens with a rise in temperature. The first stage of the thermostat is already closed, then the second stage closes and starts the compressor. The compressor was the last on and will be the first off.

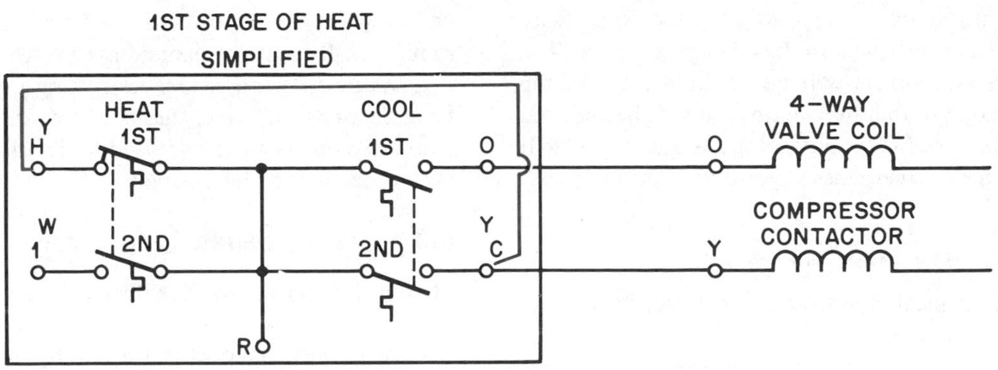

FIGURE 43–36 The space temperature drops to below the cooling set points for the first stage of heating, the compressor starts. This time, the "four-way valve" magnetic coil is deenergized and the unit will be in the heating mode.

It will operate continuously because it was the first on, and the second-stage auxiliary heat will stop and start to assist the compressor, Figure 43–37. There are as many methods of controlling the foregoing cooling and heating sequence as there are manufacturers. One common variation for controlling the heating cycle is to use outdoor thermostats to control the auxiliary heat, Figure 43–38. This keeps all of the auxiliary heat from coming on with the second stage of heating. It will come on based on outdoor temperature by the use of outdoor thermostats. These thermostats only allow the auxiliary heat to come on when needed, using the structure balance point as a guideline.

Emergency Heat Control

The emergency heat feature is only used when the heat pump cannot function. For example, if a homeowner were to have a compressor problem, such as a noise that suddenly develops, the thermostat selector switch can be changed to

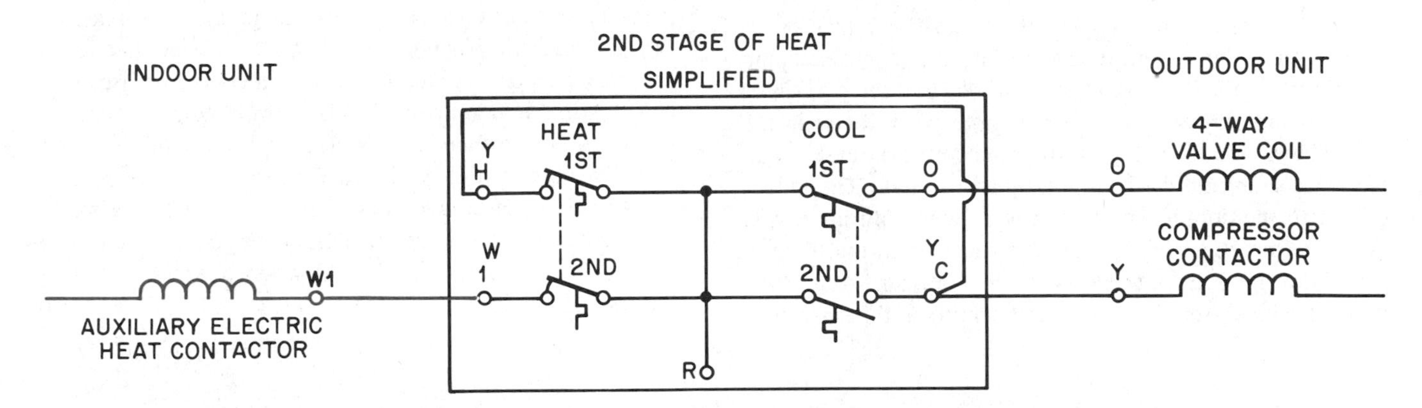

FIGURE 43–37 When the outside temperature continues to fall, it will pass the balance point of the heat pump. When the space temperature drops approximately 1.5°F, the second-stage contacts of the thermostat will close and the auxiliary heat contactors will start the auxiliary heat.

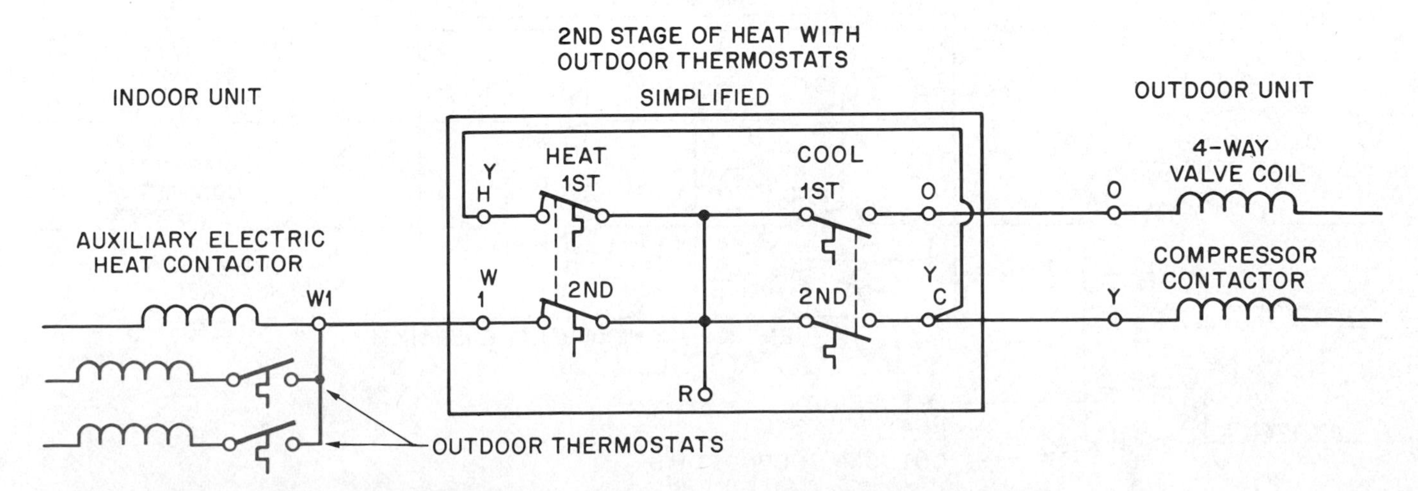

FIGURE 43–38 Wiring diagram with outdoor thermostats to control the electric auxiliary heat.

the emergency heat setting. This will stop the compressor and energize the auxiliary heat in the emergency mode. The outdoor thermostat contacts will be controlled out of the circuit for this mode so that maximum heat will be obtained from the auxiliary heat. The indoor space thermostat will control the auxiliary (emergency) heat only, Figure 43–39.

43.29 THE DEFROST CYCLE

The *defrost cycle* is used to defrost the ice from the outside coils during winter operation. The outside coils operate below freezing any time the outdoor temperature is below 45°F because the outdoor coil (evaporator) operates 20°F to 25°F colder than the outdoor air temperature. The outdoor coil is colder than the outdoor air because it has to be able to absorb heat from the outdoor air, Figure 43–22. The need for defrost varies with the outdoor air conditions and the running time of the heat pump. For example, when the outdoor temperature is 45°F, the heat pump will satisfy and shut off from time to time. The frost will melt off the coil during this off time. As the temperature falls, more running time causes more frost to form on the coil. When the outside air contains more moisture, more frost forms. During cold rainy weather, when the air temperature is 35°F to 45°F, frost will form so fast that the coil will be covered with ice between defrost cycles.

Defrost is accomplished with the air-to-air heat pump by reversing the system to the cooling mode and stopping the outdoor fan. This makes the outdoor coil the condenser with no fan, so the coil will get quite warm even in the coldest weather. Ice will melt from the coil and run down on the surface around the heat pump outdoor unit. The indoor coil is in the cooling mode and will blow cold air during defrost. One stage of the auxiliary heat is normally energized during defrost to prevent the cool air from being noticed. The system is cooling and heating at the same time and is not efficient so the defrost time must be held to a minimum. It must not operate except when needed.

Demand defrost means defrosting only when needed. Combinations of time, temperature, and pressure drop across the outdoor coil are used to determine when defrost should be started and stopped.

Instigating the Defrost Cycle

Starting the defrost cycle with the correct frost buildup is desirable. Manufacturers design the systems to start defrost as close as possible to when the coil builds frost that affects performance. Some manufacturers use time and temperature to start defrost. This is called *time and temperature instigated* and is performed with a timer and temperature-sensing device. The timer closes a set of contacts for 10 to 20 sec for a trial defrost every 90 min. The timer contacts are in series with the temperature sensor contacts so that both must be made at the same time. This means that two conditions must be met before defrost can start, Figure 43–40. Normally the timer tries for a defrost every 90 min. The sensing device contacts will close if the coil temperature is as low as 25°F, and defrost will start. Figure 43–41 shows a typical timer and a sensor. Typically, the timer runs any time the compressor runs, even in the cooling mode.

Another common method of starting defrost uses an air pressure switch that measures the air pressure drop across the outdoor coil. When the unit begins to accumulate ice, a pressure drop occurs and the air switch contacts close. This can be wired in conjunction with the timer and temperature sensor to ensure that there is an actual ice buildup on the coil. The combination of time and temperature only ensures that time has passed and the coil is cold enough to actually accumulate ice. It does not actually sense an ice thickness. It can be more efficient to defrost when there is an actual buildup of ice, Figure 43–42.

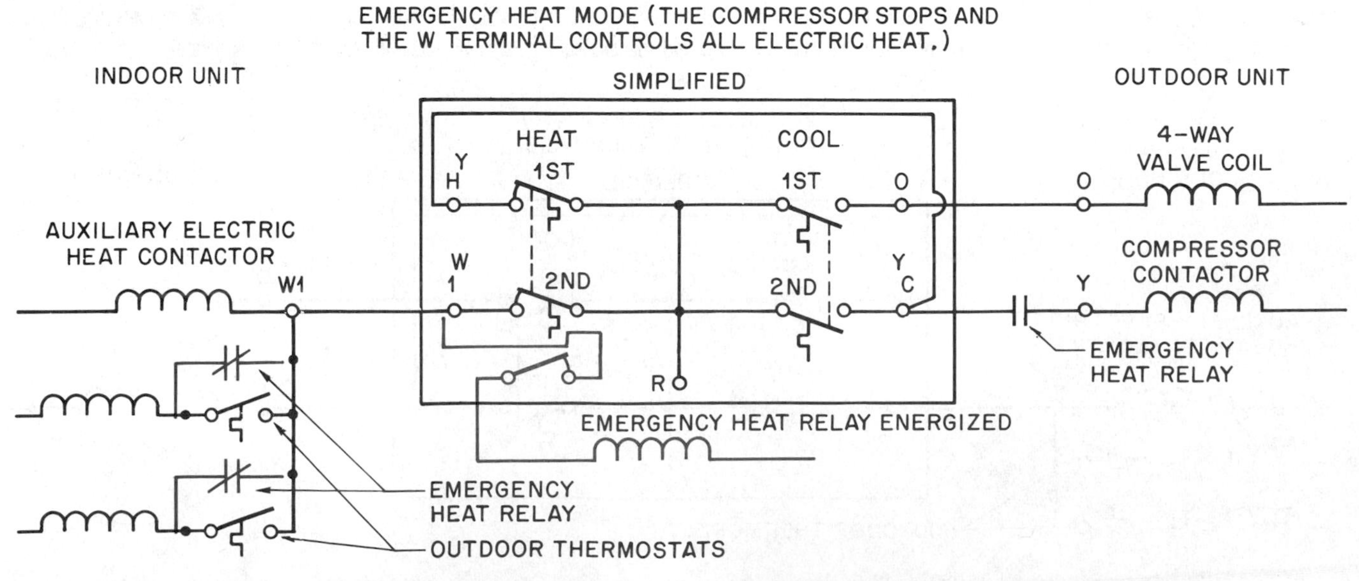

FIGURE 43–39 Wiring diagram of space temperature thermostat set to the emergency heat mode.

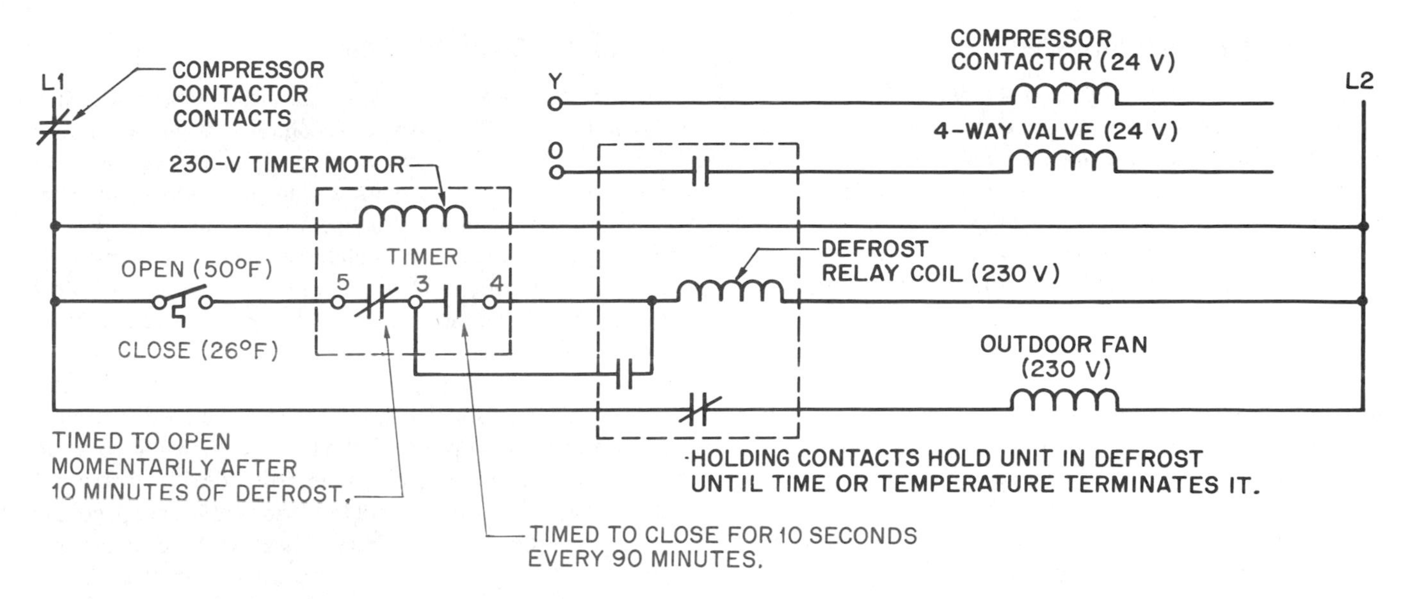

FIGURE 43–40 Wiring diagram of conditions to be met for defrost to start.

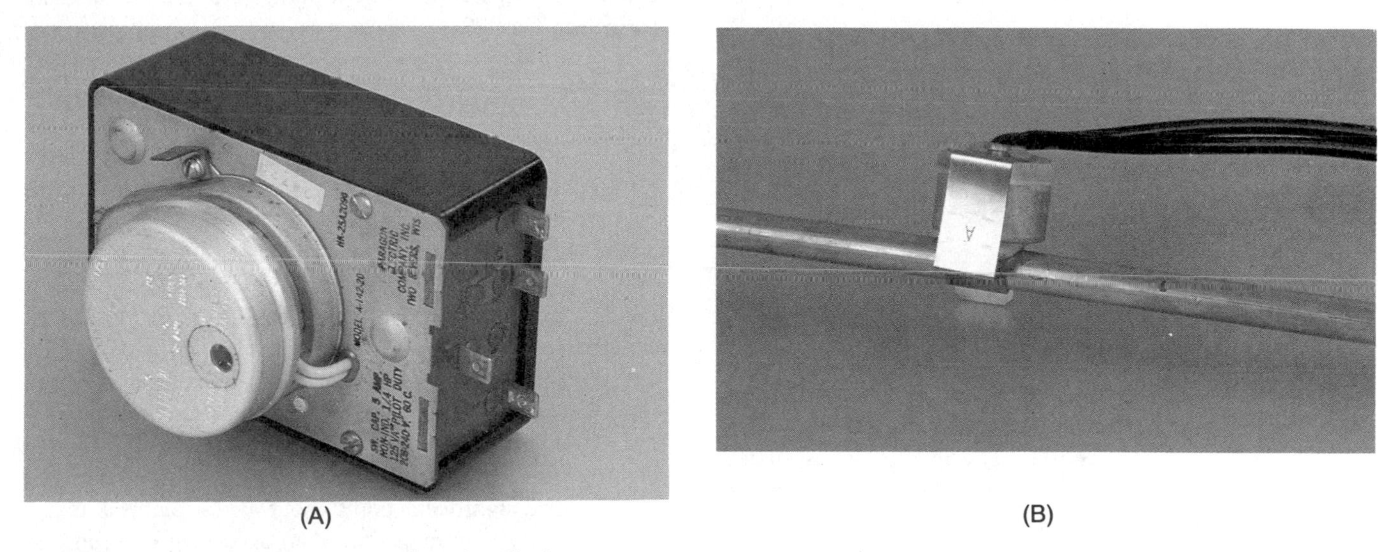

FIGURE 43–41 (A) Timer. (B) Sensor. *Photos by Bill Johnson*

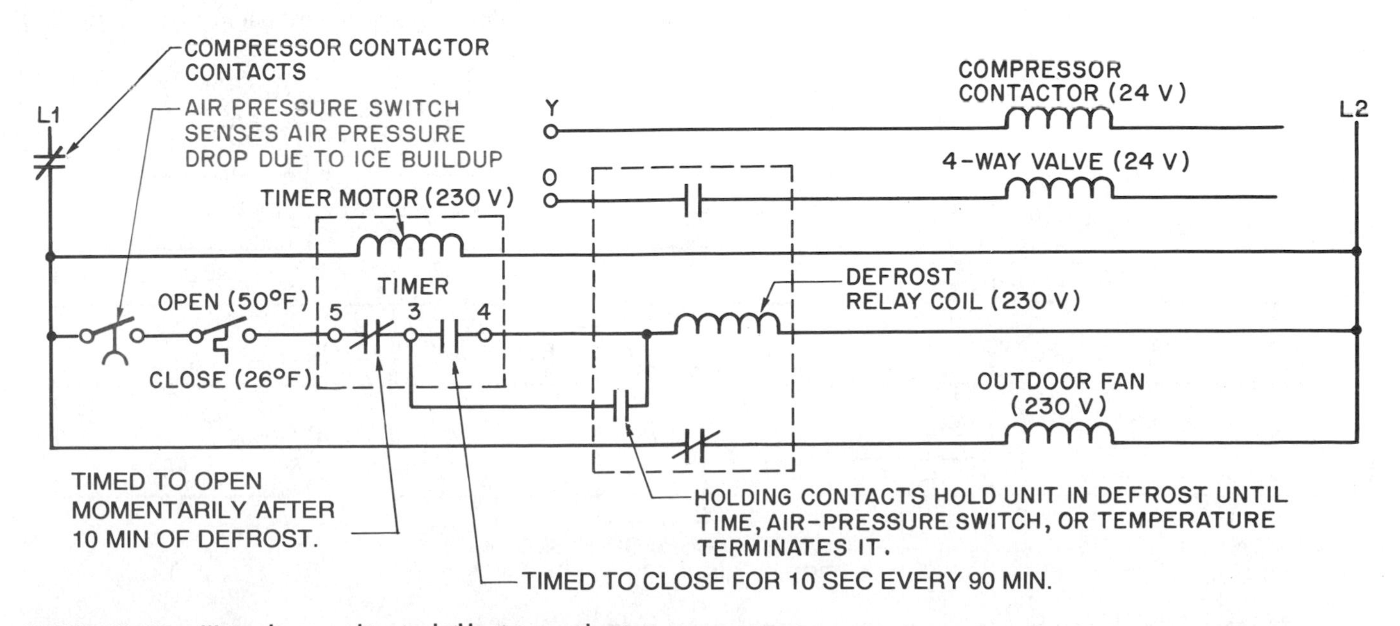

FIGURE 43–42 Wiring diagram taking ice buildup into consideration.

Terminating the Defrost Cycle

Terminating the defrost cycle at the correct time is just as important as starting defrost only when needed. This is done in several ways. *Time and temperature* was the terminology used to start defrost. *Time or temperature terminated* is the terminology used to terminate one type of defrost cycle. The difference is that two conditions must be satisfied to start defrost, and either one of two conditions can terminate defrost. Once defrost has started, time or temperature can terminate it. For example, if the defrost temperature sensor warms up to the point that it is obvious that ice is no longer on the coil and its contacts break, defrost will stop. This temperature termination is normally 50°F at the location of the temperature sensor. If the contacts do not open because it is too cold outside, the timer will terminate defrost within a predetermined time period. Normally the maximum time the timer will allow for defrost is 10 min, Figure 43–43.

Electronic Control of Defrost

Many heat pumps use electronic circuit boards for the timing portion of the defrost cycle. The electronic circuit board is also used to keep the unit from cycling too often and to monitor voltage and amperage. Consult the manufacturer for service information if you need to work on one of these boards.

43.30 INDOOR FAN MOTOR CONTROL

Starting the indoor fan motor of a heat pump differs from other heating systems. In other systems the indoor fan is started with a temperature-operated fan switch. With a heat pump the fan must start at the beginning of the cycle, which is controlled by the thermostat. The indoor fan is started with the fan terminal, sometimes labeled G. The compressor operates whenever the unit is calling for cooling or heating, so the G or fan terminal is energized whenever there is a call for cooling or heating.

43.31 AUXILIARY HEAT

During cold weather the air-to-air heat pump must have auxiliary heat. This is usually accomplished with an electric furnace with a heat pump coil for cooling and heating. Electric heat is often started with a sequencer with a time-start and time-stop feature for the electric heating element. This means that the electric heat element is energized for a time period if the unit is shut off with the thermostat on-off switch. When the thermostat is allowed to operate the system normally, the heat pump will be the first component to come on and the last to shut off. The indoor fan will start with the compressor through the fan relay circuit. The electric strip heat will operate as the last component to come on and the first to shut off. When the owner decides to shut off the heat at the thermostat and the strip heat is energized, the heating elements will continue to heat until the sequencers stop them. A temperature switch is often used to sense this heat and keep the fan running until the heaters cool. The electric furnaces used for heat pumps are like the ones described in Unit 29.

43.32 SERVICING THE AIR-TO-AIR HEAT PUMP

Servicing the air-to-air heat pump is much like servicing a refrigeration system. During the cooling season, the unit is operating as a high-temperature refrigeration system, and during the heating season it is operating as a low-temperature system with planned defrost. The servicing of the system is divided into electrical and mechanical servicing.

Servicing the electrical system is much like servicing any electrical system that is used to operate and control refrigeration equipment. The components that the manufacturer furnishes are built to last for many years of normal service. However, parts fail due to manufacturing faults and misuse. One significant point about a heat pump is that it has much more running time than a cooling only unit. A typical cooling unit in a residence in Atlanta, Georgia, will run about 120 days per year. A heat pump may well run 250

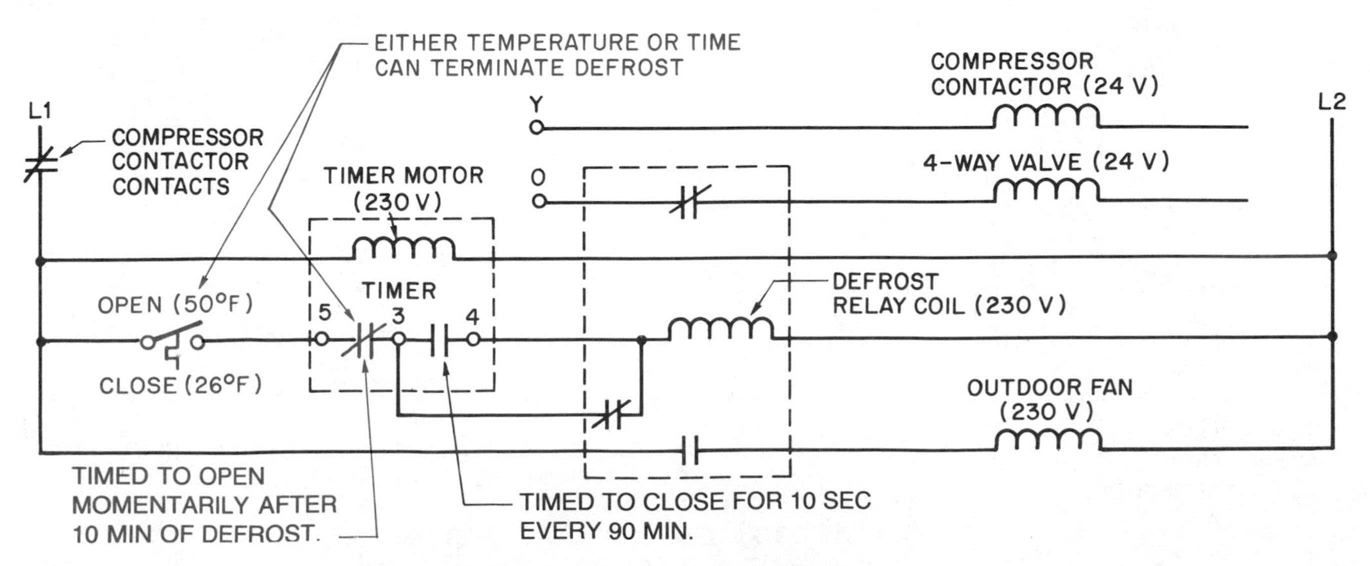

FIGURE 43–43 Wiring diagram for time- or temperature-terminated method of defrost.

days, more than twice the time. In winter the heat pump may run for days and not stop when the temperature is below the balance point of the structure.

SAFETY PRECAUTION: *The same safety precautions must be observed when troubleshooting electrical components of a heat pump as were observed when servicing an air conditioning system. When the units are located in crawl spaces and you are in contact with the ground, be careful *not* to establish a path for current to flow through your body to ground.*

43.33 TROUBLESHOOTING THE ELECTRICAL SYSTEM

Troubleshooting the electrical system of an air-to-air heat pump is similar to troubleshooting a cooling air conditioning system. Recall that there must be a power supply, a path for the electrical energy to flow on, and a load before electricity will flow. The power supply provides the electrical energy to the unit, the path is the wiring, and the loads are the electrical components that perform the work. Every circuit in the heat pump wiring can be reduced to these three items. It is important that you think in terms of power, conductor, and load, Figure 43–44.

It is usually easy to find electrical problems if some component will not function. For example, when a compressor contactor coil is open, it is evident that the problem is in the contactor. A voltmeter reading can be taken on each terminal of the contactor coil. If there is voltage and it will not energize the contacts, a resistance check of the coil will show if there is a circuit through it. A correctly operating circuit has the correct resistance. No resistance or too little resistance will create too much current flow. It sometimes helps to be able to compare a suspected bad component (eg, one with too little resistance) to a component of correct resistance, Figure 43–45.

Electrical malfunctions that cause the technician real problems are the intermittent ones, such as a unit that does not always defrost and ice builds up on the outdoor coil. The heat pump can be difficult to troubleshoot in winter because of the weather. ***If it is wet outside the emergency heat feature of the heat pump should be used until the weather improves. It is hard enough to find electrical problems in cold weather; the moisture makes it dangerous.***

Typical Electrical Problems

1. Indoor fan motor or outdoor fan motor
   A. The electrical circuit is open; it will not start; draws no current.

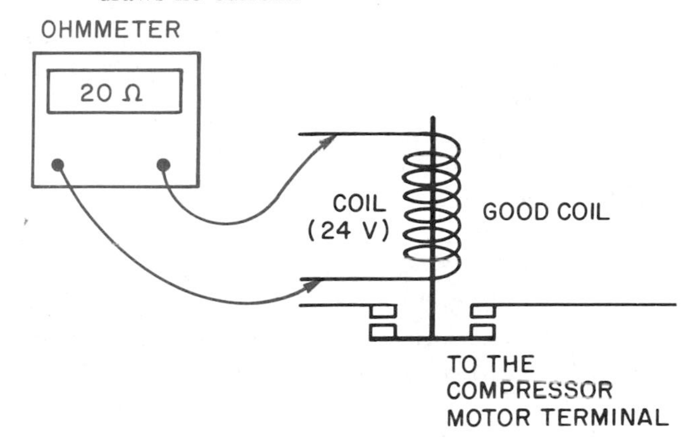

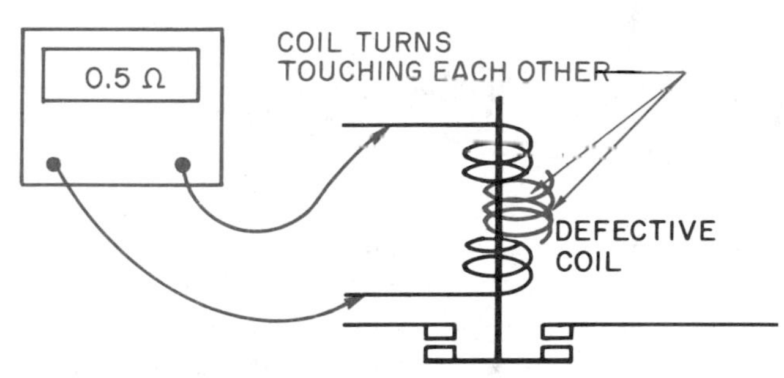

FIGURE 43–45 Magnetic holding coils used for the holding contactor. One of them has a measurable resistance of 0.5 Ω. The other one, the correct one, has a measurable resistance of 20 Ω. The one with the lower resistance is shorted and will draw too much current and place a load on the control transformer. If the load is too much, it will overload the transformer.

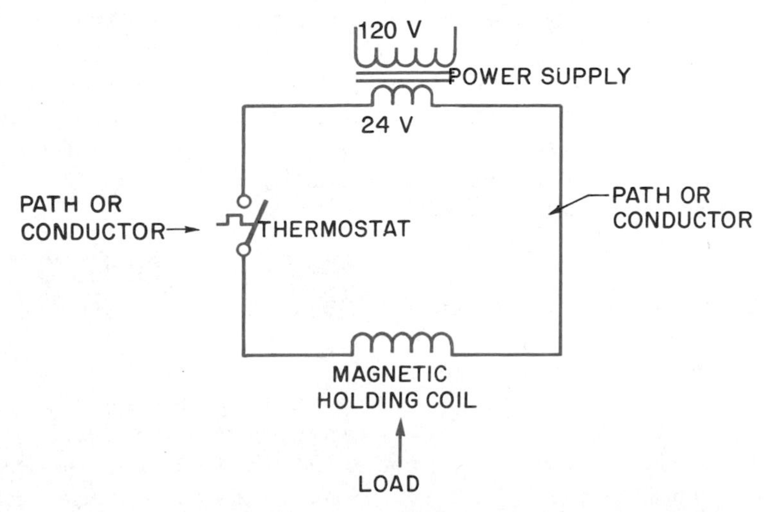

FIGURE 43–44 This is the basic example of a power supply, a path or conductor, and a load to consume power.

   B. Burned; will not start; may draw high current.
   C. Bad capacitor; fan may run slowly and draw high current.
2. Compressor contactor, indoor fan relay, or outdoor fan relay
   A. Coil winding is open; will not close contacts.
   B. Coil burned; might overload control transformer.
   C. Contacts burned; will cause local heating and burn wires.
3. Defrost relay
   A. Coil winding is open; will not close contacts.
   B. Coil burned; might overload control transformer.
   C. Contacts burned; cause defrost circuit problems; interfere with normal procedures of defrost.
4. Compressor
   A. Open winding; will not start; will draw high current if only one winding is open.
   B. Burned; high current; breaker tripped.
   C. Open internal overload; will not start; will show measurable resistance from R to S terminals.
5. Wiring problems
   A. Loose connections; local heat; insulation burned.
   B. Wires rubbing together; might cause electrical short.

43.34 TROUBLESHOOTING MECHANICAL PROBLEMS

Mechanical problems can be hard to find in a heat pump, particularly in winter operation. Summer operation of a heat pump is similar to that of a conventional cooling unit. The pressures and temperatures are the same and the weather is not as difficult to work in. Mechanical problems are solved with gage manifolds, wet and dry bulb thermometers, and air-measuring instruments.

Some Typical Mechanical Problems

1. Indoor unit
   A. Air filter dirty (winter); high head pressure; low COP.
   B. Dirty coil (winter); high head pressure; low COP.
   C. Air filter dirty (summer); low suction pressure.
   D. Dirty coil (summer); low suction pressure.
   E. Refrigerant restriction (summer); low suction.
   F. Refrigerant restriction (winter); high head pressure.
   G. Defective TXV on the indoor coil (summer); low suction pressure.
   H. Defective TXV on the outdoor coil (winter); low suction.
   I. Leaking check valve (summer); flooded indoor coil.
2. Outdoor unit
   A. Dirty coil (winter); low suction pressure.
   B. Dirty coil (summer); high head pressure.
   C. Four-way valve will not shift; stays in same mode.
   D. Four-way valve halfway; pressures will equalize while running and appear as a compressor that will not pump.
   E. Compressor efficiency down; capacity down, suction and discharge pressures too close together. Compressor will not pump pressure differential.
   F. Defective TXVs (winter); low suction pressure.
   G. Leaking check valve (winter); flooded coil.

The technician should be aware of three mechanical problems: (1) four-way valve leaking through, (2) compressor not pumping to capacity, and (3) charging the heat pump. These particular problems are different enough from commercial refrigeration that they need individual attention.

SAFETY PRECAUTION: *Be aware of the high-pressure refrigerant in the system while connecting gage lines. High-pressure refrigerant can pierce your skin and blow particles in your eyes; use eye protection. Liquid refrigerant can freeze your skin; wear gloves. The hot gas line can cause serious burns and should be avoided.*

43.35 TROUBLESHOOTING THE FOUR-WAY VALVE

The four-way valve leaking through can easily be confused with a compressor that is not pumping to capacity. The capacity of the system will not be up to normal in summer or winter cycles. This is the same symptom as small amounts of hot gas leaking from the high-pressure side of the system to the low-pressure side. When the gas is pumped around and around, work is accomplished in the compression process, but usable refrigeration is not available. When you suspect that a four-way valve is leaking from the high-pressure line to the low-pressure line, use a good-quality thermometer to check the temperature of the low-side line, the suction line from the evaporator (the indoor coil in summer or the outdoor coil in winter), and the permanent suction line between the four-way valve and the compressor. The temperature difference should not be more than about 3°F. ***Take these special precautions when recording temperatures: (1) Take the temperatures at least 5 in. from the valve body (to keep valve body temperature from affecting the reading). (2) Insulate the temperature lead that is fastened tightly to the refrigerant line, Figure 43–46.***

43.36 TROUBLESHOOTING THE COMPRESSOR

Checking a compressor in a heat pump for pumping to capacity is much like checking a refrigeration compressor except there are normally no service valves to work with. Some manufacturers furnish a chart with the unit to show what the compressor characteristic should be under different

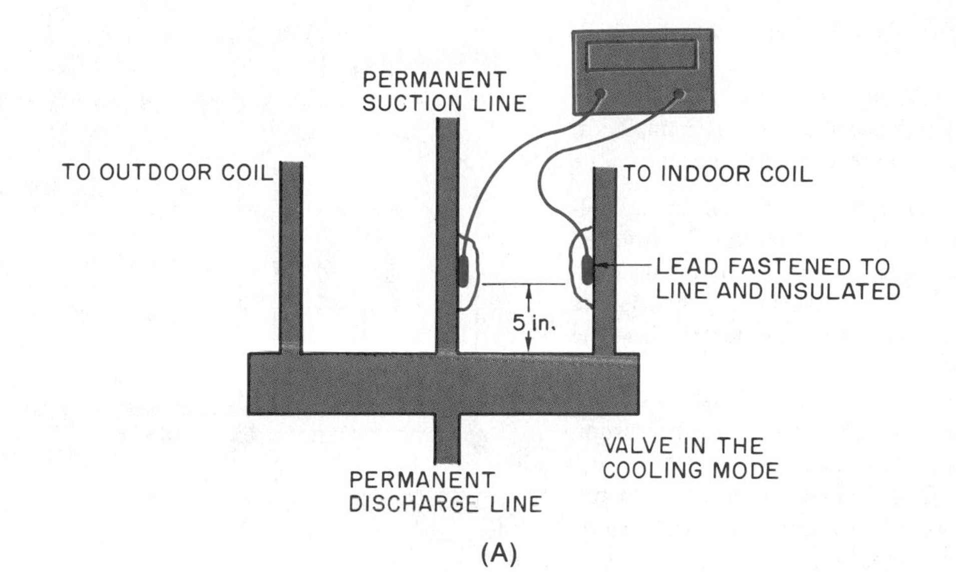

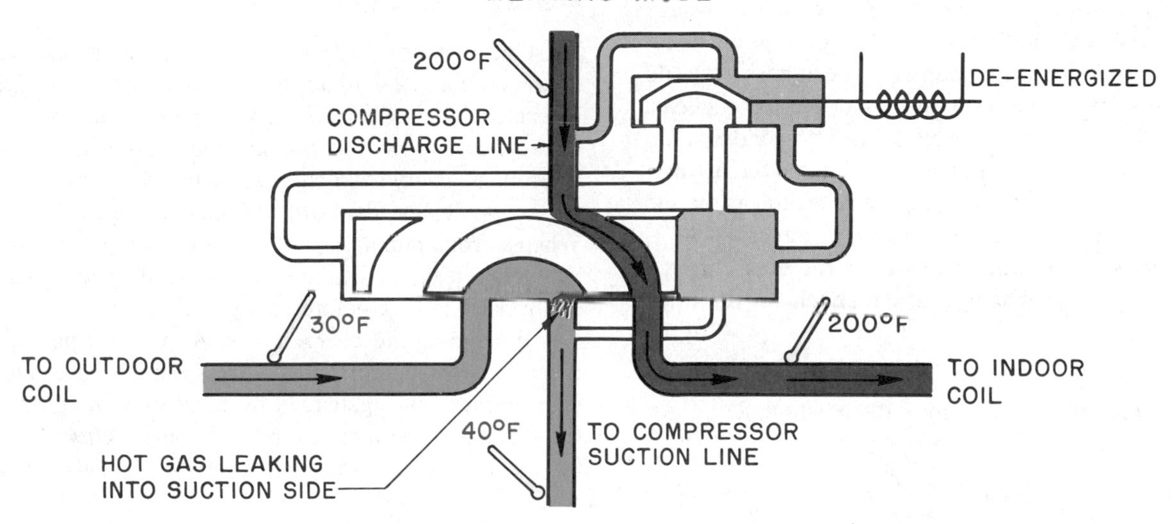

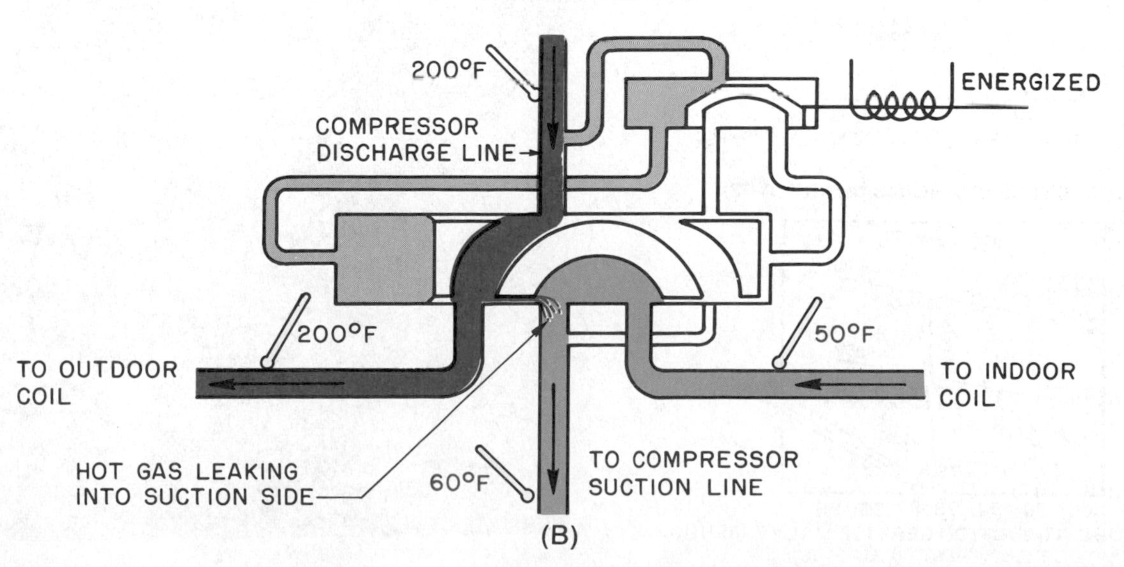

FIGURE 43–46 (A) Method to check performance of a four-way valve using temperature comparison. (B) Line diagram showing defective valve in cooling and heating.

operating conditions, Figure 43–47. If such a chart is available, use it.

The following is a reliable test, using field working conditions, that will tell if the compressor is pumping at near capacity. Any large inefficiencies will appear.

1. Whether summer or winter, operate the unit in the cooling mode. If winter the four-way valve may be switched by energizing it with a jumper or deenergizing it by disconnecting a wire to switch the unit to the summer mode. This will allow the auxiliary heat to heat the structure and keep it from getting too cold.
2. Block the condenser airflow until the head pressure is 275 psig (this simulates a 95°F day) and the suction pressure is about 70 psig.
3. The compressor amperage should be at close to full load. The combination of discharge pressure, suction pressure, and amperage should reveal if the compressor is working at near full load. If the suction pressure is high and the discharge pressure is low, the amperage will be low. This indicates that the compressor is not pumping to capacity. Sometimes you can hear a whistle when the compressor is shut off under these conditions. The suction line may also get warm immediately after shutting off the unit under these conditions. Such symptoms indicate that the compressor is leaking from the high side to the low side internally, Figure 43–48. ***If there is any doubt as to where the leak through may be, perform a temperature check on the four-way valve.***

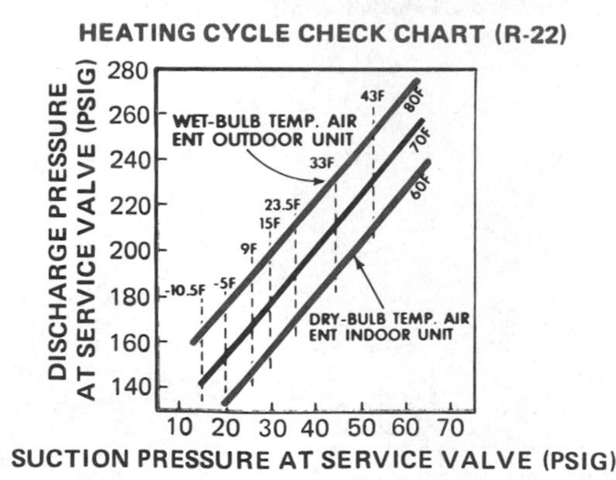

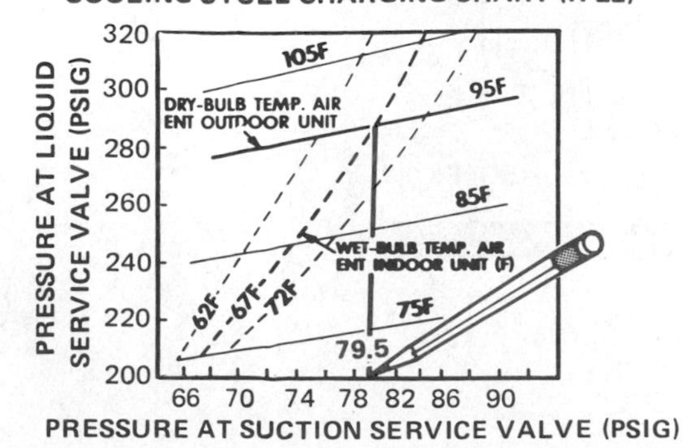

FIGURE 43–47 Performance chart for a 36,000 Btu/h heat pump system. *Reproduced courtesy of Carrier Corporation*

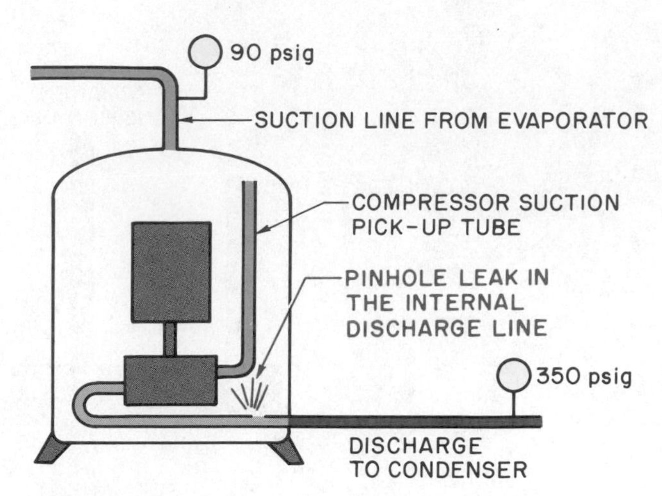

FIGURE 43–48 Compressor has vapor leaking from the high side to the low side.

43.37 CHECKING THE CHARGE

Most heat pumps have a critical refrigerant charge. The tolerance could be as close as ± 1/2 oz of refrigerant. **Therefore do not install a standard gage manifold each time you suspect a problem.** See Figure 43–49 for an example of a very short coupled system that may be used on the high-side line that will not alter the operating charge. When a heat pump has a partial charge, it is obvious that the refrigerant leaked out of the system. Leak check the system and repair the leak before charging.

When a partial charge is found in a heat pump, some manufacturers recommend that the system be evacuated to a deep vacuum and recharged by measuring the charge into the system. Some manufacturers furnish a charging procedure to allow you to add a partial charge. It is always best to

FIGURE 43–49 High-pressure gage for checking the head pressure on a heat pump with a critical charge. *Photos by Bill Johnson*

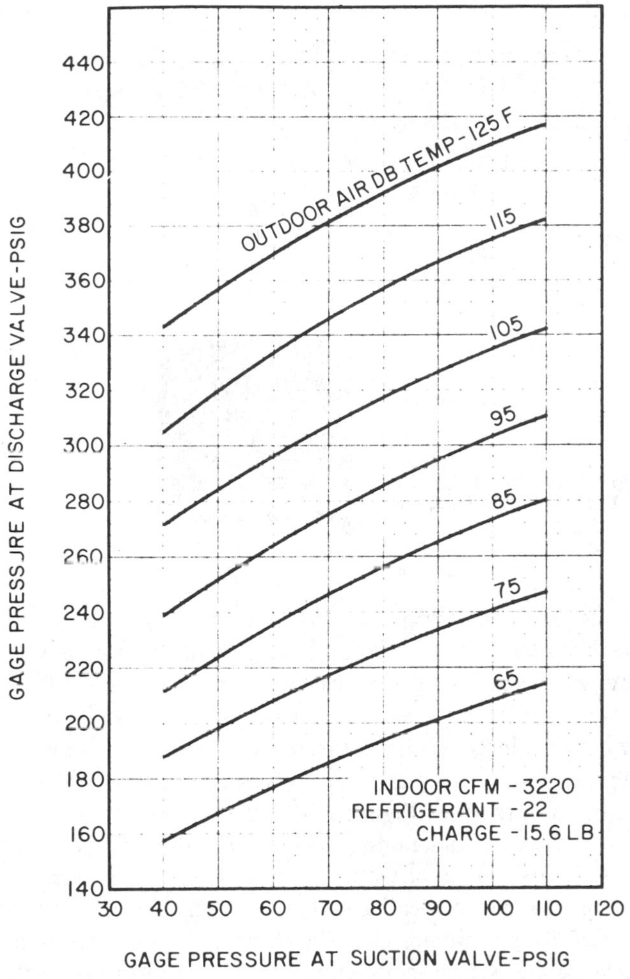

FIGURE 43–50 Heat pump performance chart.

follow the manufacturer's recommendation. Figure 43–50 is a sample performance chart for a heat pump.

If the manufacturer's recommendations are not available, the following method may be used to partially charge a system. This is specifically for systems with *fixed-bore metering devices,* not TXVs.

1. Start the unit in the cooling mode. If in the winter mode leave the electric heat where it can heat the structure. Block the condenser and cause the head pressure to rise to 275 psig. This will simulate a 95°F day operation. If the head pressure will not go up, refrigerant may have to be added.
2. Fasten a thermometer to the gas line—it will be cold while in the cooling mode. The suction pressure should be about 70 to 80 psig when fully charged, but may be lower if there is a reduced latent-heat load because of low humidity. If the system is split with a short gas line (10 to 30 ft), charge until the system has a 10°F to 15°F superheat at the line entering the outdoor unit, Figure 43–51. If the line is long (30 to 50 ft), charge until the superheat is 12°F to 18°F.

This charging procedure is close to correct for a typical heat pump that has no liquid line-to-gas line heat exchange (some manufacturers have liquid-to-gas heat exchanges to improve their particular performance). This heat exchange is normally inside the outdoor unit and may not affect these procedures.

SAFETY PRECAUTION: *There is one particular precaution when working with the charge on any heat pump that has a suction-line accumulator, and this includes most heat pumps. Part of the charge can be stored in the accumulator and will boil out later. You can heat the

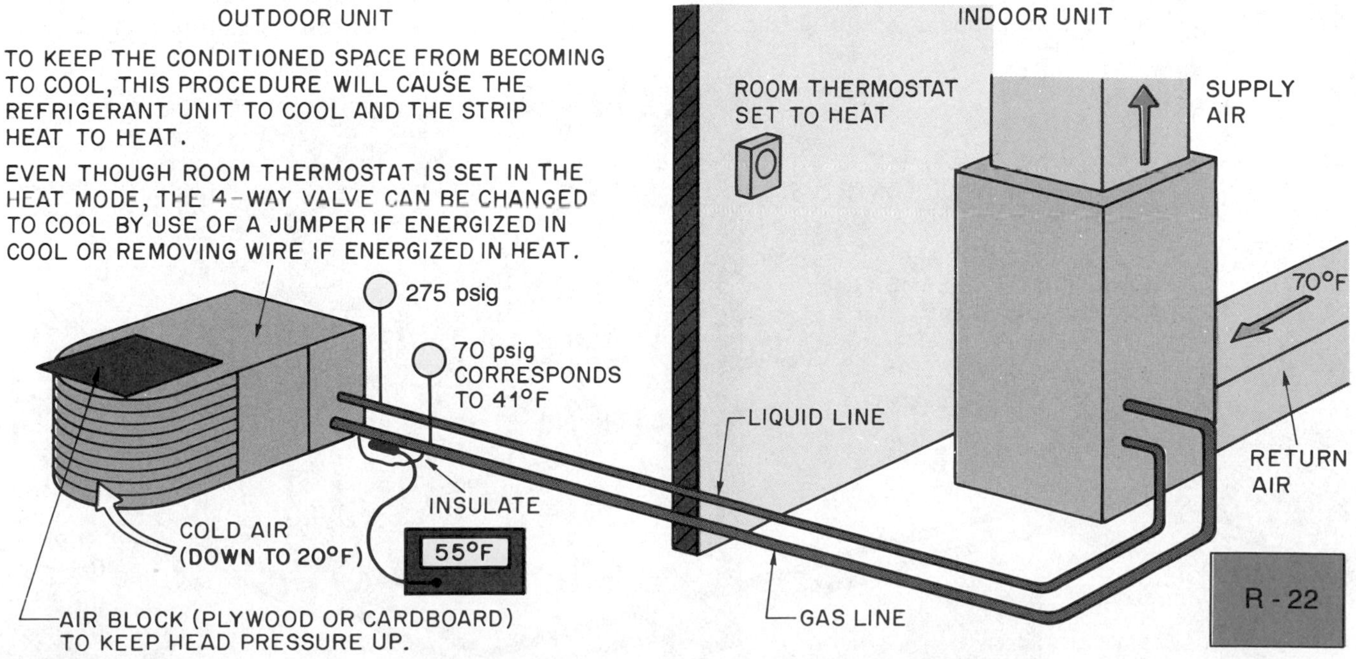

FIGURE 43–51 This illustration shows how a heat pump with a fixed-bore metering device may be charged when the manufacturer's chart or information is not available. See text for procedure.

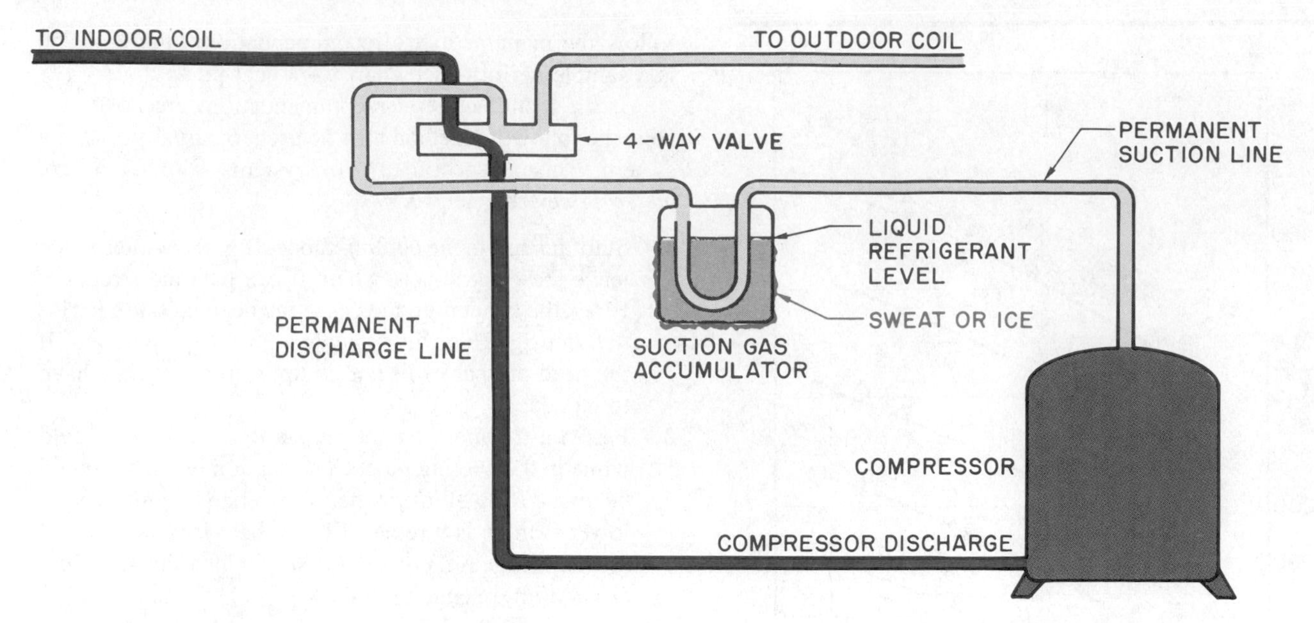

FIGURE 43–52 Suction-line accumulator that is sweating.

accumulator by running water over it to drive the refrigerant out if you are in doubt. You may also give the liquid time to boil out on its own. Often, the accumulator will frost or sweat at a particular level if liquid refrigerant is contained in it, Figure 43–52.*

43.38 SPECIAL APPLICATIONS FOR HEAT PUMPS

The use of oil or gas furnaces for auxiliary heat is a special application. Several manufacturers have systems designed to accomplish this. This type of system has several advantages. Usually, natural gas is less expensive to use as the auxiliary fuel than electricity is at a 1:1 COP. Fuel oil may be considered where a fuel oil system is already installed, where fuel oil has a price advantage, or where natural gas is not available. Both systems have similar control arrangements. The heat pump indoor coil can be used in conjunction with an oil or gas furnace, but the coil must be downstream of the oil or gas heat exchanger. The air must flow through the oil or gas heat exchanger first, then through the heat pump indoor coil. ***This means that the gas or oil furnace must not operate at the same time that the heat pump is operating, Figure 43–53.***

The control function can be accomplished with an outdoor thermostat set at the balance point of the structure to change the call for heating to the oil or gas furnace and to shut off the heat pump. This allows the heat pump to operate down to the balance point when the oil or gas furnace will take over. When defrost occurs, the oil or gas furnace will come on during defrost because the same controls are used as for a standard heat pump, and a call for auxiliary heat to warm the cool air during defrost is still part of the control

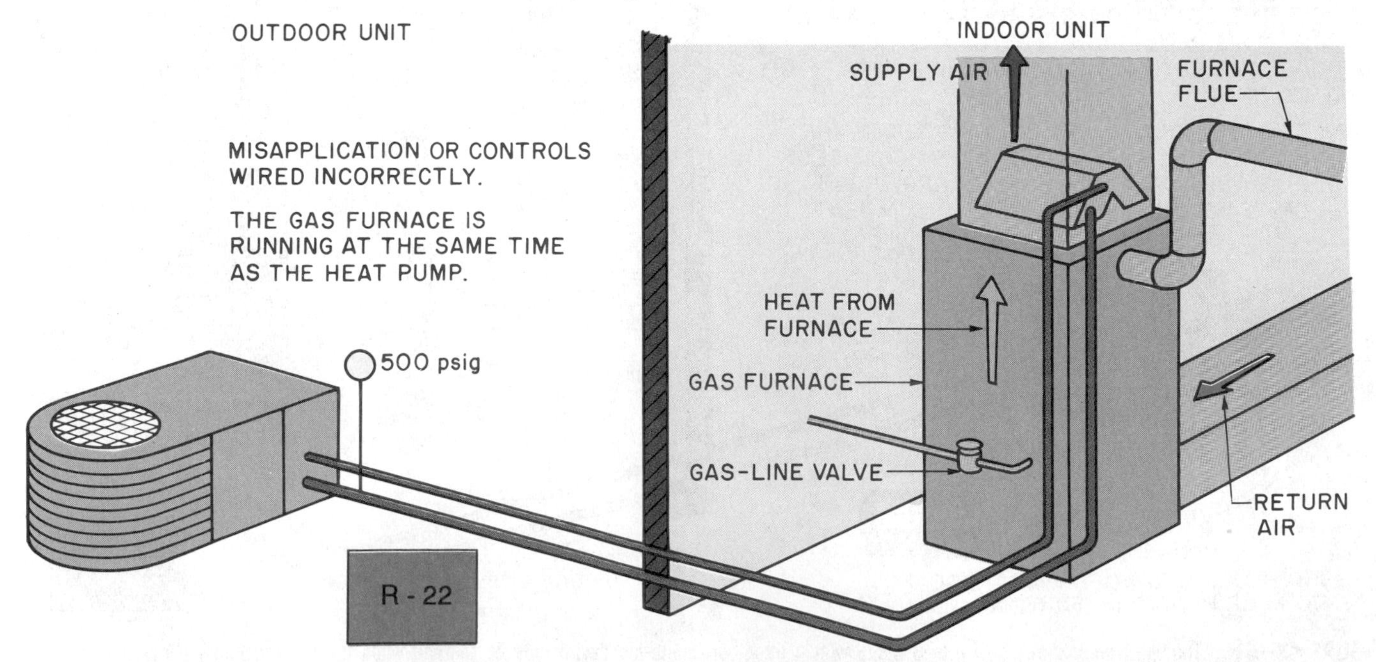

FIGURE 43–53 Operating conditions inside a heat pump coil while a gas furnace is operating at the same time that a heat pump is operating.

sequence. ***A high-pressure control can prevent the compressor from overloading if a defrost does not terminate due to a defective defrost control.***

43.39 MAXIMUM HEAT PUMP RUNNING TIME

Other control modifications allow the heat pump to run below the balance point. Such arrangements have more relays and controls. The heat pump is designed to operate whenever it can. This is accomplished by using the second-stage contacts to start the oil or gas heat. This sequence stops the heat pump until the second-stage thermostat contacts satisfy. The heat pump then starts again. This sequence repeats itself each cycle. Before considering this control sequence, a study of the system should be done to see if it is more economical to switch over from the heat pump to the auxiliary source or to restart the heat pump. The economic balance point of the structure may be determined using the cost comparison of fuels.

43.40 ADD-ON HEAT PUMP TO EXISTING ELECTRIC FURNACE

When a heat pump indoor coil is added to an existing electric furnace installation, the older furnace might not be equipped for the heat pump coil to be located before the electric heating elements. If so, a wiring configuration similar to an oil or gas installation may be used, Figure 43–54.

Many manufacturers have different methods of building a heat pump for their own performance characteristics. These methods involve many different piping configurations with special heat exchangers. This text is intended to explain the typical heat pump. If a different type of system is encountered, the manufacturer should be consulted.

43.41 HEAT PUMPS USING SCROLL COMPRESSORS

Manufacturers are continually working to make heating and cooling equipment more efficient and less expensive to purchase. Some of the new techniques involve the use of improved compressors.

Compressor pumping efficiency has been improved with the *scroll compressor*, Figure 43–55. This compressor is ideally suited for heat pump application because of its pumping characteristics. The compression takes place between two spiral-shaped forms, Figure 43–56. The scroll compressor does not lose as much capacity as the reciprocating compressor at the higher head pressures of summer or the lower suction pressures of winter operation. This is because the scroll compressor does not have a top-of-the-stroke clearance volume loss that a reciprocating compressor has. The gas trapped in the clearance volume of a reciprocating compressor must reexpand before the cylinder starts to fill. This reduces the efficiency.

The scroll compressor is about 15% more efficient than the reciprocating compressor. It has many of the same pumping characteristics as the rotary compressor. The scroll compressor does not require crankcase heat because it is not as sensitive to liquid as the reciprocating compressor. This results in a saving in power.

Scroll compressor pressures are about the same as reciprocating compressors, so the technician should not experience any difference in gage readings. They are discharge gas cooled so the technician will notice that the compressor shell is much hotter than a suction-cooled compressor. The scroll compressor runs with less vibration and a lower noise level. There are several stages of compression occurring at the same time in the scroll compressor. Each pocket of gas represents

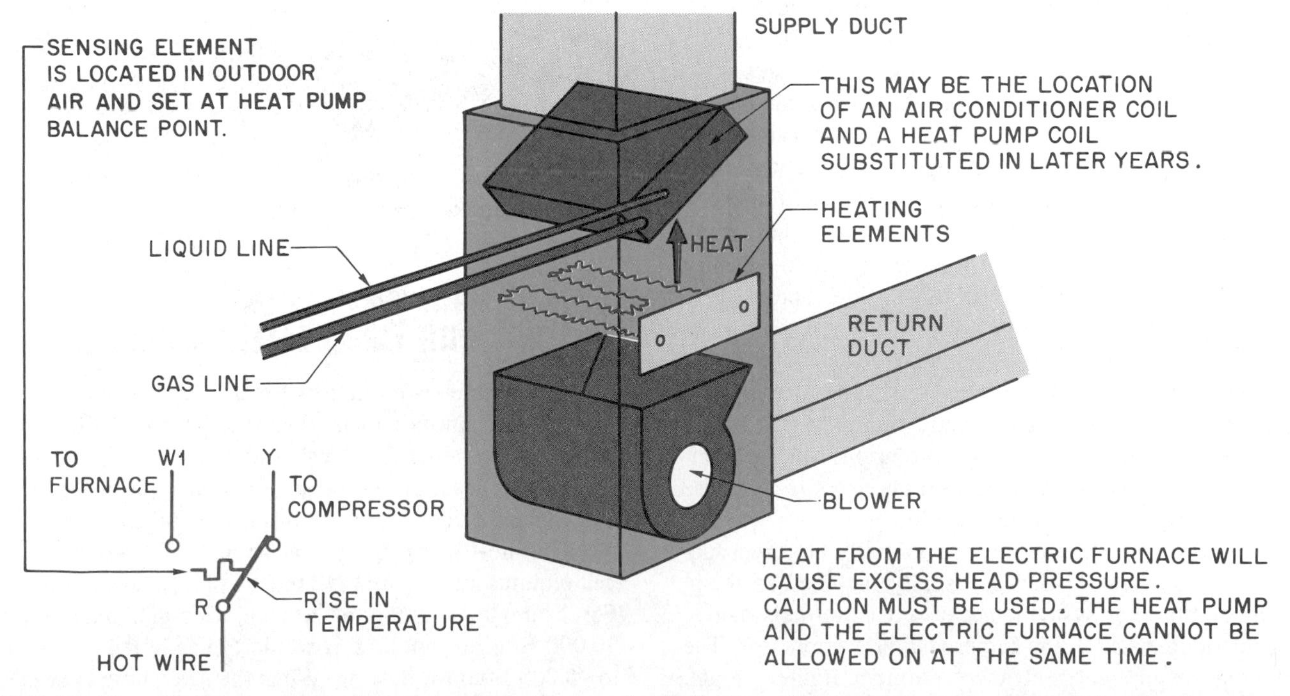

FIGURE 43–54 Electric furnace heat coils are below the heat pump coil and will cause a high head pressure if used incorrectly with a heat pump.

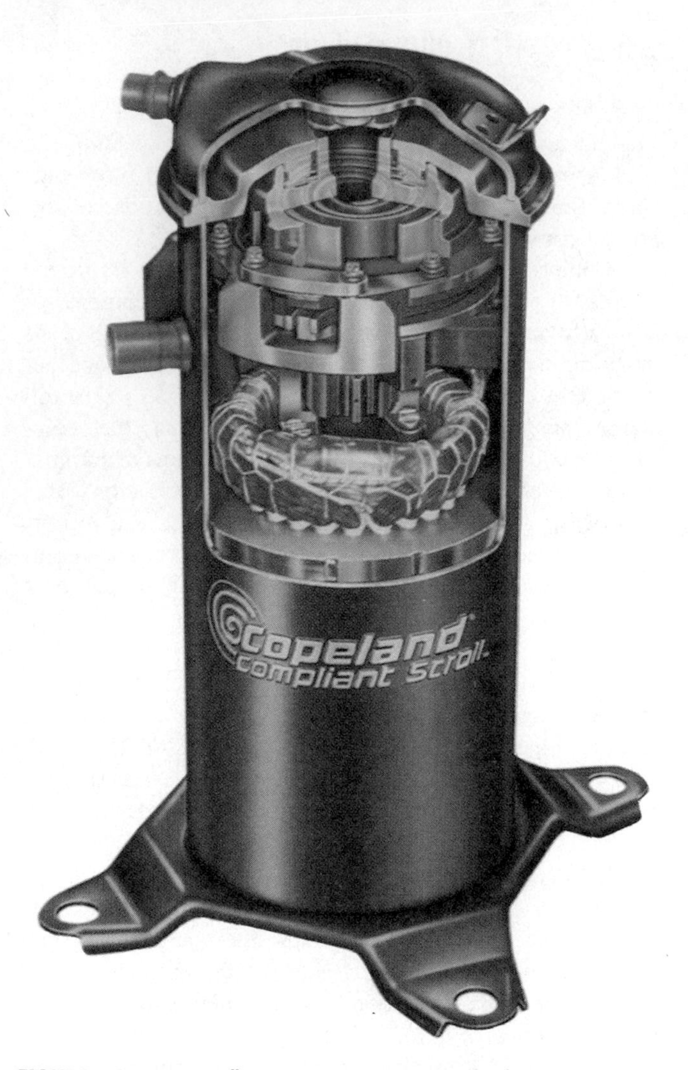

FIGURE 43–55 Scroll compressor. *Courtesy Copeland Corporation*

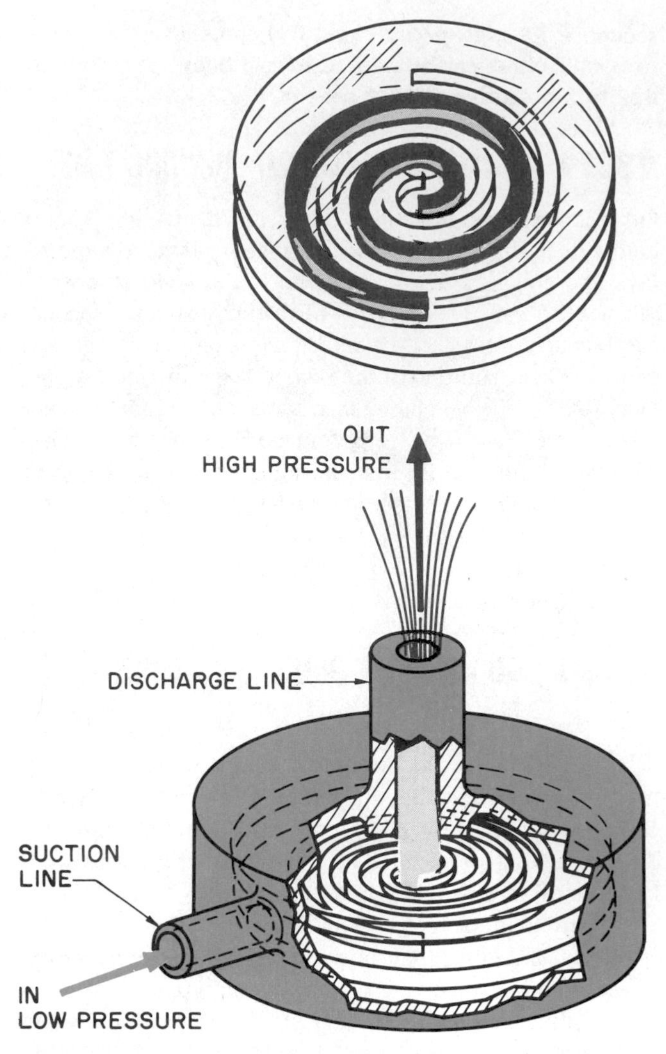

FIGURE 43–56 Orbiting action between scrolls in scroll compressor provides compression.

a stage of compression. Figure 43–57 illustrates one pocket.

The scroll compressor has a check valve in the discharge leaving the compressor to prevent pressures from equalizing back through the compressor during the off cycle. When the compressor is shut off, the gas trapped between the check valve and the compressor scroll will back up through the scroll and the compressor will make a strange sound at the moment it is shut off until these pressures equalize. The high-side pressure will then equalize through the metering device to the low side, like a reciprocating compressor system.

The equalizing of pressure between the check valve and the compressor scroll allows the compressor to start up without hard-start assistance because there is no pressure differential for the first starting revolutions of the compressor. This results in fewer start components to purchase and fewer components to cause trouble.

The scroll compressor does not normally require a suction line accumulator, which is usually included in reciprocating compressor systems, because it is not as sensitive to liquid flood back as the reciprocating compressor. The scroll compressor is more efficient, stronger, lighter weight, and requires fewer accessories for good performance.

FIGURE 43–57 Compression of one pocket of refrigerant in scroll compressor. *Courtesy Copeland Corporation*

43.42 HEAT PUMP SYSTEMS WITH VARIABLE-SPEED MOTORS

Use of variable-speed motors for the compressor and both fan motors is another method used to improve efficiency in heat pump systems. This technology allows for a different selection of heat pump capacity. Previously, heat pump system capacities have been determined by the cooling capacity of the heat pump. A typical structure in the heat pump belt requires about 2 to 2 1/2 times as much heat as cooling. For example, a home in Atlanta, Georgia, may require 30,000 Btuh for cooling. It would typically require 60,000 to 75,000 Btuh for heating. When the heat pump is selected for 30,000 Btuh, it requires considerable auxiliary heat for winter conditions below the balance point of the structure. If the heat pump is selected for the winter capacity need, it would be oversized in the summer. Poor performance and high humidity would occur, along with low efficiency. The variable-speed heat pump closes this gap because the system may be selected at closer to the heating requirement at full load and will run at part load and reduced power in warmer months. Less auxiliary heat is required, so year-round efficiency is achieved.

Variable speed is accomplished with electronics and electronically controlled motors. Each manufacturer has its own method of control and checkout procedure. Repair of a unit with variable-speed motors and electronic controls should not be attempted without the manufacturer's checklist.

PREVENTIVE MAINTENANCE

The typical heat pump requires the same maintenance as an air-cooled air conditioning system with electric heat. The air side of the system should always handle the maximum air for which it is designed. This is important because some homeowners shut registers off in unused rooms. This should never be done with a heat pump. The indoor system is the condenser in the winter cycle. Any reduction in airflow will cause high head pressure and a low COP. This will cause higher than normal power bills with less cooling. Filters should be maintained more closely in heat pump systems for the same reason. Reduced airflow cannot be allowed.

The outdoor unit should be examined for a dirty coil. The insulation on the outdoor gas line should extend all the way to the unit or heat will be lost to the outdoor air. The gas line temperature may be as high as 220°F and will transfer a lot of heat to the cold outside air. This is heat that is paid for and lost.

The contactor should be examined at least once a year for pitting and loose connections. This contactor functions about twice as many times each year as an air conditioning system contactor. Frayed wires should be repaired or replaced. Wires or capillary tubes rubbing on the cabinet or refrigerant lines will soon cause problems and they should be adjusted or isolated.

The fan motor may require lubrication. Follow the manufacturer's directions.

The customer should be quizzed about the winter performance of a heat pump to include the amount of ice buildup on the outdoor unit. If excess ice seems to be a problem, defrost problems may be causing the unit to stay in defrost. Defrost may be simulated even in hot weather with most heat pumps by disconnecting the outdoor fan and running the unit in heating mode. With the outdoor fan motor disconnected, the unit will not overload because of high evaporator temperatures. Most manufacturers have a recommended procedure for checking defrost. If the system has the same timer shown in Figure 43–41, you may run the unit until the coil and sensor are cold and covered with frost. This should be below the make point of 26°F. At this point, carefully jump from terminal 3 to 4 on the timer with an insulated jumper. The unit should immediately go into defrost if the sensor circuit is made. Should it not go into defrost at this time, wait a few more minutes and try again. The sensor is large and may take a few minutes to make. If it will not go into defrost, leave the jumper on the 3 to 4 terminals and jump the sensor. If the unit immediately goes into defrost, the sensor is either defective or not in good contact with the liquid line.

After the unit goes into defrost normally, remove the jumper from 3 to 4 and allow the unit to terminate defrost. If the ambient temperature is above freezing and there is no ice accumulation, this should take about a minute. If it takes a full 10 min, the defrost control contacts may be permanently closed. They may be checked by allowing the control to rise to ambient temperature above 26°F and check with an ohmmeter to see if the contacts remain closed. They should be open when the sensor is above 50°F. One sign of permanently closed contacts is if the unit will go into defrost every time the timer calls for it to do so, whether the sensor is cold or not, and stay in defrost the full 10 min. This is hard on the compressor if the unit does not have a high-pressure control because high pressure will occur at this time.

HVAC GOLDEN RULES

When making a service call to a residence:

- Look professional and be professional. Clean clothes and shoes should be worn.
- Wear coveralls for work under the house and remove them before entering the house.
- Ask the customer about the problem. The customer can often help you solve the problem and save you time.

Added Value to the Customer. Here are some simple, inexpensive procedures that may be included in the basic service call.

- Check line temperatures by touch in summer or winter cycle for the correct temperature.
- Clean or replace air filters as needed. This is very important for heat pumps.
- Make sure all air registers are open, again very important for heat pumps.
- Replace all panels with the correct fasteners.
- Check indoor and outdoor coils for dirt or debris.
- Ask the owners whether they have noticed any ice buildup in winter operation. This may lead to a defrost checkout for the system.
- Check gas line for proper insulation. Heat pumps may lose capacity due to loss of line insulation.

43.43 SERVICE TECHNICIAN CALLS

When reading through these service calls, keep in mind that three things are necessary for an electrical circuit to be complete and current to flow: a power supply (line voltage), a path (wire), and a load (a measurable resistance). The mechanical part of the system has four components that work together: the evaporator absorbs heat, the condenser rejects heat, the compressor pumps the heat-laden vapor, and the metering device meters the refrigerant.

SERVICE CALL 1

A homeowner calls reporting that the air conditioning system is heating, not cooling. This is the first time the unit has been operated in the cooling mode this season. *The problem is the magnetic coil winding on the four-way valve is open and will not switch the valve over into the cooling mode.*

The service technician turns the space thermostat to the cooling mode and starts the system. This is a split system with the air handler in a crawl space under the house. ***The technician goes to the outdoor unit and carefully touches the gas line (the large line); it is hot, meaning the unit is in the heating mode.*** The panel is removed and the voltage is checked at the four-way valve holding coil. 24 V are present; the coil should be energized. The unit is shut off and one side of the four-way valve coil is disconnected. A continuity check shows that the coil is open. The coil is replaced and the unit is started. It now operates correctly in the cooling mode.

SERVICE CALL 2

A store owner calls stating that the air conditioning system in the small store will not start. This is a split-system heat pump with the outdoor coil on the roof. *The problem is the four-way valve 24-V holding coil is burned and has overloaded the control transformer.* This has blown the fuse in the 24-V control circuit.

The technician begins by trying to start the unit at the space temperature thermostat. The unit will not start. When the FAN ON switch is turned to FAN ON, the fan will not start. A control voltage problem is suspected. The indoor unit is in a utility closet downstairs. ***The power is turned off and the door to the indoor unit is removed.*** The power is restored and it is found that there is no power at the low-voltage transformer secondary. The low voltage fuse is checked and found to be blown. A new fuse is in a box in the utility room. Before installing the new fuse, the technician turns off the power and then replaces the fuse. One lead is then removed from the control transformer. An ohmmeter is fastened to the two leads leaving the transformer and the resistance is 0.5 Ω. This seems very low and will likely blow another fuse if power is restored. Some 24-V component must be burned and have a shorted coil. The meter should indicate a 20-Ω resistance. This will increase the current flow enough to blow the fuse. A look at the unit diagram will show which components are 24 V. One of them must be burned. It is decided to check the components in the indoor unit first and save a trip onto the roof, if possible. Each component is checked, one at a time, by removing one lead and applying the ohmmeter. There are three sequencers for the electric heat, and one fan relay. There is no problem here. ***The technician goes to the roof and turns off the power.*** The compressor contactor is checked. It shows a resistance of 26 Ω. The four-way valve magnetic holding coil shows a resistance of 0.5 Ω. It is replaced with a new coil. The new coil has a resistance of 20 Ω. The problem has been found. The unit is put in operation with an ammeter attached to the lead leaving the control transformer to check for excessive current. An amperage multiplier made of thermostat wire wrapped 10 times around the ammeter lead is used. The current is 1 A, which is normal.

SERVICE CALL 3

A customer indicates that the heat pump has an ice bank built up on the outdoor coil. The customer is advised to turn the thermostat to the emergency heat mode until a technician can get there. *The problem is the defrost relay coil winding is open, and the unit will not go into defrost.*

The weather is below freezing, so when the technician arrives, there is still an ice bank on the outdoor coil. ***With the unit off, the technician applies a jumper wire across terminals 3 and 4 on the defrost timer, Figure 43–58.*** The unit is then started. This should force the unit into defrost if the timer contacts are not closing. This does not put the unit into defrost, so the jumper is left in place and the defrost temperature sensor is jumped. This satisfies the two conditions that must be satisfied to start defrost, but nothing happens. ***The technician turns off the unit, turns off the power, locks and tags the panel, removes the jumpers, and checks the defrost relay.*** The coil is open. The relay is changed for a new one and the unit is started again. When the jumper is again applied to terminals 3 and 4, the unit goes into

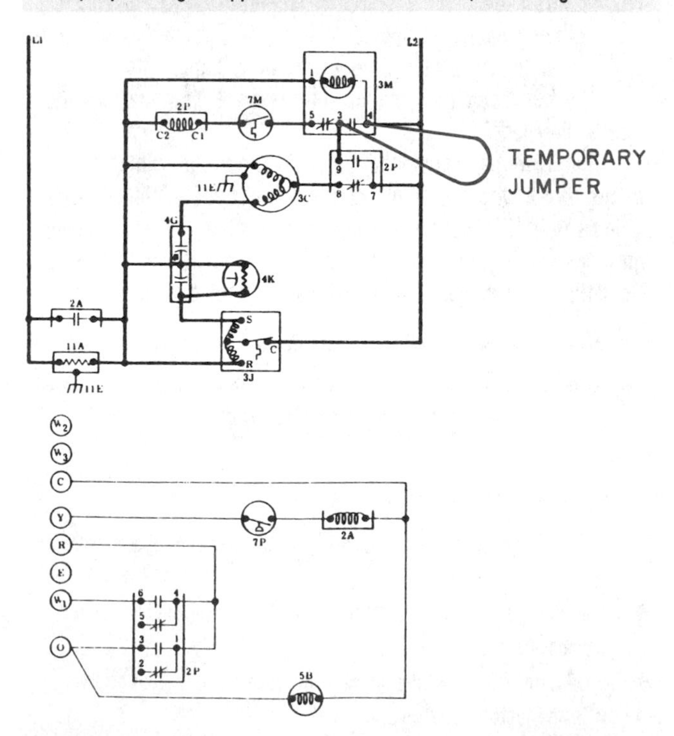

FIGURE 43–58 *With the unit off, the technician applies a jumper wire from terminals 3 to 4 to cause a defrost. This is to keep from waiting for the timer. This is a high-voltage circuit and this exercise should not be performed unless the technician is very experienced.* *Reproduced courtesy of Carrier Corporation*

defrost. Remember that two conditions, time and temperature, must be satisfied before defrost can start. The temperature condition is satisfied because the coil is frozen. When the connection from terminals 3 to 4 is made, defrost is started. This defrost would have occurred normally when the time clock advanced to a trial defrost. By jumping from 3 to 4, the technician did not have to wait for the timer. The unit is allowed to go through defrost, and the coil is still iced. The technician decides to shut off the unit and use artificial means to defrost the coil. ***The power is turned off. The panel is locked and tagged. Electrical components are protected with plastic. A water hose is pulled over to the unit and city water is used to melt the ice. City water is about 45°F. Care is taken that water is not directed on to the electrical components.*** When the ice is melted, the unit is put back in operation. A call back the next morning to the customer verifies that the unit is not icing any more.

SERVICE CALL 4

A homeowner reports that the heat pump unit is running continuously and not heating as well as it used to. *The problem is the outdoor fan is not running because the defrost relay has a set of burned contacts—the ones that furnish power to the outdoor fan.* These contacts are used to stop the outdoor fan in the defrost cycle. They should also start the fan at the end of the cycle.

The technician finds that the outdoor fan is not running during the normal running cycle. The gas line is warm, and the coil is iced up, so the unit is in the heating cycle. The fan is not running. The unit is shut off and the panel to the fan control compartment is removed. A voltmeter is fastened to the fan motor leads and the unit is started. There is no voltage going to the fan motor. The wiring diagram shows that the power supply to the fan goes through the defrost relay. ***The power is checked entering and leaving the defrost relay.*** Power is going in but not coming out. The defrost relay is changed and the unit starts and runs as it should.

SERVICE CALL 5

A residential customer calls. The heat pump serving the residence is blowing cold air for short periods of time. This is a split-system heat pump with the indoor unit in the crawl space under the house. *The problem is the contacts in the defrost relay that bring on the auxiliary heat during defrost are open and will not allow the heat to come on during defrost.*

The technician is familiar with the particular heat pump and believes that the problem is in the defrost relay or the first-stage heat sequencer. The first thing to do is to see if the sequencer is working correctly. The space temperature thermostat is turned up until the second-stage contacts are closed. The technician goes under the house with an ammeter and verifies that the auxiliary heat will operate with the space thermostat. Checking the defrost relay is the next step. This can be done by falsely energizing the relay, by waiting for a defrost cycle, or by simulating a defrost cycle. The technician decides to simulate a defrost cycle in the following manner. ***The power is turned off. To prevent the outdoor fan from running, one of the motor leads is removed.*** Then the unit is restarted. This causes the outdoor coil to form ice very quickly and make the defrost temperature thermostat contacts close. When the coil has ice on it for 5 min, the technician jumps the contacts on the defrost timer from 3 to 4, and the unit goes into defrost, Figure 43–58. The technician goes under the house and checks the electric strip heat; it is not operating. The problem must be the defrost relay. By the time the technician gets back to the outdoor unit, the unit is out of defrost. It is decided that the defrost cycle is functioning as it should, so the technician carefully energizes the defrost relay with a jumper wire. The 24-V power supply feeds terminal 4 and then should go out on terminal 6 to the auxiliary heat sequencer. These contacts are not closing. The defrost relay is changed, and the new relay is energized. The technician goes under the house and verifies that the strip heat is working. A call back later verifies that the unit is operating properly.

SERVICE CALL 6

A customer calls and says that a large amount of ice is built up on the outdoor unit of the heat pump. *The problem is the changeover contacts in the defrost relay are open.* The system is not reversing in the defrost cycle. When this happens, the 10-min maximum time causes the unit to stay in heating without defrosting the outdoor coil because the timer is still functioning.

The technician goes to the outdoor unit first because it is obvious there is a defrost problem. The coil looks like a large bank of ice. The technician removes the control box panel and jumps the timer contacts from 3 to 4; the unit fan stops, but the cycle does not reverse as it should. See Figure 43–58 for the wiring diagram. A voltage check at the four-way valve coil shows that no voltage is getting to the valve. The technician then jumps from the R terminal to the O terminal, and the four-way valve changes over to cooling, so the jumper is removed. The valve must not be getting voltage through the defrost relay. The defrost relay contacts are jumped from 1 to 3, and the valve reverses. The defrost relay is changed, and another defrost is simulated by jumping the 3 to 4 contacts. The defrost thermostat contacts are still closed, and the unit goes into defrost. The unit has so much ice accumulation that the technician uses water to defrost the coil and get the unit back to a normal running condition. A call the next day indicates that the unit is defrosting correctly.

SERVICE CALL 7

A shop owner calls reporting that the heat pump in their small shop is running constantly. The unit is blowing cold air for short periods every now and then. The power bill was excessive last month, and it was not a cold month. This is a split system with 20-kW strip heat for quick recovery, so the customer can turn the thermostat back to 60°F on the weekend and have quick recovery on Monday morning. The outdoor unit is in the rear of the shop. *The problem is the contacts in the defrost relay that change the unit to cooling in the defrost mode are stuck closed.* The unit is running in the cooling mode, and the strip heat is heating the structure. It must also make up for the cooling effect of the cooling mode.

The technician sets the thermostat to the first stage of heating. The air at the registers is cool not warm. The unit should have warm air in the first stage of heat (85°F to 100°F), depending on the outside air temperature. The indoor coil is in a closet in the back of the shop. The gas line is not hot but cold. This indicates that the unit is in the cooling mode. The technician thinks the four-way valve may be stuck in the cooling mode, but a voltage check should be performed. See Figure 43–59 for a wiring diagram. The technician removes the control voltage panel cover and finds 24 V at the C to O terminals. This means the four-way valve coil is energized and the unit should be in the cooling mode. Now, the technician must decide if the 24 V is coming from the space thermostat or the defrost relay. The field wire from the space thermostat is removed, and the valve stays energized. The voltage must be coming from the defrost relay. The wire is replaced on the O terminal and the wire on terminal 3 on the defrost relay is carefully removed. The unit changes to heating. The defrost relay is changed, and the unit operates correctly. A call back later finds the unit to be operating as expected.

SERVICE CALL 8

A store manager calls stating that the heat pump heating a small retail store has large amounts of ice built up on the outdoor unit. The unit is located outside in back of the store. *The problem is the defrost thermostat contacts are not closing when the timer advances to the defrost cycle.* This is a combination control—the timer and the sensor built into the same control.

The technician goes to the outdoor unit and finds a large bank of ice on the coil. The control panel is removed. This is a unit that has a timer that can be advanced by hand. The timer and the defrost sensor are built into one control. It is obvious that the sensor is cold enough for the contacts to make, so the dial of the timer is advanced all the way around. The unit does not go into defrost, so it must be the contacts in the timer sensor that are faulty. A new timer is installed and advanced by hand until the contacts make. The unit goes into defrost normally. One defrost cycle does not melt enough ice from the coil to leave the heat pump on its own. The technician forces the unit through three more defrosts before the ice is melted to a normal level. If a water hose had been available, it would have been used to melt the ice to save time.

SERVICE CALL 9

While on a routine service call at an apartment house complex, the technician notices a unit go into defrost when there is no ice on the coil. The unit stays in defrost for a long time, probably the full 10 min allowable. This is a split-system heat pump and should have defrosted for only 2 or 3 min with a minimum of ice buildup. The compressor was loud, as if it were pumping against a high head pressure. *The problem is the defrost thermostat contacts are shut.* The unit is going into defrost every 90 min and staying for the full 10 min. The timer is

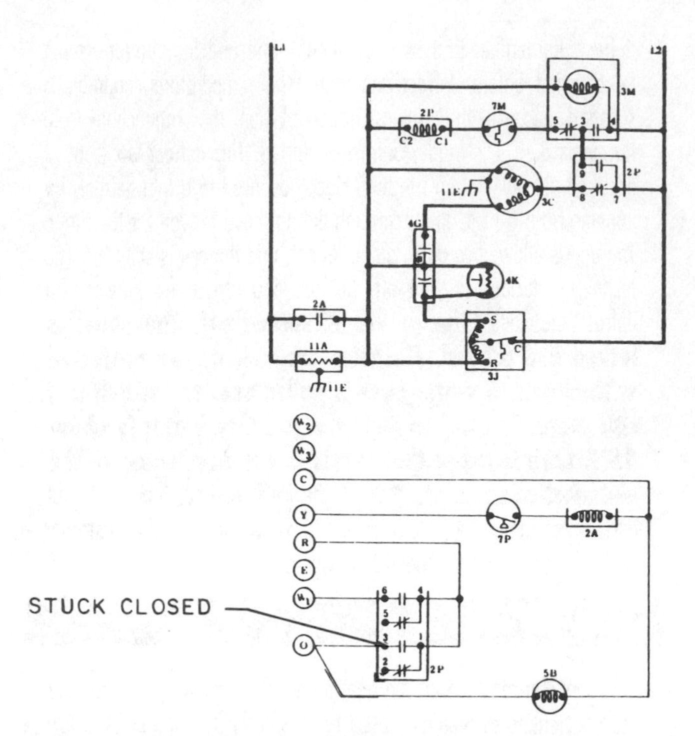

FIGURE 43–59 The contacts that change the unit to cooling for defrost are stuck closed. *Reproduced courtesy of Carrier Corporation*

terminating the defrost cycle after 10 min. Toward the end of the cycle, the head pressure is very high.

The technician shuts off the unit, removes the panel to the control box, removes one lead of the defrost thermostat, and takes a continuity check across the thermostat. The circuit should be open if the coil temperature is above 50°F but the contacts are closed. (This control should close at about 25°F and open at about 50°F. Allow plenty of time for this control to function because it is large and it takes a while for the unit's line to exchange heat from and to the control.) The technician changes the defrost thermostat, and the unit is put back in operation.

SERVICE CALL 10

A residential customer reports that there is no heat in their home. It smells like something is hot or burning. ***The customer is told to turn off the unit at the indoor breaker to stop all power to the unit.*** *The problem is the indoor fan motor has an open circuit, and the electric strip heat is coming on from time to time and then shutting off because of excessive temperature.*

The technician goes to the space thermostat and sets it to OFF. The electrical breakers are then energized. The thermostat is turned to FAN ON. The fan relay can be heard energizing, but the fan will not start. The indoor unit is under the house in the crawl space, so the technician takes tools and electrical test equipment to the unit. ***The breaker under the house is turned off and the door to the fan compartment is opened. The fan motor is checked and found to have an open winding.*** It is replaced with a new one and the unit is started. The technician uses an ammeter to check all three stages of electric heat for current draw. One stage draws no current. ***The breaker is again turned off, and the electric**

heat is checked with an ohmmeter.* The circuit in one unit is open. This happened because the fan was not running and the unit overheated. A new fuse link is installed and the heat is started again. This time it is drawing current. The technician puts the unit panels back together and leaves the job.

SERVICE CALL 11

A homeowner reports that the residential heat pump smells hot and not much air is coming out of the registers. ***The customer is told to shut off the unit at the breaker for the indoor unit.*** *The problem is the indoor fan capacitor is defective, and the fan motor is not running up to speed.* The fan motor is cutting off from an internal overload periodically.

The technician sets the thermostat to OFF and turns on the breaker. The FAN ON switch at the thermostat is set to ON. The fan relay can be heard energizing, and the fan starts. There does not seem to be enough air coming out of the registers. The fan motor is not running smoothly. The technician goes to the attic crawl space where the indoor unit is located, taking tools and electrical test instruments. ***The breaker next to the indoor unit is shut off and the panel to the control box is removed. The breaker is then turned on. It is noticed that the fan is not turning up to speed. The breaker is turned off again. The fan capacitor is removed and a new one is installed.*** The breaker is energized again, and the fan comes up to speed. A current check of the fan motor shows the fan is operating correctly. ***The power is shut off, locked and tagged and the motor is oiled.*** The technician then replaces all panels and turns on the power.

SERVICE CALL 12

A homeowner calls indicating that the unit in their residence is blowing cool air from time to time. One of the air registers is close to their television chair, and the cool air bothers them. *The problem is the breaker serving the outdoor unit is tripped, apparently for no reason.* The control voltage is supplied at the indoor unit so the electric auxiliary heat is still operable.

The technician turns the space temperature thermostat to the first stage of heat and notices that the outdoor unit does not start. When the thermostat is turned to a higher setting, the auxiliary heat comes on and heat comes out of the registers. With the thermostat in the first-stage setting, only the outdoor unit should be operating. The auxiliary heat should not be operating. In the second stage, the heat pump and the auxiliary heat should be operating. The technician goes to the outdoor unit and finds that the breaker is tripped. ***There is no way of knowing how long the breaker has been off, but it is not wise to start the unit because there has been no crankcase heat on the compressor. The breaker is switched to the off position. The panel is locked and tagged.*** The technician uses an ohmmeter and checks the resistance across the contactor load-side circuit to make sure there is a measurable resistance. It is normal, about 2 Ω (there is a compressor motor, fan motor, and defrost timer motor to read resistance through). The contactor load-side terminals are checked to ground to see if any of the power-consuming devices are grounded. The meter reads infinity, the highest resistance. It is assumed that someone may have turned the space temperature thermostat off and back on before the system pressures were equalized and the compressor motor would not start. This means that the compressor tried to start in a locked-rotor condition, and the breaker tripped before the motor overload protector turned off the compressor. A sudden power outage then back on will cause the same thing.

The technician does not reset the breaker before going to the indoor thermostat, setting it to the emergency heat mode. This prevents the compressor from starting with a cold crankcase. The technician then tells the homeowner to switch back to normal heat mode in 10 h. This gives the crankcase heater time to warm the compressor. The homeowner is instructed to see if the outdoor unit starts correctly or if the breaker trips. If the breaker trips, the thermostat should be set back to emergency heat, the breaker reset, and the technician should be called. If the technician must go back, the compressor can be started without fear of harming it. The compressor could have a starting problem, and it may not start the next time. The customer does not call back, so the unit must be operating correctly.

SERVICE CALL 13

A customer says that the heat pump is running constantly in the heating mode in mild weather. It did not do this last year; it cut off and on in the same kind of weather. *The problem is the check valve in the liquid line (parallel to the TXV) is stuck open and not forcing liquid refrigerant through the TXV.* This is overfeeding the evaporator and causing the suction pressure to be too high, hurting the efficiency of the heat pump. For example, if it is 40°F outside, the evaporator should be operating at about 15°F for the outside air to give up heat to the evaporator. If the evaporator is operating at 30°F because it is overfeeding, it will not exchange as much heat as it should. The capacity will be down.

The technician turns the room thermostat to the first stage of heating and goes to the outdoor unit at the back of the house. The unit is running. The gas line temperature is warm but not as hot as it should be. The compressor compartment door is removed so the compressor can be observed. It is sweating all over from liquid refrigerant returning to the compressor. This means that the TXV is not controlling the liquid refrigerant flow to the evaporator. A gage manifold is fastened to the high and low sides of the system. The suction pressure is higher than normal, and the head pressure is slightly lower than normal. The head pressure is down slightly because some of the charge that would normally be in the condenser is in the evaporator. A thermometer lead is attached to the suction line just before the four-way valve and the superheat is found to be 0°F. The outdoor coil is flooded. Not enough refrigerant is flooding to the compressor to make a noise because the liquid is not reaching the cylinders. It is boiling away in the compressor crankcase. The technician tries adjusting the TXV and it makes no difference. The TXV is open and not closing or else the check valve is open and bypassing the TXV. The technician goes to the supply house and gets a TXV and a check valve. **The unit charge is removed and recovered into an approved recovery cylinder.** The check valve is removed first and tested. The technician can blow nitrogen vapor through the valve in either direction, so the check valve is defective. It is changed and the TXV is left alone. The unit is pressured, leak checked, evacuated, and charged. When the unit is started in the heating mode, it operates correctly. The check valve was the problem. The superheat is checked to be sure that it is normal (8°F to 12°F) and the unit panels are assembled. A call later verifies that the unit is cycling as it should.

SERVICE CALL 14

A residential customer reports that the heat pump serving the home is not performing to capacity. The power bills are much higher than last year's bills although the heating season is much the same. The customer would like a performance check on the heat pump. *The problem is the heat pump has a two-cylinder compressor and one of the suction valves has been damaged in the compressor due to liquid floodback on start-up.* The owner had shut the disconnect off at the condensing unit last summer and left it off all summer. When the unit was started in the fall, the compressor had enough liquid in the crankcase to damage one of the suction valves. If the crankcase heat had been allowed to operate for 8 to 10 h before starting the unit, this would not have happened. Note that if one cylinder of a two-cylinder compressor is not pumping, the compressor will pump at half capacity.

The technician fastens gages to the system and starts it. The manufacturer's performance chart is used to compare pressures. The unit is running a high suction pressure and a low discharge pressure. The technician switches the unit to cooling and blocks the outdoor coil air flow to build up head pressure. The head pressure goes to 200 psi and the suction to 100 psi. The compressor current is below normal. The compressor is not pumping to capacity; it is defective. A temperature check of the gas lines at the four-way valve proves the valve is not bypassing gas.

The technician goes to a supply house and gets a new compressor. The old compressor is not in warranty, so there is no compressor exchange. **The refrigerant is recovered and the compressor is changed.** A new liquid-line filter-drier is installed. The system is pressured, leak checked, and evacuated before charging. The charge is measured into the system and the system is started. The system is operating correctly. A call to the customer later in the day proves the system is functioning correctly. The outdoor unit is satisfying the room thermostat and shutting the unit off from time to time because the weather is above the balance point of the house.

SERVICE CALL 15

A customer calls the heating and cooling service company indicating that the heat pump serving the residence must not be operating efficiently because the power bill has been abnormally high for the month of January compared to last year's bill. Both years have had about the same weather pattern. *The problem is the four-way valve in the heat pump is stuck in mid-position and will not change all the way to heat.* There is no apparent reason for the sticking valve; it must be a factory defect.

The technician turns the room thermostat to the first stage of heating and goes to the outdoor unit. The gas line leaving the unit is warm, not hot as it should be. Gages are installed on the system. The head pressure is low, and the suction pressure is high. The compressor amperage is low. All of these are signs of a compressor that is not pumping up to capacity. The outside temperature is 40°F, and the coil does not have an ice buildup, so the unit should be working at a high capacity. The technician suspects the compressor, so a compressor performance test is performed. A jumper wire is installed between the R and O terminals to energize the four-way valve coil and change the unit over to the cooling mode. The thermostat is left in the first-stage heat mode. If the house begins to cool off, the second-stage of heat will come on and keep the house from getting too cool. A plastic bag is wrapped around the condenser coil to slow the air entering the coil. The head pressure begins to climb, but the suction pressure rises with it. The compressor will not build a sufficient differential between the high and low sides. Generally, the unit should be able to build 300 psig of head pressure with 70 to 80 psig suction. This unit has a 120 psig suction at 250 psig head pressure. Before the compressor is condemned, the technician decides to check the four-way valve operation. An electronic thermometer is fastened with the two leads on the cold gas line entering and leaving the four-way valve. A difference in temperature should not be more than 3°F between these two lines, but it is 15°F. The four-way valve is leaking hot gas into the suction gas. **The four-way valve is changed after the refrigerant is recovered.** The unit is started after the pressure leak test, evacuation, and charging. The unit now performs to capacity with normal suction and head pressures.

SAFETY PRECAUTION: *Use common sense and caution while troubleshooting any piece of equipment. As a technician you are constantly exposed to *potential danger*. High-pressure refrigerant, electrical shock hazard, rotating equipment shafts, hot metal, sharp metal, and lifting heavy objects are among the most common hazards you will face. Experience and listening to experienced people will help to minimize the dangers.*

SERVICE CALL 16

An apartment house complex with 100 package heat pumps is under contract with a local company for routine service. While on a routine visit, the technician noticed that a compressor was sweating all over in the heating cycle. *The problem is one of the newer technicians serviced this unit during the last days of the cooling season and added refrigerant to the system.* It is a capillary tube metering device system. The unit now has an overcharge.

The technician knows that a compressor should never sweat all over, so a gage manifold is installed to check the pressures. A charging chart is furnished with this unit to use when adjusting the charge. When the gage readings are compared to the chart readings, both the suction and the discharge pressures are too high. This symptom calls for reducing the charge. ⊛**Refrigerant is allowed to flow into an approved empty cylinder until the pressures on the chart compare to the pressure in the machine.**⊛ An observant technician has saved a compressor.

The following service calls do not have the solutions described. The solution may be found in the Instructor's Guide.

SERVICE CALL 17

A beauty shop owner calls and says the unit serving the beauty shop is blowing cool air from time to time.

This is a split system with the outdoor unit on the roof. The technician turns the thermostat to the first stage of heat and cool air comes out of the air registers. The technician goes to the outdoor unit—the outdoor fan is running, but the compressor is not. ***The power is turned off. The panel is locked and tagged before the technician removes the cover to the compressor compartment.*** An ohm check shows that there is an open circuit between the R and C terminal and between the S and C terminal, but there is continuity between R and S. The compressor body temperature is warm, not hot.

What is the likely problem and the recommended solution?

SERVICE CALL 18

A residential customer calls and says the heat pump outdoor unit is off because of a tripped breaker. The customer reset it three times, but it would not stay reset.

The technician goes to the outdoor unit first and sees the breaker is in the tripped position. It is decided that before resetting it a motor test should be performed. ***A volt reading is taken at the compressor contactor line side and it is found that 50 V is present.*** This must be caused by the repeated resetting of the breaker. Voltage is leaking through the breaker. The main heating breaker is turned off, and the 50 V disappears. ***If the technician had taken a continuity reading first, the meter would have been damaged.*** With the main breaker off, a resistance check on the load side of the compressor contactor shows a reading of 0 Ω to ground (the compressor suction line is a convenient ground).

What is the likely problem and the recommended solution?

SERVICE CALL 19

An owner of a two-story house calls indicating that the unit serving the upstairs of the house is blowing cold air part of the time and not heating properly.

The technician turns the space temperature thermostat to the first stage of heating and notices that cool air is coming out of the air registers. The outdoor unit can be seen from a window and the fan is running. The technician goes to the outdoor unit and finds that the gas line leaving the unit is barely warm, not hot. The compressor is either not running, or it is running and not working. The compressor access panel is removed and the compressor can be touched and seen. It is hot to the touch.

What is the likely problem and the recommended solution?

SERVICE CALL 20

A dress shop manager calls to report that the heat pump serving the dress shop will not cool properly. This is the first day of operation for the system this summer. The compressor was changed about 2 weeks ago after a bad motor burn. The system has not operated much during the 2-week period because the weather has been mild.

The technician turns the thermostat to the cooling mode and starts the unit. The compressor is on the roof, and the indoor unit is in a closet. The technician goes to the indoor unit and feels the gas line; it is hot. The unit is in the heating mode with the thermostat set in the cooling mode. The technician goes to the roof and removes the low-voltage control panel. In this unit the four-way valve is energized in the cooling mode, so it should be energized at this time. A voltmeter check shows that power is at the C to O terminals (the common terminal and the four-way valve coil terminal). The technician removes the wire from the O terminal and hears the pilot solenoid valve change (with a click). Every time the wire is touched to the terminal, the valve can be heard to position. This means that the pilot valve is moving.

What is the likely problem and the recommended solution?

SERVICE CALL 21

A retail store owner reports that the heat pump serving the store is blowing cold air from time to time and not heating as it normally does. It runs constantly. The unit has a TXV.

The technician turns the thermostat to the first stage of the heating mode. The air coming out of the air registers is cool; it feels like the return air. The outdoor unit is on the roof, and the indoor unit is in an attic crawl space. The technician goes to the roof; the thermostat is set with the first stage calling for heat. The outdoor unit should be on. The outdoor unit is running, but the gas line is cool. A gage manifold is installed on the high and low sides of the system. The low side is operating in a vacuum, and the high-side pressure is 122 psig, which corresponds to the inside temperature of 70°F. This indicates that liquid refrigerant is in the condenser (the indoor coil). If the condenser pressure were 60 psig, well below the 122 psig corresponding to 70°F, it would indicate that the unit is low in refrigerant charge.

What is the likely problem and the recommended solution?

SUMMARY

- All compression cycle refrigeration systems are similar to heat pumps; however, heat pumps have the capability of pumping heat either way.
- A new component, the four-way valve enables the heat pump to reverse the refrigeration cycle and reject heat in either direction.
- Air is not the only source of heat for a heat pump, but it is the most common. The ground and large bodies of water also contain heat that may be used.
- Air at 0°F still has about 85% of the available heat contained in it compared to no heat (−460°F).
- *Water to air* is the term used to describe heat pumps that absorb heat from water and transfer it to air.
- Water-to-air heat pumps use lakes, deep wells, and water from industrial cooling systems for a heat source.
- Commercial buildings may use water to absorb heat from one part of a building and reject the same heat to another part of the same building.
- Other methods, such as solar assistance, are being developed using the reverse-cycle heat pump.
- Air-to-air heat pumps are similar to summer air conditioning equipment.
- There are two styles of equipment: split systems and package (self-contained) systems.
- New names are applied to the heat pump components because of the reverse-cycle operation.
- When applied to cooling air conditioning equipment, the coil inside the house is called an evaporator. On a heat pump unit this is called an indoor coil. Sometimes it is operated as a condenser (in winter) and sometimes as an evaporator (in summer).
- When applied to cooling air conditioning, the unit outside the house is called a condensing unit. With a heat pump this unit is a condenser in the summer and an evaporator in the winter. It is called the outdoor unit.
- The terms "indoor coil" and "outdoor coil" are used with package heat pumps also to avoid confusion.
- The refrigerant lines that connect the indoor coil with the outdoor coil are the gas line (hot gas in winter and cold gas in summer) and the liquid line.
- The liquid line is always the liquid line; the flow reverses from season to season.
- Several metering devices are used with heat pumps. The first was the TXV. There must be two of them, with a check valve piped in parallel to the valve to force the liquid refrigerant to flow through the valve.
- Capillary tube metering devices are common. There must be two of them, one for cooling and one for heating operations. They must have a check valve piped in parallel to force the liquid refrigerant to flow correctly.
- A fixed-bore metering device that will allow full flow in one direction and restricted flow in the other direction is used by many manufacturers.
- The electronic expansion valve is used by some manufacturers with close-coupled equipment because it will meter in both directions and maintain the correct superheat.
- When standard filter-driers are used with a heat pump installation, they must be placed in the circuit with the check valve to ensure correct flow. ***It must have a check valve piped in series with it, or when the flow reverses, it will back-wash and the particles will be pushed back into the system.***
- A special biflow drier may be used in the liquid line. The biflow drier is actually two driers in one shell with check valves to force the refrigerant through the correct circuit.
- The air-to-air heat pump loses capacity as the outside temperature goes down. The unit responds in reverse to the load. When the weather gets colder, the structure needs more heat and the air-to-air heat pump provides less.
- At the balance point the heat pump alone will run constantly and just heat the structure. If the outdoor temperature drops any lower, the heat pump must have help.
- Auxiliary heat is the heat that a heat pump uses as a supplement.
- Auxiliary heat is normally electric resistance heat. Oil and gas may be used in some installations.
- When the auxiliary heat is used as the only heat source, such as when the heat pump fails, it is called emergency heat.
- Emergency heat is controlled with a switch in the room thermostat. This switch turns off the heat pump and turns on the auxiliary heat.
- Coefficient of performance (COP) is determined from the heat pump's heating output divided by the input. A COP of 3:1 is common with air-to-air heat pumps at the 47°F outdoor temperature level. A COP of 1.5:1 is common at 0°F.
- The efficiency of a heat pump is a result of its being used to capture heat from the outdoors and to pump that heat indoors.
- A heat pump is installed in much the same manner as a cooling air conditioner. The air distribution requires at least 400 cfm of air per ton of cooling capacity.
- The terminal air must be distributed correctly because it is not as hot as the air in oil or gas installations. Heat pump air normally is not over 100°F with only the heat pump operating.
- ***Provision for water drainage at the outdoor unit must be provided in winter.***
- Space temperature control differs from controlling a combination cooling and heating system with electric air conditioning and gas or oil heat. There are two different heating systems: the heat pump and the auxiliary heat.
- The indoor fan must run when the compressor is operating, and the compressor operates in both summer and winter modes.
- The four-way valve determines whether the unit is in the heating or cooling mode. In one position, the compressor discharge gas is directed toward the outdoor coil

and the unit is in the cooling mode. In the other position, the hot discharge gas is directed toward the indoor coil and the unit is in the heating mode.

- The technician can tell which mode the unit is in by the temperature of the gas line. It can get very hot.
- The space temperature thermostat controls the direction of the hot gas by controlling the position of the four-way valve. Most manufacturers energize the four-way valve in the winter cycle; many also energize it in the summer cycle.
- Because the heat pump evaporator operates below freezing in the winter, frost and ice will build up on the outdoor coil. Defrost is accomplished by reversing the unit for a short time to remove this ice.
- When the system is in defrost, it is in the cooling mode with the outdoor fan off to aid in the buildup of heat. The indoor coil will blow cool air, so the auxiliary heat is usually energized during defrost.
- Defrost is normally instigated (started) by **time and temperature** and terminated by **time or temperature.**
- Servicing a heat pump involves both electrical and mechanical troubleshooting.
- The refrigerant charge is normally critical with a heat pump. The recommended charging procedures are to follow the manufacturer's charging chart or to recover the charge, evacuate the system to a deep vacuum, and measure the correct charge back into the system.
- The use of scroll compressors increases the efficiency of heat pump systems.
- Variable speed motors allow heat pump compressors to be sized closer to cold weather use, increasing efficiency.

REVIEW QUESTIONS

1. How does a heat pump resemble a refrigeration system?
2. What is the lowest temperature at which heat can be removed from a substance?
3. Name the three common sources of heat for heat pumps.
4. What component allows a heat pump to reverse its refrigerant cycle?
5. What large line connects the indoor unit to the outdoor unit?
6. What small line connects the indoor unit to the outdoor unit?
7. Why is the indoor unit not called an evaporator?
8. When is the outdoor unit an evaporator?
9. When is the indoor unit a condenser?
10. Why is it important to have drainage for the outdoor unit?
11. Name three common metering devices used with heat pumps.
12. Which metering device is most common?
13. Which metering device is most efficient?
14. Where is the only permanent suction line on a heat pump?
15. What type of drier may be used in the liquid line of a heat pump?
16. Where must a suction-line drier be placed in a heat pump after a motor burnout?
17. What controls the heat pump to determine whether it is in the heating cycle or the cooling cycle?
18. Can a heat pump switch from heating to cooling and from cooling to heating automatically?
19. What must be done when frost and ice build up on the outdoor coil of a heat pump?
20. Are all heat pumps practical anywhere in the United States? Why?
21. Why are scroll compressors more efficient than reciprocating compressors?
22. Describe why variable-speed compressor motors are more efficient than single-speed motors in heat pump systems.

UNIT 43 DIAGNOSTIC CHART FOR HEAT PUMPS IN THE HEATING MODE

The heat pump is the most complex of the heating and air conditioning systems. These are found in many homes and businesses in the south and southwest parts of the United States. This climate is best suited for heat pumps. These systems operate in both heating and cooling modes. The heating mode is discussed here.

PROBLEM	POSSIBLE CAUSE	POSSIBLE REPAIR	PARAGRAPH NUMBER
COOLING SEASON	See summer air conditioning (Unit 42)		
HEATING SEASON			
No Heat—outdoor unit will not run—indoor fan runs	Open outdoor disconnect	Close disconnect	41.9
	Open fuse of breaker	Replace fuse, reset breaker and determine problem	41.9
	Faulty wiring	Repair or replace faulty wiring or connections	41.9
No heat—indoor fan or outdoor unit will not run	Low voltage control problem		
	A. Thermostat	Repair loose connections or replace thermostat and or subbase	41.9
	B. Interconnecting, wiring or connections	Repair or replace wiring or connections	41.9
	C. Transformer	Replace if defective	41.9
		SAFETY PRECAUTION: Look out for too much current draw due to ground circuit or shorted coil	
No heating, indoor and outdoor fans running, but compressor not running	Compressor overload tripped		
	A. Low line voltage	Correct low voltage, power company or loose connections	41.9, 41.10 43.33
	B. High head pressure		
	1. Dirty indoor (condenser) coil	Clean indoor coil	43.34
	2. Indoor fan not running all the time	Check indoor fan motor and capacitor	43.34
	3. Overcharge of refrigerant	Correct charge	43.34
	4. Restricted indoor airflow	Change filters	43.34
		Open all supply registers	43.34
		Clear return air blockage	43.34
Outdoor coil (evaporator) freezes and ice will not melt	Defrost control sequence not operating	Follow manufacturers directions and correct defrost sequence	43.29
	Low charge, not enough refrigerant to perform adequate defrost	Correct charge and run unit through enough defrost cycles to clear ice off, then allow unit to run normally	43.37
Unit will not change from cooling to heating or heating to cooling	Four-way valve not changing over		
	A. Defective defrost relay of circuit board	Replace relay or circuit board	43.35
	B. Four-way valve stuck	Change four-way valve	43.35
	C. Thermostat not changing to heat in subbase	Repair or replace subbase	43.28
Excessive power bill	Compressor not running, heating off or operating on auxilliary heat	Repair compressor circuit or replace compressor	43.34
	Suction pressure too high and head pressure too low for conditions	Perform four-way valve temperature check. If valve is defective, change valve. If valve is good, change compressor.	43.35

SECTION

9 Domestic Appliances

U N I T

44 Domestic Refrigerators

OBJECTIVES

After studying this unit, you should be able to

- define refrigeration.
- describe the refrigeration cycle for household refrigerators.
- describe the types, physical characteristics, and typical locations of the evaporator, compressor, condenser, and metering device.
- explain the various defrost systems.
- describe how to dispose of the condensate.
- discuss typical refrigerator designs.
- explain the purpose of mullion and panel heaters.
- describe the electrical controls used in household refrigerators.
- discuss ice-maker operation.
- describe various service techniques used by the refrigeration technician.

SAFETY CHECKLIST

* Never use a sharp object to remove ice from an evaporator.
* Remove refrigerator doors or latch mechanisms before disposing of a refrigerator.
* Use proper equipment when moving refrigerators.
* Use a back belt brace when lifting.
* Do not raise the pressure in a refrigerator above the manufacturer's low-side specified design pressure.
* Tubing lines may contain oil that may flare up and burn when soldering. Always keep a fire extinguisher within reach when soldering.
* Use all electrical safety precautions when servicing or troubleshooting electrical circuits.

44.1 REFRIGERATION

You should have a firm understanding of Section 1 of this text before proceeding into this unit. The term refrigeration means to move heat from a place where it is not wanted to a place where it makes little or no difference. The domestic refrigerator is no exception to this statement. Heat enters the refrigerator through the walls of the box by conduction, convection, and from warm food placed inside. When the food is warmer than the box temperature, it raises the temperature in the box. Heat travels naturally from a warm to a cold substance, Figure 44–1. The refrigerator moves this heat from inside the box to the room where it makes little or no difference, Figure 44–2.

The domestic or household refrigerator is a plug-in appliance and can be moved from one location to another. Typically no license is required to install plug-in appliances. It is a package unit that is completely factory assembled and charged with refrigerant.

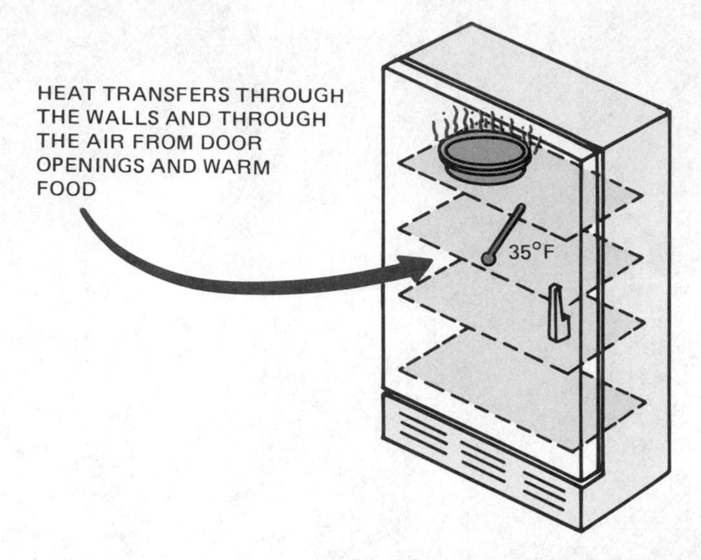

FIGURE 44–1 Warm food brings heat into the refrigerator.

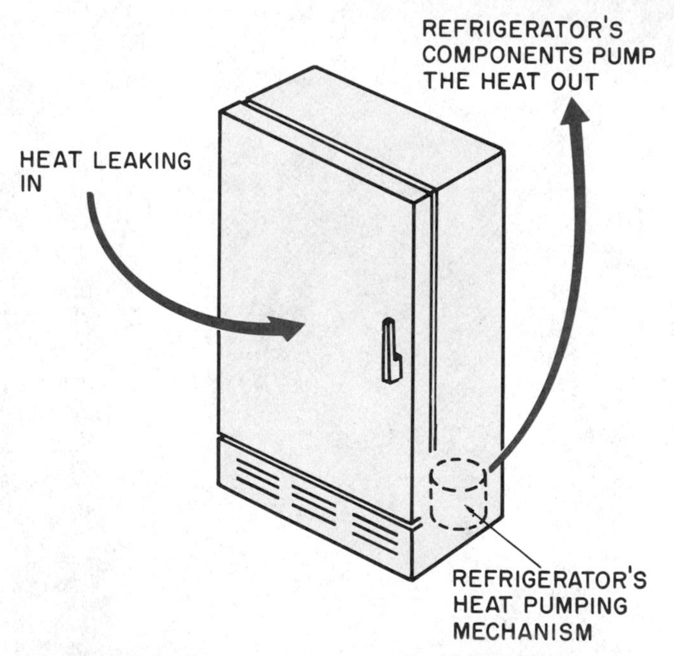

FIGURE 44–2 The refrigeration cycle moves the heat from the refrigerated box to the room, where it makes little or no difference.

The refrigeration system circulates air inside the box across a cold refrigerated coil, Figure 44–3. The air gives up sensible heat to the coil and the air temperature is lowered. It gives up latent heat (from moisture in the air) to the coil and dehumidification occurs. This causes frost to be formed on the evaporator coil. When the air has given up heat to the coil it is distributed back to the box at a much colder temperature so that it can absorb more heat and humidity, Fig-

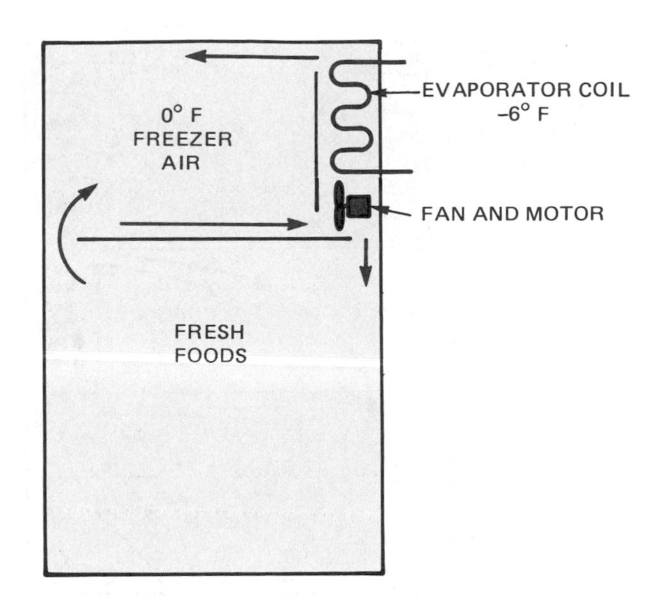

FIGURE 44–3 Air gives up heat to the cold coil.

ure 44–4. This process continues until the box temperature is reduced to the desired level. The typical domestic box inside temperature is 35°F to 40°F when the room temperature is normal. This typical box temperature is the temperature of the return air to the evaporator coil, Figure 44–4. If a thermometer were located in the center of the food, such as in a glass of water in the middle of the box, it would also register the return average air temperature. It would respond slowly to the air changes around it and react as an average of the return air temperature from the start to the end of the refrigeration cycle. These temperatures are typical of a domestic refrigerator located in the comfort conditions of a residence. The refrigerator does not perform within these temperatures if it is located in a place of extreme temperature, such as outside in the summer and winter, Figure 44–5.

44.2 THE EVAPORATOR

The household refrigerator evaporator absorbs heat into the refrigeration system. To accomplish this it must be colder than the air in the refrigerated box. In a typical commercial box application there is one box for maintaining frozen food and a separate one for fresh food such as vegetables and dairy products. The household refrigerator does both with one box. Therefore, the single compressor operates under conditions for the lowest box temperature. The freezing compartment is the lowest temperature. It is typically operated at −10°F to +5°F.

The evaporator in the household box also must operate at the low-temperature condition and still maintain the fresh-food compartment. This may be accomplished by allowing part of the air from the frozen-food compartment to flow into the fresh-food compartment, Figure 44–6. It may also be accomplished with two evaporators that are in series, one for the frozen-food compartment and the other for the fresh-food medium-temperature compartment, Figure 44–7. In either case, frost will form on the evaporator and a defrost method must be used. This is described in more detail later.

The evaporators in household refrigerators can be of two types, natural draft or forced draft, Figure 44–8. The fan improves the efficiency of the evaporator and allows for

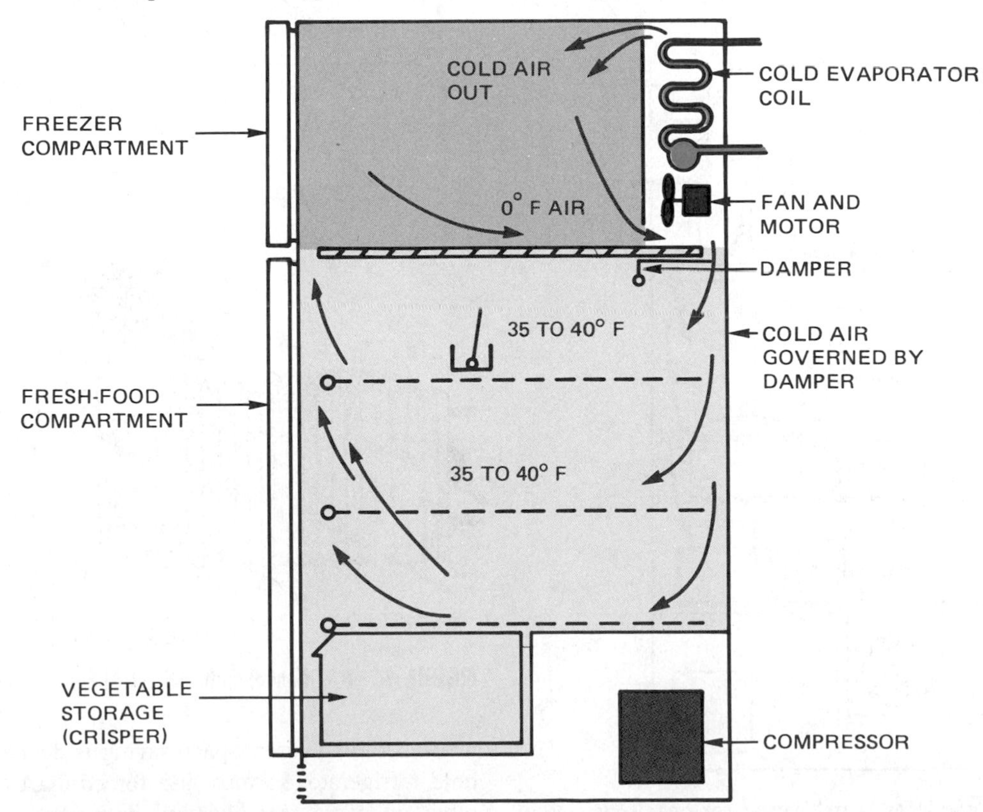

FIGURE 44–4 Cold air enters the box from the coil.

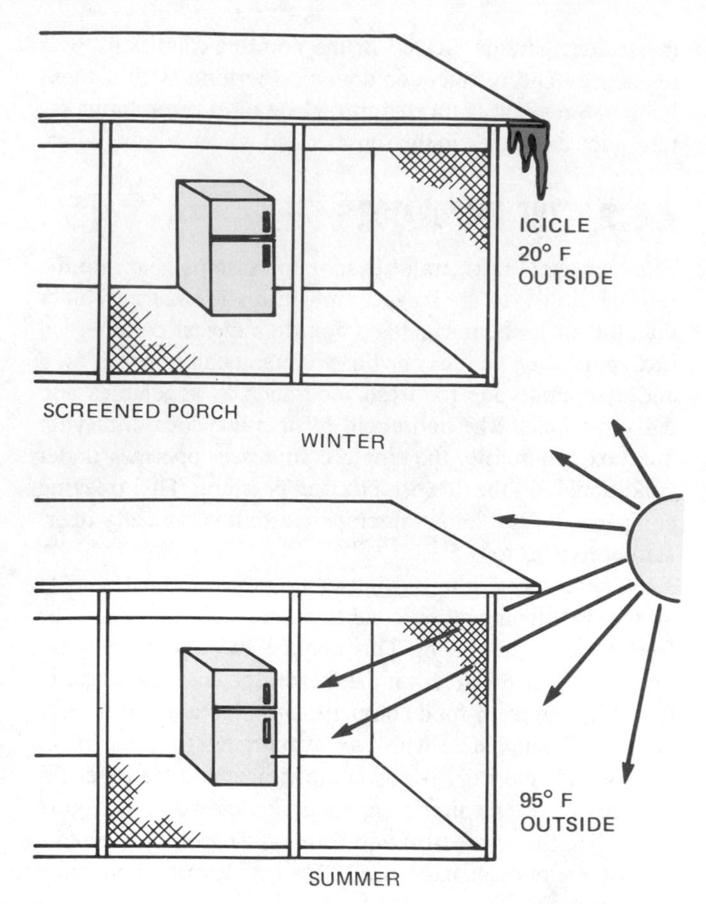

FIGURE 44–5 The ambient temperature for these refrigerators is not within their proper operating range.

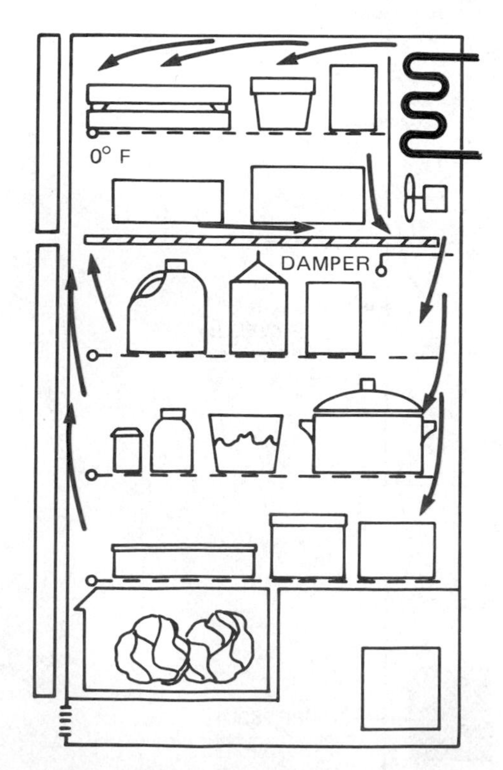

FIGURE 44–6 Air flows inside the refrigerated box from the low-temperature compartment to the medium-temperature compartment.

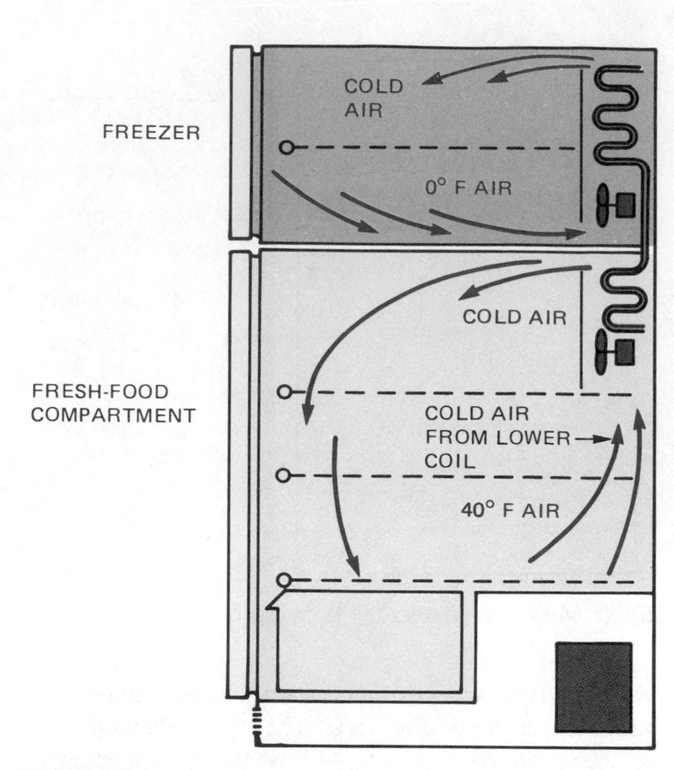

FIGURE 44–7 A two-evaporator box.

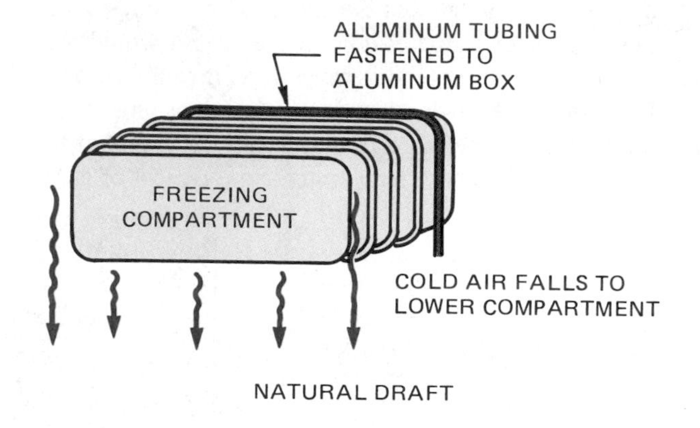

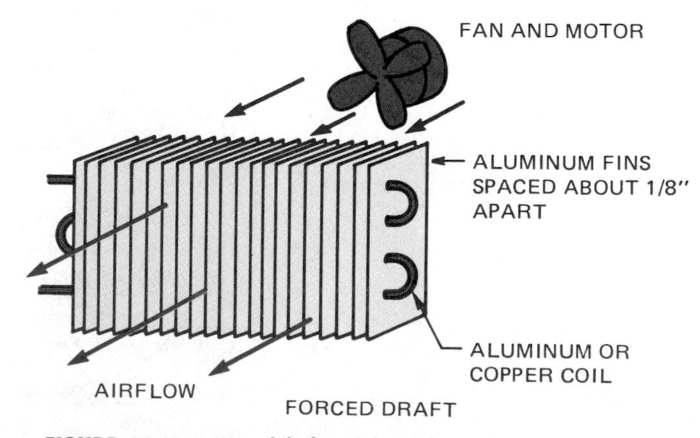

FIGURE 44–8 Natural-draft and forced-draft evaporators.

a smaller evaporator. Space saving is desirable in a household refrigerator so most use forced-draft coils. However, other units are manufactured with natural-draft coils for economy and simplicity.

44.3 NATURAL-DRAFT EVAPORATORS

These evaporators are normally the flat plate type with the refrigerant passages stamped into the plate, Figure 44–9. They are effective from a heat transfer standpoint and require natural air currents to be able to flow freely over them. The food in the frozen-food compartment may be in direct contact with the flat plate evaporator. Air from the bottom and sides may flow to the fresh-food compartment, Figure 44–10.

These natural-draft evaporators are more visible than the forced-draft evaporators and are subject to physical abuse. Models use either automatic defrost systems or manual defrost. A manual defrost system requires that the unit be shut off and the door to the compartment normally left open to accomplish the defrost. Frost is melted by room temperature, Figure 44–11. In a few instances, owners have become impatient and have used sharp objects to remove the ice. This may puncture the evaporator. ***Sharp objects should never be used around the evaporator, Figure 44–12.*** Defrost may be more quickly accomplished with a small amount of external heat, such as a hair drier or a small fan that blows room air into the box until the ice is melted. A pan of warm water may be placed under the coil. The melted ice normally drips into a pan below the evaporator, Figure 44–13.

Fan/coil-type (finned) evaporators are used to reduce the space the evaporator normally uses. The smaller the evaporator, the more internal space is available for food. The evaporator fan and coils are normally recessed in the cabinet and not exposed, Figure 44–14. Because the coils and fans are recessed, air ducts may provide the airflow direction and dampers may help control the volume of the air to the various compartments. ***Each manufacturer of refrigerators has its own method of locating the evap-**

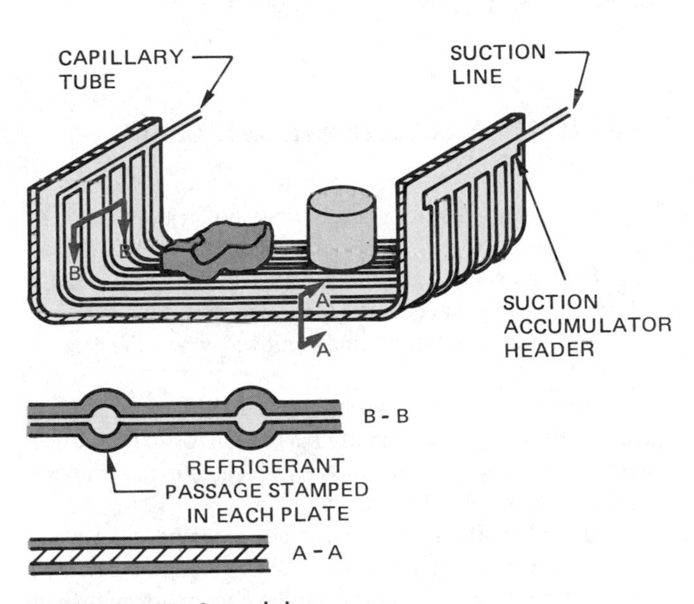

FIGURE 44–9 Stamped-plate evaporator.

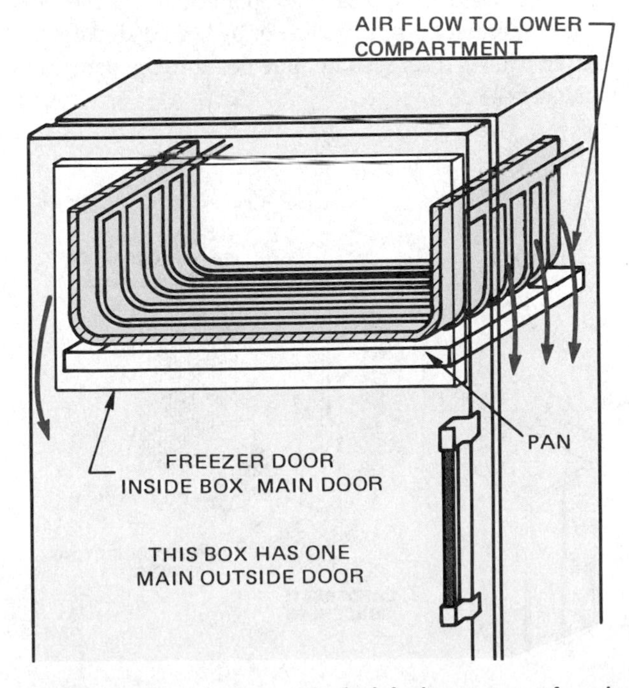

FIGURE 44–10 Air flows to the fresh-food compartment from the flat plate evaporator.

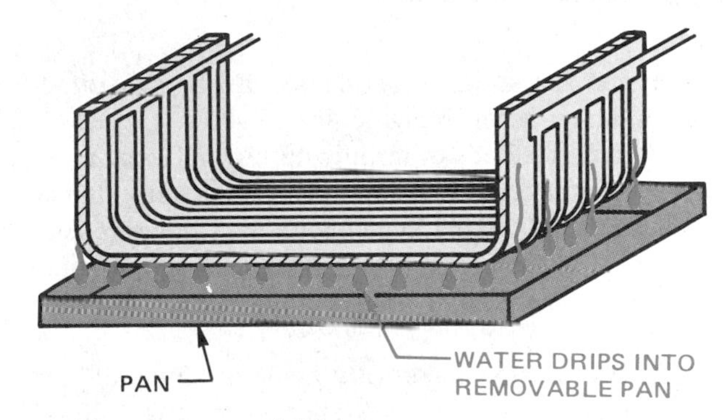

FIGURE 44–11 Manual defrost.

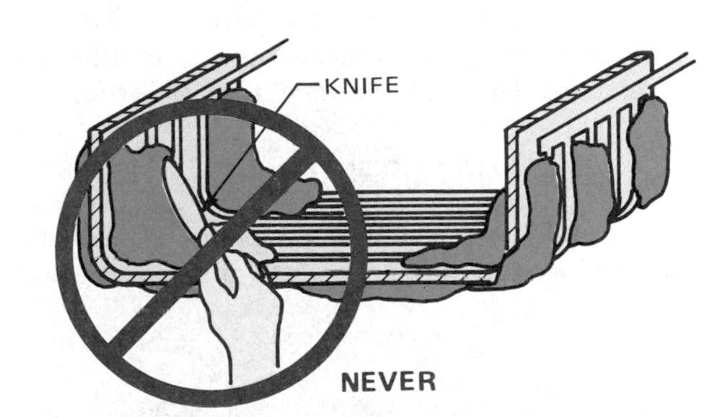

FIGURE 44–12 *Do not use sharp objects to chip frost and ice.*

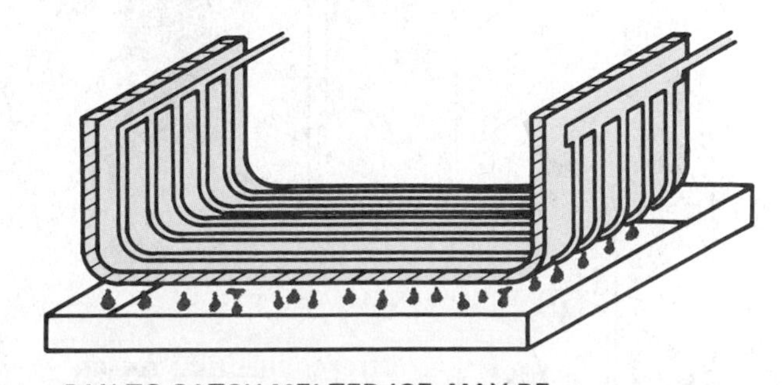

FIGURE 44–13 Melted ice (condensate) is caught in pan.

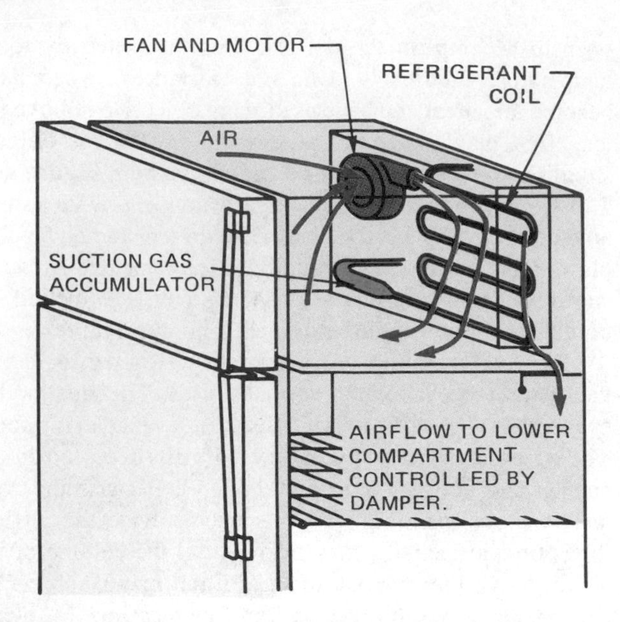

FIGURE 44–14 Forced-draft evaporator.

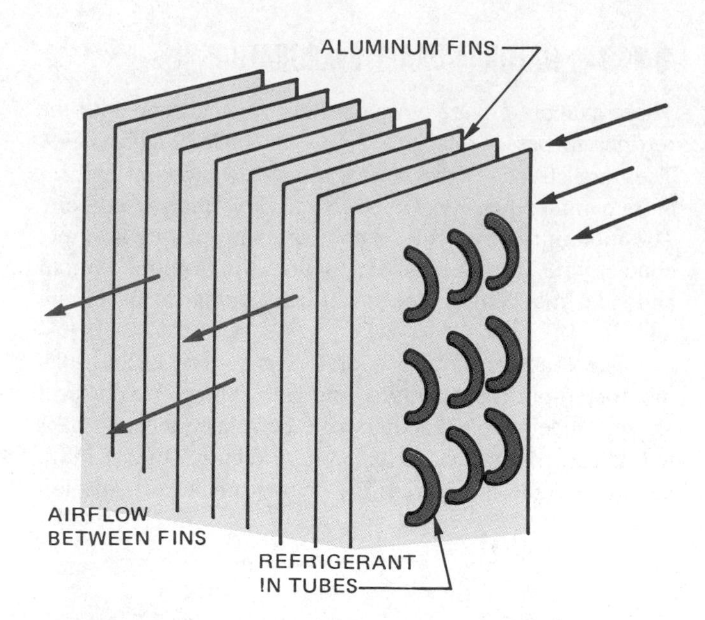

FIGURE 44–16 Forced-draft evaporator and fin spacing.

orator and fan so its literature should be consulted for specific information.* Figure 44–15 shows typical examples of some methods of manufacturing and locating evaporators. Most evaporators have an accumulator at the outlet of the evaporator. The accumulator allows the evaporator to operate as full as possible with liquid refrigerant and still protect the compressor by allowing liquid to collect and boil to a vapor.

The evaporator is normally made of aluminum tubing that may have fins to give the tubes more surface area. The fins are spaced fairly wide apart to allow for frost to build up and not block the airflow, Figure 44–16. The evaporator does not require regular maintenance because air is recirculated within the refrigerated box and it has no air filters.

44.4 EVAPORATOR DEFROST

Manual defrost is accomplished by turning off the unit, removing the food, and using room heat or a small heater. A large amount of frost may accumulate on an evaporator by the time it is defrosted. The water from this type of defrost must be disposed of manually. Units that require manual defrost normally have coils that food has touched. The shelves should be cleaned and sanitized when the frost is removed.

Automatic defrost is accomplished either with internal heat by the compressor supplying hot gas or external heat supplied by electric heating elements located in the evaporator fins, Figure 44–17.

No matter how the defrost is accomplished, the water from the coil must be dealt with. With automatic defrost a pint of water may be melted from the evaporator with each defrost. This water is typically evaporated using the heat from the compressor discharge line or by heated air from the condenser. This is discussed in more detail in the paragraph on condensers.

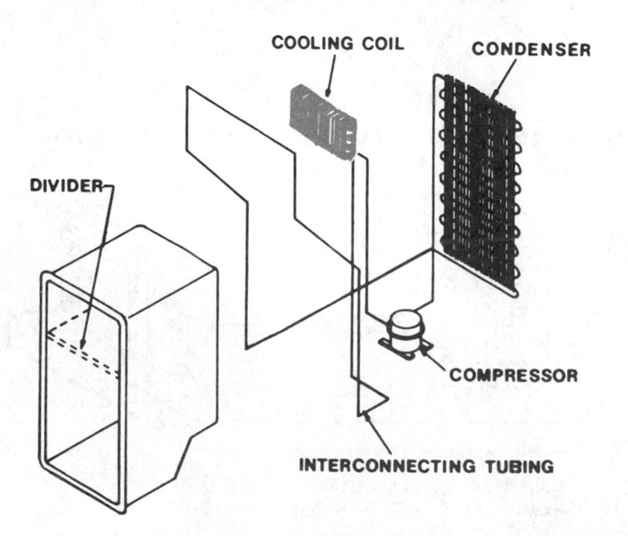

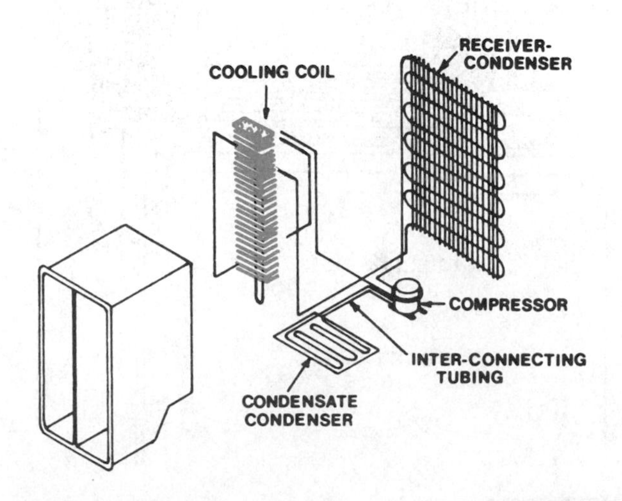

FIGURE 44–15 Typical evaporator locations in a refrigerator. *Courtesy White Consolidated Industries, Inc.*

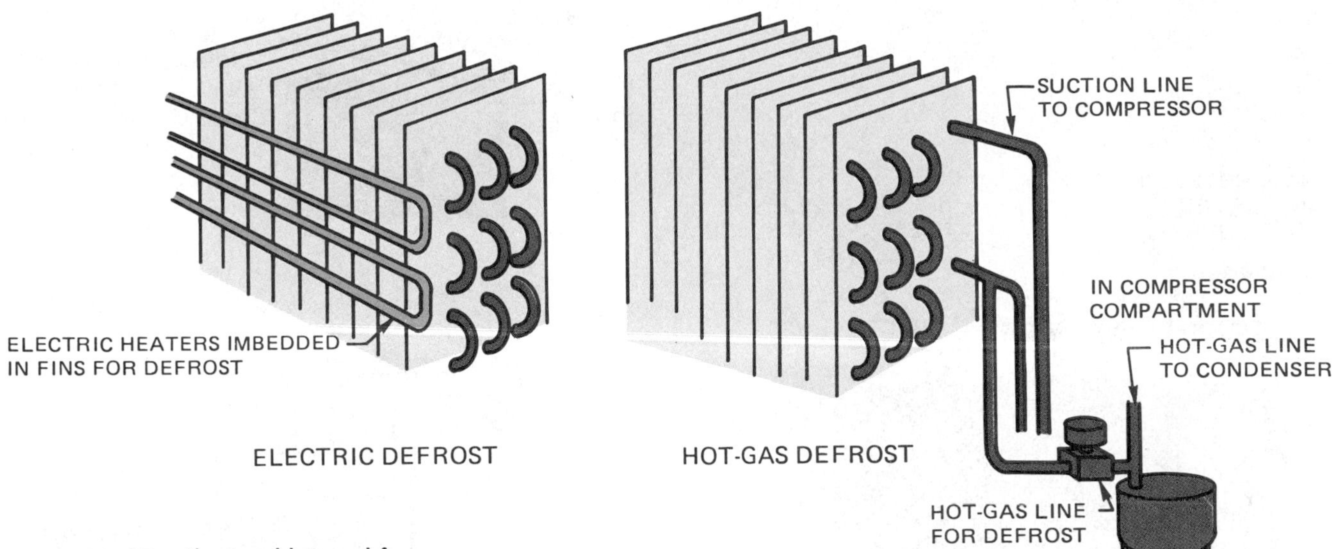

FIGURE 44–17 Electric and hot gas defrost.

44.5 THE COMPRESSOR

The compressor circulates the heat-laden refrigerant by removing it from the evaporator at a low pressure and pumping it into the condenser as a superheated vapor at a higher pressure. The compressors used in domestic refrigerators are very small in comparison to the ones used in air conditioning and commercial refrigeration systems. They are all in the fractional horsepower size ranging from about 1/10 horsepower to 1/3 horsepower, depending on the size of the box. In many of these systems it is hard to tell which is the suction and which is the discharge line. They are copper or steel and are usually the same size, 1/4, 5/16 or 3/8 in. outside diameter.

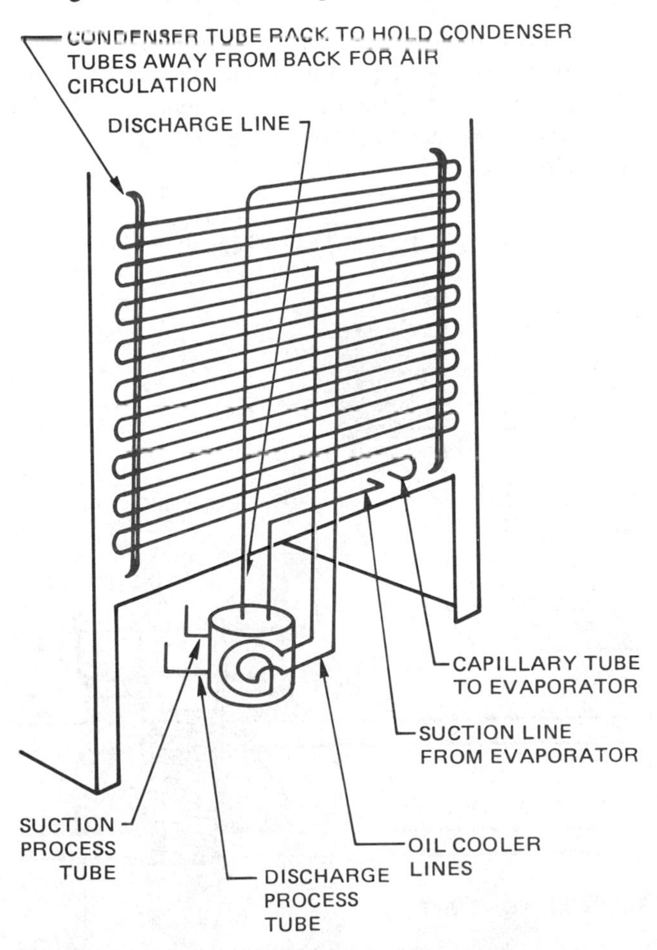

FIGURE 44–18 Illustration of compressor and related piping in refrigerator shell.

Many compressors have a suction line, a discharge line, a process tube, and two oil-cooler lines, all protruding from the shell. A shell diagram is needed to know which line to pipe to which connection when replacing a compressor with one that is not an exact replacement, Figure 44–18.

The compressors used in household refrigerators are all welded hermetically sealed types, Figure 44–19. They are positive-displacement compressors and may use either rotary or reciprocating type of pumping action, Figure 44–20. These compressors are reliable and made to last many years. A typical refrigerator may be used continuously for 20 or more years. At the end of this period of its service, it is often moved to secondary duty, traded and resold, or discarded. ***Careful disposal is vital as described later.***

Household refrigerators manufactured for the United States market operate on 120 V, 60 cycle alternating current. A power supply of 220 V, 50 cycle is used in some other

FIGURE 44–19 Welded, hermetically sealed compressor. *Photo by Bill Johnson*

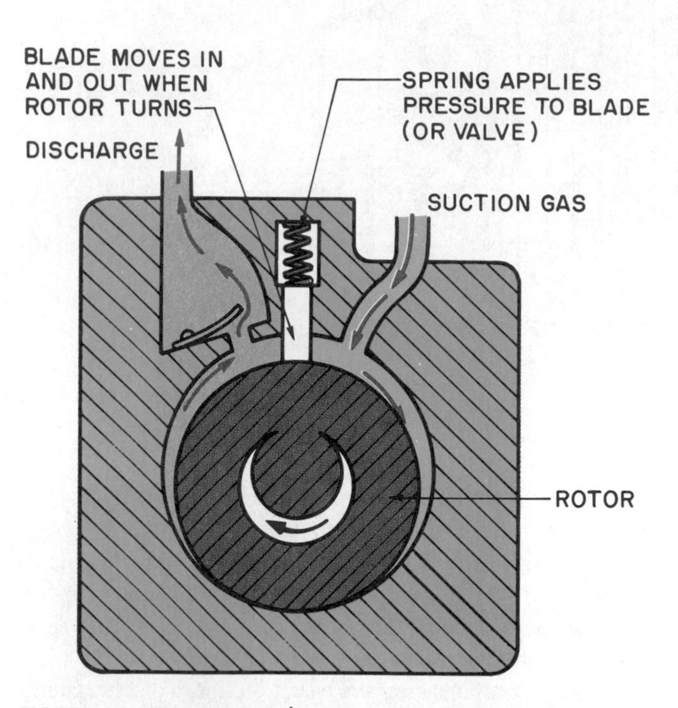

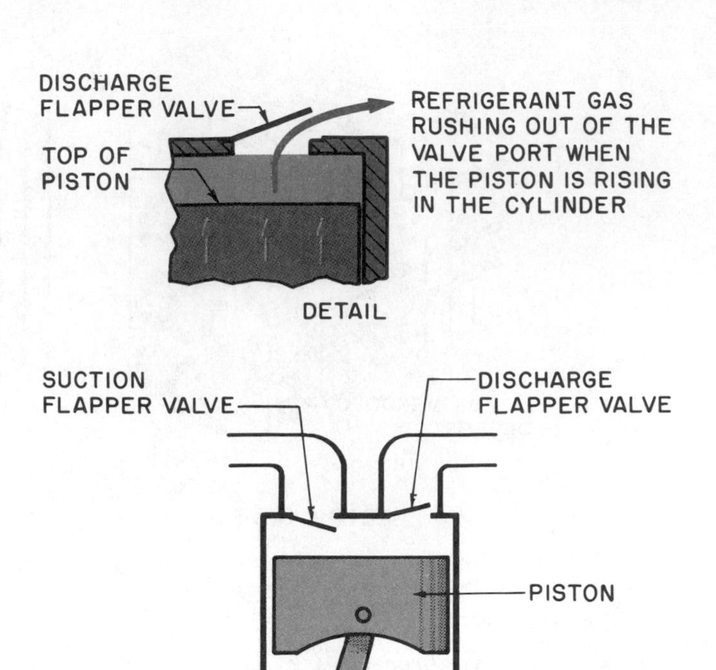

FIGURE 44–20 Rotary and reciprocating compressor action.

countries. American refrigerators may be used in foreign countries with the correct adapters.

The compressor is located at the bottom of the refrigerator and accessed at the back of the box. The refrigerator must be moved away from the wall to service the compressor, Figure 44–21. Most modern, large refrigerators have small wheels to enable you to move the box out for cleaning and to provide easy access for service, Figure 44–22.

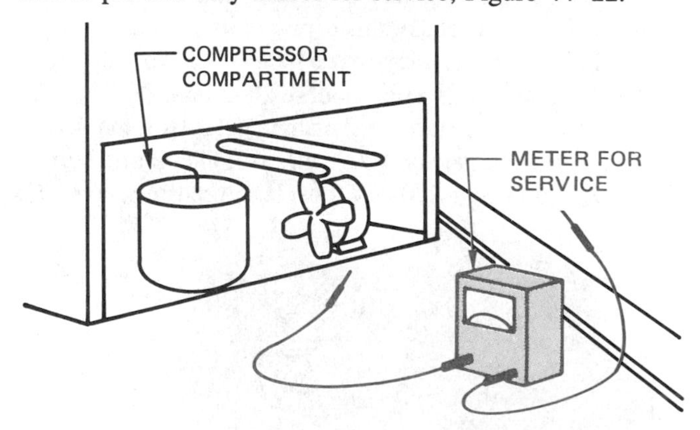

FIGURE 44–21 Refrigerator moved from wall to service compressor.

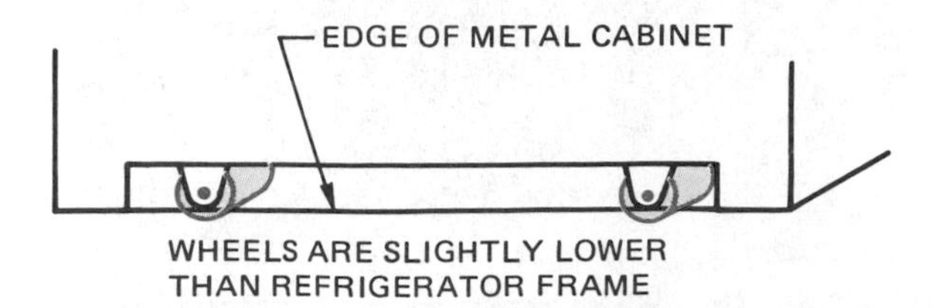

FIGURE 44–22 Wheels on modern refrigerator.

Compressors are typically mounted on internal springs and external flexible, rubberlike feet, Figure 44–23. The refrigerator is located in the living area and must produce a low noise level.

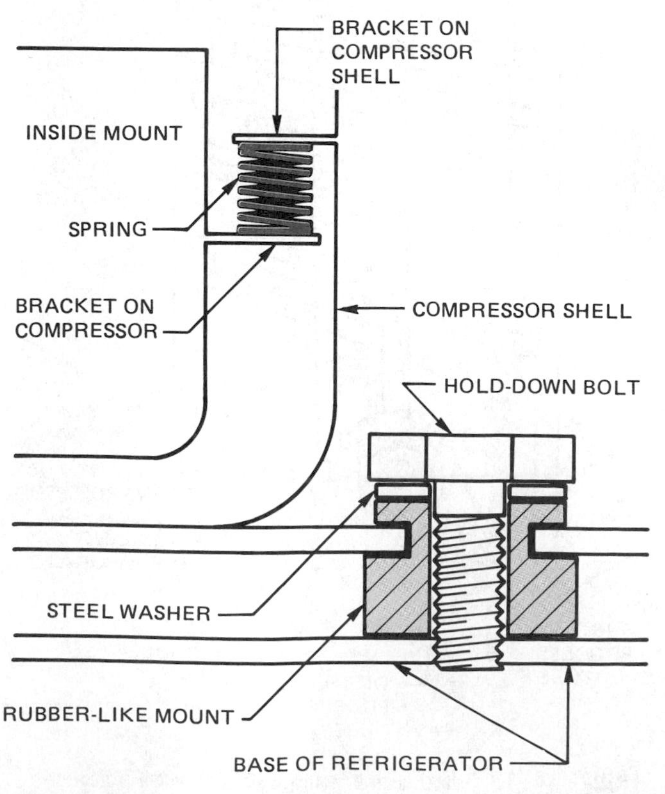

FIGURE 44–23 Compressor mount.

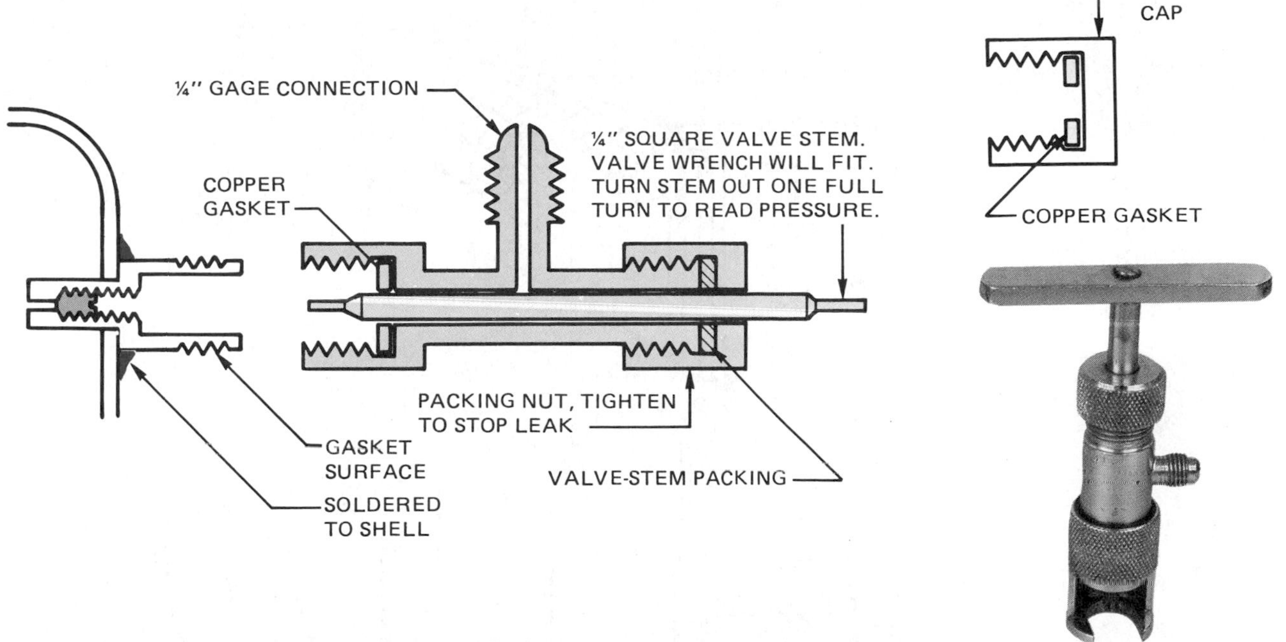

FIGURE 44–24 Special service port valve. *Photo by Bill Johnson*

The lines connecting the compressor to the refrigerant piping may be of copper or steel. Correct solder must be used when repairing any piping connections.

The household refrigerator was the first common application for the welded hermetic compressor. Service requires a skilled technician. Many units do not even have service ports for the service technician to attach gages for compressor suction and discharge readings. When the manufacturer does furnish a service port, it may be in the form of a special port that requires special connectors, Figure 44–24. When the special connectors are not furnished, line tap valves are often used. These should only be used when necessary and the manufacturer's instructions must be strictly followed or the results will be unsatisfactory.

44.6 THE CONDENSER

Condensers for domestic refrigerators are all air cooled. This permits the refrigerator to be moved to a new location or house as needed. The condensers are either cooled by natural convection or by small forced-air fans, Figure 44–25.

Natural convection condensers were the first and are more simple. They may be located on the back of the unit and care must be used when locating the refrigerator to ensure that air can flow freely over it. Many of these units have been abused by locating them under low overhanging cabinets. This causes poor air circulation and high head pressures. The unit may then have extra long running times because the capacity is reduced. In severe cases, the unit

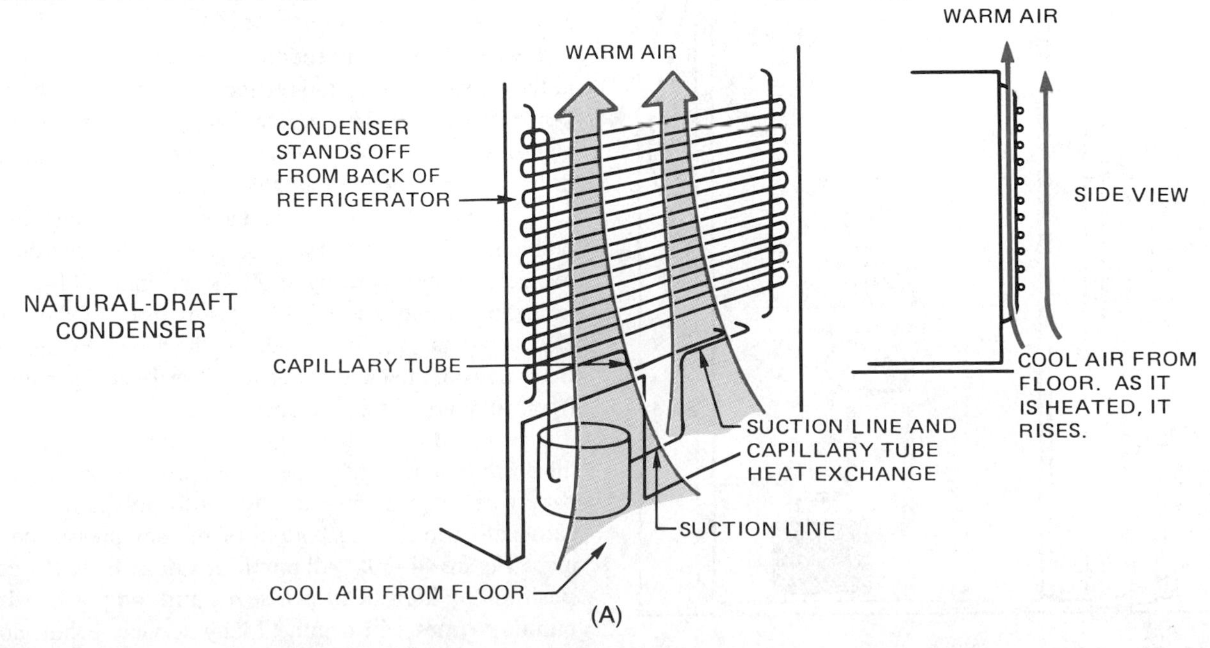

FIGURE 44–25 (A) Natural- and (B) forced-draft condensers.

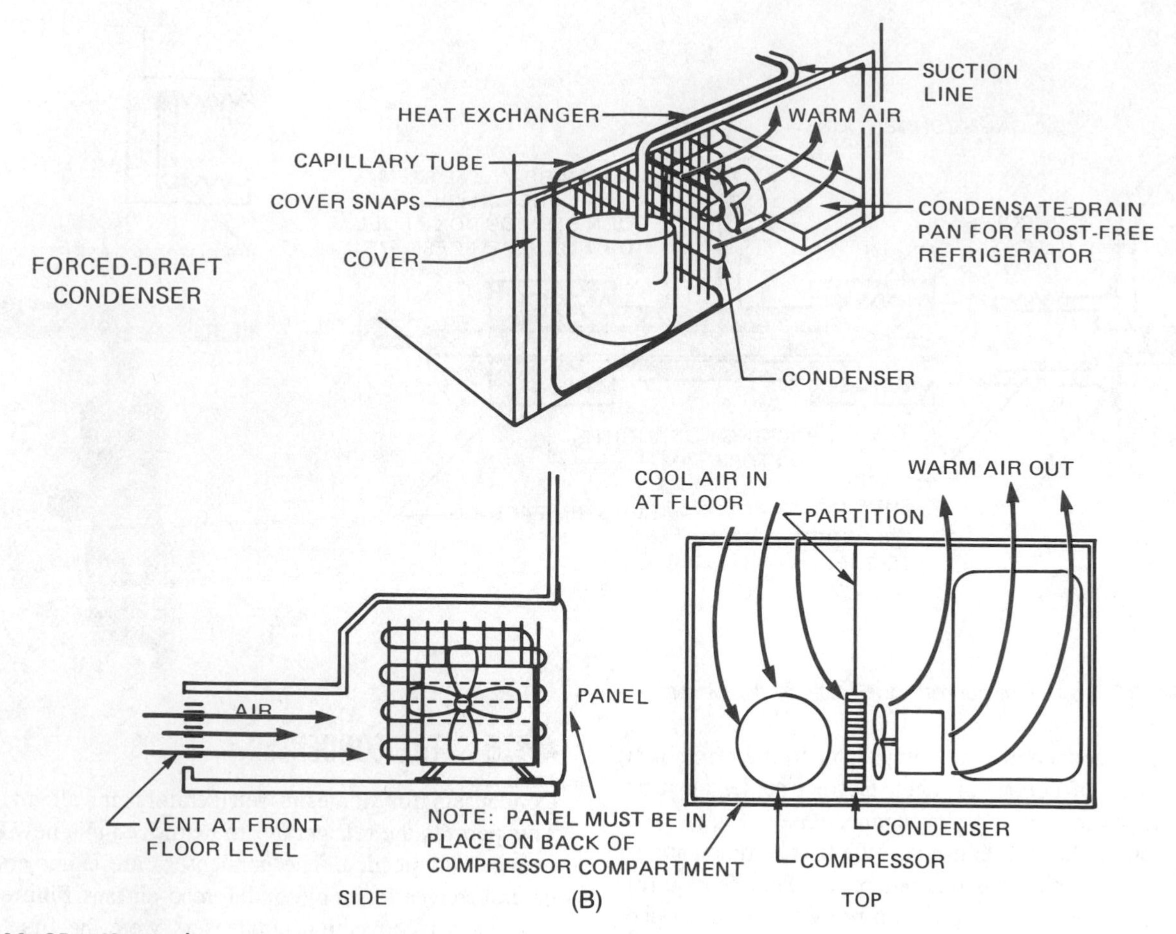

FIGURE 44–25 *(Continued)*

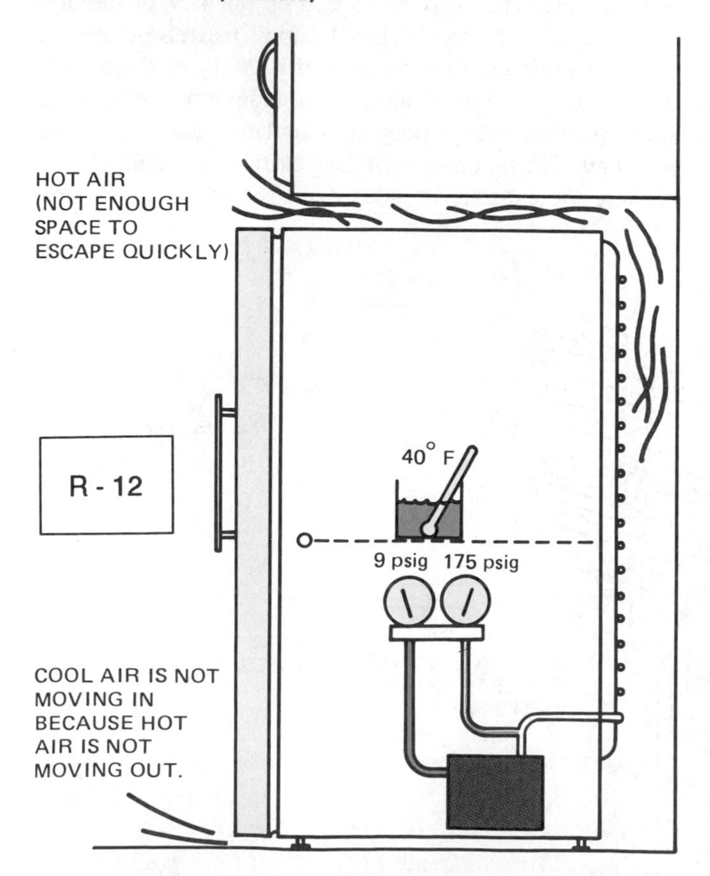

FIGURE 44–26 Unit running all the time due to poor location.

may run all the time and still not keep the food compartment cool, Figure 44–26.

Care must be used when moving any appliance, but the external condenser type of refrigerator requires special care or the condenser may be damaged. If hand trucks are used, care must be used by placing the belt under the condenser and around the unit, Figure 44–27.

Some natural convection condensers have been located in the outside wall of the refrigerator by fastening the condenser tubes to the inside of the outer sheet metal shell. This unit may work well in the open, where air can circulate over it, but will not work in an alcove.

Forced-draft condensers have solved many of these problems. The forced-draft condenser is located under the refrigerator and typically at the back, Figure 44–28. Air is taken in on one side of the bottom front of the box and discharged out the other side of the front, Figure 44–29. Because both inlet and outlet are close to the floor and at the front, they are not easily obstructed.

Forced-draft condensers must have an air pattern for the air to be forced over the finned tube condenser. This air pattern is often maintained by cardboard partitions and a cardboard back to the bottom of the compressor compartment, Figure 44–30. ***All partitions must be in the correct position, or high head pressures and long or continuous running times will occur.*** Many service technicians and homeowners have discarded the cardboard back cover thinking

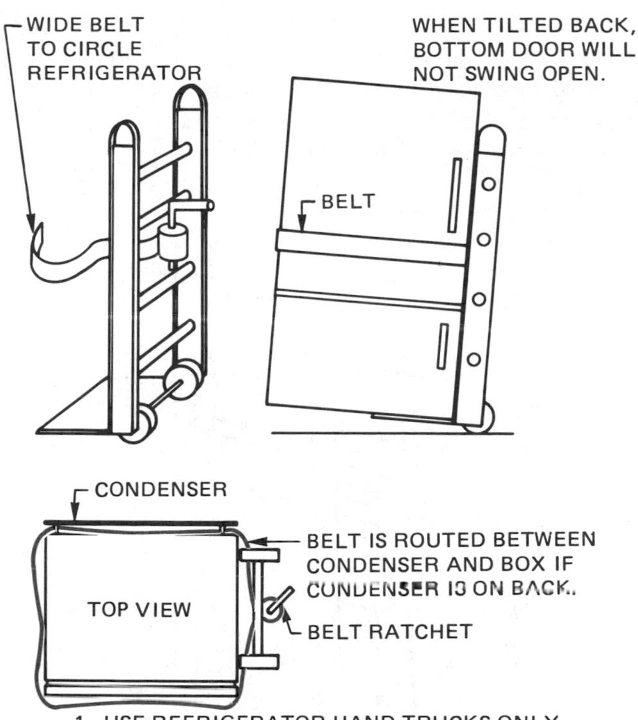

1. USE REFRIGERATOR HAND TRUCKS ONLY.
2. TRUCK REFRIGERATOR FROM SIDE SO IT WILL GO THROUGH DOORS.
3. DO **NOT** WRAP BELT AROUND NATURAL-DRAFT CONDENSER.
4. BE AWARE OF WHICH WAY THE DOOR WILL SWING IF TILTED BACK.

FIGURE 44–27 Care should be used while trucking a refrigerator.

FIGURE 44–28 Location of forced-draft condenser.

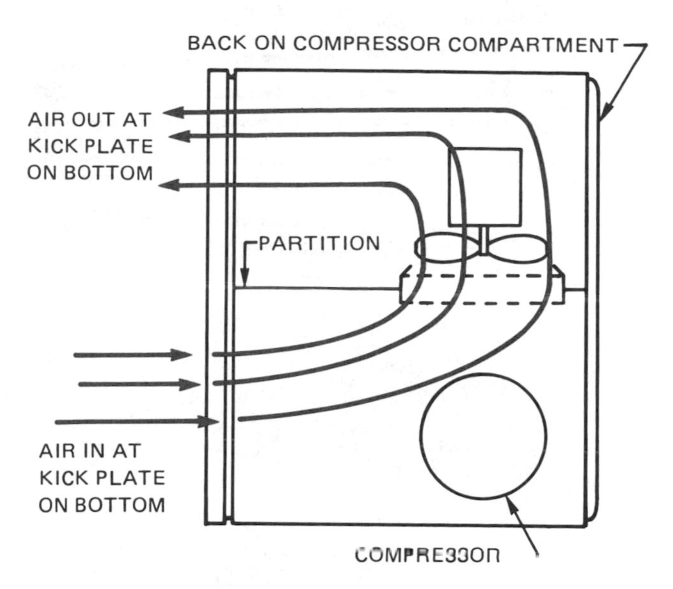

FIGURE 44–29 Airflow of forced-draft condenser.

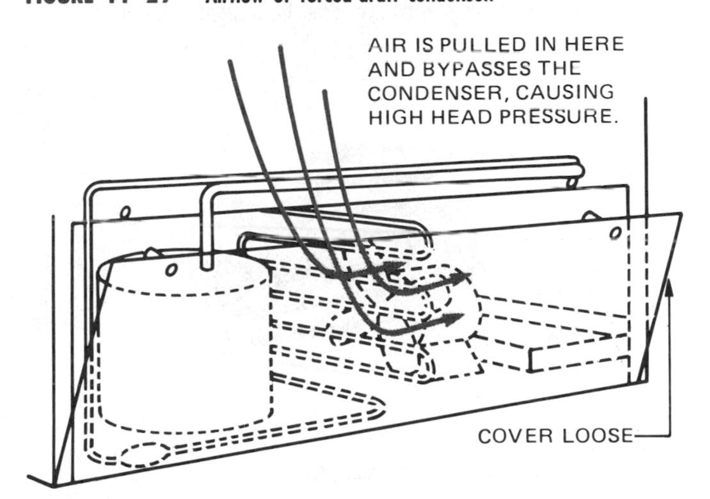

FIGURE 44–30 ***Be sure to replace all cardboard partitions on a forced-draft condenser.***

it not necessary. The refrigerator will not perform correctly without the back cover and permanent damage can result.

44.7 DEFROST CONDENSATE, AUTOMATIC DEFROST

All domestic refrigerators are low-temperature appliances and frost accumulates on the evaporator. When defrost occurs, something must be done with the water. Automatic defrost is unattended. The compressor and condenser section of the refrigerator are used to evaporate this water.

The heat available at the compressor discharge line can be used to evaporate the water when direct contact is made. Many units are designed so that the discharge line is passed through the pan that collects the defrost water, Figure 44–31. This method is used primarily on units with natural-draft condensers.

When the unit has a forced-draft condenser, warm air from the condenser may be forced over a collection pan of

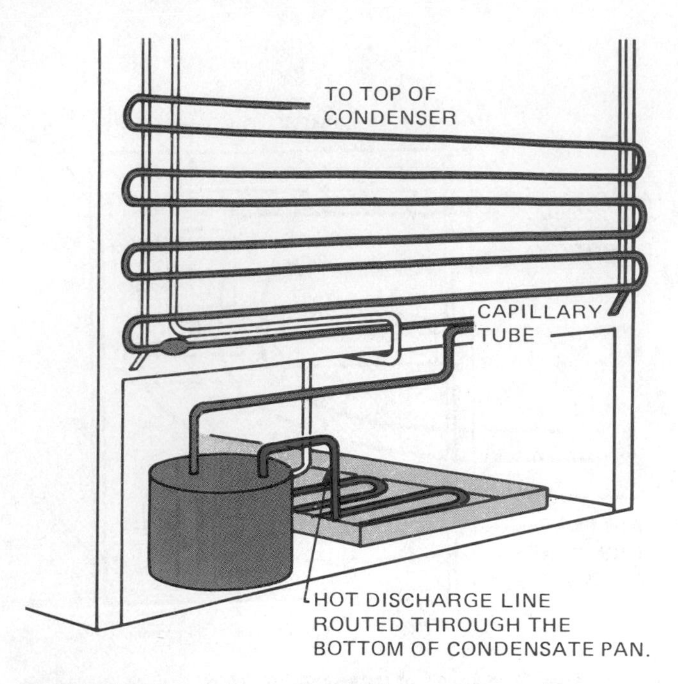

FIGURE 44–31 Compressor heat used to evaporate condensate.

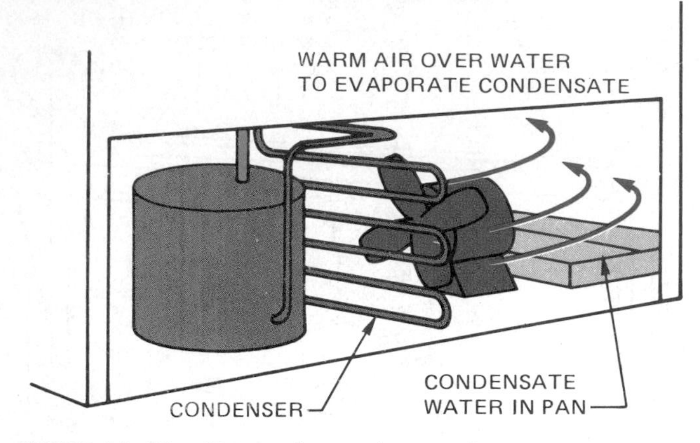

FIGURE 44–32 Warm air from condenser used to evaporate condensate.

water for the purpose of evaporation, Figure 44–32. In either case, the unit has the compressor running time from one defrost cycle to the next to evaporate the water depending on the manufacturer's design. Defrost may occur during the off cycle or as few as two or three times per 24 h.

The defrost water collection pan in the bottom of the unit is a place where lint and dirt may collect. This pan can and should be removed occasionally and cleaned or it may become unsanitary.

44.8 COMPRESSOR OIL COOLERS

The condenser may also contain an oil cooler for the compressor. This oil cooler keeps the crankcase oil at a lower temperature. This also cools the compressor assembly. It may be accomplished by routing the compressor discharge line through the condenser and removing some of the heat, then routing it back through the compressor crankcase in a closed loop to pick up heat from the oil, Figure 44–33. There are extra lines on the compressor for the oil loop.

Another method for cooling oil may be a gravity loop that allows the oil to leave the crankcase and circulate to a point where it may be cooled, Figure 44–34.

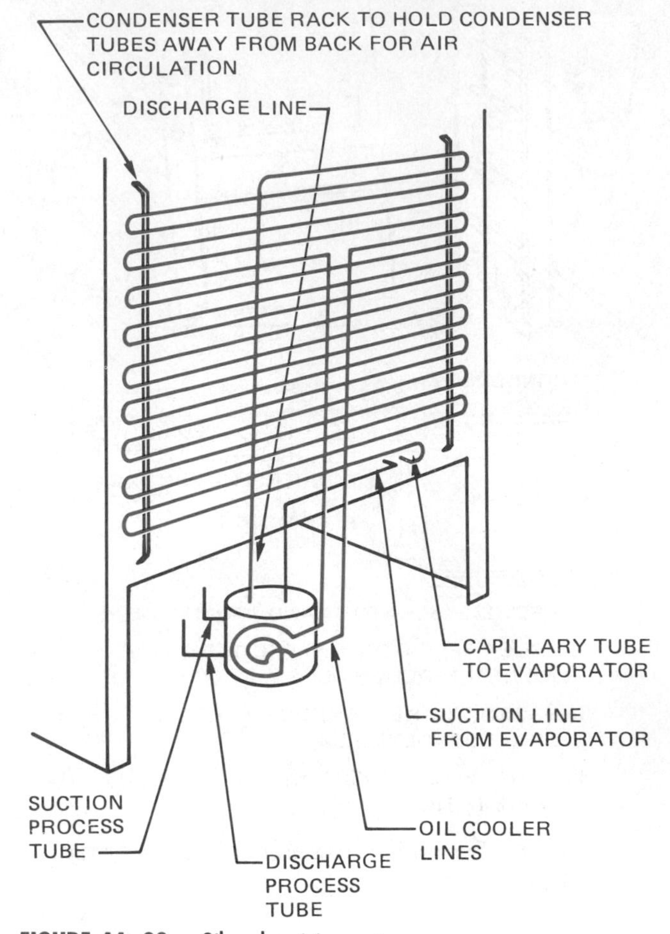

FIGURE 44–33 Oil cooler piping route.

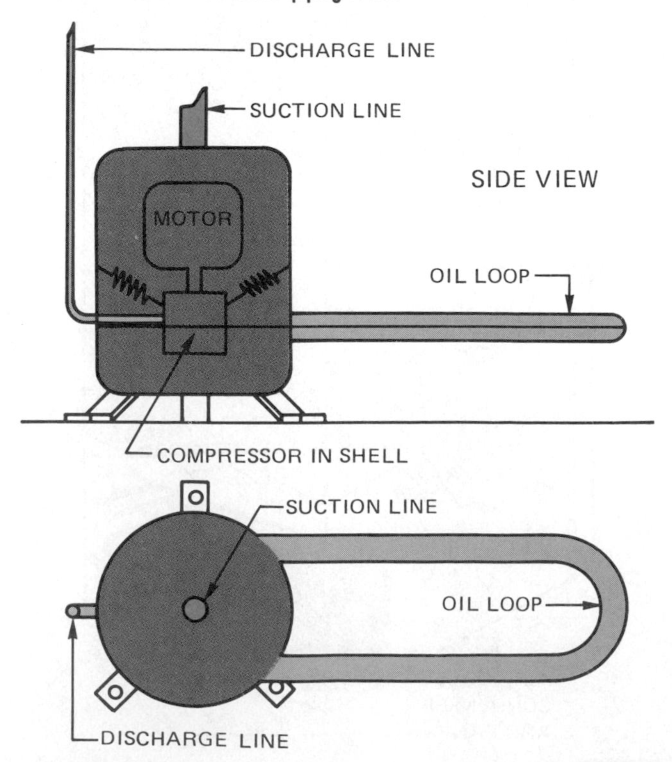

FIGURE 44–34 Loop to cool oil.

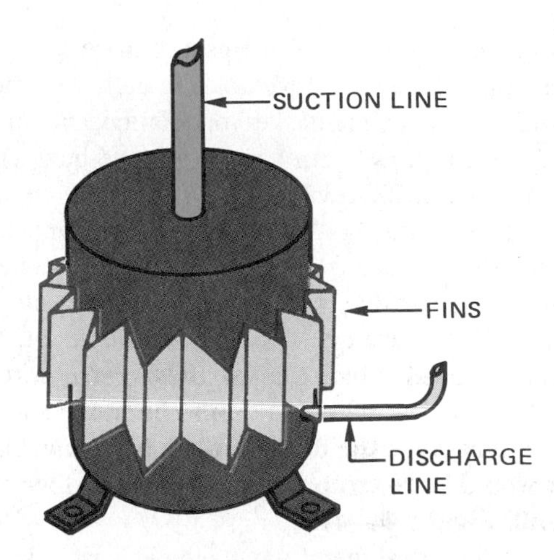

FIGURE 44–35 Some compressors have fins to cool them.

Some compressors have fins on the crankcase to accomplish oil cooling, Figure 44–35.

44.9 METERING DEVICE

Most domestic refrigerators use the capillary tube metering device. This is a fixed-bore metering device and the amount of refrigerant flow through the device is determined by the bore of the tube and the length of the tube. This is predetermined by the manufacturer. The capillary tube is usually fastened to the suction line for a heat exchange, Figure 44–36. In some cases the capillary tube may be run inside the suction line for this heat exchange, Figure 44–37. The heat exchange prevents liquid from returning to the compressor and improves the evaporator capacity by subcooling the liquid in the first part of the capillary tube, Figure 44–38.

The capillary tube may be serviced if necessary and may in some cases be changed for another one, but this is difficult and usually not practical.

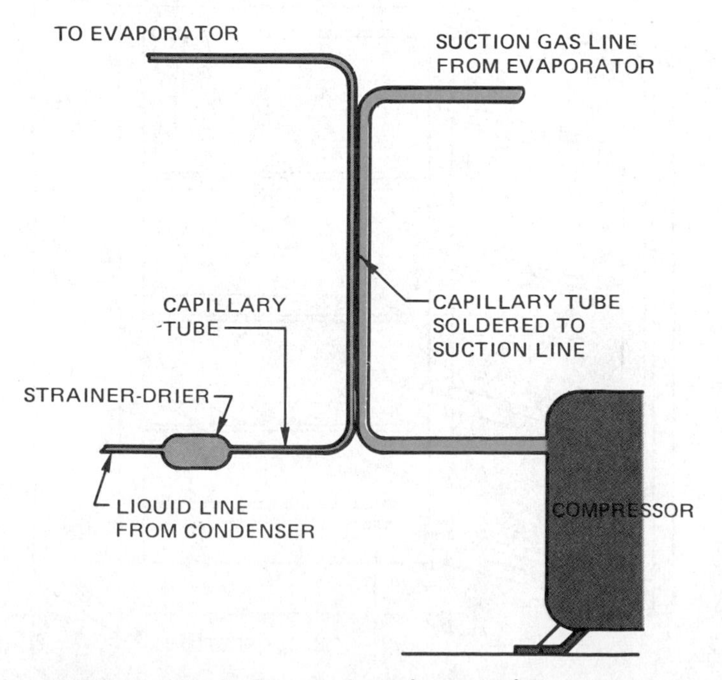

FIGURE 44–36 Capillary tube fastened to suction line.

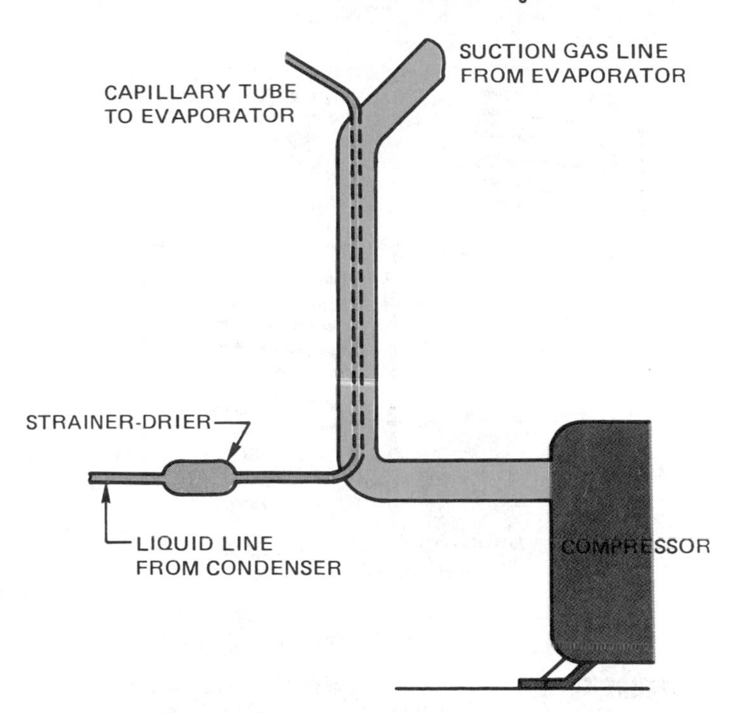

FIGURE 44–37 Capillary tube routed through the suction line.

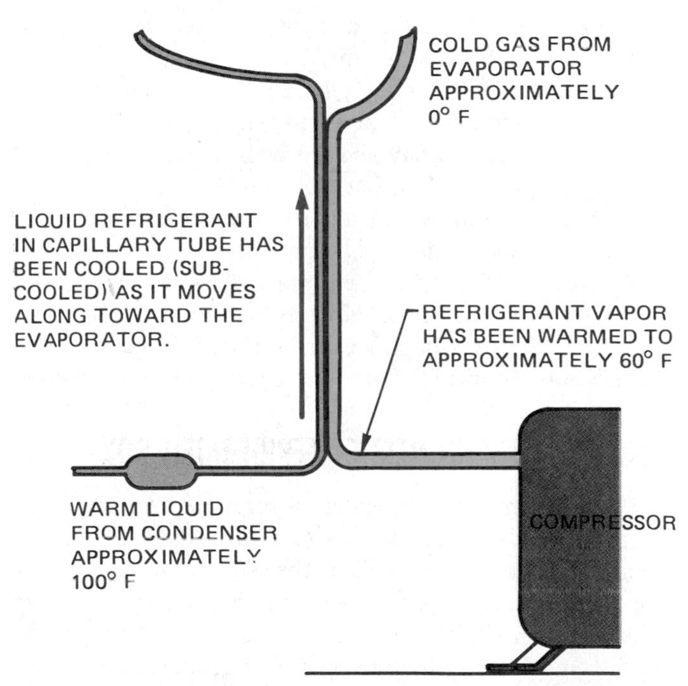

FIGURE 44–38 Heat exchange between capillary tube and suction line.

It was mentioned earlier that a domestic refrigerator would perform correctly when located in the living space. The capillary tube metering device is one reason. This device is sized to pass a certain amount of liquid refrigerant at typical living conditions. This may be considered between 65°F and 95°F. If the room ambient temperature rises above the recommended, the head pressure will climb and push more refrigerant through the capillary tube, causing the suction pressure to rise. Capacity will suffer to some extent, Figure 44–39.

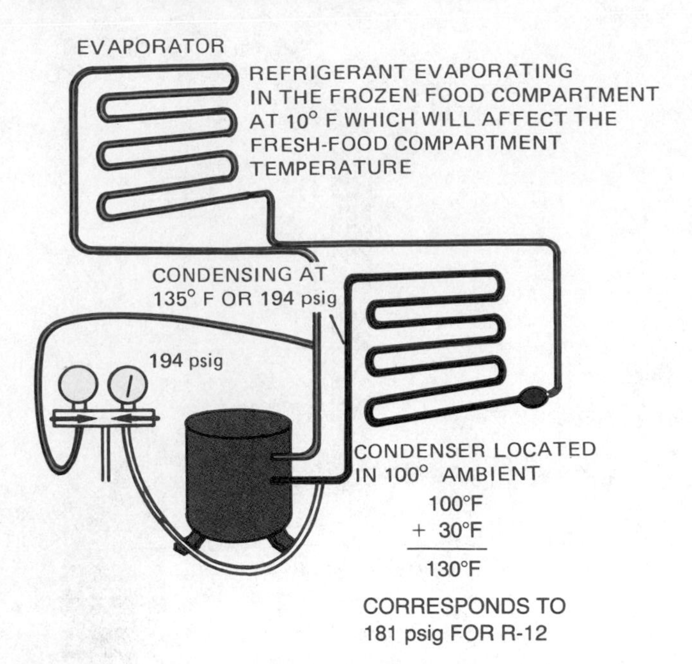

FIGURE 44–39 Capacity of refrigerator is reduced by high ambient temperature.

Most residences will not exceed the extreme temperatures for long periods of time. For example, a house temperature may rise to 100°F in the daytime if it does not have air conditioning but the house will cool at night. The refrigerator may work all day and not cool down to the correct setting, but may cycle off at night.

Domestic refrigerators and freezers are not intended to be placed outdoors or in buildings where the temperatures vary from the 65°F to 95°F temperature unless the manufacturer's literature states that they can be operated at other temperatures. Poor performance and shorter life span may be expected if the manufacturer's directions are not followed.

44.10 THE DOMESTIC REFRIGERATED BOX

The first domestic refrigerated boxes were constructed of wood. In fact, some of the earlier ice boxes could be adapted to mechanical refrigeration with the addition of the evaporator, compressor, condenser, and metering device. Later, the boxes were made of metal on a wood frame. They had an open compressor and were very heavy. Just after the turn of the century, domestic refrigeration became popular and the manufacturers began active campaigns to design more efficient systems. After World War II, foam insulation was developed and manufacturers had developed the welded hermetic compressor. Foam insulation is lighter and more efficient, Figure 44–40.

Box design has been used as a selling tool because of the possibilities of attractive design while the manufacturers have worked to make the box durable and long lasting. Many different colors have been developed. Color-coordinated kitchens are common with all major appliances the same color.

The first designs were one door to the outside and a freezer compartment door on the inside, Figure 44–41. These doors had gaskets with air pockets inside them that compressed when the door was closed, Figure 44–42. The compressibility of the gasket would eventually become ineffective with fatigue as the gasket material became old. The door would then allow room air to enter the box. This increased the load on the evaporator and caused more frost accumulation, Figure 44–43. The doors had mechanical door latches that could only be opened from the outside. If a child were to get inside with the door closed, all of the oxygen could be consumed and death could result. ***Caution must be used when one of these refrigerators is taken out of service. The door must be made safe. One method is to remove the door or door latch mechanism. Another would be to strap the door shut and then turn it to the wall, Figure 44–44.***

The modern design has a magnetic strip gasket all around the door or doors, which also has a compression-

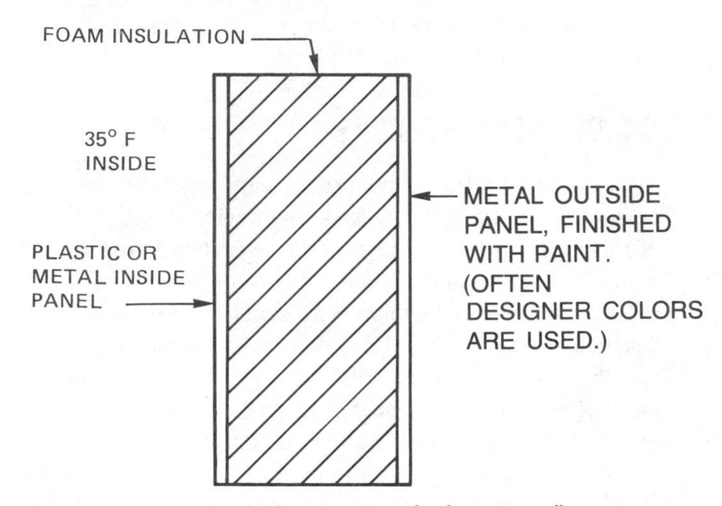

FIGURE 44–40 Sandwich construction of refrigerator wall.

FIGURE 44–41 One-door refrigerator.

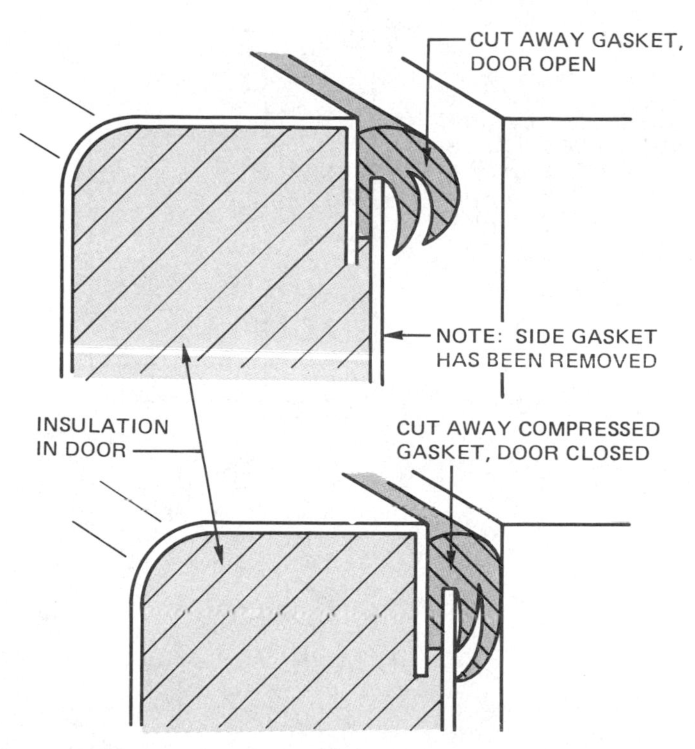

FIGURE 44–42 Early door gasket construction.

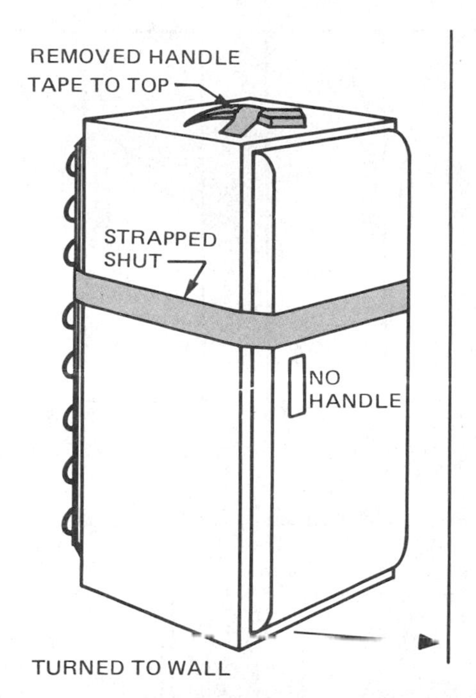

FIGURE 44–44 *Remove handle, turn to wall and strap shut old refrigerators for maximum safety.*

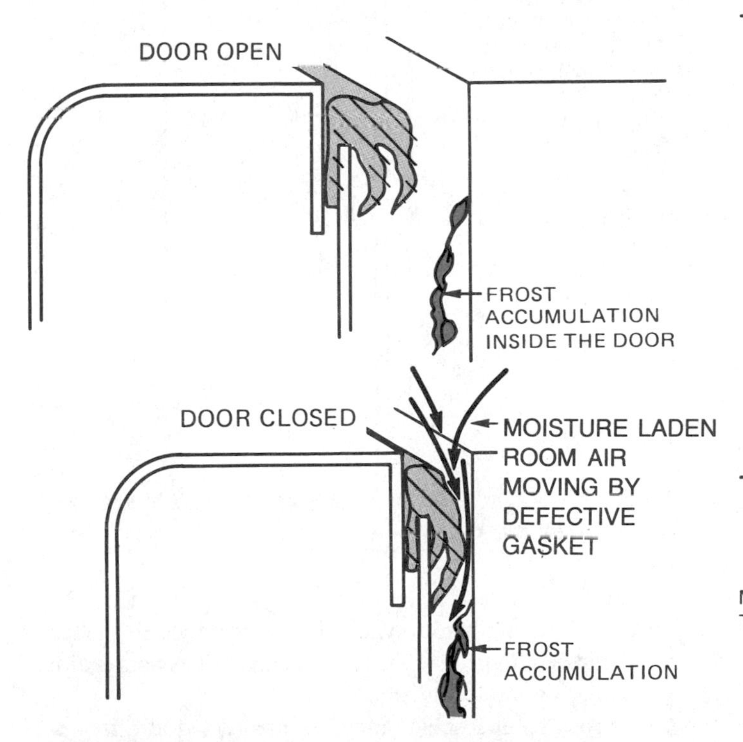

FIGURE 44–43 Older door gasket that is worn out causing frost accumulation.

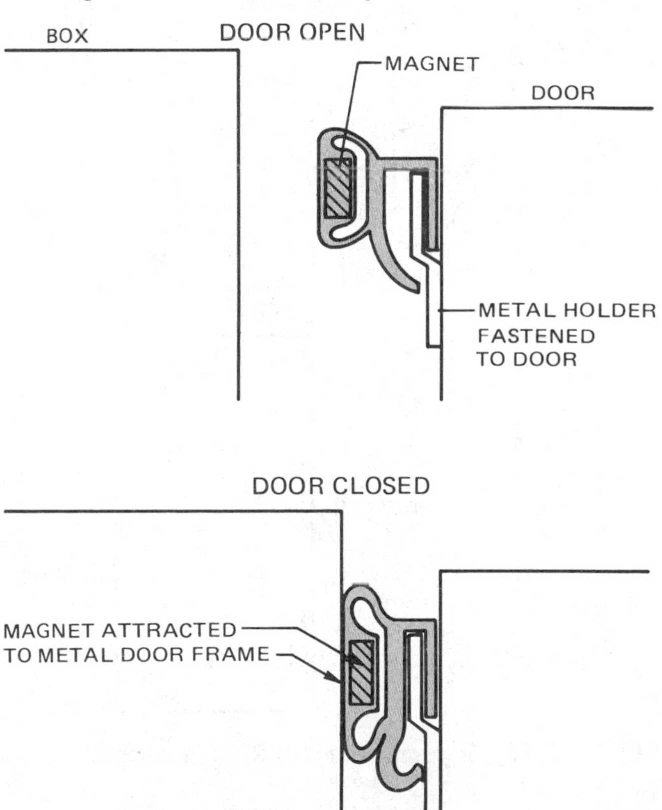

FIGURE 44–45 Modern refrigerator gasket with magnet.

type seal that maintains a good seal to keep air out, Figure 44–45. These doors can be opened from the inside. The magnetic gasket is also much easier to replace than the old type of gasket.

The typical box has two outside doors. The door combinations may be side by side or over and under. In the side-by-side styles the freezer may be either on the left or right, depending on the customer's needs, Figure 44–46. In some over-and-under and single-door styles, the door opens to the left or to the right for customer convenience, Figure 44–47.

The box must be airtight to prevent excess load and frost. If the door is shut and the air inside the box cools and shrinks, the door may be hard to open immediately after closing, Figure 44–48. A relief port may be used to allow the air to equalize when there is a pressure difference.

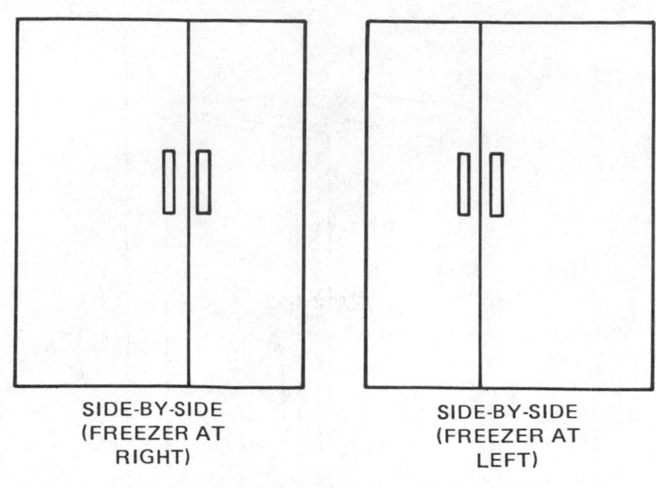
SIDE-BY-SIDE
(FREEZER AT
RIGHT)
SIDE-BY-SIDE
(FREEZER AT
LEFT)

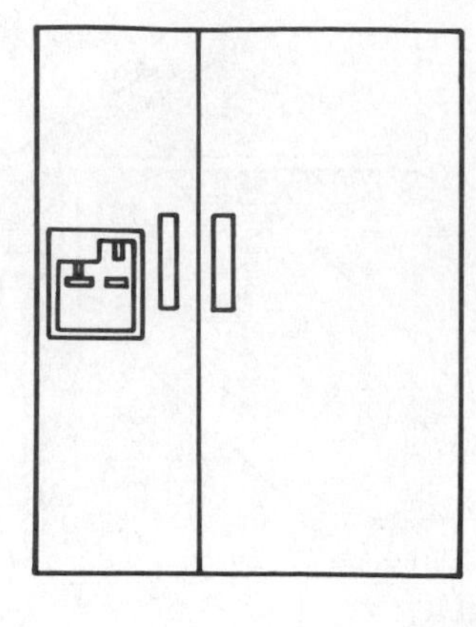
SIDE-BY-SIDE
(FREEZER AT LEFT.
NOTE ICE AND WATER
DISPENSER IN
FREEZER DOOR)

FIGURE 44–46 Different side-by-side door arrangements.

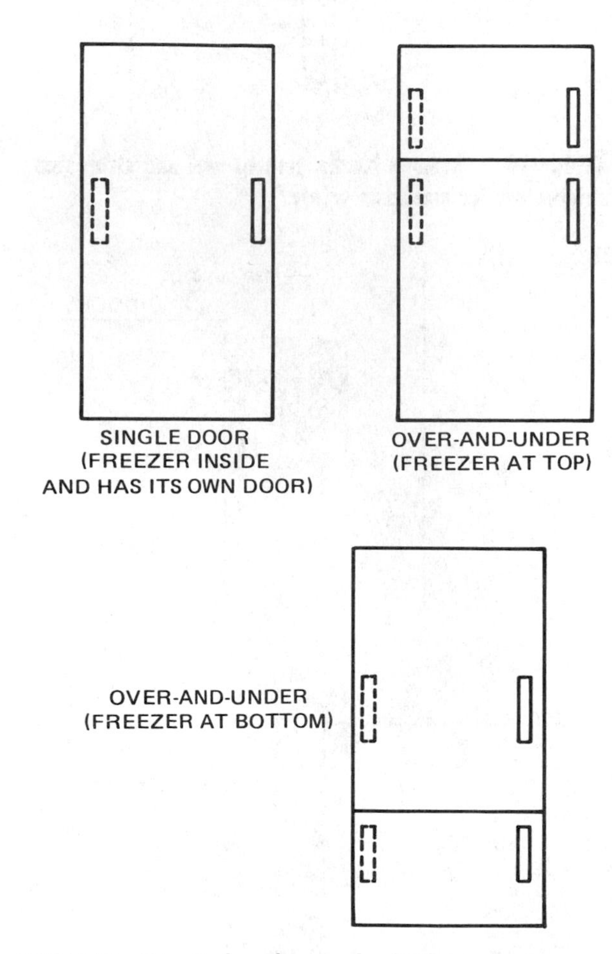
SINGLE DOOR
(FREEZER INSIDE
AND HAS ITS OWN DOOR)
OVER-AND-UNDER
(FREEZER AT TOP)
OVER-AND-UNDER
(FREEZER AT BOTTOM)

FIGURE 44–47 Single and over-and-under door arrangements.

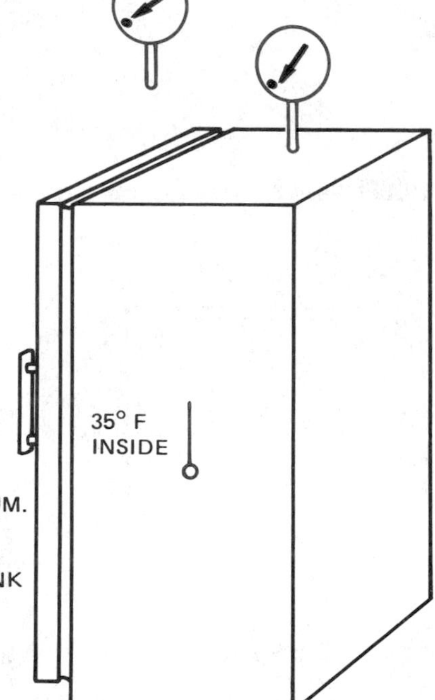
IF A HIGH QUALITY
VACUUM GAGE WERE
PLACED IN THE ROOM
IT WOULD READ 0 IN.
Hg VACUUM.
IF THEN PLACED TO
MEASURE PRESSURE
INSIDE A REFRIGERA-
TED BOX THAT HAD
BEEN OPENED THEN
CLOSED, IT WOULD
READ A SLIGHT VACUUM.
THIS IS BECAUSE THE
WARM AIR THAT EN-
TERED THE BOX SHRANK
WHEN COOLED TO BOX
TEMPERATURE.
35° F
INSIDE

FIGURE 44–48 Cooled air inside the box may make it hard to open door immediately after closing.

The condensate must be allowed to travel from the inside of the box to underneath the box to the collecting pan where it is evaporated. A trap arrangement at the bottom prevents air from traveling into the box through this opening, Figure 44–49. If a refrigerator is located outside in the winter for storage or use, the water in the drain trap may freeze, Figure 44–50.

The refrigerator may have provision for an ice maker in the frozen-food compartment. This is accomplished by furnishing a place for water to be piped to the frozen-food compartment and a bracket to which the ice maker can be fastened. A wiring harness may also be provided so the ice maker can be plugged in. This arrangement does not require any wiring by the technician.

Various compartments may be maintained at different temperatures, such as the crisper for fresh vegetables and the butter warmer. The crisper is usually maintained by enclosing it in a drawer to keep the temperature slightly higher and prevent dehydration of the food, Figure 44–51. The butter warmer may have a small heater in a closed compartment to keep the butter at a slightly higher temperature than the space temperature so it will spread more easily, Figure 44–52.

The inside surface of the modern refrigerator is usually made of plastic. This surface is easy to keep clean and will last for years if not abused. The plastic in the freezing

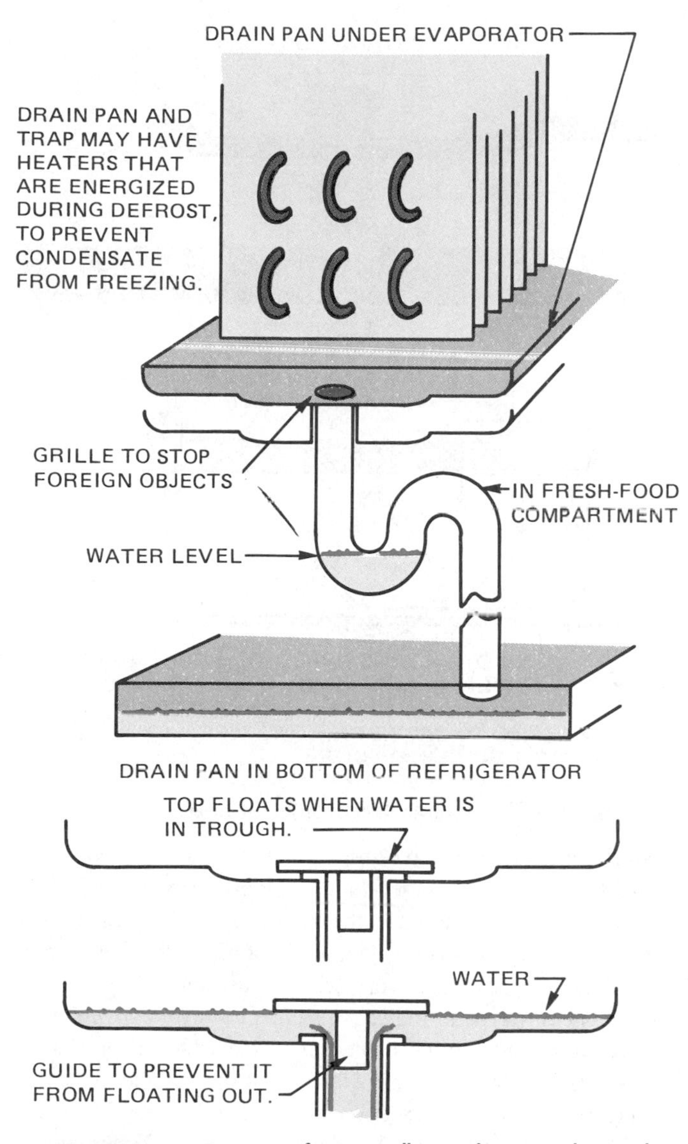

FIGURE 44–49 Two types of traps to allow condensate to drain and to prevent air from entering.

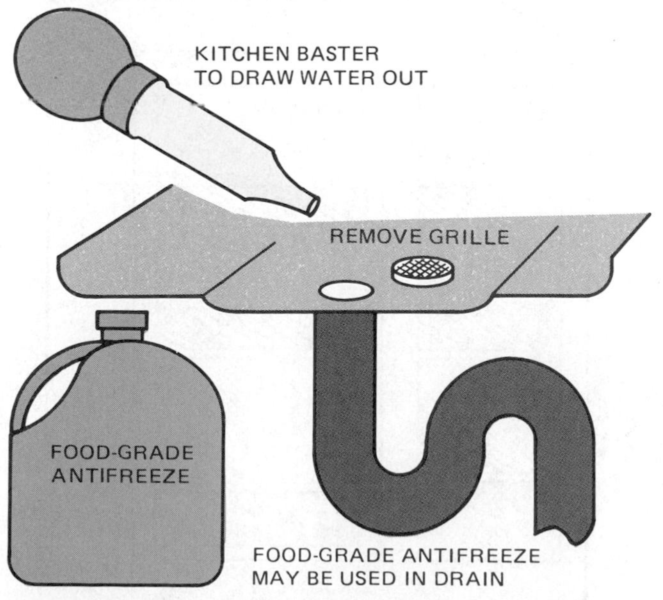

FIGURE 44–50 Service drain trap for winter storage or it may freeze.

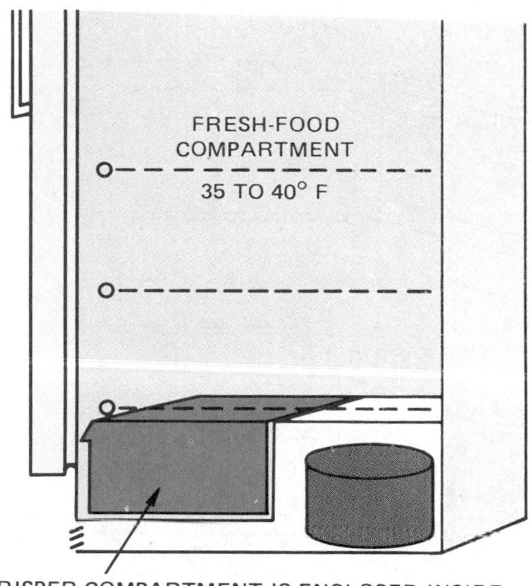

FIGURE 44–51 Crisper drawer.

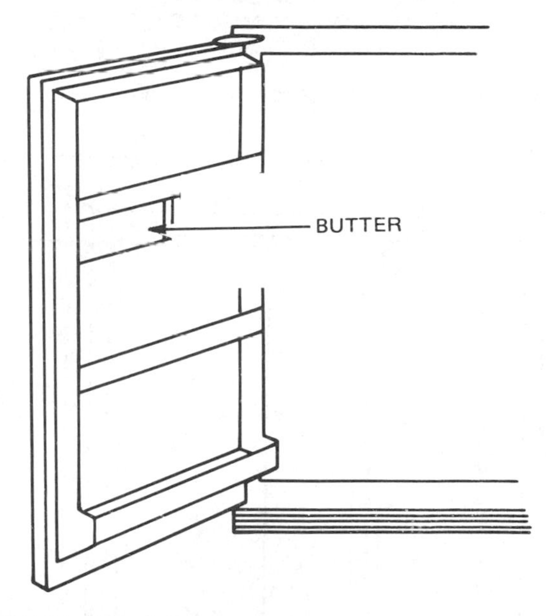

FIGURE 44–52 Butter warmer compartment.

compartment is at very cold temperatures, so care must be taken not to drop frozen food on the bottom rail or breakage may occur, Figure 44–53. If the plastic is broken, the piece should be saved. The box should be warmed to room temperature and the piece replaced with epoxy glue to prevent air from circulating in the wall of the box. The plastic may be backed with foam, which would retard air circulation.

The inside refrigerator temperature may be −5°F in the frozen-food compartment and 35°F in the fresh-food compartment. This can cause a sweating problem around the doors because the moldings may be below the dew point temperature of the room air. Special heaters, sometimes called mullion or panel heaters, are located around the doors to keep the temperature of the door facing above the dew point temperature of the room air, Figure 44–54. Some units have an energy-saver switch that may allow the owner to shut off some of the heaters when not needed, such as in the

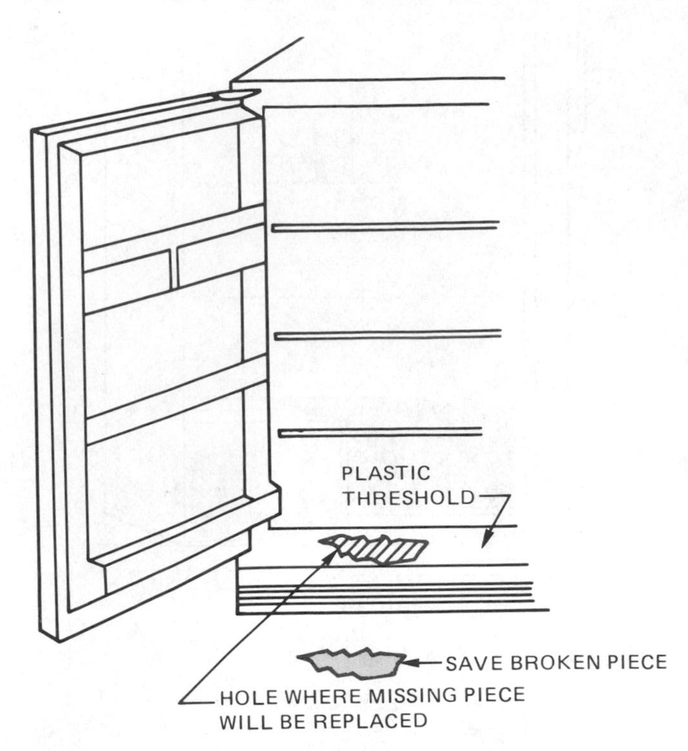

FIGURE 44–53 Broken bottom rail of refrigerator.

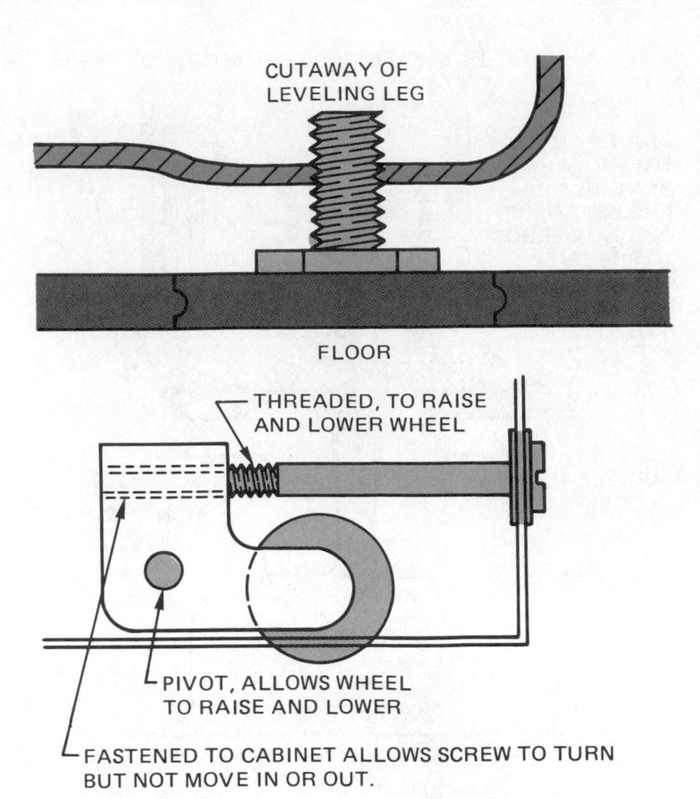

FIGURE 44–55 Leveling devices for feet or rollers.

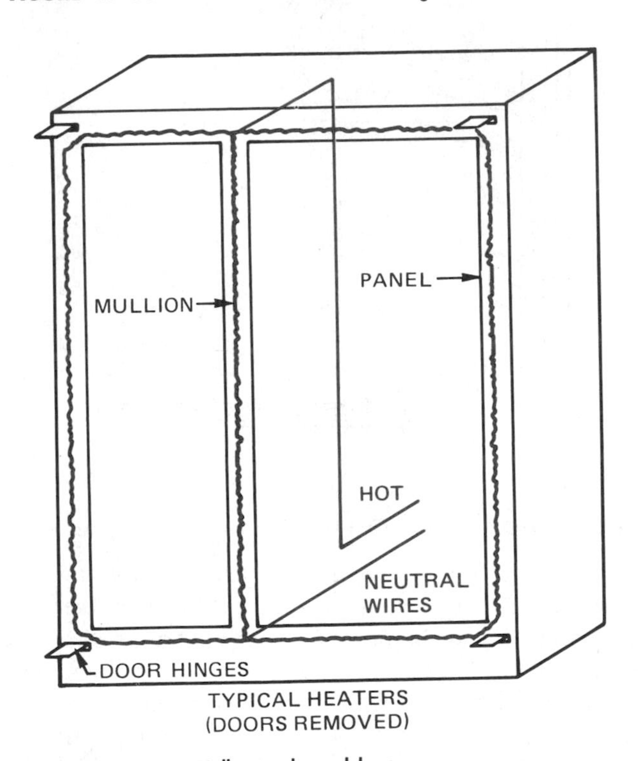

FIGURE 44–54 Mullion and panel heaters.

winter when the humidity is normally low. An explanation of the wiring for these is discussed in the paragraphs relating to controls.

Refrigerated boxes should be level. This is important if the unit has an ice maker or water may overflow the ice maker when it is automatically filled. Most refrigerators have leveling devices for the feet or rollers, Figure 44–55.

Refrigerators are manufactured in over-and-under and side-by-side door configurations. The boxes that are over-and-under seem to have more room because the compartments are typically wider than the side-by-side models. The side-by-side models have the appearance of being narrow and deep, Figure 44–56. This will not affect the function of the box but may make service more difficult.

The refrigerator may have some extras for convenience, such as an outside ice dispenser or ice water dispenser on the outside, Figure 44–57. Notice that the outside dispenser may be located in a section of the door. The wiring and water connections must be made to the door. When an ice dispenser is on the front and in the door, a chute connects

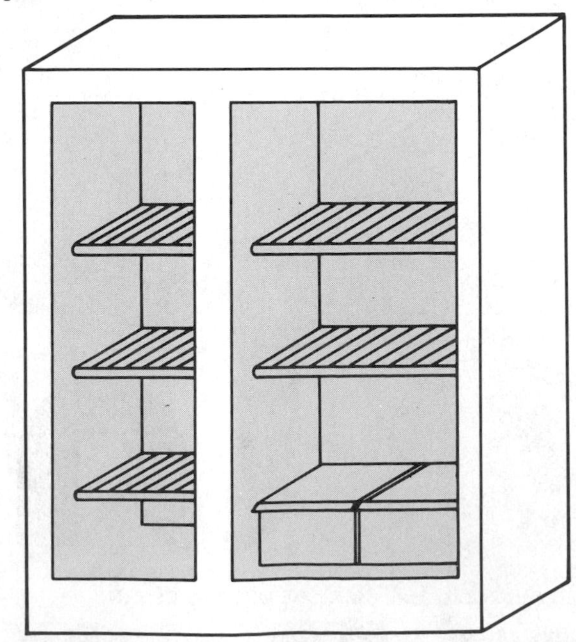

FIGURE 44–56 Side-by-side refrigerator compartments are narrower.

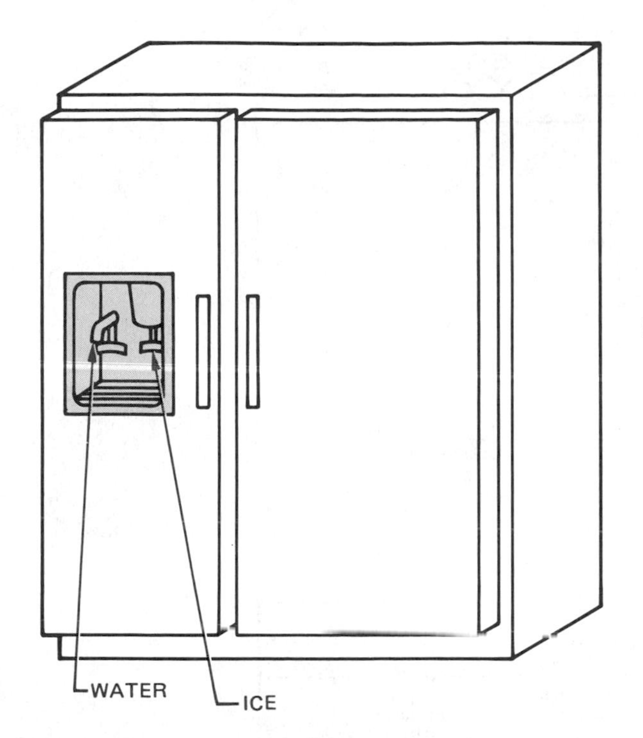

FIGURE 44–57 Water and ice dispenser on outside.

the ice maker in the freezer to the dispenser in the door. These features add to the cost and make service more complex.

44.11 WIRING AND CONTROLS

Each refrigerator should have a wiring diagram permanently fastened to the box, normally on the back. These wiring diagrams are usually of two types, pictorial and line. The pictorial diagram may show the outline of the control and the location of the control, Figure 44–58. The line diagram is used to illustrate how the circuit functions by showing all power-consuming devices between the two lines, Figure 44–59. Keep in mind that we will be dealing with 120 V only so there will be one hot wire and one neutral used to operate the equipment and a ground (green wire) for the frame ground protection.

The components of the typical refrigerator to be controlled include:

- Compressor
- Defrost components
- Various heaters, butter and panel or mullion
- Lights for the interior
- Evaporator fan
- Ice maker

44.12 COMPRESSOR CONTROLS

The space temperature thermostat controls the compressor. It is a line voltage device that passes line power to the compressor start/run circuit. It is a make-on-a-rise-in temperature device.

Several methods are used to control the compressor but all are from space temperature. For example, one manufac-

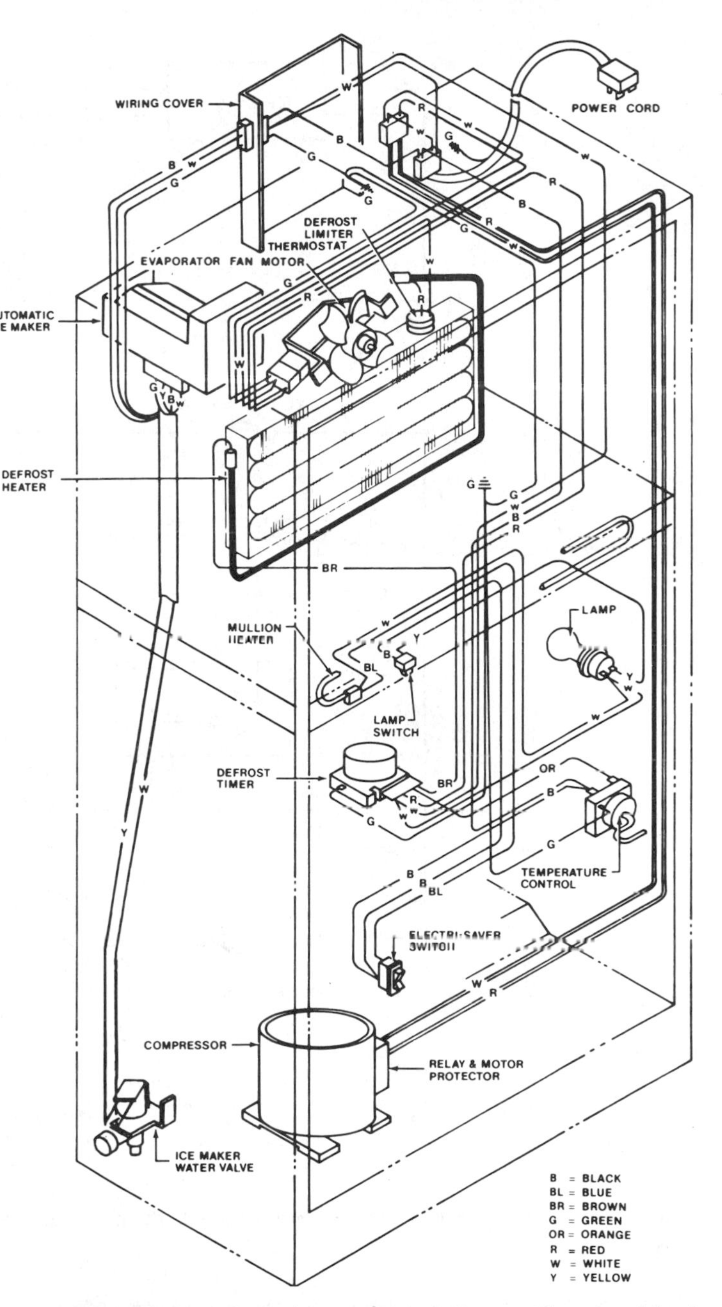

FIGURE 44–58 Pictorial wiring diagram. *Courtesy White Consolidated Industries, Inc.*

turer may use the fresh-food temperature for the control with enough heat removal from the frozen-food compartment to keep the frozen food. Another manufacturer may use the frozen-food compartment temperature to control the compressor with enough planned heat removal from the fresh-food compartment to keep it cold. Still another manufacturer may use the evaporator-plate temperature at a certain planned location as the basis for shutting off the compressor.

It does not matter which method is used, the compressor is still shut off with a thermostat based on some condition inside the refrigerated box. This is a planned condition with the intent of keeping both the frozen-food compartment and the fresh-food compartment at the correct temperature. For many years, this control has been called the thermostat or the cold control, Figure 44–60. It is adjustable

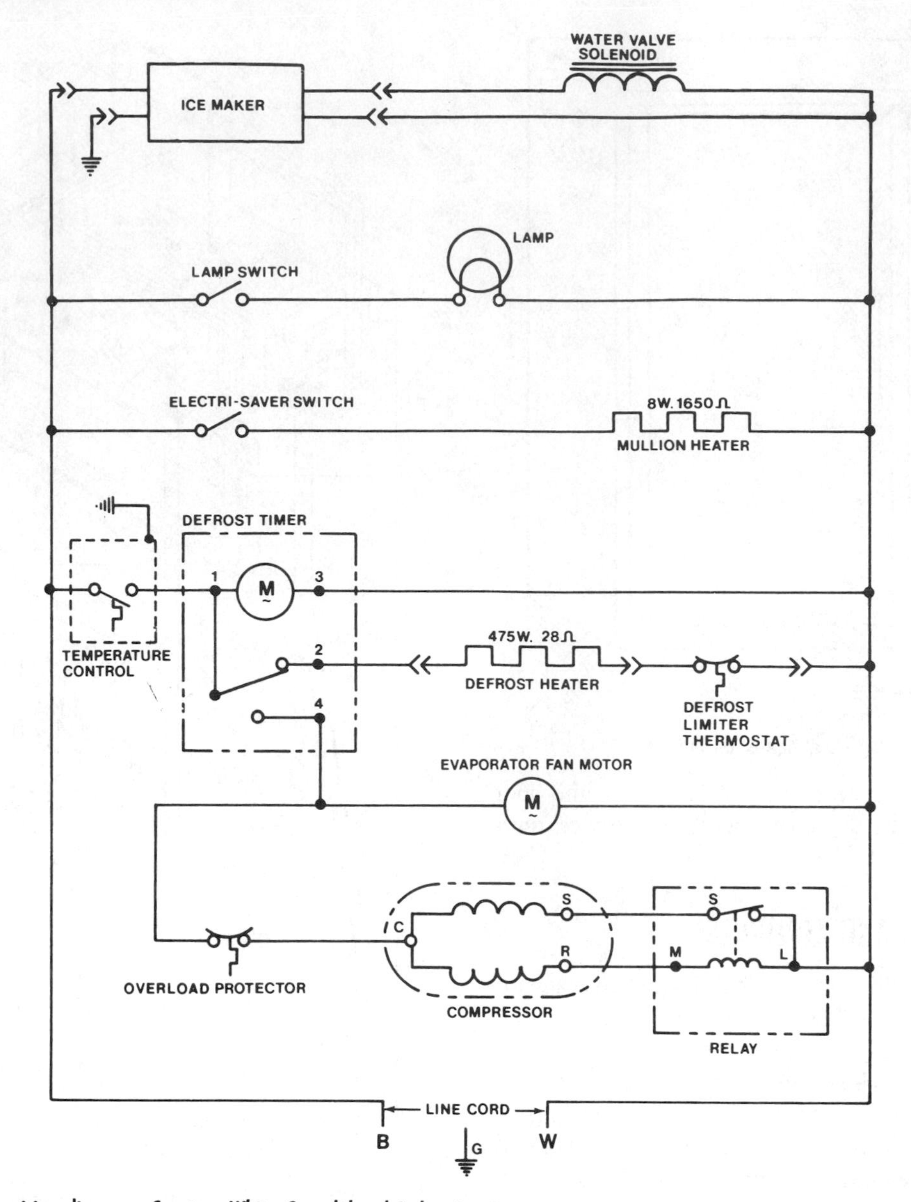

FIGURE 44–59 Line wiring diagram. *Courtesy White Consolidated Industries, Inc.*

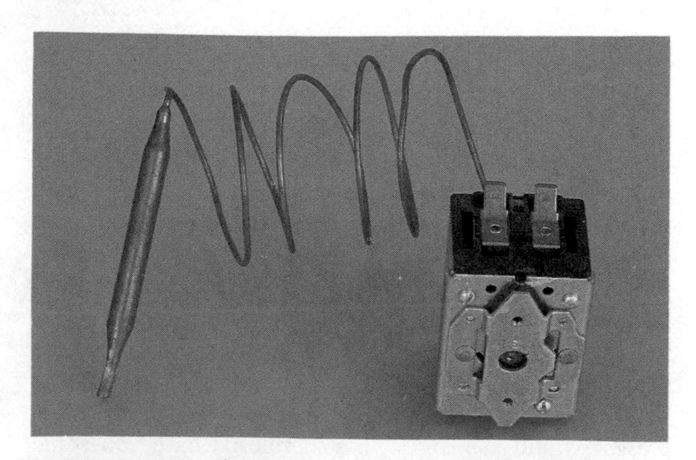

FIGURE 44–60 Thermostat or cold control. *Photo by Bill Johnson.*

and can be considered a remote bulb thermostat. These are small but usually have a large dial with graduated numbers. These numbers typically run from 1 to 10 and have no relation to the actual temperature scale. The dial may read colder with an arrow pointing in the direction where the numbers will yield colder temperatures. This control must be electrically rated with contacts that are able to stop and start the compressor for years of service. They must be reliable because food spoilage is the risk if the control were to fail.

Refrigeration thermostats have a fluid inside the sensing bulb that exerts pressure against the bottom of a diaphragm or bellows, Figure 44–61. The atmosphere is on the other side of the bellows. If the atmospheric pressure is much lower than normal, the control setting must be calibrated for a new pressure, such as when a refrigerator is located at a high altitude. Altitude adjustment varies from one manufacturer to another so the manual should be consulted. See Figure 44–62 for an example of a control with altitude adjustment. The box thermostat passes power to the compressor start circuit to start the compressor.

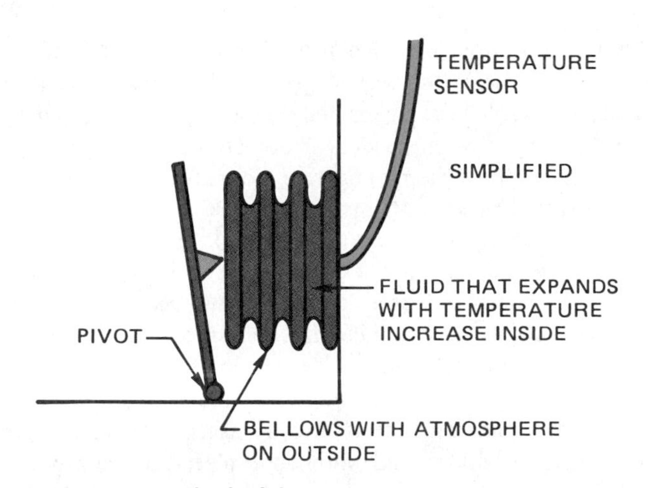

FIGURE 44–61 Sketch of thermostat.

44.13 COMPRESSOR START CIRCUIT

The compressor start circuit receives power from the thermostat circuit when its contacts close and then helps the compressor start. The compressor may need some additional help because the pressures may not totally equalize from the high to the low side between cycles through the capillary tube. The start circuit may consist of a start relay, usually a current type, and a start capacitor and circuit, Figure 44–63. Some refrigerators may use the positive temperature coefficient (PTC) device and circuit, Figure 44–64. Each device and circuit is provided to give the compressor additional starting torque. The starting components and their circuit are typically located in the vicinity of the compressor in the back.

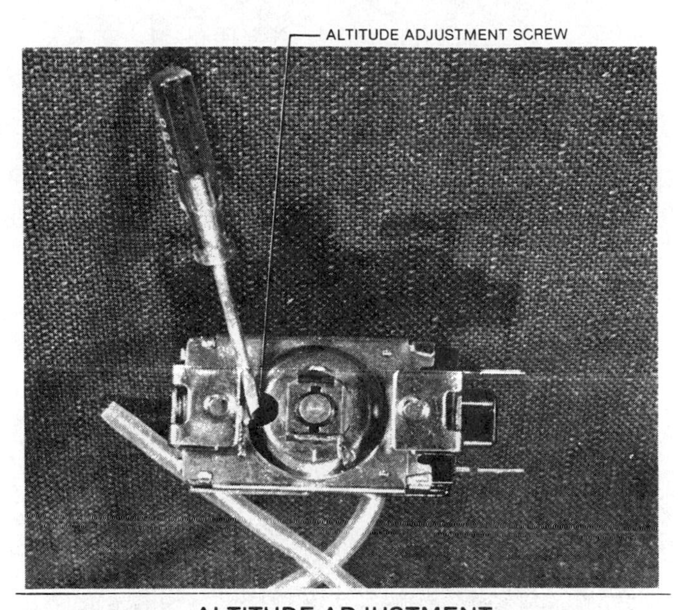

ALTITUDE ADJUSTMENT	
Altitude Above Sea Level - Feet	Range Screw Adjustment (Turns-Clockwise)
2,000	1/8
4,000	1/4
6,000	3/8
8,000	1/2
10,000	5/8

FIGURE 44–62 Altitude adjustment on a thermostat. *Courtesy White Consolidated Industries, Inc.*

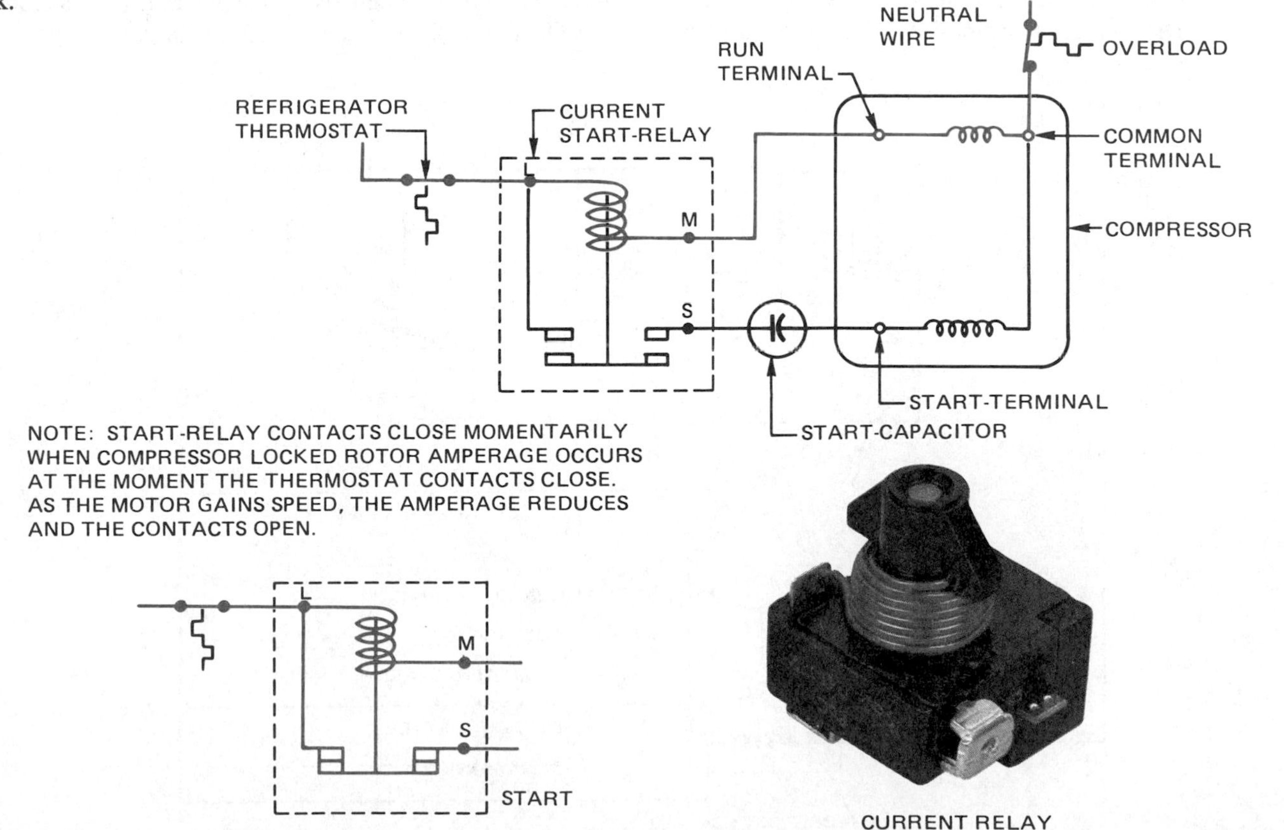

FIGURE 44–63 Refrigerator compressor start circuit using current relay. *Photo by Bill Johnson*

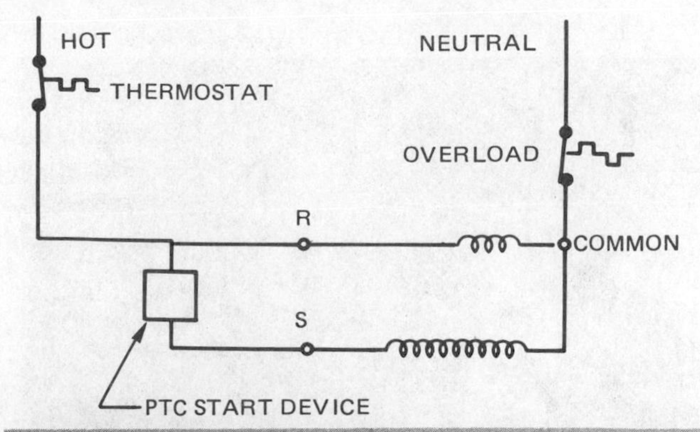

PTC START DEVICE

FIGURE 44–64 Refrigerator compressor start circuit using a PTC starting device. *Photo by Bill Johnson*

44.14 DEFROST CYCLE

All refrigerators have freezing compartments so they are low-temperature refrigeration systems. The evaporator will gather frost and it must be removed from the evaporator. Automatic defrost is called *frost free* by many manufacturers, customers, and service people. This automatic defrost is desirable because manual defrosting is a chore and is often not done when needed. Automatic defrost will help the refrigeration system to operate more efficiently because the frost will be kept off the coil.

Manufacturers use several methods to start defrost. Some manufacturers use a door switch to count the door openings to determine when to start defrost. Many manufacturers use compressor running time to start defrost by wiring a defrost timer in parallel with the compressor, where it builds time whenever the compressor runs, Figure 44–65. Compressor running time is directly associated with door openings, infiltration, and warm food placed in the box.

The defrost cycle may be terminated by two methods, time or temperature. Some units use a termination thermostat that is backed up by the timer. If the thermostat does not terminate defrost, the timer will as a safety feature.

The hot gas defrost uses the heat from the compressor to melt the ice from the coil. A solenoid valve connects the hot gas line to the inlet of the evaporator. The compressor must be running during this cycle for the heat to be available for defrost. The evaporator fan must be stopped, or heat from the evaporator will raise the temperature of the frozen-food compartment. Figure 44–66 shows a diagram of a unit with hot gas defrost.

Electric heat defrost is accomplished with electric heaters located close to the evaporator to melt the ice, Figure 44–67. When the unit calls for defrost, the compressor and evaporator fan stop and the heaters are energized. This condition is maintained until the end of the defrost cycle. Figure 44–68 shows an electric heat defrost diagram.

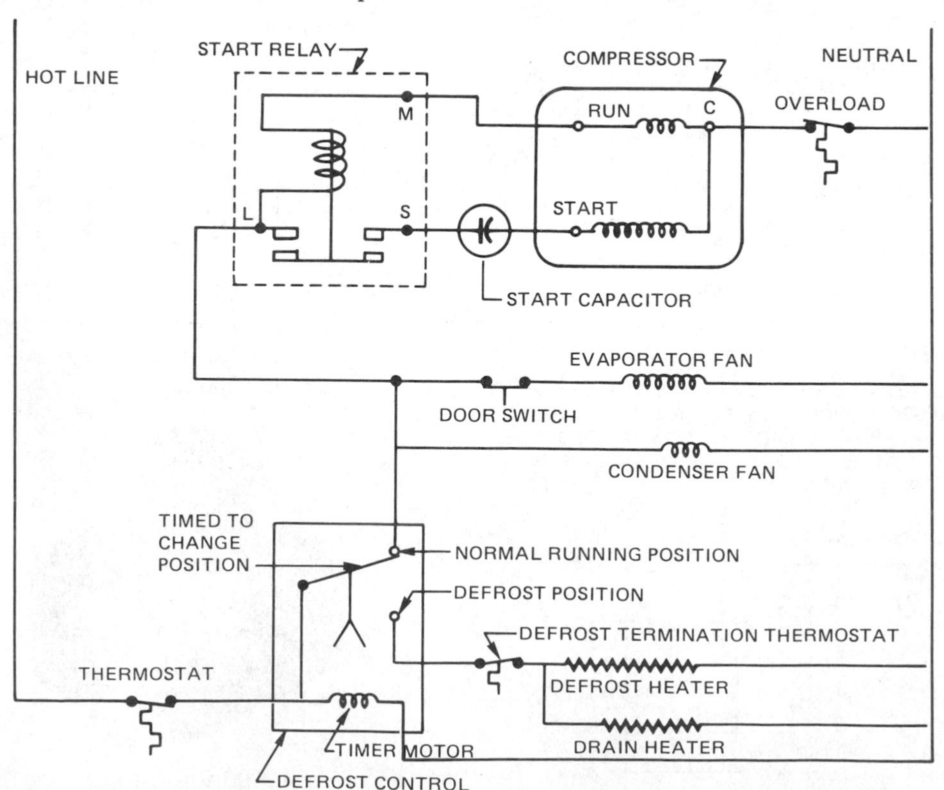

FIGURE 44–65 Timer wiring for refrigerator defrost.

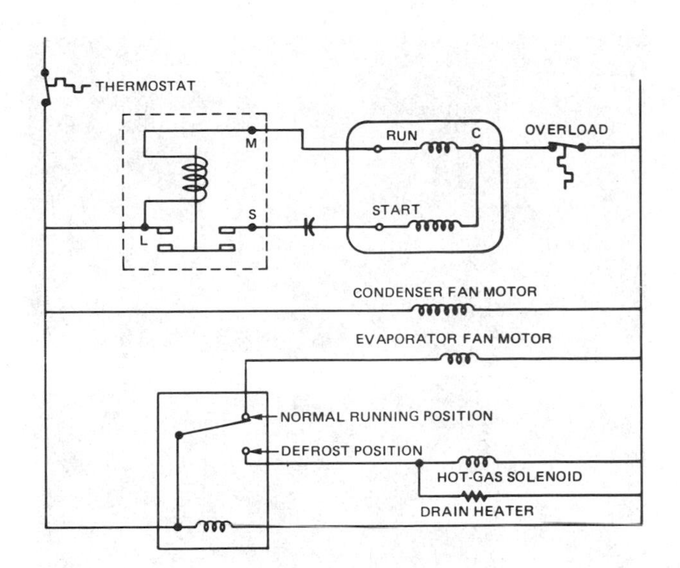

FIGURE 44–66 Wiring diagram for hot gas defrost.

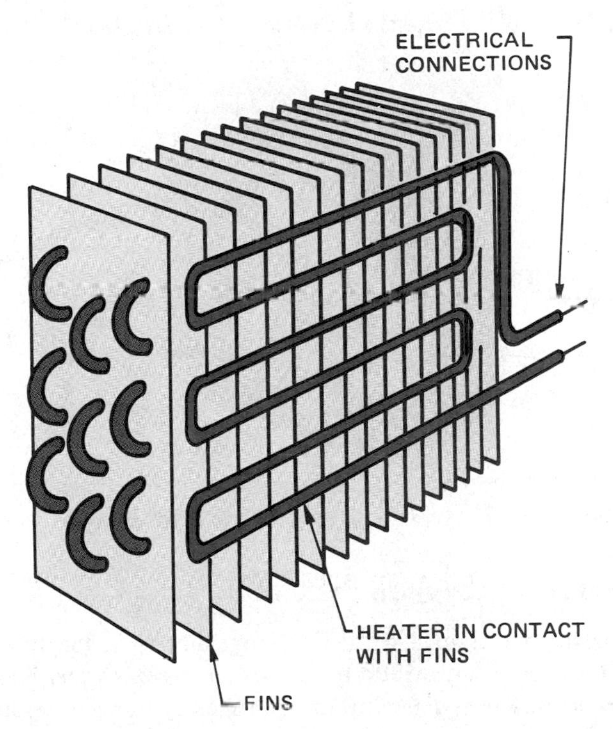

FIGURE 44–67 Electric heaters for defrost.

Drain pan heaters may be energized when either hot gas or electric heat defrost is used to keep the condensate from freezing as it leaves the drain pan.

44.15 SWEAT PREVENTION HEATERS

Most sweat prevention heaters are small electrically insulated wire heaters that are mounted against the cabinet walls at the door openings. These are to keep the outside cabinet temperature above the dew point temperature of the room so they will not sweat. Actually this condensation will not reduce capacity, but it may drop on the floor or be a nuisance on the cabinet and in some cases cause rust. Figure 44–69 shows a diagram of the various types of heaters and describes their location. Some units may have energy-saver switches to allow the owner to switch part of the heaters off

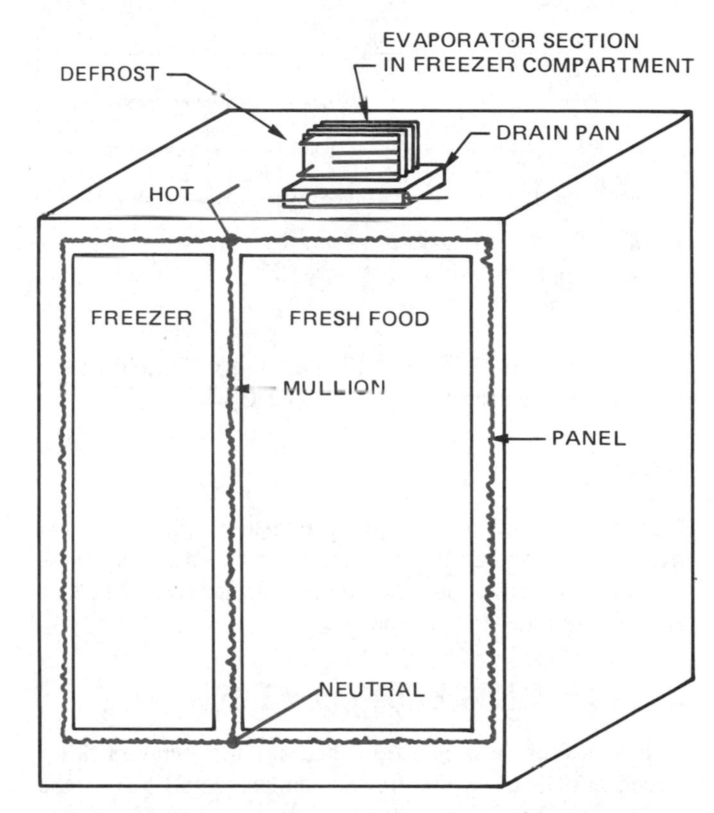

FIGURE 44–69 Heaters in a refrigerator showing their locations.

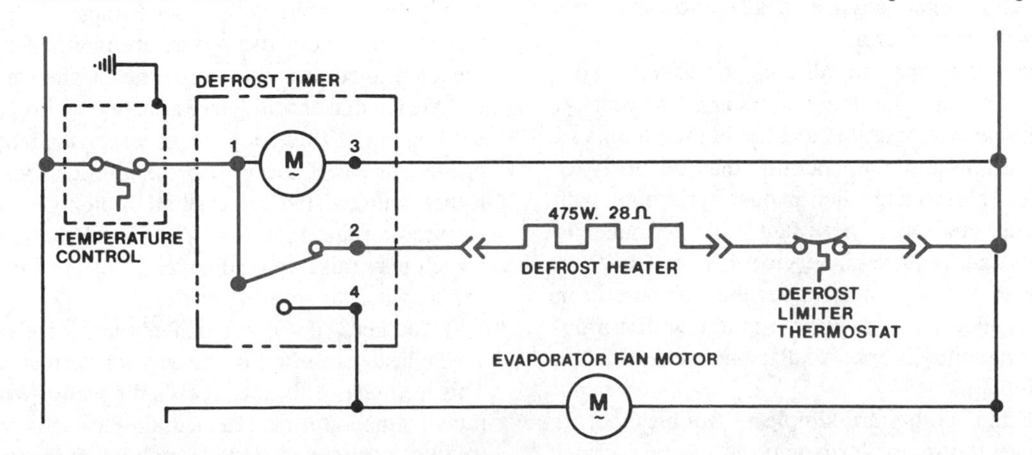

FIGURE 44–68 Electric heat defrost. *Courtesy White Consolidated Industries, Inc.*

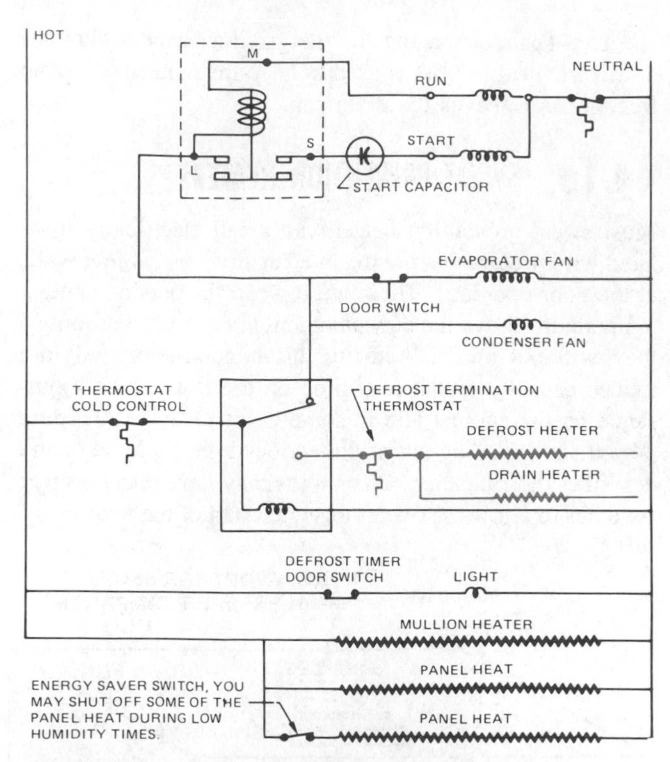

FIGURE 44–70 Wiring diagram shows position of energy saver switch.

if the room humidity is low, Figure 44–70. These switches may be switched on if sweat is noticed.

44.16 LIGHTS

Most refrigerators have lights mounted in the fresh- and frozen-food compartments so the food can be seen. These lights are typically controlled by door switches that make a circuit when the door is opened.

44.17 REFRIGERATOR FAN MOTORS

Two types of fans are furnished on refrigerators when forced draft is used: one for the condenser and one for the evaporator. These fans are usually prop-type fans with shaded-pole motors. Small squirrel cage centrifugal fans may be used for the evaporator.

The evaporator fan may run all the time except in defrost, so it may have many operating hours in a few years. It is typically a reliable device and is located in the vicinity of the evaporator, usually under a panel that may be easily removed for service. This motor is often an open-type motor with no covers over the windings. These fans have permanently lubricated motors that require no service, Figure 44–71.

The condenser fan is located under the refrigerator in the back and is typically a shaded-pole motor with a prop-type fan. It is permanently lubricated also but is covered, not open, Figure 44–72.

These small fan motors are simple to troubleshoot. If you have power to the motor leads and the motor will not turn, either the bearings are tight or the motor is defective.

FIGURE 44–71 Evaporator fan motor. *Photo by Bill Johnson*

FIGURE 44–72 Condenser fan motor. *Photo by Bill Johnson*

44.18 ICE-MAKER OPERATION

The ice maker in a domestic refrigerator is in the low-temperature compartment of the box and freezes water into ice cubes. This is generally accomplished by filling a tray in the freezing compartment with water from the home water supply. A solenoid valve opens long enough to allow the tray to fill. Time is used to determine the fill time. When a predetermined time has past, time for the water to freeze to ice, the ice maker will harvest the ice by dropping it into the holding tray. This may be accomplished by twisting the plastic tray as it turns to break the ice loose. The tray is turned and twisted by a small motor geared to have the required torque, Figure 44–73. When the bin is full, a bail switch may raise up or the weight of the ice in the bin may trip a switch to stop the cycle.

Another type of ice maker makes ice in a metal tray with a heater inserted in the tray for defrost at harvest time. This ice maker fills and freezes the water. When a predetermined temperature is reached, the heater is energized and a small gear motor is started, applying force to the bottom of the ice cubes with fingerlike devices. When they turn loose,

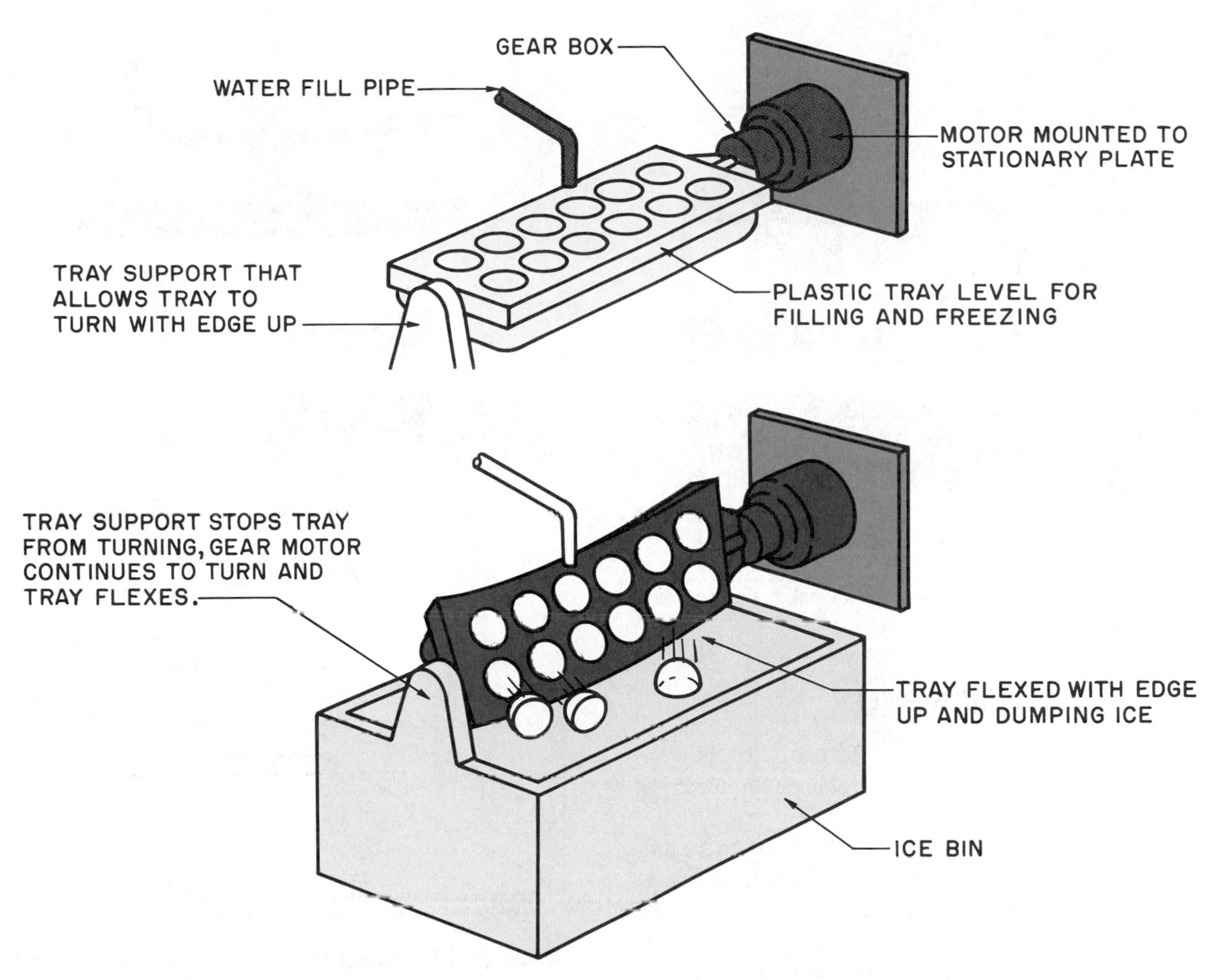

FIGURE 44–73 Water fills the tray and freezes. At the proper time, ice harvest is accomplished by twisting the plastic tray and dumping the ice into the ice bin.

the gear motor moves them to the storage area and the process starts again. This is repeated until the ice bin is full, Figure 44–74.

These domestic ice makers do not make ice like commercial ice makers, which use an auger or inverted evaporator with water flowing over it. Domestic ice makers use time more than any other method to determine the sequence of events. Some may use electronic circuits for these sequences, Figure 44–75.

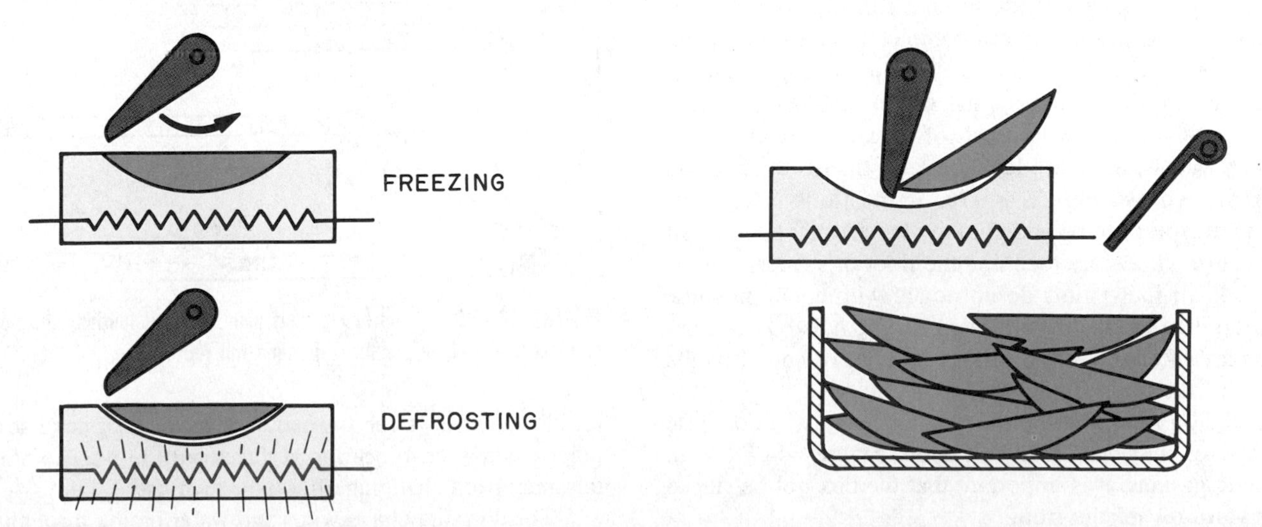

FIGURE 44–74 The water-filled tray freezes. When the control circuit determines the water is frozen, the heater energizes. Then the fingers push the ice out of the tray to the storage bin. When the bin is full, the bail switch is moved to the right and will not return; production stops.

FIGURE 44–75 Electronic circuit board for ice maker. *Courtesy White Consolidated Industries, Inc.*

44.19 REFRIGERATOR SERVICE

The technician should make every effort to separate problems into definite categories. Some problems are electrical and some mechanical.

Unit 5 describes tools and equipment used by technicians for the service of refrigeration equipment. A poorly equipped service technician works at a disadvantage. Lack of proper instruments may prevent the technician from determining the problem. Damage can be caused to the equipment or the customer's property or the technician may be injured if the correct equipment to move a refrigerated box is not used. To be a professional, you must be well equipped.

44.20 CABINET PROBLEMS

Domestic refrigerators must be level for the refrigerant to correctly circulate through the evaporator and condenser. The ice maker may overflow when filled with water if it is not level. Condensate may not completely drain during defrost if the refrigerator is not level. The leveling screws or wheels are on the bottom of the box or cabinet. Leveling feet may be adjusted with a pair of adjustable pliers. If the unit has wheels, there are leveling adjustments to raise and lower the wheels, Figure 44–76. If the floor is too low, spacers may need to be added to the lowest point so that all four feet or wheels are touching the floor or spacers, Figure 44–77. If all four points do not touch with equal pressure, vibration may cause the refrigerator to be noisy. Food containers are often glass and can create a lot of sound on the outside of the box, Figure 44–78.

If the box is not level, the door or doors will not close correctly or will tend to swing open, Figure 44–79. With magnetic gaskets, it is important that the box not be sloped downward toward the front.

The door has the most abuse of any part of the box because it is opened and closed many times. It may have a lot of weight in it due to food storage. It must have strong hinges. Many have bearing surfaces built into these hinges that may need changing after excessive use, Figure 44–80.

The door may have wires and water piping through the hinges to furnish power and water to circuits that may operate ice or water dispensers located in the door, Figure 44–81.

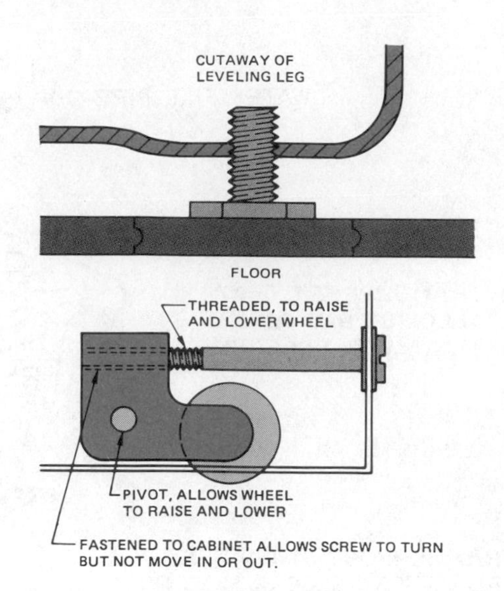

FIGURE 44–76 Refrigerator leveling devices.

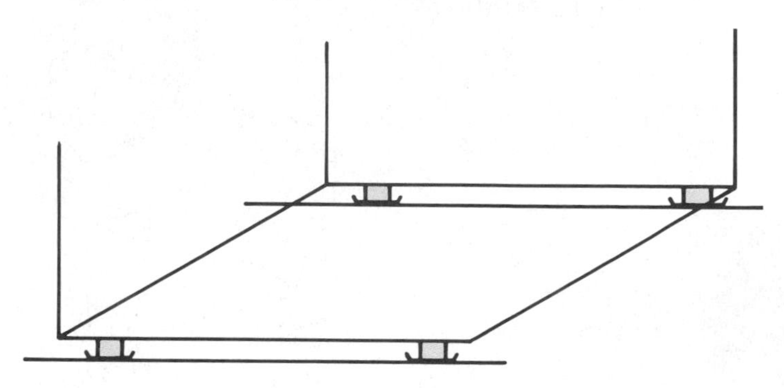

FIGURE 44–77 All four feet or wheels must be touching the floor.

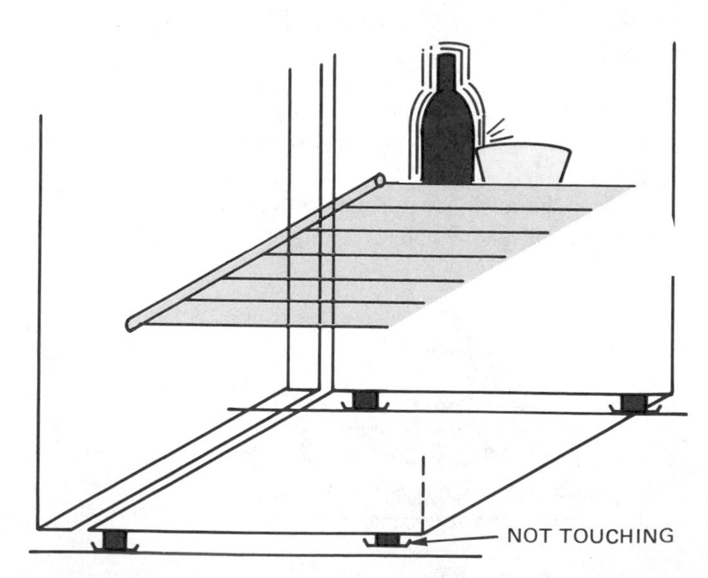

FIGURE 44–78 If all four contact points are not touching the floor or spacers with equal pressure, vibration will cause noise.

FIGURE 44–79 If the box is not level, the doors may not function correctly.

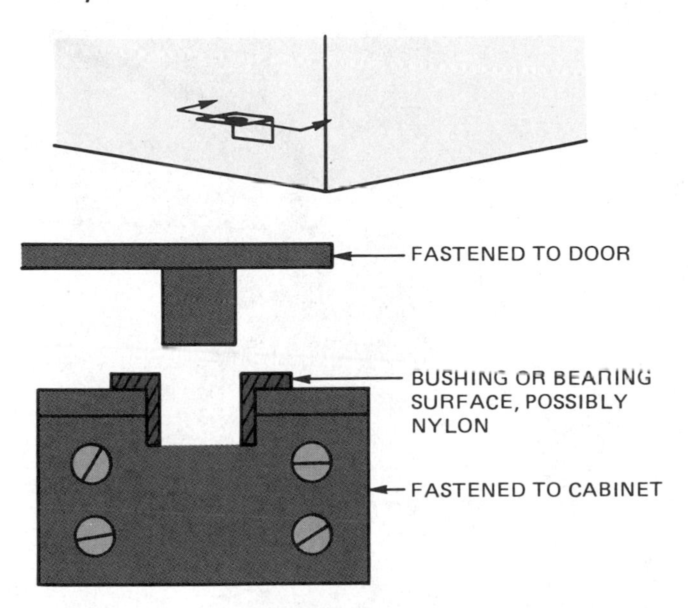

FIGURE 44–80 Hinges may have bearing surfaces to make the door opening easier.

These connections may require service when wear occurs with use.

As door gaskets age, they may need replacement. Different manufacturers use different gaskets that may require special tools. These tools help in removing the old gaskets and replacing the new ones. The manufacturer's directions should be followed when you change the gaskets. Care should be used when the new gaskets are in place to ensure that the door is properly aligned and the gaskets fit properly. A door may be distorted from use or abuse, Figure 44–82.

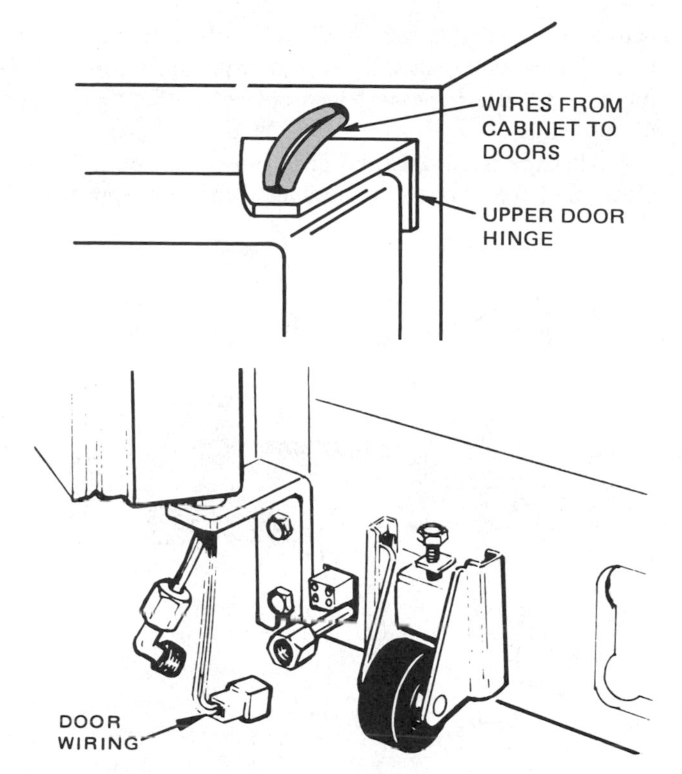

FIGURE 44–81 Power circuits and tubing for water may be connected to the door, through the hinges. *Courtesy White Consolidated Industries, Inc.*

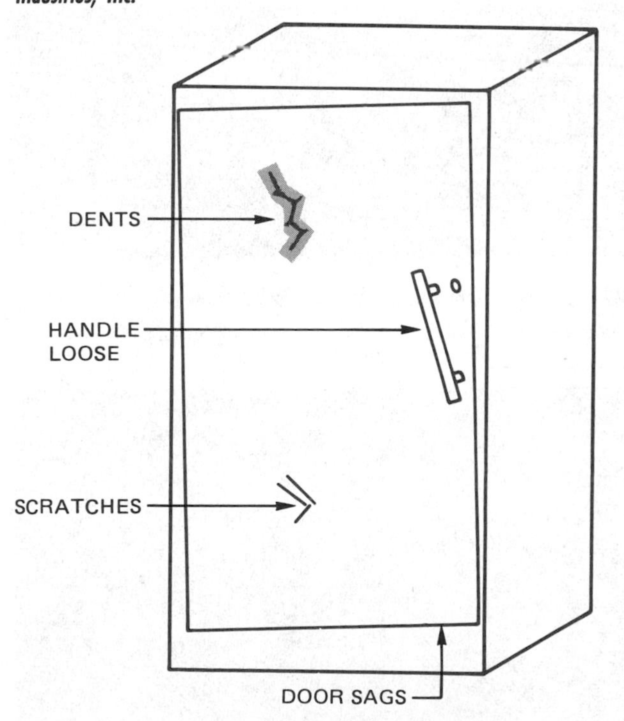

FIGURE 44–82 A door distorted from use or abuse.

44.21 GAGE CONNECTIONS

Domestic refrigerated devices do not have gage ports like commercial refrigeration or air conditioning systems. The system is hermetically sealed at the factory and the use of gages may never be needed. When a system has been charged correctly and as long as all conditions remain the same, it should never have to be adjusted. Leaks, field analysis requiring pressures, and field repair of components are the only reasons to apply gages. Many service technicians have a tendency to routinely apply gages. This can be poor practice. Taking a high-pressure reading on a high-pressure gage line full of condensed liquid refrigerant can be enough to adversely affect the operating charge. Many service technicians have started with the correct charge, but as a result of taking pressure readings have altered the charge and caused a problem. Gages should be applied as a last resort, and when applied, should be done with great care. Other methods to determine system problems without gages are discussed later.

All gage manifolds and gage lines must be leak free, clean, and free from contaminants. It is good practice to have a set of gages for each type of refrigerant you commonly use. Keep the gages under pressure with clean refrigerant from one use to the next. When you start to use a set of gages that are still under pressure from the last use, you know they are leak free, Figure 44–83. You may want to remove the Schrader valve depressors furnished in the gage lines and use special adapters for depressing Schrader valve stems, Figure 44–84. This enables you to have clear gage lines for quick evacuation of a wet or damp system.

Some manufacturers furnish a service port arrangement where an attachment may be fastened to the compressor for taking gage readings, Figure 44–85.

Other manufacturers do not furnish any service ports and field service ports may be installed in the field in the form of line tap valves, Figure 44–86. These special valves

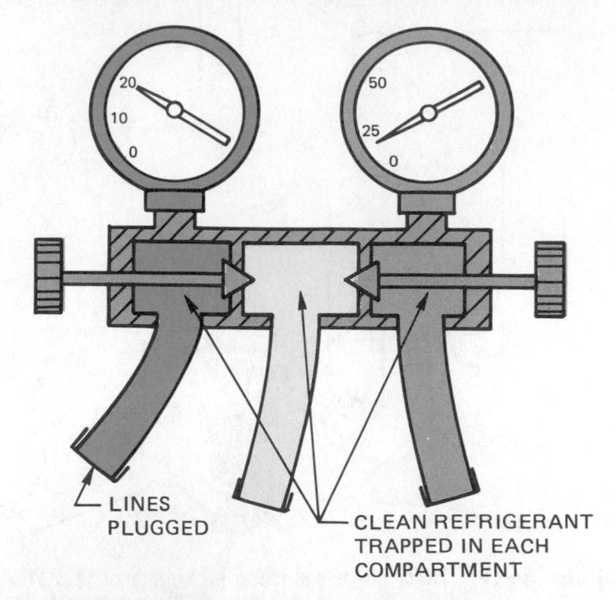

FIGURE 44–83 A clean, leak-free set of gages.

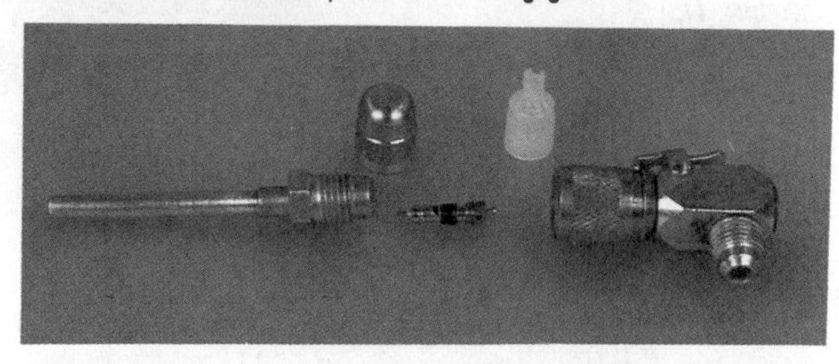

FIGURE 44–84 Remove the Schrader valve depressors from gage lines and use the special fitting on the right to depress the valve cores. *Photo by Bill Johnson*

FIGURE 44–85 Service valve assembly. *Photo by Bill Johnson*

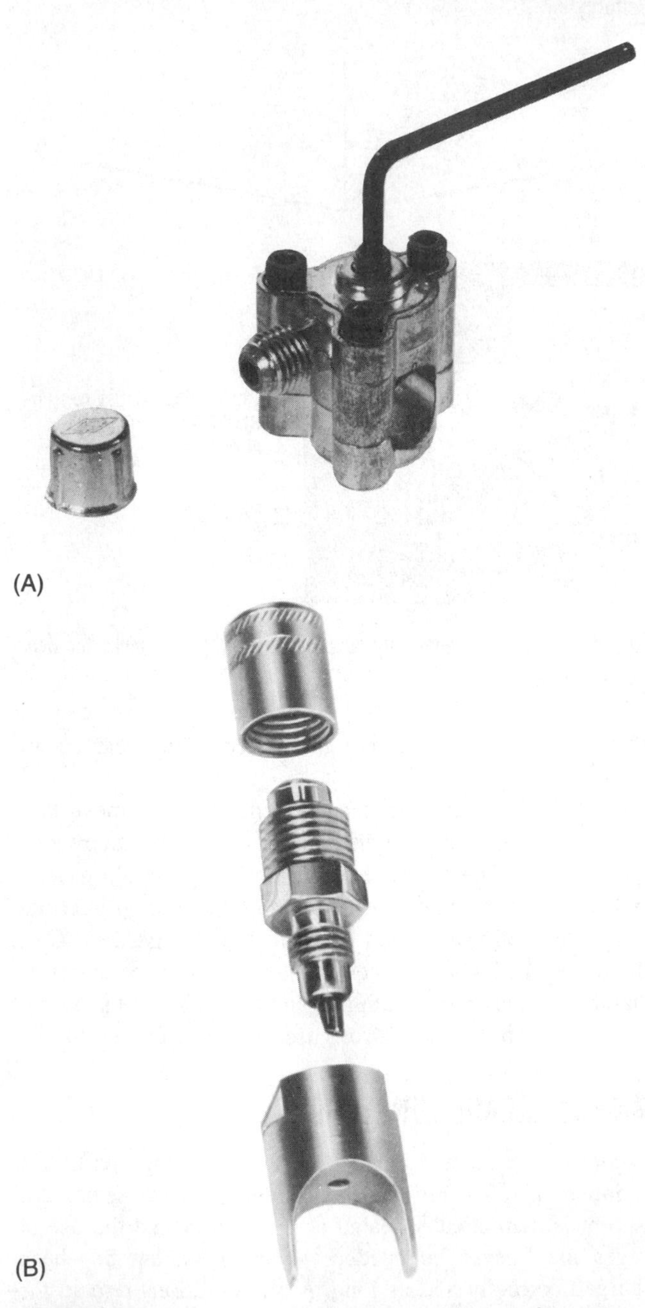

FIGURE 44–86 Line tap valve for access to refrigerant lines. *(A) Photo by Bill Johnson, (B) Courtesy J/B Industries*

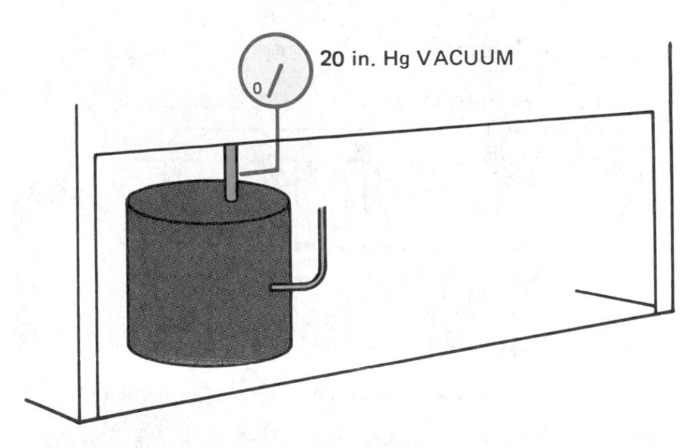

FIGURE 44–87 *If the unit is operating in a vacuum, air may be pulled into the system when gages are installed.*

should be installed using the manufacturer's instructions. Some points to remember are to always use the correct size valve based on the line size. ***If there is a chance that the system pressure may be in a vacuum, shut the unit off and let the pressures equalize before installing a low-side line tap valve or atmosphere will enter the system, Figure 44–87.*** When installing a line tap valve on the high-pressure side of the system, it is best to install it on the liquid line and not on the compressor discharge line where the gasket will be subjected to excess heat, Figure 44–88.

If repairs are made to components in the refrigerant cycle or a refrigerant charge is lost completely, it is best to use the process tubes on the compressor for pressure readings. Fittings with Schrader valves may be soldered to the process tubes for the service work and evacuation, Figure 44–89. These tubes may be capped, or they may be pinched off using a special pinch-off tool, Figure 44–90, and soldered shut. A special pinch-off tool is necessary because you cannot pinch off refrigerant lines with pliers, Figure 44–91.

44.22 LOW REFRIGERANT CHARGE

If a refrigeration unit had the correct charge when it left the factory, it will maintain that charge until a leak develops. If

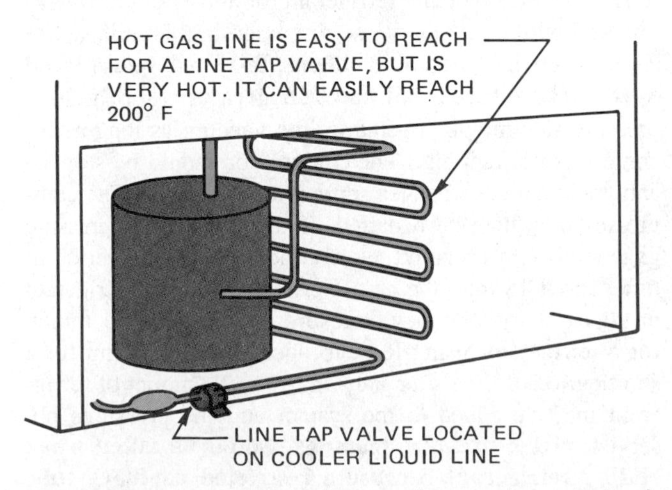

FIGURE 44–88 Locate the high-pressure line tap valve on the liquid line, not the hot gas line.

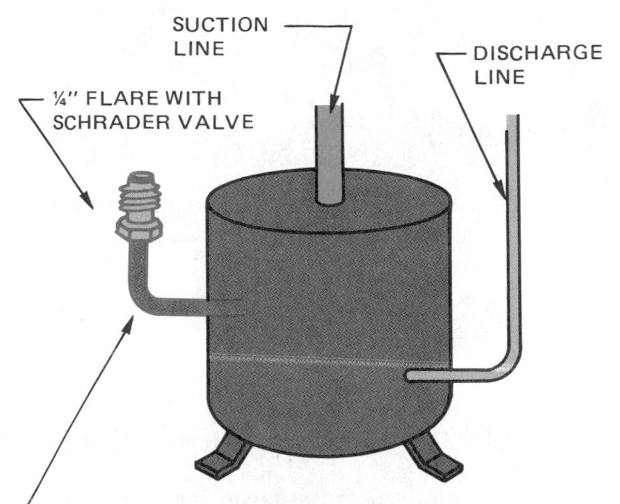

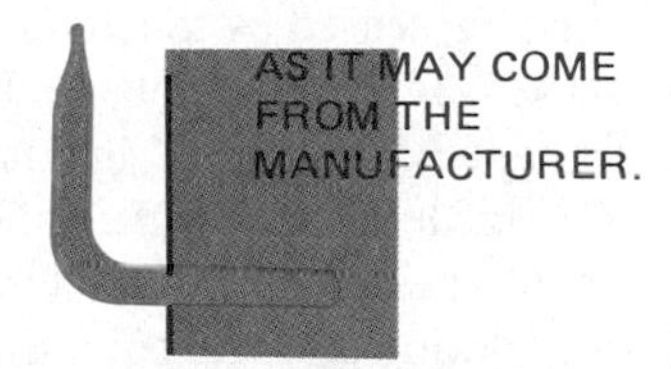

FIGURE 44–89 Fittings may be soldered on the process tube.

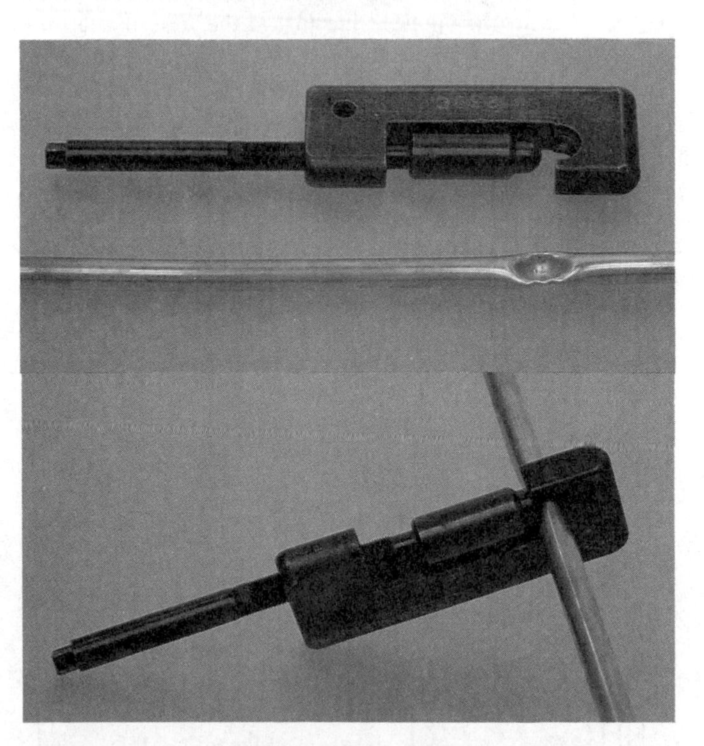

FIGURE 44–90 Special pinch-off tool. *Photos by Bill Johnson*

a unit does have a low refrigerant charge, every effort should be made to determine the cause.

One service technique that many experienced service technicians use to determine if the charge is approximately correct in a system (before connecting gages) only requires

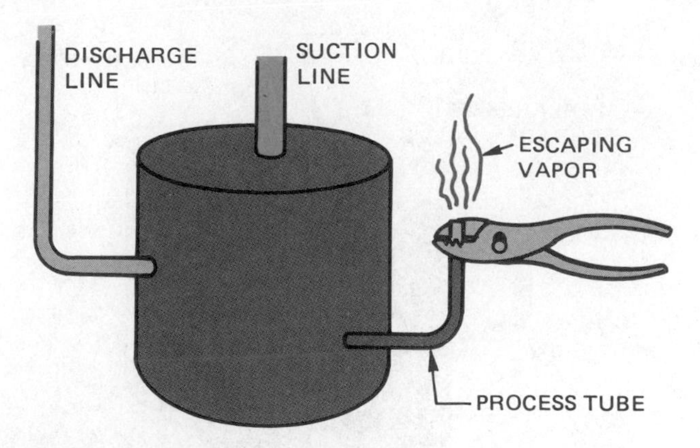

FIGURE 44–91 *You cannot use pliers to pinch off a line.*

that the unit be stopped and restarted. The unit is shut off and the pressures allowed to equalize, which takes about 5 min. Before the compressor is restarted, the technician places a hand on the suction line, where it leaves the evaporator and before any heat exchanger, in such a manner that the line temperature may be sensed by touch, Figure 44–92. The compressor is started. If this line gets cold for a short period of time, the chances are the refrigerant charge is correct, Figure 44–93. This test assumes that when the pressures equalize, much of the refrigerant in the condenser moves to the evaporator during pressure equalization. When the low-side pressure in the suction line is reduced by the compressor starting, a small amount of liquid refrigerant will move into the suction line just at the time of start-up and causes the line to become very cold for just a moment. There is such a small amount of refrigerant in a domestic refrigerator that if the charge were short, there would not be enough liquid refrigerant to leave the evaporator. This simple test has helped many experienced refrigeration technicians keep the gages in the tool bin and look for other problems. The evaporator may have ice buildup due to lack of defrost, Figure 44–94, or the evaporator fan may not be functioning.

When a technician suspects a low charge, the unit should be shut off and the pressures allowed to equalize.

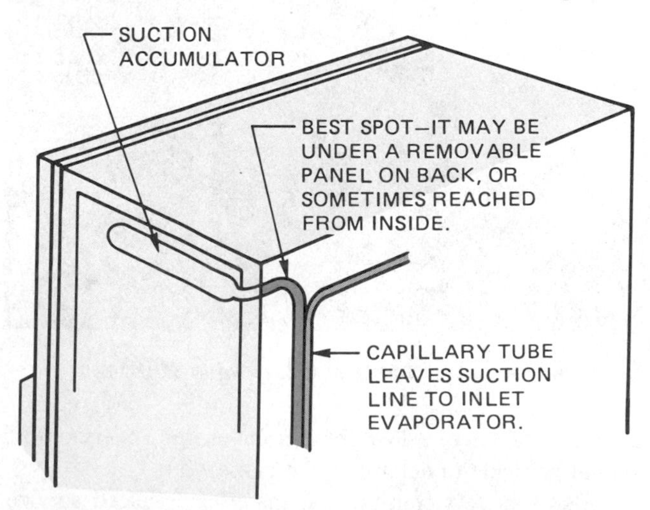

FIGURE 44–92 Location of cold spot on suction line for checking for refrigerant charge.

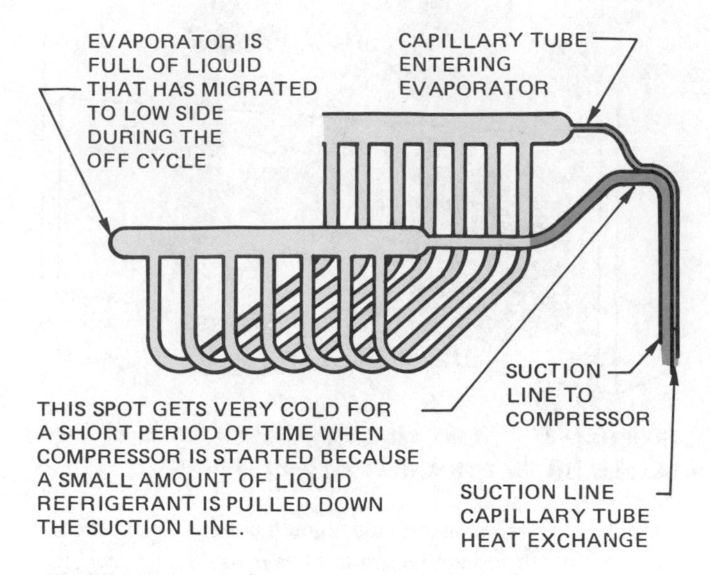

FIGURE 44–93 Performing the suction touch test for correct refrigerant charge.

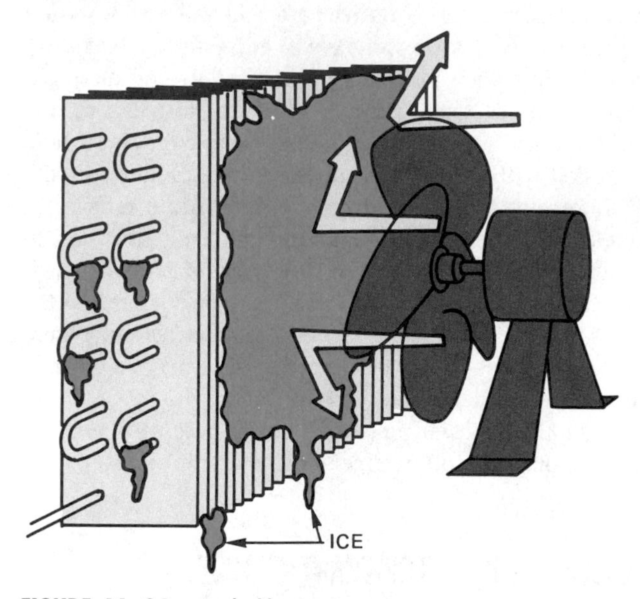

FIGURE 44–94 Ice buildup on evaporator.

R-12 is the most popular refrigerant for household refrigerators so it will be the only one discussed in this unit. **Because of environmental problems, R-12 will be replaced in the near future in all new refrigerators.** (The low-pressure side may be operating in a vacuum as long as the compressor is running.) The compressor should be stopped and uncontaminated gages applied to the system. The compressor may then be restarted. If the system low-pressure gage reads in a vacuum (below atmosphere) for a period of time, about 15 min, the unit is probably low on refrigerant or the capillary tube may be restricted. It is not uncommon for a refrigerator using R-12 to operate in a vacuum for a short period of time after start-up. A small amount of refrigerant may be added to the system and the pressures observed. ***High-pressure readings should be taken when adding refrigerant because a restricted capillary tube will cause high head pressure readings if refrigerant is added. You will not be able to determine this from a low-**

pressure reading only. If a high-pressure reading is not possible, attach a clamp-on ammeter to the compressor common wire and do not allow the compressor amperage to rise above the run-load amperage (RLA) rating of the compressor. If it does, shut it off, Figure 44–95.*

⊛Manufacturers recommend that when a low charge is found, find the leak, remove and recover the charge, repair the leak, and add refrigerant.⊛ Then leak check the repair and evacuate the system to a deep vacuum. A measured charge may then be transferred into the system.

Some experienced service technicians may successfully add a partial charge by the frost-line method. This method is used to add refrigerant while the unit is operating and works

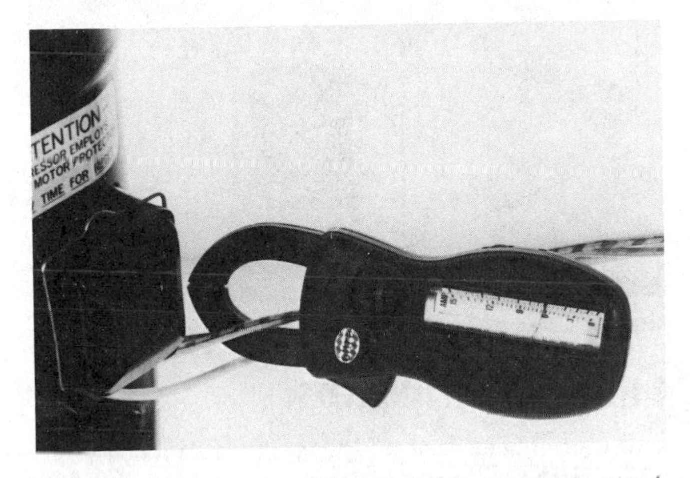

FIGURE 44–95 Taking a current reading at the compressor. *Photo by Bill Johnson*

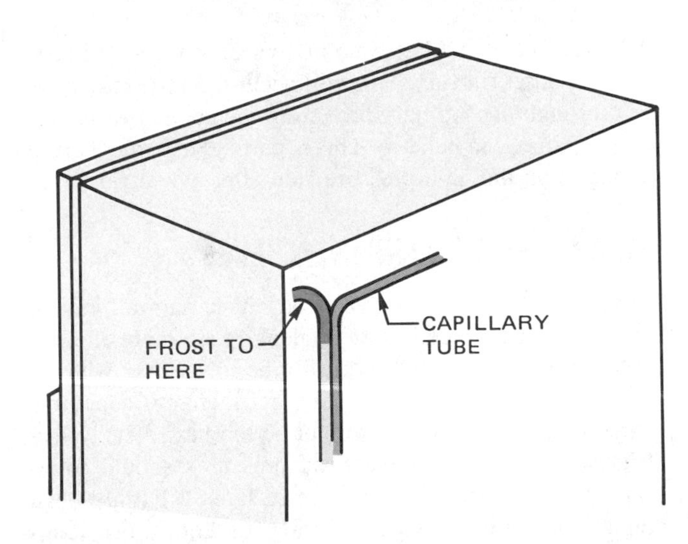

FIGURE 44–96 Checking the refrigerator charge using the frostline method.

as follows. A point on the suction line leaving the refrigerated box is located where the frost line may be observed, possibly where the suction line leaves the back of the box, Figure 44–96. Refrigerant is added very slowly by opening and closing the gage manifold low-side valve until frost appears at this point. Then add no more. The suction pressure may be 10 to 20 psig at this time and should reduce to about 2 to 5 psig just before the refrigerator thermostat shuts the compressor off. **⊛If the frost line creeps toward the compressor as the box temperature reduces, refrigerant may be recovered slowly through the low side until the frost line is correct, Figure 44–97.⊛**

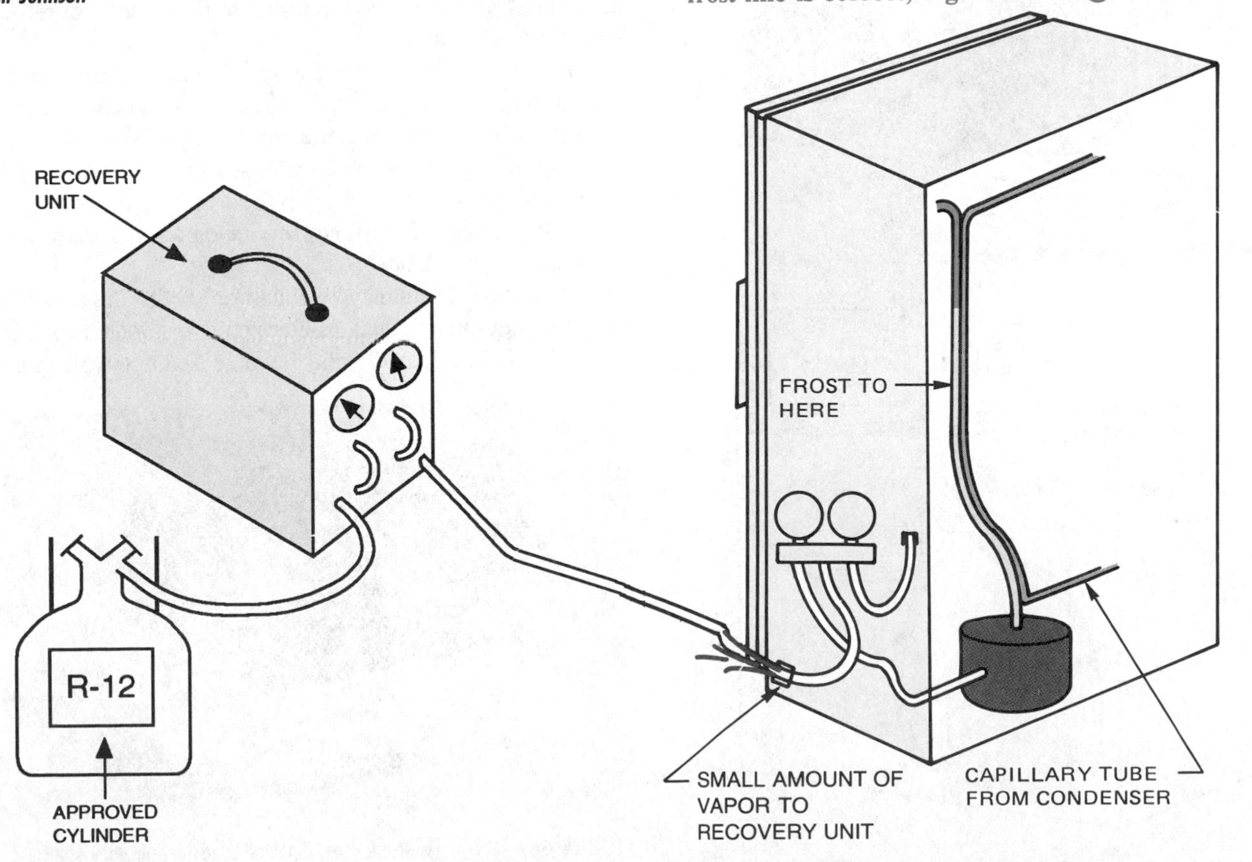

FIGURE 44–97 Frost line has moved toward the compressor; a small amount of vapor is recovered through the low-side gage.

NOTE: Many units will operate in a vacuum for a long period of time after start-up. This will occur until the refrigerant feeds through the capillary tube from the condenser and the system charge is in balance. This is particularly true if a unit is charged on the high-pressure side after evacuation.

44.23 REFRIGERANT OVERCHARGE

The refrigerator condenser may be either natural draft or forced draft. The forced-draft condensers are more efficient and the head pressure will typically be lower than with the natural-draft condenser. If too much refrigerant is added to a refrigerator, the head pressure will be too high. The typical refrigerant is R-12. Typical head pressures should correspond to a condensing temperature 25°F to 35°F higher than room temperature at design operating conditions. For example, if the room temperature at the floor is 70°F, the head pressure should be between 117 and 136 psig, Figure 44–98. Refer to the pressure/temperature chart in Unit 3 in the column for R-12. If the head pressure is higher than 136 psig, it is too high. ***Be sure to check the load on the refrigerated space before drawing any final conclusions. These head pressures will be higher during a hot pull down of the refrigerated space, Figure 44–99.*** If the compressor is sweating around the suction line, there is too much refrigerant, Figure 44–100. The suction pressure will also be too high with an overcharge of refrigerant.

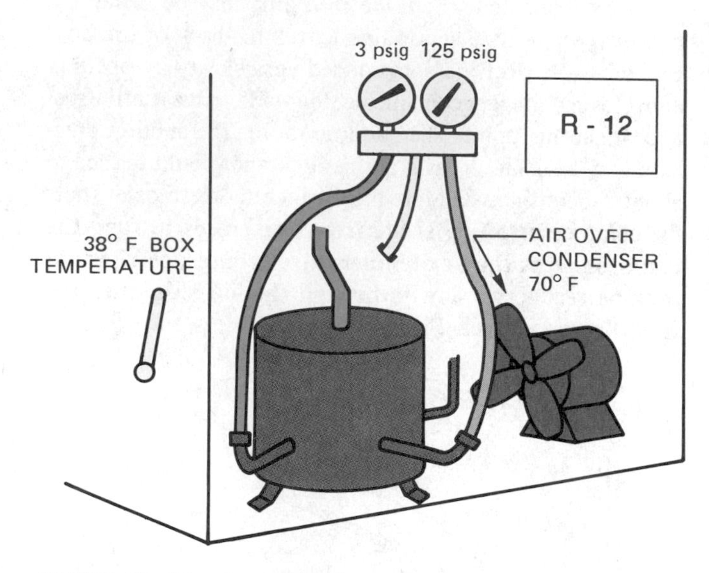

FIGURE 44–98 A typical head pressure.

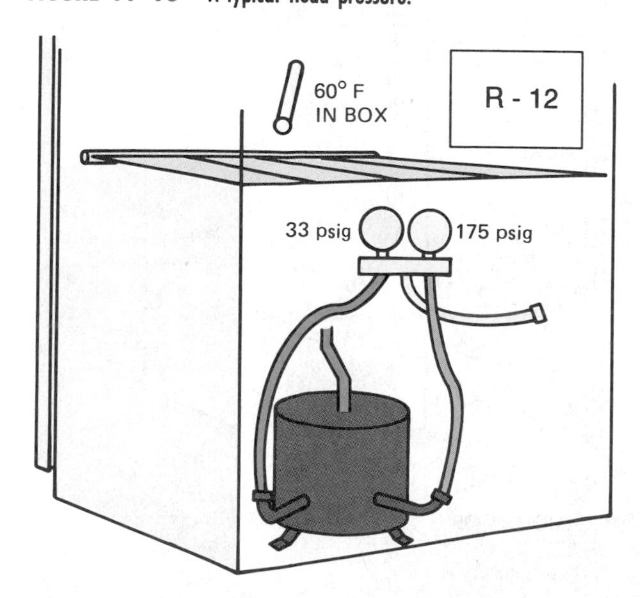

FIGURE 44–99 Head pressure under an abnormal load.

FIGURE 44–100 Compressor sweating around the suction line, too much refrigerant.

44.24 REFRIGERANT LEAKS

A very small amount of refrigerant lost from a household refrigerator will affect the performance. In a few instances a low charge that occurs from the factory due to a leak may be discovered when the refrigerator is started. It will not refrigerate from the start.

When the box has run for some period of time and a leak occurs, it may be hard to locate in the field. Many technicians prefer to move the unit to the shop and loan the customer a box. They can then repair the defective unit at their own pace when there is no food in it. In any case, the best place to perform difficult service on a refrigerator is in a shop, not a residence.

Very small leaks may be found only with the best leak detection equipment, such as electronic leak detectors, Figure 44–101. The pressure may be increased in the refrigera-

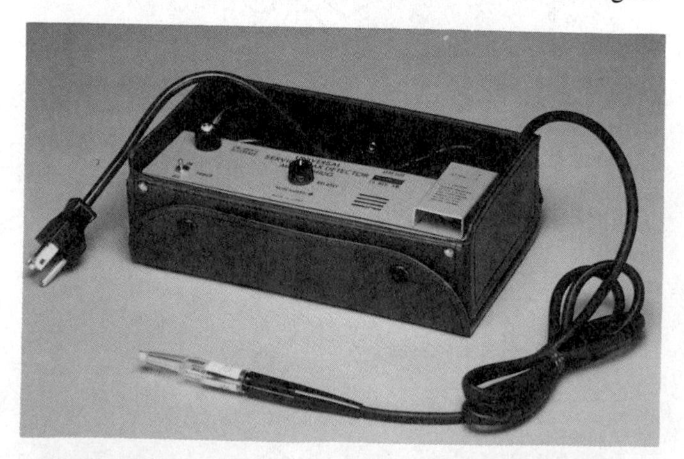

FIGURE 44–101 Electronic leak detector used to find very small leaks. *Courtesy of Yokogawa*

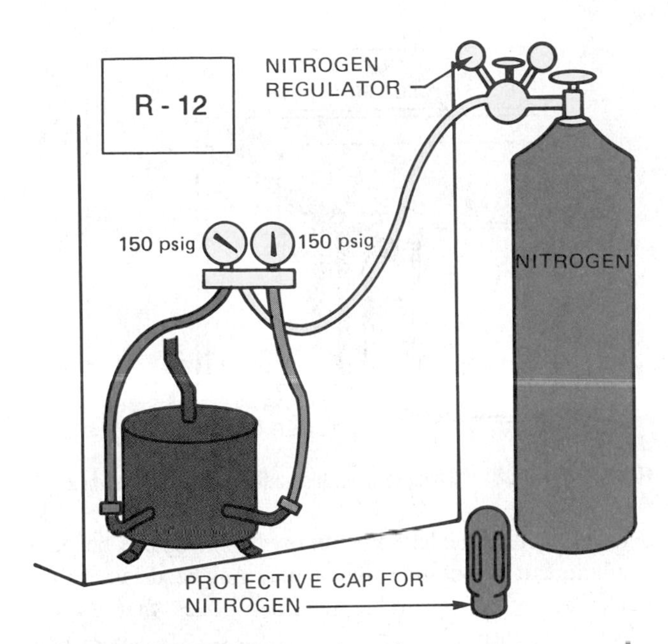

FIGURE 44–102 Nitrogen may be used to create more pressure for leak checking.

tor by using nitrogen or R-22, Figure 44–102. ***Do not raise the pressure above the manufacturer's specified low-side working pressures. This is usually 150 psig.***

44.25 EVAPORATOR LEAKS

Leaks may occur in the evaporator due to abuse from the use of a sharp tool when the unit is being defrosted manually. When this occurs in an aluminum evaporator, the evaporator may be repaired. Soldering the leak from a puncture may be difficult because of location and the contraction and expansion of the evaporator. Leaks may be repaired with the proper epoxy. Special epoxy products are available that are compatible with refrigerants.

One method is to clean the surface according to the epoxy manufacturer's directions. Apply the epoxy to the hole while the unit is in a slight vacuum, about 5 in. Hg. This will pull a small amount of epoxy into the hole and form a mushroom-shaped mound on the inside of the pipe, Figure 44–103. This mound will prevent the patch from being pushed out when the refrigerator is unplugged and the low-side pressure rises to the pressure corresponding to the room temperature. If the refrigerated box is located outside where it may reach 100°F, the pressure inside may rise to 117 psig. This may be enough to push a plain patch off the hole, Figure 44–104. ***Care should be taken not to allow too much epoxy to be drawn into the hole or a restriction may occur.***

Another method may be to use a short sheet metal (self-tapping) screw and epoxy in the hole. The vessel must have enough room for the screw inside for this to work. The puncture and the screw are cleaned according to the epoxy manufacturer's recommendations. The epoxy is applied to the hole and the screw is screwed so that the head is tight against the hole to hold the epoxy at any time high pressure occurs, Figure 44–105.

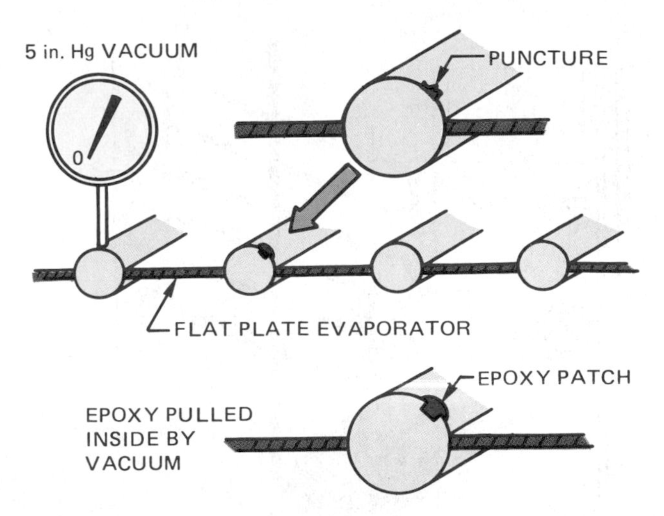

FIGURE 44–103 A slight vacuum may be used to pull a small amount of epoxy into the puncture.

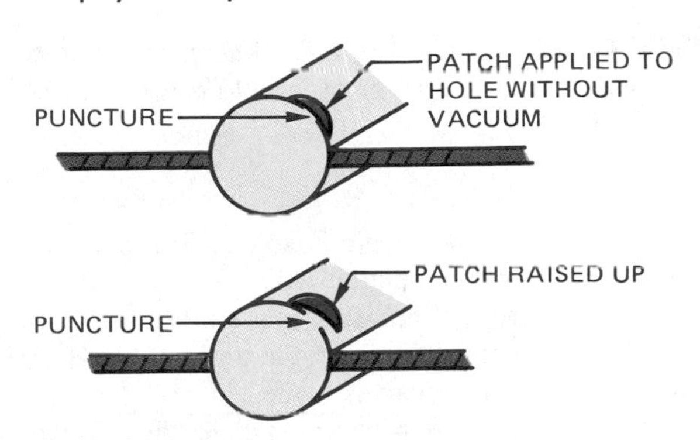

FIGURE 44–104 The pressure inside may push the patch off the puncture under some circumstances. This may occur during storage when the low-side pressure is high.

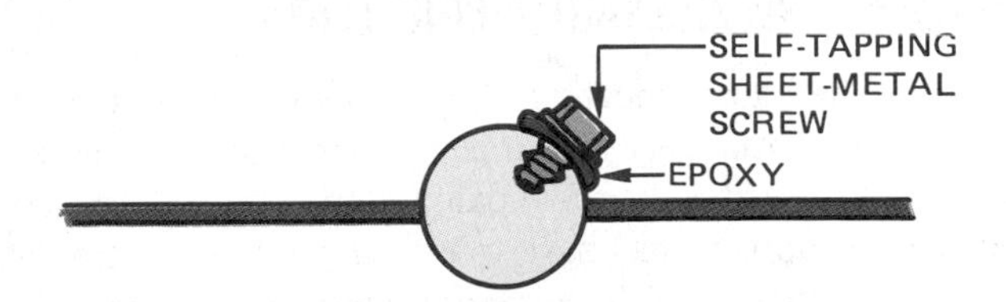

NOTE: IF SCREW IS TOO LONG, START THE THREADS AND BACK THE SCREW OUT. BLUNT THE SCREW ON GRINDER AND THEN THREAD IT IN THE PUNCTURE.

FIGURE 44–105 A screw may be used in some cases to hold the epoxy in the puncture hole.

44.26 CONDENSER LEAKS

Condensers in refrigerators are usually made of steel. Leaks usually do not occur in the middle of the tube, but at the end where connections are made or where a tube vibrated against the cabinet, causing a hole. This part of the system operates at the high-pressure condition and a small leak will lose refrigerant faster than the same size leak in the low-pressure side, Figure 44–106. Wherever the leak occurs in

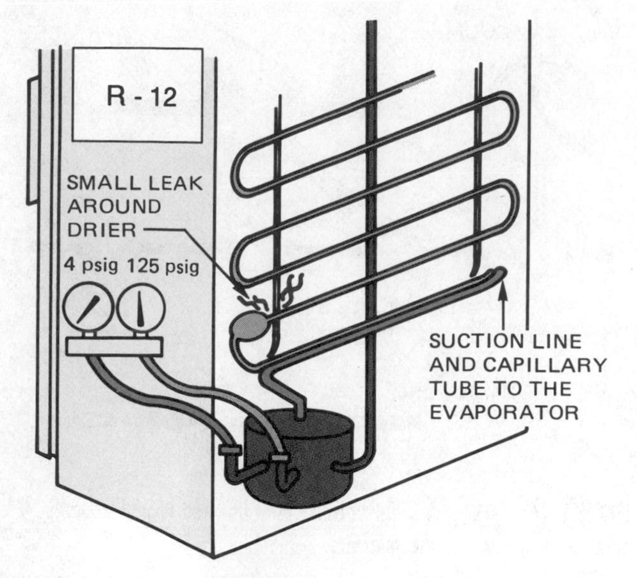

FIGURE 44–106 A small leak in the high-pressure side of the system will cause a greater refrigerant loss than a leak on the low-pressure side.

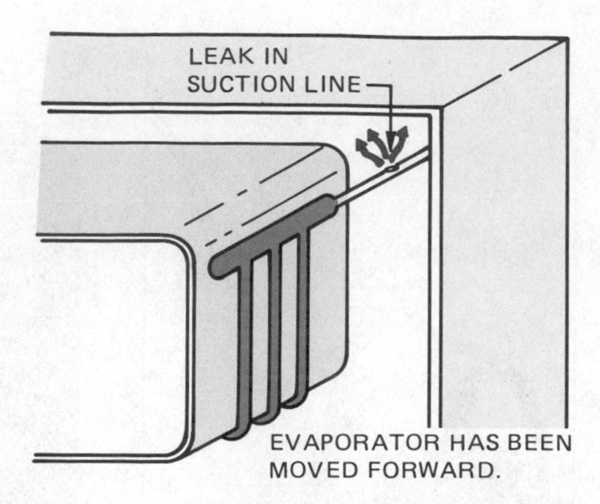

FIGURE 44–108 Piping leak on the suction line inside the box.

steel tubing, the best repair is solder. The correct solder must be used, one compatible with steel. Usually it has a high silver content. When flux is used, be sure to clean it away from the connection after the repair is made. Always leak check after the repair is made.

When the condenser is located under the refrigerator, it is often hard to gain access for leak repair. The box may be tilted to the side or back for this repair, Figure 44–107.

44.27 REFRIGERANT PIPING LEAKS

Leaks in the interconnecting piping may occur in the walls of the box. Fortunately this does not occur often because it can be difficult to make the repair and may not be economical. The evaporator may have to be removed to repair a leak in the adjacent evaporator piping, Figure 44–108. Each box has a different method for removal. The manufacturer's literature may be used. If it is not available, you may have to determine the procedure on your own. When a box has foam insulation, it is possible that it may not be disassembled. Leaks are often repairable in older boxes that used fiberglass for insulation, but it may be less expensive to purchase a new refrigerator.

When leaks occur in the wall of a fiberglass insulated box after many years, moisture in the insulation may be the cause due to electrolysis. This is caused by mild acid and current flow and usually occurs with aluminum or steel tubing. If one leak occurs due to electrolysis, more leaks will usually occur soon because the tubing is probably thin in several other places. The best repair is replacement of the box.

Leaks may occur in the connection where the copper suction line is attached to the aluminum tubing leaving the evaporator, Figure 44–109. This is a hard place for a repair particularly because of the dissimilar metals. A flare union is sometimes used in this location. Some manufacturers may have repair kits for this connection.

44.28 COMPRESSOR CHANGEOUT

⊛**Compressors may be changed in a refrigerator by recovering the refrigerant charge and removing the old compressor.**⊛ A new compressor should be ready for replacement before the old one is removed. An exact replace-

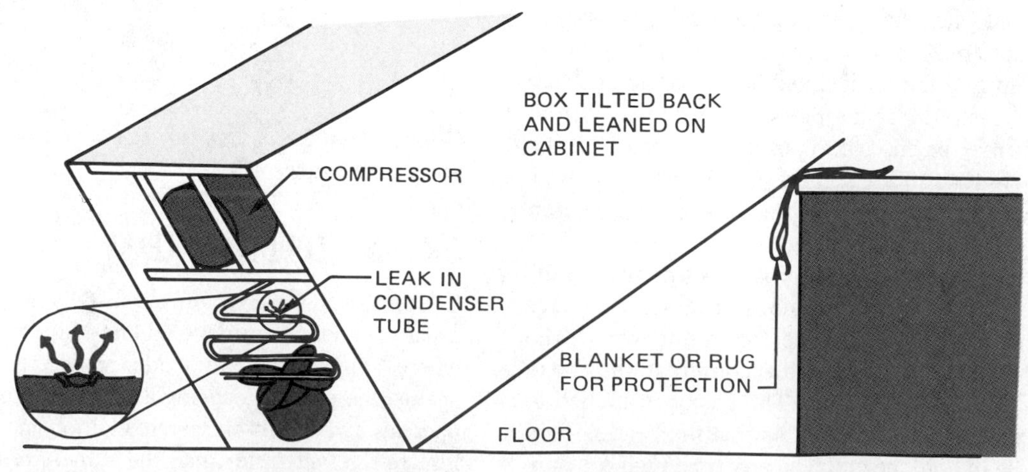

FIGURE 44–107 A refrigerator may be tilted to one side or the back to service the condenser under the box.

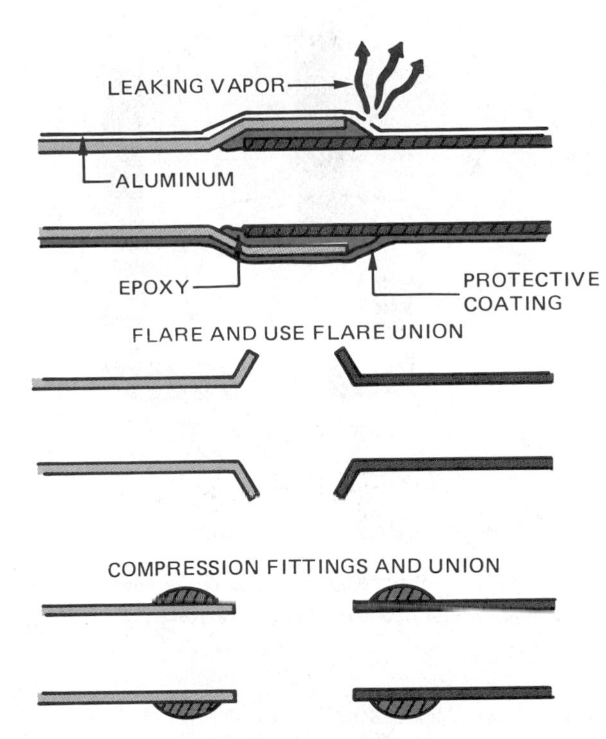

FIGURE 44–109 Procedures for repairing a leak in the connection between the copper suction line and the aluminum tubing leaving the evaporator.

ment is the best choice but may not be available. A diagram of the tube connections and the mounting should be available to help you connect it correctly. Remember, there may be several lines, including suction, discharge, suction access, discharge access, and two oil cooler lines that may all be the same size, Figure 44–110.

⊛**The best way to remove the lines from the old compressor is to recover the charge and pinch the lines off close to the compressor, Figure 44–111, or cut them using a very small tubing cutter.**⊛ If the old tubing is removed with a torch, the tubing ends should be cleaned using a file to remove excess solder, being careful not to allow filings to enter the system. ***Some of the lines may contain oil, which may flame up when separated with a torch. A fire extinguisher should always be present.*** The old tubes should be filed until they are clean and will slide inside the compressor fittings, Figure 44–112. Approved sand tape (sand tape with nonconducting abrasive) should be used to further clean the tubing ends. They must be

FIGURE 44–110 A compressor may have many lines from the shell. *Photo by Bill Johnson*

perfectly clean with no dirty pits, Figure 44–113. Dirt trapped in pits will expand into the solder connection when heated, Figure 44–114, and cause leaks.

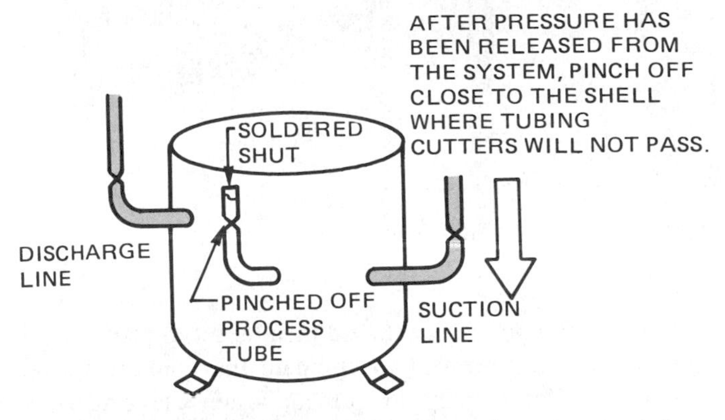

FIGURE 44–111 Pinch the old lines close to the compressor shell using side-cutting pliers.

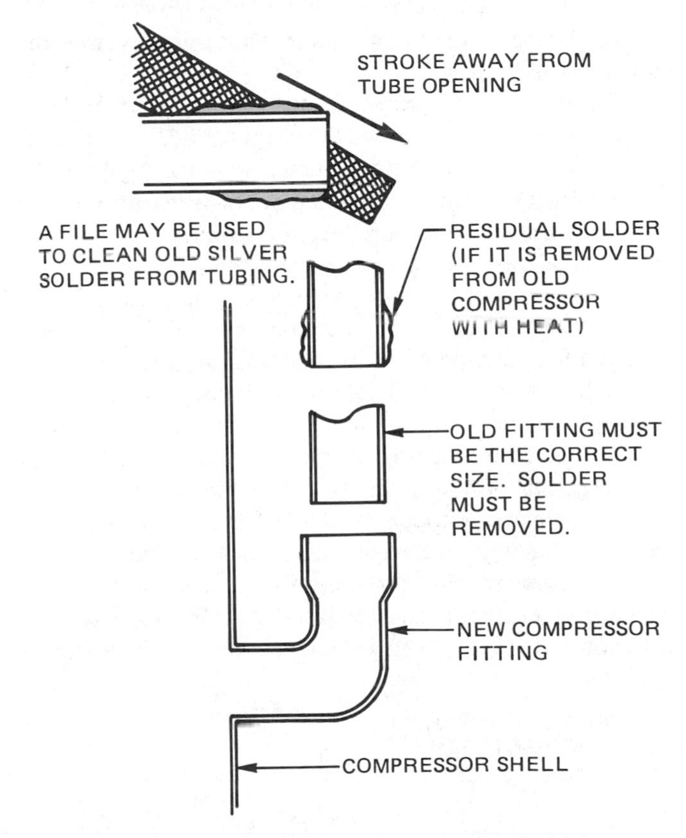

FIGURE 44–112 The old tubing must be cleaned of all old solder; a file may be needed.

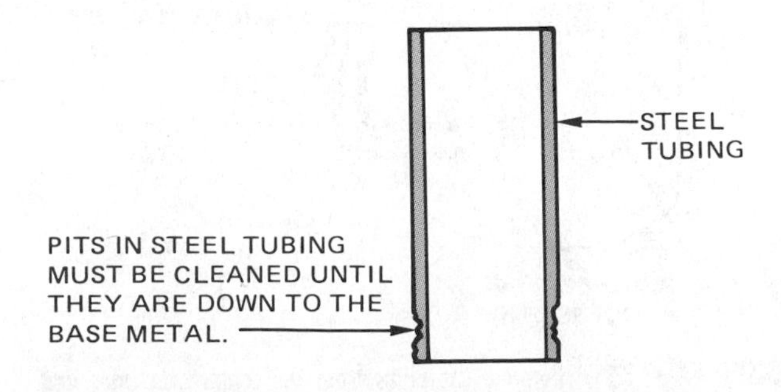

FIGURE 44–113 All pits in steel tubing must be cleaned.

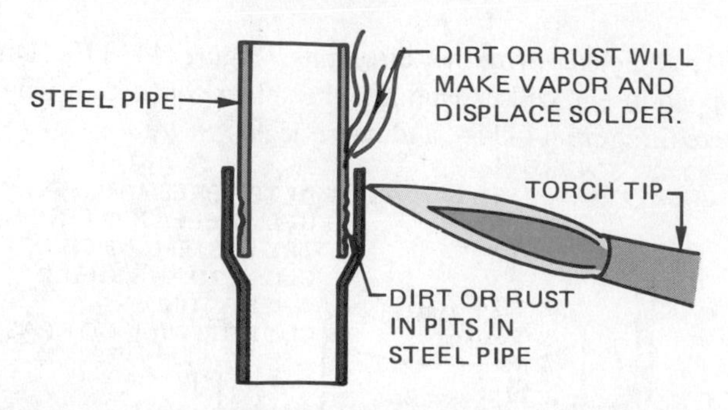

FIGURE 44–114 Dirt trapped in pits will expand when heated and cause leaks.

The new compressor should then be set in place in the compressor compartment to compare the connections on the box with those on the compressor. When it is certain that the connections line up correctly, remove the new compressor from the compartment. Remove the plugs from the compressor lines and clean the tubing ends, Figure 44–115. ***Care should be taken not to allow anything to enter the compressor lines.***

Set the new compressor in place and connect all lines. Flux may be applied if this is the last time the lines are to be connected before soldering, Figure 44–116. ***Solder the connections carefully, being particularly careful not to overheat the surrounding parts or cabinet.*** A shield of sheet metal may be used to prevent the heat of the torch flame from touching the surrounding components and cabinet, Figure 44–117. Use the minimum of heat recommended for the type of connection you are making.

While the compressor is being soldered to the lines, it is a good time to solder process tubes to the compressor. Sometimes a "tee" fitting is soldered into the suction and discharge line with a Schrader fitting, Figure 44–118.

A filter-drier should be added if the system has been open long enough to change the compressor. The refrigerator manufacturer may recommend using a liquid-line drier of the correct size. If an oversized liquid-line drier is used, additional refrigerant charge must be added. Many techni-

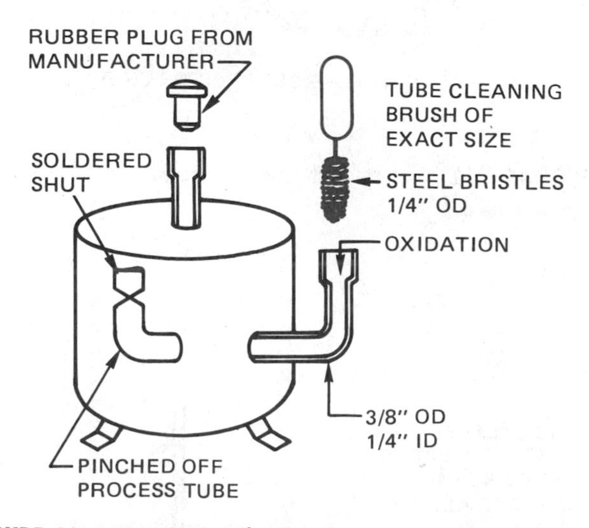

FIGURE 44–115 Remove the plugs from the compressor lines and clean the tubing ends.

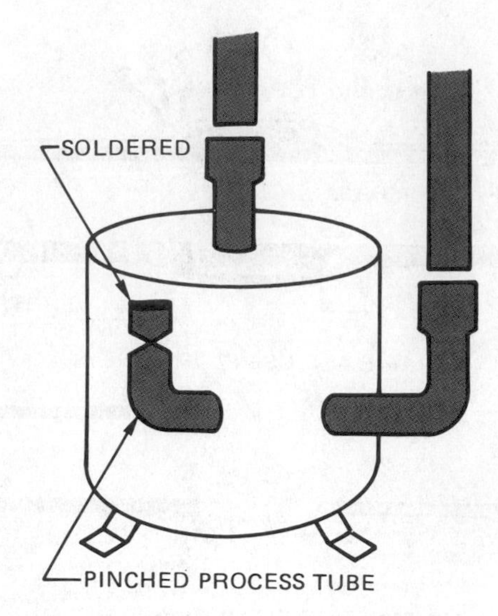

FIGURE 44–116 *Do not apply flux until the lines are being fastened together for the last time before soldering.*

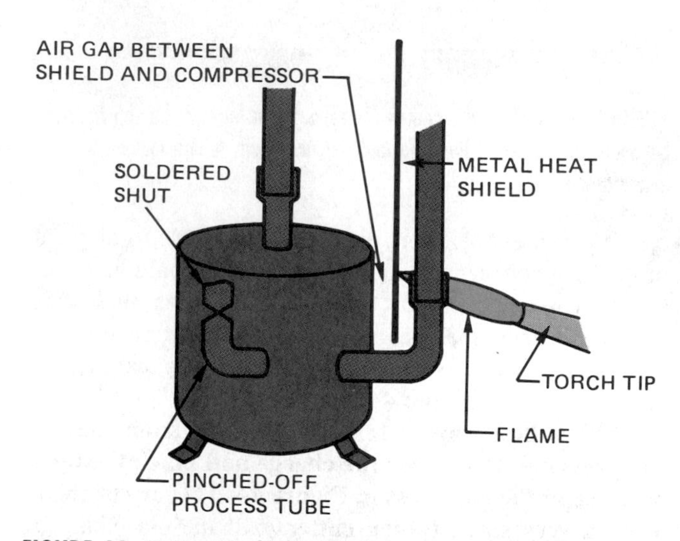

FIGURE 44–117 *A shield may be used to protect the surrounding components and cabinet from heat while soldering.*

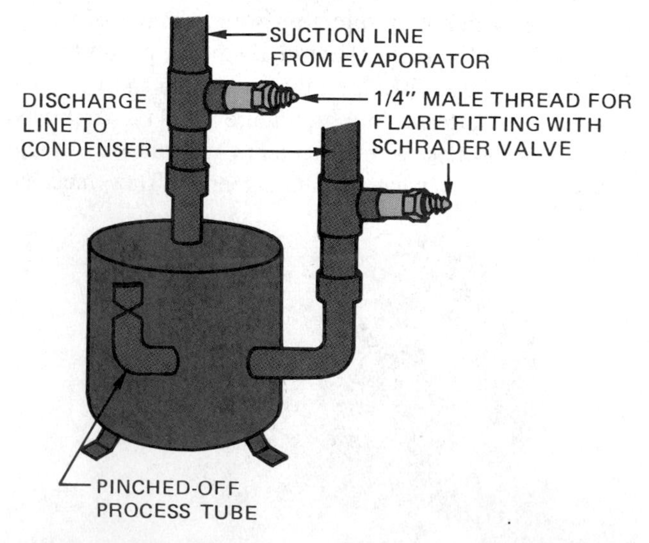

FIGURE 44–118 A "tee" fitting may be soldered in the suction and discharge line for service connection while the system is open.

cians use a suction-line drier in this case because it is in the suction vapor at low pressure and will not require added charge, Figure 44–119. The suction-line drier does not protect the capillary tube from particles or moisture, but the capillary tube will have its own strainer for particle protection and there should not be any moisture with a correct evacuation.

It is recommended that after the compressor is installed, the system be swept with nitrogen and then the drier installed. This prevents drier contamination with whatever may be in the system. This sweep may be accomplished by cutting the liquid line or the suction line, wherever the drier is going to be installed. Connect the gage manifold to the service ports and a cylinder of nitrogen. Allow vapor to flow first into the high side of the system. This will force pressure into the high side, then through the capillary tube, through the evaporator and out the loose suction line (this example is for suction-line installation), Figure 44–120. The vapor flowing from the suction line will be very slow because it is moving through the capillary tube, but it will sweep the entire system, except the compressor, which is new.

Now connect the drier by opening the whole system to the atmosphere by leaving gage valves open, so there will

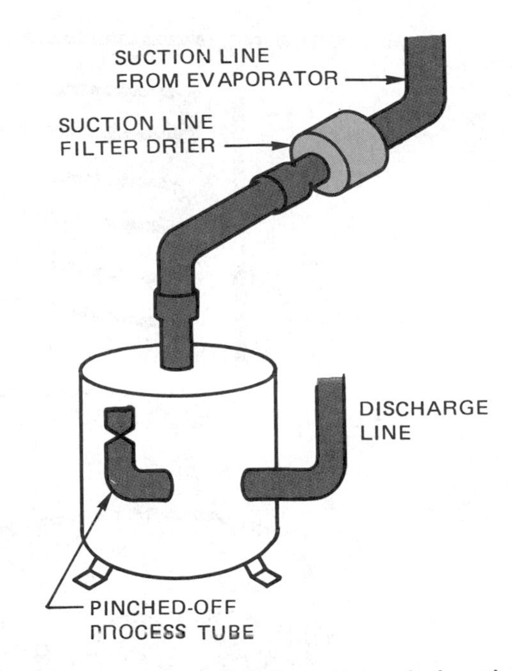
SUCTION LINE
FROM EVAPORATOR
SUCTION LINE
FILTER DRIER
DISCHARGE
LINE
PINCHED-OFF
PROCESS TUBE

FIGURE 44–119 A suction-line drier added to the low side will not require extra charge because the refrigerant is in the vapor state.

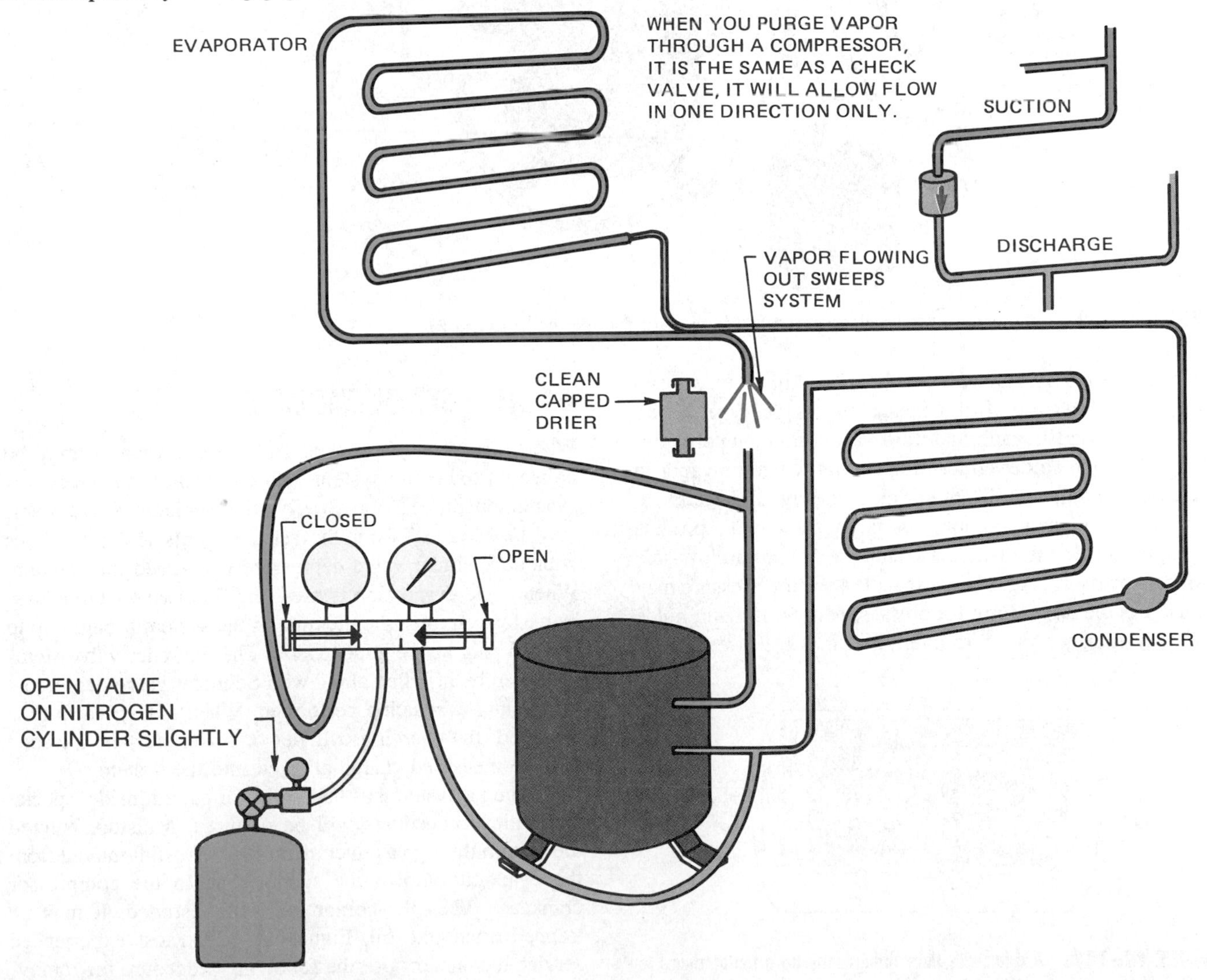
EVAPORATOR
WHEN YOU PURGE VAPOR
THROUGH A COMPRESSOR,
IT IS THE SAME AS A CHECK
VALVE, IT WILL ALLOW FLOW
IN ONE DIRECTION ONLY.
SUCTION
DISCHARGE
VAPOR FLOWING
OUT SWEEPS
SYSTEM
CLEAN
CAPPED
DRIER
CLOSED
OPEN
CONDENSER
OPEN VALVE
ON NITROGEN
CYLINDER SLIGHTLY

FIGURE 44–120 Purging or sweeping a system.

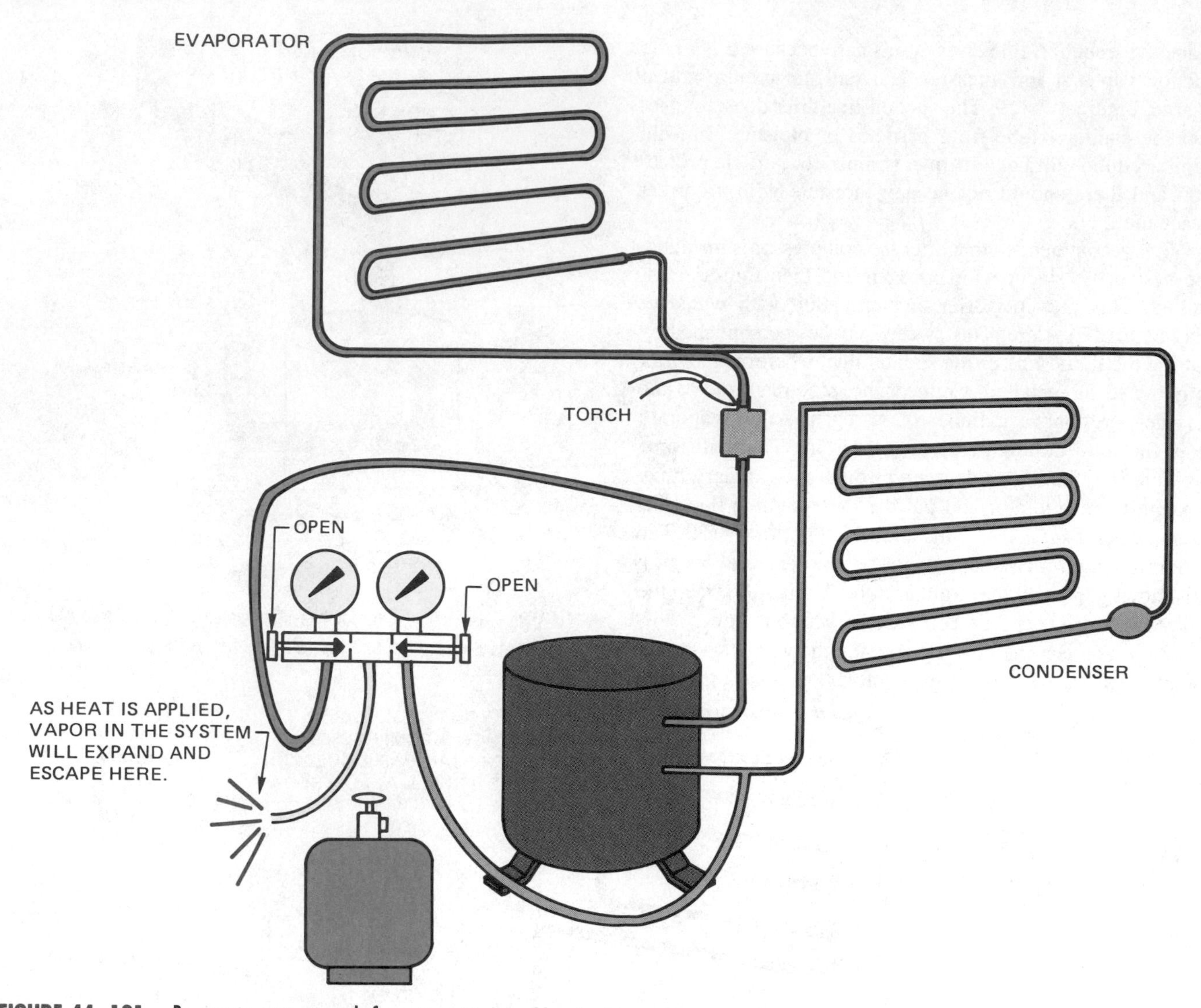
EVAPORATOR
TORCH
OPEN
OPEN
CONDENSER
AS HEAT IS APPLIED,
VAPOR IN THE SYSTEM
WILL EXPAND AND
ESCAPE HERE.

FIGURE 44–121 Be sure to open gages before attempting to solder the drier in the suction line.

be no pressure buildup, Figure 44–121. You may want to use a drier with a flare fitting instead of a solder type, Figure 44–122. Close the gage manifold valves immediately after completing the solder connection so that when the vapor in the system cools it will not shrink and draw air inside.

When the new compressor is in place with process tubes, leak check the whole assembly at maximum low-side working pressure. Again the low-side working pressure may be used as the upper limit for pressure testing. If a unit holds 150 psig nitrogen pressure overnight, it is leak free.

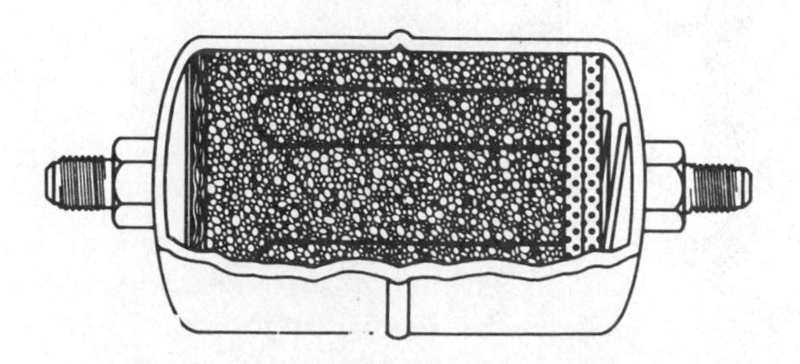

FIGURE 44–122 A drier with flare fittings may be a better choice to avoid soldering. *Courtesy Mueller Brass Co.*

44.29 SYSTEM EVACUATION

When the system is proved leak free, a vacuum may be pulled on the entire system with confidence. Unit 8 covers evacuation procedures. Briefly, the Schrader valve stems may be removed from the service stems and gage lines without Schrader valve depressors will speed the vacuum. When triple evacuation is used, the first two vacuums may be performed, then the pressure in the system brought up to about 5 psig above atmospheric. The Schrader valve stems may then be installed along with Schrader valve depressors and a final evacuation performed. When a deep vacuum is achieved, the charging cylinder or scales may be attached and the measured charge allowed into the system.

When a system has had moisture pulled inside, special evacuation procedures will be required. Moisture entered the system through a puncture during defrost. It moved from the evaporator down the suction line to the compressor crankcase when the compressor was restarted. It may be trapped under the oil, Figure 44–123. Most experienced service technicians use the following procedure for removing moisture.

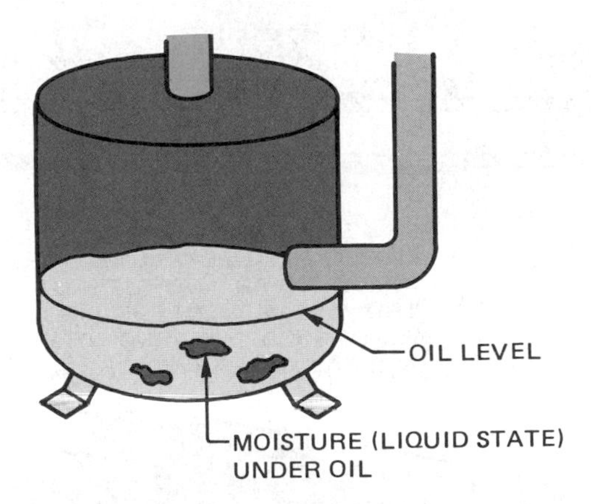

FIGURE 44–123 Moisture may be trapped under the oil in the compressor.

Install full-size gage connections, such as "tees" in the suction and discharge lines so evacuation is from both sides of the system, Figure 44–124. ***Make sure there are no Schrader depressors in the gage lines or fittings on the compressor.*** Use a two-stage rotary vacuum pump and

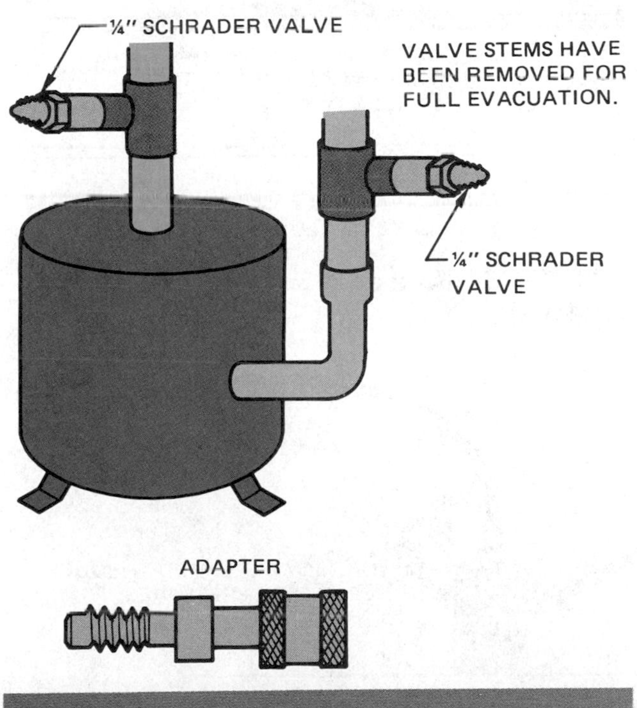

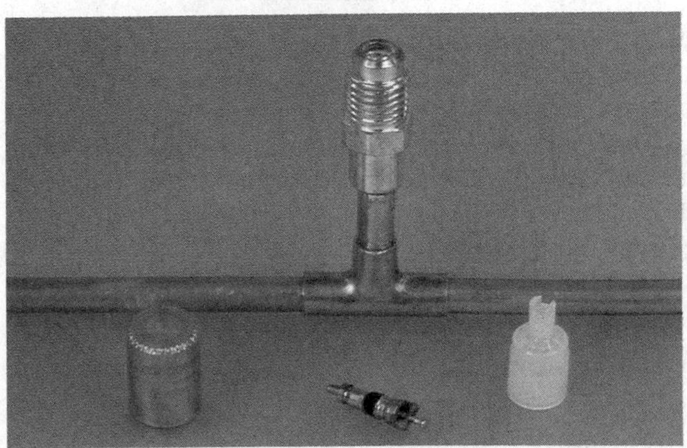

FIGURE 44–124 Full-size gage connections must be installed for the purpose of removing moisture. *Photo by Bill Johnson*

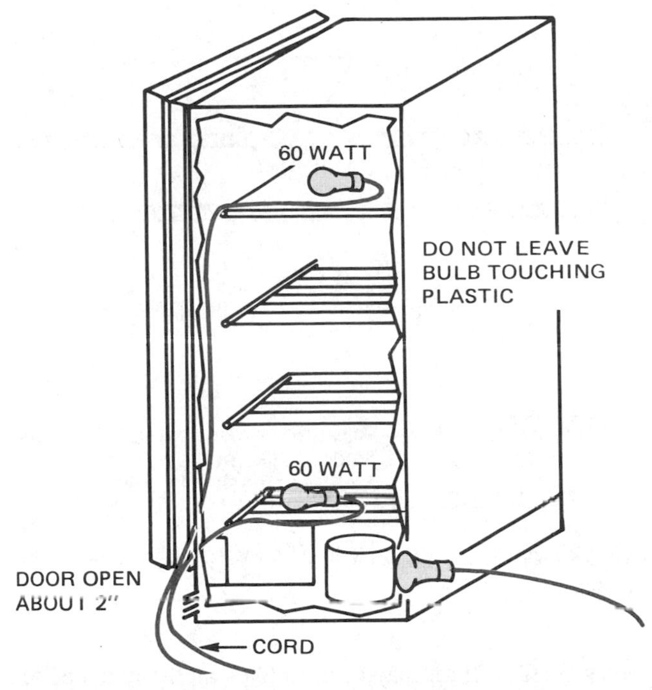

FIGURE 44–125 Placing light bulbs in a refrigerator for the purpose of heating the refrigerant circuit and not overheating the plastic.

start it. Apply heat to the compressor crankcase by placing a light bulb next to the compressor. Place a small light bulb (about 60 W) in both the low-and medium-temperature compartments and partially shut the doors, Figure 44–125. ***Do not shut the doors all the way or the heat will melt the plastic inside the refrigerator.*** Allow the vacuum pump to run for at least 8 h. Break the vacuum using the refrigerant characteristic to the system and pull another vacuum. This time monitor it with a manometer or electronic vacuum gage, see Unit 8. When a deep vacuum has been achieved, such as 1 mm Hg or 500 microns, break the vacuum with refrigerant to atmospheric pressure and remove the heat. Cut the suction line and install a suction-line filter-drier. The liquid-line drier may also be replaced if the manufacturer recommends it. It may be more trouble than it is worth because the previous evacuation procedure will remove some of the moisture from this drier giving it some capacity. The new suction-line drier will give the system enough drier capacity. Pull one more vacuum and charge the unit with a measured charge.

44.30 CAPILLARY TUBE REPAIR

Capillary tube repair may involve patching a leak in the tube because it rubbed against some other component or the cabinet. You may need to change the drier strainer at the tube inlet. Repair may consist of clearing the tube from a partial restriction or even replacing a capillary tube. Whatever the repair, the capillary tube must be handled with care because it is small and delicate. It can be pinched easily.

When a capillary tube must be cut for any reason, the following is recommended. Use a file and file the tube nearly in two, then break it the rest of the way, Figure 44–126. Examine the end and clean any particles from the end. A

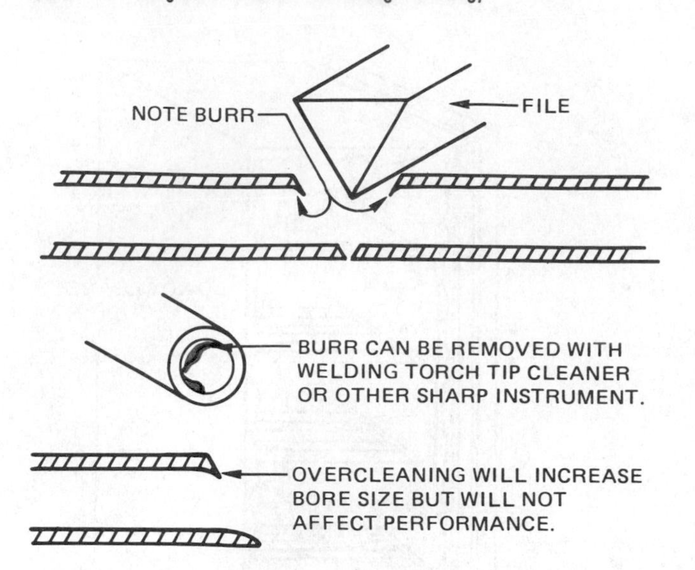

FIGURE 44–126 Use a file to cut a capillary tube to the correct length.

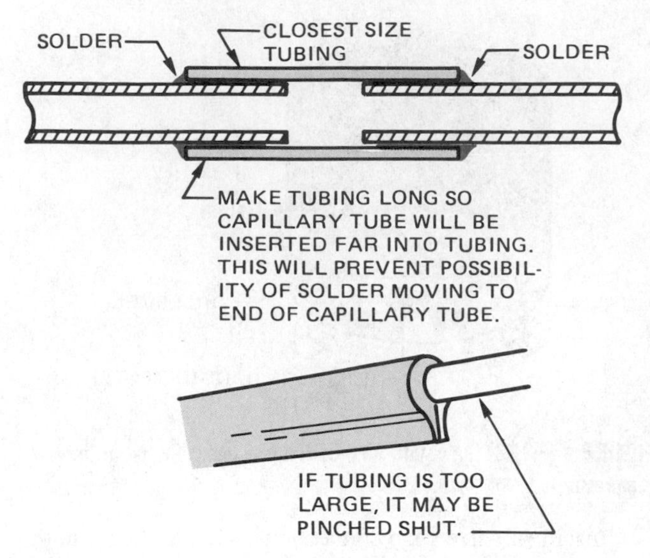

FIGURE 44–128 Repairing a broken capillary tube.

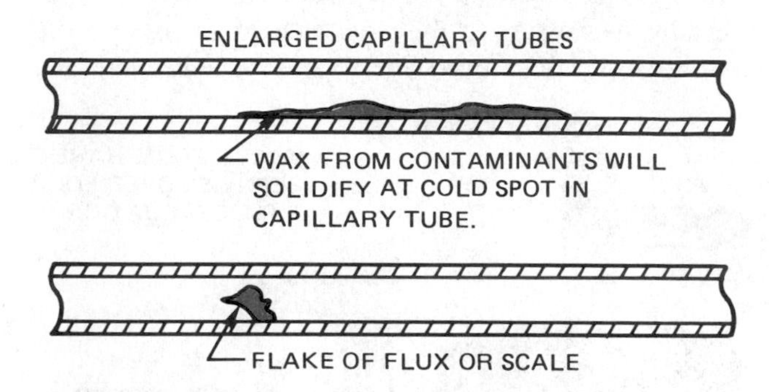

FIGURE 44–129 Wax or a small particle may enter and lodge in the capillary tube.

very small drill bit may be used to clean the tube end to the full bore. ***The tube end must have the full dimension of the inside diameter or it will cause a restriction. Proper care at this time cannot be overemphasized.***

When it is necessary to solder a capillary tube into a fitting, such as in the end of a new strainer, do *not* apply flux or clean the tube all the way to the end. Allow the outside of the end of the tube to remain dirty because the solder will not flow over the end of the tube if it is not cleaned, Figure 44–127.

A capillary tube that is broken in two may usually be repaired by cleaning both ends so the inside dimension is maintained then pieced together with a size larger tubing, Figure 44–128. When a capillary tube has a rub hole in it, it is recommended that the tube be cut in two at this point and repaired as a broken tube.

Capillary tube cleanout may sometimes be accomplished with a capillary tube pump. Usually the restriction is wax or a small particle lodged in the capillary tube, Figure 44–129. A capillary tube pump may pump an approved solvent or oil through the tube at a great pressure until the tube is clear, Figure 44–130. The only way that you will know for sure that the tube is clear is to charge the system, start it, and observe the pressures. Of course this means putting the system back to normal working order with a leak check and evacuation. This can be a lot of trouble only to find that the tube is still partially blocked.

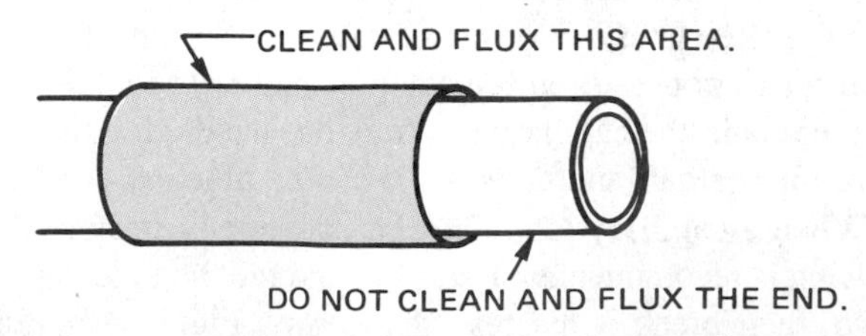

FIGURE 44–127 Application of flux to a capillary tube to prevent solder from entering the tube.

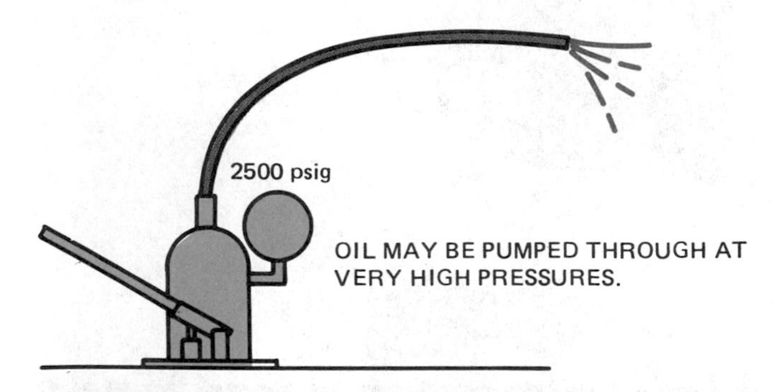

FIGURE 44–130 A capillary tube-cleaning pump may be used.

The capillary tube may be changed in some instances, but usually it is not economical because it is fastened to the suction line for a capillary tube-to-suction line heat exchange. You can never duplicate this connection exactly. You may insulate the suction line with the capillary tube under the insulation and come very close, Figure 44–131.

44.31 COMPRESSOR CAPACITY CHECK

One of the most difficult problems to diagnose is a compressor that is pumping to partial capacity. The customer may complain that the unit is running all the time. It may still have enough capacity to maintain the food compartments at reasonable temperatures. One of the first signs that

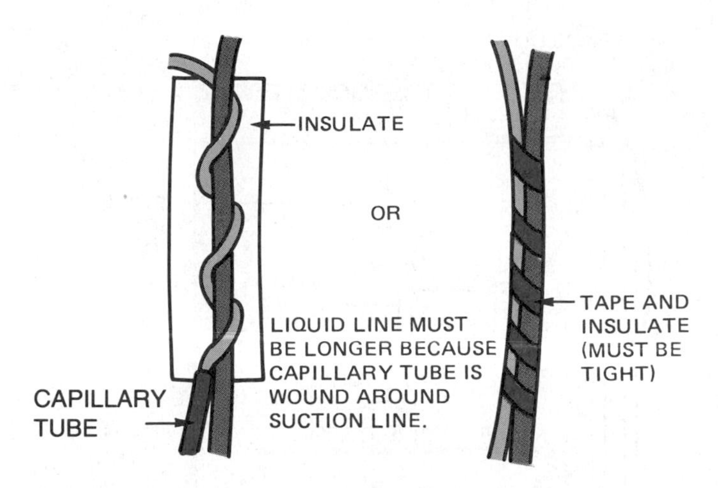

FIGURE 44–131 You may pass the capillary tube close to the suction line under insulation and get some heat exchange.

the compartment temperature is not being maintained is that the ice cream will be slightly soft, when it has not been previously. Liquids such as water or milk served from the fresh-food compartment may not seem as cool. When the above complaint of running all the time and not holding conditions is noticed, either the refrigerator has a false load or the compressor is not operating to capacity. You may have to decide which.

The first thing to check is to make sure that all door gaskets are in good condition and that the doors shut tight. Then check to make sure there is no extra load such as hot food being put in the refrigerator too often, Figure 44–132. The defrost hot gas solenoid may be leaking hot gas through to the low side of the system, Figure 44–133. The light bulb

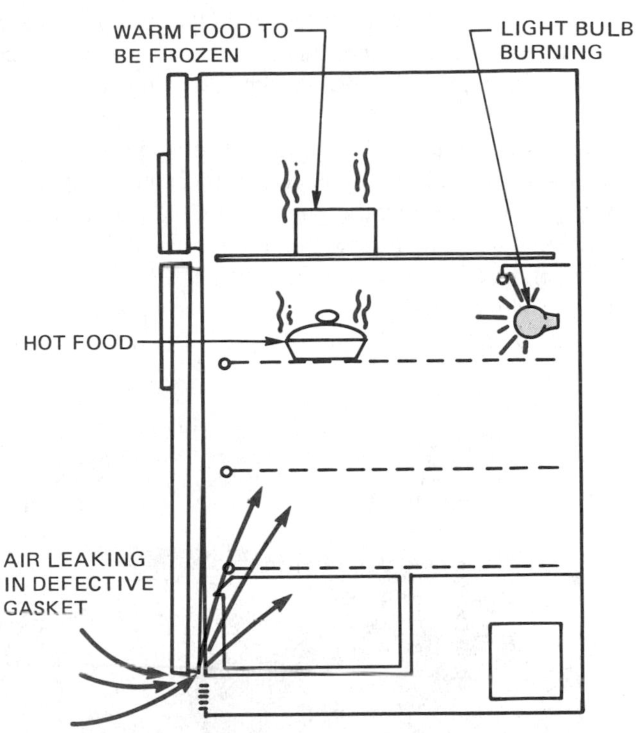

FIGURE 44–132 Extra load on a refrigerator caused by defective gaskets, hot food placed in the box, or light bulb burning all the time.

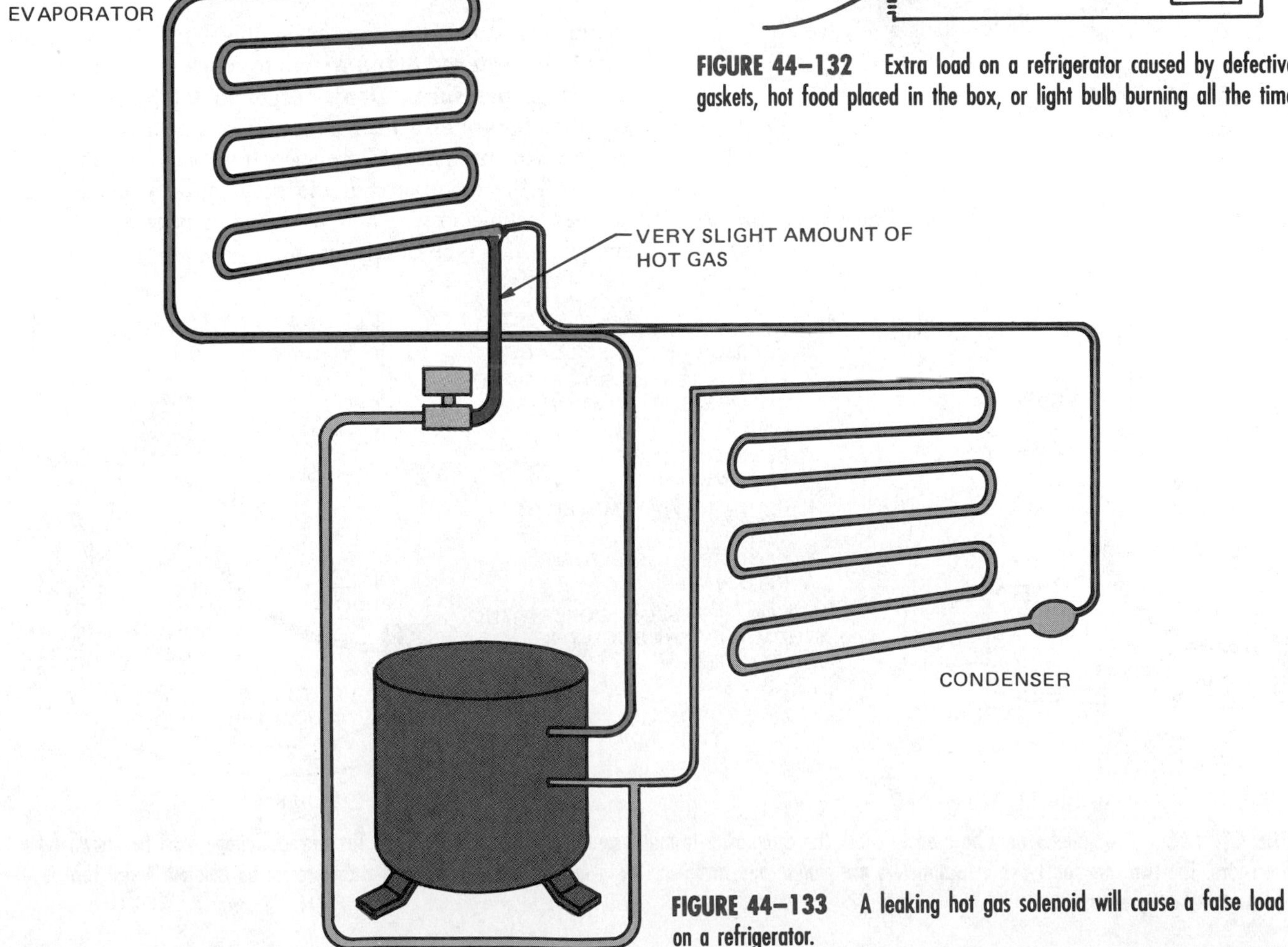

FIGURE 44–133 A leaking hot gas solenoid will cause a false load on a refrigerator.

may be on all the time in one of the compartments. Make sure that the condenser has the proper airflow, with all baffles in place for forced draft, Figure 44–134, and that it is not under a cabinet if natural draft. Make sure the unit is not in a location that is too hot for its capacity. Usually any temperature over 100°F will cause the unit to run all the time, but it should maintain conditions. A unit located outside, particularly where it is affected by the sun, will always be a problem because the unit is not designed to be an outside unit. When you have checked all of this and everything proves satisfactory, then suspect the compressor.

The manufacturer's literature is invaluable for the test. A thermometer lead should be placed in each the fresh- and frozen-food compartment, Figure 44–135. A wattmeter can be placed in the compressor common electrical line and the wattage of the compressor can be verified. If the compressor wattage for the conditions is low, the compressor is not doing all of its work, Figure 44–136. Perform the low-

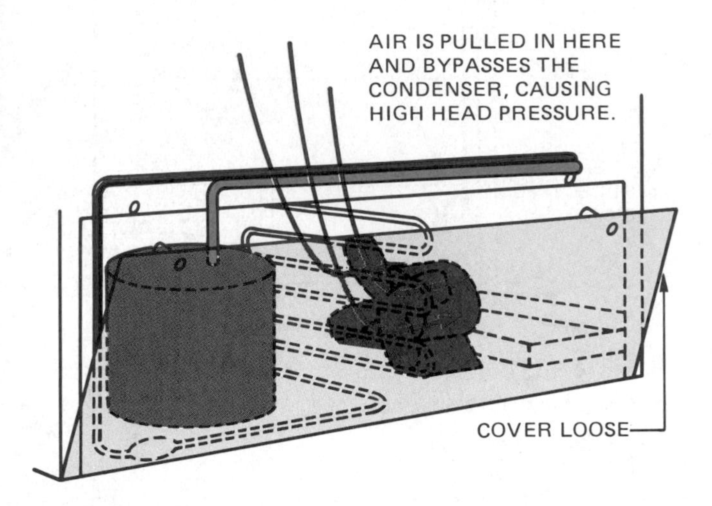

FIGURE 44–134 Make sure all baffles are in place in the condenser area for correct airflow.

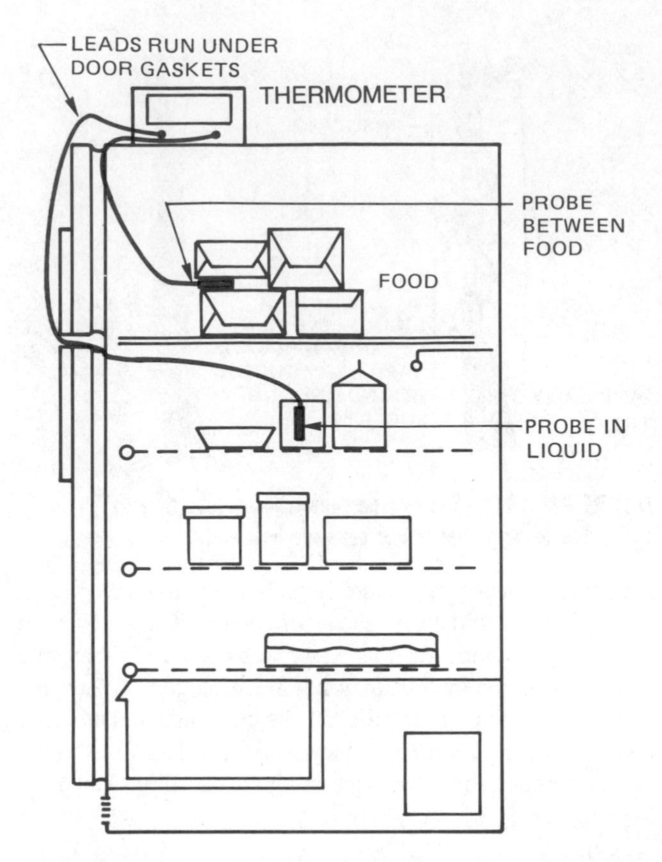

FIGURE 44–135 A thermometer with leads may be used to check the compartment temperatures.

charge test mentioned earlier in this unit. ***Now gages should be fastened to the system to check the suction and discharge pressures. Don't forget to let the pressures equalize before attaching gages.*** Use the manufacturer's literature to check the performance. If it is not available, you may call the distributor of the product, or you may call another technician who may have had considerable experience with this type of appliance.

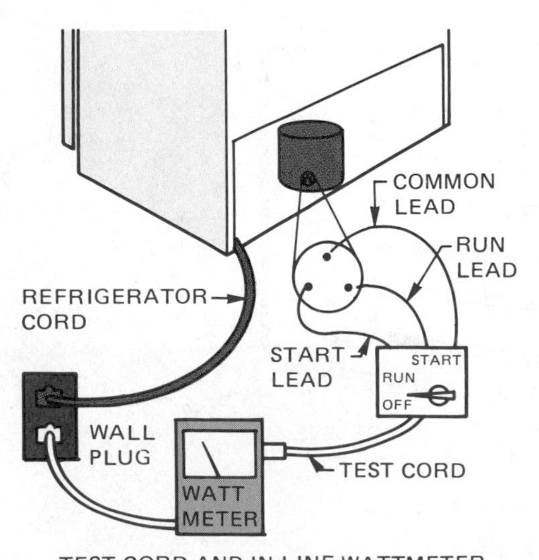

1. REMOVE WIRING FROM COMPRESSOR AND TAPE LEADS.
2. FASTEN TEST CORD LEADS TO COMPRESSOR, COMMON, RUN, AND START.
3. PLUG TEST CORD INTO WATTMETER.
4. PLUG THE WATTMETER INTO WALL.
5. PLUG REFRIGERATOR INTO WALL.

NOW COMPRESSOR IS OPERATING NORMALLY THROUGH THE WATTMETER.

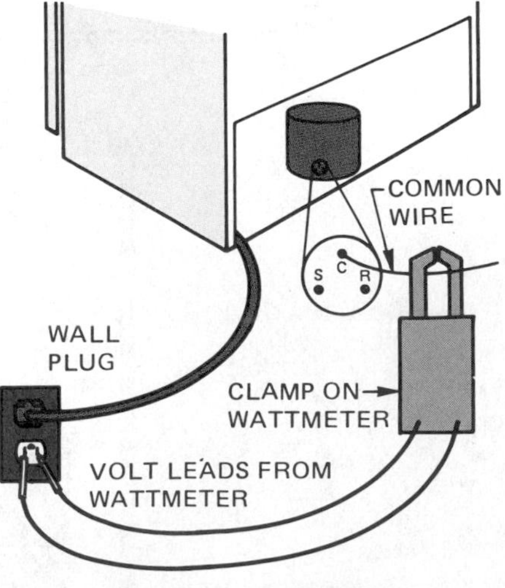

FIGURE 44–136 A wattmeter may be used to check the compressor to make sure it is working to capacity. The correct wattage must be known for this procedure. The two common types of wattmeters are plug-in and clamp-on. The plug-in type requires that the compressor be isolated. A test cord is used to start the compressor.

HVAC GOLDEN RULES

When making a service call to a residence:

- Make and keep firm appointments. Call ahead if you are delayed. The customer's time is valuable also.
- Keep customers informed if you must leave the job for parts or other reasons. Customers should not be upset if you inform them of your return schedule if it is reasonable.

Added Value to the Customer. Here are some simple, inexpensive procedures that may be included in the basic service call.

- Clean the condenser.
- Make sure the condenser fan is turning freely, where applicable.
- Check the refrigerator light and replace if needed.
- Make sure evaporator coil has no ice buildup.
- Make sure box is level.

44.32 SERVICE TECHNICIAN CALLS

SERVICE CALL 1

A customer calls and describes to the dispatcher that a new refrigerator is sweating on the outside of the cabinet between the side-by-side doors. *The problem is a defective connection at the back of the refrigerator in a mullion heater circuit.*

The technician arrives and can see from the beginning that a mullion heater is not heating. The house temperature and humidity are normal because the house has central air conditioning. The technician dreads pulling the panels off to get to the heater between the doors if it is defective because this is a difficult time-consuming job. A look at the diagram on the back of the box reveals a junction box at the back corner of the unit where the mullion heater wires are connected before running to the front. The refrigerator is unplugged and the junction box is located. The correct wires are located and the connection is checked. It seems loose, but the technician wants to be sure so the connection is taken apart. An ohm check of the heater circuit proves the heater has a complete circuit. The connection is made back in a secure manner and the refrigerator is plugged in and started. Power is checked at the connection. Power is available from the neutral wire to the hot wire going to the heater. To make sure, the technician applies the ammeter to the circuit only to find that it seems to be passing no current. A lower scale is used; the heater has a very low current draw. The wire is wrapped around the ammeter jaws to obtain a reading, Figure 44–137. When this is done it is determined that current is flowing and the technician feels confident the problem has been corrected.

SERVICE CALL 2

A customer reports that the refrigerator is running all the time and not keeping ice cream hard. The milk does not seem cool enough. *The problem is the condenser fan is not running, the unit is old, and the bearings are seized.*

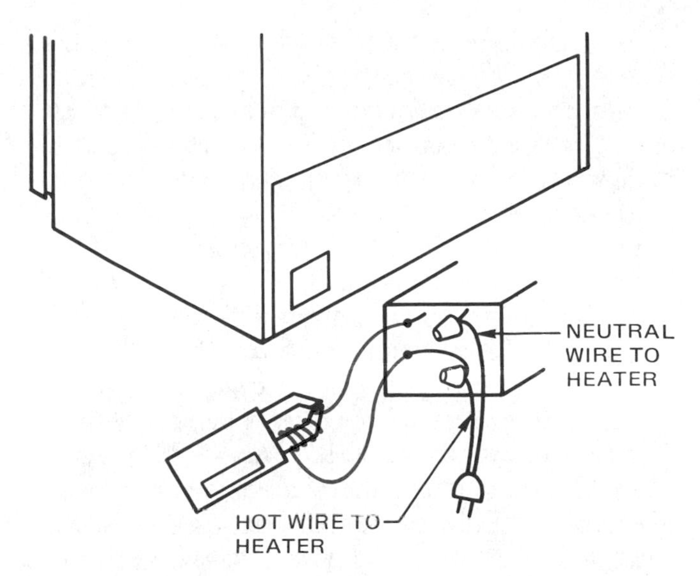

FIGURE 44–137 Wrapping the heater wire around the ammeter jaws to amplify the reading.

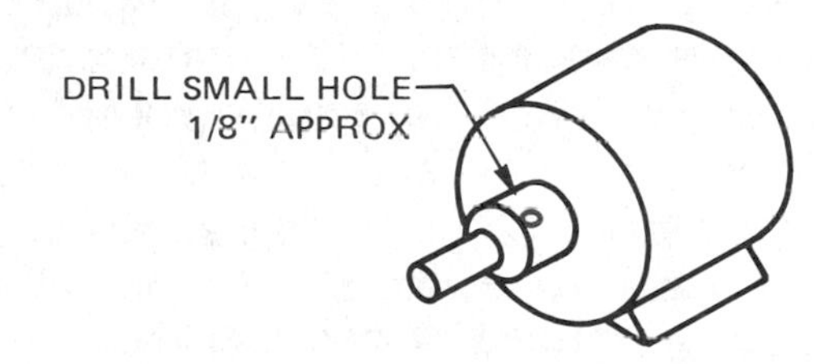

FIGURE 44–138 The motor housing may be drilled and penetrating oil forced into the bearings for a temporary repair. *The only permanent repair is to replace the motor.*

The technician observes the inside of the freezer and can tell the unit is not cold enough because of the ice cream. When the technician leans over to listen for the condenser fan, it is noticed that it is not operating. The refrigerator is pulled out from the wall. The fan can be observed from the back and it is not running. The motor is hot, so it is getting power but not turning. This is an impedance-protected motor with no overload protection. It can set with power to the leads and not turn, and it will not burn out. The technician unplugs the box and checks the fan to see if it will turn. It is very tight. To make a temporary repair, a small hole is drilled in the bearing housing and penetrating oil is forced into the hole, Figure 44–138. The fan blade is worked back and forth until it is free and will turn over easily. Motor bearing oil is applied after the penetrating oil. Aluminum foil is formed under the motor to catch any excess oil that may drip and the motor is plugged in. The fan motor starts and runs satisfactorily.

The technician explains to the owner that a new fan motor will be ordered and the next time a call close by is received the motor will be changed.

SERVICE CALL 3

A customer called to indicate that a refrigerator is not running. Something making a clicking sound from time to time can be heard.

The owner is advised to turn the unit off or unplug it until the technician arrives. *The problem is the compressor is stuck. The electrical circuit to the motor windings is shutting off because of the overload.*

The technician pulls the box from the wall and clamps an ammeter to the compressor common terminal before starting the compressor. When the unit is turned on, the compressor amperage rises to 20 A and before the technician can shut it off, it clicks and shuts off because of the overload.

The technician must now determine if the compressor will not start because of electrical problems or internal problems. A compressor starting test cord is brought in from the truck. The unit is unplugged. The three wires are removed from the motor terminals and the test cord attached to common, run, and start, Figure 44–139. The cord is plugged in and an ammeter clamped around the common wire. A voltmeter is attached to the common and run leads to ensure correct voltage is present. The test cord switch is rotated to start and the compressor hums, the amperage is still 20 A indicating a stuck compressor. The voltage is 112 V, well within the correct voltage limits, Figure 44–140. The technician places a start capacitor in the start circuit of the test cord and tries to start the compressor; again, it does not start, Figure 44–141. The run and start wires are reversed, Figure 44–142. This will allow the technician to start the compressor in reverse rotation. The compressor starts. ***It is stopped quickly because it cannot run in this mode for long without causing damage.*** The leads are reversed and the compressor is tried again. It starts correctly this time. The technician unplugs the test cord and fastens the compressor start circuit back to the compressor. The ammeter is clamped around the common wire. When the refrigerator is plugged in, the compressor starts and runs with normal amperage.

The technician has explained the procedure to the customer. The technician is quick to point out that something would not allow the

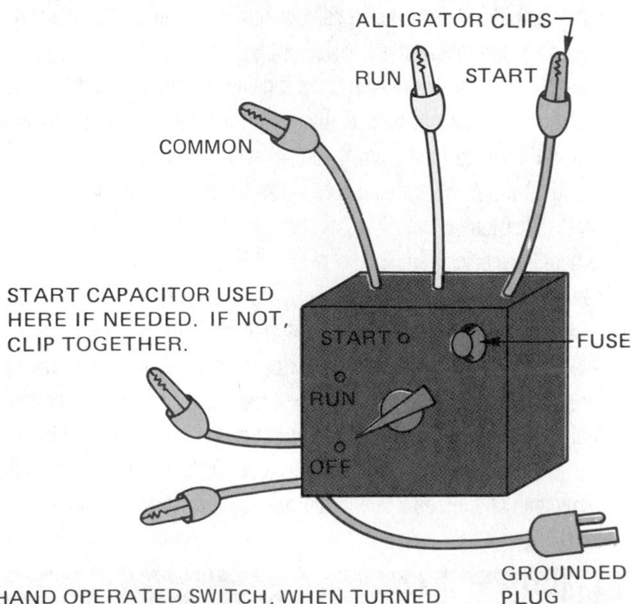

FIGURE 44–139 Hermetic starting test cord.

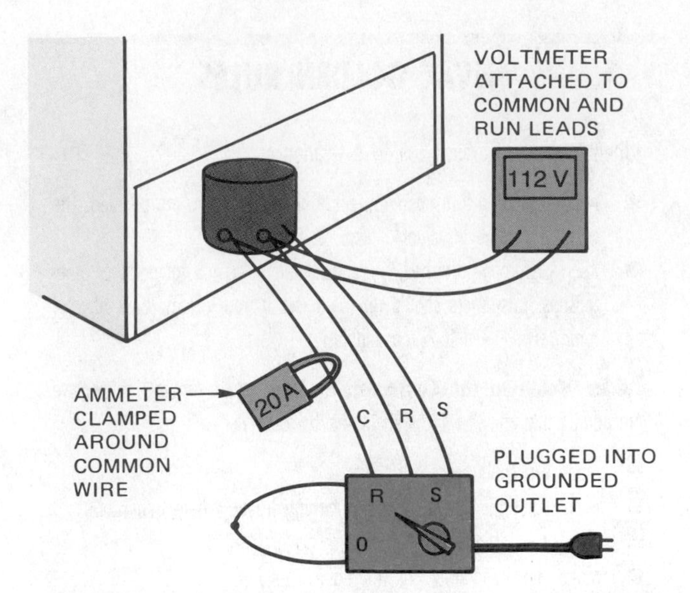

FIGURE 44–140 Voltage and amperage readings are recorded at the time the compressor is started.

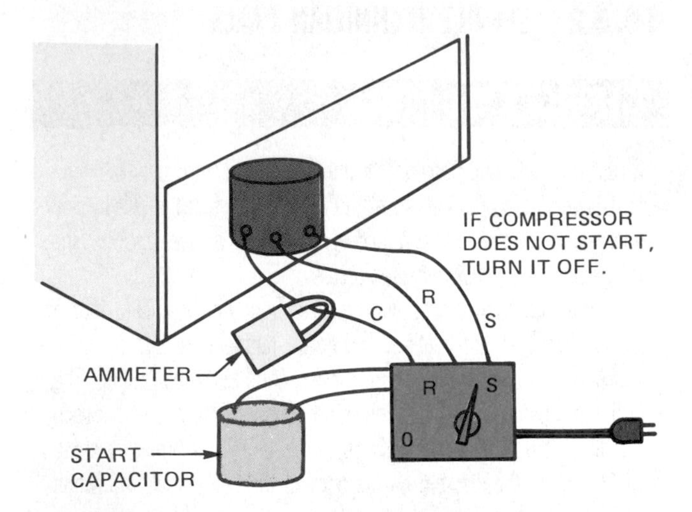

FIGURE 44–141 When stuck, a start capacitor may be placed in the start windings to give the compressor extra torque for starting.

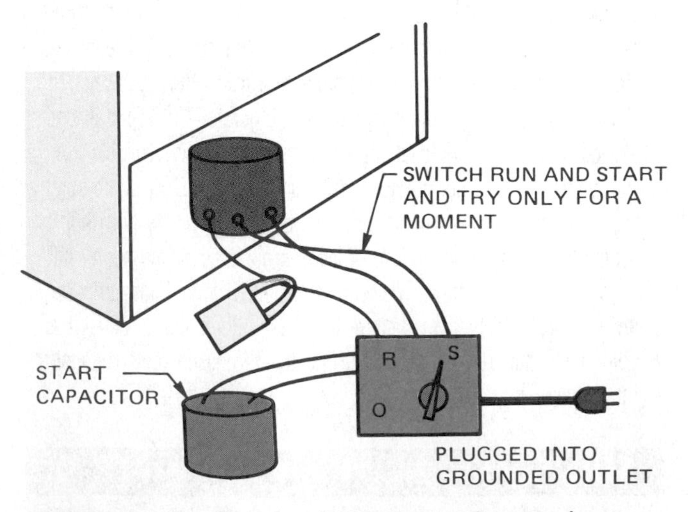

FIGURE 44–142 The run and start wires may be reversed to reverse the motor for a moment. ***The motor must not run for more than a few seconds in the reverse rotation.***

compressor to restart when it stopped. It could be a particle stuck in the cylinder or internal friction due to wear. There is no way of knowing. It may stop again. The customer is told not to put food into the box until the next day to allow the box to cycle a few times during the night. If the compressor stops again, the box should be unplugged and the technician called again.

The technician copies all data from the compressor nameplate, and draws a diagram of the compressor lines and mount in case a substitute compressor must be obtained. The customer wants this refrigerator repaired. All information is copied from the refrigerator nameplate before pushing the refrigerator back to the wall and leaving.

The next morning, the customer calls. The box is off again, just as suspected. The technician goes by the supply house and obtains an exact replacement compressor. This is best because it will fit exactly. A suction-line drier is also obtained.

When the technician arrives, the refrigerator hand truck is fastened to the refrigerator from the side so the unit will go through the door. The strap around the doors keeps the doors closed during movement. The unit is moved to the garage and the technician starts. **The charge is recovered from the system.**

A torch, fire extinguisher, gages, gage ports for the process tubes, refrigerant, scales, and wrenches are carried into the garage while the charge is being recovered. The compressor suction and discharge lines are snipped off close to the compressor connectors. The oil cooler lines are also cut. The old compressor is removed. The new compressor is set in place for a trial fit of all lines; it is a perfect fit.

The compressor is removed and the lines on the refrigerator are cut off square on the ends using a tubing cutter, Figure 44–143. They can be reached with the compressor out of the way. The ends are now cleaned with proper nonconducting sand tape. The plugs are removed from the compressor lines and it is set in place. The lines are fitted to the compressor. A process tube is attached to the high and low side of the compressor with Schrader valves in them. The valve stems are removed for soldering and are left out for a quick evacuation.

The compressor and all lines are soldered with the system open to the atmosphere. When this is finished, the technician attaches gages to the gage ports. The suction line is cut to allow for the suction-line filter-drier. The system is swept with nitrogen and the gage lines removed again in preparation for soldering the suction-line filter-drier, Figure 44–144. The filter-drier is soldered into the line. The gage lines are attached as soon as possible to prevent air from being drawn into the system. The system is practically clean at this time, but you cannot be too careful with low-temperature refrigeration.

The system is now ready for the leak check. When the technician removed the plugs from the compressor, it was noticed that a vapor holding charge was still in the compressor, so all factory connections are leak free. The system is pressured by adding a small amount of R-22 (to 5 psig) then the pressure is increased to 150 psig using nitrogen. All connections are checked with an electronic leak detector. The technician is satisfied that the connections completed during the repair are not leaking. The refrigerator had its original charge, so it does not leak. The pressure is released from the system.

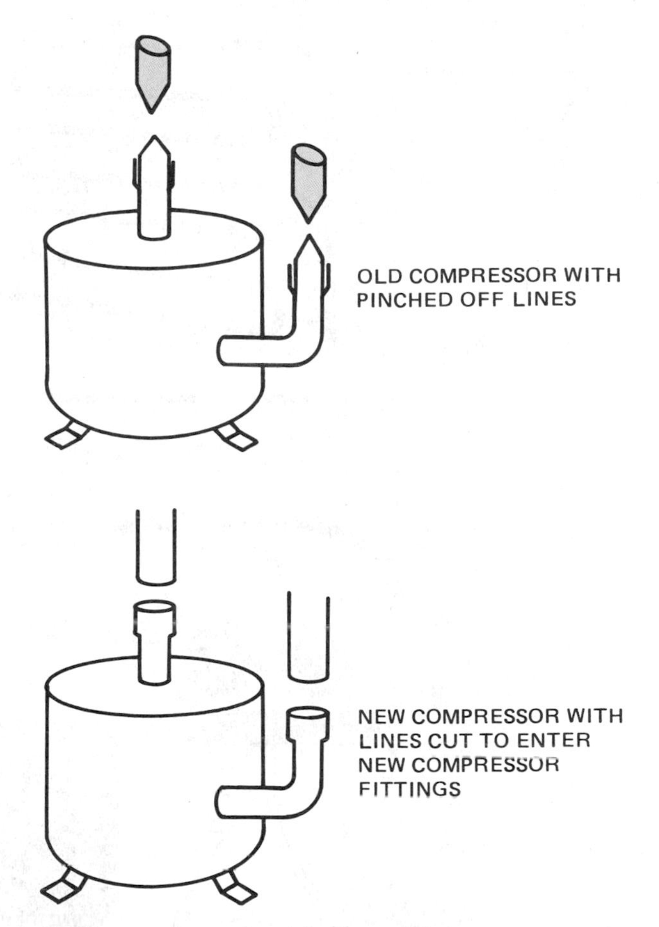

FIGURE 44–143 The pinched lines of the compressor must be cut off before inserting the new compressor connections.

The vacuum pump is attached to the system and started. Remember, the valve stems are not in the Schrader valves. The technician's vacuum gage is in the shop for repair, so sound will be used to determine the correct vacuum. After the pump has run for about 20 min, it is not making any pumping noises; see Unit 8 for how to determine this. The vacuum is broken by adding nitrogen to about 20 in. Hg and the vacuum pump is started again. When another vacuum is obtained the pump is stopped and nitrogen pressure is allowed into the system to 5 psig.

The Schrader valve stems are installed and Schrader adapters are fastened to the ends of the gage lines. The gages are fastened to the system again and a third vacuum is performed. While it is pulling down, the technician reads the correct charge from the unit nameplate and gets set up to charge the system. The electrical connections are made to the compressor. A good technician knows how to manage time and use the vacuum pump time for cleanup of details. When the vacuum is reached, the measured charge is allowed to enter the refrigerator. The high-side line is removed because the unit is about to run and the technician does not want refrigerant to condense in this line.

An ammeter is placed on the common wire to the compressor so the technician will know if the compressor starts correctly. The refrigerator is plugged in and started. It seems to run and the low-side pressure starts down. The last bit of charge is pulled into the low-pressure side of the system. The unit is shut off and trucked back into

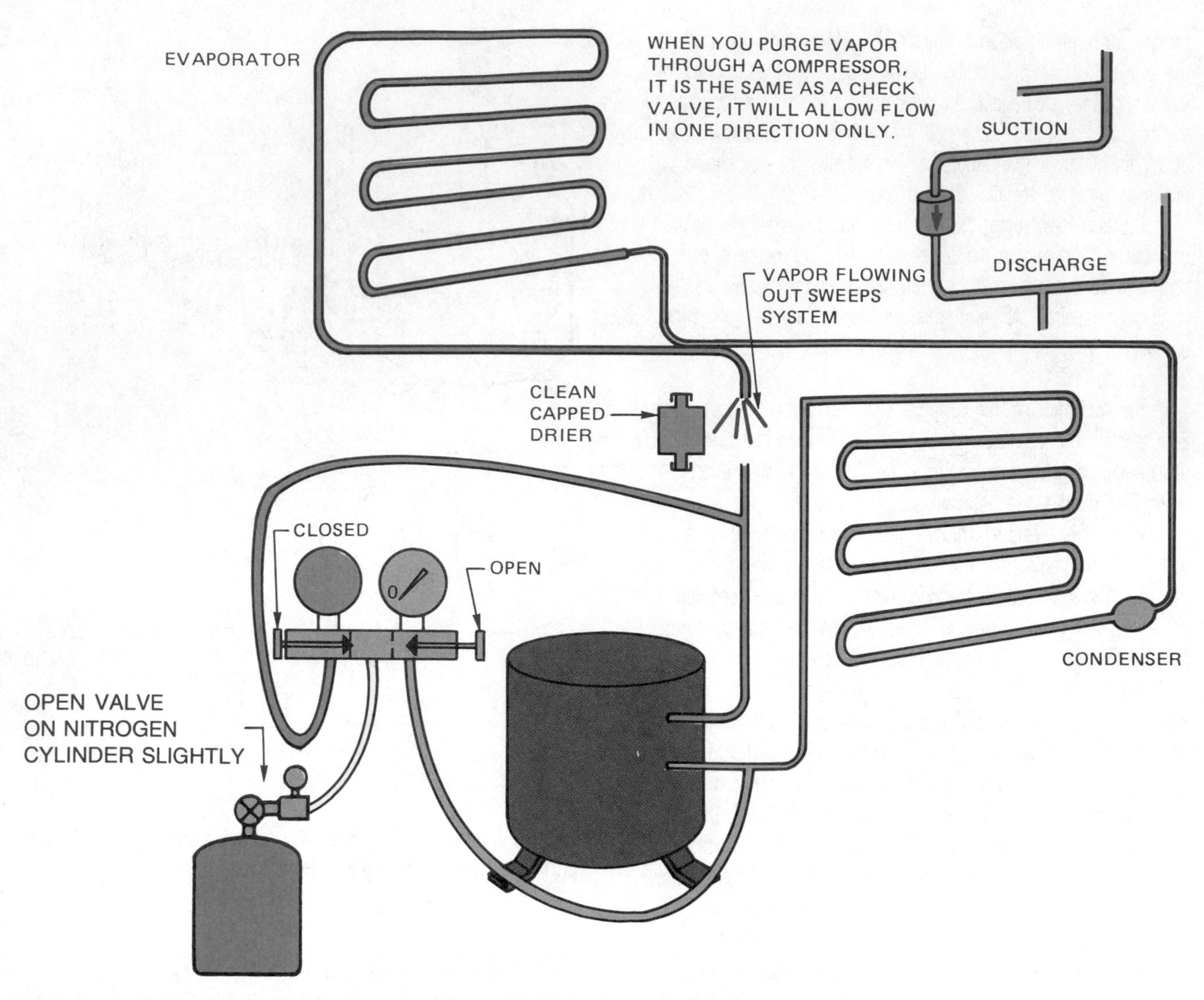

FIGURE 44–144 Sweeping or purging the system before soldering the drier in the line.

place and restarted. The technician leaves the job and calls back later in the day to learn that the refrigerator is performing correctly.

SERVICE CALL 4

A customer calls and reports the refrigerator in the lunchroom is not cooling correctly. It is running all the time and the box does not seem cool enough. *The problem is the defrost timer motor is burned out and will not advance the timer into defrost. The evaporator is frozen solid.*

The technician arrives and opens the freezing compartment inside the top door. The evaporator fan can be heard to run, but no air can be felt coming out the vents. A package of ice cream is soft, indicating the compartment is not cold enough. The technician removes the cover to the evaporator and finds it frozen, a sure sign of lack of defrost.

The compartment door is shut and the box pulled from the wall. The timer is in the back and has a small window to observe rotation of the timer motor, Figure 44–145. The timer is not turning. A voltage check of the timer terminals shows voltage to the timer motor. The winding is checked for continuity; it is open and defective.

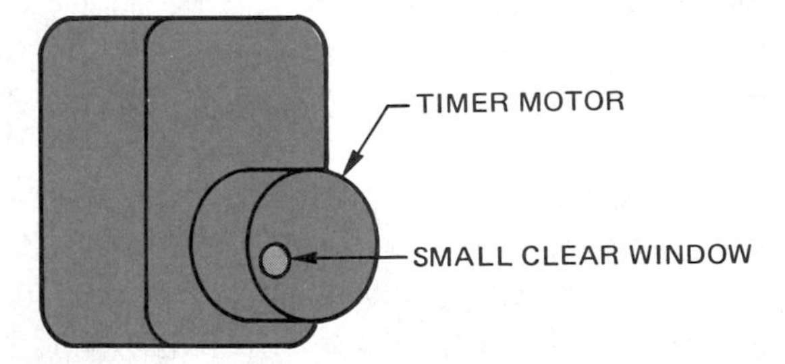

FIGURE 44–145 A small window may be provided to check a timer for rotation.

The technician opens the refrigerator door and removes the food. A small fan is placed in such a manner as to blow room air into the box for rapid defrost, Figure 44–146. The customer is told to watch for water on the floor as the frost begins to melt. The technician is going to make another service call on the way to the supply house for a defrost timer.

On returning, the defrost timer is replaced. The evaporator is thawed, but the pan underneath the unit is full and must be emptied. When emptied, the pan is cleaned and sanitized. The pan is replaced, the cover to the evaporator is replaced, and the unit is started. The coil begins to cool, so the food is replaced and the technician leaves.

FIGURE 44–146 A room fan may be used for rapid defrost.

SERVICE CALL 5

A homeowner reports that the refrigerator is running all the time and not shutting off. *The problem is the door gaskets are defective and the door is out of alignment.* The customer has four children.

The technician arrives and can see the problem easily when the refrigerator door is opened. The gaskets are worn badly and daylight may be seen under the door when looking from the side. The model number of the refrigerator is written down and the technician tells the owner that a trip to the supply house is necessary.

When the gaskets are obtained, the technician returns to the job and replaces them following the manufacturer's recommendations. The door is not closing tightly at the bottom, so the technician removes the internal shelving and adjusts the brackets in the door so that it hangs straight.

SERVICE CALL 6

A customer reports that the refrigerator is running all the time. It is cool, but never shuts off. This refrigerator is in warranty. *The problem is that a piece of frozen food has fallen and knocked the door light switch plunger off and the light in the freezer is staying on all the time. This is enough increased load to keep the compressor on all the time.*

The technician arrives and looks the refrigerator over. There does not seem to be any frost buildup on the evaporator. The owner is questioned about placing hot food in the refrigerator or leaving the door open. This does not seem to be the problem. The customer says the refrigerator is always running. It should be cycling off in the morning, no matter how much load it has.

This sounds like the compressor is not pumping, or maybe there is an additional load on the compressor. The unit has electric defrost and the evaporator coil seems to be clean of frost. The technician opens the freezer door and notices that the light switch in the freezer does not have the plunger that touches the door when it is closed, Figure 44–147. The light is staying on all the time.

The technician informs the customer that a switch will have to be obtained from the manufacturer. The light bulb in the freezer is removed until a switch is obtained.

The technician returns the next day with a switch and replaces it. The owner tells the technician that the refrigerator was shut off at breakfast, so the diagnosis was good.

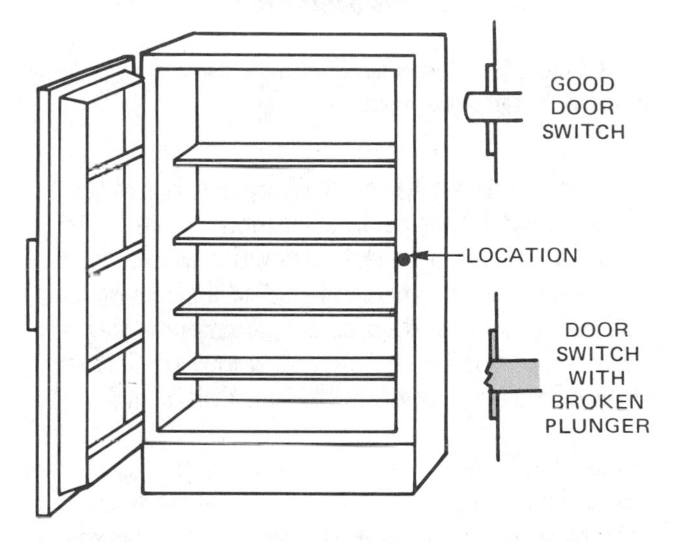

FIGURE 44–147 Plunger missing from door switch.

SERVICE CALL 7

An apartment house tenant has an old refrigerator that must be defrosted manually. The ice was not melting fast enough so an ice pick was used and the evaporator was punctured. She heard the hiss of the refrigerant leaking out, but decided to try it anyway. *The problem is the refrigerant leaked out and when the refrigerator was restarted, water was pulled into the system.* Without refrigerant the low-pressure side of the system will pull into a vacuum. When it would not cool, she called the management office. This would normally be a throw-away situation, but the apartment house has 200 refrigerators like this and a repair shop.

The technician takes a replacement refrigerator along after reading the service ticket. This has happened many times, so many that a definite procedure has been established. The food is transferred to the replacement box and the other one is trucked to the shop in the basement.

The technician solders two process tubes into the system, one in the suction line and the other in the discharge line, Figure 44–148. Pressure is added to the circuit using nitrogen and the leak is located. It is in the evaporator plate, so a patch of epoxy is going to be the repair procedure. The nitrogen pressure is allowed to escape the system and the puncture is cleaned with a nonconductive sand tape. A solvent that is recommended by the epoxy manufacturer is used to remove all dirt and grease from the puncture area. A vacuum pump is connected to the service ports. Note: The service ports have no

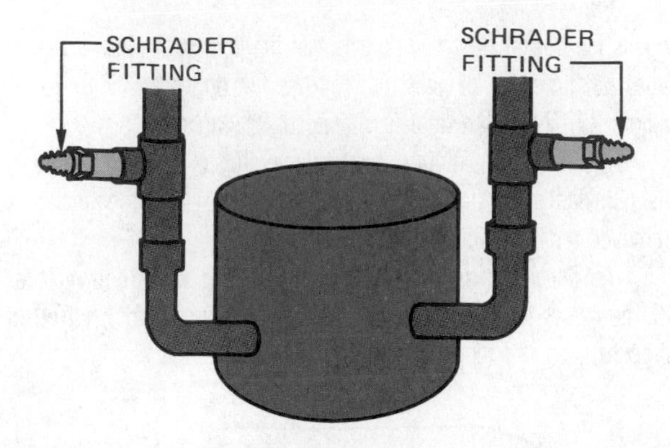

FIGURE 44–148 **Process tubes soldered in the suction and discharge lines offer the best service access.**

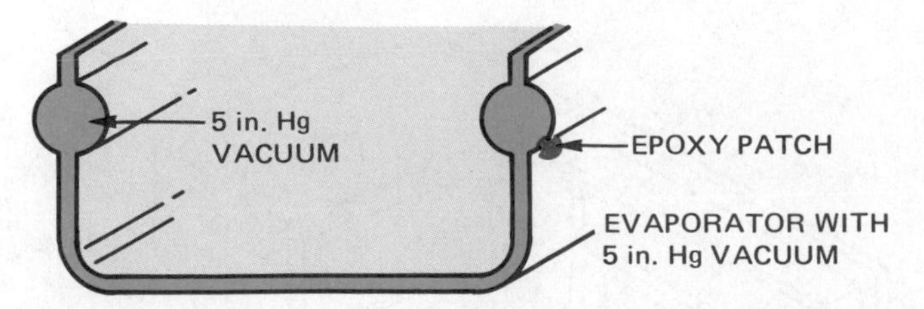

FIGURE 44–149 **Vacuum is used to pull a small amount of epoxy into the evaporator puncture.**

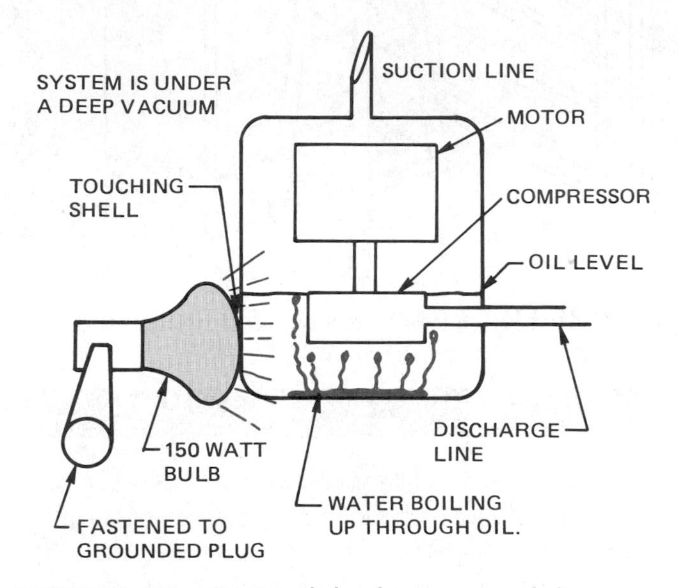

FIGURE 44–150 **Heat is applied to the compressor to boil any moisture from under the oil.**

Schrader valve plungers in them and the gage lines have no Schrader valve depressors in them; they have been removed for the time being.

The vacuum pump is started and allowed to run until 5 in. Hg is registered on the suction compound gage and the gages are valved off. Epoxy is mixed. Remember, this is a two-part mixture and it must be used very fast or it will harden, usually within 5 min. Some epoxy is spread over the puncture hole. The vacuum pulls a small amount into the hole to form the mushroom formation mentioned in the text, Figure 44–149. The vacuum causes a hole to be pulled through the center of the epoxy patch, so some more is spread over the hole. It is now becoming solid. The gages are opened to the atmosphere to equalize the pressure on each side of the patch and the epoxy is allowed to dry.

The epoxy is allowed several hours to dry while other service tasks are performed. A small amount of R-22 is used to pressure the system to 5 psig, nitrogen is then used to pressure the system to 100 psig. The epoxy patch is checked for a leak using soap bubbles. The service stems are also checked for a leak. There are no leaks. The tricky part of the service procedure comes next. This technician knows the procedure well and knows from experience that a step must not be skipped.

The pressure is allowed to escape the system and the vacuum pump is started with no stems or depressors in the Schrader fittings. This allows full bore of the gage hose. A 60-W bulb is placed in the freezer compartment and one in the fresh-food compartment. ***The doors are partially shut. A 150-W bulb is placed touching the compressor crankcase, Figure 44–150. None of these bulbs should ever be placed in contact with plastic or it will melt.*** The vacuum pump is started and allowed to run until the next day. It is not making any pumping sounds, so a good vacuum has been achieved. The technician breaks the vacuum to about 2 psig positive pressure using nitrogen and disconnects the vacuum pump.

The vacuum pump is placed on the work bench and the oil drained. There is quite a bit of moisture in the oil. The oil is replaced with special vacuum pump oil and the opening to the vacuum pump is capped to prevent atmosphere from being pumped and the pump is started. It is allowed to run for about 15 min, long enough to allow the oil to get warm. It is shut off and the oil is drained again and the special vacuum pump oil is added back to the pump. The drained oil looks good, so the pump is connected to the system.

Before turning the vacuum pump on, the technician starts the refrigerator compressor for about 30 sec. If there is any moisture trapped in the cylinder of the compressor, this will move it through the system and the next vacuum will remove it. The vacuum pump is now started and allowed to run for about 2 h; then the vacuum is again broken to about 5 psig using nitrogen. The gage lines are removed one at a time and the Schrader valve stems and depressors are added back to the gage hoses.

A suction-line filter-drier is installed in the suction line and the system is pressured back to 100 psig using nitrogen to leak check the drier connection. This may seem like a long procedure, but the service technician has tried short cuts, and they do not work. The most obvious one would be to cut the drier into the circuit at the beginning. It would become contaminated before the refrigerator is started if cut into a wet system. The liquid-line drier is not changed because it has been at least partially reactivated with the evacuation procedure and it is not blocked with particles. It takes a lot of time to change many liquid-line driers and it may not be economical. Let the suction drier do the job; it is easy to work with.

The vacuum pump is started for the third and final evacuation. It is allowed to run for about 2 h. During these waiting periods, the technician may have other service duties to perform.

The refrigerant for the refrigerator is prepared on the scales when the vacuum is complete. The high-side line is pinched off and soldered shut; the refrigerant is charged into the system, see Unit 8. The refrigerator is started to pull the last of the charge into the system. The suction line is disconnected and a cap is placed on the Schrader valve. The discharge line is soldered shut. The next time a gage reading is needed, a suction-line port is available, but no discharge port. This is common; there is less chance of a leak under the low-side pressure than under the high-side pressure.

SERVICE CALL 8

A customer calls indicating that the refrigerator is off; it tripped a breaker. He tried to reset it and it tripped again. The dispatcher advised the customer to unplug the refrigerator and reset the breaker because there may be something else on this breaker that he may need. *The problem is the compressor motor is grounded and tripping the breaker.*

The technician takes a helper, a refrigerator to loan the customer, and a refrigerator hand truck and goes to the customer's house. A bad electrical problem is suspected. This unit is in warranty and will be moved to the shop if the problem is complicated.

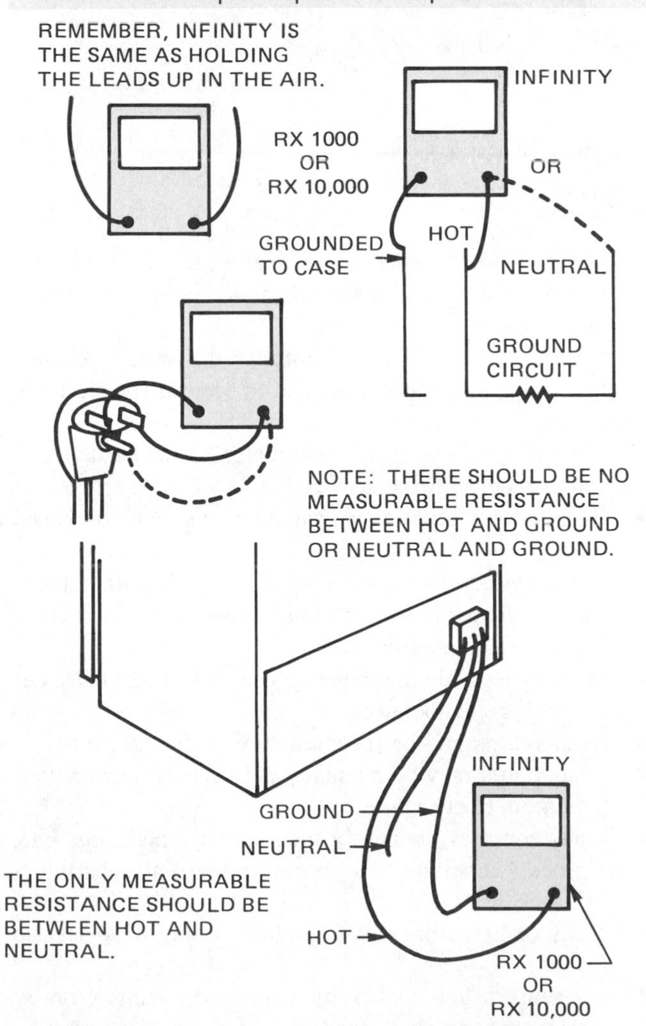
REMEMBER, INFINITY IS THE SAME AS HOLDING THE LEADS UP IN THE AIR.
RX 1000 OR RX 10,000
INFINITY
OR
GROUNDED TO CASE
HOT
NEUTRAL
GROUND CIRCUIT
NOTE: THERE SHOULD BE NO MEASURABLE RESISTANCE BETWEEN HOT AND GROUND OR NEUTRAL AND GROUND.
INFINITY
GROUND
NEUTRAL
THE ONLY MEASURABLE RESISTANCE SHOULD BE BETWEEN HOT AND NEUTRAL.
HOT
RX 1000 OR RX 10,000

FIGURE 44–151 Checking the unit for a grounded circuit using the appliance plug.

The technician uses an ohmmeter at the power cord to check for a ground. The meter is set to R × 1 and reads 0 Ω when touching one meter lead to the power cord plug and the other to the cabinet indicating a ground. ***The white or black lead must be used for this test because the green lead is grounded to the cabinet. The white or black lead will be one of the flat prongs on the plug, Figure 44–151.****

The refrigerator is moved from the wall to see if the ground can be located. It could be in the power cord and repaired on the spot. The leads are removed from the compressor and the compressor terminals are checked with one lead on a compressor terminal and the other on the cabinet or one of the refrigerant lines. The compressor shows 0 Ω to ground, Figure 44–152. It has internal problems and must be changed.

The technician and the helper move the refrigerator to the middle of the kitchen floor and put the spare refrigerator in place. The food is transferred to the spare refrigerator. It is cold because it was plugged in at the shop. It is plugged in and starts to run. The faulty refrigerator is trucked to the shop and unloaded.

The technician knows that the refrigerant in the box may be badly contaminated. **The refrigerant is then removed to a recovery cylinder used for contaminated refrigerants.** The service technician decides on a plan. Change the compressor and the liquid-line drier and add a suction-line drier.

The refrigerator is set up on a work bench to make it easy to access. The compressor is changed and process tubes soldered in place with 1/4 in. fittings on the end. **No Schrader fittings will be used on this one to demonstrate how a system is totally sealed after an evacuation.** The liquid line is cut and the old drier removed. The technician then fastens the gage lines to the process tubes. There are no Schrader stems or depressors. Nitrogen is purged through the system from the suction line back through the evaporator and capillary tube, which is loose at the drier. Nitrogen under pressure is purged through the high side and out the liquid line that is attached to the liquid-line drier, Figure 44–153. The liquid-line drier and then the suction-line drier are soldered in place.

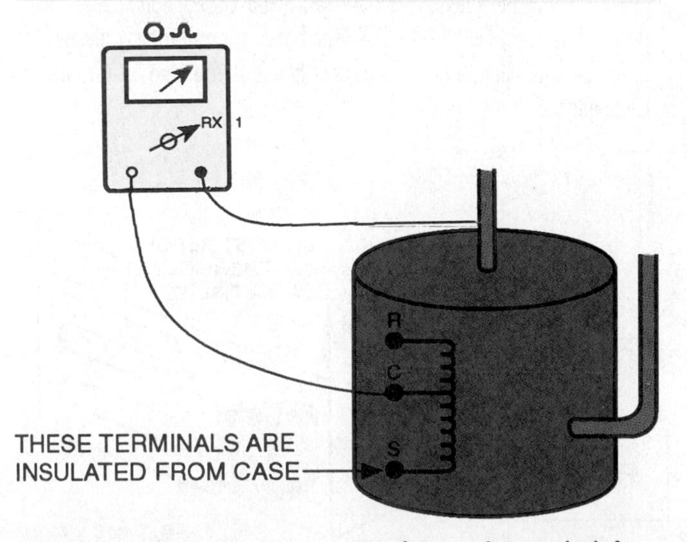
RX 1
R
C
S
THESE TERMINALS ARE INSULATED FROM CASE

FIGURE 44–152 Using the compressor frame or lines to check for an electrical ground circuit.

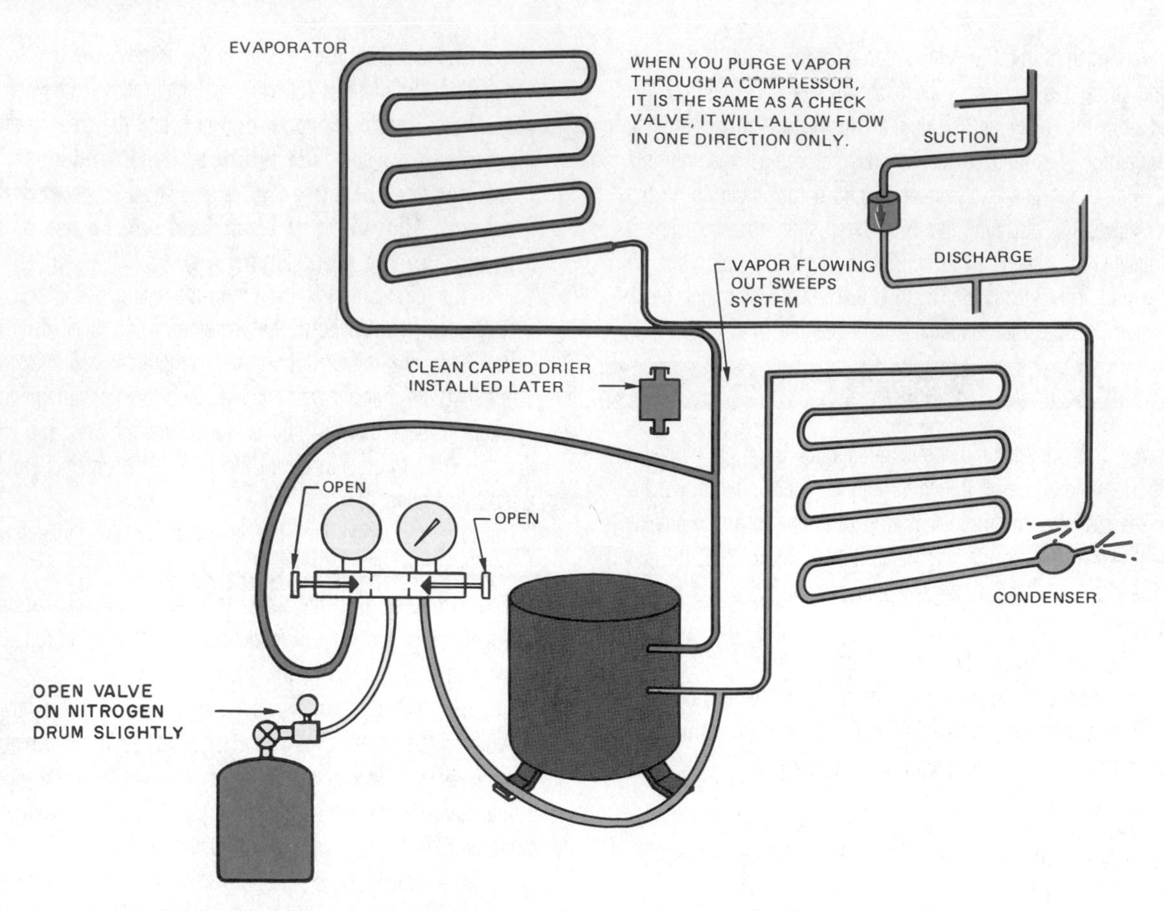

FIGURE 44–153 Purging the system at the suction-line drier connection before soldering the drier in the line.

The difference in the purge method in this system is that there is no moisture present, just contaminated refrigerant and oil, and possibly smoke from the motor ground in the system. It can be purged to remove the large particles and the driers will remove the rest.

The system is leak checked and triple evacuated. At the end of the third evacuation, the discharge service tube is pinched off using a special pinch-off tool. The system is charged with a measured charge and started.

The low-side pressure is observed to be correct during the pull down of the box, Figure 44–154. The box is allowed to stay plugged in for 24 h to operate on its own and then is transported back to the customer.

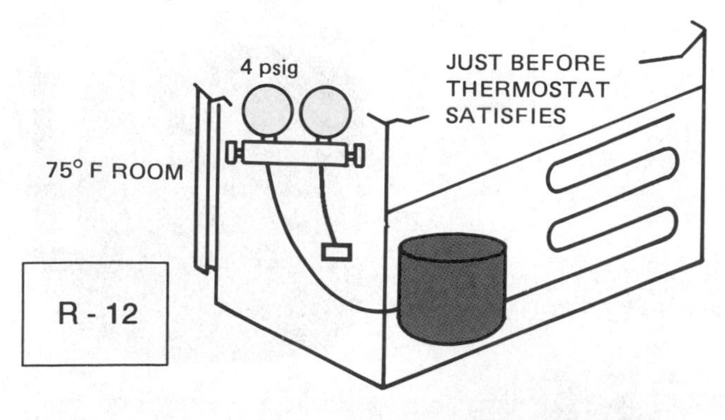

FIGURE 44–154 Typical pressure while the box is pulling down.

SUMMARY

- Refrigeration means to move heat from a place where it is not wanted to a place where it makes little or no difference.
- Heat enters the refrigerator through the walls of the box by conduction, convection, and by warm food placed in the box.
- The refrigeration system circulates air inside the box across a cold refrigerated coil.
- Evaporator compartments can be natural draft or forced draft.
- The evaporator in the household refrigerator operates at the low-temperature condition, yet maintains food in the fresh-food compartment.
- Sharp objects should never be used when manually defrosting the evaporator.
- Evaporators may be flat plate type or fan coil type.
- Evaporators may be manually defrosted or have an automatic defrost feature.
- Many compressors have a suction line, a discharge line, a process tube and two oil cooler lines, all protruding from the shell.
- Many compressors in refrigerators do not have service ports.
- Condensers are cooled by natural convection or by forced air from small fans.
- The condenser may have an oil cooler to keep the crankcase oil at a lower temperature.

- Most domestic refrigerators use the capillary tube metering device.
- These refrigerators should not be placed outdoors or in other areas where the temperature will exceed the normal temperature ranges as specified by the manufacturer.
- Current model refrigerators have a magnetic strip gasket with a compression seal around the door or doors.
- Condensate from the evaporator defrost is routed through tubing to a pan under the refrigerator.
- The space temperature thermostat controls the compressor.
- There are two types of automatic defrost, hot gas and electric heat.
- Most cabinet frost and moisture prevention heaters are small electrically insulated wire heaters that are mounted against the cabinet walls near the door openings.
- The fans used for forced-draft condensers and evaporators are typically prop-type with shaded-pole motors.
- Ice makers are located in the low-temperature compartment and freeze water into ice cubes.
- The service technician should not routinely apply gages to a domestic refrigerator system. Explore all other troubleshooting techniques before applying gages.
- A very small amount of refrigerant lost from a refrigerator will affect the performance. Very small leaks may be found only with the best leak detection equipment.
- An epoxy may be used to seal a leak in an aluminum evaporator.
- Tubing in condensers is usually made of steel. These leaks should be repaired with the appropriate solder.
- When a leak has been repaired a triple evacuation should be pulled on the entire system.
- Under certain conditions capillary tube metering devices may be repaired.
- ⊛**Refrigerant should be recovered and never exhausted into the atmosphere.**⊛

REVIEW QUESTIONS

1. Define refrigeration.
2. Describe the refrigeration cycle in a household refrigerator.
3. What is the typical temperature range in the fresh-food section of a household refrigerator?
4. What is the typical temperature range in a frozen-food compartment?
5. Describe how a single evaporator may provide the typical temperature ranges for both compartments?
6. Is more than one evaporator ever used?
7. What are two types of evaporator draft systems that may be found in a refrigerator?
8. What must be done to defrost a manual defrost refrigerator?
9. What are two methods of automatic defrost?
10. What are the approximate horsepower ranges for compressors in these refrigerators?
11. List five kinds of lines (tubes) that may protrude from a compressor.
12. Why is a piping diagram needed when replacing a compressor?
13. What voltage is used on these refrigerators in the United States?
14. What two types of draft systems are used with condensers?
15. How is the condensate typically disposed of in household refrigerators?
16. What are compressor oil coolers used for?
17. Describe three methods of cooling crankcase oil.
18. What type of metering device do these units use?
19. Is this metering device ever serviced?
20. What material is the inside surface of the modern box usually made of?
21. What are mullion and panel heaters used for?
22. What controls the compressor?
23. Describe a typical thermostat used to control a refrigerated box.
24. What are two methods that may be used to terminate automatic defrost?
25. What is the purpose of each of the two fan motors that may be used in a refrigerator?
26. Describe how one typical ice maker in a household refrigerator operates.
27. List two reasons why a refrigerator should be level.
28. Why would a refrigerator door have piping and wiring routed through the hinge on the door?
29. Why is it poor practice for a technician to routinely apply gages to a refrigerator when servicing it?
30. Why would a technician remove Schrader valve depressors from gage lines?
31. If there is a chance the system pressure is in a vacuum, what should you do before installing a low-side line tap?
32. Describe a technique for determining whether or not the system has a low-refrigerant charge without using gages.
33. Describe the frost-line method used when adding refrigerant.
34. Describe a method for repairing an evaporator refrigerant leak with epoxy.
35. When should a filter drier be added to the system?
36. How should a system that has moisture in it be evacuated?
37. Describe the procedure for soldering a capillary tube into a fitting?
38. What checks should you make when determining whether or not a compressor is operating at capacity?

UNIT 45

Domestic Freezers

OBJECTIVES

After studying this unit, you should be able to

- describe freezer burn.
- discuss the construction of typical freezer cabinets.
- identify three types of freezer evaporators.
- describe two types of freezer compressors.
- discuss two types of natural-draft condensers.
- explain the function of the capillary tube in the freezer.
- describe condenser efficiency relative to ambient air passing over it.
- explain procedures to defrost a freezer manually.
- discuss procedures to remove spoiled food odors from the box.
- describe procedures for moving upright and chest-type freezers.

SAFETY CHECKLIST

* Technicians should wear appropriate back brace belts when moving or lifting appliances.
* Never place your hands under a freezer that has been lifted. Use a stick or screwdriver to position a mat under the freezer feet.
* Do not let dry ice come in contact with skin; it will cause frostbite.
* Never allow refrigerant pressures to exceed the manufacturer's recommendations.
* Always follow all electrical safety precautions.

Many of the components and operating systems in domestic freezers are the same as or similar to those in domestic refrigerators. Unit 44 should be studied first because many descriptions and procedures are not duplicated in this unit.

45.1 THE DOMESTIC FREEZER

The domestic freezer, different from the refrigerator, is a one-temperature appliance. It operates as a low-temperature refrigeration system. Like the refrigerator, it is a stand-alone appliance, plugs into the power supply, and can be moved from one location to another. Freezers are normally not opened as often as refrigerators. Food is placed in them for long periods of time. The freezer may be located in an out-of-the-way place, such as a basement or storage room, away from the kitchen. The functions of the freezer are to hold food that has been purchased frozen and to freeze small amounts of food placed in the freezer and hold it frozen.

Food to be frozen must be packaged correctly, in airtight packages. If not, freezer burn will occur. Freezer burn is a result of dehydration of the product. This is most noticeable in products that have a tear in the package. The product looks dry and burned in the vicinity of the tear. Actually it is dry. Freezer burn does not ruin the product, but makes it unattractive and may change the flavor. When freezer burn occurs, moisture in the vicinity of the tear in the package leaves the food and transfers to the air. The moisture is in the solid state while in the food and is changed to the vapor state in the air where it collects on the coil. It is then carried off during defrost. This process of changing from a solid to a vapor is known as sublimation. Ice cubes become smaller if stored for long periods of time in the freezing compartment. This is sublimation. Moisture changes from a solid to a vapor and collects on the evaporator as ice. It is melted during defrost and leaves in the condensate.

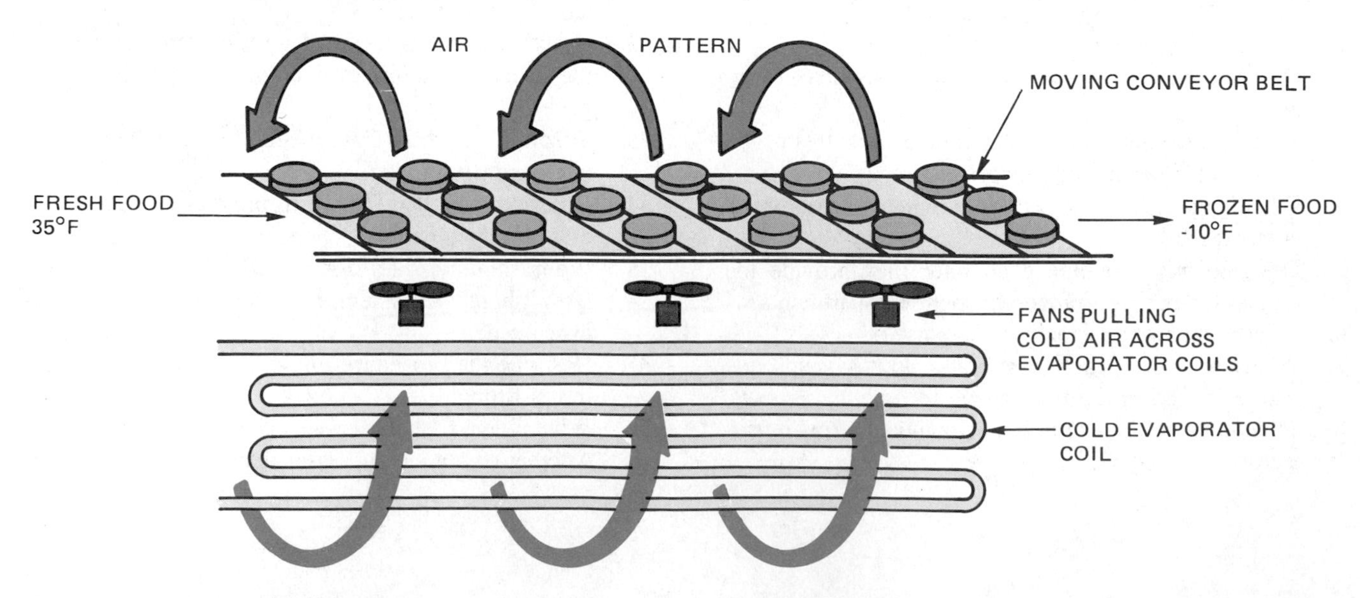

FIGURE 45–1 Commercial flash freezer.

Another problem the owner may have when freezing food in the home freezer is the amount of time it takes to freeze the product solid. This is accomplished with flash freezers in a packing plant. The flash freezer may pass the food over a flash of very cold air that will freeze it instantly, Figure 45–1. In a home freezer this is not possible. Some freezers that have forced-draft evaporators may have a quick-freeze rack that is located in the fan discharge, the coldest air in the box, Figure 45–2. The velocity of the air and the temperature will greatly speed the freezing process. This is not comparable to flash freezing done commercially. Quick freezing in a home freezer can be used for small amounts of food only. The system has very small capacity.

Many owners may purchase a quarter or half carcass of beef that has been butchered and wrapped and try to freeze the whole amount at one time in a home freezer. This causes problems. Food that is frozen slowly will have ice crystals form in the product cells, Figure 45–3. This may puncture the cells of the product. You may have noticed that a steak frozen by the store tastes different from a steak that you carried home and froze. The difference is in the time it takes to freeze the steak. You may have noticed the puddle of water and blood when you thawed the steak. This is the loss due to puncture of the cells, Figure 45–4. You may have transported the steak home in the car where it may warm up to 70°F or higher, then you may place it in the freezer where it must be lowered to 32°F before it starts to freeze. The best policy is to lower the temperature of the product to as close to 32°F as possible by placing it in the coldest place in the refrigerator for several hours, then move it to the freezer and locate it in the coldest place in the freezer. This may be a quick-freeze compartment or on the evaporator plate, Figure 45–5.

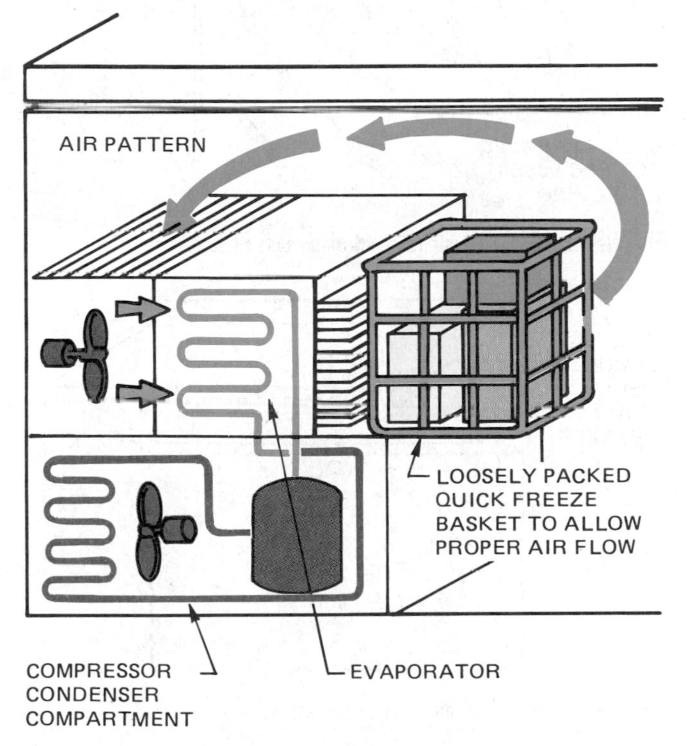
AIR PATTERN
LOOSELY PACKED
QUICK FREEZE
BASKET TO ALLOW
PROPER AIR FLOW
COMPRESSOR
CONDENSER
COMPARTMENT
EVAPORATOR

FIGURE 45–2 Quick-freeze section in freezer.

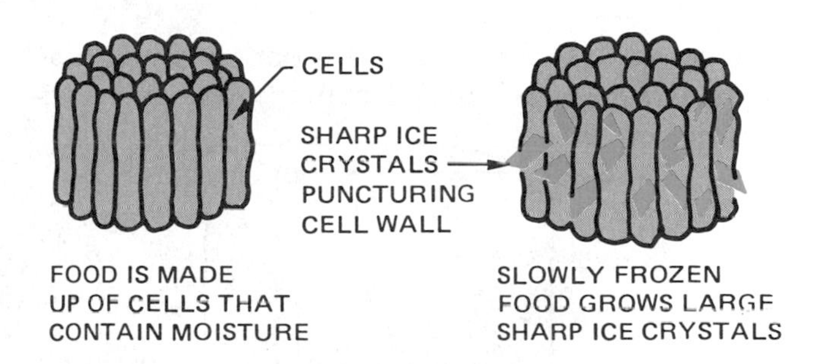
CELLS
SHARP ICE
CRYSTALS
PUNCTURING
CELL WALL
FOOD IS MADE
UP OF CELLS THAT
CONTAIN MOISTURE
SLOWLY FROZEN
FOOD GROWS LARGE
SHARP ICE CRYSTALS

FIGURE 45–3 Food frozen slowly has large ice crystals.

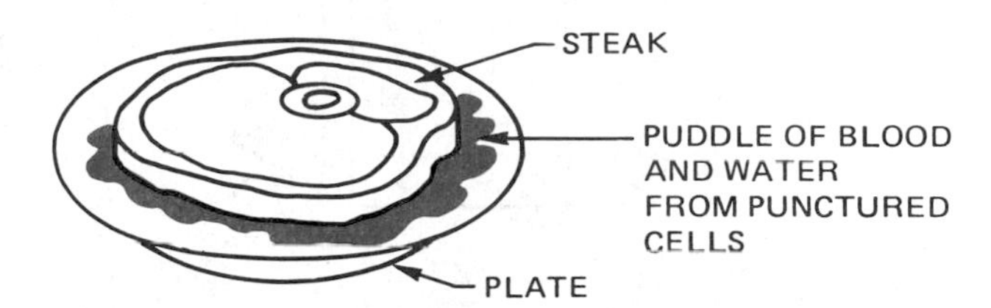
STEAK
PUDDLE OF BLOOD
AND WATER
FROM PUNCTURED
CELLS
PLATE

FIGURE 45–4 Large ice crystals may puncture the cell walls of the food.

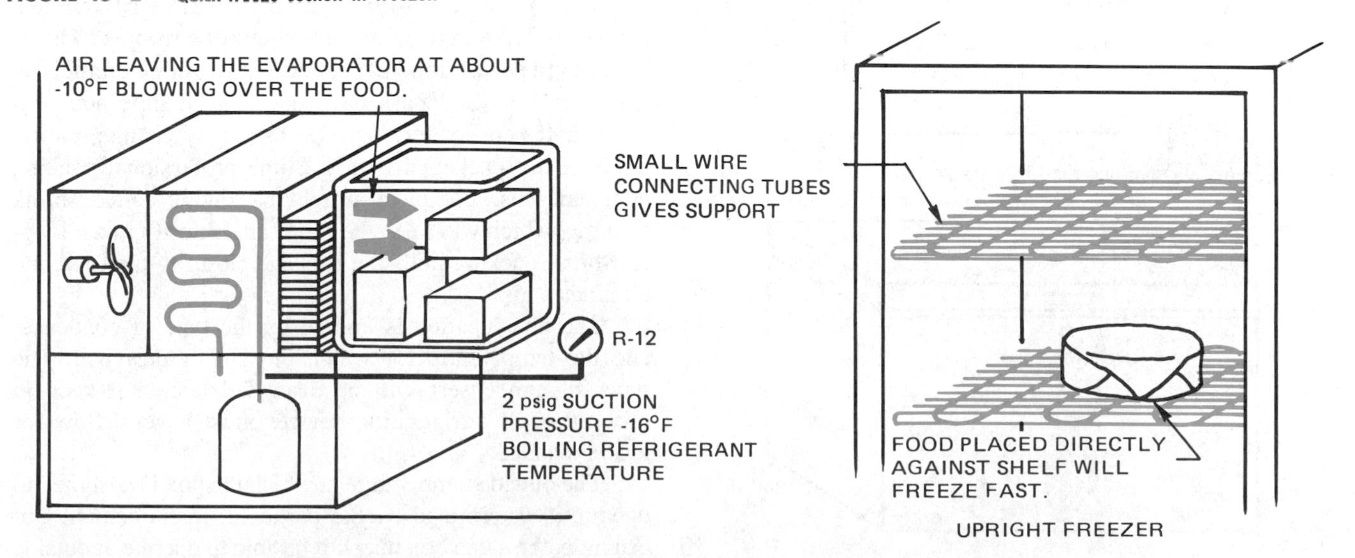
AIR LEAVING THE EVAPORATOR AT ABOUT
-10°F BLOWING OVER THE FOOD.
SMALL WIRE
CONNECTING TUBES
GIVES SUPPORT
R-12
2 psig SUCTION
PRESSURE -16°F
BOILING REFRIGERANT
TEMPERATURE
FOOD PLACED DIRECTLY
AGAINST SHELF WILL
FREEZE FAST.
UPRIGHT FREEZER

FIGURE 45–5 Two quick-freeze methods using a special compartment or placing food directly on the evaporator.

45.2 THE CABINET OR BOX

The actual box is normally made of sheet metal on the outside with metal or plastic on the inside. The outside of the box may be painted any color fashionable at the time. It may match the rest of the appliances in the kitchen. The box may be an upright type or a chest type, Figure 45–6. When upright, the door may open to the left or right, for convenience. When the freezer is a chest type, the door is called a lid and raises, Figure 45–7.

The upright box takes up much less space and is often used in the kitchen where floor space is a premium. It is probably not as efficient as a chest type, because every time the door is opened, air in the box falls out the bottom of the opening, Figure 45–8. This does not change the food temperature much, just the air temperature that must be recooled. Moisture also enters with the air and will collect on the coils. Door opening should be kept at a minimum. When a chest-type freezer is opened, the air stays in the box, Figure 45–9.

Locating food in a chest type is not as easy as in an upright freezer because the upright freezer will have multiple shelves. The chest-type freezer may have baskets that lift out so the storage on the bottom may be reached. This is important because food cannot be kept fresh indefinitely in a freezer. Fish and pork have one storage time and beef another. If kept longer, the food taste may change, particularly if not packaged tightly. Some professional packing companies use complete airtight packaging called shrink packing, which will keep the food satisfactorily much longer. Shrink packing pulls the package closer to the food and eliminates air.

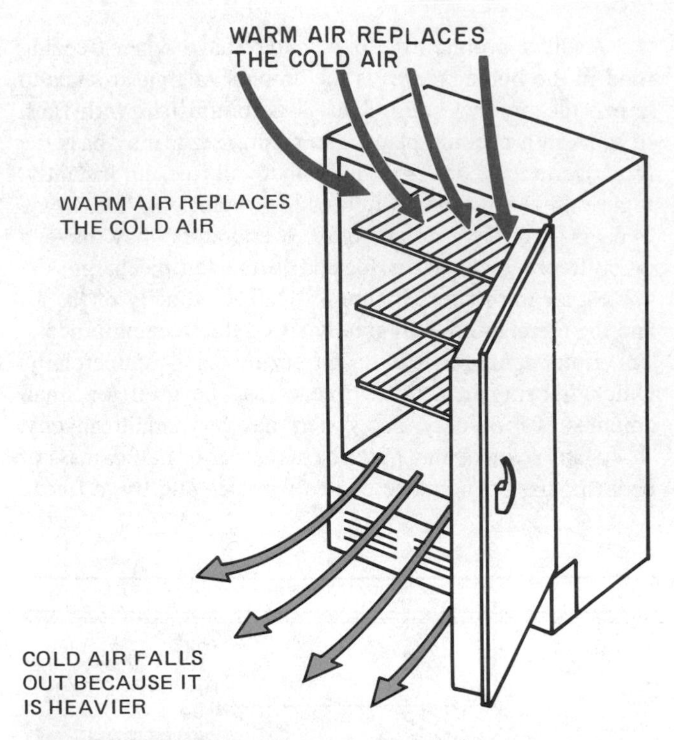
WARM AIR REPLACES THE COLD AIR
WARM AIR REPLACES THE COLD AIR
COLD AIR FALLS OUT BECAUSE IT IS HEAVIER

FIGURE 45–8 Cold air falls out of an upright freezer.

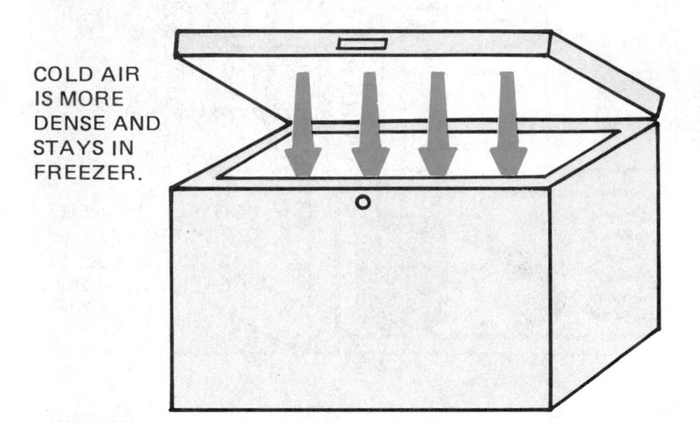
COLD AIR IS MORE DENSE AND STAYS IN FREEZER.

FIGURE 45–9 Cold air stays in a chest-type freezer.

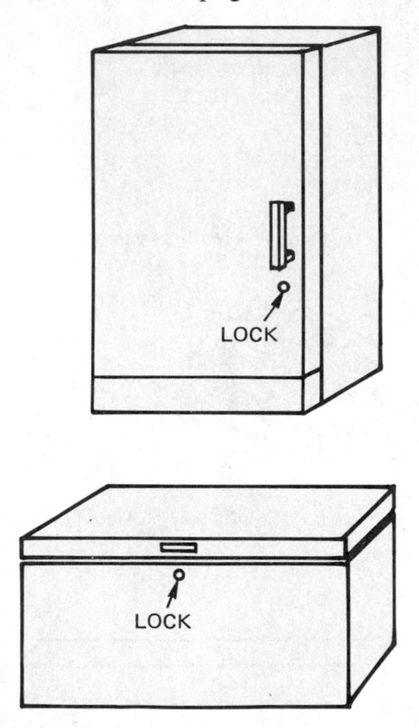
LOCK
LOCK

FIGURE 45–6 Upright and chest-type freezers.

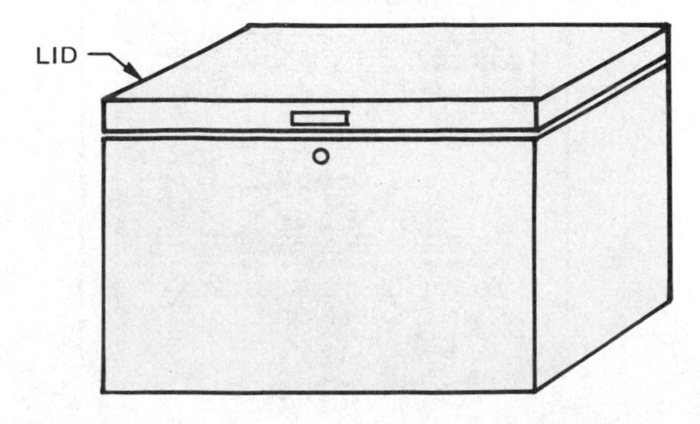
LID

FIGURE 45–7 The door is known as a lid on a chest-type freezer.

The box location is limited by the type of condenser and the temperature for which the box is designed. The types of condensers will be discussed later, but keep in mind, that all refrigeration devices must have airflow for heat exchange.

The outside temperature at which the box is designed to operate is determined by the particular manufacturer. For example, the same box may not be able to operate at outside temperatures during the summer and winter. This is partly due to the operation of the condenser in the winter. The box insulation may not be adequate for location in extreme heat in the summer.

45.3 CABINET INTERIOR

The inside of many freezers may have a lining of plastic. This plastic is strong and easy to clean and maintain. However, it may be brittle at the low temperatures inside the box, so care should be taken not to strike it. Some boxes have metal liners and plastic trim. The metal may be either painted or coated with porcelain.

The walls of the typical modern box are much like the refrigerator, a sandwich construction of metal or plastic on the inside and metal on the outside with foam insulation between. Many older boxes will have fiberglass between the walls.

Inside the Upright Box

The upright box will have door storage for small packages. This door must be strong to hold the weight of the packages and the door itself. Care must be used not to abuse the door or it may warp. It is not uncommon for a package of frozen meat to drop on the bottom rail of plastic in an upright freezer and break it, Figure 45–10. This should be repaired. Duct tape, the kind used for duct systems, may be used to seal the damaged area until the next time it is defrosted, then the spot may be repaired by fastening the piece back with epoxy glue, Figure 45–11. The glue may not be applied while the box is cold.

The upright box has many of the same gasket and door alignment features as the refrigerator. The gaskets must remain tight fitting and the door must remain in alignment or leaks will occur, Figure 45–12. Leaks will cause the compressor to run more than it was designed to and cause frost buildup. Defrost is discussed later. Many freezers are manual defrost because defrost is not needed as often with freezers as with refrigerators because the doors are not opened as often.

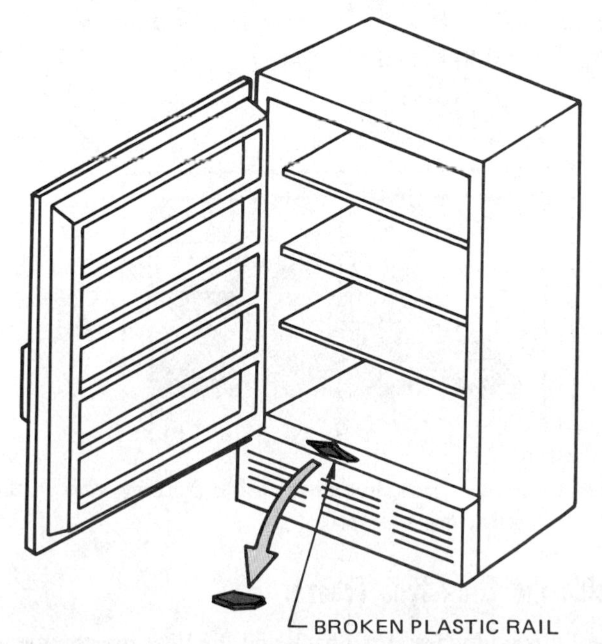

FIGURE 45–10 If frozen food is dropped, it may break the plastic rail in a freezer.

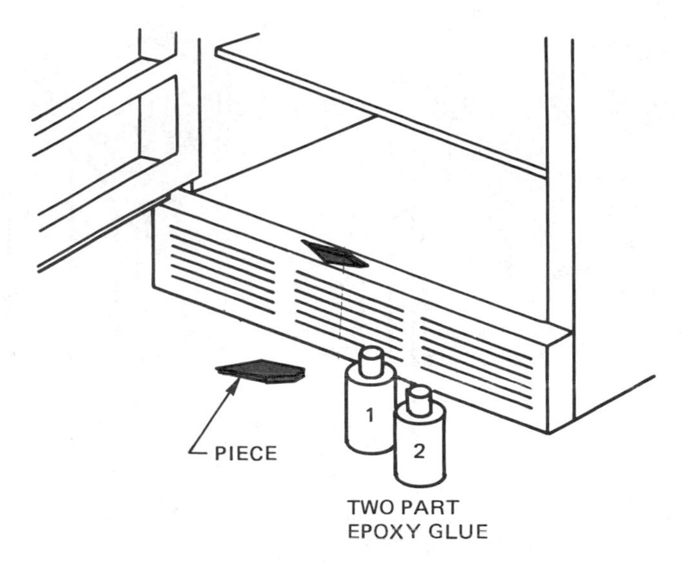

FIGURE 45–11 The piece may be glued back in place when the freezer is defrosted and cleaned.

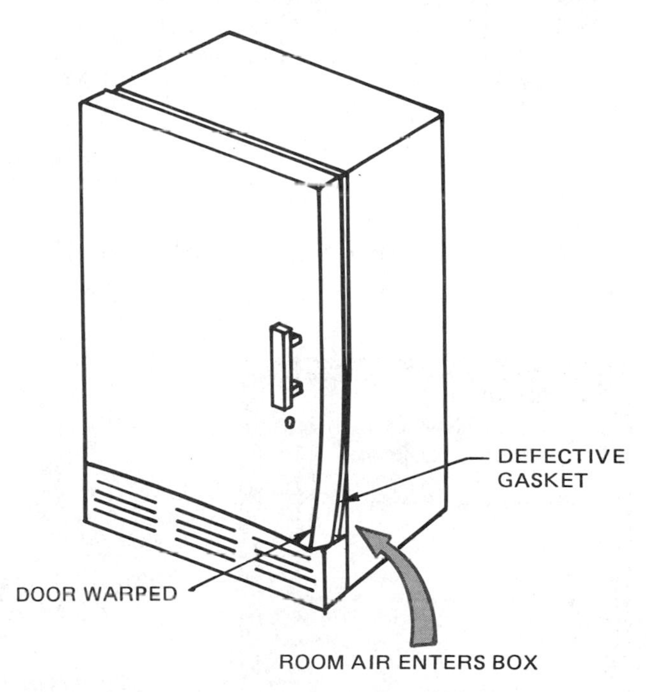

FIGURE 45–12 The door gaskets must be in good condition and the door must fit right or moisture-laden air will enter the freezer.

The evaporator, which freezes the food inside, may be one of three types—a forced air, a shelf, or a plate in the wall of the freezer, Figure 45–13. Forced-air units may have fast-freeze compartments, Figure 45–14. When plate evaporators are used, they may be the tube type with wires connecting them or the stamped type used to make a shelf, Figure 45–15. When the plate or tube evaporator is used, the fastest method for freezing food is to lay it directly on

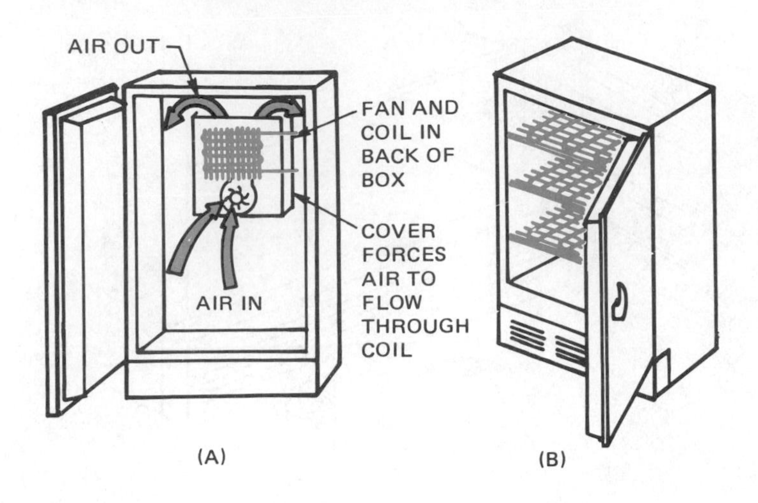

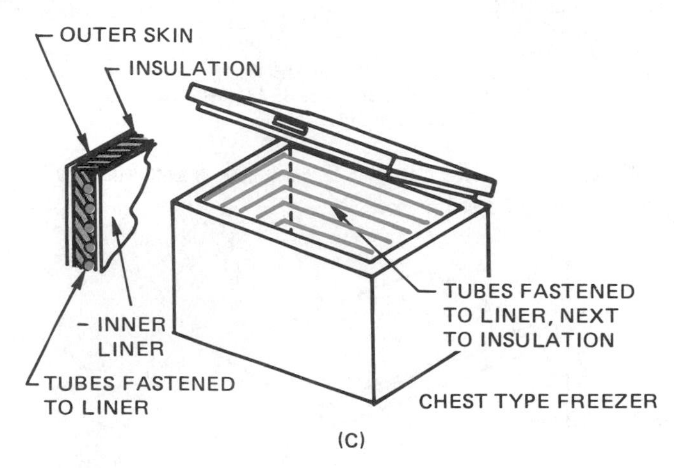

FIGURE 45–13 Three types of evaporators. (A) Forced air. (B) Shelf. (C) Plate (wall).

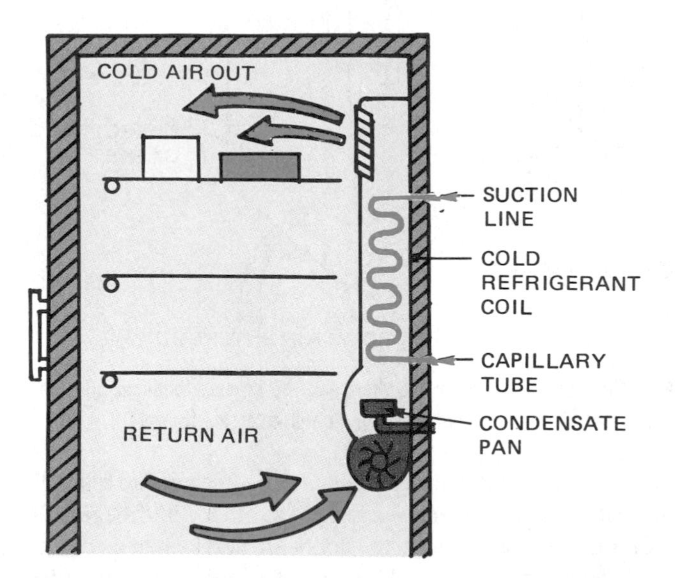

FIGURE 45–14 Forced-air type may use the discharge air from the cold evaporator for fast freezing.

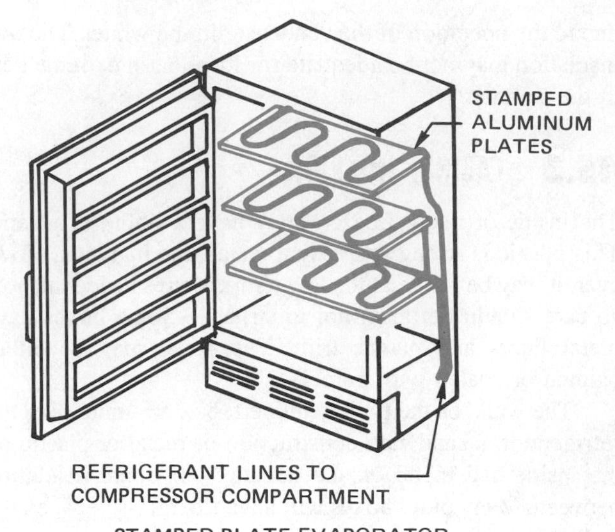

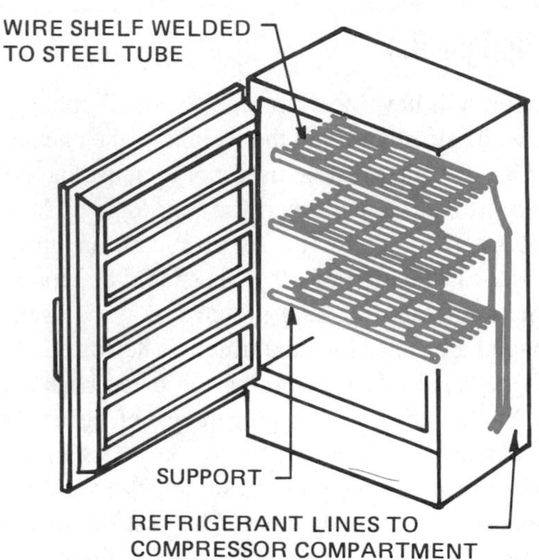

FIGURE 45–15 Plate evaporators may form the shelves, using stamped plate or steel tubing with wire.

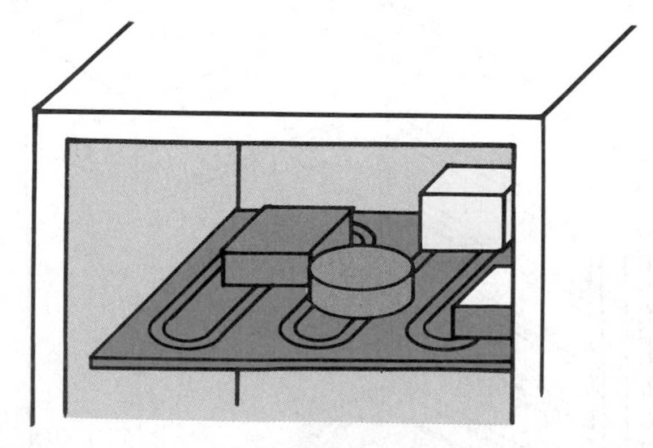

FIGURE 45–16 Placing food directly on the evaporator plate or tube will help to quickly freeze the food.

the plate, Figure 45–16. There is a limited amount of available plate surface area.

A light is normally located in the freezer to see the food. A door switch is used to turn the light on and off.

Inside the Chest-Type Freezer

The lid is on hinges at the back and the liner may be metal or plastic. Gaskets are located around the lid of the box. These gaskets must be in good condition or air with moisture will leak in. There will be a fast-frost buildup, normally right inside the lid where air is leaking in, Figure 45–17.

The chest-type freezer has a hump in one end of the compartment. The compressor is located under the hump, Figure 45–18. It may have a forced-air fan coil at the same end for fast freezing, Figure 45–19.

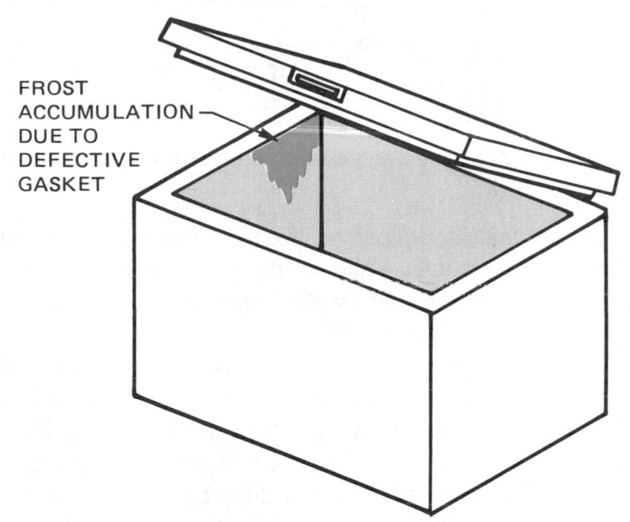

FIGURE 45–17 A leaking lid gasket will cause fast frost accumulation.

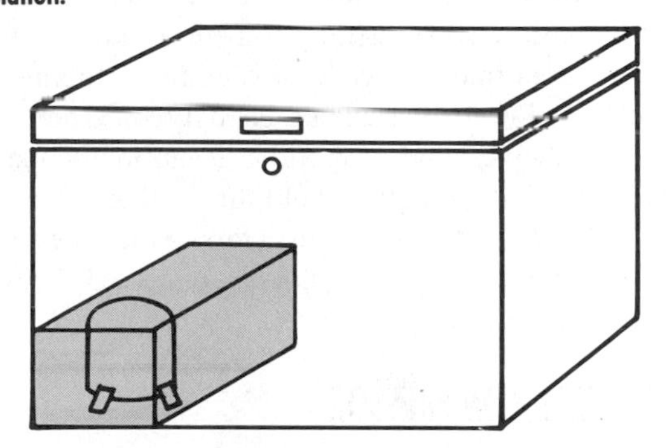

FIGURE 45–18 The compressor is located under the hump in a chest-type freezer.

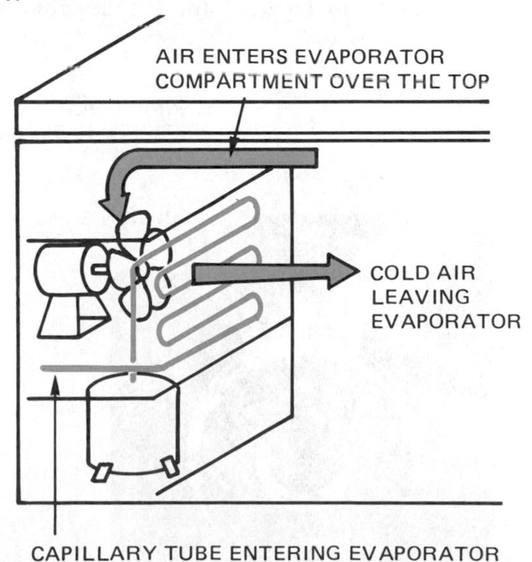

FIGURE 45–19 A small forced-draft evaporator may be located on one end and used as a fast freezer.

Care should be taken when loading a chest-type freezer or food will be hard to find at the bottom. Larger parcels should be located on the bottom and all parcels should be dated. Most freezers have baskets that may be set aside while looking for parcels at the bottom.

45.4 THE EVAPORATOR

The evaporator is the device that allows heat to be absorbed into the refrigerant and is located inside the freezer. It may be a forced-air type or a tube or plate with natural draft.

The forced-draft type is typically located behind a panel with a fan, prop or centrifugal type, passing air over it, Figure 45–20. This type may have automatic defrost because the evaporator is in a location where the thawed moisture may be drained from the cabinet interior without dripping on the food, Figure 45–21.

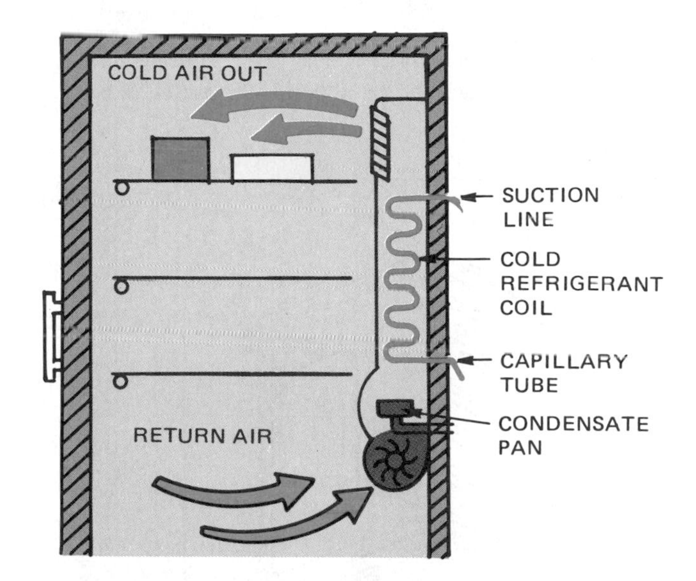

FIGURE 45–20 The forced-draft evaporator and fan are usually located behind a panel.

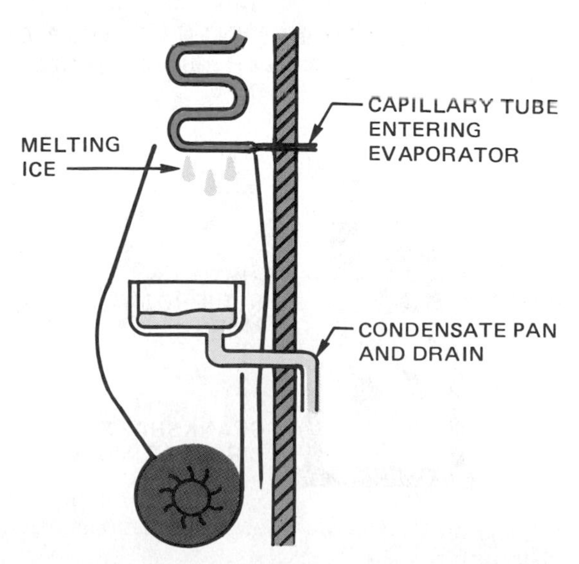

FIGURE 45–21 The evaporator must have a pan for catching condensate during defrost.

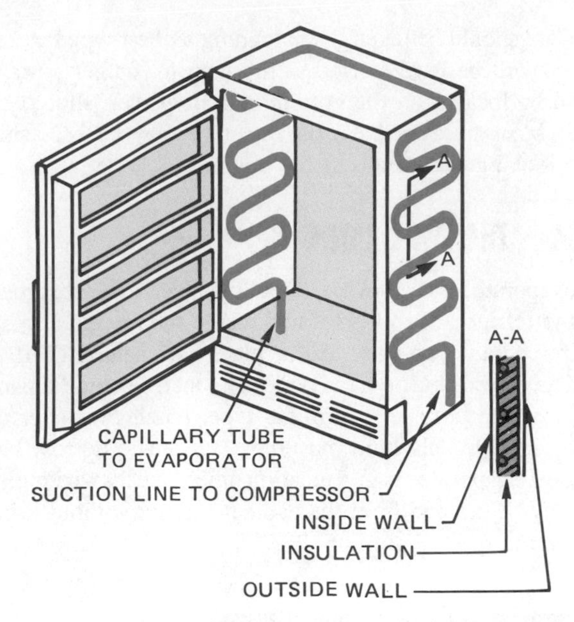

FIGURE 45–22 A plate-type evaporator using the freezer walls as the plate. The evaporator tubes are fastened to the liner of the box.

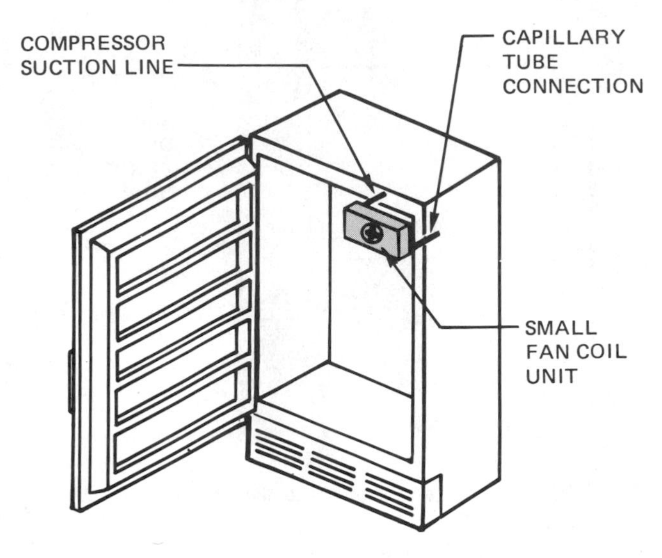

FIGURE 45–23 When a plate-type evaporator is defective, a small forced-draft evaporator may be installed and substituted if necessary.

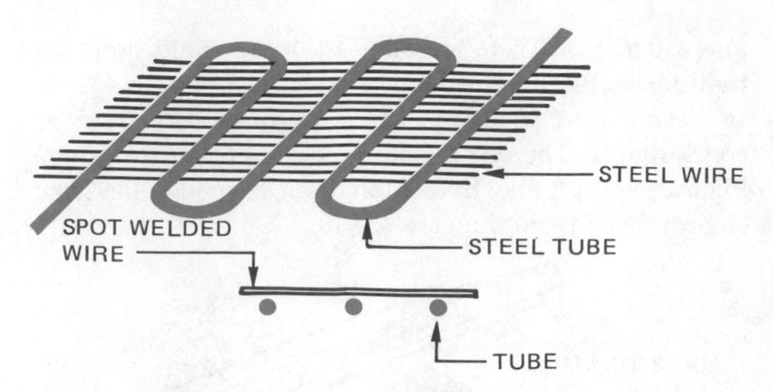

FIGURE 45–24 Shelf-type evaporators are normally steel tubing with wire to add support.

When the evaporator is a plate type or tube type, no fan is used. The evaporator may be the shelf and it must be a manual defrost because the water from thawed ice cannot be collected automatically.

Another type of plate evaporator uses the wall of the interior of the box as the evaporator plate. This is accomplished by fastening the tubing to the inner metal liner of the box, Figure 45–22. This type of evaporator is the hardest to service because the inner liner must be removed for service. If the liner has foam insulation, it is next to impossible to remove and not economical to do so. If necessary, a forced-draft evaporator could be substituted, Figure 45–23. The new evaporator must be sized by someone with experience.

Evaporators may be made of aluminum tubing when they are forced-draft or stamped shelf type. When tubing is used for shelves, the evaporator is normally made of steel. This gives it strength to hold the food and provides a method of fastening the wire connectors to the tubes, Figure 45–24. The support wires may not be connected easily with aluminum tube types.

45.5 THE COMPRESSOR

Compressors in domestic freezers are like the ones in domestic refrigerators. They may be rotary or reciprocating, Figure 45–25. Refer to Unit 44 for the description of compressors.

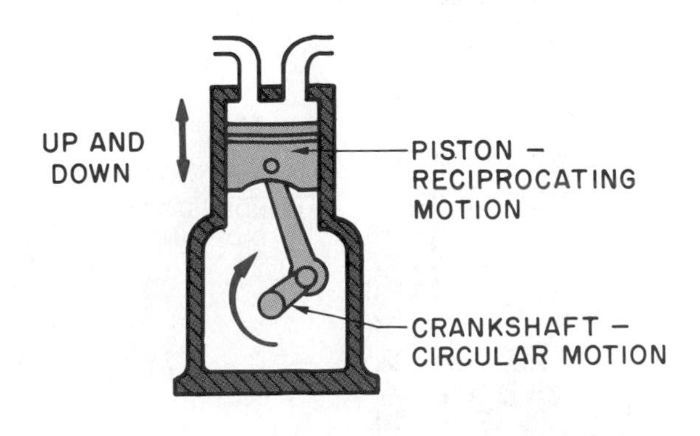

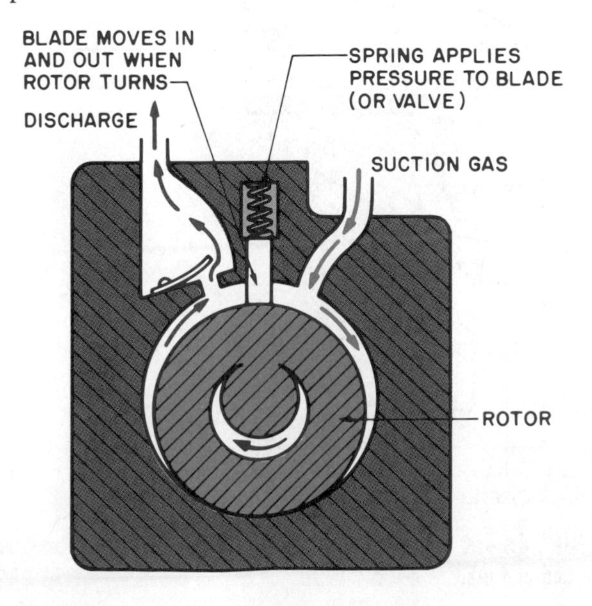

FIGURE 45–25 Freezer compressors are either reciprocating or rotary.

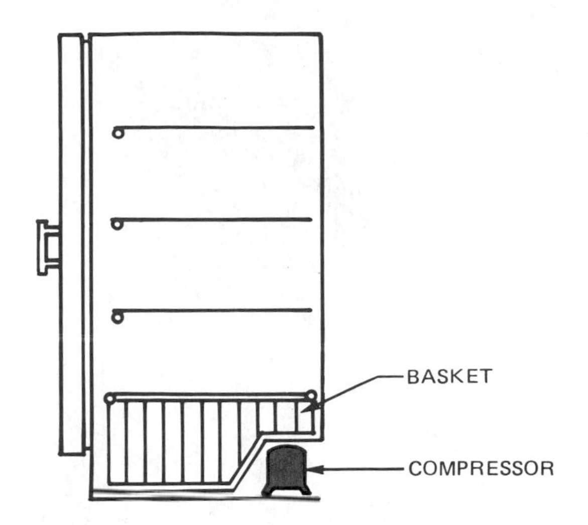

FIGURE 45–26 The compressor compartment is on the bottom and at the back on a typical upright freezer.

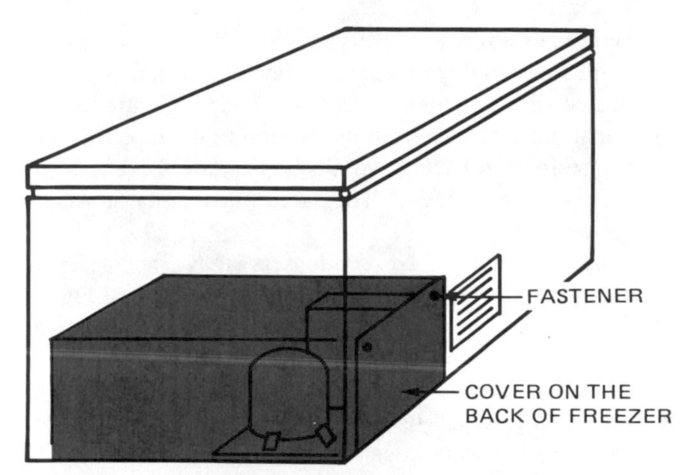

FIGURE 45–27 The chest-type compressor compartment is on the end, but usually serviced from the back.

The compressor is located at the back of the box on the bottom if the box is an upright, Figure 45–26. The box must be moved from the wall when service is needed. When it is a chest type, the compressor will be at one end or the other, but still is usually serviced from the back, Figure 45–27. Some manufacturers in the past have mounted the compressor and an air-cooled forced-draft condenser on a tray that may slide out for service. With this design the lines to the compressor must be formed in a coil to allow them to be extended for service, Figure 45–28.

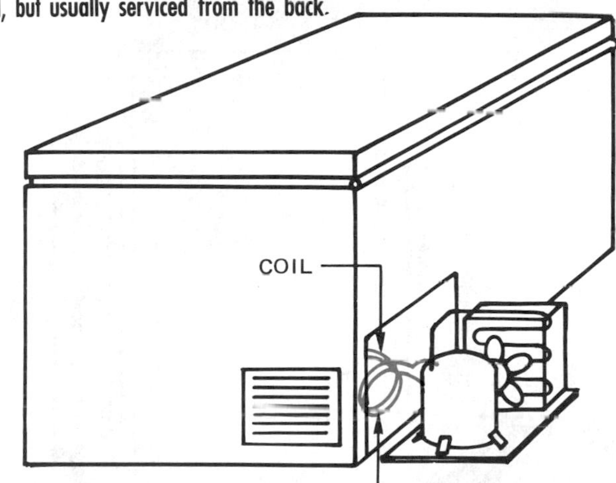

FIGURE 45–28 When the compressor is on a tray, the lines must be coiled in such a manner to allow the compressor to slide out.

45.6 THE CONDENSER

All home freezers have air-cooled condensers. They may be the chimney type located at the back of the box, in the walls

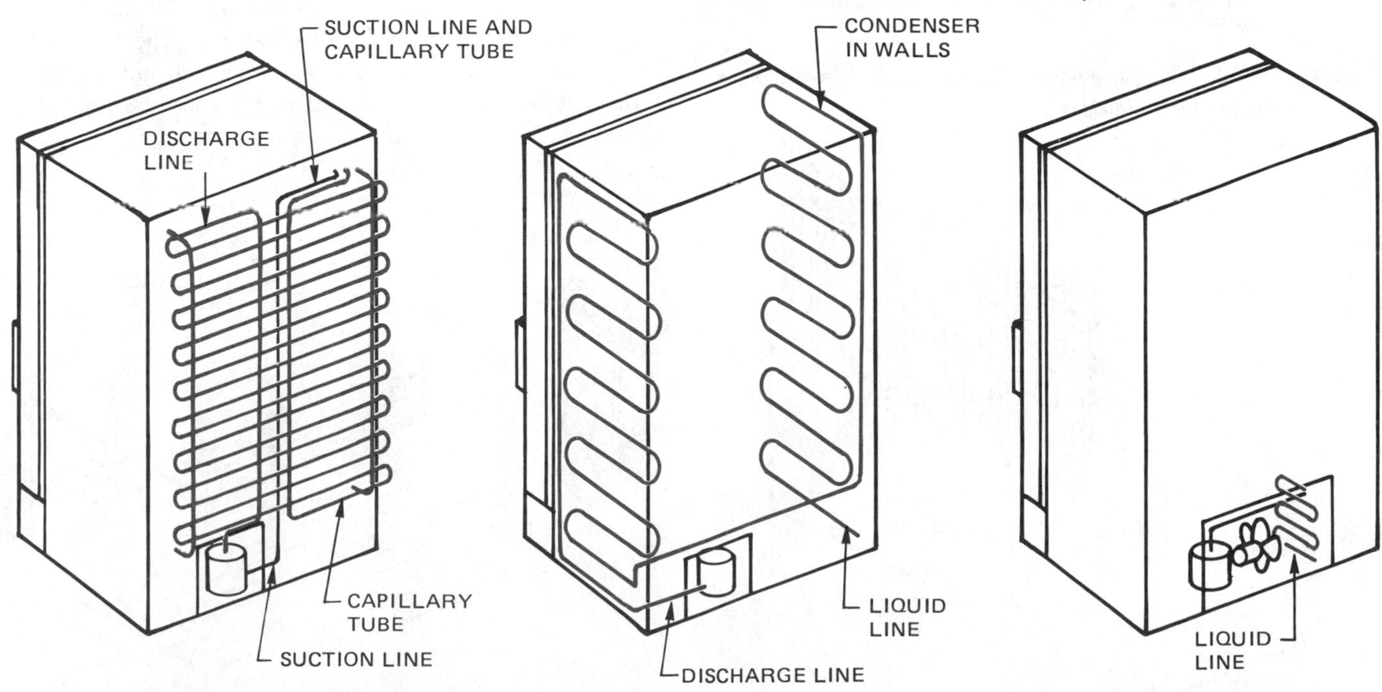

FIGURE 45–29 The condensers may be the chimney type, in the wall, or forced draft, but they must all have airflow.

of the box, or a forced-draft type with a fan, Figure 45–29.

The chimney-type located at the back of the box must be located where enough air can flow over it, Figure 45–30. Note that the air flows from the bottom to the top. This box cannot be located back in an alcove because there will not be enough air circulation. The same rules apply as with a refrigerator.

These chimney-type condensers may be made of stamped steel or steel tubes connected with wires, Figure 45–31. They normally require no routine care, except to keep lint, dust, or dirt from covering them, Figure 45–32.

When the condenser is located in the wall of the box, the tubing is fastened to the outside panel. The condenser will not be seen when located under or at the back of the appliance. The outside of the box will be very warm, Figure 45–33. Air must be allowed to circulate along the sides.

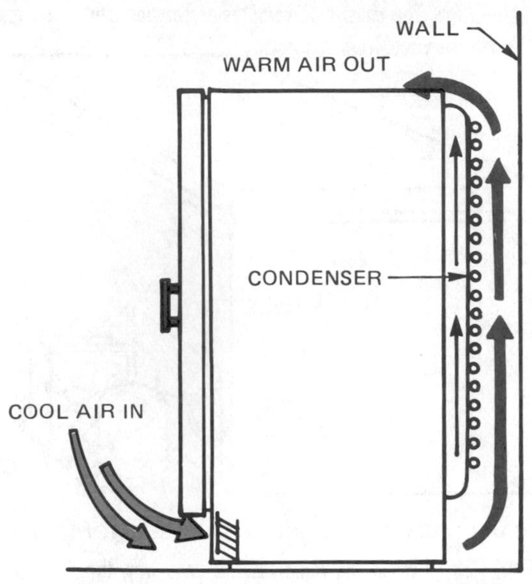

FIGURE 45–30 The chimney-type condenser must be located where it can have natural-draft circulation.

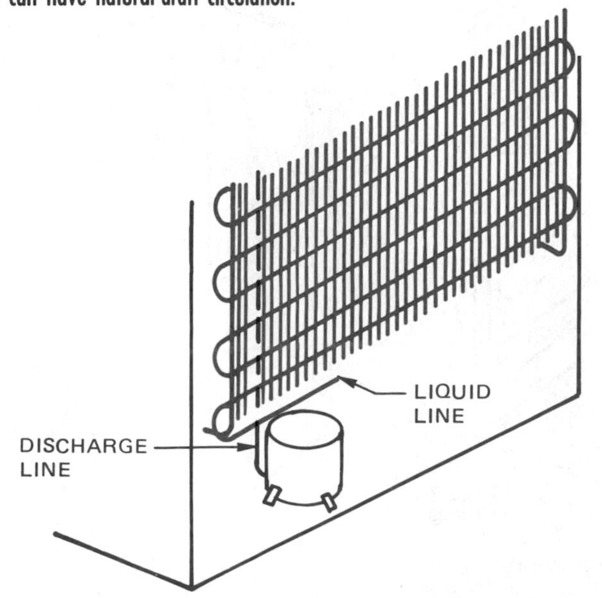

FIGURE 45–31 Some of the chimney-type condensers have steel tubes connected with wires.

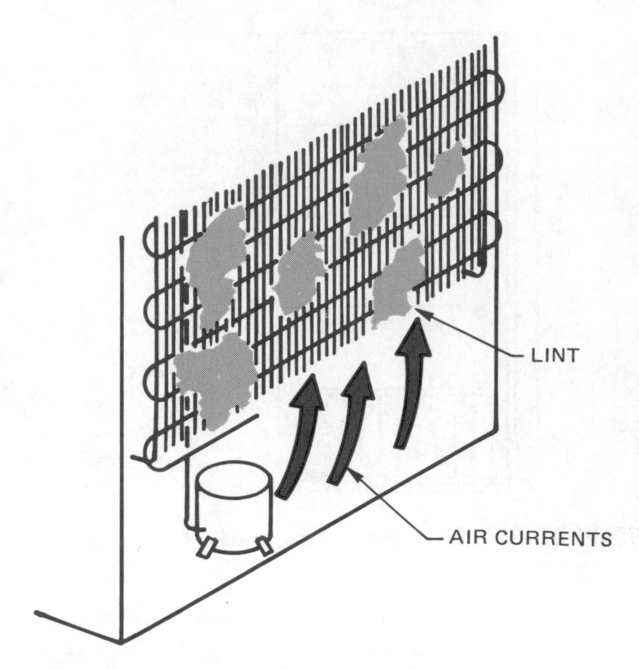

FIGURE 45–32 Condensers with wires are a natural place for lint to collect.

The forced-draft condensers are typically used with units that have automatic defrost and may require more care. They are smaller and have air forced over them. Lint or large particles may be trapped in the fins, Figure 45–34. The air in conjunction with the heat of the discharge line may be used to evaporate any water from defrost, Figure 45–35.

Many freezers with forced-draft condensers are located in laundry rooms. These rooms have more lint than other parts of the house and may be much warmer. These freezers must be more closely observed because of lint and temperature. Lint will collect in the fins of the condenser and the defrost drain pan and can cause a problem.

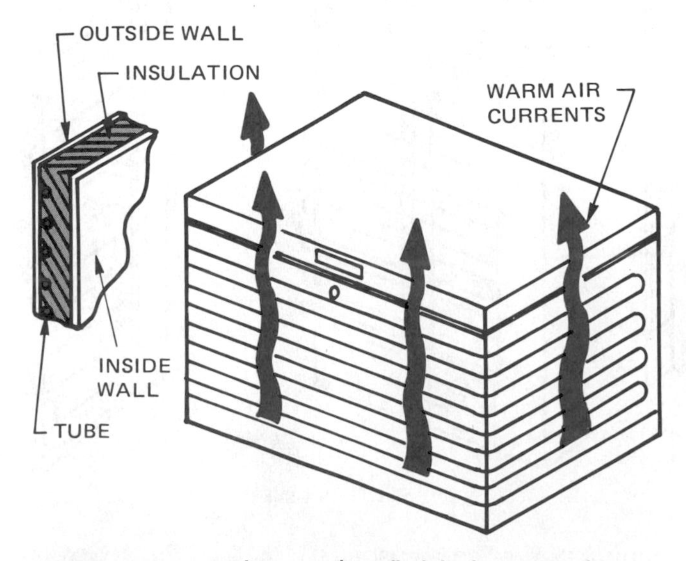

FIGURE 45–33 Condensers in the wall of the freezer must have natural air circulation.

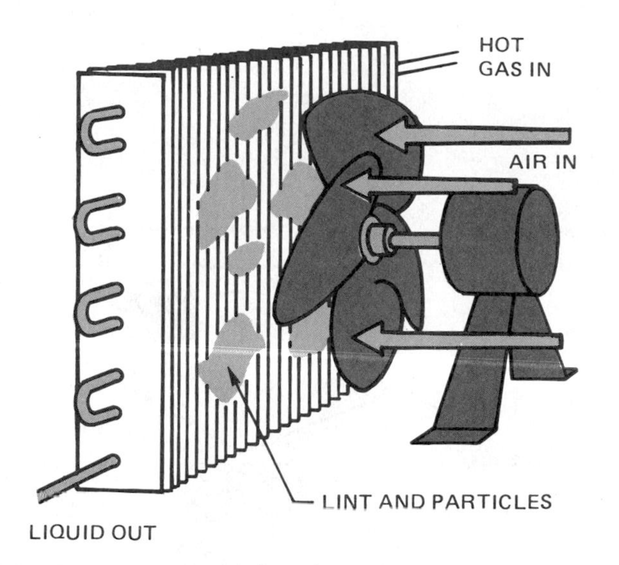

FIGURE 45–34 Forced-draft condensers have fins. They are normally close to the floor and will collect particles such as lint and dust.

45.7 THE METERING DEVICE

The capillary tube is the typical metering device for home freezers. These follow the same general design rules as for domestic refrigerators. They are fastened to the suction line for a heat exchange. See Unit 44 for information on capillary tubes. Some domestic freezers use R-22 for the refrigerant, so the technician should be sure to look at the nameplate to determine the correct refrigerant and capillary tube selection.

45.8 TYPICAL OPERATING CONDITIONS, EVAPORATOR

The home freezer must be operated with a box inside temperature cold enough to freeze ice cream hard. The following pressures and temperatures may be found in the pressure/temperature chart, Figure 3–38. Some of the pressures are fractions of a psig and are estimated between the numbers on the chart. For ice cream to be frozen hard, the temperature must be about 0°F. Some ice cream with a lot of cream and sugar content will not be frozen hard until about −10°F. These will be slightly soft at 0°F. The evaporator must operate at a temperature that will lower the air temperature to a minimum of 0°F. When the air temperature is 0°F, the coil temperature for a plate-type evaporator may be as low as −18°F. This corresponds to a pressure of 1.3 psig for R-12 and 11.3 psig for R-22, the two refrigerants used in domestic freezers. See Figure 45–36 for an illustration of both types of refrigerants.

When the evaporator is a forced-draft type, the coil temperature may be a little higher because the air is forced over the evaporator. When the air temperature is 0°F, the coil temperature may be about −11°F, which corresponds to a pressure of 4 psig for R-12 and 15.5 psig for R-22, Figure 45–37. These are temperatures and pressures at the end of the cycle, at about the time the compressor is to stop. Temperatures and pressures during a pull down will of course be higher, depending on the box temperature.

All pressures for the condenser are determined using gage readings. The freezer, like the refrigerator, may have no gage ports. Gage readings should not be taken except as a last resort. Unit 44 will explain the reason and if readings are needed, how to obtain them.

45.9 TYPICAL OPERATING CONDITIONS, CONDENSER

All freezers are air cooled because of the need for them to be transportable. The condensers are either natural- or forced-draft type. The typical freezer is designed to be located in living space so the design air temperatures are from about 65°F to 95°F. Most residences should be maintained within this temperature range most of the time. The condenser then will have air passing over it from about 65°F to 95°F. The condenser should be able to condense the refrigerant to 25°F to 35°F higher than these conditions. This is a

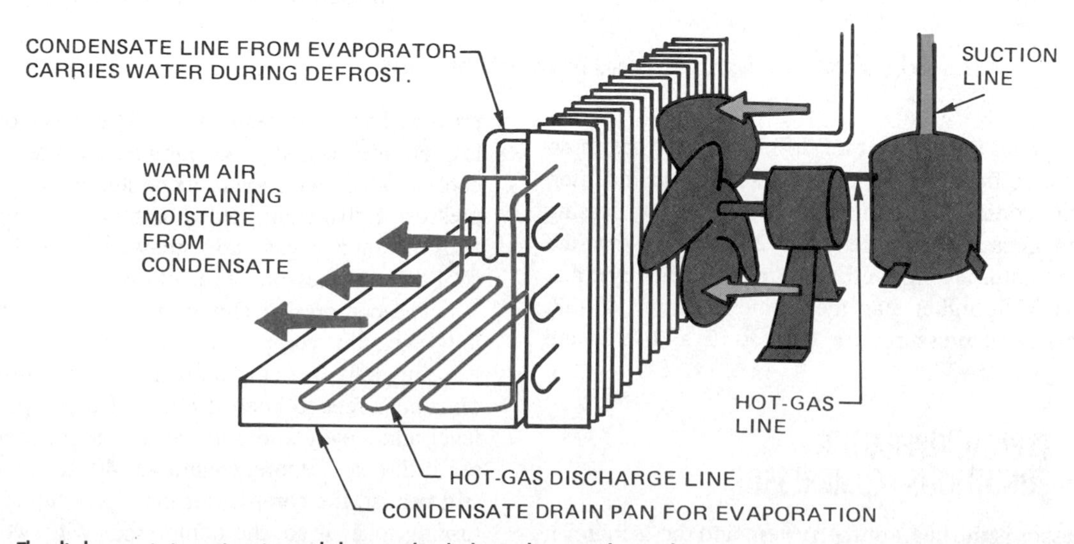

FIGURE 45–35 The discharge air in conjunction with heat in the discharge line may be used to evaporate condensate from defrost.

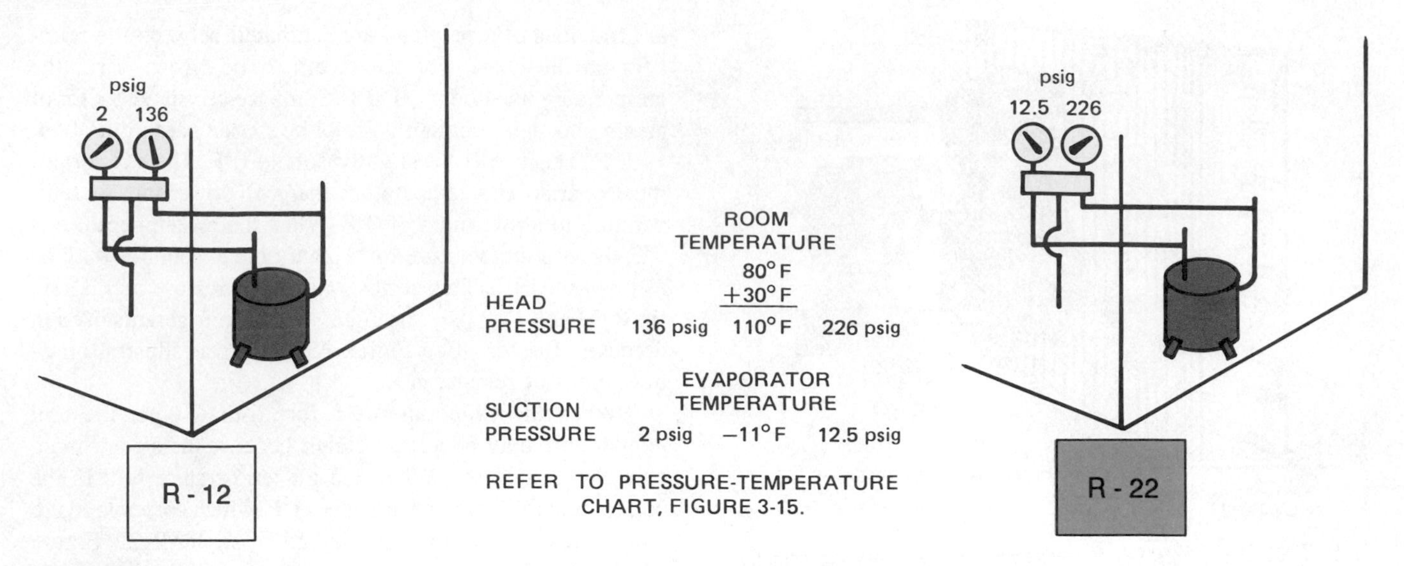

FIGURE 45–36 R-12 or R-22 may be used for the refrigerant in a domestic freezer. Be sure to read the nameplate.

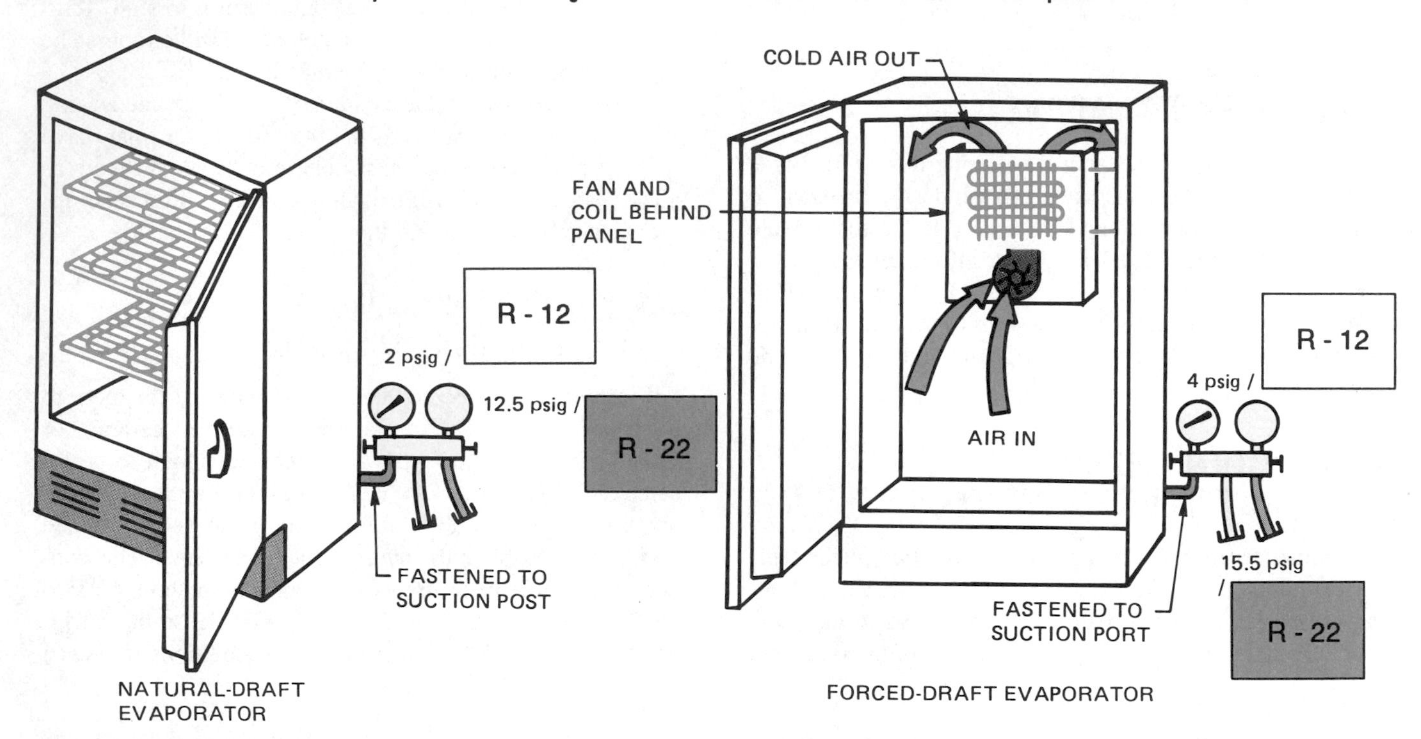

FIGURE 45–37 Some typical evaporator conditions, forced- and natural-draft evaporators.

fairly wide spread in temperature difference and has to do with condenser efficiency. The more efficient the condenser, the closer the condensing temperature is to the air passing over it. A condenser that condenses at 25°F higher than the ambient temperature is more efficient than a condenser that condenses at 35°F higher than the ambient, Figure 45–38. Some typical head pressures for both R-12 and R-22 are shown in Figure 45–39.

45.10 TYPICAL OPERATING CONDITIONS, COMPRESSOR

The compressor is the heart of the system and the technician should be able to readily recognize problems with the compressor. The compressor is hot to the touch on a freezer that is operating normally. All compressors are air or refrigerant cooled. When air cooled, there should be fins on the compressor. Refrigerant-cooled compressors may be cooled with the suction gas and it may have an oil cooler to help cool the compressor, see Unit 44.

The compressor should not be noisy. If the freezer is not level, the compressor may bind and it may vibrate. Leveling a freezer is important. The compressor is mounted on rubber feet to keep it quiet. If a compressor is setting level and steady and still has vibration, check the lines to see if they are secure, Figure 45–40. If vibration and noise still persist, the compressor may have liquid refrigerant returning to it. If so, the compressor will sweat on the side, Figure 45–41.

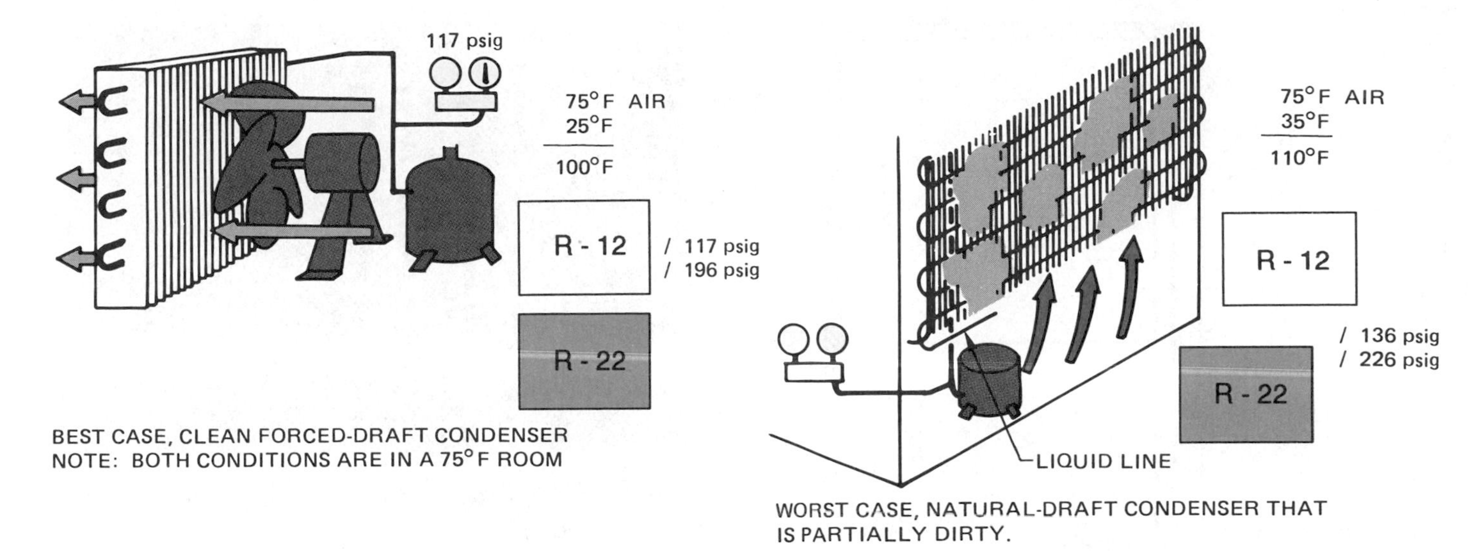

FIGURE 45–38 The refrigerant typically condenses at 25°F to 35°F higher than the ambient temperature.

	FORCED AIR CONDENSER	NATURAL-DRAFT CONDENSER
COOL ROOM	65° F + 25° F 90° F CONDENSING TEMPERATURE 100 psig R-12 168 psig R-22	65° F + 30° F 95° F CONDENSING TEMPERATURE 108 psig R-12 182 psig R-22
NORMAL ROOM	75° F + 25° F 100° F CONDENSING TEMPERATURE 117 psig R-12 196 psig R-22	75° F + 30° F 105° F CONDENSING TEMPERATURE 127 psig R-12 211 psig R-22
HOT ROOM	95° F + 25° F 120° F CONDENSING TEMPERATURE 158 psig R-12 260 psig R-22	95° F + 30° F 125° F CONDENSING TEMPERATURE 169 psig R-12 278 psig R-22
HOT ROOM DIRTY CONDENSER	95° F + 35° F 130° F CONDENSING TEMPERATURE 181 psig R-12 297 psig R-22	95° F + 35° F 130° F CONDENSING TEMPERATURE 181 psig R-12 297 psig R-22

FIGURE 45–39 Some typical R-12 and R-22 head pressure conditions.

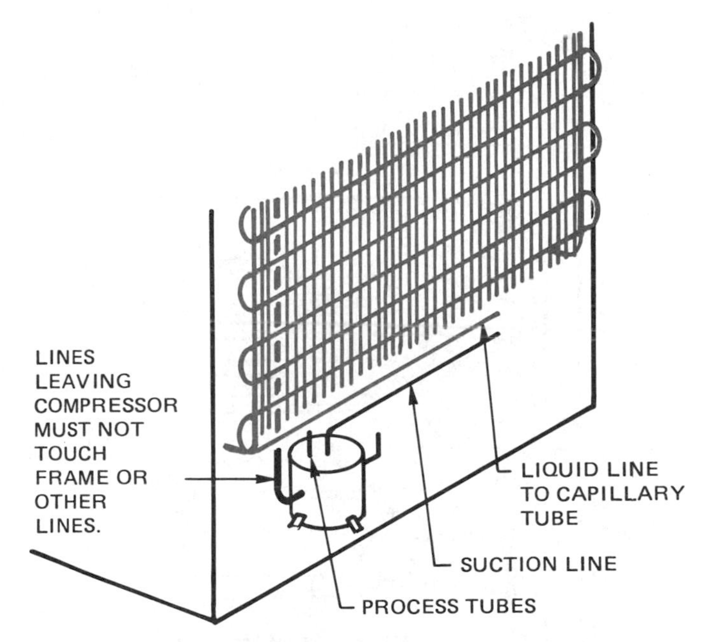

FIGURE 45–40 Check lines to make sure they are secure.

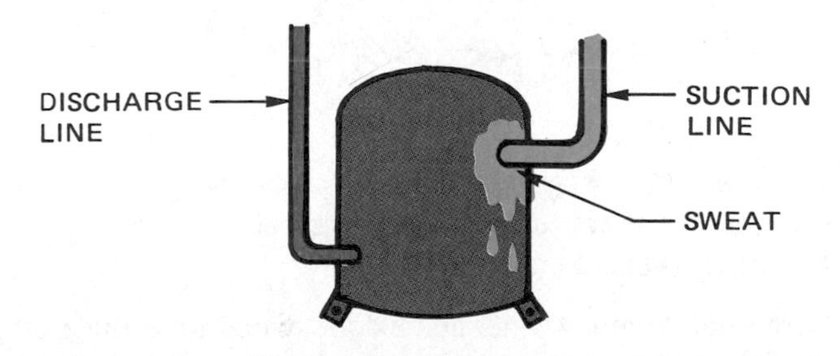

FIGURE 45–41 If liquid refrigerant is returning to the compressor, it will be sweating on the side.

If the capacity of the compressor is questioned, as when the unit is not cold enough, first suspect an extra load. A light remaining on will keep the compressor running all the time, but the space temperature may be cold enough. A gasket leak will cause an extra load. The owner may be using the box too often and opening the door too much. All conditions should be analyzed before the compressor is considered defective, Figure 45–42. The compressor may not be pumping to capacity. See Unit 44 for information on analyzing the compressor.

45.11 CONTROLS

Controls for freezers are simple except for those controlling automatic defrost. A typical freezer has only two controls, a

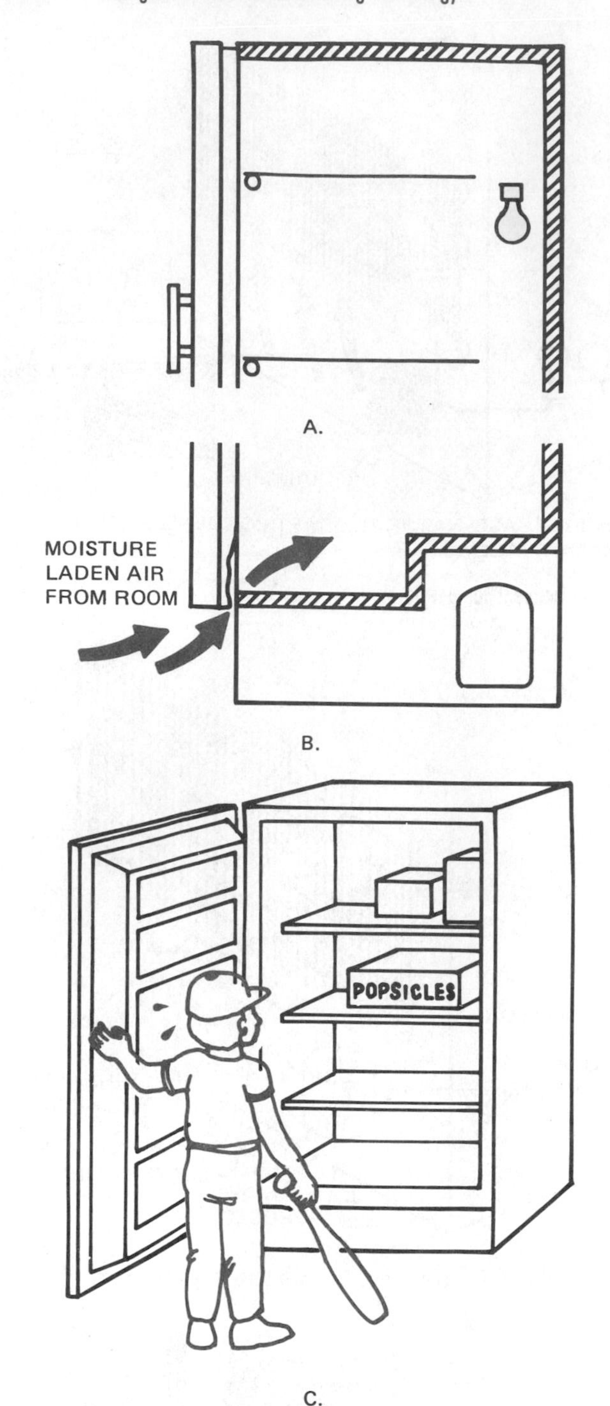

FIGURE 45–42 A light remaining on, a gasket leak, or too many door openings may keep the compressor running while the space temperature seems to be cold enough.

thermostat, Figure 45–43, and a door switch for turning off the light and possibly a fan on some models. As with any control circuitry, the technician must know what the manufacturer's intent was when the system was designed. A wiring diagram is normally fastened to the back of the box.

There are panel heaters to prevent sweat like those on refrigerators. These devices will appear on the wiring diagram, but the manufacturer's literature may have to be consulted for their exact location.

On a freezer with automatic defrost, the evaporator must be located in such a manner that the ice will melt and

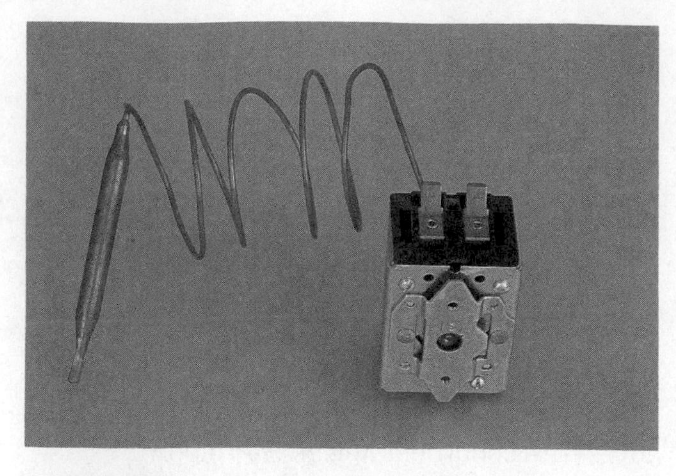

FIGURE 45–43 A cold control (thermostat) for a freezer. *Photo by Bill Johnson.*

the condensate will drain to a place where it will evaporate. This is typically a forced-draft type, Figure 45–44. The controls are similar to a refrigerator. Compressor running time may be used to instigate defrost. Electric heaters may be energized to accomplish defrost. A drain heater may be located in the drain to make sure the water does not freeze in the drain. Figure 45–45 shows a typical diagram.

45.12 SERVICING THE FREEZER

The same tools and techniques are used for servicing freezers as for refrigerators. However, when servicing freezers, the technician does not have to think about the fresh-food compartment, only the low-temperature compartment. The type of box determines the type of service. For example, a box that has a forced-draft evaporator will have a fan motor that can cause problems. These fans run many hours per year. Usually they are wired in parallel with the compressor and will be off when the compressor is off and during defrost. These fan motors are very small open-type motors with permanently lubricated bearings, Figure 45–46. The technician should remember that the fan may be controlled by a

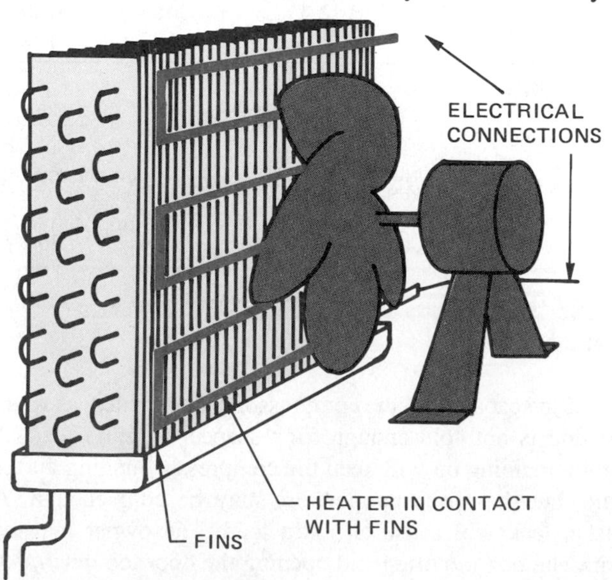

FIGURE 45–44 Forced-draft evaporator with automatic defrost.

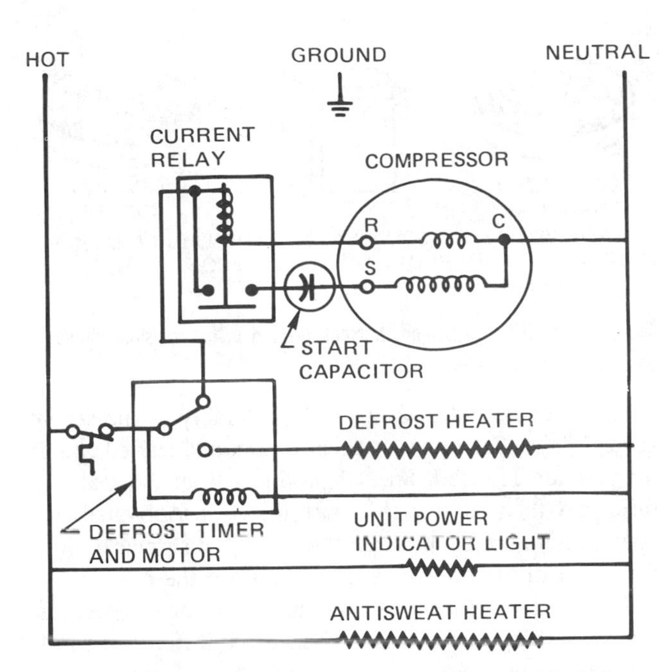

FIGURE 45–45 Typical wiring diagram with automatic defrost.

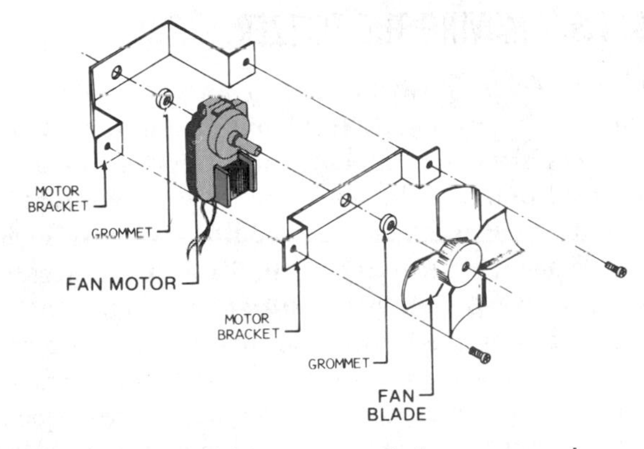

FIGURE 45–46 Small fan motors usually have permanently lubricated bearings. *Courtesy White Consolidated Industries, Inc.*

fan switch and may stop normally when the door is opened. To hear the fan with the door open, the switch must be depressed.

If the fan does not run, some food may have to be removed to be able to remove the fan panel to determine what the problem is. Typically, the fan motor may be stuck. If this is the case, a little lubrication may be added, such as penetrating oil, to get the fan motor started, Figure 45–47. It may be allowed to run temporarily until a replacement is obtained. The fan motor should be changed or it will probably fail again. ***A freezer is not considered an attended appliance. It may not be opened regularly. It must be kept reliable and no chances taken with the food.***

Proper defrost methods must be used for manual defrost. The food may be removed and external heat may be used to defrost the box. The food may be placed in ice chests and a fan used to circulate room air into the freezer compartment, Figure 45–48. Pans with hot water may be placed in the box with the door shut to aid defrost, Figure 45–49.

It is not uncommon for the owner to get impatient with slow defrost and start chipping the frost, puncturing a hole in the evaporator. ***Never use sharp instruments to hasten ice removal.*** The refrigerant will leak out. Later the owner may start the unit and water will be drawn into the system. When this has happened the evacuation procedures described in Unit 44 should be used.

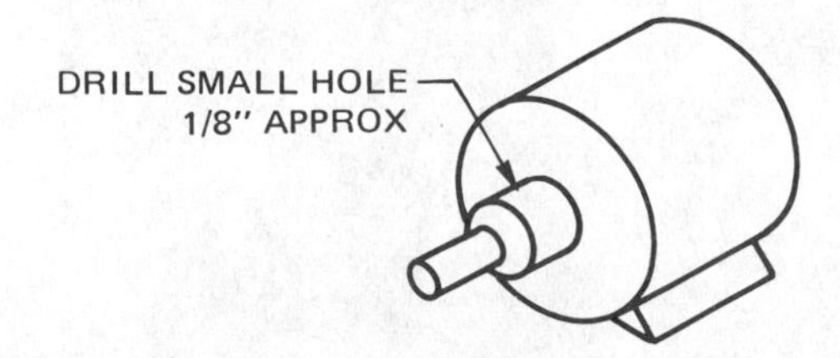

FIGURE 45–47 A fan motor may be drilled and lubricated for a temporary repair.

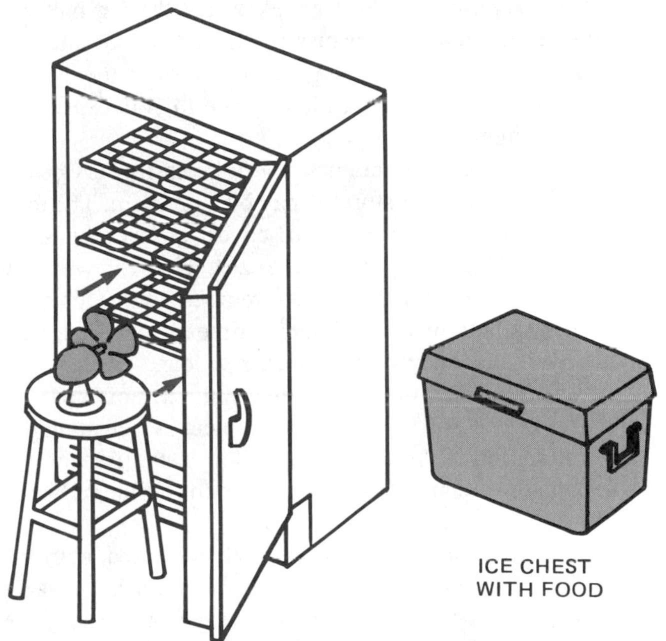

FIGURE 45–48 Food may be placed in ice chests during manual defrost. A fan may be used to circulate room air over the evaporator to defrost ice.

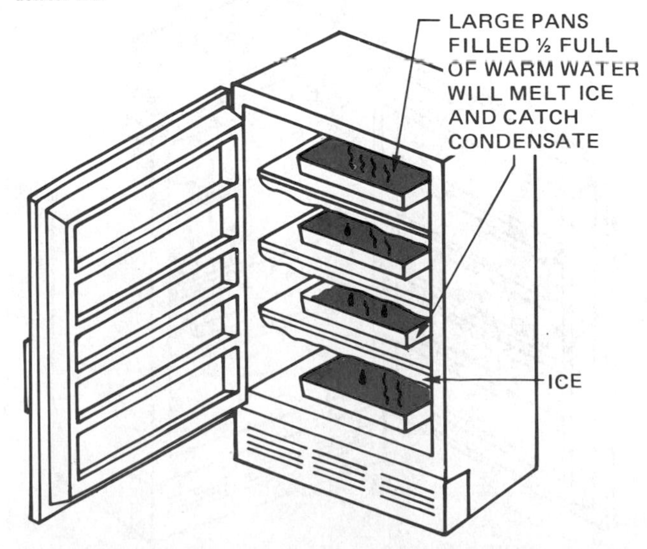

FIGURE 45–49 Pans of hot water may be used for defrost.

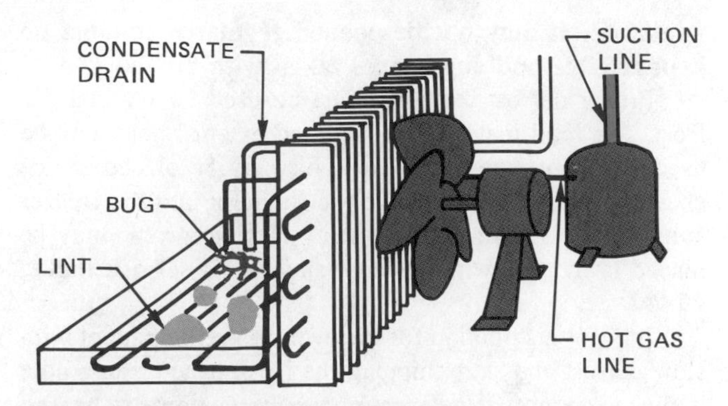

FIGURE 45–50 A condensate pan at the bottom of the unit is a likely place to collect dust, lint, and other debris.

The freezer may have a forced-draft condenser with a fan motor. Again, care must be taken that the fan motor is repaired and is reliable because the freezer is not attended regularly. The fan motor may blow air over the defrost condensate water. This is a likely place to collet dust, lint and pests, Figure 45–50.

Freezers, like refrigerators, are more easily serviced in a shop. However, food storage may be a problem. The food is frozen and when the owner discovers the freezer is not working, the food may still be frozen. If so, it will stay frozen for several hours if the door remains shut. This gives the owner and the service technician time to develop a plan. Some companies may have a freezer to loan.

If the food has thawed, it may only be soft and still cold. If so, it is still edible. The problem is consuming it before it spoils. If the food has spoiled in the freezer, the owner and the technician will have an odor problem, Figure 45–51.

When the entire box has become saturated with the odor of spoiled food, the box may or may not be salvageable. If it has fiberglass insulation, the technician may not be able to remove the odor because the insulation is saturated. One method used to salvage a unit is to clean the box,

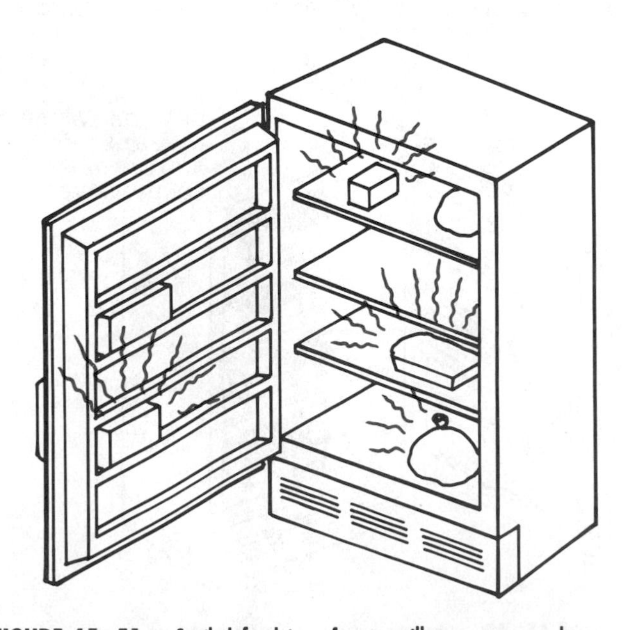

FIGURE 45–51 Spoiled food in a freezer will generate an odor.

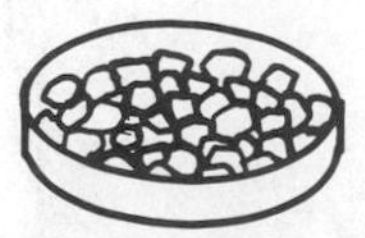

FIGURE 45–52 Activated charcoal, ground coffee, or soda powder will help to absorb odors.

open the door, and let it air out, preferably in the sun for several days. The box may then be shut and started and the temperature lowered. Special products from the manufacturer may then be used to absorb the odor. Activated charcoal, ground coffee, or soda powder in a dish placed in the freezer with the door closed may absorb the odor, Figure 45–52. If the freezer still has an odor when cooled and odor-absorbing devices have been used, it may have to be replaced.

45.13 MOVING THE FREEZER

The technician must use great care when lifting and moving any heavy object. Think out the safest method to move any object. Special tools and equipment should be used at all times. Technicians should wear appropriate back brace belts when moving and lifting objects, Figure 45–53. When it is necessary to lift, lift with your legs, not your back. Keep your back straight. A refrigerator or freezer is among the heaviest and most awkward type of equipment to move. The surroundings or the freezer may be damaged when moving the freezer. Freezers are frequently located in tight places and often need to be moved upstairs or downstairs. The correct moving devices must be used. ***Never lay a freezer on its side.***

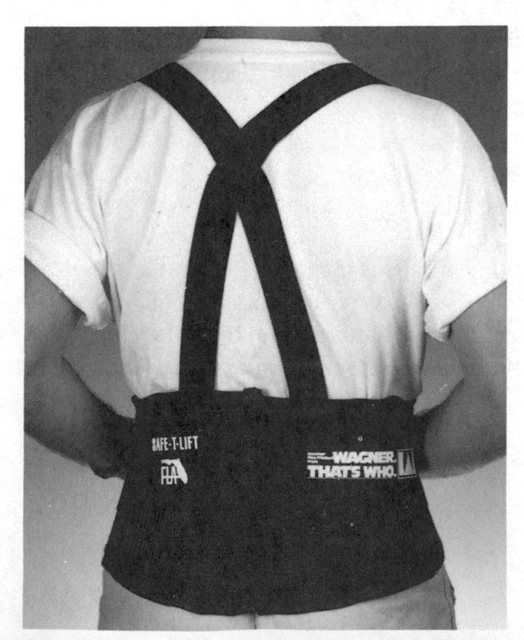

FIGURE 45–53 *Use back brace to lift objects. Use legs, not back; keep back straight.*

Moving Upright Freezers

Refrigerator hand trucks with wide belts are used for upright boxes, Figure 45–54. These must be used in such a manner as to not damage the condenser and not let the door open, Figure 45–55. The box should be trucked from the side where the door closes if possible. Freezers or refrigerators should not be trucked or moved with food in them. The combination of food and freezer for a large unit may well be over 1000 lb. When repositioned, the box must be on a solid foundation, Figure 45–56.

The freezer may be moved out from the wall enough for service without removing the food if done carefully. If the freezer is located on a floor that must be protected, the technician is responsible for knowing how to accomplish the move without damage to the floor or freezer. A small rug may be used under the legs if the box does not have wheels. A correct mat will slide across a fine-finished floor without damage to the floor.

The technician should slightly tilt the box over to one side and have someone slide the mat under the feet. Then the box must be tilted to the other side for the other feet. This method can only be used where there is room to tilt the unit, Figure 45–57. If there is no tilt room, the freezer may have to be pulled out from the wall far enough to place the mat under the front feet, then the box may be pulled out from the wall far enough for service, Figure 45–58. The front feet may be screwed upward one at a time and a mat placed under them and then screwed back out if no side or backward tilt is possible, Figure 45–59. Care should be used

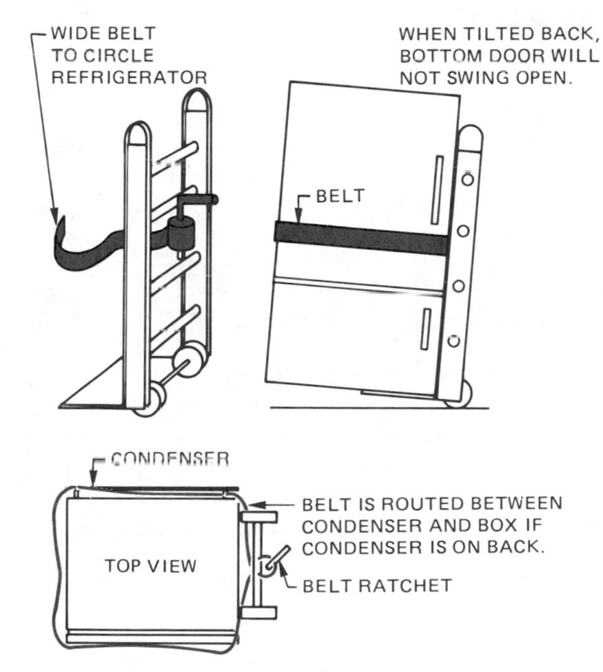

FIGURE 45–54 Refrigerator hand trucks.

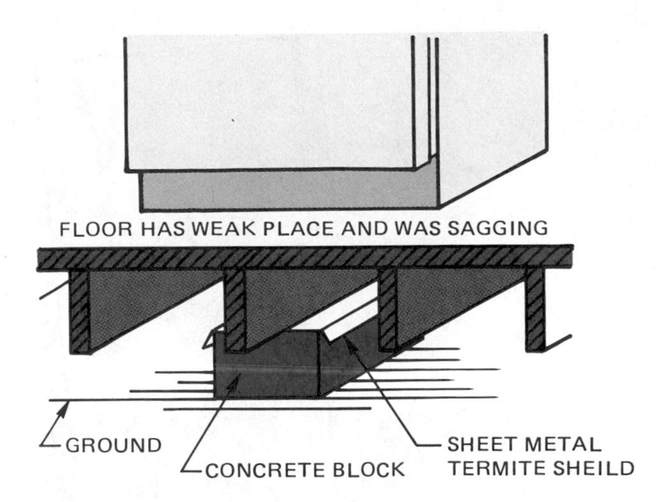

FIGURE 45–56 Freezers must be on a solid foundation.

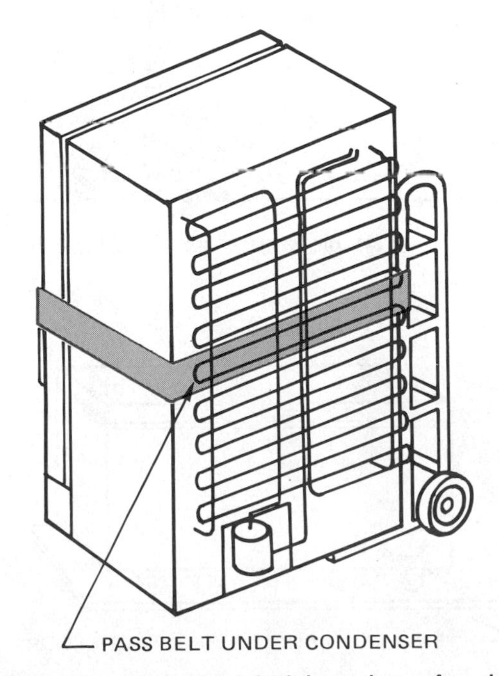

FIGURE 45–55 Pull the belt behind the condenser, if possible, when the condenser is in the back.

FIGURE 45–57 Room to tilt freezer to place a mat under the feet.

FIGURE 45–58 The freezer may be pulled out a few inches for the purpose of placing a mat under the feet.

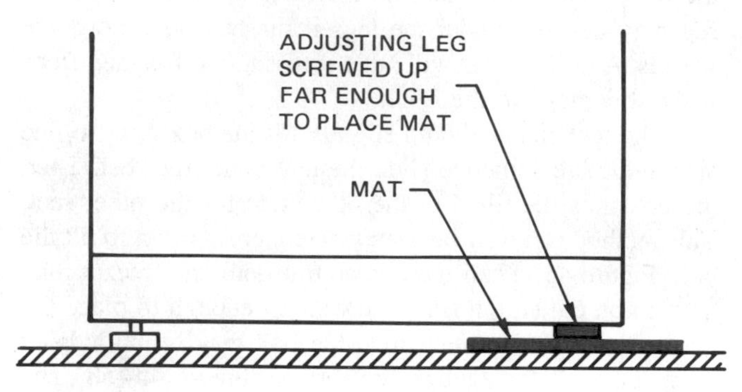

FIGURE 45–59 The feet may be screwed upward, one at a time, to work the mat under the feet.

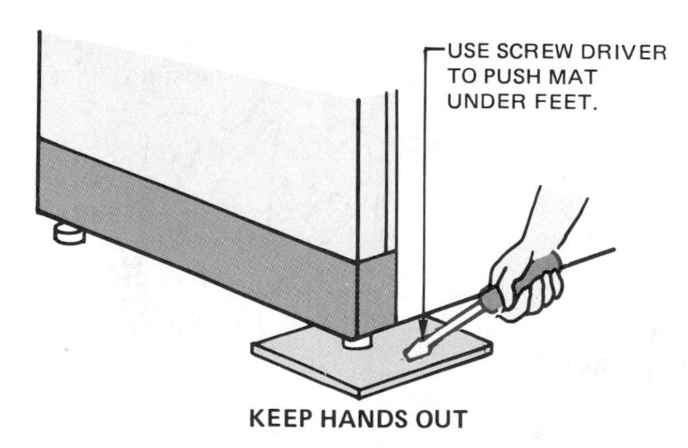

FIGURE 45–60 *Use a small stick or long screwdriver for placing the mat, not your hands.*

when placing a mat under the feet. ***Do not let the freezer fall on your hands. A screwdriver or stick may be used to slide the mat under the feet, Figure 45–60.***

Moving Chest-Type Freezers

Chest-type freezers are more difficult to move than upright freezers. They may be larger, take up more floor space, and contain more food. Refrigerator hand trucks will not work as well on chest-type freezers because the freezer is longer than it is tall, Figure 45–61.

When a chest-type freezer must be moved out from the wall for service, the same guidelines apply as for the upright freezer. Carefully using a pry bar on a piece of plywood, a mat can be placed under the feet, Figure 45–62.

When the chest-type freezer is moved from one room to another, two four-wheel dollies, Figure 45–63, may be used. The freezer must carefully be raised to the level of the dollies. This requires at least two people and caution. The freezer is manufactured in a size to enable it to be moved through standard doors, but care should be used when moving the dolly wheels over the threshold.

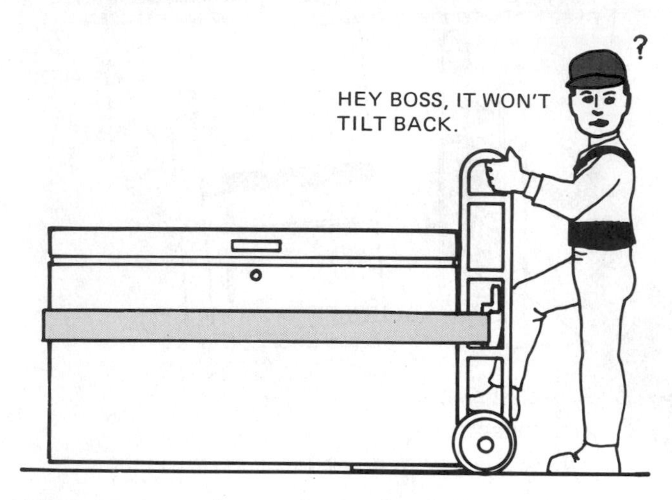

FIGURE 45–61 Refrigerator hand trucks will not work on a chest-type freezer.

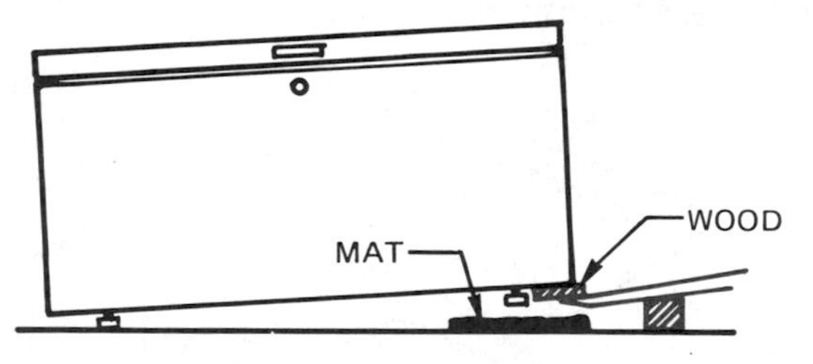

FIGURE 45–62 A mat can be placed under a chest-type freezer by using a pry bar.

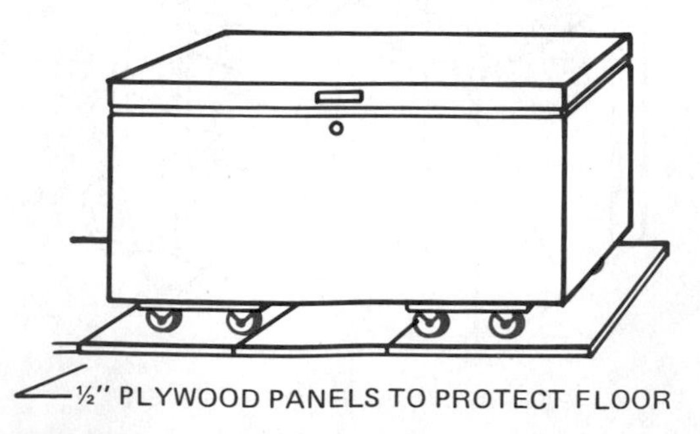

FIGURE 45–63 Use of two dollies with four wheels each is a good choice for moving a full chest-type freezer.

45.14 TEMPORARY FOOD STORAGE

The technician often must help the owner find a place in which to store food temporarily while a freezer is being repaired. If the freezer is still under warranty and a problem arises, the technician will probably be obligated to help find a place to store the food. Some appliance companies may send out a new freezer and merely change the food over if another box of the same type is available. The old box may be moved to the dealer's shop, repaired, and sold as a used freezer. If this box was originally sold to the customer as a special model, it may have to be repaired. In this case, the technician may move another box to the site and transfer the food to the loaned box until the repair is made. If this is not possible, dry ice may be used to store the food. In a chest-type freezer, the dry ice may be placed in the freezer on top of the food. ***A layer of protection is placed between the dry ice and the food, Figure 45–64. Do not shut the lid tight; leave a small gap for the CO to escape from the dry ice.*** The food in upright freezers should be emptied into boxes. The dry ice is placed on top of the food with a layer of protection between the food and dry ice.

Dry ice is very cold and will dehydrate the food if placed in contact with it. ***The technician should never allow dry ice to come in contact with skin or frostbite will occur. Use heavy gloves or better still, a scoop to handle the dry ice.*** About 20 lb of dry ice per 5 ft^3 of freezer capacity will hold the food for 24 h.

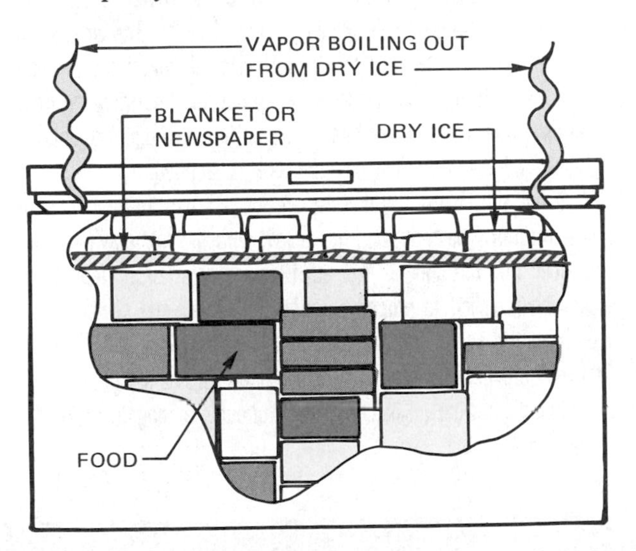

FIGURE 45–64 A layer of protection is placed between the food and the dry ice.

45.15 SERVICE TECHNICIAN CALLS

SERVICE CALL 1

A homeowner who lives in the country calls. This customer has recently purchased a freezer and has had a steer killed, butchered, quick frozen, and placed in the new freezer. *The problem is that the condenser fan motor has failed and the freezer compressor is running all the time.* The food is still frozen hard, but the temperature is beginning to rise.

The technician arrives and notices right away that the condenser fan is not running. The customer has a thermometer in the box and it reads 15°F. No time can be wasted. The box is removed from the wall for access to the fan motor. The freezer is unplugged and the technician checks the fan motor. It turns freely, so the technician removes one motor lead and an ohm check is performed. The motor has an open circuit through the windings, Figure 45–65. The technician does not have a motor and must get one. The technician asks the customer if she has a floor fan that may be used for the condenser until a new fan motor is obtained. She has one, so the technician places the fan where it can blow over the condenser coil and plugs in the fan, Figure 45–66. The fan is allowed to operate for about 10 min to remove some of the excess heat from the condenser. The freezer is plugged in and the compressor starts. The technician can then perform another call on the way to town and back and save time. The freezer will operate and be pulling down while the trip is made.

About 3 h later the technician arrives back to the job; the freezer is still running. The door is opened and the thermometer is

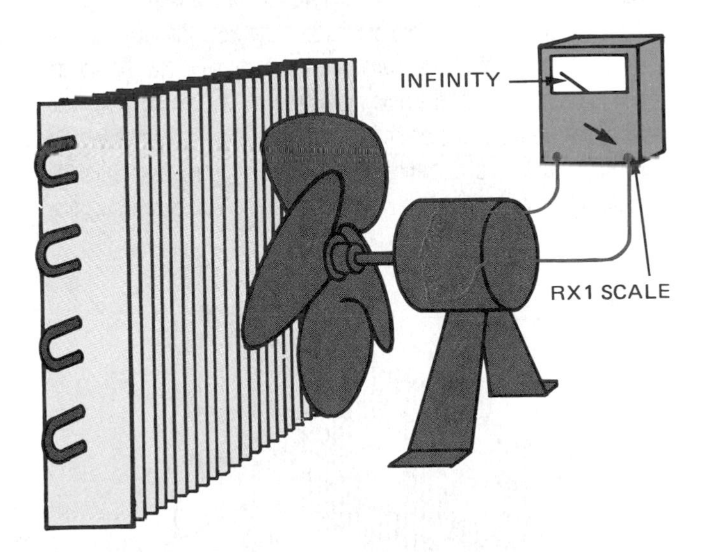

FIGURE 45–65 Motor with an open circuit through the windings.

FIGURE 45–66 Fan placed where it can temporarily pass air over the condenser.

checked. It indicates 1°F, so the freezer temperature is going down. The technician unplugs the freezer and replaces the fan motor. All panels are replaced with the proper fasteners and the freezer is plugged in. The compressor and the new condenser fan start. The freezer is moved back to the wall and the technician leaves.

SERVICE CALL 2

A customer calls to report that the freezer is running all the time. *This freezer has a vertical condenser on the back and it is covered with lint. The freezer is located in the laundry room. It is summer and the laundry room is hot from the summer heat and the clothes drier.*

The technician arrives and checks the temperature with a thermometer. It is already evident that the box is not cold enough because the ice cream is not hard. The technician places the temperature probe between two packages and closes the door. In about 5 min the instrument shows 14°F. While the temperature of the probe was pulling down, the technician has had a look around at the surroundings. It is evident that the condenser is dirty and the room is hot, Figure 45–67. The freezer is pulled out from the wall. A vacuum cleaner with a brush is used to remove the lint from the condenser. The technician then slides the box back to the wall. The technician can do no more and gives the owner the following directions. Do not dry any more clothes until the next morning. A small dial-type thermometer is sold to the customer and placed in the food compartment. The technician calls the customer the next morning and asks what the temperature indication is. It is −5°F, so the box is working fine.

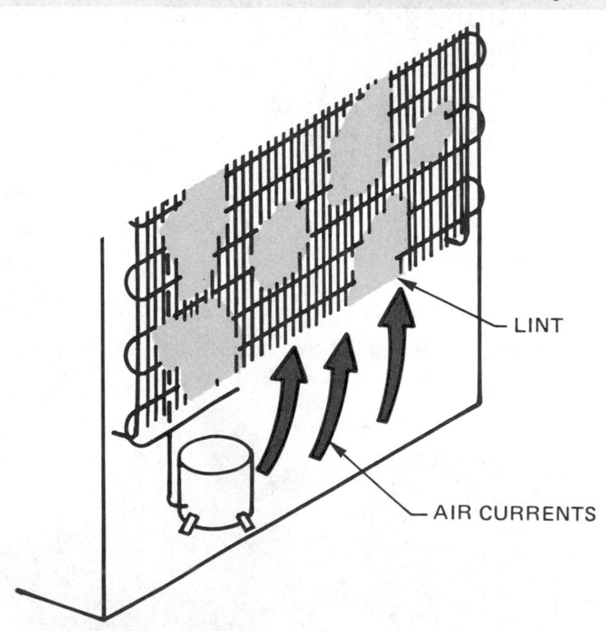

FIGURE 45–67 A dirty chimney-type condenser.

SERVICE CALL 3

A customer calls and says his freezer shocked him when he touched it and at the same time touched the water heater located beside it. *The problem is the freezer has a slight ground circuit due to moisture behind the box.* The house is old and only has a two-prong wall plug. These are ungrounded, and an adapter has been used to plug the freezer into the wall outlet. The water heater is wired straight to the electrical panel where it is grounded. The customer is advised not to touch the box. It cannot be unplugged or loss of food may result.

The technician arrives and takes in a volt-ohm-meter and tools. The technician uses the meter to check the voltage between the freezer and water heater, duplicating the situation the customer was in, Figure 45–68. The voltmeter indicates 115 V. This could be a dangerous situation. The technician removes the fuse to the freezer circuit and pulls it from the wall. When it is unplugged, the technician discovers at least half of the problem, a two-prong plug. If the unit were connected through a three-prong plug and grounded, the same type of ground that occurred would blow the fuse.

The technician then fastens one lead of the ohmmeter to the hot plug and the other to the ground (green) plug for the freezer. The meter selector switch is turned to R × 1 and the meter indicates infinity. The technician then starts turning the meter indicator switch to indicate higher resistances. When the switch is turned to R × 10,000, the meter reads 10; 10 × 10,000 means that a circuit of 100,000 Ω exists between the hot wire and the cabinet, Figure 45–69. If the cabinet had been properly grounded, the current flow through this circuit would have blown the fuse or burned a wire causing an open circuit, Figure 45–70. The technician must now find the circuit.

The compressor is disconnected while the meter is still connected, and the reading still remains, Figure 45–71. The circuits are eliminated, one at a time, until the technician discovers the junction box at the bottom of the unit to be damp due to water dripping from a very slight leak at a water heater valve. The valve packing gland is tightened by the technician and the leak is stopped.

A hair drier is used to dry the junction box and the ground is no longer present. The technician has a talk with the homeowner about the circuit and the danger involved. Before the technician leaves, an electrician is called to properly ground this circuit and several more that have the same potential problem. The technician has done all that can be expected. It would not have been satisfactory to leave the job without knowing an electrician was already coming to make the system safe.

SERVICE CALL 4

A customer reports that the beef they froze 2 weeks ago has no flavor, is dry, and seems to be tough. *The problem is this is a new freezer, sold to a person with no experience in freezing food.* The customer bought a side of beef that was not frozen and took it home to be frozen in the new freezer.

Ordinarily, the technician would not get involved in this, but this is an old customer and relations must be maintained. The technician calls the customer and asks her to thaw a steak tonight so it can be checked in the morning. The technician sees a puddle of water and blood under the steak. The technician begins to ask questions. The customer explains how the meat was frozen. She picked it up at a

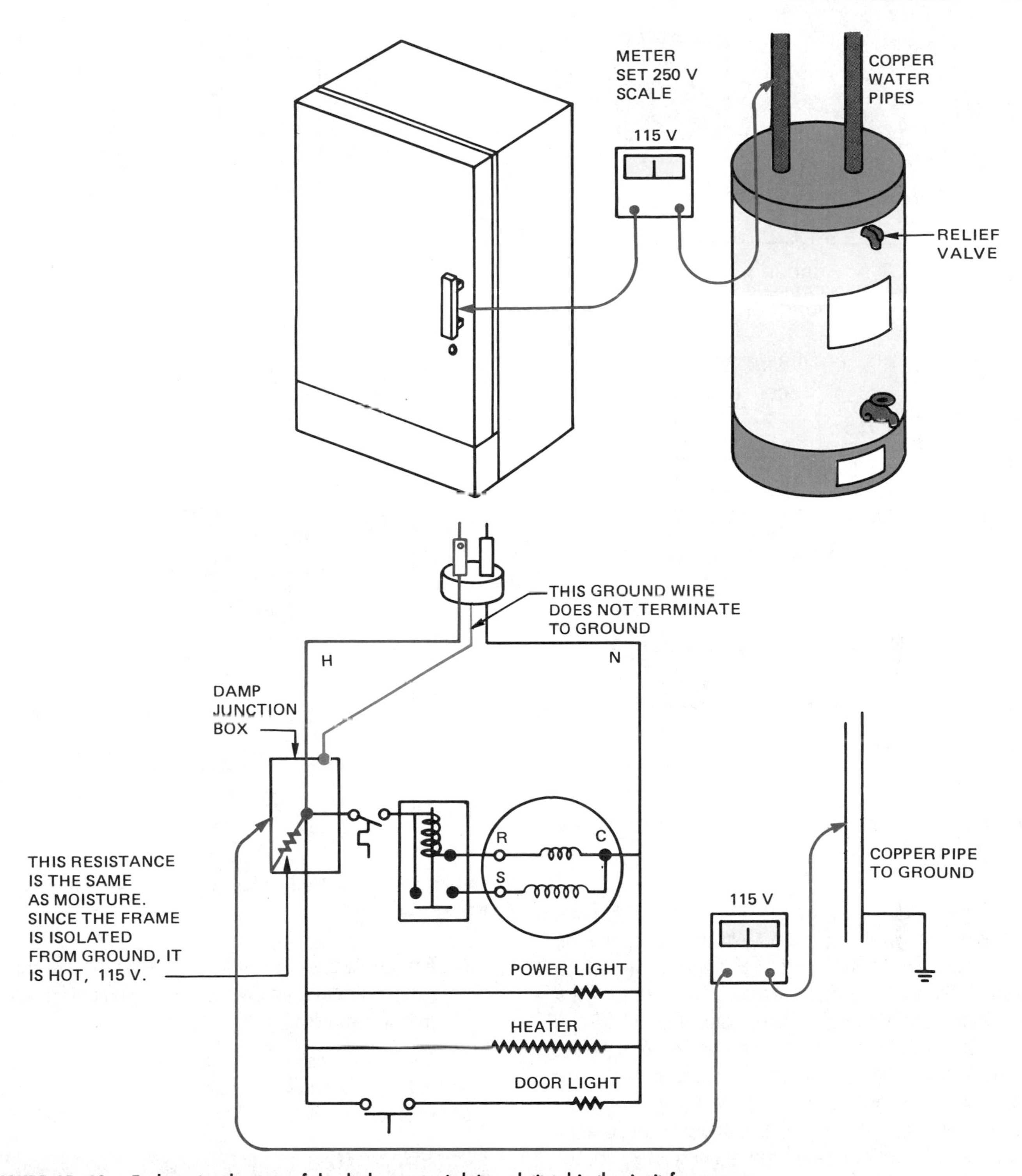

FIGURE 45–68 To determine the cause of the shock, a meter is being substituted in the circuit for a person.

packing house and made three stops before getting home; the weather was hot. The packing house clerk tried to get her to let them freeze the meat but she thought of her own new freezer and felt that it would do a better job. She had put the whole side of beef in the freezer at one time, about 200 lb, packaged in small parcels.

The technician asked to see the owner's manual. In the manual, it plainly stated that the freezer capacity to freeze meat was 40 lb for this 15 ft^3 freezer. She has a freezer full of meat that will not be of the best quality. This is partly the salesperson's fault for not explaining the freezer capabilities. The technician goes over the manual with the owner.

SERVICE CALL 5

A customer, who has been on vacation, calls. It was discovered that the freezer had warmed and thawed while they were gone. *The problem is the freezer is located in the carport storage room and the food has spoiled.*

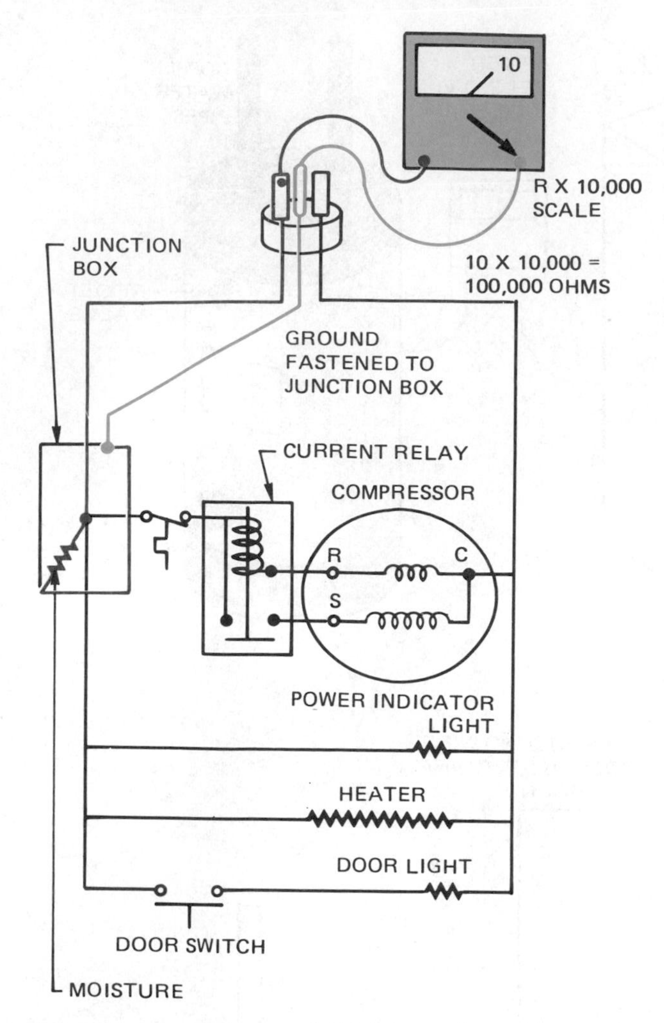

FIGURE 45–69 A resistance of 100,000 Ω is detected between the hot wire and the cabinet.

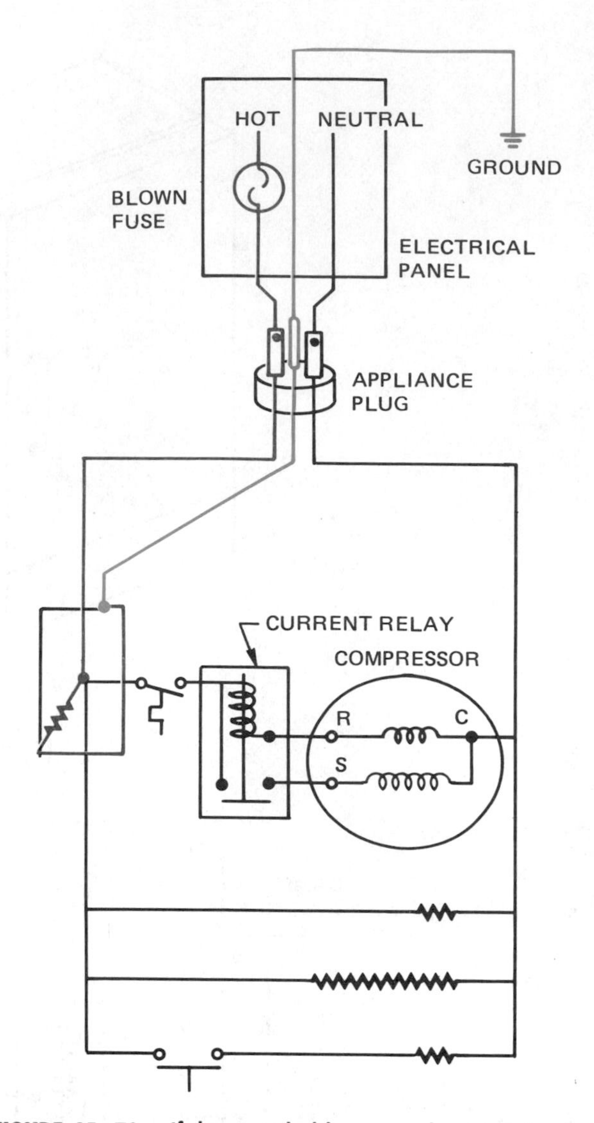

FIGURE 45–70 If this circuit had been properly grounded, a fuse may have blown.

The technician arrives with a helper. It is obvious that the freezer will have to be moved. They unload the freezer and dispose of the spoiled food. After the food is disposed of, the technician and helper move the freezer to the carport. The door is opened and the box washed with ammonia and water, several times. The door is left open for the sun to shine in. Now, the technician starts to find the problem. When the box is plugged in, the compressor does not start. A close examination shows that the common wire going to the compressor has been hot, Figure 45–72. The box is unplugged and the wire is examined closer and found to be burned in two in the junction box. The wire is replaced with proper connectors with the correct size wire. The box is plugged in again and the compressor starts. The technician can hear refrigerant circulating and feels the box is in good working order now.

This box smells so bad that the technician advises the home owner to try the following to remove the smell:

- Leave the box open for several days.
- Then close the box and place some activated charcoal, furnished by the technician, inside the box and start it.
- If odor is still present, spread ground coffee in several pans inside and leave it running for several more days.
- Then wash the freezer inside with warm soda water, about a small package to a gallon of water, and let it run with more coffee grounds inside, for several days.

If this does not remove the odor, the box may need replacing. This seems like a long procedure, but a new freezer is expensive and the old one should be saved if possible.

SERVICE CALL 6

A customer calls indicating that their freezer is not running and is making a clicking sound from time to time. *The problem is the customer has added some appliances to the circuit and the voltage is low.* A small 115-V electric heater has been added to the mother-in-law's bedroom and she is operating it all the time. The voltage is only 95 V at the wall outlet while the heater is operating. This is not enough voltage to start the freezer compressor. This is an old system with fuses. The home owner has replaced the 15-A fuse with a 20-A fuse. The owner is advised to shut the freezer off until the technician can arrive and to keep the door to the freezer shut.

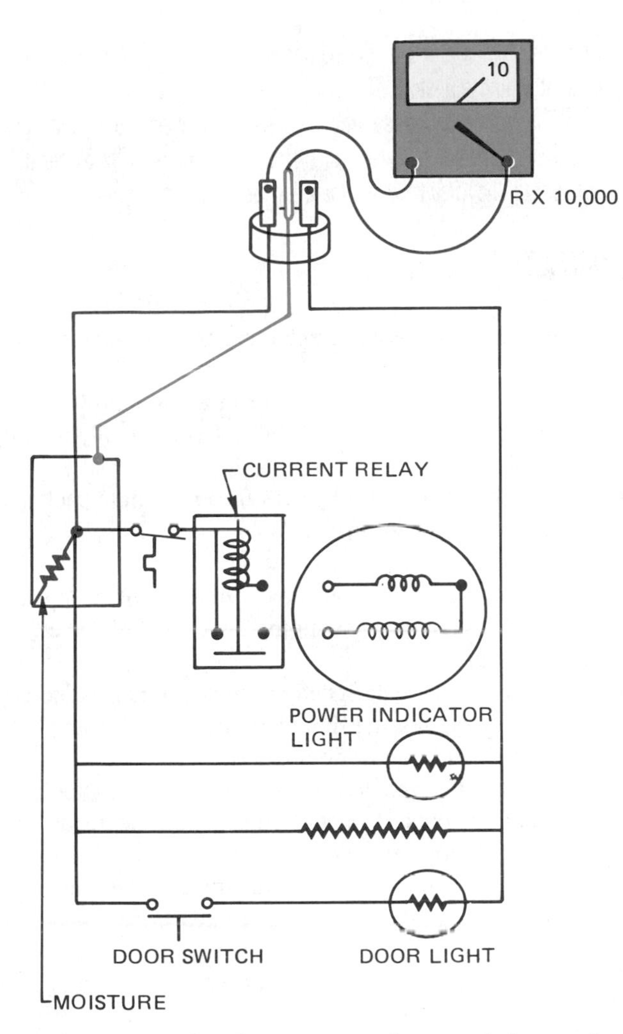

FIGURE 45–71 When the compressor is disconnected, the ground circuit is still present.

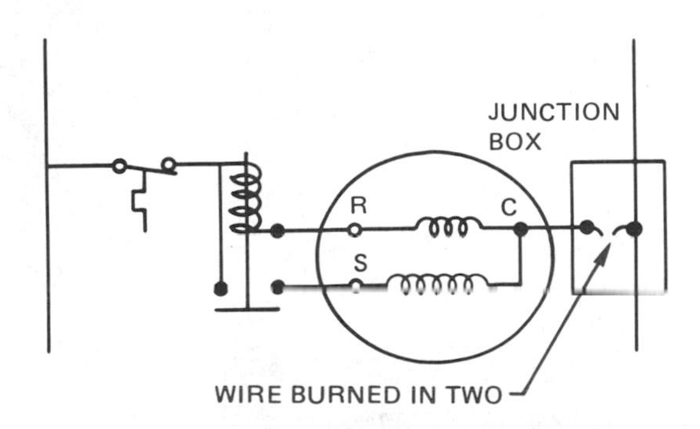

FIGURE 45–72 The insulation and wire are burned.

The technician arrives and places a temperature probe between two packages, the temperature reads 15°F after about 5 min. There is no time to spare. The technician decides to check the amperage and the voltage when starting the unit. An ammeter is fastened around the common compressor wire and a voltmeter is plugged into the other plug in the wall outlet. The voltage is 105 V; this is low to begin with, Figure 45–73. The technician plugs the box in and watches the voltmeter and the ammeter. The compressor does not start and the voltage drops to 95 V so the refrigerator is unplugged before it can trip on the overload, Figure 45–74. It is time to look further into the house circuits.

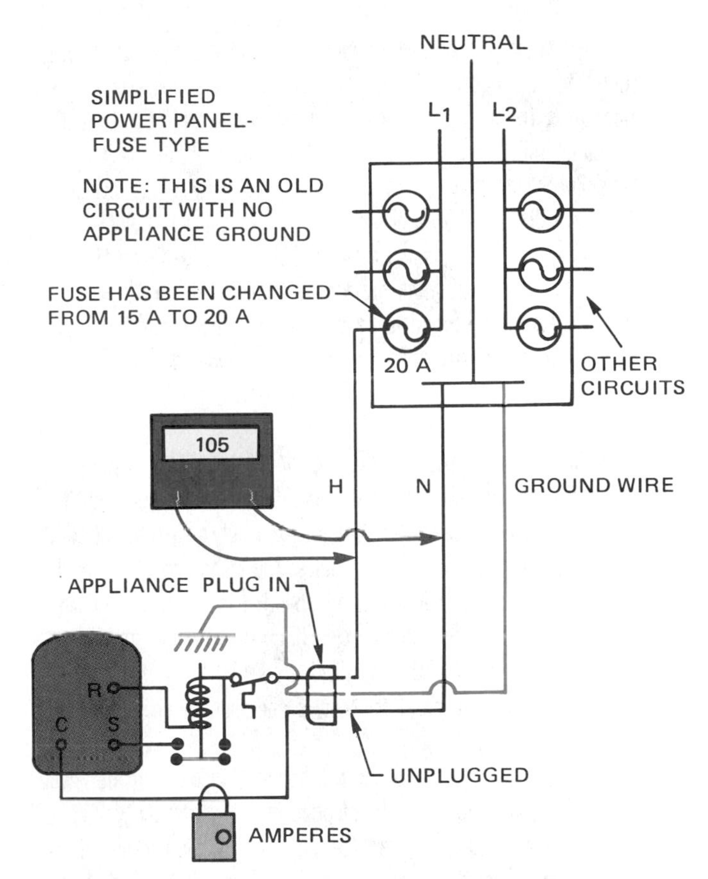

FIGURE 45–73 There are only 105 V in this circuit before the freezer is plugged in. The freezer will be an additional load on the circuit.

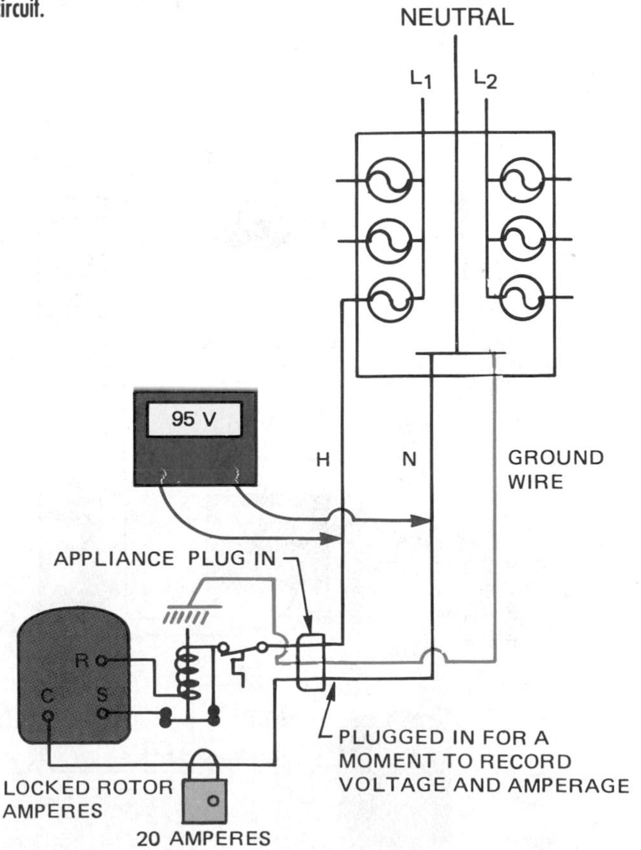

FIGURE 45–74 When the compressor is plugged in, the voltage drops to 95 V and the compressor will not start.

The technician asks the owner if any additional load has been added to the house. The heater in the mother-in-law's room is mentioned. The technician goes to the heater and finds it to be 1500-W heater. This heater should draw 13 A by itself (Watts = Amperes × Volts. Solved for amperes would be Amperes = Watts divided by Volts or Watts = 1500 divided by 115 = 13 A). However, the house only has 110 V at the main so 1500 W divided by 110 V = 13.6 A for the heater under these circumstances, Figure 45–75. The fuse is checked and found to be a 20-A fuse. The rest of the circuits with the same wire size have 15-A fuses. A voltage check at the main electrical panel shows 110 V. This circuit is losing 5 V with just the electric heater operating.

The technician unplugs the electric heater and measures the voltage at the refrigerator outlet; it is 110 V. It is obvious that the electric heater must be moved to another circuit, a circuit by itself. A smaller heater would be a better choice. The technician examines the heater and discovers that it can be operated in a lower mode that consumes 1000 W. This should be enough heat for the small room. This would pull 9.1 A, 1000 W divided by 110 V = 9.1 A. Every little bit of savings helps. This place is operating at near maximum capacity.

Another circuit is found and the heater plugged in at the lower output, 1000 W. The freezer is plugged in and starts; the voltage is 105 V while running, Figure 45–76. This is slightly above the −10% allowed for the motor voltage. The rated voltage of the box is 115 V, 115 × 0.10 = 11.5. The rated voltage of 115 − 11.5 = 103.5 V. A minimum voltage of 103.5 V can be used. Before, when the refrigerator and the heater were in the circuit at the same time, it would not start.

The technician advises the owner that an electrician should be consulted to rework the electrical system in the house. The owner did not realize the circuits were operating at near capacity.

SUMMARY

- Freezers are different from refrigerators because they operate only as low-temperature refrigeration systems.
- Functions of the household freezer are to hold food that has been purchased frozen and to freeze small amounts of food.
- Food to be frozen must be packaged in airtight packages or freezer burn will occur.
- Freezers must be level.
- Domestic freezer cabinet boxes may be of an upright design with the doors swinging out or a chest-type with a lid that raises.
- These freezers are designed to operate in normal household temperatures and should not be located where temperatures reach hot or cold extremes.
- The walls of a typical modern box are constructed with metal on the outside, metal or plastic on the inside, and foam insulation sandwiched between.
- Door gaskets must remain tight fitting or air leaks will occur, causing the compressor to run excessively and frost to build up inside the cabinet.
- Many freezers have manual defrost because the doors

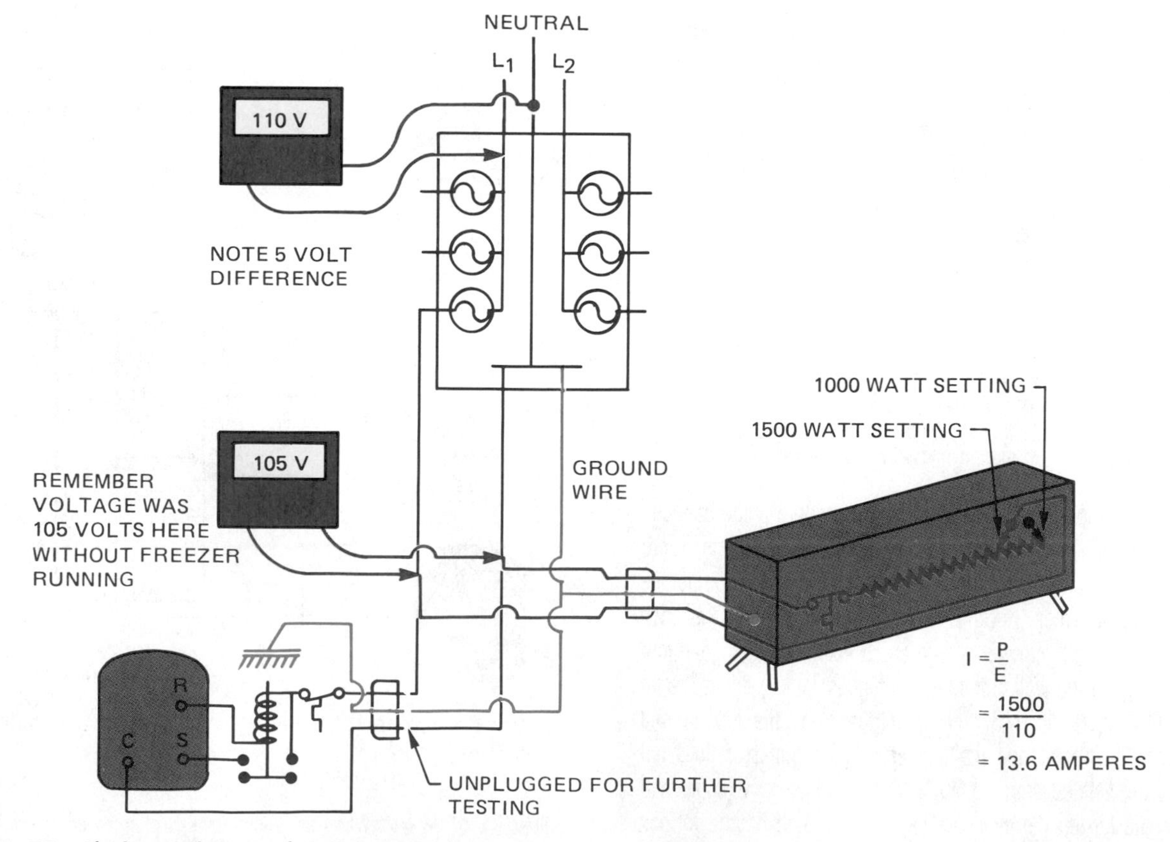

FIGURE 45–75 The home voltage is only 110 V; 1500-W heater will pull 13.6 A at 110 V.

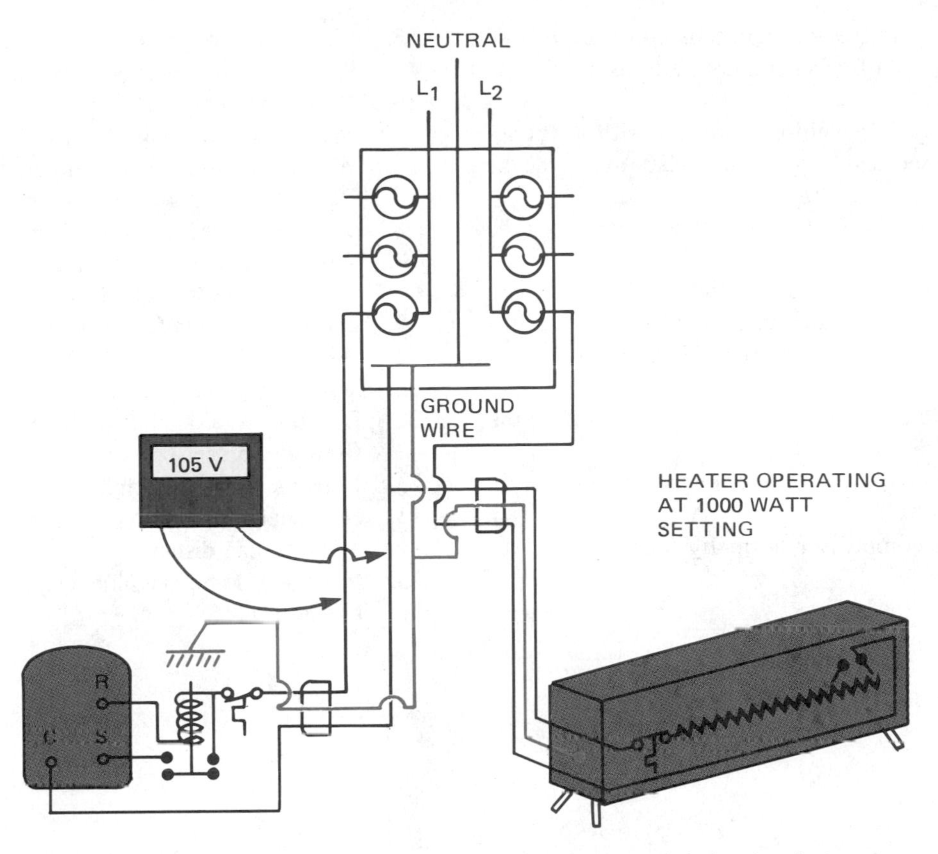

FIGURE 45–76 When the heater is moved to a new circuit, the freezer is operating at 105 V, a little low, but the best that can be done here.

are not opened as often and defrost is not needed as frequently.

- The evaporator may be a forced-air type or a tube- or plate-type with natural draft.
- Plate-type evaporator design uses the evaporator as a shelf and another type uses the wall of the interior of the box as the evaporator plate.
- Compressors can be rotary or reciprocating.
- The compressor is located at the back of the box on the bottom of an upright design and at one end or the other in a chest-type, usually serviced from the back.
- Home freezers have air-cooled condensers. These may be natural-draft chimney type, located at the back or located in the wall of the box. The freezer must be located where there is good air circulation over the condenser at the back or along the walls for this design. Forced-draft condensers, usually used with units that have automatic defrost, are located at the bottom and require more care.
- The capillary tube is the typical metering device for domestic freezers.
- The condenser should be able to condense refrigerant at 25°F to 35°F higher than the ambient air. The more efficient the condenser, the closer the condensing temperature is to the air passing over it.
- In a normally operating system, the compressor will be hot to the touch.
- Compressors are air or refrigerant cooled. An air-cooled compressor typically will have fins.
- Normally a wiring diagram is fastened to the back of the cabinet. This should be studied before you service the freezer. Controls typically consist of a thermostat and a door switch. Look for features such as automatic defrost, forced draft, drain, and panel heaters.
- When servicing the freezer, check the forced-draft evaporator and condenser fans to ensure that they are operating properly.
- If food has spoiled odor may be a problem. If the insulation is fiberglass and has been saturated, it is possible the odor cannot be eliminated.
- Refrigerator hand trucks should be used when moving upright freezers. Be careful not to damage the condenser or the door. It should be trucked from the side when possible. Extreme care should be taken not to damage the cabinet or floor.
- When a chest-type freezer must be moved from one room to another, it should be placed on two four-wheel dollies. This should not be attempted unless there are two people.
- Before a freezer is moved, the food must be removed and placed where it will stay frozen.

REVIEW QUESTIONS

1. What are the functions of the household freezer?
2. What may happen to food that is to be frozen if it is not packaged properly?

3. Describe how the freezer cabinet is constructed.
4. Why is an upright freezer not normally as efficient as the chest type?
5. When locating or installing a freezer with a natural-draft condenser inside a house, what must you be aware of?
6. How may the plastic interior of a freezer box be repaired?
7. Why must the door gaskets be tight fitting?
8. List three types of evaporators found in freezers.
9. Describe where the compressor is located in a chest-type freezer.
10. Describe a plate-type evaporator using the wall of the interior of the box.
11. What are two types of compressors used in domestic freezers?
12. Where is the compressor normally located in an upright box?
13. What are two types of condensers used in freezers?
14. What type of metering device is used on freezers?
15. Using the pressure/temperature chart, Figure 3–38, what would the psig be for R-12 at −5°F?
16. Why are household freezers air cooled?
17. Describe the temperature as it would feel to your touch of a compressor operating under normal conditions.
18. What should you suspect first if the compressor is running more than it should?
19. List two controls found on typical freezers.
20. Why do some units have panel heaters?
21. What is a typical problem with evaporator fan motors?
22. Describe how spoiled food odors may be removed from the freezer box.
23. Describe how an upright freezer should be moved.
24. How should a chest-type freezer be moved if it is to be moved some distance?
25. What are some possible ways to store frozen food temporarily?

UNIT 46

Room Air Conditioners

OBJECTIVES

After studying this unit, you should be able to

- describe the various methods of installing window air conditioning units.
- discuss the variations in the designs of window and through the wall units.
- list the major components in the refrigeration cycle of a window cooling unit.
- explain the purpose of the heat exchange between the suction line and the capillary tube.
- describe the heating cycle in the heat pump or reverse-cycle room air conditioner.
- describe the controls for room air conditioning (cooling) units.
- describe the controls for room air conditioning (cooling and heating) units.
- discuss service procedures for room air conditioners.
- list the procedures to be followed to determine whether or not to install gages.
- state the proper procedures for charging a room air conditioner.
- list the types of expansion valves that may, under some conditions, be substituted for the capillary tube.
- state the components that may require electrical service.

SAFETY CHECKLIST

* Wear back brace belts when lifting or moving any appliance.
* Lift with your legs, keeping your back straight. Whenever possible use appropriate equipment to move heavy objects.
* Observe proper electrical safety techniques when servicing any appliance or system.
* Wear goggles and gloves when transferring refrigerant into or out of a room air conditioning unit.

46.1 AIR CONDITIONING AND HEATING WITH ROOM UNITS

Room air conditioning is the process of conditioning rooms, usually one at a time, with individual units. This term applies to both heating and cooling.

Room air conditioning equipment is constructed of components that have already been discussed in detail. When a particular component or service technique is mentioned, you will be referred to the appropriate part of the text for the needed details. Systems and components characteristic to room units are discussed in this unit.

Single room air conditioning can be accomplished in several ways, but each involves the use of package (self-contained) systems of some type. The most common type is the room air conditioning window unit for cooling only, Figure 46–1. This type of unit has been expanded to include electric strip heaters in the airstream with controls for heating and cooling, Figure 46–2. Adequate air circulation between rooms allows many owners to use a room unit for more than one room.

46.2 ROOM AIR CONDITIONING, COOLING

Cooling-only units may be either window or through the wall type, Figure 46–3. They are much the same, typically having only one fan motor for the evaporator and the condenser. The capacity of these units may range from about

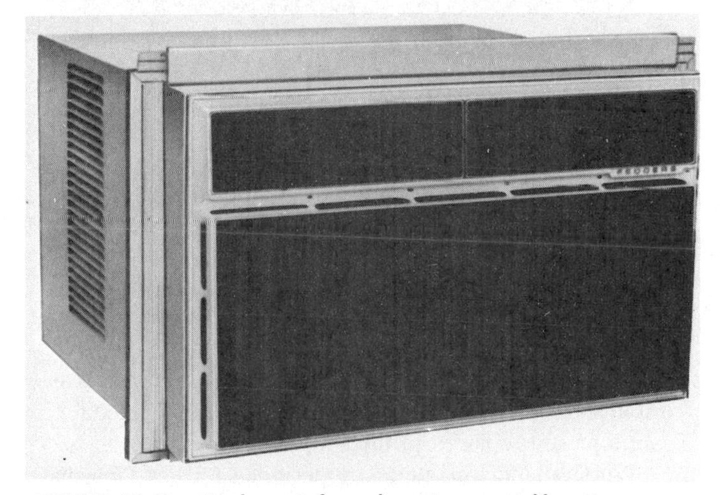

FIGURE 46–1 Window unit for cooling. *Courtesy Fedders Air Conditioning USA, Inc.*

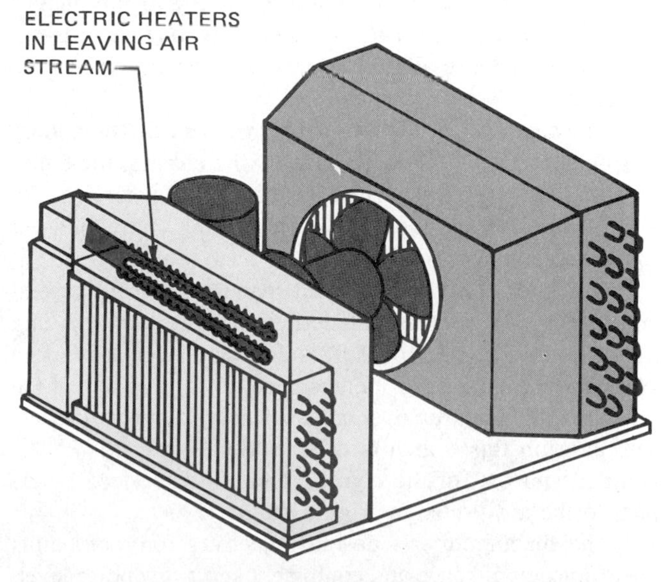

FIGURE 46–2 Window unit with heat added.

FIGURE 46–3 **Window and through the wall type units.** ***Courtesy Fedders Air Conditioning USA, Inc.***

4000 Btu/h (1/3 ton) to 24,000 Btu/h (2 tons). Some units are front-air discharge and some are top-air discharge, Figure 46–4. Top-air discharge is more common for through the wall units with the controls being located on top. These top-air discharge units are installed frequently in motels where individual room control is desirable. The controls are built in so the unit may be serviced in place or changed for a spare unit and repaired in the shop.

Window and wall units are designed for easy installation and service. Window units have two types of cases. One type is fixed to the chassis of the unit and the other is a case that fastens to the window opening and the chassis slides in and out, Figure 46–5. Older units were all slide-out type cases; in the smaller, later date units the case is built on the chassis.

The case design is important from a service standpoint. For units that have the case built on the chassis, the entire unit including the case must be removed for service. On units with slide-out chassis, the chassis may be pulled out from the case for simple service.

One special application unit manufactured by several companies is a roof-mount design for travel trailers and motor homes, Figure 46–6. This unit may be seen in some service station attendant booths because it is up out of the way. It will not get run over and does not take up wall space. The case on this unit lifts off the top, Figure 46–7. The controls and part of the control circuit are serviced inside next to the air discharge, Figure 46–8.

The manufacturer's design objectives for room units are efficiency of space and equipment and a low noise level. Most units are as compact as current manufacturing and

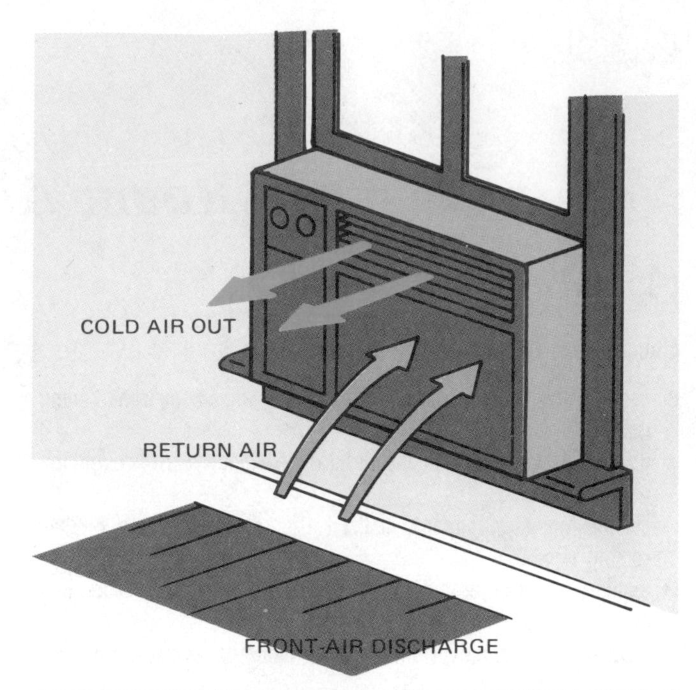

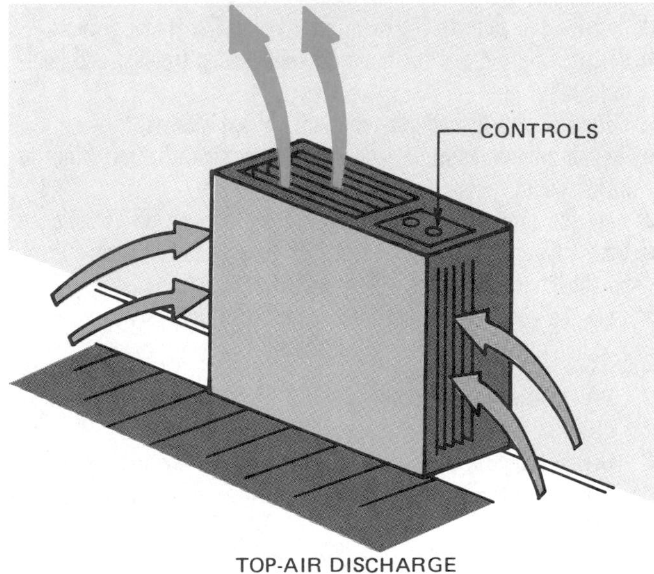

FIGURE 46–4 **Some room units are front-air discharge and some are top-air discharge.**

design standards will allow. Most components are as small and efficient as possible. The intent is to get the most capacity from the smallest unit.

46.3 THE REFRIGERATION CYCLE, COOLING

You should have a full understanding of Unit 3 before attempting to understand the following text. The most common refrigerant used for room units is R-22. This refrigerant will be the only one discussed here. When a refrigerant of another type is encountered, the pressure/temperature chart may be consulted for the difference in pressure. The operating temperatures will be the same.

The refrigeration cycle consists of the same four major components as described in Unit 3: an evaporator, to absorb heat into the system, Figure 46–9; a compressor to pump the

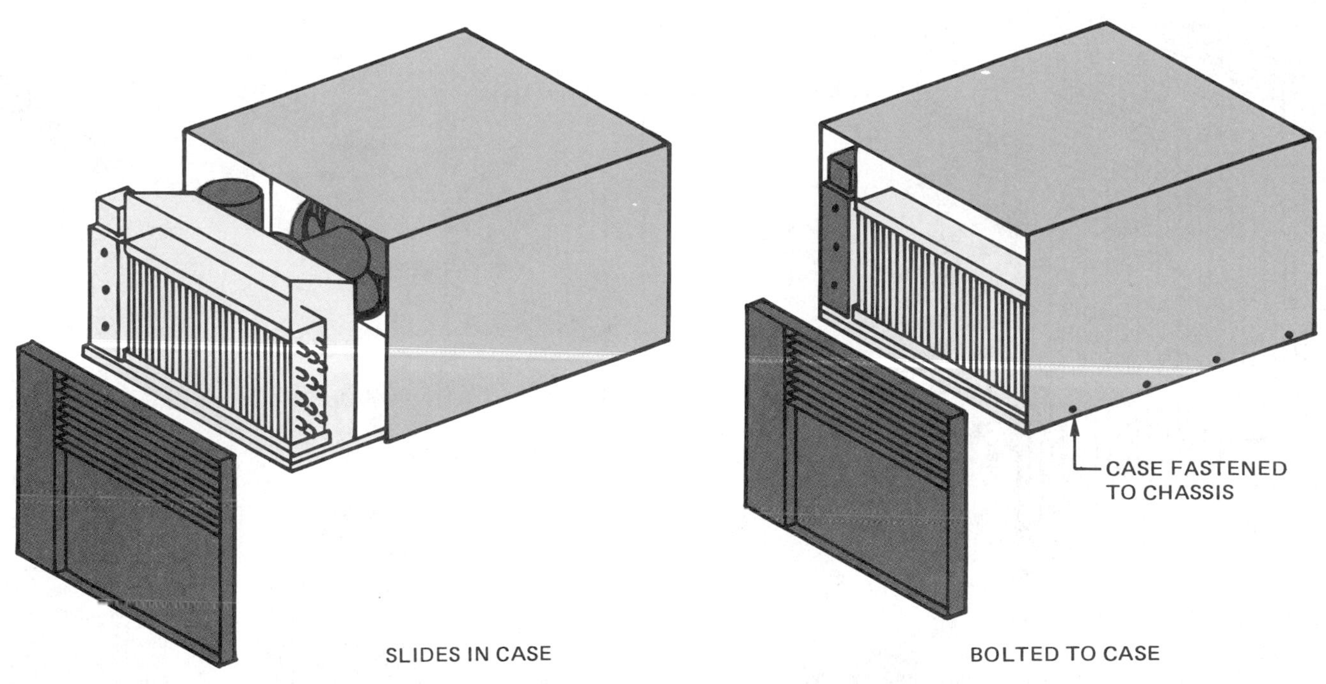

FIGURE 46–5 Some window units are of a slide-out design and in some the cover is bolted to the chassis.

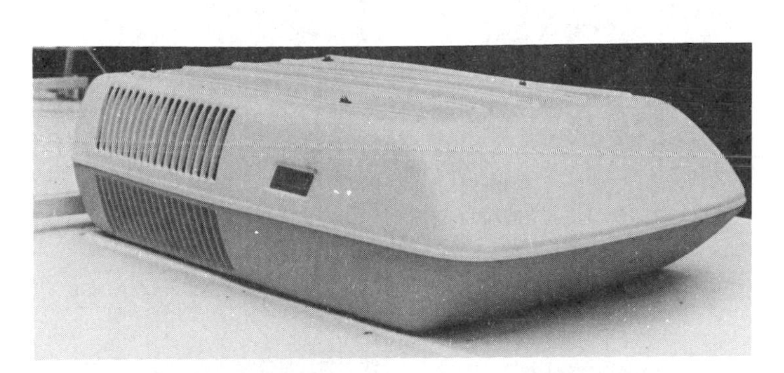

FIGURE 46–6 Roof-mount unit for recreation vehicles. Sometimes these are used in small stand-alone buildings. *Photo by Bill Johnson*

FIGURE 46–7 The case lifts off the roof-mount unit. *Photo by Bill Johnson*

FIGURE 46–8 Part of the control circuit is located inside, under the front cover. *Photo by Bill Johnson*

heat-laden refrigerant through the system, Figure 46–10; a condenser to reject the heat from the system, Figure 46–11; and an expansion device to control the flow of refrigerant, Figure 46–12. This is done by maintaining a pressure difference between the high-pressure and low-pressure sides of the system. Most units have a heat exchange between the metering device (capillary tube) and the suction line, Figure 46–13. The complete refrigeration cycle may be seen in Figure 46–14.

Cooling units are considered high-temperature refrigeration systems and must operate above freezing to prevent condensate from freezing on the coil. Typically the evaporators boil the refrigerant at about 35°F. Central air conditioners normally boil the refrigerant at 40°F. Room units boil the refrigerant much closer to freezing, so more care must be taken to prevent freezing of the evaporator.

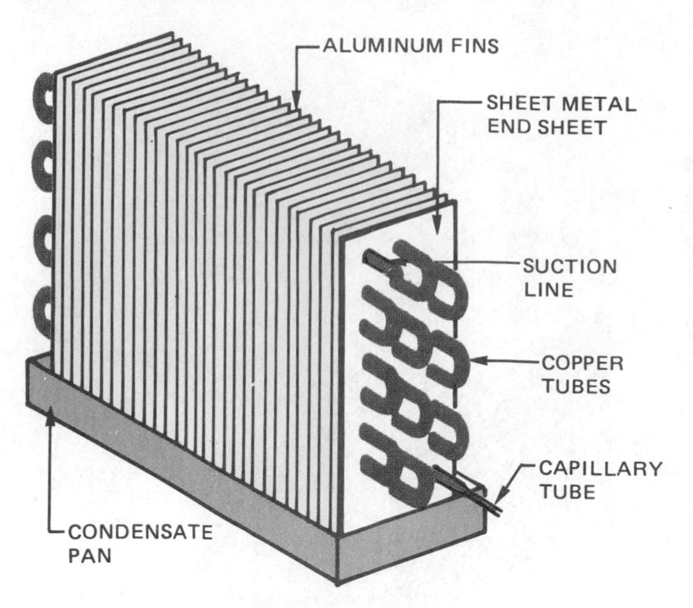

FIGURE 46–9 Room unit evaporator.

FIGURE 46–10 Room unit compressor. *Photo by Bill Johnson*

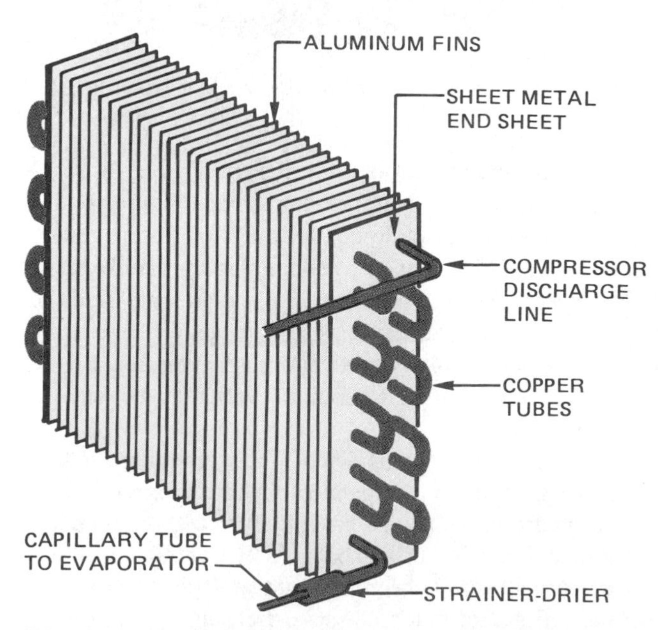

FIGURE 46–11 Room unit condenser.

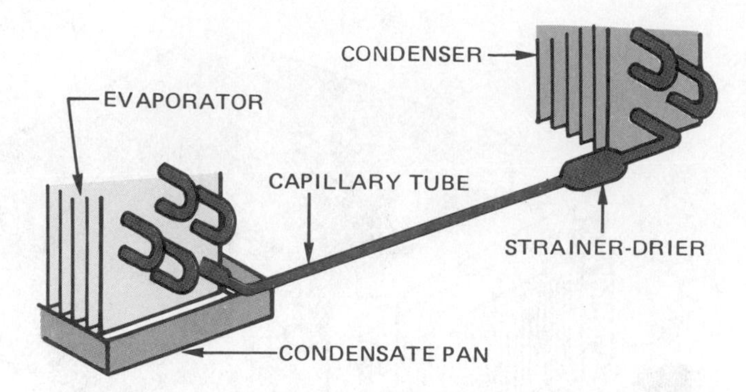

FIGURE 46–12 Room unit capillary tube.

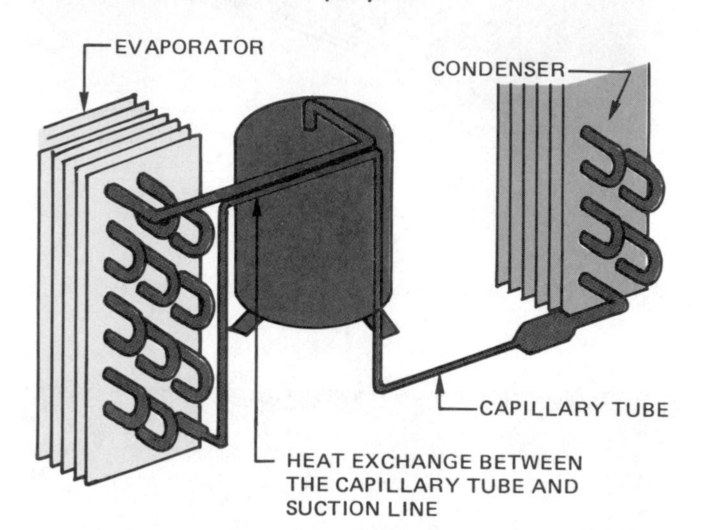

FIGURE 46–13 Suction-to-capillary tube heat exchange.

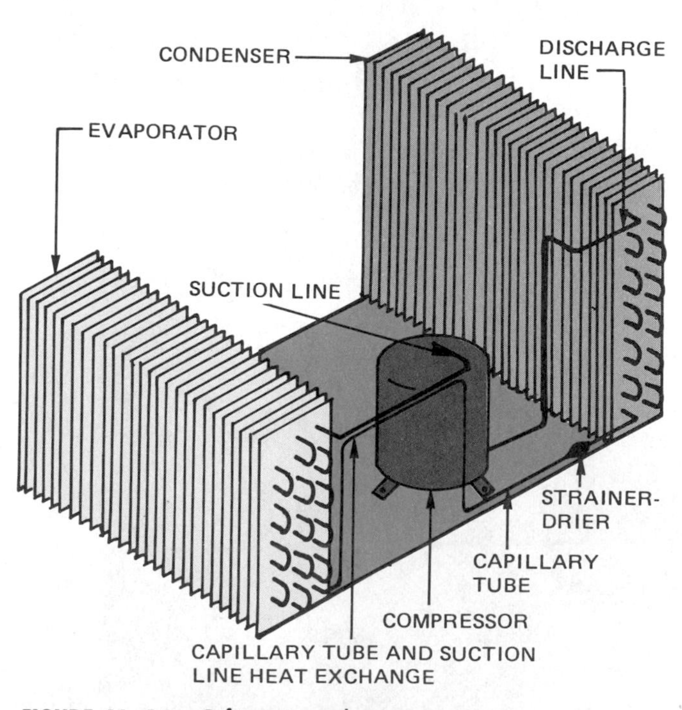

FIGURE 46–14 Refrigeration cycle components.

The evaporator is typically made of copper or aluminum tubing with aluminum fins. The fins may be straight or spine. All fin types are in close contact with the copper tubing for the best heat exchange.

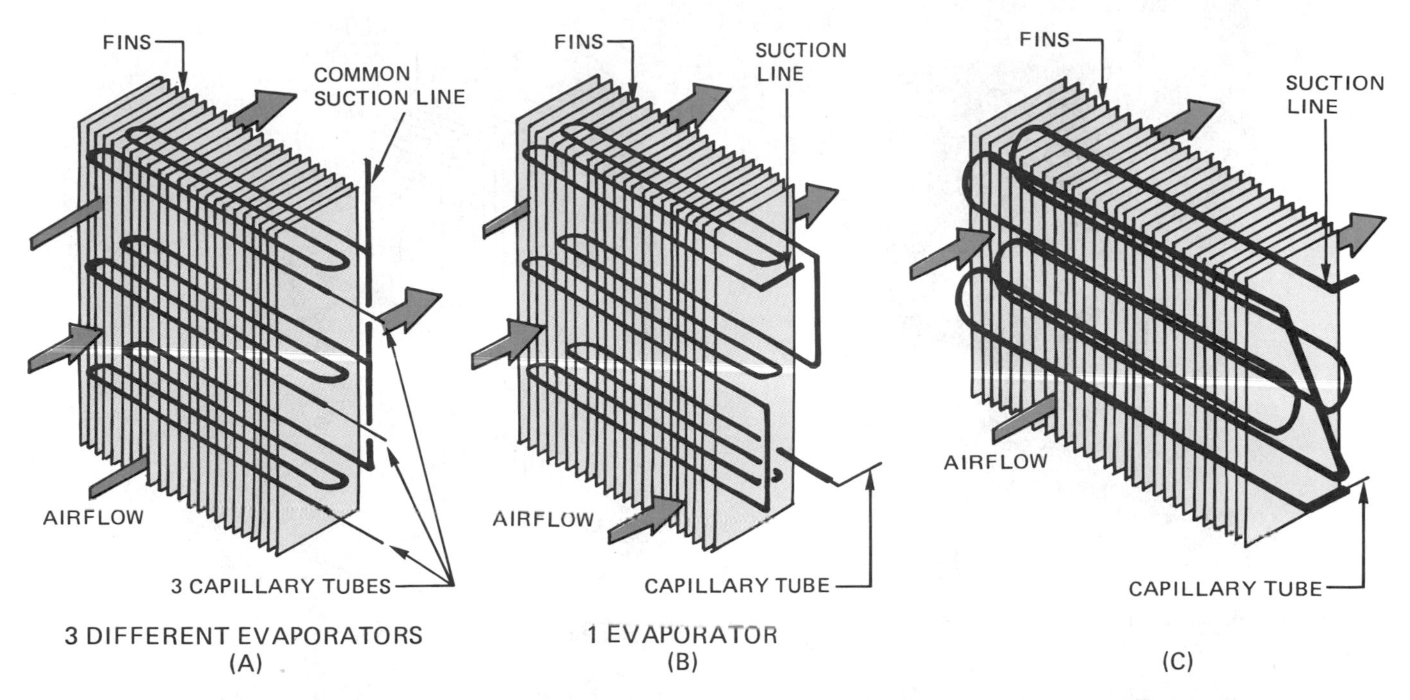

FIGURE 46–15 Some typical evaporator circuits.

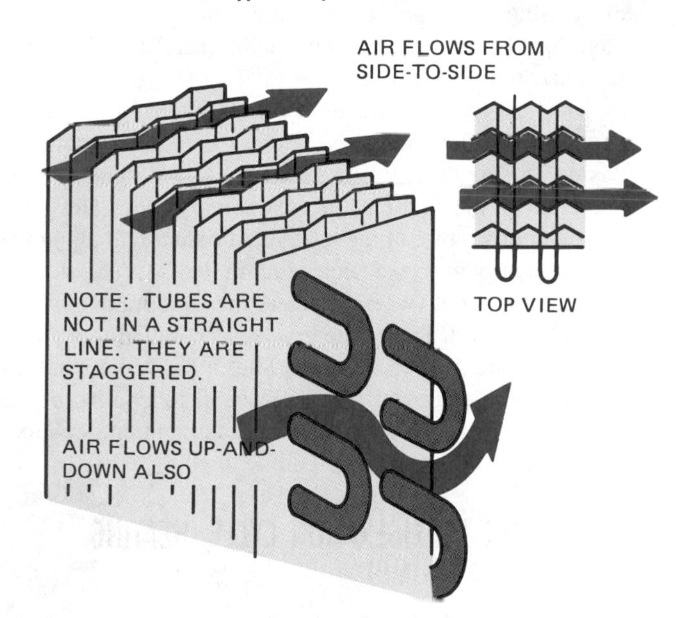

FIGURE 46–16 Some fins force the air to have contact with the fins and tubes.

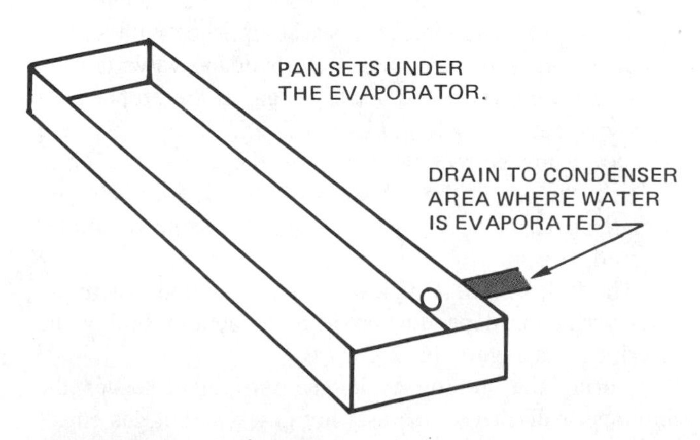

FIGURE 46–17 Condensate pan.

The evaporator may have only a few refrigerant circuits or it may have many. They may take many different paths, such as two, three, or four circuits that may be in line with the airflow (series) or that operate as multiple evaporators, Figure 46–15. Each manufacturer has designed its own and tested it to perform to its specifications.

The evaporators are small and designed for the most heat exchange obtainable. Fins that force the air to move from side to side and tubes that are staggered to force the air to pass in contact with each tube are typical, Figure 46–16.

The evaporator typically operates below the dew point temperature of the room air for the purpose of dehumidification, so condensate forms on the coil. It drains to a pan beneath the coil, Figure 46–17. The condensate is generally drained back to the condenser section and evaporated. Figure 46–18 shows a typical evaporator with temperatures and pressures.

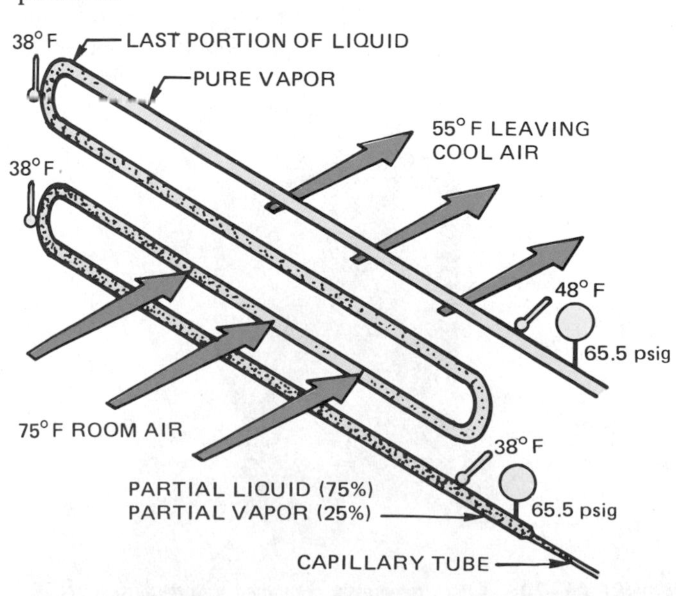

FIGURE 46–18 Typical evaporator with pressures and temperatures.

FIGURE 46–19 Hermetically sealed compressor. *Photo by Bill Johnson*

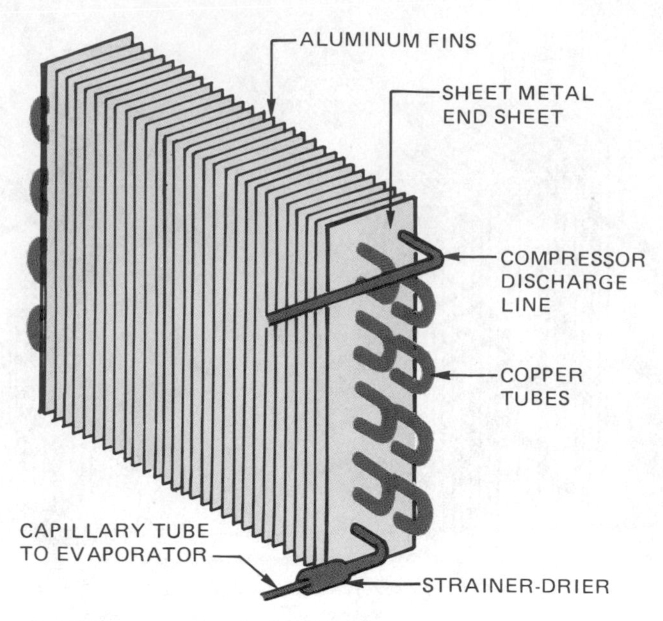

FIGURE 46–21 Finned tube condenser.

The compressor for room units is a typical hermetically sealed air conditioning compressor, Figure 46–19. The compressor may be reciprocating, Figure 46–19, or rotary, Figure 46–20. Compressors are described in detail in Unit 23.

Condensers are typically finned tube with copper or aluminum tubes and aluminum fins, similar to the evaporator, Figure 46–21. The condenser serves two purposes—it condenses the heat-laden vapor refrigerant inside the tubes and evaporates the condensate from the evaporator section. This is generally accomplished by using the heat from the discharge line and a slinger ring on the condenser fan, Figure 46–22. Evaporating the condensate keeps the unit from dripping and improves the efficiency of the condenser.

The capillary tube is the type of metering device in units manufactured during the past several years. Some early units used the automatic expansion valve, Figure 46–23. The automatic expansion valve has the advantage of controlling the pressure, which in turn controls the coil temperature. Coil freezing could be prevented by controlling the pressure. Unit 24 describes the automatic expansion device. Since there are more capillary tubes than any other expansion device for room air conditioners, it will be the one discussed in this text.

FIGURE 46–20 Rotary compressor. *Reprinted with permission of Motors and Armatures, Inc.*

Most room units are designed to exchange heat between the capillary tube and suction line. This exchange adds some superheat to the suction gas and subcools the refrigerant in the first part of the capillary tube. The pressure and temperature of the refrigerant reduces all along the tube without a heat exchange, Figure 46–24. The tube is colder at the outlet (where it enters the evaporator) than at the inlet (where it leaves the condenser). When this capillary tube is attached to a suction line, it has a net result of warming the compressor. The increase in subcooling of the capillary tube may also help the manufacturer get more capacity from the evaporator coil.

46.4 THE REFRIGERATION CYCLE, HEATING (HEAT PUMP)

Some window and through the wall units have reverse-cycle capabilities similar to heat pumps discussed in Unit 43. They can absorb heat from the outdoor air in the winter and reject heat to the indoors. This is accomplished with a four-way reversing valve. The four-way reversing valve is used to redirect the suction and discharge gas at the proper time to provide heat or cooling, Figure 46–25. Check valves are used to ensure correct flow through the correct metering device at the proper time. A study of Unit 43 will help you to understand the basic cycle because the same principles are used.

The following is a typical cycle description. There are many variations, depending on the manufacturer. Follow the description in Figure 46–26.

During the cooling cycle, the refrigerant leaves the compressor discharge line as a hot gas. The hot gas enters the four-way valve and is directed to the outdoor coil where

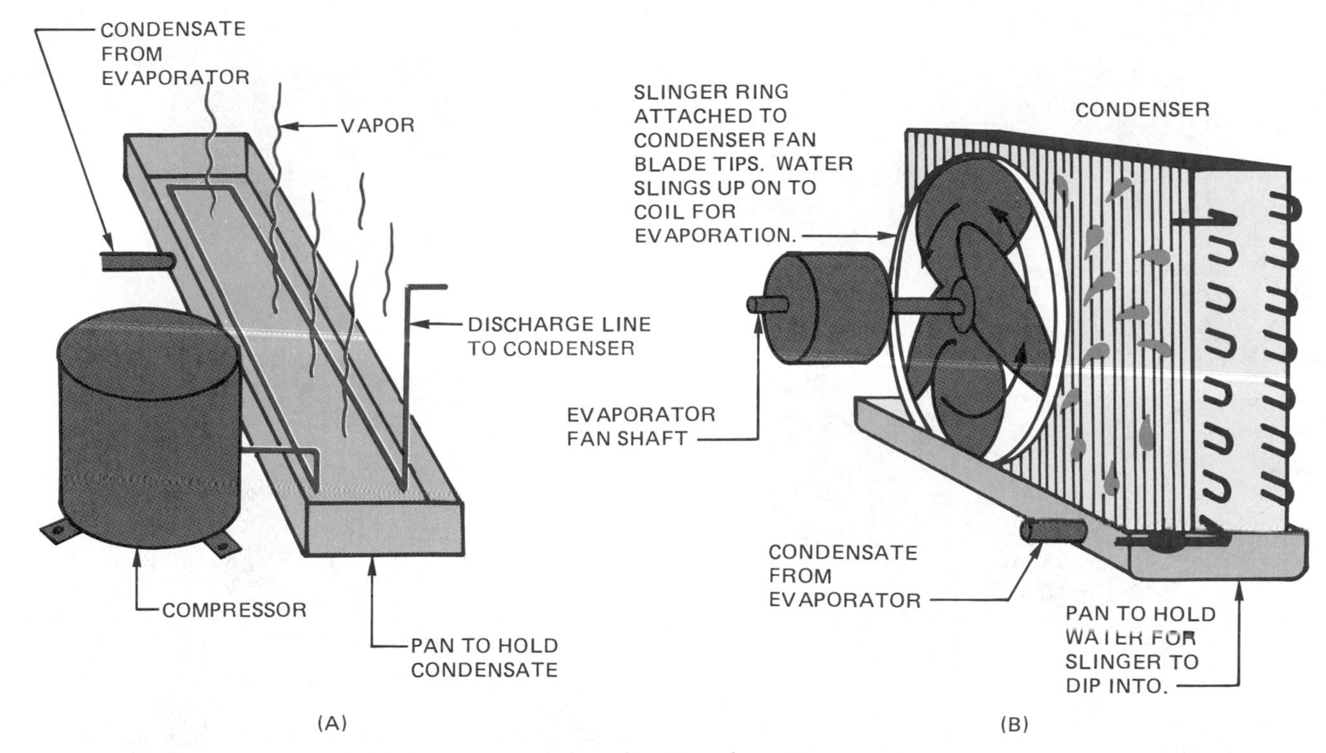

FIGURE 46–22 Moisture evaporated with the compressor discharge line (A) or slinger (B).

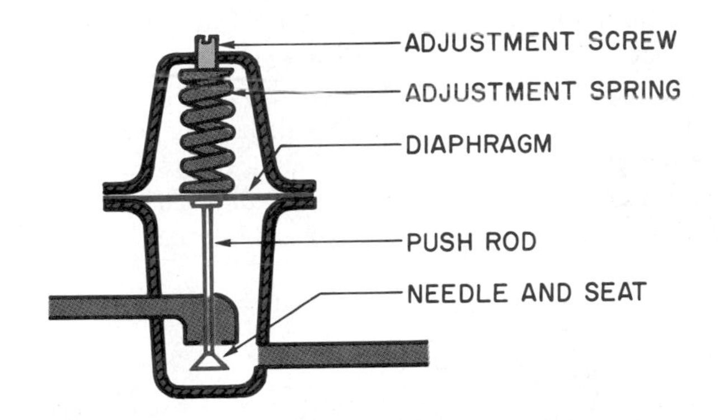

FIGURE 46–23 Some units use the automatic expansion valve.

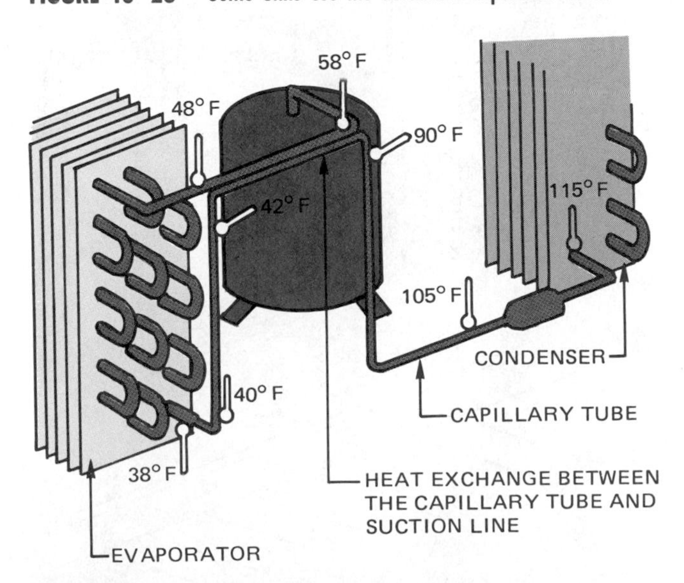

FIGURE 46–24 Typical pressures and temperatures along the capillary tube and suction-line heat exchange.

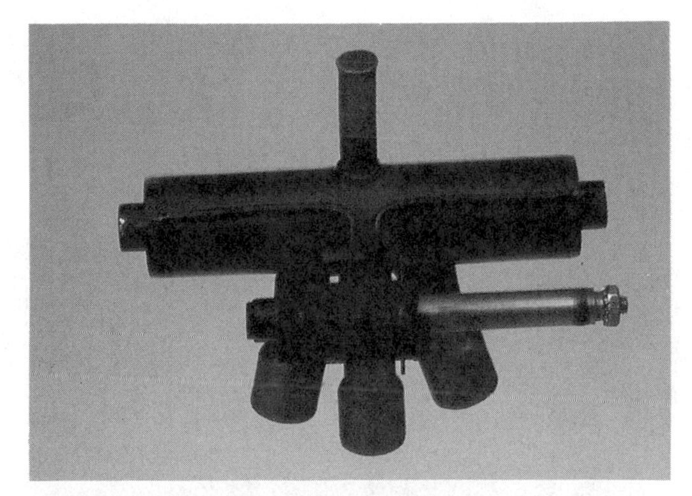

FIGURE 46–25 The four-way reversing valve for a heat pump. *Photo by Bill Johnson*

the heat is rejected to the outdoors. The refrigerant is condensed to a liquid, leaves the condenser, and flows to the indoor coil through the capillary tube. The refrigerant is expanded in the indoor coil where it boils to a vapor, just like the cooling cycle on a regular cool-only unit. The cold heat-laden vapor leaves the evaporator and enters the four-way valve body. The piston in the four-way valve directs the refrigerant to the suction line of the compressor where the refrigerant is compressed and the cycle repeats itself.

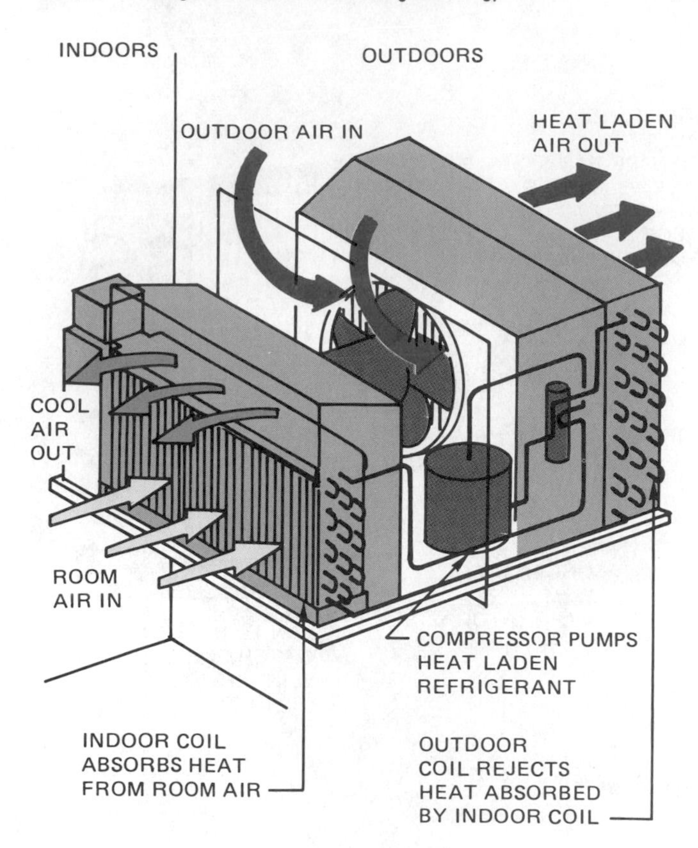

FIGURE 46–26 Refrigerant cycle and description, cooling.

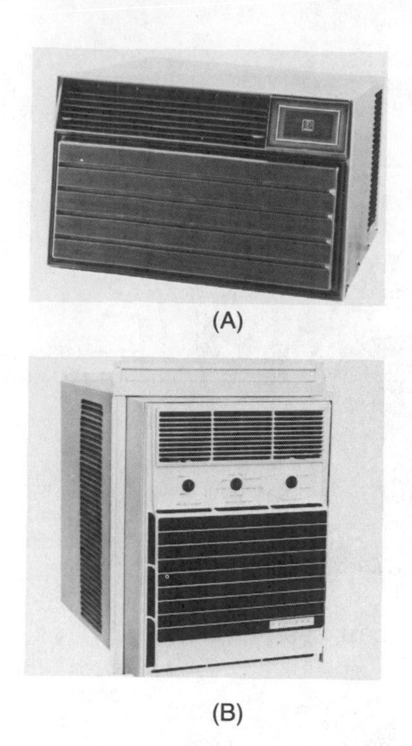

FIGURE 46–28 Window units are either for double-hung (A) or casement windows (B). *Courtesy Fedders Air Conditioning USA, Inc.*

The heating cycle may be followed in Figure 46–27. The hot gas leaves the compressor and enters the four-way valve body as in the previous example. However, the piston in the four-way valve is shifted to the heating position and the hot gas now enters the indoor coil. It now acts as the condenser and rejects heat to the conditioned space. The refrigerant condenses to a liquid and flows out the indoor

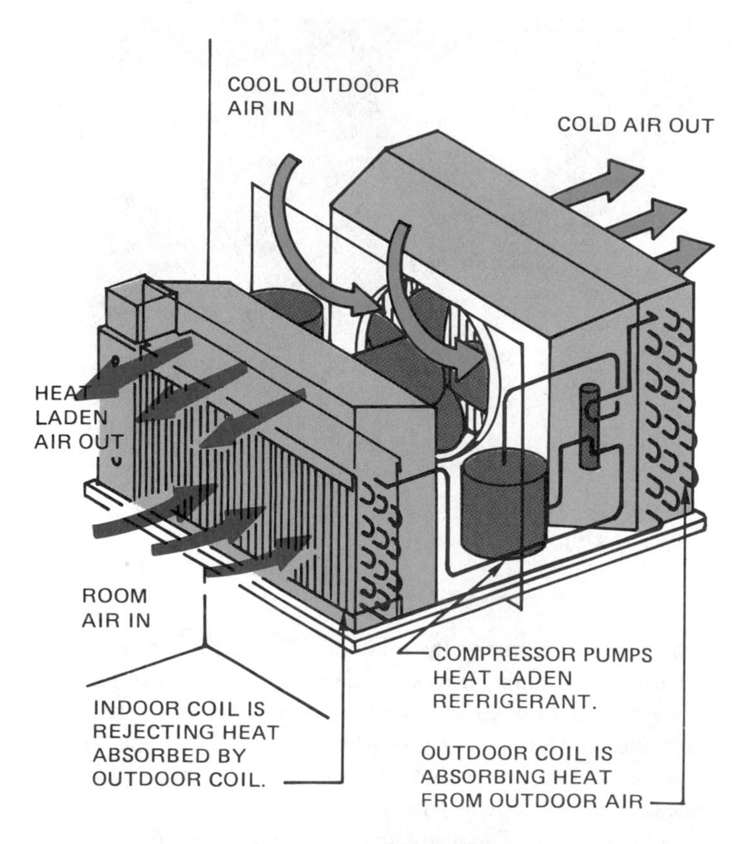

FIGURE 46–27 Refrigerant cycle and description, heating.

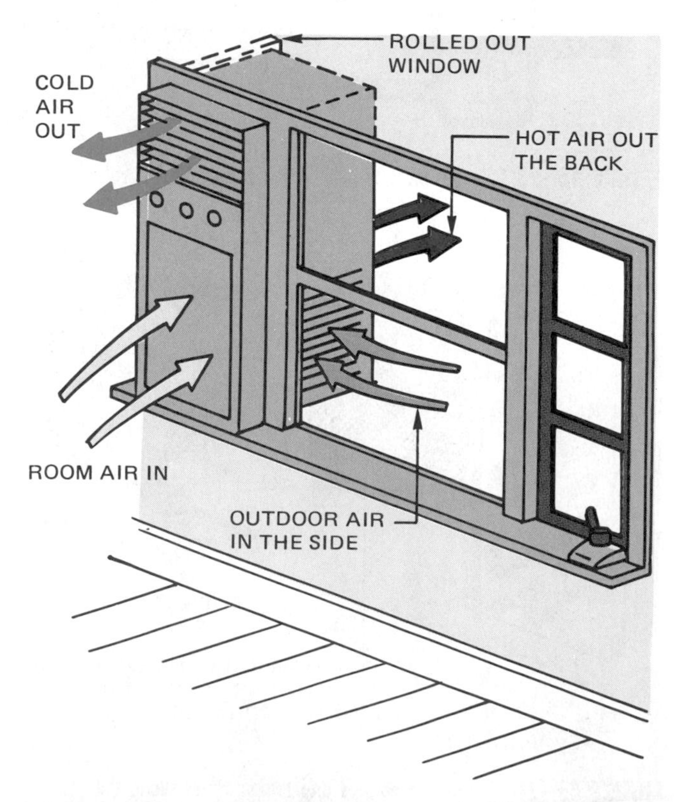

FIGURE 46–29 Window unit installed in a casement window.

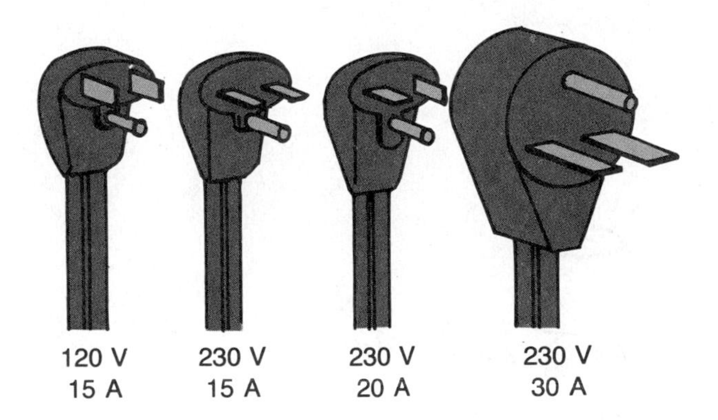

FIGURE 46–30 The proper outlet must be used to match the plug.

coil through the capillary tube to the outside coil. The refrigerant is expanded and absorbs heat while boiling to a vapor. The heat-laden vapor leaves the outdoor coil and enters the four-way valve where it is directed to the compressor suction line. The vapor is compressed and moves to the compressor discharge line to repeat the cycle.

A few things to keep in mind for any heat pump follow. The compressor is not a conventional air conditioning compressor. It is a heat pump compressor. It has a different displacement and horsepower. The compressor must have enough pumping capacity for low-temperature operation. The unit evaporator also operates at below freezing during the heating cycle and frost will build up on the evaporator. Different manufacturers will handle this frost in different ways; you will have to consult their literature. A defrost cycle may be used to defrost this ice. Another point to remember is that the capacity of a heat pump is less in colder weather so supplemental heat will probably be used. Electrical strip heat is the common heat. It may also be used to prevent the unit from blowing cold air during the defrost cycle.

46.5 INSTALLATION

There are two types of installation for room units, both cooling and cooling-heating types. These are the window

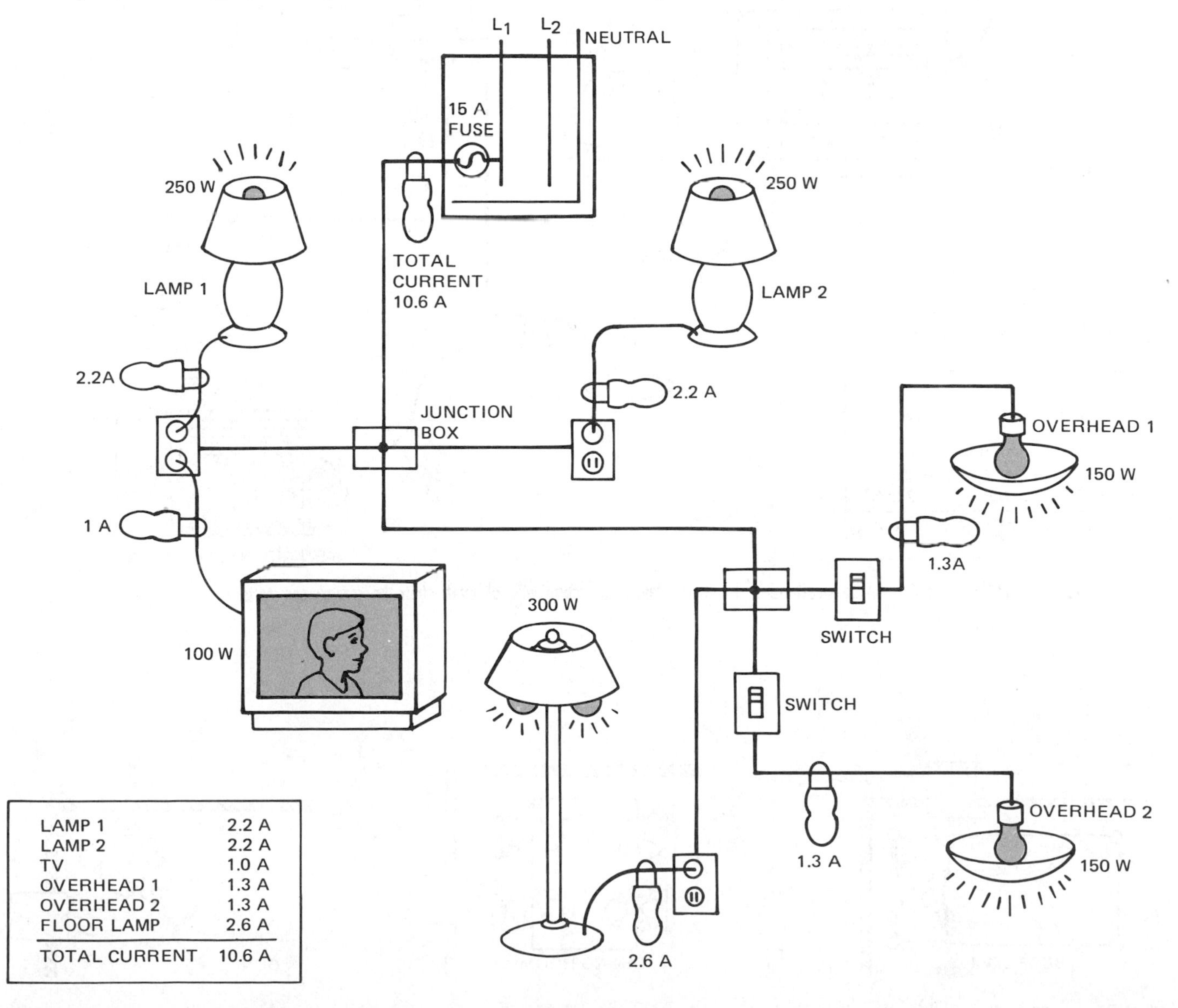

LAMP 1	2.2 A
LAMP 2	2.2 A
TV	1.0 A
OVERHEAD 1	1.3 A
OVERHEAD 2	1.3 A
FLOOR LAMP	2.6 A
TOTAL CURRENT	10.6 A

FIGURE 46–31 The circuit already has a load, without the addition of the window unit.

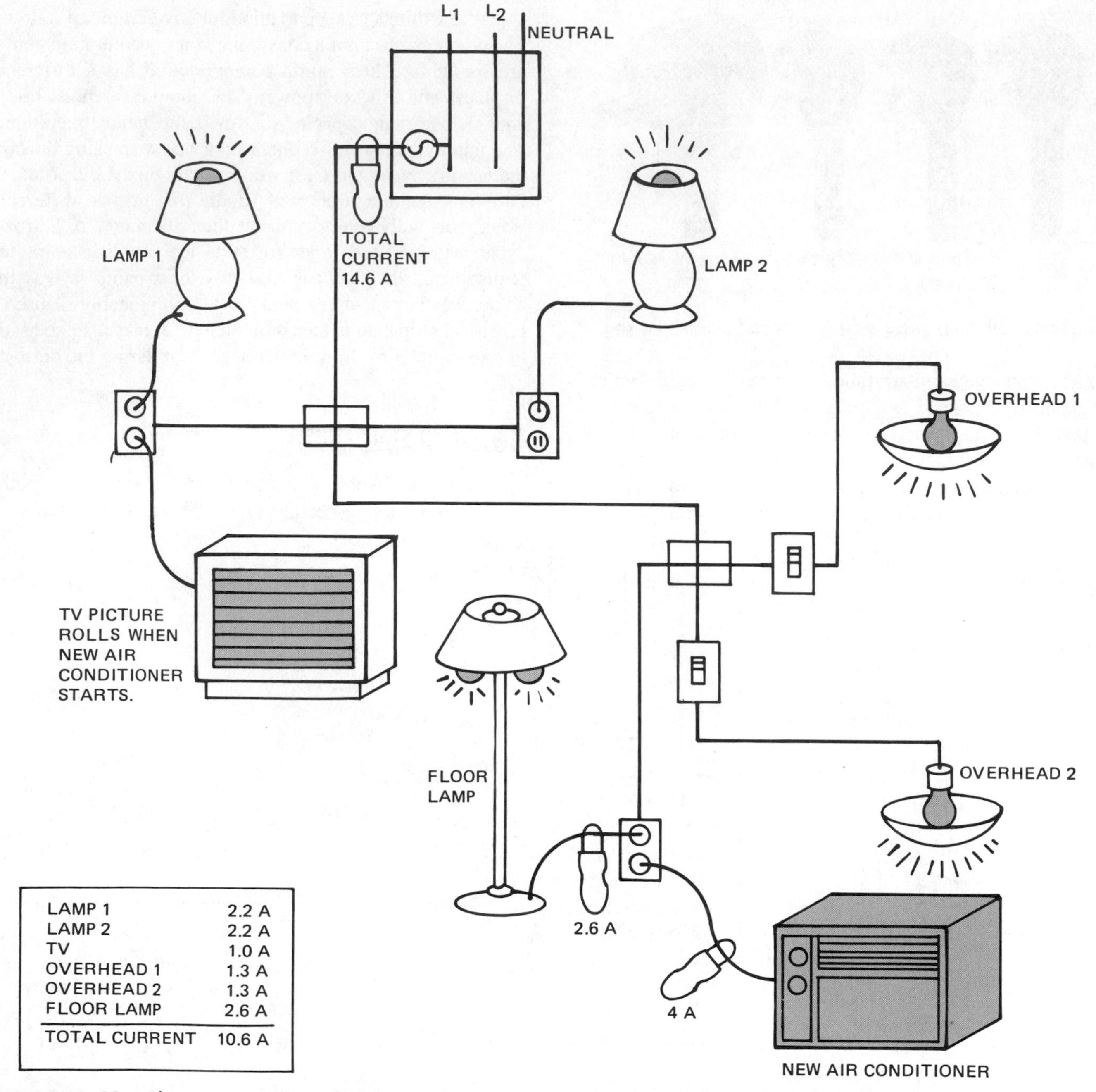

FIGURE 46–32 When a unit is on an overloaded circuit, the television picture will roll over when the compressor starts.

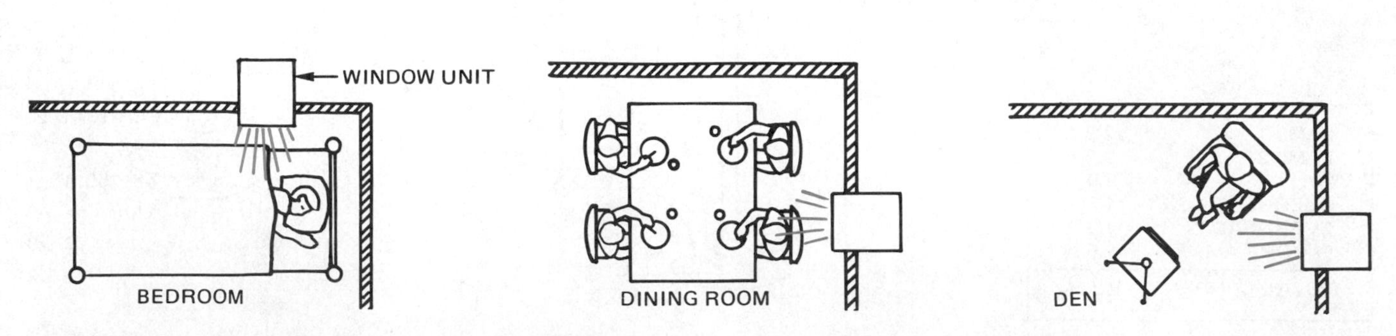

FIGURE 46–33 Some bad locations for window air conditioners.

installation and the wall installation. The window installation may be considered temporary because a unit may be removed and the window put back to its original use. The wall installation is permanent because a hole is cut in the wall and must be patched if the unit is removed. When a unit becomes old and must be changed for a new one in a window installation, a new unit to fit the window would probably be available. A new unit may not be available to fit the hole in a wall.

Window installations are either for the double-hung window or the casement window, Figure 46–28. The double-hung window is the most popular because there are more double-hung windows. Window units may be installed in other types of windows, such as picture window locations or jalousie windows, but these require a great deal of carpentry skill. A special unit is manufactured for casement window installation, Figure 46–29.

Every installation should follow some general guidelines. The inside part of the installation requires that the proper electrical outlet be located within the length of the cord furnished with the unit. Extension cords are not recommended by any manufacturer. The proper electrical outlet must match the plug on the end of the cord, Figure 46–30. Window units may be either 120- or 208/230-V operation and should be on a circuit by themselves.

When 120-V operation is chosen, it is usually to prevent having to install an electrical circuit. Often this choice is made without investigating the existing circuit as it may already have some load on it. For example, a 120-V electrical outlet may be part of the lighting circuit and have a television and several lamps on the same circuit. The addition of a window air conditioner of any size may overload the circuit, Figure 46–31.

An investigation of the electrical circuit to see what extra capacity is available is necessary. This may be accomplished by turning on all the lights in the room nearest the outlet intended for use. The adjacent rooms lights may also be turned on. Then turn off the breaker that controls the outlet. The current load of the lights may be totaled by adding them mathematically or using an ammeter on the wire to the circuit. If the circuit is already using 10 A and it is a 15-A circuit, a window unit consuming 6 A will overload the circuit. A typical 5000 Btu/h unit will pull about 7 A when 120 V is used. An inexperienced homeowner, who buys a unit at the department store and installs it, may be in trouble. If the unit will run on a slightly overloaded circuit, the television picture will roll over every time the unit compressor starts, Figure 46–32.

Other considerations on the inside of the room are the best window to use and the air direction. A window that will give the best total room circulation may not be the best choice because it may be next to the easy chair in a den, the bed in a bedroom, or behind the dining table, Figure 46–33. Some people will purchase a large window unit for several rooms. The unit must be able to circulate the air to the adjacent rooms or the thermostat will shut the unit off and cool only one room. A floor fan may be used to move the air from the room with the unit, Figure 46–34.

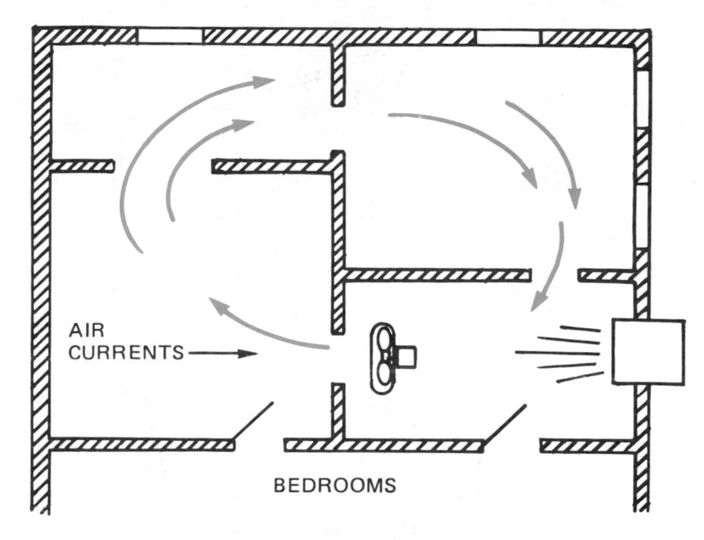

FIGURE 46–34 A floor fan may be used to circulate air to adjacent rooms.

Through the wall units in motels all seem to be located on the outside wall, under the window and where the chair or chairs and table are located. This is the best practical place for the unit, but not necessarily the best for the occupants. However, there is no other place in a small room, Figure 46–35.

A unit located where air will recirculate, for example behind drapes, will be a problem in any installation. Air will continually recirculate to the return air grille and be cold enough to satisfy the thermostat, shutting the unit off. Some small units have no thermostat, only an ON-OFF switch. These may freeze the evaporator solid with ice, Figure 46–36.

Typically, the airflow should be directed upward, as cold air will fall, Figure 46–37. This will also keep the supply air from mixing with the return air and freezing the evaporator or satisfying the thermostat.

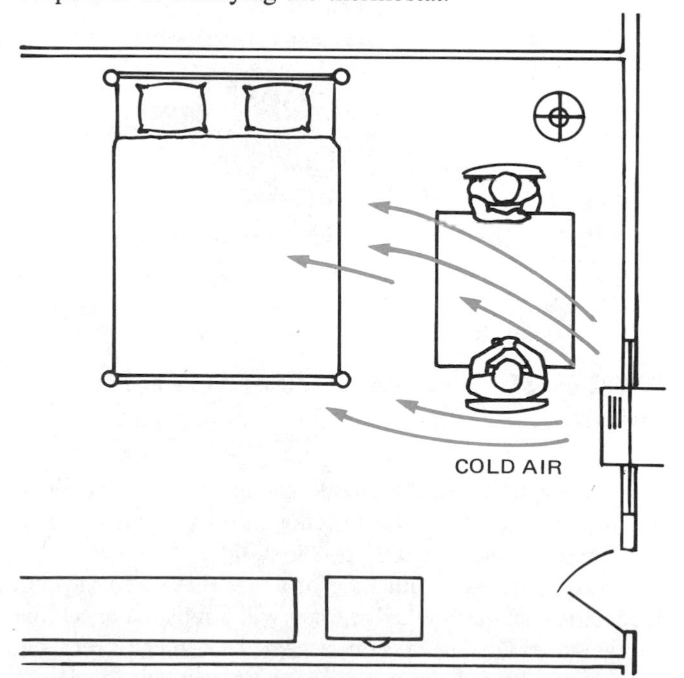

FIGURE 46–35 The outside wall may not be the best location for comfort, but it is the only place in most rooms.

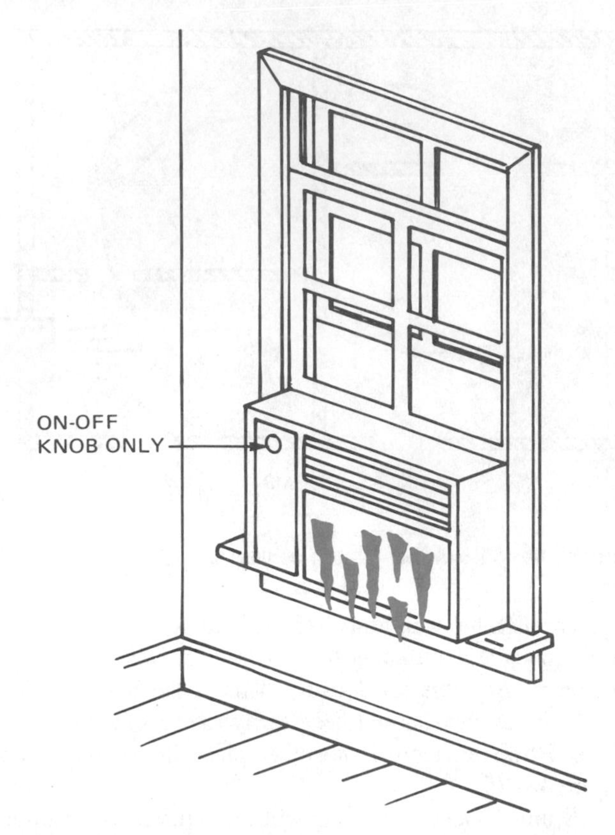

FIGURE 46–36 Some small units have no room thermostat, only an ON-OFF switch. If left on, the evaporator may freeze.

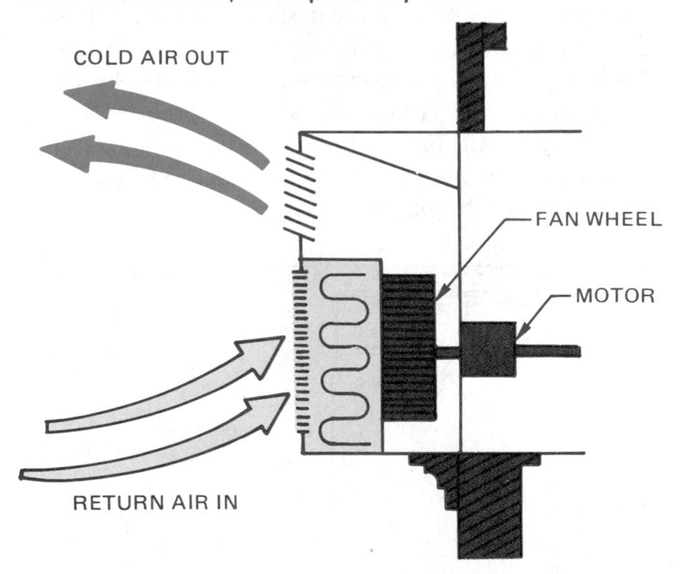

FIGURE 46–37 Cold air should be directed upward for best distribution.

The window must be wide enough for the unit to be installed in and must raise far enough for the unit to set on the window ledge, Figure 46–38.

Smaller window units are of a size that when they are placed on the window ledge, they will almost balance. The center of gravity of the unit is near the center of the unit, Figure 46–39. In larger units, the compressor is located more to the rear causing the unit to have a tendency to fall out the window unless half of the unit is extended into the room, Figure 46–40. This of course is not desirable. Most

FIGURE 46–38 The window must open wide enough for unit to fit.

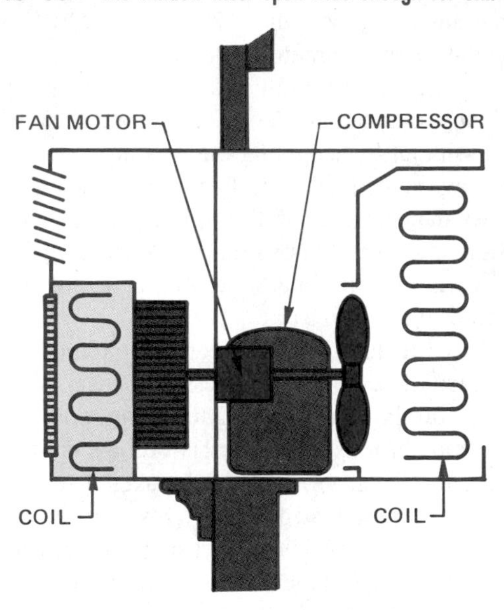

FIGURE 46–39 The center of gravity of a typical window unit is near the center.

manufacturer's provide a brace kit to support the back of the unit. The bracket takes the pressure off the upper part of the window and places the load on the bracket.

When the unit is placed in the window, it never completely fills the window. A kit is provided with each new unit to aid in neatly filling the hole. This kit may be telescoping side panels or panels to match the unit color that may be cut to fit. When the window is partly raised to accommodate the window unit, the space between the movable window parts must be insulated. Usually a foam strip is furnished for this purpose, Figure 46–41.

Directions are included with each new unit. However, in some instances the unit may be removed, the braces and window kit misplaced, and when reinstalled without braces and a proper window kit, a poor installation may result.

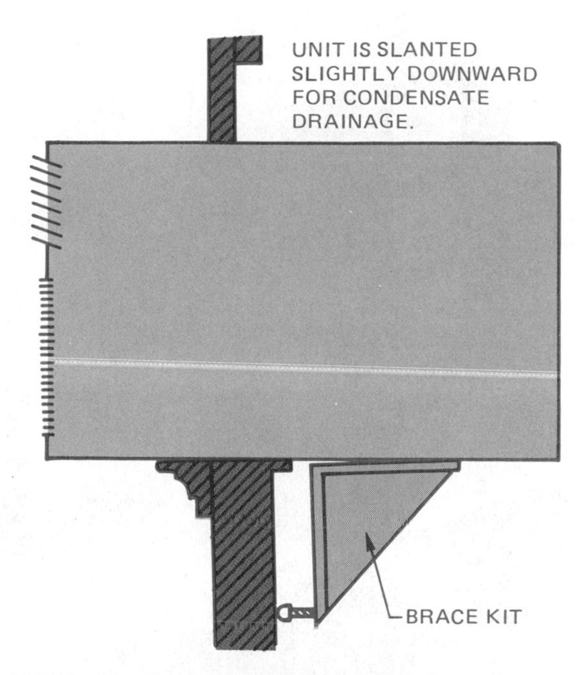

FIGURE 46–40 Larger units may be located farther to the rear so they do not protrude into the room. A brace kit should be used.

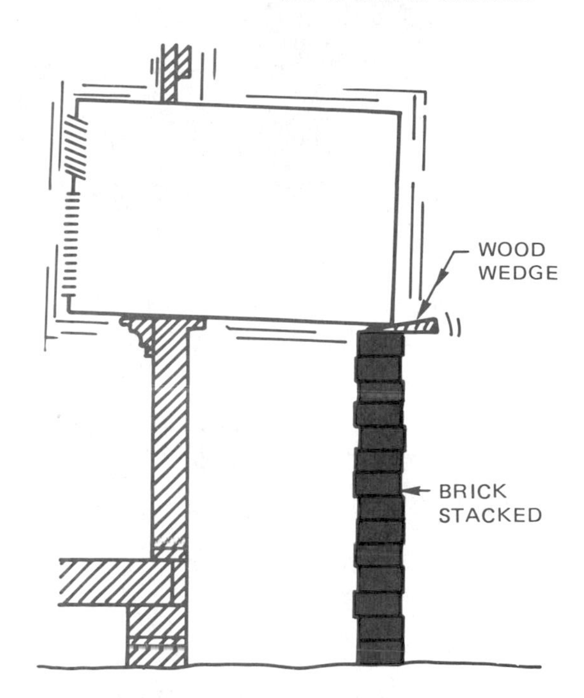

FIGURE 46–42 An awkward installation because the bracket kit is lost.

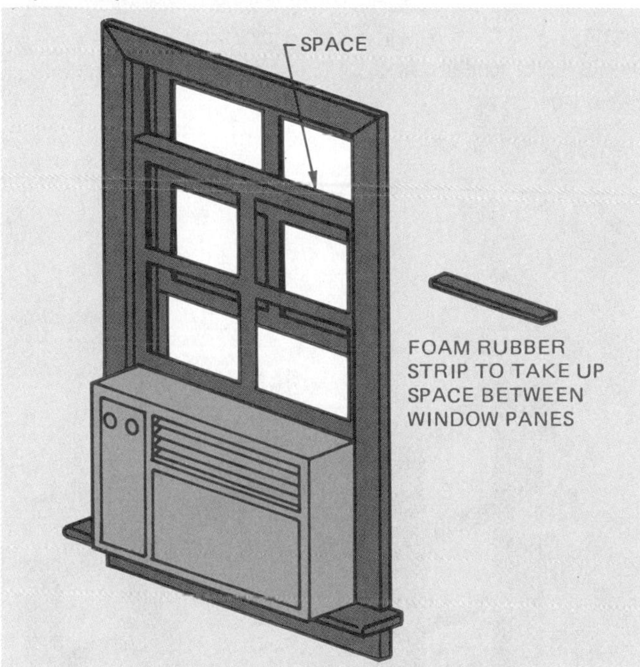

FIGURE 46–41 Foam strip for the top of the window.

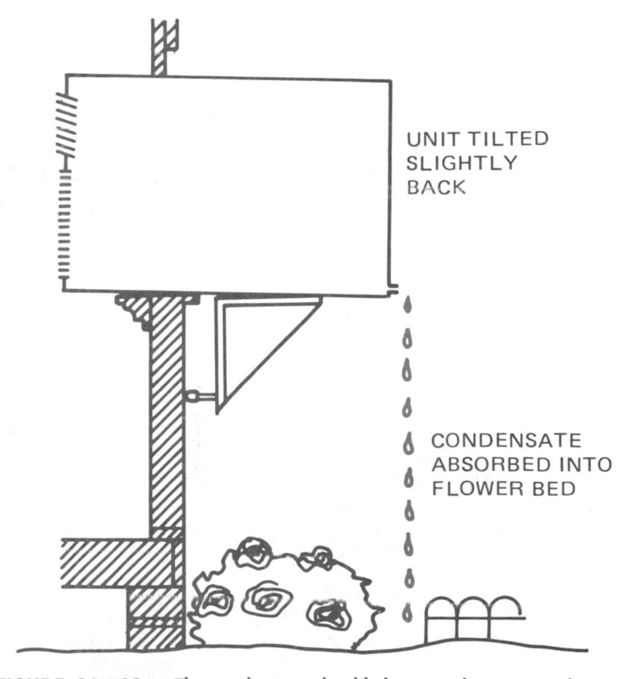

FIGURE 46–43 The condensate should drain to the proper place.

Some awkward installations can be seen because of this. Units that are installed and not properly braced will vibrate, sometimes called window shakers, Figure 46–42. The unit, properly braced, will be typically slanted slightly to the rear for proper condensate drainage. When the weight is supported by the back, it relieves the pressure on the window and reduces the vibration.

Most units evaporate the condensate, but some allow it to drain out the back. This condensate should drain to a proper location, such as a flower bed, on the grass, or in a gutter, Figure 46–43. Some are installed so that the condensate drips on the sidewalk. This is not good as algae may form and the walkway will become slick, Figure 46–44.

The unit must be installed in such a manner that air can circulate across the condenser and not recirculate. It must be allowed to escape the vicinity and not heat the surroundings. Window units have been located with the condenser in a spare room, such as in a business, Figure 46–45. This is poor practice because the unit heats the air and high operating conditions will occur.

Recirculation will cause high head pressure and high operating cost, Figure 46–46. Obstructions must not be located close to the air discharge.

Casement window installations require a special unit. This unit is narrow and tall, to fit the roll-out portion of the casement window. If the window is double, a regular window unit may be installed by cutting out part of the fixed portion of the window, Figure 46–47. This requires some

FIGURE 46–44 Condensate on a walkway can be dangerous.

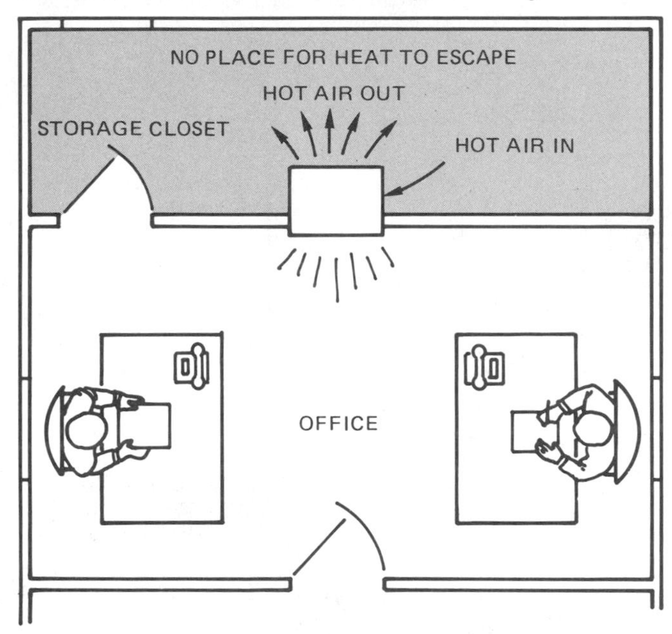

FIGURE 46–45 The heat must be able to escape from around the unit. Units with the condenser in an adjacent room will not operate efficiently.

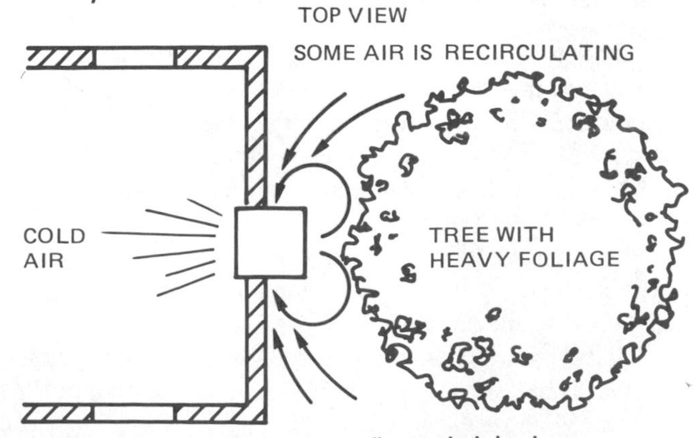

FIGURE 46–46 Recirculated air will cause high head pressure.

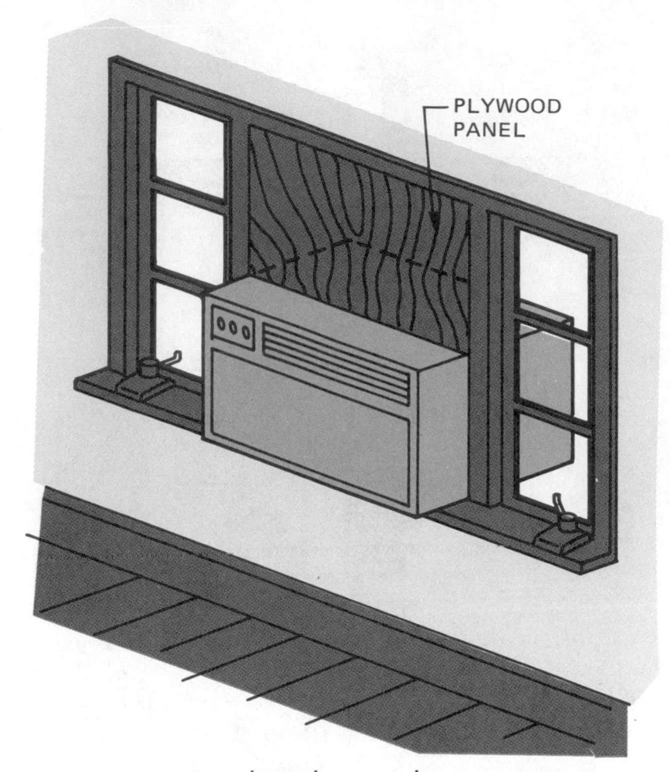

FIGURE 46–47 A regular window air conditioner is sometimes installed in the fixed portion of the casement window unit between the rollout sections.

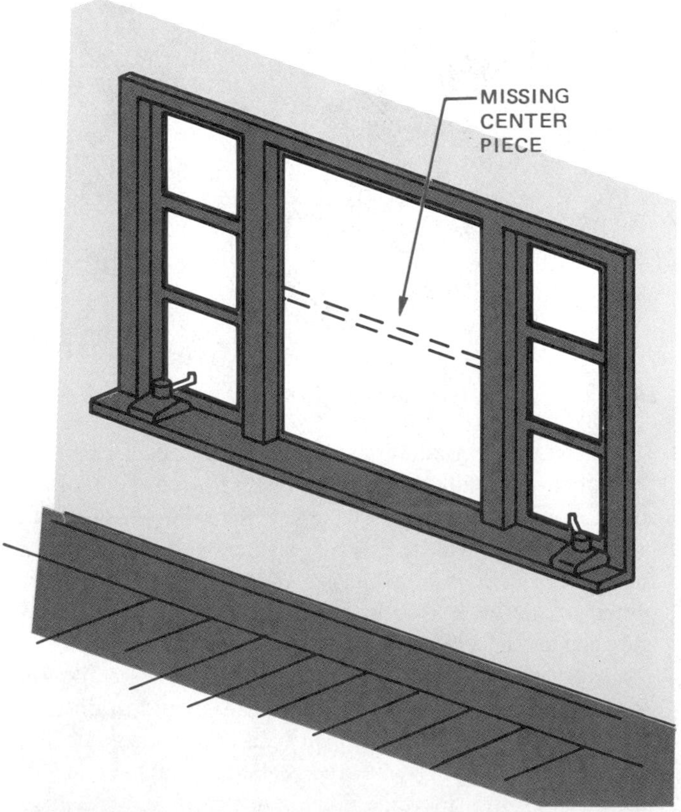

FIGURE 46–48 If the unit is moved, the window must be repaired.

FIGURE 46–49 Most of the through wall unit is located on the outside of the wall.

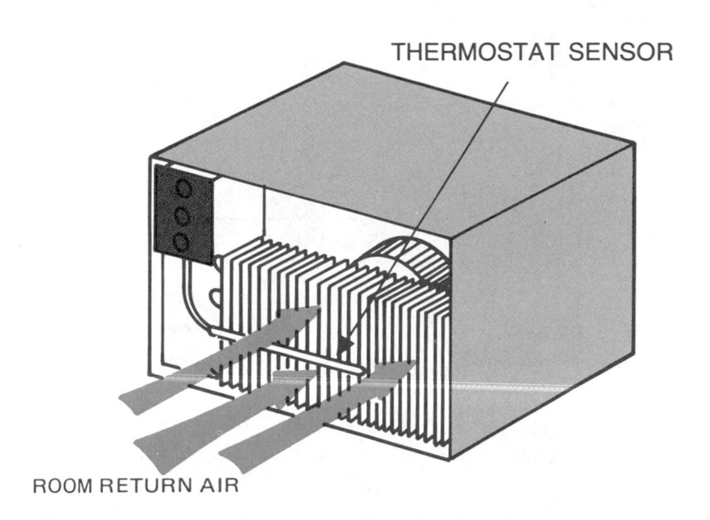

FIGURE 46–51 The thermostat is located in the return air stream.

skill. When this type of installation is performed, all stress must be on the window sill. If the unit is removed, another repair job must be performed for the window to be used, Figure 46–48.

Through the wall units typically protrude into the room only a slight amount with the bulk of the unit on the outside, Figure 46–49. When these units are installed in a motel, the outdoor portion protrudes into the walkway slightly. If condensate is not completely evaporated, it will run across the walkway and cause a hazard.

Through the wall units may be installed while the building is under construction. This is accomplished by installing wall sleeves and covering the openings until the unit is set in place, Figure 46–50. The wiring is run to the vicinity of the unit and the unit is connected when installed. It is important that the correct hole size be cut. The wall sleeve should be purchased from the manufacturer in advance to ensure this.

46.6 CONTROLS FOR ROOM UNITS, COOLING

Room units for cooling only are typically plug-in appliances furnished with a power cord. All controls are furnished with the unit. There is no wall-mounted remote thermostat as in central air conditioning. The room thermostat sensor is located in the unit return airstream, Figure 46–51. A typical room unit may have a selector switch to control the fan speed and provide power to the compressor circuit, Figure 46–52. The selector switch may be considered a power distribution center. The power cord will usually be wired straight to the selector switch for convenience. Therefore it may contain a hot, neutral, and a ground wire for a 120-V unit. A 208/230-V unit will have two hot wires and a ground, Figure 46–53. The neutral wire for 120-V service is routed straight to the power-consuming devices. On 208/230-V units there are two hot wires. One is wired through the selector switch to the power-consuming devices. The other hot wire is connected directly to the other side of the power-consuming devices. Many combinations of selector switches will allow the owner to select fan speeds, high and low cool, exhaust, and fresh air. High and low cool is accomplished

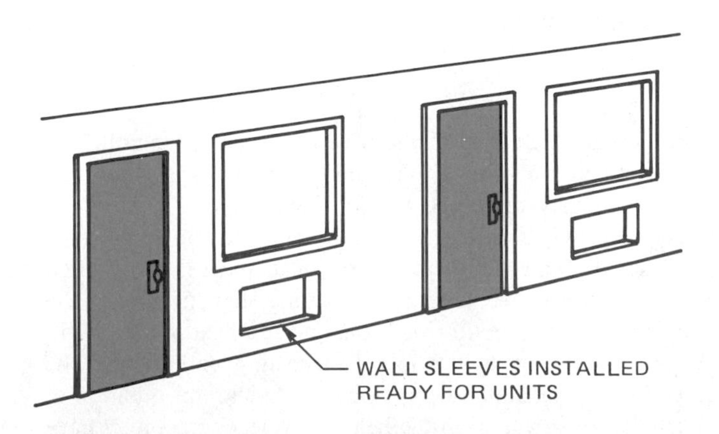

FIGURE 46–50 Wall sleeves may be installed in advance of the unit.

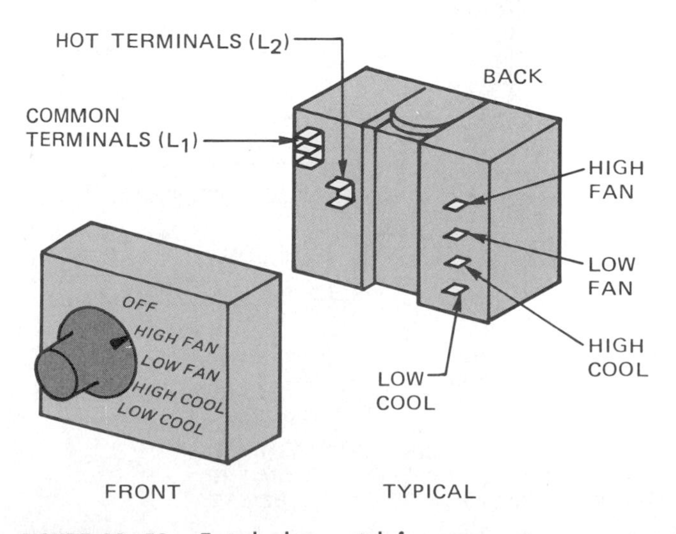

FIGURE 46–52 Typical selector switch for a room unit.

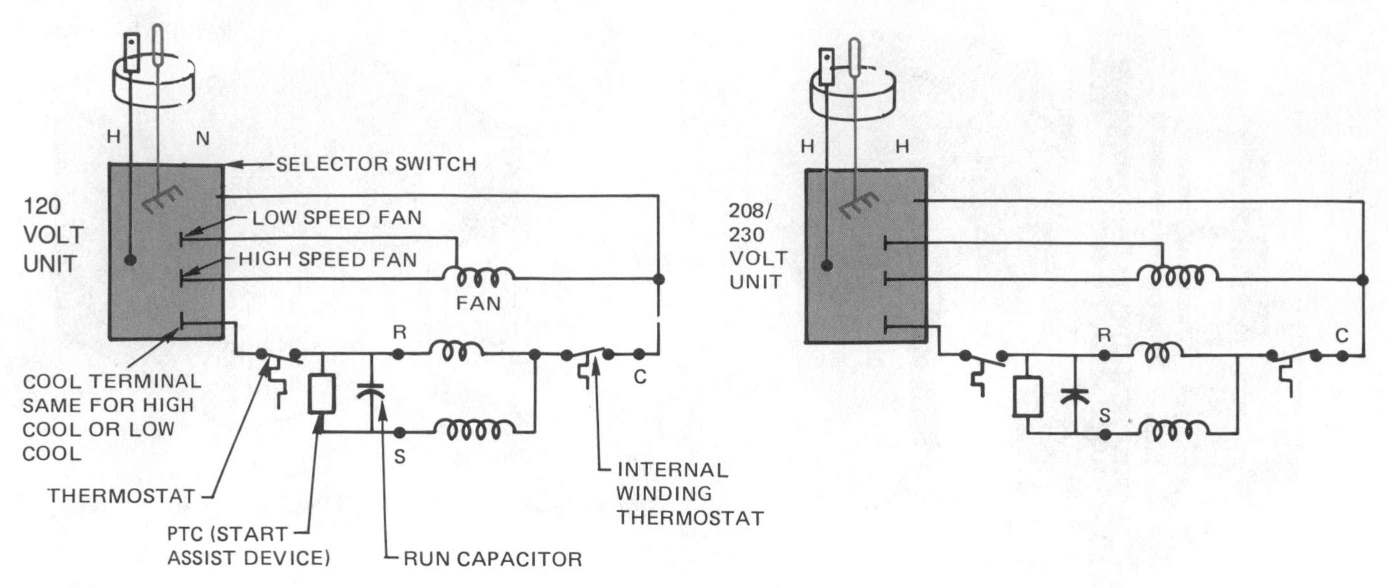

FIGURE 46–53 Typical selector switch terminal designations for 120-and 208/230-V units.

with fan speed. Because the unit has only one fan motor, a reduction in fan speed slows both the indoor and the outdoor fan. This reduces the capacity and the noise level of the unit. A slight power savings is also achieved, Figure 46–54.

The exhaust and fresh-air control is a lever that positions a damper to bring in outdoor air or exhaust indoor air, Figure 46–55.

The selector switch sends power to various circuits including the compressor start circuit to start the compressor. Motor starting was discussed in Unit 19 and will not be further discussed here. Figure 46–56 shows some selector switch and compressor diagrams with explanations.

46.7 CONTROLS, COOLING AND HEATING UNITS

Combination cooling and heating units may be either plug-in appliances or through the wall. Through the wall units usually have an electrical service furnished to the unit. The cooling controls for cooling and heating units are the same as previously discussed in the cooling cycle. Typically a selector switch changes the unit from cooling to heating. When the unit has electric strip heat, the selector switch merely directs the power to the heating element through the thermostat, Figure 46–57. If the unit is a heat pump, it will have a selector switch to select heating or cooling. The cooling cycle should be a typical cycle with the addition of the four-way valve. In the heating cycle, defrost may have to be considered.

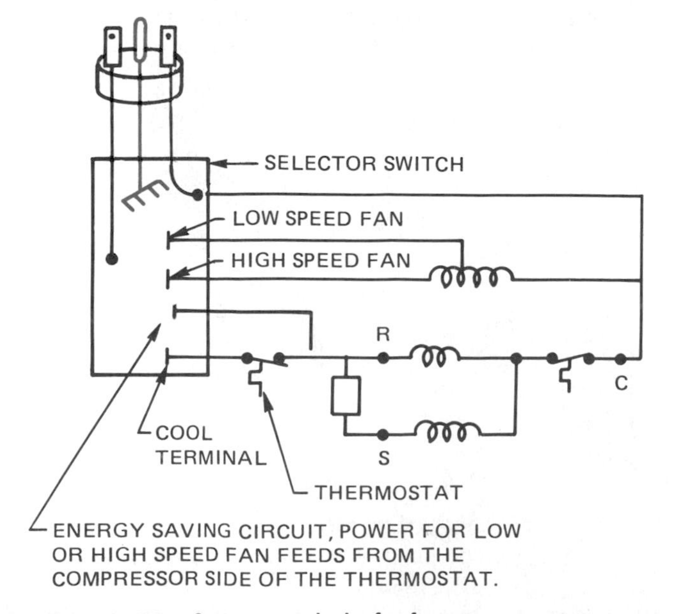

FIGURE 46–54 Some units cycle the fan for power savings.

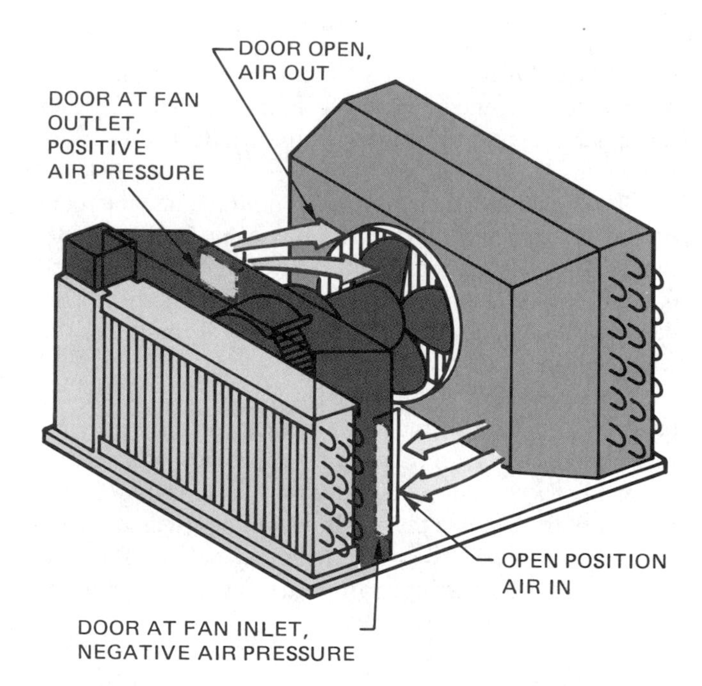

FIGURE 46–55 Exhaust and fresh air is accomplished with a damper operated from the front of the unit.

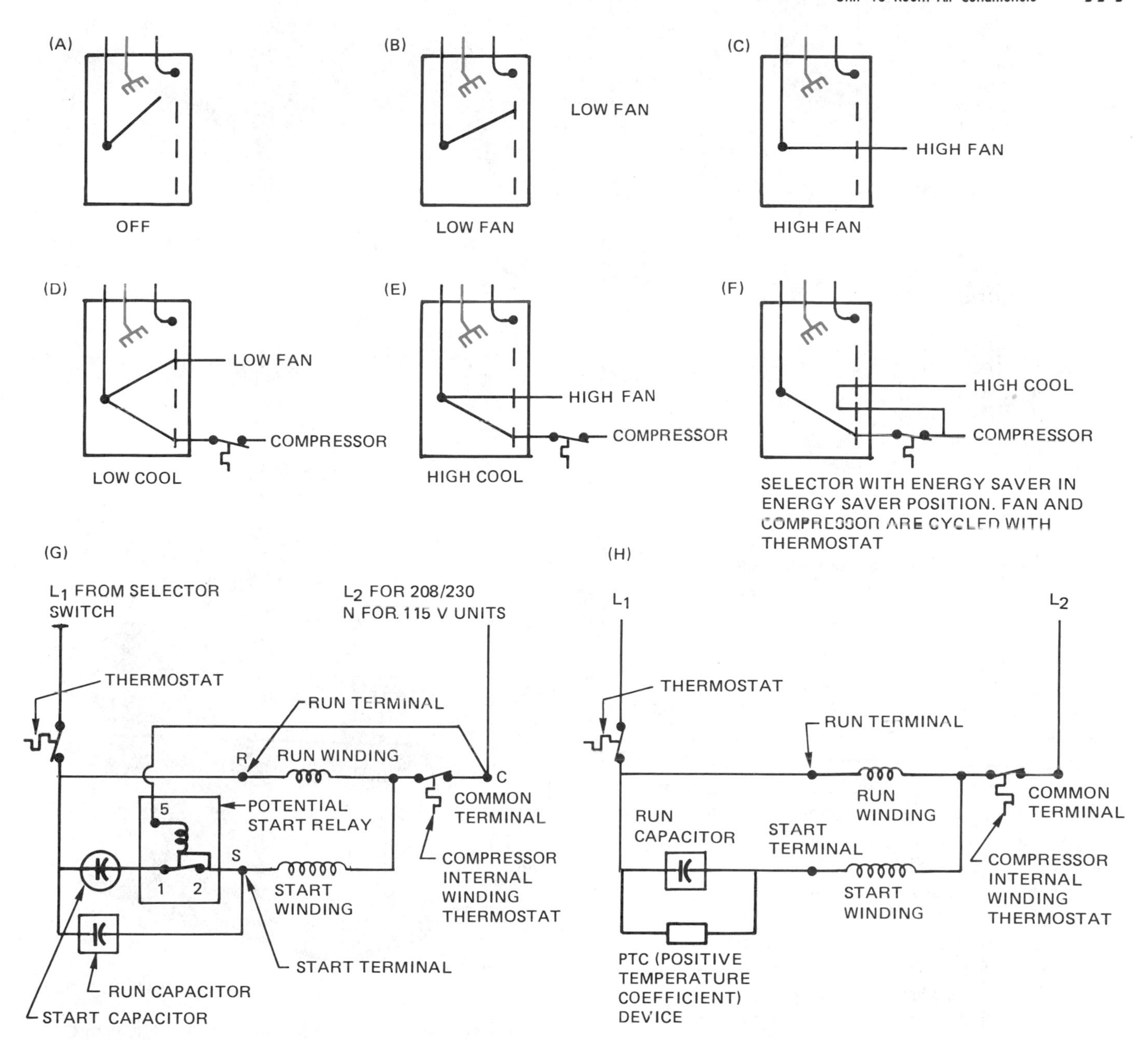

FIGURE 46–56 Some selector switch wiring diagrams and compressor starting diagrams.

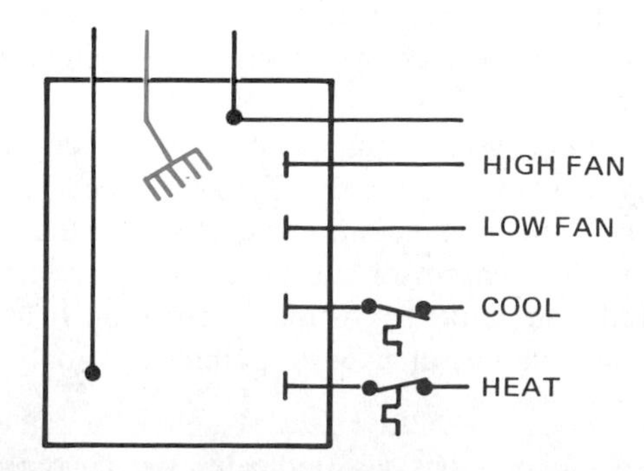

FIGURE 46–57 The selector switch directs the power to the heating or cooling circuits.

46.8 MAINTAINING AND SERVICING ROOM UNITS

Maintenance of room units basically involves keeping the filters and coils clean. The motor is typically a permanently lubricated motor in the later models. Older models may need lubrication. If the filter is not maintained, the indoor coil will become dirty and cause the unit to operate at low suction pressures. While in the cooling mode, the coil is already operating at about 35°F. If the coil temperature drops any at all, ice will begin to form. Some manufacturers provide freeze protection that will shut the compressor off and allow the indoor fan to run until the threat of freezing is over. This may be done with a thermostat located on the indoor coil or a thermostat mounted on the suction line, Figure 46–58.

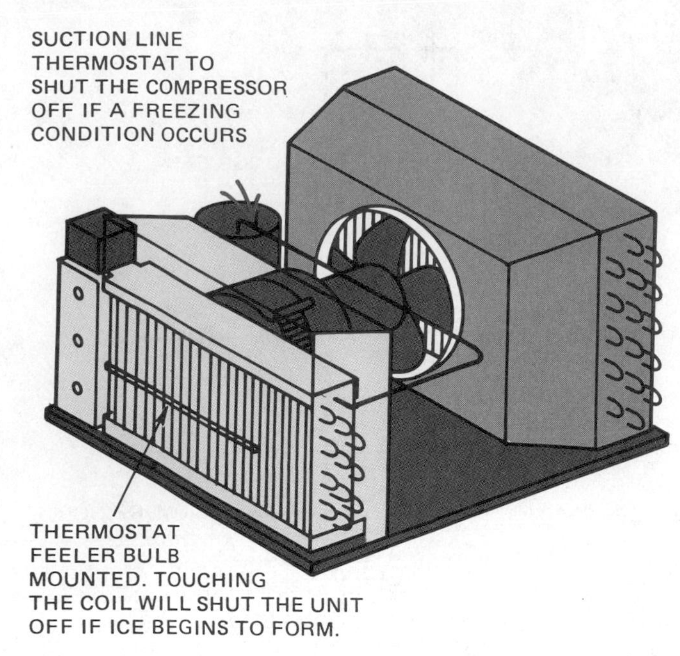
SUCTION LINE
THERMOSTAT TO
SHUT THE COMPRESSOR
OFF IF A FREEZING
CONDITION OCCURS
THERMOSTAT
FEELER BULB
MOUNTED. TOUCHING
THE COIL WILL SHUT THE UNIT
OFF IF ICE BEGINS TO FORM.

FIGURE 46–58 Protection from freezing for room unit.

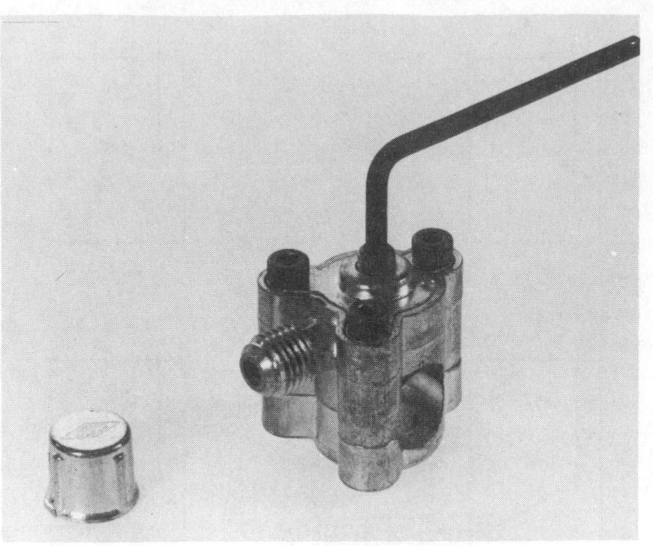

FIGURE 46–59 Line tap valve. *Photo by Bill Johnson*

Service may involve mechanical or electrical problems. Mechanical problems usually concern the fan motor and bearing or refrigerant circuit problems. All room units have a critical refrigerant charge and gages should only be installed after there is some proof that they are needed. The system will normally be sealed and require line tap valves or the installation of service ports on the process tubes for gages to be installed. Line tap valves and their use is discussed in Unit 44. Figure 46–59 shows an example of a line tap valve. ***Be sure to follow valve manufacturer's instructions.***

A bench test at the workshop is often performed on room units. The technician should perform the following test before installing gages when a low charge is suspected. Because the condenser and the evaporator are in the same temperature air, the condenser airflow may need to be reduced to move any refrigerant from the condenser due to low head pressure. A full explanation of condenser operation and charging under low ambient conditions is discussed in Unit 28. Refer to Figure 46–60 while reading the following:

- Remove the unit from its case.
- Make sure that air flows through the coils. For temporary testing, cardboard may be positioned over places where panels force air to flow through the coils.
- Start the unit in high cool.
- Let the unit run for about 5 min and observe the sweat on the suction line. It should come close to the compressor. The evaporator should be cold from bottom to top.
- Cover a portion of the condenser to force the head pressure to rise. If the room is cool, a portion of the charge may be held in the condenser. The condenser airflow may be blocked to the point that the air leaving it is hot, about 110°F.

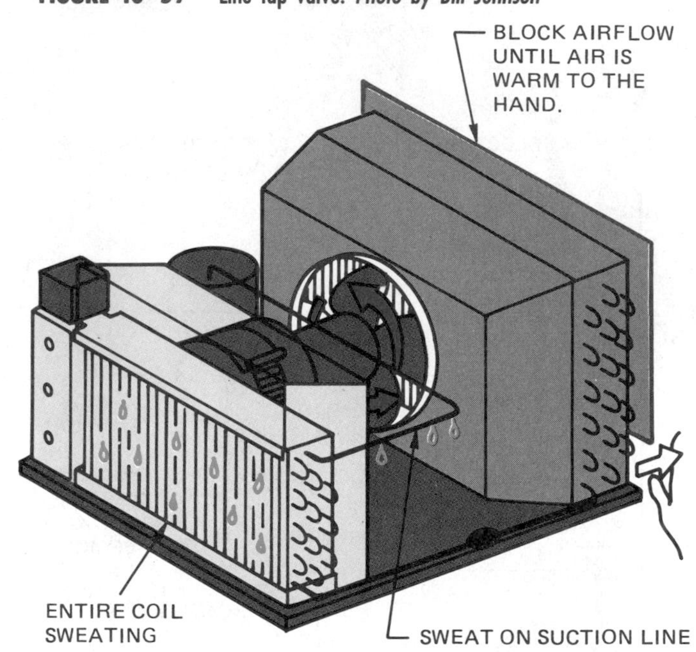
BLOCK AIRFLOW
UNTIL AIR IS
WARM TO THE
HAND.
ENTIRE COIL
SWEATING
SWEAT ON SUCTION LINE

FIGURE 46–60 Testing for a low charge before installing gages.

- With the head pressure increased, allow the unit to run for 5 min; the sweat line should move to the compressor. If the humidity is too low for the line to sweat, the suction line should be cold. The evaporator should be cold from bottom to top. If only part of the evaporator is cold (frost may form), the evaporator is starved. This may be a result of a restriction or the unit has a low charge.
- If the compressor is not pumping to capacity and the charge is correct, the coil will not be cold anywhere, only cool. An ammeter may be clamped on the common wire to the compressor and the amperage compared to the full-load amperage of the compressor. If it is very low, the compressor may be pumping on only one of two cylinders.

When the previous test indicates the charge is low, because the suction line is not cold or sweating, gages should be installed.

When the charge is low, the leak should be found. If the unit has never had a line tap or service port installed and has operated for a long period of time without a problem, it can be assumed that the leak has just occurred. It is not good practice to just add refrigerant and hope. An electronic leak detector may be used. Turn the unit off and allow the pressures to equalize. Make sure there is enough refrigerant in the unit to properly leak check, Figure 46–61. If the system pressure and temperature relationship indicates there is no liquid present, there is so little refrigerant in the system that what is left could be considered trace refrigerant for leak checking purposes. If the leak cannot be found, with the unit off, dry nitrogen may be used to increase the pressure to the lowest working pressure of the unit. This would be the evaporator working pressure and may be 150 psig, check the nameplate to be sure. When the leak check is completed, the nitrogen and trace refrigerant will have to be exhausted, the system evacuated, and a new charge added. This venting of trace refrigerant is allowed.

The correct method to charge a unit is to remove and recover the remaining charge, evacuate the unit, and measure in the correct charge printed on the nameplate. Technicians should use accurate scales similar to the one in Figure 46–62 to measure the charge. At times the technician may want to adjust the charge rather than start from the beginning. A charge may be adjusted by the following method. Figure 46–63 illustrates the procedure:

- Install suction and discharge gage lines.
- Purge the gage lines and connect to the correct refrigerant cylinder.
- Make sure that air is passing through the evaporator and the condenser coils; cardboard may be used for temporary panels.
- Start the unit in high cool.

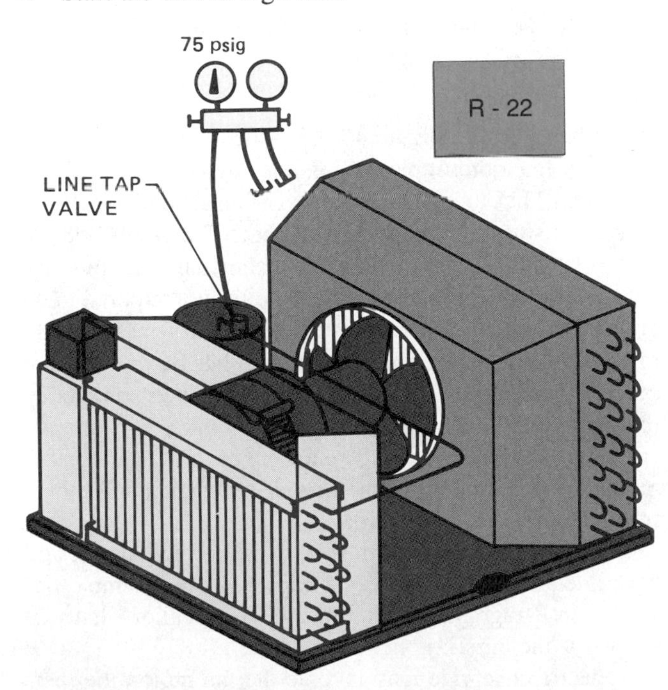

FIGURE 46–61 Make sure there is enough refrigerant in the unit for leak checking purposes.

FIGURE 46–62 Charging scales. *Courtesy Robinair*

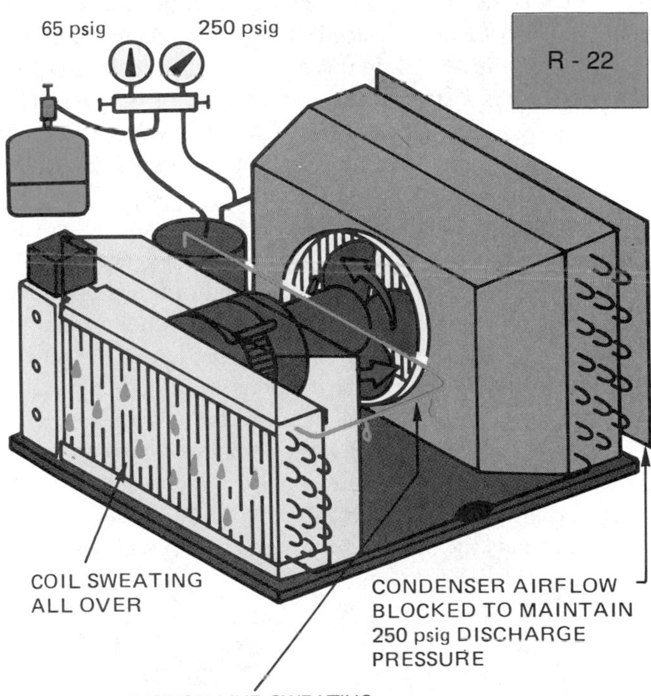

FIGURE 46–63 Charging a unit by sweat line.

- Watch the suction pressure and do not let it fall below atmospheric pressure into a vacuum; add refrigerant through the suction gage line if it falls too low.
- Adjust the airflow across the condenser until the head pressure is 260 psig, corresponding to a condensing temperature of 120°F for R-22. Refrigerant may have to be added to obtain the correct head pressure.
- Add refrigerant vapor at intervals with the head pressure of 260 psig until the compressor suction line is sweating to the compressor. As refrigerant is added, the airflow will have to be adjusted over the condenser or the pressure will rise too high. You should also note that the

evaporator coil has become colder toward the end of the coil as refrigerant is added.

- The correct charge is verified after about 15 min of operation at 260 psig and a cold (probably sweating) suction line at the compressor. The suction pressure should be about 65 psig and the compressor should be operating at near full-load amperage.

When a system must be evacuated, the same procedures are used as in Unit 8. The larger the evacuation ports, the better. Unit 44 has a complete evacuation procedure for refrigerators; this same procedure would apply here.

The capillary tube may be serviced in much the same manner as the capillary tube maintenance in Unit 44. At times a capillary tube may need to be changed. If the head pressure rises and the suction pressure does not and refrigerant will not fill the evaporator coil, there is a restriction. The capillary tube-to-suction line heat exchange for a room air conditioner presents the same problem as with a refrigerator. It may not be worth the time to make the change. Some service technicians may install a thermostatic or automatic expansion valve instead of trying to change a capillary tube, Figure 46–64. In this case you must make sure the compressor has a starting relay and capacitor or it may not start. Thermostatic and automatic expansion valves do not equalize during the off cycle. This is a workable situation for a room unit, but not a refrigerator.

Fan motor removal may be difficult in many units. The units are so compact that the fans and motor are in close proximity to the other parts. The evaporator fan is normally a centrifugal (squirrel cage design) and the condenser fan is typically a propeller type. The condenser fan shaft usually extends to a point close to the condenser coil. Often there is not room to slide the propeller fan to the end of the shaft, Figure 46–65. Either the fan motor must be raised up or the condenser coil moved, Figure 46–66. The evaporator fan wheel is normally locked in place with an Allen-head set screw that must be accessed through the fan wheel using a

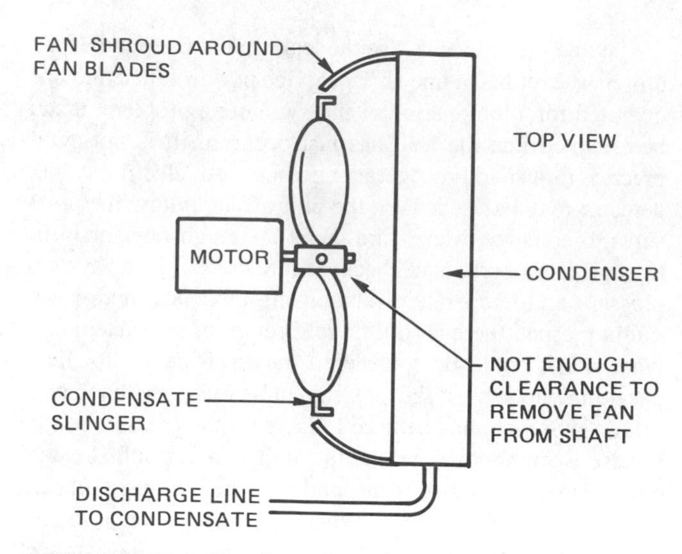

FIGURE 46–65 Often the condenser fan is hard to remove from the shaft because of the condenser.

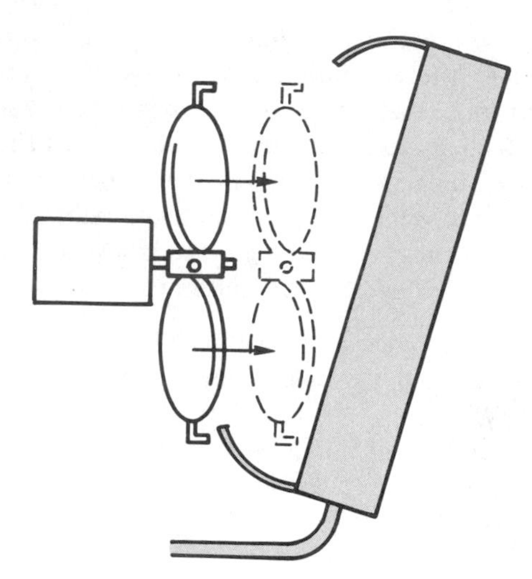

FIGURE 46–66 The condenser may be moved back or the fan motor may be raised for removal of the fan blade.

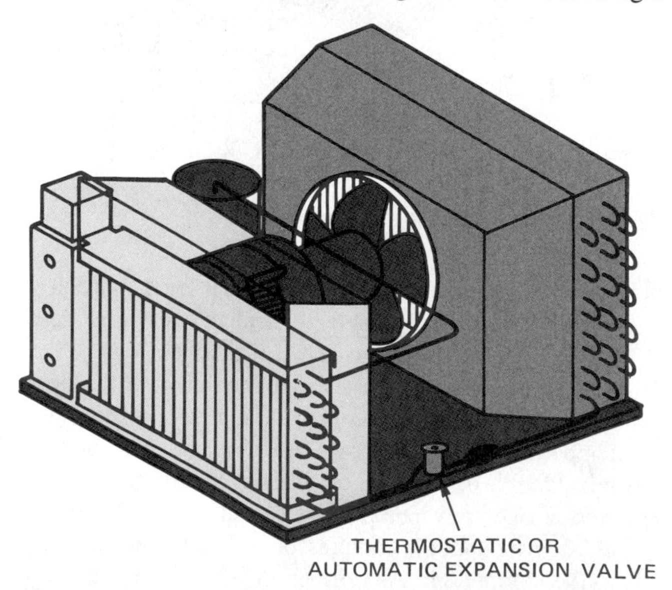

FIGURE 46–64 A thermostatic or automatic expansion valve may sometimes be installed instead of replacing a capillary tube.

long Allen wrench, Figure 46–67. When a unit is reassembled after fan motor replacement, it is often difficult to align everything back together correctly. The unit should be placed on a level surface, such as a workbench, Figure 46–68.

Electrical service requires the technician to be familiar with the electrical power supply. All units are required to be grounded to the earth. This involves a green wire to be run to every unit from the grounded neutral bus bar in the main control panel, Figure 46–69. Many units are operating today that do not have this ground and these units will operate correctly as the wire is not there to help the unit operate. It is a safety circuit to protect the service technician and others from electrical shock hazard.

The electrical service must be of the correct voltage and have adequate current-carrying capacity. The example described in Paragraph 46.5 in this unit is typical of what will happen if the circuit is not planned.

Electrical service may involve the fan motor, the thermostat, the selector switch, compressor, and the power cord. The fan motor will be either a shaded pole or a permanent

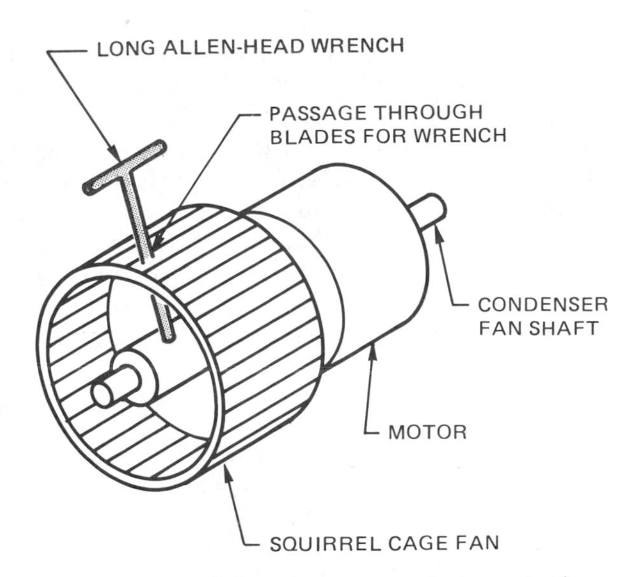

FIGURE 46–67 An Allen-head set screw may be loosened with a long Allen wrench through the blades.

FIGURE 46–68 The unit should be placed on a level surface while installing the fan and blades for proper alignment.

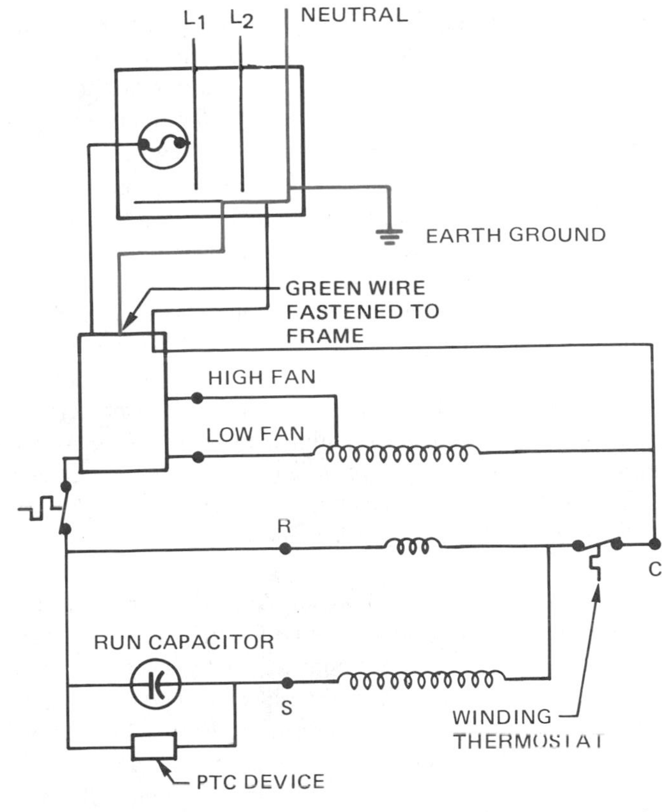

FIGURE 46–69 Wiring diagram that shows where the green ground wire terminates in the main control panel.

split-capacitor type, Figure 46–70. These motors are discussed in detail in Section 4.

The thermostat for room units is usually mounted with the control knob in the control center and the remote bulb in the vicinity of the return air. These thermostats are the line-voltage type discussed in Unit 13. They are typically slow to respond. This is somewhat overcome because most units operate with the fan running continuously anytime the se-

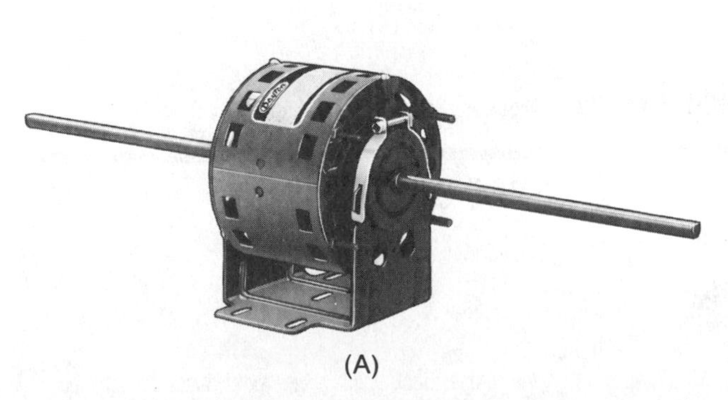

(A)

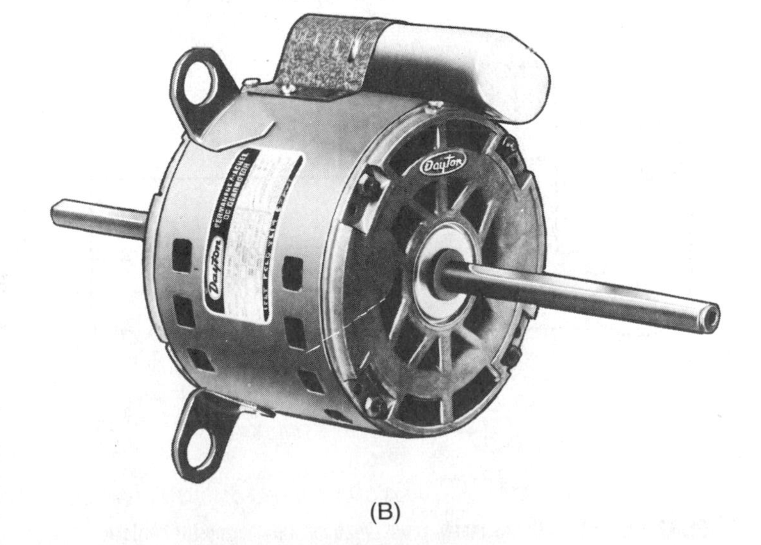

(B)

FIGURE 46–70 Fan motors. (A) Shaded pole. (B) Permanent split capacitor. *Courtesy Dayton Electric Manufacturing Company*

lector switch is set for cooling or heating. Therefore, the thermostat has a moving airstream over it at all times. Some systems may have an energy-saving feature that allows the thermostat to cycle the fan. In this case the thermostat will not provide quite as close control.

The thermostat passes power to the compressor start circuit and may be treated as a switch for troubleshooting purposes. If it is suspected that the thermostat is not passing power to the compressor start circuit, you may unplug (shut off the power and lock the circuit out if wired direct) the unit and remove the plate holding the thermostat. The plate may be moved forward far enough to apply voltmeter leads to the thermostat terminals. Turn the voltmeter selector switch to the correct setting, higher than the voltage of the unit, and plug the unit in (resume power). Turn the selector switch to the high-cool position and observe the voltmeter reading. If there is a voltage reading, the thermostat contacts are open, Figure 46–71.

The selector switch may be defective similar to the thermostat and the trouble may be found in a similar way. The panel that holds the thermostat usually has the selector switch mounted on it. If it is suspected that the selector switch is not passing power, turn the power off and remove the panel. Secure it in such a manner that the unit may be started. The selector switch is checked in a slightly different manner, using the common terminal and the power lead. For example, attach one lead of a voltmeter to the common lead and the other to the circuit to be tested, high cool, Figure 46–72. Resume the power supply and turn the selector switch to the HIGH-COOL position. Full line voltage should be indicated on the meter. If not, check the hot lead entering the selector switch. If power enters, but does not leave, the switch is defective.

Compressor electrical problems may either be in the starting circuit or the compressor itself. Room units generally use single phase. They use the potential relay or the positive temperature coefficient device (PTC) described in Unit 17 for starting the compressor. Electrical troubleshooting and capacitor checks are discussed in Unit 20. A review of these units will help you to understand motors and their problems.

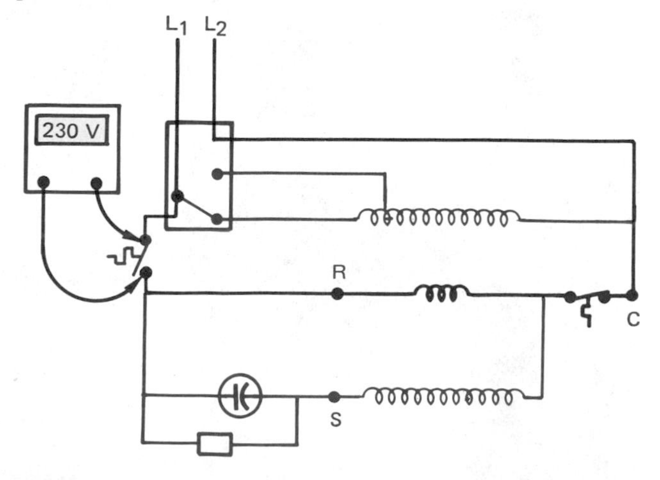

FIGURE 46–71 If the meter reads voltage, the thermostat contacts are open.

Power cord problems usually occur at the end of the cord, where it is plugged into the wall plug. Loose connections because the plugs do not fit well or the plug or power cord has been stepped on and partially unplugged are the big problems. The power cord should make the best connection possible at the wall outlet. Sometimes the cord may need support to keep the strain off the connection, Figure 46–73. If the cord becomes hot to the point of discoloration or the plug swells, the plug should be changed. The wall receptacle may need inspection to make sure damage has not occurred inside it. If the plastic is discolored or distorted, it should be changed or installing a new plug will not help and the problem will recur.

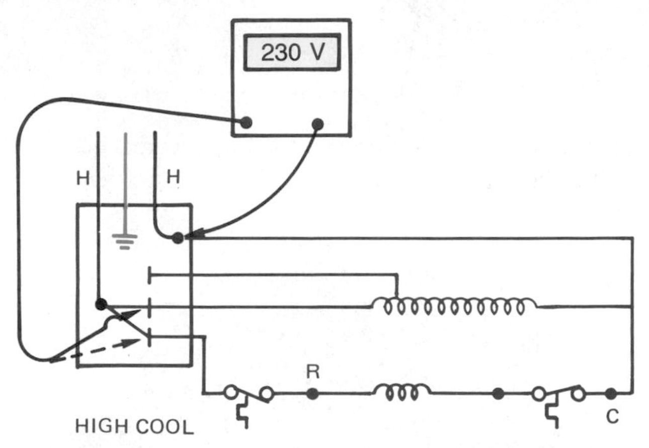

FIGURE 46–72 Using a voltmeter for checking a selector switch.

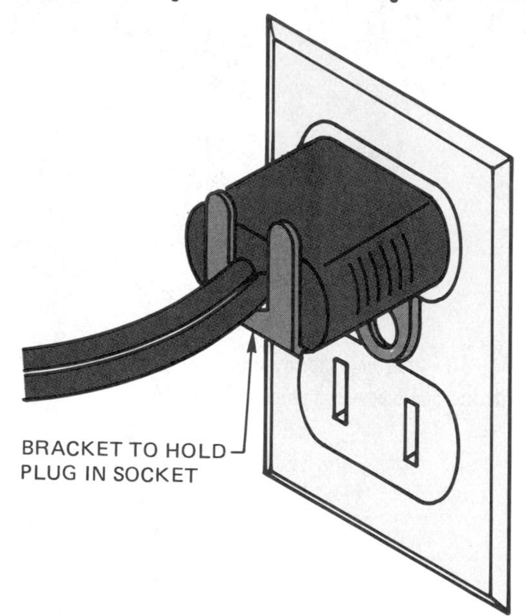

FIGURE 46–73 A power cord may need support.

HVAC GOLDEN RULES

When making a service call to a residence:

- Try to park your vehicle so that you do not block the customer's driveway.
- Always carry the proper tools and avoid extra trips through the customer's house.

- Ask the customer about the problem. The customer can often help you solve the problem and save you time.

Added Value to the Customer. Here are some simple, inexpensive procedures that may be added to the basic service call.

- Clean the air filters.
- Advise the customer about proper operation, such as do not allow conditioned air to recirculate into the return air.
- If removal of the unit is necessary, have the proper help and use dropcloths to prevent dirt from the unit from falling on the floor.

46.9 SERVICE TECHNICIAN CALLS

SERVICE CALL 1

A customer in an apartment calls and says that the window unit in the dining room, used to serve the kitchen, dining room, and living room is freezing solid with ice every night. The tenants both work during the day. They shut the unit off in the morning and turn it on when they get home from work. The air temperature is 80°F inside and outside by the time they turn the unit on. There is a high heat load in the apartment keeping the unit running. *The problem is that the head pressure is too low during the last part of the cycle, starving the evaporator and causing it to operate below freezing.*

The apartment house technician arrives late in the afternoon when the tenants arrive home to survey the situation. The tenants show the technician how the thermostat is set each night. The technician explains that the outdoor temperature is cooler than the temperature in the apartment by the time the unit is turned on. It is explained that a better approach may be to leave the unit on LOW COOL during the day, and raise the thermostat setting several degrees, letting the unit run some during the day. A call back in a few days verifies that the unit is performing well using this approach. The unit had no problem; it was lack of operator knowledge.

SERVICE CALL 2

A customer calls to indicate that the air conditioner in the guest bedroom is not cooling the room. The coil behind the filter has ice on it. This sounds like a low charge complaint. *The problem is that the drape is partially hanging in front of the unit, causing the air to recirculate. The supply air louvers are also pointing downward.*

The technician in this apartment house goes to the installation and notices that the drape hangs in front of the unit. The technician explains that the drape will have to be kept out of the airstream. The air discharge is pointed down also, causing the air to move toward the return air inlet. This is adjusted. Just to be sure that the unit does not also have a low charge, the technician pulls the unit out of the case and places it on a small stool so it can be started. The technician notices that if the unit is started, air will blow out the top of the blower compartment and bypass the indoor coil so a piece of cardboard is placed over the fan compartment to force the air through the coil. The unit is started and the condenser is partially blocked to force the head pressure higher. After the unit has run for a few minutes, the suction line starts to sweat back to the compressor. This unit has never had gages installed, so it has the original charge and it is still correct. The unit is stopped and pushed back into its case in the window. The filter is cleaned. The unit is started and left operating normally.

SERVICE CALL 3

A customer calls and reports that a new unit just purchased and installed is not cooling so it has been turned off. *The problem is a leak in the high-pressure side of the system.*

The technician arrives and starts the unit. The compressor runs, but nothing else happens. The technician turns the unit off and slides it out of its case on to a stool and restarts the compressor. It is obvious the unit has problems so the technician goes to the truck, brings back a hand truck and takes the unit to the truck and to the shop.

The unit is moved to a workbench. A line tap valve is installed on the suction process tube. When the gages are fastened to the line tap valve, no pressure registers on the gage. The system is out of refrigerant.

The technician allows R-22, the refrigerant for the system, to flow into the unit. A leak cannot be heard, so an electronic leak detector is used. A leak is found on the discharge line connection to the condenser. The refrigerant is recovered from the unit. The process tube is cut off below the line tap valve connection. A 1/4-in. flare connection is soldered to the end of the process tube and a gage manifold connected to the low side; the leak is repaired on the discharge line with high-temperature silver solder.

When the torch is removed from the connection, the gage manifold valve is closed to prevent air from being drawn into the system when the vapor inside cools and shrinks. When the connection is cooled, refrigerant is allowed into the system and the leak is checked again. It is not leaking. A small amount of refrigerant and air is purged, and the system is ready for evacuation.

The vacuum pump is attached to the system and started. While the system is being evacuated, the electronic scales are moved to the system and prepared for charging. Because the service connection is on the low-pressure side of the system, the charge will have to be allowed into the system in the vapor state. The cylinder is a 30-lb cylinder. This should be enough refrigerant to keep the pressure from dropping during charging. The unit calls for 18.5 oz of refrigerant. When the vacuum is reached, 29 in. Hg on the low-side gage, the vacuum pump is disconnected and the hose connected to the refrigerant cylinder; the gage line is purged of air. The correct charge is allowed into the system. When the correct charge is established, the unit is started. It is cooling correctly.

The technician takes the unit back to the customer and places it in the window. The unit is started and operates correctly.

SERVICE CALL 4

A customer in a warehouse office called and reported that the window unit cooling the office is not cooling to capacity. A quick fix must be performed until a new unit can be installed. *The problem is a slow leak.*

The technician arrives and looks at the unit. The unit definitely has a low charge; the evaporator coil is only cold half way to the end. The unit is pulled out of the case and set on a stool. The technician points out that the unit is old and needs replacing. The customer asks if there is anything that can be done to the unit to get them through the day.

A line tap is fastened to the suction line and a gage line connected; the suction pressure is 50 psig. The unit has R-22; this is an evaporator temperature of 26.5°F. The unit is showing signs of frost on the bottom of the coil. This is an old unit, so refrigerant is added to adjust the charge. The unit is on a stool in the room, where it is about 80°F. The condenser airflow is blocked until the leaving air is warm to the hand. Refrigerant is added until the suction line is sweating to the compressor. The gages are removed and the unit is placed back in the case. The leak was not found. The company requested that a new unit be installed tomorrow. **When the old unit is discarded, the refrigerant will have to be recovered before it is scrapped.**

SERVICE CALL 5

A new unit sold last month is brought to the shop for the technician to repair. The service repair order reads, no cooling. *The problem is the compressor is stuck and will not start.*

The technician checks to make sure the selector switch is in the OFF position and plugs the unit into the correct wall plug, 230 V. This workbench is made for servicing appliances so it has an ammeter mounted in it. The technician turns the fan to HIGH FAN; the fan runs. The selector switch is then turned to LOW FAN and the fan still runs. The amperage is normal, so the technician turns the selector switch to HIGH COOL, keeping a hand on the switch. The fan speeds up but the compressor does not start. The thermostat is not calling for cooling. The thermostat is turned to a lower setting and the compressor tries to start. The ammeter rises to a high amperage and does not fall back. The technician turns the unit off, before the compressor overload reacts. The unit is turned off and the power cord disconnected.

The technician then removes the unit from the case and looks for the compressor terminal box. The cover is removed and a test cord is installed on common-run-start, Figure 46–74. The test cord is attached to a power supply on the bench. An ammeter is still in the circuit on the test bench. The technician tries to start the compressor with the test cord by turning the selector switch to start and then releasing the selector knob, per directions. It will not start; the compressor is stuck and must be changed.

The technician installs a line tap valve on the suction process tube and starts to recover the refrigerant. While the refrigerant is being recovered slowly, a new compressor, a 1/4-in. flare fitting for the process tube, and a suction-line drier are brought from the supply room. The technician removes the compressor hold down bolts. When the refrigerant has been recovered, the technician uses an air-acetylene torch with a high-velocity tip and removes the suction and discharge lines. The old compressor is set out and the new one set in place. The suction line is altered to install the suction-line drier, but it is not installed yet. The compressor discharge line is soldered in place. The process tube was cut off below the line tap connection and the 1/4-in. flare fitting is installed on the process tube connection. The suction line is soldered to the compressor and the suction-line drier is soldered into the line. A gage line is fastened to the 1/4-in. flare connection. A small amount of R-22 is added to the system (to 5 psig), then nitrogen is used to pressure the system to 150 psig and a leak check is performed on all fittings; the unit is leak free.

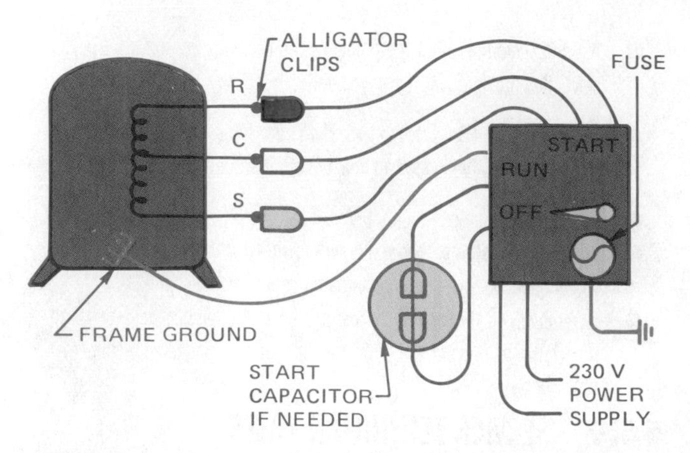

FIGURE 46–74 Test cord is fastened to common-run-start on the compressor.

The small amount of refrigerant vapor and nitrogen is purged and the vacuum pump is fastened to the gage manifold and started. When the vacuum pump has obtained a deep vacuum, according to the mercury manometer, the correct charge is weighed into the unit. The compressor wiring is reconnected while the vacuum pump operates so the unit is ready to start. The technician plugs in the cord and turns the selector switch to HIGH COOL; the compressor starts. The unit is allowed to run for 30 min and it operates satisfactorily. The process tube is pinched off and cut off below the 1/4-in. flare fitting. The process tube is closed on the end and soldered shut.

SERVICE CALL 6

A unit for an apartment is brought into the apartment workshop with a tag saying "no cooling". *The problem is the start relay is defective, a burned coil.*

The technician sets the unit on the workbench and checks the selector switch, which is in the OFF position. The unit is plugged into the correct power supply, 230 V. This workbench also has a built-in ammeter. The technician turns the selector to the HIGH FAN setting; the fan runs as it should. It is switched to LOW FAN and the fan still runs fine, at low speed. The technician then turns the selector switch to HIGH COOL, hand still on the switch. The fan speeds up to high, but the compressor does not start. The ammeter reads very high, 25 A. The technician turns the unit off before the compressor overload protector reacts.

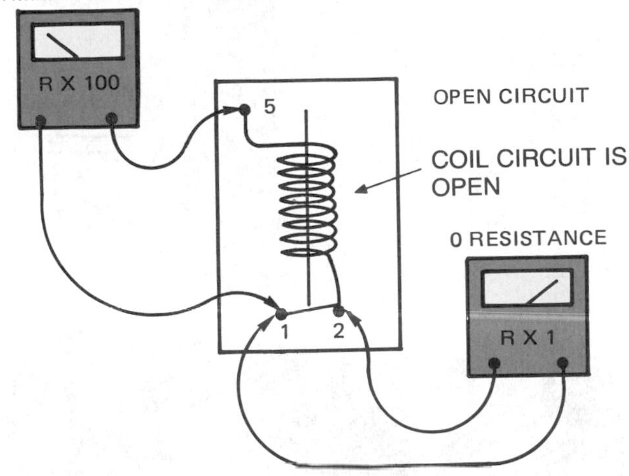

FIGURE 46–75 A start relay is checked with an ohmmeter.

The unit is turned off, unplugged, and removed from its case. The compressor motor terminal wires are removed and a test cord connected. The test cord is plugged in. The technician turns the test cord to start, then releases the knob (as per directions), and the compressor starts. This narrows the problem to the compressor start circuit. The compressor is turned off because it is running by itself, no fan.

The technician removes the terminals from the start relay and checks the coil circuit, from terminals 2 to 5; it has an open circuit, Figure 46–75. The relay coil is defective, keeping the unit in start. The capacitors are checked as explained in Unit 19 and they are satisfactory.

The relay is changed and the compressor is wired back into the circuit. The unit is plugged in again and the selector switch turned to HIGH COOL. The compressor starts and runs. The amperage is normal when compared to amperage indicated on the nameplate.

SERVICE CALL 7

A customer called to report that a window unit servicing the den and kitchen tripped the breaker. This is a fairly new unit, 6 years old. The owner is told to shut the unit off and to not reset the breaker. *The problem is the unit has a grounded compressor.*

The technician arrives and decides to give the unit an ohm test before trying to reset the breaker and start the unit. The ohmmeter selector switch is set to R × 1 and the meter is zeroed. The technician unplugs the unit and fastens one ohmmeter lead to each of the hot plugs on the 230-V plug, Figure 46–76. The unit selector switch is turned to HIGH COOL. There is a measurable resistance; all appears to be well. Before plugging the unit in, one more ohm test is necessary for the complete picture, a ground test. The ohmmeter is set at 0 on the R × 1000 scale and one lead is moved to the ground terminal. The meter reads 0 resistance; there is a ground in the unit somewhere, Figure 46–77. It could be the fan motor or the compressor. The unit is pulled from its case. The compressor terminals are disconnected. One lead of the meter is touched to the discharge line and the other to the common terminal; the meter reads 0 resistance, the compressor has the ground circuit.

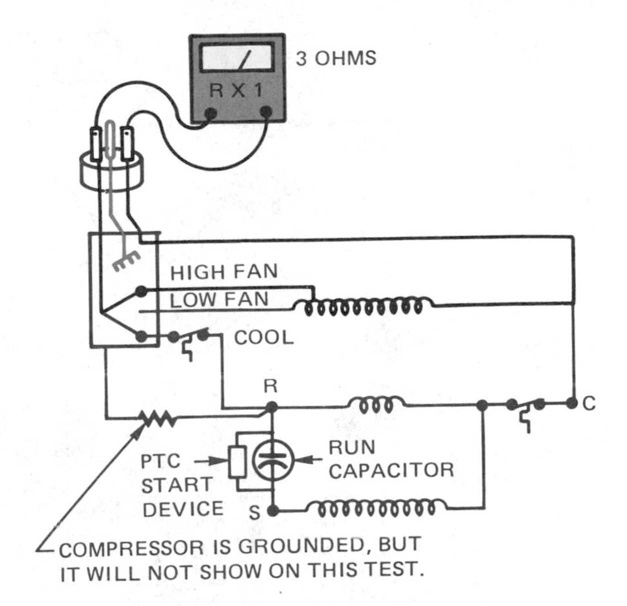

FIGURE 46–76 The unit has a ground circuit according to the ohmmeter.

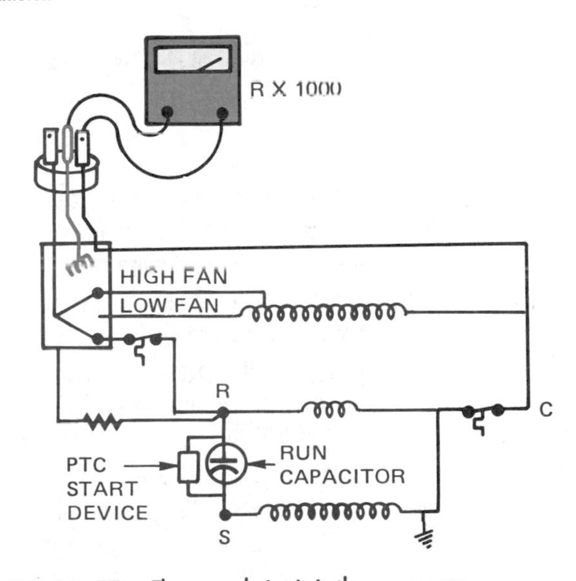

FIGURE 46–77 The ground circuit is the compressor.

The customer is informed of the problem and given a price to repair the unit. The customer decides to have it repaired because the repair cost is much less than purchasing a new unit.

The technician moves the unit to the truck and takes it to the shop. **The technician decides to recover the refrigerant from the unit to an approved cylinder for contaminated refrigerants. It may be burned and contaminated.**

The technician obtains a liquid-line drier, a suction-line drier, a 1/4-in. "tee" fitting for the liquid line, a compressor and a 1/4-in. fitting for the process tube. When the refrigerant is removed from the unit, the technician cuts the compressor discharge line at the compressor with diagonal pliers. Then the suction line is cut with a small tube cutter and the compressor is removed from the unit. A "tee" fitting is installed in the liquid line. A gage line is fastened to the "tee" fitting and nitrogen is purged through the unit, Figure 46–78. This will push much of the contamination from the unit.

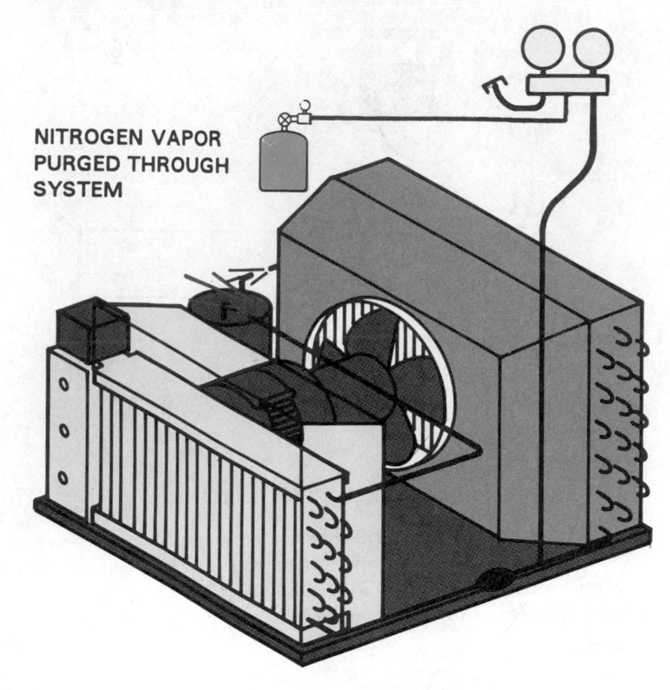

FIGURE 46–78 The system is purged using the "tee" fitting installed in the liquid line.

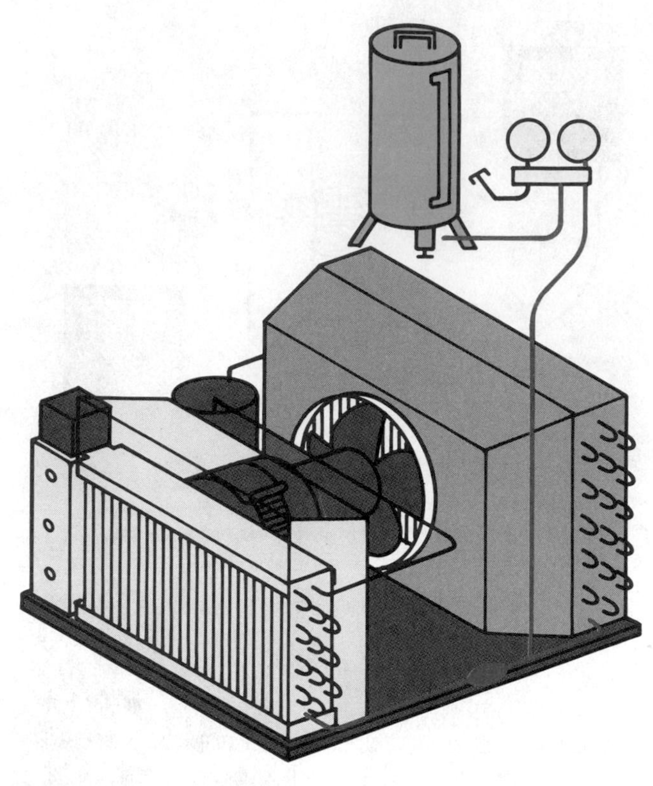

FIGURE 46–79 Liquid is charged into the system; this saves time.

The technician sets the new compressor in place and solders the discharge line to the new compressor. The suction line is cut to the correct configuration for the suction-line drier. The liquid line is now cut, just before the strainer entering the capillary tube, and the new small drier is installed. The suction-line drier and the 1/4-in. fitting for the process tube are soldered in place. The gages are fastened to the process port and the "tee" fitting in the liquid line. The unit is pressured with a small amount of R-22 (to 5 psig) and nitrogen is added to push the pressure up to 150 psig. The unit is then leak checked with an electronic leak detector. After the unit is leak checked, the trace refrigerant and nitrogen are purged from the system.

The vacuum pump is connected after the leak check and started. While the unit is being evacuated, the capacitors and relay are checked. Both capacitors, run and start, are good. The relay has a measurable resistance in the coil and the contacts look good.

The charge for the unit is 18 oz, plus the capacity of the small liquid-line drier 1.9 oz = 19.9 oz total (this information should be packed in the drier directions but all manufacturers do not furnish it). When a 500-micron vacuum is obtained according to the micron gage the charge is added in the following manner.

The charge is in a charging cylinder, such as explained in Unit 9. This charging cylinder has plenty of refrigerant and a liquid valve. The high-pressure gage line is connected to the liquid line on the unit, and the charge is allowed to enter the system through the liquid line in the liquid state, Figure 46–79. When the complete charge is inside the unit, the liquid-line valve is pinched off and soldered shut.

The unit is started and allowed to run while the technician is monitoring the suction pressure and amperage. The unit is ready to be returned to the customer.

SUMMARY

- Room air conditioners may be designed to be installed in windows or through the wall.
- These units may be designed for cooling only, cooling and heating with electric strip heat, or cooling and heating with a heat pump cycle.
- The four major components in the cooling cycle for these units are the evaporator, compressor, condenser, and metering device.
- The metering device in most units is the capillary tube.
- Most units are designed to produce a heat exchange between the capillary tube and the suction line. This adds superheat to the suction line and subcools the refrigerant in the capillary tube.
- The heat pump cycle uses the four-way valve to redirect the refrigerant.
- The heat pump compressor is different from one found in a cooling only unit. It must have enough pumping capacity for low-temperature operation.
- Window units may be installed in either double-hung or casement windows. Those installed in single casement windows must be designed for that purpose.
- Before installing a unit, the technician should make sure that the electrical service is adequate.
- When installing these units, ensure that there is no possibility of direct recirculation of air through the unit.
- The selector switch is the primary control for room air conditioning units. It switches the unit from cooling to heating, high and low cooling, high and low heating. There is also typically a control for inside or fresh air.

- The primary maintenance for these units involves keeping the filters and coils clean.
- Gages should be installed for service only when it is determined that it is absolutely necessary.
- Electrical service may involve the fan motor, thermostat, selector switch, compressor, and the power cord.

REVIEW QUESTIONS

1. Describe the difference between a window air conditioning unit and a through the wall unit.
2. Describe the two types of window unit cases.
3. What are the two methods of heating with window or through the wall units?
4. What are the four major components used in the cooling cycle?
5. Briefly describe the cooling cycle in a window unit.
6. What materials are the evaporator tubing and fins typically made of?
7. What type of metering device is generally used in these units?
8. Describe how the heat exchange occurs between the suction line and the metering device.
9. What is the component that causes the reverse cycle in the heat pump of a room air conditioner?
10. Briefly describe the heat pump heating cycle.
11. Why is the heat pump compressor different from a cooling only compressor?
12. What two types of windows can a window unit be installed in?
13. What is the problem if the window unit is installed so that the air recirculates too quickly through the evaporator?
14. What is a wall sleeve when used with a through the wall unit?
15. What does the selector switch normally control?
16. What are the two major maintenance requirements on room air conditioning units?
17. Describe the test the technician should perform when a low charge is suspected before installing gages.
18. Describe the conditions at the evaporator if the charge is correct but the compressor is not pumping to capacity.
19. When trying to detect a leak, what substance may be used to increase the refrigerant pressure?
20. What two types of expansion valves may be substituted in some cases for the capillary tube?

SECTION

10 *Chilled-Water Air Conditioning Systems*

UNIT 47

High-Pressure, Low-Pressure, and Absorption Chilled-Water Systems

OBJECTIVES

After studying this unit, you should be able to

- list different types of chilled-water air conditioning systems.
- describe how chilled-water air conditioning systems operate.
- state the types of compressors often used with high-pressure refrigerant water chillers.
- describe the operation of a centrifugal compressor in a high-pressure chiller.
- explain the difference between direct expansion and flooded chiller evaporators.
- explain what is meant by approach temperature in a water-cooled condenser.
- state two types of condensers used in chilled-water systems.
- explain subcooling.
- list the types of metering devices used in high-pressure chillers.
- list the types of refrigerants typically used in low-pressure chillers.
- state the type of compressor used in low-pressure chiller systems.
- describe the metering devices used in low-pressure chiller systems.
- explain the purge system used on a low-pressure chiller condenser.
- describe the absorption cooling system process.
- state the refrigerant generally used in large absorption chillers.
- state the compound normally used in salt solutions in large absorption chillers.
- state the type of electric motors typically used on chiller air conditioning systems.
- discuss the various start mechanisms for these motors.
- describe a load-limiting device on a chiller motor.
- discuss various motor overload protection devices and systems.

SAFETY CHECKLIST

* Wear gloves and use caution when working around hot steam pipes and other heated components.
* Use caution when working around high-pressure systems. Do not attempt to loosen fittings or connections when system is pressurized. Follow recommended procedures when using nitrogen.
* Follow all recommended safety procedures when working around electrical circuits.
* Lock and tag disconnect boxes or panels when power has been turned off to work on an electrical system. There should be one key for this lock and it should be in the technician's possession.
* Never start a motor with the door to the starter components open.
* Check pressures regularly as indicated by the manufacturer of the system to avoid excessive pressures.

Chilled-water systems are used for larger applications for central air conditioning because of the ease with which chilled water can be circulated in the system. If refrigerant were piped to all floors of a multistory building, there would be too many possibilities for leaks to occur, in addition to the expense of the refrigerant to charge the system.

The design temperature for boiling refrigerant in a coil used for cooling air is 40°F. If water can be cooled to approximately the same temperature, it can also be used to cool or condition air, Figure 47–1. This is the logic used for circulating chilled-water systems. Water is cooled to about 45°F and circulated throughout the building to air-heat exchange coils that absorb heat from the building air. When

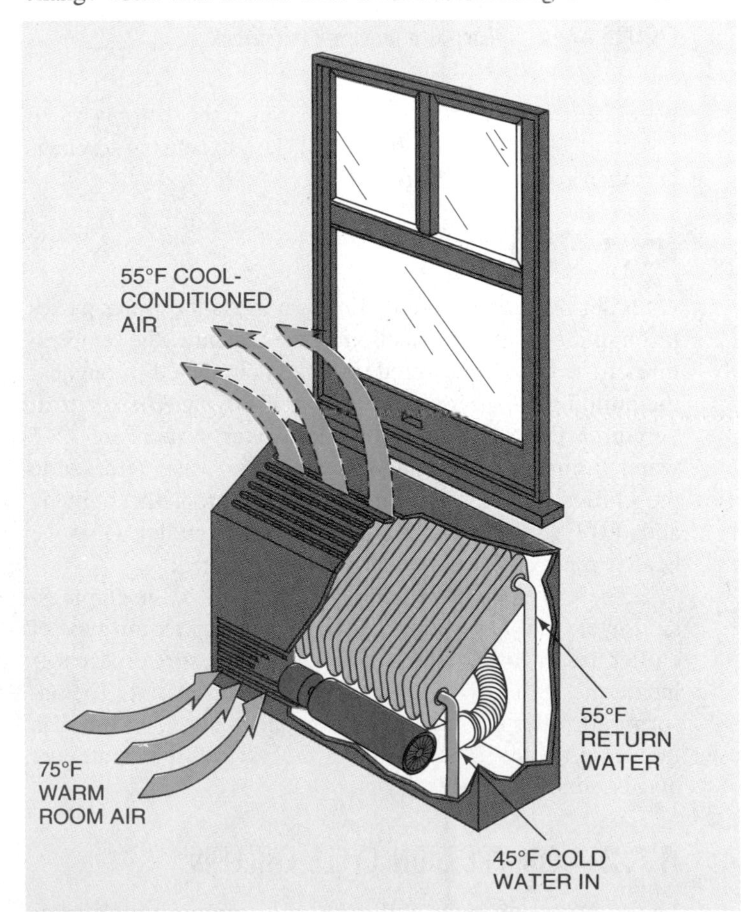

FIGURE 47–1 Water in a fan coil unit for cooling a room.

FIGURE 47–2 Water as a secondary refrigerant.

water is used for circulation in a building, the water is called a secondary refrigerant. It is much less expensive to circulate than refrigerant, Figure 47–2.

47.1 CHILLERS

A chiller refrigerates circulating water. As the water passes through the evaporator section of the machine, the temperature of the water is lowered. It is then circulated throughout the building where it picks up heat. The typical design temperatures for a circulating chilled-water system are 45°F water furnished to the building and 55°F water returned to the chiller from the building. The heat from the building adds 10°F to the water that returns to the chiller. Here the heat is removed and the water is recirculated.

There are two basic categories of chillers, the compression cycle and absorption chiller. The compression type of chiller uses a compressor to provide the pressure differences inside the chiller to boil and condense refrigerant. The absorption chiller uses a salt solution and water to accomplish the same results. These chillers are very different and are discussed separately.

47.2 COMPRESSION CYCLE CHILLERS

The compression cycle chiller has the same four basic components as a typical air conditioner: a *compressor*, an *evaporator*, a *condenser*, and a *metering device*. However, these components are generally larger to be able to handle more refrigerant and they may use a different refrigerant.

The heart of the compression cycle refrigeration system is the compressor. As mentioned in Unit 3, there are several types of compressors. The water chiller may use one of several of these types. The compressors common to water chillers are the *reciprocating*, *scroll*, *screw*, and *centrifugal*. Photos of these compressors can be found in Unit 23. The compressor can be thought of as a "**vapor pump**." The technician should think of the compressor as a component in the line that lowers the evaporator pressure to the desired boiling point of the refrigerant. Typically this is about 38°F for a chiller. It then builds the pressure in the condenser to the point that vapor will condense to a liquid for reuse in the evaporator. The typical condensing temperature is 105°F. The technician can use these temperatures to determine if a typical chiller is operating within the design parameters. The compressor must be of a design that will pump and compress the vapor to meet the needs of a particular installation.

Compression cycle chillers may be classified as either **high-pressure** or **low-pressure systems**. Following is a discussion of high-pressure refrigerant water chillers.

47.3 RECIPROCATING COMPRESSOR CHILLERS

Large reciprocating compressors used for water chillers operate the same as those for any other reciprocating compressor application with a few exceptions. A review of Unit 23 will help you to understand how compressors function. These compressors range in size from about 1/2 hp to approximately 150 hp, depending on the application. Most manufacturers have stopped using one large compressor for a large reciprocating chiller and have started using multiple small compressors. They are positive-displacement compressors and cannot pump liquid refrigerant without risk of damage to the compressor.

Several refrigerants have been used for reciprocating compressor chillers; R-500, R-502, R-12, R-134A, and R-22 are the most common. R-22 is by far the most popular.

The large reciprocating compressor will have many cylinders to produce the pumping capacity needed to move large amounts of refrigerant. Some of these compressors have as many as 12 cylinders. This becomes a machine with many moving parts and much internal friction. If one cylinder of the compressor fails, the whole system is off the line. With multiple compressors, if one compressor fails, the others can carry the load. Multiple compressors give some backup from total failure. For this reason and because of capacity control, many manufacturers use multiple compressors of a smaller size.

All large chillers must have some means for controlling the capacity or the compressor will cycle on and off. This is not satisfactory because most compressor wear occurs during start-up before oil pressure is established. A better design approach is to keep the compressor on the line and operate it at reduced capacity. Reduced capacity operation also smooths out temperature fluctuations that occur from shutting off the compressor and waiting for the water to warm up to bring the compressor back on.

47.4 CYLINDER UNLOADING

Reduced capacity for a reciprocating compressor is accomplished by *cylinder unloading*. For example, suppose a 100-ton compressor with 8 cylinders is used for the chiller for a large office building and the chiller has 12.5 tons of capacity per cylinder. When all 8 cylinders are pumping, the compressor has a capacity of 100 tons (8 × 12.5 = 100). As the cylinders are unloaded, the capacity is reduced. For example, the cylinders may unload in pairs, which would be 25 tons per unloading step. The compressor may have three steps of unloading so the compressor has three different capacities, 100 tons (8 cylinders pumping), 75 tons (6 cylinders pumping), 50 tons (4 cylinders pumping), and 25 tons (2 cylinders pumping). In the morning when the system first starts, the building may only need 25 tons of cooling. As the temperature outside rises, the chiller may need more capacity and the compressor will automatically load 2 more cylinders for 50 tons of capacity. As the temperature rises, the compressor can load up to 100% capacity or 100 tons. If the building stays open at night, such as a hotel, the compressor will start to unload as the outside temperature cools. It will unload down to 25 tons; then if this is too much capacity, the chiller will shut off. When the chiller is restarted, it will start up at the reduced capacity, lowering the starting current. A compressor cannot be unloaded to 0 pumping capacity or it would not move any refrigerant through the system to return oil that is in the system. Usually compressors will unload down to 25% to 50% of their full-load pumping capacity.

Another big advantage of cylinder unloading is that the power to operate the compressor is reduced as the capacity is reduced. The reduction in power consumption is not in direct proportion with a compressor capacity, but the power consumption is greatly reduced at part load.

Cylinder unloading is accomplished in several ways; *blocked suction* and *lifting the suction valve* are the most common.

Blocked Suction

Blocked suction is accomplished with a solenoid valve in the suction passage to the cylinder to be unloaded, Figure 47–3. If the refrigerant gas cannot reach the cylinder, no gas is pumped. If a compressor has 4 cylinders and the suction gas is blocked to one of the cylinders, the capacity of the compressor is reduced by 25% and the compressor then pumps at 75% capacity. The power consumption also goes down approximately 25%. Power consumption is related to the amperage draw of the compressor. Amperage is typically measured in the field in amperes using a clamp-on ammeter. When a compressor is running at half capacity, the amperage will be about half full-load amperage. The power consumption of the compressor is actually measured in watts. Using amperage for a measure of compressor capacity is close enough for field troubleshooting.

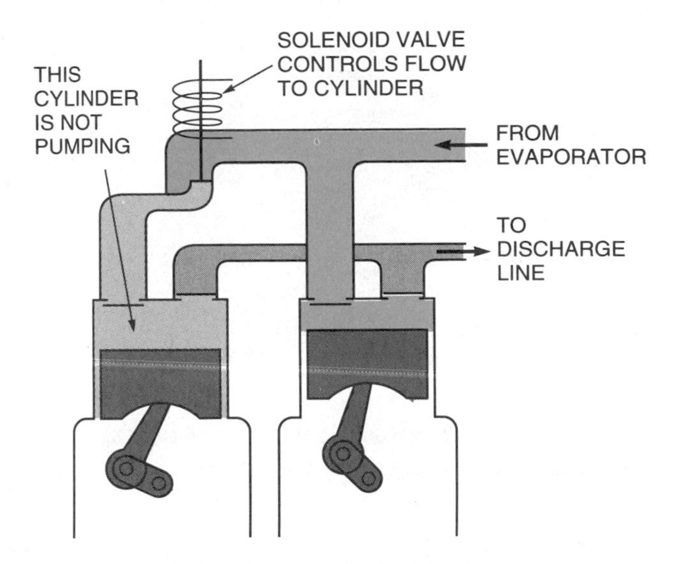

FIGURE 47–3 Blocked suction unloader for a reciprocating compressor.

Suction Valve Lift Unloading

If the suction valve is lifted off the seat of a cylinder while the compressor is pumping, this cylinder will quit pumping. Gas that enters the cylinder will be pushed back out into the suction side of the system on the upstroke. There is no resistance to pumping the refrigerant back into the suction side of the system, so it requires no energy. The power consumption will be reduced as in the example of the blocked suction. One of the advantages of lifting the suction valve is that the gas that enters the cylinder will contain oil and good cylinder lubrication will occur even while the cylinder is not pumping.

Compressor unloading could be accomplished by letting hot gas back into the cylinder, but this is not practical because power reduction will not occur. When the gas has been pumped from the low-pressure side to the high-pressure side of the system, the work has been accomplished. Also the cylinder would become overheated by compressing the hot gas again.

Except for cylinder unloading, the reciprocating chiller compressor is the same as smaller compressors. All compressors over 5 hp have pressure-lubricating systems. The pressure for lubricating the compressor is an oil pump that is typically mounted on the end of the compressor shaft and driven by the shaft, Figure 47–4. The oil pump picks up oil in the sump at evaporator pressure and delivers the oil to the bearings at about 30 to 60 psig greater than suction pressure. This is called net oil pressure, Figure 47–5. These compressors also have an oil safety shutdown in case of oil pressure failure that has a time delay of about 90 sec to allow the compressor to get started and establish oil pressure before it shuts the compressor off.

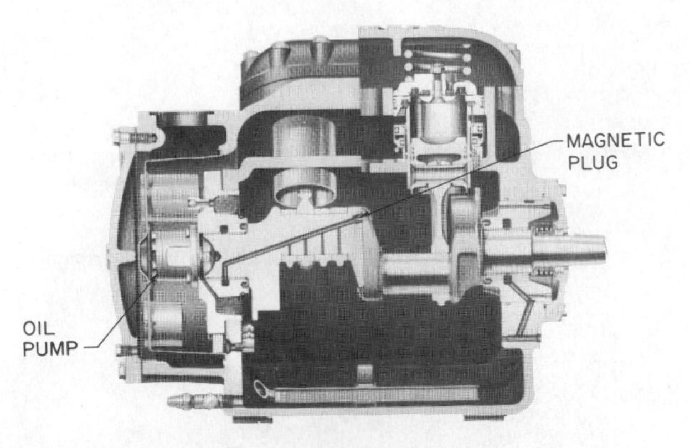

FIGURE 47–4 Pressure lubrication system for a reciprocating compressor using an oil pump. *Courtesy Trane Company*

47.5 SCROLL COMPRESSOR CHILLERS

The scroll compressor is a positive-displacement compressor. The scroll compressors applied to chillers are larger than those scroll compressors described earlier in this text. These compressors are in the 10- to 15-ton range but operate the same as the smaller compressors. See Figure 47–6 for an illustration of a scroll compressor. They are welded hermetic compressors. When these compressors are used in chillers, the capacity control of the chiller is maintained by cycling the compressors off and on in increments of 10 and 15 tons. For example, a 25-ton chiller would have a 10- and a 15-ton compressor and would have capacity control of 10 and 15 tons. A 60-ton chiller may have four 15-ton compressors and be able to operate at 100% (60 tons), 75% (45 tons), 50% (30 tons), and 25% (15 tons).

Some of the advantages of the scroll compressor:

1. Efficiency
2. Quiet
3. Fewer moving parts

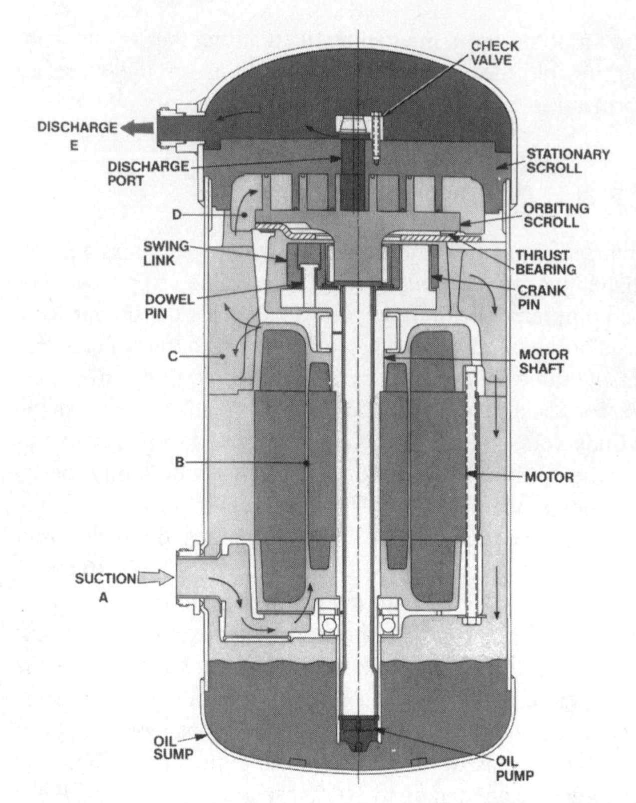

FIGURE 47–6 Check valve in a scroll compressor. *Courtesy Trane Company*

4. Size and weight
5. Can pump liquid refrigerant without compressor damage

The scroll compressor offers little resistance to refrigerant flow from the high side of the system to the low side during the off cycle. A check valve is provided to prevent backward flow when the system is shut down, Figure 47–6.

Lubrication for the scroll compressor is provided by an oil pump at the bottom of the crankshaft. The oil pump picks up oil and lubricates all moving parts on the shaft.

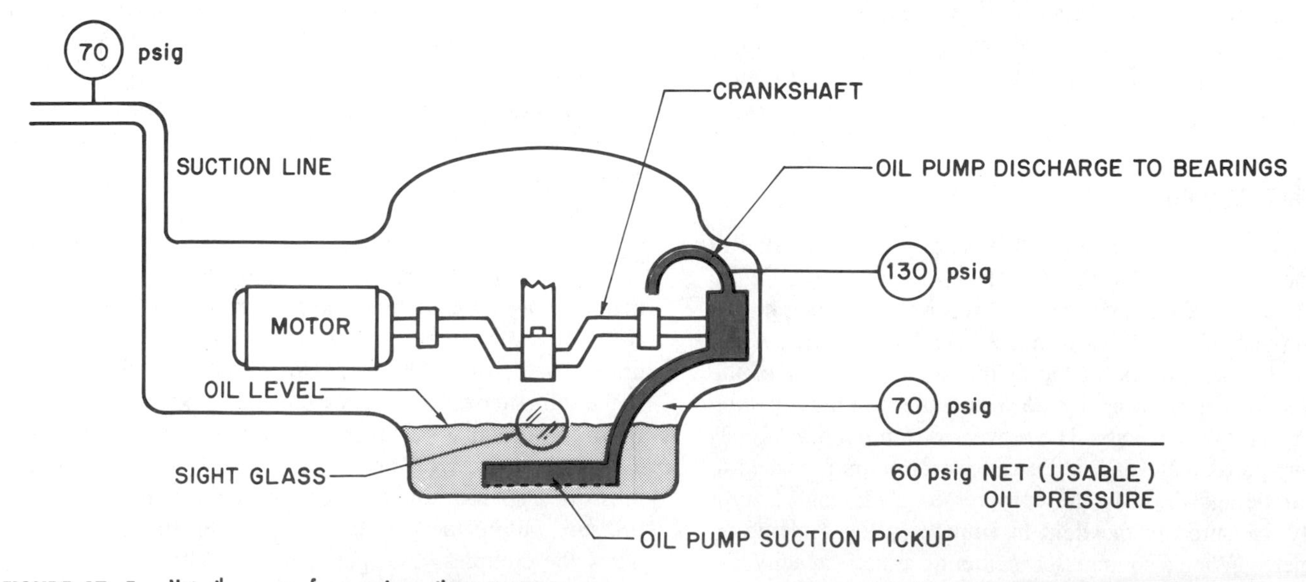

FIGURE 47–5 Net oil pressure for a reciprocating compressor.

47.6 ROTARY SCREW COMPRESSOR CHILLERS

Most all of the major manufacturers are building rotary screw compressors for larger capacity chillers. This is larger capacity equipment using high-pressure refrigerants. The rotary screw compressor is capable of handling large volumes of refrigerant with few moving parts, Figure 47–7. This type of compressor is a positive-displacement compressor that has the characteristic of being able to handle some liquid refrigerant without compressor damage. This is unlike the reciprocating compressor that cannot handle liquid refrigerant. Rotary screw compressors are manufactured in sizes from about 50 to 700 tons of capacity. These compressors are reliable and trouble free.

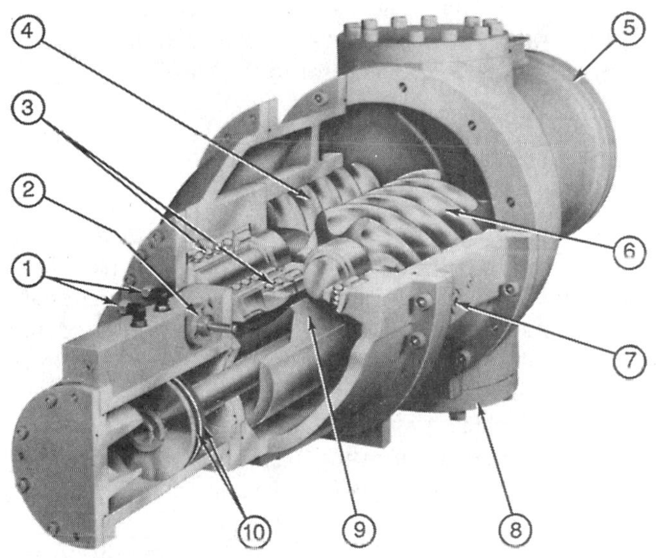

1 - Control Oil Lines
2 - Capacity Control Solenoid Valve
3 - Discharge Bearingh Assemblies
4 - Male Rotor
5 - Semi-Hermetic Motor
6 - Female Rotor
7 - Rotor Oil Injection Port
8 - Suction Inlet Flange
9 - Capacity Control Slide Valve
10 - Slide Piston Seals

FIGURE 47–7 The rotary screw compressor has few moving parts. *Reproduced courtesy of Carrier Corporation*

Manufacturers build both *semihermetic rotary screw compressors* and *open-drive compressors*. The open-drive models are direct drive and must have a shaft seal to contain the refrigerant where the shaft penetrates the compressor shell.

Capacity control is accomplished with a rotary screw compressor by means of a *slide valve* that blocks the suction gas before it enters the rotary screws in the compressor. This slide valve is typically operated by differential pressure in the system. The slide valve may be moved to the completely unloaded position before shutdown so that on start-up, the compressor is unloaded, reducing the inrush current on start-up. Most of these compressors can function from about 10% load to 100% load with sliding graduations because of the nature of the slide valve unloader. This is as opposed to the step capacity control of a reciprocating compressor, which unloads 1 or 2 cylinders at a time.

The nature of the screw compressor is to pump a great deal of oil while compressing refrigerant so these compressors typically have an oil separator to return as much oil to the compressor reservoir as possible, Figure 47–8. The oil is moved to the rotating parts of the compressor by means of pressure differential within the compressor instead of an oil pump. Oil is also accumulated in the oil reservoirs and moved to the parts that need lubrication by means of gravity. The rotating screws are close together but do not touch. The gap between the screws is sealed by oil that is pumped into the rotary screws as they turn. This oil is separated from the hot gas in the discharge line and returned to the oil pump inlet through an oil cooler.

This oil separation is necessary because if too much oil reaches the system, a poor heat exchange will occur in the evaporator and loss of capacity will occur. The natural place

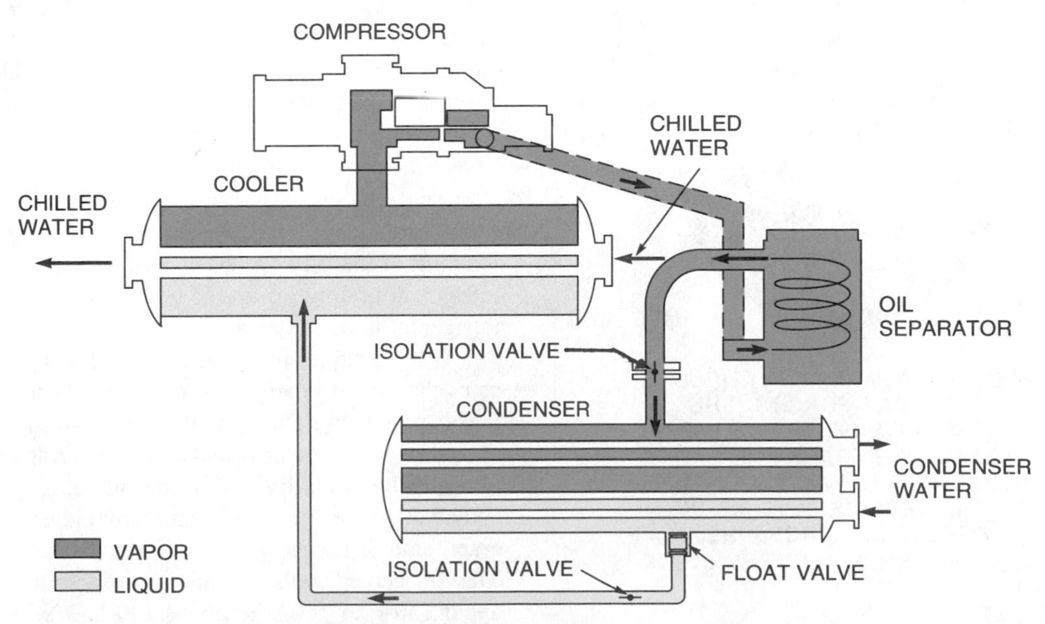

FIGURE 47–8 Oil separator for a rotary screw compressor. *Reproduced courtesy of Carrier Corporation*

to separate the oil from the refrigerant is in the discharge line. Different manufacturers use different methods but all of them have some means of oil separation.

47.7 CENTRIFUGAL COMPRESSOR CHILLERS (HIGH PRESSURE)

The centrifugal compressor uses only the centrifugal force applied to the refrigerant to move the refrigerant from the low- to the high-pressure side of the system, Figure 47–9. It is like a large fan that creates a pressure difference from one side of the compressor to the other. It does not have a great deal of force, but some companies manufacture centrifugal compressors for use in high-pressure systems. Centrifugal compressors can handle a large volume of refrigerant. To provide the pressure difference from the evaporator to the condenser for high-pressure systems the compressor is turned very fast by means of a gear box or multiple stages of compression. Typical motor speeds used for direct-drive reciprocating, scroll, and rotary screw compressors are near 3600 rpm. When faster compressor speeds are needed for centrifugal systems, a gear box is used to speed the compressor up to higher rpm levels. Speeds of about 30,000 rpm are used for some single-stage centrifugal compressors.

If the head pressure becomes too high or the evaporator pressure too low, the compressor cannot overcome the pressure difference and it quits pumping. The motor and compressor still turn, but refrigerant stops moving from the low- to the high-pressure side of the system. The compressor may make a loud whistling sound. This is called a "surge." Even though this is a loud noise, it will normally not cause damage to the compressor or motor unless it is allowed to continue for a long period of time. If damage does occur due

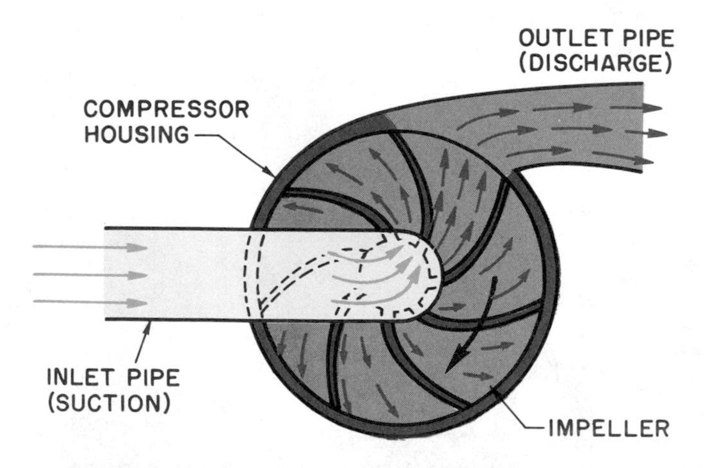

THE TURNING IMPELLER IMPARTS CENTRIFUGAL FORCE ON THE REFRIGERANT FORCING THE REFRIGERANT TO THE OUTSIDE OF THE IMPELLER. THE COMPRESSOR HOUSING TRAPS THE REFRIGERANT AND FORCES IT TO EXIT INTO THE DISCHARGE LINE. THE REFRIGERANT MOVING TO THE OUTSIDE CREATES A LOW PRESSURE IN THE CENTER OF THE IMPELLER WHERE THE INLET IS CONNECTED.

FIGURE 47–9 Centrifugal action used to compress refrigerant.

to prolonged surge times, it is likely to be to the thrust surface of the bearings or to the high-speed gear box.

The gear box used to obtain the high speeds of the centrifugal compressor adds some friction to the system that must be overcome and causes a slight loss of efficiency due to the horsepower needed to turn the gears. The gear box typically has two gears; the drive gear is the larger and the driven gear is the small gear that turns the fastest. Figure 47–10 is an illustration of a gear box.

The clearances in the impeller of a centrifugal compressor are critical, Figure 47–11. The parts must be very close together or refrigerant will bypass from the high-pressure side of the compressor to the low-pressure side.

Lubrication is accomplished with a separate motor and oil pump. This motor is a three-phase fractional horsepower (usually 1/4 hp) and is located inside the oil sump. It runs in the refrigerant and oil atmosphere. A three-phase motor is used so there will be no motor internal start switch as in a single-phase motor. This motor allows the manufacturer to start the oil pump before the centrifugal motor and establishes lubrication before the compressor and gear box start to turn. It also allows the oil pump to run during coast down of the centrifugal impeller and gear box at the end of a cycle when the machine is shut off. There is enough oil in reservoirs for a coast down in the event of a power failure to prevent bearing and gear failure. The oil pump is usually a gear type positive-displacement oil pump. It would normally furnish a net oil pressure of about 15 psig to the bearings and with some compressors to an upper sump where oil is fed by gravity to the bearings and gear box during a coast down because of power failure, Figure 47–10. There are many methods of routing the oil, but the oil must have enough pressure to reach the farthest bearings with enough pressure to circulate the oil around the bearings.

The lubrication system has a heater in the oil sump to prevent liquid refrigerant from migrating to the oil sump when the system is off, Figure 47–10. The heater typically maintains the oil sump at about 140°F. Without this heater the oil would soon become saturated with liquid refrigerant. Even with the heater operating, some liquid will migrate to the oil sump and when the compressor is started, the oil may have a tendency to foam for a few minutes immediately after start-up. The operator should keep an eye on the oil sump and run the compressor at reduced load until any foaming in the oil is reduced. Foaming oil contains liquid refrigerant boiling off and is not a good lubricant until the refrigerant is boiled away.

When a chiller is operating, the oil is heated as it lubricates the moving parts. It also absorbs heat from the bearings to cool them. The oil will become overheated if it is not cooled so an oil cooler is in the circuit during operation. The oil typically is pumped from the oil sump through a filter and an oil cooler heat exchanger with either water or refrigerant removing some of the heat from the oil, Figure 47–10. This oil cooler cools the oil from a sump temperature of about 140°F to 160°F to about 120°F. It will then pick up heat again while lubricating.

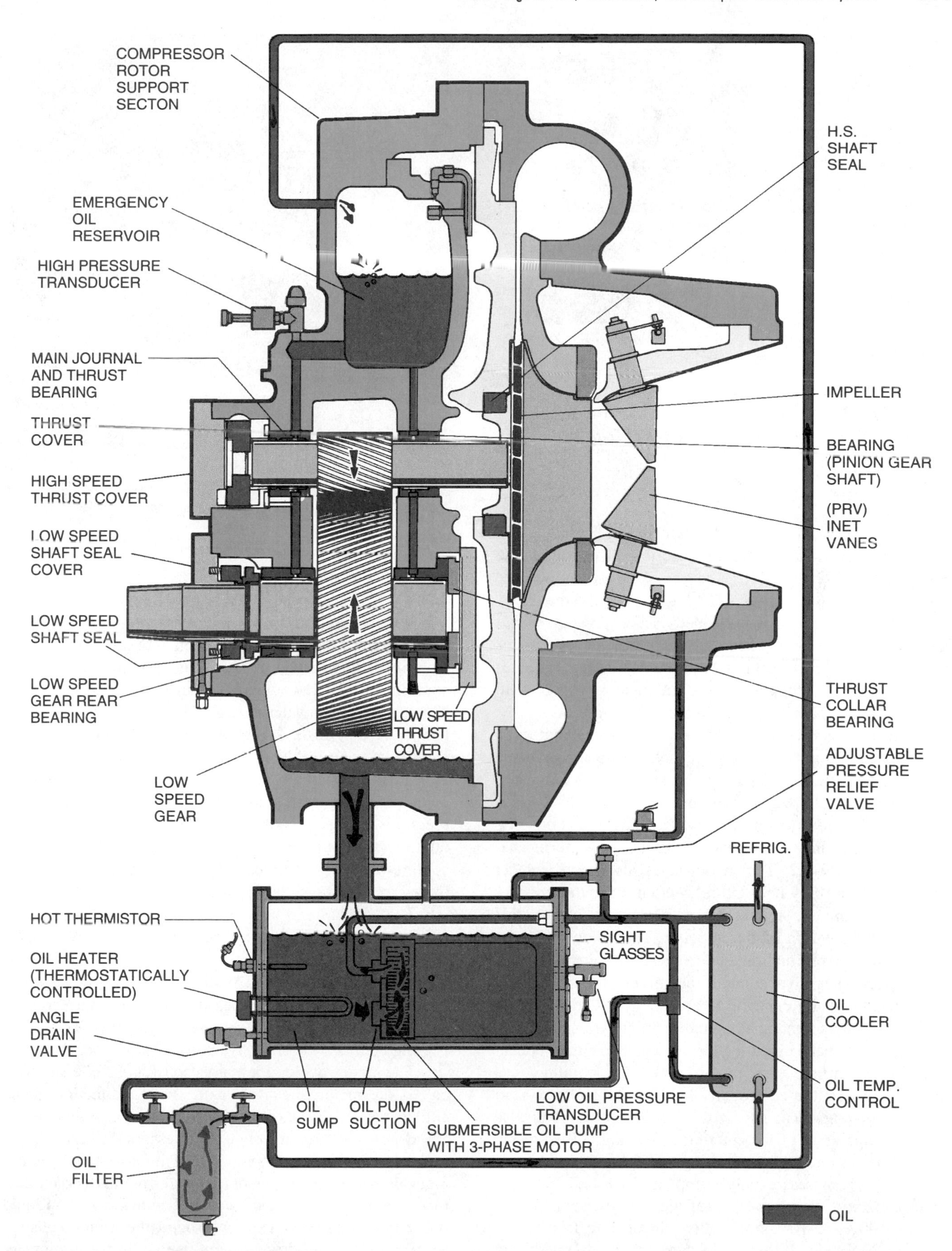

FIGURE 47–10 Gear box for a high-speed centrifugal compressor. *Courtesy York International*

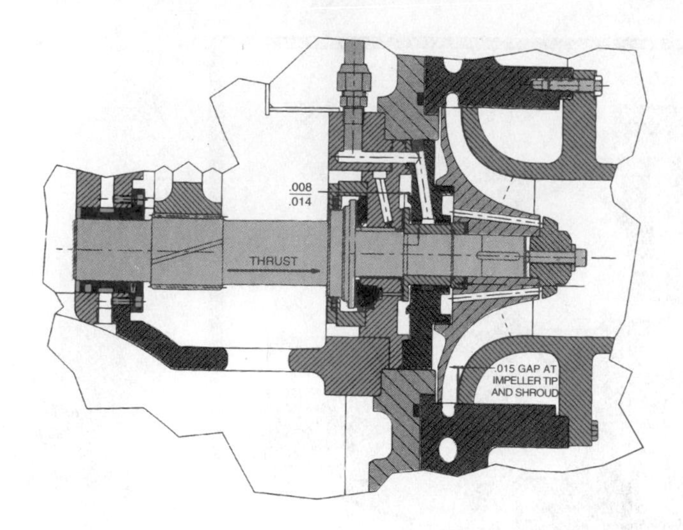

FIGURE 47–11 The parts in the high-speed centrifugal compressor have a very close tolerance. *Reproduced courtesy of Carrier Corporation*

FIGURE 47–12 Prerotation guide vanes for capacity control in a centrifugal compressor. *Courtesy York International*

The lubrication system for a centifugal chiller is a sealed system and oil is not intended to be mixed with the refrigerant and separated for recovery like the reciprocating, scroll, or rotary screw systems. If the oil gets into the refrigerant, it is difficult to remove because only vapor should leave the evaporator. When refrigerant in the evaporator becomes oil logged or saturated, the heat exchange becomes less efficient. Oil-logged refrigerant must be distilled to remove the oil. This is a long, time-consuming process. The intent of the manufacturer is to keep the two separated.

Capacity control is accomplished by means of guide vanes at the entrance to the impeller eye. These guide vanes are usually pie-shaped devices located in a circle that can rotate to allow full flow or reduced flow down to about 15% or 20%, depending on the clearances in the vane mechanism, Figure 47–12. The guide vanes also serve another purpose in that they help the refrigerant enter the impeller eye by starting the refrigerant in a rotation pattern that matches the rotation of the impeller. The guide vanes are often called the prerotation guide vanes. They are controlled using electronics or pneumatic (air-driven) motors to vary the capacity of the centrifugal compressor. The controller sends either a pneumatic air signal or an electronic signal to the inlet guide vane operating motor, which is either a pneumatic or electronic type motor. When the building load starts to satisfy, the controller tells the vane motor to start closing the vanes. The capacity control motor is controlled by the temperature of the entering or leaving chilled water, depending on the engineer's design. If the guide vanes are closed, the compressor only pumps at 15% to 20% of its rated capacity. When open all of the way, the compressor pumps 100%. A 1000-ton chiller can operate from about 150 to 200 tons to 1000 tons of capacity in any number of steps. The guide vanes can also be used for two other purposes—to prevent motor overload and to start the motor at reduced capacity to reduce the current on start-up.

When a chiller is operating at a condition where the chilled water is above the design temperature, for example, when a building is hot on Monday morning start-up, the compressor motor would run at an overloaded condition. The return chilled water may be 75°F instead of the design temperature of 55°F. The chiller control panel will have a control known as a *load limiter* that will sense the motor amperage and partially close the guide vanes to limit the compressor amperage to the full-load value. When the compressor is started, the prerotation guide vanes are closed and do not open until the motor is up to speed. Therefore, the compressor starts up unloaded. This reduces the power required to start the compressor.

The gear box and impeller have bearings to support the weight of the turning shaft. These bearings are typically a soft material known as babbitt. The babbitt is backed and supported by a steel sleeve. With a single-stage compressor, there must also be a thrust bearing to counter the thrust of the refrigerant entering the compressor impeller. This thrust bearing counters the sideways movement of the shaft.

The motor rotation for many centrifugal chillers is critical. If the motor is started in the wrong direction, some chillers can be damaged, as discussed later in this unit.

Some of these centrifugal compressors are hermetically sealed and some have open drives, Figure 47–13. The hermetically driven compressors have refrigerant-cooled motors. Figure 47–14 shows a liquid-cooled motor. Liquid refrigerant is allowed to flow around the motor housing. Systems with open-drive motors use air in the equipment room to cool the motor. The heat from a large motor is considerable and must be exhausted from the equipment room.

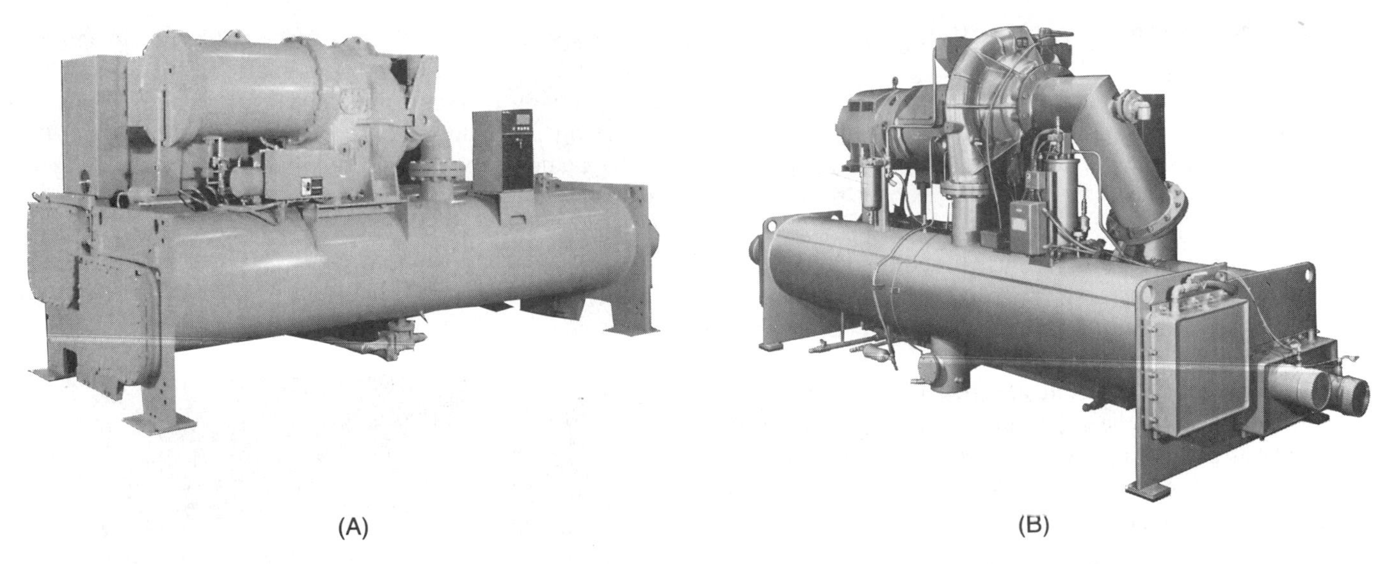

FIGURE 47–13 (A) Hermetically sealed centrifugal chiller. (B) Open-drive centrifugal chiller. *(A) Reproduced courtesy of Carrier Corporation, (B) Courtesy York International*

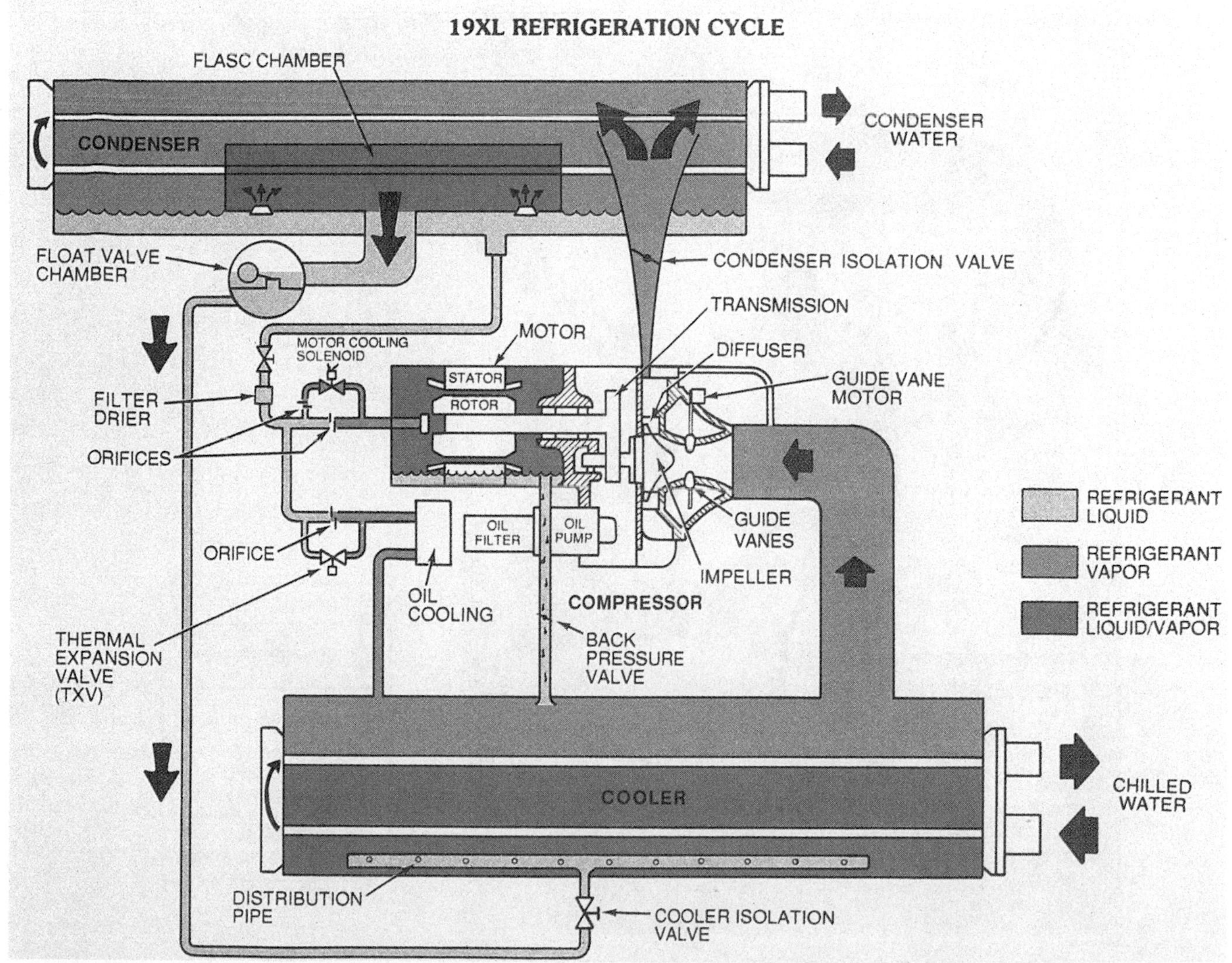

FIGURE 47–14 Refrigerant-cooled hermetic centrifugal compressor. *Reproduced courtesy of Carrier Corporation*

47.8 EVAPORATORS FOR HIGH-PRESSURE CHILLERS

The *evaporator* is the component that absorbs heat into the system. Liquid refrigerant is boiled to a vapor by the circulating water.

The heat exchange surface is typically copper for water chillers. Other materials may be used if corrosive fluids are circulated for manufacturing processes. In typical smaller air conditioning systems, air is on one side of the heat exchange process and liquid or vapor refrigerant on the other. The rate of heat exchange between air and vapor refrigerant is fair. The heat exchange between air and liquid refrigerant is better and the best heat exchange is between water and liquid refrigerant. This is where the water chiller gets part of its versatility. The heat exchange surface can be small and still produce the desired results.

The evaporators used for high-pressure chillers are either direct expansion evaporators or flooded evaporators.

47.9 DIRECT EXPANSION EVAPORATORS

Direct expansion evaporators are also known as dry type evaporators, meaning they have an established superheat at the evaporator outlet. They normally use thermostatic expansion devices for metering the refrigerant. Direct expansion evaporators are used for smaller chillers, up to approximately 150 tons on older types and to about 50 tons for more modern chillers. Direct expansion chillers introduce the refrigerant into the end of the chiller barrel with the water being introduced into the side of the shell, Figure 47–15. The water is on the outside of the tubes and baffles cause the water to be in contact with as many tubes as possible for good heat exchange. The problem with this arrangement is if the water side of the circuit ever gets

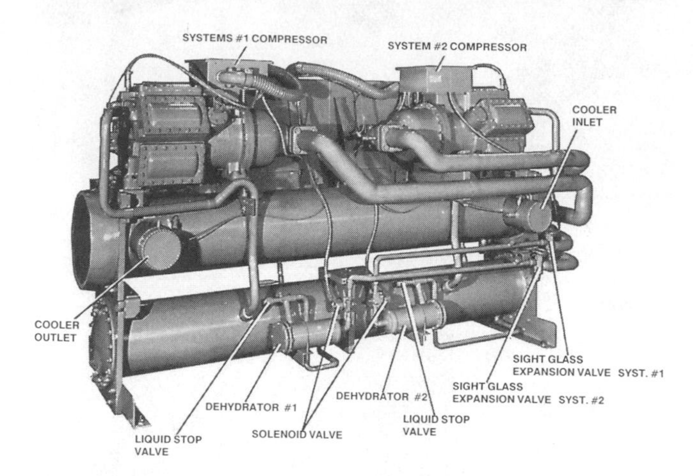

FIGURE 47–15 Direct expansion chiller with expansion device on the end of the chiller. *Courtesy York International*

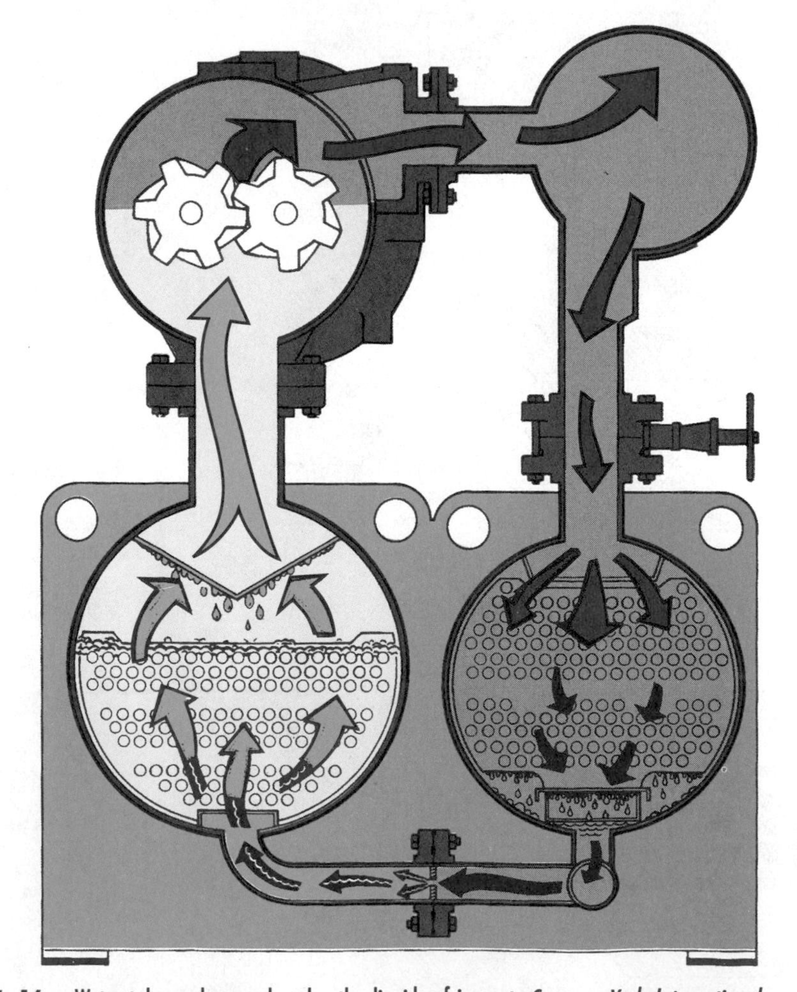

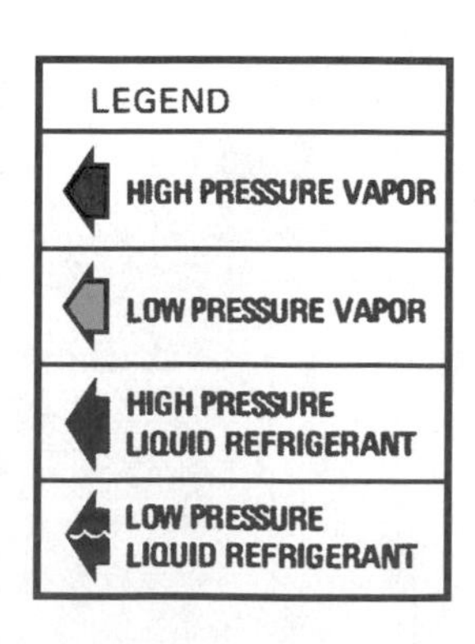

FIGURE 47–16 Water tubes submerged under the liquid refrigerant. *Courtesy York International*

fouled or dirty, the only way it can be cleaned is with the use of chemicals because the chillers cannot be cleaned with brushes. These are different from chillers where the water circulates in the tubes and the refrigerant circulates around the tubes as with flooded chillers.

47.10 FLOODED EVAPORATOR CHILLERS

Flooded chillers introduce the refrigerant at the bottom of the chiller barrel and the water circulates through the tubes. There are some advantages in this in that the tubes may be totally submerged under the refrigerant for the best heat exchange, Figure 47–16. The tubes may also be physically cleaned using several different methods. Figure 47–17 shows tubes being cleaned with a brush. The water side of a chiller should not become dirty unless the chiller is applied to an open process, such as a manufacturing operation. Most chiller systems use a closed water circuit that should remain clean unless there are many leaks and water must be continually added.

Flooded chillers use much more refrigerant charge than direct expansion chillers so leak monitoring must be part of regular maintenance with high-pressure systems.

When the water is introduced into the end of the chiller, the water is contained in water boxes. When these water boxes have removable covers they are known as marine water boxes, Figure 47–18. When the piping is attached to the cover, the piping must be removed to remove the water box cover and these are known as standard covers, Figure 47–19. The water box also is used to direct the water for different applications. For example, the water may need to

FIGURE 47–17 Cleaning water tubes. *Courtesy Goodway Tools Corporation*

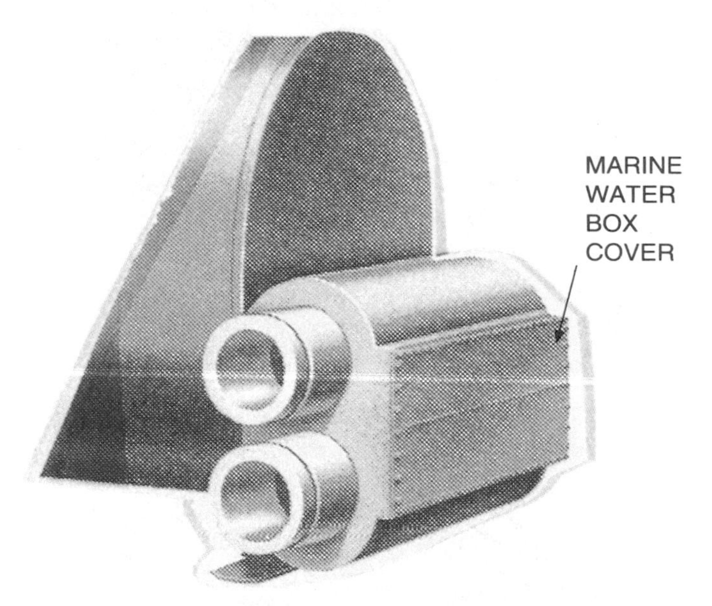

FIGURE 47–18 Marine water boxes. *Courtesy Trane Company*

FIGURE 47–19 Standard water boxes with piping connections. *Courtesy York International*

pass through the chiller one, two, three, or four times for different applications. The water box directs the water by means of partitions. This enables the manufacturer to use one chiller with combinations of water boxes and partitions for different applications.

When water passes through a one-pass chiller, it is only in contact with the refrigerant for a short period of time. Using two, three, or four passes keeps the water in contact longer but also creates more pressure drop and more pumping horsepower, Figure 47–20.

The design water temperatures are typically 55°F inlet water and 45°F outlet water for a two-pass chiller. The refrigerant is absorbing heat from the water so it is typically about 7°F cooler than the leaving water. This is called the approach temperature, Figure 47–21. **The approach temperature is very important to the technician for troubleshooting chiller performance.** If the chiller tubes be-

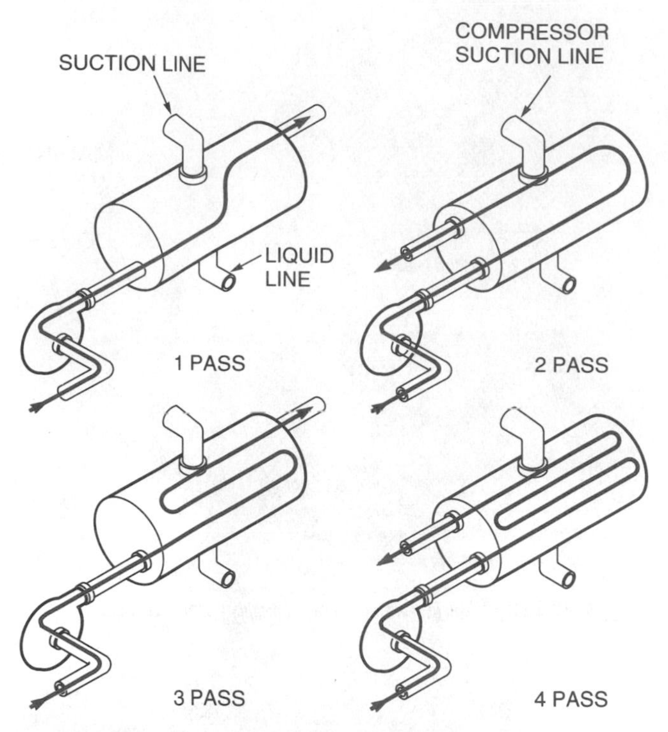

FIGURE 47–20 One-, two-, three-, and four-pass chillers.

come dirty, the approach becomes greater because the heat exchange becomes poor. The approach temperature is different for chillers with a different number of passes. Direct expansion chillers have only one pass of water (with baffles to sweep the tubes) and the approach temperature may be about 8°F. Chillers with three-pass evaporators will have an approach temperature of about 5°F and chillers with four passes may have an approach temperature of 3°F or 4°F. The longer the refrigerant is in contact with the water, the closer the approach temperature will be.

When a chiller is first started, a record should be started known as an operating performance log. This log of machine performance should be maintained when the chiller is at full-load conditions. Manufacturers rather than contractors often choose to start up larger chillers because of the cost of warranty. They will always make entries in an operating log and keep the log on file for future reference. Figure 47–22 is an example of a manufacturer's log sheet.

Flooded chillers generally have a means of measuring the actual refrigerant temperature in the evaporator either with a thermometer well or a direct readout on the control panel. Direct expansion chillers typically have a pressure gage on the low-pressure side of the system and the pressure must be converted to temperature for the evaporating refrigerant temperature.

When water is boiled in an open pot and the heat is turned up to the point that the water is boiling vigorously, water will splash out of the pot. Refrigerant in a flooded chiller acts much the same way; the compressor removes vapor from the chiller cavity, which will accelerate the process due to vapor velocity. Liquid eliminators are often placed above the tubes to prevent liquid from carrying over into the compressor suction intake at full load, Figure 47–16. The suction gas is often saturated as it enters the suction intake of a compressor on a flooded chiller. This is different from a direct expansion chiller, which will always have some superheat leaving the evaporator.

The heat exchange surface of both types of tube (direct expansion and flooded) may be improved by machining fins on the outside of the tube. Direct expansion evaporator tubes may have inserts that cause the refrigerant to sweep

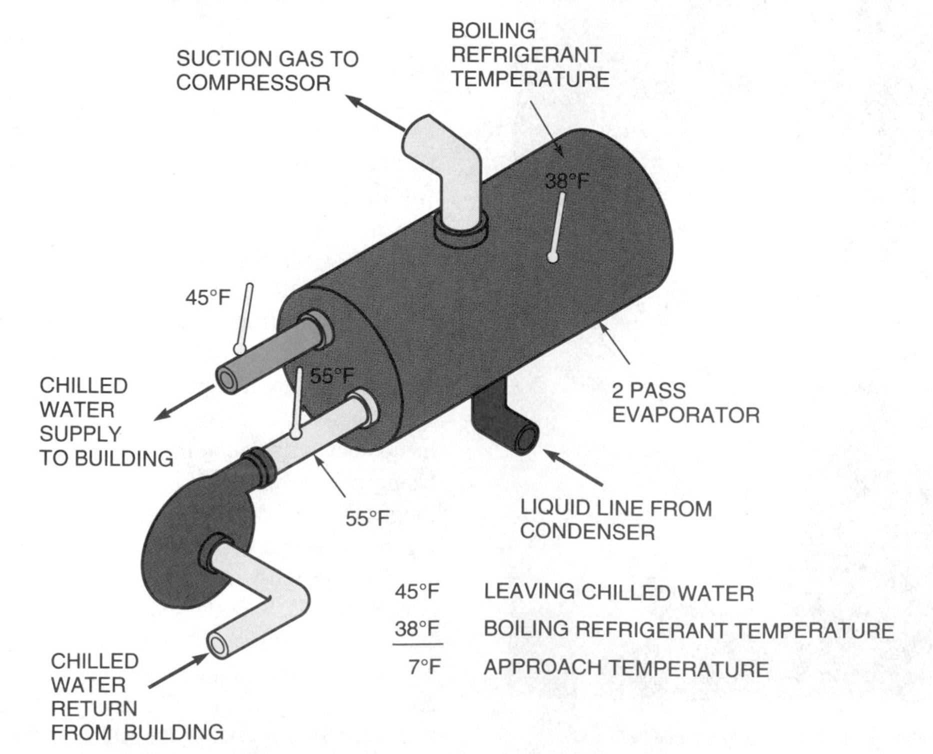

FIGURE 47–21 Approach temperature for a two-pass chiller.

YORK

ROTARY SCREW
LIQUID CHILLER LOG SHEET
125 - 400 TONS

CHILLER LOCATION ______

SYSTEM NO. ______

Date												
Time												
Hour Meter Reading												
O.A. Temperature D.B./W.B.			/	/	/	/	/	/	/	/	/	/
Compressor		Oil Level										
		Oil Pressure										
		Oil Temperature										
		Suction Temperature										
		Discharge Temperature										
		Filter PSID										
		Slide Valve Position %										
		Oil Added (gallons)										
Motor		Volts										
		Amps										
Cooler	Refrig.	Suction Pressure										
	Liquid	Inlet Temperature										
		Inlet Pressure										
		Outlet Temperature										
		Outlet Pressure										
		Flow Rate — GPM										
Condenser	Refrig.	Discharge Pressure										
		Corresponding Temperature										
		High Pressure Liquid Temperature										
		System Air — Degrees										
	Water	Inlet Temperature										
		Inlet Pressure										
		Outlet Temperature										
		Outlet Pressure										
		Flow Rate — GPM										

FORM 160.47-F6

FIGURE 47–22 Manufacturer's operating log sheet. *Courtesy York International*

the sides of the tube or the inside of the tube may be rifled like the bore of a rifle, Figure 47–23. These processes cause slight pressure drops but the results are worth the pressure drop caused.

Evaporators are constructed in a shell with end sheets to hold the tubes. Typically the tubes are made leak free in the tube sheets with a process called rolling. The tube sheet has groves cut in the hole the tube fits through. After the tube is inserted in the tube sheet hole, it is expanded with a roller into the grooves for a leak-free connection, Figure 47–24. Some chiller evaporators may have silver-soldered tubes at the tube sheet. These tubes would be hard to repair by replacement compared to the ones that are rolled in place.

The evaporator has a working pressure for the refrigerant circuit and for the water circuit. High-pressure chillers must be able to accommodate the refrigerant used, generally R-22 for newer models. The water side of the chiller may have one of two working pressures, 150 or 300 psig. The 300-psig chillers are used when the chiller is located on the bottom floor of a tall building where the water column creates a high pressure from the standing column of water.

The evaporator is normally manufactured, then leak tested. After it is proven to be leak free, it is then insulated. Many larger chillers are insulated with rubber type foam insulation that is glued directly to the shell. Smaller chillers

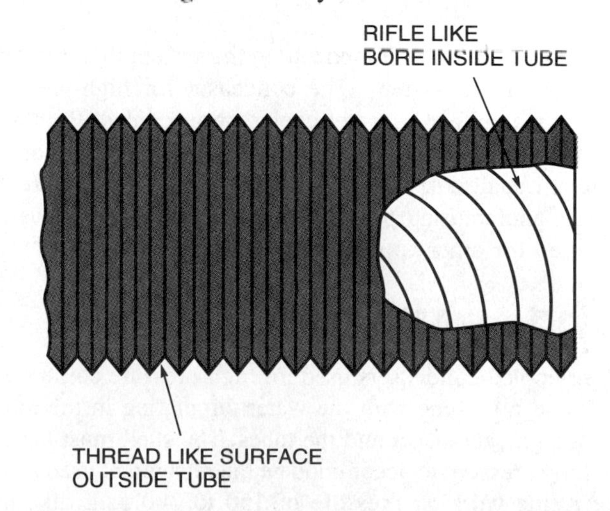

FIGURE 47–23 Fins on the outside of a chiller tube, rifle-like bore on the inside.

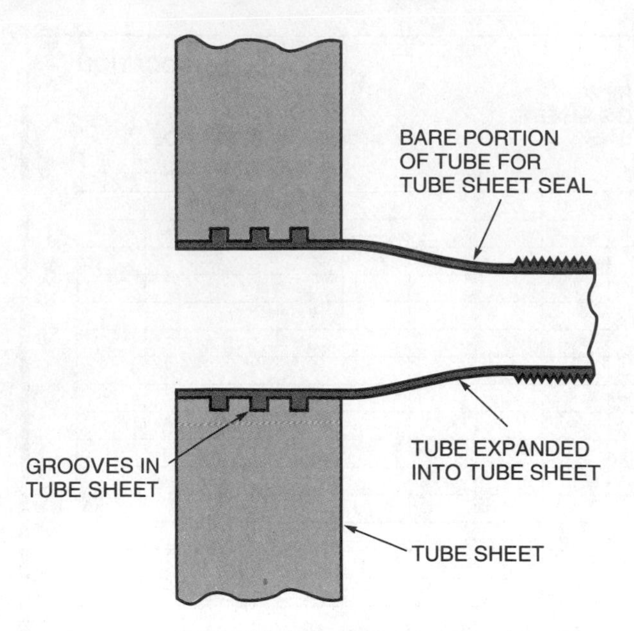

FIGURE 47–24 How a chiller tube is made leak free.

may be insulated by being placed in a skin of sheet metal and foam insulation applied between the chiller shell and the skin. If any repair is necessary, the skin will have to be removed and the foam cut away. The skin can later be replaced and foam insulation added back to insulate the chiller barrel.

Some evaporators may be located outside, as in the case of an air-cooled package chiller, so freeze protection for the water in the shell is necessary for these chillers. This can be accomplished by adding resistance heaters to the chiller barrel under the insulation. These heaters may be wired into the chiller control circuit and energized by a thermostat if the ambient temperature approaches freezing and water is not being circulated.

47.11 CONDENSERS FOR HIGH-PRESSURE CHILLERS

The condenser is the component in the system that transfers heat out of the system. The condenser for high-pressure chillers may either be water cooled or air cooled. Both will take the heat from the system and transfer the heat out of the system. Usually the heat is transferred to the atmosphere. In some manufacturing processes, the heat may be recovered and used for other applications.

47.12 WATER-COOLED CONDENSERS

Water-cooled condensers used for high-pressure chillers are shell and tube type with the water circulating in the tubes and the refrigerant around the tubes. The shell must have a working pressure to accommodate the refrigerant used and a water side working pressure of 150 to 300 psig, like the evaporators mentioned earlier. The hot discharge gas is normally discharged into the top of the condenser, Figure 47–25. The refrigerant is condensed to a liquid and drips down to the bottom of the condenser where it gathers and drains into the liquid line. As with evaporators, the heat exchange can be improved by use of extended surfaces on the refrigerant side of the condenser tube, Figure 47–23. The inside of the condenser tubes may also be grooved, as with evaporator tubes, for improved water-side heat exchange.

Condensers must also have a method for the water to be piped into the shell. There are two basic types of connections to the water box that is attached to the condenser shell. In a standard water box the piping is attached to the removable water box and some of the piping must be removed before the inspection cover can be removed on the piping end. A marine water box has a removable cover for easy access to the piping end of the machine. Tube access is more important for the condenser than the evaporator in most installations because the open cooling tower provides greater potential for getting the tubes dirty.

47.13 CONDENSER SUBCOOLING

The design condensing temperature for a water-cooled condenser is around 105°F on a day when 85°F water is supplied to the condenser and the chiller is operating at full load. The saturated liquid refrigerant would be 105°F. Many condensers have subcooling circuits that reduce the liquid temperature to below the 105°F. If the entering water at 85°F is allowed to exchange heat from the saturated refrigerant, the liquid line temperature can be reduced. If it can be reduced to 95°F, a considerable amount of capacity can be gained for the evaporator. This subcooling can increase the machine capacity about 1%/°F of subcooling. For a 300-ton chiller, 10°F subcooling would increase the capacity about 30 tons. The only expense is the extra circuit in the condenser. To subcool refrigerant, it must be isolated from the condensing process with a separate circuit, Figure 47–25.

The condenser, like the evaporator, has an approach relationship between the refrigerant condensing temperature and the leaving-water temperature. Most water-cooled condensers are two-pass condensers and are designed for 85°F entering water and 95°F leaving water with a condensing temperature of about 105°F, Figure 47–26. This is an approach temperature of 10°F and is important to the technician. Other configurations of condenser may be either one-, three-, or four-pass with different approach temperatures. As with evaporators, the longer the refrigerant is in contact with the water, the closer the approach temperature will be. The original operating log will reveal what the approach temperature was in the beginning. It should be the same for similar conditions at a later date, provided the tubes are clean.

Because the condenser rejects heat from the system, its capacity to reject heat determines the head pressure for the refrigeration process. There must be some pressure in the condenser to push the refrigerant through the expansion device. The head pressure must be controlled. There are two times when head pressure control may become too low and

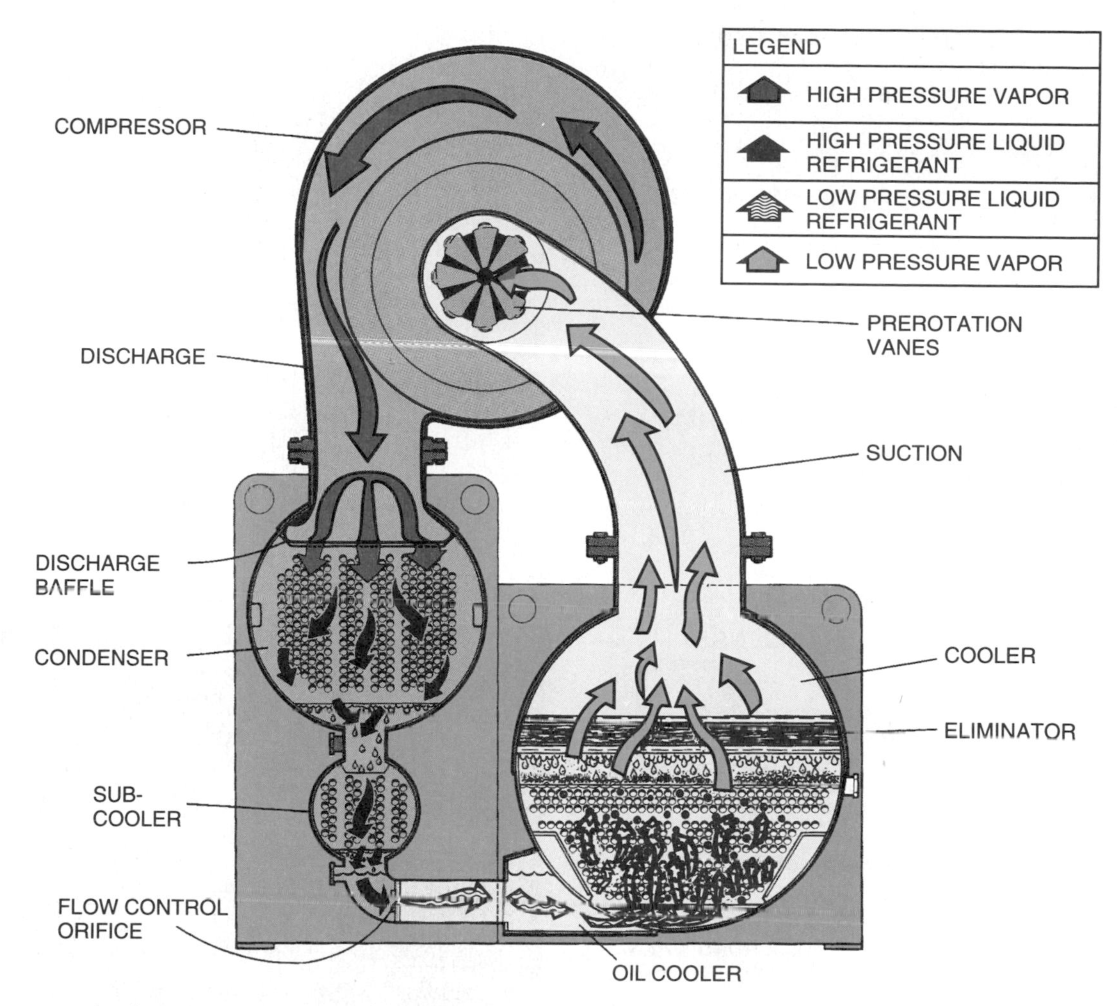

FIGURE 47–25 Subcooling circuit in a water-cooled condenser. *Courtesy York International*

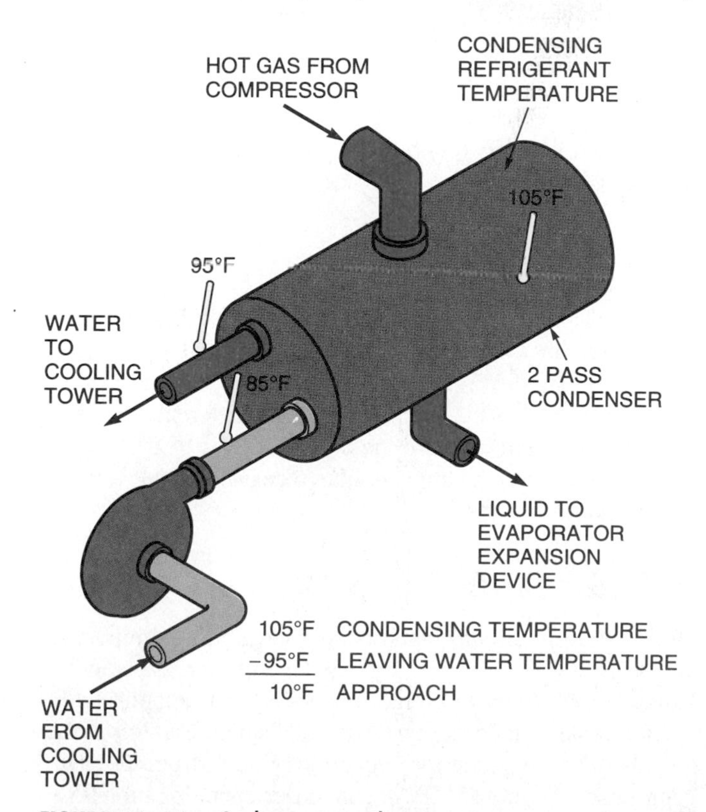

FIGURE 47–26 Condenser approach temperature.

problems may occur. These times are at start-up and very cold weather when there is still a call for air conditioning.

When there is a call to start up a water-cooled chiller and the water in the cooling tower is cold, it may cause the head pressure to become so low that the suction pressure will be low enough to either trip the low-pressure control (direct expansion chillers) or the evaporator freeze control (flooded chillers). Continued operation will also cause oil migration in many machines. A bypass valve is typically located in the condenser circuit to bypass water during start-up to prevent nuisance shutdowns due to cold condenser water. This same bypass may be used to bypass water during normal operation. This is discussed in more detail in Unit 48.

When water-cooled condensers are used, the heat from the refrigerant is transferred into the water circulating in the condenser circuit. The heat must now be transferred to the atmosphere in a typical installation. This is done by means of a cooling tower, Figure 47–27. These are discussed in Unit 48.

47.14 AIR-COOLED CONDENSERS

Air-cooled condensers are generally constructed of copper tubes with the refrigerant circulating inside and with alumi-

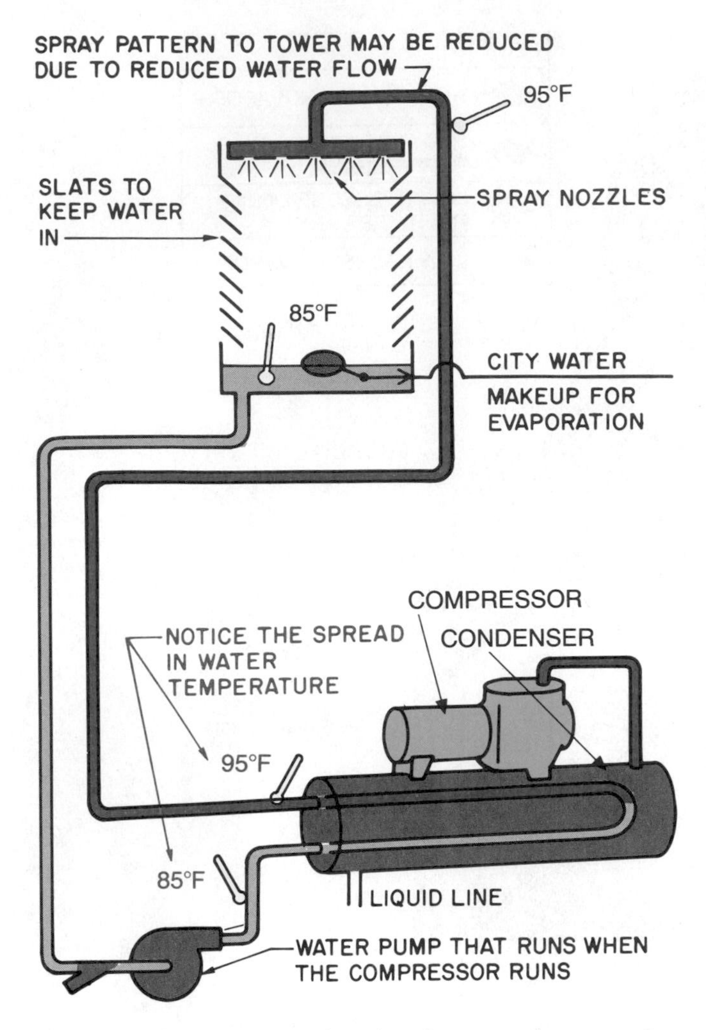

FIGURE 47–27 Heat moves from the refrigerant to the water and to the atmosphere from the cooling tower.

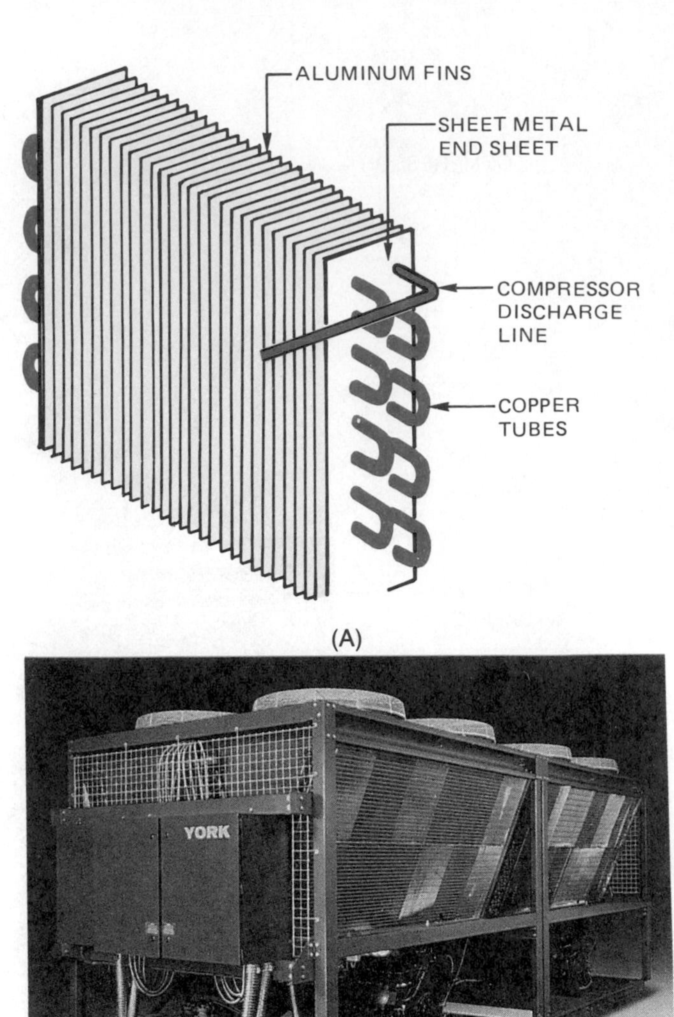

FIGURE 47–28 (A) Air-cooled condenser. (B) Air-cooled chiller. *Courtesy York International*

num fins on the outside to provide a larger heat exchange surface, Figure 47–28. These air-cooled heat exchange surfaces have been used for many years with great success. Some manufacturers furnish copper finned tubes or special coatings for the aluminum fins for locations where salt air may corrode the aluminum.

These coils are positioned differently by the manufacturer for their own design purposes. Some coils are positioned in a horizontal and some in a vertical mode.

Multiple fans are used for many of these condensers. The fans may be cycled on and off for head pressure control.

Air-cooled condensers may be in sizes from very small to several hundred tons. Air-cooled condensers eliminate the need for using water towers and the problems that occur when using water. Systems using air-cooled condensers do not operate at the low condensing temperatures and head pressures found with water-cooled condensers. Water-cooled condensers typically condense at about 105°F. Air-cooled condensers condense at 20°F to 30°F higher than the entering air temperature depending on the condenser surface area per ton of condenser. For example, an R-22 system with a water-cooled condenser would have a head pressure of about 211 psig (105°F compares to 211 psig). An air-cooled condenser on a 95°F day would have a head pressure of from 243 psig (95°F + 20°F = 115°F) to 278 psig (95°F + 30°F = 125°F). This is a considerable difference in head pressure and will increase the operating cost of the chiller. Air-cooled chillers are popular because there is less maintenance.

47.15 SUBCOOLING CIRCUIT

Subcooling in an air-cooled condenser is just as important as in a water-cooled unit. It gives the unit more capacity by about 1%/°F of subcooling. Subcooling is accomplished in an air-cooled condenser by means of a small reservoir in the condenser to separate the subcooling circuit from the main condenser, Figure 47–29. A condenser operating on a 95°F day can expect to achieve 10°F to 15°F of subcooling.

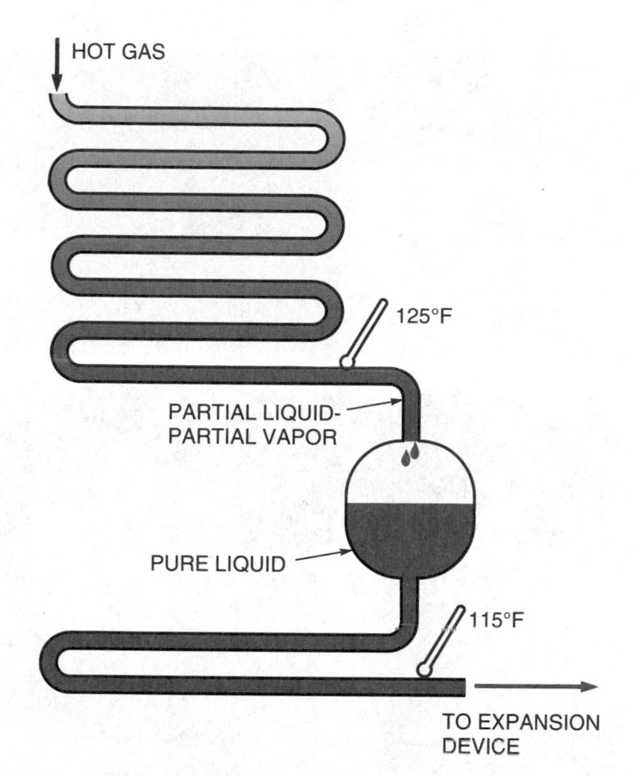

FIGURE 47–29 Subcooling circuit in an air-cooled condenser.

47.16 METERING DEVICES FOR HIGH-PRESSURE CHILLERS

The metering device holds liquid refrigerant back and meters the liquid into the evaporator at the correct rate. Four types of metering devices may be used for large chillers: thermostatic expansion valve, orifice, high- and low-side floats, and electronic expansion valve.

47.17 THERMOSTATIC EXPANSION VALVE

The thermostatic expansion valve (TXV) was discussed in detail in Unit 24. The TXVs used with chillers are the same type, only larger. The TXV maintains a constant superheat at the end of the evaporator. Usually this is 8°F to 12°F of superheat. The more superheat, the more vapor there will be in the evaporator and less heat exchange will occur. The more liquid the evaporator has, the better the heat exchange, so TXV valves are not used except on smaller chillers, up to about 150 tons.

47.18 ORIFICE

The orifice metering device is a fixed-bore metering device that is merely a restriction in the liquid line between the condenser and the evaporator, Figure 47–30. These are trouble free because they have no moving parts.

The flow of refrigerant through an orifice is a constant at a given pressure drop. When there is a greater pressure difference, more flow will occur. The flow at greater loads is accomplished because of a higher head pressure at greater loads. When the head pressure is allowed to rise, more flow

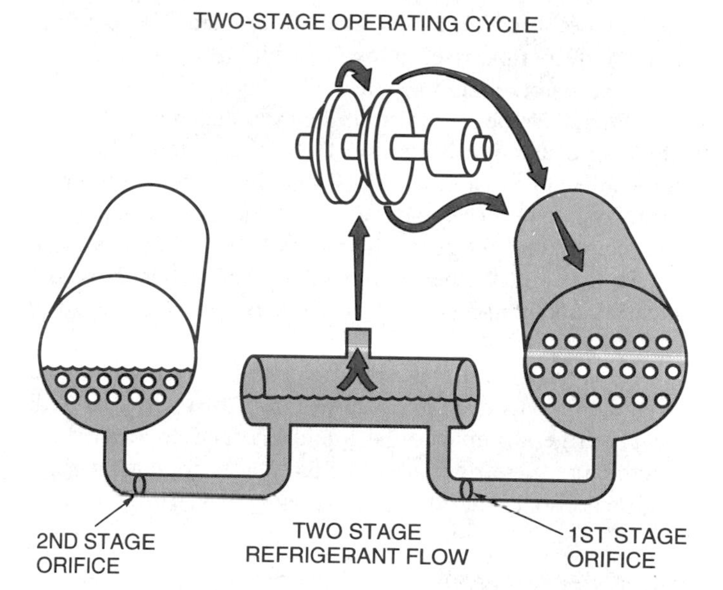

FIGURE 47–30 Orifice metering device. *Courtesy Trane Company*

will occur. Any chiller with an orifice for a metering device has a critical charge. The unit must not be overcharged or liquid refrigerant will enter the compressor inlet. This could be detrimental to a reciprocating compressor so orifices may not be used where these compressors are used. Small amounts of liquid refrigerant, often in the form of wet vapor, will not harm the scroll, rotary screw, or most centrifugal compressors.

47.19 FLOAT TYPE METERING DEVICES

There are two types of float type metering devices: the *low-side float* and the *high-side float*. The low-side float rises and throttles back the refrigerant flow when the liquid refrigerant level rises in the low side of the chiller, Figure 47–31. It is located at the correct level to produce the level of refrigerant required for the chiller.

The high-side float is located in the liquid line entering the evaporator. When the level of refrigerant is greater in the

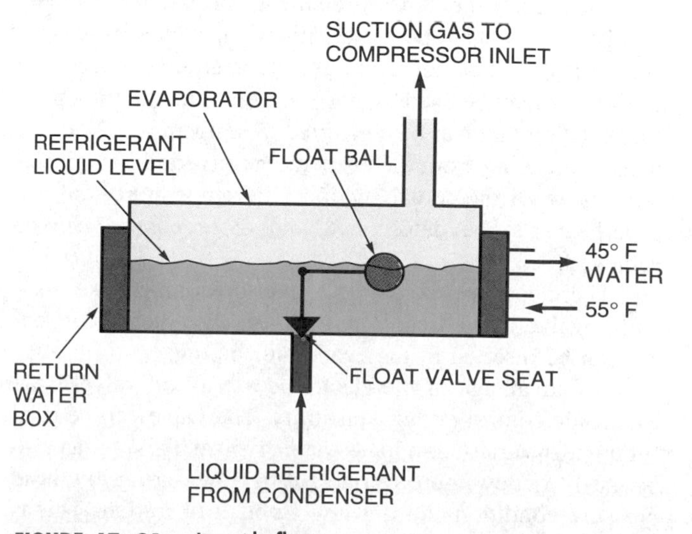

FIGURE 47–31 Low-side float.

liquid line, it is because the evaporator needs liquid refrigerant and the float rises, allowing liquid refrigerant to enter the evaporator, Figure 47–32.

Floats can be a problem if they rub the side of the float chamber causing resistance or if a hole rubs in the float. If a hole develops in the float, it will sink. In the case of the low-side float, it will open and allow full flow of refrigerant to the evaporator and no refrigerant will be held back to the liquid line and in the bottom of the condenser. If a hole wears in the high-side float, it will sink and block liquid to the evaporator.

The refrigerant charge is critical for both the low- and high-side floats. An overcharge for a low-side float will back refrigerant up into the liquid line and condenser. An overcharge of refrigerant will flood the evaporator when a high-side float is used.

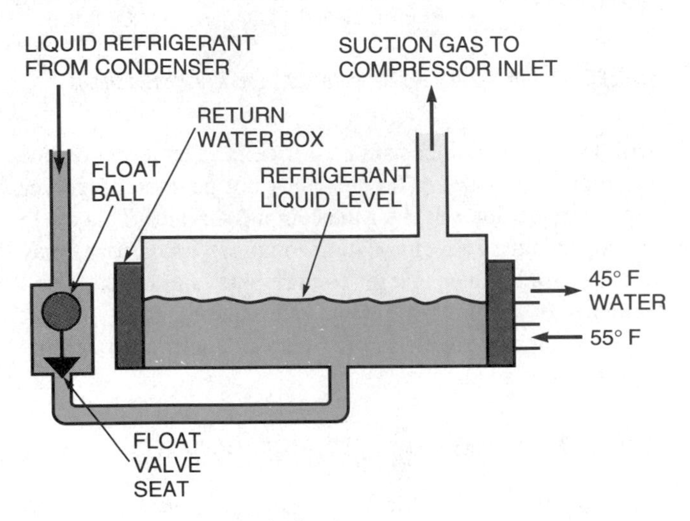

FIGURE 47–32 High-side float.

47.20 ELECTRONIC EXPANSION VALVE

The electronic expansion valve for large chillers is similar to the one explained in Unit 24. These valves operate with a thermistor to monitor the refrigerant temperature. The liquid does not reach the sensing element; an electronic circuit checks the actual temperature. This is much like a thermometer except it does not read out in temperature. A pressure transducer may be used to give the electronic circuit a pressure reading that can be converted to temperature. With the pressure reading from the evaporator converted to temperature and with the actual suction line temperature, the real superheat can be determined and close control can be achieved.

The electronic expansion valve looks much like a solenoid valve in the line, Figure 47–33. The thermistor sensor can be inserted in the evaporator piping.

An advantage of the electronic expansion valve is the electronic control circuit versatility. The valve can be used for a wider variation in loads than a typical TXV. It can also be used to allow more refrigerant to flow during low head pressure conditions during low ambient operation. This is particularly helpful for air-cooled units when the weather is cold. The large liquid handling capacity of the electronic expansion valve can allow maximum liquid refrigerant flow.

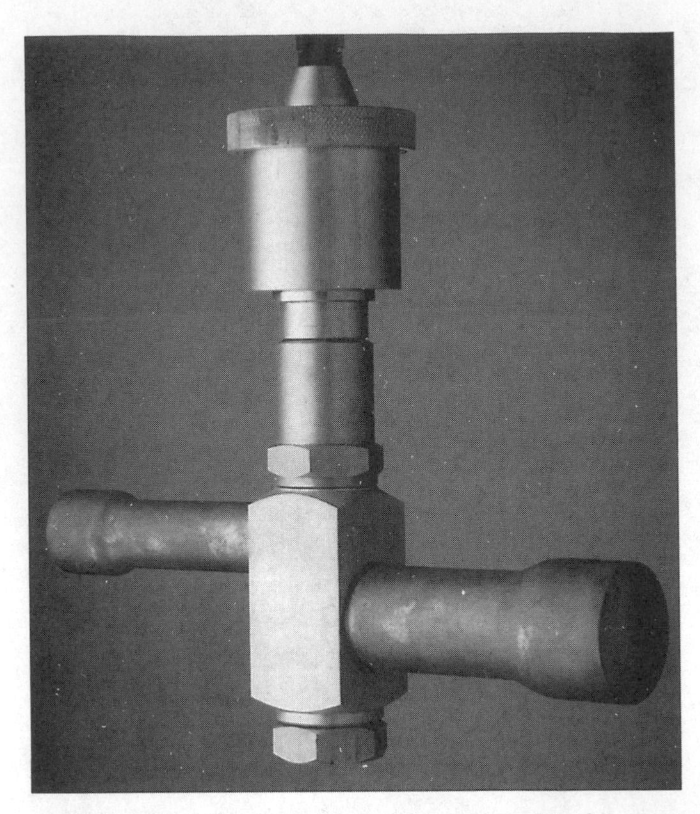

FIGURE 47–33 Electronic expansion valve. *Courtesy Trane Company*

This ability to allow maximum flow on a cold day can give the condenser much more capacity to pass on to the system, for example, if it is 40°F outside and the building is calling for cooling. The condenser can condense the refrigerant at about 60°F to 70°F. This is a head pressure of 102 to 121 psig for R-22. R-12 and R-500 would have head pressures corresponding to these temperatures. This is not enough head pressure for a typical TXV. But an electronic expansion valve (because of its flow capability) can use this low pressure to feed the evaporator. If the refrigerant is then subcooled 10°F more, the evaporator can be furnished 50°F to 60°F liquid refrigerant. The compressor is only required to pump from an evaporator pressure of approximately 65.6 psig (corresponding to 38°F evaporator boiling temperature) to 121 psig maximum head pressure. This is a big difference from a typical summer condition where the evaporator suction pressure may be 65.6 psig and the head pressure 278 psig (corresponding to a condensing temperature of 125°F) for R-22.

Each manufacturer has different design features, but all of them attempt to build a reliable, low-cost chiller that will last a long time.

47.21 LOW-PRESSURE CHILLERS

Low-pressure chillers typically use R-11, R-113, or R-123 for their refrigerants. Most small low-pressure chillers in the range up to approximately 150 tons use R-113 and chillers

VAPOR PRESSURES				
TEMP °F	113	11	114	134a
−150.0				
−140.0				29.6
−130.0				29.4
−120.0				29.1
−110.0			29.7	28.7
−100.0			29.5	28.0
−90.0		29.7	29.3	27.1
−80.0		29.6	29.0	25.7
−70.0		29.4	28.6	24.0
−60.0		29.2	28.0	21.6
−50.0	29.6	28.9	27.1	18.6
−40.0	29.5	28.4	26.1	14.7
−35.0	29.4	28.1	25.4	12.3
−30.0	29.3	27.8	24.7	9.7
−25.0	29.2	27.4	23.8	6.8
−20.0	29.0	27.0	22.9	3.6
−15.0	28.8	26.6	21.8	0.0
−10.0	28.7	26.0	20.6	2.0
−5.0	28.4	25.4	19.3	4.1
0.0	28.2	24.7	17.8	6.5
5.0	27.9	23.9	16.2	9.1
10.0	27.5	23.1	14.4	12.0
15.0	27.2	22.1	12.4	15.1
20.0	26.7	21.1	10.2	18.4
25.0	26.3	19.9	7.8	22.1
30.0	25.7	18.6	5.1	26.1
35.0	25.1	17.1	2.2	30.4
40.0	24.4	15.6	0.4	35.0
45.0	23.7	13.8	2.1	40.0
50.0	22.9	12.0	3.9	45.4
55.0	21.9	9.9	5.9	51.2
60.0	20.9	7.7	8.0	57.4
65.0	19.8	5.2	10.3	64.0
70.0	18.6	2.6	12.7	71.1
75.0	17.3	0.1	15.3	78.6
80.0	15.8	1.6	18.2	86.7
85.0	14.2	3.3	21.2	95.2
90.0	12.5	5.0	24.4	104.3
95.0	10.6	6.9	27.8	113.9
100.0	8.6	8.9	31.4	124.1
105.0	6.4	11.1	35.3	134.8
110.0	4.0	13.4	39.4	146.3
115.0	1.4	15.9	43.8	158.4
120.0	0.7	18.5	48.4	171.1
125.0	2.1	21.3	53.3	184.5
130.0	3.7	24.3	58.4	196.7
135.0	5.3	27.4	63.9	213.5
140.0	7.1	30.8	69.6	229.2
145.0	9.0	34.3	75.6	245.6
150.0	11.1	38.1	82.0	262.8

FIGURE 47–34 Pressure/temperature chart including low-pressure refrigerants.

above this range, R-11. Many of the newer chillers are using R-123. Some use R-114 for special applications. ⊛**About 1990 the trend started moving away from any refrigerant that is labeled a chlorofluorocarbon (CFC). All CFC refrigerants will be phased out in the near future. This will include R-11, R-113, and R-114, which are all low-pressure refrigerants.**⊛ Figure 47–34 shows a pressure/temperature chart for the above low-pressure refrigerants.

These chillers have all of the same components as the high-pressure chillers: a compressor, an evaporator, a condenser, and a metering device.

47.22 COMPRESSORS

All low-pressure chillers as well as some high-pressure chillers use centrifugal compressors. These low-pressure centrifugal compressors also turn from the motor speed for direct-drive compressors to very high speed (about 30,000 rpm) for gear-drive machines.

The centrifugal compressor has an impeller that turns and creates the low-pressure area in the center where the suction line is fastened to the housing, Figure 47–35. The compressed refrigerant gas is trapped in the outer shell known as the volute and guided down the discharge line to the condenser, Figure 47–36.

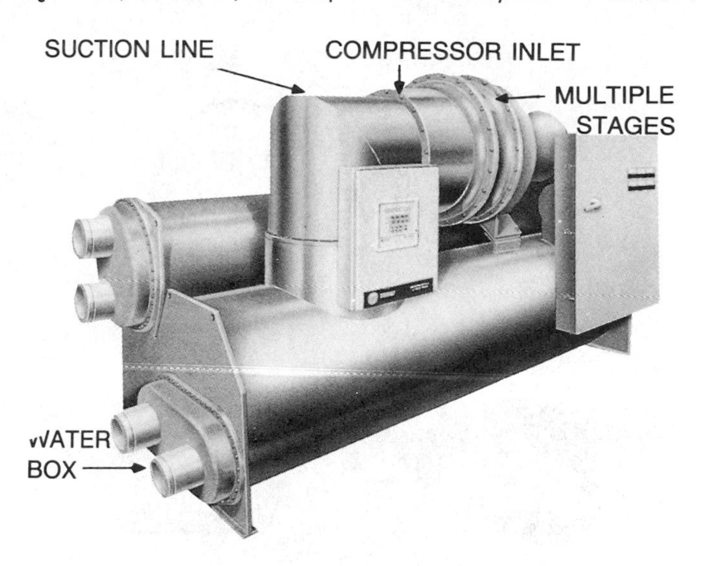

FIGURE 47–35 Multiple stages of compression. *Courtesy Trane Company*

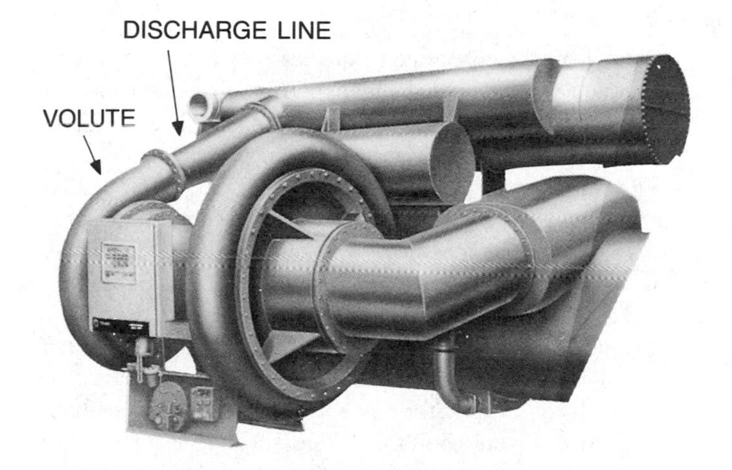

FIGURE 47–36 Hot gas entering the condenser. *Courtesy Trane Company*

Centrifugal compressors can be manufactured to produce more pressure by building more than one compressor and operating them in series, called multistage, or operating one stage of compression at a high speed. Actually with multistage compressors, the discharge from one compressor becomes the suction for the next compressor, Figure 47–37. It is common for compressors for air conditioning applications to have up to three stages of compression. Figure 47–35 shows a modern three-stage compressor that can be used for either R-11 or R-123. One of the advantages of multistage operation is that the compressor can be operated at slow speed, typically the speed of the motor, about 3600 rpm.

Refrigerant vapor can also be drawn off the liquid leaving the condenser to subcool the liquid entering the evaporator metering device. The refrigerant vapor is drawn off using the intermediate compressor stages so the liquid temperature corresponds to the pressure of the compressor intermediate suction pressure; it is less than the pressure corresponding to the condensing temperature. The subcooling

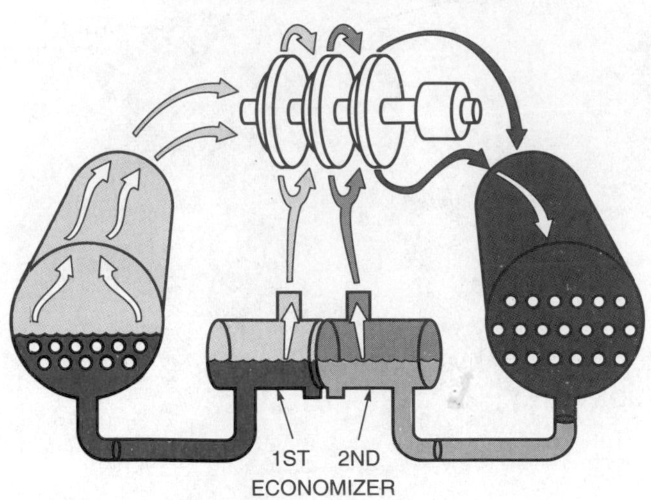

FIGURE 47–37 Three-stage centrifugal chiller with economizers. *Courtesy Trane Company*

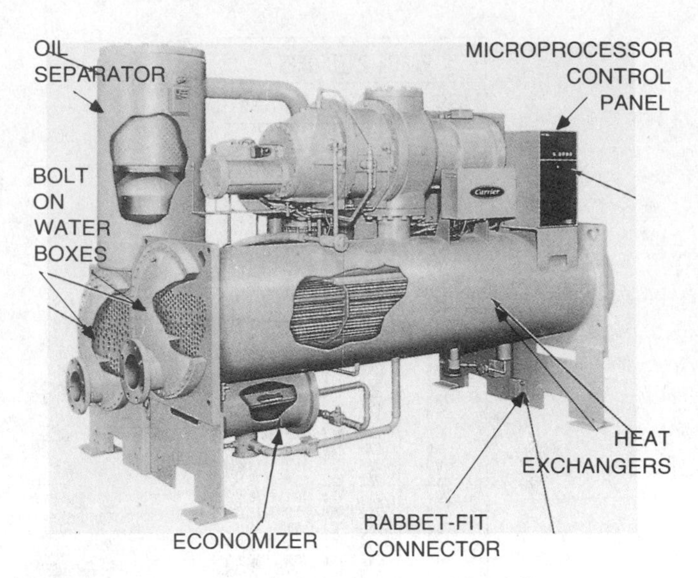

FIGURE 47–38 Cut-away of a high-pressure chiller shell. Low-pressure chiller shells are much like this. *Reproduced courtesy of Carrier Corporation*

process gives the evaporator more capacity for its respective size at no additional pumping cost. This process is called an economizer cycle, Figure 47–37.

Single-stage compression can be accomplished using a single compressor that turns at a high speed, about 30,000 rpm. This is done by means of a 3600-rpm motor and a gear box to step up the speed of the compressor, Figure 47–10. The advantage of this compressor is size and weight.

The amount of compression needed depends on the refrigerant used. For example, most centrifugal compressors used for chillers use two-pass evaporators and two-pass condensers. If the evaporator has a 7°F approach temperature, the evaporator would boil the refrigerant at 40°F for a chiller with 47°F leaving water. **NOTE:** We are using 47°F water to get an evaporator temperature of 40°F for use on a typical pressure/temperature chart. If the condenser has a 10°F approach temperature and is receiving 85°F water from the tower and has 95°F leaving water, the condenser should be condensing the refrigerant at about 105°F, Figure 47–26. ■ The compressor must overcome the following pressures to meet the requirements for this system:

REFRIGERANT TYPE	EVAP PRESS 40°F	COND PRESS 105°F
R-113	24.4 in. Hg Vac	6.4 in. Hg Vac
R-11	15.6 in. Hg Vac	11.1 psig
R-123	18.1 in. Hg Vac	8.1 psig
R-114	0.44 psig	35.3 psig
R-500	46 psig	152.2 psig
R-502	80.5 psig	231.7 psig
R-12	37 psig	126.4 psig
R-134A	35 psig	134.9 psig
R-22	68.5 psig	210.8 psig

Notice that the compressor has only to raise the pressure from 24.4 in. Hg Vac to 6.4 in. Hg Vac for R-113; this is only 18 in. Hg Vac (or 18/2.036 = 8.84 psig). The compressor that has R-502 in the system would have to raise the pressure from 80.6 psig to 231.7 psig or 151.1 psig. A different compressor would have to be used than the one for R-113. The compressors do not raise the pressure the same amount nor do they operate in the same pressure ranges. One compressor operates in a vacuum on both the high- and low-pressure side of the system and the other operates under a discharge pressure of 231.7 psig. The physical thickness of the shell of the two compressors can be very different. The above comparison will give you some idea of the different applications for centrifugals.

Many of the features and operation of low-pressure centrifugal chillers are the same or similar to those found on or with high-pressure units. Figure 47–38 shows a cut-away of a high-pressure chiller shell but looks similar to most low-pressure chillers.

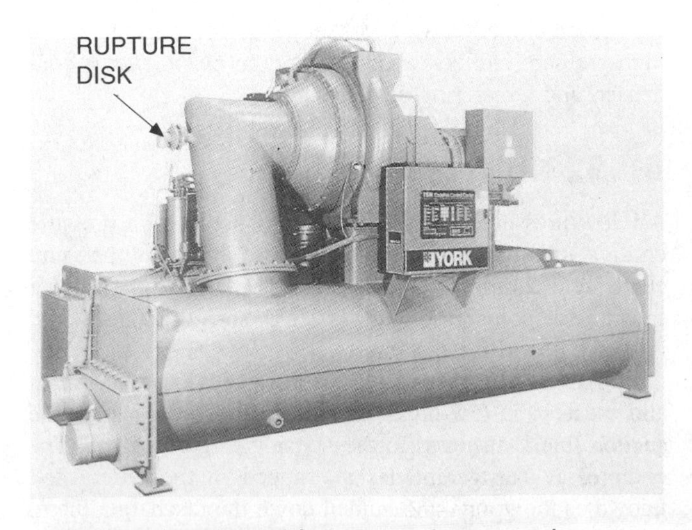

FIGURE 47–39 Rupture disk. *Courtesy York International*

The evaporator on a low-pressure system has a working pressure for the refrigerant circuit and for the water circuit. Low-pressure chillers must be able to accommodate the refrigerant used, typically R-113, R-11, or R-123. Because these are low-pressure chillers, the refrigerant side of the shell does not have to be as strong as the high-pressure chillers. These can be much lighter weight because of the pressures. The low-pressure chiller shell may have a refrigerant working pressure as low as 15 psig. The refrigerant safety device for the system is called a rupture disc; its relief pressure is 15 psig, and it is located in the low-pressure side of the system, on the evaporator section or the suction line, Figure 47–39. This device prevents excess pressure on the shell.

47.23 CONDENSERS FOR LOW-PRESSURE CHILLERS

The condenser is the component that rejects heat from the system. The condensers for low-pressure chillers are water cooled. As with high-pressure systems, the heat is usually transferred to the atmosphere. In some manufacturing processes, the heat may be recovered and used for other applications.

Water-cooled condensers used for low-pressure chillers are shell and tube type with the water circulating in the tubes and the refrigerant around the tubes. The shell must have a working pressure to accommodate the refrigerant used and a water-side working pressure of 150 to 300 psig. The hot discharge gas is discharged at the top of the condenser, Figure 47–36.

Because the condenser rejects heat from the system, its capacity to reject heat determines the head pressure for the refrigeration process. The condenser is always located above the evaporator with low-pressure chillers. The compressor lifts the refrigerant to the condenser level in the vapor state. When it is condensed to a liquid, it flows by gravity with very little pressure difference to the evaporator through the metering device. The head pressure must be controlled so that it does not drop too low below the design pressure drop.

Water-cooled condensers may also have a subcooling circuit. The purpose of the subcooling circuit is to lower the liquid refrigerant temperature to a level below the condensing temperature. It is accomplished by means of a separate chamber in the bottom of the condenser where the entering condenser water (at about 85°F) can remove heat from the condensed liquid refrigerant at about 105°F. This gives the system more efficiency, Figure 47–25.

47.24 METERING DEVICES FOR LOW-PRESSURE CHILLERS

The metering device holds liquid refrigerant back and meters the liquid into the evaporator at the correct rate. Two types of metering devices are typically used for low-pressure chillers, the orifice and the high- or low-side float. These were discussed under high-pressure chillers.

47.25 PURGE UNITS

When a centrifugal uses a low-pressure refrigerant, the low-pressure side of the system is always in a vacuum when in operation. If the system uses R-113, the complete system is in a vacuum. This means that any time the machine has a leak and it is in a vacuum, air will enter the machine, Figure 47–40. Air may cause several problems. It contains oxygen and moisture and will mix with the refrigerant, which will cause a mild acid that will eventually attack the motor windings and cause damage, or motor burn. Air will move to the condenser while the machine is running, but it will not condense. It will cause the head pressure to rise. If much air enters the system, the head pressure will rise to the point that the machine will either cause a "surge" or will shut down on high head pressure.

When the air moves to the condenser, it will collect at the top and take up condenser space. This air can be collected and removed with a purge system. The purge is a separate device that often has a compressor to collect a sample of whatever is in the top of the condenser for the purpose of trying to condense it in a separate condenser. If it can condense the product from the top of the condenser, it is refrigerant and it will return it to the system at the evaporator level, Figure 47–41. If it cannot condense the sample, the purge pressure will rise and a relief valve will allow this sample to be exhausted to the atmosphere.

There are several types of purge systems, but they all perform the same function, some more efficiently than others. Some of them use the difference in pressure from the high-pressure side of the system to the low-pressure side of

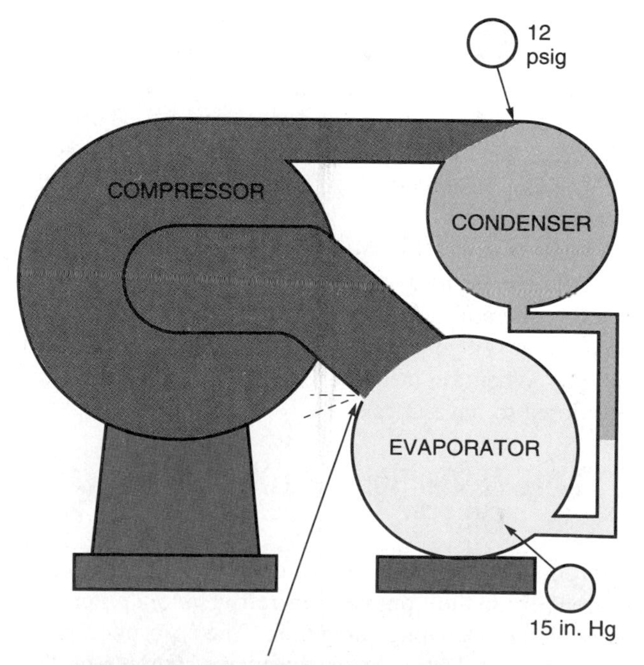
12
psig
COMPRESSOR
CONDENSER
EVAPORATOR
15 in. Hg
LEAK AT GASKET–AIR ENTERS BECAUSE EVAPORATOR IS IN A VACUUM–THE AIR WILL BE PUMPED TO THE CONDENSER BY THE COMPRESSOR WHERE IT WILL STAY BECAUSE IT CANNOT CONDENSE.

FIGURE 47–40 Low-pressure chiller with a leak.

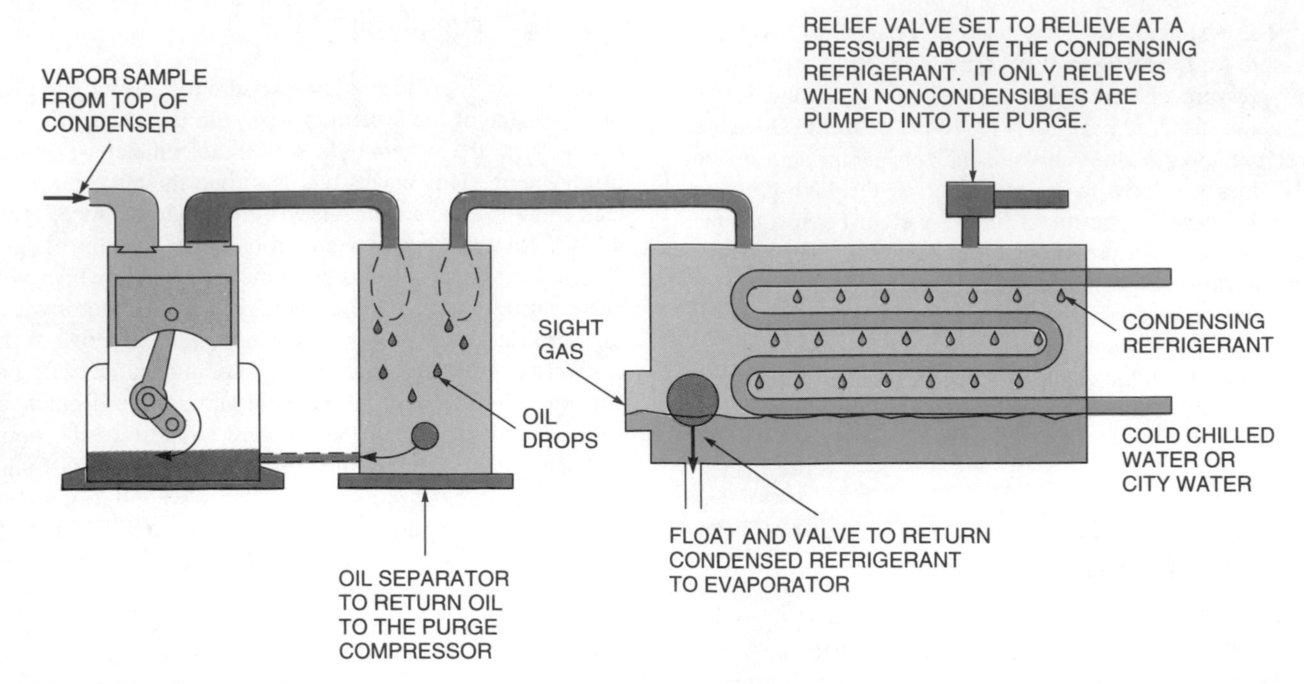

FIGURE 47–41 Purge operation.

the system to create the same pressure difference that a purge compressor will create and a condenser to condense any refrigerant from the sample.

Older purges were not efficient and a high percentage of the sample that was relieved to the atmosphere contained refrigerant. This is considered poor practice today because of environmental concerns and the price of lost refrigerant. Modern day purges must be efficient and allow only a small part of 1% of loss during the relieving process. Figure 47–42 shows one of the modern purges that releases only a very small amount of refrigerant when relieving the purge pressure. The vapor is relieved to the atmosphere only when air is present. A leak-free system will contain no air and the purge system will not have to function.

Some manufacturers relieve the purge pressure by means of a pressure switch and a solenoid valve. A counter tells the operator when the purge functions. The operator can use this as a guide to when a system has a leak and air is entering. When too many purge reliefs occur, the system is considered to have a leak.

47.26 ABSORPTION AIR CONDITIONING CHILLERS

Absorption refrigeration is a process that is considerably different from the compression refrigeration process discussed in the preceding paragraphs. The absorption process uses heat as the driving force instead of a compressor. When heat is plentiful or economical, or when it is a by-product of some other process, absorption cooling can be attractive. For example, in a manufacturing process where steam is used, it is often used at 100 psig and must be condensed before it can be reintroduced to the boiler. Condensing steam in the absorption system process is a natural choice for chilling water for air conditioning. In this case, chilled water for air conditioning can be furnished at an economical cost, basically for the cost of operating all related pumps because no compressor is involved. In other cases, natural gas is an inexpensive fuel in the summer. The gas company is looking for ways to market gas in the summer because their winter quota may depend on how much gas they can sell in the summer. Some gas companies may offer incentives to install gas appliances that consume fuel in the off-peak summer months. **Absorption refrigeration equipment also operates in a vacuum and contains non-CFC refrigerants.**

The absorption system equipment looks much like a boiler except it has chilled-water piping and condenser-water piping routed to it in addition to the piping for steam or hot water. Oil or gas burners are part of the system if it is a direct-fired chiller. Figure 47–43 shows several absorption systems by different manufacturers. The small machines are considered package machines, meaning they are completely assembled and tested at the factory and shipped as one component. They are all built in one shell or a series of chambers so the range of sizes of equipment may be limited because it must be moved as a single component. The machines are rated in tons and range in size from approximately 100 to 1700 tons. A 1700-ton system is extremely large. The large tonnage machines may be manufactured in sections for ease of rigging and moving into machine rooms, Figure 47–44. These machines are typically assembled in the factory for proper alignment except for the welds that must be accomplished in the field.

The absorption process is very different in some respects but also similar to the compression cycle refrigeration process in others. It would be helpful to understand the

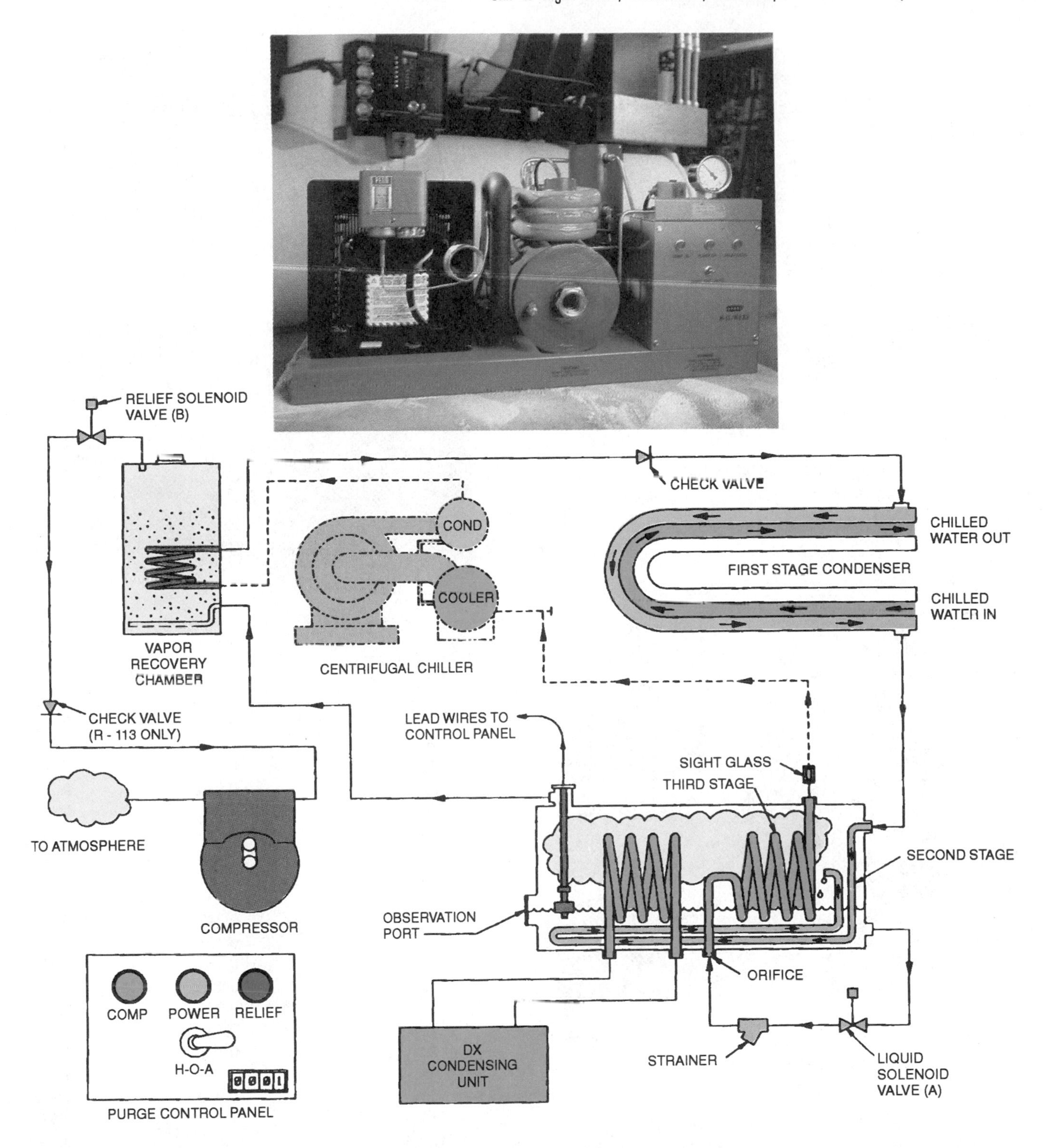

FIGURE 47–42 Modern purge with features that reduce the amount of refrigerant that escapes with the noncondensables.
Courtesy Carolina Products Inc.

compression cycle before trying to understand the absorption cycle. The boiling temperature for the refrigerant in compression cycle chillers is controlled by controlling the pressure above the boiling liquid. This boiling pressure is controlled in the compression cycle by the compressor. Figure 47–45 shows a refrigeration system using water as the refrigerant. The water must be boiled at 0.248 in. Hg absolute or 0.122 psia to boil the water at 40°F. Figure 47–46 shows a limited pressure/temperature chart for water. This refrigeration cycle would be workable using a compressor except that the volume of vapor rising from the boiling water is excessive. The compressor would have to remove 2444 ft^3 of water for every pound of water boiled at 40°F. Figure 47–47 shows a limited specific volume chart for water at different temperatures. A compressor is impractical for a refrigeration system using water as the refrigerant.

(A)

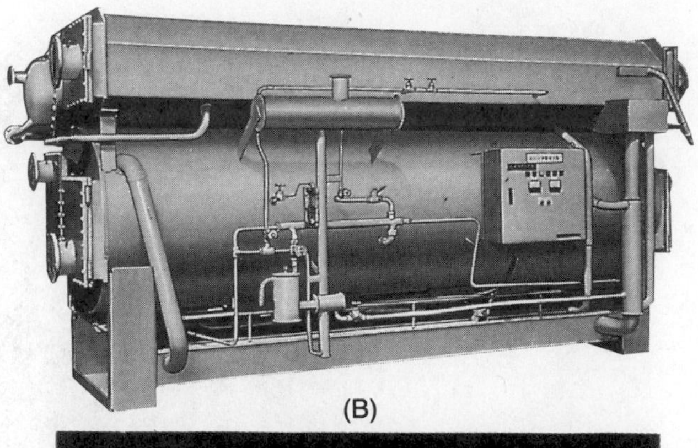

(B)

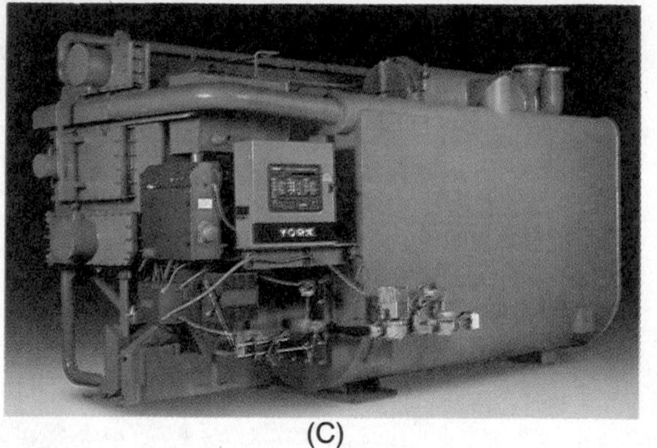

(C)

FIGURE 47–43 Absorption machines *(A) Courtesy Trane Company, (B) Reproduced courtesy of Carrier Corporation, (C) Courtesy York International*

FIGURE 47–44 This machine is assembled after being placed in the equipment room. *Reproduced courtesy of Carrier Corporation*

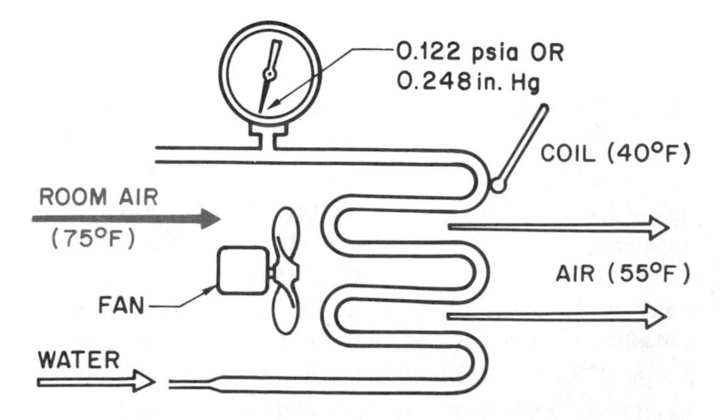

FIGURE 47–45 Using water as a refrigerant.

TEMPERATURE °F	ABSOLUTE PRESSURE lb/in.²	ABSOLUTE PRESSURE in. Hg
10	0.031	0.063
20	0.050	0.103
30	0.081	0.165
32	0.089	0.180
34	0.096	0.195
36	0.104	0.212
38	0.112	0.229
40	0.122	0.248
42	0.131	0.268
44	0.142	0.289
46	0.153	0.312
48	0.165	0.336
50	0.178	0.362
60	0.256	0.522
70	0.363	0.739
80	0.507	1.032
90	0.698	1.422
100	0.950	1.933
110	1.275	2.597
120	1.693	3.448
130	2.224	4.527
140	2.890	5.881
150	3.719	7.573
160	4.742	9.656
170	5.994	12.203
180	7.512	15.295
190	9.340	19.017
200	11.526	23.468
210	14.123	28.754
212	14.696	29.921

FIGURE 47–46 Pressure/temperature chart for water.

TEMPERATURE		SPECIFIC VOLUME OF WATER VAPOR	ABSOLUTE PRESSURE		
°C	°F	ft³/lb	lb/in.²	kPa	in. Hg
−12.2	10	9054	0.031	0.214	0.063
−6.7	20	5657	0.050	0.345	0.103
−1.1	30	3606	0.081	0.558	0.165
0.0	32	3302	0.089	0.613	0.180
1.1	34	3059	0.096	0.661	0.195
2.2	36	2837	0.104	0.717	0.212
3.3	38	2632	0.112	0.772	0.229
4.4	40	2444	0.122	0.841	0.248
5.6	42	2270	0.131	0.903	0.268
6.7	44	2111	0.142	0.978	0.289
7.8	46	1964	0.153	1.054	0.312
8.9	48	1828	0.165	1.137	0.336
10.0	50	1702	0.178	1.266	0.362
15.6	60	1206	0.256	1.764	0.522
21.1	70	867	0.363	2.501	0.739
26.7	80	633	0.507	3.493	1.032
32.2	90	468	0.698	4.809	1.422
37.8	100	350	0.950	6.546	1.933
43.3	110	265	1.275	8.785	2.597
48.9	120	203	1.693	11.665	3.448
54.4	130	157	2.224	15.323	4.527
60.0	140	123	2.890	19.912	5.881
65.6	150	97	3.719	25.624	7.573
71.1	160	77	4.742	32.672	9.656
76.7	170	62	5.994	41.299	12.203
82.2	180	50	7.512	51.758	15.295
87.8	190	41	9.340	64.353	19.017
93.3	200	34	11.526	79.414	23.468
98.9	210	28	14.123	97.307	28.754
100.0	212	27	14.696	101.255	29.921

FIGURE 47–47 Specific volume chart for water.

The absorption machine uses water for the refrigerant but does not use a compressor to create the pressure difference. It uses the fact that certain salt solutions have enough attraction for water that they may be used to create the pressure difference. You may have noticed that a grain of salt on a hot humid day will attract moisture to the point that it will become wet, Figure 47–48. This same concept is used in an absorption refrigeration system to reduce the pressure of the water to a point where it will boil at a low temperature. A type of salt solution called lithium-bromide (Li-Br) is used as the absorbent (attractant) for the water. Li-Br is a solution in the liquid form as used in absorption cooling systems. The liquid solution of Li-Br is diluted with distilled water. It is actually a mixture of about 60% Li-Br and 40% water. The term absorption means to suck up moisture.

A simplified absorption system would look like the one in Figure 47–49. This is not a complete cycle but is simplified for ease of understanding. The absorption cycle can be followed in this progressive description.

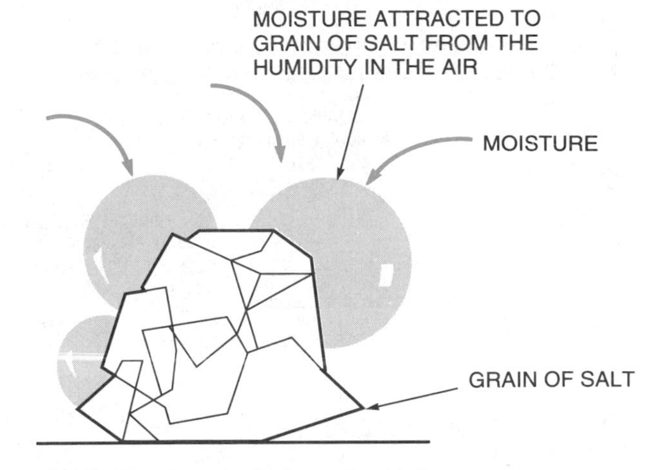

FIGURE 47–48 A grain of salt in a humid climate.

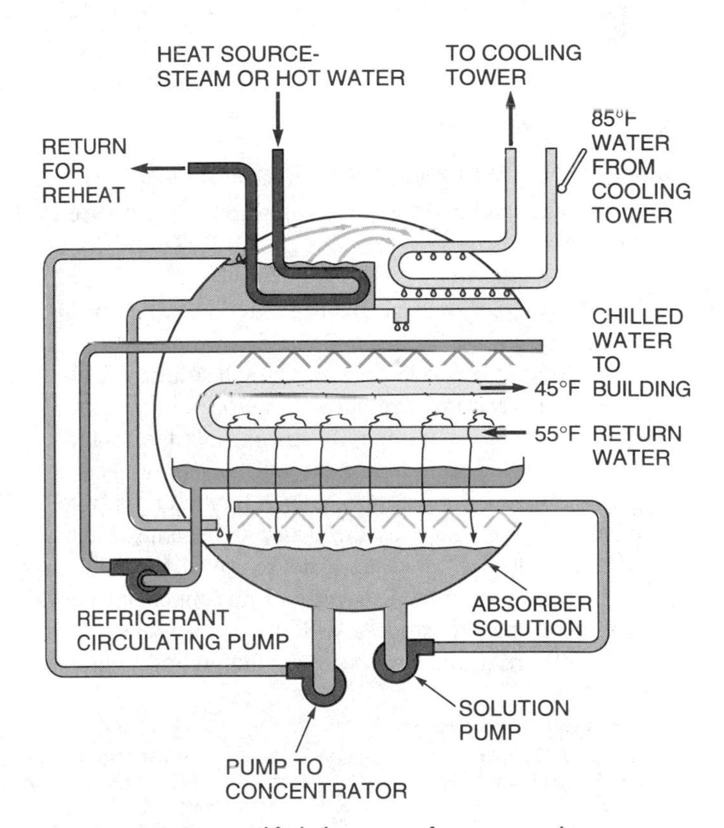

FIGURE 47–49 Simplified absorption refrigeration machine.

1. Figure 47–50 shows the equivalent of the evaporator system. Refrigerant (water) is metered into the evaporator section through a restriction (orifice). It is warm until it passes through the orifice where it flashes to a low-pressure area (about 0.248 in. Hg absolute or 0.122 psia). The reduction in pressure also reduces the temperature of the water. The water drops to the pan below the evaporator tube bundle. A refrigerant circulating pump circulates the water through spray heads to be sprayed over the evaporator tube bundle. This wets the tube bundle through which the circulating water from the system passes. The heat from the system water evaporates the refrigerant. Water is constantly being

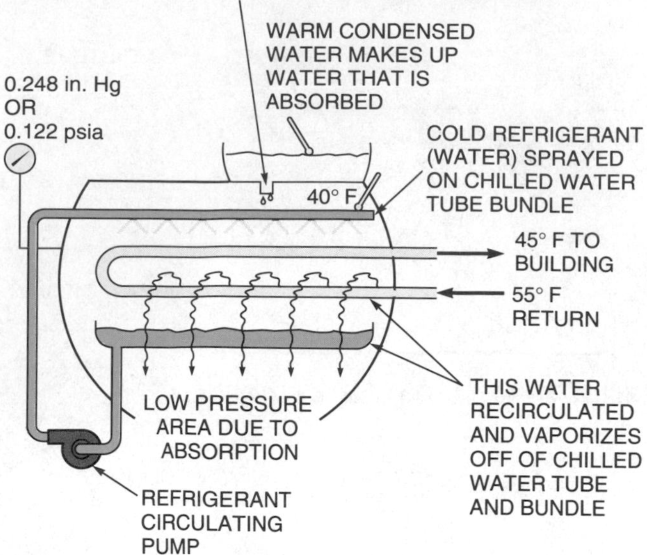

FIGURE 47–50 Evaporator section of absorption machine.

evaporated and must be made up through the orifice at the top.

2. Figure 47–51 shows the absorber section, which would be the equivalent of the suction side of the compressor in a compression cycle system. The salt solution (Li-Br) spray is a very low-pressure attraction for the evaporated water vapor so it readily absorbs into the solution. The solution is recirculated through the spray heads to give the solution more surface area to attract the water. As the solution absorbs the water, it becomes diluted with the water. If this water is not removed, the solution will become so diluted that it will no longer have any attraction and the process will stop, so another pump constantly removes some of the solution and pumps it to the next step called the concentrator. This solution that is pumped to the concentrator is called the weak solution because it contains water absorbed from the evaporator.

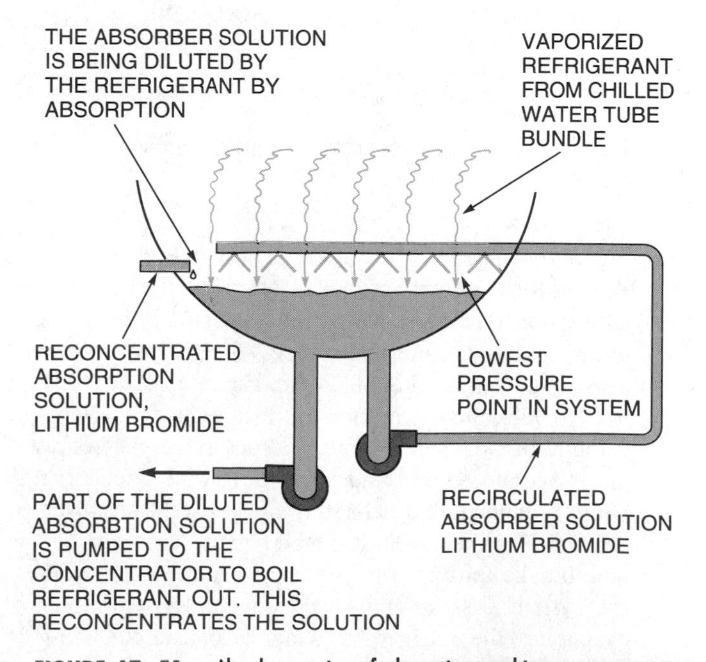

FIGURE 47–51 Absorber section of absorption machine.

3. Figure 47–52 shows the concentrator and condenser segment. The dilute solution is pumped to the concentrator where it is boiled. This boiling action changes the water to a vapor; it leaves the solution and is attracted to the condenser coils. The water is condensed to a liquid where it gathers and is metered to the evaporator section through the orifice. The heat source to boil the water is either steam or hot water in this example. Direct-fired machines are also available and are discussed later. The concentrated solution is drained back to the absorber area for circulation by the absorber pump.

As you can see from this description, the absorption process is not complicated. The only moving parts are the pump motors and impellers.

Some other features are added to the cycle to make it more efficient. One of these is shown in Figure 47–53. It is a heat exchange between the cooling tower water and the absorber solution in the absorber. This heat exchange removes heat that is generated when the water vapor is absorbed into the absorber solution.

Figure 47–54 is a heat exchanger between the dilute and the concentrated solution. This heat exchange serves two purposes; it preheats the dilute solution before it enters the concentrator and precools the concentrated solution before it enters the absorber section. It is much like the heat exchange between the suction and liquid line on a house-

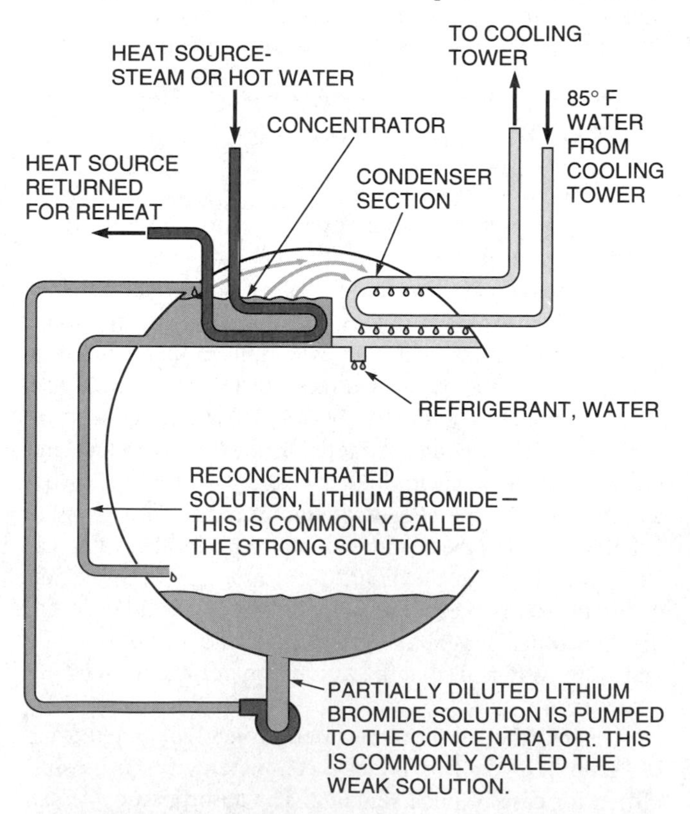

FIGURE 47–52 Concentrator and condenser section of absorption machine.

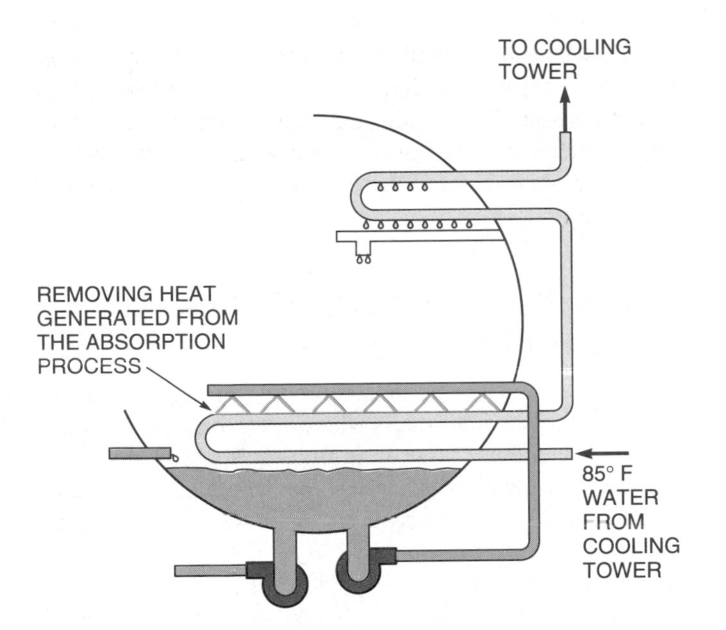

FIGURE 47–53 Heat exchange in the absorber to increase efficiency.

hold refrigerator. Without this heat exchange, the machine would be less efficient.

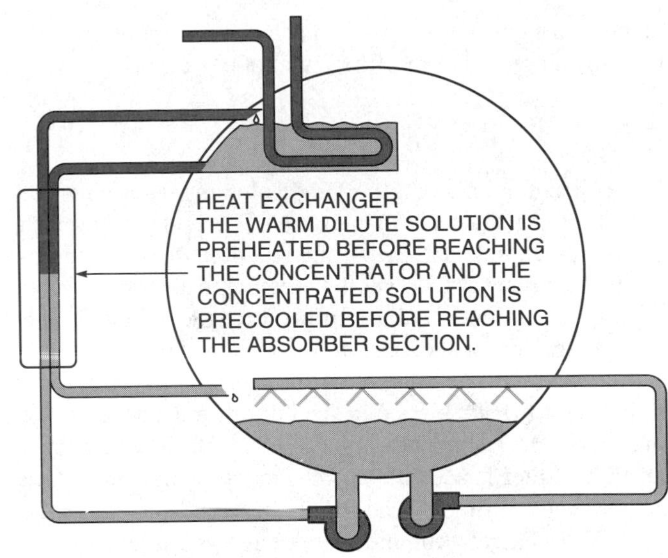

FIGURE 47–54 Heat exchanger between the solutions.

Some manufacturers have developed a two-stage absorption machine that uses a higher pressure steam or hot water, Figure 47–55. The steam pressure may be 115 psig

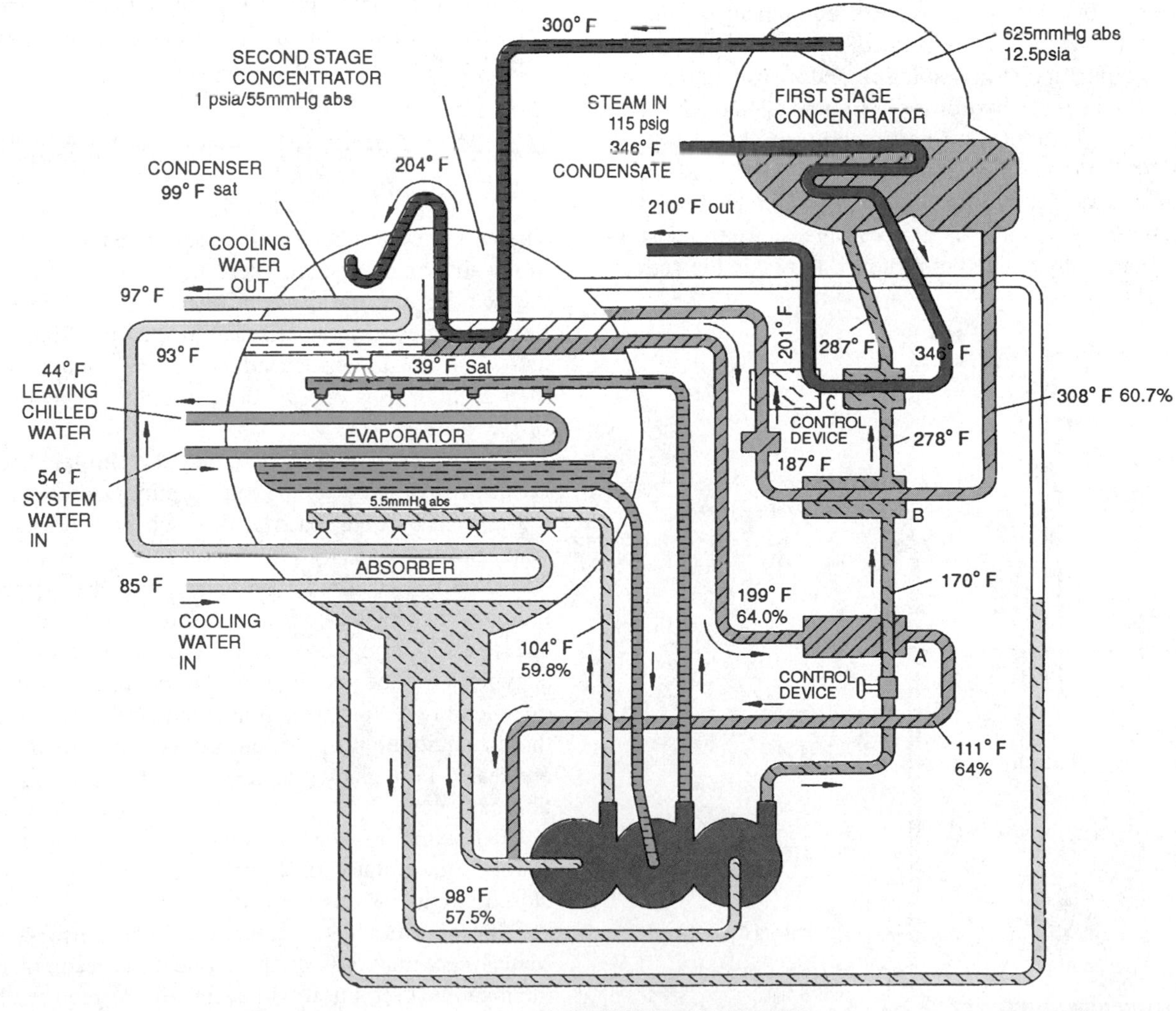

FIGURE 47–55 Two-stage absorption machine. *Courtesy Trane Company*

or hot water at 370°F. These machines are more efficient than the single-stage machine mentioned earlier.

47.27 SOLUTION STRENGTH

The concentration strength of the solutions determines the ability of the machine to perform. The wider the spread between the dilute and the strong solution, the more capacity the machine has to lower the pressure to absorb the water from the evaporator. Adjusting the strengths of these solutions is the job of the start-up technician. Some machines are shipped with no charge and the charge is added in the field. The Li-Br is shipped in steel drums to be added to the machine. The estimated amount of Li-Br is added first. Distilled water is used as the refrigerant charge and added next. When the charge is adjusted, the technician calls it the trim. You charge a compression cycle system and trim an absorption system.

Typically, a technician trims the machine with the approximate correct amount of Li-Br and water, then starts the machine. The machine is then gradually run up to full load. Full load may be determined by the temperature drop across the evaporator and full steam pressure, which is normally 12 to 14 psig (hot water or direct-fire equipment would be similar). When full load is obtained, the technician pulls a sample of dilute and strong solution and measures the specific gravity using a hydrometer, Figure 47–56. Specific gravity is the weight of a substance compared to the weight of an equal volume of water. The technician would know what these specific gravity readings should be from the manufacturer's literature or a Li-Br pressure/temperature chart. Water is either added or removed to obtain the correct specific gravity.

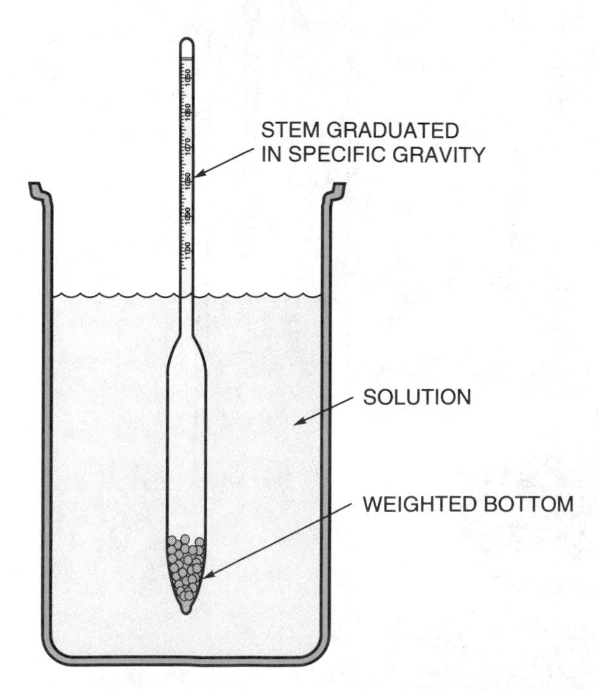

FIGURE 47–56 Hydrometer for measuring the specific gravity of absorption solutions.

This may be a relatively long process in some cases because the technician cannot always obtain full-load operation at start-up. The machine may have to be trimmed to an approximate trim and finished later when full load can be achieved. Some manufacturers furnish a trim chart for a partial load.

47.28 SOLUTIONS INSIDE THE ABSORPTION SYSTEM

The solutions inside the absorption system are corrosive. Rust is oxidation and occurs where corrosive materials and oxygen are present. The absorption machine is manufactured with materials such as steel and contains a salt water solution that will corrode if air is present. It is next to impossible to manufacture and keep a system from ever letting air with oxygen be exposed to the working parts.

These solutions must be circulated through the system and the system must be kept as clean as possible. Some of the passages that the solution must circulate through are very small, such as the spray heads in the absorber. Manufacturers use different methods of removing solid materials that may clog any small passages. Filters will stop solid particles and magnetic attraction devices will attract steel particles that may be in circulation.

47.29 CIRCULATING PUMPS FOR ABSORPTION SYSTEMS

The solutions must be circulated throughout the various parts of the absorption system. There are two distinct fluid flows: the solution flow and the refrigerant flow. Some absorption machines use two circulating pumps, Figure 47–57, and some use three, depending on the manufacturer. One manufacturer uses one motor with three pumps on the end of the shaft.

Whatever the pump type, they are similar. The pumps are centrifugal type pumps and the pump impeller and shaft must be made of a material that will not corrode or they could never be serviced after they have been operated and corrosion has occurred. The pumps must be driven with a motor and special care must be taken not to let the atmosphere enter during the pumping process. Because of this, it is typical for the motor to be hermetically sealed and to operate only within the system atmosphere. The pump is in the actual solution to be pumped, but the motor windings are sealed in their own atmosphere where no system solution can enter.

These motors must be cooled so cold refrigerant water from the evaporator or normal supply water is used in a closed circuit for this purpose.

Motors must also be serviced over a period of time. Many years may pass before servicing is recommended by the manufacturer, but it will eventually be required because there are moving parts with bearings. Different manufacturers require different procedures for motor servicing, but it is much easier to service the motor and drive if the solu-

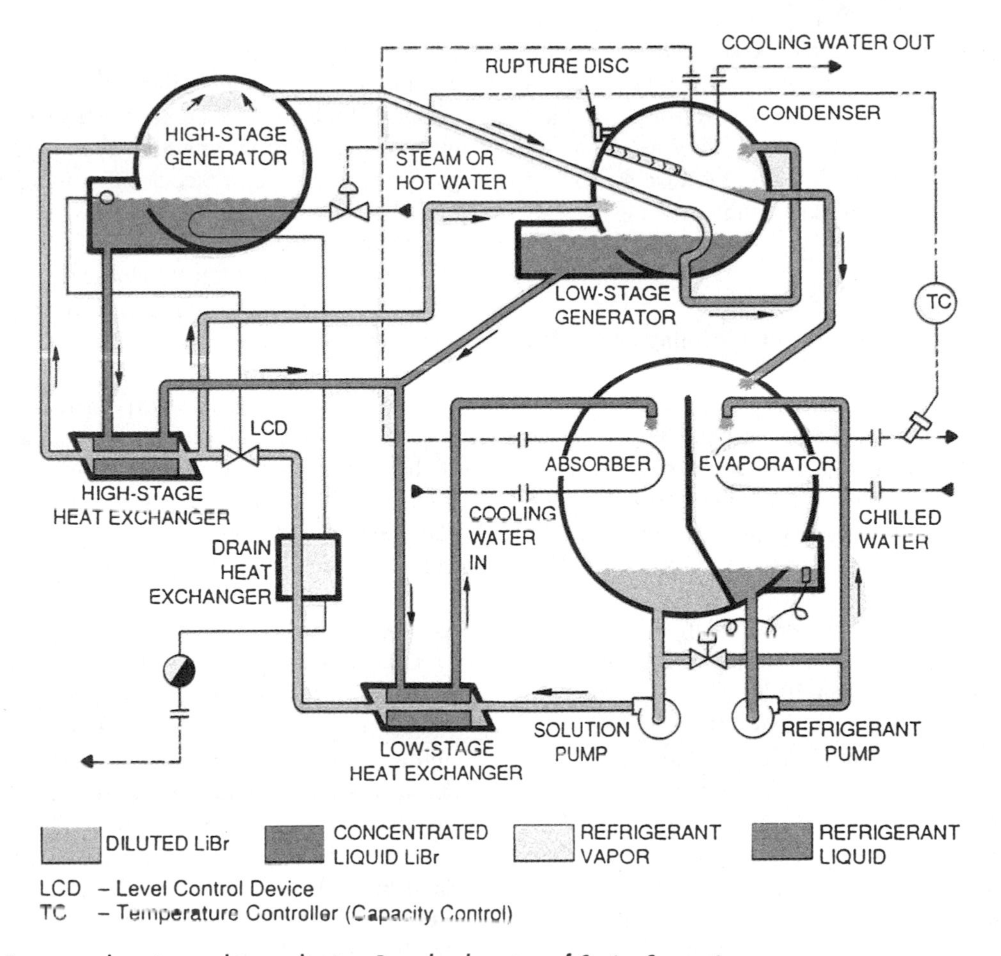

FIGURE 47–57 Two-pump absorption machine application. *Reproduced courtesy of Carrier Corporation*

tion does not have to be removed. However, if the solution must be removed, the process involves pressurizing the system with dry nitrogen to above atmospheric pressure and pushing the solution out. Once this is accomplished and service completed, the nitrogen must be removed, which is a long process. Then the system must be recharged.

47.30 CAPACITY CONTROL

The capacity can be controlled for a typical absorption system by throttling the supply of heat to the concentrator. A typical system that operates on steam uses 12 to 14 psig of steam at full load and 6 psig at half capacity. The steam can be controlled with a modulating valve, Figure 47–58. Hot

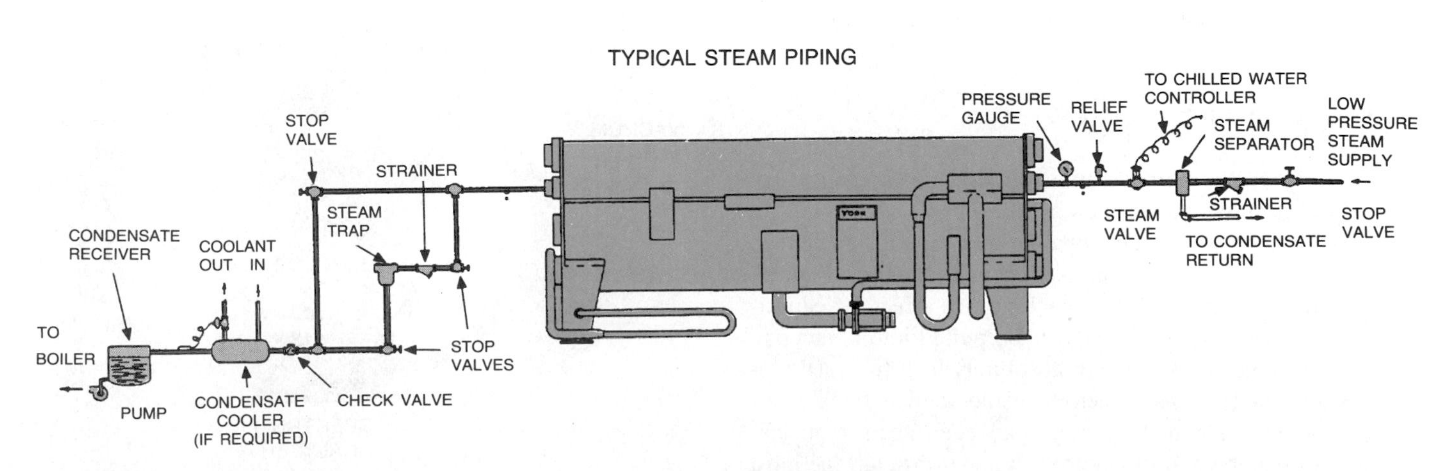

FIGURE 47–58 Capacity control using a modulating steam valve. *Courtesy York International*

water and direct-fired machines can be controlled in a similar manner. Other manufacturers may have other controls for capacity control. These may control the internal fluid flow in the machine. For example, the flow of weak (dilute) solution to the concentrator is practical because at reduced load, less solution needs to be concentrated.

47.31 CRYSTALLIZATION

The use of a salt solution for absorption cooling creates the possibility of the solution becoming too concentrated and actually turning back to rock salt. This is called *crystallization* and may occur if the machine is operated under the wrong conditions. For example, with some systems if the cooling tower water is allowed to become too cold while operating at full load, the condenser will become too efficient and remove too much water from the concentrate. This will result in a strong solution that has too little water. When this solution passes through the heat exchanger, it will turn to crystals and restrict the flow of the solution. If this is not corrected, a complete blockage will occur and the machine will stop cooling. Because this is a difficult problem to overcome when it happens, manufacturers have developed various methods to prevent this condition. One manufacturer uses pressure drop in the strong solution across the heat exchanger as a key to the problem. The action taken may be to open a valve between the refrigerant circuit and the absorber fluid circuit to make the weak solution very weak for long enough to relieve the problem. When the situation is corrected, the valve is closed and the system resumes normal operation. Another manufacturer may shut the machine down for a dilution cycle when overconcentration occurs.

Crystallization can occur for several reasons. Cold condenser water is one cause, and a shutdown of the machine due to power failure while operating at full load is another. An orderly shutdown calls for the solution pumps to operate for several minutes after shutdown to dilute the strong solution. With a power failure, the shutdown is not orderly and crystallization can occur.

Because the machine operates in a vacuum, atmospheric air can be pulled into the machine at any point where a leak occurs. Air in the system can also cause crystallization.

47.32 PURGE SYSTEM

The *purge system* removes noncondensables from the absorption machine during the operating cycle.

All absorption systems operate in a vacuum. If there is any source for a leak, the atmosphere will enter. A soft drink bottle full of air in a 500-ton machine will affect the capacity. The air will expand greatly when pulled into a vacuum. These machines must be kept absolutely leak free. All piping of the solutions have factory-welded connections wherever possible. When the typical packaged absorption machine is assembled at the factory, it is put through the most rigid of leak tests, called the mass spectrum analysis. For this test, the machine is surrounded with an envelope of helium. The system is pulled into a deep vacuum and the exhaust of the vacuum pump is analyzed with a mass spectrum analyzer for any helium, Figure 47–59. Helium is a gas with very small molecules that will leak in at any leak source.

Even with these rigid leak-check and welding procedures, the system is subject to leaks developing during shipment to the job site. Leaks may also develop after years of operation and maintenance. When a leak occurs, the noncondensables must be removed.

Absorption systems also generate small amounts of noncondensables while in normal operation. A by-product of the internal parts causes hydrogen gas and some other noncondensables to form inside during normal operation. These are held to a minimum with the use of additives, but they still occur and it is the responsibility of the purge and the operator to keep the machine free of these noncondensables.

Two kinds of purge systems are used on typical absorption cooling equipment: the nonmotorized and the motorized purge. The nonmotorized purge uses the system pumps to create a flow of noncondensable products to a chamber where they are collected and then bled off by the machine operator, Figure 47–60. This purge operates well while the machine is in operation but may not be of much use if the machine were to need service and is pressured to atmosphere using nitrogen. This machine manufacturer offers an optional motorized purge for removal of large amounts of noncondensables.

The motorized purge is essentially a two-stage vacuum pump that removes a sample of whatever gas is in the absorber and pumps it to the atmosphere. These vapors are not harmful to the atmosphere. Because the absorber operates in such a low vacuum, this gas sample would only contain noncondensables or water vapor, Figure 47–61. The motorized purge requires some maintenance because of the

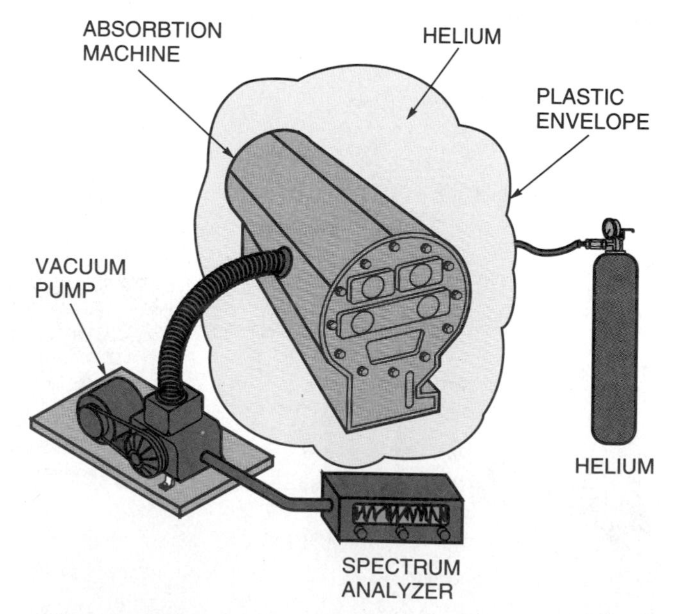

FIGURE 47–59 Leak checking an absorption machine in the factory.

MOTORLESS PURGE

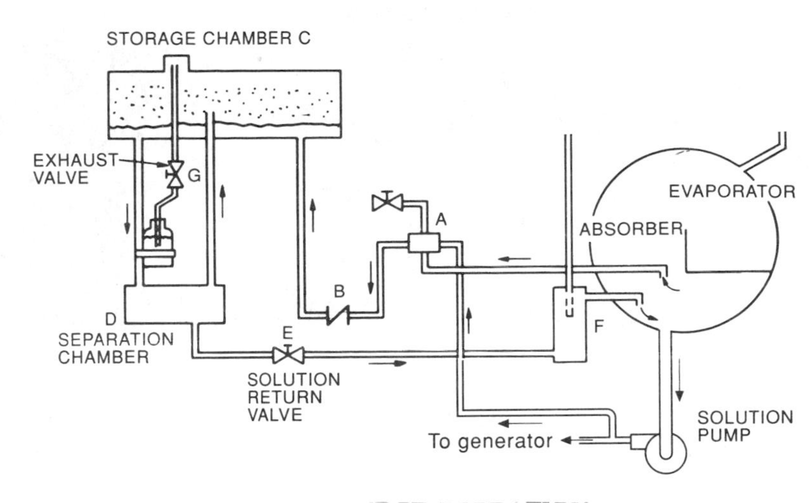

TYPICAL PURGE OPERATION SCHEMATIC

FIGURE 47–60 Collection chamber for noncondensable gases. *Reproduced courtesy of Carrier Corporation*

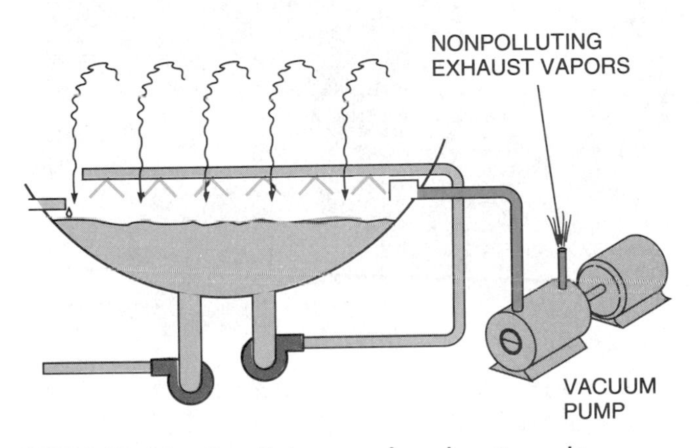

FIGURE 47–61 Nonpolluting vapors from absorption machine.

corrosive nature of the vapors that will be pulled through the vacuum pump. The vacuum pump oil must be changed on a regular basis to prevent vacuum pump failure. Vacuum pumps are expensive and must be properly maintained.

47.33 ABSORPTION SYSTEM HEAT EXCHANGERS

The absorption machine has heat exchangers as do the compression cycle chillers. There are actually more heat exchangers—the evaporator, the absorber, the concentrator, the condenser, and the first-stage heat exchanger—for two-stage systems. They are similar to compression cycle heat exchangers but have some differences. They also contain tubes. These tubes are either copper or cupronickel (a tube made of copper and nickel).

The chilled-water heat exchange tubes remove heat from the building water and add it to the refrigerant (water). The chilled-water circuit is usually a closed circuit, so the tubes rarely need maintenance except for water treatment. A flange at the end of the tube bundle allows access. These machines have either marine or standard water boxes for this access. This section of the absorption system must have some means of preventing heat from the equipment room from transferring into the cold machine parts. Some refrigeration systems are insulated in this section and some have a double-wall construction to prevent heat exchange. Like the compression cycle chillers, there is an approach temperature between the boiling refrigerant and the leaving chilled water. This approach is likely to be 2°F or 3°F for an absorption chiller because this heat exchange is good, Figure 47–62.

The absorber heat exchanger exchanges heat between the absorber solution and the water returning from the cooling tower. The design cooling tower water under full load in hot weather is 85°F. The water from this heat exchange continues to the condenser where the refrigerant vapor is condensed for reuse in the evaporator. The leaving water is likely to be from 95°F to 103°F on a hot day at full load, Figure 47–63.

The other heat exchange for standard equipment is between the heating medium: steam, hot water, or flue gases and the refrigerant. This tube bundle may have some high temperature differences because it may be at room temperature until the system is started and hot water or steam is turned into the bundle. A great deal of stress in the form of tube expansion occurs at this time. The manufacturers claim that a tube can grow in length up to 1/4 in. during this process. Different manufacturers deal with this stress in

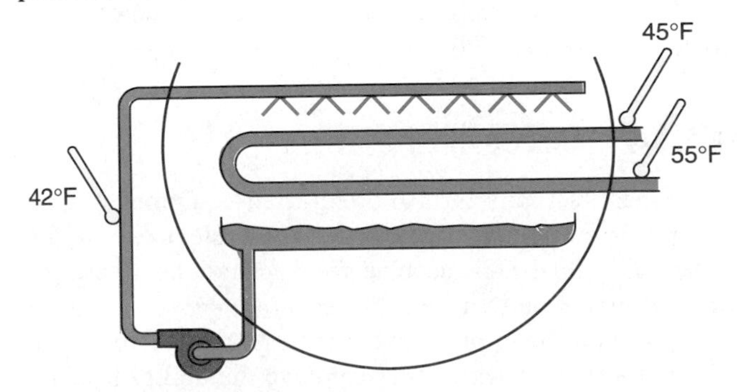

FIGURE 47–62 Chilled-water to refrigerant approach.

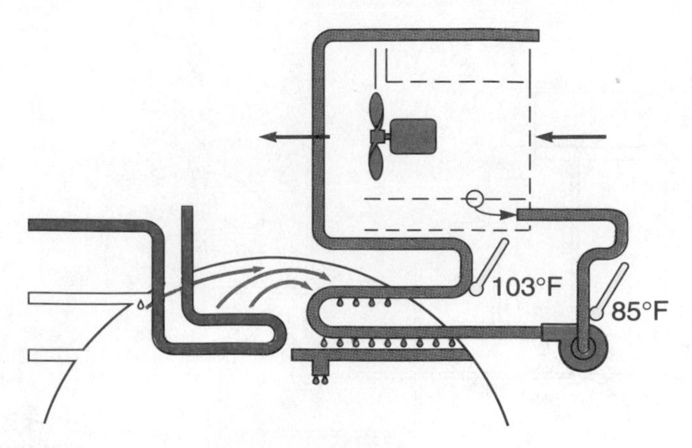

FIGURE 47–63 Cooling tower circuit.

different ways. Figure 47–64 shows a tube bundle that has the capability of floating to minimize problems.

Some absorption equipment manufacturers provide thermometer wells located at strategic spots to check the solution temperatures for assessing machine performance and to determine if the tubes are dirty. These manufacturers provide the correct temperatures and differences for their machines.

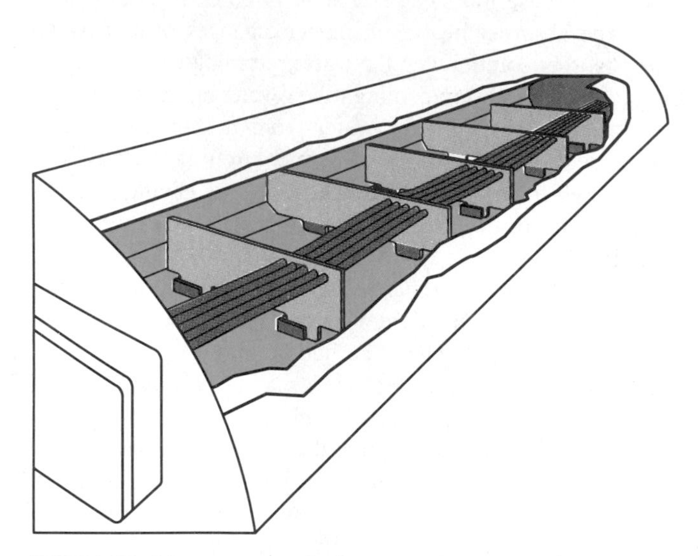

FIGURE 47–64 Expansion of tubes in an absorption machine. *Courtesy Trane Company*

47.34 DIRECT-FIRED SYSTEMS

Some manufacturers furnish direct-fired equipment. These use gas or oil for the heat source. The system can be furnished as a dual-fuel machine for applications where gas demand may require a second fuel.

These machines range in capacities from approximately 100 to 1500 tons. They also may have the ability to either heat or cool by furnishing either hot or chilled water. In some cases they may be used in an old installation to replace a boiler and an old absorption machine as one piece of equipment. Figure 47–65 shows direct-fired equipment manufactured by two different companies.

(A)

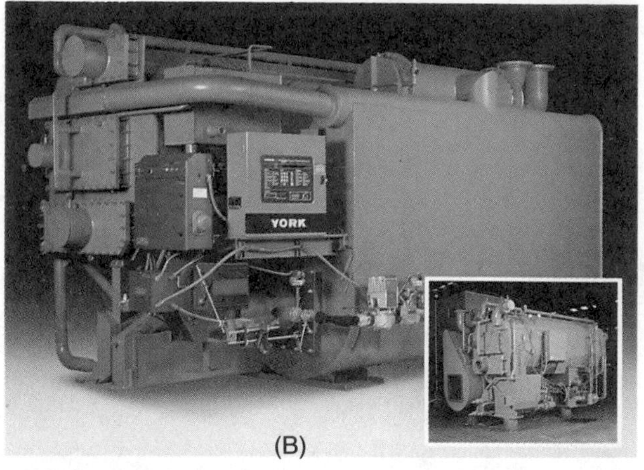

(B)

FIGURE 47–65 Direct-fired absorption machines. *Courtesy (A) Trane Company, (B) York International*

47.35 MOTORS AND DRIVES FOR COMPRESSION CYCLE CHILLERS

All of the motors that drive high-pressure and low-pressure chillers are highly efficient three-phase motors. These motors all give off some heat while in operation. This heat must be removed from the motor or it will become overheated and burn. Some manufacturers choose to keep the motor in the atmosphere for cooling. These motors are air cooled and give off heat to the equipment room, Figure 47–66. This

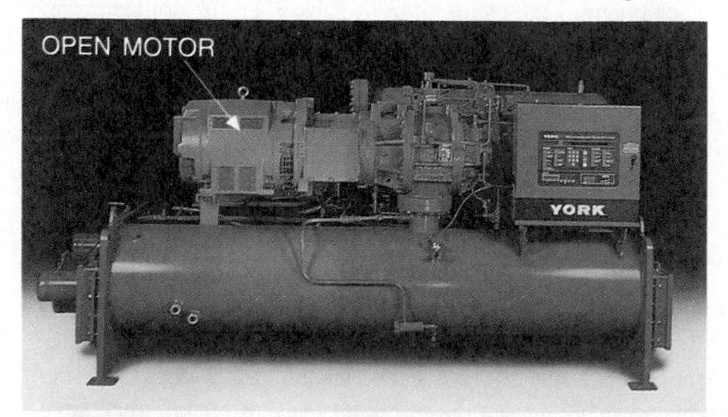

FIGURE 47–66 Air-cooled direct drive compressor motor. *Courtesy York International*

heat must be removed or the room will overheat. The heat is typically removed using an exhaust fan system. This compressor must have a shaft seal to prevent refrigerant from leaking to the atmosphere, Figure 47–67. In the past this has

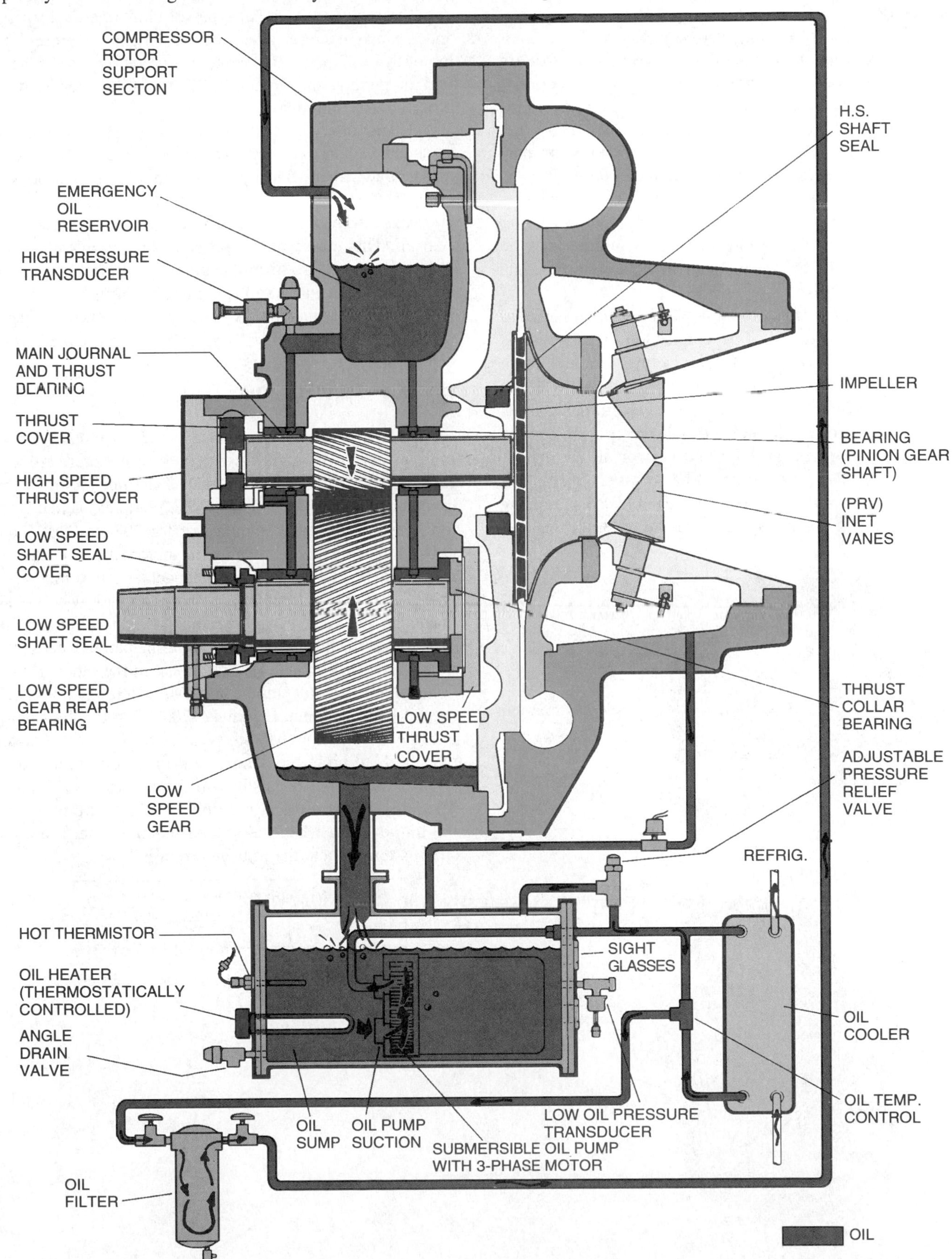

FIGURE 47–67 Shaft seal for direct-drive compressor. *Courtesy York International*

been a source of leaks, but modern seals and correct shaft alignment have improved the success with open-drive compressors.

Compressor motors may also be cooled with the vapor refrigerant leaving the evaporator. These compressors are called suction-cooled compressors and are either hermetic or semihermetic compressors. The fully hermetic compressors are in the smaller sizes and may be returned to the factory for rebuild. The semihermetic compressors may have the motor or the compressor rebuilt in the field.

Compressor motors may also be cooled with liquid refrigerant by allowing liquid to enter the motor housing around the motor. The liquid does not actually touch the motor or windings; it is contained in a jacket around the motor, Figure 47–14.

Large amounts of electrical power must be furnished to the motor through a leak-free connection. This power is often furnished through a nonconducting terminal block made of some form of phenolic or plastic with the motor terminal protruding through both sides. The motor terminals are fastened on one side and the field connections on the other with "O" rings used for the seal on the terminals, Figure 47–68. The terminal board has a gasket to seal it to the compressor housing. This terminal block is normally checked periodically for tightness of the field connections.

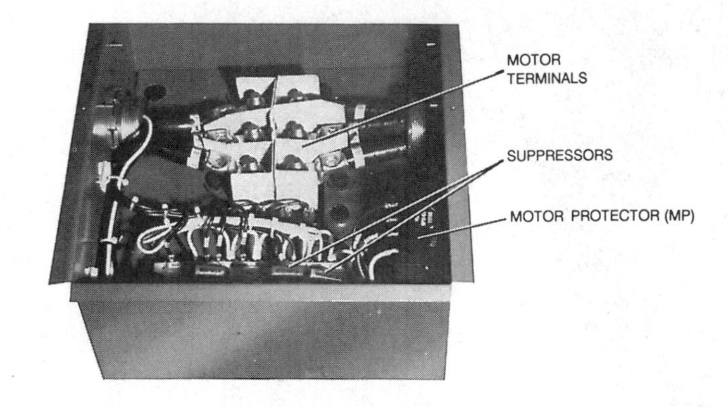

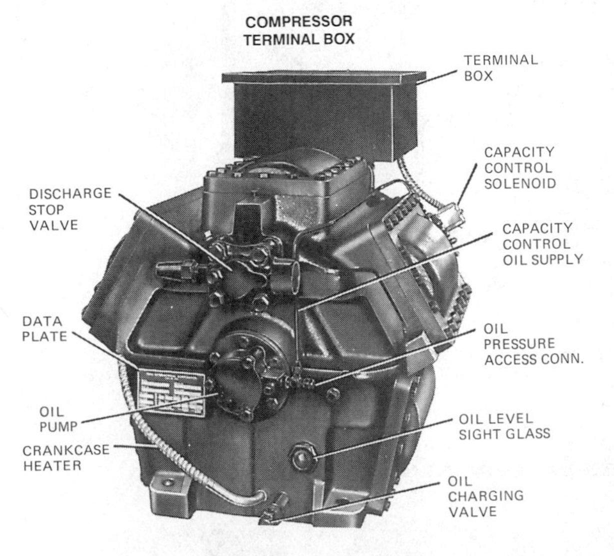

FIGURE 47–68 Terminal block for hermetic compressor motor. *Courtesy York International*

A loose connection may melt the board and cause a leak to develop.

The motors used in all types of chillers are expensive and the manufacturer goes to great lengths to protect the motor by designing it to operate in a protective atmosphere. Part of getting long motor life is to use a good start-up procedure for the motor.

A large motor is usually started with a group of components called a *starter*. There are several types of starters. A motor draws about five times as many amperes at locked rotor on start-up as at full-load amperes. If a motor draws 200 A at full load, it would draw approximately 1000 A at start-up. This inrush of current can cause problems in the electrical service so manufacturers use several different methods to start motors to minimize the inrush of current and power line fluctuations. The common ones are part-winding, autotransformer, wye-delta (often called star-delta), and electronic start.

47.36 PART-WINDING START

When compressor motors reach about 25 hp, manufacturers often use part-winding start motors. These motors are versatile and normally have nine leads. The same motor can be used for two different voltages. The compressor manufacturer can use the same compressor for 208- to 230-V and 460-V applications by changing the motor terminal arrangement, Figure 47–69. This motor is actually two motors in one. For example, a 100-hp compressor has two motors inside that are 50 hp each. When connected for 208- to 230-V application, the motors are wired in parallel and each motor is started separately, first one then the other. This is done with two motor starters and a time delay of about one second between them, Figure 47–70. When the first motor is started, the motor shaft starts turning. The second motor then starts to bring the compressor up to full speed, about 1800 or 3600 rpm depending on the motor's rated speed. This only imposes the inrush current of a 50-hp motor on the line because when the second motor is energized, the shaft is turning and the inrush current is less.

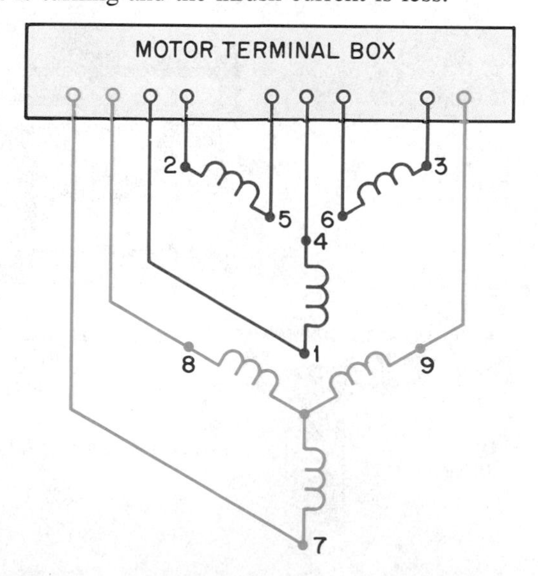

FIGURE 47–69 Nine-lead, dual-voltage motor connections.

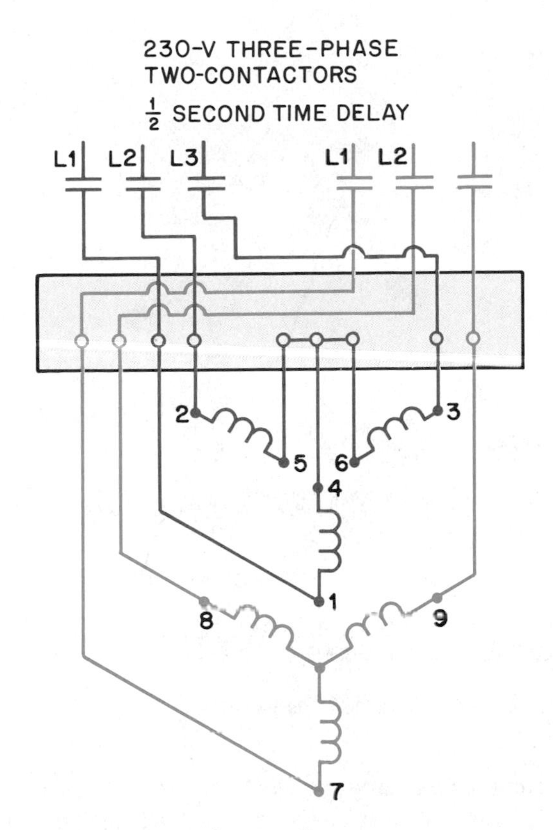

FIGURE 47–70 Part-winding start for a motor.

When the motor is used for 460-V application, the motors are wired in series and are started as one motor across the line, Figure 47–71. The higher voltage has much less inrush amperage on start-up.These motors are found on compressors up to about 150 hp.

47.37 AUTOTRANSFORMER START

An autotransformer start installation is actually a reduced voltage start. A transformer-like coil is placed between the

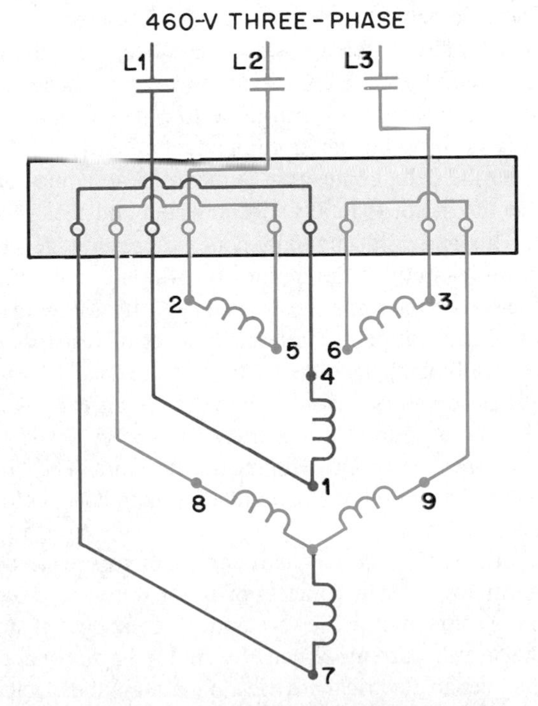

FIGURE 47–71 Motors are in series for 460-V application.

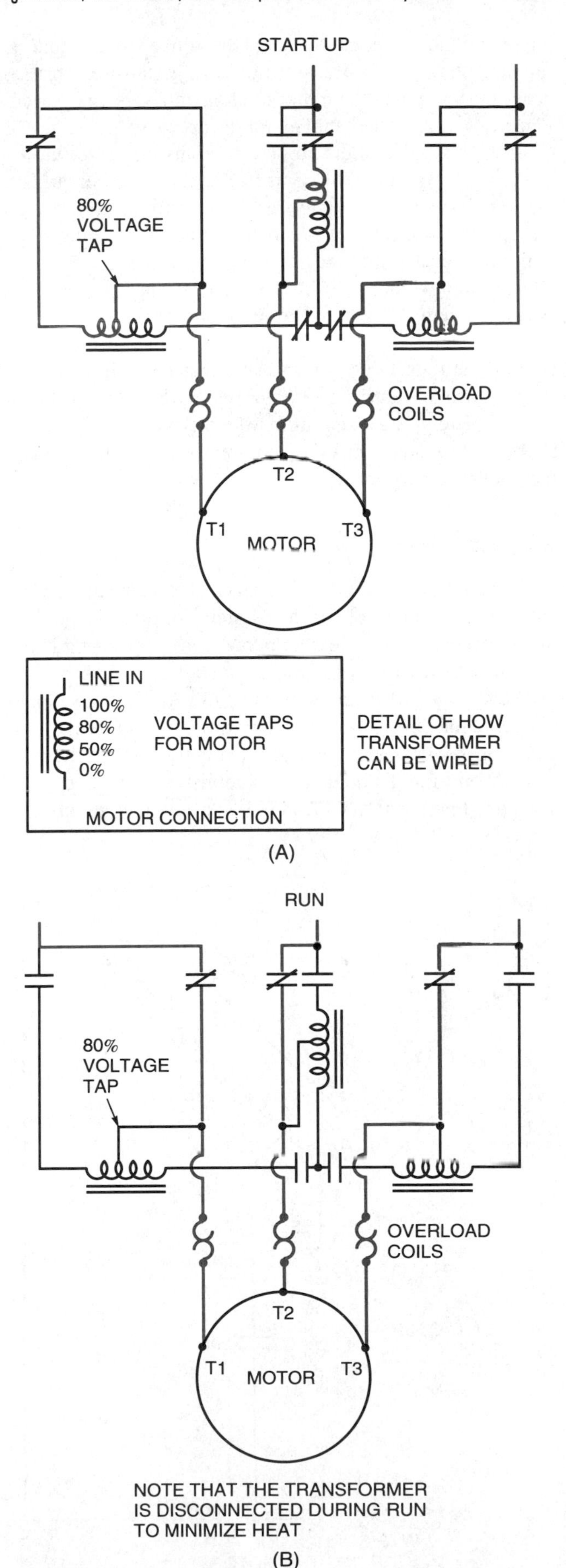

FIGURE 47–72 Autotransformer start.

motor and the starter contacts and the voltage to the motor is supplied through the transformer during start-up. This reduces the voltage to the motor until the motor is up to speed, Figure 47–72. When the motor is up to speed, a set of contacts close that short around the transformer to run the motor at full voltage. The contacts that direct power through the transformer will then open so that current does not go through the transformer when the motor is running. This minimizes heat in the area of the transformer.

The motor has very little starting torque because it is starting at reduced voltage so autotransformer start is used only in special applications. This type of motor start-up is common among large compressors that need little starting torque, such as centrifugal compressors where the pressures are completely equalized during the off cycle and compression does not start until the compressor is up to speed and the vanes start to open.

47.38 WYE-DELTA

Wye-delta start is often called star-delta start and is used with large motors with six leads and a single voltage. The motor typically draws less amperage when operating in the wye or star circuit so the motor is started in wye or star, Figure 47–73. After the motor is up to speed, a transition to a delta connection is made where the motor pulls the load and does the work more efficiently, Figure 47–74. The transition from star to delta is accomplished with a starting sequence that has three different contactors that are **electrically** and **mechanically interlocked**; two are energized in wye for start-up, then one is dropped out and the other engaged for delta operation. The time the motor operates in the wye connection depends on how long it takes the compressor to get up to speed. Large centrifugal compressors may take a minute or more to get up to speed in wye, then the transition is made to delta.

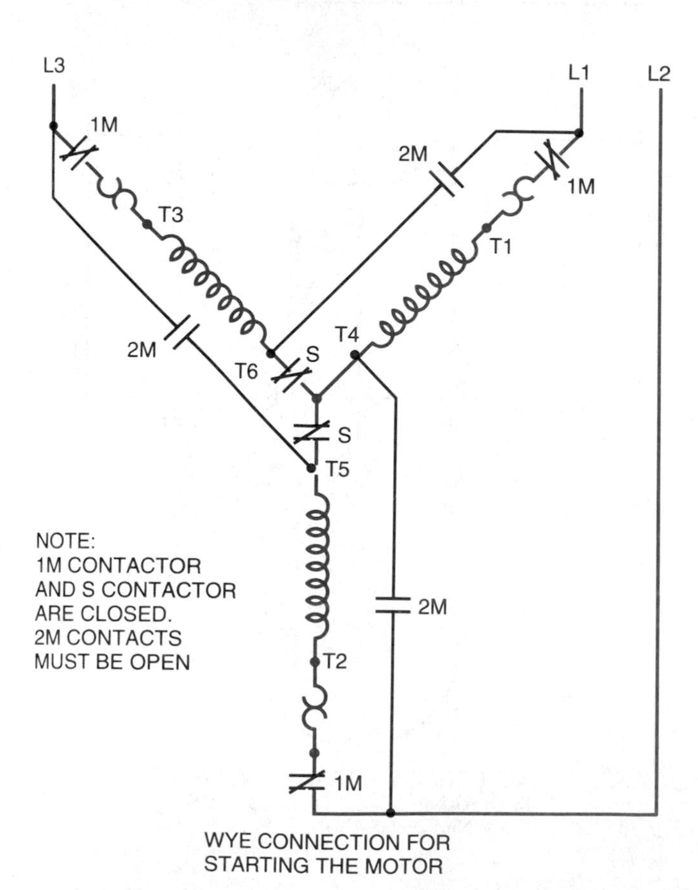

FIGURE 47–73 Starting a motor with wye circuit.

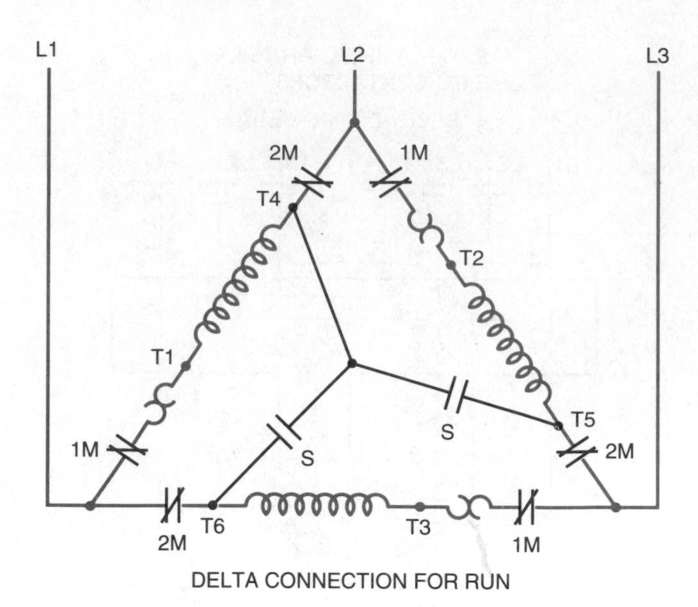

FIGURE 47–74 Delta connection for motor.

When the motor starter switches from wye to delta, the wye connection is actually disconnected electrically by means of a contactor. This disconnect is proven both electrically by the timing of the auxiliary contacts and mechanically by means of a set of levers that interlock the contactors. Then the delta connection is made. The electrical and mechanical interlock feature is necessary because if the wye connection and the delta connection were to be made at the same time, a dead short from phase to phase would occur and likely destroy the starter components, Figure 47–75.

When the delta connection is made, large amperage is drawn as the motor is totally disconnected and then reconnected. This spike of amperage could cause voltage problems in the vicinity of the motor, for example, a computer room located on the same service and close by. Many manufacturers furnish a motor starter with a set of resistors that are electrically connected and acting as the load while the transition is being made from wye to delta. This is known as a wye-delta closed transition starter. The starter would be a wye-delta open transition without the resistors. The larger the motor to be started, the more likely it is to have a closed transition starter.

All of the above starters use open contactors to start and stop the motors. When contacts of this nature are brought together to start a motor, they are under a great deal of stress with the inrush current. When the motor is disconnected from the line, an electrical arc tries to maintain the electrical path. This is much like an electrical welding arc and damage

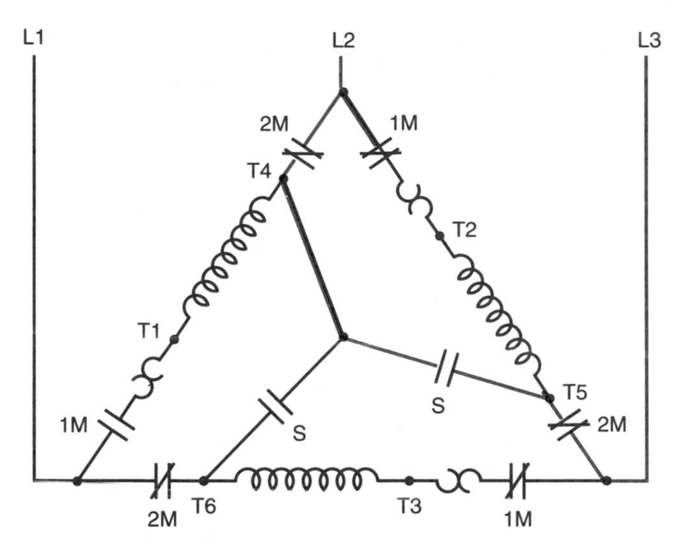
L1
L2
L3
2M
1M
T4
T2
T1
S
T5
1M
S
2M
2M
T6
T3
1M
2M AND S CONTACTS CANNOT BE CLOSED AT THE SAME TIME. THESE CONTACTS ARE KEPT SEPARATED BY ELECTRICAL INTERLOCKS AND BY LEVERS CALLED MECHANICAL INTERLOCKS.

FIGURE 47–75 Dead short for wye-delta start sequence.

to the contactor is caused each time the load is disconnected. The contacts become pitted due to this arc. The contacts will only be able to interrupt the load of a motor a limited number of times until replacement is required. Figure 47–76 shows a new and an old set of contacts for a medium-sized contactor.

The contacts in the starter will also have an arc shield to prevent the arc, when opening the contacts, from spreading to the next phase of contacts. This arc shield must be in place when starting and stopping the motor or severe damage may occur.

SAFETY PRECAUTION: *Starting and stopping a large motor using a starter requires a lot of electrical energy to be expended inside the starter cabinet. The operator should NEVER start and stop a motor with the starter door open. If something should happen inside the starter, particularly during start-up, most of this energy may be converted to heat energy and the result may be molten metal blown toward the door. Don't take the chance—shut the door.*

FIGURE 47–76 New and old set of contacts; the old ones are pitted from many starts. *Courtesy Square D Company*

47.39 ELECTRONIC STARTERS

As mentioned before for other starters, the inrush current can be five times the full-load current for a motor. This inrush current can be greatly reduced by using electronic starters. They are also called soft starters. These starters use electronic circuits to reduce the voltage and vary the frequency to the motor at start-up. The result can be a motor that would typically draw 1000 A at locked rotor to be able to start up with 100 A at locked rotor. The motor speed is then accelerated up to full speed. This start-up is much easier on all components and reduces power line voltage fluctuations.

The electronic starter also has the advantage of not having any open contacts like all of the above-mentioned starters.

47.40 MOTOR PROTECTION

The motor that drives the compressor is often the most expensive component in the system. The larger the compressor motor the more expensive and the more protection it should be afforded. Small hermetic compressors in household refrigerators have little protection. Motors used in small chillers will have motor protection like that covered in Unit 19. Here we discuss only the motor protection used for large chillers such as rotary screw or centrifugals.

The type of protection depends on the size of the system and the type of equipment used. The advances made in electronic devices for monitoring voltage, heat, and amperage have improved the protection offered today compared to the protection several years ago. However, older motor protectors are still in use and will be for many years to come.

47.41 LOAD-LIMITING DEVICES

Motors used on rotary screw and centrifugal chillers use load-limiting devices to control the motor amperage. The load-limiting device monitors the current the motor is drawing while operating and throttles the refrigerant to the suction of the compressor to prevent the motor from operating at an amperage higher than the full-load amperage. The load-limiting device controls the slide valve on a rotary screw or the prerotation vanes on a centrifugal, Figure 47–77. This device is a precision device that is adjusted at start-up and should not give any problems in the future. It is the first line of defense to prevent motor overload and is set at exactly the full-load amperage. **NOTE:** Full-load amperage may be derated to a lower value for a system that has high voltage so you cannot always go by the name plate amperage. Look at the start-up log for the derated amperage. ■

The load limiter may also have a feature that allows the operator to operate the chiller at reduced load manually.

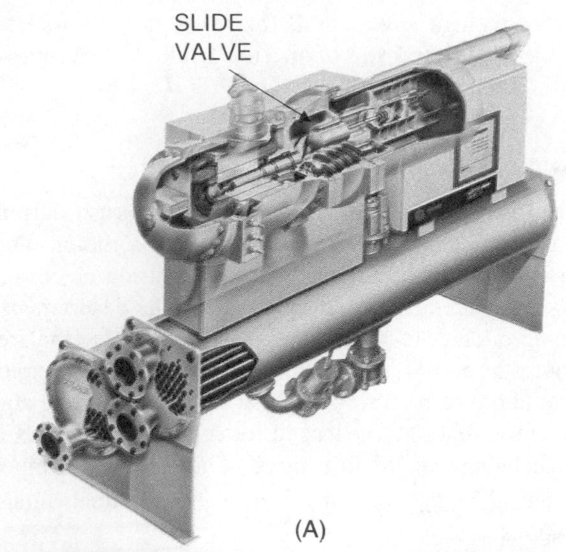

(A)

(B)

FIGURE 47–77 (A) Slide capacity control. (B) Prerotation vanes capacity control. *Courtesy (A) Trane Company, (B) York International*

Typically for rotary screw chillers this load may be varied from 10% to 100%. For centrifugal chillers it may be from 20% to 100%.

Most buildings are charged for electrical power based on the highest current draw for the month that lasted for some period of time, typically 15 to 30 min. This is called billing demand power charge with most power companies and is measured by a demand meter. Operating the equipment below the billing demand charge would be desirable whenever possible. For example, if an office building did not require cooling during the month until the last day and the operator started the chiller and allowed it to run at full load to reduce the building temperature quickly, this would take more than the required time to drive up the demand meter for the power company and the building would be charged for the whole month at a rate as though they had operated at full load all month. This excess could have been saved by an operator who started the system and kept it operating at part load and took longer to reduce the building temperature. For example, the chiller may have the feature that would allow it to run at 40% and take longer to accomplish the task.

The load limiter is set at motor full-load amperage. For example, assume a motor has a full-load amperage of 200 A. The load-limiting device will be set not to exceed full-load amperage, or 200 A.

47.42 MECHANICAL-ELECTRICAL MOTOR OVERLOAD PROTECTION

All motors must have some overload protection. A motor must not be allowed to operate at above the full-load amperage for long or damage will occur to the motor windings due to heat. Different motors have different types of protection. Overload protection for smaller motors, typically up to about 100 hp, was discussed in Unit 19.

Mechanical-electrical overload protection for large rotary screw or centrifugal compressors may be a simple dash-pot type of overload device. This device operates on the electromagnet theory that when a coil of wire is wound around a core of iron, the core of iron will move when current flows. With the dash-pot either the current from the motor or a branch of the current of the motor passing through a coil around an iron core will cause the iron core to rise, Figure 47–78.

Large motor overload protectors are typically rated to stop the motor should the current rise to 105% of the motor's full-load rating. For example, assume a motor has a rating of 200 A, the overload device would be set up to trip at 210 A (200 × 1.05 = 210). The motor should be allowed to run at this amperage for a few minutes to prevent unwanted overload trips before stopping the motor. These motors have a load-limiting device to limit the motor amper-

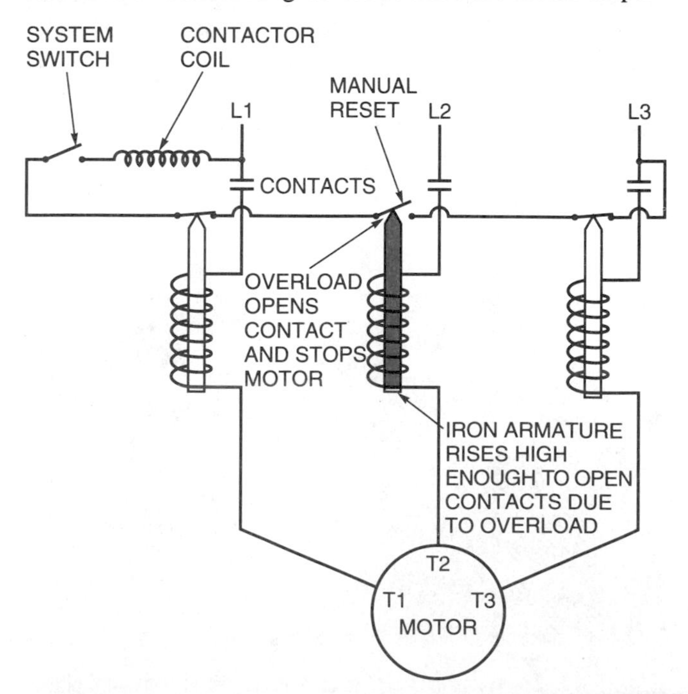

FIGURE 47–78 Iron core and current flow in a compressor overload.

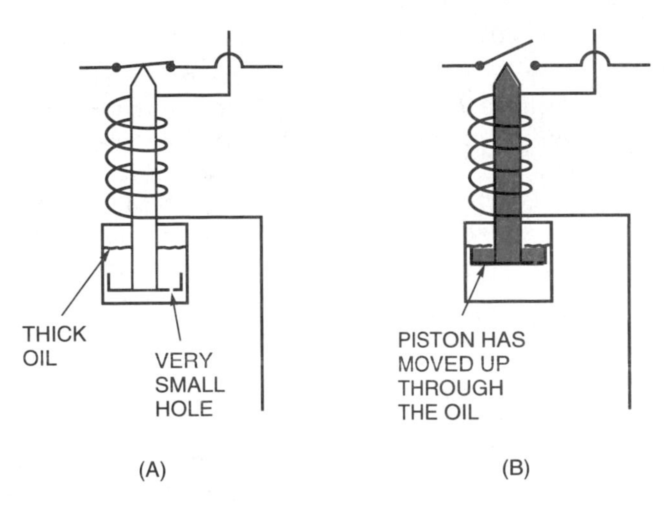

FIGURE 47–79 Dash-pot of oil gives the overload the time delay needed for motor start-up.

age, so if this device is functioning correctly, the amperage should not be excessive for longer than it takes this control to react. At this point, the amperage should be no more than full load.

The reason this overload device is called a dash-pot is because it contains a time delay to allow the motor to be able to be started. Without the time delay, the inrush current would trip the device. The time delay in the dash-pot is accomplished with a piston and a thick fluid that the piston must rise up through before tripping, Figure 47–79. The overload mechanism is wired into the main control circuit to stop the motor when it is tripped. Overload protection for large motors is a manual reset type so the operator will be aware of any problem.

47.43 ELECTRONIC SOLID-STATE OVERLOAD DEVICE PROTECTION

These protective devices are wired into the control circuit like the dash-pot but act and react as an electronic control. They have the ability to allow the motor to be started and also monitor the full-load motor amperage closely. They typically are located in the starter and are installed around the motor leads for accurate current monitoring.

47.44 ANTI-RECYCLE CONTROL

All large motors should be protected from starting too often over a specified period of time. The anti-recycle timer is a device that prevents the motor from restarting unless it has had enough running time or off time to dissipate the heat from the last start-up. Manufacturers have different ideas as to how much this time should be. Typically the larger the motor, the more time required. Many centrifugals have a 30-min time. If the motor has not been started or tried to start for 30 min, it is ready for a start. Check the manufacturer's literature for the recycle time for a specific chiller.

47.45 PHASE FAILURE PROTECTION

Large motors all use three-phase power. The typical voltages are 208, 230, 480, and 575 V. Higher voltages are used for some applications, such as 4160 V or 13,000 V. Whatever the voltage, all three phases must be furnished or the motor will overload immediately. Electronic phase protection monitors the power and ensures that all three phases are present.

47.46 VOLTAGE UNBALANCE

The voltage to a compressor must be balanced within certain limits. Usually the manufacturers use 2% voltage unbalance as the maximum.

$$\text{Voltage unbalance} = \frac{\text{Maximum deviation from average voltage}}{\text{Average voltage}}$$

For example, a technician measures the voltage on a nominal 460-V system to be:

Phase 1 to phase 2	475 V
Phase 1 to phase 3	448 V
Phase 2 to phase 3	461 V

The average voltage is:

$$\frac{475\text{ V} + 448\text{ V} + 461\text{ V}}{3} = 461.3\text{ V}$$

The maximum deviation from average is:

$$475\text{ V} - 461.3\text{ V} = 13.7\text{ V}$$

The voltage unbalance is:

$$\frac{13.7}{461.3\text{ V}} = 0.0297 \text{ or } 2.97\%$$

This is more than some manufacturers' maximum allowable. The operating technician should keep an eye on voltage unbalance because it causes the motor to overheat. Some sophisticated electronic systems have voltage unbalance protection as part of their features; otherwise, it is up to the equipment operator to monitor.

Phase unbalance is caused by the electrical load on the building being unbalanced or the power company supply voltage being out of balance.

47.47 PHASE REVERSAL

Three-phase motors turn in the direction in which they are wired to run. If the phases are reversed, the motor rotation will reverse. *This can be detrimental for many compressors.* Reciprocating compressors will normally perform in either direction because of bidirectional oil pumps. Scroll, rotary screw, and centrifugal compressors must turn in the correct direction. Phase protection is often part of the control package for these compressors. Any compressor with a separate oil pump, such as the centrifugal, will not start under these conditions because the oil pump will not pump when reversed. The scroll and rotary screw compressor will start up without protection.

SUMMARY

- A chiller refrigerates circulating water.
- There are two basic categories of chillers: the compression cycle and absorption chiller.
- The compression cycle chiller has the same four basic components as other refrigeration systems discussed previously in this text: the compressor, evaporator, condenser, and metering device.
- Compression cycle chillers may be classified as either high- or low-pressure systems.
- R-22 is the most popular refrigerant used in reciprocating compressor chillers.
- Cylinder unloading is used to control the capacity of a reciprocating compressor. This is accomplished through blocked suction or suction valve lift unloading.
- Scroll compressors used in chiller systems are usually in the 10- to 15-ton range.
- Rotary screw compressors are used for larger capacity chillers using high-pressure refrigerants.
- Centrifugal compressors may be used in high-pressure chiller single-stage systems by increasing speeds up to 30,000 rpm using a gear box.
- These compressors are lubricated with an oil pump and separate motor.
- The evaporators used for high-pressure chillers are either direct expansion or flooded evaporators.
- The refrigerant absorbs heat from the cooling water and is about 7°F cooler than the leaving water in a two-pass chiller. This is called the approach temperature.
- The condenser for high-pressure chillers may either be water or air cooled.
- Many condensers have subcooling circuits that reduce the liquid temperature.
- The condenser, like the evaporator, has an approach relationship between the refrigerant condensing temperature and the leaving-water temperature.
- Air-cooled condensers eliminate the need for using water towers.
- One of the following metering devices may be used in a high-pressure chiller: thermostatic expansion valve, orifice, float, or electronic expansion valve.
- Low-pressure chillers use centrifugal compressors that turn at a very high speed.
- These centrifugal compressors may be manufactured so that they can be operated in series called multistage operation.
- The oil pump on a centrifugal compressor is energized before the compressor is started for prelubrication of the bearings.
- The orifice or the high- or low-side float metering device is typically used for low-pressure chillers.
- When a centrifugal uses a low-pressure refrigerant, the low-pressure side is always in a vacuum. If there is a leak, air will enter the system. A purge unit will separate the refrigerant from the noncondensables and these noncondensables will be released to the atmosphere through a relief valve.
- Absorption refrigeration is a process that uses heat as the driving force rather than a compressor.
- The absorption chiller uses water as the refrigerant.
- The absorption chiller uses a salt solution consisting of lithium-bromide (Li-Br) as the attractant in the refrigeration process.
- The strength of this solution is important. When the technician adjusts the strength, it is called trimming the system.
- These solutions are corrosive and air must be kept from them.
- The absorption chillers also have a purge system.
- The motors that drive high- and low-pressure chillers are three phase.
- One of several start devices may be used: part-winding start, autotransformer start, wye-delta start, or electronic starters.
- There are many other motor controls to include motor overload protection devices, load-limiting devices, anti-recycle control, phase failure protection, voltage unbalance, and phase reversal protection.

REVIEW QUESTIONS

What medium is used for circulating the cooling throughout the building when a chiller is used?

What are the two basic categories of chillers?

What types of compressors are used with compression chilled-water systems?

What is the most popular refrigerant used with reciprocating compressor chillers?

How is the capacity of a reciprocating compressor reduced?

6. Describe what is meant by the term blocked suction.
7. Describe what is meant by the term suction valve lift unloading.
8. What approximate range of sizes of scroll compressors is used with chilled-water systems?
9. Describe how capacity control is achieved in a rotary screw compressor system.
10. At what approximate speed does a centrifugal compressor turn in a single-stage high-pressure system?
11. How is capacity control achieved in a single-stage centrifugal compressor system?
12. Describe a direct expansion evaporator.
13. Describe a flooded evaporator for a chiller.
14. What is the purpose of multiple passes in chillers?
15. What is meant by the approach temperature?
16. What type of water-cooled condenser is generally used in high-pressure chillers?
17. What is meant by condenser subcooling?
18. List the types of metering devices used in high-pressure water chillers.
19. What type of compressor is used in a low-pressure chiller system?
20. What causes a surge?
21. Describe how a centrifugal compressor is lubricated.
22. Describe the difference between a marine water box and a standard water box.
23. What is an operating performance log?
24. What types of metering devices are generally used in low-pressure chiller systems?
25. Why is a purge unit used in a low-pressure system?
26. In an absorption chiller, what is used instead of a compressor to drive the system?
27. Describe a package absorption chiller.
28. What is the compound used in most salt solutions in absorption chillers?
29. What term is used to describe the adjustment of the salt solution in an absorption chiller?
30. What refrigerant is generally used in absorption systems?
31. How can capacity control be achieved in an absorption chiller?
32. What is the term used when a salt solution becomes too concentrated and actually turns to rock salt?
33. Why is a purge system needed in absorption chillers?
34. What types of fuels are most common in direct-fired systems?
35. What type of electric motors are typically used in chilled-water systems?
36. List two ways in which compressor motors are cooled.
37. List four procedures or components used to assist electric motors to start.
38. What is a mechanical-electrical motor overload device used on large rotary screw or centrifugal compressors?
39. Describe an anti-recycle control used on large motors.
40. What do electronic phase protection devices do?

UNIT 48

Cooling Towers and Pumps

OBJECTIVES

After studying this unit, you should be able to

- describe the purpose of cooling water towers used with chilled-water systems.
- state the relationship of the cooling capacity of the water tower and the wet-bulb temperature of the outside air.
- state the means by which the cooling tower reduces water temperature.
- describe two types of cooling water towers.
- explain the various uses of fill material in cooling water towers.
- list the two types of fan drives.
- state the two types of fans used in cooling water towers.
- explain the purpose of the water tower sump.
- explain the purpose of makeup water.
- describe a centrifugal pump.
- describe water vortexing.
- explain two types of motor-pump alignment.

SAFETY CHECKLIST

* Ensure that all electrical safety precautions are observed when servicing pump motors.
* Ensure that all guards are in place when working in the area near a turning fan.
* Do not approach a turning fan that is not properly guarded.
* Be careful of your footing and balance when climbing up to or down from a cooling tower location. Do not move in a way that will cause you to lose your balance when you are working around a tower at a high elevation.
* When lifting a motor or pump, lift with your back straight. Wear a back belt support if recommended by your employer or insurance carrier.

48.1 COOLING TOWER FUNCTION

The heat that is absorbed by the chiller from the conditioned space must be rejected. The *cooling tower* is the unit in the water-cooled system that rejects this heat to the atmosphere. The water pump moves the water that contains the heat to the cooling tower. Figure 48–1 shows a basic condenser, pump, and cooling tower system.

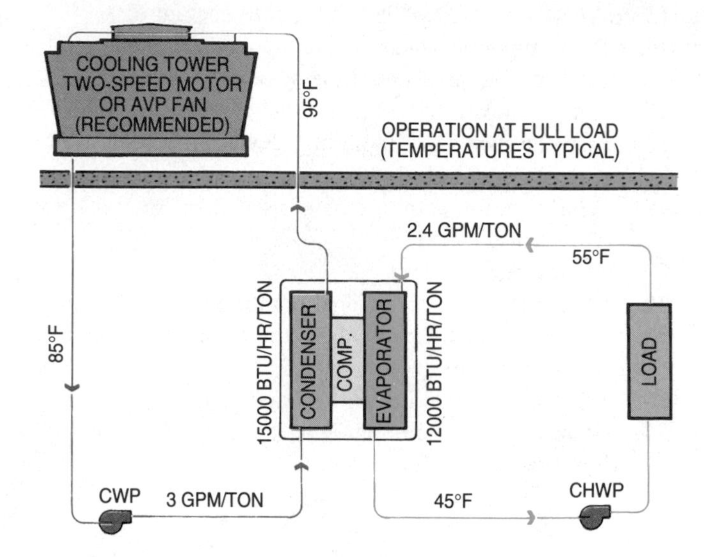

FIGURE 48–1 Example of a condenser water circuit with cooling tower. *Courtesy Marley Cooling Tower Company*

In compression systems the cooling tower must reject more heat than the chiller absorbs from the structure. The chiller absorbs the heat from the chilled-water circuit and in compression systems the compressor adds heat of compression to the hot gas pumped to the condenser, Figure 48–2. The compressor adds about 25% additional heat so the cooling tower must reject about 25% more heat than the capacity of the chiller. For example, a 1000-ton chiller would need a cooling tower that could reject about 1250 tons of heat.

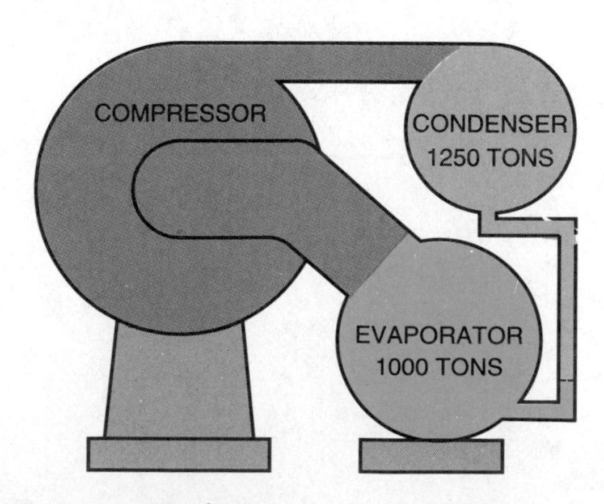

FIGURE 48–2 Heat of compression and the condenser.

The condenser must be furnished water within the design limits of the system or the system will not perform adequately. The design temperature for the water leaving the cooling tower for most refrigeration systems, including the absorption chiller is 85°F. Using a design temperature of 95°F dry bulb and 78°F wet bulb for a typical southern section of the country, the cooling tower can lower the water temperature through the tower to within 7°F of the wet-bulb temperature of the outside air. When the wet-bulb temperature is 78°F, the water temperature leaving the tower would be 85°F, Figure 48–3. This occurs even though the air temperature is 95°F.

All cooling towers work in a similar manner; they reduce the temperature of the water in the tower by means of evaporation. As the water moves through the tower, the surface area of the water is increased to enhance the evaporation. The different methods used to increase the surface area of the water is part of what makes one water tower different from another. When the water is spread out to create more surface area, the water will evaporate faster and be cooled to closer to the wet-bulb temperature of the air. The water cannot be cooled below the temperature of the

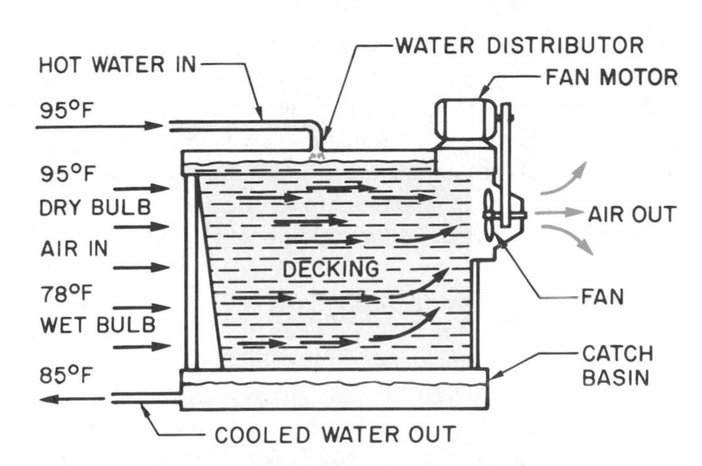

FIGURE 48–3 Typical cooling tower approach temperature.

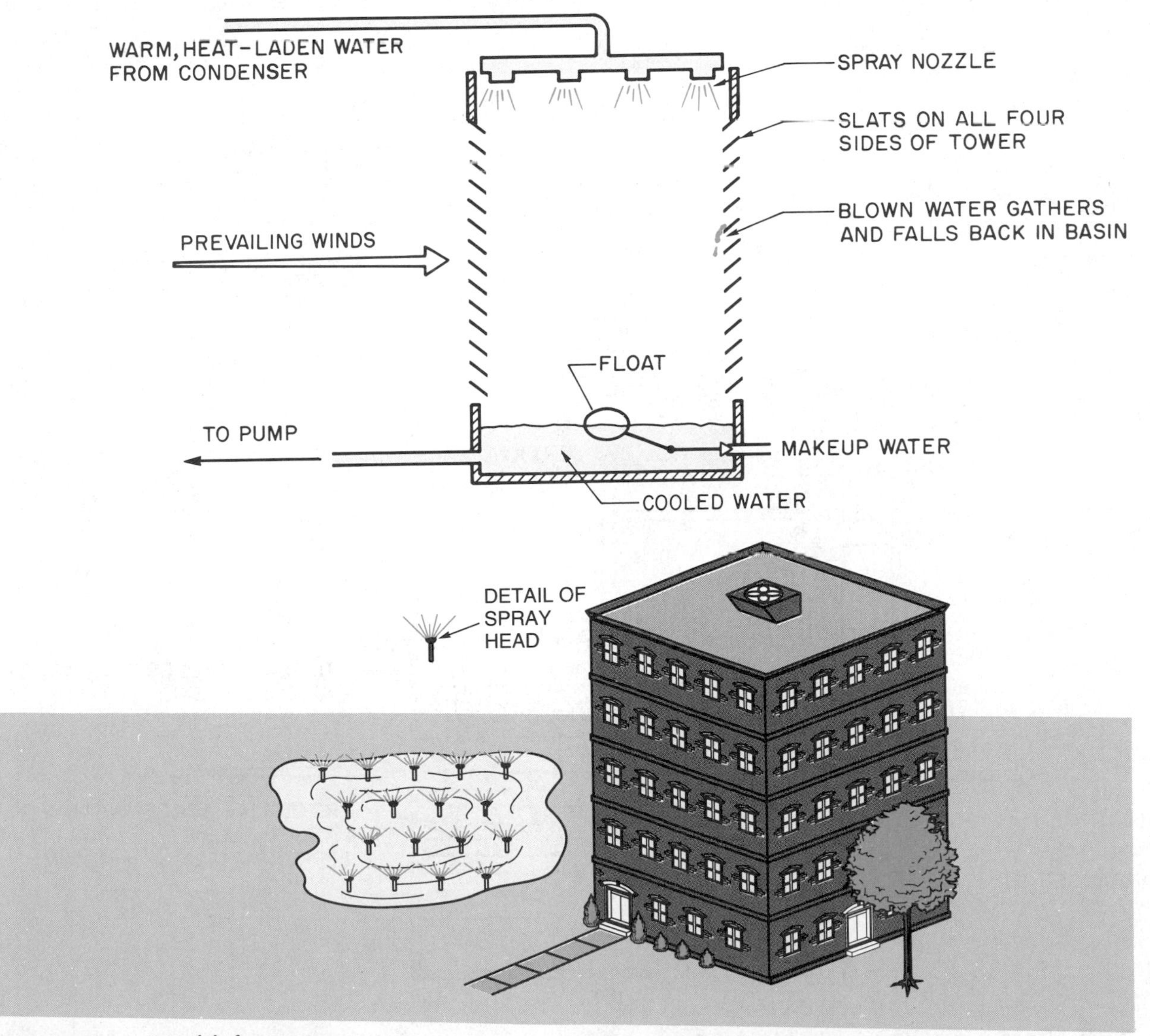

FIGURE 48–4 Natural-draft tower types.

cooling medium, which is the wet-bulb temperature. A nearly perfect cooling tower could reduce the temperature to the entering wet-bulb temperature; however, this tower would be tremendous in size. Manufacturers make towers that approach the wet-bulb temperature by 7°F for typical applications. Towers are manufactured to a 5°F approach, but they are not typically used for air conditioning applications.

48.2 TYPES OF COOLING TOWERS

Two types of cooling towers are in common use: the *natural-draft* and *forced- or induced-draft tower*. The larger the installation, the more elaborate the tower.

The natural-draft tower may be anything from a spray pond in front of a building to a tower on top of a building, Figure 48–4. The spray pond and natural-draft cooling tower rely on the prevailing winds and will cool the water to a lower temperature when the prevailing winds are greater. The natural-draft tower or spray pond application usually has an approach temperature of about 10°F because of relying only on the winds.

The spray pond can be located at ground level or it may be on the roof of a building, but wherever it is, the spray must be contained in the area of the pond. The wind can blow the spray water out of the pond and annoy people or damage property. Spray ponds on rooftops have made a comeback in the last few years because they cool the roof and reduce the solar load. If the rooftop can be cooled to 85°F, a great reduction in air conditioning load can be realized.

The spray pond and natural-draft cooling tower use pump pressure to increase the area of the water by atomizing it into droplets. Spray heads are located at strategic locations within the pond or tower and provide a spray pattern for the water. The spray nozzles are a source of restriction and must be kept clean or they will not atomize the water. Atomizing the water requires pump horsepower that may be considered an unacceptable expense because of the large volume of water used in these applications.

Forced-draft and induced-draft towers use a fan to move the air through the tower, Figure 48–5. These are the most popular applications used today because the efficiency is reliable. Centrifugal fans are used to move the air in some towers. When the tower air must be ducted to the outside of the building, the centrifugal fan must be used because it can overcome the static pressure in the ductwork.

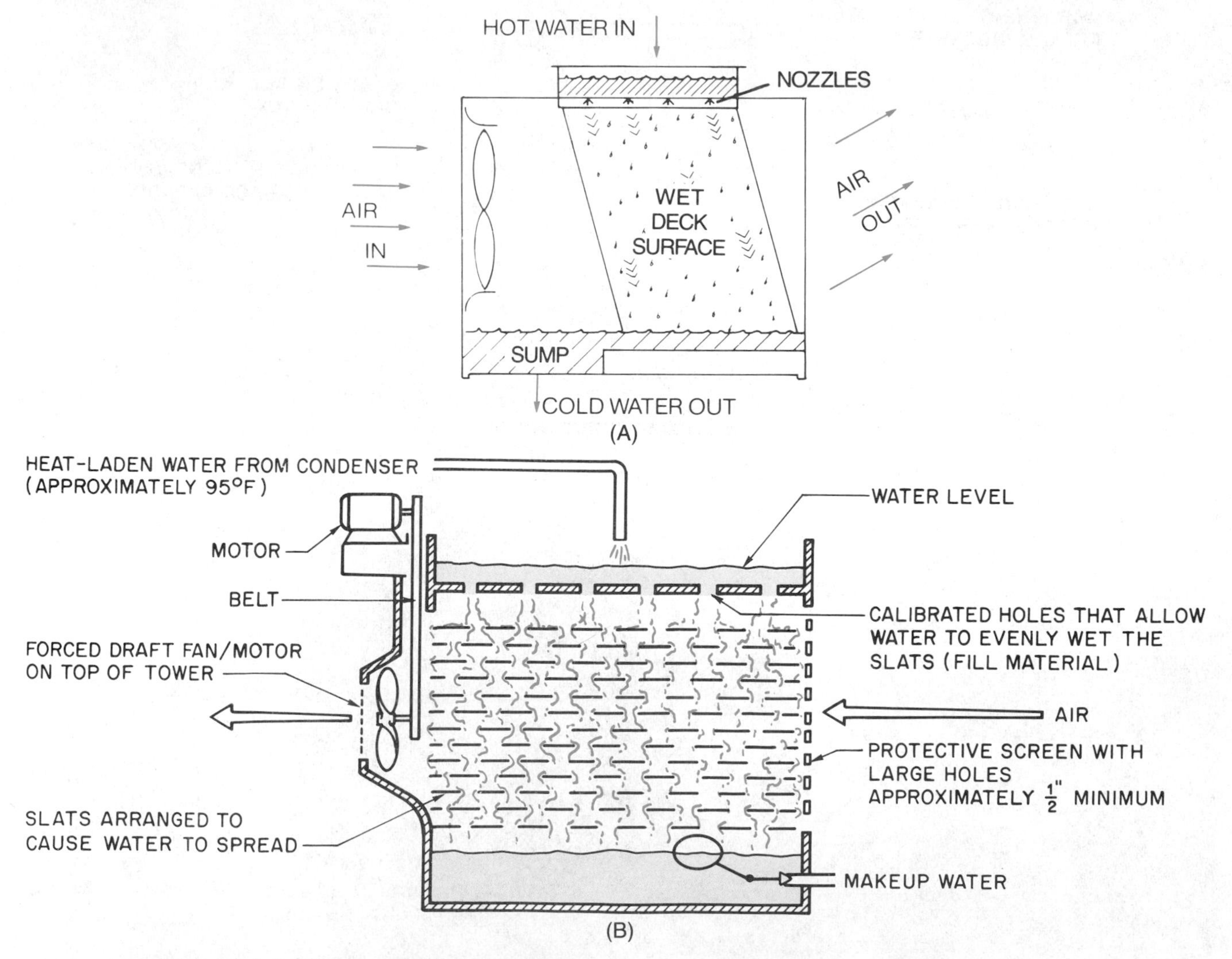

FIGURE 48–5 (A) Forced-draft cooling tower. (B) Induced-draft cooling tower.

Towers with centrifugal fans can be more compact and therefore more desirable for some applications. These towers are available up to about 500 tons capacity of heat rejection.

Larger towers use propeller type fans. These fans may either be belt driven or gear driven. Belt-driven fans will require belt maintenance on a regular basis. Gear-driven fans have a transmission that will require only lubrication.

48.3 FIRE PROTECTION

The cooling tower is a place where a fire may occur during the off season when the tower components are dry. The tower may contain materials that are flammable, such as wood or some plastics. Some fire codes and insurance companies may require that the tower be manufactured of all fireproof materials or they may require that a tower wetting system be in place. A tower wetting system may consist of sprinkler heads such as required in buildings. Some towers are controlled in such a manner that the system pump starts up on a timed cycle to wet the tower down occasionally during the off cycle. Another wetting system may be an auxiliary pumping system that keeps the tower wet all the time that the weather is above freezing. This type of system also may prevent expansion and contraction of any wood construction in the tower because it is wet all the time. Local codes and insurance requirements will dictate what method should be used to protect the tower from fire.

48.4 FILL MATERIAL

The cooling tower is designed to keep the water and the air in contact for as long as possible. Manufacturers use various methods to slow the water as it trickles down through the tower with the air moving up through the tower. Two methods are used to evaporate the water in the forced- and induced-draft tower. These are the splash method and the fill or wetted surface method. Both methods use a material in the tower often called the fill material. This material may be wood slats where the water drips down through layers of wood. Other types of materials are also used in the splash method, such as polyvinyl chloride (PVC) plastic or fiber-reinforced polyester (FRP) plastic. These have a slow burn rate and should be acceptable for use. Towers that use the splash system have a framework that supports the slats and is designed to keep them at the correct angle for proper water wetting of all slats with water flow from the top of the tower to the bottom.

The film or wetted surface type of tower uses a fill material that may be some form of plastic or fiberglass. The water is spread out on the surface while air is passed over it, Figure 48–6. This type of fill may have smaller passages for the air to travel through and is not used where particles from the surroundings can contaminate the tower and restrict the passages.

Both types of fill material rely on water to run down across the fill material, which should be kept at the angle the manufacturer recommends or the water will not take the proper path. Water running to one side or other will cause the capacity of the tower to be reduced. If the fill material is removed for maintenance, care should be taken to replace it in the correct manner for correct water movement through the tower.

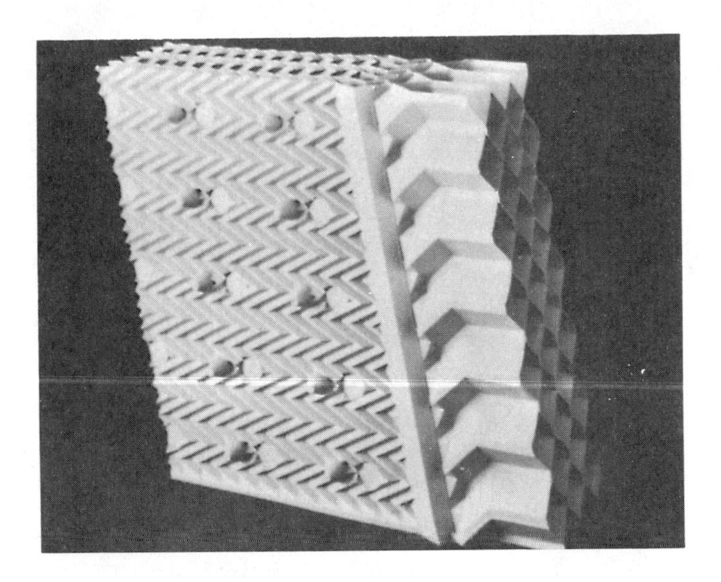

FIGURE 48–6 Wetted surface tower principle. *Courtesy Marley Cooling Tower Company*

48.5 FLOW PATTERNS

There are two distinct airflow patterns for cooling towers: *crossflow* and *counterflow*, Figure 48–7. The crossflow tower introduces the air from the side and usually pulls the air to the top of the tower for exhaust. In smaller crossflow towers the fan is on the side of the tower and the exhaust is out the side, Figure 48–5(B). When the air is exhausted out the side of the tower, care must be taken that the moisture-laden air is not exhausted to a place where it will cause a problem, such as in a walkway or parking lot where cars may be spotted with the water and chemicals used in treatment, which may be corrosive to car finishes. In the tower, the water is moving downward and the air is moving at right angles to the water.

The counterflow tower introduces the air at the bottom of the tower and exhausts the air out the top. The water is moving downward as the air moves up through it.

The water in most towers, specifically the spray type towers, has many small particles of water suspended in air. These particles are subject to being blown out of the tower from the prevailing winds. This loss of water is called drift and can be expensive and a nuisance for any surrounding areas with spray that contains chemicals. This drift of water is minimized with eliminators. Eliminators cause the spray to change direction and rub against a solid surface where the water should deposit and run back to the tower basin. The eliminators may be louvers on the side of the tower or they may be part of the fill material in some newer towers.

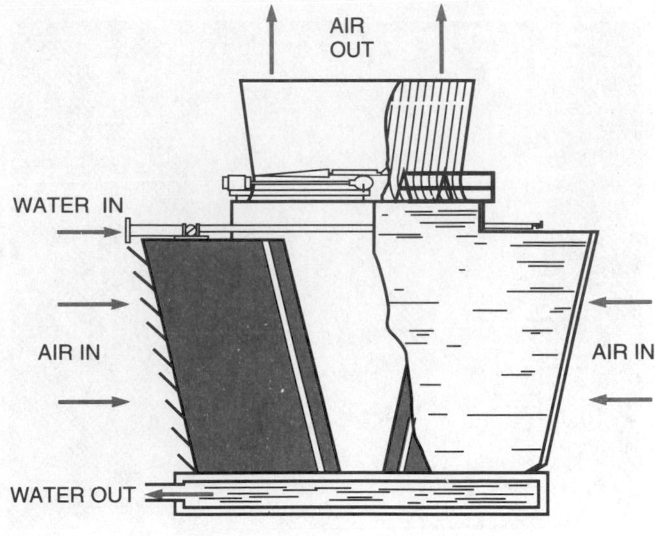

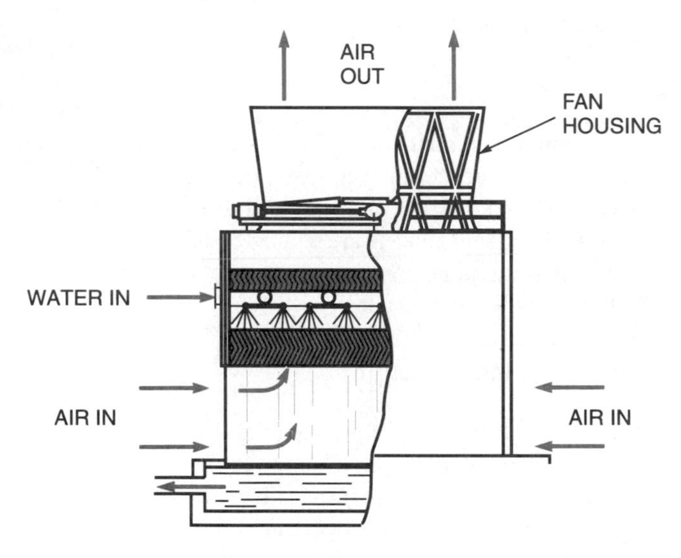

FIGURE 48–7 Crossflow and counterflow cooling towers. *Courtesy Marley Cooling Tower Company*

48.6 TOWER MATERIALS

The materials from which cooling towers are constructed must be able to withstand the environment in which the tower operates. There are towers all over the world and in different chemical environments so the materials are varied. The typical tower must be able to withstand wind, weight of tower components and related water, sun, cold, freezing weather (including any ice that may accumulate), and vibration of the fan and drive mechanisms. Towers must be carefully designed and many different materials may be used depending on the type and location of the tower. Typically, the smaller packaged towers are made of galvanized steel (for rust protection), fiberglass, or FRP. These towers are manufactured as a complete assembly and shipped to the job, Figure 48–8.

FIGURE 48–8 Packaged cooling tower. *Courtesy Baltimore Aircoil Company, Inc.*

Larger towers may have a concrete base and sump to hold the water and the sides made of other materials such as corrugated asbestos-cement panels, wood (either treated or redwood), fiberglass, or corrugated FRP. Fire prevention must be considered in the selection of materials in many cases.

48.7 FAN SECTION

The motor must have a method for turning the fan for all forced-draft towers. There are two different types of drive, the belt and gear box (transmission) drives. The belt drive is normally used for the smaller towers and includes an adjustable motor mount. This mechanism will have the greatest wear in the cooling tower and require the most maintenance so it should be located where preventive maintenance and service can be performed with as little effort as possible.

Fans for larger towers have the motor mounted out to the side and a gear box or transmission to change the direction of the motor drive shaft 90°. These motor/fan units may also be designed to change the motor-to-fan shaft speed relationship. The motor will turn at 1800 or 3600 rpm for a typical motor and the speed of the fan shaft will turn considerably slower, depending on the gear reduction. The motor is connected to the gear box using a coupling and shaft, Figure 48–9. The motor, gear box, and fan bearings must be accessible for service purposes.

The propeller fan blade is enclosed in a fan housing that improves the efficiency of the fan blades, Figure 48–7. This fan blade location is critical in most towers. The fan must be located at the correct distance from the top and sides for best performance.

48.8 TOWER ACCESS

All cooling towers will need service on a regular basis. The tower must have access doors to the fill material for cleaning and possible removal, Figure 48–10. The tower basin must also be accessible for cleaning because large amounts of sludge will accumulate. The cooling tower becomes a large filter for whatever may be airborne. Dirt, pollution, feathers, birds, plastic wrappers, and cups are among the

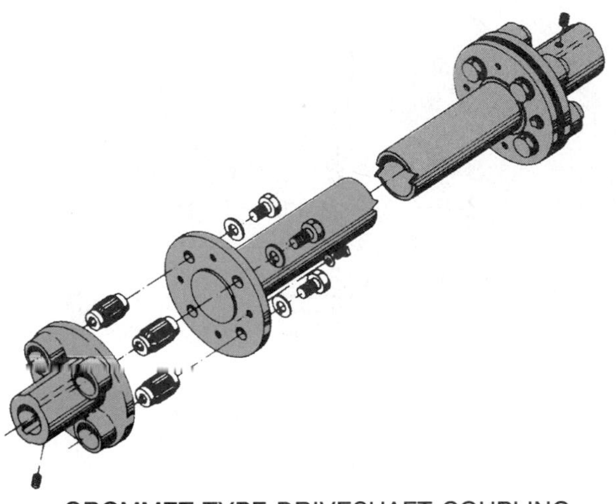

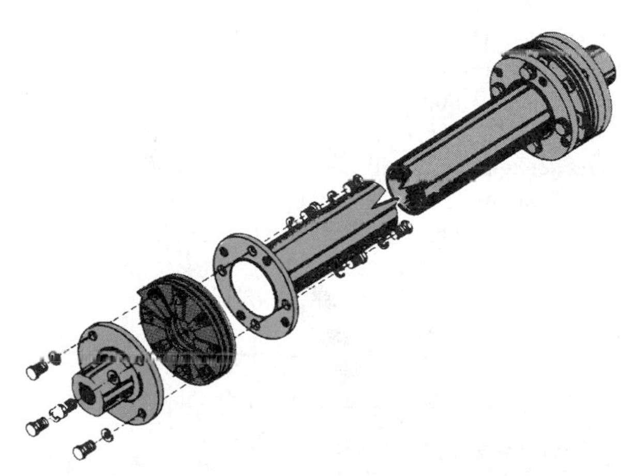

FIGURE 48–9 Coupling and shaft for gear reduction box. *Courtesy Marley Cooling Tower Company*

FIGURE 48–10 Tower access for service. *Courtesy Baltimore Aircoil Company, Inc.*

common items. These particles will gather in the sump and must be removed. There should be an adequate water supply in the vicinity of the tower for the connection of a water hose for flushing the tower sump.

FIGURE 48–11 Stairway and guardrails for large tower. *Courtesy Marley Cooling Tower Company*

When the tower is tall, it will likely have a stairway or ladder to the top for servicing the fan and drive components, Figure 48–11. Provisions should be made in the vicinity of the motor and fan to lift any component parts that may need to be removed. These parts may need to be lowered to the ground for servicing. A stair or ladder installed for servicing must meet the local codes for safety to include proper handrails and barriers to prevent stepping over the edge.

48.9 TOWER SUMP

All cooling towers must have some sort of *sump* for the water to collect in. The sump on small cooling towers may consist of a metal pan that drains to a lower point to gather the water. When this sump contains water and is located outside, it must have some means to prevent freezing such as a sump heater that is thermostatically operated, Figure 48–12. These sump heaters are used for small installations. Larger installations that would require more heat may have a circulating hot water coil or a method of using low-pressure steam to heat the tower basin water in cold weather.

Many cooling towers use underground sumps made of concrete. These sumps will not freeze in the southern climates but must be protected from the cold in northern climates unless the refrigeration system is operating and adding heat to the sump. Some sumps are located in the heated space of a building to prevent freezing.

Wherever the sump is located, it is the catch-all for all sediment and should be accessible for cleaning. A bypass filter system is often used to sweep the bottom of the sump and carry a portion of the water through the filter system for cleaning. The sump usually contains a coarse screen strainer to protect the pump, Figure 48–13. The return line may also have an in-line removable strainer.

48.10 MAKEUP WATER

Because the cooling tower operates using the principle of evaporation, water is continuously lost from the system. Several systems are used to make up water, the float valve

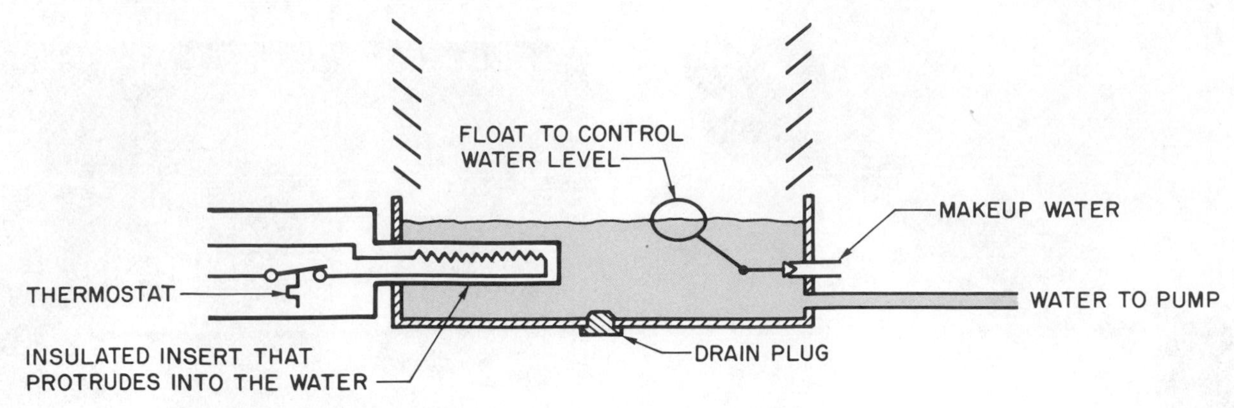

FIGURE 48–12 Freeze protection for cooling tower sump.

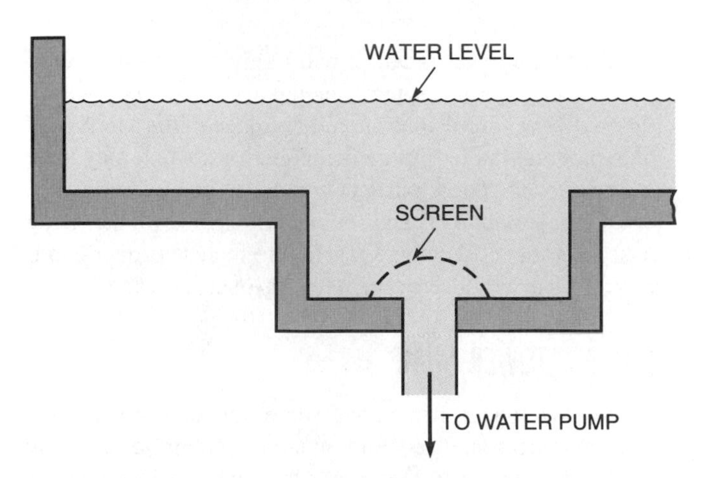

FIGURE 48–13 Filter system for cooling tower sump.

being one of the most common. When the water level drops, the float ball drops with it and opens the valve to the makeup water supply. Usually this makeup water is from the normal municipal or other similar supply water. Another method is to use float switches that fall with the water level and energize a solenoid valve to allow water to fill the sump. When the water level reaches the proper level, the float rises and shuts off the solenoid valve. Still another method uses electrodes protruding into the water that sense the water level. When the electrodes sense that the water level is too low, a solenoid valve is opened and makeup water is allowed to enter. As the water level rises, an upper electrode senses the water level and shuts off the water. Figure 48–14 shows all three level control methods.

48.11 BLOWDOWN

The water that evaporates leaves behind any solid materials that may have been in the water. These include dust particles, minerals, and algae, a low form of plant life. As the water evaporates, the water left behind becomes more and more concentrated with these particles. If this continues, the particles will drop out of the water and deposit on the surfaces of the cooling tower. These will be hard to remove because they turn to a substance much like cement. Even more important than what is happening to the cooling tower is what will happen to the condenser. Some of these particles will deposit at the hottest part of the condenser, the *outlet*. These deposits will act like an insulation to the heat exchanger. This will cause the head pressure to rise. The condenser approach will begin to spread. As mentioned in the section on condensers, the original operating log will give the technician a clue as to when this is happening.

Blowdown is the cure for part of these problems. It is merely a bleed off of the portion of the water that is being circulated. When new water is added, less sediment is present as some has gone with the water that was bled off. Fresh water is added to make up for water lost during blowdown. This fresh water dilutes the remaining water and reduces the mineral content. It is generally recognized that 3 gal/min/ton is the amount of water circulated in a water-cooled condenser used for air conditioning with a 10°F temperature rise through the condenser. Figure 48–15 shows a 30-ton application and what would happen if the cooling tower water did not have blowdown as part of the piping system. Figure 48–16 shows four methods that may be commonly used for obtaining the correct blowdown. Often, building management personnel have a difficult time understanding the purpose of blowdown. They don't understand why water is being allowed to run down the drain. **Blowdown must be managed correctly because expensive water treatment chemicals are going down the drain with the water.**

48.12 BALANCING THE WATER FLOW FOR A TOWER

The water flow for even distribution must be correct for many towers. If a tower has two or more cells where the water is distributed, the same amount of water must be fed to each cell or the tower will not perform as designed. Many towers use a distribution pan at the top to distribute the water over the fill. These pans are often a series of calibrated holes drilled into the pan. Warm return water from the condenser is poured out in the pan by the pump discharge. This discharge may be split to each side of the

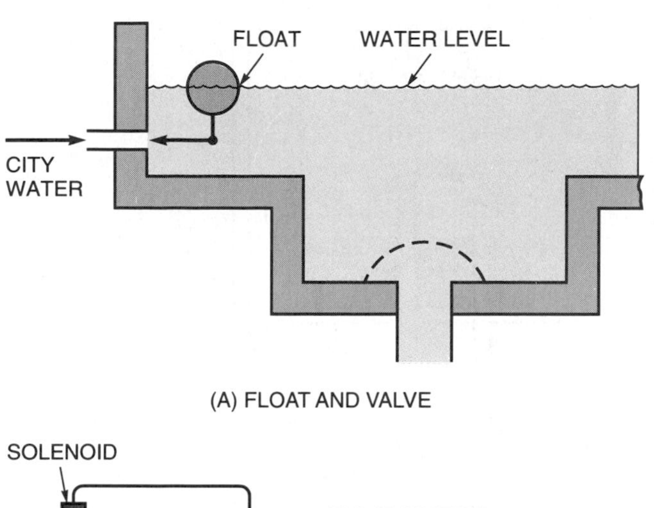

(A) FLOAT AND VALVE

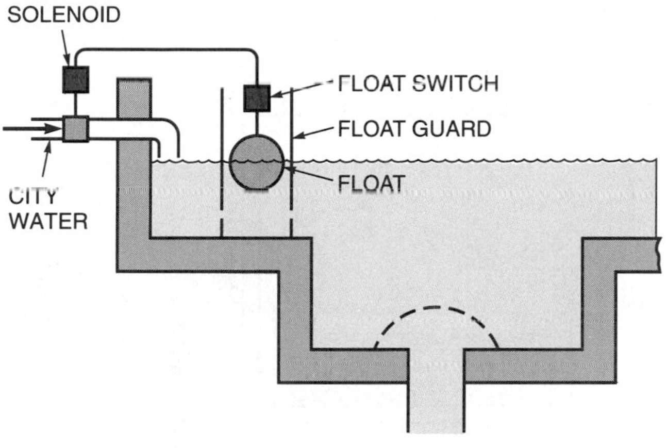

(B) FLOAT SWITCH AND SOLENOID

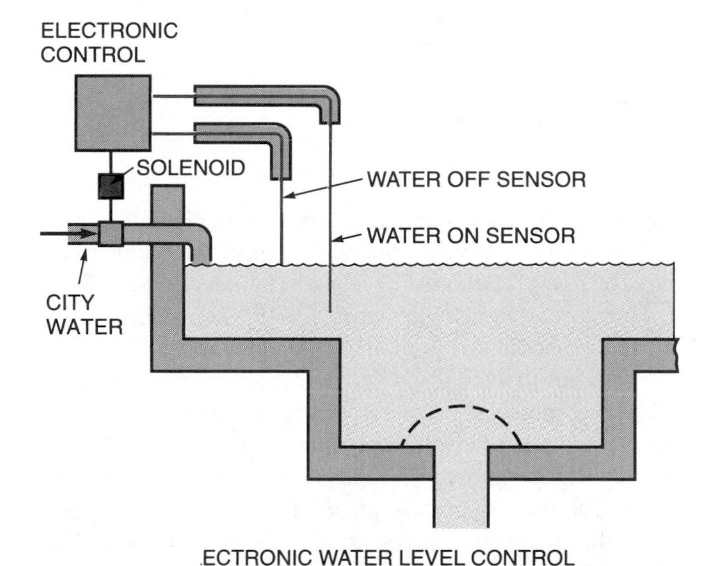

ECTRONIC WATER LEVEL CONTROL

FIGURE 48–14 **Three types of cooling tower fill methods. (A) Float and valve. (B) Float and switches. (C) Electronic sensors.**

tower. A balancing valve or valves must be set to obtain the correct flow, Figure 48–17. The calibrated holes in the top of the tower must also be clean of foreign material and be of the correct size. Figure 48–18 shows a pan where the holes are rusted out and most of the water is flowing to one side of the tower and not wetting the entire fill deck. The tower will not perform correctly when this happens. For example, the

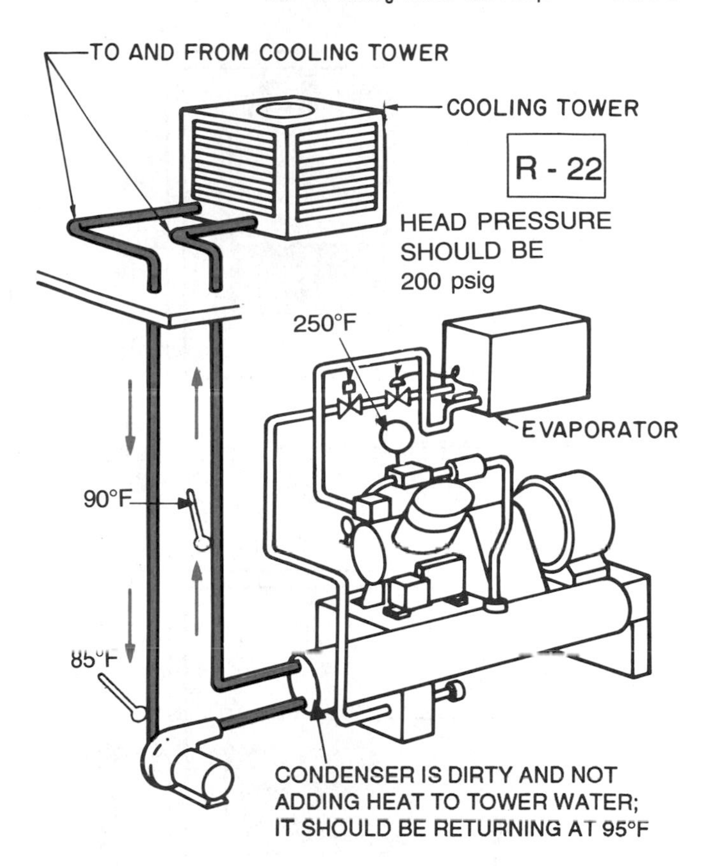

FIGURE 48–15 **Flow rate through a typical air conditioning cooling tower.**

tower may have a design water temperature of 85°F and it may be returning 90°F. This will cause high head pressure for the chiller and high operating costs.

48.13 WATER PUMPS

The condenser water pump is the device that moves the water through the condenser and cooling tower circuit. The cooling tower pump normally is located where it takes water from the cooling tower sump into the pump inlet, Figure 48–19. The water pump is the heart of the cooling tower system because it pumps the heat-laden water. This water must be pumped at the correct rate and delivered to the cooling tower at the correct pressure.

The condenser water pump is normally a *centrifugal pump*. It uses centrifugal action to impart velocity to the water that is converted to pressure. Pressure may be expressed in pounds per square inch (psig) or feet of head. One foot of head of water is 0.433 psig, or a column of water 1 ft high will cause a pressure gage to read 0.433 psig. A column of water 2.31 ft high (27.7 in.) will cause a pressure gage to read 1 psig, Figure 48–20. Pump capacities are discussed in feet of head. Centrifugal pump action is discussed in Unit 32 and a study of that unit should be helpful.

There are several types of condenser water pumps for large condenser water systems. In a *close coupled pump* the pump is located very close to the motor with the pump impeller actually mounted on the end of the motor shaft,

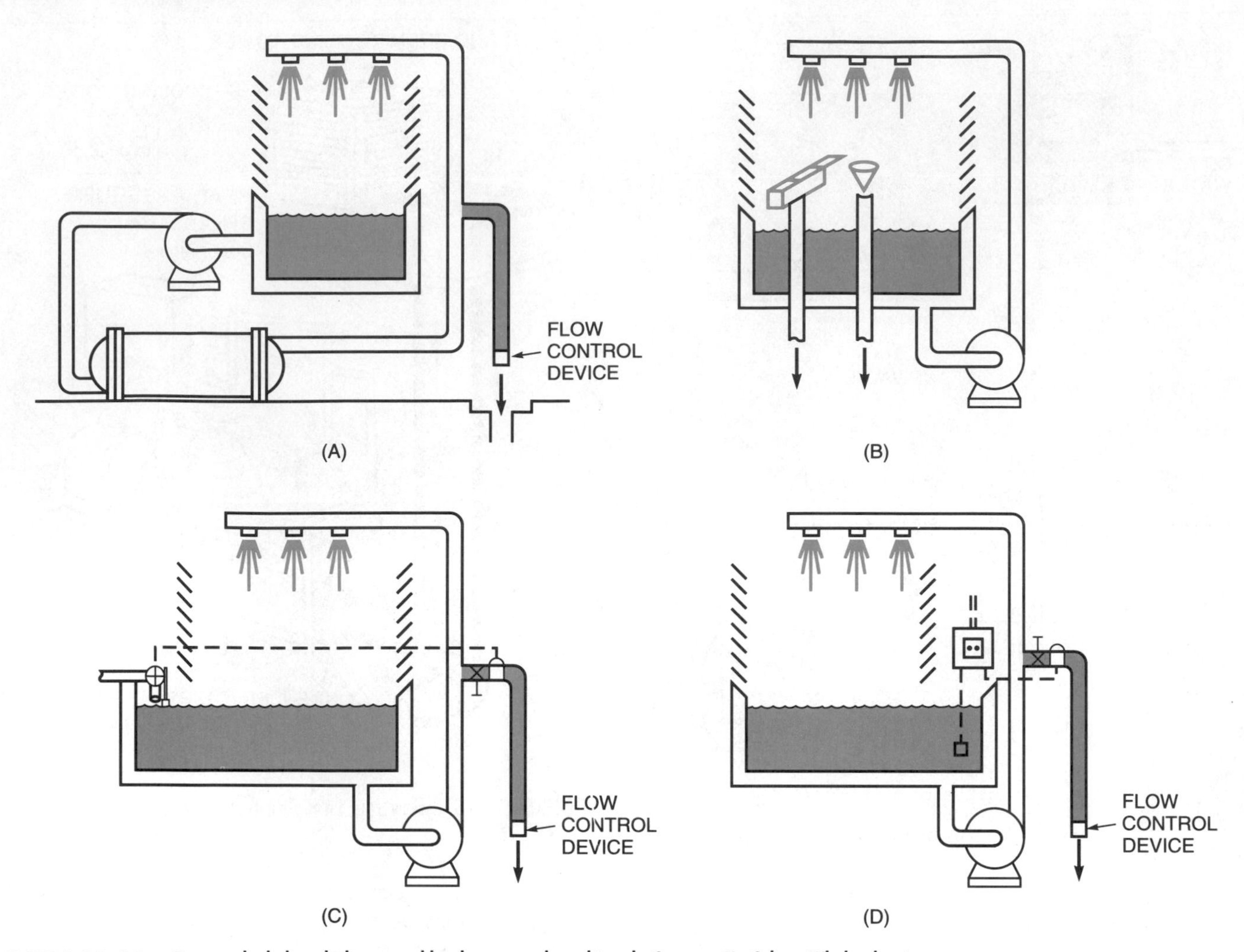

FIGURE 48–16 Four methods by which proper blowdown may be achieved. *Courtesy Nu-Calgon Wholesaler, Inc.*

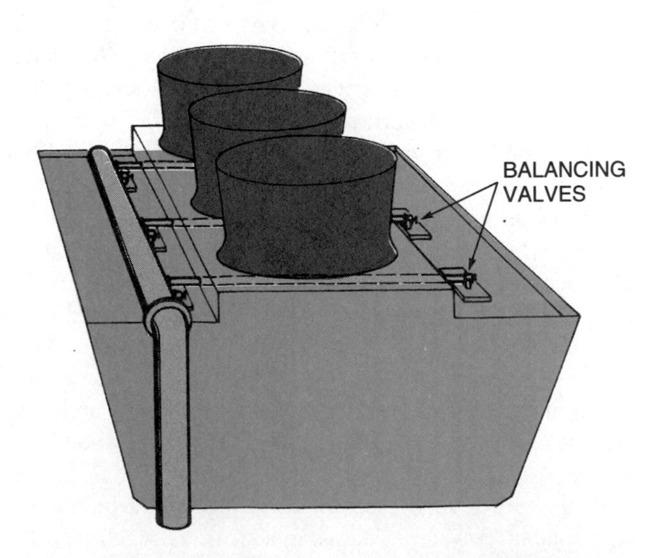

FIGURE 48–17 Balancing valves for balancing the water flow at the cooling tower. *Courtesy Marley Cooling Tower Company*

Figure 48–21. These pumps are used for small applications. Note that all of the water enters the side of the pump housing. The pump shaft must have a seal to prevent water from leaking to the atmosphere or to prevent atmosphere from leaking in should this portion of the piping be in a vacuum.

The *base-mounted* pump comes as an assembly with the pump mounted at the end of the pump shaft and a flexible coupling between the water and the pump, Figure 48–22. This pump may have a single- or double-sided impeller. The base is usually made of steel or cast iron and holds the pump and motor steady during operation. The pump base is usually fastened to the floor using cement as a filler in the open portion of the pump base. This is called grouting the pump in. The purpose is to make a more firm foundation that will last for many years. The pump and motor can still be removed from the steel or cast iron base. One of the advantages of this pump is that the motor and pump shafts are aligned at the factory and alignment is not required in the field. These pumps are used in larger applications.

As pumps become larger, the need for different designs becomes apparent. Some pumps have single-inlet impellers with all of the water entering the pump impeller on one side,

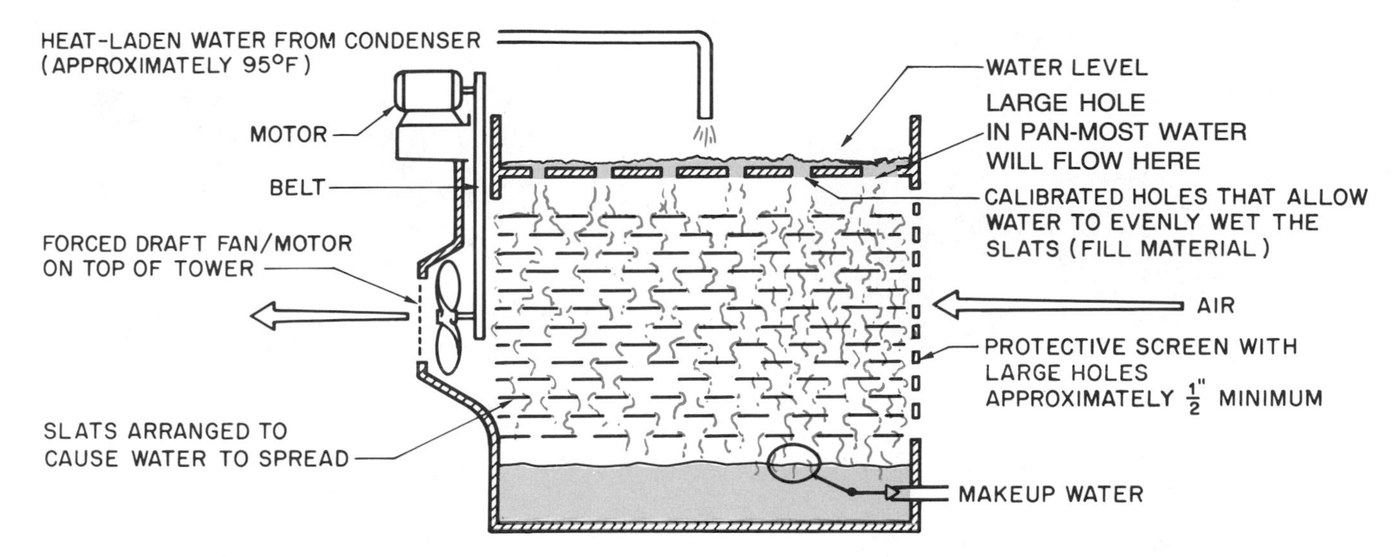

FIGURE 48–18 Calibration holes to distribute water in the tower are enlarged due to rust. Most of the water is running down the right side of tower.

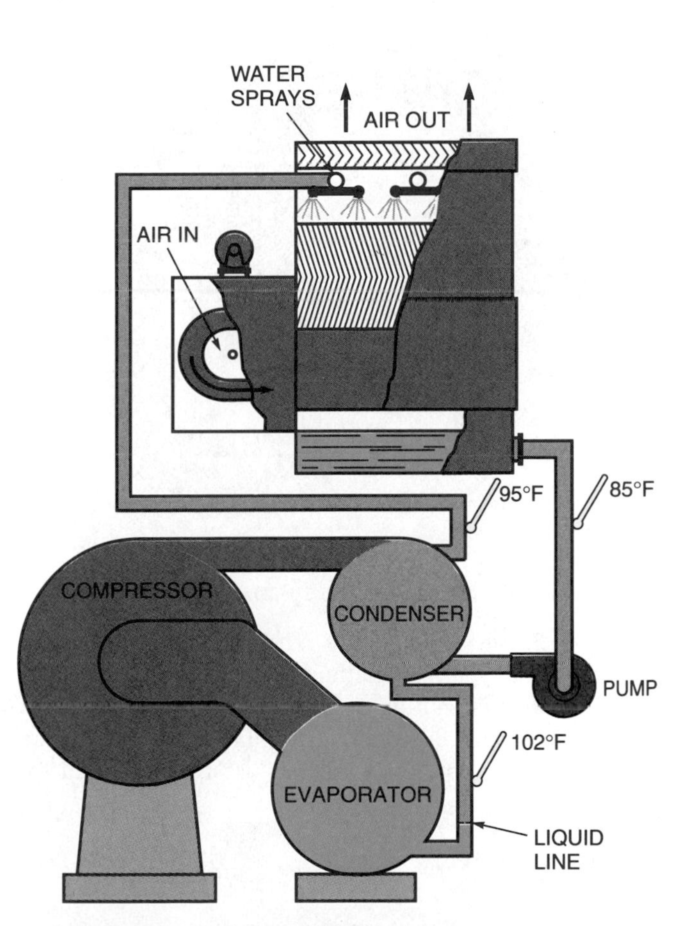

FIGURE 48–19 Cooling tower pump and system.

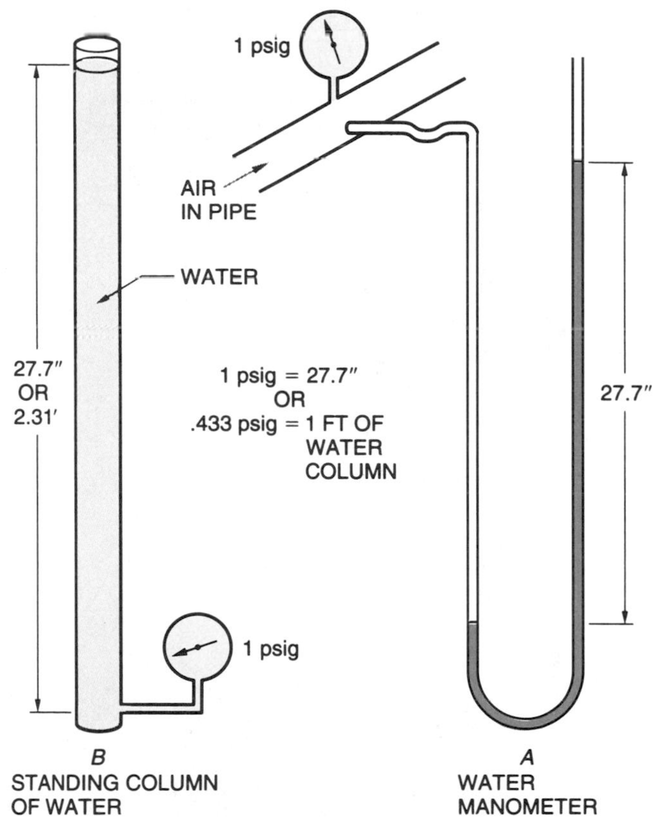

FIGURE 48–20 Standing column of water.

Figure 48–23. The double-inlet impeller allows water to flow into both sides, Figure 48–24. The pump housing must be designed differently to allow water to enter this pump. The housing at the inlet must be split so water can flow into both sides of the impeller. There are two styles of *double-inlet impeller pumps*. With one type the pump is disassembled from the ends of the pump as with the base-mounted pump and the other is a split-case pump where the

FIGURE 48–21 Close coupled pump assembly. *Courtesy ITT Bell and Gossett*

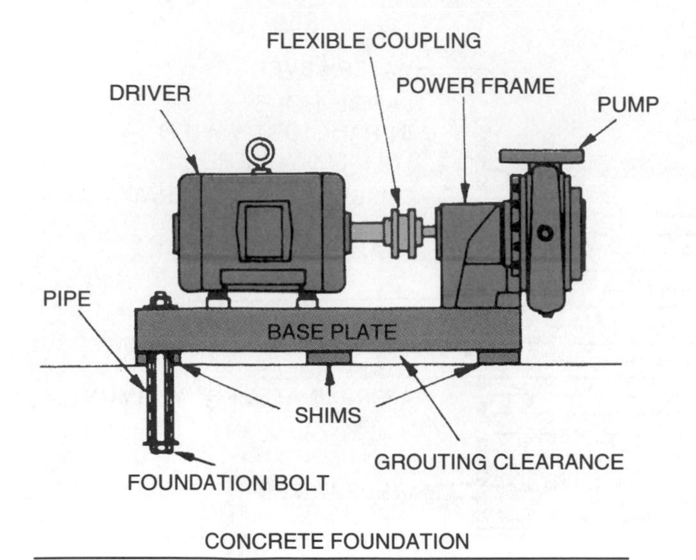

FIGURE 48–22 Flexible coupling between pump and motor. *Courtesy Amtrol, Inc.*

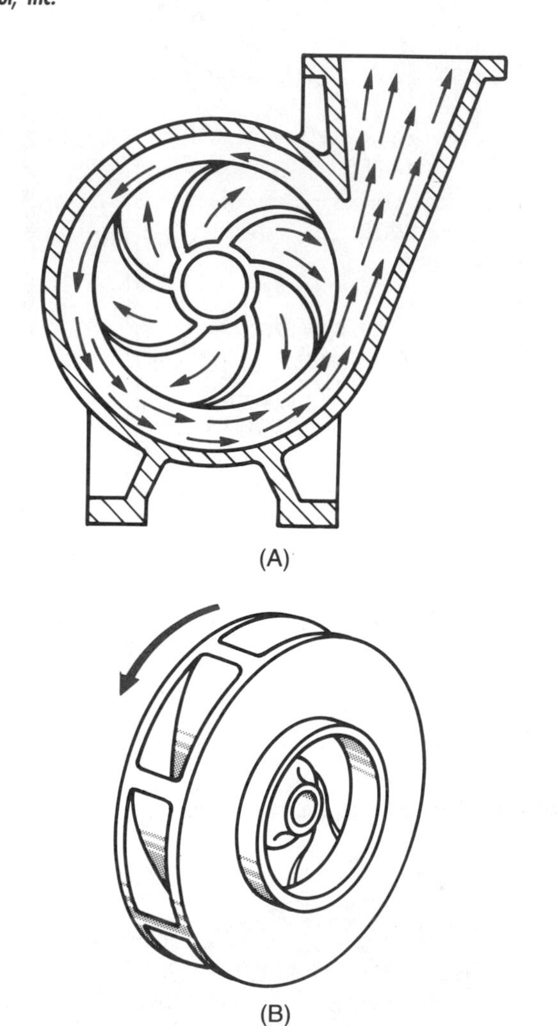

FIGURE 48–23 Single-inlet impeller.

pump is disassembled by removing the top of the pump, Figure 48–25.

These pumps must also have a shaft seal. Two kinds of seal may be used for the pumps; one is the *stuffing box* and the other the *mechanical type*. The stuffing box seal is a packing gland type seal that must be hand tightened with a wrench. When the nut on the seal is tightened, it squeezes a packing around the shaft to minimize leakage, Figure 48–26. Stuffing box types of seals are typically used for

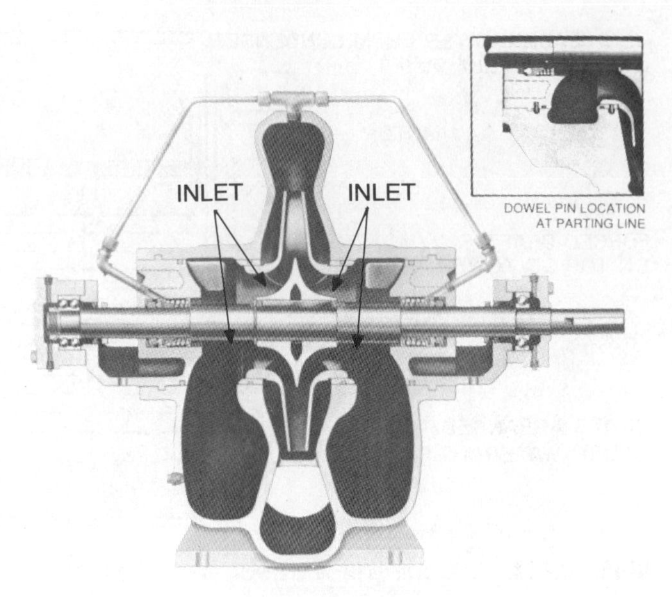

FIGURE 48–24 Double-inlet impeller pump. *Courtesy AC Pump ITT*

FIGURE 48–25 Horizontal split-case pump. *Courtesy ACC Pump ITT*

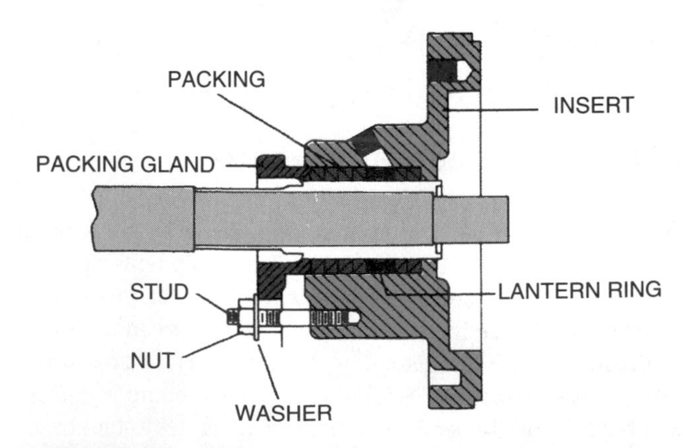

FIGURE 48–26 Stuffing box shaft seal. *Courtesy Amtrol, Inc.*

pump pressures up to about 150 psig. These seals require some regular maintenance.

The mechanical type seal may be used for higher pressures, up to about 300 psig, and usually has a carbon ring mounted on the end of a bellows that turns with the shaft. The bellows is sealed to the shaft with an "O" ring seal. The carbon ring rubs against a stationary ceramic ring while the shaft turns and seals water in the pump housing, Figure 48–27. Both of these seals are lubricated with the circulating water. The seals may have a special piping system that injects pump discharge water to the seal for split-case pumps, Figure 48–28.

Pumps may also be manufactured in a vertical configuration with the motor on top. These pumps are mounted on top of the water sump and protrude down into the sump for the pump pickup, Figure 48–29.

Most pumps for cooling towers are manufactured of cast iron. They are heavy and may be used for pressures up to 300 psig with the correct piping flanges and arrangements. Cast iron has the capacity to last many years without deterioration from rust and corrosion.

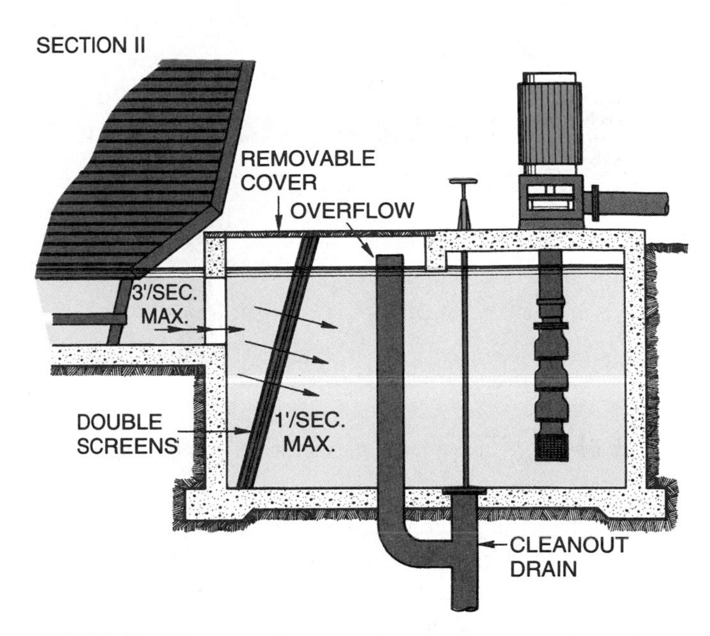

FIGURE 48–29 Vertical pump application. *Courtesy Marley Cooling Tower Company*

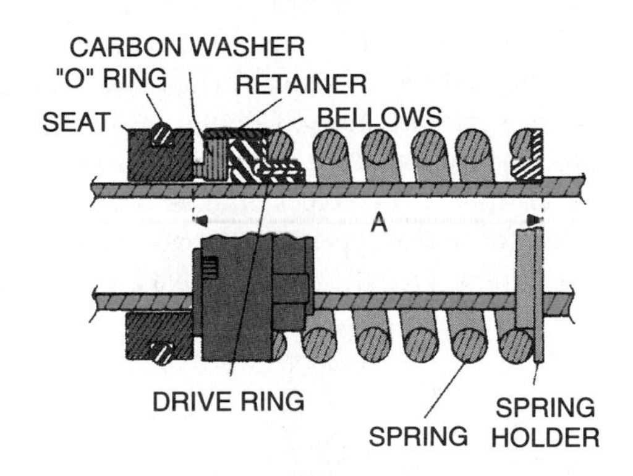

FIGURE 48–27 Mechanical shaft seal. *Courtesy Amtrol, Inc.*

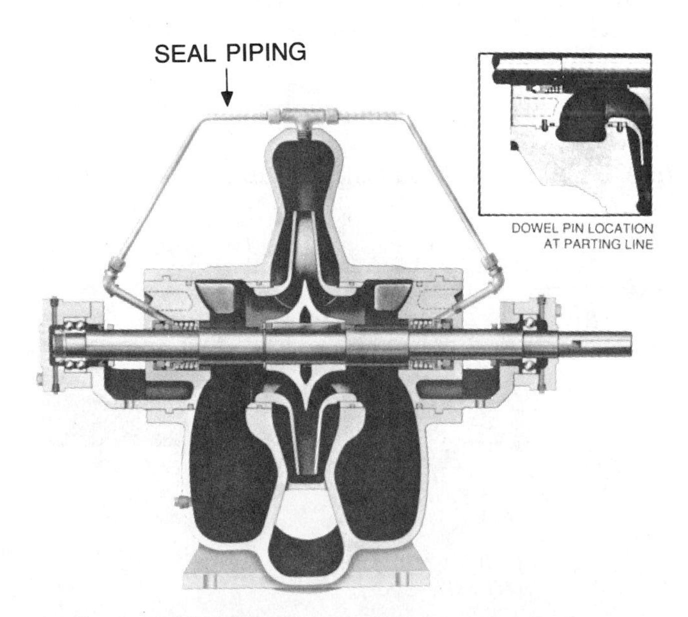

FIGURE 48–28 Lubricating a shaft seal. *Courtesy AC Pump ITT*

The impeller in the typical centrifugal pump is made of brass and is often mounted on a stainless steel shaft. This can be important when the time comes to remove the impeller from the shaft. This makes it easier to remove after many years of operation.

Impellers are furnished from the factory at a specified diameter for each pump to furnish a specific water flow against a specified pressure. Often the pump may furnish too much water flow to the point that a smaller impeller is needed. These impellers are often trimmed in the field to new specifications for the actual water flow and pressure requirements. The manufacturer can be contacted for any modifications for a specific pump.

The location of the condenser water pump is important because these pumps are normally centrifugal pumps and they must be furnished water in such a manner that the eye of the impeller is under water during start-up. Otherwise, the pump will not move any water. They **do not** have the capacity to pump air to pull water into the impeller. The pump should be located below the tower water with nothing in the pump inlet piping that will impede flow. A free flow of water to the centrifugal pump inlet is always good practice. Several things can interfere with water flow to the pump inlet, such as vortexing in the tower, fine mesh strainers, and the cooling tower bypass.

There are times when the sump to the cooling tower is located below the pump. When this is the case, special care should be taken during start-up because only air will be in the pump casing. The pump should never be started with only air inside because the pump seal will have no water for lubrication. When the sump is below the pump, a foot or check valve must be located in the line to prevent the water from flowing back into the sump during shutdown, Figure 48–30. Water must be added to the pump inlet side until the pump casing is completely full before the pump is started.

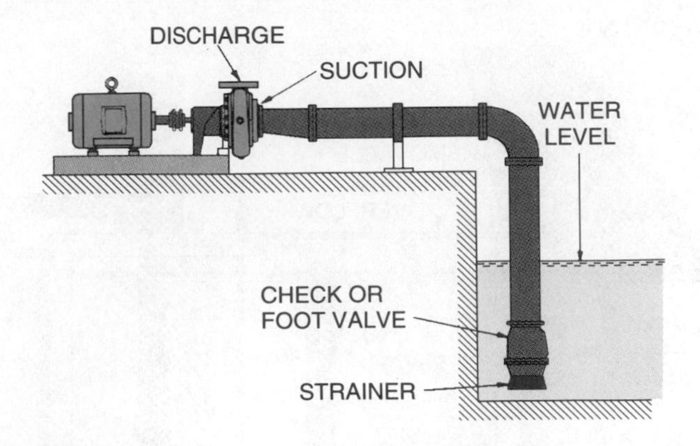

FIGURE 48–30 Cooling tower sump located below the pump.

This can be done by connecting a water hose to the pump inlet side and filling it until water escapes the bleed port on top of the pump. If the pump inlet cannot be filled, the foot or check valve is leaking and must be repaired.

Vortexing in the tower is actually a whirlpool action. This vortex interferes because it introduces air to the pump inlet. The pump is designed to pump water and air will cause problems with the pump and with the condenser at the chiller. Poor tower construction or piping practice is the cause of vortexing. The sump may not be deep enough or antivortexing design may not be in place. Vortexing can often be eliminated by means of a device placed in the sump outlet that breaks up the vortex. This may be a cross type of configuration with a plate on top. The cross configuration creates four outlets from the sump and the plate causes the water to be pulled from a further distance and from the side, Figure 48–31. Vortexing has been found in multistory buildings with the condenser water pump well below the tower and has been cured using the above methods.

There must be some sort of strainer located between the tower water and the pump inlet to prevent trash from the tower from entering the pumping circuit. Typically there is a screen type strainer in the cooling tower exit to the sump. This may be a coarse screen. Often a fine mesh screen is located before the pump inlet. If this mesh screen is too fine, problems from pressure drop will occur as the screen becomes restricted. Pressure drop causes the pump inlet to operate in a vacuum. It may pull in air while operating in a vacuum if the pump seals leak. Pressure drop can also cause pump cavitation, water turning to a vapor if the cooling tower water is warm enough and the pressure is low enough. Cavitation or air at the pump inlet would be evident from noise at the pump. It will often sound like it is pumping small rocks when this occurs. If a fine screen must be used, it may often be better applied at the pump outlet, Figure 48–32. It does less harm to throttle the pump discharge with pressure drop than the pump inlet.

The tower bypass valve helps to maintain the correct tower water temperature at the condenser during start-up and during low ambient operation. The tower bypass circuit allows water from the pump outlet to be recirculated to the pump inlet so that the cold cooling-tower water will not reach the condenser, Figure 48–33. Condenser water that is too cold in the compression cycle systems will reduce the head pressure to the point that the condenser will be so efficient that too much refrigerant will be held in the condenser and starve the evaporator. This will also cause the oil to migrate in some machines. Cooling tower water that is too cold for an absorption machine will often cause the salt solution to crystalize. Two types of three-way bypass valves may be used for tower bypass. These are the mixing valve and the diverting valve. The mixing valve has two inlets and

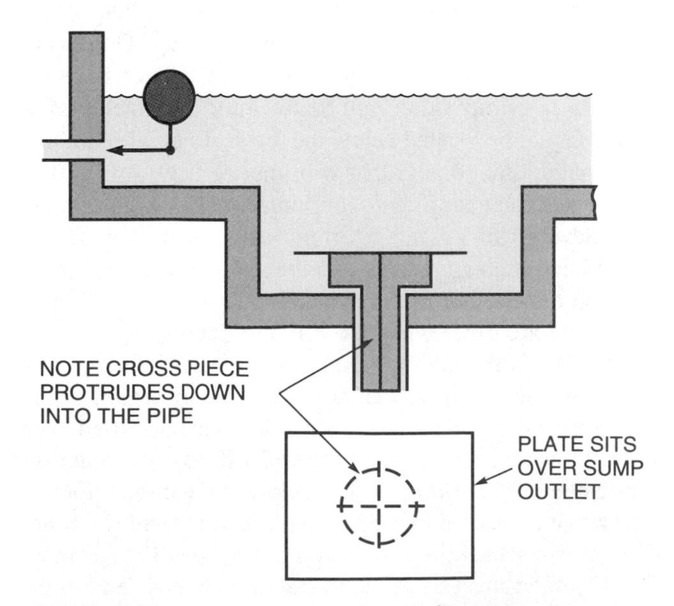

FIGURE 48–31 Vortex protection for a cooling tower.

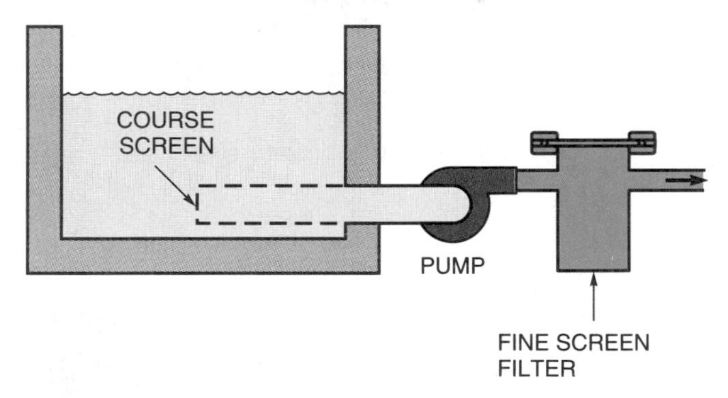

FIGURE 48–32 Best application for fine screen filter is in the pump outlet.

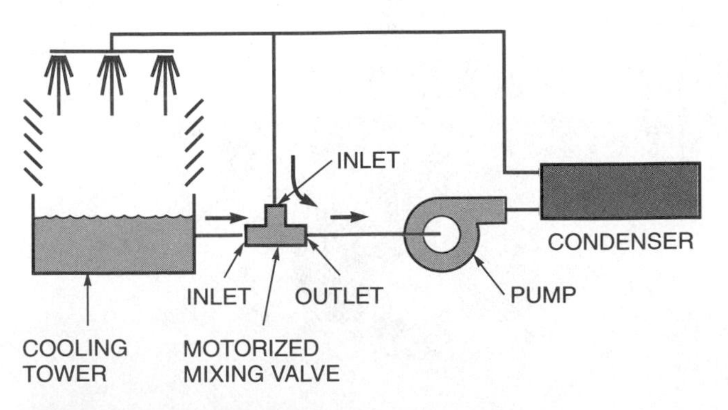

FIGURE 48–33 Mixing valve application.

one outlet and would need to be located in the pump suction line between the tower and the pump inlet, Figure 48–33. This valve can cause pressure drop in the pump suction that should be avoided if possible. The diverting valve has one inlet and two outlets, can be located in the pump discharge, and piped to the cooling tower or the pump inlet, Figure 48–34. These valves do not come in large sizes, normally a maximum of 4 in. so other arrangements may have to be made if a larger size is needed. A straight-through valve can be used in the configuration shown in Figure 48–35. Both the diverting valve and the straight-through valve allow some cooling tower water from the basin to be introduced to the pump inlet but may cause problems at start-up until the tower water becomes the correct temperature.

All pumps must have bearings that support the turning shaft while under load. Sleeve bearings are used for small pumps, and ball or roller bearings are used for larger pumps. These bearings must be lubricated on a regular basis for satisfactory service. Notice that the bearings are outside the pump on the split-case pump, Figure 48–36.

The pump must be fastened to the motor shaft in most of the above pumps in such a manner that they are within alignment tolerances. A flexible coupling may be installed between the pump and motor shaft to take out small misalignment, Figure 48–37. Because this coupling can only handle slight misalignment, the shafts must be aligned.

Two planes of alignment must be considered, angular and parallel. Angular alignment ensures that both shafts are at the same angle with each other, Figure 48–38. Parallel alignment ensures that the shafts are end to end, Figure 48–39. Correct alignment is accomplished with a dial indicator that reads to 1/1000 of an inch. It is mounted on the shaft and the shafts are rotated together through the full

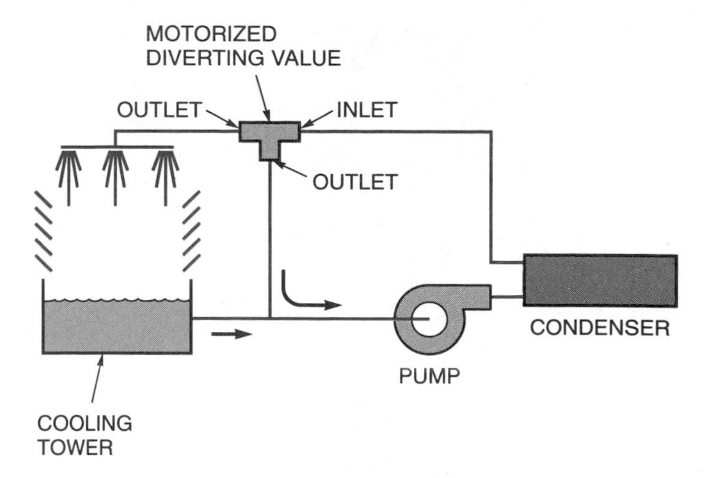

FIGURE 48–34 Diverting valve application.

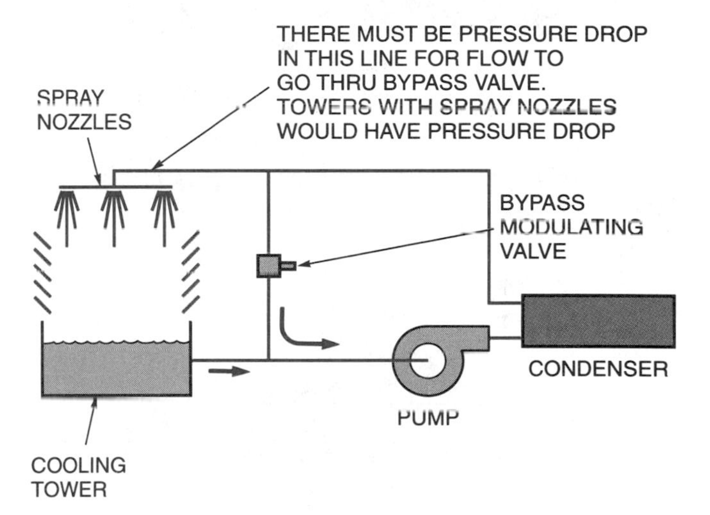

FIGURE 48–35 Straight-through valve for tower bypass.

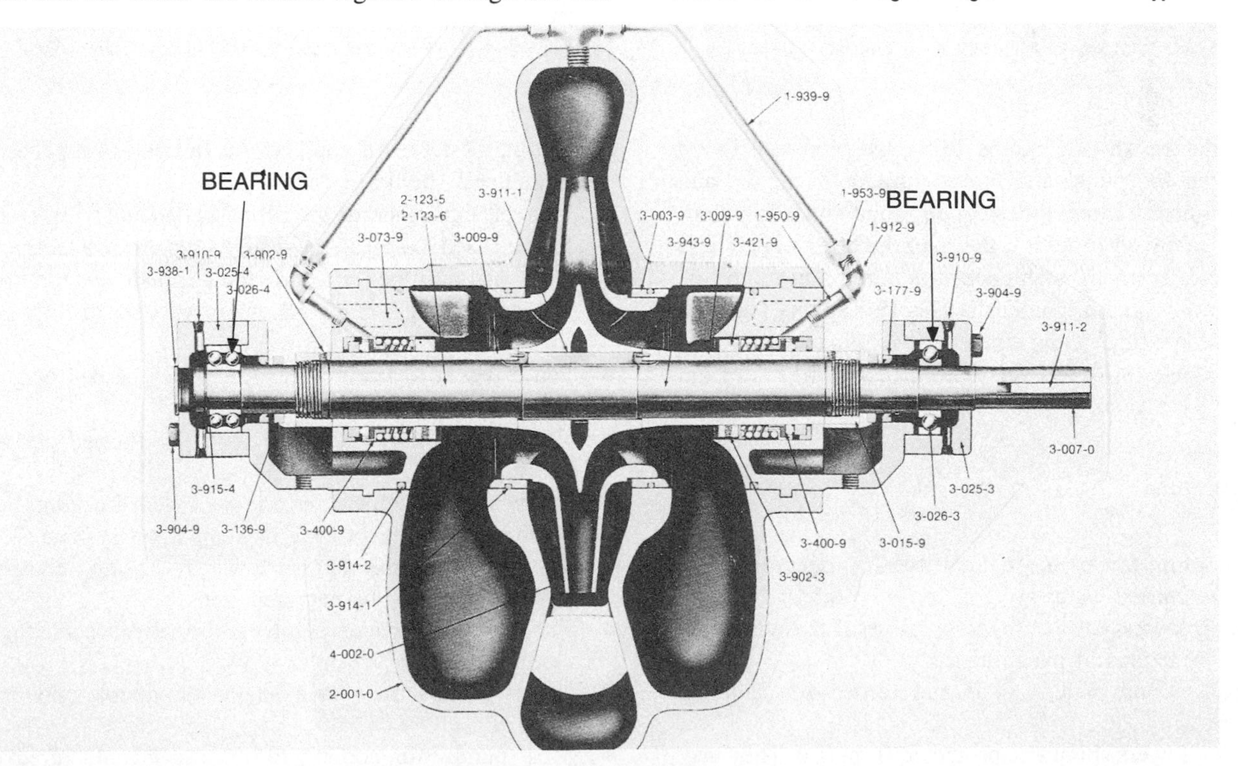

FIGURE 48–36 Bearings on a horizontal split-case pump. *Courtesy AC Pump ITT*

FIGURE 48–37 Flexible coupling. *Courtesy TB Woods & Sons*

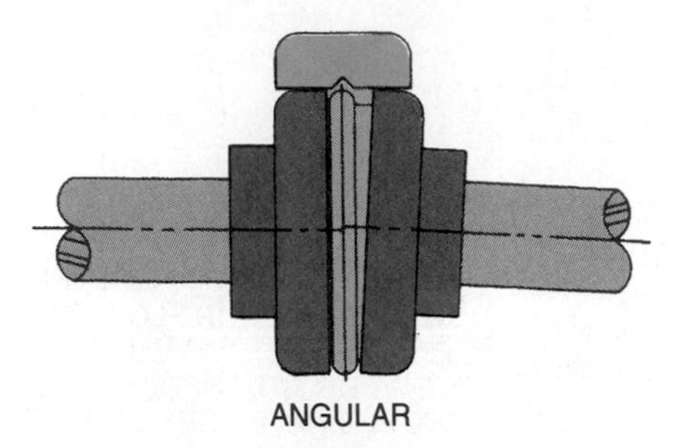

FIGURE 48–38 Angular misalignment for a coupling. *Courtesy Amtrol, Inc.*

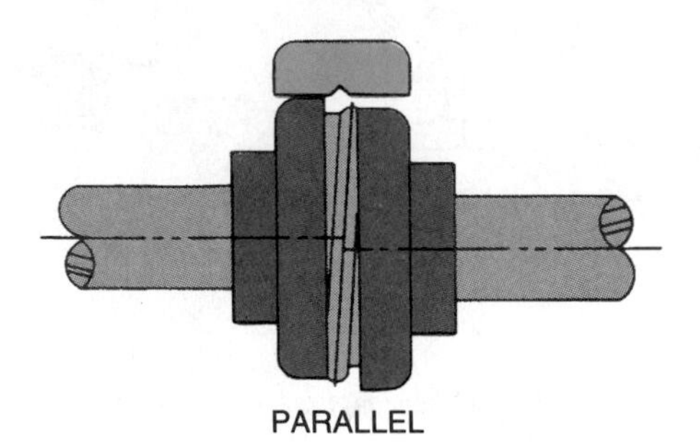

FIGURE 48–39 Parallel misalignment for a coupling. *Courtesy Amtrol, Inc.*

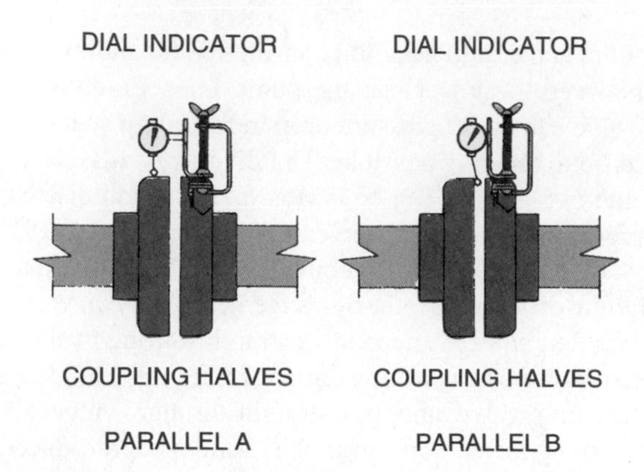

FIGURE 48–40 Dial indicators used to align coupling and shafts. *Courtesy Amtrol, Inc.*

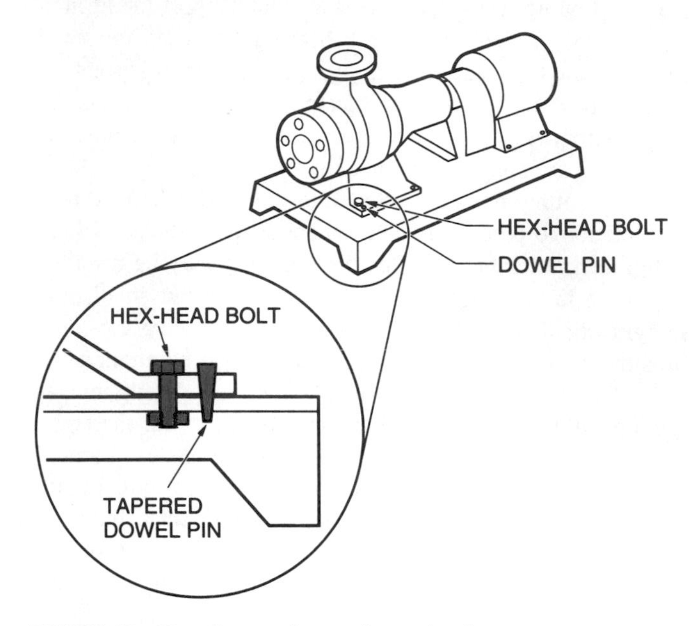

FIGURE 48–41 Pump and motor fastened to base using tapered dowel pins.

rotation, Figure 48–40. Shims (thin sheets of steel) are placed under the pump and motor until both angular and parallel alignment (sometimes called radial and axial alignment) are achieved to within the manufacturer's specifications. After alignment is attained, the pump and motor are tightened down and alignment is checked again. Then the base of both are drilled and tapered dowel pins are driven in the base, Figure 48–41. When done correctly, the motor and pump will give years of good service.

SUMMARY

- The cooling tower rejects heat that has been absorbed into the chilled-water system.
- Cooling towers lower the temperature of the water in a tower by means of evaporation.
- There are both natural-draft and forced-draft cooling towers.
- Cooling towers should be protected from fire. They may be manufactured from fireproof materials or may have a wetting system. All applicable codes and insurance requirements should be followed.
- The splash method or the fill or wetted surface method may be used to spread the water out and cause additional exposure to the air to provide adequate evaporation.
- The design of a cooling tower provides for distinct airflow patterns.
- Fan drives for forced draft towers may be belt or gear type.
- The cooling tower sump needs to be flushed periodically.
- Sumps may have heaters to protect from freezing.
- Cooling towers must have makeup water systems.
- The water flow through the tower must follow the pattern designed by the manufacturer.
- The condenser water pump is normally a centrifugal pump.
- Larger pumps may have single- or double-inlet impellers.
- Most pumps for cooling towers are manufactured of cast iron.

- Vortexing (whirlpooling) may occur in the sump. This is not good because air may be introduced into the pump. There are antivortexing designs and devices.
- A tower bypass valve helps to maintain the correct tower water temperature at the condenser.

REVIEW QUESTIONS

1. What is the purpose of the cooling tower in a chilled-water system?
2. What is the typical temperature difference between the water at the outlet of the cooling tower and outside air wet-bulb temperature?
3. Why is the water spread out as it moves through the cooling tower?
4. What are two types of water cooling towers?
5. What two types of fans are used in water towers?
6. Why must a water cooling tower have specific fire protection or be constructed of fireproof or fire-retardant materials?
7. What two methods may be used to spread the water out or slow it down as it moves through the tower?
8. What are the two types of airflow patterns in forced-draft water towers?
9. Why must there be a makeup water system in cooling towers?
10. What is meant by blowdown?
11. What type of pump is normally used for the condenser water and cooling tower water?
12. What is meant by vortexing?
13. What is the purpose of the tower bypass valve?
14. What will be the effect if the condenser water is too cold in a compression system chiller?
15. Describe the two planes of pump-motor alignment.

UNIT 49

Operation, Maintenance, and Troubleshooting of Chilled-Water Air Conditioning Systems

OBJECTIVES

After studying this unit, you should be able to

- discuss the general start-up procedures for a chilled-water air conditioning system.
- describe specific start-up procedures for a chiller using a scroll, a reciprocating, a rotary screw, and a centrifugal compressor.
- describe operating and monitoring procedures for scroll and reciprocating chilled-water systems.
- discuss preventive and other electrical maintenance and service that should be performed at least annually on chillers.
- list maintenance procedures that should be performed periodically on water-cooled chiller systems.
- describe start-up procedures for an absorption chilled-water system.
- list preventive and routine maintenance procedures for an absorption chiller.
- state preventive and periodic maintenance procedures for the purge system on an absorption system.

SAFETY CHECKLIST

* Wear gloves and use caution when working around hot steam pipes and other heated components.
* Use caution when working around high-pressure systems. Do not attempt to tighten or loosen fittings or connections when a system is pressurized. Follow recommended procedures when using nitrogen.
* Follow all recommended safety procedures when working around electrical circuits.
* Lock disconnect box or panel and tag it with your name when power is turned off to work on an electrical system. There should be only one key for this lock and it should be in the possession of the technician.
* Never start a motor with the door to the starter components open.
* Check pressures regularly as indicated by the manufacturer of the system to avoid excessive pressures.

The operation of these chillers involves starting, running, and stopping chiller systems in an orderly manner. The chiller operator should be familiar with the system and the machine before operating or monitoring any system. Most machines and systems are almost foolproof, but there is no need to take chances with this expensive equipment. It is good practice to become familiar with one piece of equipment at a time. There should be literature at the site for any major equipment installation. Study the manufacturer's literature. This literature should not leave the site or it may become lost. Many times job literature is taken home for study and never returned. Make it a practice to never let it leave the site. Manufacturers may be called on to furnish an extra copy if needed.

49.1 CHILLER START-UP

The first step in starting up a chilled-water system is to establish chilled-water flow. This is true if it is for the first time each season or the first time each day. If the system is water cooled, the next step if the condenser is water cooled is to establish condenser water flow. Air-cooled chillers have fans at the condensers and do not need the attention that water-cooled chillers do. Both the chilled water and the condenser water circuit will have some means, such as a flow switch, of proving that water is flowing. The flow switch is often a paddle that protrudes into the water and when water passes, it moves and operates a switch. Some chiller manufacturers use pressure drop controls to establish water flow. If this is the case, these are built into the water circuit by the manufacturer. The technician should know the checkpoints to make sure these switches are functioning. Because the paddle of a flow switch is in the water stream, it is not unusual for them to break off. If so, water flow may be established and not be proven with the flow switch. Water flow may be verified by the technician by means of the pressure gages. The starting point to finding out the correct water flow through a heat exchanger is to know what the pressure drop should be.

There should be a pressure test point at the inlet and outlet of any water heat exchanger, chiller or condenser, Figure 49–1. Note that two gages are used in the figure to check pressure. A better arrangement is to use one gage and two pressure connections, Figure 49–2. If two gages are used and they are at different heights, an error is automatically built in. For example, if one gage is 2.31 ft lower than the other, there is an error of 1 psig because of the difference in height. A standing column of water will have a pressure of 1 psig at the bottom, Figure 49–3. If the gages have a

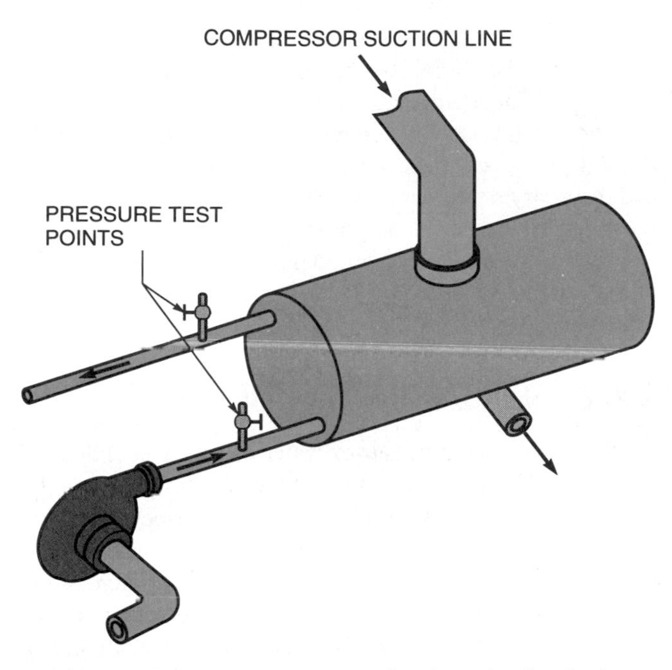

FIGURE 49–1 Pressure test points at the inlet and outlet of a heat exchanger.

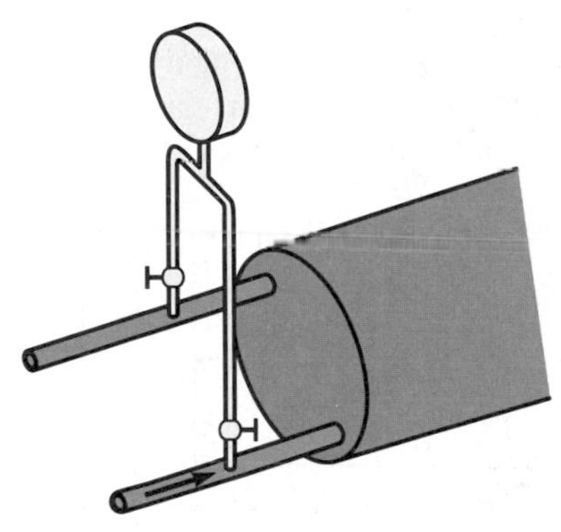

FIGURE 49–2 Using one gage and two ports for pressure checking the pressure drop through a heat exchanger.

built-in error because they are not calibrated, they cannot be expected to give correct readings. If one gage reads 1 psig when open to the atmosphere while the other gage reads 0, there is an error of 1 psig in the pressure difference. Actually, all you want to know is the pressure difference at the two points. Using one gage that is not accurate at 0 may still yield the correct pressure difference.

The technician needs to know from the pressure readings that there is a pressure drop across the heat exchanger. As water flows through the heat exchanger, the pressure drops a specific amount. The heat exchanger is a calibrated pressure drop monitoring device that may be used to determine water flow. A pressure drop chart for the heat exchanger will tell you the gallons per minute (gpm) of water flow, Figure 49–4. The original operating log sheet for the installation will show the pressure drops at start-up. This is very good information for future use. If the original log cannot be found, the manufacturer may be contacted for a copy of the log sheet or a pressure drop chart for the chiller.

The technician should also be aware of the interlock circuit through the contactors that must be satisfied before

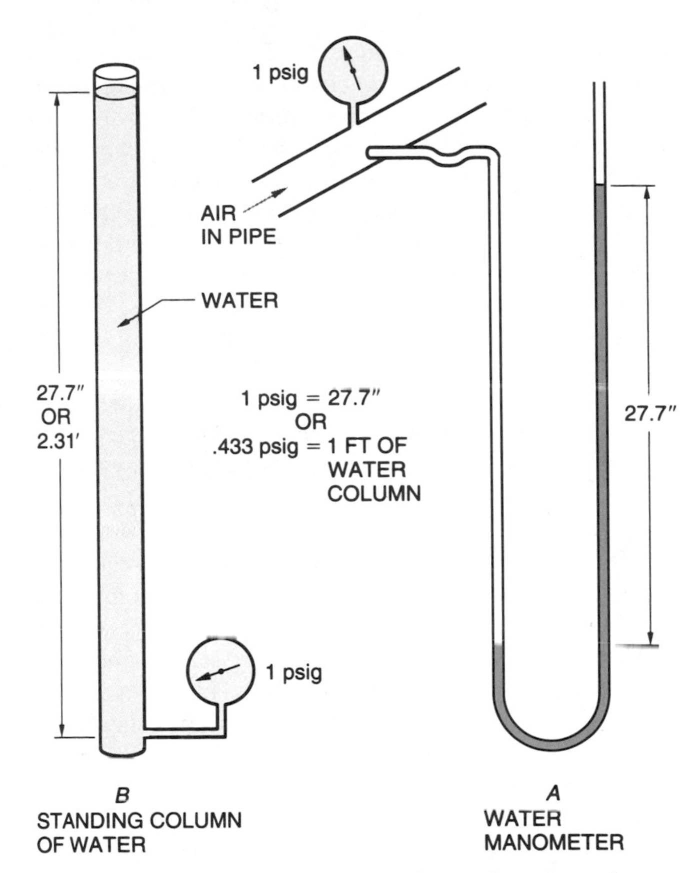

FIGURE 49–3 Pressure at the bottom of a standing column of water.

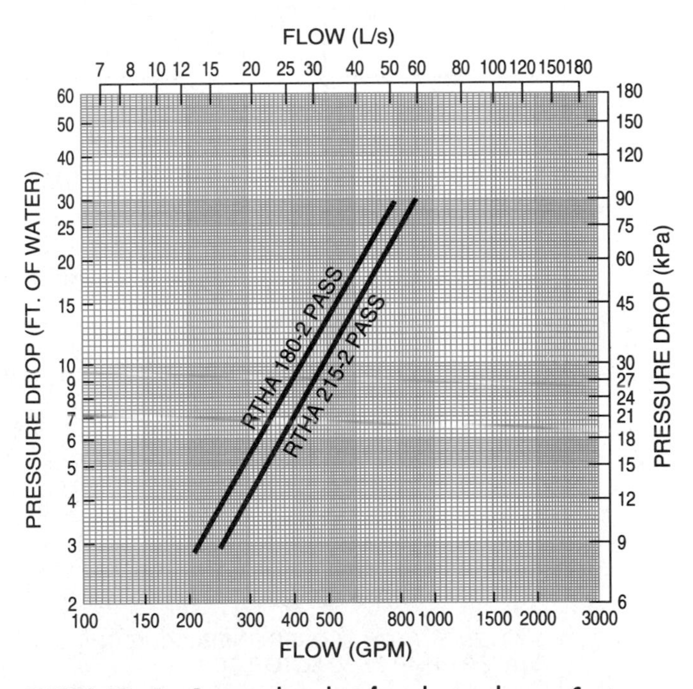

FIGURE 49–4 Pressure drop chart for a heat exchanger. *Courtesy Trane Company*

start-up. The starting sequence for most chiller systems is to start the chilled water pump first. A set of auxiliary contacts in the chilled water starter then starts the condenser water pump. When the condenser water pump starts, a set of auxiliary contacts make and pass power to the cooling tower fan and the chiller circuit, Figure 49–5. When the signal is received at the chiller control circuit, the compressor should start. This signal is often called the field control circuit because it is the circuit furnished by the contractor to the manufacturer's control circuit. If a chiller should not start, the first thing the technician should do is check the field circuit to see if it is satisfied. A technician should know where the field circuit is wired to the chiller circuit for each chiller on the job where there are several chillers. A quick check at this point will tell whether there is a field circuit problem (flow switches, pump interlock circuit, outdoor thermostat, and often the main controller) or a chiller problem. If there is no field control circuit power, there is no need to look at the chiller until there is power. Many chillers have a ready light that shows that the field control circuit is energized.

Different chillers have different starting sequences, usually depending on the type lubrication the compressor has. When the chiller has a positive-displacement compressor (scroll, reciprocating, or rotary screw), the compressor will start soon after the field control circuits are satisfied. Some manufacturers may have a time delay before the compressor starts after the field circuit is satisfied, but the compressor should start soon. Look for a time-delay circuit and make note of it and what the time delay is for any system so you won't be waiting and wondering what the problem is. The technician may think there is a problem only to have the chiller start up unexpectedly after a planned time delay. The reciprocating, scroll, and rotary screw compressors are lubricated from within and do not have a separate oil pump that must be started first. The reciprocating and the rotary screw compressors will start up unloaded and will begin to load when oil pressure is developed. The scroll compressor does not have compressor unloading capabilities and starts up under load.

49.2 SCROLL CHILLER START-UP

Scroll chillers are either air or water cooled. Regardless of the type, before attempting to start the chiller make sure the water for the chilled-water circuit is at the correct level. The system must be full. Then operate the chilled water pump and verify the water flow using pressure drop across the chiller. Make sure that all valves in the refrigerant circuit are in the correct position, usually back seated. Visually check the system for leaks. Oil on the external portion of a fitting or valve is a sure sign of a leak as oil is entrained in the refrigerant. The compressor will have crankcase heat and it must be energized for the length of time the manufacturer requires, usually 24 h. Do not start the compressor without adequate crankcase heat or compressor damage may occur.

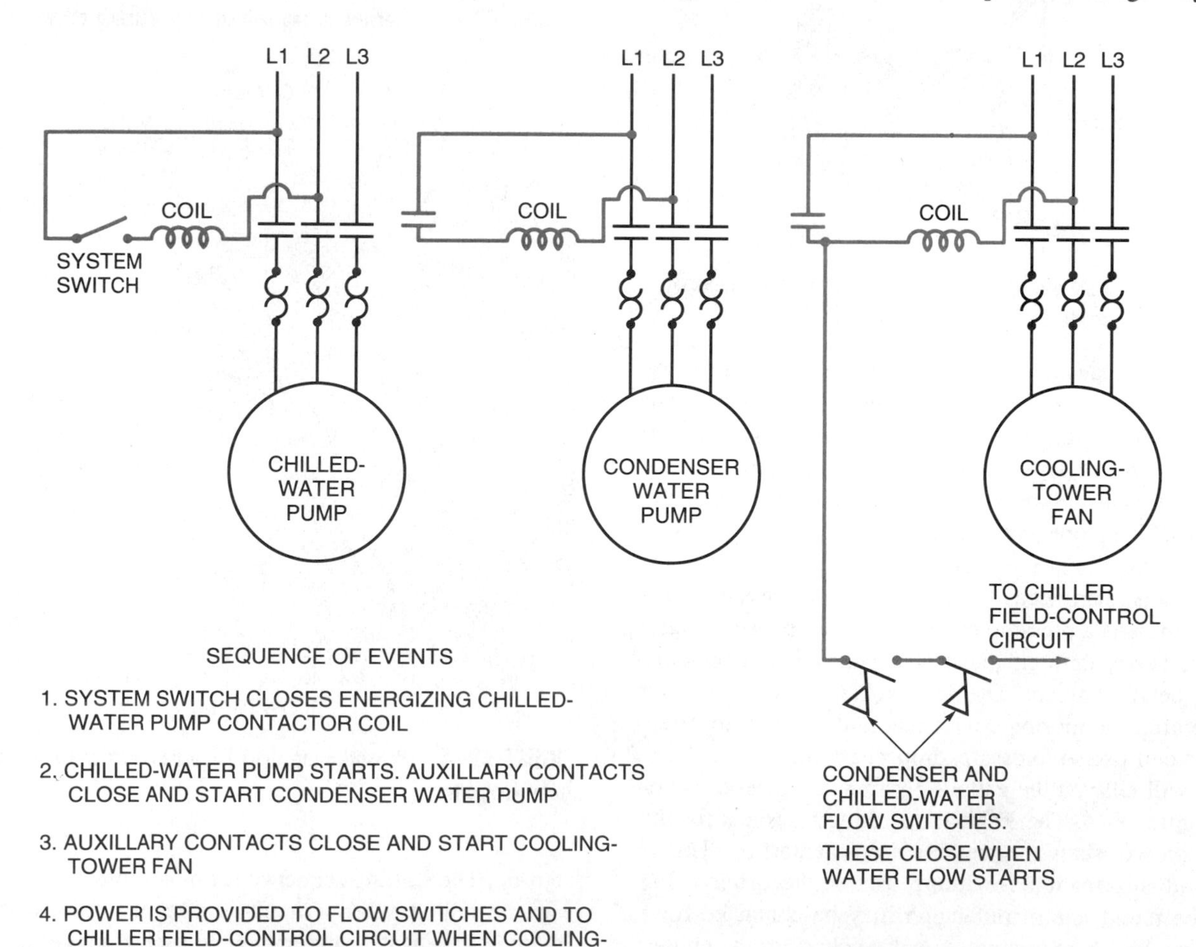
L1 L2 L3
L1 L2 L3
L1 L2 L3
COIL
COIL
COIL
SYSTEM SWITCH
CHILLED-WATER PUMP
CONDENSER WATER PUMP
COOLING-TOWER FAN
TO CHILLER FIELD-CONTROL CIRCUIT
SEQUENCE OF EVENTS
1. SYSTEM SWITCH CLOSES ENERGIZING CHILLED-WATER PUMP CONTACTOR COIL
2. CHILLED-WATER PUMP STARTS. AUXILLARY CONTACTS CLOSE AND START CONDENSER WATER PUMP
3. AUXILLARY CONTACTS CLOSE AND START COOLING-TOWER FAN
4. POWER IS PROVIDED TO FLOW SWITCHES AND TO CHILLER FIELD-CONTROL CIRCUIT WHEN COOLING-TOWER FAN IS STARTED
CONDENSER AND CHILLED-WATER FLOW SWITCHES.
THESE CLOSE WHEN WATER FLOW STARTS

FIGURE 49–5 Control circuit for a typical chiller.

When the chiller is air cooled, it will be located outside. The chiller barrel will have a heater to prevent it from freezing in the winter. Be sure this heater is wired and operable. If you don't do this at start-up, you may forget it when winter arrives. This heat strip is thermostatically controlled and will shut off when not needed.

When air cooled, make sure that the condenser fans are free to turn.

When the chiller is water cooled, the water-cooled condenser portion must be in operation. The cooling tower must be full of fresh water and the condenser water pump must be started. Make sure that water flow is established. You can look at the tower and verify this or you can check the pressure drop across the condenser.

Check to see that the field control circuit is calling for cooling. If all of the above requirements are met, the chiller should be ready to start.

When the chiller is started:

1. Observe suction pressure.
2. Observe discharge pressure.
3. Check the compressor for liquid flood back.
4. Check the entering- and leaving-water temperature on the chiller and the condenser entering- and leaving-water temperature for water cooled units.

When the chiller is operating normally, it can usually be left unattended except for the water tower when the unit is water cooled. The water tower should have regular observation and maintenance.

49.3 RECIPROCATING CHILLER START-UP

The reciprocating chiller may be either air or water cooled. When air cooled, it will be located outside and a heat strip will be applied to the chiller barrel to prevent freezing in the cold. The heat strip is thermostatically controlled and only operates when needed.

Before starting a reciprocating chiller, make sure the oil level in the sight glass is correct, usually from 1/4 to 1/2 level at the sight glass. Follow the manufacturer's instruction on oil level. When the level is high, there may be refrigerant in the oil; check for crankcase heat. Reciprocating chillers will have crankcase heaters to prevent refrigerant migration to the crankcase during the off cycle. When the crankcase heat is on, the compressor will be warm to the touch, Figure 49–6. This crankcase heat must be energized for the prescribed length of time before start-up to prevent damage to the compressor. Refrigerant in the oil will dilute the lubrication quality of the oil and can easily cause bearing damage that does not show up immediately. The safe time for crankcase heat to be energized is 24 h unless the manufacturer states otherwise. It is not good practice to ever turn the crankcase heat off, even during winter shutdown because refrigerant will migrate to the crankcase and it is not easy to boil it out to the extent that the oil is at the correct consistency for proper lubrication.

When the compressor is started, the pressures can be observed when there are panel gages. Typically, chillers

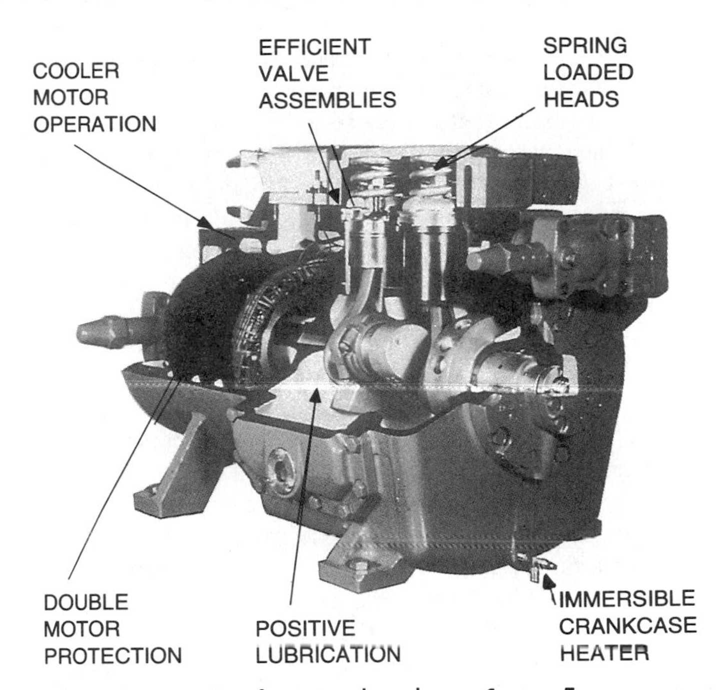

FIGURE 49–6 Heat from a crankcase heater. *Courtesy Trane Company*

over 25 tons have gages permanently installed by the manufacturer. These gages may be isolated by means of a service valve to prevent damage to the gages during long periods of running time so these valves will need to be opened for reading the gages. It is not good practice to leave the gage valves open all the time because the gage mechanism will experience considerable wear during normal operation from reciprocating compressor gas pulsations. These gages are for intermittent checking purposes. The technician should check these gages for accuracy from time to time because they cannot be relied on during the entire life of the chiller.

If a chiller has been secured for winter, the technician may have pumped the refrigerant into the receiver and the valves may need to be repositioned to the running position, which is back seated for any valve that does not have a control operating from the back seat. Figure 49–7 shows an example of a valve where the low-pressure control is fas-

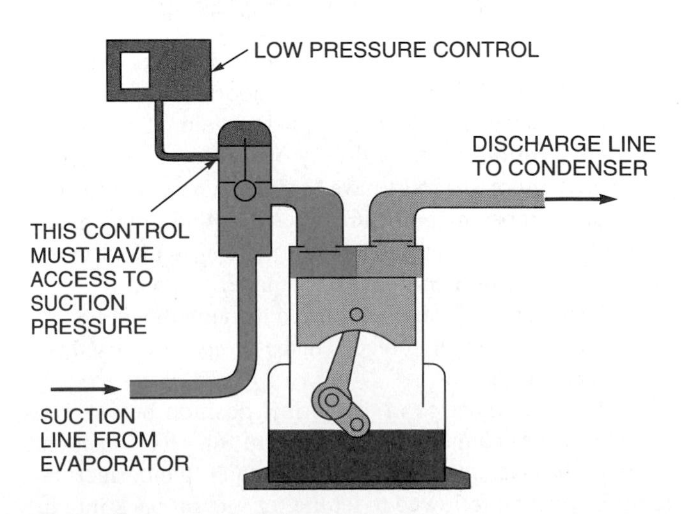

FIGURE 49–7 Service valve positions.

tened to the gage port on the back seat of the valve. If the valve is back seated, the low-pressure control will not be functional because it is isolated.

When the compressor is started, it will start up unloaded until the oil pressure builds up and loads the compressor. An ammeter can be applied to the motor leads to determine what the level of load is on the compressor. When it is fully loaded, the amperage should be close to full load.

As the system temperature begins to pull down, the technician may notice that the oil level in the compressor rises and foams. This is normal for many systems, but the oil level should reach normal level after a short period of operation, normally within 15 minutes. The compressor is suction cooled and the technician should ensure that liquid refrigerant is not flooding back into the compressor during the running cycle. This can be determined by feeling the compressor housing; the compressor motor should be cold where the suction line enters, then it should gradually become warm where the motor housing reaches the compressor housing. The compressor crankcase should never be cool to the touch after 30 min of running time or liquid refrigerant should be suspected to be flooding back, Figure 49–8.

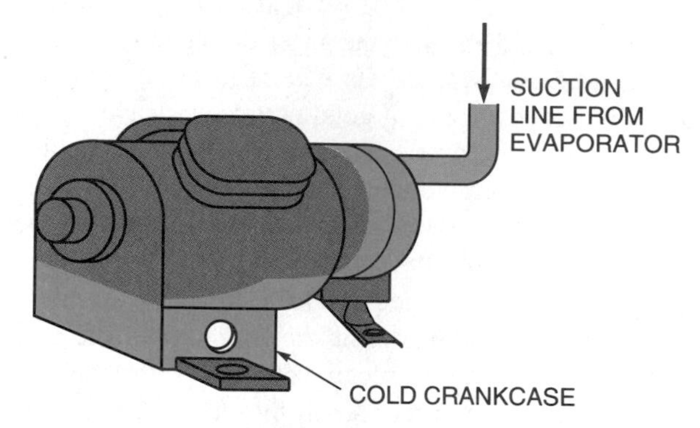

FIGURE 49–8 A cold compressor crankcase because of liquid refrigerant returning to the compressor.

49.4 ROTARY SCREW CHILLER START-UP

The rotary screw chiller may either be air or water cooled. If it is air cooled, it will be located outdoors and the heat for the chiller barrel should be verified. Again, it is thermostatically controlled and will only operate when needed.

The compressor must have crankcase heat before start-up. Again, most manufacturers require that this heat be energized for 24 h before start-up. Start-up without crankcase heat can cause compressor damage.

The water flow must be verified through the chiller and the condenser if water cooled. The best way to verify this is to use pressure gages.

All valves must be in the correct position before start-up. Check the manufacturer's literature for valve positions. If the chiller was pumped down for winter, the correct procedures must be followed to let the refrigerant back into the evaporator before start-up.

Check the field control circuit to make sure it is calling for cooling.

When all of the above is complete, the chiller is ready to start.

Start the compressor and watch for it to load up and start cooling the water. Observe the following:

1. Suction pressure
2. Discharge pressure
3. Water temperature, both chiller and condenser when water cooled
4. Look for liquid flood back to the compressor

When the chiller is operating normally, it can usually be left unattended except for the water tower when the unit is water cooled. The water tower should be observed and maintained regularly.

49.5 CENTRIFUGAL CHILLER START-UP

Centrifugal chillers often need to be started and watched for the first few minutes because of the separate oil sump system. When the chiller is a centrifugal, the oil sump should be checked before a start-up is tried. You should look for:

1. Correct oil sump temperature, from 135°F to 165°F, depending on the machine manufacturer. ***Do not attempt to start a compressor unless the oil sump temperature is within the range the compressor manufacturer recommends or serious problems may occur.*** After the machine has been operating for some time, check the bearing oil temperature to be sure that it is not overheating. If overheating occurs, check the oil cooling medium, usually water.
2. Correct level of oil in the oil sump. If the oil level is above the glass, it may be full of liquid refrigerant. Ensure that it is the correct temperature. You may start the oil pump in manual and observe what happens. If in doubt, call the manufacturer for recommendations. Unless you have experience as to what to do, do not try to start the compressor; marginal oil pressure may cause bearing damage.
3. Start the oil pump in manual and verify the oil pressure before starting the compressor, then turn it to automatic before starting the compressor.

When the chiller is started, the compressor oil pump will start and build oil pressure first. When satisfactory oil pressure is established, the compressor will start and run up to speed then change over to the run-winding configuration. This could be autotransformer or wye-delta. The centrifugal compressor starts unloaded and only begins to load up after the motor is up to speed. This is accomplished with the prerotation guide vanes mentioned earlier. When the chiller is up to speed, the prerotation vanes begin to open and will open to full load unless the machine demand limit control stops the vanes at a lower percent of full load.

When the chiller compressor starts to work, pull full- or part-load current, the technician should observe the pressure gages. The technician should look for problems such as:

1. The suction pressure, operating too low or too high. The technician should mark on the gage front or on a note pad the expected operating suction pressure for normal operation, which would be at design water temperature. Typically this would be 45°F leaving water.
2. The net oil pressure should be correct. This should be noted nearby for ready reference. If the machine has been off for a long period of time, observe the oil level in the oil sump as the compressor starts to accept the load and look for oil foaming. If the oil starts to foam, reduce the load on the compressor. This will raise the oil sump pressure and boil the refrigerant out of the oil at a slower pace. If you do not reduce the load, it is likely that the oil pressure will start to drop and the machine will shut off because of low oil pressure. If the compressor has an anti-recycle timer, you will not be able to restart the compressor until the timer completes its cycle and that may be 30 min. It is better to avoid allowing the compressor to shut off by watching the oil pressure.
3. Discharge pressure should be correct. This should be noted nearby for ready reference or marked on the pressure gage for typical operating conditions.
4. Compressor current should be correct and marked on the ammeter on the starter. The ammeter should be marked for all operating percentages, typically 40%, 60%, 80%, and 100% so the technician will be aware at all times at what load the machine is operating. The operating arm for the prerotation vanes should be marked so the travel can be measured; this is closely correlated with the current.

Chillers manufactured in the last few years all have electronic controls and a sequence of the starting events should be studied for the particular machine you are interested in. Typically they use everything from default lights on the control panel to electronic readout for what is happening. For example, a unit may have a sequence of light-emitting diode (LED) lights on the circuit board that will be lit if a particular problem has occurred. The technician must use the manufacturer's trouble chart to discover what the nature of the problem may be. Some manufacturers use the flash sequence of LED lights to describe a problem. Some manufacturers use an electronic readout that explains the problem in words or code numbers. These control sequences can be lengthy and it is beyond the scope of this text to describe them. Again, nothing beats an understanding of the equipment using the manufacturer's own description. One of the intents of these electronic controls is to make troubleshooting easier for the technician.

49.6 SCROLL AND RECIPROCATING CHILLER OPERATION

Once the chiller is on and operating, the operator should observe the chiller from time to time to make sure it is operating correctly. Small air-cooled chillers may be located in remote locations and not observed on a regular basis. These chillers are usually operated as unattended. There is not much that can go wrong if the chilled-water supply is maintained.

All water-cooled chillers require more attention because of the water circuit and cooling tower. It is good practice to check any piece of operating equipment on a periodic basis, but it is expensive so most close observation is reserved for larger systems. It is good practice to check the cooling tower water several times per day to be sure that the water level is correct. Often, a partial loss of water pressure will cause the tower water to drop due to the fact that the water is evaporating faster than the water supply can make it up. With careful observation, the technician can catch this problem before it shuts the unit off and a complaint occurs. If there is no regular technician on the job, for example, at a small office building, someone can be assigned to make this check with very little instruction. These chillers are often on top of buildings where no one goes on a regular basis. Someone should be assigned to check these installations every week.

Water-cooled chillers must have water treatment to prevent minerals and algae from forming in the water circuit. A qualified water treatment specialist is necessary for all water tower applications or problems will occur. Review the text on water towers for types of treatment needed. All towers must have blowdown, which is a percentage of the circulated water passing down the drain to prevent the tower water from becoming overconcentrated with minerals due to evaporation. Blowdown for the cooling tower is necessary but often management has a difficult time watching water that looks clean flow down the drain. The technician should know how to explain this to them.

49.7 LARGE POSITIVE-DISPLACEMENT CHILLER OPERATION

Large chillers that use reciprocating or rotary screw compressors may range in size to over 1000 tons of capacity. Chillers of this size are extremely expensive and must be observed on a regular basis or problems may occur. These problems can often be expensive to correct. These large chillers are reliable but observation can be important. The technician should look for any potential problem, as explained in the previous paragraphs. These chillers may also have pressure gages that are active all the time in that they are not valved off. If so, these may be observed as much as once per hour on some critical applications. Often an operating log is maintained on a regular basis. The frequency of the recording in the operating log depends on the importance of the job. If it is a chiller used for critical manufacturing, the entries in the log may be required hourly. If it is an office building, the log may be maintained on a daily or weekly basis. An operating log form appears earlier in the text.

49.8 CENTRIFUGAL CHILLER OPERATION

Centrifugal chillers were the largest chillers made for many years. Typically when the requirements were above 100 tons, a centrifugal chiller was considered or multiple reciprocating chillers were used. Centrifugal chillers range in size from 100 tons to about 10,000 tons for a single chiller. These were the most expensive of all chillers and also the most reliable so much attention has been given to the observation of the operation and maintenance of these chillers. For years, it has been customary for the operation of these chillers to be observed and an operating log maintained on a regular basis, usually daily. With all operating logs, slowly occurring problems can be plotted. For example, if the cooling tower water treatment has not been working and the condenser water tubes are beginning to have a mineral buildup, the condenser water approach will begin to spread. An alert operator will notice this and take corrective action. This action may be to change the water treatment and certainly to clean the water tubes because the heat exchange has been reduced.

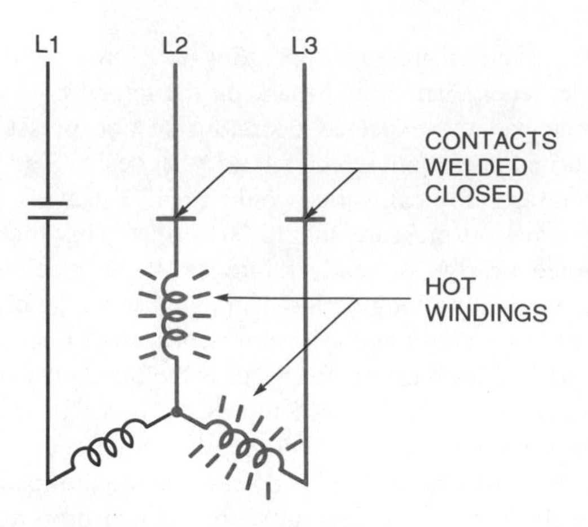

FIGURE 49–9 Welded contacts on a wye-wound motor.

49.9 AIR-COOLED CHILLER MAINTENANCE

Air-cooled chillers require little maintenance, which is one reason why they are so popular. The fan section of these chillers may require lubrication of the fan motors or the motors may have permanently lubricated bearings. It is typical for these chillers to have multiple fans and motors. The motor horsepower can be reduced in this manner and the fans can be direct drive. This eliminates belts and the need for routine lubrication in many cases. If one fan fails, the chiller can continue to operate.

The following electrical maintenance should be performed annually for a typical system:

1. Inspect the complete power wiring circuit because this is where the most current is drawn. Look for places where hot spots have occurred. For example, the wires on the compressor contactor should show no signs of heat. The insulation should not appear to have been hot. If so, this must be repaired. This can be done by cutting the wire back to clean copper. If the lead is not long enough, a splice of new wire may be needed or the wire may need to be replaced back to the next junction.
2. Inspect the motor terminal connections at the compressor. These should not show any discoloration or a repair should be made. A hot terminal block can cause refrigerant leaks so the wiring should be inspected carefully.
3. The contacts in all contactors should be inspected and replaced if pitting is excessive. If excess pitting begins to occur, it is likely that the contacts will weld shut in the future and the motor cannot be stopped using the contactor. This can lead to single phasing of a wye-wound motor if any two contacts weld shut. Motor burn will then occur because there is nothing to shut the motor off except the breaker, Figure 49–9.
4. The compressor motor should be checked for internal ground using an ohmmeter that can check ohm readings in the millions of ohms. This meter is called a megger® (megohmmeter). Figure 49–10 shows an example of a basic megger. This instrument uses about 50 V to check for leaking circuits; others may use much higher voltages and have a crank type generator to achieve the high voltage. The correct motor manufacturer's recommendations should be used for the allowable leakage of the circuit. Typically, a motor should not have less than 100 megohms circuit to ground or to another winding in the case of a delta type motor or dual-voltage motor, Figure 49–11. The required reading depends on the motor temperature because the temperature changes the requirements. You should consult the manufacturer if proper guidelines are not known. A reliable local motor shop can also give you guidelines to use.

   It is important to start a process of megging a motor when it is new and keep good records. If the motor megohm value begins to reduce, it is a sign that moisture or some other foreign matter may be getting in the system.

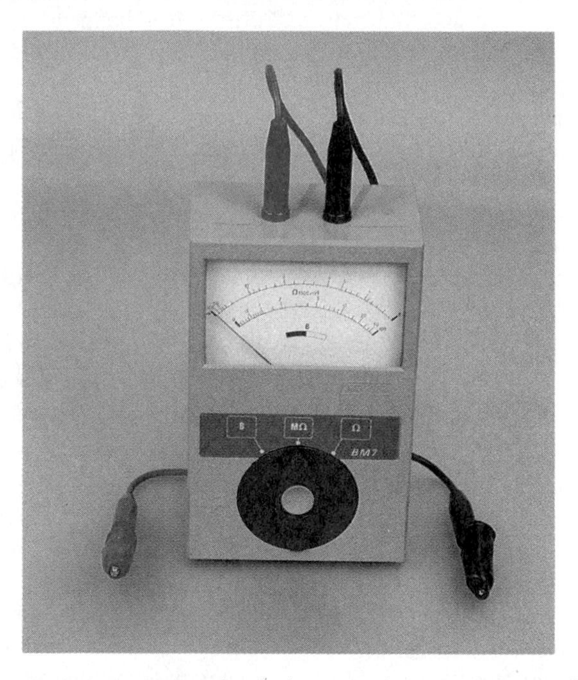

FIGURE 49–10 Megger for checking ground circuits in motors. *Photo by Bill Johnson*

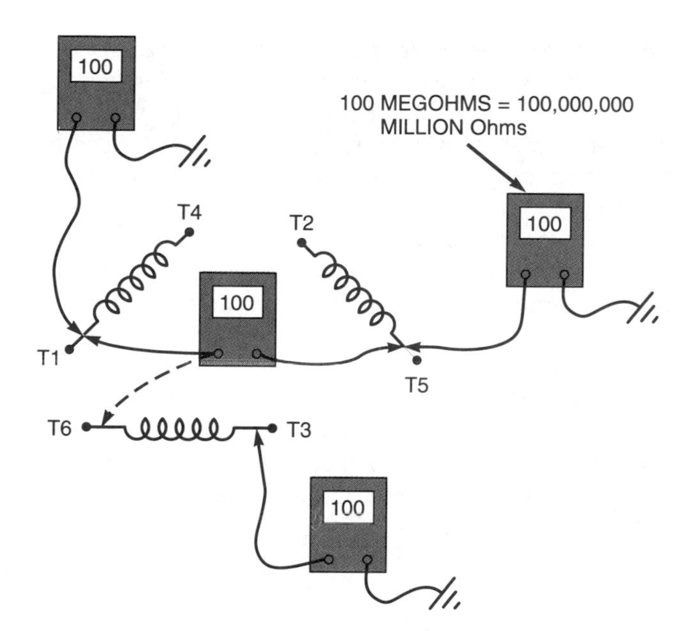

FIGURE 49–11 Using a megger to check a motor.

5. An oil sample from the compressor crankcase can be sent to an oil laboratory to determine the condition of the compressor. This test, like the megohm test, should be started when the chiller is new and performed every year. If certain elements start appearing in greater quantities in the oil, it is evident that problems are occurring. For example, if the bearings are made of babbitt and the babbitt content increases each year, at some point it is evident that the bearings need to be inspected. If small amounts of water appear, the tubes need to be inspected for leaks. This will only occur if the water pressure is greater than the refrigerant pressure. For high-pressure chillers this can occur if the chiller is at the bottom of a high column of water.
6. Inspect the condenser for deposits such as lint, dust, or dirt on the coil surface. These must be removed because they block airflow. It is good practice to clean all air-cooled condensers once a year or more often if needed. If the operating log shows an increase in head pressure compared to the entering air temperature, the coil is becoming dirty. Air-cooled coils can be deceiving in that they can be dirty and you may not notice it because the dirt may be imbedded in the interior of the coil. This cannot be seen until the coil is cleaned. A coil should be cleaned by saturating the coil with an approved detergent and then washing the coil backward, in reverse to the direction of the airflow, called back-washing. An apparently clean coil may yield a large volume of dirt from the interior of the coil.

49.10 WATER-COOLED CHILLER MAINTENANCE

Routine maintenance may involve cleaning the equipment room and keeping all pipes and pumps in good working order. The technician should ensure that the room and equipment are kept clean and in good order. It is always easier to keep a room clean than to have to clean it when the chiller breaks down and needs to be disassembled. The cooling tower should be checked to be sure the strainer is not restricted. Look at the water for signs of rust, dirt, and floating debris and clean if necessary. The water treatment should be checked on a regular basis. Some systems call for the operating technician to perform water analysis on a daily basis to check for mineral content. When the water treatment gets out of balance, the technician should either make adjustments or call the water treatment company. Someone must be in charge of keeping the water chemicals at the correct level for best performance.

Annual maintenance should involve checking the complete system and chiller to prepare it for a season of routine maintenance. Some chillers operate year around and are only shut down for annual maintenance. During this maintenance, it is often wise to bring in the factory maintenance representatives. These technicians are privileged to know what is happening across the nation or around the world with regard to any service or failure problems with their equipment. They will provide the best technical knowledge for your equipment. Often, these technicians will perform the complete start-up procedure for your equipment just as though it is for the first time. All controls may be checked at this time to be sure they will perform up to standard. It is not uncommon for a system to operate for years without having a control checkup and the system then fail only to discover that a control has failed and should have saved the system but didn't. A control checkup once a year can be good insurance against many failures.

During annual maintenance, all electrical connections should be checked; refer to the electrical maintenance for air-cooled systems stated previously. Water-cooled equipment has the same electrical symptoms.

The water-cooled condenser should be inspected on the inside every year by draining the water from the condenser and removing the heads. The tubes should be clean. You can tell when a tube is clean when you take a light and shine inside the tube and all you see is copper tube with a dull copper finish. A penlight is very good for this. The penlight can be inserted inside the tube. Any film on the inside of the tube must be removed for proper heat exchange. Many technicians believe that if the tube is open enough to allow a good water flow that it is clean. This is not true; the tube must be clean for proper efficiency.

When it is discovered that the tubes are dirty, several approaches may be used. The tubes may be brushed with a nylon brush that is the size of the inside of the tube. This is customarily done with a machine that turns the brush while flushing the tube with water. Some manufacturers recommend using a brush with fine brass bristles if the nylon will not remove the scale from the tubes. Check with the manufacturer because some do not recommend this practice. Always keep the tubes wet until it is time to brush them or the scale will become more hardened. It is good practice to be prepared to brush the tubes before removing the heads so the tubes will stay wet until they are brushed.

If brushing the tubes will not work, the tubes may be chemically cleaned using the recommended acid for the application, Figure 49–12.

SAFETY PRECAUTION: *This is a very delicate process because the tubes may be damaged. Consult the chemical manufacturer for this procedure.*

It may be good practice to let the chemical manufacturer clean the tubes, then if damage occurs, they are responsible. However the process is accomplished, make sure that the chemicals are neutralized or damage to the tubes may occur. The damage that may occur would be severe because the chemicals may eat the tube material and leaks between the water and refrigerant may occur.

After the tubes have been chemically treated, they may be clean or the chemicals may just soften the scale and the tubes may need brushing to remove the scale.

The water pump should be checked to make sure the coupling is in good condition. ***This can be done by turning off the power, locking and tagging the disconnect, and removing the cover to the pump coupling.*** If materials such as filings are found inside the coupling housing, deterioration of the coupling should be suspected. Some couplings are made of rubber and some are steel. Look for the flexible material that takes up the slack to see if it is wearing.

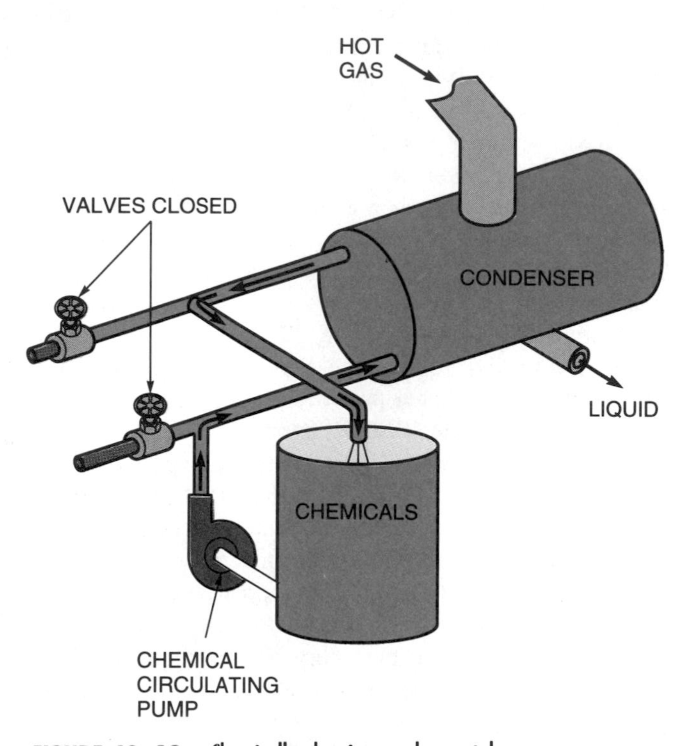

FIGURE 49–12 Chemically cleaning condenser tubes.

After the condenser has been cleaned, the condenser tubes may be checked for defects such as stress cracks, erosion, corrosion, and wear on the outside by means of an eddy current check. This test should be performed on the evaporator tubes as well. This check determines if there are irregularities in the tube. One common problem is wear on the tube's outer surface where it is supported by the tube support sheets. Each set of tubes has a different type of stress applied to the outside. In the condenser, the hot discharge gas is pulsating as it enters the condenser and has a tendency to shake the tubes. In the evaporator, the boiling refrigerant shakes the tubes. The tube support sheets are sheets of steel that the tubes pass through to help prevent this action and are evenly spaced between the end sheets. They are located in the condenser and the evaporator. When the evaporator or condenser shell is tubed, the tubes are guided in place through the support sheets. When the tube is in place, a roll mechanism is inserted into the tube and expanded to hold the tube tight in the support sheet, Figure 49–13. The tube roller may miss a tube and this tube can shake or vibrate until the tube sheet wears the tube. If this wear continues, the tube will rupture and a leak between the water and refrigerant circuit will occur. The worst case can flood the chiller on the refrigerant side with water. This may happen in particular with low-pressure chillers. The eddy current test can be used to find potential tube failures and these tubes may be pulled and replaced or plugged if there are not too many.

The eddy current test instrument has a probe that can be pushed through the clean tube while an operator watches a screen monitor, Figure 49–14. As the probe passes through the tube, the probe sends out a magnetic current signal that reacts with the tube and shows a profile on the screen monitor. When unusual profiles are noticed, the tube is marked for further study, which may involve comparing the profile to that of a known good tube. A decision may need to be made to pull or plug the tube. A tube failure in the middle of the cooling season can take days to repair and it is good practice to find these problems in advance.

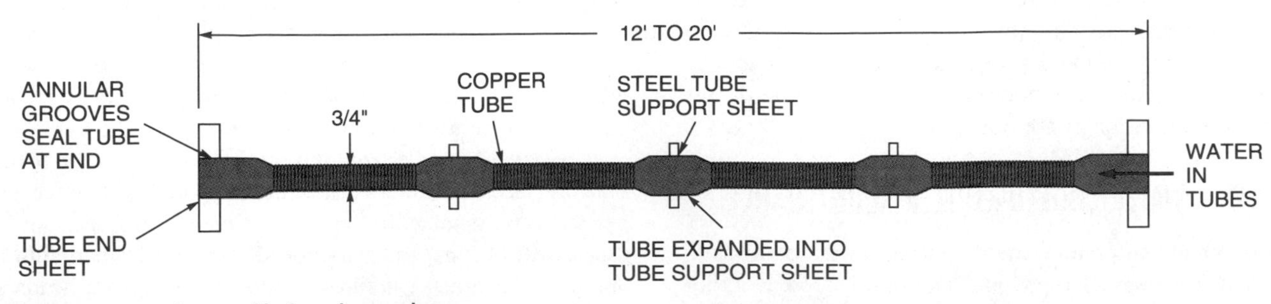

FIGURE 49–13 Tube assembly for a heat exchanger.

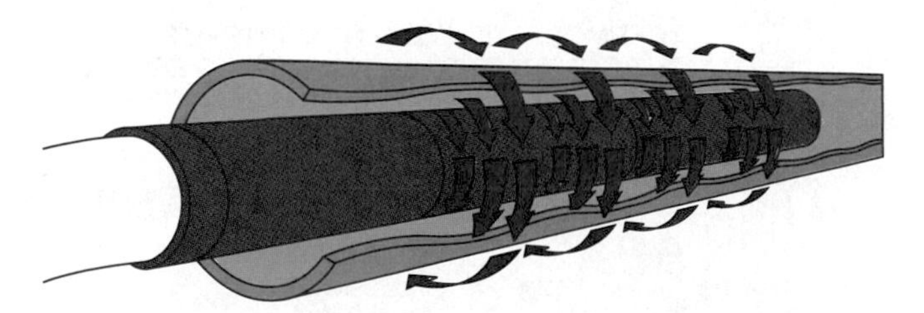

FIGURE 49–14 Probe for checking tubes using an eddy current and monitor.

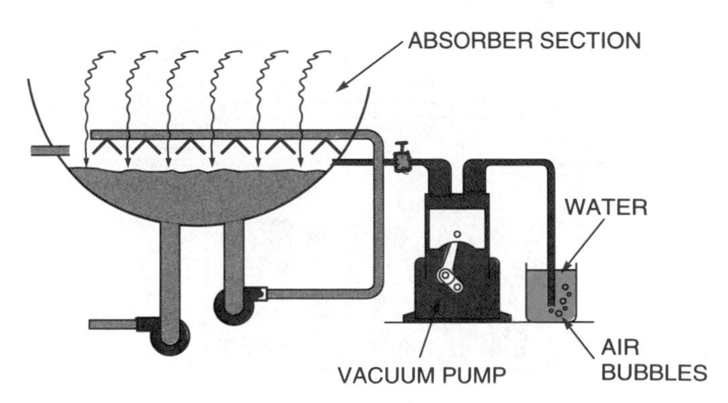

FIGURE 49–15 The vacuum pump exhaust is placed in a glass of water. When bubbles are observed, the vacuum pump is exhausting noncondensables.

Remove and clean any strainers that may be in the piping. Some systems have a strainer before or after the pump. This should be checked and cleaned.

The cooling tower should be cleaned and scrubbed out, removing any residue that may have accumulated in the basin.

If the chiller and water pumps are to stand idle during the winter months, all precautions should be taken to prevent freezing of any components such as piping, sumps, pumps, or chiller if it is located where it may freeze. Even if a chiller is in a building, it is good practice to drain it during winter because if the building loses power and the interior freezes, it would be expensive if the chiller tubes were to freeze. It is easier to just drain it and not take a chance.

49.11 ABSORPTION CHILLED-WATER SYSTEM START-UP

The absorption system start-up is similar to any chiller; chilled-water flow and condenser water flow must be established before the chiller is started. Ensure that the cooling tower water temperature is within the range of the chiller manufacturer's recommendations. If the absorption machine is started with the cooling tower water too cold, the lithium-bromide (Li-Br) may crystallize, causing a serious problem. In addition to this, the heat source must be verified whether it be steam, hot water, natural gas, or oil. Most absorption chillers are steam operated so the steam pressure must be available. If the chiller is operated from a boiler dedicated to it, the boiler must be started and operated until it is up to pressure.

The purge discharge may be monitored by placing the vacuum pump exhaust into a glass of water and looking for bubbles, Figure 49–15. If there are bubbles, it is a sure sign that noncondensables are in the chiller. It is pointless to start the chiller until the bubbles stop. Let the vacuum pump run even when the chiller is started (unless the manufacturer recommends not to) because often noncondensables will migrate to the purge pick-up point after start-up.

When all systems are ready, the chiller may be started. The fluid pumps will start to circulate the Li-Br and the refrigerant and the heat source will start. The chiller will begin to cool in a very few minutes and this can be verified by the temperature drop across the chilled-water circuit. When absorption chillers are refrigerating, they make a sound like ice cracking. When this sound begins to come from the machine, the chilled-water temperature should begin to drop.

49.12 ABSORPTION CHILLER OPERATION AND MAINTENANCE

Absorption chillers must be observed for proper operation on a more regular basis than compression cycle chillers because they can be more intricate. Chiller manufacturers have different checkpoints to determine the chiller operation. The refrigerant temperature and the absorption fluid temperatures measured against the leaving chilled-water temperatures are among a few. The manufacturer's literature must be consulted for the checkpoints and procedures. The maintenance of an operating log is highly recommended for absorption installations.

When the chiller is operating, it is good practice to pay close attention to the purge operation. If the chiller is requiring excessive purge operation, the machine may have leaks or it may be manufacturing excess hydrogen internally and need additives. The machine must be kept free of noncondensables.

The heat source may need maintenance. If it is steam, the steam valve should be checked for leaks and operating problems. When steam is used, a condensate trap will also need checking to be sure it is operating properly. The condensate trap ensures that only water (condensed steam) returns to the boiler. A typical condensate trap may contain a float. When the water level rises, the float rises and only allows water to move into the condensate return line, Figure 49–16. There are several other different types of trap; this one is shown as an example.

Condensate traps are often checked with an infrared checking device that measures the temperature on both sides of the trap, Figure 49–17.

When the chiller is operated using hot water, the hot water valve should be observed for proper operation and to ensure that there are no leaks.

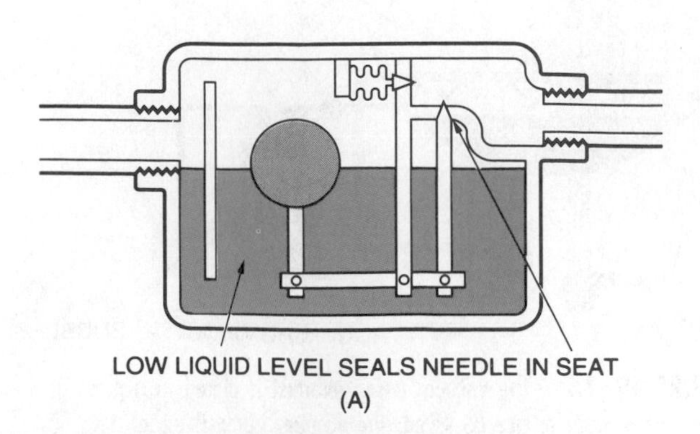
LOW LIQUID LEVEL SEALS NEEDLE IN SEAT
(A)

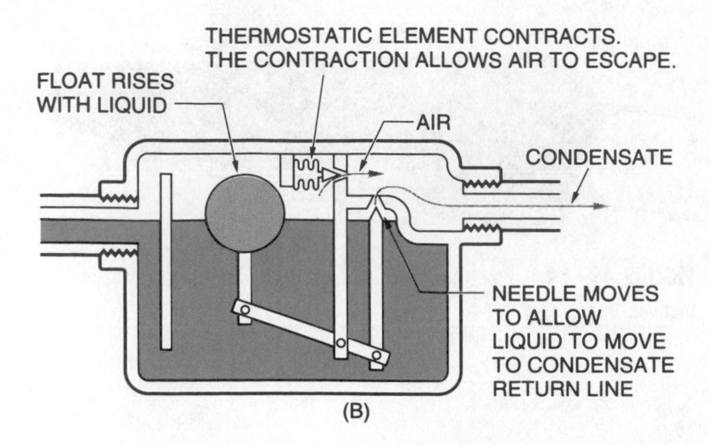
THERMOSTATIC ELEMENT CONTRACTS.
THE CONTRACTION ALLOWS AIR TO ESCAPE.
FLOAT RISES
WITH LIQUID
AIR
CONDENSATE
NEEDLE MOVES
TO ALLOW
LIQUID TO MOVE
TO CONDENSATE
RETURN LINE
(B)

FIGURE 49–16 Condensate float trap for a steam coil.

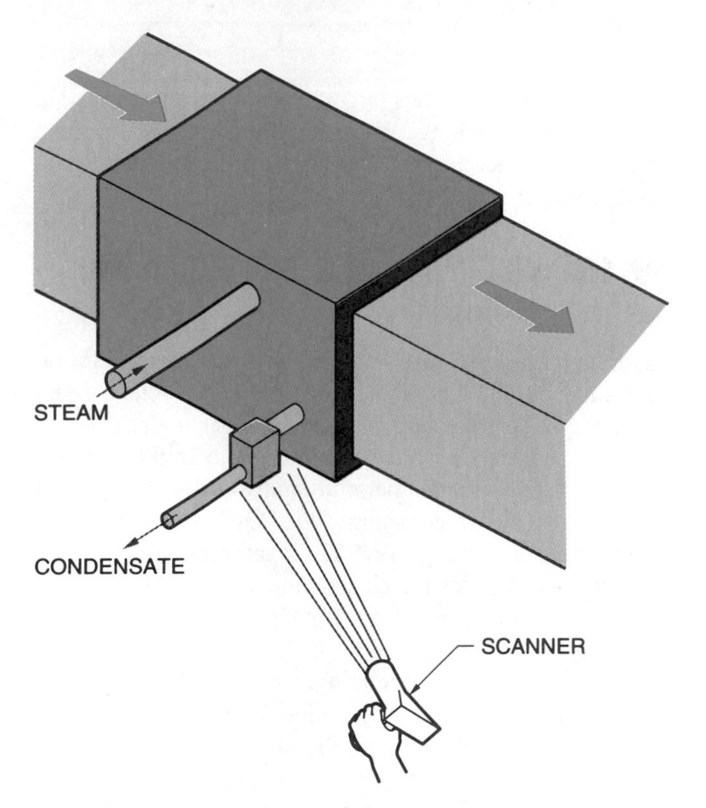
STEAM
CONDENSATE
SCANNER

FIGURE 49–17 Checking a condensate trap using an infrared scanner.

Gas- or oil-operated machines will need the typical maintenance for these fuels. Oil in particular requires maintenance of the filter systems and nozzles.

The purge system may require the most maintenance with the absorption chiller. If there is a vacuum pump with the chiller, regular oil changes will help to make it last longer. Li-Br is salt and will corrode the vacuum pump on the inside. ***It is not recommended that a vacuum pump used for a compression cycle system be used on an absorption system because of the corrosion problem, but if one is used, extra care should be taken to change the oil immediately after use.***

Any Li-Br that is spilled in the equipment room will rust any metal it is in contact with. It is good practice to keep the area where it is handled washed with fresh water to prevent this. The machine and other equipment may need to be painted periodically for additional protection in these areas.

The system solution pump or pumps may require periodic maintenance. A check with the manufacturer will detail the type and frequency of this maintenance.

The machine room must be kept above freezing or the refrigerant (water) inside the machine will freeze and cause tube rupture problems. Unlike the other chillers, this water cannot be drained and must be protected.

The water pumps need regular maintenance as with any other system.

The steam or hot water valve and condensate trap should be inspected for any needed maintenance. A strainer will be located in the vicinity of the condensate trap and should be cleaned during the off season.

The cooling tower must be maintained by cleaning and water treatment. Don't forget blowdown water. Absorption chillers operate at higher temperatures than compression cycle chillers and good water treatment maintenance is vital because the tubes will become fouled much quicker.

Absorption chillers may need the tubes probed with an eddy current probe on a more regular basis than compression cycle chillers because of the higher temperatures in the condenser section. The tubes are stressed when the heat source is applied to these tubes and they expand and then contract when cooled during the off cycle. A check with the manufacturer is a good idea.

49.13 GENERAL MAINTENANCE FOR ALL CHILLERS

Technicians who maintain chillers must be well qualified and should stay in contact with the manufacturers for the latest training schools and advice. Manufacturers receive feedback from all over the world on their equipment and know what is happening, such as with premature failures. They should share any potential problems with you to prevent future problems with your equipment. Technicians should attend factory seminars and schools and make friends with the factory technicians.

49.14 SERVICE TECHNICIAN CALLS

SERVICE CALL 1

A building manager calls and reports that the building is becoming hot. *The problem is the cooling tower fan has a broken belt and the fan is off. The chiller is off because of high head pressure.*

The technician arrives and goes to the basement where the chiller is located. It is a reciprocating water-cooled chiller. The chiller is off and the technician looks around for the problem. The control panel is opened and it is noticed that the reset button on the high-pressure control is out, signaling that the control needs resetting. The pressure drop across the condenser is normal and the water temperature is normal (because the chiller has been off the tower is able to lower the temperature without the fan). The technician resets the button and the compressor starts. As the technician watches, the compressor starts to load up. Everything runs normally for about 10 min and then the compressor sounds as though it is straining. The technician feels the condenser water line entering the condenser. It is warmer than hand temperature. The cooling tower water is too hot. The technician shuts the compressor off rather than letting it shut off with the high-pressure control.

The technician goes to the roof to look at the cooling tower, fully expecting to see it full of algae because the management does not require regular maintenance. When on the roof, it is noticed that the fan belt to the cooling tower fan is broken. The disconnect to the fan is shut off, locked out, and tagged for safety. The numbers are written down from the old belt and the technician goes to the truck for a new one. While at the truck, the technician gets a grease gun for greasing the fan and motor bearings at the cooling tower. The belt is replaced and the tower fan is turned on. It starts to turn and the technician can see from the moist air leaving the tower that the water is loaded with heat.

The technician goes back to the basement and touches the water line leading from the tower to the condenser and notices that the water has cooled back to below hand temperature. The compressor is restarted. The technician stays with the compressor for about 30 min to make sure that it will stay on and it does. This time is used lubricating pump and fan motors and in general looking for any other potential problems.

SERVICE CALL 2

A building owner reports that the building is hot and the chiller is off. The chiller is a low-pressure R-11 chiller and has an air leak. *The building maintenance personnel have been operating the purge for several days to remove air that is leaking in waiting for the weekend to leak test the machine. The purge water has been valved off and the purge condenser is not operable, causing high purge pressure and releasing refrigerant.*

The technician arrives and goes to the equipment room where the chiller is located. The signal light indicates that the chiller is off because of freeze protection. The technician looks around and notices the purge pressure is very high as the purge is still running on manual. On inspection, it is noticed that the purge relief valve is seeping refrigerant. The technician then checks the water circuit to the purge and discovers that there is no water cooling the condenser in the purge. The valve is opened to the purge water condenser. This system uses chilled water, which is now warmed up to 80°F, but will still condense the refrigerant in the purge.

The technician then starts the compressor in the 40% mode and watches the suction pressure; it is not too low so the load is advanced to 60%. The suction pressure begins to drop. The technician looks in the sight glass at the evaporator for the refrigerant level and it is low. There is some refrigerant at the site glass so the technician gets set up to add refrigerant. The refrigerant drum is piped to the charging valve and the valve is opened. Meanwhile the technician places a thermometer in the well in the evaporator to check the refrigerant temperature. The chilled-water temperature leaving the chiller is down to 55°F and the thermometer in the refrigerant reads 43°F. This is a 12°F approach and it should be 7°F according to the log sheet that was left at start-up. Two hundred pounds of refrigerant is added before the compressor is operated at full load and the refrigerant approach is lowered to 7°F. The technician is satisfied with the system charge. The chilled-water temperature pulls down to 45°F and the compressor begins to unload in about 30 min. The purge pressure is running normal and not relieving continuously as it was. The technician leaves and will return for the leak check later.

SERVICE CALL 3

A building manager reports that the chiller is off and the building is hot. *The problem is the flow switch in the condenser water circuit has a broken paddle and the field control circuit is not passing power to the chiller telling it to start.*

The technician arrives and goes straight to the chiller. This chiller has a ready light on the control panel that indicates when the field circuit is calling for cooling. The light is off. The technician first looks to see what the chilled-water temperature is. It is 80°F; the controller is calling for cooling. The technician then checks the water gages that register the pressure drop across the chiller and the condenser. The correct pressure drop registers across both so there is water flow. The technician then goes to the wiring diagram for the system and looks to see what other controls are in this circuit. The flow switches and the pump interlocks are all in the circuit. It is a matter of tracing the wiring down to see what has caused the chiller to shut down.

The technician realizes that the flow switches are likely to be a problem because there is a lot of movement of these devices since they are located in a moving water stream. The compressor switch is turned off so the compressor will not try to start during the control testing. The technician removes the cover from the chilled-water flow switch and checks voltage across it. There is no voltage.

NOTE: If the switch were open, there would be voltage. The cover is replaced and the same test is performed on the condenser water flow switch and there is voltage. This flow switch is open. The technician jumps the

switch with a jumper and the light on the chiller panel lights up. This is the problem.

The technician leaves the jumper on the flow switch and starts the chiller. Flow in the condenser water circuit is not as critical as it would be in a chilled-water circuit. If the water flow were to stop in the condenser circuit, the compressor would shut down because of high pressure. ***The chilled water flow switch should not be jumped with the compressor running because you could freeze the chiller.***

The technician explains to the building manager that the flow switch paddle must be replaced. The manager requests this be done after hours so the technician leaves and will return at 5:00 PM with the parts to repair the switch.

The technician returns to the equipment room to make the repair. The chiller is off for the day because of the energy management time clock. The condenser water pump is turned off and locked. The valves on either side of the condenser system are valved off and the water is drained from the condenser. The flow switch is removed, repaired, and replaced. The water valves are turned on and the condenser system is allowed to fill. Air is vented out of the water system at the high places in the piping. The technician then turns the water pump on and verifies that the flow switch is passing power. By leaving the meter leads on the flow switch and shutting off the water, it is verified the flow switch will also shut the system off if the water flow is stopped. The system is made ready to run the next morning when the time clock calls for cooling and the technician leaves.

SERVICE CALL 4

A building manager telephones and indicates that the centrifugal chiller in the basement of the building is making a screaming noise. He went to the door and was scared to go in. *The problem is the cooling tower strainer is stopped up and restricting the water flow to the condenser. The centrifugal has high head pressure causing a surge, which produces a loud screaming sound but is normally otherwise harmless.*

The technician arrives and from the parking lot can hear the compressor screaming. When the technician enters the room, the machine capacity control is turned down to 40% to reduce the load. The machine has high head pressure.

A look at the pressure gages across the condenser show that there is reduced water flow so the technician proceeds to the cooling tower on the roof.

A look at the cooling tower basin shows what has happened; there are tree leaves in the cooling tower basin. The technician removes the cover from the cooling tower and removes the leaves from the strainer. It can be seen immediately that the water is flowing faster.

The technician replaces the cover to the tower basin, goes to the basement, and turns the capacity control up to 100%.

The technician then talks to the manager about cleaning the cooling tower after hours when the building air conditioning is off and gets permission to do this.

That afternoon, the technician arrives and the system is off because of the time clock in the energy management system. The tower is cleaned and washed out. There are many small leaves in the tower. The technician then goes to the basement and prepares to check the condenser for leaves. The condenser circuit is valved off and drained. Then the technician removes the marine water box cover on the piping inlet end of the condenser where any trash would accumulate. There are small leaves in the condenser that have passed through the cooling tower strainer. These are cleaned out. While the water box cover is removed, the technician uses a small penlight and examines the tubes. There is no scale or fouling of the tubes, which is good news.

The water box cover is replaced and the valves are opened to the condenser. The technician waits a few minutes for the tower to refill with water after the condenser has been filled, then starts the condenser water pump in the manual mode and verifies there is water flow. This is a necessary step if the condenser is large enough to drain most of the water from the tower; the tower must have time to refill before starting the pump. Now the technician knows that when the machine starts in the morning there should be no cooling tower water problems.

SERVICE CALL 5

A building maintenance man calls and says that the chiller is off. A light is lit that says freeze control. He wants to know if it should be reset. It is a mild day without much need for air conditioning. The dispatcher tells him no, a technician will be over within 30 min. *The problem is the main controller is not calibrated and is drifting, causing the water temperature to fall too low.*

When the technician arrives the maintenance man leads the way to the chiller, which is on the roof in a penthouse. The technician checks the water flow through the chiller by comparing the pressure in and out to the log sheet and it is correct. The freeze control is reset and the chiller compressor starts. When it starts to lower the water temperature, the technician watches the controller. It is a pneumatic controller and as the water temperature drops, the air pressure should drop from the 15 psig furnished. When the water temperature reaches 43°F, the pneumatic air pressure should be 7 psig. When the water temperature reaches 41°F, the air pressure should be 3 psig and the chiller should shut off. This does not happen; the chiller continues to run and the water temperature reaches 39°F without shutting off. The technician then adjusts the controller to the correct calibration. The chiller shuts off.

The technician waits for the chiller to restart when the building water warms up; this takes about 30 min. When it is established the chiller is operating correctly, the technician leaves.

SERVICE CALL 6

A building manager calls and reports that the chiller is off and the building is getting hot. This is a centrifugal chiller. *The problem is the building technician has been adjusting the controls and has adjusted the load-limiting control, trying to get the motor to operate at more capacity to reduce the building temperature faster. The compressor was pulling too much current, which tripped the overload device.*

The technician arrives and goes to the equipment room with the building technician. The building technician does not tell the technician that the load-limiting control has been adjusted. The technician checks the field control circuit and discovers that all field controls are calling for cooling. A further check shows that an overload device has tripped. The technician knows that something has happened and caution should be used.

The technician shuts the power to the compressor starter off and checks the motor windings for a ground or a short circuit from winding to winding. The motor seems fine so power is restored. There is not much else to do except to start the motor and observe.

The technician starts the compressor in the 40% mode. When the compressor starts to load up, the current is observed. Start-up data show that the compressor should draw 600 A at full load and 240 A at 40%. The compressor is pulling 264 A; this is too much so the technician turns the load limiter back to 240 A and then turns the limiter up to 100%. The compressor begins to load up and reaches 600 A and throttles to hold this current. The technician then turns to the building maintenance technician and asks if anyone has adjusted the demand limiter. The building technician admits that it had been done. The technician is told that these controls should not be adjusted because they are reliable. The technician then uses nail polish to mark the setting on the control so it will be known if it is adjusted again.

The technician realizes that the controls belong to the building owner, but they should be adjusted only by a qualified technician.

SERVICE CALL 7

A customer calls and says the building is becoming warm. The customer says that water is still flowing over the cooling tower, but there is no cooling. The unit is a 300-ton centrifugal with a hermetic compressor. *The problem is the compressor motor is burned and will need to be replaced or rewound.*

The technician goes straight to the equipment room on the tenth floor of an office building. It is noticed that the building is hot so there is a problem with the chiller.

The technician checks the field control circuit and discovers that all signals are correct; it is calling for cooling. The technician then goes to the starter and discovers that there is no line voltage to the starter. The technician goes to the main power supply and discovers the breaker is tripped. There must be something very wrong.

The technician locks out and tags the breaker and uses a megger to check the motor leads in the starter to ground and discovers a circuit to ground. There is always a possibility that the problem may be in the wiring between the starter and the motor terminals on the motor so the technician removes the leads from the motor terminals and disconnects them. The motor is checked at the terminals without the wiring and it shows a circuit to ground.

The technician goes to the building management for advice as to the next step. It is explained that the motor will have to be removed from the equipment room for replacement or rebuild. ⊛**Because the motor is a hermetically sealed motor, the refrigerant will have to be recovered.**⊛ The technician explains that there are two possibilities for making the repair. The motor can be replaced with a factory motor, which may take several days for delivery, or the motor may be rebuilt at a local rebuild shop that is approved to work with hermetic motors. Building management gives permission to proceed with the job of a rebuild to save time because the building must have air conditioning. Like many modern buildings, no windows open and the computers rely on the air conditioning.

The technician calls the home office and requests backup help and the recommended tools for the job. This would include a hoist to lift the motor and a dolly to transport it on to the elevator and out to the truck, which is equipped with a lift gate.

While the helper is getting the tools together, the technician prepares to pull the motor. ⊛**The refrigerant is recovered from the machine using the recovery system brought by the helper. Water in the chilled water circuit and condenser water are circulated through the machine during recovery to prevent the tubes from freezing and to aid in boiling the refrigerant out of the machine. The refrigerant is stored in refrigerant drums. The machine uses R-11 and the refrigerant will be checked for acid content and recycled for reuse in the machine if needed. When the refrigerant is recovered down to 28 in. of vacuum in the machine, the technician breaks the vacuum with dry nitrogen that has been brought by the helper.**⊛ Breaking the vacuum with dry nitrogen will minimize oxidation inside the machine when it is opened to the atmosphere. The machine is made of steel inside and it will quickly rust.

The technician then removes the wires from the motor terminal box and starts to disconnect the compressor from the motor. When the motor is free to move, the hoist is set up and the motor is set on a rolling dolly to transport it to the truck. The motor is then rolled onto the lift gate, lifted to the back of the truck, and secured. The helper takes the motor to the motor shop for rebuild. This will take about 48 h. It is now 2:30 on Tuesday.

The technician goes back to the equipment room and prepares the equipment for when the motor is returned. A sample of the refrigerant and a sample of the refrigerant oil from the compressor sump are pulled and shipped by next day carrier to a chemical analysis firm. They will be analyzed and the report can be called in by Wednesday afternoon to determine what must be done about the refrigerant. The refrigerant and oil sample do not smell as though they have very much acid, but the analysis will reveal the facts.

The technician checks the contacts in the starter. There must have been quite a surge of current when the motor grounded. The contacts are good and do not require replacement.

The oil heater is turned off, locked, and tagged. The technician then cleans all gasket sealant from all compressor flanges and is now ready for the motor to be returned. Time has been saved by doing all of this before the motor is brought back.

Thursday morning the technician checks with the rebuild shop and is told the motor will be ready by 11:30. The technician arranges for the helper to pick up the motor and a compressor gasket kit. They meet at the building at 1:00 PM and proceed to install the motor. The motor is moved to the equipment room and lifted back into place. All gaskets are installed and everything is tightened. This system has a history of being very tight so the only leak checking that needs to be performed is on the fittings, including the gasketed flanges that were removed and in the area where the work was done.

When all is ready, the machine is pressured with nitrogen to 10 psig with a trace of R-22 refrigerant that is used to detect leaks. The machine is leak checked and found to be good. The nitrogen and refrigerant are exhausted and it is time to pull the vacuum. The motor leads are connected to the motor terminals.

The vacuum pump is installed and started. It will take about 5 h to pull the vacuum down to 0 in. Hg on a mercury manometer and it is 5:30 PM. The technician leaves the job and decides to come back at 10:30 PM to check the vacuum. If the vacuum can be verified this evening and the vacuum pump shut off for 8 h, the machine can be started in the morning if all is well. The technician checks the vacuum at about 11:00 PM and the manometer reads 0 in. Hg vacuum. The vacuum pump is valved off and the technician leaves to return in the morning.

When the technician arrives at 7:30 in the morning, the mercury manometer is still reading 0 in. Hg vacuum so the machine is leak free and it is time to charge the system. The refrigerant and oil report came back "OK" so no extra precautions need to be taken with the refrigerant. The technician starts the chilled-water pump and the condenser water pump in preparation to start-up. A drum of refrigerant is connected to the machine at the drum vapor valve. The valve is slowly opened to the machine and vapor is allowed to enter the machine until the pressure is above that corresponding to freezing. This is important because if liquid refrigerant is allowed to enter the machine at such a low pressure, the tubes may freeze. The chilled water is circulating, but if there is one tube that does not have circulation, this tube may freeze. The pressure in the machine must be above 17 in. Hg vacuum to prevent freezing while charging liquid.

When the machine pressure reaches 17 in. Hg vacuum, the technician then starts charging liquid refrigerant. While the technician is charging refrigerant, the helper pulls the oil in the oil sump using the machine vacuum. When the oil is charged, the oil heater is turned on to bring the oil up to temperature.

The technician is not able to charge all of the refrigerant into the machine using the vacuum because the chilled-water circuit warms the refrigerant and pressure is soon raised. The technician then changes over and pushes the remainder of the refrigerant into the machine by pressuring the drum with nitrogen on top of the liquid and drawing vapor from the bottom of the drum, Figure 49–18.

The technician starts the purge pump to pull any nitrogen that may have entered the system while charging, even though the technician was careful and could see no nitrogen in the clear charging hose while charging. The purge pressure does not rise, so the machine should be free of nitrogen. The technician has closed the main breaker and is ready to start the machine. The machine is started and quickly turned up to 100% to reduce the building temperature as quickly as possible. There is no apparent reason that the motor developed a circuit to ground; the motor rebuild shop could not find any problem so the motor failure is accounted for as a random failure.

The chiller is quickly pulling the building water down to 45°F leaving-water temperature. The technician stays with the chiller for 2 h and completes entries including the motor current in an operating log. When it is determined that the system is functioning properly the technician leaves.

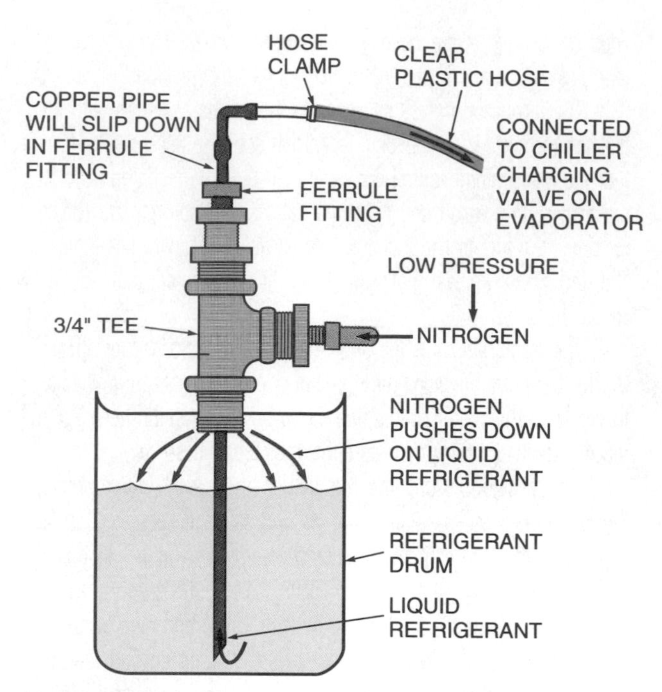

FIGURE 49–18 Using nitrogen to charge refrigerant into a low-pressure chiller.

SUMMARY

- The operation of chilled-water air conditioning systems involves starting, running, and stopping the chiller systems in an orderly manner.
- When starting a chilled-water system, first check for sufficient chilled-water flow. If water-cooled check for condenser water flow. The interlock circuit through the contactors must be satisfied; the cooling tower fan should start, followed by the compressor. There may be a time delay before the compressor starts.
- The reciprocating, scroll, and rotary screw compressors may be either air or water cooled.
- The centrifugal chiller will have a separate oil lubrication system. It should be verified that the lubrication system is functioning satisfactorily before starting the compressor.
- After the chiller is on and operating, the technician should observe the operation for a period of time to ensure that it is all operating correctly.
- Water-cooled chillers must have water treatment to prevent minerals and algae from collecting and forming.

- Inspecting and cleaning the water tower should be done on a regular basis.
- Water-cooled condenser tubes should be checked at least annually. If they are found to be not clean or to have a scale buildup they must be cleaned either with a brush or chemically. Only technicians trained in the use of the chemicals should perform this task.
- Condenser and evaporator tubes may be checked for defects with an eddy current test instrument.
- Absorption chilled-water system and compression cycle system start-up are similar in many respects. Condenser and chilled-water flow must be established before the chiller is started. The cooling tower water must not be too cold or the Li-Br may crystalize. If the chiller has been off for a long period of time, it is good practice to operate the vacuum pump purge system for several hours before starting the machine.
- Absorption chillers must be observed for proper operation on a more regular basis than compression cycle chillers because they can be more intricate.
- The purge system may require more maintenance than other systems on the absorption chiller. The vacuum pump oil should be changed regularly.
- Technicians who maintain chillers should stay in contact with the manufacturers of the equipment for the latest information, advice, and the schedule for service schools. Technicians should attend factory schools and seminars whenever possible.

REVIEW QUESTIONS

1. What is the first step that should be taken when starting up a chilled-water air conditioning system?
2. How does the lubrication system differ between a reciprocating compressor and a centrifugal compressor?
3. What types of condenser systems may a scroll chilled-water system have?
4. Why do reciprocating compressors have crankcase heaters?
5. Does the reciprocating compressor chiller start up fully loaded?
6. Does the rotary screw chiller compressor have crankcase heat?
7. Does the centrifugal compressor chiller start up fully loaded?
8. What component must start before the centrifugal compressor?
9. Why are the data in an operating log important?
10. Why should condenser tubes be kept wet until they are brushed or otherwise cleaned?
11. What is the name of the test to determine the condition of the interior of the condenser tubes?
12. If the water tower temperature is too cold when an absorption chilled-water system is started, what may happen to the Li-Br?

EPA Stratospheric Ozone Protection

Final Rule Summary

COMPLYING WITH THE REFRIGERANT RECYCLING RULE

This fact sheet provides an overview of the refrigerant recycling requirements of section 608 of the Clean Air Act, 1990, as amended (CAA), including final regulations published on May 14, 1993 (58 FR 28660), and the prohibition on venting that became effective on July 1, 1992.

APPENDIX 1 CONTENTS

Overview

Under section 608 of the CAA, EPA has established regulations that:

- Require service practices that maximize recycling of ozone-depleting compounds (both chlorofluorocarbons [CFCs] and hydrochlorofluorocarbons [HCFCs]) during the servicing and disposal of air-conditioning and refrigeration equipment.
- Set certification requirements for recycling and recovery equipment, technicians, and reclaimers.
- Restrict the sale of refrigerant to certified technicians.
- Require persons servicing or disposing of air-conditioning and refrigeration equipment to certify to EPA that they have acquired recycling or recovery equipment and are complying with the requirements of the rule.
- Require the repair of substantial leaks in air-conditioning and refrigeration equipment with a charge of greater than 50 pounds.
- Establish safe disposal requirements to ensure removal of refrigerants from

goods that enter the waste stream with the charge intact (e.g., motor vehicle air conditioners, home refrigerators, and room air conditioners).

The Prohibition on Venting

Effective July 1, 1992, section 608 of the Act prohibits individuals from knowingly venting ozone-depleting compounds used as refrigerants into the atmosphere while maintaining, servicing, repairing, or disposing of air-conditioning or refrigeration equipment. Only four types of releases are permitted under the prohibition:

1. "De minimis" quantities of refrigerant released in the course of making good faith attempts to recapture and recycle or safely dispose of refrigerant.

2. Refrigerants emitted in the course of normal operation of air-conditioning and refrigeration equipment (as opposed to during the maintenance, servicing, repair, or disposal of this equipment) such as from mechanical purging and leaks. However, EPA is requiring the repair of substantial leaks.

3. Mixtures of nitrogen and R-22 that are used as holding charges or as leak test gases, because in these cases, the ozone-depleting compound is not used as a refrigerant. However, a technician may not avoid recovering refrigerant by adding nitrogen to a charged system; before nitrogen is added, the system must be evacuated to the appropriate level in Table 1. Otherwise, the CFC or HCFC vented along with the nitrogen will be considered a refrigerant. Similarly, *pure* CFCs or HCFCs released from appliances will be presumed to be refrigerants, and their release will be considered a violation of the prohibition on venting.

4. Small releases of refrigerant which result from purging hoses or from connecting or disconnecting hoses to charge or service appliances will not be considered violations of the prohibition on venting. However, recovery and recycling equipment manufactured after November 15, 1993, must equipped with low-loss fittings.

Regulatory Requirements

Service Practice Requirements

1. Evacuation Requirements. Beginning July 13, 1993, technicians are required to evacuate air-conditioning and refrigeration equipment to established vacuum levels. If the technician's recovery or recycling equipment is manufactured any time before November 15, 1993, the air-conditioning and refrigeration equipment must be evacuated to the levels described in the first column of Table 1. If the technician's recovery or recycling equipment is manufactured on or after November 15, 1993, the air-conditioning and refrigeration equipment must be evacuated to the levels described in the second column of Table 1, and the recovery or recycling equipment must have been certified by an EPA-approved equipment testing organization (see *Equipment Certification*, below).

Technicians repairing small appliances, such as household refrigerators, household freezers, and water coolers, are required to recover 80-90 percent of the refrigerant in the system, depending on the status of the system's compressor.

TABLE 1
REQUIRED LEVELS OF EVACUATION FOR APPLIANCES
EXCEPT FOR SMALL APPLIANCES, MVACS, AND MVAC-LIKE APPLIANCES

Type of Appliance	Inches of Mercury Vacuum* Using Equipment Manufactured:	
	Before Nov. 15, 1993	On or after Nov. 15, 1993
HCFC-22 appliance** normally containing less than 200 pounds of refrigerant	0	0
HCFC-22 appliance** normally containing 200 pounds or more of refrigerant	4	10
Other high-pressure appliance** normally containing less than 200 pounds of refrigerant (CFC-12, -500, -502, -114)	4	10
Other high-pressure appliance** normally containing 200 pounds or more of refrigerant (CFC-12, -500, -502, -114)	4	15
Very High Pressure Appliance (CFC-13, -503)	0	0
Low-Pressure Appliance (CFC-11, HCFC-123)	25	25 mm Hg absolute

*Relative to standard atmospheric pressure of 29.9" Hg.
**Or isolated component of such an appliance

2. *Exceptions to Evacuation Requirements.* EPA has established limited exceptions to its evacuation requirements for 1) repairs to leaky equipment and 2) repairs that are not major and that are not followed by an evacuation of the equipment to the environment.

If, due to leaks, evacuation to the levels in Table 1 is not attainable, or would substantially contaminate the refrigerant being recovered, persons opening the appliance must:

- isolate leaking from non-leaking components wherever possible;
- evacuate non-leaking components to the levels in Table 1; and
- evacuate leaking components to the lowest level that can be attained without substantially contaminating the refrigerant. This level cannot exceed 0 psig.

If evacuation of the equipment to the environment is not to be performed when repairs are complete, and if the repair is not major, then the appliance must:

- be evacuated to at least 0 psig before it is opened if it is a high- or very high-pressure appliance; or

- be pressurized to 0 psig before it is opened if it is a low-pressure appliance. Methods that require subsequent purging (e.g., nitrogen) cannot be used.

"Major" repairs are those involving removal of the compressor, condenser, evaporator, or auxiliary heat exchanger coil.

3. Reclamation Requirement. EPA has also established that refrigerant recovered and/or recycled can be returned to the same system or other systems owned by the same person without restriction. If refrigerant changes ownership, however, that refrigerant must be reclaimed (i.e., cleaned to the ARI 700 standard of purity and chemically analyzed to verify that it meets this standard). This provision will expire in May, 1995, when it may be replaced an off-site recycling standard.

Equipment Certification

The Agency has established a certification program for recovery and recycling equipment. Under the program, EPA requires that equipment manufactured on or after November 15, 1993, be tested by an EPA-approved testing organization to ensure that it meets EPA requirements. Recycling and recovery equipment intended for use with air-conditioning and refrigeration equipment besides small appliances must be tested under the ARI 740-1993 test protocol, which is included in the final rule as Appendix B. Recovery equipment intended for use with small appliances must be tested under either the ARI 740-1993 protocol or Appendix C of the final rule. The Agency is requiring recovery efficiency standards that vary depending on the size and type of air-conditioning or refrigeration equipment being serviced. For recovery and recycling equipment intended for use with air-conditioning and refrigeration equipment besides small appliances, these standards are the same as those in the second column of Table 1. Recovery equipment intended for use with small appliances must be able to recover 90 percent of the refrigerant in the small appliance when the small appliance compressor is operating and 80 percent of the refrigerant in the small appliance when the compressor is not operating.

Equipment Grandfathering

Equipment manufactured before November 15, 1993, including home-made equipment, will be grandfathered if it meets the standards in the first column of Table 1. Third-party testing is not required for equipment manufactured before November 15, 1993, but equipment manufactured on or after that date, including home-made equipment, must be tested by a third-party (see *Equipment Certification* above).

Refrigerant Leaks

Owners of equipment with charges of greater than 50 pounds are required to repair substantial leaks. A 35 percent annual leak rate is established for the industrial process and commercial refrigeration sectors as the trigger for requiring repairs. An annual leak rate of 15 percent of charge per year is established for comfort cooling chillers and all other equipment with a charge of over 50 pounds other than industrial process and commercial refrigeration equipment. Owners of air-conditioning and refrigeration equipment with more than 50 pounds of charge must keep records of the quantity of refrigerant added to their equipment during servicing and maintenance procedures.

Mandatory Technician Certification

EPA has established a mandatory technician certification program. The Agency has developed four types of certification:

- For servicing small appliances (Type I).
- For servicing or disposing of high- or very high-pressure appliances, except small appliances and MVACs (Type II).
- For servicing or disposing of low-pressure appliances (Type III)
- For servicing all types of equipment (Universal).

Persons removing refrigerant from small appliances and motor vehicle air conditioners for purposes of disposal of these appliances do not have to be certified.

Technicians are required to pass an EPA-approved test given by an EPA-approved certifying organization to become certified under the mandatory program. Technicians must be certified by November 14, 1994. EPA expects to have approved some certifying organizations by September of this year. The Stratospheric Ozone Hotline will distribute lists of approved organizations at that time.

EPA plans to "grandfather" individuals who have already participated in training and testing programs provided the testing programs 1) are approved by EPA and 2) provide additional, EPA-approved materials or testing to these individuals to ensure that they have the required level of knowledge.

Although any organization may apply to become an approved certifier, EPA plans to give priority to national organizations able to reach large numbers of people. EPA encourages smaller training organizations to make arrangements with national testing organizations to administer certification examinations at the conclusion of their courses.

Refrigerant Sales Restrictions

Under Section 609 of the Clean Air Act, sales of CFC-12 in containers smaller than 20 pounds are now restricted to technicians certified under EPA's motor vehicle air conditioning regulations. Persons servicing appliances other than motor vehicle air conditioners may still buy containers of CFC-12 larger than 20 pounds.

After November 14, 1994, the sale of refrigerant in any size container will be restricted to technicians certified either under the program described in *Technician Certification* above or under EPA's motor vehicle air conditioning regulations.

Certification by Owners of Recycling and Recovery Equipment

EPA is requiring that persons servicing or disposing of air-conditioning and refrigeration equipment certify to EPA that they have acquired (built, bought, or leased) recovery or recycling equipment and that they are complying with the applicable requirements of this rule. This certification must be signed by the owner of the equipment or another responsible officer and sent to the appropriate EPA Regional Office by August 12, 1993. A sample form for this certification is attached. Although owners of recycling and recovery equipment are required to list the number of trucks based at their shops, they do not need to have a piece of recycling or recovery equipment for every truck.

Reclaimer Certification

Reclaimers are required to return refrigerant to the purity level specified in ARI Standard 700-1988 (an industry-set purity standard) and to verify this purity using the laboratory protocol set forth in the same standard. In addition, reclaimers must release no more than 1.5 percent of the refrigerant during the reclamation process and must dispose of wastes properly. Reclaimers must certify by August 12, 1993, to the Section 608 Recycling Program Manager at EPA headquarters that they are complying with these requirements and that the information given is true and correct. The certification must also include the name and address of the reclaimer and a list of equipment used to reprocess and to analyze the refrigerant.

EPA encourages reclaimers to participate in third-party reclaimer certification programs, such as that operated by the Air-Conditioning and Refrigeration Institute (ARI). Third-party certification can enhance the attractiveness of a reclaimer's product by providing an objective assessment of its purity.

MVAC-like Appliances

Some of the air conditioners that are covered by this rule are identical to motor vehicle air conditioners (MVACs), but they are not covered by the MVAC refrigerant recycling rule (40 CFR Part 82 Subpart B) because they are used in vehicles that are not defined as "motor vehicles." These air conditioners include many systems used in construction equipment, farm vehicles, boats, and airplanes. Like MVACs in cars and trucks, these air conditioners typically contain two or three pounds of CFC-12 and use open-drive compressors to cool the passenger compartments of vehicles. (Vehicle air conditioners utilizing HCFC-22 are not included in this group and are therefore subject to the requirements outlined above for HCFC-22 equipment.) EPA is defining these air conditioners as "MVAC-like appliances" and is applying the MVAC rule's requirements for the certification and use of recycling and recovery equipment to them. That is, technicians servicing MVAC-like appliances must "properly use" recycling or recovery equipment that has been certified to meet the standards in Appendix A to 40 CFR Part 82, Subpart B. In addition, EPA is allowing technicians who service MVAC-like appliances to be certified by a certification program approved under the MVAC rule, if they wish.

Safe Disposal Requirements

Under EPA's rule, equipment that is typically dismantled on-site before disposal (e.g., retail food refrigeration, cold storage warehouse refrigeration, chillers, and industrial process refrigeration) has to have the refrigerant recovered in accordance with EPA's requirements for servicing. However, equipment that typically enters the waste stream with the charge intact (e.g., motor vehicle air conditioners, household refrigerators and freezers, and room air conditioners) is subject to special safe disposal requirements.

Under these requirements, the final person in the disposal chain (e.g., a scrap metal recycler or landfill owner) is responsible for ensuring that refrigerant is recovered from equipment before the final disposal of the equipment. However, persons "upstream" can remove the refrigerant and provide documentation of its removal to the final person if this is more cost-effective.

The equipment used to recover refrigerant from appliances prior to their final disposal must meet the same "performance standards" as equipment used prior to servicing, but it does not need to be tested by a laboratory. This means that self-built equipment is allowed as long as it meets the performance requirements. For MVACs and MVAC-like appliances, the performance requirement is 102 mm of mercury vacuum and for small appliances, the recover equipment performance requirements are 90 percent efficiency when the appliance compressor is operational, and 80 percent efficiency when the appliance compressor is not operational.

Technician certification is not required for individuals removing refrigerant from appliances in the waste stream.

The safe disposal requirements are effective on July 13, 1993. The equipment must be registered or certified with the Agency by August 12, 1993. A sample form is attached.

Major Recordkeeping Requirements

Technicians servicing appliances that contain 50 or more pounds of refrigerant must provide the owner with an invoice that indicates the amount of refrigerant added to the appliance. Technicians must also keep a copy of their proof of certification at their place of business.

Owners of appliances that contain 50 or more pounds of refrigerant must keep servicing records documenting the date and type of service, as well as the quantity of refrigerant added.

Wholesalers who sell CFC and HCFC refrigerants must retain invoices that indicate the name of the purchaser, the date of sale, and the quantity of refrigerant purchased.

Reclaimers must maintain records of the names and addresses of persons sending them material for reclamation and the quantity of material sent to them for reclamation. This information must be maintained on a transactional basis. Within 30 days of the end of the calendar year, reclaimers must report to EPA the total quantity of material sent to them that year for reclamation, the mass of refrigerant reclaimed that year, and the mass of waste products generated that year.

Hazardous Waste Disposal

If refrigerants are recycled or reclaimed, they are not considered hazardous under federal law. In addition, used oils contaminated with CFCs are not hazardous on the condition that:

- They are not mixed with other waste.
- They are subjected to CFC recycling or reclamation.
- They are not mixed with used oils from other sources.

Used oils that contain CFCs after the CFC reclamation procedure, however, are subject to specification limits for used oil fuels if these oils are destined for burning. Individuals with questions regarding the proper handling of these materials should contact EPA's RCRA Hotline at 800-424-9346 or 703-920-9810.

Enforcement

EPA is performing random inspections, responding to tips, and pursuing potential cases against violators. Under the Act, EPA is authorized to assess fines of up to $25,000 per day for any violation of these regulations.

Planning and Acting for the Future

Observing the refrigerant recycling regulations for section 608 is essential in order to conserve existing stocks of refrigerants, as well as to comply with Clean Air Act requirements. However, owners of equipment that contains CFC refrigerants should look beyond the immediate need to maintain existing equipment in working order. **EPA urges equipment owners to act now and prepare for the phaseout of CFCs, which will be completed by January 1, 1996.** Owners are advised to begin the process of converting or replacing existing equipment with equipment that uses alternative refrigerants.

To assist owners, suppliers, technicians and others involved in comfort chiller and commercial refrigeration management, EPA has published a series of short fact sheets and expects to produce additional material. Copies of material produced by the EPA Stratospheric Protection Division are available from the Stratospheric Ozone Information Hotline (see hotline number below).

For Further Information

For further information concerning regulations related to stratospheric ozone protection, please call the Stratospheric Ozone Information Hotline: 800-296-1996. The Hotline is open between 10:00 AM and 4:00 PM, Eastern Time.

TABLE 2
MAJOR RECYCLING RULE COMPLIANCE DATES

Requirement	Date
• Date after which owners of equipment containing more than 50 pounds of refrigerant with substantial leaks must have such leaks repaired.	June 14, 1993
• Evacuation requirements go into effect. • Recovery and recycling equipment requirements go into effect.	July 13, 1993
• Owners of recycling and recovery equipment must have certified to EPA that they have acquired such equipment and that they are complying with the rule. • Reclamation requirement goes into effect.	August 12, 1993
• All newly manufactured recycling and recovery equipment must be certified by an EPA-approved testing organization to meet the requirements in the second column of Table 1.	November 15, 1993
• All technicians must be certified. • Sales restriction goes into effect.	November 14, 1994
• Reclamation requirement expires.	May 14, 1995

APPENDIX II—TEMPERATURE CONVERSION TABLE

TEMPERATURE CONVERSION TABLE

°F	Temperature to be Converted	°C	°F	Temperature to be Converted	°C
−76.0	−60	−51.1	23.0	−5	−20.6
−74.2	−59	−50.6	24.8	−4	−20.0
−72.4	−58	−50.0	26.6	−3	−19.4
−70.6	−57	−49.4	28.4	−2	−18.9
−68.8	−56	−48.9	30.2	−1	−18.3
−67.0	−55	−48.3	32.0	0	−17.8
−65.2	−54	−47.8	33.8	1	−17.2
−63.4	−53	−47.2	35.6	2	−16.7
−61.6	−52	−46.7	37.4	3	−16.1
−59.8	−51	−46.1	39.2	4	−15.6
−58.0	−50	−45.6	41.0	5	−15.0
−56.2	−49	−45.0	42.8	6	−14.4
−54.4	−48	−44.4	44.6	7	−13.9
−52.6	−47	−43.9	46.4	8	−13.3
−50.8	−46	−43.3	48.2	9	−12.8
−49.0	−45	−42.8	50.0	10	−12.2
−47.2	−44	−42.2	51.8	11	−11.7
−45.4	−43	−41.7	53.6	12	−11.1
−43.6	−42	−41.1	55.4	13	−10.6
−41.8	−41	−40.6	57.2	14	−10.0
−40.0	−40	−40.0	59.0	15	−9.4
−38.2	−39	−39.4	60.8	16	−8.9
−36.4	−38	−38.9	62.6	17	−8.3
−34.6	−37	−38.3	64.4	18	−7.8
−32.8	−36	−37.8	66.2	19	−7.2
−31.0	−35	−37.2	68.0	20	−6.7
−29.2	−34	−36.7	69.8	21	−6.1
−27.4	−33	−36.1	71.6	22	−5.6
−25.6	−32	−35.6	73.4	23	−5.0
−23.8	−31	−35.0	75.2	24	−4.4
−22.0	−30	−34.4	77.0	25	−3.9
−20.2	−29	−33.9	78.8	26	−3.3
−18.4	−28	−33.3	80.6	27	−2.8
−16.6	−27	−32.8	82.4	28	−2.2
−14.8	−26	−32.2	84.2	29	−1.7
−13.0	−25	−31.7	86.0	30	−1.1
−11.2	−24	−31.1	87.8	31	−0.6
−9.4	−23	−30.6	89.6	32	0.0
−7.6	−22	−30.0	91.4	33	0.6
−5.8	−21	−29.4	93.2	34	1.1
−4.0	−20	−28.9	95.0	35	1.7
−2.2	−19	−28.3	96.8	36	2.2
−0.4	−18	−27.8	98.6	37	2.8
1.4	−17	−27.2	100.4	38	3.3
3.2	−16	−26.7	102.2	39	3.9
5.0	−15	−26.1	104.0	40	4.4
6.8	−14	−25.6	105.8	41	5.0
8.6	−13	−25.0	107.6	42	5.6
10.4	−12	−24.4	109.4	43	6.1
12.2	−11	−23.9	111.2	44	6.7
14.0	−10	−23.3	113.0	45	7.2
15.8	−9	−22.8	114.8	46	7.8
17.6	−8	−22.2	116.6	47	8.3
19.4	−7	−21.7	118.4	48	8.9
21.2	−6	−21.1	120.2	49	9.4

°F	Temperature to be Converted	°C
122.0	50	10.0
123.8	51	10.6
125.6	52	11.1
127.4	53	11.7
129.2	54	12.2
131.0	55	12.8
132.8	56	13.3
134.6	57	13.9
136.4	58	14.4
138.2	59	15.0
140.0	60	15.6
141.8	61	16.1
143.6	62	16.7
145.4	63	17.2
147.2	64	17.8
149.0	65	18.3
150.8	66	18.9
152.6	67	19.4
154.4	68	20.0
156.2	69	20.6
158.0	70	21.1
159.8	71	21.7
161.8	72	22.2
163.4	73	22.8
165.2	74	23.3
167.0	75	23.9
168.8	76	24.4
170.6	77	25.0
172.4	78	25.6
174.2	79	26.1
176.0	80	26.7
177.8	81	27.2
179.6	82	27.8
181.4	83	28.3
183.2	84	28.9
185.0	85	29.4
186.8	86	30.0
188.6	87	30.6
190.4	88	31.1
192.2	89	31.7
194.0	90	32.2
195.8	91	32.8
197.6	92	33.3
199.4	93	33.9
201.2	94	34.4
203.0	95	35.0
204.8	96	35.6
206.6	97	36.1
208.4	98	36.7
210.2	99	37.2
212.0	100	37.8
213.8	101	38.3
215.6	102	38.9
217.4	103	39.4
219.2	104	40.0
221.0	105	40.6
222.8	106	41.1
224.6	107	41.7
226.4	108	42.2
228.2	109	42.8
230.0	110	43.3
231.8	111	43.9
233.6	112	44.4
235.4	113	45.0
237.2	114	45.6
239.0	115	46.1
240.8	116	46.6
242.6	117	47.2
244.4	118	47.7
246.2	119	48.3
248.0	120	48.8
257.0	125	51.7
266.0	130	54.4
275.0	135	57.2
284.0	140	60.0
293.0	145	62.8
302.0	150	65.6
311.0	155	68.3
320.0	160	71.1
329.0	165	73.9
338.0	170	76.7
347.0	175	79.4
356.0	180	82.2
365.0	185	85.0
374.0	190	87.8
383.0	195	90.6
392.0	200	93.3
401.0	205	96.1
410.0	210	98.9
413.6	212	100.0
428.0	220	104.4
446.0	230	110.0
464.0	240	115.6
482.0	250	121.1
500.0	260	126.7

Example 1. To find 37°F as a Celsius equivalent, find 37 in the Temperature to be Converted column and read the value in the °C column which is 2.8°C.

Example 2. To find 75°C as a Fahrenheit equivalent, find 75 in the Temperature to be Converted column and read the value in the °F column which is 167.0°F.

APPENDIX III—ELECTRICAL SYMBOLS CHART

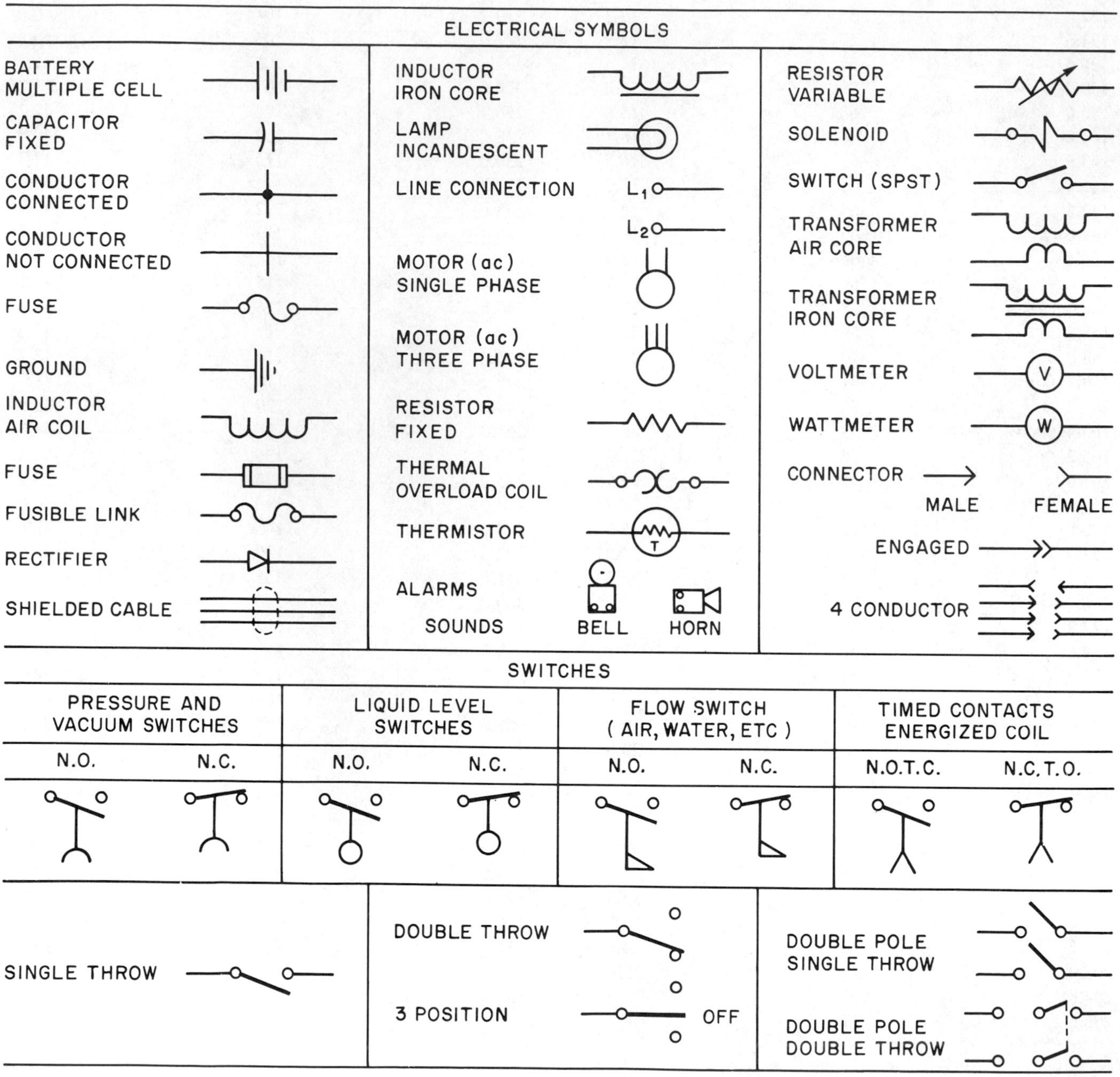

ELECTRICAL SYMBOLS

BATTERY MULTIPLE CELL	INDUCTOR IRON CORE	RESISTOR VARIABLE
CAPACITOR FIXED	LAMP INCANDESCENT	SOLENOID
CONDUCTOR CONNECTED	LINE CONNECTION L_1 L_2	SWITCH (SPST)
CONDUCTOR NOT CONNECTED	MOTOR (ac) SINGLE PHASE	TRANSFORMER AIR CORE
FUSE	MOTOR (ac) THREE PHASE	TRANSFORMER IRON CORE
GROUND	RESISTOR FIXED	VOLTMETER (V)
INDUCTOR AIR COIL	THERMAL OVERLOAD COIL	WATTMETER (W)
FUSE	THERMISTOR (T)	CONNECTOR MALE FEMALE
FUSIBLE LINK	ALARMS SOUNDS BELL HORN	ENGAGED
RECTIFIER		4 CONDUCTOR
SHIELDED CABLE		

SWITCHES

PRESSURE AND VACUUM SWITCHES		LIQUID LEVEL SWITCHES		FLOW SWITCH (AIR, WATER, ETC)		TIMED CONTACTS ENERGIZED COIL	
N.O.	N.C.	N.O.	N.C.	N.O.	N.C.	N.O.T.C.	N.C.T.O.

SINGLE THROW	DOUBLE THROW	DOUBLE POLE SINGLE THROW
	3 POSITION OFF	DOUBLE POLE DOUBLE THROW

APPENDIX IV

Add-On & Replacement
Whole House Worksheet

Customer's Name ______________________ Address ______________________

City ______________ State __________ Zip __________ Telephone Number ______________

WINTER: Inside Design Temp ________ **°F—Outside Design Temp** ________ **°F = Heating Temp Difference** ________ **°F**

SUMMER: Outside Design Temp ________ **°F—Inside Design Temp** ________ **°F = Cooling Temp Difference** ________ **°F**

HEATING		COMMON DATA SECTION		COOLING	
BTUH LOSS	HEATING FACTOR	SUBJECT	SQ. FT.	COOLING FACTOR	BTUH GAIN
	FROM TABLE E	GROSS WALL		FROM TABLE E	
		DOORS & WINDOWS (Table A or B)			
		NET WALL			
		CEILING			
		FLOORS			

Infiltration Btu/hr	=	Heating Table D	x 10 x 1.1/60 x	Volume (Cu. Ft.)	Volume (Cu. Ft)	x	1.1/60	x	ΔT x	Cooling Table D	=	Infiltration Btu/hr
	=		x 0.18333	x		x	0.01833	x	x		=	

BTUH LOSS	SUBJECT	BTUH GAIN
	SUB-TOTAL BTUH LOSS (per 10°F)	
x	ADJUSTMENT FACTOR (Table C)	
	TOTAL BTUH LOSS	
	PEOPLE______ x 300 BTUH GAIN (Assume 2 persons per bedroom)	
	APPLIANCES BTUH	1200
	SUB-TOTAL BTUH GAIN (room sensible only)	
x	DUCT LOSS/GAIN FACTOR (Table F)	x
	SUB-TOTAL BTUH (Sensible Gain)	
	MOISTURE REMOVAL (sub total x 1.3)	x 1.3
	TOTAL BTUH LOSS/GAIN	

TABLE A—HEATING—DOORS & WOOD FRAME WINDOWS (PER 10°F)

For sliding glass doors - use factors for the same type window construction.

Window & Door Types	Frames: Wood	TIM	Metal	x Area	= Btuh Loss
Single Pane Clear	9.90	10.45	11.55		
With Storm	4.75	5.25	6.50		
Double Pane Clear	5.51	6.09	7.25		
With Storm	3.41	3.85	4.90		
Triple Pane Clear	3.80	4.39	5.46		
Jalousie Single	—	—	11.0		
Single w/storm	—	—	5.0		
Skylights Single	11.07	11.69	12.92		
Double	6.65	7.35	8.75		
Door Wood Only	4.60	—	—		
Wood w/storm	3.20	—	—		
Urethane Core (R-5)	—	—	1.90		
Urethane Core (R-5) w/storm	—	—	1.70		
			TOTALS		

TABLE B — COOLING — DOORS & WINDOWS

Factors assume windows have inside shading by draperies or venetian blinds and sliding glass doors are treated as windows.

	SINGLE GLASS TEMP. DIFF.			DOUBLE GLASS TEMP. DIFF.			TRIPLE GLASS TEMP. DIFF.			X Area	= BTUH GAIN
Direction	15°	20°	25°	15°	20°	25°	15°	20°	25°		
N	18	22	26	14	16	18	11	12	13		
NE & NW	37	41	45	31	33	35	26	27	28		
E & W	52	56	60	44	46	48	38	39	40		
SE & SW	45	49	53	39	41	43	33	34	35		
S	28	32	36	23	25	27	19	20	21		
Skylights	164	168	172	141	143	145	132	136	140		
Wood ①	8.6	10.9	13.2	8.6	10.9	13.2	8.6	10.9	13.2		
Metal ②	3.5	4.5	5.4	3.5	4.5	5.4	3.5	4.5	5.4		
									TOTALS		

① For wood doors and polystyrene core metal doors

② For urethane core metal doors

TABLE C — ADJUSTMENT FACTORS — (HEATING)

°F. Temperature Diff.	30	40	50	60	70	80	90
Adjustment Factor	3	4	5	6	7	8	9

TABLE D — INFILTRATION MULTIPLIERS

Winter Air Changes Per Hour

Floor Area	900 or less	900-1500	1500-2100	over 2100
Best	0.4	0.4	0.3	0.3
Average	1.2	1.0	0.8	0.7
Poor	2.2	1.6	1.2	1.0

For each fireplace add: Best 0.1, Average 0.2, Poor 0.6

Summer Air Changes Per Hour

Floor Area	900 or less	900-1500	1500-2100	over 2100
Best	0.2	0.2	0.2	0.2
Average	0.5	0.5	0.4	0.4
Poor	0.8	0.7	0.6	0.5

Pub. No. 22-8020-9 P.I. (L)

RESIDENTIAL SURVEY FORM

1. Direction House Faces ______________________

2. Windows: ________ Glazed ________ Storm
   ________ Certified ________ Non-certified

3. Insulation: Wall R= ______ Ceiling R= ______ Floor R= ______

4. Basement: Conditioned ______ Yes, ______ No
   Insulation R= ______ Above Grade ______ Below Grade

5. Crawl Space: Conditioned ____ Yes ____ No. Insulation R= ______

6. Ceiling Height ______________________

7. Cathedral Ceiling ______________________

8. Standard Doors: Type ________ Weatherstripped ________
   Storm ______________

9. Sliding Glass Doors: ________ Glazed ________ Weatherstripped
   Storm ________ Certified ________ Non-certified ________

10. Age of House ______________

11. Furnace or Air Handler: ________ Capacity ______ Fuel ______ Age

12. Blower Motor: ______ H.P. ________ Drive ________ Wheel Size

13. Fan Relay ______ Yes, ______ No.

14. Plenum Size: ________ High ________ Wide ________ Deep

15. Ductwork Size: ________ Return Air, ________ Supply Air

16. Duct Insulation: ______ Yes, ______ No.
    Where Needed? ______________________

17. Supply Registers: Type ________ High ________ Low
    Adequate for Cooling? ______________________

18. Outside Location of Condensing Unit ______________________

19. Refrigerant Line Length Needed ______________________

20. Drain Line Length Needed ______________________

21. Is Condensate Pump Necessary ______________________

22. Electrical Service Entrance Panel ______ Volts ______ Amps

23. Distance from Entrance Panel to Condenser Site ______________

24. Is Attic Ventilation Adequate? ______ Yes ______ No

25. Has Heating and/or Cooling System been Satisfactory? ____ Yes ____ No

26. Does Customer Want: ____ Electronic Air Cleaner ____ Humidifier

27. Type of Thermostat ______________________

28. Location of Thermostat ______________________

HOUSE & DUCT LAYOUT (1 SQ. = ______ Feet)

HEAT LOSS & GAIN FACTORS

TABLE E
CONSTRUCTION FACTORS HEATING & COOLING

Heating Factor (1)	TYPE OF CONSTRUCTION	Cooling Factor (°F. Temp. Diff.) 15°	20°	25°
	WALLS (Use Sq. Ft.)			
	Walls — wood frame w/sheeting & siding, veneer or other finish			
2.71	A) No insulation, 1/2″ Gypsum Board	5.0	6.4	7.8
0.90	B) R-11 Cavity insulation + 1/2″ Gypsum Board	1.7	2.1	2.6
0.80	C) R-13 Cavity insulation + 1/2″ Gypsum Board	1.5	1.9	2.3
0.70	D) R-13 Cavity insulation + 3/4″ Bead Board (R-2.7)	1.3	1.7	2.0
0.60	E) R-19 Cavity insulation + 1/2″ Gypsum Board	1.1	1.4	1.7
0.50	F) R-19 Cavity insulation + 3/4″ Extruded Poly	0.9	1.2	1.4
	Masonry Walls			
5.10	A) Above grade No insulation	5.8	8.3	10.9
1.44	B) Above grade + R-5	1.6	2.3	3.1
0.77	C) Above grade + R-11	0.9	1.3	1.6
1.25	D) Below grade No insulation	0.0	0.0	0.0
0.74	E) Below grade + R-5	0.0	0.0	0.0
0.51	F) Below grade + R-11	0.0	0.0	0.0
	CEILINGS (Use Sq. Ft.)			
5.99	A) No insulation	17.0	19.2	21.4
1.20	B) 2″-2½″ insulation R-7	4.4	4.9	5.5
0.88	C) 3″-3½″ insulation R-11	3.2	3.7	4.1
0.53	D) 5¼″-6½″ insulation R-19	2.1	2.3	2.6
0.48	E) 6″-7″ insulation R-22	1.9	2.1	2.4
0.33	F) 8″-9½″ insulation R-30	1.3	1.5	1.6
0.26	G) 10″-12″ insulation R-38	1.0	1.1	1.3
0.23	H) 12″-13″ insulation R-44	0.9	1.0	1.1
3.08	I) Cathedral type No insulation (roof/ceiling combination)	11.2	12.6	14.1
0.72	J) Cathedral type R-11 (roof/ceiling combination)	2.8	3.2	3.5
0.49	K) Cathedral type R-19 (roof/ceiling combination)	1.9	2.2	2.4
0.45	L) Cathedral type R-22 (roof/ceiling combination)	1.8	2.0	2.2
0.40	M) Cathedral type R-26 (roof/ceiling combination)	1.6	1.8	2.0
	FLOORS (Use Sq. Ft. OR Linear Ft.)			
	Floors over unconditioned space (use sq. ft.)			
1.56	A) Over basement or enclosed crawl space (not vented)	0.0	0.0	0.0
0.40	B) Same as "A" + R-11 insulation	0.0	0.0	0.0
0.26	C) Same as "A" + R-19 insulation	0.0	0.0	0.0
3.12	D) Over vented space or garage	3.9	5.8	7.7
0.80	E) Over vented space or garage + R-11 insulation	0.8	1.3	1.7
0.52	F) Over vented space or garage + R-19 insulation	0.5	0.8	1.1
0.24	**Basement Floors (use sq. ft.)**	0.0	0.0	0.0
	Concrete slab floor unheated (use linear ft.)			
8.10	A) No edge insulation	0.0	0.0	0.0
4.10	B) 1″ edge insulation R-5	0.0	0.0	0.0
2.10	C) 2″ edge insulation R-9	0.0	0.0	0.0
	Concrete Slab floor duct in slab (use linear ft.)			
19.00	A) No edge insulation	0.0	0.0	0.0
11.40	B) 1″ edge insulation R-5	0.0	0.0	0.0
9.30	C) 2″ edge insulation R-9	0.0	0.0	0.0

(1) Heating Factor for 10° Temperature Rise

TABLE F

Calculate only if duct is located in an unconditioned space.

DUCT LOSS MULTIPLIERS

Case I - Supply Air Temperatures Below 120°F Duct Location and Insulation Value	Duct Loss Multipliers Winter Design Below 15°F	Winter Design Above 15°F
Exposed to Outdoor Ambient		
Attic, Garage, Exterior Wall, Open Crawl Space - None	1.30	1.25
Attic, Garage, Exterior Wall, Open Crawl Space - R2	1.20	1.15
Attic, Garage, Exterior Wall, Open Crawl Space - R4	1.15	1.10
Attic, Garage, Exterior Wall, Open Crawl Space - R6	1.10	1.05
Enclosed in Unheated Space		
Vented or Unvented Crawl Space or Basement - None	1.20	1.15
Vented or Unvented Crawl Space or Basement - R2	1.15	1.10
Vented or Unvented Crawl Space or Basement - R4	1.10	1.05
Vented or Unvented Crawl Space or Basement - R6	1.05	1.00
Duct Buried In or Under Concrete Slab		
No Edge Insulation	1.25	1.20
Edge Insulation R Value = 3 to 4	1.15	1.10
Edge Insulation R Value = 5 to 7	1.10	1.05
Edge Insulation R Value = 7 to 9	1.05	1.00
Case II - Supply Air Temperatures Above 120° F Duct Location and Insulation Value	**Winter Design Below 15°F**	**Winter Design Above 15°F**
Exposed to Outdoor Ambient		
Attic, Garage, Exterior Wall, Open Crawl Space - None	1.35	1.30
Attic, Garage, Exterior Wall, Open Crawl Space - R2	1.25	1.20
Attic, Garage, Exterior Wall, Open Crawl Space - R4	1.20	1.15
Attic, Garage, Exterior Wall, Open Crawl Space - R6	1.15	1.10
Enclosed in Unheated Space		
Vented or Unvented Crawl Space or Basement - None	1.25	1.20
Vented or Unvented Crawl Space or Basement - R2	1.20	1.15
Vented or Unvented Crawl Space or Basement - R4	1.15	1.10
Vented or Unvented Crawl Space or Basement - R6	1.10	1.05
Duct Buried In or Under Concrete Slab		
No Edge Insulation	1.30	1.25
Edge Insulation R Value = 3 to 4	1.20	1.15
Edge Insulation R Value = 5 to 7	1.15	1.10
Edge Insulation R Value = 7 to 9	1.10	1.05

DUCT GAIN MULTIPLIERS

Duct Location and Insulation Value	Duct Gain Multiplier
Exposed to Outdoor Ambient	
Attic, Garage, Exterior Wall, Open Crawl Space - None	1.30
Attic, Garage, Exterior Wall, Open Crawl Space - R2	1.20
Attic, Garage, Exterior Wall, Open Crawl Space - R4	1.15
Attic, Garage, Exterior Wall, Open Crawl Space - R6	1.10
Enclosed In Unconditioned Space	
Vented or Unvented Crawl Space or Basement - None	1.15
Vented or Unvented Crawl Space or Basement - R2	1.10
Vented or Unvented Crawl Space or Basement - R4	1.05
Vented or Unvented Crawl Space or Basement - R6	1.00
Duct Buried In or Under Concrete Slab	
No Edge Insulation	1.10
Edge Insulation R Value = 3 to 4	1.05
Edge Insulation R Value = 5 to 7	1.00
Edge Insulation R Value = 7 to 9	1.00

ESTIMATED PROCEDURES

1. Fill in customer information.
2. Record inside and outside design temperatures; find temp difference.
3. Measure length of each outside wall, multiply each by ceiling height. Record the total sq. ft. of exposed wall under "gross wall".
4. Using Tables A and B, determine the total area for windows & doors & enter in common data section.
5. Determine Net Wall by subtracting windows and doors from gross.
6. Measure & record total ceiling area.
7. Measure and record total floor area for floors over crawl space or basement. Total floor edge *length* (perimeter) if floor is a slab.
8. Using Table E select construction type and use the corresponding heat and cool factors on the form.
9. Determine BTUH Loss & Gain in Tables A and B by multiplying the area of glass and doors by the multiplier under the specified temperature difference. Enter total BTUH Loss/Gain on worksheet.
10. On worksheet, multiply the areas x the factors and total as instructed.

QUICK-SIZING TABLE for HEATING (FURNACE) ONLY DUCT SYSTEM (Without a Cooling Coil in Place)

OUTPUT CAPACITY (See Notes)	Min. Air Flow Req'd.	Supply Duct or Extended Plenum @ 800 FPM	Min. Sq. Inch Needed for Spec. CFM (Total Area of All Supply Duct)	Min. Number Supply Runs @ 600 FPM			MINIMUM SIZE	
				5" RUNS 80 CFM	6" RUNS 115 CFM	7" RUNS 155 CFM	Return Duct Furnace or Air Handler @ 800 FPM	Return Air Grille (or equivalent) @ Face Velocity of 500 FPM
45,000 to 55,000	500 CFM	14" x 8" or 12" round	100	7	5	4	14" x 8" or 12" round	12" x 12"
60,000 to 70,000	700 CFM	18" x 8" or 14" round	140	10	6	5	18" x 8" or 14" round	24" x 10"
75,000 to 85,000	800 CFM	22" x 8" or 14" round	170	10	7	5	22" x 8" or 14" round	24" x 12"
95,000 to 105,000	900 CFM	24" x 8" or 15" round	190	12	8	6	24" x 8" or 14" round	24" x 12"
105,000 to 115,000	1100 CFM	22" x 10" or 16" round	220	—	10	7	22" x 10" or 16" round	30" x 12"
125,000 to 150,000	1400 CFM	24" x 12" or 18" round	280	—	12	9	24" x 12" or 18" round	30" x 14"
155,000 to 160,000	1600 CFM	1-35" x 10" or 20" round or 2-22" x 8"	360	—	14	10	32" x 10" or 20" round	30" x 18"

Notes:
1. BTUH with **maximum** temperature rise.
2. Gas furnaces are rated in input capacity. Rated output capacity is 80% of input.
3. Oil and electric furnace are rated in output capacity.

QUICK-SIZING TABLE for HEATING and COOLING DUCT SYSTEM

Air conditioning systems should never be sized on the basis of floor area only, but knowledge of the approximate floor area (sq. ft.) that can be cooled with a ton of air conditioning will be of invaluable assistance to you in avoiding serious mathematical errors.

Size of O.D. Unit	Normal Air Flow Req'd @ 400 CFM per Ton	Furnace		Supply Duct or Extended Plenum @ 800 FPM	Min. Number Supply Runs @ 600 FPM				Min. Return Duct. Size at Furnace or Air Handler @ 800 FPM	Min. Return Air Grille Size (or equivalent) @ Face Velocity of 500 FPM
		Blower Motor H.P.	Blower Wheel Dia. X Width		5" RUNS 80 CFM	6" RUNS 115 CFM	7" RUNS 155 CFM	3½ X 14" 170 CFM		
1½ ton 18,000 BTUH	600 CFM	1/4 H.P.	9" X 8" 10" X 8"	16" X 8" or 12" round	8	5	4	4	16" X 8" or 12" round	24" X 8"
2 ton 24,000 BTUH	800 CFM	1/4 H.P.	9" X 9" 10" X 8"	22" X 8" or 14" round	10	7	5	5	22" X 8" or 14" round	22" X 12"
2½ ton 30,000 BTUH	1000 CFM	1/3 H.P.	10" X 8" 10" X 10" 12" X 9"	20" X 10" or 18" round	13	9	7	6	20" X 10" or 16" round	30" X 12"
3 ton 36,000 BTUH	1200 CFM	1/3 H.P.	10" X 8" 10" X 10" 12" X 9"	24" X 10" or 18" round	—	11	8	7	24" X 10" or 18" round	30" X 12"
3½ ton 42,000 BTUH	1400 CFM	1/2 H.P. 3/4 H.P.	10" X 8" 10" X 10" 12" X 9" 12" X 10"	24" X 12" or 18" round	—	12	9	8	24" X 12" or 18" round	30" X 14"
4 ton 48,000 BTUH	1600 CFM	1/2 H.P. 3/4 H.P.	10" X 10" 12" X 9" 12" X 10" 12" X 12"	32" X 10" or 20" round	—	14	11	10	28" X 12" or 20" round	30" X 18"

EQUIPMENT CAPACITY CALCULATIONS FROM WORKSHEET DATA

COOLING CAPACITY

TOTAL GAIN ____________ BTUH x [1] ________ = ________ BTUH ARI Rating

ARI Rating ____________ BTUH / 12,000 BTUH/TON = ________ TONS OF COOLING

[1] CAPACITY MULTIPLIER (BASED ON 75°F RETURN AIR TEMPERATURE)

90°F O.D.T. use 1.00
95°F O.D.T. use 1.05
100°F O.D.T. use 1.10
105°F O.D.T. use 1.15

HEATING CAPACITY

TOTAL LOSS ____________ BTUH / (Select Factor Below) = ____________ KW

ELECTRIC HEATER OR FURNACE DIVIDE USING 3413 BTUH/KW NATURAL GAS OR PROPANE FURNACE . . . SELECT OUTPUT TO BE AS CLOSE TO CALCULATED HEAT LOSS AS POSSIBLE

CALCULATION BY: ____________________

DATE: ____________________

EQUIPMENT SELECTION

Model Number(s)

HEAT PUMP .... ____________

AIR HANDLER .... ____________

SUPPLEMENTARY HEATER ... ____________

CONDENSING UNIT .... ____________

EVAPORATOR COIL .... ____________

FURNACE, ELECTRIC .... ____________

FURNACE, NATURAL GAS/PROPANE .... ____________

REFRIGERANT LINES .... ____________

THERMOSTAT .... ____________

ELECTRONIC AIR CLEANER .... ____________

HUMIDIFIER .... ____________

ACCESSORIES .... ____________

____

____

____

COST SCHEDULE(S) MATERIAL/LABOR

New Construction

Whole House Worksheet

Customer's Name ______________________ Address ______________________

City ______________ State ________ Zip ________ Telephone Number ______________

WINTER: Inside Design Temp ________ **°F—Outside Design Temp** ________ **°F = Heating Temp Difference** ________ **°F**

SUMMER: Outside Design Temp ________ **°F—Inside Design Temp** ________ **°F = Cooling Temp Difference** ________ **°F**

HEATING		COMMON DATA SECTION		COOLING	
BTUH LOSS	HEATING FACTOR	SUBJECT	SQ. FT.	COOLING FACTOR	BTUH GAIN
	FROM TABLE E	GROSS WALL		FROM TABLE E	
		DOORS & WINDOWS (Table A or B)			
		NET WALL			
		CEILING			
		FLOORS			
Infiltration Btu/hr =	Heating Table D	x 10 x 1.1/60 x Volume (Cu. Ft.)	Volume (Cu. Ft) x 1.1/60	x △T x Cooling Table D	= Infiltration Btu/hr
=		x 0.18333 x	x 0.01833	x x	=
		SUB-TOTAL BTUH LOSS (per 10°F)			
x		ADJUSTMENT FACTOR (Table C)			
		TOTAL BTUH LOSS			
		PEOPLE_____ x 300 BTUH GAIN (Assume 2 persons per bedroom)			
		APPLIANCES BTUH			1200
		SUB-TOTAL BTUH GAIN (room sensible only)			
x		DUCT LOSS/GAIN FACTOR (Table F)			x
		SUB-TOTAL BTUH (Sensible Gain)			
		MOISTURE REMOVAL (sub total x 1.3)			x 1.3
		TOTAL BTUH LOSS/GAIN			

TABLE A—HEATING—DOORS & WOOD FRAME WINDOWS (PER 10°F)

For sliding glass doors - use factors for the same type window construction.

Window & Door Types	Frames: Wood	TIM	Metal	x Area	= Btuh Loss
Single Pane Clear	9.90	10.45	11.55		
With Storm	4.75	5.25	6.50		
Double Pane Clear	5.51	6.09	7.25		
With Storm	3.41	3.85	4.90		
Triple Pane Clear	3.80	4.39	5.46		
Jalousie Single	—	—	11.0		
Single w/storm	—	—	5.0		
Skylights Single	11.07	11.69	12.92		
Double	6.65	7.35	8.75		
Door Wood Only	4.60	—	—		
Wood w/storm	3.20	—	—		
Urethane Core (R-5)	—	—	1.90		
Urethane Core (R-5) w/storm	—	—	1.70		
			TOTALS		

TABLE B — COOLING — DOORS & WINDOWS

Factors assume windows have inside shading by draperies or venetian blinds and sliding glass doors are treated as windows.

	SINGLE GLASS			DOUBLE GLASS			TRIPLE GLASS				
	TEMP. DIFF.			TEMP. DIFF.			TEMP. DIFF.			X Area	= BTUH GAIN
Direction	15°	20°	25°	15°	20°	25°	15°	20°	25°		
N	18	22	26	14	16	18	11	12	13		
NE & NW	37	41	45	31	33	35	26	27	28		
E & W	52	56	60	44	46	48	38	39	40		
SE & SW	45	49	53	39	41	43	33	34	35		
S	28	32	36	23	25	27	19	20	21		
Skylights	164	168	172	141	143	145	132	136	140		
Wood ①	8.6	10.9	13.2	8.6	10.9	13.2	8.6	10.9	13.2		
Metal ②	3.5	4.5	5.4	3.5	4.5	5.4	3.5	4.5	5.4		
									TOTALS		

① For wood doors and polystyrene core metal doors

② For urethane core metal doors

TABLE C — ADJUSTMENT FACTORS — (HEATING)

°F. Temperature Diff.	30	40	50	60	70	80	90
Adjustment Factor	3	4	5	6	7	8	9

TABLE D — INFILTRATION MULTIPLIERS

Winter Air Changes Per Hour

Floor Area	900 or less	900-1500	1500-2100	over 2100
Best	0.4	0.4	0.3	0.3
Average	1.2	1.0	0.8	0.7
Poor	2.2	1.6	1.2	1.0

For each fireplace add: Best 0.1, Average 0.2, Poor 0.6

Summer Air Changes Per Hour

Floor Area	900 or less	900-1500	1500-2100	over 2100
Best	0.2	0.2	0.2	0.2
Average	0.5	0.5	0.4	0.4
Poor	0.8	0.7	0.6	0.5

Pub. No. 22-8019-10 P.I. (L)

HOUSE & DUCT LAYOUT (1 SQ. = ____Feet)

HEAT LOSS & GAIN FACTORS

TABLE E
CONSTRUCTION FACTORS HEATING & COOLING

Heating Factor ①	TYPE OF CONSTRUCTION	Cooling Factor (°F. Temp. Diff.) 15°	20°	25°
	WALLS (Use Sq. Ft.)			
	Walls — wood frame w/sheeting & siding, veneer or other finish			
2.71	A) No insulation, 1/2″ Gypsum Board	5.0	6.4	7.8
0.90	B) R-11 Cavity insulation + 1/2″ Gypsum Board	1.7	2.1	2.6
0.80	C) R-13 Cavity insulation + 1/2″ Gypsum Board	1.5	1.9	2.3
0.70	D) R-13 Cavity insulation + 3/4″ Bead Board (R-2.7)	1.3	1.7	2.0
0.60	E) R-19 Cavity insulation + 1/2″ Gypsum Board	1.1	1.4	1.7
0.50	F) R-19 Cavity insulation + 3/4″ Extruded Poly	0.9	1.2	1.4
	Masonry Walls			
5.10	A) Above grade No insulation	5.8	8.3	10.9
1.44	B) Above grade + R-5	1.6	2.3	3.1
0.77	C) Above grade + R-11	0.9	1.3	1.6
1.25	D) Below grade No insulation	0.0	0.0	0.0
0.74	E) Below grade + R-5	0.0	0.0	0.0
0.51	F) Below grade + R-11	0.0	0.0	0.0
	CEILINGS (Use Sq. Ft.)			
5.99	A) No insulation	17.0	19.2	21.4
1.20	B) 2″-2½″ insulation R-7	4.4	4.9	5.5
0.88	C) 3″-3½″ insulation R-11	3.2	3.7	4.1
0.53	D) 5¼″-6½″ insulation R-19	2.1	2.3	2.6
0.48	E) 6″-7″ insulation R-22	1.9	2.1	2.4
0.33	F) 8″-9½″ insulation R-30	1.3	1.5	1.6
0.26	G) 10″-12″ insulation R-38	1.0	1.1	1.3
0.23	H) 12″-13″ insulation R-44	0.9	1.0	1.1
3.08	I) Cathedral type No insulation (roof/ceiling combination)	11.2	12.6	14.1
0.72	J) Cathedral type R-11 (roof/ceiling combination)	2.8	3.2	3.5
0.49	K) Cathedral type R-19 (roof/ceiling combination)	1.9	2.2	2.4
0.45	L) Cathedral type R-22 (roof/ceiling combination)	1.8	2.0	2.2
0.40	M) Cathedral type R-26 (roof/ceiling combination)	1.6	1.8	2.0
	FLOORS (Use Sq. Ft. OR Linear Ft.)			
1.56	**Floors over unconditioned space (use sq. ft.)** A) Over basement or enclosed crawl space (not vented)	0.0	0.0	0.0
0.40	B) Same as "A" + R-11 insulation	0.0	0.0	0.0
0.26	C) Same as "A" + R-19 insulation	0.0	0.0	0.0
3.12	D) Over vented space or garage	3.9	5.8	7.7
0.80	E) Over vented space or garage + R-11 insulation	0.8	1.3	1.7
0.52	F) Over vented space or garage + R-19 insulation	0.5	0.8	1.1
0.24	**Basement Floors (use sq. ft.)**	0.0	0.0	0.0
8.10	**Concrete slab floor unheated (use linear ft.)** A) No edge insulation	0.0	0.0	0.0
4.10	B) 1″ edge insulation R-5	0.0	0.0	0.0
2.10	C) 2″ edge insulation R-9	0.0	0.0	0.0
19.00	**Concrete Slab floor duct in slab (use linear ft.)** A) No edge insulation	0.0	0.0	0.0
11.40	B) 1″ edge insulation R-5	0.0	0.0	0.0
9.30	C) 2″ edge insulation R-9	0.0	0.0	0.0

① Heating Factor for 10° Temperature Rise

TABLE F

Calculate only if duct is located in an unconditioned space.

DUCT LOSS MULTIPLIERS

	Duct Loss Multipliers	
Case I - Supply Air Temperatures Below 120°F Duct Location and Insulation Value	Winter Design Below 15°F	Winter Design Above 15°F
Exposed to Outdoor Ambient		
Attic, Garage, Exterior Wall, Open Crawl Space - None	1.30	1.25
Attic, Garage, Exterior Wall, Open Crawl Space - R2	1.20	1.15
Attic, Garage, Exterior Wall, Open Crawl Space - R4	1.15	1.10
Attic, Garage, Exterior Wall, Open Crawl Space - R6	1.10	1.05
Enclosed in Unheated Space		
Vented or Unvented Crawl Space or Basement - None	1.20	1.15
Vented or Unvented Crawl Space or Basement - R2	1.15	1.10
Vented or Unvented Crawl Space or Basement - R4	1.10	1.05
Vented or Unvented Crawl Space or Basement - R6	1.05	1.00
Duct Buried In or Under Concrete Slab		
No Edge Insulation	1.25	1.20
Edge Insulation R Value = 3 to 4	1.15	1.10
Edge Insulation R Value = 5 to 7	1.10	1.05
Edge Insulation R Value = 7 to 9	1.05	1.00
Case II - Supply Air Temperatures Above 120° F Duct Location and Insulation Value	Winter Design Below 15°F	Winter Design Above 15°F
Exposed to Outdoor Ambient		
Attic, Garage, Exterior Wall, Open Crawl Space - None	1.35	1.30
Attic, Garage, Exterior Wall, Open Crawl Space - R2	1.25	1.20
Attic, Garage, Exterior Wall, Open Crawl Space - R4	1.20	1.15
Attic, Garage, Exterior Wall, Open Crawl Space - R6	1.15	1.10
Enclosed in Unheated Space		
Vented or Unvented Crawl Space or Basement - None	1.25	1.20
Vented or Unvented Crawl Space or Basement - R2	1.20	1.15
Vented or Unvented Crawl Space or Basement - R4	1.15	1.10
Vented or Unvented Crawl Space or Basement - R6	1.10	1.05
Duct Buried In or Under Concrete Slab		
No Edge Insulation	1.30	1.25
Edge Insulation R Value = 3 to 4	1.20	1.15
Edge Insulation R Value = 5 to 7	1.15	1.10
Edge Insulation R Value = 7 to 9	1.10	1.05

DUCT GAIN MULTIPLIERS

Duct Location and Insulation Value	Duct Gain Multiplier
Exposed to Outdoor Ambient	
Attic, Garage, Exterior Wall, Open Crawl Space - None	1.30
Attic, Garage, Exterior Wall, Open Crawl Space - R2	1.20
Attic, Garage, Exterior Wall, Open Crawl Space - R4	1.15
Attic, Garage, Exterior Wall, Open Crawl Space - R6	1.10
Enclosed In Unconditioned Space	
Vented or Unvented Crawl Space or Basement - None	1.15
Vented or Unvented Crawl Space or Basement - R2	1.10
Vented or Unvented Crawl Space or Basement - R4	1.05
Vented or Unvented Crawl Space or Basement - R6	1.00
Duct Buried In or Under Concrete Slab	
No Edge Insulation	1.10
Edge Insulation R Value = 3 to 4	1.05
Edge Insulation R Value = 5 to 7	1.00
Edge Insulation R Value = 7 to 9	1.00

ESTIMATED PROCEDURES

1. Fill in customer information.
2. Record inside and outside design temperatures; find temp difference.
3. Measure length of each outside wall, multiply each by ceiling height. Record the total sq. ft. of exposed wall under "gross wall".
4. Using Tables A and B, determine the total area for windows & doors & enter in common data section.
5. Determine Net Wall by subtracting windows and doors from gross.
6. Measure & record total ceiling area.
7. Measure and record total floor area for floors over crawl space or basement. Total floor edge *length* (perimeter) if floor is a slab.
8. Using Table E select construction type and use the corresponding heat and cool factors on the form.
9. Determine BTUH Loss & Gain in Tables A and B by multiplying the area of glass and doors by the multiplier under the specified temperature difference. Enter total BTUH Loss/Gain on worksheet.
10. On worksheet, multiply the areas x the factors and total as instructed.

EQUIPMENT CAPACITY CALCULATIONS FROM WORKSHEET DATA

COOLING CAPACITY

TOTAL GAIN ______________ BTUH x ①_______ = ______ BTUH ARI Rating

ARI Rating _____________ BTUH / 12,000 BTUH/TON = ________ TONS OF COOLING

① CAPACITY MULTIPLIER
(BASED ON 75°F RETURN AIR TEMPERATURE)

90°F O.D.T. use 1.00	100°F O.D.T. use 1.10
95°F O.D.T. use 1.05	105°F O.D.T. use 1.15

HEATING CAPACITY

TOTAL LOSS _________________ BTUH / (Select Factor Below) = ___________ KW

ELECTRIC HEATER OR FURNACE DIVIDE USING 3413 BTUH/KW
NATURAL GAS OR PROPANE FURNACE . . . SELECT OUTPUT TO BE AS CLOSE TO CALCULATED HEAT LOSS AS POSSIBLE

CALCULATION BY: ______________________________

DATE: ______________________________

EQUIPMENT SELECTION

Model Number(s)

HEAT PUMP................. ______________________

AIR HANDLER............... ______________________

SUPPLEMENTARY HEATER .. ______________________

CONDENSING UNIT.......... ______________________

EVAPORATOR COIL .... ______________________

FURNACE, ELECTRIC .... ______________________

FURNACE, NATURAL GAS/PROPANE .... ______________________

REFRIGERANT LINES .... ______________________

THERMOSTAT .... ______________________

ELECTRONIC AIR CLEANER .... ______________________

HUMIDIFIER .... ______________________

ACCESSORIES .... ______________________

____

____

____

COST SCHEDULE(S) MATERIAL/LABOR

____

ROOM AIRFLOW DISTRIBUTION

1. Divide conditioned space into airflow zones.
2. Record linear feet of outside wall in each zone. Obtain total of all zones.
3. Distribute CFM for windows, doors, appliances, ceiling and people per instructions. Obtain total of all zones.
4. Obtain total CFM distributed in Step 6.
5. Find total airflow of selected system.
6. Use equation below and find CFM per linear foot of outside wall.
7. Using linear feet obtained in Step 2 find CFM for linear feet of wall in each zone. (Step 2 x Step 8).
8. Obtain total CFM in each zone (Step 9 plus 3, 4, and 5).

AIRFLOW ZONES	OUTSIDE WALLS: LINEAR FEET	OUTSIDE WALLS: CFM PER FOOT	CFM: OUTSIDE WALLS	CFM: GLASS DOORS & WINDOWS (DIRECTION / CFM per sq. ft.: N 1; NE, NW, S 1½; E, SE, SW, W 2)	CFM: EXPOSED CEILING .1CFM/sq.ft.	DOORS 15 CFM per STD. 3' DOOR; APPLIANCES 60 CFM TOTAL; PEOPLE 15 CFM per person	TOTAL CFM PER ZONE
STEP 1	2	8	9	3	4	5	10
A							
B		CONSTANT					
C							
D							
E							
F							
G							
H							
I							
J							
K							
L							

STEP 6 [ ]

STEP 7 [ ] − [ ] = [ ] ÷ [ ] = [ ]

DESIGN FRICTION LOSS LIMITS FOR LONGEST (EQUIVALENT) S & R RUN

1. Layout duct system on scaled drawing and show airflow volumes to each airflow zone.
2. Select apparent longest supply and longest return — consider actual length of ducts and equivalent length of duct fittings.
3. List the air distribution devices which make-up the longest supply and longest return runs: cooling coil, duct sections (main and branch), types of duct fittings, register, grille and filter as applicable.
4. Assign the airflow volumes through each component listed in Step 3.
5. Find static pressure losses of each of the fixed distribution devices: cooling coil, register, grille, filter as applicable. Total fixed pressure losses. Consult the manufacturer's engineering data for accurate information.
6. From the duct layout and equivalent length tables find the actual or equivalent length of each item listed in Step 3.
7. Select indoor air moving equipment from manufacturer's data. Find maximum external static pressure available at the required air volume.
8. Using equation below find allowable pressure loss per 100′ of duct.

DUCT SECTIONS		PRESSURE LOSS	LENGTH	DUCT WORK SIZE
① IDENTIFICATION	CFM	① FIXED ITEMS	ACTUAL OR EQUIVALENT	MAINS, BRANCHES
3	4	5	6	9

9. Size **ALL** ducts using duct calculator. Do not exceed friction loss found in Step 8 **or** recommend air velocities.

   Main Ducts: 900 FPM
   Branch Ducts: 600 FPM
   Branch Risers: 500 FPM

NOTES

7 [   ] − [   ] = [   ] ÷ [   ] X [100] = [   ] 8

① Return Grille = .02 if Manufacturer's data is not available.

Glossary

Note: English term appears boldface followed by Spanish term set in Italic.

ABS pipe. Acrylonitrilebutadiene styrene plastic pipe used for water, drains, waste, and venting.

Tubo de acronitrilo-butadieno-estireno. Tubo plástico de acronitrilo-butadieno-estireno utilizado para el agua, los drenajes, los desperdicios y la ventilación.

Absolute pressure. Gage pressure plus the pressure of the atmosphere, normally 14.696 at sea level at 68°F.

Presión absoluta. La presión del calibrador más la presión de la atmósfera, que generalmente es 14,696 al nivel del mar a 68°F (20°C).

Absolute zero temperature. The lowest obtainable temperature where molecular motion stops, −460°F and −273°C.

Temperatura del cero absoluto. La temperatura más baja obtenible donde se detiene el movimiento molecular, −460°F and −273°C.

Absorbent (attractant). The salt solution used to attract water in an absorption chiller.

Hidrófilo. Solución salina utilizada para atraer el agua en un enfriador por absorción.

Absorber. That part of an absorption chiller where the water is absorbed by the salt solution.

Absorbedor. El lugar en el enfriador por absorción donde la solución salina absorbe el agua.

Absorption air conditioning chiller. A system using a salt substance, water, and heat to provide cooling for an air conditioning system.

Enfriador por absorción para acondicionamiento de aire. Sistema que utiliza una sustancia salina, agua y calor para proveer enfriamiento en un sistema de acondicionamiento de aire.

Accumulator. A storage tank located in the suction line of a compressor. It allows small amounts of liquid refrigerant to boil away before entering the compressor. Sometimes used to store excess refrigerant in heat pump systems during the winter cycle.

Acumulador. Tanque de almacenaje ubicado en el conducto de aspiración de un compresor. Permite que pequeñas cantidades de refrigerante líquido se evaporen antes de entrar al compresor. Algunas veces se utiliza para almacenar exceso de refrigerante en sistemas de bombas de calor durante el ciclo de invierno.

Acid-contaminated system. A refrigeration system that contains acid due to contamination.

Sistema contaminado de ácido. Sistema de refrigeración que, debido a la contaminación, contiene ácido.

ACR tubing. Air Conditioning and Refrigeration tubing that is very clean, dry, and normally charged with dry nitrogen. The tubing is sealed at the ends to contain the nitrogen.

Tubería ACR. Tubería para el acondicionamiento de aire y la refrigeración que es muy limpia y seca, y que por lo general está cargada de nitrógeno seco. La tubería se sella en ambos extremos para contener el nitrógeno.

Activated alumina. A chemical desiccant used in refrigerant driers.

Alúmina activada. Disecante químico utilizado en secadores de refrigerantes.

Active solar system. A system that uses electrical and/or mechanical devices to help collect, store, and distribute the sun's energy.

Sistema solar activo. Sistema que utiliza dispositivos eléctricos y/o mecánicos para ayudar a acumular, almacenar y distribuir la energía del sol.

Air-acetylene. A mixture of air and acetylene gas that when ignited is used for soldering, brazing, and other applications.

Aire-acetilénico. Mezcla de aire y de gas acetileno que se utiliza en la soldadura, la broncesoldadura y otras aplicaciones al ser encendida.

Air heat exchanger. A device used to exchange heat between air and another medium at different temperature levels, such as air-to-air, air-to-water, or air-to-refrigerant.

Intercambiador de aire y calor. Dispositivo utilizado para intercambiar el calor entre el aire y otro medio, como por ejemplo aire y aire, aire y agua o aire y refrigerante, a diferentes niveles de temperatura.

Air conditioner. Equipment that conditions air by cleaning, cooling, heating, humidifying, or dehumidifying it. A term often applied to comfort cooling equipment.

Acondicionador de aire. Equipo que acondiciona el aire limpiándolo, enfriándolo, calentándolo, humidificándolo o deshumidificándolo. Término comúnmente aplicado al equipo de enfriamiento para comodidad.

Air conditioning. A process that maintains comfort conditions in a defined area.

Acondicionamiento de aire. Proceso que mantiene condiciones agradables en un área definida.

Air-cooled condenser. One of the four main components of an air-cooled refrigeration system. It receives hot gas from the compressor and rejects it to a place where it makes no difference.

Condensador enfriado por aire. Uno de los cuatro componentes principales de un sistema de refrigeración enfriado por aire. Recibe el gas caliente del compresor y lo dirige a un lugar donde no afecte la temperatura.

Air gap. The clearance between the rotating rotor and the stationary winding on an open motor. Known as a vapor gap in a hermetically sealed compressor motor.

Espacio de aire. Espacio libre entre el rotor giratorio y el devandado fijo en un motor abierto. Conocido como espacio de vapor en un motor de compresor sellado herméticamente.

Air handler. The device that moves the air across the heat exchanger in a forced-air system—normally considered to be the fan and its housing.

Tratante de aire. Dispositivo que dirige el aire a través del intercambiador de calor en un sistema de aire forzado—considerado generalmente como el abanico y su alojamiento.

Air pressure control (switch). Used to detect air pressure drop across the coil in a heat pump outdoor unit due to ice buildup.

Regulador de la presión de aire (conmutador). Utilizado para detectar una caída en la presión del aire a través de la bobina en una unidad de bomba de calor para exteriores debido a la acumulación de hielo.

Air sensor. A device that registers changes in air conditions such as pressure, velocity, temperature, or moisture content.

Sensor de aire. Dispositivo que registra los cambios en las condiciones del aire, como por ejemplo cambios en presión, velocidad, temperatura o contenido de humedad.

Air, standard. Dry air at 70°F and 14.696 psi at which it has a mass density of 0.075 lb/ft^3 and a specific volume of 13.33 ft^3/lb, ASHRAE 1986.

Aire, estándar. Aire seco a 70°F (21.11°C) y 14,696 psi [libra por pulgada cuadrada]; a dicha temperatura tiene una densidad de masa de 0,075 pies/libras^3 y un volumen específico de 13,33 pies^3/libras, ASHRAE 1986.

Air vent. A fitting used to vent air manually or automatically from a system.

Válvula de aire. Accesorio utilizado para darle al aire salida manual o automática de un sistema.

Algae. A form of green or black, slimy plant life that grows in water systems.

Alga. Tipo de planta legamosa de color verde o negro que crece en sistemas acuáticos.

Allen head. A recessed hex head in a fastener.

Cabeza allen. Cabeza de concavidad hexagonal en un asegurador.

Alternating current. An electric current that reverses its direction at regular intervals.

Corriente alterna. Corriente eléctrica que invierte su dirección a intervalos regulares.

Altitude adjustment. An adjustment to a refrigerator thermostat to account for a lower than normal atmospheric pressure such as may be found at a high altitude.

Ajuste para elevación. Ajuste al termóstato de un refrigerador para regular una presión atmosférica más baja que la normal, como la que se encuentra en elevaciones altas.

Ambient temperature. The surrounding air temperature.

Temperatura ambiente. Temperatura del aire circundante.

American standard pipe thread. Standard thread used on pipe to prevent leaks.

Rosca estándar estadounidense para tubos. Rosca estándar utilizada en tubos para evitar fugas.

Ammeter. A meter used to measure current flow in an electrical circuit.

Amperímetro. Instrumento utilizado para medir el flujo de corriente en un circuito eléctrico.

Amperage. Amount of electron or current flow (the number of electrons passing a point in a given time) in an electrical circuit.

Amperaje. Cantidad de flujo de electrones o de corriente (el número de electrones que sobrepasa un punto específico en un tiempo fijo) en un circuito eléctrico.

Ampere. Unit of current flow.

Amperio. Unidad de flujo de corriente.

Anemometer. An instrument used to measure the velocity of air.

Anemómetro. Instrumento utilizado para medir la velocidad del aire.

Angle valve. Valve with one opening at a 90° angle from the other opening.

Válvula en ángulo. Válvula con una abertura a un ángulo de 90° con respecto a la otra abertura.

Anode. A terminal or connection point on a semiconductor.

Ánodo. Punto de conexión o terminal en un semiconductor.

Approach temperature. The difference in temperature between the refrigerant and the leaving water in a chilled-water system.

Temperatura de acercamiento. Diferencia en temperatura entre el refrigerante y el agua de salida en un sistema de agua enfriada.

A.S.A. Abbreviation for the American Standards Association [now known as American National Standards Institute (ANSI)].

A.S.A. Abreviatura de Asociación Estadounidense de Normas [conocida ahora como Instituto Nacional Estadounidense de Normas (ANSI)].

ASHRAE. Abbreviation for the American Society of Heating, Refrigerating, and Air Conditioning Engineers.

ASHRAE. Abreviatura de Sociedad Estadounidense de Ingenieros de Calefacción, Refrigeración y Acondicionamiento de Aire.

ASME. Abbreviation for the American Society of Mechanical Engineers.

ASME. Abreviatura de Sociedad Estadounidense de Ingenieros Mecánicos.

Aspect ratio. The ratio of the length to width of a component.

Coeficiente de alargamiento. Relación del largo al ancho de un componente.

Atmospheric pressure. The weight of the atmosphere's gases pressing down on the earth. Equal to 14.696 psi at sea level and 70°F.

Presión atmosférica. El peso de la presión ejercida por los gases de la atmósfera sobre la tierra, equivalente a 14,696 psi al nivel del mar a 70°F.

Atom. The smallest particle of an element.

Átomo. Partícula más pequeña de un elemento.

Atomize. Using pressure to change liquid to small particles of vapor.

Atomizar. Utilizar la presión para cambiar un líquido a partículas pequeñas de vapor.

Automatic combination gas valve. A gas valve for gas furnaces that incorporates a manual control, gas supply for the pilot, adjustment and safety features for the pilot, pressure regulator, the controls for and the main gas valve.

Válvula de gas de combinación automática. Válvula de gas para hornos de gas que incorpora un regulador manual, suministro de gas para la llama piloto, ajuste y dispositivos de seguridad, regulador de presión, la válvula de gas principal y los reguladores de la válvula.

Automatic control. Controls that react to a change in conditions to cause the condition to stabilize.

Regulador automático. Reguladores que reaccionan a un cambio en las condiciones para provocar la estabilidad de dicha condición.

Automatic defrost. Using automatic means to remove ice from a refrigeration coil.

Desempañador automático. La utilización de medios automáticos para remover el hielo de una bobina de refrigeración.

Automatic expansion valve. A refrigerant control valve that maintains a constant pressure in an evaporator.

Válvula de expansión automática. Válvula de regulación del refrigerante que mantiene una presión constante en un evaporador.

Back pressure. The pressure on the low-pressure side of a refrigeration system (also known as suction pressure).

Contrapresión. La presión en el lado de baja presión de un sistema de refrigeración (conocido también como presión de aspiración).

Back seat. The position of a refrigeration service valve when the stem is turned away from the valve body and seated.

Asiento trasero. Posición de una válvula de servicio de refrigeración cuando el vástago está orientado fuera del cuerpo de la válvula y aplicado sobre su asiento.

Baffle. A plate used to keep fluids from moving back and forth at will in a container.

Deflector. Placa utilizada para evitar el libre movimiento de líquidos en un recipiente.

Balanced port TXV. A valve that will meter refrigerant at the same rate when the condenser head pressure is low.

Válvula electrónica de expansión con conducto equilibrado. Válvula que medirá el refrigerante a la misma proporción cuando la presión en la cabeza del condensador sea baja.

Ball check valve. A valve with a ball-shaped internal assembly that only allows fluid flow in one direction.

Válvula de retención de bolas. Válvula con un conjunto interior en forma de bola que permite el flujo de fluido en una sola dirección.

Barometer. A device used to measure atmospheric pressure that is commonly calibrated in inches or millimeters of mercury. There are two types: mercury column and aneroid.

Barómetro. Dispositivo comúnmente calibrado en pulgadas o en milímetros de mercurio que se utiliza para medir la presión atmosférica. Existen dos tipos: columna de mercurio y aneroide.

Base. A terminal on a semiconductor.

Base. Punto terminal en un semiconductor.

Battery. A device that produces electricity from the interaction of metals and acid.

Pila. Dispositivo que genera electricidad de la interacción entre metales y el ácido.

Bearing. A device that surrounds a rotating shaft and provides a low-friction contact surface to reduce wear from the rotating shaft.

Cojinete. Dispositivo que rodea un árbol giratorio y provee una superficie de contacto de baja fricción para disminuir el desgaste de dicho árbol.

Bellows. An accordion-like device that expands and contracts when internal pressure changes.

Fuelles. Dispositivo en forma de acordeón con pliegues que se expanden y contraen cuando la presión interna sufre cambios.

Bellows seal. A method of sealing a rotating shaft or valve stem that allows rotary movement of the shaft or stem without leaking.

Cierre hermético de fuelles. Método de sellar un árbol giratorio o el vástago de una válvula que permite el movimiento giratorio del árbol o del vástago sin producir fugas.

Bending spring. A coil spring that can be fitted inside or outside a piece of tubing to prevent its walls from collapsing when being formed.

Muelle de flexión. Muelle helicoidal que puede acomodarse dentro o fuera de una pieza de tubería para evitar que sus paredes se doblen al ser formadas.

Bimetal. Two dissimilar metals fastened together to create a distortion of the assembly with temperature changes.

Bimetal. Dos metales distintos fijados entre sí para producir una distorción del conjunto al ocurrir cambios de temperatura.

Bimetal strip. Two dissimilar metal strips fastened back to back.

Banda bimetálica. Dos bandas de metales distintos fijadas entre sí en su parte posterior.

Bleeding. Allowing pressure to move from one pressure level to another very slowly.

Sangradura. Proceso a través del cual se permite el movimiento de presión de un nivel a otro de manera muy lenta.

Bleed valve. A valve with a small port usually used to bleed pressure from a vessel to the atmosphere.

Válvula de descarga. Válvula con un conducto pequeño utilizado normalmente para purgar la presión de un depósito a la atmósfera.

Blocked suction. A method of cylinder unloading. The suction line passage to a cylinder in a reciprocating compressor is blocked, thus causing that cylinder to stop pumping.

Aspiración obturada. Método de descarga de un cilindro. El paso del conducto de aspiración a un cilindro en un compresor alternativo se obtura, provocando así que el cilindro deje de bombear.

Blowdown. A system in a cooling tower whereby some of the circulating water is bled off and replaced with fresh water to dilute the sediment in the sump.

Vaciado. Sistema en una torre de refrigeración por medio del cual se purga parte del agua circulante y se reemplaza con agua fresca para diluir el sedimento en el sumidero.

Boiler. A container in which a liquid may be heated using any heat source. When the liquid is heated to the point that vapor forms and is used as the circulating medium, it is called a steam boiler.

Caldera. Recipiente en el que se puede calentar un líquido utilizando cualquier fuente de calor. Cuando se calienta el líquido al punto en que se produce vapor y se utiliza éste como el medio para la circulación, se llama caldera de vapor.

Boiling point. The temperature level of a liquid at which it begins to change to a vapor. The boiling temperature is controlled by the vapor pressure above the liquid.

Punto de ebullición. El nivel de temperatura de un líquido al que el líquido empieza a convertirse en vapor. La temperatura de ebullición se regula por medio de la presión del vapor sobre líquido.

Bore. The inside diameter of a cylinder.

Calibre. Diámetro interior de un cilindro.

Bourdon tube. C-shaped tube manufactured of thin metal and closed on one end. When pressure is increased inside, it tends to straighten. It is used in a gage to indicate pressure.

Tubo Bourdon. Tubo en forma de C fabricado de metal delgado y cerrado en uno de los extremos. Al aumentarse la presión en su interior, el tubo tiende a enderezarse. Se utiliza dentro de un calibrador para indicar la presión.

Brazing. High-temperature (above 800°F) soldering of two metals.

Broncesoldadura. Soldadura de dos metales a temperaturas altas (sobre los 800°F ó 430°C).

Breaker. A heat-activated electrical device used to open an electrical circuit to protect it from excessive current flow.

Interruptor. Dispositivo eléctrico activado por el calor que se utiliza para abrir un circuito eléctrico a fin protegerlo de un flujo excesivo de corriente.

British thermal unit. The amount (quantity) of heat required to raise the temperature of 1 lb of water 1°F.

Unidad térmica británica. Cantidad de calor necesario para elevar en 1°F (-17.56°C) la temperatura de una libra inglesa de agua.

Btu. Abbreviation for British thermal unit.

BTU. Abreviatura de unidad térmica británica.

Bulb, sensor. The part of a sealed automatic control used to sense temperature.

Bombilla sensora. Pieza de un regulador automático sellado que se utiliza para advertir la temperatura.

Burner. A device used to prepare and burn fuel.

Quemador. Dispositivo utilizado para la preparación y la quema de combustible.

Burr. Excess material squeezed into the end of tubing after a cut has been made. This burr must be removed.

Rebaba. Exceso de material introducido por fuerza en el extremo de una tubería después de hacerse un corte. Esta rebaba debe removerse.

Butane gas. A liquefied petroleum gas burned for heat.

Gas butano. Gas licuado derivado del petróleo que se quema para producir calor.

Cad cell. A device used to prove the flame in an oil burning furnace containing cadmium sulfide.

Celda de cadmio. Dispositivo utilizado para probar la llama en un horno de aceite pesado que contiene sulfuro de cadmio.

Calibrate. To adjust instruments or gages to the correct setting for conditions.

Calibrar. Ajustar instrumentos o calibradores en posición correcta para su operación.

Capacitance. The term used to describe the electrical storage ability of a capacitor.

Capacitancia. Término utilizado para describir la capacidad de almacenamiento eléctrico de un capacitador.

Capacitor. An electrical storage device used to start motors (start capacitor) and to improve the efficiency of motors (run capacitor).

Capacitador. Dispositivo de almacenamiento eléctrico utilizado para arrancar motores (capacitador de arranque) y para mejorar el rendimiento de motores (capacitador de funcionamiento).

Capacity. The rating system of equipment used to heat or cool substances.

Capacidad. Sistema de clasificación de equipo utilizado para calentar o enfriar sustancias.

Capillary attraction. The attraction of a liquid material between two pieces of material such as two pieces of copper or copper and brass. For instance, in a joint made up of copper tubing and a brass fitting, the solder filler material has a greater attraction to the copper and brass than to itself and is drawn into the space between them.

Atracción capilar. Atracción de un material líquido entre dos piezas de material, como por ejemplo dos piezas de cobre o cobre y latón. Por ejemplo, en una junta fabricada de tubería de cobre y un accesorio de latón, el material de relleno de la soldadura tiene mayor atracción al cobre y al latón que a sí mismo y es arrastrado hacia el espacio entre éstos.

Capillary tube. A fixed-bore metering device. This is a small diameter tube that can vary in length from a few inches to several feet. The amount of refrigerant flow needed is predetermined and the length and diameter of the capillary tube is sized accordingly.

Tubo capilar. Dispositivo de medición de calibre fijo. Este es un tubo de diámetro pequeño cuyo largo puede oscilar entre unas cuantas pulgadas a varios pies. La cantidad de flujo de refrigerante requerida es predeterminada y, de acuerdo a esto, se fijan el largo y el diámetro del tubo capilar.

Carbon dioxide. A by-product of natural gas combustion that is not harmful.

Bióxido de carbono. Subproducto de la combustión del gas natural que no es nocivo.

Carbon monoxide. A poisonous, colorless, odorless, tasteless gas generated by incomplete combustion.

Monóxido de carbono. Gas mortífero, inodoro, incoloro e insípido que se desprende en la combustión incompleta del carbono.

Catalytic combustor stove. A stove that contains a cell-like structure consisting of a substrate, washcoat, and catalyst producing a chemical reaction causing pollutants to be burned at much lower temperatures.

Estufa de combustor catalítico. Estufa con una estructura en forma de celda compuesta de una subestructura, una capa brochada y un catalizador que produce una reacción química. Esta reacción provoca la quema de contaminantes a temperaturas mucho más bajas.

Cathode. A terminal or connection point on a semiconductor.

Cátodo. Punto de conexión o terminal en un semiconductor.

Cavitation. A vapor formed due to a drop in pressure in a pumping system. Air at a pump inlet may be caused at a cooling tower if the pressure is low and water is turned to vapor.

Cavitación. Vapor producido como consecuencia de una caída de presión en un sistema de bombeo. El aire a la entrada de una bomba puede ser producido en una torre de refrigeración si la presión es baja y el agua se convierte en vapor.

Celsius scale. A temperature scale with 1200 graduations between water freezing (0°C) and water boiling (100°C).

Escala Celsio. Escala dividida en cien grados, con el cero marcado a la temperatura de fusión del hielo (0°C) y el cien a la de ebullición del agua (100°C).

Centigrade scale. See Celsius.

Centígrado. Véase escala Celsio.

Centrifugal compressor. A compressor used for large refrigeration systems. It is not positive displacement, but it is similar to a blower.

Compresor centrífugo. Compresor utilizado en sistemas grandes de refrigeración. No es desplazamiento positivo, pero es similar a un soplador.

Centrifugal switch. A switch that uses a centrifugal action to disconnect the start windings from the circuit.

Conmutador centrífugo. Conmutador que utiliza una acción centrífuga para desconectar los devanados de arranque del circuito.

Change of state. The condition that occurs when a substance changes from one physical state to another, such as ice to water and water to steam.

Cambio de estado. Condición que ocurre cuando una sustancia cambia de un estado físico a otro, como por ejemplo el hielo a agua y el agua a vapor.

Charge. The quantity of refrigerant in a system.

Carga. Cantidad de refrigerante en un sistema.

Charging cylinder. A device that allows the technician to accurately charge a refrigeration system with refrigerant.

Cilindro cargador. Dispositivo que le permite al mecánico cargar correctamente un sistema de refrigeración con refrigerante.

Check valve. A device that permits fluid flow in one direction only.

Válvula de retención. Dispositivo que permite el flujo de fluido en una sola dirección.

Chill factor. A factor or number that is a combination of temperature, humidity, and wind velocity that is used to compare a relative condition to a known condition.

Factor de frío. Factor o número que es una combinación de la temperatura, la humedad y la velocidad del viento utilizado para comparar una condición relativa a una condición conocida.

Chilled-water system. An air conditioning system that circulates refrigerated water to the area to be cooled. The refrigerated water picks up heat from the area, thus cooling the area.

Sistema de agua enfriada. Sistema de acondicionamiento de aire que hace circular agua refrigerada al área que será enfriada. El agua refrigerada atrapa el calor del área y la enfria.

Chiller purge unit. A system that removes air from a low-pressure chiller.

Unidad enfriadora de purga. Sistema que remueve el aire de un enfriador de baja presión.

Chimney. A vertical shaft used to convey flue gases above the rooftop.

Chimenea. Cañón vertical utilizado para conducir los gases de combustión por encima del techo.

Chimney effect. A term used to describe air or gas when it expands and rises when heated.

Efecto de chimenea. Término utilizado para describir el aire o el gas cuando se expande y sube al calentarse.

Chlorofluocarbons (CFC). Those refrigerants thought to contribute to the depletion of the ozone layer.

Cloroflurocarburos. Líquidos refrigerantes que, según algunos, han contribuido a la reducción de la capa de ozono.

Circuit. An electron or fluid-flow path that makes a complete loop.

Circuito. Electrón o trayectoria del flujo de fluido que hace un ciclo completo.

Circuit breaker. A device that opens an electric circuit when an overload occurs.

Interruptor para circuitos. Dispositivo que abre un circuito eléctrico cuando ocurre una sobrecarga.

Clamp-on ammeter. An instrument that can be clamped around one conductor in an electrical circuit and measure the current.

Amperímetro fijado con abrazadera. Instrumento que puede fijarse con una abrazadera a un conductor en un circuito eléctrico y medir la corriente.

Clearance volume. The volume at the top of the stroke in a compressor cylinder between the top of the piston and the valve plate.

Volumen de holgura. Volumen en la parte superior de una carrera en el cilindro de un compresor entre la parte superior del pistón y la placa de una válvula.

Closed circuit. A complete path for electrons to flow on.

Circuito cerrado. Circuito de trayectoria ininterrumpida que permite un flujo continuo de electrones.

Closed loop. Piping circuit that is complete and not open to the atmosphere.

Ciclo cerrado. Circuito de tubería completo y no abierto a la atmósfera.

Code. The local, state, or national rules that govern safe installation and service of systems and equipment for the purpose of safety of the public and trade personnel.

Código. Reglamentos locales, estaduales o federales que rigen la instalación segura y el servicio de sistemas y equipo con el propósito de garantizar la seguridad del personal público y profesional.

Coefficient of performance (COP). The ratio of usable output energy divided by input energy.

Coeficiente de rendimiento. Relación de la de energía de salida utilizable dividida por la energía de entrada.

CO_2 indicator. An instrument used to detect the quantity of carbon dioxide in flue gas for efficiency purposes.

Indicador del CO_2. Instrumento utilizado para detectar la cantidad de bióxido de carbono en el gas de combustión a fin de lograr un mejor rendimiento.

Cold. The word used to describe heat at lower levels of intensity.

Frío. Término utilizado para describir el calor a niveles de intensidad más bajos.

Cold anticipator. A device that anticipates a need for cooling and starts the cooling system early enough for it to reach capacity when it is needed.

Anticipador de frío. Dispositivo que anticipa la necesidad de enfriamiento y pone en marcha el sistema de enfriamiento con suficiente anticipación para que éste alcance su máxima capacidad cuando vaya a ser utilizado.

Cold junction. The opposite junction to the hot junction in a thermocouple.

Empalme frío. El empalme opuesto al empalme caliente en un termopar.

Cold trap. A device to help trap moisture in a refrigeration system.

Trampa del frío. Dispositivo utilizado para ayudar a atrapar la humedad en un sistema de refrigeración.

Cold wall. The term used in comfort heating to describe a cold outside wall and its effect on human comfort.

Pared fría. Término utilizado en la calefacción para comodidad que describe una pared exterior fría y sus efectos en la comodidad de una persona.

Collector. A terminal on a semiconductor.

Colector. Punto terminal en un semiconductor.

Combustion. A reaction called rapid oxidation or burning produced with the right combination of a fuel, oxygen, and heat.

Combustión. Reacción conocida como oxidación rápida o quema producida con la combinación correcta de combustible, oxígeno y calor.

Comfort chart. A chart used to compare the relative comfort of one temperature and humidity condition to another condition.

Esquema de comodidad. Esquema utilizado para comparar la comodidad relativa de una condición de temperatura y humedad a otra condición.

Compound gage. A gage used to measure the pressure above and below the atmosphere's standard pressure. It is a Bourdon tube sensing device and can be found on all gage manifolds used for air conditioning and refrigeration service work.

Calibrador compuesto. Calibrador utilizado para medir la presión mayor y menor que la presión estándar de la atmósfera. Es un dispositivo sensor de tubo Bourdon que puede encontrarse en todos los distribuidores de calibrador utilizados para el servicio de sistemas de acondicionamiento de aire y de refrigeración.

Compression. A term used to describe a vapor when pressure is applied and the molecules are compacted closer together.

Compresión. Término utilizado para describir un vapor cuando se aplica presión y se compactan las moléculas.

Compression ratio. A term used with compressors to describe the actual difference in the low- and high-pressure sides of the compression cycle. It is absolute discharge pressure divided by absolute suction pressure.

Relación de compresión. Término utilizado con compresores para describir la diferencia real en los lados de baja y alta presión del ciclo de compresión. Es la presión absoluta de descarga dividida por la presión absoluta de aspiración.

Compressor. A vapor pump that pumps vapor (refrigerant or air) from one pressure level to a higher pressure level.

Compresor. Bomba de vapor que bombea el vapor (refrigerante o aire) de un nivel de presión a un nivel de presión más alto.

Compressor displacement. The internal volume of a compressor, used to calculate the pumping capacity of the compressor.

Desplazamiento del compresor. Volumen interno de un compresor, utilizado para calcular la capacidad de bombeo del mismo.

Compressor shaft seal. The seal that prevents refrigerant inside the compressor from leaking around the rotating shaft.

Junta de estanqueidad del árbol del compresor. La junta de estanqueidadque evita la fuga, alrededor del árbol giratorio, del refrigerante en el interior del compresor.

Concentrator. That part of an absorption chiller where the dilute salt solution is boiled to release the water.

Concentrador. El lugar en el enfriador por absorción donde se hierve la solución salina diluida para liberar el agua.

Condensate. The moisture collected on an evaporator coil.

Condensado. Humedad acumulada en la bobina de un evaporador.

Condensate pump. A small pump used to pump condensate to a higher level.

Bomba para condensado. Bomba pequeña utilizada para bombear el condensado a un nivel más alto.

Condensation. Liquid formed when a vapor condenses.

Condensación. El líquido formado cuando se condensa un vapor.

Condense. Changing a vapor to a liquid at a particular pressure.

Condensar. Convertir un vapor en líquido a una presión específica.

Condenser. The component in a refrigeration system that transfers heat from the system by condensing refrigerant.

Condensador. Componente en un sistema de refrigeración que transmite el calor del sistema al condensar el refrigerante.

Condenser flooding. A method of maintaining a correct head pressure by adding liquid refrigerant to the condenser from a receiver to increase the head pressure.

Inundación del condensador. Método de mantener una presión correcta en la cabeza agregando refrigerante líquido al condensador de un receptor para aumentar la presión en la cabeza.

Condensing-gas furnace. A furnace with a condensing heat exchanger that condenses moisture from the flue gases resulting in greater efficiency.

Horno para condensación de gas. Horno con un intercambiador de calor para condensación que condensa la humedad de los gases de combustión. El resultado será un mayor rendimiento.

Condensing pressure. The pressure that corresponds to the condensing temperature in a refrigeration system.

Presión para condensación. La presión que corresponde a la temperatura de condensación en un sistema de refrigeración.

Condensing temperature. The temperature at which a vapor changes to a liquid.

Temperatura de condensación. Temperatura a la que un vapor se convierte en líquido.

Condensing unit. A complete unit that includes the compressor and the condensing coil.

Conjunto del condensador. Unidad completa que incluye el compresor y la bobina condensadora.

Conduction. Heat transfer from one molecule to another within a substance or from one substance to another.

Conducción. Transmisión de calor de una molécula a otra dentro de una sustancia o de una sustancia a otra.

Conductivity. The ability of a substance to conduct electricity or heat.

Conductividad. Capacidad de una sustancia de conducir electricidad o calor.

Conductor. A path for electrical energy to flow on.

Conductor. Trayectoria que permite un flujo continuo de energía eléctrica.

Contactor. A larger version of the relay. It can be repaired or rebuilt and has moveable and stationary contacts.

Contactador. Versión más grande del relé. Puede ser reparado o reconstruido. Tiene contactos móviles y fijos.

Contaminant. Any substance in a refrigeration system that is foreign to the system, particularly if it causes damage.

Contaminante. Cualquier sustancia en un sistema de refrigeración extraña a éste, principalmente si causa averías.

Control. A device to stop, start, or modulate flow of electricity or fluid to maintain a preset condition.

Regulador. Dispositivo para detener, poner en marcha o modular el flujo de electricidad o de fluido a fin de mantener una condición establecida con anticipación.

Control system. A network of controls to maintain desired conditions in a system or space.

Sistema de regulación. Red de reguladores que mantienen las condiciones deseadas en un sistema o un espacio.

Convection. Heat transfer from one place to another using a fluid.

Convección. Transmisión de calor de un lugar a otro por medio de un fluido.

Conversion factor. A number used to convert from one equivalent value to another.

Factor de conversión. Número utilizado en la conversión de un valor equivalente a otro.

Cooler. A walk-in or reach-in refrigerated box.

Nevera. Caja refrigerada donde se puede entrar o introducir la mano.

Cooling tower. The final device in many water-cooled systems, which rejects heat from the system into the atmosphere by evaporation of water.

Torre de refrigeración. Dispositivo final en muchos sistemas enfriados por agua, que dirige el calor del sistema a la atmósfera por medio de la evaporación de agua.

Copper plating. Small amounts of copper are removed by electrolysis and deposited on the ferrous metal parts in a compressor.

Encobrado. Remoción de pequeñas cantidades de cobre por medio de electrólisis que luego se colocan en las piezas de metal férreo en un compresor.

Corrosion. A chemical action that eats into or wears away material from a substance.

Corrosión Acción química que carcome o desgasta el material de una sustancia.

Counter EMF. Voltage generated or induced above the applied voltage in a single-phase motor.

Contra EMF. Tensión generada o inducida sobre la tensión aplicada en un motor unifásico.

Counterflow. Two fluids flowing in opposite directions.

Contraflujo. Dos fluidos que fluyen en direcciones opuestas.

Coupling. A device for joining two fluid-flow lines. Also the device connecting a motor drive shaft to the driven shaft in a direct-drive system.

Acoplamiento. Dispositivo utilizado para la conexión de dos conductos de flujo de fluido. Es también el dispositivo que conecta un árbol de mando del motor al árbol accionado en un sistema de mando directo.

CPVC (Chlorinated polyvinyl chloride). Plastic pipe similar to PVC except that it can be used with temperatures up to 180°F at 100 psig.

CPVC (Cloruro de polivinilo clorado). Tubo plástico similar al PVC, pero que puede utilizarse a temperaturas de hasta 180°F (82°C) a 100 psig [indicador de libras por pulgada cuadrada].

Crackage. Small spaces in a structure that allow air to infiltrate the structure.

Formación de grietas. Espacios pequeños en una estructura que permiten la infiltración del aire dentro de la misma.

Crankcase heat. Heat provided to the compressor crankcase.

Calor para el cárter del cigüeñal. Calor suministrado al cárter del cigüeñal del compresor.

Crankcase pressure regulator (CPR). A valve installed in the suction line, usually close to the compressor. It is used to keep a low-temperature compressor from overloading on a hot pull down.

Regulador de la presión del cárter del cigüeñal. Válvula instalada en el conducto de aspiración, normalmente cerca del compresor. Se utiliza para evitar la sobrecarga en un compresor de temperatura baja durante un arrastre caliente hacia abajo.

Crankshaft seal. Same as the compressor shaft seal.

Junta de estanqueidad del árbol del cigüeñal. Exactamente igual que la junta de estanqueidad del árbol del compresor.

Crankshaft throw. The off-center portion of a crankshaft that changes rotating motion to reciprocating motion.

Excentricidad del cigüeñal. Porción descentrada de un cigüeñal que cambia el movimiento giratorio a un movimiento alternativo.

Creosote. A mixture of unburned organic material found in the smoke from a wood-burning fire.

Creosota. Mezcla del material orgánico no quemado que se encuentra en el humo proveniente de un incendio de madera.

Crisper. A refrigerated compartment that maintains a high humidity and a low temperature.

Encrespador. Compartimiento refrigerado que mantiene una humedad alta y una temperatura baja.

Cross charge. A control with a sealed bulb that contains two different fluids that work together for a common specific condition.

Carga transversal. Regulador con una bombilla sellada compuesta de dos fluidos diferentes que pueden funcionar juntos para una condición común específica.

Cross liquid charge bulb. A type of charge in the sensing bulb of the TXV that has different characteristics from the system refrigerant. This is designed to help prevent liquid refrigerant from flooding to the compressor at startup.

Bombilla de carga del líquido transversal. Tipo de carga en la bombilla sensora de la válvula electrónica de expansión que tiene características diferentes a las del refrigerante del sistema. La carga está diseñada para ayudar a evitar que el refrigerante líquido se derrame dentro del compresor durante la puesta en marcha.

Cross vapor charge bulb. Similar to the vapor charge bulb but contains a fluid different from the system refrigerant. This is a special-type charge and produces a different pressure/temperature relationship under different conditions.

Bombilla de carga del vapor transversal. Similar a la bombilla de carga del vapor pero contiene un fluido diferente al del refrigerante del sistema. Esta es una carga de tipo especial y produce una relación diferente entre la presión y la temperatura bajo condiciones diferentes.

Crystallization. When a salt solution becomes too concentrated and part of the solution turns to salt.

Cristalización. Condición que ocurre cuando una solución salina se concentra demasiado y una parte de la solución se convierte en sal.

Current, electrical. Electrons flowing along a conductor.

Corriente eléctrica. Electrones que fluyen a través de un conductor.

Current relay. An electrical device activated by a change in current flow.

Relé para corriente. Dispositivo eléctrico accionado por un cambio en el flujo de corriente.

Cut-in and cut-out. The two points at which a control opens or closes its contacts based on the condition it is supposed to maintain.

Puntos de conexión y desconexión. Los dos puntos en los que un regulador abre o cierra sus contactos según las condiciones que debe mantener.

Cycle. A complete sequence of events (from start to finish) in a system.

Ciclo. Secuencia completa de eventos, de comienzo a fin, que ocurre en un sistema.

Cylinder. A circular container with straight sides used to contain fluids or to contain the compression process (the piston movement) in a compressor.

Cilindro. Recipiente circular con lados rectos, utilizado para contener fluidos o el proceso de compresión (movimiento del pistón) en un compresor.

Cylinder, compressor. The part of the compressor that contains the piston and its travel.

Cilindro del compresor. Pieza del compresor que contiene el pistón y su movimiento.

Cylinder head, compressor. The top to the cylinder on the high-pressure side of the compressor.

Culata del cilindro del compresor. Tapa del cilindro en el lado de alta presión del compresor.

Cylinder, refrigerant. The container that holds refrigerant.

Cilindro del refrigerante. El recipiente que contiene el refrigerante.

Cylinder unloading. A method of providing capacity control by causing a cylinder in a reciprocating compressor to stop pumping.

Descarga del cilindro. Método de suministrar regulación de capacidad provocando que el cilindro en un compresor alternativo deje de bombear.

Damper. A component in an air distribution system that restricts airflow for the purpose of air balance.

Desviador. Componente en un sistema de distribución de aire que limita el flujo de aire para mantener un equilibrio de aire.

Declination angle. The angle of the tilt of the earth on its axis.

Ángulo de declinación. Ángulo de inclinación de la Tierra en su eje.

Defrost. Melting of ice.

Descongelar. Convertir hielo en líquido.

Defrost cycle. The portion of the refrigeration cycle that melts the ice off the evaporator.

Ciclo de descongelación. Parte del ciclo de refrigeración que derrite el hielo del evaporador.

Defrost timer. A timer used to start and stop the defrost cycle.

Temporizador de descongelación. Temporizador utilizado para poner en marcha y detener el ciclo de descongelación.

Degreaser. A cleaning solution used to remove grease from parts and coils.

Desengrasador. Solución limpiadora utilizada para remover la grasa de piezas y bobinas.

Dehumidify. To remove moisture from air.

Deshumidificar. Remover la humedad del aire.

Dehydrate. To remove moisture from a sealed system or a product.

Deshidratar. Remover la humedad de un sistema sellado o un producto.

Density. The weight per unit of volume of a substance.

Densidad. Relación entre el peso de una sustancia y su volumen.

Desiccant. Substance in a refrigeration system drier that collects moisture.

Disecante. Sustancia en el secador de un sistema de refrigeración que acumula la humedad.

Design pressure. The pressure at which the system is designed to operate under normal conditions.

Presión de diseño. Presión a la que el sistema ha sido diseñado para funcionar bajo condiciones normales.

De-superheating. Removing heat from the superheated hot refrigerant gas down to the condensing temperature.

De sobrecalentamiento. Reducir el calor del gas caliente del refrigerante sobrecalentado hasta alcanzar la temperatura de condensación.

Detector. A device to search and find.

Detector. Dispositivo de búsqueda y detección.

Dew. Moisture droplets that form on a cool surface.

Rocío. Gotitas de humedad que se forman en una superficie fría.

Dew point. The exact temperature at which moisture begins to form.

Punto de rocío. Temperatura exacta a la que la humedad comienza a formarse.

DIAC. A semiconductor often used as a voltage-sensitive switching device.

DIAC. Semiconductor utilizado frecuentemente como dispositivo de conmutación sensible a la tensión.

Diaphragm. A thin flexible material (metal, rubber, or plastic) that separates two pressure differences.

Diafragma. Material delgado y flexible, como por ejemplo el metal, el caucho o el plástico, que separa dos presiones diferentes.

Die. A tool used to make an external thread such as on the end of a piece of pipe.

Troquel. Herramienta utilizada para formar un filete externo, como por ejemplo, en el extremo de un tubo.

Differential. The difference in the cut-in and cut-out points of a control, pressure, time, temperature, or level.

Diferencial. Diferencia entre los puntos de conexión y desconexión de un regulador, una presión, un intervalo de tiempo, una temperatura o un nivel.

Diffuser. The terminal or end device in an air distribution system that directs air in a specific direction using louvers.

Placa difusora. Punto o dispositivo terminal en un sistema de distribución de aire que dirige el aire a una dirección específica, utilizando aberturas tipo celosía.

Diode. A solid state device composed of both P-type and N-type material. When connected in a circuit one way, current will flow. When the diode is reversed, current will not flow.

Diodo. Dispositivo de estado sólido compuesto de material P y de material N. Cuando se conecta a un circuito de una manera, la corriente fluye. Cuando la dirección del diodo cambia, la corriente deja de fluir.

Direct current. Electricity in which all electron flow is continuously in one direction.

Corriente continua. Electricidad en la que todos los electrones fluyen continuamente en una sola dirección.

Direct expansion. The term used to describe an evaporator with an expansion device other than a low-side float type.

Expansión directa. Término utilizado para describir un evaporador con un dispositivo de expansión diferente al tipo de dispositivo flotador de lado bajo.

Direct-spark ignition (DSI). A system that provides direct ignition to the main burner.

Encendido de chispa directa. Sistema que le provee un encendido directo al quemador principal.

Discus compressor. A reciprocating compressor distinguished by its disc-type valve system.

Compresor de disco. Compresor alternativo caracterizado por su sistema de válvulas de tipo disco.

Discus valve. A reciprocating compressor valve design with a low clearance volume and larger bore.

Válvula de disco. Válvula de compresor alternativo diseñada con un volumen de holgura bajo y un calibre más grande.

Distributor. A component installed at the outlet of the expansion valve that distributes the refrigerant to each evaporator circuit.

Distribuidor. Componente instalado a la salida de la váluva de expansión que distribuye el refrigerante a cada circuito del evaporador.

Doping. Adding an impurity to a semiconductor to produce a desired charge.

Impurificación. La adición de una impureza para producir una carga deseada.

Double flare. A connection used on copper, aluminum, or steel tubing that folds tubing wall to a double thickness.

Abocinado doble. Conexión utilizada en tuberías de cobre, aluminio, o acero que pliega la pared de la tubería y crea un espesor doble.

Dowel pin. A pin, which may or may not be tapered, used to align and fasten two parts.

Pasador de espiga. Pasador, que puede o no ser cónico, utilizado para alinear y fijar dos piezas.

Draft gage. A gage used to measure very small pressures (above and below atmospheric) and compare them to the atmosphere's pressure. Used to determine the flow of flue gas in a chimney or vent.

Calibrador de tiro. Calibrador utilizado para medir presiones sumamente pequeñas, (mayores o menores que la atmosférica), y compararlas con la presión de la atmósfera. Utilizado para determinar el flujo de gas de combustión en una chimenea o válvula.

Drier. A device used in a refrigerant line to remove moisture.

Secador. Dispositivo utilizado en un conducto de refrigerante para remover la humedad.

Drip pan. A pan shaped to collect moisture condensing on an evaporator coil in an air conditioning or refrigeration system.

Colector de goteo. Un colector formado para acumular la humedad que se condensa en la bobina de un evaporador en un sistema de acondicionamiento de aire o de refrigeración.

Dry-bulb temperature. The temperature measured using a plain thermometer.

Temperatura de bombilla seca. Temperatura que se mide con un termómetro sencillo.

Duct. A sealed channel used to convey air from the system to and from the point of utilization.

Conducto. Canal sellado que se emplea para dirigir el aire del sistema hacia y desde el punto de utilización.

Eccentric. An off-center device that rotates in a circle around a shaft.

Excéntrico. Dispositivo descentrado que gira en un círculo alrededor de un árbol.

Eddy current test. A test with an instrument to find potential failures in evaporator or condenser tubes.

Prueba para la corriente de Foucault. Prueba que se realiza con un instrumento para detectar posibles fallas en los tubos del evaporador o del condensador.

Effective temperature. Different combinations of temperature and humidity that provide the same comfort level.

Temperatura efectiva. Diferentes combinaciones de temperatura y humedad que proveen el mismo nivel de comodidad.

Electric heat. The process of converting electrical energy, using resistance, into heat.

Calor eléctrico. Proceso de convertir energía eléctrica en calor a través de la resistencia.

Electrical power. Electrical power is measured in watts. One watt is equal to one ampere flowing with a potential of one volt. Watts = Volts × Amperes (P = E × I)

Potencia eléctrica. La potencia eléctrica se mide en watios. Un watio equivale a un amperio que fluye con una potencia de un voltio. Watios = voltios × amperios P = E × I)

Electrical shock. When an electrical current travels through a human body.

Sacudida eléctrica. Paso brusco de una corriente eléctrica a través del cuerpo humano.

Electromagnet. A coil of wire wrapped around a soft iron core that creates a magnet.

Electroimán. Bobina de alambre devanado alrededor de un núcleo de hierro blando que crea un imán.

Electron. The smallest portion of an atom that carries a negative charge.

Electrón. Partícula más pequeña de un átomo que tiene carga negativa.

Electronic air filter. A filter that charges dust particles using high-voltage direct current and then collects these particles on a plate of an opposite charge.

Filtro de aire electrónico. Filtro que carga partículas de polvo utilizando una corriente continua de alta tensión y luego las acumula en una placa de carga opuesta.

Electronic charging scale. An electronically operated scale used to accurately charge refrigeration systems by weight.

Escala electrónica para carga. Escala accionada electrónicamente que se utiliza para cargar correctamente sistemas de refrigeración por peso.

Electronic expansion valve (TXV). A metering valve that uses a thermistor as a temperature-sensing element that varies the voltage to a heat motor-operated valve.

Válvula electrónica de expansión. Válvula de medición que utiliza un termistor como elemento sensor de temperatura para variar la tensión a una válvula de calor accionada por motor.

Electronic leak detector. An instrument used to detect gases in very small portions by using electronic sensors and circuits.

Detector electrónico de fugas. Instrumento que se emplea para detectar cantidades de gases sumamente pequeñas utilizando sensores y circuitos electrónicos.

Electronics. The use of electron flow in conductors, semiconductors, and other devices.

Electrónica. La utilización del flujo de electrones en conductores, semiconductores y otros dispositivos.

Emitter. A terminal on a semiconductor.

Emisor. Punto terminal en un semiconductor.

End bell. The end structure of an electric motor that normally contains the bearings and lubrication system.

Extremo acampanado. Estructura terminal de un motor eléctrico que generalmente contiene los cojinetes y el sistema de lubrificación.

End play. The amount of lateral travel in a motor or pump shaft.

Holgadura. Amplitud de movimiento lateral en un motor o en el árbol de una bomba.

Energy. The capacity for doing work.

Energía. Capacidad para realizar un trabajo.

Energy efficiency ratio (EER). An equipment efficiency rating that is determined by dividing the output in Btuh by the input in watts. This does not take into account the startup and shutdown for each cycle.

Relación del rendimiento de engería. Clasificación del rendimiento de un equipo que se determina al dividir la salida en Btuh por la entrada en watios. Esto no toma en cuenta la puesta en marcha y la parada de cada ciclo.

Enthalpy. The amount of heat a substance contains determined from a predetermined base or point.

Entalpía. Cantidad de calor que contiene una sustancia, establecida desde una base o un punto predeterminado.

Environment. Our surroundings, including the atmosphere.

Medio ambiente. Nuestros alrededores, incluyendo la atmósfera.

Ethane gas. The fossil fuel, natural gas, used for heat.

Gas etano. Combustible fósil, gas natural, utilizados para generar calor.

Evacuation. The removal of any gases not characteristic to a system or vessel.

Evacuación. Remoción de los gases no característicos de un sistema o depósito.

Evaporation. The condition that occurs when heat is absorbed by liquid and it changes to vapor.

Evaporación. Condición que ocurre cuando un líquido absorbe calor y se convierte en vapor.

Evaporator. The component in a refrigeration system that absorbs heat into the system and evaporates the liquid refrigerant.

Evaporador. El componente en un sistema de refrigeración que absorbe el calor hacia el sistema y evapora el refrigerante líquido.

Evaporator fan. A forced convector used to improve the efficiency of an evaporator by air movement over the coil.

Abanico del evaporador. Convector forzado que se utiliza para mejorar el rendimiento de un evaporador por medio del movimiento de aire a través de la bobina.

Evaporator pressure regulator (EPR). A mechanical control installed in the suction line at the evaporator outlet that keeps the evaporator pressure from dropping below a certain point.

Regulador de presión del evaporador. Regulador mecánico instalado en el conducto de aspiración de la salida del evaporador; evita que la presión del evaporador caiga hasta alcanzar un nivel por debajo del nivel específico.

Evaporator types. Flooded—an evaporator where the liquid refrigerant level is maintained to the top of the heat exchange coil. Dry type—an evaporator coil that achieves the heat exchange process with a minimum of refrigerant charge.

Clases de evaporadores. Inundado—un evaporador en el que se mantiene el nivel del refrigerante líquido en la parte superior de la bobina de intercambio de calor. Seco—una bobina de evaporador que logra el proceso de intercambio de calor con una mínima cantidad de carga de refrigerante.

Exhaust valve. The movable component in a refrigeration compressor that allows hot gas to flow to the condenser and prevents it from refilling the cylinder on the downstroke.

Válvula de escape. Componente móvil en un compresor de refrigeración que permite el flujo de gas caliente al condensador y evita que este gas rellene el cilindro durante la carrera descendente.

Expansion (metering) device. The component between the high-pressure liquid line and the evaporator that feeds the liquid refrigerant into the evaporator.

Dispositivo de (medición) de expansión. Componente entre el conducto de líquido de alta presión y el evaporador que alimenta el refrigerante líquido hacia el evaporador.

Expansion joint. A flexible portion of a piping system or building structure that allows for expansion of the materials due to temperature changes.

Junta de expansión. Parte flexible de un sistema de tubería o de la estructura de un edificio que permite la expanión de los materiales debido a cambios de temperatura.

External drive. An external type of compressor motor drive, as opposed to a hermetic compressor.

Motor externo. Motor tipo externo de un compresor, en comparación con un compresor hermético.

External equalizer. The connection from the evaporator outlet to the bottom of the diaphragm on a thermostatic expansion valve.

Equilibrador externo. Conexión de la salida del evaporador a la parte inferior del diafragma en una válvula de expansión termostática.

Fahrenheit scale. The temperature scale that places the boiling point of water at 212°F and the freezing point at 32°F.

Escala Fahrenheit. Escala de temperatura en la que el punto de ebullición del agua se encuentra a 212°F y el punto de fusión del hielo a 32°F.

Fan. A device that produces a pressure difference in air to move it.

Abanico. Dispositivo que produce una diferencia de presión en el aire para moverlo.

Fan cycling. The use of a pressure control to turn a condenser fan on and off to maintain a correct pressure within the system.

Funcionamiento cíclico. La utilización de un regulador de presión para poner en marcha y detener el abanico de un condensador a fin de mantener una presión correcta dentro del sistema.

Fan relay coil. A magnetic coil that controls the starting and stopping of a fan.

Bobina de relé del abanico. Bobina magnética que regula la puesta en marcha y la parada de un abanico.

Farad. The unit of capacity of a capacitor. Capacitors in our industry are rated in microfarads.

Faradio. Unidad de capacidad de un capacitador. En nuestro medio, los capacitadores se clasifican en microfaradios.

Female thread. The internal thread in a fitting.

Filete hembra. Filete interno en un accesorio.

Fill or wetted-surface method. Water in a cooling tower is spread out over a wetted surface while air is passed over it to enhance evaporation.

Método de relleno o de superficie mojada. El agua en una torre de refrigeración se extiende sobre una superficie mojada mientras el aire se dirige por encima de la misma para facilitar la evaporación.

Film factor. The relationship between the medium giving up heat and the heat exchange surface (evaporator). This relates to the velocity of the medium passing over the evaporator. When the velocity is too slow, the film between the air and the evaporator becomes greater and becomes an insulator, which slows the heat exchange.

Factor de película. Relación entre el medio que emite calor y la superficie del intercambiador de calor (evaporador). Esto se refiere a la velocidad del medio que pasa sobre el evaporador. Cuando la velocidad es demasiado lenta, la película entre el aire y el evaporador se expande y se convierte en un aislador, disminuyendo así la velocidad del intercambio del calor.

Filter. A fine mesh or porous material that removes particles from passing fluids.

Filtro. Malla fina o material poroso que remueve partículas de los fluidos que pasan por él.

Fin comb. A hand tool used to straighten the fins on an air-cooled condenser.

Herramienta para aletas. Herramienta manual utilizada para enderezar las aletas en un condensador enfriado por aire.

Fixed resistor. A nonadjustable resistor. The resistance cannot be changed.

Resistor fijo. Resistor no ajustable. La resistencia no se puede cambiar.

Fixed-bore device. An expansion device with a fixed diameter that does not adjust to varying load conditions.

Dispositivo de calibre fijo. Dispositivo de expansión con un diámetro fijo que no se ajusta a las condiciones de carga variables.

Flapper valve. See reed valve.

Chapaleta. Véase válvula de lámina.

Flare. The angle that may be fashioned at the end of a piece of tubing to match a fitting and create a leak-free connection.

Abocinado. Ángulo que puede formarse en el extremo de una pieza de tubería para emparejar un accesorio y crear una conexión libre de fugas.

Flare nut. A connector used in a flare assembly for tubing.

Tuerca abocinada. Conector utilizado en un conjunto abocinado para tuberías.

Flash gas. A term used to describe the pressure drop in an expansion device when some of the liquid passing through the valve is changed quickly to a gas and cools the remaining liquid to the corresponding temperature.

Gas instantáneo. Término utilizado para describir la caída de la presión en un dispositivo de expansión cuando una parte del líquido que pasa a través de la válvula se convierte rápidamente en gas y enfria el líquido restante a la temperatura correspondiente.

Float, valve or switch. An assembly used to maintain or monitor a liquid level.

Válvula o conmutador de flotador. Conjunto utilizado para mantener o controlar el nivel de un líquido.

Flooded system. A refrigeration system operated with the liquid refrigerant level very close to the outlet of the evaporator coil for improved heat exchange.

Sistema inundado. Sistema de refrigeración que funciona con el nivel del refrigerante líquido bastante próximo a la salida de la bobina del evaporador para mejorar el intercambio de calor.

Flooding. The term applied to a refrigeration system when the liquid refrigerant reaches the compressor.

Inundación. Término aplicado a un sistema de refrigeración cuando el nivel del refrigerante líquido llega al compresor.

Flue. The duct that carries the products of combustion out of a structure for a fossil- or a solid-fuel system.

Conducto de humo. Conducto que extrae los productos de combustión de una estructura en sistemas de combustible fósil o sólido.

Flue-gas analysis instruments. Instruments used to analyze the operation of fossil—fuel-burning equipment such as oil and gas furnaces by analyzing the flue gases.

Instrumentos para el análisis del gas de combustión. Instrumentos utilizados para llevar a cabo un análisis del funcionamiento de los quemadores de combustible fósil, como por ejemplo hornos de aceite pesado o gas, a través del estudio de los gases de combustión.

Fluid. The state of matter of liquids and gases.

Fluido. Estado de la materia de líquidos y gases.

Fluid expansion device. Using a bulb or sensor, tube, and diaphragm filled with fluid, this device will produce movement at the diaphragm when the fluid is heated or cooled. A bellows may be added to produce more movement. These devices may contain vapor and liquid.

Dispositivo para la expansión del fluido. Utilizando una bombilla o sensor, un tubo y un diafragma lleno de fluido, este dispositivo generará movimiento en el diafragma cuando se caliente o enfríe el fluido. Se le puede agregar un fuelle para generar aún más movimiento. Dichos dispositivos pueden contener vapor y líquido.

Flush. The process of using a fluid to push contaminants from a system.

Descarga. Proceso de utilizar un fluido para remover los contaminantes de un sistema.

Flux. A substance applied to soldered and brazed connections to prevent oxidation during the heating process.

Fundente. Sustancia aplicada a conexiones soldadas y broncesoldadas para evitar la oxidación durante el proceso de calentamiento.

Foaming. A term used to describe oil when it has liquid refrigerant boiling out of it.

Espumación. Término utilizado para describir el aceite cuando el refrigerante líquido se derrama del mismo.

Foot-pound. The amount of work accomplished by lifting 1 lb of weight 1 ft; a unit of energy.

Libra-pie. Medida de la cantidad de energía o fuerza que se requiere para levantar una libra a una distancia de un pie; unidad de energía.

Force. Energy exerted.

Fuerza. Energía ejercida sobre un objeto.

Forced convection. The movement of fluid by mechanical means.

Convección forzada. Movimiento de fluido por medios mecánicos.

Fossil fuels. Natural gas, oil, and coal formed millions of years ago from dead plants and animals.

Combustibles fósiles. El gas natural, el petroleo y el carbón que se formaron hace millones de años de plantas y animales muertos.

Four-way valve. The valve in a heat pump system that changes the direction of the refrigerant flow between the heating and cooling cycles.

Válvula con cuatro vías. Válvula en un sistema de bomba de calor que cambia la dirección del flujo de refrigerante entre los ciclos de calentamiento y enfriamiento.

Freezer burn. The term applied to frozen food when it becomes dry and hard from dehydration due to poor packaging.

Quemadura del congelador. Término aplicado a la comida congelada cuando se seca y endurece debido a la deshidratación ocacionada por el empaque de calidad inferior.

Freeze up. Excess ice or frost accumulation on an evaporator to the point that airflow may be affected.

Congelación. Acumulación excesiva de hielo o congelación en un evaporador a tal extremo que el flujo de aire puede ser afectado.

Freezing. The change of state of water from a liquid to a solid.

Congelamiento. Cambio de estado del agua de líquido a sólido.

Freon. The trade name for refrigerants manufactured by E. I. du Pont de Nemours & Co., Inc.

Freón. Marca registrada para refrigerantes fabricados por la compañía E. I. du Pont de Nemours, S.A.

Frequency. The cycles per second (cps) of the electrical current supplied by the power company. This is normally 60 cps in the United States.

Frecuencia. Ciclos por segundo (cps), generalmente 60 cps en los Estados Unidos, de la corriente eléctrica suministrada por la empresa de fuerza motriz.

Front seated. A position on a valve that will not allow refrigerant flow in one direction.

Sentado delante. Posición en una válvula que no permite el flujo de refrigerante en una dirección.

Frost back. A condition of frost on the suction line and even the compressor body usually due to liquid refrigerant in the suction line.

Obturación por congelación. Condición de congelación que ocurre en el conducto de aspiración e inclusive en el cuerpo del compresor, normalmente debido a la presencia de refrigerante líquido en el conducto de aspiración.

Frostbite. When skin freezes.

Quemadura por frío. Congelación de la piel.

Frozen. The term used to describe water in the solid state; also used to describe a rotating shaft that will not turn.

Congelado. Término utilizado para describir el agua en un estado sólido; utilizado también para describir un árbol giratorio que no gira.

Fuel oil. The fossil fuel used for heating; a petroleum distillate.

Aceite pesado. Combustible fósil utilizado para calentar; un destilado de petróleo.

Full-load amperage (FLA). The current an electric motor draws while operating under a full-load condition. Also called the run-load amperage.

Amperaje de carga total. Corriente que un motor eléctrico consume mientras funciona en una condición de carga completa. Conocido también como amperaje de carga de funcionamiento.

Furnace. Equipment used to convert heating energy, such as fuel oil, gas, or electricity, to usable heat. It usually contains a heat exchanger, a blower, and the controls to operate the system.

Horno. Equipo utilizado para la conversión de energía calórica, como por ejemplo el aceite pesado, el gas o la electricidad, en calor utilizable. Normalmente contiene un intercambiador de calor, un soplador y los reguladores para accionar el sistema.

Fuse. A safety device used in electrical circuits for the protection of the circuit conductor and components.

Fusible. Dispositivo de segurdidad utilizado en circuitos eléctricos para la protección del conductor y de los componentes del circuito.

Fusible link. An electrical safety device normally located in a furnace that burns and opens the circuit during an overheat situation.

Cartucho de fusible. Dispositivo eléctrico de seguridad ubicado por lo general en un horno, que quema y abre el circuito en caso de sobrecalentamiento.

Fusible plug. A device (made of low-melting temperature metal) used in pressure vessels that is sensitive to low temperatures and relieves the vessel contents in an overheating situation.

Tapón de fusible. Dispositivo utilizado en depósitos en presión, hecho de un metal que tiene una temperatura de fusión baja. Este dispositivo es sensible a temperaturas bajas y alivia el contenido del depósito en caso de sobrecalentamiento.

Gage. An instrument used to detect pressure.

Calibrador. Instrumento utilizado para detectar presión.

Gage manifold. A tool that may have more than one gage with a valve arrangement to control fluid flow.

Distribuidor de calibrador. Herramienta que puede tener más de un calibrador con las válvulas arregladas a fin de regular el flujo de fluido.

Gage port. The service port used to attach a gage for service procedures.

Orificio de calibrador. Orificio de servicio utilizado con el propósito de fijar un calibrador para procedimientos de servicio.

Gas. The vapor state of matter.

Gas. Estado de vapor de una materia.

Gas-pressure switch. Used to detect gas pressure before gas burners are allowed to ignite.

Conmutador de presión del gas. Utilizado para detectar la presión del gas antes de que los quemadores de gas puedan encenderse.

Gas valve. A value used to stop, start, or modulate the flow of natural gas.

Válvula de gas. Válvula utilizada para detener, poner en marcha o modular el flujo de gas natural.

Gasket. A thin piece of flexible material used between two metal plates to prevent leakage.

Guarnición. Pieza delgada de material flexible utilizada entre dos placas de metal para evitar fugas.

Gate. A terminal on a semiconductor.

Compuerta. Punto terminal en un semiconductor.

Germanium. A substance from which many semiconductors are made.

Germanio. Sustancia de la que se fabrican muchos semiconductores.

Glow coil. A device that automatically reignites a pilot light if it goes out.

Bobina encendedora. Dispostivo que automáticamente vuelve a encender la llama piloto si ésta se apaga.

Graduated cylinder. A cylinder with a visible column of liquid refrigerant used to measure the refrigerant charged into a system. Refrigerant temperatures can be dialed on the graduated cylinder.

Cilindro graduado. Cilindro con una columna visible de refrigerante líquido utilizado para medir el refrigerante inyectado al sistema. Las temperaturas del refrigerante pueden marcarse en el cilindro graduado.

Grain. Unit of measure. One pound = 7000 grains.

Grano. Unidad de medida. Una libra equivale a 7000 granos.

Gram. Metric measurement term used to express weight.

Gramo. Término utilizado para referirse a la unidad básica de peso en el sistema métrico.

Grille. A louvered, often decorative, component in an air system at the inlet or the outlet of the airflow.

Rejilla. Componente con celosías, comúnmente decorativo, en un sistema de aire que se encuentra a la entrada o a la salida del flujo de aire.

Grommet. A rubber, plastic, or metal protector usually used where wire or pipe goes through a metal panel.

Guardaojal. Protector de caucho, plástico o metal normalmente utilizado donde un alambre o un tubo pasa a través de una base de metal.

Ground, electrical. A circuit or path for electron flow to the earth ground.

Tierra eléctrica. Circuito o trayectoria para el flujo de electrones a la puesta a tierra.

Ground wire. A wire from the frame of an electrical device to be wired to the earth ground.

Alambre a tierra. Alambre que va desde el armazón de un dispositivo eléctrico para ser conectado a la puesta a tierra.

Guide vanes. Vanes used to produce capacity control in a centrifugal compressor. Also called prerotation guide vanes.

Paletas directrices. Paletas utilizadas para producir la regulación de capacidad en un compresor centrífugo. Conocidas también como paletas directrices para prerotación.

Halide refrigerants. Refrigerants that contain halogen chemicals; R-12, R-22, R-500, and R-502 are among them.

Refrigerantes de hálido. Refrigerantes que contienen productos químicos de halógeno; entre ellos se encuentran el R-12, R-22, R-500 y R-502.

Halide torch. A torch-type leak detector used to detect the halogen refrigerants.

Soplete de hálido. Detector de fugas de tipo soplete utilizado para detectar los refrigerantes de halógeno.

Halogens. Chemical substances found in many refrigerants containing chlorine, bromine, iodine, and fluorine.

Halógenos. Sustancias químicas presentes en muchos refrigerantes que contienen cloro, bromo, yodo y flúor.

Hand truck. A two-wheeled piece of equipment that can be used for moving heavy objects.

Vagoneta para mano. Equipo con dos ruedas que puede utilizarse para transportar objetos pesados.

Hanger. A device used to support tubing, pipe, duct, or other components of a system.

Soporte. Dispositivo utilizado para apoyar tuberías, tubos, conductos u otros componentes de un sistema.

Head. Another term for pressure, usually referring to gas or liquid.

Carga. Otro término para presión, refiriéndose normalmente a gas o líquido.

Head pressure control. A control that regulates the head pressure in a refrigeration or air conditioning system.

Regulador de la presión de la carga. Regulador que controla la presión de la carga en un sistema de refrigeración o de acondicionamiento de aire.

Header. A pipe or containment to which other pipe lines are connected.

Conductor principal. Tubo o conducto al que se conectan otras conexiones.

Heat. Energy that causes molecules to be in motion and to raise the temperature of a substance.

Calor. Energía que ocasiona el movimiento de las moléculas provocando un aumento de temperatura en una sustancia.

Heat anticipator. A device that anticipates the need for cutting off the heating system prematurely so the fan can cool the furnace.

Anticipador de calor. Dispositivo que anticipa la necesidad de detener la marcha del sistema de calentamiento para que el abanico pueda enfriar el horno.

Heat coil. A device made of tubing or pipe designed to transfer heat to a cooler substance by using fluids.

Bobina de calor. Dispositivo hecho de tubos, diseñado para transmitir calor a una sustancia más fría por medio de fluidos.

Heat exchanger. A device that transfers heat from one substance to another.

Intercambiador de calor. Dispositivo que transmite calor de una sustancia a otra.

Heat of compression. That part of the energy from the pressurization of a gas or a liquid converted to heat.

Calor de compresión. La parte de la energía generada de la presurización de un gas o un líquido que se ha convertido en calor.

Heat of fusion. The heat released when a substance is changing from a liquid to a solid.

Calor de fusión. Calor liberado cuando una sustancia se convierte de líquido a sólido.

Heat of respiration. When oxygen and carbon hydrates are taken in by a substance or when carbon dioxide and water are given off. Associated with fresh fruits and vegetables during their aging process while stored.

Calor de respiración. Cuando se admiten oxígeno e hidratos de carbono en una sustancia o cuando se emiten bióxido de carbono y agua. Se asocia con el proceso de maduración de frutas y legumbres frescas durante su almacenamiento.

Heat pump. A refrigeration system used to supply heat or cooling using valves to reverse the refrigerant gas flow.

Bomba de calor. Sistema de refrigeración utilizado para suministrar calor o frío mediante válvulas que cambian la dirección del flujo de gas del refrigerante.

Heat reclaim. Using heat from a condenser for purposes such as space and domestic water heating.

Reclamación de calor. La utilización del calor de un condensador para propósitos tales como la calefacción de espacio y el calentamiento doméstico de agua.

Heat sink. A low-temperature surface to which heat can transfer.

Fuente fría. Superficie de temperatura baja a la que puede transmitírsele calor.

Heat transfer. The transfer of heat from a warmer to a colder substance.

Transmisión de calor. Cuando se transmite calor de una sustancia más caliente a una más fría.

Helix coil. A bimetal formed into a helix-shaped coil that provides longer travel when heated.

Bobina en forma de hélice. Bimetal encofrado en una bobina en forma de hélice que provee mayor movimiento al ser calentado.

Hermetic system. A totally enclosed refrigeration system where the motor and compressor are sealed within the same system with the refrigerant.

Sistema hermético. Sistema de refrigeración completamente cerrado donde el motor y el compresor se obturan dentro del mismo sistema con el refrigerante.

Hertz. Cycles per second.

Hertz. Ciclos por segundo.

Hg. Abbreviation for the element mercury.

Hg. Abreviatura del elemento mercurio.

High-pressure control. A control that stops a boiler heating device or a compressor when the pressure becomes too high.

Regulador de alta presión. Regulador que detiene la marcha del dispositivo de calentamiento de una caldera o de un compresor cuando la presión alcanza un nivel demasiado alto.

High side. A term used to indicate the high-pressure or condensing side of the refrigeration system.

Lado de alta presión. Término utilizado para indicar el lado de alta presión o de condensación del sistema de refrigeración.

High-temperature refrigeration. A refrigeration temperature range starting with evaporator temperatures no lower than 35°F, a range usually used in air conditioning (cooling).

Refrigeración a temperatura alta. Margen de la temperatura de refrigeración que comienza con temperaturas de evaporadores no menores de 35°F (2°C). Este margen se utiliza normalmente en el acondicionamiento de aire (enfriamiento).

High-vacuum pump. A pump that can produce a vacuum in the low micron range.

Bomba de vacío alto. Bomba que puede generar un vacío dentro del margen de micrón bajo.

Horsepower. A unit equal to 33,000 ft-lb of work per minute.

Potencia en caballos. Unidad equivalente a 33.000 libras-pies de trabajo por minuto.

Hot gas. The refrigerant vapor as it leaves the compressor. This is often used to defrost evaporators.

Gas caliente. El vapor del refrigerante al salir del compresor. Esto se utiliza con frecuencia para descongelar evaporadores.

Hot gas bypass. Piping that allows hot refrigerant gas into the cooler low-pressure side of a refrigeration system usually for system capacity control.

Desviación de gas caliente. Tubería que permite la entrada de gas caliente del refrigerante en el lado más frío de baja presión de un sistema de refrigeración, normalmente para la regulación de la capacidad del sistema.

Hot gas defrost. A system where the hot refrigerant gases are passed through the evaporator to defrost it.

Descongelación con gas caliente. Sistema en el que los gases calientes del refrigerante se pasan a través del evaporador para descongelarlo.

Hot gas line. The tubing between the compressor and condenser.

Conducto de gas caliente. Tubería entre el compresor y el condensador.

Hot junction. That part of a thermocouple or thermopile where heat is applied.

Empalme caliente. El lugar en un termopar o pila termoeléctrica donde se aplica el calor.

Hot pull down. The process of lowering the refrigerated space to the design temperature after it has been allowed to warm up considerably over this temperature.

Descenso caliente. Proceso de bajar la temperatura del espacio refrigerado a la temperatura de diseño luego de habérsele permitido calentarse a un punto sumamente superior a esta temperatura.

Hot water heat. A heating system using hot water to distribute the heat.

Calor de agua caliente. Sistema de calefacción que utiliza agua caliente para la distribución del calor.

Hot wire. The wire in an electrical circuit that has a voltage potential between it and another electrical source or between it and ground.

Conductor electrizado. Conductor en un circuito eléctrico a través del cual fluye la tensión entre éste y otra fuente de electricidad o entre éste y la tierra.

Humidifier. A device used to add moisture to the air.

Humedecedor. Dispositivo utilizado para agregarle humedad al aire.

Humidistat. A control operated by a change in humidity.

Humidistato. Regulador activado por un cambio en la humedad.

Humidity. Moisture in the air.

Humedad. Vapor de agua existente en el ambiente.

Hydraulics. Producing mechanical motion by using liquids under pressure.

Hidráulico. Generación de movimiento mecánico por medio de líquidos bajo presión.

Hydrocarbons. Organic compounds containing hydrogen and carbon found in many heating fuels.

Hidrocarburos. Compuestos orgánicos que contienen el hidrógeno y el carbón presentes en muchos combustibles de calentamiento.

Hydrochlorofluorocarbons (HCFC). Refrigerants thought to contribute to the depletion of the ozone layer although not to the extent of chlorofluorocarbons.

Hidrocloroflurocarburos. Líquidos refrigerantes que, según algunos, han contribuido a la reducción de la capa de ozono aunque no en tal grado como los cloroflurocarburos.

Hydrometer. An instrument used to measure the specific gravity of a liquid.

Hidrómetro Instrumento utilizado para medir la gravedad específica de un líquido.

Hydronic. Usually refers to a hot water heating system.

Hidrónico. Normalmente se refiere a un sistema de calefacción de agua caliente.

Hygrometer. An instrument used to measure the amount of moisture in the air.

Higrómetro. Instrumento utilizado para medir la cantidad de humedad en el aire.

Idler. A pulley on which a belt rides. It does not transfer power but is used to provide tension or reduce vibration.

Polea tensora. Polea sobre la que se mueve una correa. No sirve para transmitir potencia, pero se utiliza para proveer tensión o disminuir la vibración.

Ignition transformer. Provides a high-voltage current, usually to produce a spark to ignite a furnace fuel, either gas or oil.

Transformador para encendido. Provee una corriente de alta tensión, normalmente para generar una chispa a fin de encender el combustible de un horno, sea gas o aceite pesado.

Impedance. A form of resistance in an alternating current circuit.

Impedancia. Forma de resistencia en un circuito de corriente alterna.

Impeller. The rotating part of a pump that causes the centrifugal force to develop fluid flow and pressure difference.

Impulsor. Pieza giratoria de una bomba que hace que la fuerza centrífuga desarrolle flujo de fluido y una diferencia en presión.

Impingement. The condition in a gas or oil furnace when the flame strikes the sides of the combustion chamber, resulting in poor combustion efficiency.

Golpeo. Condición que ocurre en un horno de gas o de aceite pesado cuando la llama golpea los lados de la cámara de combustión. Esta condición trae como resultado un rendimiento de combustión pobre.

Inclined water manometer. Indicates air pressures in very low-pressure systems.

Manómetro de agua inclinada. Señala las presiones de aire en sistemas de muy baja presión.

Induced magnetism. Magnetism produced, usually in a metal, from another magnetic field.

Magnetismo inducido. Magnetismo generado, normalmente en un metal, desde otro campo magnético.

Inductance. An induced voltage producing a resistance in an alternating current circuit.

Inductancia. Tensión inducida que genera una resistencia en un circuito de corriente alterna.

Induction motor. An alternating current motor where the rotor turns from induced magnetism from the field windings.

Motor inductor. Motor de corriente alterna donde el rotor gira debido al magnetismo inducido desde los devanados inductores.

Inductive reactance. A resistance to the flow of an alternating current produced by an electromagnetic induction.

Reactancia inductiva. Resistencia al flujo de una corriente alterna generada por una inducción electromagnética.

Metering device. A valve or small fixed-size tubing or orifice that meters liquid refrigerant into the evaporator.

Dispositivo de medida. Válvula o tubería pequeña u orificio que mide la cantidad de refrigerante líquido que entra en el evaporador.

Methane. Natural gas is composed of 90% to 95% methane, a combustible hydrocarbon.

Metano. El gas natural se compone de un 90% a un 95% de metano, un hidrocarburo combustible.

Metric system. System International (SI)—system of measurement used by most countries in the world.

Sistema métrico. Sistema internacional; el sistema de medida utilizado por la mayoría de los países del mundo.

Micro. A prefix meaning 1/1,000,000.

Micro. Prefijo que significa una parte de un millón.

Microfarad. Capacitor capacity equal to 1/1,000,000 of a farad.

Microfaradio. Capacidad de un capacitador equivalente a 1/1.000.000 de un faradio.

Micrometer. A precision measuring instrument.

Micrómetro Instrumento de precisión utilizado para medir.

Micron. A unit of length equal to 1/1000 of a millimeter, 1/1,000,000 of a meter.

Micrón. Unidad de largo equivalente a 1/1000 de un milímetro, o 1/1.000.000 de un metro.

Micron gage. A gage used when it is necessary to measure pressure close to a perfect vacuum.

Calibrador de micrón. Calibrador utilizado cuando es necesario medir la presión de un vacío casi perfecto.

Midseated (cracked). A position on a valve that allows refrigerant flow in all directions.

Sentado en el medio (agrietado). Posición en una válvula que permite el flujo de refrigerante en cualquier dirección.

Milli. A prefix meaning 1/1000.

Mili. Prefijo que significa una parte de mil.

Modulator. A device that adjusts by small increments or changes.

Modulador. Dispositivo que se ajusta por medio de incrementos o cambios pequeños.

Moisture indicator. A device for determining moisture in a refrigerant.

Indicador de humedad. Dispositivo utilizado para determinar la humedad en un refrigerante.

Molecule. The smallest particle that a substance can be broken into and still retain its chemical identity.

Molécula. La partícula más pequeña en la que una sustancia puede dividirse y aún conservar sus propias características.

Molecular motion. The movement of molecules within a substance.

Movimiento molecular. Movimiento de moléculas dentro de una sustancia.

Monochlorodifluoromethane. The refrigerant R-22.

Monoclorodifluorometano. El refrigerante R-22.

Motor service factor. A factor above an electric motor's normal operating design parameters, indicated on the nameplate, under which it can operate.

Factor de servicio del motor. Factor superior a los parámetros de diseño normales de funcionamiento de un motor eléctrico, indicados en el marbete; este factor indica su nivel de funcionamiento.

Motor starter. Electromagnetic contactors that contain motor protection and are used for switching electric motors on and off.

Arrancador de motor. Contactadores electromagnéticos que contienen protección para el motor y se utilizan para arrancar y detener motores eléctricos.

Muffler compressor. Sound absorber at the compressor.

Silenciador del compresor. Absorbedor de sonido ubicado en el compresor.

Mullion. Stationary frame between two doors.

Parteluz. Armazón fijo entre dos puertas.

Mullion heater. Heating element mounted in mullion of a refrigerator to keep moisture from forming on it.

Calentador del parteluz. Elemento de calentamiento montado en el parteluz de un refrigerador para evitar la formación de humedad en el mismo.

Multimeter. An instrument that will measure voltage, resistance, and milliamperes.

Multímetro. Instrumento que mide la tensión, la resistencia y los miliamperios.

Multiple evacuation. A procedure for removing the refrigerant from a system. A vacuum is pulled, a small amount of refrigerant allowed into the system, and the procedure duplicated. This is often done three times.

Evacuación múltiple. Procedimiento para remover el refrigerante de un sistema. Se crea un vacío, se permite la entrada de una pequeña cantidad de refrigerante al sistema, y se repite el procedimiento. Con frecuencia esto se lleva a cabo tres veces.

National electrical code (NEC). A publication that sets the standards for all electrical installations, including motor overload protection.

Código estadounidense de electricidad. Publicación que establece las normas para todas las instalaciones eléctricas, incluyendo la protección contra la sobrecarga de un motor.

National pipe taper (NPT). The standard designation for a standard tapered pipe thread.

Cono estadounidense para tubos. Designación estándar para una rosca cónica para tubos estándar.

Natural convection. The natural movement of a gas or fluid caused by differences in temperature.

Convección natural. Movimiento natural de un gas o fluido ocacionado por diferencias en temperatura.

Natural gas. A fossil fuel formed over millions of years from dead vegetation and animals that were deposited or washed deep into the earth.

Gas natural. Combustible fósil formado a través de millones de años de la vegetación y los animales muertos que fueron depositados o arrastrados a una gran profundidad dentro la tierra.

Needlepoint valve. A device having a needle and a very small orifice for controlling the flow of a fluid.

Válvula de aguja. Dispositivo que tiene una aguja y un orificio bastante pequeño para regular el flujo de un fluido.

Negative electrical charge. An atom or component that has an excess of electrons.

Carga eléctrica negativa. Átomo o componente que tiene un exceso de electrones.

Neoprene. Synthetic flexible material used for gaskets and seals.

Neopreno. Material sintético flexible utilizado en guarniciones y juntas de estanqueidad.

Net oil pressure. Difference in the suction pressure and the compressor oil pump outlet pressure.

Presión neta del aceite. Diferencia en la presión de aspiración y la presión a la salida de la bomba de aceite del compresor.

Neutralizer. A substance used to counteract acids.

Neutralizador. Sustancia utilizada para contrarrestar ácidos.

Newton/meter^2. Metric unit of measurement for pressure. Also called a pascal.

Metro-Newton^2. Unidad métrica de medida de presión. Conocido también como pascal.

Nichrome. A metal made of nickle chromium that when formed into a wire is used as a resistance heating element in electric heaters and furnaces.

Níquel-cromio. Metal fabricado de níquel-cromio que al ser convertido en alambre, se utiliza como un elemento de calentamiento de resistencia en calentadores y hornos eléctricos.

Nitrogen. An inert gas often used to "sweep" a refrigeration system to help ensure that all refrigerant and contaminants have been removed.

Nitrógeno. Gas inerte utilizado con frecuencia para purgar un sistema de refrigeración. Esta gas ayuda a asegurar la remoción de todo el refrigerante y los contaminantes del sistema.

Nominal. A rounded-off stated size. The nominal size is the closest rounded-off size.

Nominal. Tamaño redondeado establecido. El tamaño nominal es el tamaño redondeado más cercano.

Noncondensable gas. A gas that does not change into a liquid under normal operating conditions.

Gas no condensable. Gas que no se convierte en líquido bajo condiciones de funcionamiento normales.

Nonferrous. Metals containing no iron.

No férreos. Metales que no contienen hierro.

North pole, magnetic. One end of a magnet.

Polo norte magnético. El extremo de un imán.

Nut driver. These tools have a socket head used primarily to drive hex head screws on air conditioning, heating, and refrigeration cabinets.

Extractor de tuercas. Estas herramientas tienen una cabeza hueca utilizada principalmente para darles vueltas a tornillos de cabeza hexagonal en gabinetes de acondicionamiento de aire, de calefacción y de refrigeración.

Off cycle. A period when a system is not operating.

Ciclo de apagado. Período de tiempo cuando un sistema no está en funcionamiento.

Ohm. A unit of measurement of electrical resistance.

Ohmio. Unidad de medida de la resistencia eléctrica.

Ohmmeter. A meter that measures electrical resistance.

Ohmiómetro.Instrumento que mide la resistencia eléctrica.

Ohm's law. A law involving electrical relationships discovered by Georg Ohm: E = I × R.

Ley de ohm. Ley que define las relaciones eléctricas, descubierta por Georg Ohm: E = I × R.

Oil-pressure safety control (switch). A control used to ensure that a compressor has adequate oil lubricating pressure.

Regulador de seguridad para la presión de aceite (conmutador). Regulador utilizado para asegurar que un compresor tenga la presión de lubrificación de aceite adecuada.

Oil, refrigeration. Oil used in refrigeration systems.

Aceite de refrigeración. Aceite utilizado en sistemas de refrigeración.

Oil separator. Apparatus that removes oil from a gaseous refrigerant.

Separador de aceite. Aparato que remueve el aceite de un refrigerante gaseoso.

Open compressor. A compressor with an external drive.

Compresor abierto. Compresor con un motor externo.

Operating pressure. The actual pressure under operating conditions.

Presión de funcionamiento. La presión real bajo las condiciones de funcionamiento.

Organic. Materials formed from living organisms.

Orgánico. Materiales formados de organismos vivos.

Orifice. A small opening through which fluid flows.

Orificio. Pequeña abertura a través de la cual fluye un fluido.

Overload protection. A system or device that will shut down a system if an overcurrent condition exists.

Protección contra sobrecarga. Sistema o dispositivo que detendrá la marcha de un sistema si existe una condición de sobreintensidad.

Oxidation. The combining of a material with oxygen to form a different substance. This results in the deterioration of the original substance.

Oxidación. La combinación de un material con oxígeno para formar una sustancia diferente, lo que ocasiona el deterioro de la sustancia original.

Ozone. A form of oxygen (O_3). A layer of ozone in the stratosphere that protects the earth from certain of the sun's ultraviolet wave lengths.

Ozono. Forma de oxígeno (O_3). La capa de ozono en la estratósfera que protege la tierra de ciertos rayos ultravioletas del sol.

Package unit. A refrigerating system where all major components are located in one cabinet.

Unidad completa. Sistema de refrigeración donde todos los componentes principales se encuentran en un solo gabinete.

Packing. A soft material that can be shaped and compressed to provide a seal. It is commonly applied around valve stems.

Empaquetadura. Material blando que puede formarse y comprimirse para proveer una junta de estanqueidad. Comúnmente se aplica alrededor de los vástagos de válvulas.

Parallel circuit. An electrical or fluid circuit where the current or fluid takes more than one path at a junction.

Circuito paralelo. Corriente eléctrica o fluida donde la corriente o el fluido siguen más de una trayectoria en un empalme.

Pascal. A metric unit of measurement of pressure.

Pascal. Unidad métrica de medida de presión.

Passive solar design. The use of nonmoving parts of a building to provide heat or cooling, or to eliminate certain parts of a building that cause inefficient heating or cooling.

Diseño solar pasivo. La utilización de piezas fijas de un edificio para proveer calefacción o enfriamiento, o para eliminar ciertas piezas de un edificio que causan calefacción o enfriamiento ineficientes.

PE (polyethylene). Plastic pipe used for water, gas, and irrigation systems.

Polietileno. Tubo plástico utilizado en sistemas de agua, de gas y de irrigación.

Permanent magnet. An object that has its own permanent magnetic field.

Imán permanente. Objeto que tiene su propio campo magnético permanente.

Permanent split-capacitor motor (PSC). A split-phase motor with a run capacitor only. It has a very low starting torque.

Motor permanente de capacitador separado. Motor de fase separada que sólo tiene un capacitador de funcionamiento. Su par de arranque es sumamente bajo.

Phase. One distinct part of a cycle.

Fase. Una parte específica de un ciclo.

Pilot light. The flame that ignites the main burner on a gas furnace.

Llama piloto. Llama que enciende el quemador principal en un horno de gas.

Piston. The part that moves up and down in a cylinder.

Pistón. La pieza que asciende y desciende dentro de un cilindro.

Piston displacement. The volume within the cylinder that is displaced with the movement of the piston from top to bottom.

Desplazamiento del pistón. Volumen dentro del cilindro que se desplaza de arriba a abajo con el movimiento del pistón.

Pitot tube. Part of an instrument for measuring air velocities.

Tubo Pitot. Pieza de un instrumento para medir velocidades de aire.

Planned defrost. Shutting the compressor off with a timer so that the space temperature can provide the defrost.

Descongelación proyectada. Detención de la marcha de un compresor con un temporizador para que la temperatura del espacio lleve a cabo la descongelación.

Plenum. A sealed chamber at the inlet or outlet of an air handler. The duct attaches to the plenum.

Plenum. Cámara sellada a la entrada o a la salida de un tratante de aire. El conducto se fija al plenum.

Polycyclic organic matter. By-products of wood combustion found in smoke and considered to be health hazards.

Materia orgánica policíclica. Subproductos de la combustión de madera presentes en el humo y considerados nocivos para la salud.

Polyphase. Three or more phases.

Polifase. Tres o más fases.

Porcelain. A ceramic material.

Porcelana. Material cerámico.

Portable dolly. A small platform with four wheels on which heavy objects can be placed and moved.

Carretilla portátil. Plataforma pequeña con cuatro ruedas sobre la que pueden colocarse y transportarse objetos pesados.

Positive displacement. A term used with a pumping device such as a compressor that is designed to move all matter from a volume such as a cylinder or it will stall, possibly causing failure of a part.

Desplazamiento positivo. Término utilizado con un dispositivo de bombeo, como por ejemplo un compresor, diseñado para mover toda la materia de un volumen, como un cilindro o se bloqueará, posiblemente causándole fallas a una pieza.

Positive electrical charge. An atom or component that has a shortage of electrons.

Carga eléctrica positiva. Átomo o componente que tiene una insuficiencia de electrones.

Positive temperature coefficient start device. A thermistor used to provide start assistance to a permanent split-capacitor motor.

Dispositivo de arranque de coeficiente de temperatura positiva. Termistor utilizado para ayudar a arrancar un motor permanente de capacitador separado.

Potential relay. A switching device used with hermetic motors that breaks the circuit to the start windings after the motor has reached approximately 75% of its running speed.

Relé de potencial. Dispositivo de conmutación utilizado con motores herméticos que interrupe el circuito de los devandos de arranque antes de que el motor haya alcanzado aproximadamente un 75% de su velocidad de marcha.

Potentiometer. An instrument that controls electrical current.

Potenciómetro. Instrumento que regula corriente eléctrica.

Power. The rate at which work is done.

Potencia. Velocidad a la que se realiza un trabajo.

Pressure. Force per unit of area.

Presión. Fuerza por unidad de área.

Pressure drop. The difference in pressure between two points.

Caída de presión. Diferencia en presión entre dos puntos.

Pressure/enthalpy diagram. A chart indicating the pressure and heat content of a refrigerant and the extent to which the refrigerant is a liquid and vapor.

Diagrama de presión y entalpía. Esquema que indica la presión y el contenido de calor de un refrigerante y el punto en que el refrigerante es líquido y vapor.

Pressure limiter. A device that opens when a certain pressure is reached.

Dispositivo limitador de presión. Dispositivo que se abre cuando se alcanza una presión específica.

Pressure-limiting TXV. A valve designed to allow the evaporator to build only to a predetermined temperature when the valve will shut off the flow of refrigerant.

Válvula electrónica de expansión limitadora de presión. Válvula diseñada para permitir que la temperatura del evaporador alcance un límite predeterminado cuando la válvula detenga el flujo de refrigerante.

Pressure regulator. A valve capable of maintaining a constant outlet pressure when a variable inlet pressure occurs. Used for regulating fluid flow such as natural gas, refrigerant, and water.

Regulador de presión. Válvula capaz de mantener una presión constante a la salida cuando ocurre una presión variable a la entrada. Utilizado para regular el flujo de fluidos, como por ejemplo el gas natural, el refrigerante y el agua.

Pressure switch. A switch operated by a change in pressure.

Conmutador accionado por presión. Conmutador accionado por un cambio de presión.

Pressure/temperature relationship. This refers to the pressure/temperature relationship of a liquid and vapor in a closed container. If the temperature increases, the pressure will also increase. If the temperature is lowered, the pressure will decrease.

Relación entre presión y temperatura. Se refiere a la relación entre la presión y la temperatura de un líquido y un vapor en un recipiente cerrado. Si la temperatura aumenta, la presión también aumentará. Si la temperatura baja, habrá una caída de presión.

Pressure vessels and piping. Piping, tubing, cylinders, drums, and other containers that have pressurized contents.

Depósitos y tubería con presión. Tubería, cilindros, tambores y otros recipientes que tienen un contenido presurizado.

Primary control. Controlling device for an oil burner to ensure ignition within a specific time span, usually 90 seconds.

Regulador principal. Dispositivo de regulación para un quemador de aceite pesado. El regulador principal asegura el encendido dentro de un período de tiempo específico, normalmente 90 segundos.

Propane. An LP gas used for heat.

Propano. Gas de petróleo licuado que se utiliza para producir calor.

Proton. That part of an atom having a positive charge.

Protón. Parte de un átomo que tiene carga positiva.

psi. Abbreviation for pounds per square inch.

psi. Abreviatura de libras por pulgada cuadrada.

psia. Abbreviation for pounds per square inch absolute.

psia. Abreviatura de libras por pulgada cuadrada absoluta.

psig. Abbreviation for pounds per square inch gage.

psig. Abreviatura de indicador de libras por pulgada cuadrada.

Psychrometer. An instrument for determining relative humidity.

Sicrómetro. Instrumento para medir la humedad relativa.

Psychrometric chart. A chart that shows the relationship of temperature, pressure, and humidity in the air.

Esquema sicrométrico. Esquema que indica la relación entre la temperatura, la presión y la humedad en el aire.

Pump. A device that forces fluids through a system.

Bomba. Dispositivo que introduce fluidos por fuerza a través de un sistema.

Pump down. To use a compressor to pump the refrigerant charge into the condenser and/or receiver.

Extraer con bomba. Utilizar un compresor para bombear la carga del refrigerante dentro del condensador y/o receptor.

Purge. To remove or release fluid from a system.

Purga. Remover o liberar el fluido de un sistema.

PVC (Polyvinyl chloride). Plastic pipe used in pressure applications for water and gas as well as for sewage and certain industrial applications.

Cloruro de polivinilo (PVC). Tubo plástico utilizado tanto en aplicaciones de presión para agua y gas, como en ciertas aplicaciones industriales y de aguas negras.

Quench. To submerge a hot object in a fluid for cooling.

Enfriamiento por inmersión. Sumersión de un objeto caliente en un fluido para enfriarlo.

Quick-connect coupling. A device designed for easy connecting or disconnecting of fluid lines.

Acoplamiento de conexión rápida. Dispositivo diseñado para facilitar la conexión o desconexión de conductos de fluido.

R-12. Dichlorodifluoromethane, a popular refrigerant for refrigeration systems.

R-12. Diclorodiflorometano, refrigerante muy utilizado en sistemas de refrigeración.

R-22. Monochlorodifluoromethane, a popular refrigerant for air conditioning systems.

R-22. Monoclorodiflorometano, refrigerante muy utilizado en sistemas de acondicionamiento de aire.

R-123. Dichlorotrifluoroethane, a refrigerant developed for low-pressure application.

R-123. Diclorotrifloroetano, refrigerante elaborado para aplicaciones de baja presión.

R-134a. Tetrafluoroethane, a refrigerant developed for refrigeration systems and as a possible replacement for R-12.

R-134a. Tetrafloroetano, refrigerante elaborado para sistemas de refrigeración y como posible sustituto del R-12.

R-502. An azeotropic mixture of R-22 and R-115, a popular refrigerant for low-temperature refrigeration systems.

R-502. Mezcla azeotrópica de R-22 y R-115, refrigerante muy utilizado en sistemas de refrigeración de temperatura baja.

Radiant heat. Heat that passes through air, heating solid objects that in turn heat the surrounding area.

Calor radiante. Calor que pasa a través del aire y calienta objetos sólidos que a su vez calientan el ambiente.

Radiation. Heat transfer. See radiant heat.

Radiación. Transferencia de calor. Véase calor radiante.

Random or off-cycle defrost. Defrost provided by the space temperature during the normal off cycle.

Descongelación variable o de ciclo apagado. Descongelación llevada a cabo por la temperatura del espacio durante el ciclo normal de apagado.

Rankine. The absolute Fahrenheit scale with 0 at the point where all molecular motion stops.

Rankine. Escala absoluta de Fahrenheit con el 0 al punto donde se detiene todo movimiento molecular.

Reactance. A type of resistance in an alternating current circuit.

Reactancia. Tipo de resistencia en un circuito de corriente alterna.

Reamer. Tool to remove burrs from inside a pipe after it has been cut.

Escariador. Herramienta utilizada para remover las rebabas de un tubo después de haber sido cortado.

Receiver-drier. A component in a refrigeration system for storing and drying refrigerant.

Receptor-secador. Componente en un sistema de refrigeración que almacena y seca el refrigerante.

Reciprocating. Back-and-forth motion.

Movimiento alternativo. Movimiento de atrás para adelante.

Reciprocating compressor. A compressor that uses a piston in a cylinder and a back-and-forth motion to compress vapor.

Compresor alternativo. Compresor que utiliza un pistón en un cilindro y un movimiento de atrás para adelante a fin de comprir el vapor.

Rectifier. A device for changing alternating current to direct current.

Rectificador. Dispositivo utilizado para convertir corriente alterna en corriente continua.

Reed valve. A thin steel plate used as a valve in a compressor.

Válvula con lámina. Placa delgada de acero utilizada como una válvula en un compresor.

Refrigerant. The fluid in a refrigeration system that changes from a liquid to a vapor and back to a liquid at practical pressures.

Refrigerante. Fluido en un sistema de refrigeración que se convierte de líquido en vapor y nuevamente en líquido a presiones prácticas.

Refrigerant reclaim. Recovering the refrigerant and processing it so that it can be reused.

Recuperación del refrigerante. La recuperación del refrigerante y su procesamiento para que pueda ser utilizado de nuevo.

Refrigerant reclaim. "To process refrigerant to new product specifications by means which may include distillation. It will require chemical analysis of the refrigerant to determine that appropriate product specifications are met. This term usually implies the use of processes or procedures available only at a reprocessing or manufacturing facility."

Recuperación del refrigerante. "Procesar refrigerante según nuevas especificaciones para productos a través de métodos que pueden incluir la destilación. Se requiere un análisis químico del refrigerante para asegurar el cumplimiento de las especificaciones para productos adecuadas. Por lo general este término supone la utilización de procesos o de procedimientos disponibles solamente en fábricas de reprocesamiento o manufactura."

Refrigerant recovery. "To remove refrigerant in any condition from a system and store it in an external container without necessarily testing or processing it in any way."

Recobrar refrigerante líquido. "Remover refrigerante en cualquier estado de un sistema y almacenarlo en un recipiente externo sin ponerlo a prueba o elaborarlo de ninguna manera."

Refrigerant recycling. "To clean the refrigerant by oil separation and single or multiple passes through devices, such as replaceable core filter-driers, which reduce moisture, acidity and particulate matter. This term usually applies to procedures implemented at the job site or at a local service shop."

Recirculación de refrigerante. "Limpieza del refrigerante por medio de la separación del aceite y pasadas sencillas o múltiples a través de dispositivos, como por ejemplo secadores filtros con núcleos reemplazables que disminyen la humedad, la acidez y las partículas. Por lo general este término se aplica a los procedimientos utilizados en el lugar del trabajo o en un taller de servicio local."

Refrigeration. The process of removing heat from a place where it is not wanted and transferring that heat to a place where it makes little or no difference.

Refrigeración. Proceso de remover el calor de un lugar donde no es deseado y transferirlo a un lugar donde no afecte la temperatura.

Register. A terminal device on an air distribution system that directs air but also has a damper to adjust airflow.

Registro. Dispositivo terminal en un sistema de distribución de aire que dirige el aire y además tiene un desviador para ajustar su flujo.

Relative humidity. The amount of moisture contained in the air as compared to the amount the air could hold at that temperature.

Humedad relativa. Cantidad de humedad presente en el aire, comparada con la cantidad de humedad que el aire pueda contener a dicha temperatura.

Relay. A small electromagnetic device to control a switch, motor, or valve.

Relé. Pequeño dispositivo electromagnético utilizado para regular un conmutador, un motor o una válvula.

Relief valve. A valve designed to open and release liquids at a certain pressure.

Válvula para alivio. Válvula diseñada para abrir y liberar líquidos a una presión específica.

Remote system. Often called a split system where the condenser is located away from the evaporator and/or other parts of the system.

Sistema remoto. Llamado muchas veces sistema separado donde el condensador se coloca lejos del evaporador y/u otras piezas del sistema.

Resistance. The opposition to the flow of an electrical current or a fluid.

Resistencia. Oposición al flujo de una corriente eléctrica o de un fluido.

Resistor. An electrical or electronic component with a specific opposition to electron flow. It is used to create voltage drop or heat.

Resistor. Componente eléctrico o elctrónico con una oposición específica al flujo de electrones; se utiliza para producir una caída de tensión o calor.

Restrictor. A device used to create a planned resistance to fluid flow.

Limitador. Dispositivo utilizado para producir una resistencia proyectada al flujo de fluido.

Reverse cycle. The ability to direct the hot gas flow into the indoor or the outdoor coil in a heat pump to control the system for heating or cooling purposes.

Ciclo invertido. Capacidad de dirigir el flujo de gas caliente dentro de la bobina interior o exterior en una bomba de calor a fin de regular el sistema para propósitos de calentamiento o enfriamiento.

Rod and tube. The rod and tube are each made of a different metal. The tube has a high expansion rate and the rod a low expansion rate.

Varilla y tubo. La varilla y el tubo se fabrican de un metal diferente. El tubo tiene una tasa de expansión alta y la varilla una tasa de expansión baja.

Rotary compressor. A compressor that uses rotary motion to pump fluids. It is a positive-displacement pump.

Compresor giratorio. Compresor que utiliza un movimiento giratorio para bombear fluidos. Es una bomba de desplazmiento positivo.

Rotor. The rotating or moving component of a motor, including the shaft.

Rotor. Componente giratorio o en movimiento de un motor, incluyendo el árbol.

Running time. The time a unit operates. Also called the on time.

Período de funcionamiento. El período de tiempo en que funciona una unidad. Conocido también como período de conexión.

Run winding. The electrical winding in a motor that draws current during the entire running cycle.

Devanado de funcionamiento. Devanado eléctrico en un motor que consume corriente durante todo el ciclo de funcionamiento.

Rupture disk. Safety device for a centrifugal low-pressure chiller.

Disco de ruptura. Dispositivo de seguridad para un enfriador centrífugo de baja presión.

Saddle valve. A valve that straddles a fluid line and is fastened by solder or screws. It normally contains a device to puncture the line for pressure readings.

Válvula de silleta. Válvula que está sentada a horcajadas en un conducto de fluido y se fija por medio de la soldadura o tornillos. Por lo general contiene un dispositivo para agujerear el conducto a fin de que se puedan tomar lecturas de presión.

Safety control. An electrical, mechanical, or electromechanical control to protect the equipment or public from harm.

Regulador de seguridad. Regulador eléctrico, mecánico o electromecánico para proteger al equipo de posibles averías o al público de sufrir alguna lesión.

Safety plug. A fusible plug.

Tapón de seguridad. Tapón fusible.

Sail switch. A safety switch with a lightweight sensitive sail that operates by sensing an airflow.

Conmutador con vela. Conmutador de seguridad con una vela liviana sensible que funciona al advertir el flujo de aire.

Saturated vapor. The refrigerant when all of the liquid has changed to a vapor.

Vapor saturado. El refrigerante cuando todo el líquido se ha convertido en vapor.

Saturation. A term used to describe a substance when it contains all of another substance it can hold.

Saturación. Término utilizado para describir una sustancia cuando contiene lo más que puede de otra sustancia.

Scavenger pump. A pump used to remove the fluid from a sump.

Bomba de barrido. Bomba utilizada para remover el fluido de un sumidero.

Schrader valve. A valve similar to the valve on an auto tire that allows refrigerant to be charged or discharged from the system.

Válvula Schrader. Válvula similar a la válvula del neumático de un automóvil que permite la entrada o la salida de refrigerante del sistema.

Scotch yoke. A mechanism used to create reciprocating motion from the electric motor drive in very small compressors.

Yugo escocés. Mecanismo utilizado para producir movimiento alternativo del accionador del motor eléctrico en compresores bastante pequeños.

Screw compressor. A form of positive-displacement compressor that squeezes fluid from a low-pressure area to a high-pressure area, using screw-type mechanisms.

Compresor de tornillo. Forma de compresor de desplazamiento positivo que introduce por fuerza el fluido de un área de baja presión a un área de alta presión, a través de mecanismos de tipo de tornillo.

Scroll compressor. A compressor that uses two scroll-type components to compress vapor.

Compresor espiral. Compresor que utiliza dos componentes de tipo espiral para comprimir el vapor.

Sealed unit. The term used to describe a refrigeration system, including the compressor, that is completely welded closed. The pressures can be accessed by saddle valves.

Unidad sellada. Término utilizado para describir un sistema de refrigeración, incluyendo el compresor, que es soldado completamente cerrado. Las presiones son accesibles por medio de válvulas de silleta.

Seasonal energy efficiency ratio (SEER). An equipment efficiency rating that takes into account the startup and shutdown for each cycle.

Relación del rendimiento de energía temporal. Clasificación del rendimiento de un equipo que toma en cuenta la puesta en marcha y la parada de cada ciclo.

Seat. The stationary part of a valve that the moving part of the valve presses against for shutoff.

Asiento. Pieza fija de una válvula contra la que la pieza en movimiento de la válvula presiona para cerrarla.

Semiconductor. A component in an electronic system that is considered neither an insulator nor a conductor but a partial conductor.

Semiconductor. Componente en un sistema eléctrico que no se considera ni aislante ni conductor, sino conductor parcial.

Semihermetic compressor. A motor compressor that can be opened or disassembled by removing bolts and flanges. Also known as a serviceable hermetic.

Compresor semihermético. Compresor de un motor que puede abrirse o desmontarse al removerle los pernos y bridas. Conocido también como hermético utilizable.

Sensible heat. Heat that causes a change in the level of a thermometer.

Calor sensible. Calor que produce un cambio en el nivel de un termómetro.

Sensor. A component for detection that changes shape, form, or resistance when a condition changes.

Sensor. Componente para la deteción que cambia de forma o de resistencia cuando cambia una condición.

Sequencer. A control that causes a staging of events, such as a sequencer between stages of electric heat.

Regulador de secuencia. Regulador que produce una sucesión de acontecimientos, como por ejemplo etapas sucesivas de calor eléctrico.

Series circuit. An electrical or piping circuit where all of the current or fluid flows through the entire circuit.

Circuito en serie. Circuito eléctrico o de tubería donde toda la corriente o todo el fluido fluye a través de todo el circuito.

Service valve. A manually operated valve in a refrigeration system used for various service procedures.

Válvula de servicio. Válvula de un sistema de refrigeración accionada manualmente que se utiliza en varios procedimientos de servicio.

Serviceable hermetic. See semihermetic compressor.

Compresor hermético utilizable. Véase compresor semihermético.

Shaded-pole motor. An alternating current motor used for very light loads.

Motor polar en sombra. Motor de corriente alterna utilizado en cargas sumamente livianas.

Shell and coil. A vessel with a coil of tubing inside that is used as a heat exchanger.

Coraza y bobina. Depósito con una bobina de tubería en su interior que se utiliza como intercambiador de calor.

Shell and tube. A heat exchanger with straight tubes in a shell that can normally be mechanically cleaned.

Coraza y tubo. Intercambiador de calor con tubos rectos en una coraza que por lo general puede limpiarse mecánicamente.

Short circuit. A circuit that does not have the correct measurable resistance; too much current flows and will overload the conductors.

Cortocircuito. Corriente que no tiene la resistencia medible correcta; un exceso de corriente fluye a través del circuito provocando una sobrecarga de los conductores.

Short cycle. The term used to describe the running time (on time) of a unit when it is not running long enough.

Ciclo corto. Término utilizado para describir el período de funcionamiento (de encendido) de una unidad cuando no funciona por un período de tiempo suficiente.

Shroud. A fan housing that ensures maximum airflow through the coil.

Bóveda. Alojamiento del abanico que asegura un flujo máximo de aire a través de la bobina.

Sight glass. A clear window in a fluid line.

Mirilla para observación. Ventana clara en un conducto de fluido.

Silica gel. A chemical compound often used in refrigerant driers to remove moisture from the refrigerant.

Gel silíceo. Compuesto químico utilizado a menudo en secadores de refrigerantes para remover la humedad del refrigerante.

Silicon. A substance from which many semiconductors are made.

Silicio. Sustancia de la cual se fabrican muchos semiconductores.

Silicon-controlled rectifier (SCR). A semiconductor control device.

Rectificador controlado por silicio. Dispositivo para regular un semiconductor.

Silver brazing. A high-temperature (above 800°F) brazing process for bonding metals.

Soldadura con plata. Soldadura a temperatura alta (sobre los 800°F ó 430°C) para unir metales.

Sine wave. The graph or curve used to describe the characteristics of alternating current voltage.

Onda sinusoidal. Gráfica o curva utilizada para describir las características de tensión de corriente alterna.

Single phase. The electrical power supplied to equipment or small motors, normally under 7½ hp.

Monofásico. Potencia eléctrica suministrada a equipos o motores pequeños, por lo general menor de 7½ hp.

Single phasing. The condition in a three-phase motor when one phase of the power supply is open.

Fasaje sencillo. Condición en un motor trifásico cuando una fase de la fuente de alimentación está abierta.

Sling psychrometer. A device with two thermometers, one a wet bulb and one a dry bulb, used for checking air conditions, temperature, and humidity.

Sicrómetro con eslinga. Dispositivo con dos termómetros, uno con una bombilla húmeda y otro con una bombilla seca, utilizados para revisar las condiciones del aire, de la temperatura y de la humedad.

Slip. The difference in the rated rpm of a motor and the actual operating rpm.

Deslizamiento. Diferencia entre las rpm nominales de un motor y las rpm de funcionamiento reales.

Slugging. A term used to describe the condition when large amounts of liquid enter a pumping compressor cylinder.

Relleno. Término utilizado para describir la condición donde grandes cantidades de líquido entran en el cilindro de un compresor de bombeo.

Smoke test. A test performed to determine the amount of unburned fuel in an oil burner flue-gas sample.

Prueba de humo. Prueba llevada a cabo para determinar la cantidad de combustible no quemado en una muestra de gas de combustión que se obtiene de un quemador de aceite pesado.

Snap-disc. An application of the bimetal. Two different metals fastened together in the form of a disc that provides a warping condition when heated. This also provides a snap action that is beneficial in controls that start and stop current flow in electrical circuits.

Disco de acción rápida. Aplicación del bimetal. Dos metales diferentes fijados entre sí en forma de un disco que provee una deformación al ser calentado. Esto provee también una acción rápida, ventajosa para reguladores que ponen en marcha y detienen el flujo de corriente en circuitos eléctricos.

Solar collectors. Components of a solar system designed to collect the heat from the sun, using air, a liquid, or refrigerant as the medium.

Colectores solares. Componentes de un sistema solar diseñados para acumular el calor emitido por el sol, utilizando el aire, un líquido o un refrigerante como el medio.

Solar heat. Heat from the sun's rays.

Calor solar. Calor emitido por los rayos del sol.

Soldering. Fastening two base metals together by using a third, filler metal that melts at a temperature below 800°F.

Soldadura. La fijación entre sí de dos metales bases utilizando un tercer metal de relleno que se funde a una temperatura menor de 800°F (430°C).

Solder pot. A device using a low-melting solder and an overload heater sized for the amperage of the motor it is protecting. The solder will melt, opening the circuit when there is an overload. It can be reset.

Olla para soldadura. Dispositivo que utiliza una soldadura con un punto de fusión bajo y un calentador de sobrecarga diseñado para el amperaje del motor al que provee protección. La soldadura se fundirá, abriendo así el circuito cuando ocurra una sobrecarga. Puede ser reconectado.

Solenoid. A coil of wire designed to carry an electrical current producing a magnetic field.

Solenoide. Bobina de alambre diseñada para conducir una corriente eléctrica generando un campo magnético.

Soild. Molecules of a solid are highly attracted to each other forming a mass that exerts all of its weight downward.

Sólido. Las moléculas de un sólido se atraen entre sí y forman una masa que ejerce todo su peso hacia abajo.

Specific gravity. The weight of a substance compared to the weight of an equal volume of water.

Gravedad específica. El peso de una sustancia comparada con el peso de un volumen igual de agua.

Specific heat. The amount of heat required to raise the temperature of 1 lb of a substance 1°F.

Calor específico. La cantidad de calor requerido para elevar la temperatura de una libra de una sustancia 1°F (−17°C).

Specific volume. The volume occupied by 1 lb of a fluid.

Volumen específico. Volumen que ocupa una libra de fluido.

Splash lubrication system. A system of furnishing lubrication to a compressor by agitating the oil.

Sistema de lubrificación por salpicadura. Método de proveerle lubrificación a un compresor agitando el aceite.

Splash method. A method of water dropping from a higher level in a cooling tower and splashing on slots with air passing through for more efficient evaporation.

Método de salpicaduras. Método de dejar caer agua desde un nivel más alto en una torre de refrigeración y salpicándola en ranuras, mientras el aire pasa a través de las mismas con el propósito de lograr una evaporación más eficaz.

Split-phase motor. A motor with run and start windings.

Motor de fase separada. Motor con devandos de funcionamiento y de arranque.

Split system. A refrigeration or air conditioning system that has the condensing unit remote from the indoor (evaporator) coil.

Sistema separado. Sistema de refrigeración o de acondicionamiento de aire cuya unidad de condensación se encuentra en un sitio alejado de la bobina interior del evaporador.

Spray pond. A pond with spray heads used for cooling water in water-cooled air conditioning or refrigeration systems.

Tanque de rociado. Tanque con una cabeza rociadora utilizada para enfriar el agua en sistemas de acondicionamiento de aire o de refrigeración enfriados por agua.

Squirrel cage fan. A fan assembly used to move air.

Abanico con jaula de ardilla. Conjunto de abanico utilizado para mover el aire.

Standard atmosphere or standard conditions. Air at sea level at 70°F when the atmosphere's pressure is 14.696 psia (29.92 in. Hg). Air at this condition has a volume of 13.33 ft^3/lb.

Atmósfera estándar o condiciones estándares. El aire al nivel del mar a una temperatura de 70°F (21°C) cuando la presión de la atmósfera es 14,696 psia (29,92 pulgadas Hg). Bajo esta condición, el aire tiene un volumen de 13,33 ft^3/lb (libras/pies).

Standing pilot. Pilot flame that remains burning continuously.

Piloto constante. Llama piloto que se quema de manera continua.

Start capacitor. A capacitor used to help an electric motor start.

Capacitador de arranque. Capacitador utilizado para ayudar en el arranque de un motor eléctrico.

Starting relay. An electrical relay used to disconnect the start winding in a hermetic compressor.

Relé de arranque. Relé eléctrico utilizado para desconectar el devanado de arranque en un compresor hermético.

Starting winding. The winding in a motor used primarily to give the motor extra starting torque.

Devanado de arranque. Devanado en un motor utilizado principalmente para proveerle al motor mayor par de arranque.

Starved coil. The condition in an evaporator when the metering device is not feeding enough refrigerant to the evaporator.

Bobina estrangulada. Condición que ocurre en un evaporador cuando el dispositivo de medida no le suministra suficiente refrigerante al evaporador.

Stator. The component in a motor that contains the windings; it does not turn.

Estátor. Componente en un motor que contiene los devanados y que no gira.

Steam. The vapor state of water.

Vapor. Estado de vapor del agua.

Strainer. A fine-mesh device that allows fluid flow and holds back solid particles.

Colador. Dispositivo de malla fina que permite el flujo de fluido a través de él y atrapa partículas sólidas.

Stratification. The condition where a fluid appears in layers.

Estratificación. Condición que ocurre cuando un fluido aparece en capas.

Stress crack. A crack in piping or other component caused by age or abnormal conditions such as vibration.

Grieta por tensión. Grieta que aparece en una tubería u otro componente ocasionada por envejecimiento o condiciones anormales, como por ejemplo vibración.

Subbase. The part of a space temperature thermostat that is mounted on the wall and to which the interconnecting wiring is attached.

Subbase. Pieza de un termóstato que mide la temperatura de un espacio que se monta sobre la pared y a la que se fijan los conductores eléctricos interconectados.

Subcooling. The temperature of a liquid when it is cooled below its condensing temperature.

Subenfriamiento. La temperatura de un líquido cuando se enfria a una temperatura menor que su temperatura de condensación.

Sublimation. When a substance changes from the solid state to the vapor state without going through the liquid state.

Sublimación. Cuando una sustancia cambia de sólido a vapor sin convertirse primero en líquido.

Suction gas. The refrigerant vapor in an operating refrigeration system found in the tubing from the evaporator to the compressor and in the compressor shell.

Gas de aspiración. El vapor del refrigerante en un sistema de refrigeración en funcionamiento presente en la tubería que va del evaporador al compresor y en la coraza del compresor.

Suction line. The pipe that carries the heat-laden refrigerant gas from the evaporator to the compressor.

Conducto de aspiración. Tubo que conduce el gas de refrigerante lleno de calor del evaporador al compresor.

Suction service valve. A manually operated valve with front and back seats located at the compressor.

Válvula de aspiración para servicio. Válvula accionada manualmente que tiene asientos delanteros y traseros ubicados en el compresor.

Suction valve lift unloading. The suction valve in a reciprocating compressor cylinder is lifted, causing that cylinder to stop pumping.

Descarga por levantamiento de la válvula de aspiración. La válvula de aspiración en el cilindro de un compresor alternativo se levanta, provocando que el cilindro deje de bombear.

Sump. A reservoir at the bottom of a cooling tower to collect the water that has passed through the tower.

Sumidero. Tanque que se encuentra en el fondo de una torre de refrigeración para acumular el agua que ha pasado a través de la torre.

Superheat. The temperature of vapor refrigerant above its saturation change of state temperature.

Sobrecalor. Temperatura del refrigerante de vapor mayor que su temperatura de cambio de estado de saturación.

Surge. When the head pressure becomes too great or the evaporator pressure too low, refrigerant will flow from the high- to the low-pressure side of a centrifugal compressor system, making a loud sound.

Movimiento repentino. Cuando la presión en la cabeza aumenta demasiado o la presión en el evaporador es demasiado baja, el refrigerante fluye del lado de alta presión al lado de baja presión de un sistema de compresor centrífugo. Este movimiento produce un sonido fuerte.

Swaged joint. The joining of two pieces of copper tubing by expanding or stretching the end of one piece of tubing to fit over the other piece.

Junta estampada. La conexión de dos piezas de tubería de cobre dilatando o alargando el extremo de una pieza de tubería para ajustarla sobre otra.

Swaging tool. A tool used to enlarge a piece of tubing for a solder or braze connection.

Herramienta de estampado. Herramienta utilizada para agrandar una pieza de tubería a utilizarse en una conexión soldada o broncesoldada.

Swamp cooler. A slang term used to describe an evaporative cooler.

Nevera pantanoso. Término del argot utilizado para describir una nevera de evaporación.

Sweating. A word used to describe moisture collection on a line or coil that is operating below the dew point temperature of the air.

Exudación. Término utilizado para describir la acumulación de humedad en un conducto o una bobina que está funcionando a una temperatura menor que la del punto de rocío del aire.

Tank. A closed vessel used to contain a fluid.

Tanque. Depósito cerrado utilizado para contener un fluido.

Tap. A tool used to cut internal threads in a fastener or fitting.

Macho de roscar. Herramienta utilizada para cortar filetes internos en un aparto fijador o en un accesorio.

Temperature. A word used to describe the level of heat or molecular activity, expressed in Fahrenheit, Rankine, Celsius, or Kelvin units.

Temperatura. Término utilizado para describir el nivel de calor o actividad molecular, expresado en unidades Fahrenheit, Rankine, Celsio o Kelvin.

Test light. A light bulb arrangement used to prove the presence of electrical power in a circuit.

Luz de prueba. Arreglo de bombillas utilizado para probar la presencia de fuerza eléctrica en un circuito.

Therm. Quantity of heat, 100,000 Btu.

Therm. Cantidad de calor, mil unidades térmicas inglesas.

Thermistor. A semiconductor electronic device that changes resistance with a change in temperature.

Termistor. Dispositivo eléctrico semiconductor que cambia su resistencia cuando se produce un cambio en temperatura.

Thermocouple. A device made of two unlike metals that generates electricity when there is a difference in temperature from one end to the other. Thermocouples have a hot and cold junction.

Thermopar. Dispositivo hecho de dos metales distintos que genera electricidad cuando hay una diferencia en temperatura de un extremo al otro. Los termopares tienen un empalme caliente y uno frío.

Thermometer. An instrument used to detect differences in the level of heat.

Termómetro. Instrumento utilizado para detectar diferencias en el nivel de calor.

Thermopile. A group of thermocouples connected in series to increase voltage output.

Pila termoeléctrica. Grupo de termopares conectados en serie para aumentar la salida de tensión.

Thermostat. A device that senses temperature change and changes some dimension or condition within to control an operating device.

Termostato. Dispositivo que advierte un cambio en temperatura y cambia alguna dimensión o condición dentro de sí para regular un dispositivo en funcionamiento.

Thermostatic expansion valve (TXV). A valve used in refrigeration systems to control the superheat in an evaporator by metering the correct refrigerant flow to the evaporator.

Válvula de gobierno termostático para expansión. Válvula utilizada en sistemas de refrigeración para regular el sobrecalor en un evaporador midiendo el flujo correcto de refrigerante al evaporador.

Three-phase power. A type of power supply usually used for operating heavy loads. It consists of three sine waves that are out of phase with each other.

Potencia trifásica. Tipo de fuente de alimentación normalmente utilizada en el funcionamiento de cargas pesadas. Consiste de tres ondas sinusoidales que no están en fase la una con la otra.

Throttling. Creating a restriction in a fluid line.

Estrangulamiento. Que ocasiona una restricción en un conducto de fluido.

Timers. Clock-operated devices used to time various sequences of events in circuits.

Temporizadores. Dispositivos accionados por un reloj utilizados para medir el tiempo de varias secuencias de eventos en circuitos.

Ton of refrigeration. The amount of heat required to melt a ton (2000 lb) of ice at 32°F, 288,000 Btu/24 h, 12,000 Btu/h, or 200 Btu/min.

Tonelada de refrigeración. Cantidad de calor necesario para fundir una tonelada (2000 libras) de hielo a 32°F (0°C), 288.000 Btu/24 h, 12.000 Btu/h, o 200 Btu/min.

Torque. The twisting force often applied to the starting power of a motor.

Par de torsión. Fuerza de torsión aplicada con frecuencia a la fuerza de arranque de un motor.

Torque wrench. A wrench used to apply a prescribed amount of torque or tightening to a connector.

Llave de torsión. Llave utilizada para aplicar una cantidad específica de torsión o de apriete a un conector.

Total heat. The total amount of sensible heat and latent heat contained in a substance from a reference point.

Calor total. Cantidad total de calor sensible o de calor latente presente en una sustancia desde un punto de referencia.

Transformer. A coil of wire wrapped around an iron core that induces a current to another coil of wire wrapped around the same iron core. Note: A transformer can have an air core.

Transformador. Bobina de alambre devanado alrededor de un núcleo de hierro que induce una corriente a otra bobina de alambre devanado alrededor del mismo núcleo de hierro. Nota: Un transformador puede tener un núcleo de aire.

Transistor. A semiconductor often used as a switch or amplifier.

Transistor. Semiconductor que suele utilizarse como conmutador o amplificador.

TRIAC. A semiconductor switching device.

TRIAC. Dispositivo de conmutación para semiconductores.

Tube within a tube coil. A coil used for heat transfer that has a pipe in a pipe and is fastened together so that the outer tube becomes one circuit and the inner tube another.

Bobina de tubo dentro de un tubo. Bobina utilizada en la transferencia de calor que tiene un tubo dentro de otro y se sujeta de manera que el tubo exterior se convierte en un circuito y el tubo interior en otro circuito.

Tubing. Pipe with a thin wall used to carry fluids.

Tubería. Tubo que tiene una pared delgada utilizado para conducir fluidos.

Two-temperature valve. A valve used in systems with multiple evaporators to control the evaporator pressures and maintain different temperatures in each evaporator. Sometimes called a hold-back valve.

Válvula de dos temperaturas. Válvula utilizada en sistemas con evaporadores múltiples para regular las presiones de los evaporadores y mantener temperaturas diferentes en cada uno de ellos. Conocida también como válvula de retención.

Ultraviolet. Light waves that can only be seen under a special lamp.

Ultravioleta. Ondas de luz que pueden observarse solamente utilizando una lámpara especial.

Urethane foam. A foam that can be applied between two walls for insulation.

Espuma de uretano. Espuma que puede aplicarse entre dos paredes para crear un aislamiento.

U-Tube mercury manometer. A U-tube containing mercury, which indicates the level of vacuum while evacuating a refrigeration system.

Manómetro de mercurio de tubo en U. Tubo en U que contiene mercurio y que indica el nivel del vacío mientras vacía un sistema de refrigeración.

U-Tube water manometer. Indicates natural gas and propane gas pressures. It is usually calibrated in inches of water.

Manómetro de agua de tubo en U. Indica las presiones del gas natural y del propano. Se calibra normalmente en pulgadas de agua.

Vacuum. The pressure range between the earth's atmosphere and no pressure, normally expressed in inches of mercury (in. Hg) vacuum.

Vacío. Margen de presión entre la atmósfera de la Tierra y cero presión, por lo general expresado en pulgadas de mercurio (pulgadas Hg) en vacío.

Vacuum pump. A pump used to remove some fluids such as air and moisture from a system at a pressure below the earth's atmosphere.

Bomba de vacío. Bomba utilizada para remover algunos fluidos, como por ejemplo aire y humedad de un sistema a una presión menor que la de la atmósfera de la Tierra.

Valve. A device used to control fluid flow.

Válvula. Dispositivo utilizado para regular el flujo de fluido.

Valve plate. A plate of steel bolted between the head and the body of a compressor that contains the suction and discharge reed or flapper valves.

Placa de válvula. Placa de acero empernado entre la cabeza y el cuerpo de un compresor que contiene la lámina de aspiración y de descarga o las chapaletas.

Valve seat. That part of a valve that is usually stationary. The movable part comes in contact with the valve seat to stop the flow of fluids.

Asiento de la válvula. Pieza de una válvula que es normalmente fija. La pieza móvil entra en contacto con el asiento de la válvula para detener el flujo de fluidos.

Vapor. The gaseous state of a substance.

Vapor. Estado gaseoso de una sustancia.

Vapor barrier. A thin film used in construction to keep moisture from migrating through building materials.

Película impermeable. Película delgada utilizada en construcciones para evitar que la humeded penetre a través de los materiales de construcción.

Vapor charge valve. A charge in a thermostatic expansion valve bulb that boils to a complete vapor. When this point is reached, an increase in temperature will not produce an increase in pressure.

Válvula para la carga de vapor. Carga en la bombilla de una válvula de expansión termostática que hierve a un vapor completo. Al llegar a este punto, un aumento en temperatura no produce un aumento en presión.

Vapor lock. A condition where vapor is trapped in a liquid line and impedes liquid flow.

Bolsa de vapor. Condición que ocurre cuando el vapor queda atrapado en el conducto de líquido e impide el flujo de líquido.

Vapor pump. Another term for compressor.

Bomba de vapor. Otro término para compresor.

Vapor refrigerant charging. Adding refrigerant to a system by allowing vapor to move out of the vapor space of a refrigerant cylinder and into the low-pressure side of the refrigeration system.

Carga del refrigerante de vapor. Agregarle refrigerante a un sistema permitiendo que el vapor salga del espacio de vapor de un cilindro de refrigerante y que entre en el lado de baja presión del sistema de refrigeración.

Vaporization. The changing of a liquid to a gas or vapor.

Vaporización. Cuando un líquido se convierte en gas o vapor.

Variable pitch pulley. A pulley whose diameter can be adjusted.

Polea de paso variable. Polea cuyo diámetro puede ajustarse.

Variable resistor. A type of resistor where the resistance can be varied.

Resistor variable. Tipo de resistor donde la resistencia puede variarse.

V belt. A belt that has a V-shaped contact surface and is used to drive compressors, fans, or pumps.

Correa en V. Correa que tiene una superficie de contacto en forma de V y se utiliza para accionar compresores, abanicos o bombas.

Velocity meter. A meter used to detect the velocity of fluids, air, or water.

Velocímetro. Instrumento utilizado para medir la velocidad de fluidos, aire o agua.

Velocity. The speed at which a substance passes a point.

Velocidad. Rapidez a la que una sustancia sobrepasa un punto.

Volt-ohm-milliammeter (VOM). A multimeter that measures voltage, resistance, and current in milliamperes.

Voltio-ohmio-miliamperímetro. Multímetro que mide tensión, resistencia y corriente en miliamperios.

Voltage. The potential electrical difference for electron flow from one line to another in an electrical circuit.

Tensión. Diferencia de potencial eléctrico del flujo de electrones de un conducto a otro en un circuito eléctrico.

Voltmeter. An instrument used for checking electrical potential.

Voltímetro. Instrumento utilizado para revisar la potencia eléctrica.

Volumetric efficiency. The pumping efficiency of a compressor or vacuum pump that describes the pumping capacity in relationship to the actual volume of the pump.

Rendimiento volumétrico. Rendimiento de bombeo de un compresor o de una bomba de vacío que describe la capacidad de bombeo con relación al volumen real de la bomba.

Vortexing. A whirlpool action in the sump of a cooling tower.

Acción de vórtice. Torbellino en el sumidero de una torre de refrigeración.

Walk-in cooler. A large refrigerated space used for storage of refrigerated products.

Nevera con acceso al interior. Espacio refrigerado grande utilizado para almacenar productos refrigerados.

Water box. A container or reservoir at the end of a chiller where water is introduced and contained.

Caja de agua. Recipiente o depósito al extremo de un enfriador por donde entra y se retiene el agua.

Water column (WC). The pressure it takes to push a column of water up vertically. One inch of water column is the amount of pressure it would take to push a column of water in a tube up one inch.

Columna de agua. Presión necesaria para levantar una columna de agua verticalmente. Una pulgada de columna de agua es la cantidad de presión necesaria para levantar una columna de agua a una distancia de una pulgada en un tubo.

Water-cooled condenser. A condenser used to reject heat from a refrigeration system into water.

Condensador enfriado por agua. Condensador utilizado para dirigir el calor de un sistema de refrigeración al agua.

Water-regulating valve. An operating control regulating the flow of water.

Válvula reguladora de agua. Regulador de mando que controla el flujo de agua.

Watt. A unit of power applied to electron flow. One watt equals 3.414 Btu.

Watio. Unidad de potencia eléctrica aplicada al flujo de electrones. Un watio equivale a 3,414 Btu.

Watt-hour. The unit of power that takes into consideration the time of consumption. It is the equivalent of a 1-watt bulb burning for 1 hour.

Watio hora. Unidad de potencia eléctrica que toma en cuenta la duración de consumo. Es el equivalente de una bombilla de 1 watio encendida por espacio de una hora.

Wet-bulb temperature. A wet-bulb temperature of air is used to evaluate the humidity in the air. It is obtained with a wet thermometer bulb to record the evaporation rate with an airstream passing over the bulb to help in evaporation.

Temperatura de una bombilla húmeda. La temperatura de una bombilla húmeda se utiliza para evaluar la humedad presente en el aire. Se obtiene con la bombilla húmeda de un termómetro para registrar el margen de evaporación con un flujo de aire circulando sobre la bombilla para ayudar en evaporar el agua.

Wet heat. A heating system using steam or hot water as the heating medium.

Calor húmedo. Sistema de calentamiento que utiliza vapor o agua caliente como medio de calentamiento.

Window unit. An air conditioner installed in a window that rejects the heat outside the structure.

Acondicionador de aire para la ventana. Acondicionador de aire instalado en una ventana que desvía el calor proveniente del exterior de la estructura.

Work. A force moving an object in the direction of the force. Work = Force × Distance.

Trabajo. Fuerza que mueve un objeto en la dirección de la fuerza. Trabajo = Fuerza × Distancia.

Index

A

B

D

E

G

H

I

J

K

L

M

N

O

P

R

S

W

ACE
AIR CONDITIONING
EXCELLENCE
Technician Certification Program

On the following pages, we have assembled some of the most commonly-asked questions concerning the newly-announced **AIR CONDITIONING EXCELLENCE (ACE) PROGRAM,** *(formerly the National HVACR Technician Certification Program).*

Perhaps you are an HVACR technician...or a contractor...or a vocational instructor. Whoever you are, if you are a part of the HVACR industry, and you care about the advancement of the profession and your personal career growth, you have probably heard about the development of this new certification program. We invite you to read further to find out more about the program that hopes to significantly raise the level of excellence in the HVACR industry.

What is the ACE Program?

The **Air Conditioning Excellence Program**, (ACE) is more than a technician certification program. It is part of a comprehensive ACCA-led industry initiative to:

- recognize technicians who meet high industry standards
- promote education and improve technical proficiency in the industry
- give the industry clearly defined structure, so that the consumer can easily identify qualified technicians.

This program has been developed with the help of dozens of ACCA volunteers, chapters, consultants, as well as Ferris State University, a recognized certification provider with over 52 years of HVACR education and testing experience.

Nationwide testing has begun on a Service Technician Module for Residential/Light Commercial Applications. Another module, on Installation for Residential/Light Commercial Applications, is nearing completion, and still others, on Commercial/Industrial Applications, will follow.

BUILDING A BETTER INDUSTRY THROUGH EDUCATION

Who is ACCA?

ACCA stands for the **Air Conditioning Contractors of America**, a non-profit trade association of heating, ventilation, air conditioning, and refrigeration (HVACR) contractors. ACCA represents and serves firms who design, install, service, and repair air conditioning, heating, refrigeration, humidification, dehumidification, air purification, and ventilation systems for residential and commercial customers. ACCA was formed in 1969 through the merger of Air Conditioning and Refrigeration Contractors of America (established in 1946) and the National Warm Air Heating and Air Conditioning Association (established in 1914).

ACCA represents over 4,000 small businesses with over 60 local chapters across the country. In addition, the association's membership includes manufacturers, wholesalers and distributors of HVACR equipment, and vocational and technical schools.

Why Should Technicians be Certified?

Certification helps technicians, their employers, HVACR equipment manufacturers, and consumers. How?

- Technicians who gain ACE Certification receive recognition, respect, and rewards. Certified technicians set a higher standard and have a better quality of life.
- Employers and customers know that technicians who have worked to earn professional certification care about their professions and their reputations – enough to put their skills to the toughest test.
- By unifying standards, certification programs lower training costs through focused training courses.
- Certification programs carry real weight in the marketplace, too. Potential customers tend to rely more on companies who use certified technicians. (So ... there's usually more and better work for certified technicians and the companies that employ them!)
- Manufacturers know that when their equipment has been serviced or installed by an ACE certified technician, the job has been done properly.
- Certified technicians will receive a handsome certificate, wallet card, and a uniform patch to announce their ACE Certified status to their fellow technicians, their employers, and their customers.

"The ACE Technician Certification Program will elevate this industry to an accredited profession, giving the men and women of the HVACR industry the recognition they deserve."

Phil Forner,
President
Allendale Heating Co., Inc.
Allendale, MI

What is recommended prior to taking the exam?

The ACE Program is designed to bring structure to the learning process and identify training weaknesses through a comprehensive testing process. This program is testing lifelong HVACR competencies and skills, and therefore there is no prerequisite training course prior to taking this exam.

However, technicians and proctors may wish to review prior to the exams using a variety of existing educational and training materials. Unlike the ACCA/FSU/EPA CFC Certification Program, ACCA will not directly produce test taking training materials for this program. We are, however, pleased to announce that ACCA will use this textbook, *Refrigeration & Air Conditioning Technology,* as a recommended resource for preparation of the Technician Certification exam. Additionally, instructor materials and additional study guides have been developed. These training packages will correspond to each of the subject areas in the Testing Results Detail Report (see page 13). The training courses will eventually total fifteen in all; with the first two, Physics of Refrigeration and Air Conditioning, and Electrical Fundamentals, being available by Fall 1997.

To order this text book, call ACCA at 1-888-290-2220 or Delmar Publishers at 1-800-347-7707.

Aren't There Already Certification and Training Programs Available?

It's true that many manufacturers of HVACR equipment offer courses of various kinds. But each covers different equipment, and has different standards, making them difficult to compare.

The ACE Program is a single, uniform standard. When a technician is certified in an area, it's clear he or she meets a broad, comprehensive set of expectations – not just some standards for one kind of equipment, or one model.

The formal mission statement of the program is:

"To encourage a higher level of professionalism, the ACE Program establishes a means to measure technicians' capabilities and promote education to the residential/light commercial HVACR industry."

There is, of course, The Environmental Protection Agency (EPA) certification, and it is a requirement to purchase refrigerant and work on equipment that utilizes refrigerant. The ACE Program covers certification in service and installation of heating and air conditioning equipment.

What Do the ACE Certification Tests Cover?

Tests are comprehensive and cover many skills. Take a look.

AREAS COVERED BY THE ACE PROGRAM TESTS

SERVICE TECHNICIAN MODULE

PART 1: CORE UNITS

Unit I: Safety, Tools, and Soft Skills

A. Heating safety
B. A/C safety
C. Electrical safety
D. Controls safety
E. Job safety
F. Personal safety
G. Driver safety
H. Hazardous substances
J. Customer communication
K. Customer courtesy
L. Customer conflict resolution
M. Salespersonship
N. Appearance and self-esteem
P. Service contracts
Q. Tools
R. Job clean-up

Unit II: Principles of Heating, Air Conditioning, & Refrigeration

A. Heat and matter
B. Temperature/degree
C. Pressures and gas laws
D. Manufacturers' data plates
E. Refrigeration system components and applications
F. Lubricants
G. Refrigeration cycle
H. Refrigerants (properties and applications)
J. Refrigerant cylinders
K. Refrigerant transfer
L. Leak checking
M. Recovery and recycling
N. System evacuation procedures
P. Charging procedure
Q. Brazing and silver soldering
R. Copper systems
S. Air cooled condensers
T. Evaporators
U. Defrost systems
W. Metering devices
X. Food preservation and spoilage
Y. Ice machine components and troubleshooting

Unit III: Electricity and Controls

A. Ohm's law and series circuits
B. Parallel circuits and combination circuits
C. Applications of series and parallel circuits
D. Electromagnetism and inductance
E. Inductors and capacitors in AC circuits
F. Conductors and overcurrent protection
G. AC capacitors
H. Single-phase transformers
J. Measuring devices and instrumentation
K. Single-phase motors characteristics and high-efficiency applications
L. Hermetic motors and starting relays
M. Compressor overloading devices
N. Three-phase motors
P. Motor starters and variable frequency drives
Q. Wiring diagrams and schematics
R. Control theory
S. Low-voltage controls
T. Electronic controls
U. A/C control systems
W. Flame safeguard controls
X. Gas heating controls
Y. Oil heating system controls
Z. Defrost timers
AA. HVACR system schematics and troubleshooting

"The ACE Technician Certification Program will give more structure to our service technicians' career paths, which will result in HVACR contractors delivering a better product to our customers!"

Alan J. Guzik,
Chairman/CEO
Energy Management Specialists, Inc.
Cleveland, OH

AREAS COVERED BY THE ACE PROGRAM TESTS

SERVICE TECHNICIAN MODULE (CONT.)

PART 2: SPECIALIST CERTIFICATIONS

One or both of the following units is required for certification.

Unit IV: Heating Systems

A. Combustion, combustion testing and efficiency
B. Venting/exhaust
C. Operating cycle of gas furnace
D. Gas pilot burner and thermocouple
E. Gas pressure regulators
F. Gas controllers and combination valves
G. Gas-fired systems
H. Troubleshooting
J. Oil-fired systems
K. Boilers, hydronic and steam
L. Electric heating
M. Threaded pipe systems and fittings
N. Operation of high-pressure gun-type oil burners
P. Principles of combustion and controls
Q. Heat transfer
R. Draft and combustion air

Unit V: Air Conditioning Systems

A. Properties of air and effects of humidity
B. Air measuring devices
C. Psychrometric chart applications
D. Water-cooled condensers
E. Mechanical system troubleshooting
F. Compressor construction, design and applications
G. Determination, cause and cleanup or compressor burnout
H. A/C systems
J. Economizer systems

"ACCA has a long history of providing successful education and training programs that enhance the value of the HVACR industry to our customers."

Alan L. Barnes, Sr.
President
Aircond Corporation
Atlanta, GA

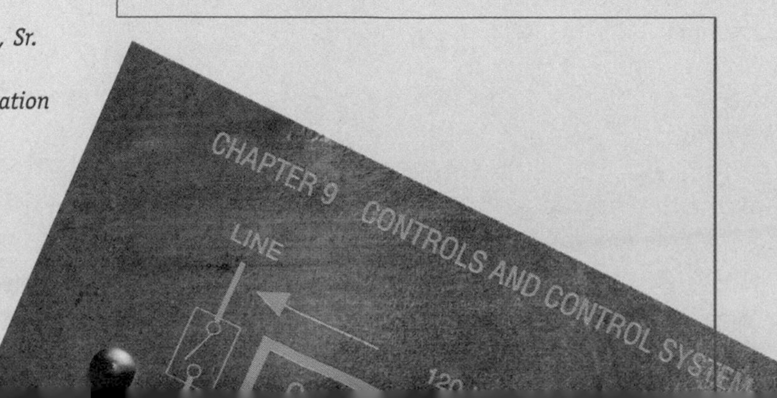

WHAT KIND OF QUESTIONS ARE ON THE TESTS?

ACCA solicited test questions from vocational school instructors, manufacturers, contractors, and other relevant parties. It received more than ten thousand possible questions. Ferris State University and the ACE Technician Certification Work Group scrutinized the questions, edited them, and organized them into test areas. Hundreds of different questions were tested and evaluated by nearly 1,000 technicians in the pilot testing that took place in 1996 across the U.S. All questions on the tests are multiple-choice. Here are some sample questions.

Answers on page 11

SAMPLE QUESTIONS ACE PROGRAM

Residential/Light Commercial Applications — Service Technician Module

1. As an oil becomes more diluted with refrigerant, its viscosity:

A. increases.
B. decreases.
C. remains the same.
D. none of the above.

2. Upon entering the condenser, the hot gas discharge first gives up its:

A. specific heat.
B. latent heat.
C. superheat.
D. vaporization.

3. The device in which the refrigerant is vaporized is called the:

A. evaporator.
B. compressor.
C. vaporizer.
D. condenser.

Hundreds of different questions were tested and evaluated by nearly 1,000 technicians in "pilot testing" that took place in 1996 across the U.S.

4. The principle reason for pulling a deep vacuum on a refrigeration system is to:

A. lower the head pressure.
B. remove refrigerant in the system.
C. remove air.
D. remove moisture.

5. If the load on a compressor increases, suction:

A. pressure will decrease, discharge pressure will increase.
B. pressure will increase, discharge pressure will decrease.
C. and discharge pressures will increase.
D. and discharge pressures will decrease.

6. If the outlet valve on a receiver is shut off (assume operating compressor), the:

A. high-pressure control will stop the system.
B. unit will pump-down.
C. suction pressure will increase.
D. valves and/or gaskets will break on the compressor (assume unit has a correctly sized receiver and proper refrigerant charge).

7. Which of the following will not cause a high discharge pressure?

A. Dirty evaporator.
B. Dirty condenser.
C. Burned out condenser fan motor.
D. Air in the system.

8. An air conditioning system is pumped down and the power is shut off. If the suction pressure rises very rapidly, this is an indication that the:

A. system is overcharged.
B. system has a restriction in the refrigerant lines.
C. suction valves are bad.
D. discharge valves are bad.

9. What are the three conditions necessary for combustion?

A. Fuel, air, and moisture.
B. Carbon, air, and pressure.
C. Fuel, heat, and oxygen.
D. Carbon, hydrogen, and oxygen.

10. Incomplete combustion can be the result of:

A. a fire that is too hot.
B. too much draft.
C. insufficient fuel supply.
D. insufficient air supply.

11. If heat is added to a saturated vapor it will become:

A. a superheated vapor.
B. a supersaturated vapor.
C. a subcooled vapor.
D. a mixture of liquid and vapor.

12. Subcooling is normally accomplished in the:

A. first part of the condenser.
B. last part of the condenser.
C. first part of the evaporator.
D. last part of the evaporator.

13. An HVACR company's fall protection work plan should contain the following elements:

A. locations on a job site where ladders may be stored at night.
B. diagrams for attaching fall arrest systems to the side of a building.
C. written emergency procedures to be followed if a worker falls off a roof.
D. written identification of all areas on the job site where a fall hazard exists.

14. Where is the marked terminal of the run capacitor connected?

A. To R terminal.
B. To C terminal.
C. To same side of the line or contactor as the R terminal.
D. To S terminal.

15. A 24 v relay coil with a reading of 30 ohms will:

A. draw about 0.4 amps.
B. draw about 0.6 amps.
C. draw about 0.8 amps.
D. cannot be determined with these readings.

16. To properly set a heat anticipator it is necessary to check:

A. the voltage rating of the gas valve.
B. the heating control circuit amperage draw.
C. the Btu output of the furnace.
D. the VA rating of the transformer.

17. You wish to determine the common, start, and run terminals on a new single phase compressor. From terminal X to terminal Y you read one ohm, Y to Z you measure five ohms, and from X to Z you read four ohms of resistance. From the data given, identify terminals X, Y, and Z.

A. X = common, Y = start, Z = run
B. X = start, Y = run, Z = common
C. X = common, Y = run, Z = start
D. X = run, Y = common, Z = start

18. When using a voltmeter to troubleshoot an inoperative circuit, first check for voltage at the:

A. first box.
B. switch.
C. last outlet.
D. last transformer.

19. Switches are always wired:

A. in series with a load.
B. in parallel with a load.
C. after a load.
D. none of the above.

20. If a resistance is 10 ohms and the voltage is 230 volts, what is the current in amperes?

A. 10
B. 15
C. 23
D. 100

21. Resistance is measured in:

A. amps.
B. volts.
C. watts.
D. ohms.

22. The thermostatic expansion valve is designed to maintain constant:

A. flow.
B. temperature.
C. pressure.
D. superheat.

23. The fin spacing on low temperature evaporators is ________ on medium temperature evaporators.

A. the same as
B. closer together than
C. farther apart than
D. none of the above

24. Does the heat loss from the condenser equal the heat gain to the evaporator?

A. It is more.
B. It is less.
C. They are equal.
D. They are equal only after running for a certain time.

25. Failure of a system to hold a vacuum concluding the evacuation process indicates that:

A. the system is ready to be charged.
B. the system has been adequately evacuated.
C. a leak in the system may exist.
D. the feed device is plugged.

26. What is the purpose of a suction line accumulator?

A. To prevent liquid flood back.
B. To protect the compressor from liquid flood back.
C. To control refrigerant flow into the evaporator.
D. To stop condenser freeze up.

27. The ideal location for a filter/drier is:

A. between the condenser and feed device.
B. midway through the condenser.
C. before the condenser.
D. between the T.E.V. and the evaporator.

28. A sight glass:

A. is used on all systems of 4 tons or larger.
B. is located just before the capillary tube.
C. is used only on systems with R-11 or R-12.
D. is used on systems which employ thermostatic expansion valves.

29. What can be used to reduce the foaming of oil on start up?

A. Anti-foam system
B. Adding new refrigerant
C. Subcool the oil
D. Crankcase heater

30. When installing a remote condensing unit that will be located outdoors, it is necessary to install a crankcase heater on the compressor to:

A. prevent migration of liquid refrigerant in oil.
B. keep the oil viscosity low for good lubrication.
C. prevent ice formation on compressor.
D. prevent sludging of the oil.

31. The suction line connects:

A. the condenser and the receiver.
B. the evaporator to the compressor.
C. two parts of the system and is the smallest size of line used.
D. the metering device to the evaporator.

32. The largest diameter line used on a vapor-compression system is the:

A. hot gas by-pass line.
B. liquid line.
C. suction line.
D. no one line, as the lines are all the same size.

ANSWERS:

1. B	5. C	9. C	13. D	17. C	21. D	25. C	29. D
2. C	6. B	10. D	14. C	18. B	22. D	26. B	30. A
3. A	7. A	11. A	15. C	19. A	23. C	27. A	31. B
4. D	8. D	12. B	16. B	20. C	24. A	28. D	32. C

How long are the tests? Can I take them all at once?

In the Service Technician Module, the three mandatory core units have a combined 260 questions. Most candidates complete these units in three to four hours.

The two specialist unit tests have 110 to 120 questions, and most candidates require two to three hours to complete them. The tests are tough, but fair. ACCA does not recommend that a candidate take both the core units and the specialist units in the same test sitting.

Does everybody get the same test questions?

NO. To insure the integrity of the test, there are several versions of the tests for each module, with all versions presenting the same level of difficulty.

Can I bring books or notes to the test?

NO. You're not allowed to bring any scratch paper, books, notes, equations, or other written material with you. All necessary materials, including a pressure/temperature (P/T) chart, are found in the test booklets.

What's the passing grade?

A candidate must pass all three of the core units, and must pass at least one of the specialist units to be ACE certified. (Both specialist units may be taken if desired.)

Each unit has its own passing grade, as follows:

Service Module—Core Units

Unit I: Safety, Tools, and Soft Skills
78% correct

Unit II: Principles of Heating, Air Conditioning, & Refrigeration
73% correct

Unit III: Electricity and Controls
67% correct

Service Module—Specialist Units

Unit IV : Heating Systems
60% correct

Unit V: Air Conditioning Systems
69% correct

Test questions and units for the Installation Module are in the late stages of development, and passing grades have not yet been specified. Completion is expected in early 1998.

Who sees the test results? Do employers see their technicians' test scores?

Candidates receive their scores in the mail, in a Testing Results Detail Report. (See box, below.)

Ferris State University tabulates the scores and mails them to candidates about two weeks after it receives the test papers from the testing site.

Test reports show a person's scores by unit, so he/she can pinpoint his/her own strengths or weaknesses.

Ferris State University will not release test scores to anyone except the person who took the test. Employees are encouraged to discuss their results with their supervisor, for their own merit and to bring structure to their continuing education.

Proctors for a test will receive a report on how the participants in a test group performed on each test unit. This feedback process lets instructors know of areas where further training might be needed.

Air Conditioning Excellence Program—Service Module

Testing Results Detail Report

Name: Pat Freemason
Test Date: 12/06/1997
ID#: 000-00-0000

Core Units	Number of Questions	% Required	% Correct	Pass?	Certified?
Safety, Tools, and Soft Skills (Unit 1)	40	78%	75%	N	
Workplace Safety			79%		
Customer Relations			71%		
Principles of HVACR (Unit 2)	120	73%	82%	Y	
Physics of refrigeration and air conditioning			67%		
Fundamentals for refrigeration & system service			85%		
Copper systems & heat exchangers			88%		
Electricity and Controls (Unit 3)	100	67%	74%	Y	N
Electrical fundamentals			59%		
Electrical diagrams, controls, & troubleshooting			88%		
Motors and starting devices			78%		
Heating Systems Specialist (Unit 4)	120	60%	55%	N	N
Combustion & venting, electric heat, piping systems			61%		
Gas-fired systems			55%		
Oil-fired systems			52%		
Troubleshooting heating systems			50%		
Air Conditioning Systems Specialist (Unit 5)	110	69%	65%	N	N
Air properties			55%		
A/C systems & heat pumps			71%		
Compressors, system service, & mechanical troubleshooting			69%		

SAMPLE

If I don't pass, can I take the test again?

YES, and those units you already passed need not be taken over, within the validity period of the certification.

Will I have to re-certify?

Not all details of the ACE Program have been worked out, and the issue of re-certification and continuing education is one of them. Because systems, standards, and equipment change, re-certification will be necessary every four years or so. ACCA will notify participants in the certification program as those decisions are reached.

Sounds good. How do I sign up for the test?

Most ACCA Chapters administer the Technician Certification exams, and have established their own testing schedules and locations. Also, many vocational schools and training centers offer testing. Candidates should not have to go very far to take the tests.

The schedule of times and locations for the tests changes frequently as new sites and dates are added. To learn the latest dates and sites, or for other information about certification, phone ACCA toll-free at:

1-888-290-2220

or contact one of the ACCA Chapters listed on the following page.